THERMAL EXPANSION
Metallic Elements and Alloys

THERMOPHYSICAL PROPERTIES OF MATTER

The TPRC Data Series

A Comprehensive Compilation of Data by the
Thermophysical Properties Research Center (TPRC), Purdue University

Y. S. Touloukian, Series Editor
C. Y. Ho, Series Technical Editor

Volume 1. Thermal Conductivity–Metallic Elements and Alloys
Volume 2. Thermal Conductivity–Nonmetallic Solids
Volume 3. Thermal Conductivity–Nonmetallic Liquids and Gases
Volume 4. Specific Heat–Metallic Elements and Alloys
Volume 5. Specific Heat–Nonmetallic Solids
Volume 6. Specific Heat–Nonmetallic Liquids and Gases
Volume 7. Thermal Radiative Properties–Metallic Elements and Alloys
Volume 8. Thermal Radiative Properties–Nonmetallic Solids
Volume 9. Thermal Radiative Properties–Coatings
Volume 10. Thermal Diffusivity
Volume 11. Viscosity
Volume 12. Thermal Expansion–Metallic Elements and Alloys
Volume 13. Thermal Expansion–Nonmetallic Solids

New data on thermophysical properties are being constantly accumulated at TPRC. Contact TPRC and use its interim updating services for the most current information.

THERMOPHYSICAL PROPERTIES OF MATTER

VOLUME 12

THERMAL EXPANSION

Metallic Elements and Alloys

Y. S. Touloukian

Director
Thermophysical Properties Research Center
and
Distinguished Atkins Professor of Engineering
School of Mechanical Engineering
Purdue University
and
Visiting Professor of Mechanical Engineering
Auburn University

R. K. Kirby

Physicist
Crystallography Section
Inorganic Materials Division
Institute for Materials Research
U.S. National Bureau of Standards

R. E. Taylor

Senior Researcher
Thermophysical Properties Research Center
and
Head
Properties Research Laboratory
School of Mechanical Engineering
Purdue University

P. D. Desai

Assistant Senior Researcher
Thermophysical Properties Research Center
Purdue University

IFI/PLENUM • NEW YORK-WASHINGTON

Library of Congress Catalog Card Number 73-129616

ISBN (13-Volume Set) 0-306-67020-8
ISBN (Volume 12) 0-306-67032-1

IFI/Plenum Data Company is a division of
Plenum Publishing Corporation
227 West 17th Street, New York, N.Y. 10011

Distributed in Europe by Heyden & Son, Ltd.
Spectrum House, Alderton Crescent
London NW4 3XX, England

Printed in the United States of America

"In this work, when it shall be found that much is omitted, let it not be forgotten that much likewise is performed..."

SAMUEL JOHNSON, A.M.

From last paragraph of Preface to his two-volume *Dictionary of the English Language*, Vol. I, page 5, 1755, London, Printed by Strahan.

Foreword

In 1957, the Thermophysical Properties Research Center (TPRC) of Purdue University, under the leadership of its founder, Professor Y. S. Touloukian, began to develop a coordinated experimental, theoretical, and literature review program covering a set of properties of great importance to science and technology. Over the years, this program has grown steadily, producing bibliographies, data compilations and recommendations, experimental measurements, and other output. The series of volumes for which these remarks constitute a foreword is one of these many important products. These volumes are a monumental accomplishment in themselves, requiring for their production the combined knowledge and skills of dozens of dedicated specialists. The Thermophysical Properties Research Center deserves the gratitude of every scientist and engineer who uses these compiled data.

The individual nontechnical citizen of the United States has a stake in this work also, for much of the science and technology that contributes to his well-being relies on the use of these data. Indeed, recognition of this importance is indicated by a mere reading of the list of the financial sponsors of the Thermophysical Properties Research Center; leaders of the technical industry of the United States and agencies of the Federal Government are well represented.

Experimental measurements made in a laboratory have many potential applications. They might be used, for example, to check a theory, or to help design a chemical manufacturing plant, or to compute the characteristics of a heat exchanger in a nuclear power plant. The progress of science and technology demands that results be published in the open literature so that others may use them. Fortunately for progress, the useful data in any single field are not scattered throughout the tens of thousands of technical journals published throughout the world. In most fields, fifty percent of the useful work appears in no more than thirty or forty journals. However, in the case of TPRC, its field is so broad that about 100 journals are required to yield fifty percent. But that other fifty percent! It is scattered through more than 3500 journals and other documents, often items not readily identifiable or obtainable. Over 78,000 references are now in the files.

Thus, the man who wants to use existing data, rather than make new measurements himself, faces a long and costly task if he wants to assure himself that he has found all the relevant results. More often than not, a search for data stops after one or two results are found—or after the searcher decides he has spent enough time looking. Now with the appearance of these volumes, the scientist or engineer who needs these kinds of data can consider himself very fortunate. He has a single source to turn to; thousands of hours of search time will be saved, innumerable repetitions of measurements will be avoided, and several billions of dollars of investment in research work will have been preserved.

However, the task is not ended with the generation of these volumes. A critical evaluation of much of the data is still needed. Why are discrepant results obtained by different experimentalists? What undetected sources of systematic error may affect some or even all measurements? What value can be derived as a "recommended" figure from the various conflicting values that may be reported? These questions are difficult to answer, requiring the most sophisticated judgment of a specialist in the field. While a number of the volumes in this Series do contain critically evaluated and recommended data, these are still in the minority. The data are now being more intensively evaluated by the staff of TPRC as an integral part of the effort of the National Standard Reference Data System (NSRDS). The task of the National Standard Reference Data System is to organize and operate a comprehensive program to prepare compilations of critically evaluated data on the properties of substances. The NSRDS is administered by the National Bureau of Standards under a directive from the Federal Council for Science and Technology, augmented by special legislation of the Congress of the United States. TPRC is one of the national resources participating in the National Standard Reference Data System in a united effort to satisfy the needs of the technical community for readily accessible, critically evaluated data.

As a representative of the NBS Office of Standard Reference Data, I want to congratulate Professor Touloukian and his colleagues on the accomplishments represented by this Series of reference data books. Scientists and engineers the world over are indebted to them. The task ahead is still an awesome one and I urge the nation's private industries and all concerned Federal agencies to participate in fulfilling this national need of assuring the availability of standard numerical reference data for science and technology.

EDWARD L. BRADY
Associate Director for Information Programs
National Bureau of Standards

Preface

Thermophysical Properties of Matter, the TPRC Data Series, is the culmination of eighteen years of pioneering effort in the generation of tables of numerical data for science and technology. It constitutes the restructuring, accompanied by extensive revision and expansion of coverage, of the original *TPRC Data Book*, first released in 1960 in loose-leaf format, 11″ × 17″ in size, and issued in June and December annually in the form of supplements. The original loose-leaf *Data Book* was organized in three volumes: (1) metallic elements and alloys; (2) nonmetallic elements, compounds, and mixtures which are solid at N.T.P., and (3) nonmetallic elements, compounds, and mixtures which are liquid or gaseous at N.T.P. Within each volume, each property constituted a chapter.

Because of the vast proportions the *Data Book* began to assume over the years of its growth and the greatly increased effort necessary in its maintenance by the user, it was decided in 1967 to change from the loose-leaf format to a conventional publication. Thus, the December 1966 supplement of the original *Data Book* was the last supplement disseminated by TPRC.

While the manifold physical, logistic, and economic advantages of the bound volume over the loose-leaf oversize format are obvious and welcome to all who have used the unwieldy original volumes, the assumption that this work will no longer be kept on a current basis because of its bound format would not be correct. Fully recognizing the need of many important research and development programs which require the latest available information, TPRC has instituted a *Data Update Plan* enabling the subscriber to inquire, by telephone if necessary, for specific information and receive, in many instances, same-day response on any new data processed or revision of published data since the latest edition. In this context, the TPRC Data Series departs drastically from the conventional handbook and giant multivolume classical works, which are no longer adequate media for the dissemination of numerical data of science and technology without a continuing activity on contemporary coverage. The loose-leaf arrangements of many works fully recognize this fact and attempt to develop a combination of bound volumes and loose-leaf supplement arrangements as the work becomes increasingly large. TPRC's *Data Update Plan* is indeed unique in this sense since it maintains the contents of the TPRC Data Series current and live on a day-to-day basis between editions. In this spirit, I strongly urge all purchasers of these volumes to complete in detail and return the *Volume Registration Certificate* which accompanies each volume in order to assure themselves of the continuous receipt of annual listing of corrigenda during the life of the edition.

The TPRC Data Series consists initially of 13 independent volumes. The first seven volumes were published in 1970, Volumes 8 and 9 in 1972, Volume 10 in 1973, and Volume 11 in 1975. Volume 12 is planned for publication in late 1975 and Volume 13 is planned for 1976. It is also contemplated that subsequent to the first edition, each volume will be revised, up-dated, and reissued in a new edition approximately every fifth year. The organization of the TPRC Data Series makes each volume a self-contained entity available individually without the need to purchase the entire Series.

The coverage of the specific thermophysical properties represented by this Series constitutes the most comprehensive and authoritative collection of numerical data of its kind for science and technology.

Whenever possible, a uniform format has been used in all volumes, except when variations in presentation were necessitated by the nature of the property or the physical state concerned. In spite of the wealth of data reported in these volumes, it should be recognized that all volumes are not of the same degree of completeness. However, as additional data are processed at TPRC on a continuing basis, subsequent editions will become increasingly more complete and up to date. Each volume in the Series basically comprises three sections, consisting of a text, the body of numerical data with source references, and a material index.

The aim of the textual material is to provide a complementary or supporting role to the body of numerical data rather than to present a treatise on the subject of the property. The user will find a basic theoretical treatment, a comprehensive presentation of selected works which constitute reviews, or compendia of empirical relations useful in estimation of the property when there exists a paucity of data or when data are completely lacking. Established major experimental techniques are also briefly reviewed.

The body of data is the core of each volume and is presented in both graphical and tabular formats for convenience of the user. Every single point of numerical data is fully referenced as to its original source and no secondary sources of information are used in data extraction. In general, it has not been possible to critically scrutinize all the original data presented in these volumes, except to eliminate perpetuation of gross errors. However, in a significant number of cases, such as for the properties of liquids and gases, the thermal conductivity and thermal diffusivity of all the elements, and the thermal expansion of most materials in all material categories, the task of full evaluation, synthesis, and correlation has been completed. It is hoped that in subsequent editions of this continuing work, not only new information will be reported but the critical evaluation will be extended to increasingly broader classes of materials and properties.

The third and final major section of each volume is the material index. This is the key to the volume, enabling the user to exercise full freedom of access to its contents by any choice of substance name or detailed alloy and mixture composition, trade name, synonym, etc. Of particular interest here is the fact that in the case of those properties which are reported in separate companion volumes, the material index in each of the volumes also reports the contents of the other companion volumes.* The sets of companion volumes are as follows:

Thermal conductivity:	Volumes 1, 2, 3
Specific heat:	Volumes 4, 5, 6
Radiative properties:	Volumes 7, 8, 9
Thermal expansion:	Volumes 12, 13

*For the first edition of the Series, this arrangement was not feasible for Volumes 7, 8, and 12 due to the sequence and the schedule of their publication. This situation will be resolved in subsequent editions.

The ultimate aims and functions of TPRC's Data Tables Division are to extract, evaluate, reconcile, correlate, and synthesize all available data for the thermophysical properties of materials with the result of obtaining internally consistent sets of property values, termed the "recommended reference values." In such work, gaps in the data often occur, for ranges of temperature, composition, etc. Whenever feasible, various techniques are used to fill in such missing information, ranging from empirical procedures to detailed theoretical calculations. Such studies are resulting in valuable new estimation methods being developed which have made it possible to estimate values for substances and/or physical conditions presently unmeasured or not amenable to laboratory investigation. Depending on the available information for a particular property and substance, the end product may vary from simple tabulations of isolated values to detailed tabulations with generating equations, plots showing the concordance of the different values, and, in some cases, over a range of parameters presently unexplored in the laboratory.

The TPRC Data Series constitutes a permanent and valuable contribution to science and technology. These constantly growing volumes are invaluable sources of data to engineers and scientists, sources in which a wealth of information heretofore unknown or not readily available has been made accessible. We look forward to continued improvement of both format and contents so that TPRC may serve the scientific and technological community with ever-increasing excellence in the years to come. In this connection, the staff of TPRC is most anxious to receive comments, suggestions, and criticisms from all users of these volumes. An increasing number of colleagues are making available at the earliest possible moment reprints of their papers and reports as well as pertinent information on the more obscure publications. I wish to renew my earnest request that this procedure become a universal practice since it will prove to be most helpful in making TPRC's continuing effort more complete and up to date.

It is indeed a pleasure to acknowledge with gratitude the multisource financial assistance received from over fifty sponsors which has made the continued generation of these tables possible. In particular, I wish to single out the sustained major support received from the Air Force Materials Laboratory–Air Force Systems Command, the Defense Supply Agency, the Office of Standard Reference Data–National Bureau of Standards, and the Office of

Advanced Research and Technology–National Aeronautics and Space Administration. TPRC is indeed proud to have been designated as a National Information Analysis Center for the Department of Defense as well as a component of the National Standard Reference Data System under the cognizance of the National Bureau of Standards.

While the preparation and continued maintenance of this work is the responsibility of TPRC's Data Tables Division, it would not have been possible without the direct input of TPRC's Scientific Documentation Division and, to a lesser degree, the Theoretical and Experimental Research Divisions. The authors of the various volumes are the Senior staff members in responsible charge of the work. It should be clearly understood, however, that many have contributed over the years and their contributions are specifically acknowledged in each volume. I wish to take this opportunity to personally thank those members of the staff, assistant researchers, graduate research assistants, and supporting graphics and technical typing personnel without whose diligent and painstaking efforts this work could not have materialized.

Y. S. TOULOUKIAN

Director
Thermophysical Properties Research Center
Distinguished Atkins Professor of Engineering

Purdue University
West Lafayette, Indiana
June 1975

Introduction to Volume 12

This volume of *Thermophysical Properties of Matter, The TPRC Data Series*, presents the data and information on the thermal expansion of metallic elements, alloys, and intermetallic compounds.

The volume comprises three major sections: the front text on *theory, estimation, and measurement* together with its bibliography, the main body of *numerical data* with its references, and the *material index*.

The text material is intended to assume a role complementary to the main body of numerical data, the presentation of which is the primary purpose of this volume. It is felt that a moderately detailed discussion of the theoretical nature of the property under consideration together with an overview of predictive procedures and recognized experimental methods and techniques will be appropriate in a major reference work of this kind. The extensive reference citations given in the text should lead the interested reader to sufficient literature for a more comprehensive study. It is hoped, however, that enough detail is presented for this volume to be self-contained for the practical user.

The main body of the volume consists of the presentation of numerical data compiled over the years in a most comprehensive and meticulous manner. The extraction of all data directly from their original sources ensures freedom from errors of transcription. Furthermore, gross errors appearing in the original source documents have been corrected. The organization and presentation of the data together with other pertinent information on the use of the tables and figures are discussed in detail in the introductory material to the section entitled *Numerical Data*. The materials covered include 64 metallic elements, 94 intermetallic compounds, 125 binary alloy systems, and 70 groups of multiple alloys. Of these 353 materials and systems, the data for 246 (i.e., 70%) have been critically evaluated, analyzed, and synthesized, and recommended reference values or provisional values have been generated and are presented in the volume together with the original experimental data. Such recommended values are those that were considered to be the most probable when assessments were made of the available data and information. It should be realized, however, that these recommended values are not necessarily the final true values and that changes directed toward this end will often become necessary as more data become available. Future editions will contain these changes.

As stated earlier, all data have been obtained from their original sources and each data set is so referenced. TPRC has in its files all data-source documents cited in this volume. Those that cannot readily be obtained elsewhere are available from TPRC in microfiche form.

We wish to gratefully acknowledge that this volume has grown out of activities made possible principally through the support of the Defense Supply Agency (DSA) under the technical monitorship of the Army Materials and Mechanics Research Center (AMMRC). The guidance and support of representatives of the sponsor have been essential to the success of the activities. Particular mention should be made of the understanding support given by Mr. Joseph L. Blue and Mr. Michael Corridore of DSA and Mr. Samuel Valencia and Dr. John J. Burke of AMMRC. Over the years, several assistant researchers have contributed to the preparation of this volume for varying periods under the authors' supervision. In chronological order of their association with TPRC, we wish to acknowledge the contributions of Messrs. P. L. Wang, B. M. Whitcomb, S. N. Vo, and T. Y. R. Lee. Mr. Lee, who is still at TPRC, not only has participated in all phases of work in the preparation of this volume but has made significant contributions to it.

Inherent in the character of this work is the fact that in the preparation of this volume we have drawn most heavily upon the scientific literature and feel a debt of gratitude to the authors of the referenced articles. While their often discordant results have caused us much difficulty in reconciling their findings, we consider this to be our challenge and our contribution to negative entropy of information, as an effort is made to create from the randomly distributed data a condensed, more orderly state.

While this volume is primarily intended as a reference work for the designer, researcher, experimentalist, and theoretician, the teacher at the graduate level may also use it as a teaching tool to point out to his students the topography of the state of knowledge on the thermal expansion of metallic elements, alloys, and intermetallic compounds. We believe there is also much food for reflection by the specialist and the academician concerning the meaning of "original" investigation and its "information content."

The authors and their contributing associates are keenly aware of the possibility of many weaknesses in a work of this scope. We hope that we will not be judged too harshly and that we will receive the benefit of suggestions regarding references omitted, additional material groups needing detailed treatment, improvements in presentation or in recommended values, and, most important, any inadvertent errors. If the *Volume Registration Certificate* accompanying this volume is returned, the reader will assure himself of receiving annually a list of corrigenda as possible errors come to our attention.

West Lafayette, Indiana
June 1975

Y. S. Touloukian
R. K. Kirby
R. E. Taylor
P. D. Desai

Contents

Numerical Data

Material Index

GROUPING OF MATERIALS AND LIST OF FIGURES AND TABLES

1. ELEMENTS

* Number marked with an asterisk indicates that recommended/provisional/typical values are not reported for this material; for others these values are reported in separate figure and table of the same number followed by the letter R.

† No original-data figure given.

‡ Only provisional values calculated from density data presented.

1. ELEMENTS (continued)

2. INTERMETALLIC COMPOUNDS

* Number marked with an asterisk indicates that recommended/provisional/typical values are not reported for this material; for others these values are reported in separate figure and table of the same number followed by the letter R.

† No original-data figure given.

‡ Only provisional values calculated from density data presented.

2. INTERMETALLIC COMPOUNDS (continued)

* Number marked with an asterisk indicates that recommended/provisional/typical values are not reported for this material; for others these values are reported in separate figure and table of the same number followed by the letter R.

† No original-data figure given.

2. INTERMETALLIC COMPOUNDS (continued)

* Number marked with an asterisk indicates that recommended/provisional/typical values are not reported for this material; for others these values are reported in separate figure and table of the same number followed by the letter R.

† No original-data figure given.

3. BINARY ALLOY SYSTEMS

* Number marked with an asterisk indicates that recommended/provisional/typical values are not reported for this material; for others these values are reported in separate figure and table of the same number followed by the letter R.

† No original-data figure given.

3. BINARY ALLOY SYSTEMS (continued)

* Number marked with an asterisk indicates that recommended/provisional/typical values are not reported for this material; for others these values are reported in separate figure and table of the same number followed by the letter R.

† No original-data figure given.

3. BINARY ALLOY SYSTEMS (continued)

* Number marked with an asterisk indicates that recommended/provisional/typical values are not reported for this material; for others these values are reported in separate figure and table of the same number followed by the letter R.

† No original-data figure given.

4. MULTIPLE ALLOYS

* Number marked with an asterisk indicates that recommended/provisional/typical values are not reported for this material; for others these values are reported in separate figure and table of the same number followed by the letter R.

† No original-data figure given.

4. MULTIPLE ALLOYS (continued)

* Number marked with an asterisk indicates that recommended/provisional/typical values are not reported for this material; for others these values are reported in separate figure and table of the same number followed by the letter R.

† No original-data figure given.

Theory, Estimation, and Measurement

Notation

A	Amplitude of vibration
a	Lattice spacing; Constant
a_1, a_2, a_3	Empirical constants
B_S	Adiabatic bulk modulus
B_T	Isothermal bulk modulus
B_0	Isothermal bulk modulus at 0 K
B_C	Isothermal bulk modulus of composite
B_d	Isothermal bulk modulus of dispersed material
B_i	Isothermal bulk modulus of *i*th material
B_{mt}	Isothermal bulk modulus of matrix material
b	Lattice spacing; Constant
C	Empirical constant; Capacitance
C_i	Heat capacity of *i*th mode
C_p	Heat capacity at constant pressure
C_V	Heat capacity at constant volume
C_{293}	Heat capacity at constant volume at 293 K
C_{el}	Heat capacity of free electrons
C_{lt}	Heat capacity of lattice
C_{mg}	Heat capacity of magnetic interaction
c	Lattice spacing
c_0, c_1	Calibration constants
c_i	Elastic constant
D_i	Thickness of *i*th material
d	Diameter; Separation of lattice planes
d_p	Diameter of *p*th fringe
E	Young's modulus
E_i	Young's modulus of *i*th material
E_D	Internal energy
E_{293}	Internal energy at 293 K
E_M	Internal energy at T_M
E_{mg}	Magnetic interaction energy
F	Free energy; Frequency
F_{th}	Free energy of thermal vibrations
f	Focal length of lens
G_C	Shear modulus of composite
G_{mt}	Shear modulus of matrix material
g	Acceleration of free fall
H	Enthalpy of formation
h	Planck's constant; Miller index
k	Boltzmann constant; Constant; Wave number; Miller index
L	Length
L_{293}	Length at 293 K
L_M	Length at T_M
L_R	Length of reference material
L_S	Length of specimen
l	Miller index
m	Mass
N	Number of particles; Order of interference
$N(E_F)$	Density of energy states
n	Number of vacancies; Index of refraction
P	Pressure
P_{st}	Static pressure
P_{th}	Thermal pressure
p	*p*th fringe
Q	Constant
R	Gas constant; Radius of a circle
S	Entropy
S_i	Sound velocity of the *i*th branch
s	Anharmonicity constant; Entropy of formation; Distance between fringes
s_{ij}	Elastic compliances
T	Temperature
T_M	Temperature at melting point
T	Mean temperature
t	Time
V	Volume
V_0	Volume at 0 K
V_{293}	Volume at 293 K
V_M	Volume at T_M
v	Fractional part of order of interference
v_d	Volume fraction of dispersed material
v_i	Volume fraction of *i*th material
W	Distance
x	Position of particle
α	Coefficient of linear thermal expansion; Angle between lattice spacings *b* and *c*

α_{293}	Coefficient of linear thermal expansion at 293 K
α_C	Coefficient of linear thermal expansion of composite
α_I	Instantaneous coefficient of linear thermal expansion
α_M	Mean coefficient of linear thermal expansion from 0 K to T_M
α_a	Coefficient of linear thermal expansion of lattice
α_b	Coefficient of linear thermal expansion of bulk material
α_d	Coefficient of linear thermal expansion of dispersed material
α_i	Coefficient of linear thermal expansion of ith material
α_m	Mean coefficient of linear thermal expansion
α_{em}	Coefficient of linear thermal expansion of electronic and magnetic interactions
α_{lt}	Coefficient of linear thermal expansion of lattice vibrations
α_{mt}	Coefficient of linear thermal expansion of matrix material
β	Coefficient of thermal expansion; Angle between lattice spacings a and c
β_{293}	Coefficient of thermal expansion at 293 K
β_C	Coefficient of thermal expansion of composite
β_M	Mean coefficient of thermal expansion from 0 K to T_M
β_m	Mean coefficient of thermal expansion
β_{mg}	Coefficient of thermal expansion of magnetic interactions
γ	Grüneisen parameter; Angle between lattice spacings a and b
γ_0	Grüneisen parameter at 0 K
γ_i	Grüneisen parameter of ith mode or ith branch
γ_j	Grüneisen parameter tensor
γ_{el}	Grüneisen parameter of free electrons
γ_{em}	Grüneisen parameter of free electrons and magnetic interactions
γ_{lt}	Grüneisen parameter of lattice
γ_{mg}	Grüneisen parameter of magnetic interactions
γ_∞	Grüneisen parameter at high temperatures
θ_D	Debye temperature
θ_0	Debye temperature at 0 K
θ	Angle of incident rays
λ	Wavelength
λ_v	Wavelength in vacuum
μ	Poisson's ratio
μ_i	Poisson's ratio of ith material
ν_0	Frequency of oscillation of a particle
ν_D	Debye frequency
ν_i	Frequency of ith mode
ρ	Density
σ	Stress
τ	Period of a pendulum
Φ	Potential
Φ_0	Potential energy of a lattice including zero point energy
ϕ	Angle
ϕ_1, ϕ_2, ϕ_3	Angles between crystallographic translation vectors
Ω	Solid angle
ω	Angle
ω_i	Angle between direction of interest and the principal crystallographic axes

Theory of Thermal Expansion of Solids

1. INTRODUCTION

"From ghoulies and ghosties, and long leggety beasties and things that go bump in the night, Good Lord deliver us" is an old Scottish prayer. Among the things that went "bump in the night" were certainly coats of armor, structural members of buildings, etc., as they cooled off. It is not difficult to visualize the consternation of our forefathers when such items suddenly emitted strange sounds. Modern homes also creak and groan when changing thermal gradients cause stresses due to unequal thermal expansions. The uses of thermal expansion in our daily lives are readily observable—mercury thermometers to indicate temperature, thermostats to control our heating and cooling systems, and even the use of hot water to "loosen" a stuck jar lid. On a hot summer day the effect of thermal expansion would also be readily apparent if not allowed for by the design engineer. To paraphrase a facetious comment that appeared in a small book, *Physics for Fools*, published in Russia in 1908: Bridges, railroad tracks, and concrete sections of highways are always made shorter than they should be since bodies always expand when heated. Even so, after an unusually hot day, newspapers will have photographs of highways that have buckled and draw bridges that have stuck. The importance to technology of accurate knowledge of thermal expansion is readily apparent when one considers the problems associated with high-performance engines, atomic reactors, re-entry vehicles, and the like.

One of the first scientific problems associated with dimensional changes in solids caused by temperature fluctuations concerned the measurement of time with pendulums. Although the period, τ,* of a pendulum is almost independent of the amplitude, it does critically depend upon its length:

$$\tau = 2\pi\sqrt{L/g} \tag{1}$$

*See *Notation*, p. 1a, for meaning of symbols used in text.

For instance, a change in temperature of 10 deg will cause a change in period of about 0.01% if the pendulum is made of brass. It is not surprising, therefore, that some of the first quantitative measurements of thermal expansion were made in order to obtain more accurate clocks. These measurements were made by Petrus van Musschenbrock [1], a professor of astronomy in Utrecht, around 1730. Professor Musschenbrock measured the thermal expansion of iron, steel, copper, brass, tin, and lead. He found that lead expanded the most and iron the least. In his own words, "Therefore it is very proper to make the Rods of Pendulums for Clocks, of Iron: They are not so good of Steel; and much worse of Brass; yet sometimes they are made of Copper, because it is not so liable to rust; but yet, that is wrong." Since this time, a large amount of information on thermal expansion has been obtained and with advances in technology the quality of the information has improved. A great deal of this information, however, is worthless, or nearly so, because the material was not sufficiently characterized and/or the experiment was not sufficiently documented. In particular, accurate data on well characterized materials are needed to understand the behavior of alloy and composite systems. Even so, within the past 20 years measurements on thermal expansion have greatly increased our knowledge in areas such as lattice dynamics, electronic and magnetic interactions, thermal defects, and phase transitions.

2. DEFINITIONS

When heat is added to a material so that there is a change in temperature, $T_1 \rightarrow T_2$, there is a corresponding change in volume, $V_1 \rightarrow V_2$. To describe this change the mean coefficient of volumetric thermal expansion of the material is defined by

$$\beta_m = \frac{V_2 - V_1}{V_1(T_2 - T_1)} \tag{2}$$

The limiting value of this ratio (at constant pressure P) as the temperature changes by a differential amount dT is defined as the true coefficient of volumetric thermal expansion, or just as the coefficient of thermal expansion, i.e.,

$$\beta = \frac{1}{V}\left(\frac{\partial V}{\partial T}\right)_P \tag{3}$$

The thermal expansion of the substance on heating from T_1 to T_2 is often expressed as percent expansion,

$$\frac{100(V_2 - V_1)}{V_1} \quad \text{or} \quad \frac{100\Delta V}{V} \tag{4}$$

The corresponding definitions for the linear (or unidirectional) case are

$$\alpha_m = \frac{L_2 - L_1}{L_1(T_2 - T_1)} \tag{5}$$

$$\alpha = \frac{1}{L}\left(\frac{\partial L}{\partial T}\right)_P \tag{6}$$

and

$$\frac{100(L_2 - L_1)}{L_1} \quad \text{or} \quad \frac{100\Delta L}{L} \tag{7}$$

where L represents the length. If the expansion is determined from measurements of density, ρ, then

$$\beta_m = \frac{\rho_1 - \rho_2}{\rho_2(T_2 - T_1)} \tag{8}$$

and

$$\beta = -\frac{1}{\rho}\left(\frac{\partial \rho}{\partial T}\right)_P \tag{9}$$

If the substance is isotropic the coefficient of thermal expansion is equal to three times the coefficient of linear thermal expansion; i.e.,

$$\beta = 3\alpha \tag{10}$$

The same is not strictly true, however, for the mean coefficients, where

$$\beta_m \simeq 3\alpha_m[1 + (T_2 - T_1)\alpha_m] \tag{11}$$

Ordinarily the coefficient of thermal expansion is not measured directly but it is either calculated directly from consecutive observations of expansion or by differentiating an equation that represents the expansion. Under certain conditions the instantaneous coefficient of linear thermal expansion will satisfactorily represent the true coefficient. The instantaneous coefficient of linear thermal expansion is defined as

$$\alpha_I = \frac{L_2 - L_1}{L_{293}(T_2 - T_1)} \quad \text{at } T_m = \frac{T_2 + T_1}{2} \tag{12}$$

where L_{293} is the length at 293 K. If the expansion over a limited temperature range can be approximated by a polynomial,

$$\frac{L - L_{293}}{L_{293}} = a_0 + a_1 T + a_2 T^2 + a_3 T^3 + \cdots \tag{13}$$

then

$$\alpha_I = \alpha_1 + 2a_2 T_m + a_3(4T_m^2 - T_2 T_1) + \cdots \tag{14}$$

and

$$\alpha = \frac{L_{293}}{L}(a_1 + 2a_2 T_m + 3a_3 T_m^2 + \cdots) \tag{15}$$

From these equations it can be shown that

$$\alpha = \frac{L_{293}}{L}\left[\alpha_I - \frac{a_3}{4}(\Delta T)^2 - \cdots\right] \tag{16}$$

If $L_{293} \simeq L$ and $\Delta T \to 0$ then $\alpha \simeq \alpha_I$. If the expansion can be represented by a quadratic, $a_3 = 0$ in Eq. (16), large temperature intervals can be used.

The coefficients needed to describe the expansion of crystals with different symmetries [2] are shown in Table I. The principal coefficients (α_1, α_2, and α_3) describe the volume change of the crystal so that

$$\beta = \alpha_1 + \alpha_2 + \alpha_3 \tag{17}$$

These coefficients are not necessarily in the direction of the crystallographic translation vectors but are orthogonal to each other. The remaining coefficients (α_4, α_5, and α_6) describe the change in shape (but not the symmetry) of the crystal. For example, a monoclinic crystal will change shape during a temperature change even if cut in the crystallographic reference system, but the interaxial angles ϕ_1 and ϕ_3, which are restricted to 90° by the symmetry, will not change.

Table I. Thermal Expansion Coefficients for Various Crystal Symmetries

Cubic	α_1	α_1	α_1	0	0	0
Hexagonal	α_1	α_1	α_3	0	0	0
Tetragonal	α_1	α_1	α_3	0	0	0
Trigonal	α_1	α_1	α_3	0	0	0
Orthorhombic	α_1	α_2	α_3	0	0	0
Monoclinic	α_1	α_2	α_3	0	α_5	0
Triclinic	α_1	α_2	α_3	α_4	α_5	α_6

The expansivity of a single crystal in a given direction is given by

$$\alpha = \sum_{i=1}^{3} \alpha_i \cos^2 \omega_i \tag{18}$$

where ω_i are the angles between the direction in question and the principal crystallographic axes. For hexagonal, tetragonal, and trigonal symmetries this relationship reduces to

$$\alpha = \alpha_1 + (\alpha_3 - \alpha_1) \cos^2 \omega_3 \tag{19}$$

3. POTENTIAL WELL—A SIMPLE MODEL

A three-dimensional array of atoms which are free to vibrate about their equilibrium positions is a reasonable model for a solid substance. The heat capacity of this model can be calculated by assuming that the vibrations are harmonic and their energies are quantized. While these assumptions are very successful in calculating the heat capacity of real solids the vibrations must be assumed to be anharmonic in order to account for thermal expansion.

This can be understood qualitatively by considering the vibrations of a single particle in the asymmetric potential well

$$\Phi = \tfrac{1}{2}Cx^2 - \tfrac{1}{3}sCx^3 \tag{20}$$

An approximate solution of the nonlinear equation of motion

$$m\frac{d^2x}{dt^2} + Cx - sCx^2 = 0 \tag{21}$$

can be shown [3] to be

$$x = A\left(\cos \nu_0 t - \frac{sA}{6} \cos 2\nu_0 t\right) + \frac{sA^2}{2} \tag{22}$$

where A is the amplitude of the motion. Since $\langle \cos \nu_0 t \rangle = \langle \cos 2\nu_0 t \rangle = 0$, the time-average position of the particle is

$$\langle x \rangle = \frac{sA^2}{2} \tag{23}$$

A shift of the average position that is proportional to the anharmonicity constant s has therefore occurred toward the softer side of the potential well (see Fig. 1). According to classical theory, A^2 is proportional to temperature so that the value of $\langle x \rangle$ can represent the thermal expansion of the system and the coefficient of thermal expansion is a constant that is proportional to s. In the case of a harmonic potential where $s = 0$, the average position of the particle does not shift and there is no corresponding thermal expansion.

4. DERIVATION OF THE GRÜNEISEN RELATION

In the quasi-harmonic approximation the effect of anharmonic interactions is simulated by treating the vibrations as harmonic but with frequencies that are volume dependent [4]. Applying statistical mechanics to this model it can be shown that

$$\frac{\partial^2 F}{\partial V \, \partial T} = -\frac{1}{V} \sum_{i=1}^{3N} \gamma_i C_i \tag{24}$$

where F is the free energy of the system and C_i is the contribution of each vibrational mode to the heat capacity at constant volume,

$$C_V = \sum_{i=1}^{3N} C_i \tag{25}$$

The term γ_i is the measure of the volume dependency of the frequency of each vibrational mode,

$$\gamma_i = -\frac{V}{\nu_i} \frac{d\nu_i}{dV} = -\frac{d \ln \nu_i}{d \ln V} \tag{26}$$

and hence indicates the effect of the anharmonic interactions. If the overall effect of the interactions is taken as the weighted average of the individual γ_i,

$$\gamma = \frac{\sum_{i=1}^{3N} \gamma_i C_i}{C_V} \tag{27}$$

then

$$\frac{\partial^2 F}{\partial V \, \partial T} = -\frac{\gamma C_V}{V} \tag{28}$$

Since it can also be shown from thermodynamics that

$$\frac{\partial^2 F}{\partial V \, \partial T} = -B_T \beta \tag{29}$$

where B_T is the isothermal bulk modulus, the Grüneisen relation can be obtained:

$$\beta = \frac{\gamma C_V}{V B_T} \tag{30}$$

where γ is known as the Grüneisen parameter.

Since both γ and the product of V and B_T are weak functions of temperature, the value of β is nearly proportional to C_V at all temperatures; that is, at high temperatures β is nearly a constant and at low

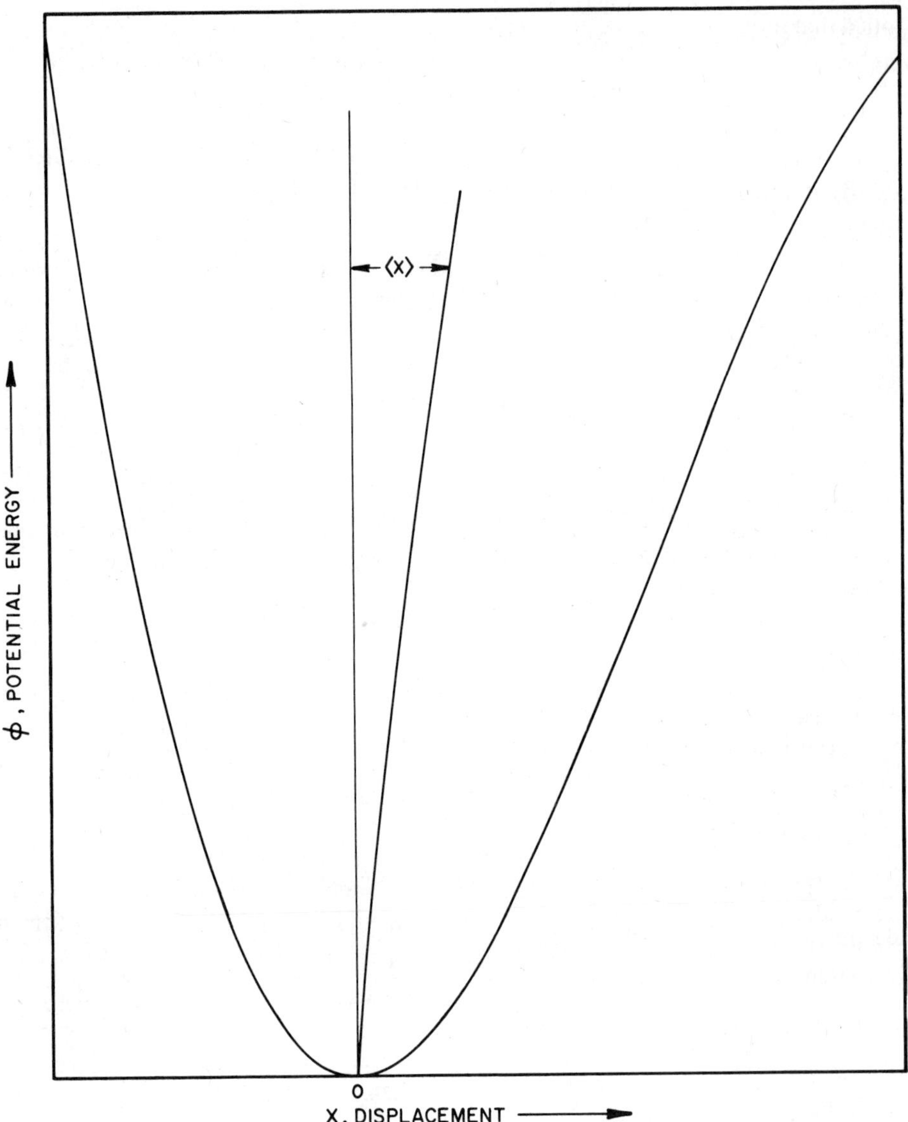

Fig. 1. Plot of asymmetric potential energy well and time-averaged position of vibrating particles.

temperatures, where the atoms must be treated as quantum oscillators, $\beta \to 0$ as $T \to 0$ K.

5. DEBYE EQUATION OF STATE

The reaction of a solid to external pressure can be divided into two parts, thermal and static. From thermodynamics it can be shown that thermal pressure in a solid is given by

$$P_{th} = -\left(\frac{\partial F_{th}}{\partial V}\right)_T \tag{31}$$

while static pressure is given by

$$P_{st} = -\frac{\partial \Phi_0}{\partial V} \tag{32}$$

where Φ_0 is the effective energy of the lattice which includes zero point energy. The total pressure is equal to the sum of the thermal and static pressures,

$$P = -\frac{\partial \Phi_0}{\partial V} - \left(\frac{\partial F_{th}}{\partial V}\right)_T \tag{33}$$

To approximate the effect of anharmonic interactions

Debye [5] assumed that the characteristic temperature,

$$\theta_D = \frac{h\nu_D}{k} \tag{34}$$

in his model of the lattice heat capacity, is a function of volume only. By differentiating Eq. (34) it can be shown with the help of Eq. (26) that

$$\frac{\partial \theta_D}{\partial V} = -\frac{\theta_D \gamma}{V} \tag{35}$$

It is obvious that in this model γ is also a function of volume only. Since it can also be shown from thermodynamics that

$$\frac{\partial F_{\text{th}}}{\partial \theta_D} = \frac{1}{\theta_D}\frac{\partial(F_{\text{th}}/T)}{\partial(1/T)} = \frac{E_D}{\theta_D} \tag{36}$$

where E_D is the Debye energy of the lattice vibrations, the Debye equation of state can be obtained:

$$P = -\frac{\partial \Phi_0}{\partial V} - \left(\frac{\partial F_{\text{th}}}{\partial \theta_D}\right)\left(\frac{\partial \theta_D}{\partial V}\right) = -\frac{\partial \Phi_0}{\partial V} + \frac{\gamma E_D}{V} \tag{37}$$

Grüneisen's relation can be obtained by differentiating P with respect to temperature at constant volume, i.e.,

$$\left(\frac{\partial P}{\partial T}\right)_V = \frac{\gamma C_V}{V} = B_T\beta \tag{38}$$

If the pressure is set equal to zero in Eq. (37), $V(\partial\Phi_0/\partial V)$ is expanded in a Taylor series in volume, and the constants of the series evaluated at zero Kelvin, the Grüneisen equation can be obtained [6]:

$$\frac{V_T - V_0}{V_0} = \frac{E_D}{Q - kE_D} \tag{39}$$

where

$$Q = \frac{V_0 B_0}{\gamma} \tag{40}$$

has a value that is large when compared to E_D. The constant k is best determined by experiment but can be related to the potential energy function.

6. GRÜNEISEN PARAMETER

As indicated in Section 4, the Grüneisen parameter, γ, is a weighted average of the γ_i, the weight for each vibrational mode being its contribution to the heat capacity. In the Debye model γ is a function of volume only:

$$\gamma = -\frac{V}{\theta_D}\frac{\partial \theta_D}{\partial V} \tag{41}$$

In a broader sense γ is also a function of temperature, and without assuming that the lattice vibrations are harmonic it can be defined by

$$\gamma(V, T) = \frac{V}{C_V}\left(\frac{\partial S}{\partial V}\right)_T \tag{42}$$

where S is the entropy of the system. In terms of the Grüneisen relation, $\gamma(V, T)$ can be expressed as

$$\gamma(V, T) = \frac{VB_T\beta}{C_V} = \frac{VB_s\beta}{C_p} \tag{43}$$

where B_s is the adiabatic bulk modulus and C_p is the heat capacity at constant pressure. At temperatures higher than the Debye characteristic temperature θ_D the value of γ is nearly constant since the whole spectrum of vibrational frequencies is excited and γ_∞ is merely the arithmetic average of the γ_i,

$$\gamma_\infty = \frac{1}{3N}\sum_{i=1}^{3N} \gamma_i \tag{44}$$

In the low-temperature limit the Debye continuum is a valid model and again γ will approach a constant value,

$$\gamma_0 = -\frac{V}{\theta_0}\frac{d\theta_0}{dV} \tag{45}$$

where θ_0 is the limiting value of θ_D at low temperatures.

If the weighting of the γ_i changes with temperature the results can be a large variation of γ with temperature, sometimes leading to negative values. This can happen in open-structure crystals, such as germanium between 16 and 40 K [7] (see Fig. 2), when the transverse modes, which can have negative γ_i, outweigh the longitudinal modes, which have positive γ_i. The result is that the coefficient of thermal expansion is also negative and the material will contract on heating. This can be explained by a simple model in which the atoms that lie in directions perpendicular to the direction of vibration are pulled further than they are forced outward by their own vibrations.

In solids which have noncubic symmetry, Grüneisen's relation can be expressed as

$$\alpha_i = \frac{C_V}{V}\sum_{j=1}^{6} s_{ij}\gamma_j, \qquad i = 1, 2, 3 \tag{46}$$

where s_{ij} are the elastic compliances [8]. This equation gives a linear relation between components of the expansion tensor and the heat capacity. The coefficient of thermal expansion is the trace of the expansion

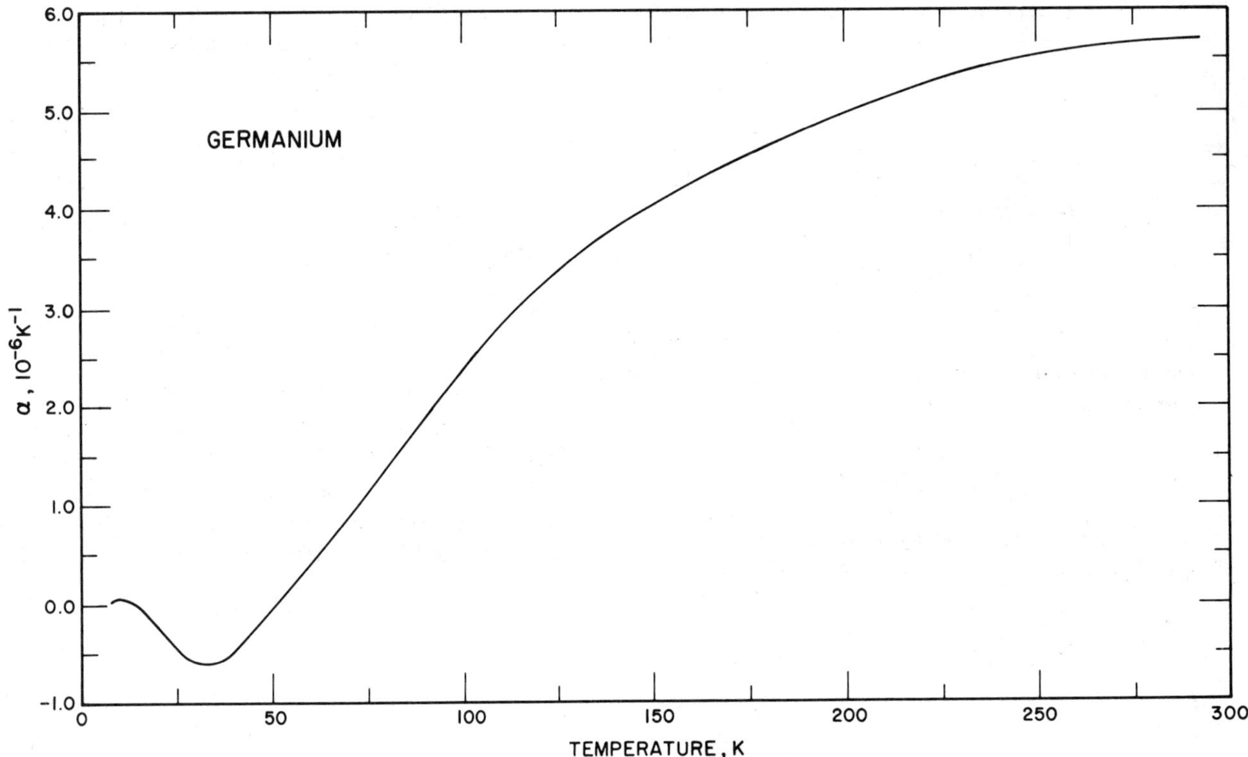

Fig. 2. Coefficient of expansion of germanium showing negative values between 16 and 40 K.

tensor so that

$$\beta = \frac{C_V}{V} \sum_{i=1}^{3} \sum_{j=1}^{6} s_{ij}\gamma_j \tag{47}$$

where $s_{ij}(j = 4, 5, 6) = 0$, as in cubic, hexagonal, orthorhombic, and some tetragonal crystals. A single value of γ can be determined by

$$\gamma = \sum_{i,j=1}^{3} s_{ij}\gamma_j \Big/ \sum_{i,j}^{3} s_{ij} \tag{48}$$

For hexagonal crystals (see Table I),

$$\alpha_1 = \frac{C_V}{V}[(s_{11} + s_{12})\gamma_1 + s_{13}\gamma_3] \tag{49}$$

$$\alpha_3 = \frac{C_V}{V}[2s_{13}\gamma_1 + s_{33}\gamma_3] \tag{50}$$

$$\beta = \frac{C_V}{V}[2(s_{11} + s_{12} + s_{13})\gamma_1 + (2s_{13} + s_{33})\gamma_3] \tag{51}$$

and

$$\gamma = \frac{2(s_{11} + s_{12} + s_{13})\gamma_1 + (2s_{13} + s_{23})\gamma_3}{2s_{11} + s_{33} + 2s_{12} + 4s_{13}} \tag{52}$$

The thermal expansion and elasticity of an ideal hexagonal close-packed structure, $c/a = 1.633$, should be isotropic. In this case $\alpha_1 = \alpha_3$ and $s_{11} + s_{12} = s_{13} + s_{33}$ (see Table II). From Table II and Eq. (46) it is obvious that in anisotropic materials the larger expansion occurs in the softer direction. For example, in the case of zinc the structure is drawn out in the c direction and compresses more easily along this axis than normal to it. The lattice vibrations in this direction, therefore, are generally of lower frequency and are excited first when the temperature is raised above zero Kelvin. This results in a rapid expansion along the c axis accompanied by a contraction normal to it (see Fig. 3).

In 1958 Sheard [9] applied an anisotropic continuum model to calculate the Grüneisen parameters

Table II. Isotropy of Hexagonal Close-Packed Crystals

Crystal	c/a	$\frac{s_{13} + s_{33}}{s_{11} + s_{12}}$	α_3/α_1
Zn	1.856	2.36	4.95
Mg	1.624	1.04	1.07
Be	1.567	0.46	0.72

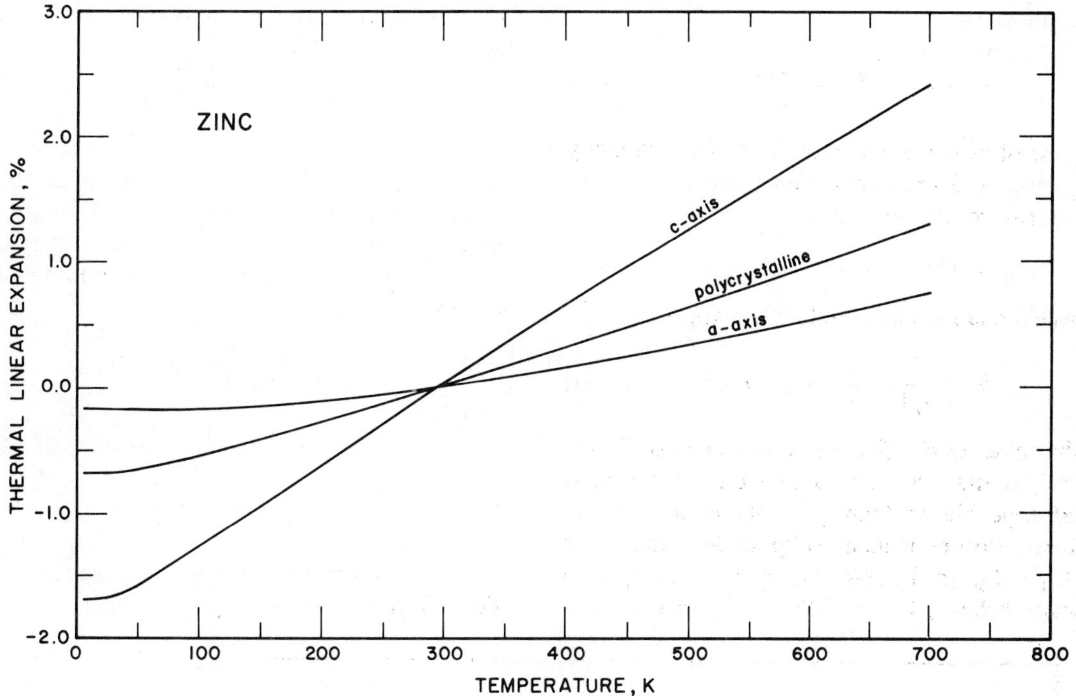

Fig. 3. Thermal linear expansion of zinc showing rapid expansion along c axis and contraction along a axis below 200 K.

at low and high temperatures from the pressure dependence of the elastic constants. Assuming that at long wavelengths for each of the three acoustic branches in cubic crystals

$$\rho S_i^2 = c_i \tag{53}$$

(S_i is the velocity of sound and c_i is the elastic constant) and

$$v_i = kS_i \tag{54}$$

where the application of cyclic boundary conditions requires that

$$kL = 0, \pm 1, \pm 2, \ldots, \pm N \tag{55}$$

(k is the wave number and L is the length of the crystal), it can be shown that

$$\gamma_i = -\frac{V}{v_i}\frac{dv_i}{dV} = \frac{1}{3} + \frac{B_T}{S_i}\frac{dS_i}{dP} = \frac{B_T}{2c_i}\frac{dc_i}{dP} - \frac{1}{6} \tag{56}$$

Thus, for high temperatures

$$\gamma_\infty = \frac{1}{3N}\sum_i^3 \gamma_i = \frac{1}{12\pi}\sum_i^3 \int \gamma_i \, d\Omega \tag{57}$$

and for low termperatures

$$\gamma_0 = \frac{\sum_i^3 \int \gamma_i S_i^{-3} \, d\Omega}{\sum_i^3 \int S_i^{-3} \, d\Omega} = \frac{\sum_i^3 \int \gamma_i c_i^{-3/2} \, d\Omega}{\sum_i^3 \int c_i^{-3/2} \, d\Omega} \tag{58}$$

where Ω is the solid angle within the Debye sphere. These integrations are usually accomplished by numerical techniques on a computer. It can be seen from Eq. (56) that when the pressure derivative of an elastic constant is sufficiently small or negative, the corresponding γ_i will be negative. If this γ_i also corresponds to a dominant mode (such as a shear or transverse mode at low temperatures) it then can cause the weighted average to be small or negative, with a resulting small or negative coefficient of thermal expansion.

In 1960 White [10] reported his measurements on the electronic contribution to the thermal expansion of metals. Because the entropy and the free energy are additive functions of state, Grüneisen's parameter can be expressed as a sum of contributions from the lattice vibrations and the conduction electrons in metals:

$$\gamma(V, T) = \frac{\gamma_{lt}C_{lt} + \gamma_{el}C_{el}}{C_{lt} + C_{el}} \tag{59}$$

In the case of nonmagnetic metals, the heat capacity of the lattice and the conduction electrons at very low temperatures can be approximated by

$$C_{lt} \simeq bT^3 \quad \text{and} \quad C_{el} \simeq aT \tag{60}$$

Grüneisen's relation can therefore be expressed as

$$\beta \simeq \frac{1}{B_T V}(\gamma_{el}aT + \gamma_{lt}bT^3) \tag{61}$$

and if the values of $B_T V\beta/T$ are plotted against T^2, the values of $\gamma_{el}a$ and $\gamma_{lt}b$ can be obtained from the intercept and slope. Since γ_{el} and γ_{lt} are of the same magnitude, these measurements must be made in the range where $C_{el} \sim C_{lt}$. In the case of copper and aluminum this means below 10 K and 20 K, respectively. The electronic Grüneisen parameter for a metal is given by

$$\gamma_{el} = 1 + \left(\frac{\partial \ln N(E_F)}{\partial \ln V}\right)_T \tag{62}$$

where $N(E_F)$ is the density of states at the Fermi surface. For the free electron model, $\gamma_{el} = 2/3$. In magnetic metals the contribution of the magnetic interaction gives rise to an additional thermal expansion term

$$\beta_{mg} = \frac{\gamma_{mg}C_{mg}}{B_T V} \tag{63}$$

where the magnetic Grüneisen parameter is given by

$$\gamma_{mg} = -\left(\frac{\partial \ln E_{mg}}{\partial \ln V}\right)_T \tag{64}$$

E_{mg} is the interaction energy. The magnetic contribution, however, is also proportional to T at low

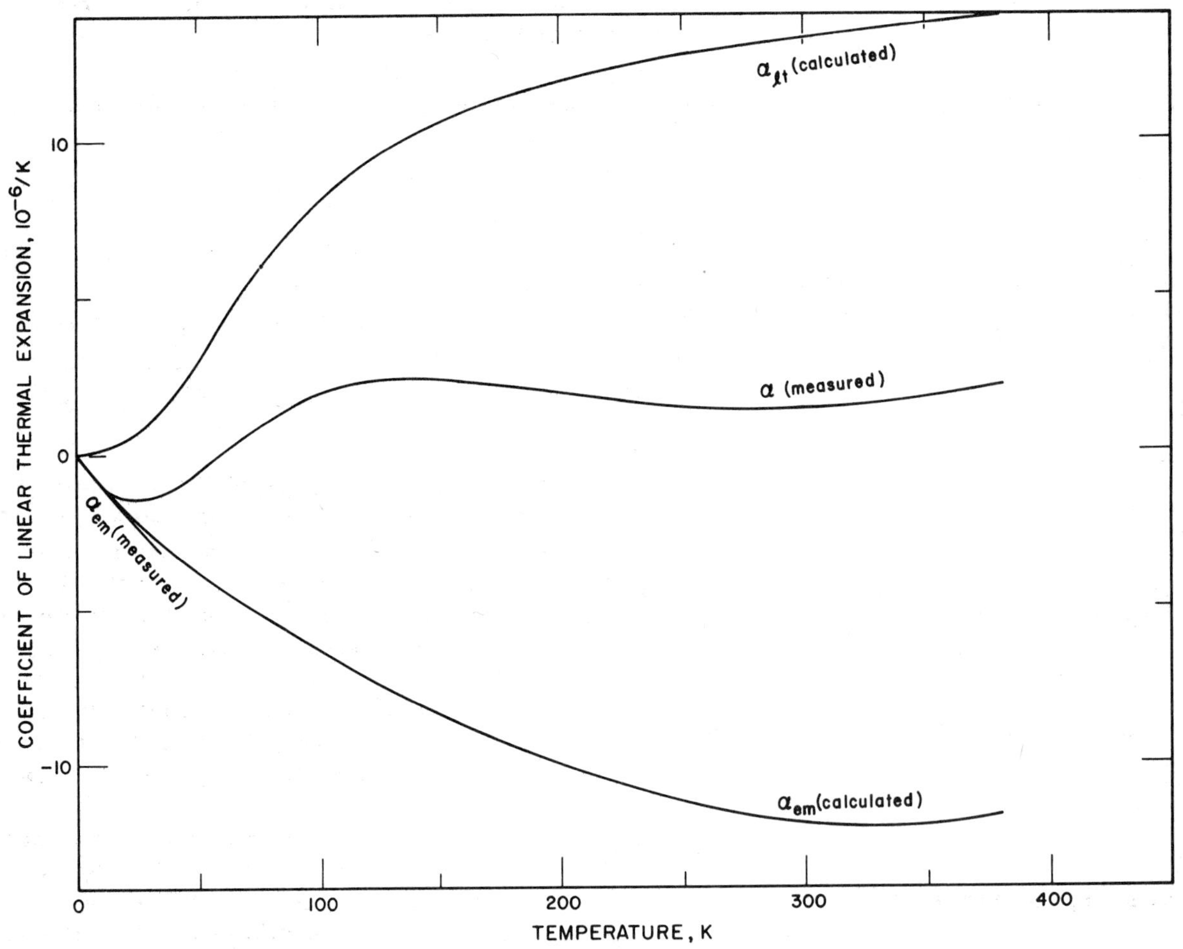

Fig. 4. Coefficients of expansion of Invar showing that the combined electronic-magnetic contribution nearly cancels the lattice contribution causing the expansion to be nearly independent of temperature.

temperatures, so that it is not possible to separate it from the electronic contribution. In this case a combination parameter, γ_{em}, is determined.

The low expansion of Invar, a face-centered cubic iron–36% nickel alloy, can be explained by a large negative value of γ_{em} (~ -20). The result is that the electronic and magnetic contribution to the expansivity, α_{em}, which is proportional to the temperature, predominates at low temperatures and almost equals the lattice contribution, α_{lt}, at room temperature (see Fig. 4).

7. SOME SEMIEMPIRICAL RELATIONSHIPS

In Section 4 it was pointed out that the changes of expansivity and heat capacity with temperature are similar, and because they are both determined as derivatives of experimental data, mathematical techniques that are used in the computation and analysis of heat capacity are useful in the computation and analysis of expansivity.

The relationship of β to the melting temperature, T_M, for crystalline materials has been known since at least 1879 [11]. On the average the total expansion of the close-packed metals on heating from 0 K to T_M is

$$\frac{V_M - V_0}{V_0} \sim 0.0681 \tag{65}$$

That this is proportional to the melting temperature can easily be shown:

$$\frac{V_M - V_0}{V_0} = \int_0^{T_M} \beta \, dT = \beta_M \int_0^{T_M} dT = T_M \beta_M \tag{66}$$

where β_M is the mean coefficient of thermal expansion from 0 K to T_M. The corresponding relationship for linear expansion is

$$\frac{L_M - L_0}{L_0} = T_M \alpha_M \sim 0.0222 \tag{67}$$

A rule of thumb for engineering materials with melting temperatures above 900 K is that these materials will expand about 2% on heating from room temperature to their melting temperature (see Fig. 5).

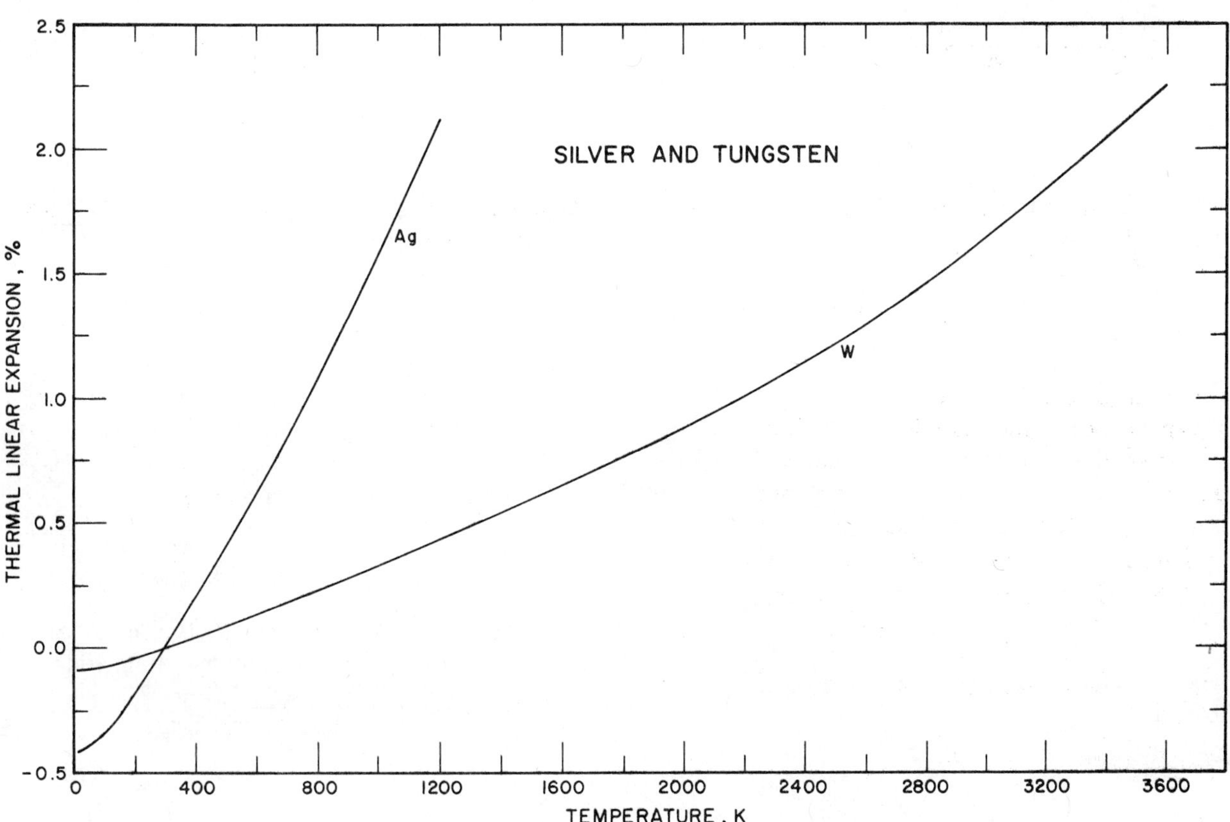

Fig. 5. Thermal linear expansion of silver and tungsten showing that both expand to about 2% at their respective melting temperatures.

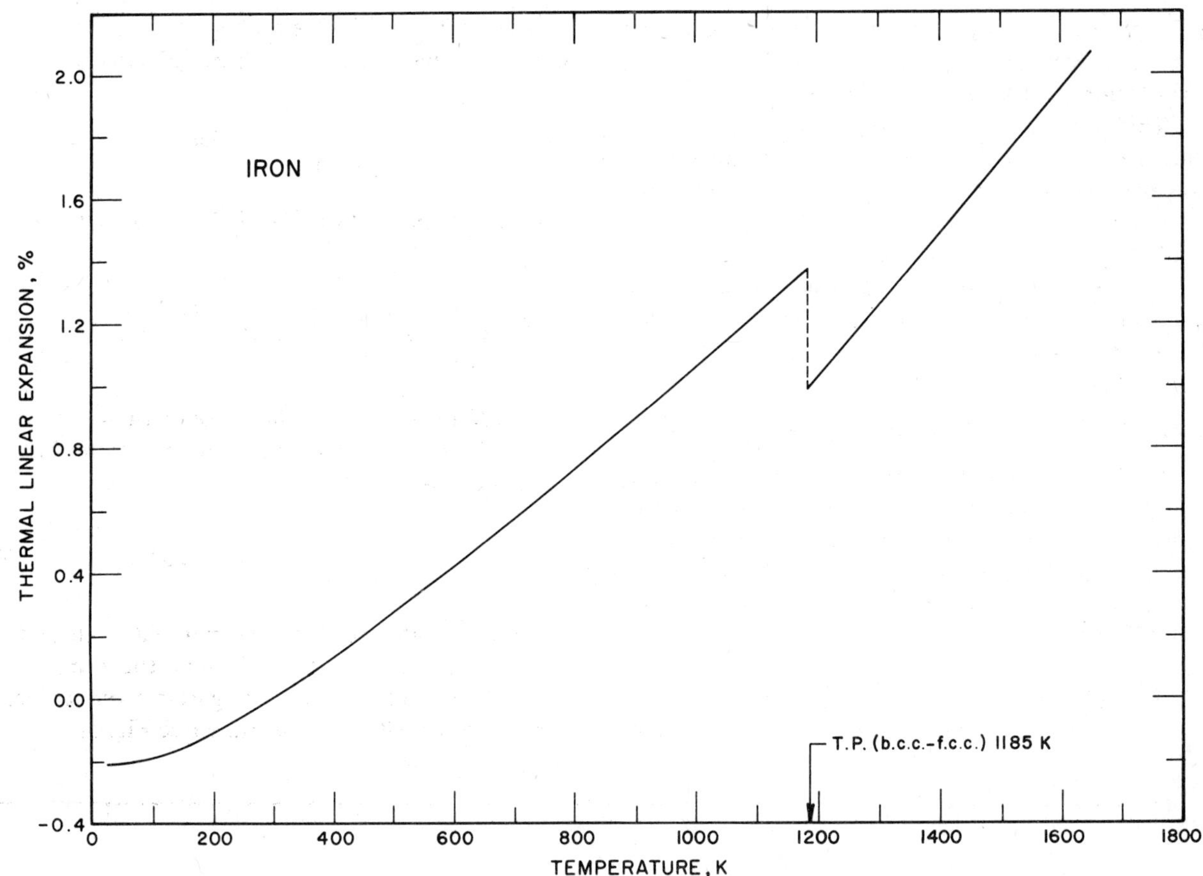

Fig. 6. Thermal linear expansion of iron showing effect of phase transition.

A slightly more useful empirical equation for fcc, bcc, and hcp crystals is

$$T_M \alpha\left(\frac{T_M}{2}\right) \sim 0.0244 \tag{68}$$

where $\alpha(T_M/2)$ is the coefficient of linear thermal expansion at half the melting temperature.

It should be noted that β is related to all parameters that are also related to the amplitude of thermal vibrations. For example, if E_M is the vibrational energy at the melting point, then in the classical limit it is equal to $3RT_M$, and from Eq. (66)

$$E_M \beta_M = 3R(0.0681) \tag{69}$$

where R is the gas constant. Also from Eq. (39) one can infer that

$$\frac{V_0 B_0 \beta_M}{\gamma} = 3R \tag{70}$$

A useful relationship for estimating the contraction from room temperature to zero Kelvin can be based on the Grüneisen equation. From Eq. (39) and its derivative we find that

$$\frac{V_{293} - V_0}{V_0} = \frac{E_{293}}{Q - kE_{293}} \tag{71}$$

and

$$\beta_{293} \simeq \frac{QC_{293}}{(Q - kE_{293})^2} \tag{72}$$

Dividing Eq. (71) by Eq. (72), we get

$$\frac{V_{293} - V_0}{V_0 \beta_{293}} = \frac{E_{293}(Q - kE_{293})}{QC_{293}} \tag{73}$$

Since $kE_{293} \ll Q$ for most engineering materials the contraction from 293 to 0 K can be estimated from

$$\frac{V_{293} - V_0}{V_0} = \frac{E(\theta_D/293)}{C(\theta_D/293)} \beta_{293} \tag{74}$$

or

$$\frac{L_{293} - L_0}{L_0} = \frac{E(\theta_D/293)}{C(\theta_D/293)} \alpha_{293} \tag{75}$$

where $E(\theta_D/293)$ and $C(\theta_D/293)$ are the values for energy and heat capacity calculated from the Debye equations [5]. Since the characteristic temperatures, θ_D, of these materials usually fall between 400 and 500 K, a good approximation [12] for the linear contraction is

$$\frac{L_{293} - L_0}{L_0} = 180\alpha_{293} \tag{76}$$

8. PHASE TRANSFORMATIONS, IMPERFECTIONS, AND POINT DEFECTS

First-order transformations in single-phase systems are characterized by a discontinuous change in volume at constant temperature and pressure. This type of transformation, in which a discontinuous reconstruction of the lattice occurs, includes the change from one crystal structure to another as in iron at 1184 K, where the change is between the body-centered cubic and the face-centered cubic forms (see Fig. 6). The change in volume may be large or very small and positive or negative. In multiphase systems such as alloys, the transition may extend over an appreciable temperature range and occur at generally higher temperatures on heating than on cooling, with hysteresis as the result.

Second-order transformations are generally characterized by the disappearance of ordered structures, with the result that a discontinuous change occurs in the coefficient of thermal expansion at constant temperature and pressure (see Fig. 7). The change in the coefficient may be large or very small and positive or negative. The ordered structures may consist of chemical compounds in alloy systems, magnetic moments as in ferromagnetic materials, and electron energy states in superconductors.

If individual crystals in a polycrystalline material are highly anisotropic, such as uranium, the material may increase in size as the result of a thermal cycle

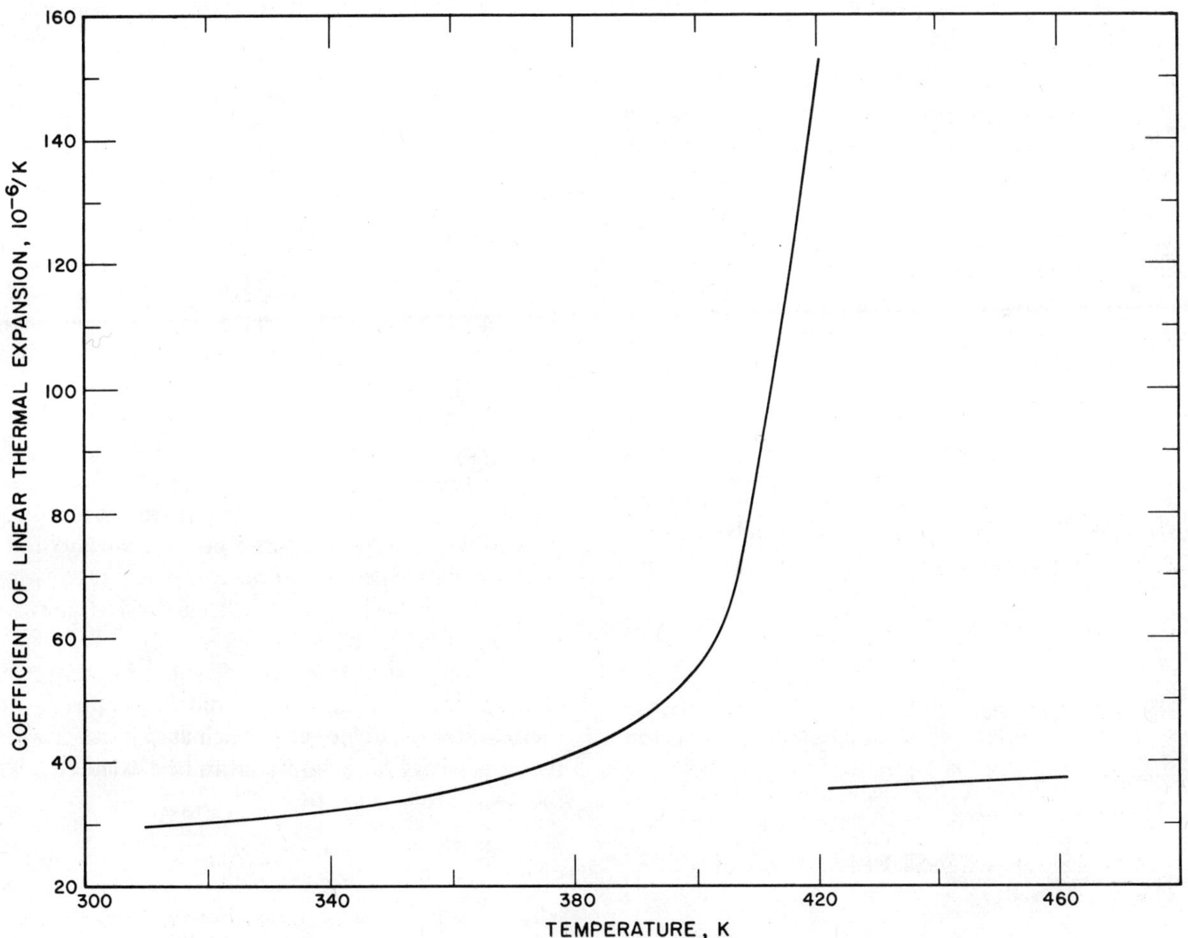

Fig. 7. Coefficient of expansion of Mg_3Cd showing effects of order–disorder transformation.

over several hundred degrees. The most probable cause of this growth is the large thermal stress between neighboring crystals that do not have their crystal axes aligned with each other [13]. Because of the differences in the axial expansivity of the crystals the stress is built up at the low end of the cycle on cooling, which results in plastic deformation, and near the upper end on heating, but in the opposite direction. If the temperature is high enough at the upper end of the cycle, the stress will be relieved. As a result, voids are created and the material grows even after more than a thousand cycles.

In general the presence of voids in a solid will not significantly affect the expansion since the hole will expand an equal amount. This will not be true, however, if the number of voids changes with temperature. For example, when a material is heated to temperatures close to its melting point a significant number of thermal vacancies, n, are generated,

$$n = N\, e^{S/k}\, e^{-H/kT} \tag{77}$$

where S is the entropy of formation and H is the enthalpy of formation per vacancy. If the vacancy is formed by an atom moving from a lattice site in the interior to a lattice site on the surface (a Schottky defect) the density of the crystal will be decreased. Since the dimensions of the crystal lattice at low vacancy concentrations will not be significantly changed, the increase in the vacancy concentration on heating from some sufficiently low temperature can be determined as

$$\frac{n}{N} = 3\left(\frac{\Delta L}{L} - \frac{\Delta a}{a}\right) \tag{78}$$

where a is the lattice parameter as measured with x rays [14]. It follows that

$$\alpha_b - \alpha_a = \frac{nH}{3NkT^2} \tag{79}$$

where α_b is the expansivity of the bulk and α_a is the expansivity of the lattice. In the case of copper at the melting point $\alpha_b - \alpha_a = 1.5 \times 10^{-6}$/K.

9. EFFECTS OF PRESSURE AND STRESS ON EXPANSION

From the definitions of the coefficient of thermal expansion and the bulk modulus it can be shown that the effect of applied pressure on the expansivity of a solid is given by

$$\left(\frac{\partial \beta}{\partial P}\right)_T = \frac{1}{B_T{}^2}\left(\frac{\partial B_T}{\partial T}\right)_P \tag{80}$$

Since $(\partial B_T/\partial T)_P$ is negative, the expansivity will decrease with increasing pressure. The following equation, which can be derived [15] from the Murnaghan equation of state, is very useful for estimating the coefficient of thermal expansion at high pressures:

$$\beta(P, T) = \frac{B_T(0, T)\beta(0, T)}{P(\partial B_T(0, T)/\partial P)_T + B_T(0, T)}. \tag{81}$$

The effect of uniaxial stress, σ, on the linear expansivity of a solid [16] is given by

$$\left(\frac{\partial \alpha}{\partial \sigma}\right)_T = -\frac{1}{E^2}\left(\frac{\partial E}{\partial T}\right)_\sigma \tag{82}$$

where E is Young's modulus (the tensile stress is taken as negative). In terms of the bulk modulus and Poisson's ratio, μ, the effect of stress is given by

$$\left(\frac{\partial \alpha}{\partial \sigma}\right)_T = \frac{1}{3B_T(1 - 2\mu)}\left[\frac{1}{B_T}\left(\frac{\partial B_T}{\partial T}\right)_P - \frac{2}{(1 - 2\mu)}\left(\frac{\partial \mu}{\partial T}\right)_\sigma\right] \tag{83}$$

10. THERMAL EXPANSION OF COMPOSITES AND MIXTURES

When thermal stresses arise in solids because of differences between the expansion of neighboring grains or substructures the expansion of the composite will be a function of the expansivity and the elastic properties of the individual grains or substructures. In general, the thermal behavior of a composite is a complex function of the thermal and elastic properties of the components but can be predicted by the use of a discrete element method of analysis with the aid of a computer [17]. Many simplified relationships have been derived, however, and some of these are cited here.

In composites where the elastic properties of the components are nearly equal and in mixtures where thermal stresses do not arise, such as in loose powders, the expansivity of the material can be calculated with the simple rule of mixtures,

$$\alpha_C = \sum_i \alpha_i v_i \tag{84}$$

where v_i is the volume fraction of the ith component.

An initially straight bimetallic strip will bend and form an arc of a circle when subjected to a temperature change ΔT [18]. The radius of the circle is given by

$$R = \frac{3(D_1 + D_2)^2 D_1 D_2 E_1 E_2 + (D_1E_1 + D_2E_2)(D_1^3E_1 + D_2^3E_2)}{6\,\Delta\alpha\,\Delta T(D_1 + D_2)D_1D_2E_1E_2} \tag{85}$$

where D_1 and D_2 are the thicknesses of the metal strips. The expansivity of a symmetrical laminate (sandwich) composite in the direction parallel to the layers [19] is

$$\alpha_C = \sum_i \frac{\alpha_i E_i D_i}{1 - \mu_i} \Big/ \sum_i \frac{E_i D_i}{1 - \mu_i} \tag{86}$$

The expansivity of a composite in which the thermal shear stresses are low, such as some alloys, is given by Turner's formula [20]

$$\alpha_C = \sum_i \alpha_i B_i v_i \Big/ \sum_i B_i v_i . \tag{87}$$

According to Kerner [21], a composite made with packed grains, none of which are continuous, will expand according to

$$\alpha_C = \left(\frac{4G_C}{B_C} + 3\right) \sum_i \frac{\alpha_i v_i}{(4G_C/B_i) + 3} \tag{88}$$

(where G_C is the shear modulus of the composite), while a composite composed of a matrix of one phase in which another phase is dispersed will expand according to

$$\alpha_C = \alpha_{mt} + \frac{(\alpha_d - \alpha_{mt})v_d}{\dfrac{12B_{mt}G_{mt}}{3B_{mt} + 4G_{mt}}\left[\dfrac{v_d}{3B_{mt}} + \dfrac{1}{4G_{mt}} + \dfrac{1 - v_d}{3B_d}\right]} \tag{89}$$

A polycrystalline specimen of an element or a compound that is anisotropic can also be considered as a composite material. In this case

$$\beta_C \neq \alpha_1 + \alpha_2 + \alpha_3 \tag{90}$$

However, for crystals with hexagonal symmetry where Young's modulus E depends on the angle ω_3 with respect to the principal crystallographic axes [2], the expansivity can be approximated by

$$\alpha_C = \frac{\int \alpha E \sin\omega_3\, d\omega_3}{\int E \sin\omega_3\, d\omega_3} \tag{91}$$

All of the approximations for the expansivity of composites depend upon the following assumptions: (1) no voids; (2) no cracking or sintering during a temperature cycle; (3) the thermal strain does not exceed the elastic limit; and (4) no chemical reaction between phases.

Methods for the Measurement of the Linear Thermal Expansion of Solids

1. INTRODUCTION

Many different methods and variations of methods for measuring the thermal expansion of solids have been developed, especially during the past fifty years. A variety of methods are useful for many different reasons. For instance: (1) high accuracy and sensitivity are necessary for measurements at temperatures approaching absolute zero where the coefficient of thermal expansion is very small; (2) at moderate temperatures an industrial application may call for automation and fast response; or (3) a knowledge of the lattice spacings is needed to understand the behavior of an anisotropic material. The choice of method may also depend upon the material to be measured, the amount of material available, the temperature range of the measurement, and the type of information required.

Broadly speaking, all methods for measuring thermal expansion may be divided into two classes: (1) relative methods in which the expansion of the material being investigated is measured relative to the expansion of another material; and (2) absolute methods in which the expansion of the material being investigated is measured directly.

Among early measurements of expansion were those of Musschenbrock in the 18th century. In his apparatus the expansion of a horizontally mounted specimen (about 15 cm in length) was magnified by a simple gear train. Heat was provided by 1, 2, 3, 4, or 5 candles placed beneath the specimen. In 1805, Lavoisier and Laplace [22] used an optical lever for magnification in an improved version of Musschenbrock's apparatus. Ramsden [22] in 1785 used a variation of the twin-telemicroscope (comparator) technique to measure thermal expansion. This technique was further developed by Callendar [23] in 1887 and by Holborn and Day [24] in 1900. Although Fizeau [25] measured the absolute expansion of a reference material that separated the optical flats in his interferometer (a very difficult measurement for him to make), he routinely measured the expansion of specimens relative to that material. Reimerdes [26] in 1896 followed by Scheel [27] in 1902 was among the first to use interferometry to measure thermal expansion in the true absolute sense.

In 1927, Becker [28] successfully used x rays to measure thermal expansion. This was particularly important because it became possible to directly measure the expansion of anisotropic crystals and to study high-temperature phases. In 1960, White [29] used a highly sensitive capacitance technique at low temperatures to measure the contribution of free electrons to the thermal expansion of a metal. Developments are continuing on methods of measuring thermal expansion with the application of modern technology. Examples are the use of stabilized lasers [30] to obtain sensitivities of 10^{-13} in expansion measurements and the use of computer-controlled data acquisition systems to measure a specimen during a rapid temperature change [31].

In the sections that follow, the major methods are briefly described. For finer details of experimental design and technique, the reader is referred to the references given for the individual methods.

2. PUSH-ROD DILATOMETERS

The push-rod dilatometer method for measuring thermal expansion is experimentally simple (see Fig. 8), reliable, and easy to automate [32]. With this method, the expansion of the specimen is transferred out of the heated zone to an extensometer by means of rods (or tubes) of some stable material. The expansion of the specimen is given by

$$\frac{\Delta L}{L_{293}} = c_0 \frac{(\Delta L)_a}{L_{293}} + c_1 \qquad (92)$$

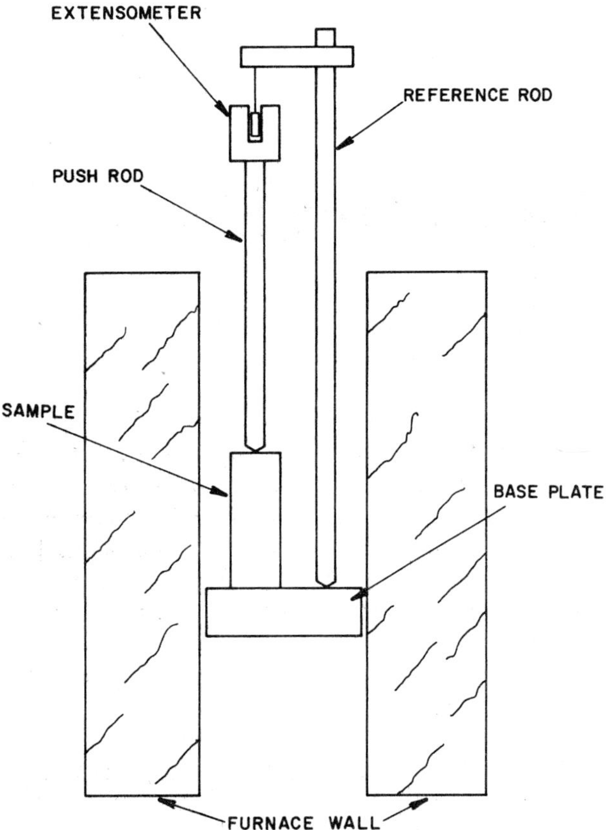

Fig. 8. Schematic diagram for push-rod dilatometer.

where $(\Delta L)_a$ is the apparent change in length as calculated from the difference between the extensometer readings at two different temperatures and c_0 and c_1 are calibration constants for the system. If the reference rod is made the same length as the push rod and a second specimen placed on the base plate, the dilatometer will measure the difference between the specimens [33]. The difference, or differential expansion, is given by

$$\frac{(\Delta L)_2}{L_{293}} - \frac{(\Delta L)_1}{L_{293}} = c_0 \frac{(\Delta L)_a}{L_{293}} + c_1 \tag{93}$$

When used this way the dilatometer can have a very high sensitivity. This technique is also very useful for quality control measurements and for studying phase transitions.

In the Baudran [34] or scissors dilatometer, the expansion of a specimen is sensed by rods that penetrate the hot zone in a direction perpendicular to the specimen axis instead of in line with it (see Fig. 9). Because the expansion of the rods does not add to the expansion of the specimen, the measurement is absolute.

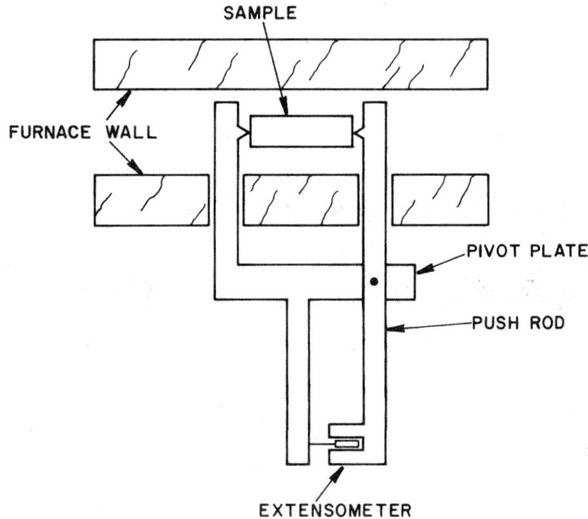

Fig. 9. Schematic diagram for Baudran dilatometer.

One of the most common sources of error in using dilatometers is the measurement of temperature. All too often the temperature that is measured is not the temperature of the specmen. This is especially true in measurements made while the temperature is changing. If a thermocouple is used, care must be taken to ensure that its junction and the specimen are at the same temperature; they can be at different temperatures even if in contact with each other.

Not only should the extensometer and thermometer be very carefully calibrated but care should be taken to avoid any drift in the calibration. If the calibrations are accurate, the value of c_0 in Eq. (92) is equal to 1. If the push rod and the reference rod in the usual type of dilatometer expand exactly the same, c_1 is the expansion of that part of the reference rod that is equal in length to the specimen. In the differential or Baudran dilatometer c_1 should be equal to zero. In normal usage, however, c_0 and c_1 will be found to vary from their ideal values. A single reference material of known expansion can be used to determine the value of c_1, or two or more reference materials can be used to determine the values of both c_0 and c_1 [35]. It can be shown that the latter method is better except when the expansion of a reference material is within two percent of the test material.

When used at temperatures below 1200 K dilatometers are usually made of vitreous silica [36]. For use at higher temperatures they are made of alumina [37], silicon carbide [38], tantalum [39], and other refractory materials. Linear variable differential transformers [32] are most frequently used for measuring expansion, but many other types of extensometers

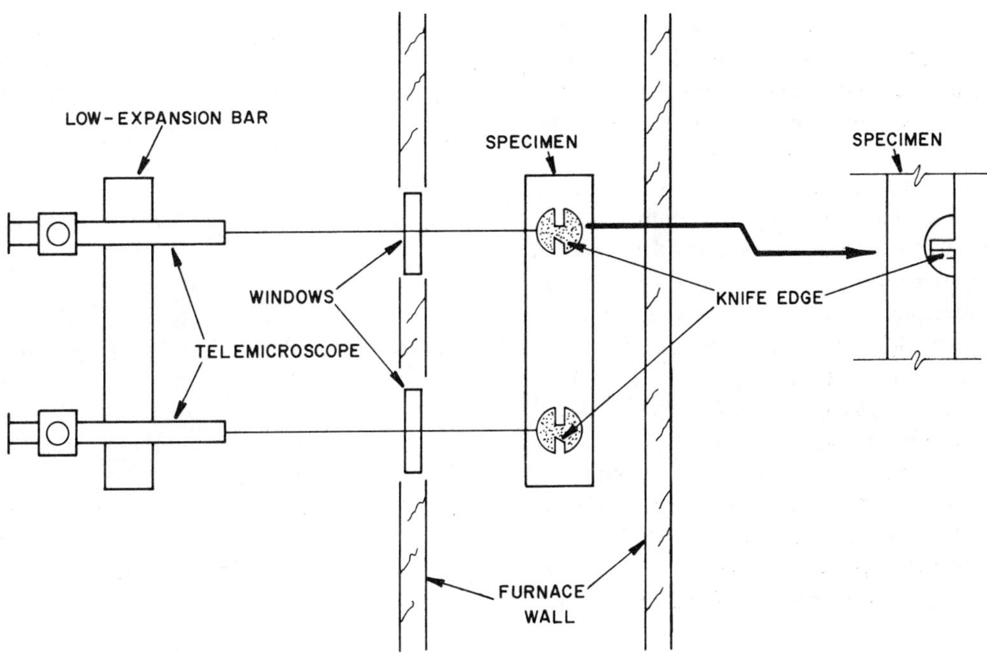

Fig. 10. Schematic diagram for telemicroscope method.

have been used. These include dial gauges [36], optical levers [40], microscopes [41], optical gratings [42], interferometers [43], and micrometers [44].

3. TWIN-TELEMICROSCOPE METHOD

The twin-telemicroscope method is most useful for measuring the absolute expansion of large specimens at high temperatures. The best results with this method are obtained when the two microscopes are rigidly mounted to a bar of low-expansion material and the length change measured with filar micrometer eyepieces [45]. Telemicroscopes (microscopes with a relay lens) are needed because of the large working distance imposed by the furnace (over 15 cm) and the necessity for a magnification of at least $50\times$ to obtain measurements with a sensitivity of 10^{-6} m. In the temperature range below 1200 K fiducial marks may be cut into or indented into flat surfaces machined on the specimen [46]. Above 1200 K the gauge length is best defined by knife edges machined into the specimen in such a way that they will form sharp images [47]. Both knife edges should face the same direction (see Fig. 10), so that vaporization or changes in lighting conditions will not affect the measurements. Small, clean holes drilled through a thin section of the specimen are also satisfactory [48].

The furnace windows may cause significant errors in length measurements if both surfaces are not flat. An argon atmosphere is sometimes necessary to keep the windows clean, especially if an optical pyrometer is used to measure temperature. When an atmosphere is used, however, care must be taken to avoid the bending of the sight paths when they pass through nonperpendicular temperature gradients [48]. Each microscope should be separately calibrated and positioned on the supporting bar so that its plane of focus coincides with that of the other. Because of the large sample size, an effort should be made to provide either a sufficiently large constant temperature zone or a zone with a known temperature gradient [49].

4. INTERFEROMETERS

These methods are based on the interference of monochromatic light reflected from two surfaces [50] that are separated by a specimen or by the combination of a specimen and a reference material. The general condition for interference is

$$2nL\cos\theta = N\lambda_v \qquad (94)$$

where n is the index of refraction of the atmosphere between the surfaces, L the distance between the two surfaces, θ the angle between the direction of the incident rays and the normal to the surfaces, N the order of interference, and λ_v the wavelength of the light in vacuum. Monochromatic light sources that may be used include cadmium, helium, mercury, and sodium low-pressure discharge lamps [51] and a stabilized He–Ne laser [52].

If slightly inclined surfaces are illuminated with collimated light and viewed at normal incidence ($\theta = 0$ for all rays), then fringes of equal thickness are observed. When the surfaces are flat the fringes will be straight; otherwise they are determined by the contour of the surfaces. This type of interference is used in the Fizeau interferometer [53]. If plane-parallel surfaces are illuminated with an extended source (θ will vary), then fringes of equal inclination are observed. This type of interference (concentric rings) is used in the Fabry–Perot interferometer [54].

When an interferometer is used to measure thermal expansion, the expansion of the specimen is given by

$$\frac{\Delta L}{L} = \frac{\lambda_v \Delta N}{2nL\cos\theta} - \frac{\Delta n}{n} \tag{95}$$

where ΔN is the number of fringes that pass a fiducial mark and Δn is the change of refractive index. A useful approximation for the refractive index is

$$n = 1 + (n_r - 1)\frac{T_r P}{P_r T} \tag{96}$$

where n_r is the index at the reference temperature T_r and the reference pressure P_r. In vacuum or in a sufficiently low-pressure atmosphere

$$\frac{\Delta L}{L} = \frac{\lambda_v \Delta N}{2L\cos\theta} \tag{97}$$

In a Fizeau interferometer (straight fringes) the fractional part of ΔN is easily determined from measurements of the position of the fiducial mark between two fringes. In a Fabry–Perot interferometer (circular fringes) the fractional part, v, is given by

$$v = p - 1 + \frac{Ld_p^2}{4f^2\lambda_v} \tag{98}$$

where d_p is the diameter of the pth fringe (counted from the center of the concentric pattern) and f is the focal length of the lens that forms the pattern. If measurements are made on the second and fourth fringes, the fractional part is

$$v = \frac{3d_2^2 - d_4^2}{d_4^2 - d_2^2} \tag{99}$$

The number of fringes that move past a reference point during the expansion of a specimen can be counted by eye or automatically by photographic [55] or photoelectric [56] techniques. Another way of determining ΔN is by finding the value of N at each temperature. This can be done by using at least three different wavelengths [51]. From Eq. (94) it can be seen that

$$(N_1 + v_1)\lambda_1 = (N_2 + v_2)\lambda_2 = (N_3 + v_3)\lambda_3 \tag{100}$$

where N_1, N_2, and N_3 are the fringe integers and v_1, v_2, and v_3 are the fringe fractions for each of the three wavelengths. In the method of exact fractions a value is guessed for N_1 ($\sim 2nL/\lambda_1$) and using the measured value of v_1 the values of $(N_2 + v_2)$ and $(N_3 + v_3)$ are calculated. This procedure is repeated with different values of N_1 until the calculated values of v_2 and v_3 agree with their measured values. If N is known, then

$$\frac{\Delta L}{L} = \frac{\Delta N}{N} - \frac{\Delta n}{n} \tag{101}$$

The Fizeau interferometer can be used to measure either the absolute or relative expansion of a specimen. In the relative method a pedestal of one material fills most of the space within a ring or cylinder of a second material (see Fig. 11). The pedestal is preferably made of a reference material. While the ring supports the optical flat, the interference is formed by reflections from the optical flat and the top surface of the pedestal. Since the reflecting surfaces are close together this method has three advantages: (1) the fringes are bright and well defined; (2) the change of the refractive index of the gas within the small space does not affect the measurements; and

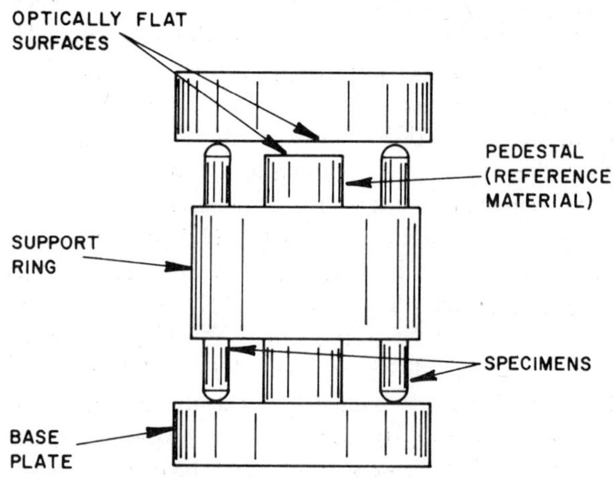

Fig. 11. Schematic diagram for Fizeau interferometer, relative method.

(3) longer specimens can be used with a corresponding increase in sensitivity. The thermal expansion is given by

$$\left(\frac{\Delta L}{L}\right)_S = \frac{\lambda_v \Delta N}{2nL_S} + \frac{L_R}{L_S}\left(\frac{\Delta L}{L}\right)_R + \frac{L_S - L_R}{L_S}\frac{\Delta n}{n} \quad (102)$$

where the subscripts refer to the specimen and reference materials. When $L_R \sim L_S$ the last term can be ignored.

If one or all three of the separators in a Fizeau interferometer expand differently, that difference will cause a rotation and/or change in the spacing of the fringes. In the first case where the two similar separators made from a reference material are exactly the same length the difference in expansion of the other material is given by

$$\left(\frac{\Delta L}{L}\right)_S - \left(\frac{\Delta L}{L}\right)_R = \frac{W\lambda_v}{2L_S}\left(\frac{1}{n_2 s_2} - \frac{1}{n_1 s_1}\right) \quad (103)$$

where W is the perpendicular distance between the specimen and the line that joins the separators that are made from the reference material and s is the distance between fringes.

When the optical flats of a Fabry–Perot interferometer are made highly reflecting the multiply reflected beams cause a great increase in sharpness of the fringes. This sharpness results in a higher sensitivity in the measurement of fringe fractions and hence in the expansion measurements. The sensitivity of a polarizing interferometer [57] is also higher because the measurement of the polarization angle is a more sensitive way of determining a fringe fraction. The laser is an extremely efficient source of radiation for this type of interferometer because it can be constructed to emit a polarized beam.

5. X-RAY METHODS

These methods are based on the diffraction of a collimated beam of monochromatic x rays that is scattered by atoms in a crystal lattice. The Bragg law

$$\lambda = 2d(hkl)\sin\theta \quad (104)$$

gives the condition for constructive reflection of the incident radiation. Here d is the separation of the lattice planes, h, k, and l are the Miller indices for the planes, and θ is the angle measured between the direction of the incident or reflected beam and the planes. Except for a small correction due to refraction, the measurement of expansion is independent of wavelength:

$$\frac{\Delta d(hkl)}{d(hkl)} = -\cot\theta\,\Delta\theta = \frac{\sin\theta_1 - \sin\theta_2}{\sin\theta_2} \quad (105)$$

The relationship between the separation of lattice planes and the symmetry of the crystal lattice is given by

$$d^2 = [1 - \cos^2\alpha - \cos^2\beta - \cos^2\gamma + 2\cos\alpha\cos\beta\cos\gamma] \Big/ \Big[\left(\frac{h}{a}\right)^2 \sin^2\alpha + \left(\frac{k}{b}\right)^2 \sin^2\beta + \left(\frac{l}{c}\right)^2 \sin^2\gamma + \frac{2hk}{ab} \times (\cos\alpha\cos\beta - \cos\gamma) + \frac{2hl}{ac}(\cos\alpha\cos\gamma - \cos\beta) + \frac{2kl}{bc}(\cos\beta\cos\gamma - \cos\alpha)\Big] \quad (106)$$

Using this equation it can be shown that the expansion of cubic crystals (where $a = b = c$ and $\alpha = \beta = \gamma = 90°$) can be obtained from any set of lattice planes (hkl):

$$\frac{\Delta a}{a} = \frac{\Delta d(hkl)}{d(hkl)} = -\cot\theta\,\Delta\theta \quad (107)$$

For monoclinic crystals, where the symmetry restrictions are $\alpha = \gamma = 90°$, only the expansion in the $[0, 1, 0]$ direction can be obtained directly:

$$\frac{\Delta b}{b} = \frac{\Delta d(0k0)}{d(0k0)} \quad (108)$$

The expansion of the interdependent quantities a, c, and β of a monoclinic crystal can be obtained from the variation of the spacings of three noncoplanar planes of the $(h0l)$ type.

The thermal expansion of crystalline materials can be accurately measured with x-ray cameras and diffractometers under conditions that preclude the use of any other method, as when the specimens are very small, weak, and/or irregular in shape. These methods are also unique in that they can easily be used to determine the principal coefficients of thermal expansion of anisotropic crystals and permit direct observation of phase changes. There is a further advantage in that measurements with x rays do not include effects that are observed in measurements on bulk specimens.

When using the camera method [58] the specimen must be either a fine-grained polycrystalline wire, a

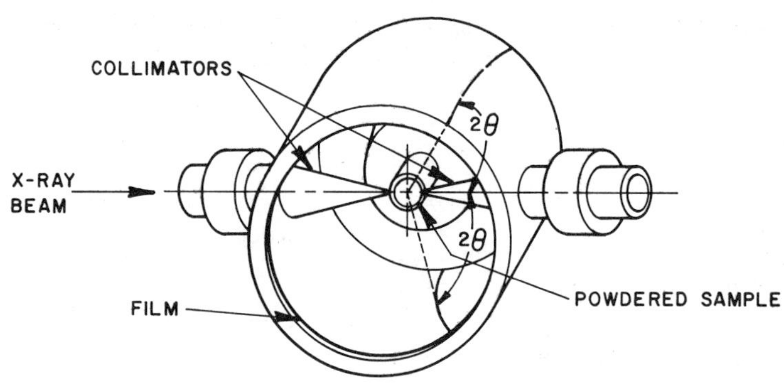

Fig. 12A. Schematic diagram for Debye–Sherrer x-ray camera method.

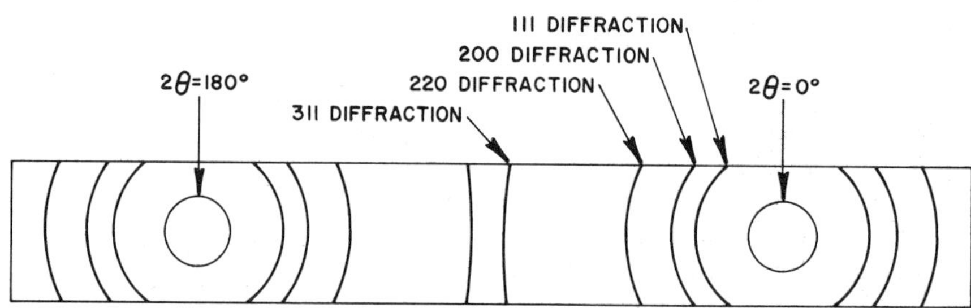

Fig. 12B. Diffraction pattern of face-centered cubic crystal.

fine powder, or a single crystal that is rotated during the exposure of the film. In powder form the specimen can be held in a thin-walled glass tube. The basic geometry of the Debye–Sherrer method is illustrated in Fig. 12A. The specimen is located at the center of the cylindrical camera and the film is placed on the inside wall. Filtered x rays enter through a collimator and either are diffracted by the crystal planes according to Eq. (104) or are scattered or pass directly through the exit. The diffracted rays are recorded on the film as sharp lines. Figure 12B shows the pattern obtained for a fcc crystal. Not all the angles possible according to Eq. (104) are diffracted, however, because of the structure factor [5]. For example, in the case of a fcc crystal reflections can occur only from those planes for which the Miller indices (*hkl*) are all even or all odd. The values of θ for the diffracted rays are determined from the position of the lines on the film. For increased accuracy the value of d is obtained using large values of θ. The accuracy also depends upon the centering of the specimen, the axial divergence of the collimated beam, corrections for absorption and refraction within the specimen, knowledge of the camera diameter, and the shrinkage of the film during development.

X-ray diffractometers, in which a diffracted beam is detected with an ionization counter and the angle determined from a wide-angle goniometer, are widely used for expansion measurements [59]. In x-ray diffractometers it must be possible to determine and to adjust the position of the specimen on the focusing circle at any temperature.

The problems associated with heating the specimen and accurately measuring its temperature in both x-ray cameras and diffractometers are discussed in several articles [58, 59, 60]. Large temperature gradients across the specimen are often encountered unless special arrangements are made [61]. At 2000 K the temperature gradient may typically be 20 K/cm and the temperature measurements accurate to only 50 K. Even larger values may occur when powder specimens are used. It is possible to use a material of known expansion to determine the temperature [58]. In many cases this reference material can be mixed with the specimen and heated with it.

Unique x-ray techniques exist for special situations. In particular, the use of the Bond technique for single crystals [62] has resulted in a sensitivity of 10^{-7} in measurements of $\Delta a/a$ [63]. In this technique the specimen is rotated between equivalent diffracting

Table III. Various Contributions to Coefficients of Linear Thermal Expansion at Low Temperatures

	Pb ($\theta_D \sim 100$ K)		Cu ($\theta_D \sim 340$ K)		Fe ($\theta_D \sim 470$ K)		Si ($\theta_D \sim 640$ K)	KCl ($\theta_D \sim 240$ K)	Al_2O_3 ($\theta_D \sim 1030$ K)
T, K	Lattice	Electronic	Lattice	Electronic	Lattice	Electro-magnetic	Lattice	Lattice	Lattice
0	0	0	0	0	0	0	0	0	0
4	4×10^{-8}	2×10^{-8}	1×10^{-9}	2×10^{-9}	8×10^{-9}	24×10^{-9}	3×10^{-11}	1×10^{-9}	2×10^{-11}
8	192×10^{-8}	6×10^{-8}	25×10^{-9}	7×10^{-9}	2×10^{-8}	10×10^{-8}	8×10^{-10}	4×10^{-8}	5×10^{-10}
12	134×10^{-7}	1×10^{-7}	13×10^{-8}	2×10^{-8}	6×10^{-8}	22×10^{-8}	4×10^{-9}	3×10^{-7}	3×10^{-9}
16	361×10^{-7}	3×10^{-7}	45×10^{-8}	3×10^{-8}	1×10^{-7}	4×10^{-7}	9×10^{-9}	1×10^{-6}	9×10^{-9}
20	740×10^{-7}	4×10^{-7}	119×10^{-8}	4×10^{-8}	3×10^{-7}	6×10^{-7}	6×10^{-9}	3×10^{-6}	2×10^{-8}
24					6×10^{-7}	9×10^{-7}	-3×10^{-8}		5×10^{-8}
28					1×10^{-6}	1×10^{-6}	-1×10^{-7}		9×10^{-8}
32					2×10^{-6}	1×10^{-6}	-3×10^{-7}		1×10^{-7}
36							-7×10^{-7}		2×10^{-7}
40							-1×10^{-6}		4×10^{-7}

orientations on either side of the incident beam. The value of θ thus obtained is unaffected by any specimen eccentricity, absorption, and zero errors, and errors due to specimen tilt and beam axial divergence are minimized.

Neutron diffraction has also been used to measure thermal expansion [64]. The use of neutrons instead of x rays offers several advantages because the scattering depends upon the nuclei of the atoms instead of the electron distribution. These include the measurement of materials that are composed of light elements such as hydrogen and carbon and of some neighboring elements such as iron and cobalt. Another advantage is that neutrons are diffracted from many deep crystal planes ($V \sim 1$ cm^3) and not from just a few near the surface. This tends to make both furnace construction and temperature measurements easier. Sources of thermal neutrons (atomic reactors) are, however, somewhat more difficult to obtain than sources of x rays.

6. HIGH-SENSITIVITY METHODS

The precise measurement of very small coefficients of linear thermal expansion requires that the sensitivity of the measurements be of the order of 10^{-10}. At low temperatures (<30 K) expansivities are very small (see Table III), but the determination of their temperature dependence contributes a great deal to the understanding of lattice dynamics. On the other hand, the behavior of low-expansion materials at ambient temperatures is of technological importance. Five methods that have been successfully used for these measurements are briefly described.

The three-terminal parallel plate capacitor technique that was developed by White [29] has been used extensively [65, 66, 67] in the range of 0 to 310 K. In this technique the capacitance of the plates is measured with a capacitance bridge at about 1000 Hz. The value of the capacitance, C, is about 10 pF and it can be measured with a sensitivity of 10^{-7} pF. As routinely used, this technique measures the relative expansion of the test material,

$$\left(\frac{\Delta L}{L}\right)_S = \left(\frac{\Delta L}{L}\right)_R + \frac{L_R - L_S}{L_S}\left(\frac{\Delta C}{C}\right) \tag{109}$$

Since $(L_R - L_S)/L_S$ is of the order of 3×10^{-3}, the relative expansion can be measured to 10^{-10}. As can be seen in Fig. 13, the capacitance is measured between electrodes (1) and (2). Electrodes (2) and (3) are both made of the reference material, which is usually copper. Electrode (3) completely surrounds the other two conductors and forms an earth shield so that the capacitance between (1) and (2) does not involve the lead wires.

Kos and Lamarche [68] have used a capacitance technique to measure the expansion of copper, silver, and gold from 4 to 15 K. In their technique a three-terminal capacitor is connected to a high-Q coil which is immersed in liquid helium to form an LC circuit. The resonant frequency, F, of this circuit is mixed with a standard frequency and the beat frequency is determined with a counter. The expansion of the specimen is given by

$$\left(\frac{\Delta L}{L}\right)_S = \frac{2(L_R - L_S)}{L_S}\left(\frac{\Delta F}{F}\right) \tag{110}$$

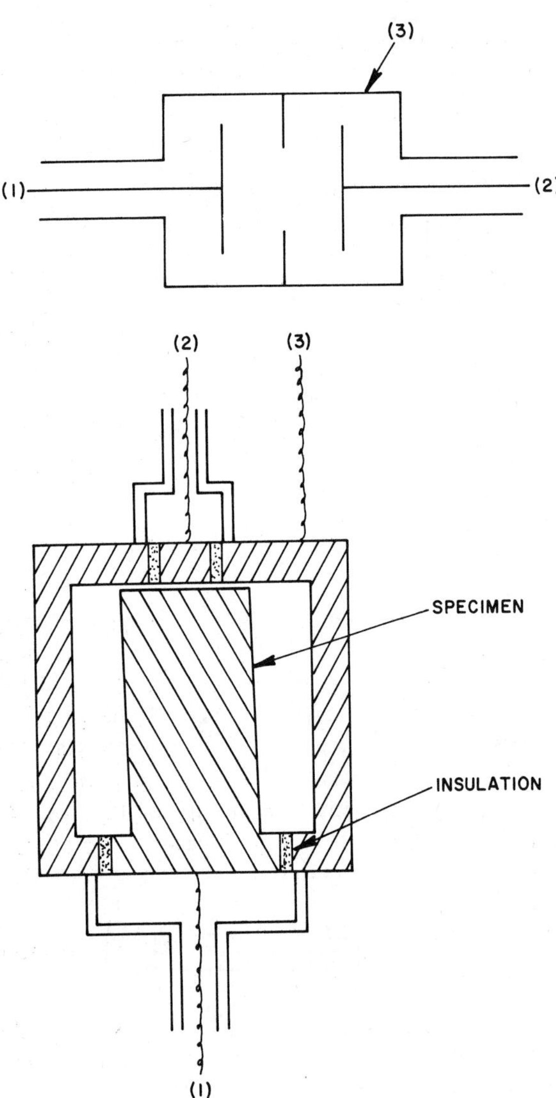

Fig. 13. Schematic diagram for three-terminal parallel plate capacitor technique.

When $\Delta F/F = 1/(4 \times 10^7) = 2.5 \times 10^{-8}$ and $(L_R - L_S)/L_S = 3 \times 10^{-4}$, the expansion can be measured to 10^{-11}.

A high-sensitivity variable differential transformer was developed by Sparks and Swenson [69] for use from 2 to 40 K. In this technique the relative displacement between the primary and secondary windings of a transformer is detected with a mutual inductance bridge. As the secondary winding moves (see Fig. 14), the change in inductance is balanced by the inductance between the primary and the pickup windings. Since the change in mutual inductance is linear to within one percent over ± 1 mm and can be measured to 10^{-10}, the expansion can be measured to 10^{-11} when the specimen's length is 10 cm.

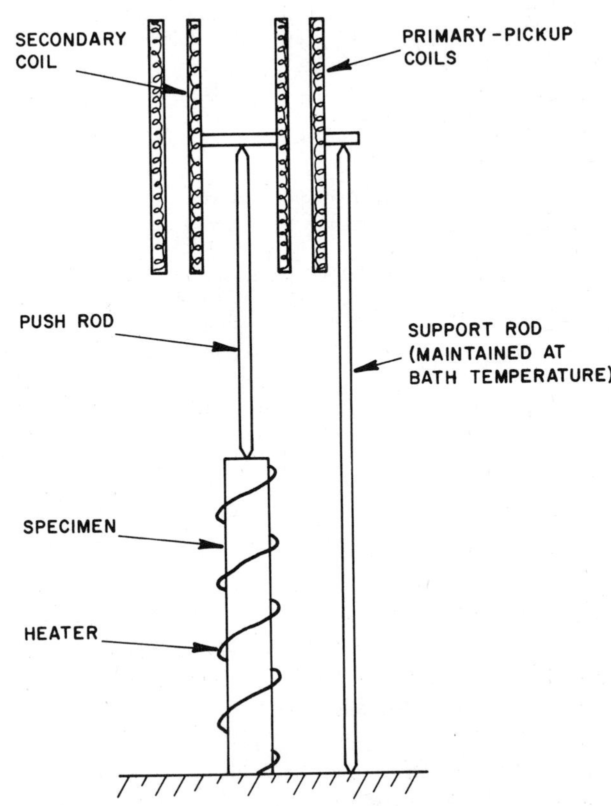

Fig. 14. Schematic diagram for variable differential transformer method.

While several mechanical-optical systems have been developed [70, 71] the most interesting is that of Pereira, Barnes, and Graham [72] for use from 2 to 30 K. In this system the expansion of a 6-cm specimen causes the rotation of a mirror which is attached to the center of a doubly twisted strip (see Fig. 15). The rotation of the mirror is detected with a photoelectric device that has a sensitivity of $\sim 10^{-8}$ rad. The system is calibrated by the piezoelectric effect of a quartz crystal which is placed in series with the specimen [73]. The response of the quartz crystal is $\sim 2 \times 10^{-10}$ cm/volt. The overall sensitivity of the system for measuring expansion is 10^{-11}.

A technique which utilizes the frequency shift of a Fabry–Perot interferometer has been developed by Foster [74] and by Jacobs [30]. The latter has used his system to measure low-expansion materials from 273 to almost 600 K with a sensitivity of 10^{-9}. The beam from a stable He–Ne laser is passed through an electro-optical modulator, an optical isolator, and a Fabry–Perot etalon whose mirrors are separated by a spacer made from the specimen. The beam is

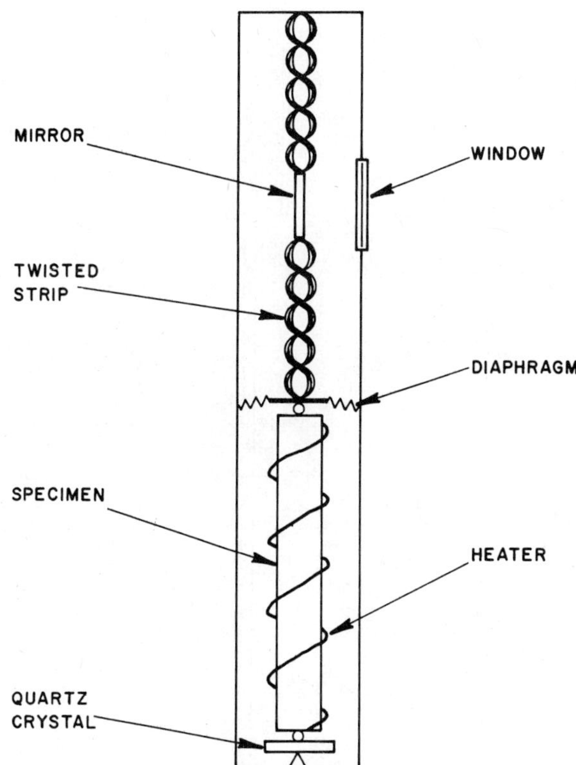

Fig. 15. Schematic diagram for mechanical–optical technique.

modulated so that frequency sidebands are impressed on it that can be tuned to coincide with a transmission peak of the etalon. At coincidence, light passes through the etalon and the change in modulation frequency is proportional to the expansion of the specimen.

7. HIGH-SPEED METHODS

High-speed methods of measuring various thermophysical properties of electrical conductors have proved to be extremely useful [75]. Although no accurate measurements of thermal expansion have yet been made with these systems, they are feasible and are of future importance. In these systems temperatures above 2000 K can be obtained in less than one second by means of pulse heating in vacuum. Temperatures are measured with high-speed optical pyrometers and the data collected with a digital acquisition system.

Cezairliyan [31] has described a system in which the expansion of a specimen is measured by detecting the change in radiance from a constant radiation source. The change in radiance is caused by the expansion of the specimen which partially blocks the radiation. The incident radiation is modulated at 600 Hz so that an optical pyrometer alternately measures the radiances from the source and specimen together and from the specimen alone. The difference between the two radiances is proportional to the width ($\sim$10 mm) of the specimen.

Ruffino *et al.* [76] have described a very complex system which utilizes two image followers and a Michelson interferometer. Each image follower tracks the movement of a hole that has been drilled through the specimen while a corner cube mounted on the follower forms part of the interferometer which measures the amount of movement.

Other Major Sources of Thermal Expansion Data

There exists in the literature a number of reference sources which, while less extensive in scope than the present volume, may nevertheless prove valuable to the reader. While it is not the intent to cite here every available review, it is felt that the following works may prove to be of some value.

Volumes I and II of the *International Critical Tables* [77] contain thermal expansion data on a variety of materials, but these data are rather out of date. Similarly, the *Smithsonian Physical Tables* [78] give data on a variety of materials. Corruccini and Gniewick [79] give data on technical solids at cryogenic temperatures. Gschneider [80], in his review article on physical properties and their interrelationships, includes data on metallic and semimetallic elements. Wood and Deem [81] include thermal expansion data in their compilation of thermophysical property data for materials useful above 1500 K. Pearson [82] gives a summary of published lattice thermal expansion data.

Touloukian [83] edited a handbook series of six volumes (nine books) entitled *Thermophysical Properties of High Temperature Solid Materials.* Volumes 1, 2, and 3 include data on elements, nonferrous alloys, and ferrous alloys, respectively. This series is an update of the five-volume series of Goldsmith, Waterman, and Hirschhorn [84] entitled *Handbook of Thermophysical Properties of Solid Materials,* which includes data on thermal expansion for materials melting above 1000 F. Data on a variety of materials are also included in the Landolt–Börnstein tables [85]. Papadakis [86] gives a compendium of engineering data for various alloys above room temperature, and a recent handbook containing evaluated thermal expansion data is the *American Institute of Physics Handbook*, Third Edition (1972) [87], but these evaluations have been taken into account in the present Volumes 12 and 13.

Comprehensive bibliographic citations (not data) are included in Volumes 1 and 3 of the six-volume *Supplement I* to *Thermophysical Properties Research Literature Retrieval Guide*, edited by Touloukian, Gerritsen, and Shafer [88].

In addition to the above works, the proceedings of the *Thermal Expansion Symposia* [89, 90, 91] represent valuable references to data, methods, and theory on thermal expansion.

References to Text

1. Desaguliers, J. T., *A Course of Experimental Philosophy*, London, 436–46, 1745.
2. Nye, J. F., *Physical Properties of Crystals*, Oxford University Press, 322 pp., 1960.
3. Kittel, C., *Mechanics—Berkeley Physics Course*, Vol. I, McGraw-Hill, New York, 480 pp., 1964.
4. Collins, J. G., and White, G. K., "Thermal Expansion of Solids," in *Progress in Low Temperature Physics* (C. J. Gorter, Editor), Vol. IV, John Wiley and Sons, New York, 450–79, 1964.
5. Kittel, C., *Introduction to Solid State Physics*, 3rd Edition, John Wiley and Sons, New York, 648 pp., 1966.
6. Wallace, D. C., *Thermodynamics of Crystals*, John Wiley and Sons, New York, 484 pp., 1972.
7. Carr, R. H., McCammon, R. D., and White, G. K., "Thermal Expansion of Germanium and Silicon at Low Temperatures," *Phil. Mag.*, **12**, 157–63, 1965.
8. Barron, T. H. K., and Munn, R. W., "Analysis of the Thermal Expansion of Anisotropic Solids, Application to Zinc," *Phil. Mag.*, **15**, 85–103, 1967.
9. Sheard, F. W., "Calculation of the Thermal Expansion of Solids from the Third-Order Elastic Constants," *Phil. Mag.*, **3**, 1381–90, 1958.
10. White, G. K., "Thermal Expansion at Low Temperatures," *Nature*, **187**, 927–9, 1960.
11. Carnelley, T., "The Relation Between the Melting Point of the Elements and the Thermal Expansion Coefficients," *Ber. Deut. Chem. Ges.*, **12**, 439–42, 1879.
12. Clark, A. F., "Low Temperature Thermal Expansion of Some Metallic Alloys," *Cryogenics*, **8**, 282–9, 1968.
13. Gittus, J. H., *Uranium*, Butterworths, 623 pp., 1963.
14. Schoknecht, W. E., and Simmons, R. O., "Thermal Vacancies and Thermal Expansion," in *AIP Conf. Proc. No. 3—Thermal Expansion*, American Institute of Physics, New York, 169–82, 1972.
15. Anderson, O. L., "Equation for Thermal Expansivity in Planetary Interiors," *J. Geophys. Res.*, **72**, 3661–8, 1967.
16. Rosenfield, A. R., and Averbach, B. L., "Effect of Stress on the Expansion Coefficient," *J. Appl. Phys.*, **27**, 154–6, 1956.
17. Fahmy, A. A., and Ragai-Ellozy, A. N., "A Discrete Element Method for the Calculation of the Thermal Expansion Coefficients of Unidirectional Fiber Composites," in *AIP Conf. Proc. No. 17—Thermal Expansion*, American Institute of Physics, New York, 231–40, 1974.
18. Hidnert, P., and Souder, W., "Thermal Expansion of Solids, Natl. Bur. Stand. Circular 48, 29 pp., 1950.
19. Gulati, S. J., and Plummer, W. A., "Influence of Poisson's Ratio on the Composite Expansion of Laminated Plates and and Cylinders," in *AIP Conf. Proc. No. 17—Thermal Expansion*, American Institute of Physics, New York, 196–206, 1974.
20. Fahmy, A. A., and Ragai, A. N., "Thermal-Expansion Behavior of Two-Phase Solids," *J. Appl. Phys.*, **41**, 5108–11, 1970.
21. Kerner, E. H., "The Elastic and Thermo-Elastic Properties of Composite Media," *Proc. Phys. Soc.*, **69**, 808–13, 1956.
22. Preston, T., *The Theory of Heat*, MacMillan and Company, London, 1904.
23. Callendar, H. L., "On the Practical Measurement of Temperature," *Phil. Trans. Roy. Soc.*, **178**, 161–230, 1887.
24. Holborn, L., and Day, A., "The Air Thermometer at High Temperature," *Ann. Phys.*, **307**, 505–45, 1900.
25. Fizeau, M. H., "Memoir on the Expansion of Solids by Heat," *Compt. Rend.*, **62**, 1101–6, 1866.
26. Reimerdes, E., University of Jena, Dissertation, 38 pp., 1896.
27. Scheel, K., "The Thermal Expansion of Quartz in the Dimension of the Main Axis," *Ann. Physik*, **9**, 837–53, 1902.
28. Becker, K., "An X-Ray Method to Determine the Thermal Expansion Coefficient at High Temperature," *Z. Physik*, **40**, 37–41, 1926.
29. White, G. K., "Measurement of Thermal Expansion at Low Temperatures," *Cryogenics*, **1**, 151–8, 1961.
30. Jacobs, S. F., Berthold, J. W., III, and Osmundsen, J., "Ultraprecise Measurement of Thermal Expansion Coefficients—Recent Progress," in *AIP Conf. Proc. No. 3—Thermal Expansion*, American Institute of Physics, New York, 1–12, 1972.
31. Cezairliyan, A., "A High Speed Method of Measuring Thermal Expansion of Electrical Conductors," *Rev. Sci. Instr.*, **42**, 540–1, 1971.
32. Clusener, G. R., "Economy Considerations for Pushrod-Type Dilatometers," in *AIP Conf. Proc. No. 3—Thermal Expansion*, American Institute of Physics, New York, 51–8, 1972.
33. Plummer, W. A., "Differential Dilatometry, A Powerful Tool," in *AIP Conf. Proc. No. 17—Thermal Expansion*, American Institute of Physics, New York, 147–8, 1974.
34. Baudran, A., "Absolute Dilatometric Recorder at High Temperature," *Bull. S.F.C.*, **27**, 13–24, 1955.
35. Hahn, T. A., and Kirby, R. K., "Thermal Expansion of a Borosilicate Glass from 80 to 680 K—Standard Reference Material 731," in *AIP Conf. Proc. No. 17—Thermal Expansion*, American Institute of Physics, New York, 93–101, 1974.
36. "ASTM Method of Test, E228, for Linear Thermal Expansion of Rigid Solids with a Vitreous Silica Dilatometer," ASTM Standards, Part 41, 1974.
37. Hyde, G. R., Domingues, L. P., and Furlong, L. R., "Improved Dilatometer," *Rev. Sci. Instr.*, **36**, 204–8, 1965.
38. Mark, S. D., and Emanuelson, R. C., "A Thermal Expansion Apparatus with a Silicon Carbide Dilatometer for Temperatures to 1500 C," *Amer. Ceram. Soc. Bull.*, **37**, 193–6, 1956.

39. Eyerly, G. B., and Lambertson, W. A., "Development of Dilatometer for Temperatures of 1000–2500 C," in *High-Temperature Technology* (J. E. Campbell and E. M. Sherwood, Editors), John Wiley and Sons, New York, 449–50, 1956.
40. Douglas, R. W., and Isard, J. O., "An Apparatus for the Measurement of Small Differential Expansions and Its Use for the Study of Fused Silica," *J. Sci. Instr.*, **29**, 13–5, 1952.
41. Seigel, S., and Quimby, S. L., "The Thermal Expansion of Crystalline Sodium between 80 K and 290 K," *Phys. Rev.*, **54**, 76–8, 1938.
42. Lieberman, A., and Crandall, W. B., "Design and Construction of a Self-Calibrating Dilatometer for High-Temperature Use," *J. Amer. Ceram. Soc.*, **35**, 304–8, 1952.
43. Meyerhoff, R. W., and Smith, J. F., "Anisotropic Thermal Expansion of Single Crystals of Thallium, Yttrium, Beryllium, and Zinc at Low Temperatures," *J. Appl. Phys.*, **33**, 219–24, 1962.
44. Kollie, T. G., McElroy, D. L., and Hutton, J. T., "A Computer Operated Fused Quartz Differential Dilatometer," in *AIP Conf. Proc. No.* 17—*Thermal Expansion*, American Institute of Physics, New York, 129–46, 1974.
45. Rothrock, B. D., and Kirby, R. K., "An Apparatus for Measuring Thermal Expansion at Elevated Temperatures," *J. Res. Natl. Bur. Stand.*, **71C**, 85–91, 1967.
46. Simmons, R. O., and Balluffii, R. W., "Measurements of Equilibrium Vacancy Concentrations in Aluminum," *Phys. Rev.*, **117**, 52–61, 1960.
47. Gaal, P. S., "Some Experimental Aspects of High-Temperature Thermal Expansion Measurements by Optical Telescopes," *High Temp.—High Pres.*, **4**, 49–57, 1972.
48. Conway, J. B., and Losekamp, A. C., "Thermal Expansion Characteristics of Several Refractory Metals to 2500 C," *Trans. Met. Soc. AIME*, **236**, 702–9, 1966.
49. Otto, J., and Thomas, W., "Thermal Expansion of Quartz Glass from 0 to 1060 C," *Z. Physik*, **175**, 337–44, 1963.
50. Jenkins, F. A., and White, H. E., *Fundamentals of Optics*, McGraw-Hill, New York, 637 pp., 1957.
51. Candler, C., *Modern Interferometers*, The University Press, Glasgow, 502 pp., 1951.
52. Plummer, W. A., "Thermal Expansion Measurements to 130 C by Laser Interferometry," in *AIP Conf. Proc. No.* 3—*Thermal Expansion*, American Institute of Physics, New York, 36–43, 1972.
53. "ASTM Method of Test, E289, for Linear Thermal Expansion of Rigid Solids with Interferometry," ASTM Standards, Part 41, 1974.
54. Fraser, D. B., and Hollis-Hallet, A. C., "The Coefficient of Thermal Expansion of Various Cubic Metals Below 100 K," *Can. J. Phys.*, **43**, 193–219, 1965.
55. Saunders, J. B., "An Apparatus for Photographing Interference Phenomena," *J. Res. Natl. Bur. Stand.*, **35**, 157–86, 1945.
56. Feder, R., and Charbnau, H. P., "Equilibrium Defect Concentration in Crystalline Sodium," *Phys. Rev.*, **149**, 464–71, 1966.
57. Dyson, J., *Interferometry as a Measuring Tool*, Hunt Barnard Printing Ltd., Aylesbury, 206 pp., 1970.
58. Merryman, R. G., and Kempter, C. P., "Precise Temperature Measurement in Debye–Scherrer Specimens at Elevated Temperatures," *J. Amer. Ceram. Soc.*, **48**, 202–5, 1965.
59. Mauer, F. A., and Bolz, L. H., "Problems in the Temperature Calibration of an X-Ray Diffractometer Furnace," in *Advances* in *X-Ray Analysis* (W. M. Mueller, Editor), Vol. 5, Plenum Press, New York, 564 pp., 1961.
60. Brand, J. A., and Goldschmidt, H. J., "Temperature Calibration of a High-Temperature X-Ray Diffraction Camera," *J. Sci. Instr.*, **33**, 41–5, 1956.
61. Ostertag, W., and Fischer, G. R., "Temperature Measurements with Metal Ribbon High Temperature X-Ray Furnaces," *Rev. Sci. Instr.*, **39**, 888–9, 1968.
62. Mauer, F. A., and Hahn, T. A., "Thermal Expansion of Some Azides by a Single Crystal X-Ray Method," in *AIP Conf. Proc. No.* 3—*Thermal Expansion*, American Institute of Physics, New York, 139–50, 1972.
63. d'Heurle, F. M., Feder, R., and Nowick, A. S., "Equilibrium Concentration of Lattice Vacancies in Lead and Lead Alloys," *J. Phys. Soc. Japan*, **18** (Suppl. II), 184–90, 1963.
64. Bowman, A. L., Krikorian, N. H., and Neveson, N. G., "The Variation of Lattice Parameters of UC-ZrC Solid Solutions with Temperature and Composition," in *AIP Conf. Proc. No.* 3—*Thermal Expansion*, American Institute of Physics, New York, 119–30, 1972.
65. White, G. K., and Collins, J. G., "Thermal Expansion of Copper, Silver, and Gold at Low Temperatures," *J. Low Temp. Phys.*, **7**, 43–75, 1972.
66. Browder, J. S., and Ballard, S. S., "Low Temperature Thermal Expansion Measurements on Optical Materials," *Appl. Opt.*, **8**, 793–8, 1969.
67. Schlosser, W. F., Graham, G. M., and Meincke, P. P. M., "The Temperature and Magnetic Field Dependence of the Forced Magnetostriction and Thermal Expansion in Invar," *J. Phys. Chem. Solids*, **32**, 927–38, 1971.
68. Kos, J. F., and Lamarche, J. L. G., "Thermal Expansion of the Noble Metals below 15 K," *Can. J. Phys.*, **47**, 2509–18, 1969.
69. Sparks, P. W., and Swenson, C. A., "Thermal Expansions from 2 to 40 K of Germanium, Silicon, and Four III–V Compounds," *Phys. Rev.*, **163**, 779–90, 1967.
70. Huzan, E., Abbis, C. P., and Jones, G. O., "Thermal Expansion of Aluminum at Low Temperatures," *Phil. Mag.*, **6**, 277–85, 1961.
71. Andres, K., "The Measurement of Thermal Expansion of Metals at Low Temperatures," *Cryogenics*, **2**, 93–7, 1961.
72. Pereira, F. N. D. D., Barnes, C. H., and Graham, G. M., "Sensitive Dilatometer for Low Temperatures and the Thermal Expansion of Copper below 10 K," *J. Appl. Phys.*, **41**, 5050–4, 1970.
73. Periera, F. N. D. D., and Graham, G. M., "Thermal Expansion of the N.B.S. Standard Copper below 30 K," in *AIP Conf. Proc. No.* 3—*Thermal Expansion*, American Institute of Physics, New York, 65–71, 1972.
74. Foster, J. D., and Finnie, I., "Method for Measuring Small Thermal Expansion with a Single Frequency Helium–Neon Laser," *Rev. Sci. Instr.*, **39**, 654–7, 1966.
75. Cezairliyan, A., and McClure, J. L., "High-Speed (Subsecond) Measurement of Heat Capacity, Electrical Resistivity, and Thermal Radiation Properties in the Range 2000 to 3600 K," *J. Res. Natl. But. Stand.*, **75A**, 283–90, 1971.
76. Ruffino, G., Rosso, A., Coslori, L., and Righini, G., "Fast Interferometric Dilatometer," in *AIP Conf. Proc. No.* 17—*Thermal Expansion*, American Institute of Physics, New York, 159–66, 1974.
77. Washburn, E. W. (Editor-in-Chief), *International Critical Tables of Numerical Data, Physics, Chemistry and Technology*, Vols. I and II, McGraw-Hill, New York, 1926 (Vol. I), 1927 (Vol. II).

78. Forsythe, W. E. (Editor), *Smithsonian Physical Tables*, Ninth Revised Edition, Smithsonian Institute (Lord Baltimore Press), 1954.

79. Corruccini, R. J., and Gniewick, J. J., "Thermal Expansion of Technical Solids at Low Temperatures," NBS Monograph 29, 22 pp., 1961.

80. Gschneider, K. A., Jr., "Physical Properties and Interrelationships of Metallic and Semimetallic Elements," in *Solid State Physics* (F. Seitz and D. Turnbull, Editors), Vol. 16, Academic Press, New York, 275–476, 1964.

81. Wood, W. D., and Deem, H. W., "Thermal Properties of High-Temperature Materials," U.S. Army Missile Command Rept. RSIC-202, 399 pp., 1964. [AD 455 069].

82. Pearson, W. B., *A Handbook of Lattice Spacings and Structures of Metals and Alloys*, Vols. 1 and 2, Pergamon Press, New York, 1958 (Vol. 1), 1967 (Vol. 2).

83. Touloukian, Y. S. (Editor), *Thermophysical Properties of High Temperature Solid Materials*, Vols. I to 6, The Macmillan Co., New York, 8549 pp., 1967.

84. Goldsmith, A., Waterman, T. E., and Hirschhorn, H. J., *Handbook of Thermophysical Properties of Solid Materials*, The MacMillan Co., New York, 1961.

85. Ebert, H., "Thermal Expansion of Solid and Liquid Technical Materials," in *Landolt–Börnstein Zahlenwerte und Funktionen aus Physik, Chemie, Astronomie, Geophysik, und Technik*, 6th Edition, Vol. IV, Part 4, Springer-Verlag, Berlin, 781–874, 1967.

86. Papadakis, E. P., "Tabulation of the Coefficients of a Quadratic Function for the Thermal Expansion of Various Alloys and Other Engineering Materials," *Mater. Sci. Eng.*, **10**, 195–203, 1972.

87. Kirby, R. K., Hahn, T. A., and Rothrock, B. D., "Thermal Expansion," in *American Institute of Physics Handbook* (D. E. Gray, Editor), Third Edition, McGraw-Hill, New York, pp. 4-119 to 4-142, 1972.

88. Touloukian, Y. S., Gerritsen, J. K., and Shafer, W. H. (Editors), *Thermophysical Properties Research Literature Retrieval Guide—Supplement I*, Vols. 1 to 6, IFI/Plenum, New York, 2225 pp., 1973.

89. Simmons, R. O., and Wallace, D. C. (Editors), "Symposium on Thermal Expansion of Solids," *J. Appl. Phys.*, **41**, 5043–54, 1970.

90. Graham, M. G., and Hagy, H. E. (Editors), *AIP Conference Proceedings No. 3, Thermal Expansion*—1971 (Third Symposium), American Institute of Physics, New York, 312 pp., 1972.

91. Taylor, R. E., and Denman, G. D. (Editors), *AIP Conference Proceedings No. 17, Thermal Expansion*—1973 (Lake of the Ozarks), American Institute of Physics, New York, 304 pp., 1974.

Numerical Data

Data Presentation and Related General Information

1. SCOPE OF COVERAGE

This volume presents the data on the percent thermal linear expansion and coefficient of thermal linear expansion for almost all metallic materials for which data are known to exist. The materials covered include 64 elements, 94 intermetallic compounds, 125 binary alloy systems, and 70 groups of multiple alloys. These data were obtained by processing over 1400 research documents dated from 1901 to 1972 and including a number of articles published more recently. Materials within each category are arranged alphabetically by name, as listed in the *Grouping of Materials and List of Figures and Tables* in the front of the volume. In all, this volume reports 4253 sets of data on 672 materials, which are listed in the *Material Index* at the end of the volume.

Of the 353 materials, systems, and groups covered, the data for 246 (i.e., 70 %) have been critically evaluated, analyzed, and synthesized, and recommended reference values or provisional values have been generated and are presented in this volume together with the original experimental data. These 246 include 62 of the 64 elements, 65 of the 94 intermetallic compounds, 69 of the 125 binary alloy systems, and 50 of the 70 groups of multiple alloys.

The temperature ranges covered by the thermal expansion data for many materials are from near absolute zero to near the melting point. The change in the thermal expansion due to transitions such as super-conducting, magnetic, and structural transitions are also taken into account.

It is important to note that, although the numerical data presented in this volume are all for thermal *linear* expansion, values for thermal *volumetric* expansion can easily be calculated from the data for thermal linear expansion by using the equations given in the text on the theory of thermal expansion. Consequently, this volume is useful for both *linear* and *volumetric* thermal expansion of metallic solids.

2. PRESENTATION OF DATA

The thermal expansion data and information for each material are presented generally in four sections arranged in the following order: (1) Recommended Values Table and Figure, (2) Original Data Plot, (3) Specification Table, and (4) Data Table.

Recommended Values for the percent thermal linear expansion of materials for which critical data evaluation and analysis is possible are presented in both tabular and graphic forms. Empirical equations in the form of a cubic polynomial which can approximately represent the recommended values are also given in the REMARKS of the table. In addition, a table of the coefficient of thermal linear expansion is also given.

The Original Data Plot is a full-page linear-scale graphic presentation of the thermal linear expansion data as a function of temperature. The data on the coefficient of thermal linear expansion for temperatures below 293 K and the data on the percent thermal linear expansion for superconducting materials below superconducting transition temperature are given as inserts. Although the recommendations made are for pure materials, this type of figure should give an excellent overall view of the quality of data available from the literature. For a number of materials for which there exists only a small number of data, the Original Data Plot is omitted. When several sets of data are too close together to be distinguishable, some of the curves (or data sets) are omitted from the plot for the sake of clarity. These are clearly indicated by asterisks in both the Specification Table and the Data Table.

The Specification Table provides in a concise form the comprehensive information on the test specimens for which data are reported. The curve numbers in the Specification Table correspond exactly to the numbers which appear in the original Data Plot and the Data Table. The Specification Table gives for each set of data the reference number which corresponds to the number in the list of References to Data Sources, the names of authors, the year of publication of the document, the method of measurement, the temperature range, the particular name of the material other than that appearing in the title of the table and the specimen designation, and the specimen composition, characterization, and the test conditions, which include any variables reported by the investigator that may be responsible for the variation in the thermal expansion data. The information of the last category, which is reported to the extent provided in the original source document, includes the following:

(1) Purity, chemical composition, carrier concentration, and type and concentration of crystal defects;
(2) Type of crystal and crystal axis orientation;
(3) Microstructure, grain size, inhomogeneity, and additional phases;
(4) Specimen shape and dimensions, and method and procedure of fabrication;
(5) Thermal history, cold work history, heat treatment, and mechanical, irradiative, and other treatments;
(6) Manufacturer and supplier, stock number, and catalog number;
(7) Test environment, degree of vacuum or pressure, strength and orientation of applied magnetic field;
(8) Pertinent physical properties such as density, porosity, and hardness;
(9) Form in which the extracted data are presented in the original source document other than raw data points;
(10) Additional information obtained directly from the author.

Unfortunately, in the majority of cases the authors do not report in their research papers all the necessary pertinent information to fully characterize and identify the materials for which their data are reported. This is particularly true for the authors of earlier investigations. Consequently, the amount of information on specimen characterization reported in the Specification Tables varies greatly from author to author.

Tabular presentation for all the data described in the Specification Table whether shown or not in the Original Data Plot is given in the Data Table. Attempts have been made to contact authors for tabular data whenever the original data are given in the research paper in a graph which is too small to allow accurate data extraction compatible with the accuracy of the measurements.

Data on percent thermal linear expansion are reported in this volume as percent elongation, given by the expression

$$\frac{\Delta L}{L_0}(\%) = \frac{L_T - L_{293\,\mathrm{K}}}{L_{293\,\mathrm{K}}} \times 100$$

where L_T and L_0 are lengths (or lattice parameters) at temperature T and at 293 K, respectively. In order to compare all the available data from the world literature on the same basis, the thermal expansion has been arbitrarily set to be zero at 293 K, whenever this is possible. In these cases the zero-point correction or the lattice parameters at the first measured temperature and the extrapolated value at 293 K are given in the Specification Table. This information is useful in deriving the original data from the presented data. In cases where it is not feasible to change the author's original data to a 293 K reference (because a large extrapolation is required or because of problems associated with phase transformations) the presented form differs from this, and the new reference temperature is specified in the Specification Table. In the cases where data are taken during the subsequent cooling or heating cycles and "permanent sets" have occurred, the data have been biased to zero expansion at 293 K and the appropriate zero-point corrections are indicated in the Specification Tables. These curves are not included in the Original Data Plot and this fact is indicated clearly on the individual curves by asterisks in the Specification Tables and Data Tables. Whenever an author reports data on both thermal linear expansion and the coefficient of thermal linear expansion, the latter is also reported in a second Data Table. In this case the same curve number is used in the Original Data Plot, Specification Table, and the two Data Tables. In this way, the information in the Specification Table for a given specimen applies to data on both the thermal linear expansion and the coefficient of thermal linear expansion for that specimen. If the author only reported coefficient of expansion data, these results are integrated to yield values of thermal expansion and this is indicated by a double dagger (‡) in the Data Table.

The coefficient of thermal linear expansion is defined as

$$\alpha = \frac{d}{dT}\left(\frac{\Delta L}{L_0}\right) = \frac{1}{L_0}\frac{dL}{dT}$$

As discussed in the text on the theory of thermal expansion of solids, there are several other commonly used definitions of the coefficient; most frequently used is the mean coefficient between two temperatures. Since the mean coefficient is applied to a range of temperatures and cannot be specified for a single temperature, such data are not reported. Instead, they have been converted to thermal linear expansion.

3. CRITERIA FOR DATA ANALYSIS

The end result of critical evaluation, analysis, and synthesis of the available data is the generation of the "most probable values." Depending upon the level of confidence these values are designated as "recommended values," "provisional values," or "typical values." Whenever possible, the reliability of experimental data is judged on the basis of a set of objective criteria. However, there are many instances when adequate information is not provided by the author to fully assess the quality of the data. In such cases, data evaluation becomes exceedingly difficult. In such instances considerable weight is given to factors such as past performance of the author and the reputation of the laboratory where the measurement was made. The criteria for arriving at recommended values are that the original data consist of concordant results from measurements carried out by several reliable and competent researchers on well-characterized materials using proven techniques, or the results of a single reliable and competent researcher on well-characterized materials using at least two independent methods. Any serious deviation from these conditions leads to designations as either "provisional values" or "typical values."

The most probable values for the majority of the elements are "recommended values." However, for rare earth elements which cannot be obtained in the pure state and for which the linear thermal expansion data scatter considerably depending upon the content and the nature of impurities, the values are designated as either "provisional values" or "typical values" depending upon the scatter. For many elements considerable scatter is observed in the thermal expansion data near phase transformations and the melting point mainly due to impurities and to the use of unsuitable heating and cooling rates. In such cases, the uncertainty assigned to the values is either increased, or they are designated as "provisional values" or "typical values."

Although most of the original data for intermetallic compounds look fairly good, they are not substantiated. Therefore, the most probable values for these materials are "provisional values" or "typical values," depending upon the scatter.

There is a vast quantity of data on binary and multiple alloys reported in the world literature. However, the majority of the data, especially for the binary alloy systems, are either conflicting or not substantiated. Due to various limitations, the extent of work involved in thoroughly analyzing all the data for the binary alloy systems would be very time consuming. Therefore, the most probable values for one or two illustrative compositions for each of the alloys, wherever possible, are reported as "provisional values." In order to show such conflicting data more clearly, the most probable values for the constituent elements are also shown in the graph. In the same way the most probable values for multiple alloys of engineering importance are also reported as "provisional values." It is worth noting that, in the case of "provisional values," it may be the lack of corroboration that has led to the assignment of a high uncertainty.

4. CLASSIFICATION OF MATERIALS

The classification scheme shown in Table IV is based on the chemical composition of the materials. This scheme is mainly for the convenience of materials grouping and data organization and is not intended to be used as the basic definitions for the various material categories. The term "binary alloy system" refers to all alloys in the full range of composition of two constituent elements, no matter which constituent element is more prevalent. Multiple alloys are, however, classified such that in each group of alloys the two elements listed are the highest and second highest constituents, respectively.

5. SYMBOLS AND ABBREVIATIONS USED IN THE TABLES AND FIGURES

In the specification tables the code designations used for the experimental methods are as follows:

D	Density method
E	Electrical capacitor method
F	Electron diffraction
I	Interferometer method
L	Dilatometer method
O	Optical method
R	Variable resistance transducer
S	Strain gauge method
T	Telemicroscope method
V	Variable-inductance transformer
X	X-ray diffraction method

Table IV. Classification of Materials

	Limits of composition (weight percent)*			
	x_1	$x_1 + x_2$	x_2	x_3
1. Elements	>99.5	–	<0.2	<0.2
2. Intermetallic compounds	>99.5	–	<0.2	<0.2
3. Binary alloy systems	–	≥99.5	–	≤0.2
4. Multiple alloys	–	≥99.5	>0.2	>0.2
	–	<99.5	≥0.2	≤0.2
	–	<99.5	>0.2	>0.2
	≤99.5	–	<0.2	<0.2

*$x_1 \geq x_2 \geq x_3$.

Other symbols and abbreviations used in the tables and/or figures are as follows:

b.c.c.	Body-centered cubic
c	Cubic
c.p.h.	Close-packed hexagonal
d	Density
f.c.c.	Face-centered cubic
f.c.t.	Face-centered tetragonal
h	Hexagonal
I.D.	Inside diameter
$\Delta L/L_0$	Percent thermal linear expansion
α	Coefficient of thermal expansion
M.P.	Melting point
monocl.	Monoclinic
NTP	Normal temperature and pressure
O.D.	Outside diameter
orthorh.	Orthorhombic
r.	Rhombohedral
s.c.	Superconducting
Subl.	Sublimation
T	Temperature
t.	Tetragonal
Temp.	Temperature
T.P.	Transition point
μ	Micro
>	Greater than
<	Less than
~	Approximately
③	Curve number for polycrystalline material if any of the following three kinds of curve number symbols appears also in the same figure; otherwise it just represents any curve number
⬡3	Curve number for measurements ∥ to *a* axis

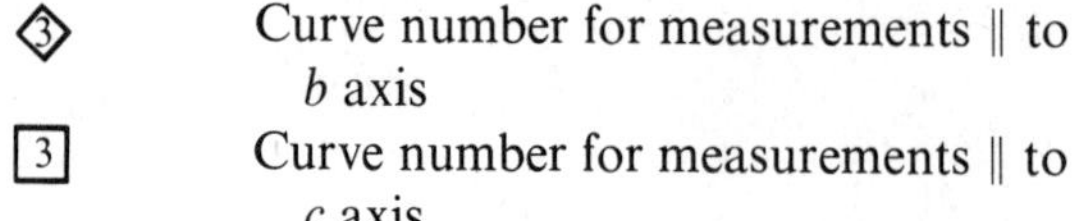

◇3	Curve number for measurements ∥ to *b* axis
[3]	Curve number for measurements ∥ to *c* axis

6. CONVENTION FOR BIBLIOGRAPHIC CITATION

For the following types of documents the bibliographic information is cited in the sequences given below.

Journal Article

a. Author(s)—The names and initials of all authors are given. The last name is written first, followed by initials.
b. Title of article.
c. Journal name—The abbreviated name of the journal as used in *Chemical Abstracts* is given.
d. Series, volume, and number—If the series is designated by a letter, no comma is used between the letter for series and the numeral for volume, and they are underlined together. If the series is also designated by a numeral, a comma is used between the numeral for series and the numeral for volume, and only the numeral representing volume is underlined. No comma is used between the numerals representing volume and number. The numeral for number is enclosed in parentheses.
e. Pages—The inclusive page numbers of the article are given.
f. Year—The year of publication.

Report

a. Author(s).
b. Title of report.
c. Name of the responsible organization.
d. Report of bulletin, circular, technical note, etc.
e. Number.
f. Part.
g. Pages.
h. Year.
i. ASTIA's AD number—This is given in square brackets whenever available.

Book

a. Author(s).
b. Title.
c. Volume.
d. Edition.
e. Publisher.
f. City, state (address of the publisher).
g. Pages.
h. Year.

7. CONVERSION FACTORS

The following conversion factors are used for the conversion of the data/values reported in the Specification Tables and Data Tables:

$$1 \text{ in.} = 0.0254 \text{ m}$$

$$1 \text{ lb} = 0.45359237 \text{ kg}$$

$$1 \text{ kX} = 1.00202 \text{ Å}$$

$$T(\text{K}) = t(°\text{C}) + 273.15$$

$$T(\text{K}) = [5/9][t(°\text{F}) + 459.67]$$

The same number of decimal places as in the original data given by the author was retained in the temperature conversion.

The kilo X unit (kX) was used in early years for relative wavelength measurements and is defined by the relation

$$1 \text{ kX} = \frac{(200) \text{ plane spacing of calcite}}{3.02945}$$

8. CRYSTAL STRUCTURES, TRANSITION TEMPERATURES, AND PERTINENT PHYSICAL CONSTANTS OF THE ELEMENTS, AND CRYSTAL STRUCTURES AND MELTING POINTS OF INTERMETALLIC COMPOUNDS COVERED IN THIS VOLUME

Table V contains information on the crystal structures, transition temperatures, and certain pertinent physical constants of the elements. Crystal structures and melting points (wherever available) of the intermetallic compounds covered in this volume are given in Table VI. This information is very useful in data analysis and synthesis. For example, the thermal expansion of a material generally changes abruptly when the material undergoes any transformation. One must therefore be extremely cautious in attempting to extrapolate the thermal expansion data across any phase, state, magnetic, or superconducting transition temperature, as given in the table. Each table has an independent series of numbered references which immediately follows the table.

Even though no attempt has been made to critically evaluate the temperatures/constants given in the tables and these should not be taken as recommended reference values, Table V for the elements is significantly different from the similar table published earlier in Volumes 1, 2, 4, 5, 7, and 10 of this Series. This revised Table V is an updated and upgraded version of the earlier table in that most of the information/data given in Table V have been taken from the most recent and reliable sources and that all temperatures have been converted to the International Practical Temperature Scale of 1968 (IPTS-68).

Table V. Crystal Structures, Transition Temperatures, and Pertinent Physical Constants of the Elements[a]

Name	Atomic Number	Atomic[b] Weight	Density[c], kg m^{-3} x 10^{-3}	Crystal Structure	Phase Transition Temp., K	Superconducting Transition Temp., K	Curie Temp., K	Néel Temp., K	Debye Temperature At 0 K, K	Debye Temperature At 298 K, K	Melting Point, K	Boiling Point, K	Critical Temp., K
Actinium	89	(227)	10.07[10][d]	f.c.c.[10]					124[3]	100[4] (at ~50 K)	1324[16]	3200 ±300[6]	
Aluminum	13	26.9815[4]	2.702[16]	f.c.c.[10]		1.196[16] 1.17[8] 1.183[16]			423 ±5[3]	390[3]	933.52[37]	2798[21]	8680[11] 7795[49]
Americium	95	(243)	11.7[12]	Double c.p.h.[10]							1269 ±4[21]	2885[75]	
Antimony	51	121.75	6.684[29]	r.[10]					150[3]	200[14]	903.89[37]	1860 ±50[21]	2994[15]
Argon	18	39.948	0.0017824[29] (at 273.2 K and 1 atm)	f.c.c.[9]						90[4] (at ~45 K)	83.81[21]	87.30[21]	151[15]
Arsenic	33	74.9216	5.73[10] (gray, at 295.7 K) 4.7[29] (black) 2.0[29] (yellow)	r.[10] (gray) orthorh.[10] (black) c.[5] (yellow)					233[3]	275[18]	1091[13] (35.8 atm) subl. 886[16]	1091[13] (35.8 atm)	
Astatine	85	(210)									573.2[19]	650[20]	
Barium	56	137.34	3.5[16]	b.c.c.[10] (α)					110.5 ±1.8[22]	116[23]	1002 ±2[21]	2174[21]	3670[15] 3928[49]
Berkelium	97	(249)											
Beryllium	4	9.01218	1.85[16]	c.p.h.[10] (α) b.c.c.[10] (β)	1530[10] (α-β)				1160[25]	1031[3]	1562[21]	2749[21]	6170[15]
Bismuth	83	208.9804	9.78[16]	r.[10]					119 ±2[3]	116 ±5[3]	544.592[37]	1839[21]	4620[27]
Boron	5	10.81	2.50[42]	Simple r.[10] (α)					1315[53]	1362[3]	2573[16]	4284[21]	
Bromine	35	79.904	3.119[29]	orthorh.[9]							265.9[21]	332.25[21]	584[15]
Cadmium	48	112.40	8.65[16]	c.p.h.[10]		0.56[16] 0.518[21]			252 ±48[3]	221[3]	594.258[37]	1038[3]	1905[15] 3567[49]
Calcium	20	40.08	1.55[16]	f.c.c.[10] (α) b.c.c.[10] (β)	720 ±2[21] (α-β)				234 ±5[3]	230[3]	1113 ±2[21]	1759[21]	3273[15]
Californium	98	(251)											
Carbon (amorphous)	6	12.011	1.8 ~ 2.1[29]								Subl. 3925-3970[5]	4473[5]	
Carbon (diamond)	6	12.011	3.52[16]	d.[9]					2240 ±5[31]	1874[3]	>3823[16]	5100[16]	
Carbon (graphite)	6	12.011	2.25[16] (α)	h.[9] (α) r.[9] (β)					402 ±11[3]	1550[3]	Subl. 3925-3970	4484[5]	

[a] All temperatures have been converted to the International Practical Temperature Scale of 1968 (IPTS-68).

[b] Atomic weights are based on ^{12}C = 12 as adopted by the International Union of Pure and Applied Chemistry in 1971; those in parentheses are the mass numbers of the isotopes of longest known half-life.

[c] Density values are given at 293.2 K unless otherwise noted.

[d] Superscript numbers designate references listed at the end of the table.

Name	Atomic Number	Atomic Weight[b]	Density[c], kg m^{-3} x 10^{-3}	Crystal Structure	Phase Transition Temp., K	Superconducting Transition Temp., K	Curie Temp., K	Néel Temp., K	Debye Temperature At 0 K, K	Debye Temperature At 298 K, K	Melting Point, K	Boiling Point, K	Critical Temp., K
Cerium	58	140.12	6.90[29]	f.c.c.[10] (α) Double c.p.h.[10] (β) f.c.c.[10] (γ) b.c.c.[10] (δ)	103 ± 5[33] (α-β) 263 ± 5[9] (β-γ) 1003[9] (γ-δ)			13[32]	146[3]	138[34]	1072 ± 3[21]	3699[16]	10440[49]
Cesium	55	132.9054	1.873[29]	b.c.c.[10]					40 ± 5[3]	43[23]	301.55[21]	944[16]	2063[46,47,48] 1900[49]
Chlorine	17	35.453	0.003214[29] (at 273.2 K)	t.[9]						115[4,36] (at ~58 K)	172.16[21]	239.10[21]	417[15]
Chromium	24	51.996	7.14[16]	b.c.c.[21]				311[37]	598 ± 32[3]	424[3]	2133 ± 20[21]	2950[21]	
Cobalt	27	58.9332	8.862[42]	c.p.h.[10] (α) f.c.c.[10] (β)	690[73] (α-β)		1395[74] (α) 1131[74] (β)		452 ± 17[3]	386[3]	1767[37]	3207[21]	
Copper	29	63.546	8.933[16]	f.c.c.[10]					342 ± 2[3]	310[3]	1357.6[37]	2840[21]	8530[11] 8310[49]
Curium	96	(247)	7[42]	Double c.p.h.[8]									
Dysprosium	66	162.50	8.556[42]	c.p.h.[10] (α) b.c.c.[10] (β)	1659[21] (α-β)		83.5[43] (ferro.- antiferromag.)	174[43]	172 ± 35[3]	158[44]	1684[21]	2835[16]	7665[49]
Einsteinium	99	(254)											
Erbium	68	167.26	9.06[42]	c.p.h.[10] (α) b.c.c.[10] (β)			19[4] (ferro.- antiferromag.)	80[4]	134 ± 10[45]	163[44]	1797[12]	3136[16]	7274[49]
Europium	63	151.96	5.245[28]	b.c.c.[10]				~90[4] 88[76]	127[3]		1091[21]	1870[16]	4611[49]
Fermium	100	(253)											
Fluorine	9	18.99840	0.001695[29] (at 273.2 K and 1 atm)	c.[75] (β-F_2)							53.48[21]	84.95[21]	144[15]
Francium	87	(223)							39[3]		300.2[19]	879[75]	
Gadolinium	64	157.25	7.87[42]	c.p.h.[10] (α) b.c.c.[10] (β)	1535[32] (α-β)		291.8[1]		170[3]	155 ± 3[3]	1587 ± 2[12]	3539[16]	8700[49]
Gallium	31	69.72	5.91[29]	orthorh.[4] (α) t.[4] (β)		1.091[16] 1.087[16]			317[3]	240[14] 125[4] (tetra at ~63 K)	302.90[21] 275.6[13] (at 8.86 x 10^6 mm Hg)	2482[21]	7645[27]
Germanium	32	72.59	5.36[29]	d.[10]					378 ± 22[3]	403[3]	1211.7[21]	3112[21]	5656[15]
Gold	79	196.9665	18.88[16]	f.c.c.[10]					165 ± 1[3]	178 ± 8[3]	1337.58[37]	3135[21]	9540[11] 8090[49]
Hafnium	72	178.49	13.3[16]	c.p.h.[10] (α) b.c.c.[10] (β)	2015 ± 20[21] (α-β)				256 ± 5[3]	213[23]	2500 ± 20[21]	4891[21]	

Name	Atomic Number	Atomic Weight[b]	Density[c], kg m^{-3} x 10^{-3}	Crystal Structure	Phase Transition Temp., K	Superconducting Transition Temp., K	Curie Temp., K	Néel Temp., K	Debye Temperature At 0 K, K	Debye Temperature At 298 K, K	Melting Point, K	Boiling Point, K	Critical Temp., K
Helium	2	4.00260	0.0001785[29] (at 273.2 K and 1 atm)	c.p.h.[16]						30[4] (at ~15 K)	3.45[29] 1.8 ±0.2[17] (at 30 atm)	4215[21]	5.3[15]
Holmium	67	164.9304	8.80[29]	c.p.h.[10] (α) b.c.c.[10] (β)	1703[12] (α-β)		17.5[72] (ferro.-antiferromag.)	133 ±2[72]	114 ±7[45]	161[44]	1745[12]	2968[16]	
Hydrogen	1	1.0079	0.00008987[29] (at 273.2 K and 1 atm)	c.p.h.[9]						116[36] (para., at ~58 K) 105[36] (ortho., at ~53 K)	13.803[21]	20.397[37]	33.3[15]
Indium	49	114.82	7.3[29]	f.c.t.[10]		3.4035[16] 3.407[16]			108.8 ±0.3[3]	129[14]	429.784[37]	2349[21]	4380[15] 7070[49]
Iodine	53	126.9045	4.94[16]	orthorh.[9]						105[4] (at ~53 K)	386.7[21] subl. 298.16[13] (at 0.31 mm Hg)	458.4[21]	785[15] 819[24]
Iridium	77	192.22	22.5[42]	f.c.c.[10]		0.105 ±0.004[21]			425 ±5[3]	228[3]	2720[37]	4712[21]	
Iron	26	55.847	7.87[28]	b.c.c.-ferromag.[10] (α) b.c.c.-paramag.[10] (β) f.c.c.[10] (β) b.c.c.[10] (δ)	1185[21] (β-γ) 1667[21] (γ-δ)		1043[40]		457 ±12[3]	373[3]	1811[21]	3140[21]	6770[30] 9440[49]
Krypton	36	83.80	0.003708[29] (at 273.2 K and 1 atm)	f.c.c.[16]						60[4] (at ~30 K)	115.78[21]	119.80[21]	209.4[15]
Lanthanum	57	138.9055	6.18[42]	Double c.p.h.[10] (α) f.c.c.[10] (β) b.c.c.[10] (γ)	583[32] (α-β) 1141[32] (β-γ)	4.80[16] (α) 4.9[16] (α) 5.91[16] (β) 6.06[16] (β)			142 ±3[52]	135 ±5[44]	1195 ±5[21]	3730[16]	10550[49]
Lawrencium	103	(257)											
Lead	82	207.2	11.342[16]	f.c.c.[10]		7.193[5]			102 ±5[3]	87 ±1[3]	600.652[37]	2026[21]	5415[27] 4770[49]
Lithium	3	6.94	0.534[16]	c.p.h.[10] (α) b.c.c.[10] (β)	80 ±10[21] (α-β)				352 ±17[3]	448[3]	453.7[21]	1617[21]	4160[11] 3727[49]
Lutetium	71	174.97	9.85[29]	c.p.h.[10] (α) b.c.c.[10] (β)	Near m.p.[50] (α-β)				210[54]	116[3]	1938[21]	3668[16]	
Magnesium	12	24.305	1.74[16]	c.p.h.[10]					396 ±54[3]	330[3]	922[21]	1364[21]	3537[49]
Manganese	25	54.9380	7.43(α)[28] 7.29(β)[28] 7.18(γ)[28]	b.c.c.[10] (α) c.[10] (β) f.c.c.[10] (γ) b.c.c.[10] (δ)	980 ±20[21] (α-β) 1360 ±20[21] (β-γ) 1411 ±20[21] (γ-δ)			95 (α) ~580[35] (β) 660[74] (γ)	418 ±32[3]	363[3]	1519 ±5[21]	2338[21]	6067[49]
Mendelevium	101	(256)											

Name	Atomic Number	Atomic Weight[b]	Density[c], kg m^{-3} x 10^{-3}	Crystal Structure	Phase Transition Temp., K	Superconducting Transition Temp., K	Curie Temp., K	Néel Temp., K	Debye Temperature At 0 K, K	Debye Temperature At 298 K, K	Melting Point, K	Boiling Point, K	Critical Temp., K
Mercury	80	200.59	13.546[29] 14.19[29] (at 234.25 K)	r.[10] (α) b.c.t.-pressure induced structure (β)	Martensitic transformation at low temp.[56]	4.16[16] (α) 4.154[16] (α) 3.949[16] (β)			~75[58]	92 ± 8[3]	234.288[37]	629.81[37]	1735[27] 1707[49]
Molybdenum	42	95.94	10.24[42]	b.c.c.[10]		0.92[16] 0.917[16]			459 ±11[3]	377[3]	2894 ±10[35]	4924[21]	17100[11] 16900[49]
Neodymium	60	144.24	7.007[29]	Double c.p.h.[10] (α) b.c.c.[10] (β)	1129[21] (α-β)			8[4] (ordinary) 19[4] (special)	159[3]	148 ± 8[3]	1290 ±5[21]	3341[21]	7930[49]
Neon	10	20.179	0.0009002[29] (at 273.2 K and 1 atm)	f.c.c.[9]						60[4] (at ~30 K)	24.553[21]	27.102[37]	44.5[15]
Neptunium	93	237.0482	20.46[42]	orthorh.[10] (α) t.[10] (β) b.c.c.[10] (γ)	551[2] (α-β) 847 ±5[26] (β-γ)				121[3]	163[3]	910 ±2[21]	4160[3]	
Nickel	28	58.71	8.90[42]	f.c.c.[10]			631[40]		427 ±14[3]	345[3]	1728[37]	3193[21]	6312[15] 11800[49]
Niobium	41	92.9064	8.57[42]	b.c.c.[10]		9.26[16] 9.17[16] 9.1[8]			241 ±13[3]	260[64]	2744[21]	5029[21]	19200[49]
Nitrogen	7	14.0067	0.0012506[29]	c.[9] (α) h.[77] (β)	35.61[21] (α-β)					70[4] (at ~35 K)	63.14[21]	77.348[37]	126.2[15]
Nobelium	102	(254)											
Osmium	76	190.2	22.48[29]	c.p.h.[10]		0.655[16] 0.65[8]			500[67]	400[68]	3288 ±10[69]	5298[21]	
Oxygen	8	15.9994	0.001429[29] (at 273.2 K and 1 atm)	b.c. orthorh.[7] (α) r.[7] (β) c.[7] (γ)	23.86[21] (α-β) 43.79[21] (β-γ)					250[4] (at ~125 K) 500[36] (at ~250 K)	54.35[21]	90.188[37]	154.8[15]
Palladium	46	106.4	12.02[28]	f.c.c.[10]					283 ±16[3]	275[14]	1827[37]	3243[21]	
Phosphorus	15	30.97376	1.82[29] (β) 2.22[29] (γ) 2.69[29] (δ)	h. ?[7] (α) b.c.c.[7] (β) c.[7] (γ) f.c. orthorh.[17] (δ)	196[71] (α-β) 298.16[13] (β-γ) 298.16[13] (β-δ)				193[3] (white) 325[3] (red)	576[3] (white) 800[3] (red)	317.3[16] (white) 1300[82] (black)	550[21]	993.8[15]
Platinum	78	195.09	21.45[29]	f.c.c.[10]					234 ±1[3]	225 ±5[3]	2045[37]	4109[21]	8310[15]

Name	Atomic Number	Atomic[b] Weight	Density[c], kg m^{-3} x 10^{-3}	Crystal Structure	Phase Transition Temp., K	Superconducting Transition Temp., K	Curie Temp., K	Néel Temp., K	Debye Temperature At 0 K, K	Debye Temperature At 298 K, K	Melting Point, K	Boiling Point, K	Critical Temp., K
Plutonium	94	(242)	19.737[29] (at 298.2 K)	Simple monocl.[10] (α) b.c. monocl.[10] (β) f.c. orthorh.[10] (γ) f.c.c.[10] (δ) b.c.t.[10] (δ') b.c.c.[10] (ε)	395 ±2[21] (α-β) 480 ±5[21] (β-γ) 588 ±5[21] (γ-δ) 730 ±5[21] (δ-δ') 753 ±5[21] (δ'-ε)			60[8,83] (α)	171[70]	176[70]	913 ±1[21]	3510[21]	
Polonium	84	(210)	9.3[29] (α) 9.5[29] (β)	Simple c.[10] (α) r.[10] (β)	327 ±1.5[66] (α-β)				81[3]		527[16]	1236[20]	2284[15]
Potassium	19	39.098	0.87[16]	b.c.c.[10]					89.4 ±0.5[3]	100[3]	336.35 ±0.5[21]	1032[21]	2454[11] 2144[49]
Praseodymium	59	140.9077	6.48[16]	Double c.p.h.[10] (α) b.c.c. (β)	1069 ±3[21] (α-β)			25[65]	85 ±1[45]	138[63]	1205 ±4[21]	3785[21]	8930[49]
Promethium	61	(145)		h.[7] (α) b.c.c.[38] (β)	1185[38] (α-β)			6[38]			1353 ±10[62]	2734[3]	
Protactinium	91	231.0359	15.37[42]	b.c.t.[10]		1.4[21]			159[3]	262[3]	1505[5]	4692[3]	
Radium	88	226.0254	5[29]						89[3]		973[16]	1900[3]	
Radon	86	(222)	0.00973[29] (at 273.2 K and 1 atm)	f.c.c.[9]						400[4] (at ~200 K)	202[16]	211[21]	377.16[15]
Rhenium	75	186.2	21.1[42]	c.p.h.[10]		1.700 ±0.002[21]			429 ±22[3]	275[23]	3459 ±20[21]	5882[21]	20100[11]
Rhodium	45	102.9055	12.45[42]	f.c.c.[10]	possible trans-formation at[57] 1373-1473 K				480 ±32[3]	350[3]	2236[37]	3978[21]	
Rubidium	37	85.4678	1.53[29]	b.c.c.[10]					54 ±4[3]	59[23]	312.64 ±0.5[21]	961[21]	2105[41,46,48]
Ruthenium	44	101.07	12.2[29]	c.p.h.[10]		0.49[16]			600[67]	415[3]	2523 ±10[69]	4433[21]	2035[49]
Samarium	62	150.4	7.54[29]	r.[10] (α) b.c.c.[32] (β)	1191[32] (α-β)		14[8] (ferro.-antiferromag.)	106[8]	116[45]	184 ±4[3]	1346[61]	2064[16]	5415[49]
Scandium	21	44.9559	3.00[42]	c.p.h.[10] (α) b.c.c.[10] (β)	1610[21] (α-β)				470 ±80[52]	476[3]	1814[12]	3547[21]	
Selenium	34	78.96	4.50[29] (α) 4.80[29] (β)	monocl.[10] (α) h. (β) amorphous[7]	304[40,60] (vitrification) 398[13] (vit.-β) 423[13] (α-β)				151.7 ±0.4[59]	89[36] (at ~45 K) 150[4] (at ~75 K)	494[21]	1009[13] (Se_6) 958.0[13] (Se_4,[37]) 1027[13] (Se_2)	1759[15]
Silicon	14	28.086	2.42[16]	d.[10]					647 ±11[3]	692[3]	1687 ±2[21]	2753[28]	5172[15]
Silver	47	107.868	10.492[16]	f.c.c.[10]					228 ±3[3]	221[3]	1235.08[37]	2440[21]	7485[11]
Sodium	11	22.98977	0.9712[16]	b.c.c.[10]	Martensitic transformation at low temp.[56]				157 ±1[3]	155 ±5[3]	371.0[21]	1157[21]	2805[11] 2405[49]

Name	Atomic Number	Atomic Weight[b]	Density[c], kg m^{-3} x 10^{-3}	Crystal Structure	Phase Transition Temp., K	Superconducting Transition Temp., K	Curie Temp., K	Néel Temp., K	Debye Temperature At 0 K, K	Debye Temperature At 298 K, K	Melting Point, K	Boiling Point, K	Critical Temp., K
Strontium	38	87.62	2.60[28]	f.c.c.[10] (α) b.c.c.[10] (γ)	830[21] (α–γ)				147 ±1[22]	148[23]	1042[21]	1652[21]	3064[15] 3818[49]
Sulfur	16	32.06	2.07[29] (α) 1.96[29] (β)	orthorh.[10] (α) monocl.[10] (β) r.[10] (γ)	368.64[21] (α–β)				200[3] (β)	527[55] (α) 250[55] (α, at 40 K)	386.0[5] (α) 392.2[5] (β) subl.368.6[13]	717.824[37]	1314[15]
Technetium	43	(99)	11.50[29]	c.p.h.[10]		8.22[16]			351[3]	422[3]	2451[16]	4550[21]	
Tellurium	52	127.60	6.24[29] (α) 6.00[5] (amorph.)	h.[10] (α)					141 ±12[3]		722.7[16]	1262[21]	2332[15]
Terbium	65	158.9254	8.25[29]	c.p.h.[10] (α) b.c.c.[10] (β)	1562[21] (α–β)		219[80] (ferro.-antiferromag.)	230[80]	150[55]	158[44]	1632[12]	3503[16]	
Thallium	81	204.37	11.85[29]	c.p.h.[10] (α) b.c.c.[10] (β)	507 ±2[21] (α–β)	2.39[16] 2.38[8]			88 ±1[3]	96[14]	577 ±2[21]	1748[21]	3225[15]
Thorium	90	232.0381	11.7[42]	f.c.c.[10] (α) b.c.c.[10] (β)	1636 ±10[21] (α–β)	1.368[16]			170[84]	100[14]	2028[21]	5073[21]	14600[49]
Thulium	69	168.9342	9.32[29]	c.p.h.[10] (α) b.c.c.[10] (β)	Near m.p.[50] (α–β)			53[81]	127 ±1[45]	167[44]	1820[12]	2220[16]	6448[49]
Tin	50	118.69	5.750[29] (α) 7.31[29] (β)	f.c.c.[10] (α) b.c.t.[10] (β)	286.2 ±3[51] (α–β)	3.722[16] (β)			236 ±24[3] (gray) 196 ±9[3] (white)	254[3] (gray) 170[14] (white)	505.1181[37]	2881[21]	8030[11] 9335[49]
Titanium	22	47.90	4.5[29]	c.p.h.[10] (α) b.c.c.[10] (β)	1156[21] (α–β)	0.39[16] 0.42[16]			426 ±5[3]	380[14]	1946 ±3[21]	3569[21]	
Tungsten	74	183.85	19.3[29]	b.c.c.[10]		0.012[16]			388 ±17[3]	312 ±3[3]	3660[37]	5844[21]	23200[11]
Uranium	92	238.029	19.07[28]	orthorh.[10] (α) t.[10] (β) b.c.c.[10] (γ)	37 ±2[39] (α_0–α) 941 ±2[21] (α–β) 1048 ±2[21] (β–γ)	0.68[16] 0.8[33] (β) 1.80[5] (γ)			200[84]	300[3]	1407.1 ±0.6[85]	4417[21]	12550[27] 12050[49]
Vanadium	23	50.941	6.1[28]	b.c.c.[10]		5.3[16] 5.37[16]			326 ±54[3]	390[14]	2202 ±6[21]	3689[21]	11250[49]
Xenon	54	131.30	0.005851[29] (at 273.2 K and 1 atm)	f.c.c.[9]							161.36[21]	165.03[21]	289.75[15]
Ytterbium	70	173.04	7.02[42]	f.c.c.[10] (α) b.c.c.[10] (β)	1033[21] (α–β)				118[78]		1098[12]	1467[21]	4430[49]
Yttrium	39	88.9059	4.47[29]	c.p.h.[10] (α) b.c.c.[10] (β)	1754[21] (α–β)				268 ±32[3]	214[79]	1800[86]	3618[21]	8983[49]
Zinc	30	65.38	7.140[29]	c.p.h.[10]		0.825[21]			316 ±20[3]	237 ±3[3]	692.73[37]	1181[21]	2172[15]
Zirconium	40	91.22	6.57[59]	c.p.h.[10] (α) b.c.c.[10] (β)	1137 ±5[21] (α–β)	0.875[16] 0.546[16]					2127[21]	4693[21]	2915[49] 2303[49]

REFERENCES

(Crystal Structures, Transition Temperatures, and Pertinent Physical Constants of the Elements)

1. Griffel, M., Skochdopole, R.E., and Spedding, F.H., "The Heat Capacity of Gadolinium from 15 to 355 K," Phys. Rev., 93, 657-61, 1954.
2. Elliott, R.P., Constitution of Binary Alloys, 1st Suppl., McGraw-Hill, 1965.
3. Gschneider, K.A., Jr., "Physical Properties and Interrelationships of Metallic and Semimetallic Elements" (Sietz, F. and Turnbull, D., Editors), Solid State Phys., 16, 275-426, 1964.
4. Gopal, E.S.R., Specific Heat at Low Temperatures, Plenum Press, 1966.
5. Weast, R.C. (Editor), Handbook of Chemistry and Physics, 47th Ed., The Chemical Rubber Co., 1966-67.
6. Foster, K.W. and Fauble, L.G., "The Volatility of Actinium," J. Phys. Chem., 64, 958-60, 1960.
7. The Institution of Metallurgists, Annual Yearbook, 68-73, 1960-61.
8. Meaden, G.T., Electrical Resistance of Metals, Plenum Press, 1965.
9. Gray, D.E. (Coordinating Editor), American Institute of Physics Handbook, McGraw Hill, 1972.
10. Pearson, W.B., A Handbook of Lattice Spacings and Structures of Metals and Alloys, Pergamon Press, 1967.
11. Grosse, A.V., "Electrical and Thermal Conductivities of Metals Over Their Entire Liquid Range," Rev. Hautes Temp. Refractaires, 3, 115-46, 1966.
12. Dennison, D.H., Gschneider, K.A., Jr., and Daane, A.H., "High-Temperature Heat Contents and Related Thermodynamic Functions of Eight Rare-Earth Metals: Sc, Gd, Tb, Dy, Ho, Er, Tm, and Lu," J. Chem. Phys., 44(11), 4273-82.
13. Rossini, F.D., Wagman, D.D., Evans, W.H., Levine, S., and Jaffe, I., "Selected Values of Chemical Thermodynamic Properties," NBS Circular 500, 537-822, 1952.
14. deLaunay, J., "Theory of Specific Heats and Lattice Vibrations," Solid State Phys., 2, 219-303, 1956.
15. Gates, D.S. and Thodos, G., AIChE J., 6(1), 50-4, 1960.
16. Weast, R.C. (Editor), Handbook of Chemistry and Physics, 7th Ed., The Chemical Rubber Co., 1973-74.
17. Sasaki, K. and Sekito, S., "Three Crystalline Modifications of Electrolytic Chromium," Trans. Electrochem. Soc., 59, 437-60, 1931.
18. Anderson, C.T., "The Heat Capacities of Arsenic, Arsenic Trioxide and Arsenic Pentoxide at Low Temperatures," J. Am. Chem. Soc., 52, 2296-300, 1930.
19. Trombe, F., "Observations on the Vapor Tensions of Rare Earth Metals, Their Separation and Their Purification," Bull. Soc. Chim. (France), 20, 1010-2, 1953.
20. Stull, D.R. and Sinke, G.C., Thermodynamic Properties of the Elements in Their Standard State, American Chemical Soc., 1956.
21. Hultgren, R., Desai, P.D., Hawkins, D.T., Gleiser, M., Kelley, K.K., and Wagman, D.D., Selected Values of the Thermodynamic Properties of the Elements, American Society for Metals, 1973.
22. Roberts, L.M., "The Atomic Heats of Calcium, Strontium and Barium Between 1.5 and 20 K," Proc. Phys. Soc. (London), B70, 738-43, 1957.
23. Zemansky, M.W., Heat and Thermodynamics, 4th Ed., McGraw-Hill, 1957.
24. Mathews, J.F., "The Critical Constants of Inorganic Substances," Chem. Rev., 72(1), 71-100, 1972.
25. Hill, R.W. and Smith, P.L., "The Specific Heat of Beryllium at Low Temperatures," Phil. Mag., 44(7), 636-44, 1953.
26. Lee, J.A., "A Review of the Physical Metallurgy of Neptunium," in Progress in Nuclear Energy, Series V. Metallurgy and Fuels, Vol. 3, Pergamon Press, New York, 453-67, 1961.
27. Grosse, A.V., Temple Univ. Research Institute Rept., 40 pp., 1960.
28. Lyman, T. (Editor), Metals Handbook, Vol. 1, 8th Ed., American Soc. for Metals, 1961.
29. Lange, N.A. (Editor), Handbook of Chemistry, Revised 10th Edition, McGraw-Hill, 1967.
30. Gross, A.V., Research Institute of Temple Univ., Report on USAEC Contract No. AT(30-1)-2082, 71 pp., 1965.
31. Burk, D.L. and Friedberg, S.A., "Atomic Heat of Diamond From 11 to 200 K," Phys. Rev., 111(5), 1275-82, 1958.
32. Spedding, F.H. and Daane, A.H. (Editors), The Rare Earths, John Wiley, 1961.
33. Matthias, B.T., Geballe, T.H., Corenzwit, E., Andres, K., and Hall, G.W., "Superconductivity of Beta-Uranium," Science, 151, 985-6, 1966.
34. Arajs, S. and Colvin, R.V., "Analysis of Specific Heats of Cerium, Neodymium, and Samarium at High Temperatures," J. Less-Common Metals, 4, 159-68, 1962.
35. Cezairliyan, A., Morse, M.S., and Beckett, C.W., "Measurement of Melting Point and Electrical Resistivity (above 2840 K) of Molybdenum by a Pulse Heating Method," Rev. Int. Hautes. Temp. Refractaires, 7(4), 382-8, 1970.

36. Rosenberg, H.M., Low Temperature Solid State Physics, Oxford at Clarendon Press, 1965.

37. Comité International des Poids et Mesures, "The International Practical Temperature Scale of 1968," Metrologia, 5(2), 35-44, 1969.

38. Williams, R.K. and McElroy, D.L., "Estimated Thermal Conductivity Values for Solid and Liquid Promethium," USAEC Rept. ORNL-TM-1424, 32 pp., 1966.

39. Fisher, E.S. and Dever, D., "Elastic Moduli and Phase Transition in Uranium at T < 43 K," Phys. Rev., 2, 170(3), 607-13, 1968.

40. Abdullaev, G.B., Mekhtieva, S.I., Abdinov, D.Sh., Aliev, G.M., and Alieva, S.G., "Thermal Conductivity of Selenium," Phys. Status Solidi, 13(2), 315-23, 1966.

41. Hochman, J.M., Silver, I.L., and Bonilla, C.F., "The Electrical and Thermal Conductivity of Liquid Rubidium to 2900 F and the Critical Point of Rubidium," USAEC Rept. CU-2660-13, 1964.

42. Touloukian, Y.S. (Editor), Thermophysical Properties of High Temperature Solid Materials, MacMillan, Vol. 1, 1967.

43. Griffel, M., Skochdopole, R.E., and Spedding, F.H., "Heat Capacity of Dysprosium from 15 to 300 K," J. Chem. Phys., 25(1), 75-9, 1956.

44. Gschneidner, K.A., Jr., Rare Earth Alloys, Van Nostrand, 1961.

45. Dreyfus, B., Goodman, B.B., Lacaze, A., and Trolliet, G., "The Specific Heats of Rare Earth Metals Between 0.5 and 4 K," Compt. Rend., 253, 1764-6, 1961.

46. Dillon, I.G., Illinois Institute of Technology, Ph.D. Thesis, June 1965.

47. Hochman, J.M. and Bonilla, C.F., Advances in Thermophysical Properties at Extreme Temperatures and Pressures (Gratch, S., Editor), ASME 3rd Symposium on Thermophysical Properties, Purdue University, 122-30, 1965.

48. Grosse, A.V., "Comparison of the Experimental Entropies of Vaporization of Rubidium and Cesium From the Melting Point to the Critical Region with Those of Mercury and the Validity of the Principle of Corresponding States for Metals," J. Inorg. Nucl. Chem., 28, 2125-9, 1966.

49. Kopp, I.Z., "An Estimate of the Critical Temperatures of the Elements," Russ. J. Phys. Chem., 41(6), 782-3, 1967.

50. Miller, A.E. and Daane, A.H., "The High-Temperature Allotropy of Some Heavy Rare-Earth Metals," Trans. AIME, 230, 568-72, 1964.

51. Raynor, G.V. and Smith, R.W., "The Transition Temperature of the Transition Between Grey and White Tin," Proc. Roy. Soc. (London), A244, 101-9, 1958.

52. Montgomery, H. and Pells, G.P., "The Low Temperature Specific Heat of Scandium and Yttrium," Proc. Phys. Soc. (London), 78, 622-5, 1961.

53. Kaufman, L. and Clougherty, E.V., ManLabs, Inc., Semi-Annual Rept. No. 2, 1963.

54. Lounasmaa, O.V., Proc. 3rd Rare Earth Conf., 1963, Gordon and Breach, New York, 1964.

55. Roach, P.R. and Lounasmaa, O.V., "Specific Heat of Terbium Between 0.37 and 4.2 K," Bull. Am. Phys. Soc., 7, 408, 1962.

56. Reed, R.P. and Breedis, J.F., ASTM STP 387, 60-132, 1966.

57. Hansen, M., Constitution of Binary Alloys, 2nd Edition, McGraw-Hill, p. 1268, 1958.

58. Smith, P.L., Conf. Phys. Basses Temp., Inst. Intern. du Froid, Paris, 281, 1956.

59. Fukuroi, T. and Muto, Y., "Specific Heat of Tellurium and Selenium at Very Low Temperatures," Tohoku Univ. Res. Inst. Sci. Rept., A8, 213-22, 1956.

60. Abdullaev, G.B., Mekhtiyeva, S.I., Abdinov, D.Sh., and Aliev, G.M., "Amorphous Selenium Thermal Conductivity Study in Softening Region," Phys. Letters, 23(3), 215-6, 1966.

61. McKeown, J.J., "The High Temperature Heat Capacity and Related Thermodynamic Functions of Some Rare Earth Metals," State University of Iowa, Ph.D. Thesis, 113 pp., 1958.

62. Weigel, F., Angew. Chem., 75, 451, 1963.

63. Murao, T., "The Magnetic Properties of the Praseodymium and the Neodymium Metals," Progr. Theoret. Phys. (Kyoto), 20(3), 277-86, 1958.

64. Morin, F.J. and Maita, J.P., "Specific Heats of Transition Metal Superconductors," Phys. Rev., 129(3), 1115-20, 1963.

65. Cable, J.W., Moon, R.M., Koehler, W.C., and Wollan, E.O., "Antiferromagnetism of Praseodymium," Phys. Rev. Letters, 12(20), 553-5, 1964.

66. Goode, J.M., "Phase Transition Temperature of Polonium," J. Chem. Phys., 26(5), 1269-71, 1957.

67. Walcott, N.M., Conf. Phys. Basses Temp., Inst. Intern. du Froid, Paris, 286, 1956.

68. White, G.K. and Woods, S.B., "Electrical and Thermal Resistivity of the Transition Elements at Low Temperatures," Phil. Trans. Roy. Soc. (London), A251(995), 273-302, 1959.

69. Douglass, R.W. and Adkins, E.F., "A Study of the Spectral Emissivities and Melting Temperatures of Osmium and Ruthenium," Trans. Met. Soc. AIME, 221, 248-9, 1961.

70. Sandenaw, T.A., Olsen, C.E., and Gibney, R.B., "Heat Capacity of Plutonium Metal Below 420 K," in Plutonium 1960, Proc. 2nd Intern. Conf. (Grison, E., Lord, W.B.H., and Fowler, R.D., Editors, 66-79, 1961.

71. Bridgman, P.W., "Two New Modifications of Phosphorus," J. Am. Chem. Soc., 36(7), 1344-63, 1914.

72. Rhodes, B.L., Legvold, S., and Spedding, F.H., "Magnetic Properties of Holmium and Thulium Metals," Phys. Rev., 109, 1547-50,

73. Lagneborg, R. and Kaplow, R., "Radial Distribution Functions in Solid Cobalt," Acta Met., 15(1), 13-24, 1967.

74. Meaden, G.T., "General Physical Properties of Manganese Metal," Met. Rev., 13(125), 97-114, 1968.

75. Samsonov, G.V. (Editor), Handbook of the Physicochemical Properties of the Elements, Plenum Press, 1968.

76. Meaden, G.T. and Sze, N.H., "Fluctuations and Critical Indices Close to the Néel Temperature of Europium Metal," presented at the International Colloquium on the Rare Earth Elements, Paris-Grenoble, May 1969.

77. Streib, W.E., Jordan, T.H., and Lipscomb, W.N., "Single-Crystal X-Ray Diffraction Study of β Nitrogen," J. Chem. Phys., 37(12), 2962-5, 1962.

78. Lounasmaa, O.V., "Specific Heat of Gadolinium and Ytterbium Metals Between 0.4 and 4 K," Phys. Rev., 129, 2460-4, 1963.

79. Jennings, L.D., Miller, R.E., and Spedding, F.H., "Lattice Heat Capacity of the Rare Earths. Heat Capacities of Yttrium and Lutetium from 15-350 K," J. Chem. Phys., 33(6), 1849-52, 1960.

80. Arajs, S. and Colvin, R.V., "Thermal Conductivity and Electrical Resistivity of Terbium Between 5 and 300 K," Phys. Rev., A136(2), 439-41, 1964.

81. Aliev, N.G. and Volkenstein, N.V., "Thermal Conductivity of Tm, Yb, and Lu at Low Temperatures," Soviet Physics-JETP, 22(5), 997-8, 1966.

82. Slack, G.A., "Thermal Conductivity of Elements with Complex Lattices: B, P, S," Phys. Rev., A139(2), 507-15, 1965.

83. Meaden, G.T., "Electronic Properties of the Actinide Metals at Low Temperatures," Proc. Roy. Soc., 276, 553-70, 1963.

84. Smith, P.L. and Walcott, N.M., Conf. Phys. Basses Temp., Inst. Intern. du Froid, 283, 1956.

85. Argonne National Laboratory, USAEC Rept. ANL-5717, 67 pp., 1957.

86. Beaudry, B.J., "The Effect of Titanium on the Melting Point and Transition Temperature of Yttrium," J. Less-Common Metals, 14(3), 370-2, 1968.

Table VI. Crystal Structures and Melting Points of Intermetallic Compounds[a]

Name	Crystal Structure	Melting Point, K	Name	Crystal Structure	Melting Point, K
AlSb	c.	~1335	Gd_2Fe_{17}	h.[3]	
Al_2Cu	b.c.t		$GdPd_3$	f.c.c.	
AlCu	Orthrh.		GaNi	b.c.c.	
$AlCu_2$			$GaAg_3$	c.p.h.	
Al_4Cu_9	c.		GeLa	Orthrh.[5]	1627
$AlCu_3$			$GeMg_2$	f.c.c.	1389[4]
Al_2Au	f.c.c.	1334	Ge_2Pr	t.[2]	1745[7]
Al_5Fe_2	Orthrh.	~1440	GePr	Orthrh.[2]	1815[7]
Al_3Fe	Monocl.	1425	Ge_3Pr_5	h.[6]	1697[7]
Al_2Fe	rhomb.		Au_4Mn	t.	
AlFe	b.c.c.		Au_2Mn	t.	
$AlFe_3$	b.c.c.		AuMn	b.c.c.	
AlNi	b.c.c.	1914	Au_4V	t.[2]	
$AlNi_3$	f.c.c.		AuZn	b.c.c.	999
$AlAg_2$	c.p.h.	1000	$HoZn_2$	Orthrh.[8]	1332[8]
Al_3U	c.[2]		In_3Pd	f.c.t.[9]	
SbGa	f.c.c.	985	In_3Pr	c.	
SbIn	f.c.c.	800	In_3Yb	f.c.c.[2]	
SbLa	f.c.c.		In_3Y	f.c.c.[2]	
β-$Be_{17}Hf_2$	h.		$Fe_{17}Lu_2$	h.[3]	
$Be_{12}Nb$	b.c.t.		$FeNi_3$	f.c.c.	
Be_2Ta	b.c.t.		FeRh	b.c.c.[2]	
$Be_{17}Ta_2$	r.		$Fe_{17}Y_2$	h.[3]	
$Be_{13}U$	f.c.c.	2280	$LaRu_2$	f.c.c.[2]	
$Be_{13}Zr$	f.c.c.	1920	$LaSn_3$	f.c.c.[2]	1410[4]
Bi_2Pt	c.		PbLi	r.	
BiPt	h.		MgAg	b.c.c.	1094
CdAu	b.c.c.	900	Mg_2Sn	f.c.c.	1045
CdLi	f.c.c.	822	MnHg	b.c.c.[2]	
Cd_3Mg	h.		MnNi	t.[2]	
CdTe	f.c.c.	1365	MnPd	t.[2]	
$CaMg_2$	h.	990	Mn_3Pt	f.c.c.[2]	
$CeIn_3$	f.c.c.[2][b]	1419[4]	MnPt	t.[2]	
$CePd_3$	f.c.c.[2]		Ni_3Nb	Orthrh.	1678
$CeRu_2$	f.c.c.		NiTi	b.c.c.	1585
$CeSn_3$	f.c.c.	1435[4]	Ni_5Y	h.[3]	
CrFe	t[2]		Nb_3Sn	c.	
Co_2Dy	f.c.c.[2]		Pd_3Sn	f.c.c.	1601
Co_2Gd	f.c.c.[2]		Pd_3Yb	f.c.c.[10]	
$Co_{17}Y_2$	h[3]		$PrRu_2$	f.c.c.[2]	
Co_5Y	h.[2]		$PrSn_3$	f.c.c.[2]	1419[4]
Cu_3AuI	f.c.c.		$SmAg_3$	t.[11]	
CuAu	f.c.c.	1184	Ag_2Tb	t.[12]	
Cu_2Mg	f.c.c.	1093	AgZn	h.	
Cu_5Zn_8	b.c.c.		NaTl	f.c.c.	578
Cu-Zn			$YbZn_2$	Orthrh.	
$GdIn_3$	f.c.c.		Zn_2Zr	f.c.c.	1454

[a] All the information given in the table is taken from ref. [1] unless otherwise noted. All temperatures have been converted to the 1968-IPTS scale.

[b] Superscript numbers designate references listed at the end of the table.

REFERENCES

(Crystal Structures and Melting Points of Intermetallic Compounds)

1. Hultgren, R., Desai, P.D., Hawkins, D.T., Gleiser, M., and Kelley, K.K., Selected Values of the Thermodynamic Properties of Binary Alloys, American Society for Metals, Ohio, 1435 pp., 1973.
2. Pearson, W.B., A Handbook of Lattice Spacings and Structures of Metals and Alloys, Pergamon Press, London, Vol. 2, 1446 pp., 1967.
3. Shah, J.S., "Thermal Lattice Expansion of Various Types of Solids", Ph.D. Thesis, University of Missouri, University Microfilms, UM-71-25, 126 pp., 1971.
4. Hansen, M., Constitution of Binary Alloys, McGraw-Hill Book Company, New York, 1305 pp., 1958.
5. Samsonov, G.V., Paderno, Yu.B., and Rud, B.M., "Physical Properties of Germanides of Ceric Group and Yttrium Rare Earth Metals", Rev. Intern. Hautes Temp. Refract., 5(2), 105-10, 1968.
6. Shunk, F.A., Constitution of Binary Alloys, Second Supplement, McGraw-Hill Book Company, New York, 720 pp., 1969.
7. Rud, B.M., Lynchak, K.A., and Paderno, Yu.B., "Some Properties of Praseodymium-Germanides", Izv. Akad. Nauk, SSSR., Neorg. Mater., 5(8), 1350-3, 1969.
8. Michel, D.J., "Some Thermal and Mechanical Properties of Intermetallic Compound $HoZn_2$", Penn. State University, Ph.D. Thesis, University Microfilms No. 69-14543, 77 pp., 1968.
9. Harris, I.R., Norman, M., and Bryant, A.W., "A Study of Some Palladium-Indium, Platinum-Indium and Platinum-Tin Alloys", J. Less-Common Metals, 16, 427-40, 1968.
10. Harris, I.R., Raynor, G.V., and Winstanley, C.J. "Rare Earth Intermediate Phases. IV. High-Temperature Lattice Spacings of Some R.E.-Pd_3 Phases", J. Less-Common Metals, 12(1), 69-74, 1967.
11. Steeb, B., Godel, D., and Lohr, C., "On the Sturcture of the Compounds Ag_3-R.E. (R.E. = Y, La, Ce, Sm, Gd, Dy, Ho, Er)", J. Less-Common Metals, 15(2), 137-41, 1968.
12. Atoji, M., "Magnetic Structure of Terbium-Silver Intermetallic", J. Chem. Phys. 48(8), 3380-3, 1968.

Numerical Data on Thermal Linear Expansion of Metallic Elements and Alloys

1. ELEMENTS

FIGURE AND TABLE NO. 1R. RECOMMENDED VALUES FOR THERMAL LINEAR EXPANSION OF ALUMINUM Al

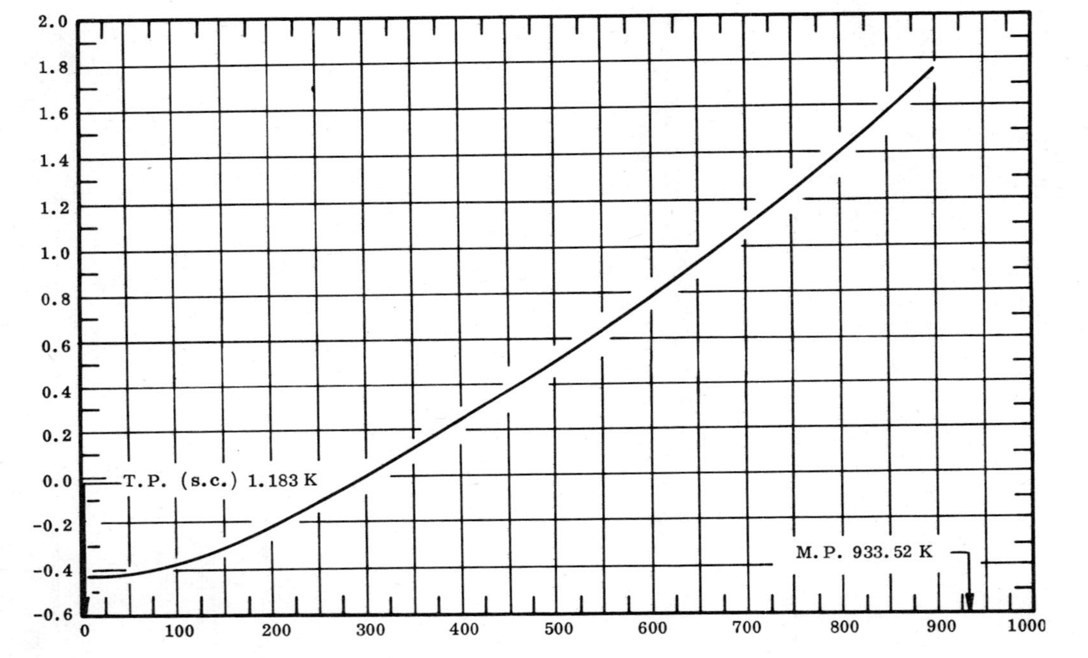

RECOMMENDED VALUES

[Temperature, T, K; Linear Expansion, $\Delta L/L_0$, %; α, K^{-1}]

T	$\Delta L/L_0$	$\alpha \times 10^6$
5	-0.418	0.001
50	-0.413	3.9
100	-0.371	12.2
200	-0.203	20.3
293	0.000	23.1
400	0.259	25.1
500	0.514	26.4
600	0.787	28.4
700	1.084	30.9
800	1.408	34.0
900	1.764	37.4

REMARKS

The tabulated values are considered accurate to within ± 3% over the entire temperature range. These values can be represented approximately by the following equations:

$$\Delta L/L_0 = -0.420 - 1.415 \times 10^{-4} (T - 5) + 8.077 \times 10^{-6} (T - 5)^2 - 8.656 \times 10^{-9} (T - 5)^3 \qquad (5 < T < 300)$$

$$\Delta L/L_0 = 0.018 + 2.364 \times 10^{-3} (T - 300) + 4.164 \times 10^{-7} (T - 300)^2 + 8.270 \times 10^{-10} (T - 300)^3 \qquad (300 < T < 900)$$

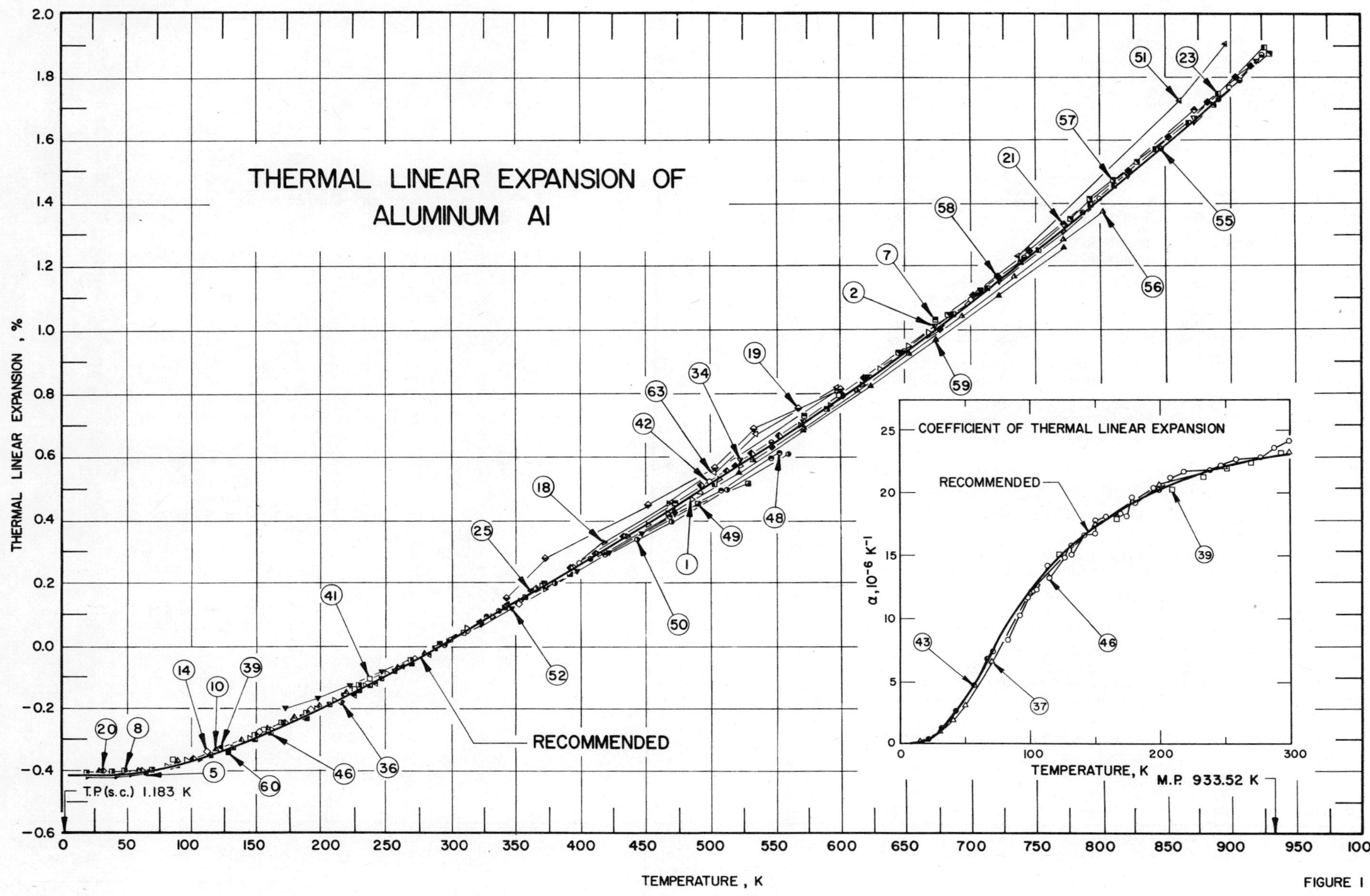

THERMAL LINEAR EXPANSION OF
ALUMINUM Al
THERMAL LINEAR EXPANSION, %
TEMPERATURE, K
RECOMMENDED
T.P. (s.c.) 1.183 K
M.P. 933.52 K
COEFFICIENT OF THERMAL LINEAR EXPANSION
RECOMMENDED
α, 10⁻⁶ K⁻¹
TEMPERATURE, K

FIGURE 1

SPECIFICATION TABLE 1. THERMAL LINEAR EXPANSION OF ALUMINUM Al

Cur. No.	Ref. No.	Author(s)	Year	Method Used	Temp. Range, K	Name and Specimen Designation	Composition (weight percent), Specifications, and Remarks
1	1	Nix, F.C. and MacNair, D.	1941	I	82-667		99.997 Al; rod specimen about 6 mm long subjected to prolonged annealing; measured in helium gas at pressure of 3 mm Hg in tests up to 440 K; measured in vacuum of 10^{-5} mm Hg above 440 K; zero-point correction of -0.047% determined by graphical interpolation.
2	2	King, D., Cornish, A.J., and Burke, J.	1966	X	523-923		99.995 Al pure, crushed specimen; expansion measured on a-axis.
3*	3	Fraser, D.B. and Hollis-Hallet, A.C.	1965	I	23-75		99.99 Al cubic crystal structure; hollow cylindrical specimen 2.54 cm O.D., 2.06 cm I.D., 5.08 cm long; expansion measured on first test; this curve is here reported using the first given temperature as reference temperature at which $\Delta L/L_0 = 0$; data on the coefficient of thermal linear expansion also reported.
4*	3	Fraser, D.B. and Hollis-Hallet, A.C.	1965	I	23-96		The above specimen; expansion measured on second test; this curve is here reported using the first given temperature as reference temperature at which $\Delta L/L_0 = 0$; data on the coefficient of thermal linear expansion also reported.
5	4	Figgins, B.F., Jones, G.O., and Riley, D.P.	1956	X	20-299		99.99 Al; wire specimen obtained from Johnson, Matthey and Co.; lattice parameter reported at 298.7 K is 4.04968 Å; 4.04921 Å at 293 K determined by graphical interpolation.
6*	5	Gerchikova, N.S., Kolobnev, N.I., Stepanova, M.G., and Fridlyander, I.N.	1963		293-573		Pressed rod made of sintered aluminum powder.
7	6	Matveyev, B.I., Kishnev, P.V., and Khanova, I.R.	1963		293-673	Soft Aluminum	99.5 Al.
8	7	Bijl, D. and Pullan, H.	1955	L	20-273		99.994 Al; supplied by Johnson, Matthey, Ltd., catalog No. JM 340; zero-point correction of -0.0420% determined by graphical extrapolation.
9*	8	Altman, H.W., Rubin, T., and Johnston, H.L.	1954	I	20-300		99.99 Al; specimen furnished by the Aluminum Co. of America; zero-point correction of 0.016% determined by graphical interpolation.
10	9	Gibbons, D.F.	1958	I	20-300		99.99 Al; single crystal; annealed; expansion measured in (111) direction; zero-point correction of -0.043% determined by graphical extrapolation.
11*	10	Wilson, A.J.C.	1942	X	273-673		99.99 Al; unannealed filings; lattice parameter reported at 273 K is 4.0385 Å; 4.04045 Å at 293 K determined by graphical interpolation.
12*	10	Wilson, A.J.C.	1942	X	273-673		99.99 Al; filings annealed at 870 K; lattice parameter reported at 273 K is 4.0391 Å; 4.0410 Å at 293 K determined by graphical interpolation.
13*	11	Taylor, C.S., Willey, L.A., Smith, D.W., and Edwards, J.D.	1938	I	293-773		99.996 Al; ring specimen 1.78 cm diameter x 0.51 cm long with short legs; cut in one piece from 14 gage aluminum sheet, bent and filed to desired shape.
14	12	Ayres, H.D.	1905	I	88-351		Expansion measured using 5460 Å line of mercury.
15*	12	Ayres, H.D.	1905	I	98-353		Similar to the above specimen.
16*	12	Ayres, H.D.	1905	I	86-373		Similar to the above specimen.
17*	13	Irmann, R.	1955		293-773	SAP (Sintered Aluminum Powder)	Atomized aluminum produced by spraying molten aluminum through nozzle; prepared to fine powder in stamp (Hametog process); further comminuted, together with fine foil shavings; flaky powder obtained, 0.01 μ thick having oxide coating less than 0.1 μ; powder contained 13% oxide; produced by High Duty Alloys, Ltd., under proprietary name Hiduminium 100.

*Not shown in figure.

SPECIFICATION TABLE 1. THERMAL LINEAR EXPANSION OF ALUMINUM Al (continued)

Cur. No.	Ref. No.	Author(s)	Year	Method Used	Temp. Range, K	Name and Specimen Designation	Composition (weight percent), Specifications, and Remarks
18	14	Palatnik, L.S., Pugachev, A.T., Boiko, B.T. and Bratsykhin, V.M.	1967	F	373-598		Underlayer of NaCl condensed on base previous to formation of film specimen; specimen 400 Å to 600 Å thick; expansion measured in vacuum 10^{-4} to 10^{-5} torrs with increasing temperature.
19	14	Palatnik, L.S., et al.	1967	F	600-343		The above specimen; expansion measured with decreasing temperature.
20	15	Rhodes, B.L., Moeller, C.E., Hopkins, V. and Marx, T.I.	1963	L	18-573		99.99 Al; 10.16 cm long, 6.2 mm diameter; temperature determined by thermocouple imbedded in side of specimen at mid-point; reported error 1.5%.
21	16	Souder, W. and Hidnert, P.	1921	T	293-873	Sample 1	99.774 Al, 0.13 Fe, 0.12 Si, 0.006 Cu; specimen cut from cast bar.
22*	16	Souder, W. and Hidnert, P.	1921	T	293-873	Sample 2	Similar to the above specimen.
23	17	Simmons, R.O. and Baluffi, R.W.	1960	T	503-929		99.995 Al; stock from Aluminum Co. of America; reported error 0.1%; average of heating and cooling measurements.
24*	17	Simmons, R.O. and Baluffi, R.W.	1960	X	503-929		The above specimen; average of heating and cooling measurements.
25	18	Zubenko, V.V. and Umansky, M.M.	1956	X	227-375		Polycrystalline specimen; lattice parameter reported at 283 K is 4.03995 kX (4.04803 Å); 4.04090 kX (4.04898 Å) at 293 K determined by graphical interpolation.
26*	19	Hidnert, P. and Krider, H.S.	1952	L	293-573	Sample 821AN	99.952 Al; annealed 1 hr at 773 K and cooled slowly; expansion measured with increasing temperature.
27*	19	Hidnert, P. and Krider, H.S.	1952	L	573-293	Sample 821AN	The above specimen; expansion measured with decreasing temperature.
28*	19	Hidnert, P. and Krider, H.S.	1952	L	293-573	Sample 821AN	The above specimen; expansion measured with increasing temperature on second test.
29*	19	Hidnert, P. and Krider, H.S.	1952	L	573-293	Sample 821AN	The above specimen; expansion measured with decreasing temperature.
30*	19	Hidnert, P. and Krider, H.S.	1952	L	293-573	Sample 828AN	Similar to the above specimen; expansion measured with increasing temperature.
31*	19	Hidnert, P. and Krider, H.S.	1952	L	573-293	Sample 828AN	The above specimen; expansion measured with decreasing temperature.
32*	19	Hidnert, P. and Krider, H.S.	1952	L	293-573	Sample 828AN	The above specimen; expansion measured with increasing temperature on second test.
33*	19	Hidnert, P. and Krider, H.S.	1952	L	573-293	Sample 828AN	The above specimen; expansion measured with decreasing temperature.
34	20	Hidnert, P.	1925	T	293-873	S821	99.952 Al, 0.019 Cu, 0.015 Fe, 0.014 Si; specimen cut from bar cast in graphite mold; specimen similar to one used for curves 26-29 except not annealed; expansion measured on several tests with increasing temperature.
35*	20	Hidnert, P.	1925	T	293-873	S828	Specimen cut from same cast bar; similar to specimen used for curves 30-33 except not annealed.
36	21	Schmidt, F.J. and Hess, I.J.	1965	O	216, 366		Specimen extrapolated onto substrate, then removed; 0.066 cm thick x 7.62 cm long; data average of two runs on each of three similar specimens; average density 2.66 g cm^{-3}; zero-point determined by assuming coefficient of expansion to be constant over given temperature range.
37*	22	Fraser, D.B. and Hollis-Hallett, A.C.	1961	I	15-300		Metal machined into form of thick-walled cylindrical tube, having ends which were carefully ground plane and parallel; annealed; reported error 20% maximum at 15 K, error decreases with increasing temperature.
38*	23	Carr, R.H.	1963	V	0-20	Sample 5 Run 17	Polycrystalline sample from high-purity stock, expansion fitted to curve $(L-L_0)/L_0 = (40.5\ T^2 + 0.765\ T^4) \times 10^{-11}$.

* Not shown in figure.

SPECIFICATION TABLE 1. THERMAL LINEAR EXPANSION OF ALUMINUM Al (continued)

Cur. No.	Ref. No.	Author(s)	Year	Method Used	Temp. Range, K	Name and Specimen Designation	Composition (weight percent), Specifications, and Remarks
39	24	Nicklow, R.M. and Young, R.A.	1963	X	124-293		Small single crystal specimen prepared from large single crystal of pure material by sawing, following by extensive etching.
40*	25	Wilson, A.J.C.	1941	X	273-923		No details given; reported error 0.01%; lattice spacing reported at 298 K is 4.0413 Å; 4.0408 Å at 293 K determined by graphical interpolation.
41	26	Buffington, R.M. and Latimer, W.M.	1926	I	87-315		Specimen turned from hand-rolled aluminum rod.
42	27	Pathak, P.D. and Vasavada, N.G.	1970	L	300-900		99.999 Al metal rod obtained from Johnson Matthey and Co., London; 10 cm long, about 0.5 cm diameter; reported error 0.5-1.0%.
43*	28	Abiss, C.P., Huzan, E. and Jones, G.O.	1961	O	16-71		May be same experiment as reported in Volumetric Expansion, Curve 2 (TPRC No. 37107); this curve is here reported using the first given temperature as reference temperature at which $\Delta L/L_0 = 0$.
44*	29	Branchereau, M., Navez, M. and Pervoux, M.	1962	L	273, 573		99.998 Al; specimen 100 mm long, 8 mm diameter.
45*	29	Branchereau, M., et al.	1962	L	273, 573		99.996 Al; 100 mm long, 8 mm diameter specimen.
46	30	Strelkov, P.G. and Novikova, S.I.	1969	L	85-343		99.99 Al; specimen 10 mm long cut from ingot; data results of several sets of tests.
47*	31	Kagan, A.S.	1964	X	295, 373		Massive sample glued to copper holder; reported error 2.0%.
48	32	Kochanovska, A.	1949	X	295-553		Polycrystalline sample, supplied by Everitt and Co., Ltd., Liverpool, as chemically pure, values of lattice constant determined from reflection (115) by $CuK\alpha_1$ radiation; lattice parameter reported at 295 K is 4.0410 Å; 4.0408 Å at 293 K determined by graphical extrapolation.
49	32	Kochanovska, A.	1949	X	295-530		The above specimen; values of lattice constant determined from reflection (024) by $CoK\alpha_1$ radiation; lattice parameter reported at 295 K is 4.0412 Å; 4.0410 Å at 293 K determined by graphical extrapolation.
50	32	Kochanovska, A.	1949	X	295-560		The above specimen; values of lattice constant determined from reflection (222) by $CrK\alpha_1$ radiation; lattice parameter reported at 295 K is 4.0410 Å; 4.0408 Å at 295 K determined by graphical extrapolation.
51	33	Richards, J.W.	1942	T	82-896		99.989 Al, 0.004 Si, 0.004 Cu, 0.003 Fe; furnished by Aluminum Co. of America; expansion measured in hydrogen or nitrogen atmosphere.
52	34	Asay, J.R.	1968	L	173-473	Al-1060	99.6 Al; density 2.697 g cm^{-3}; zero-point correction of 0.011% determined by graphical interpolation.
53*	35	Honda, K. and Okubo, Y.	1924	L	334-621	A	Pure Al, 15.025 cm long, 5 mm diameter cylindrical rod specimen.
54*	36	Otte, H.M., Montague, W.G. and Welch, D.O.	1963	X	293-312		99.99 Al polycrystalline rod 3.12 mm diameter, $CuK\alpha_1$ radiation used; temperature controlled to ± 0.5 C, extrapolated lattice parameter.
55	37	Cormish, A.J. and Burke, J.	1965	X	293-927		99.995 Al; lattice parameter at 293.2 K is 4.0488 Å.
56	38	Uffelmann, F.L.	1930	I	373-803		Cubic specimen of side 4 mm.

* Not shown in figure.

SPECIFICATION TABLE 1. THERMAL LINEAR EXPANSION OF ALUMINUM Al (continued)

Cur. No.	Ref. No.	Author(s)	Year	Method Used	Temp. Range, K	Name and Specimen Designation	Composition (weight percent), Specifications, and Remarks
57	66	Feder, R. and Nowick, A. S.	1958	L	473-931		99.997 Al single crystal grown by Bridgman Technique; macroscopic expansion studied; zero-point correction of -0.050% determined by graphical extrapolation.
58	66	Feder, R. and Nowick, A. S.	1958	X	494-928		The above specimen; microscopic expansion studied; zero-point correction of -0.05% determined by graphical extrapolation.
59	255	Bollenrath, F.	1934	L	293-773		99.87 Al; Hoopes-Aluminium.
60	342, 697	Gupta, M. L.	1968	X	110-290		99.9 pure powder form specimen; provided by Research Chemicals, Inc., Burbank, Calif.; annealed at 1373 K for 72 hr; lattice parameter reported at 290 K is 4.0490 Å; 4.0493 Å at 293 K determined by graphical extrapolation.
61*	342	Gupta, M. L.	1968	X	290-110		Similar to the above specimen; expansion measured with decreasing temperature; lattice parameter reported at 290 K is 4.0488 Å; 4.0491 Å at 293 K determined by graphical extrapolation.
62*	612	Boiko, B.T., Palatnik, L.S. and Pugachev, A.T.	1967	F	293-566		99.9$^+$ pure thin film specimen condensed in vacuum 5 x 10^{-5} torr; thickness of film in the range 600 + 10 Å; lattice parameter expansion measured with increasing temperature.
63	612	Boiko, B.T., et al.	1967	F	597-343		The above specimen; lattice parameter expansion measured with decreasing temperature.
64*	617, 701	Staumanis, M.E. and Woodard, C.L.	1971	X	40-180		99$^+$ pure specimen from Pierce Chemical Co., Rockford, Ill., particle size 8-15 μ; lattice parameter reported at 125.7 K is 4.03527 Å; this curve is here reported using the first given temperature as reference temperature at which $\Delta L/L_0 = 0$; data on the coefficient of thermal linear expansion are also reported.
65*	647	Ellwood, E.C. and Silcock, J.M.	1948	X	600-864		99.9985 pure aluminum containing 0.005 each of Si, Fe, Cu from the British Aluminum Co. Ltd; lattice parameters reported at 600 K is 4.0808 kX; this curve is here reported using the first given temperature as reference temperature at which $\Delta L/L_0 = 0$.
66*	696	Awad, F.G. and Gugan, D.	1971	E	10-80	Superaffinal	99.999 pure, larged-grained (5 mm) polycrystal samples; two runs made on the specimen, these results smoothed and only rounded values of linear thermal expansion coefficient reported; error limits of ± 2% on linear thermal expansion coefficient; this curve is here reported using the first given temperature as reference temperature at which $\Delta L/L_0 = 0$.
67*	699	McLean, K.O.	1969	V	3.8-27		Cylindrical specimen approximately 0.5 cm in diameter; obtained from Alcoa; $\ell_0 = 9.872$ cm at 0 K (estimated using the Gruneisen correlation reduced to $(\ell-\ell_0)/\ell_0 = -0.415$ at 0 K); uncertainty of ± 2% due to systematic errors.
68*	713	Esser, H. and Eusterbrock, H.	1941	O	293-673		Specimen annealed for 2 hr at 793 K and cooled slowly.
69*	716	Andres, K. and Rohrer, H.	1961	O	3.0-10.5		No details given; coefficient of linear thermal expansion also given; this curve is here reported using the first given temperature as reference temperature at which $\Delta L/L_0 = 0$.
70*	714	Andres, K.	1964	O	3.0-12		99.99 pure; this curve is here reported using the first given temperature as reference temperature at which $\Delta L/L_0 = 0$.
71*	723	Griessen, R. and Ott, H.R.	1971	E	0.3-1.18		High purity single crystal; cylindrical specimen of 36 mm length and 6 mm diameter; tabulated $\Delta L/L_0$ are defined as $L_s - L_n/L_s$ where L_s and L_n are lengths respectively in superconducting and normal states, transition temperature 1.18 K at which $L_s - L_n/L_s = 0$.

* Not shown in figure.

DATA TABLE 1. THERMAL LINEAR EXPANSION OF ALUMINUM Al

[Temperature, T, K; Linear Expansion, $\Delta L/L_0$, %]

T	$\Delta L/L_0$
CURVE 1	
82.2	-0.391
88.7	-0.386*
96.7	-0.377
103.7	-0.367*
111.7	-0.358
117.2	-0.348*
125.2	-0.339*
130.5	-0.330*
130.2	-0.321*
142.7	-0.311*
147.2	-0.302
153.2	-0.292*
158.2	-0.283*
164.2	-0.274
169.2	-0.265*
173.7	-0.255*
179.6	-0.246
184.2	-0.236*
189.2	-0.227*
193.7	-0.217*
197.9	-0.208
202.7	-0.199*
207.2	-0.189*
211.9	-0.180
216.7	-0.171*
220.7	-0.161*
225.2	-0.152*
229.7	-0.143*
235.7	-0.133
238.6	-0.124*
243.1	-0.115*
247.2	-0.105*
251.2	-0.096
256.2	-0.086*
260.0	-0.077*
264.2	-0.068*
268.2	-0.058*
273.2	-0.047*
276.7	-0.040*
280.8	-0.030*
284.2	-0.021*
288.3	-0.012*
292.4	-0.002
294.2	0.002*
296.2	0.007*
298.2	0.012*
CURVE 1 (cont.)	
301.2	0.019*
302.2	0.023*
304.2	0.027*
308.2	0.037*
312.2	0.046
316.2	0.055*
320.7	0.065*
323.9	0.074*
328.2	0.084*
332.2	0.093
336.0	0.102*
340.2	0.112*
343.5	0.121*
349.1	0.130*
352.2	0.140
356.0	0.149*
360.2	0.159*
364.2	0.168*
368.2	0.177*
370.8	0.186
376.0	0.197*
379.6	0.205*
383.7	0.215*
387.2	0.224*
391.0	0.233*
395.7	0.244
399.2	0.252*
403.2	0.262*
407.6	0.271*
411.2	0.280*
415.2	0.290
418.3	0.299*
422.2	0.308*
426.2	0.318*
430.2	0.327*
433.2	0.336*
435.7	0.342
436.2	0.346*
440.4	0.355*
444.2	0.364*
448.2	0.374*
451.8	0.383
455.2	0.392*
459.1	0.402*
463.1	0.411*
466.2	0.420
CURVE 1 (cont.)	
469.9	0.430*
474.0	0.439*
477.2	0.449*
481.2	0.458*
484.7	0.467
488.2	0.477*
492.1	0.486*
495.7	0.495*
499.1	0.505*
502.9	0.514*
506.4	0.523
510.3	0.533*
513.1	0.542*
517.4	0.551*
521.1	0.561*
524.2	0.570
527.9	0.582*
531.2	0.589*
534.3	0.598*
537.8	0.608*
541.2	0.617*
544.2	0.626*
548.2	0.636*
551.1	0.645*
554.2	0.654*
557.7	0.664*
560.7	0.673*
564.2	0.683*
567.2	0.692*
570.3	0.701
573.8	0.711*
576.9	0.720*
580.2	0.729*
583.2	0.739*
586.2	0.748*
589.2	0.758*
592.9	0.767
596.0	0.776*
599.3	0.786*
602.4	0.795*
606.2	0.804*
609.2	0.814*
612.2	0.823*
615.2	0.832
618.2	0.842*
621.2	0.851*
CURVE 1 (cont.)	
624.2	0.861*
627.3	0.870*
630.2	0.879
633.2	0.888*
636.2	0.898*
639.3	0.907*
642.6	0.917*
646.0	0.926
648.7	0.936*
652.2	0.945
654.8	0.954*
657.9	0.963*
660.7	0.973*
664.2	0.982*
667.0	0.991
CURVE 2	
523	0.575*
573	0.713
623	0.853
673	1.003
723	1.158
773	1.317
823	1.483
873	1.658
923	1.844
CURVE 3*,†	
22.6	0.0000
27.8	0.0003
32.1	0.0008
38.2	0.0018
49.6	0.0048
53.2	0.0064
58.0	0.0086
62.0	0.0102
65.5	0.0124
68.8	0.0149
71.9	0.0169
74.5	0.0192
CURVE 4*,†	
23.2	0.0000
CURVE 4 (cont.)*,†	
26.5	0.0002
30.0	0.0005
34.0	0.0010
41.7	0.0028
45.0	0.0034
49.1	0.0046
54.3	0.0066
58.7	0.0089
63.4	0.0110
67.0	0.0134
70.1	0.0153
73.4	0.0186
76.2	0.0203
79.4	0.0233
85.2	0.0289
96.0	0.0398
CURVE 5	
20.4	-0.428
22.3	-0.427
44.4	-0.425
55.1	-0.420
66.0	-0.415
75.0	-0.407
85.7	-0.397
106.2	-0.373
115.2	-0.360
125.0	-0.344*
298.7	0.0116
CURVE 6*	
293	0.0000
373	0.1920
473	0.4446
573	0.7168
CURVE 7	
293.2	0.000
373.2	0.192
473.2	0.450
573.2	0.728
673.2	1.026
CURVE 8	
20	-0.416
30	-0.415*
40	-0.414
50	-0.412
60	-0.408*
70	-0.402
80	-0.394*
90	-0.383
110	-0.357*
130	-0.327*
150	-0.294
170	-0.258
190	-0.220
210	-0.180*
230	-0.138
250	-0.0939*
270	-0.0489
273.2	-0.0420*
CURVE 9*	
20	-0.415
30	-0.414
40	-0.413
50	-0.410
60	-0.405
70	-0.399
80	-0.391
90	-0.381
100	-0.369
110	-0.357
120	-0.342
130	-0.327
140	-0.311
150	-0.294
160	-0.277
170	-0.258
180	-0.240
190	-0.220
200	-0.201
220	-0.160
240	-0.118
260	-0.074
273	-0.045
290	-0.007
298	0.011
300	0.016

* Not shown in figure.
† This curve is here reported using the first given temperature as reference temperature at which $\Delta L/L_0 = 0$.

DATA TABLE 1. THERMAL LINEAR EXPANSION OF ALUMINUM Al (continued)

T	$\Delta L/L_0$
CURVE 10	
20	-0.409*
30	-0.408
40	-0.407*
50	-0.404*
60	-0.401
70	-0.395*
80	-0.387*
90	-0.377
100	-0.366*
110	-0.353*
120	-0.339
130	-0.324*
140	-0.308
150	-0.291*
160	-0.274
170	-0.255*
180	-0.236
190	-0.217*
200	-0.196
210	-0.177*
220	-0.156
230	-0.134*
240	-0.113*
250	-0.091*
260	-0.071
270	-0.049
280	-0.028
290	-0.006*
300	0.016*
CURVE 11*	
273	-0.048
373	0.194
423	0.323
473	0.452
573	0.721
673	1.026
CURVE 12*	
273	-0.046
373	0.187
423	0.304
473	0.434
573	0.711
673	1.006
CURVE 13*	
293	0.000
373	0.191
473	0.442
573	0.713
673	1.007
773	1.329
CURVE 14	
88	-0.375*
113	-0.341
153	-0.280
193	-0.206
233	-0.128*
273	-0.043
313	0.043
351	0.132
CURVE 15*	
98	-0.368
113	-0.348
153	-0.285
193	-0.211
233	-0.131
273	-0.046
313	0.046
353	0.138
CURVE 16*	
86	-0.380
113	-0.344
153	-0.279
193	-0.208
233	-0.128
273	-0.043
313	0.043
	0.182
CURVE 17*	
293	0.00
373	0.16
473	0.378
573	0.602
673	0.836
773	1.08
CURVE 18	
373	0.172
418	0.323
468	0.453
598	0.815
CURVE 19	
600	0.817
568	0.752
534	0.690
504	0.551
451	0.449
373	0.277
343	0.153
CURVE 20	
18	-0.406*
23	-0.406*
33	-0.405
43	-0.404*
53	-0.403*
63	-0.401
73	-0.396*
83	-0.389*
93	-0.380*
103	-0.369
113	-0.356*
123	-0.341*
133	-0.325*
143	-0.308*
153	-0.292*
163	-0.274*
173	-0.255
193	-0.216*
213	-0.177*
233	-0.135*
253	-0.091*
273	-0.046*
293	0.000*
313	0.047*
333	0.094*
353	0.144*
373	0.194*
393	0.243
413	0.295
433	0.346
453	0.399
CURVE 20 (cont.)	
473	0.452
493	0.505
513	0.559
533	0.614
553	0.668
573	0.723
CURVE 21	
293	0.000*
373	0.190*
473	0.444*
573	0.717*
673	1.015*
773	1.338
873	1.697
CURVE 22*	
293	0.000
373	0.190
473	0.451
573	0.721
673	1.023
773	1.352
873	1.714
CURVE 23	
503	0.518
686	1.044
687	1.051*
713	1.125
752	1.248
754	1.258*
769	1.301*
794	1.393
806	1.433*
811	1.451
831	1.518*
844	1.567
855	1.603*
868	1.653
878	1.691*
892	1.747
918	1.852*
929	1.898
CURVE 24*	
503	0.518
548	0.648
577	0.726
581	0.739
584	0.747
590	0.766
629	0.875
631	0.884
651	0.939
659	0.964
689	1.055
709	1.114
729	1.175
741	1.215
764	1.293
772	1.319
795	1.389
806	1.427
811	1.443
830	1.508
845	1.560
856	1.599
869	1.643
882	1.691
892	1.726
905	1.777
918	1.824
929	1.870
CURVE 25	
227	-0.162
243	-0.122
263	-0.074
283	-0.025
303	0.025
323	0.073
343	0.122
363	0.170
375	0.199*
CURVE 26*	
293	0.000
373	0.188
473	0.443
573	0.714
CURVE 27*	
573	0.717
473	0.443
373	0.187
293	0.000
CURVE 28*	
293	0.000
373	0.188
473	0.443
573	0.714
CURVE 29*	
573	0.714
473	0.443
373	0.190
293	0.000
CURVE 30*	
293	0.000
373	0.187
473	0.439
573	0.708
CURVE 31*	
573	0.714
473	0.439
373	0.187
293	0.000
CURVE 32*	
293	0.000
373	0.186
473	0.439
573	0.711
CURVE 33*	
573	0.714
473	0.439
373	0.187
293	0.000

* Not shown in figure.

DATA TABLE 1. THERMAL LINEAR EXPANSION OF ALUMINUM Al (continued)

T	$\Delta L/L_0$
CURVE 34	
293	0.000*
373	0.190*
473	0.446*
523	0.582
573	0.722*
673	1.015*
773	1.330*
873	1.665
CURVE 35*	
293	0.000
373	0.190
473	0.445
523	0.591
573	0.720
673	1.015
773	1.330
873	1.670
CURVE 36	
216	-0.188
366	0.180
CURVE 37*, ‡	
15	-0.417
20	-0.417
30	-0.416
40	-0.415
50	-0.412
70	-0.403
100	-0.374
150	-0.300
200	-0.205
250	-0.098
300	0.015
CURVE 38*, ‡	
0.0	0.000
2.0	0.019
3.2	0.055
5.4	0.19
6.7	0.34
7.6	0.47
9.8	1.08
CURVE 38 (cont.)*, ‡	
10.8	1.48
11.4	1.86
11.8	2.05
12.5	2.50
13.4	3.31
14.2	3.98
14.6	4.41
15.4	5.26
15.7	5.52
17.1	7.55
17.3	7.76
18.5	9.75
19.5	11.2
19.5	11.4
CURVE 39‡	
124	-0.335
168	-0.262*
210	-0.182*
234	-0.132*
251	-0.095*
270	-0.053*
293	0.000*
CURVE 40*	
273	-0.042
298	0.012
373	0.193
423	0.322
473	0.455
573	0.725
673	1.020
773	1.334
873	1.680
923	1.866
CURVE 41	
86.5	-0.3774
89.3	-0.3733*
98.1	-0.3651*
114.5	-0.3439*
115.9	-0.3425*
124.7	-0.3299*
158.9	-0.2744
196.5	-0.2044*
CURVE 41 (cont.)	
226.5	-0.1439
239.3	-0.1174
269.2	-0.0531*
273.9	-0.0431*
276.4	-0.0375*
297.0	0.0090*
310.7	0.0402
314.5	0.0489*
315.3	0.0508*
CURVE 42‡	
300	0.017
400	0.260
500	0.520
600	0.796
700	1.092
800	1.411
900	1.767
CURVE 43*, ‡	
16	0.000×10^{-3}
20	0.113
23	0.228
26	0.375
31	0.845
32	0.978
43	3.063
43	3.089
56	7.890
56	7.940
67	14.495
70	17.023
CURVE 44*	
273	-0.051
573	0.710
CURVE 45*	
273	-0.0507
573	0.7110
CURVE 46‡	
84	-0.393
CURVE 46 (cont.)‡	
92	-0.386
99	-0.378
105	-0.370
114	-0.359
115	-0.357
122	-0.347
127	-0.340
131	-0.334
132	-0.332
140	-0.319
142	-0.317
150	-0.303
150	-0.302
159	-0.286
162	-0.282
170	-0.266
175	-0.258
178	-0.252
179	-0.249
181	-0.246
188	-0.232
195	-0.218
200	-0.209
208	-0.191
209	-0.189
210	-0.186
218	-0.169
219	-0.168
219	-0.167
228	-0.147
239	-0.125
247	-0.106
249	-0.102*
258	-0.081
278	-0.034*
280	-0.031*
288	-0.011
293	0.000*
300	0.017*
311	0.043*
331	0.094*
342	0.120
CURVE 47*	
295	0.005
373	0.195
CURVE 48	
295	0.005
328	0.087
352	0.129*
375	0.188*
417	0.287
469	0.403
473	0.411*
509	0.495
546	0.596
553	0.616
CURVE 49	
295.	0.005*
341.	0.111*
357.	0.153
390.	0.217*
392.	0.225
417.	0.284*
490.	0.450
530.	0.514
CURVE 50	
295	0.004*
337	0.101
380	0.198
444	0.339
470	0.393
512	0.494
560	0.608
CURVE 51	
82	-0.382*
300	0.018*
466	0.449*
736	1.230
861	1.721
896	1.902
CURVE 52	
173	-0.212
198	-0.179
223	-0.144
248	-0.098
273	-0.053
CURVE 52 (cont.)	
298	0.000
323	0.052
348	0.107
373	0.165
398	0.227
423	0.290
448	0.356
473	0.421
CURVE 53*, ‡	
334	0.100
373	0.195
423	0.321
472	0.447
522	0.579
572	0.714
621	0.849
CURVE 54*	
292.5	0.0000
297.4	0.0114
302.5	0.0238
312.4	0.0467
CURVE 55	
293.2	0.000
408.1	0.276
472.3	0.443
548.7	0.645
576.7	0.726
649.7	0.935
707.0	1.112
741.8	1.224
748.1	1.240
787.6	1.367
797.8	1.404
848.2	1.574
892.1	1.729
907.8	1.787
914.1	1.812
926.8	1.861
CURVE 56‡	
373	0.191

* Not shown in figure.
† This curve is here reported using the first given temperature as reference temperature at which $\Delta L/L_0 = 0$.
‡ Author's data for coefficient of thermal expansion have been integrated by TPRC to obtain $\Delta L/L_0$.

DATA TABLE 1. THERMAL LINEAR EXPANSION OF ALUMINUM Al (continued)

T	$\Delta L/L_0$
CURVE 56 (cont.) ‡	
413	0.287
453	0.385
493	0.487
533	0.591
573	0.699
613	0.810
653	0.925
693	1.041
733	1.161
773	1.283
803	1.378
CURVE 57	
473	0.443*
514	0.552*
516	0.560*
521	0.568*
522	0.576*
554	0.661*
562	0.686*
567	0.692*
568	0.699*
590	0.757
592	0.764*
611	0.816*
645	0.921
666	0.986*
668	0.993*
684	1.041
685	1.051*
709	1.120
709	1.131*
711	1.137*
739	1.217
753	1.265*
777	1.348
793	1.403
810	1.464
828	1.522
867	1.650*
887	1.714
898	1.751*
907	1.790*
925	1.851*
931	1.879
CURVE 58	
494	0.501*
502	0.517*
520	0.570
524	0.581*
527	0.597*
546	0.635
546	0.640*
572	0.717*
573	0.717*
601	0.791
614	0.832*
615	0.837*
619	0.847
671	1.003
702	1.103
705	1.112*
721	1.167
736	1.208*
745	1.245
746	1.245*
749	1.263*
770	1.327*
774	1.343*
792	1.401*
822	1.500
844	1.577*
853	1.614
863	1.645*
873	1.688*
883	1.720
893	1.750*
894	1.762*
905	1.800
915	1.833*
916	1.833*
919	1.838
922	1.855*
926	1.865*
927	1.865*
928	1.876*
CURVE 59	
293	0.0000*
323	0.0690*
373	0.1856*
CURVE 59 (cont.)	
423	0.3047*
473	0.4284*
523	0.5534
573	0.6877
623	0.8243
673	0.9652
723	1.1098
773	1.2605
CURVE 60	
110	-0.368*
130	-0.343
150	-0.306*
170	-0.267*
190	-0.230
210	-0.195*
230	-0.141
250	-0.099*
270	-0.057
290	-0.007*
CURVE 61*	
290	-0.007
270	-0.049
250	-0.101
230	-0.141
210	-0.183
190	-0.225
170	-0.259
150	-0.306
130	-0.338
110	-0.368
CURVE 62*	
293	0.000
373	0.173
417	0.320
468	0.456
529	0.699
566	0.755
CURVE 63	
597	0.820
535	0.676
503	0.551
452	0.452*
373	0.272*
343	0.153*
CURVE 64*	
40	0.0000
60	0.0079
80	0.0238
100	0.0448
126	0.082
140	0.105
180	0.175
CURVE 65*	
600	0.000
649	0.149
703	0.313
742	0.441
790	0.595
834	0.740
864	0.845
CURVE 66*, †, ‡	
10	0.000×10^{-3}
15	0.040
20	0.135
25	0.315
30	0.650
35	1.23
40	2.13
45	3.14
50	5.13
60	9.92
65	13.0
70	16.5
75	20.4
CURVE 67*, †, ‡	
3.874	0.00×10^{-5}
4.009	0.01
CURVE 67 (cont.)*, †, ‡	
4.256	0.02×10^{-5}
4.588	0.06
4.993	0.08
5.487	0.13
5.785	0.16
6.081	0.19
6.609	0.27
6.889	0.31
7.221	0.37
7.544	0.43
7.749	0.48
8.516	0.66
9.152	0.85
9.785	1.06
10.401	1.31
11.052	1.61
11.759	1.99
12.225	2.28
12.831	2.69
13.209	2.97
13.599	3.28
14.097	3.72
14.682	4.28
15.709	5.43
16.619	6.63
17.358	7.75
18.085	8.99
18.692	10.14
19.433	11.69
20.027	13.07
20.516	14.29
21.152	16.03
21.897	18.27
22.809	21.35
23.583	24.28
24.196	26.84
24.913	30.09
25.553	33.24
26.236	36.90
26.811	40.23
CURVE 68*	
293	0.000
373	0.192
473	0.443
573	0.711
673	0.999
CURVE 69*, †, ‡	
3	0.000×10^{-4}
4	0.005
5	0.012
6	0.022
7	0.035
8	0.052
9	0.075
10	0.105
10.5	0.124
CURVE 70*, †	
3.0	0.000×10^{-4}
3.0	0.002
3.1	0.001
3.6	0.002
3.6	0.002
3.8	0.005
3.8	0.006
3.9	0.006
4.5	0.008
4.6	0.007
4.7	0.011
5.5	0.015
5.6	0.017
6.8	0.033
6.8	0.034
7.9	0.052
9.3	0.082
10.0	0.121
10.6	0.126
11.8	0.187
CURVE 71*, ǂ	
0.3	0.85×10^{-6}
0.5	0.76
0.7	0.61
1.0	0.28
1.18	0.00

* Not shown in figure.
† This curve is here reported using the first given temperature as reference temperature at which $\Delta L/L_0 = 0$.
‡ Author's data for coefficient of thermal expansion have been integrated by TPRC to obtain $\Delta L/L_0$.
ǂ For explaination, see specification table.

DATA TABLE 1. COEFFICIENT OF THERMAL LINEAR EXPANSION OF ALUMINUM Al

[Temperature, T, K; Coefficient of Expansion, α, 10^{-6} K^{-1}]

T	α
CURVE 3*, ‡	
25.2	0.62
29.5	0.89
35.1	1.65
43.9	2.64
53.8	4.51
57.6	4.37
61.7	4.99
65.4	6.81
68.7	7.12
71.6	7.65
CURVE 4*	
26.6	0.71
30.2	1.1
35.8	1.94
39.5	2.18
45.4	2.48
49.6	3.45
53.9	4.48
58.8	4.79
62.8	5.35
66.7	6.39
70.2	8.17
73.1	8.14
76.4	7.86
80.7	9.53
82.3	9.55
90.6	10.1
CURVE 37‡	
15	0.18
20	0.30
30	1.0
40	1.9
50	3.1
70	6.7
100	12.2
150	17.5
200	20.6
250	22.1
300	23.3
CURVE 39‡	
124	15.0
168	18.0
210	20.4
234	21.3
251	21.9
270	22.4
293	23.2
CURVE 42*, ‡	
300	23.4
400	25.2
500	26.8
600	28.5
700	30.6
800	33.3
900	37.8
CURVE 43‡	
16.0	0.13
20.7	0.35
23.5	0.47
26.1	0.66
31.3	1.15
32.4	1.27
43.2	2.59
43.3	2.78
56.0	4.78
56.1	4.93
67.1	6.99
70.6	7.46
CURVE 46‡	
85	8.40
92	10.29
99	11.72
106	12.37
114	14.18
115	13.28
123	14.25*
128	14.96
132	15.84
133	15.07
141	16.32*
CURVE 46 (cont.)‡	
142	16.65
151	16.68
151	17.85
160	18.13
162	18.26*
171	18.48
175	18.26
179	19.28*
180	19.58
181	19.38
189	19.83*
196	20.49
200	20.35
208	21.14
209	21.14*
211	21.21*
219	21.63
219	21.21*
220	21.66
229	21.32*
239	21.95
248	22.30
249	22.28*
259	22.52
279	23.99*
280	23.56*
288	23.56
293	23.56*
300	24.08
311	24.35*
332	24.35*
343	24.84*
CURVE 53*, ‡	
334	23.7
373	25.0
423	25.4
472	26.0
522	26.8
572	27.4
621	27.8
CURVE 56*, ‡	
373	23.5
CURVE 56 (cont.)*, ‡	
413	24.3
453	25.0
493	25.7
533	26.5
573	27.4
613	28.3
653	28.8
693	29.6
733	30.1
773	31.1
803	32.3
CURVE 64	
40	2.47
50	4.00
60	6.02
70	7.74
80	9.41
110	13.89
120	15.05
140	16.76
160	17.49
CURVE 66‡	
10	0.03
15	0.13*
20	0.25*
25	0.47
30	0.87
35	1.44
40	2.15
45	2.99
50	3.90*
60	5.68
65	6.56
70	7.46*
75	8.33
CURVE 67‡	
3.874	0.00550
4.009	0.00582
4.256	0.00639
4.588	0.00742
CURVE 67 (cont.)‡	
4.993	0.00868
5.487	0.01040
5.785	0.01160
6.081	0.01280
6.609	0.01510
6.889	0.01664
7.221	0.01843
7.544	0.02022
7.749	0.02157
8.516	0.02667
9.152	0.03171
9.785	0.03727
10.401	0.04297
11.052	0.04978
11.759	0.05773
12.225	0.06375
12.831	0.07185
13.209	0.07715
13.599	0.08332
14.097	0.09164
14.682	0.10150
15.709	0.12200
16.619	0.14216
17.358	0.16040
18.085	0.18013
18.692	0.19795
19.433	0.22196
20.027	0.24121
20.516	0.26050
21.152	0.28516
21.897	0.31640
22.809	0.35930
23.583	0.39866
24.196	0.43440
24.913	0.47210
25.553	0.51330
26.236	0.55890
26.811	0.59940
CURVE 69‡	
3	0.0039
4	0.0060
5	0.0083
6	0.0113
7	0.0150
CURVE 69 (cont.)‡	
8	0.0199
9	0.0257
10	0.0349
10.5	0.0406

*Not shown in figure.

‡Author's data for coefficient of thermal expansion have been integrated by TPRC to obtain $\Delta L/L_0$.

FIGURE AND TABLE NO. 2R. RECOMMENDED VALUES FOR THERMAL LINEAR EXPANSION OF ANTIMONY Sb

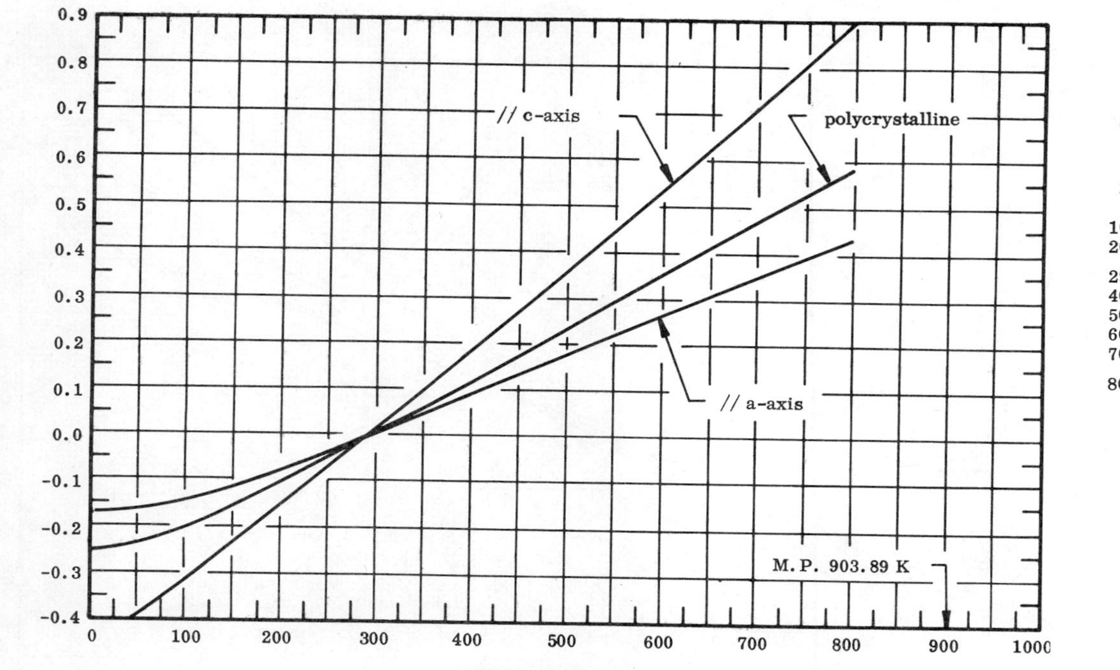

RECOMMENDED VALUES

[Temperature, T, K; Linear Expansion, $\Delta L/L_0$, %; α, K^{-1}]

T	// a-axis $\Delta L/L_0$	// c-axis $\Delta L/L_0$	polycrystalline $\Delta L/L_0$	polycrystalline $\alpha \times 10^6$
5	-0.167	-0.414	-0.249	0.02
25	-0.167	-0.409	-0.248	2.3
50	-0.164	-0.384	-0.237	5.8
100	-0.143	-0.312	-0.199	9.1
200	-0.074	-0.152	-0.100	10.5
293	0.000	0.000	0.000	11.0
400	0.093	0.179	0.122	11.5
500	0.181	0.353	0.238	11.7
600	0.266	0.531	0.354	11.7
700	0.348	0.716	0.471	11.7
800	0.432	0.900	0.588	11.7

REMARKS

The tabulated values are considered accurate to within ± 3% at temperatures below 293 K and ± 5% above. The tabulated values for polycrystalline antimony are calculated from single crystal data and are considered accurate to about ± 5% over the entire temperature range. Most of the experimental data on the polycrystalline antimony are slightly lower than the recommended values. The tabulated values can be represented approximately by the following equations:

// a-axis: $\Delta L/L_0 = -0.204 + 5.364 \times 10^{-4} T + 7.013 \times 10^{-7} T^2 - 4.781 \times 10^{-10} T^3$ $(100 < T < 800)$

// c-axis: $\Delta L/L_0 = -0.466 + 1.512 \times 10^{-3} T + 2.702 \times 10^{-7} T^2 - 3.112 \times 10^{-11} T^3$ $(100 < T < 800)$

polycrystalline: $\Delta L/L_0 = -0.291 + 8.629 \times 10^{-4} T + 5.536 \times 10^{-7} T^2 - 3.259 \times 10^{-10} T^3$ $(100 < T < 800)$

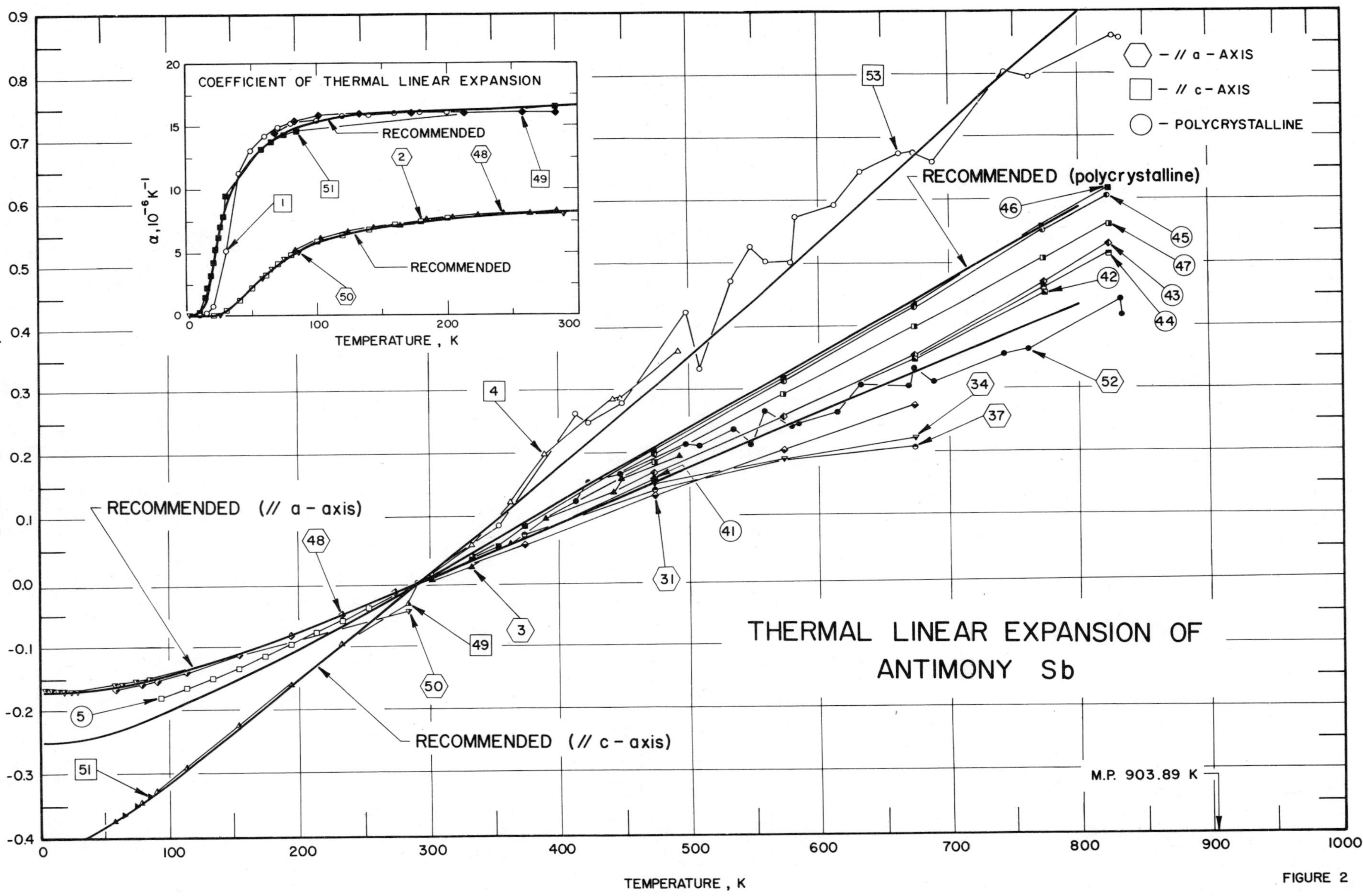
THERMAL LINEAR EXPANSION OF
ANTIMONY Sb
— // a – AXIS
— // c – AXIS
— POLYCRYSTALLINE
RECOMMENDED (polycrystalline)
RECOMMENDED (// a – axis)
RECOMMENDED (// c – axis)
M.P. 903.89 K
THERMAL LINEAR EXPANSION, %
TEMPERATURE, K
COEFFICIENT OF THERMAL LINEAR EXPANSION
RECOMMENDED
RECOMMENDED
α, 10⁻⁶ K⁻¹
TEMPERATURE, K

FIGURE 2

SPECIFICATION TABLE 2. THERMAL LINEAR EXPANSION OF ANTIMONY Sb

Cur. No.	Ref. No.	Author(s)	Year	Method Used	Temp. Range, K	Name and Specimen Designation	Composition (weight percent), Specifications, and Remarks
1*	63	Bunton, G. V. and Weintroub, S.	1969	L	10-200		Single crystal about 10 cm long, 0.55 x 0.55 cm^2 cross-section grown from precast ingot of purity higher than 99.999; expansion measured parallel to the trigonal axis; this curve is here reported using the first given temperature as reference temperature at which $\Delta L/L_0 = 0$.
2*	63	Bunton, G. V. and Weintroub, S.	1969	L	10-200		The above specimen; expansion values perpendicular to trigonal axis derived from their measurements at an angle 67 deg to trigonal axis; this curve is here reported using the first given temperature as reference temperature at which $\Delta L/L_0 = 0$.
3	64	Deshpande, V. T. and Pawar, R. R.	1969	X	301-493		Specpure metal supplied by Johnson-Matthey and Co., annealed in vacuum for about 6 hr at 723 K; expansion measured along a-axis; lattice parameter reported at 301 K is 4.3079 Å; 4.3076 Å at 293 K determined by graphical extrapolation.
4	64	Deshpande, V. T. and Pawar, R. R.	1969	X	301-493		The above specimen; expansion measured along c-axis; lattice parameter reported at 301 K is 11.2741 Å; 11.2728 Å at 293 K determined by graphical extrapolation.
5	48	Dorsey, H. G.	1907	I	93-293		Chemically pure, cast specimen; 1.3409 cm long; density 6.88 g cm^{-3}.
6*	65	Hidnert, P.	1935	L	293-673	No. 1528	Single crystal grown from melt; expansion measured 32 deg from trigonal axis with increasing temperature.
7*	65	Hidnert, P.	1935	L	293-673	No. 1528	The above specimen (32 deg); expansion measured with increasing temperature on second test.
8*	65	Hidnert, P.	1935	L	293-673	No. 1529	Similar to the above specimen; expansion measured with increasing temperature.
9*	65	Hidnert, P.	1935	L	673-293	No. 1529	The above specimen; expansion measured with decreasing temperature.
10*	65	Hidnert, P.	1935	L	293-673	No. 1529	The above specimen; expansion measured with increasing temperature on second test.
11*	65	Hidnert, P.	1935	L	293-773	No. 1526	Similar to the above specimen; expansion measured 63 deg to the trigonal axis with increasing temperature.
12*	65	Hidnert, P.	1935	L	773-293	No. 1526	The above specimen; expansion measured with decreasing temperature.
13*	65	Hidnert, P.	1935	L	293-673	No. 1526	The above specimen; expansion measured with increasing temperature on second test.
14*	65	Hidnert, P.	1935	L	293-773	No. 1526	The above specimen; expansion measured with increasing temperature on third test.
15*	65	Hidnert, P.	1935	L	773-293	No. 1526	The above specimen; expansion measured with decreasing temperature.
16*	65	Hidnert, P.	1935	L	293-673	No. 1525	Similar to the above specimen; expansion measured 69 deg to the trigonal axis with increasing temperature.
17*	65	Hidnert, P.	1935	L	673-293	No. 1525	The above specimen; expansion measured with decreasing temperature.
18*	65	Hidnert, P.	1935	L	293-673	No. 1525	The above specimen; expansion measured with increasing temperature on second test.
19*	65	Hidnert, P.	1935	L	293-673	No. 1524	Similar to the above specimen; expansion measured 72 deg to the trigonal axis with increasing temperature.
20*	65	Hidnert, P.	1935	L	673-293	No. 1524	The above specimen; expansion measured with decreasing temperature.

* Not shown in figure.

SPECIFICATION TABLE 2. THERMAL LINEAR EXPANSION OF ANTIMONY Sb (continued)

Cur. No.	Ref. No.	Author(s)	Year	Method Used	Temp. Range, K	Name and Specimen Designation	Composition (weight percent), Specifications, and Remarks
21*	65	Hidnert, P.	1935	L	293-673	No. 1524	The above specimen; expansion measured with increasing temperature on second test.
22*	65	Hidnert, P.	1935	L	293-673	No. 1527	Similar to the above specimen; expansion measured 73 deg to the trigonal axis with increasing temperature.
23*	65	Hidnert, P.	1935	L	673-293	No. 1527	The above specimen; expansion measured with decreasing temperature.
24*	65	Hidnert, P.	1935	L	293-673	No. 1527	The above specimen; expansion measured with increasing temperature on second test.
25*	65	Hidnert, P.	1935	L	293-673	No. 1531	Similar to the above specimen; expansion measured 74 deg to trigonal axis with increasing temperature.
26*	65	Hidnert, P.	1935	L	673-293	No. 1531	The above specimen; expansion measured with decreasing temperature.
27*	65	Hidnert, P.	1935	L	293-673	No. 1531	The above specimen; expansion measured with increasing temperature on second test.
28*	65	Hidnert, P.	1935	L	293-673	No. 1530	Similar to the above specimen; expansion measured 77 deg to trigonal axis with increasing temperature.
29*	65	Hidnert, P.	1935	L	673-293	No. 1530	The above specimen; expansion measured with decreasing temperature.
30*	65	Hidnert, P.	1935	L	293-673	No. 1530	The above specimen; expansion measured with increasing temperature on second test.
31	65	Hidnert, P.	1935	L	293-673	No. 1532	Similar to the above specimen; expansion measured 90 deg to trigonal axis with increasing temperature.
32*	65	Hidnert, P.	1935	L	673-293	No. 1532	The above specimen; expansion measured with decreasing temperature.
33*	65	Hidnert, P.	1935	L	293-673	No. 1532	The above specimen; expansion measured with increasing temperature on second test.
34	65	Hidnert, P.	1935	L	293-673	No. 1553	Similar to the above specimen; expansion measured with increasing temperature.
35*	65	Hidnert, P.	1935	L	673-293	No. 1553	The above specimen; expansion measured with decreasing temperature.
36*	65	Hidnert, P.	1935	L	293-673	No. 1553	The above specimen; expansion measured with increasing temperature on second test.
37	65	Hidnert, P.	1935	L	293-673	No. 1553A	Similar to the above specimen; expansion measured with increasing temperature.
38*	65	Hidnert, P.	1935	L	673-293	No. 1553A	The above specimen; expansion measured with decreasing temperature.
39*	65	Hidnert, P.	1935	L	293-673	No. 1553A	The above specimen; expansion measured with increasing temperature on second test.
40*	65	Hidnert, P.	1935	L	293-673	No. 1553A	The above specimen; expansion measured with increasing temperature on third test.
41	65	Hidnert, P.	1935	L	293-473	No. 1067	99. 8 Sb, polycrystalline cast specimen; density 6. 639 g cm^{-3}; expansion measured on first test.
42	65	Hidnert, P.	1935	L	293-773	No. 1067	The above specimen; expansion measured on second test.
43	65	Hidnert, P.	1935	L	293-823	No. 1067	The above specimen; expansion measured on third test.

* Not shown in figure.

SPECIFICATION TABLE 2. THERMAL LINEAR EXPANSION OF ANTIMONY Sb (continued)

Cur. No.	Ref. No.	Author(s)	Year	Method Used	Temp. Range, K	Name and Specimen Designation	Composition (weight percent), Specifications, and Remarks
44	65	Hidnert, P.	1935	L	293-823	No. 1067	The above specimen; expansion measured on fourth test.
45	65	Hidnert, P.	1935	L	293-823	No. 1450	99.9^+ Sb; two polycrystalline bars joined by pouring molten Sb into crack between them; density 6.656 g cm^{-3}; expansion measured on first test.
46	65	Hidnert, P.	1935	L	293-823	No. 1450	The above specimen; expansion measured on second test.
47	65	Hidnert, P.	1935	L	293-823	No. 1451	99.9^+ Sb; the above specimen melted and cast; density 6.684 g cm^{-3}.
48	710	Erfling, H.D.	1939	I	293-58		Measurements on single crystal, purity of sample checked by resistivity value; measurements along a-axis; data on the coefficients of thermal linear expansion also reported.
49	710	Erfling, H.D.	1939	I	293-58		Similar to the above specimen; measurements along c-axis; data on the coefficient of thermal linear expansion also reported.
50	711	White, G.K.	1972	L	2-85, 283	Sb1	Cylindrical specimen of 2 cm diameter and 4 cm long; single crystal grown from 5N grade material from Metals Research Limited of Cambridge; the results are extrapolated to 293 K using other available data; measured along perpendicular to trigonal axis; average values of several runs.
51	711	White, G.K.	1972	L	2-85, 283	Sb3	Single crystal grown by J. McAllan of NSL Thermometry Section from 5N grade antimony supplied by Bradley Mining Co., California; five pieces of 10 x 10 x 8 mm were spark cut and cleaned in the (111) plane and lapped by hand and stacked to form a column of 4 cm high, parallel to the trigonal axis; average values of several runs.
52	712	Klemm, W., Spitzer, H., and Niermann, H.	1960	X	293-833		Pure specimen; measured along a-axis; average of two runs; lattice parameter reported at 293 K is 4.3029 Å.
53	712	Klemm, W., et al.	1960	X	293-831		The above specimen; measured along c-axis; average of two runs; lattice parameter reported at 293 K is 11.2711 Å.

* Not shown in figure.

DATA TABLE 2. THERMAL LINEAR EXPANSION OF ANTIMONY Sb

[Temperature, T, K; Linear Expansion, $\Delta L/L_0$, %]

T	$\Delta L/L_0$
CURVE 1*, †, ‡	
10	0.00000
15	0.00007
20	0.0003
30	0.0033
40	0.011
50	0.023
60	0.037
70	0.052
80	0.067
100	0.098
120	0.129
140	0.161
160	0.193
180	0.225
200	0.258
CURVE 2*, †, ‡	
10	0.00000
15	0.00002
20	0.00003
30	0.0002
40	0.001
50	0.002
60	0.005
70	0.009
80	0.013
100	0.024
120	0.037
140	0.050
160	0.064
180	0.079
200	0.094
CURVE 3	
301	0.007
332	0.026
364	0.058
389	0.100
441	0.137
448	0.160
493	0.197

T	$\Delta L/L_0$
CURVE 4	
301	0.012
332	0.057
364	0.125
389	0.200
441	0.288
448	0.288
493	0.366
CURVE 5	
93	-0.181
113	-0.166
133	-0.150
153	-0.133
173	-0.115
193	-0.097
213	-0.078
233	-0.059
253	-0.040
273	-0.020*
293	0.000*
CURVE 6*	
293	0.000
373	0.106
473	0.243
573	0.375
673	0.505
CURVE 7*	
293	0.000
373	0.099
473	0.239
573	0.378
673	0.509
CURVE 8*	
293	0.000
373	0.101
473	0.223
573	0.330
673	0.460

T	$\Delta L/L_0$
CURVE 9*	
673	0.467
473	0.220
373	0.091
293	0.000
CURVE 10*	
293	0.000
373	0.091
473	0.209
573	0.336
673	0.460
CURVE 11*	
293	0.000
373	0.081
473	0.169
573	0.260
673	0.346
773	0.413
CURVE 12*	
773	0.461
673	0.361
473	0.162
373	0.073
293	0.000
CURVE 13*	
293	0.000
373	0.077
473	0.169
573	0.260
673	0.342
CURVE 14*	
293	0.000
373	0.076
473	0.171
573	0.266
673	0.365
773	0.442

T	$\Delta L/L_0$
CURVE 15*	
773	0.461
673	0.365
573	0.269
423	0.171
373	0.075
293	0.000
CURVE 16*	
293	0.000
373	0.073
473	0.153
573	0.241
673	0.319
CURVE 17*	
673	0.334
473	0.153
373	0.072
293	0.000
CURVE 18*	
293	0.000
373	0.072
473	0.157
573	0.246
673	0.334
CURVE 19*	
293	0.000
373	0.065
473	0.144
573	0.218
673	0.285
CURVE 20*	
673	0.312
473	0.142
373	0.059
293	0.000

T	$\Delta L/L_0$
CURVE 21*	
293	0.000
373	0.062
473	0.149
573	0.230
673	0.308
CURVE 22*	
293	0.000
373	0.070
473	0.151
573	0.227
673	0.300
CURVE 23*	
673	0.323
473	0.145
373	0.070
293	0.000
CURVE 24*	
293	0.000
373	0.060
473	0.146
573	0.232
673	0.312
CURVE 25*	
293	0.000
373	0.070
473	0.151
573	0.235
673	0.315
CURVE 26*	
673	0.327
473	0.149
373	0.064
293	0.000

T	$\Delta L/L_0$
CURVE 27*	
293	0.000
373	0.068
473	0.155
573	0.241
673	0.323
CURVE 28*	
293	0.000
373	0.063
473	0.144
573	0.218
673	0.293
CURVE 29*	
673	0.300
473	0.132
373	0.058
293	0.000
CURVE 30*	
293	0.000
373	0.062
473	0.142
573	0.224
673	0.300
CURVE 31	
293	0.000*
373	0.058
473	0.135
573	0.204
673	0.274
CURVE 32*	
673	0.304
473	0.140
373	0.059
293	0.000

* Not shown in figure.
† This curve is here reported using the first given temperature as reference temperature at which $\Delta L/L_0 = 0$.
‡ Author's data for coefficient of thermal expansion have been integrated by TPRC to obtain $\Delta L/L_0$.

DATA TABLE 2. THERMAL LINEAR EXPANSION OF ANTIMONY Sb (continued)

T	$\Delta L/L_0$
CURVE 33*	
293	0.000
373	0.066
473	0.148
573	0.235
673	0.319
CURVE 34	
293	0.000*
373	0.053*
473	0.115
573	0.154
673	0.186
CURVE 35*	
673	0.293
573	0.213
473	0.144
373	0.062
293	0.000
CURVE 36*	
293	0.000
373	0.061
473	0.142
573	0.221
673	0.304
CURVE 37	
293	0.000*
373	0.072
473	0.140
573	0.186*
673	0.208
CURVE 38*	
673	0.300
573	0.216
473	0.133
373	0.060
293	0.000
CURVE 39*	
293	0.000
373	0.066
473	0.162
573	0.244
673	0.315
CURVE 40*	
293	0.000
373	0.069
473	0.157
573	0.241
673	0.312
CURVE 41	
293	0.000*
333	0.034
373	0.067
473	0.157
CURVE 42	
293	0.000*
373	0.078*
473	0.167*
573	0.258*
673	0.350
773	0.456
CURVE 43	
293	0.000*
373	0.075*
473	0.169*
573	0.260*
673	0.353
773	0.470
823	0.530
CURVE 44	
293	0.000*
373	0.074*
473	0.167*
573	0.258*
673	0.353*
773	0.461
823	0.514
CURVE 45	
293	0.000*
333	0.043*
373	0.088*
473	0.200
573	0.316
673	0.433
773	0.552
823	0.610
CURVE 46	
293	0.000*
373	0.087
473	0.203
573	0.319
673	0.437
773	0.557
823	0.615
CURVE 47	
293	0.000*
333	0.041
373	0.082*
473	0.187
573	0.294
673	0.399
773	0.509
823	0.562
CURVE 48	
293	0.0000*
273	-0.0165
233	-0.0489
193	-0.0803
153	-0.1100
113	-0.1372
90	-0.1511
78	-0.1574
58	-0.1659
CURVE 49	
293	0.000*
283	-0.0324
233	-0.0971
193	-0.1617
CURVE 49 (cont.)	
153	-0.2261
113	-0.2904
90	-0.3268
78	-0.3454
58	-0.3747
CURVE 50‡	
2.5	-0.167*
3	-0.167*
4	-0.167*
5	-0.167
6	-0.167*
7	-0.167*
8	-0.167*
10	-0.167
12	-0.167*
14	-0.167*
16	-0.167
18	-0.167*
20	-0.167
22	-0.167*
24	-0.167*
26	-0.167
28	-0.167*
30	-0.167
57.5	-0.161
65	-0.159
75	-0.155
85	-0.150
283	-0.045
CURVE 51‡	
2.5	-0.414*
3	-0.414*
4	-0.414*
5	-0.414*
6	-0.414*
7	-0.414*
8	-0.414*
10	-0.414*
12	-0.414*
14	-0.414*
16	-0.413*
18	-0.413*
20	-0.412*
22	-0.411*
CURVE 51 (cont.)‡	
24	-0.410*
26	-0.408*
28	-0.407*
30	-0.405*
57.5	-0.374*
65	-0.364
75	-0.350
85	-0.335
283	-0.011*
CURVE 52	
293	0.000*
353	0.056
414	0.123
422	0.151
449	0.153
498	0.216
509	0.211
535	0.239
549	0.216
559	0.265
580	0.242
585	0.249
614	0.263
632	0.309
669	0.309
672	0.332
688	0.311
743	0.356
763	0.362
831	0.442
833	0.418
CURVE 53	
293	0.000
354	0.088
413	0.263
423	0.250
448	0.280
499	0.426
509	0.334
533	0.437
548	0.528
560	0.503
579	0.503
582	0.574
CURVE 53 (cont.)	
612	0.593
633	0.648
667	0.675
674	0.676
688	0.661
744	0.801
761	0.797
826	0.861
831	0.859

* Not shown in figure.

‡ Author's data for coefficient of thermal expansion have been integrated by TPRC to obtain $\Delta L/L_0$.

DATA TABLE 2. COEFFICIENT OF THERMAL LINEAR EXPANSION OF ANTIMONY Sb

[Temperature, T, K; Coefficient of Expansion, α, 10^{-6} K^{-1}]

T	α
CURVE 1‡	
10	0.04
15	0.24
20	0.78
30	5.19
40	11.36
50	13.17
60	14.29
70	14.96
80	15.29
100	15.59
120	15.86
140	15.99
160	16.11
180	16.14
200	16.19
CURVE 2‡	
10	0.01
15	0.05*
20	0.01
30	0.47
40	1.32
50	2.31
60	3.28
70	4.15
80	4.85
100	5.91
120	6.41
140	6.88
160	7.25
180	7.50
200	7.68
CURVE 48	
283	8.24
263	8.15
243	8.07
222	7.93
203	7.74
183	7.52
163	7.33
143	6.98
123	6.58
102	6.04
CURVE 48 (cont.)	
84	5.27
68	4.25*
CURVE 49	
283	16.18
253	16.17
213	16.15
173	16.11
133	16.08
102	15.81
84	15.48
68	14.66
CURVE 50‡	
2.5	-0.001
3	-0.0013*
4	-0.002*
5	-0.003*
6	-0.004*
7	-0.006*
8	-0.009*
10	-0.020*
12	-0.030*
14	-0.039*
16	-0.040*
18	-0.030*
20	0.000*
22	0.05*
24	0.115*
26	0.20
28	0.30
30	0.42
57.5	3.10*
65	3.75
75	4.50
85	5.10*
283	8.10
CURVE 51‡	
2.5	0.007
3	0.012*
4	0.027*
5	0.050*
CURVE 51 (cont.)‡	
6	0.088
7	0.15*
8	0.23*
10	0.49
12	0.90*
14	1.55
16	2.35
18	3.30
20	4.30
22	5.30
24	6.25
26	7.10
28	7.95
30	8.65
57.5	13.30
65	13.83
75	14.42
85	14.68
283	16.55

* Not shown in figure.

‡ Author's data for coefficient of thermal expansion have been integrated by TPRC to obtain $\Delta L/L_0$.

FIGURE AND TABLE NO. 3R. PROVISIONAL VALUES FOR THERMAL LINEAR EXPANSION OF BARIUM Ba

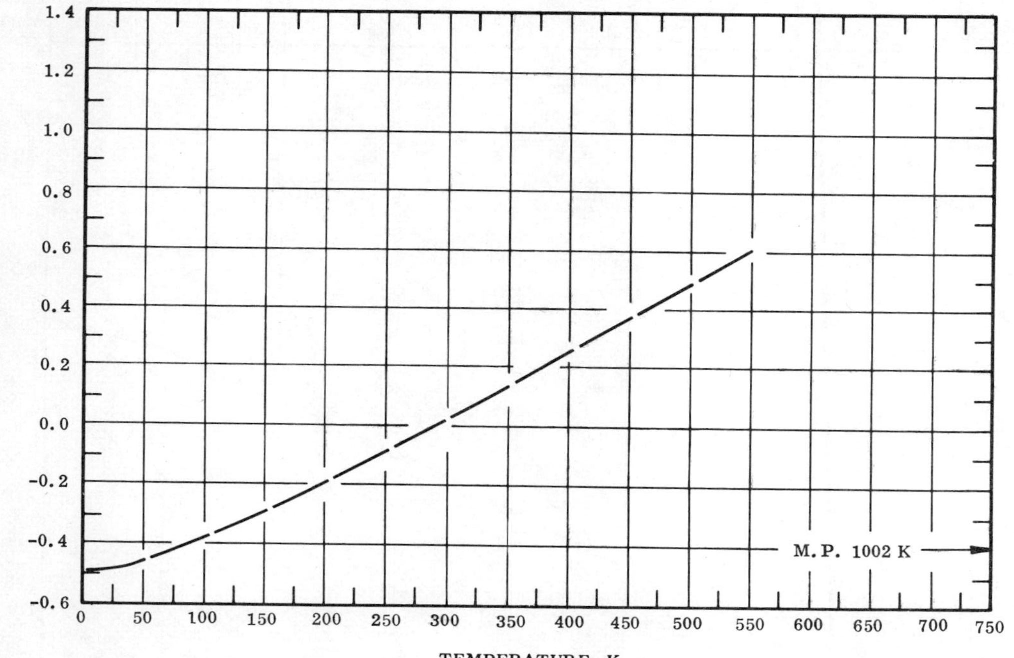

PROVISIONAL VALUES

[Temperature, T, K; Linear Expansion, $\Delta L/L_0$, %; α, K^{-1}]

T	$\Delta L/L_0$	$\alpha \times 10^6$
3	-0.494	0.2
25	-0.485	8.8
50	-0.455	14.7
100	-0.372	17.5
200	-0.186	19.5
293	0.000	20.6
400	0.227	21.8
500	0.451	22.7
550	0.567	22.7

REMARKS

The tabulated values are considered accurate to within ± 7% over the entire temperature range. These values can be represented approximately by the following equations:

$$\Delta L/L_0 = -0.496 + 3.122 \times 10^{-4} (T - 3) + 1.459 \times 10^{-5} (T - 3)^2 - 4.764 \times 10^{-8} (T - 3)^3 \quad (3 < T < 100)$$

$$\Delta L/L_0 = -0.373 + 1.798 \times 10^{-3} (T - 100) + 7.268 \times 10^{-7} (T - 100)^2 - 1.811 \times 10^{-10} (T - 100)^3 \quad (100 < T < 550)$$

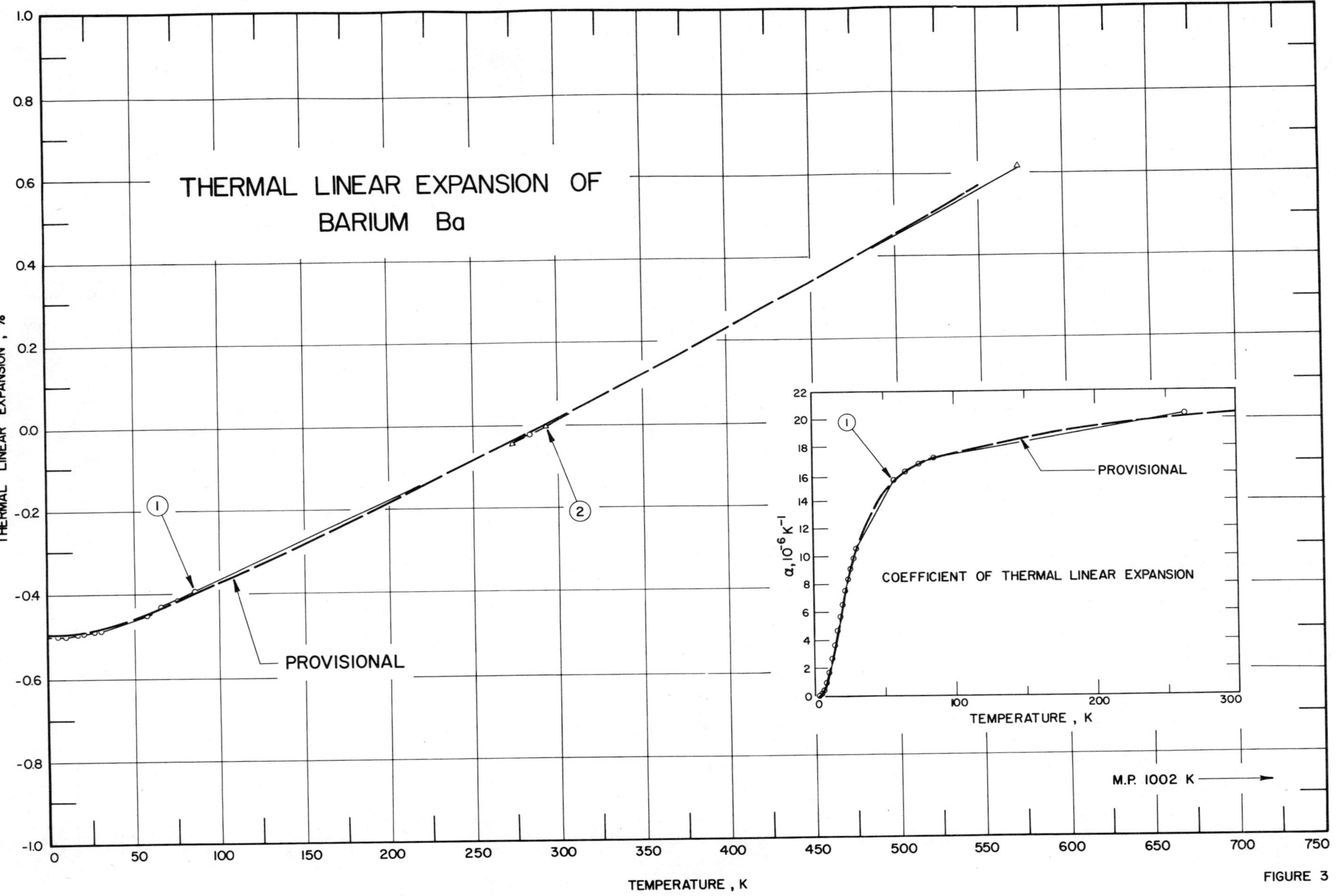
THERMAL LINEAR EXPANSION OF
BARIUM Ba
THERMAL LINEAR EXPANSION , %
1.0
0.8
0.6
0.4
0.2
0.0
-0.2
-0.4
-0.6
-0.8
-1.0
0
50
100
150
200
250
300
350
400
450
500
550
600
650
700
750
TEMPERATURE , K
PROVISIONAL
1
2
M.P. 1002 K
COEFFICIENT OF THERMAL LINEAR EXPANSION
α, 10⁻⁶ K⁻¹
22
20
18
16
14
12
10
8
6
4
2
0
100
200
300
TEMPERATURE , K
PROVISIONAL
1
FIGURE 3

SPECIFICATION TABLE 3. THERMAL LINEAR EXPANSION OF BARIUM Ba

Cur. No.	Ref. No.	Author(s)	Year	Method Used	Temp. Range, K	Name and Specimen Designation	Composition (weight percent), Specifications, and Remarks
1	726	White, G.K.	1972	E	3-85, 283		Specimen from A.D. Mackay, New York, Spectrographic analysis by the Australian Mineral Development Laboratories shows about 1.0 Sr, 0.05 Ca, 0.03 Mg, 0.02 Al, and 0.01 Fe.
2	727	Cath, P.G., and Steenis, O.L.	1936	L			No specifications given; expansion measured with increasing temperature.

DATA TABLE THERMAL LINEAR EXPANSION OF BARIUM Ba

[Temperature, T, K; Linear Expansion, $\Delta L/L_0$, %]

T	$\Delta L/L_0$	T	$\Delta L/L_0$	T	$\Delta L/L_0$
CURVE 1‡		CURVE 1 (cont.)‡		CURVE 2	
3	-0.493*	22	-0.487*	273	-0.044
4	-0.493*	24	-0.486*	293	0.000
5	-0.493	26	-0.484	573	0.616
6	-0.493*	28	-0.482*		
8	-0.493*	30	-0.480		
10	-0.493	57	-0.444		
12	-0.492*	65	-0.431		
14	-0.492*	75	-0.414		
16	-0.491	85	-0.397		
18	-0.490*	283	-0.021		
20	-0.489				

DATA TABLE COEFFICIENT OF THERMAL LINEAR EXPANSION OF BARIUM Ba

[Temperature, T, K; Coefficient of Expansion, α, 10^{-6} K^{-1}]

T	α	T	α	T	α	T	α
CURVE 1‡		CURVE 1 (cont.)‡		CURVE 1 (cont.)‡		CURVE 1 (cont.)‡	
3	0.0375	10	1.73	20	6.58	30	10.55
4	0.0962*	12	2.66	22	7.49	57	15.55
5	0.200	14	3.62	24	8.36	65	16.15
6	0.382	16	4.64	26	9.13	75	16.75
8	0.956	18	5.64	28	9.85	85	17.10
						283	20.35

* Not shown in figure.

‡ Author's data for coefficient of thermal expansion have been integrated by TPRC to obtain $\Delta L/L_0$.

FIGURE AND TABLE NO. 4R. RECOMMENDED VALUES FOR THERMAL LINEAR EXPANSION OF BERYLLIUM Be

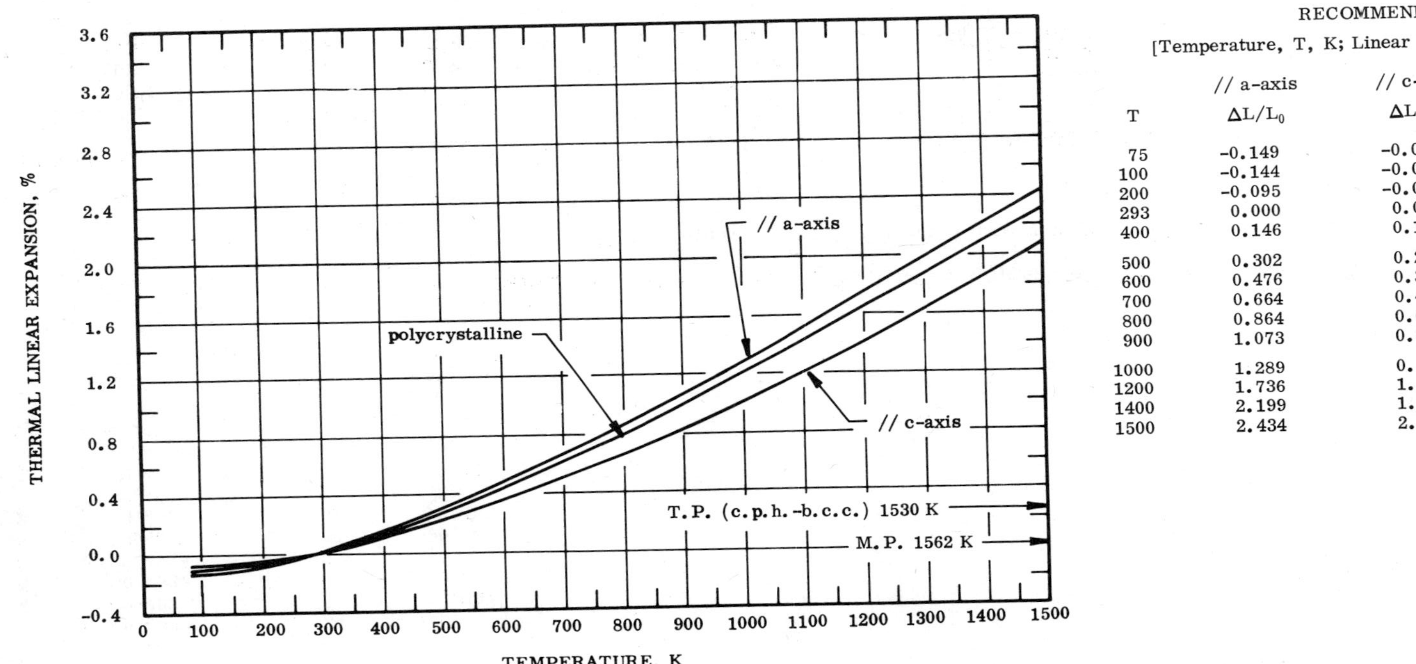

RECOMMENDED VALUES

[Temperature, T, K; Linear Expansion, $\Delta L/L_0$, %; α, K^{-1}]

	// a-axis	// c-axis	polycrystalline	
T	$\Delta L/L_0$	$\Delta L/L_0$	$\Delta L/L_0$	$\alpha \times 10^6$
75	-0.149	-0.097	-0.132	0.5
100	-0.144	-0.096	-0.128	1.3
200	-0.095	-0.066	-0.085	7.1
293	0.000	0.000	0.000	11.3
400	0.146	0.111	0.134	13.6
500	0.302	0.227	0.277	15.1
600	0.476	0.356	0.436	16.6
700	0.664	0.497	0.608	17.8
800	0.864	0.650	0.793	19.1
900	1.073	0.818	0.988	20.0
1000	1.289	0.998	1.192	20.9
1200	1.736	1.398	1.623	22.3
1400	2.199	1.841	2.080	23.3
1500	2.434	2.077	2.315	23.7

REMARKS

The tabulated values are considered accurate to within ± 3% over the entire temperature range. These values can be represented approximately by the following equations:

// a-axis: $\Delta L/L_0 = 1.242 \times 10^{-3} (T-293) + 1.144 \times 10^{-6} (T-293)^2 - 4.567 \times 10^{-10} (T-293)^3$ $(293 < T < 895)$

$\Delta L/L_0 = 1.063 + 2.123 \times 10^{-3} (T-895) + 3.197 \times 10^{-7} (T-895)^2 - 1.373 \times 10^{-10} (T-895)^3$ $(895 < T < 1500)$

// c-axis: $\Delta L/L_0 = 9.881 \times 10^{-4} (T-293) + 5.475 \times 10^{-7} (T-293)^2 + 7.916 \times 10^{-11} (T-293)^3$ $(293 < T < 895)$

$\Delta L/L_0 = 0.809 + 1.733 \times 10^{-3} (T-895) + 6.904 \times 10^{-7} (T-895)^2 - 1.519 \times 10^{-10} (T-895)^3$ $(895 < T < 1500)$

polycrystalline: $\Delta L/L_0 = 1.148 \times 10^{-3} (T-293) + 9.724 \times 10^{-7} (T-293)^2 - 2.978 \times 10^{-10} (T-293)^3$ $(293 < T < 895)$

$\Delta L/L_0 = 0.978 + 1.995 \times 10^{-3} (T-895) + 4.346 \times 10^{-7} (T-895)^2 - 1.307 \times 10^{-10} (T-895)^3$ $(895 < T < 1500)$

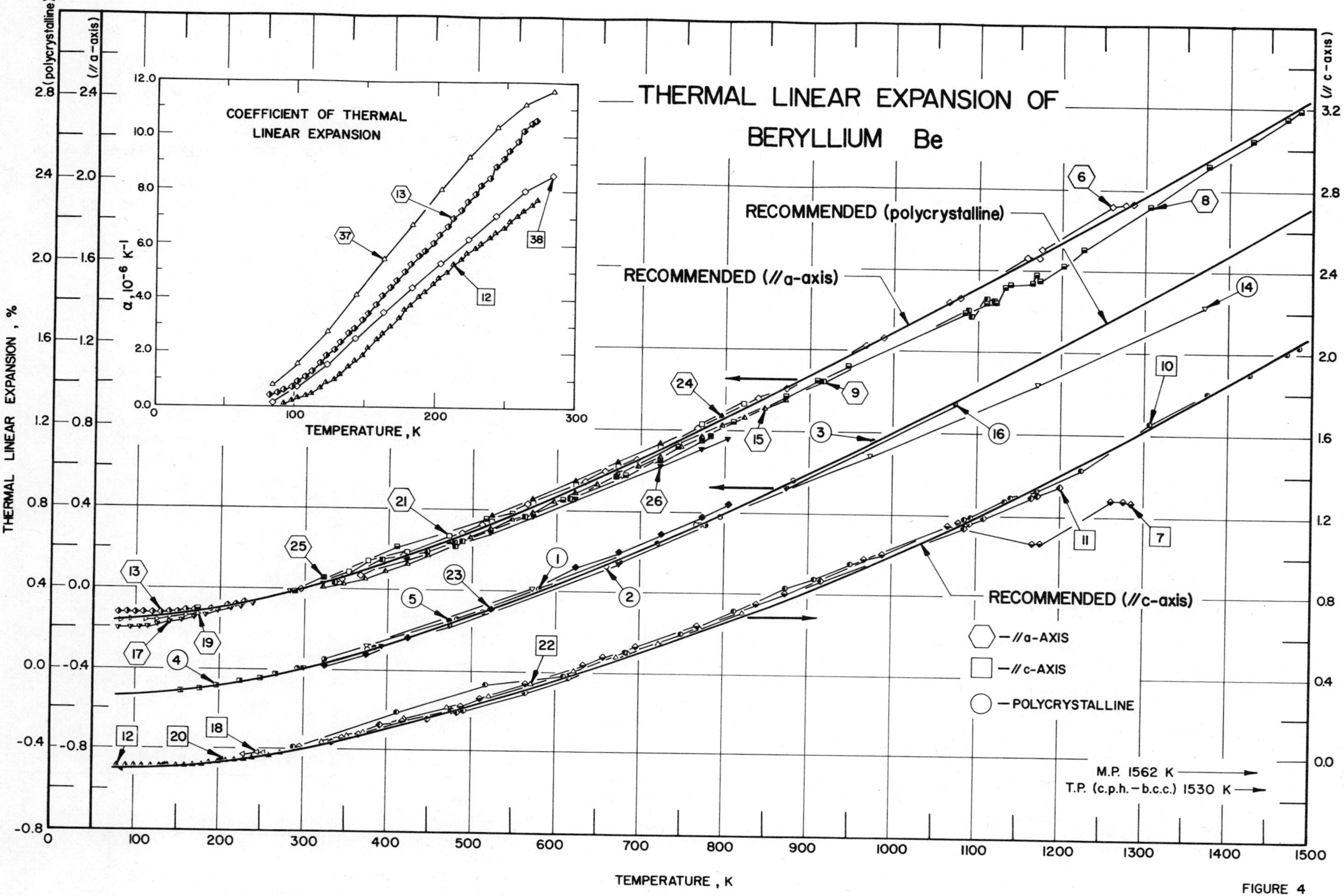
THERMAL LINEAR EXPANSION OF
BERYLLIUM Be
COEFFICIENT OF THERMAL
LINEAR EXPANSION
α, 10⁻⁶ K⁻¹
TEMPERATURE, K
RECOMMENDED (polycrystalline)
RECOMMENDED (//a-axis)
RECOMMENDED (//c-axis)
—//a-AXIS
—//c-AXIS
—POLYCRYSTALLINE
M.P. 1562 K
T.P. (c.p.h.–b.c.c.) 1530 K
THERMAL LINEAR EXPANSION, %
(polycrystalline)
(//a-axis)
(//c-axis)
TEMPERATURE, K

FIGURE 4

SPECIFICATION TABLE 4. THERMAL LINEAR EXPANSION OF BERYLLIUM Be

Cur. No.	Ref. No.	Author(s)	Year	Method Used	Temp. Range, K	Name and Specimen Designation	Composition (weight percent), Specifications, and Remarks
1	400	Hidnert, P. and Sweeney, W.T.	1927	T	293-774	Sample 1223	98.9 Be with 0.90 Fe, 0.20 Mn, 0.18 Si, 0.06 Cu; about 300 mm long x 10 mm diameter; furnished by the Beryllium Corp. of America; prepared in flux of about 90% $BaCl_2$ and 10% BaF_2; expansion measured in air furnace on first test.
2	400	Hidnert, P. and Sweeney, W.T.	1927	T	293-774	Sample 1223	The above expansion; expansion measured on second test.
3	400	Hidnert, P. and Sweeney, W.T.	1927	T	293-774	Sample 1223	The above specimen; expansion measured on third test.
4	400	Hidnert, P. and Sweeney, W.T.	1927	T	153-293	Sample 1223	The above specimen; expansion measured on fourth test.
5	400	Hidnert, P. and Sweeney, W.T.	1927	T	153-293	Sample 1223	The above specimen; expansion measured on fifth test.
6	90	Gordon, P.	1949	X	297-1286		Powder specimen (-325 mesh) from Brush Beryllium Co.; assayed only 97 Be with possible impurity of 2% oxygen and 0.094 Fe, 0.28 Al, 0.141 Si, 0.58 Mg; oxygen has little effect on the lattice parameter; measurements along a-axis; lattice parameter reported at 297 K is 2.2858 Å and the same at 293 K determined by graphical extrapolation; measurements on the several specimens; data on the coefficient of thermal linear expansion also reported.
7	90	Gordon, P.	1949	X	297-1286		Similar to the above specimen; measurements along c-axis; lattice parameter reported at 297 K is 3.5842 Å; 3.5838 Å at 293 K determined by graphical extrapolation; measurements on the several specimens; coefficient of thermal linear expansion also reported.
8	92	Martin, A.J. and Moore, A.	1959	X	75-1520		99.4 Be; 0.3 BeO, 0.1 Al, 0.1 Fe; expansion measured in 10^{-5} mm Hg vacuum; expansion measured along a-axis with increasing temperature; lattice parameter reported at 290 K is 2.2869 Å; 2.2870 Å at 293 K determined by graphical interpolation.
9	92	Martin, A.J. and Moore, A.	1959	X	1121-301		The above specimen; expansion measured with decreasing temperature; lattice parameter reported at 301 K is 2.2871 Å; 2.2868 Å at 293 K determined by graphical extrapolation.
10	92	Martin, A.J. and Moore, A.	1959	X	77-1520		The above specimen; expansion measured with increasing temperature along c-axis; lattice parameter reported at 290 K is 3.5834 Å; 3.5836 Å at 293 K determined by graphical interpolation.
11	92	Martin, A.J. and Moore, A.	1959	X	1201-290		The above specimen; expansion measured with decreasing temperature; lattice parameter reported at 290 K is 3.5834 Å; 3.5835 Å at 293 K determined by graphical interpolation.
12	93	Meyerhoff, R.W. and Smith, J.F.	1962	L	80-273		99.5^+ Be; slight impurities Fe, Al, Ca, Cr, Cu, Ni, Mn; expansion measured parallel to c-axis; zero-point correction of -0.0198% determined by graphical extrapolation; data on the coefficient of thermal linear expansion also reported.
13	93	Meyerhoff, R.W. and Smith, J.F.	1962	L	80-273		Similar to the above specimen; expansion measured perpendicular to the c-axis; zero-point correction of -0.023% determined by graphical extrapolation; data on the coefficient of thermal linear expansion also reported.
14	94	Chirkin, V.S.	1966		293-1373		98.736 Be; impurity predominately Mg and Al; hot-pressed specimen.
15	95	Brett, N.H. and Russell, L.E.	1960	L	298-873		Specimen cut from extruded rod; expansion measured parallel to extrusion.

* Not shown in figure.

SPECIFICATION TABLE 4. THERMAL LINEAR EXPANSION OF BERYLLIUM Be (continued)

Cur. No.	Ref. No.	Author(s)	Year	Method Used	Temp. Range, K	Name and Specimen Designation	Composition (weight percent), Specifications, and Remarks
16	96	Watrous, J.D.	1961	L	302-1073		Reactor-grade hot-pressed and extruded beryllium; from Brush Beryllium Co., Cleveland, Ohio; mid-values recorded from the range given by author; zero-point correction of 0.015% determined by graphical extrapolation.
17	97	Finkel, V.A. and Papirov, I.I.	1968	X	79-296	Sample 1	99.0 Be single crystal, prepared by zone melting; lattice parameter measured with accuracy > 0.0001Å; margin of error in temperature measurement < 0.2 K; expansion measured along a-axis; lattice parameter reported at 293 K is 2.2858 Å.
18	97	Finkel, V.A. and Papirov, I.I.	1968	X	80-299	Sample 1	The above specimen; expansion measured along c-axis; lattice parameter reported at 293 K is 3.5865 Å.
19	97	Finkel, V.A. and Papirov, I.I.	1968	X	79-296	Sample 2	99.9 Be; similar to the above specimen; expansion measured along a-axis; lattice parameter reported at 293 K is 2.2824 Å.
20	97	Finkel, V.A. and Papirov, I.I.	1968	X	79-299	Sample 2	The above specimen; expansion measured along c-axis; lattice parameter reported at 293 K is 3.5834 Å.
21	98	Owen, E.A. and Richards, T.L.	1936	X	293-823		99.8 Be; foil 0.25 mm thick; expansion measured along base or a-axis; lattice parameter reported at 293 K is 2.2813 Å.
22	98	Owen, E.A. and Richards, T.L.	1936	X	293-823		The above specimen; expansion measured along height or c-axis; lattice parameter reported at 293 K is 3.5773 Å.
23	99	Treco, R.M.	1950	L	301-805	Sample 4	99.28 Be, 0.179 O_2, 0.170 Fe, 0.140 Al, 0.086 Si, 0.080 C, 0.020 Mn, 0.020 Cu, 0.014 Mg, 0.007 Ni, and 0.007 Ca; obtained from Brush Beryllium Co.; remelted, cast in high-vacuum induction furnace; specimen machined to 0.250 in. diameter, 3.000 in. long; surrounded by argon gas; accuracy of temperature measurement ±0.25 C; annealed after extrusion; expansion measured longitudinally; zero-point correction of 0.010% determined by graphical extrapolation.
24	99	Treco, R.M.	1950	L	297-797	Sample 5	Similar to the above specimen except specimen tested as extruded; zero-point correction of 0.005% determined by graphical extrapolation.
25	99	Treco, R.M.	1950	L	298-784	Sample 6	Similar to the above specimen except annealed after extrusion; expansion measured transversely; zero-point correction of 0.007% determined by graphical extrapolation.
26	99	Treco, R.M.	1950	L	302-804	Sample 7	Similar to the above specimen except specimen tested as extruded; zero-point correction of 0.010% determined by graphical extrapolation.
27*	651	Goggin, W.R. and Paguin, R.A.	1970	L	275-311	FP-17	Specimen from Stanford Research Institute; values reported for mutually perpendicular cube axes; unidentified with regard to pressing direction.
28*	651	Goggin, W.R. and Paguin, R.A.	1970	L	275-311	FP-17	The above measurements; second test.
29*	651	Goggin, W.R. and Paguin, R.A.	1970	L	275-311	FP-40	Specimen from Stanford Research Institute.
30*	651	Goggin, W.R. and Paguin, R.A.	1970	L	275-311	I-400	Specimen from Brush Beryllium Co.; values reported for mutually perpendicular cube axes; unidentified with regard to pressing direction.
31*	651	Goggin, W.R. and Paguin, R.A.	1970	L	275-311	I-400	The above measurements; second test.
32*	651	Goggin, W.R. and Paguin, R.A.	1970	L	275-311	I-400	The above measurements; third test.

*Not shown in figure.

SPECIFICATION TABLE 4. THERMAL LINEAR EXPANSION OF BERYLLIUM Be (continued)

Cur. No.	Ref. No.	Author(s)	Year	Method Used	Temp. Range, K	Name and Specimen Designation	Composition (weight percent), Specifications, and Remarks
33*	651	Goggin, W.R. and Paguin, R.A.	1970	L	275-311	I-400	The above specimen; measurements parallel to pressing direction.
34*	651	Goggin, W.R. and Paguin, R.A.	1970	L	275-311	I-400	The abpve specimen; measurements perpendicular to pressing direction.
35*	651	Goggin, W.R. and Paguin, R.A.	1970	L	275-311	SP	Specimen from Kawacki-Berylco Industries, Inc.
36*	91	Zelikman, A.N., Kislyakov, I.P., and Bal'shin, M.Yu.	1961		298-373		99.96 pure specimen.
37*	712	Erfling, H.D.	1939	I	293-58		Single crystal from the Degussa, Frankfürt; sample purity checked by resistivity value; measurements along a-axis; data on the coefficient of thermal linear expansion also reported.
38*	712	Erfling, H.D.	1939	I	293-58		Similar to the above specimen; measurements along c-axis; data on the coefficient of thermal linear expansion also reported.

* Not shown in figure.

DATA TABLE 4. THERMAL LINEAR EXPANSION OF BERYLLIUM Be

[Temperature, T, K; Linear Expansion, $\Delta L/L_0$, %]

T	$\Delta L/L_0$
CURVE 1	
293	0.000*
328	0.045*
372	0.100*
483	0.254
580	0.401
676	0.563*
777	0.746
CURVE 2	
293	0.000*
377	0.104
511	0.289
659	0.532
769	0.741
882	0.959
CURVE 3	
293	0.000*
382	0.107
720	0.645
794	0.773
979	1.158
CURVE 4	
155	-0.112
177	-0.103
197	-0.088
224	-0.065
249	-0.045
266	-0.025
293	0.000*
CURVE 5	
293	0.000*
324	0.036
373	0.097*
476	0.246
CURVE 6	
297	0.000*
298	0.000*
CURVE 6 (cont.)	
299.7	0.000*
347	0.057*
368	0.074
420	0.184*
488	0.284
512	0.324
566	0.437
633	0.542
658	0.599
696	0.656
769	0.809
840.7	0.962
874.5	1.019
968	1.234
989	1.269
1066.7	1.448
1080.7	1.470
1168	1.676
1174.5	1.676
1177.5	1.711
1262.7	1.925
1277	1.934
1286	1.938
CURVE 7	
297	0.011
298	0.014*
299.7	0.017*
347	0.061
368	0.075
420	0.145
488	0.223
512	0.257
566	0.338
633	0.435
658	0.477
696	0.527
769	0.639
840.7	0.745
874.5	0.806
968	0.985
989	1.007
1066.7	1.152
1080.7	1.163
1168	1.066
CURVE 7 (cont.)	
1174.5	1.066*
1177.5	1.085
1262.7	1.281
1277	1.281
1286	1.272
CURVE 8	
75	-0.115*
290	-0.005*
412	0.213
519	0.358
549	0.379
607	0.454
682	0.576
746	0.720*
811	0.843
873	0.974
910	1.053
947	1.123
1086	1.402
1112	1.451
1121	1.451
1134	1.525
1140	1.538
1167	1.547
1171	1.586
1174	1.564*
1203	1.634
1228	1.713
1306	1.923
1377	2.128
1430	2.255
1471	2.360
1487	2.399
1520	2.474
CURVE 9	
1121	1.460
1112	1.460
1094	1.382
1090	1.408
948	1.123*
918	1.049
873	0.983*
CURVE 9 (cont.)	
811	0.852*
746	0.730
682	0.585*
616	0.459
564	0.367
489	0.240*
481	0.240
481	0.218
448	0.196*
394	0.148
345	0.069*
337	0.039
301	0.013*
CURVE 10	
77	-0.091*
290	-0.005
411	0.182
519	0.308
611	0.372
684	0.495
750	0.598
813	0.710
875	0.824
908	0.877
951	0.952
1086	1.184
1094	1.192
1135	1.273
1142	1.287
1173	1.324
1227	1.430
1308	1.653
1378	1.801
1428	1.904
1472	2.004
1488	2.049
1520	2.113
CURVE 11	
1201	1.345
1174	1.300
1168	1.289
1110	1.194
CURVE 11 (cont.)	
1093	1.163
1087	1.174
916	0.865
875	0.826*
813	0.711*
750	0.599*
684	0.496
616	0.365
566	0.290
492	0.200
484	0.186
479	0.198
449	0.153
393	0.103
347	0.050*
335	0.033
290	-0.002*
CURVE 12	
80	-0.088
85	-0.088*
90	-0.088
95	-0.088*
100	-0.087
105	-0.087*
110	-0.087
115	-0.087*
120	-0.086
125	-0.086*
130	-0.086
135	-0.085*
140	-0.084
145	-0.083*
150	-0.082
155	-0.081*
160	-0.080
165	-0.079*
170	-0.077
175	-0.075*
180	-0.074
185	-0.072*
190	-0.070
195	-0.068*
200	-0.065
205	-0.063*
CURVE 12 (cont.)	
210	-0.060
215	-0.058*
220	-0.055
225	-0.052*
230	-0.049
235	-0.046*
240	-0.043
245	-0.040*
250	-0.036*
255	-0.033*
260	-0.029
265	-0.026*
270	-0.023
273.2	-0.020
CURVE 13	
80	-0.121
85	-0.120*
90	-0.120
95	-0.120*
100	-0.120
105	-0.119*
110	-0.118
115	-0.118*
120	-0.117
125	-0.116*
130	-0.115
135	-0.114*
140	-0.112
145	-0.111*
150	-0.110
155	-0.108*
160	-0.106
165	-0.104*
170	-0.102
175	-0.099*
180	-0.097*
185	-0.094*
190	-0.091
195	-0.088*
200	-0.085*
205	-0.082*
210	-0.079
215	-0.075*
220	-0.072

* Not shown in figure.

DATA TABLE 4. THERMAL LINEAR EXPANSION OF BERYLLIUM Be (continued)

T	$\Delta L/L_0$
CURVE 13 (cont.)	
225	-0.068*
230	-0.064
235	-0.060*
240	-0.056*
245	-0.051*
250	-0.046*
255	-0.042*
260	-0.037*
265	-0.032*
270	-0.026*
273.2	-0.023*
CURVE 14‡	
293	0.000*
373	0.113
573	0.416
773	0.743
973	1.087
1173	1.451
1373	1.834
CURVE 15	
298	0.000*
323	0.010
348	0.025
373	0.052
398	0.087
423	0.127
448	0.175
473	0.217*
498	0.262
523	0.312
548	0.357
573	0.392
598	0.442
623	0.487
648	0.530
673	0.585
698	0.625
723	0.672
748	0.720
773	0.767
798	0.825
823	0.865
848	0.910
873	0.957

T	$\Delta L/L_0$
CURVE 16	
302	0.015
393	0.117
473	0.220*
673	0.548
873	0.936
1073	1.354
CURVE 17	
79	-0.210
86	-0.204*
93	-0.204*
96	-0.199
100	-0.206*
103	-0.196*
106	-0.200
109	-0.191*
114	-0.193*
120	-0.187
126	-0.194*
126	-0.185
129	-0.179*
133	-0.195*
133	-0.172*
136	-0.178*
140	-0.172
146	-0.166*
153	-0.166*
157	-0.160
161	-0.156*
166	-0.150*
170	-0.150
175	-0.142*
180	-0.139*
183	-0.139
188	-0.129*
194	-0.127*
199	-0.123
199	-0.116*
203	-0.112*
208	-0.114*
210	-0.112*
214	-0.106
220	-0.094*
225	-0.094*
229	-0.081
232	-0.087*
236	-0.080*

T	$\Delta L/L_0$
CURVE 17 (cont.)	
236	-0.067*
241	-0.070
244	-0.063*
249	-0.063*
251	-0.051*
255	-0.048*
260	-0.042*
266	-0.038*
270	-0.028*
272	-0.020*
278	-0.021*
281	-0.015*
287	-0.010
289	-0.002*
293	0.000*
296	0.000*
296	0.006*
CURVE 18	
80	-0.080*
83	-0.078*
89	-0.080*
90	-0.075*
94	-0.079*
98	-0.071*
100	-0.077*
102	-0.074*
107	-0.073*
113	-0.075*
118	-0.070*
122	-0.070*
126	-0.068*
132	-0.068*
137	-0.068*
141	-0.064*
146	-0.064*
152	-0.060*
158	-0.062*
163	-0.059*
168	-0.055*
171	-0.062*
174	-0.057*
180	-0.053*
185	-0.053*
192	-0.048*
199	-0.050*
202	-0.045*

T	$\Delta L/L_0$
CURVE 18 (cont.)	
207	-0.046*
210	-0.040*
215	-0.043*
219	-0.039*
221	-0.036*
229	-0.037*
230	-0.035
231	-0.032*
236	-0.030*
240	-0.029*
241	-0.024*
247	-0.023
254	-0.023*
258	-0.018
264	-0.018*
268	-0.013*
273	-0.010*
278	-0.007*
284	-0.006*
286	-0.001*
289	-0.003*
293	0.000*
295	0.002*
299	0.004*
CURVE 19	
79	-0.167*
84	-0.167
87	-0.162*
94	-0.167
94	-0.157*
100	-0.159*
106	-0.159*
111	-0.161
112	-0.152*
116	-0.160*
118	-0.149*
124	-0.146*
126	-0.153
130	-0.144*
132	-0.151*
137	-0.148*
139	-0.142
144	-0.144*
148	-0.139*
154	-0.134*
160	-0.134

T	$\Delta L/L_0$
CURVE 19 (cont.)	
164	-0.130*
169	-0.125*
173	-0.130
175	-0.119*
180	-0.124*
182	-0.116*
188	-0.116*
193	-0.107
197	-0.104*
204	-0.104*
207	-0.095*
214	-0.093*
217	-0.082*
224	-0.086*
228	-0.072*
234	-0.076*
238	-0.064*
247	-0.062*
250	-0.050*
256	-0.050*
260	-0.043*
263	-0.034*
267	-0.033*
271	-0.023*
277	-0.028*
278	-0.013*
283	-0.011*
284	-0.003*
288	-0.007*
293	0.000*
293	0.000*
296	0.014*
CURVE 20	
79	-0.102*
82	-0.100*
87	-0.100
89	-0.102*
89	-0.096*
95	-0.096*
99	-0.098*
103	-0.093*
108	-0.096*
110	-0.091*
116	-0.087*
120	-0.089*
125	-0.085*

T	$\Delta L/L_0$
CURVE 20 (cont.)	
133	-0.085*
134	-0.081*
137	-0.080
139	-0.085*
142	-0.081*
147	-0.077*
151	-0.077*
157	-0.076*
161	-0.069*
165	-0.072*
167	-0.068*
170	-0.066*
174	-0.069*
178	-0.067*
183	-0.058*
184	-0.063*
189	-0.061*
192	-0.058*
199	-0.054*
202	-0.054
208	-0.053*
212	-0.050*
216	-0.050*
219	-0.043*
223	-0.050*
229	-0.041*
234	-0.040*
238	-0.035*
238	-0.040*
243	-0.030*
249	-0.030*
252	-0.027*
258	-0.024*
261	-0.020*
266	-0.021*
269	-0.016*
270	-0.012*
273	-0.016*
277	-0.014*
280	-0.008*
285	-0.005*
288	-0.005*
290	0.000*
293	0.000*
299	0.005*

* Not shown in figure.
‡ Author's data for coefficient of thermal expansion have been integrated by TPRC to obtain $\Delta L/L_0$.

DATA TABLE 4. THERMAL LINEAR EXPANSION OF BERYLLIUM Be (continued)

T	ΔL/L₀
CURVE 21	
293	0.000*
323	0.039*
353	0.083
373	0.114
423	0.184
473	0.263
523	0.342
573	0.430
623	0.522
673	0.618
723	0.714
773	0.824
823	0.929
CURVE 22	
293	0.000*
323	0.031
353	0.067
373	0.089
423	0.148
473	0.207
523	0.268
573	0.338
623	0.400
673	0.470
723	0.542
773	0.618
823	0.699
CURVE 23	
301	0.010
323	0.038*
373	0.098
423	0.170*
473	0.244
523	0.333*
573	0.415
623	0.510
673	0.605*
723	0.699
773	0.793
805	0.858
CURVE 24	
297	0.005*
CURVE 24 (cont.)	
323	0.040*
373	0.099
423	0.179
473	0.265
523	0.355
573	0.447
623	0.538
673	0.635
723	0.730
773	0.824*
797	0.868
CURVE 25	
298	0.007*
323	0.044
373	0.111*
423	0.180*
473	0.252*
523	0.327*
573	0.407*
623	0.488*
673	0.578
723	0.666
773	0.754
784	0.771
CURVE 26	
302	0.010*
323	0.035
373	0.095*
423	0.161
473	0.228*
523	0.299
573	0.386
623	0.461
673	0.546*
723	0.629
773	0.710
804	0.764
CURVE 27*	
275	-0.010
293	0.000
311	0.020
CURVE 28*	
275	-0.019
293	0.000
311	0.019
CURVE 29*	
275	-0.021
293	0.000
311	0.021
CURVE 30*	
275	-0.019
293	0.000
311	0.019
CURVE 31*	
275	-0.018
293	0.000
311	0.018
CURVE 32*	
275	-0.020
293	0.000
311	0.020
CURVE 33*	
275	-0.019
293	0.000
311	0.019
CURVE 34*	
275	-0.019
293	0.000
311	0.019
CURVE 35*	
275	-0.019
293	0.000
311	0.019
CURVE 36*	
298	0.006
348	0.066
398	0.127
448	0.192
CURVE 37*	
293	0.000
273	-0.023
253	-0.046
233	-0.066
213	-0.085
193	-0.101
173	-0.114
153	-0.125
133	-0.134
113	-0.139
90	-0.143
78	-0.144
CURVE 38*	
293	0.000
273	-0.017
253	-0.033
233	-0.048
213	-0.060
193	-0.071
173	-0.080
153	-0.087
133	-0.092
113	-0.095
90	-0.096
57	-0.097

* Not shown in figure.

DATA TABLE 4. COEFFICIENT OF THERMAL LINEAR EXPANSION OF BERYLLIUM Be

[Temperature, T, K; Coefficient of Expansion, α, 10^{-6} K^{-1}]

T	α
CURVE 6*	
323	10.7
343	11.7
373	12.6
423	13.9
473	14.7
523	15.1
573	15.7
623	16.1
673	16.5
723	16.8
773	17.2
823	17.5
873	17.8
923	18.0
973	18.3
1023	18.5
1073	18.7
1123	19.0
1173	19.2
1223	19.5
1273	19.7
CURVE 7*	
323	8.7
348	9.4
373	9.6
423	10.4
473	11.0
523	11.6
573	12.1
623	12.5
673	12.8
723	13.1
773	13.3
823	13.6
873	13.8
923	14.1
973	14.3
1023	14.5
1073	14.8
1123	15.0
1173	15.3
1223	15.5
1273	15.8
CURVE 12	
271.6	7.7
268	7.5
262	7.3
258	7.1
252	6.9
248	6.7
242	6.5
238	6.3
232	6.1
228	5.9
222	5.7
218	5.5
212	5.3
208	5.0
202	4.8
198	4.6
192	4.3
188	4.1
182	3.8
178	3.6
172	3.3
168	3.0
162	2.7
158	2.5
152	2.2
148	1.9
142	1.7
138	1.5
132	1.2
128	1.0
122	0.9
118	0.7
112	0.5
108	0.4
102	0.3
98	0.2
92	0.1
CURVE 13	
271.6	10.6
268	10.5
262	10.2
258	9.8
252	9.5
248	9.2
CURVE 13 (cont.)	
242	8.9
238	8.5
232	8.2
228	7.9
222	7.6
218	7.3
212	7.0
208	6.7
202	6.4
198	6.1
192	5.8
188	5.6
182	5.3
178	5.0
172	4.7
168	4.4
162	4.1
158	3.8
152	3.5
148	3.2
142	2.9
138	2.7
132	2.4
128	2.1
122	1.9
118	1.6
112	1.3
108	1.1
102	0.9
98	0.7
92	0.6
88	0.5
82	0.4
CURVE 14*, ‡	
293	13.9
373	14.3
573	16.0
773	16.7
973	17.7
1173	18.7
1373	19.6
CURVE 37	
283	11.7
263	11.16
243	10.30
223	9.24
203	8.06
183	6.75
163	5.43
143	4.12
123	2.78
102	1.58
84	0.79
CURVE 38	
283	8.59
263	8.03
243	7.13
223	6.24
203	5.37
183	4.43
163	3.50
143	2.56
123	1.57
102	0.73
74	0.13

* Not shown in figure.
‡ Author's data for coefficient of thermal expansion have been integrated by TPRC to obtain $\Delta L/L_0$.

FIGURE AND TABLE NO. 5R. RECOMMENDED VALUES FOR THERMAL LINEAR EXPANSION OF BISMUTH Bi

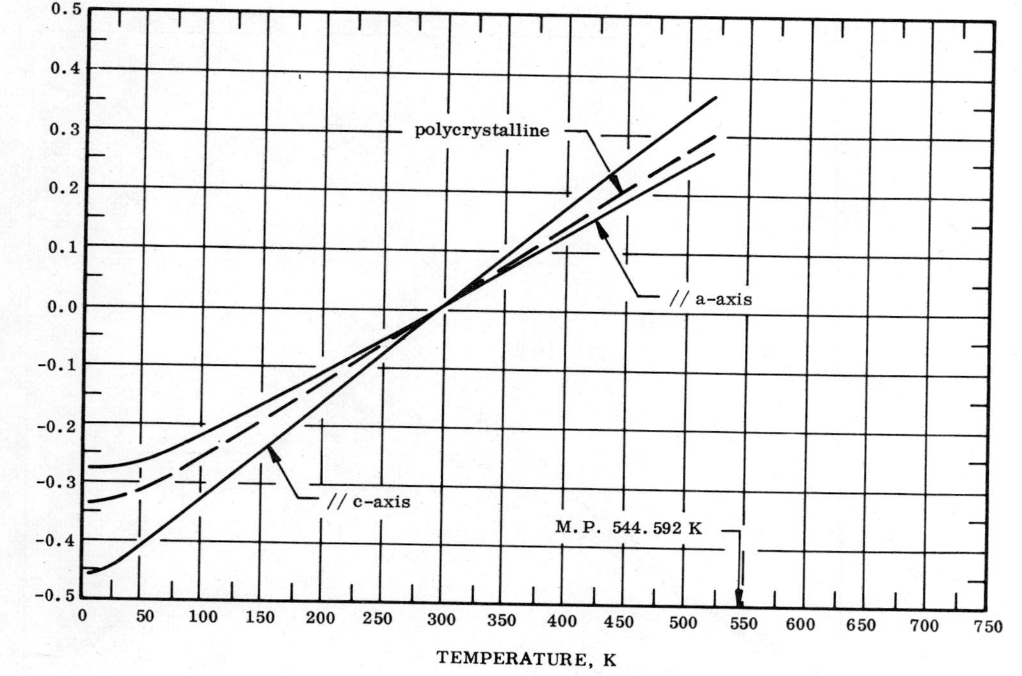

RECOMMENDED VALUES

[Temperature, T, K; Linear Expansion, $\Delta L/L_0$, %; α, K^{-1}]

	// a-axis	// c-axis	polycrystalline	
T	$\Delta L/L_0$	$\Delta L/L_0$	$\Delta L/L_0$	$\alpha \times 10^6$
2	-0.274	-0.455	-0.334†	0.01
25	-0.272	-0.444	-0.329†	5.3
50	-0.262	-0.406	-0.310†	10.0
100	-0.219	-0.326	-0.255†	12.3
200	-0.108	-0.157	-0.124†	13.1
293	0.000	0.000	0.000	13.4
400	0.125	0.172	0.141†	13.3
500	0.242	0.334	0.273†	12.7
525	0.273	0.374	0.307†	12.4

† Provisional values.

REMARKS

The tabulated values are considered accurate to within ± 3% at temperatures below 293 K and ± 5% above. Since no experimental data on the polycrystalline bismuth were located in the literature, the tabulated values are calculated from single crystal data. These values are considered accurate to within ± 10% over the entire temperature range. These values can be represented approximately by the following equations:

// a-axis: $\Delta L/L_0 = -0.310 + 8.414 \times 10^{-4} T + 1.019 \times 10^{-6} T^2 - 9.825 \times 10^{-10} T^3$ $(50 < T < 525)$

// c-axis: $\Delta L/L_0 = -0.490 + 1.637 \times 10^{-3} T + 2.141 \times 10^{-7} T^2 - 3.843 \times 10^{-10} T^3$ $(50 < T < 525)$

polycrystalline: $\Delta L/L_0 = -0.370 + 1.103 \times 10^{-3} T + 7.676 \times 10^{-7} T^2 - 8.058 \times 10^{-10} T^3$ $(50 < T < 525)$

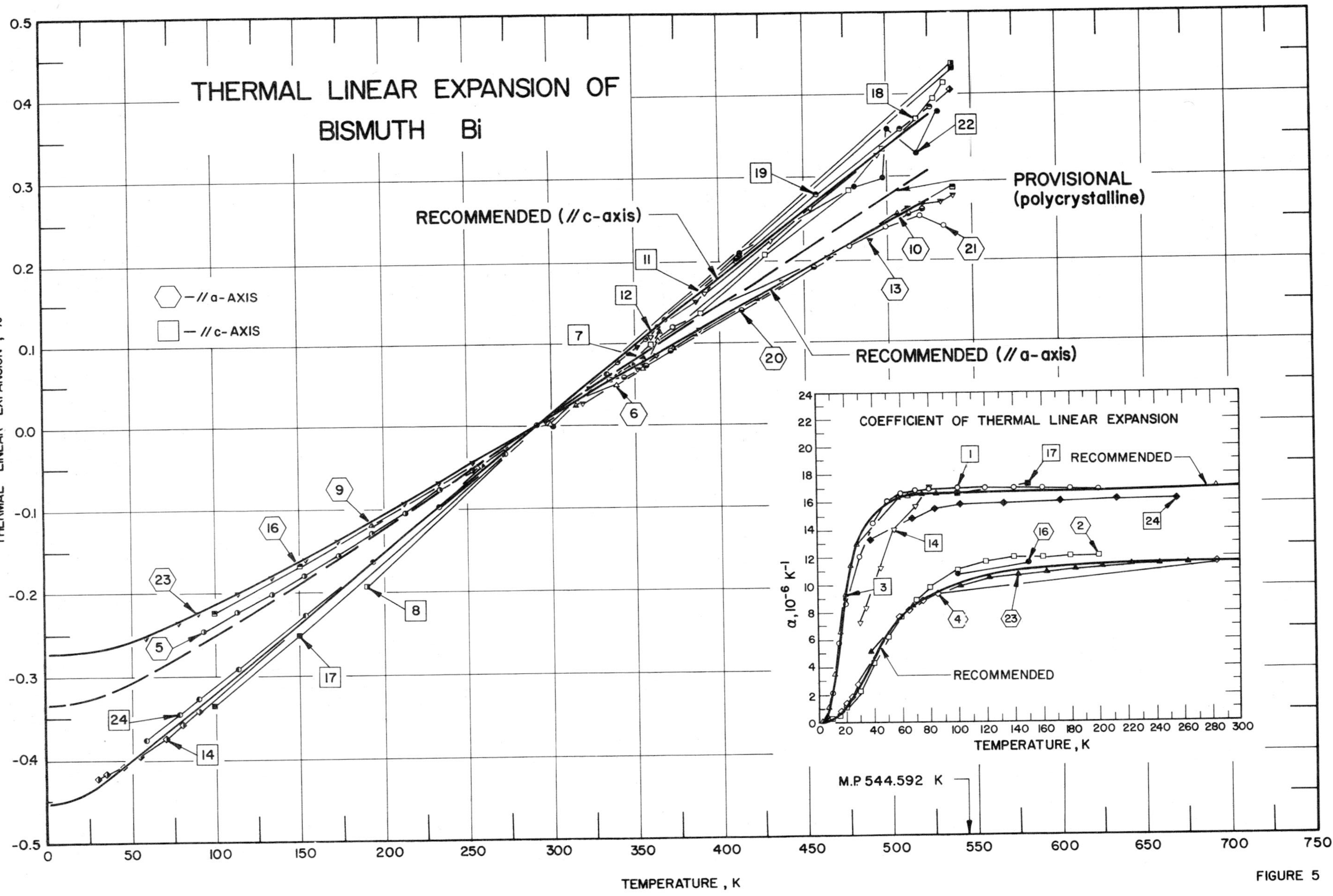

THERMAL LINEAR EXPANSION OF
BISMUTH Bi
— // a-AXIS
— // c-AXIS
PROVISIONAL
(polycrystalline)
RECOMMENDED (// c-axis)
RECOMMENDED (// a-axis)
M.P. 544.592 K
THERMAL LINEAR EXPANSION, %
TEMPERATURE, K
0.5
0.4
0.3
0.2
0.1
0.0
-0.1
-0.2
-0.3
-0.4
-0.5
0
50
100
150
200
250
300
350
400
450
500
550
600
650
700
750
COEFFICIENT OF THERMAL LINEAR EXPANSION
RECOMMENDED
RECOMMENDED
α, 10⁻⁶ K⁻¹
TEMPERATURE, K
0
20
40
60
80
100
120
140
160
180
200
220
240
260
280
300
0
2
4
6
8
10
12
14
16
18
20
22
24
1
2
3
4
5
6
7
8
9
10
11
12
13
14
16
17
18
19
20
21
22
23
24

FIGURE 5

SPECIFICATION TABLE 5. THERMAL LINEAR EXPANSION OF BISMUTH Bi

Cur. No.	Ref. No.	Author(s)	Year	Method Used	Temp. Range, K	Name and Specimen Designation	Composition (weight percent), Specifications, and Remarks
1*	63	Bunton, G. V. and Weintroub, S.	1969	L	10-200		99.9999$^+$ Bi single crystal supplied by Kawecki-Billiton Ltd.; expansion measured parallel to trigonal axis; this curve is here reported using the first given temperature as reference temperature at which $\Delta L/L_0 = 0$.
2*	63	Bunton, G. V. and Weintroub, S.	1969	L	10-200		The above specimen; expansion measured perpendicular to trigonal axis; this curve is here reported using the first given temperature as reference temperature at which $\Delta L/L_0 = 0$.
3*	100	White, G. K.	1969	E	2.5-85, 283		Single crystal; expansion measured parallel to trigonal axis; this curve is here reported using the first given temperature as reference temperature at which $\Delta L/L_0 = 0$.
4*	100	White, G. K.	1969	E	2.5-85, 283		Single crystal; expansion measured perpendicular to trigonal axis; this curve is here reported using the first given temperature as reference temperature at which $\Delta L/L_0 = 0$.
5	48	Dorsey, H. G.	1907	I	93-293		Specimen marked free from arsenic; cast; length 1.3409 cm; density 9.84 g cm^{-3}.
6	18	Zubenko, V.V. and Umansky, M.M.	1956	X	294-365		Polycrystalline specimen; expansion measured along a-axis; lattice parameter reported at 293 K is 4.5373 kX (4.5464 Å).
7	18	Zubenko, V.V. and Umansky, M.M.	1956	X	293-365		The above specimen; expansion measured along c-axis; lattice parameter reported at 293 K is 11.8390 kX (11.8627 Å).
8	101	Cave, E. F.	1958	X	150, 540		Single crystal, grown by Czorchralski technique; expansion measured along c-axis; reported error 2% in the coefficient of expansion.
9	101	Cave, E. F.	1958	X	193, 373		Powder specimen; (500) (hexagonal indices) reflection used to determine average expansion coefficient; reported error 3% in the coefficient of expansion.
10	102	Roberts, J. K.	1924	O	315-506		Expansion measured perpendicular to axis.
11	102	Roberts, J. K.	1924	O	322-495		Expansion measured parallel to axis.
12	103	Jay, A. H.	1933	X	291-414		99.995 Bi, 0.004 Ag sample supplied by A. Hilger Ltd.; filings from rod of sample; expansion measured parallel to principal axis; lattice spacing reported at 293 K is 11.836 kX.
13	103	Jay, A. H.	1933	X	292-541		Similar to the above specimen; expansion measured perpendicular to principal axis; lattice spacing reported at 293 K is 3.9287 kX.
14	256	Jacobs, R. B. and Goetz, A.	1937	X	30-538		[111] interplanar spacing measured; expansion measured parallel to principal axis; changes in coefficient of expansion at 258 K and 348 K interpreted as second order phase changes by author.
15*	670	Cave, E. F. and Holroyd, L. V.	1960	O	333-373		Single crystal, ~0.0001 Cu, ~0.0001 Fe, prepared by pulling from melt of purified Bi maintained under vacuum; sample oriented for a-axis measurement about 10 cm long; this curve is here reported using the first given temperature as reference temperature at which $\Delta L/L_0 = 0$.
16	670	Cave, E. F. and Holroyd, L. V.	1960	O	100-540		Similar to the above specimen; expansion measured along a-axis.
17	670	Cave, E. F. and Holroyd, L. V.	1960	O	100-540		The above specimen; expansion measured along c-axis.

* Not shown in figure.

SPECIFICATION TABLE 5. THERMAL LINEAR EXPANSION OF BISMUTH Bi (continued)

Cur. No.	Ref. No.	Author(s)	Year	Method Used	Temp. Range, K	Name and Specimen Designation	Composition (weight percent), Specifications, and Remarks
18	703	Goetz, A. and Hergenrother, R.C.	1932	X	294-535		Single crystal; expansion measured along the [111] plane; lattice parameter reported at 294 K is 3.9453 Å; 3.9452 Å at 293 K determined by graphical extrapolation.
19	700	Gachkovskii, V.F. and Strelkov, P.G.	1937	L	288-527		Single crystal; expansion measured parallel to the axis.
20	700	Gachkovskii, V.F. and Strelkov, P.G.	1937	L	286-514		Single crystal; expansion measured perpendicular to the axis.
21	712	Klemm, W., Spitzer, H., and Niermann, H.	1960	X	293-537		Pure specimen; measured along a-axis.
22	712	Klemm, W., et al.	1960	X	293-532		The above specimen; measured along c-axis.
23	710	Erfling, H.D.	1939	I	293-58		Crystal, sample purity checked by resistivity value; measurements along a-axis; data on the coefficients of thermal linear expansion also reported.
24	710	Erfling, H.D.	1939	I	293-58		Similar to the above specimen; measurements along c-axis; data on the coefficients of thermal linear expansion also reported.

DATA TABLE 5. THERMAL LINEAR EXPANSION OF BISMUTH Bi

[Temperature, T, K; Linear Expansion, $\Delta L/L_0$, %]

T	$\Delta L/L_0$
CURVE 1*, †, ‡	
10	0.000
15	0.002
20	0.005
30	0.016
40	0.029
50	0.044
60	0.061
70	0.078
80	0.094
100	0.128
120	0.162
140	0.196
160	0.230
180	0.264
200	0.298
CURVE 2*, †, ‡	
10	0.0000
15	0.0002
20	0.0006
30	0.002
40	0.005
50	0.011
60	0.018
70	0.026
80	0.035
100	0.056
120	0.079
140	0.103
160	0.127
180	0.151
200	0.175
CURVE 3*, †, ‡	
2.5	0.000×10^{-3}
3	0.002
4	0.011
6	0.063
8	0.205
10	0.515
12	1.06
16	3.07
20	6.25
24	10.04
CURVE 3 (cont.)*, †, ‡	
28	15.26×10^{-3}
58	59.19
65	70.67
75	87.15
85	103.7
CURVE 4*, †, ‡	
2.5	0.000×10^{-3}
3	0.0004
4	0.002
6	0.007
8	0.021
10	0.051
12	0.106
16	0.325
20	0.726
24	1.342
28	2.256
58	17.87
65	23.42
75	31.93
85	41.03
CURVE 5	
93	-0.025
113	-0.022
133	-0.020
153	-0.018
173	-0.015
193	-0.013
213	-0.010
233	-0.008
253	-0.005
273	-0.002
293	0.000*
CURVE 6	
293	0.000*
300	0.003
315	0.033
339	0.051
355	0.070
364	0.103
CURVE 7	
293	0.000*
300	0.006*
316	0.033*
339	0.061
355	0.086
365	0.117
CURVE 8	
190	-0.182
540	0.437
CURVE 9	
193	-0.118
373	0.094
CURVE 10‡	
315.0	0.026
355.3	0.074
386.2	0.111
424.0	0.157
469.3	0.212
506.4	0.257
CURVE 11‡	
322.5	0.047
360.8	0.109
392.5	0.161
432.2	0.225
455.7	0.263
495.1	0.327
CURVE 12	
291	-0.002*
293	0.000*
298	0.005*
351	0.097
362	0.115
387	0.151
414	0.206
CURVE 13	
292	0.000*
293	0.000*
299	0.005*
319	0.027
341	0.051*
352	0.069
362	0.083
387	0.115
413	0.140*
437	0.173
465	0.203
476	0.221*
481	0.222
504	0.251*
512	0.264
519	0.272*
521	0.267
525	0.268*
532	0.270
538	0.274*
541	0.279
CURVE 14‡	
30	-0.423
35	-0.419
45	-0.409
55	-0.396
70	-0.374
80	-0.358
90	-0.341
258	-0.048
348	0.076
538	0.407
CURVE 15*, †	
333	0.000
343	0.246
353	0.484
363	0.739
373	0.994
CURVE 16‡	
100	-0.223
CURVE 16 (cont.)‡	
150	-0.167
540	0.289
CURVE 17‡	
100	-0.334
150	-0.250
540	0.431
CURVE 18	
294	0.003*
329	0.058
360	0.101
390	0.137
429	0.208
469	0.284
499	0.335
519	0.370
529	0.395
535	0.416
CURVE 19	
288.1	-0.009*
288.2	-0.008*
289.6	-0.006*
290.2	-0.005*
293.2	0.000*
339.8	0.080
349.8	0.097*
356.2	0.104
368.4	0.124*
368.4	0.129
410.8	0.197*
410.8	0.201
458.5	0.280
509.5	0.359
526.6	0.385
CURVE 20	
286.3	-0.008*
287.0	-0.007*
287.2	-0.007*
287.6	-0.007*
CURVE 20 (cont.)	
287.9	-0.006*
293.2	0.000*
343.4	0.057*
343.9	0.058
354.8	0.072*
355.3	0.074
371.0	0.090
413.8	0.138*
413.9	0.140
413.9	0.142*
413.9	0.135*
456.1	0.191*
456.4	0.191
513.6	0.257
520.8	0.261
CURVE 21	
293	0.000*
373	0.118
478	0.218
500	0.240
520	0.254
537	0.241
CURVE 22	
293	0.00*
301	0.00
369	0.12
413	0.21
481	0.29
499	0.30
501	0.36
519	0.33
532	0.38
CURVE 23	
293	0.000*
273	-0.023*
253	-0.046
233	-0.069
213	-0.092
193	-0.115*
173	-0.137

* Not shown in figure.
† This curve is here reported using the first given temperature as reference temperature at which $\Delta L/L_0 = 0$.
‡ Author's data for coefficient of thermal expansion have been integrated by TPRC to obtain $\Delta L/L_0$.

DATA TABLE 5. THERMAL LINEAR EXPANSION OF BISMUTH Bi (continued)

T	$\Delta L/L_0$
CURVE 23 (cont.)	
153	-0.159
133	-0.180
113	-0.201
90	-0.224
78	-0.235
58	-0.253
20	-0.272
CURVE 24	
333	0.065
293	0.000*
273	-0.032
233	-0.097
193	-0.162
153	-0.226
113	-0.290
90	-0.326
78	-0.345
58	-0.374
20	-0.425

* Not shown in figure.

DATA TABLE 5. COEFFICIENT OF THERMAL LINEAR EXPANSION OF BISMUTH Bi

[Temperature, T, K; Coefficient of Expansion, α, 10^{-6} K^{-1}]

T	α
CURVE 1‡	
10	2.04
15	5.71
20	8.60
30	12.11
40	14.51
50	16.09
60	16.65
70	16.84
80	16.91
100	17.00
120	17.01
140	16.98
160	16.94
180	16.87
200	16.85
CURVE 2‡	
10	0.20
15	0.52
20	1.02
30	2.20
40	4.33
50	6.24
60	7.72
70	8.90
80	9.81
100	11.10
120	11.72
140	12.01
160	12.04
180	12.10
200	12.11
CURVE 3‡	
2.5	0.0319
3	0.0540
4	0.1185*
6	0.400
8	1.027
10	2.070
12	3.435
16	6.58
20	9.33
24	11.40
28	12.92

T	α
CURVE 3 (cont.)‡	
58	16.37
65	16.40
75	16.57
85	16.62
283	17.05
CURVE 4‡	
2.5	0.0063
3	0.0100*
4	0.0170*
6	0.0379*
8	0.098*
10	0.200*
12	0.353*
16	0.744
20	1.26
24	1.82
28	2.75
58	7.66
65	8.19
75	8.84
85	9.36
283	11.61
CURVE 10*, ‡	
315.0	11.8
355.3	12.0
386.2	12.1
424.0	12.2
469.3	12.0
506.4	12.1
CURVE 11*, ‡	
322.5	16.2
360.8	16.1
392.5	16.3
432.2	16.3
455.7	16.2
495.1	16.1
CURVE 14‡	
30	7.2
35	8.3
45	11.2
55	14.0

T	α
CURVE 14 (cont.)‡	
70	15.6
80	17.0
CURVE 16‡	
100	10.8
150	11.6
540	11.8*
CURVE 17‡	
100	16.6
150	17.3
540	17.6*
CURVE 23	
283	11.6
263	11.56
243	11.50
223	11.44
203	11.32
183	11.14
163	10.93
143	10.72
123	10.46
102	9.89
84	9.25*
68	8.53
38	5.11
CURVE 24	
303	16.2
253	16.2
213	16.2
173	16.08
133	15.94
102	15.86
84	15.55
68	14.88
38	13.28

* Not shown in figure.

‡ Author's data for coefficient of thermal expansion have been integrated by TPRC to obtain $\Delta L/L_0$.

FIGURE AND TABLE NO. 6R. RECOMMENDED VALUES FOR THERMAL LINEAR EXPANSION OF CADMIUM Cd

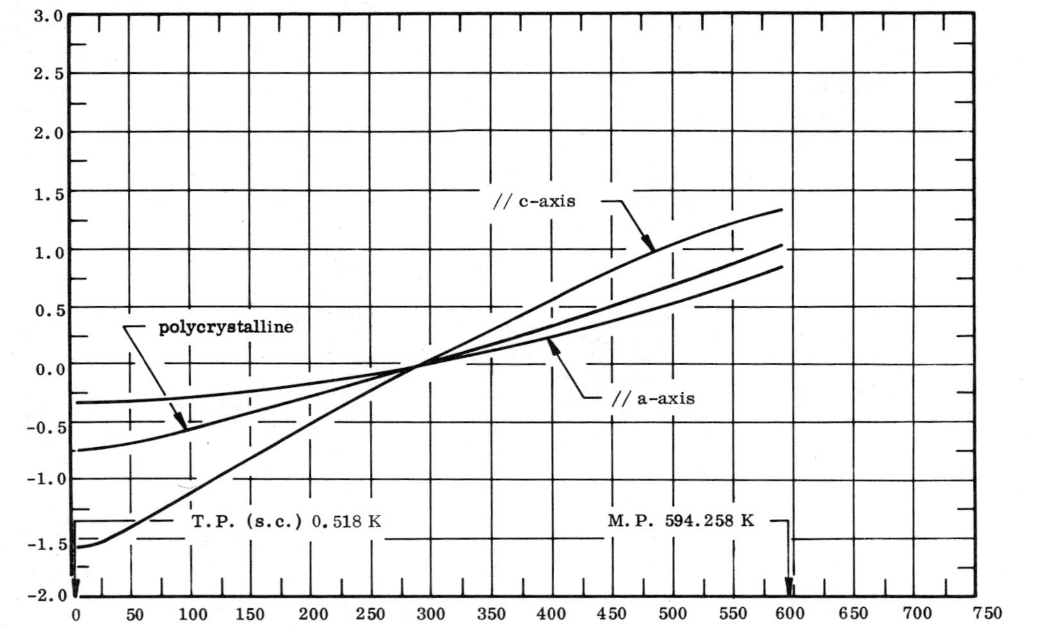

RECOMMENDED VALUES

[Temperature, T, K; Linear Expansion, $\Delta L/L_0$, %; α, K^{-1}]

	// a-axis	// c-axis	polycrystalline	
T	$\Delta L/L_0$	$\Delta L/L_0$	$\Delta L/L_0$	$\alpha \times 10^6$
5	-0.332	-1.569	-0.744	0.05
25	-0.336	-1.530	-0.734	11.8
50	-0.338	-1.398	-0.691	20.9
100	-0.302	-1.100	-0.568	26.9
200	-0.166	-0.517	-0.283	29.8
293	0.000	0.000	0.000	30.8
400	0.231	0.551	0.338	32.8
500	0.512	1.027	0.684	36.0
590	0.862	1.359	1.028	40.0

REMARKS

The tabulated values are considered accurate to within ± 3% over the entire temperature range. Since no experimental data on the polycrystalline cadmium were located in the literature, the tabulated values are calculated from single crystal data. These values are considered accurate to within ± 5% over the entire temperature range. These values can be represented approximately by the following equations:

// a-axis:

$\Delta L/L_0 = -0.302 + 1.063 \times 10^{-3}(T-100) + 3.305 \times 10^{-6}(T-100)^2 - 3.578 \times 10^{-9}(T-100)^3$ $(100 < T < 293)$

$\Delta L/L_0 = 1.939 \times 10^{-3}(T-293) + 1.233 \times 10^{-6}(T-293)^2 + 6.742 \times 10^{-9}(T-293)^3$ $(293 < T < 590)$

// c-axis:

$\Delta L/L_0 = -1.100 + 6.030 \times 10^{-3}(T-100) - 2.088 \times 10^{-6}(T-100)^2 + 1.705 \times 10^{-9}(T-100)^3$ $(100 < T < 293)$

$\Delta L/L_0 = 5.415 \times 10^{-3}(T-293) - 1.100 \times 10^{-6}(T-293)^2 - 5.594 \times 10^{-9}(T-293)^3$ $(293 < T < 590)$

polycrystalline:

$\Delta L/L_0 = -0.568 + 2.724 \times 10^{-3}(T-100) + 1.450 \times 10^{-6}(T-100)^2 - 1.675 \times 10^{-9}(T-100)^3$ $(100 < T < 293)$

$\Delta L/L_0 = 3.096 \times 10^{-3}(T-293) + 4.795 \times 10^{-7}(T-293)^2 + 2.549 \times 10^{-9}(T-293)^3$ $(293 < T < 590)$

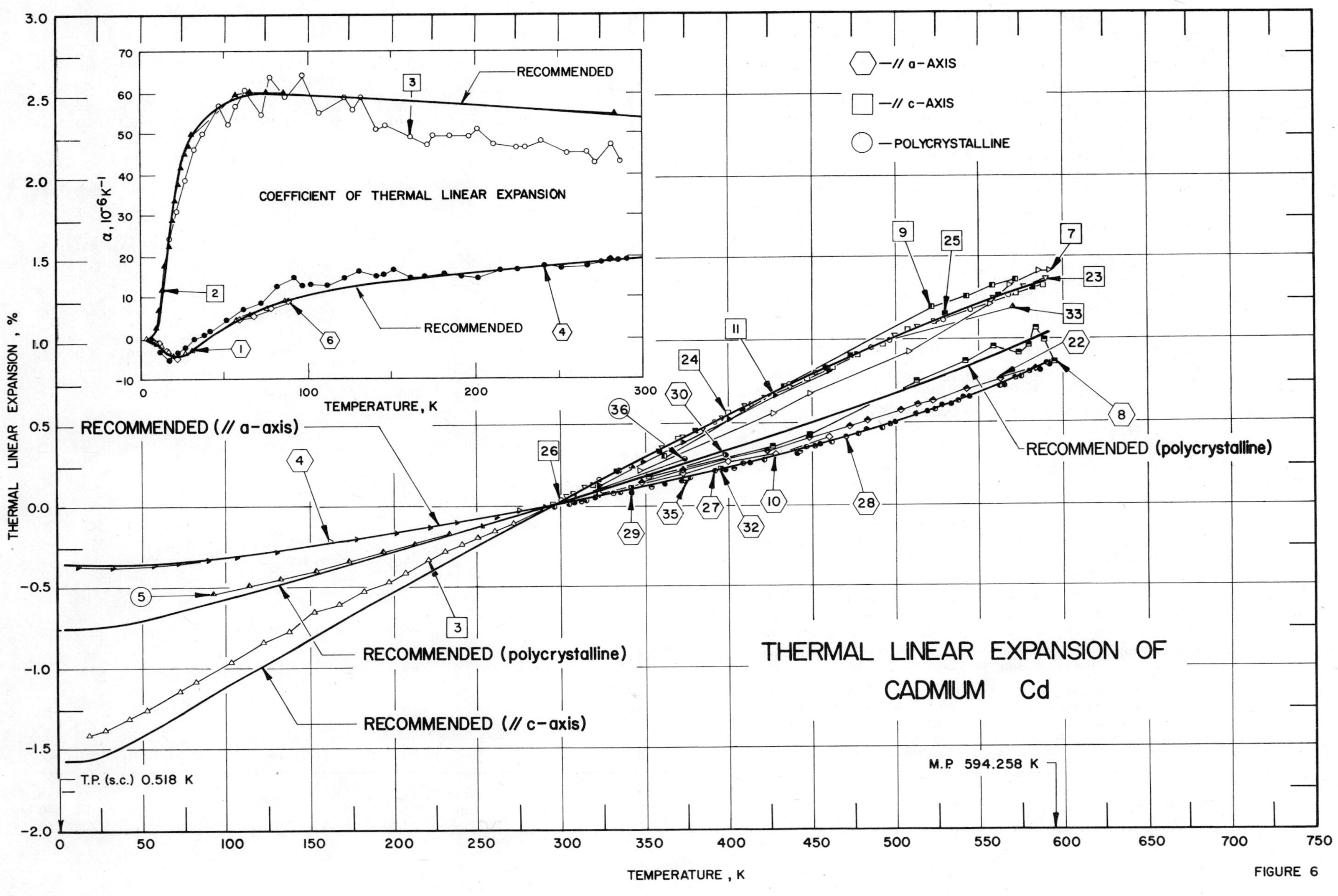
—// a-AXIS
—// c-AXIS
—POLYCRYSTALLINE
COEFFICIENT OF THERMAL LINEAR EXPANSION
RECOMMENDED
RECOMMENDED
α, 10⁻⁶ K⁻¹
TEMPERATURE, K
RECOMMENDED (// a-axis)
RECOMMENDED (polycrystalline)
RECOMMENDED (polycrystalline)
RECOMMENDED (// c-axis)
THERMAL LINEAR EXPANSION OF
CADMIUM Cd
M.P 594.258 K
T.P. (s.c.) 0.518 K
THERMAL LINEAR EXPANSION, %
TEMPERATURE, K

FIGURE 6

SPECIFICATION TABLE 6. THERMAL LINEAR EXPANSION OF CADMIUM Cd

Cur. No.	Ref. No.	Author(s)	Year	Method Used	Temp. Range, K	Name and Specimen Designation	Composition (weight percent), Specifications, and Remarks
1*	104	McCammon, R. D. and White, G. K.	1965	E	3-283	1 and 2	Specimens grown from 99.999 Cd; cylindrical crystals 5 cm long, 2 cm diameter oriented perpendicular to hexad axis; 2 identical crystals supplied by Metals Research Ltd., Cambridge, U. K.; cylinders spark machined, ends fine sparked; data given is average of two specimens; this curve is here reported using the first given temperature as reference temperature at which $\Delta L/L_0 = 0$.
2*	104	McCammon, R. D. and White, G. K.	1965	E	3-283	3	Similar to the above specimen except diameter oriented parallel to hexad axis; this curve is here reported using the first given temperature as reference temperature at which $\Delta L/L_0 = 0$.
3	105	Madaiah, N. and Graham, G. M.	1964	E	17-286		Specimen obtained by seeding melt of spectroscopically pure material to grow crystal 5 to 6 cm long, with length parallel to hexad axis of crystal; reported error in the coefficient of expansion $<5\%$ in temperature range 50 to 240 K.
4	105	Madaiah, N. and Graham, G. M.	1964	E	11-290		Similar to the above specimen but with length perpendicular to hexad axis of crystal.
5	48	Dorsey, H. G.	1907	I	93-293		Pure specimen from Eimer and Amend; cast, length 1.1659 cm; density 8.62 g cm^{-3}.
6*	106	White, G. K.	1963	E	7-281		Expansion measured normal to hexad axis; this curve is here reported using the first given temperature as reference temperature at which $\Delta L/L_0 = 0$.
7	107	Kotel'nikov, V. A. and Petrov, Yu. I.	1969	X	346-592		Specimen particles ~200 Å diameter deposited from metallic vapor in Ar atmosphere; expansion measured in vacuum 10^{-4} mm of Hg along c-axis of hexagonal crystals; zero-point correction of 0.1% determined by graphical extrapolation.
8	107	Kotel'nikov, V. A. and Petrov, Yu. I.	1969	X	352-594		The above specimen; expansion measured along a-axis of basal plane; zero-point correction of 0.1% determined by graphical extrapolation.
9	107	Kotel'nikov, V. A. and Petrov, Yu. I.	1969	X	362-572		Similar to the above specimen; particle diameter ~1500 Å; expansion measured along c-axis.
10	108	Owen, E. A. and Roberts, E. W.	1936	X	291-460		>99.9 Cd, specimen prepared by sublimation on clean surface of copper backing sheet, deposited layer 0.05 mm thick; expansion measured perpendicular to hexagonal axis, Ni radiation; lattice parameter reported at 291 K is 2.9723 Å; 2.97235 Å at 293 K determined by graphical extrapolation.
11	108	Owen, E. A. and Roberts, E. W.	1936	X	291-460		The above specimen; expansion measured along hexagonal axis; lattice parameter reported at 322 K is 5.6153 Å; 5.6080 Å at 293 K determined by graphical extrapolation.
12*	108	Owen, E. A. and Roberts, E. W.	1936	X	358, 396		Similar to the above specimen except deposited layer 0.12 mm thick; expansion measured perpendicular to hexagonal axis; this curve is here reported using the first given temperature as reference temperature at which $\Delta L/L_0 = 0$; lattice parameter reported at 358 K is 2.9771 Å.
13	108	Owen, E. A. and Roberts, E. W.	1936	X	358, 396		The above specimen; expansion measured along hexagonal axis; this curve is here reported using the first given temperature as reference temperature at which $\Delta L/L_0 = 0$; lattice parameter reported at 358 K is 5.6251 Å.
14*	108	Owen, E. A. and Roberts, E. W.	1936	X	382, 449		Similar to the above specimen except deposited layer 0.03 mm thick; expansion measured perpendicular to hexagonal axis; this curve is here reported using the first given temperature as reference temperature at which $\Delta L/L_0 = 0$; lattice parameter reported at 382 K is 2.9784 Å.

* Not shown in figure.

SPECIFICATION TABLE 6. THERMAL LINEAR EXPANSION OF CADMIUM Cd (continued)

Cur. No.	Ref. No.	Author(s)	Year	Method Used	Temp. Range, K	Name and Specimen Designation	Composition (weight percent), Specifications, and Remarks
15*	108	Owen, E. A. and Roberts, E. W.	1936	X	382, 449		The above specimen; expansion measured along hexagonal axis; lattice parameter reported at 382 K is 5.6315 Å; this curve is here reported using the first given temperature as reference temperature at which $\Delta L/L_0 = 0$.
16*	108	Owen, E. A. and Roberts, E. W.	1936	X	481-552		Similar to the above specimen except sublimated on Cd backing sheet, chamber filled with nitrogen; expansion measured perpendicular to hexagonal axis; lattice parameter reported at 481 K is 2.9876 Å; this curve is here reported using the first given temperature as reference temperature at which $\Delta L/L_0 = 0$.
17*	108	Owen, E. A. and Roberts, E. W.	1936	X	481-552		The above specimen; expansion measured along hexagonal axis; lattice parameter reported at 481 K is 5.6597 Å; this curve is here reported using the first given temperature as reference temperature at which $\Delta L/L_0 = 0$.
18*	109	Hachkovsky, W. F. and Strelkov, P. G.	1937		579-591		Monocrystal specimen; expansion measured perpendicular to hexagonal axis with increasing temperature; this curve is here reported using the first given temperature as reference temperature at which $\Delta L/L_0 = 0$.
19*	109	Hachkovsky, W. F. and Strelkov, P. G.	1937		591-579		The above specimen; expansion measured with decreasing temperature; this curve is here reported using the first given temperature as reference temperature at which $\Delta L/L_0 = 0$.
20*	109	Hachkovsky, W. F. and Strelkov, P. G.	1937		579-591		The above specimen; expansion measured with increasing temperature; this curve is here reported using the first given temperature as reference temperature at which $\Delta L/L_0 = 0$.
21*	109	Hachkovsky, W. F. and Strelkov, P. G.	1937		591-579		The above specimen; expansion measured with decreasing temperature; this curve is here reported using the first given temperature as reference temperature at which $\Delta L/L_0 = 0$.
22	594	Gertsriken, S. D. and Slysar, B. F.	1958	L	273-583		99.99 pure Cd; specimen in cylindrical rods 3.5 mm diameter, 20-25 mm long; specimens annealed 2-3 hr at fairly high temperatures and slowly furnace cooled to room temperature; zero-point correction of -0.054% determined by graphical interpolation.
23	625	Feder, R. and Norwick, A. S.	1972	I	302-590		99.999 pure specimen, Bridgman grown; expansion measured along c-axis.
24	625	Feder, R. and Norwick, A. S.	1972	I	295-588		The above specimen, cut from Bridgman grown boule, angle between specimen axis and c-axis is 18°; expansion measured along c-axis.
25	625	Feder, R. and Norwick, A. S.	1972	I	334-589		The above specimen; seeded, horizontal boat; expansion measured along c-axis.
26	625	Feder, R. and Norwick, A. S.	1972	X	297-591		The above specimen; expansion measured along c-axis; lattice parameter 5.61792 Å at 299 K.
27	625	Feder, R. and Norwick, A. S.	1972	I	297-590		The above specimen; large crystal cut to get angle between specimen and c-axis equal to 90°; expansion measured along a-axis.
28	625	Feder, R. and Norwick, A. S.	1972	I	298-580		The above specimen; Bridgman grown, cut to get angle between specimen and c-axis equal to 85°; expansion measured along a-axis.
29	625	Feder, R. and Norwick, A. S.	1972	X	296-592		The above specimen; expansion measured along a-axis; lattice parameter 2.97910 Å at 299 K.
30	665	Balasundaram, L. J. and Sinha, A. N.	1971	L	293-398		Specimen is chemically pure metal.
31*	700	Gachkovskii, V. F. and Strelkov, P. G.	1937	L	297-573		Single crystal sample, specimen containing 0.02 Pb, 0.005 Cu, 0.0005 Ag, $\ell_0 = 58.71$ mm; expansion measured in direction 75.5° with respect to axis of crystal.

* Not shown in figure.

SPECIFICATION TABLE 6. THERMAL LINEAR EXPANSION OF CADMIUM Cd (continued)

Cur. No.	Ref. No.	Author(s)	Year	Method Used	Temp. Range, K	Name and Specimen Designation	Composition (weight percent), Specifications, and Remarks
32	704	Edwards, D.A., Wallace, W.E., and Craig, R.S.	1952	X	298-569	Cd	"Pure cadmium", sample prepared by filing bulk alloys under argon or helium atmosphere, size of filings were smaller than 200 mesh; obtained from Anaconda Copper Mining Co.; strain annealed for 30 hr at 498 K; alloys examined spectroscopically and metallic contaminants appeared only in traces; average measurement error of 0.20%; expansion measured along the a-axis; lattice parameter reported at 298 K is 2.9793 Å; 2.9790 Å at 293 K by graphical extrapolation.
33	704	Edwards, D.A., et al.	1952	X	298-569	Cd	The above specimen; expansion measured along c-axis; lattice parameter reported at 298 K is 5.6185 Å; 5.6170 Å at 293 K by graphical extrapolation.
34*	715	Shinoda, G.	1934	X	294, 373		Expansion measured along the c-axis.
35	715	Shinoda, G.	1934	X	294, 373		Expansion measured perpendicular to the c-axis.
36	715	Shinoda, G.	1934	X	294, 373		Similar to the above specimen except polycrystalline.
37*	714	Andres, K.	1964	O	1.9-6.1	Probe I	99.999 pure, polycrystalline sample; this curve is here reported using the first given temperature as reference temperature at which $\Delta L/L_0 = 0$.
38*	714	Andres, K.	1964	O	2.2-6.4	Probe II	99.999 pure, polycrystalline sample; specimen melted in high vacuum and then quenched; this curve is here reported using the first given temperature as reference temperature at which $\Delta L/L_0 = 0$.
39*	714	Andres, K.	1964	O	2.2-12	Probe II	The above specimen; this curve is here reported using the first given temperature as reference temperature at which $\Delta L/L_0 = 0$.

* Not shown in figure.

DATA TABLE 6. THERMAL LINEAR EXPANSION OF CADMIUM Cd

[Temperature, T, K; Linear Expansion, $\Delta L/L_0$, %]

T	$\Delta L/L_0$
CURVE 1*, †, ‡	
3	0.0000 x 10^{-4}
4	0.0037
5	0.0045
6	-0.0039
7	-0.017
8	-0.477
10	-1.924
12	-4.774
14	-9.174
16	-14.97
18	-21.18
20	-29.22
22	-36.97
24	-44.67
26	-52.12
28	-59.12
30	-65.42
57.5	-43.42
65	-38.62
75	64.64
85	148.9
CURVE 2*, †, ‡	
3	0.000 x 10^{-4}
4	0.058
5	0.243
6	0.758
7	1.958
8	4.308
10	14.56
12	33.76
14	63.76
16	105.4
18	158.3
20	221.6
22	293.5
24	373.5
26	460.5
28	553.0
30	650.5
57.5	2163.0
65	2615.3
75	3220.3
85	3823.8
CURVE 3‡	
17	-1.408
22	-1.394*
26	-1.379
32	-1.354*
37	-1.330*
42	-1.305
47	-1.278*
52	-1.250
57	-1.223*
63	-1.188*
67	-1.164*
72	-1.136
77	-1.106*
82	-1.074
86	-1.050*
92	-1.013*
96	-0.988*
102	-0.950
106	-0.927*
111	-0.899*
116	-0.870*
121	-0.840
126	-0.812*
131	-0.783*
136	-0.754
141	-0.727*
147	-0.696*
150	-0.680*
156	-0.650
161	-0.625*
166	-0.601
171	-0.577*
175	-0.558*
181	-0.528
185	-0.508*
191	-0.478*
196	-0.453
201	-0.428*
206	-0.403
211	-0.378*
216	-0.355*
220	-0.336
225	-0.313*
230	-0.289
235	-0.266*
240	-0.242
CURVE 3 (cont.)‡	
245	-0.218*
250	-0.195
255	-0.172*
260	-0.150
266	-0.123*
271	-0.100
275	-0.083*
281	-0.055*
286	-0.033*
CURVE 4‡	
11	-0.362
17	-0.364*
22	-0.366*
26	-0.367*
32	-0.368
37	-0.367*
41	-0.366*
46	-0.365*
52	-0.363*
56	-0.361
61	-0.358*
66	-0.354*
71	-0.350
77	-0.344*
81	-0.339*
88	-0.330
91	-0.326*
96	-0.319*
101	-0.312*
107	-0.304
111	-0.299*
115	-0.294*
121	-0.285*
126	-0.277*
130	-0.271
136	-0.262*
141	-0.254*
145	-0.248*
151	-0.238*
156	-0.230*
161	-0.222
165	-0.216*
170	-0.208*
176	-0.199
CURVE 4 (cont.)‡	
182	-0.189*
185	-0.185*
191	-0.176*
196	-0.168*
201	-0.160
205	-0.154*
210	-0.147*
215	-0.139*
221	-0.129
225	-0.122*
230	-0.114*
236	-0.104
241	-0.095*
245	-0.088*
251	-0.078*
254	-0.073*
260	-0.062
266	-0.051*
270	-0.044*
275	-0.035
280	-0.025*
285	-0.016*
290	-0.006
CURVE 5	
93	-0.540
113	-0.494
133	-0.446
153	-0.395
173	-0.342
193	-0.289
213	-0.234
233	-0.176
253	-0.117
273	-0.059
293	0.000
CURVE 6*, †, ‡	
7	0.000 x 10^{-4}
8	-0.095
9	-0.75
11	-6.69
13	-11.19
15	-17.10
CURVE 6 (cont.)*, †, ‡	
17	-23.97 x 10^{-4}
19	-31.53
21	-39.63
23	-47.72
25	-55.35
27	-58.89
28	-65.21
30	-43.46
59	-24.28
63	-7.65
66	59.85
76	142.85
CURVE 7	
346	0.215
427	0.563
449	0.685
507	0.946
553	1.239
567	1.346
579	1.383*
585	1.428
592	1.428
CURVE 8	
352	0.175
363	0.188*
426	0.354
434	0.402*
449	0.438
512	0.752
522	0.780*
542	0.876
552	0.935*
559	0.951
566	0.951*
574	0.927
580	0.973
584	1.017
589	1.000
594	0.864
CURVE 9	
362	0.302
433	0.747
521	1.215
543	1.278
559	1.346
572	1.378
CURVE 10	
291	-0.002*
322	0.075
349	0.133*
373	0.200
379	0.210*
400	0.270
428	0.334
460	0.422
CURVE 11	
322	0.130
349	0.258
373	0.399
379	0.404*
400	0.536
428	0.657
460	0.804
CURVE 12*, †	
358	0.000
396	0.080
CURVE 13†	
358	0.000
396	0.527
CURVE 14*, †	
382	0.000
449	0.171

*Not shown in figure.

†This curve is here reported using the first given temperature as reference temperature at which $\Delta L/L_0 = 0$.

‡Author's data for coefficient of thermal expansion have been integrated by TPRC to obtain $\Delta L/L_0$.

DATA TABLE 6. THERMAL LINEAR EXPANSION OF CADMIUM Cd (continued)

T	$\Delta L/L_0$
CURVE 15*, †	
382	0.000
449	0.311
CURVE 16*, †	
481	0.000
502	0.037
527	0.140
536	0.194
552	0.274
CURVE 17*, †	
481	0.000
502	0.063
527	0.173
536	0.238
552	0.249
CURVE 18*, †, ‡	
579	0.000
584	0.033
588	0.071
591	0.153
CURVE 19*, †, ‡	
591	0.146
587	0.046
583	0.026
579	0.000
CURVE 20*, †, ‡	
579	0.000
584	0.0302
588	0.0496
591	0.185
CURVE 21*, †, ‡	
591	0.000
588	-0.106
579	-0.163
CURVE 22	
273	-0.054*
298	0.013
323	0.079*
348	0.145
373	0.212
398	0.279*
423	0.346
448	0.414
473	0.490
483	0.519
493	0.551*
503	0.580
513	0.612
523	0.643
533	0.675*
543	0.708
553	0.741*
563	0.776
573	0.811*
583	0.849
CURVE 23	
302	0.051
314	0.111
335	0.223
360	0.351
372	0.414
385	0.478
397	0.541
410	0.601
423	0.666
436	0.727
448	0.789*
461	0.852*
487	0.969*
500	1.028
512	1.079*
513	1.081
526	1.134*
537	1.181*
548	1.221*
559	1.263
568	1.294*
577	1.322
587	1.350*
590	1.357
CURVE 24	
295	0.011
307	0.078
319	0.138
328	0.183*
348	0.288*
358	0.340*
370	0.404
385	0.479*
400	0.553
415	0.628*
446	0.774
462	0.848
477	0.919
492	0.988
507	1.056
523	1.119
538	1.177*
558	1.250*
573	1.298
579	1.318*
588	1.339
CURVE 25	
334	0.216
359	0.345
380	0.453
409	0.596
460	0.843
474	0.911
486	0.964*
508	1.061*
530	1.153
538	1.181*
563	1.276
573	1.312*
574	1.312*
575	1.314*
583	1.338
586	1.349*
588	1.352*
589	1.356*
CURVE 26	
297	0.026
300	0.042
CURVE 26 (cont.)	
311	0.098*
317	0.130*
320	0.146*
323	0.161
341	0.252*
361	0.358*
371	0.411*
382	0.464*
392	0.515
403	0.571*
414	0.622
425	0.678*
430	0.702*
436	0.730
441	0.755*
447	0.784*
452	0.806
463	0.855*
468	0.881*
474	0.907*
479	0.928*
485	0.957
490	0.979*
496	1.000
501	1.023*
507	1.050*
511	1.067*
512	1.071*
523	1.114*
529	1.139
534	1.156*
540	1.178*
545	1.197
551	1.216*
556	1.234*
562	1.253*
567	1.271
570	1.277*
577	1.299*
579	1.303*
584	1.315*
588	1.328*
591	1.334*
CURVE 27	
297	0.009
305	0.025
315	0.044
322	0.058*
326	0.065*
335	0.084
350	0.115*
377	0.176
384	0.192*
391	0.209
397	0.225*
409	0.253
410	0.254*
412	0.260*
417	0.274*
422	0.286
438	0.327*
446	0.353
453	0.371*
461	0.393
464	0.401*
466	0.407*
477	0.441
479	0.445*
487	0.469*
492	0.486
512	0.552*
524	0.596
532	0.623*
536	0.638*
537	0.643
552	0.698*
562	0.735
562	0.739*
575	0.795
584	0.836*
586	0.842*
590	0.865
CURVE 28	
298	0.012*
305	0.025*
313	0.038
321	0.053*
331	0.075
342	0.098*
CURVE 28 (cont.)	
354	0.122
367	0.153*
379	0.181*
392	0.212*
404	0.241
418	0.274*
430	0.306*
443	0.341
451	0.363
456	0.379*
457	0.383*
469	0.417*
470	0.420
482	0.453*
483	0.458*
489	0.479*
490	0.482*
495	0.496*
496	0.500
499	0.512*
502	0.516*
502	0.520*
503	0.525*
509	0.545*
511	0.551
517	0.570*
519	0.578*
520	0.581*
525	0.598*
527	0.605
532	0.627*
534	0.634*
540	0.653
541	0.661*
548	0.683*
556	0.716*
563	0.748*
572	0.785
580	0.823
CURVE 29	
296	0.009*
308	0.033
310	0.036*
313	0.041*
320	0.053

* Not shown in figure.
† This curve is here reported using the first given temperature as reference temperature at which $\Delta L/L_0 = 0$.
‡ Author's data for coefficient of thermal expansion have been integrated by TPRC to obtain $\Delta L/L_0$.

DATA TABLE 6. THERMAL LINEAR EXPANSION OF CADMIUM Cd (continued)

T	$\Delta L/L_0$
CURVE 29 (cont.)	
320	0.054*
325	0.063*
332	0.078*
342	0.098
344	0.103*
352	0.121*
354	0.122*
362	0.141
363	0.147*
372	0.165
373	0.170*
382	0.186*
392	0.212*
399	0.226
402	0.235*
403	0.238*
413	0.264
420	0.278*
424	0.289*
434	0.316*
435	0.319*
441	0.336
445	0.343*
446	0.348*
455	0.375
462	0.394*
467	0.407*
477	0.437*
478	0.442*
483	0.456
488	0.472*
497	0.500*
500	0.508
505	0.525*
519	0.573
522	0.582*
527	0.597*
527	0.601*
534	0.622
540	0.646*
542	0.653*
544	0.660*
545	0.664
549	0.678*
555	0.705*
563	0.734*
565	0.740
571	0.768*
576	0.791*
578	0.797*
586	0.831
588	0.843*
589	0.846*
592	0.859
CURVE 30	
293	0.000*
398	0.313
CURVE 31*	
376.7	0.198
416.7	0.297
459.2	0.464
502.4	0.593
553.2	0.768
572.9	0.845
CURVE 32	
298	0.01*
341	0.11
396	0.23
495	0.49*
569	0.76*
CURVE 33	
298	0.027*
341	0.250
396	0.539*
495	1.030*
569	1.210
CURVE 34*	
294	0.005
373	0.400
CURVE 35	
294	0.002*
373	0.143
CURVE 36	
294	0.003*
373	0.229
CURVE 37*, †	
1.9	0.000 x 10^{-4}
1.9	0.000
2.3	0.001
2.3	0.002
2.3	0.002
2.9	0.005
2.9	0.005
3.2	0.009
3.2	0.010
3.6	0.016
3.6	0.017
4.0	0.025
4.0	0.027
4.5	0.046
4.5	0.047
4.7	0.063
5.0	0.078
5.0	0.079
5.2	0.100
5.3	0.115
5.4	0.126
5.4	0.127
5.6	0.152
5.7	0.167
5.9	0.201
5.9	0.202
5.9	0.203
6.1	0.233
6.1	0.237
CURVE 38*, †	
2.2	0.000 x 10^{-4}
2.2	0.000
2.8	0.002
2.8	0.003
3.1	0.006
4.0	0.021
4.2	0.031
4.4	0.037
4.4	0.038
4.9	0.070
5.2	0.092
CURVE 38 (cont.)*, †	
5.4	0.121 x 10^{-4}
5.6	0.143
5.7	0.164
5.9	0.191
6.0	0.221
6.2	0.261
6.4	0.311
CURVE 39*, †	
2.2	0.000 x 10^{-4}
2.8	0.002
3.1	0.006
3.9	0.021
4.3	0.036
4.9	0.068
5.1	0.091
5.4	0.118
5.7	0.164
6.0	0.217
6.4	0.310
6.6	0.374
6.8	0.457
7.1	0.559
7.4	0.687
7.6	0.808
7.9	1.00
8.2	1.27
8.6	1.61
9.1	2.05
9.5	2.64
10.1	3.46
10.7	4.50
11.0	5.19
11.4	5.96
11.8	6.90
12.2	8.02
12.2	8.06

* Not shown in figure.
† This curve is here reported using the first given temperature as reference temperature at which $\Delta L/L_0 = 0$.

DATA TABLE 6. COEFFICIENT OF THERMAL LINEAR EXPANSION OF CADMIUM Cd

[Temperature, T, K; Coefficient of Expansion, α, 10^{-6} K^{-1}]

T	α
CURVE 1‡	
3	0.039
4	0.035*
5	-0.01*
6	-0.07*
7	-0.20*
8	-0.40*
10	-1.05
12	-1.8
14	-2.6*
16	-3.2
18	-3.6*
20	-3.85
22	-3.9*
24	-3.8*
26	-3.65
28	-3.35*
30	-2.95
57.5	4.55
65	6.0
75	7.7
85	9.15
283	19.3
CURVE 2‡	
3	0.026
4	0.090*
5	0.28*
6	0.75
7	1.65*
8	3.05
10	7.2
12	12.0
14	18.0
16	23.7
18	29.2
20	34.0
22	38.0
24	42.0
26	45.0
28	47.5
30	50.0
57.5	60.0
65	60.6
75	60.4
85	60.3
283	54.8
CURVE 3‡	
17	24.3
22	31.9
26	39.2
32	46.2
37	50.0
42	50.2*
47	57.4
52	52.4
57	57.0
63	61.0
67	56.5*
72	55.0
77	64.4
82	63.2*
86	59.8
92	62.3*
96	64.3
102	61.6*
106	55.3
111	57.0*
116	58.5*
121	59.1
126	56.2
131	59.0
136	57.1*
141	51.4
147	52.1
150	51.1*
156	49.5*
161	49.5
166	47.4*
171	47.4
175	49.6
181	49.9*
185	49.8
191	49.8*
196	49.1
201	51.2
206	50.8*
211	47.5
216	46.8*
220	47.0*
225	46.4
230	46.3
235	47.2*
CURVE 3 (cont.)‡	
240	48.0
245	47.1*
250	46.8*
255	45.2
260	44.5*
266	45.3
271	43.2
275	44.7*
281	47.3
286	43.3
CURVE 4‡	
11	-2.90
17	-4.30
22	-3.57
26	-2.02
32	0.00
37	1.52
41	2.00
46	2.92*
52	4.78
56	5.67*
61	7.14
66	7.66*
71	8.22
77	11.55*
81	12.86
88	13.98*
91	14.44
96	12.99
101	13.20
107	13.05*
111	12.73
115	14.20*
121	14.73
126	15.37*
130	16.10
136	15.61*
141	15.61
145	15.97
151	16.37
156	15.89*
161	15.04
165	15.50*
CURVE 4 (cont.)‡	
170	15.33
176	15.73*
182	15.94
185	15.28*
191	15.18
196	15.42*
201	14.95
205	15.05*
210	15.61*
215	16.41
221	16.41*
225	16.84
230	16.91*
236	16.91*
241	17.34
245	17.25*
251	17.09
254	17.46*
260	17.71*
266	17.83
270	18.16*
275	18.85
280	19.50
285	18.75
290	19.87
CURVE 6‡	
7	0.00
8	-0.19*
9	-0.48*
11	-1.12
13	-1.85*
15	-2.65
17	-3.26*
19	-3.61*
21	-3.95*
23	-4.15
25	-3.94*
27	-3.69*
28	-3.40*
30	-2.92*
59	4.42
63	5.17*
66	5.92
CURVE 6 (cont.)‡	
76	7.57
86	9.04
281	19.33
CURVE 18*,‡	
579	52.00
584	74.00
588	122.00
591	426.00
CURVE 19*,‡	
591	444.00
587	68.00
584	55.00
579	52.00
CURVE 20*,‡	
579	52.00
584	62.00
588	35.00
591	896.00
CURVE 21*,‡	
591	613.00
588	73.00
579	52.00

*Not shown in figure.
‡ Author's data for coefficient of thermal expansion have been integrated by TPRC to obtain $\Delta L/L_0$.

FIGURE AND TABLE NO. 7R. RECOMMENDED VALUES FOR THERMAL LINEAR EXPANSION OF CALCIUM Ca

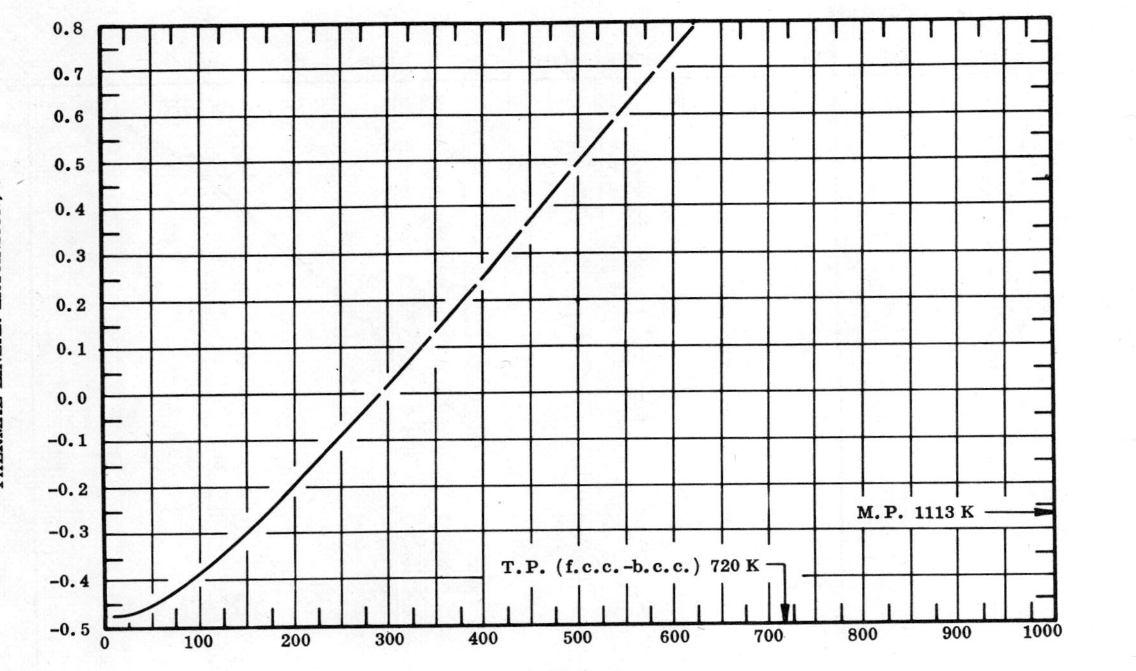

RECOMMENDED VALUES

[Temperature, T, K; Linear Expansion, $\Delta L/L_0$, %; α, K^{-1}]

T	$\Delta L/L_0$	$\alpha \times 10^6$
5	-0.474	0.003
25	-0.473	1.9
50	-0.459	9.5
100	-0.391	16.6
200	-0.200	20.6
293	0.000	22.3
400	0.243†	23.4
500	0.480†	24.0
600	0.723†	24.5
625	0.785†	24.5

† Provisional values.

REMARKS

The tabulated values are considered accurate to within ±3% at temperatures below 293 K and ±10% above. These values can be represented approximately by the following equation:

$$\Delta L/L_0 = -0.560 + 1.543 \times 10^{-3}\, T + 1.499 \times 10^{-6}\, T^2 - 8.363 \times 10^{-10}\, T^3 \qquad (100 < T < 625)$$

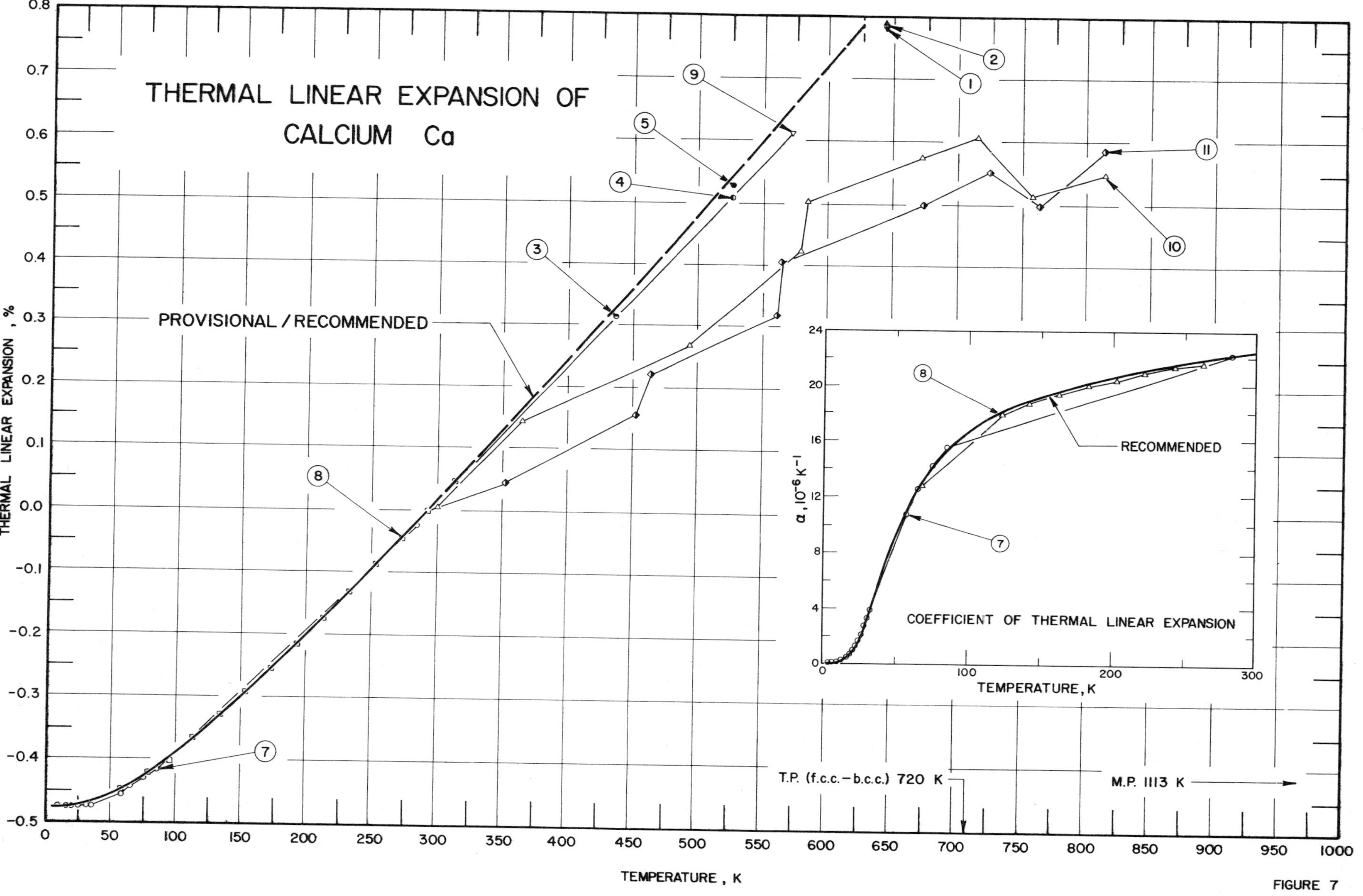
THERMAL LINEAR EXPANSION OF
CALCIUM Ca
PROVISIONAL / RECOMMENDED
THERMAL LINEAR EXPANSION, %
TEMPERATURE, K
T.P. (f.c.c.–b.c.c.) 720 K
M.P. 1113 K
COEFFICIENT OF THERMAL LINEAR EXPANSION
RECOMMENDED
TEMPERATURE, K
α, 10⁻⁶ K⁻¹

FIGURE 7

SPECIFICATION TABLE 7. THERMAL LINEAR EXPANSION OF CALCIUM Ca

Cur. No.	Ref. No.	Author(s)	Year	Method Used	Temp. Range, K	Name and Specimen Designation	Composition (weight percent), Specifications, and Remarks
1	110	Bernstein, B.T. and Smith, J.F.	1959	X	299, 644		Bar specimen maintained under inert atm of purified helium, data average of five to nine experimental points; f.c.c. structure; transition temperature 723-733 K expansion measured in (511) direction; reported error ±2% in the coefficient of expansion.
2	110	Bernstein, B.T. and Smith, J.F.	1959	X	299, 644		The above specimen, expansion in (531) direction.
3	110	Bernstein, B.T. and Smith, J.F.	1959	X	299, 437		The above specimen, expansion in (600) direction; reported error ±5% in the coefficient of expansion.
4	110	Bernstein, B.T. and Smith, J.F.	1959	X	299, 526		The above specimen, expansion in (620) direction; reported error ±3% in the coefficient of expansion.
5	110	Bernstein, B.T. and Smith, J.F.	1959	X	299, 526		The above specimen, expansion in (533) direction; reported error ±4% in the coefficient of expansion.
6*	110	Bernstein, B.T. and Smith, J.F.	1959	X	740, 876		Similar to the above specimen, except b.c.c. structure; expansion measured in (211) direction; reported error ±5% in the coefficient of expansion.
7	726	White, G.K.	1972	E	3-85, 273		Specimen from the A.D. Mackay, New York, spectrographic analysis by the Australian Mineral Development Laboratories shows 0.5 Mg and Al, less than 0.01 Fe and Mn, average of experiments on two specimen; values referred to 293 K using other available data.
8	728	Erfling, H.D.	1942	I	68-313		Pure specimen from the Merck Co., cylindrical specimen of 1 cm long and 1.5 mm diameter tempered at 673 K in vacuum, treated in CO_2 atmosphere; data on the coefficient of thermal linear expansion also reported.
9	727	Cath, P.G. and Steenis, O.L.	1936	L	273-573		No specifications given; specimen heated to 573 K and allowed to cool; expansion measured with increasing temperature.
10	729	Schulze, A.	1935	L	293-814		99.9 Ca; expansion measured with increasing temperature; zero-point correction is 0.006%.
11	729	Schulze, A.	1935	L	814-293		The above specimen; expansion measured with decreasing temperature; zero-point correction is 0.044%.

* Not shown in figure.

DATA TABLE 7. THERMAL LINEAR EXPANSION OF CALCIUM Ca

[Temperature, T, K; Linear Expansion, $\Delta L/L_0$, %]

T	$\Delta L/L_0$
CURVE 1	
299	0.0134
644	0.7862
CURVE 2	
299	0.0135
644	0.7898
CURVE 3	
299	0.0131
437	0.3154
CURVE 4	
299	0.0131
526	0.5103
CURVE 5	
299	0.0137
526	0.5312
CURVE 6*	
740	1.5019
876	1.9589
CURVE 7‡	
3	-0.474*
4	-0.474*
5	-0.474
6	-0.474*
8	-0.474*
10	-0.474
12	-0.474*
14	-0.474*
16	-0.474
CURVE 7 (cont.)‡	
18	-0.474*
20	-0.474
22	-0.473*
24	-0.473*
26	-0.473
28	-0.472*
30	-0.472
32	-0.471*
57	-0.452
65	-0.442
75	-0.429
85	-0.414
283	-0.022
CURVE 8	
313	0.045
293	0.000
273	-0.044
253	-0.087
CURVE 8 (cont.)	
233	-0.130
213	-0.172
193	-0.213
173	-0.252
153	-0.291
133	-0.328
113	-0.364
90	-0.403
78	-0.421
57	-0.447
CURVE 9	
273	-0.044*
293	0.000*
573	0.616
CURVE 10	
293	0.000*
CURVE 10 (cont.)	
300	0.006
365	0.144
493	0.269
579	0.426
583	0.506
672	0.577
715	0.611
757	0.515
814	0.549
CURVE 11	
293	0.000*
352	0.044
454	0.157
462	0.222
561	0.320
564	0.409
673	0.501
CURVE 11 (cont.)	
725	0.554
763	0.500
814	0.587

DATA TABLE 7. COEFFICIENT OF THERMAL LINEAR EXPANSION OF CALCIUM Ca

[Temperature, T, K; Coefficient of Expansion, α, 10^{-6} K^{-1}]

T	α
CURVE 7‡	
3	0.0026
4	0.0055*
5	0.09
6	0.019*
8	0.0465*
10	0.094
12	0.173*
14	0.295
16	0.47*
18	0.7
20	1.00
22	1.36
24	1.78
26	2.22
28	2.73
30	3.3
CURVE 7 (cont.)‡	
32	3.9
57	10.75
65	12.45
75	14.16
85	15.5
283	22.15
CURVE 8	
303	22.52
283	22.12*
263	21.58
243	21.32
223	20.88
203	20.39
183	19.93
CURVE 8 (cont.)	
163	10.32
143	18.70
123	17.95
102	16.76
84	15.32*
67	12.61

* Not shown in figure.
‡ Author's data for coefficient of thermal expansion have been integrated by TPRC to obtain $\Delta L/L_0$.

FIGURE AND TABLE NO. 8R. RECOMMENDED VALUES FOR THERMAL LINEAR EXPANSION OF CERIUM Ce

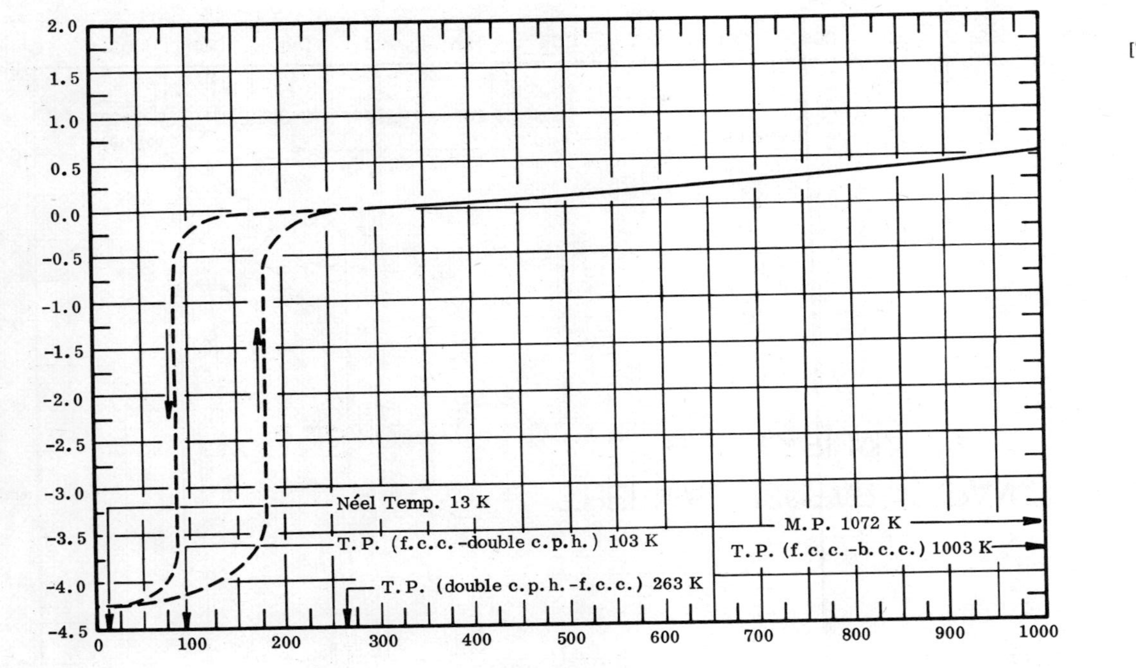

RECOMMENDED VALUES

[Temperature, T, K; Linear Expansion, $\Delta L/L_0$, %; α, K^{-1}]

T	$\Delta L/L_0$	$\alpha \times 10^6$
293	0.000	5.2
400	0.059	5.9
500	0.121	6.4
600	0.188	7.0
700	0.259	7.6
800	0.337	8.2
900	0.421	8.8
1000	0.512	9.4

REMARKS

The tabulated values are considered accurate to within ±3% over the entire temperature range. Several investigations for temperatures below 293 K indicated hysteresis and metastable transitions possibly due to unsuitable heating and cooling rates. The values above 293 K can be represented approximately by the following equation:

$$\Delta L/L_0 = -0.141 + 4.297 \times 10^{-4}\, T + 1.581 \times 10^{-7}\, T^2 + 6.604 \times 10^{-11}\, T^3$$

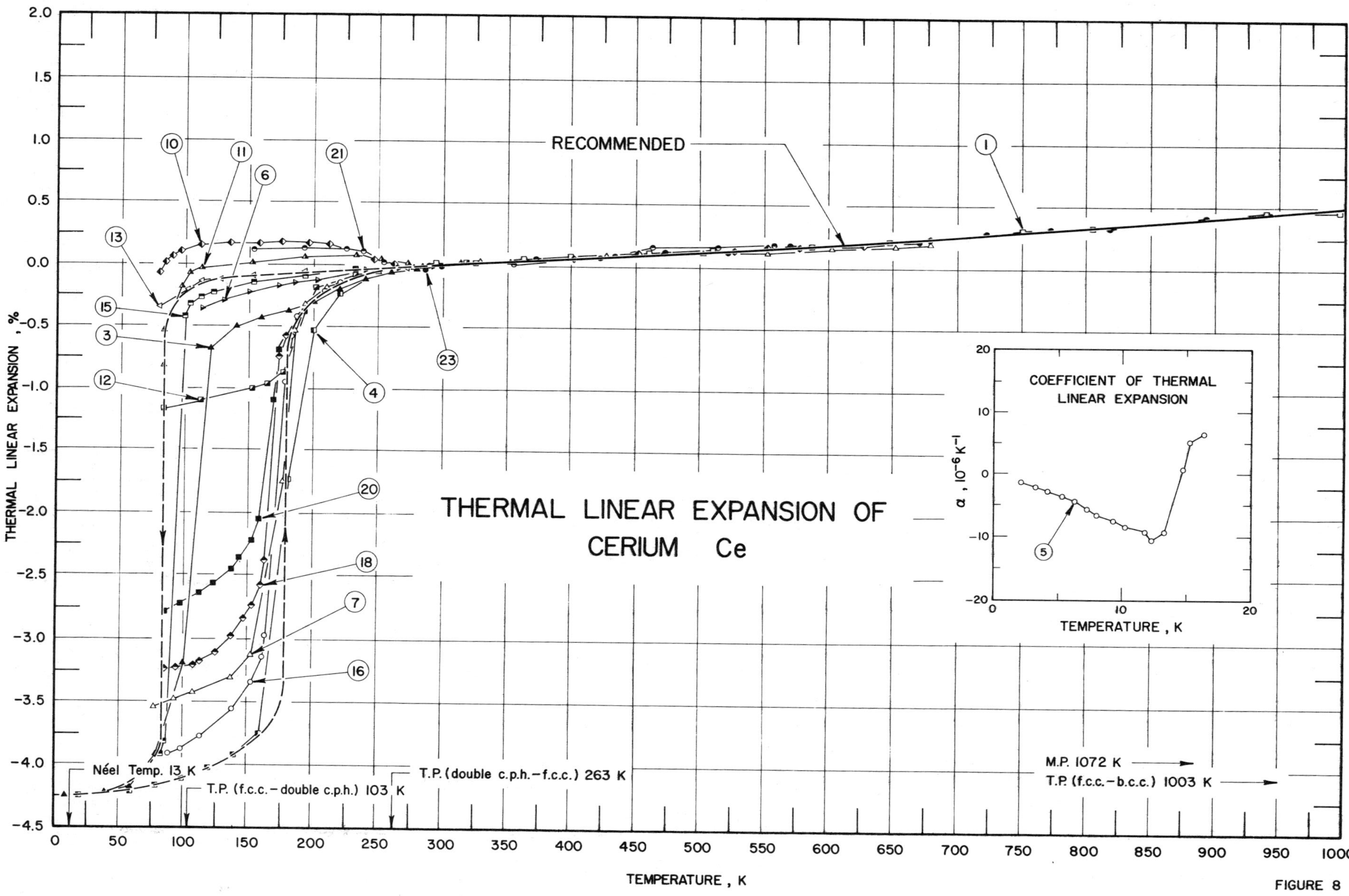
THERMAL LINEAR EXPANSION OF
CERIUM Ce
RECOMMENDED
THERMAL LINEAR EXPANSION, %
TEMPERATURE, K
COEFFICIENT OF THERMAL
LINEAR EXPANSION
α, 10^{-6} K^{-1}
TEMPERATURE, K
Néel Temp. 13 K
T.P. (f.c.c.-double c.p.h.) 103 K
T.P. (double c.p.h.-f.c.c.) 263 K
M.P. 1072 K
T.P. (f.c.c.-b.c.c.) 1003 K

FIGURE 8

SPECIFICATION TABLE 8. THERMAL LINEAR EXPANSION OF CERIUM Ce

Cur. No.	Ref. No.	Author(s)	Year	Method Used	Temp. Range, K	Name and Specimen Designation	Composition (weight percent), Specifications, and Remarks
1	78	Barson, F., Legvold, S. and Spedding, F.H.	1956	L	294-996		99.5 Ce; specimen 0.6 cm diameter x 5 cm long; prepared from fluorides; separated and purified by ion exchange method; possible impurities include other rare earths, C, N; expansion measured with increasing temperature; zero-point correction of 0.004% determined by graphical extrapolation.
2*	78	Barson, F., et al.	1956	L	986-300		The above specimen; expansion measured with decreasing temperature; zero-point correction of 0.065% determined by graphical extrapolation.
3	111	Smith, R.D. and Morrice, E.	1963	L	10-280		99.83 pure cerium metal electrowon from its oxide dissolved in cerous fluoride-lithium fluoride-barium fluoride (CeF_3-LiF-BaF_2) electrolyte: 0.25 in. diameter x 2.5 in. long; annealed at 725 K under vacuum of 10^{-6} torr for 1 hr; $\alpha\rightarrow\gamma$ transition temperature 120 ± 5 K; expansion measured with increasing temperature; reported error ±3%; zero-point correction of 0.03% determined by graphical extrapolation.
4	111	Smith, R.D. and Morrice, E.	1963	L	280-10		The above specimen; $\gamma\rightarrow\alpha$ transition temperature 165 ± 5 K; expansion measured with decreasing temperature; zero-point correction of 0.02% determined by graphical extrapolation.
5*	80	Andres, K.	1963	O	2.0-16		99.5^{+} pure, specimen melted into tantalum form; this curve is here reported using the first given temperature as reference temperature at which $\Delta L/L_0 = 0$.
6	112	Pavlov, V.S. and Finkel, V.A.	1967	X	114-299	γ-Ce	0.045 O, 0.025 C, <<0.01 Si, <0.01 Al, <<0.007 Fe, 0.002 N, 0.002 H, <0.001 Cu, <<0.001 Mg; specimen produced by zone melting combined with electron transfer; specimen deformed 50%, annealed for 30 min. at 500 C, cooled to room temperature at rate 1°/min., completely recrystallized; expansion measured in atmospheric pressure; lattice constant reported at 293 K is 5.15545 Å.
7	113	Trombe, F. and Foex, M.	1943	L	77-679		99.6 Ce, 15 cm long specimen; α-state transforms to γ-state at higher temperatures; expansion measured with increasing temperature; zero-point correction is 0.238%.
8*	113	Trombe, F. and Foex, M.	1943	L	679-77		The above specimen; expansion measured with decreasing temperature; γ-state transforms to α-state by fast cooling; zero-point correction is 0.263%.
9*	113	Trombe, F. and Foex, M.	1943	L	80-679		Similar to the above specimen; β-state transforms to γ-state; expansion measured with increasing temperature; zero-point correction is 0.03%.
10	113	Trombe, F. and Foex, M.	1943	L	679-80		The above specimen; β-state obtained from γ-state by slow cooling; expansion measured with decreasing temperature; zero-point correction is 0.263%.
11	114	Rashid, M.S. and Altstetter, C.J.	1966	L	323-84		99.9 pure specimens from three sources tested; annealed at 723 K; expansion measured on first thermal cycle with decreasing temperature; zero-point correction is -0.004%.
12	114	Rashid, M.S. and Altstetter, C.J.	1966	L	84-317		The above specimen; expansion measured with increasing temperature; zero-point correction is -0.339%.
13	114	Rashid, M.S. and Altstetter, C.J.	1966	L	325-84		The above specimen; expansion measured on fourth thermal cycle with decreasing temperature; zero-point correction is -0.399%.
14*	114	Rashid, M.S. and Altstetter, C.J.	1966	L	84-473		The above specimen; expansion measured with increasing temperature; zero-point correction is -0.52%.
15	114	Rashid, M.S. and Altstetter, C.J.	1966	L	293-82		Similar to the above specimen; expansion measured on fourth thermal cylce with decreasing temperature.

* Not shown in figure.

SPECIFICATION TABLE 8. THERMAL LINEAR EXPANSION OF CERIUM Ce (continued)

Cur. No.	Ref. No.	Author(s)	Year	Method Used	Temp. Range, K	Name and Specimen Designation	Composition (weight percent), Specifications, and Remarks
16	114	Rashid, M. S. and Altstetter, C. J.	1966	L	82-293		The above specimen; expansion measured with increasing temperature.
17*	114	Rashid, M. S. and Altstetter, C. J.	1966	L	293-85		The above specimen; expansion measured on sixth thermal cycle with decreasing temperature.
18	114	Rashid, M. S. and Altstetter, C. J.	1966	L	85-293		The above specimen; expansion measured with increasing temperature.
19*	114	Rashid, M. S. and Altstetter, C. J.	1966	L	293-85		The above specimen; expansion measured on tenth thermal cycle with decreasing temperature.
20	114	Rashid, M. S. and Altstetter, C. J.	1966	L	85-293		The above specimen; expansion measured with increasing temperature.
21	114	Rashid, M. S. and Altstetter, C. J.	1966	L	557-153		Similar to the above specimen; expansion measured with decreasing temperature; zero-point correction is 0.007%.
22*	114	Rashid, M. S. and Altstetter, C. J.	1966	L	153-557		The above specimen; expansion measured with increasing temperature; zero-point correction is -0.244%.
23	82	Hanak, J. J.	1959	X	286-892		0.1 Si, <0.1 Ta, <0.03 Ca, ≤0.02 each La, Pr, Nd, 0.012 H, 0.01 Mg, <0.01 each any other impurity; wire specimen cut from sheet prepared at Ames Laboratory; face-centered cubic structure; lattice parameter reported at 293 K is 5.161 Å.
24*	82	Hanak, J. J.	1959	X	1015-1052		Similar to the above specimen except body-centered cubic structure; lattice parameter reported at 1015 K is 4.11 Å; this curve is here reported using the first given temperature as reference temperature at which $\Delta L/L_0 = 0$.

* Not shown in figure.

DATA TABLE 8. THERMAL LINEAR EXPANSION OF CERIUM Ce

[Temperature, T, K; Linear Expansion, $\Delta L/L_0$, %]

T	$\Delta L/L_0$
CURVE 1	
294	0.001
307	0.012*
325	0.030
337	0.039*
348	0.049
363	0.059
381	0.070*
400	0.082
419	0.090*
439	0.105*
460	0.119
485	0.133*
510	0.149
536	0.163*
562	0.179*
586	0.198
616	0.212*
646	0.231
674	0.250*
698	0.267*
725	0.290*
750	0.309
777	0.328*
805	0.347
835	0.370*
862	0.393*
892	0.416*
916	0.438*
940	0.459
971	0.485*
992	0.497*
996	0.489
CURVE 2*	
986	0.514
977	0.501
952	0.477
938	0.463
913	0.435
885	0.411
857	0.383
829	0.364
804	0.345
780	0.325
752	0.305
CURVE 2 (cont.)*	
725	0.286
700	0.265
676	0.250
653	0.234
629	0.218
602	0.202
576	0.183
553	0.171
530	0.152
508	0.142
485	0.126
462	0.114
442	0.102
422	0.088
403	0.075
384	0.063
369	0.053
354	0.042
340	0.032
325	0.023
312	0.013
300	0.005
CURVE 3	
10	-4.25
20	-4.24*
40	-4.23
60	-4.17
80	-3.92
100	-3.18
120	-0.68
140	-0.50
160	-0.43
180	-0.36
200	-0.29
220	-0.19
240	-0.10
260	-0.05
280	-0.02
CURVE 4	
280	-0.03*
260	-0.07*
240	-0.12*
CURVE 4 (cont.)	
220	-0.21
200	-0.53
180	-1.74
160	-3.75
140	-3.93
120	-4.03
100	-4.11
80	-4.17
60	-4.21
40	-4.23*
20	-4.25
10	-4.26*
CURVE 5 *, †, ‡	
2.01	0.000 x 10^{-1}
3.04	-0.002
4.01	-0.004
5.04	-0.008
6.05	-0.012
7.04	-0.017
8.00	-0.023
9.03	-0.030
10.11	-0.039
11.08	-0.048
12.02	-0.057
13.01	-0.067
14.06	-0.072
15.12	-0.069
16.06	-0.064
CURVE 6	
114	-0.356
117	-0.345*
118	-0.326*
121	-0.309*
124	-0.280*
126	-0.279*
130	-0.281
136	-0.258*
136	-0.235*
140	-0.242*
141	-0.230*
144	-0.222*
145	-0.213*
CURVE 6 (cont.)	
151	-0.212
151	-0.196*
155	-0.208*
157	-0.198*
161	-0.184*
166	-0.176*
172	-0.168
173	-0.158*
177	-0.158*
178	-0.152*
181	-0.153*
185	-0.145
185	-0.140*
187	-0.130*
192	-0.120*
195	-0.109*
198	-0.114*
202	-0.115
212	-0.093*
220	-0.079*
221	-0.085*
225	-0.068*
228	-0.074*
240	-0.045
241	-0.051*
243	-0.053*
257	-0.036*
265	-0.026
275	-0.011*
281	-0.013*
289	-0.003*
293	0.000
299	0.004*
CURVE 7	
77	-3.541
94	-3.487
107	-3.432
123	-3.368*
138	-3.302
153	-3.149
164	-2.963*
169	-2.701*
176	-1.758
179	-1.025*
CURVE 7 (cont.)	
185	-0.536
193	-0.319
210	-0.180
226	-0.108*
250	-0.044
277	-0.004
293	0.000*
297	0.010*
326	0.026
351	0.041
377	0.061*
401	0.074*
426	0.088
452	0.080*
476	0.089*
502	0.111
527	0.119*
552	0.127
577	0.135*
602	0.151
627	0.168*
652	0.181
679	0.205
CURVE 8*	
679	0.230
652	0.206
627	0.193
602	0.176
577	0.160
552	0.152
527	0.144
502	0.136
476	0.114
452	0.105
426	0.091
401	0.073
376	0.059
351	0.041
326	0.030
297	0.014
293	0.000
277	-0.005
254	-0.015
227	-0.031
CURVE 8 (cont.)*	
201	-0.069
173	-0.114
156	-0.148
135	-0.191
125	-0.234
118	-0.348
114	-0.646
108	-1.643
106	-2.354
99	-2.835
90	-3.293
77	-3.516
CURVE 9*	
80	-0.350
100	-0.338
120	-0.307
134	-0.286
148	-0.250
164	-0.225
175	-0.181
188	-0.134
209	-0.085
231	-0.042
252	-0.030
277	-0.007
293	0.000
296	0.007
326	0.023
352	0.043
377	0.038
402	-0.010
416	-0.094
427	-0.129
451	-0.148
476	-0.150
502	-0.157
527	-0.149
552	-0.141
577	-0.133
602	-0.117
627	-0.100
652	-0.087
679	-0.063

* Not shown in figure.
† This curve is here reported using the first given temperature as reference temperature at which $\Delta L/L_0 = 0$.
‡ Author's data for coefficient of thermal expansion have been integrated by TPRC to obtain $\Delta L/L_0$.

DATA TABLE 8. THERMAL LINEAR EXPANSION OF CERIUM Ce (continued)

CURVE 10

T	$\Delta L/L_0$
679	0.230
652	0.206*
627	0.193
602	0.176*
577	0.160
552	0.152
527	0.144
502	0.136*
476	0.114*
452	0.105
426	0.091*
401	0.073
376	0.059*
351	0.041*
326	0.030*
297	0.014*
293	0.000*
277	0.005*
261	0.001
246	0.050
224	0.131*
211	0.170
196	0.181
175	0.190
158	0.186
135	0.182
113	0.154
96	0.112
90	0.075
85	0.017
80	-0.057

CURVE 11

T	$\Delta L/L_0$
323	0.032*
313	0.024
293	0.000*
284	-0.004
273	0.018
252	0.065
233	0.081
193	0.067
153	0.023
113	-0.028
104	-0.053
96	-0.177
84	-0.547
84	-0.821

CURVE 12

T	$\Delta L/L_0$
84	-1.156
113	-1.110
153	-1.009
165	-0.962
176	-0.876
193	-0.372
197	-0.254*
201	-0.180
212	-0.116*
233	-0.070
258	-0.034*
273	-0.017*
293	0.000*
317	0.013

CURVE 13

T	$\Delta L/L_0$
325	0.016*
313	0.010*
293	0.000*
273	-0.008
233	-0.027
193	-0.050
153	-0.089
128	-0.111
113	-0.141
102	-0.200
84	-0.353

CURVE 14*

T	$\Delta L/L_0$
84	-0.474
113	-0.442
153	-0.387
164	-0.363
174	-0.326
181	-0.266
193	-0.143
204	-0.094
222	-0.065
233	-0.054
273	-0.018
293	0.000
313	0.019
353	0.045
393	0.045
403	0.015
406	-0.009

CURVE 14 (cont.)*

T	$\Delta L/L_0$
419	-0.009
419	-0.107
433	-0.107
446	-0.136
473	-0.125

CURVE 15

T	$\Delta L/L_0$
293	0.000*
273	-0.010*
233	-0.030
193	-0.080
154	-0.140
122	-0.220
113	-0.260
104	-0.320
100	-0.410
85	-3.810
82	-3.930

CURVE 16

T	$\Delta L/L_0$
82	-3.930*
89	-3.910
99	-3.870
113	-3.780
138	-3.550
153	-3.340
161	-3.140
164	-2.960
179	-0.950
182	-0.570
186	-0.420
193	-0.330*
209	-0.200
220	-0.140
233	-0.080*
250	-0.040*
273	-0.010*
293	0.000*

CURVE 17*

T	$\Delta L/L_0$
293	0.000
273	-0.010
233	-0.030
193	-0.080
153	-0.150

CURVE 17 (cont.)*

T	$\Delta L/L_0$
121	-0.240
113	-0.290
106	-0.410
88	-3.120
85	-3.240

CURVE 18

T	$\Delta L/L_0$
85	-3.240
94	-3.230
108	-3.200
113	-3.170
125	-3.100
136	-2.970
147	-2.840
153	-2.740
160	-2.570
163	-2.370
174	-0.740
177	-0.570
186	-0.420*
193	-0.330*
209	-0.200*
220	-0.140*
233	-0.080*
250	-0.040*
273	-0.010*
293	0.000*

CURVE 19*

T	$\Delta L/L_0$
293	0.000
273	-0.010
233	-0.030
193	-0.080
177	-0.110
153	-0.170
139	-0.210
124	-0.280
113	-0.360
108	-0.470
87	-2.660
85	-2.780

CURVE 20

T	$\Delta L/L_0$
85	-2.780
98	-2.730

CURVE 20 (cont.)

T	$\Delta L/L_0$
113	-2.640
124	-2.560
136	-2.450
144	-2.350
153	-2.210
159	-2.050
169	-1.090
173	-0.680
177	-0.570*
186	-0.420*
193	-0.330*
209	-0.200*
220	-0.140*
233	-0.080*
250	-0.040*
273	-0.010*
293	0.000*

CURVE 21

T	$\Delta L/L_0$
557	0.183
513	0.157
433	0.096
354	0.036
293	0.000*
273	-0.016*
264	-0.025*
254	0.041
238	0.113
225	0.147
193	0.147
153	0.124

CURVE 22*

T	$\Delta L/L_0$
153	-0.127
193	-0.076
273	-0.012
293	0.000
353	0.035
375	0.035
433	-0.055
474	-0.119
513	-0.094
557	-0.068

CURVE 23

T	$\Delta L/L_0$
286	-0.019
289	0.000
298	0.000
328	0.019*
370	0.058*
372	0.058
423	0.078
473	0.116
486	0.116*
521	0.136
570	0.194
575	0.194*
577	0.174*
628	0.194*
670	0.233
724	0.291
773	0.310
819	0.349
892	0.407

CURVE 24*, †

T	$\Delta L/L_0$
1015	0.342
1021	0.586
1022	0.830
1029	0.586
1032	0.342
1034	0.098
1036	0.830
1052	0.586

* Not shown in figure.
† This curve is here reported using the first given temperature as reference temperature at which $\Delta L/L_0 = 0$.

DATA TABLE COEFFICIENT OF THERMAL LINEAR EXPANSION OF CERIUM Ce

[Temperature, T, K; Coefficient of Expansion, α, 10^{-6} K^{-1}]

T	α
CURVE 5‡	
2.01	-1.41
3.04	-2.07
4.01	-2.92
5.04	-3.62
6.05	-4.65
7.04	-5.92
8.00	-6.77
9.03	-7.56
10.11	-8.35
11.08	-9.21
12.02	-10.59
13.01	-9.15
14.06	-0.89
15.12	5.00
16.06	6.26

‡ Author's data for coefficient of thermal expansion have been integrated by TPRC to obtain $\Delta L/L_0$.

FIGURE AND TABLE NO. 9R. PROVISIONAL VALUES FOR THERMAL LINEAR EXPANSION OF CESIUM Cs

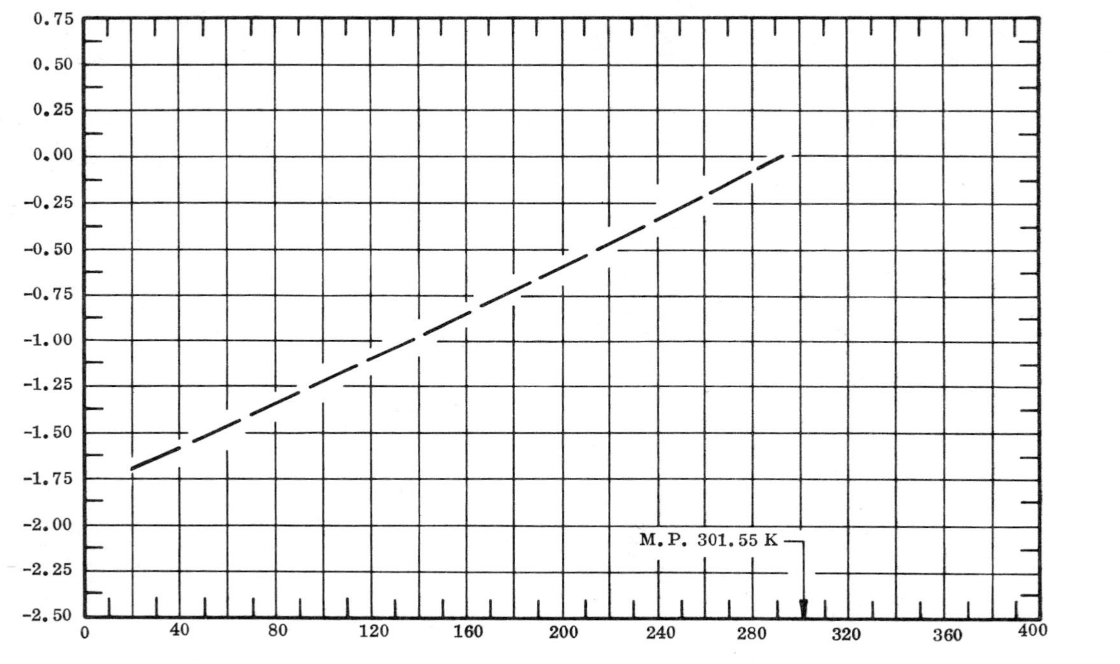

PROVISIONAL VALUES

[Temperature, T, K; Linear Expansion, $\Delta L/L_0$, %]

T	$\Delta L/L_0$ ‡
20	-1.695
30	-1.642
40	-1.588
50	-1.532
60	-1.476
70	-1.418
80	-1.359
90	-1.299
100	-1.238
120	-1.113
140	-0.986
160	-0.857
180	-0.727
200	-0.596
220	-0.466
240	-0.336
260	-0.207
280	-0.081
293	0.000

‡ Values under zero Kbar pressure.

REMARKS

Since no experimental data were located in the literature, the tabulated values are calculated from the density data. The tabulated values are considered accurate to within ±10% over the entire temperature range and can be represented approximately by the following equation:

$$\Delta L/L_0 = -1.795 + 4.879 \times 10^{-3}\, T + 8.416 \times 10^{-6}\, T^2 - 1.418 \times 10^{-6}\, T^3$$

FIGURE AND TABLE NO. 10R. RECOMMENDED VALUES FOR THERMAL LINEAR EXPANSION OF CHROMIUM Cr

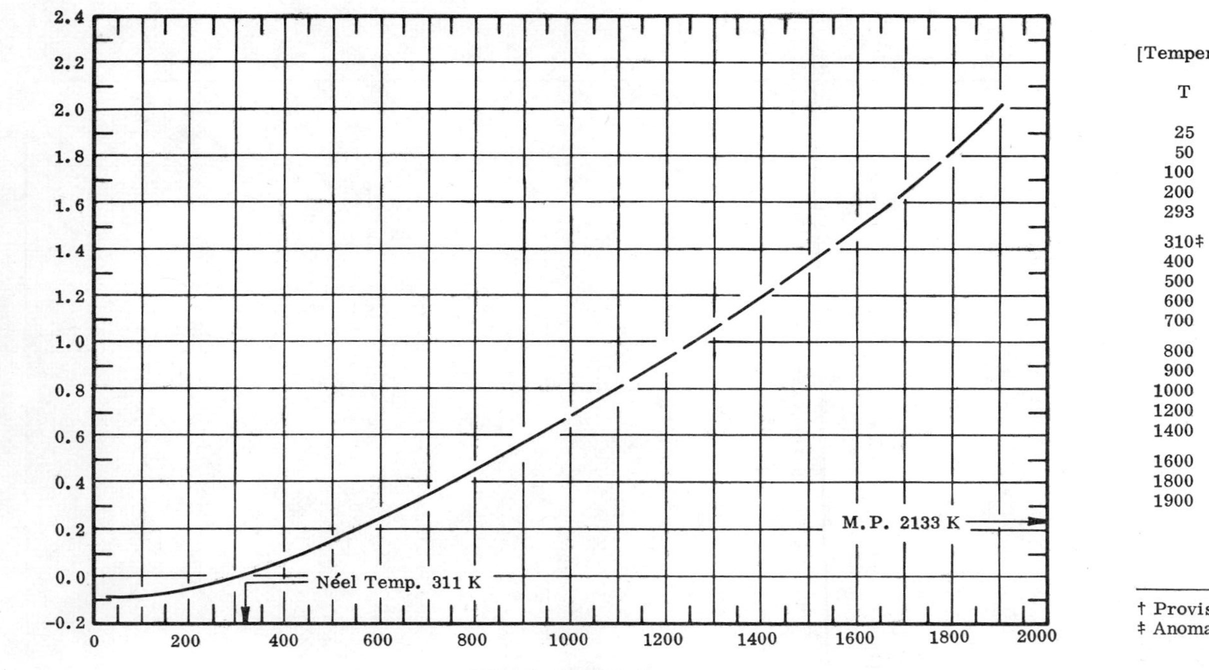

RECOMMENDED VALUES

[Temperature, T, K; Linear Expansion, $\Delta L/L_0$, %; α, K^{-1}]

T	$\Delta L/L_0$	$\alpha \times 10^6$
25	-0.098	0.1
50	-0.098	0.7
100	-0.091	2.3
200	-0.051	5.3
293	0.000	4.9
310‡	0.008	4.0
400	0.075	7.8
500	0.159	8.8
600	0.254	9.6
700	0.354	10.2
800	0.460	10.8
900	0.570	11.3
1000	0.685	11.8
1200	0.931†	12.8
1400	1.198†	14.0
1600	1.499†	15.8
1800	1.837†	17.9
1900	2.020†	19.0

† Provisional values.
‡ Anomaly due to Néel temperature.

REMARKS

The tabulated values are considered accurate to within ± 3% for temperatures below 1000 K, ± 7% below 1400 K and ± 12% at higher temperature. These values can be represented approximately by the following equations:

$$\Delta L/L_0 = 6.390 \times 10^{-4} (T-293) + 7.331 \times 10^{-7} (T-293)^2 - 3.839 \times 10^{-10} (T-293)^3 \quad (293 < T < 795)$$

$$\Delta L/L_0 = 0.456 + 1.085 \times 10^{-3} (T-795) + 1.550 \times 10^{-7} (T-795)^2 + 1.363 \times 10^{-10} (T-795)^3 \quad (795 < T < 1295)$$

$$\Delta L/L_0 = 1.053 + 1.342 \times 10^{-3} (T-1295) + 3.594 \times 10^{-7} (T-1295)^2 + 1.091 \times 10^{-10} (T-1295)^3 \quad (1295 < T < 1900)$$

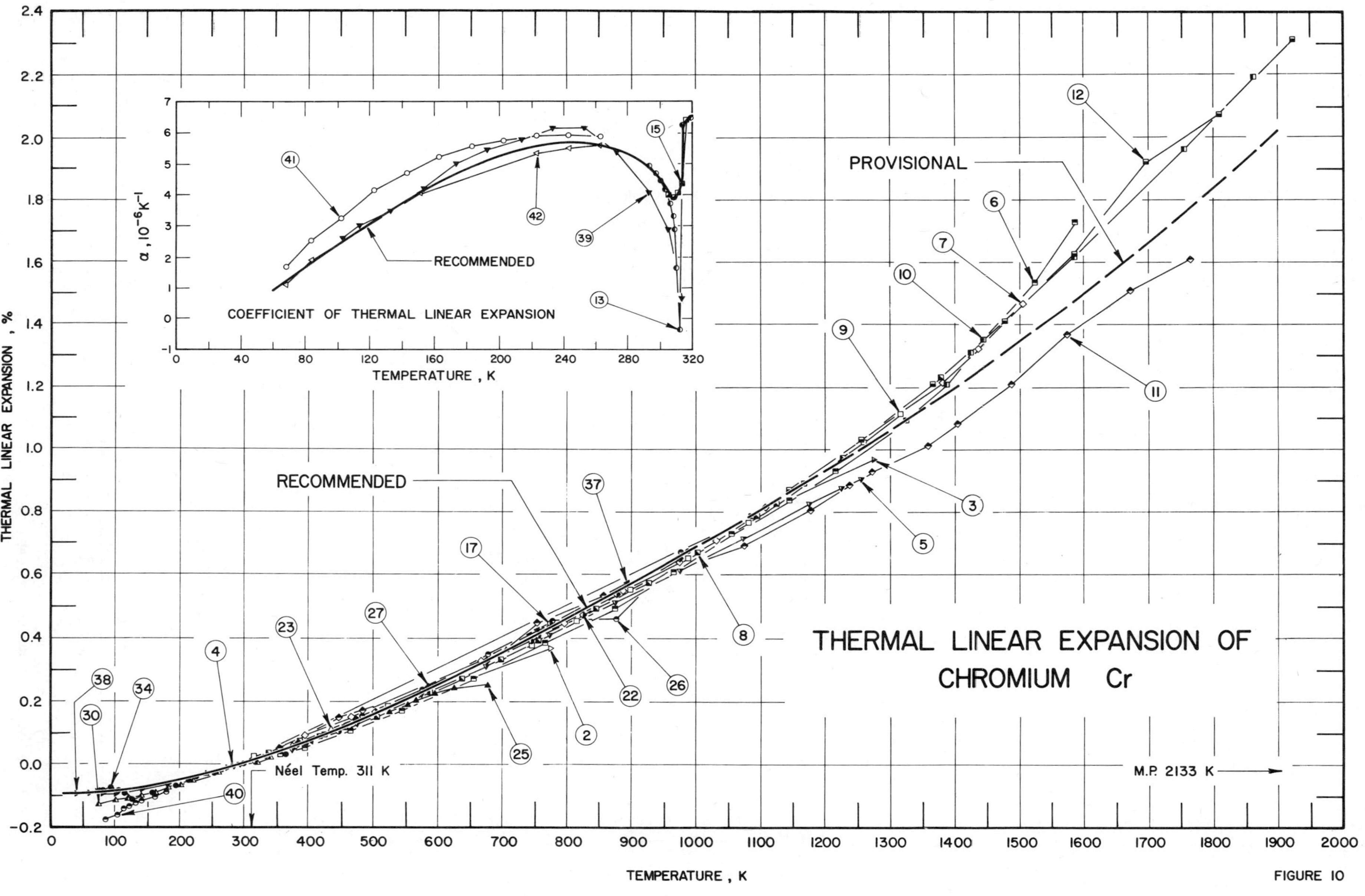
THERMAL LINEAR EXPANSION OF CHROMIUM Cr
THERMAL LINEAR EXPANSION , %
TEMPERATURE , K
PROVISIONAL
RECOMMENDED
M.P. 2133 K
Néel Temp. 311 K
COEFFICIENT OF THERMAL LINEAR EXPANSION
α , 10⁻⁶K⁻¹
TEMPERATURE , K
RECOMMENDED

FIGURE 10

SPECIFICATION TABLE 10. THERMAL LINEAR EXPANSION OF CHROMIUM Cr

Cur. No.	Ref. No.	Author(s)	Year	Method Used	Temp. Range, K	Name and Specimen Designation	Composition (weight percent), Specifications, and Remarks
1*	116	Ishikawa, Y., Hosikina, S., and Endoh, Y.	1967	S	245-323	S-1	Pure Cr; from Johnson Matthey; Neel temperature 311 K; zero-point correction is 0.0087%.
2	117	Yaggee, F. L. and Styles, J. W.	1966		273, 773		99.97 Cr; density 7.16 ±0.01 g cm^{-1}.
3	117	Yaggee, F. L. and Styles, J. W.	1966		273, 1273		The above specimen.
4	118	Straumanis, M. E. and Weng, C. C.	1955	X	283-333		99.95^{+} Cr, 0.01 Sb, 0.019 N, 0.0088 O, 0.005 C, 0.0001 H; small quantities of sintered electrolytic metal ground, passed through 325 mesh sieve; powder placed in silica bulb, heated while evacuating bulb; heated to 1020 K for 2.5 hr; powder recrystallized; change in slope of curve at 305.7 K; reported error 0.0005% in measurement of lattice constant; lattice parameter reported at 293 K is 2.879108 Å.
5	119	Yaggee, F. L., Gilbert, E. R., and Styles, J. W.	1969	L	273-1252		99.94^{+} Cr; in shape of rod 0.64 cm diameter supplied by Bureau of Mines, annealed in a vacuum of 2 x 10^{-6} mm of Hg at 1270 K for 1 hr; density at 298 K is 7.166 ±0.002 g cm^{-3}; zero-point correction of -0.013% determined by graphical interpolation.
6	120	Shevlin, T. S. and Hauck, C. A.	1954	L	297-1589		Electrolytic, 99% minimum chromium content, -325 mesh, pressed into bar; expansion measured with increasing temperature on first run; zero-point correction of -0.009% determined by graphical extrapolation.
7	120	Shevlin, T. S. and Hauck, C. A.	1954	L	1503-300		The above specimen; expansion measured with decreasing temperature; zero-point correction of -0.033% determined by graphical extrapolation.
8	120	Shevlin, T. S. and Hauck, C. A.	1954	L	300-1378		The above specimen; expansion measured with increasing temperature on second test; zero-point correction of 0.009% determined by graphical extrapolation.
9	120	Shevlin, T. S. and Hauck, C. A.	1954	L	1314-300		The above specimen; expansion measured with decreasing temperature; zero-point correction of -0.004% determined by graphical extrapolation.
10	121	Lucks, C. F. and Deem, H. W.	1956	L	293-1866		Expansion measured in vacuum of 5 x 10^{-5} mm Hg with temperature increasing at 2.8 C/min.
11	122	Müller, S. and Dünner, Ph.	1965	X	288-1761		99.4 Cr, 0.5 O, 0.02 Si, 0.018 C, 0.018 P, 0.01 Mn, 0.005 S; lattice parameter reported at 288 K is 2.8840 Å; 2.8841 Å at 293 K determined by graphical interpolation.
12	123	Lucks, C. F. and Deem, H. W.	1958	L	477-1922		Ductile, c.p.; cylindrical specimen 0.375 in. diameter and 3 in. long; material from Bureau of Mines, Albany, Ore.; heating rate 2.8 K per min., measured under pressure 5 x 10^{-5} mm of Hg.
13*	124	White, G. K.	1961	E	307-315	Cr 1	0.03 O_2, 0.0008 N_2; machined from extruded bar; annealed at 1073 K for 4 hr; sample from Aeronautics Research Laboratory, Melbourne; zero-point correction of 0.0067 obtained from integration of coefficient of thermal linear expansion from same author; data on the coefficient of thermal linear expansion also reported.
14*	124	White, G. K.	1961	E	308-314	Cr 2	Similar to the above specimen; zero-point correction assumed to be the same as above.
15*	124	White, G. K.	1961	E	308-314	Cr 3	0.03 O_2, 0.004 N_2; machined from arc-cast alab which had been press forged at 1323 K; sample from Aeronatucis Research Laboratory, Melbourne; zero-point correction of 0.0067 obtained from integration of coefficient of thermal linear expansion from same author; data on the coefficient of thermal linear expansion also reported.

* Not shown in figure.

SPECIFICATION TABLE 10. THERMAL LINEAR EXPANSION OF CHROMIUM Cr (continued)

Cur. No.	Ref. No.	Author(s)	Year	Method Used	Temp. Range, K	Name and Specimen Designation	Composition (weight percent), Specifications, and Remarks
16*	124	White, G. K.	1961	E	308-315	Cr 3a	The above specimen after annealing at 1323 K and relapping; zero-point correction assumed to be the same as above.
17	125	Hidnert, P.	1941	T	771-298	1359A	99.3 Cr, 0.002 Fe, 0.002 Si; electrolytic; hollow cylindrical specimen 300 mm long, 9 mm O.D., 8 mm I.D.; prepared at N.B.S. by G.E. Rentro; similar sample previously found containing 0.1% H_2; expansion measured with decreasing temperature in air atmosphere on first test; zero-point correction of 1.113% determined by graphical extrapolation.
18*	125	Hidnert, P.	1941	T	296-770	1359A	The above specimen; expansion measured with increasing temperature on second test; zero-point correction of 0.002% determined by graphical extrapolation.
19*	125	Hidnert, P.	1941	T	763-299	1359A	The above specimen; expansion measured with decreasing temperature on second test; zero-point correction of 0.022% determined by graphical extrapolation.
20*	125	Hidnert, P.	1941	T	294-770	1359A	The above specimen; expansion measured with increasing temperature on third test.
21*	125	Hidnert, P.	1941	T	736-295	1359A	The above specimen; expansion measured with decreasing temperature on third test; zero-point correction of 0.009% determined by graphical extrapolation.
22	125	Hidnert, P.	1941	T	296-971	1359A	The above specimen; expansion measured with increasing temperature on tenth test; zero-point correction of -0.197% determined by graphical extrapolation.
23	125	Hidnert, P.	1941	T	906-296	1359A	The above specimen; expansion measured with decreasing temperature on tenth test; zero-point correction of -0.193% determined by graphical extrapolation.
24*	125	Hidnert, P.	1941	I	295-571	1328I	98.7 Cr, 0.003 Fe; electrolytic; three pieces, each 8 mm in length; sample prepared by Union Carbide Carbon Research Laboratories, Inc., New York, N.Y.; expansion measured with increasing temperature on first test; zero-point correction of 0.001% determined by graphical extrapolation.
25	125	Hidnert, P.	1941	I	297-677	1328I	The above specimen; expansion measured with increasing temperature on second test; zero-point correction of 0.002% determined by graphical extrapolation.
26	125	Hidnert, P.	1941	I	298-987	1328I	The above specimen; expansion measured with increasing temperature on fourth test; zero-point correction of 0.003% determined by graphical extrapolation.
27	125	Hidnert, P.	1941	I	298-985	1328I	The above specimen; expansion measured with increasing temperature on fifth test; zero-point correction of 0.003% determined by graphical extrapolation.
28*	126	James, W. E., Straumanis, M. E. and Rao, P. B.	1970	X	283-333	Sample I	0.05 O, 0.01 Fe; 325 mesh powder obtained from duPont; microscopic examination showed well-defined, fine crystals; lattice parameter reported at 293 K is 2.879103 kX.
29*	126	James, W. E., et al.	1970	X	293-333	Sample II	Similar to the above specimen, from the same batch at duPont except examination showed long fibrous crystals, matted together; lattice parameter reported at 293 K is 2.879161 kX.
30	257	Bolef, D. I. and deKlerk, J.	1963	L	74-360		Single crystal grown from 99.95 pure U.S. Bureau of Mines arc-cast Cr by annealing several hrs at ~ 1770 K; 2 specimens cut, one oriented along [100] crystal axis, second along [110] axis; density 7.200 g cm^{-3} at 293 K; results from specimens given; zero-point correction is -0.0167%.

* Not shown in figure.

SPECIFICATION TABLE 10. THERMAL LINEAR EXPANSION OF CHROMIUM Cr (continued)

Cur. No.	Ref. No.	Author(s)	Year	Method Used	Temp. Range, K	Name and Specimen Designation	Composition (weight percent), Specifications, and Remarks
31*	258	Matsumoto, T. and Mitsui, T.	1969	S	305-315		Single crystal cut from ingot melted in plasma furnace, annealed at 1470 K; zero-point correction of 0.0044% determined by graphical extrapolation.
32*	595	Sully, A. H., Brandes, E. A. and Mitchell, K. W.	1953	X	293-493		Sample purified by hydrogen reduction at 1720 K; powder finer than 200 mesh; parameter calculated from films taken with chromium radiation; lattice parameter reported at 293 K is 2.8848 Å.
33*	626	Stebler, B., Anderson, C-G. and Kristensson, O.	1970	X	298-324		99.996 pure crystal from Materials Research Corp., Orangeburg, N. Y.; phase change at 311 K.
34	627	Gordienko, V.A. and Nikolaev, V.I.	1971	X	81-399		< 0.010 O, < 0.008 N; sample was relieved of mechanical strain by electrochemical etching, magnetic anomaly near 130 K and phase transformation near 300 K; lattice parameter reported at 293 K is 2.8838 Å.
35*	707	Pridantseva, K.S. and Solov'yeva, N.A.	1965	L	293-973		Specimen prepared by powdered metallurgy method; impurities mainly 1% oxygen, 1% measurement error.
36*	707	Pridantseva, K.S. and Solov'yeva, N.A.	1965	L	293-973		Similar to the above specimen; annealed at 980 K for 240 hr; 1% measurement error.
37	708	Ageev, N.V. and Medel, M.S.	1963	X	282-893		99.96 pure, specimen made from electrolytic chromium; expansion measured at 10^{-4} mm Hg; accuracy ±0.0001 Å; lattice parameter at 293 K is 2.8851 Å, determined by graphical interpolation.
38	709	Shah, J.S.	1971	X	40-180		99.3 pure, powder sample of particle size 8-10 μ from Koch Light Labs, Inc., Colnbrook, England; lattice parameter not corrected for refraction; lattice parameter reported as 2.8834 Å at 180 K, the values referred to 293 K using other available data; zero-point correction is -0.061%.
39*	724	Fine, M.E., Greiner, E.S., and Ellis, W.C.	1951	L	98-673		Wrought electrolytic chromium; data on the coefficient of thermal linear expansion also reported.
40	730	Kondorsky, E.I., Kostina, T.I., and Ekonomova, L.N.	1971	S	86-334		Polycrystalline sample; anomaly observed between 77 and 85 K on fast heating which disappeared on very slow cooling; author reports first order phase transition at 122 K.
41*	710	Erfling, H.D.	1939	I	273-67	Cr-I	Pure specimen; zero-point correction of 0.012% determined by graphical extrapolation; data on the coefficient of thermal linear expansion also reported.
42*	710	Erfling, H.D.	1939	I	273-67	Cr-II	Relatively more pure than Cr-I specimen; zero-point correction of 0.011% determined by graphical extrapolation; data on the coefficient of thermal linear expansion also reported.

* Not shown in figure.

DATA TABLE 10. THERMAL LINEAR EXPANSION OF CHROMIUM Cr

[Temperature, T, K; Linear Expansion, $\Delta L/L_0$, %]

T	$\Delta L/L_0$
CURVE 1*	
245	-0.0355
248	-0.0342
252	-0.0305
255	-0.0284
257	-0.0264
259	-0.0245
261	-0.0228
264	-0.0207
266	-0.0190
268	-0.0172
271	-0.0155
273	-0.0139
275	-0.0119
277	-0.0106
279	-0.0092
282	-0.0074
284	-0.0060
286	-0.0047
288	-0.0030
291	-0.0016
293	0.0000
295	0.0015
297	0.0031
299	0.0048
302	0.0060
304	0.0079
306	0.0090
308	0.0107
310	0.0118
313	0.0134
315	0.0152
317	0.0166
320	0.0184
321	0.0200
323	0.0221
CURVE 2	
273	-0.016
773	0.374
CURVE 3	
273	-0.020*
1273	0.965
CURVE 4	
283	-0.0035
285	-0.0030*
293	0.0000
303	0.0053*
313	0.01199
323	0.0193*
333	0.0269
CURVE 5	
273	-0.013*
373	0.053
473	0.134
573	0.221
673	0.313
773	0.410
873	0.508
973	0.611
1073	0.717
1173	0.823
1223	0.875
1252	0.909
CURVE 6	
297	0.002*
336	0.025*
355	0.034
394	0.059
464	0.102
544	0.171
586	0.211*
655	0.276
711	0.329*
769	0.389
816	0.435*
875	0.495
966	0.601
1008	0.659*
1055	0.728
1144	0.837
1214	0.924
1325	1.093*
1386	1.218*
1441	1.344*
1525	1.538
CURVE 6 (cont.)	
1589	1.724
CURVE 7	
1503	1.468
1436	1.328
1380	1.217
1150	0.855*
1108	0.819
1030	0.707
972	0.640
886	0.544
800	0.448
666	0.325
469	0.151
391	0.094
300	0.007*
CURVE 8	
300	0.009*
308	0.012*
339	0.037
386	0.070
419	0.091
453	0.120
505	0.157
639	0.275
697	0.339
753	0.395
847	0.491
925	0.579
1008	0.674
1091	0.785
1122	0.824
1226	0.976
1254	1.019*
1314	1.110*
1378	1.228
CURVE 9	
1319	1.116
1258	1.021
1169	0.888
1080	0.761
CURVE 9 (cont.)	
989	0.653
900	0.550
811	0.453
744	0.380
575	0.225*
525	0.181
458	0.129*
414	0.096*
355	0.051*
316	0.021
300	0.007*
CURVE 10	
293	0.000*
422	0.093
589	0.230
755	0.386*
922	0.568*
1089	0.785
1255	1.028*
1422	1.305
1589	1.620
1755	1.960
1866	2.193
CURVE 11	
288	-0.004*
352	0.051
449	0.145
484	0.176*
577	0.231*
673	0.308*
754	0.446
859	0.533
1073	0.696
1177	0.804
1237	0.883
1270	0.925
1358	1.015
1401	1.081
1488	1.209
1571	1.369
1620	1.393*
1670	1.507
CURVE 11 (cont.)	
1761	1.601
CURVE 12	
478	0.136*
589	0.229*
700	0.331*
811	0.444*
922	0.568*
1033	0.706*
1144	0.862
1255	1.030
1366	1.211
1478	1.407
1589	1.621
1700	1.920
1811	2.071
1866	2.192*
1922	2.316
CURVE 13*	
307.38	0.0065
308.14	0.0067
309.02	0.0070
309.37	0.0071
310.05	0.0073
310.76	0.0073
310.85	0.0073
310.97	0.0073
311.28	0.0073
311.54	0.0072
311.68	0.0072
311.95	0.0072
312.30	0.0074
313.15	0.0078
314.27	0.0085
315.25	0.0091
CURVE 14*	
307.97	0.0067
309.44	0.0071
310.38	0.0073
310.87	0.0075
311.25	0.0075
CURVE 14 (cont.)*	
311.40	0.0075
311.58	0.0075
311.64	0.0075
311.78	0.0076
311.97	0.0076
312.65	0.0079
313.09	0.0081
313.82	0.0085
CURVE 15*	
307.93	0.0067
309.40	0.0072
309.88	0.0074
310.49	0.0076
311.05	0.0078
311.44	0.0079
311.79	0.0081
312.56	0.0084
312.98	0.0087
313.70	0.0090
CURVE 16*	
307.78	0.0066
308.95	0.0069
309.85	0.0071
310.17	0.0072
310.56	0.0072
310.77	0.0073
311.12	0.0073
311.68	0.0075
312.58	0.0079
313.53	0.0084
314.97	0.0091
CURVE 17	
771	0.442
764	0.418*
755	0.402
742	0.388
503	0.156*
470	0.127*
446	0.107
418	0.083*

* Not shown in figure.

DATA TABLE 10. THERMAL LINEAR EXPANSION OF CHROMIUM Cr (continued)

T	$\Delta L/L_0$
CURVE 17 (cont.)	
374	0.051*
298	0.003
CURVE 18*	
296	0.002
333	0.026
366	0.051
477	0.140
570	0.226
669	0.322
728	0.321
756	0.405
770	0.416
CURVE 19*	
763	0.423
577	0.234
533	0.191
458	0.123
392	0.071
299	0.004
CURVE 20*	
294	0.000
326	0.019
387	0.067
479	0.144
572	0.231
709	0.365
760	0.414
770	0.422
CURVE 21*	
736	0.394
523	0.186
455	0.124
413	0.089
295	0.002
CURVE 22	
296	0.003*
368	0.049
477	0.138*
501	0.161
CURVE 22 (cont.)	
574	0.229*
677	0.331
759	0.413*
821	0.477
877	0.540
971	0.642
CURVE 23	
706	0.361
438	0.108
372	0.055
296	0.002*
CURVE 24*	
295	0.001
333	0.027
373	0.056
471	0.128
571	0.175
CURVE 25	
297	0.003*
337	0.030*
377	0.059*
479	0.134*
527	0.175
551	0.196
566	0.206
583	0.215*
593	0.223
624	0.241
677	0.254
CURVE 26	
298	0.003*
377	0.058*
481	0.152
579	0.239
679	0.345
778	0.451
878	0.460
975	0.667
987	0.671
CURVE 27	
298	0.003*
339	0.029*
381	0.062
474	0.140
583	0.240
676	0.340*
778	0.441*
874	0.549*
974	0.671*
985	0.678*
CURVE 28*	
283	-0.0049
293	0.0000
313	0.0084
323	0.0149
333	0.0198
CURVE 29*	
293.0	0.0000
294.0	0.0005
301.5	0.0029
312.9	0.0087
323.0	0.0164
333.0	0.0234
CURVE 30	
74	-0.1263
80	-0.1246
100	-0.1184
120	-0.1104
140	-0.1013
160	-0.0908
180	-0.0777
200	-0.0655
220	-0.0527
240	-0.0388
260	-0.0242
280	-0.0085*
290	-0.0014*
293	0.0000*
300	0.0020*
310	0.0031*
320	0.0065
340	0.0250
360	0.0503*
CURVE 31*	
305.37	0.0036
305.50	0.0037
306.00	0.0038
306.50	0.0040
307.00	0.0041
307.50	0.0042
308.00	0.0044
308.25	0.0044
308.50	0.0045
308.75	0.0046
309.00	0.0046
309.25	0.0047
309.50	0.0047
309.75	0.0047
310.00	0.0048
310.25	0.0049
310.50	0.0045
310.75	0.0045
311.00	0.0044
311.25	0.0044
311.50	0.0046
311.75	0.0047
312.00	0.0049
312.50	0.0051
313.00	0.0054
313.50	0.0057
314.00	0.0059
314.50	0.0062
315.00	0.0060
315.25	0.0067
CURVE 32*	
293	0.000
299	0.006
308	0.006
319	0.023
332	0.020
341	0.034
353	0.034
368	0.041
379	0.051
393	0.061
397	0.069
422	0.082
433	0.099
448	0.097
455	0.115
475	0.121
493	0.131
CURVE 33*	
298.2	0.0000
303.7	-0.0005
308.4	-0.0086
310.9	-0.0029
311.0	0.0063
311.6	0.0233
312.0	0.0351
312.5	0.0327
313.3	0.0304
320.0	0.0222
322.5	0.0222
324.2	0.0234
CURVE 34	
80	-0.081
86	-0.097*
88	-0.081*
92	-0.079
97	-0.084*
99	-0.087
101	-0.084*
103	-0.087
104	-0.091*
107	-0.088*
108	-0.093*
108	-0.097*
112	-0.088*
113	-0.093*
114	-0.096
117	-0.089*
117	-0.100*
119	-0.109*
122	-0.094*
122	-0.102*
122	-0.104*
124	-0.116
127	-0.101*
129	-0.105*
132	-0.110*
133	-0.104*
139	-0.105*
142	-0.104*
148	-0.097*
157	-0.091
162	-0.089*
172	-0.080*
177	-0.077*
187	-0.069*
CURVE 34 (cont.)	
192	-0.065
196	-0.064*
199	-0.061*
201	-0.059*
207	-0.055*
211	-0.051
212	-0.053*
217	-0.050*
222	-0.048*
227	-0.409
237	-0.033*
242	-0.032*
248	-0.023*
252	-0.024
257	-0.020*
262	-0.015*
272	-0.007*
278	-0.003*
282	-0.009*
287	-0.000*
288	-0.005*
293	0.000*
293	0.000*
295	-0.001*
299	0.002*
302	-0.004*
306	-0.006*
307	-0.007*
308	-0.006*
313	-0.003*
316	0.001*
318	0.003*
327	0.013*
337	0.021*
362	0.037
399	0.066
CURVE 35*	
293	0.000
573	0.238
773	0.446
973	0.655
CURVE 36*	
293	0.000
573	0.234
773	0.455
973	0.670

* Not shown in figure.

† This curve is here reported using the first given temperature as reference temperature at which $\Delta L/L_0 = 0$.

DATA TABLE 10. THERMAL LINEAR EXPANSION OF CHROMIUM Cr (continued)

T	$\Delta L/L_0$
CURVE 37	
282	-0.007*
293	0.000*
404	0.076
591	0.254
752	0.424
893	0.570
CURVE 38	
40	-0.097
60	-0.095
80	-0.094
100	-0.091
140	-0.080
180	-0.061
CURVE 39*	
93	-0.097
123	-0.088
173	-0.069
223	-0.043
273	-0.013
283	-0.004
293	0.000
303	0.003
313	0.018
323	0.011
373	0.045
423	0.087
473	0.126
523	0.175
573	0.220
623	0.267
673	0.315
CURVE 40	
86	-0.175
94	-0.169*
102	-0.161
104	-0.158*
108	-0.153*
112	-0.147
118	-0.137*
120	-0.137
126	-0.130*
128	-0.131*

T	$\Delta L/L_0$
CURVE 40 (cont.)	
130	-0.127
138	-0.120*
140	-0.118
148	-0.112*
152	-0.108*
160	-0.102
168	-0.095*
178	-0.086
186	-0.080*
202	-0.066*
204	-0.065*
224	-0.050*
244	-0.036*
246	-0.035*
262	-0.023*
264	-0.022*
284	-0.007*
286	-0.006*
300	0.007*
308	0.015*
312	0.017*
318	0.021*
334	0.031*
CURVE 41*	
273	-0.012
253	-0.023
233	-0.035
213	-0.047
193	-0.059
173	-0.070
153	-0.080
133	-0.090
113	-0.098
90	-0.106
78	-0.109
57.14	-0.112
CURVE 42*	
273	-0.011
253	-0.022
233	-0.033
213	-0.044
90	-0.094
78	-0.096
57.34	-0.099

* Not shown in figure.

† This curve is here reported using the first given temperature as reference temperature at which $\Delta L/L_0 = 0$.

DATA TABLE 10. COEFFICIENT OF THERMAL LINEAR EXPANSION OF CHROMIUM Cr

[Temperature, T, K; Coefficient of Expansion, α, 10^{-6} K^{-1}]

T	α
CURVE 13	
293	4.96
294	4.94*
295	4.83*
296	4.74*
297	4.68
298	4.63*
299	4.49*
300	4.44
301	4.38*
302	4.17
303	4.10*
304	4.00*
305	3.86*
306	3.68
307	3.54*
308	3.29
309	2.86
310	1.66
312	0.34
313	4.39*
314	6.30
315	6.39*
316	6.43*
317	6.47*
318	6.46*
319	6.51
CURVE 15	
300	4.40
301	4.36*
302	4.25*
303	4.12*
304	4.07*
305	4.00
306	3.97*
307	3.92*
308	3.92
309	3.92*
310	3.97*
311	4.03
312	4.19*
313	4.39
CURVE 39	
103	2.6
113	3.0
133	3.5
153	4.2
173	5.0
193	5.5
213	5.8
233	6.2
253	6.2
273	5.4
293	4.1
303	2.9
313	0.7
323	6.5*
333	6.8*
353	7.2*
373	7.7*
393	8.0*
413	8.5*
433	8.8*
453	9.0*
473	9.1*
CURVE 40*	
90	7.7
94	9.1
99	10.5
103	12.0
107	13.8
110	15.4
113	16.4
118	16.6
123	5.1
124	6.5
125	10.1
128	7.4
133	8.0
139	8.4
145	8.4
150	8.4
154	7.9
160	8.4
164	9.0
173	8.7
182	8.1
CURVE 40 (cont.)*	
195	8.3
203	7.2
213	7.2
224	7.2
234	7.2
245	7.2
254	7.2
263	7.4
274	7.2
285	7.9
293	8.8
300	9.6
304	10.2
308	10.2
310	5.3
315	6.3
326	6.3
CURVE 41	
263	5.92
243	5.97
223	5.94
203	5.78
183	5.59
163	5.22
143	4.73
123	4.15
102	3.26
84	2.56
68	1.72
CURVE 42	
263	5.61
243	5.55
223	5.39
151.65	4.05
84	1.94
67.76	1.14

* Not shown in figure.

FIGURE AND TABLE NO. 11R. RECOMMENDED VALUES FOR THERMAL LINEAR EXPANSION OF COBALT Co

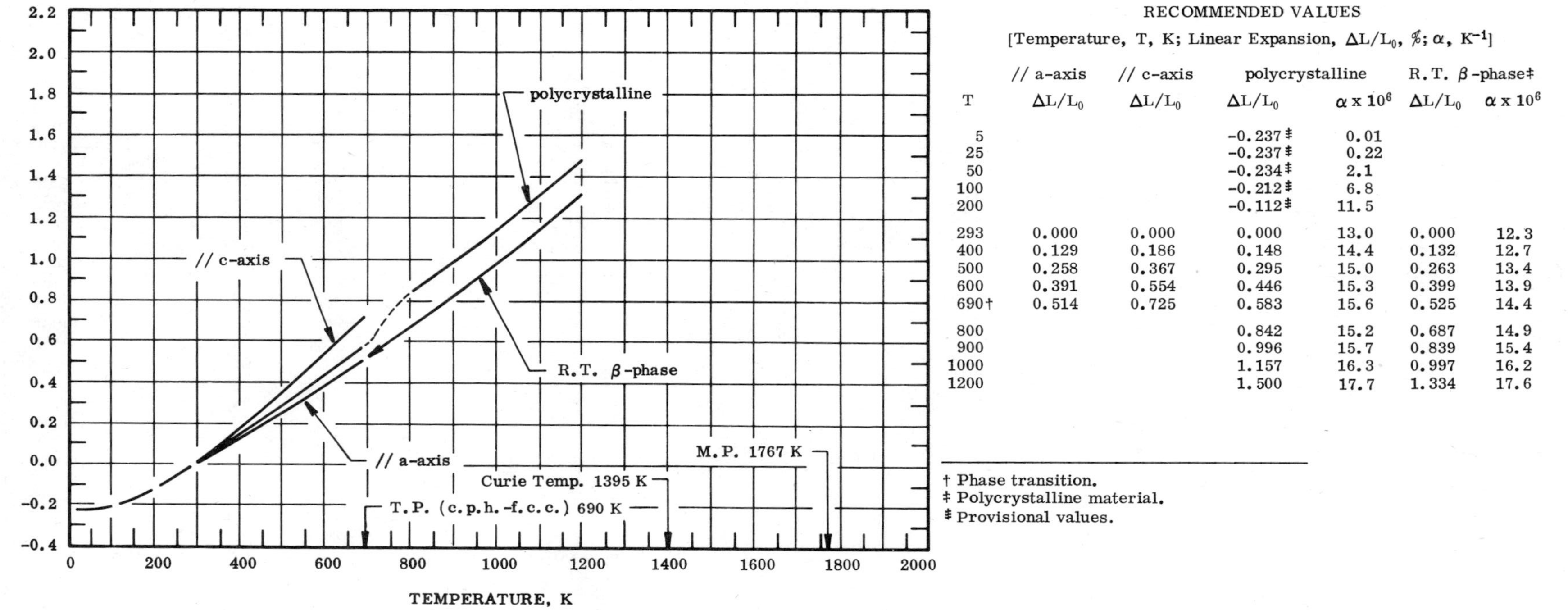

RECOMMENDED VALUES

[Temperature, T, K; Linear Expansion, $\Delta L/L_0$, %; α, K^{-1}]

T	// a-axis $\Delta L/L_0$	// c-axis $\Delta L/L_0$	polycrystalline $\Delta L/L_0$	polycrystalline $\alpha \times 10^6$	R.T. β-phase‡ $\Delta L/L_0$	R.T. β-phase‡ $\alpha \times 10^6$
5			-0.237#	0.01		
25			-0.237#	0.22		
50			-0.234#	2.1		
100			-0.212#	6.8		
200			-0.112#	11.5		
293	0.000	0.000	0.000	13.0	0.000	12.3
400	0.129	0.186	0.148	14.4	0.132	12.7
500	0.258	0.367	0.295	15.0	0.263	13.4
600	0.391	0.554	0.446	15.3	0.399	13.9
690†	0.514	0.725	0.583	15.6	0.525	14.4
800			0.842	15.2	0.687	14.9
900			0.996	15.7	0.839	15.4
1000			1.157	16.3	0.997	16.2
1200			1.500	17.7	1.334	17.6

† Phase transition.
‡ Polycrystalline material.
# Provisional values.

REMARKS

The tabulated values are considered accurate to within ± 3% at temperatures below 293 K and ± 5% above for polycrystalline cobalt. The values along a-axis and c-axis are considered accurate to within ± 5% for temperatures above 293 K and ± 7% below. Hysteresis affect in the thermal expansion probably due to impurities and unsuitable heating and cooling rates are observed near the phase transition. The tabulated values can be represented approximately by the following equations:

// a-axis: $\Delta L/L_0 = -0.326 + 1.035 \times 10^{-3} T + 2.645 \times 10^{-7} T^2 + 5.717 \times 10^{-12} T^3$

// c-axis: $\Delta L/L_0 = -0.479 + 1.556 \times 10^{-3} T + 2.566 \times 10^{-7} T^2 + 2.948 \times 10^{-11} T^3$

polycrystalline: $\Delta L/L_0 = -0.283 + 5.481 \times 10^{-4} T + 1.721 \times 10^{-6} T^2 - 1.013 \times 10^{-9} T^3$ $(100 < T < 690)$

$\Delta L/L_0 = -0.044 + 6.509 \times 10^{-4} T + 6.532 \times 10^{-7} T^2 - 1.029 \times 10^{-10} T^3$ $(800 < T < 1200)$

R.T. β-phase: $\Delta L/L_0 = -0.343 + 1.105 \times 10^{-3} T + 1.930 \times 10^{-7} T^2 + 4.179 \times 10^{-11} T^3$

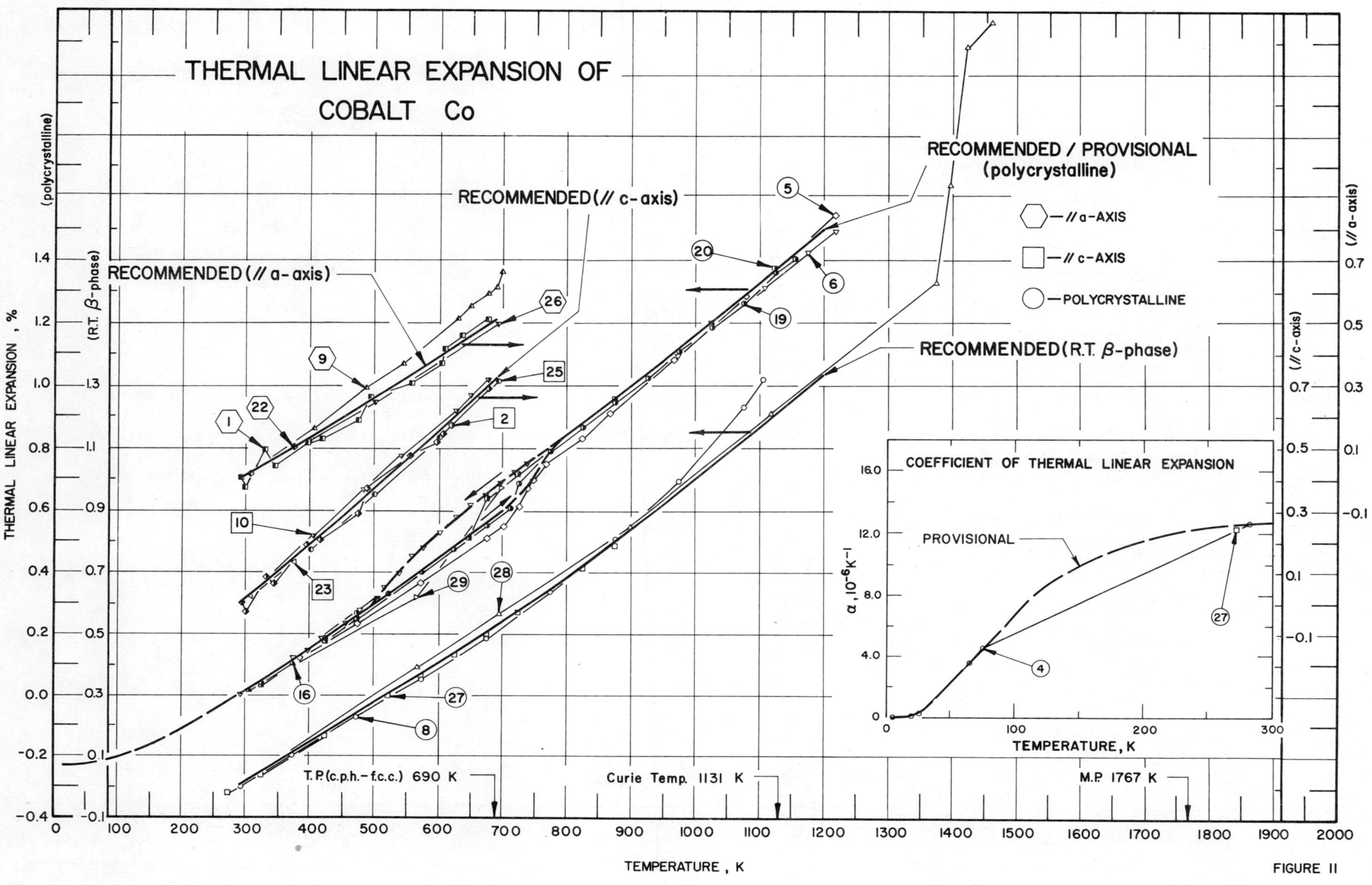

THERMAL LINEAR EXPANSION OF
COBALT Co
RECOMMENDED / PROVISIONAL
(polycrystalline)
RECOMMENDED (// c-axis)
RECOMMENDED (// a-axis)
—// a-AXIS
—// c-AXIS
—POLYCRYSTALLINE
RECOMMENDED (R.T. β-phase)
COEFFICIENT OF THERMAL LINEAR EXPANSION
PROVISIONAL
α, 10⁻⁶K⁻¹
TEMPERATURE, K
T.P. (c.p.h.–f.c.c.) 690 K
Curie Temp. 1131 K
M.P. 1767 K
THERMAL LINEAR EXPANSION, %
(polycrystalline)
(R.T. β-phase)
(// c-axis)
(// a-axis)
TEMPERATURE, K

FIGURE 11

SPECIFICATION TABLE 11. THERMAL LINEAR EXPANSION OF COBALT Co

Cur. No.	Ref. No.	Author(s)	Year	Method Used	Temp. Range, K	Name and Specimen Designation	Composition (weight percent), Specifications, and Remarks
1	127	Bolgov, I. S., Smirnov, Yu. N. and Finkel, V. A.	1964	X	293-675		99.98 Co, 80 x 8 x 2 mm bar; annealed in vacuum at 1573 K; expansion measured along hexagonal crystal a-axis below transformation temperature; lattice parameter reported at 293 K is 2.5023 Å.
2	127	Bolgov, I. S., et al.	1964	X	293-674		The above specimen; expansion measured along hexagonal crystal c-axis below transformation; lattice parameter reported at 293 K is 4.0674 Å.
3*	127	Bolgov, I. S., et al.	1964	X	677-1593		The above specimen; expansion measured along cubic a-axis above transformation temperature of 676 K; this curve is here reported using the first given temperature as reference temperature at which $\Delta L/L_0 = 0$.
4*	41	White, G. K.	1965	E	4-85, 283		99.99$^+$ Co; from Johnson Matthey; specimen 5 cm long and 2 cm in diameter; annealed in vacuo for some hrs after machining at 1273 K; this curve is here reported using the first given temperature as reference temperature at which $\Delta L/L_0 = 0$.
5	128	Brenner, A., Burkhead, P. and Seegmiller, E.	1947	C	294-1217		Specimen electrodeposited on copper rod, rod then removed leaving cylinder 30 cm long x 5 mm diameter; expansion measured with increasing temperature; zero-point correction of 0.001% determined by graphical extrapolation.
6	128	Brenner, A., et al.	1947	C	1217-293		The above specimen; expansion measured with decreasing temperature; zero-point correction is -0.055%.
7	129	Masumoto, H.	1934	L	293, 333		0.13 Fe, 0.074 Al, 0.05 C, 0.007 P, 0.002 Si, trace Ni, S, Mn; electrolytic cobalt; supplied by Sugibayashi & Co., 10 cm long.
8	130	Petrov, U. E.	1965	X	293-1103		Spheroid aerosol particles obtained by condensation of metal vapor under argon atmosphere; expansion measured in vacuum of 10^{-4} mm Hg; β-phase from 293 K.
9	131	Kulesko, G. I. and Seryugim, A. L.	1968	X	293-696	α-cobalt	99.99 Co polycrystalline specimen with grain size 0.1-0.3 mm mechanically then electrochemically polished in mixture of equal quantities of HCl and ethyl alcohol; vacuum annealed at 1573 K for several hr; flat specimen put into furnace; expansion measured along a-axis in α-phase; lattice parameter reported at 293 K is 2.5059 Å.
10	131	Kulesko, G. I. and Seryugim, A. L.	1968	X	293-673	α-cobalt	The above specimen; expansion measured along c-axis; lattice parameter reported at 293 K is 4.0691 Å.
11	131	Kulesko, G. I. and Seryugim, A. L.	1968	X	293-701	β-cobalt	Similar to the above specimen; expansion measured along a-axis in β-phase; lattice parameter reported at 293 K is 3.5440 Å.
12*	55	Richter, F. and Lotter, V.	1969	L	1173-1486		0.028 Cu, 0.02 Si, 0.01 each Fe, Ca, < 0.01 each Al, C, Mg, Mn, Ni, 0.002 S; cylindrical specimen, 6 mm diameter, 35 mm long, one side being of hemi-spherical shape; supplied by Ges. für Elektrometallurgie, Düsseldorf; heating rate 20 K/min; volume magnetostriction $\omega_M \times 10^4 = -2.5 \pm 0.3$; Curie temperature 1390 ± 4 K; expansion measured under constant magnetic field; this curve is here reported using the first given temperature as reference temperature at which $\Delta L/L_0 = 0$.
13*	133	Masumoto, H., Saito, H. and Kikuchi, M.	1967	L	173-635	No. 1	Single crystal of hexagonal structure, 6 to 11 cm long, 2 mm diameter; expansion measurements along 2.06 degrees from c-axis; zero-point correction is 0.042%.
14*	133	Masumoto, H., et al.	1967	L	173-634	No. 4	Similar to the above specimen; expansion measurements along 43.40 degrees from the c-axis; zero-point correction is -0.034%.
15*	133	Masumoto, H., et al.	1967	L	173-636	No. 7	Similar to the above specimen; expansion measurements along 82.80 degrees from the c-axis; zero-point correction is -0.030%.

* Not shown in figure.

SPECIFICATION TABLE 11. THERMAL LINEAR EXPANSION OF COBALT Co (continued)

Cur. No.	Ref. No.	Author(s)	Year	Method Used	Temp. Range, K	Name and Specimen Designation	Composition (weight percent), Specifications, and Remarks
16	56	Masumoto, H. and Nara, S.	1927	L	303, 373		Pure Co.
17*	59	Masumoto, H.	1929	L	303, 373		Electrolytic Co, circular rod specimen, 4 mm thick, 10 cm long; expansion measured in vacuum.
18*	301	Arbuzov, M. P. and Zelenkov, I. A.	1965	L	423-1024		Pure metal; this curve is here reported using the first given temperature as reference temperature at which $\Delta L/L_0 = 0$.
19	555	Fine, M. E. and Ellis, W. C.	1948	L	293-1152	Electrolytic Cobalt	99.9 Co; major impurities, 0.05 Fe, 0.02 Si, 0.04 Ni; specimen was prepared from cast ingots by first hot-swaging followed by cold-swaging to 0.18 in. diameter rod 2.5 in. long with flat-ground and polished end; before measurement they are annealed in hydrogen for 1 hr at 1173 K; zero-point correction is 0.008%.
20	555	Fine, M. E. and Ellis, W. C.	1948	L	1152-293	Electrolytic Cobalt	The above specimen; expansion measured with decreasing temperature; zero-point correction is 0.018%.
21*	628	Krajewski, W., Kruger, J. and Winterhager, H.	1970	L	373-1073		99.99 Co; cylindrical specimen 5 mm diameter, 30 mm long; polished and annealed in vacuum.
22	715	Shinoda, G.	1934	X	306, 373		Expansion measured perpendicular to the c-axis.
23	715	Shinoda, G.	1934	X	306, 373		Expansion measured along the c-axis.
24*	715	Shinoda, G.	1934	X	306, 373		Similar to the above specimen except for polycrystalline.
25	721	Bibring, H. and Sebilleau, F.	1955	X, L	293-690		99.5^+ pure electrolytic cobalt, α-phase; principal impurities 0.10 Ni, 0.10 Fe, 0.02 Cu; sample was degased at 1473 K for 100 hr, melted in furnace, machined and electrolytically polished; expansion measured along the c-axis; expansion calculated from $\alpha_{//} = 15.4 \times 10^{-6} + 12.8 \times 10^{-9}\,T(C)$.
26	721	Bibring, H. and Sebilleau, F.	1955	X, L	293-690		The above specimen; expansion measured perpendicular to the c-axis; expansion calculated from $\alpha_{\perp} = 11 \times 10^{-6} + 7.5 \times 10^{-9}\,T(C)$.
27	722	Owen, E.A. and Jones, D.M.	1954	X	293-873		Cobalt sponge, powder sample from Johnson Matthey and Co., annealed for 12 hr at temperature between 793 and 823 K; quenched in ice water, face-centered cubic β-phase, lattice parameter reported as 3.5441 ±0.0002 Å at 291 K.
28	725	Marick, L.	1936	X	293-1460		99.9^+ Co, <0.01 Fe, <0.01 Ni, Kahlbaum cobalt; specimen 8 cm long, 8.71 gm/cc density, face-centered cubic β-phase; lattice parameter reported as 3.559 ± 0.002 Å at 293 K.
29	725	Marick, L.	1936	X	293-693		The above specimen; close-packed, hexagonal α-phase; lattice parameter reported as 2.519 ±0.002 Å at 293 K.

* Not shown in figure.

DATA TABLE 11. THERMAL LINEAR EXPANSION OF COBALT Co

[Temperature, T, K; Linear Expansion, $\Delta L/L_0$, %]

T	$\Delta L/L_0$
CURVE 1	
293	0.000
299	-0.036
328	0.072
342	0.036
395	0.108
418	0.124
474	0.184
492	0.260
556	0.304
601	0.368
609	0.416
632	0.460
675	0.508
CURVE 2	
293	0.000
299	-0.030
329	0.084
342	0.064
392	0.187
415	0.202
473	0.290
488	0.369
552	0.477
596	0.514
603	0.546
617	0.570
674	0.691
CURVE 3*	
677	0.000
685	0.034
708	0.028
730	0.070
774	0.124
828	0.219
882	0.306
918	0.326
938	0.413
967	0.427
1038	0.503
1038	0.548
1073	0.610
1092	0.627
CURVE 3 (cont.)*	
1119	0.655
1163	0.723
1167	0.798
1218	0.821
1225	0.911
1259	0.922
1277	0.917
1300	1.049
1300	1.113
1319	1.125
1325	1.226
1350	1.293
1354	1.352
1379	1.380
1411	1.428
1489	1.580
1502	1.642
1524	1.667
1564	1.684
1593	1.754
CURVE 4*, †, ‡	
4	0.000 x 10^{-4}
5	0.012
6	0.028
8	0.070
10	0.130
12	0.210
14	0.313
16	0.445
18	0.611
20	0.821
22	1.095
24	1.453
26	1.905
65	76.005
75	116.255
85	166.255
CURVE 5	
294	0.001*
384	0.121
473	0.232
571	0.365
CURVE 5 (cont.)	
672	0.508
700	0.547
724	0.610
738	0.668
746	0.694
766	0.747
821	0.830
865	0.910
965	1.085
1076	1.289
1217	1.553
CURVE 6	
1217	1.497
1173	1.424
1107	1.313
970	0.994
736	0.745
718	0.719
694	0.682
670	0.646
648	0.617
624	0.576
600	0.528
574	0.475
558	0.447
538	0.397
515	0.349
505	0.317
477	0.275
452	0.234
417	0.185
396	0.145
373	0.120
293	0.000
CURVE 7	
293	0.000
333	0.048
CURVE 8	
293	0.000
373	0.100
CURVE 8 (cont.)	
473	0.225
573	0.350
673	0.485
773	0.635
873	0.805
973	0.990
1073	1.230
1103	1.320
CURVE 9	
293	0.000*
404	0.160
488	0.295
544	0.371
627	0.515
649	0.555
673	0.594
687	0.618
696	0.666
CURVE 10	
293	0.000*
403	0.216
484	0.361
540	0.469
623	0.617
649	0.664
673	0.713
CURVE 11	
293	0.000*
406	0.152
499	0.293
541	0.356
623	0.485
650	0.511
673	0.564
868	0.567*
701	0.587
CURVE 12*, †, ‡	
1173	0.000
CURVE 12 (cont.)*, †, ‡	
1184	0.021
1194	0.040
1200	0.052
1218	0.086
1221	0.092
1231	0.111
1242	0.133
1282	0.210
1315	0.276
1379	0.404
1385	0.417
1390	0.429
1420	0.495
1453	0.567
1462	0.586
1470	0.604
1473	0.610
1486	0.639
CURVE 13*	
173	-0.042
192	-0.035
214	-0.028
234	-0.021
254	-0.014
272	-0.007
293	0.000
316	0.007
336	0.014
355	0.021
375	0.028
395	0.035
415	0.042
436	0.049
455	0.056
475	0.063
495	0.071
515	0.078
535	0.085
555	0.092
575	0.099
596	0.107
615	0.114
635	0.121
CURVE 14*	
173	-0.034
213	-0.023
234	-0.017
255	-0.011
274	-0.005
293	0.000
296	0.001
315	0.007
333	0.013
355	0.019
375	0.025
394	0.031
414	0.036
433	0.043
454	0.049
474	0.055
495	0.061
516	0.067
534	0.073
557	0.080
576	0.085
596	0.092
617	0.098
634	0.104
CURVE 15*	
173	-0.030
194	-0.025
216	-0.020
235	-0.014
253	-0.009
274	-0.004
293	0.000
296	0.001
314	0.006
334	0.012
354	0.017
375	0.022
393	0.027
415	0.033
434	0.038
453	0.044
475	0.050
495	0.054
515	0.060

* Not shown in figure.

† This curve is here reported using the first given temperature as reference temperature at which $\Delta L/L_0 = 0$.

‡ Author's data for coefficient of thermal expansion have been integrated by TPRC to obtain $\Delta L/L_0$.

DATA TABLE 11. THERMAL LINEAR EXPANSION OF COBALT Co (continued)

T	$\Delta L/L_0$
CURVE 15 (cont.)*	
534	0.066
554	0.070
575	0.076
595	0.082
614	0.087
636	0.093
CURVE 16	
303	0.013
373	0.102
CURVE 17*	
303	0.012
373	0.097
CURVE 18*, †, ‡	
423	0.000
476	0.066
524	0.129
575	0.198
626	0.270
673	0.339
699	0.379
721	0.516
746	0.467
752	0.478
770	0.509
799	0.553
823	0.589
873	0.664
921	0.739
975	0.824
1024	0.909
CURVE 19	
293	0.000*
323	0.040
373	0.111*
423	0.184
473	0.264
523	0.329
573	0.399
623	0.474
673	0.549
CURVE 19 (cont.)	
710	0.605
723	0.681
773	0.785
823	0.863
873	0.946
923	1.022
973	1.103
1023	1.183
1073	1.264
1123	1.365
1152	1.408
CURVE 20	
1152	1.418*
1123	1.374
1073	1.283*
1023	1.193
973	1.113*
923	1.032*
873	0.956
823	0.873*
773	0.795*
723	0.714
673	0.639
648	0.510
623	0.473*
573	0.399*
523	0.332*
473	0.250
423	0.179
373	0.107*
323	0.033
293	0.000*
CURVE 21*	
373	0.109
673	0.517
1073	1.233
CURVE 22	
306	0.016
373	0.101
CURVE 23	
306	0.021
373	0.129
CURVE 24*	
306	0.018
373	0.110
CURVE 25‡	
293	0.000*
400	0.172
500	0.346
600	0.533
690	0.712
CURVE 26‡	
293	0.000*
400	0.122
500	0.244
600	0.373*
690	0.495
CURVE 27‡	
273	-0.025
323	0.037
373	0.100*
423	0.164
473	0.229*
523	0.295
573	0.362
623	0.430
673	0.499
723	0.569
773	0.640*
823	0.712
873	0.785
CURVE 28	
293	0.000*
568	0.393
693	0.562
898	0.843
1116	1.21
1372	1.63
CURVE 28 (cont.)	
1394	1.94
1421	2.39
1460	2.47
CURVE 29	
293	0.000*
568	0.318
693	0.675

* Not shown in figure.
† This curve is here reported using the first given temperature as reference temperature at which $\Delta L/L_0 = 0$.
‡ Author's data for coefficient of thermal expansion have been integrated by TPRC to obtain $\Delta L/L_0$.

DATA TABLE 11. COEFFICIENT OF THERMAL LINEAR EXPANSION OF COBALT Co

[Temperature, T, K; Coefficient of Expansion, α, 10^{-6} K^{-1}]

T	α
CURVE 4‡	
4	0.011
5	0.014
6	0.017*
8	0.025*
10	0.035*
12	0.045*
14	0.058*
16	0.074*
18	0.092*
20	0.118
22	0.156*
24	0.202*
26	0.250
65	3.55
75	4.55
85	5.50
283	12.7
CURVE 12*, ‡	
1173	19.71
1184	18.97
1194	18.97
1200	19.22
1218	19.22
1221	19.40
1231	19.40
1242	19.11
1282	19.76
1315	20.05
1379	20.05
1385	22.52
1390	22.80
1420	21.72
1453	21.48
1462	22.09
1470	22.09
1473	21.79
1486	21.90
CURVE 18*, ‡	
423	12.26 x 10^{-6}
476	12.80
524	13.25
575	13.76
CURVE 18 (cont.)*, ‡	
626	14.45
673	14.99
699	15.70
721	18.07
746	22.75
752	17.56
770	15.66
799	14.99
823	14.73
873	15.28
921	15.84
975	16.61
1024	17.04
CURVE 25*, ‡	
293	15.7
400	17.0
500	18.3
600	19.6
690	20.7
CURVE 26*, ‡	
293	11.2
400	12.0
500	12.7
600	13.5
690	14.1
CURVE 27‡	
273	12.3 x 10^{-6}
323	12.5*
373	12.7*
423	12.9*
473	13.1*
523	13.3*
573	13.5*
623	13.7*
673	13.9*
723	14.1*
773	14.3*
823	14.5*
873	14.75*

* Not shown in figure.
‡ Author's data for coefficient of thermal expansion have been integrated by TPRC to obtain $\Delta L/L_0$.

FIGURE AND TABLE NO. 12R. RECOMMENDED VALUES FOR THERMAL LINEAR EXPANSION OF COPPER Cu

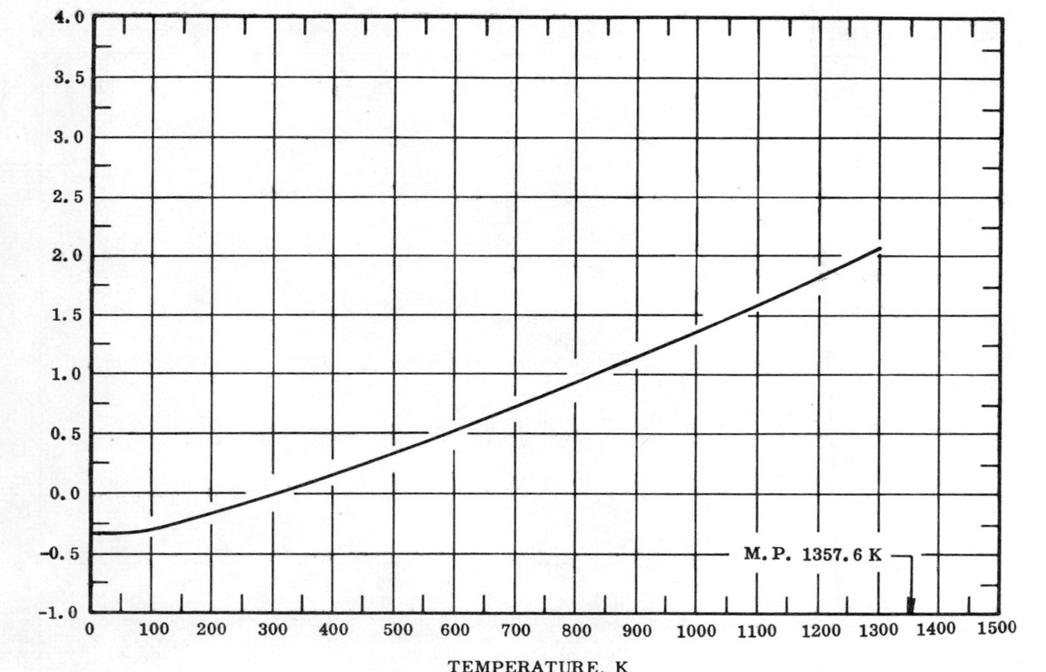

RECOMMENDED VALUES

[Temperature, T, K; Linear Expansion, $\Delta L/L_0$, %; α, K^{-1}]

T	$\Delta L/L_0$	$\alpha \times 10^6$
1	-0.324	0.0003
5	-0.324	0.005
25	-0.324	0.63
50	-0.318	3.87
100	-0.282	10.3
200	-0.148	15.2
293	0.000	16.5
400	0.182	17.6
500	0.362	18.3
600	0.549	18.9
700	0.741	19.5
800	0.939	20.3
900	1.147	21.3
1000	1.366	22.4
1200	1.838	24.9
1300	2.095	25.8

REMARKS

The tabulated values are considered accurate to within ± 3% over the entire temperature range. These values can be represented approximately by the following equations:

$$\Delta L/L_0 = -0.281 + 1.073 \times 10^{-3}(T - 100) + 2.904 \times 10^{-6}(T - 100)^2 - 4.548 \times 10^{-9}(T - 100)^3 \quad (100 < T < 293)$$

$$\Delta L/L_0 = 1.685 \times 10^{-3}(T - 293) + 2.702 \times 10^{-7}(T - 293)^2 + 1.149 \times 10^{-10}(T - 293)^3 \quad (293 < T < 1300)$$

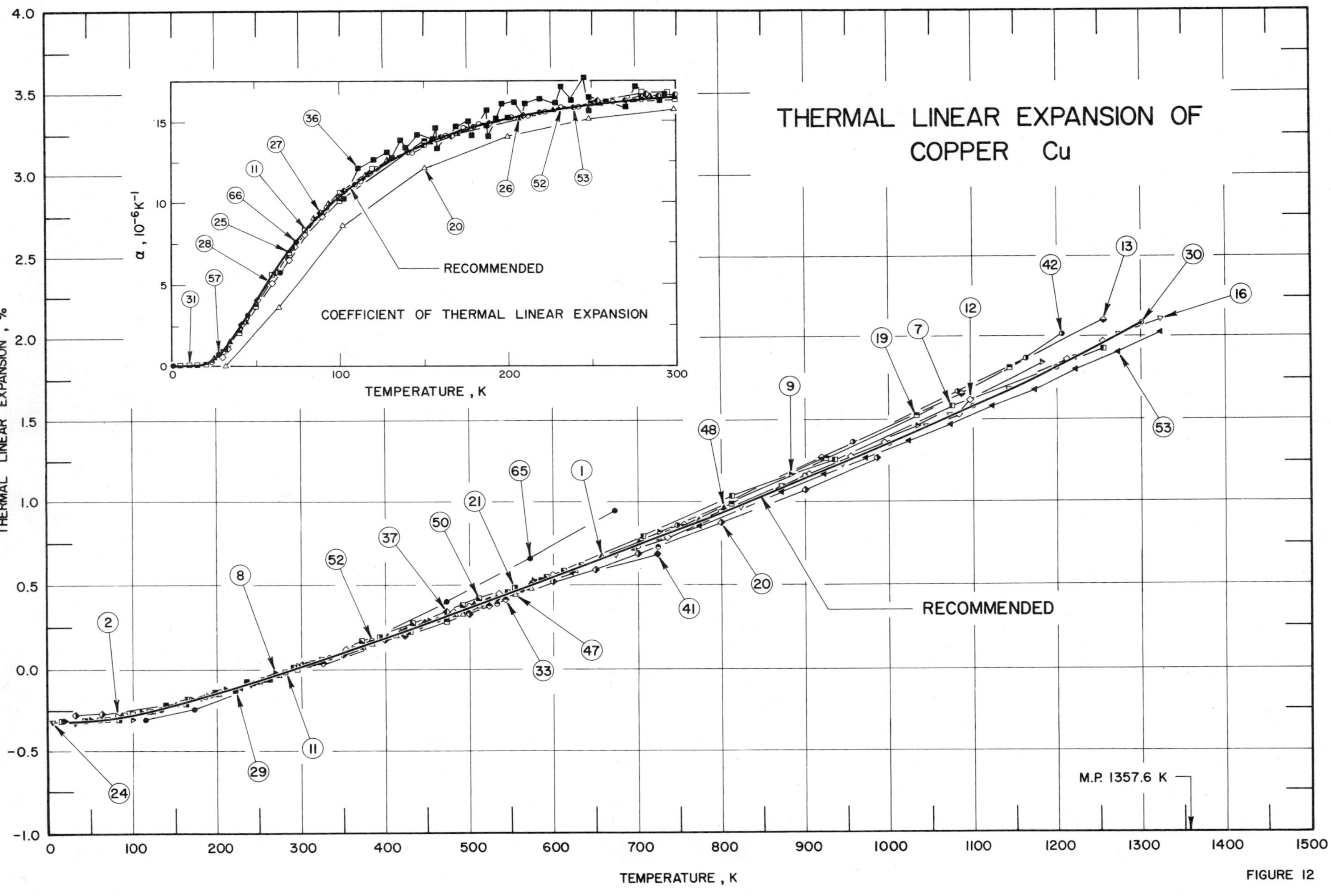

THERMAL LINEAR EXPANSION OF
COPPER Cu
THERMAL LINEAR EXPANSION , %
TEMPERATURE , K
RECOMMENDED
M.P. 1357.6 K
COEFFICIENT OF THERMAL LINEAR EXPANSION
α , 10⁻⁶K⁻¹
TEMPERATURE , K
RECOMMENDED

FIGURE 12

SPECIFICATION TABLE 12. THERMAL LINEAR EXPANSION OF COPPER Cu

Cur. No.	Ref. No.	Author(s)	Year	Method Used	Temp. Range, K	Name and Specimen Designation	Composition (weight percent), Specifications, and Remarks
1	1	Nix, F.C. and MacNair, D.	1941	I	88-773		99.979 Cu, 0.02 O; rod specimen about 6 mm long; obtained from Adam Hilger; low temperature portion of test made at cooling rate of 20 C per hr from room temperature with system containing helium gas at pressure of 3 mm Hg; region above room temperature, gas at pressure of 3 mm Hg up to 420 K with heating rate of 15 C per hr; at 420 K, system evacuated to a pressure of 10^{-5} mm Hg; zero-point correction of -0.035% determined by graphical interpolation.
2	133	Ullrich, H.-J.	1967	X	83-363		99.93 Cu; reported error 5%; measurements made with respect to a-axis.
3*	3	Fraser, D.B. and Hollis-Hallett, A.C.	1965	I	17-100		99.9^+ Cu; cold drawn, oxygen-free, high conductivity; vacuum annealed for 4 hr at 573 K; this curve is here reported using the first given temperature as reference temperature at which $\Delta L/L_0 = 0$; zero-point correction is 0.295%; data on the coefficient of thermal linear expansion also reported.
4*	134	Simmons, R.O. and Balluffi, R.W.	1957	X	16-103		99.996 Cu single crystal specimen; expansion measured in vacuum on first test; this curve is here reported using the first given temperature as reference temperature at which $\Delta L/L_0 = 0$; zero-point correction is 0.0001%.
5*	134	Simmons, R.O. and Balluffi, R.W.	1957	X	9-77		The above specimen; expansion measured on second test; this curve is here reported using the first given temperature as reference temperature at which $\Delta L/L_0 = 0$; zero-point correction is 0.0004%.
6*	134	Simmons, R.O. and Balluffi, R.W.	1957	X	8-90		The above specimen; expansion measured on third test; this curve is here reported using the first given temperature as reference temperature at which $\Delta L/L_0 = 0$; zero-point correction is 0.0006%.
7	135	Paine, R.M., Stonehouse, A.J., and Beaver, W.M.	1959		290-1256		Electrolytic tough pitch copper.
8	7	Bijl, D. and Pullan, H.	1955	L	20-273		99.98 Cu, supplied by Johnson Matthey Ltd., specification Bss 1433; zero-point correction of -0.03% determined by graphical extrapolation.
9	136	Gehlen, P.C.	1969	X	100-1183		Single crystal cut to a (111) face; molybdenum radiation used; lattice parameter reported at 292 K is 3.6139 Å; 3.6139 Å at 293 K determined by graphical interpolation.
10*	137	Carr, R.H., McCammon, R.D., and White, G.K.	1964	E	5-30		99.999 Cu, either Fe, Sb or Cr <1 ppm and each of As and Te <2 ppm; supplied by the American Smelting and Refining Co., New Jersey; after machining annealed at 770 K for several hr; this curve is here reported using the first given temperature as reference temperature at which $\Delta L/L_0 = 0$.
11	137	Carr, R.H., et al.	1964	E	60-295		Similar to the above specimen.
12	138	Fieldhouse, I.B.	1956	T	351-1257		Data average of three separate runs; zero-point correction of 0.01% determined by graphical extrapolation.
13	121	Lucks, C.F. and Deem, H.W.	1956	L	293-1255		Expansion measured in vacuum of 5×10^{-5} mm Hg with temperature increasing at 2.8 C/min.
14*	139	Bunton, G.V. and Weintroub, S.	1968	L	10-200		Grade 1 copper with a total metallic impurity level of less than 10 ppm, supplied by Johnson Matthey Ltd., machined into a right-angled cylinder of 6 cm length and 1 cm^2 cross-sectional area; annealed at 820 K in argon for 3 hr and slowly cooled down to room temperature; this curve is here reported using the first given temperature as reference temperature at which $\Delta L/L_0 = 0$.

* Not shown in figure.

SPECIFICATION TABLE 12. THERMAL LINEAR EXPANSION OF COPPER Cu (continued)

Cur. No.	Ref. No.	Author(s)	Year	Method Used	Temp. Range, K	Name and Specimen Designation	Composition (weight percent), Specifications, and Remarks
15*	140	Beenakker, Y.Y.M. and Swenson, C.A.	1955	L	4.2-300		99.999 Cu; zero-point correction of -0.034% determined by graphical interpolation.
16	141	Simmons, R.O. and Balluffi, R.W.	1963	X	323-1323		99.999^+ Cu grade A-58 material from American Smelting and Refining Co.; no impurity greater than 0.0002; specimen prepared by melting and controlled directional solidification to produce grain size >1.0 cm in prepurified nitrogen atmosphere; bar trimmed to 1.27 x 1.27 x 53 cm; large crystal at center of bar studied using $NiK\alpha_1$ x-radiation (λ = 1.65784 Å); data from two thermal cycles smoothed by author.
17	141	Simmons, R.O. and Balluffi, R.W.	1963	T	323-1323		The above specimen.
18	142	Aoyama, S. and Ito, T.	1939	L	77-273	Electrolytic	99.968 Cu, 0.015 Sb, 0.01 Fe, 0.007 S; round bar 5 mm diameter, 100 mm long; annealed at 380 C for 4 hr.
19	123	Lucks, C.F. and Deem, H.W.	1958	L	293-1255		0.375 in. diameter, 3 in. long; cold drawn, electrolytic tough pitch; Federal specification QQ-C-502; from Williams and Co., Columbus, Ohio (Revere Copper and Brass, Inc.); heating rate 2.2 C per min; vertical type quartz tube dilatometer.
20	143	Nasekovs'kii, A.P.	1967		31-985		Electrolytic specimen.
21	15	Rhodes, B.L., Moeller, C.E., Hopkins, V., and Marx, T.I.	1963	L	18-573		Specimen 10.16 cm long by 6.2 mm; temperature determined by thermocouple imbedded in the side of specimen at its midpoint; reported error <1.5%.
22	144	Keesom, W.H., Agt, V., and Jansen, A.F.J.	1926		20-374		Length of inner vacuum vessel, 947.96 mm; constructed from electrolytic copper by L. Ouwerkerk; expansion measured under atmospheric pressure.
23	48	Dorsey, H.G.	1907	I	93-293		99.9 Cu; commercial seamless tubing; 0.9712 cm long; density 8.91 g cm^{-3}.
24	145	Sandenaw, T.A.	1960	V	4-300		Oxygen-free, high conductivity (OFHC) specimen; zero-point correction of -0.031% determined by graphical extrapolation.
25	22	Fraser, D.B. and Hollis-Hallett, A.C.	1961	I	20-300		Metal machined into the form of thick-walled cylindrical tubes having ends carefully ground plane and parallel, annealed; reported error 20% maximum at 10 K.
26	146	Rubin, T., Altman, H.W., and Johnston, H.L.	1956	I	25-286	OFHC polycrystalline	Polycrystalline copper sample cut from large bar of oxygen-free high conductivity (OFHC) copper, obtained from American Brass Co., slice of proper thickness cut from bar and two surfaces perpendicular to axis of original bar made as parallel as possible, ground flat and smooth; polished to length 0.4948±0.0005 cm; reported error 0.3 to 0.7%.
27*	146	Rubin, T., et al.	1956	I	24-284	Single crystal	Single crystal copper sample cut from small etched annealed rod of single crystal copper obtained from General Electric Co. cut to proper length in hardened plaster matrix, and quartered, then adjusted to uniform length by method used for above specimen; polished to length 0.4953±0.0005 cm.
28*	147	Novikova, S.I. and Strelkov, P.G.	1960	L	30-105		This curve is here reported using the first given temperature as reference temperature at which $\Delta L/L_0 = 0$.
29	26	Buffington, R.M. and Latimer, W.M.	1926	I	109-312		Specimen turned from hard-rolled copper rod.
30	27	Pathak, P.D. and Vasavada, N.G.	1970	L	300-1300		99.999 Cu; rod 10 cm long, about 0.5 cm diameter; obtained from Johnson Matthey and Co., London; reported error 0.5 to 1.0%.

*Not shown in figure.

SPECIFICATION TABLE 12. THERMAL LINEAR EXPANSION OF COPPER Cu (continued)

Cur. No.	Ref. No.	Author(s)	Year	Method Used	Temp. Range, K	Name and Specimen Designation	Composition (weight percent), Specifications, and Remarks
31*	148	Kos, J. F. and Lamarche, J. L. G.	1969	E	4-15		0.0004 Ag, 0.0002 Fe, 0.0001 Cd, 0.0001 Mg, 0.0001 Pb, 0.0001 Si; polycrystalline specimen 7 mm diameter, 10 cm long; from Johnson, Matthey and Co.; electron Grüneisen $\gamma_e = 0.67 \pm 0.03$; low temperature lattice Grüneisen $\gamma_0 = 1.74 \pm 0.02$; expansion measured on first test; reported error 5% at 5 K, 10% at 10 K; this curve is here reported using the first given temperature as reference temperature at which $\Delta L/L_0 = 0$.
32*	148	Kos, J. F. and Lamarche, J. L. G.	1969	E	4-15		The above specimen; second test; this curve is here reported using the first given temperature as reference temperature at which $\Delta L/L_0 = 0$.
33	149	Burger, E. E.	1934	L	298-544		Cylindrical specimen 0.25 in. diameter, 12 in. long; zero-point correction of 0.008% determined by graphical extrapolation.
34	150	Eisenstein, A.	1946	X	293, 1148		Cubic crystal structure; reported error 0.2% in measurement of lattice constant, 15% in determination of coefficient of expansion.
35*	151	Kasai, N.	1968	E	0-60		Cylindrical specimen about 2 cm long, 8 mm diameter; this curve is here reported using the first given temperature as reference temperature at which $\Delta L/L_0 = 0$.
36	30	Strelkov, P. G. and Novikova, S. I.	1969	L	100-345		Specimen cut from copper rail 19 mm long; data are combined results of seven sets of measurements.
37	31	Kagan, A. S.	1964	X	295, 473		Massive specimen, glued to copper specimen holder; reported error 3.0%.
38*	31	Kagan, A. S.	1964	X	295, 473		Powdered specimen, particle diameter $\sim 20\ \mu$; glued to copper holder.
39*	31	Kagan, A. S.	1964	X	295, 473		Powdered specimen, particle diameter $\sim 2\ \mu$; glued to copper holder.
40	152	Eppelsheimer, D. S. and Penman, R. R.	1950		291-1043		99.999$^+$ Cu, produced by National Research Corp., Cambridge, Mass.; specimen prepared by swaging down to 0.050 in., annealed at 470 K for 4 hr in cast iron chips, further reduced down by etching in concentrated HNO_3; expansion measured in (024) plane; lattice parameter reported at 291 K is 5.6059 Å; 5.6060 Å at 293 K determined by graphical interpolation.
41	152	Eppelsheimer, D. S. and Penman, R. R.	1950		291-1043		The above specimen; expansion measured in (331) plane; lattice parameter reported at 291 K is 3.6056 Å; 3.6057 Å at 293 K determined by graphical interpolation.
42	33	Richards, J. W.	1942	T	83-1207		Bus-bar copper, high purity; expansion measured in hydrogen atmosphere.
43	153	Masumoto, H., Saito, H., and Sawaya, S.	1969	L	273, 313	No. 21	Electrolytic 99.900 Cu; 0.009 Si, 0.008 Fe, 0.005 Sb, 0.005 S, 0.001 Mn; rod specimen 2 mm diameter, 10 cm long cooled at rate 570 K per hr, after heating in vacuum at 1170 K for 30 min; density 8.93 g cm^{-3} at 293 K.
44*	154	Shapiro, J. M., Taylor, D. R., and Graham, G. M.	1964	L	7.5-54		Specimen prepared from rod of Johnson-Matthey spectroscopically pure copper, 8.14 cm long, 0.50 cm diameter; vacuum annealed at 620 K for 48 hr to relieve strains produced by machining ends; reported error ±2%; this curve is here reported using the first given temperature as reference temperature at which $\Delta L/L_0 = 0$.
45*	154	Shapiro, J. M., et al.	1964	L	6.4-55		Similar to the above specimen; another run; this curve is here reported using the first given temperature as reference temperature at which $\Delta L/L_0 = 0$.
46	155	Hume-Rothery, W. and Andrews, K. W.	1942	X	291-1144		99.998 Cu; lattice spacing reported at 291 K is 3.6074 Å; 3.60745 Å at 293 K determined by graphical interpolation.
47	38	Uffelmann, F. L.	1930	I	383-573		Cubic specimen of side 4 mm.

*Not shown in figure.

SPECIFICATION TABLE 12. THERMAL LINEAR EXPANSION OF COPPER Cu (continued)

Cur. No.	Ref. No.	Author(s)	Year	Method Used	Temp. Range, K	Name and Specimen Designation	Composition (weight percent), Specifications, and Remarks
48	156	Mitra, G.B. and Mitra, S.K.	1962	X	303-803		Spectroscopically pure rod supplied by Johnson Matthey and Co. Ltd., London; 0.0003 Ni, Pb, 0.0001 Si, Fe, Li, <0.0001 Na, K, Mg, Ca, 0.00005 Ag, Mn; drawn into thin wire about 0.5 mm diameter; lattice parameter reported at 303 K is 3.6147 Å; 3.6141 Å at 293 K determined by graphical extrapolation.
49	161	Hahn, T.A.	1970	I	20-800	Standard Reference Material 736	OFHC polycrystalline copper rod 0.25 in. diameter; annealed in vacuum at 811 K; 99.99 at.% Cu determined from resistivity ratio $\rho_{273}/\rho_4 = 62.53$; reported data are summary of certified values for the reference material.
50	254	Rosenbohm, E.	1938	L	293-633		Specimen length 40.90 mm at 293 K; reported error ~1%.
51	393	Pavese, F., Righini, F., and Ruffino, G.	1970	L	293-873		99.999 Cu; specimen obtained from Leico Industries with specified impurities 0.0002 Ag, 0.00005 Fe, 0.00003 Si, and 0.00001 each Ca and Mg; annealed at 1073 K in inert atmosphere; expansion measured using 0.54607 μm line of mercury; data obtained by author from quadratic fit: $L_t = L_0[1 + (15.9079 + 0.0040\, t^2) \times 10^{-6}]$ which best fit experimental data before correcting for expansion of silica, given by manufacturer as $\alpha = 0.54 \times 10^{-6}\ C^{-1}$; averaged from measurements with both increasing and decreasing temperature.
52	426	Lifanov, I.I. and Shertyukov, N.G.	1968	L	88-573		Reported data are smoothed values obtained by authors from experiments on eleven high purity specimens; some specimens annealed, some cylindrical; no significant differences observed between specimens.
53	452	Leksina, I.E. and Novikova, S.I.	1963	L	90-1323		99.99, oxygen-free Cu annealed in vacuum at 1070 K; specimen <25 mm long.
54	611	Kurnakow, W.S. and Ageew, N.W.	1931	L	293-873		Electrolytic copper; expansion measured with increasing temperature.
55	611	Kurnakow, W.S. and Ageew, N.W.	1931	L	773-293		The above specimen; expansion measured with decreasing temperature.
56*	614	McLean, K.O., Swenson, C.A., and Case, C.R.	1972	V	0-26	Cu-1	99.999^+ Cu; pure specimen from Nuclear Elements Corp., vacuum cast; this curve is here reported using the first given temperature as reference temperature at which $\Delta L/L_0 = 0$.
57*	614	McLean, K.O., et al.	1972	V	0-31	Cu-3	99.999^+ Cu; pure specimen from Asarco Lab was vacuum cast in graphite into a 2 cm diameter rod and was swaged into 0.8 cm rod; this curve is here reported using the first given temperature as reference temperature at which $\Delta L/L_0 = 0$.
58*	615	White, G.K. and Collins, J.G.	1972	L	2.7-29.6	Cu2 (Si1a)	99.999^+ Cu; spectroanalysis quoted <0.0001 Fe, Sb, or Cr; <0.0002 As and Te; cylindrical specimen 5.09 cm long, 1.9 cm diameter Asarco from American Smelting and Refining Co., N.Y.; specimen annealed at 770 K in vacuo after machining; measurements with Si1a specimen of the following specification in the cell: cylinder 5.08 cm long, 2.2 cm diameter; 100 Ωcm n-type from Merck, Sharp, and Dohme, plated with Ca, 0.002 cm Ni, and later with layer of evaporated Ag; results corrected for the expansion of Si; this curve is here reported using the first given temperature as reference temperature at which $\Delta L/L_0 = 0$.
59*	615	White, G.K. and Collins, J.G.	1972	L	2.5-33.9	Cu2 (Si1b)	The above Cu2 specimen; measurements with the above Si1a specimen with addition layer of evaporated gold; corrected for thermal expansion of Si; this curve is here reported using the first given temperature as reference temperature at which $\Delta L/L_0 = 0$.
60*	615	White, G.K. and Collins, J.G.	1972	L	2.6-29.3	Cu2 (Si1c)	The above Cu2 specimen; measurements with the above Si1a specimen, after cleaning, relapping, electroless Ni plating and coating with final film of gold; corrected for thermal expansion of Si; this curve is here reported using the first given temperature as reference temperature at which $\Delta L/L_0 = 0$.

* Not shown in figure.

SPECIFICATION TABLE 12. THERMAL LINEAR EXPANSION OF COPPER Cu (continued)

Cur. No.	Ref. No.	Author(s)	Year	Method Used	Temp. Range, K	Name and Specimen Designation	Composition (weight percent), Specifications, and Remarks
61*	615	White, G.K. and Collins, J.G.	1972	L	2.5-33.9	Cu2 (Si2a)	The above Cu2 specimen; measurements with 4.48 cm long, 1.9 cm diameter cylinder with an 0.2 cm hole along central axis and a Cu cap 0.60 cm thick; 100 Ωcm n-type specimen from duPont de Nemours; corrected for thermal expansion of Si; this curve is here reported using the first given temperature as reference temperature at which $\Delta L/L_0 = 0$.
62*	615	White, G.K. and Collins, J.G.	1972	L	2.4-29.2	Cu2 (Si2b)	The above Cu2 specimen; measurements with the above Si2a specimen, however, cut into two cylinders, ground and lapped to lengths 2.56 cm and 1.73 cm; mounted together with a Cu end piece 0.80 cm long; corrected for thermal expansion of Si; this curve is here reported using the first given temperature as reference temperature at which $\Delta L/L_0 = 0$.
63*	615	White, G K. and Collins, J.G.	1972	L	2.7-28.7	Cu2 (LiF)	The above Cu2 specimen; measurements with LiF specimen of 5.08 cm long, 2.2 cm diameter cylinder, optical quality crystal from Isomet Corp., N.J., ground and coated with an evaporated film of Ag; corrected for thermal expansion of LiF; this curve is here reported using the first given temperature as reference temperature at which $\Delta L/L_0 = 0$.
64*	639	Fitzer, E.	1971	L	293-1273		99.999 pure specimen; average of seven participants.
65	664	Masumoto, H., Sawaya, S., and Kikuchi, M.	1972	L	115-673		Electrolytic specimen with 0.008 Fe, 0.001 Mn, 0.005 S; rod specimen; average of heating and cooling experiments; zero-point correction is -0.313%.
66*	696	Awad, F.C. and Gugan, D.	1971	E	10-80		99.9 pure, sample from commercial "high purity" polycrystalline copper; results based on three data runs, these results were smoothed and rounded values of thermal linear expansion coefficient were obtained; error limits of ±2% on linear thermal expansion coefficient given; this curve is here reported using the first given temperature as reference temperature at which $\Delta L/L_0 = 0$.
67*	709	Shah, J.S.	1971	X	40-180		99.7 pure, powder sample of particle size 8-10 μ from Koch Light Labs, Inc., Colnbrook, England; lattice parameter not corrected for refraction; lattice parameter reported at 40 K is 3.60306 Å; this curve is here reported using the first given temperature as reference temperature at which $\Delta L/L_0 = 0$.
68*	738	Kantola, M. and Tokola, E.	1967	X	296-843		Sample prepared using MD 154 copper shot from Metals Disintegrating Corp., N.J.; sample powdered and sifted through sieve of diameter equal to 0.060 mm; lattice parameter at 296 K reported as 3.61443 Å, 3.6140 Å at 293 K determined by graphical extrapolation.

* Not shown in figure.

DATA TABLE 12. THERMAL LINEAR EXPANSION OF COPPER Cu

[Temperature, T, K; Linear Expansion, $\Delta L/L_0$, %]

T	$\Delta L/L_0$
CURVE 1	
87.7	-0.295
94.2	-0.289*
100.2	-0.284*
104.7	-0.279*
108.7	-0.274*
113.7	-0.269*
119.2	-0.264*
121.2	-0.259
126.2	-0.253*
130.7	-0.249*
135.7	-0.243*
139.2	-0.238*
142.0	-0.233*
150.2	-0.223
154.0	-0.217*
157.2	-0.213*
161.2	-0.207*
165.2	-0.202*
168.7	-0.197*
172.7	-0.192*
176.0	-0.187*
179.7	-0.181*
182.7	-0.176
187.1	-0.171*
190.0	-0.166*
193.0	-0.161*
196.2	-0.156*
200.0	-0.150
203.2	-0.136*
207.0	-0.140*
211.2	-0.135*
213.0	-0.130*
216.6	-0.125*
220.2	-0.119*
224.2	-0.115*
226.7	-0.110*
229.1	-0.105*
233.7	-0.099*
237.1	-0.094*
240.1	-0.089*
243.6	-0.084*
246.2	-0.079*
250.0	-0.074
252.9	-0.068*
255.9	-0.063*
258.7	-0.058*
CURVE 1 (cont.)	
262.2	-0.053*
265.2	-0.048*
268.7	-0.043*
271.7	-0.038*
273.2	-0.035
274.7	-0.032*
277.4	-0.027*
280.7	-0.022*
284.2	-0.017*
286.6	-0.012*
289.2	-0.007*
292.4	-0.002*
295.3	0.004*
298.2	0.008*
301.7	0.013*
304.9	0.018*
310.2	0.027*
315.5	0.037*
320.3	0.046
327.2	0.056*
332.8	0.066*
338.7	0.076*
344.2	0.085*
350.2	0.095*
356.2	0.104*
362.2	0.114*
365.2	0.119*
367.7	0.124*
373.9	0.138
377.2	0.142*
380.2	0.147*
382.1	0.151*
385.2	0.157*
387.8	0.161*
390.2	0.166*
393.7	0.171*
395.8	0.176*
399.2	0.180*
402.0	0.185
405.2	0.190*
407.9	0.195*
411.2	0.200*
413.3	0.204*
419.2	0.214
424.6	0.223*
431.0	0.233*
CURVE 1 (cont.)	
435.7	0.242*
441.2	0.252*
446.6	0.262*
451.2	0.271*
456.3	0.281*
462.2	0.290*
467.0	0.300*
473.0	0.309*
477.2	0.319*
483.0	0.328
488.6	0.338*
494.0	0.347*
499.4	0.357*
505.8	0.366*
516.7	0.386*
522.2	0.395
532.2	0.414*
536.4	0.424*
541.3	0.433*
546.7	0.443*
552.2	0.453*
557.0	0.462*
562.0	0.471*
566.2	0.481*
572.0	0.490
576.9	0.500*
585.1	0.514*
587.2	0.519*
592.0	0.529*
596.9	0.538*
602.0	0.548*
611.4	0.567*
617.1	0.576*
626.4	0.595
635.8	0.615*
645.3	0.633*
657.2	0.653
667.3	0.671*
677.0	0.691*
686.0	0.710*
695.2	0.729
705.7	0.748*
715.2	0.767*
725.3	0.786
735.2	0.805*
744.4	0.824*
CURVE 1 (cont.)	
754.2	0.843*
762.9	0.862*
773.3	0.882
CURVE 2	
83	-0.295
188	-0.162
298	0.009
363	0.127
CURVE 3*,†	
17.4	0.0000
20.7	0.0001
22.5	0.0002
25.55	0.0003
32.4	0.0010
33.7	0.0014
34.4	0.0015
37.0	0.0018
42.2	0.0030
43.8	0.0034
45.5	0.0039
46.7	0.0042
49.5	0.0053
51.45	0.0060
53.2	0.0068
54.7	0.0074
57.0	0.0086
59.4	0.0098
60.85	0.0106
62.6	0.0116
67.0	0.0142
70.0	0.0162
73.3	0.0182
76.7	0.0207
80.55	0.0231
86.9	0.0286
94.5	0.0353
99.8	0.0412
CURVE 4*,†	
16	0.000×10^{-3}
19	-0.400
CURVE 4 (cont.)*,†	
27	-0.400×10^{-3}
33	1.30
36	1.70
40	2.9
44	3.7
50	6.9
63	13.7
69	18.5
73	20.9
92	37.0
102	48.9
103	48.3
CURVE 5*,†	
9	0.0×10^{-3}
14	0.3
16	-0.2
19	1.1
23	0.3
28	0.7
39	2.6
41	3.2
44	2.9
46	3.9
48	3.8
49	5.2
52	6.8
70	18.9
71	18.8
73	19.7
76	21.9
77	22.4
CURVE 6*,†	
8	0.0×10^{-3}
9	1.3
12	1.0
13	0.3
47	5.0
57	9.2
79	25.0
83	27.9
90	35.9
CURVE 7	
293	0.000
430	0.220
548	0.445
709	0.791
813	1.030
935	1.256
1077	1.582
1256	1.946
CURVE 8	
20	-0.321
30	-0.320*
40	-0.319*
50	-0.317
60	-0.313*
70	-0.307*
80	-0.299*
90	-0.289*
110	-0.268
130	-0.244*
150	-0.219*
170	-0.190
190	-0.161*
210	-0.130
230	-0.0993*
250	-0.0677*
270	-0.0352
273	-0.0300*
CURVE 9	
100	-0.302
195	-0.158
292	0.000*
295	0.003*
452	0.279
577	0.509
728	0.808
882	1.151
1033	1.469
1183	1.859
CURVE 10*,†	
5	0.000×10^{-3}

* Not shown in figure.
† This curve is here reported using the first given temperature as reference temperature at which $\Delta L/L_0 = 0$.

DATA TABLE 12. THERMAL LINEAR EXPANSION OF COPPER Cu (continued)

T	$\Delta L/L_0$
CURVE 10 (cont.)*,†	
6	0.0055 x 10^{-3}
7	0.014
8	0.028
9	0.047
10	0.071
12	0.153
14	0.284
16	0.499
18	0.814
20	1.254
22	1.864
24	2.694
26	3.834
28	5.244
30	7.044
CURVE 11‡	
60	-0.309
70	-0.303
80	-0.295*
90	-0.286
100	-0.276
110	-0.265*
120	-0.253
280	-0.022
295	0.003
CURVE 12	
352	0.104
482	0.335
537	0.447
555	0.455*
713	0.765*
737	0.796
816	0.981*
904	1.165
954	1.285
994	1.366*
1000	1.405*
1082	1.545
1096	1.620
1211	1.866
1257	1.973
CURVE 13	
293	0.000*
422	0.200
589	0.51*
755	0.86
922	1.27
1089	1.68
1255	2.11
CURVE 14*,†,‡	
10	0.00000
15	0.00003
20	0.00012
25	0.00031
30	0.00065
40	0.00208
50	0.0048
60	0.0094
70	0.0155
80	0.0230
90	0.0319
100	0.0418
110	0.0528
120	0.0645
130	0.077
140	0.090
150	0.104
160	0.118
170	0.132
180	0.146
190	0.161
200	0.176
CURVE 15*	
4.2	-0.327
25	-0.327
50	-0.322
75	-0.306
100	-0.283
125	-0.255
150	-0.222
175	-0.186
200	-0.150
250	-0.070
300	0.011
CURVE 16	
323	0.050*
373	0.135*
423	0.222*
473	0.311
523	0.402*
573	0.495*
623	0.590*
673	0.686
723	0.784*
773	0.884*
823	0.987
873	1.091
923	1.198*
944	1.244
973	1.307*
1023	1.419*
1044	1.467
1073	1.533
1123	1.650*
1144	1.699
1173	1.769*
1223	1.891
1273	2.018
1323	2.149
CURVE 17*	
323	0.050
373	0.135
423	0.222
473	0.311
523	0.402
573	0.495
623	0.590
673	0.686
723	0.784
773	0.884
823	0.987
873	1.091
923	1.198
973	1.307
1023	1.419
1073	1.534
1123	1.651
1173	1.770
1223	1.893
1273	2.021
1323	2.154
CURVE 18*	
77	-0.304
91	-0.291
133	-0.242
153	-0.215
173	-0.187
193	-0.157
233	-0.095
273	-0.032
CURVE 19	
293	0.000*
366	0.110*
477	0.296
589	0.504
700	0.736*
812	0.991
922	1.264
1033	1.538
1144	1.820
1255	2.106*
CURVE 20‡	
31	-0.284
64	-0.278
103	-0.254*
150	-0.205*
200	-0.140*
249	-0.068*
300	0.011
350	0.091
400	0.172
450	0.254*
500	0.337
550	0.422*
600	0.507
650	0.595
700	0.684
750	0.776*
800	0.871
850	0.970*
900	1.074
950	1.184*
985	1.265
CURVE 21	
18	-0.328
23	-0.328*
33	-0.328*
43	-0.326
53	-0.324*
63	-0.319*
73	-0.313*
83	-0.306
93	-0.296*
103	-0.286*
113	-0.275*
123	-0.264*
133	-0.251*
143	-0.238*
153	-0.225*
163	-0.211
173	-0.196*
193	-0.166*
213	-0.134*
233	-0.102*
253	-0.068*
273	-0.034*
293	0.000*
313	0.035
333	0.071*
353	0.108*
373	0.144*
393	0.180
413	0.216*
433	0.253*
453	0.288*
473	0.326*
493	0.362*
513	0.399*
533	0.436*
553	0.474
573	0.510*
CURVE 22*	
20.31	-0.323
87.68	-0.290
170.28	-0.190
273.20	-0.032
374.09	0.131
CURVE 23*	
93	-0.288
113	-0.267
133	-0.243
153	-0.217
173	-0.188
193	-0.159
213	-0.129
233	-0.097
253	-0.065
273	-0.033
293	0.000
CURVE 24	
4	-0.336
7	-0.336*
12	-0.335*
24	-0.330*
34	-0.330
52	-0.321*
82	-0.299*
119	-0.257*
151	-0.214*
165	-0.191
184	-0.166*
240	-0.080
300	0.011*
CURVE 25*,‡	
20	-0.326
30	-0.325
40	-0.324
50	-0.321
70	-0.310
100	-0.283
150	-0.221
200	-0.149
250	-0.070
300	0.011
CURVE 26*,‡	
25.25	-0.325
34.53	-0.324
43.49	-0.322
51.55	-0.320
60.38	-0.315

* Not shown in figure.
† This curve is here reported using the first given temperature as reference temperature at which $\Delta L/L_0 = 0$.
‡ Author's data for coefficient of thermal expansion have been integrated by TPRC to obtain $\Delta L/L_0$.

DATA TABLE 12. THERMAL LINEAR EXPANSION OF COPPER Cu (continued)

T	$\Delta L/L_0$
CURVE 26 (cont.)*, ‡	
69.48	-0.310
79.65	-0.302
85.67	-0.297
90.00	-0.293
94.75	-0.288
105.43	-0.277
116.82	-0.264
129.32	-0.249
142.08	-0.232
153.70	-0.216
169.40	-0.194
187.12	-0.168
207.11	-0.138
229.81	-0.103
255.64	-0.062
285.64	-0.012
CURVE 27*, ‡	
24.25	-0.325
32.16	-0.325
41.21	-0.323
51.03	-0.320
61.95	-0.314
73.29	-0.307
81.35	-0.300
89.49	-0.293
103.96	-0.278
119.59	-0.261
141.94	-0.232
166.55	-0.199
169.18	-0.195
189.39	-0.165
210.02	-0.134
229.86	-0.103
254.66	-0.063
284.34	-0.015
CURVE 28*, †, ‡	
30	0.000 x 10^{-3}
40	1.425
50	4.230
60	8.565
70	14.46
80	21.80
90	30.50
CURVE 28 (cont.)*, †, ‡	
100	40.38 x 10^{-3}
105	45.35
CURVE 29	
108.6	-0.271*
139.4	-0.2331
170.8	-0.1901*
198.9	-0.1490
223.5	-0.1351
237.8	-0.0891
261.4	-0.0520
285.0	-0.0135*
295.0	0.0180*
304.0	0.0180*
310.9	0.0308
312.0	0.0319*
CURVE 30‡	
300	0.012*
400	0.185*
500	0.364*
600	0.550*
700	0.743*
800	0.943*
900	1.151
1000	1.367
1100	1.594
1200	1.832
1300	2.090
CURVE 31*, †, ‡	
4	0.00000 x 10^{-3}
5	0.00029
6	0.00085
7	0.00172
8	0.00303
9	0.00489
10	0.00744
11	0.0109
12	0.0153
13	0.0210
14	0.0280
15	0.0368
CURVE 32*, †, ‡	
4	0.00000 x 10^{-3}
5	0.00029
6	0.00084
7	0.00169
8	0.00297
9	0.00479
10	0.00728
11	0.0106
12	0.0149
13	0.0205
14	0.0274
15	0.0359
CURVE 33	
298	0.008*
323	0.047
348	0.090*
373	0.128*
398	0.170*
423	0.213*
448	0.254
473	0.298*
498	0.336*
523	0.376
544	0.408
CURVE 34	
293	0.000
1148	1.377
CURVE 35*, †	
0	0.0 x 10^{-3}
20	0.0
40	2.0
60	10.0
CURVE 36*, ‡	
100	-0.289
104	-0.284
111	-0.276
114	-0.272
121	-0.264
122	-0.263
129	-0.254
CURVE 36 (cont.)*, ‡	
132	-0.250
137	-0.243
140	-0.240
146	-0.231
155	-0.218
157	-0.216
158	-0.215
168	-0.200
176	-0.188
179	-0.183
187	-0.172
189	-0.168
193	-0.163
196	-0.159
205	-0.144
208	-0.139
211	-0.135
219	-0.123
221	-0.119
230	-0.105
234	-0.097
239	-0.089
246	-0.077
249	-0.071
259	-0.056
270	-0.038
281	-0.019
291	-0.003
293	-0.000
301	0.014
303	0.016
313	0.033
324	0.053
335	0.070
345	0.088
CURVE 37	
295	0.004
473	0.320
CURVE 38*	
295	0.004
473	0.315
CURVE 39*	
295	0.003
473	0.308
CURVE 40	
291	-0.003*
533	0.371
723	0.723
1043	1.239
CURVE 41	
291	-0.004*
533	0.424
723	0.684
1043	1.222*
CURVE 42	
83	-0.293*
301	0.013*
496	0.341*
749	0.852
958	1.364
1081	1.671
1163	1.875
1207	2.018
CURVE 43*	
273	-0.031
313	0.031
CURVE 44*, †, ‡	
7.5	0.0000 x 10^{-3}
7.8	0.0005
8.0	0.0009
8.5	0.002
8.6	0.002
9.1	0.003
9.8	0.005
10.4	0.007
11.1	0.010
11.4	0.0116
11.9	0.014
12.2	0.016
12.5	0.018
13.3	0.024
CURVE 44 (cont.)*, †, ‡	
13.4	0.024 x 10^{-3}
13.5	0.024
14.9	0.025
16.3	0.038
17.2	0.0549
17.6	0.068
18.5	0.075
18.7	0.093
19.3	0.112
20.9	0.156
21.4	0.171
22.5	0.212
22.8	0.224
23.2	0.241
23.5	0.255
24.2	0.288
24.9	0.324
25.7	0.371
26.1	0.400
26.5	0.422
27.1	0.465
27.7	0.510
28.1	0.540
32.0	0.934
35.7	1.456
37.3	1.741
38.6	2.006
39.6	2.222
42.5	2.930
44.8	3.614
50.8	5.851
52.9	6.746
54.4	7.419
CURVE 45*, †, ‡	
6.4	0.0000 x 10^{-3}
6.9	0.0006
7.1	0.0009
7.4	0.0013
7.7	0.0017
8.6	0.0034
9.4	0.0053
9.6	0.0059
9.7	0.0062
9.9	0.0068
10.0	0.0071
10.3	0.0082

*Not shown in figure.
†This curve is here reported using the first given temperature as reference temperature at which $\Delta L/L_0 = 0$.
‡Author's data for coefficient of thermal expansion have been integrated by TPRC to obtain $\Delta L/L_0$.

DATA TABLE 12. THERMAL LINEAR EXPANSION OF COPPER Cu (continued)

T	$\Delta L/L_0$
CURVE 45 (cont.)*, †, ‡	
10.5	0.0090 x 10^{-3}
10.7	0.0098
10.8	0.010
11.1	0.012
11.4	0.013
12.3	0.018
15.6	0.047
16.2	0.054
16.4	0.057
23.8	0.287
24.1	0.300
24.8	0.336
25.1	0.354
26.3	0.430
31.2	0.863
40.3	2.395
43.5	3.243
47.5	4.555
49.1	5.151
50.8	5.818
54.6	7.526
CURVE 46*	
291	-0.004
573	0.509
773	0.894
944	1.249
1044	1.463
1144	1.684
CURVE 47‡	
383	0.150
393	0.167*
413	0.200
433	0.234*
453	0.267*
473	0.301*
493	0.335
513	0.370
523	0.387*
533	0.404
553	0.439
573	0.474
CURVE 48	
303	0.017*
403	0.192*
503	0.374
603	0.557*
703	0.751
803	0.955
CURVE 49*	
20	-0.3239
40	-0.3217
60	-0.3141
80	-0.3004
100	-0.2817
130	-0.2470
160	-0.2068
200	-0.1480
250	-0.0700
293	0.0000
300	0.0116
400	0.1828
500	0.3624
600	0.5487
700	0.7407
800	0.9385
CURVE 50	
293	0.000*
373	0.166
393	0.198*
413	0.231*
433	0.264
453	0.299*
473	0.333*
493	0.367
513	0.405
533	0.436*
553	0.471*
573	0.505*
593	0.541
613	0.575
633	0.607
CURVE 51*	
293	0.000
373	0.135
473	0.312
573	0.496
673	0.689
773	0.889
873	1.098
CURVE 52‡	
88	-0.295*
93	-0.290*
103	-0.279*
113	-0.268*
123	-0.257*
133	-0.244
143	-0.231*
153	-0.217*
163	-0.203*
173	-0.189*
183	-0.174*
193	-0.159*
203	-0.144
213	-0.129*
223	-0.113*
233	-0.098*
243	-0.082
253	-0.066*
263	-0.049*
273	-0.033*
283	-0.017*
293	0.000*
303	0.017*
313	0.033*
323	0.050*
333	0.067
343	0.084*
353	0.101*
363	0.118*
373	0.135*
383	0.152
393	0.169*
403	0.186*
413	0.204*
423	0.221*
CURVE 52 (cont.)‡	
433	0.238*
443	0.255
453	0.273*
463	0.290
473	0.308*
483	0.325*
493	0.343*
503	0.361*
513	0.378*
523	0.396*
533	0.415*
543	0.433*
553	0.452*
563	0.470
573	0.489*
CURVE 53‡	
90	-0.292*
100	-0.282*
110	-0.272*
120	-0.260*
130	-0.248*
140	-0.235*
150	-0.221*
160	-0.207*
170	-0.193*
180	-0.179
190	-0.164*
200	-0.149*
210	-0.133*
220	-0.118*
230	-0.102
240	-0.086*
250	-0.070*
260	-0.054*
270	-0.038*
280	-0.022*
290	-0.005*
300	0.012*
310	0.028*
320	0.045*
330	0.062*
373	0.136*
423	0.222
473	0.310*
523	0.400
CURVE 53 (cont.)‡	
573	0.491*
623	0.584
773	0.869
823	0.966*
873	1.064
923	1.165
973	1.267
1023	1.372
1073	1.479
1123	1.588
1173	1.699
1223	1.813
1273	1.929
1323	2.048
CURVE 54*	
293	0.00
373	0.16
473	0.32
573	0.48
673	0.66
773	0.83
873	1.00
CURVE 55*	
773	0.84
673	0.66
573	0.47
473	0.31
373	0.15
293	0.00
CURVE 56*, †, ‡	
0	0.00000 x 10^{-3}
1	0.00001
2	0.00007
3	0.0062
4	0.0004
5	0.0008
6	0.0015
7	0.0025
8	0.0040
9	0.0092
10	0.0091
CURVE 56 (cont.)*, †, ‡	
12	0.0182 x 10^{-3}
14	0.0329
15	0.0430
16	0.0554
18	0.0887
20	0.1356
22	0.1995
24	0.2854
25	0.3381
26	0.3984
CURVE 57*, †, ‡	
0	0.00000 x 10^{-3}
1	0.00001
2	0.00005
3	0.00015
4	0.0003
5	0.0007
6	0.0013
7	0.0021
8	0.0034
9	0.0055
10	0.0079
12	0.0159
14	0.0293
15	0.0385
16	0.0499
18	0.0810
20	0.126
22	0.187
24	0.274
25	0.326
26	0.386
28	0.730
30	0.711
CURVE 58*, †, ‡	
2.714	0.0000 x 10^{-3}
3.549	0.0001
4.060	0.0002
4.440	0.0004
4.871	0.0005
5.289	0.0007
5.672	0.0009
6.033	0.0012
6.466	0.0016

*Not shown in figure.

†This curve is here reported using the first given temperature as reference temperature at which $\Delta L/L_0 = 0$.

‡Author's data for coefficient of thermal expansion have been integrated by TPRC to obtain $\Delta L/L_0$.

DATA TABLE 12. THERMAL LINEAR EXPANSION OF COPPER Cu (continued)

CURVE 58 (cont.)*, †, ‡

T	$\Delta L/L_0$
6.931	0.0020 x 10^{-3}
7.376	0.0026
7.914	0.0034
8.553	0.0045
9.246	0.0061
9.988	0.0081
10.76	0.011
11.786	0.015
13.076	0.023
14.335	0.033
15.576	0.046
16.974	0.060
18.575	0.095
20.225	0.135
21.767	0.184
23.558	0.256
25.265	0.344
26.684	0.435
28.159	0.584
29.606	0.678

CURVE 59*, †, ‡

T	$\Delta L/L_0$
2.466	0.0000 x 10^{-4}
3.091	0.00008
3.649	0.00019
4.065	0.00028
4.418	0.00039
4.827	0.00055
5.250	0.00075
5.623	0.00097
5.966	0.0012
6.350	0.0015
6.767	0.0019
7.226	0.0024
7.781	0.0032
8.422	0.0043
9.109	0.0058
9.829	0.0078
10.660	0.011
11.676	0.015
12.837	0.022
14.028	0.031
15.225	0.043
16.551	0.060
18.079	0.086
19.744	0.124
21.424	0.175

CURVE 59 (cont.)*, †, ‡

T	$\Delta L/L_0$
23.021	0.236 x 10^{-4}
24.706	0.318
26.476	0.427
28.020	0.542
29.220	0.647
30.255	0.749
31.428	0.878
32.669	1.031
33.904	1.202

CURVE 60*, †, ‡

T	$\Delta L/L_0$
2.567	0.00000 x 10^{-3}
2.978	0.00004
3.387	0.00010
3.898	0.00021
4.411	0.00036
4.823	0.00051
5.249	0.00070
5.626	0.00092
5.969	0.0011
6.354	0.0015
6.771	0.0019
7.228	0.0024
7.780	0.0031
8.426	0.0042
9.119	0.0057
9.834	0.0076
10.659	0.0104
11.689	0.0149
12.874	0.0218
14.068	0.0311
15.269	0 0432
16.605	0.0608
18.137	0.0873
19.738	0.124
21.362	0.172
23.027	0.236
24.748	0.320
26.478	0.427
28.034	0.544
29.307	0.657

CURVE 61*, †, ‡

T	$\Delta L/L_0$
2.478	0.00000 x 10^{-3}
3.110	0.00007
3.659	0.00017

CURVE 61 (cont.)*, †, ‡

T	$\Delta L/L_0$
4.069	0.00026 x 10^{-3}
4.410	0.00037
4.815	0.00052
5.248	0.00071
5.630	0.00092
5.974	0.0011
6.350	0.0014
6.762	0.0018
7.226	0.0023
7.780	0.0031
8.415	0.0041
9.098	0.0056
9.811	0.0074
10.642	0.0102
11.675	0.0146
12.842	0.0213
14.036	0.0304
15.230	0.0423
16.560	0.0596
18.088	0.0857
19.676	0.122
21.345	0.171
23.006	0.235
24.705	0.318
26.488	0.428
28.026	0.544
29.228	0.650
30.324	0.759
31.523	0.893
32.785	1.050
33.988	1.219

CURVE 62*, †, ‡

T	$\Delta L/L_0$
2.444	0.00000 x 10^{-3}
3.065	0.00007
3.629	0.00017
4.060	0.00028
4.425	0.00040
4.828	0.00056
5.247	0.00076
5.621	0.00097
5.993	0.0012
6.374	0.0015
6.762	0.0019
7.221	0.0024
7.772	0.0032
8.410	0.0042
9.093	0.00567
9.806	0.00756

CURVE 62 (cont.)*, †, ‡

T	$\Delta L/L_0$
10.639	0.0103 x 10^{-3}
11.680	0.0148
12.855	0.0216
14.051	0.0308
15.244	0.0428
16.588	0.0604
18.127	0.0869
19.719	0.123
21.357	0.172
22.988	0.235
24.638	0.316
26.418	0.42566
27.949	0.541
29.170	0.649

CURVE 63*, †, ‡

T	$\Delta L/L_0$
2.696	0.00000 x 10^{-4}
3.450	0.00012
3.987	0.00024
4.429	0.00038
4.730	0.00050
5.278	0.00077
5.654	0.00099
5.998	0.0012
6.386	0.0015
6.893	0.0021
7.559	0.0029
8.356	0.0042
9.164	0.0059
9.919	0.0080
10.700	0.0106
11.769	0.0152
13.052	0.0228
14.278	0.0326
15.511	0.0454
16.895	0.0643
18.478	0.0931
20.135	0.133
21.831	0.187
23.706	0.266
25.611	0.369
27.308	0.485
28.751	0.603

CURVE 64*

T	$\Delta L/L_0$
293	0.000
573	0.477
673	0.657
773	0.859
873	1.066
973	1.279
1073	1.508
1173	1.746
1273	1.935

CURVE 65

T	$\Delta L/L_0$
115	-0.313
173	-0.224
273	-0.047*
293	0.000*
373	0.158*
473	0.395
573	0.658
673	0.921

CURVE 66*, †, ‡

T	$\Delta L/L_0$
10	0.000 x 10^{-3}
12	0.008
14	0.021
15	0.030
16	0.041
18	0.072
20	0.12
22	0.17
24	0.26
25	0.30
26	0.37
28	0.52
30	0.71
35	1.39
40	2.41
45	3.81
50	5.59
60	10.35
65	13.35
70	16.70
75	20.37
80	24.33

CURVE 67*, †

T	$\Delta L/L_0$
40	0.000
60	0.009
80	0.022
100	0.041
140	0.095
180	0.147

CURVE 68

T	$\Delta L/L_0$
296	0.012
298	0.017
344	0.097
388	0.180
449	0.296
451	0.296
520	0.426
550	0.501
580	0.545
581	0.548
607	0.600
649	0.700
718	0.822
779	0.949
843	1.074

*Not shown in figure.
†This curve is here reported using the first given temperature as reference temperature at which $\Delta L/L_0 = 0$.
‡Author's data for coefficient of thermal expansion have been integrated by TPRC to obtain $\Delta L/L_0$.

DATA TABLE 12. COEFFICIENT OF THERMAL LINEAR EXPANSION OF COPPER Cu

[Temperature, T, K; Coefficient of Expansion, α, 10^{-6} K^{-1}]

T	α
CURVE 3	
19.9	0.32
23.1	0.52
27.4	0.89
29.6	1.25
34.7	1.63
38.3	1.91
40.4	2.30
43.8	2.58
45.2	2.88
47.5	3.42
49.0	3.77
51.3	4.10
53.0	4.22
55.1	4.84
57.0	5.17
58.9	5.10
61.0	5.59
63.9	5.80
66.3	6.22
70.0	6.35
73.3	6.75
76.7	6.80
81.7	7.65
87.5	8.73
93.3	9.72
CURVE 11‡	
60	5.650
70	7.100*
80	8.400
90	9.600*
100	10.650
110	11.500*
120	12.200
280	16.700
295	16.799
CURVE 14‡	
10	0.02
15	0.10
20	0.27
25	0.50
30	0.86
40	2.00
CURVE 14 (cont.)‡	
50	3.62
60	5.34
70	6.88
80	8.24
90	9.40
100	10.48
110	11.42
120	12.18
130	12.82
140	13.30
150	13.74
160	14.08
170	14.40
180	14.68
190	14.92
200	15.12
CURVE 20‡	
31	0.00
64	3.51
103	8.75
150	12.15
200	14.03
249	15.09
300	15.81
350	16.21*
400	16.35*
450	16.49*
500	16.78*
550	16.95*
600	17.30*
650	17.67*
700	18.12*
750	18.70*
800	19.39*
850	20.19*
900	21.28*
950	22.66*
985	23.69*
CURVE 25‡	
20	0.2
30	1.0
40	2.2
CURVE 25 (cont.)‡	
50	3.7
70	7.1
100	10.8
150	13.8
200	15.3
250	16.1
300	16.4
CURVE 26‡	
25.25	0.55
34.53	1.58
43.49	2.78
51.55	4.12*
60.38	5.58*
69.48	6.92
79.65	8.32*
85.67	9.12
90.00	9.40*
94.75	10.00
105.43	10.95*
116.82	11.87
129.32	12.53
142.08	13.33
153.70	13.85
169.40	14.37
187.12	14.88
207.11	15.37
229.81	15.72
255.64	16.16
285.64	16.60
CURVE 27‡	
24.25	0.59
32.16	1.22
41.21	2.52
51.03	4.02
61.95	5.86
73.29	7.45
81.35	8.42*
89.49	9.43
103.96	10.87
119.59	11.93
141.94	13.26
166.55	14.18
CURVE 27 (cont.)‡	
169.18	14.35*
189.39	14.96
210.02	15.42
229.86	15.81*
254.66	16.18
284.34	16.65
CURVE 28‡	
30	0.75
40	2.10*
50	3.51*
60	5.16
70	6.63
80	8.05
90	9.36
100	10.40
105	10.75
CURVE 30*, ‡	
300	17.0
400	17.6
500	18.3
600	18.9
700	19.6
800	20.4
900	21.2
1000	22.1
1100	23.2
1200	24.5
1300	27.0
CURVE 31‡	
4	0.00160
5	0.00424
6	0.00688*
7	0.01062*
8	0.01549*
9	0.02169*
10	0.02940
11	0.03878*
12	0.05000*
13	0.06322*
14	0.07864*
15	0.09637
CURVE 32*, ‡	
4	0.00161
5	0.00418
6	0.00675
7	0.01041
8	0.01515
9	0.02119
10	0.02870
11	0.03784
12	0.04876
13	0.06164
14	0.07663
15	0.09390
CURVE 36‡	
100	10.51
104	10.23
111	12.25
114	11.19*
121	12.64
122	12.27*
129	13.15
132	12.86
137	13.91
140	13.40
146	14.19
155	13.97
157	14.56
158	13.43
168	14.69
176	15.00
179	14.02
187	15.77
189	14.02
193	15.33
196	16.01
205	16.37
208	15.11*
211	16.19
219	15.54
221	16.55
230	16.18
234	17.26
239	16.42
246	17.71
249	15.67
CURVE 36 (cont.)‡	
259	16.32
270	15.95
281	17.28
291	16.48
293	16.78
301	16.07*
303	16.69*
313	16.61*
324	17.20*
335	17.20*
345	17.20*
CURVE 44*, ‡	
7.5	0.016
7.8	0.017
8.0	0.020
8.5	0.020
8.6	0.021
9.1	0.025
9.8	0.030
10.4	0.038
11.1	0.047
11.4	0.050
11.9	0.058
12.2	0.062
12.5	0.064
13.3	0.077
13.4	0.094
13.5	0.082
14.9	0.097
16.3	0.146
17.2	0.154
17.6	0.182
18.5	0.224
18.7	0.215
19.3	0.240
20.9	0.309
21.4	0.327
22.5	0.402
22.8	0.414
23.2	0.429
23.5	0.474
24.2	0.474
24.9	0.553
25.7	0.594

* Not shown in figure.
‡ Author's data for coefficient of thermal expansion have been integrated by TPRC to obtain $\Delta L/L_0$.

DATA TABLE 12. COEFFICIENT OF THERMAL LINEAR EXPANSION OF COPPER Cu (continued)

CURVE 44 (cont.)*, ‡

T	α
26.1	0.655
26.5	0.646
27.1	0.716
27.7	0.760
28.1	0.787
32.0	1.233
35.7	1.589
37.3	1.968
38.6	2.113
39.6	2.193
42.5	2.692
44.8	3.258
50.8	4.198
52.9	4.325
54.4	4.656

CURVE 45*, ‡

T	α
6.4	0.010
6.9	0.013
7.1	0.015
7.4	0.014
7.7	0.015
8.6	0.022
9.4	0.027
9.6	0.029
9.7	0.030
9.9	0.031
10.0	0.034
10.3	0.037
10.5	0.038
10.7	0.044
10.8	0.047
11.1	0.049
11.4	0.051
12.3	0.060
15.6	0.113
16.2	0.136
16.4	0.151
23.8	0.470
24.1	0.442
24.8	0.577
25.1	0.604
26.3	0.671
31.2	1.096
40.3	2.270
43.5	3.034
47.5	3.524

CURVE 45 (cont.)*, ‡

T	α
49.1	3.926
50.8	3.926
54.6	4.875

CURVE 47*, ‡

T	α
383	16.63
393	16.68
413	16.80
433	16.87
453	16.93
473	16.99
493	17.07
513	17.18
523	17.22
533	17.35
553	17.40
573	17.52

CURVE 52‡

T	α
88	9.440*
93	9.870
103	10.70
113	11.48
123	12.19
133	12.81*
143	13.35
153	13.81*
163	14.20
173	14.53
183	14.80
193	15.03*
203	15.25
213	15.45
223	15.64
233	15.82
243	15.99
253	16.15*
263	16.30
273	16.43*
283	16.54
293	16.63
303	16.71*
313	16.78*
323	16.84*
333	16.89*
343	16.94*

CURVE 52 (cont.)‡

T	α
353	16.99*
363	17.04*
373	17.08*
383	17.11*
393	17.13*
403	17.16*
413	17.19*
423	17.22*
433	17.26*
443	17.31*
453	17.37*
463	17.44*
473	17.52*
483	17.61*
493	17.71*
503	17.82*
513	17.95*
523	18.10*
533	18.27*
543	18.46*
553	18.66*
563	18.86*
573	19.06*

CURVE 53‡

T	α
90	9.38*
100	10.40*
110	11.27
120	12.01*
130	12.64*
140	13.18*
150	13.65
160	14.06
170	14.41*
180	14.71
190	14.97*
200	15.20*
210	15.40*
220	15.58
230	15.75*
240	15.91
250	16.07
260	16.22*
270	16.36
280	16.49
290	16.61
300	16.72

CURVE 53 (cont.)‡

T	α
310	16.82*
320	16.90*
330	16.97*
373	17.18*
423	17.45*
473	17.75*
523	18.10*
573	18.45*
623	18.80*
773	19.15*
823	19.52*
873	19.91*
923	20.31*
973	20.72*
1023	21.13*
1073	21.56*
1123	22.01*
1173	22.50*
1223	23.02*
1273	23.54*
1323	24.10*

CURVE 56*, ‡

T	α
0	0.00000
1	0.00031
2	0.00080
3	0.0016
4	0.0030
5	0.00517
6	0.00820
7	0.0123
8	0.0178
9	0.0249
10	0.0335
12	0.0573
14	0.0901
15	0.1117
16	0.1364
18	0.1963
20	0.273
22	0.366
24	0.493
25	0.562
26	0.643

CURVE 57‡

T	α
0	0.00000
1	0.00023*
2	0.00062*
3	0.00133*
4	0.00252*
5	0.00436*
6	0.00701*
7	0.1066*
8	0.1550*
9	0.0217*
10	0.0296*
12	0.0511*
14	0.0822*
15	0.1020*
16	0.1259*
18	0.1851*
20	0.2633*
22	0.3637*
24	0.4872
25	0.5578*
26	0.6345*
28	0.8074
30	1.007*

CURVE 58*, ‡

T	α
2.714	0.0012
3.549	0.002
4.060	0.0026
4.440	0.0034
4.871	0.0044
5.289	0.0056
5.672	0.0066
6.033	0.0073
6.466	0.0092
6.931	0.011
7.376	0.0133
7.914	0.0158
8.553	0.0199
9.246	0.0247
9.988	0.0307
10.76	0.0381
11.786	0.0497
13.076	0.0699
14.335	0.092
15.576	0.1201
16.974	0.1558
18.575	0.2107

CURVE 58 (cont.)*, ‡

T	α
20.225	0.2763
21.767	0.3524
23.558	0.461
25.265	0.571
26.684	0.703
28.159	0.832
29.606	0.971

CURVE 59*, ‡

T	α
2.466	0.0010
3.091	0.0014
3.649	0.0023
4.065	0.0028
4.418	0.0034
4.827	0.0043
5.250	0.0053
5.623	0.0063
5.966	0.0075
6.350	0.0088
6.767	0.0105
7.226	0.0126
7.781	0.0154
8.422	0.0193
9.109	0.024
9.829	0.0298
10.660	0.0382
11.676	0.049
12.837	0.0668
14.028	0.0880
15.225	0.113
16.551	0.148
18.079	0.195
19.744	0.260
21.424	0.339
23.021	0.432
24.706	0.544
26.476	0.683
28.020	0.815
29.220	0.934
30.255	1.033
31.428	1.167
32.669	1.292
33.904	1.480

* Not shown in figure.
‡ Author's data for coefficient of thermal expansion have been integrated by TPRC to obtain $\Delta L/L_0$.

DATA TABLE 12. COEFFICIENT OF THERMAL LINEAR EXPANSION OF COPPER Cu (continued)

T	α
CURVE 60*, ‡	
2.567	0.0010
2.978	0.0011
3.387	0.0018
3.898	0.0025
4.411	0.0033
4.823	0.0040
5.249	0.0052
5.626	0.0063
5.969	0.0072
6.354	0.0087
6.771	0.0103
7.228	0.0124
7.780	0.0153
8.426	0.0191
9.119	0.0237
9.834	0.0299
10.659	0.0379
11.689	0.0492
12.874	0.0669
14.068	0.0883
15.269	0.1141
16.605	0.1486
18.137	0.1972
19.738	0.2582
21.362	0.3368
23.027	0.4317
24.748	0.5480
26.478	0.6840
28.034	0.8230
29.307	0.5490
CURVE 61*, ‡	
2.478	0.0010
3.110	0.0013
3.659	0.0021
4.069	0.0026
4.410	0.0035
4.815	0.0039
5.248	0.0050
5.630	0.0061
5.974	0.0070
6.350	0.0085
6.762	0.0095
7.226	0.0122
7.780	0.0149
8.415	0.0187
9.098	0.0232

T	α
CURVE 61 (cont.)*, ‡	
9.811	0.0293
10.642	0.0374
11.675	0.0482
12.842	0.0661
14.036	0.0871
15.230	0.1125
16.560	0.1468
18.088	0.1948
19.676	0.2589
21.345	0.3347
23.006	0.4326
24.705	0.547
26.488	0.687
28.026	0.822
29.228	0.938
30.324	1.051
31.523	1.178
32.785	1.326
33.988	1.475
CURVE 62*, ‡	
2.444	0.0009
3.065	0.0013
3.629	0.0023
4.060	0.0030
4.425	0.0037
4.828	0.0040
5.247	0.0055
5.621	0.0058
5.993	0.0078
6.374	0.0085
6.762	0.0104
7.221	0.0124
7.772	0.0151
8.410	0.0181
9.093	0.0238
9.806	0.0291
10.639	0.0378
11.680	0.0485
12.855	0.0668
14.051	0.0873
15.244	0.114
16.588	0.148
18.127	0.197
19.719	0.260
21.357	0.338
22.988	0.434

T	α
CURVE 62 (cont.)*, ‡	
24.638	0.544
26.418	0.689
27.949	0.824
29.170	0.937
CURVE 63*, ‡	
2.696	0.0013
3.450	0.0020
3.987	0.0025
4.429	0.0036
4.730	0.0043
5.278	0.0055
5.654	0.0066
5.998	0.0076
6.386	0.0087
6.893	0.0111
7.559	0.0143
8.356	0.0185
9.164	0.0238
9.919	0.0303
10.700	0.0370
11.769	0.0497
13.052	0.0689
14.278	0.0908
15.511	0.1173
16.895	0.1553
18.478	0.2081
20.135	0.2756
21.831	0.3615
23.706	0.4758
25.611	0.612
27.308	0.754
28.751	0.885
CURVE 66‡	
10	0.03
12	0.05*
14	0.08*
15	0.10*
16	0.13*
18	0.18*
20	0.25*
22	0.34*
24	0.47*
25	0.55*
26	0.69*

T	α
CURVE 66 (cont.)‡	
28	0.84*
30	1.04*
35	1.69*
40	2.39*
45	3.20
50	3.91*
60	5.62*
65	6.36
70	7.05*
75	7.62
80	8.25*

* Not shown in figure.
‡ Author's data for coefficient of thermal expansion have been integrated by TPRC to obtain $\Delta L/L_0$.

FIGURE AND TABLE NO. 13R. RECOMMENDED VALUES FOR THERMAL LINEAR EXPANSION OF DYSPROSIUM Dy

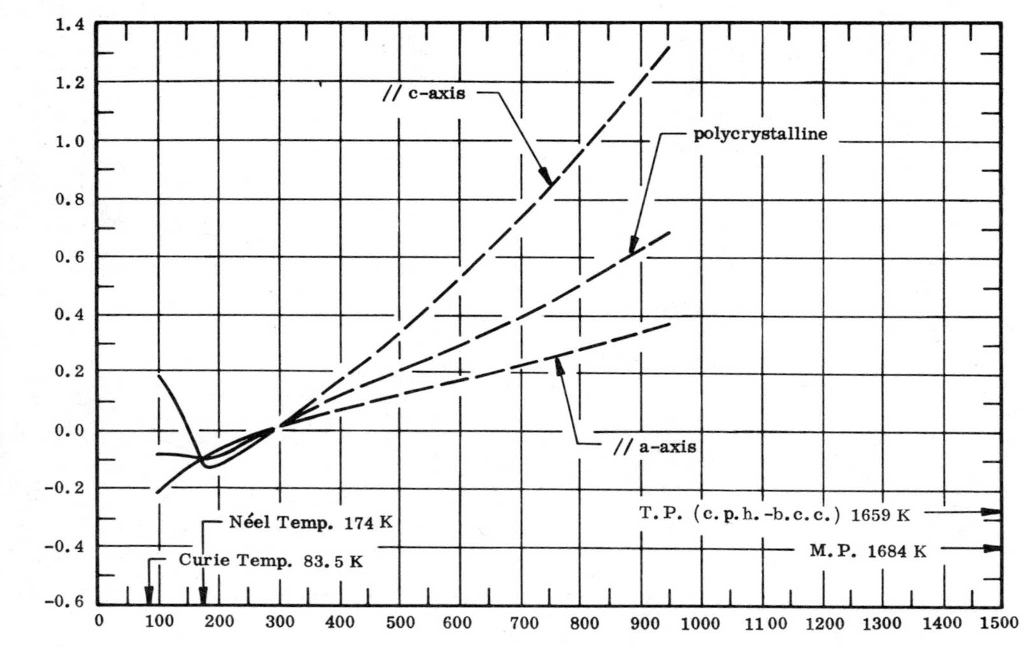

RECOMMENDED VALUES

[Temperature, T, K; Linear Expansion, $\Delta L/L_0$, %; α, K^{-1}]

T	// a-axis $\Delta L/L_0$	// c-axis $\Delta L/L_0$	polycrystalline $\Delta L/L_0$	polycrystalline $\alpha \times 10^6$
100	-0.229	0.185	-0.091	
150	-0.143	0.005	-0.097	
174†	-0.110	-0.112	-0.111	
200	-0.077	-0.128	-0.094	
293	0.000	0.000	0.000	9.6
400	0.069‡	0.170‡	0.103‡	9.2
500	0.120‡	0.339‡	0.193‡	9.0
600	0.166‡	0.520‡	0.284‡	9.3
700	0.212‡	0.720‡	0.381‡	10.4
800	0.266‡	0.939‡	0.490‡	11.8
900	0.332‡	1.182‡	0.615‡	13.8
950	0.372‡	1.314‡	0.686‡	14.7

† Anomaly due to Neel temperature.
‡ Provisional values.

REMARKS

The tabulated values are considered accurate to within ± 3% at temperatures below 293 K and ± 7% above. The tabulated values for polycrystalline dysprosium are calculated from single crystal data and are considered to about ± 5% below 293 K and about ± 10%. Experimental data for polycrystalline clysprosium are in fair agreement below 293 K and slightly lower above. These values can be represented approximately by the following equations:

// a-axis: $\Delta L/L_0 = -0.320 + 1.546 \times 10^{-3} T - 1.855 \times 10^{-6} T^2 + 1.047 \times 10^{-9} T^3$ $(293 < T < 950)$

// c-axis: $\Delta L/L_0 = -0.442 + 1.520 \times 10^{-3} T - 2.114 \times 10^{-7} T^2 + 5.852 \times 10^{-10} T^3$ $(293 < T < 950)$

polycrystalline: $\Delta L/L_0 = -0.360 + 1.538 \times 10^{-3} T - 1.307 \times 10^{-6} T^2 + 8.931 \times 10^{-10} T^3$ $(293 < T < 950)$

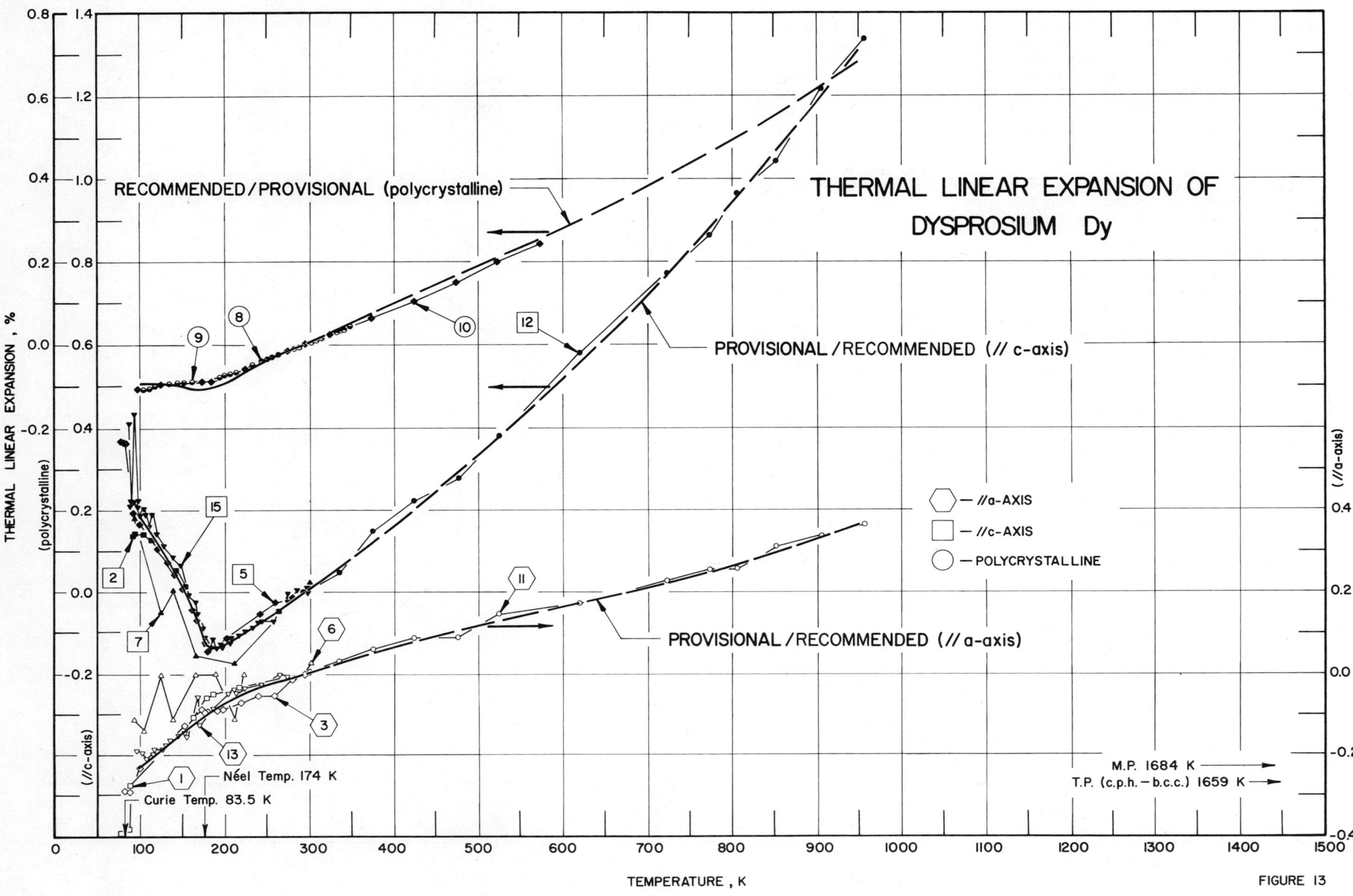
THERMAL LINEAR EXPANSION OF
DYSPROSIUM Dy
RECOMMENDED / PROVISIONAL (polycrystalline)
PROVISIONAL / RECOMMENDED (// c-axis)
PROVISIONAL / RECOMMENDED (// a-axis)
— //a-AXIS
— //c-AXIS
— POLYCRYSTALLINE
M.P. 1684 K
T.P. (c.p.h. – b.c.c.) 1659 K
Néel Temp. 174 K
Curie Temp. 83.5 K
THERMAL LINEAR EXPANSION, %
(polycrystalline)
(// c-axis)
(// a-axis)
TEMPERATURE, K

FIGURE 13

SPECIFICATION TABLE 13. THERMAL LINEAR EXPANSION OF DYSPROSIUM Dy

Cur. No.	Ref. No.	Author(s)	Year	Method Used	Temp. Range, K	Name and Specimen Designation	Composition (weight percent), Specifications, and Remarks
1	157	Clark, A.E., De Savage, B.F., and Bozorth, R.	1965	S	79-273	1	Single crystal plate cut from large crystal loaned by Dr. F.H. Spedding and Dr. S. Legvold of Iowa State Univ; plate contains a [110] axis and b[1$\bar{1}$0] axis; expansion measured along a-axis; reported error ±15%; zero-point correction of -0.01% determined by graphical extrapolation.
2	157	Clark, A.E., et al.	1965	S	91-272	2	Specimen similar to above, except contains a[110]-axis and c[001]-axis; expansion measured along c-axis; zero-point correction of -0.031% determined by graphical extrapolation.
3	158	Finkel', V.A. and Vorob'ev, V.V.	1967	X	73-300		99.2 Dy; polycrystalline specimen; expansion measured along a-axis; lattice parameter reported at 293 K is 3.5913 Å.
4*	158	Finkel', V.A. and Vorob'ev, V.V.	1967	X	76-300		The above specimen; expansion measured along b-axis; lattice parameter reported at 293 K is 6.2191 Å.
5	158	Finkel', V.A. and Vorob'ev, V.V.	1967	X	78-300		The above specimen; expansion measured along c-axis; lattice parameter reported at 293 K is 5.6506 Å.
6	159	Banister, J.R., Legvold, S., and Spedding, F.H.	1954	X	49-301		0.07 Y, 0.01 Ho, Ca detectable spectrographically, traces of Fe, Ta, and Mg; expansion measured along a-axis; lattice parameter reported at 293 K is 3.595 ± 0.002 Å.
7	159	Banister, J.R., et al.	1954	X	49-301		The above specimen; expansion measured along c-axis; lattice parameter reported at 293 K is 5.6557 Å.
8	79	Barson, F., Legvold, S., and Spedding, F.H.	1953	L	341-104		Prepared from the salt by bomb reduction with calcium; cast in tantalum crucible and turned to cylinder about 0.3 cm diameter; ends of rod machined flat, subjected to prolonged annealing at 823 K; strong spectroscopic lines of calcium and weak lines of magnesium, Ho < 0.1%, Y < 0.07%, and traces of tantalum; expansion measured with decreasing temperature on run 1; zero-point correction of -0.015% determined by graphical interpolation.
9	79	Barson, F., et al.	1953	L	103-349		The above specimen; expansion measured with increasing temperature on run 2; zero-point correction of -0.0155% determined by graphical interpolation.
10	81	Trombe, F. and Foex, M.	1953	L	97-573		Distilled specimen containing 0.5 to 1.0 other rare earths; expansion measured in hydrogen atmosphere below room temperature; in argon atmosphere above room temperature; zero-point correction of -0.02% determined by graphical interpolation.
11	82	Hanak, J.J.	1959	X	293-958		≤0.1 each Ta, Tb, < 0.05 Ca, < 0.03 Si, ≤0.02 each Ho, Er, < 0.01 Ni; wire cut from sheet prepared by Ames Laboratory; hexagonal close packed structure; expansion measured along a-axis; lattice parameter of 3.5928 Å at 293 K taken from their smooth values.
12	82	Hanak, J.J.	1959	X	293-958		The above specimen; expansion measured along c-axis; lattice parameter of 5.656 Å at 293 K taken from their smooth values.
13	160	Darnell, F.T.	1963	X	29-296		Single crystal containing total of 0.1 impurities; platelet approximately 0.1 mm thick and 1 mm^2 area, face the (001) plane; below 86 K structure is orthorhombic, above 86 K hexagonal close-packed; measured along a-axis; reported error in measurement of lattice constant ±0.001 Å; lattice parameter reported at 283 K is 3.5943 Å; 3.5944 Å at 293 K determined by graphical interpolation.

* Not shown in figure.

SPECIFICATION TABLE 13. THERMAL LINEAR EXPANSION OF DYSPROSIUM Dy (continued)

Cur. No.	Ref. No.	Author(s)	Year	Method Used	Temp. Range, K	Name and Specimen Designation	Composition (weight percent), Specifications, and Remarks
14*	160	Darnell, F. T.	1963	X	29-85		The above specimen; expansion measured along b-axis; this curve is here reported using the first given temperature as reference temperature at which $\Delta L/L_0 = 0$; lattice parameter reported at 293 K is 6.1773 Å.
15	160	Darnell, F. T.	1963	X	9-298		The above specimen; measured along c-axis; lattice parameter reported at 293 K is 5.6540 Å.

* Not shown in figure.

DATA TABLE 13. THERMAL LINEAR EXPANSION OF DYSPROSIUM Dy

[Temperature, T, K; Linear Expansion, $\Delta L/L_0$, %]

T	$\Delta L/L_0$
CURVE 1	
79	-0.393
87	-0.381
88	-0.274
145	-0.152*
162	-0.108
179	-0.058
186	-0.048
218	-0.030
273	-0.007*
CURVE 2	
91	0.140
94	0.148
104	0.143
114	0.132
134	0.086*
143	0.055
153	0.017
163	-0.031*
172	-0.091*
176	-0.029*
181	-0.136*
191	-0.135*
199	-0.129*
209	-0.121*
218	-0.109*
226	-0.098*
236	-0.085*
244	-0.062*
254	-0.058*
264	-0.046
272	-0.031*
CURVE 3	
73	-0.846*
76	-0.858*
81	-0.872*
82	-0.287
89	-0.292
100	-0.234
113	-0.206*
123	-0.181*
132	-0.159*
139	-0.153*
CURVE 3 (cont.)	
151	-0.125
160	-0.111*
168	-0.100*
172	-0.089
176	-0.095
180	-0.092
185	-0.092*
191	-0.089
199	-0.084
220	-0.070
240	-0.056
259	-0.050
280	-0.014
293	0.000
300	0.008
CURVE 4*	
76	-0.156
82	-0.117
82	-0.257
87	-0.254
100	-0.219
109	-0.191
120	-0.174
129	-0.153
140	-0.130
150	-0.109
159	-0.088
166	-0.082
172	-0.077
180	-0.074
186	-0.079
191	-0.068
200	-0.063
220	-0.048
241	-0.032
260	-0.019
281	-0.011
293	0.000
300	0.002
CURVE 5	
78	0.370
81	0.366
CURVE 5 (cont.)	
83	0.363
84	0.211*
91	0.195
100	0.168
110	0.136*
120	0.106
131	0.078
140	0.042
150	0.003
161	-0.041
166	-0.067
171	-0.097*
176	-0.133*
180	-0.143
192	-0.126*
201	-0.111
220	-0.089*
241	-0.057
260	-0.023
280	-0.011
293	0.000*
300	0.000*
CURVE 6	
	Å
49	3.584 (orth.)
76	3.583
	$\Delta L/L_0$
93	-0.111
105	-0.139
123	0.000
139	-0.111
166	0.000
190	0.000
211	-0.111
233	0.000
300	0.000*
301	0.028*
CURVE 7	Å
49	5.668* (orth.)
76	5.673*
	$\Delta L/L_0$
93	0.181
105	0.146*
123	-0.049
139	0.005
166	-0.155
190	-0.225*
211	-0.172
233	-0.190*
300	0.022
301	-0.119
CURVE 8	
341.4	0.0372
339.9	0.0364*
338.7	0.0355
337.4	0.0346*
336.0	0.0337*
334.0	0.0320*
331.8	0.0302
329.7	0.0285*
327.5	0.0268*
325.3	0.0250*
323.1	0.0233*
320.8	0.0216*
317.5	0.0190*
314.2	0.0164
311.0	0.0138*
307.7	0.0112
304.3	0.0086*
300.9	0.0061
297.4	0.0035*
294.1	0.0009*
290.9	-0.0017*
287.5	-0.0042
284.2	-0.0068*
281.0	-0.0094
277.5	-0.0119*
274.3	-0.0145*
270.9	-0.0170*
267.6	-0.0196*
CURVE 8 (cont.)	
264.2	-0.0221
260.8	-0.0246*
257.3	-0.0271
253.9	-0 0297*
250.5	-0.0322
247.1	-0.0347*
243.8	-0.0372
240.5	-0.0397*
237.0	-0.0422*
233.5	-0.0447*
230.1	-0.0472
226.6	-0.0497*
223.3	-0.0521*
220.1	-0.0546*
216.7	-0.0571*
213.7	-0.0595*
210.1	-0.0620
206.8	-0.0644*
203.5	-0.0669*
200.2	-0.0693*
196.8	-0.0718*
193.5	-0.0742*
190.1	-0.0766*
186.6	-0.0790*
183.0	-0.0814*
179.1	-0.0838*
169.6	-0.0854*
158.6	-0.0852
151.6	-0.0859*
147.7	-0.0867*
144.1	-0.0873*
141.6	-0.0880*
138.7	-0.0888*
136.4	-0.0895*
133.9	-0.0902*
129.7	-0.0917*
127.5	-0.0924
123.6	-0.0939*
119.4	-0.0953*
116.0	-0.0968*
112.2	-0.0982*
110.1	-0.0989*
108.4	-0.0996*
106.4	-0.1003*
104.2	-0.1010
CURVE 9	
103.3	-0.1053
104.5	-0.1046*
105.7	-0.1038*
107.2	-0.1031*
108.6	-0.1024*
111.4	-0.1009
114.7	-0.0995*
118.7	-0.0981
122.4	-0.0966*
126.3	-0.0951*
130.5	-0.0937*
134.9	-0.0922
140.1	-0.0908*
143.3	-0.0900
146.7	-0.0893*
149.1	-0.0886*
150.9	-0.0878
155.6	-0.0871*
161.1	-0.0864
169.9	-0.0865
172.0	-0.0863*
175.5	-0.0873*
177.9	-0.0866*
179.3	-0.0858*
181.9	-0.0842*
184.3	-0.0826*
186.5	-0.0810*
189.6	-0.0786*
193.9	-0.0753
197.3	-0.0729*
200.5	-0.0705
203.9	-0.0681*
207.2	-0.0656
210.4	-0.0632*
213.6	-0.0607
217.0	-0.0582*
220.2	-0.0558*
223.5	-0.0533*
226.8	-0.0508*
229.9	-0.0483*
233.4	-0.0458
236.7	-0.0433*
240.2	-0.0408*
243.5	-0.0384*
246.6	-0.0358*
250.0	-0.0333*

* Not shown in figure.

DATA TABLE 13. THERMAL LINEAR EXPANSION OF DYSPROSIUM Dy (continued)

T	$\Delta L/L_0$
CURVE 9 (cont.)	
253.3	-0.0308*
256.6	-0.0283*
259.9	-0.0258*
263.3	-0.0233*
269.7	-0 0182*
272.9	-0.0157*
276.2	-0.0131*
279.5	-0.0106*
282.8	-0.0080*
286.0	-0.0055*
289.3	-0.0029*
292.6	-0.0003*
295.9	0.0023*
299.2	0.0048*
303.6	0.0082*
306.9	0.0107*
310.2	0.0134*
313.4	0.0160*
316.8	0.0186*
320.0	0.0212*
323.2	0.0237*
328.1	0.0281*
332.0	0.0315*
335.0	0.0341*
337.7	0.0367*
340.7	0.0393*
343.5	0.0419*
346.6	0.0444*
349.2	0.0470
CURVE 10	
97	-0.102*
123	-0.091*
173	-0.083*
184	-0.083*
223	-0.054*
273	-0.017*
293	0.000*
323	0.024*
373	0.065
423	0.106
473	0.155
523	0.201
573	0.243

T	$\Delta L/L_0$
CURVE 11	
293	0.000*
333	0.032
375	0.060
423	0.088
477	0.088
525	0.144
620	0.172
722	0.227
773	0.255
806	0.255
851	0.311
906	0.339
958	0.366
CURVE 12	
293	0.000
333	0.049
375	0.155
423	0.225
477	0.278
525	0.384
620	0.579
722	0.773
773	0.862
806	0.968
851	1.039
906	1.215
958	1.339
CURVE 13	
	Å
29	3.602 (orth.)
31	3.601
40	3.600
51	3.600
55	3.599
61	3.598
77	3.597
77	3.596
84	3.596
	$\Delta L/L_0$
96	-0.186
98	-0.189*
101	-0.195

T	$\Delta L/L_0$
CURVE 13 (cont.)	
109	-0.209
115	-0.198
116	-0.186
127	-0.186
130	-0.175
134	-0.161
136	-0.161*
144	-0.150
145	-0.142*
151	-0.142
155	-0.153
169	-0.056
170	-0.125
186	-0.081
203	-0.047
211	-0.036
213	-0.045
224	-0.033
242	-0.025
265	0.000
274	-0.003
283	-0.003
296	0.000*
CURVE 14*	
	Å
29	6.177 (orth.)
31	6.176
42	6.176
52	6.178
55	6.180
61	6.182
79	6.182
79	6.184
85	6.183
CURVE 15	
9	5.676 (orth.)
16	5.676
23	5.676
26	5.677
32	5.676
32	5.676
34	5.676
39	5.677
51	5.677

T	Å
CURVE 15 (cont.)	
59	5.678
72	5.677
76	5.678
76	5.677
81	5.677
81	5.666
	$\Delta L/L_0$
88	0.412
88	0.216
89	0.225
92	0.437
92	0.225
96	0.211
98	0.223
100	0.190
104	0.205
107	0.193
110	0.161
113	0.188*
114	0.193
120	0.144
128	0.119
139	0.087
149	0.064
149	0.059*
158	-0.005
166	-0.023
169	-0.051
174	-0.086
177	-0.122
178	-0.116
183	-0.131
186	-0.115
190	-0.131
197	-0.127
198	-0.132
206	-0.122
208	-0.111
219	-0.104
224	-0.097
234	-0.083
240	-0.076
245	-0.070
259	-0.053
274	-0.001
275	-0.023
286	0.002

T	$\Delta L/L_0$
CURVE 15 (cont.)	
298	-0.001
298	0.011

* Not shown in figure.

FIGURE AND TABLE NO. 14R. RECOMMENDED VALUES FOR THERMAL LINEAR EXPANSION OF ERBIUM Er

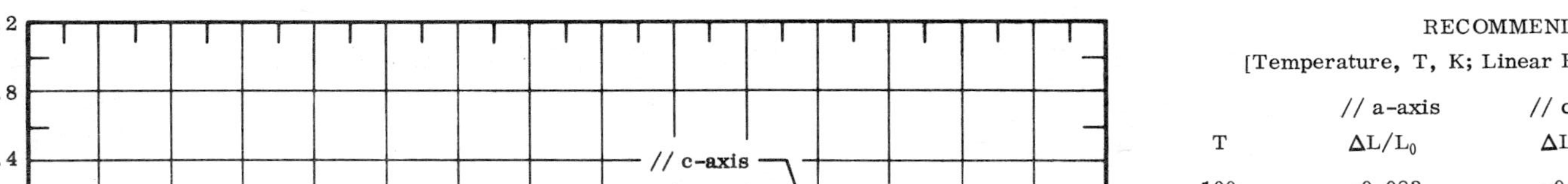

RECOMMENDED VALUES

[Temperature, T, K; Linear Expansion, $\Delta L/L_0$, %; α, K^{-1}]

T	// a-axis $\Delta L/L_0$	// c-axis $\Delta L/L_0$	polycrystalline $\Delta L/L_0$	polycrystalline $\alpha \times 10^6$
100	-0.083	-0.360	-0.175	8.2
200	-0.040	-0.160	-0.080	8.9
250	-0.025	-0.080	-0.043	9.2
293	0.000	0.000	0.000	9.4
400	0.063†	0.181†	0.102†	9.9
500	0.126†	0.362†	0.205†	10.5
600	0.196†	0.550†	0.314†	11.2
700	0.274†	0.741†	0.430†	11.9
800	0.364†	0.931†	0.553†	12.7
900	0.466†	1.125†	0.686†	14.0
1000	0.584†	1.332†	0.833†	15.7
1100	0.720†	1.563†	1.001†	17.8
1200	0.879†	1.828†	1.195†	20.9

† Provisional values.

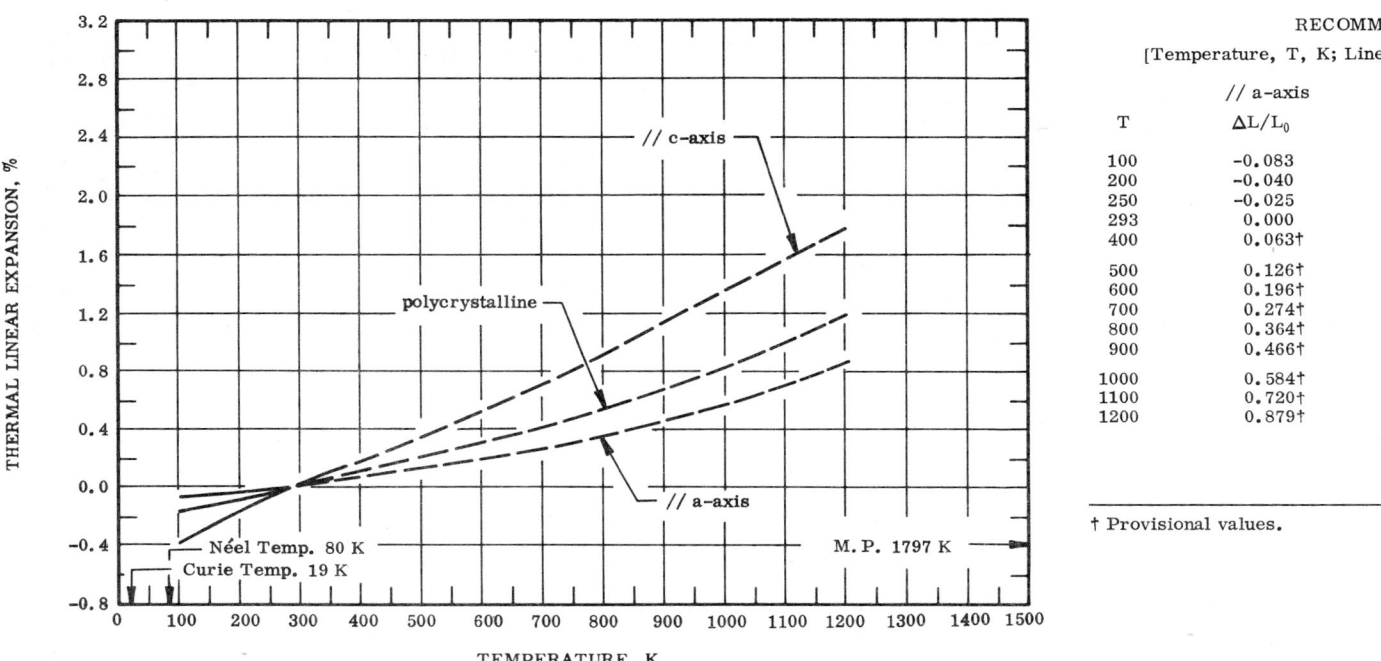

REMARKS

The tabulated values are considered accurate to within ± 3% at temperatures below 293 K and ± 7% above. These values can be represented approximately by the following equations:

// a-axis: $\Delta L/L_0 = 5.674 \times 10^{-4} (T - 293) + 1.286 \times 10^{-7} (T - 293)^2 + 3.279 \times 10^{-10} (T - 293)^3$ $(293 < T < 745)$

$\Delta L/L_0 = 0.313 + 8.846 \times 10^{-4} (T - 745) + 5.732 \times 10^{-7} (T - 745)^2 + 4.700 \times 10^{-10} (T - 745)^3$ $(745 < T < 1200)$

// c-axis: $\Delta L/L_0 = 1.614 \times 10^{-3} (T - 293) + 7.812 \times 10^{-7} (T - 293)^2 - 6.777 \times 10^{-10} (T - 293)^3$ $(293 < T < 745)$

$\Delta L/L_0 = 0.826 + 1.904 \times 10^{-3} (T - 745) - 1.377 \times 10^{-7} (T - 745)^2 + 1.730 \times 10^{-9} (T - 745)^3$ $(745 < T < 1200)$

polycrystalline: $\Delta L/L_0 = 9.161 \times 10^{-4} (T - 293) + 3.519 \times 10^{-7} (T - 293)^2 - 1.804 \times 10^{-11} (T - 293)^3$ $(293 < T < 745)$

$\Delta L/L_0 = 0.484 + 1.223 \times 10^{-3} (T - 745) + 3.274 \times 10^{-7} (T - 745)^2 + 9.188 \times 10^{-10} (T - 745)^3$ $(745 < T < 1200)$

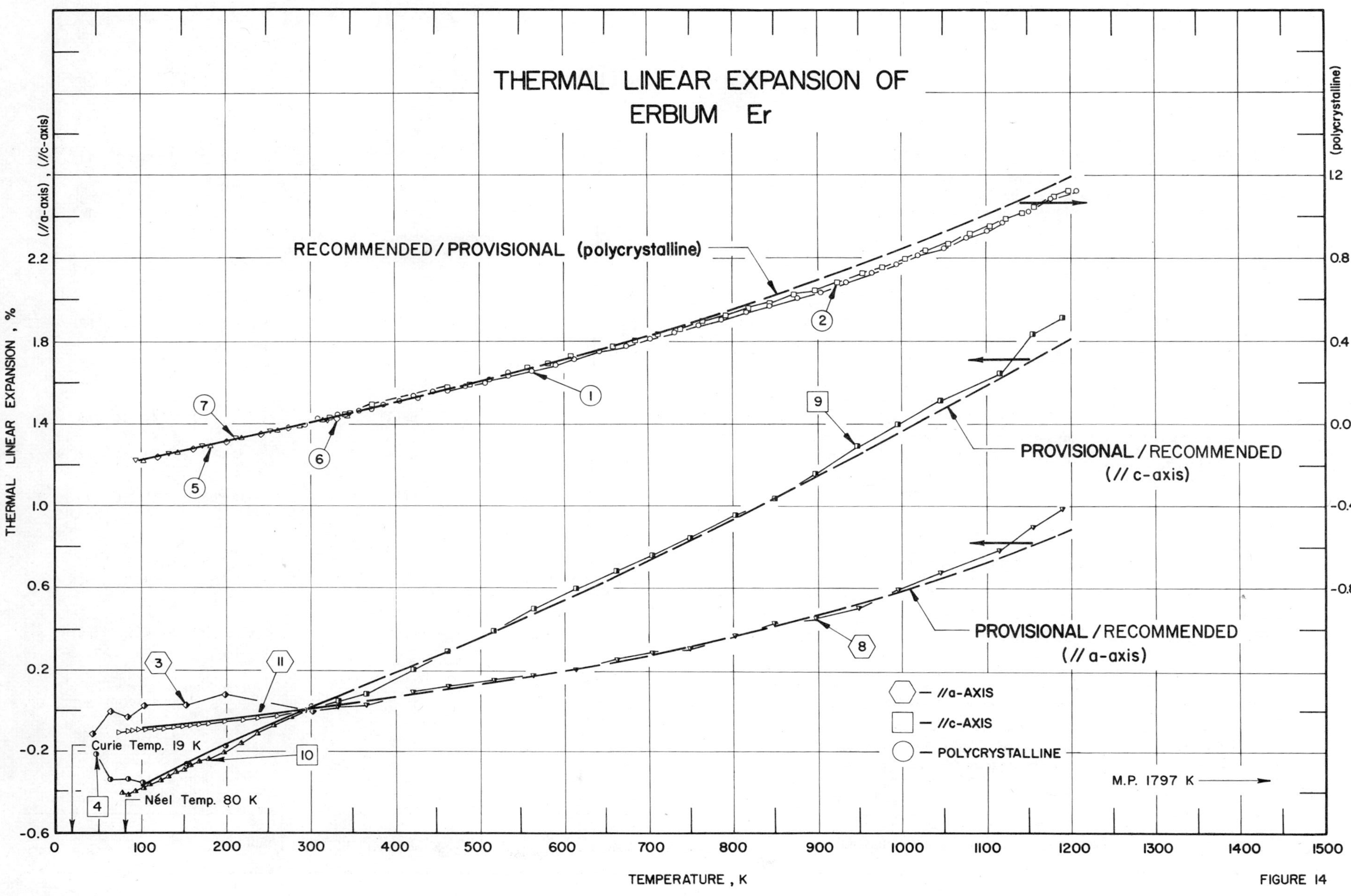
THERMAL LINEAR EXPANSION OF
ERBIUM Er
RECOMMENDED/PROVISIONAL (polycrystalline)
PROVISIONAL/RECOMMENDED
(// c-axis)
PROVISIONAL/RECOMMENDED
(// a-axis)
//a-AXIS
//c-AXIS
POLYCRYSTALLINE
Curie Temp. 19 K
Néel Temp. 80 K
M.P. 1797 K
THERMAL LINEAR EXPANSION, %
(//a-axis), (//c-axis)
(polycrystalline)
TEMPERATURE, K
1
2
3
4
5
6
7
8
9
10
11

FIGURE 14

SPECIFICATION TABLE 14. THERMAL LINEAR EXPANSION OF ERBIUM Er

Cur. No.	Ref. No.	Author(s)	Year	Method Used	Temp. Range, K	Name and Specimen Designation	Composition (weight percent), Specifications, and Remarks
1	78	Barson, F., Legvold, S. and Spedding, F. H.	1956	L	299-1207		0.07 Ca, 0.04 Si, 0.03 Fe, 0.01 each Ho, Yb, 0.0095 C, 0.005 Dy, 0.003 N_2, 0.002 Th; specimen 0.6 cm diameter, 6 cm long; fluorides of erbium bomb reduced with calcium metal to produce compact metallic sample, then vacuum cast, turned to shape; expansion measured with increasing temperature.
2	78	Barson, F., et al.	1956	L	1199-324		The above specimen; expansion measured with decreasing temperature.
3	159	Banister, J.R., Legvold, S. and Spedding, F. H.	1954	X	43-301		< 0.02 each Ho, Y, traces of Ca, Fe, Mg and Si; Cu-α radiation employed; hexagonal, closed packed structure; expansion measured along a -axis; lattice parameter reported at 301 K is 3.562 Å; 3.562 Å at 293 K determined by graphical interpolation.
4	159	Banister, J.R., et al.	1954	X	43-301		The above specimen; expansion measured along c -axis; lattice parameter reported at 301 K is 5.602 Å; 5.601 Å at 293 K determined by graphical interpolation.
5	79	Barson, F., Legvold, S. and Spedding, F. H.	1953	L	344-102		< 0.01 Y, traces of Mg and Si; specimen prepared from salt by bomb reduction with calcium, cast in tantalum crucibles and turned to cylinder about 0.3 cm diameter, ends machined flat; subjected to prolonged annealing at about 823 K; expansion measured with decreasing temperature on test 1; zero-point correction of -0.018% determined by graphical interpolation.
6	79	Barson, F., et al.	1953	L	103-349		The above specimen; expansion measured with increasing temperature on test 2; zero-point correction of -0.020% determined by graphical interpolation.
7	79	Barson, F., et al.	1953	L	94-348		The above specimen; expansion measured with increasing temperature on test 3; zero-point correction of -0.0177% determined by graphical interpolation.
8	82	Hanak, J. J.	1959	X	293-1190		≤ 0.008 Ho, ≤ 0.002 Tm; wire specimen cut from sheet prepared at Ames Laboratory; hexagonal close packed structure; expansion measured along a-axis; lattice parameter reported at 293 K is 3.560 Å.
9	82	Hanak, J. J.	1959	X	293-1190		The above specimen; expansion measured along c-axis; lattice parameter reported at 293 K is 5.596 Å.
10	733	Finkel', V.A. and Palatnik, M.I.	1971	X	79-300		99.4 Er; polycrystalline sample, annealed for 10 hr at ~ 10^{-6} Torr and 1273 K; expansion measured along c-axis; lattice parameter reported at 293 K is 5.5836 Å; coefficient of thermal linear expansion also reported.
11	733	Finkel', V.A. and Palatnik, M.I.	1971	X	72-293		The above specimen; expansion measured along a-axis; lattice parameter reported at 293 K is 3.55502 Å; coefficient of thermal linear expansion also reported.
12	771	Darnell, F.J.	1963	X	11-297		99.9 pure single crystal formed during reduction of metal chloride, major impurities consist of other heavy rare earths; thermal expansion measured along a-axis; lattice parameter reported at 293 K is 3.5628 Å.
13	771	Darnell, F.J.	1963	X	7-294		The above specimen; thermal expansion measured along c-axis; lattice parameter reported at 293 K is 5.5994 Å.

DATA TABLE 14. THERMAL LINEAR EXPANSION OF ERBIUM Er

[Temperature, T, K; Linear Expansion, $\Delta L/L_0$, %]

CURVE 1

T	$\Delta L/L_0$
293	0.000
310	0.026
333	0.041*
357	0.063
373	0.074*
387	0.092
407	0.108
423	0.138
429	0.128*
446	0.157
448	0.147*
465	0.162
484	0.184*
509	0.207
514	0.217*
537	0.237*
537	0.248
565	0.260
591	0.286
612	0.313*
642	0.349
674	0.378*
701	0.411
731	0.440
760	0.471
788	0.506*
817	0.540
845	0.572
878	0.611*
905	0.652
935	0.686*
965	0.731
994	0.775*
1020	0.813
1051	0.856*
1079	0.905*
1101	0.938
1120	0.977*
1151	1.038
1178	1.090*
1207	1.131

CURVE 2

T	$\Delta L/L_0$
1199	1.123
1181	1.098
1159	1.056*
1143	1.027*
1125	0.995
1104	0.956
1081	0.920*
1057	0.875
1029	0.836*
1004	0.798
979	0.755*
954	0.725
925	0.686
899	0.645*
874	0.620
846	0.588
820	0.559*
793	0.525
765	0.492*
738	0.459
712	0.435*
686	0.403
659	0.375
634	0.350*
610	0.328
584	0.295
559	0.272
489	0.197
465	0.179
373	0.092
341	0.055*
324	0.035*

CURVE 3

T	$\Delta L/L_0$
43	-0.112
65	0.000
85	-0.028
103	0.028
152	0.028
200	0.084
301	0.000

CURVE 4

T	$\Delta L/L_0$
43	-0.196
65	-0.339
85	-0.339
103	-0.357
152	-0.268
200	-0.179
301	0.018

CURVE 5

T	$\Delta L/L_0$
343.9	0.0487
343.2	0.0481*
342.3	0.0470*
341.0	0.0460*
339.5	0.0441*
337.9	0.0424*
335.6	0.0401*
333.1	0.0378*
330.8	0.0355*
328.5	0.0333*
326.1	0.0310*
323.7	0.0287*
321.3	0.0264*
318.9	0.0241*
316.5	0.0219
314.0	0.0196*
311.7	0.0173*
309.3	0.0150*
306.8	0.0128*
304.3	0.0105*
301.9	0.0082*
299.6	0.0060*
297.1	0.0037*
294.5	0.0014*
292.1	-0.0008*
289.7	-0.0031*
287.2	-0.0054
284.8	-0.0076*
282.8	-0.0099*
280.0	-0.0121*
277.5	-0.0144*
275.1	-0.0167*
272.6	-0.0188*
269.5	-0.0217*
267.0	-0.0239*
264.0	-0.0267*

CURVE 5 (cont.)

T	$\Delta L/L_0$
261.5	-0.0290
259.2	-0.0312*
256.7	-0.0334*
254.3	-0.0356*
252.1	-0.0378*
249.7	-0.0401*
247.6	-0.0423*
245.4	-0.0445*
243.0	-0.0467*
240.7	-0.0489*
238.0	-0.0511*
235.7	-0.0533*
233.1	-0.0555*
230.7	-0.0577*
228.3	-0.0600*
226.0	-0.0622*
223.5	-0.0644*
220.9	-0.0665
218.7	-0.0687*
215.8	-0.0709*
213.4	-0.0731*
210.9	-0.0753*
208.6	-0.0775*
206.2	-0.0796*
203.7	-0.0818*
201.3	-0.0840*
198.8	-0.0861*
196.6	-0.0883*
194.1	-0.0905*
191.2	-0.0932*
188.9	-0.0953*
186.6	-0.0975*
184.6	-0.0996*
182.0	-0.1018
179.5	-0.1039*
177.1	-0.1060*
174.5	-0.1082*
172.3	-0.1103*
169.9	-0.1124*
167.6	-0.1146*
165.2	-0.1167*
162.7	-0.1188*
160.2	-0.1209*
157.0	-0.1235*
154.6	-0.1256*
152.1	-0.1278*

CURVE 5 (cont.)

T	$\Delta L/L_0$
149.6	-0.1298*
147.0	-0.1319*
145.1	-0.1335*
142.4	-0.1356
139.7	-0.1377*
137.2	-0.1397*
134.8	-0.1418*
132.3	-0.1439*
129.8	-0.1460*
127.4	-0.1481*
124.9	-0.1501*
122.6	-0.1522*
120.0	-0.1542*
117.5	-0.1563*
114.9	-0.1583*
112.4	-0.1603*
110.5	-0.1618*
107.8	-0.1638*
105.5	-0.1659*
103.3	-0.1679*
101.9	-0.1689

CURVE 6

T	$\Delta L/L_0$
102.6	-0.1695*
104.2	-0.1685*
105.4	-0.1675*
106.5	-0.1665*
108.8	-0.1645*
111.2	-0.1624*
113.5	-0.1604*
115.7	-0.1583*
118.2	-0.1563*
120.4	-0.1543
122.3	-0.1527*
124.6	-0.1507*
127.1	-0.1486*
129.4	-0.1465*
131.8	-0.1444*
134.2	-0.1424*
136.7	-0.1403*
139.2	-0.1382*
141.7	-0.1361*
144.2	-0.1341*
146.9	-0.1320*
149.6	-0.1299*

CURVE 6 (cont.)

T	$\Delta L/L_0$
152.2	-0.1278*
154.7	-0.1257*
157.3	-0.1235*
159.3	-0.1220*
162.0	-0.1199
164.7	-0.1178*
167.3	-0.1157*
169.9	-0.1135*
172.7	-0.1114*
175.4	-0.1093*
177.9	-0.1071*
180.3	-0.1050*
182.7	-0.1029*
185.0	-0.1007*
187.3	-0.0986*
189.7	-0.0964*
192.1	-0.0943*
194.6	-0.0921*
196.7	-0.0899*
199.6	-0.0872*
201.9	-0.0851
204.3	-0.0829*
206.7	-0.0807*
209.2	-0.0786*
211.5	-0.0764*
213.9	-0.0742*
216.4	-0.0720*
219.4	-0.0693*
222.0	-0.0671*
224.6	-0.0649*
227.1	-0.0627*
229.5	-0.0605*
231.9	-0.0583*
234.4	-0.0561*
236.9	-0.0539*
239.3	-0.0517*
241.6	-0.0495
244.1	-0.0472*
246.5	-0.0450*
248.9	-0.0429*
251.4	-0.0416*
253.7	-0.0384*
256.1	-0.0361*
258.5	-0.0339*
260.9	-0.0317*
263.2	-0.0295*

*Not shown in figure.

DATA TABLE 14. THERMAL LINEAR EXPANSION OF ERBIUM Er (continued)

T	$\Delta L/L_0$
CURVE 6 (cont.)	
265.6	-0.0272*
267.9	-0.0250*
270.4	-0.0227*
272.6	-0.0205*
275.0	-0.0183
277.9	-0.0155*
280.2	-0.0132*
282.5	-0.0110*
284.5	-0.0087*
286.7	-0.0065*
288.8	-0.0042*
291.1	-0.0020*
293.8	0.0008*
296.2	0.0031*
298.4	0.0053*
301.5	0.0082*
303.8	0.0105*
306.2	0.0127*
308.1	0.0144*
310.4	0.0167*
312.7	0.0190*
315.2	0.0213*
318.4	0.0241*
321.0	0.0264*
323.4	0.0287*
325.8	0.0309*
329.1	0.0338*
331.8	0.0361
334.0	0.0384*
336.8	0.0407*
340.2	0.0442*
342.8	0.0465*
345.3	0.0485*
347.5	0.0511*
348.7	0.0522*
CURVE 7	
94.2	-0.1722
96.8	-0.1702*
99.5	-0.1682*
102.3	-0.1662*
104.8	-0.1642*
107.7	-0.1616*
109.9	-0.1596*
111.9	-0.1581*
114.8	-0.1561*

T	$\Delta L/L_0$
CURVE 7 (cont.)	
117.1	-0.1541*
119.5	-0.1520*
122.0	-0.1499*
124.5	-0.1479*
126.9	-0.1458*
129.6	-0.1438*
131.5	-0.1422
133.5	-0.1407*
136.0	-0.1386*
138.3	-0.1365*
141.3	-0.1344*
144.3	-0.1324*
146.7	-0.1303*
149.3	-0.1282*
151.8	-0.1261*
154.4	-0.1240*
156.9	-0.1219*
159.1	-0.1198*
161.9	-0.1171*
165.4	-0.1145*
168.0	-0.1124*
170.5	-0.1103*
172.9	-0.1081
175.5	-0.1060*
178.1	-0.1039*
180.6	-0.1017*
183.1	-0.0996*
185.5	-0.0974*
187.8	-0.0953*
190.4	-0.0931*
193.2	-0.0904*
195.6	-0.0883*
198.1	-0.0861*
200.5	-0.0839*
202.8	-0.0818*
205.1	-0.0796*
207.6	-0.0774*
210.0	-0.0753*
212.5	-0.0731
214.7	-0.0709*
218.2	-0.0681*
220.6	-0.0660*
223.1	-0.0616*
225.5	-0.0594*
227.8	-0.0572*
230.2	-0.0550*
232.6	-0.0550*

T	$\Delta L/L_0$
CURVE 7 (cont.)	
235.0	-0.0528*
237.4	-0.0506*
239.9	-0.0484*
242.2	-0.0461*
244.7	-0.0439*
246.9	-0.0417*
249.9	-0.0389*
252.3	-0.0367
254.8	-0.0345*
257.5	-0.0323*
259.9	-0.0301*
262.4	-0.0278*
264.8	-0.0256*
267.4	-0.0234*
269.7	-0.0211*
272.0	-0.0189*
274.6	-0.0166*
277.0	-0.0144*
279.5	-0.0121*
282.1	-0.0099*
284.3	-0.0076*
287.2	-0.0054*
290.3	-0.0025*
292.8	-0.0002*
295.3	0.0020*
297.8	0.0043*
300.2	0.0066*
302.7	0.0088*
305.1	0.0111*
307.5	0.0134*
310.0	0.0157*
312.4	0.0179*
314.9	0.0202*
317.5	0.0225*
320.1	0.0248*
322.7	0.0271
325.1	0.0293*
327.5	0.0316*
330.0	0.0339*
332.7	0.0362*
335.6	0.0385*
338.1	0.0409*
340.7	0.0432*
343.1	0.0455*
345.5	0.0477*
347.8	0.0500

T	$\Delta L/L_0$
CURVE 8	
293	0.000
332	0.028*
367	0.028*
423	0.084
463	0.112*
518	0.140
565	0.169*
615	0.197
662	0.253*
705	0.281*
750	0.309
802	0.365*
850	0.421
899	0.449*
949	0.506
996	0.590*
1047	0.674*
1116	0.787
1156	0.899*
1190	0.983
CURVE 9	
293	0.000*
332	0.054*
367	0.089
423	0.214*
463	0.304
518	0.411*
565	0.518
615	0.608*
662	0.697
705	0.768*
750	0.858
802	0.965*
850	1.054
899	1.179*
949	1.305
996	1.412*
1047	1.537*
1116	1.662
1156	1.858*
1190	1.930

T	$\Delta L/L_0$
CURVE 10	
79	-0.403
85	-0.410
91	-0.399
96	-0.390*
101	-0.385
110	-0.364
122	-0.346
131	-0.331
140	-0.310
150	-0.293
159	-0.274
169	-0.258
179	-0.242
199	-0.201
219	-0.161
238	-0.118
258	-0.073
278	-0.032
293	0.000*
300	0.007*
CURVE 11	
72	-0.116*
75	-0.109
82	-0.103*
86	-0.103
90	-0.100
98	-0.098*
105	-0.094
115	-0.089
126	-0.083
138	-0.089
149	-0.076
158	-0.070
168	-0.065
178	-0.059
199	-0.050
220	-0.039
238	-0.028
260	-0.021
279	-0.006
293	0.000*
CURVE 12*	
11	-0.255

T	$\Delta L/L_0$
CURVE 12 (cont.)*	
13	-0.247
14	-0.255
16	-0.264
17	-0.250
19	-0.244
22	-0.146
28	-0.140
30	-0.135
33	-0.121
35	-0.121
40	-0.093
41	-0.087
61	-0.056
63	-0.073
69	-0.062
72	-0.062
78	-0.053
85	-0.053
89	-0.053
101	-0.042
115	-0.059
117	-0.053
123	-0.042
127	-0.048
142	-0.048
145	-0.056
155	-0.053
164	-0.053
171	-0.037
190	-0.025
293	0.000
297	0.006
CURVE 13*	
7	0.111
8	0.121
10	0.118
13	0.109
14	0.116
16	0.111
19	0.102
20	-0.246
24	-0.257
28	-0.277
32	-0.300
35	-0.300

*Not shown in figure.

DATA TABLE 14. THERMAL LINEAR EXPANSION OF ERBIUM Er (continued)

T	$\Delta L/L_0$	T	$\Delta L/L_0$	T	$\Delta L/L_0$
CURVE 13 (cont.)*		CURVE 13 (cont.)*		CURVE 13 (cont.)*	
38	-0.323	137	-0.300	248	-0.111
40	-0.323	143	-0.275	263	-0.043
49	-0.345	147	-0.272	266	-0.071
50	-0.348	150	-0.259	278	-0.016
52	-0.361	157	-0.266	293	0.000
54	-0.346	161	-0.255	294	0.002
55	-0.350	164	-0.255		
65	-0.359	166	-0.255		
68	-0.368	174	-0.248		
81	-0.373	179	-0.248		
99	-0.366	179	-0.241		
100	-0.366	189	-0.229		
103	-0.350	196	-0.204		
106	-0.347	198	-0.200		
111	-0.332	211	-0.187		
120	-0.322	224	-0.169		
126	-0.322	237	-0.148		
136	-0.284	244	-0.107		

DATA TABLE 14. COEFFICIENT OF THERMAL LINEAR EXPANSION OF ERBIUM Er

[Temperature, T, K; Coefficient of Expansion, α, $10^{-6}\ K^{-1}$]

T	α
CURVE 10*	
85	-11
85	18
100	18
200	18
293	18
300	18
CURVE 11*	
78	17.4
83	13.2
83	4.6
100	5.1
150	5.6
200	6.1
250	6.1
293	6.1
300	6.1

* Not shown in figure.

FIGURE AND TABLE NO. 15R. PROVISIONAL VALUES FOR THERMAL LINEAR EXPANSION OF EUROPIUM Eu

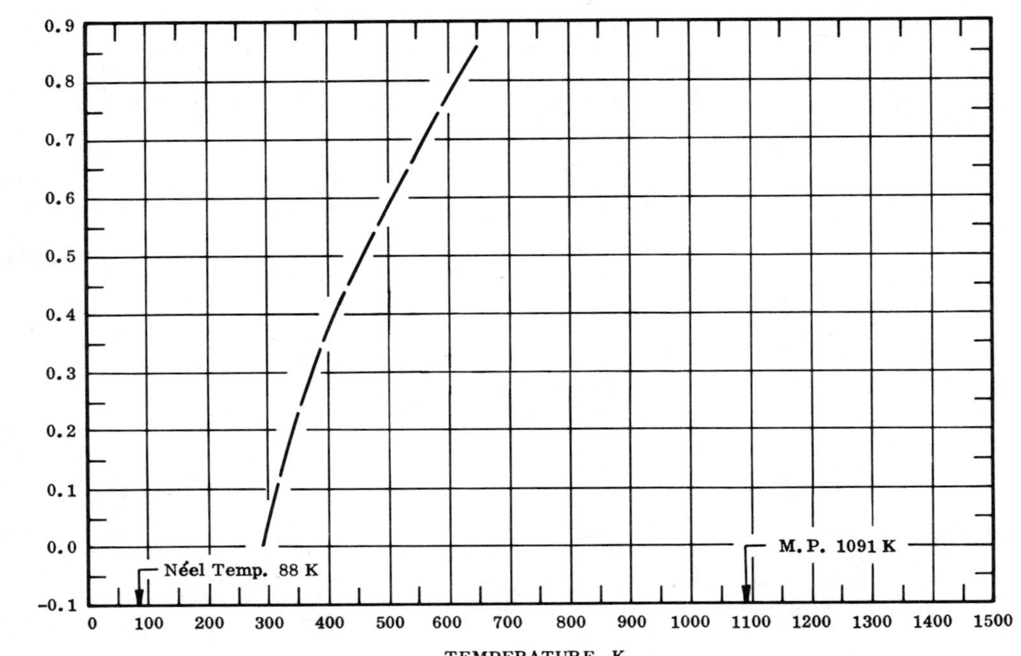

PROVISIONAL VALUES

[Temperature, T, K; Linear Expansion, $\Delta L/L_0$, %; α, K^{-1}]

T	$\Delta L/L_0$	$\alpha \times 10^6$
293	0.000	41
400	0.355	26
500	0.577	19
600	0.763	19
650	0.863	18

REMARKS

The tabulated values are considered accurate to within ±10% over the entire temperature range. These values can be represented approximately by the following equation:

$$\Delta L/L_0 = -2.231 + 1.209 \times 10^{-2}\, T - 1.851 \times 10^{-5}\, T^2 + 1.112 \times 10^{-8}\, T^3$$

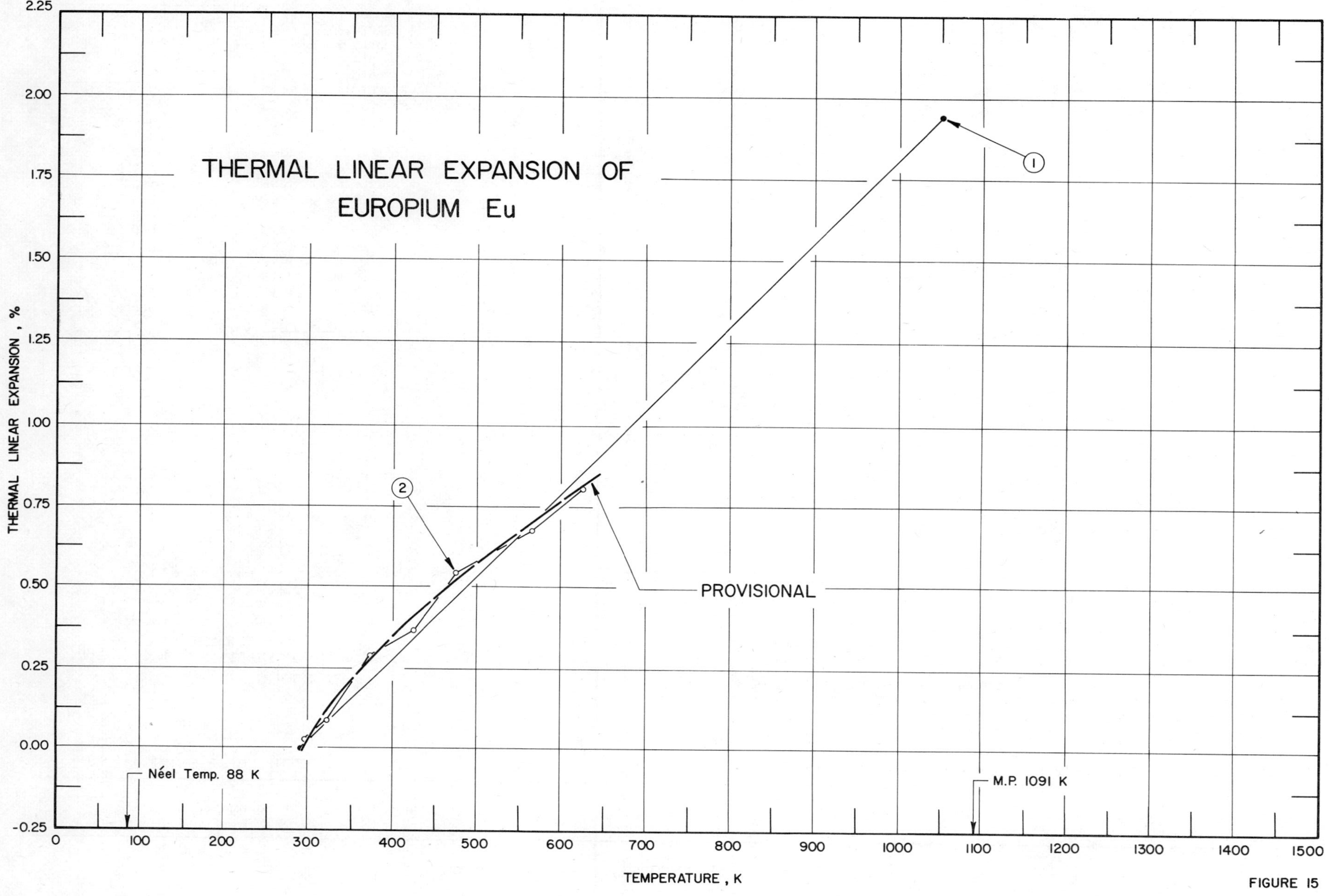
THERMAL LINEAR EXPANSION OF
EUROPIUM Eu
1
2
PROVISIONAL
Néel Temp. 88 K
M.P. 1091 K
THERMAL LINEAR EXPANSION, %
2.25
2.00
1.75
1.50
1.25
1.00
0.75
0.50
0.25
0.00
-0.25
0
100
200
300
400
500
600
700
800
900
1000
1100
1200
1300
1400
1500
TEMPERATURE, K
FIGURE 15

SPECIFICATION TABLE 15. THERMAL LINEAR EXPANSION OF EUROPIUM Eu

Cur. No.	Ref. No.	Author(s)	Year	Method Used	Temp. Range, K	Name and Specimen Designation	Composition (weight percent), Specifications, and Remarks
1	162	Spedding, F.H., Hanak, J.J., and Daane, A.H.	1958	X	293, 1053		High purity specimen prepared by heating Eu_2O_3 with La metal in vacuo at 1473 K.
2	82	Hanak, J.J.	1959	X	293-625		<0.1 Yb, 0.045 Sm, <0.01 Gd; wire specimen cut from sheet prepared by Ames Lab; body-centered cubic structure; lattice parameter reported at 293 K is 4.582 Å; however the value of 4.581 Å was taken from the smooth curve and used in the calculations.
3*	731	Cohen, R.L., Hufner, S., and West, K.W.	1969	E	81-97		Nominally 99.9 pure specimen, contains substantial impurities confirmed by chemical analysis; this curve is here reported using the first given temperature as reference temperature at which $\Delta L/L_0 = 0$.

DATA TABLE 15. THERMAL LINEAR EXPANSION OF EUROPIUM Eu

[Temperature, T, K; Linear Expansion, $\Delta L/L_0$, %]

T	$\Delta L/L_0$	T	$\Delta L/L_0$	T	$\Delta L/L_0$
CURVE 1		CURVE 3 (cont.)*, †		CURVE 3 (cont.)*, †	
293	0.000	83.2	0.0014	89.4	0.0110
1053	1.976	84.0	0.0022	89.7	0.0109
		84.1	0.0023	90.1	0.0109
CURVE 2		84.6	0.0029	90.5	0.0109
		84.6	0.0032	90.9	0.0109
293	0.000*	85.0	0.0034	91.2	0.0110
298	0.022	85.4	0.0049	91.6	0.0108
327	0.087	85.9	0.0049	92.1	0.0108
375	0.284	86.4	0.0061	92.8	0.0109
425	0.371	86.7	0.0078	93.7	0.0109
475	0.546	86.8	0.0087	94.6	0.0111
578	0.677	87.0	0.0091	95.6	0.0112
625	0.808	87.0	0.0096	96.6	0.0112
		87.3	0.0098		
CURVE 3*, †		87.4	0.0102		
		87.8	0.0106		
80.7	0.0000	88.3	0.0107		
81.6	0.0003	88.7	0.0110		
82.4	0.0008	89.0	0.0109		

*Not shown in figure.
† This curve is here reported using the first given temperature as reference temperature at which $\Delta L/L_0 = 0$.

FIGURE AND TABLE NO. 16R. RECOMMENDED VALUES FOR THERMAL LINEAR EXPANSION OF GADOLINIUM Gd

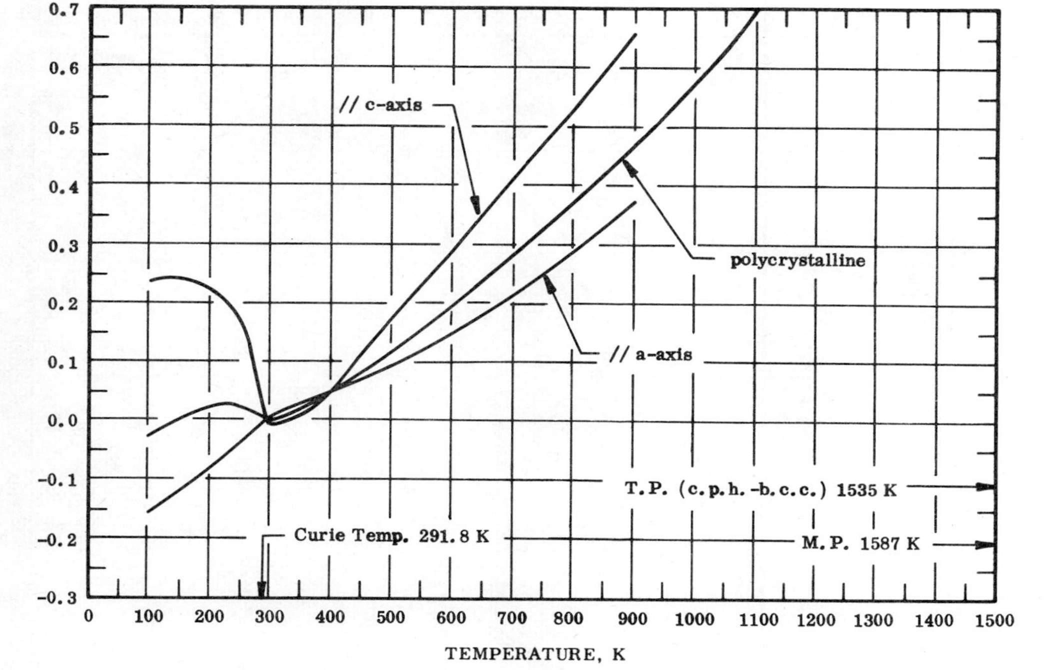

RECOMMENDED VALUES

[Temperature, T, K; Linear Expansion, $\Delta L/L_0$, %; α, K^{-1}]

T	// a-axis $\Delta L/L_0$	// c-axis $\Delta L/L_0$	polycrystalline $\Delta L/L_0$	polycrystalline $\alpha \times 10^6$
100	-0.156	0.239	-0.024	
150	-0.120	0.241	0.000	
200	-0.082	0.229	0.022	
293	0.000	0.000	0.000	
350	0.270	0.070	0.020	
400	0.051	0.055	0.052	6.4
500	0.093	0.171	0.119	7.3
600	0.150	0.291	0.197	8.2
700	0.216	0.410	0.281	8.7
800	0.291	0.531	0.371	9.2
900	0.373	0.661	0.469	10.4
1000			0.575	11.0
1200			0.854	16.8

REMARKS

The tabulated values are considered accurate to within ±5% over the entire temperature range. These values can be represented approximately by the following equations:

// a-axis: $\Delta L/L_0 = 0.018 - 2.484 \times 10^{-4}\,T + 9.063 \times 10^{-7}\,T^2 - 2.130 \times 10^{-10}\,T^3$ $(400 < T < 900)$

// c-axis: $\Delta L/L_0 = -0.475 + 1.526 \times 10^{-3}\,T - 6.679 \times 10^{-7}\,T^2 + 4.167 \times 10^{-10}\,T^3$ $(400 < T < 900)$

polycrystalline: $\Delta L/L_0 = 0.051 + 6.381 \times 10^{-4}\,(T-400) + 5.446 \times 10^{-7}\,(T-400)^2 - 3.157 \times 10^{-10}\,(T-400)^3$ $(400 < T < 800)$

$\Delta L/L_0 = 0.373 + 9.222 \times 10^{-4}\,(T-800) + 1.657 \times 10^{-7}\,(T-800)^2 + 1.347 \times 10^{-9}\,(T-800)^3$ $(800 < T < 1200)$

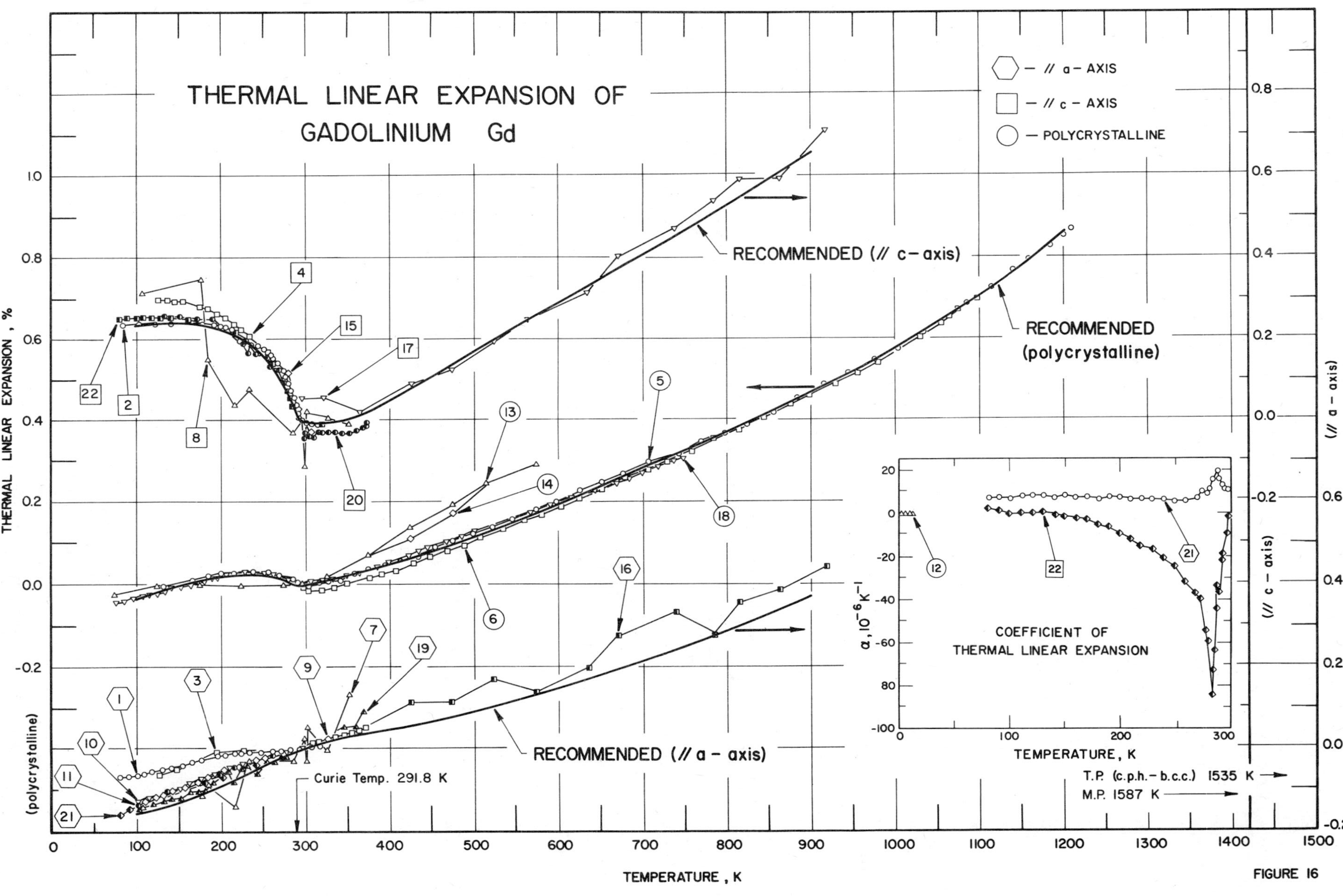
THERMAL LINEAR EXPANSION OF
GADOLINIUM Gd
— // a – AXIS
— // c – AXIS
— POLYCRYSTALLINE
RECOMMENDED (// c – axis)
RECOMMENDED
(polycrystalline)
RECOMMENDED (// a – axis)
Curie Temp. 291.8 K
COEFFICIENT OF
THERMAL LINEAR EXPANSION
α, 10⁻⁶ K⁻¹
TEMPERATURE, K
T.P. (c.p.h. – b.c.c.) 1535 K
M.P. 1587 K
THERMAL LINEAR EXPANSION, %
(polycrystalline)
(// c – axis)
(// a – axis)
TEMPERATURE, K

FIGURE 16

SPECIFICATION TABLE 16. THERMAL LINEAR EXPANSION OF GADOLINIUM Gd

Cur. No.	Ref. No.	Author(s)	Year	Method Used	Temp. Range, K	Name and Specimen Designation	Composition (weight percent), Specifications, and Remarks
1	163	Ergin, Y.V.	1965	S	79-316		Single crystal grown in the laboratory at E.M. Savitskii; expansion measured along a-axis; zero-point correction of -0.145% determined by graphical interpolation.
2	163	Ergin, Y.V.	1965	S	83-315		Similar to the above; expansion measured along c-axis; zero-point correction of -0.035% determined by graphical interpolation.
3	164	Vorob'ev, V.V., Smirnov, Yu.N., and Finkel', V.A.	1966	X	125-369		99.7 pure; specimen annealed 15 min at 1320 K, cooled slowly; expansion measured along a-axis; lattice parameter reported at 297 K is 3.6330 Å; 3.6329 Å at 293 K determined by graphical interpolation.
4	164	Vorob'ev, V.V., et al.	1966	X	124-369		The above specimen; expansion measured along c-axis; lattice parameter at 293 K is 5.77727 Å.
5	78	Barson, F., Legvold, S., and Spedding, F.H.	1956	L	129-1208		0.3 Ta, <0.04 Ca, <0.01 Fe, <0.01 Mg, <0.01 Si; specimen 0.6 cm diameter, 6 cm long; Curie temperature about 289 K; expansion measured with increasing temperature; data below 223 K unreliable.
6	78	Barson, F., et al.	1956	L	1098-295		The above specimen; expansion measured with decreasing temperature.
7	159	Banister, J.R., Legvold, S., and Spedding, F.H.	1954	X	106-349		0.1 Sm, 0.05 Nd, 0.02 each Ca, Tb; powder specimen; hexagonal, close-packed structure at room temperature; expansion measured along a -axis; lattice parameter reported at 298 K is 3.635 Å; 3.6342 Å at 293 K determined by graphical interpolation.
8	159	Banister, J.R., et al.	1954	X	106-349		The above specimen; ferromagnetic Curie point 289 K; expansion measured along c -axis; lattice parameter reported at 285 K is 5.781 Å; 5.778 Å at 293 K determined by graphical interpolation.
9	79	Barson, F., Legvold, S., and Spedding, F.H.	1953	L	344-109		<0.15 Sn, <0.03 Ca, ~0.02 Fe, <0.02 Mg, traces of Tb; prepared from salt by bomb reduction with Ca, cast and then rolled into cylinder about 0.3 cm diameter; ends machined flat; subjected to prolonged annealing at about 820 K; expansion measured with decreasing temperature on test 1; zero-point correction of -0.0113% determined by graphical interpolation.
10	79	Barson, F., et al.	1953	L	100-348		The above specimen; expansion measured with increasing temperature on test 2; zero-point correction of -0.0124% determined by graphical interpolation.
11	79	Barson, F., et al.	1953	L	349-99		The above specimen; expansion measured with decreasing temperature on test 3; zero-point correction of -0.0123% determined by graphical interpolation.
12*	80	Andres, K.	1963	O	1.9-13		Polycrystalline specimen, melted into Ta form; this curve is here reported using the first given temperature as reference temperature at which $\Delta L/L_0$ = 0.
13	81	Trombe, F. and Foex, M.	1953	L	73-573		0.1 to 0.2 Mg, trace Fe; Curie temperature 289 ± 2 K determined magnetically; expansion measured with increasing temperature in hydrogen atmosphere below room temperature; in argon atmosphere above; zero-point correction is -0.006%.
14*	81	Trombe, F. and Foex, M.	1953	L	573-73		The above specimen; expansion measured with decreasing temperature; hysteresis due to slight transformation, but not detected in magnetic experiments; zero-point correction is -0.006%

* Not shown in figure.

SPECIFICATION TABLE 16. THERMAL LINEAR EXPANSION OF GADOLINIUM Gd (continued)

Cur. No.	Ref. No.	Author(s)	Year	Method Used	Temp. Range, K	Name and Specimen Designation	Composition (weight percent), Specifications, and Remarks
15	165	Cadieu, F.J. and Douglass, D.H.	1968	C	279-306		Single crystal prepared from metal sponge from Lunex Co.; sponge contained traces of Ti and other rare earths; ferromagnetic transition observed at 292.50 ± 0.02 K; expansion measured along c-axis.
16	82	Hanak, J.J.	1959	X	295-917		< 0.1 Ta, ≤ 0.05 each La, Sm, < 0.04 Ca, < 0.03 Si, < 0.023 C, ≤ 0.02 Tb, 0.013 N, < 0.01 each any other impurity; wire specimen cut from sheet prepared at Ames Laboratory; hexagonal close-packed structure; expansion measured along a-axis; lattice parameter reported at 293 K is 3.634 Å.
17	82	Hanak, J.J.	1959	X	295-917		The above specimen; expansion measured along c-axis; lattice parameter reported at 293 K is 5.784 Å.
18	166	Birss, R.R.	1960		76-746		< 0.005 Fe, < 0.1 other rare earths; high purity polycrystalline specimen from Johnson, Matthey; 6.88 mm long; zero-point correction is 0.0015%.
19	160	Darnell, F.J.	1963	X	101-366		0.1 total impurity; single crystal, platelet face (001) plane, approximately 0.1 mm thick and 1 mm^2 in area; expansion measured along a-axis; lattice parameter reported at 296 is 3.6309 Å; 3.6307 Å at 293 K determined by graphical interpolation.
20	160	Darnell, F.J.	1963	X	131-371		The above specimen; expansion measured along c-axis; lattice parameter reported at 291 K is 5.7822 Å; 5.7812 Å at 293 K determined by graphical interpolation.
21	718	Vorob'ev, V.V., Palatnik, M.I. and Finkel, V.A.	1971	X	81-313		99.7 pure specimen; lattice parameter reported at 293 K is 3.6316 Å; measurements along a-axis; data on the coefficient of linear expansion also reported.
22	718	Vorob'ev, V.V., et al.	1971	X	79-319		The above specimen; lattice parameter reported at 293 K is 5.7773 Å; measurements along c-axis; data on the coefficient of linear expansion also reported.

DATA TABLE 16. THERMAL LINEAR EXPANSION OF GADOLINIUM Gd

[Temperature, T, K; Linear Expansion, $\Delta L/L_0$, %]

T	$\Delta L/L_0$
CURVE 1	
79	-0.071
88	-0.070
94	-0.068*
100	-0.067
108	-0.064
117	-0.059
128	-0.055
138	-0.051
151	-0.043
164	-0.036
174	-0.030
189	-0.024*
196	-0.020
203	-0.019
209	-0.015*
216	-0.014
223	-0.014
231	-0.013
246	-0.012
259	-0.010
271	-0.008
280	-0.007
286	-0.004*
291	-0.001
297	0.002*
301	0.003*
305	0.004
312	0.004*
316	0.003
CURVE 2	
83	0.238
101	0.239
121	0.240
139	0.240
159	0.240
177	0.239
191	0.238
198	0.235
204	0.231
212	0.223
221	0.210
226	0.203
231	0.196
243	0.184
CURVE 2 (cont.)	
250	0.177
253	0.174*
257	0.166
260	0.155
266	0.141
271	0.123
274	0.108
278	0.090
282	0.072
285	0.054
287	0.036
289	0.017
290	0.007
291	0.003*
294	-0.001
296	-0.004*
300	-0.008
302	-0.009*
306	-0.011*
309	-0.012
315	-0.014
CURVE 3	
125	-0.068
133	-0.062*
143	-0.051
153	-0.044*
172	-0.029*
183	-0.021*
193	-0.016
203	-0.012*
213	-0.010*
223	-0.009
232	-0.009*
242	-0.009*
255	-0.010*
262	-0.011*
272	-0.013*
277	-0.013*
283	-0.010*
287	-0.003*
297	0.002*
303	0.008*
307	0.010
313	0.012
CURVE 3 (cont.)	
320	0.018*
332	0.024
343	0.029
353	0.038
362	0.042
369	0.047
CURVE 4	
124	0.300
135	0.297
143	0.294
153	0.293
173	0.281
184	0.275
195	0.261
204	0.252
213	0.237
223	0.222
233	0.205
244	0.186
253	0.164
263	0.137*
272	0.103*
277	0.076*
281	0.048*
287	0.014*
291	0.000*
295	0.000*
301	0.0002*
306	-0.005*
311	-0.007*
320	-0.004*
330	-0.002*
340	0.003*
351	0.008*
359	0.014*
369	0.021*
CURVE 5	
129	-0.007
165	0.008
184	0.020
199	0.026
213	0.024
CURVE 5 (cont.)	
236	0.029
253	0.027
266	0.020
283	0.013
290	0.001
303	-0.002
314	0.004
324	0.005
340	0.008
358	0.022
373	0.035
392	0.041
410	0.055
432	0.067
452	0.083
474	0.102
499	0.119
521	0.136
545	0.157
573	0.179
597	0.196
624	0.223
650	0.242
675	0.264
706	0.293
740	0.314
768	0.343
797	0.367
823	0.393
853	0.418
883	0.450
914	0.487
943	0.517
973	0.546
1001	0.574
1033	0.617
1086	0.686
1114	0.725
1139	0.767
1159	0.793
1183	0.828
1199	0.852
1208	0.865
CURVE 6	
1098	0.697
1075	0.667
1063	0.656
1054	0.636
1028	0.604
979	0.541
953	0.515
928	0.487
898	0.460
875	0.435
842	0.405
813	0.372
784	0.350
756	0.320
728	0.296
705	0.277
677	0.255
650	0.227
624	0.207
602	0.186
579	0.166
559	0.153
533	0.129
507	0.110
487	0.089
467	0.079
446	0.064
426	0.047
406	0.030
388	0.023
375	0.015
348	-0.002
332	-0.011
318	-0.016
303	-0.018
295	-0.012
CURVE 7	
106	-0.144
175	-0.116
183	-0.089
216	-0.144
231	-0.034
283	-0.034
285	-0.034
CURVE 7 (cont.)	
298	0.021
299	-0.034
300	0.049
324	-0.006
349	0.131
CURVE 8	
106	0.313
175	0.347
183	0.140
216	0.036
231	0.071
283	-0.033
285	0.053*
298	-0.033*
299	-0.120
300	0.019
324	0.001
249	-0.016
CURVE 9	
343.9	0.0312*
340.8	0.0298*
338.2	0.0284*
335.7	0.0269*
333.2	0.0255*
330.9	0.0241*
328.6	0.0228*
326.4	0.0214*
324.4	0.0200
322.2	0.0187*
320.0	0.0173*
317.9	0.0159*
315.8	0.0145*
311.6	0.0118*
309.5	0.0104*
307.4	0.0091*
305.4	0.0077*
303.3	0.0064*
301.1	0.0050*
299.1	0.0036*
296.8	0.0023*
294.6	0.0009*
292.2	-0.0005*

*Not shown in figure.

DATA TABLE 16. THERMAL LINEAR EXPANSION OF GADOLINIUM Gd (continued)

T	$\Delta L/L_0$
CURVE 9 (cont.)	
289.8	-0.0019*
287.4	-0.0032*
285.1	-0.0046*
282.9	-0.0059*
280.5	-0.0073*
277.9	-0.008*
275.6	-0.0100*
273.1	-0.0113*
270.8	-0.0127*
267.2	-0.0149*
263.6	-0.0167*
260.1	-0.0187*
256.6	-0.0207*
253.5	-0.0227*
250.2	-0.0247
246.9	-0.0266*
243.0	-0.0286*
240.5	-0.0306
237.5	-0.0325*
234.5	-0.0345*
231.3	-0.0364*
226.5	-0.0397*
223.6	-0.0416*
220.8	-0.0436
216.3	-0.0468*
213.7	-0.0487*
211.3	-0.0506*
208.8	-0.0525*
205.7	-0.0551*
202.4	-0.0576
199.2	-0.0601*
196.0	-0.0627*
192.5	-0.0652*
188.7	-0.0677*
185.5	-0.0702*
182.5	-0.0727
179.5	-0.0753*
176.3	-0.0778*
173.2	-0.0803*
170.2	-0.0827
167.2	-0.0852*
164.2	-0.0877*
161.1	-0.0902*
158.1	-0.0927*
155.2	-0.0951
152.2	-0.0976*
149.2	-0.1000*
CURVE 9 (cont.)	
146.2	-0.1025*
143.2	-0.1049*
140.1	-0.1074
137.0	-0.1098*
133.9	-0.1122*
130.8	-0.1146
127.6	-0.1170*
124.3	-0.1194*
121.8	-0.1213
119.4	-0.1230*
117.1	-0.1248*
114.9	-0.1266*
112.5	-0.1284*
110.1	-0.1301*
108.8	-0.1313
CURVE 10	
100.2	-0.1330
101.5	-0.1324*
102.5	-0.1318*
104.4	-0.1307*
106.8	-0.1289*
109.3	-0.1272*
112.0	-0.1254
114.4	-0.1237*
116.9	-0.1219*
119.0	-0.1201*
121.7	-0.1183
124.3	-0.1165*
126.7	-0.1147*
129.2	-0.1129*
131.4	-0.1111*
133.9	-0.1093*
136.3	-0.1075*
138.7	-0.1057*
141.2	-0.1039
143.6	-0.1020*
146.0	-0.1002*
148.4	-0.0984*
150.9	-0.0965
153.2	-0.0947*
155.5	-0.0928*
157.9	-0.0910*
160.4	-0.0892
162.9	-0.0873*
165.3	-0.0854*
CURVE 10 (cont.)	
167.6	-0.0836*
170.0	-0.0817*
172.3	-0.0799*
174.8	-0.0780
177.2	-0.0761*
179.5	-0.0743*
181.9	-0.0724*
184.3	-0.0705*
186.8	-0.0686*
189.1	-0.0667*
191.6	-0.0648
194.0	-0.0629*
196.5	-0.0610*
199.7	-0.0585*
202.2	-0.0566*
204.8	-0.0547*
207.4	-0.0528*
209.9	-0.0509
212.5	-0.0489*
215.2	-0.0470*
217.8	-0.0451*
220.4	-0.0432*
223.4	-0.0412*
226.4	-0.0393
229.3	-0.0374*
232.3	-0.0354*
235.6	-0.0334*
239.0	-0.0315*
242.0	-0.0295*
245.5	-0.0275
249.1	-0.0256*
252.6	-0.0236*
254.9	-0.0223*
257.6	-0.0209*
260.0	-0.0196*
262.4	-0.0182*
264.8	-0.0169*
267.3	-0.0156*
269.8	-0.0142*
272.3	-0.0129*
275.9	-0.0109
278.2	-0.0095*
280.6	-0.0081*
282.7	-0.0068*
284.8	-0.0055*
286.9	-0.0041*
288.9	-0.0027*
CURVE 10 (cont.)	
290.8	-0.0014*
292.8	0.0000*
294.7	0.0013*
296.7	0.0027*
298.7	0.0040*
300.7	0.0054*
303.6	0.0074*
305.7	0.0087*
307.6	0.0101*
309.5	0.0114*
311.5	0.0128*
313.5	0.0142*
315.4	0.0155*
317.4	0.0169*
319.3	0.0183*
321.4	0.0196*
323.4	0.0210*
325.3	0.0223*
327.3	0.0237*
329.2	0.0251*
331.1	0.0264*
333.0	0.0278*
334.9	0.0292*
336.7	0.0305*
338.7	0.0319*
340.7	0.0333*
342.6	0.0347*
344.4	0.0361*
346.2	0.0374*
347.9	0.0388
CURVE 11	
348.7	0.0374*
347.7	0.0366*
345.9	0.0353*
344.0	0.0339*
342.0	0.0326*
340.2	0.0312*
338.1	0.0298*
336.2	0.0284*
333.1	0.0264*
330.1	0.0243*
327.1	0.0222*
325.0	0.0209*
323.0	0.0195*
321.1	0.0182*
CURVE 11 (cont.)	
319.1	0.0168*
317.1	0.0154*
315.0	0.0141*
313.0	0.0127*
311.0	0.0113*
309.0	0.0100*
307.0	0.0086*
304.9	0.0073*
302.9	0.0059*
300.6	0.0046*
298.6	0.0032*
296.4	0.0019*
293.9	0.0005*
291.6	-0.0009*
289.4	-0.0023*
287.1	-0.0036*
285.0	-0.0050*
282.9	-0.0063*
280.8	-0.0077*
278.6	-0.0090*
276.5	-0.0104*
274.2	-0.0117*
270.8	-0.0137*
268.7	-0.0150*
266.5	-0.0164*
264.4	-0.0177*
262.3	-0.0191*
260.1	-0.0204
258.0	-0.0217*
255.7	-0.0231*
253.7	-0.0244*
251.9	-0.0257*
249.6	-0.0270*
247.6	-0.0283*
244.6	-0.0303*
241.6	-0.0323*
238.6	-0.0342*
235.4	-0.0361*
232.2	-0.0380*
229.2	-0.0400*
226.5	-0.0420*
223.9	-0.0439*
221.3	-0.0459*
218.5	-0.0478*
215.6	-0.0497
212.4	-0.0523*
209.9	-0.0542*
CURVE 11 (cont.)	
207.6	-0.0561*
204.2	-0.0586*
201.7	-0.0605*
199.1	-0.0625*
197.2	-0.0644
194.8	-0.0663*
192.4	-0.0681*
189.9	-0.0700*
187.6	-0.0719*
185.2	-0.0738*
183.0	-0.0757*
180.5	-0.0776*
178.2	-0.0795*
175.1	-0.0820*
172.9	-0.0838*
171.2	-0.0857*
168.2	-0.0876*
165.4	-0.0894*
163.7	-0.0913*
161.3	-0.0931*
158.9	-0.0950*
156.6	-0.0968*
154.2	-0.0987*
151.9	-0.1005*
149.6	-0.1024*
147.4	-0.1042*
144.6	-0.1060*
142.1	-0.1078*
139.9	-0.1097*
137.5	-0.1115*
135.3	-0.1133
133.0	-0.1151*
130.6	-0.1169*
128.3	-0.1187*
125.2	-0.1211*
123.0	-0.1229*
120.6	-0.1247*
118.2	-0.1265*
115.7	-0.1283*
113.3	-0.1301*
110.9	-0.1318*
108.5	-0.1336*
106.0	-0.1354*
103.6	-0.1371*
101.1	-0.1389*
99.2	-0.1400

*Not shown in figure.

DATA TABLE 16. THERMAL LINEAR EXPANSION OF GADOLINIUM Gd (continued)

T	$\Delta L/L_0$
CURVE 12*, †, ‡	
1.95	0.000 x 10^{-6}
2.92	2.357
3.43	3.788
3.96	4.569
4.90	3.752
5.91	1.191
6.94	-2.512
7.96	-6.449
8.93	-9.950
9.98	-13.221
11.22	-15.001
12.00	-14.186
12.85	-11.882
13.05	-11.184
CURVE 13	
73	-0.026
123	-0.008
172	-0.006
223	-0.008
273	-0.006
293	0.000
323	0.018
373	0.068
423	0.133
473	0.190
523	0.240
573	0.287
CURVE 14	
573	0.287*
523	0.240*
473	0.170*
423	0.109*
373	0.063*
323	0.018*
293	0.000*
273	-0.006*
223	-0.008
173	-0.006
123	-0.008*
73	-0.026*
CURVE 15	
278	0.1078
282	0.0863*
285	0.0642*
286	0.0582*
286	0.0515*
288	0.0430*
289	0.0341*
290	0.0254
290	0.0215*
290	0.0194*
291	0.0168*
291	0.0144*
291	0.0123*
291	0.0097*
292	0.0082*
292	0.0071*
292	0.0067*
292	0.0061*
292	0.0050*
292	0.0043*
292	0.0040*
292	0.0036*
292	0.0032*
292	0.0028*
292	0.0021*
292	0.0013*
292	0.0012*
292	0.0008*
292	0.0004*
292	0.0002*
293	-0.0006*
293	-0.0007*
293	-0.0018*
293	-0.0020*
293	-0.0028*
293	-0.0038*
294	-0.0048*
295	-0.0084*
297	-0.0145*
302	-0.0276*
303	-0.0301*
303	-0.0306*
305	-0.0343
CURVE 16	
295	0.000*
320	0.055*
CURVE 16 (cont.)	
361	0.028*
424	0.110
472	0.110
522	0.165
563	0.138
632	0.193
670	0.275
738	0.330
784	0.275
815	0.358
862	0.385
917	0.440
CURVE 17	
295	0.052
320	0.052
361	0.017
424	0.086
472	0.121
522	0.190
563	0.242
632	0.311
670	0.399
738	0.467
784	0.536
815	0.588
862	0.588
917	0.709
CURVE 18	
76	-0.047
85	-0.042
95	-0.037
106	-0.032
114	-0.027
123	-0.023
131	-0.018
138	-0.012
150	-0.007
162	-0.002
170	0.002
178	0.006
187	0.012
195	0.017
203	0.021
CURVE 18 (cont.)	
210	0.024*
218	0.026*
228	0.028
240	0.028
250	0.025*
259	0.022*
264	0.021*
271	0.018*
273	0.014*
278	0.012
283	0.007*
287	0.004*
293	0.000*
296	-0.001*
305	0.002
312	0.004*
320	0.007
330	0.010
338	0.014*
348	0.020
362	0.024
370	0.033*
384	0.040
396	0.050
409	0.058*
421	0.067
432	0.075
444	0.083
454	0.091*
467	0.100
482	0.110
500	0.123
524	0.139*
544	0.155*
567	0.171
590	0.188
615	0.205
641	0.213
668	0.240
681	0.255
699	0.266
717	0.280
735	0.294
746	0.304
CURVE 19	
101	-0.148
109	-0.148*
118	-0.140
130	-0.134
140	-0.123
151	-0.123
161	-0.109
171	-0.109
179	-0.104
188	-0.085
199	-0.068*
214	-0.082
221	-0.054
230	-0.046
241	-0.063
252	-0.038
262	-0.038
268	-0.021
274	-0.021
278	-0.013*
282	-0.030*
288	-0.010*
289	-0.005*
296	0.004*
298	-0.002*
298	0.006*
301	-0.002*
309	0.004*
317	0.020*
332	0.015*
342	0.050
358	0.050
366	0.089
CURVE 20	
131	0.258
150	0.258*
172	0.246
190	0.246
194	0.240*
196	0.244*
200	0.235
200	0.239*
202	0.244*
205	0.240*
207	0.235*
CURVE 20 (cont.)	
209	0.228*
212	0.218
218	0.204*
224	0.208*
231	0.199*
235	0.187
241	0.164
245	0.183
249	0.163*
253	0.157
256	0.138*
261	0.138*
267	0.106
268	0.131
268	0.125*
271	0.111*
271	0.107*
273	0.095
275	0.099*
275	0.093*
275	0.086*
277	0.093*
278	0.057*
280	0.086
283	0.081*
284	0.028
286	0.057*
288	0.038*
291	0.017*
295	-0.017*
298	-0.029*
298	-0.036*
298	-0.043
301	-0.043
304	-0.040*
307	-0.040*
309	-0.043
316	-0.036
319	-0.036
327	-0.036
334	-0.036
342	-0.036
350	-0.036
353	-0.028*
359	-0.028*
368	-0.019
369	-0.010*
371	-0.010
371	-0.019

*Not shown in figure.
†This curve is here reported using the first given temperature as reference temperature at which $\Delta L/L_0 = 0$.
‡Author's data for coefficient of thermal linear expansion have been integrated by TPRC to obtain $\Delta L/L_0$.

DATA TABLE 16. THERMAL LINEAR EXPANSION OF GADOLINIUM Gd (continued)

T	$\Delta L/L_0$	T	$\Delta L/L_0$	T	$\Delta L/L_0$	T	$\Delta L/L_0$
CURVE 21		CURVE 21 (cont.)		CURVE 22 (cont.)		CURVE 22 (cont.)	
80	-0.161	260	-0.027*	100	0.252	274	0.086*
90	-0.154	268	-0.020*	107	0.253	278	0.071
100	-0.148*	273	-0.017*	118	0.253	281	0.058
108	-0.140*	277	-0.014*	128	0.253	283	0.049*
119	-0.130*	285	-0.008*	139	0.251	285	0.036
128	-0.124*	289	-0.005*	150	0.251*	288	0.023*
139	-0.114*	293	-0.000*	159	0.250*	290	0.009*
150	-0.107*	296	-0.005*	169	0.248*	293	0.000*
158	-0.099*	300	-0.008*	178	0.244*	296	-0.002*
169	-0.090*	305	0.011*	189	0.240*	300	-0.006*
181	-0.083	309	0.011*	200	0.232*	305	-0.007*
189	-0.077*	314	0.015*	210	0.224*	309	-0.008*
200	-0.069	319	0.018*	218	0.213	314	-0.008*
210	-0.061*			229	0.198	319	-0.006
218	-0.055*	CURVE 22		239	0.181*		
229	-0.048*			250	0.163*		
240	-0.040	79	0.248	259	0.139		
251	-0.033*	88	0.251	269	0.140*		

DATA TABLE 16. COEFFICIENT OF THERMAL LINEAR EXPANSION OF GADOLINIUM Gd

[Temperature, T, K; Coefficient of Expansion, α, 10^{-6} K^{-1}]

T	α	T	α	T	α	T	α	T	α
CURVE 12‡		CURVE 21 (cont.)		CURVE 21 (cont.)		CURVE 22 (cont.)		CURVE 22 (cont.)	
1.95	0.0220	101	6.9	279	9.3	121	0.5	283	-84.1
2.92	0.0266*	112	8.2	281	11.0	131	0.8	284	-72.4
3.43	0.0295*	121	8.2	285	15.2	142	-0.3	286	-63.4
3.96	0.0000*	131	8.2	289	19.3	150	-1.2	288	-43.8
4.90	-0.0174*	141	7.5	290	16.0	161	-2.2	288	-33.6
5.91	-0.0333	150	8.5	293	13.1	170	-2.8	290	-36.6
6.94	-0.0386*	160	7.7	294	11.3	180	-5.3	293	-21.6
7.96	-0.0386*	170	7.7	299	5.4	190	-6.0	294	-18.8
8.93	-0.0336*	181	6.6	302	5.4*	200	-9.1	298	-9.4
9.98	-0.0287	191	7.7	304	3.9*	210	-12.2	299	-1.4
11.22	0.0000*	200	7.7	308	5.1*	219	-14.7	303	-2.2*
12.00	0.0209*	210	6.7	313	5.1*	230	-16.6	309	-0.7*
12.85	0.0333*	220	6.7			239	-20.4	313	1.1*
13.05	0.0365	230	6.7	CURVE 22		250	-24.2		
		240	6.7			259	-31.2		
CURVE 21		250	5.8	81	2.4	268	-36.2		
		259	5.8	91	1.5	273	-39.2		
82	7.4	270	7.3	100	0.0	278	-53.9		
91	7.4	275	10.0	111	0.5	280	-58.9		

* Not shown in figure.
‡ Author's data for coefficient of thermal expansion have been integrated by TPRC to obtain $\Delta L/L_0$.

SPECIFICATION TABLE 17. THERMAL LINEAR EXPANSION OF GALLIUM Ga

Cur. No.	Ref. No.	Author(s)	Year	Method Used	Temp. Range, K	Name and Specimen Designation	Composition (weight percent), Specifications, and Remarks
1	779	Powell, R.W.	1949		273-293		Single crystal rod of 0.45 cm diameter and 15 cm long grown approximately in the direction of a-axis.
2	779	Powell, R.W.	1949		273-293		Similar to the above specimen except grown approximately in the direction of b-axis.
3	779	Powell, R.W.	1949		273-293		Similar to the above specimen except grown approximately in the direction of c-axis.

DATA TABLE 17. THERMAL LINEAR EXPANSION OF GALLIUM Ga

[Temperature, T, K; Linear Expansion, $\Delta L/L_0$, %]

T	$\Delta L/L_0$
CURVE 1	
273	-0.033
293	0.000
CURVE 2	
273	-0.023
293	0.000
CURVE 3	
273	-0.062
293	0.000

FIGURE AND TABLE NO. 18R. RECOMMENDED VALUES FOR THERMAL LINEAR EXPANSION OF GERMANIUM Ge

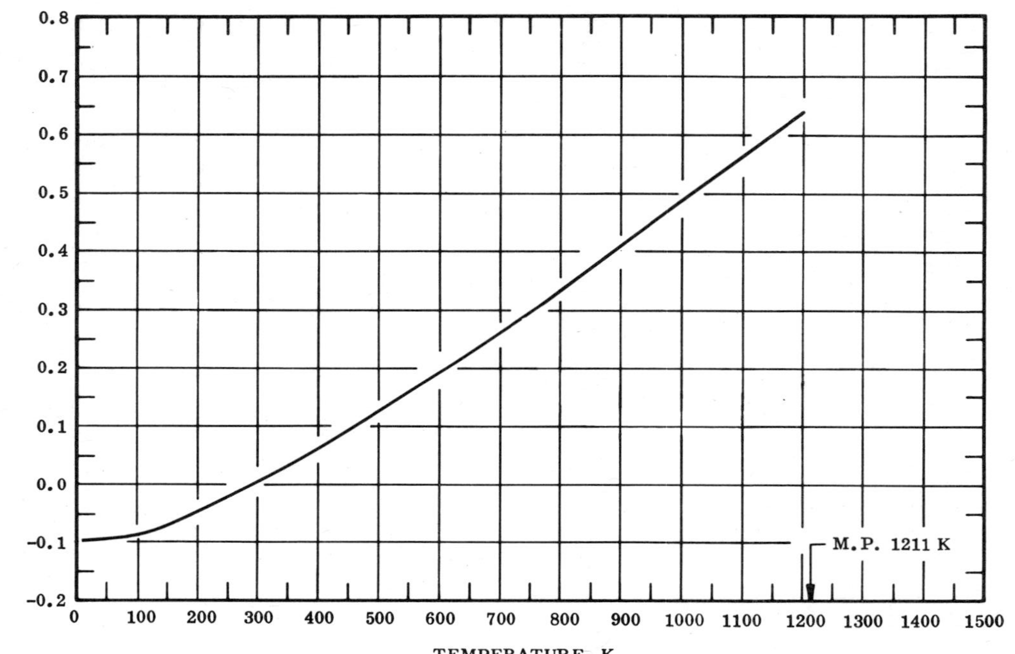

RECOMMENDED VALUES

[Temperature, T, K; Linear Expansion, $\Delta L/L_0$, %; α, K^{-1}]

T	$\Delta L/L_0$	$\alpha \times 10^6$
10	-0.094	0.06
25	-0.094	-0.57
50	-0.095	-0.62
100	-0.089	2.4
200	-0.050	4.9
293	0.000	5.7
400	0.064	6.2
500	0.127	6.5
600	0.193	6.7
700	0.262	7.0
800	0.333	7.2
900	0.406	7.4
1000	0.481	7.6
1200	0.636	8.0

REMARKS

The tabulated values are considered accurate to within ± 3% over the entire temperature range. These values can be represented approximately by the following equations:

$$\Delta L/L_0 = -0.089 + 2.626 \times 10^{-4}(T-100) + 1.463 \times 10^{-6}(T-100)^2 - 2.221 \times 10^{-9}(T-100)^3 \quad (100 < T < 293)$$

$$\Delta L/L_0 = 5.790 \times 10^{-4}(T-293) + 1.768 \times 10^{-7}(T-293)^2 - 4.562 \times 10^{-11}(T-293)^3 \quad (293 < T < 1200)$$

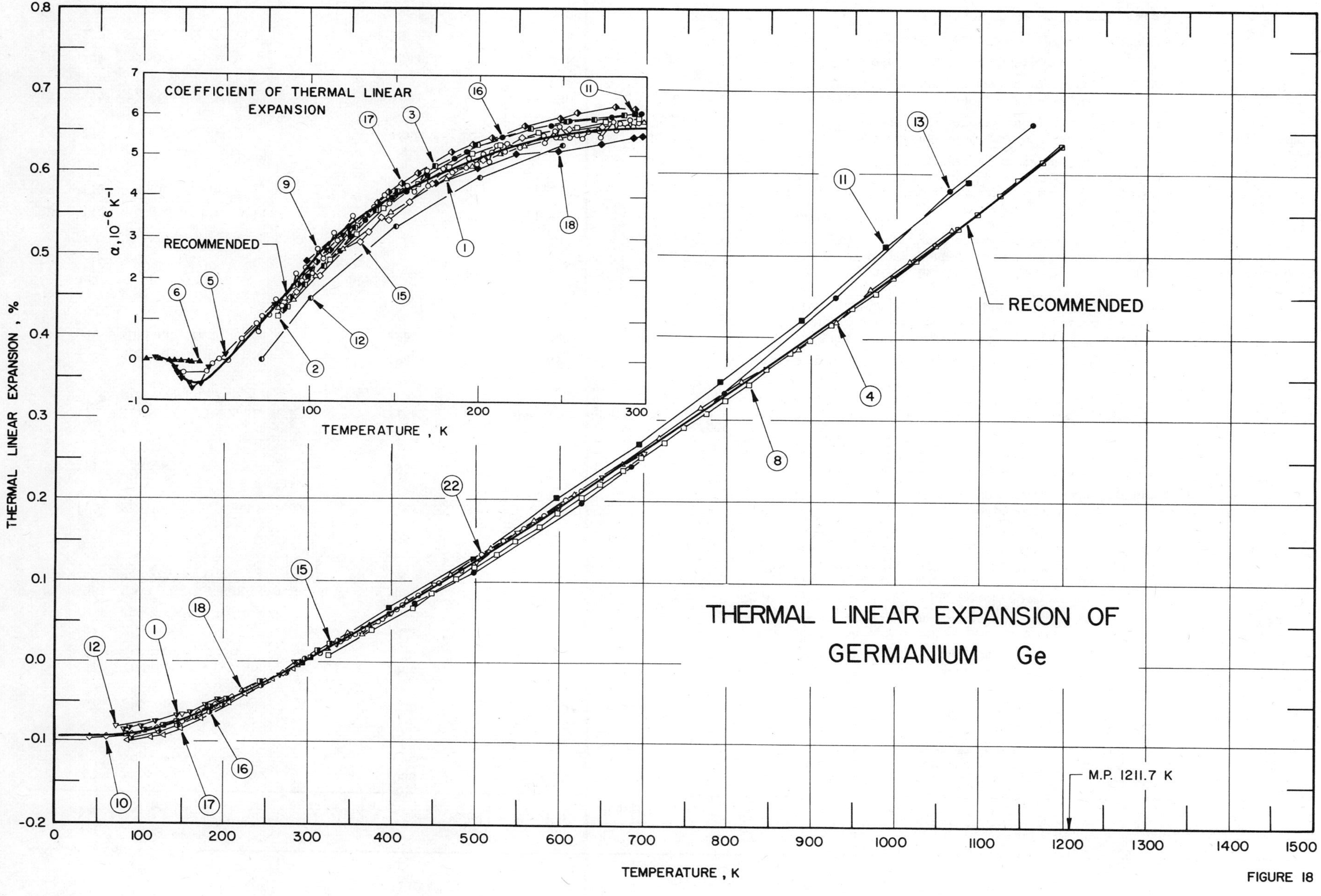
THERMAL LINEAR EXPANSION OF
GERMANIUM Ge
THERMAL LINEAR EXPANSION , %
TEMPERATURE , K
0.8
0.7
0.6
0.5
0.4
0.3
0.2
0.1
0.0
-0.1
-0.2
0
100
200
300
400
500
600
700
800
900
1000
1100
1200
1300
1400
1500
RECOMMENDED
M.P. 1211.7 K
1
4
8
10
11
12
13
15
16
17
18
22
COEFFICIENT OF THERMAL LINEAR
EXPANSION
α, 10⁻⁶ K⁻¹
TEMPERATURE , K
7
6
5
4
3
2
1
0
-1
0
100
200
300
RECOMMENDED
1
2
3
5
6
9
11
12
15
16
17
18

FIGURE 18

SPECIFICATION TABLE 18. THERMAL LINEAR EXPANSION OF GERMANIUM Ge

Cur. No.	Ref. No.	Author(s)	Year	Method Used	Temp. Range, K	Name and Specimen Designation	Composition (weight percent), Specifications, and Remarks
1	167	Zhdanova, V.V. and Kontorova, T.A.	1966	L	81-350	n-type Ge	n-type; doped with antimony, with carrier concentration 1.14 x 10^{14} cm^{-3}; specimen cut from single crystal block oriented along [111], length between 0.8 to 1.6 cm, and cross sectional area between 0.4 to 0.8 cm^2; reported error 2%.
2*	167	Zhdanova, V.V. and Kontorova, T.A.	1966	L	81-350	n-type Ge	Similar to the above specimen, with carrier concentration 2.4 x 10^{18} cm^{-3}.
3*	167	Zhdanova, V.V. and Kontorova, T.A.	1966	L	81-350	n-type Ge	Similar to the above specimen, doped with arsenic with carrier concentration 3.0 x 10^{19} cm^{-3}.
4	168	Zhdanova, V.V., Kekua, M.G. and Samadashvili, T.Z.	1967	L	289-1067		Pure; specimen with cross section area between 1.0 and 2.7 mm^2, and length between 0.1 and 0.8 mm; measured in vacuum; reported error 1 to 3%.
5*	169	Carr, R.H., McCammon, R.D. and White, G.K.	1965	E	2.5-20		Cylindrical specimen having less than 2 x 10^{15} donor impurities cm^{-3}; crystal supplied is GE 1K from Amalgamated Wireless Valve Company of Sydney; reported error ± 1%; this curve is here reported using the first given temperature as reference temperature at which $\Delta L/L_0 = 0$; the data on the coefficients of thermal linear expansion also reported.
6*	170	Sparks, P.W. and Swenson, C.A.	1967	V	2-34		Specimen from Mono-Silicon, tested as received; this curve is here reported using the first given temperature as reference temperature at which $\Delta L/L_0 = 0$.
7*	171	Sparks, P.W.	1966	V	1.9-32		Specimen from Mono-Silicon; diameter 1.2 cm, length 13.46 cm at 4 K; this curve is here reported using the first given temperature as reference temperature at which $\Delta L/L_0 = 0$.
8	172	Janot, C., Bianchi, G. and George, B.	1968	L	323-1198		High purity single crystal specimen of n-type; expansion observed along [111] axis; reported error <3.0%; zero-point correction of -0.005% determined by graphical extrapolation.
9*	173	Novikova, S.I.	1960	L	25-349		Data compilation of 16 series of tests on two crystalline specimens.
10	9	Gibbons, D.F.	1958	I	40-300		Single crystal, concentration of electrically active impurities 10^{14} cm^{-3}; expansion measured along <100> direction; zero-point correction of -0.0111% determined by graphical interpolation.
11	174	Singh, H.P.	1968	X	293-885		Specpure powder filled in quartz capillary of 0.5 mm inner diameter and annealed; lattice parameter reported at 293 K is 5.6575 Å; data on the coefficient of thermal linear expansion are also reported.
12	143	Nasekovskii, A.P.	1967		71-700		High purity specimen.
13	175	Shaw, N. and Liu, Y.H.	1965	X	291-1163		99.9999 Ge powder, annealed at 1170 K for 48 hr before high-temperature investigation; lattice parameter reported at 291 K is 5.6462 Å; 5.6463 Å at 293 K determined by graphical interpolation.
14*	176	Baldwin, T.O.	1968	X	299-773	FR-2	Crystal specimen; expansion affected by annealing of damage caused by neutron irradiation; reference temperature not reported by author.
15	177	Zhdanova, V.V.	1964	L	77-340	p-type	Single crystal p-type specimen 1.2 to 2.0 cm long x 0.3 to 0.8 cm^2 cross-section 1.15 x 10^{15} cm^{-3}; expansion measured under 200 gm load; reported error ±2%.
16	177	Zhdanova, V.V.	1964	L	77-340	p-type	Similar to the above specimen; carrier concentration 1.9 x 10^{18} cm^{-3}.
17	177	Zhdanova, V.V.	1964	L	77-340	p-type	Similar to the above specimen; carrier concentration 5.2 x 10^{19} cm^{-3}.
18	178	Fine, M.E.	1953	L	98-548		99.99^+ Ge single crystal, specimen cut with lengths parallel to <111> and <100> direction.

* Not shown in figure.

SPECIFICATION TABLE 18. THERMAL LINEAR EXPANSION OF GERMANIUM Ge (continued)

Cur. No.	Ref. No.	Author(s)	Year	Method Used	Temp. Range, K	Name and Specimen Designation	Composition (weight percent), Specifications, and Remarks
19*	179	Straumanis, M.E. and Aka, E.Z.	1952	X	283-323		99.99 pure Ge obtained from Eagle-Picher Co., Joplin, Mo., density 5.3234 g cm^{-3} at 298 K; lattice parameter reported at 293 K is 5.64589 kX.
20*	179	Straumanis, M.E. and Aka, E.Z.	1952	X	283-323		99.999 pure Ge, obtained from Eagle-Picher Co., Joplin, Mo., density 5.3234 g cm^{-3} at 298 K; lattice parameter reported at 283 K is 5.645518 kX; 5.64589 kX at 293 K determined by graphical interpolation.
21*	732	Bhalla, A.S. and White, E.W.	1971	X	302-314		No details given; zero-point correction of -0.003% determined by graphical extrapolation.
22	733	Feder, R. and Light, T.B.	1972	I	298-625		Pure specimen of a right circular cylinder 1 cm in diameter and 3.0774 cm length; zero-point correction of 0.03% obtained by graphical extrapolation; data on the coefficient of thermal linear expansion also reported.

* Not shown in figure.

DATA TABLE 18. THERMAL LINEAR EXPANSION OF GERMANIUM Ge

[Temperature, T, K; Linear Expansion, $\Delta L/L_0$, %]

T	$\Delta L/L_0$
CURVE 1‡	
81	-0.084
86	-0.075*
90	-0.083
92	-0.083*
97	-0.082*
104	-0.080
112	-0.079*
118	-0.077*
120	-0.077
124	-0.076*
136	-0.072*
147	-0.068
163	-0.062
180	-0.055
195	-0.048
214	-0.041*
228	-0.036*
248	-0.025*
270	-0.013
284	-0.005*
299	0.004*
315	0.013*
329	0.022*
339	0.028
CURVE 2*, ‡	
81	-0.093
86	-0.092
90	-0.092
103	-0.089
112	-0.087
128	-0.083
144	-0.077
163	-0.070
173	-0.065
182	-0.061
195	-0.055
214	-0.045
237	-0.032
258	-0.021
275	-0.011
295	0.001
308	0.009
341	0.029
CURVE 3*, ‡	
81	-0.096
86	-0.096
90	-0.095
96	-0.094
100	-0.093
108	-0.091
118	-0.089
126	-0.086
138	-0.082
150	-0.077
162	-0.072
173	-0.067
186	-0.061
199	-0.054
212	-0.047
231	-0.036
251	-0.025
268	-0.015
286	-0.004
306	0.008
322	0.018
338	0.028
CURVE 4‡	
289	-0.002
348	0.032
367	0.044
418	0.076
470	0.110
519	0.142
570	0.177
618	0.209
671	0.246
718	0.279
769	0.315
817	0.349
871	0.388
918	0.422
970	0.460
1017	0.495
1067	0.532
CURVE 5*, †, ‡	
2.51	0.00 x 10^{-6}
4.24	0.03
6.01	0.19
6.91	0.29
8.11	0.71
9.24	1.24
9.55	1.49
11.75	2.53
12.20	2.80
12.37	2.84
14.43	3.64
16.48	3.13
18.03	1.56
18.97	0.06
19.99	0.07
CURVE 6*, †, ‡	
2	0.000 x 10^{-3}
4	0.001
6	0.003
8	0.008
10	0.017
12	0.025
14	0.030
16	0.028
18	0.011
20	-0.025
22	-0.087
24	-0.171
26	-0.272
28	-0.386
30	-0.509
32	-0.638
34	-0.765
CURVE 7*, †	
1.89	0.00000 x 10^{-6}
3.86	0.0413
4.58	0.0839
4.76	0.1014
7.05	0.4914
7.84	0.7094
8.75	1.1304
CURVE 7 (cont.)*, †	
9.35	1.4704 x 10^{-6}
9.51	1.5404
10.40	1.9004
11.57	2.4704
12.63	2.8504
13.94	3.1804
14.66	3.2104
15.75	3.1104
16.09	3.0304
17.92	1.9704
18.29	1.2600
18.03	0.6384
18.17	0.5254
18.39	0.3664
20.21	-1.3196
20.50	-1.7096
20.58	-1.8096
22.12	-4.4096
22.41	-4.9896
22.48	-5.1396
23.96	-8.8296
24.10	-9.2296
24.31	-9.8296
25.82	-14.896
26.15	-15.9496
27.76	-22.3496
27.95	-23.1496
28.08	-23.4496
29.61	-30.6496
29.90	-38.6496
31.52	-38.6496
31.69	-39.3496
31.80	-39.8496
CURVE 8	
323.2	0.010
373.2	0.040
423.2	0.068
448.2	0.087
473.2	0.103
498.2	0.119
523.2	0.136
548.2	0.152
573.2	0.169
CURVE 8 (cont.)	
598.2	0.185
623.2	0.202
648.2	0.220
673.2	0.237
698.2	0.254
723.2	0.272
748.2	0.290
773.2	0.308
798.2	0.324
823.2	0.344
848.2	0.362
873.2	0.381
898.2	0.399
923.2	0.418
948.2	0.437
973.2	0.456
998.2	0.475
1023.2	0.495
1048.2	0.514
1073.2	0.534
1098.2	0.554
1123.2	0.574
1148.2	0.594
1173.2	0.614
1198.2	0.635
CURVE 9*, ‡	
25	-0.0095
28	-0.0095
33	-0.0096
39	-0.0096
42	-0.0095
46	-0.0096
52	-0.0096
53	-0.0096
54	-0.0096
60	-0.0096
63	-0.0095
69	-0.0095
70	-0.0095
71	-0.0095
76	-0.0094
77	-0.0094
80	-0.0094
CURVE 9 (cont.)*, ‡	
84	-0.0094
85	-0.0093
90	-0.0092
91	-0.0092
95	-0.0091
97	-0.0091
101	-0.0090
103	-0.0089
105	-0.0089
108	-0.0088
112	-0.0087
114	-0.0086
116	-0.0086
122	-0.0084
125	-0.0083
128	-0.0082
131	-0.0081
136	-0.0079
138	-0.0078
144	-0.0076
146	-0.0075
148	-0.0074
155	-0.0072
157	-0.0071
161	-0.0069
162	-0.0069
168	-0.0066
169	-0.0065
171	-0.0065
178	-0.0061
184	-0.0060
188	-0.0057
192	-0.0055
193	-0.0054
198	-0.0052
200	-0.0051
203	-0.0049
206	-0.0048
207	-0.0047
209	-0.0046
211	-0.0045
212	-0.0045
216	-0.0043
217	-0.0042
225	-0.0038

* Not shown in figure.
† This curve is here reported using the first given temperature as reference temperature at which $\Delta L/L_0 = 0$.
‡ Author's data for coefficient of thermal expansion have been integrated by TPRC to obtain $\Delta L/L_0$.

CURVE 9 (cont.)*

T	$\Delta L/L_0$
226	-0.0038
233	-0.0034
238	-0.0031
240	-0.0030
241	-0.0029
245	-0.0027
247	-0.0026
248	-0.0025
249	-0.0025
253	-0.0023
254	-0.0022
256	-0.0021
259	-0.0019
262	-0.0018
263	-0.0017
265	-0.0016
267	-0.0015
271	-0.0013
272	-0.0012
275	-0.0010
278	-0.0009
280	-0.0008
283	-0.0006
284	-0.0005
288	-0.0003
290	-0.0002
292	-0.00006
295	0.0001
301	0.0005
304	0.0006
307	0.0008
313	0.0012
316	0.0014
320	0.0018
323	0.0018
324	0.0018
331	0.0023
334	0.0023
337	0.0026
338	0.0027
339	0.0027
347	0.0032
349	0.0033

CURVE 10

T	$\Delta L/L_0$
40	-0.0922
50	-0.0921*
60	-0.0919
70	-0.0914*
80	-0.0905*
90	-0.0893
100	-0.0874*
110	-0.0850*
120	-0.0819
130	-0.0786*
140	-0.0749*
150	-0.0720
160	-0.0670*
170	-0.0627
180	-0.0583*
190	-0.0536
200	-0.0487*
210	-0.0439
220	-0.0390*
230	-0.0340
240	-0.0289*
250	-0.0237
260	-0.0183*
270	-0.0129
280	-0.0074*
290	-0.0017*
300	0.0040*

CURVE 11

T	$\Delta L/L_0$
293	0.000
397	0.069
496	0.129
595	0.205
692	0.270
790	0.348
889	0.422
987	0.514
1085	0.590

CURVE 12‡

T	$\Delta L/L_0$
71	-0.082
100	-0.080*
150	-0.068
200	-0.048*

CURVE 12 (cont.)‡

T	$\Delta L/L_0$
250	-0.024*
300	0.004*
350	0.033
400	0.064
450	0.096
500	0.127
550	0.160
600	0.192
650	0.225
700	0.258

CURVE 13

T	$\Delta L/L_0$
291	-0.001*
427	0.073
496	0.112
625	0.199
683	0.242
796	0.332
929	0.451
1062	0.580
1163	0.665

CURVE 14*

T	$\Delta L/L_0$
299	0.0105
353	0.0107
383	0.0096
405	0.0085
438	0.0056
469	0.0027
498	0.0037
523	0.0035
673	0.0027
773	0.00095

CURVE 15‡

T	$\Delta L/L_0$
85	-0.090
93	-0.089*
103	-0.087*
106	-0.086
113	-0.084*
122	-0.082*
130	-0.080
135	-0.078*

CURVE 15 (cont.)‡

T	$\Delta L/L_0$
142	-0.076*
146	-0.075
158	-0.070*
172	-0.064*
185	-0.058
190	-0.056*
203	-0.050*
207	-0.048
226	-0.038*
233	-0.034*
246	-0.027
254	-0.022*
273	-0.011*
292	-0.000*
293	0.000*
303	0.006
315	0.014
329	0.022

CURVE 16‡

T	$\Delta L/L_0$
99	-0.092
112	-0.089*
118	-0.088*
125	-0.086
133	-0.083
139	-0.081
147	-0.078
156	-0.074*
158	-0.073*
168	-0.069
178	-0.064*
185	-0.061
187	-0.060*
192	-0.057*
200	-0.053
215	-0.045*
224	-0.040*
244	-0.029
252	-0.024*
254	-0.023*
273	-0.012
278	-0.009*
293	0.000*
298	0.004

CURVE 16 (cont.)‡

T	$\Delta L/L_0$
306	0.008*
326	0.021*
337	0.028

CURVE 17‡

T	$\Delta L/L_0$
85	-0.097
89	-0.097*
93	-0.096*
105	-0.094*
113	-0.092
117	-0.091*
123	-0.089*
126	-0.088*
127	-0.087
134	-0.085*
140	-0.083*
146	-0.081
148	-0.080
154	-0.077*
161	-0.074*
163	-0.073*
171	-0.069
184	-0.063*
190	-0.060*
196	-0.057*
209	-0.050
228	-0.039
232	-0.037*
249	-0.027*
257	-0.022*
260	-0.020
263	-0.018*
277	-0.010*
281	-0.007*
282	-0.006
292	-0.000*
293	0.000*
310	0.011
329	0.023*

CURVE 18‡

T	$\Delta L/L_0$
98	-0.087*
123	-0.080*
148	-0.071*

CURVE 18 (cont.)‡

T	$\Delta L/L_0$
173	-0.060*
198	-0.049*
223	-0.037
248	-0.024*
273	-0.011*
293	0.000*
298	0.003*
323	0.017
348	0.031*
373	0.045
398	0.060
423	0.074*
448	0.089*
473	0.104*
498	0.119*
523	0.134*
548	0.149*

CURVE 19*

T	$\Delta L/L_0$
283	-0.0074
293	0.000
303	0.0051
313	0.0120
323	0.0195

CURVE 20*

T	$\Delta L/L_0$
283	-0.0058
303	0.0060
323	0.0179

CURVE 21*

T	$\Delta L/L_0$
302	0.005
306	0.007
310	0.010
314	0.012

CURVE 22

T	$\Delta L/L_0$
298	0.003
312	0.012
338	0.026
345	0.030
357	0.036
363	0.042

*Not shown in figure.
‡Author's data for coefficient of thermal expansion have been integrated by TPRC to obtain $\Delta L/L_0$.

DATA TABLE 18. THERMAL LINEAR EXPANSION OF GERMANIUM Ge (continued)

T	$\Delta L/L_0$
CURVE 22 (cont.)	
381	0.052
388	0.056
407	0.068
413	0.072
432	0.084
438	0.088*
458	0.100
464	0.104*
483	0.117
489	0.121*
509	0.135
515	0.138*
534	0.151
538	0.154*
559	0.168
564	0.171*
581	0.182
589	0.188*
608	0.201
613	0.204*
625	0.212

* Not shown in figure.

DATA TABLE 18. COEFFICIENT OF THERMAL LINEAR EXPANSION OF GERMANIUM Ge

[Temperature, T, K; Coefficient of Expansion, α, $10^{-6}\,K^{-1}$]

T	α
CURVE 1‡	
81	1.30
86	1.40*
90	1.53
92	1.68*
97	1.77*
104	2.06*
112	2.26*
118	2.55*
120	2.78
124	2.82*
136	3.22*
147	3.63
163	4.05*
180	4.48
195	4.78
214	5.02
228	5.26
248	5.49
270	5.67
284	5.75
299	5.83
315	5.90*
329	5.95*
339	5.98*
CURVE 2‡	
81	1.13
86	1.34*
90	1.53*
103	2.15*
112	2.50
128	3.10
144	3.73
163	4.34
173	4.56*
182	4.75
195	5.04
214	5.29
237	5.60
258	5.74
275	5.89
295	5.96
308	6.02*
341	6.11*
CURVE 3‡	
81	1.13
86	1.34
90	1.53*
96	1.86
100	2.12*
108	2.36
118	2.84*
126	3.21
138	3.68
150	4.13
162	4.48*
173	4.76
186	5.03*
199	5.25
212	5.44*
231	5.67
251	5.84
268	5.96
286	6.05
306	6.13*
322	6.16*
338	6.22*
CURVE 4*, ‡	
289	5.76
348	6.03
367	6.14
418	6.38
470	6.57
519	6.65
570	6.76
618	6.88
671	6.91
718	7.03
769	7.12
817	7.17
871	7.24
918	7.27
970	7.38
1017	7.45
1067	7.47
CURVE 5	
8	0.0035
10	0.0045*
CURVE 5 (cont.)	
12	0.0060*
14	0.0025*
16	-0.0045*
18	-0.013*
20	-0.023
22	-0.036
24	-0.049
26	-0.056*
28	-0.061
30	-0.071
35	-0.062
40	-0.024
50	0.165
60	0.523*
70	0.918*
80	1.38
90	1.85*
100	2.28
110	2.69
120	3.07
130	3.41*
CURVE 6‡	
2	0.000062
4	0.00050*
6	0.0016*
8	0.0040*
10	0.0045
12	0.0040*
14	0.0012*
16	-0.0040
18	-0.0125*
20	-0.0240
22	-0.0375*
24	-0.0470
26	-0.0540*
28	-0.0595
30	-0.0640
32	-0.0650*
34	-0.0615
CURVE 9‡	
25	-0.36*
28	-0.38*
33	-0.24*
CURVE 9 (cont.) ‡	
39	-0.31
42	-0.10
46	-0.03
52	-0.01
53	0.34*
54	0.16*
60	0.58
63	0.63*
69	0.95
70	0.73
71	1.10
76	1.10
77	1.27*
80	1.44
84	1.33
85	1.52
90	1.80
91	2.15
95	1.99
97	2.25
101	2.15
103	2.49*
105	2.77
108	2.53
112	2.75*
114	3.15
116	2.95
122	3.15*
125	3.53
128	3.35
131	3.55*
136	3.57*
138	3.78
144	4.00
146	3.81
148	3.99*
155	4.12*
157	4.28
161	4.16
162	4.36*
168	4.65
169	4.29
171	4.53*
178	4.72
184	4.64
188	4.75*
192	4.75
CURVE 9 (cont.) ‡	
193	4.98
198	4.80
200	4.91*
203	5.01
206	4.87
207	5.09
209	4.91*
211	5.26
212	5.04*
216	5.13
217	5.31
225	5.22
226	5.40*
233	5.44*
238	5.53
240	5.35
241	5.57*
245	5.61
247	5.44
248	5.77
249	5.64*
253	5.64*
254	5.43
256	5.58*
259	5.57
262	5.66*
263	5.44
265	5.62
267	5.47*
271	5.65
272	5.47
275	5.75*
278	5.75*
280	5.86
283	5.63
284	5.80
288	5.72
290	5.91
292	5.72
295	5.84
301	6.00*
304	5.87*
307	5.90*
313	6.00*
316	5.87*
320	6.13*
323	6.10*
CURVE 9 (cont.) ‡	
324	5.96
331	5.91*
334	5.87*
337	5.90*
338	6.13*
339	5.93*
347	6.10*
349	5.89*
CURVE 11	
293	6.12
397	6.49*
496	6.83*
595	7.17*
692	7.49*
790	7.81*
889	8.13*
987	8.44*
1085	8.74*
CURVE 12‡	
71	0.00
100	1.58
150	3.31
200	4.51
250	5.27
300	5.81*
350	6.04*
400	6.24*
450	6.35*
500	6.40*
550	6.48*
600	6.52*
650	6.56*
700	6.59*
CURVE 15‡	
85	1.42
93	1.66
103	2.05*
106	2.08
113	2.41*
122	2.71*
130	2.94

*Not shown in figure.
‡Author's data for coefficient of thermal expansion have been integrated by TPRC to obtain $\Delta L/L_0$.

DATA TABLE 18. COEFFICIENT OF THERMAL LINEAR EXPANSION OF GERMANIUM Ge (continued)

T	α
CURVE 15 (cont.)‡	
135	3.18
142	3.53
146	3.47
158	3.89
172	4.39
185	4.63
190	4.78*
203	4.92
207	5.10*
226	5.45
233	5.47*
246	5.60*
254	5.68
273	5.78
292	5.88
293	5.95*
303	6.02*
315	5.99*
329	6.12*
CURVE 16‡	
99	2.01
112	2.50*
118	2.72
125	3.03
133	3.41
139	3.62*
147	3.94
156	4.14
158	4.30*
168	4.59
178	4.72*
185	4.97
187	4.95*
192	5.11
200	5.23*
215	5.45
224	5.54*
244	5.77
252	5.86*
254	5.87
273	5.95*
278	6.00
293	6.09*
298	6.07
306	6.15*
326	6.17*
337	6.21*
CURVE 17‡	
85	1.23
89	1.59
93	1.81
105	2.40
113	2.70
117	2.93*
123	3.16
126	3.30*
127	3.25*
134	3.58
140	3.81
146	4.10
148	4.06*
154	4.35
161	4.58*
163	4.60
171	4.77*
184	5.10
190	5.24*
196	5.27
209	5.44
228	5.75
232	5.84*
249	5.97
257	6.13
260	6.10*
263	6.11*
277	6.17*
281	6.21
282	6.14*
292	6.18*
293	6.20
310	6.24*
329	6.30*
CURVE 18‡	
98	2.46
123	3.25
148	4.00
173	4.39
198	4.77
223	5.01
248	5.18
273	5.35
293	5.48
298	5.55
323	5.61*
348	5.73*
CURVE 18 (cont.)‡	
373	5.78*
398	5.80*
423	5.87*
448	5.92*
473	5.96*
498	5.98*
523	6.00*
548	6.00*
CURVE 22*	
293	5.49
360	5.98
386	6.17
408	6.26
435	6.39
462	6.51
486	6.61
509	6.65
532	6.74
559	6.76
584	6.76
607	6.80

*Not shown in figure.
‡Author's data for coefficient of thermal expansion have been integrated by TPRC to obtain $\Delta L/L_0$.

FIGURE AND TABLE NO. 19R. RECOMMENDED VALUES FOR THERMAL LINEAR EXPANSION OF GOLD Au

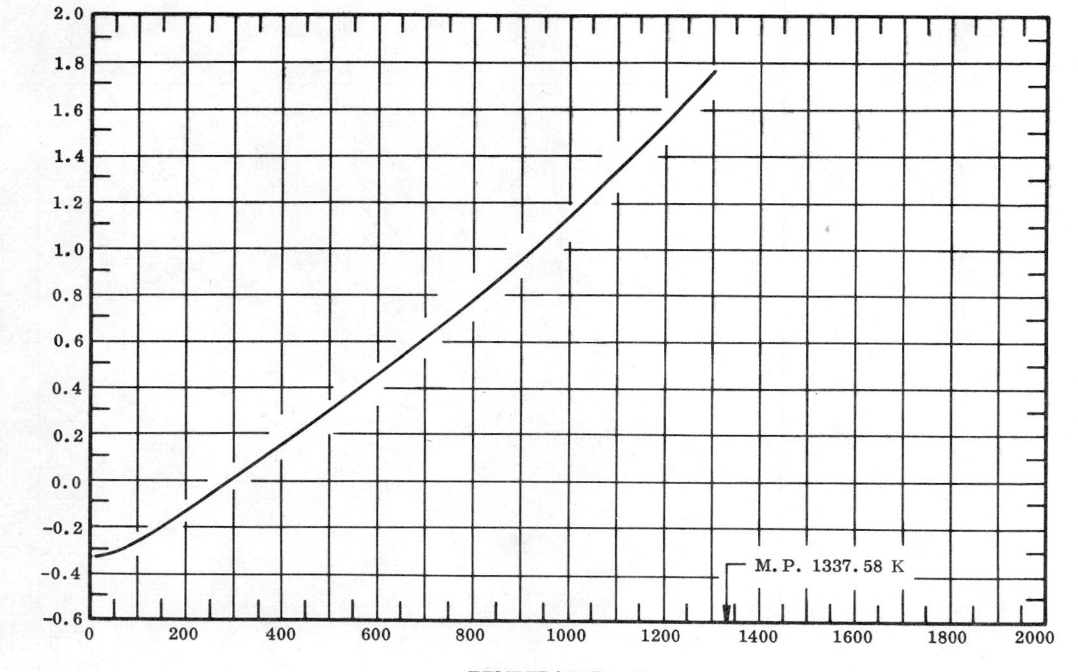

RECOMMENDED VALUES

[Temperature, T, K; Linear Expansion, $\Delta L/L_0$, %; α, K^{-1}]

T	$\Delta L/L_0$	$\alpha \times 10^6$
2	-0.327	0.002
5	-0.327	0.03
25	-0.325	2.8
50	-0.311	7.7
100	-0.260	11.8
200	-0.130	13.7
293	0.000	14.2
400	0.155	14.8
500	0.306	15.4
600	0.463	15.9
700	0.624	16.4
800	0.790	17.0
900	0.963	17.7
1000	1.146	18.6
1200	1.542	20.8
1300	1.757	22.1

REMARKS

The tabulated values are considered accurate to within ±3% over the entire temperature range. These values can be represented approximately by the following equations:

$$\Delta L/L_0 = -0.260 + 1.167 \times 10^{-3} (T - 100) + 1.347 \times 10^{-6} (T-100)^2 - 2.117 \times 10^{-9} (T-100)^3 \quad (100 < T < 293)$$

$$\Delta L/L_0 = 1.451 \times 10^{-3} (T - 293) + 1.213 \times 10^{-7} (T - 293)^2 + 1.681 \times 10^{-10} (T - 293)^3 \quad (293 < T < 1300)$$

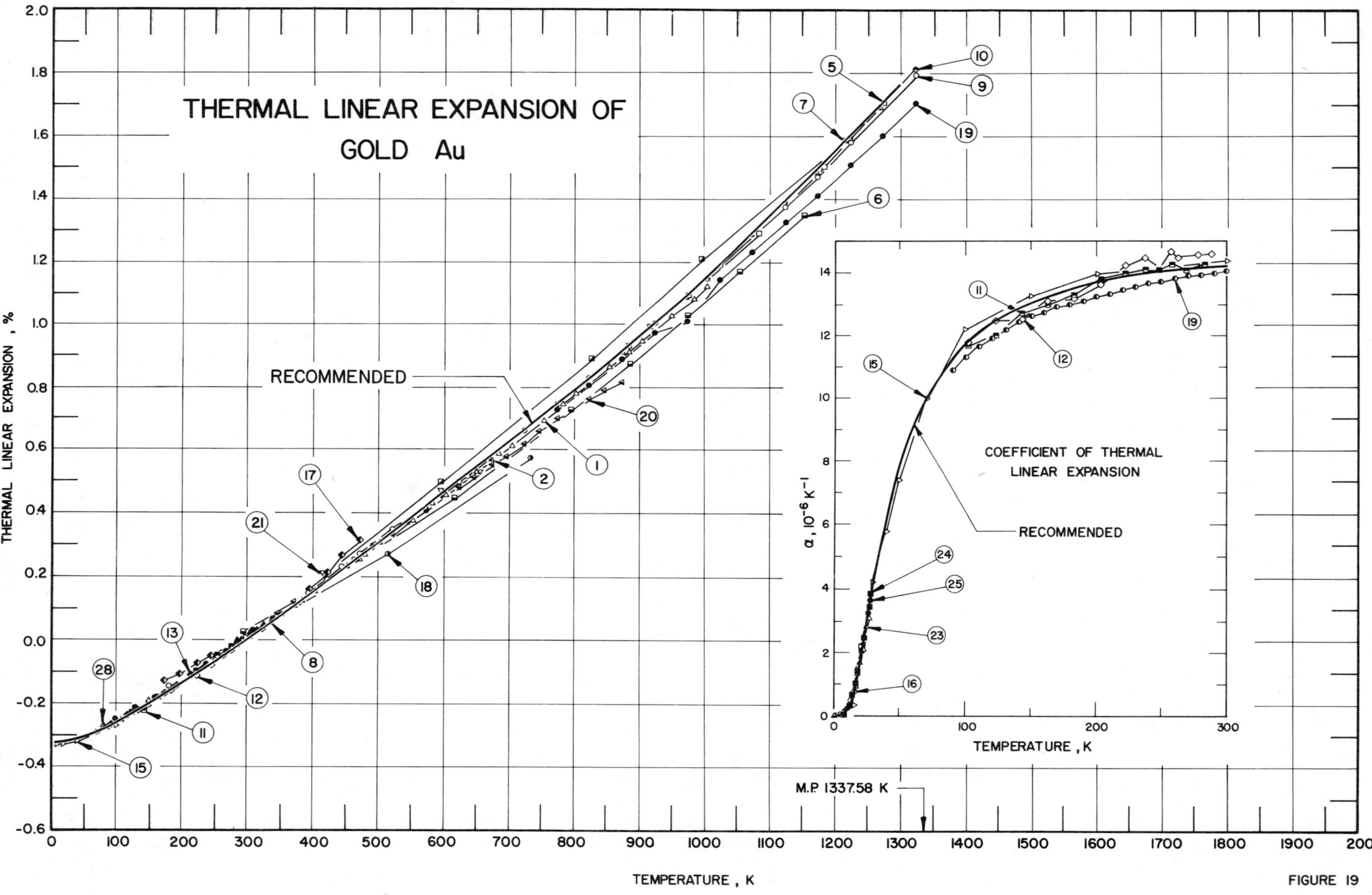
THERMAL LINEAR EXPANSION OF
GOLD Au
RECOMMENDED
THERMAL LINEAR EXPANSION , %
TEMPERATURE , K
M.P. 1337.58 K
COEFFICIENT OF THERMAL
LINEAR EXPANSION
RECOMMENDED
$\alpha, 10^{-6} K^{-1}$
TEMPERATURE , K

FIGURE 19

SPECIFICATION TABLE 19. THERMAL LINEAR EXPANSION OF GOLD Au

Cur. No.	Ref. No.	Author(s)	Year	Method Used	Temp. Range, K	Name and Specimen Designation	Composition (weight percent), Specifications, and Remarks
1	1	Nix, F.C. and MacNair, D.	1941	I	86-1003		99.99$^+$ Au with traces of Ag, Cu, and Pd; rod specimen about 6 mm long; low temperature portion of test made at cooling rate of 20 K per hr from room temperature with system containing helium gas at 3 mm Hg; region above room temperature the gas at a pressure of 3 mm Hg up to 423 K at heating rate of 15 K per hr, at which point system evacuated to pressure of 10^{-5} mm Hg; zero-point correction of -0.0275% determined by graphical interpolation.
2	181	Merryman, R.G. and Kempter, C.P.	1965	X	293-1270		Wire specimen.
3*	3	Fraser, D.B. and Hollis-Hallet, A.C.	1965	I	7.9-60		Cubic crystal structure; hollow cylinder 5.08 cm long, 2.54 cm O.D., 2.06 I.D., expansion measured on first test; this curve is here reported using the first given temperature as reference temperature at which $\Delta L/L_0 = 0$; zero-point correction is 0.3%; data on the coefficient of thermal linear expansion also reported.
4*	3	Fraser, D.B. and Hollis-Hallet, A.C.	1965	I	8.5-100		The above specimen; expansion measured on second test; this curve is here reported using the first given temperature as reference temperature at which $\Delta L/L_0 = 0$; zero-point correction is 0.3%; data on the coefficient of thermal linear expansion also reported.
5	182	Merryman, R.G.	1964	X	298-1275		<0.1 each Zn, Rh, Ir; <0.05 each K, V, Co, Sr, Cd; <0.01 each Na, Al, Ca, Cr, Fe, Ni, Ru, Pd, Ag, Ba, Pt; <0.004 each B, Mn, Cu, Pb; <0.002 Li, and <0.001 each Be, Mg; expansion measured in vacuo in the range 1 x 10^{-3} to 5 x 10^{-3} mm Hg.
6	183	Dutta, B.N. and Dayal, B.	1963	X	298-1151		99.9 Au, uniform cylindrical specimen of 0.04 cm diameter, supplied by Vin Vish Corp., Bombay, India; lattice parameter reported at 298 K is 4.0780 Å; 4.0770 Å at 293 K determined by graphical extrapolation.
7	184	Vermaak, J.S. and Kuhlmann-Wilsdorf, D.	1968	F	298-1213		99.999 Au evaporated onto amorphous carbon substrates, heated to 1023 K expansion determined by comparison with thallous chloride as standard; lattice parameter reported at 298 K is 4.0739 Å; 4.07357 Å at 293 K determined by graphical extrapolation.
8	185	Patel, V.K.	1967	X	288-338		99.999 Au filed from metal; powder annealed at 1023 K; expansion measured using CoK_{α_1} x-radiation; lattice parameter reported at 298 K is 4.07868 Å; 4.07833 Å at 293 K determined by graphical interpolation.
9	186	Simmons, R.O. and Balluffi, R.W.	1962	X	323-1323		99.999 Au from Sigmond Cohn and Co.; 0.0001 to 0.0009 Cu and Ag, no other impurities; specimen prepared by melting and controlled solidification to give bar 50 cm long x 1.27 cm square; cast in purified nitrogen atmosphere; average grain size 1.5 cm, nucleated along both sides of sample; grain used for measurement [211] direction inclined 1° to surface normal; expansion measured using NiK_{α_1} x-radiation (λ = 1.65784 Å), film exposure time varied from 2 min at 288 K to 12 min at 1330 K; data smoothed by author from measurements on two thermal cycles.
10	186	Simmons, R.O. and Balluffi, R.W.	1962	I	323-1323		The above specimen; expansion measured after annealing 6 hr at 1293 K with different technique.
11	70	Dorsey, H.G.	1908	I	103-288		99.998 Au, 0.002 Ag; obtained from U.S. Philadelphia Mint; specimen 1.3523 cm long; results average of several trials.

* Not shown in figure.

SPECIFICATION TABLE 19. THERMAL LINEAR EXPANSION OF GOLD Au (continued)

Cur. No.	Ref. No.	Author(s)	Year	Method Used	Temp. Range, K	Name and Specimen Designation	Composition (weight percent), Specifications, and Remarks
12	70	Dorsey, H. G.	1908	I	103-288		Specimen purified by solution method leaving any noble metals and traces of copper and silver; surface definitely crystalline; 0.8856 cm long.
13	48	Dorsey, H. G.	1907	I	93-293		99.0 Au, containing traces of platinum; cast; 1.1261 cm long; density 19.49 g cm^{-3}.
14*	48	Dorsey, H. G.	1907	I	193-295		The above specimen; expansion measured on second test.
15	22	Fraser, D. B. and Hollis-Hallett, A. C.	1961	I	10-300		Metal machined into form of thick-walled cylindrical tube having ends carefully ground plane and parallel to each other; annealed; reported error decreases from max of 20% at 10 K; data on the coefficient of thermal linear expansion also reported.
16*	148	Kos, J. F. and Lamarche, J. L. G.	1969	E	5-15		0.03 Cu, 0.03 Ag; polycrystalline sample 7 mm diameter, 10 cm long; from Johnson Matthey and Co.; temperature accurate to ±0.1 K; this curve is here reported using the first given temperature as reference temperature at which $\Delta L/L_0 = 0$.
17	34	Asay, J. R	1968		173-473		99.0 Au, density 18.959 g cm^{-3}; zero-point correction is -0.005%.
18	261	Shinoda, G.	1934	X	288-732		Expansion measured in (422) plane using CuKα x-radiation; zero-point correction of -0.006% determined by graphical interpolation.
19	452	Leksina, I. E. and Novikova, S.I.	1963	L	90-1323		99.99 Au, remelted in vacuum; specimen <25 mm long.
20	611	Kurnakow, N. S. and Ageew, N. W.	1931	L	293-873		Chemically purified sample; expansion measured with increasing temperature.
21	612	Boiko, B. T., Palatnik, L. S., and Pugachev, A. T.	1967	F	676-293		99.9$^+$ pure, thin film specimen condensed in vacuum 5 x 10^{-5} torr; thickness of film in the range 600 + 10 Å; lattice parameter expansion measured with decreasing temperature.
22*	613	Fitzer, E. and Weisenburger, S.	1972	L, O, X	293-1173		High purity specimen; average values from 16 laboratories.
23*	614	McLean, K. O., Swenson, C. A., and Case, C. R.	1972	V	1-26	Cu-1	99.999$^+$ pure, specimen from Nuclear Element Corp was vacuum cast; this curve is here reported using the first given temperature as reference temperature at which $\Delta L/L_0 = 0$; ±2% accuracy, specimen 0.6 cm diameter and 8.819 cm long at 4.2 K.
24*	615	White, G. K. and Collins, J. G.	1972	L	2-29, 57-85, 283	Au 1	99.9$^+$ Au, chemical analysis at Australian Mineral Development Laboratories Ltd.; showed 0.0070 Fe, <0.0010 Mn; cylindrical specimen 5.09 cm long, 1.9 cm diameter from Matthey Garrett Ltd.; this curve is here reported using the first given temperature as reference temperature at which $\Delta L/L_0 = 0$.
25*	615	White, G. K. and Collins, J. G.	1972	L	2-29	Au 2	99.95$^+$ pure, cylindrical specimen 5.09 cm long, 1.6 cm diameter from Matthey Garrett Ltd., cast in graphite (ATJ grade); this curve is here reported using the first given temperature as reference temperature at which $\Delta L/L_0 = 0$.
26*	616	Straumanis, M. E.	1971	X	288-338		99.999 pure specimen recovered from cold work (filing) occurred at room temperature; lattice parameter reported at 287.7 K is 4.07799 Å; 4.07835 Å at 293 K determined by graphical interpolation.
27*	639	Fitzer, E.	1971	L	293-1173		99.6 pure specimen by Wohlwill Electrolytic Refining Process; refined to 99.999 by several processes, mostly liquid-liquid extraction method; remelted in graphite crucible in a high frequency induction process and vacuum 10^{-3} torr, poured into slit graphite mold, cold-rolled to rolling sections of 12 to 6 mm and then cold-drawn to 3-10 mm diameter; average of 10 participants.

* Not shown in figure.

SPECIFICATION TABLE 19. THERMAL LINEAR EXPANSION OF GOLD Au (continued)

Cur. No.	Ref. No.	Author(s)	Year	Method Used	Temp. Range, K	Name and Specimen Designation	Composition (weight percent), Specifications, and Remarks
28	654	Nagender Naidu, S.V. and Houska, C.R.	1971	X	80-298		Prepared from 99.97 Au from Engelhard Industries Inc. by melting in induction furnace under hydrogen atmosphere; homogenized for 1 week at 1173 K; annealed at 873 K in evacuated quartz tubes after flushing 3 times with argon; lattice parameter reported at 298 K is 4.0779 Å; 4.0777 Å at 293 K determined by graphical interpolation.
29*	657	Vest, R.W.	1971	L	293-1072		Specimen prepared from 99.9 pure metal; data on the coefficient of thermal linear expansion also reported.
30*	767	Straumanis, M.E. and Patel, V.K.	1972	X	288-338		99.999 Au from American Smelting and Refining Co. containing from 0.0001 to 0.0005 Cu, Mg, Fe, Si, Pb, and Ag.

* Not shown in figure.

DATA TABLE 19. THERMAL LINEAR EXPANSION OF GOLD Au

[Temperature, T, K; Linear Expansion, $\Delta L/L_0$, %]

T	$\Delta L/L_0$
CURVE 1	
86.2	-0.2723 *
88.2	-0.2680 *
94.2	-0.2622 *
98.2	-0.2565 *
103.2	-0.2507
108.7	-0.2449 *
113.2	-0.2402 *
118.5	-0.2434 *
123.2	-0.2276 *
128.2	-0.2219 *
132.7	-0.2161 *
136.2	-0.2106 *
142.1	-0.2048 *
146.7	-0.1988 *
150.7	-0.1930
155.9	-0.1874 *
160.5	-0.1815 *
165.2	-0.1757 *
169.2	-0.1699 *
173.7	-0.1642 *
178.2	-0.1584 *
182.7	-0.1527
187.0	-0.1469 *
191.2	-0.1411 *
194.7	-0.1353 *
199.4	-0.1296 *
204.2	-0.1238*
208.2	-0.1181 *
213.0	-0.1123 *
216.7	-0.1065 *
221.2	-0.1007 *
225.0	-0.0950 *
228.6	-0.0892 *
233.2	-0.0835 *
237.2	-0.0780 *
242.0	-0.0719 *
246.2	-0.0661 *
250.2	-0.0604
254.2	-0.0546 *
258.2	-0.0488 *
262.2	-0.0434 *
267.1	-0.0373 *
270.5	-0.0315 *
273.2	-0.0275 *
CURVE 1 (cont.)	
275.2	-0.0262 *
277.2	-0.0223 *
283.2	-0.0142 *
287.2	-0.0085 *
291.2	-0.0027 *
295.2	0.0034 *
298.2	0.0074 *
298.2	0.0074 *
302.9	0.0143
307.2	0.0201 *
311.2	0.0258 *
315.8	0.0316 *
319.0	0.0374 *
323.3	0.0434 *
327.2	0.0489 *
331.7	0.0550 *
335.2	0.0605 *
339.2	0.0662
343.2	0.0720 *
347.4	0.0778 *
352.0	0.0835
356.2	0.0896 *
360.2	0.0951 *
364.2	0.1008 *
367.2	0.1054 *
373.2	0.1137 *
377.4	0.1189 *
381.7	0.1247
385.5	0.1304 *
390.2	0.1362 *
394.2	0.1420 *
398.0	0.1480 *
402.7	0.1546
406.2	0.1593 *
410.8	0.1659 *
414.2	0.1713 *
418.2	0.1771 *
422.7	0.1829 *
426.1	0.1881 *
429.7	0.1938 *
434.2	0.2002 *
437.2	0.2054 *
441.3	0.2114 *
445.7	0.2172 *
448.4	0.2226 *
CURVE 1 (cont.)	
453.2	0.2284
456.4	0.2342*
461.4	0.2402*
464.2	0.2457*
469.8	0.2517*
472.2	0.2572*
477.2	0.2630*
480.0	0.2690
484.2	0.2748*
488.2	0.2811*
491.4	0.2861*
495.6	0.2918*
499.5	0.2976*
504.0	0.3039
507.8	0.3097*
511.7	0.3149*
516.4	0.3212*
519.4	0.3264*
523.0	0.3322*
527.2	0.3379*
530.8	0.3437*
534.5	0.3494*
538.1	0.3552*
542.0	0.3610*
545.9	0.3627*
549.3	0.3728*
552.8	0.3788
557.0	0.3846*
560.2	0.3904*
564.0	0.3958*
567.2	0.4019*
570.9	0.4077*
574.6	0.4134*
578.3	0.4192*
581.2	0.4247
584.8	0.4301*
588.8	0.4362*
592.0	0.4417*
595.4	0.4477*
599.2	0.4532*
603.2	0.4592
607.1	0.4647*
610.6	0.4705*
614.2	0.4765*
618.0	0.4820*
CURVE 1 (cont.)	
621.0	0.4878*
624.5	0.4935*
628.0	0.4993*
631.4	0.5051*
635.7	0.5108*
639.2	0.5172*
642.0	0.5224*
646.2	0.5281*
650.2	0.5339
653.5	0.5396*
657.6	0.5454*
660.3	0.5512*
664.4	0.5569*
668.3	0.5627*
672.0	0.5685*
676.3	0.5742*
679.1	0.5800*
683.1	0.5858
686.2	0.5915*
690.2	0.5973*
693.7	0.6030*
697.2	0.6088*
701.2	0.6146
704.8	0.6203*
708.5	0.6261*
712.9	0.6319*
716.2	0.6382*
719.2	0.6434*
723.0	0.6492*
726.9	0.6549*
730.2	0.6607*
731.0	0.6664*
734.0	0.6722*
740.8	0.6780*
744.8	0.6837*
748.3	0.6895*
752.1	0.6953
762.2	0.7126*
765.2	0.7183*
779.4	0.7408*
783.7	0.7483
788.0	0.7543*
790.5	0.7592*
803.0	0.7803
830.2	0.8299*
CURVE 1 (cont.)	
832.2	0.8328*
841.2	0.8484*
842.3	0.8507*
845.9	0.8559*
848.2	0.8605*
849.2	0.8616*
852.2	0.8674
856.3	0.8740*
859.2	0.8795*
862.2	0.8847*
865.5	0.8905*
869.2	0.8968*
872.2	0.8970*
876.2	0.9078*
879.2	0.9135
883.0	0.9193*
887.0	0.9256*
889.7	0.9314*
893.0	0.9366*
896.9	0.9429*
901.2	0.9481
903.2	0.9539*
906.2	0.9596*
910.0	0.9654*
913.7	0.9720*
917.2	0.9774*
920.2	0.9827*
923.7	0.9884*
927.0	0.9942*
930.7	1.0000*
934.2	1.0057*
936.2	1.0115*
941.2	1.0173*
945.2	1.0236*
948.2	1.0288*
950.2	1.0345
954.2	1.0403*
958.0	1.0461*
961.2	1.0518*
964.2	1.0576*
967.5	1.0639*
970.8	1.0691*
974.2	1.0755*
978.2	1.0815*
981.4	1.0864
CURVE 1 (cont.)	
984.2	1.0922*
987.2	1.0979*
990.2	1.1043*
993.7	1.1095*
997.0	1.1152*
999.2	1.1210*
1003.0	1.1268
CURVE 2	
293	0.000
298	0.006*
373	0.105
474	0.253
673	0.570
771	0.741
872	0.915
970	1.094
1071	1.288
1170	1.482
1270	1.690
CURVE 3*, †	
7.9	0.0000
9.5	0.0001
10.4	0.0001
11.9	0.0002
13.6	0.0003
17.4	0.0006
18.8	0.0008
20.0	0.0011
21.7	0.0011
24.8	0.0025
26.6	0.0031
29.2	0.0040
31.6	0.0053
34.5	0.0063
35.9	0.0075
38.3	0.0086
41.7	0.0103
46.0	0.0130
49.4	0.0156
53.0	0.0185
56.6	0.0217
59.9	0.0244

* Not shown in figure.

† This curve is here reported using the first given temperature as reference temperature at which $\Delta L/L_0 = 0$.

DATA TABLE 19. THERMAL LINEAR EXPANSION OF GOLD Au (continued)

T	$\Delta L/L_0$
CURVE 4*, †	
8.5	0.0000
10.4	0.0001
11.5	0.0001
12.7	0.0002
14.4	0.0002
16.8	0.0004
20.2	0.0010
23.0	0.0017
25.6	0.0026
27.8	0.0033
32.7	0.0057
35.9	0.0071
37.1	0.0079
39.4	0.0091
41.6	0.0103
45.7	0.0128
49.5	0.0157
53.0	0.0185
56.5	0.0217
60.3	0.0249
63.4	0.0278
66.7	0.0309
70.0	0.0342
73.2	0.0374
76.2	0.0407
79.3	0.0440
82.4	0.0473
85.7	0.0509
88.2	0.0539
91.2	0.0571
94.0	0.0606
99.8	0.0669
CURVE 5	
298	0.0071*
385.5	0.1317*
469	0.2508
585.5	0.4330*
775	0.7436*
883	0.9307
918	0.9942
973	1.0979
1081	1.1290
1183	1.5096
1275	1.7094

T	$\Delta L/L_0$
CURVE 6	
298	0.024
618	0.442
798	0.724
888	0.878
973	1.028
1058	1.172
1151	1.347
CURVE 7	
298	0.008*
598	0.499
828	0.892
998	1.216
1213	1.582
CURVE 8	
288.0	-0.0082*
298.2	0.0087*
308.5	0.0221
318.4	0.0319
328.2	0.0464
337.6	0.0579
CURVE 9	
323	0.043*
348	0.079
373	0.116*
395	0.148
423	0.189
447	0.225
473	0.265
518	0.334*
523	0.342
573	0.419
582	0.433*
623	0.498
636	0.519
673	0.578*
674	0.580
723	0.660
773	0.745*
823	0.829
873	0.915*
923	1.004
973	1.094

T	$\Delta L/L_0$
CURVE 9 (cont.)	
1023	1.187
1073	1.281*
1123	1.377
1173	1.477
1223	1.580
1273	1.685*
1323	1.793
CURVE 10	
323	0.043*
373	0.116
423	0.189
472	0.263
473	0.265
523	0.342
573	0.419
574	0.421
623	0.498
672	0.577
673	0.578
723	0.660
773	0.745
779	0.755
823	0.829
872	0.913
873	0.915
923	1.004
973	1.095
974	1.097*
1003	1.151
1023	1.189*
1073	1.285*
1123	1.384
1173	1.485*
1223	1.591
1273	1.701*
1323	1.815
CURVE 11 ‡	
103	-0.26
123	-0.23
143	-0.21
163	-0.18
183	-0.15*
203	-0.13*
223	-0.10

T	$\Delta L/L_0$
CURVE 11 (cont.) ‡	
238	-0.08*
243	-0.07*
248	-0.06
258	-0.05
263	-0.04*
268	-0.04*
278	-0.02
283	-0.01*
288	-0.007
CURVE 12 ‡	
103	-0.26*
123	-0.24*
143	-0.22*
163	-0.19*
183	-0.16
203	-0.14*
223	-0.11
238	-0.09*
243	-0.08
248	-0.07*
258	-0.06*
263	-0.05
268	-0.04*
278	-0.03
283	-0.02*
288	-0.01*
CURVE 13	
93	-0.268
113	-0.244
133	-0.220
153	-0.195*
173	-0.169
193	-0.143*
213	-0.116
233	-0.086
253	-0.059*
273	-0.029
293	0.000*
CURVE 14*	
193	-0.142
201.3	-0.129
213	-0.113

T	$\Delta L/L_0$
CURVE 14 (cont.)*	
223	-0.097
233	-0.084
243	-0.071
253	-0.056
263	-0.041
273	-0.027
284.8	-0.010
294.5	0.002
CURVE 15 ‡	
10	-0.330
15	-0.330*
20	-0.329
30	-0.326*
40	-0.321
50	-0.315*
70	-0.297
100	-0.264
150	-0.200
200	-0.132*
250	-0.061*
300	0.010
CURVE 16*, †, ‡	
5	0.000 x 10^{-4}
6	0.036
7	0.097
8	0.193
9	0.335
10	0.534
11	0.804
12	1.159
13	1.611
14	2.174
15	2.868
CURVE 17	
173	-0.125
198	-0.102
223	-0.077
248	-0.050
273	-0.022*
298	0.005*
323	0.037
348	0.070*

T	$\Delta L/L_0$
CURVE 17 (cont.) ‡	
373	0.109*
398	0.157
423	0.208
448	0.263
473	0.317
CURVE 18	
288	-0.006*
515	0.268
732	0.572
CURVE 19 ‡	
90	-0.264
100	-0.252
110	-0.241*
120	-0.229*
130	-0.217
140	-0.205*
150	-0.192*
160	-0.180
170	-0.167*
180	-0.154
190	-0.141*
200	-0.127*
210	-0.114*
220	-0.101*
230	-0.087*
240	-0.073
250	-0.060*
260	-0.046*
270	-0.032
280	-0.018*
290	-0.004*
300	0.010*
310	0.024
320	0.038*
330	0.052*
340	0.067
373	0.114*
423	0.187*
473	0.260*
523	0.335*
573	0.411
623	0.489*
773	0.725
823	0.805

* Not shown in figure.

† This curve is here reported using the first given temperature as reference temperature at which $\Delta L/L_0 = 0$.

‡ Author's data for coefficient of thermal expansion have been integrated by TPRC to obtain $\Delta L/L_0$.

DATA TABLE 19. THERMAL LINEAR EXPANSION OF GOLD Au (continued)

T	$\Delta L/L_0$
CURVE 19 (cont.) ‡	
873	0.887
923	0.970
973	1.055
1023	1.142
1073	1.231
1123	1.321
1173	1.414
1223	1.509
1273	1.606
1323	1.705
CURVE 20	
293	0.00*
323	0.04
348	0.08
373	0.12
398	0.15*
423	0.19*
448	0.23*
473	0.27*
498	0.30
523	0.33
548	0.37
573	0.41
598	0.45
623	0.48
648	0.51
673	0.55
698	0.58
723	0.62
748	0.66
773	0.70
798	0.73*
823	0.76
848	0.79
873	0.82
CURVE 21	
676	0.565*
651	0.528
598	0.466
548	0.367*
492	0.291
419	0.205
373	0.103*
297	0.013*
CURVE 22*	
293	0.000
573	0.417
773	0.744
973	1.096
1173	1.480
CURVE 23*, †, ‡	
0	0.000 x 10^{-4}
1	0.0002
2	0.0016
3	0.0063
4	0.0176
5	0.0405
6	0.0810
7	0.146
8	0.244
9	0.387
10	0.587
12	1.22
14	2.26
15	2.98
16	3.85
18	6.11
20	9.13
22	13.0
24	17.7
25	20.4
26	23.4
CURVE 24*, †, ‡	
2.458	0.000 x 10^{-4}
2.829	0.002
3.072	0.004
3.373	0.007
3.621	0.009
3.772	0.011
4.058	0.016
4.083	0.016
4.444	0.022
4.882	0.033
5.311	0.046
5.694	0.061
6.033	0.077
6.416	0.098
6.846	0.126
7.319	0.165
CURVE 24 (cont.)*, †, ‡	
7.908	0.224 x 10^{-4}
8.587	0.310
9.291	0.424
10.027	0.574
10.859	0.788
11.741	1.07
11.859	1.12
13.055	1.66
14.310	2.41
15.544	3.36
16.917	4.71
18.564	6.77
20.235	9.40
21.857	12.5
23.368	15.9
24.897	19.9
26.369	24.3
27.787	28.9
29.394	34.8
CURVE 24*, †, ‡	
1.193	0.000 x 10^{-4}
2.539	0.002
3.012	0.004
3.392	0.008
3.758	0.012
4.070	0.016
4.427	0.023
4.854	0.033
5.282	0.046
5.659	0.060
6.009	0.076
6.402	0.098
6.824	0.126
7.290	0.163
7.855	0.219
8.504	0.300
9.198	0.410
9.929	0.557
10.704	0.753
11.739	1.09
12.984	1.64
14.197	2.36
15.425	3.30
16.777	4.62
18.324	6.52
20.028	9.15
CURVE 25 (cont.)*, †, ‡	
21.759	12.4 x 10^{-4}
23.403	16.1
24.967	20.2
26.726	25.5
28.512	31.7
CURVE 26*	
287.7	-0.0088
298.2	0.0081
308.5	0.0216
318.2	0.0314
328.2	0.0458
337.6	0.0574
CURVE 27*	
293	0.000
573	0.418
673	0.577
773	0.745
873	0.919
973	1.097
1073	1.265
1173	1.479
CURVE 28	
80	-0.270
195	-0.128
298	0.005*
CURVE 29*	
293	0.000
373	0.091
473	0.225
573	0.365
673	0.511
773	0.671
873	0.834
973	1.005
1072	1.169
CURVE 30*	
288	-0.007
338	0.060

* Not shown in figure.
† This curve is here reported using the first given temperature as reference temperature at which $\Delta L/L_0 = 0$.
‡ Author's data for coefficient of thermal expansion have been integrated by TPRC to obtain $\Delta L/L_0$.

DATA TABLE 19. COEFFICIENT OF THERMAL LINEAR EXPANSION OF GOLD Au

[Temperature, T, K; Coefficient of Expansion, α, 10^{-6} K^{-1}]

T	α
CURVE 3	
11.5	0.51
13.9	0.83
15.8	1.11
16.8	1.40
19.5	2.25
22.4	2.95
24.1	3.22
27.0	3.45
29.1	4.38
31.8	4.34
33.7	4.98
36.4	6.03
38.8	4.86
42.1	5.77
45.5	6.92
49.5	7.74
53.0	8.43
56.4	8.59
60.0	8.62
CURVE 4	
10.6	0.39
12.4	0.47
14.7	0.69
17.3	1.36
19.9	2.06
22.9	2.79
25.4	3.34
29.6	4.39
31.8	4.69
34.9	4.95
37.6	5.54
39.3	5.49
42.5	5.94
45.5	6.82
49.3	7.74
53.0	8.50
56.6	8.74
59.9	8.86
63.5	9.27
66.7	9.61
69.9	10.09
73.1	10.5
76.2	10.7
79.3	10.6
CURVE 4 (cont.)	
82.5	10.8
85.3	11.3
88.4	11.2
CURVE 11‡	
103	11.70
123	12.08
143	12.79
163	13.03
183	13.39
203	13.80
223	14.04
238	14.18
243	14.15*
248	14.16
258	14.35
263	14.25*
268	14.16
278	14.26*
283	14.29
288	14.40
CURVE 12‡	
103	11.80
123	12.49
143	12.54
163	13.11
183	13.21
203	13.67
223	14.29
238	14.51
243	14.31*
248	14.12*
258	14.70
263	14.55
268	14.54*
278	14.60
283	14.62*
288	14.60
CURVE 15‡	
10	0.2
15	0.3
CURVE 15 (cont.)‡	
20	2.2
30	4.2
40	5.8
50	7.4
70	10.0
100	12.2
150	13.3
200	14.0
250	14.2*
300	14.4
CURVE 16‡	
5	0.0260
6	0.0464*
7	0.0751*
8	0.1163*
9	0.168*
10	0.231
11	0.309*
12	0.400*
13	0.504
14	0.622*
15	0.767
CURVE 19‡	
90	10.94*
100	11.32
110	11.66
120	11.95
130	12.20
140	12.42
150	12.61
160	12.78
170	12.93
180	13.06
190	13.18
200	13.29
210	13.39
220	13.49
230	13.58
240	13.67
250	13.75
260	13.82
270	13.88
CURVE 19 (cont.)‡	
280	13.94
290	14.00
300	14.06
310	14.12*
320	14.18*
330	14.23*
340	14.28*
373	14.43*
423	14.66*
473	14.87*
523	15.10*
573	15.34*
623	15.60*
773	15.88*
823	16.18*
873	16.50*
923	16.83*
973	17.18*
1023	17.54*
1073	17.92*
1123	18.32*
1173	18.74*
1223	19.18*
1273	19.65*
1323	20.15*
CURVE 23‡	
1	0.00044*
2	0.00239*
3	0.0069*
4	0.0158*
5	0.0300*
6	0.0509*
7	0.0795*
8	0.1176*
9	0.168*
10	0.231*
12	0.404*
14	0.643*
15	0.788*
16	0.950
18	1.310
20	1.716
22	2.150
24	2.61*
CURVE 23 (cont.)‡	
25	2.84
26	3.08
CURVE 24‡	
2.458	0.0048*
2.829	0.0067*
3.072	0.0078*
3.373	0.0101*
3.621	0.0124
3.772	0.0133*
4.058	0.0170*
4.083	0.0156*
4.444	0.0204*
4.882	0.0276*
5.311	0.0351*
5.694	0.0425*
6.033	0.0497
6.416	0.0596*
6.846	0.0731*
7.319	0.0886*
7.908	0.1128*
8.587	0.1418*
9.291	0.1805*
10.027	0.2267*
10.859	0.2891*
11.741	0.3677
11.859	0.3804*
13.055	0.5186*
14.310	0.6784
15.544	0.866*
16.917	1.100
18.564	1.407
20.235	1.740*
21.857	2.101
23.368	2.441
24.897	2.794*
26.369	3.133*
27.787	3.464
29.394	3.834
57.5	8.75§
65	9.58§
75	10.41§
85	11.03§
283	14.08§
CURVE 25‡	
1.973	0.0020*
2.539	0.0044*
3.012	0.0071*
3.392	0.0099*
3.758	0.0127*
4.070	0.0159*
4.427	0.0207*
4.854	0.0269*
5.382	0.0343*
5.659	0.0415*
6.009	0.0499*
6.402	0.0602*
6.824	0.0716*
7.290	0.0883*
7.855	0.1104*
8.504	0.1397*
9.198	0.1773*
9.929	0.2227*
10.704	0.2834*
11.739	0.3729*
12.984	0.5133*
14.197	0.6725*
15.425	0.8561*
16.777	1.0960*
18.324	1.367*
20.028	1.714*
21.759	2.086*
23.403	2.456*
24.967	2.813*
26.726	3.219
28.512	3.622
CURVE 29*	
373	12.77
473	13.48
573	14.16
673	14.86
773	15.58
873	16.21
973	16.94
1073	17.60

* Not shown in figure.
‡ Author's data for coefficient of thermal expansion have been integrated by TPRC to obtain $\Delta L/L_0$.
§ Average of two runs.

FIGURE AND TABLE NO. 20R. RECOMMENDED VALUES FOR THERMAL LINEAR EXPANSION OF HAFNIUM Hf

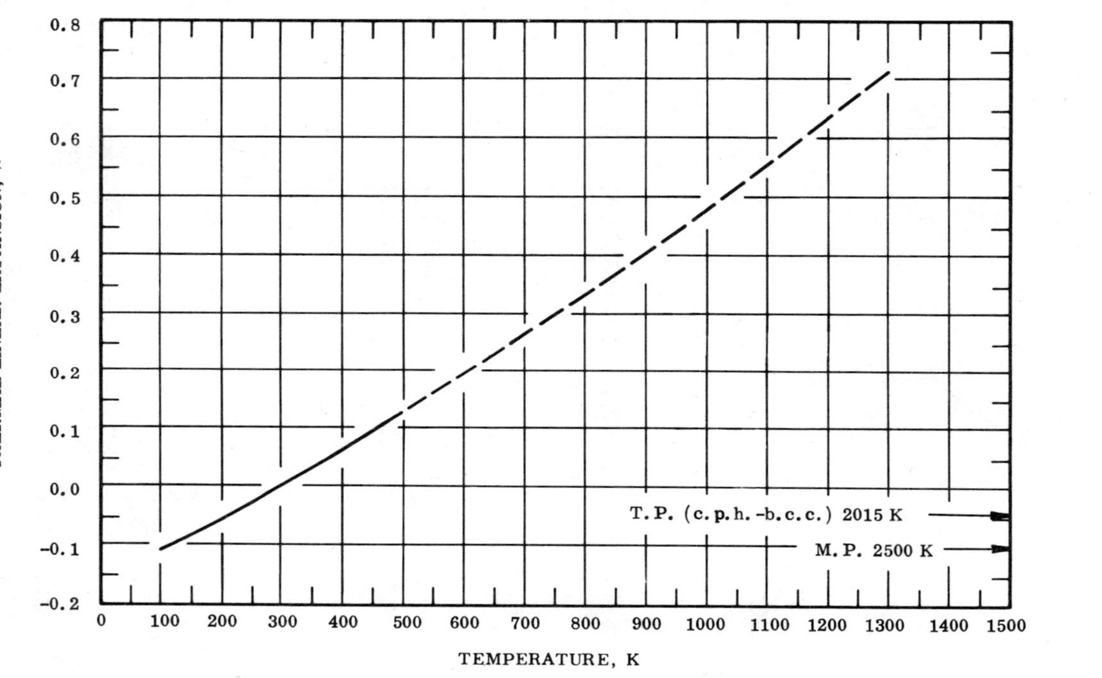

RECOMMENDED VALUES

[Temperature, T, K; Linear Expansion, $\Delta L/L_0$, %; α, K^{-1}]

T	$\Delta L/L_0$	$\alpha \times 10^6$
100	-0.110	5.3
200	-0.054	5.7
293	0.000	5.9
400	0.065	6.2
500	0.128	6.4
600	0.193†	6.6
700	0.260†	6.8
800	0.330†	7.0
900	0.402†	7.2
1000	0.475†	7.5
1200	0.630†	8.1
1300	0.712†	8.4

† Provisional values.

REMARKS

The tabulated values are considered accurate to within ± 5% at temperatures below 500 K and ±10% above. These values can be represented approximately by the following equation:

$$\Delta L/L_0 = -0.164 + 5.365 \times 10^{-4}\, T + 9.631 \times 10^{-8}\, T^2 + 7.281 \times 10^{-12}\, T^3$$

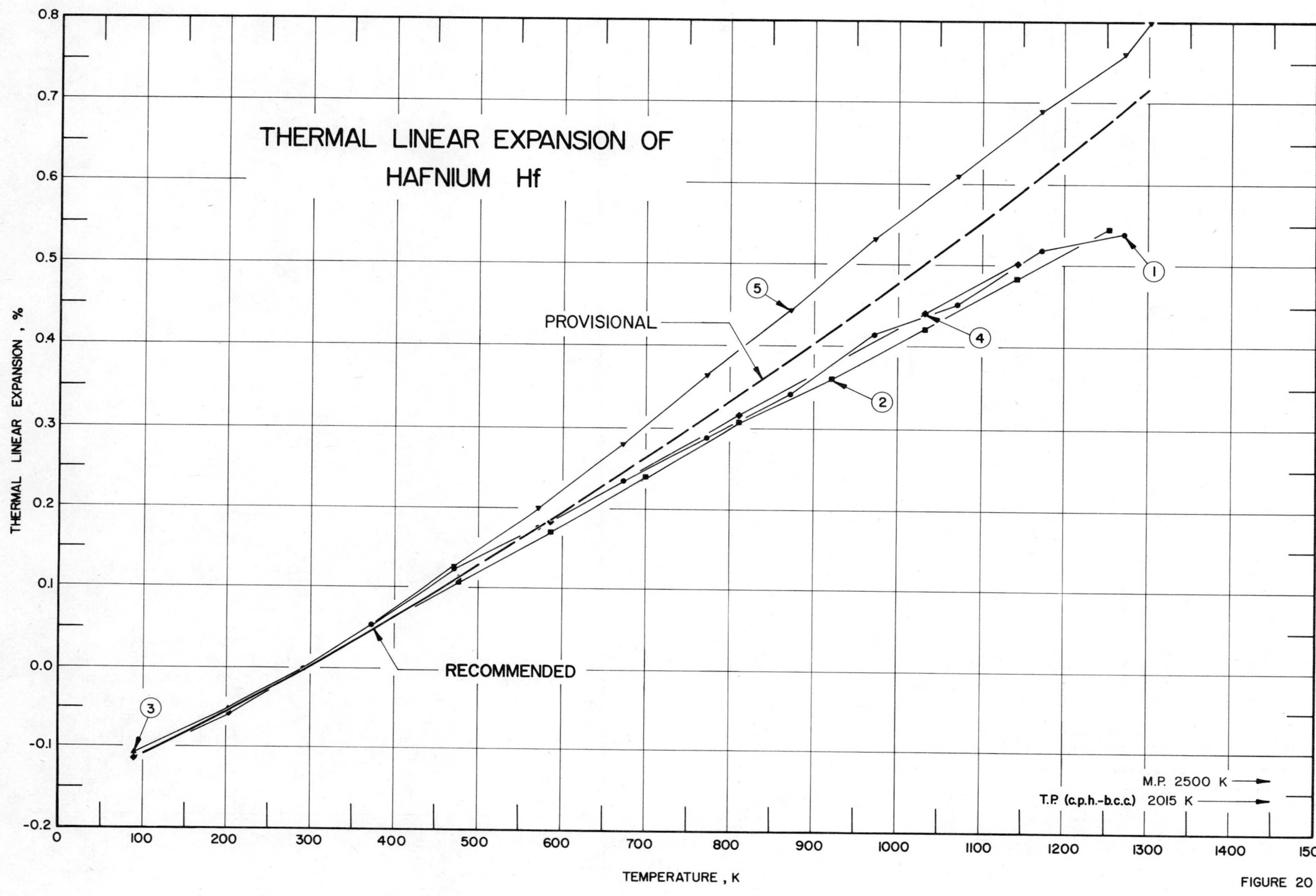
THERMAL LINEAR EXPANSION OF
HAFNIUM Hf
PROVISIONAL
RECOMMENDED
1
2
3
4
5
M.P. 2500 K
T.P. (c.p.h.-b.c.c.) 2015 K
THERMAL LINEAR EXPANSION, %
0.8
0.7
0.6
0.5
0.4
0.3
0.2
0.1
0.0
-0.1
-0.2
TEMPERATURE, K
0
100
200
300
400
500
600
700
800
900
1000
1100
1200
1300
1400
1500

FIGURE 20

SPECIFICATION TABLE 20. THERMAL LINEAR EXPANSION OF HAFNIUM Hf

Cur. No.	Ref. No.	Author(s)	Year	Method Used	Temp. Range, K	Name and Specimen Designation	Composition (weight percent), Specifications, and Remarks
1	188	Baldwin, E. E.	1954	L	293-1258		Prepared from a billet, arc melted by WAPD, extruded at 1373 K at KAPL, swaged cold to the final size in several passes and annealed at 1023 K after each swaging pass; data on the coefficient of thermal linear expansion also reported.
2	189	Adenstedt, H. K.	1952	L	300-1335	No. 775	99^+ Hf, cold swaged bar obtained from Foot Mineral Co.; expansion measured in vacuum with temperature increasing 0.55 K/min.
3	189	Adenstedt, H. K.	1952	L	94-487	No. 773	Similar to the above specimen.
4	189	Adenstedt, H. K.	1952	L	94-1153	No. 783	Similar to the above specimen except annealed at 1273 K.
5	735	Golutvin, Y. M., and Maslennikov, E. G.	1970	L	298-1302		Polycrystalline specimen obtained by iodine reaction; impurities: 0.79 Zr, 0.12 Mo, <0.005 each Fe, Si, and Ti, <0.001 each Mn, and Ca, 0.0014 Ni, <0.002 Cr, 0.013 C, and 0.003 N.

DATA TABLE 20. THERMAL LINEAR EXPANSION OF HAFNIUM Hf

[Temperature, T, K; Linear Expansion, $\Delta L/L_0$, %]

T	$\Delta L/L_0$	T	$\Delta L/L_0$	T	$\Delta L/L_0$	T	$\Delta L/L_0$	T	$\Delta L/L_0$
CURVE 1		CURVE 2		CURVE 3		CURVE 4 (cont.)		CURVE 5 (cont.)	
293	0.000	293	0.000*	90	-0.109	589	0.184	673	0.280
373	0.053	477	0.108	203	-0.051	811	0.316	773	0.366
473	0.121	589	0.170	293	0.000*	1033	0.442	873	0.446
573	0.176	700	0.240	477	0.112	1144	0.502	973	0.532
673	0.232	811	0.307					1073	0.609
773	0.288	922	0.361	CURVE 4		CURVE 5		1173	0.689
873	0.342	1033	0.422					1273	0.762
973	0.415	1144	0.484	90	-0.115	293	0.000*	1302	0.800
1073	0.452	1255	0.547	203	-0.053	373	0.053*		
1173	0.519			293	0.000*	473	0.123		
1273	0.539			477	0.115*	573	0.200		

DATA TABLE 20. COEFFICIENT OF THERMAL LINEAR EXPANSION OF HAFNIUM Hf

[Temperature, T, K; Coefficient of Expansion, α, $10^{-6}\,K^{-1}$]

T	α
CURVE 1*	
373	6.8
473	6.3
573	5.8
673	5.5
773	5.8
873	5.5
973	5.5
1073	5.5
1173	5.5
1273	5.5

*Not shown in figure.

FIGURE AND TABLE NO. 21R. PROVISIONAL VALUES FOR THERMAL LINEAR EXPANSION OF HOLMIUM Ho

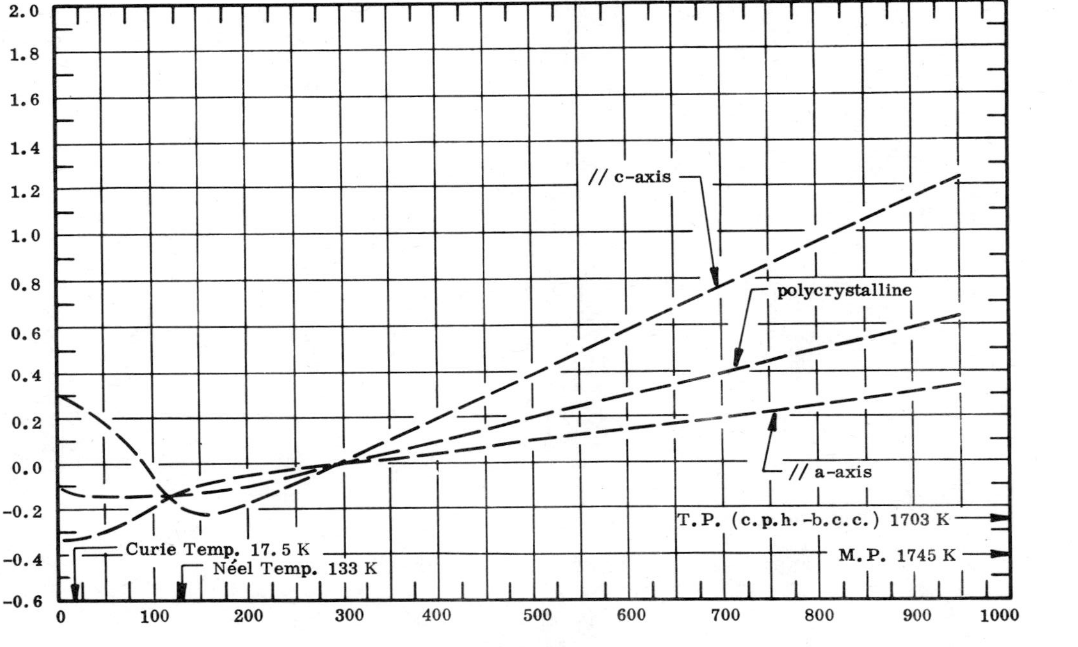

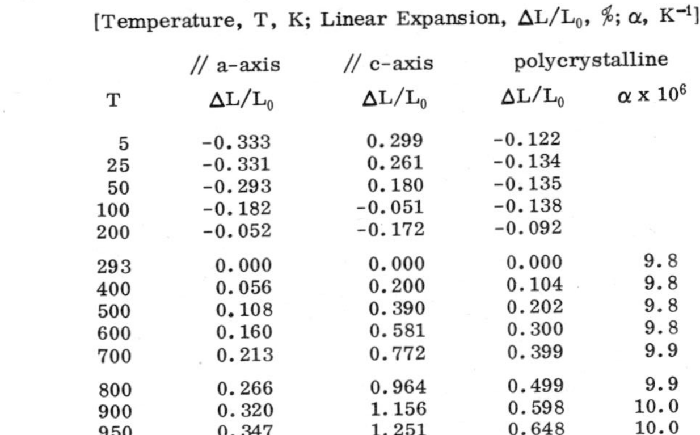

PROVISIONAL VALUES

[Temperature, T, K; Linear Expansion, $\Delta L/L_0$, %; α, K^{-1}]

T	// a-axis $\Delta L/L_0$	// c-axis $\Delta L/L_0$	polycrystalline $\Delta L/L_0$	polycrystalline $\alpha \times 10^6$
5	-0.333	0.299	-0.122	
25	-0.331	0.261	-0.134	
50	-0.293	0.180	-0.135	
100	-0.182	-0.051	-0.138	
200	-0.052	-0.172	-0.092	
293	0.000	0.000	0.000	9.8
400	0.056	0.200	0.104	9.8
500	0.108	0.390	0.202	9.8
600	0.160	0.581	0.300	9.8
700	0.213	0.772	0.399	9.9
800	0.266	0.964	0.499	9.9
900	0.320	1.156	0.598	10.0
950	0.347	1.251	0.648	10.0

REMARKS

The tabulated values are considered accurate to within ±10% at temperatures below 133 K and ±7% above. Since no experimental data on polycrystalline holmium were located in the literature, the tabulated values are calculated from single crystal data and are considered accurate to about ±10% over the entire temperature range. These values can be represented approximately by the following equations:

// a-axis: $\Delta L/L_0 = -0.156 + 5.432 \times 10^{-4}\, T - 5.148 \times 10^{-8}\, T^2 + 3.912 \times 10^{-11}\, T^3$ $(293 < T < 950)$

// c-axis: $\Delta L/L_0 = -0.538 + 1.791 \times 10^{-3}\, T + 1.663 \times 10^{-7}\, T^2 - 7.355 \times 10^{-11}\, T^3$ $(293 < T < 950)$

polycrystalline: $\Delta L/L_0 = -0.281 + 9.474 \times 10^{-4}\, T + 3.955 \times 10^{-8}\, T^2 - 7.359 \times 10^{-12}\, T^3$ $(293 < T < 950)$

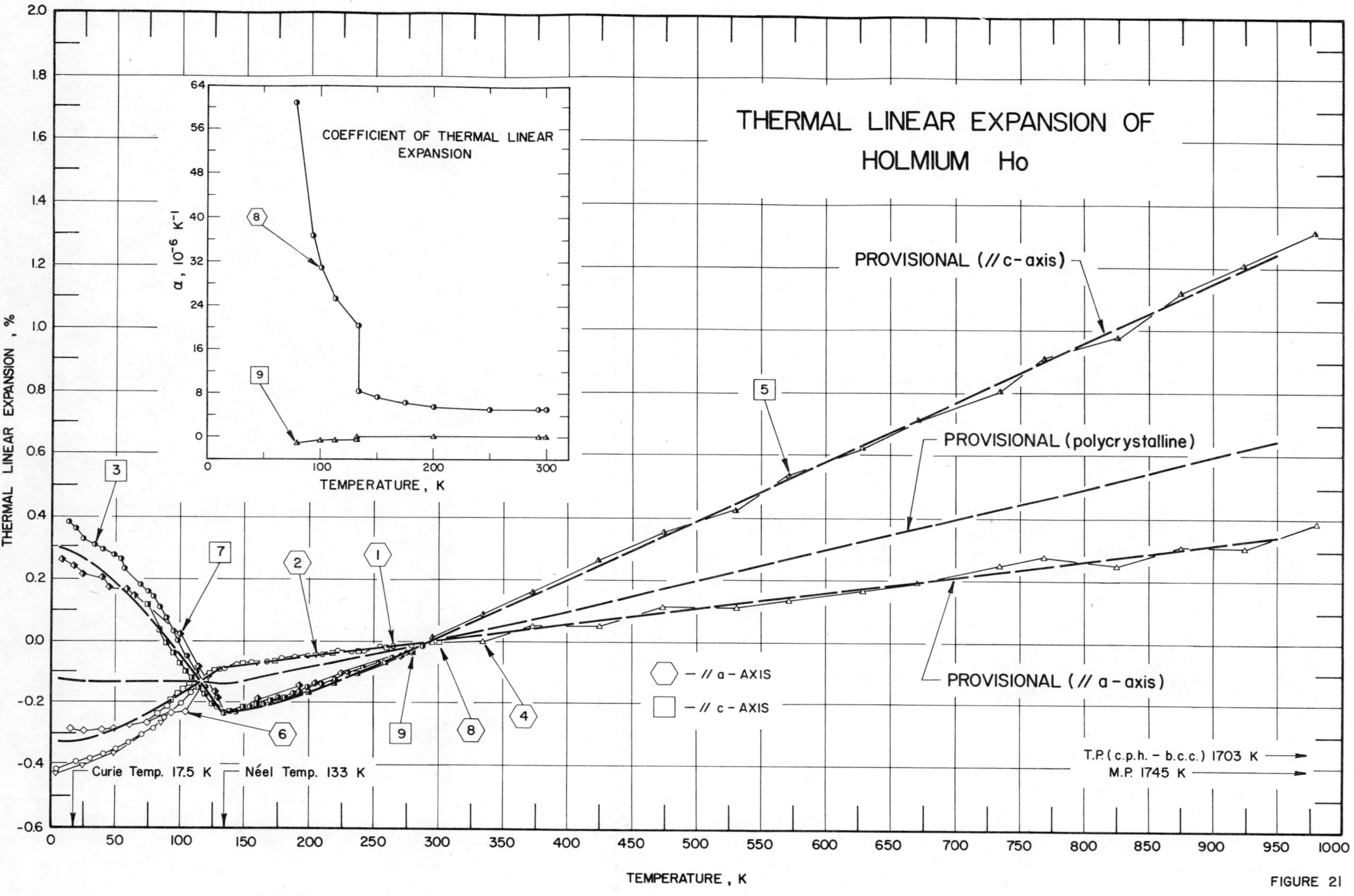
THERMAL LINEAR EXPANSION OF
HOLMIUM Ho
COEFFICIENT OF THERMAL LINEAR
EXPANSION
α, 10⁻⁶ K⁻¹
TEMPERATURE, K
PROVISIONAL (// c-axis)
PROVISIONAL (polycrystalline)
PROVISIONAL (// a-axis)
— // a – AXIS
— // c – AXIS
T.P. (c.p.h. – b.c.c.) 1703 K
M.P. 1745 K
Curie Temp. 17.5 K
Néel Temp. 133 K
THERMAL LINEAR EXPANSION, %
TEMPERATURE, K

FIGURE 21

SPECIFICATION TABLE 21. THERMAL LINEAR EXPANSION OF HOLMIUM Ho

Cur. No.	Ref. No.	Author(s)	Year	Method Used	Temp. Range, K	Name and Specimen Designation	Composition (weight percent), Specifications, and Remarks
1	190	Rhyne, J.J., Legvold, S., and Rodine, E.T.	1967	S	6-277		Hexagonal, close-packed single crystal grown by thermal strain anneal method; basal-plane and b-c-axis plane disk shaped specimen about 9 mm in diameter and 1.5 mm thick, spark-cut from the bulk crystal; expansion measured parallel to a-axis and no magnetic field applied; zero-point correction of 0.003% determined by graphical extrapolation.
2	190	Rhyne, J.J., et al.	1967	S	5-299		Similar to the above specimen except expansion measured parallel to b-axis and without magnetic field; results are identical, within experimental error to the above curve measured along a-axis; zero-point correction of 0.002% determined by graphical extrapolation.
3	190	Rhyne, J.J., et al.	1967	S	13-287		Similar to the above specimen except expansion measured parallel to c-axis and without magnetic field; zero-point correction of 0.015% determined by graphical extrapolation.
4	82	Hanak, J.J.	1959	X	295-981		≤0.04 Dy, <0.02 Er, <0.01 Tm, <0.005 Cr; wire specimen cut from sheet prepared at Ames Laboratory; hexagonal close-packed structure; expansion measured along a-axis; lattice parameter at 293 K is 3.577 Å.
5	82	Hanak, J.J.	1959	X	295-981		The above specimen; expansion measured along c-axis; lattice parameter at 293 K is 5.619 Å.
6	160	Darnell, F.J.	1963	X	16-299		Single crystal contain total of 0.1 impurities; platelet face (001) plane, approximately 0.1 mm thick, 1 mm^2 area; expansion measured along a-axis; reported error ±0.1 Å in measurement of lattice parameter; lattice parameter reported at 283 K is 3.5826 Å; 3.5828 Å at 293 K determined by graphical interpolation, magnetic transition near 132 K.
7	160	Darnell, F.J.	1963	X	8-299		The above specimen; expansion measured along c-axis; lattice parameter reported at 299 K is 5.6283 Å; 5.6277 Å at 293 K determined by graphical interpolation, magnetic transition near 132 K.
8	733	Finkel', V.A., Palatnik, M.I.	1971	X	76-300		99.5 Ho; polycrystalline sample, annealed for 10 hr at ~ 10^{-6} Torr and 1273 K; expansion measured along c-axis; lattice parameter reported at 293 K is 5.6129 Å; data on the coefficient of thermal linear expansion also reported.
9	733	Finkel', V.A., Palatnik, M.I.	1971	X	78-300		The above specimen; expansion measured along a-axis; lattice parameter reported at 293 K is 3.56626 Å; data on the coefficient of thermal linear expansion also reported.

DATA TABLE 21. THERMAL LINEAR EXPANSION OF HOLMIUM Ho

[Temperature, T, K; Linear Expansion, $\Delta L/L_0$, %]

T	$\Delta L/L_0$
CURVE 1	
6	-0.416
11	-0.415*
14	-0.412*
17	-0.407*
19	-0.406*
20	-0.395
22	-0.393*
32	-0.380
41	-0.367
44	-0.363*
52	-0.349
60	-0.331
69	-0.310*
71	-0.306
77	-0.289*
79	-0.283
87	-0.256
101	-0.201
103	-0.192*
108	-0.173
113	-0.152*
120	-0.128
125	-0.111*
131	-0.094
134	-0.088
145	-0.078
156	-0.071
167	-0.063
178	-0.056
188	-0.051
200	-0.043
211	-0.040
222	-0.030
233	-0.027
243	-0.022
255	-0.016
265	-0.011
277	-0.006
CURVE 2	
5	-0.422
11	-0.422*
14	-0.420*
17	-0.415*
17	-0.409*
CURVE 2 (cont.)	
19	-0.408*
20	-0.403
23	-0.400*
25	-0.395*
29	-0.389*
34	-0.384*
40	-0.370*
47	-0.365*
49	-0.360
58	-0.340*
67	-0.319*
78	-0.292*
86	-0.263
89	-0.247*
98	-0.214*
105	-0.188*
111	-0.161*
117	-0.143*
120	-0.129*
122	-0.126*
128	-0.103*
133	-0.091*
137	-0.087*
145	-0.080*
157	-0.071
169	-0.063*
182	-0.054*
195	-0.047*
206	-0.041
217	-0.034*
231	-0.028*
243	-0.022*
255	-0.017*
269	-0.009*
279	-0.006*
290	-0.001*
299	0.001*
CURVE 3	
13	0.381
15	0.375*
16	0.371*
18	0.361
20	0.349*
21	0.345*
CURVE 3 (cont.)	
22	0.340*
25	0.330
27	0.323*
34	0.310
40	0.299
49	0.279
55	0.263
62	0.237
69	0.192
75	0.159
79	0.144
84	0.110
89	0.074
95	0.034
98	0.002
106	-0.048
113	-0.101
120	-0.156
129	-0.216
133	-0.230
144	-0.224
155	-0.213
166	-0.201
176	-0.186
187	-0.172
199	-0.154
210	-0.139
221	-0.123
231	-0.104
242	-0.086*
254	-0.068*
265	-0.049*
276	-0.029*
287	-0.010
CURVE 4	
295	0.000
334	0.000
372	0.056
424	0.056
473	0.112
531	0.112
571	0.140
628	0.168
670	0.196
CURVE 4 (cont.)	
734	0.252
768	0.280
825	0.252
875	0.308
924	0.308
981	0.391
CURVE 5	
295	0.018
334	0.089
372	0.160
424	0.267
473	0.356
531	0.427
571	0.534
628	0.623
670	0.712
734	0.801
768	0.908
825	0.979
875	1.121
924	1.210
981	1.317
CURVE 6	
16	-0.284
21	-0.290*
21	-0.276*
26	-0.290
28	-0.273*
32	-0.293*
37	-0.287
45	-0.262*
46	-0.293*
49	-0.293*
49	-0.282
52	-0.282*
61	-0.268
66	-0.268*
75	-0.265
77	-0.273*
77	-0.234*
85	-0.248*
93	-0.237
CURVE 6 (cont.)	
95	-0.237*
100	-0.229*
104	-0.229
116	-0.167*
122	-0.189*
127	-0.117*
128	-0.128*
135	-0.111*
136	-0.100*
139	-0.100*
142	-0.100*
147	-0.067*
151	-0.092*
151	-0.083*
156	-0.069*
160	-0.078*
168	-0.069*
177	-0.069*
181	-0.061*
185	-0.067*
202	-0.047*
209	-0.053*
222	-0.053*
228	-0.042*
231	-0.042*
235	-0.036*
253	-0.025*
269	-0.011*
283	-0.005*
299	0.003*
CURVE 7	
8	0.264
8	0.256*
10	0.265*
10	0.255*
12	0.248*
13	0.253*
14	0.249*
17	0.240
17	0.235*
25	0.217
29	0.219*
32	0.208*
40	0.208*
CURVE 7 (cont.)	
46	0.187*
46	0.176
54	0.164*
59	0.164*
59	0.153*
63	0.153*
66	0.146
71	0.146*
76	0.121*
77	0.114*
79	0.127*
79	0.104*
84	0.104*
84	0.098*
85	0.102*
90	0.088*
90	0.079*
96	0.041*
99	0.036*
100	0.021
103	0.006*
104	0.015*
106	-0.015*
107	0.008*
107	-0.026*
111	-0.037*
112	-0.047*
113	-0.056*
114	-0.081*
114	-0.092
114	-0.097*
115	-0.072*
116	-0.099*
118	-0.120*
119	-0.113*
120	-0.090*
123	-0.142*
123	-0.147*
126	-0.156*
127	-0.166*
127	-0.174
128	-0.174*
129	-0.191
129	-0.204*
131	-0.222*
134	-0.245*

*Not shown in figure.

DATA TABLE 21. THERMAL LINEAR EXPANSION OF HOLMIUM Ho (continued)

T	$\Delta L/L_0$
CURVE 7 (cont.)	
135	-0.236*
135	-0.225*
138	-0.236*
143	-0.236*
148	-0.222*
150	-0.214*
161	-0.197
163	-0.197*
166	-0.197*
177	-0.190*
177	-0.175
185	-0.177*
185	-0.166*
192	-0.163
192	-0.149
204	-0.138
209	-0.129*
217	-0.133*
217	-0.101*
225	-0.106
230	-0.101*
231	-0.095*
235	-0.099*
241	-0.090*
244	-0.081*
254	-0.069*
271	-0.049
277	-0.044*
280	-0.024*
299	0.011*
CURVE 8	
78	-0.286
83	-0.236
89	-0.211
93	-0.190
98	-0.172
103	-0.153
107	-0.139
113	-0.125
117	-0.112*
123	-0.102
127	-0.091
130	-0.085*
135	-0.081*
143	-0.077*
149	-0.073
157	-0.067*

T	$\Delta L/L_0$
CURVE 8 (cont.)	
168	-0.061
179	-0.057*
189	-0.050*
199	-0.046*
219	-0.035
239	-0.028
260	-0.016
281	-0.007*
293	0.000*
300	0.003
CURVE 9	
76	0.119
84	0.039
90	-0.002
95	-0.036
100	-0.071
104	-0.100
110	-0.127
114	-0.150
120	-0.173
126	-0.201
130	-0.219*
133	-0.230*
139	-0.223
144	-0.219*
150	-0.214
160	-0.207
171	-0.198
181	-0.187
190	-0.176*
200	-0.162
219	-0.137
237	-0.109
258	-0.071
277	-0.032
293	0.000*
300	0.009*

* Not shown in figure.

FIGURE AND TABLE NO. 22R. RECOMMENDED VALUES FOR THERMAL LINEAR EXPANSION OF INDIUM In

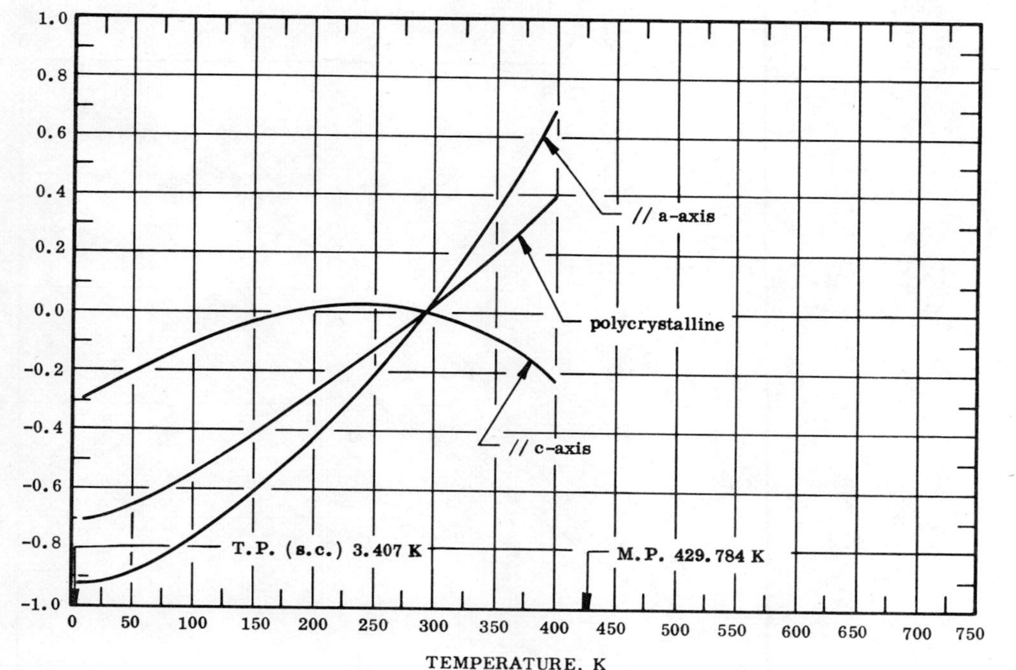

RECOMMENDED VALUES

[Temperature, T, K; Linear Expansion, $\Delta L/L_0$, %; α, K^{-1}]

T	// a-axis $\Delta L/L_0$	// c-axis $\Delta L/L_0$	polycrystalline $\Delta L/L_0$	polycrystalline $\alpha \times 10^6$
5	-0.918	-0.287	-0.708	
25	-0.914	-0.264	-0.697	
50	-0.883	-0.211	-0.659	
100	-0.765	-0.106	-0.545	25.4
200	-0.425	0.015	-0.278	28.1
293	0.000	0.000	0.000	32.1
400	0.686	-0.224	0.383	40.3

LOW-TEMPERATURE THERMAL EXPANSION VALUES

	Normal State $\Delta L/L \times 10^4$			Superconducting State $\Delta L/L \times 10^4$ †		
T	// a-axis $\Delta L/L_0$	// c-axis $\Delta L/L_0$	polycrystalline $\Delta L/L_0$	// a-axis $\Delta L/L_0$	// c-axis $\Delta L/L_0$	polycrystalline $\Delta L/L_0$
0	0.00	0.00	0.00	-0.04	0.16	0.03
1.5	0.03	-0.05	0.01	0.00	0.10	0.03
2	0.06	-0.10	0.02	0.03	0.03	0.03
2.5	0.14	-0.24	0.04	0.12	-0.14	0.03
3	0.29	-0.49	0.09	0.28	-0.44	0.04
3.4	0.46	-0.75	0.17	0.46	-0.75	0.06

† Superconducting state value is set equal to that of normal state value at the transition temperature, $T_c = 3.4$ K.

REMARKS

The tabulated values are considered accurate to within ±3% over the entire temperature range, with the exception of the values above 293 K along c-axis, which are considered accurate to within ±5%. The tabulated values for polycrystalline are calculated from single crystal data and agree well at the temperature below 293 K and slightly lower above are considered accurate to within ±5% over the entire temperature range. These values can be represented approximately by the following equations:

// a-axis: $\Delta L/L_0 = -1.049 + 2.766 \times 10^{-3} T - 3.779 \times 10^{-7} T^2 + 1.076 \times 10^{-6} T^3$ $(100 < T < 400)$

// c-axis: $\Delta L/L_0 = -0.317 + 2.399 \times 10^{-3} T - 1.969 \times 10^{-6} T^2 - 8.608 \times 10^{-9} T^3$ $(100 < T < 400)$

polycrystalline: $\Delta L/L_0 = -0.805 + 2.640 \times 10^{-3} T - 8.968 \times 10^{-7} T^2 + 4.294 \times 10^{-9} T^3$ $(100 < T < 400)$

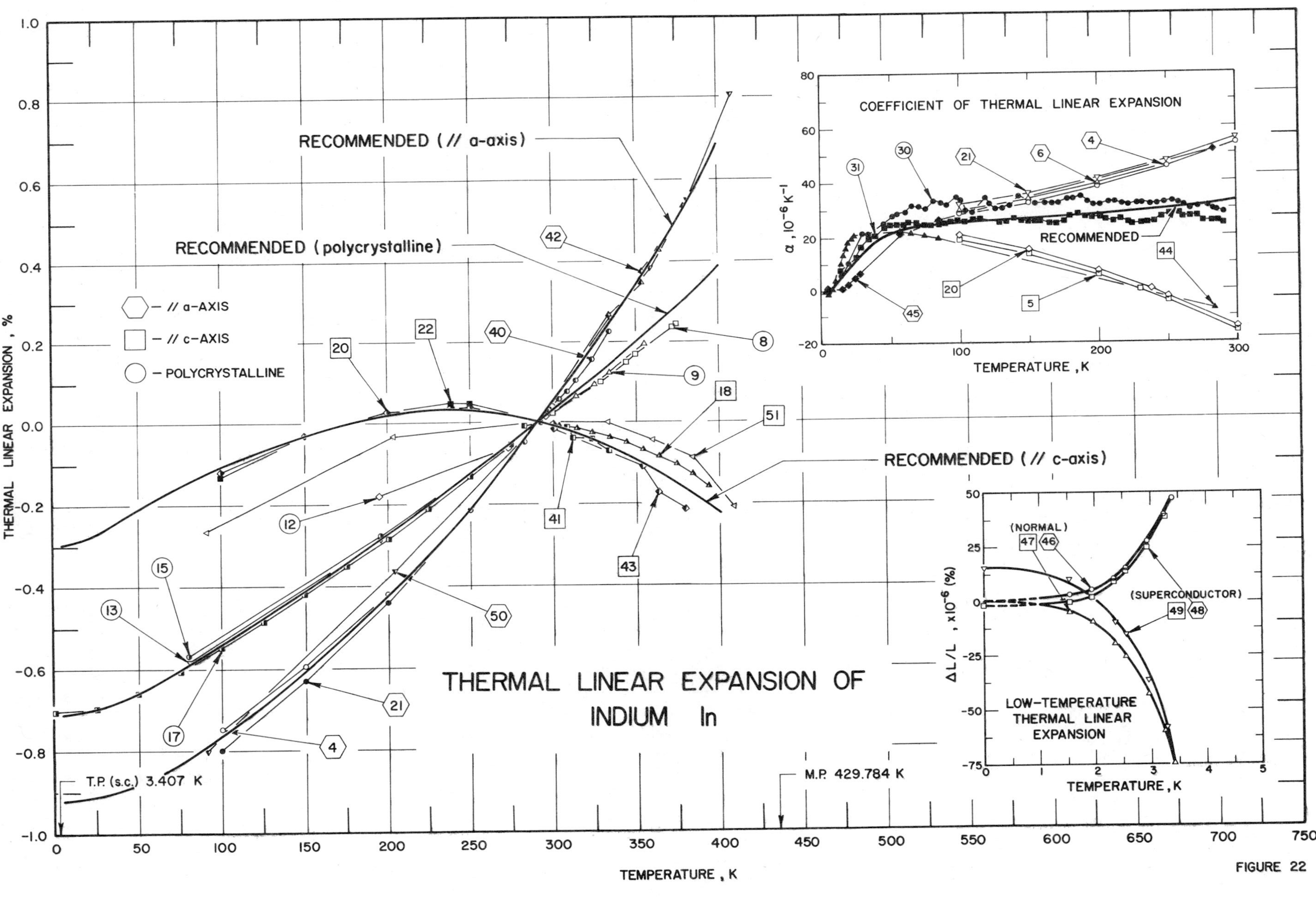
THERMAL LINEAR EXPANSION OF
INDIUM In
THERMAL LINEAR EXPANSION, %
TEMPERATURE, K
RECOMMENDED (// a-axis)
RECOMMENDED (polycrystalline)
RECOMMENDED (// c-axis)
// a-AXIS
// c-AXIS
POLYCRYSTALLINE
T.P. (s.c.) 3.407 K
M.P. 429.784 K
COEFFICIENT OF THERMAL LINEAR EXPANSION
RECOMMENDED
α, 10^{-6} K^{-1}
TEMPERATURE, K
LOW-TEMPERATURE THERMAL LINEAR EXPANSION
(NORMAL)
(SUPERCONDUCTOR)
ΔL/L, x10^{-6} (%)
TEMPERATURE, K

FIGURE 22

SPECIFICATION TABLE 22. THERMAL LINEAR EXPANSION OF INDIUM In

Cur. No.	Ref. No.	Author(s)	Year	Method Used	Temp. Range, K	Name and Specimen Designation	Composition (weight percent), Specifications, and Remarks
1*	191	Pollock, D.B.	1969	L	77, 293		Specimen 3.2 cm long.
2*	192	Shinoda, G.	1933	X	296, 360		Material from Dr. Peters and Dr. Rost; f.c.t.; determination of thermal expansion coefficient carried out with regard to planes (351), (244), (153); expansion measured parallel to axis.
3*	192	Shinoda, G.	1933	X	296, 360		The above specimen; expansion measured perpendicular to axis.
4	193	Smith, J.F. and Schneider, V.L.	1964	X	100-300		99.9999 In, supplied by Consolidated Mining and Smelting Co. of Canada; expansion measured along a-axis.
5*	193	Smith, J.F. and Schneider, V.L.	1964	X	100-300		The above specimen; expansion measured along c-axis.
6*	193	Smith, J.F. and Schneider, V.L.	1964	X	100-300		99.999 In, supplied by Consolidated Mining and Smelting Co. of Canada; expansion measured along a-axis.
7*	194	Olsen, J.L. and Rohrer, H.	1957	N	1.6-3.2		Change of length measured at normal-superconducting transition; data average of 10-15 trials on each of two polycrystalline specimens.
8	195	Hidnert, P. and Blair, M.G.	1943	T	300-374		Rod of cast material obtained from Indium Corp. of America, estimated 99.9 pure; expansion measured with increasing temperature; zero-point correction of 0.021% determined by graphical extrapolation.
9	195	Hidnert, P. and Blair, M.G.	1943	T	355-300		The above specimen; expansion measured with decreasing temperature; zero-point correction of 0.033% determined by graphical extrapolation.
10*	195	Hidnert, P. and Blair, M.G.	1943	T	298-378		The above specimen; expansion measured with increasing temperature on second test; zero-point correction of 0.015% determined by graphical extrapolation.
11*	195	Hidnert, P. and Blair, M.G.	1943	T	361-300		The above specimen; expansion measured with decreasing temperature; zero-point correction of 0.015% determined by graphical extrapolation.
12	195	Hidnert, P. and Blair, M.G.	1943	L	298-195		The above specimen; expansion measured with decreasing temperature in low-temperature expansion apparatus on third test; zero-point correction of 0.015% determined by graphical extrapolation.
13	195	Hidnert, P. and Blair, M.G.	1943	L	273-81		The above specimen; expansion measured with decreasing temperature on fourth test; zero-point correction of -0.059% determined by graphical extrapolation.
14*	195	Hidnert, P. and Blair, M.G.	1943	L	201-299		The above specimen; expansion measured with increasing temperature; zero-point correction of 0.04% determined by graphical interpolation.
15	195	Hidnert, P. and Blair, M.G.	1943	L	300-80		The above specimen; expansion measured with decreasing temperature on fifth test; zero-point correction of 0.020% determined by graphical interpolation.
16*	195	Hidnert, P. and Blair, M.G.	1943	L	196-300		The above specimen; expansion measured with increasing temperature; zero-point correction of 0.031% determined by graphical interpolation.
17	196	Swenson, C.A.	1955	L	0-300		99.9 In specimen from A. D. Mackay; specimen machined from ingot cast under vacuum; zero-point correction of -0.063% determined by graphical interpolation.
18	197	Vernon, E.V. and Weintroub, S.	1953	I	303-393		Single crystal about 2 cm long; expansion measured parallel to c-axis.
19*	197	Vernon, E.V. and Weintroub, S.	1953	I	303-393		Similar to the above specimen; expansion measured perpendicular to c-axis.

* Not shown in figure.

SPECIFICATION TABLE 22. THERMAL LINEAR EXPANSION OF INDIUM In (continued)

Cur. No.	Ref. No.	Author(s)	Year	Method Used	Temp. Range, K	Name and Specimen Designation	Composition (weight percent), Specifications, and Remarks
20	193	Smith, J.F. and Schneider, V.L.	1964	X	80-300		99.9999 In, from Consolidated Mining and Smelting Co. of Canada; CuKα x-radiation used; expansion measured parallel to the unique axis; reported error ± 5%.
21	193	Smith, J.F. and Schneider, V.L.	1964	X	100-300		99.97 pure, from the Consolidated Mining and Smelting Co. of Canada; CuKα radiation used; expansion measured perpendicular to the unique axis; reported error ± 5%.
22	193	Smith, J.F. and Schneider, V.L.	1964	X	100-300		The above specimen; expansion measured parallel to the unique axis.
23*	193	Smith, J.F. and Schneider, V.L.	1964	X	100-300		99.83 In, 0.17 Tl (0.1 at %); dilute solid solution of thallium prepared from the 99.9999% indium by casting in sealed pyrex containers; expansion measured perpendicular to the unique axis; reported error ± 5%.
24*	193	Smith, J.F. and Schneider, V.L.	1964	X	100-300		The above specimen; expansion measured parallel to the unique axis.
25*	193	Smith, J.F. and Schneider, V.L.	1964	X	100-300		99.81 In, 0.19 Cd (0.2 at %); solid solution prepared from 99.9999% indium by casting in sealed pyrex containers; expansion measured perpendicular to the unique axis; reported error ± 5%.
26*	193	Smith, J.F. and Schneider, V.L.	1964	X	100-300		The above specimen; expansion measured parallel to the unique axis.
27*	193	Smith, J.F. and Schneider, V.L.	1964	X	100-300		99.80 In, 0.2 Sn (0.2 at %); solid solution prepared from 99.9999% indium by casting in sealed pyrex containers; expansion measured perpendicular to the unique axis; CuKα radiation employed; reported error ± 5%.
28*	193	Smith, J.F. and Schneider, V.L.	1964	X	100-300		The above specimen; linear coefficient parallel to the unique axis.
29*	105	Madaiah, N. and Graham, G.M.	1964	E	9-290	Specimen 1	Crystal specimen grown from melt of spectroscopically pure material; specimen 5 to 6 cm long; tetrad axis at 35° to rod axis; expansion measured along rod axis; reported error < 5% in range 50 K to 240 K.
30*	105	Madaiah, N. and Graham, G.M.	1964	E	14-291	Specimen 2	Similar to the above specimen except polycrystalline, grown by quenching from the melt.
31*	105	Madaiah, N. and Graham, G.M.	1964	E	14-291	Specimen 3	Similar to the above specimen except single crystal, destroyed before orientation determined.
32*	612	Boiko, B.T., Palatnik, L.S., and Pugachev, A.T.	1967	F	353-328		> 99.9 pure thin film condensed in vacuum 5 x 10^{-5} torr; thickness of film in the range 600 + 10 Å; thermal expansion measured in the direction of (113) plane with decreasing temperature.
33*	612	Boiko, B.T., et al.	1967	F	293-392		The above specimen; thermal expansion measured in the direction of (113) plane with increasing temperature.
34*	612	Boiko, B.T., et al.	1967	F	353-329		The above specimen; thermal expansion measured in the direction of (022) plane with decreasing temperature.
35*	612	Boiko, B.T., et al.	1967	F	293-392		The above specimen; thermal expansion measured in the direction of (022) plane with increasing temperature.
36*	612	Boiko, B.T., et al.	1967	F	353-328		The above specimen; thermal expansion measured in the direction of (111) plane with decreasing temperature.
37*	612	Boiko, B.T., et al.	1967	F	293-393		The above specimen; thermal expansion measured in the direction of (111) plane with increasing temperature.

* Not shown in figure.

SPECIFICATION TABLE 22. THERMAL LINEAR EXPANSION OF INDIUM In (continued)

Cur. No.	Ref. No.	Author(s)	Year	Method Used	Temp. Range, K	Name and Specimen Designation	Composition (weight percent), Specifications, and Remarks
38*	612	Boiko, B.T., et al.	1967	F	352-328		The above specimen; thermal expansion measured in the direction of (220) plane with decreasing temperature.
39*	612	Boiko, B.T., et al.	1967	F	293-391		The above specimen; thermal expansion measured in the direction of (220) plane with increasing temperature.
40	618	Straumanis, M.E., Rao, P.B., and James, W.J.	1971	X	283-333		99.999 pure specimen from American Smelting and Refining Co.; lattice parameters measured along a-direction; lattice parameter reported at 293 K is 4.5981 Å.
41	618	Straumanis, M.E., et al.	1971	X	283-333		The above specimen; lattice parameters measured along c-axis; lattice parameter reported at 293 K is 4.9509 Å.
42	619	Deshpande, V.T. and Pawar, R.R.	1969	X	300-379		Pure specimen; measurements along a-axis; lattice parameter reported at 300 K is 4.6002 Å; 4.5982 Å at 293 K determined by graphical extrapolation; data on the coefficient of thermal expansion are also reported.
43	619	Deshpande, V.T. and Pawar, R.R.	1969	X	300-379		The above specimen; measurements along c-axis; lattice parameter at 300 K is 4.9463 Å; 4.9463 Å at 293 K determined by graphical extrapolation; data on the coefficient of thermal expansion are also reported.
44*	633	Collins, J.G., Cowan, J.G., and White, G.K.	1967	E	2-284		99.99 pure single crystal of 0.75 in. diameter and 2 in. length from Metals Research Ltd., Cambridge were reduced to required length of 5.08 cm; cylinder axis was within 2 deg of the tetrad axis.
45*	633	Collins, J.G., et al.	1967	E	2-284		The above specimen; cylinder axis was within 2 deg of a plane perpendicular to tetrad axis.
46	633	Collins, J.G., et al.	1967	E	1.5-3.4		The above specimen; measurements along a-axis in the magnetic field of about 10^3 Oe provided by a superconducting solenoid; superconducting transition temperature is 3.4 K; this curve is here reported using the first given temperature as reference temperature at which $\Delta L/L_0 = 0$.
47	633	Collins, J.G., et al.	1967	E	1.5-3.4		The above specimen; measurements along c-axis in the magnetic field; this curve is here reported using the first given temperature as reference temperature at which $\Delta L/L_0 = 0$.
48	633	Collins, J.G., et al.	1967	E	1.5-3.4		The above specimen; measurements along a-axis for superconductors without magnetic applied field; authors report $(\ell s - \ell n)/\ell n$ values where ℓn and ℓs respectfully are lengths of specimen with and without applied magnetic field; this curve referred to the normal state value of curve of 46 at the critical temperature, $T_c = 3.4$ K.
49	633	Collins, J.G., et al.	1967	E	1.5-3.4		The above specimen; measurements along c-axis for superconductor without applied magnetic field; authors report $(\ell s - \ell n)/\ell n$ values; this curve referred to the normal state value of curve 47 at the critical temperature, $T_c = 3.4$ K.
50	644	Graham, J., Moore, A., and Raynor, G.V.	1967	E	91-408		99.97 pure specimen; lattice parameter expansion measured along a-axis; lattice parameter reported at 293 K is 4.5893 kX; data on the coefficients of thermal linear expansion also reported.
51	644	Graham, J., et al.	1967	E	91-408		The above specimen; measurements along c-axis; lattice parameter reported at 293 K is 4.9372 kX; data on the coefficients of thermal linear expansion also reported.

* Not shown in figure.

DATA TABLE 22. THERMAL LINEAR EXPANSION OF INDIUM In

[Temperature, T, K; Linear Expansion, $\Delta L/L_0$, %]

T	$\Delta L/L_0$
CURVE 1*	
77.2	-0.588
293.2	0.000
CURVE 2*	
296	0.000
360	0.302
CURVE 3*	
296	0.000
360	0.078
CURVE 4‡	
100	-0.749
150	-0.597
200	-0.421
250	-0.213
300	0.038
CURVE 5*, ‡	
100	-0.088
150	-0.008
200	0.039
230	0.048
250	0.044
300	-0.011
CURVE 6*, ‡	
100	-0.771
150	-0.610
200	-0.427
250	-0.214
300	0.038
CURVE 7*, †, ǂ	
1.60	0.0230 x 10^{-6}
1.75	0.0226
1.81	0.0213
1.86	0.0222
1.94	0.0206
1.98	0.0219
CURVE 7 (cont.)*, †, ǂ	
2.05	0.0204 x 10^{-6}
2.06	0.0185
2.13	0.0177
2.23	0.0182
2.45	0.0155
2.57	0.0132
2.66	0.0129
2.71	0.0116
2.86	0.0106
2.97	0.0083
3.04	0.0070
3.17	0.0061
3.40	0.0000
CURVE 8	
300	0.021
328	0.103
344	0.148
350	0.167
372	0.234
374	0.240
CURVE 9	
355	0.193
333	0.123
324	0.097
313	0.063
300	0.021*
CURVE 10*	
298	0.015
316	0.068
321	0.082
327	0.102
344	0.153
351	0.175
369	0.233
373	0.246
CURVE 11*	
361	0.213
332	0.119
CURVE 11 (cont.)*	
322	0.089
315	0.066
310	0.052
300	0.021
CURVE 12	
298	0.015*
296	0.012
273	-0.060
195	-0.181
CURVE 13	
273	-0.059*
196	-0.284
81	-0.582
CURVE 14*	
201	-0.263
205	-0.251
273	-0.059
276	-0.051
299	0.015
CURVE 15	
300	0.020*
273	-0.059*
196	-0.279
80	-0.569
CURVE 16*	
196	-0.262
273	-0.054
300	0.018
CURVE 17	
6	-0.702
17	-0.700
31	-0.688
45	-0.667
59	-0.642
CURVE 17 (cont.)	
66	-0.626
73	-0.609
76	-0.601
78	-0.596
98	-0.551
108	-0.527
125	-0.484
152	-0.413
172	-0.363
184	-0.322
219	-0.229
232	-0.189
252	-0.129
274	-0.062
293	0.000
303	0.029
CURVE 18‡	
303	-0.008
313	-0.016
323	-0.026
333	-0.038
343	-0.051
353	-0.067
363	-0.085
373	-0.107
383	-0.131
393	-0.160
CURVE 19*, ‡	
303	0.050
313	0.102
323	0.156
333	0.212
343	0.271
353	0.333
363	0.398
373	0.466
383	0.538
393	0.163
CURVE 20‡	
100	-0.119
CURVE 20 (cont.)‡	
150	-0.030
200	0.026
238	0.039
250	0.038
300	-0.006*
CURVE 21‡	
100	-0.800
150	-0.632
200	-0.441
250	-0.221*
300	0.036*
CURVE 22‡	
100	-0.136
150	-0.034*
200	0.029*
238	0.045
250	0.043
300	-0.007*
CURVE 23*, ‡	
100	-0.779
150	-0.621
200	-0.438
250	-0.222
300	0.036
CURVE 24*, ‡	
100	-0.099
150	-0.021
200	0.027
235	0.038
250	0.036
300	-0.006
CURVE 25*, ‡	
100	-0.923
150	-0.739
200	-0.524
250	-0.266
300	0.043
CURVE 26*, ‡	
100	-0.106
150	-0.027
200	0.023
238	0.035
250	0.034
300	-0.005
CURVE 27*, ‡	
100	-0.809
150	-0.654
200	-0.468
250	-0.240
300	0.039
CURVE 28*, ‡	
100	-0.174
150	-0.079
200	-0.012
250	0.016
257	0.016
300	-0.003
CURVE 29*, ‡	
9	-0.431
13	-0.427
19	-0.420
25	-0.413
30	-0.405
34	-0.398
40	-0.387
44	-0.380
49	-0.371
55	-0.360
60	-0.350
64	-0.342
70	-0.330
74	-0.322
79	-0.313
83	-0.305
88	-0.295
94	-0.283
99	-0.274
104	-0.264

* Not shown in figure.
† This curve is here reported using the first given temperature as reference temperature at which $\Delta L/L_0 = 0$.
‡ Author's data for coefficient of thermal expansion have been integrated by TPRC to obtain $\Delta L/L_0$.
ǂ For explaination, see specification table.

DATA TABLE 22. THERMAL LINEAR EXPANSION OF INDIUM In (continued)

T	$\Delta L/L_0$
CURVE 29 (cont.)*, ‡	
110	-0.253
114	-0.246
120	-0.235
124	-0.228
129	-0.219
133	-0.212
139	-0.201
144	-0.192
150	-0.181
154	-0.174
159	-0.165
163	-0.158
170	-0.147
176	-0.137
180	-0.131
184	-0.124
191	-0.113
195	-0.107
200	-0.100
205	-0.093
210	-0.086
214	-0.081
220	-0.073
226	-0.065
230	-0.060
235	-0.054
239	-0.049
245	-0.043
250	-0.037
256	-0.031
261	-0.027
265	-0.023
271	-0.018
275	-0.015
281	-0.010
287	-0.005
290	-0.003
CURVE 30*, ‡	
14	-0.838
20	-0.833
30	-0.817
35	-0.807
45	-0.784
52	-0.765
55	-0.757
CURVE 30 (cont.)*, ‡	
61	-0.740
66	-0.724
71	-0.709
76	-0.694
82	-0.674
91	-0.645
98	-0.622
101	-0.612
105	-0.599
110	-0.584
119	-0.556
122	-0.546
126	-0.534
131	-0.518
136	-0.503
143	-0.480
148	-0.463
153	-0.446
155	-0.440
162	-0.417
167	-0.401
172	-0.385
178	-0.366
182	-0.352
188	-0.332
193	-0.315
197	-0.302
202	-0.286
208	-0.267
212	-0.254
218	-0.234
222	-0.222
228	-0.203
232	-0.190
238	-0.171
243	-0.155
248	-0.139
254	-0.119
258	-0.106
262	-0.094
267	-0.078
272	-0.063
279	-0.042
283	-0.030
286	-0.021
291	-0.006
CURVE 31*, ‡	
14	-0.688
19	-0.683
25	-0.676
29	-0.671
35	-0.660
39	-0.652
46	-0.636
50	-0.627
55	-0.615
60	-0.602
63	-0.595
69	-0.580
75	-0.565
80	-0.553
85	-0.541
90	-0.528
94	-0.518
99	-0.505
104	-0.492
110	-0.476
115	-0.463
129	-0.427
134	-0.414
140	-0.399
145	-0.386
150	-0.373
155	-0.360
159	-0.350
165	-0.335
170	-0.323
176	-0.308
180	-0.297
186	-0.280
191	-0.267
196	-0.253
202	-0.237
206	-0.226
211	-0.213
217	-0.198
222	-0.185
225	-0.177
231	-0.163
236	-0.150
241	-0.138
246	-0.125
252	-0.108
259	-0.089
CURVE 32 (cont.)*, ‡	
261	-0.083
267	-0.067
272	-0.054
276	-0.043
282	-0.028
287	-0.015
291	-0.005
CURVE 32*	
353	0.079
328	0.052
CURVE 33*	
293	0.000
319	0.015
343	0.053
363	0.053
392	0.089
CURVE 34*	
353	0.160
329	0.096
CURVE 35*	
293	0.000
319	0.076
343	0.146
362	0.207
392	0.248
CURVE 36*	
353	0.238
328	0.133
293	0.000
CURVE 37*	
293	0.000
318	0.097
343	0.192
362	0.260
393	0.422
CURVE 38*	
352	0.338
328	0.163
293	0.000
CURVE 39*	
293	0.000
318	0.139
343	0.246
362	0.340
391	0.471
CURVE 40	
283	-0.047
293	0.000
298	0.028
303	0.052
308	0.078
313	0.106
323	0.158
333	0.221
CURVE 41	
283	0.008
293	0.000*
298	-0.022*
303	-0.014*
308	-0.016
313	-0.040
323	-0.040
333	-0.069
CURVE 42	
300	0.043*
301	0.048
323	0.148*
333	0.263
353	0.348
353	0.372
363	0.428
379	0.535
CURVE 43	
300	-0.002
301	-0.010*
323	-0.024*
333	-0.067*
353	-0.111
353	-0.119*
363	-0.174
379	-0.214
CURVE 44*, †, ‡	
2	0.000 x 10^{-3}
2.5	-0.014
3	-0.039
3.5	-0.075
4	-0.122
5	-0.237
6	-0.350
7	-0.415
8	-0.393
10	0.085
12	1.216
14	2.993
15	4.116
16	5.385
18	8.326
20	11.735
22	15.515
24	19.579
25	21.696
26	23.859
28	28.298
57	94.009
65	110.25
75	131.09
85	151.05
284	268.45
CURVE 45*, †, ‡	
2	0.000 x 10^{-3}
2.5	0.008
3	0.024
3.5	0.047
4	0.078
5	0.163
6	0.270

* Not shown in figure.

† This curve is here reported using the first given temperature as reference temperature at which $\Delta L/L_0 = 0$.

‡ Author's data for coefficient of thermal expansion have been integrated by TPRC to obtain $\Delta L/L_0$.

DATA TABLE 22. THERMAL LINEAR EXPANSION OF INDIUM In (continued)

T	$\Delta L/L_0$
CURVE 45 (cont.)*, †, ‡	
7	0.386×10^{-3}
8	0.501
10	0.707
12	0.878
14	1.036
15	1.123
16	1.223
18	1.495
20	1.894
22	2.456
24	3.209
25	3.664
26	4.173
28	5.359
57	44.727
65	60.590
75	83.740
85	108.89
284	880.00
CURVE 46‡	
1.53	3×10^{-6}
1.94	5
2.35	11
2.55	16
2.93	27
3.26	40
3.40	47
CURVE 47‡	
1.53	-5×10^{-6}
1.94	-10
2.35	-19
2.55	-26
2.93	-43
3.26	-65
3.40	-75
CURVE 48‡	
0.00	-4×10^{-6}
1.53	-0.4
1.94	2
2.35	9

T	$\Delta L/L_0$
CURVE 48 (cont.)‡	
2.55	14×10^{-6}
2.93	25
3.26	39
3.40	47
CURVE 49‡	
0.00	16×10^{-6}
1.53	10
1.94	4*
2.35	-10
2.55	-16
2.93	-37
3.26	-63
3.40	-75
CURVE 50	
91	-0.800
204	-0.368
293	0.000
295	0.009
334	0.259
359	0.381
382	0.551
408	0.804
CURVE 51	
91	-0.263
204	-0.033
293	0.000
295	0.000
333	0.000
360	-0.044
383	-0.087
408	-0.209

* Not shown in figure.
† This curve is here reported using the first given temperature as reference temperature at which $\Delta L/L_0 = 0$.
‡ Author's data for coefficient of thermal expansion have been integrated by TPRC to obtain $\Delta L/L_0$.
‡ For explaination see, specification table.

DATA TABLE 22. COEFFICIENT OF THERMAL LINEAR EXPANSION OF INDIUM In

[Temperature, T, K; Coefficient of Expansion, α, 10^{-6} K^{-1}]

T	α
CURVE 4‡	
100	28.4
150	32.4
200	38.0
250	45.2
300	54.0
CURVE 5‡	
100	18.6
150	13.2
200	5.6
230	0.0
250	-4.2
300	-16.1
CURVE 6‡	
100	30.3
150	34.0
200	39.2
250	45.8*
300	53.9*
CURVE 18*, ‡	
303	-7.4
313	-8.8
323	-11.3
333	-13.0
343	-13.1
353	-17.4
363	-19.9
373	-23.0
383	-26.2
393	-31.0
CURVE 19*, ‡	
303	49.9
313	52.7
323	55.4
333	57.8
343	60.7
353	63.2
363	66.3
373	69.9
CURVE 19 (cont.)*, ‡	
383	72.9
393	78.7
CURVE 20‡	
100	20.6
150	15.0
200	7.3
238	0.0
250	-2.7
300	-14.9
CURVE 21‡	
100	31.9
150	35.5
200	40.7
250	47.3
300	55.5
CURVE 22*, ‡	
100	23.4
150	17.1
200	8.3
238	0.0
250	-3.1
300	-16.9
CURVE 23*, ‡	
100	29.6
150	33.7
200	39.5
250	47.0
300	56.1
CURVE 24*, ‡	
100	18.0
150	13.1
200	6.1
235	0.0
250	-2.8
300	-13.8
CURVE 25*, ‡	
100	34.2
150	39.4
200	46.8
250	56.2
300	67.7
CURVE 26*, ‡	
100	18.3
150	13.4
200	6.5
238	0.0
250	-2.4
300	-13.2
CURVE 27*, ‡	
100	28.4
150	33.6
200	40.8
250	50.2
300	61.6
CURVE 28*, ‡	
100	21.4
150	16.6
200	9.9
250	1.4
257	0.0
300	-9.1
CURVE 29*, ‡	
9	8.15
13	10.48
19	11.68
25	13.60
30	16.87
34	17.95
40	18.15
44	16.55
49	18.75
55	20.02
60	20.28
64	19.54
CURVE 29 (cont.)*, ‡	
70	19.36
74	18.80
79	20.44
83	19.64
88	18.93
94	19.45
99	19.17
104	18.81
110	19.24
114	17.67
120	18.31
124	17.44
129	17.05
133	17.61
139	17.73
144	18.51
150	18.79
154	17.75
159	17.13
163	16.57
170	16.40
176	16.18
180	16.38
184	15.85
191	15.53
195	14.81
200	13.58
205	14.01
210	14.09
214	13.30
220	13.04
226	12.90
230	12.09
235	11.85
239	11.50
245	10.80
250	10.05
256	9.70
261	9.31
265	8.48
271	8.53
275	8.23
281	8.07
287	7.96
290	8.03
CURVE 30‡	
14	6.02
20	10.29
30	21.40
35	21.13
45	24.97
52	27.96
55	28.37
61	28.89
66	31.40
71	30.98
76	30.41
82	33.32
87	32.70
91	31.85
98	34.43
101	33.46
105	29.52
110	28.86
119	34.06
122	31.98
126	30.56
131	30.57
136	31.32
143	34.45
148	33.56*
153	32.39*
155	32.39
162	32.28
167	32.06
172	32.06
178	33.50
182	33.82
188	34.45
193	32.73
197	31.66
202	31.81
208	32.38
212	32.79
218	31.91
222	31.57
228	32.18
232	32.02
238	31.79
243	31.79
248	32.18
254	32.70
CURVE 30 (cont.)‡	
258	31.93
262	31.26
267	29.98
272	30.57
279	30.68
283	29.60
286	29.25
291	28.65
CURVE 31‡	
14	7.74
19	10.77*
25	12.77
29	16.40
35	19.50
39	20.53
46	23.58
50	24.54
55	24.30
60	24.30
63	25.18
69	25.08
75	24.50
80	24.18
85	24.88
90	24.89
94	25.54
99	26.26
104	25.71
110	26.22
115	26.36
129	25.62
134	25.62
140	26.12
145	25.72
150	25.38
155	25.66
159	25.26
165	24.74
170	24.66
176	25.52
180	27.18
186	28.15
191	27.27
196	26.82

* Not shown in figure.
‡ Author's data for coefficient of thermal expansion have been integrated by TPRC to obtain $\Delta L/L_0$.

DATA TABLE 22. COEFFICIENT OF THERMAL LINEAR EXPANSION OF INDIUM In (continued)

T	α
CURVE 31 (cont.)‡	
202	27.13
206	26.32
211	25.62
217	26.12
222	25.14
225	24.44
231	24.72
236	24.78
241	25.26
246	26.50
252	27.93
259	28.15
261	26.97
267	27.03
272	26.72
276	25.68
282	25.22
287	25.26
291	24.25
CURVE 44‡	
2	-0.20
2.5	-0.38*
3	-0.60*
3.5	-0.83*
4	-1.05*
5	-1.26
6	-0.99*
7	-0.32*
8	0.77
10	4.01
12	7.30*
14	10.47
15	11.98*
16	13.41
18	16.00
20	18.09
22	19.71
24	20.93
25	21.41*
26	21.84*
28	22.55*
57	22.0
65	21.3
75	20.4
85	19.5
284	-7.7

T	α
CURVE 45‡	
2	0.12
2.5	0.24*
3	0.38*
3.5	0.54*
4	0.70*
5	0.99*
6	1.15
7	1.18*
8	1.12*
10	0.94*
12	0.77*
14	0.81*
15	0.92
16	1.09*
18	1.63*
20	2.36
22	3.26*
24	4.27*
25	4.82
26	5.37*
28	6.49
57	20.2*
65	22.1
75	24.2*
85	26.1
284	51.4
CURVE 50*	
100	29.35
200	39.70
300	60.55
400	84.70
CURVE 51*	
100	22.45
200	12.10
300	-1.55
400	-25.70

* Not shown in figure.
‡ Author's data for coefficient of thermal expansion have been integrated by TPRC to obtain $\Delta L/L_0$.

FIGURE AND TABLE NO. 23R. RECOMMENDED VALUES FOR THERMAL LINEAR EXPANSION OF IRIDIUM Ir

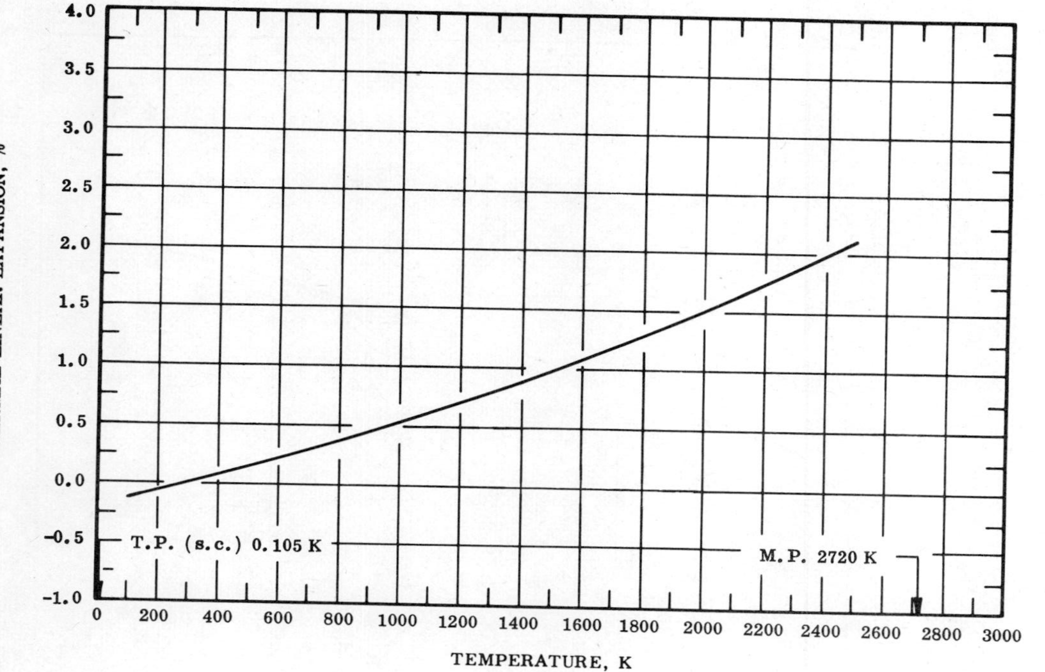

RECOMMENDED VALUES

[Temperature, T, K; Linear Expansion, $\Delta L/L_0$, %; α, K^{-1}]

T	$\Delta L/L_0$	$\alpha \times 10^6$
100	-0.111	4.4
200	-0.058	5.9
293	0.000	6.4
400	0.070	6.7
500	0.139	7.2
600	0.212	7.6
700	0.289	7.9
800	0.370	8.1
900	0.452	8.2
1000	0.535	8.4
1200	0.705	8.6
1400	0.880	9.0
1600	1.068	9.8
1800	1.275	10.9
2000	1.505	12.1
2200	1.758	13.2
2400	2.032	14.2
2500	2.178	14.7

REMARKS

The tabulated values are considered accurate to within ± 3% over the entire temperature range. These values can be represented approximately by the following equations:

$$\Delta L/L_0 = 6.005 \times 10^{-4} (T - 293) + 3.184 \times 10^{-7} (T - 293)^2 - 1.410 \times 10^{-10} (T - 293)^3 \qquad (293 < T < 1025)$$

$$\Delta L/L_0 = 0.557 + 8.400 \times 10^{-4} (T - 1025) + 8.762 \times 10^{-9} (T - 1025)^2 + 1.332 \times 10^{-10} (T - 1025)^3 \qquad (1025 < T < 1760)$$

$$\Delta L/L_0 = 1.232 + 1.069 \times 10^{-3} (T - 1760) + 3.025 \times 10^{-7} (T - 1760)^2 - 2.528 \times 10^{-11} (T - 1760)^3 \qquad (1760 < T < 2500)$$

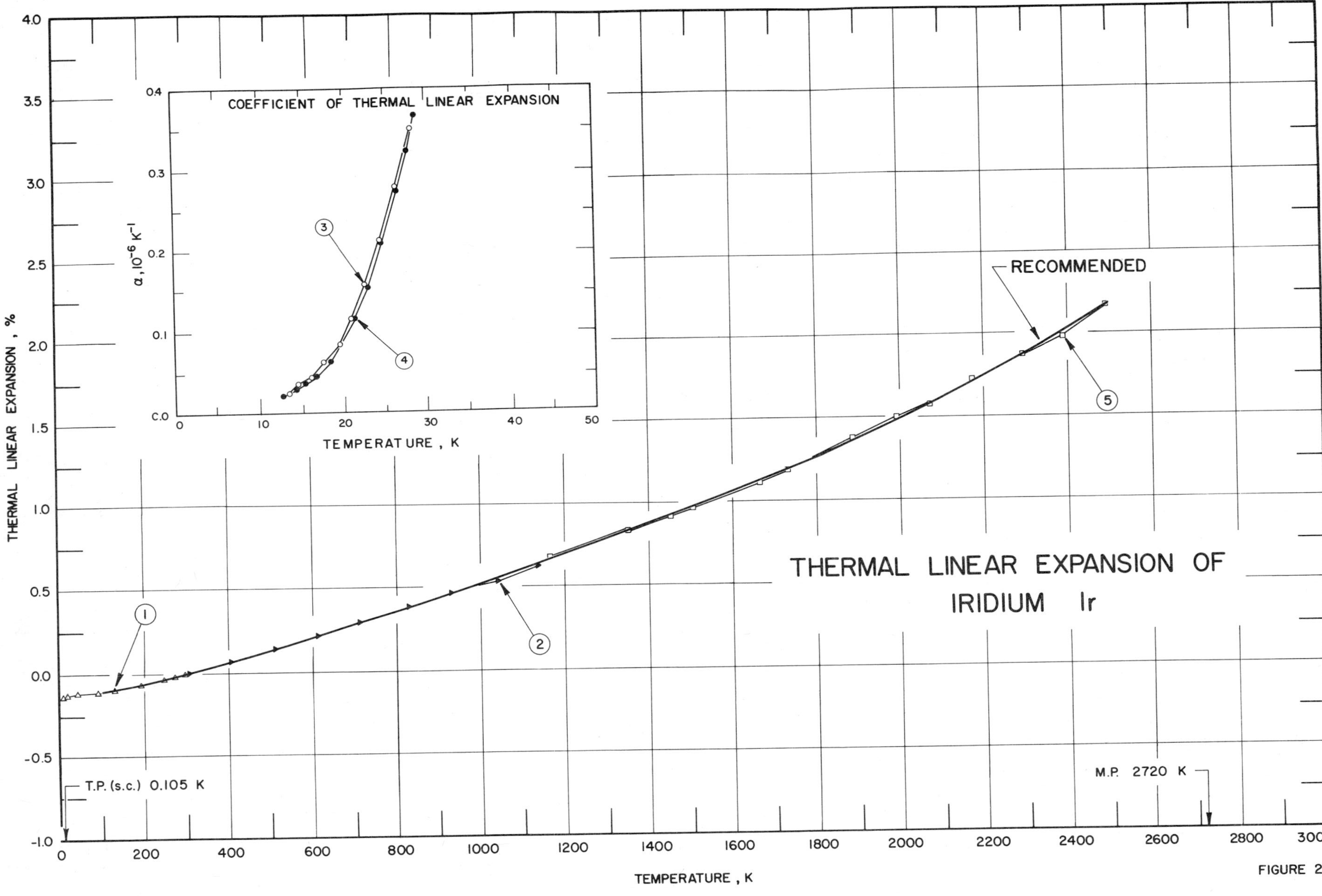
THERMAL LINEAR EXPANSION OF
IRIDIUM Ir
RECOMMENDED
M.P. 2720 K
T.P. (s.c.) 0.105 K
THERMAL LINEAR EXPANSION , %
TEMPERATURE , K
COEFFICIENT OF THERMAL LINEAR EXPANSION
α, 10⁻⁶ K⁻¹
TEMPERATURE , K
1
2
3
4
5

FIGURE 23

SPECIFICATION TABLE 23. THERMAL LINEAR EXPANSION OF IRIDIUM Ir

Cur. No.	Ref. No.	Author(s)	Year	Method Used	Temp. Range, K	Name and Specimen Designation	Composition (weight percent), Specifications, and Remarks
1	198, 652	Schaake, H. F.	1968	X	4.2-298		Specimen in the form of a 0.2 mm thick layer of iridium powder < 250 mesh, supplied as iridium sponge by Engelhard Industries, and used in the "as received" condition; pressed into copper specimen mount; lattice parameter reported at 271 K is 3.83893 Å; 3.83946 Å at 293 K determined by graphical interpolation.
2	174	Singh, H. P.	1968	X	303-1138		Specpure powder filled in quartz capillary of 0.5 mm inner diameter and annealed; lattice parameter reported at 303 K is 3.8390 Å; 3.8387 Å at 293 K determined by graphical extrapolation; data on the coefficient of thermal linear expansion also reported.
3*	598	White, G.K. and Pawlowicz, A.K.	1970	E	13-28		Purity > 99.9; specimen 1.4 cm diameter x 2.5 cm long from International Nickel Co.; made from powder by sintering and hot forging; annealed at 1270 K in vacuum; this curve is here reported using the first given temperature as reference temperature at which $\Delta L/L_0 = 0$.
4*	598	White, G.K. and Pawlowicz, A.K.	1970	E	13-29		The above specimen; this curve is here reported using the first given temperature as reference temperature at which $\Delta L/L_0 = 0$.
5	624, 653	Halvorsen, J.J.	1971	I	1164-2494		Commercially pure wire specimen of 0.635 cm diameter from Engelhard Industries, N. J.; impurities as follows: 0.0087 Fe, 0.0054 each Rh and Ru, 0.0030 O, 0.015 each Pt and N, 0.009 Al, and <0.0010 each Sb, As, Te, Zn.

* Not shown in figure.

DATA TABLE 23. THERMAL LINEAR EXPANSION OF IRIDIUM Ir

[Temperature, T, K; Linear Expansion, $\Delta L/L_0$, %]

T	$\Delta L/L_0$
CURVE 1	
4.2	-0.132
20.0	-0.130
42.5	-0.128
61.0	-0.125*
77.4	-0.122*
90.0	-0.120
101.0	-0.110*
131.0	-0.0979
162.0	-0.0656*
194.5	-0.0599
241.5	-0.0336
271.0	-0.0138
298.0	0.0044
CURVE 2	
303	0.006
406	0.068
508	0.144
611	0.204
713	0.290
825	0.373
928	0.452
1040	0.537
1138	0.629
CURVE 3*, †, ‡	
13.40	0.000×10^{-4}
CURVE 3 (cont.)*, †, ‡	
14.57	0.038×10^{-4}
16.05	0.099
17.57	0.181
19.28	0.306
21.03	0.481
22.84	0.728
24.58	1.049
26.36	1.485
28.11	2.034
CURVE 4*, †, ‡	
12.90	0.000×10^{-4}
CURVE 4*, †, ‡ (cont.)	
14.09	0.032×10^{-4}
15.27	0.071
16.64	0.129
18.09	0.209
19.73	0.33
21.36	0.495
23.03	0.722
24.80	1.046
26.53	1.462
27.96	1.889
28.96	2.234
CURVE 5	
1164	0.67
1350	0.834
1450	0.919
1465	0.923*
1502	0.951
1509	0.984*
1667	1.106
1733	1.190
1750	1.256*
1888	1.382
1991	1.500
2071	1.575
2178	1.739
CURVE 5 (cont.)	
2298	1.875
2334	1.940
2386	1.992
2494	2.174

DATA TABLE 23. COEFFICIENT OF THERMAL LINEAR EXPANSION OF IRIDIUM Ir

[Temperature, T, K; Coefficient of Expansion, α, 10^{-6} K^{-1}]

T	α
CURVE 2*	
303	6.21
406	6.48
508	6.77
611	7.08
713	7.39
825	7.76
928	8.11
1040	8.51
1138	8.87
CURVE 3‡	
13.40	0.0279
14.57	0.037
16.05	0.046
17.57	0.061
19.28	0.085
21.03	0.115
CURVE 3 (cont.)‡	
22.84	0.157
24.58	0.212
26.36	0.277
28.11	0.350
CURVE 4‡	
12.90	0.022
14.09	0.030
15.27	0.037
16.64	0.047
18.09	0.063
19.73	0.085*
21.36	0.116
23.03	0.155
24.80	0.210
26.53	0.271
27.96	0.325
28.96	0.365

*Not shown in figure.

†This curve is here reported using the first given temperature as reference temperature at which $\Delta L/L_0 = 0$.

‡Author's data for coefficient of thermal expansion have been integrated by TPRC to obtain $\Delta L/L_0$.

FIGURE AND TABLE NO. 24R. RECOMMENDED VALUES FOR THERMAL LINEAR EXPANSION OF IRON Fe

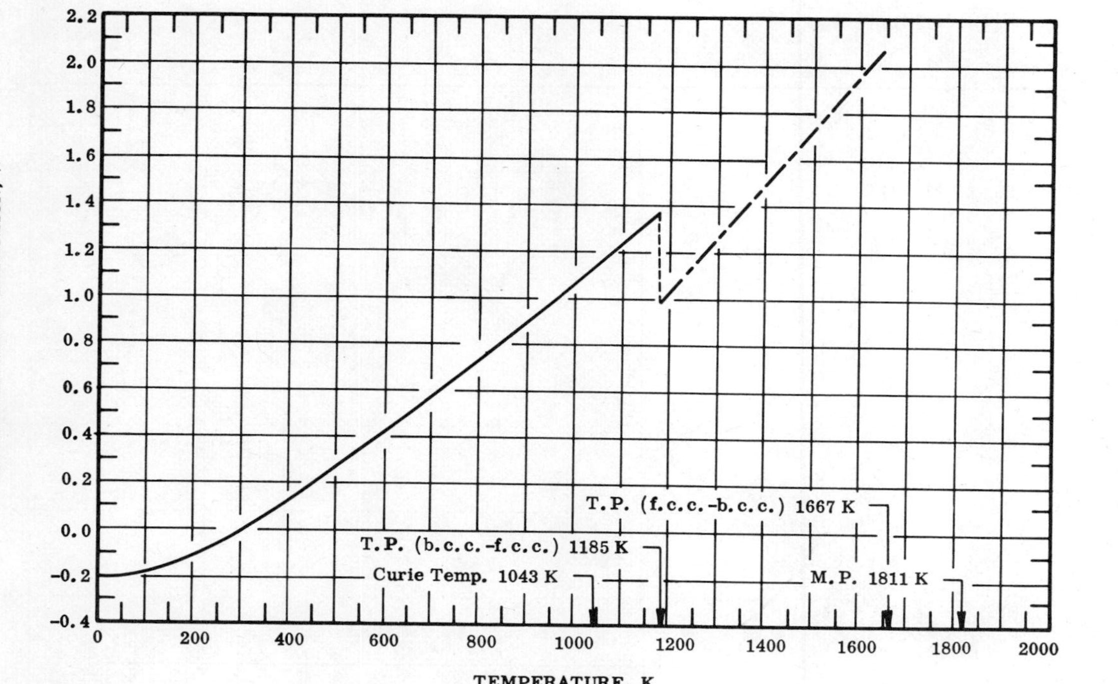

RECOMMENDED VALUES

[Temperature, T, K; Linear Expansion, $\Delta L/L_0$, %; α, K^{-1}]

T	$\Delta L/L_0$	$\alpha \times 10^6$
5	-0.204	0.01
25	-0.203	0.20
50	-0.203	1.3
100	-0.184	5.6
200	-0.102	10.1
293	0.000	11.8
400	0.134	13.4
500	0.274	14.4
600	0.421	15.1
700	0.575	15.7
800	0.735	16.2
900	0.899	16.4
1000	1.065	16.6
1100	1.230	16.7
1185‡	1.370	16.8
1185‡	0.993†	23.3
1200	1.028†	23.3
1400	1.494†	23.3
1600	1.960†	23.3
1650	2.077†	23.3

† Typical values.
‡ Phase transition.

REMARKS

The tabulated values are considered accurate to within ± 3% at temperatures below 900 K, ± 5% below 1185 K, and ± 20% above. Hysteresis effects in the thermal expansion probably due to unsuitable heating and cooling rates are observed near the α-β transition near 1185 ± 50 K. The tabulated values can be represented approximately by the following equations:

$$\Delta L/L_0 = -0.206 - 5.905 \times 10^{-5} (T - 5) + 3.650 \times 10^{-6} (T - 5)^2 - 3.389 \times 10^{-9} (T - 5)^3 \quad (5 < T < 300)$$

$$\Delta L/L_0 = 0.007 + 1.210 \times 10^{-3} (T - 300) + 6.504 \times 10^{-7} (T - 300)^2 - 3.140 \times 10^{-10} (T - 300)^3 \quad (300 < T < 1185)$$

$$\Delta L/L_0 = -1.810 + 2.435 \times 10^{-3} T - 8.100 \times 10^{-8} T^2 + 2.057 \times 10^{-11} T^3 \quad (1185 < T < 1650)$$

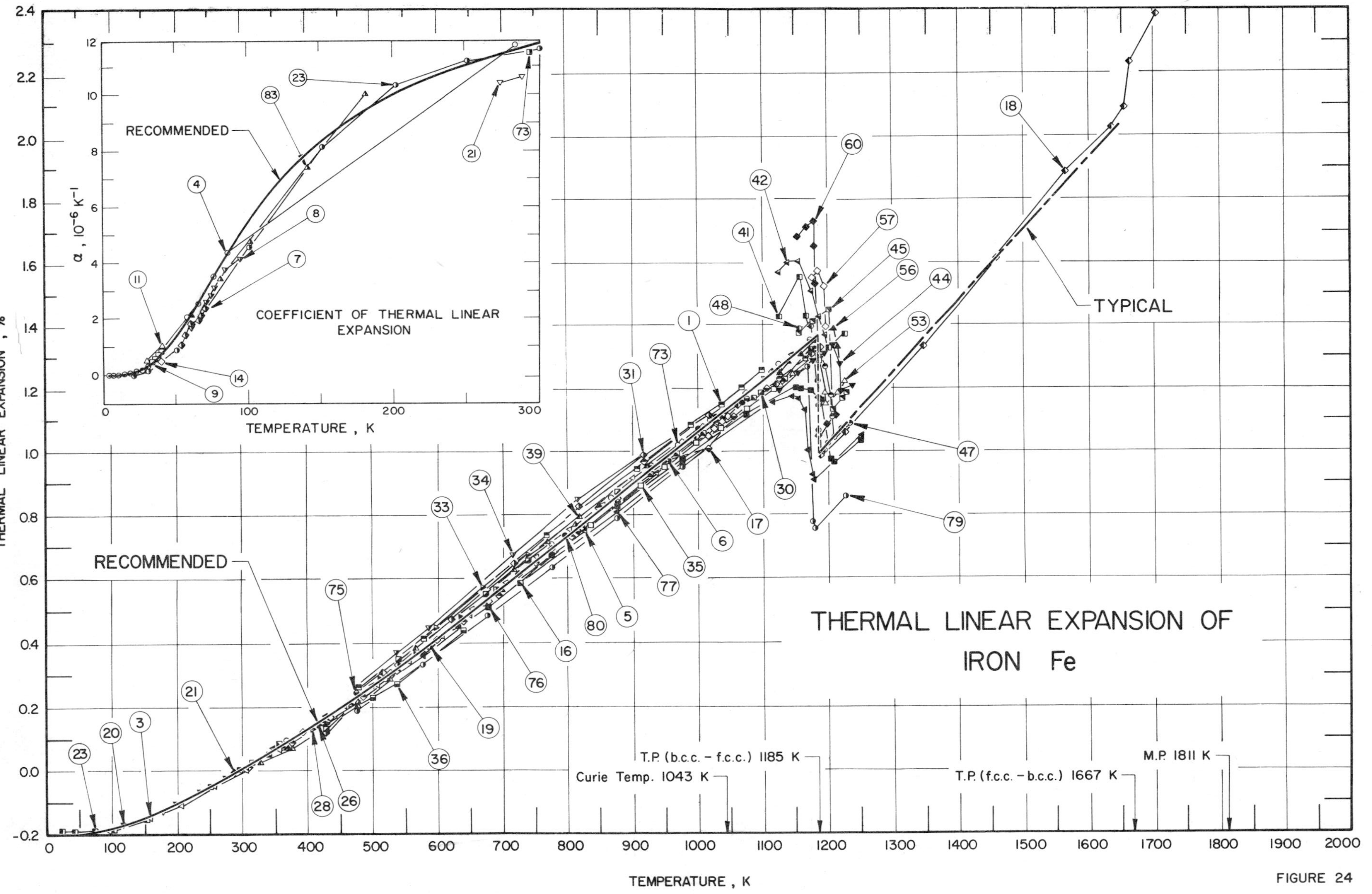
THERMAL LINEAR EXPANSION OF
IRON Fe
TYPICAL
RECOMMENDED
M.P. 1811 K
T.P. (f.c.c. – b.c.c.) 1667 K
T.P. (b.c.c. – f.c.c.) 1185 K
Curie Temp. 1043 K
TEMPERATURE , K
THERMAL LINEAR EXPANSION , %
COEFFICIENT OF THERMAL LINEAR EXPANSION
α , 10⁻⁶ K⁻¹
RECOMMENDED
TEMPERATURE , K

FIGURE 24

SPECIFICATION TABLE 24. THERMAL LINEAR EXPANSION OF IRON Fe

Cur. No.	Ref. No.	Author(s)	Year	Method Used	Temp. Range, K	Name and Specimen Designation	Composition (weight percent), Specifications, and Remarks
1	40	Gorton, A.T., Bitsianes, G., and Joseph, T.L.	1965	X	295-1098	α	99.67 Fe; Buel electrolytic iron powder, hydrogen annealed at 738 K magnetic transformation temperature range 988-1098 K; α-phase; lattice parameter reported at 295 K is 2.8663 Å; 2.8663 Å at 293 K determined by graphical extrapolation.
2*	40	Gorton, A.T., et al.	1965	X	1193-1343	γ	The above specimen in γ-phase; this curve is here reported using the first given temperature as reference temperature at which $\Delta L/L_0 = 0$; lattice parameter reported at 1193 K is 3.6477 Å.
3	1	Nix, F.C. and MacNair, D.	1941	I	92-958		99.992 Fe, 0.0023 S, 0.002 Cu, $<$0.001 Be, 0.001 C, 0.001 Si, 0.0005 P, 0.0003 O_2, 0.0002 H_2, and 0.0002 N_2; furnished by Dr. J.G. Thompson of NBS; specimen 6 mm in length; heated in He gas at pressure of 3 mm Hg up to 370 K at a rate of 10 C per hr and then in a vacuum of 10^{-5} mm Hg up to 1070 K; zero-point correction of -0.025% determined by graphical interpolation.
4*	41	White, G.K.	1965	E	3-85		99.95$^+$ Fe, 0.1 Si, 0.1 Mn, and 0.008 C; vacuum-remelted; specimen 5 cm long and 2 cm diameter; annealed in vacuo for some hrs after machining at 870 K; this curve is here reported using the first given temperature as reference temperature at which $\Delta L/L_0 = 0$.
5	42	Stuart, H. and Ridley, N.	1966	X	348-1148		99.998 Fe; powder ground to 200 Mesh, sealed in vacuum in 0.5 mm diameter silica tubes and stress relief annealed at 870 K for 15 min formed into wire 0.5 mm diameter, sealed in silica tubes, then cooled with graphite; lattice parameter reported at 293 K is 2.8662 Å.
6	43	Fassiska, E.J. and Zwell, L.	1967	X	296-951		99.5$^+$ Fe; crushed for analysis; reported error 2%; lattice parameter reported at 196 K is 2.8664 Å; 1.8663 Å at 293 K determined by graphical extrapolation.
7*	3	Fraser, D.B. and Hollis-Hallet, A.C.	1965	I	20-69	α	Cubic crystal structure in α-phase; hollow cylindrical specimen 2.54 cm O.D., 2.06 cm I.D., 5.0 cm long; expansion measured on first test; this curve is here reported using the first given temperature as reference temperature at which $\Delta L/L_0 = 0$; data on the coefficient of thermal linear expansion also reported.
8*	3	Fraser, D.B. and Hollis-Hallet, A.C.	1965	I	20-121		The above specimen; expansion on second test; this curve is here reported using the first given temperature as reference temperature at which $\Delta L/L_0 = 0$, data on the coefficient of thermal linear expansion also reported.
9*	44	Krishnan, K.S.	1966	E	5-39		Zone refined specimen obtained from Battelle Institute; this curve is here reported using the first given temperature as reference temperature at which $\Delta L/L_0 = 0$.
10*	44	Krishnan, K.S.	1966	E	8-39		Floating zone refined by author at Bell Telephone Laboratories; this curve is here reported using the first given temperature as reference temperature at which $\Delta L/L_0 = 0$.
11*	44	Krishnan, K.S.	1966	E	5-40		Zone melted iron with 0.02 Mn added; Fe obtained from American Iron and Steel Institute; this curve is here reported using the first given temperature as reference temperature at which $\Delta L/L_0 = 0$.
12*	44	Krishnan, K.S.	1966	E	5-39		Vacuum melted Fe with 0.025 Mn added; Fe obtained from American Iron and Steel Institute; this curve is here reported using the first given temperature as reference temperature at which $\Delta L/L_0 = 0$.

* Not shown in figure.

SPECIFICATION TABLE 24. THERMAL LINEAR EXPANSION OF IRON Fe (continued)

Cur. No.	Ref. No.	Author(s)	Year	Method Used	Temp. Range, K	Name and Specimen Designation	Composition (weight percent), Specifications, and Remarks
13*	44	Krishnan, K.S.	1966	E	5-38		Similar to the above specimen but with 0.15 Mn added; this curve is here reported using the first given temperature as reference temperature at which $\Delta L/L_0 = 0$.
14*	44	Krishnan, K.S.	1966	E	5-39		Vacuum melted Fe with 0.015 Mo added; Fe obtained from American Iron and Steel Inst.; Mo from General Electric Co.; this curve is here reported using the first given temperature as reference temperature at which $\Delta L/L_0 = 0$.
15*	44	Krishnan, K.S.	1966	E	5-39		Similar to the above specimen except 0.039 Mo added; this curve is here reported using the first given temperature as reference temperature at which $\Delta L/L_0 = 0$.
16	45	Ridley, N. and Stuart, H.	1968	X	298-1172		99.98 Fe from Johnson Matthey; specimen filings from bulk material as supplied; Cu Kα x-radiation used; lattice parameter reported at 293 K is 2.8622 Å.
17	45	Ridley, N. and Stuart, H.	1968	X	292-1106		99.90 Fe, Japanese iron; similar to the above specimen; lattice parameter at 293K is 2.8622 Å.
18	46	Basinski, Z.S., Hume-Rothery, W., and Sutton, A.L.	1955	X	293-1807		Two specimens used; for data 293-1273 K, filing of high purity iron bar (42AF1) from National Physical Laboratory of 99.97$^+$ purity; for data 1273 K up, filings taken from 25 S.W.G. high purity iron wire (AHN2) from British Iron and Steel Research Assn., of 99.98$^+$ purity; expansion in 293-1273 K measured with filings in vacuum in silica container; above 1273 K specimen in hydrogen atmosphere at pressure of 1.5 atmosphere; lattice parameter reported at 293 K is 2.4772 kX (2.4822 Å).
19	47	Allen, R.D.	1959	L	293-922	Armco Iron	No details given.
20	48	Dorsey, H.G.	1907	I	93-293		0.058 C, 0.008 Si, 0.071 Mn, trace P; rolled rod from Pennsylvania Steel Co.; length 2.0129 cm; density 7.77 g cm^{-3}.
21	49	Owen, E.A. and Yates, E.L.	1937	X	287-913		99.966 Fe; supplied by the Mond Nickel Co., Ltd.; α-phase; lattice parameter reported at 291 K is 2.8605 Å; 2.8606 Å at 293 K determined by graphical interpolation; data on the coefficient of thermal linear expansion also reported.
22	50	Jaeger, F.M. and Rosenbohm, E.	1938	I	373-1073		Specimen 41 mm long at 273 K; zero-point correction of -0.02% determined by graphical extrapolation.
23	22	Fraser, D.B. and Hollis-Hallett, A.C.	1961	L	20-300		Metal machined into form of thick-walled cylindrical tube having ends which were carefully ground plane and parallel to each other; annealed; reported error decreases from maximum of 20% at 20 K.
24*	51	Souder, W. and Hidnert, P.	1922	T	298-973		0.02 C, 0.007 S, 0.006 Si, 0.014 total of Co, Cu, Ni; annealed specimen.
25*	52	National Physical Laboratory	1936		273-523	Batch 5	0.0045 C, 0.002 Mn, 0.0015 B, 0.0010 P; total of all other impurities <0.0025.
26	53	Batchelder, F.W.V. and Raeuchle, R.F.	1954	X	298-634		Expansion measured in 310 plane; Co x-radiation used; reported error ±0.02; lattice constant reported at 298 K is 2.8665 Å; 2.8664 Å at 293 K determined by graphical extrapolation.
27*	53	Batchelder, F.W.V. and Raeuchle, R.F.	1954	X	298-636		The above specimen, expansion measured in 211 plane; Cr x-radiation lattice constant reported at 298 K is 2.8665 Å; 2.8664 Å at 293 K determined by graphical extrapolation.
28	53	Batchelder, F.W.V. and Raeuchle, R.F.	1954	X	304-621		The above specimen, expansion measured in 220 plane; Fe x-radiation; lattice constant reported at 304 K is 2.8666 Å; 2.8663 Å at 293 K determined by graphical extrapolation.

*Not shown in figure.

SPECIFICATION TABLE 24. THERMAL LINEAR EXPANSION OF IRON Fe (continued)

Cur. No.	Ref. No.	Author(s)	Year	Method Used	Temp. Range, K	Name and Specimen Designation	Composition (weight percent), Specifications, and Remarks
29*	54	Straumanis, M.E. and Kim, D.C.	1969	X	283-332		Main impurities 0.0020 Ni, 0.0015 Al, 0.0010 Ca, 0.0010 Si, 0.0010 V, 0.0009 P, and 0.0009 C; zone-refined iron from Battelle Memorial Institute; filing from inner portion of rod taken; vacuum annealed at 770 K for 30 min to relieve stress; 300 mesh; lattice constant reported at 293 K is 2.8661 Å.
30	54	Straumanis, M.E. and Kim, D.C.	1969	X	282-1097		Similar to the above specimen; measured in vacuum; lattice constant reported at 293 K is 2.8664 Å.
31	54	Straumanis, M.E. and Kim, D.C.	1969	X	293-1113		Similar to the above specimen; measured in H_2, pressures change with temperatures, at 300 K, 1 atm., at 520 K, 1.7 atm., at 770 K, 2.6 atm., at 970 K, 3.2 atm., at 1120 K, 3.7 atm.; lattice parameter reported at 293 K is 2.8664 Å.
32*	54	Straumanis, M.E. and Kim, D.C.	1969	X	288-329		Single crystal whisker grown by reducing $FeCl_2$ in dry hydrogen stream at 100 K for 2 hr in tube furnace, molten $FeCl_2$ cooled with hydrogen; 1 to 2 cm long, about 0.1 mm thick, and frequently curved; lattice constant reported at 293 K is 2.8661 Å.
33	54	Straumanis, M.E. and Kim, D.C.	1969	X	282-1119		Similar to the above specimen; whisker packed into quartz capillary, sealed under vacuum, measured in vacuum; lattice constant reported at 293 K is 2.8664 Å.
34	54	Straumanis, M.E. and Kim, D.C.	1969	X	293-1103		Similar to the above specimen; measured in H_2, pressure changes with temperature, 300 K - 1 atm., 520 K - 1.7 atm., 770 K - 2.6 atm., 970 K - 3.2 atm., 1120 K - 3.7 atm.; lattice parameter reported at 293 K is 2.8664 Å.
35	55	Richter, F. and Lotter, V.	1969	V	831-1115		0.01 O, 0.0002 each Mg, Mn, Si, 0.0001 each Cu, Ag; cylindrical specimen 6 mm diameter and 35 mm long, one side being of hemispherical shape; supplied by HEK, Lübeck; heating rate 20 C per min; volume magnetostriction $\omega_M = 4.0 \pm 0.4 \times 10^{-4}$; Curie temperature 1036 ± 3 K; expansion measured with constant field H.
36	32	Kachanooska, A.	1949	X	295-639		Polycrystalline chemically pure sample, supplied by Everitt Co., Ltd., Liverpool; expansion measured from (013) reflection using Co Kα x-radiation; lattice parameter reported at 295 K is 2.8607 Å; 2.8606 Å at 293 K determined by graphical extrapolation.
37*	32	Kachanooska, A.	1949	X	295-612		The above specimen; expansion measured from (112) reflection using Cr Kα x-radiation; lattice parameter reported at 295 K is 2.8609 Å; 2.8608 Å at 293 K determined by graphical extrapolation.
38*	56	Masumoto, H. and Nara, S.	1927	L	303, 373		Electrolytic Fe.
39	57	Miller, I.G.	1958		310-1123		Cylindrical specimen; 6.5 mm diameter, 65 mm long; reported error in measurement of ΔL, 2 μ; zero-point correction of 0.016% determined by graphical extrapolation.
40	58	Austin, J.B. and Pierce, R.H.H.	1934	I	273-1073	Sample 1	0.07 H_2, 0.013 Si, 0.012 C, and 0.004 P; Burgess double-refined electroytic iron, used without treatment; specimen approximately 5 mm height, 2 mm diameter; heating rate 3.0 to 3.5 C per min; accuracy of length measurement $\pm 3 \times 10^{-6}$ cm.
41	58	Austin, J.B. and Pierce, R.H.H.	1934	I	1123-1223	Sample 1	The above specimen; expansion measured during α-γ-phase transformation with increasing temperature; zero-point correction is -0.0234%.

SPECIFICATION TABLE 24. THERMAL LINEAR EXPANSION OF IRON Fe (continued)

Cur. No.	Ref. No.	Author(s)	Year	Method Used	Temp. Range, K	Name and Specimen Designation	Composition (weight percent), Specifications, and Remarks
42	58	Austin, J.B. and Pierce, R.H.H.	1934	I	1223-1123	Sample 1	The above specimen; expansion measured with decreasing temperature; zero-point correction is -0.023%.
43*	58	Austin, J.B. and Pierce, R.H.H.	1934	I	273-1073	Sample 2	0.035 Si, 0.003 O, and 0.0009 N; Westinghouse iron (A185) vacuum-melted electrolytic iron; specimen approximately 5 mm height, 2 mm diameter; heating rate 3.0 to 3.5 C per min; accuracy of length measured ±3 x 10^{-6} cm.
44	58	Austin, J.B. and Pierce, R.H.H.	1934	I	1123-1231	Sample 2	The above specimen; expansion measured during α-γ-phase transformation with increasing temperature; zero-point correction is -0.023%.
45	58	Austin, J.B. and Pierce, R.H.H.	1934	I	1231-1154	Sample 2	The above specimen; expansion measured with decreasing temperature; zero-point correction is -0.023%.
46*	58	Austin, J.B. and Pierce, R.H.H.	1934	I	273-1073	Sample 3	0.03 Si, 0.012 P, 0.006 S, 0.002 C, and trace Mn; Westinghouse (A160) hydrogen-melted electrolytic iron; specimen approximately 5 mm high, 2 mm diameter, accuracy of length measurement ±3 x 10^{-6} cm; heating rate 3.0 to 3.5 C per min.
47	58	Austin, J.B. and Pierce, R.H.H.	1934	I	1123-1230	Sample 3	The above specimen; expansion measured during α-γ-phase transformation with increasing temperature; zero-point correction is -0.023%.
48	58	Austin, J.B. and Pierce, R.H.H.	1934	I	1230-1156	Sample 3	The above specimen; expansion measured with decreasing temperature; zero-point correction is -0.023%.
49*	58	Austin, J.B. and Pierce, R.H.H.	1934	I	273-1073	Sample 4	Open-hearth ingot iron melted in hydrogen; specimen approximately 5 mm high, 2 mm diameter, accuracy of length measurement ±3 x 10^{-6} cm; heating rate 3.0 to 3.5 c per min.
50*	58	Austin, J.B. and Pierce, R.H.H.	1934	I	1123-1216	Sample 4	The above specimen; expansion measured during α-γ-phase transformation with increasing temperature; zero-point correction is -0.022%.
51*	58	Austin, J.B. and Pierce, R.H.H.	1934	I	1216-1137	Sample 4	The above specimen; expansion measured with decreasing temperature; zero-point correction is -0.022%.
52*	58	Austin, J.B. and Pierce, R.H.H.	1934	I	273-1073	Sample 5	0.03 Mn, 0.01 Si, 0.005 C, 0.004 P, 0.003 Si, 0.002 O, and 0.0001 N; Armco iron held in moist H_2 at 1770 K for 18 hr; specimen approximately 5 mm height, 2 mm diameter, accuracy of length measurement ±3 x 10^{-6} cm; heating rate 3.0 to 3.5 C per min.
53	58	Austin, J.B. and Pierce, R.H.H.	1934	I	1123-1221	Sample 5	The above specimen; expansion measured during α-γ-phase transformation with increasing temperature; zero-point correction is -0.023%.
54*	58	Austin, J.B. and Pierce, R.H.H.	1934	I	1221-1156	Sample 5	The above specimen; expansion measured with decreasing temperature; zero-point correction is -0.023%.
55	58	Austin, J.B. and Pierce, R.H.H.	1934	I	273-1073	Sample 6	0.012 C, 0.004 O, Walters (V53) vacuum-melted electrolytic iron; specimen approximately 5 mm high, 2 mm diameter, accuracy of length measurement ±3 x 10^{-6} cm; heating rate 3.0 to 3.5 C per min.
56	58	Austin, J.B. and Pierce, R.H.H.	1934	I	1123-1226	Sample 6	The above specimen; expansion measured during α-γ-phase transformation with increasing temperature; zero-point correction is -0.022%.
57	58	Austin, J.B. and Pierce, R.H.H.	1934	I	1226-1177	Sample 6	The above specimen; expansion measured with decreasing temperature; zero-point correction is -0.022%.

*Not shown in figure.

SPECIFICATION TABLE 24. THERMAL LINEAR EXPANSION OF IRON Fe (continued)

Cur. No.	Ref. No.	Author(s)	Year	Method Used	Temp. Range, K	Name and Specimen Designation	Composition (weight percent), Specifications, and Remarks
58*	58	Austin, J.B. and Pierce, R.H.H.	1934	I	273-1073	Sample 7	Powdered carbonyl iron melted in moist hydrogen; specimen approximately 5 mm high, 2 mm diameter, accuracy of length measurement $\pm 3 \times 10^{-6}$ cm; heating rate 3.0 to 3.5 C per min.
59	58	Austin, J.B. and Pierce, R.H.H.	1934	I	1123-1210	Sample 7	The above specimen; expansion measured during α-γ-phase transformation with increasing temperature; zero-point correction is -0.024%.
60	58	Austin, J.B. and Pierce, R.H.H.	1934	I	1210-1151	Sample 7	The above specimen; expansion measured with decreasing temperature; zero-point correction is -0.024%.
61*	58	Austin, J.B. and Pierce, R.H.H.	1934	I	273-1073	Sample 8	Powdered carbonyl iron, sintered in moist H_2 at 1770 K for 18 hr; specimen approximately 5 mm high, 2 mm diameter, accuracy of length measurement $\pm 3 \times 10^{-6}$ cm; heating rate 3.0 to 3.5 C per min.
62*	58	Austin, J.B. and Pierce, R.H.H.	1934	I	1123-1230	Sample 8	Similar to the above specimen; expansion measured during α-γ-phase transformation with increasing temperature; zero-point correction is -0.024%.
63*	58	Austin, J.B. and Pierce, R.H.H.	1934	I	1230-1183	Sample 8	The above specimen; expansion measured with decreasing temperature; zero-point correction is -0.024%.
64*	58	Austin, J.B. and Pierce, R.H.H.	1934	I	273-1073	Sample 9	0.0168 N, 0.012 O; carbonyl iron plate as received; specimen approximately 5 mm high, 2 mm diameter, accuracy of length measurement $\pm 3 \times 10^{-6}$ cm; heating rate 3.0 to 3.5 C per min.
65*	58	Austin, J.B. and Pierce, R.H.H.	1934	I	1123-1214	Sample 9	Similar to the above specimen; expansion measured during α-γ-phase transformation; zero-point correction is -0.023%.
66*	58	Austin, J.B. and Pierce, R.H.H.	1934	I	1123-1214	Sample 9	The above specimen; expansion measured with decreasing temperature; zero-point correction is -0.023%.
67*	58	Austin, J.B. and Pierce, R.H.H.	1934	I	273-1073	Sample 10	Similar to the above specimen except held in moist H_2 at 1770 K for 47 hr; heating.
68*	58	Austin, J.B. and Pierce, R.H.H.	1934	I	1123-1218	Sample 10	The above specimen; expansion measured during α-γ-phase transformation with increasing temperature; zero-point correction is -0.023%.
69*	58	Austin, J.B. and Pierce, R.H.H.	1934	I	1218-1181	Sample 10	The above specimen; expansion measured with decreasing temperature; zero-point correction is -0.023%.
70*	59	Masumoto, H.	1929	L	303, 373	Electrolytic Fe	Circular rod specimen 4 mm thick, 10 cm long; expansion measured in vacuum.
71*	59	Masumoto, H.	1929	L	303, 373	Armco Fe	Similar to the above specimen.
72*	60	Holborn, L. and Day, A.L.	1901	T	273-1023		Wrought iron; zero-point correction of -0.026% determined by graphical interpolation.
73*	61	Austin, J.B. and Pierce, R.H.H.	1933	I	293-1125	Sample A	Specimen specially purified carbonyl iron sintered 18 hr at 1773 K in hydrogen; α to γ transformation at 1188 ± 2 K.
74*	61	Austin, J.B. and Pierce, R.H.H.	1933	I	293-1173	Sample B	Specimen specially purified carbonyl iron melted in hydrogen; α to γ transformation at 1188 ± 2 K.
75	62	Yaggee, F.L. and Styles, J.W.	1969	L	273-1173	Armco Iron	α to γ transformation temperature 1183 K on heating, 1175 K on cooling; reported error $< \pm 2$%.

* Not shown in figure.

SPECIFICATION TABLE 24. THERMAL LINEAR EXPANSION OF IRON Fe (continued)

Cur. No.	Ref. No.	Author(s)	Year	Method Used	Temp. Range, K	Name and Specimen Designation	Composition (weight percent), Specifications, and Remarks
76	67	Richter, F.	1970	L	293-1250	Armco Iron	0.042 O_2, 0.020 each C, Mn, Ni, 0.010 each Cr, Cu, P, <0.010 Si, 0.006 S, 0.003 Al, 0.002 Sn; expansion measured with increasing temperature.
77	67	Richter, F.	1970	L	1250-293	Armco Iron	The above specimen; expansion measured with decreasing temperature; zero-point correction of 0.01% determined by graphical extrapolation.
78*	67	Richter, F.	1970	L	293-1223		0.005 C, 0.003 S, 0.002 each Mn, N, O, P, Si, 0.001 Al; expansion measured with increasing temperature.
79	67	Richter, F.	1970	L	1223-293		The above specimen; expansion measured with decreasing temperature; zero-point correction is -0.189%.
80	301	Arbuzov, M.P. and Zelenko, I.A.	1964	L	293-1069		Armco iron; data on the coefficient of thermal linear expansion are also reported.
81*	389	Owen, E.A., Yates, E.L., and Sully, A.H.	1937	X	288-873		Lump material heat-treated, then filings taken; heated 14 hr at 870 K; lattice parameter reported at 288 K is 2.86050 Å; 2.86067 Å at 293 K determined by graphical interpolation.
82	631	Masumoto, H., and Kobayashi, T.	1965	L	273-313		99.99 pure electrolytic pure specimen.
83	709	Shah, J.S.	1971	X	40-180		99.9 pure, powder sample of particle size 8-10M from Koch Light Labs, Inc., Colnbrook, England, lattice parameter not corrected for refraction, lattice parameter reported as 2.86086 Å at 40 K, curve reported using first givent temperature as reference temperature at which $\Delta L/L_0 = 0$, coefficient of thermal linear expansion also reported.
84	709	Shah, J.S.	1971	X	40-180		Similar to the above specimen, 99.999 pure, lattice parameter reported as 2.86077 Å at 40 K; lattice parameter not corrected for refraction, curve reported using first given temperature as reference temperature at which $\Delta L/L_0 = 0$.
85		Sinha, A.N. and Bulasundaram, L.	1970	L	273-673		Pure iron.
86	769	Gen, Ya.M., and Petror, Yu.I.	1968	X	293-1019		Measurements on aerosol particles, averaged of two heating runs, lattice parameter reported at 293 K is 2.88647 Å; zero-point correction is -0.01%.
87	693	Ryabov, V.R., Lozovskaya, A.V., and Vasil'yev, V.G.	1969	L	293-773		99.97 pure specimen.
88		Lysak, L.I., and Andrushchik, L.O.	1969	L	93-293		No details given; specimen suddenly cooled in liquid nitrogen.
89		Andreeva, L.P., and Gel'd, P.V.	1971	L	293-1273		99.95 pure specimen.
90*	555	Fine, M.E., and Ellis, W.C.	1948	L	303-1073		0.02 C; electrolytic iron.

* Not shown in figure.

DATA TABLE 24. THERMAL LINEAR EXPANSION OF IRON Fe

[Temperature, T, K; Linear Expansion, $\Delta L/L_0$, %]

T	$\Delta L/L_0$
CURVE 1	
295	0.000*
298	0.024*
378	0.094*
478	0.262
573	0.412
613	0.467*
673	0.558*
768	0.736
798	0.764*
903	0.942
923	0.980*
988	1.082
1023	1.116*
1038	1.148
1048	1.162*
1068	1.204
1073	1.221*
1098	1.259
CURVE 2*, †	
1193	0.000
1208	0.025
1323	0.255
1343	0.282
CURVE 3	
91.7	-0.191
97.2	-0.186*
107.2	-0.177*
114.2	-0.172*
125.2	-0.167*
131.2	-0.162*
139.2	-0.158*
146.2	-0.152*
152.2	-0.147
158.7	-0.142*
163.7	-0.137*
169.7	-0.133*
175.2	-0.127*
180.7	-0.122*
186.2	-0.117*
192.0	-0.112*
196.2	-0.107*
200.2	-0.103
CURVE 3 (cont.)	
206.1	-0.097*
210.4	-0.092*
215.4	-0.087*
219.9	-0.083*
225.2	-0.077*
231.0	-0.072*
235.0	-0.067*
240.0	-0.062*
244.2	-0.057*
248.2	-0.052*
252.7	-0.048
257.2	-0.042*
262.2	-0.037*
266.5	-0.032*
270.5	-0.027*
273.2	-0.025*
275.7	-0.022*
279.0	-0.017*
283.7	-0.012*
287.2	-0.007*
291.2	-0.002*
295.2	0.003*
301.2	0.008
305.7	0.010*
309.2	0.017*
313.5	0.022*
318.3	0.027*
322.0	0.032*
326.2	0.037*
330.2	0.042*
334.7	0.047*
338.2	0.052*
342.1	0.057*
346.2	0.062*
351.0	0.067
354.2	0.072*
358.4	0.077*
363.2	0.082*
366.2	0.087*
370.2	0.092*
376.1	0.097*
379.4	0.102*
384.8	0.107*
388.4	0.112*
393.1	0.118*
396.2	0.122*
CURVE 3 (cont.)	
400.2	0.127
405.2	0.132*
409.2	0.138*
414.2	0.143*
417.2	0.147*
420.2	0.153*
424.2	0.157*
428.3	0.163*
434.8	0.172*
442.4	0.182*
449.7	0.193*
456.2	0.202
463.2	0.212*
470.2	0.223*
478.2	0.232
485.2	0.243*
491.3	0.252*
499.3	0.263*
505.5	0.272
512.8	0.282*
519.4	0.292*
526.1	0.302*
533.2	0.312*
539.2	0.322*
545.9	0.332*
552.2	0.342
558.4	0.352*
565.3	0.362*
571.3	0.372*
577.7	0.382*
584.0	0.393*
590.2	0.403*
597.0	0.413*
603.0	0.423
609.2	0.433*
615.2	0.442*
621.2	0.453*
627.3	0.463*
633.7	0.473*
635.4	0.482*
646.2	0.493*
653.0	0.503
658.3	0.513*
665.2	0.523*
671.7	0.533*
678.4	0.543*
CURVE 3 (cont.)	
683.9	0.553*
690.3	0.563*
696.4	0.573*
703.2	0.583
708.5	0.593*
714.5	0.603*
720.2	0.613*
726.2	0.623*
732.2	0.633*
738.2	0.643*
744.2	0.653*
750.2	0.663
756.7	0.673*
762.0	0.683*
768.2	0.693*
774.2	0.703*
780.2	0.713*
786.2	0.723*
792.2	0.733*
798.2	0.743*
804.7	0.753
810.2	0.763*
816.2	0.773*
822.2	0.784*
828.4	0.794*
834.2	0.803*
840.2	0.813*
846.2	0.823*
852.2	0.833
856.0	0.839*
857.4	0.843*
863.7	0.853*
871.1	0.863*
876.0	0.873*
882.2	0.883*
889.0	0.893*
895.3	0.903*
900.7	0.913
908.4	0.923*
914.4	0.933*
921.7	0.944*
928.2	0.954*
933.2	0.963*
940.2	0.973*
947.5	0.984*
951.2	0.988*
CURVE 3 (cont.)	
954.2	0.994*
958.0	1.004
CURVE 4*, †, ‡	
3	0.000 x 10^{-4}
4	0.012
5	0.028
6	0.047
8	0.098
10	0.168
12	0.259
14	0.373
16	0.514
18	0.687
20	0.898
22	1.152
24	1.455
26	1.822
28	2.274
30	2.834
57.5	35.421
65	52.896
75	83.647
85	123.396
CURVE 5	
348	0.070
428	0.171
497	0.269
649	0.488
821	0.750
933	0.932
1028	1.089
1148	1.301
CURVE 6	
296	0.004
687	0.562
951	0.966
CURVE 7*, †	
20.3	0.00 x 10^{-3}
22.8	0.01
28.9	0.18
32.2	0.37
37.3	1.37
49.5	1.64
52.7	2.12
56.5	2.71
60.0	3.35
63.20	4.01
66.6	4.68
CURVE 8*, †	
20.2	0.00 x 10^{-3}
22.7	-0.01
31.7	0.13
32.8	0.17
50.1	1.36
53.1	1.64
56.5	2.03
60.0	2.54
63.2	3.12
66.7	3.86
70.1	4.65
73.0	5.50
76.3	6.43
79.1	7.43
88.1	10.82
88.2	10.94
98.5	15.15
98.6	15.24
121.3	27.79
CURVE 9*, †, ‡	
5	0.000 x 10^{-4}
6	0.018
7	0.068
8	0.168
9	0.293
10	0.413
11	0.518
12	0.613
13	0.708

* Not shown in figure.
† This curve is here reported using the first given temperature as reference temperature at which $\Delta L/L_0 = 0$.
‡ Author's data for coefficient of thermal expansion have been integrated by TPRC to obtain $\Delta L/L_0$.

DATA TABLE 24. THERMAL LINEAR EXPANSION OF IRON Fe (continued)

T	$\Delta L/L_0$
CURVE 9 (cont.)*, †, ‡	
14	0.808 x 10^{-4}
15	0.903
16	1.003
17	1.118
18	1.238
19	1.368
20	1.518
21	1.688
22	1.883
23	2.103
24	2.343
25	2.608
26	2.893
27	3.188
28	3.498
29	3.828
30	4.178
31	4.563
32	4.988
33	5.453
34	5.983
35	6.573
36	7.203
37	7.878
38	8.613
39	9.408
CURVE 10*, †	
8	0.000 x 10^{-4}
9	0.040
10	0.080
11	0.120
12	0.160
13	0.205
14	0.260
15	0.315
16	0.380
17	0.465
18	0.565
19	0.685
20	0.830
21	1.005
22	1.205
23	1.435
24	1.690
25	1.965

T	$\Delta L/L_0$
CURVE 10 (cont.)*, †	
26	2.265 x 10^{-4}
27	2.575
28	2.905
29	3.275
30	3.680
31	4.125
32	4.610
33	5.135
34	5.700
35	6.275
36	6.865
37	7.520
38	8.290
39	9.180
CURVE 11*, †, ‡	
5	0.000 x 10^{-4}
6	0.025
7	0.070
8	0.125
9	0.180
10	0.225
11	0.250
12	0.265
13	0.290
14	0.325
15	0.370
16	0.440
17	0.535
18	0.645
19	0.775
20	0.930
21	1.115
22	1.320
23	1.545
24	1.795
25	2.075
26	2.395
27	2.750
28	3.150
29	3.590
30	4.070
31	4.610
32	5.200
33	5.820
34	6.455

T	$\Delta L/L_0$
CURVE 11 (cont.)*, †, ‡	
35	7.100 x 10^{-4}
36	7.765
37	8.465
38	9.260
39	10.180
40	11.200
CURVE 12*, †, ‡	
5	0.000 x 10^{-4}
6	-0.030
7	-0.070
8	-0.110
9	-0.145
10	-0.165
11	-0.165
12	-0.145
13	-0.090
14	-0.005
15	0.090
16	0.200
17	0.325
18	0.455
19	0.590
20	0.740
21	0.905
22	1.080
23	1.280
24	1.515
25	1.775
26	2.055
27	2.355
28	2.680
29	3.030
30	3.400
31	3.810
32	4.270
33	4.780
34	5.355
35	5.980
36	6.645
37	7.350
38	8.095
39	8.885

T	$\Delta L/L_0$
CURVE 13*, †, ‡	
5	0.000 x 10^{-4}
6	0.055
7	0.115
8	0.165
9	0.200
10	0.225
11	0.245
12	0.265
13	0.290
14	0.320
15	0.345
16	0.375
17	0.415
18	0.460
19	0.515
20	0.585
21	0.680
22	0.800
23	0.950
24	1.130
25	1.340
26	1.585
27	1.850
28	2.145
29	2.475
30	2.830
31	3.215
32	3.625
33	4.055
34	4.460
35	4.825
36	5.215
37	5.695
38	6.300
CURVE 14*, †, ‡	
5	0.000 x 10^{-4}
6	0.005
7	0.020
8	0.025
9	0.040
10	0.065
11	0.095
12	0.130
13	0.180
14	0.235

T	$\Delta L/L_0$
CURVE 14 (cont.)*, †, ‡	
15	0.285 x 10^{-4}
16	0.340
17	0.405
18	0.480
19	0.560
20	0.640
21	0.725
22	0.810
23	0.895
24	0.985
25	1.090
26	1.215
27	1.350
28	1.505
29	1.685
30	1.895
31	2.140
32	2.415
33	2.715
34	3.015
35	3.340
36	3.750
37	4.225
38	4.715
39	5.215
CURVE 15*, †, ‡	
5	0.000 x 10^{-4}
6	0.005
7	0.015
8	0.020
9	0.020
10	0.020
11	0.030
12	0.055
13	0.095
14	0.150
15	0.205
16	0.270
17	0.355
18	0.450
19	0.555
20	0.675
21	0.815
22	0.975
23	1.150

T	$\Delta L/L_0$
CURVE 15 (cont.)*, †, ‡	
24	1.340 x 10^{-4}
25	1.555
26	1.790
27	2.035
28	2.300
29	2.595
30	2.925
31	3.305
32	3.725
33	4.160
34	4.595
35	5.045
36	5.595
37	6.240
38	6.860
39	7.440
CURVE 16	
298	0.007*
354	0.084
433	0.178*
497	0.269*
651	0.482*
721	0.583
821	0.750*
926	0.921
933	0.936*
940	0.967*
950	0.957*
961	0.979*
963	0.991
977	1.01*
986	1.03*
989	1.04*
991	1.04*
1008	1.06*
1009	1.06
1018	1.09*
1020	1.09*
1024	1.09*
1026	1.10*
1038	1.11
1048	1.11*
1066	1.13*
1075	1.13*
1083	1.16*

* Not shown in figure.
† This curve is here reported using the first given temperature as reference temperature at which $\Delta L/L_0 = 0$.
‡ Author's data for coefficient of thermal expansion have been integrated by TPRC to obtain $\Delta L/L_0$.

DATA TABLE 24. THERMAL LINEAR EXPANSION OF IRON Fe (continued)

T	$\Delta L/L_0$
CURVE 16 (cont.)	
1085	1.17
1096	1.17*
1105	1.18*
1106	1.18*
1115	1.21*
1131	1.23
1133	1.25*
1145	1.26*
1160	1.29*
1172	1.32
CURVE 17	
292	-0.001*
813	0.746
906	0.903*
929	0.951
971	1.02*
973	1.02*
1005	1.07*
1012	1.09
1015	1.09*
1029	1.11*
1033	1.13*
1053	1.13
1056	1.14*
1072	1.15*
1094	1.18*
1096	1.20*
1106	1.20
CURVE 18	
293	0.000*
513	0.307
623	0.468*
722	0.617
822	0.771*
921	0.924*
994	1.046
1026	1.094*
1033	1.098*
1043	1.110*
1060	1.142
1120	1.219
1175	1.312
1189	1.328
1189	0.982
CURVE 18 (cont.)	
1196	0.991*
1223	1.060
1347	1.343
1457	1.615
1565	1.881
1635	2.026
1653	2.092
1661	2.095*
1662	2.278
1667	2.305*
1705	2.389
1775	2.551*
1807	2.615*
CURVE 19	
293.2	0.000*
366.5	0.070
477.6	0.220
588.7	0.390
699.8	0.575
810.9	0.767
922.1	0.955*
CURVE 20	
93	-0.1849*
113	-0.1737
133	-0.1603*
153	-0.1450*
173	-0.1278
193	-0.1093
213	-0.0894*
233	-0.0680
253	-0.0462*
273	-0.0234
293	0.0000*
CURVE 21	
287	-0.007
288	-0.003*
289	-0.007*
291	-0.003*
295	0.003*
309	0.014
371	0.091
CURVE 21 (cont.)	
420	0.157
468	0.220*
481	0.238
534	0.318
567	0.353*
588	0.395*
639	0.468
664	0.507*
690	0.552
749	0.647
761	0.671*
803	0.738
849	0.825
913	0.968
CURVE 22	
373	0.096*
383	0.108*
393	0.121
403	0.134*
413	0.147*
418	0.154*
423	0.161*
433	0.174
443	0.188*
453	0.201*
463	0.215*
473	0.230*
498	0.265*
523	0.302
548	0.326*
573	0.376*
598	0.414
623	0.453*
648	0.491*
673	0.531
698	0.570*
723	0.610*
748	0.651*
773	0.691
798	0.732*
823	0.773*
848	0.815*
873	0.856
898	0.898*
923	0.939*
CURVE 22 (cont.)	
948	0.981*
973	1.022
983	1.039*
993	1.056*
999	1.066*
1003	1.068*
1004	1.068*
1005	1.068
1006	1.063*
1007	1.052*
1008	1.047*
1009	1.047*
1010	1.047*
1011	1.048*
1012	1.050
1013	1.053*
1018	1.067*
1023	1.077
1028	1.088*
1033	1.098
1043	1.119
1053	1.140*
1063	1.161*
1073	1.182
CURVE 23‡	
20	-0.197
30	-0.197*
40	-0.197
50	-0.196*
70	-0.193
100	-0.182
150	-0.150
200	-0.104*
250	-0.050
300	0.008*
CURVE 24*	
298	0.006
373	0.096
473	0.226
573	0.371
673	0.524
773	0.683
873	0.851
973	1.025
CURVE 25*	
273	-0.0224
373	0.0896
423	0.1534
473	0.2160
523	0.2852
CURVE 26	
298	0.006*
349	0.067*
418	0.154
473	0.235
526	0.314*
575	0.379
634	0.475
CURVE 27*	
298	0.005
348	0.056
407	0.139
462	0.212
518	0.293
574	0.371
636	0.474
CURVE 28	
304	0.013*
360	0.079
404	0.132
464	0.206
510	0.292
564	0.367
621	0.451
CURVE 29*	
283	-0.0108
293	0.0000
303	0.0130
313	0.0241
323	0.0352
332	0.0452
CURVE 30	
282	-0.009*
293	0.000*
427	0.138
539	0.352
670	0.551
739	0.659*
809	0.775*
874	0.868*
955	0.992
1023	1.094*
1097	1.187
CURVE 31	
293	0.000*
427	0.123
539	0.337
618	0.479
713	0.644
816	0.827
911	0.988
1015	1.117
1113	1.254
CURVE 32*	
288	-0.0080
293	0.0000
298	0.0059
307	0.0152
317	0.0282
329	0.0445
CURVE 33	
280	0.003*
293	0.000*
423	0.153
597	0.446
666	0.565
737	0.673
802	0.753*
874	0.872*
949	0.966
1062	1.144*
1116	1.214

* Not shown in figure.
‡ Author's data for coefficient of thermal expansion have been integrated by TPRC to obtain $\Delta L/L_0$.

DATA TABLE 24. THERMAL LINEAR EXPANSION OF IRON Fe (continued)

T	$\Delta L/L_0$
CURVE 34	
293	0.000*
425	0.163*
533	0.373
584	0.446
711	0.670
814	0.841
911	0.973*
1016	1.106*
1102	1.238
CURVE 35‡	
831	0.768
873	0.834*
910	0.893
926	0.918*
939	0.937*
949	0.953
970	0.984*
994	1.022
1014	1.051*
1026	1.069*
1031	1.076
1043	1.092*
1049	1.100*
1056	1.111*
1063	1.121*
1073	1.137
1115	1.200
CURVE 36	
295	0.002*
314	0.027
339	0.058
427	0.149*
499	0.230
533	0.275
639	0.440
CURVE 37*	
295	0.002
357	0.068
395	0.110
448	0.170
CURVE 37 (cont.)*	
496	0.222
512	0.257
573	0.331
612	0.390
CURVE 38*	
303	0.0122
373	0.0854
CURVE 39	
310	0.016*
322	0.027
372	0.076
420	0.136*
473	0.204
523	0.291
567	0.381
620	0.468*
673	0.551*
718	0.633
768	0.718
817	0.798
873	0.879*
920	0.958
974	1.031*
1021	1.116*
1073	1.184*
1123	1.264
CURVE 40*	
273	-0.0234
373	0.0936
473	0.2206
573	0.3666
673	0.5286
773	0.6916
873	0.8526
973	1.0056
1073	1.2006
CURVE 41	
1123	1.422
1128	1.463*
CURVE 41 (cont.)	
1136	1.509*
1147	1.557*
1157	1.558
1163	1.510*
1165	1.467*
1167	1.423
1172	1.369*
1178	1.333*
1186	1.285
1200	1.323
1204	1.335*
1223	1.365
CURVE 42	
1223	1.365*
1204	1.335*
1200	1.323*
1197	1.334*
1191	1.368
1184	1.425
1179	1.467*
1172	1.514
1161	1.557*
1151	1.600
1139	1.600
1123	1.566
CURVE 43*	
273	-0.0224
373	0.0896
473	0.222
573	0.371
673	0.529
773	0.693
873	0.854
973	1.007
1073	1.146
CURVE 44	
1123	1.197
1134	1.214*
1154	1.246
1166	1.265*
1178	1.283
CURVE 44 (cont.)	
1186	1.295*
1193	1.308
1200	1.323*
1209	1.335
1213	1.330*
1216	1.278
1218	1.190
1225	1.192*
1231	1.206
CURVE 45	
1154	1.375
1165	1.395*
1176	1.415
1191	1.439*
1200	1.453
1207	1.134
1213	1.149*
1220	1.167
1227	1.188
1231	1.206*
CURVE 46*	
273	-0.0232
373	0.0928
473	0.225
573	0.367
673	0.525
773	0.697
873	0.859
973	1.019
1073	1.169
CURVE 47	
1123	1.234
1138	1.259*
1164	1.298
1180	1.322
1187	1.300*
1194	1.262
1199	1.222*
1201	1.180
1203	1.142*
1205	1.101
CURVE 47 (cont.)	
1206	1.061*
1209	1.040
1217	1.061*
1230	1.087
CURVE 48	
1230	1.087
1220	1.064
1209	1.040*
1199	1.022
1192	1.003
1189	1.022*
1186	1.062
1183	1.141
1181	1.221
1178	1.300*
1176	1.342*
1172	1.395
1156	1.381
CURVE 49*	
273	-0.0224
373	0.0896
473	0.2178
573	0.3640
673	0.5206
773	0.6816
873	0.8352
973	0.9928
1073	1.1466
CURVE 50*	
1123	1.207
1133	1.220
1148	1.245
1167	1.271
1185	1.292
1191	1.229
1199	1.141
1203	1.081
1210	1.098
1216	1.105
CURVE 51*	
1216	1.105
1199	1.058
1187	1.037
1183	1.162
1181	1.222
1177	1.257
1171	1.258
1159	1.238
1146	1.218
1137	1.207
CURVE 52*	
273	-0.023
373	0.092
473	0.223
573	0.364
673	0.517
773	0.682
873	0.841
973	0.999
1073	1.154
CURVE 53	
1123	1.215*
1127	1.220*
1148	1.251
1168	1.283*
1186	1.314
1190	1.319*
1192	1.190
1194	1.153
1211	1.187
1221	1.221
CURVE 54*	
1221	1.221
1206	1.187
1184	1.131
1183	1.488
1173	1.472
1156	1.442

*Not shown in figure.

DATA TABLE 24. THERMAL LINEAR EXPANSION OF IRON Fe (continued)

T	$\Delta L/L_0$
CURVE 55*	
273	-0.022
373	0.089
473	0.219
573	0.361
673	0.517
773	0.672
873	0.835
973	0.993
1073	1.170
CURVE 56	
1123	1.251
1127	1.262*
1139	1.285
1159	1.315
1180	1.349
1201	1.381*
1207	1.389
1210	1.365*
1211	1.331
1212	1.268*
1214	1.206*
1215	1.180*
1226	1.195*
CURVE 57	
1226	1.195*
1209	1.159*
1200	1.142*
1197	1.266*
1195	1.395
1193	1.461*
1191	1.524
1188	1.567*
1183	1.575
1177	1.554
CURVE 58*	
273	-0.024
373	0.096
473	0.228
573	0.369
673	0.524
773	0.696
CURVE 58 (cont.)*	
873	0.858
973	1.012
1073	1.160
CURVE 59	
1123	1.218
1132	1.233
1151	1.260
1170	1.293
1189	1.321
1190	1.266
1191	1.201
1192	1.138
1192	1.073
1210	1.112
CURVE 60	
1210	1.112
1199	1.083
1183	1.004
1183	1.110*
1182	1.174
1181	1.293*
1180	1.413*
1180	1.535
1179	1.654
1179	1.738
1168	1.719
1151	1.687
CURVE 61*	
273	-0.024
373	0.096
473	0.228
573	0.369
673	0.524
773	0.696
873	0.858
973	1.026
1073	1.184
CURVE 62*	
1123	1.247
1140	1.279
1163	1.318
1185	1.354
1203	1.383
1204	1.307
1206	1.236
1207	1.160
1207	1.084
1208	1.012
1221	1.041
1230	1.064
CURVE 63*	
1230	1.064
1217	1.021
1199	0.982
1197	1.055
1196	1.132
1196	1.206
1195	1.278
1195	1.352
1195	1.501
1195	1.595
1192	1.643
1188	1.632
1183	1.620
CURVE 64*	
273	-0.023
373	0.090
473	0.229
573	0.376
673	0.533
773	0.702
873	0.865
973	1.020
1073	1.169
CURVE 65*	
1123	1.216
1137	1.243
1155	1.229
1170	1.200
CURVE 65 (cont.)*	
1178	1.132
1186	1.069
1192	1.031
1205	1.073
1214	1.103
CURVE 66*	
1214	1.103
1206	1.069
1176	0.999
1166	1.039
1162	1.102
1158	1.165
1153	1.224
1148	1.260
1144	1.292
1139	1.296
1123	1.278
CURVE 67*	
273	-0.023
373	0.094
473	0.225
573	0.369
673	0.529
773	0.697
873	0.865
973	1.019
1073	1.169
CURVE 68*	
1123	1.205
1129	1.214
1146	1.244
1165	1.274
1183	1.301
1199	1.324
1202	1.308
1204	1.162
1205	1.046
1206	0.965
1218	0.994
CURVE 69*	
1218	0.994
1202	0.963
1198	0.959
1196	1.041
1196	1.124
1195	1.235
1194	1.351
1194	1.467
1193	1.576
1193	1.691
1192	1.788
1188	1.779
1181	1.768
CURVE 70*	
303	0.0119
373	0.0954
CURVE 71*	
303	0.0123
373	0.0984
CURVE 72*	
273	-0.026
523	0.299
648	0.487
773	0.690
898	0.896
1023	1.097
CURVE 73‡	
293	0.000*
318	0.030*
368	0.091
417	0.155*
470	0.226
519	0.294
572	0.370*
622	0.445*
669	0.521
723	0.612*
775	0.701
821	0.777*
CURVE 73 (cont.)‡	
869	0.856
922	0.942*
970	1.020
1020	1.100
1071	1.183*
1125	1.273
CURVE 74*, ‡	
293	0.000
419	0.157
520	0.295
723	0.612
824	0.780
871	0.855
927	0.944
974	1.016
1023	1.090
1070	1.160
1123	1.240
1173	1.317
CURVE 75	
273	-0.03*
473	0.24
673	0.53*
873	0.84
1073	1.17
1173	1.35
CURVE 76	
293	0.000*
373	0.109*
473	0.226*
573	0.375*
673	0.514
773	0.674
873	0.824
973	0.976
1073	1.118
1121	1.178*
1150	1.200
1160	1.200
1173	1.193
1189	1.160

* Not shown in figure.

‡ Author's data for coefficient of thermal expansion have been integrated by TPRC to obtain $\Delta L/L_0$.

DATA TABLE 24. THERMAL LINEAR EXPANSION OF IRON Fe (continued)

T	$\Delta L/L_0$
CURVE 76 (cont.)	
1197	1.091*
1203	0.977
1209	0.962
1250	1.035
CURVE 77	
1250	1.045
1177	0.909
1173	0.920
1169	1.001
1162	1.133
1154	1.168
1142	1.172
1124	1.169*
1112	1.158
1073	1.108*
973	0.965
873	0.811
773	0.668*
673	0.502*
573	0.364
473	0.223*
373	0.108*
300	0.010*
CURVE 78*	
293	0.000
373	0.080
473	0.210
573	0.363
673	0.502
773	0.651
873	0.819
973	0.973
1073	1.122
1173	1.288
1184	1.310
1186	0.996
1191	0.981
1223	1.048
CURVE 79	
1223	0.859
1177	0.756
1173	0.772
1169	1.263

T	$\Delta L/L_0$
CURVE 79 (cont.)	
1073	1.105*
973	0.956
873	0.790
773	0.633
673	0.482
573	0.339
473	0.193
373	0.074*
293	0.000*
CURVE 80‡	
293	0.000*
374	0.098*
423	0.162*
473	0.230*
519	0.295*
574	0.376*
628	0.458
674	0.530*
725	0.614*
770	0.689*
799	0.738
824	0.780*
874	0.862*
925	0.946*
974	1.026*
1004	1.073
1027	1.104
1069	1.157
CURVE 81*	
288	-0.006
373	0.095
473	0.228
573	0.369
673	0.522
773	0.687
873	0.869
CURVE 82*	
273	-0.023
293	0.000
313	0.023

T	$\Delta L/L_0$
CURVE 83*, †	
40	0.000
60	0.003
80	0.008
100	0.017
140	0.041
180	0.076
CURVE 84*, †	
40	0.000
60	0.005
80	0.010
100	0.017
140	0.035
180	0.074
CURVE 85*	
273	-0.025
293	0.000
673	0.480
CURVE 86*	
293	0.000
449	0.179
548	0.334
564	0.361
583	0.378
593	0.407
613	0.402
619	0.396
710	0.558
734	0.568
739	0.591
759	0.580
772	0.628
794	0.610
813	0.697
829	0.744
833	0.712
860	0.793
869	0.768
873	0.780
873	0.819
908	0.829
919	0.870
936	0.911
936	0.938

T	$\Delta L/L_0$
CURVE 86 (cont.)*	
949	0.911
959	0.968
981	0.968
981	0.980
981	1.003
1011	1.003
1019	1.066
CURVE 87*	
293	0.000
373	0.094
473	0.233
573	0.384
673	0.538
773	0.687
CURVE 88*	
93	-1.345
111	-1.231
122	-1.133
134	-1.058
154	-0.904
171	-0.835
193	-0.659
217	-0.511
237	-0.362
262	-0.217
284	-0.064
293	0.000
CURVE 89*, ‡	
293	0.000
363	0.085
473	0.226
573	0.363
673	0.504
773	0.648
873	0.796
973	0.946
1138	1.198
1206	1.305
CURVE 90*	
303	0.013
473	0.235

T	$\Delta L/L_0$
CURVE 90 (cont.)*	
673	0.535
873	0.851
1073	1.163

* Not shown in figure.
† This curve is here reported using the first given temperature as reference temperature at which $\Delta L/L_0 = 0$.
‡ Author's data for coefficient of thermal expansion have been integrated by TPRC to obtain $\Delta L/L_0$.

DATA TABLE 24. COEFFICIENT OF THERMAL LINEAR EXPANSION OF IRON Fe

[Temperature, T, K; Coefficient of Expansion, α, 10^{-6} K^{-1}]

CURVE 4‡

T	α
3	0.011
4	0.014*
5	0.017*
6	0.021*
8	0.030*
10	0.040
12	0.051*
14	0.063*
16	0.078*
18	0.095*
20	0.116
22	0.138*
24	0.165*
26	0.202*
28	0.250*
30	0.310
57.5	2.06
65	2.60
75	3.55
85	4.40
283	11.85

CURVE 7*, †

T	α
24.6	0.13
27.5	0.18*
33.1	0.30*
53.0	1.08
56.3	1.46
59.9	1.83
63.3	1.97
66.2	2.14

CURVE 8*, †

T	α
26.0	0.11
27.7	0.18*
53.3	1.03*
56.6	1.30*
59.8	1.66
63.4	1.96*
66.6	2.22*
69.9	2.60
73.2	2.85
76.0	3.14
83.6	3.77

CURVE 8 (cont.)*, †

T	α
93.2	4.15
93.4	4.16

CURVE 9‡

T	α
5	0.017
6	0.02*
7	0.08*
8	0.12*
9	0.13*
10	0.11*
11	0.10*
12	0.09*
13	0.10*
14	0.10*
15	0.09
16	0.11*
17	0.12*
18	0.12*
19	0.14*
20	0.16*
21	0.18*
22	0.21*
23	0.23*
24	0.25*
25	0.28
26	0.29*
27	0.30*
28	0.32*
29	0.34*
30	0.36*
31	0.41*
32	0.44*
33	0.49*
34	0.57*
35	0.61
36	0.65*
37	0.70*
38	0.77*
39	0.82

CURVE 10*, ‡

T	α
8	0.04
9	0.04
10	0.04

CURVE 10 (cont.)*, ‡

T	α
11	0.04
12	0.04
13	0.05
14	0.06
15	0.05
16	0.08
17	0.09
18	0.11
19	0.13
20	0.16
21	0.19
22	0.21
23	0.25
24	0.26
25	0.29
26	0.31
27	0.31
28	0.35
29	0.39
30	0.42
31	0.47
32	0.50
33	0.55
34	0.58
35	0.57
36	0.61
37	0.70
38	0.84
39	0.94

CURVE 11‡

T	α
5	0.01
6	0.04*
7	0.05*
8	0.06*
9	0.05*
10	0.04*
11	0.01*
12	0.02*
13	0.03*
14	0.04*
15	0.05*
16	0.09*
17	0.10*
18	0.12*

CURVE 11 (cont.)‡

T	α
19	0.14
20	0.17*
21	0.20*
22	0.21*
23	0.24*
24	0.26*
25	0.30*
26	0.34*
27	0.37*
28	0.43*
29	0.45*
30	0.51
31	0.57*
32	0.61*
33	0.63*
34	0.64*
35	0.65*
36	0.68*
37	0.72*
38	0.87*
39	0.97*
40	1.07

CURVE 12*, ‡

T	α
5	-0.02
6	-0.04
7	-0.04
8	-0.04
9	-0.03
10	-0.01
11	0.01
12	0.03
13	0.08
14	0.09
15	0.10
16	0.12
17	0.13
18	0.13
19	0.14
20	0.16
21	0.17
22	0.18
23	0.22
24	0.25
25	0.27

CURVE 12 (cont.)*, ‡

T	α
26	0.29
27	0.31
28	0.34
29	0.36
30	0.38
31	0.44
32	0.48
33	0.54
34	0.61
35	0.64
36	0.69
37	0.72
38	0.77
39	0.81

CURVE 13*, ‡

T	α
5	0.05
6	0.06
7	0.06
8	0.04
9	0.03
10	0.02
11	0.02
12	0.02
13	0.03
14	0.03
15	0.02
16	0.04
17	0.04
18	0.05
19	0.06
20	0.08
21	0.11
22	0.13
23	0.17
24	0.19
25	0.23
26	0.26
27	0.27
28	0.32
29	0.34
30	0.37
31	0.40
32	0.42
33	0.44

CURVE 13 (cont.)*, ‡

T	α
35	0.37
35	0.36
36	0.42
37	0.54
38	0.67

CURVE 14‡

T	α
5	-0.02
6	0.03*
7	0.00*
8	0.01*
9	0.02*
10	0.03*
11	0.03*
12	0.04*
13	0.06*
14	0.05*
15	0.05*
16	0.06*
17	0.07*
18	0.08*
19	0.08*
20	0.08*
21	0.09*
22	0.08*
23	0.09*
24	0.09*
25	0.12*
26	0.13*
27	0.14*
28	0.17*
29	0.19*
30	0.23*
31	0.26*
32	0.29*
33	0.31*
34	0.29*
35	0.36*
36	0.46*
37	0.49*
38	0.49*
39	0.51

* Not shown in figure.
† This curve is here reported using the first given temperature as reference temperature at which $\Delta L/L_0 = 0$.
‡ Author's data for coefficient of thermal expansion have been integrated by TPRC to obtain $\Delta L/L_0$.

T	α
CURVE 15*, ‡	
5	0.00
6	0.01
7	0.01
8	0.00
9	0.00
10	0.00
11	0.02
12	0.03
13	0.05
14	0.06
15	0.05
16	0.08
17	0.09
18	0.10
19	0.11
20	0.13
21	0.15
22	0.17
23	0.18
24	0.20
25	0.23
26	0.24
27	0.25
28	0.28
29	0.31
30	0.35
31	0.41
32	0.43
33	0.44
34	0.43
35	0.47
36	0.63
37	0.66
38	0.58
39	0.58
CURVE 23‡	
20	0.03
30	0.15
40	0.45*
50	0.90
70	2.4
100	4.6
150	8.2
200	10.4
250	11.3
300	11.7

T	α
CURVE 35*, ‡	
831	15.87
873	15.87
910	15.87
926	15.57
939	15.82
949	15.82
970	15.61
994	15.44
1014	15.09
1026	14.73
1031	14.26
1043	16.71
1049	16.54
1056	16.78
1063	16.39
1073	16.04
1115	16.04
CURVE 73*, ‡	
293	11.58
318	12.03*
368	12.73*
417	13.21*
470	13.61*
519	14.01*
572	14.63*
622	15.55*
669	16.72*
723	17.21*
775	16.80*
821	16.49*
369	16.33*
922	16.20*
970	16.11*
1020	16.02*
1071	16.49*
1125	16.84*
CURVE 74*, ‡	
293	11.58
419	13.32
520	14.12
723	17.03
824	16.24
871	15.93
927	15.62

T	α
CURVE 74 (cont.)*, ‡	
974	15.26
1023	14.83
1070	14.94
1123	15.29
1173	15.73
CURVE 80*, ‡	
293	11.65
374	12.62
423	13.23
473	13.96
519	14.44
574	14.90
628	15.50
674	16.08
725	16.61
770	17.01
799	16.81
824	16.38
874	16.52
925	16.40
974	16.17
1004	14.99
1027	12.50
1069	12.50
CURVE 83*, †	
40	0.79
60	2.11
80	3.45
100	4.78
140	7.44
180	10.11
CURVE 89*, ‡	
293	11.9
363	12.4
473	13.3
573	14.0
673	14.3
773	14.5
873	15.0
973	15.0
1138	15.6
1206	15.8

* Not shown in figure.
† This curve is here reported using the first given temperature as reference temperature at which $\Delta L/L_0 = 0$.
‡ Author's data for coefficient of thermal expansion have been integrated by TPRC to obtain $\Delta L/L_0$.

FIGURE AND TABLE NO. 25R. PROVISIONAL VALUES FOR THERMAL LINEAR EXPANSION OF LANTHANUM La

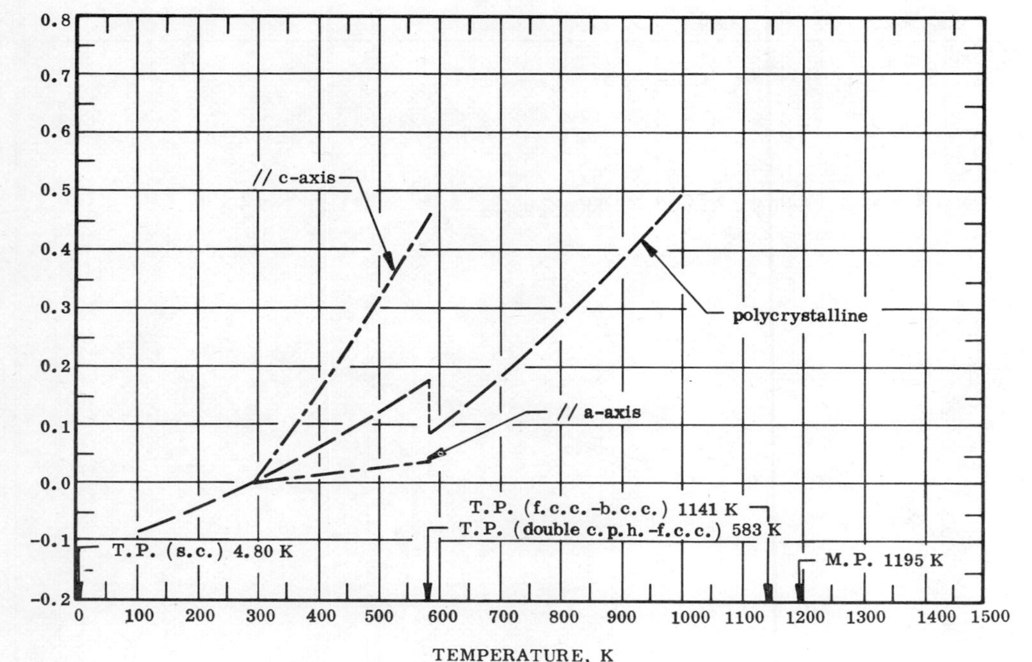

PROVISIONAL VALUES

[Temperature, T, K; Linear Expansion, $\Delta L/L_0$, %; α, K^{-1}]

T	// a-axis $\Delta L/L_0$	// c-axis $\Delta L/L_0$	polycrystalline $\Delta L/L_0$	polycrystalline $\alpha \times 10^6$
100			-0.083	3.5
200			-0.043	4.4
293	0.000‡	0.000‡	0.000	5.2
400	0.014‡	0.146‡	0.059	5.9
500	0.027‡	0.315‡	0.121	6.6
583†	0.036‡	0.460‡	0.178	7.3
583†			0.088	7.6
600			0.100	7.2
700			0.183	9.4
800			0.280	10.2
900			0.385	10.8
1000			0.497	11.3

† Phase transition.
‡ Typical values.

REMARKS

The tabulated values are considered accurate to within ± 7% at temperatures below 300 K and ± 10% above. Hysteresis effects in the thermal linear expansion are observed near transition and near melting. The tabulated values can be represented approximately by the following equations:

// a-axis: $\Delta L/L_0 = -0.016 - 4.828 \times 10^{-5}\, T + 4.640 \times 10^{-7}\, T^2 - 3.924 \times 10^{-10}\, T^3$

// c-axis: $\Delta L/L_0 = 0.039 - 1.742 \times 10^{-3}\, T + 6.760 \times 10^{-6}\, T^2 - 4.349 \times 10^{-9}\, T^3$

polycrystalline: $\Delta L/L_0 = -0.115 + 2.786 \times 10^{-4}\, T + 4.094 \times 10^{-7}\, T^2 - 4.481 \times 10^{-11}\, T^3$ $(100 < T < 583)$

$\Delta L/L_0 = 0.066 - 8.563 \times 10^{-4}\, T + 1.871 \times 10^{-6}\, T^2 - 5.843 \times 10^{-10}\, T^3$ $(583 < T < 1000)$

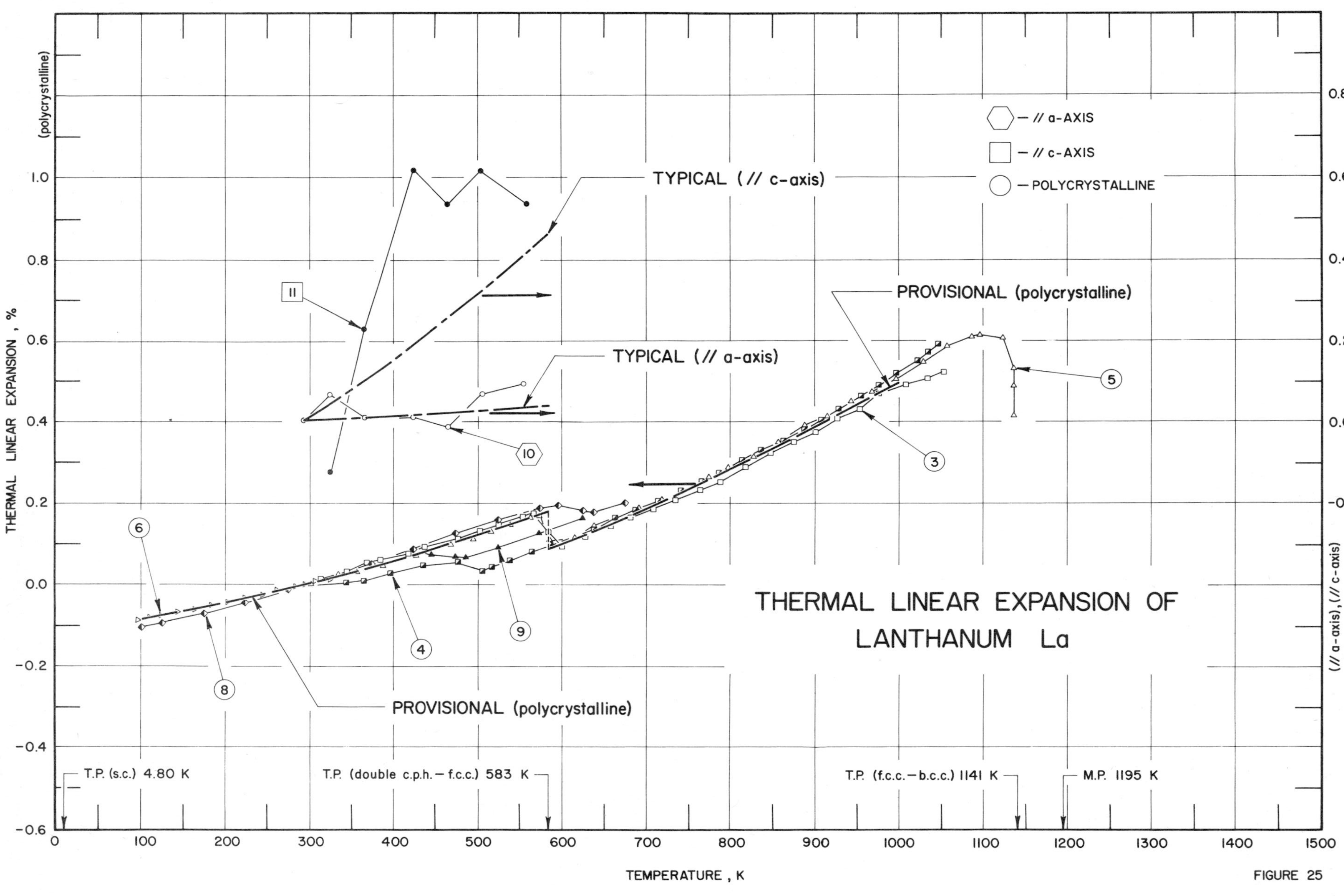
THERMAL LINEAR EXPANSION OF
LANTHANUM La
— // a-AXIS
— // c-AXIS
— POLYCRYSTALLINE
TYPICAL (// c-axis)
TYPICAL (// a-axis)
PROVISIONAL (polycrystalline)
PROVISIONAL (polycrystalline)
T.P. (s.c.) 4.80 K
T.P. (double c.p.h. – f.c.c.) 583 K
T.P. (f.c.c. – b.c.c.) 1141 K
M.P. 1195 K
THERMAL LINEAR EXPANSION , %
(polycrystalline)
(// a-axis), (// c-axis)
TEMPERATURE , K
0
100
200
300
400
500
600
700
800
900
1000
1100
1200
1300
1400
1500
-0.6
-0.4
-0.2
0.0
0.2
0.4
0.6
0.8
1.0
-0.2
0.0
0.2
0.4
0.6
0.8
3
4
5
6
8
9
10
11
FIGURE 25

SPECIFICATION TABLE 25. THERMAL LINEAR EXPANSION OF LANTHANUM La

Cur. No.	Ref. No.	Author(s)	Year	Method Used	Temp. Range, K	Name and Specimen Designation	Composition (weight percent), Specifications, and Remarks
1	77	Andres, K.	1968	E	1.4-24	La Specimen 1	99.9 La, 0.2 Nd, 0.1 Si, 0.02 Cr, 0.01 Mn, 0.01 Ce; f.c.c. structure; cylindrical specimens 3.78 cm long at 6.0 K; subjected to 20 kbar hydrostatic pressure at 473 K for about 10 min, cooled to 170 K under pressure and stored in liquid nitrogen; data below 24 K obtained by heating above helium bath temperature; data above 24 K obtained during helium transfer while cooling from 77 K; superconducting transition observed at 5-95 K; measurements under magnetic field of 6900 Oe; this curve is here reported using the first given temperature as reference temperature at which $\Delta L/L_0 = 0$; zero-point correction is 0.016 x 10^{-3}%.
2	77	Andres, K.	1968	E	1.4-5.95		The above specimen; the curve is here reported by equalling the value of curve 1 at the superconducting transition temperature, 5.95 K; zero-point correction is 0.0158 x 10^{-3}%.
3	78	Barson, F., Legvold, S., and Spedding, F.H.	1956	L	312-1052		99.9 La, 0.04 Ca, 0.04 N_2, 0.0175 C; specimen 0.6 cm diameter, 6 cm long; fluorides of lanthanum bomb reduced with calcium metal to produce compact metallic sample; vacuum cast, turned to shape in a lathe; expansion with increasing temperature; first test, zero-point correction of 0.017%.
4	78	Barson, F., et al.	1956	L	1048-342		The above specimen; expansion with decreasing temperature; zero-point correction of 0.1% determined by graphical extrapolation.
5	78	Barson, F., et al.	1956	L	300-1137		The above specimen; expansion with increasing temperature; softening occurred at high temperatures; second test.
6	79	Barson, F., Legvold, S., and Spedding, F.H.	1953	L	321-98		<1.0 Mg, <0.025 Ca, <0.0086 Fe, <0.01 others; prepared from the salt by bomb reduction with calcium; extruded and turned into cylindrical rod about 0.3 cm diameter, the ends machined flat; subjected to prolonged annealing at about 820 K; expansion measured with decreasing temperature; first test; zero-point correction of -0.01% determined by graphical interpolation.
7	79	Barson, F., et al.	1953	L	99-321		The above specimen, expansion with increasing temperature, second test; zero-point correction of -0.0094% determined by graphical interpolation.
8	81	Trombe, F. and Foex, M.	1953	L	100-673		Expansion measured with increasing temperature in hydrogen atmosphere below room temperature, in argon atmosphere above room temperature; zero-point correction is 0.02%.
9	81	Trombe, F. and Foex, M.	1953	L	673-100		The above specimen; expansion measured with decreasing temperature; zero-point correction is 0.02%.
10	82	Hanak, J.J.	1959	X	293-557		<0.1 Ta; <0.005 each all other impurities; thin wire specimen cut from sheet prepared at Ames Laboratory; c.p.h. sturcture; expansion measured along a-axis; lattice parameter reported at 293 K is 3.771 Å.
11	82	Hanak, J.J.	1959	X	293-557		The above specimen; expansion measured along c-axis; lattice parameter reported at 293 K is 12.13 Å.
12*	82	Hanak, J.J.	1959	X	294-871		Similar to the above specimen except f.c.c. structure; expansion measured along a-axis; lattice parameter reported at 293 K is 5.305 Å
13*	82	Hanak, J.J.	1959	X	1151-1175		Similar to the above specimen except b.c.c. structure; this curve is here reported using the first given temperature as reference temperature at which $\Delta L/L_0 = 0$; lattice parameter reported at 1151 K is 4.26 Å.

* Not shown in figure.

DATA TABLE 25. THERMAL LINEAR EXPANSION OF LANTHANUM La

[Temperature, T, K; Linear Expansion, $\Delta L/L_0$, %]

CURVE 1†

T	$\Delta L/L_0$
1.63	0.0000 x 10^{-3}
1.91	0.0008
2.08	0.0012
2.55	0.0015
2.77	0.0014
3.05	0.0005
3.65	-0.0006
4.00	-0.0019
4.23	-0.0022
4.49	-0.0034
4.73	-0.0048
4.98	-0.0063
5.21	-0.0083
5.48	-0.0099
5.95	-0.0153
6.03	-0.0162
6.07	-0.0169
6.13	-0.0178
6.16	-0.0185
6.22	-0.0193
6.29	-0.0202
6.35	-0.0208
6.38	-0.0219
6.46	-0.0230
6.71	-0.0264
6.96	-0.0319
7.20	-0.0374
7.47	-0.0428
7.91	-0.0606
8.72	-0.0844
9.27	-0.103
9.89	-0.121
10.80	-0.163
11.78	-0.222
12.83	-0.489
13.87	-0.349
14.89	-0.431
16.84	-0.603
19.76	-0.883
24.00	-1.243

CURVE 2‡

T	$\Delta L/L_0$
1.45	-0.0639 x 10^{-3}
3.02	-0.0588
3.77	-0.0517
4.00	-0.0475
4.49	-0.0388
4.96	-0.0302
5.47	-0.0191
5.88	-0.003
5.95	-0.0153

CURVE 3

T	$\Delta L/L_0$
312	0.017
324	0.028*
343	0.036
355	0.044*
369	0.052
383	0.060
399	0.070*
418	0.078
437	0.091
455	0.103*
476	0.115
501	0.132
525	0.148
551	0.166
568	0.177
577	0.153*
581	0.126
583	0.104*
589	0.094*
600	0.098
626	0.120
657	0.141
680	0.163
707	0.184
734	0.207
761	0.233
789	0.259
819	0.290
846	0.321
875	0.350
900	0.377
927	0.408
952	0.435
977	0.466
1009	0.495
1032	0.514
1052	0.523

CURVE 4

T	$\Delta L/L_0$
1048	0.594
1034	0.575
1022	0.553
998	0.522
975	0.491
954	0.461
929	0.430
909	0.403
888	0.380
863	0.353
838	0.332
814	0.303
788	0.276
765	0.253
740	0.229
712	0.205
686	0.181
661	0.162
637	0.138
612	0.12*
588	0.10
564	0.08
539	0.06
516	0.042
505	0.036
492	0.052*
477	0.059
452	0.056*
433	0.044
413	0.037*
397	0.028
380	0.018*
365	0.012
351	0.01*
342	0.007

CURVE 5

T	$\Delta L/L_0$
300	0.005
310	0.011*
331	0.021
355	0.034
369	0.040*
386	0.047
406	0.061*
427	0.073
447	0.087*
469	0.100
491	0.115
515	0.133
539	0.149
562	0.164
573	0.166
579	0.154*
583	0.134*
585	0.117
589	0.103*
615	0.119
639	0.143
664	0.167*
690	0.187
718	0.210
742	0.235*
773	0.261
798	0.287
826	0.318
858	0.354
886	0.389
914	0.417
942	0.449
967	0.478
997	0.511
1027	0.552
1056	0.587
1086	0.612
1098	0.615
1111	0.607*
1122	0.602
1127	0.587*
1130	0.570*
1135	0.537
1136	0.519*
1136	0.449*
1136	0.433*
1136	0.418
1137	0.487
1137	0.468*

CURVE 6

T	$\Delta L/L_0$
321.1	0.014
318.1	0.013*
316.0	0.012*
314.5	0.012*
313.2	0.011*
310.6	0.010*
308.0	0.008*
305.5	0.007
303.0	0.006*
300.4	0.004*
297.8	0.003*
295.1	0.002*
292.5	0.000
290.0	-0.001*
287.3	-0.002*
284.7	-0.004*
282.1	-0.005
279.5	-0.006*
276.6	-0.007*
271.3	-0.010*
268.6	-0.012*
265.9	-0.013*
263.0	-0.014*
260.2	-0.016
257.4	-0.017*
254.5	-0.018*
251.8	-0.019*
249.1	-0.021*
246.2	-0.022*
242.0	-0.024
239.3	-0.025*
236.3	-0.026*
233.4	-0.028*
230.6	-0.029*
227.7	-0.030*
225.0	-0.032*
222.2	-0.033
219.4	-0.034*
216.1	-0.035*
213.0	-0.037*
210.0	-0.038*
207.5	-0.039*
204.7	-0.040*
202.0	-0.042
199.2	-0.043*
196.8	-0.044*
193.9	-0.045*
191.5	-0.047*
188.6	-0.048*
184.6	-0.050*
182.2	-0.051
179.6	-0.052*
176.7	-0.053*
173.4	-0.054*
170.7	-0.056*
168.0	-0.057*
164.0	-0.058*
161.4	-0.060
158.2	-0.061*
155.4	-0.062*
152.8	-0.063*
150.6	-0.064*
147.3	-0.065*
144.5	-0.066*
141.7	-0.068
139.4	-0.069*
136.5	-0.070*
133.8	-0.071*
131.0	-0.072*
128.1	-0.073*
125.3	-0.074*
122.7	-0.076
119.7	-0.077*
116.8	-0.078*
114.0	-0.079*
110.9	-0.080
107.8	-0.081*
104.5	-0.082*
101.3	-0.083*
97.6	-0.084

CURVE 7*

T	$\Delta L/L_0$
99.3	-0.082
102.4	-0.081
104.1	-0.081
106.0	-0.080
107.4	-0.080
109.0	-0.079
111.7	-0.078
114.6	-0.077
117.1	-0.076
119.6	-0.075
122.4	-0.074

* Not shown in figure.

† This curve is here reported using the first given temperature as reference temperature at which $\Delta L/L_0 = 0$.

‡ For explaination, see specification table.

DATA TABLE 25. THERMAL LINEAR EXPANSION OF LANTHANUM La (continued)

T	$\Delta L/L_0$
CURVE 7 (cont.)*	
125.1	-0.073
127.4	-0.072
130.2	-0.070
132.8	-0.069
135.1	-0.068
137.8	-0.067
140.0	-0.066
142.9	-0.065
145.6	-0.064
148.3	-0.062
150.9	-0.061
154.0	-0.060
157.3	-0.059
160.5	-0.058
163.5	-0.057
166.5	-0.056
169.3	-0.054
172.0	-0.053
175.0	-0.052
177.8	-0.051
180.7	-0.050
183.6	-0.048
186.3	-0.047
189.2	-0.046
192.2	-0.045
195.1	-0.044
198.0	-0.042
200.9	-0.041
203.7	-0.040
206.7	-0.039
209.7	-0.037
212.6	-0.036
215.5	-0.035
218.4	-0.034
221.4	-0.032
224.3	-0.031
227.3	-0.030
230.2	-0.028
233.2	-0.027
236.1	-0.026
239.3	-0.025
242.2	-0.023
245.1	-0.022
248.0	-0.021
250.9	-0.020
253.7	-0.018
256.7	-0.017
259.6	-0.016

T	$\Delta L/L_0$
CURVE 7 (cont.)*	
262.5	-0.014
265.3	-0.013
268.2	-0.012
271.1	-0.010
274.1	-0.009
276.7	-0.008
279.6	-0.006
282.5	-0.005
285.2	-0.004
287.9	-0.002
290.8	-0.001
293.6	0.000
296.4	0.002
299.1	0.003
301.9	0.004
304.3	0.006
307.6	0.007
310.3	0.008
313.0	0.010
315.8	0.011
318.6	0.012
321.3	0.014
CURVE 8	
100	-0.101
123	-0.094
173	-0.070
223	-0.042
273	-0.014
293	0.000*
323	0.016*
373	0.051
423	0.084
473	0.121
523	0.157
573	0.189
597	0.192
623	0.180
639	0.179
673	0.200
CURVE 9	
673	0.200*
623	0.167
573	0.127
523	0.094

T	$\Delta L/L_0$
CURVE 9 (cont.)	
485	0.066
473	0.066
445	0.079
423	0.082*
373	0.051*
323	0.016*
293	0.000*
273	-0.014*
223	-0.042*
173	-0.070*
123	-0.094*
100	-0.101*
CURVE 10	
293	0.000
323	0.064
366	0.011
421	0.011
462	-0.016
504	0.064
557	0.091
CURVE 11	
293	0.000
323	-0.124
366	0.288
421	0.618
462	0.537
504	0.618
557	0.537
CURVE 12	
294	0.000*
341	0.000
387	0.000
523	0.094*
552	0.226
580	0.170
586	0.226
611	0.189
642	0.226
661	0.245
669	0.207
679	0.226
734	0.283

T	$\Delta L/L_0$
CURVE 12 (cont.)	
783	0.358
825	0.339
871	0.434
CURVE 13*, †	
1151	0.000
1154	-0.238
1175	0.000

* Not shown in figure.

† This curve is here reported using the first given temperature as reference temperature at which $\Delta L/L_0 = 0$.

FIGURE AND TABLE NO. 26R. RECOMMENDED VALUES FOR THERMAL LINEAR EXPANSION OF LEAD Pb

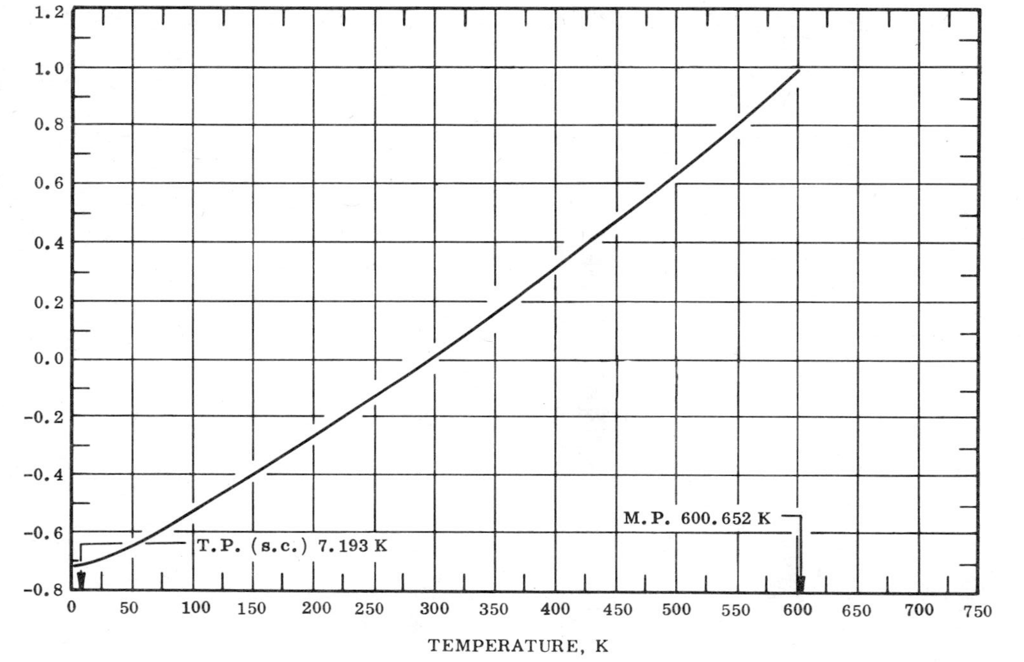

RECOMMENDED VALUES

[Temperature, T, K; Linear Expansion, $\Delta L/L_0$, %; α, K^{-1}]

T	$\Delta L/L_0$	$\alpha \times 10^6$
5	-0.708	0.25
25	-0.694	14.4
50	-0.647	21.6
100	-0.526	25.6
200	-0.261	27.5
293	0.000	28.9
400	0.317	30.6
500	0.638	33.3
600	0.988	36.7

LOW-TEMPERATURE THERMAL EXPANSION VALUES

	Normal State	Superconducting State
T	$\Delta L/L \times 10^4$	$\Delta L/L$† $\times 10^4$
0	0.00	
1	0.005	
2	0.015	0.18
3	0.046	0.21
4	0.12	0.25
5	0.28	0.38
6	0.62	0.71
7	1.32	1.34
7.2	1.47	1.47

† Superconducting state value is set equal to that of normal state value at the transition temperature, $T_c = 7.2$ K.

REMARKS

The tabulated values are considered accurate to within ± 3% over the entire temperature range. These values can be represented approximately by the following equation:

$$\Delta L/L_0 = -0.786 + 2.572 \times 10^{-3}\, T + 1.147 \times 10^{-7}\, T^2 + 8.770 \times 10^{-10}\, T^3 \qquad (100 < T < 600)$$

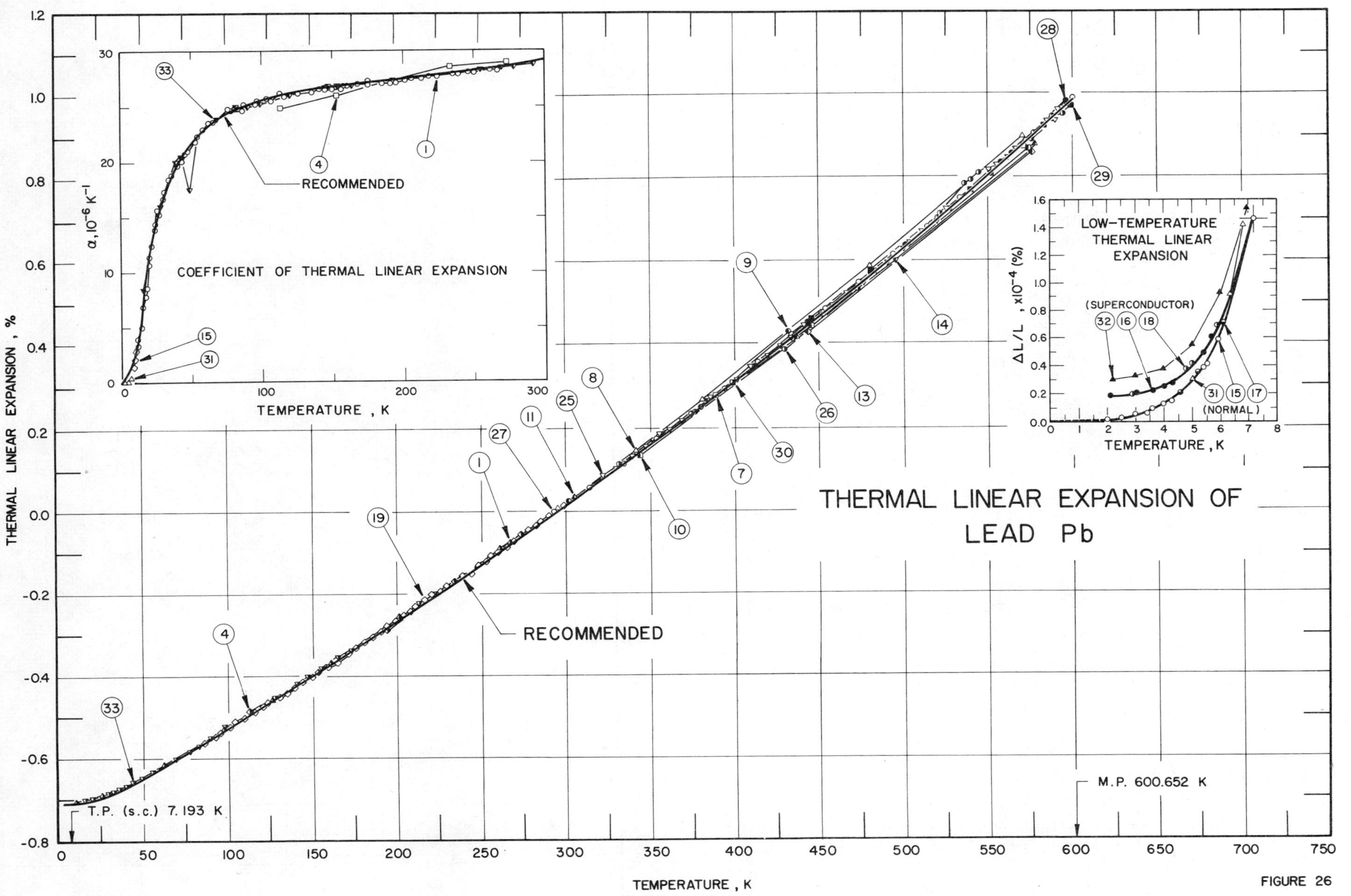
THERMAL LINEAR EXPANSION OF
LEAD Pb
THERMAL LINEAR EXPANSION, %
TEMPERATURE, K
RECOMMENDED
T.P. (s.c.) 7.193 K
M.P. 600.652 K
LOW-TEMPERATURE
THERMAL LINEAR
EXPANSION
(SUPERCONDUCTOR)
(NORMAL)
ΔL/L, x10⁻⁴ (%)
TEMPERATURE, K
COEFFICIENT OF THERMAL LINEAR EXPANSION
RECOMMENDED
α, 10⁻⁶ K⁻¹
TEMPERATURE, K

FIGURE 26

SPECIFICATION TABLE 26. THERMAL LINEAR EXPANSION OF LEAD Pb

Cur. No.	Ref. No.	Author(s)	Year	Method Used	Temp. Range, K	Name and Specimen Designation	Composition (weight percent), Specifications, and Remarks
1	68	Channing, D.A. and Weintroub, S.	1965	I	11-267		99.995 Pb polycrystalline hollow cylinder; specimen 10 mm long, 9 mm I.D., 16 mm O.D.; machined from lead rod; three small aquidistant projecting feet milled at each end; data from 8 overlapping runs.
2	610	Stokes, A.R. and Wilson, A.J.C.	1941	X	286-594		99.997 Pb, 0.0009 Cu, 0.0007 Sb, 0.0006 Ag, 0.0002 Bi; lattice parameter reported at 291.8 K is 4.9393 Å; 4.9395 Å at 293 K determined by graphical interpolation.
3*	71	Dheer, P.N. and Surange, S.L.	1958	E	4.7-10		Spectroscopically pure specimen from Johnson Matthey Ltd. in form of cylinder 4.87 cm x 0.7 cm diameter; this curve is here reported using the first given temperature as reference temperature at which $\Delta L/L_0=0$; correction -0.1×10^{-4}%.
4	70	Dorsey, H.G.	1908	I	118-273		Specimen supposedly pure, cast; length 1.5320 cm.
5*	72	Olsen, J.S. and Rohrer, H.	1957	N	2-7	V-Pb-2	Change of length measured below normal-superconducting transtion; data are result of 10-15 trials; tabulated $\Delta L/L_0$ are defined as (Ls-Ln)/Ls where Ls is length of specimen in superconducting state and Ln in normal state; transition temperature 7.2 K at which Ls - Ln = 0.
6*	72	Olsen, J.L. and Rohrer, H.	1957	N	2-7	V-Pb-4	Similar to the above specimen; tabulated $\Delta L/L_0$ are defined as (Ls-Ln)/Ls where Ls length of specimen in superconducting state and Ln in normal state, transition temperature 7.2 K at which Ls - Ln = 0.
7	73	Hidnert, P. and Sweeney, W.	1932	T	570-294	1001	Specimen cast in preheated steel mold; expansion measured with decreasing temperature; zero-point correction is -0.02%.
8	73	Hidnert, P. and Sweeney, W.	1932	T	295-570	1001	The above specimen; expansion measured with increasing temperature; zero-point correction of 0.007% determined by graphical extrapolation.
9	73	Hidnert, P. and Sweeney, W.	1932	T	575-301	1144	99.9 Pb, cast in sand mold; expansion measured with decreasing temperature; zero-point correction of -0.08% determined by graphical extrapolation.
10	73	Hidnert, P. and Sweeney, W.	1932	T	299-575	1144	The above specimen; expansion measured with increasing temperature; zero-point correction of -0.085% determined by graphical extrapolation.
11	73	Hidnert, P. and Sweeney, W.	1932	T	577-305	1215	99.8 Pb, cast in sand mold; expansion measured with decreasing temperature; zero-point correction of -0.177% determined by graphical extrapolation.
12	73	Hidnert, P. and Sweeney, W.	1932	T	301-577	1215	The above specimen; expansion measured with increasing temperature; zero-point correction of -0.177% determined by graphical extrapolation.
13	73	Hidnert, P. and Sweeney, W.	1932	T	576-303	1215	The above specimen; expansion measured with decreasing temperature on second test; zero-point correction of -0.290% determined by graphical extrapolation.
14	73	Hidnert, P. and Sweeney, W.	1932	T	304-576	1215	The above specimen; expansion measured with increasing temperature; zero-point correction of -0.265% determined by graphical extrapolation.
15	74	White, G.K.	1962	E	2.0-7.2	Pb1	99.99 Pb, machined from "melting point" lead from National Bureau of Standards; expansion of normal state in applied magnetic field of 900 Oe; this curve referred to 0 K, at which $\Delta L/L_0 = 0$; reported error is ±1%; data on the coefficient of thermal linear expansion also reported.
16	74	White, G.K.	1962	E	2.1-7.2	Pbl	The above specimen; this curve referred to the normal state value of Curve 15 at the critical temperature, $T_c = 7.2$ K; reported error is ±1%; data on the coefficient of thermal linear expansion also reported.

* Not shown in figure.

SPECIFICATION TABLE 26. THERMAL LINEAR EXPANSION OF LEAD Pb (continued)

Cur. No.	Ref. No.	Author(s)	Year	Method Used	Temp. Range, K	Name and Specimen Designation	Composition (weight percent), Specifications, and Remarks
17	74	White, G.K.	1962	E	1.8-7.2	Pb2	99.9999 Pb, machined from "Tadanac" (Consolidated Mining and Smelting Co. of Canada); expansion of normal state in magnetic field of 900 Oe; this curve referred to 0 K, at which $\Delta L/L_0 = 0$; reported error is ±1%.
18	74	White, G.K.	1962	E	1.3-7.2	Pb2	The above specimen; expansion of superconducting state; this curve referred to the normal state value of Curve 17 at the critical temperature, $T_c = 7.2$ K; reported error is ±1%.
19	75	Nix, F.C. and MacNair, D.	1942	I	85-298		99.99 Pb specimen from National Lead Co.; zero-point correction of -0.058% determined by graphical interpolation.
20	76	Van Duijn, J. and Van Galen, J.	1957	X	313-560		Total impurity < 0.01; polycrystalline specimen supplied by Johnson and Matthey Ltd., London.
21	76	Van Duijn, J. and Van Galen, J.	1957	L	323-572		Similar to the above specimen; cylindrical specimen, 50 mm long, 7 mm diameter.
22	33	Richards, J.W.	1942	T	83-598		0.0001 each Cu, Ag, Fe; Doe Run lead furnished by Electric Storage Battery Co.; expansion measured in hydrogen or nitrogen atmosphere.
23	34	Asay, J.R.	1968		173-473		99.0 Pb, block; density 11.171 g cm^{-3}.
24	38	Uffelmann, F.L.	1930	I	353-553		Cubic specimen of side 4 mm.
25	66	Feder, R. and Nowick, A.S.	1958	L	293-592		99.998 Pb single crystal grown by Bridgman technique; macroscopic expansion studied; zero-point correction is -0.06%.
26	66	Feder, R. and Nowick, A.S.	1958	X	293-595		The above specimen; microscopic expansion studied; zero-point correction is -0.06%.
27	69	Feder, R. and Nowick, A.S.	1967	I	293-600		99.999 Pb; cast and cold worked; machined into hollow cylinder 2.5 in. long, 1.0 in. O.D., 0.625 in. I.D.; annealed in vacuo at 583 K for several days; measurement used He-Ne laser as light source.
28	742	d'Heurle, F.M., Feder, R., and Norwick, A.S.	1963	L	435-600		99.999 Pb; sample cut from bar after growth of large crystals.
29	742	d'Heurle, et al.	1963	X	433-599		Similar to the above specimen.
30	665	Balasundaram, L.J., and Sinha, A.N.	1971	L	293-398		Specimen of chemically pure Pb.
31	714, 89	Andres, K.	1964, 1961	O	2.2-7.6		99.99 Pb, normal state; this curve referred to 0 K, at which $\Delta L/L_0 = 0$; data on the coefficient of thermal linear expansion also reported.
32	714, 89	Andres, K.	1964, 1961	O	2.2-7.6		The above specimen, superconducting state; this curve referred to the normal state value of Curve 31 at the critical temperature, $T_c = 7.2$ K, data on the coefficient of thermal linear expansion also reported.
33	737	Rubin, T., Johnson, H.L., and Altman, H.W.	1961	I	16-292		99.99 Pb from National Bureau of Standards; cut to three pillars, filed to same length within 1/5 wavelength of sodium - D radiation; pillars served to separate interference plates.

* Not shown in figure.

DATA TABLE 26. THERMAL LINEAR EXPANSION OF LEAD Pb

[Temperature, T, K; Linear Expansion, $\Delta L/L_0$, %]

T	$\Delta L/L_0$
CURVE 1‡	
11	-0.705
11	-0.705
12	-0.705*
14	-0.704*
15	-0.704*
17	-0.702*
18	-0.701*
20	-0.699*
20	-0.699*
22	-0.697*
24	-0.694*
24	-0.694*
26	-0.691
27	-0.690*
30	-0.685*
31	-0.683*
34	-0.678*
36	-0.674*
40	-0.666*
43	-0.660*
45	-0.656*
47	-0.652*
53	-0.639*
54	-0.637*
58	-0.628*
62	-0.619
65	-0.612*
76	-0.586*
78	-0.581*
86	-0.561*
87	-0.558*
96	-0.536
97	-0.533*
104	-0.515*
107	-0.508*
113	-0.492*
116	-0.484*
125	-0.461
135	-0.435*
141	-0.419*
145	-0.408*
150	-0.395*
156	-0.379*
158	-0.374*
160	-0.368
168	-0.347*
CURVE 1 (cont.)‡	
175	-0.328*
175	-0.328*
185	-0.301*
191	-0.284*
195	-0.274*
201	-0.257
206	-0.244*
213	-0.224*
218	-0.210*
224	-0.194*
236	-0.161
239	-0.152*
246	-0.133*
250	-0.121*
261	-0.090
267	-0.073
CURVE 2	
285.8	-0.016
287.7	-0.016*
287.8	-0.014*
291.8	-0.003*
331.4	0.112
333.3	0.122
371.1	0.235*
372.5	0.241
423.8	0.397*
431.6	0.422
469.9	0.547*
581.4	0.584
523.0	0.729*
528.0	0.744
573.8	0.897*
575.5	0.908
594.0	0.976
CURVE 3*, †	
4.69	0.00 x 10^{-4}
5.09	0.13
5.55	0.23
6.03	0.51
6.06	0.45
6.34	0.60
6.60	0.78
CURVE 3 (cont.)*, †	
6.80	0.95 x 10^{-4}
7.03	1.19
7.25	1.30
7.25	1.42
7.52	1.62
7.61	1.70
7.73	1.95
7.93	2.22
8.11	2.46
8.16	2.56
8.39	3.05
8.65	3.41
8.85	3.83
9.09	4.29
9.41	5.20
9.41	5.28
9.59	5.52
9.63	5.77
9.83	6.24
9.96	6.96
CURVE 4‡	
113	-0.491
153	-0.390
193	-0.284
233	-0.172
273	-0.057
CURVE 5*, ǂ	
1.7	0.155 x 10^{-6}
2.1	0.152
2.4	0.150
2.7	0.148
3.0	0.146
3.1	0.144
3.5	0.141
3.7	0.135
4.0	0.127
7.2	0.000
CURVE 6*, ǂ	
1.7	0.145 x 10^{-6}
2.0	0.144
CURVE 6 (cont.)*, ǂ	
2.2	0.144 x 10^{-6}
2.5	0.143
2.6	0.142
3.0	0.139
3.1	0.138
3.3	0.134
3.6	0.131
3.7	0.128
3.8	0.126
4.0	0.122
7.2	0.000
CURVE 7	
570	0.874
388	0.279
294	0.000*
CURVE 8	
295	0.007
313	0.058*
341	0.143
381	0.265
479	0.586
570	0.901
CURVE 9	
575	0.884*
437	0.427
379	0.251
301	0.025
CURVE 10	
299	0.015*
343	0.138
387	0.266*
444	0.441
490	0.587*
573	0.873
575	0.879*
CURVE 11	
577	0.879
436	0.416
354	0.173*
305	0.032
CURVE 12*	
301	0.023
337	0.129
372	0.230
480	0.557
577	0.879
CURVE 13	
576	0.860
443	0.423
303	0.007*
CURVE 14	
304	0.035*
383	0.264
494	0.600
576	0.885*
CURVE 15ǂ	
2.0	0.015 x 10^{-4}
2.5	0.023
3.0	0.046
3.4	0.068
3.6	0.088
3.7	0.096
3.8	0.108
4.0	0.122
4.3	0.151
5.4	0.392
5.5	0.416
5.9	0.593
7.2	1.470
CURVE 16ǂ	
2.1	0.182 x 10^{-4}
2.2	0.184
CURVE 16 (cont.)ǂ	
2.6	0.191 x 10^{-4}
2.9	0.198
3.0	0.204
3.1	0.197
3.4	0.217
3.6	0.229
3.7	0.235
3.8	0.241
4.0	0.252
4.3	0.278
5.3	0.465
5.4	0.496
5.6	0.545
5.7	0.611
7.2	1.470
CURVE 17ǂ	
1.8	0.011 x 10^{-4}
2.3	0.014
2.6	0.027
3.1	0.053
3.1	0.058
3.6	0.089
3.9	0.116
4.1	0.133
4.6	0.200
5.2	0.349
5.4	0.409
5.5	0.426
6.1	0.717
7.2	1.470
CURVE 18ǂ	
1.3	0.181 x 10^{-4}
1.5	0.182
1.9	0.185
2.3	0.192
2.9	0.208
3.0	0.214
3.1	0.211
3.4	0.227
3.5	0.241
3.7	0.243
3.9	0.258

* Not shown in figure.
† This curve is here reported using the first given temperature as reference temperature at which $\Delta L/L_0 = 0$.
‡ Author's data for coefficient of thermal expansion have been integrated by TPRC to obtain $\Delta L/L_0$.
ǂ For explaination, see specification table.

DATA TABLE 26. THERMAL LINEAR EXPANSION OF LEAD Pb (continued)

T	$\Delta L/L_0$
CURVE 18 (cont.)‡	
4.0	0.271 x 10^{-4}
4.2	0.284
4.8	0.382
4.9	0.395
4.9	0.410
5.0	0.429
5.2	0.446
5.7	0.554
5.9	0.689
7.2	1.470
CURVE 19	
85.2	-0.562
92.2	-0.549
95.7	-0.539
100.2	-0.525
104.7	-0.513
109.7	-0.502
115.7	-0.490
120.2	-0.476
123.7	-0.464
129.7	-0.453
134.2	-0.442
138.2	-0.427
143.7	-0.415
148.2	-0.402
152.6	-0.390*
158.2	-0.378
163.0	-0.366
170.2	-0.346
174.2	-0.335*
175.2	-0.329
180.7	-0.317
185.2	-0.304
190.0	-0.292
193.7	-0.280
198.0	-0.267
203.5	-0.255
207.2	-0.243
211.2	-0.231
216.2	-0.218
220.7	-0.206
225.1	-0.194
229.6	-0.181
234.2	-0.169*
238.6	-0.157

T	$\Delta L/L_0$
CURVE 19 (cont.)	
243.2	-0.154
247.4	-0.132
252.2	-0.120
255.8	-0.108
260.7	-0.095
264.2	-0.083
268.7	-0.071
273.2	-0.058*
277.4	-0.046
282.1	-0.034
285.4	-0.022
290.2	-0.009
294.4	0.003*
298.2	0.015
CURVE 20	
293	0.000*
313	0.059
339	0.117*
349	0.186
368	0.205*
378	0.245*
400	0.294
415	0.340*
417	0.339*
426	0.376
437	0.404*
452	0.457*
460	0.538
491	0.580*
514	0.653
523	0.680*
545	0.762*
559	0.808
CURVE 21	
293	0.000*
323	0.084
341	0.135*
360	0.190
386	0.270*
400	0.304
412	0.346*
429	0.391
458	0.481*

T	$\Delta L/L_0$
CURVE 21 (cont.)	
478	0.542*
494	0.588*
499	0.599*
532	0.712
551	0.775*
565	0.823*
572	0.868
CURVE 22	
83	-0.503
300	0.020
376	0.235
433	0.410
496	0.584
582	0.902
598	0.995
CURVE 23	
173	-0.276
198	-0.224*
223	-0.169*
248	-0.111
273	-0.050*
298	0.012*
323	0.079*
348	0.149
373	0.221*
398	0.274
423	0.295*
448	0.371
473	0.386
CURVE 24‡	
353	0.174
373	0.232*
393	0.291*
413	0.350
433	0.410*
443	0.440
453	0.471*
463	0.501
473	0.532*
493	0.595
513	0.659*

T	$\Delta L/L_0$
CURVE 24 (cont.)‡	
523	0.691*
533	0.724
553	0.791
CURVE 25	
293	0.000*
312	0.058*
321	0.082
329	0.104
338	0.130
347	0.153
355	0.182
361	0.199
368	0.218
386	0.270
390	0.283
395	0.294*
401	0.315
409	0.345
419	0.370
427	0.394
433	0.414
441	0.442
451	0.469
457	0.493
466	0.520
475	0.548
480	0.562
483	0.573
490	0.596
500	0.629
503	0.645
511	0.670
519	0.698
525	0.727
527	0.727
534	0.749*
551	0.815
555	0.822
562	0.849
566	0.868
569	0.877*
573	0.888*
574	0.897*
585	0.932
588	0.937
592	0.961

T	$\Delta L/L_0$
CURVE 26	
293	0.000*
297	0.011*
331	0.111
376	0.239
428	0.386
472	0.536
520	0.707
521	0.714
541	0.785
545	0.796
549	0.809
553	0.819*
568	0.870*
571	0.886*
575	0.890
577	0.907
587	0.944
593	0.953
595	0.965
CURVE 27	
293	0.000
313	0.058
333	0.116
353	0.176
373	0.236
393	0.296
413	0.358
433	0.421
453	0.484
473	0.548
493	0.615
513	0.681
533	0.750
553	0.819
573	0.890
593	0.963
600	0.989
CURVE 28	
435	0.420*
442	0.450*
443	0.451*
443	0.453
455	0.487*

T	$\Delta L/L_0$
CURVE 28 (cont.)	
455	0.490*
466	0.527*
466	0.528*
483	0.582*
483	0.584*
484	0.583*
498	0.631*
501	0.638
541	0.777*
557	0.835
564	0.859
572	0.887*
572	0.888*
573	0.890*
573	0.893*
585	0.934*
593	0.967*
596	0.975*
596	0.975*
596	0.979
599	0.988*
600	0.989*
CURVE 29	
433	0.416*
446	0.461
452	0.481*
473	0.547*
481	0.576
493	0.613*
494	0.617*
501	0.636*
512	0.673*
512	0.676*
540	0.771
550	0.809*
552	0.814
565	0.859*
569	0.875*
572	0.881*
574	0.890*
577	0.900*
577	0.904*
583	0.923*
584	0.925
585	0.931*

* Not shown in figure.
‡ Author's data for coefficient of thermal expansion have been integrated by TPRC to obtain $\Delta L/L_0$.
‡ For explaination, see specification table.

DATA TABLE 26. THERMAL LINEAR EXPANSION OF LEAD Pb (continued)

T	$\Delta L/L_0$
CURVE 29 (cont.)	
590	0.948*
594	0.961*
594	0.963*
597	0.973
598	0.979*
599	0.982*
CURVE 30	
293	0.000*
398	0.308
CURVE 31‡	
2.8	0.023 x 10^{-4}
2.8	0.026
3.3	0.045
3.7	0.079
3.8	0.084
3.9	0.092
3.9	0.091
3.9	0.097
4.0	0.104
4.0	0.114
4.2	0.136
4.4	0.159
4.6	0.205
5.0	0.294
5.0	0.296
5.4	0.418
5.8	0.612
6.3	0.912
6.8	1.420
7.2	1.820
7.3	1.890
CURVE 32‡	
2.2	0.308 x 10^{-4}
2.5	0.313
2.6	0.314
2.7	0.318
2.9	0.320
2.9	0.321
3.0	0.327
3.2	0.331
3.2	0.333
3.5	0.347
3.5	0.353
CURVE 32 (cont.)‡	
3.7	0.358 x 10^{-4}
3.8	0.370
3.8	0.371
4.0	0.387
4.0	0.384
4.1	0.303
4.2	0.415
4.4	0.437
4.8	0.496
5.0	0.558
5.2	0.610
5.3	0.612
6.0	0.934
6.9	1.540
7.2	1.820
7.6	2.360
CURVE 33	
16.873	-0.701
20.96	-0.697
24.91	-0.692*
28.59	-0.686
32.04	-0.681
35.30	-0.675
38.98	-0.668
43.50	-0.659
48.89	-0.648
55.32	-0.635
61.091	-0.622*
68.14	-0.606
76.05	-0.586
82.817	-0.570
83.67	-0.568*
90.62	-0.550
92.47	-0.545*
99.87	-0.527
102.78	-0.519*
109.99	-0.501*
115.02	-0.488*
120.88	-0.473*
127.95	-0.455
133.62	-0.440*
140.77	-0.421
147.03	-0.404
151.84	-0.391*
154.68	-0.384
162.62	-0.362*
164.24	-0.358
CURVE 33 (cont.)	
173.413	-0.333
184.895	-0.302*
197.52	-0.267*
213.157	-0.225
215.12	-0.219*
222.10	-0.200
234.07	-0.167*
247.90	-0.128*
255.73	-0.106*
269.86	-0.066*
278.32	-0.042*
292.12	-0.003*

* Not shown in figure.
‡ For explaination, see specification table.

DATA TABLE 26. COEFFICIENT OF THERMAL LINEAR EXPANSION OF LEAD Pb

[Temperature, T, K; Coefficient of Expansion, α, 10^{-6} K^{-1}]

T	α	T	α	T	α	T	α
CURVE 1‡		CURVE 1 (cont.)‡		CURVE 24*,‡		CURVE 33 (cont.)	
11	3.3	168	27.0	353	28.9	38.98	19.91
11	3.8*	175	27.0*	373	29.1	43.50	20.58
12	4.0*	175	27.3	393	29.5	48.89	17.55
14	5.0*	185	27.1*	413	29.9	55.32	22.52
15	6.8	191	27.1	433	30.2	61.091	23.27
17	7.8*	195	27.1*	443	30.3	68.14	23.85
18	8.6*	201	27.3*	453	30.5	76.05	24.66
20	10.7*	206	27.5*	463	30.9	82.817	24.97
20	11.4	213	27.5*	473	31.2	83.67	24.89
22	12.4*	218	27.7	493	31.6	90.62	25.21
24	13.9*	224	27.7*	513	32.0	92.47	25.15
24	14.4*	236	27.9	523	32.3	99.87	25.36
26	15.7*	239	27.8*	533	33.0	102.78	25.47
27	15.3*	246	28.1	553	34.3	109.99	25.64
30	16.7	250	28.0*			115.02	25.75
31	17.3*	261	28.3*	CURVE 31		120.88	25.94
34	18.5*	267	28.2			127.95	26.13
36	18.8*			2.0	0.0135	133.62	26.46
40	19.7*	CURVE 4‡		2.5	0.0239*	140.77	26.40
43	20.1*			3.0	0.0394*	147.03	26.82
45	20.7*	113	24.8	3.5	0.0698*	151.84	26.66
47	21.0*	153	25.9	4.0	0.1153	154.68	26.73
53	21.8	193	27.2	4.5	0.1853*	162.62	27.17
54	22.1*	233	28.5	5.0	0.2795*	164.24	26.84
58	22.7*	273	28.9	5.5	0.4222	173.413	27.02
62	23.3*					184.895	27.24
65	23.3*	CURVE 15		CURVE 32*		197.52	27.37
76	24.8					213.157	27.50
78	24.7*	3	0.052*	2.0	0.0103	215.12	27.66
86	24.7*	4	0.110*	2.5	0.0189	222.10	27.44
87	25.2*	5	0.240*	3.0	0.0342	234.07	27.87
96	25.2	6	0.480*	3.5	0.0580	247.90	28.11
97	25.5*	7	0.900*	4.0	0.0981	255.73	28.22
104	25.7*	8	1.440	4.5	0.1490	269.86	28.45
107	25.5*	9	2.120	5.0	0.2330	278.32	28.49
113	26.2	10	2.880	5.5	0.4120	292.12	28.68
116	25.8*	11	3.800*				
125	26.2*			CURVE 33			
135	26.3*	CURVE 16*					
141	26.6			16.873	8.40		
145	26.5*	3	0.035	20.96	11.44		
150	26.5*	4	0.080	24.91	13.96		
156	26.5	5	0.220	28.59	15.80		
158	26.8*	6	0.440	32.04	17.25		
160	26.8*	7	0.880	35.30	18.41		

* Not shown in figure.
‡ Author's data for coefficient of thermal expansion have been integrated by TPRC to obtain $\Delta L/L_0$.

FIGURE AND TABLE NO. 27R. RECOMMENDED VALUES FOR THERMAL LINEAR EXPANSION OF LITHIUM Li

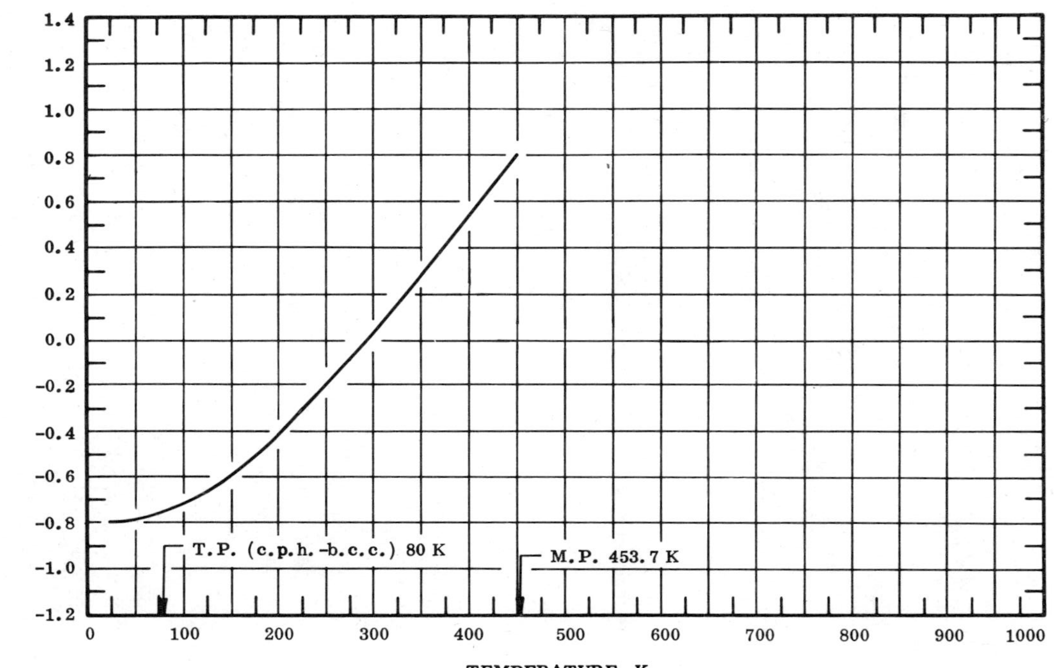

RECOMMENDED VALUES

[Temperature, T, K; Linear Expansion, $\Delta L/L_0$, %; α, K^{-1}]

T	$\Delta L/L_0$	$\alpha \times 10^6$
20	-0.792	1
25	-0.792	1
50	-0.789	5
100	-0.721	21
200	-0.404	39
293	0.000	46
400	0.530	53
450	0.804	56

REMARKS

The tabulated values are considered accurate to within ± 3% at temperatures above 293 K and ± 5% below. These values can be represented approximately by the following equations:

$$\Delta L/L_0 = -0.792 - 5.316 \times 10^{-4} (T - 20) + 1.951 \times 10^{-5} (T - 20)^2 - 2.578 \times 10^{-8} (T - 20)^3 \quad (20 < T < 235)$$

$$\Delta L/L_0 = -0.259 + 4.283 \times 10^{-3} (T - 235) + 2.880 \times 10^{-6} (T - 235)^2 + 8.687 \times 10^{-10} (T - 235)^3 \quad (235 < T < 450)$$

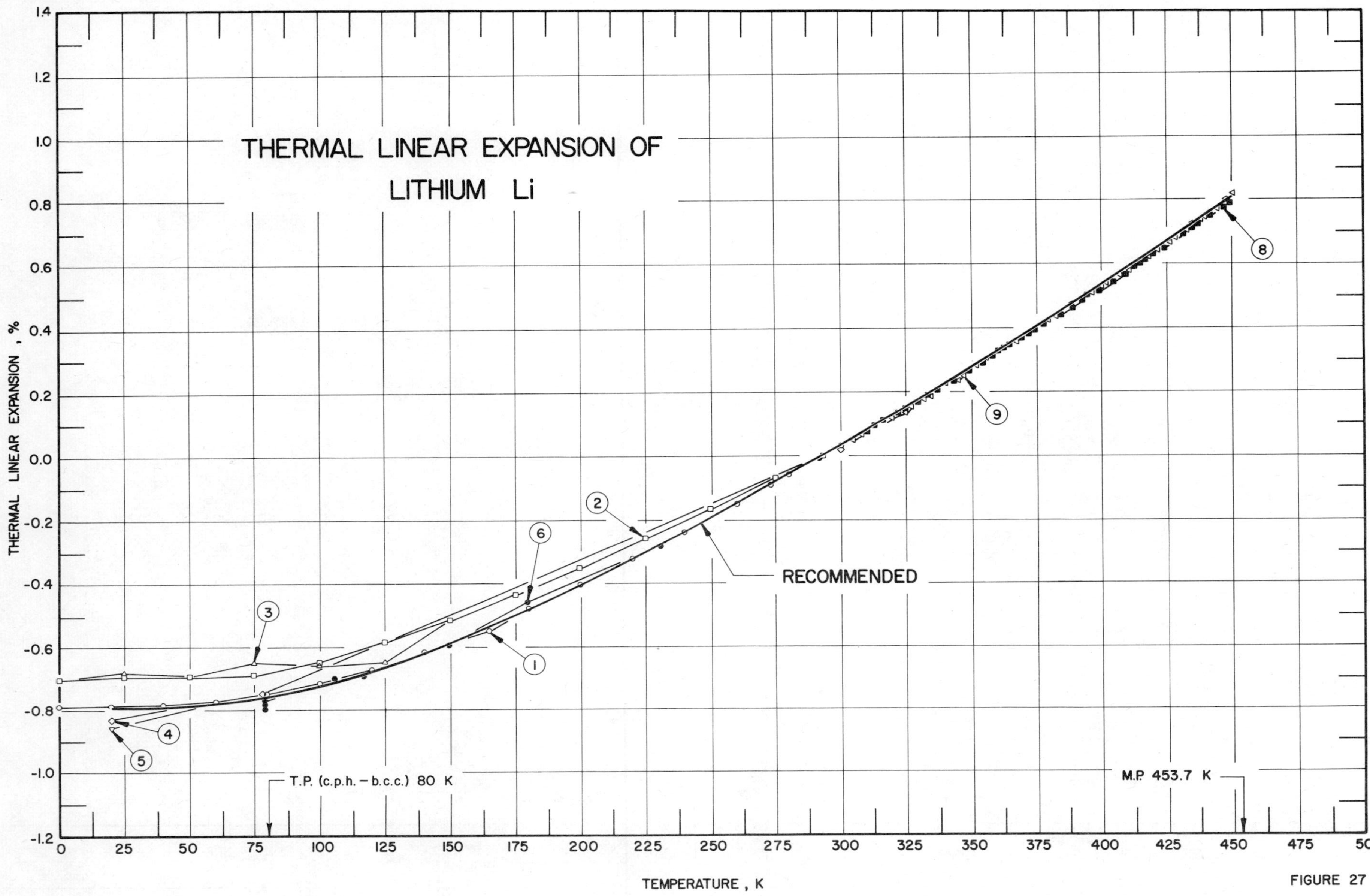

THERMAL LINEAR EXPANSION OF
LITHIUM Li
THERMAL LINEAR EXPANSION, %
TEMPERATURE, K
RECOMMENDED
M.P. 453.7 K
T.P. (c.p.h. – b.c.c.) 80 K
1.4
1.2
1.0
0.8
0.6
0.4
0.2
0.0
-0.2
-0.4
-0.6
-0.8
-1.0
-1.2
0
25
50
75
100
125
150
175
200
225
250
275
300
325
350
375
400
425
450
475
500
1
2
3
4
5
6
8
9

FIGURE 27

SPECIFICATION TABLE 27. THERMAL LINEAR EXPANSION OF LITHIUM Li

Cur. No.	Ref. No.	Author(s)	Year	Method Used	Temp. Range, K	Name and Specimen Designation	Composition (weight percent), Specifications, and Remarks
1	83	Laquer, H. L.	1952	L	0-300		Data averaged by author from two specimens; both diameter 0.50 cm and length of one 1.976 cm, other 2.806 cm; zero-point correction of -0.09% determined by graphical interpolation.
2	84	Snyder, D. D., Zimmerman, W. B. and Kuhlman, H.	1964	V	300-0		Data average of results on several runs; expansion measured with temperature decreasing at 265 K per minute; zero-point correction of 0.018% determined by graphical interpolation.
3	84	Snyder, D. D., et al.	1964	V	0-300		The above specimen; expansion measured with increasing temperature; zero-point correction of 0.018% determined by graphical interpolation.
4	85	Kogan, V. S. and Khotkevish, V. I.	1962	X	20-300		Light isotope (Li^6), low temperature run made in cryostat, where sample and film immersed completely in coolant for run in liquid hydrogen or partially for run in liquid nitrogen; cell diameter 57.3 mm; reported error in measurement of lattice parameter ± 0.1%; lattice parameter reported at 300 K is 3.510 Å; 3.509 Å at 293 K determined by graphical interpolation.
5	85	Kogan, V. S. and Khotkevish, V. I.	1962	X	20-300		Similar to the above specimen, heavy isotope (Li^7); lattice parameter reported at 300 K is 3.509 Å; 3.5082 Å at 293 K determined by graphical interpolation.
6	87	Owen, E. A. and Williams, G. J.	1954	X	79-294		99.92 Li, specimen prepared from lump of material, impurities as Na, K, Mg each not more than 0.01, Ca 0.05, flat disc form about 6 mm diameter, about 1 mm thick; lattice parameter reported at 293 K is 3.5025 kX (3.5095 Å).
7*	86	Lynch, R. W. and Edwards, L. R.	1970	L	293		99.95 pure from Foote Mineral Co., measurements on 0.63 cm diameter and 2.5 cm long rod; authors report a single value of the coefficient of thermal linear expansion at 293 K, i.e. 46×10^{-6} K^{-1}.
8	734	Feder, R.	1970	X	304-450		Single crystal of high purity zone refined Li in a grease coated stainless steel mold grown by Bridgman technique; average of heating and cooling.
9	734	Feder, R.	1970	L	304-451		Similar to the above specimen; average of heating and cooling.

*Not shown in figure.

DATA TABLE 27. THERMAL LINEAR EXPANSION OF LITHIUM Li

[Temperature, T, K; Linear Expansion, $\Delta L/L_0$, %]

T	$\Delta L/L_0$
CURVE 1	
0	-0.788
20	-0.787
40	-0.782
60	-0.772
80	-0.748
100	-0.713
120	-0.668
140	-0.612
160	-0.548
180	-0.477
200	-0.401
220	-0.320
240	-0.236
260	-0.149
273	-0.090
280	-0.059
300	0.032
CURVE 2	
293	0.000
275	-0.071
250	-0.166
225	-0.259
200	-0.353
175	-0.436
150	-0.514
125	-0.582
100	-0.645
75	-0.686
50	-0.691
25	-0.691
0	-0.700
CURVE 3	
0	-0.700*
25	-0.679
50	-0.682*
75	-0.648
100	-0.652
125	-0.642
150	-0.514*
175	-0.436*
200	-0.353*
CURVE 3 (cont.)	
225	-0.259*
250	-0.166*
275	-0.071*
293	0.000*
CURVE 4	
20	-0.830
78	-0.745
300	0.024
CURVE 5	
20	-0.859
78	-0.745*
300	0.024*
CURVE 6	
79	-0.797
79	-0.779
79	-0.766
80	-0.763*
106	-0.699
117	-0.690
118	-0.690*
149	-0.593
179	-0.458
231	-0.280
292	-0.006
292	0.007
293	0.000*
294	0.001*
CURVE 7	
x	
CURVE 8	
310	0.078
313	0.096
314	0.101*
316	0.107*
316	0.110
317	0.114*
CURVE 8 (cont.)	
322	0.139
324	0.146*
325	0.149
329	0.170
333	0.191
334	0.194*
337	0.207
343	0.238
349	0.268
354	0.292
358	0.312
359	0.317*
361	0.329
363	0.340
365	0.351
369	0.372
372	0.385
373	0.391*
374	0.393
378	0.414
384	0.445*
384	0.445*
385	0.446
385	0.447*
389	0.466
393	0.490
395	0.504
399	0.520
400	0.524*
405	0.548
409	0.570
413	0.593
415	0.604
417	0.614
420	0.633
424	0.654
427	0.666
432	0.695
435	0.716
437	0.728
442	0.753
443	0.757*
447	0.779
449	0.790
450	0.795*
CURVE 9	
304	0.052
304	0.054*
305	0.058*
308	0.071
309	0.077*
311	0.083*
312	0.089*
313	0.093*
317	0.115*
319	0.120
319	0.124*
320	0.129*
321	0.131*
322	0.136*
325	0.139
325	0.149*
326	0.148*
327	0.155
328	0.159*
329	0.166*
330	0.171*
331	0.173*
332	0.181
334	0.186
335	0.196*
336	0.198*
341	0.205*
341	0.229
342	0.228*
344	0.236
345	0.245*
347	0.251*
347	0.256*
347	0.257
348	0.259*
349	0.266*
350	0.270*
350	0.273*
352	0.286
354	0.289*
356	0.303
356	0.305*
358	0.315*
359	0.322*
363	0.340*
364	0.343*
CURVE 9 (cont.)	
367	0.360
369	0.371*
370	0.373*
370	0.378*
372	0.385*
375	0.400*
377	0.408*
378	0.417*
380	0.428
383	0.441
384	0.446*
384	0.449*
385	0.452*
386	0.456*
390	0.475*
390	0.480
391	0.484*
392	0.486*
393	0.493*
395	0.503*
396	0.508*
397	0.513*
398	0.519*
400	0.531*
400	0.533*
401	0.536
402	0.541
402	0.543*
404	0.552*
407	0.569
408	0.572*
409	0.578*
411	0.586
411	0.590*
412	0.593*
413	0.599*
414	0.606*
415	0.612*
418	0.627
421	0.641*
422	0.646
423	0.654*
423	0.656*
427	0.677
428	0.681*
428	0.683*
CURVE 9 (cont.)	
431	0.703*
431	0.703
433	0.714*
434	0.716*
435	0.726*
436	0.728
438	0.741
439	0.743*
440	0.749
442	0.759*
442	0.766
443	0.771*
444	0.776
445	0.781*
445	0.783*
446	0.788*
447	0.794*
448	0.800
448	0.803*
449	0.805
450	0.812
451	0.820

* Not shown in figure.

x Author reports coefficient of thermal linear expansion only at 293 K, see specification table.

FIGURE AND TABLE NO. 28R. PROVISIONAL VALUES FOR THERMAL LINEAR EXPANSION OF LUTETIUM Lu

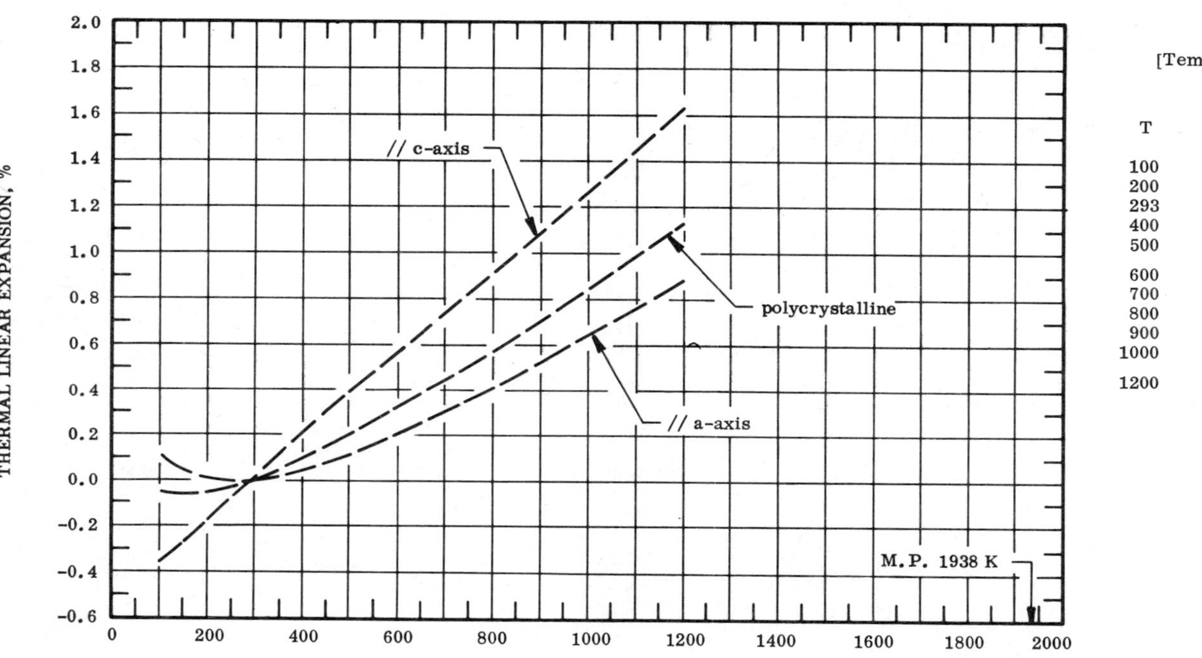

PROVISIONAL VALUES

[Temperature, T, K; Linear Expansion, $\Delta L/L_0$, %; α, K^{-1}]

T	// a-axis $\Delta L/L_0$	// c-axis $\Delta L/L_0$	polycrystalline $\Delta L/L_0$	polycrystalline $\alpha \times 10^6$
100	0.106	-0.351	-0.046	
200	0.004	-0.172	-0.055	
293	0.000	0.000	0.000	8.2
400	0.048	0.195	0.097	10.2
500	0.118	0.381	0.206	11.3
600	0.205	0.564	0.325	12.0
700	0.301	0.740	0.447	12.7
800	0.407	0.913	0.576	13.2
900	0.521	1.085	0.709	13.6
1000	0.641	1.261	0.848	14.0
1200	0.891	1.622	1.135	14.8

REMARKS

The tabulated values are considered accurate to within ±10% at temperatures below 293 K and ±7% above. Since no experimental data on the polycrystalline lutetium were located in the literature, the tabulated values are calculated from single crystal data. These values are considered accurate to within ±7% over the entire temperature range. These values can be represented approximately by the following equation:

$$\Delta L/L_0 = -0.215 + 5.254 \times 10^{-4}\, T + 7.313 \times 10^{-7}\, T^2 - 1.941 \times 10^{-10}\, T^3 \qquad (293 < T < 1200)$$

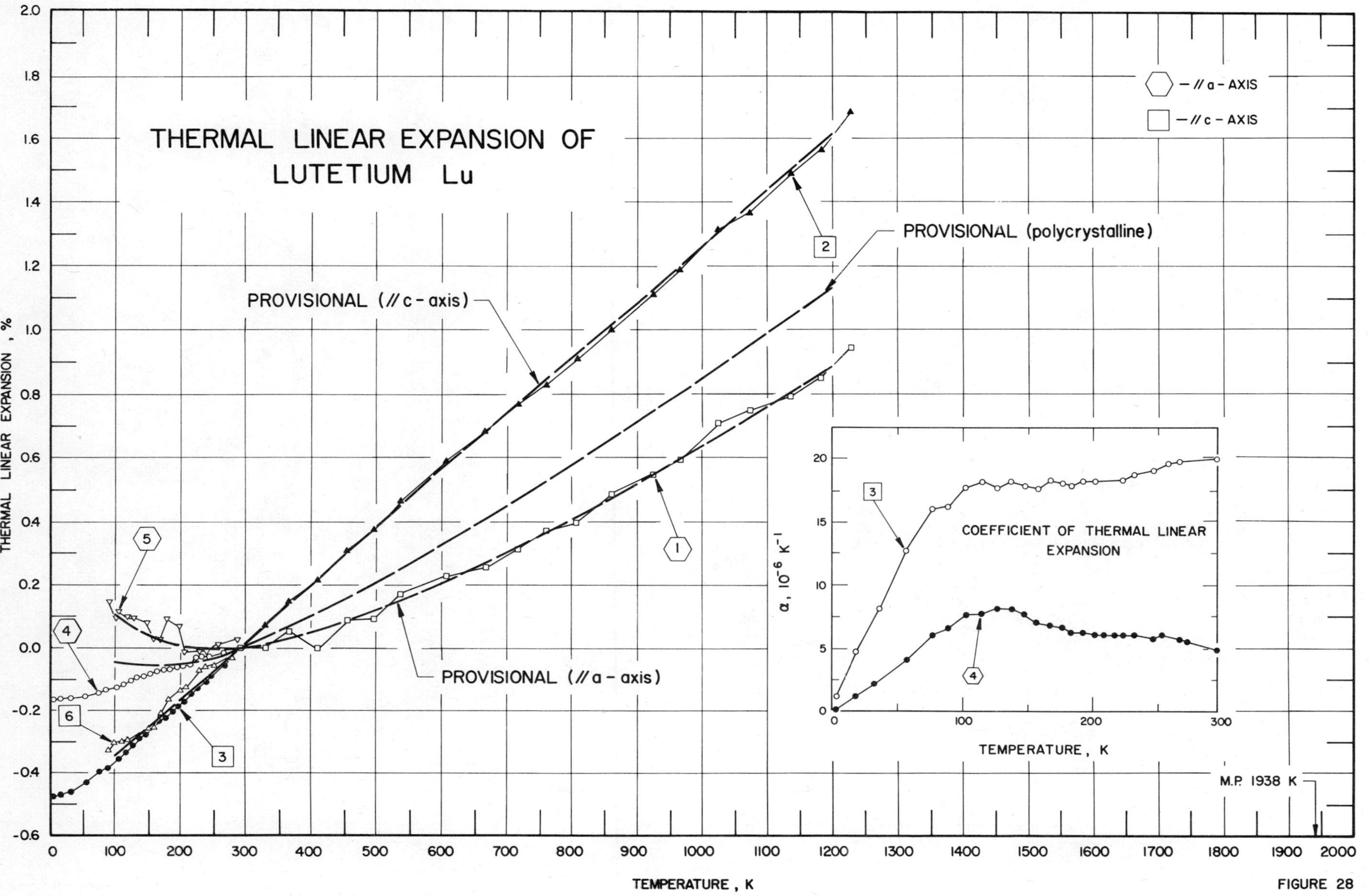

THERMAL LINEAR EXPANSION OF
LUTETIUM Lu
— // a - AXIS
— // c - AXIS
PROVISIONAL (// c - axis)
PROVISIONAL (polycrystalline)
PROVISIONAL (// a - axis)
THERMAL LINEAR EXPANSION, %
TEMPERATURE, K
COEFFICIENT OF THERMAL LINEAR
EXPANSION
α, 10⁻⁶ K⁻¹
TEMPERATURE, K
M.P. 1938 K

FIGURE 28

SPECIFICATION TABLE 28. THERMAL LINEAR EXPANSION OF LUTETIUM Lu

Cur. No.	Ref. No.	Author(s)	Year	Method Used	Temp. Range, K	Name and Specimen Designation	Composition (weight percent), Specifications, and Remarks
1	82	Hanak, J.J.	1959	X	291-1229		< 0.05 Ca, < 0.03 each Mg, Si, < 0.02 Cu, < 0.01 Cr, ≤ 0.005 Yb; wire specimen cut from sheet prepared at Ames Laboratory; hexagonal close packed structure; expansion measured along a-axis; lattice parameter reported at 293 K is 3.505 Å.
2	82	Hanak, J.J.	1959	X	291-1229		The above specimen; expansion measured along c-axis; lattice parameter reported at 293 K is 5.551 Å.
3	635	Tonnies, J.J.	1971	L	300-4		Impurities consisted of 0.007 O, 0.004 N, 0.0014 Ta, 0.001 C, and 0.002 total other impurities; measurements on the single crystals along c-axis; zero-point correction of -0.01% determined by graphical interpolation; data on the coefficients of thermal linear expansion also reported.
4	635	Tonnies, J.J.	1971	L	300-4		Similar to the above specimen, along a-axis; zero-point correction of -0.038% determined by graphical interpolation; data on the coefficients of thermal linear expansion also reported.
5	736	Singh, S., Khanduri, N.C., and Tsang, T.	1971	X	90-301		Specimen from Research Chemicals, Phoenix, Arizona, impurity levels 0.0001 percent, reduced to filing of 250 mesh, annealed at temperature below melting point for 12 hr in argon atmosphere, measurement along a-axis, lattice parameter reported at 293 K is 3.5357 Å.
6	736	Singh, S., et al.	1971	X	91-301		Similar to the above specimen, measurements along c-axis, lattice parameter reported at 293 K is 5.5575 Å.

DATA TABLE 28. THERMAL LINEAR EXPANSION OF LUTETIUM Lu

[Temperature, T, K; Linear Expansion, $\Delta L/L_0$, %]

T	$\Delta L/L_0$
CURVE 1	
291	0.000
330	0.000
365	0.057
410	0.000
454	0.086
498	0.086
537	0.171
608	0.228
666	0.257
716	0.314
760	0.371
809	0.399
860	0.485
924	0.542
968	0.599
1021	0.713
1073	0.742
1136	0.799
1181	0.856
1229	0.942
CURVE 2	
291	0.000*
330	0.072
365	0.144
410	0.216
454	0.306
498	0.378
537	0.468
608	0.594
666	0.684
716	0.775
760	0.829
809	0.919
860	1.009
924	1.117
965	1.189
1021	1.315
1073	1.369
1136	1.495
1131	1.567
1229	1.693
CURVE 3	
300.8	0.014
273.2	-0.038*
266.6	-0.050
259.7	-0.090*
252.9	-0.077*
246.0	-0.090
239.0	-0.104
231.7	-0.118*
224.4	-0.131
216.8	-0.145
209.1	-0.165*
201.2	-0.175
193.0	-0.189
184.6	-0.205
176.0	-0.221
167.0	-0.237
157.6	-0.255*
147.8	-0.272
137.6	-0.291
126.7	-0.311
115.2	-0.332
102.7	-0.355
89.1	-0.380
77.2	-0.399
56.1	-0.429
33.0	-0.459
18.8	-0.468
4.2	-0.471
CURVE 4	
300.8	0.024
273.2	-0.010
266.4	-0.014
259.7	-0.018
252.9	-0.022
246.0	-0.026
239.0	-0.030
231.7	-0.025
224.4	-0.029
216.8	-0.044
209.1	-0.048
201.2	-0.054
193.0	-0.059
184.6	-0.064
176.0	-0.070
167.0	-0.076
157.6	-0.084
147.8	-0.090
137.6	-0.098
126.7	-0.106
115.2	-0.116
102.7	-0.125
89.1	-0.135
77.2	-0.142
56.1	-0.154
33.0	-0.160
18.8	-0.163
4.2	-0.164
CURVE 5	
90	0.141
101	0.096
109	0.113
120	0.099
130	0.093
150	0.079
160	0.034
169	0.028
180	0.082
200	0.068
209	-0.011
230	-0.003
241	-0.003
251	0.000
261	0.008
180	-0.008
290	0.025
293	0.000
301	0.000
CURVE 6	
91	-0.324
101	-0.306
110	-0.306
120	-0.297
130	-0.250
150	-0.259
160	-0.257
170	-0.212
180	-0.162
200	-0.138
209	-0.126
231	-0.065
241	-0.056
251	-0.065
260	-0.016
280	-0.031
290	0.007*
293	0.000
301	0.007*

DATA TABLE 28. COEFFICIENT OF THERMAL LINEAR EXPANSION OF LUTETIUM Lu

[Temperature, T, K; Coefficient of Expansion, α, 10^{-6} K^{-1}]

T	α
CURVE 3	
300.8	20.01
273.2	19.76*
266.6	19.85
259.7	19.63
252.9	19.46*
246.0	19.06
239.0	18.90*
231.7	18.85
224.4	18.44
216.8	18.43*
209.1	18.10*
201.2	18.12
193.0	18.32
184.6	17.98
166.0	18.01
167.0	18.34
157.6	17.69
147.8	17.79
137.6	18.31
126.7	17.62
115.2	18.16
102.7	17.76
89.1	16.21
77.2	16.01
56.1	12.78
33.0	8.08
18.3	4.96
4.2	1.12
CURVE 4	
300.8	4.82
273.2	5.59
266.4	5.75
259.7	6.02*
252.9	6.15
246.0	5.98
239.0	6.10*
231.7	6.06
224.4	6.19
216.8	6.14
209.1	6.28
201.2	6.20
193.0	6.40
184.6	6.46
176.0	6.70
167.0	6.96
157.6	7.18
147.8	7.74
137.6	8.10
126.7	8.10
115.2	7.69
102.7	7.51
89.1	6.63
77.2	6.03
56.1	4.15
33.0	2.20
18.8	1.22
4.2	0.25

* Not shown in figure.

FIGURE AND TABLE NO. 29R. RECOMMENDED VALUES FOR THERMAL LINEAR EXPANSION OF MAGNESIUM Mg

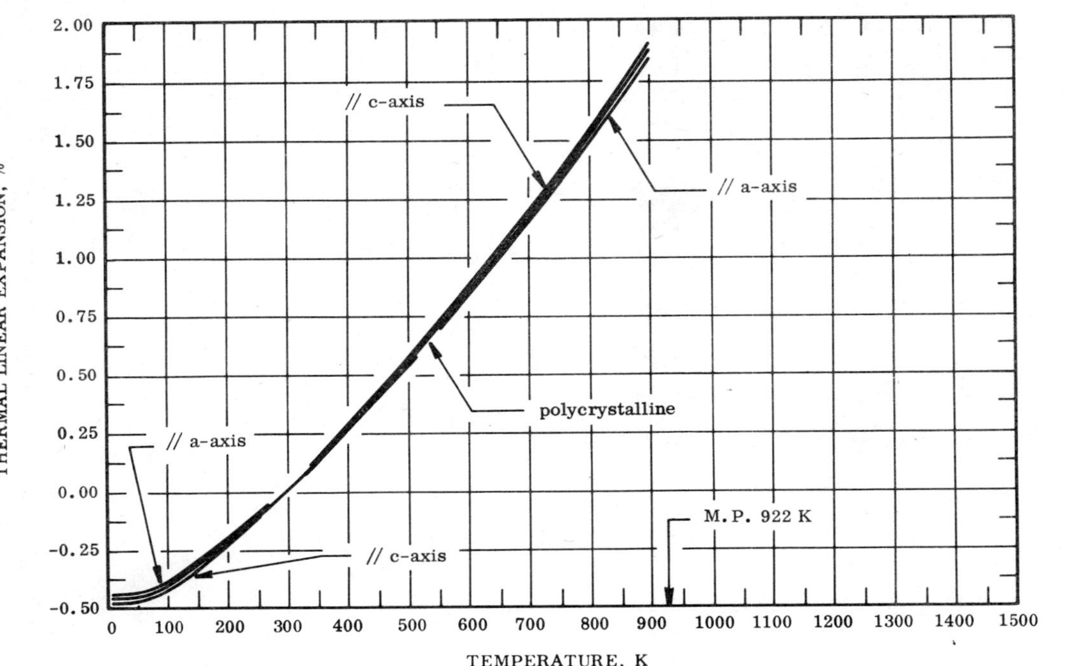

RECOMMENDED VALUES

[Temperature, T, K; Linear Expansion, $\Delta L/L_0$, %; α, K^{-1}]

T	// a-axis $\Delta L/L_0$	// c-axis $\Delta L/L_0$	polycrystalline $\Delta L/L_0$	polycrystalline $\alpha \times 10^6$
5	-0.448	-0.480	-0.459	0.01
25	-0.448	-0.480	-0.459	0.7
50	-0.440	-0.473	-0.451	5.8
100	-0.387	-0.416	-0.397	14.6
200	-0.209	-0.222	-0.213	21.2
293	0.000	0.000	0.000	24.8
400	0.277	0.285	0.281	27.3
500	0.556	0.571	0.561	29.1
600	0.851	0.877	0.860	30.9
700	1.167	1.200	1.178	33.0
800	1.509	1.544	1.521	35.4
900	1.876	1.906	1.886	37.6

REMARKS

The tabulated values are considered accurate to within ± 5% at temperatures below 293 K and ± 3% above. The tabulated values for polycrystalline magnesium calculated from single crystal data are considered accurate to within ± 5% over the entire temperature range. These values can be represented approximately by the following equations:

// a-axis:

$$\Delta L/L_0 = -0.384 + 1.271 \times 10^{-3} (T - 100) + 5.294 \times 10^{-6} (T - 100)^2 - 7.606 \times 10^{-9} (T - 100)^3 \quad (100 < T < 293)$$

$$\Delta L/L_0 = 2.465 \times 10^{-3} (T - 293) + 8.906 \times 10^{-7} (T - 293)^2 + 2.118 \times 10^{-10} (T - 293)^3 \quad (293 < T < 900)$$

// c-axis:

$$\Delta L/L_0 = -0.416 + 1.515 \times 10^{-3} (T - 100) + 4.507 \times 10^{-6} (T - 100)^2 - 6.177 \times 10^{-9} (T - 100)^3 \quad (100 < T < 293)$$

$$\Delta L/L_0 = 2.564 \times 10^{-3} (T - 293) + 9.311 \times 10^{-7} (T - 293)^2 + 3.270 \times 10^{-11} (T - 293)^3 \quad (293 < T < 900)$$

polycrystalline:

$$\Delta L/L_0 = -0.395 + 1.372 \times 10^{-3} (T - 100) + 4.878 \times 10^{-6} (T - 100)^2 - 6.813 \times 10^{-9} (T - 100)^3 \quad (100 < T < 293)$$

$$\Delta L/L_0 = 2.493 \times 10^{-3} (T - 293) + 9.334 \times 10^{-7} (T - 293)^2 + 1.152 \times 10^{-10} (T - 293)^3 \quad (293 < T < 900)$$

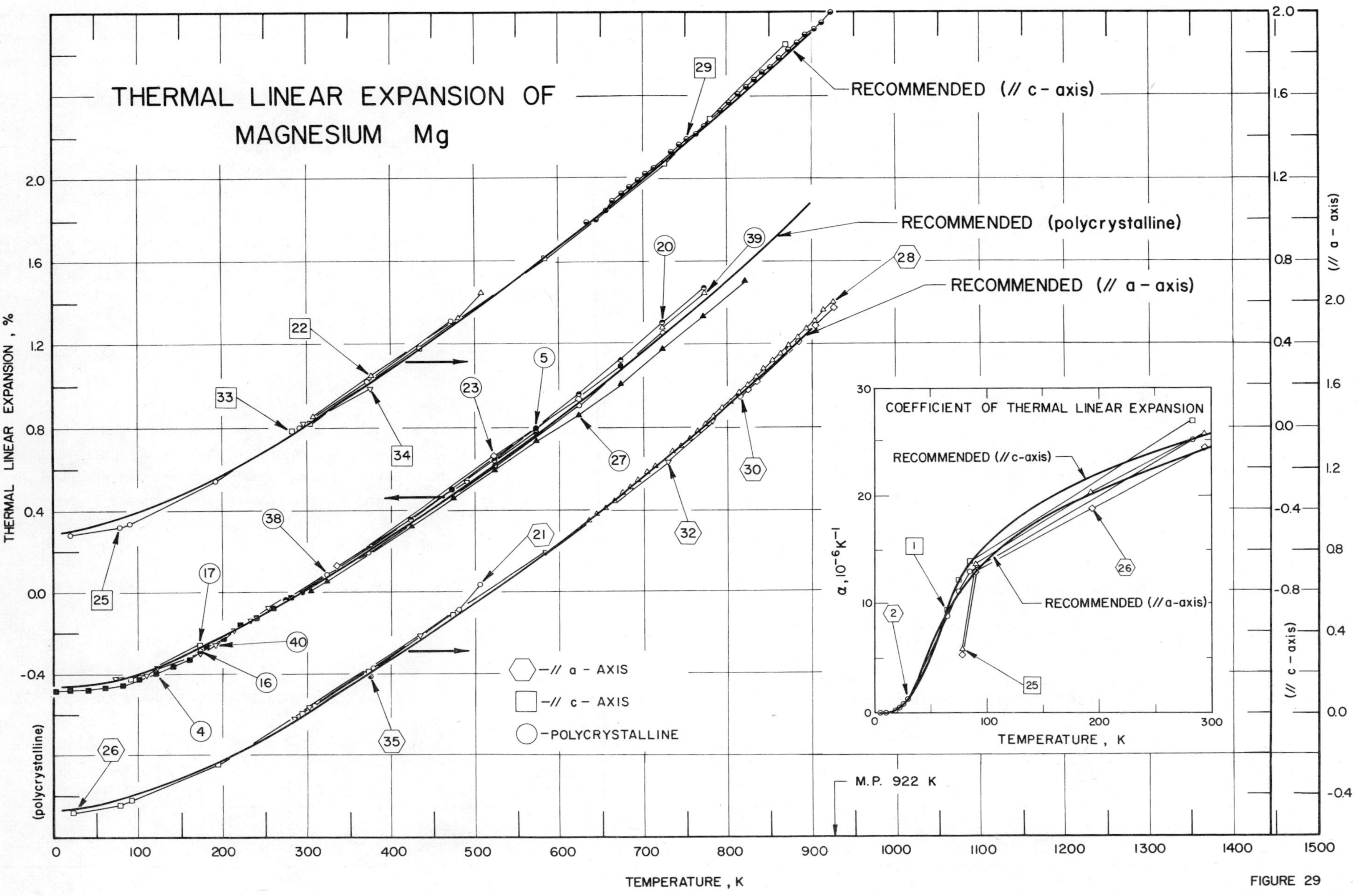
THERMAL LINEAR EXPANSION OF
MAGNESIUM Mg
RECOMMENDED (// c - axis)
RECOMMENDED (polycrystalline)
RECOMMENDED (// a - axis)
-// a - AXIS
-// c - AXIS
-POLYCRYSTALLINE
M.P. 922 K
TEMPERATURE , K
THERMAL LINEAR EXPANSION , %
(polycrystalline)
(// a - axis)
(// c - axis)
COEFFICIENT OF THERMAL LINEAR EXPANSION
RECOMMENDED (//c-axis)
RECOMMENDED (//a-axis)
$\alpha, 10^{-6} K^{-1}$
TEMPERATURE , K

FIGURE 29

SPECIFICATION TABLE 29. THERMAL LINEAR EXPANSION OF MAGNESIUM Mg

Cur. No.	Ref. No.	Author(s)	Year	Method Used	Temp. Range, K	Name and Specimen Designation	Composition (weight percent), Specifications, and Remarks
1*	104	McCammon, R. D. and White, G. K.	1965	E	5-85, 283	1	Specimen grown from 99.98^{+} Mg; cylindrical crystal 5 cm long, 2 cm diameter oriented perpendicular to hexad axis; crystal supplied by Metals Research Ltd., Cambridge, U. K.; cylinder spark machined, ends fine sparked; this curve is here reported using the first given temperature as reference temperature at which $\Delta L/L_0 = 0$.
2*	104	McCammon, R. D. and White, G. K.	1965	E	5-85, 283	2	Similar to the above specimen except oriented parallel to hexad axis; this curve is here reported using the first given temperature as reference temperature at which $\Delta L/L_0 = 0$.
3*	104	McCammon, R. D. and White, G. K.	1965	E	1.7-10	3	Specimen No. 2 shortened 0.2 cm and relapped to determine effect of surface damage; this curve is here reported using the first given temperature as reference temperature at which $\Delta L/L_0 = 0$.
4	83	Laquer, H. L.	1952	L	0-300		Specimen turned from stock rod to 2.54 cm long; zero-point correction of -0.052% determined by graphical interpolation.
5	199	Bell, I. P.	1954	L	293-673		Pure specimen.
6*	200	Hidnert, P. and Sweeney, W. T.	1928	T	293-573	Sample 1269	99.99 Mg, 0.005 Fe, 0.002 Cu; specimen cast in vacuum furnace at 938 K.
7*	200	Hidnert, P. and Sweeney, W. T.	1928	T	173-573	Sample 1269	The above specimen, second test.
8*	200	Hidnert, P. and Sweeney, W. T.	1928	L	90, 293	Sample 1269	The above specimen, third test after heating twice to 773 K.
9*	200	Hidnert, P. and Sweeney, W. T.	1928	L	90-293	Sample 1269	The above specimen, fourth test.
10*	200	Hidnert, P. and Sweeney, W. T.	1928	I	293-773	Sample 1269I	Similar to the above specimen.
11*	200	Hidnert, P. and Sweeney, W. T.	1928	T	293-773	Sample 1269A	Similar to the above specimen; expansion measured in air.
12*	200	Hidnert, P. and Sweeney, W. T.	1928	T	293-773	Sample 1269A	The above specimen, second test.
13*	200	Hidnert, P. and Sweeney, W. T.	1928	T	293-773	Sample 1268A	99.996 Mg, 0.002 Si, < 0.001 Fe, Cu; specimen extruded at 683 K; expansion measured in air.
14*	200	Hidnert, P. and Sweeney, W. T.	1928	T	293-773	Sample 1268A	The above specimen, second test.
15*	200	Hidnert, P. and Sweeney, W. T.	1928	T	293-573	Sample 1270	Similar to the above specimen.
16	200	Hidnert, P. and Sweeney, W. T.	1928	T	173-573	Sample 1270	The above specimen, second test.
17	200	Hidnert, P. and Sweeney, W. T.	1928	L	90-293	Sample 1270	The above specimen; tested after heating to 773 K.
18*	200	Hidnert, P. and Sweeney, W. T.	1928	T	293-773	Sample 1270A	Similar to the above specimen; expansion measured in air.
19*	200	Hidnert, P. and Sweeney, W. T.	1928	T	293-773	Sample 1270A	The above specimen, second test.
20	201	Takahasi, K. and Kikuti, R.	1936		273-773		Rod specimen, about 16 cm long, 5 mm diameter, heating rate 2 C/min.
21	202	Hanawalt, J. P. and Frevel, L. K.	1938	X	309-506		Thin disc specimen; expansion measured in a-axis.
22	202	Hanawalt, J. P. and Frevel, L. K.	1938	X	309-506		The above specimen; expansion measured in c-axis.
23	35	Honda, K. and Okubo, Y.	1924	L	334-621		Pure Mg, cylindrical rod specimen, 14.940 cm long, 5 mm thick.
24	255	Bollenrath, Fr.	1934	L	293-673		99.96 Mg, 0.034 Fe, 0.004 Mn, 0.003 each Cu, Si.
25	262	Goens, E. and Schmid, E.	1936	X	21-473		99.95 Mg crystal specimen; expansion measured parallel to crystallographic axis.

* Not shown in figure.

SPECIFICATION TABLE 29. THERMAL LINEAR EXPANSION OF MAGNESIUM Mg (continued)

Cur. No.	Ref. No.	Author(s)	Year	Method Used	Temp. Range, K	Name and Specimen Designation	Composition (weight percent), Specifications, and Remarks
26	262	Goens, E. and Schmid, E.	1936	X	21-473		The above specimen; expansion measured perpendicular to axis.
27	263	Grube, G. and Vosskühler, H.	1934	L	303-823		0.052 Fe, 0.018 Si, and traces of Cu and Al.
28	264	Janot, C., Mallejac, D. and George, B.	1970	L	633-924		99.99$^+$ Mg; hexagonal specimen measured perpendicular to principal axis.
29	264	Janot, C., et al.	1970	L	633-924		The above specimen; expansion measured parallel to principal axis.
30	264	Janot, C., et al.	1970	X	633-924		The above specimen; microscopic expansion measured perpendicular to axis.
31*	264	Janot, C., et al.	1970	X	633-924		The above specimen; expansion measured parallel to axis.
32	392	Raynor, G. V. and Hume-Rothery, W.	1939	X	283-870		99.9 Mg starting material in form of rod, turned to < 0.6 cm diameter; annealed 20 hr at 570 K and recrystallized during annealing; expansion measured along a-axis; reported probable error in measurement of lattice constant ± 0.0002 Å; lattice parameter reported at 283 K is 3.2017 Å; 3.2025 Å at 293 K determined by graphical interpolation.
33	392	Raynor, G. V. and Hume-Rothery, W.	1939	X	283-870		The above specimen; expansion measured along c-axis; lattice parameter reported at 283 K is 5.1986 Å; 5.1999 Å at 293 K determined by graphical interpolation.
34	715	Shinoda, G.	1934	X	297, 373		Expansion measured along c-axis.
35	715	Shinoda, G.	1934	X	297, 373		Expansion measured perpendicular to c-axis.
36*	715	Shinoda, G.	1934	X	297, 373		Similar to the above specimen except polycrystalline.
37*	714	Andres, K.	1964	O	2.2-14		99.99 pure, polycrystalline specimen; this curve is here reported using the first given temperature as reference temperature at which $\Delta L/L_0 = 0$.
38	719	Austin, J. B.	1932	I	273-773		99.99 pure polycrystalline specimen cut from 0.25 in. extruded rod, turned down to 0.125 in. diameter and 4 mm length; zero-point correction of 0.048% determined by graphical interpolation, data on the coefficients of thermal linear expansion also reported.
39	719	Austin, J. B.	1932	I	273-773		Similar to the above specimen; measured after first heating; zero-point correction of 0.052% determined by graphical interpolation, data on the coefficients of thermal linear expansion also reported.
40	720	Schuerch, H. U.	1972	L	73-490		No specification given.

* Not shown in figure.

DATA TABLE 29. THERMAL LINEAR EXPANSION OF MAGNESIUM Mg

[Temperature, T, K; Linear Expansion, $\Delta L/L_0$, %]

T	$\Delta L/L_0$
CURVE 1*, †, ‡	
5	0.0000 x 10^{-3}
6	0.0009
7	0.0020
8	0.0036
10	0.0086
12	0.0170
14	0.0306
16	0.0522
18	0.0854
20	0.1364
22	0.2114
24	0.3154
26	0.4564
28	0.6424
30	0.8824
65	19.8699
75	30.5699
85	43.5199
CURVE 2*, †, ‡	
5	0.0000 x 10^{-3}
6	0.0013
7	0.0032
8	0.0056
10	0.0134
12	0.0267
14	0.0476
16	0.0782
18	0.1222
20	0.1842
22	0.2692
24	0.3832
26	0.5332
28	0.7272
30	0.9712
65	19.4337
75	29.6337
85	41.7837
CURVE 3*, †	
1.7	0.0000 x 10^{-4}
2.8	0.0070
3.0	0.0021
3.3	0.0044
CURVE 3 (cont.)*, †	
3.6	0.0057
4.3	0.0082
5.5	0.0170
5.9	0.0211
7.1	0.0344
7.7	0.0421
8.4	0.0560
9.3	0.0769
10.1	0.0987
CURVE 4	
0	-0.487
20	-0.486
40	-0.482
60	-0.475
80	-0.458
100	-0.432
120	-0.400
140	-0.363
160	-0.323
180	-0.279
200	-0.233
220	-0.184
240	-0.134
260	-0.084
273.2	-0.052*
280	-0.034
300	0.016*
CURVE 5	
293	0.000
373	0.219
473	0.502
573	0.792
673	1.094
CURVE 6*	
293	0.000
373	0.209
473	0.486
573	0.787
CURVE 7*	
173	-0.299
293	0.000
373	0.208
473	0.486
573	0.781
CURVE 8*	
90	-0.445
293	0.000
CURVE 9*	
90	-0.432
173	-0.277
293	0.000
CURVE 10*	
293	0.000
473	0.490
573	0.784
673	1.094
773	1.430
CURVE 11*	
293	0.000
373	0.213
473	0.486
573	0.787
673	1.094
773	1.421
CURVE 12*	
293	0.000
373	0.208
473	0.486
573	0.781
673	1.102
773	1.445
CURVE 13*	
293	0.000
CURVE 13 (cont.)*	
373	0.211
473	0.481
573	0.770
673	1.087
773	1.421
CURVE 14*	
293	0.000
373	0.210
473	0.484
573	0.781
673	1.098
773	1.435
CURVE 15*	
293	0.000
373	0.208
473	0.484
573	0.778
CURVE 16	
173	-0.282
293	0.000*
373	0.208*
473	0.488*
573	0.781
CURVE 17	
90	-0.432
173	-0.278
293	0.000*
CURVE 18*	
293	0.000
373	0.210
473	0.481
573	0.767
673	1.064
773	1.392
CURVE 19*	
293	0.000
373	0.211
473	0.486
573	0.778
673	1.098
773	1.435
CURVE 20	
273	-0.0508
323	0.0763*
373	0.2083*
423	0.3446
473	0.4877*
523	0.6352
573	0.7891*
623	0.9484
673	1.1127
723	1.2830
773	1.4600
CURVE 21	
309	0.043
377	0.227
480	0.499
506	0.634
CURVE 22	
309	0.046
377	0.244
480	0.533
506	0.658
CURVE 23‡	
334	0.112
373	0.219*
423	0.359*
471	0.494*
523	0.643
572	0.787*
621	0.933
CURVE 24*	
293	0.0000
323	0.0731
373	0.1990
423	0.3310
473	0.4691
523	0.6127
573	0.7616
623	0.9164
673	1.075
CURVE 25	
21	-0.517
78	-0.483
90	-0.466
193	-0.257
293	0.000
373	0.220
473	0.511
CURVE 26	
21	-0.483
78	-0.451
90	-0.436
193	-0.243
293	0.000
373	0.209
473	0.487
CURVE 27	
303	0.0251
323	0.0501
373	0.2004*
423	0.3296
473	0.4621
523	0.5914
573	0.7233
623	0.8551
673	1.0000
723	1.1610
773	1.3193
823	1.4802

* Not shown in figure.
† This curve is here reported using the first given temperature as reference temperature at which $\Delta L/L_0 = 0$.
‡ Author's data for coefficient of thermal expansion have been integrated by TPRC to obtain $\Delta L/L_0$.

DATA TABLE 29. THERMAL LINEAR EXPANSION OF MAGNESIUM Mg (continued)

T	$\Delta L/L_0$
CURVE 28	
633	0.9511
643	0.9829
653	1.0150
663	1.0473
673	1.0799
683	1.1127
693	1.1457
703	1.1790
713	1.2126
723	1.2465
733	1.2806
743	1.3150
753	1.3498
763	1.3849
773	1.4203
783	1.4561
793	1.4922
803	1.5287
813	1.5657
823	1.6030
833	1.6408
843	1.6790
853	1.7176
863	1.7569
873	1.7965
883	1.8367
893	1.8774
903	1.9187
913	1.9605
923	2.0030*
924	2.0073
CURVE 29	
633	0.9915
643	1.0239
653	1.0563
663	1.0888
673	1.1216
683	1.1545
693	1.1872
703	1.2209
713	1.2545
723	1.2882*
733	1.3221
743	1.3562
753	1.3907
763	1.4253
CURVE 29 (cont.)	
773	1.4600
783	1.4952*
793	1.5305
803	1.5661
813	1.6020
823	1.6383
833	1.6747
843	1.7115
853	1.7486
863	1.7866
873	1.8238
883	1.8620
893	1.9005
903	1.9394
913	1.9787
923	2.0184*
924	2.0224
CURVE 30	
633	0.9510*
643	0.9825*
653	1.0142*
663	1.0462*
673	1.0783*
683	1.1105*
693	1.1432*
703	1.1759*
713	1.2088*
723	1.2419*
733	1.2754*
743	1.3091*
753	1.3431*
763	1.3773*
773	1.4118*
783	1.4466*
793	1.4817*
803	1.5172*
813	1.5531*
823	1.5892
833	1.6257
843	1.6627
853	1.7000
863	1.7379
873	1.7761
883	1.8148
893	1.8539
903	1.8936
CURVE 30 (cont.)	
913	1.9339
923	1.9747*
924	1.9789
CURVE 31*	
633	0.9911
643	1.0234
653	1.0558
663	1.0883
673	1.1211
683	1.1540
693	1.1872
703	1.2204
713	1.2540
723	1.2876
733	1.3214
743	1.3555
753	1.3896
763	1.4240
773	1.4585
783	1.4934
793	1.5283
803	1.5635
813	1.5987
823	1.6343
833	1.6701
843	1.7060
853	1.7422
863	1.7786
873	1.8152
883	1.8520
893	1.8890
903	1.9263
913	1.9638
923	2.0015
924	2.0053
CURVE 32	
283	-0.026
300.9	0.020
303	0.030*
304.6	0.033*
433	0.379
583.7	0.789
728	1.232
780.0	1.435
870	1.791*
CURVE 33	
283	-0.026*
300.9	0.020
303	0.030*
304.6	0.034*
433	0.391
583.7	0.814
728	1.276
780.0	1.485
870	1.847
CURVE 34	
296.7	0.008
373	0.190
CURVE 35	
296.7	0.008*
373	0.188
CURVE 36*	
296.7	0.008
373	0.189
CURVE 37*,†	
2.2	0.00 x 10^{-3}
2.2	0.00
2.7	0.1
2.7	0.1
2.7	0.1
3.3	0.3
3.3	0.3
3.3	0.3
3.4	0.5
3.4	0.5
3.9	0.8
4.2	1.0
4.2	1.0
4.6	1.4
4.6	1.5
5.2	2.1
6.0	3.1
6.3	3.5
6.3	3.6
6.7	4.4
7.2	5.4
CURVE 37 (cont.)*,†	
7.5	6.1 x 10^{-3}
7.8	7.1
8.1	8.0
8.5	9.2
8.8	10.7
9.3	12.3
9.3	12.8
9.8	14.5
10.3	17.5
11.0	21.5
11.0	21.9
11.4	24.8
11.7	27.1
12.1	31.3
12.5	34.7
13.5	45.5
CURVE 38	
273	-0.048*
293	0.000*
323	0.072
373	0.197
423	0.326*
473	0.462*
523	0.604
573	0.753*
623	0.912
673	1.077*
723	1.257
773	1.442*
CURVE 39	
273	-0.052*
293	0.000*
323	0.078*
373	0.208*
423	0.342*
473	0.483*
523	0.683*
573	0.785*
623	0.943*
673	1.101*
723	1.263
773	1.438
CURVE 40	
73	-0.438
123	-0.375
173	-0.300
193	-0.237
213	-0.192
233	-0.146
253	-0.098
273	-0.049*
293	0.000*
323	0.075*
373	0.206*
423	0.341*
473	0.484*
490	0.533

* Not shown in figure.

† This curve is here reported using the first given temperature as reference temperature at which $\Delta L/L_0 = 0$.

DATA TABLE 29. COEFFICIENT OF THERMAL LINEAR EXPANSION OF MAGNESIUM Mg

[Temperature, T, K; Coefficient of Expansion, α, 10^{-6} K^{-1}]

T	α
CURVE 1	
5	0.075*
6	0.010*
7	0.013*
8	0.018*
10	0.032*
12	0.052*
14	0.084*
16	0.132*
18	0.20*
20	0.31*
22	0.44*
24	0.60*
26	0.81*
28	1.05*
30	1.35*
65	9.50
75	11.90
85	14.00
283	26.80
CURVE 2	
5	0.011
6	0.016*
7	0.021*
8	0.028*
10	0.050
12	0.083*
14	0.126*
16	0.18*
18	0.26
20	0.36*
22	0.49
24	0.65*
26	0.85
28	1.09*
30	1.35
65	9.20
75	11.20
85	13.10
283	25.10
CURVE 23*, ‡	
334	27.0
373	27.8
CURVE 23 (cont.)*, ‡	
423	28.0
471	28.4
523	28.9
572	29.7
621	29.9
CURVE 25	
78	5.8
90	13.8
193	20.3
293	25.7
373	27.5*
473	29.1*
CURVE 26	
78	5.3
90	13.0
193	18.8
293	24.3
373	26.1*
473	27.8*
CURVE 38	
293	24.5
323	25.4
373	26.5
423	27.8
473	29.1
523	30.8
573	32.4
623	34.5
673	56.5
723	38.6
CURVE 39	
293	25.0
323	26.0
373	26.9
423	27.8
473	28.8
523	29.8
573	30.7
CURVE 39 (cont.)	
623	31.6
673	32.6
723	33.5
773	34.5

* Not shown in figure.
‡ Author's data for coefficient of thermal expansion have been integrated by TPRC to obtain $\Delta L/L_0$.

FIGURE AND TABLE NO. 30R. RECOMMENDED VALUES FOR THERMAL LINEAR EXPANSION OF MANGANESE Mn

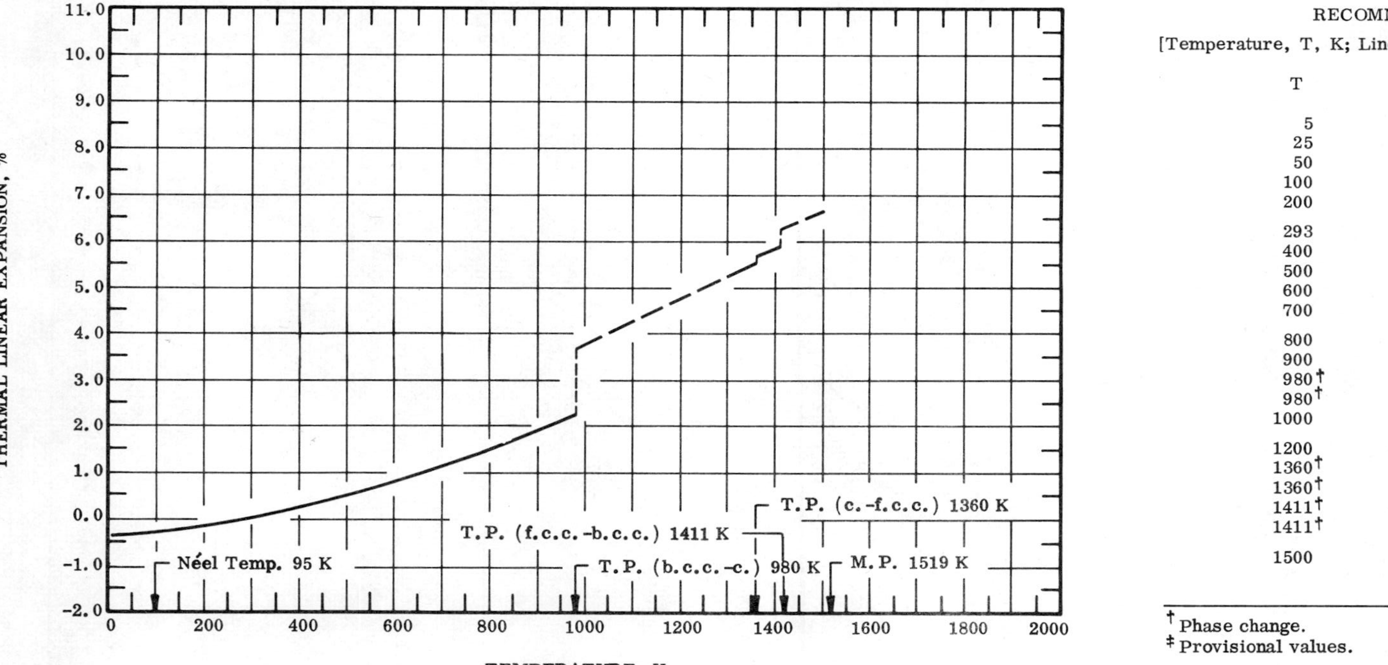

RECOMMENDED VALUES

[Temperature, T, K; Linear Expansion, $\Delta L/L_0$, %; α, K^{-1}]

T	$\Delta L/L_0$	$\alpha \times 10^6$
5	-0.354	-0.15
25	-0.355	-0.70
50	-0.357	-0.80
100	-0.342	10.2
200	-0.188	18.5
293	0.000	21.7
400	0.247	24.5
500	0.502	27.0
600	0.791	30.3
700	1.116	34.5
800	1.448	39.5
900	1.912	45.5
980†	2.294	50.5
980†	3.676‡	
1000	3.772‡	
1200	4.738‡	
1360†	5.510‡	
1360†	5.635‡	
1411†	5.892‡	
1411†	6.214‡	
1500	6.604‡	

† Phase change.
‡ Provisional values.

REMARKS

The tabulated values are considered accurate to within ±3% at temperatures below 900 K and ±7% above. These values can be represented approximately by the following equation:

$$\Delta L/L_0 = -0.582 + 1.897 \times 10^{-3}\, T - 1.315 \times 10^{-8}\, T^2 + 1.090 \times 10^{-9}\, T^3 \qquad (200 < T < 980)$$

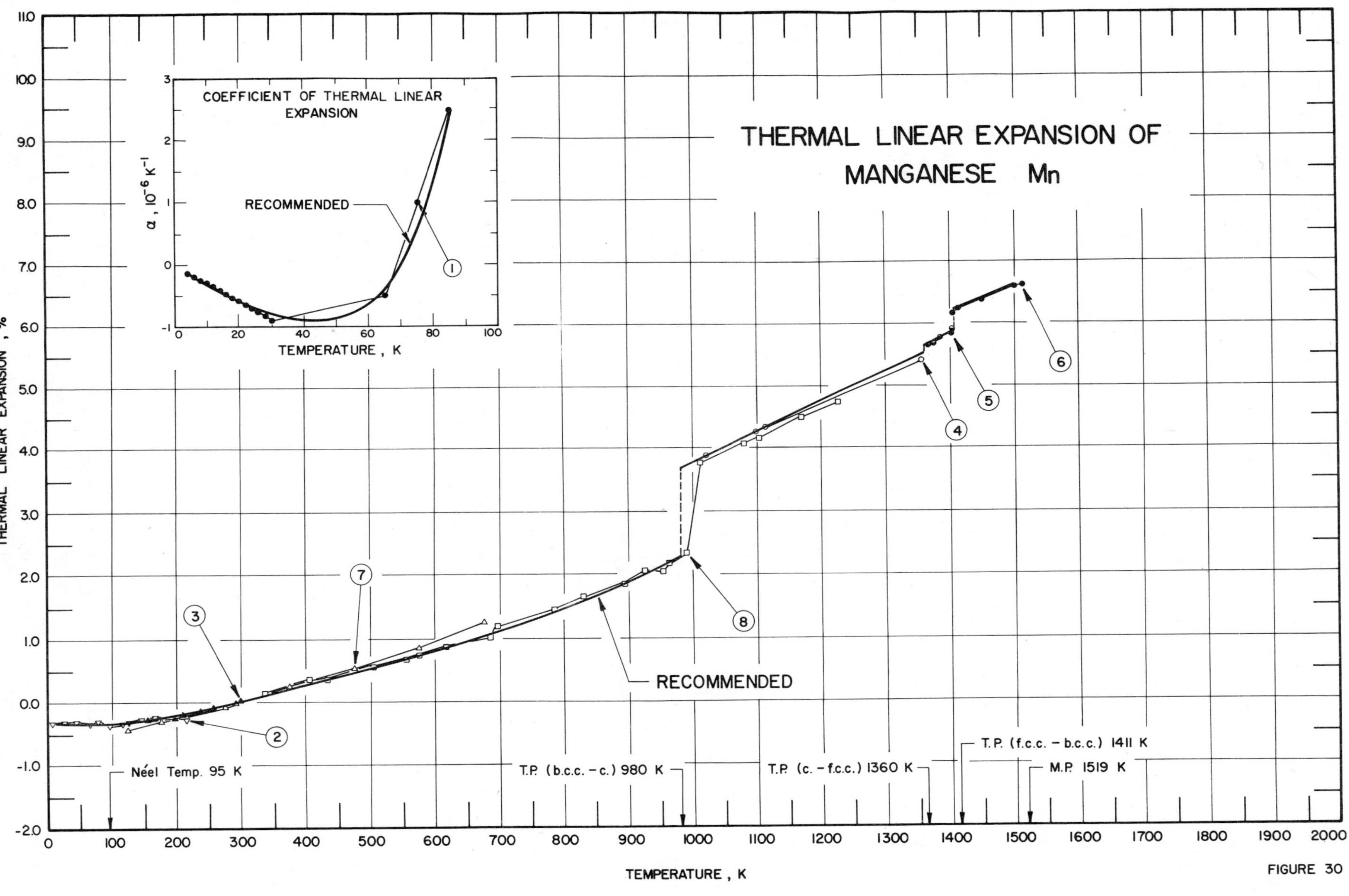
THERMAL LINEAR EXPANSION OF
MANGANESE Mn
THERMAL LINEAR EXPANSION, %
TEMPERATURE, K
RECOMMENDED
Néel Temp. 95 K
T.P. (b.c.c. – c.) 980 K
T.P. (c. – f.c.c.) 1360 K
T.P. (f.c.c. – b.c.c.) 1411 K
M.P. 1519 K
COEFFICIENT OF THERMAL LINEAR EXPANSION
RECOMMENDED
α, 10⁻⁶ K⁻¹
TEMPERATURE, K

FIGURE 30

SPECIFICATION TABLE 30. THERMAL LINEAR EXPANSION OF MANGANESE Mn

Cur. No.	Ref. No.	Author(s)	Year	Method Used	Temp. Range, K	Name and Specimen Designation	Composition (weight percent), Specifications, and Remarks
1*	41	White, G.K.	1965	E	4-85	α-Mn	Prepared by melting 99.9$^+$ manganese in vacuo and chill-casting; specimen 6 cm long and 2 cm diameter; bulk density ~7.5 g cm^{-3}; measured as received condition; this curve is here reported using the first given temperature as reference temperature at which $\Delta L/L_0 = 0$.
2	203	Marples, J.A.C.	1967	X	5-299	α-Mn	0.0005 Mg, 0.0025 Si, and 0.0001 Cu; plate ground flat and etched with hydrochloric acid; lattice parameter reported at 299 K is 8.9104 Å; 8.9094 Å at 293 K determined by graphical interpolation.
3	204	Gazzara, C.P., Middleton, R.M., Weiss, R.J. and Halls, E.O.	1967	X	5-299		Specimen (1) pure, carbon free, electrolytic ground to -400 mesh powder; Specimen (2) electrolytic plate, ground with mortar and pestle, and sifted to -400 mesh powder; Specimen (3) sample 2 annealed at 710 K, 90 hr in a vacuum; Specimen (4) Johnson-Matthey H.S. Manganese flake, lightly pickled in hydrochloric acid to remove surface oxide, washed, dried, and crushed; all specimens contained as impurity <0.6 oxygen; author averaged data from above four specimens; lattice parameter reported at 299 K is 8.9121 Å; 8.9107 Å at 293 K determined by graphical interpolation.
4*	205	Basinski, Z.S. and Christian, J.W.	1954	X	293-1353	β-Mn	Complex cubic structure (20 atoms per unit cell); cylindrical rod specimen about 1 mm in diameter, 1 m in length obtained by cutting rectangular rod from selected flat of α-Mn with glass knife, carefully grinding in jig, finally etching away in dilute nitric acid; expansion measured under several atmospheres of purified hydrogen. Lattice parameter referred to α-phase by multiplication by $(n_\alpha/n_\beta)^{1/3}$, where n_α is number of atoms per unit cell in α-phase (58) and n_β is number of atoms per unit cell in β-phase; α-phase lattice parameter at 293 K taken as 8.9139 Å.
5*	205	Basinski, Z.S. and Christian, J.W.	1954	X	1368-1405	γ-Mn	Similar to the above specimen, f.c.c. structure (4 atoms per unit cell); lattice parameter referred to α-phase by multiplication by $(n_\alpha/n_\gamma)^{1/3}$, where n_α is number of atoms per unit cell in α-phase (58) and n_γ is number of atoms per unit cell in γ-phase α-phase lattice parameter at 293 K taken as 8.9139 Å.
6	205	Basinski, Z.S. and Christian, J.W.	1954	X	1406-1512	δ-Mn	Similar to the above specimen, b.c.c. structure (2 atoms per unit cell); lattice parameter referred to α-phase by multiplication by $(n_\alpha/n_\delta)^{1/3}$, where n_α is number of atoms per unit cell in α-phase (58) and n_δ is number of atoms per unit cell in δ-phase; α-phase lattice parameter at 293 K taken as 8.9139 Å.
7	663, 664	Masumoto, H., Sawaya, S., and Kikuchi, M.	1971, 1972	L	115-673		Electrolytic rod specimen of 2 mm diameter and 11 cm length with 0.007 Fe, 0.008 Si, 0.007 C, and 0.030 S; average of heating and cooling experiments; zero-point correction is -0.423%.
8	776	Schmitz-Pranghe, N., and Dunner, P.	1968	X	85-1221		99.87 Mn, <0.0001 Cu, 0.0005 Mg, Si, 0.009 C, 0.02 H_2, 0.0025 N_2, and 0.1 O_2 polycrystalline specimen; discs of 7-8 mm diameter, and 1/2 - 2 mm thickness; average of heating and cooling; below 1273 K, measurements in vacuum and above that temperature, measurements in Helium atmosphere, lattice parameter reported at 295 K is 8.910 Å.

* Not shown in figure.

DATA TABLE 30. THERMAL LINEAR EXPANSION OF MANGANESE Mn

[Temperature, T, K; Linear Expansion, $\Delta L/L_0$, %]

T	$\Delta L/L_0$
CURVE 1*, ‡	
4	0.000 x 10^{-3}
5	-0.013
6	-0.028
8	-0.068
10	-0.119
12	-0.181
14	-0.255
16	-0.341
18	-0.439
20	-0.549
22	-0.671
24	-0.805
26	-0.951
28	-1.11
30	-1.28
65	-3.73
75	-3.48
85	-1.73
CURVE 2	
5	-0.320
26	-0.321
45	-0.323
61	-0.330*
63	-0.324*
77	-0.329
86	-0.330*
94	-0.337*
97	-0.336
100	-0.331*
113	-0.315*
123	-0.304
144	-0.273
164	-0.231
196	-0.173
222	-0.128
299	0.011
CURVE 3	
5	-0.33
42	-0.34
102	-0.34
109	-0.34*
CURVE 3 (cont.)	
116	-0.34
151	-0.29
174	-0.25
191	-0.23
207	-0.20
235	-0.14
255	-0.099
273	-0.052
299	0.016
CURVE 4	
293	1.025
298	1.047
1023	3.862
1102	4.245
1110	4.280
1119	4.319
1353	5.416
CURVE 5*, †	
1368	5.659
1372	5.666
1388	5.769
1402	5.801
1405	5.859
CURVE 6*, †	
1406	6.165
1408	6.199*
1410	6.169*
1414	6.204
1422	6.224*
1452	6.384
1499	6.596
1512	6.612
CURVE 7	
115	-0.423
173	-0.299
273	-0.056
293	0.000
CURVE 7 (cont.)	
373	0.215
473	0.534
573	0.895
673	1.281
CURVE 8	
85	-0.438
111	-0.404
134	-0.337
165	-0.269
183	-0.236
216	-0.168
237	-0.101
284	-0.011
295	0.000
339	0.123
371	0.213*
406	0.337
434	0.370
505	0.572
553	0.685
573	0.752
617	0.887
686	1.021
696	1.201
785	1.448
830	1.683
895	1.874
923	2.065
952	2.076
961	2.188
990	2.334
1008	3.744
1016	3.776
1081	4.064
1108	4.160
1169	4.480
1221	4.753

* Not shown in figure.

† This curve is here reported using the first given temperature as reference temperature at which $\Delta L/L_0 = 0$.

‡ Author's data for coefficient of thermal expansion have been integrated by TPRC to obtain $\Delta L/L_0$.

DATA TABLE 30. COEFFICIENT OF THERMAL LINEAR EXPANSION OF MANGANESE Mn

[Temperature, T, K; Coefficient of Expansion, α, 10^{-6} K^{-1}]

T	α
CURVE 1‡	
4	-1.1
5	-1.4
6	-1.7
8	-2.3
10	-2.8
12	-3.4
14	-4.0
16	-4.6
18	-5.2
20	-5.8
22	-6.4
24	-7.0
26	-7.6
28	-8.2
30	-9.0
65	-5.0
75	10.0
85	25.0

‡ Author's data for coefficient of thermal expansion have been integrated by TPRC to obtain $\Delta L/L_0$.

FIGURE AND TABLE NO. 31R. PROVISIONAL VALUES FOR THERMAL LINEAR EXPANSION OF MERCURY Hg

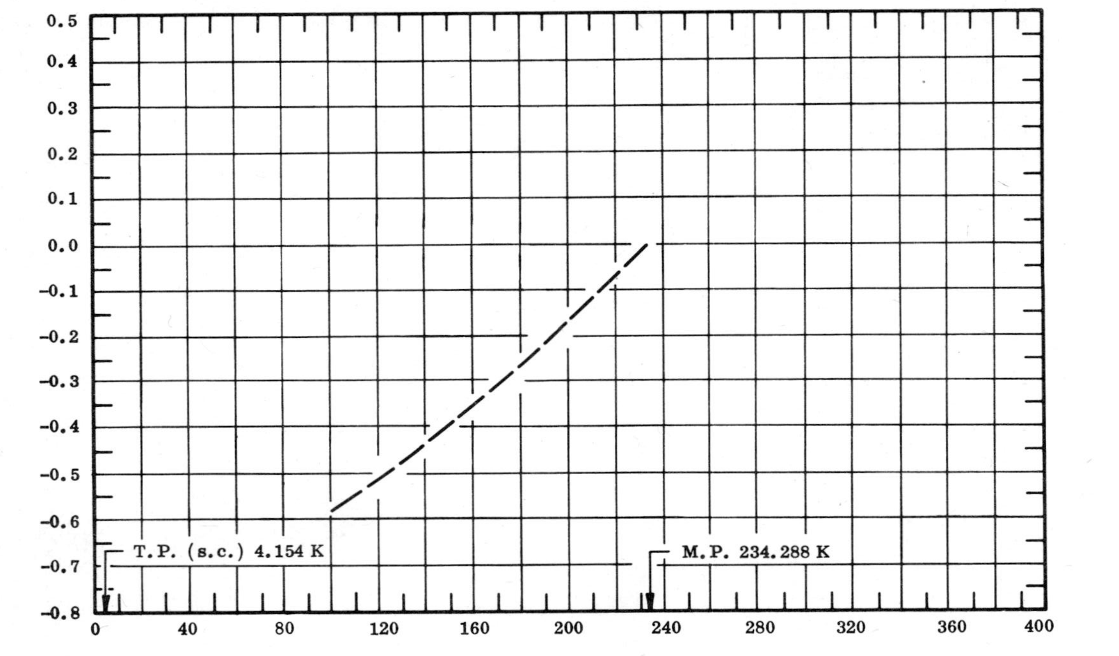

PROVISIONAL VALUES

[Temperature, T, K; Linear Expansion, $\Delta L/L_0$, %; α, K^{-1}]

T	$\Delta L/L_0$	$\alpha \times 10^6$
100	-0.587	34
150	-0.397	42
200	-0.170	49
234	0.000	52

REMARKS

The tabulated values are considered accurate to within ± 10% over the entire temperature range. Because the freezing point of mercury is below room temperature, 234 K, the thermal linear expansion data are referred to 234 K as reference temperature at which $\Delta L/L_0 = 0$. These values can be represented approximately by the following equation:

$$\Delta L/L_0 = -0.858 + 1.875 \times 10^{-3}\, T + 8.626 \times 10^{-6}\, T^2 - 4.197 \times 10^{-9}\, T^3$$

SPECIFICATION TABLE 31. THERMAL LINEAR EXPANSION OF MERCURY Hg

Cur. No.	Ref. No.	Author(s)	Year	Method Used	Temp. Range, K	Name and Specimen Designation	Composition (weight percent), Specifications, and Remarks
1*	206	Hill, D. M.	1935	L	83-158		Single crystal specimen grown slowly in mold; expansion measured parallel to trigonal axis; this curve is here reported using the first given temperature as reference temperature at which $\Delta L/L_0 = 0$.
2*	206	Hill, D. M.	1935	L	83-158		Similar to the above specimen; expansion measured perpendicular to trigonal axis; this curve is here reported using the first given temperature as reference temperature at which $\Delta L/L_0 = 0$.

DATA TABLE THERMAL LINEAR EXPANSION OF MERCURY Hg

[Temperature, T, K; Linear Expansion, $\Delta L/L_0$, %]

T	$\Delta L/L_0$	T	$\Delta L/L_0$	T	$\Delta L/L_0$	T	$\Delta L/L_0$
CURVE 1*, †, ‡		CURVE 1 (cont.)*, †, ‡		CURVE 2*, †, ‡		CURVE 2 (cont.)*, †, ‡	
83	0.000	138	0.246	83	0.000	138	0.192
113	0.131	143	0.270	113	0.103	143	0.211
118	0.154	148	0.295	118	0.121	148	0.229
123	0.177	153	0.319	123	0.138	153	0.248
128	0.200	158	0.344	128	0.156	158	0.267
133	0.223			133	0.174		

DATA TABLE COEFFICIENT OF THERMAL LINEAR EXPANSION OF MERCURY Hg

[Temperature, T, K; Coefficient of Expansion, α, $10^{-6}\ K^{-1}$]

T	α	T	α	T	α	T	α
CURVE 1‡		CURVE 1 (cont.)‡		CURVE 2‡		CURVE 2 (cont.)‡	
83	42.6	138	47.5	83	33.4	138	36.8
113	44.9	143	48.0	113	35.2	143	36.9
118	45.5	148	48.8	118	35.4	148	37.2
123	46.0	153	49.6	123	35.5	153	37.5
128	46.4	158	49.8	128	35.6	158	37.7
133	46.6			133	36.0		

* Not shown in figure.
† This curve is here reported using the first given temperature as reference temperature at which $\Delta L/L_0 = 0$.
‡ Author's data for coefficient of thermal expansion have been integrated by TPRC to obtain $\Delta L/L_0$.

FIGURE AND TABLE NO. 32R. RECOMMENDED VALUES FOR THERMAL LINEAR EXPANSION OF MOLYBDENUM Mo

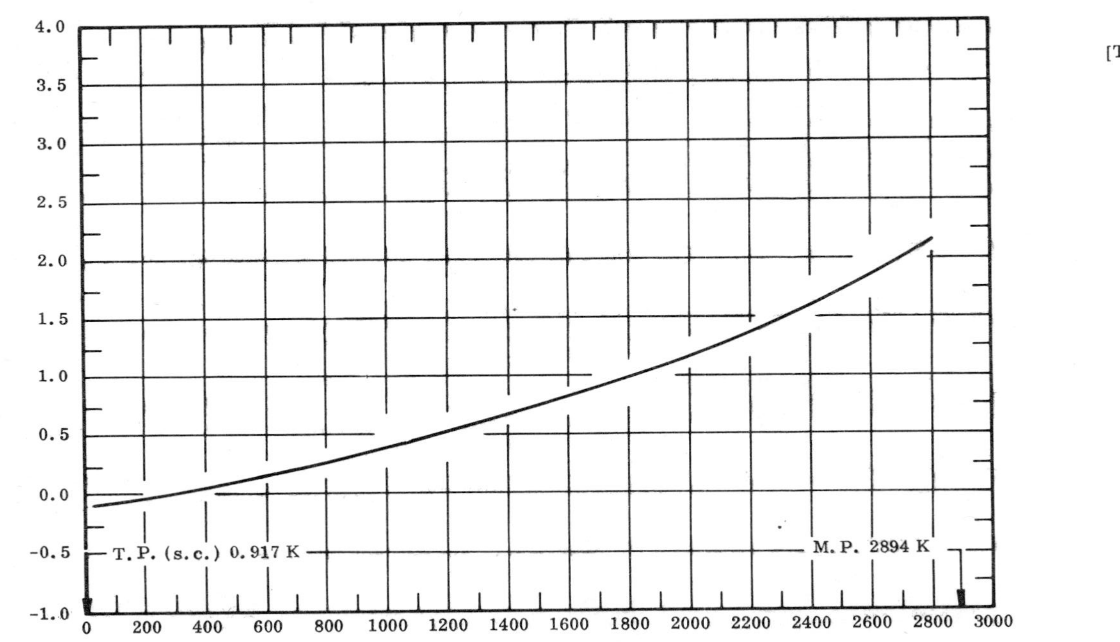

RECOMMENDED VALUES

[Temperature, T, K; Linear Expansion, $\Delta L/L_0$, %; α, K^{-1}]

T	$\Delta L/L_0$	$\alpha \times 10^6$
20	-0.092	0.3
25	-0.092	0.4
50	-0.091	1.0
100	-0.083	2.8
200	-0.045	4.6
293	0.000	4.8
400	0.052	4.9
500	0.103	5.1
600	0.155	5.3
700	0.209	5.5
800	0.265	5.7
900	0.324	6.0
1000	0.385	6.2
1200	0.514	6.7
1400	0.653	7.2
1600	0.802	7.8
1800	0.964	8.4
2000	1.145	9.6
2200	1.347	10.6
2400	1.577	12.4
2600	1.842	14.2
2800	2.150	16.5

REMARKS

The tabulated values are considered accurate to within ± 3% at temperatures below 1500 K and ± 5% above. These values can be represented approximately by the following equations:

$$\Delta L/L_0 = 4.697 \times 10^{-4} (T - 293) + 9.756 \times 10^{-8} (T - 293)^2 + 9.403 \times 10^{-12} (T - 293)^3 \quad (293 < T < 1545)$$

$$\Delta L/L_0 = 0.760 + 7.583 \times 10^{-4} (T - 1545) + 1.329 \times 10^{-7} (T - 1545)^2 + 1.149 \times 10^{-10} (T - 1545)^3 \quad (1545 < T < 2800)$$

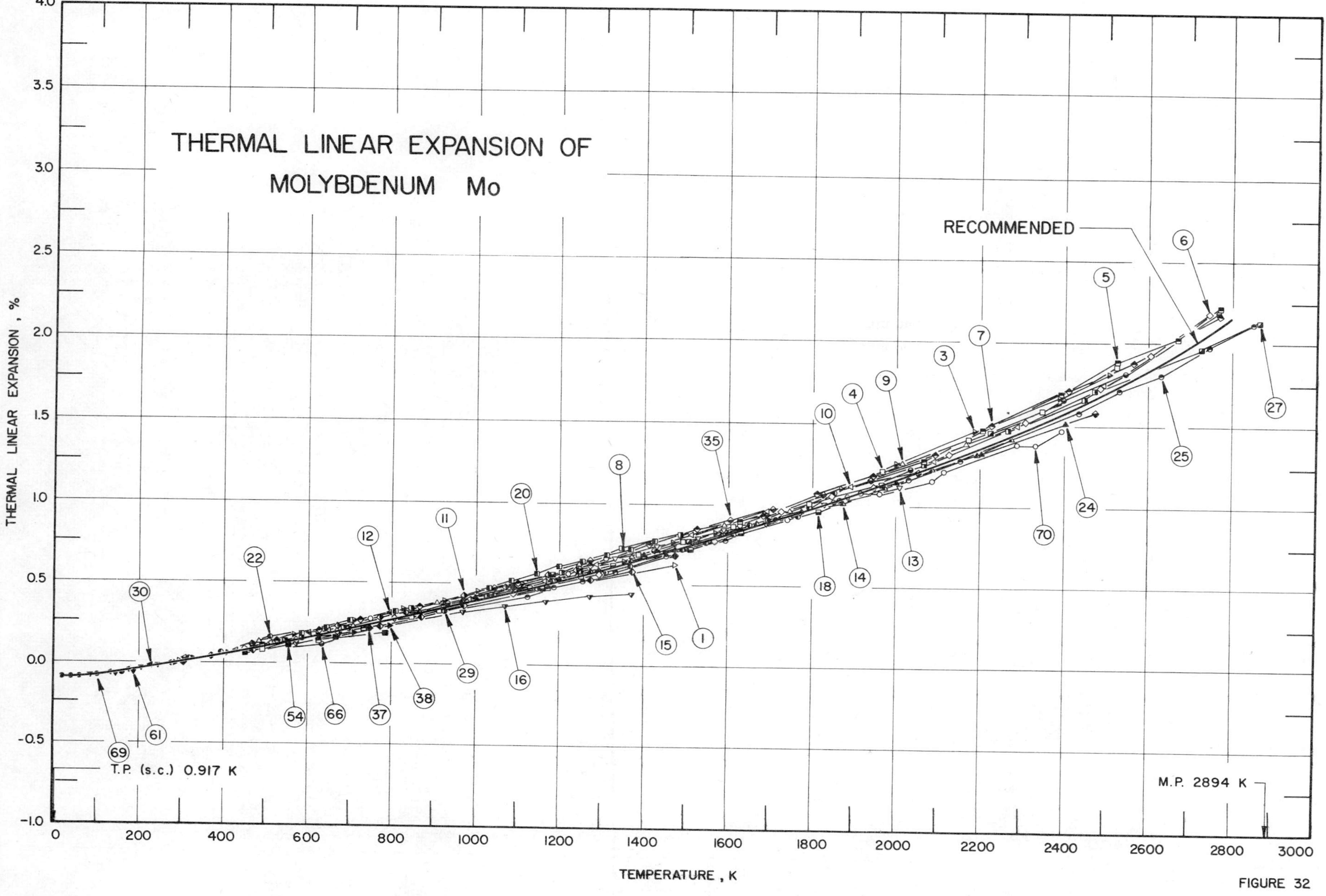

THERMAL LINEAR EXPANSION OF
MOLYBDENUM Mo
RECOMMENDED
T.P. (s.c.) 0.917 K
M.P. 2894 K
TEMPERATURE , K
THERMAL LINEAR EXPANSION , %

FIGURE 32

SPECIFICATION TABLE 32. THERMAL LINEAR EXPANSION OF MOLYBDENUM Mo

Cur. No.	Ref. No.	Author(s)	Year	Method Used	Temp. Range, K	Name and Specimen Designation	Composition (weight percent), Specifications, and Remarks
1	208	Frantsevich, I. N., Zhurakovskii, E. A., and Lyashchenko, A. B.	1967	X	293, 1473		Sample prepared from powder by vacuum sintering the pressed block, sintering temperature between 1470 and 2870 K.
2*	209	Shodhan, R. P.	1965	X	288-338		99.9935 Mo obtained from Hilger Co. in form of fine powder; placed in dry quartz tube, evacuated, and sealed; heated 144 hrs at 1370 K, then quenched in water; lattice parameter reported at 298 K is 3.14700 Å; 3.14692 Å at 293 K determined by graphical interpolation.
3	210	Conway, J. B. and Losekamp, A. C.	1966	T	467-2506	A	99.87^{+} Mo; sheet 20 mil thick; formed by arc-casting; expansion measured in helium atmosphere on first test.
4	210	Conway, J. B. and Losekamp, A. C.	1966	T	490-2524	A	The above specimen; expansion measured on second test.
5	210	Conway, J. B. and Losekamp, A. C.	1966	T	497-2773		99.9^{+} Mo rod 0.5 in. diameter formed by powder metallurgy techniques; expansion measured in helium atmosphere on first test.
6	210	Conway, J. B. and Losekamp, A. C.	1966	T	452-2742		The above specimen; data on second test.
7	210	Conway, J. B. and Losekamp, A. C.	1966	T	528-2772		The above specimen; data measured on third test.
8	210	Conway, J. B. and Losekamp, A. C.	1966	T	452-2473		99.90^{+} Mo; sheet 20 mil thick formed by powder metallurgy techniques; expansion in helium atmosphere on first test.
9	210	Conway, J. B. and Losekamp, A. C.	1966	T	486-2511		The above specimen; expansion measured on second test.
10	210	Conway, J. B. and Losekamp, A. C.	1966	T	514-2491		The above specimen; expansion measured on third test.
11	211	Yaggee, F. L. and Styles, J. W.	1965	L	288-1240		No details given; zero-point correction is -0.003%.
12	212	Pawar, R. R.		X	309-799		Specpure specimen from John Matthey and Co.; annealed 6 hr at 1470 K; specimen in powder form; lattice parameter reported at 309 K is 3.1462 Å; 3.1459 Å at 293 K determined by graphical extrapolation.
13	187	Ross, R. G. and Hume-Rothery, W.	1963	X	295-2006		No details given; lattice parameter reported at 294.7 K is 3.1470 Å; 3.1470 Å at 293 K determined by graphical interpolation.
14	213	Amonenko, V. M., V'yugov, P. N., and Gumenyuk, V. S.	1964	L	552-2273		Specimen normal commercial rod material; annealed in tungsten furnace for 1.5-2.0 hr above 2270 K before measurement started under vacuum; reported error 1%.
15	214	Mulyakaev, L. M., Dubinin, G. N., Ryumin, V. P., and Golubeva, A. S.	1967	L	293-1374		99.90^{+} Mo, <0.1 impurities.
16	214	Mulyakaev, L. M., et al.	1967	L	293-1374		Similar to the above specimen; surface chrome plated from the vapor phase in mixture of 50% Cr + 43% Al_2O_3 + 7% NH_4Cl at 1470 K.
17*	214	Mulyakaev, L. M., et al.	1967	L	293-1374		Similar to the above specimen; chrome plated from vapor phase in the mixture of 30% $Cr\ Cl_2$ + 70% Al_2O_3 at 1470 K.
18	215	Edwards, J. W., Speiser, R., and Johnston, H. L.	1951	X	291-2073		99.9 Mo specimen furnished by Climax Molybdenum Co.; specimen prepared by Fansteel Metallurgical Corp; lattice parameter reported at 291 K is 3.1473 Å.
19*	217	Totskii, E. E.	1964		273-1373		No details given.
20	218	Levingstein, M. A.	1961		297-1422		High purity; 0.003 O, 0.0005 N; wire formed by spraying onto mandrel; mandrel removed by leaching; induction heat-treated in wet hydrogen for 1 hr at 3870 K, then machined into 0.19 in. diameter by 2.5 in. long specimen; reported error 2%; zero-point correction of 0.003% determined by graphical extrapolation.

* Not shown in figure.

SPECIFICATION TABLE 32. THERMAL LINEAR EXPANSION OF MOLYBDENUM Mo (continued)

Cur. No.	Ref. No.	Author(s)	Year	Method Used	Temp. Range, K	Name and Specimen Designation	Composition (weight percent), Specifications, and Remarks
21*	219	Denman, G. L.	1962	L	355-1366		Pure Mo; zero-point correction of 0.003% determined by graphical extrapolation.
22	138	Fieldhouse, I. B., et al.	1956	T	319-1850		Data averaged by author from three separate runs; zero-point correction of 0.002% determined by graphical interpolation.
23*	121	Lucks, C. F. and Deem, H. W.	1956	L	293-1866		Expansion measured in vacuum of 5 x 10^{-5} mm Hg with temperature increasing at rate of 2.8 C/min.
24	220	Worthing, A. G.	1926	T	300-2400		Well seasoned specimen mounted as filament in lamp.
25	221	Rasor, N. S. and McClelland, J. D.	1957	T	308-2852		After thermal testing, impurities 0.25 Fe, 0.063 Si, 0.021 Ti, 0.013 Cu, and 0.008 C; specimen obtained as arc-melted swaged rods 5.08 cm long x 0.95 cm diameter from Climax Molybdenum Co.; expansion measured with increasing temperature on first test.
26*	221	Rasor, N. S. and McClelland, J. D.	1957	T	2759-311		The above specimen; expansion measured with decreasing temperature.
27	221	Rasor, N. S. and McClelland, J. D.	1957	T	301-2864		The above specimen; expansion measured with increasing temperature.
28*	221	Rasor, N. S. and McClelland, J. D.	1957	T	2804-1143		The above specimen; expansion measured with decreasing temperature.
29	123	Lucks, C. F. and Deem, H. W.	1958	L	477-1866		Cylincrical specimen 0.375 in. in diameter by 3 in. long; arc melted, unalloyed from Climax Molybdenum Co.; measured under pressure of 5 x 10^{-5} mm Hg; heating and cooling rate 5 F per min.
30	222	Schad, L. W. and Hidnert, P.	1919	T	290-131	Test 1	99.85 Mo, 0.12 Fe, 0.015 Si, and 0.002 S; specimen obtained from Pfanstiehl Co.; after sintering, heated to 1323 K in hydrogen atmosphere and swaged to size; expansion measured with decreasing temperature; reported error ±0.05%.
31*	222	Schad, L. W. and Hidnert, P.	1919	T	131-578	Test 1	The above specimen; expansion measured with increasing temperature.
32*	222	Schad, L. W. and Hidnert, P.	1919	T	578-295	Test 1	The above specimen; expansion measured with decreasing temperature.
33*	222	Schad, L. W. and Hidnert, P.	1919	T	289-578	Test 2	The above specimen; expansion measured with increasing temperature on second test.
34*	222	Schad, L. W. and Hidnert, P.	1919	T	496-293	Test 2	The above specimen; expansion measured with decreasing temperature.
35	223	Valentich, J.	1969	L	371-2476		99.9 Mo; arc-cast specimen; expansion measured with increasing temperature.
36*	223	Valentich, J.	1969	L	2476-396		The above specimen; expansion measured with decreasing temperature; zero-point correction of 0.052% determined by graphical extrapolation.
37	224	Clark, D. and Knight, D.	1965	X	294-1477		Expansion measured for purpose of comparison; lattice parameter reported at 294 K is 3.1471 Å.
38	225	Valentich, J.	1965	L	293-799		Expansion measured in vacuum.
39*	226	Hidnert, P. and Gore, W. B.	1924	T	301-782	S515	0.04 Fe and 0.003 Si; hexagonal ingot 1.59 cm diameter; prepared from coarse grained powder; expansion measured with increasing temperature; zero-point correction of 0.003% determined by graphical extrapolation.
40*	226	Hidnert, P. and Gore, W. B.	1924	T	782-303	S515	The above specimen; expansion measured with decreasing temperature; zero-point correction of 0.013% determined by graphical extrapolation.
41*	226	Hidnert, P. and Gore, W. B.	1924	T	297-774	S540	Similar to the above specimen; expansion measured with increasing temperature; zero-point correction of 0.002% determined by graphical extrapolation.

* Not shown in figure.

SPECIFICATION TABLE 32. THERMAL LINEAR EXPANSION OF MOLYBDENUM Mo (continued)

Cur. No.	Ref. No.	Author(s)	Year	Method Used	Temp. Range, K	Name and Specimen Designation	Composition (weight percent), Specifications, and Remarks
42*	226	Hidnert, P. and Gore, W. B.	1924	T	774-298	S540	The above specimen; expansion measured with decreasing temperature; zero-point correction of 0.0385% determined by graphical extrapolation.
43*	226	Hidnert, P. and Gore, W. B.	1924	T	291-773	S540A	The above specimen; swaged to 0.635 cm diameter; expansion measured with increasing temperature; zero-point correction is -0.101%.
44*	226	Hidnert, P. and Gore, W. B.	1924	T	773-291	S540A	The above specimen; expansion measured with decreasing temperature; zero-point correction is -0.110 %.
45*	226	Hidnert, P. and Gore, W. B.	1924	T	289-778	S540B	0.11 Fe, 0.014 Cu, 0.003 Si; the above specimen; swaged to 0.445 cm diameter; expansion measured with increasing temperature; zero-point correction is -0.202%.
46*	226	Hidnert, P. and Gore, W. B.	1924	T	778-286	S540B	The above specimen; expansion measured with decreasing temperature; zero-point correction is -0.219%.
47*	226	Hidnert, P. and Gore, W. B.	1924	T	293-773	S540C	0.03 Fe, 0.006 Cu, 0.004 Si; the above specimen; swaged to 0.254 cm diameter; expansion measured with increasing temperature; zero-point correction is -0.3%.
48*	226	Hidnert, P. and Gore, W. B.	1924	T	773-393	S540C	The above specimen; expansion measured with decreasing temperature; zero-point correction is -0.311 %.
49*	226	Hidnert, P. and Gore, W. B.	1924	T	293-769	S603A	0.10 Fe, 0.017 Si, <0.005 Ca; swaged to 0.635 cm diameter from 1.59 cm diameter hexagonal ingot prepared from coarse grained powder; expansion measured with increasing temperature.
50*	226	Hidnert, P. and Gore, W. B.	1924	T	769-284	S603A	The above specimen; expansion measured with decreasing temperature; zero-point correction is -0.0258%.
51*	226	Hidnert, P. and Gore, W. B.	1924	T	293-773	S603B	The above specimen; swaged to 0.445 cm diameter; expansion measured with increasing temperature; zero-point correction is -0.1%.
52*	226	Hidnert, P. and Gore, W. B.	1924	T	773-297	S603B	The above specimen; expansion measured with decreasing temperature; zero-point correction of -0.102% determined by graphical extrapolation.
53*	226	Hidnert, P. and Gore, W. B.	1924	T	301-783	S603C	0.11 Fe, 0.010 Ca, 0.007 Si, 0.005 Cu; the above specimen; swaged to 0.254 cm diameter; expansion measured with increasing temperature; zero-point correction of -0.195% determined by graphical interpolation.
54	226	Hidnert, P. and Gore, W. B.	1924	T	783-298	S603C	The above specimen; expansion measured with decreasing temperature; zero-point correction of -0.241% determined by graphical extrapolation.
55*	226	Hidnert, P. and Gore, W. B.	1924	T	298-573	S1052	0.015 Fe, 0.002 Si; cold worked specimen prepared from fine-grained powder; 0.254 cm diameter.
56*	226	Hidnert, P. and Gore, W. B.	1924	T	298-773	S1052	The above specimen, second test.
57*	226	Hidnert, P. and Gore, W. B.	1924	T	298-773	S1052	Similar to the above specimen except annealed at 1773 K.
58*	226	Hidnert, P. and Gore, W. B.	1924	T	298-573	S1053	0.010 Fe, 0.004 Si; cold worked specimen prepared from coarse grained powder; 0.254 cm diameter.
59*	226	Hidnert, P. and Gore, W. B.	1924	T	298-773	S1053	The above specimen, second test.
60*	226	Hidnert, P. and Gore, W. B.	1924	T	298-773	S1053	Similar to the above specimen except annealed at 1773 K.
61	75	Nix, F. C. and Mac Nair, D.	1942	I	86-298		99.95 Mo; specimen from Westinghouse Lamp Co.; zero-point correction of -0.010% determined by graphical interpolation.

* Not shown in figure.

SPECIFICATION TABLE 32. THERMAL LINEAR EXPANSION OF MOLYBDENUM Mo (continued)

Cur. No.	Ref. No.	Author(s)	Year	Method Used	Temp. Range, K	Name and Specimen Designation	Composition (weight percent), Specifications, and Remarks
62*	227	Davidson, D. L.	1968	X	80-373		Specimen cut from single crystal grown from material obtained from Chase Copper and Brass Co.; specimen 50 mm long; zero-point correction is -0.016 %.
63*	149	Burger, E. E.	1934	L	298-823		Specimen in rod form 0.25 in. diameter, 12 in. long; zero-point correction of 0.003% determined by graphical extrapolation.
64*	228	Baskin, L. M., et al.	1965	L	291, 673		Specimen prepared from powder; sintered at 1973 K, annealed 5 hr at 1223 K and quenched; porosity ~ 1% by volume; reported error 2.5%.
65*	301	Arbuzov, M. P. and Zelenkov, I. A.	1964	L	293-873		Pure metal.
66	596	Conway, J. B. and Flagella, P. N.	1967		498-2765		No details given; zero-point correction is ± 0.01%; the original data reported using 298 K as reference temperature.
67*	596	Conway, J. B. and Flagella, P. N.	1967	T	501-2778		No details given; the original data reported using 298 K as reference temperature.
68*	596	Conway, J. B. and Flagella, P. N.	1967	T	406-2748		No details given; the original data reported using 298 K as reference temperature.
69*	617, 701	Straumanis, M. E. and Woodard, C. L.	1971	X	20-180		Mo powder embedded in a thin In layer which adhered well the Mo sample holder, lattice parameter reported at 180 K is 3.1453 Å, the results have been referred to 293 K using other available information, zero-point correction is -0.053%.
70	623	Mochalov, G. A. and Ivanov, O. S.	1969	L	563-2394		99.9 pure specimen, below 1273 K using quartz dilatometer and above that with cathetometer.
71*	707	Pridantseva, K. S. and Solov'yeva, N. A.	1965	L	293, 1073		Specimen prepared by powder metallurgy method, 1% measurement error.
72*	707	Pridantseva, K. S. and Solov'yeva, N. A.	1965	L	293-973		Similar to the above specimen, 1% measurement error.
73*	714	Andres, K.	1964	O	4.2-13.3		99.95 pure, the results have been referred to 293 K using other available information; zero-point correction is -0.09280%.
74*	748	Hubbell, W. C. and Brotzen, F. R.	1972	L	83, 373		Powder specimen prepared from reactor grade Mo mixed and compacted into 6.2 mm rounds approximately 200 mm long, sample melted by electron-beam floating zone in vacuum of 10^{-6} torn, cylinders about 10 mm long spark sliced from sample and end faces polished, crystal axes orientation determined by Lave back-reflection method, $\pm 1.5 \times 10^{-6}$ K^{-1} uncertainty in α, 10.223 gm · cm^{-3} at 298 K.
75	739	Petukhov, V. A., and Chekhouskoi, V. Ya.	1972	O	293-2819		99.95 Mo, 0.004 O_2, 0.0008 H_2, 0.004 N_2; measurements in helium atmosphere.
76	747	Lisovskii, Yu. A.	1973	E	55-300		99.97 pure polycrystalline specimen of 10 cm length.
77	777	Chekhouskoi, V. Ya, and Petukhov, V. A.	1970	O	1185-2470		Average values between 99.9 pure specimen manufactured by powder metallurgy and 99.97 pure specimen manufactured by the zone melting.

* Not shown in figure.

DATA TABLE 32. THERMAL LINEAR EXPANSION OF MOLYBDENUM Mo

[Temperature, T, K; Linear Expansion, $\Delta L/L_0$, %]

T	$\Delta L/L_0$
CURVE 1	
293	0.000*
1473	0.637
CURVE 2*	
288.2	-0.0025
298.2	0.0025
308.2	0.0076
318.2	0.0121
328.2	0.0168
338.2	0.0219
CURVE 3	
467	0.10
562	0.17
716	0.26
828	0.34
928	0.39
1005	0.45
1116	0.50
1247	0.60
1350	0.65
1464	0.76
1572	0.83
1693	0.94
1864	1.11
2008	1.26
2184	1.46
2384	1.66
2506	1.83
CURVE 4	
490	0.08
614	0.17
845	0.29
914	0.33
1128	0.47
1254	0.55
1383	0.69
1495	0.77
1612	0.87
1794	1.03
1969	1.22
CURVE 4 (cont.)	
2172	1.42
2346	1.59
2524	1.86
CURVE 5	
497	0.10
626	0.14
737	0.22
921	0.35
1174	0.48
1436	0.70
1631	0.90
2062	1.28
2204	1.46
2385	1.68
2527	1.87
2676	2.03
2773	2.23
CURVE 6	
452	0.08
575	0.14
668	0.19
804	0.28
1093	0.44
1293	0.56
1565	0.78
1730	0.96
1843	1.06
1941	1.12
2123	1.33
2308	1.53
2486	1.74
2605	1.94
2742	2.19
CURVE 7	
528	0.14
628	0.20
722	0.26
866	0.35
1002	0.44
CURVE 7 (cont.)	
1102	0.50
1274	0.61
1415	0.75
1527	0.85
1688	0.94
1819	1.09
1943	1.19
2094	1.33
2228	1.50
2410	1.71
2563	1.89
2772	2.17
CURVE 8	
452	0.09*
545	0.14
681	0.22
848	0.34
1028	0.41
1205	0.57
1348	0.73
1481	0.81
1629	0.92*
1848	1.07*
2064	1.26
2265	1.47
2451	1.66
2473	1.71
CURVE 9	
486	0.13
714	0.23
912	0.37
1104	0.49*
1243	0.62
1371	0.67
1466	0.76*
1593	0.88
1744	0.97
1864	1.11*
2019	1.27
2162	1.39
2305	1.54*
CURVE 9 (cont.)	
2417	1.71*
2511	1.81*
CURVE 10	
514	0.12
649	0.20
876	0.35*
981	0.40
1165	0.52
1281	0.63
1512	0.81
1698	0.96
1893	1.13
2093	1.29
2284	1.50
2491	1.75
CURVE 11	
288	-0.003
373	0.046
473	0.103
573	0.162*
673	0.223
773	0.288
873	0.352*
973	0.418
1073	0.486
1174	0.552
1240	0.597
CURVE 12	
309	0.011
426	0.088
512	0.133*
602	0.185
698	0.225*
799	0.301
CURVE 13	
294.7	0.001*
1173.2	0.541
CURVE 13 (cont.)	
1333.2	0.659
1432.2	0.741
1528.2	0.792
1594.2	0.840
1698.2	0.888
1826.2	1.03*
1921.2	1.05*
2006.2	1.13
CURVE 14	
552	0.15*
672	0.21*
773	0.27*
865	0.32
969	0.39
1071	0.46*
1165	0.52*
1273	0.58
1368	0.65
1479	0.72
1581	0.81
1670	0.89
1778	0.96
1879	1.05
1975	1.13
2088	1.24
2193	1.32
2273	1.43
CURVE 15	
293	0.000*
373	0.037*
473	0.085
573	0.136*
673	0.189*
773	0.244
873	0.300
973	0.357
1073	0.416*
1173	0.474*
1273	0.532
1374	0.583
CURVE 16	
293	0.000*
373	0.034*
473	0.079*
573	0.126
673	0.176
773	0.228*
873	0.275*
973	0.320
1073	0.360
1173	0.395
1273	0.424
1374	0.446
CURVE 17*	
293	0.000
373	0.037
473	0.085
573	0.136
673	0.189
773	0.245
873	0.301
973	0.359
1073	0.414
1173	0.469
1273	0.516
1374	0.559
CURVE 18	
291	0.000*
1126	0.467*
1162	0.473
1327	0.616
1510	0.737
1639	0.839
1819	0.979
1968	1.14
2073	1.26*
CURVE 19*	
273	0.000
373	0.041
473	0.094

*Not shown in figure.

DATA TABLE 32. THERMAL LINEAR EXPANSION OF MOLYBDENUM Mo (continued)

T	$\Delta L/L_0$
CURVE 19 (cont.)*	
573	0.148
673	0.205
773	0.263
873	0.323
973	0.385
1073	0.448
1173	0.514
1273	0.581
1373	0.650
CURVE 20	
297	0.003*
367	0.047*
422	0.082*
478	0.115*
533	0.149*
589	0.183
644	0.216
700	0.251
755	0.287*
811	0.323
867	0.358*
922	0.399*
978	0.438*
1034	0.478
1089	0.518
1144	0.560
1200	0.601
1255	0.643
1311	0.686
1367	0.728
1422	0.763
CURVE 21*	
355	0.038
444	0.093
538	0.128
622	0.178
710	0.229
790	0.278
877	0.328
966	0.383
1055	0.443
1144	0.498
CURVE 21 (cont.)*	
1233	0.549
1322	0.623
1366	0.653
CURVE 22	
319	0.012
397	0.057
428	0.064*
483	0.092*
500	0.127*
511	0.152
553	0.112*
623	0.174*
735	0.242*
803	0.287*
839	0.312
889	0.334*
978	0.388*
1047	0.427
1184	0.529
1194	0.522*
1205	0.540*
1289	0.595*
1305	0.570
1358	0.647*
1366	0.627
1391	0.650*
1480	0.735*
1495	0.732
1544	0.777
1558	0.775*
1571	0.786*
1573	0.802*
1616	0.817
1625	0.832*
1657	0.882
1673	0.871*
1707	0.891*
1729	0.927
1748	0.931*
1762	0.944
1783	0.979*
1797	0.993
1850	1.072
CURVE 23*	
293	0.000
422	0.070
589	0.154
755	0.240
922	0.337
1089	0.441
1255	0.560
1422	0.684
1589	0.814
1755	0.950
1866	1.047
CURVE 24	
300	0.00*
1000	0.41
1200	0.54
1400	0.68
1600	0.83
1800	0.99*
2000	1.15
2200	1.33
2400	1.51
CURVE 25	
308	0.007*
1123	0.403
1189	0.487
1259	0.534
1332	0.578
1452	0.689
1599	0.790
1716	0.907
1806	0.967*
1889	1.029*
2030	1.169
2159	1.298
2273	1.475*
2439	1.585
2530	1.712
2637	1.817
2689	1.913*
2748	1.998
2779	2.103*
2852	2.114
CURVE 26*	
2759	2.053
2683	1.960
2592	1.839
2418	1.623
2264	1.425
2116	1.249
1931	1.094
1735	0.942
1604	0.844
1472	0.710
1332	0.626
1187	0.537
1149	0.518
311	0.043
CURVE 27	
301	0.005*
1090	0.485*
1184	0.559
1301	0.632*
1403	0.722*
1507	0.786
1618	0.876*
1734	0.966*
1854	1.068
1915	1.132*
2055	1.285*
2221	1.457
2401	1.588*
2624	1.845*
2724	1.983
2762	2.079*
2819	2.101*
2864	2.143
CURVE 28*	
2804	2.086
2760	2.022
2615	1.803
2533	1.741
2448	1.611
2319	1.475
2079	1.243
1907	1.069
1775	0.946
CURVE 28 (cont.)*	
1604	0.821
1444	0.720
1331	0.601
1143	0.496
CURVE 29	
477	0.098*
588	0.153
699	0.210
810	0.272*
922	0.337
1033	0.407*
1144	0.481
1255	0.560
1366	0.640*
1477	0.725*
1588	0.814*
1699	0.904
1810	0.999*
1866	1.048
CURVE 30	
291.0	-0.0003*
290.2	-0.0004*
273.4	-0.009
249.7	-0.028*
223.2	-0.034*
198.0	-0.045*
173.4	-0.055
130.7	-0.071
CURVE 31*	
130.7	-0.072
150.1	-0.064
196.0	-0.045
248.8	-0.020
291.9	-0.007
322.4	0.015
374.8	0.043
424.2	0.068
472.0	0.094
522.3	0.12
578.4	0.16
578.5	0.16
CURVE 32*	
578.5	0.156
497.1	0.109
414.7	0.062
356.6	0.033
295.6	0.0012
CURVE 33*	
289.2	-0.002
374.5	0.041
445.5	0.080
499.0	0.109
577.7	0.154
CURVE 34*	
496.4	0.109
416.3	0.064
416.0	0.064
293.0	0.0000
CURVE 35	
371	0.048*
929	0.392*
1044	0.476
1156	0.562*
1270	0.641
1376	0.727*
1494	0.813*
1601	0.902
1711	0.987
1822	1.076
1940	1.162
2050	1.203
2274	1.412*
2476	1.598
CURVE 36*	
2476	1.629
2386	1.527
2273	1.428
2158	1.332
2046	1.232
1823	1.000

*Not shown in figure.

DATA TABLE 32. THERMAL LINEAR EXPANSION OF MOLYBDENUM Mo (continued)

T	$\Delta L/L_0$
CURVE 36 (cont.)*	
1708	0.971
1600	0.885
1496	0.793
1374	0.708
1264	0.623
1158	0.537
816	0.296
711	0.224
597	0.156
481	0.081
376	0.034
CURVE 37	
294	0.000*
304	-0.009
495	0.108*
744	0.271
916	0.366*
1099	0.474
1290	0.595
1477	0.696
CURVE 38	
293	0.000*
328	0.012
405	0.053
518	0.107*
623	0.154
704	0.193
799	0.240
CURVE 39*	
301	0.003
318	0.010
373	0.031
424	0.060
500	0.098
531	0.109
599	0.151
662	0.187
718	0.218
782	0.255
CURVE 40*	
782	0.265
664	0.197
545	0.130
481	0.095
387	0.046
303	0.006
CURVE 41*	
297	0.002
319	0.013
377	0.043
428	0.067
477	0.092
524	0.118
572	0.143
625	0.169
677	0.195
722	0.213
774	0.226
CURVE 42*	
774	0.2629
646	0.1820
568	0.1370
473	0.0901
386	0.0568
298	0.0032
CURVE 43*	
291	-0.001
293	0.000
326	0.016
380	0.042
439	0.071
470	0.085
519	0.111
582	0.144
617	0.166
654	0.192
729	0.239
773	0.274
CURVE 44*	
773	0.274
667	0.213
550	0.147
491	0.111
389	0.055
293	0.009
291	0.009
CURVE 45*	
289	-0.002
293	0.000
323	0.013
377	0.038
452	0.070
530	0.110
582	0.135
626	0.157
648	0.169
728	0.216
778	0.252
CURVE 46*	
778	0.252
571	0.135
477	0.089
378	0.049
293	0.017
286	0.015
CURVE 47*	
293	0.000
334	0.017
389	0.043
429	0.065
474	0.078
531	0.102
585	0.128
633	0.154
694	0.202
718	0.219
773	0.256
CURVE 48*	
773	0.256
638	0.180
612	0.169
561	0.143
492	0.106
384	0.054
293	0.011
CURVE 49	
293	0.000*
318	0.013*
379	0.042*
433	0.070
483	0.093*
534	0.120*
578	0.143*
629	0.175
657	0.195
725	0.240*
769	0.272
CURVE 50*	
769	0.272
662	0.212
570	0.166
474	0.188
356	0.057
293	0.026
284	0.021
CURVE 51*	
293	0.000
319	0.013
380	0.044
427	0.068
473	0.091
531	0.120
550	0.133
628	0.175
677	0.206
734	0.243
773	0.269
CURVE 52*	
773	0.2674
666	0.2036
569	0.1452
478	0.093
377	0.041
297	0.002
CURVE 53*	
301	0.0050
328	0.0142
389	0.0398
452	0.0696
543	0.1174
572	0.1300
642	0.1611
669	0.1766
717	0.2031
783	0.2413
CURVE 54	
783	0.1953
666	0.1555
556	0.1103
457	0.0685
298	0.0019*
CURVE 55*	
298	0.002
373	0.040
473	0.093
573	0.146
CURVE 56*	
298	0.002
373	0.034
473	0.085
573	0.138
673	0.193
773	0.254
CURVE 57*	
298	0.003
373	0.042
CURVE 57 (cont.)*	
473	0.096
573	0.151
673	0.207
773	0.266
CURVE 58*	
298	0.002
373	0.038
473	0.088
573	0.138
CURVE 59*	
298	0.002
373	0.038
473	0.088
573	0.139
673	0.188
773	0.246
CURVE 60*	
298	0.003
373	0.042
473	0.098
573	0.154
673	0.212
773	0.279
CURVE 61	
86.2	-0.090
90.7	-0.089*
101.2	-0.086*
107.7	-0.084*
116.7	-0.081*
123.7	-0.079*
130.2	-0.076*
137.2	-0.074*
143.7	-0.072
149.7	-0.069*
156.2	-0.067*
161.7	-0.064*
166.2	-0.062*

*Not shown in figure.

DATA TABLE 32. THERMAL LINEAR EXPANSION OF MOLYBDENUM Mo (continued)

T	$\Delta L/L_0$
CURVE 61 (cont.)	
172.5	-0.060*
178.4	-0.057*
183.2	-0.055*
188.2	-0.052
193.2	-0.050*
198.3	-0.048*
203.7	-0.045
209.2	-0.043*
213.7	-0.040*
218.5	-0.038*
224.2	-0.036*
228.7	-0.033*
233.7	-0.031*
238.2	-0.028*
243.3	-0.026
248.2	-0.023*
253.2	-0.021*
257.6	-0.019*
262.5	-0.016*
266.4	-0.014*
271.7	-0.011*
273.2	-0.011*
276.2	-0.009*
281.7	-0.007*
285.2	-0.004*
290.3	-0.002*
294.2	0.001*
298.2	0.003*
CURVE 62*	
73	-0.105
98	-0.101
123	-0.093
148	-0.083
173	-0.070
198	-0.057
223	-0.045
248	-0.031
273	-0.016
293	0.000
323	0.016
348	0.033
373	0.050

T	$\Delta L/L_0$
CURVE 63*	
298	0.003
323	0.019
348	0.033
373	0.047
398	0.061
423	0.077
448	0.092
473	0.105
498	0.120
523	0.133
548	0.151
573	0.164
598	0.177
623	0.192
648	0.207
673	0.223
698	0.237
723	0.250
748	0.267
773	0.281
798	0.295
823	0.308
CURVE 64*	
291	-0.001
673	0.205
CURVE 65*, ‡	
293	0.000
473	0.711
567	1.187
673	1.667
719	1.910
772	2.171
873	2.623
CURVE 66	
498	0.102*
636	0.124
751	0.233*
938	0.375

T	$\Delta L/L_0$
CURVE 66 (cont.)	
1171	0.500*
1425	0.707
1609	0.885*
1853	1.036
2037	1.247
2208	1.451*
2393	1.663
2547	1.835
1678	2.011*
2765	2.205
CURVE 67*	
501	0.142
573	0.172
711	0.279
858	0.333
1000	0.434
1101	0.500
1253	0.614
1403	0.757
1506	0.841
1673	0.950
1825	1.077
1937	1.168
2070	1.302
2243	1.501
2421	1.712
2555	1.893
2778	2.169
CURVE 68*	
406	0.080
549	0.174
815	0.298
1067	0.431
1253	0.534
1547	0.785
1723	0.936
1941	1.105
2187	1.314
2298	1.500
2487	1.727

T	$\Delta L/L_0$
CURVE 68 (cont.)*	
2599	1.902
2748	2.178
CURVE 69	
20	-0.091
40	-0.091
60	-0.089
80	-0.086
90	-0.085*
100	-0.082
120	-0.076*
140	-0.070*
160	-0.067
180	-0.053*
CURVE 70	
563	0.146*
641	0.199
751	0.272
873	0.343*
966	0.402*
1056	0.460
1170	0.531*
1220	0.565
1293	0.618*
1312	0.665
1361	0.645*
1513	0.744*
1630	0.889
1699	0.889*
1699	0.942*
1740	0.922
1883	1.051
1916	1.017*
1916	1.088
1960	1.077
2084	1.166
2116	1.221
2289	1.396
2330	1.381
2394	1.476

T	$\Delta L/L_0$
CURVE 71*	
293	0.000
1073	0.434
CURVE 72*	
293	0.000
573	0.136
773	0.261
973	0.399
CURVE 73*	
4.2	-0.09281
4.7	-0.09281
6.0	-0.09281
6.0	-0.09281
6.2	-0.09281
6.2	-0.09281
7.3	-0.09281
9.0	-0.09281
10.3	-0.09280
11.2	-0.09280
12.3	-0.09280
12.3	-0.09280
12.9	-0.09280
13.3	-0.09280
CURVE 74*	
83	-0.105
373	0.040
CURVE 75	
2023	1.184
2141	1.294
2178	1.327
2278	1.428
2332	1.483
2453	1.626
2548	1.747
2563	1.765
2603	1.814
2609	1.822

T	$\Delta L/L_0$
CURVE 75 (cont.)	
2630	1.860
2639	1.866
2671	1.915
2718	1.990
2728	1.991
2748	2.032
2775	2.082
2790	2.115
2814	2.155
2819	2.164
CURVE 76‡	
55	0.089
60	-0.088
70	-0.087
80	-0.085
90	-0.083
100	-0.080
110	-0.077
120	-0.074
130	-0.071
140	-0.067
150	-0.063
160	-0.060
170	-0.056
180	-0.051
190	-0.047
200	-0.043
210	-0.039
220	-0.034
230	-0.030
240	-0.025
250	-0.020
260	-0.016
270	-0.011
280	-0.006
290	-0.001
293	0.000
300	0.003

*Not shown in figure.
‡Author's data for coefficient of thermal expansion have been integrated by TPRC to obtain $\Delta L/L_0$.

DATA TABLE 32. THERMAL LINEAR EXPANSION OF MOLYBDENUM Mo (continued)

T	$\Delta L/L_0$	T	$\Delta L/L_0$	T	$\Delta L/L_0$
CURVE 77		CURVE 77 (cont.)		CURVE 77 (cont.)	
1185	0.697	1902	1.255	2425	1.796
1209	0.713	1962	1.309	2470	1.851
1283	0.768	2025	1.372		
1311	0.786	2033	1.371		
1498	0.927	2047	1.391		
1526	0.949	2128	1.467		
1608	1.010	2160	1.450		
1633	1.032	2201	1.539		
1665	1.058	2216	1.552		
1669	1.056	2267	1.616		
1713	1.101	2298	1.644		
1766	1.135	2333	1.686		
1771	1.144	2365	1.717		
1800	1.167	2390	1.748		
1875	1.233	2403	1.765		

DATA TABLE 32. COEFFICIENT OF THERMAL LINEAR EXPANSION OF MOLYBDENUM Mo

[Temperature, T, K; Coefficient of Expansion, α, 10^{-6} K^{-1}]

T	α	T	α
CURVE 65*, ‡		CURVE 76 (cont.) ‡	
293	3.51	140	3.65
473	4.39	150	3.80
576	4.85	160	3.94
673	5.06	170	4.05
719	5.48	180	4.16
772	4.39	190	4.27
873	4.56	200	4.35
		210	4.41
CURVE 76‡		220	4.48
		230	4.56
55	1.10	240	4.61
60	1.32	250	4.66
70	1.65	260	4.71
80	2.10	270	4.73
90	2.42	280	4.76
100	2.76	290	4.78
110	3.06	293	4.79
120	3.27	300	4.82
130	3.47		

* Not shown in figure.
‡ Author's data for coefficient of thermal expansion have been integrated by TPRC to obtain $\Delta L/L_0$.

FIGURE AND TABLE NO. 33R. RECOMMENDED VALUES FOR THERMAL LINEAR EXPANSION OF NEODYMIUM Nd

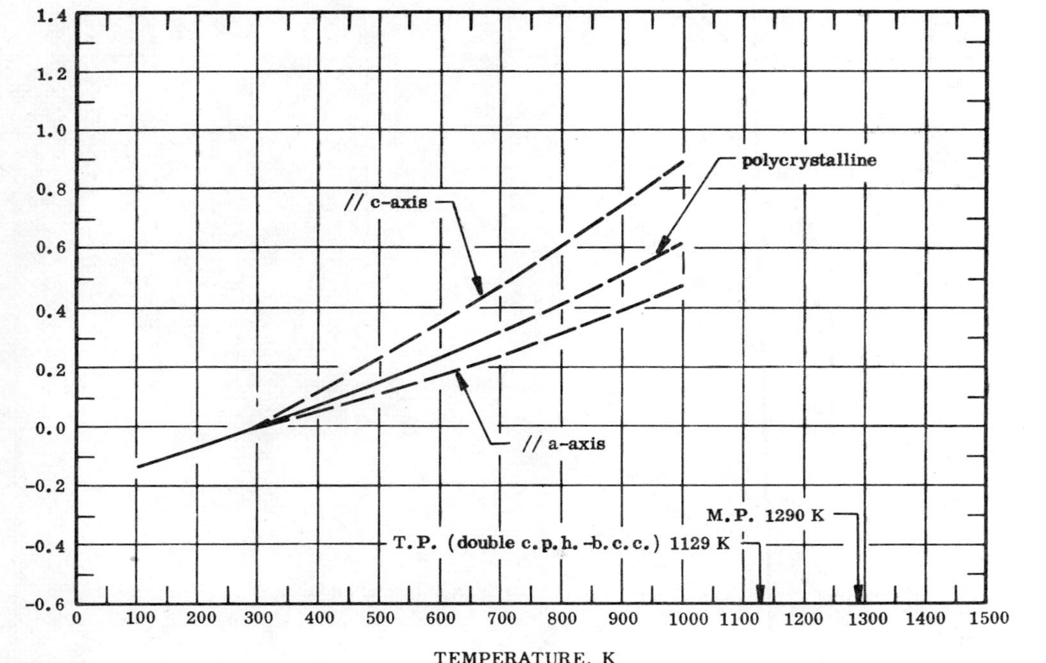

RECOMMENDED VALUES

[Temperature, T, K; Linear Expansion, $\Delta L/L_0$, %; α, K^{-1}]

T	// a-axis $\Delta L/L_0$	// c-axis $\Delta L/L_0$	polycrystalline $\Delta L/L_0$	polycrystalline $\alpha \times 10^6$
100			-0.130	6.5
200			-0.064	6.7
293	0.000	0.000	0.000	6.9
400	0.051†	0.120†	0.074	7.2
500	0.106†	0.233†	0.148	7.7
600	0.168†	0.348†	0.228	8.2
700	0.236†	0.470†	0.314	8.9
800	0.309†	0.600†	0.406†	9.6
900	0.389†	0.740†	0.506†	10.3
1000	0.473†	0.894†	0.614†	11.0

† Provisional values.

REMARKS

The tabulated values for polycrystalline neodymium are considered accurate to within ± 3% at temperatures below 700 K and ± 5% above. The tabulated values along a-axis and c-axis are considered accurate to within ± 10% over the entire temperature range. These values can be represented approximately by the following equations:

// a-axis: $\Delta L/L_0 = -0.951 + 2.043 \times 10^{-4} T + 4.287 \times 10^{-7} T^2 - 6.489 \times 10^{-11} T^3$

// c-axis: $\Delta L/L_0 = -0.350 + 1.299 \times 10^{-3} T - 4.787 \times 10^{-7} T^2 + 4.238 \times 10^{-10} T^3$

polycrystalline: $\Delta L/L_0 = -0.194 + 6.442 \times 10^{-4} T + 3.916 \times 10^{-9} T^2 + 1.605 \times 10^{-10} T^3$

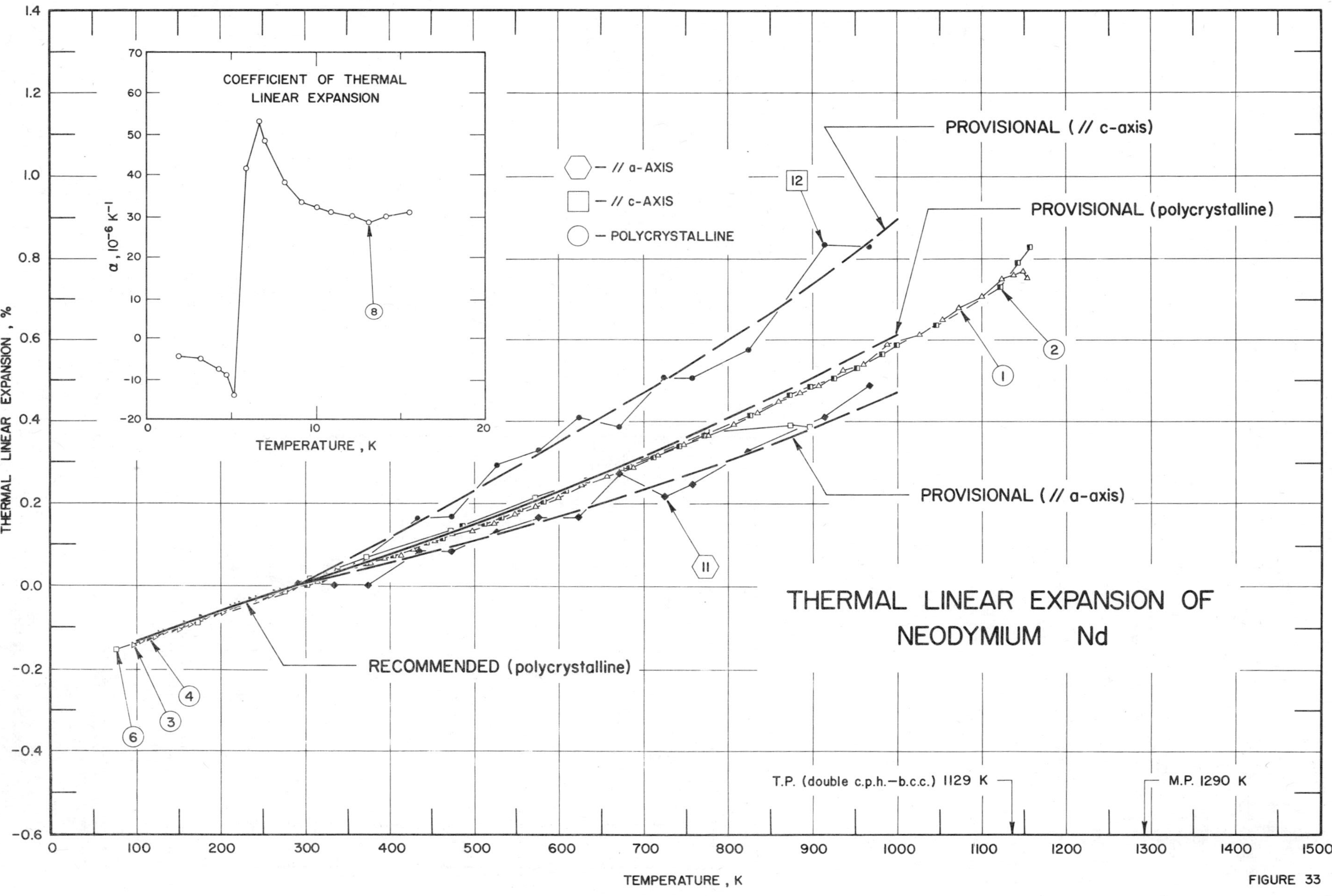
COEFFICIENT OF THERMAL
LINEAR EXPANSION
α, 10⁻⁶ K⁻¹
TEMPERATURE , K
— // a-AXIS
— // c-AXIS
— POLYCRYSTALLINE
PROVISIONAL (// c-axis)
PROVISIONAL (polycrystalline)
PROVISIONAL (// a-axis)
RECOMMENDED (polycrystalline)
THERMAL LINEAR EXPANSION OF
NEODYMIUM Nd
T.P. (double c.p.h.–b.c.c.) 1129 K
M.P. 1290 K
THERMAL LINEAR EXPANSION , %
TEMPERATURE , K

FIGURE 33

SPECIFICATION TABLE 33. THERMAL LINEAR EXPANSION OF NEODYMIUM Nd

Cur. No.	Ref. No.	Author(s)	Year	Method Used	Temp. Range, K	Name and Specimen Designation	Composition (weight percent), Specifications, and Remarks
1	78	Barson, F., Legvold, S. and Spedding, F. H.	1956	L	295-1155		99.7 Nd, 0.08 Pr, 0.06 Sm, 0.06 N_2, 0.05 Ta, 0.025 Si, 0.0175 C, 0.01 Ca; rod specimen 0.6 cm diameter, 6 cm long; fluorides of cerium, bomb reduced with calcium metal to produce compact metallic sample; vacuum cast; turned to shape on a lathe; anomaly observed at 1140 ± 2 K; expansion measured with increasing temperature; zero-point correction of 0.001% determined by graphical extrapolation.
2	78	Barson, F., et al.	1956	L	1158-301		The above specimen; expansion measured with decreasing temperature; zero-point correction of 0.114% determined by graphical extrapolation.
3	79	Barson, F., Legvold, S. and Spedding, F. H.	1953	L	341-99		< 1 Mg, < 0.25 Ca, < 178 ppm Fe, < 0.01 others; prepared from the salt by bomb reduction with calcium; cast in tantalum crucibles and then turned into cylindrical rod 0.3 cm in diameter, ends machined flat; subjected to prolonged annealing at about 550 C; expansion measured with decreasing temperature on first test; zero-point correction of -0.0138% determined by graphical interpolation.
4	79	Barson, F., et al.	1953	L	117-342		The above specimen; expansion measured with increasing temperature on second test; zero-point correction of -0.0136% determined by graphical interpolation.
5*	79	Barson, F., et al.	1953	L	103-349		The above specimen; expansion measured with increasing temperature on third test; zero-point correction of -0.0143% determined by graphical interpolation.
6	229	Trombe, F. and Foex, M.	1951	L	78-898		< 0.01 impurities, mainly Si; low temperature measurements made in hydrogen atmosphere, high temperature in argon; expansion measured with increasing temperature; zero-point correction of -0.016% determined by graphical interpolation.
7*	229	Trombe, F. and Foex, M.	1951	L	898-78		The above specimen; expansion measured with decreasing temperature; zero-point correction of -0.003% determined by graphical interpolation.
8*	80	Andres, K.	1963	O	2.0-15		Polycrystalline specimen melted into Ta form; this curve is here reported using the first given temperature as reference temperature at which $\Delta L/L_0 = 0$.
9*	230	Jaeger, F. M., Bottema, J. A. and Rosenbohm, E.	1938	L	293-1039		99.5 Nd, traces of Fe, Si, Al; transformation from α to δ phase begins at 733 K, complete at 1019 K; expansion measured with increasing temperature; data as reported by authors appear to be in error, corrected by TPRC by dividing by 41 mm, the length of the specimen in other experiments using this apparatus. [See Ref. No. 254].
10*	230	Jaeger, F. M., et al.	1938	L	1033-873		The above specimen; expansion measured with decreasing temperature.
11	82	Hanak, J. J.	1959	X	293-969		0.19 O, 0.1 Ta, < 0.06 Sm, < 0.05 Ca, < 0.025 Si, < 0.01 each any other impurities; wire specimen cut from sheet prepared at Ames Laboratory; hexagonal-close-packed structure; expansion measured along a-axis; lattice parameter reported at 293 K is 3.660 Å.
12	82	Hanak, J. J.	1959	X	293-969		The above specimen; expansion measured along c-axis; lattice parameter reported at 293 K is 11.807 Å.
13*	82	Hanak, J. J.	1959	X	1123-1168		Similar to the above specimen except b.c.c. structure; this curve is here reported using the first given temperature as reference temperature at which $\Delta L/L_0 = 0$; lattice parameter reported at 1123 K is 4.12 Å.

* Not shown in figure.

DATA TABLE 33. THERMAL LINEAR EXPANSION OF NEODYMIUM Nd

[Temperature, T, K; Linear Expansion, $\Delta L/L_0$, %]

T	$\Delta L/L_0$
CURVE 1	
295	0.001
313	0.017
333	0.028
359	0.048
376	0.058
392	0.069
411	0.076
429	0.089
451	0.111
473	0.127
497	0.139
523	0.157
549	0.179
573	0.199
600	0.218
627	0.243
657	0.265
689	0.291
719	0.320
748	0.344
777	0.365
806	0.396
835	0.420
860	0.450
885	0.473
907	0.494
936	0.525
960	0.549
986	0.587
1029	0.617
1053	0.655
1074	0.681
1100	0.708
1123	0.750
1138	0.758
1150	0.765
1155	0.756
CURVE 2	
301	0.009
309	0.018
321	0.027*
342	0.038
CURVE 2 (cont.)	
357	0.047*
373	0.057
389	0.066*
403	0.078
424	0.090*
444	0.105
464	0.119
488	0.142
512	0.152
531	0.168
555	0.187
581	0.209
609	0.230
632	0.255
656	0.267*
683	0.292
711	0.318
741	0.340
771	0.368
801	0.394*
827	0.419
849	0.435*
874	0.461
899	0.489
925	0.508
956	0.536
981	0.563
1001	0.587
1027	0.614*
1048	0.640
1074	0.675*
1101	0.711*
1121	0.734
1131	0.751*
1136	0.769*
1141	0.790
1148	0.811*
1158	0.833
CURVE 3	
98.6	-0.141
99.4	-0.141*
101.1	-0.139*
CURVE 3 (cont.)	
102.7	-0.138*
104.2	-0.137
106.8	-0.135*
109.1	-0.133*
111.6	-0.131*
114.1	-0.129*
116.6	-0.127*
119.2	-0.125*
121.4	-0.124
124.7	-0.121*
127.3	-0.119*
129.8	-0.117*
132.3	-0.115
134.7	-0.113*
137.1	-0.111*
139.6	-0.109*
142.3	-0.107*
144.9	-0.105*
147.5	-0.103*
150.1	-0.102
152.8	-0.100*
155.5	-0.098*
158.1	-0.096*
161.6	-0.093
164.4	-0.091*
167.0	-0.089*
169.7	-0.087*
172.5	-0.085*
174.6	-0.083*
178.0	-0.081*
181.7	-0.078
184.5	-0.076*
187.3	-0.074*
191.1	-0.072*
194.0	-0.070*
196.9	-0.068*
199.8	-0.066*
202.7	-0.064
205.6	-0.062*
208.7	-0.059*
211.4	-0.057
214.6	-0.055*
217.5	-0.053*
220.5	-0.051*
CURVE 3 (cont.)	
223.4	-0.049
226.3	-0.047*
229.2	-0.045*
232.1	-0.043*
235.1	-0.041*
238.1	-0.039*
241.1	-0.037
244.0	-0.034*
247.0	-0.032*
249.8	-0.030*
252.8	-0.028*
255.8	-0.026*
258:9	-0.024
261.9	-0.022*
264.9	-0.020*
267.9	-0.018*
270.8	-0.016
273.9	-0.013*
276.9	-0.011*
280.1	-0.009
283.1	-0.007*
285.9	-0.005*
288.9	-0.003*
292.1	-0.001*
295.9	0.002*
298.1	0.004*
301.3	0.006*
304.1	0.008*
307.1	0.010*
310.0	0.012*
313.1	0.015*
316.1	0.017*
319.1	0.019*
322.3	0.021
325.4	0.023*
238.7	0.026*
331.4	0.028
334.3	0.030*
336.4	0.031*
338.5	0.033*
340.7	0.034
CURVE 4	
116.8	-0.123
118.7	-0.122*
120.4	-0.121*
122.3	-0.120*
123.9	-0.118*
126.5	-0.116*
129.2	-0.114
131.8	-0.113*
134.4	-0.111*
137.1	-0.109
139.7	-0.107*
142.2	-0.105
144.8	-0.103*
147.3	-0.101*
149.8	-0.099*
152.5	-0.097*
155.2	-0.095*
157.5	-0.093
160.2	-0.092*
163.2	-0.090*
166.0	-0.088*
168.8	-0.086*
171.5	-0.084*
174.3	-0.082*
177.1	-0.080
179.9	-0.078*
182.8	-0.076*
185.6	-0.074*
188.4	-0.072*
191.3	-0.070
194.2	-0.068*
197.1	-0.066*
199.9	-0.064*
202.8	-0.062*
205.8	-0.060*
208.7	-0.058*
211.7	-0.056*
214.7	-0.053*
217.7	-0.051*
220.7	-0.049
223.9	-0.047*
226.8	-0.045*
229.9	-0.043*
232.9	-0.041*
CURVE 4 (cont.)	
236.0	-0.039
239.3	-0.037*
242.3	-0.035*
245.4	-0.033*
248.5	-0.030*
251.5	-0.028
254.5	-0.026*
257.6	-0.024*
260.8	-0.022*
263.8	-0.020
266.9	-0.018*
270.1	-0.016*
273.2	-0.014*
276.3	-0.012*
279.3	-0.009*
282.6	-0.007*
285.7	-0.005*
288.6	-0.003
292.0	-0.001*
295.0	0.002
298.1	0.004
301.4	0.006*
304.5	0.008*
307.6	0.010*
310.6	0.012*
313.9	0.014*
316.9	0.017*
320.1	0.019
323.2	0.021*
326.6	0.023*
329.7	0.026*
332.8	0.028*
336.1	0.030*
329.3	0.032*
342.4	0.034*
CURVE 5*	
102.5	-0.141
104.9	-0.140
106.7	-0.139
108.8	-0.137
111.2	-0.136
113.7	-0.134

* Not shown in figure.

DATA TABLE 33. THERMAL LINEAR EXPANSION OF NEODYMIUM Nd (continued)

T	$\Delta L/L_0$
CURVE 5 (cont.)*	
116.1	-0.132
118.5	-0.130
120.9	-0.128
123.3	-0.126
125.5	-0.124
127.9	-0.122
130.4	-0.120
132.7	-0.118
135.2	-0.116
137.5	-0.114
139.9	-0.112
142.4	-0.110
145.1	-0.108
147.6	-0.106
150.1	-0.104
152.6	-0.102
155.3	-0.101
157.8	-0.099
160.4	-0.097
163.0	-0.095
167.2	-0.093
169.8	-0.091
172.6	-0.089
175.3	-0.087
177.9	-0.085
180.4	-0.083
183.0	-0.081
185.5	-0.079
188.1	-0.077
190.8	-0.075
193.5	-0.073
196.1	-0.071
198.8	-0.069
201.4	-0.066
204.1	-0.064
206.9	-0.062
209.5	-0.060
212.3	-0.058
215.0	-0.056
217.9	-0.054
220.7	-0.052
223.4	-0.050
226.3	-0.048
229.1	-0.046
232.0	-0.044
234.9	-0.042
237.9	-0.040
CURVE 5 (cont.)*	
241.8	-0.037
244.7	-0.035
247.6	-0.033
250.5	-0.031
253.4	-0.028
256.3	-0.026
259.4	-0.024
262.3	-0.022
265.2	-0.020
268.1	-0.018
271.1	-0.016
274.1	-0.014
277.0	-0.012
280.0	-0.010
282.9	-0.007
285.8	-0.005
288.7	-0.003
291.7	-0.001
294.7	0.001
298.0	0.004
301.0	0.006
304.1	0.010
310.2	0.012
313.2	0.014
316.1	0.016
319.0	0.019
322.3	0.021
325.3	0.023
328.2	0.025
332.2	0.028
334.6	0.030
338.4	0.033
341.1	0.035
343.9	0.037
346.9	0.039
349.2	0.041
CURVE 6	
78	-0.152
173	-0.086
273	-0.016
373	0.064
473	0.134
573	0.214
673	0.289
773	0.373
CURVE 6 (cont.)	
873	0.390
898	0.384
CURVE 7*	
78	-0.153
173	-0.086
273	-0.013
373	0.062
473	0.132
573	0.212
673	0.277
773	0.372
873	0.390
898	0.397
CURVE 8*, †, ‡	
1.98	0.00 x 10^{-4}
3.02	-0.48
4.03	-1.13
4.54	-1.55
5.02	-2.10
5.97	-0.77
6.58	2.13
6.99	4.21
8.04	8.75
9.03	12.31
10.02	15.58
10.99	18.67
12.09	22.09
13.03	24.87
14.04	27.85
15.05	30.98
CURVE 9*	
293	0.000
373	0.042
393	0.055
413	0.068
433	0.082
453	0.098
473	0.114
493	0.131
513	0.149
533	0.167
CURVE 9 (cont.)*	
553	0.185
573	0.203
593	0.220
613	0.237
633	0.255
653	0.274
673	0.291
693	0.316
713	0.343
733	0.381
753	0.395
773	0.419
793	0.440
813	0.464
833	0.489
853	0.516
873	0.541
893	0.568
913	0.591
933	0.612
953	0.633
973	0.663
993	0.685
1013	0.713
1017	0.717
1018	0.716
1019.7	0.702
1021.7	0.698
1027	0.698
1028	0.701
1033	0.711
1039	0.721
CURVE 10*	
1033	0.724
1023	0.714
1013	0.702
993	0.680
988	0.678
981	0.681
977	0.692
973	0.693
969	0.692
953	0.674
933	0.668
923	0.668
CURVE 10 (cont.)*	
913	0.662
893	0.637
873	0.611
853	0.585
833	0.557
CURVE 11	
293	0.000
331	0.000
374	0.000
431	0.082
474	0.082
529	0.137
577	0.164
621	0.164
670	0.273
724	0.219
759	0.246
822	0.328
912	0.410
969	0.492
CURVE 12	
293	0.000*
331	0.034*
374	0.068*
431	0.161
474	0.169
529	0.296
577	0.330
621	0.407
670	0.381
724	0.508
759	0.508
822	0.576
912	0.838
969	0.830
CURVE 13*, †	
1123	0.000
1163	0.000
1168	0.485
1168	0.485

* Not shown in figure.

† This curve is here reported using the first given temperature as reference temperature at which $\Delta L/L_0 = 0$.

‡ Author's data for coefficient of thermal expansion have been integrated by TPRC to obtain $\Delta L/L_0$.

DATA TABLE 33. COEFFICIENT OF THERMAL LINEAR EXPANSION OF NEODYMIUM Nd

[Temperature, T, K; Coefficient of Expansion, α, 10^{-6} K^{-1}]

T	α
CURVE 8[‡]	
1.98	-0.44
3.02	-0.50
4.03	-0.78
4.54	-0.87
5.02	-1.40
5.97	4.19
6.58	5.34
6.99	4.84
8.04	3.80
9.03	3.39
10.02	3.22
10.99	3.16
12.09	3.05
13.03	2.87
14.04	3.03
15.05	3.17

‡ Author's data for coefficient of thermal expansion have been integrated by TPRC to obtain $\Delta L/L_0$.

FIGURE AND TABLE NO. 34R. RECOMMENDED VALUES FOR THERMAL LINEAR EXPANSION OF NICKEL Ni

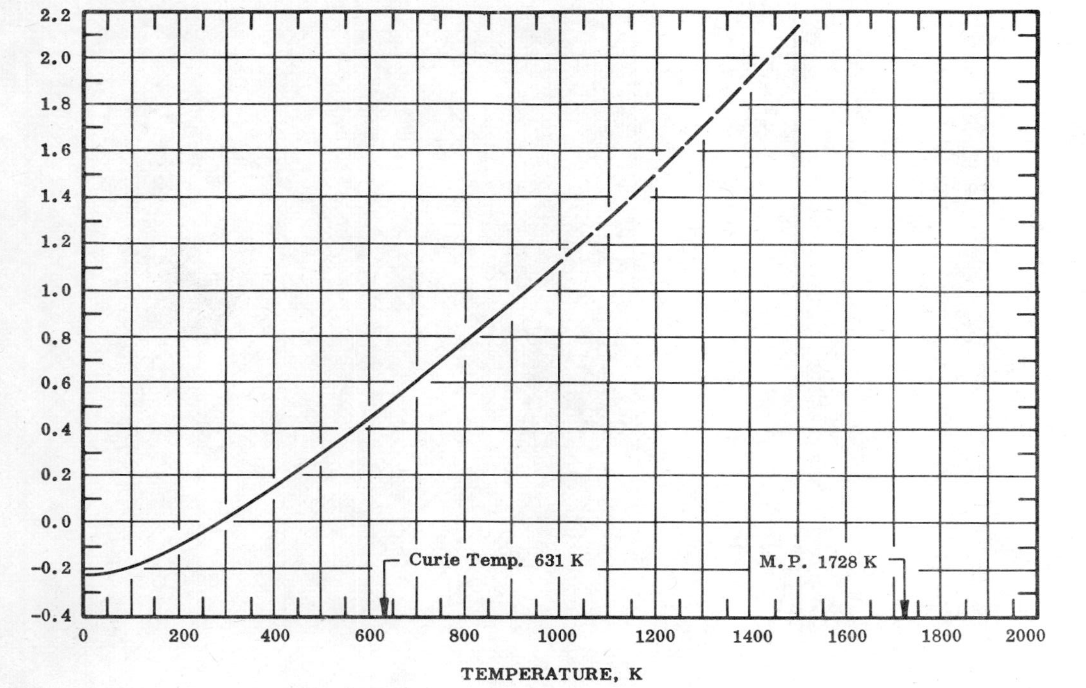

RECOMMENDED VALUES

[Temperature, T, K; Linear Expansion, $\Delta L/L_0$, %; α, K^{-1}]

T	$\Delta L/L_0$	$\alpha \times 10^6$
5	-0.231	0.02
25	-0.230	0.25
50	-0.229	1.5
75	-0.222	4.3
100	-0.208	6.6
200	-0.114	11.3
293	0.000	13.4
400	0.150	14.5
500	0.299	15.3
600	0.455	15.9
700	0.617	16.4
800	0.783	16.8
900	0.953	17.1
1000	1.126†	17.4
1100	1.302†	17.8
1200	1.483†	18.3
1300	1.669†	18.9
1400	1.861†	19.5
1500	2.060†	20.3

† Provisional values.

REMARKS

The tabulated values are considered accurate to within ±3% at temperatures below 1000 K and ±7% above. These values can be represented approximately by the following equations:

$$\Delta L/L_0 = 1.362 \times 10^{-3}\,(T-293) + 4.544 \times 10^{-7}\,(T-293)^2 - 1.806 \times 10^{-10}\,(T-293)^3 \qquad (293 < T < 895)$$

$$\Delta L/L_0 = 0.944 + 1.713 \times 10^{-3}\,(T-895) + 1.283 \times 10^{-7}\,(T-895)^2 + 1.447 \times 10^{-10}\,(T-895)^3 \qquad (895 < T < 1500)$$

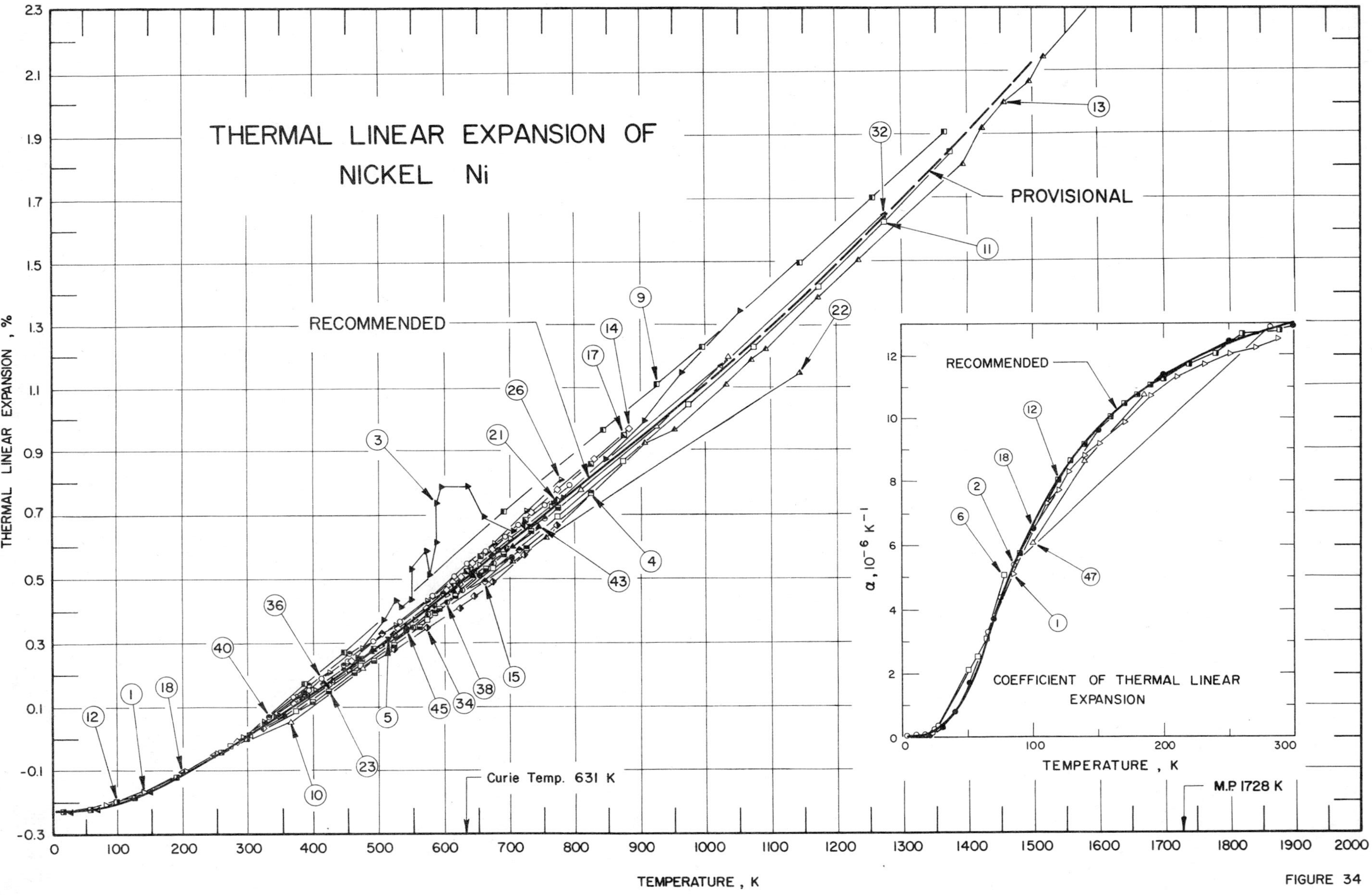
THERMAL LINEAR EXPANSION OF
NICKEL Ni
PROVISIONAL
RECOMMENDED
Curie Temp. 631 K
M.P. 1728 K
THERMAL LINEAR EXPANSION, %
TEMPERATURE, K
RECOMMENDED
COEFFICIENT OF THERMAL LINEAR
EXPANSION
α, 10⁻⁶ K⁻¹
TEMPERATURE, K

FIGURE 34

SPECIFICATION TABLE 34. THERMAL LINEAR EXPANSION OF NICKEL Ni

Cur. No.	Ref. No.	Author(s)	Year	Method Used	Temp. Range, K	Name and Specimen Designation	Composition (weight percent), Specifications, and Remarks
1	1	Nix, F.C. and MacNair, D.	1941	I	81-762		99.90 Ni, 0.05 C, 0.03 Fe, 0.011 SiO_2, 0.0043 Al, 0.0035 Mg, <0.001 Mn, and 0.0005 Cu; specimen obtained from International Nickel Co.; rod specimen ~6 cm long; expansion below room temperature measured in helium atmosphere at pressure of 3 mm Hg with temperature decreasing at 20 K/hr; above room temperature expansion measured in helium atmosphere with temperature increasing at 15 K/hr; expansion above 423 K measured in vacuum of 10^{-5} mm Hg; zero-point correction is -0.025%.
2*	41	White, G.K.	1965	E	4-85, 283		99.99^+ Ni; obtained from Johnson Matthey; annealed at 1023 K in vacuo for some hours after machining.
3	231	Petrov, Yu.I.	1965	X	296-1054		Small nickel particle ~5 x 10^{-6} cm in diameter, prepared by condensing metal vapor in rarefied argon atmosphere, stored for over a year in hermetically sealed container; expansion measured under vacuum of ~10^{-4} torr using Ni filtered Cu x-radiation; smoothed values for the initial heating of six specimens; zero-point correction of -0.004% determined by graphical extrapolation.
4	231	Petrov, Yu.I.	1965	X	586-825		One of the above specimens; quenched from 1053 K; expansion measured with increasing temperature on second test.
5	231	Petrov, Yu.I.	1965	X	390-693		Similar to the above specimen except quenched from 703 K.
6*	3	Fraser, D.B. and Hollis-Hallet, A.C.	1965	I	45-84	First test	Cubic crystal structure; external diameter 2.54 cm, internal diameter 2.06 cm, 5.08 cm long; expansion measured on first test; this curve is here reported using the first given temperature as reference temperature at which $\Delta L/L_0 = 0$; zero-point correction is 0.2%; data on the coefficient of thermal linear expansion also reported.
7*	3	Fraser, D.B. and Hollis-Hallet, A.C.	1965	I	29-83	Second test	The above specimen; second test; this curve is here reported using the first given temperature as reference temperature at which $\Delta L/L_0 = 0$; zero-point correction is 0.201%; data on the coefficient of thermal linear expansion also reported.
8*	3	Fraser, D.B. and Hollis-Hallet, A.C.	1965	I	31-78	Third test	The above specimen; third test; this curve is here reported using the first given temperature as reference temperature at which $\Delta L/L_0 = 0$; zero-point correction is 0.201%; data on the coefficient of thermal linear expansion also reported.
9	232	Shevlin, T.S., Newkirk, H.W., Stevens, E.G., and Greenhouse, H.M.	1956	X	302-1366		Pure wire, 0.006 in. diameter, obtained from Driver-Harris Co., Harrison, N.J.; zero-point correction of 0.017% determined by graphical extrapolation.
10	47	Allen, R.D.	1959		293-1033	"A" nickel	Average of 10 runs; standard deviation 0.17 x 10^{-6} per degree F.
11	217	Totskii, E.E.	1964		273-1373	Electrolytic Ni	No details given; zero-point correction of -0.021% determined by graphical interpolation.
12	8	Altman, H.W., Rubin, T., and Johnston, H.L.	1954	I	20-300	L-Nickel	99.6 Ni furnished by the International Nickel Co.; zero-point correction of 0.0088% determined by graphical interpolation; data on the coefficient of thermal linear expansion also reported.
13	138	Fieldhouse, I.B., Hedge, J.C., Lang, J.I., Takata, A.N., and Waterman, T.E.	1956	L	386-1658		Data average of three separate runs.
14	233	Owen, E.A. and Yates, E.L.	1936	X	285-881		99.98 Ni; powder specimens prepared from four different samples of annealed filings; specimens 1 and 2 annealed 14 hr at 873 K and cooled slowly; specimen 3 annealed 1 hr at 603 K and quenched in ice water; specimen 4 annealed 1 hr at 713 K and quenched in ice water; all specimens gave same results; data reported are results from all specimens; lattice parameter reported at 289 K is 3.51718 Å; 3.51738 Å at 293 K determined by graphical interpolation.

* Not shown in figure.

SPECIFICATION TABLE 34. THERMAL LINEAR EXPANSION OF NICKEL Ni (continued)

Cur. No.	Ref. No.	Author(s)	Year	Method Used	Temp. Range, K	Name and Specimen Designation	Composition (weight percent), Specifications, and Remarks
15	14	Palatnik, L. S., Pugachev, A. T., Boiko, B. T. and Bratsykhin, V. M.	1967	F	293-715		Underlayer of NaCl condensed on base prior to formation of film; film specimen 400 to 600 Å thick; expansion measured under vacuum of 10^{-4} to 10^{-5} torr with increasing temperature.
16*	14	Palatnik, L. S., et al.	1967	F	705-357		The above specimen; expansion measured with decreasing temperature.
17	235	Owen, E. A. and Yates, E. L.	1937	X	284-875		99.98 Ni; specimen provided by the Mond Nickel Co., Ltd.; lattice parameter reported at 290 K is 3.5172 Å; 3.5173 Å at 293 K determined by graphical interpolation.
18	22	Fraser, D. B. and Hollis-Hallett, A. C.	1961	I	20-300		Metal machined into form of thick-walled cylindrical tube with ends carefully ground plane and parallel to each other, annealed; reported error decreases from maximum of 20% at 20 K.
19*	236	Davis, M., Densem, C. E. and Rendall, J. H.	1955		293, 593		0.07 C, 0.016 Si, 0.013 Fe, 0.01 to 0.2 O, 0.003 S, 0.0005 Mn, 0.0003 Mg; grade A carbonyl nickel powder supplied by the Mond Nickel Co., Ltd.; annealed.
20*	237	Jordan, L. and Swanger, W. H.	1930	T	298-1173	B-2	99.94 Ni, 0.03 Fe, 0.016 Co, 0.006 each Cu, Si, 0.005 C, 0.004 S; specimen annealed and cold swaged to 0.35 cm diameter, 31.1 cm long; annealed; expansion measured by P. Hidnert.
21	81	Trombe, F. and Foex, M.	1953	L	273-773		Expansion measured in hydrogen atmosphere below room temperature, in argon atmosphere above room temperature; zero-point correction is -0.023%.
22	150	Eisenstein, A.	1946	X	293, 1148		Cubic crystal structure; reported error in measurement of lattice constant 0.2%.
23	55	Richter, F. and Lotter, U.	1969	V	373-773	Nickel "S"	0.02 C, 0.01 each Cu, Fe, Mg, Mn, Si; cylindrical specimen 6 mm diameter, 35 mm long; one side being of hemispherical shape; obtained from Vakuumschmelze Hanau; heating rate 20 K/min; volume magnetostriction $\omega_m \times 10^4 = -2.7 \pm 0.3$; Curie temperature $\theta_C = 628 \pm 3$ K; expansion measured under constant magnetic field H.
24*	31	Kagan, A. S.	1964	X	295,473		Massive specimen, glued to copper holder; reported error 2%.
25*	31	Kagan, A. S.	1964	X	295, 473		Powdered specimen, particle diameter ~2 μ, glued to copper holder; reported error 2%.
26	238	Masumoto, H., Saito, H., Marakami, Y. and Kikuchi, M.	1968	L	273-780	No. 1	99.8 Ni, 0.190 Co, 0.019 C, 0.014 Fe, 0.011 Al, 0.004 Si, 0.001 S; commercial electrolytic Ni; melted in vacuum, annealed; single crystal rod 12 cm long, 3 mm diameter; transition point 633 K; zero-point correction is -0.025%.
27*	238	Masumoto, H., et al.	1968	L	273-780	No. 5	Similar to the above specimen; zero-point correction is -0.173%.
28*	238	Masumoto, H., et al.	1968	L	273-780	No. 7	Similar to the above specimen; zero-point correction is -0.281%.
29*	56	Masumoto, H. and Nara, S.	1927	L	303, 373		Pure Ni.
30*	59	Masumoto, H.	1929	L	303, 373	Electrolytic Ni	Circular rod specimen, 4 mm diameter, 10 cm long; expansion measured in vacuum.
31*	59	Masumoto, H.	1929	L	303, 373	Mond Ni	Raw material supplied by Sugibayashi and Co., refined by electrolysis, circular rod specimen, 4 mm thick, 10 cm long; expansion measured in vacuum.
32	60	Holborn, L. and Day, A. L.	1901	T	273-1273		No details given; zero-point correction of -0.028% determined by graphical interpolation.

* Not shown in figure.

SPECIFICATION TABLE 34. THERMAL LINEAR EXPANSION OF NICKEL Ni (continued)

Cur. No.	Ref. No.	Author(s)	Year	Method Used	Temp. Range, K	Name and Specimen Designation	Composition (weight percent), Specifications, and Remarks
33*	240	Pathak, P. D., Gupta, M. C. and Trivedi, J. M.	1969		273-773		99.99 Ni obtained from Messrs Johnson Matthey and Co. Ltd., London; rod of 10 cm long, 0.5 cm diameter, transition temperature 630 K.
34	153	Masumoto, H., Saito, H. and Sawaya, S.	1969	L	273-773		99.980 Ni, 0.180 Co, 0.002 S, 0.001 Cu, 0.001 Fe, electrolytic Ni; rod specimen 2 mm diameter, 10 cm long, cooled at rate 300 K/hr after heating in vacuum at 1173 K for 30 min, density 8.90 g cm^{-3} at 293 K; zero-point correction is -0.028%.
35*	241	Masumoto, H., Sawaya, S. and Kadowaki, S.	1969	L	273, 313		Electrolytic Ni, 0.180 Co, 0.002 S, 0.002 C, 0.001 Cu, 0.001 Fe, rod specimen 2 mm diameter, 10 cm long, cooled at rate 300 K/hr after heating in vacuum at 1273 K for 1 hr, density 8.90 g cm^{-3} at 293 K.
36	254	Rosenbohm, E.	1938	L	293-793		Specimen 41.00 mm long at 293 K; reported error 1%.
37*	301	Arbuzov, M. P. and Zelenkov, I. A.	1964	L	295-974		Pure metal.
38	388	Nakamura, K.	1936	L	288-678		Rod specimen 2.30 mm diameter, 8.00 cm long; density 8.904 g cm^{-3}.
39*	389	Owen, E. A., Yates, E. L. and Sully, A. H.	1937	X	288-873		Lump specimen heat treated, then filings taken, heated 14 hr at 973 K; lattice parameter reported at 298 K is 3.51713 Å; 3.51738 Å at 293 K determined by graphical interpolation.
40	415	Saur, E. and Wenkowitsch, V.	1958	X	291-753	Probe 1	0.1 total impurity; specimen obtained from Vacuumschmelze AG, Hanau; expansion measured using CuKα λ = 1.5374 Å x-radiation.
41*	415	Saur, E. and Wenkowitsch, V.	1958	I	291-744	Probe 1	The above specimen; macroscopic expansion measured using red line of He λ = 0.6680 μ; specimen 4.661 mm long at 291 K.
42*	415	Saur, E. and Wenkowitsch, V.	1958	X	305-675	Probe 2	0.13 Mg, 0.10 Fe, 0.05 Si, 0.02 each C, Cu, 0.002 S; obtained from Mond Nickel Co., London; expansion measured using CuKα λ = 1.5374 Å x-radiation.
43	415	Saur, E. and Wenkowitsch, V.	1958	I	291-745	Probe 2	The above specimen; macroscopic expansion measured using red line of He λ = 0.6680 μ; specimen 4.468 mm long at 291 K.
44*	612	Boiko, B. T., Palatnik, L. S. and Pugachev, A. T.	1967	F	293-713		>99.9 pure thin film specimen condensed in vacuum 5 x 10^{-5} torr; thickness of the film in the range 600 + 10 Å; lattice parameter expansion measured with increasing temperature.
45	612	Boiko, B. T., et al.	1967	F	704-351		The above specimen; lattice parameter expansion measured with decreasing temperature.
46*	658	Masumoto, H. and Sawaya, S.	1970	L	293-773		Electrolytic sample; zero-point correction is -0.030%.
47*	709	Shah, J. S.	1971	X	40-180		99.9 pure, powder sample of particle size 8-10 μ from Koch Light Labs, Inc., Colnbrook, England; lattice parameter not corrected for refraction; lattice parameter reported as 3.51594 Å at 40 K; this curve is here reported using the first given temperature as reference temperature at which $\Delta L/L_0 = 0$; data on the coefficient of thermal linear expansion also reported.
48*	740	Gurevich, M. Ye. and Larikov, L. N.	1970	L	293-841		High purity Ni; no other details given; authors' values for thermal linear expansion do not agree with their data on coefficient of thermal linear expansion.
49*	741	Tanji, Y.	1971	L	273-1073		99.97 Ni, 0.012 Fe, 0.007 Si, 0.008 Cu, 0.003 Mn; specimen prepared with electrolytic Ni; melted in high frequency induction furnace; forged, rolled, faced to finished flat rectangular bar of length 12.000 cm, width 1.1989 cm, and thickness 0.1351 cm; annealed in vacuum at 1273 K for 10 hr and cooled in furnace; density 8.904 g cm^{-3} at room temperature.

* Not shown in figure.

SPECIFICATION TABLE 34. THERMAL LINEAR EXPANSION OF NICKEL Ni (continued)

Cur. No.	Ref. No.	Author(s)	Year	Method Used	Temp. Range, K	Name and Specimen Designation	Composition (weight percent), Specifications, and Remarks
50*	738	Kantola, M. and Tokola, E.	1967	X	298-1063		Alloy prepared using MD 101 nickel powder from Metals Disintegrating Corp., N.J.; sifted through sieve of diameter equal to 0.060 mm; lattice parameter at 298 K reported as 3.5236 Å; 3.5240 Å at 293 K determined by graphical extrapolation.

* Not shown in figure.

DATA TABLE 34. THERMAL LINEAR EXPANSION OF NICKEL Ni

[Temperature, T, K; Linear Expansion, $\Delta L/L_0$, %]

T	$\Delta L/L_0$
CURVE 1	
81.2	-0.21311
82.2	-0.21060*
88.7	-0.20608*
99.2	-0.20056*
107.7	-0.19553*
115.0	-0.19051*
121.2	-0.18549*
128.2	-0.18097*
133.4	-0.16720*
140.1	-0.17043
145.2	-0.16541*
150.7	-0.16039*
155.7	-0.15561*
161.2	-0.15034*
166.2	-0.14607*
172.2	-0.14030*
177.2	-0.13528*
181.2	-0.13026*
176.2	-0.12524*
181.2	-0.12072*
195.2	-0.11570*
199.2	-0.11017*
204.2	-0.10515
208.6	-0.10063*
213.6	-0.09511*
217.6	-0.09009*
222.0	-0.08507*
226.2	-0.08055*
230.9	-0.07502*
235.0	-0.07050*
239.3	-0.06498*
243.5	-0.06021*
247.8	-0.05494*
252.1	-0.04941
256.2	-0.04490*
260.2	-0.03987*
264.2	-0.03485*
268.7	-0.02983*
273.2	-0.02481*
275.2	-0.02130*
281.3	-0.01477*
285.2	-0.00975*
288.3	-0.00472*
293.3	0.00030

T	$\Delta L/L_0$
CURVE 1 (cont.)	
296.4	0.00532*
301.2	0.01109*
304.4	0.01536*
309.2	0.02038*
312.6	0.02540*
316.2	0.03042*
320.2	0.03545*
324.2	0.04047*
328.2	0.04549*
332.2	0.05051*
335.4	0.05553*
339.2	0.06055*
343.0	0.06557*
347.2	0.07059*
350.9	0.07562
354.7	0.08064*
358.6	0.08566*
362.3	0.09068*
366.6	0.09570*
370.3	0.10072*
372.2	0.10449*
374.2	0.10762*
377.2	0.11203*
380.2	0.11693*
384.3	0.12182*
388.7	0.12721*
391.5	0.13161*
395.8	0.13675*
389.3	0.14141*
402.8	0.14655
407.0	0.15169*
410.2	0.15609*
413.7	0.16123*
417.2	0.16662*
420.7	0.17078*
424.2	0.17592*
427.2	0.18057*
431.0	0.18620*
433.3	0.19075*
436.7	0.19526*
440.2	0.20065*
443.2	0.20505*
446.6	0.22995*
449.4	0.21509*

T	$\Delta L/L_0$
CURVE 1 (cont.)	
453.2	0.21999
456.0	0.22464*
459.2	0.23002*
462.3	0.23443*
465.8	0.23981*
469.4	0.24446*
472.6	0.24912*
476.2	0.25450*
479.2	0.25891*
482.4	0.26405*
486.2	0.26943*
489.2	0.27433*
492.2	0.27849*
495.9	0.28388*
499.7	0.28902*
502.8	0.29342
506.5	0.29881*
509.2	0.30332*
512.2	0.30811*
516.2	0.31325*
519.2	0.31790*
522.2	0.32304*
525.3	0.32818*
528.2	0.33528*
531.5	0.33724*
534.5	0.34287*
537.7	0.34703*
541.2	0.35242*
543.5	0.35683*
546.8	0.36197*
552.4	0.37176
555.5	0.37665*
558.2	0.38130*
562.0	0.38645*
564.4	0.39110*
567.2	0.39599*
570.2	0.40162*
572.2	0.40627*
576.2	0.41093*
578.9	0.41558*
581.2	0.42047*
584.8	0.42537*
587.2	0.43026*
590.2	0.43516*

T	$\Delta L/L_0$
CURVE 1 (cont.)	
593.4	0.44006*
596.0	0.44520*
599.2	0.45034*
602.2	0.45499
605.0	0.46013*
608.0	0.46527*
610.3	0.46992*
613.7	0.47482*
616.2	0.47922*
617.9	0.48167*
619.2	0.48412*
620.7	0.48657*
621.5	0.48901*
622.8	0.49146*
624.1	0.49391*
626.0	0.49711*
627.2	0.49881*
628.2	0.50125*
630.3	0.50419*
631.3	0.50615*
632.7	0.50909*
634.0	0.51105*
635.2	0.51349*
636.6	0.51594*
637.2	0.51839*
639.4	0.52084*
641.0	0.52353*
642.2	0.52573*
643.8	0.52818*
645.6	0.53136*
646.9	0.53332*
648.2	0.53553*
649.9	0.53797*
651.2	0.54091
652.3	0.54287*
654.4	0.54532*
655.7	0.54801*
658.5	0.55266*
661.2	0.55756*
664.8	0.56245*
667.8	0.56735*
671.0	0.57225*
674.0	0.57714*
677.5	0.58253*

T	$\Delta L/L_0$
CURVE 1 (cont.)	
680.2	0.58742*
682.2	0.59183*
686.2	0.59672*
688.7	0.60187*
691.7	0.60652*
695.2	0.61166*
697.2	0.61631*
700.3	0.62120
703.2	0.62659*
707.2	0.63149*
715.9	0.64568*
727.2	0.66527*
740.4	0.68485*
750.7	0.70443*
759.4	0.72010*
761.7	0.72402
CURVE 2*, †, ‡	
4	0.000 x 10^{-2}
5	0.002
6	0.004
8	0.010
10	0.019
12	0.030
14	0.045
16	0.063
18	0.086
20	0.114
22	0.148
24	0.191
26	0.247
65	7.296
75	11.146
85	16.046
CURVE 3	
296	0.004*
343	0.069
404	0.156*
451	0.228*
469	0.253
501	0.293*

T	$\Delta L/L_0$
CURVE 3 (cont.)	
504	0.374
526	0.437
531	0.417
550	0.434
550	0.532
573	0.583
577	0.509
588	0.618
588	0.739
595	0.781
631	0.789
635	0.733*
660	0.698
661	0.700*
674	0.669*
704	0.654
768	0.735
782	0.754
845	0.877
905	0.999
964	1.148
1054	1.348
CURVE 4	
586	0.393
729	0.597
825	0.767
CURVE 5	
390	0.136
515	0.314
585	0.418
593	0.439*
612	0.480
640	0.523
693	0.597
CURVE 6*, †	
45.0	0.000 x 10^{-3}
45.4	0.034
58.0	2.822

* Not shown in figure.
† This curve is here reported using the first given temperature as reference temperature at which $\Delta L/L_0 = 0$.
‡ Author's data for coefficient of thermal expansion have been integrated by TPRC to obtain $\Delta L/L_0$.

DATA TABLE 34. THERMAL LINEAR EXPANSION OF NICKEL Ni (continued)

T	$\Delta L/L_0$
CURVE 6 (cont.)*, †	
69.8	6.249 x 10^{-3}
70.3	6.707
70.6	6.782
72.0	6.884
84.1	13.637
CURVE 7*, †	
29.0	0.000 x 10^{-3}
30.7	0.020
34.8	0.145
40.4	0.462
45.0	0.992
54.8	3.114
65.1	5.889
83.4	13.463
CURVE 8*, †	
31.3	0.000 x 10^{-3}
33.5	0.023
39.0	0.385
48.8	1.820
52.7	2.561
61.7	4.615
74.7	9.089
78.4	10.563
CURVE 9	
302	0.017
388	0.172
449	0.272
690	0.711
846	0.966
927	1.108
994	1.236
1148	1.506
1255	1.703
1366	1.916
CURVE 10	
293	0.000*
366	0.054
477	0.220
588	0.400*
699	0.590*
CURVE 10 (cont.)	
810	0.780
922	0.980
1033	1.200
CURVE 11	
273	-0.021
373	0.084
473	0.225
573	0.377
673	0.536
773	0.699
873	0.869
973	1.046
1073	1.232
1173	1.427
1273	1.632
1373	1.847
CURVE 12	
20	-0.2301
30	-0.2299*
40	-0.2293*
50	-0.2281*
60	-0.2259
70	-0.2226*
80	-0.2182*
90	-0.2129*
100	-0.2067
110	-0.1996*
120	-0.1919*
130	-0.1836
140	-0.1748*
150	-0.1654*
160	-0.1556*
170	-0.1454*
180	-0.1349*
190	-0.1240
200	-0.1129*
220	-0.0900*
240	-0.0662*
260	-0.0417
273	-0.0254*
290	-0.0041*
298	0.0064*
300	0.0088
CURVE 13	
386	0.125
430	0.181
586	0.410*
703	0.552
714	0.591*
758	0.629
908	0.921
952	0.969
1030	1.118
1070	1.193
1091	1.225
1175	1.390
1236	1.512
1394	1.810
1422	1.929
1458	2.012
1498	2.070
1519	2.146
1658	2.483*
CURVE 14	
285	-0.011
286	-0.011*
287	-0.011*
288	-0.009*
289	-0.006*
371	0.109
460	0.243
530	0.345
548	0.365*
574	0.413
604	0.461
607	0.466*
609	0.466*
611	0.469*
614	0.481*
615	0.482
617	0.479*
618	0.488*
619	0.493*
621	0.492*
622	0.494*
625	0.502
629	0.516*
632	0.521*
635	0.529*
636	0.530*
CURVE 14 (cont.)	
638	0.535*
643	0.552
646	0.550*
647	0.552*
651	0.561*
661	0.584*
671	0.596
678	0.605*
683	0.609*
684	0.610*
702	0.650*
733	0.706
774	0.780
830	0.876
876	0.950*
881	0.975
CURVE 15	
293	0.000*
418	0.176
507	0.331
606	0.462*
663	0.497
715	0.591
CURVE 16*	
705	0.602
651	0.526
605	0.436
546	0.346
496	0.284
435	0.183
389	0.147
357	0.078
CURVE 17	
284	-0.007*
290	-0.004*
373	0.110*
473	0.257
541	0.357
576	0.411*
625	0.493*
635	0.522
646	0.547*
CURVE 17 (cont.)	
655	0.567
666	0.584*
676	0.601*
727	0.687
775	0.772*
827	0.860
875	0.948
CURVE 18 ‡	
20	-0.230*
30	-0.229
40	-0.229*
50	-0.228*
70	-0.222
100	-0.207*
150	-0.167
200	-0.114
250	-0.055
300	0.009*
CURVE 19*	
293	0.000
593	0.399
CURVE 20*	
298	0.007
373	0.106
473	0.250
573	0.404
623	0.490
673	0.572
773	0.731
873	0.900
973	1.071
1073	1.248
1173	1.434
CURVE 21	
273	-0.023*
293	0.000*
323	0.038
373	0.106*
423	0.177
473	0.249*
523	0.327
CURVE 21 (cont.)	
573	0.406*
623	0.494
673	0.579
723	0.665
773	0.741
CURVE 22	
293	0.000*
1148	1.154
CURVE 23 ‡	
373	0.089*
400	0.119
425	0.151
462	0.202
492	0.245
524	0.294
561	0.353
592	0.405
617	0.448
623	0.459*
666	0.536
690	0.578
732	0.650
773	0.721
CURVE 24*	
295	0.003
473	0.254
CURVE 25*	
295	0.003
473	0.241
CURVE 26	
273	-0.025*
293	0.000*
324	0.053
379	0.131
427	0.205
478	0.282
529	0.356
575	0.432
633	0.525*
676	0.608
728	0.708
780	0.806

* Not shown in figure.
† This curve is here reported using the first given temperature as reference temperature at which $\Delta L/L_0 = 0$.
‡ Author's data for coefficient of thermal expansion have been integrated by TPRC to obtain $\Delta L/L_0$.

DATA TABLE 34. THERMAL LINEAR EXPANSION OF NICKEL Ni (continued)

T	$\Delta L/L_0$
CURVE 27*	
273	-0.028
293	0.000
328	0.050
376	0.122
427	0.199
477	0.274
526	0.345
577	0.423
633	0.511
677	0.599
729	0.698
780	0.799
CURVE 28*	
273	-0.027
293	0.000
328	0.049
376	0.123
427	0.200
477	0.275
526	0.348
577	0.424
633	0.512
677	0.596
730	0.698
780	0.793
CURVE 29*	
303	0.0136
373	0.1086
CURVE 30*	
303	0.0128
373	0.1021
CURVE 31*	
303	0.0129
373	0.1032
CURVE 32	
273	-0.028*
523	0.320*
648	0.523
773	0.727*
1023	1.170
1273	1.649
CURVE 33*, ‡	
273	-0.026
373	0.105
473	0.248
523	0.324
573	0.406
593	0.443
643	0.547
650	0.563
673	0.608
723	0.691
773	0.778
CURVE 34	
273	-0.028*
293	0.000*
323	0.037*
373	0.099*
423	0.158*
473	0.223*
523	0.284
573	0.350
623	0.409
648	0.450
673	0.490
723	0.576
773	0.667
CURVE 35*	
273	-0.027
313	0.027
CURVE 36	
293	0.0000*
373	0.1302
393	0.1554
413	0.1861
433	0.2156*
453	0.2437
473	0.2731*
493	0.3039
513	0.3341*
533	0.3651
553	0.3976*
573	0.4300*
583	0.4464
CURVE 36 (cont.)	
593	0.4629*
603	0.4795*
608	0.4880*
613	0.4965*
618	0.5053
623	0.5145*
628	0.5244*
633	0.5344*
638	0.5441
643	0.5532*
648	0.5616*
653	0.5697*
663	0.5858
673	0.6020*
693	0.6351
713	0.6677
733	0.7006*
753	0.7339
773	0.7671*
793	0.7995
CURVE 37*, ‡	
295	0.003
373	0.106
425	0.177
473	0.245
523	0.319
571	0.395
622	0.483
633	0.503
674	0.572
725	0.657
774	0.741
794	0.774
824	0.823
874	0.907
922	0.990
974	1.082
CURVE 38	
288	-0.006*
293	0.000*
303	0.018*
318	0.033
328	0.053*
353	0.075*
CURVE 38 (cont.)	
378	0.111*
403	0.145*
428	0.178*
453	0.210*
478	0.248*
503	0.281*
528	0.329*
553	0.353
578	0.391
603	0.424
628	0.463
653	0.563*
678	0.541
CURVE 39*	
288	-0.007
373	0.110
473	0.256
573	0.409
673	0.598
773	0.774
878	0.952
CURVE 40	
334	0.069
394	0.170
448	0.230
456	0.265
582	0.452*
600	0.458
626	0.515*
659	0.562*
681	0.589
730	0.669
753	0.696
CURVE 41*	
291	-0.0028
385	0.1287
446	0.2145
492	0.2758
553	0.3665
612	0.4592
624	0.4789
628	0.4852
CURVE 41 (cont.)*	
642	0.5084
656	0.5303
671	0.5548
706	0.6037
744	0.6605
CURVE 42*	
305	0.023
335	0.061
409	0.167
498	0.301
556	0.382
608	0.485
638	0.538
674	0.575
675	0.575
CURVE 43	
291	-0.0028*
383	0.1270*
449	0.2200
490	0.2762
554	0.3686*
613	0.4622
619	0.4708*
626	0.4847*
641	0.5093
653	0.5273*
674	0.5572*
705	0.6011
745	0.6610
CURVE 44*	
293	0.000
414	0.166
504	0.323
604	0.450
659	0.492
713	0.590
CURVE 45	
704	0.569
653	0.513
604	0.431*
CURVE 45 (cont.)	
542	0.341
493	0.277*
430	0.178*
387	0.143
351	0.074
CURVE 46*	
293	0.000
323	0.039
373	0.103
423	0.169
473	0.230
523	0.291
573	0.352
623	0.416
673	0.491
723	0.574
773	0.648
CURVE 47*, †	
40	0.000
60	0.004
80	0.012
100	0.022
140	0.052
180	0.091
CURVE 48*, ‡	
293	0.000
373	0.103
398	0.136
423	0.171
448	0.206
473	0.243
498	0.281
523	0.320
548	0.362
573	0.405
598	0.449
623	0.494
648	0.537
673	0.579
698	0.619
723	0.660
748	0.700

* Not shown in figure.
† This curve is here reported using the first given temperature as reference temperature at which $\Delta L/L_0 = 0$.
‡ Author's data for coefficient of thermal expansion have been integrated by TPRC to obtain $\Delta L/L_0$.

DATA TABLE 34. THERMAL LINEAR EXPANSION OF NICKEL Ni (continued)

T	$\Delta L/L_0$
CURVE 48 (cont.)*, ‡	
773	0.741
798	0.781
823	0.821
841	0.862
CURVE 49*, ‡	
273	-0.024
293	0.000
323	0.037
373	0.101
423	0.167
473	0.236
483	0.250
493	0.264
503	0.278
513	0.292
523	0.306
533	0.321
543	0.336
553	0.351
563	0.366
573	0.381
583	0.397
593	0.412
603	0.428
613	0.445
623	0.462
633	0.480
643	0.498
653	0.515
663	0.531
673	0.547
723	0.627
773	0.708
803	0.758
823	0.792
873	0.878
923	0.966
973	1.056
1023	1.148
1073	1.242

T	$\Delta L/L_0$
CURVE 50*	
298	-0.011
381	0.128
401	0.150
466	0.238
505	0.295
523	0.321
544	0.349
594	0.426
626	0.471
646	0.499
702	0.536
689	0.582
711	0.616
728	0.644
757	0.698
775	0.732
773	0.726
833	0.826
892	0.928
961	1.041
1063	1.234

* Not shown in figure.
‡ Author's data for coefficient of thermal expansion have been integrated by TPRC to obtain $\Delta L/L_0$.

DATA TABLE 34. COEFFICIENT OF THERMAL LINEAR EXPANSION OF NICKEL Ni

[Temperature, T, K; Coefficient of Expansion, α, 10^{-6} K^{-1}]

CURVE 1

T	α
79	4.84*
84	5.10
102	6.57*
111	7.29
120	7.78
128	8.29
140	8.81
151	9.19
170	9.86
190	10.69
209	11.30
209	13.24*
231	11.76
250	12.05
271	12.24
287	12.49
312	12.69*
325	12.89*
353	13.32*
372	13.47*
394	13.72*
414	13.99*
432	14.21*
474	14.53*
513	15.13*
534	15.68*
552	15.96*
573	16.29*
594	16.76*
609	17.23*
618	17.68*
626	18.06*
629	17.65*
633	17.07*
640	16.64*
646	16.52*
659	16.35*
673	16.23*
695	16.10*
715	16.02*
736	15.98*
752	15.93*
786	15.95*

CURVE 2‡

T	α
4	0.017
5	0.020*
6	0.025*
8	0.036*
10	0.050
12	0.065*
14	0.082*
16	0.102
18	0.125*
20	0.155*
22	0.190
24	0.240
26	0.315
65	3.300
75	4.400
85	5.400
283	12.900

CURVE 6

T	α
51.5	2.16
57.6	2.54
64.1	3.16
77.3	5.06

CURVE 7*

T	α
31.9	0.25
32.7	0.35
37.6	0.56
42.7	1.15
49.9	2.16
60.0	2.69
74.2	4.13

CURVE 8*

T	α
35.2	0.50
43.9	1.46
50.8	1.90
57.2	2.28
68.2	3.44
76.6	4.00

CURVE 12

T	α
30	0.36*
40	0.87*
50	1.73*
60	2.73*
70	3.83*
80	4.84*
90	5.77
100	6.61*
110	7.37*
120	8.05
130	8.65
140	9.18
150	9.65*
160	10.07
170	10.44
180	10.77
190	11.06
200	11.31
220	11.72
240	12.05
260	12.67
290	12.79
300	12.9*

CURVE 18‡

T	α
20	0.05
30	0.30
40	0.80
50	1.7
70	3.7
100	6.5
150	9.6
200	11.4
250	12.4
300	12.9

CURVE 23*,‡

T	α
373	10.31
400	12.04
425	13.19
462	14.21
492	14.90
524	15.61
561	16.50

CURVE 23 (cont.)*,‡

T	α
592	16.87
617	17.42
623	18.33
666	17.82
690	17.09
732	17.20
773	17.11

CURVE 33*,‡

T	α
273	12.5
373	13.7
473	14.9
523	15.5
573	17.4
593	19.1
643	22.6
650	23.5
673	16.1
723	17.0
773	17.6

CURVE 37*,‡

T	α
295	12.950
373	13.500
425	13.810
473	14.430
523	15.500
571	16.160
622	18.330
633	17.140
674	16.670
725	16.730
774	17.220
794	16.380
824	16.480
874	17.050
922	17.490
974	17.850

CURVE 47

T	α
40	1.25*
60	2.96*
80	4.57*

CURVE 47 (cont.)

T	α
100	6.05
140	8.67
180	10.8

CURVE 48*,‡

T	α
293	11.41
373	12.82
398	13.26
423	13.92
448	14.36
473	14.63
498	15.05
523	15.86
548	16.66
573	17.22
598	17.68
623	17.98
648	17.29
673	16.59
698	16.25
723	16.16
748	16.16
773	16.16
798	16.16
823	16.16
841	16.19

CURVE 49*,‡

T	α
273	12.10
293	12.31
323	12.56
373	13.05
423	13.41
473	13.86
483	14.00
493	14.07
503	14.19
513	14.33
523	14.54
533	14.65
543	14.76
553	15.06
563	15.23
573	15.42

CURVE 49 (cont.)*,‡

T	α
583	15.66
593	15.89
603	16.18
613	16.76
623	17.48
633	18.29
643	17.24
653	16.73
663	16.09
673	15.68
723	16.16
773	16.56
803	16.76
823	17.03
873	17.42
923	17.83
973	18.14
1023	18.63
1073	19.03

* Not shown in figure.

‡ Author's data for coefficient of thermal expansion have been integrated by TPRC to obtain $\Delta L/L_0$.

FIGURE AND TABLE NO. 35R. RECOMMENDED VALUES FOR THERMAL LINEAR EXPANSION OF NIOBIUM Nb

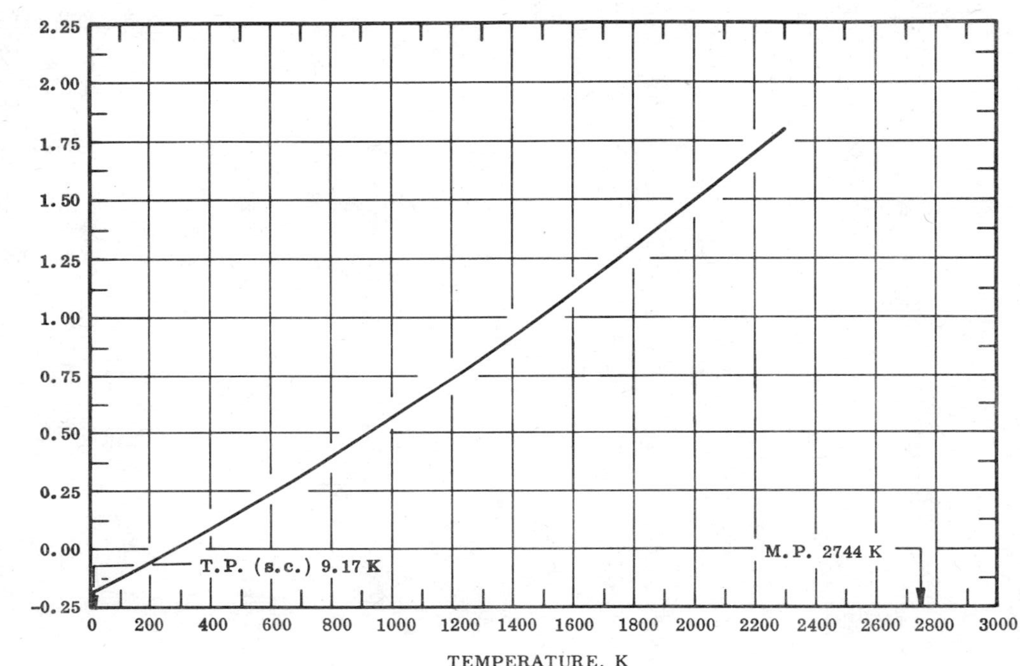

RECOMMENDED VALUES

[Temperature, T, K; Linear Expansion, $\Delta L/L_0$, %; α, K^{-1}]

T	$\Delta L/L_0$	$\alpha \times 10^6$
5	-0.147	0.01
25	-0.147	0.39
50	-0.144	2.0
100	-0.126	5.2
200	-0.064	6.8
293	0.000	7.3
400	0.078	7.4
500	0.155	7.8
600	0.232	8.0
700	0.312	8.1
800	0.393	8.2
900	0.477	8.4
1000	0.561	8.6
1200	0.737	8.8
1400	0.916	9.2
1600	1.102	9.4
1800	1.292	9.6
2000	1.488	9.8
2200	1.687	10.0
2300	1.788	10.1

LOW-TEMPERATURE THERMAL EXPANSION VALUES

T	Normal State $\Delta L/L \times 10^6$	Superconducting State $\Delta L/L^{\dagger} \times 10^6$
0	0.00	-
1	0.11	9.1
2	0.45	8.9
3	1.0	8.7
4	1.9	8.7
5	3.1	8.7
6	4.7	8.9
7	6.7	9.6
8	9.2	10.8
9	12.5	12.7
9.26	13.5	13.5

† Superconducting state value is set equal to that of normal state at the transition temperature, $T_c = 9.26$ K

REMARKS

The tabulated values are considered accurate to within ± 3% at temperatures below 1200 K and ± 5% above. These values can be represented approximately by the following equations:

$$\Delta L/L_0 = -0.125 + 5.164 \times 10^{-4} (T-100) + 9.858 \times 10^{-7} (T-100)^2 - 1.525 \times 10^{-9} (T-100)^3 \quad (100 < T < 293)$$

$$\Delta L/L_0 = 7.265 \times 10^{-4} (T-293) + 1.026 \times 10^{-7} (T-293)^2 - 1.032 \times 10^{-11} (T-293)^3 \quad (293 < T < 2300)$$

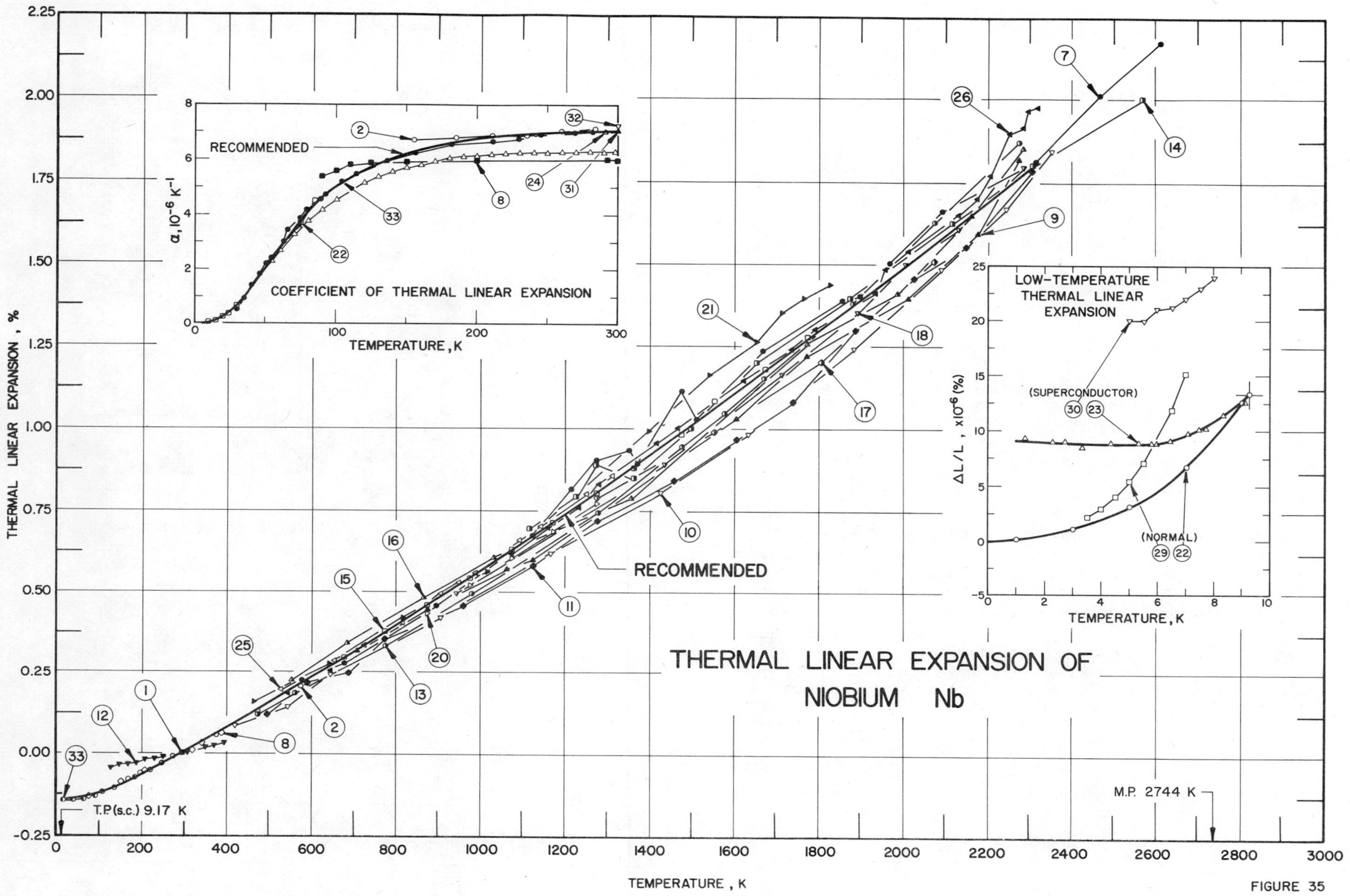
THERMAL LINEAR EXPANSION OF
NIOBIUM Nb
THERMAL LINEAR EXPANSION, %
TEMPERATURE, K
RECOMMENDED
M.P. 2744 K
T.P.(s.c.) 9.17 K
COEFFICIENT OF THERMAL LINEAR EXPANSION
α, 10⁻⁶ K⁻¹
TEMPERATURE, K
RECOMMENDED
LOW-TEMPERATURE
THERMAL LINEAR
EXPANSION
ΔL/L, x10⁻⁶ (%)
TEMPERATURE, K
(SUPERCONDUCTOR)
(NORMAL)

FIGURE 35

SPECIFICATION TABLE 35. THERMAL LINEAR EXPANSION OF NIOBIUM Nb

Cur. No.	Ref. No.	Author(s)	Year	Method Used	Temp. Range, K	Name and Specimen Designation	Composition (weight percent), Specifications, and Remarks
1	208	Frantsevich, I.N., Zhurakovskii, E.A., and Lyashchenko, A.B.	1967	X	293, 1473		Sample prepared from powder by vacuum sintering of pressed block, sintering temperature between 1473 and 2873 K.
2	242	Hidnert, P. and Krider, H.S.	1933	I	138-573		0.93 Sn and 0.26 Fe; specimen 300 mm long and 4.6 mm in diameter; cut from rod, prepared by pressing powder into 0.625 in. square and 16 in. long, heated in vacuum for hours, then hummered and heated again in vacuum, then swaged and heated in vacuum, and swaged to 4.6 mm diameter; test 2; zero-point correction of -0.013% determined by graphical interpolation; data on the coefficient of thermal linear expansion also reported.
3*	242	Hidnert, P. and Krider, H.S.	1933	I	290-573		The above specimen; test 3; zero-point correction of -0.013% determined by graphical interpolation.
4*	242	Hidnert, P. and Krider, H.S.	1933	I	297-574		The above specimen; test 4; zero-point correction of -0.013% determined by graphical extrapolation.
5*	242	Hidnert, P. and Krider, H.S.	1933	I	295-475		The above specimen; test 5; zero-point correction of -0.013% determined by graphical extrapolation.
6*	242	Hidnert, P. and Krider, H.S.	1933	I	292-575		The above specimen; test 6; zero-point correction of -0.013% determined by graphical interpolation.
7	243	Vasyutinskiy, B.M., Kartmazov, G.N., Smirnov, Yu.M., and Finkel, V.A.	1966	X	293-2616		99.95 Nb, gas impurities; specimen 0.1 to 0.3 mm thick; heated in vacuum of 2 x 10^{-5} mm Hg; expansion measured using Cu-Kα x-radiation with (330, 411) diffraction lines focused at large angles; lattice parameter reported at 293 K is 3.3000 Å.
8	244	Harris, B. and Peacock, D.E.	1966	L	77-387		Pure Nb; recrystallized; measurements made in baths of liquid nitrogen, liquid bromoethane, and liquid polyalkylene glycol; reported error ± 1%; zero-point correction is -0.13%; data on the coefficient of thermal linear expansion also reported.
9	210	Conway, J.B. and Losekamp, A.C.	1966	T	578-2272		99.94^{+} Nb, arc-cast sheet 20 mil thick; expansion measured in helium atmosphere on first test.
10	210	Conway, J.B. and Losekamp, A.C.	1966	T	421-2355		The above specimen; second test.
11	210	Conway, J.B. and Losekamp, A.C.	1966	T	497-2305		The above specimen; third test.
12	246	Smirnov, Yu.M. and Finkel, V.A.	1966	X	128-396		99.95 Nb polycrystalline specimens, in form of plates and strips 0.1 to 0.3 mm thick; average data from several specimens; lattice parameter reported at 293 K is 3.30074 Å.
13	213	Amonenko, V.M., V'yugov, P.N. and Gumenyuk, V.S.	1964	L	473-2273		Nb melted in an arc furnace and subsequently subjected to pressure working; specimens heated in tungsten furnace; annealed for 1.5-2.0 hr above 2273 K before measurements started under vacuum; reported error 1%.
14	215	Edwards, J.W., Speiser, R., and Johnston, H.L.	1951	X	291-2470		99.8 Nb; specimen obtained from Fansteel Metallurgical Corp.; lattice parameter reported at 291 K is 3.3004 Å.
15	247	Johnson, P.M., Lincoln, R.L., and McClure, E.R.	1968	L	473-1773		High purity (electron-beam-melted) Nb; specimen 2 in. cylinder, ends made parallel to 500 microinches per 0.25 in., lapped flat to 10 microinches per 0.25 in., and polished to 60% reflectivity; helium-neon continuous gas laser used as light source.

* Not shown in figure.

SPECIFICATION TABLE 35. THERMAL LINEAR EXPANSION OF NIOBIUM Nb (continued)

Cur. No.	Ref. No.	Author(s)	Year	Method Used	Temp. Range, K	Name and Specimen Designation	Composition (weight percent), Specifications, and Remarks
16	248	Conway, J. B., Fincel, R. M., Jr., and Losekamp, A. C.	1965	T	556-2281		Arc-cast niobium; specimen in the form of 20-mil sheet formed into a sharp U shape about 0.3 in. wide, 0.3 in. depth; expansion measured in helium atmosphere with increasing temperature on first test.
17	248	Conway, J. B., et al.	1965	T	1224-2074		The above specimen; expansion measured with decreasing temperature; first test.
18	248	Conway, J. B., et al.	1965	T	484-2281		The above specimen; expansion measured with increasing temperature; second test.
19*	248	Conway, J. B., et al.	1965	T	298-2064		The above specimen; expansion measured with increasing temperature; second test.
20	249	Tottle, C.R.	1957	L	291-1273		99.5 Nb; expansion measured in 10^{-4} mm Hg vacuum.
21	250	Fieldhouse, I. B., Hedge, J. C., and Lang, J. I.	1958	T	300-1824		No details given; zero-point correction of 0.007% determined by graphical extrapolation.
22	251	White, G.K.	1962	E	6-85		Specimen machined from ingot supplied by Temescal Metallurgical Corp.; originally material from Wah Chang Corp., approximately 99.8 pure, then electron beam melted; sample stress-relieved 2 hr in vacuum at 773 K; end ground parallel; Rockwell hardness R_E 60; expansion of normal state in magnetic field; values at and below T_c = 9.26 K, are referred to 0 K as reference temperature at which $\Delta L/L_0 = 0$.
23	251	White, G.K.	1962	E	6-9.26		The above specimen; expansion measured in superconducting state; this curve referred to the normal state value of curve 22 at the critical temperature, T_c = 9.26 K; zero-point correction is 4.4×10^{-6}%.
24*	252	Heal, J.T.	1958		291-1173		0.004 N, 0.002 C, 0.001 O, 0.0002 H; small specimen purified by electron bombardment melting.
25	253	Ul'yanov, R.A., Tarasov, N.D., and Mikhayolov, Ya.D.	1964	L	299-1307		Starting material 98.7 pure, vacuum refined to 0.1 C, <0.1 any other impurities; expansion measured in vacuum of 10^{-5} mm Hg.
26	623	Mochalov, G.A. and Ivanov, O.S.	1969	L	543-2417		98.9 pure specimen melted in an arc furnace with a nonconsumable tungsten electrode; quartz dilatometer below 1273 K and above that cathetometer.
27*	707	Pridantseva, K.S. and Solov'yeva, N.A.	1965	L	293, 1073		Specimen prepared by powder metallurgy method; 1% measurement error.
28*	707	Pridantseva, K.S. and Solov'yeva, N.A.	1965	L	293-1073		Similar to the above specimen.
29	89	Andres, K.	1961	O	3.0-7.0		No details given; expansion of normal state in magnetic field; this curve is here referred to 0 K as reference temperature for which $\Delta L/L_0 = 0$.
30	89	Andres, K.	1961	O	5.0-8.5		No details given; expansion of superconducting state; author found abnormal behavior of unexplained nature at low temperatures; this curve referred to the normal state value of curve 29 at the critical temperature, T_c = 9.26 K; zero-point correction is 20×10^{-6}%.
31*	747	Lisovskii, Yu.A.	1973	E	300		99.97 pure polycrystalline specimen; data on the coefficient of thermal linear expansion only at 300 K is reported.
32*	744	Linkoano, M. and Rantavuori, E.	1970	X	300		99.8 pure specimen; data on the coefficient of thermal linear expansion only at 300 K is reported.
33	745	Lebedev, V.P., Mamalui, A., Pervakov, V.A., Petrenko, N.S., Popov, V.P., and Khotkevich, V.I.	1969	L	7.6-282.3		99.8 pure specimen.

* Not shown in figure.

SPECIFICATION TABLE 35. THERMAL LINEAR EXPANSION OF NIOBIUM Nb (continued)

Cur. No.	Ref. No.	Author(s)	Year	Method Used	Temp. Range, K	Name and Specimen Designation	Composition (weight percent), Specifications, and Remarks
34*	746	Straumanis, M. E. and Zyszezynski, S.	1970	X	288-338		Rod specimen from the Battelle Memorial Institute; impurities: 0.1 Ta, 0.01 O, 0.004 each C and N, 0.003 each Mo, Ti, and Zr, 0.001 Ca, 0.0001 H; lattice parameter reported as 3.30020 Å at 288 K, 3.30028 Å at 293 K determined by graphical interpolation.
35*	748	Hubbell, W. C. and Brotzen, F. R.	1972	L	83, 373		99.98 Nb metallurgical grade niobium rod 0.25 in. diameter; single crystal grown by electron beam melting, cylinders about 10 mm long spark sliced from rod and end faces polished, 8.569 g cm^{-3} at 298 K; $\pm 1.5 \times 10^{-6}$ K^{-1} uncertainty in α.
36	747	Lisovskii, Yu. A.	1972	E	55-300		99.97 pure polycrystalline 100 mm long specimen.

* Not shown in figure.

DATA TABLE 35. THERMAL LINEAR EXPANSION OF NIOBIUM Nb

[Temperature, T, K; Linear Expansion, $\Delta L/L_0$, %]

T	$\Delta L/L_0$
CURVE 1	
293	0.000
1473	0.956*
CURVE 2	
138	-0.105
168	-0.085
196	-0.067
220	-0.051
245	-0.034
272	-0.014
294	0.001*
375	0.059
473	0.134
573	0.212
CURVE 3*	
290	-0.002
376	0.060
475	0.136
573	0.212
CURVE 4*	
297	0.003
373	0.060
476	0.137
574	0.211
CURVE 5*	
295	0.002
338	0.033
372	0.058
424	0.097
475	0.135
CURVE 6*	
292	0.000
373	0.059
477	0.139
575	0.214
CURVE 7	
293	0.000*
579	0.227
675	0.279
894	0.461
1121	0.673
1217	0.815
1270	0.909
1351	0.936
1362	0.897
1473	1.115
1504	1.027
1664	1.239
1855	1.391
1894	1.409
1966	1.506
2097	1.688
2314	1.818
2467	2.015
2616	2.176
CURVE 8	
77	-0.130
142	-0.089*
153	-0.086
168	-0.078*
168	-0.077*
180	-0.067*
210	-0.052
219	-0.047*
228	-0.039*
289	-0.003*
293	0.000*
296	0.005*
319	0.013
340	0.026*
340	0.031
364	0.043*
387	0.060
CURVE 9	
578	0.22*
CURVE 9 (cont.)	
705	0.32
868	0.45
1062	0.57
1380	0.77
1604	1.03
1774	1.22
2010	1.39
2178	1.64
2272	1.82
CURVE 10	
421	0.08
545	0.14
642	0.24
904	0.42
1164	0.62
1425	0.81
1633	0.98
1885	1.24
2091	1.48
2243	1.67
2355	1.84
CURVE 11	
497	0.12
587	0.19
685	0.25
773	0.36
960	0.46
1122	0.58
1274	0.72
1458	0.84
1604	0.97
1739	1.13
1885	1.30
2025	1.46
2152	1.56
2305	1.78
CURVE 12	
128	-0.042
132	-0.040*
135	-0.041*
135	-0.040*
142	-0.038*
145	-0.038*
150	-0.037
157	-0.035*
160	-0.035*
165	-0.034
172	-0.031*
179	-0.030*
185	-0.030
189	-0.028*
194	-0.027*
199	-0.025*
205	-0.023
210	-0.022*
219	-0.019*
230	-0.017
234	-0.016*
239	-0.014*
244	-0.012*
250	-0.012
252	-0.011*
259	-0.009*
264	-0.009*
267	-0.007*
273	-0.007*
274	-0.005*
282	-0.003*
285	-0.003*
290	-0.003*
293	0.000*
297	0.006*
300	0.006*
305	0.003*
309	0.003*
310	0.005
316	0.005*
318	0.007*
320	0.008*
324	0.007*
CURVE 12 (cont.)	
324	0.008*
326	0.009*
329	0.009*
330	0.012*
334	0.012*
336	0.011*
338	0.013*
338	0.013*
342	0.015*
347	0.016*
348	0.015*
350	0.018
355	0.018*
360	0.019*
364	0.021*
371	0.024
380	0.027*
388	0.030*
396	0.033
CURVE 13	
473	0.12
560	0.18
674	0.24*
773	0.33
872	0.40*
978	0.49
1071	0.56*
1273	0.74
1372	0.85*
1475	0.94
1578	1.04*
1670	1.16
1778	1.27*
1880	1.38
1975	1.50*
2076	1.63
2186	1.76*
2273	1.87
CURVE 14	
293	0.000*
1115	0.697
1225	0.794
1270	0.891
1355	0.933*
1360	0.857
1360	0.882
1465	1.10*
1495	1.02
1655	1.20
1855	1.37*
1870	1.39
1965	1.50*
2116	1.63
2309	1.83
2470	2.00
CURVE 15	
473	0.140*
573	0.218*
673	0.297
773	0.378
873	0.458
973	0.542
1073	0.629
1173	0.716*
1273	0.804
1373	0.896*
1473	0.988
1573	1.081*
1673	1.179
1773	1.276
CURVE 16	
556	0.227
682	0.347
865	0.478
1033	0.610
1359	0.784
1572	1.040*
1773	1.263

* Not shown in figure.

DATA TABLE 35. THERMAL LINEAR EXPANSION OF NIOBIUM Nb (continued)

T	$\Delta L/L_0$
CURVE 16 (cont.)	
1985	1.412
2161	1.656
2281	1.868
CURVE 17	
1274	0.797
1555	0.996
1802	1.210
2074	1.513
CURVE 18	
484	0.149*
577	0.208*
665	0.276*
774	0.357*
942	0.491
1112	0.618*
1274	0.729*
1438	0.891
1717	1.168
1892	1.354
2006	1.500
2135	1.609
2281	1.793
CURVE 19*	
298	0.050
1191	0.699
1457	1.002
1823	1.295
2064	1.542
CURVE 20	
291.2	0.000*
573.2	0.205*
673.2	0.281*
773.2	0.359*
873.2	0.438
973.2	0.520
1073.2	0.602
1173.2	0.686
1273.2	0.772

T	$\Delta L/L_0$
CURVE 21	
300	0.007*
462	0.163
560	0.239*
642	0.279
819	0.423
898	0.487*
1018	0.569
1105	0.648*
1147	0.693*
1223	0.782*
1306	0.908*
1399	0.996
1476	1.089*
1541	1.166
1594	1.208*
1655	1.268
1712	1.356
1762	1.396
1824	1.434
CURVE 22‡	
1	0.11 x 10^{-6}§
3	1.03 x 10^{-6}§
5	3.09 x 10^{-6}§
7	6.71 x 10^{-6}§
9	12.52 x 10^{-6}§
9.26	13.48 x 10^{-6}§
10	-0.1471
11	-0.1471
12	-0.1471
14	-0.1471
15	-0.1471
16	-0.1471
18	-0.1470
20	-0.1470
22	-0.1469
24	-0.1469
26	-0.1468
28	-0.1467
30	-0.1466
75	-0.1367
85	-0.1326
CURVE 23‡, ǂ	
1.43	9.25 x 10^{-6}
1.51	9.0*

T	$\Delta L/L_0$
CURVE 23 (cont.) ‡, ǂ	
2.26	9.0 x 10^{-6}
2.46	8.7*
2.51	9.0
3.39	8.5
3.49	8.9*
4.22	8.8
5.25	8.8
5.80	8.8
5.96	8.9
6.47	9.1
6.64	9.2*
7.09	9.7
7.43	10.1
7.62	10.5*
7.70	10.3
8.32	11.4
8.41	11.2*
9.08	12.6
9.13	12.9
CURVE 24*, ‡	
291	-0.002
323	0.021
373	0.057
473	0.130
573	0.205
673	0.282
773	0.361
873	0.443
973	0.526
1073	0.612
1173	0.699
CURVE 25	
299	0.003*
337	0.038*
373	0.063*
428	0.110*
473	0.142*
530	0.190
565	0.209*
573	0.218*
588	0.228*
653	0.281
673	0.295*
701	0.318*

T	$\Delta L/L_0$
CURVE 25 (cont.)	
725	0.338
772	0.374*
813	0.412
862	0.451*
873	0.471*
889	0.479*
903	0.492
916	0.496*
949	0.529
973	0.553*
983	0.559
996	0.575*
1030	0.602*
1073	0.634*
1098	0.662
1110	0.673*
1146	0.702
1173	0.731*
1249	0.804
1273	0.824*
1307	0.854
CURVE 26	
543	0.185
643	0.253
708	0.314*
768	0.360*
864	0.445*
980	0.536*
1076	0.621
1160	0.701
1229	0.762
1281	0.830
1357	0.883*
1417	0.959
1465	1.009
1496	0.998*
1623	1.146
1739	1.246
1789	1.313
1935	1.417
1959	1.484
2010	1.538
2134	1.657
2187	1.708
2206	1.770
2351	1.906

T	$\Delta L/L_0$
CURVE 26 (cont.)	
2381	1.916
2399	1.970
2417	1.975
CURVE 27*	
293	0.000
1073	0.611
CURVE 28*	
293	0.000
573	0.217
1173	0.683
CURVE 29‡	
3.0	1.23*
3.5	2.09
4.0	2.96
4.5	4.07
5.0	5.47
5.5	7.21
6.0	9.36*
6.5	11.98
7.0	15.13
CURVE 30‡, ǂ	
5.0	20 x 10^{-6}
5.5	20
6.0	21
6.5	21
7.0	22
7.5	23
8.0	24
8.5	26*
CURVE 31*	
⧖	
CURVE 32*	
⧖	
CURVE 33‡	
7.6	-0.147*
15	-0.146
20	-0.146*
25	-0.146*
30	-0.146*
35	-0.146*
40	-0.145

T	$\Delta L/L_0$
CURVE 33 (cont.) ‡	
45	-0.144*
50	-0.143*
53	-0.143*
63	-0.140*
65	-0.139
75	-0.136*
78.9	-0.134*
88.3	-0.130*
93.0	-0.128
104.2	-0.122
114.4	-0.117*
135.7	-0.105*
155.6	-0.093*
182.1	-0.076
211.3	-0.056*
229.3	-0.044*
244.9	-0.034*
268.9	-0.017*
282.3	-0.008*
CURVE 34*	
288	-0.002
298	0.004
308	0.011
318	0.016
328	0.027
338	0.035
CURVE 35*	
83	-0.181
373	0.069
CURVE 36*	
55	-0.131
60	-0.129
70	-0.126
80	-0.123
90	-0.119
100	-0.115
110	-0.110
120	-0.105
130	-0.100
140	-0.094
150	-0.088
160	-0.083
169	-0.077
180	-0.071
190	-0.065

* Not shown in figure.
‡ Author's data for coefficient of thermal expansion have been integrated by TPRC to obtain $\Delta L/L_0$.
ǂ For explanation see specification table.
§ These are not percent expansion values, see specification table.
⧖ Author reports coefficient of thermal linear expansion only at 300 K.

DATA TABLE 35. THERMAL LINEAR EXPANSION OF NIOBIUM Nb (continued)

T	$\Delta L/L_0$
CURVE 36 (cont.)*	
200	-0.058
210	-0.052
220	-0.046
230	-0.040
240	-0.033
250	-0.027
260	-0.021
270	-0.015
280	-0.008
290	-0.002
293	0.000
300	0.004

DATA TABLE 35. COEFFICIENT OF THERMAL LINEAR EXPANSION OF NIOBIUM Nb

[Temperature, T, K; Coefficient of Expansion, α, 10^{-6} K^{-1}]

T	α
CURVE 2	
155	6.7
185	6.8
210	6.9
235	6.9
260	7.0
283	7.1
CURVE 8	
90	5.35
100	5.56
110	5.77
125	5.85
150	5.87
200	5.93
293	5.96
300	5.98
400	5.99*
CURVE 22‡	
6	0.018*
7	0.023*
8	0.029*
9	0.036*
10	0.044
11	0.065*

T	α
CURVE 22 (cont.)‡	
12	0.090*
14	0.10*
15	0.12
16	0.17*
18	0.22*
20	0.28
22	0.35*
24	0.39
26	0.44*
28	0.53*
30	0.64
75	3.70
85	4.50
CURVE 24*, ‡	
291	7.02
323	7.08
373	7.19
473	7.40
573	7.62
673	7.83
773	8.04
873	8.25
973	8.46
1073	8.63
1173	8.89

T	α
CURVE 29*, ‡	
3.0	0.011
3.5	0.015
4.0	0.020
4.5	0.025
5.0	0.031
5.5	0.039
6.0	0.047
6.5	0.057
7.0	0.069
CURVE 30*, ‡	
5.0	0.0039
5.5	0.0065
6.0	0.0101
6.5	0.0141
7.0	0.0180
7.5	0.0233
8.0	0.0324
8.5	0.0430
CURVE 31*	
300	7.08

T	α
CURVE 32*	
300	7.2
CURVE 33‡	
7.6	0.02*
15	0.10*
20	0.20*
25	0.40*
30	0.60
35	0.95
40	1.40
45	1.80
50	2.20
53	2.40
63	3.00
65	3.40
75	3.80
78.9	4.18
88.3	4.51
93.0	4.79
104.2	5.14
114.4	5.48
135.7	5.82
155.6	6.26
182.1	6.52
211.3	6.75
229.3	6.74

T	α
CURVE 33 (cont.)‡	
244.9	6.91
268.9	6.97
282.3	7.07
CURVE 36*	
55	2.29
60	2.67
70	3.28
80	3.76
90	4.19
100	4.53
110	4.88
120	5.18
130	5.42
140	5.60
150	5.74
160	5.85
169	5.96
180	6.07
190	6.12
200	6.18
210	6.22
220	6.25
230	6.28
240	6.29
250	6.29

T	α
CURVE 36 (cont.)*	
260	6.31
270	6.31
280	6.31
290	6.31
293	6.30
300	6.28

* Not shown in figure.

‡ Author's data for coefficient of thermal expansion have been integrated by TPRC to obtain $\Delta L/L_0$.

FIGURE AND TABLE NO. 36R. PROVISIONAL VALUES FOR THERMAL LINEAR EXPANSION OF OSMIUM Os

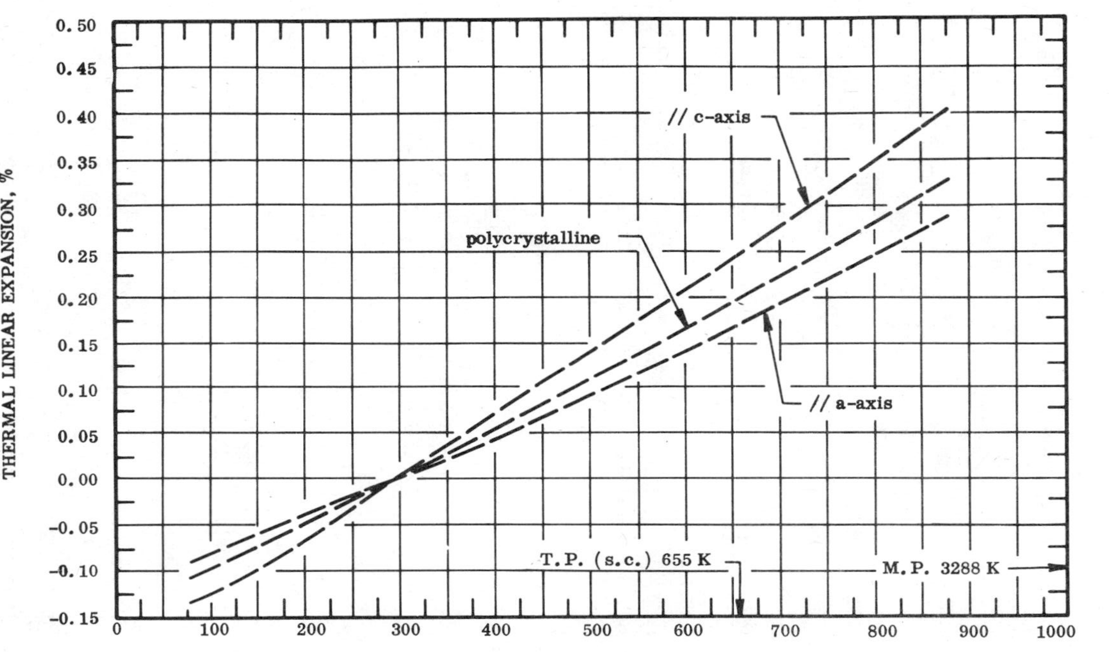

PROVISIONAL VALUES

[Temperature, T, K; Linear Expansion, $\Delta L/L_0$, %; α, K^{-1}]

T	// a-axis $\Delta L/L_0$	// c-axis $\Delta L/L_0$	polycrystalline $\Delta L/L_0$	polycrystalline $\alpha \times 10^6$
75	-0.094	-0.138	-0.109	4.8
100	-0.083	-0.136	-0.101	5.1
200	-0.042	-0.065	-0.050	5.1
293	0.000	0.000	0.000	5.1
400	0.045	0.073	0.054	5.2
500	0.091	0.142	0.108	5.4
600	0.140	0.210	0.163	5.6
700	0.191	0.277	0.220	5.8
800	0.245	0.346	0.279	5.9
875	0.287	0.397	0.324	5.9

REMARKS

The tabulated values are considered accurate to within ± 7% over the entire temperature range. Since no experimental data on the polycrystalline osmium were located in the literature, the tabulated values are calculated from single crystal data. These values are considered accurate to within ± 10% over the entire temperature range. These values can be represented approximately by the following equations:

// a-axis: $\Delta L/L_0 = -0.125 + 4.100 \times 10^{-4}\, T + 9.412 \times 10^{-9}\, T^2 + 6.855 \times 10^{-11}\, T^3$

// c-axis: $\Delta L/L_0 = -0.193 + 6.246 \times 10^{-4}\, T + 1.286 \times 10^{-7}\, T^2 - 8.279 \times 10^{-11}\, T^3$ $(150 < T < 875)$

polycrystalline: $\Delta L/L_0 = -0.147 + 4.846 \times 10^{-4}\, T + 3.946 \times 10^{-8}\, T^2 + 2.579 \times 10^{-11}\, T^3$

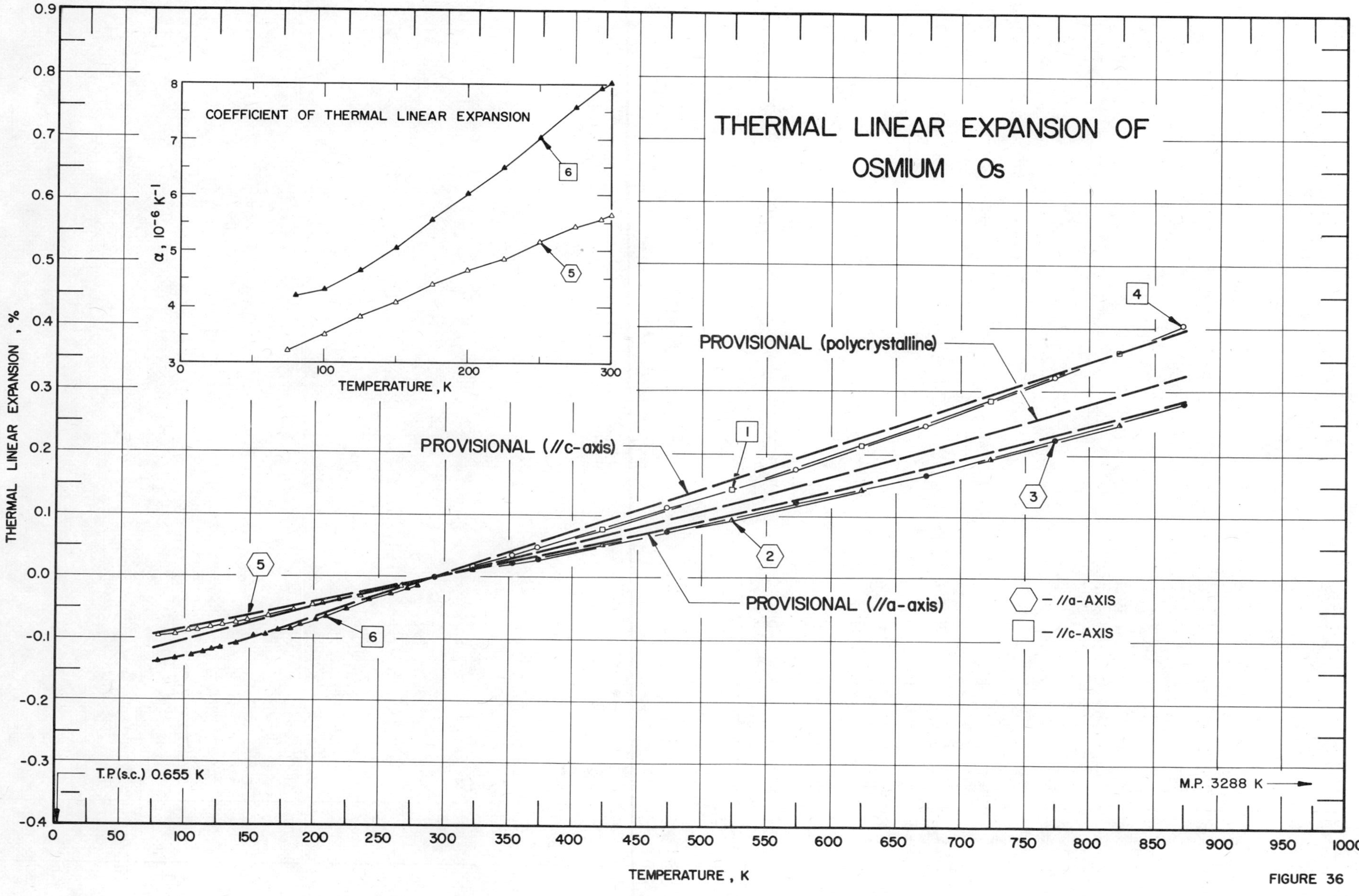

THERMAL LINEAR EXPANSION OF
OSMIUM Os
COEFFICIENT OF THERMAL LINEAR EXPANSION
α, 10⁻⁶ K⁻¹
TEMPERATURE, K
PROVISIONAL (polycrystalline)
PROVISIONAL (//c-axis)
PROVISIONAL (//a-axis)
— //a-AXIS
— //c-AXIS
T.P.(s.c.) 0.655 K
M.P. 3288 K
THERMAL LINEAR EXPANSION, %
TEMPERATURE, K

FIGURE 36

SPECIFICATION TABLE 36. THERMAL LINEAR EXPANSION OF OSMIUM Os

Cur. No.	Ref. No.	Author(s)	Year	Method Used	Temp. Range, K	Name and Specimen Designation	Composition (weight percent), Specifications, and Remarks
1	108	Owen, E. A. and Roberts, E. W.	1936	X	323-823		99.8^+ Os; small lump crushed and ground into powder; annealed 5 hr at 1273 K to remove effect of cold work; expansion measured parallel to hexagonal axis.
2	108	Owen, E. A. and Roberts, E. W.	1936	X	323-823		The above specimen; expansion measured perpendicular to hexagonal axis.
3	265	Owen. E. A. and Roberts, E. W.	1937	X	293-873		99.8^+ Os; powder annealed 5 hr at 1273 K; expansion measured in base of hexagonal crystal; base side reported at 293 K is 2.7298 Å.
4	265	Owen, E. A. and Roberts, E. W.	1937	X	293-873		The above specimen; expansion measured along height of hexagonal crystal; height reported at 293 K is 4.3104 Å.
5	660	Finkel, V. A., Palatnik, M. I., and Kovtun, G. P.	1971	X	79-300		Single crystal grown by zone recrystallization; measurements along a-axis; lattice parameter reported at 292 K is 2.7344 Å; 2.7344 Å at 293 K determined by interpolation; data on the coefficient of thermal linear expansion also reported.
6	660	Finkel, V. A., et al.	1971	X	80-300		Single crystal grown by zone recrystallization; measurements along c-axis; lattice parameter reported at 293 K is 4.3172 Å; data on the coefficient of thermal linear expansion also reported.

DATA TABLE 36. THERMAL LINEAR EXPANSION OF OSMIUM Os

[Temperature, T, K; Linear Expansion, $\Delta L/L_0$, %]

T	$\Delta L/L_0$
CURVE 1‡	
323	0.018
423	0.079
523	0.143
623	0.211
723	0.285
823	0.364
CURVE 2‡	
323	0.012
423	0.053
523	0.097
623	0.144
723	0.195
823	0.250
CURVE 3	
293	0.000
323	0.011*
353	0.022
373	0.029
423	0.051*
473	0.073
523	0.099*
573	0.121
623	0.143*
673	0.168
723	0.194*
773	0.223
823	0.249*
873	0.282
CURVE 4	
293	0.000*
323	0.016*
353	0.035
373	0.048
423	0.078*
473	0.111
523	0.143*
573	0.176
623	0.213*
673	0.248
723	0.285*
773	0.324
823	0.366*
873	0.408
CURVE 5	
79	-0.095
83	-0.094*
88	-0.094*
91	-0.092
97	-0.091
103	-0.088
109	-0.085
120	-0.080
130	-0.078
134	-0.076*
139	-0.073
149	-0.070
153	-0.069*
165	-0.064
167	-0.061*
172	-0.061*
178	-0.058*
185	-0.055*
188	-0.052*
194	-0.052*
199	-0.049
207	-0.045
219	-0.038
224	-0.037*
228	-0.033*
236	-0.030
241	-0.030*
242	-0.027*
249	-0.024*
257	-0.020*
262	-0.019
267	-0.016*
271	-0.012*
277	-0.009*
285	-0.006
289	-0.003*
292	-0.0003*
300	0.0007*
CURVE 6	
80	-0.136
88	-0.134*
92	-0.131
97	-0.131*
105	-0.126
109	-0.123*
114	-0.121
120	-0.118
128	-0.115
140	-0.109
145	-0.104*
154	-0.099
163	-0.095
171	-0.089
177	-0.086
183	-0.081*
184	-0.079
193	-0.075*
201	-0.070
205	-0.068*
210	-0.064
216	-0.063*
218	-0.058*
225	-0.052
228	-0.053*
234	-0.048*
238	-0.044*
243	-0.043*
244	-0.039
247	-0.038*
255	-0.032*
259	-0.029
265	-0.025*
267	-0.021*
272	-0.020
275	-0.014*
280	-0.013
288	-0.004*
293	0.000
294	0.004*
300	0.004*

DATA TABLE 36. COEFFICIENT OF THERMAL LINEAR EXPANSION OF OSMIUM Os

[Temperature, T, K; Coefficient of Expansion, α, 10^{-6} K^{-1}]

T	α
CURVE 1*, ‡	
323	5.9
423	6.2
523	6.6
623	7.1
723	7.6
823	8.3
CURVE 2*, ‡	
323	4.0
423	4.2
523	4.5
623	4.9
723	5.3
823	5.8
CURVE 5	
75	3.23
100	3.51
125	3.82
150	4.08
175	4.39
200	4.64
225	4.85
250	5.16
275	5.44
293	5.57
300	5.65
CURVE 6	
80	4.20
100	4.31
125	4.65
150	5.05
175	5.56
200	6.03
225	6.50
250	7.04
275	7.59
293	7.92
300	8.02

* Not shown in figure.

‡ Author's data for coefficient of thermal expansion have been integrated by TPRC to obtain $\Delta L/L_0$.

FIGURE AND TABLE NO. 37R. RECOMMENDED VALUES FOR THERMAL LINEAR EXPANSION OF PALLADIUM Pd

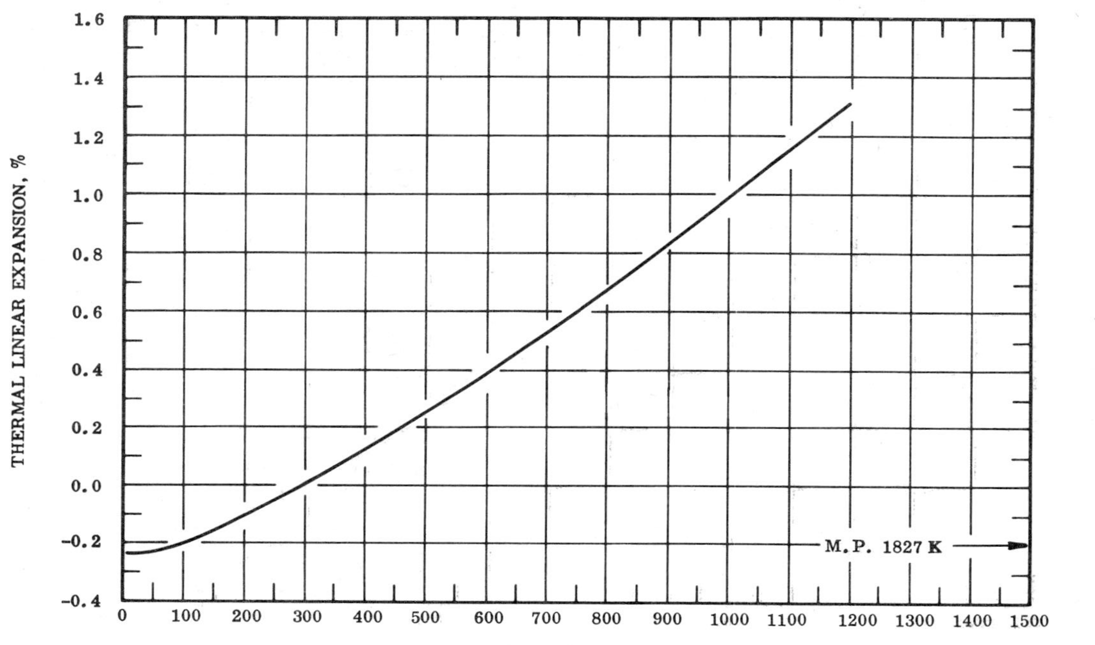

RECOMMENDED VALUES

[Temperature, T, K; Linear Expansion, $\Delta L/L_0$, %; α, K^{-1}]

T	$\Delta L/L_0$	$\alpha \times 10^6$
5	-0.237	0.02
25	-0.237	0.79
50	-0.232	3.5
100	-0.200	8.0
200	-0.104	10.7
293	0.000	11.8
400	0.131	12.6
500	0.261	13.2
600	0.395	13.6
700	0.532	14.1
800	0.674	14.5
900	0.821	15.0
1000	0.975	15.6
1200	1.302	16.9

REMARKS

The tabulated values are considered accurate to within ± 3% at temperatures below 800 K and ± 5% above. These values can be represented approximately by the following equations:

$$\Delta L/L_0 = -0.200 + 7.400 \times 10^{-4} (T - 100) + 2.237 \times 10^{-6} (T - 100)^2 - 3.519 \times 10^{-9} (T - 100)^3 \quad (100 < T < 293)$$

$$\Delta L/L_0 = 1.210 \times 10^{-3} (T - 293) + 2.000 \times 10^{-7} (T - 293)^2 + 4.859 \times 10^{-11} (T - 293)^3 \quad (293 < T < 1200)$$

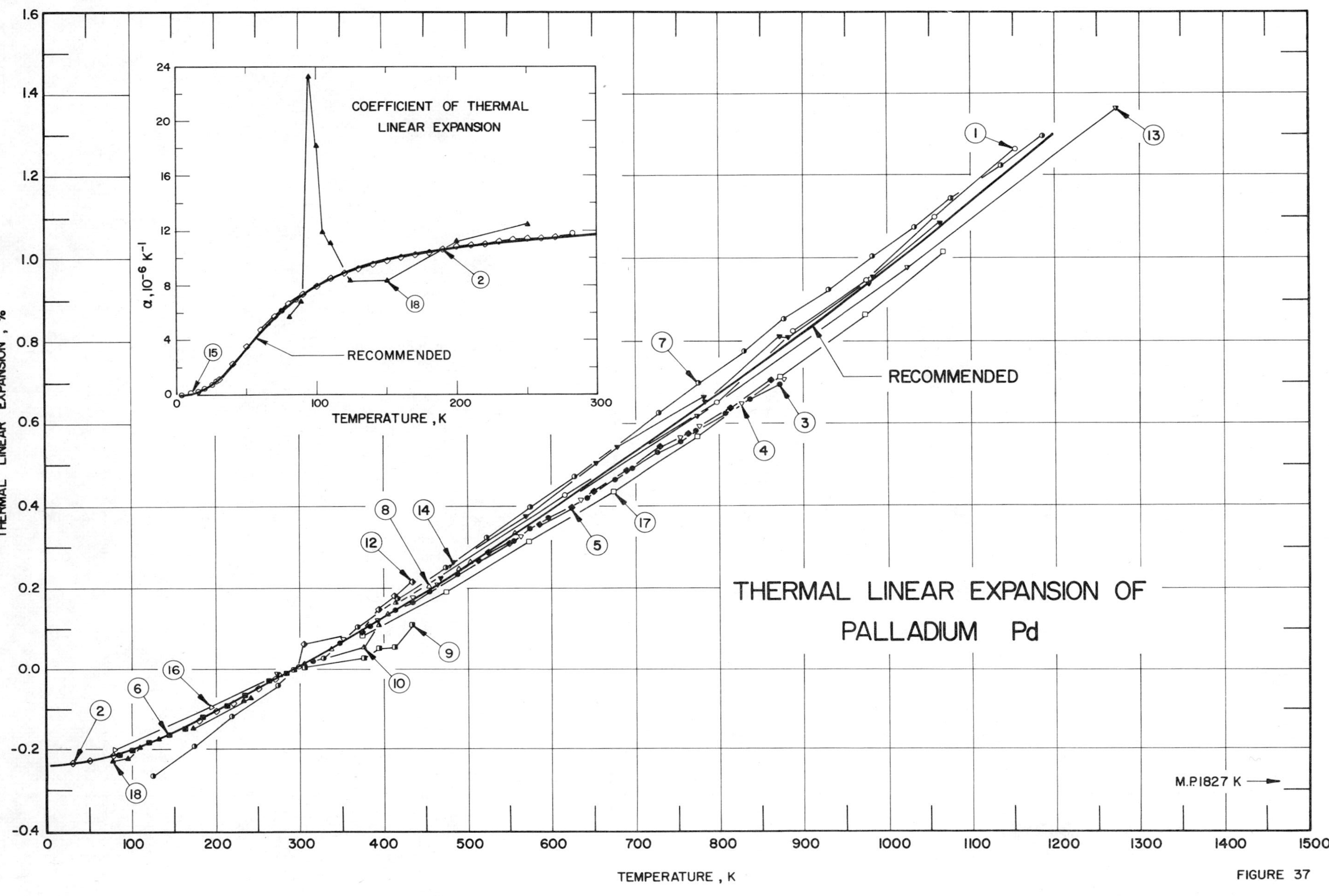
THERMAL LINEAR EXPANSION OF
PALLADIUM Pd
THERMAL LINEAR EXPANSION, %
TEMPERATURE, K
RECOMMENDED
M.P. 1827 K
COEFFICIENT OF THERMAL
LINEAR EXPANSION
α, 10⁻⁶ K⁻¹
TEMPERATURE, K
RECOMMENDED

FIGURE 37

SPECIFICATION TABLE 37. THERMAL LINEAR EXPANSION OF PALLADIUM Pd

Cur. No.	Ref. No.	Author(s)	Year	Method Used	Temp. Range, K	Name and Specimen Designation	Composition (weight percent), Specifications, and Remarks
1	268	Dutta, B. N. and Dayal, B.	1963	X	298-1151		99.9 Pd, specimen in the form of a wire; obtained from Vin Vish Corp., Bombay, India; lattice parameter reported at 298 K is 3.8899 Å; 3.88987 Å at 293 K determined by graphical extrapolation.
2	269	Waterhouse, N. and Yates, B.	1968	I	30-270		0.1 Al, 0.05 Si, 0.03 Fe, 0.003 each Pt, Mn, 0.002 each Ti, Ni, 0.001 each Cu, Rh, Ag, < 0.001 Mg; supplied by Johnson, Matthey and Co. Ltd.; hollow cylinder 2.6 cm O.D., 2.3 cm I.D., 1 cm long; annealed in vacuum for between 30 and 60 min. at 1173 K, cooled over a period of 24 hr.
3	270	Masumoto, H., Saito, H. and Kadowaki, S.	1968	L	315-871	Specimen No. 1	99.91 Pd; 3 mm diameter, 10 to 11 cm long; cubic single crystal; zero-point correction of -0.233% determined by graphical extrapolation.
4	270	Masumoto, H., et al.	1968	L	338-878	Specimen No. 2	Similar to the above specimen; zero-point correction of -0.124% determined by graphical extrapolation.
5	270	Masumoto, H., et al.	1968	L	296-862	Specimen No. 11	Similar to the above specimen; zero-point correction of -0.033% determined by graphical extrapolation.
6	75	Nix, F. C. and MacNair, D.	1942	I	86-299		99.99$^+$ Pd, specimen obtained from International Nickel Co.; zero-point correction of -0.0224% determined by graphical interpolation.
7	271	Masumoto, H. and Sawaya, S.	1969	L	125-1184		99.9 Pd; cylindrical specimen 2 mm diameter, 100 mm long.
8	272	Owen, E. A. and Jones, J. I.	1937	X	283-559		99.9 Pd; annealed in vacuo 3.5 hr at 723 K; lattice parameter reported at 284 K is 3.8822 Å; 3.8826 Å at 293 K determined by graphical interpolation.
9	272	Owen, E. A. and Jones, J. I.	1937	X	293-433		Similar to the above specimen; lattice parameter reported at 293 K is 3.8852 Å.
10	272	Owen, E. A. and Jones, J. I.	1937	X	293-413		Similar to the above specimen except 7.5 atomic % hydrogen; lattice parameter reported at 293 K is 3.8889 Å.
11*	272	Owen, E. A. and Jones, J. I.	1937	X	293-433		Similar to the above specimen except 5.0 atomic % hydrogen; lattice parameter reported at 293 K is 3.8882 Å.
12	272	Owen, E. A. and Jones, J. I.	1937	X	293-433		Similar to the above specimen except 3.5 atomic % hydrogen; lattice parameter reported at 293 K is 3.8854 Å.
13	60	Holborn, L. and Day, A. L.	1901	T	273-1273		No details given; zero-point correction of -0.023% determined by graphical interpolation.
14	648	Rao, C.N. and Rao, K.K.	1964	X	296-1063		Spec pure specimen from Johnson Matthey; arc melted under inert atmosphere; homogenized for a week at 1273 K in vacuum; filed powder of -240 mesh vacuum sealed in quartz capillary tube, annealed at 923 K to remove stress; lattice parameter reported at 296 K and 298 K are respectively 3.8907 Å and 3.8905 Å; 3.8904 Å at 293 K determined by extrapolation.
15*	598	White, G.K. and Pawlowicz, A.T.	1970	E	3-85, 283		Average of data on the 99.98 Pd from the Malthey Garret of Sydney Australia; and from the Bell Telephone Labs; this curve here reported is referred to 293 K as reference temperature using other available data.
16	654	Nagender Naidu, S.V., and Houska, C.R.	1971	X	80-298		Prepared from 99.97 Pd from Engelhard Industries Inc. by melting in induction furnace under hydrogen atm; homogenized for one week at 1173 K; annealed at 873 K in evaluated quartz tubes after flushing 3 times with argon; lattice parameter reported at 298 K is 3.8900 Å; 3.8897 Å at 293 K determined by graphical interpolation.

* Not shown in figure.

SPECIFICATION TABLE 37. THERMAL LINEAR EXPANSION OF PALLADIUM Pd (continued)

Cur. No.	Ref. No.	Author(s)	Year	Method Used	Temp. Range, K	Name and Specimen Designation	Composition (weight percent), Specifications, and Remarks
17	657	Vest, R.W.	1971	L	293-1068		Specimen prepared from 99.9 pure metal of -352 mesh; data on the coefficient of thermal linear expansion also reported.
18	656	Smirnov, Yu.N., and Timoshenko, V.N.	1972	X	76-298		99.98 (at.%) polycrystalline specimen of 0.13 mm thickness plate; annealed in vacuum at 1300 K for 10 hours; lattice parameters indicate a kink near 95 K, the temperature at which paramagnons became disorder; also indicate a bend near 130 K; lattice parameter reported at 293 K is 3.8892 Å; data on coefficient of thermal linear expansion also reported.

DATA TABLE 37. THERMAL LINEAR EXPANSION OF PALLADIUM Pd

[Temperature, T, K; Linear Expansion, $\Delta L/L_0$, %]

T	$\Delta L/L_0$
CURVE 1	
298	0.00077
618	0.428
798	0.695
888	0.823
973	0.957
1058	1.104
1151	1.268
CURVE 2‡	
30	-0.236
40	-0.234*
50	-0.232
60	-0.227*
70	-0.222*
80	-0.216
90	-0.209*
100	-0.201*
110	-0.193*
120	-0.184*
130	-0.175
140	-0.166*
150	-0.156*
160	-0.146*
170	-0.136*
180	-0.126
190	-0.115*
200	-0.105*
210	-0.094*
220	-0.083*
230	-0.072
240	-0.061*
250	-0.049
260	-0.038*
270	-0.026
CURVE 3	
315	0.026
347	0.064
383	0.103
414	0.144
434	0.171
453	0.193
489	0.234
554	0.315
CURVE 3 (cont.)	
573	0.343
599	0.371
641	0.420
675	0.463
696	0.491
727	0.530
753	0.555
771	0.581
808	0.624
838	0.660
871	0.699
CURVE 4	
338	0.056*
351	0.075
393	0.120
434	0.175
464	0.207
490	0.242*
563	0.328
635	0.415
675	0.465*
724	0.523*
751	0.562
777	0.593
802	0.624*
828	0.544
838	0.666*
878	0.709
CURVE 5	
296	0.004*
352	0.078*
373	0.094
394	0.117*
414	0.143*
434	0.173*
453	0.195*
486	0.236*
513	0.268
525	0.287
550	0.311
560	0.324*
587	0.358
CURVE 5 (cont.)	
622	0.399
650	0.436
689	0.486
739	0.546
763	0.577
777	0.590*
812	0.638
862	0.709
CURVE 6	
86.2	-0.216
90.2	-0.212*
94.2	-0.208*
100.2	-0.203*
106.7	-0.199*
109.7	-0.194*
117.2	-0.189*
120.2	-0.185
126.2	-0.180*
133.4	-0.174*
142.2	-0.166*
145.2	-0.163
152.0	-0.158*
156.2	-0.153
161.2	-0.148
167.7	-0.141*
173.2	-0.134*
178.0	-0.130*
182.2	-0.125*
187.2	-0.120
190.9	-0.115*
195.2	-0.110*
200.2	-0.106*
203.2	-0.101*
208.2	-0.096*
211.7	-0.092
215.6	-0.087*
220.6	-0.083*
224.5	-0.078*
228.7	-0.073*
233.0	-0.069*
236.7	-0.064
240.7	-0.059*
244.7	-0.055*
249.1	-0.050*
CURVE 6 (cont.)	
252.9	-0.045*
256.6	-0.041*
260.6	-0.036
264.6	-0.032*
268.7	-0.027*
273.2	-0.022*
276.3	-0.018*
281.3	-0.013*
285.3	-0.008
289.2	-0.004*
294.2	0.001*
299.2	0.007*
CURVE 7	
125	-0.264
172	-0.192
219	-0.119
274	-0.044
293	0.000*
318	0.028
368	0.102
416	0.174
473	0.251
521	0.326
575	0.399
627	0.472
674	0.541*
727	0.626
775	0.701
830	0.778
877	0.854
930	0.924
982	1.003
1031	1.078
1076	1.142
1136	1.222
1184	1.298
CURVE 8	
283	-0.016
284	-0.010*
337	0.051
383	0.110*
405	0.136
CURVE 8 (cont.)	
454	0.203
490	0.247
504	0.262
559	0.337
CURVE 9	
293	0.000
305	0.005
373	0.031
393	0.056
413	0.057
433	0.108
CURVE 10	
293	0.000*
305	0.013
373	0.056
393	0.114
413	0.163
CURVE 11*	
293	0.000
305	0.012
373	0.049
393	0.112
413	0.148
433	0.194
CURVE 12	
293	0.000*
305	0.064
373	0.095*
393	0.148
413	0.180
433	0.216
CURVE 13	
273	-0.023*
523	0.282*
773	0.616
1023	0.975
1273	1.362
CURVE 14	
296	0.008*
298	0.002*
468	0.221
483	0.262
570	0.373
573	0.375*
663	0.504
678	0.542
783	0.663
783	0.658
873	0.812
881	0.807
978	0.946
981	0.954
1063	1.087
CURVE 15*†	
3	-0.2379
4	-0.2379
5	-0.2379
6	-0.2379
8	-0.2379
10	-0.2379
12	-0.2379
15	-0.2379
18	-0.2379
20	-0.2379
22	-0.2378
25	-0.2376
28	-0.2374
30	-0.237
65	-0.226
75	-0.220
85	-0.214
283	-0.012
CURVE 16	
80	-0.200
195	-0.095
298	0.008*
CURVE 17	
293	0.000*
373	0.081
473	0.190
573	0.311

* Not shown in figure.
† This curve is here reported using the first given temperature as reference temperature at which $\Delta L/L_0 = 0$.
‡ Author's data for coefficient of thermal expansion have been integrated by TPRC to obtain $\Delta L/L_0$.

DATA TABLE 37. THERMAL LINEAR EXPANSION OF PALLADIUM Pd (continued)

T	$\Delta L/L_0$	T	$\Delta L/L_0$	T	$\Delta L/L_0$	T	$\Delta L/L_0$
CURVE 17 (cont.)		CURVE 18 (cont.)		CURVE 18 (cont.)		CURVE 18 (cont.)	
673	0.438	102	-0.213*	190	-0.131*	241	-0.069
773	0.570	107	-0.203*	192	-0.126*	244	-0.064*
873	0.714	110	-0.198	202	-0.110*	250	-0.051*
973	0.868	114	-0.195*	204	-0.116*	253	-0.051*
1068	1.019	117	-0.193*	205	-0.113*	257	-0.041*
		125	-0.188*	208	-0.116*	258	-0.046*
CURVE 18		128	-0.182*	208	-0.105*	261	-0.038*
		134	-0.177*	211	-0.108*	264	-0.036*
77	-0.229	138	-0.177*	213	-0.105*	269	-0.028*
77	-0.229	140	-0.175*	216	-0.100*	270	-0.023*
79	-0.229*	151	-0.170*	221	-0.095*	271	-0.028*
80	-0.229*	152	-0.164*	221	-0.090*	278	-0.020*
84	-0.226*	158	-0.159*	231	-0.080	284	-0.005*
86	-0.226*	159	-0.157*	233	-0.080*	293	0.000*
87	-0.226*	160	-0.162*	234	-0.080*	298	0.005*
87	-0.226*	165	-0.154	234	-0.077*		
90	-0.221*	171	-0.149*	235	-0.077*		
92	-0.221*	174	-0.142*	236	-0.077*		
94	-0.221	177	-0.141*	238	-0.075*		
97	-0.213*	182	-0.134*	238	-0.069*		

DATA TABLE 37. COEFFICIENT OF THERMAL LINEAR EXPANSION OF PALLADIUM Pd

[Temperature, T, K; Coefficient of Expansion, α, 10^{-6} K^{-1}]

T	α	T	α	T	α	T	α
CURVE 2‡		CURVE 2 (cont.)‡		CURVE 15 (cont.)		CURVE 17 (cont.)*	
30	1.12	210	10.9	18	0.330*	873	14.48
40	2.25	220	11.0	20	0.435	973	15.32
50	3.58	230	11.2	22	0.560*	1073	16.12
60	4.76	240	11.3	25	0.785		
70	5.78	250	11.4	28	1.05	CURVE 18	
80	6.65	260	11.4	30	1.24*		
90	7.38	270	11.5	65	5.23	80	5.7
100	7.99			75	6.21	89	6.8
110	8.50	CURVE 15		85	7.00	94	23.3
120	8.88			283	11.78	100	18.1
130	9.22	3	0.0135*			104	11.9
140	9.53	4	0.019*	CURVE 17*		110	11.1
150	9.79	5	0.0255*			124	8.3
160	10.0	6	0.0335*	373	10.28	150	8.3
170	10.2	8	0.055*	473	11.14	200	11.2
180	10.4	10	0.085	573	11.95	250	12.4
190	10.6	12	0.126*	673	12.78		
200	10.8	15	0.213	773	13.63		

* Not shown in figure.

‡ Author's data for coefficient of thermal expansion have been integrated by TPRC to obtain $\Delta L/L_0$.

FIGURE AND TABLE NO. 38R. RECOMMENDED VALUES FOR THERMAL LINEAR EXPANSION OF PLATINUM Pt

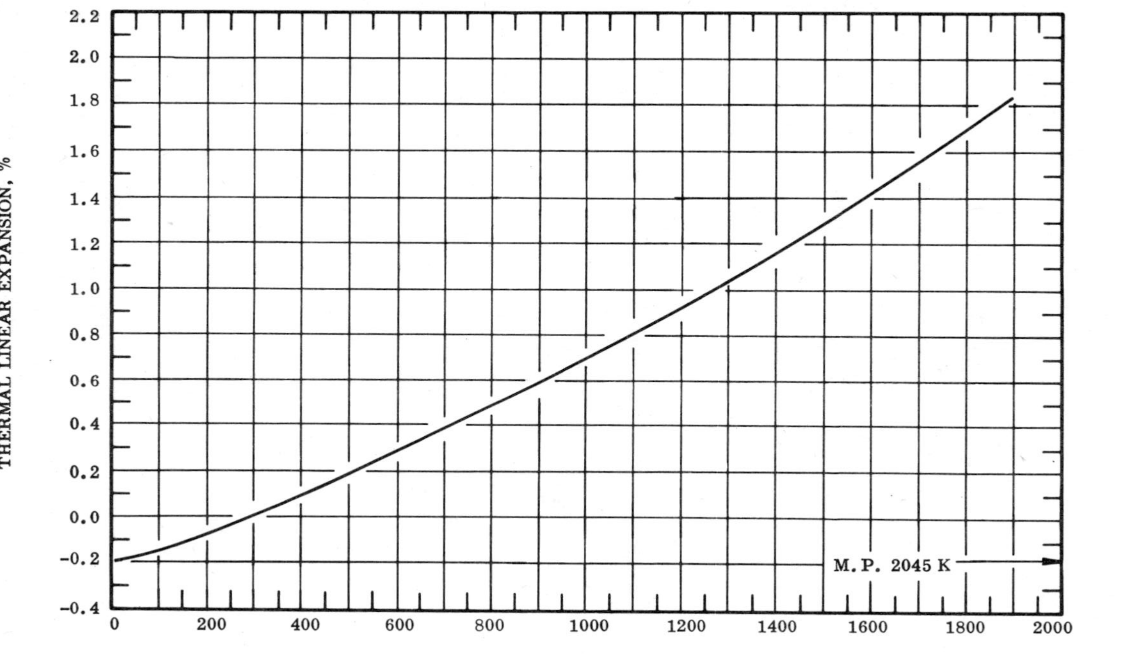

RECOMMENDED VALUES

[Temperature, T, K; Linear Expansion, $\Delta L/L_0$, %; α, K^{-1}]

T	$\Delta L/L_0$	$\alpha \times 10^6$
5	-0.192	0.02
25	-0.191	0.93
50	-0.186	3.6
100	-0.157	6.6
200	-0.081	8.5
293	0.000	8.8
400	0.096	9.2
500	0.189	9.6
600	0.288	9.8
700	0.388	10.1
800	0.490	10.3
900	0.593	10.5
1000	0.699	10.8
1200	0.920	11.4
1400	1.157	12.3
1600	1.414	13.3
1800	1.690	14.4
1900	1.837	14.9

REMARKS

The tabulated values are considered accurate to within ± 3% over the entire temperature range. These values can be represented approximately by the following equations:

$$\Delta L/L_0 = -0.157 + 6.335 \times 10^{-4} (T - 100) + 1.369 \times 10^{-6} (T - 100)^2 - 2.235 \times 10^{-9} (T - 100)^3 \quad (100 < T < 293)$$

$$\Delta L/L_0 = 9.122 \times 10^{-4} (T - 293) + 7.467 \times 10^{-8} (T - 293)^2 + 4.258 \times 10^{-11} (T - 293)^3 \quad (293 < T < 1900)$$

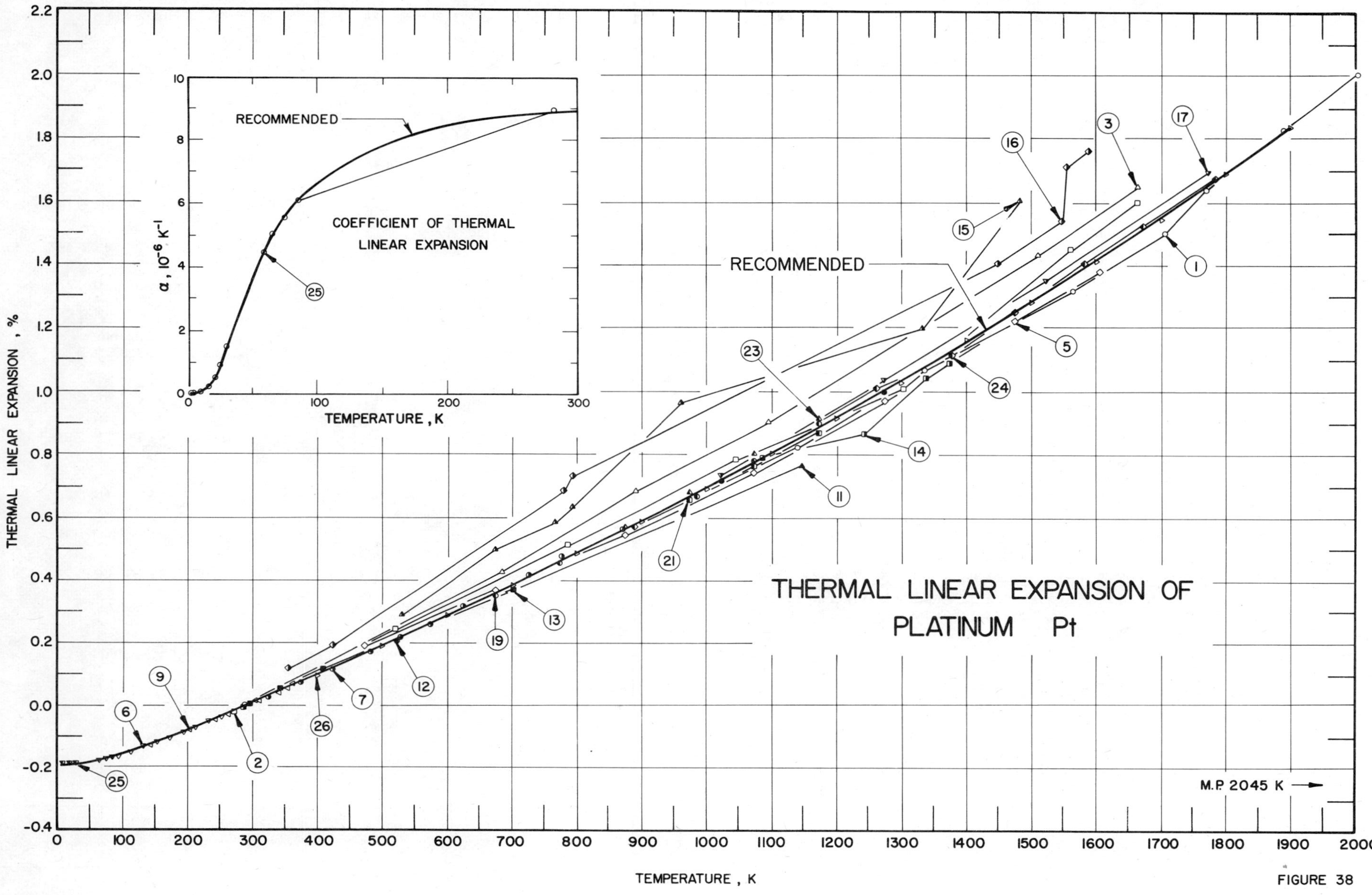
THERMAL LINEAR EXPANSION OF
PLATINUM Pt
THERMAL LINEAR EXPANSION , %
TEMPERATURE , K
RECOMMENDED
M.P. 2045 K
COEFFICIENT OF THERMAL
LINEAR EXPANSION
α, 10^{-6} K^{-1}
TEMPERATURE , K
RECOMMENDED

FIGURE 38

SPECIFICATION TABLE 38. THERMAL LINEAR EXPANSION OF PLATINUM Pt

Cur. No.	Ref. No.	Author(s)	Year	Method Used	Temp. Range, K	Name and Specimen Designation	Composition (weight percent), Specifications, and Remarks
1	215	Edwards, J.W., Speiser, R., and Johnston, H. L.	1951	X	291-2055		99.95 Pt; specimen obtained from American Platinum Works; this curve is here reported using the first given temperature as reference temperature at which $\Delta L/L_0 = 0$; lattice parameter reported at 291 K is 3.924 Å.
2	273	Mauer, F. A. and Bole, L. H.	1955	X	273-1663		High purity powder prepared by Dr. R. Gilchrist of N. B. S. by thermal decomposition of chloroplatinic acid; sintered 30 min at 1573 K; expansion measured in helium atmosphere; zero-point correction of -0.021% determined by graphical interpolation.
3	273	Mauer, F. A. and Bole, L. H.	1955	X	1663-295		The above specimen; expansion measured with decreasing temperature; zero-point correction of 0.028% determined by graphical extrapolation.
4*	291	Andres, K.	1963	O	3.3-12		Specimen polycrystalline rod; this curve is here reported using the first given temperature as reference temperature at which $\Delta L/L_0 = 0$; correction -13.57 x 10^{-7}%.
5	274	Brand, J. A. and Goldschmidt, H. J.	1956	X	273-1607		No details given; lattice parameter reported at 273 K is 3.9137 kX; 3.9145 kX (3.9224 Å) at 293 K determined by graphical interpolation.
6	48	Dorsey, H. G.	1907	I	93-293		Commercially pure; hammered specimen from J. Bishop and Co.; 1.5112 cm long; density 22.15 g cm^{-3}.
7	275	Owen, E. A. and Yates, E. L.	1934	X	287-873		Powder fixed directly to surface of copper sheet; reported error 0.01% in determination of lattice constant; lattice parameter reported at 287 K is 3.9161 Å; 3.91625 Å at 293 K determined by graphical interpolation.
8*	276	Andres, K.	1963	O	2.0-13		99.99 Pt; specimen 7.4 cm long; this curve is here reported using the first given temperature as reference temperature at which $\Delta L/L_0 = 0$; correction -4.62 x 10^{-7}%.
9	75	Nix, F. C. and Mac Nair, D.	1942	I	85-368		99.99^+ Pt; specimen refined in Bell Telephone Laboratories; zero-point correction of -0.018% determined by graphical interpolation.
10*	29	Branchereau, M., Navez, M., and Perroux, M.	1962	L	273-573		99.995 Pt; specimen 8 mm diameter, 100 mm long.
11	150	Eisenstein, A.	1946	X	293, 1148		Cubic crystal structure; reported error 0.3%.
12	60	Holborn, L. and Day, A. L.	1901	T	273-1273		No details given; zero-point correction of -0.018% determined by graphical interpolation.
13	277	Liu, L. G., Takahashi, T., and Bassett, W. A.	1970	X	286-701		Pt used as internal standard for temperature calibration; measured along a-axis of lattice; zero-point correction of 0.005% determined by graphical interpolation.
14	261	Shinoda, G.	1934	X	288-1373		Expansion measured in (422) plane using CuKα x-radiation; zero-point correction of 0.005% determined by graphical interpolation.
15	629	Vertogradskii, V. A.	1969	T	530-1479		99.9 pure specimen of 0.25 mm diameter; zero-point correction 0.055% determined by graphical extrapolation.
16	629	Vertogradskii, V. A.	1969	T	356-1589		99.9 pure specimen of 0.4 mm diameter; zero-point correction of 0.055% determined by graphical extrapolation.
17	630	Cowder, L. R., Zocher, R. W., Kerrisk, J. F., and Lyon, L. L.	1970		293-1773		No details of the specimen given.
18*	631	Masumoto, H. and Kobayashi, T.	1965	L	293-313		99.99 pure specimen.
19	632	Weisenburger, S.	1970	L	573-1173	Participant 1	99.999 pure specimen; measurements on dilatometer manufactured by Leitz, Wetzlar, W. Germany; quartz used for measuring mechanical expansion.

SPECIFICATION TABLE 38. THERMAL LINEAR EXPANSION OF PLATINUM Pt (continued)

Cur. No.	Ref. No.	Author(s)	Year	Method Used	Temp. Range, K	Name and Specimen Designation	Composition (weight percent), Specifications, and Remarks
20	632	Weisenburger, S.	1970	L	573-1173	Participant 2	The above specimen; measurements on dilatometer manufactured by Linseis, Selb, W. Germany; quartz used for measuring electromagnetic induction expansion.
21	632	Weisenburger, S.	1970	L	573-1173	Participant 3	The above specimen; measurements on dilatometer manufactured by Netsch Selb, W. Germany; Al_2O_3 used for measuring electromagnetic induction expansion.
22	632	Weisenburger, S.	1970	L	573-1173	Participant 4	The above specimen; measurements on dilatometer developed by participant; Al_2O_3 used for measuring electromagnetic induction expansion.
23	632	Weisenburger, S.	1970	L	573-1173	Participant 5	The above specimen; measurement on dilatometer manufactured by Adamel, Paris, France; Al_2O_3 used for measuring electormagnetic induction expansion.
24	636	Evans, D. L. and Fischer, G. R.	1972	X	293-1783		99.99 pure; cold-rolled to 0.13 mm thickness; data was obtained at randomly chosen temperature in addition to continuous heating and cooling runs; consistent results obtained.
25	637	White, G. K.	1972	L	3-85, 283		Thermopure grade; 3.7 cm long and 1.0 cm diameter rod specimen from Engelhard Industries; Fe, Mn, Ni, and Cu impurities less than 0.00001.
26	638	Hahn, T. A. and Kirby, R. K.	1972	I, T	293-1900		Measurements on 3 samples, 99.99^+; specimen of 6 mm diameter from J. Bishop and Co.; 99.9992 specimen of 8 mm diameter from Sigmund Cahn, and 99.999 specimen of 6 mm diameter from Heraeus; smoothed values tabulated; data on the coefficient of thermal linear expansion also reported.
27	639	Fitzer, E.	1971	L	293-1273		99.999 pure specimen from Heraeus, W. Germany; arc melted in ceramic crucible in Ar atmosphere and cast in a water cooled crucible; 2 mm of block machined off and block pickled again, drawn in special pure dies, being pickled again after about 10 passes; average of 10 participants.
28*	719	Austin, J. B.	1973	I	273-1173		99.995 pure specimen containing spectroscopic traces of Fe, Ca, and Mg, heated to 1173 K and cooled in the furnace, average of two runs, zero-point correction of -0.0180% determined by graphical interpolation, data on the coefficients of thermal linear expansion also reported.

* Not shown in figure.

DATA TABLE 38. THERMAL LINEAR EXPANSION OF PLATINUM Pt

[Temperature, T, K; Linear Expansion, $\Delta L/L_0$, %]

T	$\Delta L/L_0$	
CURVE 1†		
291	0.000	
1140	0.826	
1335	1.07	
1565	1.32	
1705	1.50	
1770	1.62*	
1770	1.64	
1890	1.83	
2005	2.00	
CURVE 2		
273	-0.021	
298	0.005*	
531	0.244	
788	0.515	
1045	0.787	
1303	1.09	
1561	1.45	
1663	1.60	
CURVE 3		
1663	1.648	
1509	1.431	
1097	0.903	
891	0.689	
685	0.426	
295	0.002*	
CURVE 4*, †		
3.26	0.0000	x 10^{-3}
4.34	0.0012	
4.53	0.0015	
4.87	0.0019	
5.76	0.0040	
5.97	0.0043	
6.23	0.0050	
6.28	0.0053	
7.13	0.0079	
7.13	0.0077	
7.86	0.0110	
8.04	0.0115	
8.95	0.0167	
9.99	0.0242	
10.72	0.0312	

T	$\Delta L/L_0$	
CURVE 4 (cont.)*, †		
11.36	0.0375	x 10^{-3}
11.78	0.0431	
12.27	0.0496	
CURVE 5		
273	-0.021*	
473	0.193	
673	0.367	
873	0.546	
1073	0.742	
1273	0.977	
1473	1.225	
1607	1.389	
CURVE 6		
93	-0.163	
113	-0.149	
133	-0.135	
153	-0.120	
173	-0.104	
193	-0.086	
213	-0.071	
233	-0.053	
253	-0.036	
273	-0.018*	
293	0.000*	
CURVE 7		
287	-0.004	
327	0.022	
375	0.065	
375	0.073	
422	0.114	
481	0.175	
529	0.216	
575	0.257	
624	0.315	
676	0.374*	
726	0.417	
778	0.474	
812	0.489	
816	0.502*	
869	0.566	
873	0.555*	

T	$\Delta L/L_0$	
CURVE 8*, †		
2.0	0.0000	x 10^{-3}
3.1	0.0008	
4.3	0.0021	
4.5	0.0024	
4.8	0.0028	
5.7	0.0048	
5.9	0.0053	
6.2	0.0059	
6.3	0.0063	
7.1	0.0086	
7.0	0.0087	
7.8	0.0120	
8.0	0.0122	
8.9	0.0176	
10.0	0.0253	
10.7	0.0321	
11.3	0.0385	
11.7	0.0443	
12.2	0.0508	
12.8	0.0605	
CURVE 9		
85.2	-0.171*	
87.2	-0.170*	
92.2	-0.166*	
99.2	-0.162*	
104.7	-0.157*	
112.2	-0.152*	
118.3	-0.148*	
124.7	-0.143*	
131.2	-0.138*	
137.2	-0.134*	
143.2	-0.129*	
148.7	-0.124	
155.6	-0.120*	
161.2	-0.115*	
167.1	-0.110*	
172.1	-0.106*	
178.3	-0.101*	
183.2	-0.096*	
189.3	-0.092*	
193.7	-0.087*	
199.3	-0.083*	
205.2	-0.078	
210.2	-0.073*	

T	$\Delta L/L_0$
CURVE 9 (cont.)	
216.2	-0.070*
221.2	-0.064*
226.7	-0.059*
232.1	-0.055*
237.2	-0.050*
242.9	-0.046
247.9	-0.041*
253.0	-0.036*
258.1	-0.032*
263.2	-0.027
268.2	-0.022*
273.2	-0.018*
275.2	-0.016*
279.1	-0.013*
283.7	-0.008*
289.2	-0.004*
294.2	0.001*
299.4	0.006*
301.2	0.008*
306.2	0.012
311.2	0.017
316.1	0.021*
321.2	0.026*
326.2	0.030*
331.5	0.035
336.2	0.040*
341.7	0.044
346.7	0.049*
351.9	0.054*
357.2	0.058
362.7	0.063
368.2	0.068*
CURVE 10*	
273	-0.0185
573	0.2594
CURVE 11	
293	0.000*
1148	0.769

T	$\Delta L/L_0$
CURVE 12	
273	-0.018*
523	0.212
773	0.457
1023	0.721
1273	1.002
CURVE 13	
285.7	-0.003*
298.2	0.005
342.7	0.046
410.2	0.117
527.2	0.219*
701.2	0.370
CURVE 14	
288	-0.005
1088	0.795
1231	0.870
1338	1.044
1373	1.093
CURVE 15	
530	0.290
672	0.501
767	0.597
793	0.641
960	0.968
1330	1.203
1479	1.610
CURVE 16	
356	0.119
423	0.192
779	0.689
793	0.737
1448	1.403
1547	1.540
1554	1.714
1589	1.761

T	$\Delta L/L_0$
CURVE 17	
293	0.000*
523	0.209*
773	0.462*
1023	0.737
1273	1.040
1523	1.357
1773	1.696
CURVE 18*	
293	0.000
313	0.012
CURVE 19	
573	0.256*
673	0.353
773	0.454
873	0.558*
973	0.671
1073	0.781
1173	0.902
CURVE 20*	
573	0.260
673	0.356
773	0.455
873	0.557
973	0.662
1073	0.778
1173	0.900
CURVE 21	
573	0.264*
673	0.366*
773	0.466*
873	0.554*
973	0.660
1073	0.770
1173	0.876

* Not shown in figure.
† This curve is here reported using the first given temperature as reference temperature at which $\Delta L/L_0$.

DATA TABLE 38. THERMAL LINEAR EXPANSION OF PLATINUM Pt (continued)

T	$\Delta L/L_0$
CURVE 22*	
573	0.263
673	0.359
773	0.464
873	0.570
973	0.680
1073	0.791
1173	0.905
CURVE 23	
573	0.269*
673	0.368*
773	0.469*
873	0.572
973	0.683
1073	0.802
1073	0.920
CURVE 24	
293	0.000*
889	0.577
1261	1.014
1378	1.119
1473	1.251
1583	1.404
1671	1.522
1783	1.677
CURVE 25‡	
3	-0.192*
4	-0.192*
5	-0.192*
6	-0.192*
8	-0.192*
10	-0.192
CURVE 25 (cont.)‡	
15	-0.192*
20	-0.192
25	-0.191*
30	-0.191
58	-0.182*
65	-0.179
75	-0.174
85	-0.168
CURVE 26	
293	0.0000*
400	0.0971
500	0.1910
600	0.2873
700	0.3858
800	0.4867
CURVE 26 (cont.)	
900	0.5902
1000	0.6969
1100	0.8069
1200	0.9205
1300	1.037
1400	1.159
1500	1.284
1600	1.413
1700	1.548
1800	1.689
1900	1.839
CURVE 27*	
293	0.000
573	0.262
673	0.363
CURVE 27 (cont.)*	
773	0.464
873	0.596
973	0.678
1073	0.786
1173	0.901
1273	1.010
CURVE 28*	
273	-0.0180
373	0.0719
473	0.1645
573	0.2595
673	0.3580
773	0.4590
873	0.5620
973	0.6665
CURVE 28 (cont.)*	
1073	0.7675
1173	0.8765

DATA TABLE 38. COEFFICIENT OF THERMAL LINEAR EXPANSION OF PLATINUM Pt

[Temperature, T, K; Coefficient of Expansion, α, 10^{-6} K^{-1}]

T	α
CURVE 25‡	
3	0.007
4	0.0107*
5	0.0155
6	0.022*
8	0.041*
10	0.071
15	0.22
20	0.51
25	0.93
30	1.43
58	4.43
65	5.01
75	5.58
85	6.09
283	8.92
CURVE 26*	
293	8.88
330	9.05
330	9.18
345	9.00
CURVE 26 (cont.)*	
360	9.18
420	9.28
432	9.46
473	9.49
504	9.58
517	9.44
521	9.57
622	9.85
669	9.89
712	10.10
720	10.00
754	10.22
815	10.31
849	10.32
866	10.50
900	10.49
910	10.67
955	10.74
1008	10.93
1127	11.30
1150	11.60
1171	11.60
CURVE 26 (cont.)*	
1186	11.86
1189	11.20
1200	11.70
1221	11.60
1290	11.89
1299	12.34
1315	11.91
1330	12.16
1355	12.34
1404	12.29
1424	12.22
1444	12.55
1481	12.74
1510	14.74
1562	13.04
1581	12.97
1634	13.50
1667	13.75
1740	14.16
1754	14.16
1814	14.60
1841	14.86
CURVE 28*	
373	9.13
473	9.39
573	9.66
673	9.93
773	10.19
873	10.46
973	10.72
1073	10.97
1173	11.27

* Not shown in figure.

‡ Author's data for coefficient of thermal expansion have been integrated by TPRC to obtain $\Delta L/L_0$.

FIGURE AND TABLE NO. 39R. RECOMMENDED VALUES FOR THERMAL LINEAR EXPANSION OF PLUTONIUM Pu

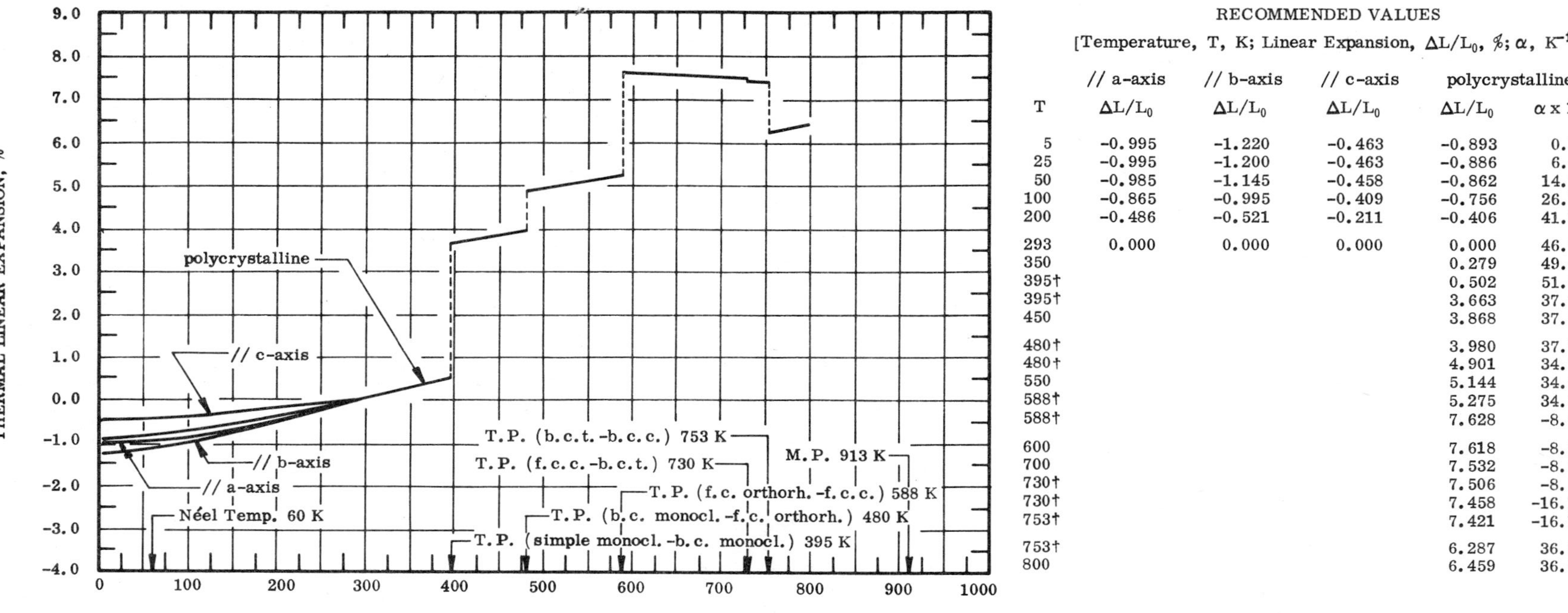

RECOMMENDED VALUES

[Temperature, T, K; Linear Expansion, $\Delta L/L_0$, %; α, K^{-1}]

T	// a-axis $\Delta L/L_0$	// b-axis $\Delta L/L_0$	// c-axis $\Delta L/L_0$	polycrystalline $\Delta L/L_0$	polycrystalline $\alpha \times 10^6$
5	-0.995	-1.220	-0.463	-0.893	0.1
25	-0.995	-1.200	-0.463	-0.886	6.4
50	-0.985	-1.145	-0.458	-0.862	14.1
100	-0.865	-0.995	-0.409	-0.756	26.7
200	-0.486	-0.521	-0.211	-0.406	41.0
293	0.000	0.000	0.000	0.000	46.7
350				0.279	49.4
395†				0.502	51.0
395†				3.663	37.3
450				3.868	37.3
480†				3.980	37.3
480†				4.901	34.6
550				5.144	34.6
588†				5.275	34.6
588†				7.628	-8.6
600				7.618	-8.6
700				7.532	-8.6
730†				7.506	-8.6
730†				7.458	-16.1
753†				7.421	-16.1
753†				6.287	36.6
800				6.459	36.6

† Phase transformation.

REMARKS

The tabulated values are considered accurate to within ± 3% at temperatures below 480 K. The tabulated values for polycrystalline plutonium are calculated from single crystal data and are considered accurate to about ± 3% at temperatures below 395 K and ± 5% above. The reliable experimental data on polycrystalline plutonium agree well with tabulated values at temperatures below 395 K and considerably lower (curve 44) at higher temperatures. The tabulated values for polycrystalline plutonium can be represented approximately by the following equations:

$$\Delta L/L_0 = -0.893 - 3.180 \times 10^{-5} (T - 5) + 1.763 \times 10^{-5} (T - 5)^2 - 2.407 \times 10^{-8} (T - 5)^3 \quad (5 < T < 200)$$

$$\Delta L/L_0 = -0.407 + 4.099 \times 10^{-3} (T - 200) + 3.552 \times 10^{-6} (T - 200)^2 - 3.360 \times 10^{-9} (T - 200)^3 \quad (200 < T < 395)$$

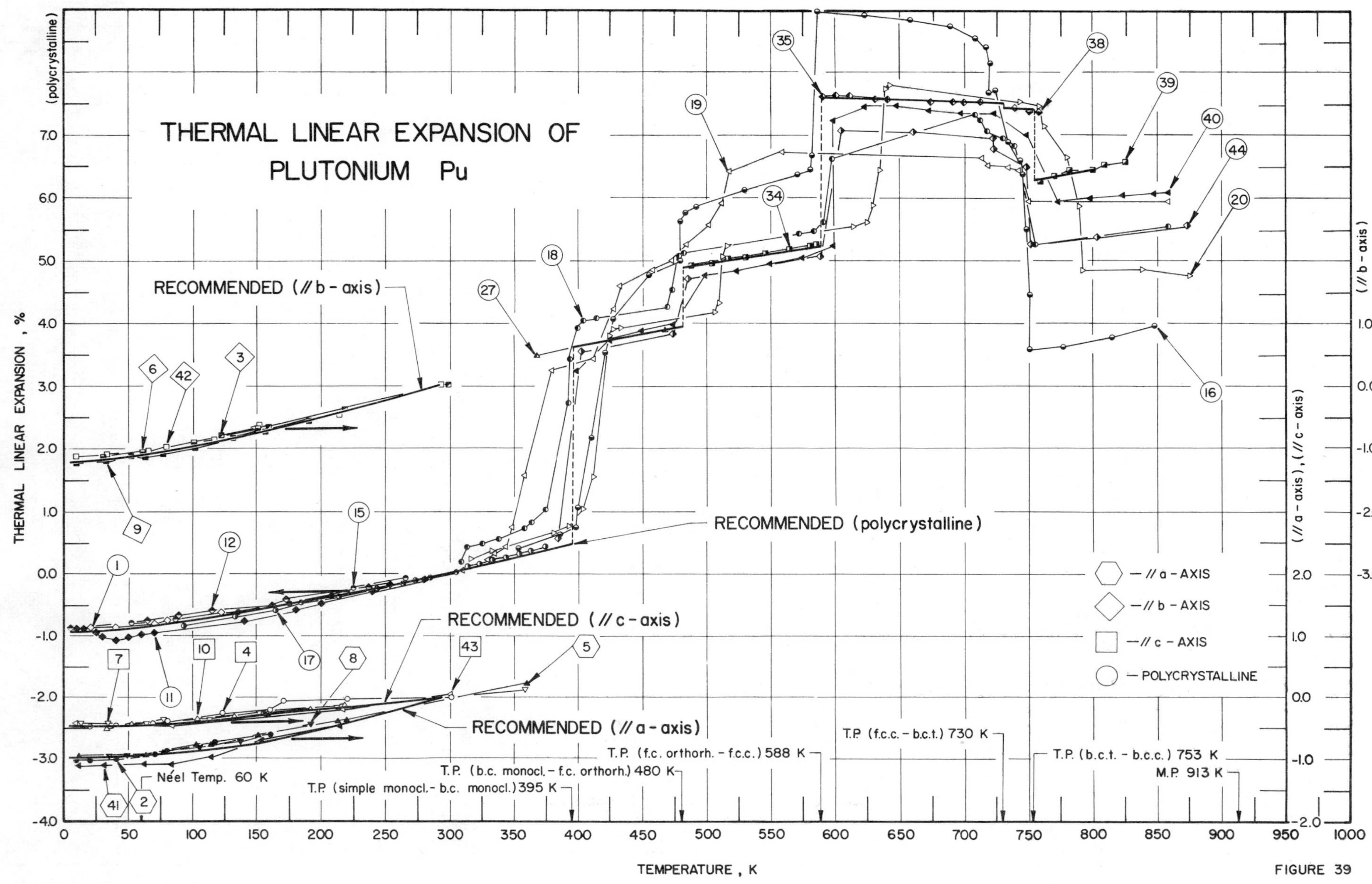
THERMAL LINEAR EXPANSION OF
PLUTONIUM Pu
THERMAL LINEAR EXPANSION, %
(polycrystalline)
(// a - axis), (// c - axis)
(// b - axis)
TEMPERATURE, K
RECOMMENDED (// b - axis)
RECOMMENDED (polycrystalline)
RECOMMENDED (// c - axis)
RECOMMENDED (// a - axis)
— // a - AXIS
— // b - AXIS
— // c - AXIS
— POLYCRYSTALLINE
Neel Temp. 60 K
T.P. (simple monocl.- b.c. monocl.) 395 K
T.P. (b.c. monocl.- f.c. orthorh.) 480 K
T.P. (f.c. orthorh. - f.c.c.) 588 K
T.P. (f.c.c. - b.c.t.) 730 K
T.P. (b.c.t. - b.c.c.) 753 K
M.P. 913 K

FIGURE 39

SPECIFICATION TABLE 39. THERMAL LINEAR EXPANSION OF PLUTONIUM Pu

Cur. No.	Ref. No.	Author(s)	Year	Method Used	Temp. Range, K	Name and Specimen Designation	Composition (weight percent), Specifications, and Remarks
1	278	Solente, P.	1964	S	20-204	α-Pu	Homogeneous, polycrystalline specimen in α-phase; zero-point correction is ±0.000%.
2	278	Solente, P.	1964	X	20-300	α-Pu	The above specimen; expansion measured along a-axis of orthogonal crystal structure; lattice parameter reported at 300 K is 6.183 Å; 6.180 Å at 293 K determined by graphical interpolation.
3	278	Solente, P.	1964	X	20-300	α-Pu	The above specimen; expansion measured along b-axis; lattice parameter reported at 300 K is 4.822 Å; 4.8201 Å at 293 K determined by graphical interpolation.
4	278	Solente, P.	1964	X	20-300	α-Pu	The above specimen; expansion measured along c-axis; lattice parameter reported at 300 K is 10.963 Å; 10.9626 Å at 293 K determined by graphical interpolation.
5	279	Lee, J.A., Marples, J.A.C., Mendelssohn, K., and Sutcliffe, P.W.	1967	X	11-360		97.58 Pu 239, 2.34 Pu 240, 0.08 Pu 241 in atomic percent; P< 0.01, Th< 0.005, 0.0025 Al, Ca< 0.0025, Ta< 0.0025, W< 0.0025, 0.0023 Fe, 0.0016 Si, 0.0012 Ga, 0.0011 Ni, 0.001 As, 0.001 Zr, 0.001 Cr; high purity plutonium prepared by USAEC; specimen as a plate ~2 x 1 x 0.1 cm; measurement made within six months of purification; expansion measured along a-axis; lattice parameter reported at 293 K is 6.1819 Å.
6	279	Lee, J.A., et al.	1967	X	11-360		The above specimen; expansion measured along b-axis; lattice parameter reported at 293 K is 4.8219 Å.
7	279	Lee, J.A., et al.	1967	X	13-359		The above specimen; expansion measured along c-axis; lattice parameter reported at 293 K is 10.9644 Å.
8	279	Lee, J.A., et al.	1967	X	8-300		Isotopic analysis 97.02 Pu 239, 2.87 Pu 240, 0.11 Pu 241; P< 0.01, W< 0.0056, Ca< 0.0025, 0.0024 Fe, 0.0021 Si, 0.0019 Ta, 0.0015 Al, As< 0.001, Zr< 0.001, 0.0005 Cr; high purity plutonium prepared by USAEC; specimen as a plate ~2 x 1 x 0.1 cm; measurements made within three months of purification; expansion measured along a-axis; lattice parameter reported at 296 K is 6.1842 Å; 6.1833 Å at 293 K determined by graphical interpolation.
9	279	Lee, J.A., et al.	1967	X	9-301		The above specimen; expansion measured along b-axis; lattice parameter reported at 297 K is 4.8242 Å; 4.8231 Å at 293 K determined by graphical interpolation.
10	279	Lee, J.A., et al.	1967	X	8-300		The above specimen; expansion measured along c-axis; lattice parameter reported at 299 K is 10.9669 Å; 10.9656 Å at 293 K determined by graphical interpolation.
11	280	Lallement, R.	1963	L	10-280		Cooled rapidly to 77 K and then to 4.2 K, heated up to 77 K in about 8 hr; also cooled rapidly to 77 K and then heated to 300 K in about 8 hr.
12	145	Sandenaw, T.A.	1960	V	298-5		Normal purity specimen; expansion measured with decreasing temperature during rapid cooling; zero-point correction of -0.081% determined by graphical interpolation.
13*	145	Sandenaw, T.A.	1960	V	298-48		Similar to the above; expansion measured during slow cooling; zero-point correction of -0.081% determined by graphical interpolation.
14*	145	Sandenaw, T.A.	1960	V	2-60		The above specimen; expansion measured with increasing temperature; this curve is here reported using the first given temperature as reference temperature at which $\Delta L/L_0 = 0$; zero point correction 0.794%.

*Not shown in figure.

SPECIFICATION TABLE 39. THERMAL LINEAR EXPANSION OF PLUTONIUM Pu (continued)

Cur. No.	Ref. No.	Author(s)	Year	Method Used	Temp. Range, K	Name and Specimen Designation	Composition (weight percent), Specifications, and Remarks
15	145	Sandenaw, T. A.	1960	V	4-300		Similar to the above specimen except 0.06 Fe added; zero-point correction of -0.065% determined by graphical interpolation.
16	281	Grove, G.R.	1966	L	301-846		Disc specimen, 0.157 cm thick; zero-point correction of 0.056% determined by graphical extrapolation.
17	282	Elliott, R.O. and Tate, R.E.	1952	L	93-373		0.068 Fe, 0.027 C, 0.05 Ni, 0.03 Al, 0.03 Cr, 0.02 Bi, 0.02 V; α-Pu; specimen 0.5 in. in diameter, 0.947 in. long; plutonium melting stock RJ 1213, cooled from 598 K under 50 000 psi, heating rate 1 C per min; dial indicator accurate to 0.00002 in.; density 19.60 g cm^{-3}.
18	283	Abramson, R., Boucher, R., Fabre, R., Monti, H., Pascard, R., Anselin, F., and Grison, E.	1958	L	308-859	Sample B	99.99 pure, 0.0035 Fe, 0.0025 C, 0.0020 Ni, Cu; obtained from USAEC; expansion measured under vacuum of 10^{-5} mm Hg with increasing temperature; observed phase transition temperatures α→β at 398 K, β→γ at 486 K, γ→δ at 587 K, δ→δ' at 737 K, δ'→ε at 758 K; zero-point correction is ±0.20%.
19	283	Abramson, R., et al.	1958	L	859-310	Sample B	The above specimen; expansion measured with decreasing temperature; phase transition temperatures; ε→δ' at 766 K, δ'→δ at 742 K, δ→γ at ~535 K, β→α at 364 K; zero-point correction of 0.58% determined by graphical extrapolation.
20	283	Abramson, R., et al.	1958	L	315-875	Sample C	Specimen extracted from a bar taken from a reactor at Saclay, France; expansion measured with increasing temperature; zero-point correction of 0.12% determined by graphical extrapolation.
21*	283	Abramson, R., et al.	1958	L	875-395	Sample C	The above specimen; expansion measured with decreasing temperature; this curve is here reported using the first given temperature as reference temperature at which $\Delta L/L_0 = 0$.
22	266	Smith, C.S.	1954		298-774		Data measured at Los Alamos in 1945; expansion measured with increasing temperature.
23	266	Smith, C.S.	1954		793-298		The above specimen; expansion with decreasing temperature; zero-point correction of -0.004% determined by graphical extrapolation.
24*	284	Zachariasen, W.H. and Ellinger, F.H.	1959	X	366-463	β-Pu	Pure; body-centered monoclinic crystal structure in β-phase with 34 atoms per unit cell; density 17.70 g cm^{-3} at 463 K; expansion measured along a-axis; reported error ±0.03% in measurement of lattice constant; this curve is here reported using the first given temperature as the reference temperature at which $\Delta L/L_0 = 0$.
25*	284	Zachariasen, W.H. and Ellinger, F.H.	1959	X	366-463	β-Pu	The above specimen; expansion measured along b-lattice; reported error ±0.04% in measurement of lattice constant; this curve is here reported using the first given temperature as reference temperature at which $\Delta L/L_0 = 0$; lattice parameter reported at 366 K is 10.449 A.
26*	284	Zachariasen, W.H. and Ellinger, F.H.	1959	X	366-463	β-Pu	The above specimen; expansion measured along c-lattice; reported error ±0.04% in measurement of lattice constant; this curve is here reported using the first given temperature as reference temperature at which $\Delta L/L_0 = 0$; unit cell reported at 366 K is 7.824 A.
27	284	Zachariasen, W.H. and Ellinger, F.H.	1959	X	366-463	β-Pu	The values reported in this curve calculated from curves 24, 25, 26 and other available data.

*Not shown in figure.

SPECIFICATION TABLE 39. THERMAL LINEAR EXPANSION OF PLUTONIUM Pu (continued)

Cur. No.	Ref. No.	Author(s)	Year	Method Used	Temp. Range, K	Name and Specimen Designation	Composition (weight percent), Specifications, and Remarks
28*	284	Zachariasen, W.H. and Ellinger, F.H.	1959	X	406-475	β-Pu	99.42 Pu, 0.58 Ce; expansion measured along a-axis; this curve is here reported using the first given temperature as reference temperature at which $\Delta L/L_0 = 0$.
29*	284	Zachariasen, W.H. and Ellinger, F.H.	1959	X	406-475	β-Pu	The above specimen; expansion measured along b-axis; this curve is here reported using the first given temperature as reference temperature at which $\Delta L/L_0 = 0$.
30*	284	Zachariasen, W.H. and Ellinger, F.H.	1959	X	406-475	β-Pu	The above specimen; expansion measured along c-axis; this curve is here reported using the first given temperature as reference temperature at which $\Delta L/L_0 = 0$.
31*	285	Zachariasen, W.H. and Ellinger, F.H.	1955	X	486-585	γ-Pu	Powder specimen in γ-phase prepared from 99.97 Pu metal; small amount of pure Ag powder added for temperature calibration; specimen placed in evacuated silica capillary; expansion measured in [100] direction; reported error 5.1% in measurement of lattice constant; lattice parameter reported at 486 K is 3.16052 Å; this curve is here reported using the first given temperature as reference temperature at which $\Delta L/L_0 = 0$.
32*	285	Zachariasen, W.H. and Ellinger, F.H.	1955	X	486-585	γ-Pu	The above specimen; expansion measured in [010] direction; reported error 1.4% in measurement of lattice constant; this curve is here reported using the first given temperature as reference temperature at which $\Delta L/L_0 = 0$; lattice constant reported at 486 K is 5.76275 Å.
33*	285	Zachariasen, W.H. and Ellinger, F.H.	1955	X	486-585	γ-Pu	The above specimen; expansion measured in [001] direction; reported error 1.9% in measurement of lattice constant; this curve is here reported using the first given temperature as reference temperature at which $\Delta L/L_0 = 0$; lattice parameter reported at 486 K is 10.1442 Å.
34	285	Zachariasen, W.H. and Ellinger, F.H.	1955	X	486-585	γ-Pu	The values reported in this curve calculated from curves 31, 32, 33 and other available data.
35*	286	Ellinger, F.H.	1957	X	590-713	δ-Pu	99.97 Pu metal filings; close-packed-cubic structure in δ-phase; specimen contained in evacuated vitreous silica capillary tube; silver added for temperature calibration; expansion measured using CuK x-radiation; (stable only 592 ± 5 K to 724 ± 4 K); lattice constant reported at 590 K is 4.63668 Å.
36*	286	Ellinger, F.H.	1957	X	738-757	δ'-Pu	The above specimen except b.c.t. structure in δ'-phase; structure stable only between 738 K and 757 K; expansion measured along a-axis using CuK x-radiation; lattice parameter reported at 738 K is 3.3245 Å; this curve is here reported using the first given temperature as reference temperature at which $\Delta L/L_0 = 0$.
37*	286	Ellinger, F.H.	1957	X	738-757	δ'-Pu	The above specimen; expansion measured along c-axis; this curve is here reported using the first given temperature as reference temperature at which $\Delta L/L_0 = 0$; lattice parameter reported at 738 K is 4.4889 Å.
38	286	Ellinger, F.H.	1957	X	738-757	δ'-Pu	The values reported in this curve calculated from curves 36, 37 and other available data.
39*	286	Ellinger, F.H.	1957	X	760-825	ϵ-Pu	The above specimen except b.c.c. structure in ϵ-phase; structure stable only from 758 K to 912 K; expansion measured using CuK x-radiation; lattice parameter reported at 760 K is 3.63498 Å.
40	772	Goldberg, A., Rose, R.L., and Matlock, D.K.	1970	L	293-859		High purity electrorefined specimen with <0.02 impurities including Am., zero-point correction is 3.35%.

* Not shown in figure.

SPECIFICATION TABLE 39. THERMAL LINEAR EXPANSION OF PLUTONIUM Pu (continued)

Cur. No.	Ref. No.	Author(s)	Year	Method Used	Temp. Range, K	Name and Specimen Designation	Composition (weight percent), Specifications, and Remarks
41	773	Fournier, J.M.	1972	X	10-300		No details given; anomaly near 80 K; measurements along a-axis; lattice parameter reported at 293 K is 5.181 Å.
42	773	Fournier, J.M.	1972	X	11-298		Similar to the above specimen; measurements along b-axis; lattice parameter reported at 293 K is 4.822 Å.
43	773	Fournier, J.M.	1972	X	10-300		Similar to the above specimen; measurements along c-axis; lattice parameter reported at 293 K is 10.966 Å.
44	778	Zette, E.R.	1955	L	308-873		99.97 pure specimen; zero-point correction is -0.08%.

DATA TABLE 39. THERMAL LINEAR EXPANSION OF PLUTONIUM Pu

[Temperature, T, K; Linear Expansion, $\Delta L/L_0$, %]

T	$\Delta L/L_0$
CURVE 1	
19.6	-0.887
39.8	-0.879
68.8	-0.799
80.8	-0.760
122.7	-0.637
161.5	-0.519
220.4	-0.326*
293.6	0.005*
CURVE 2	
20	-1.035
40	-1.003
69	-0.938
81	-0.890
123	-0.744
160	-0.615
220	-0.372
300	0.048*
CURVE 3	
20	-1.102*
40	-1.081*
69	-0.998*
81	-0.956
123	-0.790
160	-0.666
220	-0.396
300	0.039
CURVE 4	
20	-0.498
40	-0.479
69	-0.425
81	-0.397
123	-0.288
160	-0.206
220	-0.033
300	0.004
CURVE 5	
11	-0.990
12	-0.958*

T	$\Delta L/L_0$
CURVE 5 (cont.)	
32	-0.951*
63	-0.967
77	-0.932
77	-0.851*
103	-0.788
115	-0.780
151	-0.603
213	-0.396
293	0.000*
299	0.005*
300	0.057
360	0.233
CURVE 6	
11	-1.109*
11	-1.180*
32	-1.116
63	-1.056
76	-0.948*
103	-0.902
113	-0.711
151	-0.641
213	-0.458*
293	0.000
298	0.050
360	0.288
CURVE 7	
13	-0.405
13	-0.428*
34	-0.428
64	-0.447
76	-0.455*
77	-0.376
103	-0.346*
114	-0.384
152	-0.279
214	-0.183
293	0.000*
300	0.003*
300	0.069*
359	0.111

T	$\Delta L/L_0$
CURVE 8	
8	-1.024
35	-0.992
50	-0.993
65	-0.940*
78	-0.922*
105	-0.815
131	-0.741
159	-0.626*
192	-0.466
219	-0.356*
296	0.014*
300	0.055*
CURVE 9	
9	-1.217
34	-1.209
51	-1.134
64	-1.134
77	-1.097
103	-0.999
131	-0.847
158	-0.718
191	-0.565
218	-0.407*
297	0.022*
301	0.047*
CURVE 10	
8	-0.480*
8	-0.491*
34	-0.501
51	-0.475
64	-0.484*
77	-0.460
104	-0.346
131	-0.335
158	-0.271
192	-0.210
219	-0.145
299	0.012*
300	0.040*

T	$\Delta L/L_0$
CURVE 11	
10	-0.85
15	-0.88
20	-0.90*
25	-0.96
30	-1.03
40	-1.08
50	-1.03
60	-0.99
70	-0.96
80	-0.95*
100	-0.90*
120	-0.86*
140	-0.77
160	-0.70*
180	-0.58
200	-0.48
240	-0.29
280	-0.08
CURVE 12	
298	0.019*
251	-0.150
236	-0.202
171	-0.416
114	-0.597
88	-0.671
65	-0.743
5	-0.849
CURVE 13*	
298	0.019
261	-0.134
253	-0.182
201	-0.414
180	-0.513
166	-0.584
159	-0.620
151	-0.619
137	-0.617
121	-0.651
75	-0.769
64	-0.794
48	-0.835

T	$\Delta L/L_0$
CURVE 14*, †	
2	0.000
14	0.006
24	0.014
46	0.056
60	0.114
CURVE 15	
4	-0.846*
7	-0.844
24	-0.832*
29	-0.824
35	-0.815*
39	-0.805*
53	-0.770
62	-0.750*
68	-0.732*
74	-0.722*
79	-0.716*
86	-0.703
90	-0.696*
112	-0.656*
128	-0.625*
135	-0.611
156	-0.551*
166	-0.525*
175	-0.497
181	-0.475*
190	-0.451*
197	-0.404*
206	-0.350
210	-0.325*
220	-0.265*
225	-0.243
236	-0.197*
240	-0.184*
251	-0.140*
265	-0.087
300	0.023*
CURVE 16	
301	0.056
352	0.410
385	0.669

T	$\Delta L/L_0$
CURVE 16 (cont.)	
395	0.766
399	1.088
411	2.185
420	3.540
426	4.088
454	4.298
478	4.491
479	5.604
483	5.766
490	5.879
529	6.104
570	6.362
580	6.443
581	6.653
586	9.000
623	8.959
658	8.895
688	8.750
706	8.572
717	8.411
719	8.137
719	7.717
724	7.754
733	6.862
741	6.588
744	6.362
746	5.491
750	4.427
750	3.556
776	3.604
814	3.766
846	3.911
CURVE 17‡	
93	-0.814
103	-0.782*
113	-0.750*
123	-0.716*
133	-0.682
143	-0.646*
153	-0.610*
163	-0.573
173	-0.535*
183	-0.495
193	-0.455*

*Not shown in figure.
†This curve is here reported using the first given temperature as reference temperature at which $\Delta L/L_0 = 0$.
‡Author's data for coefficient of thermal expansion have been integrated by TPRC to obtain $\Delta L/L_0$.

DATA TABLE 39. THERMAL LINEAR EXPANSION OF PLUTONIUM Pu (continued)

T	$\Delta L/L_0$
CURVE 17 (cont.)‡	
203	-0.414*
213	-0.372
223	-0.329*
233	-0.284*
243	-0.240
253	-0.194*
263	-0.147
273	-0.099
283	-0.050
293	0.000*
303	0.051
313	0.102
323	0.155
333	0.209
343	0.264
353	0.319
363	0.376
373	0.433
CURVE 18	
308	0.20
312	0.32
325	0.48
337	0.58
357	0.73
364	0.82
374	1.04
391	2.73
394	3.43
400	3.91
403	4.01
413	4.08
468	4.26
472	4.53
477	5.08
481	5.14
572	5.43
584	5.47
590	5.61
598	6.60
708	7.32
711	7.22
717	7.09
730	6.95
737	6.83
753	5.24
859	5.57
CURVE 19	
859	5.95
750	5.94
739	6.41
734	6.46
718	6.53
713	6.61
558	6.73
516	6.42
511	5.91
502	5.59
485	5.28
471	5.02
457	4.88
433	4.61
427	4.42
425	3.73*
423	3.59*
419	3.52*
411	3.44
378	3.26
357	1.59
346	0.75
339	0.45
334	0.31
329	0.21
322	0.14*
310	0.07
CURVE 20	
315	0.23
331	0.37
380	0.67
391	0.77
402	1.04
411	1.55
424	3.80
426	3.90
432	3.93
505	4.19
508	4.32
512	5.08
515	5.21
612	5.53
623	5.61
628	5.87
634	6.41
CURVE 20 (cont.)	
638	7.71
641	7.78
741	7.53
757	7.47
760	7.40
762	7.16
780	6.61
788	5.87
791	4.85
838	4.87
875	4.77
CURVE 21*, †	
875	0.00
793	-0.43
788	0.03
786	0.22
764	0.35
758	0.45
607	0.63
569	0.29
566	-0.12
557	-0.71
551	-1.00
543	-1.20
525	-1.48
505	-1.76
472	-1.99
462	-2.20
457	-2.67
454	-2.88
449	-2.96
434	-3.09
411	-3.21
395	-4.36
CURVE 22	
298	0.00*
373	0.39
402	0.60*
409	0.79
427	2.51*
431	2.65
440	2.76*
484	2.96
CURVE 22 (cont.)	
492	3.13*
497	3.68
504	3.80*
573	4.01
583	4.22*
594	6.01*
602	6.08
615	6.08*
657	5.94
695	5.94*
726	5.81
740	5.61*
755	5.04*
774	5.07
CURVE 23	
793	4.85
743	4.76
735	5.03
722	5.23
673	5.32
573	5.41
522	5.47
493	5.44
484	5.36
484	5.14
473	5.00
473	4.80
451	4.56
444	4.36
427	4.16
394	3.58
385	3.27
361	2.96
337	2.73
329	1.62
316	0.54
298	0.12*
CURVE 24*, †	
366	0.00
463	0.618
CURVE 25*, †	
366	0.000
463	0.134
CURVE 26*, †	
366	0.000
463	0.447
CURVE 27	
366	3.489
463	3.913
CURVE 28*, †	
406	0.000
475	0.380
CURVE 29*, †	
406	0.000
475	0.145
CURVE 30*, †	
406	0.000
475	0.228
CURVE 31*, †	
486	0.000
503	-0.052
506	-0.045
511	-0.046
512	-0.063
515	-0.065
516	-0.081
519	-0.066
531	-0.136
537	-0.130
545	-0.129
562	-0.173
565	-0.163
569	-0.158
577	-0.177
580	-0.196
582	-0.203
585	-0.207
CURVE 32*, †	
486	0.000
503	0.000
503	0.069
506	0.086
511	0.085
512	0.113
515	0.127
516	0.133
519	0.140
531	0.190
537	0.216
545	0.238
562	0.302
565	0.310
569	0.314
577	0.367
580	0.379
582	0.392
585	0.399
CURVE 33*, †	
486	0.000
503	0.113
506	0.171
511	0.239
512	0.260
515	0.245
516	0.245
519	0.280
531	0.386
537	0.432
545	0.492
562	0.609
565	0.650
569	0.637
577	0.787
580	0.790
582	0.828
585	0.836
CURVE 34	
486	4.917
503	4.963
506	4.991*
511	4.989*
512	5.018*

*Not shown in figure.
†This curve is here reported using the first given temperature as reference temperature at which $\Delta L/L_0 = 0$.

DATA TABLE 39. THERMAL LINEAR EXPANSION OF PLUTONIUM Pu (continued)

T	$\Delta L/L_0$
CURVE 34 (cont.)	
515	5.030
516	5.021*
519	5.041*
531	5.071
537	5.098*
545	5.127
562	5.175*
565	5.195
569	5.194*
577	5.258*
580	5.252
582	5.274*
585	5.276
CURVE 35	
590	7.612
591	7.618*
591	7.631*
595	7.622*
595	7.620*
596	7.622*
596	7.620*
598	7.622*
598	7.616*
599	7.605*
599	7.608*
600	7.626
602	7.622*
602	7.618*
603	7.615*
604	7.622*
605	7.620*
605	7.607*
610	7.615
611	7.608*
612	7.613*
612	7.594*
618	7.603*
618	7.590*
622	7.620*
622	7.585*
623	7.613*
623	7.595*
623	7.592*
623	7.587*
623	7.574*
626	7.592*
CURVE 35 (cont.)	
627	7.595*
627	7.585*
630	7.579
635	7.578*
636	7.569*
639	7.594*
640	7.592
649	5.577*
651	7.562*
653	7.553*
657	7.553*
658	7.574*
658	7.562*
658	7.548*
661	7.548*
667	7.532*
669	7.546*
671	7.550*
673	7.557*
674	7.537
679	7.540*
681	7.546*
681	7.541*
688	7.518*
688	7.516*
689	7.541*
689	7.532
695	7.530*
700	7.530
701	7.575*
707	7.516*
708	7.509*
711	7.534*
712	7.533
CURVE 36*, †	
738	0.000
739	-0.009
743	0.229
743	0.388
744	0.334
744	0.397
744	0.280
744	0.241
749	0.126
749	0.409
749	0.457
CURVE 36 (cont.)*, †	
750	0.301
750	0.424
750	0.481
751	0.415
751	0.511
753	0.532
753	0.575
755	0.508
757	0.502
CURVE 37*, †	
738	0.000
739	0.062
743	-0.423
743	-0.488
743	-0.559
743	-0.813
744	-0.780
745	-0.784
745	-0.646
745	-0.472
749	-0.960
750	-0.735
750	-0.991
750	-1.145
751	-0.971
751	-1.105
753	-1.096
753	-1.254
755	-1.239
757	-1.172
CURVE 38	
738	7.447
740	7.463*
743	7.435*
744	7.419*
745	7.438*
749	7.338*
750	7.390
751	7.405*
753	7.419*
755	7.364*
757	7.384
CURVE 39†	
759	6.292
759	6.313*
761	6.323*
767	6.335*
768	6.322*
768	6.370*
771	6.351
776	6.369*
782	6.396*
784	6.433
788	6.423*
792	6.433*
799	6.461
803	6.480*
805	6.486*
806	6.500*
808	6.500*
810	6.514
818	6.558*
819	6.534*
825	6.552
825	6.563*
CURVE 40	
293	0.00
303	0.06*
323	0.14*
348	0.28
373	0.40*
398	3.26
423	3.75
448	3.87
473	3.97
498	4.76
523	4.83
548	4.93
573	5.01
598	5.21
598	7.21
623	7.45
648	7.43
673	7.37
698	7.35
723	7.33
748	6.98
773	5.93
798	5.98
CURVE 40 (cont.)	
823	6.02
848	6.07
859	6.09
CURVE 41	
10	-1.12
32	-1.12
63	-1.10
78	-1.10
115	-0.93
153	-0.71
215	-0.46
293	0.00*
300	0.02*
CURVE 42	
11	-1.14
34	-1.10
66	-1.04
79	-0.97
116	-0.85
153	-0.62
215	-0.46
293	0.00
298	0.02*
CURVE 43	
10	-0.42
32	-0.44*
64	-0.45*
77	-0.46
115	-0.38*
152	-0.27*
216	-0.18
293	0.00
300	0.03
CURVE 44	
308	0.08*
383	0.58
402	3.57
473	3.82
485	4.70
588	5.08
CURVE 44 (cont.)	
602	7.08
660	7.03
722	6.93
722	6.77
746	6.47
751	5.27
804	5.38
873	5.54

* Not shown in figure.

† Author's data for coefficient of thermal expansion have been integrated by TPRC to obtain $\Delta L/L_0$.

DATA TABLE 39. COEFFICIENT OF THERMAL LINEAR EXPANSION OF PLUTONIUM Pu

[Temperature, T, K; Coefficient of Expansion, α, 10^{-6} K^{-1}]

T	α
CURVE 17‡	
93	31.13
103	32.09
113	33.05
123	34.01
133	34.97
143	35.93
153	36.68
163	37.84
173	38.80
183	39.76
193	40.72
203	41.68
213	42.64
223	43.60
233	44.56
243	45.51
253	46.47
263	47.43
273	48.39
283	49.35
293	50.31
303	51.27*
313	52.23*
323	53.18*
333	54.14*
343	55.10*
353	56.06*
363	57.02*
373	57.97*

* Not shown in figure.

‡ Author's data for coefficient of thermal expansion have been integrated by TPRC to obtain $\Delta L/L_0$.

SPECIFICATION TABLE 40. THERMAL LINEAR EXPANSION OF POLONIUM Po

Cur. No.	Ref. No.	Author(s)	Year	Method Used	Temp. Range, K	Name and Specimen Designation	Composition (weight percent), Specifications, and Remarks
1*	287	Brocklehurst, R.E., Goode, J.M. and Vasamillet, L.F.	1957	X	179-297		Sample prepared and purified in vacuum by fractional volatilization, distilled into fused silica capillary 0.3 to 0.5 mm dia, 15 to 30 μ wall thickness; 11 μg sample used for measurement; reported error ±6.8% in determination of coefficient of expansion.

DATA TABLE 40. THERMAL LINEAR EXPANSION OF POLONIUM Po

[Temperature, T, K; Linear Expansion, $\Delta L/L_0$, %]

T	$\Delta L/L_0$
CURVE 1*	
179	-0.251
297	0.009

* No figure given.

FIGURE AND TABLE NO. 41R. PROVISIONAL VALUES FOR THERMAL LINEAR EXPANSION OF POTASSIUM K

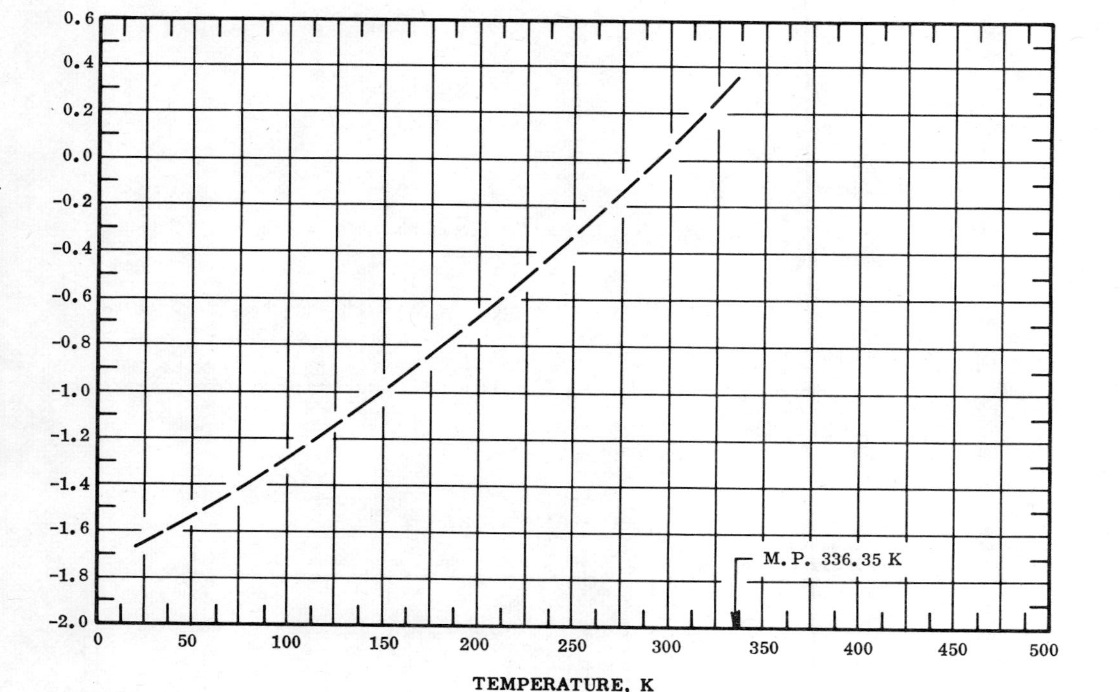

PROVISIONAL VALUES

[Temperature, T, K; Linear Expansion, $\Delta L/L_0$, %]

T	$\Delta L/L_0$ ‡
20	-1.665
25	-1.643
50	-1.527
75	-1.404
100	-1.273
150	-0.989
200	-0.674
250	-0.326
293	0.000
336.35†	0.354

† Melting point at 336.35 K.

‡ Values under zero Kbar pressure.

REMARKS

Since no experimental data were located in the literature, the tabulated values are calculated from the volumetric expansion data. The tabulated values are considered accurate to within ±10% over the entire temperature range and can be represented approximately by the following equation:

$$\Delta L/L_0 = -1.752 + 4.205 \times 10^{-3}\, T + 5.678 \times 10^{-6}\, T^2 + 1.263 \times 10^{-9}\, T^3$$

FIGURE AND TABLE NO. 42R. PROVISIONAL VALUES FOR THERMAL LINEAR EXPANSION OF PRASEODYMIUM Pr

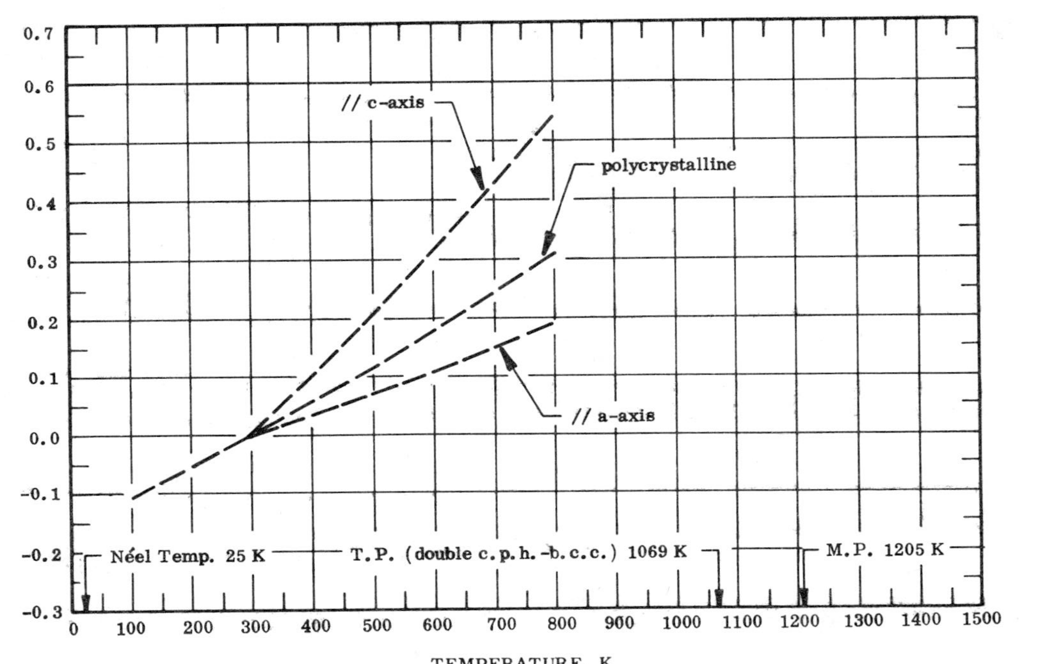

PROVISIONAL VALUES

[Temperature, T, K; Linear Expansion, $\Delta L/L_0$, %; α, K^{-1}]

T	// a-axis $\Delta L/L_0$	// c-axis $\Delta L/L_0$	polycrystalline $\Delta L/L_0$	polycrystalline $\alpha \times 10^6$
100			-0.106	5.8
200			-0.050	5.4
293	0.000	0.000	0.000	5.4
400	0.036	0.107	0.059	5.6
500	0.070	0.210	0.117	5.7
600	0.108	0.317	0.178	6.3
700	0.150	0.428	0.243	6.8
800	0.198	0.546	0.314	7.3

REMARKS

The tabulated values are considered accurate to within ±7% over the entire temperature range. These values can be represented approximately by the following equations:

// a-axis: $\Delta L/L_0 = -0.103 + 3.875 \times 10^{-4}\ T - 1.952 \times 10^{-7}\ T^2 + 2.242 \times 10^{-10}\ T^3$

// c-axis: $\Delta L/L_0 = -0.300 + 1.050 \times 10^{-3}\ T - 1.743 \times 10^{-7}\ T^2 + 2.287 \times 10^{-10}\ T^3$

polycrystalline: $\Delta L/L_0 = -0.168 + 6.552 \times 10^{-4}\ T - 3.427 \times 10^{-7}\ T^2 + 3.475 \times 10^{-10}\ T^3$

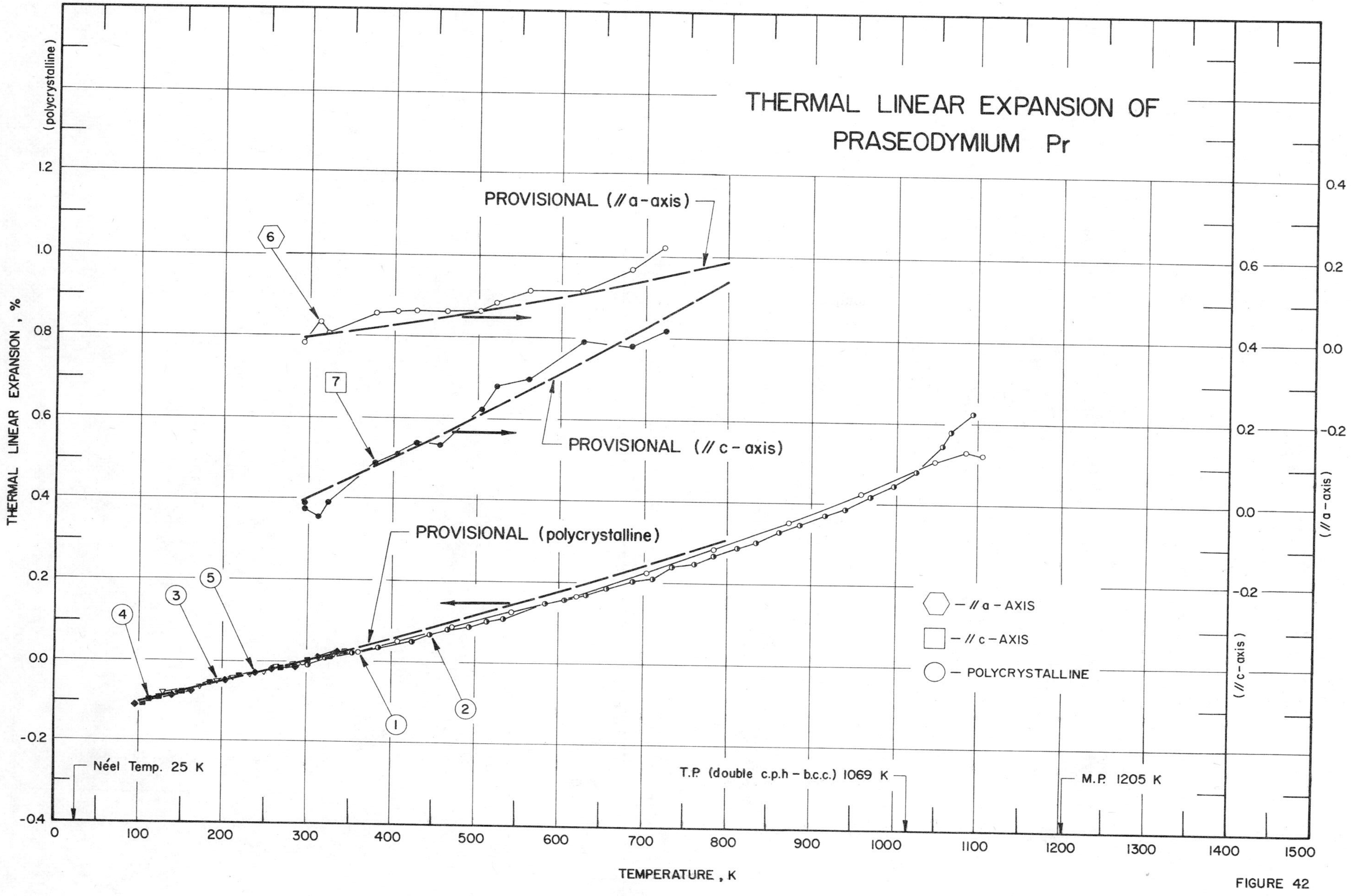
THERMAL LINEAR EXPANSION OF
PRASEODYMIUM Pr
PROVISIONAL (// a-axis)
PROVISIONAL (// c - axis)
PROVISIONAL (polycrystalline)
— // a - AXIS
— // c - AXIS
— POLYCRYSTALLINE
Néel Temp. 25 K
T.P. (double c.p.h - b.c.c.) 1069 K
M.P. 1205 K
TEMPERATURE, K
THERMAL LINEAR EXPANSION, %
(polycrystalline)
(// c - axis)
(// a - axis)

FIGURE 42

SPECIFICATION TABLE 42. THERMAL LINEAR EXPANSION OF PRASEODYMIUM Pr

Cur. No.	Ref. No.	Author(s)	Year	Method Used	Temp. Range, K	Name and Specimen Designation	Composition (weight percent), Specifications, and Remarks
1	78	Barson, F., Legvold, S., and Spedding, F.H.	1956	L	294-1106		<0.2 Nd, 0.1 Ce, 0.047 N_2, 0.0309 C, 0.03 Ta, 0.03 Si, 0.01 La, 0.006 Fe; specimen 0.6 cm diameter, 6 cm long; fluorides of Pr bomb reduced with Ca metal to produce compact metallic sample; vacuum cast, turned to shape in lathe; expansion measured with increasing temperature; anomaly masked by creeping at high temperature.
2	78	Barson, F., et al.	1956	L	1091-302		The above specimen; expansion measured with decreasing temperature; anomaly observed at 1063-1066 K; zero-point correction of 0.174% determined by graphical extrapolation.
3	79	Barson, F., Legvold, S., and Spedding, F.H.	1953	L	345-131		Prepared from the salt by a bomb reduction with calcium; cast in tantalum crucibles and then turned into a cylindrical rod about 0.3 cm in diameter; ends machined flat; subjected to prolonged annealing at about 823 K; expansion measured with decreasing temperature on first test; zero-point correction of -0.0106% determined by graphical interpolation.
4	79	Barson, F., et al.	1953	L	107-347		The above specimen; expansion measured with increasing temperature on second test; zero-point correction of -0.0106% determined by graphical interpolation.
5	79	Barson, F., et al.	1953	L	346-99		The above specimen; expansion measured with decreasing temperature on third test; zero-point correction of -0.0108% determined by graphical interpolation.
6	82	Hanak, J.J.	1959	X	295-722		0.1 Ta, <0.1 each La, Nd; 0.08 O, 0.013 N, 0.01 Mg, <0.01 any other impurity; wire specimen cut from sheet prepared at Ames Laboratory; hexagonal close-packed structure; expansion measured along a-axis; lattice parameter reported at 295 K is 3.670 Å; 3.6705 Å at 293 K determined by graphical extrapolation.
7	82	Hanak, J.J.	1959	X	295-722		The above specimen; expansion measured along c-axis; lattice parameter reported at 295 K is 11.830 Å; 11.8306 Å at 293 K determined by graphical extrapolation.
8*	82	Hanak, J.J.	1959	X	1087-1105		Similar to the above specimen except b.c.c. structure; the values referred to α-phase using atomic volumes and values of a and c parameters at 293 K for α-phase from curves 6 and 7; lattice parameters reported at 1087 K is 4.13 Å.

*Not shown in figure.

DATA TABLE 42. THERMAL LINEAR EXPANSION OF PRASEODYMIUM Pr

[Temperature, T, K; Linear Expansion, $\Delta L/L_0$, %]

T	$\Delta L/L_0$
CURVE 1	
294	0.000
303	0.006*
320	0.017*
343	0.027*
361	0.033
373	0.042*
389	0.052*
409	0.059
430	0.067*
448	0.083*
471	0.091
492	0.107*
519	0.112*
546	0.129
573	0.148*
596	0.159*
621	0.175
649	0.189*
674	0.210*
705	0.230
731	0.249*
756	0.263*
784	0.289
814	0.308*
847	0.336*
875	0.358
901	0.379*
930	0.404*
960	0.435
988	0.456*
1017	0.481*
1049	0.515
1062	0.527*
1079	0.534*
1087	0.534
1106	0.528*
CURVE 2	
1091	0.627
1069	0.583
1063	0.560*
1057	0.543
1045	0.515*
1036	0.499*
1026	0.485
CURVE 2 (cont.)	
999	0.454
970	0.423
940	0.399
916	0.377
889	0.351
863	0.332
837	0.306
811	0.291
785	0.275
762	0.254
735	0.242
711	0.219
687	0.207
658	0.190
633	0.174
609	0.160
582	0.146
559	0.126*
536	0.116
515	0.104
491	0.090
469	0.085
447	0.072
425	0.058
403	0.051*
386	0.040
369	0.036*
355	0.029
341	0.025*
321	0.018
302	-0.005
CURVE 3	
344.7	0.028*
343.3	0.027*
340.8	0.026*
337.3	0.024*
334.9	0.022*
331.0	0.020
326.8	0.018*
322.8	0.016*
318.4	0.014*
314.9	0.012*
311.5	0.010*
307.9	0.008*
CURVE 3 (cont.)	
304.4	0.006*
300.7	0.004*
297.0	0.002*
293.4	0.000*
289.9	-0.002*
286.1	-0.004
282.5	-0.006*
279.0	-0.008*
275.3	-0.010*
271.6	-0.012*
267.9	-0.014
264.4	-0.015*
261.0	-0.017*
257.5	-0.019*
253.6	-0.021*
250.2	-0.023
246.8	-0.025*
243.0	-0.027*
239.5	-0.029*
235.8	-0.031*
232.2	-0.033
228.5	-0.034*
225.0	-0.036*
221.7	-0.038*
217.9	-0.040*
214.8	-0.042
211.4	-0.044*
207.7	-0.046*
204.4	-0.047*
200.6	-0.049*
197.0	-0.051*
193.5	-0.053
189.9	-0.055*
186.2	-0.057*
183.2	-0.058*
179.0	-0.060*
174.8	-0.062
171.8	-0.064*
168.6	-0.065*
165.6	-0.067*
162.4	-0.069*
157.6	-0.071*
155.2	-0.072
152.0	-0.074*
147.8	-0.076*
143.0	-0.078*
135.3	-0.080*
130.7	-0.081
CURVE 4	
106.8	-0.104
108.1	-0.103*
109.1	-0.103*
109.9	-0.102*
111.7	-0.101*
113.6	-0.100
115.2	-0.099*
117.8	-0.097*
120.6	-0.095*
123.4	-0.094*
126.0	-0.092
129.0	-0.090*
131.8	-0.088*
134.6	-0.087*
137.5	-0.085*
140.3	-0.083*
143.0	-0.081*
145.8	-0.080*
148.9	-0.078*
151.9	-0.076*
155.0	-0.075
158.2	-0.073*
161.3	-0.071*
164.4	-0.069*
167.6	-0.067*
170.8	-0.066*
175.8	-0.064*
177.0	-0.062*
180.2	-0.060*
183.5	-0.058*
186.6	-0.057
189.8	-0.055*
194.0	-0.053*
197.2	-0.051*
200.2	-0.049*
203.5	-0.048*
206.8	-0.046*
210.2	-0.044*
213.7	-0.042*
217.4	-0.040*
220.8	-0.038
CURVE 4 (cont.)	
224.2	-0.037*
227.5	-0.035*
230.8	-0.033*
234.4	-0.031*
238.1	-0.029*
241.5	-0.027*
245.1	-0.025*
247.9	-0.023*
252.4	-0.021*
255.9	-0.020*
259.6	-0.018*
263.4	-0.016*
267.1	-0.014*
270.9	-0.012
274.5	-0.010*
278.3	-0.008*
282.0	-0.006*
285.6	-0.004*
289.3	-0.002*
293.1	0.000*
296.9	0.002*
300.8	0.004
304.5	0.006*
308.2	0.008*
311.8	0.010*
315.3	0.012*
319.6	0.014*
323.3	0.016*
327.0	0.018
330.4	0.020*
353.7	0.022*
337.1	0.024*
340.6	0.026*
343.8	0.028*
347.3	0.030
CURVE 5	
346.4	0.029*
342.4	0.028*
339.9	0.026
337.3	0.025*
334.9	0.023*
332.6	0.022*
330.4	0.021*
326.9	0.019*
CURVE 5 (cont.)	
323.2	0.017*
319.8	0.015*
316.1	0.013
312.6	0.011*
309.1	0.009*
305.5	0.007*
301.9	0.005*
298.2	0.003*
294.6	0.001*
289.8	-0.002
286.4	-0.004*
282.8	-0.006*
279.1	-0.008*
275.6	-0.010*
271.9	-0.012*
268.3	-0.014*
264.9	-0.015*
261.2	-0.017
257.8	-0.019*
254.3	-0.021*
250.7	-0.023*
247.3	-0.025*
243.6	-0.027*
240.3	-0.029
236.7	-0.031*
233.3	-0.033*
229.3	-0.035*
225.0	-0.037*
221.9	-0.039*
218.6	-0.041*
214.9	-0.043*
210.8	-0.045*
207.2	-0.047*
204.2	-0.049
200.8	-0.050*
197.7	-0.052*
195.4	-0.054*
192.3	-0.056*
189.1	-0.058*
185.9	-0.060*
182.8	-0.061*
179.7	-0.063*
175.5	-0.066*
172.5	-0.067*
169.4	-0.069*
166.4	-0.071*

*Not shown in figure.

DATA TABLE 42. THERMAL LINEAR EXPANSION OF PRASEODYMIUM Pr (continued)

T	$\Delta L/L_0$
CURVE 5 (cont.)	
163.4	-0.073
160.4	-0.074*
156.5	-0.077*
153.5	-0.078*
150.5	-0.080*
147.6	-0.082*
144.6	-0.084*
141.9	-0.085
139.1	-0.087*
136.3	-0.089*
133.6	-0.091*
130.9	-0.092*
128.1	-0.094*
125.2	-0.096*
122.6	-0.099*
120.0	-0.099*
117.2	-0.101*
114.4	-0.103*
111.8	-0.104*
109.2	-0.106*
106.6	-0.108*
104.0	-0.109*
101.5	-0.111*
98.7	-0.113
CURVE 6	
295	-0.015
295	-0.015
315	0.040
322	0.012
380	0.067
404	0.067
428	0.067
463	0.067
509	0.067
525	0.094
565	0.121
628	0.121
682	0.176
722	0.230
CURVE 7	
295	-0.026
295	-0.001
315	-0.043

T	$\Delta L/L_0$
CURVE 7 (cont.)	
322	-0.001
380	0.092
404	0.118
428	0.143
463	0.135
509	0.228
525	0.287
565	0.304
628	0.397
682	0.388
722	0.422
CURVE 8	
1087	0.00
1088	0.49
1091	0.00
1097	-0.25
1098	-0.49
1105	0.24

*Not shown in figure.

FIGURE AND TABLE NO. 43R. PROVISIONAL VALUES FOR THERMAL LINEAR EXPANSION OF PROTACTINIUM Pa

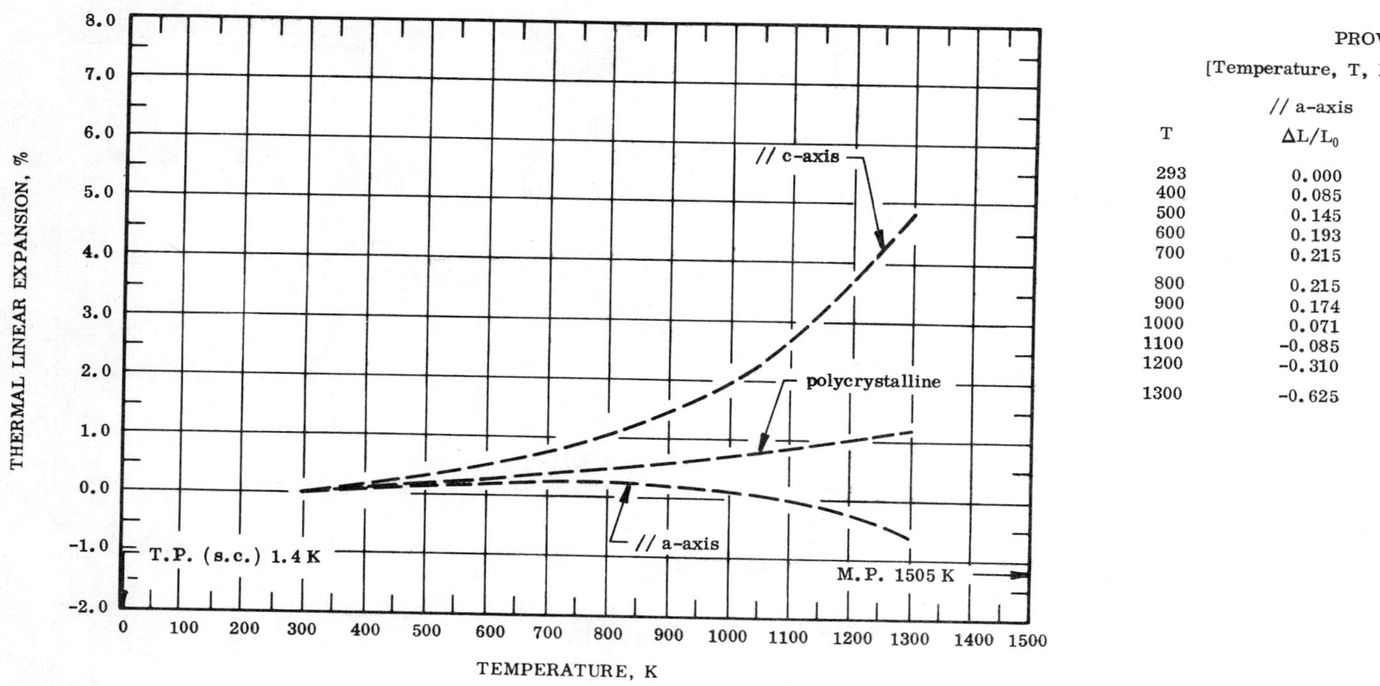

PROVISIONAL VALUES

[Temperature, T, K; Linear Expansion, $\Delta L/L_0$, %]

T	// a-axis $\Delta L/L_0$	// c-axis $\Delta L/L_0$	polycrystalline $\Delta L/L_0$
293	0.000	0.000	0.000
400	0.085	0.140	0.103
500	0.145	0.300	0.197
600	0.193	0.502	0.296
700	0.215	0.738	0.389
800	0.215	1.039	0.490
900	0.174	1.441	0.596
1000	0.071	1.995	0.712
1100	-0.085	2.720	0.850
1200	-0.310	3.690	1.023
1300	-0.625	4.870	1.207

REMARKS

The tabulated values are considered accurate to within ±10% over the entire temperature range. Since no experimental data on the polycrystalline protactinium were located in the literature, the tabulated values are calculated from single crystal data. These values are considered accurate to within ±10% over the entire temperature range.

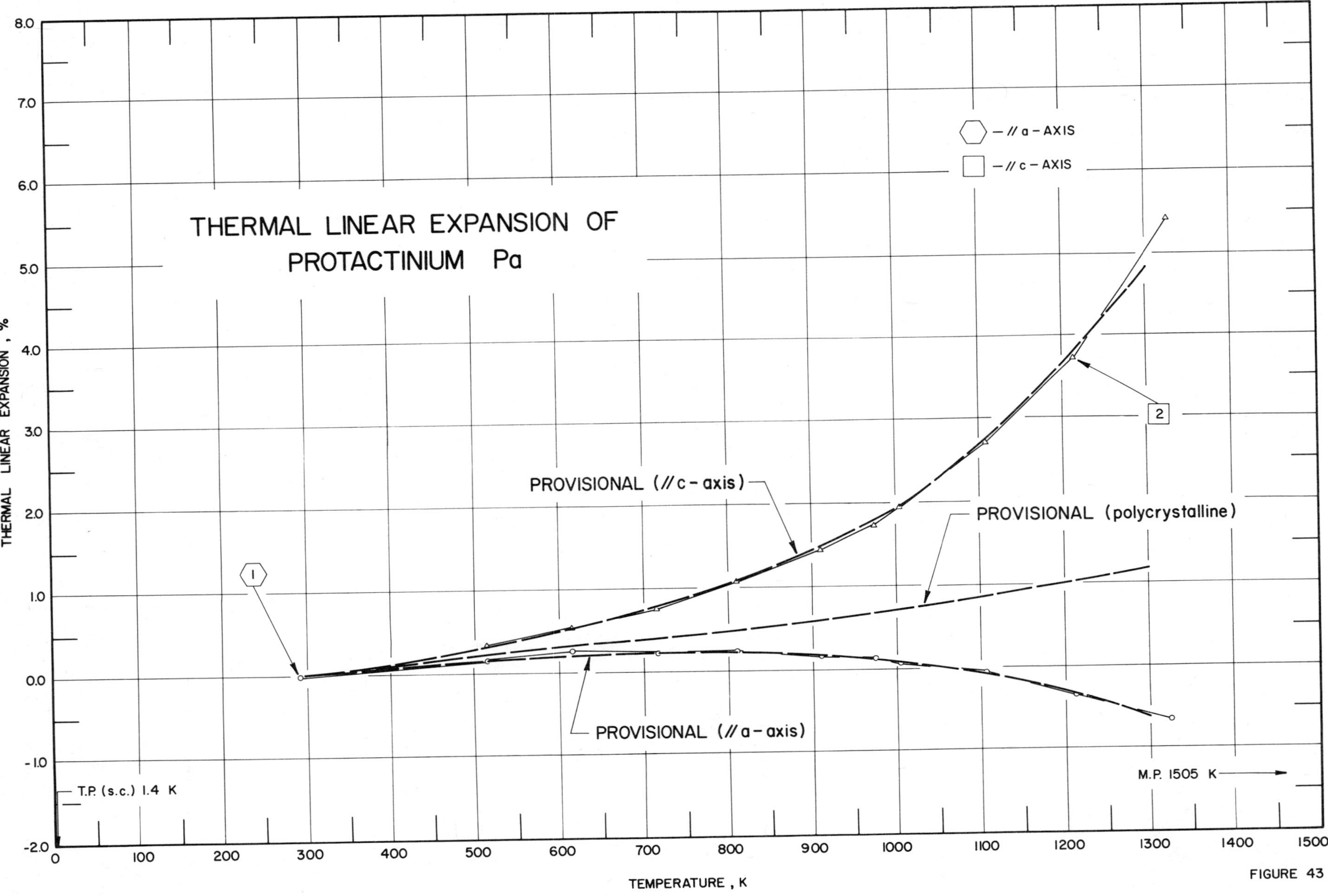
THERMAL LINEAR EXPANSION OF
PROTACTINIUM Pa
— // a – AXIS
— // c – AXIS
PROVISIONAL (// c – axis)
PROVISIONAL (polycrystalline)
PROVISIONAL (// a – axis)
2
1
M.P. 1505 K
T.P. (s.c.) 1.4 K
THERMAL LINEAR EXPANSION, %
TEMPERATURE, K
8.0
7.0
6.0
5.0
4.0
3.0
2.0
1.0
0.0
-1.0
-2.0
0
100
200
300
400
500
600
700
800
900
1000
1100
1200
1300
1400
1500
FIGURE 43

SPECIFICATION TABLE 43. THERMAL LINEAR EXPANSION OF PROTACTINIUM Pa

Cur. No.	Ref. No.	Author(s)	Year	Method Used	Temp. Range, K	Name and Specimen Designation	Composition (weight percent), Specifications, and Remarks
1	290	Marples, J. A. C.	1965	X	291-1328		80 mg specimen prepared by reducing PaF_4 with Ca and arc-melting resulting metal in CaF_2; body-centered tetragonal structure; expansion measured along a-axis; lattice parameter reported at 291 K is 3.9246 Å; 3.92465 Å at 293 K determined by graphical interpolation.
2	290	Marples, J. A. C.	1965	X	291-1328		The above specimen; expansion measured along c-axis; lattice parameter reported at 291 K is 3.2406 Å; 3.24064 Å at 293 K determined by graphical interpolation.

DATA TABLE 43. THERMAL LINEAR EXPANSION OF PROTACTINIUM Pa

[Temperature, T, K; Linear Expansion, $\Delta L/L_0$, %]

T	$\Delta L/L_0$	T	$\Delta L/L_0$
CURVE 1		CURVE 2	
291	-0.001	291	-0.003
515	0.140	515	0.329
615	0.255	615	0.523
713	0.219	713	0.767
810	0.214	810	1.076
911	0.191	911	1.458
971	0.138	971	1.739
1004	0.087	1004	1.961
1109	-0.051	1109	2.708
1211	-0.318	1211	3.735
1328	-0.652	1328	5.442

FIGURE AND TABLE NO. 44R. RECOMMENDED VALUES FOR THERMAL LINEAR EXPANSION OF RHENIUM Re

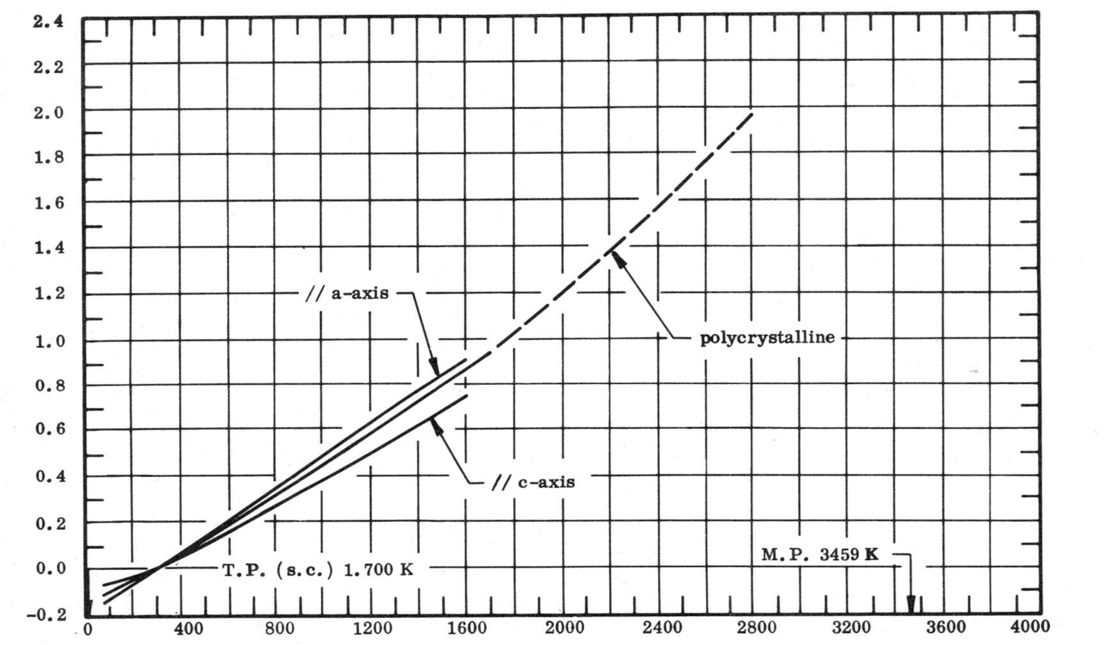

RECOMMENDED VALUES

[Temperature, T, K; Linear Expansion, $\Delta L/L_0$, %; α, K^{-1}]

	// a-axis	// c-axis	polycrystalline	
T	$\Delta L/L_0$	$\Delta L/L_0$	$\Delta L/L_0$	$\alpha \times 10^6$
75	-0.144	-0.066	-0.118	
100	-0.130	-0.052	-0.108	
200	-0.062	-0.036	-0.053	
293	0.000	0.000	0.000	6.2
400	0.073	0.055	0.067	6.3
500	0.142	0.107	0.130	6.4
600	0.211	0.160	0.194	6.4
700	0.281	0.215	0.259	6.5
800	0.352	0.271	0.325	6.7
900	0.423	0.330	0.392	6.8
1000	0.494	0.393	0.460	6.9
1200	0.636	0.528	0.599	7.1
1400	0.777	0.679	0.744	7.4
1600	0.918	0.852	0.896	7.7
1800			1.053†	8.0
2000			1.216†	8.3
2200			1.386†	8.7
2400			1.563†	9.1
2600			1.748†	9.4
2800			1.941†	9.8

† Provisional values.

REMARKS

The tabulated values are considered accurate to within ± 3% at temperatures below 293 K and ± 5% above except for polycrystalline rhenium the values are considered accurate to above ± 10% at the temperatures above 1600 K. These values can be represented approximately by the following equations:

// a-axis: $\Delta L/L_0 = -0.195 + 6.513 \times 10^{-4}\,T + 5.412 \times 10^{-8}\,T^2 - 1.652 \times 10^{-11}\,T^3$ $(293 < T < 1600)$

// c-axis: $\Delta L/L_0 = -0.149 + 5.237 \times 10^{-4}\,T - 5.667 \times 10^{-8}\,T^2 + 7.560 \times 10^{-11}\,T^3$ $(293 < T < 1600)$

polycrystalline: $\Delta L/L_0 = -0.177 + 5.948 \times 10^{-4}\,T + 3.410 \times 10^{-8}\,T^2 + 8.447 \times 10^{-12}\,T^3$ $(293 < T < 2800)$

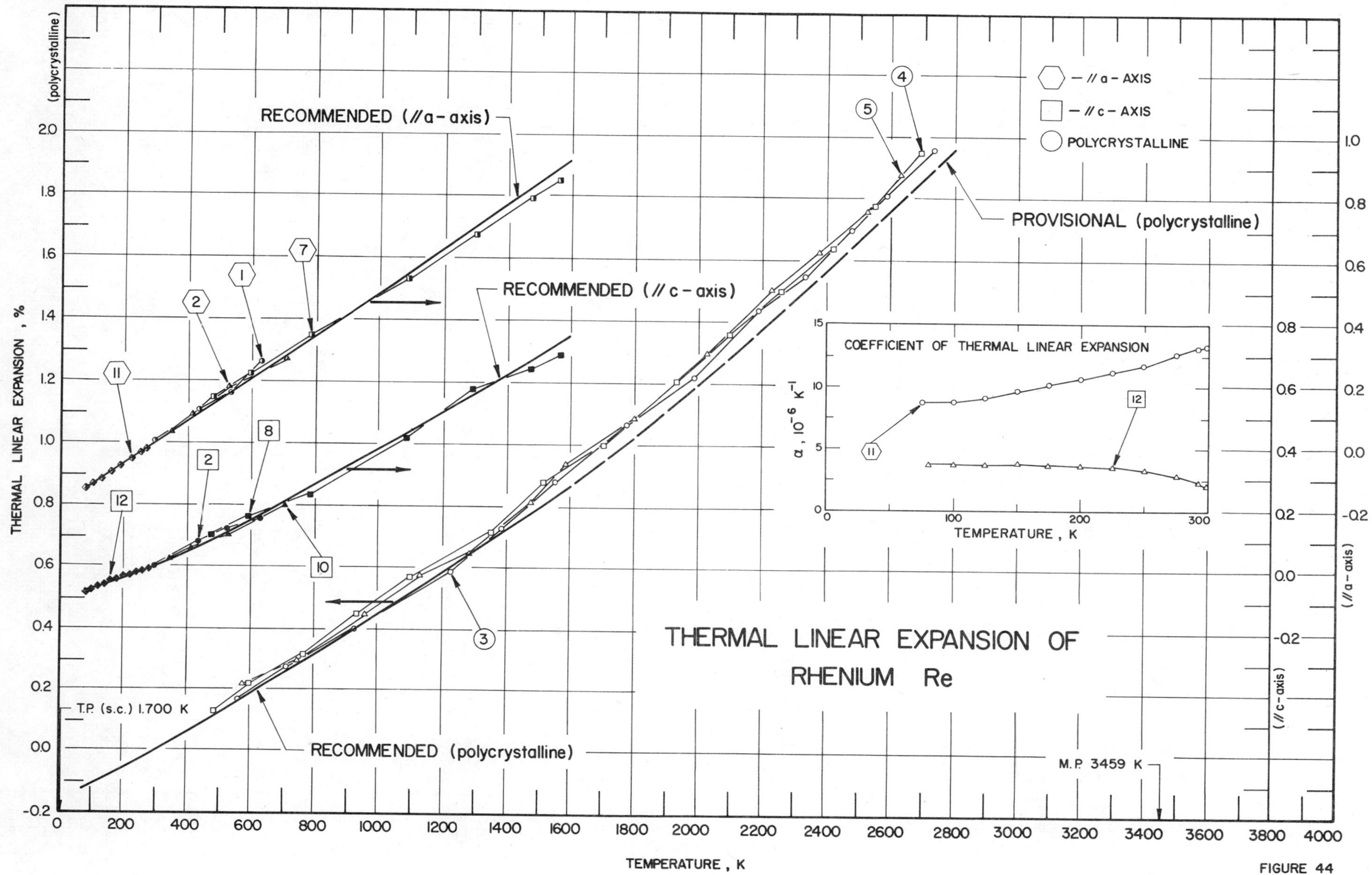
THERMAL LINEAR EXPANSION OF
RHENIUM Re
— // a – AXIS
— // c – AXIS
POLYCRYSTALLINE
RECOMMENDED (//a-axis)
RECOMMENDED (// c-axis)
RECOMMENDED (polycrystalline)
PROVISIONAL (polycrystalline)
T.P. (s.c.) 1.700 K
M.P. 3459 K
COEFFICIENT OF THERMAL LINEAR EXPANSION
α, 10⁻⁶ K⁻¹
TEMPERATURE, K
THERMAL LINEAR EXPANSION, %
(polycrystalline)
(// c-axis)
(// a-axis)
TEMPERATURE, K

FIGURE 44

SPECIFICATION TABLE 44. THERMAL LINEAR EXPANSION OF RHENIUM Re

Cur. No.	Ref. No.	Author(s)	Year	Method Used	Temp. Range, K	Name and Specimen Designation	Composition (weight percent), Specifications, and Remarks
1	292	Power, R. R. and Deshpande, V. I.	1967	X	298-627		Sample annealed 6 hr at 1473 K; five sharp reflections from (21. 1), (11. 4), (10. 5), (20. 4) and (30. 0) planes, recorded in the back-reflection region, used in evaluating the lattice parameters; expansion measured along a-axis; lattice parameter reported at 298 K is 2. 7606 Å; 2. 7605 Å at 293 K determined by graphical extrapolation.
2	292	Power, R. R. and Deshpande, V. I.	1967	X	298-627		The above specimen; expansion measured along c-axis; lattice parameter reported at 298 K is 4. 4578 Å; 4. 4397 Å at 293 K determined by graphical interpolation.
3	210	Conway, J.B. and Losekamp, A.C.	1966	T	562-2736		99. 0^+ Re, < 0. 5 Fe; sheet 20 mil thick formed by powder metallurgy techniques; expansion measured in helium atmosphere on first test; zero-point correction is ± 0. 000%.
4	210	Conway, J.B. and Losekamp, A.C.	1966	T	489-2690		The above specimen; expansion measured on second test.
5	210	Conway, J.B. and Losekamp, A.C.	1966	T	580-2632		The above specimen; expansion measured on third test.
6*	273	Andres, K.	1963	O	2. 0-13		Polycrystalline specimen; 7. 4 cm long; this curve reported using first given temperature as reference temperature at which $\Delta L/L_0 = 0$; correction -1.85×10^{-7}%.
7	293	Wasilewski, R. J.	1961	X	298-1561		Polycrystalline specimen; designated spectrographic standard by Johnson Matthey; vacuum annealed at 1273 K; expansion measured in crystalline basal plane; lattice parameter reported at 298 K is 2. 7609 Å; 2. 76078 Å at 293 K determined by graphical extrapolation.
8	293	Wasilewski, R. J.	1961	X	298-1561		The above specimen; expansion measured parallel to c-axis; lattice parameter reported at 298 K is 4. 4576 Å; 4. 45745 Å at 293 K determined by graphical interpolation.
9	277	Liu, L. G., Takahashi, T. and Bassett, W. A.	1970	X	286-701		99^+ Re powdered specimen, mixed with Pt for temperature calibration; expansion measured in helium atmosphere under pressure of 1 bar; expansion measured along a-axis; reported error ± 0. 15% in measurement of lattice parameter; zero-point correction of 0. 003% determined by graphical interpolation.
10	277	Liu, L. G., et al.	1970	X	286-701		The above specimen; expansion measured along c-axis; zero-point correction of 0. 001% determined by graphical interpolation.
11	660	Finkel', V.A., Paltnik, M.I. and Kovtun G.P.	1971	X	81-300		Single crystal grown by zone recrystallization; measurements along a-axis; lattice parameter reported at 293 K is 2. 7614 Å; data on the coefficient of thermal linear expansion also reported.
12	660	Finkel', V.A., et al.	1971	X	80-300		Single crystal grown by zone recrystallization; measurements along c-axis; lattice parameter reported at 293 K is 4. 4566 Å; data on the coefficient of thermal linear expansion also reported.

*Not shown in figure.

DATA TABLE 44. THERMAL LINEAR EXPANSION OF RHENIUM Re

[Temperature, T, K; Linear Expansion, $\Delta L/L_0$, %]

T	$\Delta L/L_0$
CURVE 1	
298	0.004
434	0.105
531	0.163
627	0.261
CURVE 2	
298	0.003
434	0.081
531	0.126
627	0.160
CURVE 3	
562	-0.17
712	0.28
927	0.40
1227	0.59
1384	0.73
1552	0.88
1774	1.07
1983	1.22
2184	1.44
2333	1.55
2477	1.70
2588	1.81
2736	1.96
CURVE 4	
489	0.13
599	0.22
766	0.32
933	0.45
1098	0.57
1350	0.72
1514	0.88
1704	1.00
1930	1.21
2096	1.36
2259	1.50
2420	1.64
2550	1.78
2690	1.95
CURVE 5	
580	0.22
746	0.30
960	0.45
1131	0.58
1283	0.65
1477	0.82
1586	0.94
1799	1.09
2026	1.30
2228	1.51
2376	1.63
2525	1.76
2632	1.86
CURVE 6*, †	
2.0	0.00 x 10^{-7}
4.4	7.30
4.4	8.83
5.3	11.77
5.8	16.12
6.1	17.87
6.1	18.42
6.5	21.07
7.9	32.82
8.5	39.12
9.8	57.72
10.4	65.54
10.8	72.62
10.8	75.47
11.2	79.30
11.2	82.85
12.6	114.55
CURVE 7	
298	0.004*
475	0.149
593	0.225
783	0.352
1083	0.533
1297	0.674
1473	0.797
1561	0.855
CURVE 8	
298	0.003*
475	0.106
593	0.162
783	0.236
1083	0.420
1297	0.584
1473	0.651
1561	0.685
CURVE 9	
285.7	-0.004*
298.2	0.003*
342.7	0.039
410.2	0.094
527.2	0.184
701.2	0.271
CURVE 10	
285.7	-0.001*
298.2	0.001*
342.7	0.026
410.2	0.055
527.2	0.115
701.2	0.201
CURVE 11	
81	-0.144
85	-0.143*
88	-0.142*
95	-0.140*
102	-0.133
107	-0.132*
110	-0.127*
116	-0.126*
121	-0.123*
130	-0.118
134	-0.114*
140	-0.112*
147	-0.107*
151	-0.104*
156	-0.101*
CURVE 11 (cont.)	
160	-0.097
164	-0.097*
173	-0.091*
173	-0.087*
178	-0.086*
181	-0.083*
185	-0.082*
188	-0.080*
191	-0.077
196	-0.074*
199	-0.070*
204	-0.068*
205	-0.066*
214	-0.062*
218	-0.057*
223	-0.055
230	-0.049*
236	-0.044*
244	-0.038*
250	-0.033*
253	-0.031*
255	-0.027
260	-0.026*
263	-0.021*
270	-0.016
274	-0.014*
278	-0.011*
285	-0.006*
292	-0.001*
292	0.001*
193	0.000*
300	0.003*
CURVE 12	
80	-0.082
90	-0.077*
100	-0.075
108	-0.068*
120	-0.063
130	-0.061*
140	-0.055
149	-0.047*
159	-0.044
CURVE 12 (cont.)	
169	-0.043*
179	-0.041
190	-0.033*
199	-0.030
211	-0.027*
220	-0.026
229	-0.020*
240	-0.016
250	-0.013*
259	-0.010
269	-0.007*
279	-0.004
288	0.0004*
293	0.000*
300	0.003*

* Not shown in figure.
† This curve is here reported using the first given temperature as reference temperature at which $\Delta L/L_0$ = 0.

DATA TABLE 44. COEFFICIENT OF THERMAL LINEAR EXPANSION OF RHENIUM Re

[Temperature, T, K; Coefficient of Expansion, α, 10^{-6} K^{-1}]

T	α
CURVE 11	
75	8.80
100	8.82
125	9.18
150	9.68
175	10.12
200	10.65
225	11.24
250	11.93
275	12.72
293	13.25
300	13.41
CURVE 12	
80	3.87
100	3.87
125	3.86
150	3.87
175	3.83
200	3.70
225	3.52
250	3.30
275	2.88
293	2.41
300	2.17

FIGURE AND TABLE NO. 45R. RECOMMENDED VALUES FOR THERMAL LINEAR EXPANSION OF RHODIUM Rh

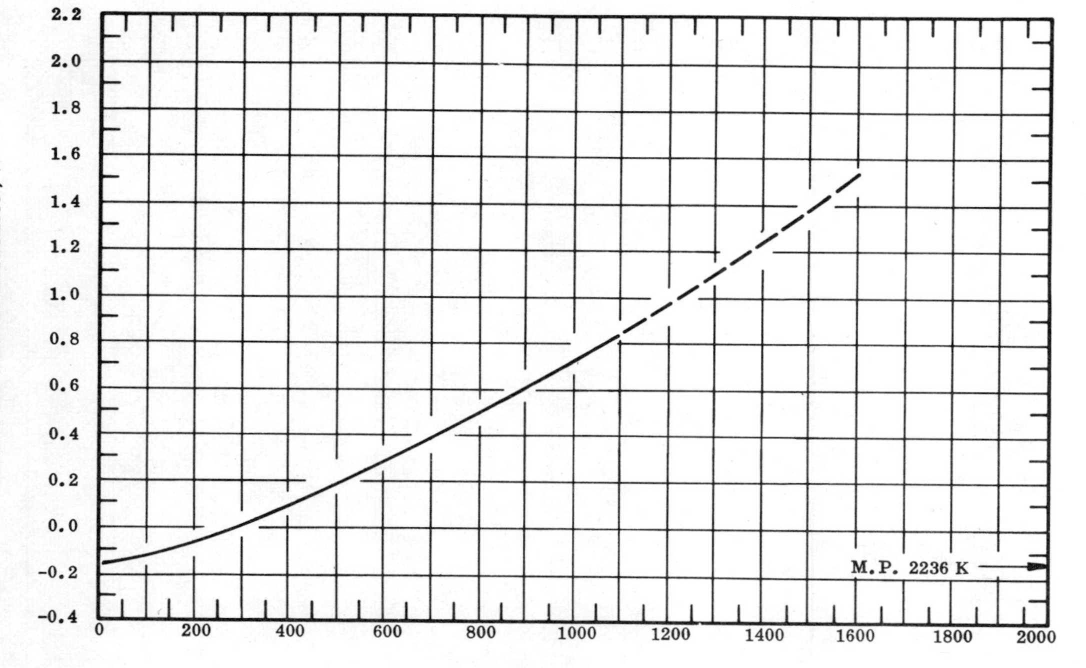

RECOMMENDED VALUES

[Temperature, T, K; Linear Expansion, $\Delta L/L_0$, %; α, K^{-1}]

T	$\Delta L/L_0$	$\alpha \times 10^6$
5	-0.149	0.01
25	-0.149	0.2
50	-0.148	1.8
100	-0.135	5.0
200	-0.073	7.3
293	0.000	8.2
400	0.091	8.8
500	0.182	9.3
600	0.278	9.8
700	0.378	10.3
800	0.484	10.8
900	0.595	11.3
1000	0.712	11.9
1200	0.961†	13.0
1400	1.232†	14.2
1600	1.526†	15.4

† Provisional values.

REMARKS

The tabulated values are considered accurate to within ± 3% at temperatures below 1100 K and ± 7% above. These values can be represented approximately by the following equations:

$$\Delta L/L_0 = -0.135 + 4.867 \times 10^{-4} (T - 100) + 1.528 \times 10^{-6} (T - 100)^2 - 2.218 \times 10^{-9} (T - 100)^3 \quad (100 < T < 293)$$

$$\Delta L/L_0 = 8.286 \times 10^{-4} (T - 293) + 2.435 \times 10^{-7} (T - 293)^2 + 1.218 \times 10^{-11} (T - 293)^3 \quad (293 < T < 1600)$$

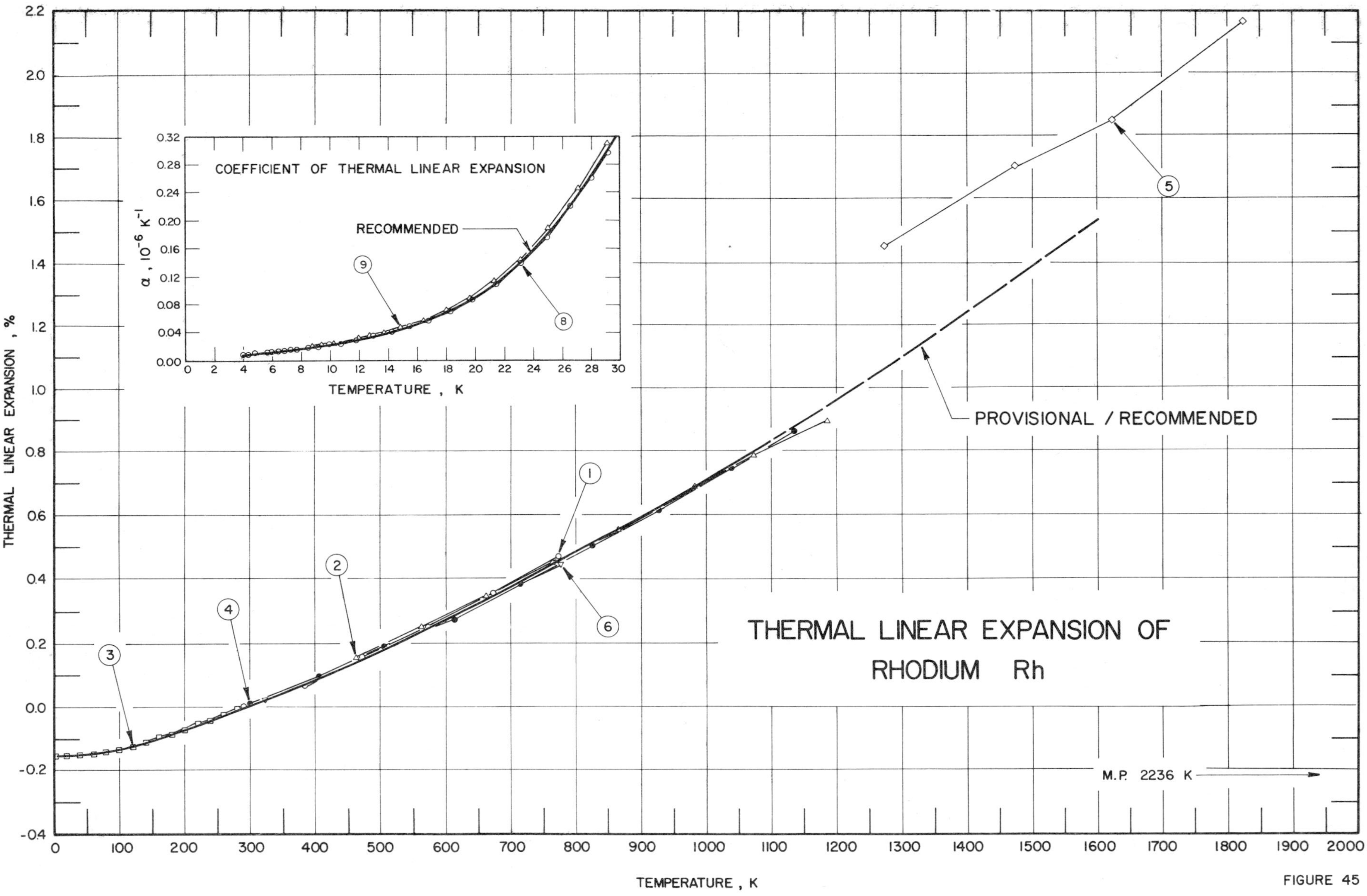
THERMAL LINEAR EXPANSION OF
RHODIUM Rh
PROVISIONAL / RECOMMENDED
M.P. 2236 K
THERMAL LINEAR EXPANSION, %
TEMPERATURE, K
COEFFICIENT OF THERMAL LINEAR EXPANSION
RECOMMENDED
α, 10⁻⁶ K⁻¹
TEMPERATURE, K

FIGURE 45

SPECIFICATION TABLE 45. THERMAL LINEAR EXPANSION OF RHODIUM Rh

Cur. No.	Ref. No.	Author(s)	Year	Method Used	Temp. Range, K	Name and Specimen Designation	Composition (weight percent), Specifications, and Remarks
1	294	Swanger, H.W.	1929	I	293-773		No impurities found by spectrographic examination except for traces of iridium in some instances; average data.
2	187	Ross, R.G. and Hume-Rothery, W.	1963	X	296-1168		No details given; lattice parameter reported at 296 K is 3.8032 Å; 3.8031 Å at 293 K determined by graphical extrapolation.
3	83	Laquer, H.L.	1952	L	0-300		Specimen prepared by slotting two pieces of sheet 5.08 cm x 0.51 cm x 0.0381 cm thick halfway down, inserting them in each other and silver-soldering the assembly; zero-point correction of -0.017% determined by graphical interpolation.
4	174	Singh, H.P.	1968	X	303-1138		Specpure powder filled in quartz capillary of 0.5 mm inner diameter; annealed; lattice parameter reported at 303 K is 3.8034 Å; 3.8031 Å at 293 K determined by graphical extrapolation; data on the coefficient of thermal linear expansion are also reported.
5	295	Bale, E.S.	1958	X	298-1819		Specimen containing less than 0.001 metallic impurities produced by reduction of the oxide; pressed, sintered in oxygen-free hydrogen and cold rolled with intermediate anneals to sheet 0.060 in. thick; lattice parameter reported at 298 K is 3.8035 Å; 3.8032 Å at 293 K determined by graphical extrapolation.
6	387	Swanger, W.H.	1930	I	293-773		Trace of Ir only impurity remaining after sequence of chemical purification; reported error 1% in determination of coefficient of expansion.
7	597	Pawar, R.R.	1968	X	301-860		Pure rhodium was used to obtain the powder pictures at different temperature; lattice parameter reported at 301 K is 3.8034 Å; 3.8033 Å at 293 K determined by graphical extrapolation.
8*	598	White, G.K. and Pawlowicz, A.T.	1970	E	4-29		Specimen 1.4 cm diameter x 2.5 cm long from International Nickel Co.; greater than 99.9% purity made from powder by sintering and hot forging; annealed at 1273 K in vacuum; this curve is here reported using the first given temperature as reference temperature at which $\Delta L/L_0 = 0$.
9*	598	White, G.K. and Pawlowicz, A.T.	1970	E	6-29		The above specimen; this curve is here reported using the first given temperature as reference temperature at which $\Delta L/L_0 = 0$.

*Not shown in figure.

DATA TABLE 45. THERMAL LINEAR EXPANSION OF RHODIUM Rh

[Temperature, T, K; Linear Expansion, $\Delta L/L_0$, %]

T	$\Delta L/L_0$
CURVE 1	
293	0.000
323	0.024*
373	0.066
473	0.153
573	0.249
673	0.353
773	0.461
CURVE 2	
295.8	0.002
467.0	0.152
565.7	0.247
664.0	0.342
767.0	0.444
867.8	0.549
983.1	0.681
1071.6	0.783
1187.5	0.896
CURVE 3	
0	-0.159
20	-0.158*
40	-0.154*
60	-0.148
80	-0.141*
100	-0.132*
120	-0.122
140	-0.110*
160	-0.098*
180	-0.085
200	-0.071*
220	-0.057
240	-0.042*
260	-0.027
273.2	-0.017*
280	-0.011
300	0.005*
CURVE 4	
303	0.008
406	0.092
508	0.190
611	0.276
CURVE 4 (cont.)	
713	0.382
825	0.503
928	0.616
1040	0.742
1138	0.865
CURVE 5	
298	0.007*
1273	1.454
1473	1.703
1622	1.858
1819	2.166
CURVE 6	
293	0.000*
323	0.024
373	0.066*
473	0.153*
573	0.249*
773	0.446
CURVE 7	
301	0.002*
443	0.128
517	0.210
602	0.289
695	0.381
785	0.475
860	0.565
CURVE 8*, †, ‡	
4.00	0.000×10^{-4}
4.38	0.003
4.85	0.008
5.67	0.018
5.98	0.022
6.44	0.028
6.83	0.034
7.21	0.040
7.77	0.049
8.49	0.063
9.16	0.076
CURVE 8 (cont.)*, †, ‡	
9.97	0.094×10^{-4}
10.74	0.11
11.72	0.14
12.97	0.18
14.19	0.23
15.42	0.28
16.76	0.36
18.27	0.45
19.86	0.58
21.46	0.74
23.14	0.95
24.97	1.24
26.60	1.57
28.03	1.92
29.19	2.24
CURVE 9*, †, ‡	
6.29	0.000×10^{-4}
7.01	0.011
7.61	0.021
8.69	0.042
9.33	0.055
10.18	0.075
11.00	0.097
11.92	0.12
12.69	0.15
13.70	0.19
14.82	0.24
16.40	0.32
18.00	0.42
19.67	0.56
21.35	0.73
23.16	0.96
25.03	1.28
27.12	1.73
29.15	2.30

* Not shown in figure.
† This curve is here reported using the first given temperature as reference temperature at which $\Delta L/L_0 = 0$.
‡ Author's data for coefficient of thermal expansion have been integrated by TPRC to obtain $\Delta L/L_0$.

DATA TABLE 45. COEFFICIENT OF THERMAL LINEAR EXPANSION OF RHODIUM Rh

[Temperature, T, K; Coefficient of Expansion, α, $10^{-6}\ K^{-1}$]

T	α
CURVE 4*	
303	7.89
406	8.44
508	9.02
611	9.62
713	10.21
825	10.88
928	11.50
1040	12.19
1138	12.80
CURVE 8‡	
4.00	0.008
4.38	0.009
4.85	0.010
5.67	0.012
5.98	0.013
6.44	0.014
6.83	0.015
7.21	0.016
7.77	0.016
8.49	0.020
9.16	0.020
9.97	0.024
10.74	0.025
11.72	0.031
12.97	0.036
14.19	0.043
15.42	0.050
16.76	0.059
18.27	0.071
19.86	0.089
21.46	0.11
23.14	0.14
24.97	0.18
26.60	0.22
28.03	0.26
29.19	0.30
CURVE 9‡	
6.29	0.014 *
7.01	0.016 *
7.61	0.017 *
8.69	0.020
9.33	0.022
10.18	0.0246
CURVE 9 (cont.)‡	
11.00	0.028
11.92	0.033
12.69	0.037
13.70	0.041
14.82	0.048
16.40	0.058
18.00	0.072
19.67	0.090
21.35	0.11
23.16	0.14
25.03	0.19
27.12	0.25
29.15	0.31

* Not shown in figure.
‡ Author's data for coefficient of thermal expansion have been integrated by TPRC to obtain $\Delta L/L_0$.

FIGURE AND TABLE NO. 46R. RECOMMENDED VALUES FOR THERMAL LINEAR EXPANSION OF RUTHENIUM Ru

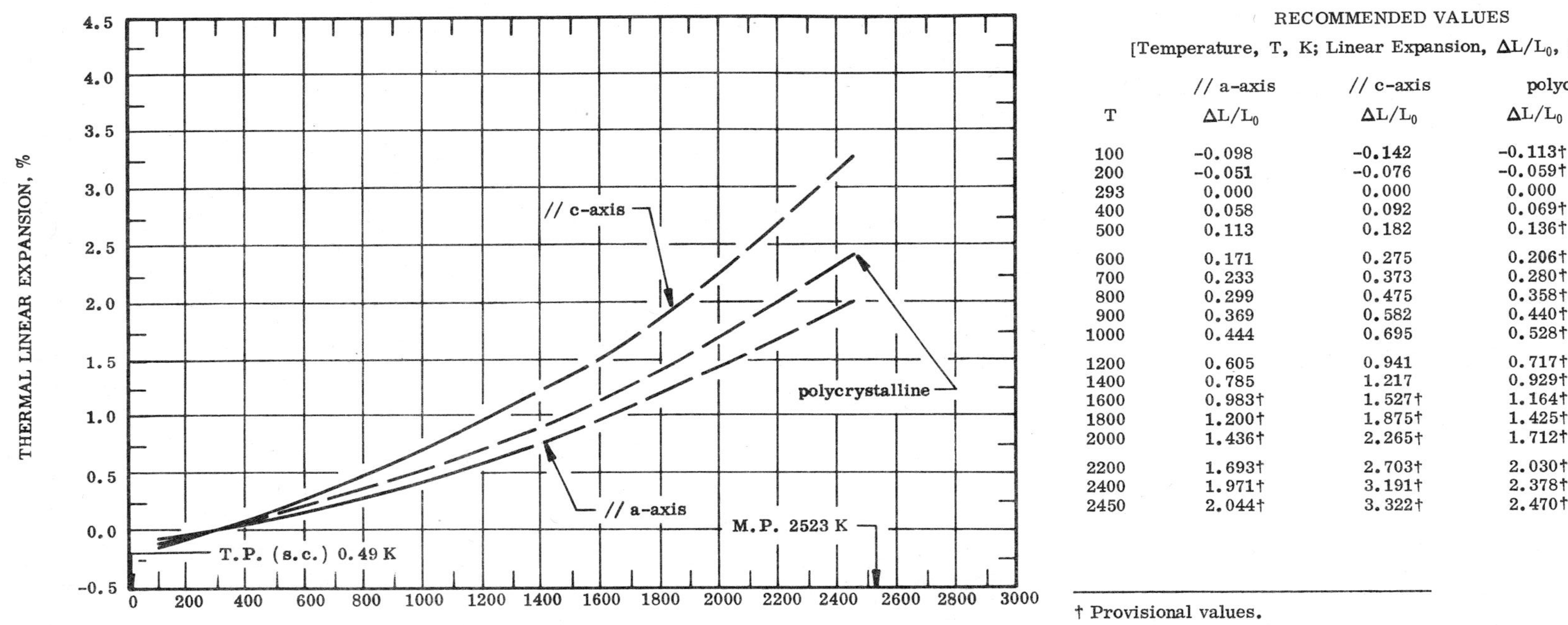

RECOMMENDED VALUES

[Temperature, T, K; Linear Expansion, $\Delta L/L_0$, %; α, K^{-1}]

T	// a-axis $\Delta L/L_0$	// c-axis $\Delta L/L_0$	polycrystalline $\Delta L/L_0$	$\alpha \times 10^6$
100	-0.098	-0.142	-0.113†	5.2
200	-0.051	-0.076	-0.059†	5.9
293	0.000	0.000	0.000	6.4
400	0.058	0.092	0.069†	6.6
500	0.113	0.182	0.136†	6.9
600	0.171	0.275	0.206†	7.2
700	0.233	0.373	0.280†	7.6
800	0.299	0.475	0.358†	8.0
900	0.369	0.582	0.440†	8.4
1000	0.444	0.695	0.528†	8.9
1200	0.605	0.941	0.717†	10.0
1400	0.785	1.217	0.929†	11.1
1600	0.983†	1.527†	1.164†	12.4
1800	1.200†	1.875†	1.425†	13.7
2000	1.436†	2.265†	1.712†	15.1
2200	1.693†	2.703†	2.030†	16.6
2400	1.971†	3.191†	2.378†	18.0
2450	2.044†	3.322†	2.470†	18.4

† Provisional values.

REMARKS

The tabulated values are considered accurate to within ± 5% at temperatures below 1400 K and ± 7% above. Since no experimental data on the polycrystalline ruthenium were located in the literature, the tabulated values are calculated from single crystal data. These values are considered accurate to within ± 10% over the entire temperature range. These values can be represented approximately by the following equations:

// a-axis: $\Delta L/L_0 = -0.122 + 3.792 \times 10^{-4}\,T + 1.729 \times 10^{-7}\,T^2 + 1.354 \times 10^{-11}\,T^3$ $(293 < T < 2450)$

// c-axis: $\Delta L/L_0 = -0.240 + 7.971 \times 10^{-4}\,T + 4.874 \times 10^{-8}\,T^2 + 8.952 \times 10^{-11}\,T^3$ $(293 < T < 2450)$

polycrystalline: $\Delta L/L_0 = -0.163 + 5.267 \times 10^{-4}\,T + 1.246 \times 10^{-7}\,T^2 + 4.048 \times 10^{-11}\,T^3$ $(293 < T < 2450)$

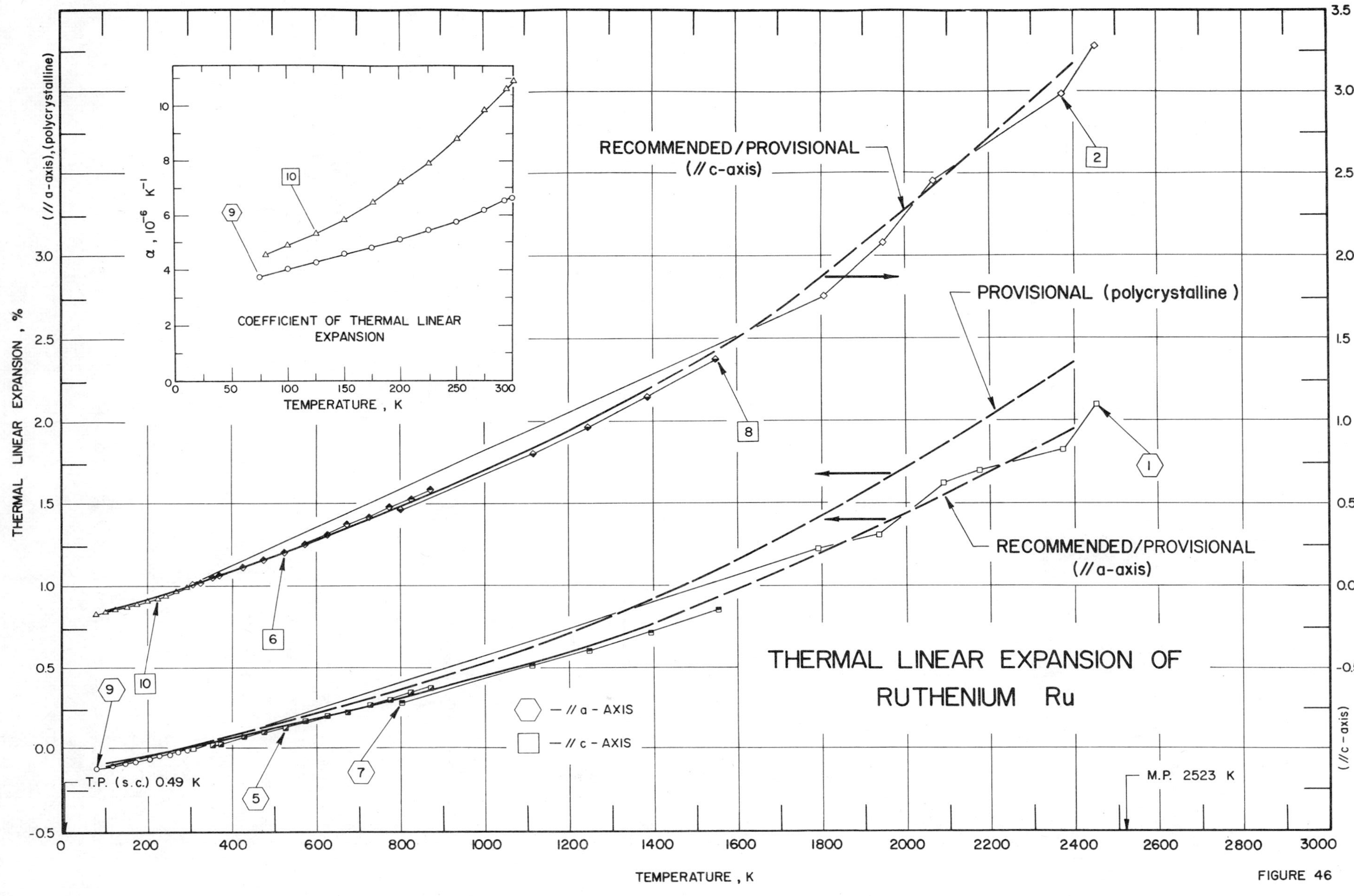
THERMAL LINEAR EXPANSION OF
RUTHENIUM Ru
RECOMMENDED/PROVISIONAL
(// c-axis)
PROVISIONAL (polycrystalline)
RECOMMENDED/PROVISIONAL
(// a-axis)
— // a - AXIS
— // c - AXIS
T.P. (s.c.) 0.49 K
M.P. 2523 K
COEFFICIENT OF THERMAL LINEAR
EXPANSION
TEMPERATURE, K
α, 10⁻⁶ K⁻¹
THERMAL LINEAR EXPANSION, %
(// a-axis), (polycrystalline)
(// c-axis)
TEMPERATURE, K

FIGURE 46

SPECIFICATION TABLE 46. THERMAL LINEAR EXPANSION OF RUTHENIUM Ru

Cur. No.	Ref. No.	Author(s)	Year	Method Used	Temp. Range, K	Name and Specimen Designation	Composition (weight percent), Specifications, and Remarks
1	187	Ross, R.G. and Hume-Rothery, W.	1963	X	303-2453		Expansion measured along a-axis; lattice parameter reported at 303 K is 3.7042 Å; 2.7040 Å at 293 K determined by graphical extrapolation.
2	187	Ross, R.G. and Hume-Rothery, W.	1963	X	303-2453		Expansion measured along c-axis; lattice parameter reported at 303 K is 4.2800 Å; 4.2795 Å at 293 K determined by graphical extrapolation.
3*	108	Owen, E.A. and Roberts, E.W.	1936	X	323-823		Expansion measured along hexagonal axis.
4*	108	Owen, E.A. and Roberts, E.W.	1936	X	323-823		Expansion measured perpendicular to hexagonal axis.
5	265	Owen, E.A. and Roberts, E.W.	1937	X	293-873		Chemically pure powder; annealed 3 hr in vacuo at 1273 K; expansion measured along a-axis; lattice parameter reported at 293 K is 2.6984 Å.
6	265	Owen, E.A. and Roberts, E.W.	1937	X	293-873		The above specimen; expansion measured along c-axis; lattice parameter reported at 293 K is 4.2730 Å.
7	599	Hall, E.O. and Crangle, J.	1957	X	293-1557		High-purity powder ruthenium from Mond Nickel Co. Ltd.; specimen mounted in thin silica capillaries and examined in vacuo in a Unicam S.150 high temperature camera; expansion measured along a-axis; lattice parameter reported at 293 K is 2.7056 Å.
8	599	Hall, E.O. and Crangle, J.	1957	X	293-1557		The above specimen; expansion measured along c-axis; lattice parameter reported at 293 K is 4.2805 Å.
9	660	Finkel, V.A., Palatnik, M.I., and Kovtun, G.P.	1971	X	80-300		Single crystal grown by zone recrystallization; measurements along a-axis; lattice parameter reported at 293 K is 2.7060 Å; data on the coefficient of thermal linear expansion also reported.
10	660	Finkel, V.A., et al.	1971	X	80-300		Single crystal grown by zone recrystallization; measurements along c-axis; lattice parameter reported at 293 K is 4.2813 Å; data on the coefficient of thermal linear expansion also reported.

* Not shown in figure.

DATA TABLE 46. THERMAL LINEAR EXPANSION OF RUTHENIUM Ru

[Temperature, T, K; Linear Expansion, $\Delta L/L_0$, %]

T	$\Delta L/L_0$
CURVE 1	
303	0.0074
1793	1.243
1933	1.428
2083	1.631
2173	1.709
2373	1.845
2453	2.112
CURVE 2	
303	0.0117
1803	1.774
1943	2.190
2073	2.465
2173	2.638
2373	2.991
2453	3.283
CURVE 3*, ‡	
323	0.027
423	0.117
523	0.213
623	0.314
823	0.535
CURVE 4*, ‡	
232	0.018
423	0.078
523	0.140
623	0.206
823	0.350
CURVE 5	
293	0.0000*
323	0.0185*
353	0.0371
373	0.0482
423	0.0778
473	0.1075
523	0.1408
573	0.1742
623	0.2075
673	0.2409
CURVE 5 (cont.)	
723	0.2779
773	0.3150
823	0.3521
873	0.3891
CURVE 6	
293	0.0000*
323	0.0257
353	0.0538
373	0.0725
423	0.1170
473	0.1662
523	0.2129
573	0.2645
623	0.3159
673	0.3698
723	0.4236
773	0.4798
823	0.5359
873	0.5968
CURVE 7	
293	0.000*
799	0.298
1115	0.511
1247	0.610
1391	0.721
1557	0.867
CURVE 8	
293	0.000*
799	0.465
1115	0.811
1247	0.969
1391	1.154
1557	1.397
CURVE 9	
80	-0.112
86	-0.109
96	-0.106
99	-0.102
CURVE 9 (cont.)	
111	-0.099
117	-0.093
128	-0.091
134	-0.087
142	-0.085
153	-0.078
163	-0.074
170	-0.072
177	-0.066*
183	-0.064*
190	-0.062*
198	-0.059*
202	-0.053
214	-0.049*
217	-0.045*
222	-0.045*
228	-0.039
234	-0.034*
242	-0.031*
247	-0.028*
251	-0.027
257	-0.022*
265	-0.018*
268	-0.017*
271	-0.012
284	-0.005*
289	-0.002*
293	0.000
300	0.001*
CURVE 10	
80	-0.156
87	-0.153*
94	-0.152*
103	-0.149
108	-0.144*
118	-0.138*
123	-0.136
128	-0.134*
137	-0.129*
140	-0.126*
144	-0.124*
154	-0.119
159	-0.116*
162	-0.113*
CURVE 10 (cont.)	
169	-0.110*
173	-0.105*
178	-0.105
186	-0.097*
191	-0.093*
197	-0.092*
200	-0.087
204	-0.084*
208	-0.082*
215	-0.076*
225	-0.069
228	-0.065*
234	-0.062*
236	-0.057*
245	-0.050
248	-0.044*
259	-0.037*
267	-0.027*
273	-0.024
277	-0.017*
285	-0.009*
292	-0.007*
293	-0.002*
293	0.000
300	0.005*

* Not shown in figure.

‡ Author's data for coefficient of thermal expansion have been integrated by TPRC to obtain $\Delta L/L_0$.

DATA TABLE 46. COEFFICIENT OF THERMAL LINEAR EXPANSION OF RUTHENIUM Ru

[Temperature, T, K; Coefficient of Expansion, α, $10^{-6}\ K^{-1}$]

T	α
CURVE 3*, ‡	
323	8.8
423	9.3
523	9.8
623	10.4
823	11.7
CURVE 4*, ‡	
323	5.9
423	6.1
523	6.4
623	6.8
823	7.6
CURVE 9	
75	3.79
100	4.06
125	4.31
150	4.64
175	4.89
200	5.19
225	5.51
250	5.82
275	6.20
293	6.60
300	6.70
CURVE 10	
80	4.58
100	4.93
125	5.35
150	5.89
175	6.52
200	7.24
225	7.98
250	8.87
275	9.93
293	10.74
300	11.02

* Not shown in figure.
‡ Author's data for coefficient of thermal expansion have been integrated by TPRC to obtain $\Delta L/L_0$.

FIGURE AND TABLE NO. 47R. TYPICAL VALUES FOR THERMAL LINEAR EXPANSION OF SCANDIUM Sc

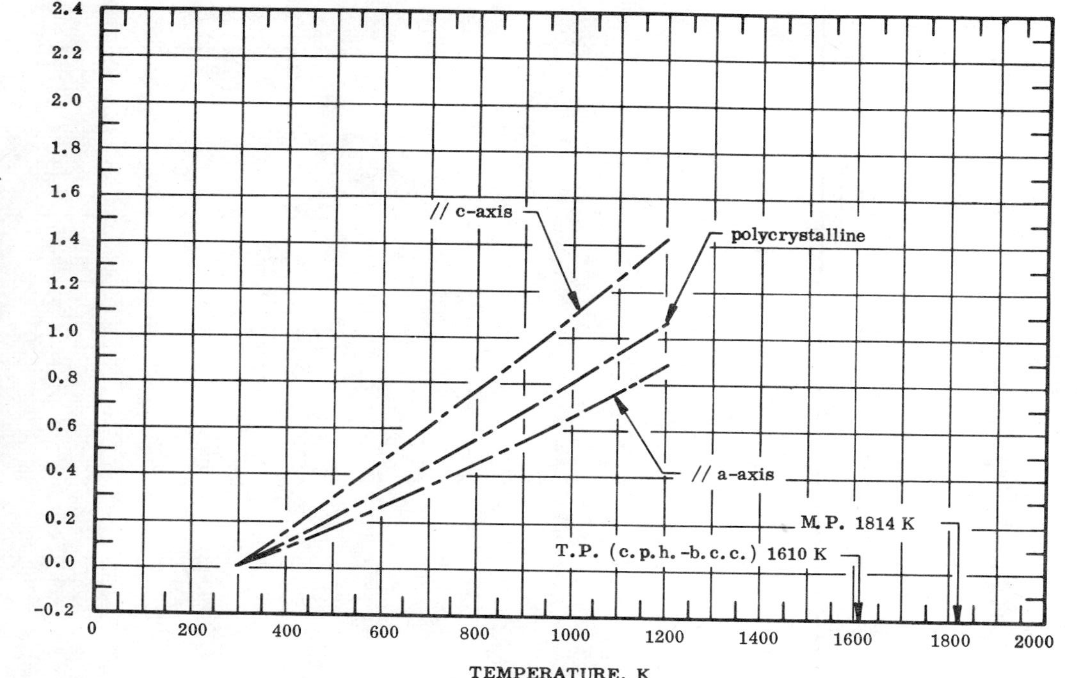

TYPICAL VALUES

[Temperature, T, K; Linear Expansion, $\Delta L/L_0$, %; α, K^{-1}]

	// a-axis	// c-axis	polycrystalline	
T	$\Delta L/L_0$	$\Delta L/L_0$	$\Delta L/L_0$	$\alpha \times 10^6$
293	0.000	0.000	0.000	10.0
400	0.084	0.163	0.110	10.5
500	0.167	0.314	0.216	10.8
600	0.255	0.465	0.325	11.1
700	0.348	0.616	0.437	11.4
800	0.445	0.769	0.553	11.8
900	0.547	0.925	0.673	12.0
1000	0.654	1.085	0.797	12.6
1200	0.882	1.420	1.060	13.7

REMARKS

The tabulated values are considered accurate to within ± 20% over the entire temperature range. Since no experimental data on the polycrystalline scandium were located in the literature, the tabulated values are calculated from single crystal data. These values are considered accurate to within ± 20% over the entire temperature range. These values can be represented approximately by the following equations:

// a-axis: $\Delta L/L_0 = -0.206 + 6.373 \times 10^{-4}\ T + 2.127 \times 10^{-7}\ T^2 + 9.550 \times 10^{-12}\ T^3$

// c-axis: $\Delta L/L_0 = -0.472 + 1.685 \times 10^{-3}\ T - 3.175 \times 10^{-7}\ T^2 + 1.898 \times 10^{-10}\ T^3$

polycrystalline: $\Delta L/L_0 = -0.289 + 9.631 \times 10^{-4}\ T + 6.492 \times 10^{-8}\ T^2 + 5.747 \times 10^{-11}\ T^3$

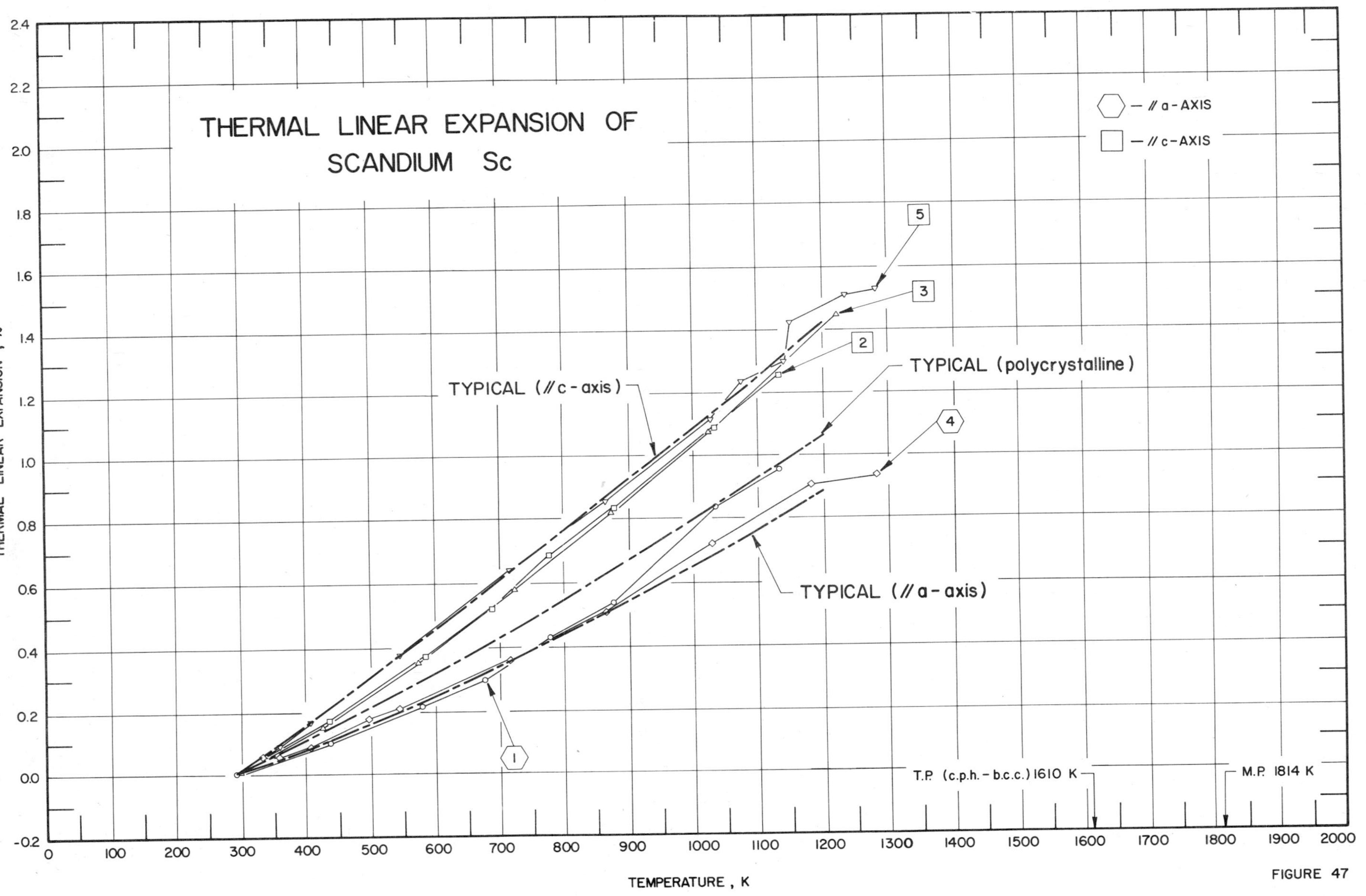
THERMAL LINEAR EXPANSION OF
SCANDIUM Sc
— // a-AXIS
— // c-AXIS
5
3
2
TYPICAL (polycrystalline)
TYPICAL (// c-axis)
4
TYPICAL (// a-axis)
1
T.P. (c.p.h. – b.c.c.) 1610 K
M.P. 1814 K
THERMAL LINEAR EXPANSION, %
2.4
2.2
2.0
1.8
1.6
1.4
1.2
1.0
0.8
0.6
0.4
0.2
0.0
-0.2
0
100
200
300
400
500
600
700
800
900
1000
1100
1200
1300
1400
1500
1600
1700
1800
1900
2000
TEMPERATURE, K

FIGURE 47

SPECIFICATION TABLE 47. THERMAL LINEAR EXPANSION OF SCANDIUM Sc

Cur. No.	Ref. No.	Author(s)	Year	Method Used	Temp. Range, K	Name and Specimen Designation	Composition (weight percent), Specifications, and Remarks
1	296	Mardon, P. G., Nichols, J. L., Pearce, J. H. and Poole, D. M.	1961	X	290-1131		< 0.1 Ta, 0.02 Cu, 0.01 Ag, 0.01 rare earths, 0.002 Fe; expansion measured in nitrogen under pressure of atmosphere; expansion measured along a-axis; lattice parameter reported at 293 K is 3.3085 Å.
2	296	Mardon, P. G., et al.	1961	X	290-1131		The above specimen; expansion measured along c-axis; lattice parameter reported at 293 K is 5.2702 Å.
3	297	Giselman, D.	1961	L	298-1223		0.9 O_2, 0.11 Fe, 0.08 F_2, 0.05 Ta, 0.05 Th, 0.03 Y, 0.02 H_2, 0.01 G, 0.01 N_2; specimen 2 in. long, 0.25 in. diameter; produced at Metal Research Laboratories by Ca reduction of ScF_3 and subsequent distillation; density 3.05 g cm^{-3}; expansion measured in argon atmosphere; zero-point correction of 0.006% determined by graphical extrapolation.
4	82	Hanak, J. J.	1959	X	293-1282		< 0.1 Fe, < 0.06 Cr, < 0.02 each Ca, Mg; wire specimen cut from sheet prepared at Ames Laboratory; hexagonal close packed structure; expansion measured along a-axis; lattice parameter reported at 293 K is 3.308 Å.
5	82	Hanak, J. J.	1959	X	293-1252		The above specimen; expansion measured along c-axis; lattice parameter reported at 293 K is 5.267 Å.

DATA TABLE 47. THERMAL LINEAR EXPANSION OF SCANDIUM Sc

[Temperature, T, K; Linear Expansion, $\Delta L/L_0$, %]

T	$\Delta L/L_0$	T	$\Delta L/L_0$	T	$\Delta L/L_0$	T	$\Delta L/L_0$	T	$\Delta L/L_0$
CURVE 1		CURVE 3		CURVE 3 (cont.)		CURVE 4 (cont.)		CURVE 5 (cont.)	
290	-0.024*	298	0.006	1173	1.352*	917	0.544*	602	0.475*
293	0.000	323	0.038*	1223	1.446	975	0.665*	671	0.570*
439	0.106	373	0.088*			1029	0.726	715	0.646
578	0.218	423	0.152	CURVE 4		1077	0.726*	771	0.740*
678	0.308	473	0.217*			1141	0.786*	816	0.778*
776	0.438	523	0.284*	293	0.000*	1183	0.907	865	0.854
874	0.544	573	0.360	331	0.060*	1236	0.998*	917	0.949*
1032	0.831	623	0.430*	360	0.060*	1282	0.937	975	1.044*
1131	0.955	673	0.510*	408	0.091			1029	1.120
CURVE 2		723	0.582	447	0.151*	CURVE 5		1077	1.234
290	-0.009*	773	0.660*	495	0.181*			1141	1.291
293	0.000*	823	0.743*	544	0.212	293	0.000*	1153	1.424
437	0.176	873	0.820	602	0.272*	331	0.057*	1236	1.519
582	0.374	923	0.903*	671	0.333*	360	0.095*	1282	1.538
678	0.524	973	0.988*	715	0.363	408	0.171		
778	0.687	1023	1.080	771	0.423*	447	0.247*		
877	0.829	1073	1.173*	816	0.484*	495	0.304*		
1031	1.097	1123	1.262*	865	0.514	544	0.380		
1131	1.256								

* Not shown in figure.

FIGURE AND TABLE NO. 48R. RECOMMENDED VALUES FOR THERMAL LINEAR EXPANSION OF SILVER Ag

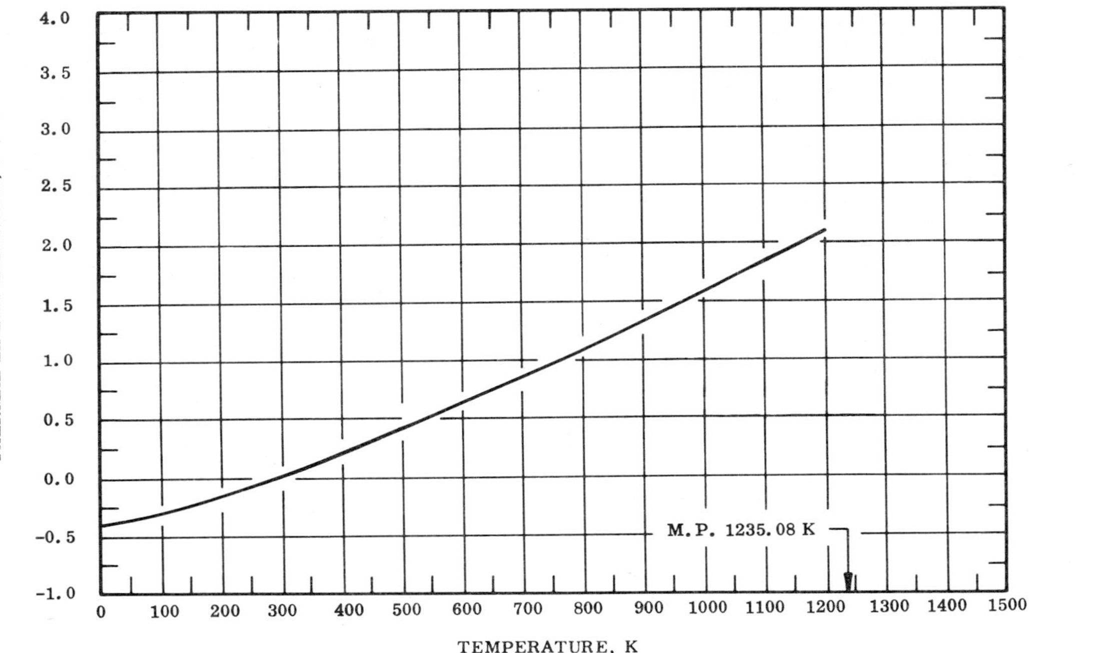

RECOMMENDED VALUES

[Temperature, T, K; Linear Expansion, $\Delta L/L_0$, %; α, K^{-1}]

T	$\Delta L/L_0$	$\alpha \times 10^6$
5	-0.410	0.015
25	-0.409	1.9
50	-0.396	8.2
100	-0.336	14.2
200	-0.171	17.8
293	0.000	18.9
400	0.207	19.7
500	0.409	20.6
600	0.620	21.5
700	0.842	22.6
800	1.072	23.7
900	1.312	24.8
1000	1.568	25.9
1200	2.110	28.4

REMARKS

The tabulated values are considered accurate to within ± 3% over the entire temperature range. These values can be represented approximately by the following equation:

$$\Delta L/L_0 = -0.515 + 1.647 \times 10^{-3}\, T + 3.739 \times 10^{-7}\, T^2 + 6.283 \times 10^{-11}\, T^3 \qquad (200 < T < 1200)$$

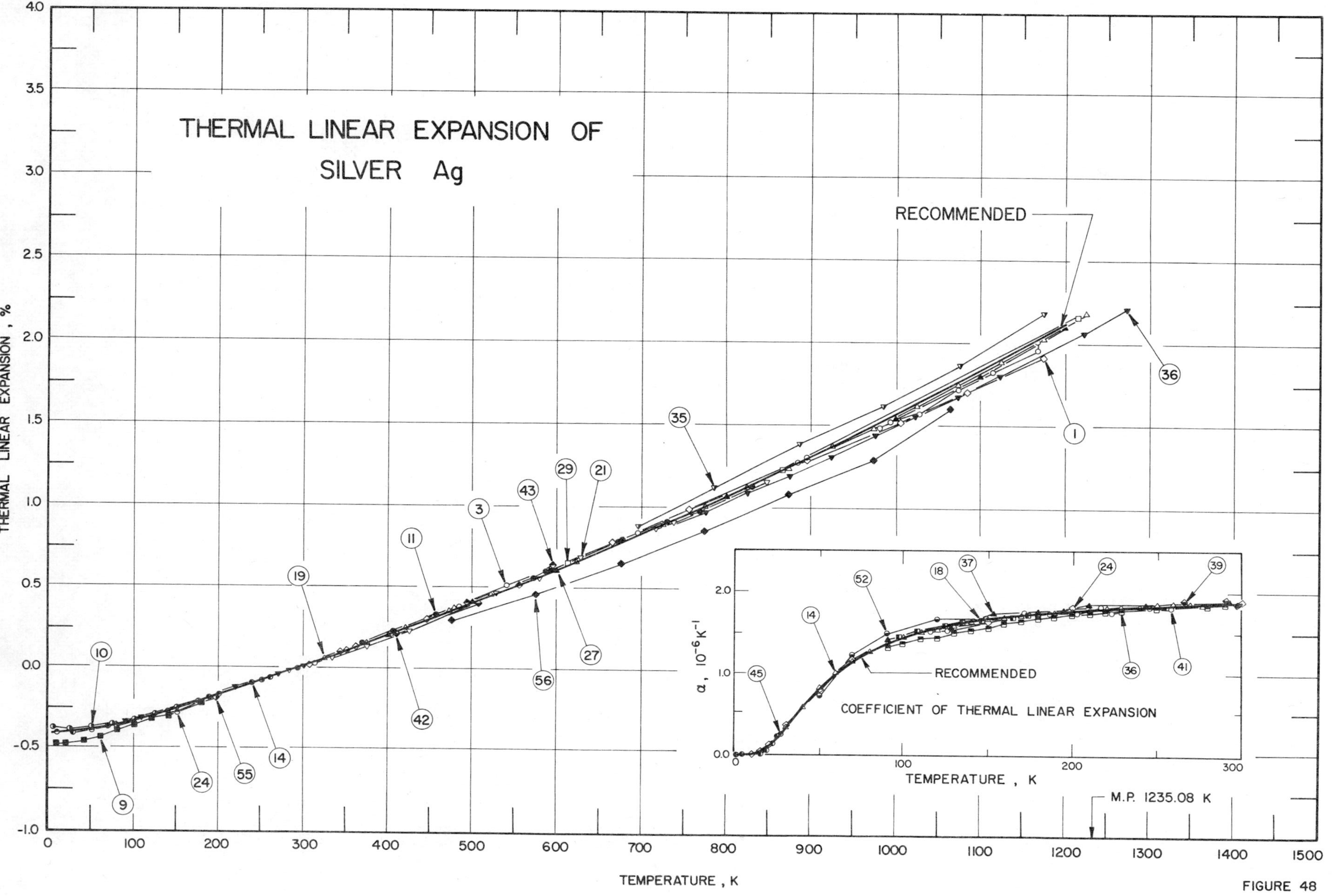
THERMAL LINEAR EXPANSION OF
SILVER Ag
RECOMMENDED
THERMAL LINEAR EXPANSION, %
TEMPERATURE, K
COEFFICIENT OF THERMAL LINEAR EXPANSION
α, $10^{-6}K^{-1}$
RECOMMENDED
TEMPERATURE, K
M.P. 1235.08 K

FIGURE 48

SPECIFICATION TABLE 48. THERMAL LINEAR EXPANSION OF SILVER Ag

Cur. No.	Ref. No.	Author(s)	Year	Method Used	Temp. Range, K	Name and Specimen Designation	Composition (weight percent), Specifications, and Remarks
1	42	Stuart, H. and Ridley, N.	1966	X	303-1177		99.999 Ag; powder ground to 200 mesh; sealed in vacuum in 0.5 mm diameter tube and stress annealed 15 min at 870 K; formed into wire 0.5 mm diameter, polished and sealed in silica tube, then coated with graphite; lattice parameter reported at 293 K is 4.0844 Å.
2*	298	Straumanis, M.E. and Riad, S.M.	1965	X	283-337	ASARCO	99.999$^+$ Ag from ASARCO; expansion measured using copper Kα x-rays in vacuum; lattice parameter reported at 293 K is 4.08587 Å.
3	298	Straumanis, M.E. and Riad, S.M.	1965	X	312-1168		The above specimen; lattice parameter reported at 312 K is 4.0877 Å; 4.0861 Å at 293 K determined by graphical extrapolation.
4*	3	Fraser, D.B. and Hallet-Hollis, A.C.	1965	I	17-100		Cubic crystal structure; hollow cylindrical specimen 2.54 cm O.D., 2.06 cm I.D., 5.08 cm long; this curve is here reported using the first given temperature as reference temperature at which $\Delta L/L_0 = 0$; correction is 0.3778%.
5	299	Petrov, Yu.I. and Fedorov, Yu.I.	1967	X	573-1260		Spherical aerosol particles of silver, $\approx 1.8 \times 10^{-6}$ cm diameter, produced by condensation of metallic vapor in argon atmosphere at pressure of 0.1 mm Hg; finished sample diameter < 0.3 mm; powder glued with bakelite lacquer to platinum filament, heated by electric current.
6	299	Petrov, Yu.I. and Fedorov, Yu.I.	1967	X	699-1153		Similar to the above specimen; rapid heating of sample to 1273 K and maintained for 4 hr.
7	273	Mauer, F.A. and Bolz, L.H.	1955	X	273-1149		99.999 Ag powder from A.D. Mackay, Inc.; zero-point correction of -0.037% determined by graphical interpolation.
8	300	Michel, D.Y.	1968	X	293-1119		High purity silver powder placed in Be foil capsule; CuKα (λ = 1.54178 Å) x-radiation used; lattice parameter reported at 293 K is 4.087.
9	280	Lallement, R.	1963	L	10-240		Specimen cooled rapidly to 77 K and then to 4.2 K, heated up to 77 K in about 8 hr; also cooled rapidly to 77 K and then heated to 300 K in about 8 hr.
10	302	Beenakker, Y.Y.M. and Swenson, C.A.	1955	L	4.2-300		German silver.
11	303	Owen, E.A. and Roberts, E.W.	1938	X	291-830		99.95 Ag; results of several different runs combined; no systematic differences noted; lattice parameter reported at 291 K is 4.0772 Å; 4.0774 Å at 293 K determined by graphical interpolation.
12	12	Ayres, H.D.	1905	I	123-373		Expansion measured using 5460 Å line of mercury.
13	12	Ayres, H.D.	1905	I	89-373		Similar to the above specimen.
14	269	Waterhouse, N. and Yates, B.	1968	I	30-270		0.0005 Cu, 0.0003 Fe, < 0.0001 Mg; hollow cylinder 2.6 cm O.D., 2.3 cm I.D., and 1 cm long; supplied by Johnson, Matthey and Co., Ltd.; annealed in vacuum for between 30 and 60 min at 770 K; cooled over a period of 24 hrs.
15	14	Palatnik, L.S., Pugachev, A.T., Boiko, B.T. and Bratsyhkin, V.M.	1967	F	293-654		Specimen condensed as film 400 to 600 Å thick on underlayer of NaCl on base; expansion measured under vacuum of 10^{-4} to 10^{-5} torr with increasing temperature; zero-point correction is -0.006%.
16	14	Palatnik, L.S., et al.	1967	F	583-351		The above specimen; expansion measured with decreasing temperature.
17	274	Brand, J.A. and Goldschmidt, H.J.	1956	X	273-1173		No details given; lattice parameter reported at 273 K is 4.0761 kX (4.0843 Å); 4.0778 kX (4.0859 Å) at 293 K determined by graphical interpolation.

* Not shown in figure.

SPECIFICATION TABLE 48. THERMAL LINEAR EXPANSION OF SILVER Ag (continued)

Cur. No.	Ref. No.	Author(s)	Year	Method Used	Temp. Range, K	Name and Specimen Designation	Composition (weight percent), Specifications, and Remarks
18*	304	Evans, D. J. and Winstanley, C. J.	1966	L	98-174		No details given; this curve is here reported using the first given temperature as reference temperature at which $\Delta L/L_0 = 0$.
19	305	Simmons, R. O. and Balluffi, R. W.	1960	X	323-1223		Specimen prepared from material of initial purity, 99.999$^+$ Ag; < 0.0003 O_2, < 0.0001 each Fe, Mg, Si, Cu; bar-shaped specimen 50 cm long, 1.27 cm square; where silver melted and cast horizontally in spectrographically pure graphite mold; solidified by rapid cooling from one end in prepurified nitrogen atmosphere, cooling rate adjusted to obtain polycrystal with average grain size of about 1.5 cm; reported standard deviation from smooth curve 10×10^{-6}.
20	48	Dorsey, H. G.	1907	I	93-293		Pure specimen from U. S. Mint; cast; length 1.2805 cm; density 10.43 g cm^{-3}.
21	275	Owen, E. A. and Yates, E. L.	1934	X	282-847	Method 1	Powder fixed to thin copper foil (0.05 mm thick) and this was backed by the thick copper sheet; reported error 0.01% in measurement of lattice parameter; lattice parameter reported at 282 K is 4.0765 Å; 4.0772 Å at 293 K determined by graphical interpolation.
22*	275	Owen, E. A. and Yates, E. L.	1934	X	289-873	Method 2	Powder fixed directly to surface of copper sheet; lattice parameter reported at 289 K is 4.0771 Å; 4.0774 Å at 293 K determined by graphical interpolation.
23	281	Grove, G. R.	1966	L	424-922		Spectroscopic grade specimen.
24	22	Fraser, D. B. and Hallet-Hollis, A. C.	1961	I	10-300		Metal machined into form of thick walled cylindrical tubes; ends ground plane and parallel to each other; annealed; reported error decreases from maximum of 20% at 10 K.
25*	75	Nix, F. C. and MacNair, D.	1942	I	87-298		99.999 Ag; specimen from Handy and Harmon; zero-point correction is -0.03838%.
26*	26	Buffington, R. M. and Latimer, W. M.	1926	I	87-310		Specimen turned from rod cast in graphite and hammered.
27	27	Pathak, P. D. and Vasavada, N. G.	1970	L	300-1200		99.999 Ag; obtained from Johnson Matthey and Co., London; rod 10 cm long, about 0.5 cm diameter; reported error 0.5 to 1%; data on the coefficient of thermal linear expansion are also reported.
28*	148	Kos, J. F. and Lamarche, J. L. G.	1969	E	5-15		0.0005 Fe, 0.0002 Cu, 0.0001 each Ca, Cd, Mg, Sn; polycrystalline specimen, 7 mm in diameter, 10 cm long; specimen obtained from Johnson Matthey and Co.; temperature accurate to ± 0.01 K.
29	306	Hume-Rothery, W. and Reynolds, P. W.	1938	X	293-1216		Chemically pure assay silver; supplied by Messrs Johnson, Matthey and Co., Ltd.; filings sealed in evacuated specimen tube, annealed, reported data obtained from several experiments; lattice parameter reported at 293 K is 4.0774 Å.
30*	60	Holborn, L. and Day, A. L.	1901	T	273-1023		No details given; zero-point correction of -0.039% determined by graphical interpolation.
31*	307	Spreadborough, J. and Christian, J. W.	1959	X	289-1119		Specimen contained in high vacuum chamber; lattice parameter reported at 289 K is 4.0773 kX (4.0854 Å); 4.0776 kX (4.0857 Å) at 293 K determined by graphical interpolation.
32*	444	Riad, S. M.	1964	X	313-1031		Similar to the specimen used for curve 3; expansion measured in oxygen atmosphere; lattice parameter reported at 313 K is 4.07917 Å; 4.0780 Å at 293 K determined by graphical extrapolation.

*Not shown in figure.

SPECIFICATION TABLE 48. THERMAL LINEAR EXPANSION OF SILVER Ag (continued)

Cur. No.	Ref. No.	Author(s)	Year	Method Used	Temp. Range, K	Name and Specimen Designation	Composition (weight percent), Specifications, and Remarks
33*	444	Riad, S. M.	1964	X	313-1173		Similar to the above specimen; expansion measured in nitrogen atmosphere; lattice parameter reported at 313 K is 4.07917 Å; 4.0776 Å at 293 K determined by graphical extrapolation.
34*	444	Riad, S. M.	1964	X	313-1128		Similar to the above specimen; expansion measured in hydrogen atmosphere; lattice parameter reported at 313 K is 4.07917 Å; 4.0776 Å at 293 K determined by graphical extrapolation.
35	444	Riad, S. M.	1964	X	313-1173		Similar to the above specimen; expansion measured in atmosphere of air; lattice parameter reported at 213 K is 4.07917 kX (4.0873 Å); 4.0776 kX (4.085 Å) at 293 K determined by graphical interpolation.
36	452	Leksina, I. E. and Novikova, S. I.	1963	L	90-1273		99.99 Ag, remelted in vacuum; specimen <25 mm long; data on the coefficient of thermal linear expansion are also reported.
37*	609	Chistov, S. F., Chernov, A. P. and Dombovskii, S. A.	1968	L	90-430		Spectrally pure; 4 mm diameter, 7.96 mm high; specimen cooled to temperature of liquid nitrogen; heating rate 1.6 deg. per minute.
38*	609	Chistov, S. F., et al.	1968	L	293-540		Similar to the above specimen.
39*	609	Chistov, S. F., et al.	1968	L	266-484		Similar to the above specimen.
40*	609	Chistov, S. F., et al.	1968	L	293-519		Similar to the above specimen.
41*	609	Chistov, S. F., et al.	1968	L	116-464		Similar to the above specimen.
42	609	Chistov, S. F., et al.	1968	L	271-505		Similar to the above specimen.
43	612	Boiko, B. T., Palatnik, L. S., and Pugachev, A. T.	1967	F	293-595		99.9$^+$ pure, thin film specimen condensed in vacuum 5 x 10^{-5} torr thickness of film in the range 600 + 10 Å; lattice parameter expansion measured with increasing temperature.
44*	612	Boiko, T. T., et al.	1967	F	584-352		The above specimen; lattice parameter measured with decreasing temperature.
45*	614	McLean, K. O., Swenson, C. A., and Case, C. R.	1972	V	0-26		99.999$^+$ pure, specimen from Nuclear Elements Corp.; vacuum cast; this curve is here reported using the first given temperature as reference temperature at which $\Delta L/L_0 = 0$.
46*	615	White, G. K. and Collins, J. G.	1972	L	2-30	Ag 1b	99.999$^+$ Ag; spectroanalysis by Australian Mineral Development Laboratory showed <0.0002 Fe, <0.0001 Mn; gas analysis at Defence Standards Lab. Victoria showed 0.00004 O, 0.000005 N; neutron activation analysis by Gulf General Atomics of Calif. showed 0.00002 O; cylindrical specimen 5.09 cm long, 2.5 cm diameter; the specimen after relapping from Consolidated Mining and Smelting Co. of Canada; this curve is here reported using the first given temperature as reference temperature at which $\Delta L/L_0 = 0$.
47*	615	White, G. K. and Collins, J. G.	1972	L	4-33	Ag 1c	The Ag 1b specimen after degassing by pumping for 1 day at 1070 K and relapping; this curve is here reported using the first given temperature as reference temperature at which $\Delta L/L_0 = 0$.
48*	615	White, G. K. and Collins, J. G.	1972	L	2-8	Ag 1c	The above specimen; measurements on second test; this curve is here reported using the first given temperature as reference temperature at which $\Delta L/L_0 = 0$.

* Not shown in figure.

SPECIFICATION TABLE 48. THERMAL LINEAR EXPANSION OF SILVER Ag (continued)

Cur. No.	Ref. No.	Author(s)	Year	Method Used	Temp. Range, K	Name and Specimen Designation	Composition (weight percent), Specifications, and Remarks
49*	615	White, G.K. and Collins, J.G.	1972	L	2-23	Ag 2	99.999 pure Tadanac "oxygen free" cylindrical specimen from Consolidated Mining and Smelting Co. of Canada; annealed at 770 K for 2 hr, after machining; this curve is here reported using the first given temperature as reference temperature at which $\Delta L/L_0 = 0$.
50*	615	White, G.K. and Collins, J.G.	1972	L	3-28	Ag 3 "Specpure" JM 24757	Cylindrical rod 5.09 cm long, 1.7 cm diameter cast in graphite from John Matthey; degassed at 1073 K for some hrs before melting; this curve is here reported using the first given temperature as reference temperature at which $\Delta L/L_0 = 0$.
51*	616	Straumanis, M.E.	1971	X	283-338		99.999 pure specimen recovered from the cold work (filing) occurred at room temperature in 5 day; lattice parameter reported at 293 K is 4.08574 Å.
52*	617	Straumanis, M.E. and Woodard, C.L.	1971	X	40-180		99.7 pure specimen; probable impurities Cu and Mg; fine powder embedded in a In layer which adhered well the Ag sample holder; lattice parameter reported at 180 K is 4.07786 Å; data on the coefficient of thermal expansion are also reported; this curve reported here is referred to 293 K as reference temperature using other available data; zero-point correction is 0.206%.
53*	617	Straumanis, M.E. and Woodard, C.L.	1971	X	40-180		The above sample embedded in vaseline adhering to another Ag sample holder; lattice parameter reported at 180 K is 4.0780 Å; this curve reported here is referred to 293 K as reference temperature using other available data; zero-point correction is 0.206%.
54*	617	Straumanis, M.E. and Woodard, C.L.	1971	X	40-180		99.99 pure fine grained sample holder; lattice parameter reported at 180 K is 4.07899 Å; this curve reported here is referred to 293 K as reference temperature using other available data; zero-point correction is 0.206%.
55	654	Nagender Naidu, S.V. and Houska, C.R.	1971	X	80-298		Prepared from 99.99 Ag from Engelhard Industries Inc. by melting in induction furnace under hydrogen atmosphere; homogenized for 1 wk. at 1173 K; annealed at 873 K in evacuated quartz tubes after flushing with argon 3 times; lattice parameter reported at 298 K is 4.0864 Å; 4.0680 Å at 293 K determined by graphical interpolation.
56	657	Vest, R.W.	1971	L	293-1067		Specimen prepared from 99.9 pure metal of -352 mesh; data on the coefficient of thermal linear expansion also reported.

*Not shown in figure.

DATA TABLE 48. THERMAL LINEAR EXPANSION OF SILVER Ag

[Temperature, T, K; Linear Expansion, $\Delta L/L_0$, %]

T	$\Delta L/L_0$
CURVE 1	
293	0.000
303	0.022
327	0.076*
361	0.147
445	0.311
581	0.619*
666	0.786
755	0.970
895	1.280
1008	1.533
1084	1.714
1177	1.934
CURVE 2*	
283	-0.018
293	0.000
309	0.027
323.3	0.057
338.1	0.085
CURVE 3	
312	0.039
484	0.379
540	0.501
695	0.839
883	1.275
895	1.302
980	1.492
991	1.524
1029	1.622
1074	1.727
1114	1.830
1167	1.960
1168	2.009
CURVE 4*, †	
17.2	0.000
21.8	0.001
23.8	0.001
25.6	0.001
26.7	0.002
30.3	0.002
32.6	0.004
CURVE 4 (cont.)*, †	
34.7	0.004
36.9	0.006
40.0	0.007
41.6	0.008
43.8	0.010
45.7	0.011
48.1	0.013
49.4	0.014
51.3	0.015
53.0	0.017
55.8	0.019
56.5	0.020
58.5	0.022
59.9	0.023
62.0	0.025
65.0	0.029
68.5	0.032
71.6	0.036
74.1	0.040
77.8	0.044
81.3	0.048
83.9	0.052
86.7	0.056
90.1	0.060
94.3	0.066
99.9	0.075
CURVE 5	
537	0.442
748	0.838
830	1.084*
857	1.167
956	1.464
1079	1.742
1166	1.974
1201	1.770
1229	1.970
1233	1.894
1260	2.032
CURVE 6	
699	0.755
1020	1.44
1153	1.832
CURVE 7	
273	-0.037*
294	0.002*
325	0.063
376	0.174
425	0.259*
479	0.357
531	0.472*
582	0.580
634	0.685*
685	0.803
788	1.033
886	1.283
994	1.540
1097	1.810
1149	1.961
CURVE 8	
292.8	0.000*
610.7	0.587
732.2	0.880
869.2	1.174
997.2	1.468
1119.2	1.786
CURVE 9*	
10	-0.485
20	-0.488
40	-0.460
60	-0.435
80	-0.395
100	-0.354
120	-0.331
140	-0.305
180	-0.225
240	-0.110*
CURVE 10	
4.2	-0.376
25	-0.376
50	-0.369
75	-0.350
100	-0.324
125	-0.291
CURVE 10 (cont.)	
150	-0.253
175	-0.214
200	-0.171*
250	-0.079
300	0.013*
CURVE 11	
291	-0.004
370	-0.0157
408	0.230
458	0.335
480	0.375
572	0.569
586	0.600
591	0.608
599	0.628*
671	0.778
674	0.794
730	0.911
759	0.974
830	1.128
CURVE 12	
123	-0.310
173	-0.229
223	-0.135*
273	-0.039
323	0.058*
372.5	0.151
CURVE 13	
89	-0.360
123	-0.308*
173	-0.224*
223	-0.134
273	-0.038*
323	0.057*
372.5	0.150*
CURVE 14‡	
30	-0.408
40	-0.404*
CURVE 14 (cont.)‡	
50	-0.397*
60	-0.388*
70	-0.377
80	-0.365*
90	-0.352*
100	-0.337*
110	-0.323*
120	-0.307*
130	-0.291*
140	-0.275
150	-0.259*
160	-0.242*
170	-0.225*
180	-0.207*
190	-0.190
200	-0.172*
210	-0.154*
220	-0.136*
230	-0.118*
240	-0.100
250	-0.081*
260	-0.063
270	-0.044*
CURVE 15	
293	0.000*
376	0.153*
422	0.249*
491	0.411
553	0.511*
593	0.647
654	0.765
CURVE 16	
583	0.619
554	0.556
507	0.463
462	0.355
404	0.226*
351	0.125
CURVE 17	
273	-0.042*
CURVE 17 (cont.)	
373	0.167*
473	0.397
573	0.628
673	0.851
773	1.045
873	1.260
973	1.488
1073	1.739
1173	2.000
CURVE 18*, †, ‡	
98	0.000
108	0.015
112	0.021
125	0.091
135	0.058
144	0.073
146	0.076
165	0.108
169	0.115
174	0.123
CURVE 19	
323	0.057
373	0.154
423	0.254
473	0.355
523	0.458
573	0.564*
623	0.671
673	0.781*
723	0.893
773	1.008
823	1.125
873	1.244
923	1.367
973	1.493
1023	1.623
1073	1.755
1123	1.893
1173	2.036
1223	2.186

*Not shown in figure.
†This curve is here reported using the first given temperature as reference temperature at which $\Delta L/L_0 = 0$.
‡Author's data for coefficient of thermal expansion have been integrated by TPRC to obtain $\Delta L/L_0$.

DATA TABLE 48. THERMAL LINEAR EXPANSION OF SILVER Ag (continued)

T	$\Delta L/L_0$
CURVE 20	
93	-0.348*
113	-0.319
133	-0.289*
153	-0.256
173	-0.222*
193	-0.187
213	-0.151*
233	-0.114
253	-0.076
273	-0.038
293	0.000*
CURVE 21	
282	-0.017
332	0.063
375	0.134
375	0.144*
426	0.242
480	0.355*
529	0.473
580	0.566
627	0.686
673	0.775
718	0.860
737	0.902
779	1.015*
797	1.066*
847	1.150
CURVE 22*	
289	0.007
330	0.069
377	0.162
423	0.265
481	0.380
532	0.471
577	0.574
624	0.687
675	0.807
725	0.917
777	1.013
809	1.101
814	1.113
865	1.234
873	1.238
CURVE 23	
424	0.241*
572	0.580
665	0.757
771	1.006
820	1.134*
870	1.216
922	1.363
CURVE 24‡	
10	-0.413
15	-0.412*
20	-0.412*
30	-0.410*
40	-0.405*
50	-0.398
70	-0.378*
100	-0.339*
150	-0.262
200	-0.175
250	-0.082*
300	0.013*
CURVE 25*	
87.2	-0.357
92.2	-0.349
97.2	-0.340
103.2	-0.332
110.2	-0.324
114.2	-0.315
121.2	-0.306
126.2	-0.298
132.2	-0.289
137.2	-0.281
141.2	-0.273
145.7	-0.265
151.7	-0.257
157.7	-0.248
161.7	-0.239
166.3	-0.231
171.2	-0.223
176.2	-0.214
181.2	-0.206
186.2	-0.197
190.0	-0.189
194.7	-0.181
CURVE 25 (cont.)*	
200.2	-0.172
205.5	-0.164
210.2	-0.155
214.6	-0.147
219.2	-0.138
223.6	-0.130
228.6	-0.122
233.0	-0.114
237.2	-0.105
242.7	-0.096
246.4	-0.088
251.2	-0.080
255.7	-0.071
260.0	-0.063
264.2	-0.054
269.2	-0.046
273.2	-0.038
278.0	-0.029
282.1	-0.021
286.7	-0.012
291.2	-0.004
293.0	0.000
295.0	0.004
297.7	0.008
CURVE 26*	
86.8	-0.354
92.4	-0.346
101.2	-0.334
134.9	-0.280
179.7	-0.206
189.0	-0.175
192.7	-0.184
211.6	-0.149
212.1	-0.148
228.3	-0.120
228.9	-0.119
241.9	-0.095
250.0	-0.080
260.3	-0.061
265.4	-0.051
267.8	-0.045
274.3	-0.037
277.2	-0.029
280.1	-0.024
295.0	0.004
306.3	0.025
310.3	0.033
CURVE 27‡	
300	0.013*
400	0.205
500	0.406
600	0.615
700	0.834*
800	1.061
900	1.299*
1000	1.548
1100	1.811
1200	2.101
CURVE 28*, †, ‡	
5	0.000 x 10^{-4}
6	0.022
7	0.053
8	0.100
9	0.169
10	0.265
11	0.395
12	0.569
13	0.796
14	1.086
15	1.454
CURVE 29	
293	0.000*
610.7	0.652
732	0.920*
869	1.236
997	1.553*
1119	1.884
1216	2.163
CURVE 30*	
273	-0.039
523	0.448
773	0.993
1023	1.602
CURVE 31*	
289.2	-0.007
295.2	0.003
407.2	0.253
530.2	0.511
CURVE 31 (cont.)*	
672.2	0.800
825.2	1.114
990.2	1.519
1119.2	1.877
CURVE 32*	
313	0.029
493	0.417
645	0.738
794	1.110
983	1.560
1031	1.665
CURVE 33*	
313	0.038
491	0.409
594	0.649
697	0.850
785	1.093
891	1.346
983	1.608
1081	1.871
1173	2.153
CURVE 34*	
313	0.038
492	0.343
593	0.559
793	1.027
889	1.218
983	1.522
1081	1.792
1128	1.863
CURVE 35	
313	0.038*
493	0.411*
593	0.632*
695	0.885
785	1.118
886	1.380
984	1.633
1079	1.876
1173	2.182
CURVE 36‡	
90	-0.339
100	-0.326*
110	-0.312
120	-0.297*
130	-0.283*
140	-0.267*
150	-0.252*
160	-0.236*
170	-0.219*
180	-0.203*
190	-0.185*
200	-0.168*
210	-0.151*
220	-0.133
230	-0.115*
240	-0.097*
250	-0.079*
260	-0.061*
270	-0.043
280	-0.024*
290	-0.006*
300	0.013
310	0.032*
320	0.051*
330	0.070*
340	0.089*
373	0.152*
423	0.250*
473	0.349*
523	0.449*
573	0.550*
623	0.654*
773	0.973
823	1.083
873	1.195
923	1.310
973	1.429
1023	1.550
1073	1.675
1123	1.803
1173	1.934*
1223	2.068
1273	2.205
CURVE 37*, ‡	
90	-0.354
102	-0.337

*Not shown in figure.
†This curve is here reported using the first given temperature as reference temperature at which $\Delta L/L_0 = 0$.
‡Author's data for coefficient of thermal expansion have been integrated by TPRC to obtain $\Delta L/L_0$.

DATA TABLE 48. THERMAL LINEAR EXPANSION OF SILVER Ag (continued)

CURVE 37 (cont.)*, ‡

T	$\Delta L/L_0$
129	-0.297
141	-0.277
153	-0.257
166	-0.234
180	-0.209
195	-0.182
210	-0.155
243	-0.094
277	-0.030
283	-0.019
293	-0.000
298	-0.010
313	-0.038
328	-0.066
344	-0.096
357	-0.121
372	-0.149
401	-0.204
415	-0.230
430	-0.259

CURVE 38*, ‡

T	$\Delta L/L_0$
293	0.000
324	0.060
329	0.070
339	0.090
455	0.319
528	0.464
540	0.488

CURVE 39*, ‡

T	$\Delta L/L_0$
266	-0.051
293	0.000
298	0.010
345	0.101
379	0.169
442	0.293
463	0.334
484	0.375

CURVE 40*, ‡

T	$\Delta L/L_0$
293	0.000
361	0.131
368	0.145

CURVE 40 (cont.)*, ‡

T	$\Delta L/L_0$
395	0.198
401	0.210
415	0.237
438	0.284
475	0.359
497	0.404
519	0.450

CURVE 41*, ‡

T	$\Delta L/L_0$
116	-0.310
127	-0.293
160	-0.240
172	-0.219
189	-0.189
203	-0.165
216	-0.141
224	-0.127
245	-0.089
258	-0.065
293	0.000
298	0.009
316	0.043
328	0.065
342	0.091
385	0.169
406	0.208
425	0.244
447	0.285
464	0.316

CURVE 42‡

T	$\Delta L/L_0$
271	-0.042*
293	0.000*
314	0.041*
332	0.077*
344	0.101
350	0.113*
373	0.159*
396	0.203*

CURVE 42 (cont.)‡

T	$\Delta L/L_0$
410	0.231
421	0.252*
505	0.416

CURVE 43

T	$\Delta L/L_0$
293	0.000
378	0.157*
424	0.255*
493	0.414
555	0.515
595	0.649

CURVE 44*

T	$\Delta L/L_0$
584	0.622
555	0.561
508	0.469
465	0.362
403	0.233
352	0.134

CURVE 45*, †, ‡

T	$\Delta L/L_0$
0	0.000 x 10^{-4}
1	0.000
2	0.001
3	0.004
4	0.010
5	0.022
6	0.043
7	0.077
8	0.128
9	0.203
10	0.308
12	0.644
14	1.211
15	1.608
16	2.100
18	3.436
20	5.328
22	7.881
24	11.18
25	13.15
26	15.34

CURVE 46*, †, ‡

T	$\Delta L/L_0$
2.383	0.000 x 10^{-4}
3.000	0.003
3.611	0.007
4.035	0.011
4.417	0.015
4.834	0.022
5.257	0.030
5.633	0.039
5.975	0.048
6.373	0.062
6.807	0.079
7.268	0.100
7.824	0.132
8.467	0.178
9.158	0.240
9.884	0.320
10.705	0.435
11.711	0.618
12.884	0.904
14.092	1.30
15.290	1.82
16.630	2.57
18.178	3.72
19.762	5.25
21.387	7.24
23.108	9.88
24.881	13.22
26.557	17.00
28.187	21.30
29.603	25.54

CURVE 47*, †, ‡

T	$\Delta L/L_0$
4.406	0.000 x 10^{-4}
4.822	0.005
5.250	0.012
5.625	0.020
5.966	0.028
6.354	0.039
6.773	0.053
7.229	0.072
7.785	0.010
8.428	0.141
9.115	0.196
9.834	0.269
10.660	0.376
11.686	0.551

CURVE 47 (cont.)*, †, ‡

T	$\Delta L/L_0$
12.869	0.825 x 10^{-4}
14.080	1.207
15.294	1.714
16.662	2.465
18.205	3.585
19.794	5.095
21.425	7.081
23.043	9.542
26.984	17.86
30.484	28.23
31.471	31.70
32.704	36.38

CURVE 48*, †, ‡

T	$\Delta L/L_0$
2.290	0.000 x 10^{-4}
3.013	0.002
3.586	0.004
4.043	0.007
4.435	0.011
4.849	0.016
5.251	0.022
5.625	0.029
5.973	0.037
6.361	0.048
6.778	0.061
7.239	0.080
7.793	1.070

CURVE 49*, †, ‡

T	$\Delta L/L_0$
2.428	0.000 x 10^{-4}
3.051	0.002
3.629	0.005
4.055	0.008
4.414	0.011
4.831	0.017
5.254	0.023
5.628	0.031
5.978	0.039
6.370	0.050
6.790	0.065
7.251	0.084
7.810	0.113
8.451	0.155
9.135	0.212
9.857	0.287

CURVE 49 (cont.)*, †, ‡

T	$\Delta L/L_0$
10.676	0.396 x 10^{-4}
11.691	0.574
12.868	0.854
14.069	1.242
15.268	1.752
16.657	2.530
18.248	3.720
19.846	5.290
21.477	7.338
23.112	9.893

CURVE 50*, †, ‡

T	$\Delta L/L_0$
2.473	0.000 x 10^{-4}
3.085	0.002
3.630	0.005
4.059	0.008
4.419	0.011
4.816	0.016
5.237	0.022
5.616	0.029
5.964	0.037
6.354	0.048
6.775	0.062
7.235	0.080
7.792	0.108
8.434	0.147
9.124	0.202
9.841	0.274
10.649	0.377
11.675	0.551
12.860	0.824
14.061	1.200
15.256	1.695
16.587	2.417
18.114	3.515
19.761	5.071
21.443	7.119
23.053	9.568
24.722	12.66
26.499	16.62
28.162	20.97

CURVE 51*

T	$\Delta L/L_0$
283	-0.018
294	0.000

*Not shown in figure.
†This curve is here reported using the first given temperature as reference temperature at which $\Delta L/L_0 = 0$.
‡Author's data for coefficient of thermal expansion have been integrated by TPRC to obtain $\Delta L/L_0$.

DATA TABLE 48. THERMAL LINEAR EXPANSION OF SILVER Ag (continued)

T	$\Delta L/L_0$
CURVE 51 (cont.)	
305	0.030
323	0.056
338	0.084
CURVE 52*, †	
40	-0.405
60	-0.391
80	-0.366
100	-0.335
140	-0.267
180	-0.206
CURVE 53*, †	
40	-0.400
60	-0.389
80	-0.363
100	-0.333
140	-0.260
180	-0.206
CURVE 54*, †	
40	-0.417
60	-0.398
80	-0.367
100	-0.336
140	-0.269
180	-0.206
CURVE 55	
80	-0.367
195	-0.186
298	0.009
CURVE 56	
293	0.000
373	0.128*
473	0.289
573	0.467
673	0.657
773	0.861
873	1.073
973	1.299
1067	1.518

*Not shown in figure.

†This curve is here reported using the first given temperature as reference temperature at which $\Delta L/L_0 = 0$.

DATA TABLE 48. COEFFICIENT OF THERMAL LINEAR EXPANSION OF SILVER Ag

[Temperature, T, K; Coefficient of Expansion, α, 10^{-6} K^{-1}]

T	α
CURVE 14‡	
30	3.20
40	5.80
50	8.21
60	10.1
70	11.5
80	12.7
90	13.7
100	14.5
110	15.1
120	15.6
130	16.0
140	16.4
150	16.7
160	17.0
170	17.2
180	17.4
190	17.6
200	17.8
210	18.0
220	18.1
230	18.3
240	18.4
250	18.6
260	18.8
270	18.9
CURVE 18‡	
98	14.61
108	15.30
112	15.41*
125	16.02
135	16.41
144	16.62
146	16.71
165	17.10
169	17.00
174	17.21
CURVE 24‡	
10	0.1
15	0.6
20	1.4
30	3.5
40	5.8*
CURVE 24 (cont.)‡	
50	8.0
70	11.7
100	14.4*
150	16.6
200	18.2
250	18.9
300	19.0
CURVE 27*, ‡	
300	18.8
400	19.6
500	20.5
600	21.4
700	22.3
800	23.2
900	24.3
1000	25.5
1100	27.2
1200	30.8
CURVE 28*, ‡	
5	0.0177
6	0.0255
7	0.0378
8	0.0563
9	0.0804
10	0.1114
11	0.1499
12	0.1978
13	0.2560
14	0.3250
15	0.4110
CURVE 36‡	
90	13.27
100	13.71
110	14.15
120	14.59
130	15.02
140	15.45
150	15.86
160	16.24
170	16.59
CURVE 36 (cont.)‡	
180	16.91
190	17.20
200	17.42
210	17.60
220	17.75
230	17.88
240	18.01*
250	18.14
260	18.26*
270	18.38*
280	18.50
290	18.62
300	18.73*
310	18.83*
320	18.93*
330	19.02*
340	19.09*
373	19.45*
423	19.63*
473	19.85*
523	20.13*
573	20.50*
623	21.00*
773	21.58*
823	22.17*
873	22.77*
923	23.38*
973	23.99*
1023	24.60*
1073	25.22*
1123	25.86*
1173	26.50*
1223	27.17*
1273	27.85*
CURVE 37‡	
90	14.12
102	14.17*
129	15.77
141	16.65*
153	17.48
166	17.43*
180	17.87
195	17.87
210	18.56
CURVE 37 (cont.)‡	
243	18.74
277	18.54
283	18.91*
293	19.08
298	18.98*
313	18.78*
328	18.96*
344	18.89*
357	18.98*
372	18.74*
401	18.67*
415	19.14*
430	19.14*
CURVE 38*, ‡	
293	19.08
324	19.59
329	20.13
339	19.56
455	20.00
528	19.59
540	21.12
CURVE 39‡	
266	19.06
293	19.08
298	18.96*
345	20.09*
379	19.83*
442	19.48*
463	19.54*
484	19.33*
CURVE 40*, ‡	
293	19.08
361	19.53
368	19.26
395	19.98
401	19.18
415	20.55
438	20.30
475	20.20
497	20.69
519	20.88
CURVE 41‡	
116	15.20
127	15.52
160	16.91*
172	17.46
189	17.46
203	17.62
216	18.40
224	17.74
245	18.31
258	18.18
293	19.08*
298	18.49*
316	18.54*
328	18.29*
342	18.45*
385	18.21*
406	19.01*
425	18.64*
447	18.34*
464	18.71*
CURVE 42*, ‡	
271	19.26
293	19.08
314	19.64
332	20.73
344	19.25
350	20.09
373	19.71
396	19.08
410	20.06
421	19.48
505	19.54
CURVE 45*, ‡	
0	0.000
1	0.00034
2	0.0013
3	0.0038
4	0.00831
5	0.0156
6	0.0264
7	0.0415
8	0.0616
9	0.0880
CURVE 45 (cont.)*, ‡	
10	0.121
12	0.215
14	0.352
15	0.441
16	0.544
18	0.792
20	1.10
22	1.45
24	1.85
25	2.07
26	2.31
CURVE 46*, ‡	
2.383	0.0031
3.000	0.0053
3.611	0.0082
4.035	0.0109
4.417	0.0131
4.834	0.0178
5.257	0.0214
5.633	0.0256
5.975	0.0299
6.373	0.0361
6.807	0.0428
7.268	0.0514
7.824	0.0632
8.467	0.0797
9.158	0.0988
9.884	0.1225
10.705	0.1571
11.711	0.2059
12.884	0.2820
14.092	0.3757
15.290	0.4873
16.630	0.6381
18.178	0.8445
19.762	1.083
21.387	1.367
23.108	1.698
24.881	2.072
26.557	2.443
28.187	2.835
29.603	3.153

* Not shown in figure.

‡ Author's data for coefficient of thermal expansion have been integrated by TPRC to obtain $\Delta L/L_0$.

DATA TABLE 48. COEFFICIENT OF THERMAL LINEAR EXPANSION OF SILVER Ag (continued)

T	α
CURVE 47*,	
4.406	0.011
4.822	0.014
5.250	0.018
5.625	0.022
5.966	0.026
6.354	0.031
6.773	0.037
7.229	0.045
7.785	0.056
8.428	0.072
9.115	0.090
9.834	0.113
10.660	0.145
11.686	0.195
12.869	0.269
14.080	0.362
15.294	0.473
16.662	0.625
18.205	0.828
19.794	1.07
21.425	1.36
23.043	1.68
26.984	2.54
30.484	3.39
31.471	3.64
32.704	3.95
CURVE 48*, ‡	
2.290	0.0014
3.013	0.0034
3.586	0.0057
4.043	0.0079
4.435	0.0103
5.251	0.0169
5.625	0.0208
5.973	0.0249
6.361	0.0298
6.778	0.0360
7.239	0.0438
7.793	0.0547
CURVE 49*, ‡	
2.428	0.0021
3.051	0.0039
3.629	0.0062

T	α
CURVE 49 (cont.)*, ‡	
4.055	0.0088
4.414	0.0107
4.831	0.0141
5.254	0.0178
5.628	0.0219
5.978	0.0254
6.370	0.0315
6.790	0.0384
7.251	0.0455
7.810	0.0576
8.451	0.0730
9.135	0.0924
9.857	0.1164
10.676	0.1504
11.691	0.1993
12.868	0.2768
14.069	0.3696
15.268	0.4812
16.657	0.6392
18.248	0.8560
19.846	1.109
21.477	1.403
23.112	1.722
CURVE 50*, ‡	
2.473	0.0022
3.085	0.0040
3.630	0.0061
4.059	0.0082
4.419	0.0105
4.816	0.0132
5.237	0.0169
5.616	0.0208
5.964	0.0249
6.354	0.0299
6.775	0.0358
7.235	0.0440
7.792	0.0548
8.434	0.0694
9.124	0.0883
9.841	0.1116
10.649	0.1448
11.675	0.1935
12.860	0.2674
14.061	0.3594
15.256	0.4682

T	α
CURVE 50 (cont.)*, ‡	
16.587	0.6179
18.114	0.8194
19.761	1.070
21.443	1.366
23.053	1.676
24.722	2.032
26.499	2.421
28.162	2.281
CURVE 52	
50	7.65
70	12.16
90	15.03
120	16.98
160	16.99
CURVE 56*	
373	16.03
473	17.02
573	18.07
673	19.12
773	20.12
873	21.19
973	22.21
1073	23.24

*Not shown in figure.
‡Author's data for coefficient of thermal expansion have been integrated by TPRC to obtain $\Delta L/L_0$.

FIGURE AND TABLE NO. 49R. RECOMMENDED VALUES FOR THERMAL LINEAR EXPANSION OF SODIUM Na

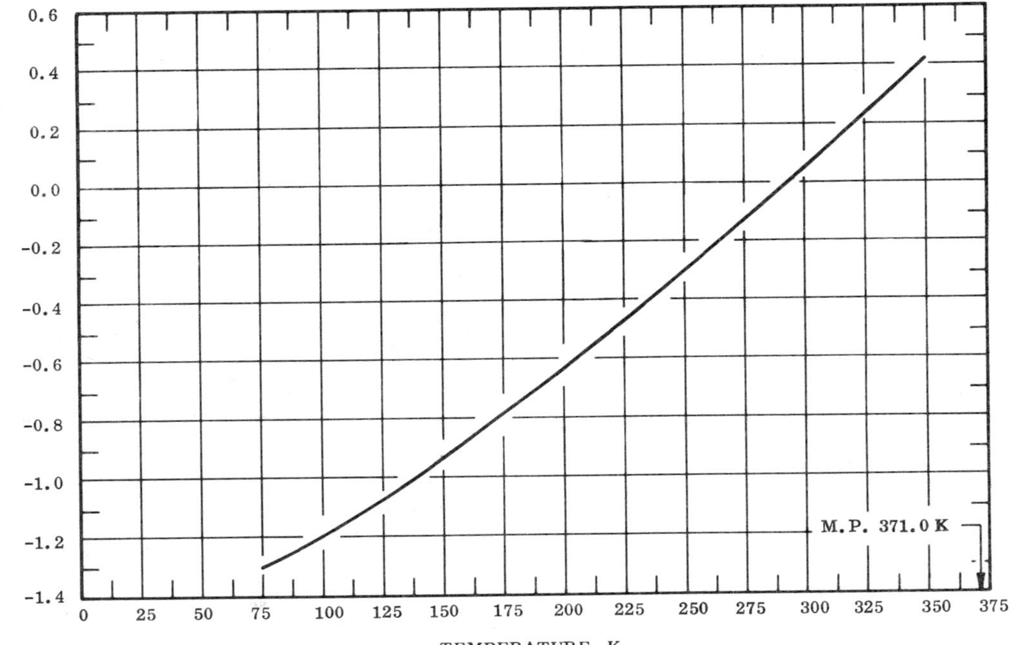

RECOMMENDED VALUES

[Temperature, T, K; Linear Expansion, $\Delta L/L_0$, %; α, K^{-1}]

T	$\Delta L/L_0$	$\alpha \times 10^6$
75	-1.308	36
100	-1.200	45
200	-0.636	65
293	0.000	71
350	0.417	74

REMARKS

The tabulated values are considered accurate to within ± 3% over the entire temperature range. These values can be represented approximately by the following equation:

$$\Delta L/L_0 = -1.560 + 2.307 \times 10^{-3}\ T + 1.419 \times 10^{-5}\ T^2 - 1.328 \times 10^{-8}\ T^3$$

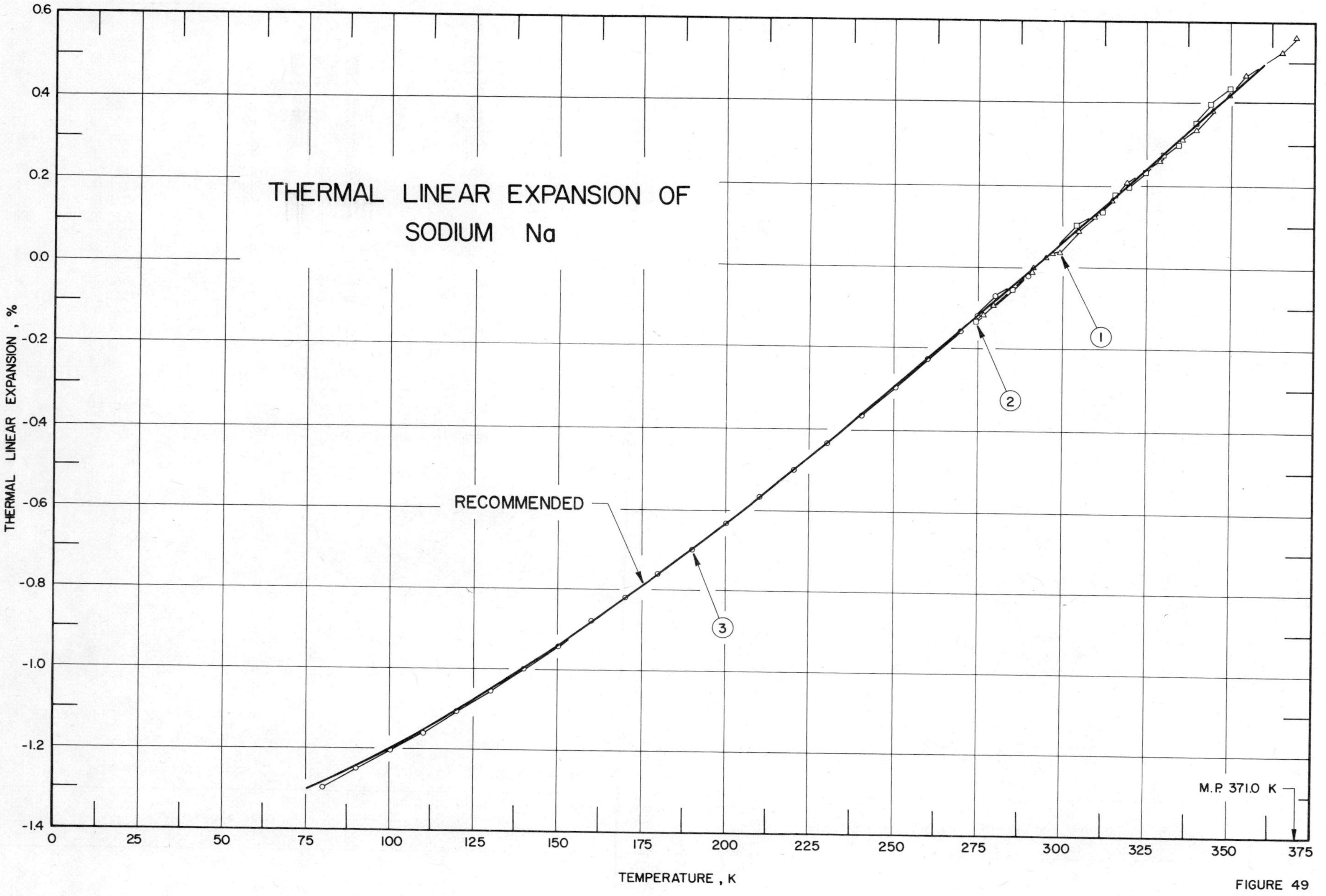
THERMAL LINEAR EXPANSION OF
SODIUM Na
RECOMMENDED
1
2
3
M.P. 371.0 K
0.6
0.4
0.2
0.0
-0.2
-0.4
-0.6
-0.8
-1.0
-1.2
-1.4
THERMAL LINEAR EXPANSION, %
0 25 50 75 100 125 150 175 200 225 250 275 300 325 350 375
TEMPERATURE, K

FIGURE 49

SPECIFICATION TABLE 49. THERMAL LINEAR EXPANSION OF SODIUM Na

Cur. No.	Ref. No.	Author(s)	Year	Method Used	Temp. Range, K	Name and Specimen Designation	Composition (weight percent), Specifications, and Remarks
1	308	Feder, R. and Charbnau	1966	X	276-370		Crystal specimens grown by Bridgman technique in form of rod, cut into wafers 0.5 cm thick; change in lattice parameter measured; zero-point correction of -0.137% determined by graphical interpolation.
2	308	Feder, R. and Charbnau	1966	L	274-369		Crystal specimen grown by Bridgman technique in form of cylinder, length maintained at 5.0 cm; change in overall length measured; zero-point correction of -0.139% determined by graphical interpolation.
3	309	Siegel, S. and Quimby, S.I.	1938	L	80-290		Three single crystal specimens 4 cm x 0.47 cm dia grown from triply distilled Mallinchrodt sodium; data given as mean of observations on three crystals; zero-point correction of -0.137% determined by graphical interpolation.

DATA TABLE 49. THERMAL LINEAR EXPANSION OF SODIUM Na

[Temperature, T, K; Linear Expansion, $\Delta L/L_0$, %]

T	$\Delta L/L_0$	T	$\Delta L/L_0$	T	$\Delta L/L_0$	T	$\Delta L/L_0$	T	$\Delta L/L_0$	T	$\Delta L/L_0$
CURVE 1		CURVE 1 (cont.)		CURVE 2 (cont.)		CURVE 2 (cont.)		CURVE 2 (cont.)		CURVE 3	
276	-0.117	338	0.330*	280	-0.087*	325	0.232*	353	0.452	80	-1.292
279	-0.093	340	0.339	285	-0.053	327	0.252*	354	0.466	90	-1.248
283	-0.067	342	0.360*	287	-0.044*	330	0.272	356	0.474	100	-1.208
291	-0.011	345	0.383	290	-0.020*	332	0.291*	358	0.499	110	-1.159
292	0.000	346	0.387*	292	-0.007*	332	0.291*	358	0.494	120	-1.107
295	0.023	350	0.421	295	0.022*	333	0.292*	358	0.493	130	-1.054
297	0.033	350	0.422*	297	0.029*	334	0.301*	360	0.507	140	-0.997
298	0.036	350	0.424*	302	0.065*	335	0.310	361	0.516	150	-0.938
299	0.048*	354	0.454*	307	0.102*	335	0.311*	362	0.527	160	-0.879
302	0.072*	357	0.470	307	0.103	335	0.313*	363	0.539	170	-0.819
305	0.091	362	0.513*	309	0.121*	338	0.331*	365	0.549	180	-0.758
310	0.127	363	0.517*	312	0.137	340	0.352	365	0.550	190	-0.694
313	0.146*	364	0.528	312	0.139*	343	0.392*	365	0.550	200	-0.630
315	0.163	368	0.560*	312	0.139*	346	0.392*	365	0.553	210	-0.565
321	0.205	370	0.571	315	0.159*	347	0.401	366	0.560	220	-0.498
321	0.207*			317	0.176	348	0.410*	367	0.571	230	-0.431
322	0.210*	CURVE 2		320	0.196	348	0.413*	367	0.571	240	-0.363
328	0.259*			321	0.205*	348	0.410*	368	0.575	250	-0.295
329	0.260	274	-0.133	322	0.215*	350	0.432	369	0.583	260	-0.227
331	0.279*	274	-0.131*	325	0.231	351	0.432			270	-0.159
336	0.317	275	-0.124	325	0.234*	351	0.434			280	-0.090
										290	-0.021

* Not shown in figure.

FIGURE AND TABLE NO. 50R. PROVISIONAL VALUES FOR THERMAL LINEAR EXPANSION OF STRONTIUM Sr

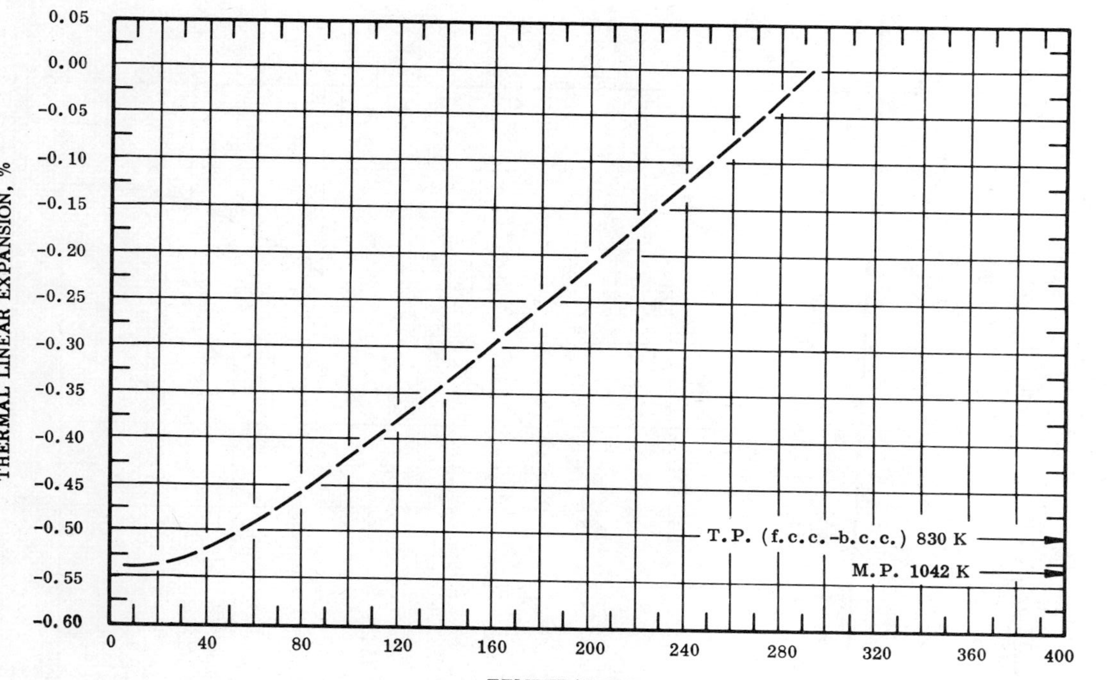

PROVISIONAL VALUES

[Temperature, T, K; Linear Expansion, $\Delta L/L_0$, %; α, K^{-1}]

T	$\Delta L/L_0$	$\alpha \times 10^6$
5	-0.536	-0.04
25	-0.533	6.0
50	-0.505	14.5
100	-0.418	19.3
200	-0.208	22.1
293	0.000	22.5

REMARKS

The tabulated values are considered accurate to within ±7% over the entire temperature range. These values can be represented approximately by the following equation:

$$\Delta L/L_0 = -0.561 + 8.831 \times 10^{-4}\, T + 6.424 \times 10^{-6}\, T^2 - 9.957 \times 10^{-9}\, T^3$$

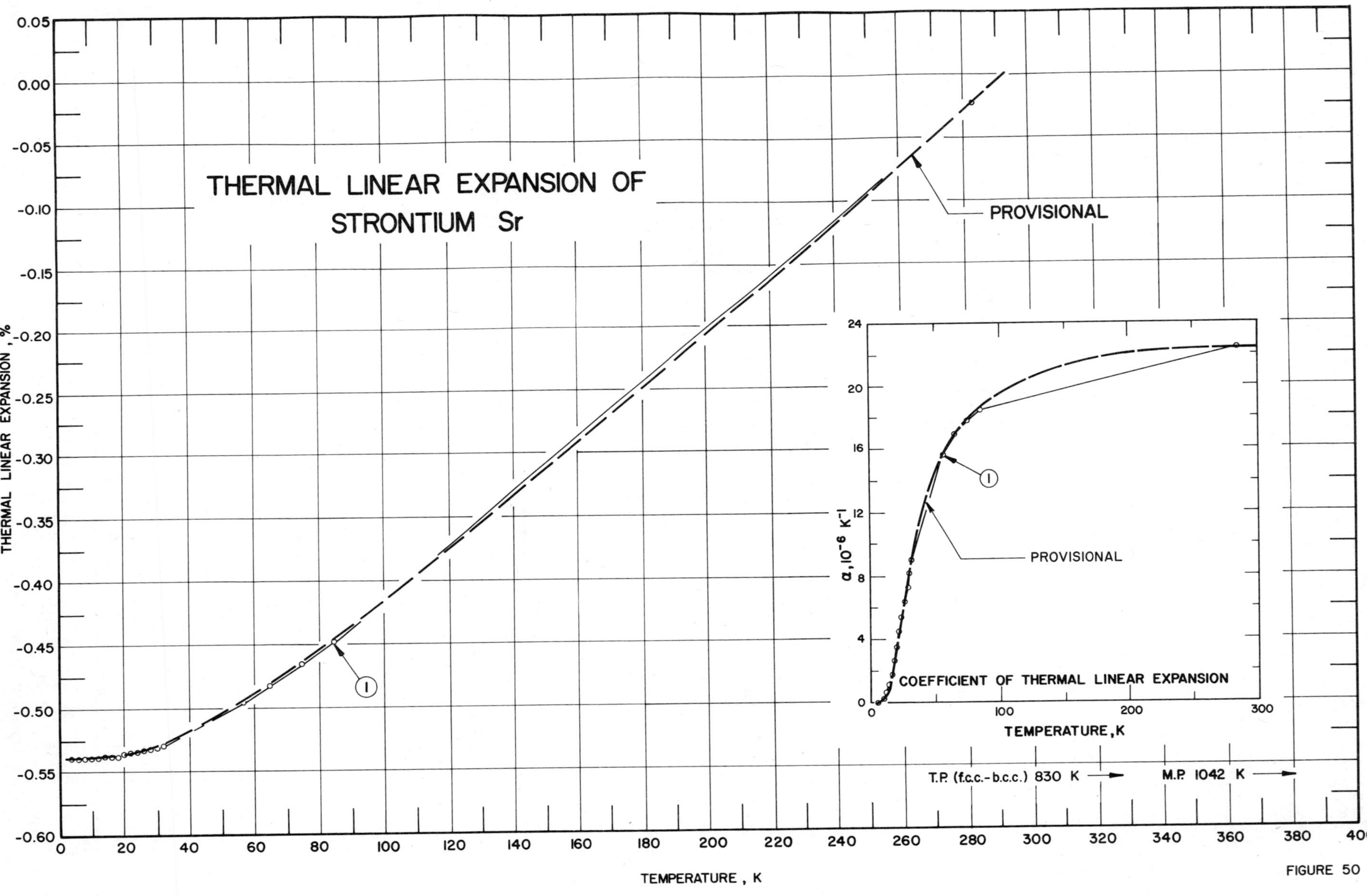
THERMAL LINEAR EXPANSION OF
STRONTIUM Sr
PROVISIONAL
1
THERMAL LINEAR EXPANSION, %
TEMPERATURE, K
COEFFICIENT OF THERMAL LINEAR EXPANSION
α, 10⁻⁶ K⁻¹
TEMPERATURE, K
PROVISIONAL
1
T.P. (f.c.c.-b.c.c.) 830 K →
M.P. 1042 K →

FIGURE 50

SPECIFICATION TABLE 50. THERMAL LINEAR EXPANSION OF STRONTIUM Sr

Cur. No.	Ref. No.	Author(s)	Year	Method Used	Temp. Range, K	Name and Specimen Designation	Composition (weight percent), Specifications, and Remarks
1	726	White, G.K.	1972	E	3-85, 283		Specimen from A.D. Mackay, New York; spectrographic analysis by Australian Mineral Development Laboratories shows about 1.0 Ca, 0.3 Al, 0.2 Si, 0.1 Ba, and 0.05 Fe.

DATA TABLE 50. THERMAL LINEAR EXPANSION OF STRONTIUM Sr

[Temperature, T, K; Linear Expansion, $\Delta L/L_0$, %]

T	$\Delta L/L_0$	T	$\Delta L/L_0$	T	$\Delta L/L_0$	T	$\Delta L/L_0$	T	$\Delta L/L_0$
CURVE 1‡		CURVE 1 (cont.)‡		CURVE 1 (cont.)‡		CURVE 1 (cont.)‡		CURVE 1 (cont.)‡	
3	-0.537*	10	-0.537	20	-0.535	30	-0.529	85	-0.447
4	-0.537	12	-0.537	22	-0.534	32	-0.527	283	-0.022
5	-0.537*	14	-0.536	24	-0.533	57	-0.495		
6	-0.537	16	-0.536	26	-0.532	65	-0.482		
8	-0.537	18	-0.536	28	-0.531	75	-0.465		

DATA TABLE 50. COEFFICIENT OF THERMAL LINEAR EXPANSION OF STRONTIUM Sr

[Temperature, T, K; Coefficient of Expansion, α, 10^{-6} K^{-1}]

T	α	T	α	T	α	T	α	T	α
CURVE 1‡		CURVE 1 (cont.)‡		CURVE 1 (cont.)‡		CURVE 1 (cont.)‡		CURVE 1 (cont.)‡	
3	-0.034*	10	0.27	20	3.50	30	8.20	85	18.45
4	-0.039*	12	0.64	22	4.45	32	9.00	283	22.45
5	-0.037	14	1.15	24	5.40	57	15.60		
6	-0.020*	16	1.80	26	6.40	65	16.90		
8	0.062*	18	2.60	28	7.30	75	17.80		

*Not shown in figure.
‡Author's data for coefficient of thermal expansion have been integrated by TPRC to obtain $\Delta L/L_0$.

FIGURE AND TABLE NO. 51R. RECOMMENDED VALUES FOR THERMAL LINEAR EXPANSION OF TANTALUM Ta

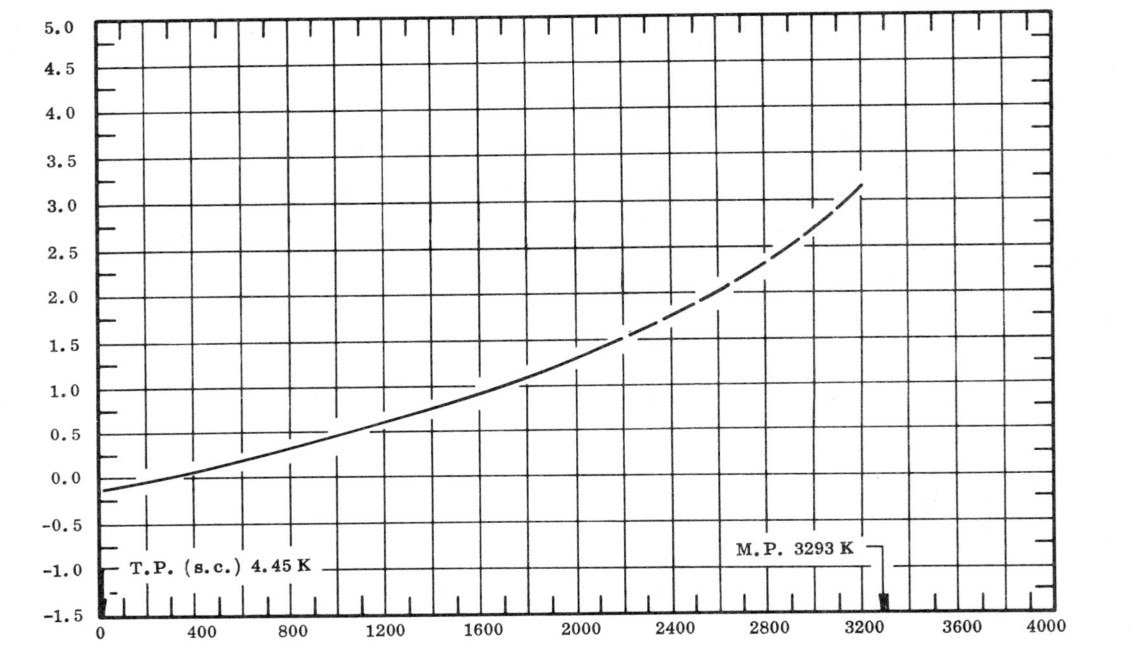

RECOMMENDED VALUES

[Temperature, T, K; Linear Expansion, $\Delta L/L_0$, %; α, K^{-1}]

T	$\Delta L/L_0$	α
5	-0.136	0.01
25	-0.135	0.61
50	-0.131	2.6
100	-0.112	4.8
200	-0.057	6.0
293	0.000	6.3
400	0.069	6.6
500	0.136	6.8
600	0.205	6.9
700	0.273	7.1
800	0.343	7.2
900	0.413	7.2
1000	0.485	7.3
1200	0.633	7.4
1400	0.784	7.7
1600	0.941	8.0
1800	1.107	8.6
2000	1.287	9.4
2200	1.488	10.7
2400	1.718†	12.3
2600	1.985†	14.4
2800	2.300†	17.1
3000	2.675†	20.4
3200	3.126†	24.4

† Provisional values.

REMARKS

The tabulated values are considered accurate to within ±3% at temperatures below 1100 K, ±5% below 2100 K, and ±10% above. These values can be represented approximately by the following equations:

$$\Delta L/L_0 = 6.308 \times 10^{-4} (T-293) + 9.706 \times 10^{-8} (T-293)^2 - 2.545 \times 10^{-11} (T-293)^3 \quad (293 < T < 1260)$$

$$\Delta L/L_0 = 0.678 + 7.471 \times 10^{-4} (T-1260) + 2.320 \times 10^{-8} (T-1260)^2 + 1.061 \times 10^{-10} (T-1260)^3 \quad (1260 < T < 2230)$$

$$\Delta L/L_0 = 1.522 + 1.092 \times 10^{-3} (T-2230) + 3.320 \times 10^{-7} (T-2230)^2 + 2.547 \times 10^{-10} (T-2230)^3 \quad (2230 < T < 3200)$$

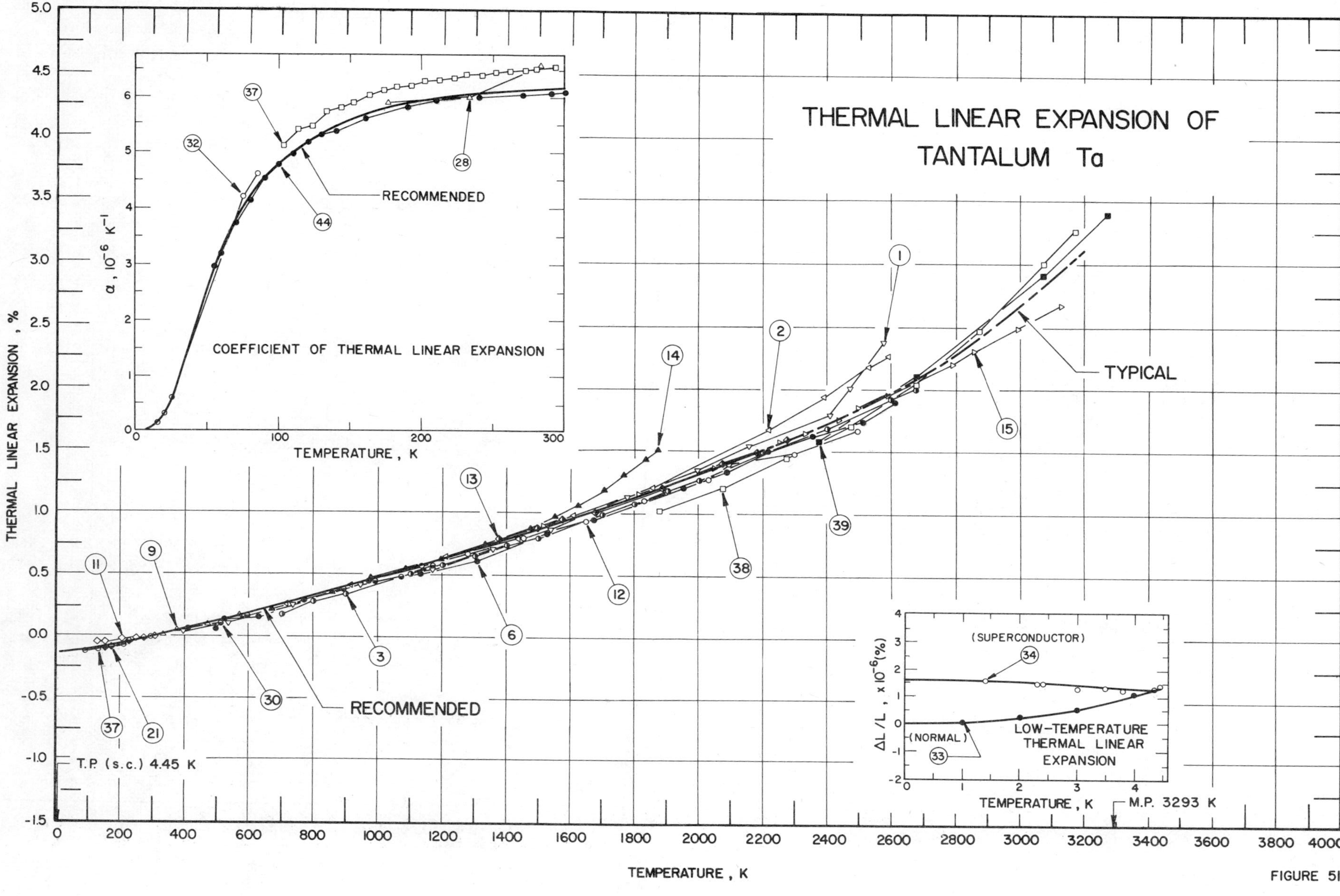
THERMAL LINEAR EXPANSION OF
TANTALUM Ta
THERMAL LINEAR EXPANSION, %
TEMPERATURE, K
TYPICAL
RECOMMENDED
1
2
3
6
9
11
12
13
14
15
21
30
37
38
39
T.P. (s.c.) 4.45 K
M.P. 3293 K
COEFFICIENT OF THERMAL LINEAR EXPANSION
α, 10⁻⁶ K⁻¹
TEMPERATURE, K
RECOMMENDED
28
32
37
44
LOW-TEMPERATURE
THERMAL LINEAR
EXPANSION
(SUPERCONDUCTOR)
(NORMAL)
34
33
ΔL/L, x 10⁻⁶(%)
TEMPERATURE, K

FIGURE 51

SPECIFICATION TABLE 51. THERMAL LINEAR EXPANSION OF TANTALUM Ta

Cur. No.	Ref. No.	Author(s)	Year	Method Used	Temp. Range, K	Name and Specimen Designation	Composition (weight percent), Specifications, and Remarks
1	310	Conway, J.B., Finel, R.M., Jr., and Losekamp, A.C.	1965	T	293-2584		Arc-cast in helium; 20 mil sheet specimen of U-shape about 0.3 in. wide, 0.3 in. deep, and 2.5 in. long; expansion measured in helium atm with increasing temperature.
2	310	Conway, J.B., et al.	1965	T	293-2580		The above specimen; second state.
3	310	Conway, J.B., et al.	1965	T	293-2673		The above specimen; specimen holding fixture was surrounded by several layers of tantalum to act as a "getter" for the cover-gas impurities to minimize permanent deformation.
4*	210	Conway, J.B., and Losekamp, A.C.	1966	T	552-2595		99.87$^+$ Ta, <0.5 W; arc-cast sheet 20 mil thick; expansion measured in helium atm.
5*	210	Conway, J.B., and Losekamp, A.C.	1966	T	533-2653		The above specimen; expansion measured on the second test.
6*	210	Conway, J.B., and Losekamp, A.C.	1966	T	497-2615		The sbove specimen; expansion measured on the third test.
7*	596	Conway, J.B., and Flagella, P.N.	1967	T	508-2588		No details given.
8*	596	Conway, J.B., and Flagella, P.N.	1967	T	383-2493		The above specimen.
9*	596	Conway, J.B., and Flagella, P.N.	1967	T	493-2643		The above specimen.
10*	311	V'yugov, P.N. and Gumenyuk, V.S.	1965	L	453-544		Specimen 2 mm diameter, between 200 and 240 mm long; one end attached to the water-cooled upper lead and the other dipped into molten tin or lead in a recess in the water-cooled lower electrode; measured at a pressure not exceeding 10^{-6} mm Hg; reported error 0.3 to 0.5%.
11	246	Smirnov, Yu.M. and Finkel, V.A.	1966	X	127-396		99.99 Ta; polycrystalline specimens in form of strips and plates 0.1 to 0.3 mm thick; data average of several specimens; lattice parameter reported at 294 K is 3.30188 Å; 3.30187 Å at 293 K determined by graphical interpolation.
12	215	Edwards, J.W., Speiser, R., and Johnston, H.L.	1951	X	291-2495		99.9 Ta; specimen obtained from Fansteel Metallurgical Corp; lattice parameter at 293 K is 3.3029 Å determined by graphical interpolation.
13	213	Amonenko, V.M., V'yngov, P.N., and Gumenyuk, V.S.	1964	L	293-2273		Commercial rod specimen; annealed in tungsten furnace 1.5-2.0 hr above 2270 K before measurement started under vacuum.
14	138	Fieldhouse, I.B., et al.	1956	T	343-1871		Sintered specimen; data average of three separate runs.
15	221	Rasor, N.S. and McClelland, J.D.	1957	T	302-3125		After heating in experiments, impurities 0.015 C, 0.013 Si; specimen prepared by pressing and sintering powdered metal, prepared by Fansteel Metallurgical Corp; expansion measured with increasing temperature on first test.
16*	221	Rasor, N.S. and McClelland, J.D.	1957	T	2992-1135		The above specimen; expansion measured with decreasing temperature.
17*	221	Rasor, N.S. and McClelland, J.D.	1957	T	1906-3065		The above specimen; expansion measured with increasing temperature on second test; data include permanent elongation of unspecified magnitude resulting from first test.
18*	221	Rasor, N.S. and McClelland, J.D.	1957	T	1918-1324		The above specimen; expansion measured with decreasing temperature.
19*	312	Hidnert, P.	1929	T	293-573		99.9 Ta, 0.019 Mg, 0.01 Fe, <0.005 C, <0.001 Mo; after last anneal or heat treatment sample worked in vacuum; expansion measured on first test.
20*	312	Hidnert, P.	1929	T	293-573		The above specimen; expansion measured on second test.
21	312	Hidnert, P.	1929	T	148-573		The above specimen; expansion measured on third test.
22*	312	Hidnert, P.	1929	T	293-773		The above specimen; expansion measured on fourth test.
23*	312	Hidnert, P.	1929	L	83-293		The above specimen; expansion measured on eight test.

* Not shown in figure.

SPECIFICATION TABLE 51. THERMAL LINEAR EXPANSION OF TANTALUM Ta (continued)

Cur. No.	Ref. No.	Author(s)	Year	Method Used	Temp. Range, K	Name and Specimen Designation	Composition (weight percent), Specifications, and Remarks
24*	312	Hidnert, P.	1929	L	83-293		The above specimen; expansion measured on ninth test.
25*	312	Hidnert, P.	1929	T	293-773		99.9 Ta, 0.014 Mg, <0.005 C, <0.001 Mo; metal in completely annealed state; heated to melting point in a vacuum furnace; expansion measured on first test.
26*	312	Hidnert, P.	1929	T	293-573		The above specimen; expansion measured on third test.
27*	312	Hidnert, P.	1929	T	293-573		The above specimen; expansion measured on fourth test.
28*	312	Hidnert, P.	1929	L	83-293		Similar to the above specimen; cut from same bar; expansion measured on second test.
29*	194	Olsen, J.L. and Rohrer, H.	1957	N	1.7-4.4		Rod specimen of spectroscopically pure material tested as received from Johnson, Matthey and Co.; expansion observed below normal-superconducting transitions; data result from 10-15 separate trials; data reported using $\Delta L/L_0$ defined as $(L_s-L_n)/L_n$ where L_s is length in superconducting state and L_n in normal state; this curve is here reported using the first given temperature as reference temperature at which $\Delta L/L_0 = 0$.
30*	313	Makin, S.M., Standring, S., and Hunter, P.M.	1953	I	407-1171		Slightly cold-worked specimen; expansion measured with increasing temperature.
31*	313	Makin, S.M., et al.	1953	I	1067-473		The above specimen; expansion measured with decreasing temperature.
32*	251	White, G.K.	1962	E	5-85		99.8$^+$ Ta from Wah Chang Corp; analysis showed <0.05 Nb, <0.03 W, <0.015 O_2 and < 0.01 any other impurity; machined until ends parallel and rod 5.09 cm long; this curve is here reported using the first given temperature as reference temperature at which $\Delta L/L_0 = 0$.
33*	251	White, G.K.	1962	E	1.3-11.1		Similar to the above specimen; measurements in the magnetic field; this curve is referred to 0 K, at which $\Delta L/L_0 = 0$.
34*	251	White, G.K.	1962	E	1.4-4.4		Similar to the above specimen; this curve referred to the normal state value of curve 33 at the critical temperature, T_c = 4.44 K.
35*	276, 714	Andres, K.	1963	O	4.44-12.6		High purity 7.4 cm long specimen; measurements in the magnetic field; this curve referred to 0 K, at which $\Delta L/L_0 = 0$.
36*	276, 714	Andres, K.	1963	O	1.6-8.5		The above specimen; this curve referred to the normal state value of curve 35 at the critical temperature, T_c = 4.44 K.
37*	75	Nix, F.C., and MacNair, D.	1942	I	92-301		99.9 Ta, 0.01 Fe, 0.003 Ca from Fansteel Metallurgical corporation; zero-point correction of -0.013% determined by graphical interpolation; data on the coefficients of thermal linear expansion also given.
38	314	Saldinger, I.L., and Glasier, L.F.	1959	T	1873-3173		Specimen from Fansteel Metallurgical Corporation containing 0.02 Si, 0.010 N_2 and O_2, 0.005 Fe, and 0.05 other impurities; data smoothed by author from several tests.
39	314	Saldinger, I.L., and Glasier, L.F.	1959	T	2373-3273		Similar to the above specimen except carburized by the American Metal Products.
40*	89	Andres, K.	1961	O	2.5-5.5		No details given; this curve is here reported using the first temperature as reference temperature at which $\Delta L/L_0 = 0$.
41*	747	Lisovskii, Yu.A.	1972	E	55-300		99.97 pure polycrystalline specimen.

* Not shown in figure.

DATA TABLE 51. THERMAL LINEAR EXPANSION OF TANTALUM Ta

[Temperature, T, K; Linear Expansion, $\Delta L/L_0$, %]

T	$\Delta L/L_0$
CURVE 1	
293	0.00
529	0.13
736	0.26
941	0.42
1164	0.55
1365	0.72
1539	0.87
1773	1.13
1956	1.36
2153	1.55
2306	1.81
2467	2.02
2577	2.37
CURVE 2	
293	0.00*
588	0.18*
713	0.27
919	0.42
1205	0.65
1434	0.81
1612	0.98
1843	1.22
2041	1.50
2213	1.68
2384	1.95
2522	2.18
2588	2.27
CURVE 3	
293	0.00*
650	0.19
800	0.29
900	0.36
1000	0.44*
1100	0.51
1200	0.58
1300	0.66
1400	0.74
1500	0.81
1600	0.91
1700	0.98
1800	1.09
1900	1.18
2000	1.27
CURVE 3 (cont.)	
2200	1.48
2400	1.69
2600	1.92
2673	2.00
CURVE 4*	
552	0.18
708	0.31
897	0.44
1069	0.51
1295	0.68
1559	0.91
1739	1.02
1982	1.26
2171	1.42
2341	1.59
2499	1.80
2595	1.93
CURVE 5*	
533	0.14
609	0.18
737	0.29
875	0.40
1173	0.57
1430	0.73
1644	0.92
1864	1.11
2102	1.35
2260	1.49
2444	1.66
2588	1.87
2653	1.92
CURVE 6*	
497	0.07
520	0.15
624	0.17
774	0.30
972	0.46
1136	0.51
1309	0.62
1524	0.84
1672	0.95
CURVE 6 (cont.)*	
1836	1.11
1959	1.22
2083	1.33
2216	1.50
2360	1.63
2510	1.73
2615	1.90
CURVE 7*	
510	0.214
706	0.335
895	0.518
1066	0.574
1277	0.716
1557	0.909
1726	1.064
1964	1.288
2166	1.446
2347	1.603
2493	1.816
2588	1.916
CURVE 8*	
383	0.098
608	0.251
778	0.313
953	0.496
1143	0.548
1273	0.684
1508	0.858
1638	0.981
1843	1.132
1948	1.209
2088	1.335
2213	1.531
2358	1.665
2493	1.780
CURVE 9*	
493	0.191
573	0.241
758	0.319
873	0.470
1163	0.606
CURVE 9 (cont.)*	
1428	0.743
1638	0.930
1863	1.132
2098	1.335
2238	1.540
2443	1.654
2578	1.867
2623	1.942
2643	1.942
CURVE 10*	
453	0.138
454	0.142
460	0.147
471	0.157
478	0.164
488	0.172
490	0.175
492	0.180
503	0.188
519	0.205
531	0.218
542	0.226
544	0.234
CURVE 11	
127	-0.046
127	-0.045*
130	-0.044*
131	-0.045*
133	-0.043*
137	-0.042*
138	-0.039*
150	-0.039
166	-0.035*
180	-0.031*
192	-0.029*
194	-0.028*
199	-0.027*
202	-0.027
209	-0.023*
210	-0.024*
215	-0.021*
221	-0.021*
224	-0.019*
CURVE 11 (cont.)	
229	-0.018*
233	-0.018*
237	-0.016*
242	-0.015*
246	-0.014*
251	-0.012
264	-0.008*
268	-0.006*
275	-0.005*
281	-0.005*
284	-0.002*
289	-0.001*
294	0.0002*
297	0.001*
303	0.004
308	0.004*
317	0.009*
326	0.012*
332	0.013*
339	0.016*
343	0.016*
344	0.015*
348	0.018*
351	0.020
355	0.021*
360	0.023*
361	0.024*
365	0.024*
370	0.026*
375	0.030*
381	0.030*
388	0.031*
389	0.034*
395	0.033*
396	0.035
CURVE 12	
293	0.000
1135	0.572
1281	0.681
1454	0.805
1641	0.948
1826	1.11
2030	1.28
2091	1.40
2299	1.48
2495	1.67
CURVE 13	
293	0.00*
1176	0.64*
1280	0.71*
1373	0.80
1495	0.88
1576	0.96
1674	1.05
1776	1.14*
1885	1.23
1968	1.32*
2073	1.41
2181	1.50
2273	1.60
CURVE 14	
343	0.027*
375	0.035*
478	0.105*
597	0.195*
669	0.245
860	0.375
914	0.425*
980	0.480
1085	0.555
1102	0.565*
1198	0.635
1250	0.685*
1335	0.765
1361	0.795*
1428	0.863*
1446	0.875*
1472	0.890
1525	0.940*
1555	0.990
1595	1.025*
1623	1.070
1653	1.100*
1703	1.200
1729	1.245*
1765	1.325
1808	1.390*
1833	1.440
1871	1.525

* Not shown in figure.

DATA TABLE 51. THERMAL LINEAR EXPANSION OF TANTALUM Ta (continued)

T	$\Delta L/L_0$
CURVE 15	
302	0.000*
1516	0.904
1809	1.163
2044	1.384
2254	1.587
2329	1.665
2434	1.776
2495	1.861
2589	1.960
2787	2.208
2855	2.321
2990	2.499
3125	2.672
CURVE 16*	
2992	2.565
2794	2.395
2648	2.201
2416	1.988
2151	1.711
1835	1.438*
1815	1.424*
1782	1.379
1525	1.122
1135	0.820
CURVE 17*	
1906	1.450
2297	1.871
2700	2.298
2851	2.488
2988	2.650
3065	2.757
CURVE 18*	
1918	1.571
1687	1.341
1324	1.046
CURVE 19*	
293	0.000
333	0.027
373	0.054
473	0.122
573	0.190
CURVE 20*	
293	0.000
333	0.026
373	0.053
473	0.119
573	0.188
CURVE 21	
148	-0.090
173	-0.076
223	-0.046
273	-0.014
293	0.000*
333	0.026*
373	0.052*
473	0.117*
573	0.185*
CURVE 22*	
293	0.000
373	0.050
473	0.113
573	0.179
673	0.243
773	0.307
CURVE 23*	
83	-0.145
273	-0.014
293	0.000
CURVE 24*	
83	-0.139
273	-0.014
293	0.000
CURVE 25*	
293	0.000
373	0.053
473	0.121
573	0.185
673	0.247
773	0.317
CURVE 26*	
293	0.000
333	0.027
373	0.054
473	0.122
573	0.190
CURVE 27*	
293	0.000
333	0.026
373	0.053
473	0.120
573	0.190
CURVE 28	
83	-0.125
193	-0.061
273	-0.013
293	0.000
CURVE 29*, †	
1.70	0.000 x 10^{-6}
1.94	-2.640
2.15	0.05
2.47	1.93
2.73	3.67
2.79	4.11
2.99	9.61
3.03	7.10
3.05	7.30
3.25	7.06
3.45	8.23
3.50	7.84
3.61	8.16
3.70	7.82
3.83	8.06
3.89	8.30
3.91	8.43
3.99	8.19
4.04	8.06
4.14	8.71
4.15	8.51
4.15	8.34
4.15	8.08
4.17	8.26
4.37	7.81
CURVE 30*	
407	0.065
514	0.126
589	0.180
773	0.304
882	0.382
988	0.452
1076	0.496
1171	0.577
CURVE 31*	
1067	0.504
980	0.449
740	0.283
620	0.200
473	0.102
CURVE 32*, †, ‡	
5	0.000 x 10^{-3}
6	0.001
7	0.002
8	0.005
9	0.008
10	0.012
12	0.025
14	0.044
15	0.057
16	0.072
18	0.11
20	0.16
22	0.24
24	0.34
25	0.39
75	12.42
85	16.82
CURVE 33*	
1	0.06 x 10^{-6}
2	0.23
3	0.59
4	1.10
4.39	1.38
CURVE 34*	
1.4	1.58 x 10^{-6}
2.3	1.49
CURVE 34 (cont.)*	
2.4	1.47 x 10^{-6}
3.0	1.37
3.1	1.29
3.5	1.33
3.8	1.21
4.39	1.38
CURVE 35*	
4.44	1.26 x 10^{-6}
5.1	1.94
5.2	1.97
5.3	2.21
5.4	2.26
5.5	2.30
5.9	2.85
6.4	3.61
6.5	3.52
6.9	4.31
7.4	5.23
8.1	6.69
8.5	7.95
8.9	9.16
9.4	10.41
9.4	10.69
9.9	12.23
10.4	14.56
11.0	17.82
11.8	22.62
12.6	28.67
CURVE 36*	
1.6	1.69 x 10^{-6}
2.3	1.60
3.2	1.51
3.7	1.45
3.7	1.40
4.0	1.51
4.0	1.40
4.4	1.39
4.44	1.26
5.1	1.88
5.2	2.00
5.4	2.26
5.4	2.29
5.9	2.76
5.9	2.82
6.4	3.41
CURVE 36 (cont.)*	
6.4	3.58 x 10^{-6}
6.8	4.19
6.8	4.27
7.4	5.18
8.1	6.68
8.5	7.88
CURVE 37*	
91.7	-0.122
101.2	-0.117
110.7	-0.112
119.2	-0.107
129.2	-0.102
137.2	-0.097
145.7	-0.093
153.7	-0.088
162.2	-0.083
170.4	-0.078
178.2	-0.073
186.0	-0.068
193.8	-0.063
202	-0.058
210.2	-0.053
217.5	-0.048
225.9	-0.043
233.2	-0.038
241.2	-0.034
248.7	-0.029
256.2	-0.024
264.1	-0.019
272.2	-0.014
273.2	-0.013
279.2	-0.009
286.7	-0.004
294.2	0.001
301.2	0.006
CURVE 38	
1873	1.04
2073	1.23
2273	1.45
2473	1.71
2673	2.04
2873	2.47
3073	2.99
3173	3.26

* Not shown in figure.
† This curve is here reported using the first given temperature as reference temperature at which $\Delta L/L_0 = 0$.
‡ Author's data for coefficient of thermal expansion have been integrated by TPRC to obtain $\Delta L/L_0$.

DATA TABLE 51. THERMAL LINEAR EXPANSION OF TANTALUM Ta (continued)

T	$\Delta L/L_0$	T	$\Delta L/L_0$	T	$\Delta L/L_0$
CURVE 39		CURVE 41*, ‡		CURVE 41 (cont.)*, ‡	
2373	1.58	55	-0.130	220	-0.044
2473	1.74	60	-0.127	230	-0.038
2673	2.11	70	-0.123	240	-0.032
2873	2.50*	80	-0.119	250	-0.026
3073	2.93	90	-0.115	260	-0.020
3273	3.40	100	-0.110	270	-0.014
		110	-0.105	280	-0.008
CURVE 40*, ‡		120	-0.100	290	-0.002
		130	-0.095	293	0.000
2.5	0.0 x 10^{-6}	140	-0.090	300	0.004
3.0	0.2	150	-0.084		
3.5	0.6	160	-0.079		
4.0	1.0	170	-0.073		
4.5	1.6	180	-0.067		
5.0	2.4	190	-0.062		
5.5	3.5	200	-0.056		
		210	-0.050		

DATA TABLE 51. COEFFICIENT OF THERMAL LINEAR EXPANSION OF TANTALUM Ta

[Temperature, T, K; Coefficient of Expansion, α, 10^{-6} K^{-1}]

T	α	T	α	T	α	T	α
CURVE 32‡		CURVE 37		CURVE 37 (cont.)		CURVE 41 (cont.)‡	
5	0.0097	103	5.13	272	6.50	150	5.51
6	0.0142	113	5.40	281	6.53	160	5.60
7	0.0200*	123	5.49	291	6.58	170	5.67
8	0.0280	133	5.71	293	6.55	180	5.75
9	0.0375*	143	5.80			190	5.80
10	0.0485	153	5.92	CURVE 41‡		200	5.86
12	0.075	163	6.01			210	5.90
14	0.116	173	6.11	55	2.94	220	5.94
15	0.142	182	6.16	60	3.18	230	5.97
26	0.170	193	6.20	70	3.70	240	6.00
18	0.235	202	6.25	80	4.12	250	6.02
20	0.320	212	6.30	90	4.50	260	6.04
22	0.425	222	6.35	100	4.75	270	6.05
24	0.550	231	6.41	110	4.99	280	6.06
25	0.610	242	6.39	120	5.18	290	6.07
75	4.20	252	6.45	130	5.31	293	6.07
85	4.60	262	6.48	140	5.38	300	6.08

* Not shown in figure.
‡ Author's data for coefficient of thermal expansion have been integrated by TPRC to obtain $\Delta L/L_0$.

FIGURE AND TABLE NO. 52R. PROVISIONAL VALUES FOR THERMAL LINEAR EXPANSION OF TERBIUM Tb

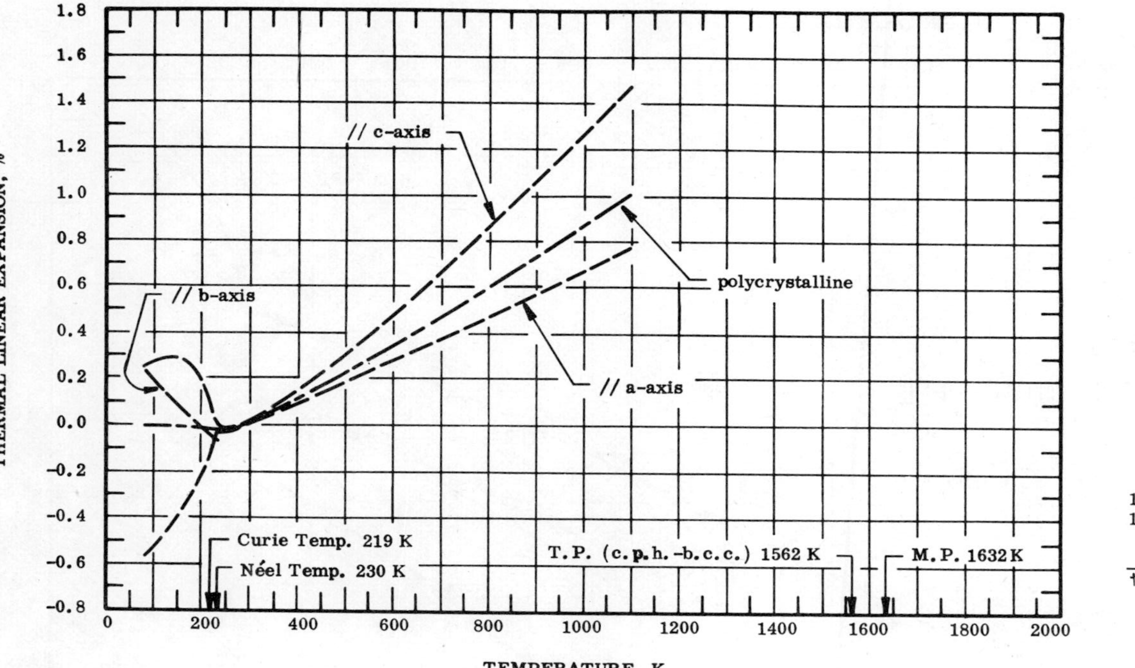

PROVISIONAL VALUES

[Temperature, T, K; Linear Expansion, $\Delta L/L_0$, %; α, K^{-1}]

T	// a-axis $\Delta L/L_0$	// b-axis $\Delta L/L_0$	// c-axis $\Delta L/L_0$	polycrystalline $\Delta L/L_0$	polycrystalline $\alpha \times 10^6$
80	-0.573	0.236	0.246	-0.030†	
100	-0.536	0.184	0.260	-0.031†	
150	-0.397	0.077	0.286	-0.011†	
170	-0.332	0.037	0.272	-0.008†	
200	-0.230	-0.018	0.192	-0.019†	
210	-0.146	-0.036	0.123	-0.020†	
230	-0.049		-0.007	-0.035†	
250	-0.037		-0.032	-0.036†	
270	-0.017		-0.024	-0.019†	
293	0.000		0.000	0.000	9.4
400	0.086		0.146	0.106†	10.5
500	0.173		0.302	0.216†	11.4
600	0.265		0.475	0.335†	12.2
700	0.361		0.661	0.461†	12.9
800	0.462		0.850	0.591†	13.4
900	0.564		1.063	0.729†	13.8
1000	0.669		1.266	0.869†	13.9
1100	0.773		1.471	1.006†	13.9

† Typical values.

REMARKS

The tabulated values are considered accurate to within ±7% at temperatures below 293 K and ±10% above. The tabulated values for polycrystalline terbium are calculated from single crystal data and are considered accurate to within ±15% over the entire temperature range. The experimental data on the polycrystalline terbium are considerably lower than the tabulated values. The tabulated values can be represented approximately by the following equations:

// a-axis: $\Delta L/L_0 = -0.190 + 5.123 \times 10^{-4}\, T + 5.087 \times 10^{-7}\, T^2 - 1.622 \times 10^{-10}\, T^3$ $(293 < T < 1100)$

// c-axis: $\Delta L/L_0 = -0.240 + 3.707 \times 10^{-4}\, T + 1.712 \times 10^{-6}\, T^2 - 5.783 \times 10^{-10}\, T^3$ $(293 < T < 1100)$

polycrystalline: $\Delta L/L_0 = -0.208 + 4.751 \times 10^{-4}\, T + 8.947 \times 10^{-7}\, T^2 - 2.941 \times 10^{-10}\, T^3$ $(293 < T < 1100)$

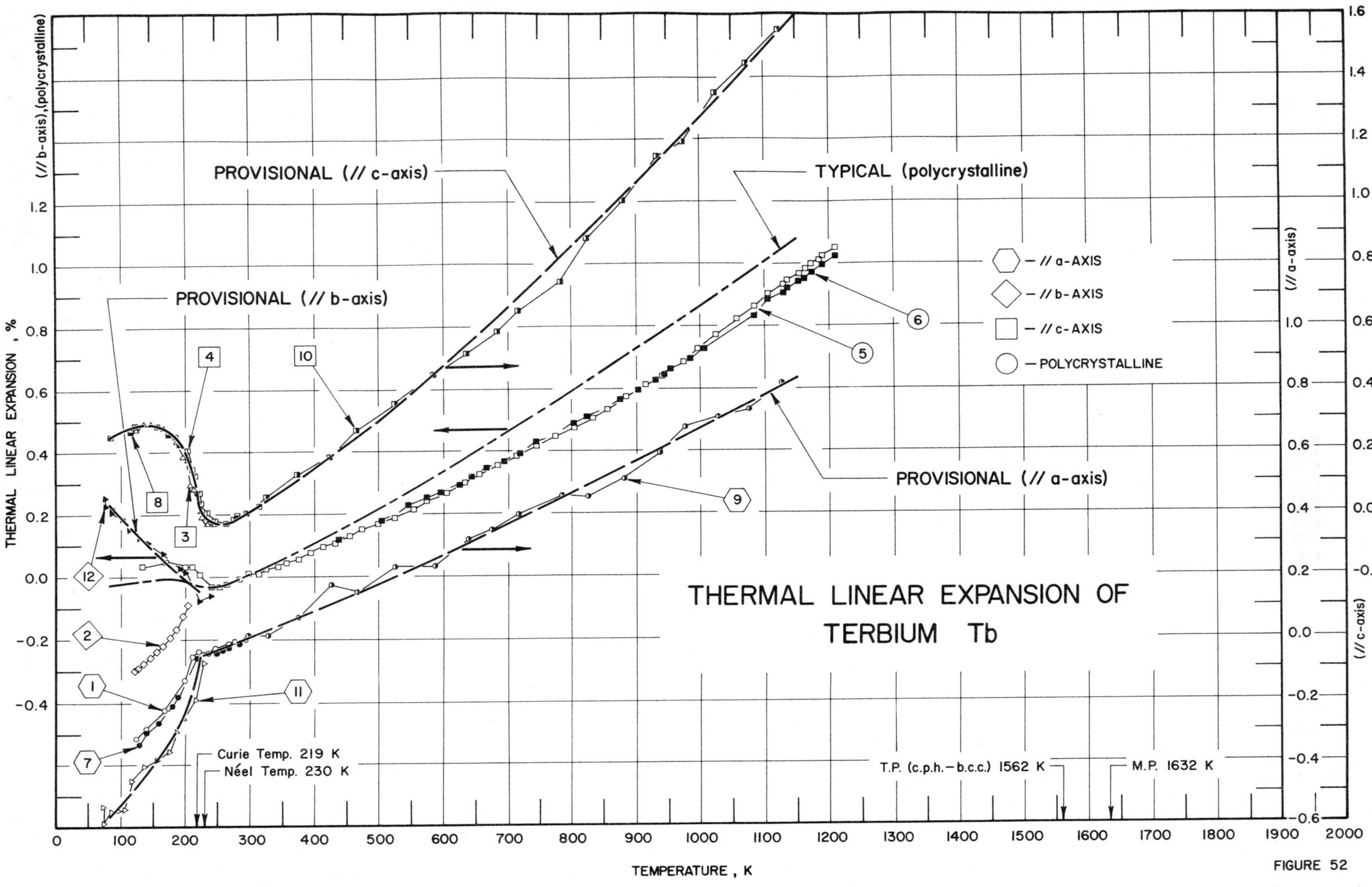
THERMAL LINEAR EXPANSION OF
TERBIUM Tb
THERMAL LINEAR EXPANSION, %
(// b-axis),(polycrystalline)
(// a-axis)
(// c-axis)
TEMPERATURE, K
PROVISIONAL (// c-axis)
TYPICAL (polycrystalline)
PROVISIONAL (// b-axis)
PROVISIONAL (// a-axis)
— // a-AXIS
— // b-AXIS
— // c-AXIS
— POLYCRYSTALLINE
Curie Temp. 219 K
Néel Temp. 230 K
T.P. (c.p.h.–b.c.c.) 1562 K
M.P. 1632 K
1
2
3
4
5
6
7
8
9
10
11
12

FIGURE 52

SPECIFICATION TABLE 52. THERMAL LINEAR EXPANSION OF TERBIUM Tb

Cur. No.	Ref. No.	Author(s)	Year	Method Used	Temp. Range, K	Name and Specimen Designation	Composition (weight percent), Specifications, and Remarks
1	315	Finkel', V.A., Smirnov, Yu.N., and Vorob'ev, V.V.	1967	X	121-295		99.5 Tb crystalline specimen; rhombic distorted hexagonal-close-packed structure; expansion measured along a-axis; transition point observed at 223 K corresponding to transformation of antiferromagnetic helicoidal structure into ferromagnetic structure with collinear ordering; lattice parameter reported at 293 K is 3.609 Å.
2	315	Finkel', V.A., et al.	1967	X	121-293		The above specimen; expansion measured along b-axis; at 223 K, b-axis becomes indistinguishable from a-axis indicating absence of rhombic distortion in hexagonal-close-packed structure; lattice parameter reported at 293 K is 6.2509 Å.
3	315	Finkel', V.A., et al.	1967	X	121-294		The above specimen; expansion measured along c-axis; lattice parameter reported at 293 K is 5.6970 Å.
4	316	Alberts, L. and de V. du Plessis, P.	1968	X	84-315		Single crystal; expansion measured with zero magnetic field along <0001> axis; negative anomaly observed at 240 K in agreement with positive forced magnetostriction.
5	78	Barson, F., Legvold, S., and Spedding, F.H.	1956	L	134-1210		99.5 Tb; specimen 0.6 cm diameter, 6 cm long; fluorides of Tb bomb reduced with Ca metal to produce compact metallic sample; vacuum cast; turned to shape; expansion measured with increasing temperature; discrepancy appeared from 973 K to 1073 K; zero-point correction of 0.001% determined by graphical interpolation.
6	78	Barson, F., et al.	1956	L	1210-300		The above specimen; expansion measured with decreasing temperature; curie temperature about 233 K; zero-point correction of -0.025% determined by graphical extrapolation.
7	317	Finkel', V.A. and Vorob'ev, V.V.	1968	X	117-294		99.5 Tb; material arc-melted in 0.5 to 0.6 argon atmosphere; specimen 10 x 20 x 2 mm cut from casting and annealed 10-50 hrs under 2 x 10^{-6} mm Hg vacuum at 1473 K; expansion measured along a-axis; lattice parameter reported at 294 K is 3.6096 Å; 2.60955 Å at 293 K determined by graphical interpolation.
8	317	Finkel', V.A. and Vorob'ev, V.V.	1968	X	118-300		The above specimen; expansion measured along c-axis; lattice parameter reported at 291 K is 5.6972 Å; 5.69727 Å at 293 K determined by graphical interpolation.
9	82	Hanak, J.J.	1959	X	297-1125		≤0.02 Dy, ≤0.01 Gd, ≤0.005 La; wire specimen cut from sheet prepared at Ames Laboratory, hexagonal close-packed structure; expansion measured along a-axis; lattice parameter reported at 297 K is 3.600 Å; 3.5997 Å at 293 K determined by graphical extrapolation.
10	82	Hanak, J.J.	1959	X	297-1125		The above specimen; expansion measured along c-axis; lattice parameter reported at 297 K is 5.697 Å; 5.6967 Å at 293 K determined by graphical extrapolation.
11	771	Darnell, F.J.	1963	X	77-296		99.9 pure single crystal cut from large crystal consisting of two or three grains which had been grown by annealing an arc-melted button, measurements along a-axis, lattice parameter reported at 293 K is 3.6075 Å.
12	771	Darnell, F.J.	1963	X	76-294		Similar to the above specimen, measurements along b-axis; lattice parameter reported at 293 K is 6.2481 Å.
13*	771	Darnell, F.J.	1963	X	78-298		Similar to the above specimen, measurements along c-axis; lattice parameter reported at 293 K is 5.7001 Å.

* Not shown in figure.

DATA TABLE 52. THERMAL LINEAR EXPANSION OF TERBIUM Tb

[Temperature, T, K; Linear Expansion, $\Delta L/L_0$, %]

T	$\Delta L/L_0$
CURVE 1	
121	-0.316
128	-0.305*
138	-0.285
147	-0.266*
157	-0.249*
167	-0.227
177	-0.205*
187	-0.172*
198	-0.130
202	-0.097*
207	-0.069*
210	-0.058
212	-0.050*
215	-0.047*
217	-0.044*
220	-0.039
221	-0.042*
222	-0.042*
224	-0.042*
226	-0.042*
229	-0.042
231	-0.042*
234	-0.042*
237	-0.042
247	-0.030*
257	-0.022*
267	-0.014
277	-0.008*
287	-0.003*
293	-0.000*
295	0.008
CURVE 2	
121	-0.299
128	-0.286
138	-0.272
148	-0.256
158	-0.235
167	-0.216
177	-0.198
187	-0.166
198	-0.126
203	-0.094
207	-0.067*
212	-0.058*
CURVE 2 (cont.)	
215	-0.050*
218	-0.048*
220	-0.046*
222	-0.042
222	-0.045*
224	-0.043*
226	-0.043*
229	-0.042*
231	-0.043*
234	-0.043*
238	-0.042*
248	-0.034*
258	-0.026*
268	-0.018*
278	-0.011*
288	-0.015*
293	0.000*
CURVE 3	
121	0.276
128	0.283*
138	0.290
148	0.291
158	0.290*
168	0.279
178	0.262
188	0.235
198	0.181
203	0.177
208	0.109
210	0.090
212	0.081
215	0.054*
218	0.037*
220	0.018*
222	0.011*
222	-0.009*
224	-0.012*
226	-0.019*
229	-0.021*
231	-0.028*
232	-0.033
234	-0.039*
235	-0.037*
237	-0.033*
CURVE 3 (cont.)	
242	-0.033*
247	-0.032
257	-0.026*
267	-0.021*
278	-0.012*
287	-0.004*
293	0.000*
294	0.002*
CURVE 4‡	
84	0.258
124	0.284*
158	0.284
182	0.257*
201	0.205
214	0.142
217	0.122
223	0.070
227	0.037
234	0.004
247	-0.019
261	-0.021
277	-0.013
294	0.001*
315	0.021
CURVE 5	
134	0.038
142	0.043*
170	0.044
181	0.043*
200	0.032
211	0.017
221	0.001
228	-0.017*
233	-0.031*
241	-0.032
253	-0.029
267	-0.021
284	-0.009
300	0.005
313	0.013
327	0.025
343	0.037
CURVE 5 (cont.)	
357	0.045
374	0.059
394	0.077
412	0.094
432	0.109
453	0.129
475	0.150
500	0.173
523	0.192
550	0.217
576	0.240
604	0.268
631	0.301
658	0.327
684	0.354
712	0.382
742	0.413
773	0.446
802	0.476
831	0.509
857	0.539
886	0.576
915	0.612
943	0.644
973	0.686
999	0.726
1026	0.771
1059	0.822
1082	0.864
1107	0.903*
1108	0.915*
1130	0.936
1135	0.951
1151	0.969
1162	0.981
1171	1.000
1183	1.019
1190	1.025
1210	1.054
CURVE 6	
1210	1.028
1190	0.999
1183	0.993*
1171	0.974
CURVE 6 (cont.)	
1162	0.955
1151	0.943
1135	0.925
1130	0.910
1108	0.889
1107	0.877*
1082	0.838
1009	0.729
984	0.698
955	0.660
930	0.628
903	0.594
878	0.564
852	0.537*
827	0.510
801	0.483
775	0.453*
747	0.425
719	0.394
692	0.366
669	0.343
642	0.318
624	0.296
597	0.271
573	0.247
549	0.222
526	0.198*
502	0.178
479	0.158*
456	0.138*
437	0.117
417	0.099*
397	0.084*
380	0.071*
365	0.060*
352	0.046*
330	0.027*
314	0.017*
300	0.005*
CURVE 7	
117	-0.339
129	-0.312*
138	-0.292
147	-0.276*
CURVE 7 (cont.)	
158	-0.265
168	-0.237*
179	-0.215
188	-0.182
198	-0.137*
203	-0.107*
211	-0.076*
213	-0.060*
216	-0.060
218	-0.060*
221	-0.054*
223	-0.054*
225	-0.051*
227	-0.051*
230	-0.051*
232	-0.048*
238	-0.048*
249	-0.043
259	-0.038
269	-0.026
275	-0.021
281	-0.015
294	0.001*
CURVE 8	
118	0.271
129	0.281*
140	0.292*
150	0.292*
161	0.287*
171	0.280
180	0.262
189	0.227
200	0.171*
207	0.125*
212	0.080*
221	0.034*
223	0.013*
223	-0.006*
228	-0.020*
232	-0.034*
235	-0.042*
240	-0.040*
252	-0.033*
261	-0.026*

*Not shown in figure.

‡Author's data for coefficient of thermal expansion have been integrated by TPRC to obtain $\Delta L/L_0$.

DATA TABLE 52. THERMAL LINEAR EXPANSION OF TERBIUM Tb (continued)

T	ΔL/L$_0$
CURVE 8 (cont.)	
271	-0.019*
282	-0.012*
291	-0.001
300	0.004*
CURVE 9	
297	0.008
328	0.008
372	0.064
425	0.175
469	0.147
524	0.230
587	0.230
637	0.314
682	0.342
717	0.397
781	0.453
822	0.453
880	0.508
932	0.592
973	0.675
1027	0.703
1072	0.731
1125	0.814
CURVE 10	
297	0.005
328	0.058
372	0.128

T	ΔL/L$_0$
CURVE 10 (cont.)	
425	0.181
467	0.268
524	0.356
587	0.444
637	0.514
682	0.584
717	0.655
781	0.742
822	0.883
880	1.006
932	1.146
973	1.199
1027	1.357
1072	1.445
1125	1.550
CURVE 11	
77	-0.593
77	-0.540
87	-0.554
93	-0.510
102	-0.507
106	-0.485*
114	-0.455
117	-0.441*
123	-0.441*
125	-0.441*
131	-0.427*
133	-0.407

T	ΔL/L$_0$
CURVE 11 (cont.)	
142	-0.366*
153	-0.382
165	-0.352*
169	-0.360
178	-0.266*
185	-0.294
194	-0.255
197	-0.208*
204	-0.194
214	-0.175*
221	-0.139*
229	-0.072
242	-0.072*
242	-0.061*
278	-0.036*
279	-0.019*
282	-0.019*
288	-0.011*
288	0.003*
291	0.019*
293	0.000*
296	0.000*
296	0.011*
CURVE 12	
76	0.253
76	0.234
85	0.222
87	0.208

T	ΔL/L$_0$
CURVE 12 (cont.)	
92	0.203*
93	0.195*
101	0.187
105	0.182*
113	0.152
115	0.160*
121	0.144*
123	0.136*
129	0.125
131	0.115*
140	0.112
151	0.099*
163	0.078*
167	0.075
172	0.056*
176	0.037*
191	0.030
194	0.000*
202	0.019
211	-0.040
218	-0.059
226	-0.083
239	-0.061
276	-0.010*
280	-0.010*
285	0.000*
293	0.000*
294	0.000*

T	ΔL/L$_0$
CURVE 13*	
78	0.221
80	0.221
116	0.272
124	0.286
137	0.286
143	0.286
152	0.288
162	0.270
168	0.246
169	0.258
181	0.219
186	0.223
195	0.216
198	0.203
206	0.144
206	0.160
214	0.133
215	0.098
223	0.000
224	0.026
227	0.000
232	-0.026
236	-0.025
238	-0.025
238	-0.035
239	-0.025
244	-0.033
249	-0.021
250	-0.039

T	ΔL/L$_0$
CURVE 13 (cont.)*	
256	-0.019
263	-0.025
268	0.000
274	0.000
277	-0.012
282	-0.002
293	0.000
296	0.007
298	0.007

DATA TABLE 52. COEFFICIENT OF THERMAL LINEAR EXPANSION OF TERBIUM Tb

[Temperature, T, K; Coefficient of Expansion, α, 10^{-6} K^{-1}]

T	α
CURVE 4‡	
84	8.6
124	4.5
158	-4.9
182	-16.9
201	-37.8
214	-60.4
217	-69.6
223	-103.4
227	-63.3

T	α
CURVE 4 (cont.)‡	
234	-30.0
247	-5.9
261	3.4
277	6.8
294	8.9
315	10.4*

* Not shown in figure.
‡ Author's data for coefficient of thermal expansion have been integrated by TPRC to obtain ΔL/L$_0$.

FIGURE AND TABLE NO. 53R. PROVISIONAL VALUES FOR THERMAL LINEAR EXPANSION OF THALLIUM Tl

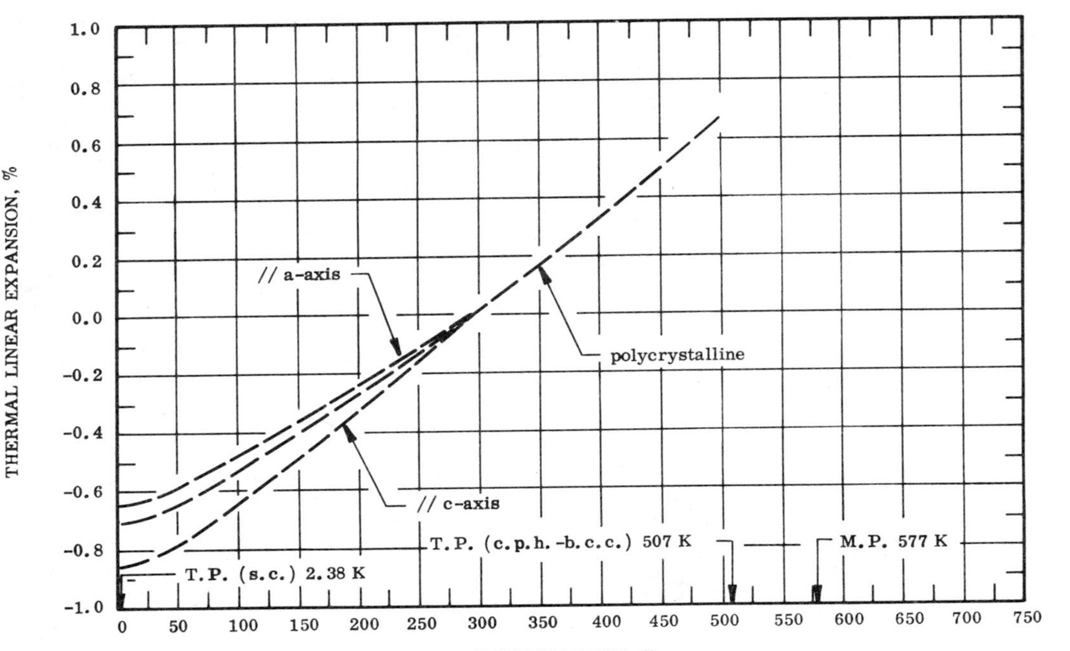

PROVISIONAL VALUES

[Temperature, T, K; Linear Expansion, $\Delta L/L_0$, %; α, K^{-1}]

T	// a-axis $\Delta L/L_0$	// c-axis $\Delta L/L_0$	polycrystalline $\Delta L/L_0$	polycrystalline $\alpha \times 10^6$
5	-0.645	-0.860	-0.717	1.0
25	-0.633	-0.833	-0.700	16.1
50	-0.591	-0.780	-0.654	21.5
100	-0.481	-0.647	-0.536	25.2
200	-0.238	-0.328	-0.268	28.0
293	0.000	0.000	0.000	29.9
400			0.331	31.9
500			0.665	34.7

REMARKS

The tabulated values are considered accurate to within ±7% over the entire temperature range. The tabulated values for polycrystalline thallium below 293 K are calculated from single crystal data and are considered accurate to within ±10% in that temperature range and to within ±7% above. These values can be represented approximately by the following equations:

// a-axis: $\Delta L/L_0 = -0.673 + 1.476 \times 10^{-3}\, T + 5.013 \times 10^{-6}\, T^2 - 7.536 \times 10^{-9}\, T^3$

// c-axis: $\Delta L/L_0 = -0.882 + 1.741 \times 10^{-3}\, T + 6.929 \times 10^{-6}\, T^2 - 8.886 \times 10^{-9}\, T^3$

polycrystalline: $\Delta L/L_0 = -0.700 + 1.941 \times 10^{-3}\,(T - 25) + 3.729 \times 10^{-6}\,(T - 25)^2 - 4.501 \times 10^{-9}\,(T - 25)^3$ $(25 < T < 260)$

$\Delta L/L_0 = -0.096 + 2.948 \times 10^{-3}\,(T - 260) + 5.553 \times 10^{-7}\,(T - 260)^2 + 1.619 \times 10^{-9}\,(T - 260)^3$ $(260 < T < 500)$

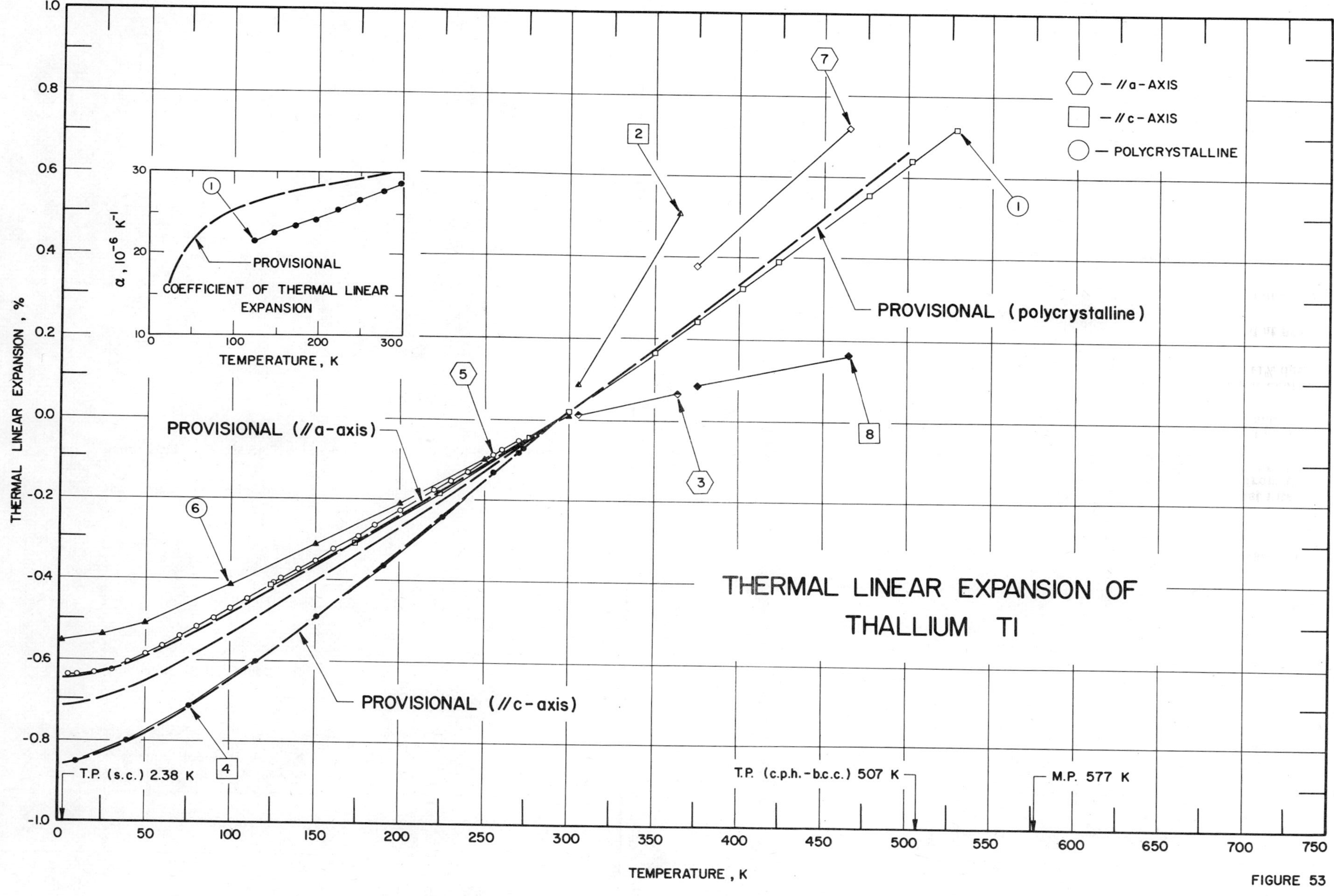
THERMAL LINEAR EXPANSION OF
THALLIUM Tl
— //a-AXIS
— //c-AXIS
— POLYCRYSTALLINE
PROVISIONAL (polycrystalline)
PROVISIONAL (//a-axis)
PROVISIONAL (//c-axis)
T.P. (s.c.) 2.38 K
T.P. (c.p.h.-b.c.c.) 507 K
M.P. 577 K
THERMAL LINEAR EXPANSION, %
TEMPERATURE, K
1.0
0.8
0.6
0.4
0.2
0.0
-0.2
-0.4
-0.6
-0.8
-1.0
0
50
100
150
200
250
300
350
400
450
500
550
600
650
700
750
1
2
3
4
5
6
7
8
PROVISIONAL
COEFFICIENT OF THERMAL LINEAR EXPANSION
α, 10⁻⁶ K⁻¹
TEMPERATURE, K
30
20
10
0
100
200
300

FIGURE 53

SPECIFICATION TABLE 53. THERMAL LINEAR EXPANSION OF THALLIUM Tl

Cur. No.	Ref. No.	Author(s)	Year	Method Used	Temp. Range, K	Name and Specimen Designation	Composition (weight percent), Specifications, and Remarks
1	318	Lanikov, L. N., Fal'chenko, V. M. and Koblova, E. A.	1966	L	124-529		99.9995 Tl.
2	192	Shinoda, G.	1933	X	305-364	α-Tl	α-Tl, hexagonal close-packed structure; specimen 0.5 mm thick; expansion measured parallel to crystal axis.
3	192	Shinoda, G.	1933	X	305-364	α-Tl	The above specimen; expansion measured perpendicular to axis.
4	93	Meyerhoff, R. W. and Smith, J. F.	1962	I	5-273		99.95^{+} Tl; supplied by the American Smelting and Refining Co.; cast into cylinders; annealed at 488 K for 3 to 7 days; specimen 0.5730 in. long cut from the grain; expansion measured parallel to the c-axis; zero-point correction of -0.0718% determined by graphical extrapolation.
5	93	Meyerhoff, R. W. and Smith, J. F.	1962	I	5-273		Similar to the above specimen; 1.0000 in. long; expansion measured perpendicular to the c-axis; zero-point correction of -0.041% determined by graphical extrapolation.
6	196	Swenson, C. A.	1955	I	0-300		99.9 Tl; obtained from A. D. Mackay; specimen machined from ingot cast in vacuum; data smoothed by author; zero-point correction of -0.041% determined by graphical interpolation.
7	690	Schneider, A., and Heymer, G.	1956	X	291-465		Pure metal, measurements along a-axis; lattice parameter reported at 291 K is 3.456 Å.
8	690	Schneider, A., and Heymer, G.	1956	X	291-465		Pure metal, measurements along c-axis; lattice parameter reported at 291 K is 5.530 Å.

DATA TABLE 53. THERMAL LINEAR EXPANSION OF THALLIUM Tl

[Temperature, T, K; Linear Expansion, $\Delta L/L_0$, %]

T	$\Delta L/L_0$
CURVE 1‡	
124	-0.419
149	-0.364*
173	-0.309
198	-0.250*
224	-0.185
249	-0.120*
276	-0.047
300	0.020
323	0.088*
350	0.166
375	0.242
402	0.324
424	0.392
450	0.472*
476	0.554
502	0.640
505	0.640*
529	0.720
CURVE 2	
305	0.086
364	0.511
CURVE 3	
305	0.011

T	$\Delta L/L_0$
CURVE 3 (cont.)	
364	0.064
CURVE 4	
5	-0.862*
10	-0.860
15	-0.854*
20	-0.846*
25	-0.837*
30	-0.827*
35	-0.816*
40	-0.805
45	-0.793*
50	-0.781*
55	-0.769*
60	-0.756*
65	-0.743*
70	-0.730*
75	-0.716
85	-0.688*
90	-0.674*
95	-0.660*
105	-0.631*
110	-0.616*
115	-0.601
130	-0.556*
135	-0.540*

T	$\Delta L/L_0$
CURVE 4 (cont.)	
145	-0.509*
150	-0.493
155	-0.477*
165	-0.445*
175	-0.412*
180	-0.396*
190	-0.363
200	-0.329*
210	-0.295*
220	-0.261*
225	-0.243
230	-0.226*
235	-0.208*
240	-0.191*
250	-0.155*
255	-0.138
260	-0.120*
270	-0.083
273	-0.072
CURVE 5	
5	-0.636
10	-0.635
15	-0.634*
20	-0.632
25	-0.628*

T	$\Delta L/L_0$
CURVE 5 (cont.)	
30	-0.622
35	-0.613*
40	-0.603
45	-0.593*
50	-0.582
55	-0.572*
60	-0.561
65	-0.549*
70	-0.538
75	-0.527*
80	-0.515
85	-0.504*
90	-0.492
95	-0.481*
100	-0.469
105	-0.458
110	-0.446
115	-0.437*
120	-0.422*
125	-0.411
130	-0.399
135	-0.387*
140	-0.375
145	-0.363*
150	-0.351
155	-0.339
160	-0.327

T	$\Delta L/L_0$
CURVE 5 (cont.)	
165	-0.315*
170	-0.302*
175	-0.290
180	-0.278*
185	-0.265*
190	-0.253*
195	-0.241*
200	-0.228
205	-0.216*
210	-0.203*
215	-0.191*
220	-0.178
225	-0.165
230	-0.153
235	-0.140*
240	-0.127
245	-0.114*
250	-0.101*
255	-0.088
260	-0.075
265	-0.062*
270	-0.049
273.2	-0.041
CURVE 6	
5	-0.552

T	$\Delta L/L_0$
CURVE 6 (cont.)	
21	-0.543
39	-0.524
56	-0.498
65	-0.486
69	-0.477
78	-0.458
103	-0.413
145	-0.332
175	-0.266
209	-0.193
242	-0.122
262	-0.074
274	-0.045
293	0.000
301	0.105
CURVE 7	
291	-0.009
375	0.367
405	0.715
CURVE 8	
291	-0.009
375	0.089
465	0.161

DATA TABLE 53. COEFFICIENT OF THERMAL LINEAR EXPANSION OF THALLIUM Tl

[Temperature, T, K; Coefficient of Expansion, α, 10^{-6} K^{-1}]

T	α
CURVE 1‡	
124	21.7
149	22.5
173	23.3
198	24.1
224	25.5
249	26.5
276	27.6
300	28.6
323	29.0*

T	α
CURVE 1 (cont.)‡	
350	29.8*
375	30.0*
402	30.3*
424	30.9*
450	31.3*
476	31.4*
502	31.4*
505	32.4*
529	32.5*

* Not shown in figure.

‡ Author's data for coefficient of thermal expansion have been integrated by TPRC to obtain $\Delta L/L_0$.

FIGURE AND TABLE NO. 54R. RECOMMENDED VALUES FOR THERMAL LINEAR EXPANSION OF THORIUM Th

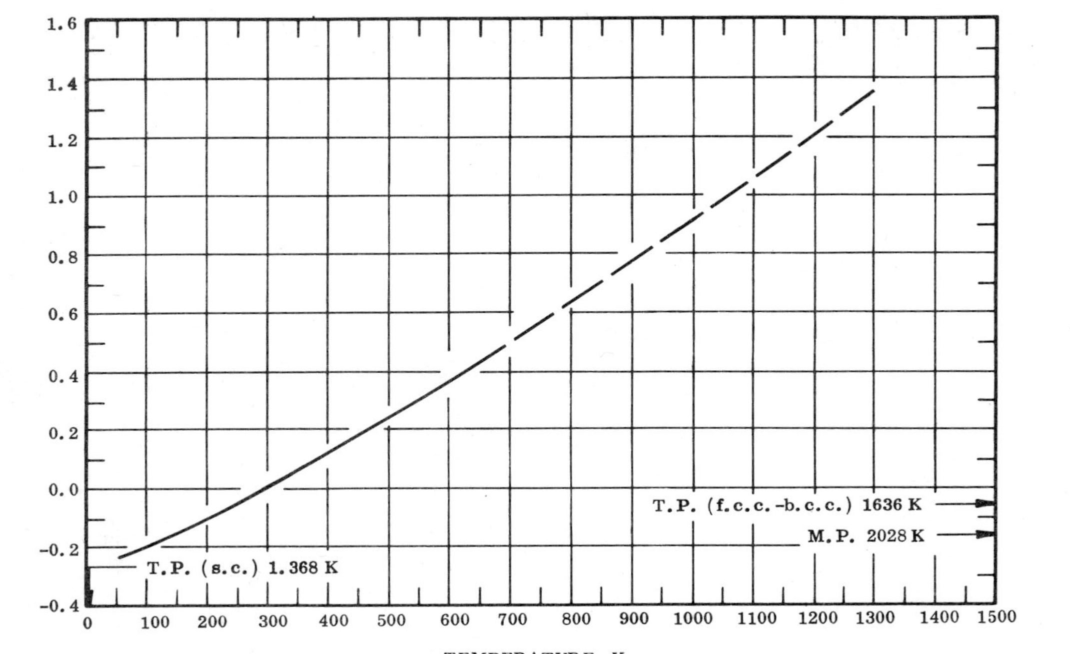

RECOMMENDED VALUES

[Temperature, T, K; Linear Expansion, $\Delta L/L_0$, %; α, K^{-1}]

T	$\Delta L/L_0$	$\alpha \times 10^6$
50	-0.236	7.9
100	-0.193	8.9
200	-0.098	10.0
293	0.000	11.0
400	0.122	11.9
500	0.245	12.6
600	0.373	13.1
700	0.507	13.5
800	0.643†	13.8
900	0.782†	14.0
1000	0.922†	14.5
1200	1.214†	14.7
1300	1.368†	14.7

† Provisional values.

REMARKS

The tabulated values are considered accurate to within ±3% at temperatures below 700 K and ±10% above. These values can be represented approximately by the following equation:

$$\Delta L/L_0 = -0.280 + 8.190 \times 10^{-4}\ T + 5.286 \times 10^{-7}\ T^2 - 1.432 \times 10^{-10}\ T^3$$

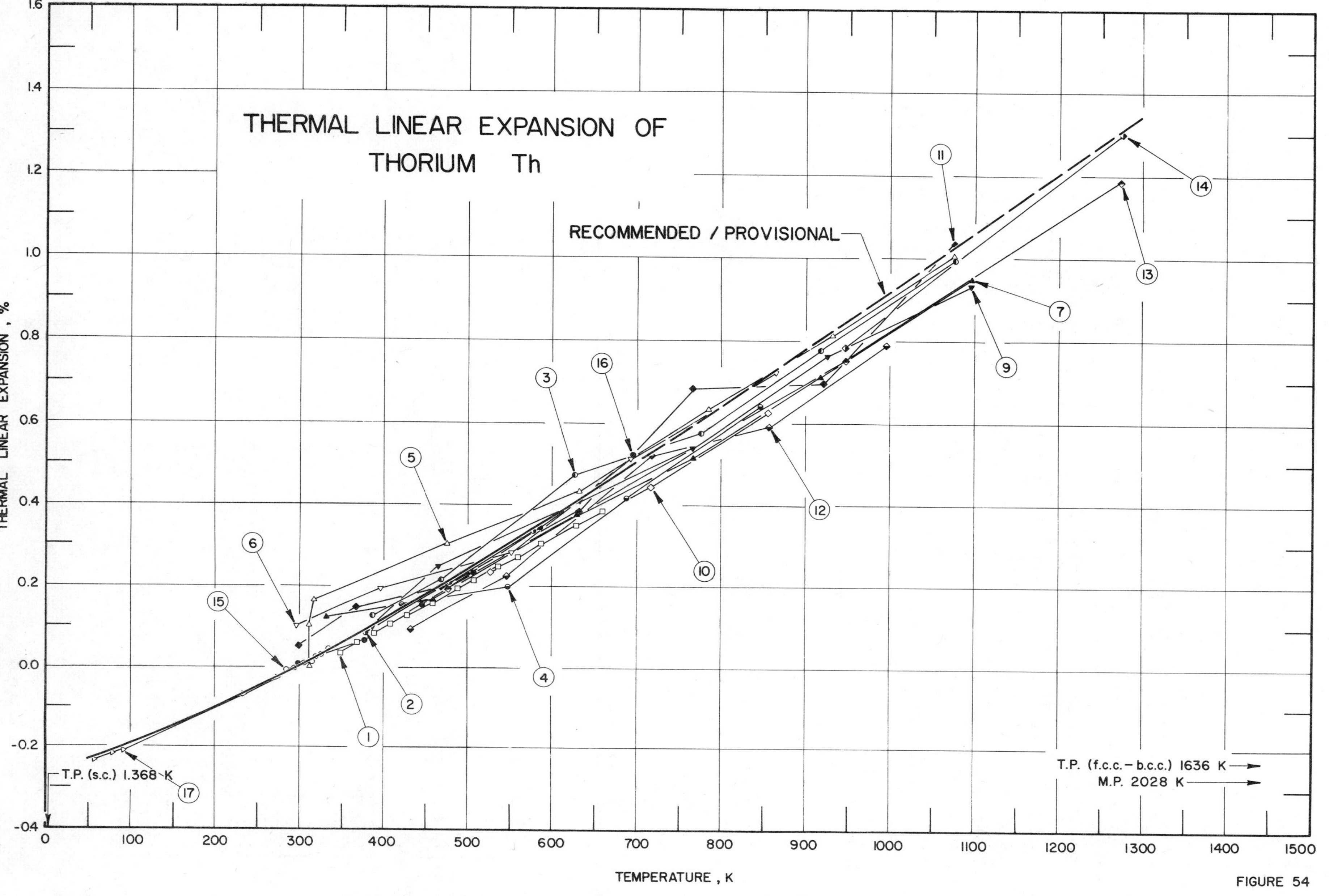
THERMAL LINEAR EXPANSION OF
THORIUM Th
RECOMMENDED / PROVISIONAL
THERMAL LINEAR EXPANSION, %
TEMPERATURE, K
1.6
1.4
1.2
1.0
0.8
0.6
0.4
0.2
0.0
-0.2
-0.4
0
100
200
300
400
500
600
700
800
900
1000
1100
1200
1300
1400
1500
T.P. (s.c.) 1.368 K
T.P. (f.c.c. – b.c.c.) 1636 K
M.P. 2028 K
1
2
3
4
5
6
7
9
10
11
12
13
14
15
16
17

FIGURE 54

SPECIFICATION TABLE 54. THERMAL LINEAR EXPANSION OF THORIUM Th

Cur. No.	Ref. No.	Author(s)	Year	Method Used	Temp. Range, K	Name and Specimen Designation	Composition (weight percent), Specifications, and Remarks
1	319	Armstrong, P. E. and Carlson, O. N.	1959	S	347-655		Single crystals grown by strain-anneal technique; specimen showed cubic symmetry; polycrystalline specimen measured; zero-point correction of 0.259% determined by graphical extrapolation.
2	320	Thompson, J. G.	1933		293-573		Swaged rod, 5 mm diameter.
3	321	Danielson, G. C., Murphy, G., Peterson, D. and Rogers, B. A.	1952	T	1074-383	Billet A-458	Raw material bomb-reduced and cast in graphite, specimen cold swaged from billet croppings into rod, machined to 0.635 cm diameter; heated to 1073 K before testing; expansion measured with decreasing temperature; zero-point correction is 0.096%.
4	321	Danielson, G. C., et al.	1952	T	420-846	Billet A-458	The above specimen; expansion measured with increasing temperature; zero-point correction is 0.095%.
5	321	Danielson, G. C., et al.	1952	T	1077-310	Billet A-496	Similar to the above specimen; expansion measured with decreasing temperature; zero-point correction is 0.125%.
6	321	Danielson, G. C., et al.	1952	T	295-861	Billet A-496	The above specimen; expansion measured with increasing temperature; zero-point correction is 0.125%.
7	321	Danielson, G. C., et al.	1952	T	1098-328	Billet A-499	Similar to the above specimen except heated to 973 K before hot swaging; expansion measured with decreasing temperature.
8	321	Danielson, G. C., et al.	1952	T	318-842	Billet A-499	The above specimen; expansion measured with increasing temperature.
9	321	Danielson, G. C., et al.	1952	T	1096-301	Billet A-496	Similar to the above specimen; expansion measured with decreasing temperature.
10	321	Danielson, G. C., et al.	1952	T	297-851	Billet A-496	The above specimen; expansion measured with increasing temperature.
11	321	Danielson, G. C., et al.	1952	T	1087-298	Billet MX-569	Similar to the above specimen; expansion measured with decreasing temperature; zero-point correction of 0.137% determined by graphical extrapolation.
12	321	Danielson, G. C., et al.	1952	T	430-996	Billet MX-569	The above specimen; expansion measured with increasing temperature; zero-point correction of 0.137% determined by graphical extrapolation.
13	322	Smith, J. F.	1958		298-1273		Specimen from Ames; experiment performed by J. G. Thompson at National Bureau of Standards.
14	322	Smith, J. F.	1958		298-1273		Similar to the above except specimen from Westinghouse.
15	394	James, W. J. and Straumanis, M. E.	1956	X	283-333		0.020 C, 0.01 O, 0.01 N, approximately 0.1 total metallic impurities; specimen prepared by van Arkel method; sample in shape of small cylinder containing quasi-crystalline aggregates oriented radially; specimen was single chip removed from sample; unfiltered CuKα_1 x-radiation used to measure lattice parameter using strong α_1 (533) line; reported error in lattice parameter ± 0.00008 kX; possible discontinuity in coefficient of expansion suggested at ~311 K; lattice parameter reported at 293 K is 5.07363 kX (5.08378 Å).
16	634	Harris, I. R. and Raynor, G. V.	1964	X	298-692		Iodide metal from Atomic Energy Research Establishment, Harwell, <0.1 metallic impurities, 0.02 C, 0.01 N, and 0.01 O; lattice parameter reported at 298 K is 5.0758 kX; 5.0753 kX at 293 K determined by graphical extrapolation.
17	728	Erfling, H. D.	1942		303-57.12		99.8 Th

DATA TABLE 54. THERMAL LINEAR EXPANSION OF THORIUM Th

[Temperature, T, K; Linear Expansion, $\Delta L/L_0$, %]

T	$\Delta L/L_0$
CURVE 1	
347	0.038
356	0.050*
366	0.060
376	0.071*
385	0.082
396	0.093*
405	0.104
415	0.114*
425	0.126
435	0.136*
444	0.147*
455	0.159
464	0.169*
475	0.181*
484	0.192
494	0.202*
503	0.214
514	0.225*
524	0.236*
533	0.247
544	0.259*
555	0.270
564	0.281*
574	0.292*
583	0.303
593	0.314*
604	0.326*
614	0.337*
624	0.350
633	0.360*
645	0.371*
655	0.383
CURVE 2	
293	0.000*
333	0.044*
373	0.090
473	0.209
573	0.339
CURVE 3	
1074	0.993
915	0.774
774	0.578
CURVE 3 (cont.)	
621	0.473
468	0.214
383	0.124
CURVE 4	
420	0.158
549	0.198
686	0.414
846	0.641
CURVE 5	
1077	1.030
929	0.813
784	0.638
630	0.438
472	0.304
314	0.166
310	0.102
CURVE 6	
295	0.102
392	0.193
550	0.280
691	0.519
861	0.724
CURVE 7	
1098	0.952
917	0.717
767	0.513
626	0.378
458	0.165
328	0.121
CURVE 8*	
318	0.043
453	0.150
682	0.411
842	0.481
CURVE 9	
1096	0.934
923	0.766
764	0.540
630	0.410
462	0.247
385	0.108
301	0.022*
CURVE 10	
297	0.007*
527	0.236
713	0.441
851	0.628
CURVE 11	
1087	1.036
920	0.702
765	0.683
629	0.382
469	0.200
368	0.146
298	0.057
CURVE 12	
430	0.098
545	0.221
715	0.521
853	0.590
996	0.791
CURVE 13	
298	0.005*
473	0.194
948	0.755
1273	1.190
CURVE 14	
298	0.006*
473	0.202*
948	0.786
1273	1.303
CURVE 15	
283	-0.008
293	0.000
303	0.010
308	0.015*
313	0.018
318	0.029
323	0.034
333	0.048
CURVE 16	
298	0.010
303	0.002*
377	0.067
447	0.154
504	0.232
582	0.345
651	0.447
692	0.528
CURVE 17	
303	0.011
293	0.000
273	-0.022
233	-0.064
90	-0.207
78	-0.217
57.12	-0.232

* Not shown in figure.
† This curve is here reported using the first given temperature as reference temperature at which $\Delta L/L_0 = 0$.

FIGURE AND TABLE NO. 55R. PROVISIONAL VALUES FOR THERMAL LINEAR EXPANSION OF THULIUM Tm

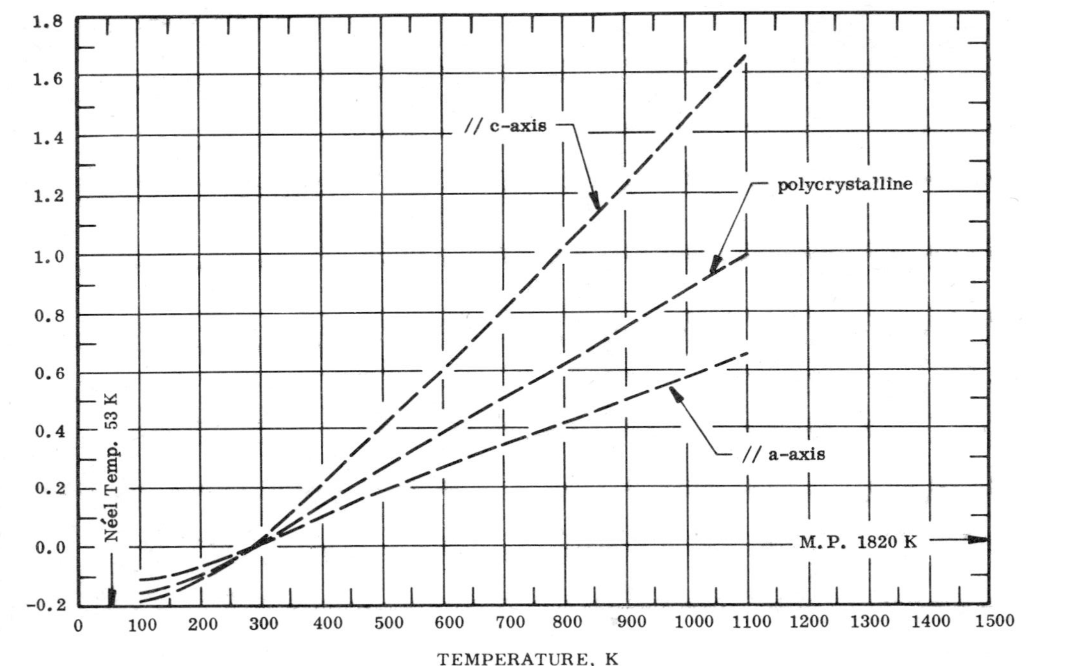

PROVISIONAL VALUES

[Temperature, T, K; Linear Expansion, $\Delta L/L_0$, %]

T	// a-axis $\Delta L/L_0$	// c-axis $\Delta L/L_0$	polycrystalline $\Delta L/L_0$
100	-0.115	-0.191	-0.140
200	-0.072	-0.106	-0.083
293	0.000	0.000	0.000
400	0.108	0.209	0.142
500	0.191	0.416	0.266
600	0.268	0.621	0.386
700	0.346	0.823	0.505
800	0.424	1.028	0.625
900	0.503	1.227	0.744
1000	0.584	1.423	0.864
1100	0.668	1.619	0.985

REMARKS

The tabulated values are considered accurate to within ±10% over the entire temperature range.

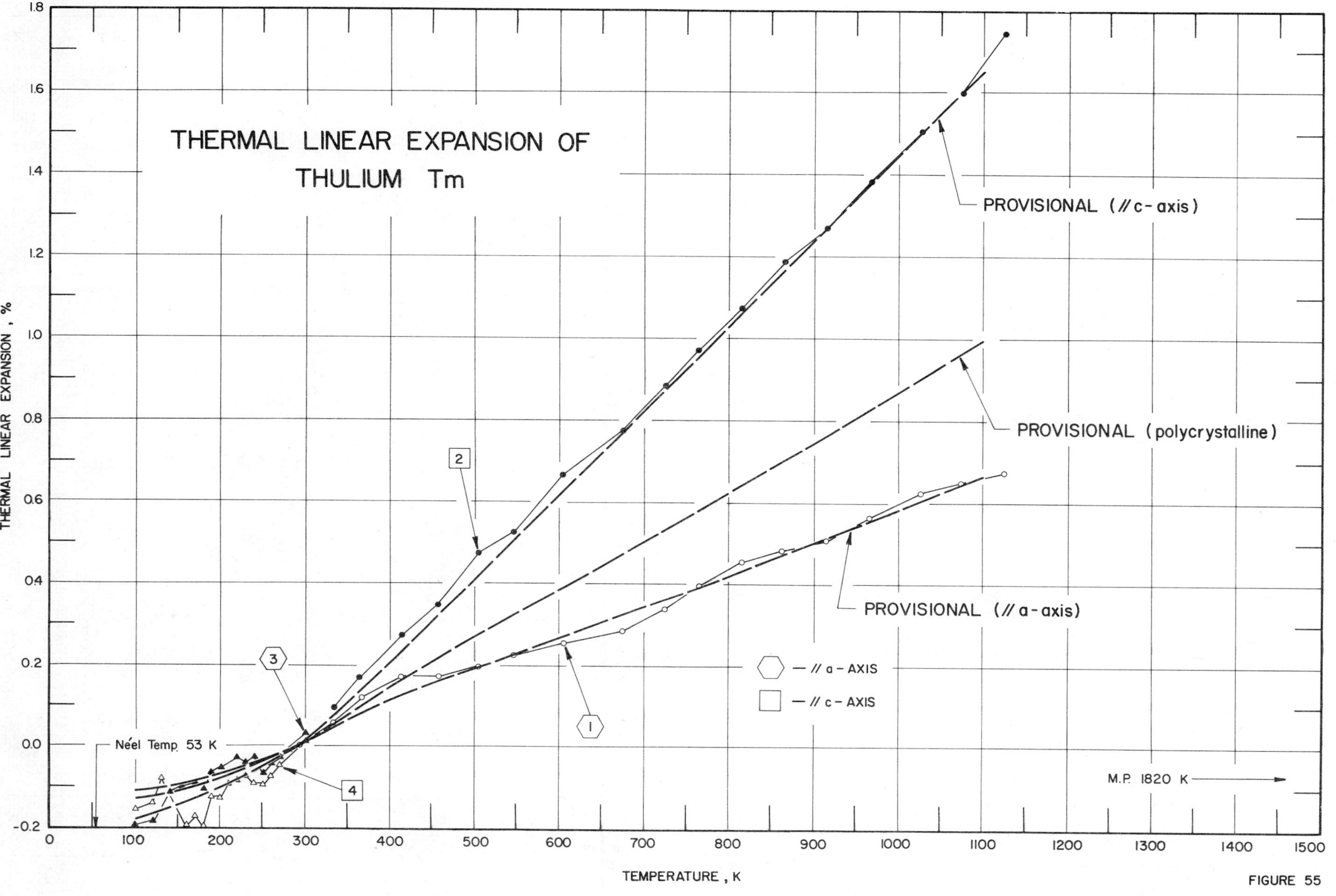
THERMAL LINEAR EXPANSION OF
THULIUM Tm
PROVISIONAL (// c- axis)
PROVISIONAL (polycrystalline)
PROVISIONAL (// a- axis)
— // a - AXIS
— // c - AXIS
Néel Temp. 53 K
M.P. 1820 K
1
2
3
4
THERMAL LINEAR EXPANSION, %
TEMPERATURE, K
1.8
1.6
1.4
1.2
1.0
0.8
0.6
0.4
0.2
0.0
-0.2
0
100
200
300
400
500
600
700
800
900
1000
1100
1200
1300
1400
1500
FIGURE 55

SPECIFICATION TABLE 55. THERMAL LINEAR EXPANSION OF THULIUM Tm

Cur. No.	Ref. No.	Author(s)	Year	Method Used	Temp. Range, K	Name and Specimen Designation	Composition (weight percent), Specifications, and Remarks
1	82	Hanak, J.J.	1959	X	293-1126		0.03 Ca, < 0.03 Fe, Mg, Si, < 0.02 Er, < 0.001 Yb; wire specimen cut from sheet prepared at Ames Laboratory; hexagonal close-packed structure; expansion measured along a-axis; lattice parameter reported at 293 K is 3.537 Å.
2	82	Hanak, J.J.	1959	X	293-1126		The above specimen, expansion measured along c-axis; lattice parameter reported at 293 K is 5.564 Å.
3	736	Singh, S., Khanduri, N.C., and Tsang, T.	1971	X	100-300		Specimen from the Research Chemicals, Phoenix, Arizona, metallic impurity 0.01 percent, filings of 250 mesh, annealed at a temperature just below melting point in argon atmosphere, measurements along a-axis; lattice parameter reported at 293 K is 3.5167 Å.
4	736	Singh, S., et al.	1971	X	100-300		Similar to the above specimen; measurements along c-axis, lattice parameters reported at 293 K is 5.5732 Å.

DATA TABLE 55. THERMAL LINEAR EXPANSION OF THULIUM Tm

[Temperature, T, K; Linear Expansion, $\Delta L/L_0$, %]

T	$\Delta L/L_0$	T	$\Delta L/L_0$	T	$\Delta L/L_0$	T	$\Delta L/L_0$	T	$\Delta L/L_0$
CURVE 1		CURVE 2		CURVE 3		CURVE 3 (cont.)		CURVE 4 (cont.)	
293	0.000	293	0.000	100	-0.205	269	-0.028	200	-0.129
331	0.057	331	0.090	111	-0.176*	280	0.011	210	-0.095
366	0.113	366	0.162	119	-0.188	293	0.000	219	-0.081
414	0.170	414	0.270	130	-0.119*	300	0.031	230	-0.077
459	0.170	459	0.341	139	-0.114			239	-0.091
505	0.198	505	0.467	147	-0.105*	CURVE 4		250	-0.109
548	0.226	548	0.521	152	-0.105			260	-0.075
606	0.254	606	0.665	158	-0.088*	100	-0.160	269	-0.032*
671	0.283	671	0.773	169	-0.088	109	-0.160*	280	-0.023
721	0.339	721	0.881	179	-0.105	119	-0.140	293	0.000
765	0.396	765	0.971	189	-0.063	130	-0.079	300	0.011
816	0.452	816	1.078	200	-0.057	140	-0.239*		
866	0.481	866	1.186	209	-0.057*	146	-0.215*		
914	0.509	914	1.276	219	-0.028	152	-0.201		
969	0.565	969	1.384	229	-0.040	159	-0.190*		
1026	0.622	1026	1.510	239	-0.028	169	-0.178		
1076	0.650	1076	1.600	249	-0.068	179	-0.208		
1126	0.679	1126	1.743	259	-0.043	189	-0.122		

* Not shown in figure.

FIGURE AND TABLE NO. 56R. PROVISIONAL VALUES FOR THERMAL LINEAR EXPANSION OF TIN Sn

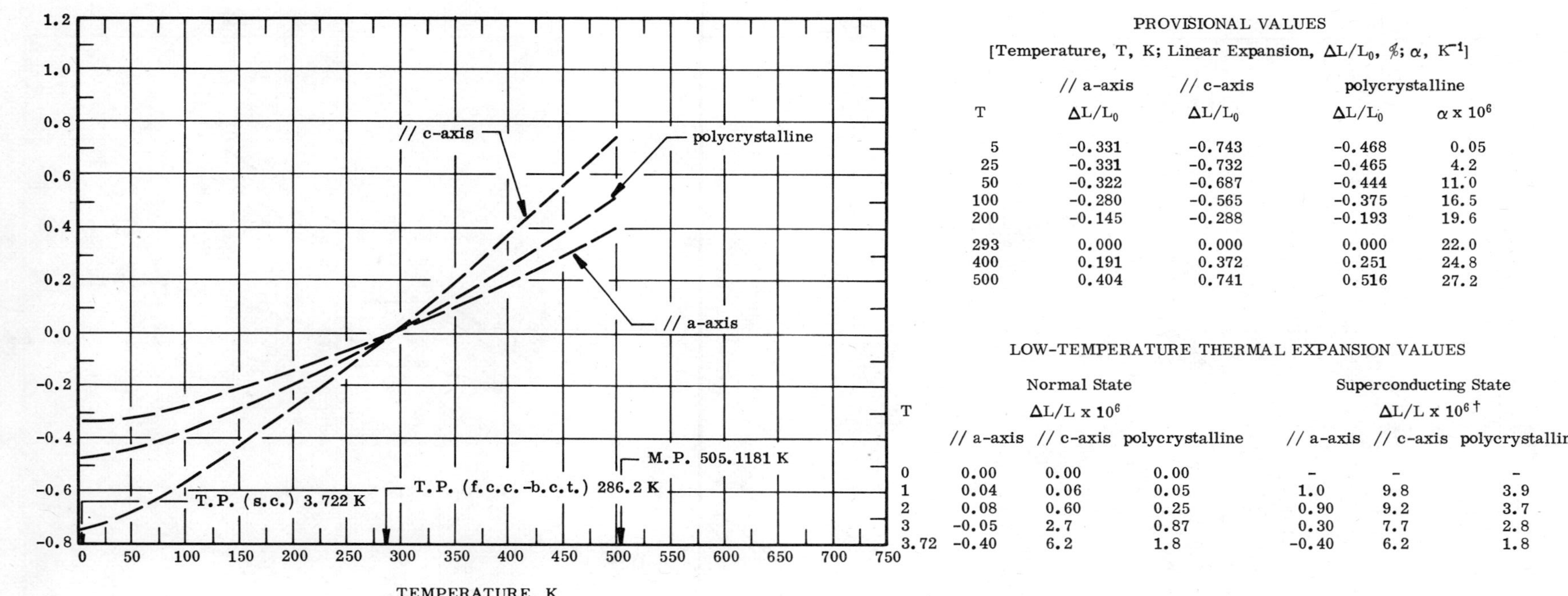

PROVISIONAL VALUES

[Temperature, T, K; Linear Expansion, $\Delta L/L_0$, %; α, K^{-1}]

T	// a-axis $\Delta L/L_0$	// c-axis $\Delta L/L_0$	polycrystalline $\Delta L/L_0$	polycrystalline $\alpha \times 10^6$
5	-0.331	-0.743	-0.468	0.05
25	-0.331	-0.732	-0.465	4.2
50	-0.322	-0.687	-0.444	11.0
100	-0.280	-0.565	-0.375	16.5
200	-0.145	-0.288	-0.193	19.6
293	0.000	0.000	0.000	22.0
400	0.191	0.372	0.251	24.8
500	0.404	0.741	0.516	27.2

LOW-TEMPERATURE THERMAL EXPANSION VALUES

T	Normal State $\Delta L/L \times 10^6$ // a-axis	// c-axis	polycrystalline	Superconducting State $\Delta L/L \times 10^6$ † // a-axis	// c-axis	polycrystalline
0	0.00	0.00	0.00	-	-	-
1	0.04	0.06	0.05	1.0	9.8	3.9
2	0.08	0.60	0.25	0.90	9.2	3.7
3	-0.05	2.7	0.87	0.30	7.7	2.8
3.72	-0.40	6.2	1.8	-0.40	6.2	1.8

† Superconducting state value is set equal to that of normal state value at the transition temperature, $T_c = 3.72$ K.

REMARKS

The tabulated values are considered accurate to within ±7% over the entire temperature range. The tabulated values for polycrystalline tin are calculated from the single crystal data and are considered accurate to within ±10% over the entire temperature range. These values can be represented approximately by the following equations:

// a-axis: $\Delta L/L_0 = -0.402 + 1.168 \times 10^{-3}\,T + 3.858 \times 10^{-7}\,T^2 + 9.973 \times 10^{-10}\,T^3$ $(100 < T < 500)$

// c-axis: $\Delta L/L_0 = -0.792 + 2.027 \times 10^{-3}\,T + 2.679 \times 10^{-6}\,T^2 - 1.188 \times 10^{-9}\,T^3$ $(100 < T < 500)$

polycrystalline: $\Delta L/L_0 = -0.525 + 1.354 \times 10^{-3}\,T + 1.587 \times 10^{-6}\,T^2 - 2.896 \times 10^{-10}\,T^3$ $(100 < T < 500)$

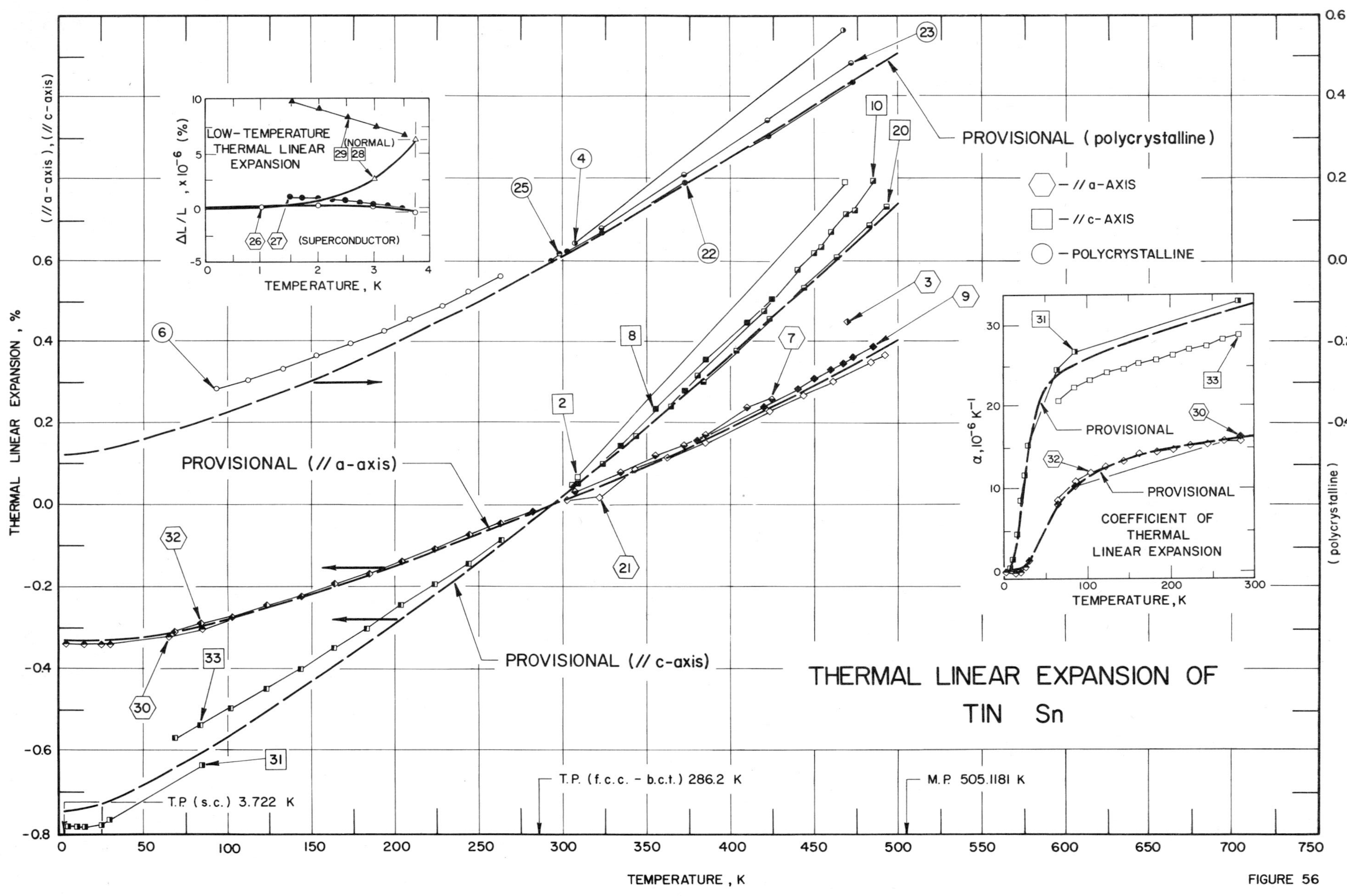
THERMAL LINEAR EXPANSION OF
TIN Sn
PROVISIONAL (polycrystalline)
— // a - AXIS
— // c - AXIS
— POLYCRYSTALLINE
PROVISIONAL (// a-axis)
PROVISIONAL (// c-axis)
THERMAL LINEAR EXPANSION , %
(// a - axis) , (// c - axis)
(polycrystalline)
TEMPERATURE , K
T.P. (s.c.) 3.722 K
T.P. (f.c.c. - b.c.t.) 286.2 K
M.P. 505.1181 K
LOW-TEMPERATURE THERMAL LINEAR EXPANSION
(NORMAL)
(SUPERCONDUCTOR)
ΔL/L , x 10⁻⁶ (%)
TEMPERATURE , K
α, 10⁻⁶ K⁻¹
PROVISIONAL
PROVISIONAL
COEFFICIENT OF THERMAL LINEAR EXPANSION
TEMPERATURE, K

FIGURE 56

SPECIFICATION TABLE 56. THERMAL LINEAR EXPANSION OF TIN Sn

Cur. No.	Ref. No.	Author(s)	Year	Method Used	Temp. Range, K	Name and Specimen Designation	Composition (weight percent), Specifications, and Remarks
1*	323	Novikova, S.I.	1961	L	24-217	α-Sn (gray tin)	α-Sn powder pressed in liquid nitrogen under 1500 atm for first specimen and 2500 atm for second specimen; recorded data compiled from 9 series of measurements on two specimens; this curve is here reported using the first given temperature as reference temperature at which $\Delta L/L_0 = 0$.
2	192	Shinoda, G.	1933	X	307-467	β-Sn	Pure granule; double body-centered tetragonal structure; expansion measured along c-axis.
3	192	Shinoda, G.	1933	X	307-470	β-Sn	The above specimen; expansion measured along a-axis.
4	192	Shinoda, G.	1933	X	307-467	β-Sn	The above specimen; mean expansion as determined by author.
5*	194	Olsen, J.L. and Rohrer, H.	1957	N	2-4	V-SN-3	Change of length measured below normal-superconducting transition; data are result of 10-15 trials; data reported using $\Delta L/L_0$ defined as $(L_s-L_n)/L_s$ where L_s is length in superconducting state and L_n in normal state; this curve is here reported using the first given temperature as reference temperature at which $\Delta L/L_0 = 0$; zero-point correction is -0.1793×10^{-6}%.
6	48	Dorsey, H.G.	1907	I	93-293		Chemically pure specimen from Eimer and Amend; 1.4332 cm long; density 7.32 g cm^{-3}.
7	324	Deshpande, V.T. and Sirdeshmukh, D.B.	1961	X	308-425	β-Sn	Tetragonal crystal structure; metal obtained from Johnsons of Hendon; melted and fine filings taken from cooled block; annealed at 423 K for 16 hr; expansion measured along a-axis; lattice parameter reported at 308 K is 5.8327 Å; 5.8313Å at 293 K determined by graphical extrapolation.
8	324	Deshpande, V.T. and Sirdeshmukh, D.B.	1961	X	308-425	β-Sn	The above specimen; expansion measured along c-axis; lattice parameter reported at 308 K is 3.1825 Å; 3.1809 Å at 293 K determined by graphical extrapolation.
9	325	Deshpande, V.T. and Sirdeshmukh, D.B.	1962	X	306-485	β-Sn	Similar to the above specimen; expansion measured along a-axis; reported error ± 0.0017 Å in measurement of lattice parameter; lattice parameter reported at 306 K is 5.8326 Å; 5.8312 Å at 293 K determined by graphical extrapolation; data on the coefficient of thermal linear expansion also reported.
10	325	Deshpande, V.T. and Sirdeshmukh, D.B.	1962	X	306-485	β-Sn	The above specimen; expansion measured along c-axis; reported error ± 0.009 Å in measurement of lattice parameter; lattice parameter reported at 306 K is 3.1812 Å; 3.1807 Å at 293 K determined by graphical extrapolation; data on the coefficient of thermal linear expansion also reported.
11	326	Thewlis, J. and Davey, A.R.	1954	X	143-296	α-Sn	99.99 Sn with main impurities Pb, Sb; grey tin supplied by Tin Research Inst.; lattice parameter reported at 293 K is 6.48923 Å.
12*	197	Vernon, E.V. and Weintroub, S.	1953	I	301-455		Single crystal about 2 cm long, with orientation $\Psi = 8°$.
13*	197	Vernon, E.V. and Weintroub, S.	1953	I	301-475		Similar to the above specimen, with orientation $\Psi = 65.5°$.
14*	87	Childs, B.G. and Weintroub, S.	1950	L	304-496		Single crystal produced from "Chempur" Sn, in form of rod 6 cm long, 3 mm diameter using horizontal gradient furnace of Kapitza type, with orientation $\Psi = 30°$.
15*	87	Childs, B.G. and Weintroub, S.	1950	L	304-496		Similar to the above specimen, with orientation $\Psi = 31.5°$.
16*	87	Childs, B.G. and Weintroub, S.	1950	L	304-497		Similar to the above specimen, with orientation $\Psi = 44.5°$.
17*	87	Childs, B.G. and Weintroub, S.	1950	L	304-497		Similar to the above specimen, with orientation $\Psi = 43.5°$.
18*	87	Childs, B.G. and Weintroub, S.	1950	L	326-453		Similar to the above specimen, with orientation $\Psi = 79.5°$.

* Not shown in figure.

SPECIFICATION TABLE 56. THERMAL LINEAR EXPANSION OF TIN Sn (continued)

Cur. No.	Ref. No.	Author(s)	Year	Method Used	Temp. Range, K	Name and Specimen Designation	Composition (weight percent), Specifications, and Remarks
19*	87	Childs, B.G. and Weintroub, S.	1950	L	303-496		Similar to the above specimen, with orientation $\Psi = 86.5°$.
20	87	Childs, B.G. and Weintroub, S.	1950	L	303-493		Similar to the above specimen, values along c-axis calculated from curves 14-19.
21	87	Childs, B.G. and Weintroub, S.	1950	L	303-493		Similar to the above specimen, values along a-axis calculated from curves 14-19.
22	263	Grube, G. and Vasskühler, H.	1934	L	303-473		Cylindrical specimen 8 cm long x 5 mm diameter.
23	263	Grube, G. and Vasskühler, H.	1934	L	303-473		0.21 Mg added, similar to the above specimen.
24*	263	Grube, G. and Vasskühler, H.	1934	L	303-473		0.4 Mg added, similar to the above specimen.
25	665	Balasundaram, L.J. and Sinha, A.N.	1971	L	293-398		Specimen is pure metal.
26	774	White, G.K.	1964	E	1.50-4.45		Cylindrical specimen of tetragonal structure prepared by Metals Research Ltd., Cambridge, England; measurements along a-axis in the magnetic field of 500 Oe; this curve is here reported using the first given temperature as reference temperature at which $\Delta L/L_0 = 0$.
27	774	White, G.K.	1964	E	1.54-4.45		The above specimen; measurements along a-axis without magnetic field; the data here referred to normal state value of curve 26 at the critical temperature, 3.72 K; zero-point correction is -0.93×10^{-6}%.
28	774	White, G.K.	1964	E	1.54-4.14		The above specimen; measurements along c-axis in the magnetic field of 500 Oe; this curve is here reported using the first given temperature as reference temperature at which $\Delta L/L_0 = 0$.
29	774	White, G.K.	1964	E	1.52-4.00		The above specimen; measurements along c-axis without magnetic field; the data here referred to normal state value of curve 28 at the critical temperature, 3.72 K; zero-point correction is 10.23×10^{-6}%.
30	774	White, G.K.	1964	E	4-85, 283		The above specimen; measurements along a-axis.
31	774	White, G.K.	1964	E	4-85, 283		The above specimen; measurements along c-axis.
32	710	Erfling, H.D.	1939	I	68-283		Tetragonal structure, 10.3 mm specimen; values along a-axis derived from the values along c-axis and along 11° to a-axis.
33	710	Erfling, H.D.	1939	I	68-283		Similar to the above specimen; values along c-axis.

* Not shown in figure.

DATA TABLE 56. THERMAL LINEAR EXPANSION OF TIN Sn

T	$\Delta L/L_0$
CURVE 1*, †, ‡	
24	0.00 x 10^{-3}
25	-0.08
28	-0.34
29	-0.43
35	-0.82
38	-0.96
39	-1.01
43	-1.14
44	-1.15
47	-1.14
48	-1.14
52	-1.02
53	-0.94
56	-0.64
57	-0.55
62	-0.05
66	0.53
71	1.53
74	2.20
75	2.43
80	3.70
84	4.75
86	5.29
87	5.58
92	7.18
94	7.81
97	8.80
99	9.52
101	10.20
102	10.54
106	11.91
108	12.63
110	13.38
111	13.78
116	15.72
119	16.94
125	19.45
128	20.73
133	23.00
137	24.91
138	25.38
139	25.84
145	38.64
152	32.10
153	32.60
154	33.09
161	36.62
163	37.63
CURVE 1 (cont.)*, †, ‡	
172	42.08
178	45.11
181	46.65
188	50.18
190	51.21
198	55.38
200	56.40
208	60.58
209	61.12
217	65.45
CURVE 2	
307	0.064
467	0.797
CURVE 3	
307	0.036*
470	0.447
CURVE 4	
307	0.045
467	0.564
CURVE 5*, †	
1.76	0.000 x 10^{-6}
1.95	-0.093
2.02	-0.131
2.11	-0.186
2.30	-0.306
2.37	-0.314
2.48	-0.443
2.67	-0.606
2.81	-0.662
2.94	-0.839
3.00	-0.856
3.11	-1.002
3.30	-1.230
3.53	-1.502
3.74	-1.793
CURVE 6	
93	-0.319
113	-0.292
133	-0.264
CURVE 6 (cont.)	
153	-0.234
173	-0.205
193	-0.174
213	-0.142
233	-0.107
253	-0.072
273	-0.037
293	0.000
CURVE 7	
308	0.024*
333	0.075
355	0.120
371	0.142
385	0.170
410	0.237
424	0.254
CURVE 8	
308	0.050
333	0.141
355	0.229
371	0.276
385	0.352
410	0.450
424	0.509
CURVE 9	
306	0.024*
379	0.156
421	0.236
439	0.282
451	0.308
453	0.312*
459	0.329
467	0.348
473	0.360
485	0.389
CURVE 10	
306	0.044
379	0.317
421	0.474
439	0.578
CURVE 10 (cont.)	
451	0.619
453	0.635
459	0.672
467	0.717
473	0.726
485	0.798
CURVE 11	
143	-0.073
159	-0.059
171	-0.057
181	-0.054*
197	-0.042
208	-0.039
221	-0.040
235	-0.023*
246	-0.021
260	-0.018
273	-0.006
293	-0.001*
293	0.000
294	0.000*
296	0.002
CURVE 12‡	
301	0.025
313	0.064
333	0.130
353	0.198
373	0.268
393	0.340
413	0.414
433	0.490
453	0.568*
455	0.576
CURVE 13‡	
301	0.015
313	0.038
333	0.077*
353	0.118
373	0.161
393	0.204
413	0.249
433	0.294
CURVE 13 (cont.)‡	
453	0.340
473	0.386*
475	0.390
CURVE 14‡	
304	0.032*
328	0.101
353	0.176
378	0.254
403	0.334
428	0.418
453	0.503
478	0.591
496	0.656
CURVE 15*, ‡	
304	0.031
328	0.098
353	0.171
378	0.246
403	0.325
428	0.406
453	0.490
478	0.576
496	0.640
CURVE 16‡	
304	0.028*
328	0.088
353	0.153
378	0.221
403	0.292
428	0.365
453	0.439
478	0.515
497	0.573
CURVE 17*, ‡	
304	0.027
328	0.086
353	0.150
378	0.217
403	0.287
428	0.359
453	0.433
CURVE 17 (cont.)*, ‡	
478	0.508
497	0.566
CURVE 18‡	
326	0.058
328	0.062*
353	0.107
378	0.153*
403	0.201
428	0.251
453	0.301
CURVE 19‡	
303	0.017
328	0.059
353	0.102*
378	0.147*
403	0.194
428	0.241
453	0.290
478	0.340
496	0.377
CURVE 20	
303	0.032*
323	0.097
343	0.164
363	0.234
383	0.305
403	0.379
423	0.455
443	0.532
463	0.611
483	0.692
493	0.733
CURVE 21	
303	0.016
323	0.019
343	0.083
363	0.118
383	0.154
403	0.191
423	0.229

* Not shown in figure.
† This curve is here reported using the first given temperature as reference temperature at which $\Delta L/L_0 = 0$.
‡ Author's data for coefficient of thermal expansion have been integrated by TPRC to obtain $\Delta L/L_0$.

DATA TABLE 56. THERMAL LINEAR EXPANSION OF TIN Sn (continued)

T	$\Delta L/L_0$
CURVE 21 (cont.)	
443	0.267
463	0.306
483	0.346
493	0.366
CURVE 22	
303	0.023
323	0.069
373	0.190
423	0.309
473	0.434
CURVE 23	
303	0.0267*
323	0.080
373	0.212
423	0.344
473	0.487
CURVE 24*	
303	0.0250
323	0.0751
373	0.2014
423	0.3310
473	0.4683
CURVE 25	
293	0.000
398	0.012
CURVE 26	
1	0.04×10^{-6}
2	0.08
3	-0.05
3.72	-0.4
CURVE 27	
1.50	0.93×10^{-6}
1.75	0.90
2.00	0.84
2.25	0.75
2.50	0.63
2.75	0.45
CURVE 27 (cont.)	
3.00	0.27
3.25	0.06
3.50	-0.16
3.72	-0.4
CURVE 28	
1	0.06×10^{-6}*
2	0.6*
3	2.7
3.72	6.16
CURVE 29	
1.52	9.81×10^{-6}
1.54	9.55*
2.05	9.31
2.56	8.23
2.63	8.33*
3.05	7.43
3.10	7.52*
3.39	6.67*
3.57	6.60
3.56	6.32*
3.72	6.16
CURVE 30‡	
283	-0.016
85	-0.306
65	-0.324
30	-0.341
25	-0.341
20	-0.341
15	-0.341
10	-0.341*
8	-0.341*
6	-0.341*
4	-0.341
CURVE 31‡	
283	-0.010*
85	-0.640
65	-0.700
30	-0.771
25	-0.781
20	-0.785*
15	-0.789
CURVE 31 (cont.)‡	
10	-0.790
8	-0.791*
6	-0.791*
4	-0.791
CURVE 32	
293	0.000
283	-0.016*
263	-0.048
243	-0.079
223	-0.109
203	-0.139
183	-0.168
163	-0.197
143	-0.224
123	-0.250
101.7	-0.277
84	-0.297
68	-0.312
CURVE 33	
293	0.000
283	-0.029*
263	-0.086
243	-0.142
223	-0.197
203	-0.250
183	-0.302
163	-0.354
143	-0.404
123	-0.453
101.7	-0.503
84	-0.544
68	-0.578

* Not shown in figure.

‡ Author's data for coefficient of thermal expansion have been integrated by TPRC to obtain $\Delta L/L_0$.

DATA TABLE 56. COEFFICIENT OF THERMAL LINEAR EXPANSION OF TIN Sn

[Temperature, T, K; Coefficient of Expansion, α, 10^{-6} K^{-1}]

T	α
CURVE 1‡	
24	-0.95
25	-0.77*
28	-0.95*
29	-0.79*
35	-0.52
38	-0.40*
39	-0.67*
43	0.03
44	-0.12*
47	0.15*
48	-0.05
52	0.66*
53	0.90
56	1.08
57	0.76
62	1.24*
66	1.68
71	2.30
74	2.18*
75	2.42
80	2.67*
84	2.57*
86	2.81
87	3.01*
92	3.37
94	3.00
97	3.56*
99	3.64
101	3.17*
102	3.57*
106	3.29
108	3.93
110	3.62*
111	3.85*
116	4.01
119	4.14
125	4.22
128	4.33*
133	4.71
137	4.86
138	4.61*
139	4.52*
145	4.81
152	5.09
153	4.83
154	4.98
CURVE 1 (cont.)‡	
161	5.12
163	4.96
172	4.93
178	5.16
181	5.10
188	4.98
190	5.40
198	5.00
200	5.30
208	5.14
209	5.61
217	5.22
CURVE 6	
103	13.61
123	13.95
143	14.88
163	14.73
183	15.54
203	16.09
223	17.29
243	17.46
263	17.80
283	18.32
CURVE 9*	
303	16.5
323	17.0
343	17.8
363	18.7
383	19.7
403	20.2
423	21.6
433	22.3
443	22.8
463	24.3
473	25.4
483	26.2
CURVE 10*	
303	32.4
323	33.9
343	34.9
CURVE 10 (cont.)*	
363	36.5
383	39.0
403	41.2
423	43.7
433	45.9
443	47.8
463	51.2
473	53.7
483	56.9
CURVE 12*, ‡	
301	31.93
313	32.51
333	33.53
353	34.55
373	35.53
393	36.46
413	37.46
433	38.44
453	39.43
455	39.54
CURVE 13*, ‡	
301	18.81
313	19.37
333	20.21
353	20.89
373	21.51
393	21.99
413	22.40
433	22.75
453	22.92
473	23.03
475	23.03
CURVE 14*, ‡	
304	28.29
328	29.39
353	30.57
378	31.70
403	32.78
428	33.79
453	34.73
CURVE 14 (cont.)*, ‡	
478	35.59
496	36.24
CURVE 15*, ‡	
304	27.48
328	28.57
353	29.66
378	30.81
403	31.94
428	32.95
453	34.07
478	35.05
496	35.77
CURVE 16*, ‡	
304	24.63
328	25.55
353	26.59
378	27.82
403	28.79
428	29.55
453	30.12
478	30.55
497	30.73
CURVE 17*, ‡	
304	24.10
328	25.00
353	26.14
378	27.39
403	28.52
428	29.30
453	29.89
478	30.48
497	30.66
CURVE 18*, ‡	
326	17.58
328	17.66
353	18.22
378	18.88
403	19.54
CURVE 18 (cont.)*, ‡	
428	20.17
453	20.54
CURVE 19‡	
303	16.46
328	17.06
353	17.70
378	18.29
403	18.93
428	29.36
453	29.79
478	20.27
496	20.43
CURVE 30‡	
283	16.3
85	10.3
65	8.0
30	1.2
25	0.43
20	-0.37
15	-0.15
10	-0.10*
8	-0.07*
6	-0.04*
4	-0.01
CURVE 31‡	
283	33.4
85	26.7
65	24.7
30	15.2
25	11.7
20	8.4
15	4.75
10	1.65
8	0.84
6	0.28
4	0.075
CURVE 32	
283	15.83
CURVE 32 (cont.)	
263	15.76
243	15.44
223	15.16
203	14.84
183	14.41
163	14.03
143	13.38
123	12.77
101.7	11.79
84	10.66
68	8.58
CURVE 33	
283	28.99
363	28.36
243	27.59
223	27.01
203	26.33
183	25.92
163	25.46
143	24.78
123	24.09
101.7	23.23
84	22.45
68	20.57

* Not shown in figure.
‡ Author's data for coefficient of thermal expansion have been integrated by TPRC to obtain $\Delta L/L_0$.

FIGURE AND TABLE NO. 57R. RECOMMENDED VALUES FOR THERMAL LINEAR EXPANSION OF TITANIUM Ti

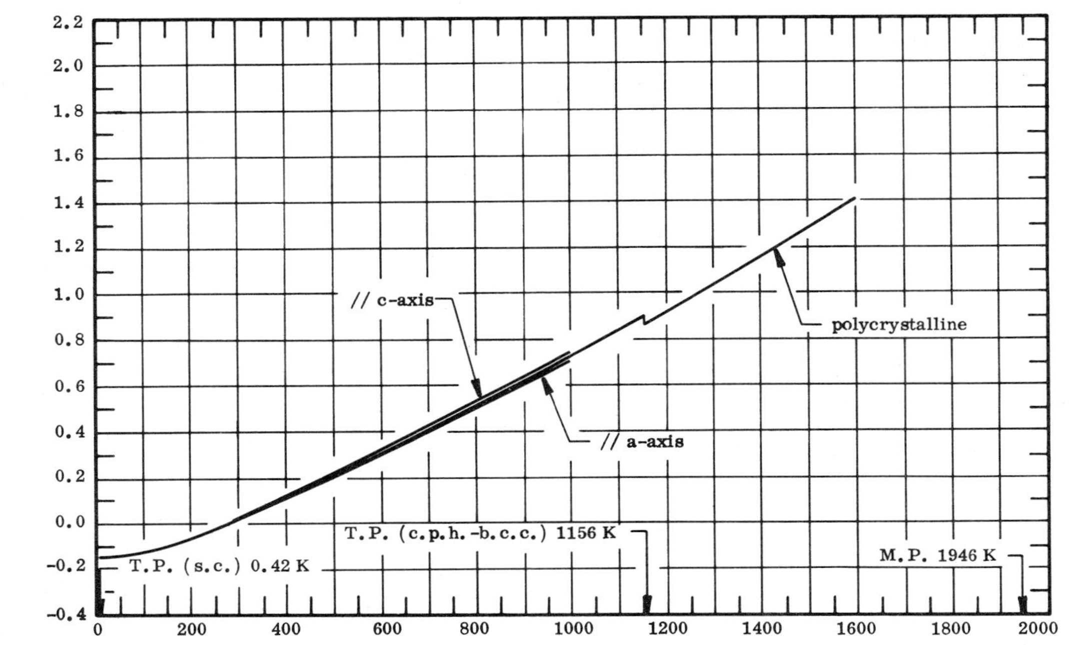

RECOMMENDED VALUES

[Temperature, T, K; Linear Expansion, $\Delta L/L_0$, %; α, K^{-1}]

T	// a-axis $\Delta L/L_0$	// c-axis $\Delta L/L_0$	polycrystalline $\Delta L/L_0$	polycrystalline α
5			-0.153	0.01
25			-0.153	0.14
50			-0.151	1.2
100			-0.136	4.5
200			-0.074	7.4
293	0.000	0.000	0.000	8.6
400	0.095	0.100	0.096	9.4
500	0.191	0.198	0.193	9.9
600	0.293	0.299	0.295	10.4
700	0.399	0.404	0.401	10.8
800	0.509	0.511	0.510	11.1
900	0.623	0.620	0.622	11.3
1000	0.740	0.730	0.736	11.5
1156†			0.918	11.8
1156†			0.868	11.0
1200			0.917	11.3
1400			1.153	12.3
1600			1.411	13.5

† Phase transition.

REMARKS

The tabulated values for polycrystalline material are considered accurate to within ± 3% at temperatures below 1156 K and ± 5% above. The tabulated values along a-axis and c-axis are considered accurate to within ± 5% over the entire temperature range. These values can be represented approximately by the following equations:

// a-axis: $\Delta L/L_0 = -0.215 + 6.081 \times 10^{-4}\,T + 4.750 \times 10^{-7}\,T^2 - 1.279 \times 10^{-10}\,T^3$

// c-axis: $\Delta L/L_0 = -0.247 + 7.617 \times 10^{-4}\,T + 2.937 \times 10^{-7}\,T^2 - 7.831 \times 10^{-11}\,T^3$

polycrystalline: $\Delta L/L_0 = -0.198 + 5.072 \times 10^{-4}\,T + 6.500 \times 10^{-7}\,T^2 - 2.203 \times 10^{-10}\,T^3$ $(150 < T < 1156)$

$\Delta L/L_0 = 0.328 - 3.555 \times 10^{-4}\,T + 8.844 \times 10^{-7}\,T^2 - 1.496 \times 10^{-10}\,T^3$ $(1156 < T < 1600)$

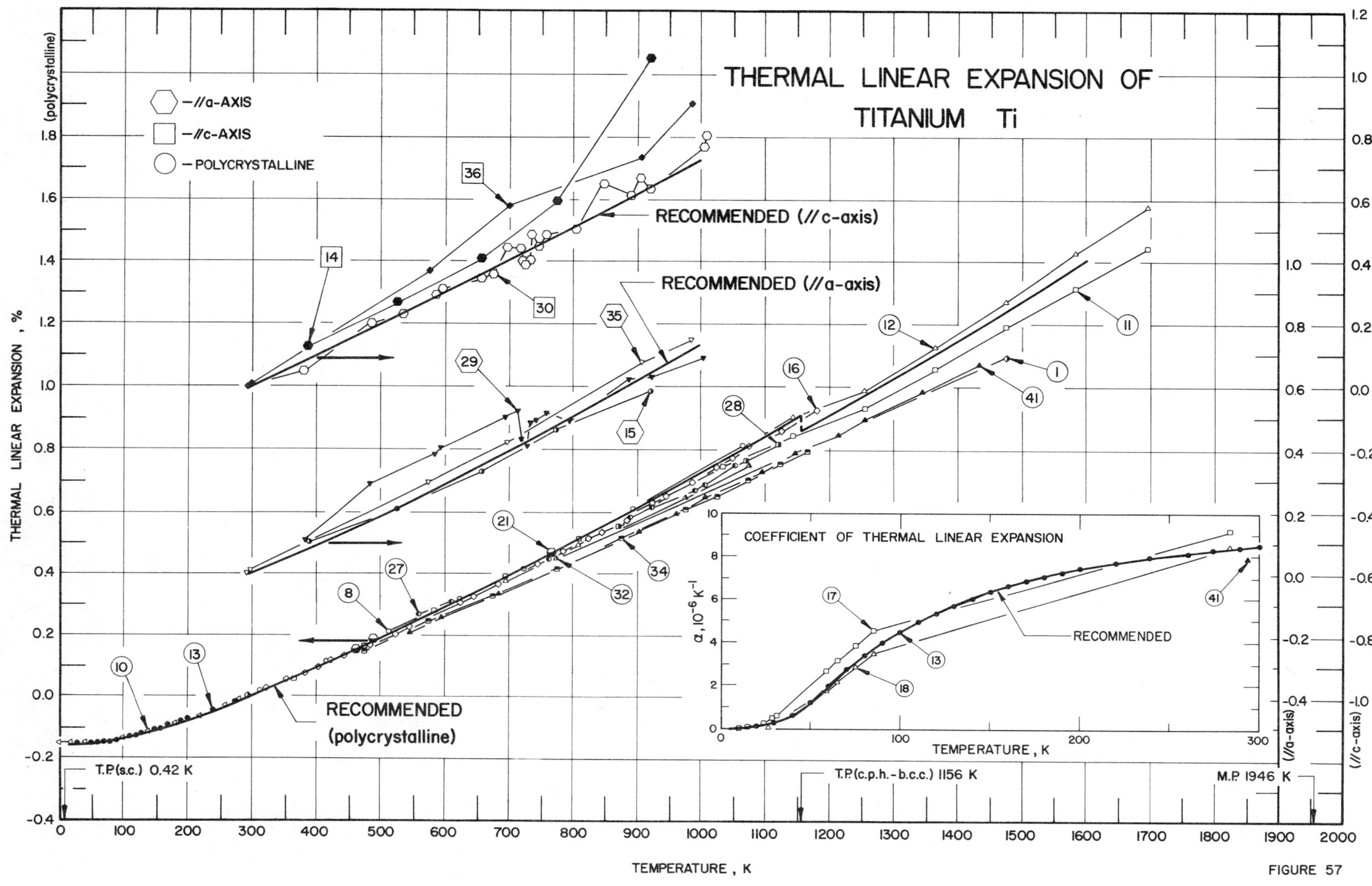

THERMAL LINEAR EXPANSION OF
TITANIUM Ti
—//a-AXIS
—//c-AXIS
—POLYCRYSTALLINE
RECOMMENDED (//c-axis)
RECOMMENDED (//a-axis)
RECOMMENDED
(polycrystalline)
THERMAL LINEAR EXPANSION, %
(polycrystalline)
(//a-axis)
(//c-axis)
TEMPERATURE, K
T.P.(s.c.) 0.42 K
T.P.(c.p.h.-b.c.c.) 1156 K
M.P. 1946 K
COEFFICIENT OF THERMAL LINEAR EXPANSION
RECOMMENDED
α, 10^{-6} K^{-1}
TEMPERATURE, K

FIGURE 57

SPECIFICATION TABLE 57. THERMAL LINEAR EXPANSION OF TITANIUM Ti

Cur. No.	Ref. No.	Author(s)	Year	Method Used	Temp. Range, K	Name and Specimen Designation	Composition (weight percent), Specifications, and Remarks
1	208	Frantsevich, I. N., Zhurakovskii, E. A. and Lyashchenko, A. B.	1967	X	293, 1473		Sample prepared from powder by vacuum sintering of pressed block.
2*	329	Kornilov, I. I. and Boriskina, N.G.	1966		330-1273		(Cr + Fe + Si + B) < 0. 015, (Al + Mb + Se) < 0. 01; specimen forged at 1323 to 1373 K; zero-point correction of 0.01% determined by graphical extrapolation; data on the coefficient of thermal linear expansion also reported.
3*	330	Kovtun, S. F. and Ul'yanov, R. A.	1963	L	74-1294		Technical grade Ti tested; zero-point correction of -0. 025% determined by graphical interpolation.
4*	331	Margolin, H. and Portisch, H.	1968	X	435-513		Bureau of Mines titanium samples used; 0. 108 H; subjected to hydrogenation at 1173 to 873 K; expansion measured along a-axis; this curve is here reported using the first given temperature as reference temperature at which $\Delta L/L_0 = 0$; lattice parameter reported at 435 K is 2. 9555 Å.
5*	331	Margolin, H. and Portisch, H.	1968	X	437-519		The above specimen; expansion measured along c-axis; this curve is here reported using the first given temperature as reference temperature at which $\Delta L/L_0 = 0$; lattice parameter reported at 437 K is 4. 6918 Å.
6*	117	Yaggee, F. L. and Styles, J. W.	1966	L	273-773		Density 4. 51 ± 0. 02 g cm^{-3}.
7*	117	Yaggee, F. L. and Styles, J. W.	1966	L	273-1273		The above specimen.
8	332	Greiner, E. S. and Ellis, W. C.	1948	L	311-1064		0. 12 Mg, 0. 06 Mn, < 0. 03 each Ca, Cu, Fe, Sn, V, < 0. 01 Si, < 0. 005 Al; cylindrical specimen, 0. 2 in. diameter, 2. 4 in. long; titanium furnished by U. S. Bureau of Mines; specimen made from annealed sintered compact 0. 38 in. thick; sections of compact cold-swaged with intermediate anneals in vacuum at 1073 K; expansion measured in slowly circulating helium atmosphere; zero-point correction of 0. 017% determined by graphical extrapolation.
9*	332	Greiner, E. S. and Ellis, W. C.	1948	L	311-1100		0. 04 Mg, 0. 03 each Mn, Si, < 0. 005 each Al, Cu, Fe, Na, < 0. 001 Ca; cylindrical specimen 0. 2 in. diameter, 2. 4 in. long; titanium furnished by U. S. Bureau of Mines; specimen made from rod of titanium swaged to size with intermediate anneals; expansion measured in slowly circulating helium atmosphere; zero-point correction of 0. 017% determined by graphical extrapolation.
10	83	Laquer, H. L.	1952	L	0-300		Specimen of unknown origin, machined to 5. 1 cm long; zero-point correction of -0. 018% determined by graphical interpolation.
11	333	Williams, D. N.	1961		293-1699	A-55 grade	Rod specimen 1. 6 cm diameter; expansion measured under 3 x 10^{-4} mm Hg vacuum with increasing temperature.
12	333	Williams, D. N.	1961		1699-293		The above specimen; expansion measured with decreasing temperature.
13	8	Altman, H. W., Rubin, T. and Johnston, H. L.	1954	L	20-300		Sample obtained from "as deposited" metal, cold swaged and unannealed; obtained from Foote Mineral Co.; zero-point correction of 0. 0045% determined by graphical interpolation; data on the coefficient of thermal linear expansion also reported.
14	334	Willens, R. H.	1961	X	293-920		Remelted iodide titanium, supplied by U. S. Ordnance Corps., Watertown Arsenal Labs. in form of hot rolled plates about 4 mm thick; small trapezoids cut from plates and milled on both sides to thickness of about 2. 5 mm, sealed in quartz tube filled with helium, annealed 323 K above α-β transformation temperature for 15 min., and then quickly returned to room temperature, riveted to a titanium specimen wheel with titanium rivets; segments on front face of specimen wheel made coplanar and normal to wheel axis by grinding; cold worked layer produced by grinding removed by acid etch of equal parts of nitric acid, hydrofluoric acid and glycerin; placed in vacuum system; after 3 days of pumping annealed at 973 K for 2 hr; expansion measured along c-axis; lattice parameter reported at 293 K is 4. 6822 Å.

* Not shown in figure.

SPECIFICATION TABLE 57. THERMAL LINEAR EXPANSION OF TITANIUM Ti (continued)

Cur. No.	Ref. No.	Author(s)	Year	Method Used	Temp. Range, K	Name and Specimen Designation	Composition (weight percent), Specifications, and Remarks
15	334	Willens, R. H.	1961	X	293-920		The above specimen; expansion measured along a-axis; lattice parameter reported at 293 K is 2. 9496 Å.
16	119	Yaggee, F. L., Gilbert, E. R. and Styles, Y. W.	1969	L	294-1180		(C, O, N, H) 0. 0917% interstitially in commercial crystal bar; rod specimen 0. 64 cm diameter; annealed in vacuum of 2 x 10^{-6} mm Hg at 1273 K for 1 hr; density at 298 K 4.512 ± 0. 003 g cm^{-3}; zero-point correction of -0. 022% determined by graphical interpolation.
17*	335	Cowan, Y. A., Pawlowicz, A. T., and White, G. K.	1968	E	10-85, 283	Ti_1	ICI iodide-titanium of Vickers hardness 100 melted and hot forged (773-873 K); pickled; vacuum annealed, machined and lapped to a cylinder 2 cm diameter and 5 cm long; this curve is here reported using the first given temperature as reference temperature at which $\Delta L/L_0 = 0$.
18*	335	Cowan, Y. A., et al.	1968	E	10-85, 283	Ti_2	99. 95 Ti; electron-beam melted material supplied by Japan Vacuum Engineering Co., Tokyo; 2 cm diameter and 2. 60 cm long; this curve is here reported using the first given temperature as reference temperature at which $\Delta L/L_0 = 0$.
19*	336	Hidnert, P.	1943	T	295-79		Specimen cut from plate furnished by Titanium Alloy Mfg. Co.; chemical analysis supplied with material 97. 2 Ti, 1. 05 Si, 1. 11 Fe, 0. 22 C, 0. 2 Nb, 0. 17 V; specimen cut to bar 15 cm long x 14 x 9 mm; expansion measured with decreasing temperature; zero-point correction of 0. 002% determined by graphical interpolation.
20*	336	Hidnert, P.	1943	T	295-79		The above specimen; expansion measured with increasing temperature; zero-point correction of 0. 002% determined by graphical interpolation.
21	336	Hidnert, P.	1943	T	295-773		Similar to the above specimen except 20 x 10 mm cross section; expansion measured with increasing temperature; zero-point correction of 0. 002% determined by graphical extrapolation.
22*	336	Hidnert, P.	1943	T	562-296		The above specimen; expansion measured with decreasing temperature; zero-point correction of 0. 012% determined by graphical extrapolation.
23*	336	Hidnert, P.	1943	T	295-973		The above specimen; expansion measured on second test with increasing temperature; zero-point correction of 0. 002% determined by graphical extrapolation.
24*	336	Hidnert, P.	1943	T	552-325		The above specimen; expansion measured with decreasing temperature; zero-point correction of -0. 012% determined by graphical extrapolation.
25*	336	Hidnert, P.	1943	T	295-973		The above specimen; expansion measured on third test with increasing temperature; zero-point correction of 0. 002% determined by graphical extrapolation.
26*	336	Hidnert, P.	1943	T	486-316		The above specimen; expansion measured with decreasing temperature; zero-point correction of 0. 01% determined by graphical extrapolation.
27	313	Makin, S. M., Standring, J. and Hunter, P. M.	1953	I	566-1054		Fully softened specimen; expansion measured with increasing temperature.
28	313	Makin, S. M., et al.	1953	I	1119-381		The above specimen; expansion measured with decreasing temperature.
29	337	Wasilewski, R. J.	1961	X	293-1007		0. 0065 O_2, 0. 008 N_2 present as interstitials in high purity titanium powder of -325 mesh; CoKα x-radiation used; expansion measured along a-axis; reported error less than ± 0. 0005 Å in measurement of lattice parameter; lattice parameter reported at 293 K is 2. 95085 Å.

* Not shown in figure.

SPECIFICATION TABLE 57. THERMAL LINEAR EXPANSION OF TITANIUM Ti (continued)

Cur. No.	Ref. No.	Author(s)	Year	Method Used	Temp. Range, K	Name and Specimen Designation	Composition (weight percent), Specifications, and Remarks
30	337	Wasilewski, R. J.	1961	X	292-1008		The above specimen; expansion measured along c-axis; lattice parameter reported at 292 K is 4.6825 Å; 4.68385 Å at 293 K determined by graphical interpolation.
31*	276	Andres, K.	1963	O	2.0-13		99.999 Ti; polycrystalline specimen 7.4 cm long; this curve reported using the first given temperature as reference temp. at which $\Delta L/L_0 = 0$; correction -9.72×10^{-7}%.
32	338	Pridantseva, K. S.	1964		293-1073		100 Ti; iodide titanium remelted in arc furnace.
33*	338	Pridantseva, K. S.	1964		293-1073		Commercial titanium containing 0.1 each Si, Zr.
34	338	Pridantseva, K. S.	1964		381-1169		Pure titanium.
35	339	Berry, R. L. P. and Raynor, G. V.	1953	X	293-983		0.16 C, 0.08 Fe, 0.013 N and no other impurity exceeding 0.02; specimen filings taken from metal prepared by iodide process; expansion measured along a-axis; lattice parameter reported at 293 K is 2.9437 Kx.
36	339	Berry, R. L. P. and Raynor, G. V.	1953	X	293-985		The above specimen; expansion measured along c-axis; lattice parameter reported at 293 K is 4.6725 kX; contamination with oxygen and nitrogen suspected.
37*	340	Greiner, E. S. and Ellis, W. C.	1949	L	303-1073		0.12 Mg, 0.06 Mn, < 0.03 each Ca, Cu, Fe, Sn, V, < 0.01 Si, 0.005 Al; sample made from annealed sintered compact, cylindrical specimen, approximately 0.2 in. diameter, 2.4 in. long.
38*	340	Greiner, E. S. and Ellis, W. C.	1949	L	303-1073		0.04 Mg, 0.03 Mn, 0.03 Si, < 0.005 each Al, Cu, Fe, Na, < 0.001 Ca; sample made from rod of Ti, cold-swaged with intermediate anneals; cylindrical specimen, approximately 0.2 in. diameter, 2.4 in. long.
39*	707	Pridantseva, K. S. and Solov'yeva, N. A.	1965	L	293, 473		Cast sample, 1% measurement error.
40*	707	Pridantseva, K. S. and Solov'yeva, N. A.	1965	L	293, 473		Powder sample, 1% measurement error.
41	757	Mal'ko, P. I., Arensburger, D. S., Pugin, V. S., Nemchenko, V. F. and L'vov, S. N.	1970	O	293-1430		Electrolytic powder specimen; sintered for 2 hr in thoroughly dried argon at 1373 K; transformation reported at 1155 K.

* Not shown in figure.

DATA TABLE 57. THERMAL LINEAR EXPANSION OF TITANIUM Ti

[Temperature, T, K; Linear Expansion, $\Delta L/L_0$, %]

T	$\Delta L/L_0$
CURVE 1	
293	0.000
1473	1.097
CURVE 2*	
330	0.029
373	0.067
420	0.105
479	0.152
532	0.203
581	0.227
620	0.263
676	0.306
727	0.352
778	0.399
829	0.444
873	0.507
927	0.546
973	0.597
1026	0.658
1073	0.718
1118	0.791
1175	0.858
1222	0.918
1273	1.011
CURVE 3*	
74	-0.166
173	-0.099
275	-0.025
294	0.002
375	0.070
475	0.169
576	0.263
675	0.362
776	0.468
875	0.570
975	0.672
1072	0.774
1173	0.883
1268	0.968
1294	0.996
CURVE 4*, †	
435	0.000
474	0.017
513	0.125
CURVE 5*, †	
437	0.000
519	0.100
CURVE 6*	
273	-0.191
773	0.459
CURVE 7*	
273	-0.0212
1273	1.040
CURVE 8	
311	0.017
417	0.117
513	0.211
623	0.317
721	0.417
891	0.616
1064	0.817
CURVE 9*	
311	0.017
417	0.119
513	0.211
623	0.318
721	0.415
902	0.605
1100	0.817
CURVE 10	
0	-0.150
20	-0.150
40	-0.149
60	-0.148*
80	-0.144*
CURVE 10 (cont.)	
100	-0.136
120	-0.126*
140	-0.115
160	-0.103*
180	-0.089
200	-0.076*
220	-0.061
240	-0.045*
260	-0.029
273	-0.018*
280	-0.011
300	0.006*
CURVE 11	
293	0.000*
366	0.057
477	0.163
588	0.277
699	0.393
810	0.511
922	0.629
1033	0.745
1144	0.845
1255	0.933
1366	1.057
1477	1.189
1588	1.319
1699	1.443
CURVE 12	
1699	1.597
1588	1.431
1477	1.273
1366	1.127
1255	0.989
1144	0.905
1033	0.753*
922	0.621*
810	0.497
699	0.379
588	0.263*
477	0.153*
366	0.053*
293	0.000*
CURVE 13	
20	-0.1537*
30	-0.1537
40	-0.1534*
50	-0.1526*
60	-0.1511
70	-0.1489*
80	-0.1457*
90	-0.1420*
100	-0.1377*
110	-0.1330*
120	-0.1279
130	-0.1223*
140	-0.1165*
150	-0.1103*
160	-0.1037
170	-0.0970*
180	-0.0900*
190	-0.0829*
200	-0.0755*
220	-0.0604*
240	-0.0448
260	-0.0287*
273.16	-0.0201*
290	-0.0040*
298.16	0.0029*
300	0.0045*
CURVE 14	
293	0.000*
386	0.128
523	0.263
656	0.406
774	0.598
920	1.051
CURVE 15	
293	0.000*
386	0.102
523	0.210
656	0.329
774	0.468
920	0.590
CURVE 16	
273	-0.022*
294	0.001*
323	0.023
345	0.042*
365	0.059
383	0.078
404	0.096
425	0.115
444	0.134
485	0.170
525	0.208
545	0.228*
585	0.266*
625	0.305
646	0.327
686	0.367
725	0.409
746	0.429*
785	0.473
824	0.516*
846	0.536
885	0.578
927	0.625*
947	0.653
986	0.698
1027	0.744
1046	0.767
1085	0.815
1126	0.864
1180	0.927
CURVE 17*, †, ‡	
10	0.000 x 10^{-3}
12	0.0079
14	0.0194
16	0.0352
18	0.0564
20	0.0844
22	0.1214
24	0.1699
26	0.2334
28	0.3154
30	0.4174
58	4.967
65	7.022
CURVE 17 (cont.)*, †, ‡	
75	10.532 x 10^{-3}
85	14.712
CURVE 18*, †, ‡	
12	0.000 x 10^{-3}
14	0.0039
16	0.0085
18	0.0140
20	0.0206
22	0.0288
24	0.0397
26	0.0550
28	0.0770
30	0.107
58	2.795
65	4.160
75	6.655
85	9.815
CURVE 19*	
295	0.002
273	-0.015
195	-0.081
79	-0.144
CURVE 20*	
79	-0.144
195	-0.081
273	-0.015
295	0.002
CURVE 21	
295	0.002*
330	0.036*
365	0.067*
374	0.070*
416	0.108*
463	0.155
489	0.188
571	0.260*
579	0.270*
674	0.375*

* Not shown in figure.
† This curve is here reported using the first given temperature as reference temperature at which $\Delta L/L_0 = 0$.
‡ Author's data for coefficient of thermal expansion have been integrated by TPRC to obtain $\Delta L/L_0$.

DATA TABLE 57. THERMAL LINEAR EXPANSION OF TITANIUM Ti (continued)

T	$\Delta L/L_0$
CURVE 21 (cont.)	
767	0.471
773	0.480*
CURVE 22*	
562	0.264
470	0.171
421	0.120
377	0.077
330	0.034
296	0.003
CURVE 23*	
295	0.002
368	0.068
375	0.073
474	0.165
479	0.172
485	0.177
551	0.245
630	0.327
778	0.487
862	0.589
973	0.732
CURVE 24*	
552	0.256
464	0.169
454	0.158
409	0.116
364	0.072
333	0.045
325	0.036
CURVE 25*	
295	0.002
373	0.075
383	0.082
409	0.112
482	0.182
495	0.193
560	0.262
679	0.378
896	0.616
973	0.712
CURVE 26*	
486	0.198
428	0.138
418	0.126
369	0.080
328	0.041
316	0.027
CURVE 27	
566	0.263
613	0.306
692	0.383*
721	0.412*
767	0.445
890	0.584
924	0.620
989	0.672
1004	0.688
1054	0.748
CURVE 28	
1119	0.819
1070	0.768
968	0.656*
869	0.555
821	0.505*
476	0.177*
381	0.088*
CURVE 29	
293	0.000*
381	0.107
485	0.290
585	0.391
595	0.405
697	0.507
716	0.503
720	0.432
730	0.415
732	0.486
743	0.496
760	0.513
799	0.496
886	0.625
921	0.632
1007	0.690
CURVE 30	
292	-0.029*
292	0.029*
380	0.050
488	0.200
531	0.230
586	0.296
595	0.313
653	0.345
675	0.364
699	0.442
719	0.441
720	0.405
724	0.392
731	0.402
735	0.490
746	0.456
759	0.494
804	0.505
847	0.657
890	0.618
901	0.674
919	0.637
1004	0.776
1008	0.804
CURVE 31*, †	
2.00	0.000 x 10^{-7}
2.41	2.77
2.41	3.28
2.88	9.14
3.08	11.86
3.29	13.13
3.33	14.47
3.51	15.67
3.91	25.07
3.91	25.73
3.96	27.79
4.37	35.39
4.93	49.06
4.93	51.71
5.08	54.88
5.48	69.15
5.51	68.16
6.09	87.68
6.09	88.93
6.20	95.48
6.74	116.88
CURVE 31 (cont.)*, †	
6.84	123.28 x 10^{-7}
7.55	155.78
7.90	170.68
8.14	191.18
8.38	204.18
8.57	211.68
9.07	247.98
9.24	257.38
9.59	284.18
9.94	306.58
10.43	344.1
10.51	351.18
10.82	385.58
10.86	390.78
11.20	427.28
11.60	454.88
11.76	469.48
12.06	510.28
12.06	517.98
12.19	520.28
12.27	521.58
13.01	604.68
CURVE 32	
293	0.000*
573	0.252*
773	0.449
973	0.646
1073	0.757
CURVE 33*	
293	0.000
573	0.246
773	0.449
973	0.656
1073	0.725
CURVE 34‡	
381	0.073*
476	0.152
578	0.241
674	0.327
774	0.419
877	0.515
978	0.612
CURVE 34 (cont.)‡	
1024	0.656
1074	0.705
1123	0.753
1169	0.797
CURVE 35	
293	0.000
298	0.003
574	0.296
697	0.421
909	0.679
983	0.747
CURVE 36	
293	0.000
298	0.008
575	0.372
700	0.580
909	0.738
985	0.910
CURVE 37*	
303	0.009
473	0.158
673	0.361
873	0.586
1073	0.811
CURVE 38*	
303	0.009
473	0.164
673	0.357
873	0.563
1073	0.871
CURVE 39*	
293	0.000
473	0.162
CURVE 40*	
293	0.000
473	0.153
CURVE 41‡	
293	0.000*
365	0.056*
365	0.056*
473	0.148*
477	0.152*
523	0.193*
547	0.214
567	0.232*
595	0.256
672	0.326*
686	0.339
774	0.420*
871	0.510*
871	0.510*
902	0.540
961	0.598
972	0.609*
1008	0.645
1094	0.728
1127	0.761*
1147	0.781
1212	0.847
1256	0.893
1341	0.982
1431	1.079

* Not shown in figure.
† This curve is here reported using the first given temperature as reference temperature at which $\Delta L/L_0 = 0$.
‡ Author's data for coefficient of thermal expansion have been integrated by TPRC to obtain $\Delta L/L_0$.

DATA TABLE 57. COEFFICIENT OF THERMAL LINEAR EXPANSION OF TITANIUM Ti

[Temperature, T, K; Coefficient of Expansion, α, 10^{-6} K^{-1}]

T	α
CURVE 2*	
377	6.45
473	6.51
581	6.80
682	7.36
780	7.91
873	8.14
967	8.77
1073	9.05
1176	9.51
1265	9.92
CURVE 13	
20	0.08
30	0.26
40	0.60
50	1.18
60	1.96
70	2.72
80	3.39
90	3.98
100	4.48
110	4.92
120	5.32
130	5.69
140	6.03
150	6.35
160	6.63
170	6.87
180	7.07
190	7.26
200	7.43
220	7.70
240	7.92
260	8.11
273.16	8.30
290	8.39
300	8.48
CURVE 17‡	
10	0.03
12	0.04*
14	0.06*
16	0.09
18	0.12*
CURVE 17 (cont.)	
20	0.16*
22	0.21*
24	0.27
26	0.36*
28	0.46*
30	0.56
58	2.6
65	3.1
75	3.8
85	4.5
283	9.1
CURVE 18‡	
12	0.01*
14	0.02*
16	0.02*
18	0.03*
20	0.03*
22	0.04*
24	0.06*
26	0.09
28	0.13*
30	0.17*
58	1.75
65	2.1
75	2.8
85	3.4
283	8.4
CURVE 34*,‡	
381	8.07
476	8.57
578	8.83
674	9.13
774	9.29
877	9.41
978	9.64
1024	9.76
1074	9.82
1123	9.75
1169	9.49
CURVE 41‡	
293	7.90
365	7.79*
365	8.72*
473	8.32*
477	8.83*
523	8.83*
547	9.10*
567	8.43*
595	8.95*
672	9.31*
686	9.13*
774	9.14*
871	9.48*
871	10.07*
902	9.62*
961	9.77*
972	10.46*
1008	9.65*
1094	9.71*
1127	10.24*
1147	10.01*
1212	10.30*
1256	10.46*
1341	10.51*
1431	10.93*

* Not shown in figure.
‡ Author's data for coefficient of thermal expansion have been integrated by TPRC to obtain $\Delta L/L_0$.

FIGURE AND TABLE NO. 58R. RECOMMENDED VALUES FOR THERMAL LINEAR EXPANSION OF TUNGSTEN W

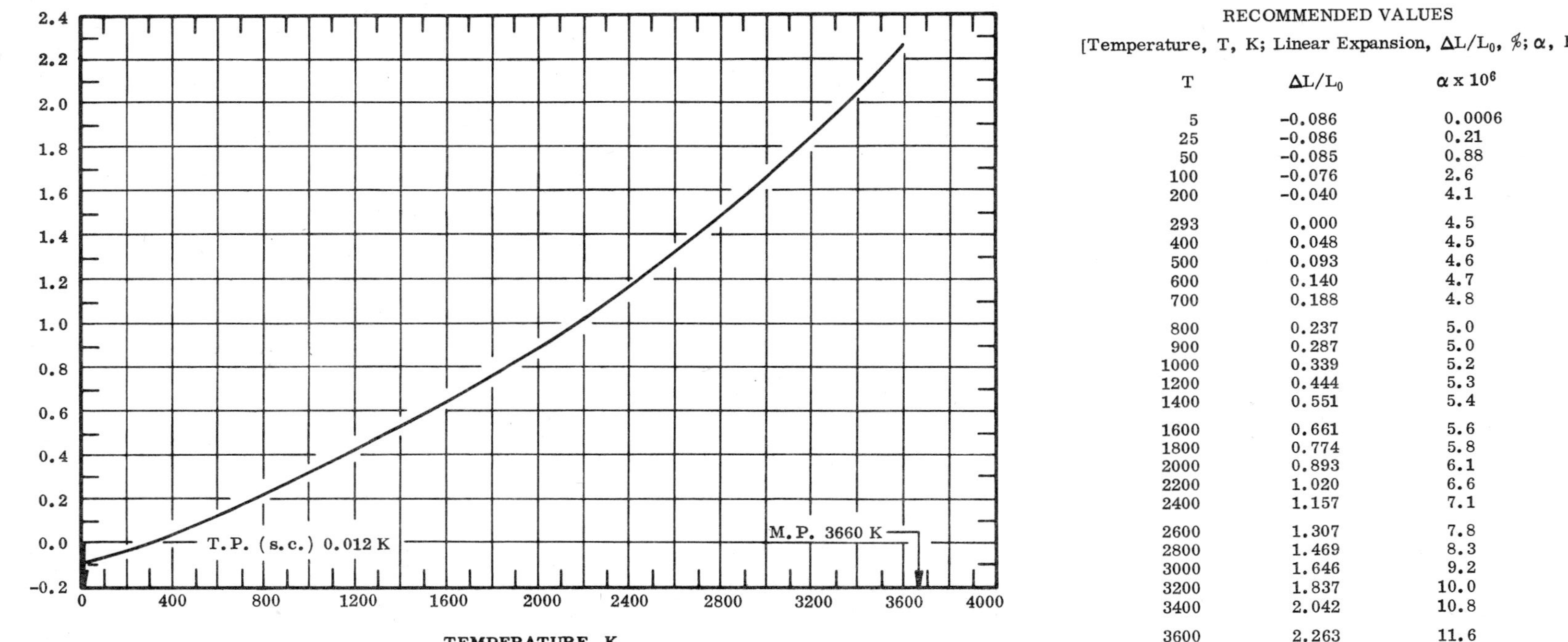

RECOMMENDED VALUES

[Temperature, T, K; Linear Expansion, $\Delta L/L_0$, %; α, K^{-1}]

T	$\Delta L/L_0$	$\alpha \times 10^6$
5	-0.086	0.0006
25	-0.086	0.21
50	-0.085	0.88
100	-0.076	2.6
200	-0.040	4.1
293	0.000	4.5
400	0.048	4.5
500	0.093	4.6
600	0.140	4.7
700	0.188	4.8
800	0.237	5.0
900	0.287	5.0
1000	0.339	5.2
1200	0.444	5.3
1400	0.551	5.4
1600	0.661	5.6
1800	0.774	5.8
2000	0.893	6.1
2200	1.020	6.6
2400	1.157	7.1
2600	1.307	7.8
2800	1.469	8.3
3000	1.646	9.2
3200	1.837	10.0
3400	2.042	10.8
3600	2.263	11.6

REMARKS

The tabulated values are considered accurate to within ± 3% over the entire temperature range. These values can be represented approximately by the following equations:

$$\Delta L/L_0 = 4.266 \times 10^{-4} (T - 293) + 8.479 \times 10^{-8} (T - 293)^2 - 1.974 \times 10^{-11} (T - 293)^3 \quad (293 < T < 1395)$$

$$\Delta L/L_0 = 0.548 + 5.416 \times 10^{-4} (T - 1395) + 1.952 \times 10^{-8} (T - 1395)^2 + 4.422 \times 10^{-11} (T - 1395)^3 \quad (1395 < T < 2495)$$

$$\Delta L/L_0 = 1.226 + 7.451 \times 10^{-4} (T - 2495) + 1.654 \times 10^{-7} (T - 2495)^2 + 7.568 \times 10^{-12} (T - 2495)^3 \quad (2495 < T < 3600)$$

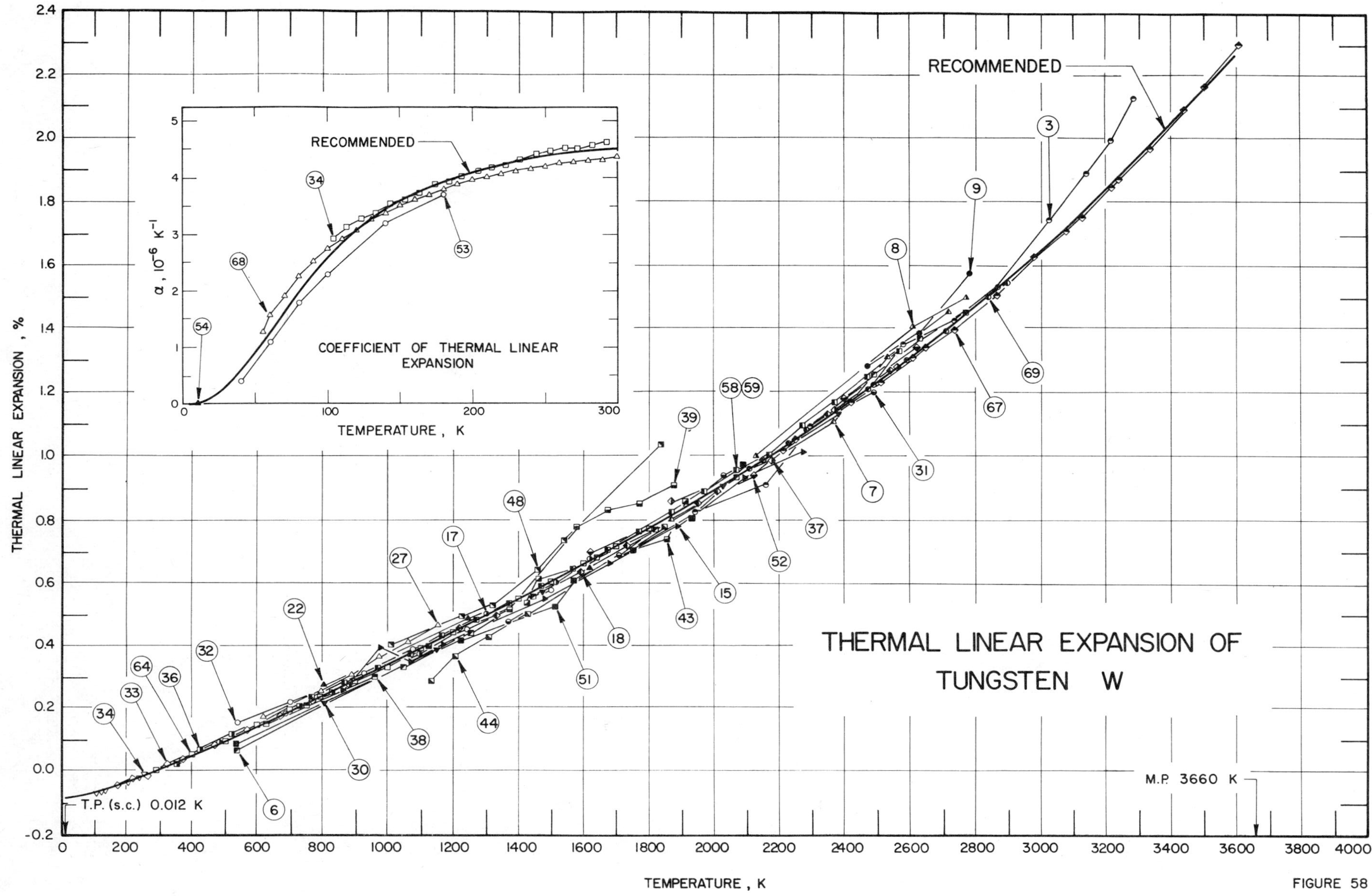
THERMAL LINEAR EXPANSION OF
TUNGSTEN W
THERMAL LINEAR EXPANSION, %
TEMPERATURE, K
RECOMMENDED
M.P. 3660 K
T.P. (s.c.) 0.012 K
COEFFICIENT OF THERMAL LINEAR EXPANSION
α, 10⁻⁶ K⁻¹
TEMPERATURE, K
RECOMMENDED

FIGURE 58

SPECIFICATION TABLE 58. THERMAL LINEAR EXPANSION OF TUNGSTEN W

Cur. No.	Ref. No.	Author(s)	Year	Method Used	Temp. Range, K	Name and Specimen Designation	Composition (weight percent), Specifications, and Remarks
1*	208	Frantsevich, I.N., Zhurakovskii, E.A., and Lyashchenko, A.B.	1967		293, 1473		No details given.
2*	341	Houska, C.R.	1963	X	298-2050		99.95 W; pressed from Wah Chang powder (<200 mesh) at 16 000 psi and annealed in hydrogen atmosphere at 1673 K for 2 hr; expansion measured in (222) and (321) directions; zero-point correction of 0.002% determined by graphical extrapolation.
3	311	V'yugov, P.N. and Gumenyuk, V.S.	1965	T	1937-3322		Technically pure; electrical resistivity ratio $\rho(273K)/\rho(4.2K) = 9$; specimen 2 mm in diameter by 200 to 240 mm long; measured in vacuum ($<10^{-6}$ mm Hg); reported error in expansion 0.3 to 0.5%; zero-point correction of -0.01% determined by graphical extrapolation.
4*	343	Takamori, T. and Tomozawa, M.	1965	L	848-293		Degreased and heated at 1373 K for 15 min; expansion measured during cooling; dilatometer used for measurements has a precision of ±10 ppm.
5*	210	Conway, J.B. and Losekamp, A.C.	1966	T	529-2747		99.95^+ W, <0.01 Ca, <0.01 Fe, <0.01 Mg, <0.01 Mo, 0.0053 C, 0.0003 N, 0.0002 O, and 0.00006 H; sample 0.635 cm in diameter by 6.35 cm long formed by powder metallurgy techniques; measured in helium atmosphere; data could not be read accurately enough from reported graph to warrant correction with zero-point correction of 0.0019% as determined from authors' least squares fit of data; first test.
6	210	Conway, J.B. and Losekamp, A.C.	1966	T	541-2772		Above specimen; second test.
7	210	Conway, J.B. and Losekamp, A.C.	1966	T	532-2723		99.0^+ W, 0.01-0.10 Mo, <0.01 Fe, <0.01 Si, 0.0032 O, 0.0018 C, 0.0003 N, and 0.0002 H; sheet 0.508 mm thick prepared by powder metallurgy techniques; measured in helium atmosphere; data could not be read accurately enough from reported graph to warrant correction with zero-point correction of 0.0019% as determined from authors' least squares fit of data; first test.
8	210	Conway, J.B. and Losekamp, A.C.	1966	T	513-2766		Above specimen; second test.
9	210	Conway, J.B. and Losekamp, A.C.	1966	T	490-2771		Curve 7 specimen; third test.
10*	344	Illinois Inst. of Technology	1966		291-2255		Tungsten plasma sprayed onto removable mandrel at a distance of 6.35 cm and 400 amp; first test, increasing temperature.
11*	344	Illinois Inst. of Technology	1966		2255-1273		Above specimen; decreasing temperature.
12*	344	Illinois Inst. of Technology	1966		1370-2255		Above specimen; second test, increasing temperature.
13*	344	Illinois Inst. of Technology	1966		2255-292		Above specimen; decreasing temperature.
14	211	Yaggee, F.L. and Styles, J.W.	1965	L	298-1263		Zero-point correction of 0.002% determined by graphical extrapolation.
15	213	Amonenko, V.M., V'yugov, P.N., and Gumenyuk, V.S.	1964	L	293-2273		Commercial rod material annealed for 1.5 to 2.0 hr above 2273 K; measured in vacuum ($<2 \times 10^{-5}$ mm Hg); errors 3% for low temperature and less than 1% at high temperature.
16*	345	Brizes, W.F.	1968	T	782-2321		High purity rod 7.62 cm long by 0.64 cm diameter; measured in helium atmosphere (15 μm); overall accuracy ±2% at 2273 K; this curve reported here is referred to 293 K as reference temperature using other available data; zero-point correction is -0.235%.

*Not shown in figure.

SPECIFICATION TABLE 58. THERMAL LINEAR EXPANSION OF TUNGSTEN W (continued)

Cur. No.	Ref. No.	Author(s)	Year	Method Used	Temp. Range, K	Name and Specimen Designation	Composition (weight percent), Specifications, and Remarks
17	273	Mauer, F.A. and Bolz, L.H.	1955	X	273-1613		99.88 W, 0.05 K, 0.04 SiO_2, 0.01 Al_2O_3, 0.01 Fe, and 0.01 Mo; specimen ground to pass a 325 mesh sieve and pressed into cavity; powder supplied by Westinghouse Electric Corp; sintered at 1673 K in hydrogen atmosphere; measured in helium atmosphere; zero-point correction of -0.0085% determined by graphical interpolation.
18	352	Worthing, A.G.	1917	T	563-2670		Tungsten filaments approximately 18 cm long mounted in evacuated tubular glass bulbs; zero-point correction of 0.003% determined by graphical extrapolation.
19*	347	Neel, D.S., Pears, C.D., and Oglesby, S., Jr.	1962	L	294-2283		0.003 Fe, 0.0026 Si, 0.001 each P, Ni, Cu, H, and N; arc-cast specimen from Union Carbide Co.; initial length 6.772 cm; density 18.87 g cm^{-3}; expansion measured at increasing temperatures in helium atmosphere; reported error 2%; expansion values which they report seem to be very low.
20*	347	Neel, D.S., et al.	1962	L	2283-294		Same as above except expansion measured at decreasing temperatures; zero-point correction is 0.05%.
21*	347	Neel, D.S., et al.	1962	L	294-3025		Similar to curve 19 specimen and conditions; specimen melted at 3039 K; zero-point correction is ±0.000%.
22	348	Anthony, F.M. and Pearl, H.A.	1960	L	300-1616		Specimen prepared by compacting and sintering powder into ingots, forging, and stress relieving at 1353 K prior to machining into 7.62 cm long by 0.94 cm diameter rod; powder obtained from Lamp Div. of General Electric; measured in argon atmosphere; reported error ± 5%; zero-point correction of 0.003% determined by graphical extrapolation.
23*	217	Totskii, E.E.	1964	L	273-1373		Annealed tungsten; reported error <1.2%; zero-point correction of -0.009% determined by graphical interpolation.
24*	218	Levinstein, M.A.	1961	L	297-1422		High purity W; 0.002 O_2 and 0.0006 N_2; plasma-sprayed onto Al mandrel using a tungsten wire in plasma gun; mandrel removed by leaching; heat-treated for 2 hr at 2478 K in vacuum; zero-point correction of 0.002% determined by graphical extrapolation.
25*	218	Levinstein, M.A.	1961	L	297-1422		High purity W; 0.042-0.113 O_2 and 0.0017 N_2; plasma-sprayed onto Al mandrel using tungsten powder in plasma gun; mandrel removed by leaching; zero-point correction of 0.002% determined by graphical extrapolation.
26*	219	Denman, G.L.	1962	L	297-1366		Pure W; 0.03 Mo. 0.03 Mo, 0.009 C, 0.005 Fe, 0.005 Ni, 0.004 Ca, and 0.003 O_2; reported probable errors -5% at 394 K and <1% at 1366 K; zero-point correction of 0.002% determined by graphical extrapolation.
27	268	Dutta, B.N. and Dayal, B.	1963	X	298-1151		99.96 W; analar sample in powder form; lattice parameter reported at 298 K is 3.1649 Å; 3.1648 Å at 293 K determined by graphical extrapolation.
28*	349	Fulkerson, S.D.	1959	L	293-1573		Rod 2.54 cm long by 0.67 cm diameter.
29*	274	Brand, J.A. and Goldschmidt, H.J.	1956	X	273-1573		Lattice parameter reported at 273 K is 3.1584 Å; 3.1587 Å at 293 K determined by graphical interpolation.
30	223	Valentich, J.	1969	L	293-2489		99.9 W; arc-cast; measured in vacuum (<5 x 10^{-6} mm Hg) with increasing temperature.
31	223	Valentich, J.	1969	L	2489-293		Same as above except measured with decreasing temperature.

* Not shown in figure.

SPECIFICATION TABLE 58. THERMAL LINEAR EXPANSION OF TUNGSTEN W (continued)

Cur. No.	Ref. No.	Author(s)	Year	Method Used	Temp. Range, K	Name and Specimen Designation	Composition (weight percent), Specifications, and Remarks
32	224	Clark, D. and Knight, D.	1965	X	300-1499		Lattice parameter reported at 300 K is 3.1647 Å; 3.1646 Å at 293 K determined by graphical extrapolation.
33	353	Hidnert, P. and Sweeney, W.T.	1925	L	173-773		99.98 W, 0.015 Mo, 0.005 Cu, and < 0.002 As (analysis after expansion tests); rod specimen 300 mm long by 4.5 mm diameter; density 19.211 g cm^{-3} at 298 K; results smoothed by author from seven independent tests.
34	75	Nix, F.C. and MacNair, D.	1942	I	102-301		99.95^{+} W from Westinghouse Lamp Corp; zero-point correction of -0.009% determined by graphical interpolation; data on the coefficient of thermal linear expansion also reported.
35*	354	Baun, W.L.	1959	X	291-1246		Lattice parameter reported at 293 K is 3.16562 Å.
36	149	Burger, E.E.	1934	L	298-823		Zero-point correction of 0.002% determined by graphical extrapolation.
37	355	Knibbs, R.H.	1969	O	1942-2558		Rod specimen 7.62 cm long by 0.64 cm diameter; density 18.6 g cm^{-3}; this curve reported here is extrapolated to 293 K using recommended value of 0.861 for $\Delta L/L_0$ at 1942 K.
38	356	Goucher, F.S.	1924	T	283-1197		Rod specimen 1 mm in diameter; expansion measured by comparison with fused silica.
39	357	Benedicks, C., Berlin, and Phragmen, G.	1924	T	288-1873		Wire specimen 140 mm long by 2.3 mm diameter; forged in hammering machine; expansion measured with increasing temperature; zero-point correction of -0.002% determined by graphical interpolation; first test.
40*	357	Benedicks, C., et al.	1924	T	288-1873		Same as above; second test.
41*	357	Benedicks, C., et al.	1924	T	288-1873		Same as above; third test.
42*	357	Benedicks, C., et al.	1924	T	288-1993		Similar to curve 40 specimen and conditions; first test.
43	357	Benedicks, C., et al.	1924	T	1972-288		Same as above except expansion measured with decreasing temperature.
44	357	Benedicks, C., et al.	1924	T	288-1843		Same as curve 42; third test.
45*	357	Benedicks, C., et al.	1924	T	288-1363		Similar to curve 40 specimen and conditions; first test.
46*	357	Benedicks, C., et al.	1924	T	288-1973		Same as above; second test.
47*	357	Benedicks, C., et al.	1924	T	288-1973		Same as above; third test.
48	357	Benedicks, C., et al.	1924	T	288-1893		Similar to curve 40 specimen and conditions.
49*	261	Shinoda, G.	1934	X	288-1328		Expansion measured in (321) plane using CuKα radiation; zero-point correction of -0.002% determined by graphical interpolation.
50*	261	Shinoda, G.	1934	X	288-1258		Same as above except expansion measured in (400) plane.
51	596	Conway, J.B. and Flagella, P.N.	1967	T	533-2748		No details given.
52	596	Conway, J.B. and Flagella, P.N.	1967	T	548-2623		No details given.
53*	709	Shah, J.S.	1971	X	40-180		99.9 pure; powder sample of particle size 8-10 μ from Koch Light Labs, Inc., Colnbrook, England, lattice parameter not corrected for refraction; lattice parameter reported as 3.16273 Å at 40 K; lattice parameter reported using the first given temperature as reference temperature at which $\Delta L/L_0 = 0$; coefficient of linear expansion also reported.

* Not shown in figure.

SPECIFICATION TABLE 58. THERMAL LINEAR EXPANSION OF TUNGSTEN W (continued)

Cur. No.	Ref. No.	Author(s)	Year	Method Used	Temp. Range, K	Name and Specimen Designation	Composition (weight percent), Specifications, and Remarks
54*	291, 89	Andres, K.	1961	O	4-10		99.99 pure specimen; this curve is here reported using the first given temperature as reference temperature at which $\Delta L/L_0 = 0$.
55*	714	Andres, K.	1964	O	6.6-14		99.99 pure; this curve is here reported using the first given temperature as reference temperature at which $\Delta L/L_0 = 0$.
56*	624	Fitzer, E.	1972	L	293-1573	Participant 14	99.98 pure; sintered specimen prepared by Metallwerk Plansee, Reutte, Austria; powder compacted under 3.65 tons cm^{-2} to yield bars 16 x 16 x 600 mm; swaged to diameter of 3.3 - 6.8 mm; ground to final dimension, resultant porosity practically zero.
57*	624	Fitzer, E.	1972	L	873-1773	Participant 33	Similar to the above specimen.
58	624	Fitzer, E.	1972	O	293-2573	Participant 46	Similar to the above specimen.
59*	624	Fitzer, E.	1972	X	293-1173	Participant 40	Similar to the above specimen.
60*	624	Fitzer, E.	1972	L	573-1273	Participant 6	Arc-melted (99.95 pure) specimen with major impurity 0.3 Mo, supplied by the Thermo Electron Corporation and distributed by AFML, Wright Patterson AFB; material from heat 100W8600 and stress relieved at 1673 K for 30 min.
61*	624	Fitzer, E.	1972	L	293-1573	Participant 8	Similar to the above specimen.
62*	624	Fitzer, E.	1972	L	293-1473	Participant 14	Similar to the above specimen.
63*	624	Fitzer, E.	1972	L	293-1973	Participant 14	The above specimen; second sample.
64	180	Kirby, R.K.	1972	T	293-1800		99.96 pure sintered specimen from General Electric Company.
65*	180	Kirby, R.K.	1972	T	1000-1700		99.97 pure sintered specimen from Metallwerk Plansee, Austria.
66*	180	Kirby, R.K.	1972	T	1000-1700		99.95 pure arc-cast specimen from Thermo Electron Corporation, USA.
67	739	Petukhov, V.A. and Chekhovskoi	1972	O	293-3335		99.96 W, 0.001 O_2, 0.0003 H_2, 0.02 N_2, measurements in helium atmosphere.
68	747	Lisovskii, Yu.A.	1972	E	55-300		99.97 pure polycrystalline 100 mm long specimen.
69	758	Kraftmakher, Ya.A.	1972	L	2050-2897		Wire specimen of 0.05 mm diameter, no other specification given; the curve reported here is extrapolated to 293 K using the recommended value of 0.923 for $\Delta L/L_0$ at 2050 K.

*Not shown in figure.

DATA TABLE 58. THERMAL LINEAR EXPANSION OF TUNGSTEN W

[Temperature, T, K; Linear Expansion, $\Delta L/L_0$, %]

T	$\Delta L/L_0$
CURVE 1*	
293	0.000
1473	0.543
CURVE 2*	
298	0.002
1027	0.346
1282	0.477
1546	0.627
1795	0.765
2050	0.921
CURVE 3	
1937	0.824
2161	0.908
2209	1.019*
2309	1.002
2343	1.117*
2363	1.157*
2490	1.226
2500	1.269*
2582	1.351
2618	1.320*
2618	1.376*
2647	1.405*
2739	1.424
2739	1.501*
2885	1.650
2927	1.650*
3026	1.743*
3146	1.886*
3146	1.941*
3206	1.993*
3253	2.057*
3285	2.129*
3322	2.133*
CURVE 4*	
848	0.262
704	0.187
563	0.123
293	0.000
CURVE 5*	
529	0.11
714	0.21
994	0.33
1212	0.42
1403	0.53
1569	0.62
1727	0.72
1900	0.82
2080	0.93
2197	1.01
2354	1.14
2465	1.21
2599	1.32
2701	1.40
2747	1.47
CURVE 6	
541	0.06
773	0.21*
1051	0.33
1250	0.44
1445	0.55
1639	0.68
1817	0.77
1971	0.88*
2073	0.94
2227	1.06*
2382	1.17*
2493	1.25
2627	1.35
2772	1.45
CURVE 7	
532	0.12*
707	0.20*
980	0.33*
1248	0.48*
1466	0.59*
1633	0.65*
1853	0.81
2201	1.04*
2370	1.16
2526	1.32
2723	1.46
CURVE 8	
513	0.11*
607	0.14*
821	0.24*
1074	0.33*
1364	0.50*
1542	0.59*
1773	0.79*
1962	0.89*
2133	1.00
2309	1.13*
2483	1.25*
2614	1.41
2766	1.50
CURVE 9	
490	0.06*
596	0.12*
733	0.20*
1012	0.33*
1171	0.42*
1326	0.50*
1504	0.62*
1687	0.74*
1938	0.89*
2119	0.99*
2273	1.13*
2472	1.28
2484	1.24*
2625	1.38
2771	1.58
CURVE 10*	
291	0.000
1171	0.416
1273	0.453
1371	0.512
1469	0.577
1568	0.641
1666	0.686
1767	0.744
1870	0.772
1969	0.782
2066	0.848
2164	0.908
2255	0.926
CURVE 11*	
2255	0.926
2070	0.784
1870	0.641
1672	0.548
1473	0.451
1273	0.347
CURVE 12*	
1370	0.413
1475	0.467
1577	0.516
1665	0.571
1667	0.536
1765	0.677
1867	0.679
1965	0.728
2067	0.839
2165	0.922
2255	0.933
CURVE 13*	
2255	0.933
2067	0.802
1869	0.666
1667	0.531
1474	0.492
293	0.000
CURVE 14*	
298	0.002
373	0.032
473	0.074
573	0.121
673	0.169
773	0.219
873	0.270
973	0.323
1073	0.377
1173	0.432
1263	0.481
CURVE 15	
293	0.00*
473	0.09*
568	0.11*
675	0.17*
773	0.22*
872	0.26
978	0.31
1070	0.35
1168	0.40
1273	0.45*
1364	0.50*
1482	0.55
1565	0.60*
1676	0.66
1783	0.72*
1887	0.78
1982	0.85
2082	0.93
2179	1.01*
2273	1.08
CURVE 16*, †, ‡	
782	0.235
1058	0.579
1159	0.647
1491	0.813
1821	0.973
2122	1.116
2321	1.214
CURVE 17	
273	-0.008*
294	0.0005*
479	0.080
685	0.178*
891	0.276
1097	0.378
1303	0.494
1509	0.602
1613	0.678
CURVE 18	
563	0.121*
660	0.168*
800	0.234*
1230	0.445
1332	0.497
1500	0.587*
1591	0.634
1735	0.719
1814	0.766*
1952	0.853
2005	0.888
2155	0.991
2194	1.020*
2347	1.129
2366	1.148*
2531	1.271
2670	1.384*
CURVE 19*	
293	0.00
966	0.17
1022	0.18
1116	0.21
1216	0.26
1377	0.30
1488	0.34
1605	0.37
1727	0.42
1822	0.47
1883	0.48
1899	0.48
2049	0.52
2166	0.56
2283	0.59
CURVE 20*	
2283	0.640
294	0.000
CURVE 21*	
293	0.00
616	0.07
1016	0.16

*Not shown in figure.
†This curve is here reported using the first given temperature as reference temperature at which $\Delta L/L_0 = 0$.
‡ Author's data for coefficient of thermal expansion have been integrated by TPRC to obtain $\Delta L/L_0$.

DATA TABLE 58. THERMAL LINEAR EXPANSION OF TUNGSTEN W (continued)

T	$\Delta L/L_0$
CURVE 21 (cont.)*	
1085	0.19
1366	0.28
1533	0.35
1649	0.40
1705	0.42
1833	0.48
2010	0.52
2099	0.56
2227	0.57
2288	0.58
2394	0.57
2577	0.61
2727	0.65
2797	0.66
2900	0.65
2955	0.57
2977	0.50
2988	0.44
3025	0.39
CURVE 22	
300	0.003*
811	0.256
1253	0.483
1616	0.656
CURVE 23*	
273	-0.009
373	0.035
473	0.080
573	0.126
673	0.172
773	0.220
873	0.268
973	0.316
1073	0.364
1173	0.415
1273	0.467
1373	0.519
CURVE 24*	
297	0.002
366	0.040

T	$\Delta L/L_0$
CURVE 24 (cont.)*	
422	0.070
477	0.098
533	0.126
588	0.155
644	0.186
699	0.217
755	0.249
810	0.280
866	0.311
922	0.342
977	0.374
1033	0.404
1088	0.435
1144	0.467
1199	0.498
1255	0.529
1310	0.560
1366	0.592
1422	0.624
CURVE 25*	
297	0.002
366	0.037
422	0.063
477	0.089
533	0.115
588	0.143
644	0.171
699	0.200
755	0.227
810	0.258
866	0.288
922	0.321
977	0.351
1033	0.387
1088	0.414
1144	0.445
1199	0.477
1255	0.509
1310	0.544
1366	0.609
1422	0.609

T	$\Delta L/L_0$
CURVE 26*	
297	0.002
394	0.042
489	0.087
578	0.122
668	0.165
759	0.217
850	0.262
942	0.312
1035	0.363
1125	0.408
1222	0.472
1322	0.527
1366	0.557
CURVE 27	
298.2	0.003*
618.2	0.167
798.2	0.250
888.2	0.306
973.2	0.366
1058.2	0.413
1151.2	0.464
CURVE 28*	
293	0.000
473	0.090
673	0.181
873	0.277
1073	0.374
1273	0.479
1373	0.530
1473	0.581
1573	0.638
CURVE 29*	
273	-0.009
373	0.041
473	0.095
573	0.142
673	0.187
773	0.228
873	0.269
973	0.316

T	$\Delta L/L_0$
CURVE 29 (cont.)*	
1073	0.364
1173	0.412
1273	0.459
1373	0.510
1473	0.564
1573	0.617
CURVE 30	
293	0.000*
811	0.209
918	0.272*
1037	0.333*
1149	0.389
1260	0.440*
1364	0.501*
1482	0.575
1593	0.641*
1705	0.707*
2034	0.905
2152	0.969*
2261	1.053*
2377	1.127
2489	1.208*
CURVE 31	
2489	1.208*
2379	1.153
2265	1.078
2151	1.002
2034	0.933
1928	0.855
1818	0.782
1703	0.694
1600	0.625
1482	0.553
1375	0.479
293	0.000*
CURVE 32	
300	0.005*
541	0.147
700	0.216
932	0.308

T	$\Delta L/L_0$
CURVE 32 (cont.)	
1077	0.393
1241	0.453
1499	0.580
CURVE 33	
173	-0.051
223	-0.030
273	-0.009*
323	0.013
373	0.035
473	0.080
573	0.126
673	0.173
773	0.221
CURVE 34	
102.2	-0.075
110.7	-0.073*
118.4	-0.070
127.2	-0.068*
134.2	-0.066
141.2	-0.063*
147.4	-0.061*
154.7	-0.058*
161.9	-0.056*
166.4	-0.054*
168.2	-0.053*
174.2	-0.051*
178.5	-0.049*
185.2	-0.047*
191.0	-0.044*
197.5	-0.042*
202.7	-0.040
209.7	-0.037*
215.2	-0.035*
220.7	-0.032*
223.2	-0.031*
226.2	-0.030*
228.2	-0.029*
232.4	-0.027*
237.2	-0.025
243.2	-0.022*
249.1	-0.020*
255.5	-0.018*

T	$\Delta L/L_0$
CURVE 34 (cont.)	
260.0	-0.015
265.2	-0.013*
271.1	-0.010*
273.2	-0.009*
276.1	-0.007*
282.7	-0.005*
287.2	-0.003*
292.4	-0.000*
301.7	0.004*
CURVE 35*	
291	-0.001
293	0.000
659	0.169
987	0.334
1164	0.422
1246	0.462
CURVE 36	
298	0.002*
323	0.012*
348	0.025
373	0.036*
398	0.048*
423	0.061
448	0.073*
473	0.085*
498	0.097*
523	0.109
548	0.120*
573	0.131*
598	0.144*
623	0.155*
648	0.167*
673	0.180*
698	0.191*
723	0.203
748	0.214*
773	0.227*
798	0.237*
823	0.249

*Not shown in figure.

DATA TABLE 58. THERMAL LINEAR EXPANSION OF TUNGSTEN W (continued)

T	$\Delta L/L_0$
CURVE 37	
1942	0.861*
1973	0.882
2073	0.949*
2173	1.018
2273	1.089*
2373	1.162
2473	1.237*
2558	1.303
CURVE 38	
283	-0.002*
293	0.000*
380	0.036*
418	0.058*
485	0.086*
566	0.125*
624	0.152
699	0.185*
750	0.207
815	0.231*
888	0.271*
957	0.304
1042	0.346*
1197	0.434*
CURVE 39	
288	-0.002*
1123	0.398
1273	0.476*
1373	0.519
1473	0.569*
1573	0.784
1673	0.834
1773	0.855
1873	0.912
CURVE 40*	
288	-0.002
1073	0.326
1173	0.334
1273	0.391
1373	0.441
1473	0.498
CURVE 40 (cont.)*	
1573	0.584
1673	0.662
1773	0.762
1873	0.776
CURVE 41*	
288	-0.002
1323	0.434
1473	0.541
1573	0.698
1673	0.748
1773	0.798
1873	0.848
CURVE 42*	
288	-0.002
1133	0.405
1173	0.426
1233	0.462
1373	0.541
1538	0.662
1993	0.898
CURVE 43	
1972	0.898*
1913	0.855
1853	0.741
1458	0.619
1423	0.541
1353	0.505*
1258	0.462*
1133	0.391*
288	-0.002*
CURVE 44	
288	-0.002*
1143	0.291
1213	0.369
1308	0.426
1368	0.462*
1423	0.505
1473	0.548*
CURVE 44 (cont.)	
1583	0.641*
1843	0.784
CURVE 45*	
288	-0.002
1103	0.441
1143	0.484
1173	0.505
1223	0.526
1253	0.548
1273	0.562
1333	0.576
1363	0.619
CURVE 46*	
288	-0.002
1218	0.426
1363	0.548
1473	0.584
1548	0.662
1785	0.769
1953	1.040
1973	1.055
CURVE 47*	
288	-0.002
1208	0.484
1333	0.548
1423	0.605
1458	0.662
1848	0.834
1973	0.969
CURVE 48	
288	-0.002*
1013	0.405
1233	0.484
1318	0.526
1403	0.562*
1458	0.641
1538	0.741
1808	0.955*
1833	1.034
1893	1.069*
CURVE 49*	
288	-0.002
515	0.101
899	0.321
1168	0.575
1228	0.617
1258	0.648
1328	0.743
CURVE 50*	
288	-0.002
1168	0.479
1258	0.654
CURVE 51	
533	0.079
723	0.204*
1003	0.331*
1223	0.416
1503	0.525
1563	0.600
1743	0.712
1928	0.813
2078	0.948
2183	1.002*
2368	1.136*
2468	1.200*
2598	1.301*
2688	1.390*
2748	1.438*
CURVE 52	
548	0.061*
798	0.197*
1063	0.351*
1258	0.440*
1448	0.550*
1623	0.690
1808	0.786*
1963	0.896*
2123	0.943
2233	1.053*
2393	1.179
2498	1.253*
2623	1.355
CURVE 53*	
40	0.000
60	0.001
80	0.004
100	0.009
140	0.020
180	0.034
CURVE 54*, †, ‡	
4	0.00×10^{-4}
6	0.25
8	0.95
10	2.52
CURVE 55*, †	
6.6	0.000×10^{-4}
7.5	0.003
7.5	0.003
8.2	0.004
8.8	0.005
8.8	0.006
9.4	0.009
9.6	0.009
10.2	0.011
10.2	0.012
11.1	0.018
11.8	0.022
12.2	0.027
12.8	0.033
13.6	0.041
CURVE 56*	
293	0.000
373	0.035
473	0.078
573	0.123
673	0.169
773	0.218
873	0.270
973	0.325
1073	0.380
1173	0.434
1273	0.487
1373	0.536
1473	0.585
1573	0.636
CURVE 57*	
873	0.310
973	0.365
1073	0.423
1173	0.481
1273	0.539
1373	0.600
1473	0.661
1573	0.726
CURVE 58	
293	0.000*
373	0.037*
473	0.083*
573	0.131*
673	0.179*
773	0.228
873	0.277
973	0.327
1073	0.378
1173	0.430
1273	0.483
1373	0.537
1473	0.592
1573	0.649
1673	0.706
1773	0.766
1873	0.827
1973	0.891
2073	0.957
2173	1.026
2273	1.097
2373	1.172
2473	1.251
2573	1.333
CURVE 59*	
293	0.000
373	0.037
473	0.080
573	0.120
673	0.170
773	0.220
873	0.266
973	0.320
1073	0.370
1173	0.426

*Not shown in figure.

†This curve is here reported using the first given temperature as reference temperature at which $\Delta L/L_0 = 0$.

‡Author's data for coefficient of thermal expansion have been integrated by TPRC to obtain $\Delta L/L_0$.

DATA TABLE 58. THERMAL LINEAR EXPANSION OF TUNGSTEN W

T	$\Delta L/L_0$
CURVE 60*	
573	0.127
673	0.174
773	0.224
897	0.276
973	0.327
1073	0.380
1173	0.440
1273	0.499
CURVE 61*	
293	0.000
373	0.033
473	0.080
573	0.130
673	0.176
773	0.230
873	0.287
973	0.349
1073	0.409
1173	0.464
1273	0.521
1373	0.578
1473	0.639
1573	0.691
CURVE 62*	
293	0.000
373	0.037
473	0.083
573	0.130
673	0.180
773	0.232
873	0.288
973	0.343
1073	0.402
1173	0.460
1273	0.517
1373	0.571
1473	0.626
1573	0.680
CURVE 63*	
293	0.000
373	0.038
473	0.084
CURVE 63 (cont.)*	
573	0.130
673	0.179
773	0.230
873	0.285
973	0.342
1073	0.399
1173	0.457
1273	0.513
1373	0.567
1473	0.621
1573	0.675
CURVE 64	
293	0.0000
400	0.0489
500	0.0951
600	0.1420
700	0.1896
800	0.2380
900	0.2873
1000	0.3375
1100	0.3890
1200	0.4410
1300	0.4940
1400	0.5490
1500	0.6050
1600	0.6610
1700	0.7190
1800	0.7780
CURVE 65*	
1000	0.337
1100	0.387
1200	0.438
1300	0.490
1400	0.543
1500	0.598
1600	0.654
1700	0.713
CURVE 66*	
1000	0.335
1100	0.386
1200	0.438
1300	0.491
CURVE 66 (cont.)*	
1400	0.545
1500	0.600
1600	0.656
1700	0.713
CURVE 67	
293	0.000*
2213	1.017
2252	1.046
2419	1.165
2509	1.228
2607	1.305
2651	1.342
2738	1.399
2775	1.442*
2848	1.504
2884	1.536*
2992	1.630
3080	1.709
3128	1.750
3219	1.845
3243	1.871
3337	1.972
3438	2.087
3503	2.163
3549	2.219*
3593	2.279*
3598	2.290*
3608	2.299
CURVE 68*	
55	-0.083
60	-0.082
70	-0.081
80	-0.079
90	-0.076
100	-0.074
110	-0.071
120	-0.068
130	-0.064
140	-0.061
150	-0.058
160	-0.054
170	-0.050
180	-0.047
190	-0.043
CURVE 68 (cont.)*	
200	-0.039
210	-0.035
220	-0.031
230	-0.027
240	-0.023
250	-0.018
260	-0.014
270	-0.010
280	-0.006
290	-0.001
293	0.000
300	0.003
CURVE 69	
2050	1.156
2108	1.194
2171	1.236
2230	1.275
2289	1.316
2345	1.355
2408	1.400
2477	1.449
2534	1.491
2591	1.534
2648	1.577
2712	1.628
2775	1.679
2839	1.733
2897	1.783

* Not shown in figure.

DATA TABLE 58. COEFFICIENT OF THERMAL LINEAR EXPANSION OF TUNGSTEN W

[Temperature, T, K; Coefficient of Expansion, α, 10^{-6} K^{-1}]

T	α
CURVE 16*, ‡	
782	4.52
1058	4.80
1159	5.13
1491	4.91
1821	4.75
2122	4.74
2321	5.13
CURVE 34	
103	2.92
113	3.14
123	3.28
133	3.37
143	3.54
153	3.62
163	3.72
174	3.90
184	3.93
193	4.04
204	4.13
213	4.18
223	4.23
233	4.33
244	4.42
254	4.47
264	4.52
273	4.52
283	4.58
293	4.64
CURVE 37*, ‡	
1942	6.60
1973	6.65
2073	6.82
2173	7.00
2273	7.19
2373	7.41
2473	7.62
2558	7.82
CURVE 53*	
40	0.4
60	1.1
CURVE 53 (cont.)*	
80	1.8
100	2.3
140	3.2
180	3.7
CURVE 54*, ‡	
4	0.0006
6	0.0019
8	0.0051
10	0.0106
CURVE 68*	
55	1.28
60	1.58
70	1.91
80	2.27
90	2.54
100	2.76
110	2.94
120	3.09
130	3.27
140	3.38
150	3.52
160	3.62
170	3.72
180	3.82
190	3.90
200	3.97
210	4.02
220	4.07
230	4.12
240	4.16
250	4.22
260	4.27
270	4.29
280	4.30
290	4.33
293	4.36
300	4.37
CURVE 69	
2050	6.53
2108	6.63
CURVE 69 (cont.)	
2171	6.60
2230	6.80
2289	6.94
2345	6.99
2408	7.16
2477	7.23
2534	7.43
2591	7.51
2648	7.83
2712	8.01
2775	8.25
2839	8.53
2897	8.78

* Not shown in figure.

‡ Author's data for coefficient of thermal expansion have been integrated by TPRC to obtain $\Delta L/L_0$.

FIGURE AND TABLE NO. 59R. RECOMMENDED VALUES FOR THERMAL LINEAR EXPANSION OF URANIUM U

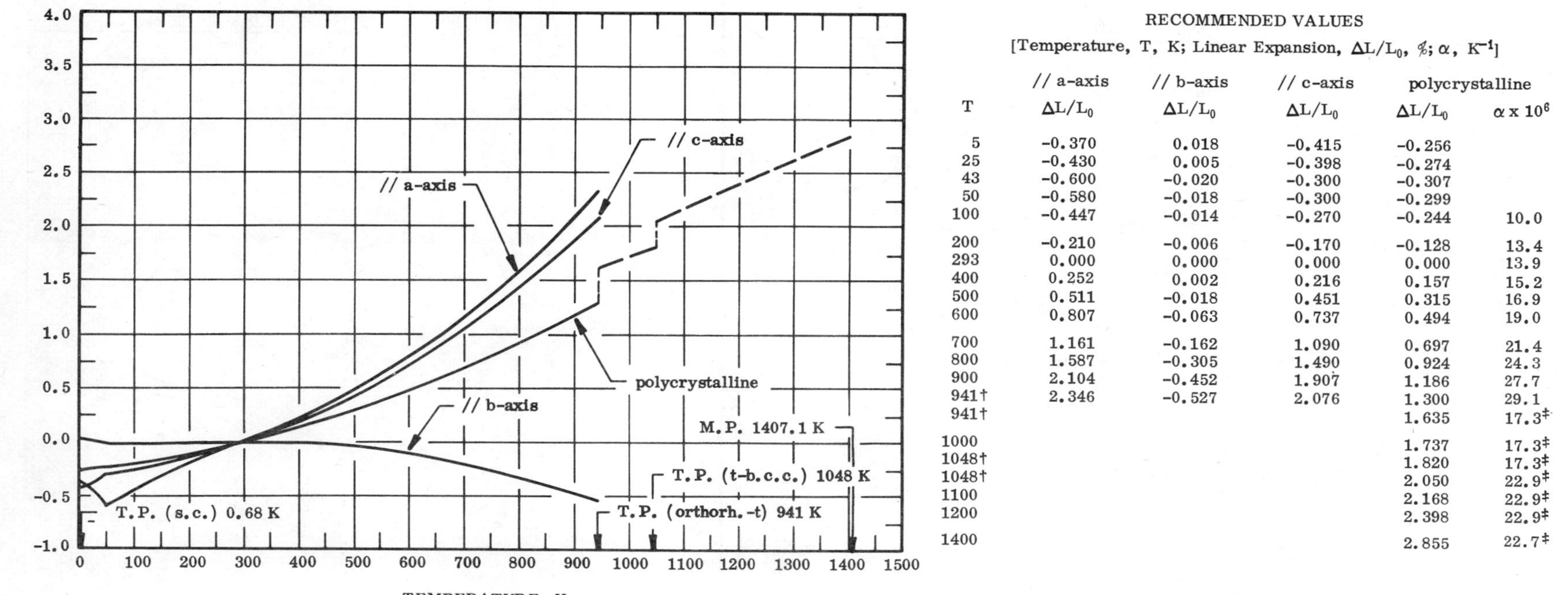

RECOMMENDED VALUES

[Temperature, T, K; Linear Expansion, $\Delta L/L_0$, %; α, K^{-1}]

T	// a-axis $\Delta L/L_0$	// b-axis $\Delta L/L_0$	// c-axis $\Delta L/L_0$	polycrystalline $\Delta L/L_0$	polycrystalline $\alpha \times 10^6$
5	-0.370	0.018	-0.415	-0.256	
25	-0.430	0.005	-0.398	-0.274	
43	-0.600	-0.020	-0.300	-0.307	
50	-0.580	-0.018	-0.300	-0.299	
100	-0.447	-0.014	-0.270	-0.244	10.0
200	-0.210	-0.006	-0.170	-0.128	13.4
293	0.000	0.000	0.000	0.000	13.9
400	0.252	0.002	0.216	0.157	15.2
500	0.511	-0.018	0.451	0.315	16.9
600	0.807	-0.063	0.737	0.494	19.0
700	1.161	-0.162	1.090	0.697	21.4
800	1.587	-0.305	1.490	0.924	24.3
900	2.104	-0.452	1.907	1.186	27.7
941†	2.346	-0.527	2.076	1.300	29.1
941†				1.635	17.3‡
1000				1.737	17.3‡
1048†				1.820	17.3‡
1048†				2.050	22.9‡
1100				2.168	22.9‡
1200				2.398	22.9‡
1400				2.855	22.7‡

† Phase transition.
‡ Provisional values.

REMARKS

The tabulated values are considered accurate to within ± 5% at temperatures below 941 K and ± 7% above. The tabulated values for polycrystalline uranium can be represented approximately by the following equations:

$$\Delta L/L_0 = -0.379 + 1.264 \times 10^{-3}\, T - 8.982 \times 10^{-8}\, T^2 + 6.844 \times 10^{-10}\, T^3 \qquad (293 < T < 941)$$

$$\Delta L/L_0 = -0.149 + 1.775 \times 10^{-3}\, T + 4.382 \times 10^{-7}\, T^2 - 1.239 \times 10^{-10}\, T^3 \qquad (1048 < T < 1400)$$

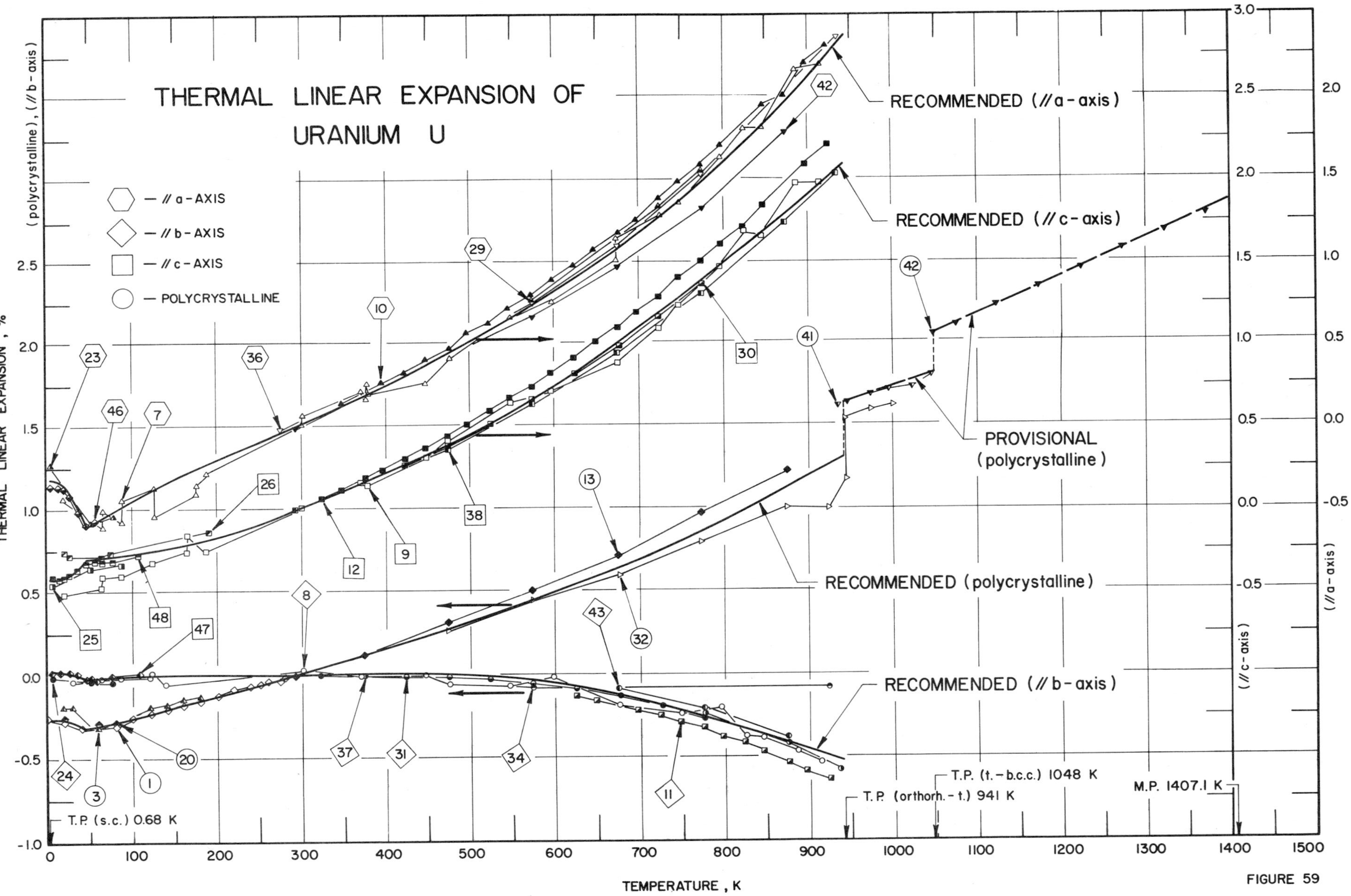
THERMAL LINEAR EXPANSION OF
URANIUM U
— // a-AXIS
— // b-AXIS
— // c-AXIS
— POLYCRYSTALLINE
RECOMMENDED (// a-axis)
RECOMMENDED (// c-axis)
PROVISIONAL
(polycrystalline)
RECOMMENDED (polycrystalline)
RECOMMENDED (// b-axis)
T.P. (s.c.) 0.68 K
T.P. (orthorh.-t.) 941 K
T.P. (t.-b.c.c.) 1048 K
M.P. 1407.1 K
TEMPERATURE, K
THERMAL LINEAR EXPANSION, %
(polycrystalline), (// b-axis)
(// c-axis)
(// a-axis)

FIGURE 59

SPECIFICATION TABLE 59. THERMAL LINEAR EXPANSION OF URANIUM U

Cur. No.	Ref. No.	Author(s)	Year	Method Used	Temp. Range, K	Name and Specimen Designation	Composition (weight percent), Specifications, and Remarks
1	83	Laquer, H. L.	1952	L	0-300	A-1	Specimen cut from sample annealed for 6 hr in the β-phase and slowly cooled cast ring, axis of specimen radial to ring; rod specimen 2.54 cm long, 0.50 cm diameter; zero-point correction of -0.022% determined by graphical interpolation.
2*	83	Laquer, H. L.	1952	L	0-380	A-2	Cut from same ring as above, except axis in plane of ring perpendicular to radius; zero-point correction of -0.022% determined by graphical interpolation.
3	83	Laquer, H. L.	1952	L	0-300	A-3	Cut from same ring as above, except axis is perpendicular to plane of ring and to radius; zero-point correction of -0.022% determined by graphical interpolation.
4*	83	Laquer, H. L.	1952	L	0-300	C-1	Specimen similar to A-1, except 0.50 cm long, 1.27 cm diameter; zero-point correction of -0.032% determined by graphical interpolation.
5*	83	Laquer, H. L.	1952	L	0-300	C-2	Specimen similar to A-2, except dimensions as the above specimen; zero-point correction of -0.035% determined by graphical interpolation.
6*	83	Laquer, H. L.	1952	L	0-300	C-3	Specimen similar to A-3, except 0.50 cm long, 1.27 cm diameter; zero-point correction of -0.035% determined by graphical interpolation.
7	360	Bridge, J.R., Schwartz, C.M., and Vaughan, D.A.	1956	X	20-913	α-U	Specimen filings from pure U bar stock; cleaned with magnet, pickled in nitric acid; annealed at 873 K in evacuated silica tube; pickled in concentrated nitric acid, washed in water, alcohol, quickly dried and sealed in evacuated thin-walled silica capillary; expansion measured along a-axis; data given are compilation of results on several samples; cell dimension reported at 188 K is 2.845 Å; 2.853 Å at 293 K determined by graphical interpolation.
8	360	Bridge, J.R., et al.	1956	X	20-913	α-U	The above specimen; expansion measured along b-axis; unit cell reported at 188 K is 5.864 Å; 5.867 Å at 293 K determined by graphical interpolation.
9	360	Bridge, J.R., et al.	1956	X	20-913	α-U	The above specimen; expansion measured along c-axis; unit cell reported at 188 K is 4.942 Å; 4.955 Å at 293 K determined by graphical interpolation.
10	361	Lloyd, L.T.	1959	L	293-923	α-V	α-uranium single crystal, obtained by grain-coarsening technique; size ranged up to 0.20 in diameter, 0.5 in length; sample prepared in the form of pseudo-unit cells with six flat surfaces parallel to principal crystallographic axis; expansion measured in [100] direction.
11	361	Lloyd, L.T.	1959	L	293-923	α-V	The above specimen; expansion measured in [010] direction.
12	361	Lloyd, L.T.	1959	L	293-923	α-V	The above specimen; expansion measured in [001] direction.
13	362	Heal, T.J. and McIntosh, A.B.	1958		291-873		99.8$^+$ U, 0.14 C, 0.003 Si; purified by magnesium reduction process.
14*	363	Andres, K.	1968	E	0.0-15	α-U	Crystalline specimen from the Argonne National Lab.; typically 2 x 3 x 5 mm; expansion measured along a-axis with increasing temperature after fast cooling; this curve reported here referred to 293 K using the recommended value of -0.374% at 15 K giving zero-point correction of -0.381%.
15*	363	Andres, K.	1968	E	15-0.0	α-U	The above specimen; expansion measured along a-axis with slowly decreasing temperature; this curve reported here referred to 293 K using the value used for curve 14.
16*	363	Andres, K.	1968	E	0.0-15	α-U	The above specimen; expansion measured along b-axis with increasing temperature after fast cooling; this curve reported here referred to 293 K using the recommended value of 0.014% at 15 K, giving zero-point correction of 0.005%.

* Not 'hown in figure.

SPECIFICATION TABLE 59. THERMAL LINEAR EXPANSION OF URANIUM U (continued)

Cur. No.	Ref. No.	Author(s)	Year	Method Used	Temp. Range, K	Name and Specimen Designation	Composition (weight percent), Specifications, and Remarks
17*	363	Andres, K.	1968	E	15-0.0	α-U	The above specimen; expansion measured along b-axis with slowly decreasing temperature; this curve reported here referred to 293 K using the value used for curve 16.
18*	363	Andres, K.	1968	E	0.0-15	α-U	The above specimen; expansion measured along c-axis with increasing temperature after fast cooling; this curve reported here referred to 293 K using the recommended value of -0.405% at 15 K, giving zero-point correction of -0.410%.
19*	363	Andres, K.	1968	E	15-0.0	α-U	The above specimen; expansion measured along c-axis with slowly decreasing temperature; this curve reported here referred to 293 K using the value used for curve 18.
20	364	Laquer, H. L. and Schuch, A. F.	1952	L	20-300		Data of average of three specimens cut from 5 cm diameter casting which was annealed 6 hr in β-phase and cooled slowly; one specimen cut radially, second specimen cut axially near periphery, and third specimen cut axially near center; no difference observed; zero-point correction of 0.009% determined by graphical interpolation.
21*	365	Thewlis, J.	1952	X	293,993	β-U	β-uranium, unit-cell dimensions deduced; lattice parameter reported at 993 K is 759 ± 0.001; expansion measured along a-axis.
22*	365	Thewlis, J.	1952	X	293,993	β-U	The above specimen; lattice parameter reported at 993 K is 5.656 Å; expansion measured along c-axis.
23	366	Barrett, C. S., Mueller, M. H., and Hitterman, R. L.	1963	X	4.2-298	α-U	Single-crystal sample, rectangular block 1 to 3 mm on a side; furnished by Lloyd, L. T.; grown from high-purity uranium by the grain-coarsening techniques; isolated from other grains and formed into rectangular blocks by metallographic procedures; rate of heating was 1 to 2 C/min.; expansion measured along a-axis; lattice parameter reported at 298 K is 2.8537 Å; 2.8534 Å at 293 K determined by graphical interpolation.
24	366	Barrett, C. S., et al.	1963	X	4.2-298	α-U	The above specimen; expansion measured along b-axis; lattice parameter reported at 298 K is 5.8695 Å; 5.8694 Å at 293 K determined by graphical interpolation.
25	366	Barrett, C. S., et al.	1963	X	4.2-298	α-U	The above specimen; expansion measured along c-axis; lattice parameter reported at 298 K is 4.9548 Å; 4.9544 Å at 293 K determined by graphical interpolation.
26	367	Schuch, A. F. and Laquer, H. L.	1952	L	20-192		Coarse grained specimen cut radially from rod.
27*	367	Schuch, A. F. and Laquer, H. L.	1952	L	20,76		Coarse grained specimen cut axially from periphery of the above rod.
28	367	Schuch, A. F. and Laquer, H. L.	1952	L	20-75		Coarse grained specimen cut axially from center of the above rod.
29	368	Ibrahim, E. F.	1964	L	323-773		Single crystal specimen; expansion measured in <100> direction.
30	368	Ibrahim, E. F.	1964	L	323-773		The above specimen; expansion measured in <001> direction.
31	368	Ibrahim, E. F.	1964	L	323-773		The above specimen; expansion measured in <010> direction.
32	369	Saller, H. A., Rough, F. A., and Chubb, W.	1956	L	293-1000		Typical data resulting from pure uranium or U + 0.076 Cr or U + 0.18 Si or U + 0.10 Ti; each specimen measured twice; no noticeable difference between measurements.
33*	370	Battelle Memorial Institute	1944	X	298-928	α-U	Specimen in α-phase; expansion measured along a-axis; data termed "tentative data" by authors; lattice parameter reported at 298 K is 2.8519 Å; 2.8516 Å at 293 K determined by graphical extrapolation.

* Not shown in figure.

SPECIFICATION TABLE 59. THERMAL LINEAR EXPANSION OF URANIUM U (continued)

Cur. No.	Ref. No.	Author(s)	Year	Method Used	Temp. Range, K	Name and Specimen Designation	Composition (weight percent), Specifications, and Remarks
34	370	Battelle Memorial Institute	1944	X	300-928	α-U	The above specimen; expansion measured along b-axis; lattice parameter reported at 300 K is 5.8590 Å; 5.8591 Å at 293 K determined by graphical extrapolation.
35*	370	Battelle Memorial Institute	1944	X	300-928	α-U	The above specimen; expansion measured along c-axis; lattice parameter reported at 300 K is 4.9443 Å; 4.9434 Å at 293 K determined by graphical extrapolation.
36	371	Klepfer, H.A. and Chiotti, P.	1957	X	273-935	α-U	Specimen in alpha (orthorhombic) phase; expansion measured along a-axis; data smoothed by author; lattice parameter reported at 298 K is 2.8535 Å; 2.8532 Å at 293 K determined by graphical interpolation.
37	371	Klepfer, H.A. and Chiotti, P.	1957	X	273-935	α-U	The above specimen; expansion measured along b-axis; lattice parameter reported at 298 K is 5.8648 Å; 5.86485 Å at 293 K determined by graphical interpolation.
38	371	Klepfer, H.A. and Chiotti, P.	1957	X	273-935	α-U	The above specimen; expansion measured along c-axis; lattice parameter reported at 298 K is 4.9543 Å; 4.9539 Å at 293 K determined by graphical interpolation.
39	371	Klepfer, H.A. and Chiotti, P.	1957	X	935-1045	β-U	Specimen in beta (tetragonal) phase; expansion measured along a-axis; data smoothed by author; this curve is here reported using the first given temperature as reference temperature at which $\Delta L/L_0 = 0$; lattice parameter reported at 935 K is 10.7444 Å.
40	371	Klepfer, H.A. and Chiotti, P.	1957	X	935-1045	β-U	The above specimen; expansion measured along c-axis; this curve is here reported using the first given temperature as reference temperature at which $\Delta L/L_0 = 0$; lattice parameter reported at 935 K is 5.6515 Å.
41	371	Klepfer, H.A. and Chiotti, P.	1957	X	935-1045	β-U	Values calculated for polycrystalline specimen from curves 39 and 40; they were referred to α-phase using corresponding atomic volumes (at 293 K for α-phase and 935-1045 K for β-phase).
42	371	Klepfer, H.A. and Chiotti, P.	1957	X	1045-1373	γ-U	Specimen in gamma (bcc) phase; data smoothed by author; values referred to α-phase with the aid of atomic volumes using corresponding lattice parameters (at 293 K for α-phase and 1045-1373 K for γ-phase); lattice parameter reported at 1045 K is 3.5321 Å.
43	414	Langeron, J.-P. and Lehr, P.	1955	L	293-873	α-U	Monocrystal 50 mm long, 3 mm by 4 mm cross section; expansion determined along a-axis in α-phase.
44	414	Langeron, J.-P. and Lehr, P.	1955	L	293-873	α-U	The above specimen; expansion determined along b-axis.
45*	414	Langeron, J.-P. and Lehr, P.	1955	L	293-873	α-U	The above specimen; expansion determined along c-axis.
46	641	Marples, J.A.C.	1970	X	4-107	α-U	Specimen in the form of polycrystalline buttons made by arc melting; examined in the as-cast state; strain free surface obtained by grinding and electro-polishing; impurities 0.004 out of which 0.0025 Si; measurements along a-axis; lattice parameter reported at 107 K is 2.8395 Å, value of 2.8533 Å at 293 K is used for calculation; anomaly observed near 43 K.
47	641	Marples, J.A.C.	1970	X	4-107	α-U	The above specimen; measurements along b-axis; lattice parameter reported at 107 K is 5.8650 Å, value of 5.8648 Å at 293 K is used in calculation.
48	641	Marples, J.A.C.	1970	X	4-107	α-U	The above specimen; measurements along c-axis; lattice parameter reported at 107 K is 4.9398 Å, value of 4.9539 Å at 293 K is used in calculation.

* Not shown in figure.

DATA TABLE 59. THERMAL LINEAR EXPANSION OF URANIUM U

[Temperature, T, K; Linear Expansion, $\Delta L/L_0$, %]

T	$\Delta L/L_0$
CURVE 1	
0	-0.263
20	-0.267
40	-0.312
60	-0.306
80	-0.302
100	-0.259
120	-0.233
140	-0.206
160	-0.179
180	-0.159
200	-0.123
220	-0.095
240	-0.068
260	-0.040
273	-0.022
280	-0.013
300	0.014
CURVE 2*	
0	-0.291
20	-0.288
40	-0.277
60	-0.262
80	-0.245
100	-0.226
120	-0.206
140	-0.185
160	-0.162
180	-0.139
200	-0.116
220	-0.091
240	-0.066
260	-0.040
273	-0.022
280	-0.013
300	0.014
320	0.041
340	0.068
360	0.095
380	0.122
CURVE 3	
0	-0.180
20	-0.184
40	-0.257
CURVE 3 (cont.)	
60	-0.258*
80	-0.237*
100	-0.215*
120	-0.192
140	-0.170
160	-0.148
180	-0.126
200	-0.104*
220	-0.081*
240	-0.059*
260	-0.037*
273	-0.022*
280	-0.015*
300	0.008*
CURVE 4*	
0	-0.292
20	-0.299
40	-0.335
60	-0.334
80	-0.313
100	-0.288
120	-0.261
140	-0.233
160	-0.205
180	-0.175
200	-0.145
220	-0.115
240	-0.084
260	-0.053
273	-0.032
280	-0.021
300	0.010
CURVE 5*	
0	-0.299
20	-0.304
40	-0.359
60	-0.364
80	-0.344
100	-0.317
120	-0.289
140	-0.256
160	-0.224
180	-0.192
CURVE 5 (cont.)*	
200	-0.158
220	-0.125
240	-0.091
260	-0.057
273.2	-0.034
280	-0.023
300	0.012
CURVE 6*	
0	-0.286
20	-0.291
40	-0.307
60	-0.303
80	-0.286
100	-0.256
120	-0.242
140	-0.218
160	-0.193
180	-0.167
200	-0.141
220	-0.113
240	-0.084
260	-0.055
273.2	-0.035
280	-0.024
300	0.007
CURVE 7	
20	-0.436
63	-0.611
63	-0.506
88	-0.576
88	-0.436
123	-0.366
123	-0.541
173	-0.401
173	-0.366
188	-0.296
300	0.020
300	0.055
300	-0.015
370	0.230
373	0.160
375	0.265
375	0.195
CURVE 7 (cont.)	
448	0.265
473	0.405
548	0.651
598	0.756
673	1.141
673	1.001
724	1.281
748	1.351
794	1.632
823	1.807
845	1.807
885	2.157
913	2.193
CURVE 8	
20	-0.036
63	-0.019
88	-0.019
88	-0.002*
123	0.015
123	-0.019
173	-0.002*
173	0.032*
188	-0.053
300	0.032
300	0.015*
300	-0.036*
370	-0.002
373	-0.019*
375	-0.019*
375	-0.036*
448	-0.002
473	-0.053
548	-0.070
598	-0.036
673	-0.087
673	-0.190
724	-0.207
748	-0.241
794	-0.207
823	-0.377
845	-0.394
885	-0.496
913	-0.548
CURVE 9	
20	-0.507
63	-0.406
63	-0.487
88	-0.406
88	-0.346
123	-0.325
123	-0.265
173	-0.265
173	-0.204
188	-0.265
300	-0.023*
300	0.038
300	0.038*
370	0.159
373	0.139
375	0.240
375	0.260*
448	0.320
473	0.421
548	0.643
598	0.704
673	0.885
673	0.986
724	1.188
748	1.249
794	1.471
823	1.695
845	1.652
885	1.975
913	1.975
CURVE 10	
323	0.074
348	0.137
373	0.202*
398	0.270*
423	0.339
448	0.411
473	0.486
498	0.562
523	0.642
548	0.724
573	0.809*
598	0.897*
623	0.987
648	1.079
CURVE 10 (cont.)	
673	1.178
698	1.276
723	1.380
748	1.483
773	1.594
798	1.707
823	1.829*
848	1.948
873	2.071
898	2.202
923	2.331
CURVE 11	
323	0.002
348	0.001
373	-0.002
398	-0.004
423	-0.009
448	-0.017
473	-0.025
498	-0.037
523	-0.051
548	-0.066
573	-0.081
598	-0.104
623	-0.129
648	-0.153
673	-0.182
698	-0.215
723	-0.249
748	-0.281
773	-0.326
798	-0.374
823	-0.419
848	-0.472
873	-0.528
898	-0.587
923	-0.649
CURVE 12	
323	0.063
348	0.118
373	0.176
398	0.237
423	0.300

*Not shown in figure.

DATA TABLE 59. THERMAL LINEAR EXPANSION OF URANIUM U (continued)

T	$\Delta L/L_0$
CURVE 12 (cont.)	
448	0.367
473	0.437
498	0.508
523	0.584
548	0.663
573	0.744
598	0.829
623	0.917
648	1.008
673	1.098
698	1.199
723	1.298
748	1.401
773	1.507
798	1.621
823	1.733
848	1.848
873	1.972
898	2.093
923	2.224
CURVE 13	
291	-0.0032
373	0.127
473	0.306
573	0.506
673	0.728
773	0.972
873	1.238
CURVE 14*	
2	-0.367
4	-0.367
5	-0.367
6	-0.367
7	-0.368
11	-0.368
12	-0.368
13	-0.369
14	-0.372
15	-0.374
CURVE 15*	
15	0.344
14	0.370
12	0.366

T	$\Delta L/L_0$
CURVE 15 (cont.)*	
11	0.364
10	0.363
8	0.362
7	0.362
4	0.361
2	0.361
CURVE 16*	
0	0.014
7	0.014
9	0.014
10	0.014
11	0.014
12	0.014
14	0.014
15	0.014
CURVE 17*	
15	0.014
14	0.014
12	0.015
11	0.015
9	0.015
6	0.015
0	0.015
CURVE 18*	
0	-0.405
10	-0.405
12	-0.405
14	-0.405
15	-0.405
CURVE 19*	
15	-0.405
14	-0.406
12	-0.406
11	-0.406
9	-0.407
8	-0.407
7	-0.407
0	-0.407

T	$\Delta L/L_0$
CURVE 20	
20	-0.265
60	-0.296
80	-0.280
100	-0.258*
120	-0.234*
140	-0.207*
160	-0.180*
180	-0.153*
200	-0.126*
220	-0.099*
240	-0.072*
260	-0.045*
280	-0.018*
300	0.009*
CURVE 21*	
993	1.61
CURVE 22*	
993	0.322
CURVE 23	
4.2	-0.243
50.0	-0.594
78.0	-0.548
298.0	0.013*
CURVE 24	
4.2	-0.009
50.0	-0.049
78.0	-0.038
298.0	0.001*
CURVE 25	
4.2	-0.460
50.0	-0.366
78.0	-0.337
298.0	0.008*
CURVE 26	
20	-0.259
26	-0.282

T	$\Delta L/L_0$
CURVE 26 (cont.)	
62	-0.288
63	-0.286*
74	-0.277
75	-0.278*
191	-0.134
192	-0.135*
CURVE 27*	
20.3	-0.276
75.6	-0.291
CURVE 28*	
20.0	-0.280
75.4	-0.294
75.4	-0.306
CURVE 29	
323	0.072*
373	0.195*
423	0.326*
473	0.466*
523	0.617*
573	0.779
623	0.954*
673	1.142*
723	1.345
773	1.563
CURVE 30	
323	0.061*
373	0.165*
423	0.275
473	0.392
523	0.519
573	0.658
623	0.811
673	0.978
723	1.162
773	1.367
CURVE 31	
323	-0.001
373	-0.002

T	$\Delta L/L_0$
CURVE 31 (cont.)	
423	-0.007
473	-0.016
523	-0.032
573	-0.056
623	-0.090
673	-0.134
723	-0.189
773	-0.260
CURVE 32	
293	0.000*
373	0.118*
473	0.268
573	0.424
673	0.594
773	0.780
873	1.000
942	1.168
942	1.515
973	1.577
1000	1.629
CURVE 33*	
298	0.012
573	0.646
928	1.783
CURVE 34	
300	-0.002*
574	-0.096
928	-0.081
CURVE 35*	
300	0.019
575	0.479
928	1.432
CURVE 36	
273	-0.034
298	0.008*
373	0.177*
473	0.447*
573	0.751

T	$\Delta L/L_0$
CURVE 36 (cont.)	
673	1.119
773	1.529
873	2.013*
935	2.353
CURVE 37	
273	0.001
298	-0.000
373	-0.002
473	-0.012
573	-0.036
673	-0.096
773	-0.225
873	-0.425
935	-0.577
CURVE 38	
273	-0.034*
298	0.0085
373	0.148
473	0.370
573	0.630
673	0.941
773	1.308
873	1.742
935	2.035
CURVE 39*, †	
935	0.000
948	0.034
973	0.089
998	0.149
1023	0.159
1045	0.256
CURVE 40*, †	
935	0.000
948	0.005
973	0.018
998	0.035
1023	0.050
1045	0.062

* Not shown in figure.

† This curve is here reported using the first given temperature as reference temperature at which $\Delta L/L_0 = 0$.

DATA TABLE 59. THERMAL LINEAR EXPANSION OF URANIUM U (continued)

T	$\Delta L/L_0$
CURVE 41	
935	1.618
948	1.643
973	1.685
998	1.731
1023	1.743
1045	1.813
CURVE 42†	
1045	2.061
1073	2.116
1123	2.232
1173	2.347
1223	2.457
1273	2.572
1323	2.679
1373	2.786
CURVE 43	
293	0.000*
473	0.403*
573	0.666
673	0.969
773	1.330
873	1.769
CURVE 44	
293	0.000*
473	0.0036*
573	-0.028*
673	-0.095
773	-0.211
873	-0.389
CURVE 45*	
293	0.000
473	0.387
573	0.664
673	0.996
773	1.387
873	1.868
CURVE 46	
4	-0.372
9	-0.364*
CURVE 46 (cont.)	
13	-0.372
16	-0.375*
20	-0.382
24	-0.410
30	-0.470*
35	-0.522
40	-0.571*
45	-0.596
50	-0.592
55	-0.585*
61	-0.564
76	-0.547
107	-0.484
CURVE 47	
4	0.019
9	0.017*
13	0.017*
13	0.019*
17	0.022
20	0.010*
24	0.015
30	0.008*
35	0.007
40	0.002*
45	-0.010
50	-0.012*
56	-0.022
61	-0.014*
76	-0.012
107	0.003
CURVE 48	
4	-0.412
9	-0.410*
14	-0.414
17	-0.411*
20	-0.408
24	-0.398
30	-0.388*
35	-0.377
40	-0.361*
45	-0.347
50	-0.337*
56	-0.337
62	-0.331
76	-0.317
107	-0.285

* Not shown in figure.

† This curve is here reported using the first given temperature as reference temperature at which $\Delta L/L_0 = 0$.

FIGURE AND TABLE NO. 60R. RECOMMENDED VALUES FOR THERMAL LINEAR EXPANSION OF VANADIUM V

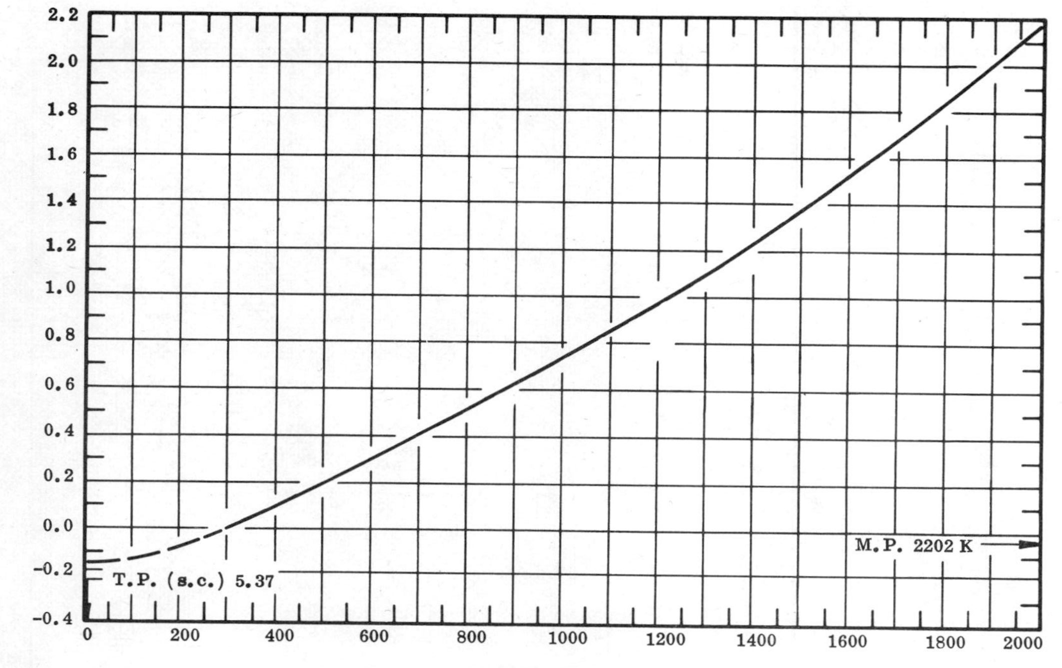

RECOMMENDED VALUES

[Temperature, T, K; Linear Expansion, $\Delta L/L_0$, %; α, K^{-1}]

T	$\Delta L/L_0$	$\alpha \times 10^6$
5	-0.155‡	0.02
25	-0.155‡	0.2
50	-0.153‡	1.3
100	-0.136‡	5.1
200	-0.072‡	7.1
293	0.000	8.4
400	0.102	9.6
500	0.200	9.9
600	0.300	10.2
700	0.405	10.5
800	0.512	10.9
900	0.622	11.2
1000	0.737	11.6
1200	0.978	12.5
1400	1.239	13.6
1600	1.521	14.7
1800	1.828	16.0
2000	2.160	17.2

‡ Provisional values.

LOW-TEMPERATURE THERMAL EXPANSION VALUES

T	Normal State $\Delta L/L \times 10^6$	Superconducting State $\Delta L/L^{\dagger} \times 10^6$
0	0.00	
1	0.20	-2.7
2	0.78	-2.3
3	1.8	-1.3
4	3.2	1.1
5	5.0	4.8
5.1	5.2	5.2

† Superconducting state value is set equal to that of normal state value at the transition temperature, $T_c = 5.1$ K.

REMARKS

The tabulated values are considered accurate to within ± 5% above 293 K and ± 10% below. Recent investigators have reported phase transition near 230 K which can very well be due to impurities. The tabulated values can be represented approximately by the following equation:

$$\Delta L/L_0 = -0.269 + 8.971 \times 10^{-4}\, T + 5.822 \times 10^{-8}\, T^2 + 5.036 \times 10^{-11}\, T^3 \qquad (293 < T < 2000)$$

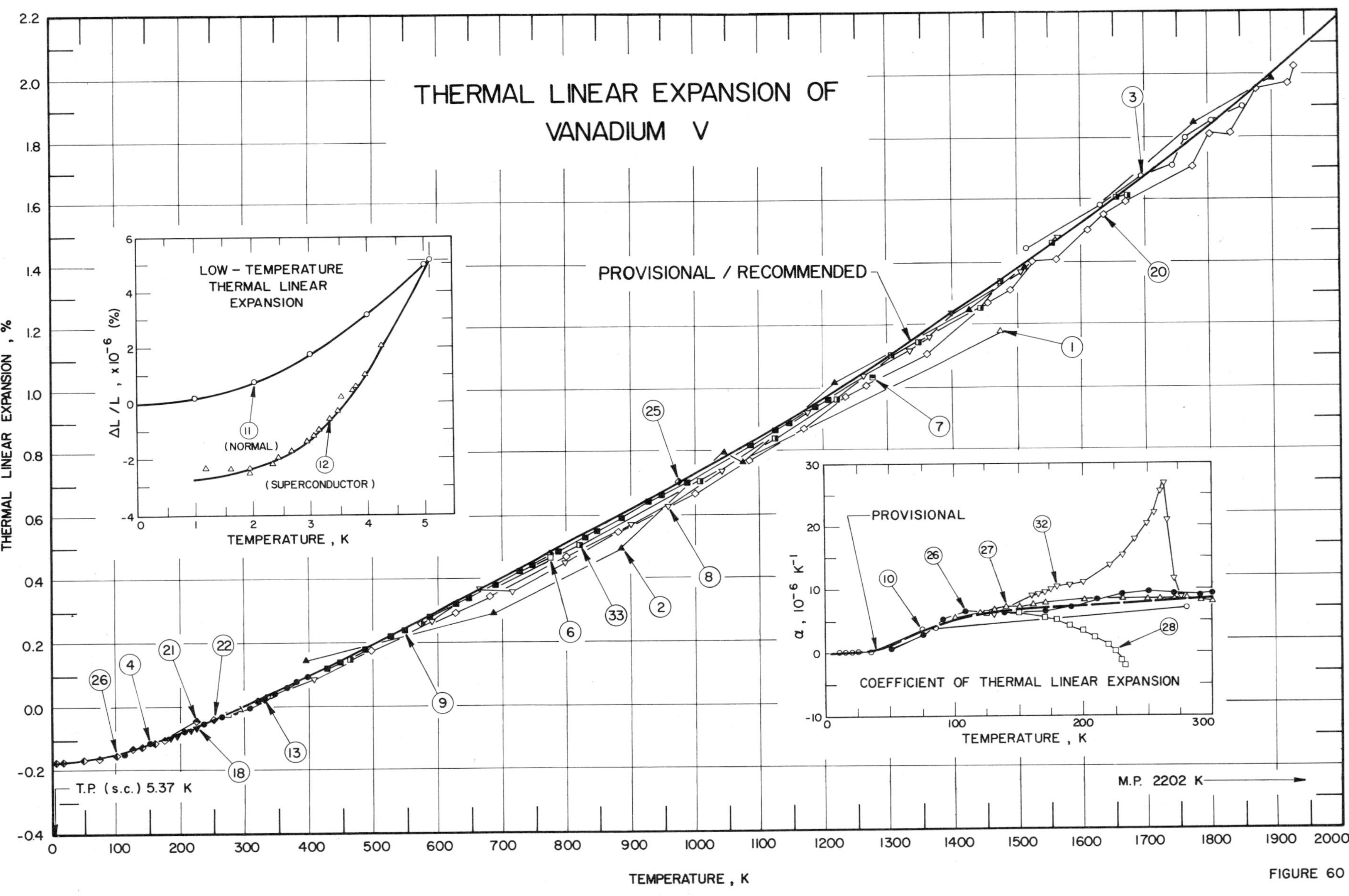
THERMAL LINEAR EXPANSION OF
VANADIUM V
PROVISIONAL / RECOMMENDED
LOW – TEMPERATURE
THERMAL LINEAR
EXPANSION
(NORMAL)
(SUPERCONDUCTOR)
TEMPERATURE , K
ΔL /L , x10⁻⁶ (%)
PROVISIONAL
COEFFICIENT OF THERMAL LINEAR EXPANSION
TEMPERATURE , K
α , 10⁻⁶ K⁻¹
T.P. (s.c.) 5.37 K
M.P. 2202 K
THERMAL LINEAR EXPANSION , %
TEMPERATURE , K

FIGURE 60

SPECIFICATION TABLE 60. THERMAL LINEAR EXPANSION OF VANADIUM V

Cur. No.	Ref. No.	Author(s)	Year	Method Used	Temp. Range, K	Name and Specimen Designation	Composition (weight percent), Specifications, and Remarks
1	208	Frantsevich, I.N., Zhurakovskii, E.A., and Lyashohenko, A.B.	1967	X	293, 1473		Sample prepared from powder by vacuum sintering of pressed block.
2	243	Vasyutinskiy, B.M., Kartmazov, G.N., Smirnov, Yu.M., and Finkel, V.A.	1966	X	293-1900		99.74 V, 0.1 O, 0.04 C, 0.005 N, and <0.063 metallic impurities; specimen 0.1 to 0.3 mm thick; heated in vacuum of 2×10^{-5} mm Hg; (321) line focused using Cu x-radiation; lattice parameter reported at 293 K is 3.0289 Å.
3	243	Vasyutinskiy, B.M., et al.	1966	X	1517-1854		The above specimen; (211) line focused using V K_2 x-radiation; lattice parameter reported at 293 K is 3.0289 Å.
4	246	Smirnov, Yu.M. and Finkel, V.A.	1966	X	112-398		99.74 V, 0.1 O, 0.04 C, 0.005 N, 0.01 H; polycrystalline samples in form of plates and strips 0.1 to 0.3 mm thick; data average of several specimens; expansion measured using CuKα x-radiation; lattice parameter reported at 293 K is 3.030758 Å.
5*	246	Smirnov, Yu.M. and Finkel, V.A.	1966	X	112-396		99.20 V, 0.61 O, 0.05 C, 0.08 N, 0.03 H; similar to the above specimen; lattice parameter reported at 293 K is 3.03064 Å.
6	117	Yaggee, F.L. and Styles, J.W.	1966	L	273, 773		Density 6.08 ± 0.1 g cm^{-1}.
7	117	Yaggee, F.L. and Styles, J.W.	1966	L	273, 1273		The above specimen.
8	358	Fieldhouse, I.B. and Lang, J.I.	1961	T	294-1735		99.74 V, 0.073 O, 0.048 Fe, 0.043 N, 0.042 C; hot-rolled and annealed; density 6.05 g cm^{-3}; measured expansion includes unspecified amount of creep at temperatures above approximately 1600 K; 0.1544 total of C, O, N, H interstitially.
9	119	Yaggee, F.L., Gilbert, E.R., and Styles, Y.W.	1969	L	297-1205		Commercial arc-cast in form of rod 0.64 cm diameter; annealed in vacuum of 2×10^{-6} mm Hg at 1273 K for 1 hr; density 6.089 ± 0.002 g cm^{-3} at 298 K; zero-point correction of -0.018% determined by graphical extrapolation.
10	251	White, G.K.	1962	E	6-85, 280		Electro-refined raw material supplied by U.S. Bureau of Mines; analysis revealed 0.09 Fe, 0.04 O_2, 0.014 Si, and 0.01 C; arc-melted into solid bar, machined to cylindrical shape; stress-relieved 2 hr at 773 K, machined to 5.09 cm; this curve reported here is extrapolated to 293 K using other available data.
11	251	White, G.K.	1962	E	1-5.1		The above specimen; measurements in the magnetic field of 1600 Oe; this curve is here reported using 0 K as the reference temperature at which $\Delta L/L_0 = 0$.
12	252	White, G.K.	1962	E	1-5.1		The above specimen; this curve is here referred to the normal state value of Curve 11 at the critical temperature, $T_c = 5.1$ K; zero-point correction is -0.05×10^{-6}%.
13	359	James, W.J. and Straumanis, M.E.	1961	X	283-330	Sample I	99.972 V, 0.015 O, 0.005 C, 0.002 N, 0.006 total metallic impurities; sample prepared at Battelle Memorial Institute by Van Ankel method; specimen about 0.1 mm diameter; lattice parameter reported at 293 K is 3.01787 kX (3.023905 Å).
14*	359	James, W.J. and Straumanis, M.E.	1961	X	284-334	Sample II	99.87 V, 0.055 O, 0.042 N, 0.031 C, 0.0026 H; specimen from Union Carbide Metals Co.; wire specimen 0.09 mm diameter; lattice parameter reported at 293 K is 3.02549 kX (3.03154 Å).

* Not shown in figure.

SPECIFICATION TABLE 60. THERMAL LINEAR EXPANSION OF VANADIUM V (continued)

Cur. No.	Ref. No.	Author(s)	Year	Method Used	Temp. Range, K	Name and Specimen Designation	Composition (weight percent), Specifications, and Remarks
15*	390	Suzuki, H. and Miyahara, S.	1966	X	173-293		Specimen cleaned from single crystal; anomalous transformation at about 240 K below which three distinct axes exist; distortion from cubic structure explained as the crystal distortion follows the distortion of the Fermi surface of conduction electrons causing a change of the Fermi energy, and if the energy decreases the crystal deforms as in the Jahn-Teller distortion; microscopic expansion reported as the spacing of the (200) plane; lattice parameter reported at 283 K is 1.51797 Å; 1.51825 Å at 293 K determined by graphical extrapolation.
16*	390	Suzuki, H. and Miyahara, S.	1966	X	173-283		The above specimen; expansion reported as the spacing of the (110) plane; lattice parameter reported at 283 K is 2.1400 Å; 2.15022 Å at 293 K determined by graphical extrapolation.
17*	390	Suzuki, H. and Miyahara, S.	1966	S	185-300		The above specimen; macroscopic expansion measured along one axis; zero-point correction of 0.0056% determined by graphical interpolation.
18*	390	Suzuki, H. and Miyahara, S.	1966	S	185-227		The above specimen; expansion measured along second axis; data indistinguishable from the above at temperatures above the transition point; zero-point correction of 0.0056% determined by graphical extrapolation.
19*	390	Suzuki, H. and Miyahara, S.	1966	S	185-235		The above specimen; expansion measured along third axis; data indistinguishable from curve 15 above the transition point; zero-point correction of 0.0056% determined by graphical extrapolation.
20	623	Mochalov, G.A. and Ivanov, O.S.	1969	O	293-1873		Commercial specimen were remelted in arc furnace with nonconsumable tungsten electrode; measurements below 1273 K using quartz dilatometer and above 1273 K using cathetometer.
21	622	Westlake, D.G. and Ockers, S.T.	1970	I	73-548		99.99^+ pure; major impurities 0.025 Si, 0.002 Ti, 0.0005 Fe, and 0.0003 N.
22*	620	Finkel, V.A., Glamazda, V.I., and Kovtun, Z.	1970	X	80-297		Single crystal in form of rod with [110] growth axis obtained by zone refining; tetragonal structures with second order transition at 230 K; measured along a-axis; lattice parameter reported at 76 and 293 K are 3.0248 and 3.0294 Å, respectively.
23*	620	Finkel, V.A., et al.	1970	X	80-230		The above specimen; b.c.c. above 230 K; measured along c-axis; lattice parameter reported at 293 K is 3.0294 Å.
24*	707	Pridantseva, K.S. and Solov'yeva, N.A.	1965	L	293, 1073		Powder metallurgy sample, 1% measurement error.
25	707	Pridantseva, K.S. and Solov'yeva, N.A.	1965	L	293-973		Similar to the above specimen except annealed at 1273 K for 24 hr; 1% measurement error.
26	759	Bolef, D.I., Smith, R.E., and Miller, J.G.	1971	L	4-300		99.82 V, 0.012 H_2, 0.003 N_2, 0.086 O_2, 0.01 Si, 0.04 Al, 0.025 Fe; single crystal cut from single ingot (grown by Verneui process using an arc as heat source) from Linde Company, anomaly near 180 K; measurements along [100] crystal axis, zero-point correction of 0.017% determined by graphical interpolation; data on the coefficient of thermal linear expansion also reported.
27*	759	Bolef, D.I., et al.	1971	L	4-300		Similar to the above specimen; expansion measured along [110] crystal axis; zero-point correction of 0.017% determined by graphical interpolation; data on the coefficient of thermal linear expansion also reported.
28*	760	Bolef, D.I., et al.	1971	L	80-296		Similar to the above specimen; obtained by cutting perpendicular to [110] axis about 1/3 off the length of above specimen; phase change in this and in the following curves near 260 K possibly due to hydrogen precipitating measurements along [100] crystal axis; zero-point correction of 0.014% obtained by graphical interpolation.

* Not shown in figure.

SPECIFICATION TABLE 60. THERMAL LINEAR EXPANSION OF VANADIUM V (continued)

Cur. No.	Ref. No.	Author(s)	Year	Method Used	Temp. Range, K	Name and Specimen Designation	Composition (weight percent), Specifications, and Remarks
29*	760	Bolef, D.I., et al.	1971	L	80-296		The above specimen; measurements along [110] direction; zero-point correction of 0.014% determined by graphical interpolation.
30*	760	Bolef, D.I., et al.	1971	L	75-362		99.76 V, 0.035 H_2, 0.067 N_2, 0.087 O_2, 0.01 Si, 0.04 Al, and 0.02 Fe specimen; obtained by cutting a piece parallel to [110] axis from the above specimen; measurements along [100] axis; zero-point correction of 0.021% determined by graphical interpolation.
31*	760	Bolef, D.I., et al.	1971	L	80-366		Similar to the above specimen; measurements along [101] axis; zero-point correction of 0.017% determined by graphical interpolation.
32*	760	Bolef, D.I., et al.	1971	L	80-357		Similar to the above specimen; measurements along [110] axis, these values are very close to the values along [101] axis; zero-point correction of 0.014% determined by graphical interpolation.
33*	761	Brodskiy, B.R. and Neymark, B.E.	1971	L	342-1672		99.82 V, 0.07 O, 0.00 N, <0.001 H, <0.005 Al, <0.01 Fe, 0.003 Si, and 0.02 Ni, obtained by electron beam melting in vacuum of pressed powder grade VEL-2.

* Not shown in figure.

DATA TABLE 60. THERMAL LINEAR EXPANSION OF VANADIUM V

[Temperature, T, K; Linear Expansion, $\Delta L/L_0$, %]

T	$\Delta L/L_0$
CURVE 1	
293	0.00
1473	1.18
CURVE 2	
293	0.000*
391	0.148
685	0.297
885	0.498
1045	0.802
1075	0.772
1217	1.020
1427	1.25
1517	1.390
1778	1.848
1900	1.987
CURVE 3	
1517	1.446
1630	1.581
1695	1.677
1743	1.710
1762	1.796
1807	1.852
1854	1.911
CURVE 4	
112	-0.142
112	-0.140*
116	-0.140*
118	-0.136*
123	-0.136*
126	-0.132*
130	-0.129*
136	-0.126*
139	-0.122*
143	-0.118*
150	-0.116
157	-0.107*
166	-0.104*
168	-0.102*
172	-0.098*
176	-0.094*
180	-0.091
CURVE 4 (cont.)	
185	-0.087*
189	-0.081*
193	-0.079*
199	-0.077*
205	-0.077
212	-0.072*
218	-0.066*
223	-0.062*
228	-0.058*
233	-0.056*
237	-0.053
241	-0.049*
245	-0.045*
249	-0.041*
253	-0.038*
256	-0.036*
259	-0.034*
262	-0.030
263	-0.030*
264	-0.026*
266	-0.029*
268	-0.025*
270	-0.026*
272	-0.022*
274	-0.019*
275	-0.016*
277	-0.019*
278	-0.016*
285	-0.011
285	-0.009*
294	0.009*
294	0.001*
298	0.001*
303	0.006
304	0.008*
308	0.011*
311	0.014*
316	0.018*
320	0.022*
322	0.023*
326	0.026*
329	0.031*
332	0.033*
335	0.036*
338	0.041*
342	0.044
CURVE 4 (cont.)	
345	0.047*
347	0.049*
351	0.053*
353	0.057*
358	0.060*
363	0.062*
367	0.066*
369	0.068*
372	0.070*
374	0.074*
380	0.081
384	0.080*
388	0.083*
390	0.084*
393	0.088*
394	0.091*
396	0,092*
398	0.094
CURVE 5*	
112	-0.135
114	-0.131
120	-0.127
121	-0.130
126	-0.119
128	-0.121
133	-0.119
135	-0.120
137	-0.116
138	-0.119
140	-0.116
144	-0.116
145	-0.113
147	-0.112
151	-0.106
155	-0.104
156	-0.107
161	-0.102
163	-0.098
168	-0.096
173	-0.095
173	-0.089
175	-0.093
178	-0.086
181	-0.081
CURVE 5 (cont.)	
187	-0.076
192	-0.071
193	-0.073
196	-0.072
198	-0.071
203	-0.068
208	-0.063
210	-0.063
214	-0.058
216	-0.060
218	-0.054
222	-0.054
229	-0.052
233	-0.052
239	-0.050
243	-0.045
244	-0.043
248	-0.043
253	-0.038
256	-0.034
260	-0.033
263	-0.030
266	-0.030
268	-0.023
271	-0.021
274	-0.020
275	-0.017
277	-0.015
282	-0.012
284	-0.013
285	-0.009
287	-0.005
290	-0.001
291	-0.002
292	-0.003
293	0.000
294	0.004
297	0.008
299	0.011
300	0.013
301	0.010
302	0.013
305	0.016
308	0.018
312	0.024
313	0.024
CURVE 5 (cont.)	
316	0.024
318	0.030
319	0.027
323	0.035
325	0.031
327	0.039
330	0.046
333	0.048
336	0.046
337	0.052
340	0.053
341	0.049
345	0.055
348	0.058
353	0.059
357	0.067
359	0.070
362	0.069
364	0.072
370	0.079
372	0.085
376	0.085
379	0.089
381	0.088
385	0.095
386	0.093
389	0.099
391	0.097
395	0.108
396	0.101
CURVE 6	
273	-0.0196
773	0.471
CURVE 7	
273	-0.0212*
1273	1.038
CURVE 8	
294	0.01*
339	0.04
359	0.06*
CURVE 8 (cont.)	
409	0.09
590	0.27
663	0.37
713	0.36
798	0.45
899	0.57
957	0.63
1040	0.74
1066	0.77
1176	0.92
1251	1.04
1313	1.12
1362	1.16
1399	1.24
1507	1.37
1565	1.48
1608	1.58*
1644	1.73*
1675	1.88*
1707	2.10*
1739	2.37*
1755	2.50*
1782	2.71*
1797	2.94*
CURVE 9	
297	0.004*
327	0.032*
348	0.051*
368	0.067*
386	0.088*
406	0.107*
426	0.127
447	0.143
486	0.184
528	0.224
547	0.242
587	0.282
627	0.323
647	0.343
687	0.382
727	0.425
748	0.446
787	0.488
827	0.531

*Not shown in figure.

DATA TABLE 60. THERMAL LINEAR EXPANSION OF VANADIUM V (continued)

T	ΔL/L₀
CURVE 9 (cont.)	
847	0.552
887	0.597
926	0.641
946	0.664
986	0.707
1027	0.754
1046	0.777*
1065	0.799*
1085	0.823
1125	0.870
1145	0.894
1166	0.915*
1184	0.941
1205	0.963
CURVE 10‡	
6	-0.155
7	-0.155
8	-0.155
9	-0.155
10	-0.155
12	-0.155
14	-0.155
15	-0.155
16	-0.155
18	-0.155
20	-0.155
22	-0.155
24	-0.155
25	-0.155
26	-0.155
28	-0.155
30	-0.155
75	-0.116
85	-0.120
280	-0.010
CURVE 11	
0	0.00×10^{-6}
1	0.20
2	0.78
3	1.77
4	3.16
5	4.98
5.1	5.19
CURVE 12	
1.38	-2.39
1.63	-2.39
1.92	-2.39
1.92	-2.25
2.24	-1.94
2.31	-2.18
2.45	-1.90
2.68	-1.70
2.94	-1.38
3.03	-1.15
3.13	-0.95
3.31	-0.55
3.49	-0.23
3.52	0.21
3.74	0.44
3.79	0.60
3.93	0.93
3.93	1.07
4.25	2.09
5.10	5.19
CURVE 13	
283.2	-0.0127*
293.2	0.000*
303.2	0.0082*
310.2	0.0185*
320.2	0.0241
330.2	0.0331
CURVE 14*	
284	-0.010
293	0.000
294	0.004
304	0.011
314	0.023
324	0.044
334	0.048
CURVE 15*	
173	-0.255
183	-0.237
193	-0.218
203	-0.203
313	-0.178
CURVE 15 (cont.)*	
223	-0.155
233	-0.132
235	-0.105
238	-0.085
240	-0.085
243	-0.091
245	-0.079
248	-0.067
250	-0.070
253	-0.062
263	-0.051
273	-0.037
283	-0.019
CURVE 16*	
173	-0.225
183	-0.216
193	-0.179
203	-0.159
213	-0.147
223	-0.121
233	-0.079
235	-0.079
238	-0.072
240	-0.073
243	-0.062
245	-0.059
248	-0.055
253	-0.052
263	-0.046
273	-0.020
283	-0.010
CURVE 17*	
185	-0.121
192	-0.112
198	-0.103
204	-0.096
211	-0.087
217	-0.077
220	-0.074
223	-0.071
226	-0.065
227	-0.063
229	-0.061
CURVE 17 (cont.)*	
231	-0.059
232	-0.057
232	-0.056
233	-0.055
235	-0.054
236	-0.051
238	-0.050
239	-0.049
240	-0.047
240	-0.046
242	-0.046
243	-0.045
243	-0.044
244	-0.044
246	-0.042
249	-0.040
252	-0.038
252	-0.037
257	-0.032
257	-0.031
263	-0.029
263	-0.026
268	-0.023
268	-0.022
273	-0.019
273	-0.017
278	-0.014
278	-0.012
283	-0.010
283	-0.008
288	-0.004
288	-0.002
294	0.001
300	0.006
CURVE 18	
185	-0.097
192	-0.092
199	-0.088*
204	-0.083*
211	-0.076*
217	-0.073*
219	-0.070*
220	-0.070
222	-0.069*
223	-0.067*
CURVE 18 (cont.)	
225	-0.066
227	-0.064*
CURVE 19*	
185	-0.072
192	-0.068
198	-0.065
204	-0.063
211	-0.060
217	-0.057
220	-0.057
223	-0.056
226	-0.055
229	-0.054
232	-0.053
235	-0.052
CURVE 20	
390	0.095*
494	0.182
628	0.292
679	0.349
796	0.468
880	0.546
1001	0.671
1083	0.774
1170	0.875
1232	0.974
1268	1.010
1359	1.113
1457	1.272
1489	1.315
1525	1.401
1563	1.411
1612	1.502
1635	1.553
1669	1.593
1774	1.702
1802	1.817
1839	1.817
1874	1.958
1928	1.971
1934	2.023
CURVE 21	
73	-0.160
123	-0.130
173	-0.100
223	-0.060
273	-0.020
293	0.000
323	0.026
348	0.049
CURVE 22*	
76	-0.152
84	-0.149
92	-0.145
101	-0.142
110	-0.139
120	-0.135
128	-0.129
138	-0.125
144	-0.119
150	-0.112
160	-0.106
169	-0.099
177	-0.089
183	-0.082
186	-0.079
191	-0.075
197	-0.069
204	-0.066
209	-0.063
213	-0.059
221	-0.056
225	-0.056
228	-0.053
234	-0.050
239	-0.043
244	-0.036
251	-0.033
257	-0.026
263	-0.020
269	-0.016
272	-0.013
279	-0.010
284	-0.003
293	0.000
297	0.007

* Not shown in figure.
‡ Author's data for coefficient of thermal expansion have been integrated by TPRC to obtain $\Delta L/L_0$.

DATA TABLE 60. THERMAL LINEAR EXPANSION OF VANADIUM V (continued)

T	$\Delta L/L_0$
CURVE 23*	
77	-0.188
87	-0.188
98	-0.182
108	-0.178
115	-0.175
125	-0.168
134	-0.162
144	-0.158
153	-0.152
163	-0.142
172	-0.135
178	-0.129
181	-0.125
186	-0.119
190	-0.112
197	-0.106
200	-0.099
203	-0.092
208	-0.086
212	-0.079
217	-0.073
221	-0.063
225	-0.056
CURVE 24*	
293	0.000
1073	0.761
CURVE 25	
293	0.000
573	0.261
773	0.480
973	0.712
CURVE 26	
4.2	-0.168
20	-0.168
30	-0.168*
40	-0.168*
50	-0.167
60	-0.166*
70	-0.164*
80	-0.161*
90	-0.155*
100	-0.150
110	-0.145*
CURVE 26 (cont.)	
120	-0.140*
130	-0.133*
140	-0.126*
150	-0.119*
160	-0.113
170	-0.107*
180	-0.101*
190	-0.094*
200	-0.086*
210	-0.077*
220	-0.068*
230	-0.059*
240	-0.050*
250	-0.040*
260	-0.031*
270	-0.021*
280	-0.010*
290	-0.003*
CURVE 27*	
4.2	-0.176
20	-0.176
30	-0.175
40	-0.174
50	-0.172
60	-0.170
70	-0.167
80	-0.165
90	-0.160
100	-0.153
110	-0.147
120	-0.141
130	-0.134
140	-0.127
150	-0.120
160	-0.112
170	-0.103
180	-0.095
190	-0.087
200	-0.028
210	-0.020
220	-0.063
230	-0.055
240	-0.045
250	-0.036
260	-0.027
270	-0.020
280	-0.012
CURVE 27 (cont.)*	
290	-0.002
300	0.010
CURVE 28*	
80	-0.123
90	-0.117
100	-0.111
110	-0.106
120	-0.100
130	-0.095
140	-0.089
150	-0.081
160	-0.076
170	-0.070
180	-0.065
190	-0.061
200	-0.056
210	-0.054
220	-0.052
230	-0.052
240	-0.044
250	-0.037
260	-0.027
267	-0.022
277	-0.014
289	-0.004
296	0.002
CURVE 29*	
80	-0.161
90	-0.157
100	-0.152
110	-0.146
120	-0.140
130	-0.134
140	-0.128
150	-0.120
160	-0.111
170	-0.101
180	-0.093
190	-0.084
200	-0.075
210	-0.066
220	-0.058
230	-0.052
240	-0.044
250	-0.037
CURVE 29 (cont.)*	
260	-0.027
267	-0.022
277	-0.014
289	-0.004
296	0.002
CURVE 30*	
75	-0.087
85	-0.081
95	-0.076
100	-0.073
120	-0.066
140	-0.053
160	-0.042
180	-0.028
200	-0.021
220	-0.016
240	-0.016
250	-0.019
260	-0.025
263	-0.026
278	-0.015
287	-0.011
293	0.000
302	0.004
309	0.011
316	0.016
326	0.022
338	0.031
345	0.038
348	0.040
362	0.044
362	0.057
CURVE 31*	
80	-0.230
100	-0.220
120	-0.212
140	-0.198
160	-0.180
180	-0.163
200	-0.145
221	-0.116
241	-0.082
250	-0.066
260	-0.041
263	-0.022
CURVE 31 (cont.)*	
271	-0.016
291	-0.002
301	0.004
309	0.009
324	0.023
334	0.033
354	0.046
366	0.055
CURVE 32*	
80	-0.225
100	-0.230
120	-0.211
140	-0.199
160	-0.176
180	-0.159
200	-0.135
219	-0.107
226	-0.096
231	-0.088
234	-0.081
237	-0.077
241	-0.069
246	-0.060
248	-0.054
255	-0.041
260	-0.028
263	-0.019
275	-0.010
288	-0.001
295	0.001
313	0.018
331	0.031
357	0.057
CURVE 33	
342	0.043*
365	0.063*
376	0.075*
445	0.058*
462	0.158
502	0.209*
547	0.247*
570	0.261*
656	0.349*
683	0.370*
738	0.428*
CURVE 33 (cont.)	
773	0.483*
791	0.486*
821	0.509
909	0.617*
918	0.627*
1004	0.708
1012	0.723*
1118	0.830*
1125	0.841*
1211	0.977
1219	0.965
1306	1.102
1312	1.083*
1334	1.146*
1441	1.253
1475	1.338
1531	1.391*
1558	1.467
1654	1.591*
1659	1.609
1672	1.619

* Not shown in figure.

DATA TABLE 60. COEFFICIENT OF THERMAL LINEAR EXPANSION OF VANADIUM V

[Temperature, T, K; Coefficient of Expansion, α, 10^{-6} K^{-1}]

T	α	T	α	T	α	T	α
CURVE 10‡		CURVE 26 (cont.)		CURVE 27 (cont.)*		CURVE 30 (cont.)*	
6	0.025	200	8.16*	313	7.92*	250	4.56
7	0.0285*	210	8.59	320	7.92*	255	-5.96
8	0.034*	220	8.91*			260	-7.63
9	0.039*	230	9.14	CURVE 28		261	-7.94
10	0.045*	240	9.51*			262	8.46
12	0.0575*	250	9.79	120	6.12*	270	8.43
14	0.072*	260	9.43*	125	6.15	280	8.23
15	0.08	270	9.50	130	6.21*	293	7.89
16	0.089*	280	9.12*	140	6.33*	300	7.66
18	0.108*	290	9.26	150	6.31*		
20	0.131	300	9.23	160	6.25*	CURVE 32	
22	0.157*	310	9.22*	170	5.73		
24	0.188*			180	5.26	120	6.13
25	0.208	CURVE 27		190	4.52	130	6.50
26	0.230*			200	3.78	140	6.95*
28	0.282*	20	0.41*	210	2.54	150	7.72*
30	0.350	30	0.64*	220	1.10	160	8.59
75	3.700	40	0.97*	225	0.29	165	9.20
85	4.110	50	1.31*	230	-1.02	170	9.70
280	7.750	60	2.13*	233	-1.75	175	10.01
		65	2.80*	233	9.27	180	10.21
CURVE 26		70	3.55*	240	9.24	190	10.60
		80	4.43*	250	9.22*	200	11.19
20	0.02*	90	5.11*	260	9.14*	210	11.98*
27	-0.02*	100	5.57	270	8.95	220	13.46
35	0.09*	110	6.15*	280	8.63*	230	15.38
43	0.37*	120	6.47	293	8.27*	240	17.62
50	0.96	130	6.79	298	8.04*	250	20.33
60	1.76*	140	7.28			255	22.17
68	2.52*	150	7.69	CURVE 30*		260	25.85
74	3.16	160	7.86			261	26.61
78	3.78*	170	8.01	120	5.67	265	20.95
83	4.49*	180	8.11*	130	5.92	270	11.76
90	5.11	190	8.15*	140	6.03	275	8.86
98	5.56*	200	8.31	150	6.01	280	7.70*
108	5.94	210	8.52*	160	5.93	293	7.40*
118	6.10*	220	8.64*	170	5.64	300	7.53*
130	6.22*	230	8.90	180	5.17		
138	6.28	240	8.90*	190	4.45		
148	6.37*	250	8.91	200	3.66		
160	6.47*	260	8.91	210	2.56		
170	6.63	270	8.91	220	1.24		
178	6.97*	280	8.58	230	-0.36		
184	7.24*	290	8.35	240	2.35		
190	7.48	300	8.05	245	3.44		

* Not shown in figure.
‡ Author's data for coefficient of thermal expansion have been integrated by TPRC to obtain $\Delta L/L_0$.

FIGURE AND TABLE NO. 61R. PROVISIONAL VALUES FOR THERMAL LINEAR EXPANSION OF YTTERBIUM Yb

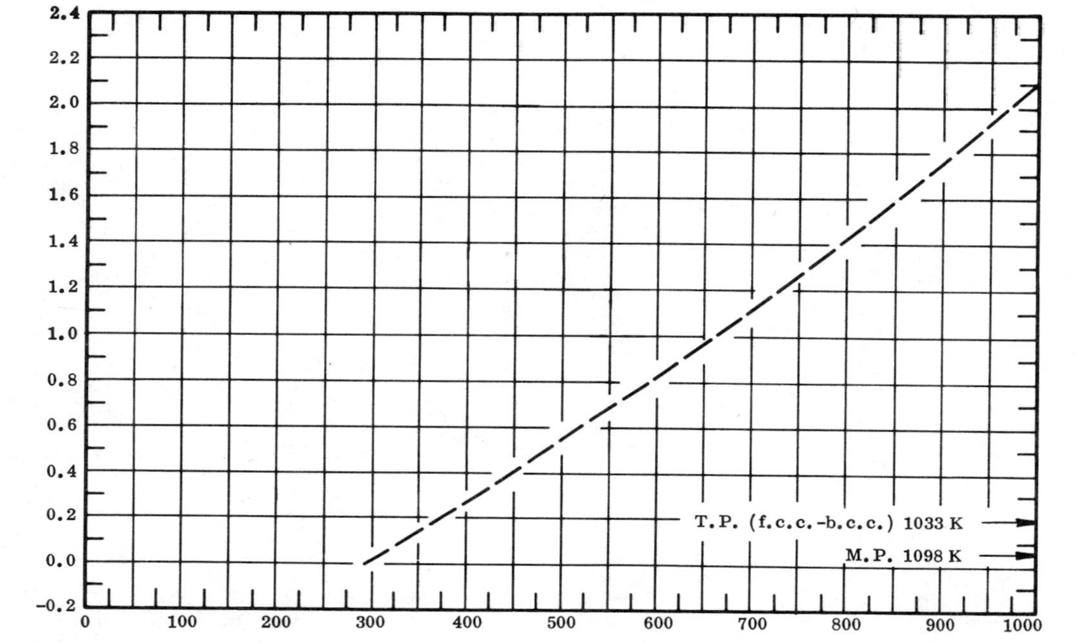

PROVISIONAL VALUES

[Temperature, T, K; Linear Expansion, $\Delta L/L_0$, %; α, K^{-1}]

T	$\Delta L/L_0$	$\alpha \times 10^6$
293	0.000	25.1
400	0.272	26.0
500	0.539	27.1
600	0.821	29.1
700	1.119	30.7
800	1.432	32.2
900	1.760	33.3
1000	2.100	34.2

REMARKS

The tabulated values are considered accurate to within ±7% over the entire temperature range. These values can be represented approximately by the following equation:

$$\Delta L/L_0 = -0.657 + 2.071 \times 10^{-3}\, T + 6.421 \times 10^{-7}\, T^2 + 4.957 \times 10^{-11}\, T^3$$

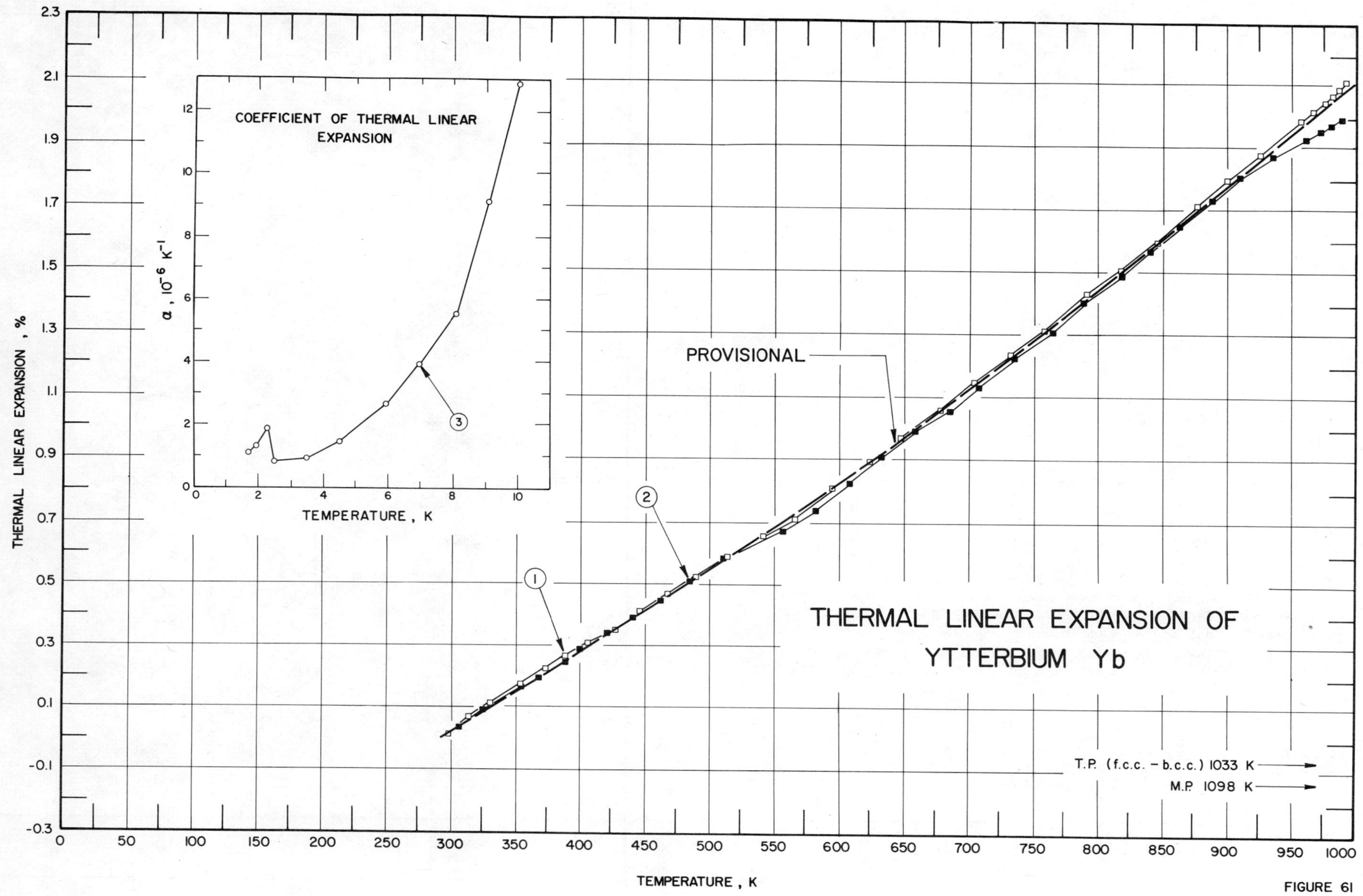
COEFFICIENT OF THERMAL LINEAR
EXPANSION
α, 10⁻⁶ K⁻¹
TEMPERATURE, K
PROVISIONAL
1
2
3
THERMAL LINEAR EXPANSION OF
YTTERBIUM Yb
T.P. (f.c.c. – b.c.c.) 1033 K
M.P. 1098 K
THERMAL LINEAR EXPANSION, %
TEMPERATURE, K

FIGURE 61

SPECIFICATION TABLE 61. THERMAL LINEAR EXPANSION OF YTTERBIUM Yb

Cur. No.	Ref. No.	Author(s)	Year	Method Used	Temp. Range, K	Name and Specimen Designation	Composition (weight percent), Specifications, and Remarks
1	78	Barson, F., Legvold, S. and Spedding, F. H.	1956	L	298-1047		0.5 Ca, 0.05 Fe, 0.05 Si, 0.03 Ta, 0.01 Tm, < 0.01 Er, 0.005 Lu, 0.0645 C, 0.01 N_2; metallic lanthanum reacted with ytterbium oxide to yield metallic ytterbium; rod specimen 0.6 cm diameter, 6 cm long; expansion measured with increasing temperature; zero-point correction of 0.013% determined by graphical extrapolation.
2	78	Barson, F., et al.	1956	L	1035-306		The above specimen; expansion measured with decreasing temperature; zero-point correction of 0.026% determined by graphical extrapolation.
3	80	Andres, K.	1963	O	1.69-10		Polycrystalline specimen; melted into tantalum form; this curve is here reported using the first given temperature as reference temperature at which $\Delta L/L_0 = 0$.
4*	82	Hanak, J. J.	1959	X	541-685		0.03 Ca, < 0.02 each Mg, Si, < 0.01 each La, Er, Tm, Lu, Fe, < 0.005 Cr; wire specimen cut from sheet prepared at Ames Laboratory; hexagonal close packed structure; expansion measured along a-axis; this curve is here reported using the first given temperature as reference temperature at which $\Delta L/L_0 = 0$; lattice parameter reported at 541 K is 3.91 Å.
5*	82	Hanak, J. J.	1959	X	541-685		The above specimen; expansion measured along c-axis; this curve is here reported using the first given temperature as reference temperature at which $\Delta L/L_0 = 0$; lattice parameter reported at 541 K is 6.403 Å.
6*	82	Hanak, J. J.	1959	X	296-849		Similar to the above specimen; f.c.c. structure; lattice parameter reported at 296 K is 5.484 Å; 5.485 Å at 293 K determined by graphical extrapolation.
7*	82	Hanak, J. J.	1959	X	993-1080		Similar to the above specimen, except b.c.c. structure; this curve is here reported using the first given temperature at reference temperature at which $\Delta L/L_0 = 0$.

* Not shown in figure.

DATA TABLE 61. THERMAL LINEAR EXPANSION OF YTTERBIUM Yb

[Temperature, T, K; Linear Expansion, $\Delta L/L_0$, %]

CURVE 1

T	$\Delta L/L_0$
298	0.013
312	0.068
330	0.108
354	0.177
372	0.221
387	0.263
405	0.302
427	0.350
445	0.413
466	0.465
488	0.521
513	0.586
540	0.657
565	0.712
594	0.806
622	0.893
647	0.963
676	1.055
704	1.141
731	1.235
759	1.316
791	1.429
817	1.512
845	1.598
877	1.714
902	1.796
928	1.878
958	1.987
967	2.019
979	2.041
983	2.064
989	2.089
994	2.118
999	2.159
1006	2.178*
1024	2.180*
1041	2.159*
1047	2.139*

CURVE 2

T	$\Delta L/L_0$
1035	2.111*
1021	2.079*
1013	2.059*
1005	2.056*
1001	2.017
992	1.995

CURVE 2 (cont.)

T	$\Delta L/L_0$
982	1.973
973	1.954
962	1.927
939	1.873
910	1.809
888	1.734
864	1.646
840	1.574
819	1.489
789	1.401
765	1.310
735	1.225
709	1.133
685	1.058
658	0.989
633	0.909
607	0.825
581	0.740
556	0.673
510	0.581
484	0.505
461	0.446
440	0.393
420	0.340
400	0.291
389	0.244
367	0.199
354	0.161
325	0.088
306	0.035

CURVE 3*, ‡

T	$\Delta L/L_0$
1.69	0.000×10^{-3}
1.97	0.0035
2.21	0.0074
2.45	0.0106
3.48	0.0198
4.49	0.0318
5.97	0.0618
6.93	0.0934
8.01	0.145
9.01	0.218
10.00	0.327

CURVE 4*, †

T	$\Delta L/L_0$
541	0.000
556	0.000
583	0.026
685	0.384

CURVE 5*, †

T	$\Delta L/L_0$
541	0.000
556	0.000
583	0.047
685	0.156

CURVE 6*

T	$\Delta L/L_0$
296	0.018
301	0.055
323	0.109
344	0.182
376	0.182
412	0.347
424	0.365
460	0.456
466	0.529
481	0.547
506	0.638
527	0.693
556	0.748
570	0.821
575	0.802
590	0.875
611	0.967
621	1.021
631	1.003
651	1.076
668	1.149
670	1.149*
681	1.240
710	1.258
722	1.295
725	1.404
765	1.532
822	1.696
849	1.733

CURVE 7*, †

T	$\Delta L/L_0$
993	0.000
1036	0.450
1051	0.225
1073	-0.225
1080	0.000

*Not shown in figure.
†This curve is here reported using the first given temperature as reference temperature at which $\Delta L/L_0 = 0$.
‡Author's data for coefficient of thermal expansion have been integrated by TPRC to obtain $\Delta L/L_0$.

DATA TABLE 61. COEFFICIENT OF THERMAL LINEAR EXPANSION OF YTTERBIUM Yb

[Temperature, T, K; Coefficient of Expansion, α, 10^{-6} K^{-1}]

T	α
CURVE 3[‡]	
1.69	0.117
1.97	0.134
2.21	0.189
2.45	0.082
3.48	0.095
4.49	0.143
5.97	0.262
6.93	0.397
8.01	0.557
9.01	0.915
10.00	1.285

[‡] Author's data for coefficient of thermal expansion have been integrated by TPRC to obtain $\Delta L/L_0$.

FIGURE AND TABLE NO. 62R. TYPICAL VALUES FOR THERMAL LINEAR EXPANSION OF YTTRIUM Y

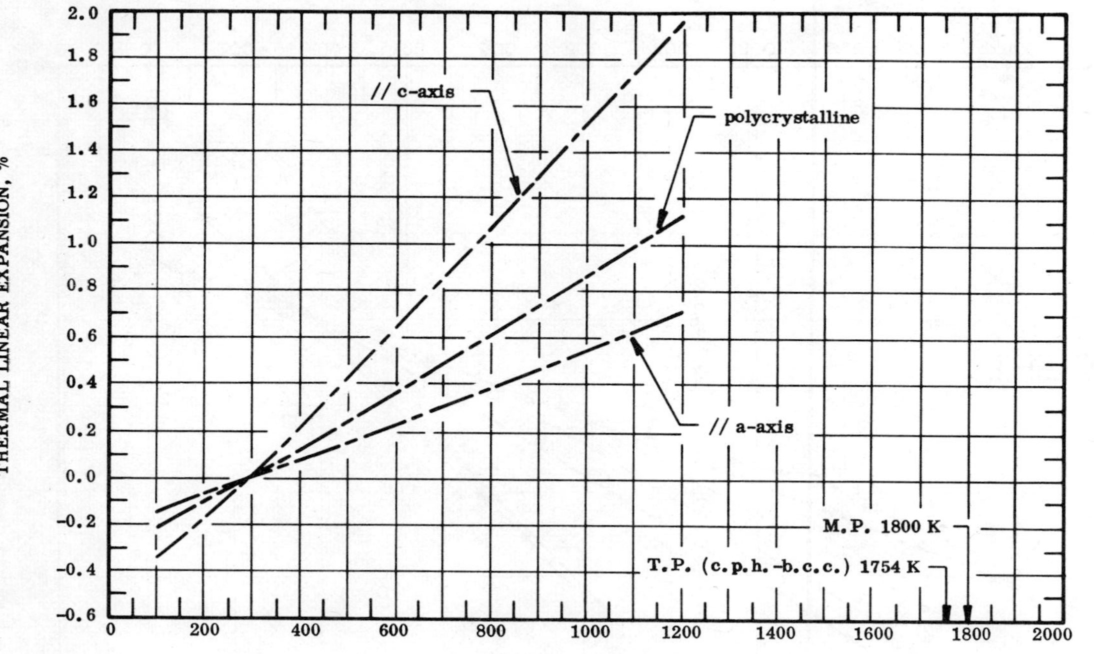

TYPICAL VALUES

[Temperature, T, K; Linear Expansion, $\Delta L/L_0$, %; α, K^{-1}]

	// a-axis	// c-axis	polycrystalline	
T	$\Delta L/L_0$	$\Delta L/L_0$	$\Delta L/L_0$	$\alpha \times 10^6$
100	-0.141	-0.345	-0.209	10.2
200	-0.067	-0.175	-0.103	10.7
293	0.000	0.000	0.000	11.3
400	0.081	0.207	0.123	11.6
500	0.156	0.411	0.241	12.0
600	0.231	0.623	0.362	12.2
700	0.307	0.841	0.485	12.5
800	0.384	1.063	0.610	12.7
900	0.462	1.288	0.737	12.8
1000	0.541	1.513	0.865	12.8
1200	0.701	1.958	1.120	12.8

REMARKS

The tabulated values are considered accurate to within ± 15% below 293 K and ± 20% above. The tabulated values for polycrystalline yttrium are calculated from single crystal data and are considered accurate to within ± 20% over the entire temperature range. Experimental data on polycrystalline yttrium are lower than the typical values. The tabulated values can be represented approximately by the following equations:

// a-axis: $\Delta L/L_0 = -0.214 + 7.328 \times 10^{-4}\, T + 9.009 \times 10^{-8}\, T^2 + 1.330 \times 10^{-11}\, T^3$

// c-axis: $\Delta L/L_0 = -0.494 + 1.461 \times 10^{-3}\, T + 8.471 \times 10^{-7}\, T^2 - 3.018 \times 10^{-10}\, T^3$

polycrystalline: $\Delta L/L_0 = -0.308 + 9.762 \times 10^{-4}\, T + 2.877 \times 10^{-7}\, T^2 - 9.160 \times 10^{-11}\, T^3$

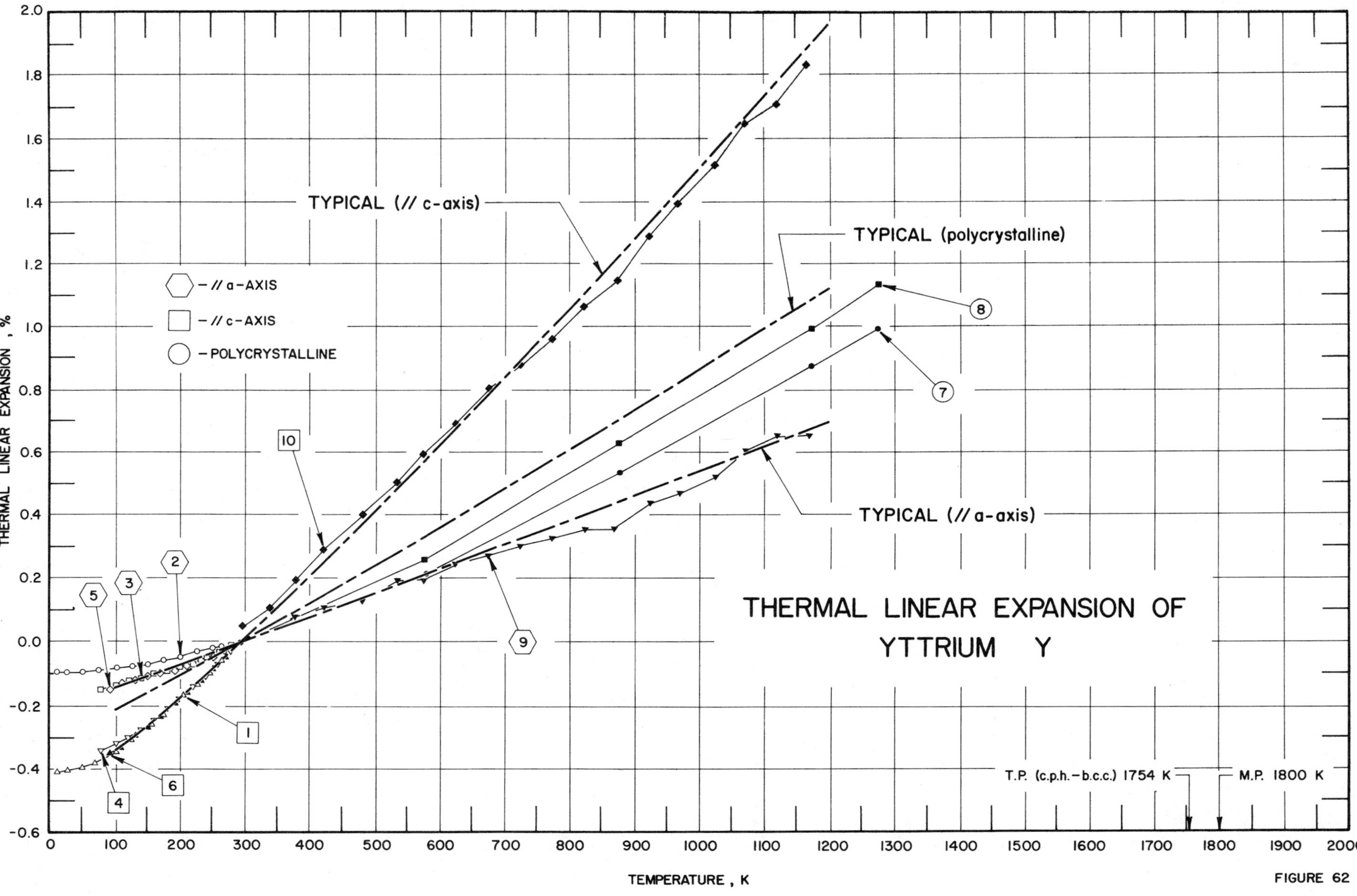
THERMAL LINEAR EXPANSION OF
YTTRIUM Y
THERMAL LINEAR EXPANSION, %
TEMPERATURE, K
TYPICAL (// c-axis)
TYPICAL (polycrystalline)
TYPICAL (// a-axis)
- // a-AXIS
- // c-AXIS
- POLYCRYSTALLINE
T.P. (c.p.h.-b.c.c.) 1754 K
M.P. 1800 K
FIGURE 62

SPECIFICATION TABLE 62. THERMAL LINEAR EXPANSION OF YTTRIUM Y

Cur. No.	Ref. No.	Author(s)	Year	Method Used	Temp. Range, K	Name and Specimen Designation	Composition (weight percent), Specifications, and Remarks
1	93	Meyerhoff, R.W. and Smith, Y.F.	1962	L	5-273		99.0+ Y, 0.5 Zr, 0.250 O, 0.0268 Ti, 0.0212 C, 0.0160 Ni, 0.0148 N, 0.0125 Si, 0.0112 Fe, 0.0040 Ca, <0.0030 Mg, <0.0010 B, <0.100 all rare earths combined; yttrium metal are melted under helium atmosphere into rod; sections of rod placed in tantalum holders and vacuum annealed 24 hr at 1573 K; after aligning the grains, specimen cut and ends ground parallel and etched; 0.6742 in. long; expansion measured parallel to c-axis; zero-point correction is -0.383%.
2	93	Meyerhoff, R.W. and Smith, Y.F.	1962	L	5-273		Similar to the above specimen; 0.5469 in. long; expansion measured perpendicular to c-axis; zero-point correction is -0.0089%.
3	317	Finkel', V.A. and Vorobev, V.V.	1968	X	79-299		99.8 Y; material melted in arc furnace under 0.5 to 0.6 atmosphere; specimen 10 x 20 x 2 mm cut from casting, annealed 10-50 hr in 2×10^{-6} mm Hg vacuum at 1473 K; expansion measured along a-axis; lattice parameter reported at 299 K is 3.6505 Å; 3.6503 Å at 293 K determined by graphical interpolation.
4	317	Finkel', V.A. and Vorobev, V.V.	1968	X	79-300		The above specimen; expansion measured along c-axis; lattice parameter reported at 300 K is 5.7411 Å; 5.7403 Å at 293 K determined by graphical interpolation.
5	372	Finkel', V.A. and Vorobev, V.V.	1968	X	80-300		99.8+ Y; polycrystalline specimen prepared from vacuum distilled material; rolled with high degree of deformation; recrystallized and annealed at 1423 K; expansion measured along a-axis; anomaly of unexplained nature observed at 180 K; lattice parameter reported at 300 K is 3.65070 Å; 3.65040 Å at 293 K determined by graphical interpolation.
6	372	Finkel', V.A. and Vorobev, V.V.	1968	X	80-300		The above specimen; expansion measured along c-axis; lattice parameter reported at 300 K is 5.74126 Å; 5.74046 Å at 293 K determined by graphical interpolation.
7	373	Nolting, H.J., Simmons, C.R., and Klingenberg, J.J.	1960	L	298-1273		99.9+ Y; metal prepared by lithium reduction of YCl_3; rod specimen 5.08 cm long, 0.64 cm diameter.
8	373	Nolting, H.J., et al.	1960	L	298-1273		99.9+ Y; metal prepared through intermediate alloy process; rod specimen 5.08 cm long, 0.64 cm diameter.
9	82	Hanak, J.J.	1959	X	296-1170		1.47 insoluble Br_2, 0.4 Ta, 0.1 Ca, 0.06 each Mg, Ni, 0.02 Fe, <0.01 each any other impurity; wire specimen cut from sheet prepared at Ames Laboratory; hexagonal-close-packed structure; expansion measured along a-axis; lattice parameter reported at 296 K is 3.648 Å; 3.645 Å at 293 K determined by graphical extrapolation.
10	82	Hanak, J.J.	1959	X	296-1170		The above specimen; expansion measured along c-axis; lattice parameter reported at 296 K is 5.735 Å; 5.732 Å at 293 K determined by graphical extrapolation.

DATA TABLE 62. THERMAL LINEAR EXPANSION OF YTTRIUM Y

[Temperature, T, K; Linear Expansion, $\Delta L/L_0$, %]

T	$\Delta L/L_0$
CURVE 1	
5	-0.409*
10	-0.409
15	-0.408*
20	-0.408*
25	-0.408
30	-0.406*
35	-0.405*
40	-0.403*
45	-0.400*
50	-0.397
55	-0.394*
60	-0.393*
65	-0.385*
70	-0.380
75	-0.375*
80	-0.369*
85	-0.363*
90	-0.357*
95	-0.350*
100	-0.343
105	-0.336*
110	-0.329*
115	-0.322*
120	-0.314*
125	-0.306
130	-0.299*
135	-0.291*
140	-0.282*
145	-0.274*
150	-0.266*
155	-0.257
160	-0.249*
165	-0.240*
170	-0.231*
175	-0.222
180	-0.213*
185	-0.204*
190	-0.295*
195	-0.286*
200	-0.277*
205	-0.168
210	-0.158*
215	-0.149*
220	-0.140*
225	-0.130
230	-0.121*
CURVE 1 (cont.)	
235	-0.111*
240	-0.102*
245	-0.092
250	-0.083*
255	-0.073*
260	-0.064*
265	-0.054
270	-0.45*
273.2	-0.038*
CURVE 2	
5	-0.091*
10	-0.091
15	-0.091*
20	-0.091*
25	-0.091*
30	-0.091*
35	-0.090*
40	-0.090*
45	-0.090*
50	-0.089
55	-0.089*
60	-0.088*
65	-0.088*
70	-0.087*
75	-0.086
80	-0.085*
85	-0.043*
90	-0.082*
95	-0.081*
100	-0.079
105	-0.078*
110	-0.076*
115	-0.075*
120	-0.073*
125	-0.071*
130	-0.069*
135	-0.068*
140	-0.066*
145	-0.064*
150	-0.062*
155	-0.060*
160	-0.058*
165	-0.056*
170	-0.054*
CURVE 2 (cont.)	
175	-0.052
180	-0.050*
185	-0.048*
190	-0.045*
195	-0.043*
200	-0.041
205	-0.039*
210	-0.037*
215	-0.035*
220	-0.033*
225	-0.030
230	-0.028*
235	-0.026*
240	-0.024*
245	-0.022*
250	-0.019
255	-0.017*
260	-0.015*
265	-0.013
270	-0.010*
273.2	-0.009*
CURVE 3	
79	-0.143
90	-0.135*
100	-0.132
109	-0.124*
119	-0.118
129	-0.113*
140	-0.107
149	-0.099*
160	-0.096
169	-0.091*
180	-0.082
190	-0.077*
200	-0.072
209	-0.066*
220	-0.061
229	-0.055*
240	-0.044
250	-0.036*
260	-0.031
270	-0.017*
280	-0.012
289	-0.003*
299	0.005*
CURVE 4	
79	-0.339
90	-0.331*
100	-0.318
110	-0.306*
119	-0.296
130	-0.283*
140	-0.271
149	-0.262*
161	-0.245
171	-0.224*
179	-0.208
189	-0.191*
198	-0.175
209	-0.156*
220	-0.137
230	-0.118*
238	-0.102
249	-0.085*
259	-0.064
270	-0.046*
280	-0.026
289	-0.008*
300	0.014*
CURVE 5	
80.5	-0.143*
89.2	-0.137
100.0	-0.130*
110.5	-0.124
119.9	-0.119*
129.7	-0.114
140.0	-0.108*
150.6	-0.103
160.3	-0.097*
169.8	-0.094
180.2	-0.088*
190.3	-0.082
200.1	-0.076*
210.2	-0.070
220.0	-0.063*
229.8	-0.056
239.5	-0.048*
250.0	-0.041
260.1	-0.032*
270.1	-0.024
CURVE 5 (cont.)	
280.0	-0.014*
288.7	-0.005*
300.0	0.007*
CURVE 6	
80.5	-0.360*
90.1	-0.346
100.4	-0.332*
109.4	-0.322
118.1	-0.307*
129.0	-0.293
140.1	-0.280*
150.2	-0.263
160.7	-0.246*
169.4	-0.229
180.0	-0.211*
189.6	-0.189
199.8	-0.170*
209.0	-0.151
220.1	-0.133*
229.1	-0.116
239.6	-0.097*
249.0	-0.078
259.8	-0.057*
271.3	-0.040
280.5	-0.022*
290.8	-0.005*
300.0	0.014*
CURVE 7	
298	0.000*
573	0.213
873	0.539
1173	0.871
1273	0.990
CURVE 8	
298	0.005*
573	0.260
873	0.632
1173	0.994
1273	1.137
CURVE 9	
296	0.082
340	0.055
378	0.082
421	0.110
480	0.137
532	0.192
573	0.192
625	0.247
673	0.274
725	0.302
774	0.329
821	0.357
870	0.357
922	0.439
969	0.466
1023	0.521
1073	0.604
1120	0.658
1170	0.658
CURVE 10	
296	0.052
340	0.105
378	0.192
421	0.297
480	0.401
532	0.506
573	0.593
625	0.698
673	0.803
725	0.872
774	0.959
821	1.082
870	1.151
922	1.291
969	1.396
1023	1.518
1073	1.657
1120	1.710
1170	1.832

* Not shown in figure.

FIGURE AND TABLE NO. 63R. RECOMMENDED VALUES FOR THERMAL LINEAR EXPANSION OF ZINC Zn

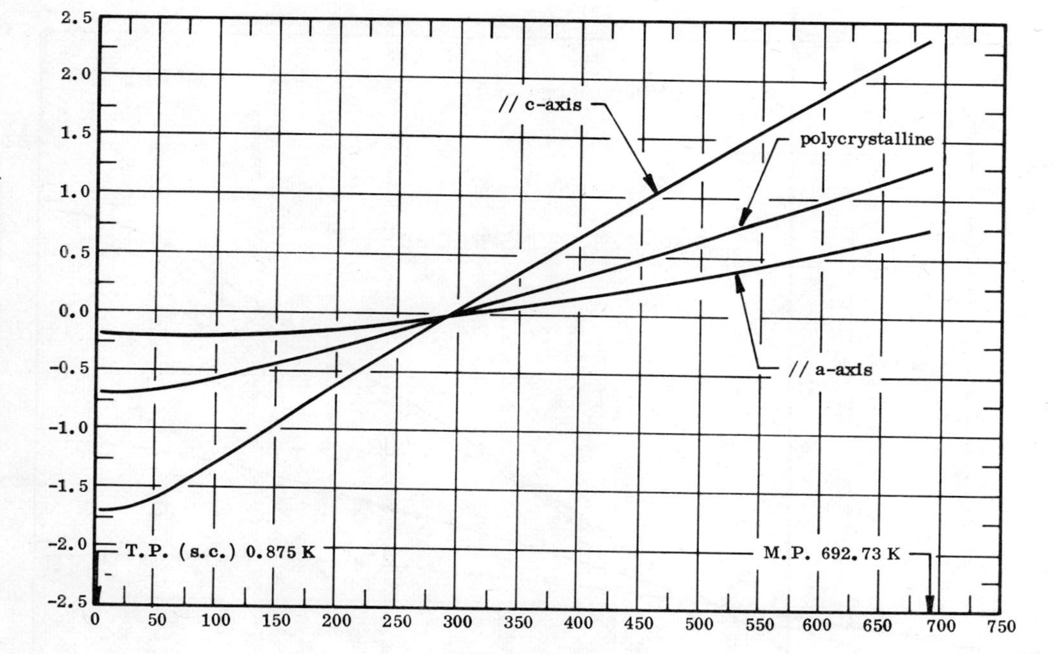

RECOMMENDED VALUES

[Temperature, T, K; Linear Expansion, $\Delta L/L_0$, %; α, K^{-1}]

	// a-axis	// c-axis	polycrystalline	
T	$\Delta L/L_0$	$\Delta L/L_0$	$\Delta L/L_0$	$\alpha \times 10^6$
5	-0.177	-1.692	-0.682	0.02
25	-0.179	-1.678	-0.679	5.6
50	-0.189	-1.575	-0.651	15.7
100	-0.187	-1.262	-0.545	24.5
200	-0.108	-0.604	-0.273	28.6
293	0.000	0.000	0.000	30.2
400	0.157	0.673	0.329	31.6
500	0.336	1.284	0.652	32.8
600	0.542	1.872	0.985	33.6
690	0.747	2.380	1.291	34.0

REMARKS

The tabulated values are considered accurate to within ± 3% over the entire temperature range. The tabulated values for polycrystalline are calculated from single crystal data. These values can be represented approximately by the following equations:

// a-axis: $\Delta L/L_0 = -0.228 + 1.987 \times 10^{-4}\, T + 2.119 \times 10^{-6}\, T^2 - 5.203 \times 10^{-10}\, T^3$ $(100 < T < 690)$

// c-axis: $\Delta L/L_0 = -1.929 + 6.714 \times 10^{-3}\, T - 2.747 \times 10^{-7}\, T^2 - 5.876 \times 10^{-10}\, T^3$ $(100 < T < 690)$

polycrystalline: $\Delta L/L_0 = -0.796 + 2.375 \times 10^{-3}\, T + 1.301 \times 10^{-6}\, T^2 - 5.249 \times 10^{-10}\, T^3$ $(100 < T < 690)$

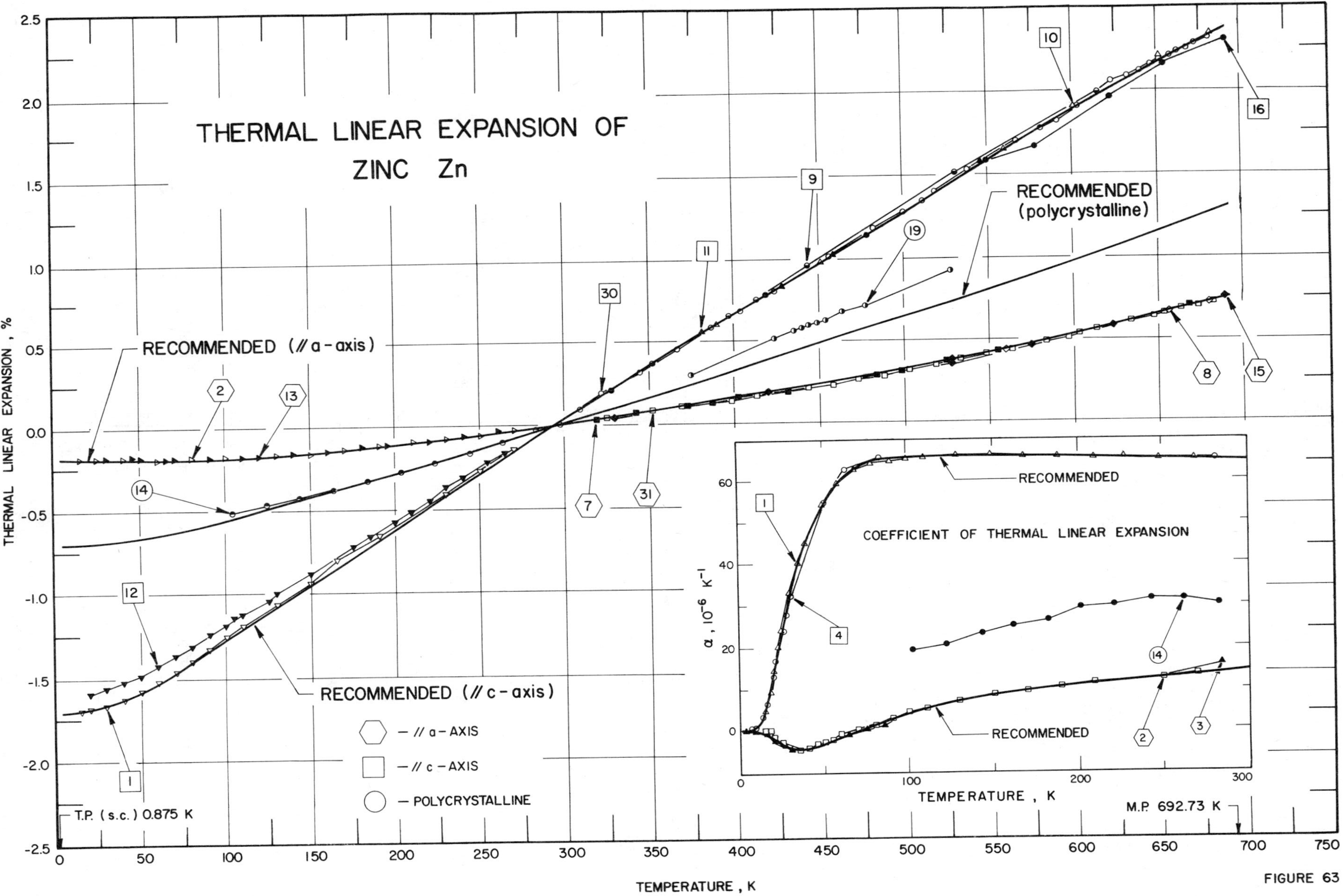
THERMAL LINEAR EXPANSION OF
ZINC Zn
THERMAL LINEAR EXPANSION, %
TEMPERATURE, K
RECOMMENDED (// a-axis)
RECOMMENDED (// c-axis)
RECOMMENDED (polycrystalline)
— // a-AXIS
— // c-AXIS
— POLYCRYSTALLINE
T.P. (s.c.) 0.875 K
M.P. 692.73 K
COEFFICIENT OF THERMAL LINEAR EXPANSION
α, 10^{-6} K^{-1}
RECOMMENDED
RECOMMENDED
TEMPERATURE, K

FIGURE 63

SPECIFICATION TABLE 63. THERMAL LINEAR EXPANSION OF ZINC Zn

Cur. No.	Ref. No.	Author(s)	Year	Method Used	Temp. Range, K	Name and Specimen Designation	Composition (weight percent), Specifications, and Remarks
1	374	Channing, D.A. and Weintroub, S.	1965	I	15-270		Single crystal grown from 99.995 pure Zn supplied by National Smelting Co.; hollow cylinder specimen 8 mm high, 2 mm wall thickness and 10 mm bore; crystal formed the spacer which separated two optically flat fused quartz plates; spacer annealed in argon at 50 cm Hg for 15 hr at 573 K; measured along direction 10.8 degrees to the hexad axis.
2	374	Channing, D.A. and Weintroub, S.	1965	I	20-270		The above specimen; values along a-axis calculated from measurements along directions 63.9 degrees to the hexad axis.
3*	104	McCammon, R.D. and White, G.K.	1965	E	4-85, 283	1	Specimen grown from 99.99$^+$ Zn; cylindrical crystals 5 cm long, 2 cm diameter oriented perpendicular hexad axis; crystal supplied by Metals Research Ltd., Cambridge, U.K.; cylinder spark machined, ends fine sparked; this curve is here reported using the first given temperature as reference temperature at which $\Delta L/L_0 = 0$.
4*	104	McCammon, R.D. and White, G.K.	1965	E	4-85, 283	2	Similar to the above specimen except oriented parallel to hexad axis; this curve is here reported using the first given temperature as reference temperature at which $\Delta L/L_0 = 0$.
5*	104	McCammon, R.D. and White, G.K.	1965	E	1.7-9.9	3	Specimen no. 1 shortened 0.2 cm and relapped to determined effects of surface damage; this curve is here reported using the first given temperature as reference temperature at which $\Delta L/L_0 = 0$.
6	83	Laquer, H.L.	1952	L	0-300		Specimen polycrystalline sample 2.54 cm long; author claims results valid for this specimen only; zero-point correction of -0.054% determined by graphical interpolation.
7	375	Gilder, H.M. and Wallmark, G.N.	1969	L	668-319		Crystal with dimensions approximately 1 x 1 x 15 cm; crystallographic and body axes aligned to within 0.5°, supplied by Monocrystals, Cleveland, Ohio; oxidized in air prior to an actual run; measurements made along a-axis; expansion measured with increasing temperature; zero-point correction of 0.036% determined by graphical extrapolation.
8	375	Gilder, H.M. and Wallmark, G.N.	1969	L	319-683		Similar to the above specimen; expansion measured with increasing temperature; zero-point correction of 0.036% determined by graphical extrapolation.
9	375	Gilder, H.M. and Wallmark, G.N.	1969	L	672-321		Similar to the above specimen; measurement made along c-axis; expansion measured with decreasing temperature; zero-point correction of 0.180% determined by graphical interpolation.
10	375	Gilder, H.M. and Wallmark, G.N.	1969	L	321-682		Similar to the above specimen; expansion measured with increasing temperature; zero-point correction of 0.179% determined by graphical extrapolation.
11	192	Shinoda, G.	1933	X	304, 391		Hexagonal-close-packed structure; expansion measured in (006) and (106) planes.
12	93	Myerhoff, R.W. and Smith, Y.F.	1962	L	5-273		99.9975 Zn, 0.0010 Pb, 0.0007 Cu, 0.0003 Fe, 0.0001 Cd, 0.00003 Sb, 0.00001 As; obtained from Bunker Hill; single crystal grown in MgO crucibles; 0.7900 in. long; expansion measured parallel to c-axis; zero-point correction of -0.1214% determined by graphical extrapolation.
13	93	Myerhoff, R.W. and Smith, Y.F.	1962	L	5-273		Similar to the above specimen; 1.0315 in. long; expansion measured perpendicular to c-axis; zero-point correction of -0.0224% determined by graphical extrapolation.

* Not shown in figure.

SPECIFICATION TABLE 63. THERMAL LINEAR EXPANSION OF ZINC Zn (continued)

Cur. No.	Ref. No.	Author(s)	Year	Method Used	Temp. Range, K	Name and Specimen Designation	Composition (weight percent), Specifications, and Remarks
14	70	Dorsey, H.G.	1908	I	103-283		Specimen obtained from Eimer and Amend, marked chemically pure; cast; length 1.6985 cm; data average of several trials on two specimens.
15	275	Olsen, E.A. and Yates, E.L.	1934	X	290-689		99.9 Zn; slightly annealed; powder mounted directly on copper sheet; expansion measured along base side of unit cell; base side reported at 293 K is 2.65906 Å.
16	275	Olsen, E.A. and Yates, E.L.	1934	X	287-692		The above specimen; expansion measured along height of unit cell; height reported at 293 K is 4.9347 Å.
17	376	Olsen, E.A. and Iball, J.	1933	X	293-673		Fine filings of Zn prepared and annealed in vacuo at 573 K for about 1 hr; filings mounted in layer 0.07 mm thick on glass fiber 0.60 mm in diameter; expansion measured along base side of unit cell; base side reported at 293 K is 2.659 Å.
18	376	Olsen. E.A. and Iball, J.	1933	X	293-673		The above specimen; expansion measured along height of unit cell; height reported at 293 K is 4.93 Å.
19	377	Freeman, J.R. and Brandt, P.F.	1926	T	293-526		0.003 Cd, 0.003 Fe, and 0.001 Pb; cast specimen 30 cm long, 0.714 cm diameter; expansion measured with increasing temperature; measurement performed by P. Hidnert and W. Sweeney at NBS.
20	377	Freeman, J.R. and Brandt, P.F.	1926	T	293-584		The above specimen; expansion measured on second test.
21	378	Claus, K. and Lohberg, K.	1955	L	293-673		Pure, extruded specimen; data measured with increasing and decreasing temperatures undifferentiable.
22*	109	Gachkovskii, V.F. and Strelkov, P.G.	1937	L	679-690		Single crystal of Zn; expansion measured parallel to hexagonal axis; this curve is here reported using the first given temperature as reference temperature at which $\Delta L/L_0 = 0$.
23*	109	Gachkovskii, V.F. and Strelkov, P.G.	1937	L	679-691		The above specimen; expansion measured perpendicular to hexagonal axis; this curve is here reported using the first given temperature as reference temperature at which $\Delta L/L_0 = 0$.
24*	700	Gachkovskii, V.F. and Strelkov, P.G.	1937	L	289-686	Sample No. 1	Monocrystalline zinc distilled in vacuum; specimen containing 0.002 Pb, 0.008 Cu, 0.0005 Fe, $\leq$0.0005 Cd, $\leq$0.0005 Ga, $\ell_0 = 65.0$ mm; expansion measured in a direction 9.5° with respect to the axis of crystal.
25*	700	Gachkovskii, V.F. and Strelkov, P.G.	1937	L	552-683	Sample 2a	Similar to the above specimen except that $\ell_0 = 63.4$ mm; expansion measured in a direction 6° with respect to the axis of crystal.
26*	700	Gachkovskii, V.F. and Strelkov, P.G.	1937	L	478-691	Sample 2b	Similar to the above specimen except that $\ell_0 = 47.4$ mm; expansion measured in a direction 6° with respect to the axis of crystal.
27*	700	Gachkovskii, V.F. and Strelkov, P.G.	1937	L	287-681	Sample 2b	Similar to the above specimen except that $\ell_0 = 58.8$ mm; expansion measured in a direction 76° with respect to the axis of crystal.
28*	700	Gachkovskii, V.F. and Strelkov, P.G.	1937	L	294-686	Sample 4	Similar to the above specimen except that $\ell_0 = 57.57$ mm; expansion measured in a direction 79° with respect to the axis of crystal.
29*	762	Apostolou, S.F.	1970	X	305-687		99.999 pure single crystal grown by Bridgman technique; preannealed for days at a temperature near melting point; measurements along c-axis; cooling curve zero-point correction of 0.17% determined by graphical extrapolation.
30	762	Apostolou, S.F.	1970	X	293-681		Similar to the above specimen; along c-axis; cooling curve; zero-point correction of 0.180% determined by graphical extrapolation.

* Not shown in figure.

SPECIFICATION TABLE 63. THERMAL LINEAR EXPANSION OF ZINC Zn (continued)

Cur. No.	Ref. No.	Author(s)	Year	Method Used	Temp. Range, K	Name and Specimen Designation	Composition (weight percent), Specifications, and Remarks
31	763	Wallmark, G.N.	1969	T	323-683		Single crystal from Mono Crystals, Cleveland, Ohio; measurements along a-axis; average of heating and cooling; zero-point correction of 0.036% determined by graphical extrapolation; data on the coefficient of thermal linear expansion also reported.
32*	763	Wallmark, G.N.	1969	T	323-683		Similar to the above specimen except treated further to ensure that c-axis varied by no more than $\frac{1}{2}°$ foam the body axis; zero-point correction of 0.17% determined by graphical extrapolation; data on the coefficient of thermal expansion also reported.
33*	765	Willemsen, H.W., Vittoratos, E., and Meincke, P.P.M.		E	3.7-9.7	Cominco 69 grade	Large single crystal grown from pure Zn; relatively free of magnetic impurities; data on the coefficient of thermal linear expansion also given; Mn impurity of the order 0.004 changes the thermal expansion considerably.

* Not shown in figure.

DATA TABLE 63. THERMAL LINEAR EXPANSION OF ZINC Zn

[Temperature, T, K; Linear Expansion, $\Delta L/L_0$, %]

T	$\Delta L/L_0$
CURVE 1‡	
15	-1.691
18	-1.688*
20	-1.686
23	-1.681*
25	-1.676*
28	-1.668
30	-1.662*
33	-1.651*
35	-1.644*
38	-1.631*
40	-1.622
45	-1.598*
50	-1.572
55	-1.544*
60	-1.515*
70	-1.454*
80	-1.391
90	-1.326*
100	-1.261*
110	-1.196
130	-1.064*
150	-0.932
170	-0.800*
190	-0.668*
210	-0.537
230	-0.407*
250	-0.277*
270	-0.148
CURVE 2‡	
20	-0.178*
23	-0.179*
25	-0.179
28	-0.180*
30	-0.181*
33	-0.183*
35	-0.183*
38	-0.185*
40	-0.186
45	-0.188*
50	-0.189*
55	-0.191*
60	-0.192
CURVE 2 (cont.)‡	
70	-0.192*
80	-0.191*
90	-0.189*
100	-0.186
110	-0.181*
130	-0.169
150	-0.154
170	-0.137
190	-0.118
210	-0.097
230	-0.075
250	-0.052
270	-0.028
CURVE 3*, †, ‡	
4	0.000 x 10^{-4}
5	-0.002
6	-0.002
7	0.000
8	0.000
10	-0.065
12	-0.3
14	-1.1
16	-2.5
18	-5.0
20	-8.6
22	-13.3
24	-19.1
26	-26.0
28	-33.9
30	-42.6
65	-149.4
75	-156.9
85	-148.9
CURVE 4*, †, ‡	
4	0.000 x 10^{-4}
5	0.004
6	0.010
7	0.020
8	0.038
10	0.137
12	0.403
14	0.973
CURVE 4 (cont.)*, †, ‡	
16	1.983 x 10^{-4}
18	3.553
20	5.783
22	8.723
24	12.373
26	16.773
28	21.973
30	28.013
65	194.438
75	257.988
85	323.088
CURVE 5*, †, ‡	
1.72	0.00 x 10^{-6}
2.12	0.14
2.23	-0.04
2.72	-0.32
3.01	-0.20
3.01	-0.42
3.21	-0.52
3.44	-0.65
3.88	-0.79
5.30	-1.09
6.12	-1.04
6.81	-0.92
7.49	-0.68
8.03	-0.71
9.15	-1.77
9.88	-5.26
CURVE 6	
0	-0.554*
20	-0.553
40	-0.545*
60	-0.526*
80	-0.497
100	-0.461*
120	-0.421*
140	-0.378
160	-0.333*
180	-0.287*
200	-0.240
220	-0.191
240	-0.141*
CURVE 6 (cont.)	
260	-0.089
273.2	-0.054*
280	-0.036*
300	-0.018
CURVE 7	
668.4	0.704
661.5	0.685*
655.0	0.667*
648.4	0.650*
641.9	0.632*
634.7	0.614*
623.9	0.586*
613.7	0.561*
598.9	0.527*
584.5	0.494*
569.9	0.460*
555.2	0.430
540.5	0.399
525.5	0.369
511.4	0.342
497.9	0.317
484.1	0.291
471.1	0.267*
458.3	0.245*
445.5	0.223
431.3	0.199
416.3	0.176
401.0	0.152
386.4	0.130
371.3	0.108
356.9	0.088*
342.0	0.070
327.6	0.047*
319.4	0.036
CURVE 8	
319.4	0.036
349.2	0.077*
371.4	0.108*
393.7	0.143*
414.7	0.172
437.7	0.210*
458.1	0.244
CURVE 8 (cont.)	
478.4	0.280*
496.5	0.314*
517.6	0.353
539.3	0.396*
560.9	0.441
582.7	0.489*
603.7	0.538
622.9	0.584*
636.8	0.619*
647.8	0.648*
656.9	0.673
667.6	0.703*
674.5	0.723*
682.6	0.747
CURVE 9	
671.5	2.294*
664.4	2.260*
659.4	2.233*
653.2	2.199*
647.3	2.170*
641.4	2.139*
635.8	2.109*
625.3	2.051*
615.7	1.999
599.8	1.905*
586.3	1.829*
573.0	1.749*
560.6	1.671*
545.9	1.588
532.4	1.505
517.7	1.415*
502.9	1.325*
487.6	1.230*
473.2	1.136*
458.3	1.048
443.1	0.952
427.8	0.855*
426.7	0.848*
412.4	0.758
396.9	0.660
380.9	0.558*
366.5	0.470*
352.1	0.378
336.8	0.282*
321.1	0.180*
CURVE 10	
321.1	0.179
361.0	0.432
390.1	0.614
427.8	0.851
450.6	0.995
473.3	1.135
495.7	1.277
517.6	1.410
539.2	1.542
560.2	1.668
579.2	1.784
600.6	1.908
620.9	2.025
635.9	2.107
644.4	2.153
653.8	2.202
662.5	2.245
671.3	2.287
676.2	2.311
681.6	2.337
CURVE 11	
304	0.071
391	0.632
CURVE 12	
5	-1.599
10	-1.597*
15	-1.592
20	-1.583*
25	-1.572*
30	-1.557
35	-1.541*
40	-1.521*
45	-1.500*
50	-1.476
55	-1.450*
60	-1.423*
65	-1.395*
70	-1.366*
75	-1.337*
80	-1.307
85	-1.277*
90	-1.247*
95	-1.216*

*Not shown in figure.

†This curve is here reported using the first given temperature as reference temperature at which $\Delta L/L_0 = 0$.

‡Author's data for coefficient of thermal expansion have been integrated by TPRC to obtain $\Delta L/L_0$.

DATA TABLE 63. THERMAL LINEAR EXPANSION OF ZINC Zn (continued)

T	$\Delta L/L_0$
CURVE 12 (cont.)	
100	-1.186
105	-1.155*
110	-1.124*
115	-1.093*
120	-1.062*
125	-1.031*
130	-0.999
135	-0.969*
140	-0.938*
145	-0.907*
150	-0.876
155	-0.845*
160	-0.814*
165	-0.783*
170	-0.752*
175	-0.721*
180	-0.690*
185	-0.660*
190	-0.629
195	-0.598*
200	-0.568*
205	-0.537*
210	-0.506*
215	-0.476*
220	-0.445*
225	-0.415*
230	-0.384*
235	-0.354*
240	-0.323
245	-0.293*
250	-0.262*
255	-0.232*
260	-0.202*
265	-0.171*
270	-0.141*
273.2	-0.121*
CURVE 13	
5	-0.170*
10	-0.170
15	-0.170*
20	-0.170*
25	-0.171*
30	-0.171*
35	-0.172*
40	-0.173*

T	$\Delta L/L_0$
CURVE 13 (cont.)	
45	-0.174*
50	-0.175*
55	-0.176*
60	-0.177
65	-0.177*
70	-0.178*
75	-0.178*
80	-0.177
85	-0.177*
90	-0.176*
95	-0.175*
100	-0.173*
105	-0.171
110	-0.169*
115	-0.167*
120	-0.164*
125	-0.161*
130	-0.158
135	-0.155*
140	-0.152*
145	-0.148*
150	-0.145
155	-0.141*
160	-0.137*
165	-0.133*
170	-0.129*
175	-0.124*
180	-0.120*
185	-0.115*
190	-0.111
195	-0.106*
200	-0.101*
205	-0.096*
210	-0.091*
215	-0.086*
220	-0.081
225	-0.075*
230	-0.070
235	-0.065*
240	-0.059
245	-0.054*
250	-0.048
255	-0.043*
260	-0.037
265	-0.032*
270	-0.026
273.2	-0.022*

T	$\Delta L/L_0$
CURVE 14‡	
103	-0.508
123	-0.469
143	-0.425
163	-0.377
183	-0.327
203	-0.272
223	-0.213
243	-0.152
263	-0.091
283	-0.030
CURVE 15	
290	-0.007
293	0.000*
296	-0.008*
296	0.003*
328	0.045
373	0.110*
420	0.184
478	0.290*
527	0.357*
528	0.380
551	0.418*
575	0.451*
577	0.470*
595	0.523*
623	0.567
628	0.587*
652	0.654*
655	0.672*
658	0.687*
680	0.744
685	0.748*
689	0.760
CURVE 16	
287	0.000
293	0.000*
296	0.018*
328	0.217
372	0.504*
419	0.794
478	1.143
527	1.425*
529	1.479*

T	$\Delta L/L_0$
CURVE 16 (cont.)	
550	1.587
574	1.686*
579	1.716
597	1.836*
624	1.960
627	1.982*
655	2.128*
655	2.156
659	2.193*
682	2.280*
685	2.286*
692	2.310
CURVE 17*	
293	0.000
523	0.338
623	0.564
673	0.601
CURVE 18*	
293	0.000
523	1.217
623	1.825
673	2.231
CURVE 19	
293.4	0.002*
373.5	0.318
424.3	0.521
435.3	0.564
440.0	0.584
443.2	0.597
448.3	0.617
453.1	0.636
463.8	0.678
476.3	0.728
526.0	0.925
CURVE 20	
293.3	0.001*
375.5	0.320
423.2	0.510
450.7	0.618

T	$\Delta L/L_0$
CURVE 20 (cont.)	
474.5	0.714
524.9	0.914
570.9	1.093
583.6	1.148
CURVE 21	
293	0.000*
298	0.007*
310	0.019
323	0.032*
340	0.059*
348	0.066*
360	0.083
373	0.103*
385	0.126*
398	0.142*
410	0.157
423	0.169*
435	0.191*
448	0.213*
460	0.233
473	0.259*
485	0.279*
498	0.299*
510	0.330
523	0.352*
535	0.377*
548	0.412*
560	0.431
573	0.461*
585	0.497*
598	0.526*
610	0.555
623	0.589*
635	0.643*
648	0.659
664	0.707*
673	0.732
CURVE 22*, †, ‡	
679	0.000
680	0.005
680	0.009
681	0.011
681	0.013

T	$\Delta L/L_0$
CURVE 22 (cont.)*, †, ‡	
683	0.028
685	0.039
686	0.043
688	0.059
688	0.061
688	0.063
689	0.065
690	0.140
690	0.154
690	0.154
690	0.158
690	0.158
690	0.165
CURVE 23*, †, ‡	
679	0.000
681	0.009
683	0.014
686	0.023
687	0.026
687	0.028
689	0.053
689	0.057
691	0.091
CURVE 24*	
288.6	-0.028
293.2	0.000
379.2	0.525
432.4	0.840
478.6	1.112
516.6	1.333
592.4	1.762
645.0	2.041
657.3	2.104
660.3	2.119
669.5	2.163
674.1	2.182
675.3	2.187
677.1	2.194
681.8	2.202
681.8	2.220
686.1	2.224
686.2	2.225

* Not shown in figure.
† This curve is here reported using the first given temperature as reference temperature at which $\Delta L/L_0 = 0$.
‡ Author's data for coefficient of thermal expansion have been integrated by TPRC to obtain $\Delta L/L_0$.

DATA TABLE 63. THERMAL LINEAR EXPANSION OF ZINC Zn (continued)

T	$\Delta L/L_0$
CURVE 25*, †	
552.2	0.000
590.1	0.222
655.4	0.571
672.8	0.656
679.6	0.690
683.1	0.711
CURVE 26*, †	
478.1	0.000
560.4	0.496
639.4	0.942
650.4	1.001
652.6	1.011
660.3	1.051
669.5	1.096
670.1	1.095
676.2	1.129
678.8	1.141
678.9	1.140
681.4	1.15
683.4	1.161
685.4	1.18
690.1	1.21
690.5	1.21
CURVE 27*	
286.8	-0.011
293.2	0.000
401.9	0.179
496.0	0.383
575.9	0.564
630.8	0.689
632.9	0.694
638.7	0.708
642.0	0.716
679.9	0.833
681.2	0.837
CURVE 28*	
293.2	0.000
418.3	0.218
463.3	0.305
574.6	0.552
619.1	0.663
CURVE 28 (cont.)*	
670.1	0.806
677.8	0.830
685.6	0.856
CURVE 29*	
304.9	0.071
313.2	0.122
321.2	0.170
335.6	0.266
353.2	0.377
365.8	0.459
386.3	0.587
401.0	0.672
422.0	0.804
447.6	0.975
469.5	1.110
482.3	1.182
498.7	1.281
512.8	1.368
536.9	1.515
552.4	1.608
564.9	1.685
581.1	1.779
600.7	1.890
619.2	2.003
635.0	2.087
644.2	2.135
649.2	2.156
655.4	2.188
661.1	2.233
666.1	2.247
669.9	2.267
676.3	2.298
682.0	2.325
687.2	2.352
CURVE 30	
292.7	0.000
307.9	0.099
321.1	0.180
329.3	0.230
344.2	0.323
366.2	0.457
387.7	0.596
403.7	0.699
CURVE 30 (cont.)	
423.8	0.820
450.0	0.980
467.1	1.090
481.9	1.180
499.1	1.288
511.7	1.365
518.5	1.412
538.4	1.532
551.8	1.611
566.8	1.704
581.4	1.787
590.7	1.839
605.8	1.921
618.9	2.000
628.1	2.055
634.3	2.090
640.8	2.122
646.7	2.155
652.2	2.183
657.2	2.212
662.9	2.244
668.2	2.266
673.8	2.295
681.3	2.331
CURVE 31	
323	0.041
338	0.061
353	0.082
368	0.103
383	0.125
398	0.147
413	0.171
428	0.194
443	0.219
458	0.244
473	0.271
488	0.298
503	0.326
518	0.355
533	0.384
548	0.414
563	0.446
573	0.468
583	0.490
593	0.513
CURVE 31 (cont.)	
603	0.536
613	0.560
623	0.584
633	0.609
643	0.635
653	0.662
663	0.690
673	0.719
683	0.748
CURVE 32*	
323	0.184
338	0.281
353	0.376
368	0.470
383	0.563
398	0.657
413	0.751
428	0.846
443	0.941
458	1.035
473	1.139
488	1.223
503	1.315
518	1.408
533	1.499
548	1.589
563	1.679
573	1.738
583	1.798
593	1.858
603	1.917
613	1.974
623	2.029
633	2.084
643	2.138
653	2.190
663	2.241
673	2.289
683	2.336
CURVE 33*	
3.7	0.49×10^{-6}
4.0	0.29
4.4	0.00
CURVE 33 (cont.)*	
4.6	-0.17×10^{-6}
4.8	-0.22
5.2	-0.22
5.5	-0.09
5.7	0.00
6.0	0.22
6.8	0.83
7.6	1.50
7.9	1.67
8.5	1.67
8.8	1.44
9.0	1.05
9.2	0.62
9.3	0.00
9.5	-0.93
9.7	-2.54

* Not shown in figure.
† This curve is here reported using the first given temperature as reference temperature at which $\Delta L/L_0 = 0$.

DATA TABLE 63. COEFFICIENT OF THERMAL LINEAR EXPANSION OF ZINC Zn

[Temperature, T, K; Coefficient of Expansion, α, 10^{-6} K^{-1}]

CURVE 1‡

T	α
15	5.0
18	9.9
20	14.1
23	20.1
25	24.2
28	30.1*
30	33.4
33	37.6*
35	40.1
38	43.4*
40	45.4
45	50.0*
50	54.1
55	57.3*
60	59.6
70	62.6
80	64.1
90	64.9
100	65.4
110	65.7
130	66.0
150	66.0
170	65.9
190	65.7
210	65.4
230	65.1
250	64.7
270	64.1

CURVE 2‡

T	α
20	-1.9
23	-2.8
25	-3.3*
28	-4.1*
30	-4.3*
33	-4.5*
35	-4.5
38	-4.3*
40	-4.2
45	-3.6
50	-3.0
55	-2.3
60	-1.5
70	0.0
80	1.6
90	2.9
100	4.1
110	5.1
130	6.8
150	8.1
170	9.2
190	10.0
210	10.7
230	11.4
250	11.8
270	12.3

CURVE 3‡

T	α
4	-0.003
5	-0.0015*
6	0.0015*
7	0.0032*
8	-0.0025*
10	-0.063
12	-0.23*
14	-0.52*
16	-0.97
18	-1.50*
20	-2.05
22	-2.65*
24	-3.2*
26	-3.7
28	-4.15*
30	-4.6
65	-1.5
75	0.0
85	1.6
283	12.19

CURVE 4‡

T	α
4	0.034*
5	0.050*
6	0.075*
7	0.12
8	0.23*
10	0.76
12	1.90*
14	3.80
16	6.35
18	9.4*
20	12.9
22	16.5
24	20.0*
26	24.0
28	28.0
30	32.4
65	62.7
75	64.4
85	65.8
283	64.5

CURVE 14‡

T	α
103	19.0
123	20.4
143	23.3
163	24.6
183	26.0
203	29.1
223	29.7
243	30.8
263	30.8
283	29.7

CURVE 22*, ‡

T	α
679.0	54
680.0	54
680.6	65
681.0	40
681.3	64
683.9	51
685.7	76
686.2	76
688.0	101
688.3	50
688.6	61
689.0	71
690.5	935
690.7	342
690.7	399
690.8	584
690.8	616
690.9	764

CURVE 23*, ‡

T	α
679.0	30
681.9	31
683.7	31
686.4	31
687.3	38
687.6	71
689.5	197
689.8	73
691.0	486

CURVE 33

T	α
4.0	-1.01
4.4	-0.51
5.0	0.00
5.4	0.33
6.0	0.71
6.3	0.90
6.7	0.99
7.3	0.99
7.7	0.79
7.9	0.49
8.2	0.00
8.4	-0.31
8.6	-0.74
8.8	-1.34
9.0	-1.96
9.2	-2.90
9.3	-4.43
9.4	-5.63

* Not shown in figure.

‡ Author's data for coefficient of thermal expansion have been integrated by TPRC to obtain $\Delta L/L_0$.

FIGURE AND TABLE NO. 64R. RECOMMENDED VALUES FOR THERMAL LINEAR EXPANSION OF ZIRCONIUM Zr

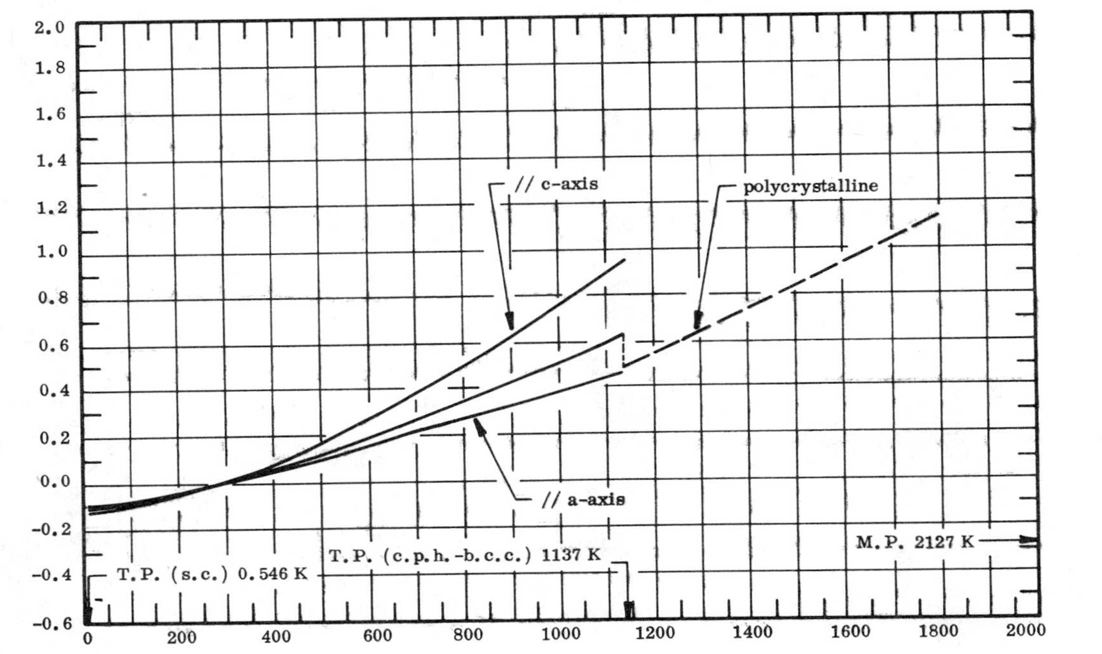

RECOMMENDED VALUES

[Temperature, T, K; Linear Expansion, $\Delta L/L_0$, %; α, K^{-1}]

T	// a-axis $\Delta L/L_0$	// c-axis $\Delta L/L_0$	polycrystalline $\Delta L/L_0$	polycrystalline $\alpha \times 10^6$
10	-0.108	-0.132	-0.116	0.82
25	-0.107	-0.131	-0.115	0.2
50	-0.106	-0.130	-0.114	1.6
100	-0.087	-0.120	-0.098	4.0
200	-0.044	-0.065	-0.051	5.2
293	0.000	0.000	0.000	5.7
400	0.051	0.079	0.060	5.9
500	0.102	0.165	0.123	6.6
600	0.156	0.263	0.192	7.1
700	0.212	0.372	0.265	7.6
800	0.269	0.491	0.343	7.9
900	0.325	0.617	0.422	8.0
1000	0.382	0.750	0.505	8.2
1100	0.436	0.887	0.586	8.2
1137†	0.455	0.940	0.617	8.2
1137†			0.482‡	9.0
1200			0.539‡	9.1
1400			0.725‡	9.5
1600			0.923‡	10.3
1800			1.139‡	11.3

† Phase transition.
‡ Provisional values

REMARKS

The tabulated values are considered accurate to within ± 3% at temperature below 1137 K and ± 10% above. Hysteresis in thermal expansion is observed at ± 100 degrees around transitional temperature, 1137 K. The tabulated values can be represented approximately by the following equations:

// a-axis: $\Delta L/L_0 = -0.107 + 2.724 \times 10^{-4}\, T + 3.658 \times 10^{-7}\, T^2 - 1.499 \times 10^{-10}\, T^3$ $(293 < T < 1137)$

// c-axis: $\Delta L/L_0 = -0.121 + 1.598 \times 10^{-4}\, T + 9.387 \times 10^{-7}\, T^2 - 2.273 \times 10^{-10}\, T^3$ $(293 < T < 1137)$

polycrystalline: $\Delta L/L_0 = -0.111 + 2.325 \times 10^{-4}\, T + 5.595 \times 10^{-7}\, T^2 - 1.768 \times 10^{-10}\, T^3$ $(293 < T < 1137)$

$\Delta L/L_0 = -0.759 + 1.474 \times 10^{-3}\, T - 5.140 \times 10^{-7}\, T^2 + 1.559 \times 10^{-10}\, T^3$ $(1137 < T < 1800)$

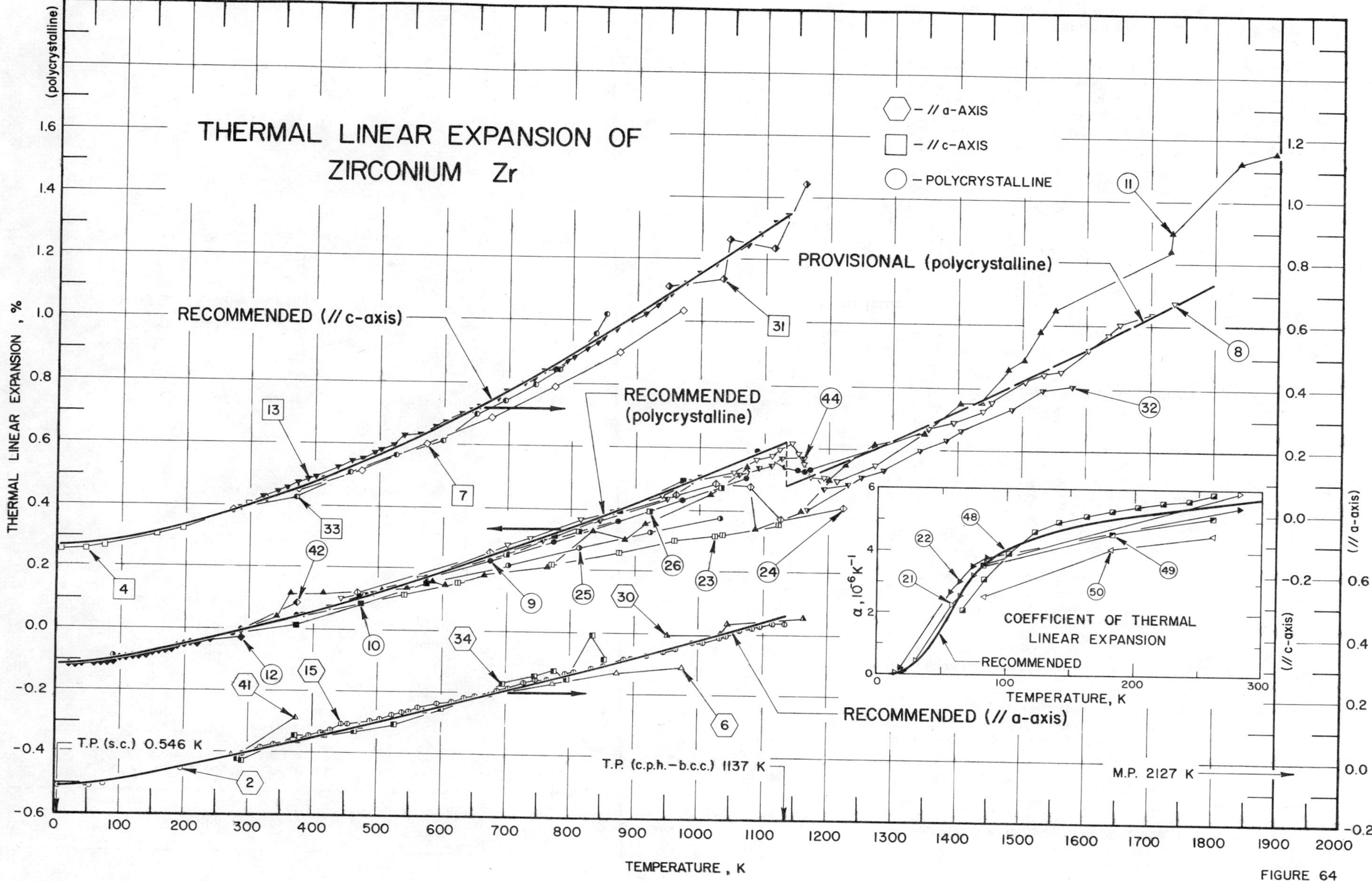
THERMAL LINEAR EXPANSION OF
ZIRCONIUM Zr
– // a-AXIS
– // c-AXIS
– POLYCRYSTALLINE
RECOMMENDED (// c-axis)
RECOMMENDED
(polycrystalline)
PROVISIONAL (polycrystalline)
RECOMMENDED (// a-axis)
T.P. (s.c.) 0.546 K
T.P. (c.p.h.–b.c.c.) 1137 K
M.P. 2127 K
THERMAL LINEAR EXPANSION, %
(polycrystalline)
(// c-axis)
(// a-axis)
TEMPERATURE, K
COEFFICIENT OF THERMAL
LINEAR EXPANSION
RECOMMENDED
α, 10^{-6} K^{-1}
TEMPERATURE, K

FIGURE 64

SPECIFICATION TABLE 64. THERMAL LINEAR EXPANSION OF ZIRCONIUM Zr

Cur. No.	Ref. No.	Author(s)	Year	Method Used	Temp. Range, K	Name and Specimen Designation	Composition (weight percent), Specifications, and Remarks
1	208	Frantsevich, I. N., Zhurakovskii, E. A. and Lyaschenko, A. B.	1967	X	293, 1473		Sample prepared from powder by vacuum sintering of pressed block.
2	379	Goldak, J., Lloyd, L. T. and Barrett, C. S.	1966	X	4.2-297		Single crystals isolated from large grains grown in iodide crystal; raw material Zr bar from Westinghouse, Hf-free, grade I; expansion measured along a-axis with Ni $K\beta_1$ (1.50010 Å) x-radiation; lattice parameter reported at 292 K is 3.23302 Å; 3.23310 Å at 293 K determined by graphical interpolation.
3	379	Goldak, J., et al.	1966	X	4.2-297		The above specimen; expansion measured along a-axis with W $L\alpha_1$ (1.47635 Å) x-radiation; lattice parameter reported at 292 K is 3.23296 Å; 3.23298 Å at 293 K determined by graphical interpolation.
4	379	Goldak, J., et al.	1966	X	4.2-300		The above specimen; expansion measured along c-axis with Ni $K\alpha$ (1.65784 Å) x-radiation; lattice parameter reported at 300.3 K is 5.14897 Å; 5.14868 Å at 293 K determined by graphical interpolation.
5	379	Goldak, J., et al.	1966	X	4.2-300		The above specimen; expansion measured along c-axis with Ni $K\alpha_1$ (1.66169 Å) x-radiation; lattice parameter reported at 300.3 K is 5.14893 Å; 5.14844 Å at 293 K determined by graphical interpolation.
6	380	Couterne, J. C. and Cizeron, G.	1966	X	273-976		99.91 Zr, 0.05 Fe, 0.02 Cu, 0.0115 Ni, 0.004 Cr, 0.0025 Al, 0.0025 Si; specimens taken from plate drawn from cylindrical billet, then cut parallel to drawing direction; expansion measured along a-axis; lattice parameter reported at 273 K is 3.2312 Å; 3.2315 Å at 293 K determined by graphical interpolation.
7	380	Couterne, J. C. and Cizeron, G.	1966	X	273-974		The above specimen; expansion measured along c-axis; lattice parameter reported at 273 K is 5.1475 Å; 5.1481 Å at 293 K determined by graphical interpolation.
8	213	Amonenko, V. M., V'yngov, P. N. and Gumenyuk, V. S.	1964	L	444-1740		Specimen prepared by the iodide method; heated in tungsten furnace; annealed 1.5-2.0 hr at 1720 K before measurements started under vacuum.
9	381	Baluffi, R. W., Resnick, R. and Timper, A. J.	1952	T	273-1353		High purity powder sintered to bar 3.81 cm long x 0.635 cm square; zero-point correction of -0.016% determined by graphical interpolation.
10	382	Toy, S. M. and Vetrano, J. B.	1960		298-973	Crystal bar	No details given.
11	358	Fieldhouse, I. B. and Lang, J. I.	1961		294-1899		99.95 Zr, 0.029 Fe, 0.017 C, 0.0045 Hf, <0.031 all other impurities; density 6.49 g cm^{-3}.
12	8	Altman, H. W., Rubin, T. and Johnston, H. L.	1954	I	20-300		Sample obtained from "as deposited" metal, cold swaged and unannealed; furnished by the Foote Mineral Co.; zero-point correction of 0.00325% determined by graphical interpolation.
13	383	Lloyd, L. T.	1963	I	290-1135		Single crystal prepared by subjecting Westinghouse, Hf-free, grade I, iodide crystal bar Zr to grain growth treatment; after grinding, faces etched in solution of 7 v/o 48% HF in 50:50 mixture of concentrated HNO_3 and water. followed by a second etch in 50:50 HNO_3 and water; heated to 1070 K for 24 hr in vacuum of at least 2 x 10^{-6} mm Hg; expansion measured along c-axis on sixth test; zero-point correction of -0.00795% determined by graphical interpolation.
14*	383	Lloyd, L. T.	1963	I	293-1134		Similar to the above specimen; expansion measured on seventh test; zero-point correction of -0.00214% determined by graphical interpolation.

* Not shown in figure.

SPECIFICATION TABLE 64. THERMAL LINEAR EXPANSION OF ZIRCONIUM Zr (continued)

Cur. No.	Ref. No.	Author(s)	Year	Method Used	Temp. Range, K	Name and Specimen Designation	Composition (weight percent), Specifications, and Remarks
15	383	Lloyd, L. T.	1963	I	291-1135		Similar to the above specimen except expansion measured perpendicular to c-axis on second test; zero-point correction of -0.00180% determined by graphical interpolation.
16*	383	Lloyd, L. T.	1963	I	292-1121		Similar to the above specimen; expansion measured perpendicular to c-axis on fifth test; zero-point correction of -0.00240% determined by graphical interpolation.
17*	383	Lloyd, L. T.	1963	I	292-1130		Similar to the above specimen; expansion measured perpendicular to c-axis on sixth test; zero-point correction of -0.00505% determined by graphical interpolation.
18*	383	Lloyd, L. T.	1963	I	292-1124		Similar to the above specimen; expansion measured perpendicular to c-axis on seventh test; zero-point correction of -0.00332% determined by graphical interpolation.
19*	384	Zwikker, C.	1926		1000-1500		Expansion measured with increasing temperature; this curve is here reported using the first given temperature as reference temperature at which $\Delta L/L_0 = 0$.
20*	384	Zwikker, C.	1926		1500-1000		The above specimen; expansion measured with decreasing temperature; hysteresis observed; explained by the author as being due to a modification of the material; this curve is here reported using the first given temperature as reference temperature at which $\Delta L/L_0 = 0$.
21	335	Cowan, Y. A., Pawlowicz, A. T. and White, G. K.	1968	E	12-85, 283	Zr_1	Arc-melted from iodide-bar supplied by Foote Mineral Co. and then heated at 1070 K in vacuum of 10^{-6} torr for 20 hr to remove hydrogen; 0.007 Hf and 0.005 to 0.006 each O_2, Si and Mg in starting material; specimen 2 cm diameter and 5 cm long; this curve is here reported using the first given temperature as reference temperature at which $\Delta L/L_0 = 0$.
22	335	Cowan, Y. A., et al.	1968	E	10-85, 283		Rod supplied by the Department of Metallurgy, University of New South Wales containing 1 at % Hf; annealed at 970 K in vacuum after machining into cylinder 2 cm diameter and 5 cm long; this curve is here reported using the first given temperature as reference temperature at which $\Delta L/L_0 = 0$.
23	313	Makin, S. M., Standring, J. and Hunter, P. M.	1953	I	293-1123		Fully softened specimen; expansion measured with increasing temperature.
24	385	Squire, C. F. and Kaufmann, A. R.	1941		472-1225		No details given except reported error $<3.0\%$; zero-point correction of -0.0014% determined by graphical extrapolation.
25	189	Adenstedt, H. K.	1952	L	94-1033	No. 755	Cold worked specimen from Foote Mineral Co.; expansion measured in vacuum with temperature increasing at 0.55 K per minute.
26	189	Adenstedt, H. K.	1952	L	481-1033	No. 755	The above specimen, annealed.
27*	189	Adenstedt, H. K.	1952	L	478-1033	No. 660	Similar to the above cold-worked specimen.
28*	189	Adenstedt, H. K.	1952	L	479-1033	No. 660	The above specimen, annealed.
29*	189	Adenstedt, H. K.	1952	L	483-1033	No. 684	Similar to the above cold-worked specimen.

* Not shown in figure.

SPECIFICATION TABLE 64. THERMAL LINEAR EXPANSION OF ZIRCONIUM Zr (continued)

Cur. No.	Ref. No.	Author(s)	Year	Method Used	Temp. Range, K	Name and Specimen Designation	Composition (weight percent), Specifications, and Remarks
30	386	Skinner, G.B. and Johnston, H.L.	1953	X	950-1164	α-zirconium	Specimen containing about 2% Hf prepared by iodide method; rod about 0.05 cm diameter; density 6.501 ± 0.005 g cm^{-3}; expansion measured along a-axis of c.p.h. crystal structure; transition (c.p.h. to b.c.c.) observed near 1165 K; reported error 10% maximum; this curve is here reported using the first given temperature as reference temperature at which $\Delta L/L_0 = 0$; lattice parameter reported at 950 K is 3.2393 kX (3.2458 Å).
31	386	Skinner, G.B. and Johnston, H.L.	1953	X	950-1164	α-zirconium	The above specimen; expansion measured along c-axis; reported error 14% maximum; this curve is here reported using the first given temperature as reference temperature at which $\Delta L/L_0 = 0$; lattice parameter at 950 K is 5.1741 kX.
32	386	Skinner, G.B. and Johnston, H.L.	1953	X	1166-1584	β-zirconium	Specimen containing about 2% Hf prepared by iodide method; rod 25.4 cm long, 0.62 cm diameter; specimen in β-phase, having b.c.c. structure; reported error 7% maximum.
33	387	Russell, R.B.	1954	X	283-852		<0.0001 Hf; crystal bar; machined to 0.471 in. diameter, dehydrogenated at 1670 K for 1 hr in vacuum of 10^{-6} mm Hg, swaged to 0.189 in. diameter; chemical analysis taken, and then rolled to 0.012 in. thick sheet, cut into strip about 1 x 0.4 in.; prick-punched with Burgess "Vibro-Tool"; annealed at 770 K in packet in quartz tube sealed off at pressure of 5 x 10^{-6} mm Hg; expansion measured along c-axis; lattice parameter reported at 292.2 K is 5.14831 Å; 5.14828 Å at 293 K determined by graphical interpolation; data on the coefficient of thermal linear expansion also given.
34	387	Russell, R.B.	1954	X	283-852		The above specimen; expansion measured along a-axis; lattice parameter reported at 292.2 K is 3.23178 Å; 3.23180 Å at 293 K determined by graphical interpolation; data on the coefficient of thermal linear expansion also given.
35*	338	Pridantseva, K.S.	1969		293-1083	Alloy No. 1	Iodide Zr, remelted in arc furnace; containing 0.03 Hf.
36*	338	Pridantseva, K.S.	1969		293-1083	Alloy No. 4	Similar to the above specimen; 0.1 Be, 0.03 Hf.
37*	338	Pridantseva, K.S.	1969		293-1083	Alloy No. 5	Similar to the above specimen; 0.1 Re, 0.03 Hf.
38	707	Pridantseva, K.S. and Solov'yeva, N.A.	1965	L	293, 473		Caste sample; 1% measurement error.
39	707	Pridantseva, K.S. and Solov'yeva, N.A.	1965	L	293, 473		Powder sample; 1% measurement error.
40	715	Shinoda, G.	1934	X	288-373		Expansion measured along the c-axis.
41	715	Shinoda, G.	1934	X	288-373		Expansion measured perpendicular to the c-axis.
42	715	Shinoda, G.	1934	X	288-373		Similar to the above specimen, except polycrystalline.
43*	761	Brodskiy, B.R., and Neymark, B.E.	1971	L	359-967		99.8 pure iodide zirconium containing 0.03 Hf 0.05 (Fe + Ni + (r); 0.020, and 0.02 C; expansion in the original state.
44	761	Brodskiy, B.R., and Neymark, B.E.	1971	L	386-1230		The above specimen after annealing at 973 K in the α-region; measurements on heating curve.
45*	761	Brodskiy, B.R., and Neymark, B.E.	1971	L	1212-1127		The above specimen; measurements on cooling curve.

* Not shown in figure.

SPECIFICATION TABLE 64. THERMAL LINEAR EXPANSION OF ZIRCONIUM Zr (continued)

Cur. No.	Ref. No.	Author(s)	Year	Method Used	Temp. Range, K	Name and Specimen Designation	Composition (weight percent), Specifications, and Remarks
46*	761	Brodskiy, B.R., and Neymark, B.E.	1971	L	1369-1230		Similar to the above specimen except annealed at 1223 K in the β region; measurements on heating curve.
47*	761	Brodskiy, B.R., and Neymark, B.E.	1971	L	1211-1098		The above specimen; measurements on cooling curve.
48	710	Erfling, H.D.	1939	I	273-59		7.7 mm long specimen; data on the coefficient of thermal linear expansion also reported.
49	728	Erfling, H.D.	1939	I	293-78		Wire specimen of 0.5 mm diameter; data on the coefficient of thermal linear expansion also reported.
50	728	Erfling, H.D.	1939	I	293-78		Rod specimen of 1.4 mm diameter which was rolled from thick piece; data on the thermal linear expansion also reported.

* Not shown in figure.

DATA TABLE 64. THERMAL LINEAR EXPANSION OF ZIRCONIUM Zr

[Temperature, T, K; Linear Expansion, $\Delta L/L_0$, %]

T	$\Delta L/L_0$
CURVE 1	
293.2	0.000
1473.2	0.956
CURVE 2	
4.2	-0.112
4.2	-0.114*
51.0	-0.113
77.4	-0.103
77.4	-0.105*
194.9	-0.056
195.2	-0.053*
292.0	-0.002
297.3	0.000*
CURVE 3*	
4.2	-0.112
4.2	-0.112
51.0	-0.112
71.4	-0.101
71.4	-0.100
194.9	-0.053
195.2	-0.050
292.0	-0.001
297.3	0.003
CURVE 4	
4.2	-0.141
4.2	-0.142*
47.2	-0.141
77.4	-0.135
154.3	-0.094
194.6	-0.072
196.2	-0.069*
196.6	-0.071*
300.3	0.006
CURVE 5*	
4.2	-0.135
4.2	-0.136
47.2	-0.136
77.4	-0.130
154.3	-0.089
194.3	-0.065
CURVE 5 (cont.)*	
196.2	-0.064
196.6	-0.066
300.3	0.009
CURVE 6	
273	-0.009
376	0.038
474	0.087
574	0.140
671	0.192
774	0.232
871	0.270
976	0.291
CURVE 7	
273	-0.012
326	0.021
376	0.052
474	0.118
575	0.202
672	0.293
773	0.392
874	0.503
974	0.639
CURVE 8	
444	0.100
567	0.172
701	0.279
739	0.299
819	0.362
943	0.442
990	0.488
1053	0.520
1085	0.549
1113	0.575
1128	0.598
1147	0.614
1155	0.581
1161	0.548
1174	0.507
1198	0.507
1217	0.499
CURVE 8 (cont.)	
1273	0.543
1341	0.595
1358	0.664
1397	0.688
1441	0.721
1454	0.755
1509	0.820
1531	0.844
1564	0.849
1604	0.924
1639	0.972
1655	1.001
1709	1.037
1740	1.078
CURVE 9	
273	-0.011
373	0.044
473	0.106
573	0.166
673	0.227
773	0.292
873	0.360
973	0.429
1073	0.500
1131	0.538
1152	0.530
1161	0.527
1173	0.534
1273	0.611
1353	0.670
CURVE 10	
298	0.000*
373	0.018
473	0.083
573	0.158
673	0.233*
773	0.310
873	0.398
973	0.496
CURVE 11	
294	0.00*
CURVE 11 (cont.)	
344	0.04
418	0.05*
465	0.12
509	0.13
546	0.14
583	0.16
609	0.15
669	0.18
763	0.21
835	0.33
882	0.30
919	0.35
1021	0.45
1073	0.54
1088	0.34
1128	0.37
1160	0.40
1201	0.50
1228	0.55
1272	0.62
1350	0.65
1404	0.75
1440	0.75
1480	0.86
1504	0.89
1530	0.98
1555	1.05
1733	1.24
1738	1.30
1844	1.53
1899	1.56
CURVE 12	
20	-0.1166
30	-0.1163
40	-0.1154
50	-0.1139*
60	-0.1117
70	-0.1089*
80	-0.1056*
90	-0.1020
100	-0.0981
110	-0.0940
120	-0.0895
130	-0.0851
140	-0.0805
CURVE 12 (cont.)	
150	-0.0757
160	-0.0709
170	-0.0659
180	-0.0609
190	-0.0558
200	-0.0507
220	-0.0402
240	-0.0296
260	-0.0188
273.16	-0.0116*
290	-0.0023
298.16	0.0022*
300	0.0033*
CURVE 13	
289.5	-0.0040
293.5	0.0006*
297.6	0.0051*
301.8	0.0097*
306.4	0.0143
311.2	0.0188*
316.5	0.0233*
321.6	0.0279
326.8	0.0324*
331.8	0.0369*
337.1	0.0415*
342.4	0.0460
347.8	0.0506*
353.0	0.0550*
358.4	0.0596
363.6	0.0642*
369.2	0.0687*
374.6	0.0732
385.1	0.0823*
390.8	0.0869
396.1	0.0914*
401.5	0.0959
407.1	0.1001*
412.5	0.1051*
417.8	0.1091*
423.1	0.114*
428.4	0.119*
433.7	0.123*
439.0	0.128
449.8	0.137*
454.6	0.141*
CURVE 13 (cont.)	
460.4	0.146
465.8	0.150*
471.2	0.155*
476.2	0.159
481.8	0.164*
487.0	0.168*
492.5	0.173
498.0	0.178*
503.2	0.182
508.7	0.187*
513.4	0.191*
518.7	0.196*
523.8	0.200
529.0	0.205*
534.0	0.209*
539.2	0.214
544.3	0.218*
549.2	0.223
554.4	0.227*
559.4	0.232*
564.4	0.237
574.2	0.246*
579.0	0.250*
583.7	0.255*
588.6	0.259
593.4	0.264*
598.2	0.268*
602.8	0.273
607.8	0.277*
612.2	0.282*
617.0	0.286*
621.4	0.291*
626.0	0.295*
630.4	0.300
635.0	0.305*
639.5	0.309*
644.0	0.314*
648.4	0.318*
653.0	0.323*
657.3	0.327
661.8	0.332*
665.8	0.336*
670.3	0.341*
674.5	0.345*
679.0	0.350*
683.1	0.354
687.2	0.359*

* Not shown in figure.

DATA TABLE 64. THERMAL LINEAR EXPANSION OF ZIRCONIUM Zr (continued)

T	ΔL/L$_0$	T	ΔL/L$_0$	T	ΔL/L$_0$	T	ΔL/L$_0$	T	ΔL/L$_0$	T	ΔL/L$_0$
CURVE 13 (cont.)		CURVE 13 (cont.)		CURVE 13 (cont.)		CURVE 14 (cont.)*		CURVE 14 (cont.)*		CURVE 14 (cont.)*	
691.5	0.364*	871.2	0.577	1026.2	0.790	336.1	0.0387	586.2	0.257	791.2	0.474
695.7	0.368*	874.6	0.581*	1029.0	0.795*	341.8	0.0433	590.9	0.261	795.1	0.479
700.1	0.373	878.0	0.586*	1032.8	0.799*	347.0	0.0478	600.7	0.270	798.9	0.484
704.2	0.377*	881.4	0.590*	1036.0	0.804*	352.1	0.0524	605.2	0.275	802.6	0.488
708.6	0.382*	885.0	0.595*	1039.1	0.808*	357.4	0.0569	610.3	0.279	806.5	0.493
712.6	0.386*	888.2	0.600*	1042.5	0.813*	368.2	0.0660	614.6	0.284	810.2	0.497
716.6	0.391*	891.6	0.604	1045.5	0.817*	373.7	0.0705	619.1	0.288	814.0	0.502
720.8	0.395*	894.8	0.609*	1048.8	0.822*	379.1	0.0750	623.9	0.293	817.7	0.506
724.8	0.400	898.3	0.613*	1052.2	0.826*	384.8	0.0796	628.6	0.297	821.4	0.511
728.9	0.404*	902.0	0.618	1055.3	0.831*	390.3	0.0841	633.2	0.302	825.0	0.515
732.9	0.409*	905.2	0.622*	1058.6	0.835*	395.6	0.0887	637.8	0.307	828.8	0.520
737.0	0.413*	908.6	0.627*	1061.5	0.840*	401.1	0.0934	642.2	0.311	832.5	0.524
741.0	0.418	912.1	0.631	1064.7	0.845*	406.5	0.0977	647.0	0.316	835.9	0.529
745.0	0.423*	915.4	0.636*	1068.2	0.849*	411.8	0.102	651.2	0.320	839.6	0.533
749.0	0.427*	918.8	0.640*	1071.3	0.854	417.1	0.107	655.9	0.325	843.2	0.538
753.1	0.432*	922.2	0.645*	1074.3	0.858*	422.5	0.111	660.2	0.329	847.0	0.543
757.2	0.436*	925.1	0.649*	1077.6	0.863*	427.8	0.116	664.6	0.334	850.4	0.547
760.8	0.441	928.8	0.654*	1080.9	0.867*	433.2	0.121	669.1	0.338	853.9	0.552
764.8	0.445*	931.9	0.659	1084.2	0.872*	438.4	0.125	673.4	0.343	857.4	0.556
768.6	0.450*	935.2	0.663*	1087.3	0.876*	443.5	0.130	681.8	0.352	861.2	0.561
772.4	0.454	938.5	0.668*	1090.5	0.881*	448.8	0.134	686.3	0.356	864.7	0.565
776.5	0.459*	941.6	0.672*	1093.4	0.885*	454.1	0.139	690.4	0.361	868.1	0.570
780.3	0.463*	944.8	0.677	1096.8	0.890	459.4	0.143	694.6	0.366	871.5	0.574
784.2	0.468*	948.3	0.681*	1099.8	0.894*	464.4	0.148	698.8	0.370	875.0	0.579
788.0	0.472*	951.5	0.686*	1102.9	0.899	469.8	0.152	703.2	0.375	878.6	0.583
791.8	0.477	954.6	0.690	1106.2	0.904*	475.0	0.157	708.0	0.379	882.3	0.588
795.3	0.482*	958.1	0.695*	1109.5	0.908*	480.3	0.161	711.4	0.384	885.5	0.592
799.2	0.486*	961.1	0.699*	1112.5	0.913*	485.3	0.166	715.7	0.388	888.9	0.597
803.0	0.491	964.3	0.704*	1115.5	0.917*	490.7	0.170	719.8	0.393	892.4	0.602
806.8	0.495*	967.6	0.708*	1118.8	0.922	496.0	0.175	724.0	0.397	895.9	0.606
810.6	0.500*	970.8	0.713*	1121.9	0.926*	501.1	0.180	727.9	0.402	899.3	0.611
814.3	0.504*	974.0	0.718*	1125.2	0.931*	506.4	0.184	731.9	0.406	902.7	0.615
818.0	0.509*	977.4	0.722*	1128.4	0.935*	511.4	0.189	736.1	0.411	906.1	0.620
821.5	0.513	980.6	0.727	1131.4	0.940*	516.3	0.193	740.0	0.415	909.5	0.624
825.1	0.518*	983.8	0.731*	1134.6	0.944	521.8	0.198	744.1	0.420	912.9	0.629
828.9	0.522*	987.2	0.736			526.6	0.202	748.1	0.425	916.2	0.633
832.4	0.527*	990.4	0.740*	CURVE 14*		531.8	0.207	752.0	0.429	919.7	0.638
836.0	0.531*	993.7	0.745*			536.7	0.211	756.1	0.434	923.1	0.642
839.6	0.536	996.9	0.749*	292.6	-0.0021	541.9	0.216	759.9	0.438	926.4	0.647
843.2	0.541*	1000.1	0.754	294.9	0.0024	547.0	0.220	764.0	0.443	929.8	0.651
846.8	0.545*	1003.5	0.758*	299.5	0.0069	552.0	0.225	767.9	0.447	933.2	0.656
850.3	0.550	1006.5	0.765*	304.2	0.0115	556.9	0.229	771.8	0.452	936.4	0.661
853.7	0.554*	1009.8	0.767*	309.5	0.0161	562.1	0.234	775.6	0.456	939.7	0.665
857.3	0.559*	1013.2	0.772*	314.8	0.0206	566.8	0.239	779.6	0.461	943.2	0.670
860.9	0.563*	1016.4	0.776*	320.2	0.0251	571.6	0.243	783.5	0.465	946.3	0.674
864.3	0.568*	1019.7	0.781*	325.4	0.0297	576.5	0.248	787.4	0.470	949.4	0.679
867.9	0.572*	1022.8	0.786*	330.4	0.0342	581.5	0.252			952.7	0.683

* Not shown in figure.

DATA TABLE 64. THERMAL LINEAR EXPANSION OF ZIRCONIUM Zr (continued)

T	$\Delta L/L_0$	T	$\Delta L/L_0$	T	$\Delta L/L_0$	T	$\Delta L/L_0$	T	$\Delta L/L_0$	T	$\Delta L/L_0$
CURVE 14 (cont.)*		CURVE 14 (cont.)*		CURVE 15 (cont.)		CURVE 15 (cont.)		CURVE 16 (cont.)*		CURVE 16 (cont.)*	
955.9	0.688	1115.5	0.915	618.0	0.172	1032.6	0.392	607.5	0.164	1068.7	0.39
959.1	0.692	1118.8	0.919	626.8	0.177*	1040.8	0.397	617.1	0.169	1078.4	0.402
962.4	0.697	1122.8	0.924	636.3	0.181	1048.9	0.401*	626.0	0.174	1089.1	0.406
965.7	0.701	1125.2	0.928	644.5	0.186*	1057.4	0.406*	636.0	0.178	1098.7	0.411
968.9	0.706	1128.0	0.933	653.8	0.191	1065.8	0.411	645.5	0.183	1109.2	0.416
972.2	0.710	1130.9	0.937	662.6	0.195*	1073.4	0.415*	655.4	0.188	1121.2	0.421
975.6	0.715	1134.0	0.942	672.6	0.200*	1084.0	0.420*	664.8	0.193		
978.9	0.720			681.0	0.204*	1094.5	0.424	674.5	0.197	CURVE 17*	
982.1	0.724	CURVE 15		689.5	0.209	1104.4	0.429	683.7	0.202		
985.4	0.729			696.4	0.213*	1116.0	0.434	693.4	0.207	292.2	-0.0051
992.0	0.738	290.7	-0.0018*	704.4	0.218	1134.5	0.438	702.6	0.212	292.6	-0.0003
995.2	0.742	297.2	0.0028*	712.8	0.222*			712.2	0.216	299.3	0.0045
998.6	0.747	304.4	0.0074*	720.4	0.227*	CURVE 16*		721.9	0.221	307.4	0.0915
1001.7	0.751	312.6	0.0120*	730.5	0.232			731.4	0.226	316.1	0.0140
1005.0	0.756	320.5	0.0165	739.1	0.236*	291.8	-0.0024	740.9	0.231	324.9	0.0187
1008.4	0.760	328.9	0.0211*	748.4	0.241	295.0	0.0024	761.5	0.240	333.1	0.0235
1011.5	0.767	339.6	0.0257	757.2	0.246*	304.3	0.0071	771.4	0.245	341.6	0.0282
1014.8	0.769	345.4	0.0303*	765.6	0.250	314.1	0.0118	781.4	0.250	350.2	0.0330
1018.1	0.774	353.8	0.0349*	774.1	0.255*	323.5	0.0166	791.7	0.254	358.7	0.0377
1021.2	0.778	362.1	0.0395	782.8	0.259*	332.4	0.0213	801.1	0.259	367.7	0.0425
1024.5	0.783	370.6	0.0440*	791.4	0.264	341.4	0.0261	810.9	0.264	376.6	0.0472
1030.7	0.792	378.9	0.0486*	799.9	0.268	349.7	0.0308	820.8	0.269	385.0	0.0520
1033.7	0.797	386.6	0.0532*	807.9	0.273	359.0	0.0356	830.8	0.273	395.2	0.0567
1037.2	0.801	395.0	0.0578	816.0	0.278*	368.3	0.0403	840.8	0.278	404.6	0.0615
1040.1	0.806	403.8	0.0674*	824.4	0.282*	377.2	0.0451	849.8	0.283	414.1	0.0662
1043.5	0.810	412.8	0.0669	832.6	0.287	385.5	0.0498	859.2	0.288	423.8	0.0710
1046.4	0.815	421.2	0.0715*	842.0	0.291*	395.8	0.0546	868.9	0.292	432.8	0.0710
1049.7	0.819	430.2	0.0761	851.0	0.297	404.5	0.0593	878.5	0.297	432.4	0.0757
1052.9	0.824	438.6	0.0867*	859.8	0.301*	413.5	0.0641	888.2	0.302	441.7	0.0805
1056.2	0.828	447.3	0.0853	868.2	0.305*	423.7	0.0681	897.6	0.307	450.9	0.0852
1059.4	0.833	456.0	0.0899	877.5	0.310*	440.7	0.0783	907.3	0.311	460.3	0.0900
1062.4	0.837	462.4	0.0945*	885.7	0.314	449.8	0.0831	917.1	0.316	469.6	0.0947
1065.7	0.842	473.1	0.0992*	894.6	0.319	459.1	0.0878	926.9	0.321	478.3	0.0990
1069.1	0.847	481.3	0.103*	903.0	0.323	467.9	0.0926	936.2	0.326	487.6	0.104
1072.1	0.851	490.0	0.108	911.8	0.328*	477.1	0.0973	945.6	0.330	496.8	0.109
1075.1	0.856	498.0	0.113*	920.4	0.333	486.5	0.102	955.4	0.335	506.4	0.114
1078.4	0.860	506.8	0.117	929.1	0.337*	495.1	0.107	964.4	0.340	515.6	0.118
1081.4	0.865	515.4	0.122*	938.0	0.342*	503.8	0.112	973.7	0.345	525.3	0.123
1084.7	0.869	523.8	0.126	946.4	0.346	513.1	0.117	982.9	0.349	534.0	0.128
1088.0	0.874	532.5	0.131*	954.8	0.351*	522.6	0.121	992.1	0.354	543.5	0.133
1090.9	0.878	540.8	0.136	963.6	0.356	532.2	0.126	1000.6	0.359	552.6	0.137
1094.0	0.883	549.0	0.140	972.0	0.360*	541.4	0.131	1009.3	0.364	562.2	0.142
1097.2	0.887	557.5	0.145*	981.1	0.365*	551.2	0.136	1018.0	0.368	571.7	0.147
1100.3	0.892	565.8	0.149	989.7	0.369	560.0	0.140	1026.5	0.373	581.3	0.152
1103.2	0.896	574.5	0.154*	998.3	0.374*	569.7	0.145	1034.9	0.378	590.7	0.156
1106.5	0.901	583.3	0.158*	1006.8	0.378	579.1	0.150	1043.1	0.383	600.5	0.161
1109.7	0.906	593.0	0.163	1015.5	0.383*	588.8	0.155	1051.7	0.387	607.5	0.166
1112.6	0.910	606.8	0.168	1024.1	0.388*	598.2	0.159	1060.1	0.392	619.4	0.171

* Not shown in figure.

DATA TABLE 64. THERMAL LINEAR EXPANSION OF ZIRCONIUM Zr (continued)

T	$\Delta L/L_0$
CURVE 17 (cont.)*	
629.4	0.175
638.9	0.180
648.9	0.185
658.2	0.190
667.5	0.194
676.9	0.199
688.8	0.204
698.0	0.209
707.0	0.213
716.5	0.218
726.0	0.223
735.3	0.228
744.2	0.232
762.9	0.242
772.3	0.247
781.5	0.251
791.4	0.256
800.1	0.261
809.8	0.266
818.9	0.270
827.4	0.275
835.8	0.280
844.2	0.285
852.7	0.289
860.7	0.294
868.6	0.299
876.6	0.304
883.9	0.308
891.8	0.313
899.7	0.318
907.4	0.323
915.2	0.327
923.3	0.332
936.0	0.337
944.6	0.342
953.0	0.346
961.5	0.351
969.7	0.356
977.8	0.361
986.2	0.365
994.5	0.370
1002.8	0.375
1010.8	0.380
1020.0	0.384
1028.3	0.389
1036.3	0.394
1044.6	0.399
CURVE 17 (cont.)*	
1052.6	0.403
1061.4	0.408
1070.2	0.413
1080.7	0.417
1091.6	0.422
1102.4	0.427
1114.3	0.432
1129.6	0.437
CURVE 18*	
291.5	-0.0033
294.0	0.0014
299.9	0.0062
307.1	0.0109
313.8	0.0157
321.4	0.0204
329.5	0.0252
337.6	0.0299
345.6	0.0349
354.0	0.0394
362.9	0.0442
371.6	0.0489
381.3	0.0537
390.0	0.0584
398.7	0.0632
407.2	0.0679
416.3	0.0727
424.7	0.0774
433.6	0.0822
442.4	0.0869
454.7	0.0917
464.0	0.0964
473.4	0.101
481.4	0.106
493.0	0.111
503.2	0.116
512.6	0.120
522.6	0.125
531.8	0.130
541.3	0.135
550.2	0.139
559.1	0.144
568.0	0.149
576.2	0.158
594.7	0.163
603.4	0.168
CURVE 18 (cont.)*	
612.6	0.173
621.4	0.177
630.2	0.182
640.6	0.187
651.0	0.192
659.3	0.196
667.4	0.201
678.1	0.206
687.9	0.211
696.9	0.215
706.0	0.220
715.8	0.225
725.9	0.230
735.8	0.234
745.9	0.239
754.9	0.244
764.1	0.249
772.5	0.253
781.6	0.258
792.4	0.263
800.0	0.268
810.2	0.272
819.1	0.277
829.2	0.282
839.1	0.287
847.7	0.291
857.7	0.291
857.7	0.296
867.7	0.301
876.8	0.306
885.6	0.310
899.0	0.315
908.3	0.320
915.8	0.325
926.8	0.329
937.2	0.334
947.1	0.339
955.9	0.344
964.7	0.348
973.9	0.353
982.8	0.358
992.8	0.363
1002.2	0.367
1012.3	0.372
1021.4	0.377
1031.2	0.382
1040.3	0.386
CURVE 18 (cont.)*	
1050.2	0.391
1058.8	0.396
1069.8	0.401
1087.0	0.405
1099.7	0.410
1110.9	0.415
1124.3	0.420
CURVE 19*, †	
1000	0.000
1100	0.065
1200	0.159
1300	0.259
1400	0.339
1500	0.404
CURVE 20*, †	
1000	0.000
1100	0.065
1200	0.181
1300	0.267
1400	0.344
1500	0.404
CURVE 21*, †, ‡	
12	0.0000 x 10^{-3}
14	0.0055
16	0.0104
18	0.0209
20	0.0374
22	0.0624
24	0.0974
26	0.144
28	0.204
30	0.281
58	3.963
65	5.626
75	8.501
85	11.901
CURVE 22*, †, ‡	
10	0.0000 x 10^{-3}
12	0.0055
14	0.0162
CURVE 22 (cont.)*, †, ‡	
14	0.0162 x 10^{-3}
16	0.0342
18	0.0617
58	5.692
65	7.67
75	10.89
85	14.509
CURVE 23	
293	0.000*
541	0.114
625	0.153
771	0.215
879	0.258
963	0.295
1028	0.316
1038	0.320*
1123	0.357
CURVE 24	
472	0.120
672	0.254
875	0.387*
961	0.446
1029	0.482
1081	0.472
1127	0.362
1225	0.412
CURVE 25	
94	-0.0901
204	-0.0420*
479	0.0937*
590	0.154*
702	0.215
813	0.274
923	0.325
1033	0.375
CURVE 26	
481	0.107*
588	0.175*
700	0.251
812	0.325
922	0.398
1033	0.473
CURVE 27*	
478	0.0912
588	0.148
699	0.203
813	0.252
921	0.301
1033	0.364
CURVE 28*	
479	0.105
589	0.173
700	0.244
811	0.312
921	0.371
1033	0.428
CURVE 29*	
483	0.110
588	0.178
700	0.255
810	0.332
922	0.415
1033	0.483
CURVE 30	
950	0.397
1032	0.397
1042	0.437
1119	0.419
1164	0.459
CURVE 31	
950	0.720
1032	0.740
1042	0.866
1119	0.841
1164	0.944
CURVE 32	
1166	0.40
1251	0.51
1289	0.53
1387	0.63
1408	0.66

* Not shown in figure.
† This curve is here reported using the first given temperature as reference temperature at which $\Delta L/L_0 = 0$.
‡ Author's data for coefficient of thermal expansion have been integrated by TPRC to obtain $\Delta L/L_0$.

DATA TABLE 64. THERMAL LINEAR EXPANSION OF ZIRCONIUM Zr (continued)

T	$\Delta L/L_0$
CURVE 32 (cont.)	
1492	0.73
1533	0.79
1584	0.80
CURVE 33	
283.2	-0.0168*
290.4	-0.0184*
290.7	-0.004*
292.2	0.0005*
370.6	0.0275
417.6	0.0899
458.7	0.107
528.9	0.163
599.2	0.219
651.1	0.300
695.1	0.342
743.4	0.394
777.7	0.445
796.9	0.450*
836.9	0.560
852.2	0.630
CURVE 34	
283.2	-0.0204*
290.4	-0.0235
290.7	-0.0111*
292.2	-0.0006*
370.6	0.0058
417.6	0.0591*
458.7	0.0696
528.9	0.0971
599.2	0.146
651.1	0.187*
695.1	0.228
743.4	0.259
777.7	0.278
796.9	0.244
836.9	0.391
852.2	0.328
CURVE 35*	
293	0.000
573	0.176
773	0.304
973	0.441
1073	0.512
CURVE 36*	
293	0.000
573	0.170
773	0.287
973	0.415
1073	0.484
CURVE 37*	
293	0.000
573	0.171
773	0.291
973	0.410
1073	0.458
CURVE 38*	
293	0.000
473	0.109
CURVE 39*	
293	0.000
473	0.107
CURVE 40*	
288	-0.001
373	0.020
CURVE 41	
288	-0.007*
373	0.114
CURVE 42	
288	-0.005
373	0.083
CURVE 43*	
359	0.033
458	0.091
640	0.203
751	0.267
957	0.390
967	0.403
CURVE 44	
386	0.046
424	0.070
492	0.119
502	0.126*
589	0.178*
673	0.239*
753	0.295
846	0.362
949	0.435
1020	0.486*
1053	0.504
1067	0.515
1095	0.532
1113	0.541
1140	0.570
1153	0.576*
1167	0.562
1175	0.534*
1177	0.528*
1177	0.513*
1178	0.500*
1178	0.487*
1193	0.471
1219	0.472*
1230	0.481
CURVE 45*	
1212	0.456
1189	0.445
1179	0.464
1154	0.546
1127	0.534
CURVE 46*	
369	0.049
518	0.125
581	0.165
660	0.225
779	0.305
881	0.385
971	0.464
1039	0.513
1062	0.531
1072	0.543
1089	0.552
1111	0.577
1127	0.591
1137	0.595
CURVE 46 (cont.)*	
1164	0.571
1200	0.496
1215	0.512
1230	0.524
CURVE 47*	
1211	0.502
1190	0.491
1179	0.500
1163	0.487
1163	0.513
1142	0.540
1130	0.565
1121	0.557
1098	0.551
CURVE 48	
273	-0.012
253	-0.024
233	-0.035
213	-0.046
193	-0.057
173	-0.068
153	-0.078
133	-0.088
113	-0.097
90	-0.106
78	-0.110
59	-0.113
CURVE 49	
293	0.000
273	-0.010
90	-0.094
78	-0.098
CURVE 50	
293	0.000
273	-0.009
90	-0.083
78	-0.087

* Not shown in figure.

DATA TABLE 64. COEFFICIENT OF THERMAL LINEAR EXPANSION OF ZIRCONIUM Zr

[Temperature, T, K; Coefficient of Expansion, α, 10^{-6} K^{-1}]

T	α	T	α
CURVE 21‡		CURVE 34	
12	0.014	273	5.599
14	0.025	293	5.644
16	0.040	298	5.656*
18	0.065	323	5.712
20	0.10	373	5.825*
22	0.15	423	5.937*
24	0.20	473	6.050*
26	0.265	523	6.162*
28	0.34	573	6.274*
30	0.43	623	6.386*
58	2.2	673	6.498*
65	2.55	723	6.610*
75	3.20	773	6.722*
85	3.60	823	6.833*
283	5.95	873	6.945*
CURVE 22‡		CURVE 48	
10	0.01	263	5.83
12	0.03	243	5.68
14	0.07	223	5.52
16	0.11	203	5.46
18	0.16	183	5.34
58	2.65	163	5.14
65	3.0	143	4.95
75	3.4	123	4.60
85	3.78	102	3.89
283	5.40	84	3.04
		68	2.08
CURVE 33			
		CURVE 49	
273	6.106		
293	6.389	263	5.01
298	6.459*	182	4.59
323	6.812*	84	3.56
373	7.517*		
423	8.220*	CURVE 50	
473	8.923*		
523	9.625*	263	4.56
573	10.32*	182	4.05
623	11.02*	84	3.47
673	11.72*		
723	12.42*		
773	13.12*		
823	13.81*		
873	14.50*		

* Not shown in figure.
‡ Author's data for coefficient of thermal expansion have been integrated by TPRC to obtain $\Delta L/L_0$.

2. INTERMETALLIC COMPOUNDS

FIGURE AND TABLE NO. 65R. PROVISIONAL VALUES FOR THERMAL LINEAR EXPANSION OF AlSb INTERMETALLIC COMPOUND

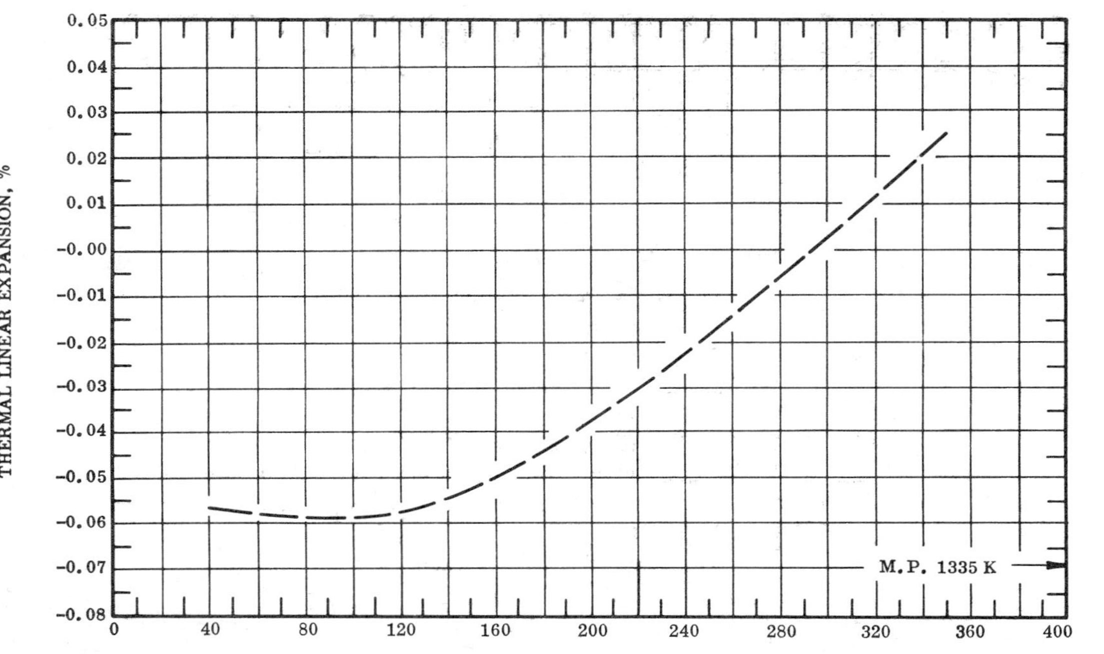

PROVISIONAL VALUES

[Temperature, T, K; Linear Expansion, $\Delta L/L_0$, %; α, K^{-1}]

T	$\Delta L/L_0$	$\alpha \times 10^6$
40	-0.056	-1.31
50	-0.058	-0.92
100	-0.059	0.82
200	-0.037	3.3
293	0.000	4.4
350	0.025	4.5

REMARKS

The tabulated values are considered accurate to within ± 7% over the entire temperature range. These values can be represented approximately by the following equation:

$$\Delta L/L_0 = -0.049 - 2.997 \times 10^{-4}\, T + 2.243 \times 10^{-6}\, T^2 - 2.234 \times 10^{-9}\, T^3$$

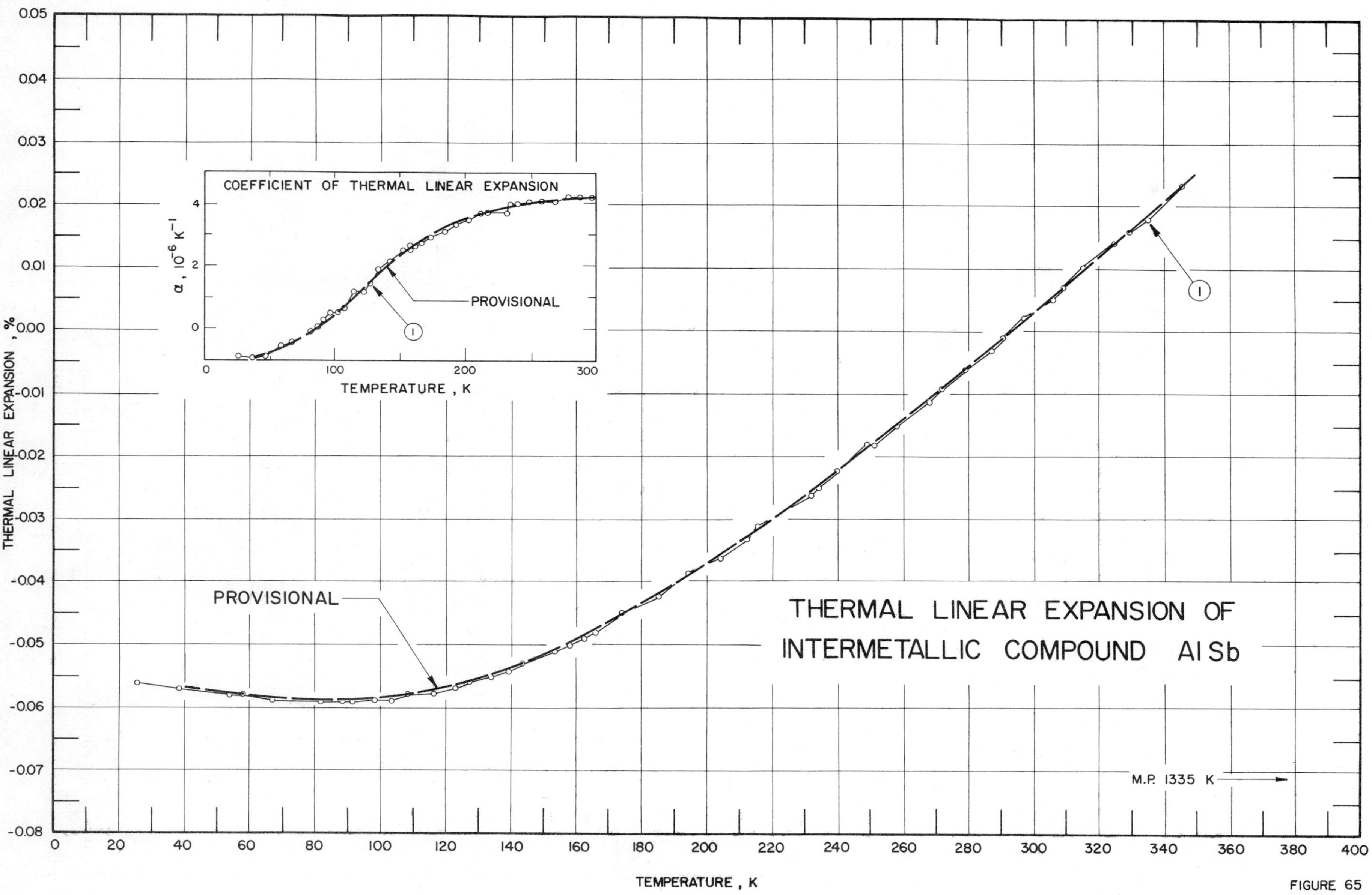

THERMAL LINEAR EXPANSION OF
INTERMETALLIC COMPOUND AlSb
THERMAL LINEAR EXPANSION, %
0.05
0.04
0.03
0.02
0.01
0.00
-0.01
-0.02
-0.03
-0.04
-0.05
-0.06
-0.07
-0.08
TEMPERATURE, K
0
20
40
60
80
100
120
140
160
180
200
220
240
260
280
300
320
340
360
380
400
PROVISIONAL
1
M.P. 1335 K
COEFFICIENT OF THERMAL LINEAR EXPANSION
α, 10⁻⁶ K⁻¹
0
2
4
TEMPERATURE, K
0
100
200
300
PROVISIONAL
1

FIGURE 65

SPECIFICATION TABLE 65. THERMAL LINEAR EXPANSION OF AlSb INTERMETALLIC COMPOUND

Cur. No.	Ref. No.	Author(s)	Year	Method Used	Temp. Range, K	Name and Specimen Designation	Composition (weight percent), Specifications, and Remarks
1	559	Novikova, S.I. and Abrikosov, N. Kh.	1964	L	26-345		Prepared from quite pure components, 99.9998 Al, 99.999 Sb; specimen coarsely crystalline and irregular form, longest dimension 16.55 mm, values of the coefficient of thermal linear expansion becomes negative at 85 K.

DATA TABLE 65. THERMAL LINEAR EXPANSION OF AlSb INTERMETALLIC COMPOUND

[Temperature, T, K; Linear Expansion, $\Delta L/L_0$, %]

T	$\Delta L/L_0$	T	$\Delta L/L_0$	T	$\Delta L/L_0$	T	$\Delta L/L_0$	T	$\Delta L/L_0$
CURVE 1‡		CURVE 1 (cont.)‡		CURVE 1 (cont.)‡		CURVE 1 (cont.)‡		CURVE 1 (cont.)‡	
26	-0.056	116	-0.058	174	-0.045	258	-0.015	330	0.016
38	-0.057	123	-0.057	185	-0.042	268	-0.011	335	0.018
47	-0.058	127	-0.056	194	-0.039	272	-0.009	345	0.023
59	-0.058	134	-0.055	204	-0.036	279	-0.006		
67	-0.059	139	-0.054	212	-0.033	287	-0.003		
82	-0.059	143	-0.053	216	-0.031	291	-0.001		
88	-0.059	153	-0.051	231	-0.026	297	0.002		
91	-0.059	158	-0.050	234	-0.025	306	0.005		
98	-0.059	158	-0.050	240	-0.022	310	0.007		
103	-0.059	162	-0.049	249	-0.018	316	0.010		
108	-0.058	166	-0.048	251	-0.018	325	0.014		

DATA TABLE 65. COEFFICIENT OF THERMAL LINEAR EXPANSION OF AlSb INTERMETALLIC COMPOUND

[Temperature, T, K; Coefficient of Expansion, α, 10^{-6} K^{-1}]

T	α	T	α	T	α	T	α	T	α	T	α
CURVE 1‡		CURVE 1 (cont.)‡		CURVE 1 (cont.)‡		CURVE 1 (cont.)‡		CURVE 1 (cont.)‡		CURVE 1 (cont.)‡	
26	-0.86	98	0.49	143	2.16	194	3.37	251	4.19	297	4.25
38	-0.97	103	0.58	153	2.50	204	3.47	258	4.11	306	4.36*
47	-0.93	108	0.68	158	2.50	212	3.71	268	4.11	310	4.46*
59	-0.57	116	1.20	158	2.68	216	3.77	272	4.19*	316	4.30*
67	-0.41	123	1.20	162	2.62	231	3.74	279	4.22	325	4.32*
82	-0.11	127	1.44	166	2.72	234	4.00	287	4.22	330	4.53*
88	0.07	134	1.88*	174	2.93	240	4.00	291	4.22*	335	4.29*
91	0.29	139	1.88*	185	3.11	249	4.13			345	4.40*

* Not shown in figure.

‡ Author's data for coefficient of thermal expansion have been integrated by TPRC to obtain $\Delta L/L_0$.

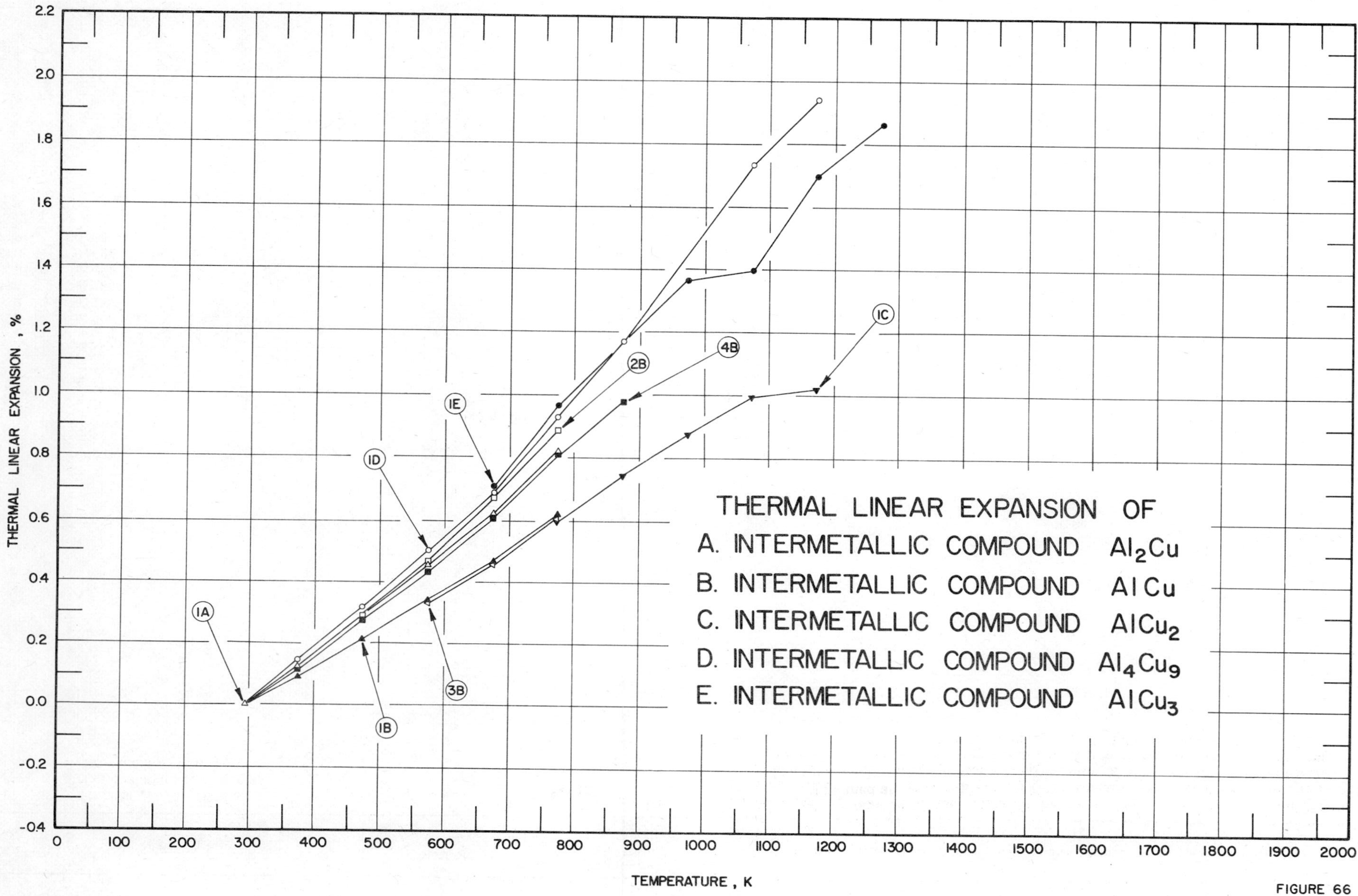

THERMAL LINEAR EXPANSION OF
A. INTERMETALLIC COMPOUND Al_2Cu
B. INTERMETALLIC COMPOUND AlCu
C. INTERMETALLIC COMPOUND $AlCu_2$
D. INTERMETALLIC COMPOUND Al_4Cu_9
E. INTERMETALLIC COMPOUND $AlCu_3$
1A
1B
1C
1D
1E
2B
3B
4B
THERMAL LINEAR EXPANSION , %
2.2
2.0
1.8
1.6
1.4
1.2
1.0
0.8
0.6
0.4
0.2
0.0
-0.2
-0.4
0
100
200
300
400
500
600
700
800
900
1000
1100
1200
1300
1400
1500
1600
1700
1800
1900
2000
TEMPERATURE , K

FIGURE 66

SPECIFICATION TABLE 66A. THERMAL LINEAR EXPANSION OF Al_2Cu INTERMETALLIC COMPOUND

Cur. No.	Ref. No.	Author(s)	Year	Method Used	Temp. Range, K	Name and Specimen Designation	Composition (weight percent), Specifications, and Remarks
1	698	Rabkin, D.M., Ryabov, V.R., Lozovskaya, A.V., and Dorzhenko, V.A.	1970	L	293-773	θ phase	Alloy containing 46.70 Al and 53.30 Cu prepared from electrolytic copper (99.99 pure) and AVOOO grade aluminum (99.99 pure) were used; specimen produced by vacuum melting (maximum vacuum of 10^{-3} mm Hg); the melting space was evacuated and then filled with argon or helium, the molten metal was maintained at 100-200 K above T_{melt}; cylindrical specimen 4 mm in diameter and 50 mm long.

DATA TABLE 66A. THERMAL LINEAR EXPANSION OF Al_2Cu INTERMETALLIC COMPOUND

[Temperature, T, K; Linear Expansion, $\Delta L/L_0$, %]

T	$\Delta L/L_0$
CURVE 1	
293	0.000
373	0.127
473	0.290*
573	0.458
673	0.624
773	0.823

* Not shown in figure.

SPECIFICATION TABLE 66B. THERMAL LINEAR EXPANSION OF AlCu INTERMETALLIC COMPOUND

Cur. No.	Ref. No.	Author(s)	Year	Method Used	Temp. Range, K	Name and Specimen Designation	Composition (weight percent), Specifications, and Remarks
1	698	Rabkin, D. M., Ryalov, V.R., Lozovskaya, A.V., and Dovzhenko, V.A.	1970	L	293-773	η-phase	Alloy containing 30.53 Al and 69.46 Cu prepared from electrolytic copper (99.99 pure) and AVOOO grade aluminum (99.99 pure) were used; specimen produced by vacuum melting (maximum vacuum of 10^{-3} mm Hg); the melting space was evacuated and then filled with argon or helium, the molten metal was maintained at 100-200 K above T_{melt}; cylindrical specimen 4 mm in diameter and 50 mm long; this η-phase has two-phase structure.
2	698	Rabkin, D. M., et al.	1970	L	293-773	ξ-phase	Similar to the above specimen with composition 26.00 Al and 74.00 Cu; this ξ-phase is characterized by the pressure of turns and a banded microstructure.
3	698	Rabkin, D. M., et al.	1970		293-773	ϵ-phase	Similar to the above specimen with composition 23.35 Al and 76.65 Cu; this ϵ-phase has polygonization lines intersecting grain boundaries.
4	698	Rabkin, D. M., et al.	1970	L	293-873	δ-phase	Similar to the above specimen with composition 22.66 Al and 77.34 Cu; this δ-phase has a two-phase structure.

DATA TABLE 66B. THERMAL LINEAR EXPANSION OF AlCu INTERMETALLIC COMPOUND

[Temperature, T, K; Linear Expansion, $\Delta L/L_0$, %]

T	$\Delta L/L_0$	T	$\Delta L/L_0$
CURVE 1		CURVE 3	
293	0.000*	293	0.000*
373	0.091	373	0.087*
473	0.214	473	0.207*
573	0.343	573	0.333
673	0.466	673	0.459
773	0.624	773	0.606
CURVE 2		CURVE 4	
293	0.000*	293	0.000
373	0.126*	373	0.119
473	0.290	473	0.272
573	0.469	573	0.434
673	0.676	673	0.608
773	0.888	773	0.811
		873	0.980

* Not shown in figure.

SPECIFICATION TABLE 66C. THERMAL LINEAR EXPANSION OF $AlCu_2$ INTERMETALLIC COMPOUND

Cur. No.	Ref. No.	Author(s)	Year	Method Used	Temp. Range, K	Name and Specimen Designation	Composition (weight percent), Specifications, and Remarks
1	698	Rabkin, D.M., Ryalov, V.R., Lozovskaya, A.V., and Dovzhenko, V.A.	1970	L	293-1173	γ_2	Alloy containing 18.13 Al and 81.87 Cu prepared from electrolytic copper (99.99 pure) and AVOOO grade aluminum (99.99 pure) were used; specimen produced by vacuum melting (maximum vacuum of 10^{-3} mm Hg); the melting space was evacuated and then filled with argon or helium, the molten metal was maintained at 100-200 K above T_{melt}; cylindrical specimen 4 mm in diameter and 50 mm long.

DATA TABLE 66C. THERMAL LINEAR EXPANSION OF $AlCu_2$ INTERMETALLIC COMPOUND

[Temperature, T, K; Linear Expansion, $\Delta L/L_0$, %]

T	$\Delta L/L_0$
CURVE 1	
293	0.000*
373	0.090*
473	0.203*
573	0.326*
673	0.456*
773	0.600
873	0.741
973	0.879
1073	0.995
1173	1.021

* Not shown in figure.

SPECIFICATION TABLE 66D. THERMAL LINEAR EXPANSION OF Al_4Cu_9 INTERMETALLIC COMPOUND

Cur. No.	Ref. No.	Author(s)	Year	Method Used	Temp. Range, K	Name and Specimen Designation	Composition (weight percent), Specifications, and Remarks
1	698	Rabkin, D.M., Ryabov, V.R., Lozovskaya, A.V., and Dovzhenko, V.A.	1970	L	293-1173	X	Alloy containing 15.37 Al and 84.63 Cu prepared from electrolytic copper (99.99 pure) and AVOOO grade aluminum (99.99 pure) were used; specimen produced by vacuum melting (maximum vacuum of 10^{-3} mm Hg); the melting space was evacuated and then filled with argon or helium, the molten metal was maintained at 100-200 K above T_{melt}; cylindrical specimen 4 mm in diameter and 50 mm long.

DATA TABLE 66D. THERMAL LINEAR EXPANSION OF Al_4Cu_9 INTERMETALLIC COMPOUND

[Temperature, T, K; Linear Expansion, $\Delta L/L_0$, %]

T	$\Delta L/L_0$
CURVE 1	
293	0.000*
373	0.150
473	0.318
573	0.501
673	0.690
773	0.933
873	1.169
973	1.370*
1073	1.739
1173	1.945

* Not shown in figure.

SPECIFICATION TABLE 66E. THERMAL LINEAR EXPANSION OF $AlCu_3$ INTERMETALLIC COMPOUND

Cur. No.	Ref. No.	Author(s)	Year	Method Used	Temp. Range, K	Name and Specimen Designation	Composition (weight percent), Specifications, and Remarks
1	698	Rabkin, D.M., Ryabov, V.R., Lozovskaya, A.V., and Dovzhenko, V.A.	1970	L	293-773	ρ	Alloy containing 13.88 Al and 86.12 Cu prepared from electrolytic copper (99.99 pure) and AVOOO grade aluminum (99.99 pure) were used; specimen produced by vacuum melting (maximum vacuum of 10^{-3} mm Hg); the melting space was evacuated and then filled with argon or helium, the molten metal was maintained at 100-200 K above T_{melt}; cylindrical specimen 4 mm in diameter and 50 mm long.

DATA TABLE 66E. THERMAL LINEAR EXPANSION OF $AlCu_3$ INTERMETALLIC COMPOUND

[Temperature, T, K; Linear Expansion, $\Delta L/L_0$, %]

T	$\Delta L/L_0$
CURVE 1	
293	0.000*
373	0.130*
473	0.297*
573	0.472*
673	0.711
773	0.969
873	1.170*
973	1.367
1073	1.404
1173	1.703
1273	1.867

* Not shown in figure.

FIGURE AND TABLE NO. 67R. PROVISIONAL VALUES FOR THERMAL LINEAR EXPANSION OF Al_2Au INTERMETALLIC COMPOUND

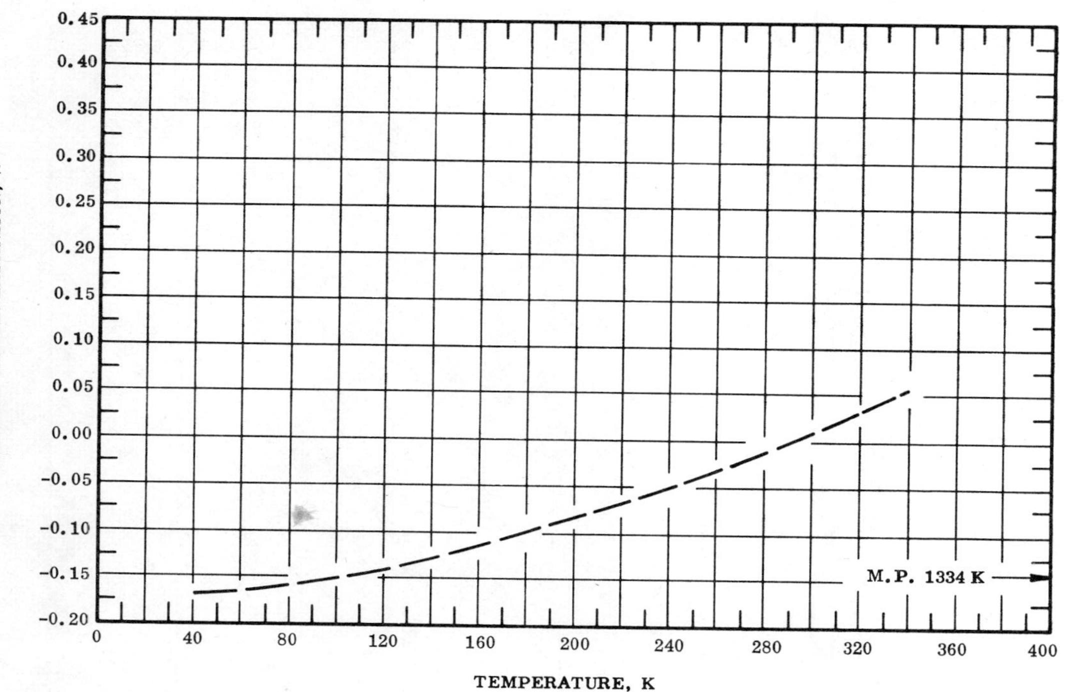

PROVISIONAL VALUES

[Temperature, T, K; Linear Expansion, $\Delta L/L_0$, %; α, K^{-1}]

T	$\Delta L/L_0$	$\alpha \times 10^6$
40	-0.168	2.3
100	-0.149	4.6
200	-0.084	7.9
293	0.000	10.5
340	0.053	11.6

REMARKS

The tabulated values are considered accurate to within ±7% over the entire temperature range. These values can be represented approximately by the following equation:

$$\Delta L/L_0 = -0.175 + 6.870 \times 10^{-5}\, T + 2.102 \times 10^{-6}\, T^2 - 9.819 \times 10^{-10}\, T^3$$

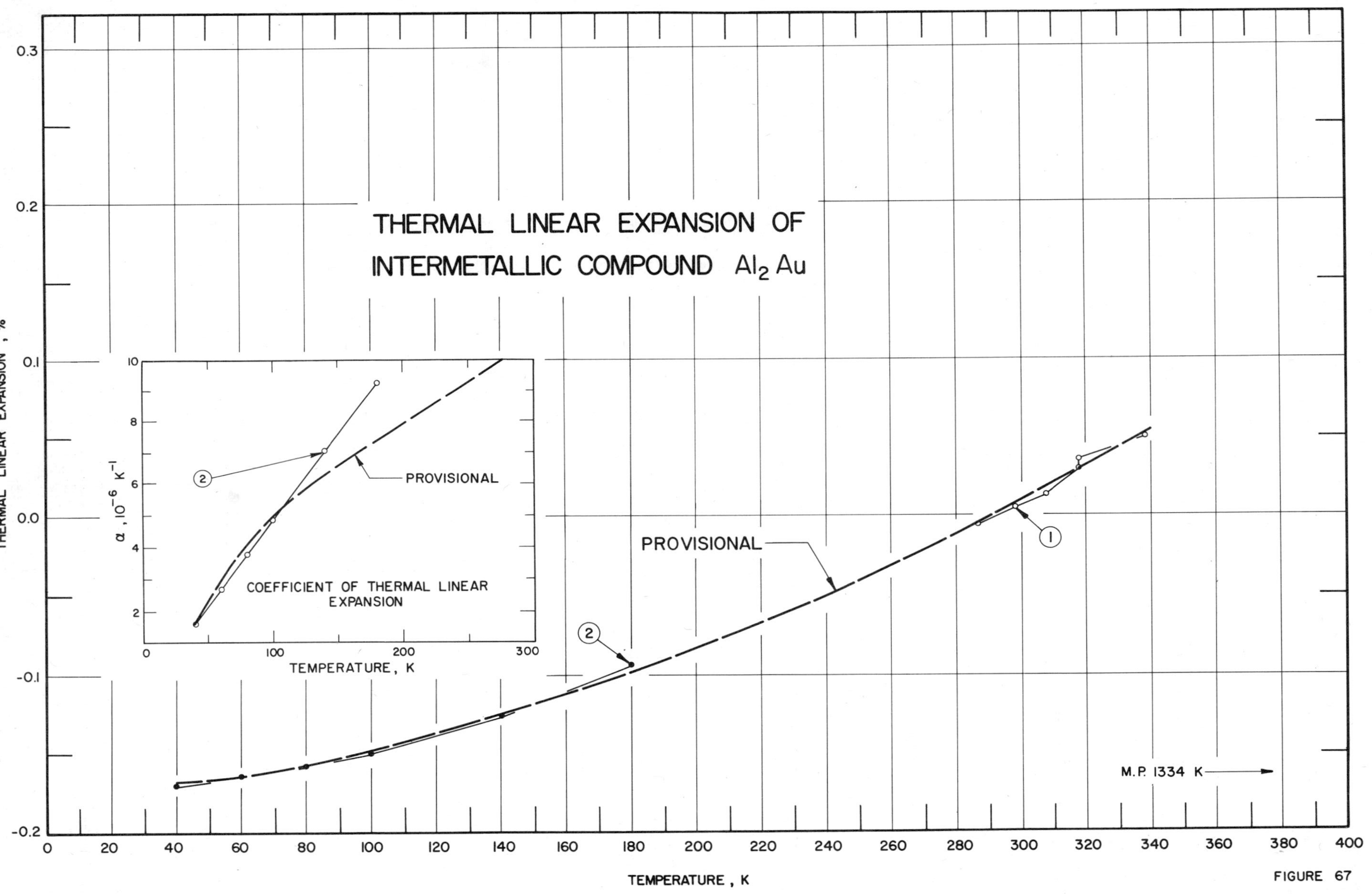
THERMAL LINEAR EXPANSION OF
INTERMETALLIC COMPOUND Al_2Au
THERMAL LINEAR EXPANSION, %
0.3
0.2
0.1
0.0
-0.1
-0.2
$\alpha, 10^{-6}\ K^{-1}$
10
8
6
4
2
2
PROVISIONAL
COEFFICIENT OF THERMAL LINEAR EXPANSION
0
100
200
300
TEMPERATURE, K
PROVISIONAL
1
2
M.P. 1334 K
0
20
40
60
80
100
120
140
160
180
200
220
240
260
280
300
320
340
360
380
400
TEMPERATURE, K
FIGURE 67

SPECIFICATION TABLE 67. THERMAL LINEAR EXPANSION OF Al_2Au INTERMETALLIC COMPOUND

Cur. No.	Ref. No.	Author(s)	Year	Method Used	Temp. Range, K	Name and Specimen Designation	Composition (weight percent), Specifications, and Remarks
1*	687	Straumanis, M. E. and Chopra, K. S.	1964	X	287-338	$Al_{1.95}Au$	Specimen with composition 78.94 Au prepared from 0.0003 Fe, 0.00013 Cu, Al from AIAG Co., Neuhausen, Switzerland, and <0.0001 Mg, 0.0001 Pb, 0.0003 Ag, ASARCO Au- by heating calculated quantities in alundum crucible; homogenized; lattice parameters reported at 287 K is 5.9963 Å; 5.9966 Å at 293 K determined by graphical extrapolation.
2	709	Shah, J. S.	1971	X	40-180		Finely divided powder sample, intermetallic compound, lattice parameter not corrected for refraction, lattice parameter reported as 5.98643 Å at 40 K, values calculated using lattice parameter at 293 K as 5.9966 Å, data on the coefficient of thermal linear expansion also reported.

DATA TABLE 67. THERMAL LINEAR EXPANSION OF Al_2Au INTERMETALLIC COMPOUND

[Temperature, T, K; Linear Expansion, $\Delta L/L_0$, %]

T	$\Delta L/L_0$	T	$\Delta L/L_0$
CURVE 1*		CURVE 2	
287	-0.005	40	-0.170
298	0.006	60	-0.165
308	0.017	80	-0.158
318	0.030	100	-0.150
318	0.039	140	-0.125
338	0.050	180	-0.093

DATA TABLE 67. COEFFICIENT OF THERMAL LINEAR EXPANSION OF Al_2Au INTERMETALLIC COMPOUND

[Temperature, T, K; Coefficient of Expansion, α, 10^{-6} K^{-1}]

T	α
CURVE 2	
40	1.6
60	2.7
80	3.8
100	4.9
140	7.1
180	9.3

* Not shown in figure.

FIGURE AND TABLE NO. 68AR. PROVISIONAL VALUES FOR THERMAL LINEAR EXPANSION OF Al_5Fe_2 INTERMETALLIC COMPOUND

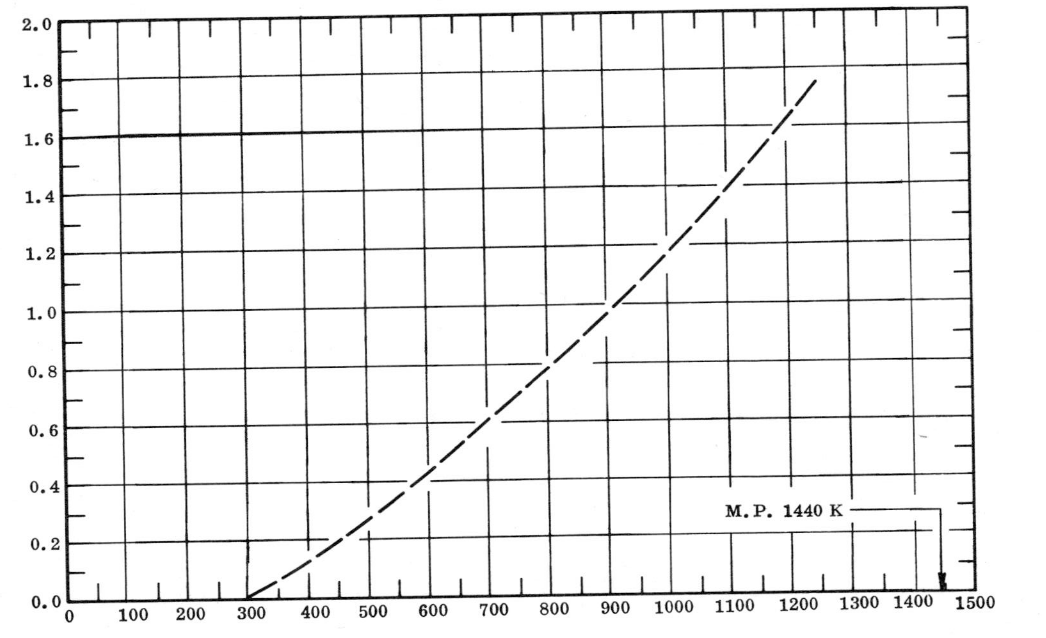

PROVISIONAL VALUES

[Temperature, T, K; Linear Expansion, $\Delta L/L_0$, %; α, K^{-1}]

T	$\Delta L/L_0$	$\alpha \times 10^6$
293	0.000	12.5
400	0.140	13.7
500	0.283	14.9
600	0.439	16.2
700	0.606	17.4
800	0.786	18.6
900	0.979	19.8
1000	1.184	21.1
1100	1.401	22.3
1200	1.631	23.6
1250	1.750	24.2

REMARKS

The tabulated values are considered accurate to within ± 7% over the entire temperature range. These values can be represented approximately by the following equation:

$$\Delta L/L_0 = -0.315 + 9.011 \times 10^{-4}\, T + 5.848 \times 10^{-7}\, T^2 + 1.275 \times 10^{-11}\, T^3$$

FIGURE AND TABLE NO. 68BR. PROVISIONAL VALUES FOR THERMAL LINEAR EXPANSION OF Al_3Fe INTERMETALLIC COMPOUND

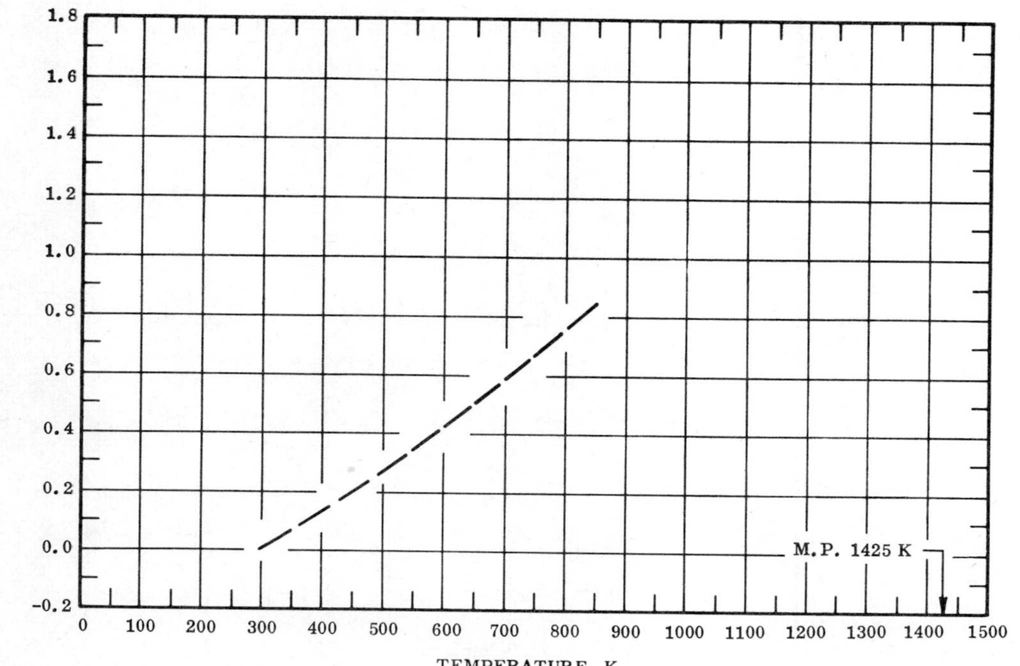

PROVISIONAL VALUES

[Temperature, T, K; Linear Expansion, $\Delta L/L_0$, %; α, K^{-1}]

T	$\Delta L/L_0$	$\alpha \times 10^6$
293	0.000	11.4
400	0.133	13.2
500	0.271	14.6
600	0.424	15.8
700	0.588	16.7
800	0.759	17.4
850	0.846	17.6

REMARKS

The tabulated values are considered accurate to within ±7% over the entire temperature range. These values can be represented approximately by the following equation:

$$\Delta L/L_0 = -0.243 + 4.857 \times 10^{-4}\,T + 1.305 \times 10^{-6}\,T^2 - 4.342 \times 10^{-10}\,T^3$$

FIGURE AND TABLE NO. 68CR. PROVISIONAL VALUES FOR THERMAL LINEAR EXPANSION OF Al_2Fe INTERMETALLIC COMPOUND

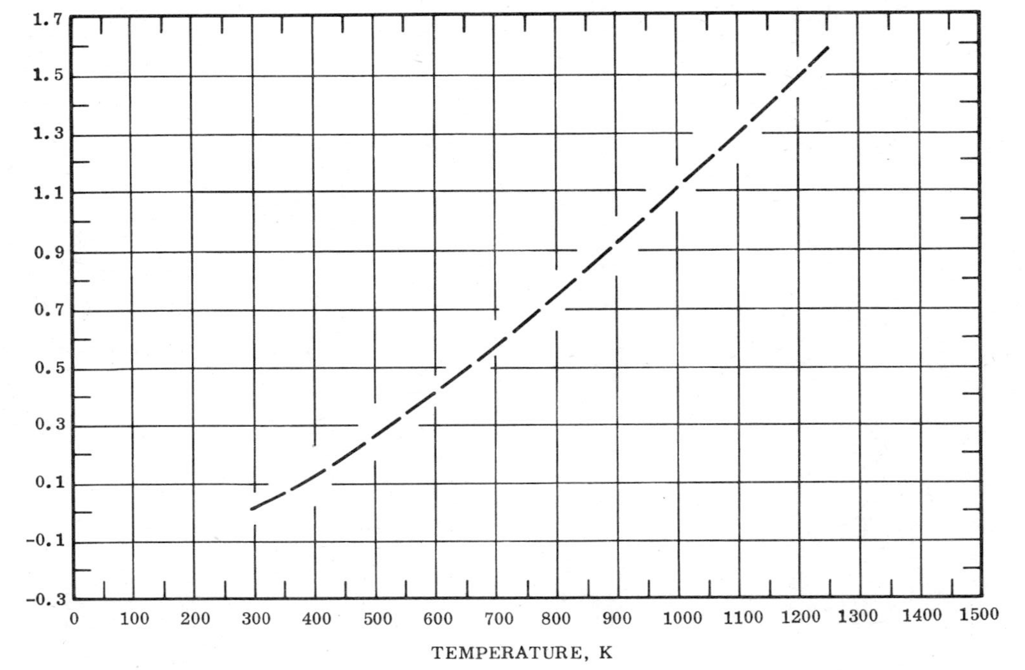

PROVISIONAL VALUES

[Temperature, T, K; Linear Expansion, $\Delta L/L_0$, %; α, K^{-1}]

T	$\Delta L/L_0$	$\alpha \times 10^6$
293	0.000	11.2
400	0.129	12.8
500	0.264	14.2
600	0.412	15.4
700	0.573	16.5
800	0.742	17.4
900	0.921	18.1
1000	1.105	18.7
1200	1.488	19.4
1250	1.585	19.4

REMARKS

The tabulated values are considered accurate to within ±7% over the entire temperature range. These values can be represented approximately by the following equation:

$$\Delta L/L_0 = -0.249 + 5.620 \times 10^{-4}\, T + 1.066 \times 10^{-6}\, T^2 - 2.739 \times 10^{-10}\, T^3$$

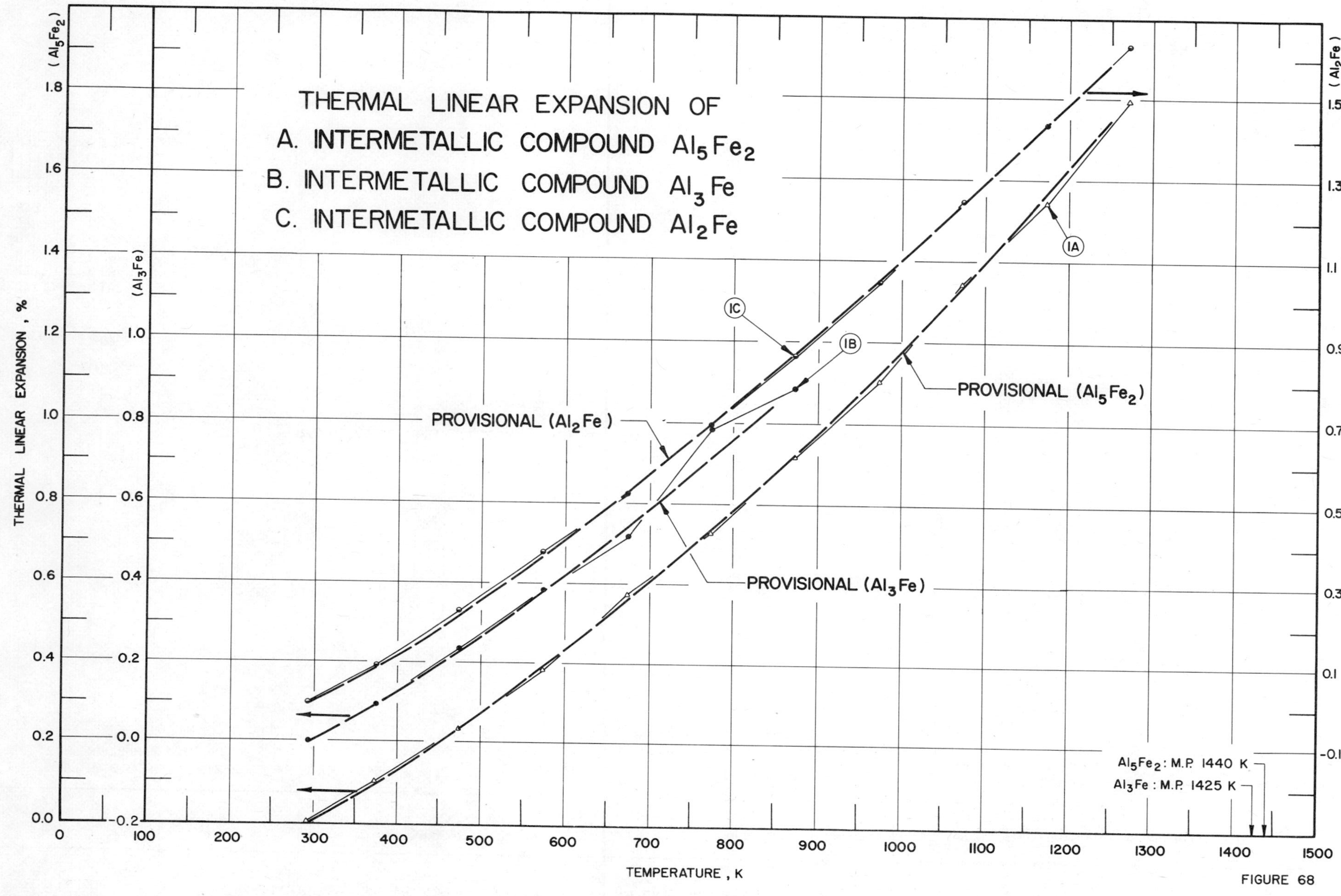

THERMAL LINEAR EXPANSION OF
A. INTERMETALLIC COMPOUND Al_5Fe_2
B. INTERMETALLIC COMPOUND Al_3Fe
C. INTERMETALLIC COMPOUND Al_2Fe
1C
1B
1A
PROVISIONAL (Al_2Fe)
PROVISIONAL (Al_5Fe_2)
PROVISIONAL (Al_3Fe)
Al_5Fe_2 : M.P. 1440 K
Al_3Fe : M.P. 1425 K
THERMAL LINEAR EXPANSION, %
(Al_5Fe_2)
1.8
1.6
1.4
1.2
1.0
0.8
0.6
0.4
0.2
0.0
(Al_3Fe)
1.0
0.8
0.6
0.4
0.2
0.0
-0.2
(Al_2Fe)
1.5
1.3
1.1
0.9
0.7
0.5
0.3
0.1
-0.1
0
100
200
300
400
500
600
700
800
900
1000
1100
1200
1300
1400
1500
TEMPERATURE, K

FIGURE 68

SPECIFICATION TABLE 68A. THERMAL LINEAR EXPANSION OF Al_5Fe_2 INTERMETALLIC COMPOUND

Cur. No.	Ref. No.	Author(s)	Year	Method Used	Temp. Range, K	Name and Specimen Designation	Composition (weight percent), Specifications, and Remarks
1	693	Ryabov, V. R., Lozovskaya, A. V., and Vasil'yev, V. G.	1968	L	293-1273		Specimen prepared by melting exact amount of 99.99 Al and 99.97 Fe in vacuum induction furnace in magnetic crucible; annealed at 1273 K for 10 days and air cooled, analysis after that indicates 46.2 Fe; melting point of this compound is 1323 K.

DATA TABLE 68A. THERMAL LINEAR EXPANSION OF Al_5Fe_2 INTERMETALLIC COMPOUND

[Temperature, T, K; Linear Expansion, $\Delta L/L_0$, %]

T	$\Delta L/L_0$
CURVE 1	
293	0.000
373	0.102
473	0.235
573	0.385
673	0.537
773	0.730
873	0.915
973	1.107
1073	1.343
1173	1.545
1273	1.799

SPECIFICATION TABLE 68B. THERMAL LINEAR EXPANSION OF Al_3Fe INTERMETALLIC COMPOUND

Cur. No.	Ref. No.	Author(s)	Year	Method Used	Temp. Range, K	Name and Specimen Designation	Composition (weight percent), Specifications, and Remarks
1	693	Ryabov, V.R., Lozovskaya, A.V. and Vasilyev, V.G.	1968	L	293-873		Specimen prepared by melting exact amount of 99.99 Al and 99.97 Fe in vacuum induction furnace in magnetic crucible; annealed at 873 K for 10 days and water cooled, analysis after that indicates 41.16 Fe.

DATA TABLE 68B. THERMAL LINEAR EXPANSION OF Al_3Fe INTERMETALLIC COMPOUND

[Temperature, T, K; Linear Expansion, $\Delta L/L_0$, %]

T	$\Delta L/L_0$
CURVE 1	
293	0.000
373	0.095
473	0.235
573	0.381
673	0.520
773	0.785
873	0.882

SPECIFICATION TABLE 68C. THERMAL LINEAR EXPANSION OF Al_2Fe INTERMETALLIC COMPOUND

Cur. No.	Ref. No.	Author(s)	Year	Method Used	Temp. Range, K	Name and Specimen Designation	Composition (weight percent), Specifications, and Remarks
1	693	Ryabov, V.R., Lozovskaya, A. V., and Vasil' yer, V. G.	1968	L	293-1273		Specimen prepared by melting exact amount of 99.99 Al and 99.97 Fe in vacuum induction furnace in magnetic crucible; annealed at 1273 K for 10 days and air cooled, analysis after that indicates 51.24 Fe; melting point of this compound is 1338 K.

DATA TABLE 68C. THERMAL LINEAR EXPANSION OF Al_2Fe INTERMETALLIC COMPOUND

[Temperature, T, K; Linear Expansion, $\Delta L/L_0$, %]

T	$\Delta L/L_0$
CURVE 1	
293	0.000
373	0.092
473	0.231
573	0.378
673	0.521
773	0.694
873	0.866
973	1.049
1073	1.248
1173	1.437
1273	1.639

FIGURE AND TABLE NO. 69AR. PROVISIONAL VALUES FOR THERMAL LINEAR EXPANSION OF AlFe INTERMETALLIC COMPOUND

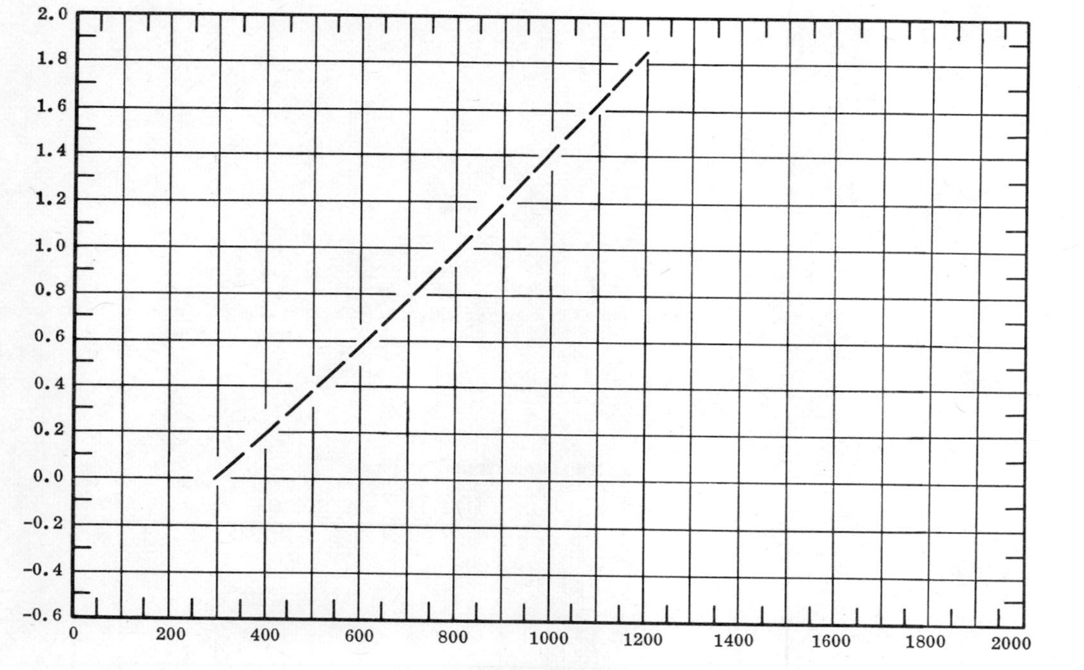

PROVISIONAL VALUES

[Temperature, T, K; Linear Expansion, $\Delta L/L_0$, %; α, K^{-1}]

T	$\Delta L/L_0$	$\alpha \times 10^6$
293	0.000	17.3
400	0.191	18.3
500	0.380	19.3
600	0.576	20.0
700	0.780	20.7
800	0.989	21.1
900	1.202	21.4
1000	1.418	21.6
1100	1.634	21.7
1200	1.850	21.8

REMARKS

The tabulated values are considered accurate to within ±7% over the entire temperature range. These values can be represented approximately by the following equation:

$$\Delta L/L_0 = -0.449 + 1.324 \times 10^{-3}\,T + 7.898 \times 10^{-7}\,T^2 - 2.472 \times 10^{-10}\,T^3$$

FIGURE AND TABLE NO. 69BR. PROVISIONAL VALUES FOR THERMAL LINEAR EXPANSION OF $AlFe_3$ INTERMETALLIC COMPOUND

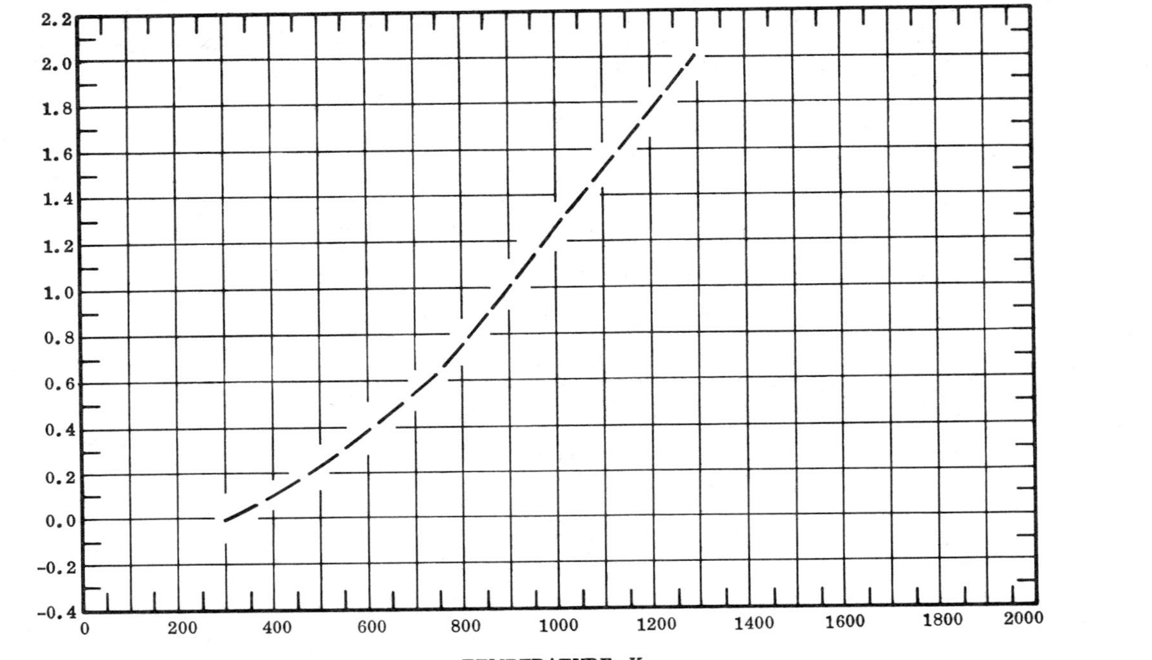

PROVISIONAL VALUES

[Temperature, T, K; Linear Expansion, $\Delta L/L_0$, %]

T	$\Delta L/L_0$
293	0.000
400	0.108
500	0.242
600	0.395
700	0.554
800	0.761
900	1.022
1000	1.263
1100	1.503
1200	1.754
1300	2.002

REMARKS

The tabulated values are considered accurate to within ± 7% over the entire temperature range.

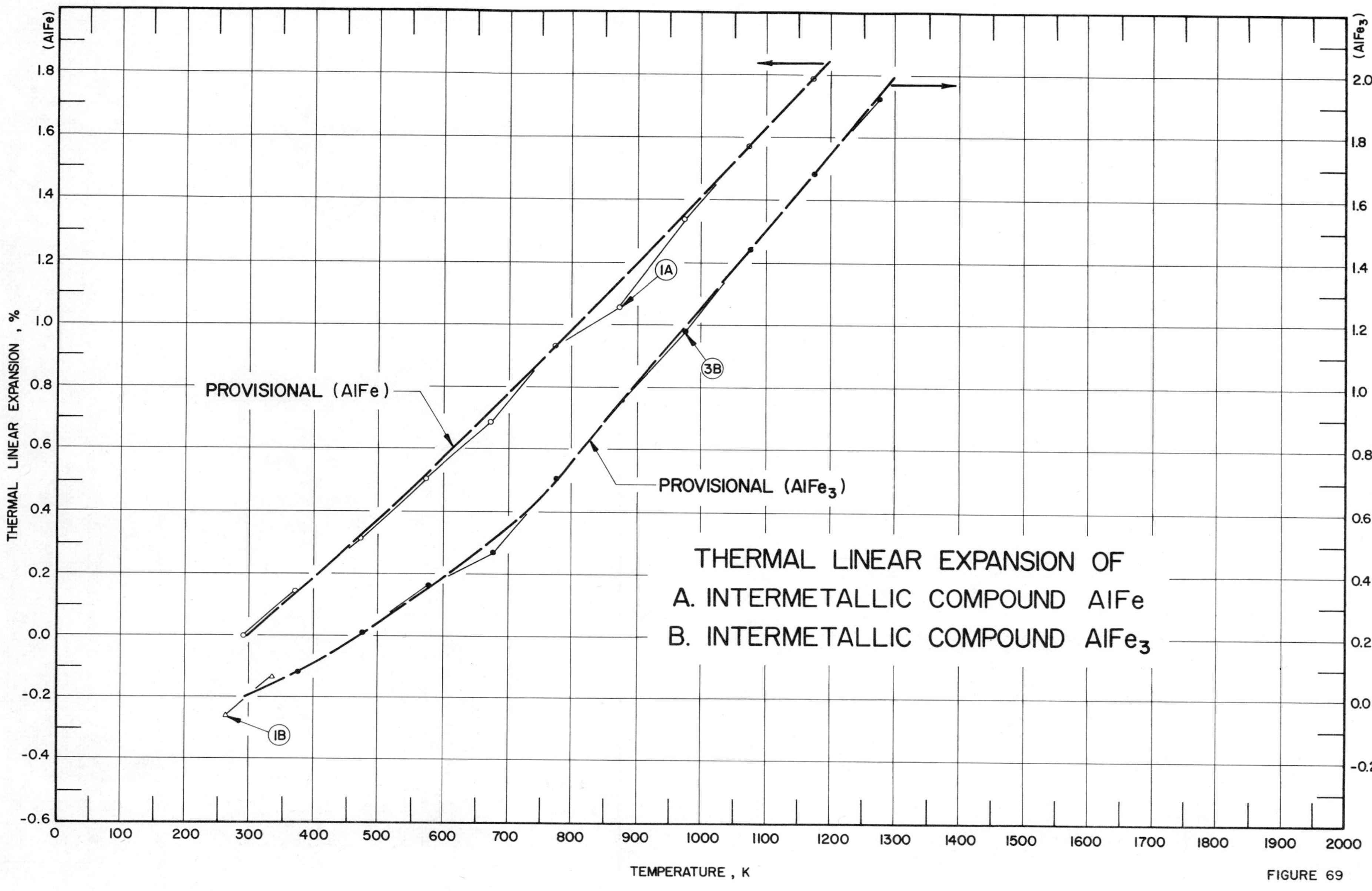

THERMAL LINEAR EXPANSION OF
A. INTERMETALLIC COMPOUND AlFe
B. INTERMETALLIC COMPOUND $AlFe_3$
PROVISIONAL (AlFe)
PROVISIONAL ($AlFe_3$)
1A
3B
1B
THERMAL LINEAR EXPANSION, %
(AlFe)
($AlFe_3$)
TEMPERATURE, K
0
100
200
300
400
500
600
700
800
900
1000
1100
1200
1300
1400
1500
1600
1700
1800
1900
2000
-0.6
-0.4
-0.2
0.0
0.2
0.4
0.6
0.8
1.0
1.2
1.4
1.6
1.8
2.0

FIGURE 69

SPECIFICATION TABLE 69A. THERMAL LINEAR EXPANSION OF AlFe INTERMETALLIC COMPOUND

Cur. No.	Ref. No.	Author(s)	Year	Method Used	Temp. Range, K	Name and Specimen Designation	Composition (weight percent), Specifications, and Remarks
1	693	Ryabov, V. R., Lozovskaya, A. V. and Vasil'yev, V. G.	1968	L	293-1173		Specimen prepared by melting exact amount of 99.99 Al and 99.97 Fe in vacuum induction furnace in magnetic crucible; annealed at 1273 K for 10 days and air cooled, analysis after that indicates 67.20 Fe; phase transformation starts at 740 K and end between 861 and 895 K, 1816 K.

DATA TABLE 69A. THERMAL LINEAR EXPANSION OF AlFe INTERMETALLIC COMPOUND

[Temperature, T, K; Linear Expansion, $\Delta L/L_0$, %]

T	$\Delta L/L_0$
CURVE 1	
293	0.000
373	0.141
473	0.317
573	0.505
673	0.686
773	0.931
873	1.058
973	1.340
1073	1.572
1173	1.795

SPECIFICATION TABLE 69B. THERMAL LINEAR EXPANSION OF $AlFe_3$ INTERMETALLIC COMPOUND

Cur. No.	Ref. No.	Author(s)	Year	Method Used	Temp. Range, K	Name and Specimen Designation	Composition (weight percent), Specifications, and Remarks
1	133	Ullrich, H.J.	1967	X	263-333		No details given; lattice parameter reported at 293 K is 2.8969 Å.
2	693	Ryabov, V.R., Lozovskaya, A.V., and Vasil'yev, V.G.	1968	L	293-1273		Specimen prepared by melting exact amount of 99.99 Al and 99.97 Fe in vacuum induction furnace annealed at 1273 K for 10 days and air cooled, analysis after that indicates 86.20 Fe in magnetic crucible; melting point 1723 K.

DATA TABLE 69B. THERMAL LINEAR EXPANSION OF $AlFe_3$ INTERMETALLIC COMPOUND

[Temperature, T, K; Linear Expansion, $\Delta L/L_0$, %]

T	$\Delta L/L_0$
CURVE 1	
263	-0.055
293	0.000
333	0.073
CURVE 2	
293	0.000
373	0.083
473	0.217
573	0.365
673	0.477
773	0.710
873	0.957
973	1.183
1073	1.441
1173	1.685
1273	1.926

FIGURE AND TABLE NO. 70R. PROVISIONAL VALUES FOR THERMAL LINEAR EXPANSION OF AlNi INTERMETALLIC COMPOUND

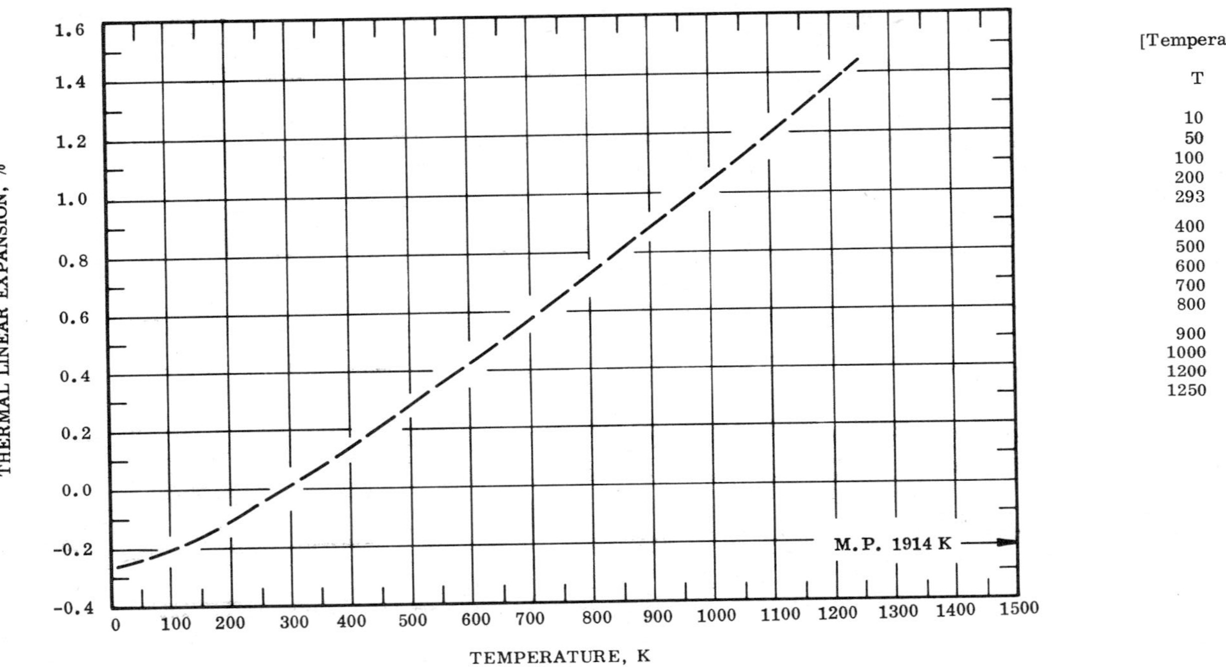

PROVISIONAL VALUES

[Temperature, T, K; Linear Expansion, $\Delta L/L_0$, %; α, K^{-1}]

T	$\Delta L/L_0$	$\alpha \times 10^6$
10	-0.258	
50	-0.251	8.7
100	-0.206	9.4
200	-0.106	10.8
293	0.000	11.9
400	0.144	13.0
500	0.279	13.9
600	0.421	14.6
700	0.570	15.2
800	0.724	15.6
900	0.881	15.8
1000	1.040	15.9
1200	1.357	16.0
1250	1.435	16.0

REMARKS

The tabulated values are considered accurate to within ±7% over the entire temperature range. These values can be represented approximately by the following equation:

$$\Delta L/L_0 = -0.294 + 7.928 \times 10^{-4}\, T + 7.893 \times 10^{-7}\, T^2 - 2.591 \times 10^{-10}\, T^3 \quad (50 < T < 1250)$$

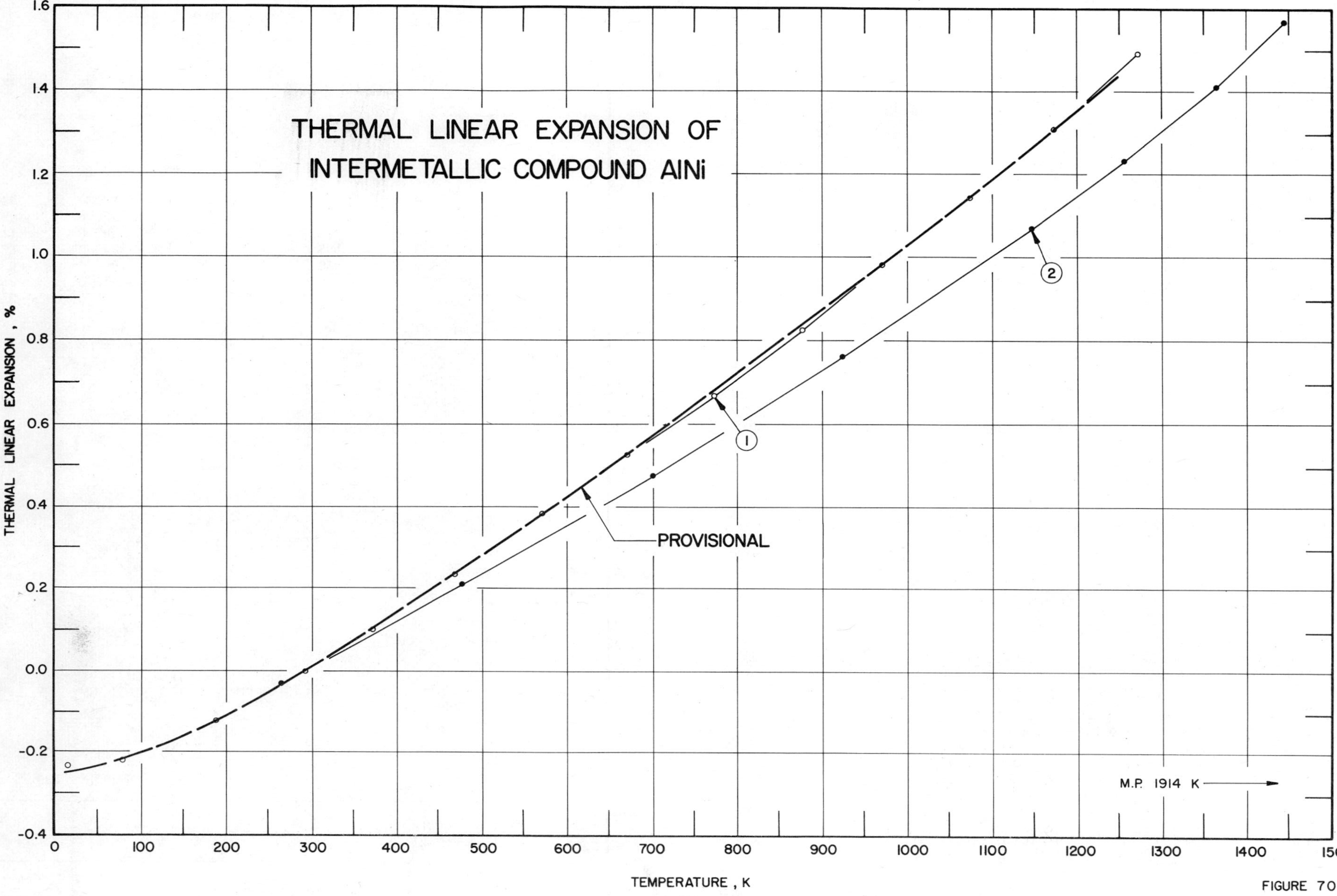
THERMAL LINEAR EXPANSION OF
INTERMETALLIC COMPOUND AlNi
THERMAL LINEAR EXPANSION , %
1.6
1.4
1.2
1.0
0.8
0.6
0.4
0.2
0.0
-0.2
-0.4
0
100
200
300
400
500
600
700
800
900
1000
1100
1200
1300
1400
1500
TEMPERATURE , K
PROVISIONAL
1
2
M.P. 1914 K

FIGURE 70

SPECIFICATION TABLE 70. THERMAL LINEAR EXPANSION OF AlNi INTERMETALLIC COMPOUND

Cur. No.	Ref. No.	Author(s)	Year	Method Used	Temp. Range, K	Name and Specimen Designation	Composition (weight percent), Specifications, and Remarks
1	582	Wasilewski, R.J.	1967	X, L	12-1274		Specimen with compositions close to equiatomic (50.4 at.% Al) prepared from high purity elements in both polycrystalline and single crystal form; average of three dilatometer and six x-ray experiments on the specimen between composition range 48.2 - 51.4 at.% Al; lattice parameter reported at 298 K is 2.887 Å.
2	583	Singleton, R. H., Wallace, A. V., and Miller, D. G.	1966	L	266-1447		99.99 pure Al added to molten, deoxidized high purity electrolytic Ni under pressure of 1 mm of Hg in Ar atmosphere; poured into copper mold, comminuted to particle size 325 mesh by crushing and dry milling; powder consolidated to 98% of theoretical density by hot-pressing under vacuum; zero-point correction of -0.029% determined by graphical interpolation.

DATA TABLE 70. THERMAL LINEAR EXPANSION OF AlNi INTERMETALLIC COMPOUND

[Temperature, T, K; Linear Expansion, $\Delta L/L_0$, %]

T	$\Delta L/L_0$	T	$\Delta L/L_0$	T	$\Delta L/L_0$
CURVE 1		CURVE 1 (cont.)		CURVE 2	
12	-0.235	680	0.529	266	-0.029
78	-0.219	773	0.670	478	0.210
189	-0.120	877	0.824	700	0.476
293	0.000	980	0.981	922	0.762
371	0.108	1074	1.143	1144	1.071
469	0.238	1171	1.318	1255	1.234
574	0.384	1274	1.499	1367	1.416
				1447	1.571

FIGURE AND TABLE NO. 71R. PROVISIONAL VALUES FOR THERMAL LINEAR EXPANSION OF $AlNi_3$ INTERMETALLIC COMPOUND

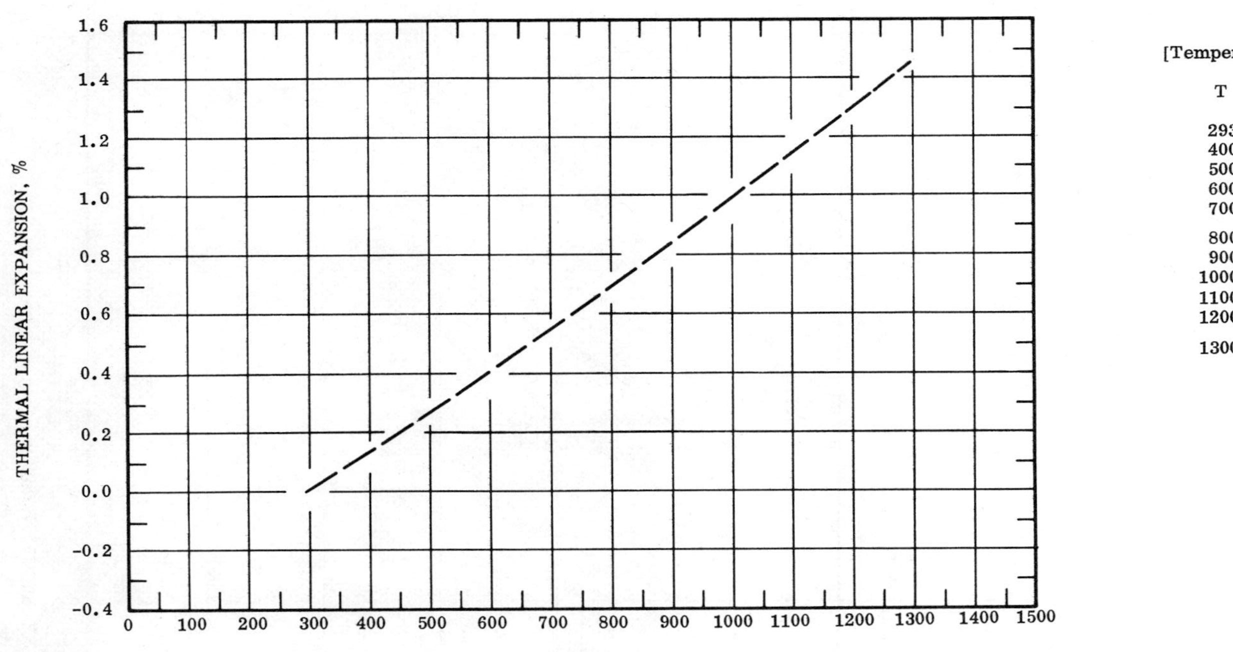

PROVISIONAL VALUES

[Temperature, T, K; Linear Expansion, $\Delta L/L_0$, %; α, K^{-1}]

T	$\Delta L/L_0$	$\alpha \times 10^6$
293	0.000	12.3
400	0.135	12.9
500	0.267	13.4
600	0.403	13.9
700	0.544	14.3
800	0.689	14.6
900	0.837	14.9
1000	0.989	15.2
1100	1.140	15.4
1200	1.294	15.5
1300	1.450	15.6

REMARKS

The tabulated values are considered accurate to within ±10% over the entire temperature range. These values can be represented approximately by the following equation:

$$\Delta L/L_0 = -0.333 + 1.034 \times 10^{-3}\ T + 3.770 \times 10^{-7}\ T^2 - 9.007 \times 10^{-11}\ T^3$$

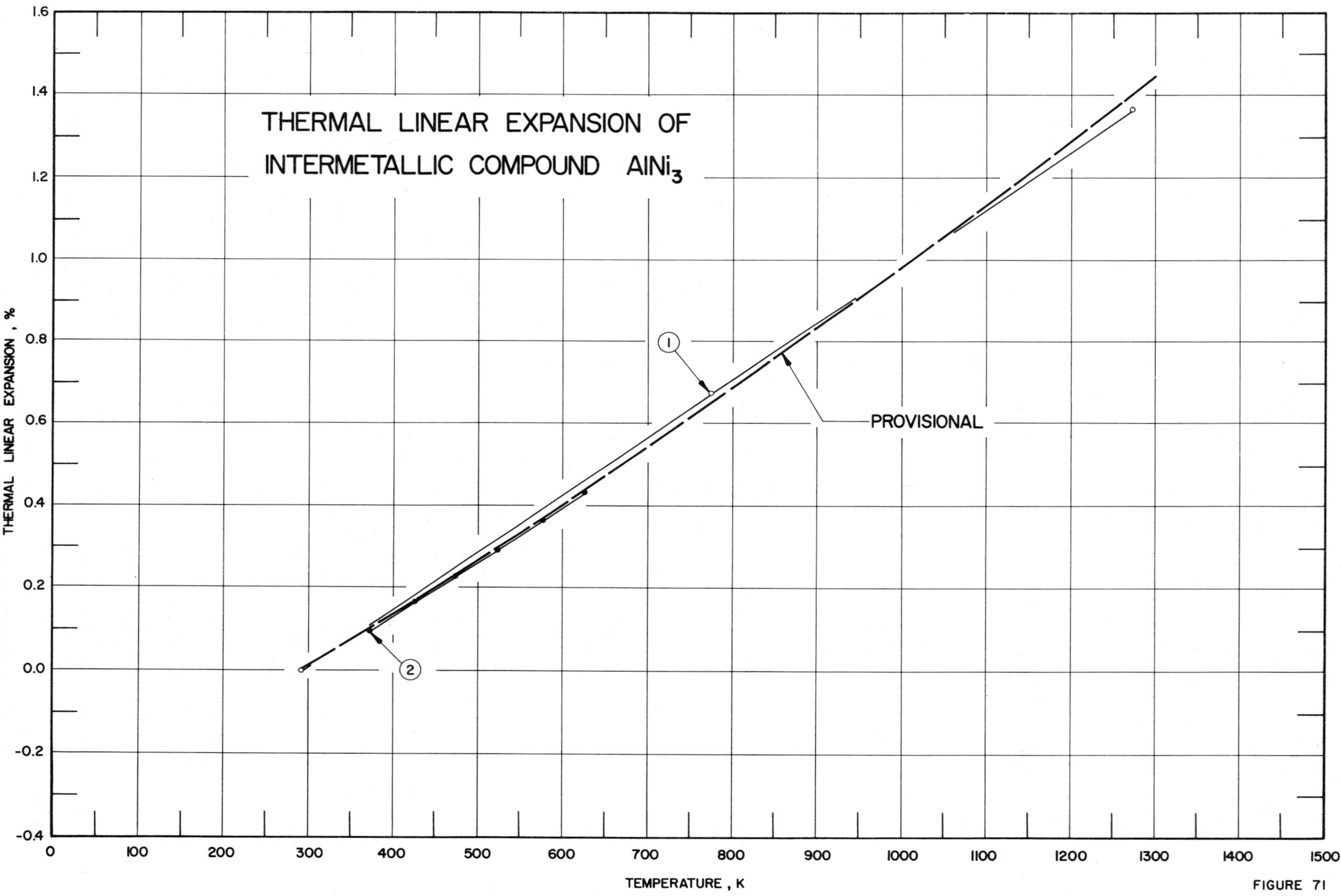
THERMAL LINEAR EXPANSION OF
INTERMETALLIC COMPOUND AlNi3
1
PROVISIONAL
2
THERMAL LINEAR EXPANSION, %
1.6
1.4
1.2
1.0
0.8
0.6
0.4
0.2
0.0
-0.2
-0.4
0
100
200
300
400
500
600
700
800
900
1000
1100
1200
1300
1400
1500
TEMPERATURE, K

FIGURE 71

SPECIFICATION TABLE 71. THERMAL LINEAR EXPANSION OF $AlNi_3$ INTERMETALLIC COMPOUND

Cur. No.	Ref. No.	Author(s)	Year	Method Used	Temp. Range, K	Name and Specimen Designation	Composition (weight percent), Specifications, and Remarks
1	673	Taylor, A. and Floyd, R.W.	1952	L	74-1273		Specimen of 5 kg heated in Leitz dilatometer to 1270 K at rate of 3 K per min, subsequently cooled slowly to room temperature overnight, both on heating and cooling, the final and initial lengths are the same, and then cooled in liquid nitrogen to 77 K; no anomalous expansion or contraction observed.
2	301	Arbazov, M. P. and Zelenkov, I. A.	1964	L	293-974		Specimen prepared from pure metals, 86.7 Ni and 13.3 Al.

DATA TABLE 71. THERMAL LINEAR EXPANSION OF $AlNi_3$ INTERMETALLIC COMPOUND

[Temperature, T, K; Linear Expansion, $\Delta L/L_0$, %]

T	$\Delta L/L_0$	T	$\Delta L/L_0$	T	$\Delta L/L_0$
CURVE 1		CURVE 2‡		CURVE 2 (cont.)‡	
74	-0.301	293	0.000*	675	0.510
293	0.000	372	0.096	724	0.585
773	0.672	427	0.166	771	0.659
1273	1.564	477	0.231	793	0.693
		523	0.293	825	0.743
		576	0.367	875	0.822
		625	0.437	974	0.986

DATA TABLE 71. COEFFICIENT OF THERMAL LINEAR EXPANSION OF $AlNi_3$ INTERMETALLIC COMPOUND

[Temperature, T, K; Coefficient of Expansion, α, 10^{-6} K^{-1}]

T	α	T	α
CURVE 2*, ‡		CURVE 2 (cont.)*, ‡	
293	11.80	675	14.96
372	12.53	724	15.47
427	12.89	771	15.91
477	13.19	793	15.46
523	13.64	825	15.69
576	14.19	875	16.01
625	14.44	974	16.99

* Not shown in figure.
‡ Author's data for coefficient of thermal expansion have been integrated by TPRC to obtain $\Delta L/L_0$.

FIGURE AND TABLE NO. 72R. PROVISIONAL VALUES FOR THERMAL LINEAR EXPANSION OF $AlAg_2$ INTERMETALLIC COMPOUND

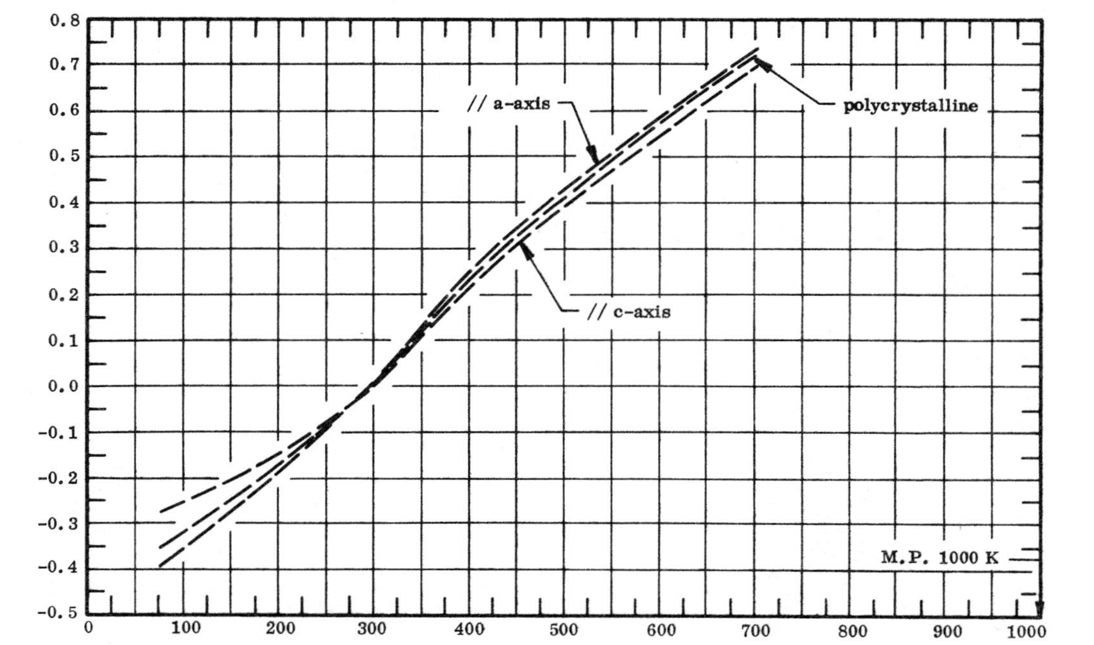

PROVISIONAL VALUES

[Temperature, T, K; Linear Expansion, $\Delta L/L_0$, %; α, K^{-1}]

	// a-axis	// c-axis	polycrystalline	
T	$\Delta L/L_0$	$\Delta L/L_0$	$\Delta L/L_0$	$\alpha \times 10^6$
75	-0.392	-0.275	-0.353	11
100	-0.352	-0.252	-0.319	12
200	-0.181	-0.148	-0.170	14
293	0.000	0.000	0.000	23
350	0.144	0.125	0.138	23
400	0.252	0.221	0.242	18
500	0.432	0.394	0.419	17
600	0.589	0.551	0.576	16
700	0.738	0.700	0.725	15

REMARKS

The tabulated values are considered accurate to within ±7%. The values for polycrystalline material in the absence of experimental data are calculated from the single crystal data.

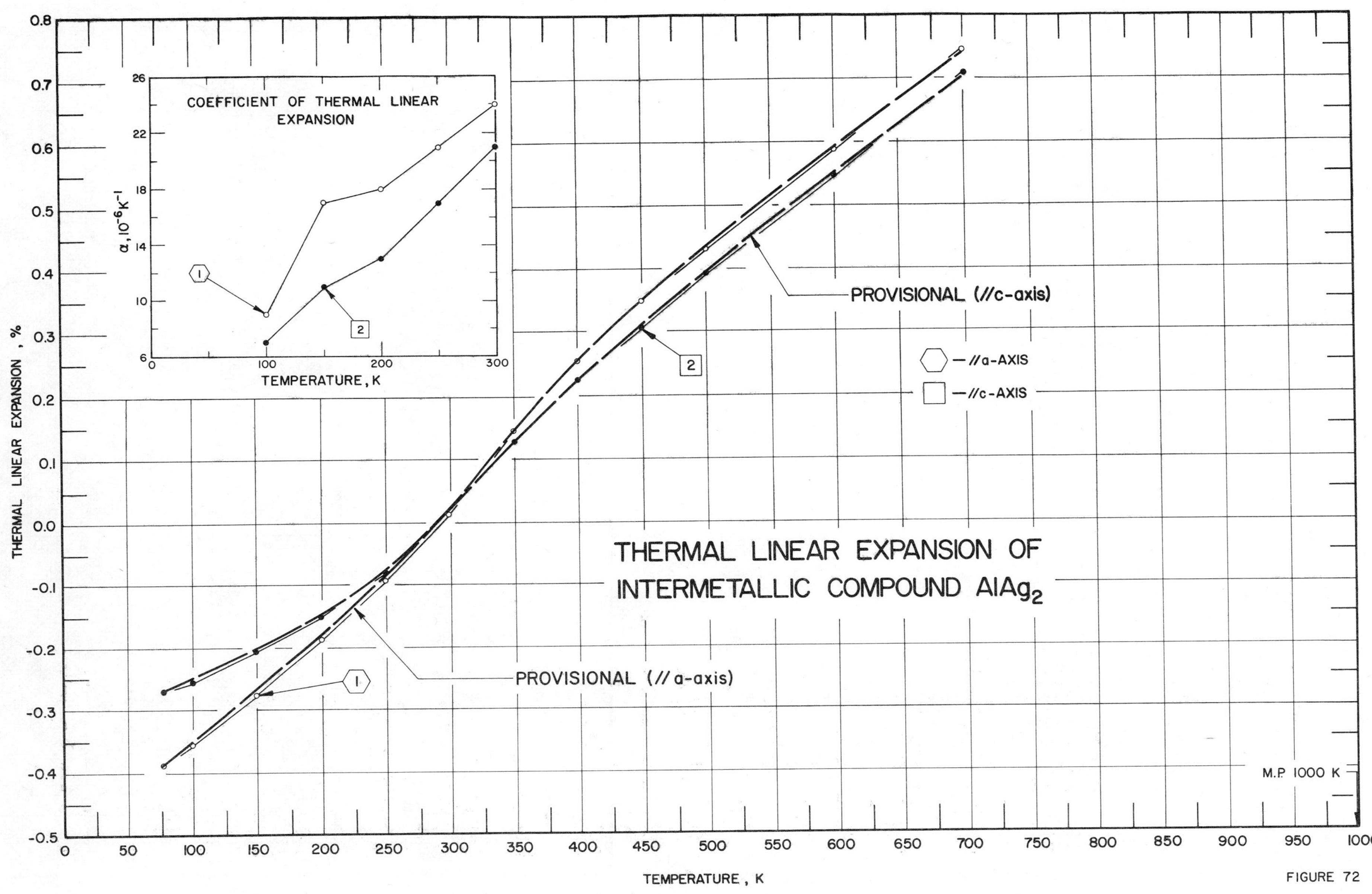
COEFFICIENT OF THERMAL LINEAR EXPANSION
$\alpha, 10^{-6}K^{-1}$
26
22
18
14
10
6
0
100
200
300
TEMPERATURE, K
1
2
PROVISIONAL (//c-axis)
— //a-AXIS
— //c-AXIS
THERMAL LINEAR EXPANSION OF
INTERMETALLIC COMPOUND AlAg2
PROVISIONAL (// a-axis)
M.P. 1000 K
THERMAL LINEAR EXPANSION, %
0.8
0.7
0.6
0.5
0.4
0.3
0.2
0.1
0.0
-0.1
-0.2
-0.3
-0.4
-0.5
0
50
100
150
200
250
300
350
400
450
500
550
600
650
700
750
800
850
900
950
1000
TEMPERATURE, K

FIGURE 72

SPECIFICATION TABLE 72. THERMAL LINEAR EXPANSION OF $AlAg_2$ INTERMETALLIC COMPOUND

Cur. No.	Ref. No.	Author(s)	Year	Method Used	Temp. Range, K	Name and Specimen Designation	Composition (weight percent), Specifications, and Remarks
1	586	Neumann, J. P. and Chang, Y. A.	1968	X	77-700		After homogenizing 1 day at 825 K, filings with size less than 40 μ prepared; samples subjected to different anneal-quenching and expansion observed along a-axis using CuKα x-radiation variation in heat-treatment made no observable difference; data smoothed by author; lattice parameter reported at 298 K is 2.8779 Å; 2.8775 Å at 293 K determined by graphical interpolation.
2	586	Neumann, J. P. and Chang, Y. A.	1968	X	77-700		The above specimen; expansion along c-axis; lattice parameter reported at 298 K is 4.6225 Å; 4.6219 Å at 293 K determined by graphical interpolation.

DATA TABLE 72. THERMAL LINEAR EXPANSION OF $AlAg_2$ INTERMETALLIC COMPOUND

[Temperature, T, K; Linear Expansion, $\Delta L/L_0$, %]

T	$\Delta L/L_0$	T	$\Delta L/L_0$	T	$\Delta L/L_0$	T	$\Delta L/L_0$
CURVE 1		CURVE 1 (cont.)		CURVE 2		CURVE 2 (cont.)	
77	-0.389	350	0.145	77	-0.270	350	0.127
100	-0.357	400	0.253	100	-0.257	400	0.225
150	-0.278	450	0.350	150	-0.209	450	0.309
200	-0.187	500	0.434	200	-0.151	500	0.391
250	-0.093	600	0.590	250	-0.080	600	0.547
298	0.013	700	0.743	298	0.012*	700	0.707

DATA TABLE 72. COEFFICIENT OF THERMAL LINEAR EXPANSION OF $AlAg_2$ INTERMETALLIC COMPOUND

[Temperature, T, K; Coefficient of Expansion, α, $10^{-6}\ K^{-1}$]

T	α	T	α	T	α	T	α
CURVE 1		CURVE 1 (cont.)		CURVE 2		CURVE 2 (cont.)	
100	9	400	21	100	7	400	18
150	17	450	18	150	11	450	17
200	18	500	17	200	13	500	16
250	21	600	15	250	17	600	16
298	24	700	15	298	21	700	16
350	24			350	21		

* Not shown in figure.

FIGURE AND TABLE NO. 73R. PROVISIONAL VALUES FOR THERMAL LINEAR EXPANSION OF Al_3U INTERMETALLIC COMPOUND

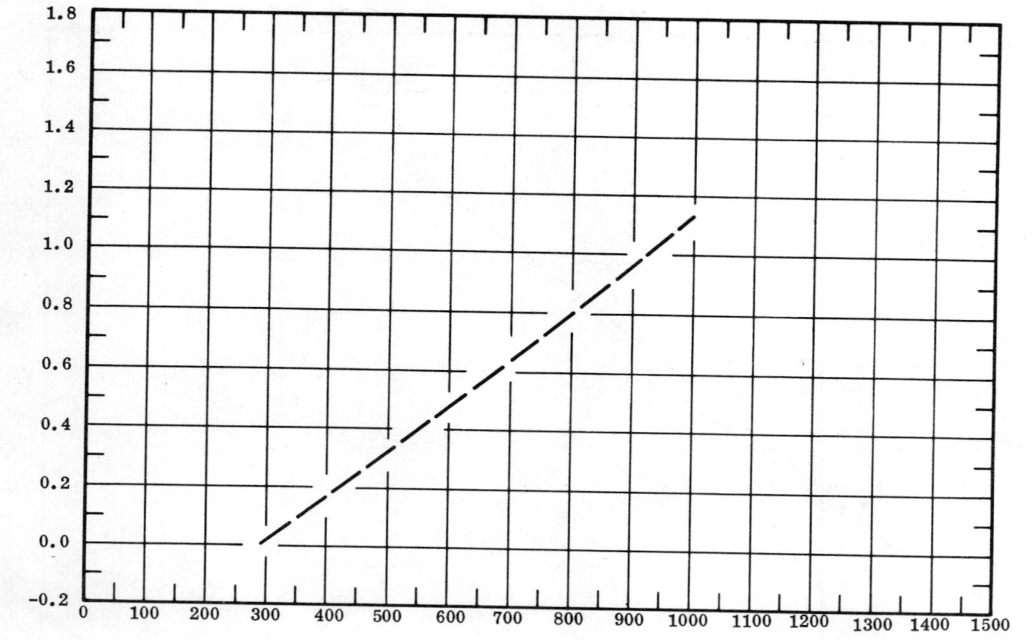

PROVISIONAL VALUES

[Temperature, T, K; Linear Expansion, $\Delta L/L_0$, %; α, K^{-1}]

T	$\Delta L/L_0$	$\alpha \times 10^6$
293	0.000	15.0
400	0.162	15.3
500	0.316	15.5
600	0.472	15.8
700	0.630	16.1
800	0.792	16.3
900	0.957	16.5
1000	1.123	16.9

REMARKS

The tabulated values are considered accurate to within ±10% over the entire temperature range. These values can be represented approximately by the following equation:

$$\Delta L/L_0 = -0.426 + 1.441 \times 10^{-3}\, T + 8.825 \times 10^{-8}\, T^2 + 2.413 \times 10^{-11}\, T^3$$

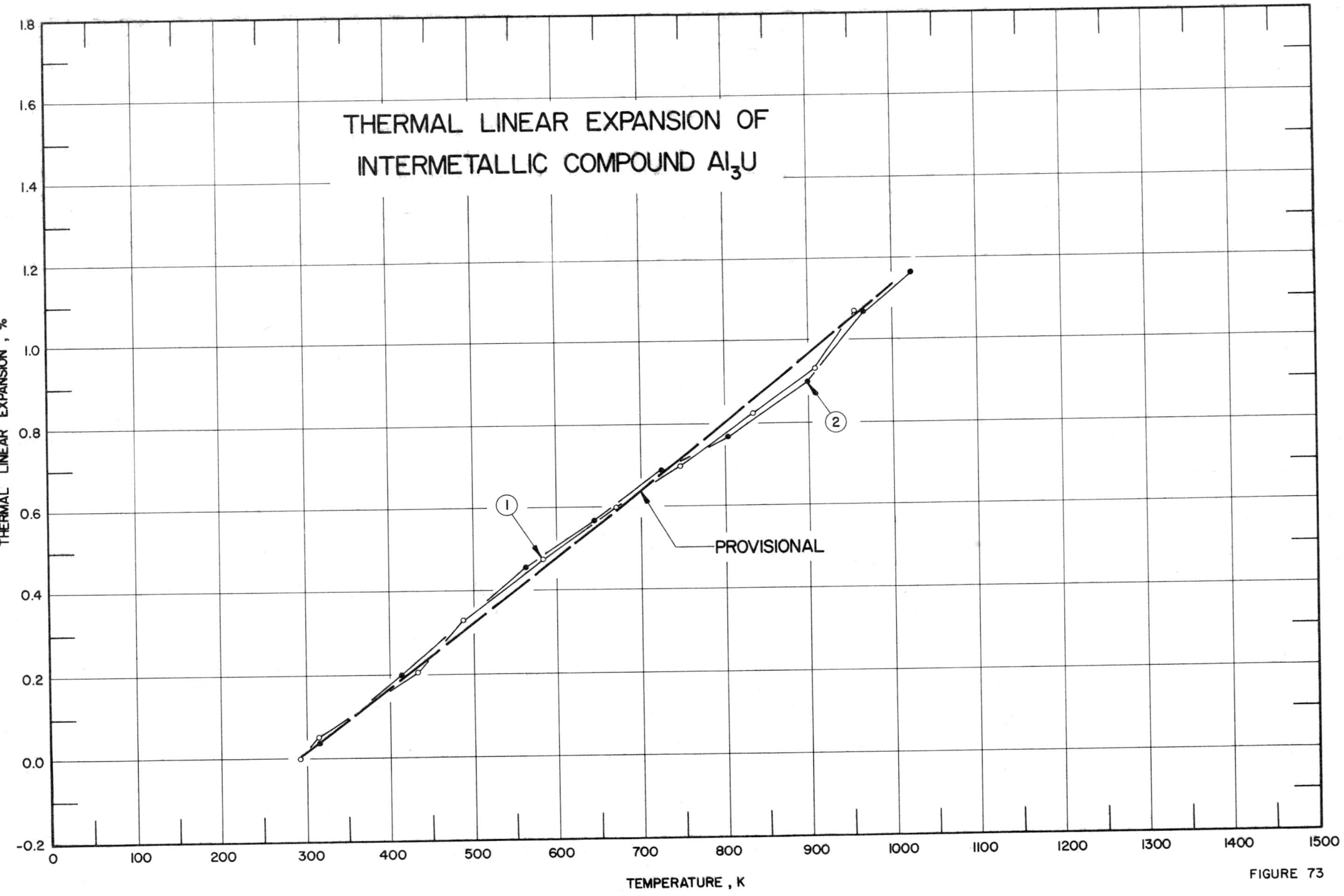
THERMAL LINEAR EXPANSION OF
INTERMETALLIC COMPOUND Al3U
1
2
PROVISIONAL
THERMAL LINEAR EXPANSION, %
-0.2 0.0 0.2 0.4 0.6 0.8 1.0 1.2 1.4 1.6 1.8
TEMPERATURE, K
0 100 200 300 400 500 600 700 800 900 1000 1100 1200 1300 1400 1500

FIGURE 73

SPECIFICATION TABLE 73. THERMAL LINEAR EXPANSION OF Al_3U INTERMETALLIC COMPOUND

Cur. No.	Ref. No.	Author(s)	Year	Method Used	Temp. Range, K	Name and Specimen Designation	Composition (weight percent), Specifications, and Remarks
2	694	Pearce, R.J.	1965	X	293-956		High purity powder specimen prepared by vapor phase reaction between reactor grade U and AlCl (prepared in situ by heating Al powder and KCl under vacuum at 873 K); sample sealed in 0.3 mm bore quartz capillary tube under high vacuum; lattice parameter reported at 293 K is 4.2651 Å before heating; expansion measured on first series.
2	694	Pearce, R.J.	1965	X	293-1021		The above specimen; expansion measured on second series.

DATA TABLE 73. THERMAL LINEAR EXPANSION OF Al_3U INTERMETALLIC COMPOUND

[Temperature, T, K; Linear Expansion, $\Delta L/L_0$, %]

T	$\Delta L/L_0$	T	$\Delta L/L_0$
CURVE 1		CURVE 2 (cont.)	
293	0.000	561	0.457
296	0.012*	641	0.567
315	0.052	726	0.687
432	0.213	804	0.767
488	0.328	899	0.903
582	0.476	967	1.071
670	0.596	1021	1.163
747	0.699		
833	0.821		
907	0.933		
956	1.074		
CURVE 2			
293	0.000		
315	0.040		
412	0.199		

* Not shown in figure.

FIGURE AND TABLE NO. 74R. PROVISIONAL VALUES FOR THERMAL LINEAR EXPANSION OF SbGa INTERMETALLIC COMPOUND

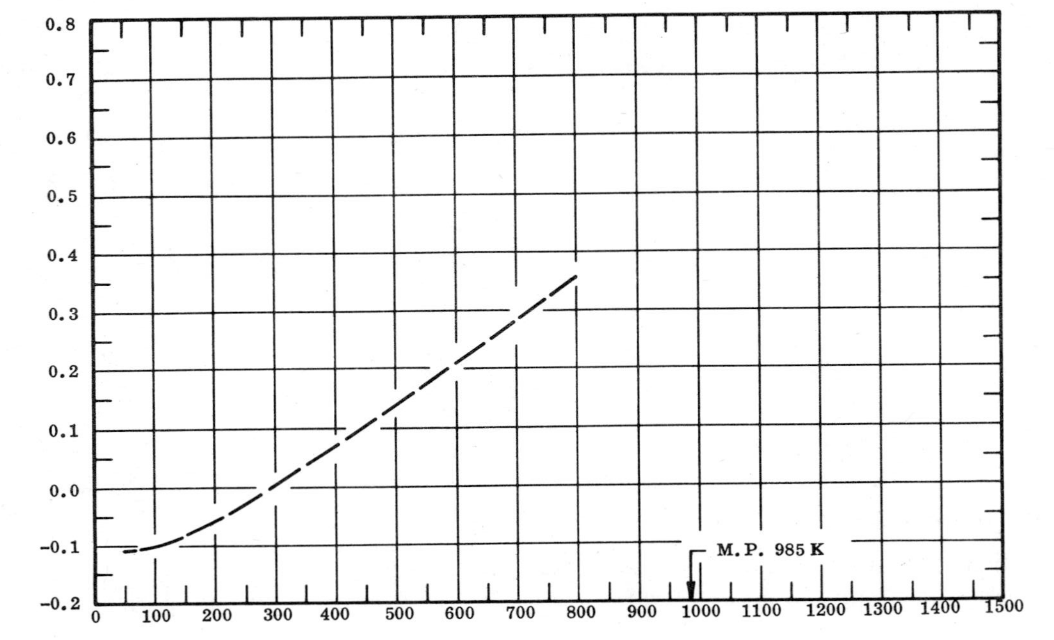

PROVISIONAL VALUES

[Temperature, T, K; Linear Expansion, $\Delta L/L_0$, %; α, K^{-1}]

T	$\Delta L/L_0$	$\alpha \times 10^6$
50	-0.107	
100	-0.100	3.0
200	-0.054	5.3
293	0.000	6.1
400	0.068	6.7
500	0.137	7.1
600	0.209	7.3
700	0.281	7.3
800	0.357	7.3

REMARKS

The tabulated values are considered accurate to within ± 7% over the entire temperature range. These values can be represented approximately by the following equation:

$$\Delta L/L_0 = -0.138 + 3.051 \times 10^{-4} T + 6.602 \times 10^{-7} T^2 - 3.380 \times 10^{-10} T^3 \qquad (100 < T < 800)$$

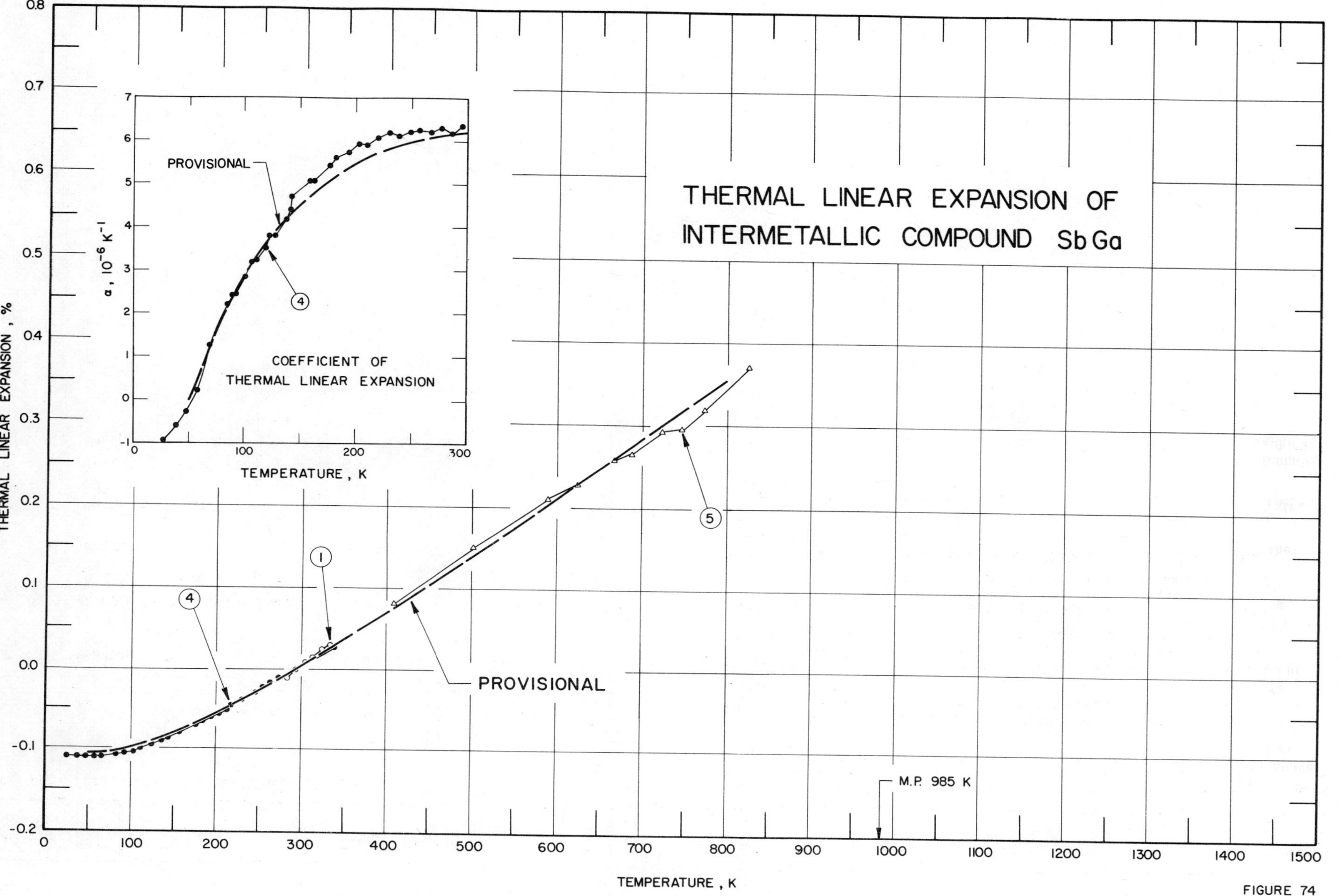

THERMAL LINEAR EXPANSION OF
INTERMETALLIC COMPOUND SbGa
THERMAL LINEAR EXPANSION, %
TEMPERATURE, K
M.P. 985 K
PROVISIONAL
1
4
5
COEFFICIENT OF
THERMAL LINEAR EXPANSION
α, 10^{-6} K^{-1}
TEMPERATURE, K
PROVISIONAL
4

FIGURE 74

SPECIFICATION TABLE 74. THERMAL LINEAR EXPANSION OF SbGa INTERMETALLIC COMPOUND

Cur. No.	Ref. No.	Author(s)	Year	Method Used	Temp. Range, K	Name and Specimen Designation	Composition (weight percent), Specifications, and Remarks
1	561	Straumanis, M. E. and Kim, C. D.	1965	X	283-333		99.999 pure elements mixed in sealed, evacuated quartz tube; heated 15 hr above melting point of each element, cooled slowly; compounds ground to powder; data is average of several runs; lattice parameter reported at 293 K is 6.09576 Å.
2*	170	Sparks, P. W. and Swenson, C. A.	1967	V	2-32		Specimen received from Texas Instruments; tested as received; this curve is here reported using the first given temperature as reference temperature at which $\Delta L/L_0 = 0$.
3*	171	Sparks, P. W.	1966	V	2-31		Specimen from Texas Instruments; diameter 0.6 cm, length 10.06 cm at 4 K; sample ends polished normal to its axis; this curve is here reported using the first given temperature as reference temperature at which $\Delta L/L_0 = 0$.
4	559	Novikova, S. I. and Abrikosov, N. Kh.	1964	L	27-337		Specimen prepared from 99.99 Ga; 99.999 Sb pure components, coarsely crystalline and irregular form; longest dimension 15.32 mm, values of the coefficients of thermal linear expansion become negative at 52 K.
5	756	Woolley, J. C.	1965	X	298-824		No details given; zero-point correction of 0.004% determined by graphical extrapolation.
6*	770	Bernstein, L., and Beals, R. J.	1961	I	310-700		Single crystal containing impurities of 10^{17} cm^{-3} grown by the Czochalski technique in a (111) direction; zero-point correction of 0.011% determined by graphical extrapolation.

*Not shown in figure.

DATA TABLE 74. THERMAL LINEAR EXPANSION OF SbGa INTERMETALLIC COMPOUND

[Temperature, T, K; Linear Expansion, $\Delta L/L_0$, %]

CURVE 1

T	$\Delta L/L_0$
283	-0.010
293	0.000
303	0.007
313	0.014
323	0.023
333	0.029

CURVE 2*, †, ‡

T	$\Delta L/L_0$
2	0.000 x 10^{-4}
6	0.004
8	0.005
10	-0.022
12	-0.12
14	-0.31
16	-0.62
18	-1.0
20	-1.6
22	-2.2
24	-2.9
26	-3.5
28	-4.2
30	-4.9
32	-5.5

CURVE 3*, †

T	$\Delta L/L_0$
1.89	0.00000 x 10^{-4}
2.10	0.00002
2.30	0.00006
2.51	0.00008
3.68	0.00054
4.58	0.0012
4.76	0.0015
5.44	0.0023
5.60	0.0026
6.25	0.0041
7.19	0.0063
7.97	0.0066
8.72	0.0032
9.02	0.00031
9.17	-0.0017
9.46	-0.0066
9.82	-0.022
10.19	-0.032
10.93	-0.068

CURVE 3 (cont.)*, †

T	$\Delta L/L_0$
11.29	-0.085
12.02	-0.15
12.38	-0.18
13.07	-0.28
13.43	-0.31
14.17	-0.44
14.53	-0.49
15.30	-0.67
15.66	-0.73
16.39	-0.94
16.76	-1.0
17.52	-1.3
17.89	-1.4
18.51	-1.64
18.88	-1.63
19.60	-2.0
19.96	-2.2
20.69	-2.5
21.05	-2.6
21.78	-3.0
22.14	-3.1
22.83	-3.5
23.18	-3.7
22.89	-4.1
24.24	-4.2
24.90	-4.7
25.24	-4.8
25.91	-5.3
26.24	-5.4
26.88	-5.9
27.20	-6.0
27.84	-6.5
28.16	-6.6
28.78	-7.1
29.08	-7.2
29.67	-7.7
29.97	-7.8
30.55	-8.2
30.84	-8.3

CURVE 4‡

T	$\Delta L/L_0$
27	-0.110
39	-0.111
48	-0.111
58	-0.111

CURVE 4 (cont.)‡

T	$\Delta L/L_0$
68	-0.110
84	-0.107
88	-0.106*
92	-0.105
103	-0.103
107	-0.101*
111	-0.100
119	-0.097*
123	-0.096
128	-0.094*
138	-0.090
142	-0.088
143	-0.088*
159	-0.080
162	-0.078*
177	-0.070
183	-0.067
194	-0.061
203	-0.056
211	-0.051
220	-0.045
231	-0.039
239	-0.034*
249	-0.028
257	-0.023
268	-0.016
277	-0.010
286	-0.005*
295	0.001*
303	0.006*
313	0.013*
329	0.023*
337	0.028

CURVE 5

T	$\Delta L/L_0$
298	0.004*
410	0.081
501	0.151
590	0.211
625	0.229
667	0.258
688	0.266
722	0.292
743	0.297
773	0.320
824	0.375

CURVE 6*

T	$\Delta L/L_0$
310	0.01
370	0.06
470	0.12
570	0.19
600	0.21
610	0.22
620	0.23
630	0.23
640	0.25
650	0.26
660	0.28
670	0.32
690	0.40
700	0.45

* Not shown in figure.

† This curve is here reported using the first given temperature as reference temperature at which $\Delta L/L_0 = 0$.

‡ Author's data for coefficient of thermal expansion have been integrated by TPRC to obtain $\Delta L/L_0$.

DATA TABLE 74. COEFFICIENT OF THERMAL LINEAR EXPANSION OF SbGa INTERMETALLIC COMPOUND

[Temperature, T, K; Coefficient of Expansion, α, 10^{-6} K^{-1}]

T	α	T	α
CURVE 2*, ‡		CURVE 4 (cont.) ‡	
2	0.000084	107	3.23
4	0.00067	111	3.28*
6	0.0022	119	3.54
8	-0.001	123	3.83*
10	-0.026	128	3.83
12	-0.070	138	4.20
14	-0.125	142	4.42*
16	-0.183	143	4.72*
18	-0.242	159	5.08
20	-0.290	162	5.08*
22	-0.323	177	5.42*
24	-0.338	183	5.62
26	-0.342	194	5.74
28	-0.335	203	5.95
30	-0.315	211	5.91
32	-0.285	220	6.09
		231	6.21
CURVE 3		239	6.13
		249	6.21
2	0.000084	257	6.26
4	0.00067	268	6.21
6	0.00226	277	6.32
8	-0.0018	286	6.18
10	-0.024	295	6.36
12	-0.071	313	6.29
14	-0.129	329	6.55
16	-0.188	337	6.55
18	-0.23		
20	-0.28		
22	-0.34		
24	-0.38		
26	-0.42		
28	-0.43		
30	-0.41		
CURVE 4‡			
27	-0.96		
39	-0.60		
48	-0.28		
58	0.23		
68	1.29		
84	2.24		
88	2.45		
92	2.49		
103	2.88		

* Not shown in figure.
‡ Author's data for coefficient of thermal expansion have been integrated by TPRC to obtain $\Delta L/L_0$.

FIGURE AND TABLE NO. 75R. PROVISIONAL VALUES FOR THERMAL LINEAR EXPANSION OF SbIn INTERMETALLIC COMPOUND

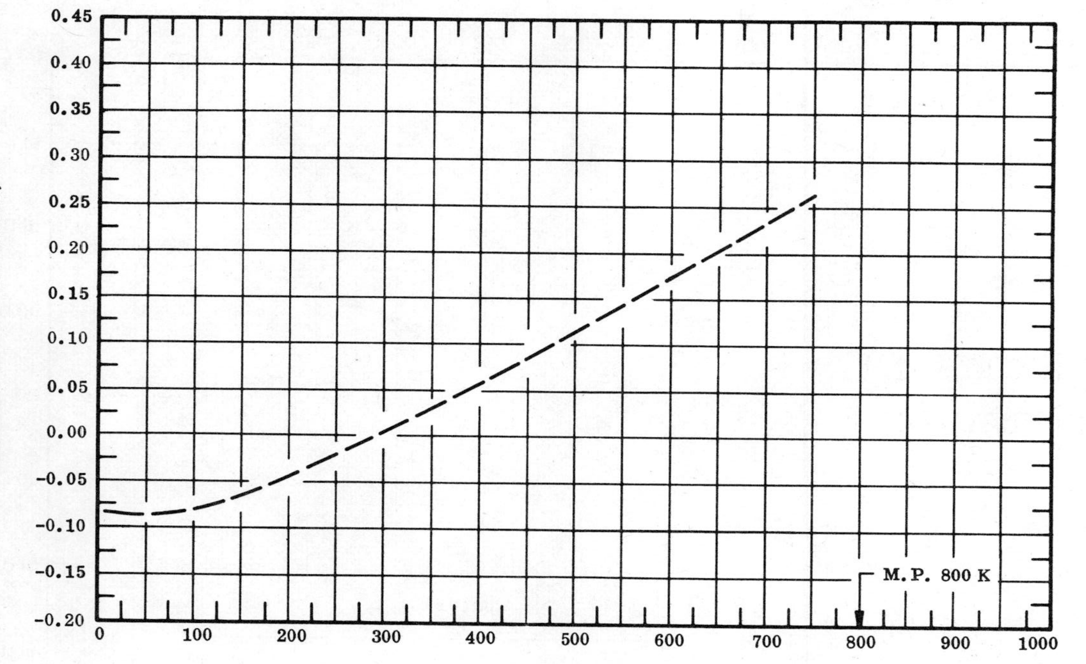

PROVISIONAL VALUES

[Temperature, T, K; Linear Expansion, $\Delta L/L_0$, %; α, K^{-1}]

T	$\Delta L/L_0$	$\alpha \times 10^6$
5	-0.088	-0.07
25	-0.090	-1.3
50	-0.092	-0.4
100	-0.086	2.8
200	-0.043	4.1
293	0.000	5.0
400	0.058	5.7
500	0.116	6.1
600	0.178	6.1
700	0.238	6.1
750	0.267	6.2

REMARKS

The tabulated values are considered accurate to within ±7% over the entire temperature range. These values can be represented approximately by the following equation:

$$\Delta L/L_0 = -0.099 + 1.249 \times 10^{-4}\,T + 8.773 \times 10^{-7}\,T^2 - 5.260 \times 10^{-10}\,T^3 \qquad (50 < T < 750)$$

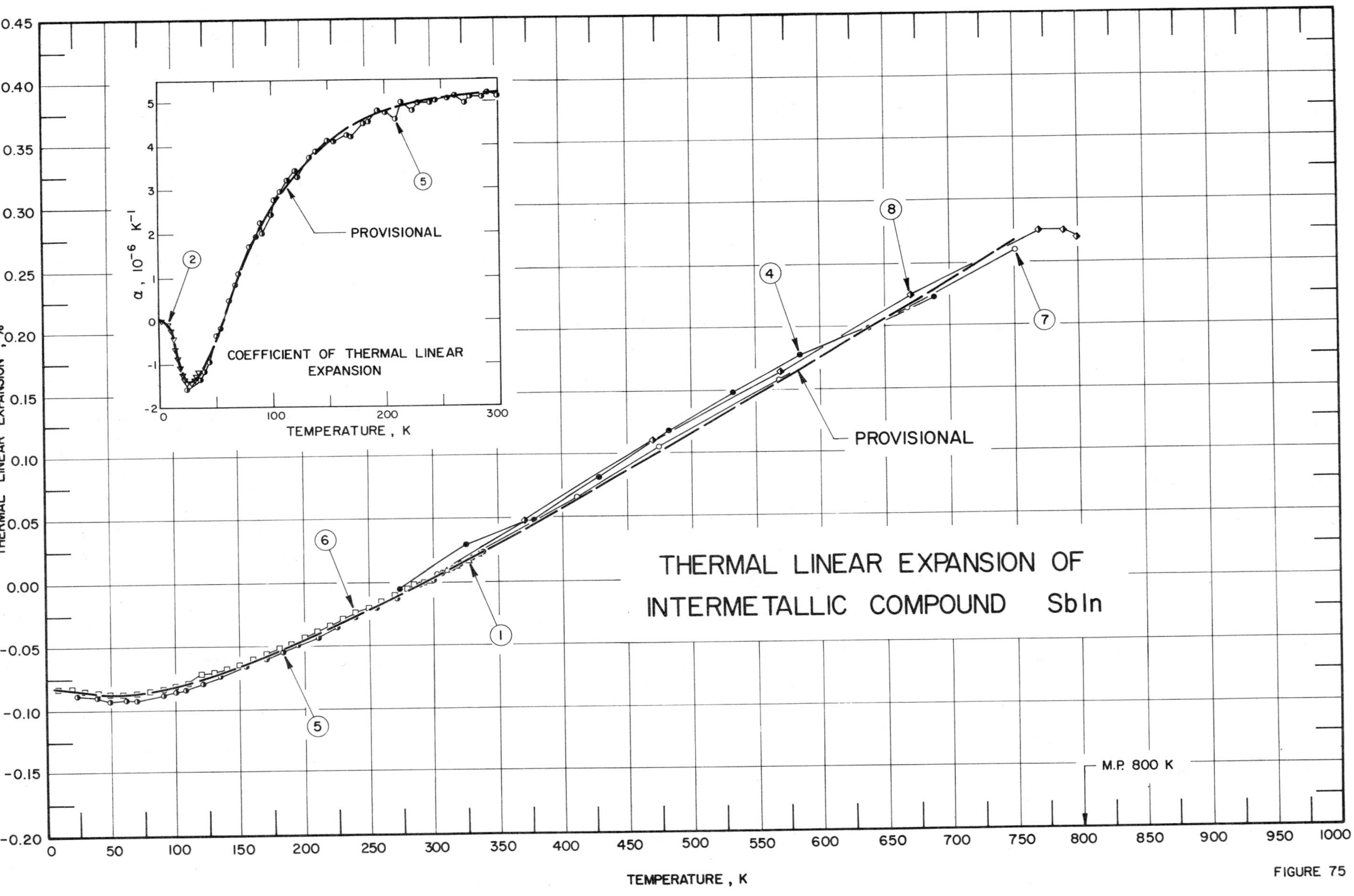
THERMAL LINEAR EXPANSION OF
INTERMETALLIC COMPOUND SbIn
PROVISIONAL
M.P. 800 K
THERMAL LINEAR EXPANSION, %
TEMPERATURE, K
0.45
0.40
0.35
0.30
0.25
0.20
0.15
0.10
0.05
0.00
-0.05
-0.10
-0.15
-0.20
0
50
100
150
200
250
300
350
400
450
500
550
600
650
700
750
800
850
900
950
1000
1
4
5
6
7
8
COEFFICIENT OF THERMAL LINEAR
EXPANSION
PROVISIONAL
α, 10⁻⁶ K⁻¹
TEMPERATURE, K
2
5
-2
-1
0
1
2
3
4
5
0
100
200
300

FIGURE 75

SPECIFICATION TABLE 75. THERMAL LINEAR EXPANSION OF SbIn INTERMETALLIC COMPOUND

Cur. No.	Ref. No.	Author(s)	Year	Method Used	Temp. Range, K	Name and Specimen Designation	Composition (weight percent), Specifications, and Remarks
1	561	Straumanis, M.E. and Kim, C.D.	1965	X	290-335		99.999 pure elements mixed in sealed, evacuated quartz tubes; heated 15 hr above melting point of each element and cooled slowly; compound ground to powder; data is average of several runs; lattice parameter reported at 290 K is 6.47912 Å; 6.47922 Å at 293 K determined by graphical interpolation.
2*	170	Sparks, P.W. and Swenson, C.A.	1967	V	2-34		Specimen from Ohio Semiconductors; tested as received; this curve is here reported using the first given temperature as reference temperature at which $\Delta L/L_0 = 0$.
3*	171	Sparks, P.W.	1966	V	1.89-31.38		Specimen from Ohio Semiconductors; diameter 1.0 cm, length 10.12 cm at 4 K; sample ends polished normal to its axis; this curve is here reported using the first given temperature as reference temperature at which $\Delta L/L_0 = 0$.
4	273	Mauer, F.A. and Bolz, L.H.	1955	X	273-689		Prepared by D. E. Roberts at N.B.S.; 10^{17} to 10^{18} impurities/cm^{-3}; zero-point correction of -0.005% determined by graphical interpolation.
5	323	Novikova, S.I.	1961	L	24-337		Specimen obtained from Semiconductor Laboratory of the Metallurgical Inst. of the Academy of Sciences of the USSR, and prepared from spectrographically pure In and Sb; data are results of 4 series of measurements on cylinder 11.48 mm long.
6	9	Gibbons, D.F.	1958	I	10-300		Two crystals obtained by crystal pulling technique; electrically active impurity concentration 10^{14} cm^{-3} and 10^{16} cm^{-3}; expansion measured along <100> axis; data results from both specimens; zero-point correction of -0.0098% determined by graphical interpolation.
7	235	Shaw, N. and Liu, Y.H.	1965	X	303-751		Prepared from component metals of spectroscopical purity in stoichiometric proportions; further purification was made by repeated zone-refining; the powder taken from the alloy ingots were annealed at 770 K for 48 hr; lattice parameter reported at 303 K is 6.4653 Å; 6.4650 Å at 293 K determined by graphical extrapolation.
8	770	Bernstein, L. and Beals, R.J.	1961	I	310-800		Single crystal containing impurities of 10^{17} cm^{-3}, grown by the Czochalski technique in a (111) direction; zero-point correction of 0.011% determined by graphical extrapolation.

* Not shown in figure.

DATA TABLE 75. THERMAL LINEAR EXPANSION OF SbIn INTERMETALLIC COMPOUND

[Temperature, T, K; Linear Expansion, $\Delta L/L_0$, %]

T	$\Delta L/L_0$
CURVE 1	
290	-0.001
298	0.001
308	0.008
318	0.012
328	0.017
335	0.022
CURVE 2*, †, ‡	
2	0.000 x 10^{-3}
4	0.0013
6	-0.0014
8	-0.0097
10	-0.043
12	-0.110
14	-0.218
16	-0.370
18	-0.565
20	-0.796
22	-1.05
24	-1.33
26	-1.61
28	-1.90
30	-2.17
32	-2.44
34	-2.68
CURVE 3*, †	
1.89	0.00000 x 10^{-3}
3.68	0.000051
4.04	0.000053
4.76	0.000018
4.93	-0.000005
5.61	-0.00037
6.03	-0.0019
6.43	-0.0028
7.34	-0.0067
7.73	-0.0094
8.57	-0.0203
8.95	-0.0254
9.75	-0.0450
10.12	-0.0538
10.86	-0.0831
11.22	-0.0955
12.06	-0.145
CURVE 3 (cont.)*, †	
12.42	-0.158
13.13	-0.213
13.49	-0.234
14.27	-0.321
14.63	-0.347
15.73	-0.458
16.10	-0.487
17.35	-0.638
17.72	-0.676
18.72	-0.822
19.09	-0.865
20.08	-1.027
20.30	-1.054
20.45	-1.072
21.35	-1.234
21.71	-1.283
23.85	-1.703
24.20	-1.759
24.99	-1.886
25.32	-1.944
26.15	-2.144
26.35	-2.179
26.48	-2.202
27.29	-2.407
27.42	-2.430
27.61	-2.465
28.30	-2.658
28.49	-2.692
28.61	-2.714
29.25	-2.899
29.56	-2.953
30.17	-3.132
30.46	-3.186
31.10	-3.359
31.21	-3.380
31.32	-3.401
31.38	-3.411
CURVE 4	
273	-0.005
299	0.003*
325	0.029
377	0.049
429	0.080
481	0.118
CURVE 4 (cont.)	
532	0.143
585	0.176
637	0.199
689	0.223
CURVE 5‡	
24	-0.089
28	-0.090*
36	-0.091*
39	-0.091
44	-0.092*
50	-0.092
54	-0.092
62	-0.092
67	-0.092*
70	-0.091
80	-0.090
86	-0.089*
89	-0.088
91	-0.088*
98	-0.086
104	-0.085*
107	-0.084
114	-0.082*
121	-0.079
123	-0.079*
133	-0.075*
139	-0.073
149	-0.069*
155	-0.066
167	-0.061*
170	-0.060
182	-0.055
186	-0.053*
194	-0.049
201	-0.046*
210	-0.042
215	-0.039*
225	-0.034
230	-0.032*
241	-0.026
245	-0.024*
256	-0.019
262	-0.016*
271	-0.011
CURVE 5 (cont.)‡	
276	-0.009*
286	-0.004*
291	-0.001*
300	0.004
306	0.007
314	0.011
321	0.014
330	0.019
337	0.023
CURVE 6	
10	-0.084
20	-0.084
30	-0.086
40	-0.087
50	-0.088
60	-0.088
70	-0.088
80	-0.086
90	-0.085
100	-0.082
110	-0.079
120	-0.076
130	-0.072
140	-0.069
150	-0.066
160	-0.061
170	-0.057
180	-0.053
190	-0.049
200	-0.044
210	-0.039
220	-0.035
230	-0.030
240	-0.026
250	-0.021
260	-0.016
270	-0.011*
280	-0.006
290	-0.001*
300	0.003*
CURVE 7	
303	0.005
CURVE 7 (cont.)	
411	0.065
475	0.104
569	0.156
667	0.213
751	0.260
CURVE 8	
310	0.011*
370	0.048
470	0.110
570	0.165
670	0.225
770	0.275
790	0.275
800	0.270

* Not shown in figure.
† This curve is here reported using the first given temperature as reference temperature at which $\Delta L/L_0 = 0$.
‡ Author's data for coefficient of thermal expansion have been integrated by TPRC to obtain $\Delta L/L_0$.

DATA TABLE 75. COEFFICIENT OF THERMAL LINEAR EXPANSION OF SbIn INTERMETALLIC COMPOUND

[Temperature, T, K; Coefficient of Expansion, α, 10^{-6} K^{-1}]

T	α
CURVE 2‡	
2	0.00013
4	0.00000*
6	-0.015*
8	-0.094
10	-0.24
12	-0.432
14	-0.65
16	-0.87
18	-1.08
20	-1.23
22	-1.34
24	-1.42
26	-1.43*
28	-1.41
30	-1.35
32	-1.26
34	-1.17
CURVE 5‡	
24	-0.157
28	-0.148*
36	-0.132
39	-0.114
44	-0.094
50	-0.032
54	-0.016
62	0.049
67	0.085
70	0.111
80	0.173
86	0.194
89	0.226
91	0.203
98	0.243
104	0.276
107	0.296*
114	0.322
121	0.345
123	0.329
133	0.374
139	0.387
149	0.413
155	0.410
167	0.428
170	0.422

T	α
CURVE 5 (cont.)‡	
182	0.453
186	0.455
194	0.483
201	0.475
210	0.461
215	0.499
225	0.481
230	0.497
241	0.497
245	0.505
256	0.508
262	0.513
271	0.497
276	0.511
286	0.511
291	0.522
300	0.511
306	0.511*
314	0.521*
321	0.521*
330	0.526*
337	0.531*

* Not shown in figure.
‡ Author's data for coefficient of thermal expansion have been integrated by TPRC to obtain $\Delta L/L_0$.

SPECIFICATION TABLE 76. THERMAL LINEAR EXPANSION OF SbLa INTERMETALLIC COMPOUND

Cur. No.	Ref. No.	Author(s)	Year	Method Used	Temp. Range, K	Name and Specimen Designation	Composition (weight percent), Specifications, and Remarks
1	566	Goncharova, E.V., Zhuze, V.P., Zhdanova, V.V., Zhukova, T.B., Smirnov, I.A., and Shadrichev, E.V.	1968	L	300-951		Synthesized from 99.8$^+$ La and 99.99 Sb; measurements on specimen cut from coarse grain ingot made in induction furnace; measurements in argon atmosphere, phase transition reported between 630-720 K, reported error in the coefficient of thermal linear expansion measurement is ±3%.
2	566	Goncharova, E.V., et al.	1968	L	295-925		Similar to the above specimen, except cut from different ingot; results based on this and above curve indicate probable homogenity range.

DATA TABLE 76. THERMAL LINEAR EXPANSION OF SbLa INTERMETALLIC COMPOUND

[Temperature, T, K; Linear Expansion, $\Delta L/L_0$, %]

T	$\Delta L/L_0$	T	$\Delta L/L_0$	T	$\Delta L/L_0$	T	$\Delta L/L_0$	T	$\Delta L/L_0$	T	$\Delta L/L_0$
CURVE 1‡		CURVE 1 (cont.)‡		CURVE 1 (cont.)‡		CURVE 2 (cont.)‡		CURVE 2 (cont.)‡		CURVE 2 (cont.)‡	
300	0.005	675	0.300	807	0.604	368	0.079*	584	0.311	797	0.438
316	0.017	686	0.313	839	0.689	380	0.091	599	0.327	820	0.475
374	0.060	700	0.334	873	0.779	400	0.112	629	0.357	846	0.516
404	0.082	708	0.350	895	0.838	431	0.145	638	0.365	878	0.567
436	0.106	713	0.361	926	0.921	442	0.157*	671	0.378	902	0.606
463	0.126	718	0.373	951	0.988	456	0.172	675	0.377*	925	0.642
506	0.159	718	0.373			463	0.179*	682	0.373		
535	0.182	726	0.393	CURVE 2‡		480	0.198	708	0.355*		
573	0.212	742	0.435			496	0.215	725	0.351		
595	0.229	750	0.456	295	0.002	523	0.244	744	0.361		
613	0.244	769	0.505	320	0.028	532	0.254*	762	0.384		
653	0.279	791	0.563	356	0.066	556	0.280	781	0.413		

DATA TABLE 76. COEFFICIENT OF THERMAL LINEAR EXPANSION OF SbLa INTERMETALLIC COMPOUND

[Temperature, T, K; Coefficient of Expansion, α, 10^{-6} K^{-1}]

T	α	T	α	T	α	T	α	T	α	T	α
CURVE 1*,‡		CURVE 1 (cont.)*,‡		CURVE 1 (cont.)*,‡		CURVE 2*,‡		CURVE 2 (cont.)*,‡		CURVE 2 (cont.)*,‡	
300	7.40	653	9.14	769	26.06	295	10.16	496	10.71	708	-5.80
316	7.33	675	10.68	791	26.14	320	10.55	523	10.97	725	0.75
374	7.45	686	12.24	807	26.23	356	10.61	532	10.73	744	10.33
404	7.45	700	17.81	839	26.49	368	10.40	556	11.01	762	15.42
436	7.52	708	21.10	873	26.60	380	10.40	584	11.01	781	15.28
463	7.62	713	22.73	895	26.71	400	10.55	599	10.77	797	15.89
506	7.75	718	25.28	926	26.71	431	10.83	629	9.19	820	15.89
535	7.82	718	25.59	951	26.93	442	10.55	638	8.17	846	15.89
573	7.80	726	26.06			456	10.73	671	-0.48	878	16.09
595	8.06	742	26.08			463	10.55	675	-3.59	902	15.69
613	8.26	750	25.94			480	10.91	682	-7.94	925	15.69

*Not shown in figure.

‡Author's data for coefficient of thermal expansion have been integrated by TPRC to obtain $\Delta L/L_0$.

FIGURE AND TABLE NO. 77R. PROVISIONAL VALUES FOR THERMAL LINEAR EXPANSION OF β-$Be_{17}Hf_2$ INTERMETALLIC COMPOUND

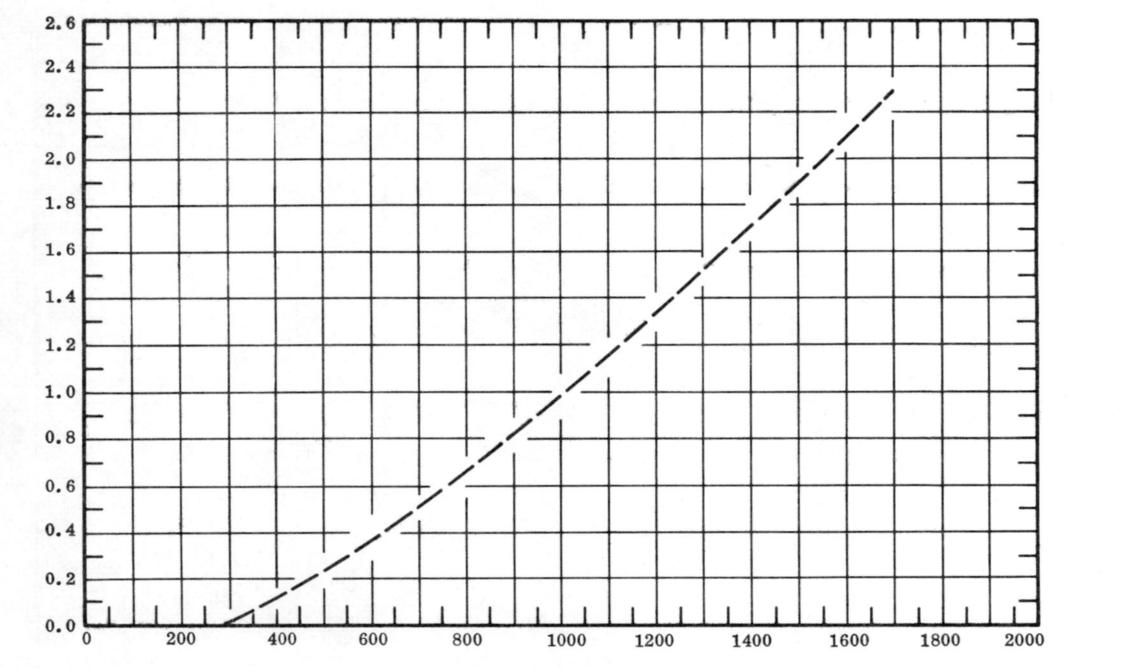

PROVISIONAL VALUES

[Temperature, T, K; Linear Expansion, $\Delta L/L_0$, %; α, K^{-1}]

T	$\Delta L/L_0$	$\alpha \times 10^6$
293	0.000	10.3
400	0.117	11.5
500	0.238	12.6
600	0.369	13.6
700	0.511	14.5
800	0.661	15.4
900	0.819	16.2
1000	0.985	16.9
1100	1.157	17.5
1200	1.335	18.0
1300	1.518	18.5
1400	1.704	18.9
1500	1.894	19.2
1600	2.087	19.4
1700	2.282	19.5

REMARKS

The tabulated values are considered accurate to within ±10% over the entire temperature range. These values can be represented approximately by the following equation:

$$\Delta L/L_0 = -0.246 + 6.385 \times 10^{-4}\ T + 7.264 \times 10^{-7}\ T^2 - 1.336 \times 10^{-10}\ T^3$$

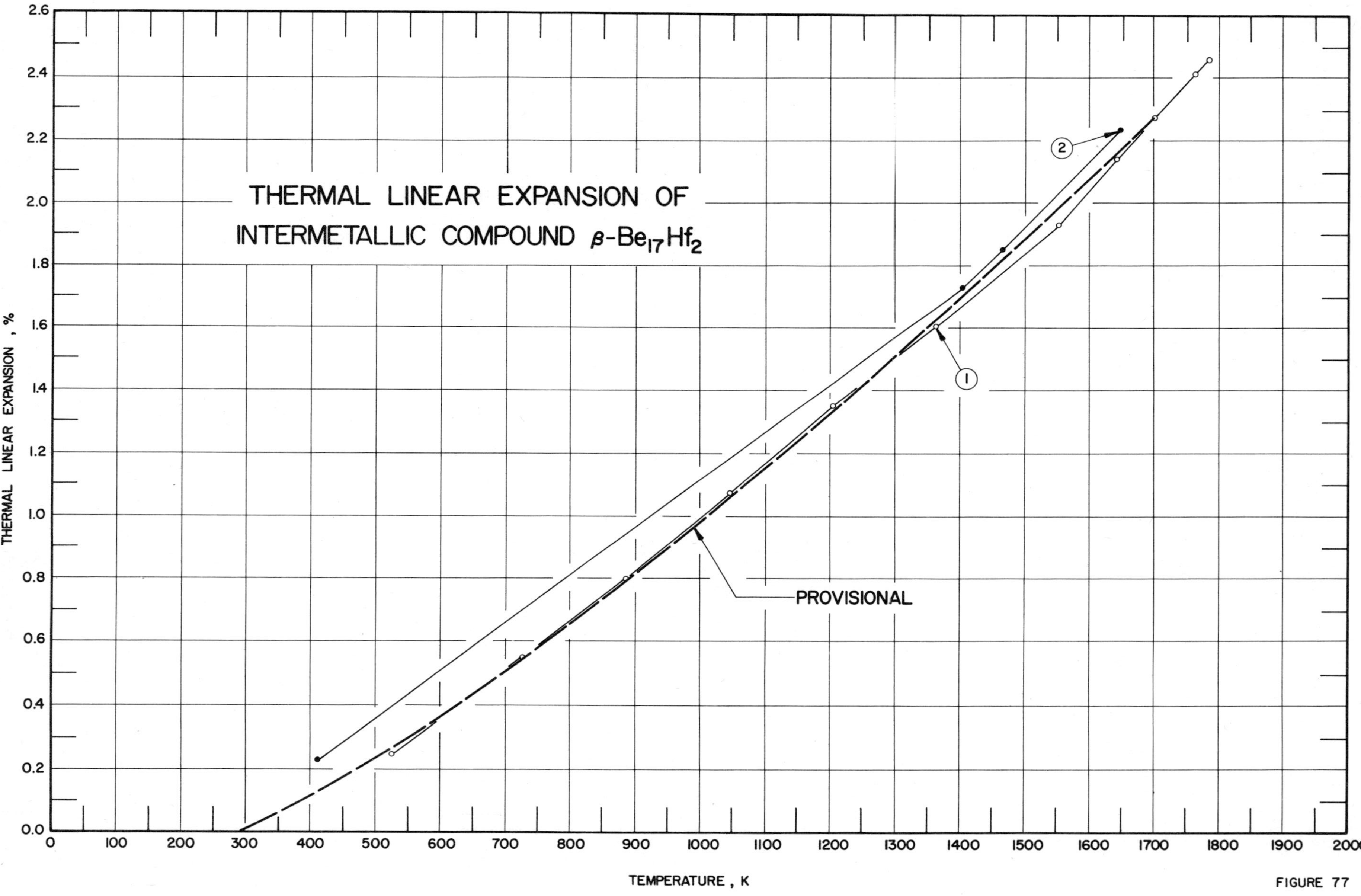
THERMAL LINEAR EXPANSION OF
INTERMETALLIC COMPOUND β-$Be_{17}Hf_2$
2
1
PROVISIONAL
THERMAL LINEAR EXPANSION, %
0.0 0.2 0.4 0.6 0.8 1.0 1.2 1.4 1.6 1.8 2.0 2.2 2.4 2.6
0 100 200 300 400 500 600 700 800 900 1000 1100 1200 1300 1400 1500 1600 1700 1800 1900 2000
TEMPERATURE, K

FIGURE 77

SPECIFICATION TABLE 77. THERMAL LINEAR EXPANSION OF β-$Be_{17}Hf_2$ INTERMETALLIC COMPOUND

Cur. No.	Ref. No.	Author(s)	Year	Method Used	Temp. Range, K	Name and Specimen Designation	Composition (weight percent), Specifications, and Remarks
1	564	Booker, J., Paine, R.M., and Stonehouse, A.J.	1960	T	523-1785		Bar specimen 3.25 in. long by 0.50 by 0.25 in.
2	564	Booker, J., et al.	1960	T	1649-410		The above specimen; expansion measured with decreasing temperature; zero-point correction of 0.200% determined by graphical extrapolation.

DATA TABLE 77. THERMAL LINEAR EXPANSION OF β-$Be_{17}Hf_2$ INTERMETALLIC COMPOUND

[Temperature, T, K; Linear Expansion, $\Delta L/L_0$, %]

T	$\Delta L/L_0$
CURVE 1	
523	0.250
724	0.552
886	0.800
1047	1.071
1204	1.349
1363	1.602
1556	1.929
1641	2.142
1703	2.281
1761	2.416
1785	2.452
CURVE 2	
1649	2.237
1467	1.855
1401	1.735
410	0.227

FIGURE AND TABLE NO. 78R. PROVISIONAL VALUES FOR THERMAL LINEAR EXPANSION OF $Be_{12}Nb$ INTERMETALLIC COMPOUND

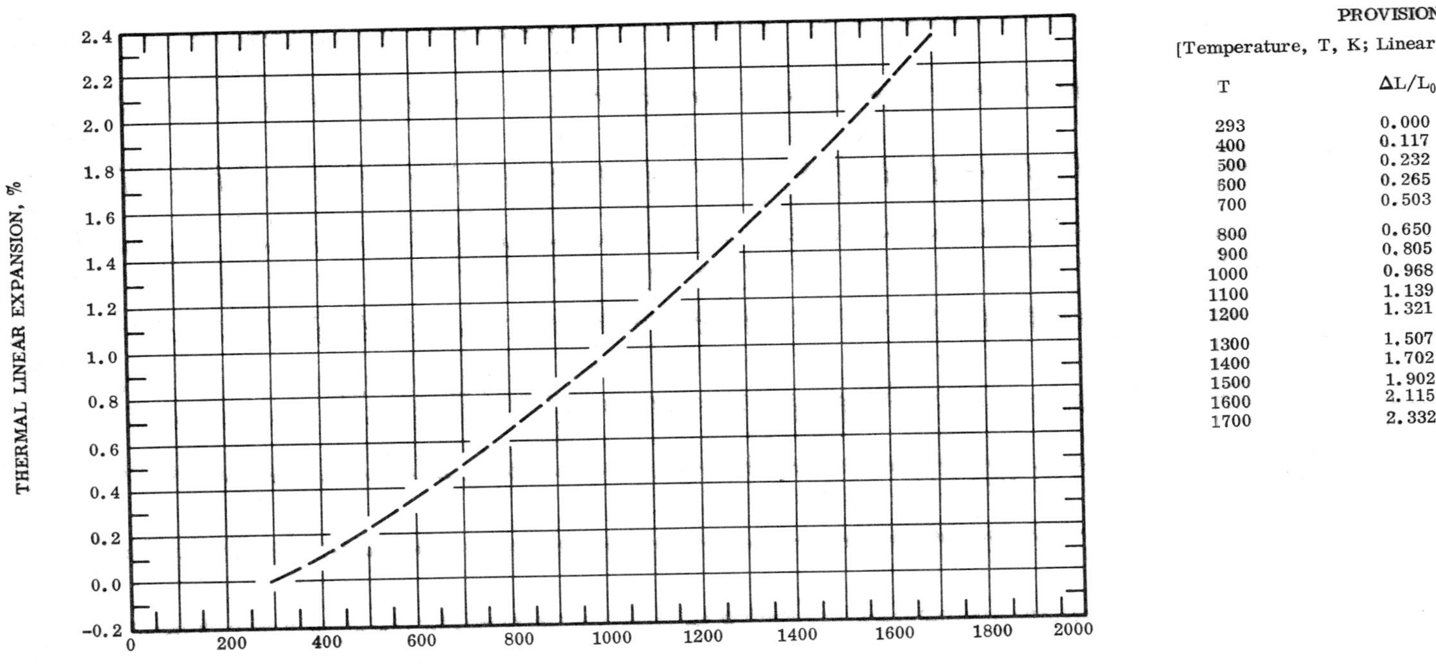

PROVISIONAL VALUES

[Temperature, T, K; Linear Expansion, $\Delta L/L_0$, %; α, K^{-1}]

T	$\Delta L/L_0$	$\alpha \times 10^6$
293	0.000	10.5
400	0.117	11.5
500	0.232	12.4
600	0.265	13.3
700	0.503	14.2
800	0.650	15.1
900	0.805	15.9
1000	0.968	16.8
1100	1.139	17.6
1200	1.321	18.4
1300	1.507	19.2
1400	1.702	19.9
1500	1.902	20.6
1600	2.115	21.4
1700	2.332	22.1

REMARKS

The tabulated values are considered accurate to within ±10% over the entire temperature range. These values can be represented approximately by the following equation:

$$\Delta L/L_0 = -0.265 + 7.590 \times 10^{-4}\, T + 5.065 \times 10^{-7}\, T^2 - 3.180 \times 10^{-11}\, T^3$$

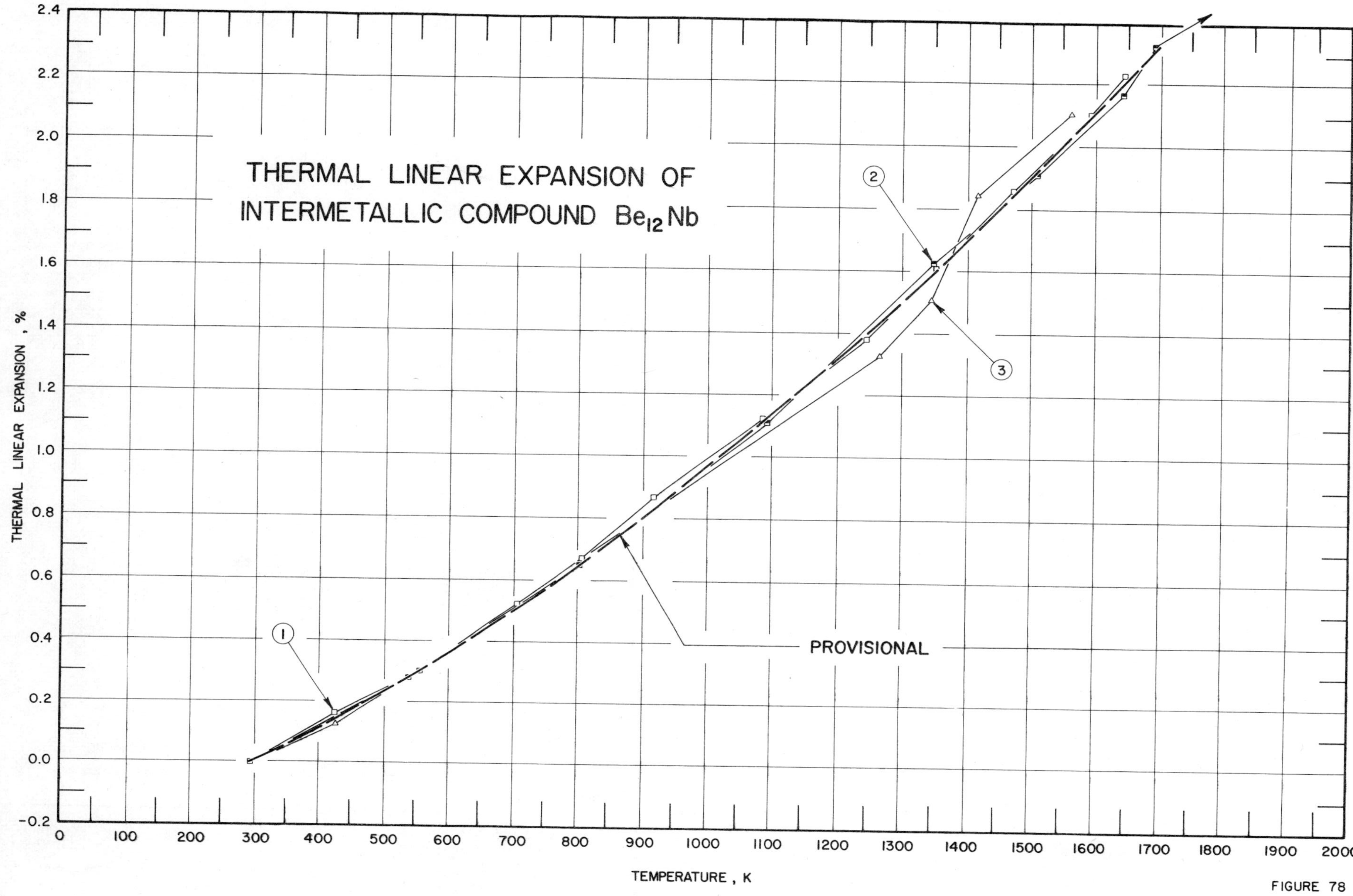
THERMAL LINEAR EXPANSION OF
INTERMETALLIC COMPOUND $Be_{12}Nb$
PROVISIONAL
1
2
3
THERMAL LINEAR EXPANSION, %
-0.2
0.0
0.2
0.4
0.6
0.8
1.0
1.2
1.4
1.6
1.8
2.0
2.2
2.4
TEMPERATURE, K
0
100
200
300
400
500
600
700
800
900
1000
1100
1200
1300
1400
1500
1600
1700
1800
1900
2000

FIGURE 78

SPECIFICATION TABLE 78. THERMAL LINEAR EXPANSION OF $Be_{12}Nb$ INTERMETALLIC COMPOUND

Cur. No.	Ref. No.	Author(s)	Year	Method Used	Temp. Range, K	Name and Specimen Designation	Composition (weight percent), Specifications, and Remarks
1	135	Paine, R.M., Stonehouse, A.J., and Beaver, W.M.	1959	T	293-1644		As pressed; expansion measured with increasing temperature.
2	135	Paine, R.M., et al.	1959	T	293-1751		The above specimen; expansion measured with increasing temperature.
3	135	Paine, R.M., et al.	1959	T	1562-423		The above specimen except expansion measured with decreasing temperature.

DATA TABLE 78. THERMAL LINEAR EXPANSION OF $Be_{12}Nb$ INTERMETALLIC COMPOUND

[Temperature, T, K; Linear Expansion, $\Delta L/L_0$, %]

T	$\Delta L/L_0$	T	$\Delta L/L_0$	T	$\Delta L/L_0$
CURVE 1		CURVE 2		CURVE 3	
293	0.000	293	0.000*	1562	2.107
422	0.161	538	0.283	1417	1.845
555	0.307	803	0.658	1349	1.515
704	0.528	1092	1.120	1268	1.333
809	0.679	1350	1.625	423	0.132
916	0.873	1510	1.910		
1089	1.129	1644	2.165		
1247	1.380	1692	2.310		
1353	1.611	1751	2.440*		
1473	1.860				
1591	2.105				
1644	2.299				

*Not shown in figure.

FIGURE AND TABLE NO. 79AR. PROVISIONAL VALUES FOR THERMAL LINEAR EXPANSION OF $Be_{12}Ta$ INTERMETALLIC COMPOUND

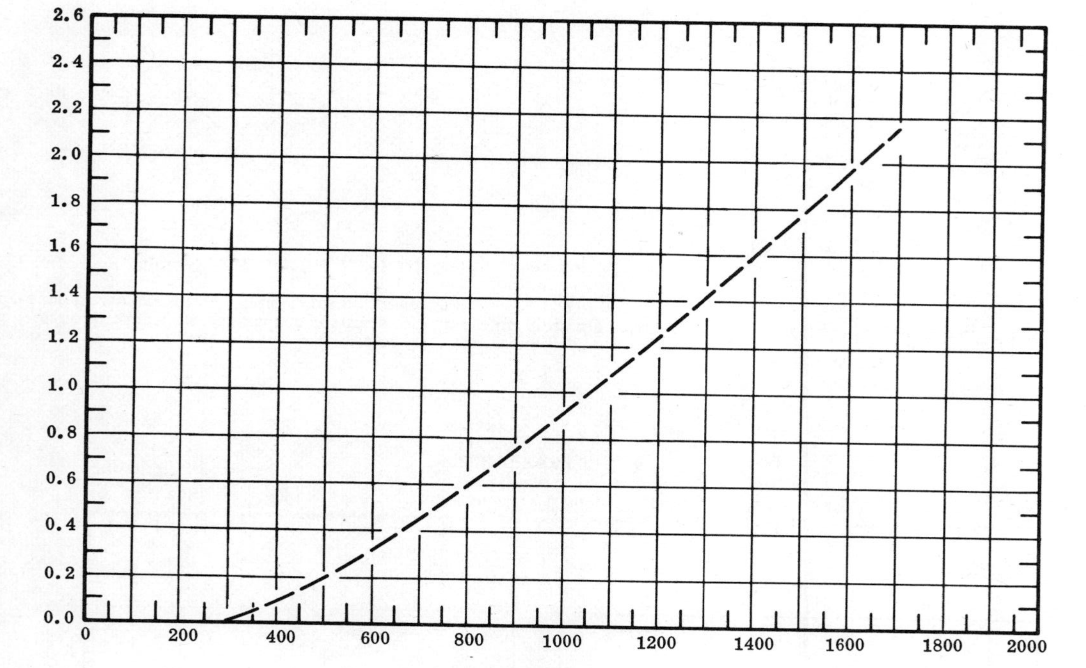

PROVISIONAL VALUES

[Temperature, T, K; Linear Expansion, $\Delta L/L_0$, %; α, K^{-1}]

T	$\Delta L/L_0$	$\alpha \times 10^6$
293	0.000	8.4
400	0.098	10.0
500	0.205	11.4
600	0.326	12.6
700	0.456	13.7
800	0.598	14.6
900	0.749	15.5
1000	0.907	16.2
1100	1.073	16.8
1200	1.244	17.3
1300	1.419	17.7
1400	1.599	18.0
1500	1.780	18.3
1600	1.965	18.4
1700	2.147	18.5

REMARKS

The tabulated values are considered accurate to within ±10% over the entire temperature range. These values can be represented approximately by the following equation:

$$\Delta L/L_0 = -0.182 + 3.699 \times 10^{-4}\, T + 9.071 \times 10^{-7}\, T^2 - 1.872 \times 10^{-10}\, T^3$$

FIGURE AND TABLE NO. 79BR. PROVISIONAL VALUES FOR THERMAL LINEAR EXPANSION OF $Be_{17}Ta_2$ INTERMETALLIC COMPOUND

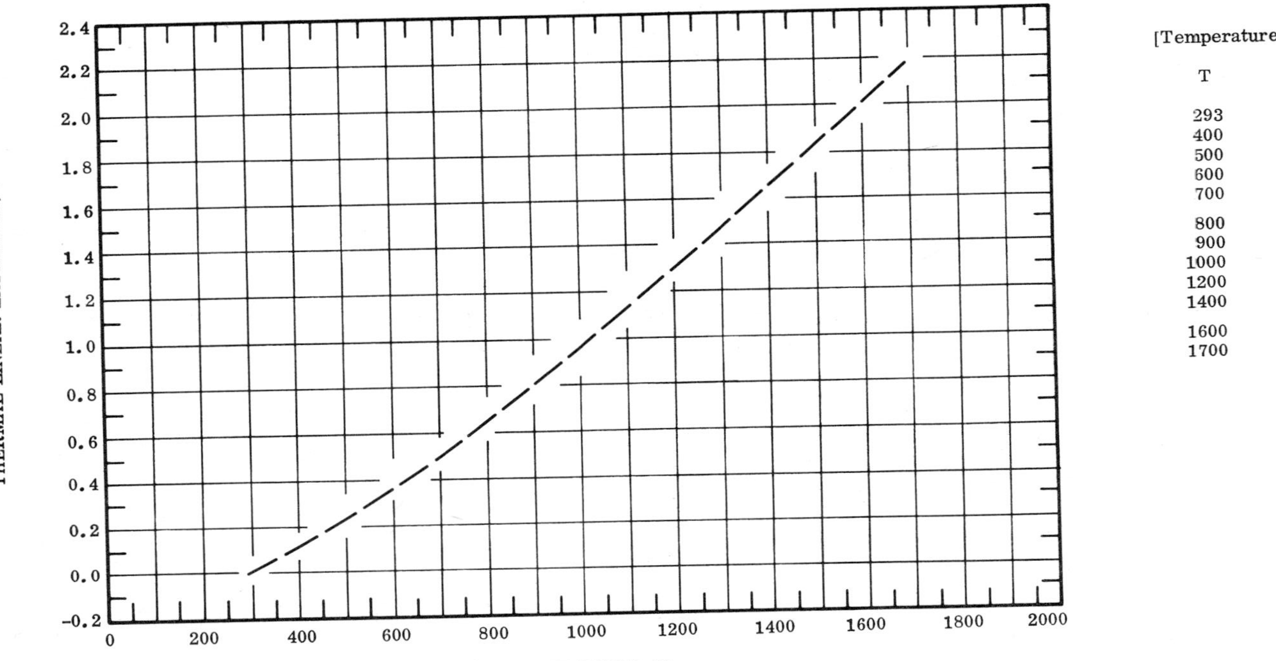

PROVISIONAL VALUES

[Temperature, T, K; Linear Expansion, $\Delta L/L_0$, %; α, K^{-1}]

T	$\Delta L/L_0$	$\alpha \times 10^6$
293	0.000	10.1
400	0.114	11.3
500	0.233	12.3
600	0.361	13.3
700	0.499	14.2
800	0.644	15.0
900	0.797	15.6
1000	0.957	16.2
1200	1.291	17.1
1400	1.640	17.7
1600	1.996	17.8
1700	2.175	17.8

REMARKS

The tabulated values are considered accurate to within ±10% over the entire temperature range. These values can be represented approximately by the following equation:

$$\Delta L/L_0 = -0.240 + 6.153 \times 10^{-4}\,T + 7.351 \times 10^{-7}\,T^2 - 1.539 \times 10^{-10}\,T^3$$

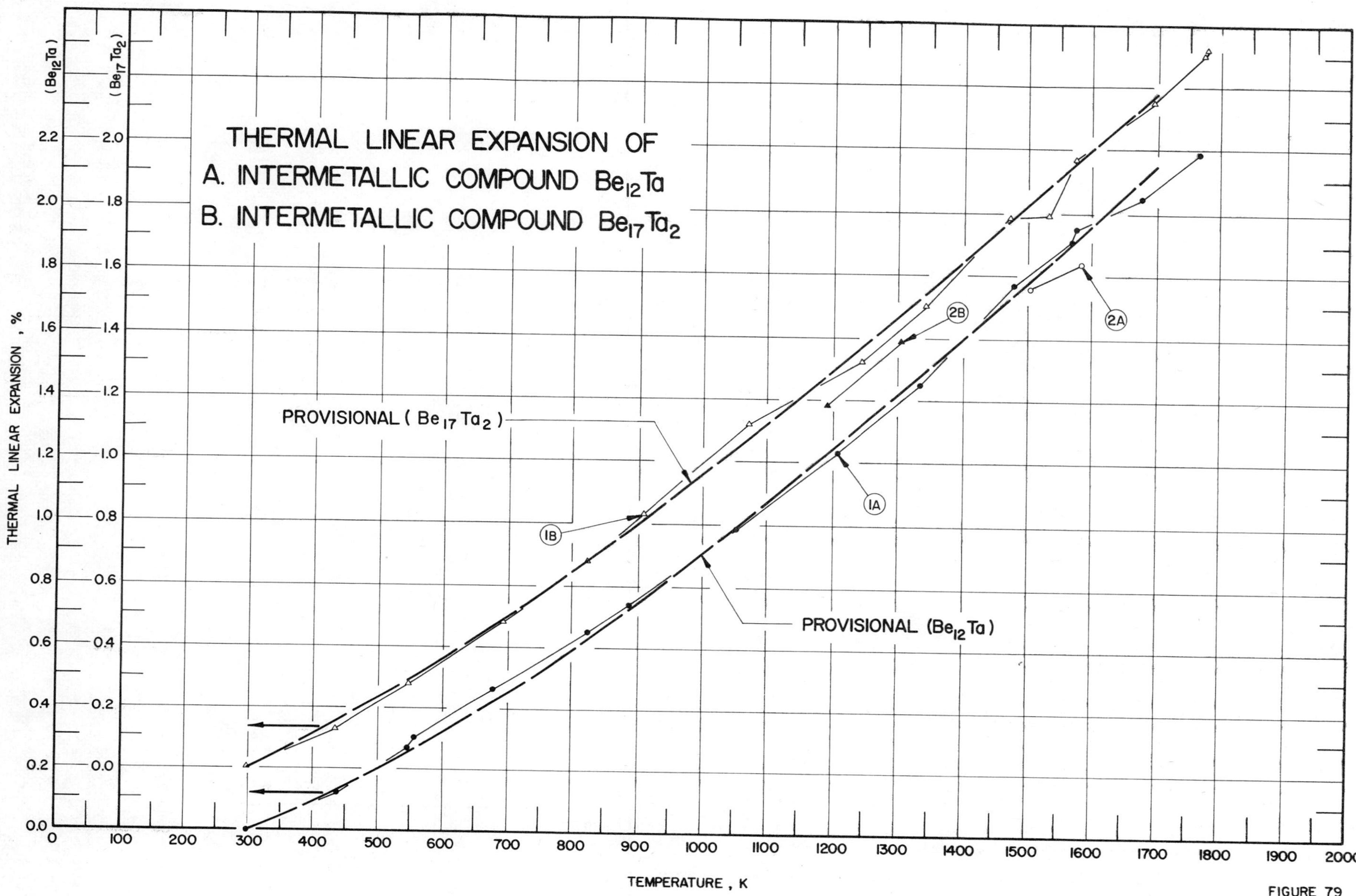

THERMAL LINEAR EXPANSION OF
A. INTERMETALLIC COMPOUND $Be_{12}Ta$
B. INTERMETALLIC COMPOUND $Be_{17}Ta_2$
PROVISIONAL ($Be_{17}Ta_2$)
PROVISIONAL ($Be_{12}Ta$)
1A
2A
1B
2B
THERMAL LINEAR EXPANSION, %
($Be_{12}Ta$)
($Be_{17}Ta_2$)
TEMPERATURE, K

FIGURE 79

SPECIFICATION TABLE 79A. THERMAL LINEAR EXPANSION OF $Be_{12}Ta$ INTERMETALLIC COMPOUND

Cur. No.	Ref. No.	Author(s)	Year	Method Used	Temp. Range, K	Name and Specimen Designation	Composition (weight percent), Specifications, and Remarks
1	589	Booker, J., Paine, R. M., and Stonehouse, A. J.	1961		299-1767	H. P. 222	Dimension 2.5 x 3.5 x 1.625 in. bar; hot pressing maximum temperature 1799.8 K, maximum pressure 2000 psi; pre-oxidized at 1755 K; 99.2% absolute density; expansion measured with increasing temperature; measured in air (>1477.6 K) and in argon atmosphere (<1477.6 K).
2	589	Booker, J., et al.	1961		1583, 1503	H. P. 222	Similar to the above specimen except expansion measured with decreasing temperature.

DATA TABLE THERMAL LINEAR EXPANSION OF $Be_{12}Ta$ INTERMETALLIC COMPOUND

[Temperature, T, K; Linear Expansion, $\Delta L/L_0$, %]

T	$\Delta L/L_0$	T	$\Delta L/L_0$
CURVE 1		CURVE 2	
299	0.006*	1583	1.84
435	0.137*	1503	1.76
544	0.275		
549	0.261*		
554	0.303		
678	0.472		
822	0.658		
887	0.742		
1054	0.991		
1208	1.233		
1335	1.477		
1480	1.771		
1570	1.916		
1577	1.959		
1677	2.046		
1767	2.194		

* Not shown in figure.

SPECIFICATION TABLE 79B. THERMAL LINEAR EXPANSION OF $Be_{17}Ta_2$ INTERMETALLIC COMPOUND

Cur. No.	Ref. No.	Author(s)	Year	Method Used	Temp. Range, K	Name and Specimen Designation	Composition (weight percent), Specifications, and Remarks
1	589	Booker, J., Paine, R. M., and Stonehouse, A. J.	1961		299-1777	H. P. 227	Dimension 2.5 x 3.5 x 1.625 in. bar; 99.2% absolute density; hot pressing maximum temperature 1747.1 K, maximum pressure 2000 psi; pre-oxidized at 1755.4 K; measured in air (>1477.6 K) and in argon atmosphere (<1477.6 K), expansion measured with increasing temperature.
2	589	Booker, J., et al.	1961		1302, 1190	H. P. 227	Similar to the above specimen except expansion measured with decreasing temperature.

DATA TABLE 79B. THERMAL LINEAR EXPANSION OF $Be_{17}Ta_2$ INTERMETALLIC COMPOUND

[Temperature, T, K; Linear Expansion, $\Delta L/L_0$, %]

T	$\Delta L/L_0$	T	$\Delta L/L_0$
CURVE 1		CURVE 2	
299	0.007	1302	1.388
433	0.136	1190	1.182
549	0.283		
692	0.486		
827	0.686		
910	0.833		
1077	1.130		
1249	1.325		
1341	1.505		
1472	1.786		
1532	1.796		
1579	1.969		
1696	2.156		
1772	2.308*		
1777	2.323		

* Not shown in figure.

FIGURE AND TABLE NO. 80R. PROVISIONAL VALUES FOR THERMAL LINEAR EXPANSION OF $Be_{13}U$ INTERMETALLIC COMPOUND

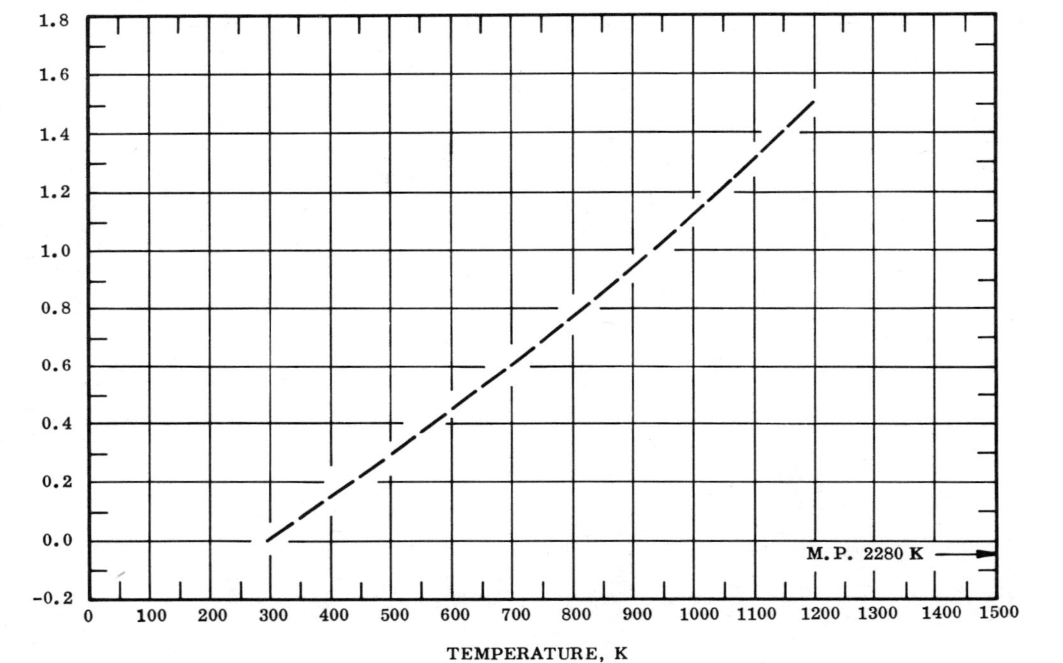

PROVISIONAL VALUES

[Temperature, T, K; Linear Expansion, $\Delta L/L_0$, %; α, K^{-1}]

T	$\Delta L/L_0$	$\alpha \times 10^6$
293	0.000	13.1
400	0.146	14.1
500	0.291	15.0
600	0.444	15.8
700	0.606	16.5
800	0.775	17.2
900	0.950	17.7
1000	1.129	18.2
1100	1.313	18.7
1200	1.502	19.0

REMARKS

The tabulated values are considered accurate to within ±7% over the entire temperature range. These values can be represented approximately by the following equation:

$$\Delta L/L_0 = -0.334 + 9.698 \times 10^{-4}\,T + 6.254 \times 10^{-7}\,T^2 - 1.323 \times 10^{-10}\,T^3$$

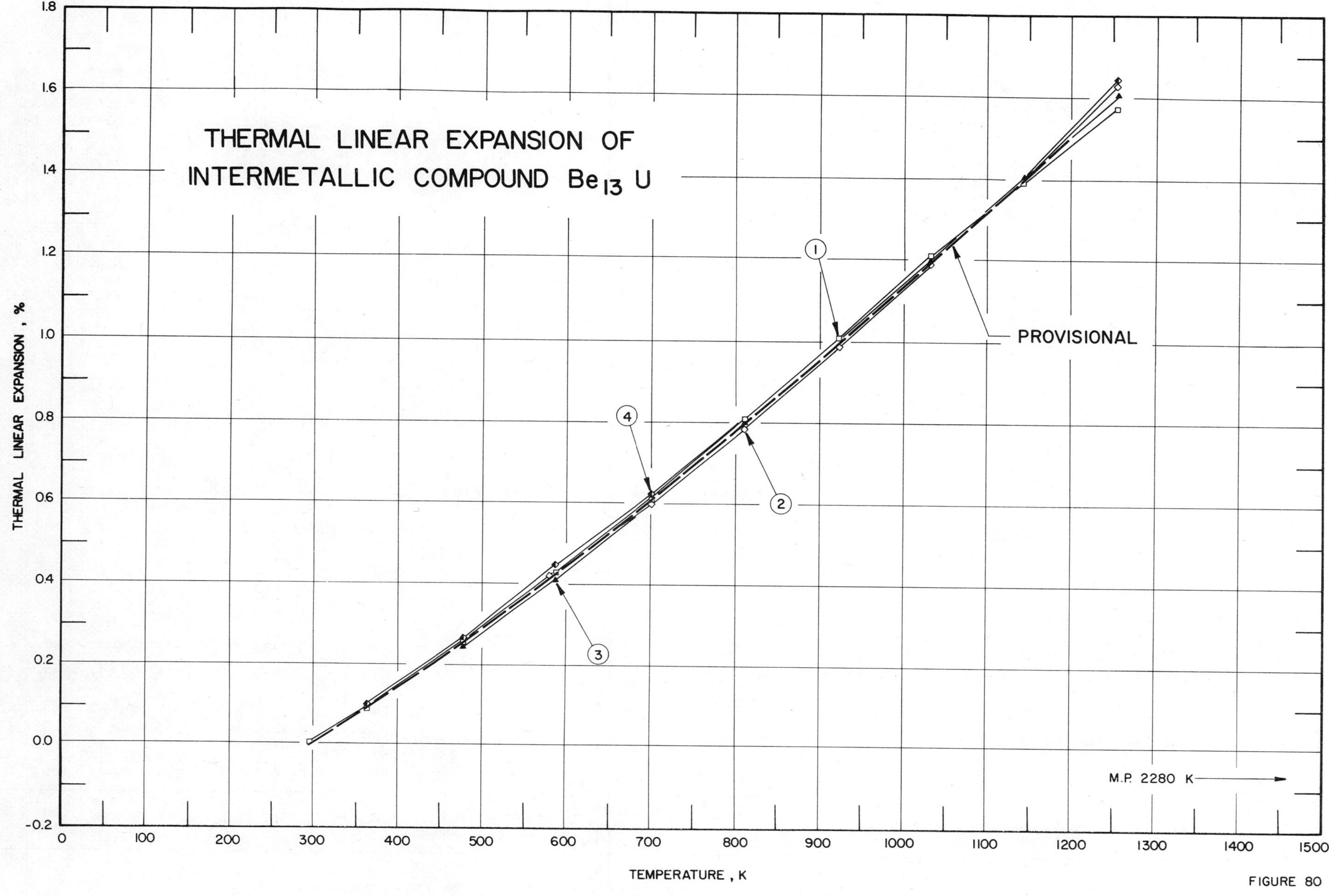

THERMAL LINEAR EXPANSION OF
INTERMETALLIC COMPOUND $Be_{13}U$
PROVISIONAL
1
2
3
4
M.P. 2280 K
THERMAL LINEAR EXPANSION, %
1.8
1.6
1.4
1.2
1.0
0.8
0.6
0.4
0.2
0.0
-0.2
0
100
200
300
400
500
600
700
800
900
1000
1100
1200
1300
1400
1500
TEMPERATURE, K

FIGURE 80

SPECIFICATION TABLE 80. THERMAL LINEAR EXPANSION OF $Be_{13}U$ INTERMETALLIC COMPOUND

Cur. No.	Ref. No.	Author(s)	Year	Method Used	Temp. Range, K	Name and Specimen Designation	Composition (weight percent), Specifications, and Remarks
1	590	Tripler, A.B., Snyder, M.J., and Duckworth, W.H.	1959	L	366-1255		Sintered specimen; expansion measured in 5 x 10^{-5} mm Hg vacuum with temperature increasing 2.8 K per min.
2	590	Tripler, A.B., et al.	1959	L	1255-366		The above specimen; expansion measured with decreasing temperature.
3	590	Tripler, A.B., et al.	1959	L	366-1255		The above specimen; expansion measured with increasing temperature on second test.
4	590	Tripler, A.B., et al.	1959	L	1255-366		The above specimen; expansion measured with decreasing temperature.
5	689	Snyder, M.J. and Tripler, A.B., Jr.	1960	L	293, 1273		Specimen formed by heating hydrostatically compacted mixture of UH_3 and Be metal in proportion in induction furnace at about 1823 K; reaction carried out under pressure of argon slightly above 1 atmosphere to minimize volatilization of Be; expansion measured under vacuum; heating and cooling rate of about 276 K per min.

DATA TABLE 80. THERMAL LINEAR EXPANSION OF $Be_{13}U$ INTERMETALLIC COMPOUND

[Temperature, T, K; Linear Expansion, $\Delta L/L_0$, %]

T	$\Delta L/L_0$	T	$\Delta L/L_0$	T	$\Delta L/L_0$
CURVE 1		CURVE 3		CURVE 5	
366	0.095	366	0.094*	293.2	0.000
477	0.252	477	0.246	1273.2	1.637
588	0.426	588	0.415		
700	0.615	700	0.608*		
811	0.811	811	0.802		
922	1.007	922	1.007*		
1033	1.212	1033	1.199		
1144	1.394	1144	1.409		
1255	1.576	1255	1.611		
CURVE 2		CURVE 4			
1255	1.628	1255	1.645		
1144	1.394*	1144	1.409*		
1033	1.185	1033	1.212*		
922	0.984	922	1.007*		
811	0.782	811	0.810*		
700	0.600	700	0.622		
588	0.420	588	0.441		
477	0.252*	477	0.265		
366	0.096*	366	0.101		

*Not shown in figure.

FIGURE AND TABLE NO. 81R. PROVISIONAL VALUES FOR THERMAL LINEAR EXPANSION OF $Be_{13}Zr$ INTERMETALLIC COMPOUND

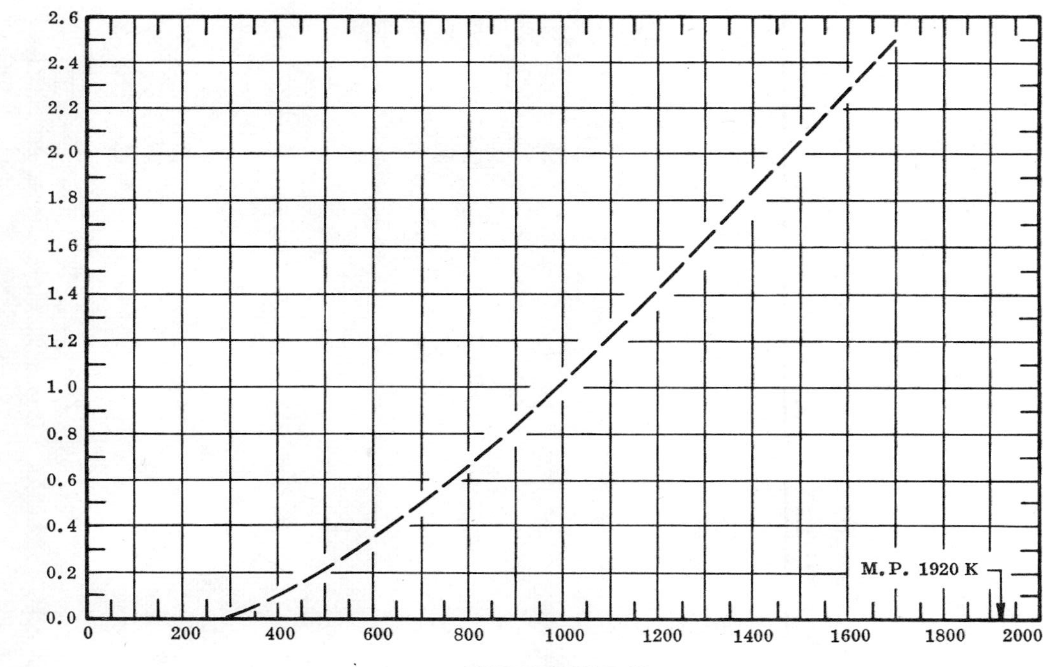

PROVISIONAL VALUES

[Temperature, T, K; Linear Expansion, $\Delta L/L_0$, %; α, K^{-1}]

T	$\Delta L/L_0$	$\alpha \times 10^6$
293	0.000	8.5
400	0.102	10.5
500	0.219	12.3
600	0.350	13.9
700	0.497	15.3
800	0.657	16.7
900	0.830	17.9
1000	1.014	19.0
1200	1.411	20.6
1400	1.834	21.6
1600	2.272	22.1
1700	2.494	22.1

REMARKS

The tabulated values are considered accurate to within ±10% over the entire temperature range. These values can be represented approximately by the following equation:

$$\Delta L/L_0 = -0.157 + 2.042 \times 10^{-4}\, T + 1.207 \times 10^{-6}\, T^2 - 2.416 \times 10^{-10}\, T^3$$

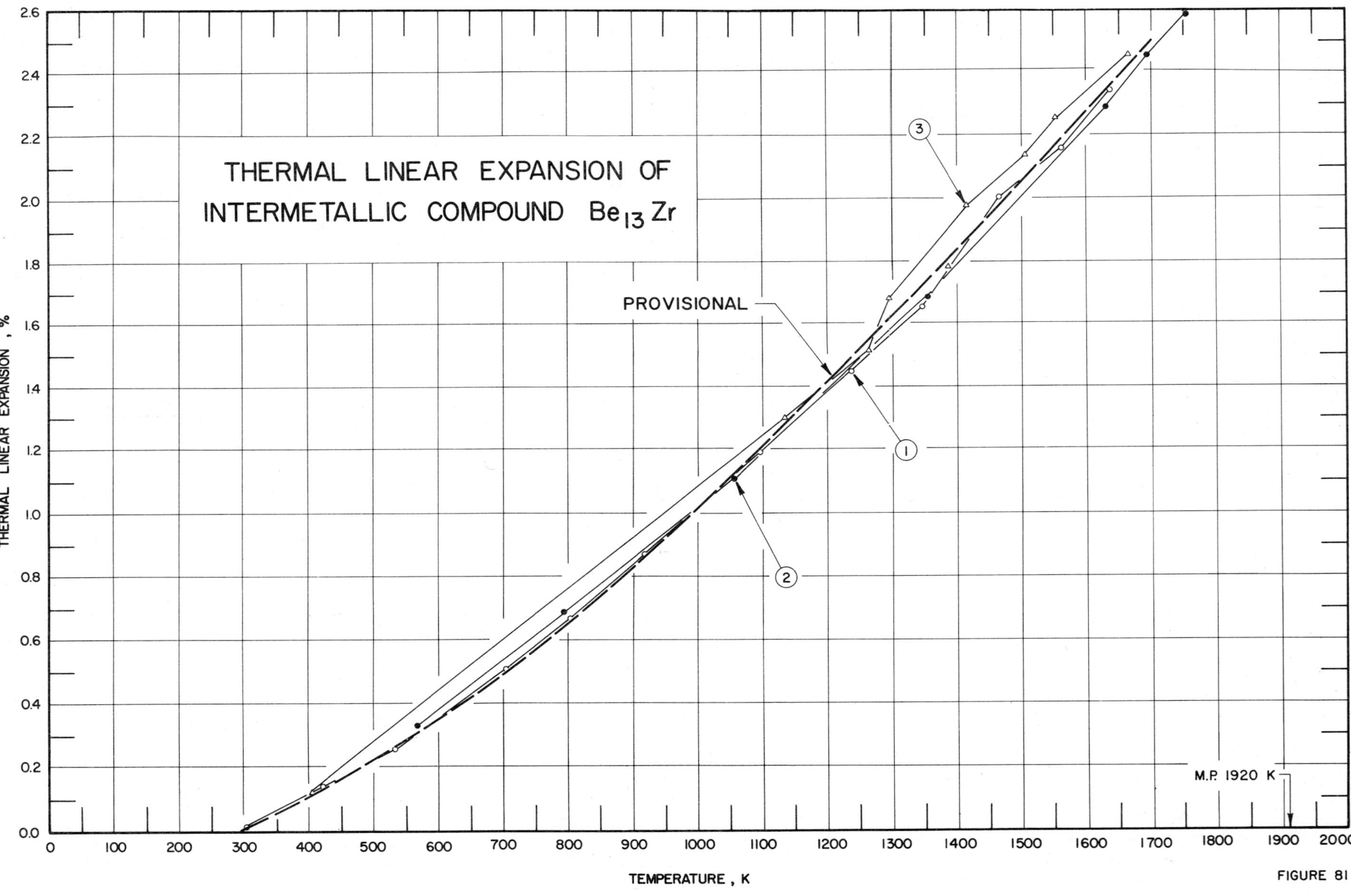
THERMAL LINEAR EXPANSION OF
INTERMETALLIC COMPOUND $Be_{13}Zr$
PROVISIONAL
M.P. 1920 K
1
2
3
THERMAL LINEAR EXPANSION, %
TEMPERATURE, K
0.0
0.2
0.4
0.6
0.8
1.0
1.2
1.4
1.6
1.8
2.0
2.2
2.4
2.6
0
100
200
300
400
500
600
700
800
900
1000
1100
1200
1300
1400
1500
1600
1700
1800
1900
2000

FIGURE 81

SPECIFICATION TABLE 81. THERMAL LINEAR EXPANSION OF $Be_{13}Zr$ INTERMETALLIC COMPOUND

Cur. No.	Ref. No.	Author(s)	Year	Method Used	Temp. Range, K	Name and Specimen Designation	Composition (weight percent), Specifications, and Remarks
1	135	Paine, R.M., Stonehouse, A.J., and Beaver, W.M.	1959		304-1638		As pressed; measured with increasing temperature.
2	135	Paine, R.M., et al.	1959		304-1792		As pressed; measured with increasing temperature.
3	135	Paine, R.M., et al.	1959		1668-304		Expansion measured with decreasing temperature ; zero-point correction of 0.022% determined by graphical extrapolation.

DATA TABLE 81. THERMAL LINEAR EXPANSION OF $Be_{13}Zr$ INTERMETALLIC COMPOUND

[Temperature, T, K; Linear Expansion, $\Delta L/L_0$, %]

T	$\Delta L/L_0$	T	$\Delta L/L_0$	T	$\Delta L/L_0$
CURVE 1		CURVE 2		CURVE 3 (cont.)	
304	0.012	304	0.014*	1388	1.781
422	0.134	569	0.334	1299	1.686
532	0.255	791	0.684	1263	1.516
704	0.506	1057	1.106	1138	1.298
807	0.662	1351	1.695	408	0.216
920	0.862	1630	2.285	304	0.022
1097	1.195	1695	2.446		
1238	1.452	1752	2.587		
1349	1.657	1792	2.664*		
1469	2.002				
1565	2.160	CURVE 3			
1638	2.342				
		1668	2.454		
		1555	2.256		
		1502	2.136		
		1418	1.980		

*Not shown in figure.

FIGURE AND TABLE NO. 82AR, PROVISIONAL VALUES FOR THERMAL LINEAR EXPANSION OF Bi_2Pt INTERMETALLIC COMPOUND

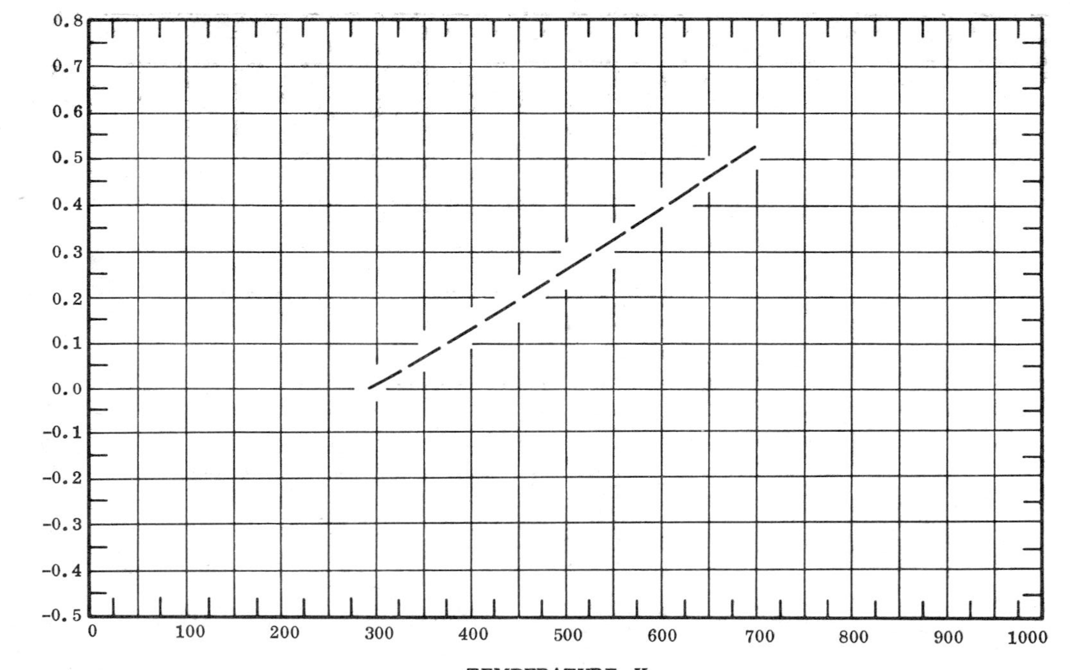

PROVISIONAL VALUES

[Temperature, T, K; Linear Expansion, $\Delta L/L_0$, %; α, K^{-1}]

T	$\Delta L/L_0$	$\alpha \times 10^6$
293	0.000	11.0
400	0.126	12.4
500	0.255	13.1
600	0.388	13.4
700	0.523	13.4

REMARKS

The tabulated values are considered accurate to within ±10% over the entire temperature range. These values can be represented approximately by the following equation:

$$\Delta L/L_0 = -0.255 + 5.626 \times 10^{-4}\, T + 1.233 \times 10^{-6}\, T^2 - 6.424 \times 10^{-10}\, T^3$$

FIGURE AND TABLE NO. 82BR. PROVISIONAL VALUES FOR THERMAL LINEAR EXPANSION OF BiPt INTERMETALLIC COMPOUND

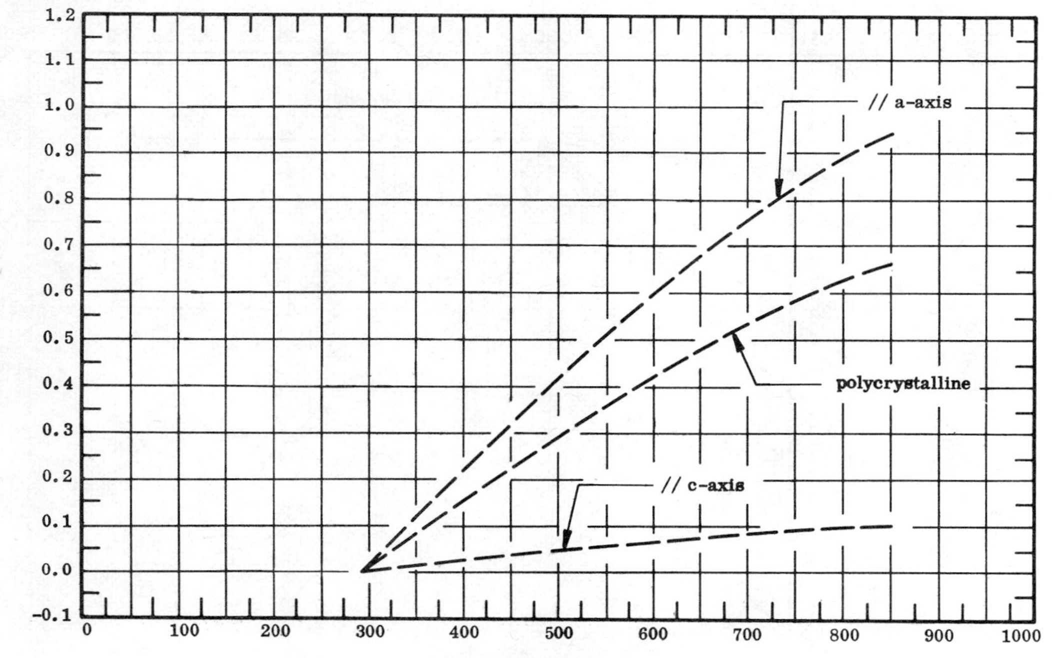

PROVISIONAL VALUES

[Temperature, T, K; Linear Expansion, $\Delta L/L_0$, %; α, K^{-1}]

	// a-axis	// c-axis	polycrystalline	
T	$\Delta L/L_0$	$\Delta L/L_0$	$\Delta L/L_0$	$\alpha \times 10^6$
293	0.000	0.000	0.000	14.6
400	0.221	0.028	0.157	14.4
500	0.419	0.049	0.296	13.6
600	0.602	0.069	0.424	12.1
700	0.762	0.085	0.536	10.1
800	0.891	0.098	0.627	7.7
850	0.941	0.103	0.661	6.2

REMARKS

The tabulated values are considered accurate to within ±10%. The values for polycrystalline material in the absence of experimental data are calculated from the single crystal data. These and other values can be represented approximately by the following equations:

// a-axis: $\Delta L/L_0 = -0.561 + 1.659 \times 10^{-3}\,T + 1.281 \times 10^{-6}\,T^2 - 1.359 \times 10^{-9}\,T^3$

// c-axis: $\Delta L/L_0 = -0.085 + 3.231 \times 10^{-4}\,T - 8.745 \times 10^{-8}\,T^2 - 3.777 \times 10^{-11}\,T^3$

polycrystalline: $\Delta L/L_0 = -0.409 + 1.240 \times 10^{-3}\,T + 7.962 \times 10^{-7}\,T^2 - 9.088 \times 10^{-10}\,T^3$

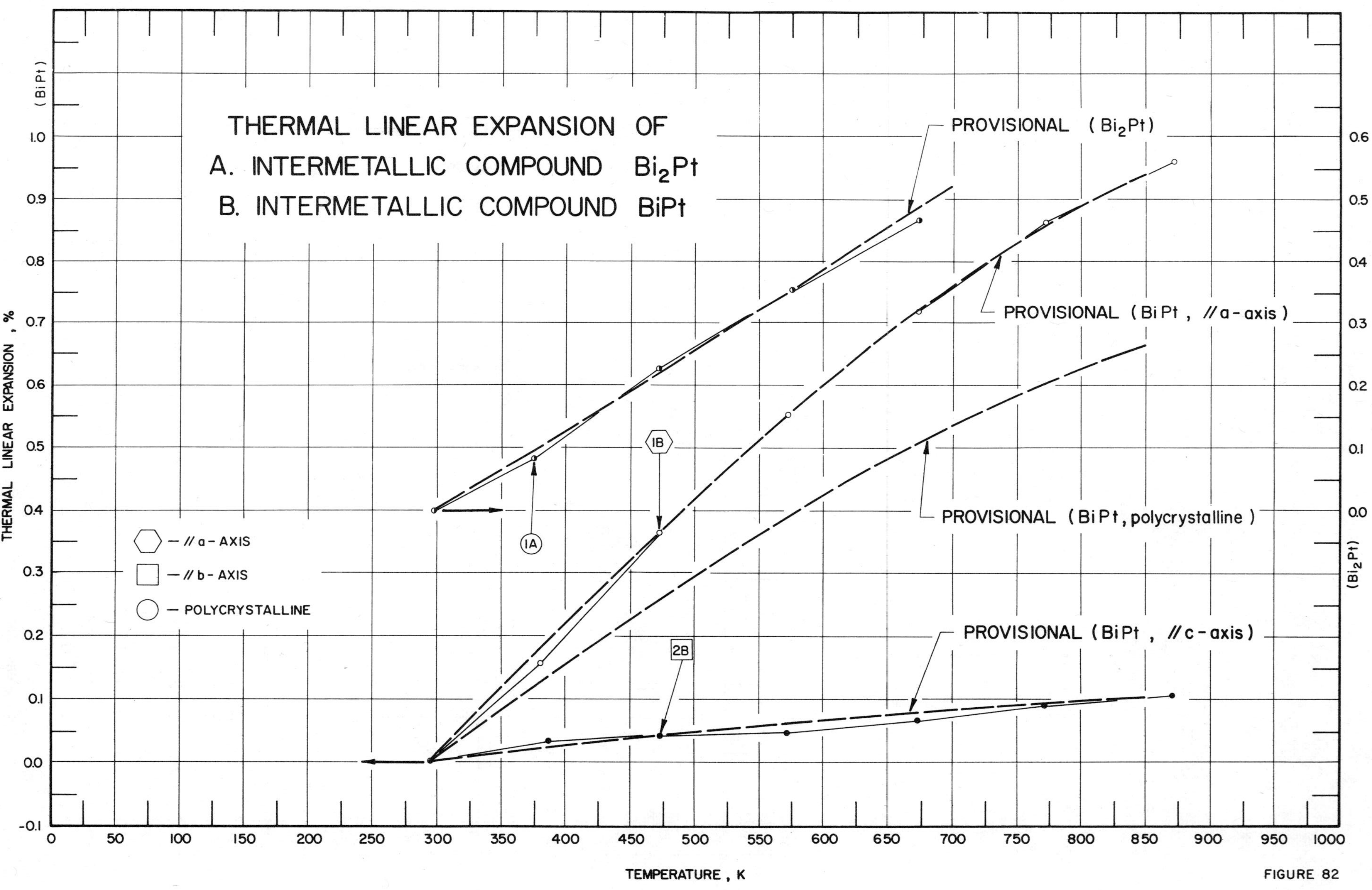
THERMAL LINEAR EXPANSION OF
A. INTERMETALLIC COMPOUND Bi_2Pt
B. INTERMETALLIC COMPOUND BiPt
— //a - AXIS
— //b - AXIS
— POLYCRYSTALLINE
PROVISIONAL (Bi_2Pt)
PROVISIONAL (BiPt , //a - axis)
PROVISIONAL (BiPt , polycrystalline)
PROVISIONAL (BiPt , //c - axis)
1A
1B
2B
THERMAL LINEAR EXPANSION , %
(BiPt)
(Bi_2Pt)
TEMPERATURE , K
0
50
100
150
200
250
300
350
400
450
500
550
600
650
700
750
800
850
900
950
1000
-0.1
0.0
0.1
0.2
0.3
0.4
0.5
0.6
0.7
0.8
0.9
1.0
0.0
0.1
0.2
0.3
0.4
0.5
0.6
FIGURE 82

SPECIFICATION TABLE 82A. THERMAL LINEAR EXPANSION OF Bi_2Pt INTERMETALLIC COMPOUND

Cur. No.	Ref. No.	Author(s)	Year	Method Used	Temp. Range, K	Name and Specimen Designation	Composition (weight percent), Specifications, and Remarks
1	679	Zhuravlev, N.N. and Stepanova, A.A.	1962	X	293-673	α-Bi_2Pt	Specimen prepared from homogenized alloys corresponding to compositions; beads of alloy ground in agate mortar, sieved, and then packed into quartz capillary with walls a few hundredths of a millimeter thick and set up in camera; specimen crystallized in cubic system with lattice constant 6.683 Å.

DATA TABLE 82A. THERMAL LINEAR EXPANSION OF Bi_2Pt INTERMETALLIC COMPOUND

[Temperature, T, K; Linear Expansion, $\Delta L/L_0$, %]

T	$\Delta L/L_0$
CURVE 1	
293	0.000
375	0.082
472	0.227
575	0.357
673	0.469

SPECIFICATION TABLE 82B. THERMAL LINEAR EXPANSION OF BiPt INTERMETALLIC COMPOUND

Cur. No.	Ref. No.	Author(s)	Year	Method Used	Temp. Range, K	Name and Specimen Designation	Composition (weight percent), Specifications, and Remarks
1	679	Zhuravlev, N.N. and Stepanova, A. A.	1962	X	293-873		Specimen prepared from homogenized alloys corresponding to compositions; beads of alloy ground in agate mortar, sieved, and then packed into quartz capillary with walls a few hundredths of a millimeter thick and set in camera; specimen of hexagonal system and nickel arsenide type structure; lattice constant of a-axis reported at room temperature is 4.315 Å; expansion measured along a-axis.
2	679	Zhuravlev, N.N. and Stepanova, A. A.	1962	X	293-870		The above specimen; lattice constant of c-axis reported at room temperature is 5.490 Å; expansion measured along c-axis.

DATA TABLE 82B. THERMAL LINEAR EXPANSION OF BiPt INTERMETALLIC COMPOUND

[Temperature, T, K; Linear Expansion, $\Delta L/L_0$, %]

T	$\Delta L/L_0$
CURVE 1	
293	0.000
380	0.159
473	0.366
572	0.556
674	0.720
772	0.866
873	0.961
CURVE 2	
293	0.000
386	0.033
474	0.044
573	0.049
673	0.069
772	0.091
870	0.109

FIGURE AND TABLE NO. 83R. PROVISIONAL VALUES FOR THERMAL LINEAR EXPANSION OF CdAu INTERMETALLIC COMPOUND

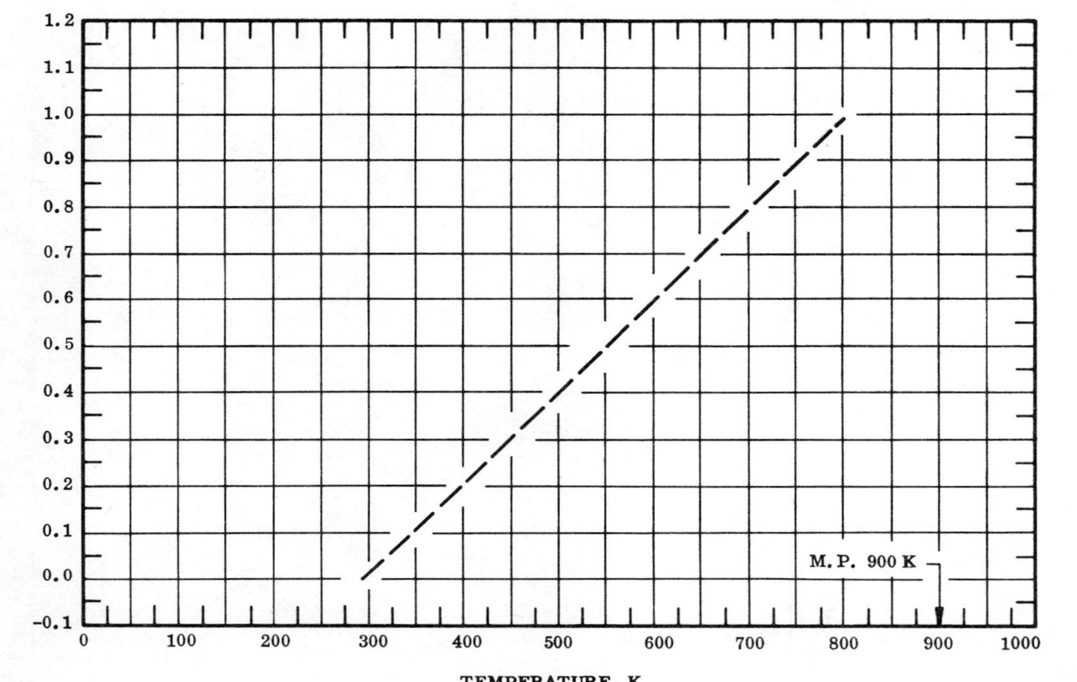

PROVISIONAL VALUES

[Temperature, T, K; Linear Expansion, $\Delta L/L_0$, %; α, K^{-1}]

T	$\Delta L/L_0$	$\alpha \times 10^6$
293	0.000	18.9
400	0.204	19.1
500	0.396	19.4
600	0.592	19.7
700	0.792	20.1
800	0.996	20.6

REMARKS

The tabulated values are considered accurate to within ±10% over the entire temperature range. These values can be represented approximately by the following equation:

$$\Delta L/L_0 = -0.552 + 1.890 \times 10^{-3}\, T - 5.466 \times 10^{-8}\, T^2 + 1.370 \times 10^{-10}\, T^3$$

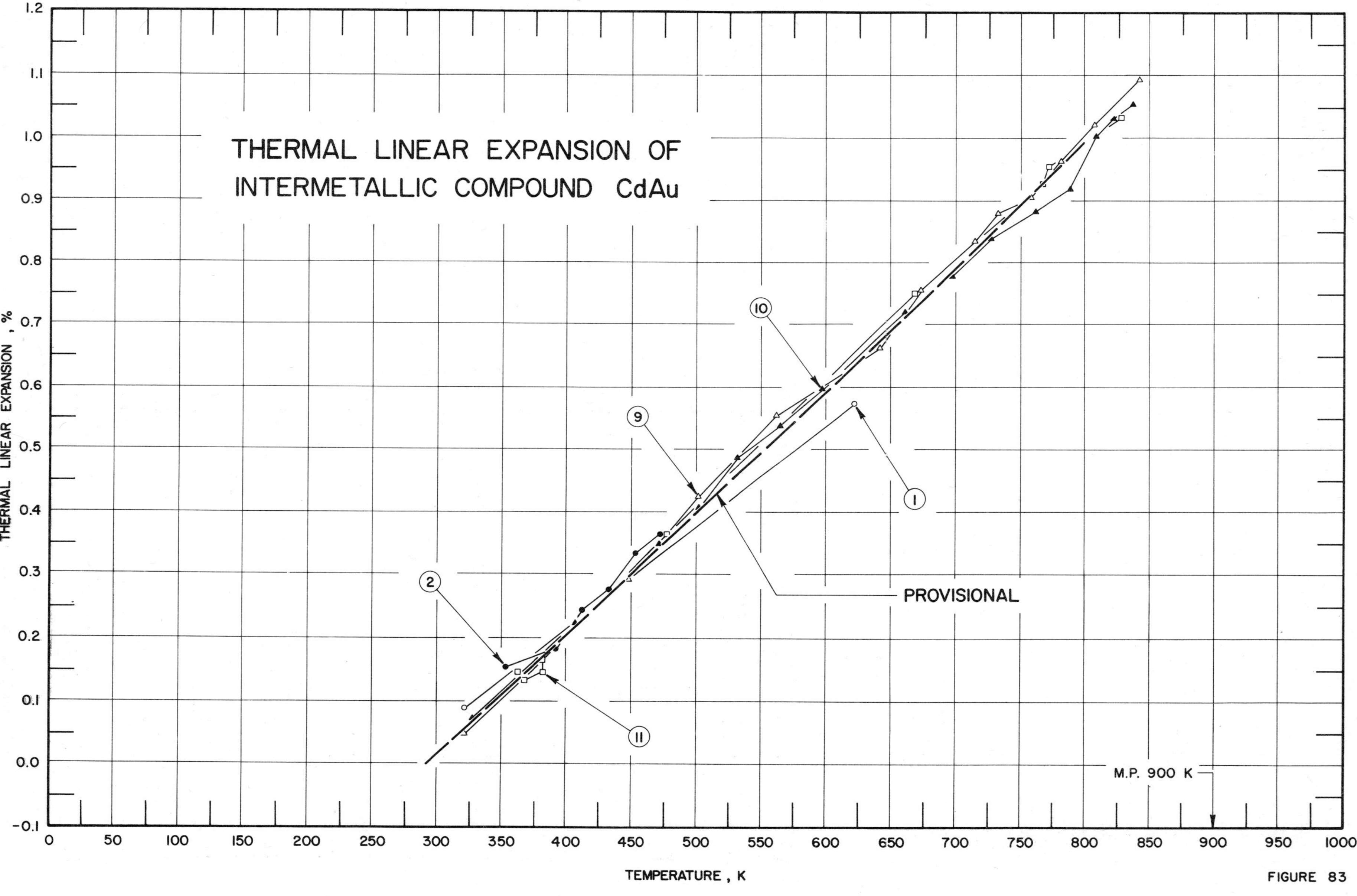

THERMAL LINEAR EXPANSION OF
INTERMETALLIC COMPOUND CdAu
PROVISIONAL
M.P. 900 K
1
2
9
10
11
TEMPERATURE, K
THERMAL LINEAR EXPANSION, %
0
50
100
150
200
250
300
350
400
450
500
550
600
650
700
750
800
850
900
950
1000
-0.1
0.0
0.1
0.2
0.3
0.4
0.5
0.6
0.7
0.8
0.9
1.0
1.1
1.2

FIGURE 83

SPECIFICATION TABLE 83. THERMAL LINEAR EXPANSION OF CdAu INTERMETALLIC COMPOUND

Cur. No.	Ref. No.	Author(s)	Year	Method Used	Temp. Range, K	Name and Specimen Designation	Composition (weight percent), Specifications, and Remarks
1	667	Koster, W. and Schneider, A.	1940	X	323,623	β-CdAu	36.3 Cd specimen from W.C. Heraeus of Hanau; powder specimen was heat treated in quartz tube in argon atm; lattice parameter reported at 323 K is 3.324 ± 0.004 Å; 3.321 Å at 293 K was used to calculate tabulated values.
2	684	Zirinsky, S.	1956	X	353-473		50.25 (at. % Cd) specimen prepared by melting accurately weighed 99.99 Au from Baker and Co., and 99.99 Cd from N.J. Zinc Co. in evacuated fused quartz tubes; lattice parameter reported at 353 K is 3.3226 kX, 3.319 kX at 293 K was used to calculate tabulated values.
3*	684	Zirinsky, S.	1956	L	353-473		The above specimen; this curve is here reported using the first given temperature as reference temperature at which $\Delta L/L_0 = 0$.
4*	691	Chang, L.C.	1951	X	343,378	β_1-CdAu	Single crystal 47.5 at. % Cd; cubic structure; this curve is here reported using the first given temperature as reference temperature at which $\Delta L/L_0 = 0$.
5*	691	Chang, L.C.	1951	X	293,318	β'-CdAu	Similar to the above specimen; orthorhombic structure; expansion measured perpendicular to (100) plane.
6*	691	Chang, L.C.	1951	X	293,318	β'-CdAu	The above specimen; expansion measured perpendicular to (010) plane.
7*	691	Chang, L.C.	1951	X	293,318	β'-CdAu	The above specimen; expansion measured perpendicular to (001) plane.
8*	691	Chang, L.C.	1951	X	293,318	β'-CdAu	The above specimen; expansion measured perpendicular to (111) plane.
9	678	Warlimont, H.	1959	X	322-842		No details given; continuous measurements; lattice parameter reported at 322 K is 3.3223 kX; 3.3207 kX at 293 K determined by graphical extrapolation.
10	678	Warlimont, H.	1959	X	329-837		No details given; continuous measurements; lattice parameter reported at 329 K is 3.3231 kX; 3.3207 kX at 293 K determined by graphical extrapolation.
11	678	Warlimont, H.	1959	X	364-829		No details given; isothermal measurements; lattice parameter reported at 364 K is 3.3256 kX; 3.3207 kX at 293 K determined by graphical extrapolation.

* Not shown in figure.

DATA TABLE 83. THERMAL LINEAR EXPANSION OF CdAu INTERMETALLIC COMPOUND

[Temperature, T, K; Linear Expansion, $\Delta L/L_0$, %]

T	$\Delta L/L_0$	T	$\Delta L/L_0$
CURVE 1		CURVE 9	
323	0.090	322	0.048
623	0.572	448	0.292
		501	0.425
CURVE 2		561	0.554
		641	0.714
353	0.120	673	0.759
373	0.151	715	0.834
393	0.181	733	0.879
413	0.241	759	0.906
433	0.271	781	0.961
453	0.331	808	1.021
473	0.361	842	1.093
CURVE 3*, †		CURVE 10	
353	0.000	329	0.072
473	0.262	406	0.223
		473	0.349
CURVE 4*, †		501	0.406
		533	0.488
343.2	0.000	565	0.539
378.2	0.115	598	0.599
		661	0.720
CURVE 5		696	0.780
		728	0.840
293.2	0.000	763	0.882
318.2	0.540	779	0.918
		808	1.006
CURVE 6		822	1.033
		837	1.057
293.2	0.000		
318.2	-0.030	CURVE 11	
CURVE 7		364	0.148
		368	0.133
293.2	0.000	377	0.145
318.2	-0.185	377	0.166
		478	0.364
CURVE 8		669	0.750
		768	0.928
293.2	0.000	773	0.952
318.2	0.247	829	1.033

* Not shown in figure.
† This curve is here reported using the first given temperature as reference temperature at which $\Delta L/L_0 = 0$.

FIGURE AND TABLE NO. 84AR. PROVISIONAL VALUES FOR THERMAL LINEAR EXPANSION OF CdLi INTERMETALLIC COMPOUND

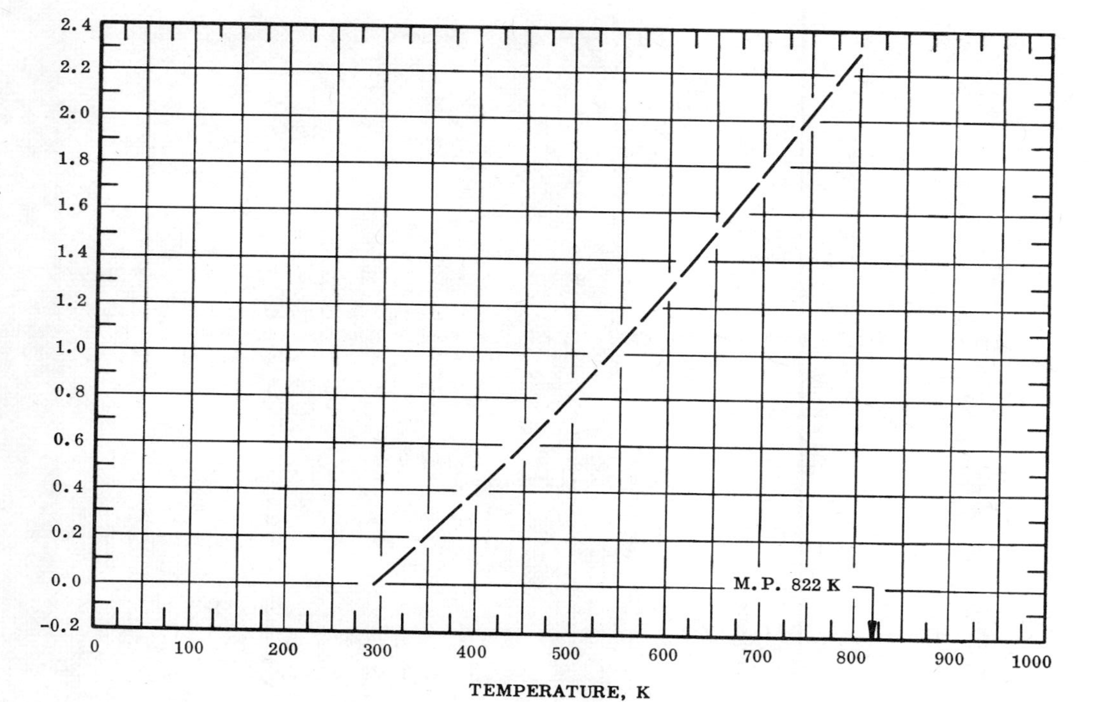

PROVISIONAL VALUES

[Temperature, T, K; Linear Expansion, $\Delta L/L_0$, %; α, K^{-1}]

T	$\Delta L/L_0$	$\alpha \times 10^6$
293	0.000	38
400	0.415	40
500	0.832	43
600	1.282	47
700	1.772	51
800	2.307	56

REMARKS

The tabulated values are considered accurate to within ±10% over the entire temperature range. These values can be represented approximately by the following equation:

$$\Delta L/L_0 = -1.025 + 3.334 \times 10^{-3}\ T + 2.938 \times 10^{-7}\ T^2 + 9.331 \times 10^{-10}\ T^3$$

FIGURE AND TABLE NO. 84CR. PROVISIONAL VALUES FOR THERMAL LINEAR EXPANSION OF CdTe INTERMETALLIC COMPOUND

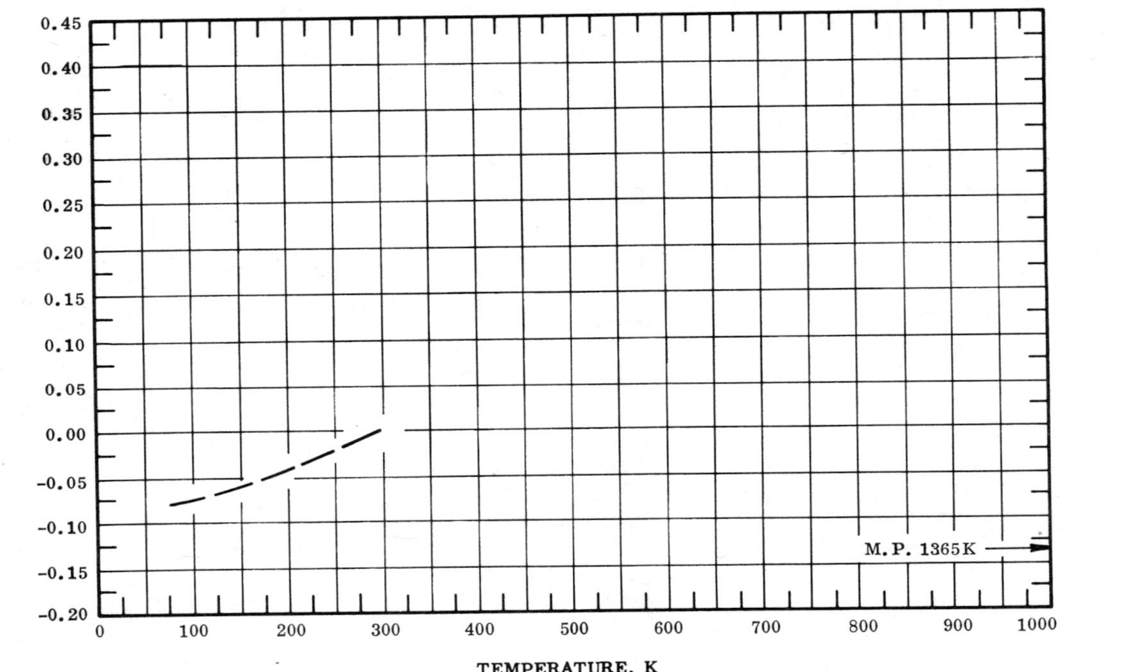

PROVISIONAL VALUES

[Temperature, T, K; Linear Expansion, $\Delta L/L_0$, %; α, K^{-1}]

T	$\Delta L/L_0$	$\alpha \times 10^6$
75	-0.079	2.2
100	-0.072	2.6
200	0.040	3.9
293	0.000	4.6

REMARKS

The tabulated values are considered accurate to within ± 7% over the entire temperature range. These values can be represented approximately by the following equation:

$$\Delta L/L_0 = -0.090 + 6.183 \times 10^{-5}\, T + 1.157 \times 10^{-6}\, T^2 - 1.102 \times 10^{-9}\, T^3$$

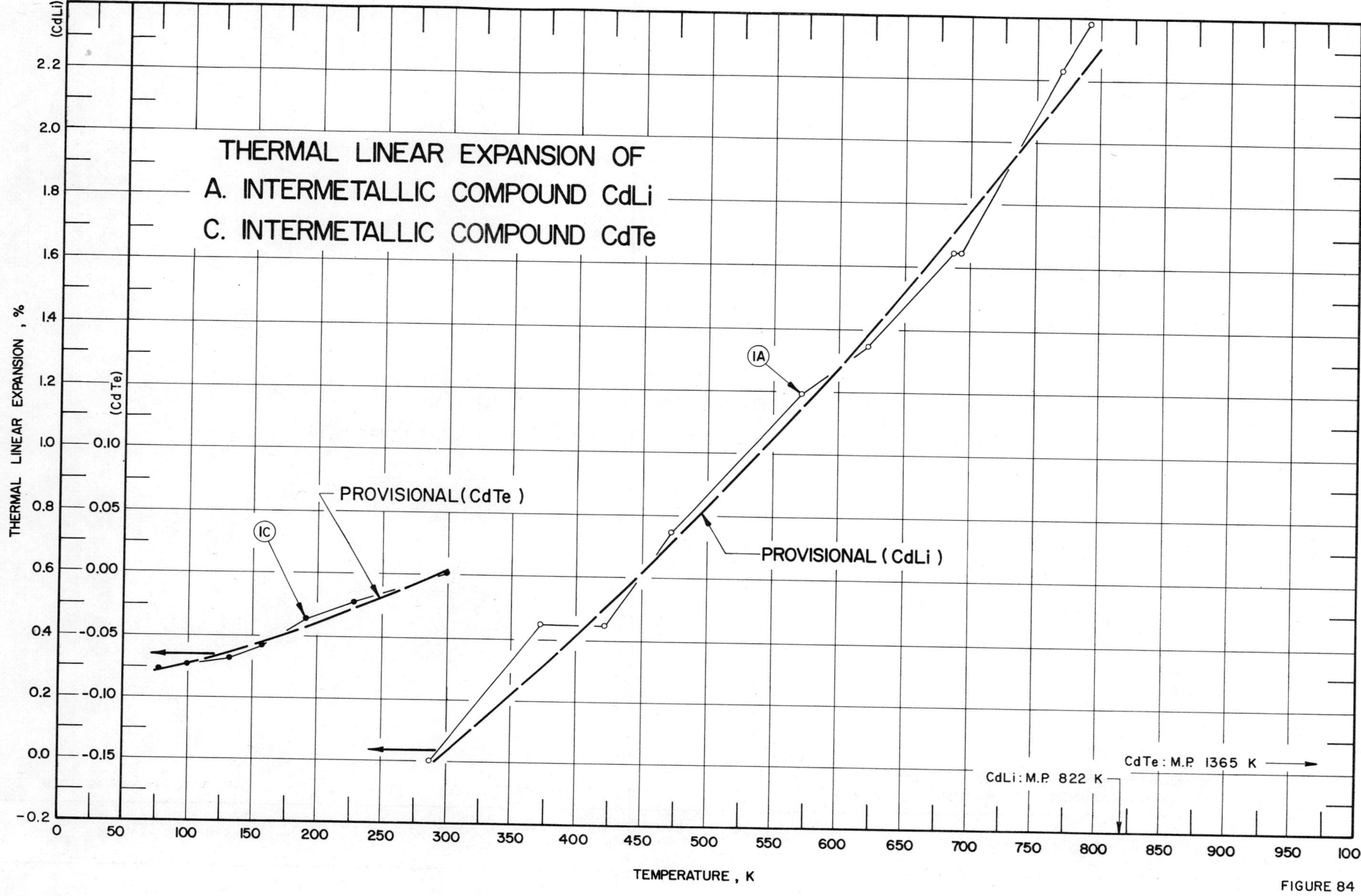
THERMAL LINEAR EXPANSION OF
A. INTERMETALLIC COMPOUND CdLi
C. INTERMETALLIC COMPOUND CdTe
THERMAL LINEAR EXPANSION, %
(CdLi)
(CdTe)
2.2
2.0
1.8
1.6
1.4
1.2
1.0
0.8
0.6
0.4
0.2
0.0
-0.2
0.10
0.05
0.00
-0.05
-0.10
-0.15
0
50
100
150
200
250
300
350
400
450
500
550
600
650
700
750
800
850
900
950
1000
TEMPERATURE, K
PROVISIONAL (CdTe)
PROVISIONAL (CdLi)
IA
IC
CdLi: M.P. 822 K
CdTe: M.P. 1365 K

FIGURE 84

SPECIFICATION TABLE 84A. THERMAL LINEAR EXPANSION OF CdLi INTERMETALLIC COMPOUND

Cur. No.	Ref. No.	Author(s)	Year	Method Used	Temp. Range, K	Name and Specimen Designation	Composition (weight percent), Specifications, and Remarks
1	690	Schneider, V.A. and Heymer, G.	1956	X	289-791		Specimen prepared from pure metals; lattice parameter reported at 289 K is 6.70 Å; 6.70 Å at 293 K determined by graphical interpolation.

DATA TABLE 84A. THERMAL LINEAR EXPANSION OF CdLi INTERMETALLIC COMPOUND

[Temperature, T, K; Linear Expansion, $\Delta L/L_0$, %]

T	$\Delta L/L_0$
CURVE 1	
289	0.000
373	0.448
423	0.448
473	0.746
573	1.194
623	1.343
663	1.642
667	1.642
763	2.239
791	2.388

SPECIFICATION TABLE 84B. THERMAL LINEAR EXPANSION OF Cd_3Mg INTERMETALLIC COMPOUND

Cur. No.	Ref. No.	Author(s)	Year	Method Used	Temp. Range, K	Name and Specimen Designation	Composition (weight percent), Specifications, and Remarks
1*	704	Edwards, D.A., Wallace, W.E., and Craig, R.S.	1952	X	298-569		Samples prepared by filing bulk alloy under Ar or He atmospheres; size of filings less than 200 mesh; spectroscopic analysis showed only traces of metallic contaminants; Mg from National Lead Co., Cd from Anaconda Copper Mining Co.; heat treated for 12 hr at 548 K and aged at room temperature for 9 months; sample becomes disordered at 348 K; expansion measured along a-axis; lattice parameter at 298 K reported as 6.2209 kX, 6.2217 kX at 293 K determined by graphical extrapolation.
2*	704	Edwards, D.A., et al.	1952	X	298-569		Similar to the above specimen; expansion measured along c-axis; lattice parameter at 298 K reported as 5.0348 kX, 5.0340 kX at 293 K determined by graphical extrapolation.

DATA TABLE 84B. THERMAL LINEAR EXPANSION OF Cd_3Mg INTERMETALLIC COMPOUND

[Temperature, T, K; Linear Expansion, $\Delta L/L_0$, %]

T	$\Delta L/L_0$	T	$\Delta L/L_0$
CURVE 1*		CURVE 2 (cont.)*	
298	-0.013	348	4.799
323	-0.056	359	4.338
339	-0.051	367	4.384
348	-0.064	414	4.619
348	-1.760	470	4.829
359	-1.545	518	5.000
367	-1.484	569	5.087
414	-1.368		
470	-1.236		
518	-0.966		
569	-0.776		
CURVE 2*			
298	0.016		
323	0.274		
339	0.526		
340	0.733		

* No figure given.

SPECIFICATION TABLE 84C. THERMAL LINEAR EXPANSION OF CdTe INTERMETALLIC COMPOUND

Cur. No.	Ref. No.	Author(s)	Year	Method Used	Temp. Range, K	Name and Specimen Designation	Composition (weight percent), Specifications, and Remarks
1	652	Schaake, H. F.	1969	X	77-300		Specimen was ground to <325 mesh, mixed with Ir and then sprinkled onto specimen holder; lattice parameter reported at 300 K is 6.4810 Å; 6.4809 Å at 293 K determined by graphical interpolation.

DATA TABLE 84C. THERMAL LINEAR EXPANSION OF CdTe INTERMETALLIC COMPOUND

[Temperature, T, K; Linear Expansion, $\Delta L/L_0$, %]

T	$\Delta L/L_0$
CURVE 1	
77	-0.076
100	-0.074
129	-0.067
159	-0.057
193	-0.036
229	-0.024
300	0.002

FIGURE AND TABLE NO. 85R. PROVISIONAL VALUES FOR THERMAL LINEAR EXPANSION OF $CaMg_2$ INTERMETALLIC COMPOUND

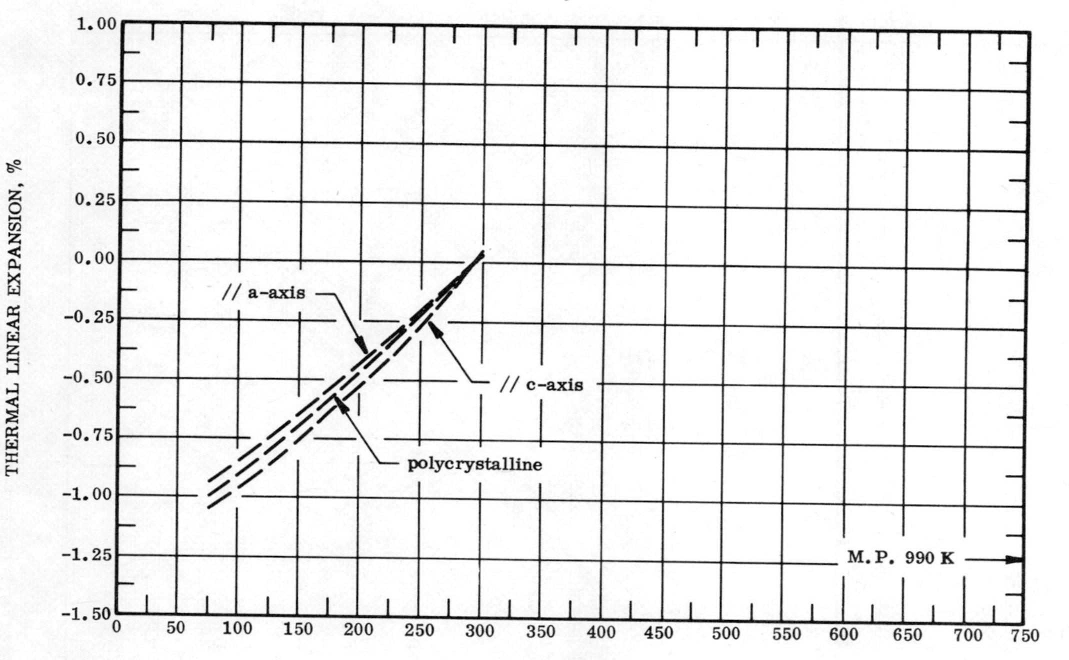

PROVISIONAL VALUES

[Temperature, T, K; Linear Expansion, $\Delta L/L_0$, %; α, K^{-1}]

T	// a-axis $\Delta L/L_0$	// c-axis $\Delta L/L_0$	polycrystalline $\Delta L/L_0$	polycrystalline $\alpha \times 10^6$
75	-0.960	-1.052	-0.990	37
100	-0.863	-0.963	-0.897	39
200	-0.444	-0.520	-0.469	47
293	0.000	0.000	0.000	54

REMARKS

The tabulated values are considered accurate to within ±10%. The values for polycrystalline material in the absence of experimental data are calculated from the single crystal data. These and other values can be represented approximately by the following equations:

// a-axis: $\Delta L/L_0 = -1.236 + 3.554 \times 10^{-3} T + 1.535 \times 10^{-6} T^2 + 2.510 \times 10^{-9} T^3$

// c-axis: $\Delta L/L_0 = -1.265 + 2.251 \times 10^{-3} T + 8.003 \times 10^{-6} T^2 - 3.240 \times 10^{-9} T^3$

polycrystalline: $\Delta L/L_0 = -0.125 + 3.099 \times 10^{-3} T + 3.840 \times 10^{-6} T^2 + 2.942 \times 10^{-10} T^3$

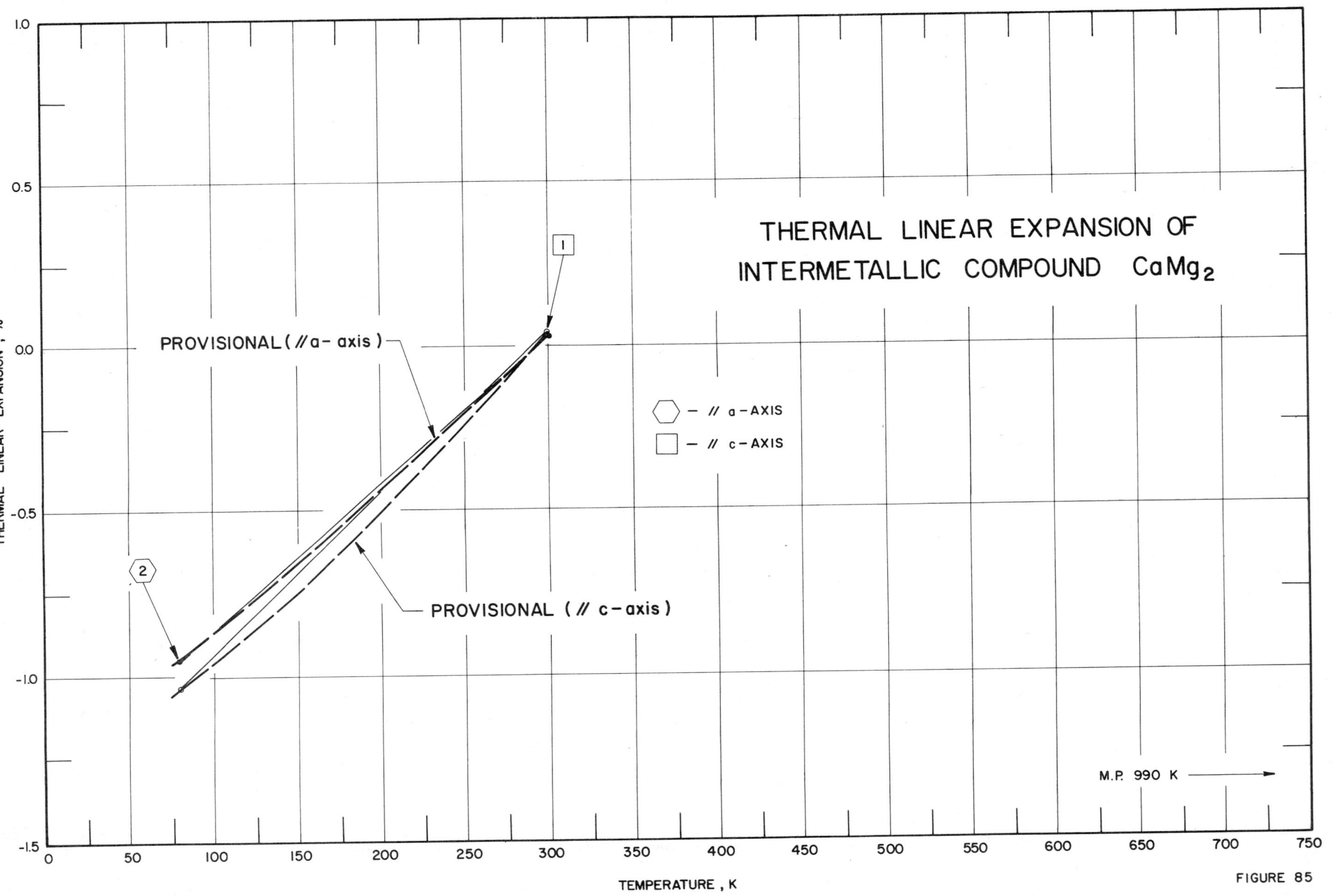
THERMAL LINEAR EXPANSION OF
INTERMETALLIC COMPOUND $CaMg_2$
— // a − AXIS
— // c − AXIS
PROVISIONAL (// a − axis)
PROVISIONAL (// c − axis)
1
2
M.P. 990 K
THERMAL LINEAR EXPANSION, %
1.0
0.5
0.0
-0.5
-1.0
-1.5
0
50
100
150
200
250
300
350
400
450
500
550
600
650
700
750
TEMPERATURE, K
FIGURE 85

SPECIFICATION TABLE 85. THERMAL LINEAR EXPANSION OF $CaMg_2$ INTERMETALLIC COMPOUND

Cur. No.	Ref. No.	Author(s)	Year	Method Used	Temp. Range, K	Name and Specimen Designation	Composition (weight percent), Specifications, and Remarks
1	688	Smith, J.F. and Ogren, J.R.	1958	X	80-300		Single crystal of hexagonal specimen; grown by Bridgman method; expansion measured along c-axis.
2	688	Smith, J.F. and Ogren, J.R.	1958	X	80-300		The above specimen; expansion measured perpendicular to c-axis.

DATA TABLE 85. THERMAL LINEAR EXPANSION OF $CaMg_2$ INTERMETALLIC COMPOUND

[Temperature, T, K; Linear Expansion, $\Delta L/L_0$, %]

T	$\Delta L/L_0$
CURVE 1	
80	-1.029
300	0.034
CURVE 2	
80	-0.946
300	0.031

FIGURE AND TABLE NO. 86AR. PROVISIONAL VALUES FOR THERMAL LINEAR EXPANSION OF $CeIn_3$ INTERMETALLIC COMPOUND

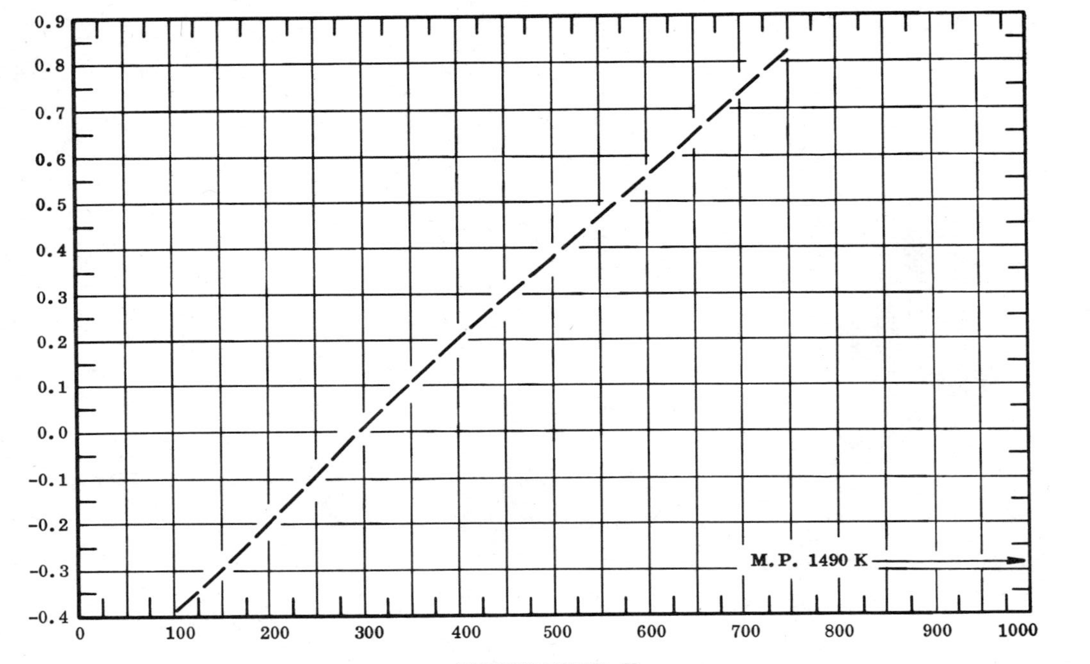

PROVISIONAL VALUES

[Temperature, T, K; Linear Expansion, $\Delta L/L_0$, %]

T	$\Delta L/L_0$
100	-0.386
200	-0.196
293	0.000
400	0.204
500	0.377
600	0.553
700	0.736
750	0.830

REMARKS

The tabulated values are considered accurate to within ±10% over the entire temperature range.

FIGURE AND TABLE NO. 86BR. PROVISIONAL VALUES FOR THERMAL LINEAR EXPANSION OF $CePd_3$ INTERMETALLIC COMPOUND

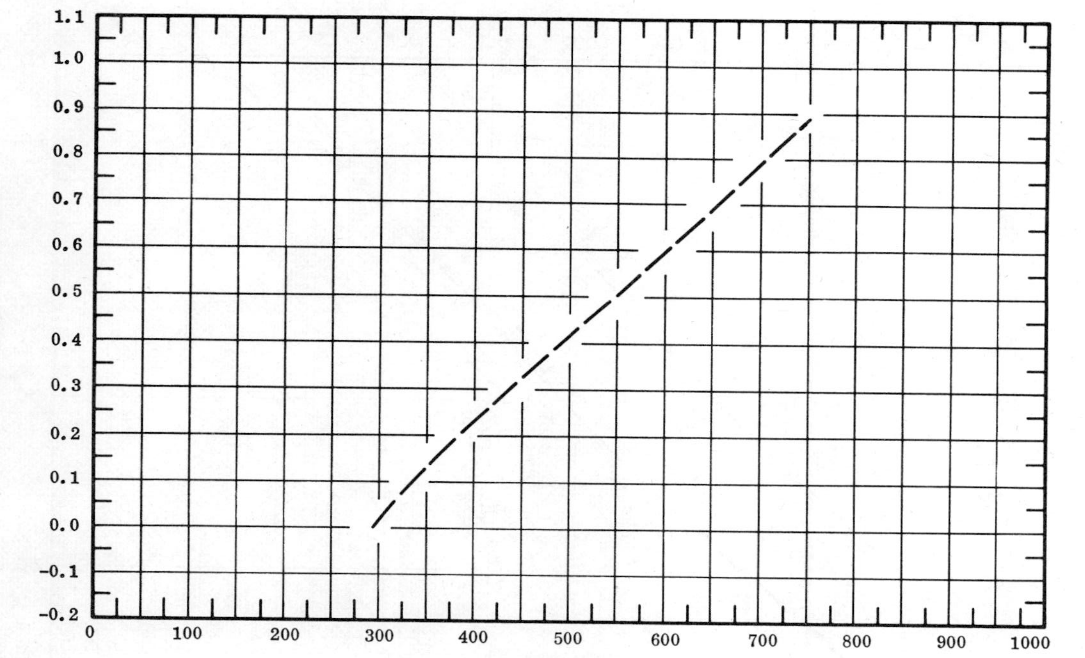

PROVISIONAL VALUES

[Temperature, T, K; Linear Expansion, $\Delta L/L_0$, %; α, K^{-1}]

T	$\Delta L/L_0$	$\alpha \times 10^6$
293	0.000	23.1
400	0.228	19.8
500	0.418	18.2
600	0.598	18.0
700	0.782	19.2
750	0.880	20.3

REMARKS

The tabulated values are considered accurate to within ± 7% over the entire temperature range. These values can be represented approximately by the following equation:

$$\Delta L/L_0 = -0.900 + 4.032 \times 10^{-3}\, T - 3.963 \times 10^{-6}\, T^2 + 2.335 \times 10^{-9}\, T^3$$

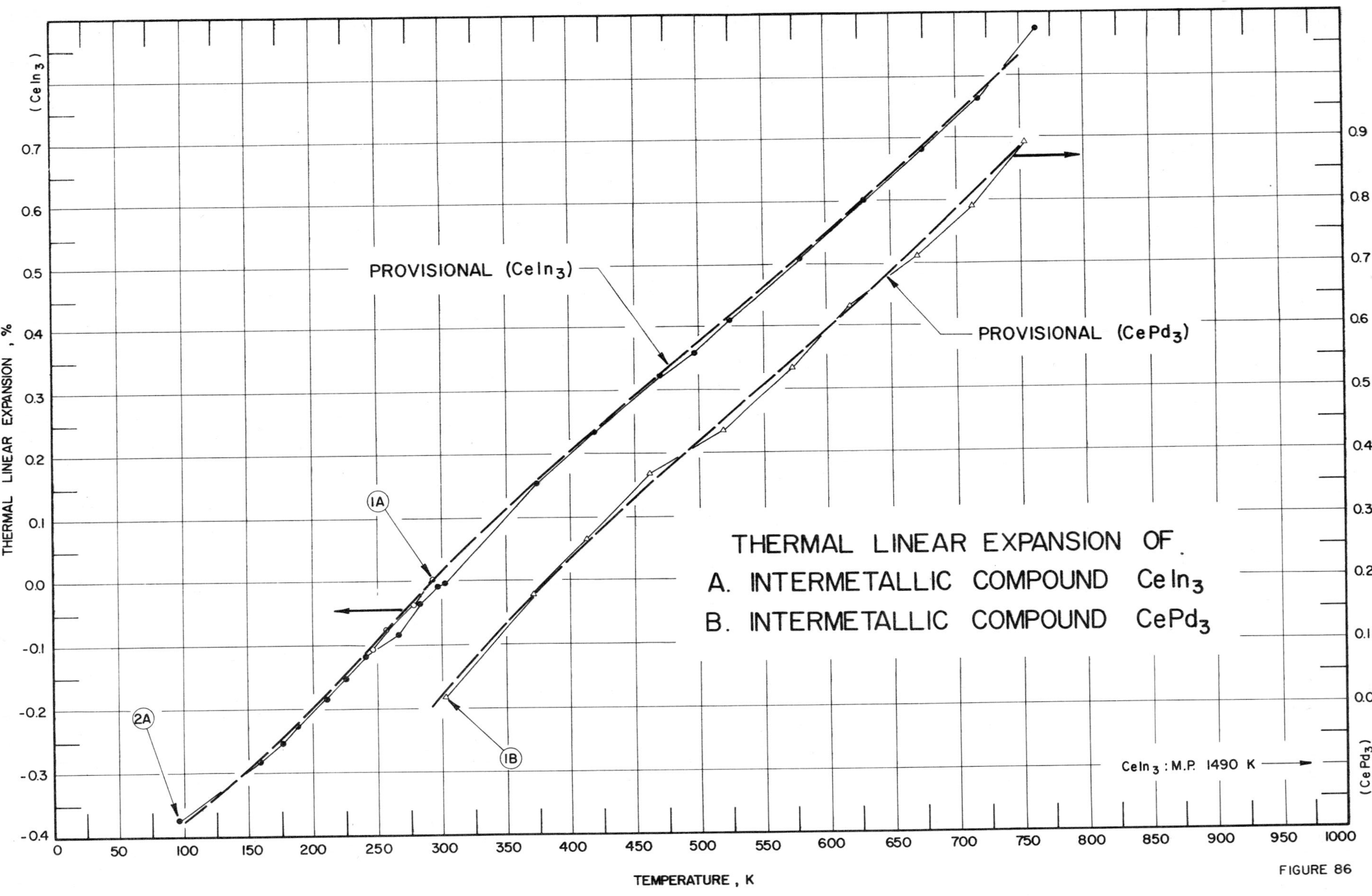

THERMAL LINEAR EXPANSION OF
A. INTERMETALLIC COMPOUND CeIn3
B. INTERMETALLIC COMPOUND CePd3
PROVISIONAL (CeIn3)
PROVISIONAL (CePd3)
CeIn3 : M.P. 1490 K
1A
1B
2A
THERMAL LINEAR EXPANSION , %
(CeIn3)
0.7
0.6
0.5
0.4
0.3
0.2
0.1
0.0
-0.1
-0.2
-0.3
-0.4
(CePd3)
0.9
0.8
0.7
0.6
0.5
0.4
0.3
0.2
0.1
0.0
TEMPERATURE , K
0
50
100
150
200
250
300
350
400
450
500
550
600
650
700
750
800
850
900
950
1000

FIGURE 86

SPECIFICATION TABLE 86A. THERMAL LINEAR EXPANSION OF $CeIn_3$ INTERMETALLIC COMPOUND

Cur. No.	Ref. No.	Author(s)	Year	Method Used	Temp. Range, K	Name and Specimen Designation	Composition (weight percent), Specifications, and Remarks
1	682	Harris, I.R. and Raynor, G.V.	1965	X	246-293		Specimen prepared from In containing 0.002 Pb and 0.001 Cd from Johnson, Matthey and Co. Ltd., and Ce containing < 0.001 other rare metals, and 0.02 other base metals; prepared by melting weighed 15 mm Hg argon wrapped in Ta foil and sealed in degassed silica tube at 2 x 10^{-6} mm Hg pressure; homogenized, annealed for 3 days at 973 K and quenched; powdered; annealed at 973 K for stress relief; slight abnormality at 513 K; lattice parameter reported at 95 K is 4.6620 kX; 4.6796 kX at 293 K determined by graphical extrapolation.
2	682	Harris, I.R. and Raynor, G.V.	1965	X	762-95		The above specimen; measurements with decreasing temperature; lattice parameter reported at 293 K is 4.6785 kX; slight abnormality observed at 143 K.

DATA TABLE 86A. THERMAL LINEAR EXPANSION OF $CeIn_3$ INTERMETALLIC COMPOUND

[Temperature, T, K; Linear Expansion, $\Delta L/L_0$, %]

T	$\Delta L/L_0$	T	$\Delta L/L_0$
CURVE 1		CURVE 2 (cont.)	
246	-0.107	302	-0.004
263	-0.079	296	-0.008
277	-0.036	278	-0.034
293	0.000	266	-0.083
		242	-0.115
CURVE 2		225	-0.152
		210	-0.186
762	0.874	189	-0.226
718	0.763	176	-0.252
673	0.684	158	-0.282
629	0.603	95	-0.376
578	0.511		
525	0.412		
496	0.359		
470	0.325		
419	0.235		
373	0.154		

SPECIFICATION TABLE 86B. THERMAL LINEAR EXPANSION OF $CePd_3$ INTERMETALLIC COMPOUND

Cur. No.	Ref. No.	Author(s)	Year	Method Used	Temp. Range, K	Name and Specimen Designation	Composition (weight percent), Specifications, and Remarks
1	557	Harris, L.R., Raynor, G.V., and Winstanley, C.J.	1967	X	301-754		Ce 99.95$^+$ pure, Pd 99.95$^+$ pure; obtained from Johnson, Matthey and Co.; CuKα radiation used for analysis; lattice spacing reported at 301 K is 4.1198 Å; 4.1190 Å at 293 K determined by graphical extrapolation.

DATA TABLE 86B. THERMAL LINEAR EXPANSION OF $CePd_3$ INTERMETALLIC COMPOUND

[Temperature, T, K; Linear Expansion, $\Delta L/L_0$, %]

T	$\Delta L/L_0$
CURVE 1	
301	0.019
373	0.180
413	0.265
461	0.359
518	0.439
574	0.537
619	0.634
670	0.716
713	0.794
754	0.893

FIGURE AND TABLE NO. 87R. PROVISIONAL VALUES FOR THERMAL LINEAR EXPANSION OF $CeRu_2$ INTERMETALLIC COMPOUND

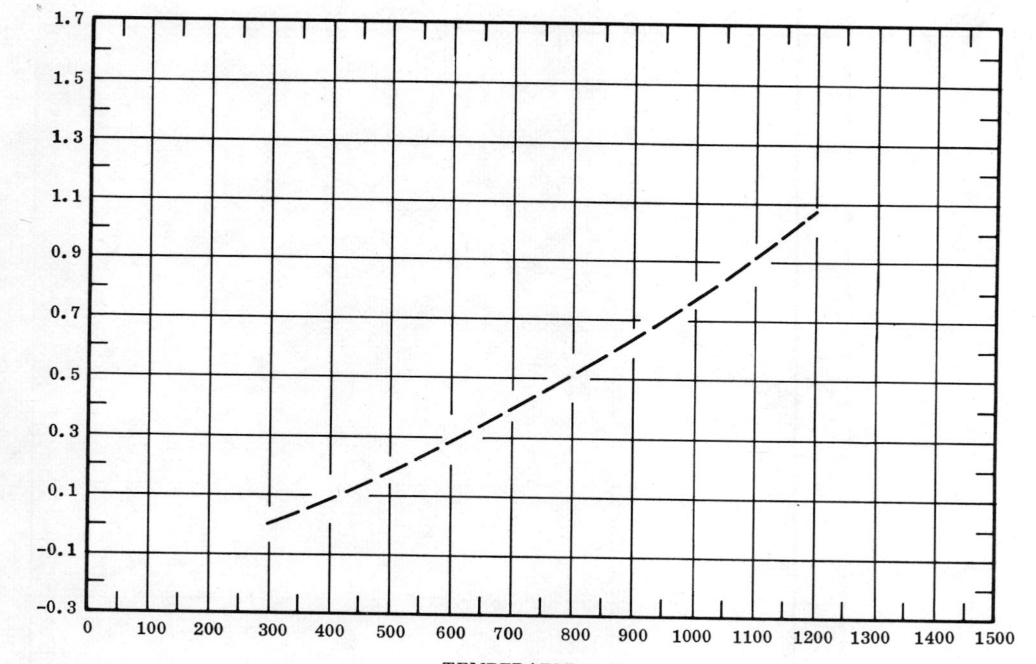

PROVISIONAL VALUES

[Temperature, T, K; Linear Expansion, $\Delta L/L_0$, %; α, K^{-1}]

T	$\Delta L/L_0$	$\alpha \times 10^6$
293	0.000	7.9
400	0.090	8.8
500	0.182	9.6
600	0.282	10.5
700	0.391	11.3
800	0.509	12.2
900	0.636	13.2
1000	0.773	14.2
1100	0.920	15.2
1200	1.077	16.2

REMARKS

The tabulated values are considered accurate to within ± 7% over the entire temperature range. These values can be represented approximately by the following equation:

$$\Delta L/L_0 = -0.202 + 5.901 \times 10^{-4}\, T + 3.294 \times 10^{-7}\, T^2 + 5.616 \times 10^{-11}\, T^3$$

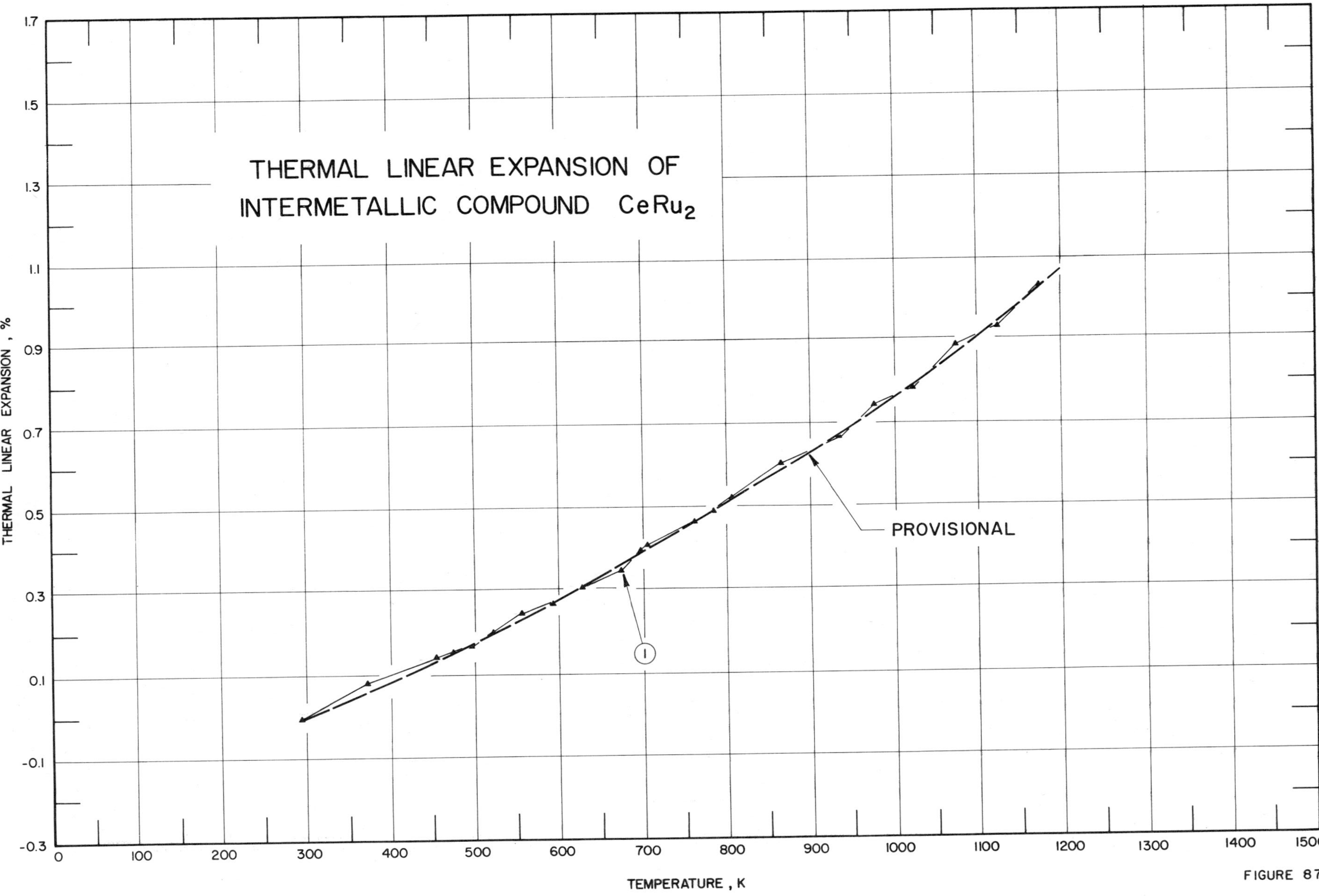
THERMAL LINEAR EXPANSION OF
INTERMETALLIC COMPOUND CeRu2
PROVISIONAL
1
THERMAL LINEAR EXPANSION , %
TEMPERATURE , K

FIGURE 87

SPECIFICATION TABLE 87. THERMAL LINEAR EXPANSION OF $CeRu_2$ INTERMETALLIC COMPOUND

Cur. No.	Ref. No.	Author(s)	Year	Method Used	Temp. Range, K	Name and Specimen Designation	Composition (weight percent), Specifications, and Remarks
1	558	Gschneidner, K.A., Jr., Elliott, R.O., and Cromer, D.T.	1965	X	272-1173		Prepared by arc-melting components; filings annealed at 723 K for 15 min; CuKα radiation used; lattice parameter reported at 293 K is 7.5361 Å.

DATA TABLE 87. THERMAL LINEAR EXPANSION OF $CeRu_2$ INTERMETALLIC COMPOUND

[Temperature, T, K; Linear Expansion, $\Delta L/L_0$, %]

T	$\Delta L/L_0$	T	$\Delta L/L_0$
CURVE 1		CURVE 1 (cont.)	
272	0.0026	808	0.5214
273	0.0066*	810	0.5254*
293	0.0000	868	0.6037
374	0.0809	931	0.6674
454	0.1433	973	0.7417
472	0.1592	1025	0.7882
493	0.1751	1073	0.8903
524	0.2016	1123	0.9394
557	0.2415	1173	1.0370
558	0.2627*		
594	0.2680		
628	0.3078		
674	0.3463		
698	0.3994		
705	0.4100		
761	0.4604		
783	0.4976		

*Not shown in figure.

FIGURE AND TABLE NO. 88R. PROVISIONAL VALUES FOR THERMAL LINEAR EXPANSION OF $CeSn_3$ INTERMETALLIC COMPOUND

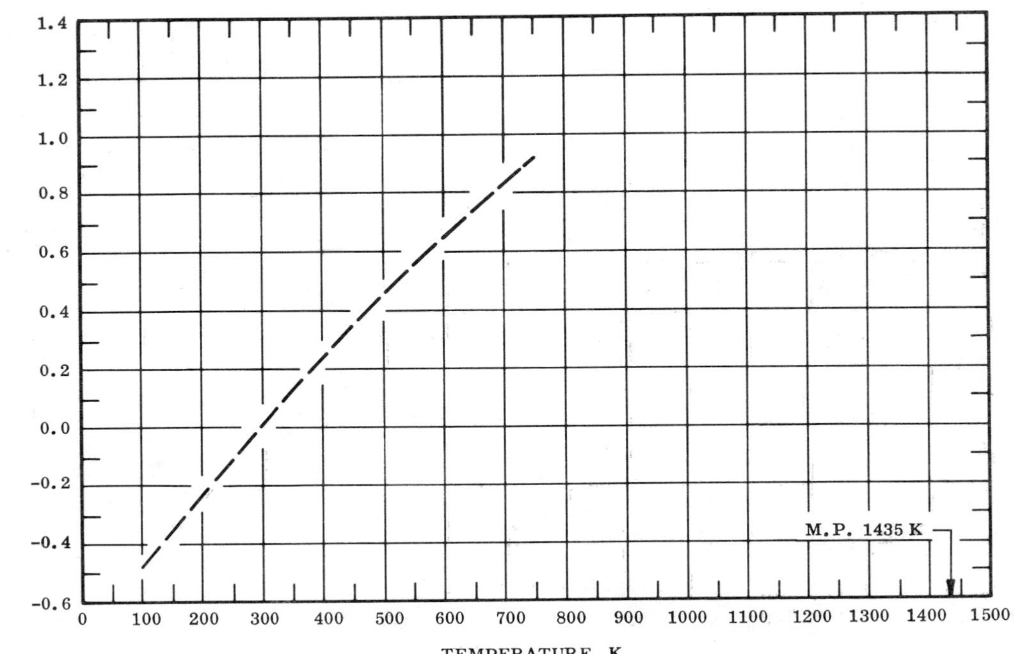

PROVISIONAL VALUES

[Temperature, T, K; Linear Expansion, $\Delta L/L_0$, %; α, K^{-1}]

T	$\Delta L/L_0$	$\alpha \times 10^6$
100	-0.493	28.0
200	-0.225	25.3
293	0.000	23.3
400	0.237	21.3
500	0.444	20.0
600	0.640	19.1
700	0.829	18.7
750	0.922	18.6

REMARKS

The tabulated values are considered accurate to within ± 7% over the entire temperature range. These values can be represented approximately by the following equation:

$$\Delta L/L_0 = -0.788 + 3.114 \times 10^{-3} T - 1.673 \times 10^{-6} T^2 + 7.475 \times 10^{-10} T^3$$

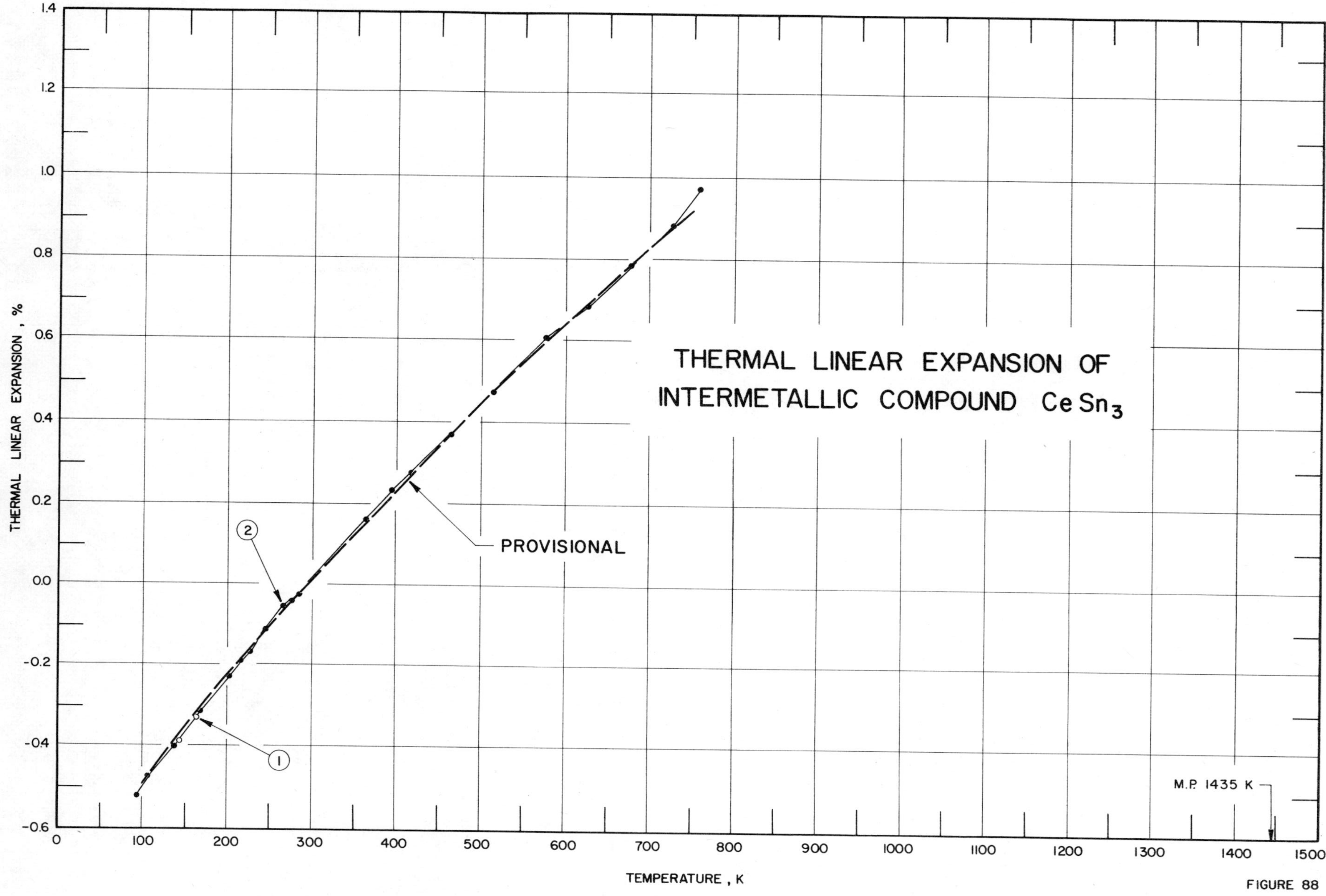

THERMAL LINEAR EXPANSION OF
INTERMETALLIC COMPOUND $CeSn_3$
PROVISIONAL
M.P. 1435 K
1
2
THERMAL LINEAR EXPANSION, %
TEMPERATURE, K
1.4
1.2
1.0
0.8
0.6
0.4
0.2
0.0
-0.2
-0.4
-0.6
0
100
200
300
400
500
600
700
800
900
1000
1100
1200
1300
1400
1500

FIGURE 88

SPECIFICATION TABLE 88. THERMAL LINEAR EXPANSION OF $CeSn_3$ INTERMETALLIC COMPOUND

Cur. No.	Ref. No.	Author(s)	Year	Method Used	Temp. Range, K	Name and Specimen Designation	Composition (weight percent), Specifications, and Remarks
1	682	Harris, I.R. and Raynor, G.V.	1965	X	142-293		Specimen prepared from In containing 0.002 Pb and 0.001 Cd from Messrs. Johnson, Matthey and Co. Ltd., and 99.999 Sn from Messrs. Capper, Pass Ltd. by melting weighed quantities in non-consumable electrode-arc furnace in 15 mm Hg argon; wrapped in Ta foil and sealed in degassed silica tubes at 2 x 10^{-6} mm Hg pressure; homogenized, annealed for 3 days at 973 K and quenched, powdered, annealed at 973 K for stress relief; measurements with increasing temperature; lattice parameter reported at 164 K is 4.6961 kX; 4.7116 kX at 293 K determined by graphical extrapolation.
2	682	Harris, I.R. and Raynor, G.V.	1965	X	759-93		The above specimen; measurements with decreasing temperature; lattice parameter reported at 293 K is 4.7118 kX.

DATA TABLE 88. THERMAL LINEAR EXPANSION OF $CeSn_3$ INTERMETALLIC COMPOUND

[Temperature, T, K; Linear Expansion, $\Delta L/L_0$, %]

T	$\Delta L/L_0$	T	$\Delta L/L_0$
CURVE 1		CURVE 2 (cont.)	
142	-0.388	288	-0.021
164	-0.329	281	-0.023
293	0.000	275	-0.044
		267	-0.059
CURVE 2		245	-0.112
		227	-0.166
759	0.968	218	-0.185
724	0.885	202	-0.227
673	0.783	193	-0.250
623	0.686	169	-0.314
579	0.607	161	-0.335
517	0.473	137	-0.401
465	0.373	108	-0.475
415	0.274	93	-0.524
396	0.235		
364	0.165		
300	0.017		

SPECIFICATION TABLE 89. THERMAL LINEAR EXPANSION OF CrFe INTERMETALLIC COMPOUND

Cur. No.	Ref. No.	Author(s)	Year	Method Used	Temp. Range, K	Name and Specimen Designation	Composition (weight percent), Specifications, and Remarks
1*	707	Pridantseva, K. S. and Solov'yeva, N. A.	1965	L	293-973		55 Fe, balance Cr, powder sample, impurities mainly 1% oxygen, samples melted in induction furnace in argon atmosphere, homogeneous non-magnetic σ-phase; annealed at 983 K for 10 days 1% measurement error.

DATA TABLE 89. THERMAL LINEAR EXPANSION OF CrFe INTERMETALLIC COMPOUND

[Temperature, T, K; Linear Expansion, $\Delta L/L_0$, %]

T	$\Delta L/L_0$
CURVE 1*	
293	0.000
573	0.357
773	0.631
973	0.945

* No figure given.

FIGURE AND TABLE NO. 90AR. PROVISIONAL VALUES FOR THERMAL LINEAR EXPANSION OF Co_2Dy INTERMETALLIC COMPOUND

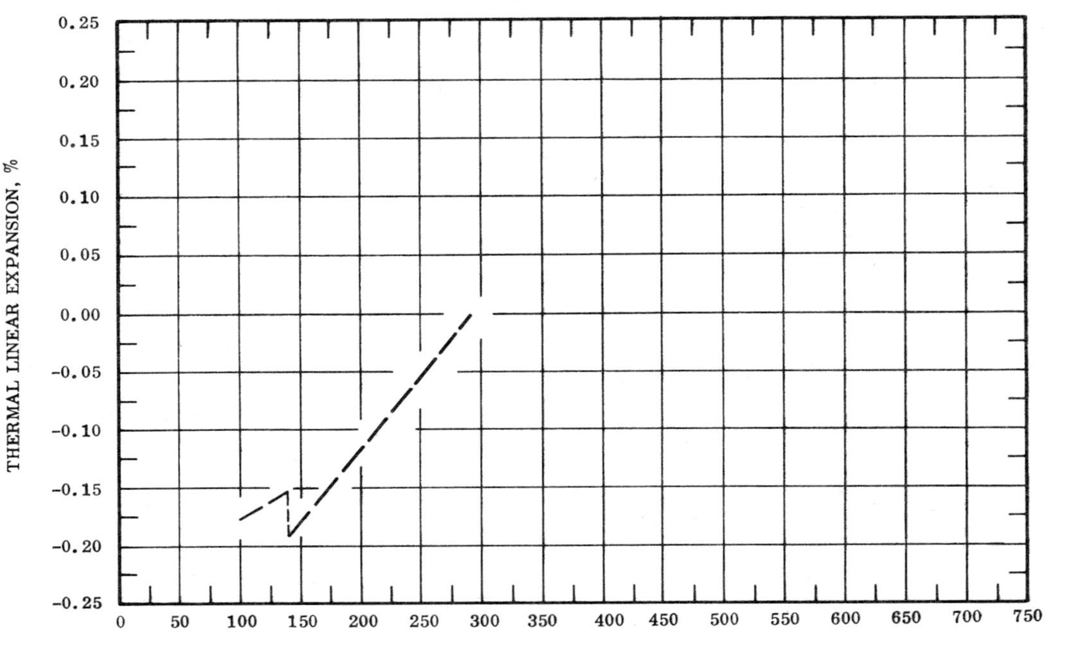

PROVISIONAL VALUES

[Temperature, T, K; Linear Expansion, $\Delta L/L_0$, %; α, K^{-1}]

T	$\Delta L/L_0$	$\alpha \times 10^6$
100	-0.179	
140†	-0.152	
140†	-0.192	11.8
150	-0.181	11.9
200	-0.119	12.5
293	0.000	13.1

† Magnetic anomaly.

REMARKS

The tabulated values are considered accurate to within ± 7% over the entire temperature range. An anomaly due to magnetic transition is observed at 140 K. These values can be represented approximately by the following equation:

$$\Delta L/L_0 = -0.343 + 9.552 \times 10^{-4}\, T + 1.020 \times 10^{-6}\, T^2 - 9.543 \times 10^{-10}\, T^3 \qquad (140 < T < 293)$$

FIGURE AND TABLE NO. 90BR. PROVISIONAL VALUES FOR THERMAL LINEAR EXPANSION OF Co_2Gd INTERMETALLIC COMPOUND

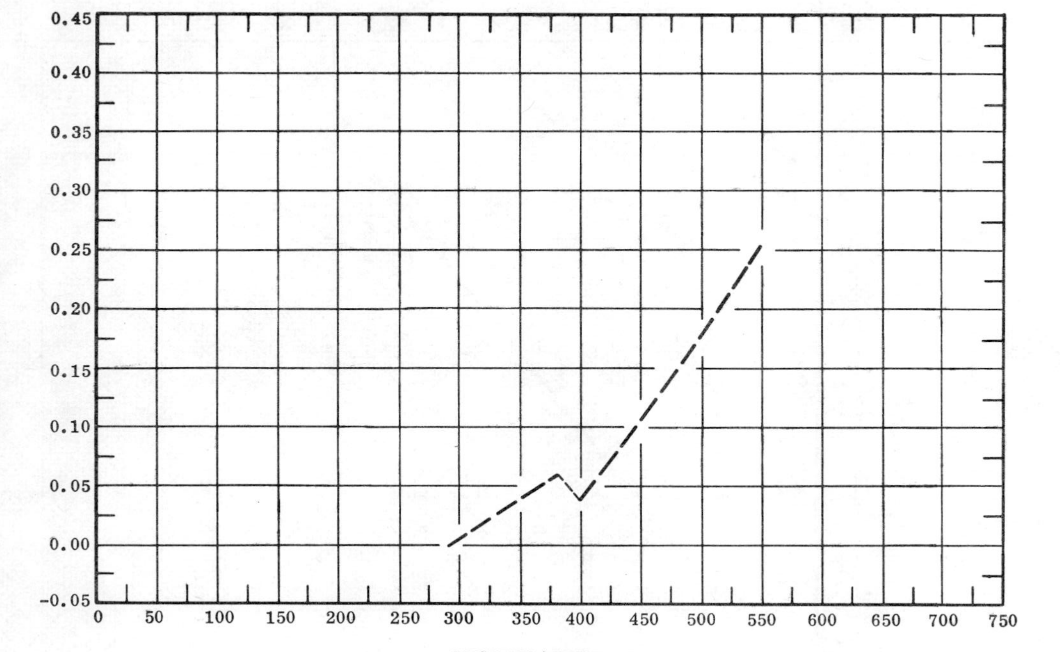

PROVISIONAL VALUES

[Temperature, T, K; Linear Expansion, $\Delta L/L_0$, %; α, K^{-1}]

T	$\Delta L/L_0$	$\alpha \times 10^6$
293	0.000	
350	0.038	6.7
383†	0.060	
400	0.039	
500	0.178	15.0
550	0.254	15.2

† Anomaly due to magnetic transformation.

REMARKS

The tabulated values are considered accurate to within ±7% over the entire temperature range. These values can be represented approximately by the following equation:

$$\Delta L/L_0 = 0.243 - 3.097 \times 10^{-3} T + 8.600 \times 10^{-6} T^2 - 5.333 \times 10^{-9} T^3 \quad (400 < T < 550)$$

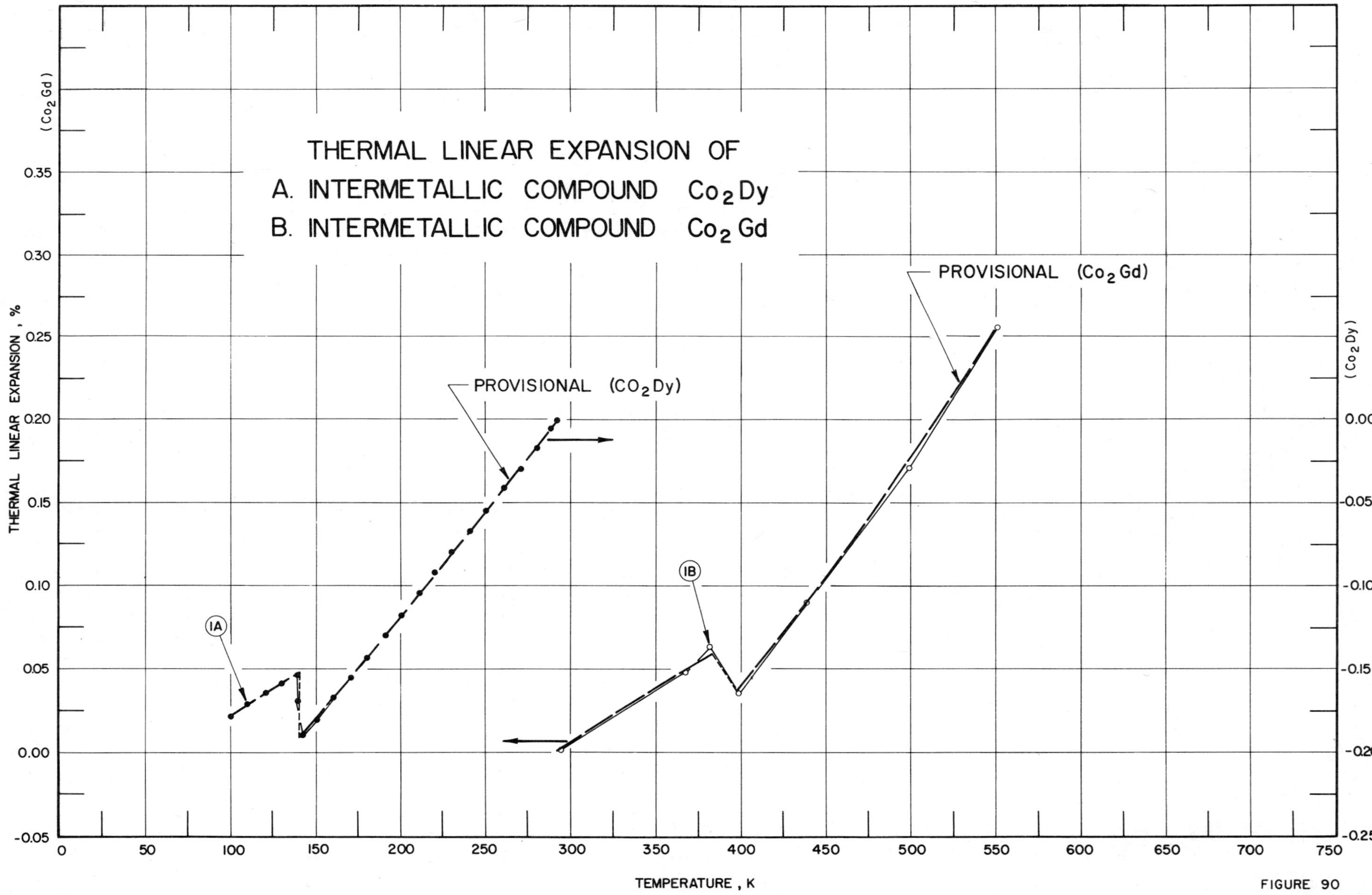
THERMAL LINEAR EXPANSION OF
A. INTERMETALLIC COMPOUND Co$_2$Dy
B. INTERMETALLIC COMPOUND Co$_2$Gd
PROVISIONAL (Co$_2$Gd)
PROVISIONAL (CO$_2$Dy)
1A
1B
THERMAL LINEAR EXPANSION, %
(Co$_2$Gd)
0.35
0.30
0.25
0.20
0.15
0.10
0.05
0.00
-0.05
(Co$_2$Dy)
0.00
-0.05
-0.10
-0.15
-0.20
-0.25
0
50
100
150
200
250
300
350
400
450
500
550
600
650
700
750
TEMPERATURE, K

FIGURE 90

SPECIFICATION TABLE 90A. THERMAL LINEAR EXPANSION OF Co_2Dy INTERMETALLIC COMPOUND

Cur. No.	Ref. No.	Author(s)	Year	Method Used	Temp. Range, K	Name and Specimen Designation	Composition (weight percent), Specifications, and Remarks
1	750	Chatterjee, D., and Taylor, K.N.R.	1971	S	100-293		Polycrystalline specimen prepared by melting stoichiometric quantities in argon atmosphere; flat surfaces were cut on the arc melted "buttons" using 0.01 inch carborundum saw and polished and further cleaned with trichloroethylene; magnetic transition reported near 140 K.

DATA TABLE 90A. THERMAL LINEAR EXPANSION OF Co_2Dy INTERMETALLIC COMPOUND

[Temperature, T, K; Linear Expansion, $\Delta L/L_0$, %]

T	$\Delta L/L_0$	T	$\Delta L/L_0$
CURVE 1		CURVE 1 (cont.)	
100	-0.179	200	-0.117
110	-0.171	210	-0.105
120	-0.164	220	-0.092
130	-0.158	230	-0.080
139	-0.153	240	-0.067
140	-0.169	250	-0.055
142	-0.190	260	-0.042
150	-0.181	270	-0.030
160	-0.167	280	-0.017
170	-0.155	288	-0.005
180	-0.143	293	0.000
190	-0.130		

SPECIFICATION TABLE 90B. THERMAL LINEAR EXPANSION OF Co_2Gd INTERMETALLIC COMPOUND

Cur. No.	Ref. No.	Author(s)	Year	Method Used	Temp. Range, K	Name and Specimen Designation	Composition (weight percent), Specifications, and Remarks
1	560	Mansey, R.C., Raynor, G.V., and Harris, I.R.	1968	X	295-551		Gd from Johnson, Matthey and Co., 99.98$^+$ pure; 99.998 pure Co from Koch-Light Laboratories, Inc.; homogenized 3 days at 773 K, rapidly cooled and brittle part pulverized under paraffin to obtain specimens; annealing found unnecessary; discontinuity in expansion found near Curie point; expansion measured using $K\alpha$ x-radiation from Cu, Co, and Cr; contraction between 383-393 K due to Curie temperature; lattice parameter reported at 295 K is 7.2417 kX (7.2562 Å); 7.2416 kX (7.2561 Å) at 293 K determined by graphical extrapolation.

DATA TABLE 90B. THERMAL LINEAR EXPANSION OF Co_2Gd INTERMETALLIC COMPOUND

[Temperature, T, K; Linear Expansion, $\Delta L/L_0$, %]

T	$\Delta L/L_0$
CURVE 1	
295	0.001
366	0.048
378	0.062
398	0.036
438	0.090
499	0.171
551	0.257

FIGURE AND TABLE NO. 91AR. TYPICAL VALUES FOR THERMAL LINEAR EXPANSION OF $Co_{17}Y_2$ INTERMETALLIC COMPOUND

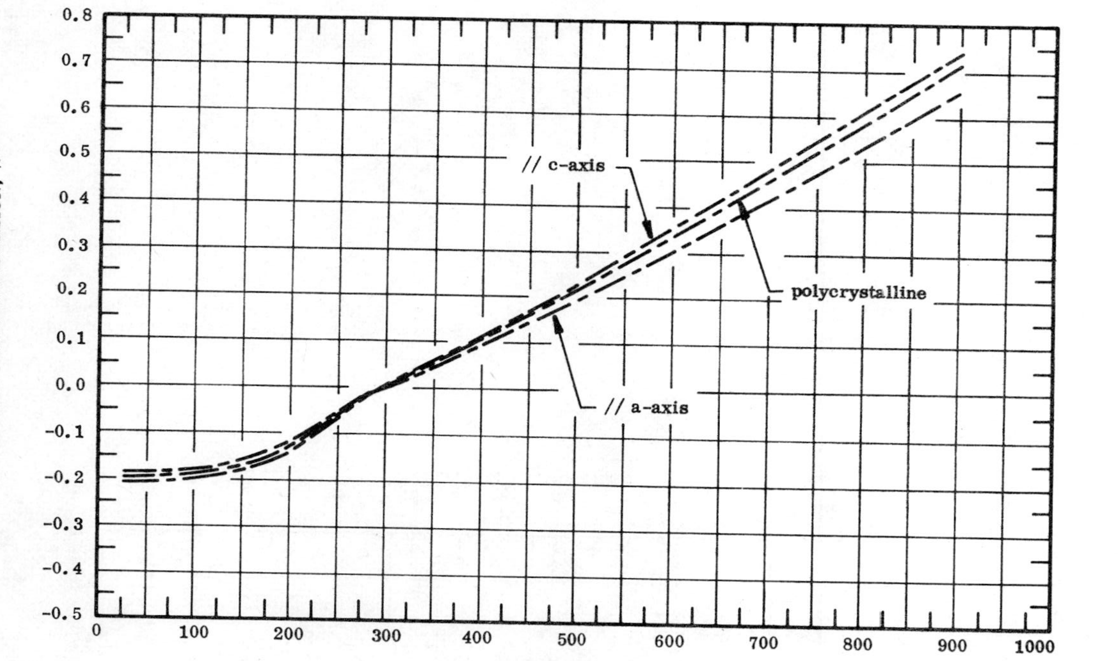

TYPICAL VALUES

[Temperature, T, K; Linear Expansion, $\Delta L/L_0$, %]

T	// a-axis $\Delta L/L_0$	// c-axis $\Delta L/L_0$	polycrystalline $\Delta L/L_0$
25	-0.205	-0.181	-0.197
50	-0.203	-0.180	-0.195
100	-0.202	-0.176	-0.193
200	-0.135	-0.104	-0.125
293	0.000	0.000	0.000
400	0.111	0.092	0.105
500	0.227	0.194	0.216
600	0.348	0.301	0.332
700	0.476	0.416	0.456
800	0.607	0.530	0.581
900	0.739	0.651	0.710

REMARKS

The tabulated values are considered accurate to within ±15%. The values for polycrystalline material in the absence of experimental data are calculated from the single crystal data.

FIGURE AND TABLE NO. 91BR. TYPICAL VALUES FOR THERMAL LINEAR EXPANSION OF Co_5Y INTERMETALLIC COMPOUND

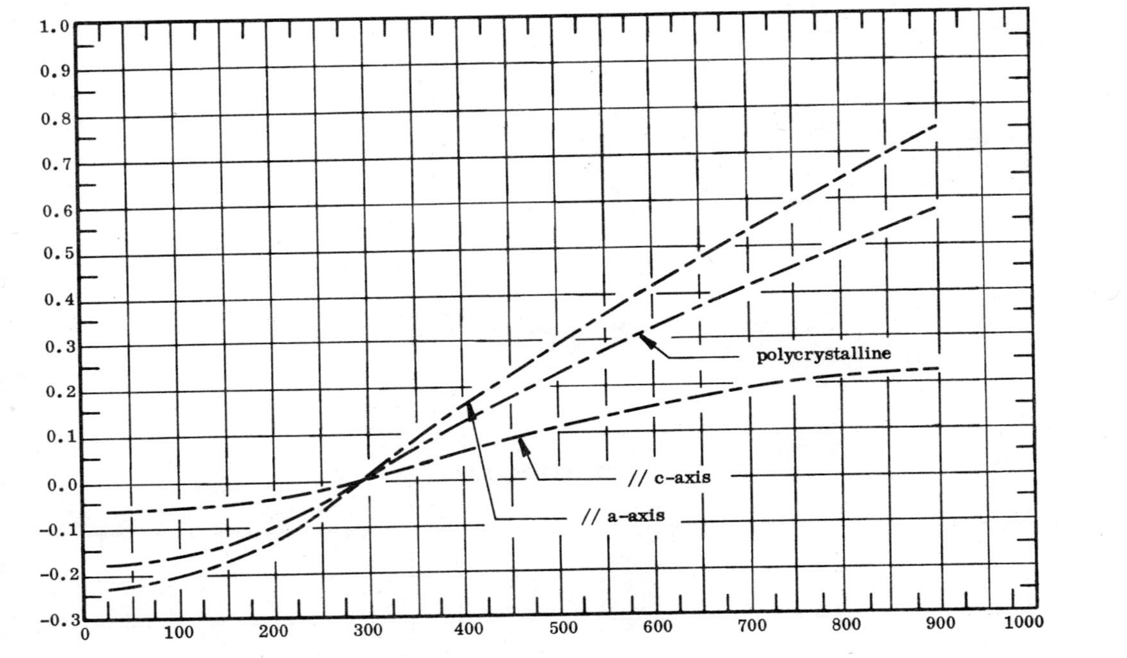

TYPICAL VALUES

[Temperature, T, K; Linear Expansion, $\Delta L/L_0$, %]

T	// a-axis $\Delta L/L_0$	// c-axis $\Delta L/L_0$	polycrystalline $\Delta L/L_0$
25	-0.235	-0.069	-0.180
50	-0.228	-0.069	-0.175
100	-0.212	-0.067	-0.164
200	-0.133	-0.040	-0.102
293	0.000	0.000	0.000
400	0.159	0.059	0.126
500	0.286	0.108	0.227
600	0.413	0.153	0.326
700	0.527	0.189	0.414
800	0.645	0.211	0.500
900	0.752	0.227	0.577

REMARKS

The tabulated values are considered accurate to within ±15%. The values for polycrystalline material in the absence of experimental data are calculated from the single crystal data.

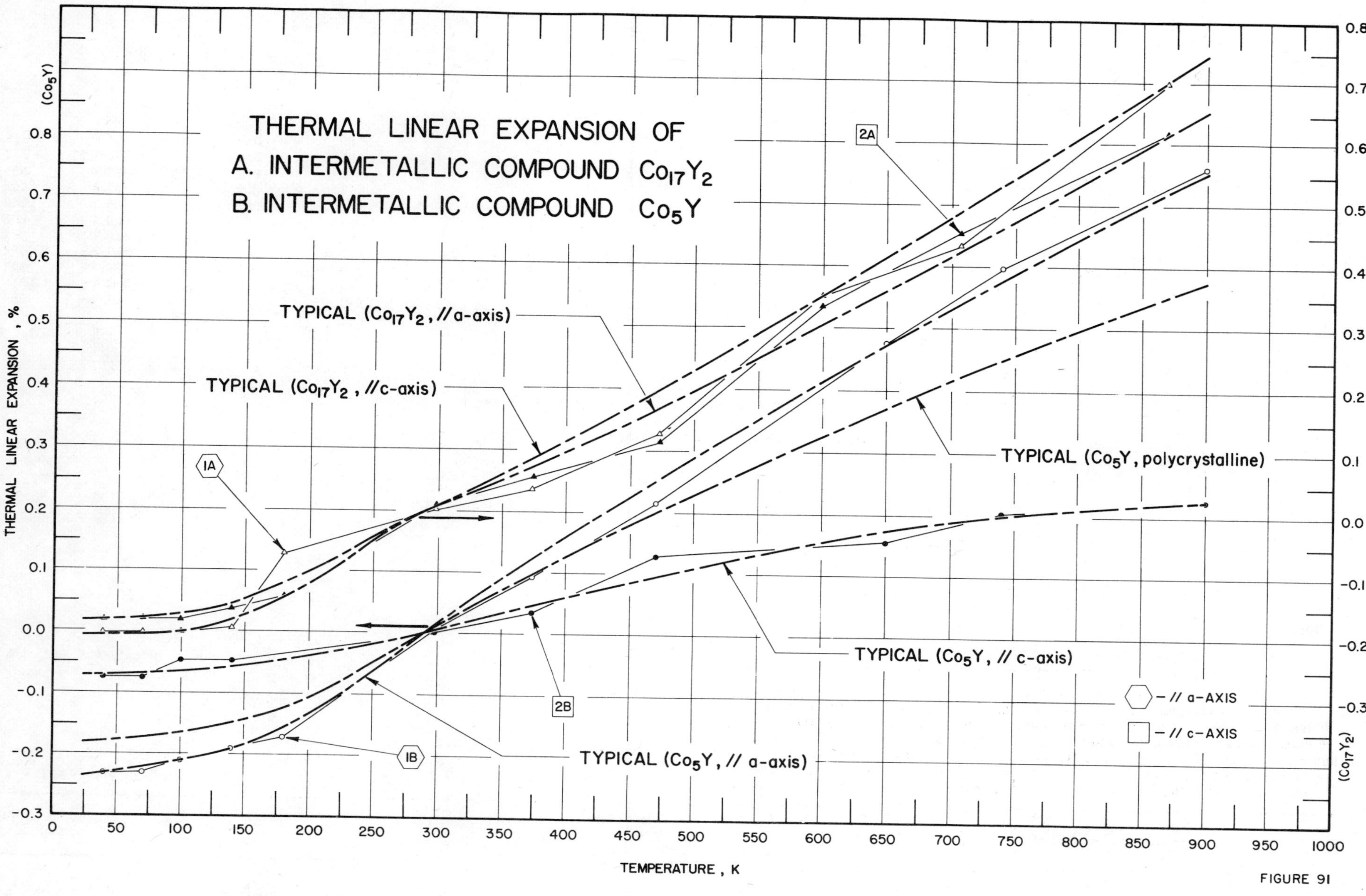
THERMAL LINEAR EXPANSION OF
A. INTERMETALLIC COMPOUND $Co_{17}Y_2$
B. INTERMETALLIC COMPOUND Co_5Y
TYPICAL ($Co_{17}Y_2$, //a-axis)
TYPICAL ($Co_{17}Y_2$, //c-axis)
TYPICAL (Co_5Y, polycrystalline)
TYPICAL (Co_5Y, // c-axis)
TYPICAL (Co_5Y, // a-axis)
2A
1A
2B
1B
– // a–AXIS
– // c–AXIS
THERMAL LINEAR EXPANSION, %
(Co_5Y)
($Co_{17}Y_2$)
TEMPERATURE, K

FIGURE 91

SPECIFICATION TABLE 91A. THERMAL LINEAR EXPANSION OF $Co_{17}Y_2$ INTERMETALLIC COMPOUND

Cur. No.	Ref. No.	Author(s)	Year	Method Used	Temp. Range, K	Name and Specimen Designation	Composition (weight percent), Specifications, and Remarks
1	709	Shah, J.S.	1971	X	40-870		Samples from Dr. R. Lemaire, C.N.R.S., Grenoble, France; samples prepared by melting elements in high-frequency levitation furnace then homogenizing at 1373 K for 3 days; powders were ground in a boron nitride mortar then sieved through a special 20 μ sieve; intermetallic compound powder samples were mounted on special quartz fiber; sample structure hexagonal Th_2Ni_{17}-type; lattice parameter reported as 8.3291 Å at 40 K; expansion measured along a-axis; lattice parameter not corrected for refraction.
2	709	Shah, J.S.	1971	X	40-870		The above specimen; lattice parameter reported as 8.1190 Å at 40 K; expansion measured along c-axis; lattice parameter not corrected for refraction.

DATA TABLE 91A. THERMAL LINEAR EXPANSION OF $Co_{17}Y_2$ INTERMETALLIC COMPOUND

[Temperature, T, K; Linear Expansion, $\Delta L/L_0$, %]

T	$\Delta L/L_0$	T	$\Delta L/L_0$
CURVE 1		CURVE 2	
40	-0.201	40	-0.181
70	-0.201	70	-0.181
100	-0.201	100	-0.181
140	-0.196	140	-0.165
180	-0.071	180	-0.144
298	0.003	298	0.006
373	0.037	373	0.053
473	0.128	473	0.114
600	0.352	600	0.376
713	0.433	713	0.455
870	0.700	870	0.620

SPECIFICATION TABLE 91B. THERMAL LINEAR EXPANSION OF Co_5Y INTERMETALLIC COMPOUND

Cur. No.	Ref. No.	Author(s)	Year	Method Used	Temp. Range, K	Name and Specimen Designation	Composition (weight percent), Specifications, and Remarks
1	709	Shah, J.S.	1971	X	40-900		Samples from Dr. R. Lemaire, C.N.R.S., Grenoble, France; compounds prepared by melting elements in high-frequency levitation furnace then homogenizing at 1373 K for 3 days; powders were ground in a boron nitride mortar then sieved through a special 20 μ sieve; intermetallic compound powder samples were mounted on special quartz fiber; sample structure hexagonal $CaCu_5$-type; expansion measured along the a-axis; lattice parameter reported as 4.9360 Å at 40 K; 4.9474 Å at 293 K determined by graphical interpolation; lattice parameter not corrected for refraction.
2	709	Shah, J.S.	1971	X	40-900		The above specimen; expansion measured along the c-axis; lattice parameter reported as 3.9730 Å at 40 K; 3.97596 Å at 293 K determined by graphical interpolation; lattice parameter not corrected for refraction.

DATA TABLE 91B. THERMAL LINEAR EXPANSION OF Co_5Y INTERMETALLIC COMPOUND

[Temperature, T, K; Linear Expansion, $\Delta L/L_0$, %]

T	$\Delta L/L_0$	T	$\Delta L/L_0$
CURVE 1		CURVE 2	
40	-0.230	40	-0.074
70	-0.230	70	-0.074
100	-0.210	100	-0.049
140	-0.190	140	-0.049
180	-0.170	180	-0.024
298	0.008	298	0.001
373	0.093	373	0.036
470	0.214	470	0.127
650	0.477	650	0.152
740	0.598	740	0.202
900	0.760	900	0.227

FIGURE AND TABLE NO. 92AR. PROVISIONAL VALUES FOR THERMAL LINEAR EXPANSION OF Cu_3AuI INTERMETALLIC COMPOUND

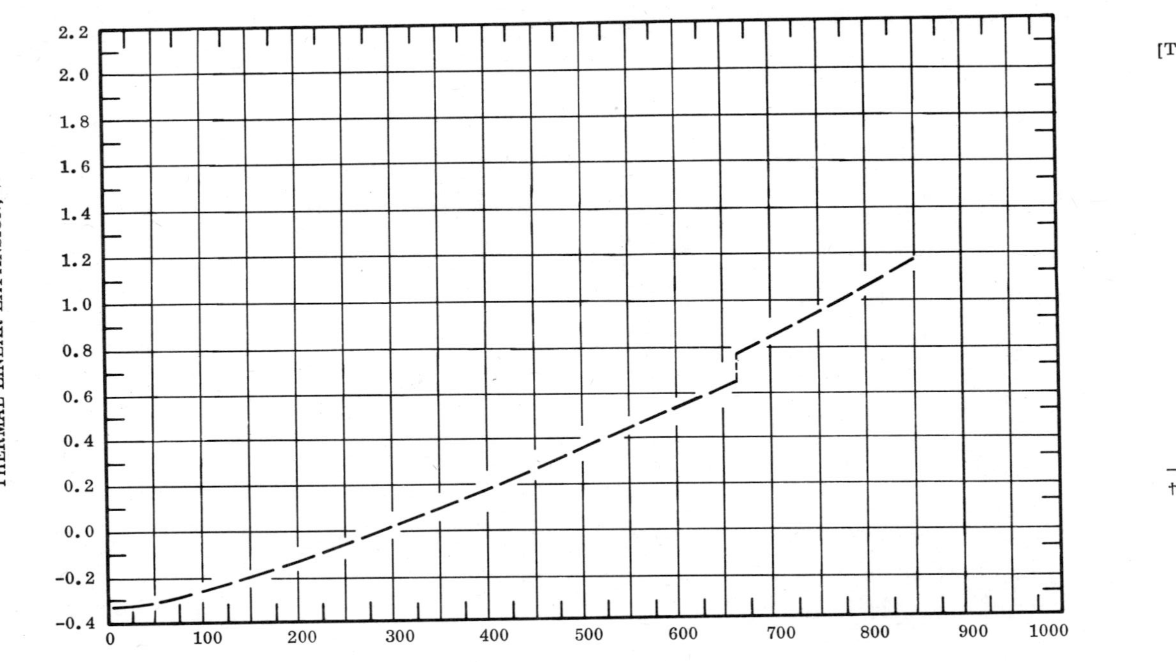

PROVISIONAL VALUES

[Temperature, T, K; Linear Expansion, $\Delta L/L_0$, %; α, K^{-1}]

T	$\Delta L/L_0$	$\alpha \times 10^6$
5	-0.330	0.1
50	-0.312	7.3
100	-0.263	10.7
200	-0.140	13.8
293	0.000	15.9
400	0.178	17.3
500	0.353	17.6
600	0.526	17.0
663†	0.630	17.0
663†	0.761	19.2
700	0.833	19.9
800	1.038	20.9
850	1.143	20.9

† Phase change.

REMARKS

The tabulated values are considered accurate to within ± 7% over the entire temperature range. These values can be represented approximately by the following equations:

$$\Delta L/L_0 = -0.350 + 6.638 \times 10^{-4}\, T + 2.265 \times 10^{-6}\, T^2 - 1.560 \times 10^{-9}\, T^3 \quad (50 < T < 663)$$

$$\Delta L/L_0 = 0.514 - 2.059 \times 10^{-3}\, T + 4.992 \times 10^{-6}\, T^2 - 2.000 \times 10^{-9}\, T^3 \quad (663 < T < 850)$$

FIGURE AND TABLE NO. 92BR. PROVISIONAL VALUES FOR THERMAL LINEAR EXPANSION OF CuAu INTERMETALLIC COMPOUND

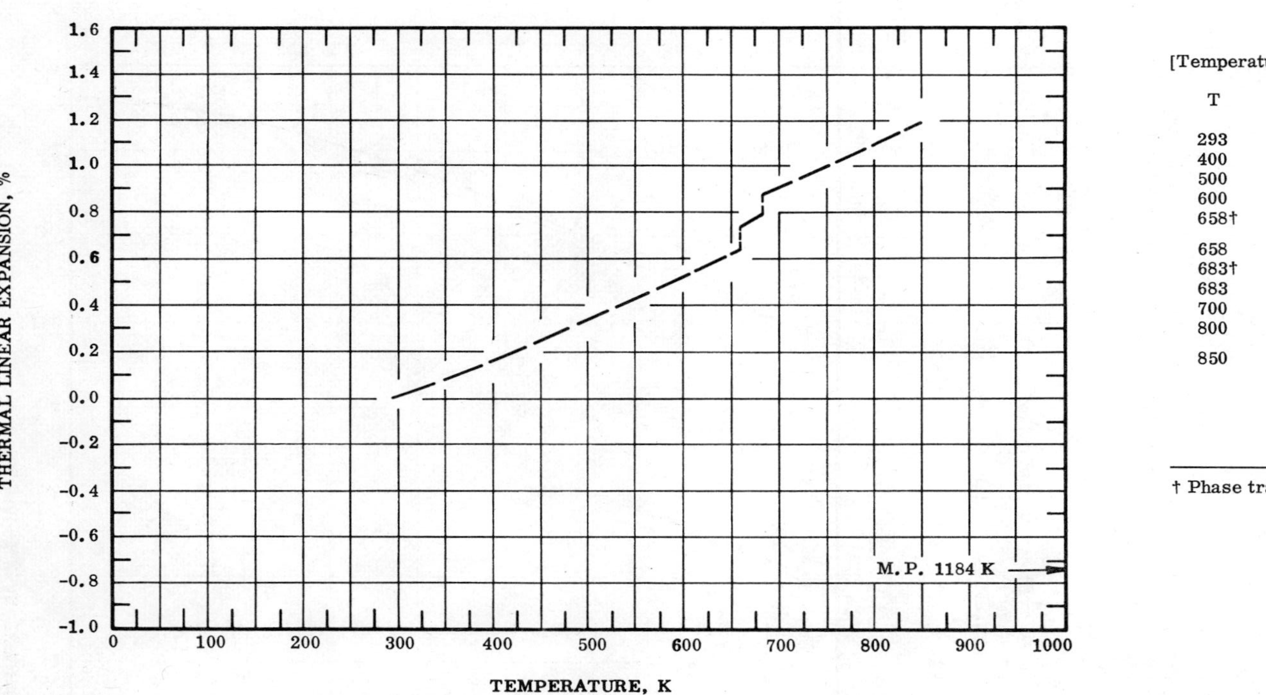

PROVISIONAL VALUES

[Temperature, T, K; Linear Expansion, $\Delta L/L_0$, %; α, K^{-1}]

T	$\Delta L/L_0$	$\alpha \times 10^6$
293	0.000	13.3
400	0.158	16.2
500	0.333	18.2
600	0.520	19.6
658†	0.636	20.0
658	0.722	
683†	0.781	
683	0.875	16.5
700	0.900	16.9
800	1.082	19.4
850	1.184	20.9

† Phase transformation.

The tabulated values are considered accurate to within ±7% over the entire temperature range. These values can be represented approximately by the following equations:

$$\Delta L/L_0 = -0.231 + 1.385 \times 10^{-4}\, T + 2.540 \times 10^{-6}\, T^2 - 1.137 \times 10^{-9}\, T^3 \quad (293 < T < 658)$$

$$\Delta L/L_0 = -0.174 + 1.995 \times 10^{-3}\, T - 1.523 \times 10^{-6}\, T^2 + 1.241 \times 10^{-9}\, T^3 \quad (683 < T < 850)$$

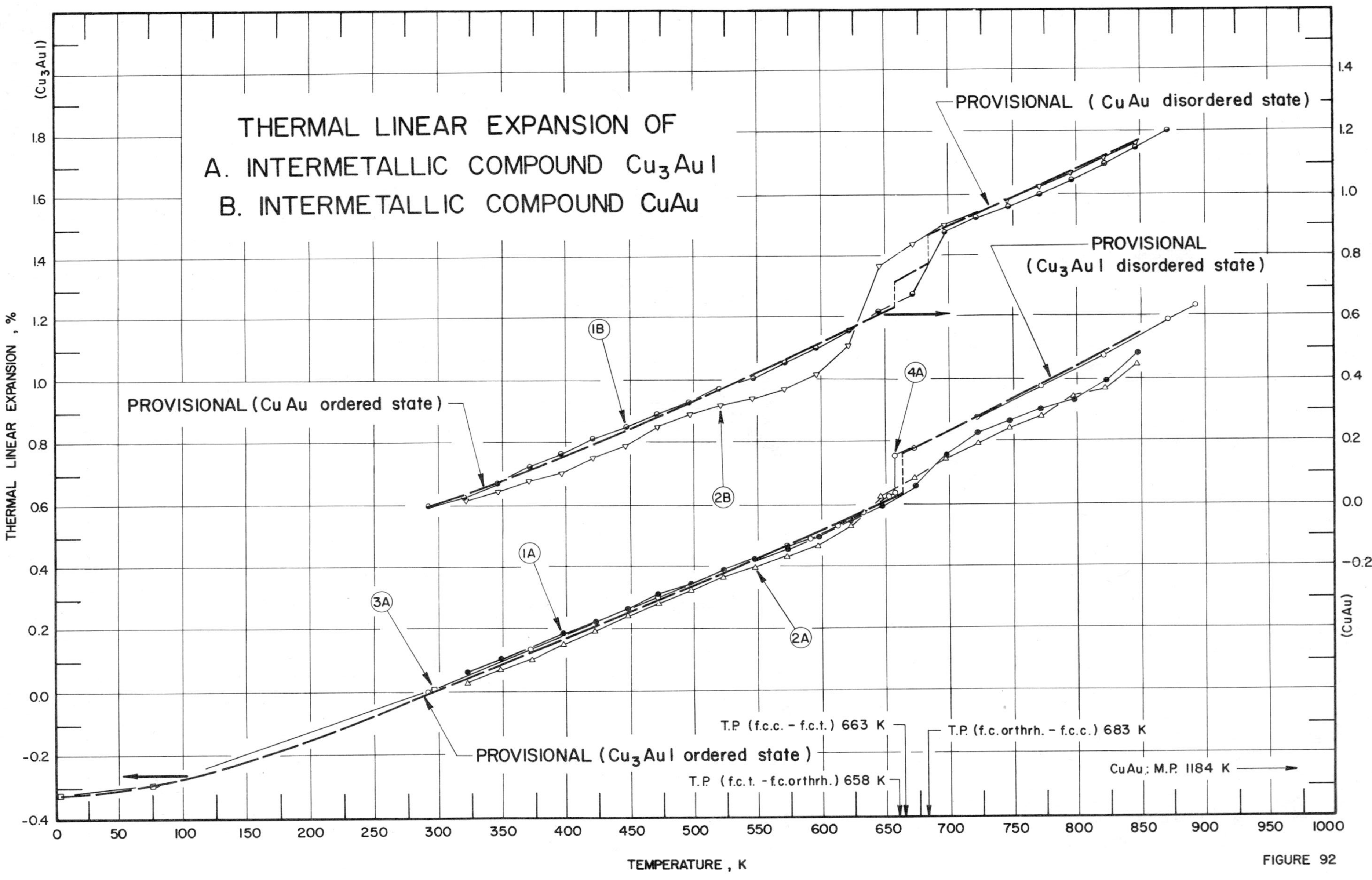
THERMAL LINEAR EXPANSION OF
A. INTERMETALLIC COMPOUND Cu_3Au I
B. INTERMETALLIC COMPOUND CuAu
PROVISIONAL (CuAu disordered state)
PROVISIONAL
(Cu_3Au I disordered state)
PROVISIONAL (CuAu ordered state)
PROVISIONAL (Cu_3Au I ordered state)
1B
2B
1A
2A
3A
4A
T.P. (f.c.c. - f.c.t.) 663 K
T.P. (f.c. orthrh. - f.c.c.) 683 K
T.P. (f.c.t. - f.c.orthrh.) 658 K
CuAu; M.P. 1184 K
THERMAL LINEAR EXPANSION, %
(Cu_3Au I)
(CuAu)
TEMPERATURE, K
0
50
100
150
200
250
300
350
400
450
500
550
600
650
700
750
800
850
900
950
1000
-0.4
-0.2
0.0
0.2
0.4
0.6
0.8
1.0
1.2
1.4
1.6
1.8
-0.2
0.0
0.2
0.4
0.6
0.8
1.0
1.2
1.4

FIGURE 92

SPECIFICATION TABLE 92A. THERMAL LINEAR EXPANSION OF Cu_3AuI INTERMETALLIC COMPOUND

Cur. No.	Ref. No.	Author(s)	Year	Method Used	Temp. Range, K	Name and Specimen Designation	Composition (weight percent), Specifications, and Remarks
1	611	Kurnakow, N. S. and Ageew, J.	1931	L	293-848	Cu_3Au I	Alloy of composition 50.70 Au and 49.30 Cu prepared from chemically pure gold and electrolytic copper; annealed at 1073 K; expansion measured with increasing temperature.
2	611	Kurnakow, N. S. and Ageew, J.	1931	L	848-293	Cu_3Au I	Similar to the above specimen; expansion measured with decreasing temperature; zero-point correction is -0.01%.
3	674	Flinn, P. A., McManus, G. M., and Rayne, J. A.	1960	X	4-298	Cu_3Au I	Single crystal with composition 50.66 Au and 49.34 Cu of disordered alloy prepared by melting appropriate composition using Bridgman method with cooling rate 2°/hr; ingot was etched electrolytically in silver cyamide and potassium cyamide solution and oriented with x-rays by Laue back reflection technique; lattice parameter reported at 298 K is 3.7490 Å; 3.7488 Å at 293 K determined by graphical interpolation.
4	692	Owen, E. A. and Liu, Y. H.	1947	X	291-893	Cu_3Au I	1 gm ingot specimen prepared free from segregation by process of alternate annealing and cold working; carried out over a period of about one month; lattice parameter reported at 291 K is 3.7402 kX; 3.7403 kX at 293 K determined by graphical interpolation.

DATA TABLE 92A. THERMAL LINEAR EXPANSION OF Cu_3AuI INTERMETALLIC COMPOUND

[Temperature, T, K; Linear Expansion, $\Delta L/L_0$, %]

CURVE 1

T	$\Delta L/L_0$
293	0.00
323	0.06
348	0.10
373	0.14*
398	0.18
423	0.22
448	0.26
473	0.30
498	0.34
523	0.38
548	0.42
573	0.45
598	0.49
623	0.54
648	0.59
673	0.65
698	0.75
723	0.82
748	0.86
773	0.89
798	0.92
823	0.98
848	1.07

CURVE 2

T	$\Delta L/L_0$
848	1.04
823	0.97
798	0.93
773	0.88
748	0.84
723	0.79
698	0.74
673	0.68
648	0.62
623	0.53
598	0.46
573	0.43
548	0.39
523	0.36
498	0.32
473	0.28
448	0.24
423	0.19
398	0.15
373	0.10
348	0.07
323	0.03
293	0.00

CURVE 3

T	$\Delta L/L_0$
4.2	-0.323
77	-0.299
298	0.005

CURVE 4

T	$\Delta L/L_0$
291	-0.003
373	0.134
473	0.294
573	0.457
593	0.489
613	0.529
633	0.572
653	0.623
657	0.636
659	0.754
673	0.778
723	0.871
773	0.973
823	1.077
873	1.187
893	1.232

* Not shown in figure.

SPECIFICATION TABLE 92B. THERMAL LINEAR EXPANSION OF CuAu INTERMETALLIC COMPOUND

Cur. No.	Ref. No.	Author(s)	Year	Method Used	Temp. Range, K	Name and Specimen Designation	Composition (weight percent), Specifications, and Remarks
1	611	Kurnakow, N.S. and Ageew, J.	1931	L	293-873		Alloy with composition 75.89 Au and 24.11 Cu prepared from pure gold and electrolytic copper; annealed at 1073 K; expansion measured with increasing temperature.
2	611	Kurnakow, N.S. and Ageew, J.	1931	L	873-293		Similar to the above specimen; expansion measured with decreasing temperature; zero-point correction is -0.01%.

DATA TABLE 92B. THERMAL LINEAR EXPANSION OF CuAu INTERMETALLIC COMPOUND

[Temperature, T, K; Linear Expansion, $\Delta L/L_0$, %]

T	$\Delta L/L_0$	T	$\Delta L/L_0$	T	$\Delta L/L_0$
CURVE 1		CURVE 1 (cont.)		CURVE 2 (cont.)	
293	0.00	723	0.93	698	0.89
323	0.03	748	0.96	673	0.84
348	0.07	773	1.00	648	0.77
373	0.12	798	1.05	623	0.51
398	0.16	823	1.10	598	0.42
423	0.21	848	1.15	573	0.37
448	0.25	873	1.21	548	0.34
473	0.29			523	0.32
498	0.33	CURVE 2		498	0.29
523	0.37			473	0.25
548	0.41	873	1.20	448	0.19
573	0.46	848	1.16	423	0.15
598	0.51	823	1.12	398	0.10
623	0.56	798	1.07	373	0.08
648	0.62	773	1.03	348	0.04
673	0.68	748	0.98	323	0.02
698	0.88	723	0.93	293	0.00

SPECIFICATION TABLE 93. THERMAL LINEAR EXPANSION OF Cu_2Mg INTERMETALLIC COMPOUND

Cur. No.	Ref. No.	Author(s)	Year	Method Used	Temp. Range, K	Name and Specimen Designation	Composition (weight percent), Specifications, and Remarks
1*	133	Ullrich, H.J.	1967	X	283-333		No details given; lattice parameter reported at 293 K is 7.0365 Å.
2*	675	Cheng, C.H.	1967		273, 523		Single crystal of stoichiometric composistions received from Dr. V.B. Kurfman of Metallurgical Laboratory of Dow Chemical Co.; grown by Bridgman method in graphite molds from high purity metals; density 5.76 g cm^{-3} at room temperature.
3*	688	Smith, J.F. and Ogren, J.R.	1958	X	80-300		Polycrystal of cubic specimen.

DATA TABLE 93. THERMAL LINEAR EXPANSION OF Cu_2Mg INTERMETALLIC COMPOUND

[Temperature, T, K; Linear Expansion, $\Delta L/L_0$, %]

T	$\Delta L/L_0$
CURVE 1*	
283	-0.029
293	0.000
333	0.143
CURVE 2*	
273	-0.037
523	0.430
CURVE 3*	
80	-0.639
300	0.021

* No figure given.

SPECIFICATION TABLE 94. THERMAL LINEAR EXPANSION OF Cu_5Zn_8 INTERMETALLIC COMPOUND

Cur. No.	Ref. No.	Author(s)	Year	Method Used	Temp. Range, K	Name and Specimen Designation	Composition (weight percent), Specifications, and Remarks
1*	668	Schubert, K. and Wall, E.	1949	X	293, 1023	γ-phase $Cu_{36.7}Zn_{63.3}$	Specimen prepared from 99.998 Zn and electrolytic copper by mixing the melts in vacuum; lattice parameter reported at 293 K is 8.85 kX.

DATA TABLE 94. THERMAL LINEAR EXPANSION OF Cu_5Zn_8 INTERMETALLIC COMPOUND

[Temperature, T, K; Linear Expansion, $\Delta L/L_0$, %]

T	$\Delta L/L_0$
CURVE 1*	
293	0.000
1023	1.964

* No figure given.

SPECIFICATION TABLE 95. THERMAL LINEAR EXPANSION OF Cu-Zn INTERMETALLIC COMPOUND

Cur. No.	Ref. No.	Author(s)	Year	Method Used	Temp. Range, K	Name and Specimen Designation	Composition (weight percent), Specifications, and Remarks
1*	668	Schubert, K. and Wall, E.	1949	X	293, 1023	ϵ-phase $Cu_{21}Zn_{79}$	Specimen prepared from 99.998 Zn and electrolytic copper by mixing the melts in vacuum; measurements along a-axis; lattice parameter reported at 293 K is 2.73 kX.
2*	668	Schubert, K. and Wall, E.	1949	X	293, 1023	ϵ-phase $Cu_{21}Zn_{79}$	The above specimen; measurements along c-axis; lattice parameter reported at 293 K is 4.26 kX.

DATA TABLE 95. THERMAL LINEAR EXPANSION OF Cu-Zn INTERMETALLIC COMPOUND

[Temperature, T, K; Linear Expansion, $\Delta L/L_0$, %]

T	$\Delta L/L_0$
CURVE 1*	
293	0.000
1023	2.336
CURVE 2*	
293	0.000
1023	1.606

* No figure given.

FIGURE AND TABLE NO. 96AR. PROVISIONAL VALUES FOR THERMAL LINEAR EXPANSION OF $GdIn_3$ INTERMETALLIC COMPOUND

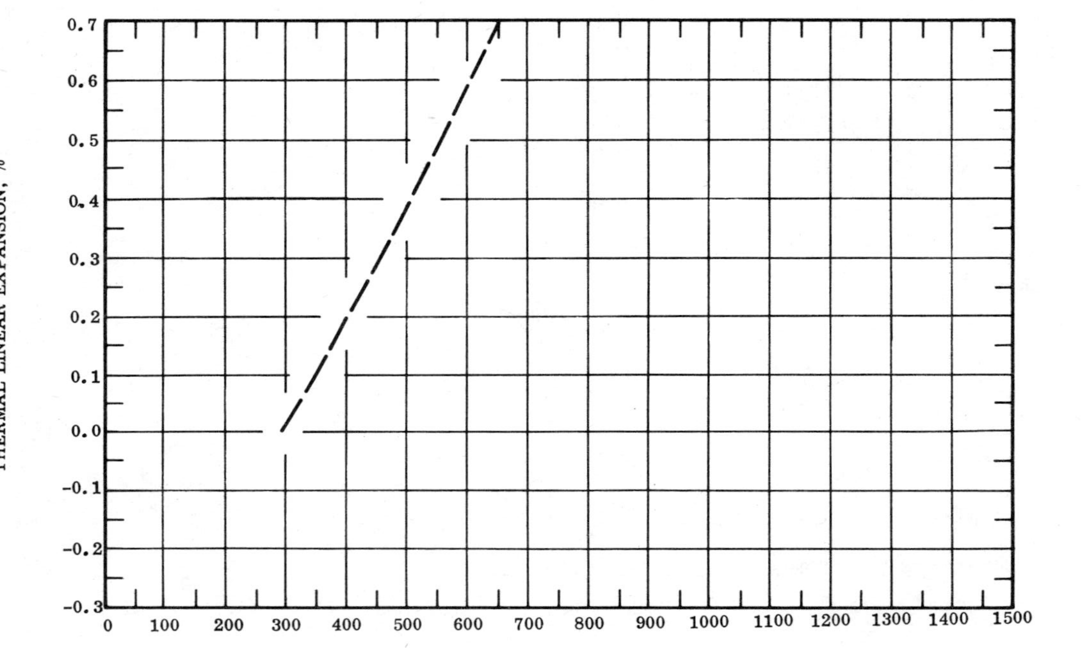

PROVISIONAL VALUES

[Temperature, T, K; Linear Expansion, $\Delta L/L_0$, %; α, K^{-1}]

T	$\Delta L/L_0$	$\alpha \times 10^6$
293	0.000	16.8
400	0.192	18.9
500	0.388	20.1
600	0.590	20.5
650	0.694	20.5

REMARKS

The tabulated values are considered accurate to within ±7% over the entire temperature range. These values can be represented approximately by the following equation:

$$\Delta L/L_0 = -0.365 + 7.038 \times 10^{-4}\,T + 2.199 \times 10^{-6}\,T^2 - 1.194 \times 10^{-9}\,T^3$$

FIGURE AND TABLE NO. 96BR. TYPICAL VALUES FOR THERMAL LINEAR EXPANSION OF Gd_2Fe_{17} INTERMETALLIC COMPOUND

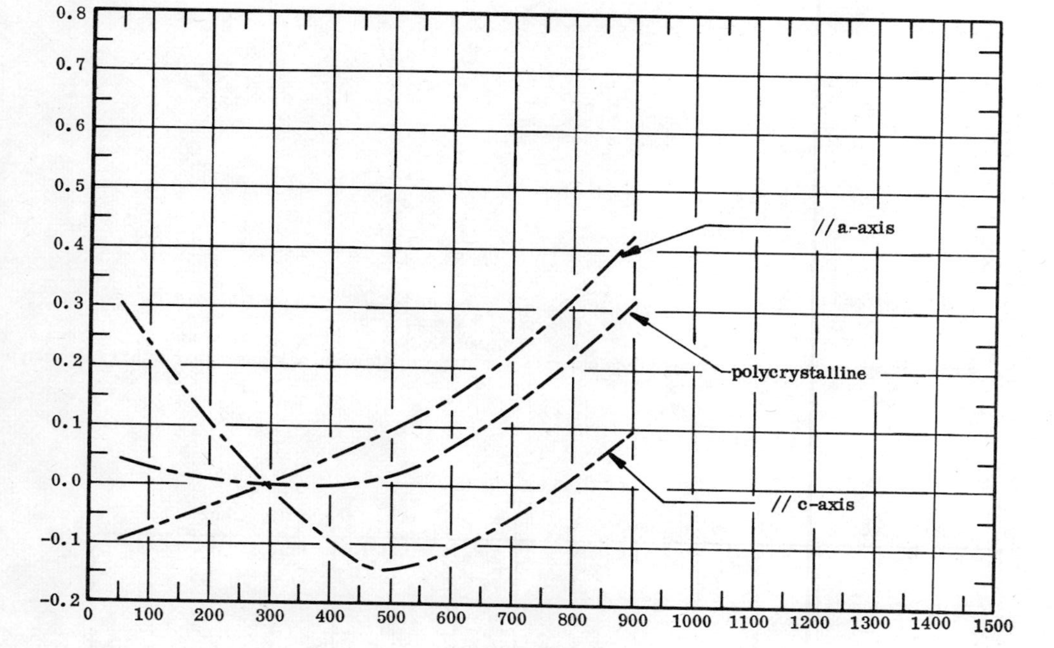

TYPICAL VALUES

[Temperature, T, K; Linear Expansion, $\Delta L/L_0$, %]

T	// a-axis $\Delta L/L_0$	// c-axis $\Delta L/L_0$	polycrystalline $\Delta L/L_0$
50	-0.096	0.305	0.038
100	-0.077	0.231	0.026
200	-0.039	0.103	0.008
293	0.000	0.000	0.000
400	0.043	-0.093	-0.002
500	0.091	-0.147	0.012
600	0.147	-0.112	0.061
700	0.223	-0.050	0.132
800	0.321	0.018	0.220
900	0.426	0.091	0.314

REMARKS

The tabulated values are considered accurate to within ±15%. The values for polycrystalline material in the absence of experimental data are calculated from the single crystal data. Anomolous behavior observed near Curie temperature at about 479 K.

FIGURE AND TABLE NO. 96CR. PROVISIONAL VALUES FOR THERMAL LINEAR EXPANSION OF $GdPd_3$ INTERMETALLIC COMPOUND

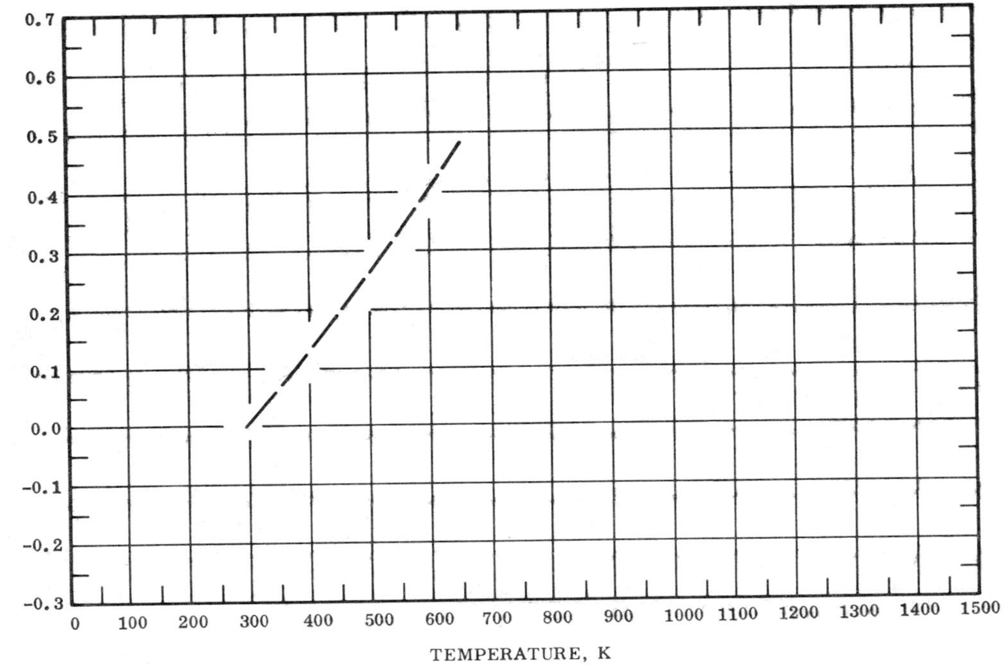

PROVISIONAL VALUES

[Temperature, T, K; Linear Expansion, $\Delta L/L_0$, %; α, K^{-1}]

T	$\Delta L/L_0$	$\alpha \times 10^6$
293	0.000	11.3
400	0.130	12.9
500	0.264	13.8
600	0.406	14.6
650	0.480	15.0

REMARKS

The tabulated values are considered accurate to within ±7% over the entire temperature range. These values can be represented approximately by the following equation:

$$\Delta L/L_0 = -0.268 + 6.466 \times 10^{-4}\, T + 1.030 \times 10^{-6}\, T^2 - 3.928 \times 10^{-10}\, T^3$$

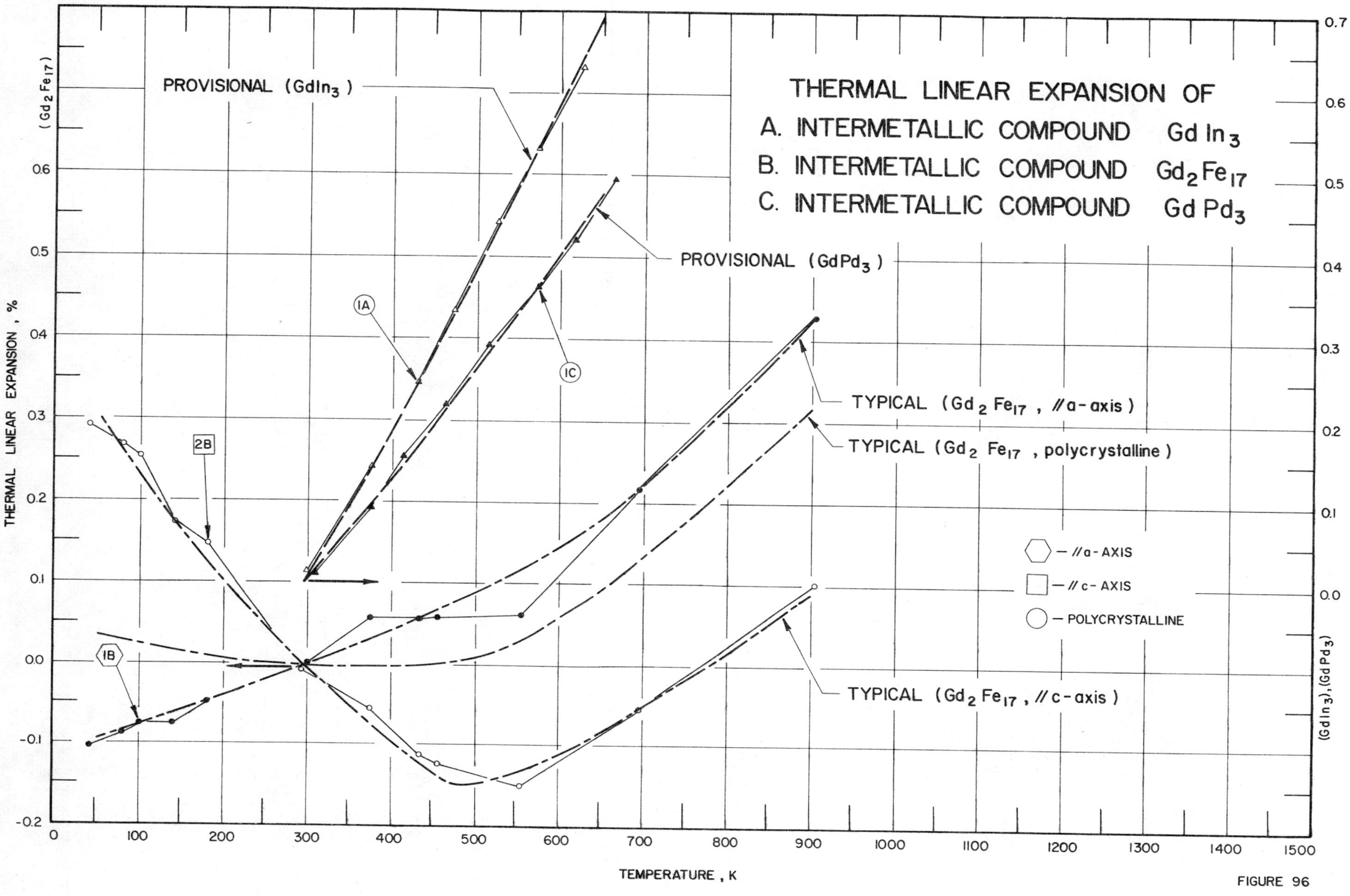
THERMAL LINEAR EXPANSION OF
A. INTERMETALLIC COMPOUND Gd In3
B. INTERMETALLIC COMPOUND Gd2 Fe17
C. INTERMETALLIC COMPOUND Gd Pd3
PROVISIONAL (GdIn3)
PROVISIONAL (GdPd3)
1A
1C
2B
1B
TYPICAL (Gd2 Fe17 , //a-axis)
TYPICAL (Gd2 Fe17 , polycrystalline)
TYPICAL (Gd2 Fe17 , //c-axis)
— //a-AXIS
— //c-AXIS
— POLYCRYSTALLINE
THERMAL LINEAR EXPANSION, %
(Gd2 Fe17)
(GdIn3), (GdPd3)
TEMPERATURE, K
-0.2
-0.1
0.0
0.1
0.2
0.3
0.4
0.5
0.6
0.7
0
100
200
300
400
500
600
700
800
900
1000
1100
1200
1300
1400
1500

FIGURE 96

SPECIFICATION TABLE 96A. THERMAL LINEAR EXPANSION OF $GdIn_3$ INTERMETALLIC COMPOUND

Cur. No.	Ref. No.	Author(s)	Year	Method Used	Temp. Range, K	Name and Specimen Designation	Composition (weight percent), Specifications, and Remarks
1	682	Harris, I.R. and Raynor, G.V.	1965	X	297-625		Specimen prepared from In containing 0.002 Pb and 0.001 Cd from Messrs. Johnson, Matthey and Co. Ltd., and Gd containing <0.10 other rare metals, 0.10 Ta, and 0.02 other base metals by melting weighed quantities in non-consumable electrode arc furnace in 15 mm Hg argon; wrapped in Ta foil and sealed in degassed silica tubes at 2 x 10^{-6} mm Hg pressure; homogenized, annealed for 3 days at 973 K and quenched; powdered, annealed at 973 K for stress relief; lattice parameter reported at 297 K is 4.5968 kX; 4.5960 kX at 293 K determined by graphical extrapolation.

DATA TABLE 96A. THERMAL LINEAR EXPANSION OF $GdIn_3$ INTERMETALLIC COMPOUND

[Temperature, T, K; Linear Expansion, $\Delta L/L_0$, %]

T	$\Delta L/L_0$
CURVE 1	
297	0.017
373	0.148
430	0.250
472	0.337
524	0.442
574	0.533
625	0.635

SPECIFICATION TABLE 96B. THERMAL LINEAR EXPANSION OF Gd_2Fe_{17} INTERMETALLIC COMPOUND

Cur. No.	Ref. No.	Author(s)	Year	Method Used	Temp. Range, K	Name and Specimen Designation	Composition (weight percent), Specifications, and Remarks
1	709	Shah, J.S.	1971	X	40-903	α-Gd_2Fe_{17}	Samples from Dr. R. Lemaire, C.N.R.S., Grenoble France; compounds prepared by melting elements in high-frequency levitation furnace then homogenizing at 1373 K for 3 days; powders were ground in a boron nitride mortar then sieved through a special 20 μ sieve; α-phase intermetallic powder samples were mounted on special quartz fiber; sample structure hexagonal Th_2Ni_{17}-type; expansion measured along a-axis; lattice parameter reported as 8.4945 Å at 298 K; 8.4943 Å at 293 K by graphical interpolation; lattice parameter not corrected for refraction.
2	709	Shah, J.S.	1971	X	40-903	α-Gd_2Fe_{17}	The above specimen; expansion measured along c-axis; lattice parameter reported as 8.3409 Å at 298 K; 8.3415 Å at 293 K by graphical interpolation; lattice parameter not corrected for refraction.

DATA TABLE 96B. THERMAL LINEAR EXPANSION OF Gd_2Fe_{17} INTERMETALLIC COMPOUND

[Temperature, T, K; Linear Expansion, $\Delta L/L_0$, %]

T	$\Delta L/L_0$	T	$\Delta L/L_0$
CURVE 1		CURVE 2 (cont.)	
40	-0.104	140	0.174
80	-0.089	180	0.150
100	-0.075	298	-0.007
140	-0.075	373	-0.054
180	-0.049	433	-0.114
298	0.002	453	-0.123
373	0.058	553	-0.150
433	0.058	693	-0.053
453	0.058	903	0.102
553	0.062		
693	0.217		
903	0.430		
CURVE 2			
40	0.294		
80	0.270		
100	0.246		

SPECIFICATION TABLE 96C. THERMAL LINEAR EXPANSION OF $GdPd_3$ INTERMETALLIC COMPOUND

Cur. No.	Ref. No.	Author(s)	Year	Method Used	Temp. Range, K	Name and Specimen Designation	Composition (weight percent), Specifications, and Remarks
1	557	Harris, I. R., Raynor, G. V., Winstanley, C.J.	1967	X	304-664		99.95⁺ Gd, 99.95⁺ Pd; obtained from Johnson Matthey and Co.; copper Kα radiation used for analysis; lattice parameter reported at 304 K is 4.0851 Å; 4.0847 Å at 293 K determined by graphical extrapolation.

DATA TABLE 96C. THERMAL LINEAR EXPANSION OF $GdPd_3$ INTERMETALLIC COMPOUND

[Temperature, T, K; Linear Expansion, $\Delta L/L_0$, %]

T	$\Delta L/L_0$
CURVE 1	
304	0.010
374	0.093
413	0.157
462	0.218
516	0.294
572	0.367
619	0.421
664	0.495

FIGURE AND TABLE NO. 97R. PROVISIONAL VALUES FOR THERMAL LINEAR EXPANSION OF GaNi INTERMETALLIC COMPOUND

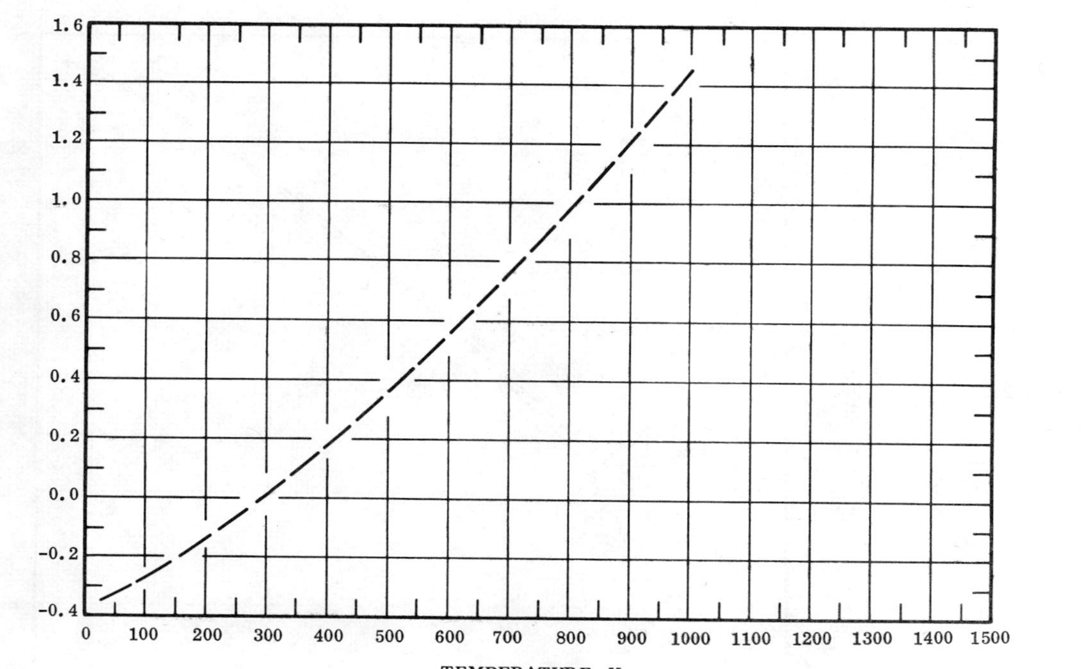

PROVISIONAL VALUES

[Temperature, T, K; Linear Expansion, $\Delta L/L_0$, %; α, K^{-1}]

T	$\Delta L/L_0$	$\alpha \times 10^6$
25	-0.351	10.3
50	-0.325	10.9
100	-0.268	11.9
200	-0.138	14.0
293	0.000	15.7
400	0.178	17.4
500	0.360	18.9
600	0.555	20.2
700	0.763	21.3
800	0.980	22.2
900	1.207	23.0
1000	1.440	23.5

REMARKS

The tabulated values are considered accurate to within ±10% over the entire temperature range. These values can be represented approximately by the following equation:

$$\Delta L/L_0 = -0.376 + 9.787 \times 10^{-4}\ T + 1.136 \times 10^{-6}\ T^2 - 2.980 \times 10^{-10}\ T^3$$

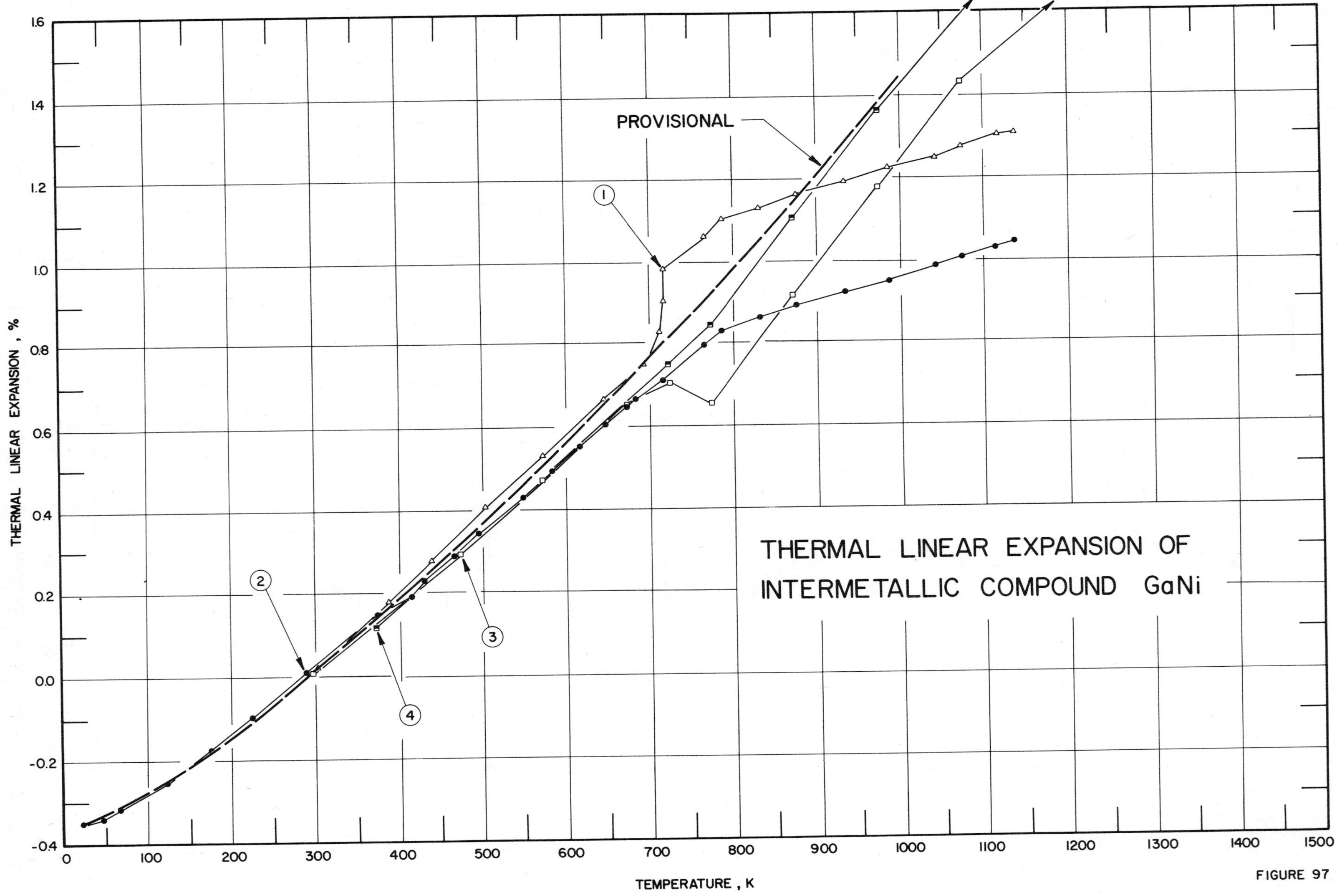
THERMAL LINEAR EXPANSION OF
INTERMETALLIC COMPOUND GaNi
PROVISIONAL
1
2
3
4
THERMAL LINEAR EXPANSION, %
TEMPERATURE, K
1.6
1.4
1.2
1.0
0.8
0.6
0.4
0.2
0.0
-0.2
-0.4
0
100
200
300
400
500
600
700
800
900
1000
1100
1200
1300
1400
1500

FIGURE 97

SPECIFICATION TABLE 97. THERMAL LINEAR EXPANSION OF GaNi INTERMETALLIC COMPOUND

Cur. No.	Ref. No.	Author(s)	Year	Method Used	Temp. Range, K	Name and Specimen Designation	Composition (weight percent), Specifications, and Remarks
1	584	Wasolewski, R.J., Butler, S.R., and Hanlon, J.E.	1968	X	301-1138		Prepared by heating together in stoichiometric mixture 99.98⁺Ni and 99.999 Ga; cooled; arc-melted in evacuated silica tube; heated to above 773 K; powder specimen; expansion measured on initial increase of temperature using (310) reflection with cobalt $K\alpha$ x-radiation; average of heating and cooling; zero-point correction of 0.277% determined by graphical extrapolation.
2	584	Wasolewski, R.J., et al.	1968	X	23-1138		Similar to the above specimen except quenched; expansion on runs after the first; zero-point correction of 0.008% determined by graphical extrapolation.
3	584	Wasolewski, R.J., et al.	1968	L	298-1173		Similar to the above specimen; chill-cast bar 5 cm long; expansion measured with initial increase of temperature; zero-point correction of -0.192% determined by graphical extrapolation.
4	584	Wasolewski, R.J., et al.	1968	L	298-1173		The above specimen; expansion measured on runs after the first; average of heating and cooling; zero-point correction of 0.008% determined by graphical extrapolation.

DATA TABLE 97. THERMAL LINEAR EXPANSION OF GaNi INTERMETALLIC COMPOUND

[Temperature, T, K; Linear Expansion, $\Delta L/L_0$, %]

T	$\Delta L/L_0$	T	$\Delta L/L_0$	T	$\Delta L/L_0$	T	$\Delta L/L_0$
CURVE 1		CURVE 2		CURVE 2 (cont.)		CURVE 3 (cont.)	
301	0.020	23	-0.354	787	0.835	973	1.180
386	0.179	49	-0.340	833	0.864	1073	1.438
436	0.279	69	-0.319	879	0.898	1173	1.710*
504	0.409	124	-0.256	933	0.926		
572	0.530	176	-0.176	985	0.958	CURVE 4	
646	0.669	227	-0.098	1043	0.989		
691	0.750	290	0.008	1073	1.011	298	0.008*
712	0.834	377	0.141	1114	1.035	373	0.113
719	0.906	412	0.188	1138	1.047	473	0.294*
719	0.982	430	0.226			573	0.474*
767	1.066	467	0.285	CURVE 3		673	0.654*
787	1.104	496	0.342			723	0.743
833	1.133	550	0.433	298	0.006	773	0.842
879	1.167	583	0.499	373	0.109*	873	1.104
933	1.195	616	0.552	473	0.292	973	1.364
985	1.227	649	0.604	573	0.470	1073	1.622*
1043	1.258	671	0.643	673	0.651	1173	1.894*
1073	1.280	684	0.666	723	0.702*		
1114	1.304	719	0.713	773	0.658		
1138	1.316	767	0.797	873	0.920		

*Not shown in figure.

SPECIFICATION TABLE 98. THERMAL LINEAR EXPANSION OF $GaAg_3$ INTERMETALLIC COMPOUND

Cur. No.	Ref. No.	Author(s)	Year	Method Used	Temp. Range, K	Name and Specimen Designation	Composition (weight percent), Specifications, and Remarks
1*	686	Hume-Rothery, W. and Andrews, K.W.	1959	X	723-852	ζ-phase	23.3 (at.%) Ga specimen prepared from 99.99 pure Ag from Messrs. Johnson, Matthey and Co. Ltd., and high purity Ga from Kaliwerke, Ascheysleben, Germany; lattice parameter reported at 723 K is 4.7292 kX; measurements along a-axis; this curve is here reported using the first given temperature as reference temperature at which $\Delta L/L_0 = 0$.
2*	686	Hume-Rothery, W. and Andrews, K.W.	1959	X	723-852	ζ-phase	The above specimen; average value of lattice parameters reported at 723 K is 4.7302 kX; measurements along b-axis; this curve is here reported using the first given temperature as reference temperature at which $\Delta L/L_0 = 0$.

DATA TABLE 98. THERMAL LINEAR EXPANSION OF $GaAg_3$ INTERMETALLIC COMPOUND

[Temperature, T, K; Linear Expansion, $\Delta L/L_0$, %]

T	$\Delta L/L_0$
CURVE 1*	
723	0.000
773	0.082
852	0.218
CURVE 2*	
723	0.000
773	0.106
852	0.304

*No figure given.

SPECIFICATION TABLE 99. THERMAL LINEAR EXPANSION OF GeLa INTERMETALLIC COMPOUND

Cur. No.	Ref. No.	Author(s)	Year	Method Used	Temp. Range, K	Name and Specimen Designation	Composition (weight percent), Specifications, and Remarks
1*	567	Samsonov, G. V., Paderno, Yu. B., and Rud', B. M.	1968	L	406-1236		65.91 La, 34.08 Ge; specimen prepared by alloying of elements in electro-arched furnace with unexpended wolframic electrode and water cooled copper bottom in surroundings of Ar and then annealed in Ar by temperature 1272 K for 10-12 hrs; structure of this compound is orthorhombic and melting point is 1625 ± 20 K.

DATA TABLE 99. THERMAL LINEAR EXPANSION OF GeLa INTERMETALLIC COMPOUND

[Temperature, T, K; Linear Expansion, $\Delta L/L_0$, %]

T	$\Delta L/L_0$	T	$\Delta L/L_0$
CURVE 1*		CURVE 1 (cont.)*	
406	0.013	940	0.425
457	0.038	964	0.459
492	0.063	994	0.469
543	0.096	1016	0.501
593	0.131	1042	0.501
638	0.168	1063	0.533
695	0.212	1092	0.560
713	0.245	1120	0.580
761	0.279	1158	0.602
775	0.300	1200	0.624
801	0.322	1236	0.653
833	0.348		
854	0.369		
874	0.386		
892	0.398		
916	0.421		

*No figure given.

SPECIFICATION TABLE 100. THERMAL LINEAR EXPANSION OF $GeMg_2$ INTERMETALLIC COMPOUND

Cur. No.	Ref. No.	Author(s)	Year	Method Used	Temp. Range, K	Name and Specimen Designation	Composition (weight percent), Specifications, and Remarks
1*	568	Chung, P. L., Whitten, W.B., and Danielson, G.C.	1965	S	89-293		Starting material 99.99 percent Mg from the Dow Chemical Co. and Ge with a resistivity of 60 ohm cm at 300 K from Eagle-Picher Co.; single crystal prepared by Bridgman method; annealed 10 hr at 1123K; specimen about 1.5 cm in diameter and 1 cm in length; spark machined with faces perpendicular to the [100] and [111] axis.

DATA TABLE 100. THERMAL LINEAR EXPANSION OF $GeMg_2$ INTERMETALLIC COMPOUND

[Temperature, T, K; Linear Expansion, $\Delta L/L_0$, %]

T	$\Delta L/L_0$
CURVE 1*	
89	-0.262
115	-0.237
135	-0.214
175	-0.167
208	-0.123
235	-0.085
261	-0.049*
293	0.000*

* No figure given.

FIGURE AND TABLE NO. 101AR. PROVISIONAL VALUES FOR THERMAL LINEAR EXPANSION OF Ge_2Pr INTERMETALLIC COMPOUND

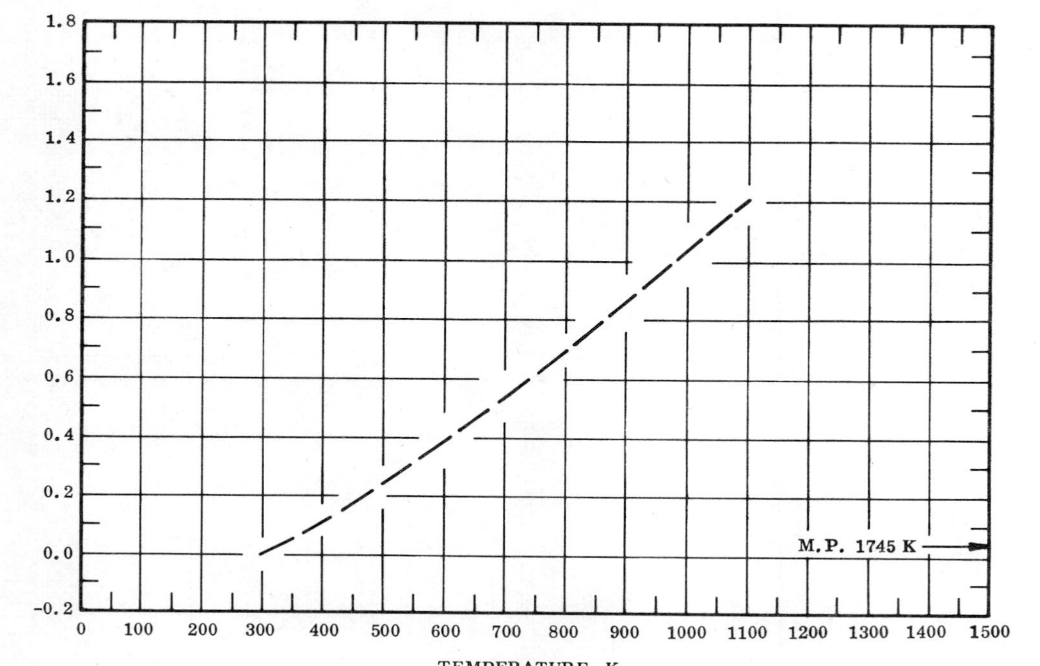

PROVISIONAL VALUES

[Temperature, T, K; Linear Expansion, $\Delta L/L_0$, %; α, K^{-1}]

T	$\Delta L/L_0$	$\alpha \times 10^6$
293	0.000	9.8
400	0.119	12.0
500	0.247	13.6
600	0.390	14.9
700	0.543	15.7
800	0.705	16.3
900	0.872	16.8
1000	1.042	17.0
1100	1.211	17.2

REMARKS

The tabulated values are considered accurate to within ±10% over the entire temperature range. These values can be represented approximately by the following equation:

$$\Delta L/L_0 = -0.197 + 2.983 \times 10^{-4}\, T + 1.422 \times 10^{-6}\, T^2 - 4.813 \times 10^{-10}\, T^3$$

FIGURE AND TABLE NO. 101BR. PROVISIONAL VALUES FOR THERMAL LINEAR EXPANSION OF GePr INTERMETALLIC COMPOUND

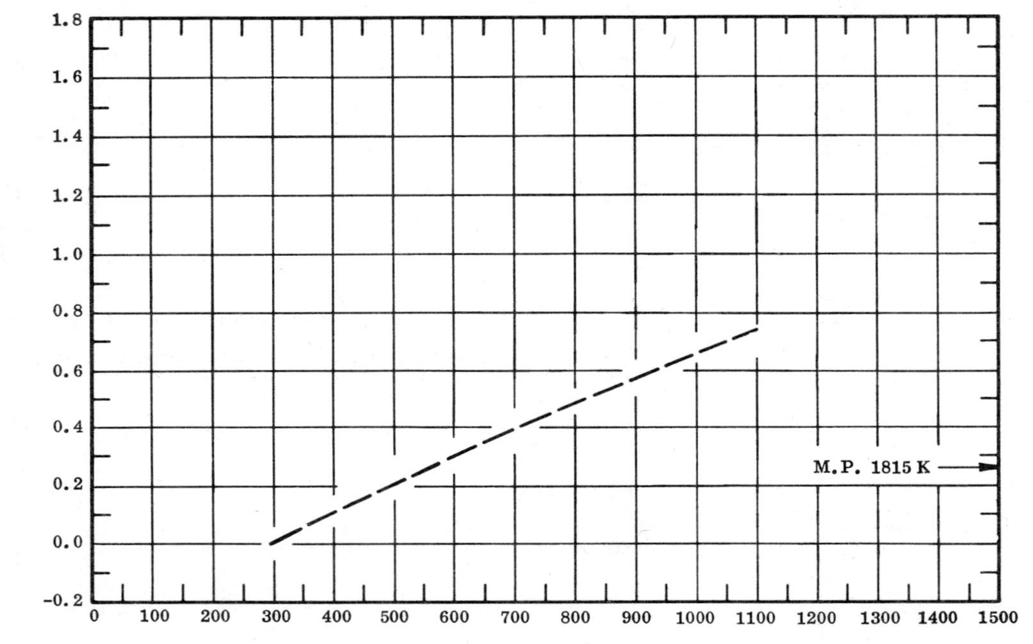

PROVISIONAL VALUES

[Temperature, T, K; Linear Expansion, $\Delta L/L_0$, %; α, K^{-1}]

T	$\Delta L/L_0$	$\alpha \times 10^6$
293	0.000	10.9
400	0.113	10.3
500	0.214	9.7
600	0.309	9.3
700	0.400	9.0
800	0.489	8.8
900	0.576	8.7
1000	0.663	8.7
1100	0.751	8.7

REMARKS

The tabulated values are considered accurate to within ±10% over the entire temperature range. These values can be represented approximately by the following equation:

$$\Delta L/L_0 = -0.356 + 1.355 \times 10^{-3}\, T - 5.228 \times 10^{-7}\, T^2 + 1.874 \times 10^{-10}\, T^3$$

FIGURE AND TABLE NO. 101CR. PROVISIONAL VALUES FOR THERMAL LINEAR EXPANSION OF Ge_3Pr_5 INTERMETALLIC COMPOUND

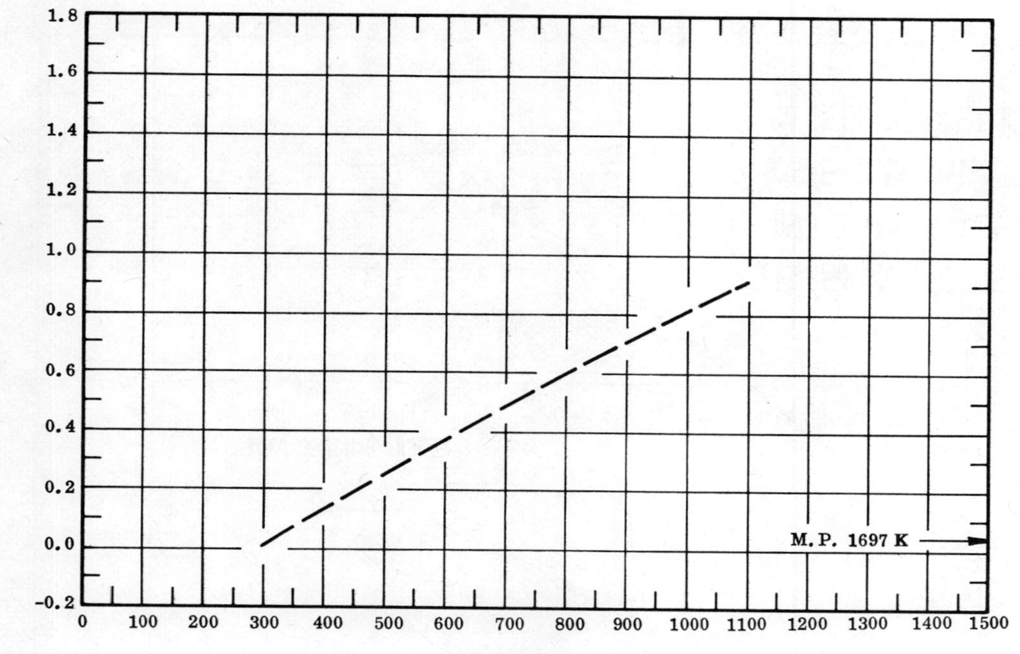

PROVISIONAL VALUES

[Temperature, T, K; Linear Expansion, $\Delta L/L_0$, %; α, K^{-1}]

T	$\Delta L/L_0$	$\alpha \times 10^6$
293	0.000	11.56
400	0.123	11.55
500	0.239	11.55
600	0.354	11.54
700	0.470	11.54
800	0.585	11.53
900	0.700	11.53
1000	0.816	11.52
1100	0.932	11.52

REMARKS

The tabulated values are considered accurate to within ±10% over the entire temperature range. These values can be represented approximately by the following equation:

$$\Delta L/L_0 = -0.340 + 1.166 \times 10^{-3}\, T - 1.973 \times 10^{-8}\, T^2 + 1.038 \times 10^{-11}\, T^3$$

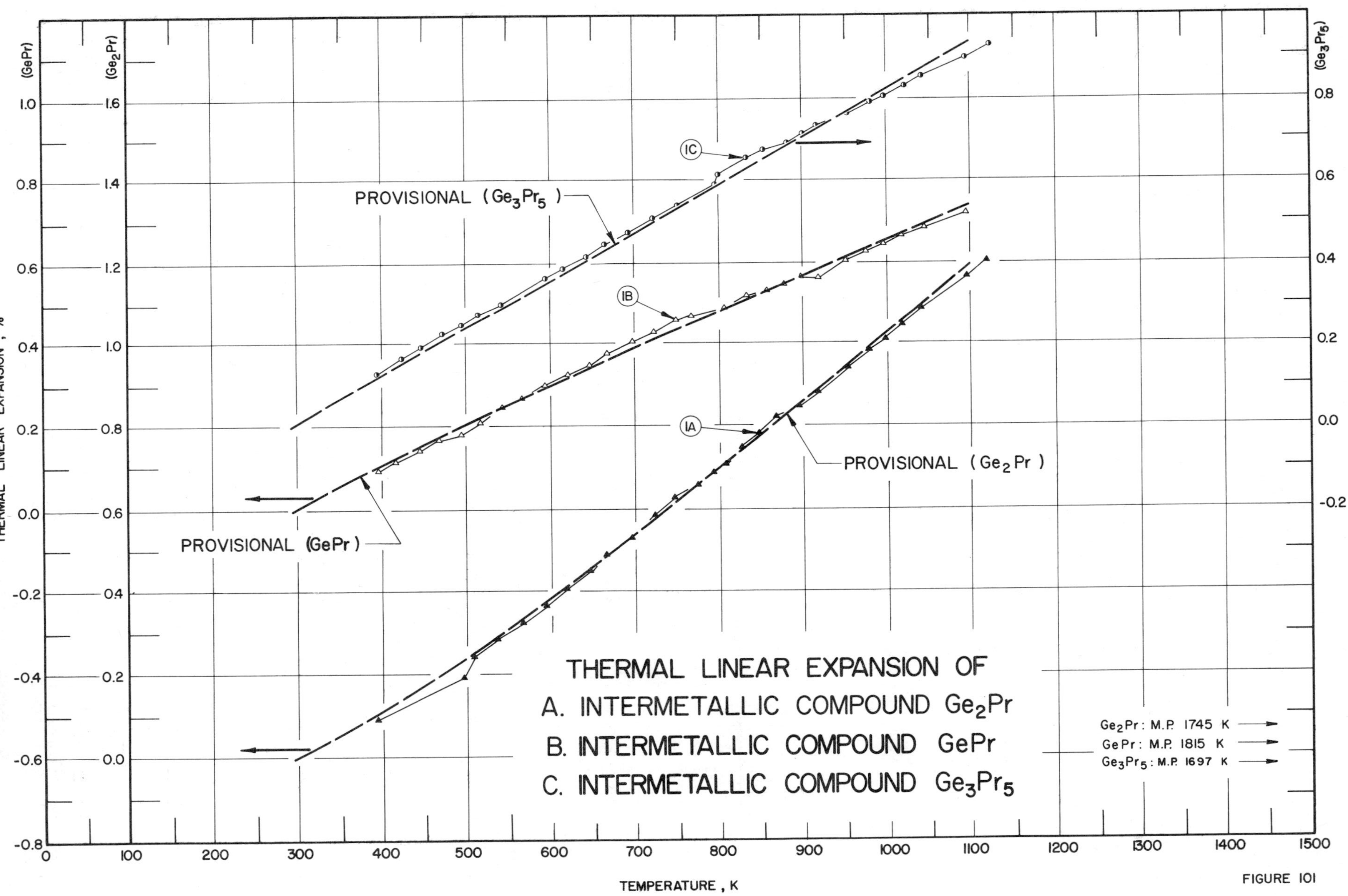

THERMAL LINEAR EXPANSION OF
A. INTERMETALLIC COMPOUND Ge_2Pr
B. INTERMETALLIC COMPOUND GePr
C. INTERMETALLIC COMPOUND Ge_3Pr_5
Ge_2Pr : M.P. 1745 K
GePr : M.P. 1815 K
Ge_3Pr_5 : M.P. 1697 K
PROVISIONAL (Ge_3Pr_5)
PROVISIONAL (GePr)
PROVISIONAL (Ge_2Pr)
1C
1B
1A
THERMAL LINEAR EXPANSION, %
(GePr)
(Ge_2Pr)
(Ge_3Pr_5)
TEMPERATURE, K

FIGURE 101

SPECIFICATION TABLE 101A. THERMAL LINEAR EXPANSION OF Ge_2Pr INTERMETALLIC COMPOUND

Cur. No.	Ref. No.	Author(s)	Year	Method Used	Temp. Range, K	Name and Specimen Designation	Composition (weight percent), Specifications, and Remarks
1	585	Rud', B.M., Lynchak, K.A., and Paderno, Yu.B.	1969	L	392-1119	Ge_2Pr	Specimen prepared by fusing 99.36 pure Pr and 99.99 pure single crystal Ge in argon atmosphere; annealed 12 hr at 1273 K and cooled over 6 hr; single phase of specimen verified by metallographic analysis; melting point reported is 1743 K anomaly of unexplained nature found near 500 K; zero-point correction of 0.041% determined by graphical extrapolation.

DATA TABLE 101A. THERMAL LINEAR EXPANSION OF Ge_2Pr INTERMETALLIC COMPOUND

[Temperature, T, K; Linear Expansion, $\Delta L/L_0$, %]

T	$\Delta L/L_0$	T	$\Delta L/L_0$
CURVE 1		CURVE 1 (cont.)	
392	0.091	772	0.656
417	0.119*	790	0.687
443	0.143*	808	0.714
469	0.172*	829	0.751
491	0.199	848	0.781
509	0.246	869	0.822
536	0.287	891	0.853
566	0.329	914	0.883
592	0.369	950	0.943
616	0.408	978	0.990
644	0.452	995	1.018
664	0.492	1018	1.052
694	0.531	1040	1.090
722	0.581	1093	1.172
749	0.622	1119	1.216

*Not shown in figure.

SPECIFICATION TABLE 101B. THERMAL LINEAR EXPANSION OF GePr INTERMETALLIC COMPOUND

Cur. No.	Ref. No.	Author(s)	Year	Method Used	Temp. Range, K	Name and Specimen Designation	Composition (weight percent), Specifications, and Remarks
1	585	Rud', B. M., Lynchak, K. A., and Paderno, Yu. B.	1969	L	395-1119	GePr	Specimen prepared by fusing 99.36 pure Pr and 99.99 pure single crystal Ge in argon atmosphere; annealed 12 hr at 1273 K and cooled over 6 hr; single phase of specimen verified by metallographic analysis; melting point reported is 1813 K, anomaly of unexplained nature found near 800 K; zero-point correction of 0.02% determined by graphical extrapolation.

DATA TABLE 101B. THERMAL LINEAR EXPANSION OF GePr INTERMETALLIC COMPOUND

[Temperature, T, K; Linear Expansion, $\Delta L/L_0$, %]

T	$\Delta L/L_0$	T	$\Delta L/L_0$
CURVE 1		CURVE 1 (cont.)	
395	0.094	769	0.473
415	0.120	809	0.499
444	0.143	834	0.522
469	0.172	859	0.537
491	0.201	878	0.522
519	0.216	897	0.570
542	0.245	918	0.568
568	0.274	950	0.609
593	0.301	974	0.631
647	0.330	993	0.647
669	0.355	1019	0.667
696	0.381	1045	0.689
721	0.410	1093	0.724
750	0.437	1119	0.745

SPECIFICATION TABLE 101C. THERMAL LINEAR EXPANSION OF Ge_3Pr_5 INTERMETALLIC COMPOUND

Cur. No.	Ref. No.	Author(s)	Year	Method Used	Temp. Range, K	Name and Specimen Designation	Composition (weight percent), Specifications, and Remarks
1	585	Rud', B.M., Lynchak, K.A., and Paderno, Yu. B.	1969	L	395-1124	Ge_3Pr_5	Specimen prepared by fusing 99.36 pure Pr and 99.99 pure single crystal Ge in argon atmosphere; annealed 12 hr at 1273 K and cooled over 6 hr; single phase of specimen verified by metallographic analysis; melting point reported is 1695 K, anomaly of unexplained nature found near 800 K; zero-point correction of 0.120% determined by graphical extrapolation.

DATA TABLE 101C. THERMAL LINEAR EXPANSION OF Ge_3Pr_5 INTERMETALLIC COMPOUND

[Temperature, T, K; Linear Expansion, $\Delta L/L_0$, %]

T	$\Delta L/L_0$	T	$\Delta L/L_0$
CURVE 1		CURVE 1 (cont.)	
395	0.132	855	0.670
422	0.167	881	0.690
449	0.196	900	0.716
471	0.225	918	0.734
497	0.246	954	0.762
518	0.273	980	0.787
543	0.300	997	0.801
568	0.332*	1021	0.826
595	0.361	1044	0.851
617	0.389	1095	0.900
643	0.416	1124	0.925
666	0.442		
691	0.472		
724	0.505		
752	0.539		
797	0.586		
800	0.614		
839	0.653		

* Not shown in figure.

SPECIFICATION TABLE 102. THERMAL LINEAR EXPANSION OF Au_4Mn INTERMETALLIC COMPOUND

Cur. No.	Ref. No.	Author(s)	Year	Method Used	Temp. Range, K	Name and Specimen Designation	Composition (weight percent), Specifications, and Remarks
1*	562	Kaneko, T. and Matsumoto, M.	1969	X	292-421		Au, 99.99 purity and electrolytic Mn 99.9 purity; weighed in desired proportion, melted in sealed silica tube; heating in vacuum at 1073 K for 48 hr, kept at 623 K for 170 hr to develop ordered arrangement of atoms; 20 mm diameter. 15 mm long; temperature measured in error ±1.5; measurements along a-axis; Curie point reported 371 K; lattice parameter reported at 293 K is 4.0792 Å.
2*	562	Kaneko, T. and Matsumoto, M.	1969	X	291-421		The above specimen; measurements along c-axis; lattice parameter reported at 293 K is 4.0213 Å.

DATA TABLE 102. THERMAL LINEAR EXPANSION OF Au_4Mn INTERMETALLIC COMPOUND

[Temperature, T, K; Linear Expansion, $\Delta L/L_0$, %]

T	$\Delta L/L_0$	T	$\Delta L/L_0$
CURVE 1*		CURVE 2*	
292	0.022	291	-0.010
310	-0.012	309	0.094
325	0.042	325	0.164
336	0.061	335	0.147
349	0.061	348	0.139
358	0.069	357	0.194
367	0.145	367	0.159
375	0.125	374	0.204
399	0.154	398	0.293
409	0.147	410	0.313
421	0.145	421	0.378

* No figure given.

FIGURE AND TABLE NO. 103R. TYPICAL VALUES FOR THERMAL LINEAR EXPANSION OF Au_2Mn INTERMETALLIC COMPOUND

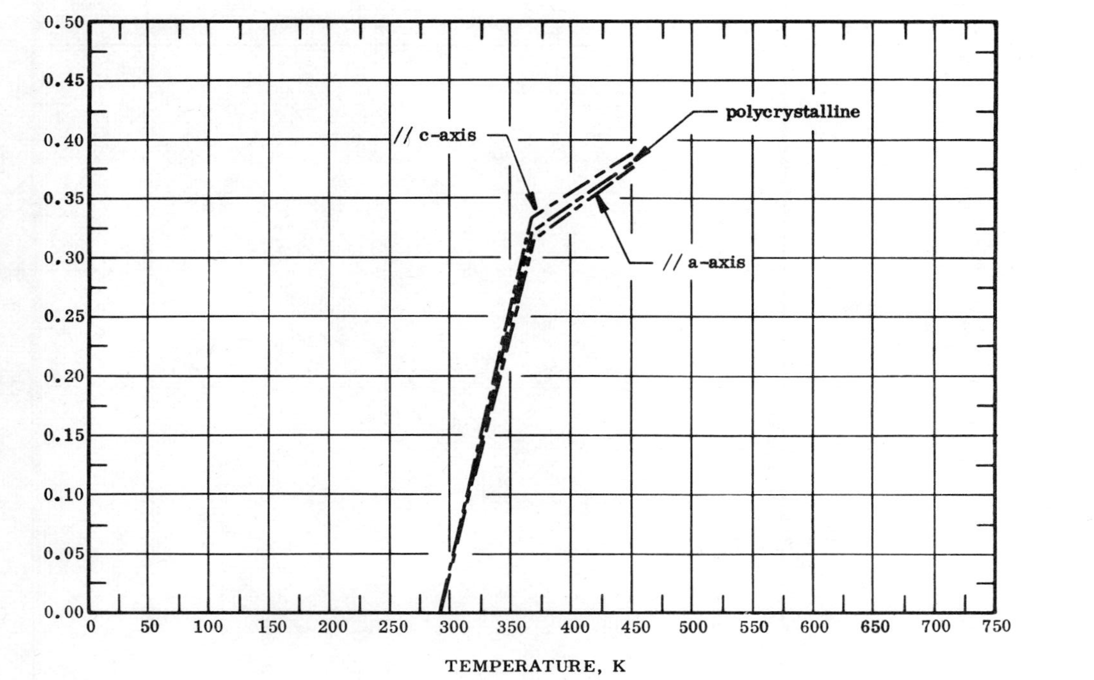

TYPICAL VALUES

[Temperature, T, K; Linear Expansion, $\Delta L/L_0$, %]

T	// a-axis $\Delta L/L_0$	// c-axis $\Delta L/L_0$	polycrystalline $\Delta L/L_0$
293	0.000	0.000	0.000
367†	0.315	0.333	0.321
400	0.338	0.356	0.344
450	0.377	0.385	0.380

† Neel temperature.

REMARKS

The tabulated values are considered accurate to within ±10% over the entire temperature range.

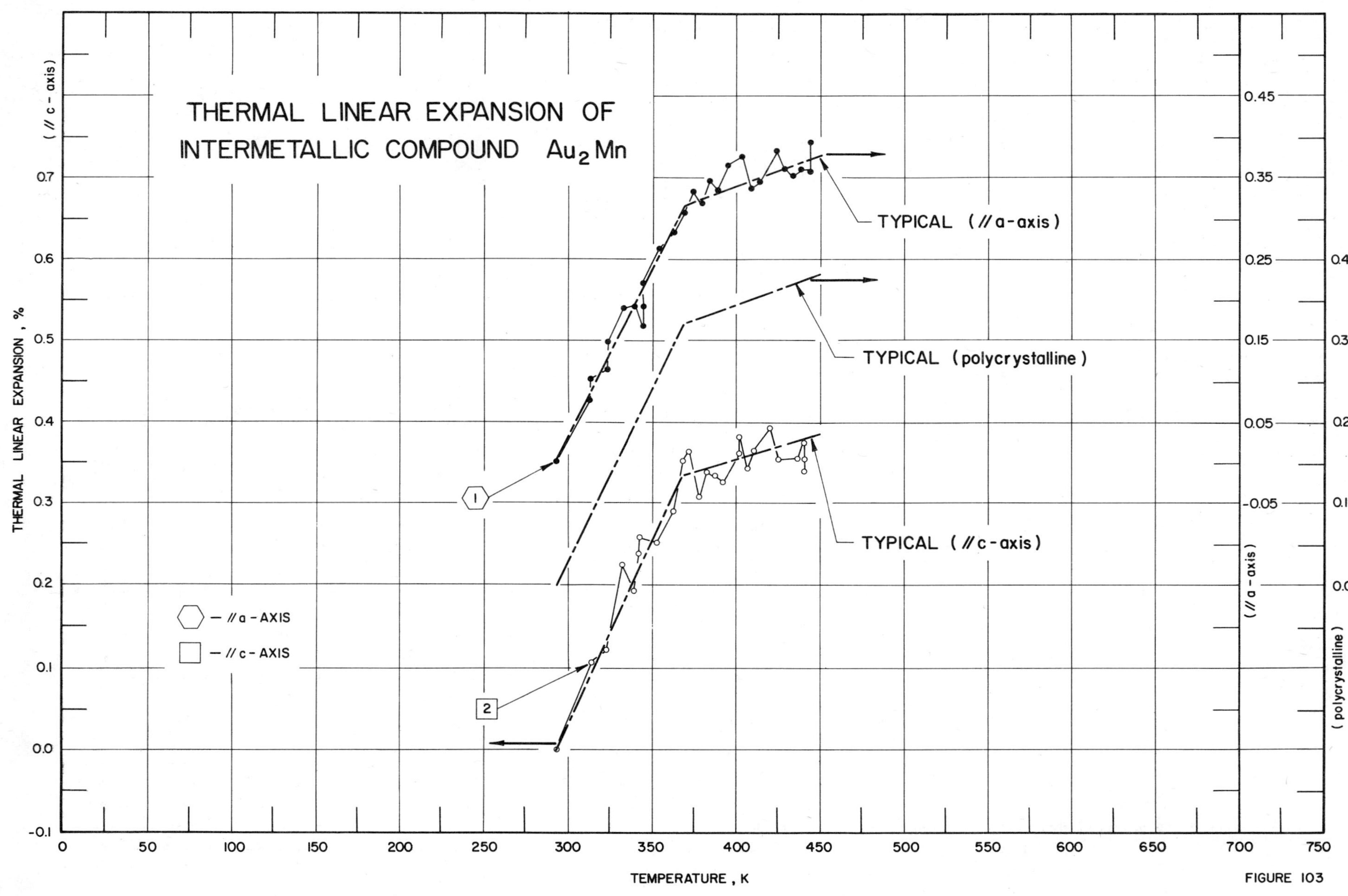
THERMAL LINEAR EXPANSION OF
INTERMETALLIC COMPOUND Au_2Mn
TYPICAL (//a-axis)
TYPICAL (polycrystalline)
TYPICAL (//c-axis)
1
2
— //a-AXIS
— //c-AXIS
THERMAL LINEAR EXPANSION, %
(// c - axis)
-0.1
0.0
0.1
0.2
0.3
0.4
0.5
0.6
0.7
(// a - axis)
-0.05
0.05
0.15
0.25
0.35
0.45
(polycrystalline)
0.0
0.1
0.2
0.3
0.4
TEMPERATURE, K
0
50
100
150
200
250
300
350
400
450
500
550
600
650
700
750

FIGURE 103

SPECIFICATION TABLE 103. THERMAL LINEAR EXPANSION OF Au_2Mn INTERMETALLIC COMPOUND

Cur. No.	Ref. No.	Author(s)	Year	Method Used	Temp. Range, K	Name and Specimen Designation	Composition (weight percent), Specifications, and Remarks
1	570	Gordienko, V. A. and Nikolaev, V. I.	1968	X	293-441		Neel temperature ~367 K; expansion measured along a-axis using iron K_α radiation; expansion affected by spontaneous magneto-striction at Neel temperature; lattice parameter reported at 293 K is 3.3629 Å.
2	570	Gordienko, V. A. and Nikolaev, V. I.	1968	X	293-440		The above specimen; expansion measured along c-axis; Neel temperature is 367 K; lattice parameter reported at 293 K is 8.5927 Å.

DATA TABLE 103. THERMAL LINEAR EXPANSION OF Au_2Mn INTERMETALLIC COMPOUND

[Temperature, T, K; Linear Expansion, $\Delta L/L_0$, %]

T	$\Delta L/L_0$	T	$\Delta L/L_0$	T	$\Delta L/L_0$	T	$\Delta L/L_0$
CURVE 1		CURVE 1 (cont.)		CURVE 2 (cont.)		CURVE 2 (cont.)	
293	0.000	393	0.333	332	0.223	425	0.355
313	0.077	393	0.363	338	0.192	436	0.355
313	0.101	402	0.378	342	0.239	440	0.340
322	0.113	408	0.336	342	0.257	440	0.356*
322	0.149	412	0.345	352	0.251	440	0.376
332	0.190	421	0.381	362	0.290		
338	0.190	427	0.360	367	0.351		
342	0.190	432	0.351	371	0.363		
342	0.220	437	0.360	376	0.307*		
343	0.169	441	0.354	382	0.339		
353	0.262	441	0.395	386	0.334*		
362	0.282			391	0.327		
367	0.309	CURVE 2		401	0.363		
373	0.333			401	0.382		
377	0.318	293	0.000*	406	0.343		
382	0.348	313	0.105*	410	0.365		
388	0.333	322	0.121	420	0.393		

* Not shown in figure.

SPECIFICATION TABLE 104. THERMAL LINEAR EXPANSION OF AuMn INTERMETALLIC COMPOUND

Cur. No.	Ref. No.	Author(s)	Year	Method Used	Temp. Range, K	Name and Specimen Designation	Composition (weight percent), Specifications, and Remarks
1*	563	Matsumoto, M., Kaneko, T., and Kamigaki, K.	1968	O	303-522		Specimen prepared from 99.99 Au and 99.9 Mn; homogenized by vacuum heating at 723 K for 19 hours and quenched to room temperature; crystal tetragonal form with little deformation; zero-point correction of -0.006% determined by graphical extrapolation; Néel temperature reported at 513 K.
2*	685	Smith J.H., and Gaunt, P.	1961	X	78-319		47 at. % Au; small block specimen about 0.3 x 0.4 x 3 cm^3; lattice parameter along a-axis of tetragonal phase calculated from [222] and [321] diffraction maxima; no volume change accompanying transformation; reported error ±0.1%; lattice parameter reported at 293 K is 3.1562 Å.
3*	685	Smith J.H., and Gaunt, P.	1961	X	77-320		The above specimen; lattice parameter along c-axis for $c/a < 1$ reported at 293 K is 3.2698 Å.
4*	685	Smith J.H., and Gaunt, P.	1961	X	344-426		Similar to the above specimen; lattice parameter along a-axis for $c/a > 1$ reported at 344 K is 3.2497 Å.
5*	685	Smith J.H., and Gaunt, P.	1961	X	343-425		The above specimen; lattice parameter along c-axis for $c/a > 1$ reported at 343 K is 3.1111 Å.
6*	685	Smith J.H., and Gaunt, P.	1961	X	442, 450		Similar to the above specimen except cubic structure stable above 442 K; lattice parameter reported at 442 K is 3.2071 Å.

* No figure given.

DATA TABLE 104. THERMAL LINEAR EXPANSION OF AuMn INTERMETALLIC COMPOUND

[Temperature, T, K; Linear Expansion, $\Delta L/L_0$, %]

T	$\Delta L/L_0$
CURVE 1*	
303	0.012
313	0.034
323	0.058
333	0.076
338	0.087
343	0.099
348	0.114
352	0.123
359	0.133
362	0.142
367	0.154
372	0.167
378	0.185
382	0.200
388	0.213
392	0.234
398	0.237
402	0.247
407	0.260
411	0.275
413	0.277
415	0.280
417	0.285
420	0.287
422	0.295
425	0.305
427	0.312
429	0.310
431	0.320
433	0.324
435	0.329
437	0.334
440	0.338
442	0.343
451	0.362
452	0.371
458	0.379
462	0.393
467	0.410
473	0.418
474	0.428
476	0.430
478	0.438
481	0.443
482	0.450
484	0.452
CURVE 1 (cont.)*	
487	0.462
489	0.470
492	0.467
494	0.470
497	0.477
498	0.478
501	0.482
503	0.482
505	0.490
506	0.492
508	0.496
511	0.499
512	0.507
514	0.510
516	0.514
520	0.520
522	0.527
CURVE 2*	
78	-1.155
293	0.000
308	0.219
319	0.413
CURVE 3*	
77	1.603
293	0.000
310	-0.021
320	-0.015
CURVE 4*	
	Å
344	3.2497 (t, c/a>1)
369	3.2384
379	3.2348
397	3.2286
420	3.2250
426	3.2250
CURVE 5*	
	Å
343	3.1111 (t, c/a>1)
369	3.1111
378	3.1338
398	3.1538
419	3.1663
425	3.1663
CURVE 6*	
	Å
442	3.2071 (cubic)
450	3.2075

* No figure given.

SPECIFICATION TABLE 105. THERMAL LINEAR EXPANSION OF Au_4V INTERMETALLIC COMPOUND

Cur. No.	Ref. No.	Author(s)	Year	Method Used	Temp. Range, K	Name and Specimen Designation	Composition (weight percent), Specifications, and Remarks
1*	751	Kasai, N. and Ogawa, S.	1971	E	4-70		Corresponding weights of 99.9 V from Showa Chemicals Co., and 99.999 Au from Sumitomo Metals and Mining Ltd, melted in BeO crucible in vacuum using induction furnace; sealed in quartz tube in vacuum of 10^{-6} torr, annealed at 1273 K for two days, homogenized and quenched in water, sample size is 19.51 mm length and 7.83 mm.
2*	751	Kasai, N. and Ogawa, S.	1971	E	5-100		The above specimen in the disordered state further annealed in vacuum at 773 K for three weeks and at 723 K for five days to get ordered state; average of values on ordered state with different heat treatment; Curie temperature reported at 52 ±1 K.

* No figure given.

DATA TABLE 105. THERMAL LINEAR EXPANSION OF Au_4V INTERMETALLIC COMPOUND

[Temperature, T, K; Linear Expansion, $\Delta L/L_0$, %]

T	$\Delta L/L_0$	T	$\Delta L/L_0$	T	$\Delta L/L_0$
CURVE 1*, †, ‡		CURVE 2*, †, ‡		CURVE 2 (cont.)*, †, ‡	
4	0.000 x 10^{-3}	5	0.000 x 10^{-3}	55	18.400 x 10^{-3}
6	0.026	7	0.064	60	21.800
10	0.094	10	0.175	70	29.300
12	0.147	14	0.449	80	37.800
14	0.222	16	0.640	90	47.100
16	0.325	20	1.160	100	57.100
20	0.651	25	2.140		
30	2.430	30	3.580		
40	5.820	35	5.620		
50	10.700	40	8.310		
60	16.800	45	11.500		
70	24.100	50	14.900		

DATA TABLE 105. COEFFICIENT OF THERMAL LINEAR EXPANSION OF Au_4V INTERMETALLIC COMPOUND

[Temperature, T, K; Coefficient of Expansion, α, 10^{-6} K^{-1}]

T	α	T	α	T	α
CURVE 1*, ‡		CURVE 2*, ‡		CURVE 2 (cont.)*, ‡	
4	0.13	5	0.25	55	6.76
6	0.13	7	0.36	60	7.02
10	0.21	10	0.53	70	7.99
12	0.32	14	0.84	80	8.94
14	0.43	16	1.07	90	9.60
16	0.60	20	1.55	100	10.47
20	1.03	25	2.35		
30	2.53	30	3.43		
40	4.24	35	4.73		
50	5.52	40	6.02		
60	6.70	45	6.67		
70	7.79	50	7.12		

* Not shown in figure.
† This curve is here reported using the first given temperature as reference temperature at which $\Delta L/L_0 = 0$.
‡ Author's data for coefficient of thermal expansion have been integrated by TPRC to obtain $\Delta L/L_0$.

FIGURE AND TABLE NO. 106R. PROVISIONAL VALUES FOR THERMAL LINEAR EXPANSION OF AuZn INTERMETALLIC COMPOUND

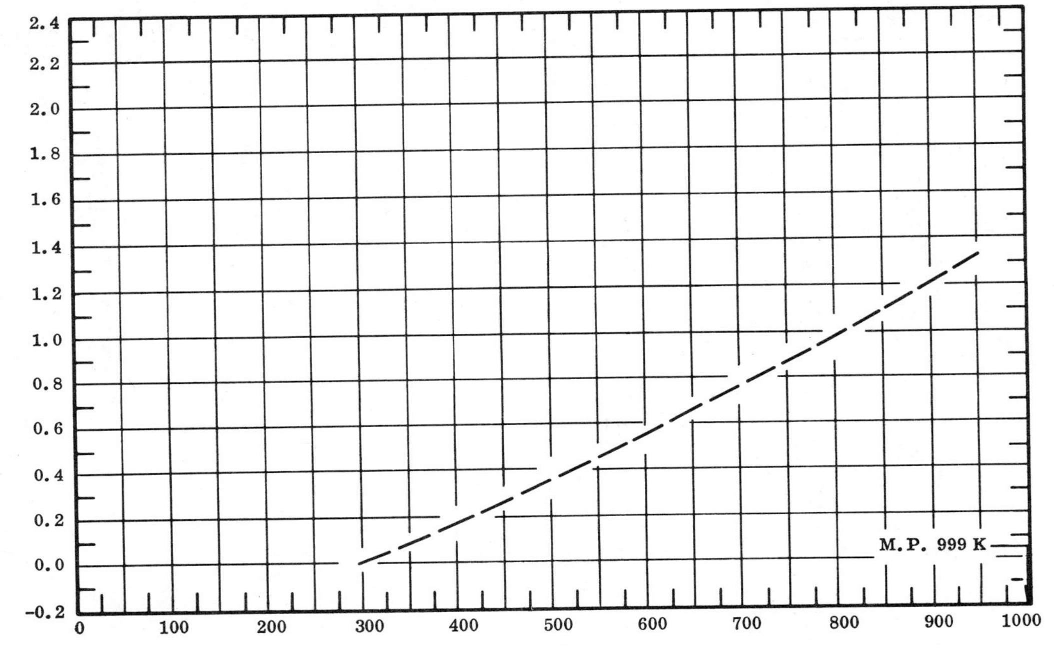

PROVISIONAL VALUES

[Temperature, T, K; Linear Expansion, $\Delta L/L_0$, %; α, K^{-1}]

T	$\Delta L/L_0$	$\alpha \times 10^6$
293	0.000	16.4
400	0.185	17.8
500	0.369	19.0
600	0.564	20.1
700	0.770	21.0
800	0.985	21.9
900	1.207	22.6
950	1.321	22.9

REMARKS

The tabulated values are considered accurate to within ±7% over the entire temperature range. These values can be represented approximately by the following equation:

$$\Delta L/L_0 = -0.417 + 1.192 \times 10^{-3}\, T + 8.545 \times 10^{-7} \times T^2 - 1.938 \times 10^{-10}\, T^3$$

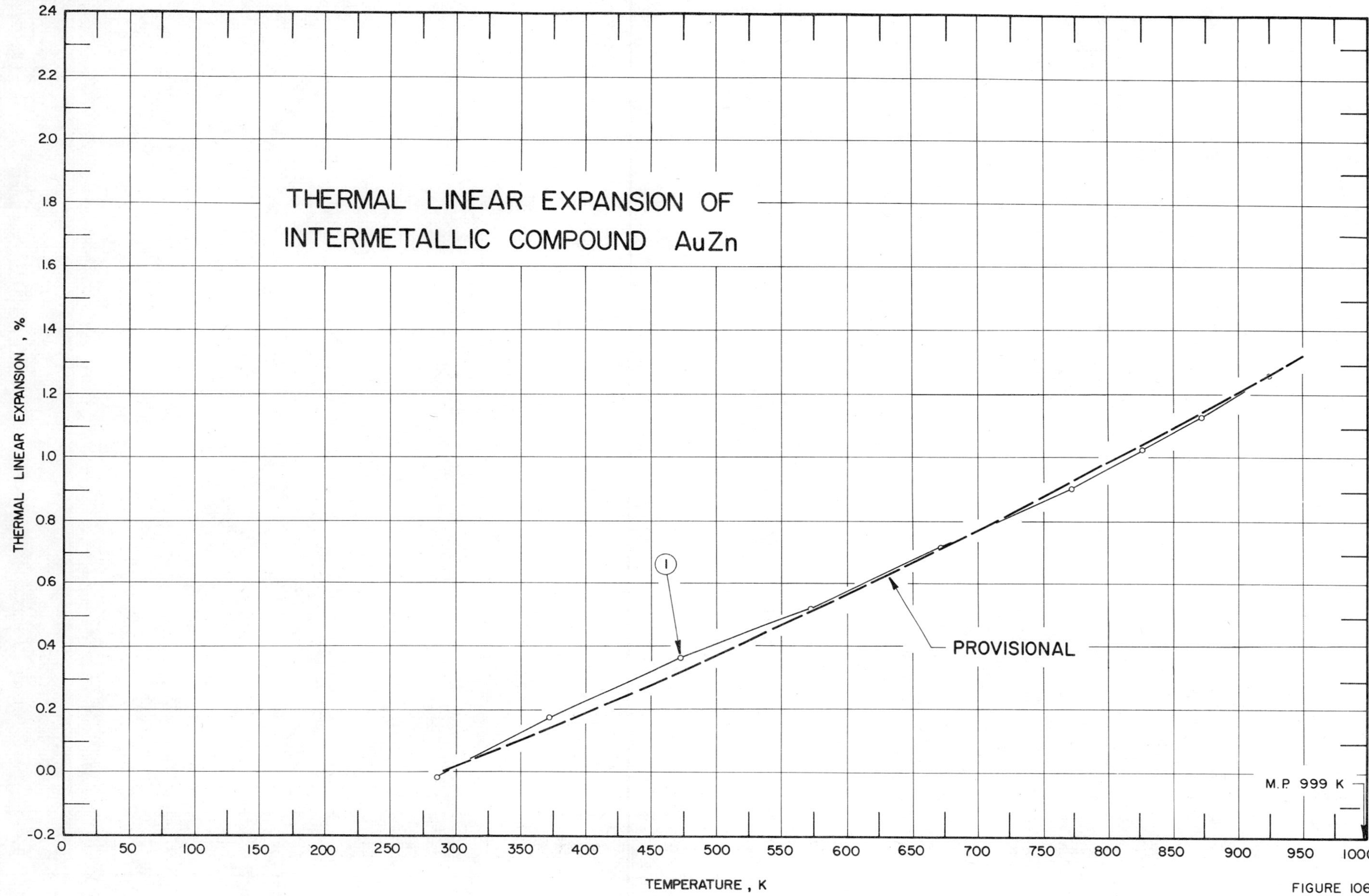
THERMAL LINEAR EXPANSION OF
INTERMETALLIC COMPOUND AuZn
THERMAL LINEAR EXPANSION , %
2.4
2.2
2.0
1.8
1.6
1.4
1.2
1.0
0.8
0.6
0.4
0.2
0.0
-0.2
0
50
100
150
200
250
300
350
400
450
500
550
600
650
700
750
800
850
900
950
1000
TEMPERATURE , K
1
PROVISIONAL
M.P. 999 K

FIGURE 106

SPECIFICATION TABLE 106. THERMAL LINEAR EXPANSION OF AuZn INTERMETALLIC COMPOUND

Cur. No.	Ref. No.	Author(s)	Year	Method Used	Temp. Range, K	Name and Specimen Designation	Composition (weight percent), Specifications, and Remarks
1	418	Iwasaki, H. and Ussugi, T.	1968	X	286-925		Alloy with composition 75.45 Au and 24.55 Zn prepared by melting 99.99 pure materials in sealed, evacuated quartz tube; single crystal grown from the melt; plate specimen cut parallel [100] plane, 1.7 mm thick, 11 by 16 mm^2, annealed and etched; m.p. = 998 K; lattice parameter reported at 286 K is 3.1513 Å; 3.1518 Å at 293 K determined by graphical interpolation.

DATA TABLE 106. THERMAL LINEAR EXPANSION OF AuZn INTERMETALLIC COMPOUND

[Temperature, T, K; Linear Expansion, $\Delta L/L_0$, %]

T	$\Delta L/L_0$
CURVE 1	
286	-0.016
373	0.178
473	0.368
574	0.520
673	0.714
773	0.904
826	1.028
873	1.133
925	1.256

FIGURE AND TABLE NO. 107R. TYPICAL VALUES FOR THERMAL LINEAR EXPANSION OF $HoZn_2$ INTERMETALLIC COMPOUND

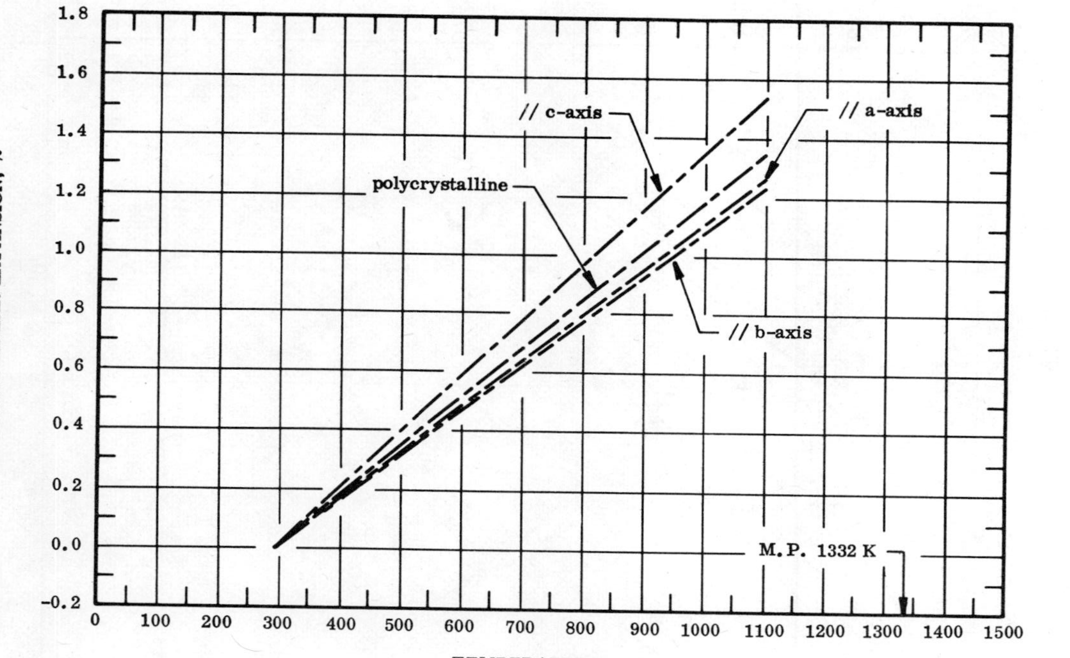

TYPICAL VALUES

[Temperature, T, K; Linear Expansion, $\Delta L/L_0$, %; α, K^{-1}]

T	//a-axis $\Delta L/L_0$	//b-axis $\Delta L/L_0$	//c-axis $\Delta L/L_0$	polycrystalline $\Delta L/L_0$
293	0.000	0.000	0.000	0.000
400	0.166	0.165	0.208	0.180
500	0.323	0.320	0.399	0.347
600	0.481	0.474	0.501	0.515
700	0.639	0.628	0.783	0.683
800	0.796	0.782	0.975	0.851
900	0.952	0.934	1.167	1.018
1000	1.107	1.087	1.358	1.184
1100	1.261	1.238	1.548	1.349

// a-axis: $\alpha = 15.7 \times 10^{-6}$ (293-1100)

// b-axis: $\alpha = 15.4 \times 10^{-6}$ (293-1100)

// c-axis: $\alpha = 19.0 \times 10^{-6}$ (293-1100)

polycrystalline: $\alpha = 16.7 \times 10^{-6}$ (293-1100)

REMARKS

The tabulated values are considered accurate to within ±20%. The values for polycrystalline material in the absence of experimental data are calculated from the single crystal data. These and other values can be represented approximately by the following equations:

// a-axis: $\Delta L/L_0 = -0.451 + 1.504 \times 10^{-3}\, T + 1.217 \times 10^{-7}\, T^2 - 6.768 \times 10^{-11}\, T^3$

// b-axis: $\Delta L/L_0 = -0.451 + 1.536 \times 10^{-3}\, T + 2.285 \times 10^{-8}\, T^2 - 2.076 \times 10^{-11}\, T^3$

// c-axis: $\Delta L/L_0 = -0.545 + 1.824 \times 10^{-3}\, T + 1.356 \times 10^{-7}\, T^2 - 6.284 \times 10^{-11}\, T^3$

polycrystalline: $\Delta L/L_0 = -0.482 + 1.632 \times 10^{-3}\, T + 7.743 \times 10^{-8}\, T^2 - 4.278 \times 10^{-11}\, T^3$

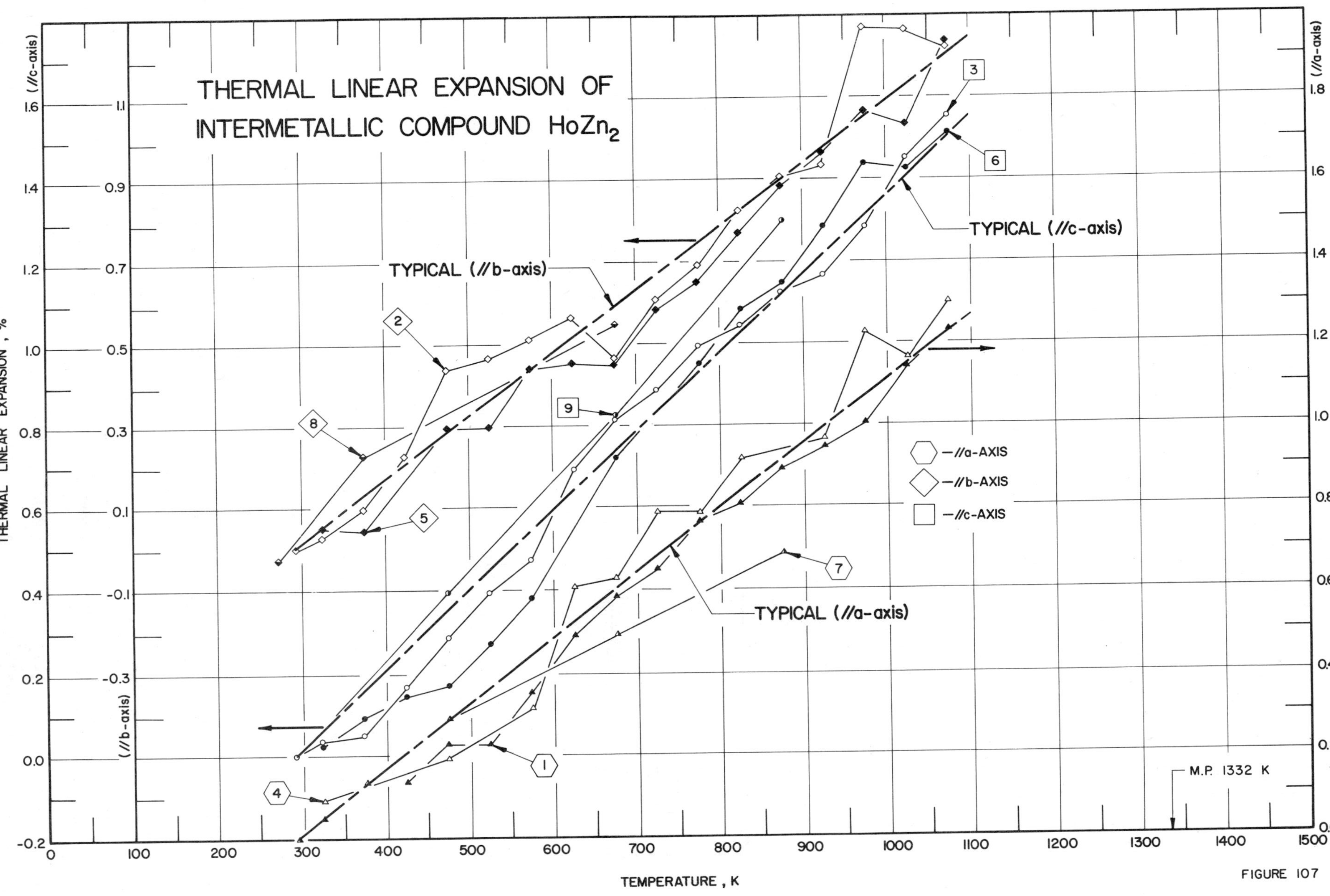
THERMAL LINEAR EXPANSION OF
INTERMETALLIC COMPOUND $HoZn_2$
TYPICAL (//b-axis)
TYPICAL (//c-axis)
TYPICAL (//a-axis)
—//a-AXIS
—//b-AXIS
—//c-AXIS
M.P. 1332 K
THERMAL LINEAR EXPANSION, %
TEMPERATURE, K
(//c-axis)
(//b-axis)
(//a-axis)
1
2
3
4
5
6
7
8
9
-0.2
0.0
0.2
0.4
0.6
0.8
1.0
1.2
1.4
1.6
-0.3
-0.1
0.1
0.3
0.5
0.7
0.9
1.1
0.0
0.2
0.4
0.6
0.8
1.0
1.2
1.4
1.6
1.8
0
100
200
300
400
500
600
700
800
900
1000
1100
1200
1300
1400
1500

FIGURE 107

SPECIFICATION TABLE 107. THERMAL LINEAR EXPANSION OF $HoZn_2$ INTERMETALLIC COMPOUND

Cur. No.	Ref. No.	Author(s)	Year	Method Used	Temp. Range, K	Name and Specimen Designation	Composition (weight percent), Specifications, and Remarks
1	300	Michel, D. J. and Ryba, E.	1968	X	293-1073		99.5$^+$% pure holmium metal supplied by the Research Div. of the Nuclear Corp. of America and Bernard Ring, Inc.; 99.999% pure zinc supplied by American Smelting and Refining Co.; $HoZn_2$ alloy reduced to powder and then encapsulated in Be foil; CuK_α (λ = 1.54178 Å) radiation used; expansion measured along a-axis; lattice parameter reported at 293 K is 4.456 Å.
2	300	Michel, D. J. and Ryba, E.	1968	X	293-1073		The above specimen; expansion measured along b-axis; lattice parameter reported at 293 K is 7.039 Å.
3	300	Michel, D. J. and Ryba, E.	1968	X	293-1073		The above specimen; expansion measured along c-axis; lattice parameter reported at 293 K is 7.641 Å.
4	300	Michel, D. J. and Ryba, E.	1968	X	293-1073		Specimen similar to the above prepared from the same $HoZn_2$ alloy; expansion measured along a-axis; lattice parameter reported at 293 K is 4.456 Å.
5	300	Michel, D. J. and Ryba, E.	1968	X	293-1073		The above specimen; expansion measured along b-axis; lattice parameter reported at 293 K is 7.039 Å.
6	300	Michel, D. J. and Ryba, E.	1968	X	293-1073		The above specimen; expansion measured along c-axis; lattice parameter reported at 293 K is 7.641 Å.
7	565	Ryba, E.	1967	X	273-873		Specimen prepared from metal powders encapsulated in Be-foil by heating; expansion measured along a-axis; lattice parameter reported at 273 K is 4.4534 Å; 4.4552 Å at 293 K determined by graphical interpolation.
8	565	Ryba, E.	1967	X	273-874		The above specimen; expansion measured along b-axis; lattice parameter reported at 273 K is 7.0399 Å; 7.0423 Å at 293 K determined by graphical interpolation.
9	565	Ryba, E.	1967	X	273-874		The above specimen; expansion measured along c-axis; lattice parameter reported at 273 K is 7.6098 Å; 7.6135 Å at 293 K determined by graphical interpolation.

DATA TABLE 107. THERMAL LINEAR EXPANSION OF $HoZn_2$ INTERMETALLIC COMPOUND

[Temperature, T, K; Linear Expansion, $\Delta L/L_0$, %]

T	$\Delta L/L_0$
CURVE 1	
293	0.000
323	0.045
373	0.135
423	0.135
473	0.224
523	0.224
573	0.359
623	0.494
673	0.583
723	0.651
773	0.763
823	0.808
873	0.897
923	0.943
973	1.010
1023	1.144
1073	1.234
CURVE 2	
293	0.000
323	0.028
373	0.099
423	0.227
473	0.440
523	0.467
573	0.511
623	0.568
673	0.483
723	0.611
773	0.696
823	0.824
873	0.909
923	0.938
973	1.278
1023	1.264
1073	1.221
CURVE 3	
293	0.000
323	0.039
373	0.052
423	0.170
473	0.289
523	0.393
CURVE 3 (cont.)	
573	0.471
623	0.694
673	0.811
723	0.890
773	0.995
823	1.046
873	1.126
923	1.165
973	1.283
1023	1.453
1073	1.557
CURVE 4	
293	0.000
323	0.090
373	0.135
423	0.135
473	0.190
523	0.224
573	0.314
623	0.606
673	0.628
723	0.785
773	0.785
823	0.920
873	0.965
923	1.055
973	1.234
1023	1.167
1073	1.302
CURVE 5	
293	0.000
323	0.057
373	0.043
423	0.227
473	0.298
523	0.298
573	0.440
623	0.455
673	0.469
723	0.582
773	0.654
823	0.796
CURVE 5 (cont.)	
873	0.881
923	0.966
973	1.065
1023	1.037
1073	1.236
CURVE 6	
293	0.000
323	0.026
373	0.092
423	0.144
473	0.170
523	0.275
573	0.380
623	0.694*
673	0.720
723	0.890*
773	0.955
823	1.086
873	1.152
923	1.283
973	1.440
1023	1.426
1073	1.518
CURVE 7	
273	-0.040
473	0.296
673	0.496
873	0.689
CURVE 8	
273	-0.034
473	0.231
673	0.551
874	0.720
CURVE 9	
273	-0.049
474	0.394
673	0.825
874	1.303

*Not shown in figure.

FIGURE AND TABLE NO. 108AR. PROVISIONAL VALUES FOR THERMAL LINEAR EXPANSION OF In_3Pd INTERMETALLIC COMPOUND

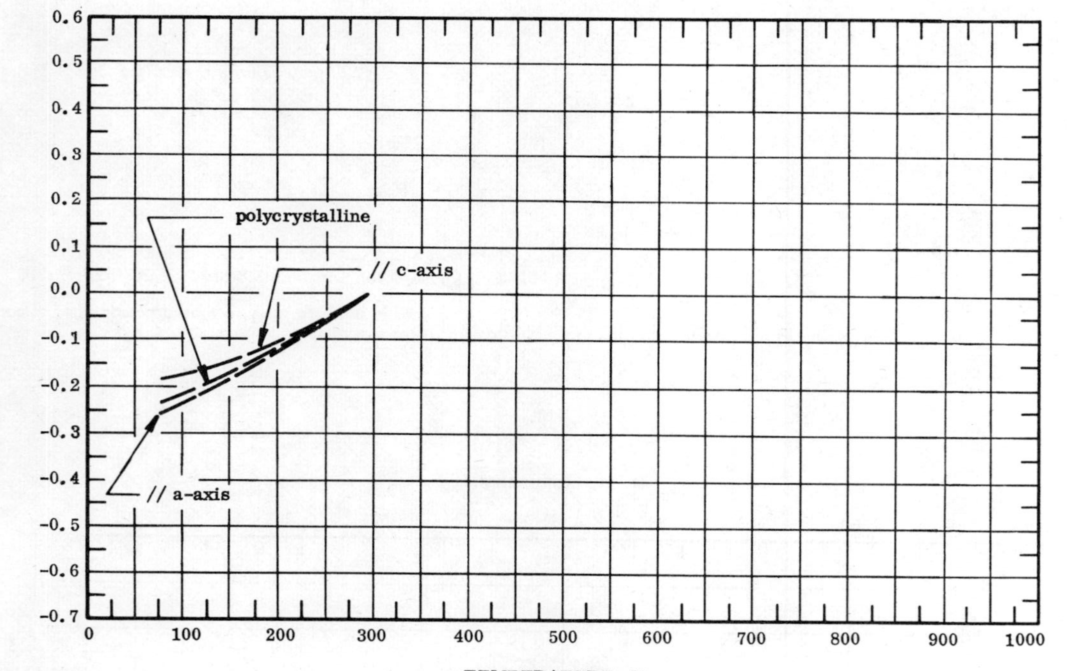

PROVISIONAL VALUES

[Temperature, T, K; Linear Expansion, $\Delta L/L_0$, %; α, K^{-1}]

	// a-axis	// c-axis	polycrystalline	
T	$\Delta L/L_0$	$\Delta L/L_0$	$\Delta L/L_0$	$\alpha \times 10^6$
75	-0.257	-0.184	-0.233	6.3
100	-0.232	-0.178	-0.214	8.2
200	-0.119	-0.101	-0.113	12.0
293	0.000	0.000	0.000	12.4

REMARKS

The tabulated values are considered accurate to within ±7%. The values for polycrystalline material in the absence of experimental data are calculated from the single crystal data.

FIGURE AND TABLE NO. 108BR. PROVISIONAL VALUES FOR THERMAL LINEAR EXPANSION OF In_3Pr INTERMETALLIC COMPOUND

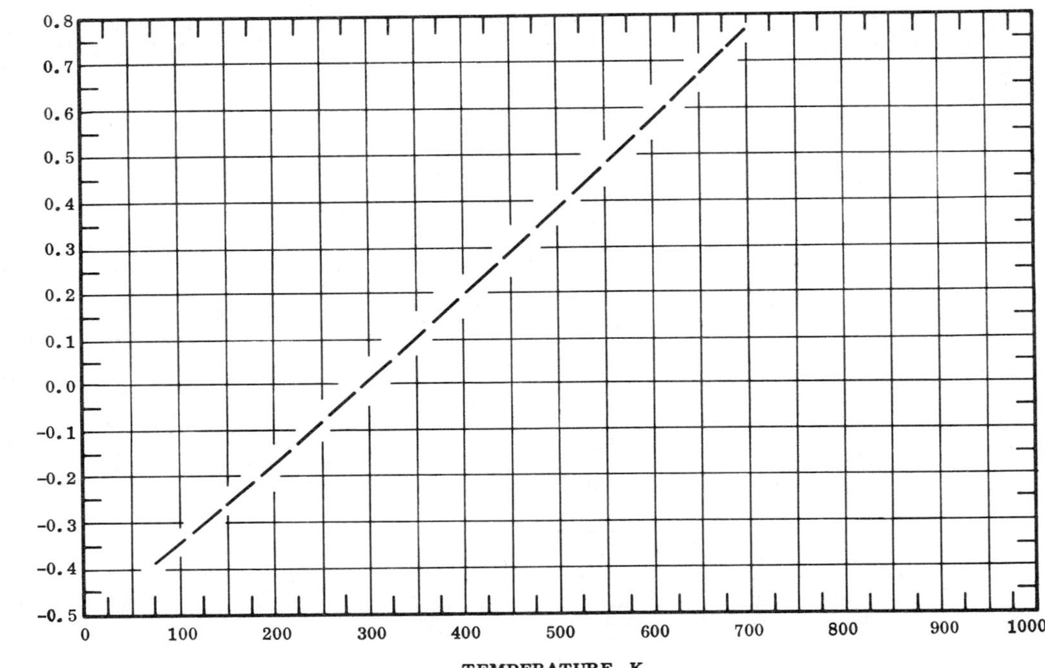

PROVISIONAL VALUES

[Temperature, T, K; Linear Expansion, $\Delta L/L_0$, %; α, K^{-1}]

T	$\Delta L/L_0$	$\alpha \times 10^6$
75	-0.386	16.9
100	-0.344	17.1
200	-0.169	17.9
293	0.000	18.4
400	0.199	18.7
500	0.387	18.8
600	0.574	18.8
700	0.760	18.8

REMARKS

The tabulated values are considered accurate to within ± 7% over the entire temperature range. These values can be represented approximately by the following equation:

$$\Delta L/L_0 = -0.511 + 1.617 \times 10^{-3}\, T + 5.439 \times 10^{-7}\, T^2 - 3.764 \times 10^{-10}\, T^3$$

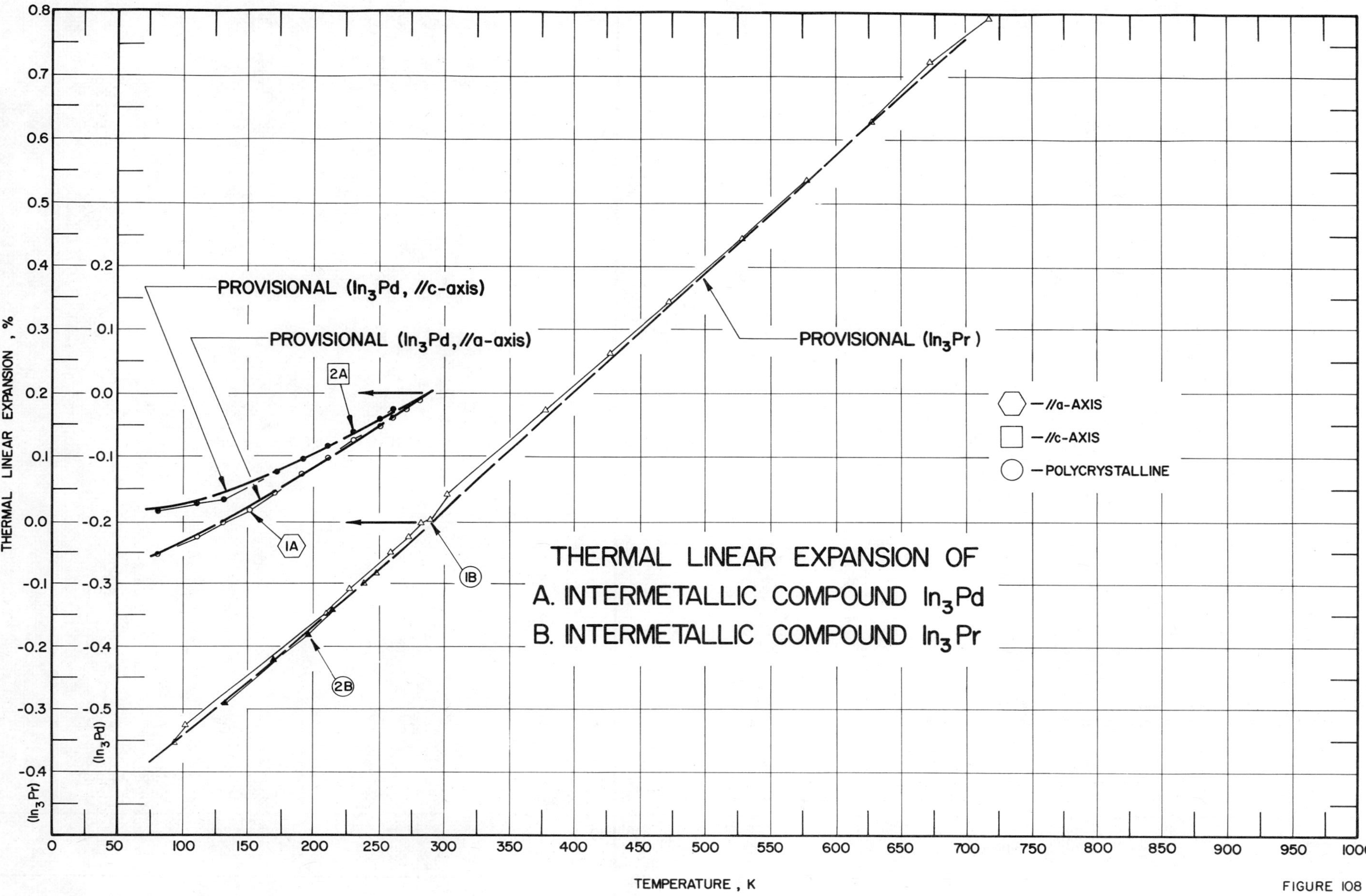

THERMAL LINEAR EXPANSION OF
A. INTERMETALLIC COMPOUND In_3Pd
B. INTERMETALLIC COMPOUND In_3Pr
PROVISIONAL (In_3Pd, //c-axis)
PROVISIONAL (In_3Pd, //a-axis)
PROVISIONAL (In_3Pr)
2A
1A
1B
2B
—//a-AXIS
—//c-AXIS
—POLYCRYSTALLINE
THERMAL LINEAR EXPANSION, %
(In_3Pr)
(In_3Pd)
TEMPERATURE, K

FIGURE 108

SPECIFICATION TABLE 108A. THERMAL LINEAR EXPANSION OF In_3Pd INTERMETALLIC COMPOUND

Cur. No.	Ref. No.	Author(s)	Year	Method Used	Temp. Range, K	Name and Specimen Designation	Composition (weight percent), Specifications, and Remarks
1	683	Harris, I.R., Norman, M., and Bryant, A.W.	1968	X	83-295		Specimen prepared by arc- or induction-melting nominal composition of Pd from Messrs. Johnson Matthey and Co., Ltd., and In from Koch-Light Laboratories Ltd.; measurements along a-axis; lattice parameter reported at 282 K is 4.0566 kX; 4.0570 K at 293 K determined by graphical extrapolation.
2	683	Harris, I.R., et al.	1968	X	83-295		The above specimen; measurements along c-axis; lattice parameter reported at 298 K is 3.7905 kX; 3.7909 kX determined by graphical extrapolation.

DATA TABLE 108A. THERMAL LINEAR EXPANSION OF In_3Pd INTERMETALLIC COMPOUND

[Temperature, T, K; Linear Expansion, $\Delta L/L_0$, %]

T	$\Delta L/L_0$	T	$\Delta L/L_0$
CURVE 1		CURVE 2	
83	-0.251	83	-0.185
113	-0.224	112	-0.171
133	-0.202	133	-0.158
153	-0.182	173	-0.124
172	-0.155	193	-0.106
192	-0.128	213	-0.082
213	-0.103	232	-0.063
232	-0.076	253	-0.042
252	-0.052	263	-0.029
253	-0.052	273	-0.024
263	-0.039	282	-0.010
273	-0.025	295	0.003
282	-0.010		
295	0.002		

SPECIFICATION TABLE 108B. THERMAL LINEAR EXPANSION OF In_3Pr INTERMETALLIC COMPOUND

Cur. No.	Ref. No.	Author(s)	Year	Method Used	Temp. Range, K	Name and Specimen Designation	Composition (weight percent), Specifications, and Remarks
1	682	Harris, I.R. and Raynor, G.V.	1965	X	717-93		Specimen prepared from In containing 0.002 Pb and 0.001 Cd from Messrs. Johnson, Matthey and Co., Ltd., and Pr containing <0.10 other rare earth metals, 0.10 Ta, and 0.02 other base metals by the method similar to the one described for $CeIn_3$; measurements with decreasing temperature; lattice parameter of 4.6607 kX at 293 K is used in the calculation.
2	682	Harris, I.R. and Raynor, G.V.	1965	X	132-215		The above specimen; measurements with increasing temperature; lattice parameter reported at 215 K is 4.6541 kX; 4.6607 kX at 293 K determined by graphical extrapolation.

DATA TABLE 108B. THERMAL LINEAR EXPANSION OF In_3Pr INTERMETALLIC COMPOUND

[Temperature, T, K; Linear Expansion, $\Delta L/L_0$, %]

T	$\Delta L/L_0$	T	$\Delta L/L_0$
CURVE 1		CURVE 1 (cont.)	
717	0.796	210	-0.148
671	0.721	101	-0.326
627	0.629	93	-0.358
579	0.538		
529	0.446	CURVE 2	
472	0.345		
426	0.262	132	-0.292
377	0.174	169	-0.221
301	0.043	195	-0.182
293	0.000	215	-0.142
288	0.006		
281	-0.004		
273	-0.026		
258	-0.049		
249	-0.082		
238	-0.101		
226	-0.109		

FIGURE AND TABLE NO. 109AR. PROVISIONAL VALUES FOR THERMAL LINEAR EXPANSION OF In_3Yb INTERMETALLIC COMPOUND

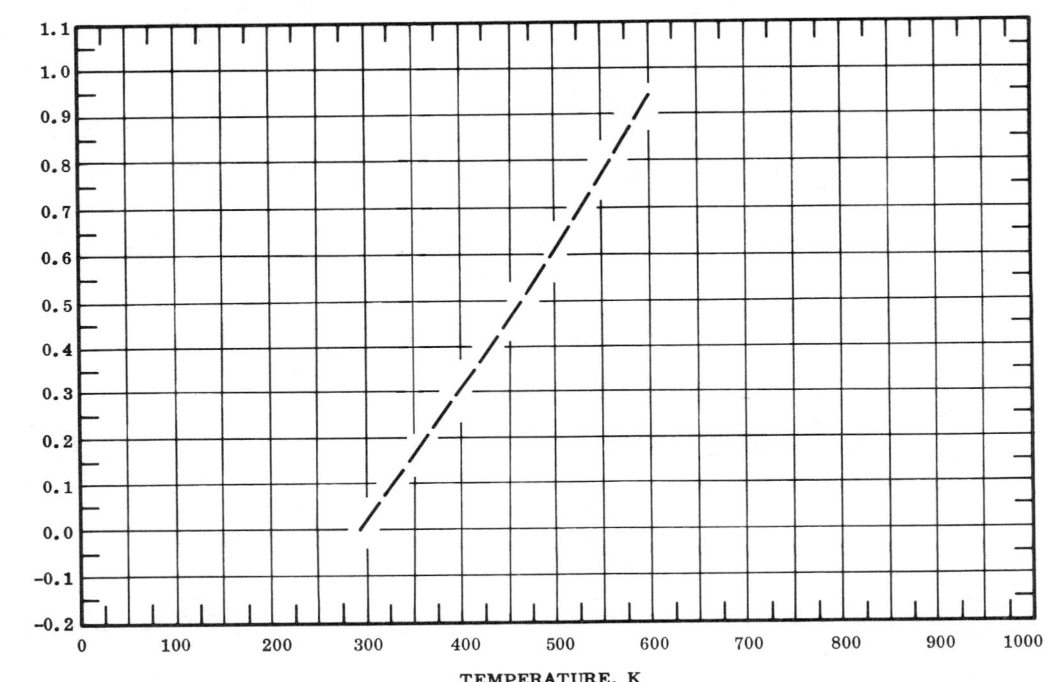

PROVISIONAL VALUES

[Temperature, T, K; Linear Expansion, $\Delta L/L_0$, %; α, K^{-1}]

T	$\Delta L/L_0$	$\alpha \times 10^6$
293	0.000	26.8
400	0.306	30.1
500	0.618	31.9
600	0.941	32.5

REMARKS

The tabulated values are considered accurate to within ± 7% over the entire temperature range. These values can be represented approximately by the following equation:

$$\Delta L/L_0 = -0.568 + 1.013 \times 10^{-3}\, T + 3.783 \times 10^{-6}\, T^2 - 2.136 \times 10^{-9}\, T^3$$

FIGURE AND TABLE NO. 109BR. PROVISIONAL VALUES FOR THERMAL LINEAR EXPANSION OF In_3Y INTERMETALLIC COMPOUND

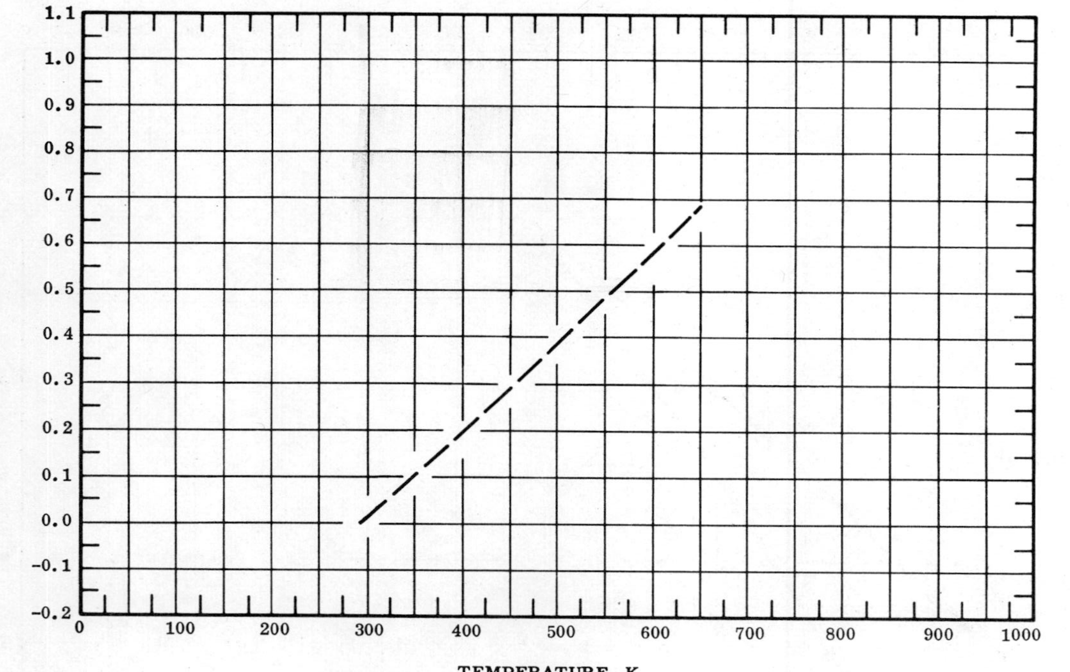

PROVISIONAL VALUES

[Temperature, T, K; Linear Expansion, $\Delta L/L_0$, %; α, K^{-1}]

T	$\Delta L/L_0$	$\alpha \times 10^6$
293	0.000	18.2
400	0.199	18.9
500	0.391	19.3
600	0.584	19.4
650	0.681	19.4

REMARKS

The tabulated values are considered accurate to within ± 7% over the entire temperature range. These values can be represented approximately by the following equation:

$$\Delta L/L_0 = -0.486 + 1.457 \times 10^{-3}\, T + 8.374 \times 10^{-7}\, T^2 - 4.869 \times 10^{-10}\, T^3$$

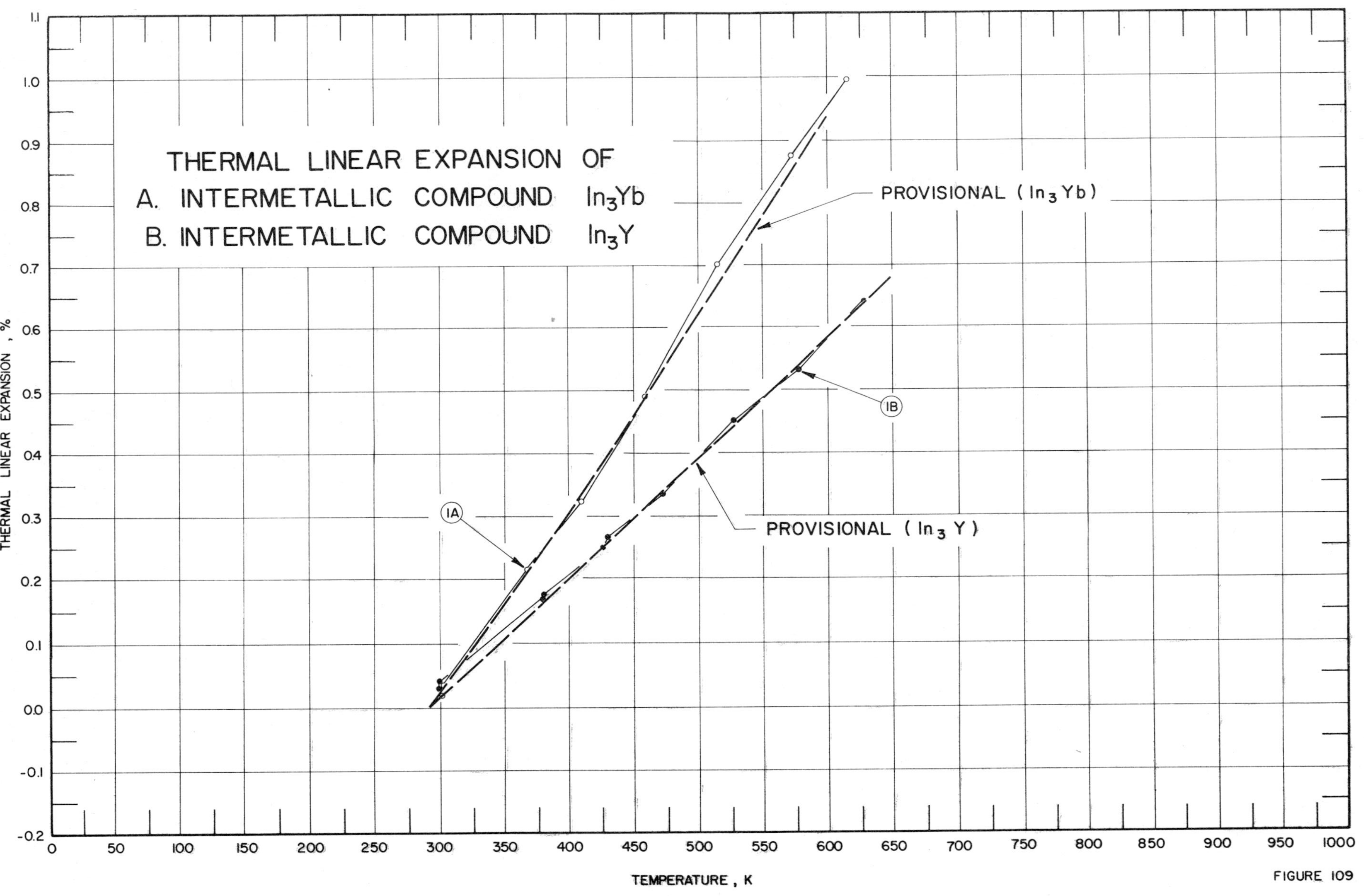
THERMAL LINEAR EXPANSION OF
A. INTERMETALLIC COMPOUND In_3Yb
B. INTERMETALLIC COMPOUND In_3Y
PROVISIONAL (In_3Yb)
PROVISIONAL (In_3Y)
1A
1B
THERMAL LINEAR EXPANSION, %
1.1
1.0
0.9
0.8
0.7
0.6
0.5
0.4
0.3
0.2
0.1
0.0
-0.1
-0.2
0
50
100
150
200
250
300
350
400
450
500
550
600
650
700
750
800
850
900
950
1000
TEMPERATURE, K
FIGURE 109

SPECIFICATION TABLE 109A. THERMAL LINEAR EXPANSION OF In_3Yb INTERMETALLIC COMPOUND

Cur. No.	Ref. No.	Author(s)	Year	Method Used	Temp. Range, K	Name and Specimen Designation	Composition (weight percent), Specifications, and Remarks
1	557	Harris, I.R., Raynov, G.U., and Winslamley, C.J.	1967	X	301-621		Yb 99.95$^+$ pure, In 99.95$^+$ pure; obtained from Johnson Matthey and Co.; copper Kα radiation used for analysis; lattice parameter reported at 301 K is 4.6050 Å; 4.6040 Å at 293 K determined by graphical extrapolation.

DATA TABLE 109A. THERMAL LINEAR EXPANSION OF In_3Yb INTERMETALLIC COMPOUND

[Temperature, T, K; Linear Expansion, $\Delta L/L_0$, %]

T	$\Delta L/L_0$
CURVE 1	
301	0.021
304	0.036
369	0.215
410	0.323
459	0.490
515	0.703
573	0.879
621	0.999

SPECIFICATION TABLE 109B. THERMAL LINEAR EXPANSION OF In_3Y INTERMETALLIC COMPOUND

Cur. No.	Ref. No.	Author(s)	Year	Method Used	Temp. Range, K	Name and Specimen Designation	Composition (weight percent), Specifications, and Remarks
1	682	Harris, I.R. and Raynor, G.V.	1967	X	300-629		Specimen prepared from In containing 0.002 Pb and 0.001 Cd from Messrs. Johnson, Matthey and Co., Ltd., and Y containing <0.010 other rare earth metals and 0.02 other base metals by the method similar to the one reported for $CeIn_3$; lattice parameter reported at 300 K is 4.5851 kX; 4.5844 kX at 293 K determined by graphical extrapolation.

DATA TABLE 109B. THERMAL LINEAR EXPANSION OF In_3Y INTERMETALLIC COMPOUND

[Temperature, T, K; Linear Expansion, $\Delta L/L_0$, %]

T	$\Delta L/L_0$
CURVE 1	
300	0.011
300	0.020
380	0.168
380	0.174
426	0.249
430	0.268
473	0.336
527	0.451
577	0.534
629	0.641

FIGURE AND TABLE NO. 110R. TYPICAL VALUES FOR THERMAL LINEAR EXPANSION OF $Fe_{17}Lu_2$ INTERMETALLIC COMPOUND

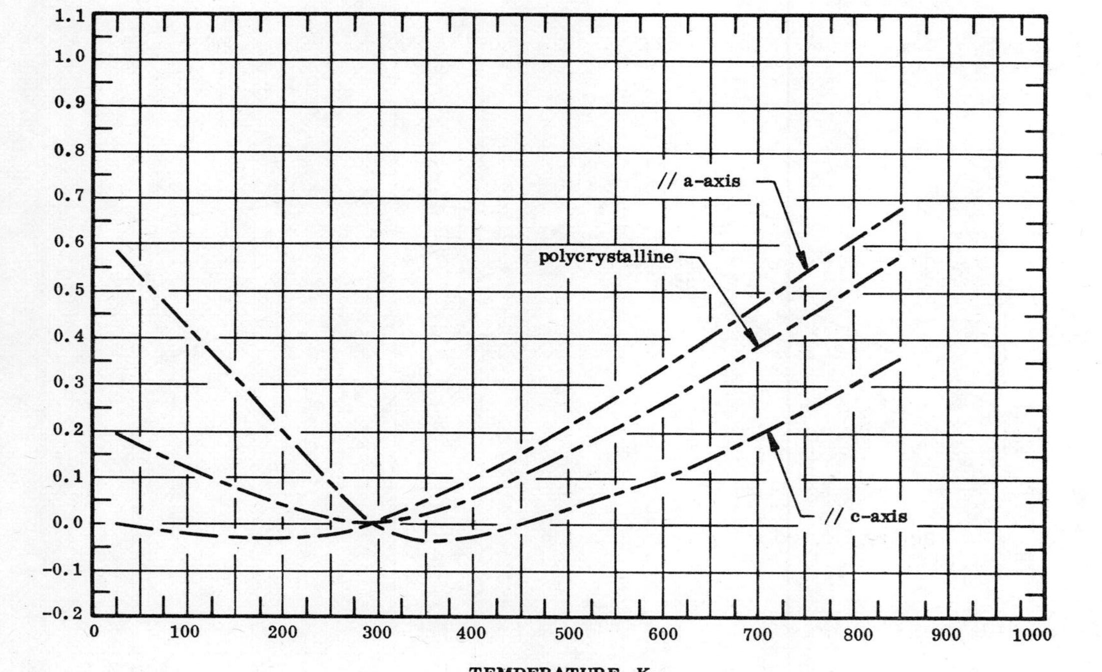

TYPICAL VALUES

[Temperature, T, K; Linear Expansion, $\Delta L/L_0$, %]

T	// a-axis $\Delta L/L_0$	// c-axis $\Delta L/L_0$	polycrystalline $\Delta L/L_0$
25	-0.001	0.588	0.195
50	-0.009	0.529	0.170
100	-0.021	0.416	0.125
200	-0.031	0.192	0.043
293	0.000	0.000	0.000
400	0.101	-0.024	0.059
500	0.215	0.031	0.154
600	0.341	0.101	0.261
700	0.475	0.196	0.382
800	0.610	0.309	0.510
850	0.680	0.359	0.573

REMARKS

The tabulated values are considered accurate to within ±15% over the entire temperature range. The values for polycrystalline material in the absence of experimental data are based on the single crystal data. An anomaly is observed at about 270 K due to magnetic transformation.

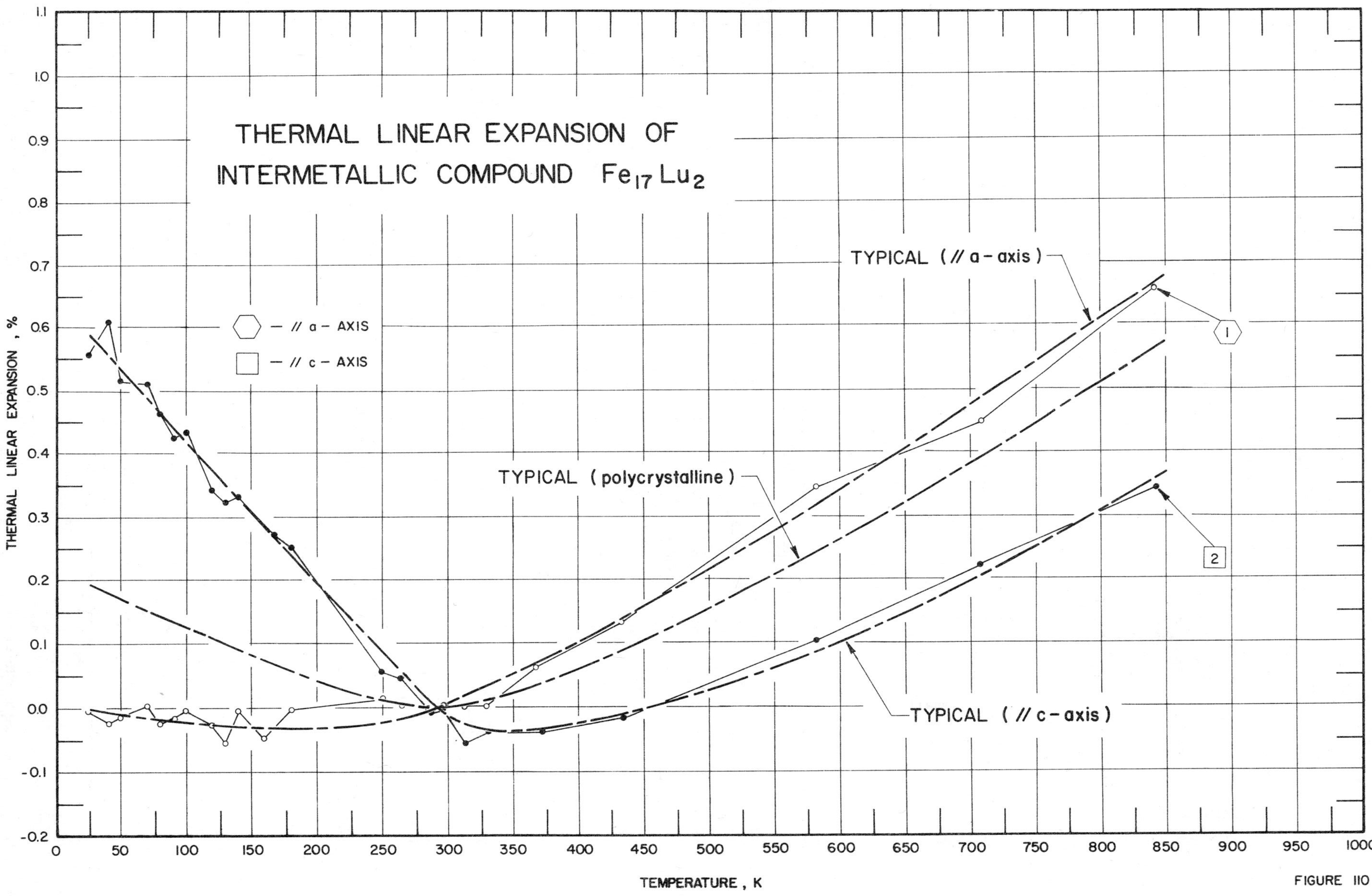

THERMAL LINEAR EXPANSION OF
INTERMETALLIC COMPOUND Fe17 Lu2
— // a – AXIS
— // c – AXIS
TYPICAL (// a – axis)
TYPICAL (polycrystalline)
TYPICAL (// c – axis)
1
2
THERMAL LINEAR EXPANSION, %
1.1
1.0
0.9
0.8
0.7
0.6
0.5
0.4
0.3
0.2
0.1
0.0
-0.1
-0.2
0
50
100
150
200
250
300
350
400
450
500
550
600
650
700
750
800
850
900
950
1000
TEMPERATURE, K

FIGURE 110

SPECIFICATION TABLE 110. THERMAL LINEAR EXPANSION OF $Fe_{17}Lu_2$ INTERMETALLIC COMPOUND

Cur. No.	Ref. No.	Author(s)	Year	Method Used	Temp. Range, K	Name and Specimen Designation	Composition (weight percent), Specifications, and Remarks
1	709	Shah, J.S.	1971	X	25-843		Samples from Dr. R. Lemaire, C.N.R.S., Grenoble, France; compounds prepared by melting elements in high-frequency levitation furnace then homogenizing at 1373 K for 3 days; powders were ground in a boron nitride mortar then sieved through a special 20 μ sieve; intermetallic compound powder samples were mounted on special quartz fiber; sample structure hexagonal Th_2Ni_{17}-type; expansion measured along a-axis; lattice parameter reported as 8.3787 Å at 25 K; 8.3792 Å at 293 K determined by graphical interpolation; lattice parameter not corrected for refraction.
2	709	Shah, J.S.	1971	X	25-843		The above specimen; expansion measured along c-axis; lattice parameter reported as 8.3377 Å at 25 K; 8.2913 Å at 293 K determined by graphical interpolation; lattice parameter not corrected for refraction, magnetic transformation reported for this and above curve at 270 K.

DATA TABLE 110. THERMAL LINEAR EXPANSION OF $Fe_{17}Lu_2$ INTERMETALLIC COMPOUND

[Temperature, T, K; Linear Expansion, $\Delta L/L_0$, %]

T	$\Delta L/L_0$	T	$\Delta L/L_0$	T	$\Delta L/L_0$
CURVE 1		CURVE 1 (cont.)		CURVE 2 (cont.)	
25	-0.006	373	0.063	160	0.277
40	-0.024	433	0.134	180	0.203
50	-0.012	583	0.348	250	0.054
70	0.004	708	0.450	264	0.048
80	-0.024	843	0.661	287	-0.004
90	-0.013			293	0.000
100	-0.001	CURVE 2		313	-0.052
120	-0.025			330	-0.039
130	-0.055	25	0.558	373	-0.017
140	-0.007	40	0.604	433	-0.013
160	-0.047	50	0.514	583	0.105
180	-0.007	70	0.510	708	0.223
250	0.014	80	0.462	843	0.341
264	0.008	90	0.427		
287	-0.006	100	0.439		
298	0.005	120	0.341		
313	0.001	130	0.322		
330	0.007	140	0.335		

SPECIFICATION TABLE 111. THERMAL LINEAR EXPANSION OF $FeNi_3$ INTERMETALLIC COMPOUND

Cur. No.	Ref. No.	Author(s)	Year	Method Used	Temp. Range, K	Name and Specimen Designation	Composition (weight percent), Specifications, and Remarks
1*	662	Ibragimov, E.A., Selisskiy, Ya.P., and Sorokin, M.N.	1966	X	523-867		0.15-0.3 Si, <0.005 C, <0.002 S, P; wire specimen of 1 mm diameter prepared in induction furnace from electrolytic Ni, Mn, and commercial Fe; prior to measurements specimen was subjected to various heat treatments and finally cold drawn and annealed at 723 K between 15 minutes to 256 hours; phase transition at 779 K; this curve here is reported using the first measured temperature as reference temperature at which $\Delta L/L_0 = 0$.

DATA TABLE 111. THERMAL LINEAR EXPANSION OF $FeNi_3$ INTERMETALLIC COMPOUND

[Temperature, T, K; Linear Expansion, $\Delta L/L_0$, %]

T	$\Delta L/L_0$
CURVE 1*, †, ‡	
523	0.000
703	0.272
779	0.413
792	0.437
867	0.557

DATA TABLE 111. COEFFICIENT OF THERMAL LINEAR EXPANSION OF $FeNi_3$ INTERMETALLIC COMPOUND

[Temperature, T, K; Coefficient of Expansion, α, 10^{-6} K^{-1}]

T	α
CURVE 1*, ‡	
523	14.2
703	16.0
779	21.1
792	16.3
867	15.7

* No figure given.
† This curve is here reported using the first given temperature as reference temperature at which $\Delta L/L_0 = 0$.
‡ Author's data for coefficient of thermal expansion have been integrated by TPRC to obtain $\Delta L/L_0$.

SPECIFICATION TABLE 112. THERMAL LINEAR EXPANSION OF FeRh INTERMETALLIC COMPOUND

Cur. No.	Ref. No.	Author(s)	Year	Method Used	Temp. Range, K	Name and Specimen Designation	Composition (weight percent), Specifications, and Remarks
1*	681	De Bergevin, F. and Muldawer, L.	1961	X	288, 338		Specimen prepared by melting; lattice parameter reported at 288 K is 2.987 Å; 2.988 Å at 293 K determined by interpolation.
2*	681	De Bergevin, F. and Muldawer, L.	1961	X	293, 363		Specimen prepared chemically; lattice parameter reported at 293 K is 2.987 Å.

DATA TABLE 112. THERMAL LINEAR EXPANSION OF FeRh INTERMETALLIC COMPOUND

[Temperature, T, K; Linear Expansion, $\Delta L/L_0$, %]

T	$\Delta L/L_0$
CURVE 1*	
288	-0.033
338	0.301
CURVE 2*	
293	0.000
363	0.368

* No figure given.

FIGURE AND TABLE NO. 113R. TYPICAL VALUES FOR THERMAL LINEAR EXPANSION OF $Fe_{17}Y_2$ INTERMETALLIC COMPOUND

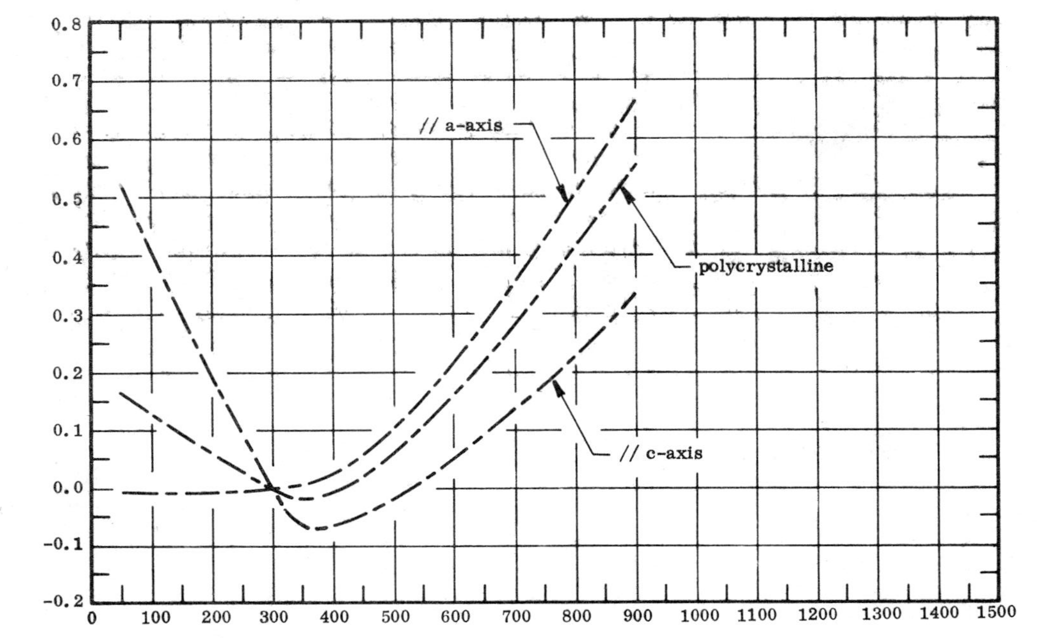

TYPICAL VALUES

[Temperature, T, K; Linear Expansion, $\Delta L/L_0$, %]

T	// a-axis $\Delta L/L_0$	// c-axis $\Delta L/L_0$	polycrystalline $\Delta L/L_0$
50	-0.008	0.519	0.168
100	-0.008	0.407	0.130
200	-0.007	0.193	0.060
293	0.000	0.000	0.000
400	0.026	-0.064	-0.004
500	0.102	-0.017	0.062
600	0.218	0.052	0.163
700	0.357	0.143	0.286
800	0.504	0.226	0.411
900	0.668	0.332	0.556

REMARKS

The tabulated values are considered accurate to within ±15% over the entire temperature range. The values for polycrystalline material in the absence of experimental data are based on the single crystal data. An anomaly is observed at about 310 K due to magnetic transformation.

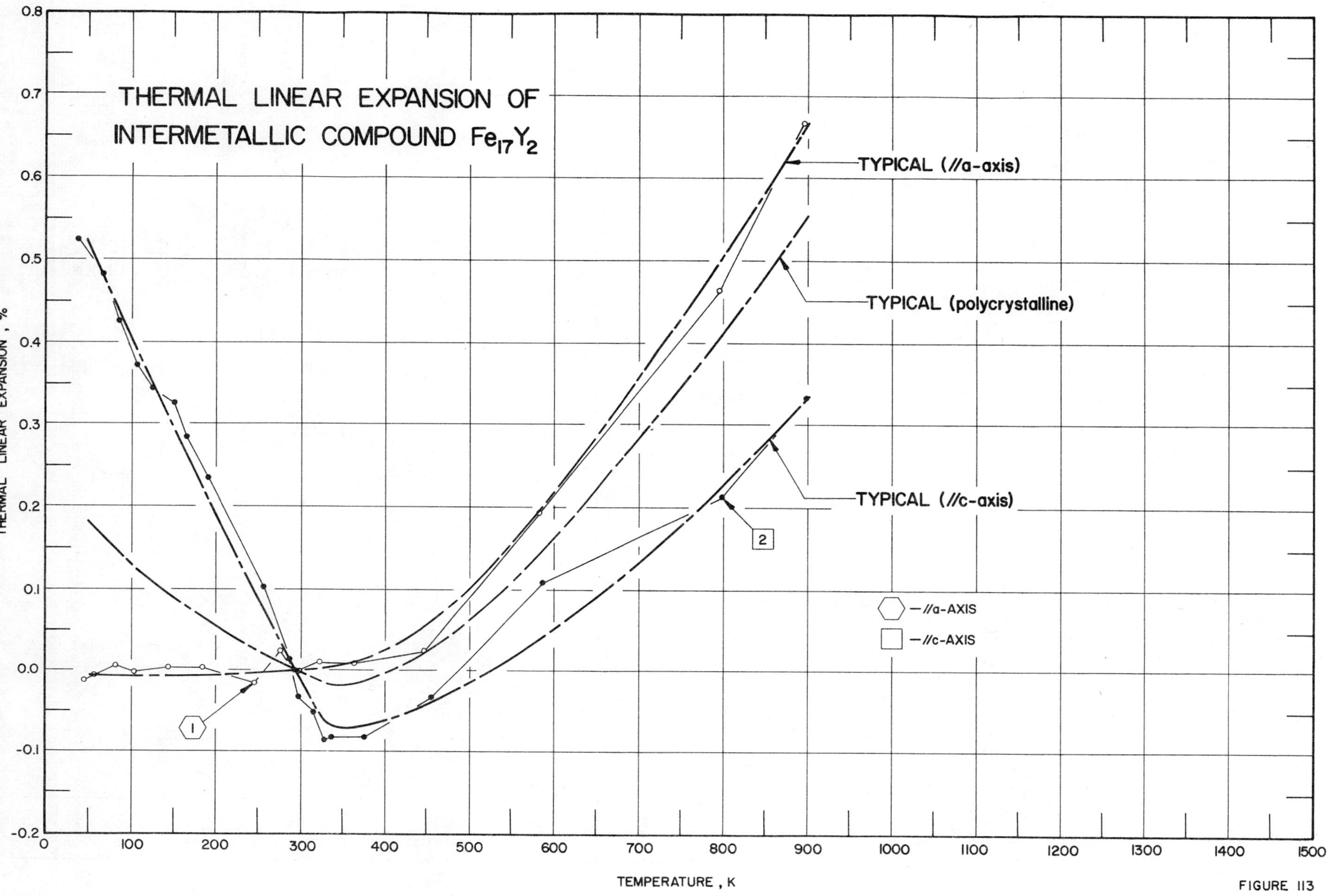
THERMAL LINEAR EXPANSION OF
INTERMETALLIC COMPOUND $Fe_{17}Y_2$
TYPICAL (//a-axis)
TYPICAL (polycrystalline)
TYPICAL (//c-axis)
—//a-AXIS
—//c-AXIS
1
2
THERMAL LINEAR EXPANSION, %
0.8
0.7
0.6
0.5
0.4
0.3
0.2
0.1
0.0
-0.1
-0.2
TEMPERATURE, K
0
100
200
300
400
500
600
700
800
900
1000
1100
1200
1300
1400
1500

FIGURE 113

SPECIFICATION TABLE 113. THERMAL LINEAR EXPANSION OF $Fe_{17}Y_2$ INTERMETALLIC COMPOUND

Cur. No.	Ref. No.	Author(s)	Year	Method Used	Temp. Range, K	Name and Specimen Designation	Composition (weight percent), Specifications, and Remarks
1	709	Shah, J.S.	1971	X	45-896		Samples from Dr. R. Lemaire, C.N.R.S., Grenoble France; compounds prepared by melting elements in high-frequency levitation furnace then homogenizing at 1373 K for 3 days; powders ground in boron nitride mortar then sieved through a special 20 μ sieve; intermetallic compound samples were mounted on special quartz fiber; sample structure Th_2Ni_{17}-type; expansion measured along a-axis; lattice parameter reported as 8.4620 Å at 40 K; 8.4625 Å at 293 K determined by graphical interpolation; lattice parameter not corrected for refraction.
2	709	Shah, J.S.	1971	X	38-899		The above specimen; expansion measured along c-axis; lattice parameter reported as 8.3477 Å at 40 K; 8.3065 Å at 293 K determined by graphical interpolation; lattice parameter not corrected for refraction, magnetic transformation for this and above curve reported at 310 K.

DATA TABLE 113. THERMAL LINEAR EXPANSION OF $Fe_{17}Y_2$ INTERMETALLIC COMPOUND

[Temperature, T, K; Linear Expansion, $\Delta L/L_0$, %]

T	$\Delta L/L_0$	T	$\Delta L/L_0$	T	$\Delta L/L_0$
CURVE 1		CURVE 2		CURVE 2 (cont.)	
45	-0.013	38	0.526	585	0.010
59	-0.008	68	0.482	798	0.212
81	0.006	81	0.426	899	0.335
104	-0.001	104	0.371		
144	0.005	123	0.344		
182	0.005	150	0.327		
247	-0.018	163	0.288		
279	0.026	190	0.233		
293	0.000	253	0.102		
299	0.000	287	0.011		
321	0.009	293	0.000		
366	0.009	298	-0.035		
449	0.027	311	-0.051		
581	0.196	324	-0.089		
795	0.466	334	-0.082		
896	0.668	375	-0.082		
		452	-0.035		

FIGURE AND TABLE NO. 114AR. PROVISIONAL VALUES FOR THERMAL LINEAR EXPANSION OF $LaRu_2$ INTERMETALLIC COMPOUND

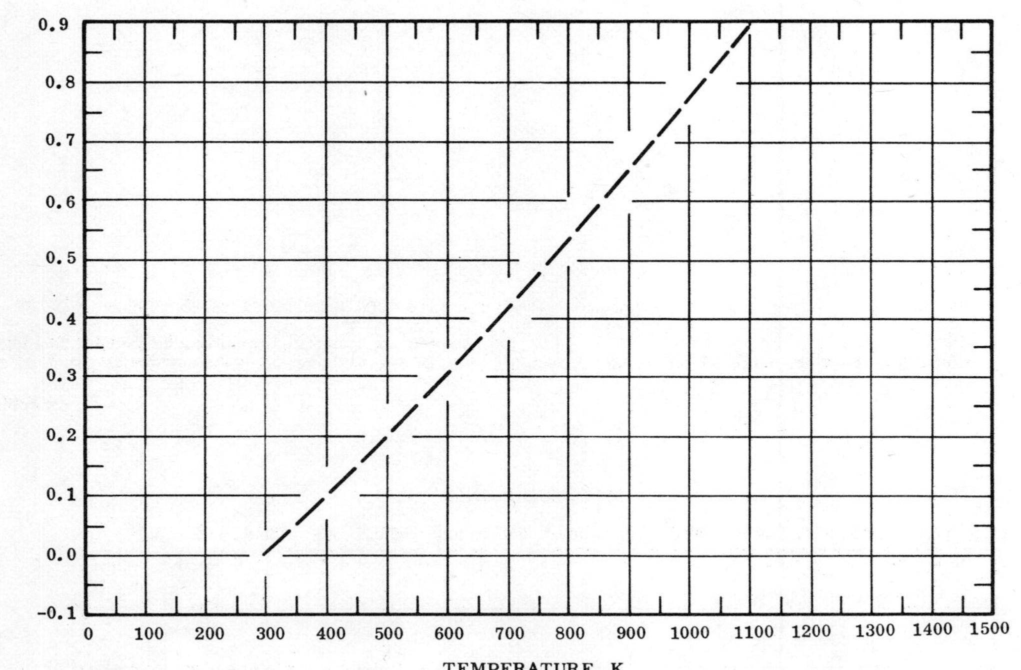

PROVISIONAL VALUES

[Temperature, T, K; Linear Expansion, $\Delta L/L_0$, %; α, K^{-1}]

T	$\Delta L/L_0$	$\alpha \times 10^6$
293	0.000	8.8
400	0.099	9.7
500	0.200	10.4
600	0.307	11.0
700	0.419	11.4
800	0.535	11.8
900	0.654	11.9
1000	0.773	11.9
1100	0.892	11.9

REMARKS

The tabulated values are considered accurate to within ± 10% over the entire temperature range. These values can be represented approximately by the following equation:

$$\Delta L/L_0 = -0.210 + 5.355 \times 10^{-4}\, T + 6.848 \times 10^{-7}\, T^2 - 2.369 \times 10^{-10}\, T^3$$

FIGURE AND TABLE NO. 114BR. PROVISIONAL VALUES FOR THERMAL LINEAR EXPANSION OF $LaSn_3$ INTERMETALLIC COMPOUND

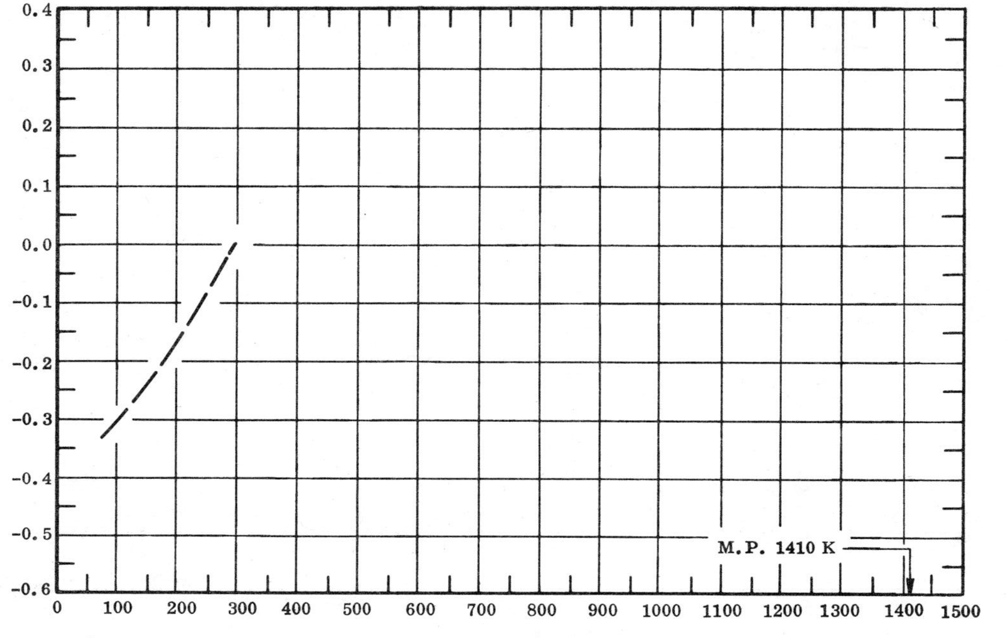

PROVISIONAL VALUES

[Temperature, T, K; Linear Expansion, $\Delta L/L_0$, %; α, K^{-1}]

T	$\Delta L/L_0$	$\alpha \times 10^6$
75	-0.332	9.5
100	-0.305	11.2
200	-0.165	16.3
293	0.000	18.8

REMARKS

The tabulated values are considered accurate to within ± 7% over the entire temperature range. These values can be represented approximately by the following equation:

$$\Delta L/L_0 = -0.381 + 3.497 \times 10^{-4}\ T + 4.496 \times 10^{-6}\ T^2 - 4.271 \times 10^{-9}\ T^3$$

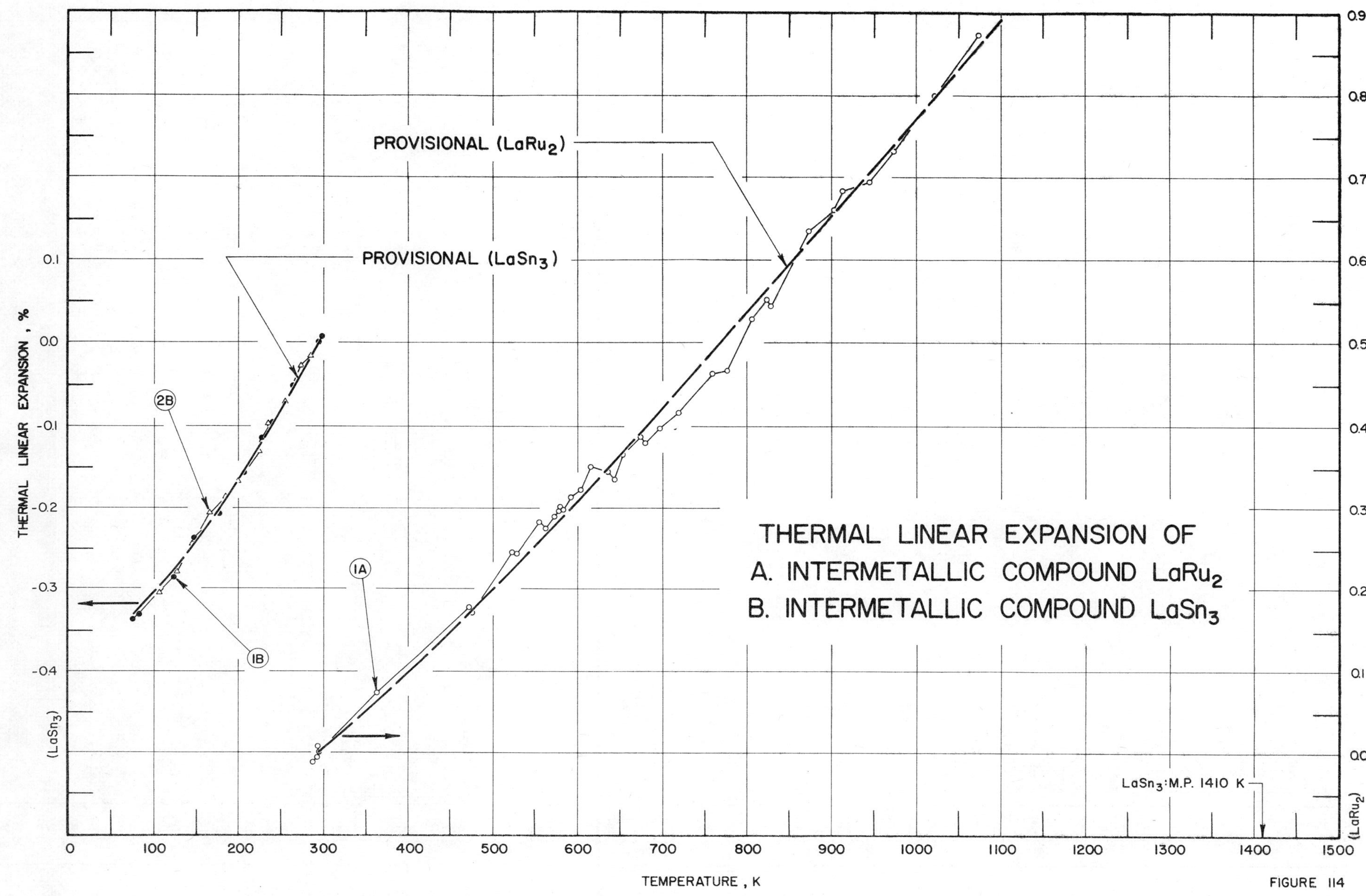
THERMAL LINEAR EXPANSION OF
A. INTERMETALLIC COMPOUND $LaRu_2$
B. INTERMETALLIC COMPOUND $LaSn_3$
PROVISIONAL ($LaRu_2$)
PROVISIONAL ($LaSn_3$)
2B
1A
1B
$LaSn_3$: M.P. 1410 K
THERMAL LINEAR EXPANSION, %
($LaSn_3$)
0.1
0.0
-0.1
-0.2
-0.3
-0.4
($LaRu_2$)
0.9
0.8
0.7
0.6
0.5
0.4
0.3
0.2
0.1
0.0
0
100
200
300
400
500
600
700
800
900
1000
1100
1200
1300
1400
1500
TEMPERATURE, K
FIGURE 114

SPECIFICATION TABLE 114A. THERMAL LINEAR EXPANSION OF $LaRu_2$ INTERMETALLIC COMPOUND

Cur. No.	Ref. No.	Author(s)	Year	Method Used	Temp. Range, K	Name and Specimen Designation	Composition (weight percent), Specifications, and Remarks
1	558	Gschneidner, K.A., Jr., Elliott, R.O., and Cromer, D.T.	1965	X	289-1073		Prepared by arc-melting components; filings for x-ray examination annealed at 723 K for 15 min; CuKα radiation used; lattice parameter reported at 293 K is 7.7035 Å.

DATA TABLE 114A. THERMAL LINEAR EXPANSION OF $LaRu_2$ INTERMETALLIC COMPOUND

[Temperature, T, K; Linear Expansion, $\Delta L/L_0$, %]

T	$\Delta L/L_0$	T	$\Delta L/L_0$	T	$\Delta L/L_0$
CURVE 1		CURVE 1 (cont.)		CURVE 1 (cont.)	
289	-0.012	604	0.321	945	0.696
291	-0.006	617	0.350	973	0.732
291	0.008	631	0.345	1023	0.800
292	0.000	641	0.335	1073	0.873
363	0.074	652	0.366		
473	0.179	673	0.387		
476	0.175	680	0.380		
521	0.245	698	0.399		
529	0.240	720	0.416		
557	0.282	760	0.463		
561	0.276	778	0.466		
563	0.278*	803	0.539		
573	0.289	823	0.553		
578	0.296	829	0.548		
580	0.301	873	0.636		
582	0.299	903	0.661		
593	0.314	913	0.685		

*Not shown in figure.

SPECIFICATION TABLE 114B. THERMAL LINEAR EXPANSION OF $LaSn_3$ INTERMETALLIC COMPOUND

Cur. No.	Ref. No.	Author(s)	Year	Method Used	Temp. Range, K	Name and Specimen Designation	Composition (weight percent), Specifications, and Remarks
1	682	Harris, I.R. and Raynor, G.V.	1965	X	296-85		Specimen prepared from 99.999 Sn from Messrs. Capper, Pass, Ltd., and La containing <0.01 other rare metals and 0.02 other base metals, by melting weighed quantities in non-consumable electrode arc furnace in 15 mm Hg argon; wrapped in Ta foil and sealed in degassed silica tube at 2 x 10^{-6} mm Hg pressure; homogenized, annealed for 3 days at 973 K and quenched; powdered; annealed at 973 K for stress relief; measurements with decreasing temperatures; lattice parameter reported at 293 K is 4.7587 kX.
2	682	Harris, I.R. and Raynor, G.V.	1965	X	107-286		The above specimen; measurement with increasing temperature; lattice parameter reported at 286 K is 4.758 kX; 4.7587 at 293 K determined by graphical extrapolation.

DATA TABLE 114B. THERMAL LINEAR EXPANSION OF $LaSn_3$ INTERMETALLIC COMPOUND

[Temperature, T, K; Linear Expansion, $\Delta L/L_0$, %]

T	$\Delta L/L_0$	T	$\Delta L/L_0$
CURVE 1		CURVE 2 (cont.)	
296	0.006	169	-0.206
293	0.000	187	-0.185
263	-0.050	200	-0.166
240	-0.096	224	-0.130
228	-0.113	238	-0.097
203	-0.158	254	-0.071
172	-0.208	263	-0.050
149	-0.237	272	-0.038
121	-0.284	286	-0.015
92	-0.330		
85	-0.336		
CURVE 2			
107	-0.305		
127	-0.279		
146	-0.244		

SPECIFICATION TABLE 115. THERMAL LINEAR EXPANSION OF PbLi INTERMETALLIC COMPOUND

Cur. No.	Ref. No.	Author(s)	Year	Method Used	Temp. Range, K	Name and Specimen Designation	Composition (weight percent), Specifications, and Remarks
1*	666	Zalkin, A. and Ramsey, W.J.	1957	X	159-629		Specimen prepared by melting in argon, weighed amounts of reagent grades Pb and Li; heated to 873 K, further heated tc 1023 K and held at 823 K for 3 hr; this compound transforms from rhombohedral to b.c.c. at 487 K; lattice parameter reported at 159 K is 3.5278 Å; 3.5404 Å at 293 K determined by graphical interpolation.

DATA TABLE 115. THERMAL LINEAR EXPANSION OF PbLi INTERMETALLIC COMPOUND

[Temperature, T, K; Linear Expansion, $\Delta L/L_0$, %]

T	$\Delta L/L_0$
CURVE 1*	
159	-0.356
299	0.019
343	0.158
469	0.596
	a (Å)
502	3.5675 b.c.c.
509	3.5663
514	3.5693
544	3.5712
582	3.5795
629	3.5867

* No figure given.

FIGURE AND TABLE NO. 116R. PROVISIONAL VALUES FOR THERMAL LINEAR EXPANSION OF MgAg INTERMETALLIC COMPOUND

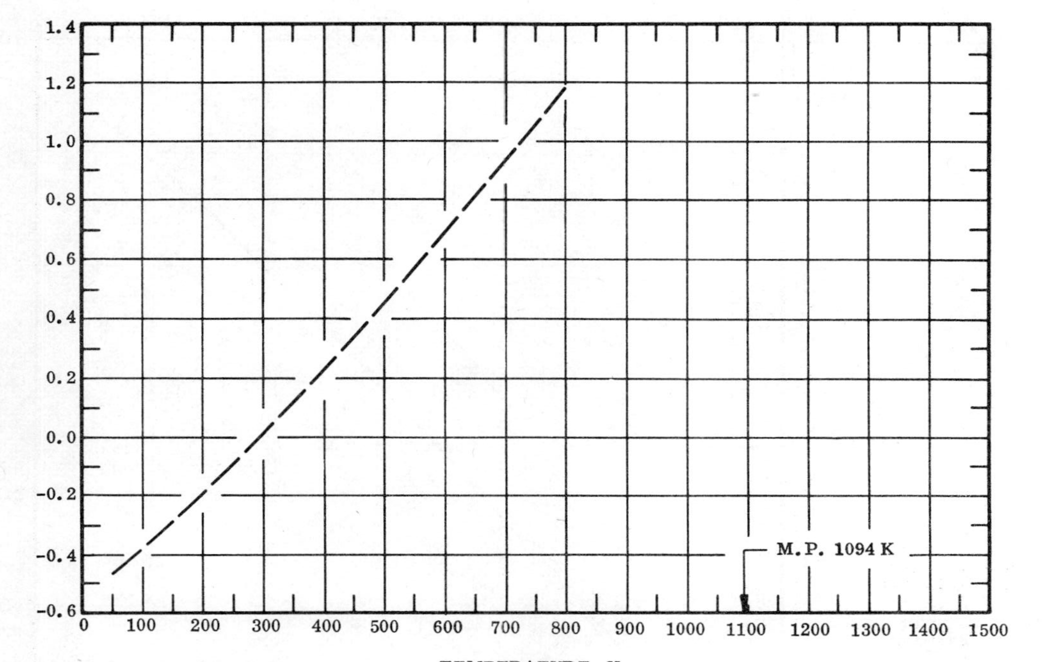

PROVISIONAL VALUES

[Temperature, T, K; Linear Expansion, $\Delta L/L_0$, %; α, K^{-1}]

T	$\Delta L/L_0$	$\alpha \times 10^6$
50	-0.466	17.1
100	-0.378	18.0
200	-0.190	19.6
293	0.000	21.0
400	0.231	22.3
500	0.461	23.3
600	0.698	24.0
700	0.942	24.6
800	1.187	24.8

REMARKS

The tabulated values are considered accurate to within ± 7% over the entire temperature range. These values can be represented approximately by the following equation:

$$\Delta L/L_0 = -0.548 + 1.597 \times 10^{-3}\, T + 1.056 \times 10^{-6}\, T^2 - 4.245 \times 10^{-10}\, T^3$$

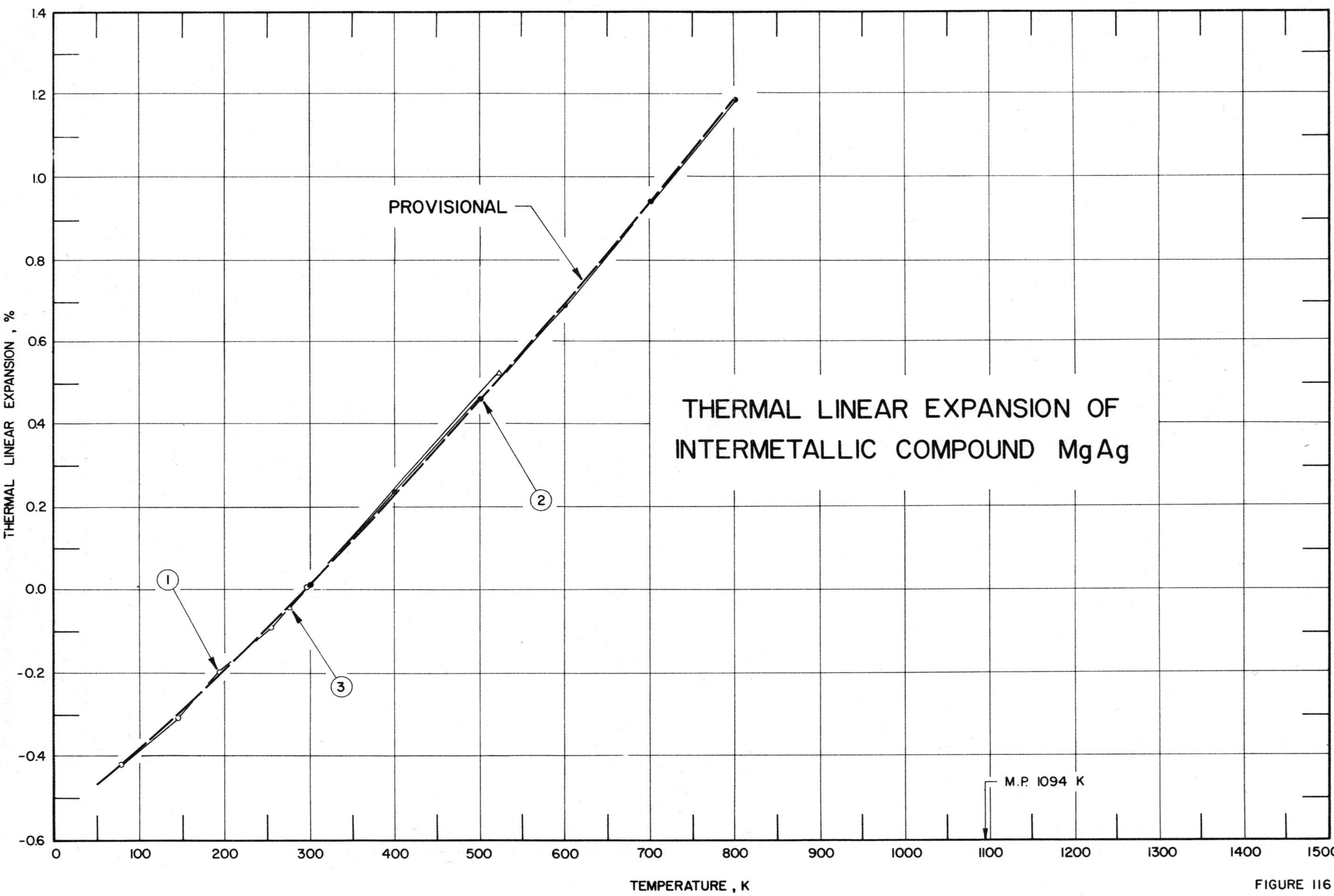
THERMAL LINEAR EXPANSION OF
INTERMETALLIC COMPOUND MgAg
PROVISIONAL
1
2
3
M.P. 1094 K
THERMAL LINEAR EXPANSION, %
1.4
1.2
1.0
0.8
0.6
0.4
0.2
0.0
-0.2
-0.4
-0.6
0
100
200
300
400
500
600
700
800
900
1000
1100
1200
1300
1400
1500
TEMPERATURE, K

FIGURE 116

SPECIFICATION TABLE 116. THERMAL LINEAR EXPANSION OF MgAg INTERMETALLIC COMPOUND

Cur. No.	Ref. No.	Author(s)	Year	Method Used	Temp. Range, K	Name and Specimen Designation	Composition (weight percent), Specifications, and Remarks
1	443	Neumann, J. P. and Chang, Y. A.	1967	X	77-295	β'-AgMg	Specimen with composition 17.34 Mg and 82.66 Ag; ingot produced by melting and chill-casting silver and magnesium (both 99.99 pure) in helium atmosphere; filings with a particle size less than 40 μ; annealed in vacuum for one day at 700 K, and cooled to room temperature with rate of 1 K per min; lattice parameter expansion measured; lattice parameter reported at 295 K is 3.3102 Å; 3.3100 Å at 293 K determined by graphical interpolation.
2	443	Neumann, J. P. and Chang, Y. A.	1967	D	298-800	β'-AgMg	Similar to above except specimen ~ 4 mm in diameter and 50 mm long; machined from ingot.
3	675	Cheng, C. H.	1967		273, 523		Single crystal of stoichiometric compositions; received from Dr. V. B. Kurfman of Metallurgical Laboratory of Dow Chemical Co.; grown by Bridgman method in graphite molds from high purity metals; density 6.042 g cm^{-3} at room temperature.

DATA TABLE 116. THERMAL LINEAR EXPANSION OF MgAg INTERMETALLIC COMPOUND

[Temperature, T, K; Linear Expansion, $\Delta L/L_0$, %]

T	$\Delta L/L_0$	T	$\Delta L/L_0$
CURVE 1		CURVE 3	
77	-0.422	273	-0.045
143	-0.304	523	0.524
196	-0.198		
251	-0.089		
295	0.004		
CURVE 2			
298	0.011		
400	0.234		
500	0.460		
600	0.697		
700	0.940		
800	1.188		

FIGURE AND TABLE NO. 117R. PROVISIONAL VALUES FOR THERMAL LINEAR EXPANSION OF Mg_2Sn INTERMETALLIC COMPOUND

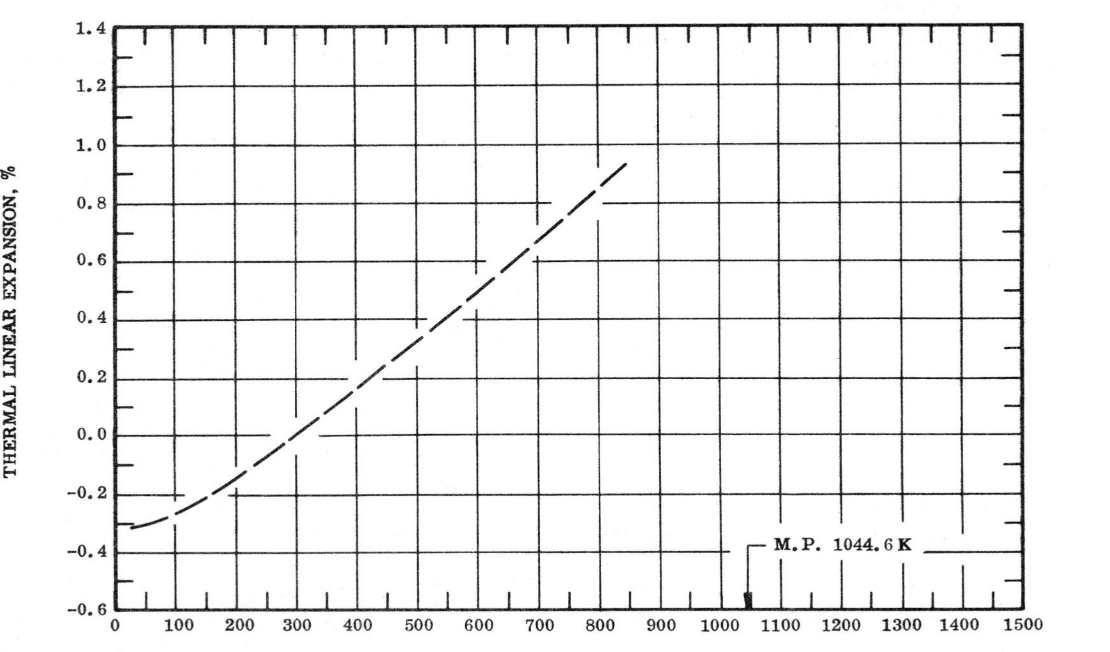

PROVISIONAL VALUES

[Temperature, T, K; Linear Expansion, $\Delta L/L_0$, %; α, K^{-1}]

T	$\Delta L/L_0$	$\alpha \times 10^6$
25	-0.320	
50	-0.319	
100	-0.262	11.9
200	-0.134	13.7
293	0.000	15.0
400	0.167	16.2
500	0.333	16.9
600	0.505	17.4
700	0.680	17.5
800	0.853	17.8
850	0.940	17.9

REMARKS

The tabulated values are considered accurate to within ±7% over the entire temperature range. These values can be represented approximately by the following equation:

$$\Delta L/L_0 = -0.372 + 9.910 \times 10^{-4}\, T + 1.106 \times 10^{-6}\, T^2 - 5.367 \times 10^{-10}\, T^3 \qquad (50 < T < 850)$$

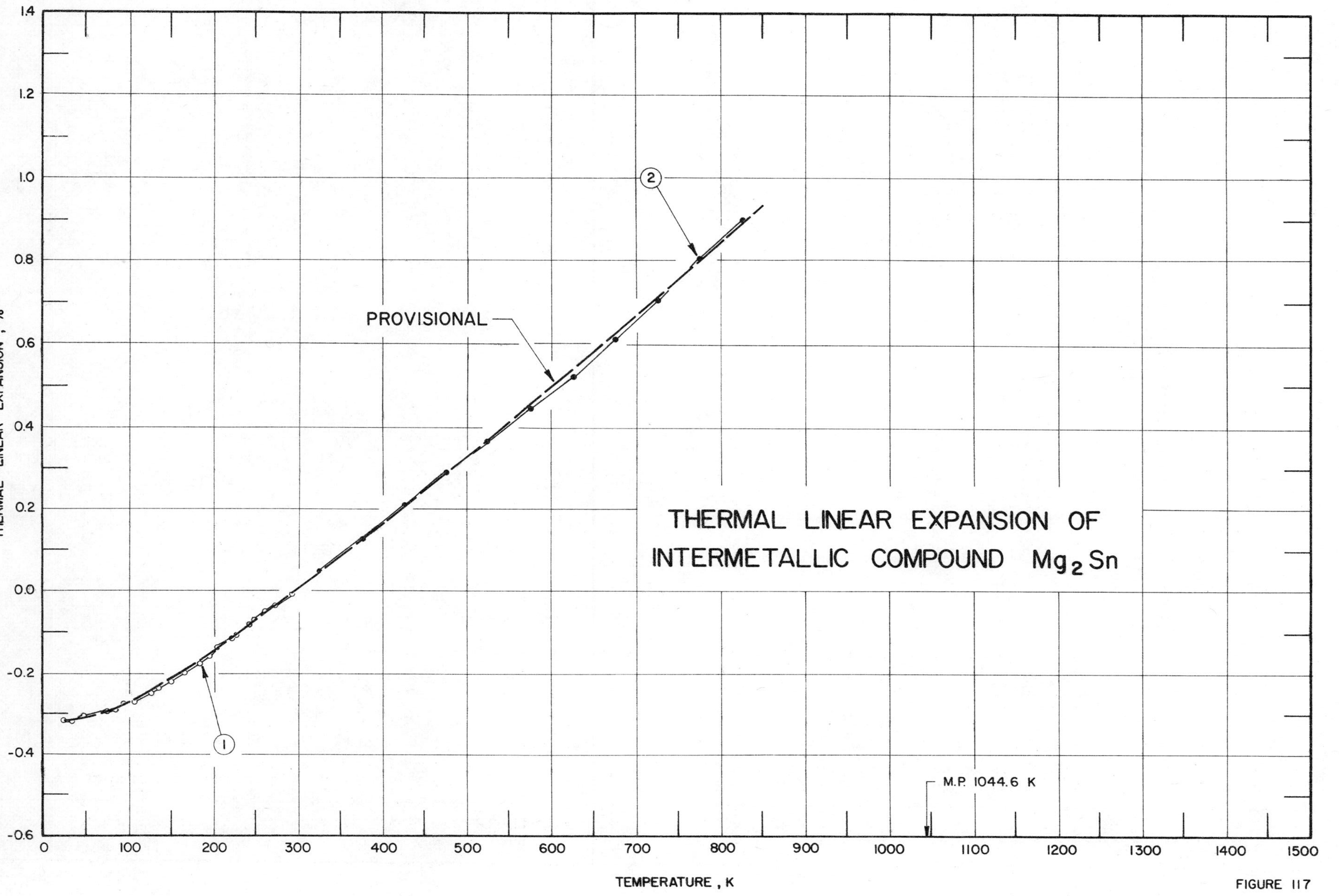

THERMAL LINEAR EXPANSION OF
INTERMETALLIC COMPOUND Mg_2Sn
PROVISIONAL
M.P. 1044.6 K
1
2
THERMAL LINEAR EXPANSION, %
1.4
1.2
1.0
0.8
0.6
0.4
0.2
0.0
-0.2
-0.4
-0.6
TEMPERATURE, K
0
100
200
300
400
500
600
700
800
900
1000
1100
1200
1300
1400
1500

FIGURE 117

SPECIFICATION TABLE 117. THERMAL LINEAR EXPANSION OF Mg_2Sn INTERMETALLIC COMPOUND

Cur. No.	Ref. No.	Author(s)	Year	Method Used	Temp. Range, K	Name and Specimen Designation	Composition (weight percent), Specifications, and Remarks
1	569	Novikova, S.I.	1969	L	26-295		Polycrystalline specimen prepared from 99.998 pure Mg, 99.999 pure Sn; data compiled from four series of measurements.
2	263	Grube, G. and Vosskühler	1934	L	303-823		Cylindrical specimen (70.9 Sn) of 10 cm length and 5 mm diameter at 303 K.

DATA TABLE 117. THERMAL LINEAR EXPANSION OF Mg_2Sn INTERMETALLIC COMPOUND

[Temperature, T, K; Linear Expansion, $\Delta L/L_0$, %]

T	$\Delta L/L_0$	T	$\Delta L/L_0$	T	$\Delta L/L_0$	T	$\Delta L/L_0$	T	$\Delta L/L_0$	T	$\Delta L/L_0$
CURVE 1‡		CURVE 1 (cont.)‡		CURVE 1 (cont.)‡		CURVE 1 (cont.)‡		CURVE 1 (cont.)‡		CURVE 2	
25	-0.313	85	-0.288	128	-0.245*	174	-0.185*	223	-0.110	303	0.016
33	-0.312	88	-0.285*	134	-0.237*	181	-0.174	228	-0.102	323	0.049
37	-0.311	96	-0.278*	135	-0.236	188	-0.164*	241	-0.083	373	0.130
41	-0.310*	96	-0.278	142	-0.227*	189	-0.162*	249	-0.069	423	0.211
49	-0.308	103	-0.272*	150	-0.217	196	-0.152	262	-0.050	473	0.295
58	-0.305*	108	-0.267	151	-0.216*	201	-0.145*	273	-0.032	523	0.370
66	-0.301*	111	-0.264*	158	-0.207*	205	-0.137	285	-0.013*	573	0.450
75	-0.295	118	-0.256*	165	-0.196	212	-0.128*	295	0.003	623	0.527
80	-0.291*	127	-0.246	169	-0.190*	217	-0.119*			673	0.614
										723	0.712
										773	0.807
										823	0.900

DATA TABLE 117. COEFFICIENT OF THERMAL LINEAR EXPANSION OF Mg_2Sn INTERMETALLIC COMPOUND

[Temperature, T, K; Coefficient of Expansion, α, $10^{-6}\ K^{-1}$]

T	α	T	α	T	α	T	α	T	α
CURVE 1‡		CURVE 1 (cont.)‡		CURVE 1 (cont.)‡		CURVE 1 (cont.)‡		CURVE 1 (cont.)‡	
25	0.60	85	7.48	128	12.32*	174	14.08	223	15.87
33	1.77	88	8.28	134	12.62	181	14.69	228	15.57*
37	1.45*	96	8.73	135	13.15*	188	14.69*	241	15.35
41	1.85	96	9.20*	142	13.03	189	15.01	249	15.78*
49	3.34	103	9.67	150	13.03	196	14.67	262	15.73
58	4.25	108	9.94*	151	12.71	201	15.33*	273	15.99
66	5.88	111	10.28	158	13.81	205	15.59	285	15.91
75	7.16	118	10.82*	165	14.09	212	15.75	295	16.58
80	7.96	127	11.29	169	14.00*	217	14.96		

*Not shown in figure.
‡Author's data for coefficient of thermal expansion have been integrated by TPRC to obtain $\Delta L/L_0$.

SPECIFICATION TABLE 118. THERMAL LINEAR EXPANSION OF MnHg INTERMETALLIC COMPOUND

Cur. No.	Ref. No.	Author(s)	Year	Method Used	Temp. Range, K	Name and Specimen Designation	Composition (weight percent), Specifications, and Remarks
1*	680	Nakagawa, Y. and Hori, T.	1962	X	93-194		Powdered specimen carefully annealed at 573 K for few days in order to obtain sharp diffraction lines; crystal structure of b.c.c. (CsCl type) at room temperature, crystal distortion from cubic to tetragonal symmetry at low temperature; temperature dependence of lattice parameter determined from (310) lines; b.c.t. unit cell reported at 93 K is 3.288 ± 0.002 Å; expansion measured along a-axis of b.c.t.; this curve is here reported using the first given temperature as reference temperature at which $\Delta L/L_0 = 0$.
2*	680	Nakagawa, Y. and Hori, T.	1962	X	188-291		The above specimen; b.c.c. unit cell reported at 291 K is 3.314 ± 0.001 Å; expansion measured along a-axis of b.c.c.
3*	680	Nakagawa, Y. and Hori, T.	1962	X	93-194		The above specimen; b.c.t. unit cell reported at 93 K is 3.313 ± 0.002 Å; expansion measured along c-axis of b.c.t.; this curve is here reported using the first given temperature as reference temperature at which $\Delta L/L_0 = 0$.

DATA TABLE 118. THERMAL LINEAR EXPANSION OF MnHg INTERMETALLIC COMPOUND

[Temperature, T, K; Linear Expansion, $\Delta L/L_0$, %]

T	$\Delta L/L_0$	T	$\Delta L/L_0$	T	$\Delta L/L_0$
CURVE 1*, †		CURVE 2*		CURVE 3 (cont.)*	
93	0.000	188	-0.272	145	0.060
95	0.000	195	-0.272	154	0.091
99	0.030	200	-0.241	163	0.060
110	0.061	206	-0.241	169	0.060
125	0.091	219	-0.181	175	0.030
136	0.152	241	-0.151	183	0.030
145	0.182	266	-0.091	185	0.030
154	0.243	291	0.000	187	0.030
163	0.243			188	0.000
168	0.274	CURVE 3*, †		194	0.000
174	0.304				
182	0.304	93	0.000		
188	0.335	96	0.000		
194	0.335	99	0.030		
		111	0.060		
		125	0.060		
		137	0.060		

*No figure given.

†This curve is here reported using the first given temperature as reference temperature at which $\Delta L/L_0 = 0$.

SPECIFICATION TABLE 119. THERMAL LINEAR EXPANSION OF MnNi INTERMETALLIC COMPOUND

Cur. No.	Ref. No.	Author(s)	Year	Method Used	Temp. Range, K	Name and Specimen Designation	Composition (weight percent), Specifications, and Remarks
1*	422	Pál, L., Krén, E., Kadar, G., Szabo, P., and Tarnoczi, T.	1968	X	447-962		Powdered sample; alloy (48.34 Mn, 51.66 Ni) metal of 99.9 purity prepared in a induction furnace under argon atmosphere; homogenized at 1073K for 24 hrs; expansion measured in atmosphere of 30 H_2 and 70 N_2 along a-axis; lattice parameter reported at 293 K is 3.743 Å.
2*	422	Pál, L., et al.	1968	X	293-965		The above specimen; expansion measured along c-axis; lattice parameter reported at 293 K is 3.531 Å.

DATA TABLE 119. THERMAL LINEAR EXPANSION OF MnNi INTERMETALLIC COMPOUND

[Temperature, T, K; Linear Expansion, $\Delta L/L_0$, %]

T	$\Delta L/L_0$	T	$\Delta L/L_0$
CURVE 1*		CURVE 2 (cont.)*	
288	0.000	869	-0.255
293	0.000	963	-0.765
446	0.080		
572	0.294		
674	0.588		
773	1.122		
872	1.763		
963	2.592		
CURVE 2*			
289	0.000		
293	0.000		
445	0.057		
572	0.057		
677	0.057		
774	0.000		

* No figure given.

SPECIFICATION TABLE 120. THERMAL LINEAR EXPANSION OF MnPd INTERMETALLIC COMPOUND

Cur. No.	Ref. No.	Author(s)	Year	Method Used	Temp. Range, K	Name and Specimen Designation	Composition (weight percent), Specifications, and Remarks
1*	422	Pál, L., Krén, E., Kádár, G., Szabo, P., and Tarnoczi, T.	1968	X	294-910		Powdered sample with composition 34.05 Mn and 65.95 Pd; alloy metal 99.9 purity placed in induction furnace under argon atmosphere and homogenized at 1073 K for 24 hr; for high temperatures up to 1173 K expansion measured in atmosphere of 30.0 hydrogen and 70.0 nitrogen gas; expansion measured along c-axis; lattice parameter reported at 293 K is 3.587 Å.
2*	422	Pál, L., et al.	1968	X	293-927		The above specimen; expansion measured along a-axis; lattice parameter reported at 293 K is 4.070 Å.

DATA TABLE 120. THERMAL LINEAR EXPANSION OF MnPd INTERMETALLIC COMPOUND

[Temperature, T, K; Linear Expansion, $\Delta L/L_0$, %]

T	$\Delta L/L_0$	T	$\Delta L/L_0$	T	$\Delta L/L_0$
CURVE 1*		CURVE 1 (cont.)*		CURVE 2 (cont.)*	
286	-0.028	859	-3.401	724	1.941
293	0.000	895	-3.624	754	2.137
371	-0.167	923	-3.959	779	2.555
421	-0.139			795	2.825
471	-0.334	CURVE 2*		812	3.268
542	-0.641			835	3.440
558	-0.390	290	-0.025	857	3.661
575	-0.613	293	0.000	897	3.931
623	-0.613	371	0.147	913	4.128
677	-0.920	421	0.393		
708	-1.032	467	0.516		
723	-1.477	536	0.835		
753	-1.673	555	0.934		
773	-1.951	575	0.983		
791	-2.175	619	1.155		
810	-2.816	672	1.499		
835	-3.122	705	1.744		

* No figure given.

SPECIFICATION TABLE 121. THERMAL LINEAR EXPANSION OF Mn_3Pt INTERMETALLIC COMPOUND

Cur. No.	Ref. No.	Author(s)	Year	Method Used	Temp. Range, K	Name and Specimen Designation	Composition (weight percent), Specifications, and Remarks
1*	571	Kren, E., Kadar, G. and Szabo, P.	1968	X	293-675	$Mn_{3.09}\,Pt_{0.91}$	3.0 excess Mn forming compound $Mn_{3.09}\,Pt_{0.91}$; lattice parameter reported at 293 K is 3.8277 Å; Neel temperature 455 ± 10 K.

DATA TABLE 121. THERMAL LINEAR EXPANSION OF Mn_3Pt INTERMETALLIC COMPOUND

[Temperature, T, K; Linear Expansion, $\Delta L/L_0$, %]

T	$\Delta L/L_0$
CURVE 1*	
293	0.000
342	0.157
362	0.264
372	0.259
392	0.342
413	0.447
421	0.562
439	0.585
446	1.105
452	1.134
472	1.199
496	1.319
522	1.395
547	1.440
572	1.534
608	1.654
642	1.693
675	1.766

* No figure given.

SPECIFICATION TABLE 122. THERMAL LINEAR EXPANSION OF MnPt INTERMETALLIC COMPOUND

Cur. No.	Ref. No.	Author(s)	Year	Method Used	Temp. Range, K	Name and Specimen Designation	Composition (weight percent), Specifications, and Remarks
1*	422	Pál, L., Krén, E., Kádár, G., Szabo, P., and Tarnoczi, T.	1968	X	293-1120	Manganese Platinum	Powdered sample with composition 22.18 Mn and 77.82 Pt; alloy prepared in an induction furnace under argon atmosphere; homogenized at 1070 K for 24 hr; expansion measured in atmosphere of 70 N, 30 H; expansion measured along a-axis; transition near 710 K; lattice parameter at 293 K is 4.003 Å.
2*	422	Pál, L., et al.	1968	X	296-1126	Manganese Platinum	The above specimen; expansion measured along the c-axis; lattice parameter at 293 K is 3.671 Å.

DATA TABLE 122. THERMAL LINEAR EXPANSION OF MnPt INTERMETALLIC COMPOUND

[Temperature, T, K; Linear Expansion, $\Delta L/L_0$, %]

T	$\Delta L/L_0$	T	$\Delta L/L_0$	T	$\Delta L/L_0$
CURVE 1*		CURVE 2*		CURVE 2 (cont.)*	
288	-0.075	290	0.000	1055	-0.872
293	0.000	293	0.000	1125	-0.981
372	0.125	368	0.000		
432	0.249	381	0.000		
476	0.249	432	0.000		
573	0.349	469	0.000		
624	0.499	494	0.000		
681	0.674	573	0.000		
742	0.724	626	0.000		
780	0.625	693	0.000		
821	0.724	745	0.000		
877	1.024	775	0.000		
921	1.224	822	-0.136		
968	1.799	874	-0.299		
1001	1.874	925	-0.299		
1053	1.948	961	-0.681		
1121	2.148	1002	-0.763		

* No figure given.

SPECIFICATION TABLE 123. THERMAL LINEAR EXPANSION OF Ni_3Nb INTERMETALLIC COMPOUND

Cur. No.	Ref. No.	Author(s)	Year	Method Used	Temp. Range, K	Name and Specimen Designation	Composition (weight percent), Specifications, and Remarks
1*	579	Arbuzov, M.P. and Chuprina, V.G.	1966	L	373-1173		Specimen prepared in arc furnace remelted in air in an induction furnace homogenized for 70 hours at 1423 K in vacuum anomaly of unexplained nature observed near 773-823 K; 4 mm diameter, 50 mm long.

DATA TABLE 123. THERMAL LINEAR EXPANSION OF Ni_3Nb INTERMETALLIC COMPOUND

[Temperature, T, K; Linear Expansion, $\Delta L/L_0$, %]

T	$\Delta L/L_0$	T	$\Delta L/L_0$
CURVE 1*, ‡		CURVE 1 (cont.)*, ‡	
373	0.101	823	0.735
423	0.164	873	0.815
473	0.228	923	0.897
523	0.295	973	0.978
573	0.363	1023	1.061
623	0.434	1073	1.146
673	0.507	1123	1.233
723	0.582	1173	1.322
773	0.658		

DATA TABLE 123. COEFFICIENT OF THERMAL LINEAR EXPANSION OF Ni_3Nb INTERMETALLIC COMPOUND

[Temperature, T, K; Coefficient of Expansion, α, 10^{-6} K^{-1}]

T	α	T	α
CURVE 1*, ‡		CURVE 1 (cont.)*, ‡	
373	12.75	823	15.70
423	13.10	873	16.20
473	13.40	923	16.60
523	13.80	973	17.00
573	14.35	1023	17.45
623	14.75	1073	17.85
673	15.10	1123	18.25
723	15.60	1173	18.60
773	15.85		

*No figure given.
‡Author's data for coefficient of thermal expansion have been integrated by TPRC to obtain $\Delta L/L_0$.

SPECIFICATION TABLE 124. THERMAL LINEAR EXPANSION OF NiTi INTERMETALLIC COMPOUND

Cur. No.	Ref. No.	Author(s)	Year	Method Used	Temp. Range, K	Name and Specimen Designation	Composition (weight percent), Specifications, and Remarks
1*	677	Buehler, W.J. and Wiley, R.C.	1962	X	297, 1173		55.1 Ni of stoichiometric composition specimen.

DATA TABLE 124. THERMAL LINEAR EXPANSION OF NiTi INTERMETALLIC COMPOUND

[Temperature, T, K; Linear Expansion, $\Delta L/L_0$, %]

T	$\Delta L/L_0$
CURVE 1*	
297	0.004
1173	0.915

*No figure given.

FIGURE AND TABLE NO. 125R. TYPICAL VALUES FOR THERMAL LINEAR EXPANSION OF Ni_5Y INTERMETALLIC COMPOUND

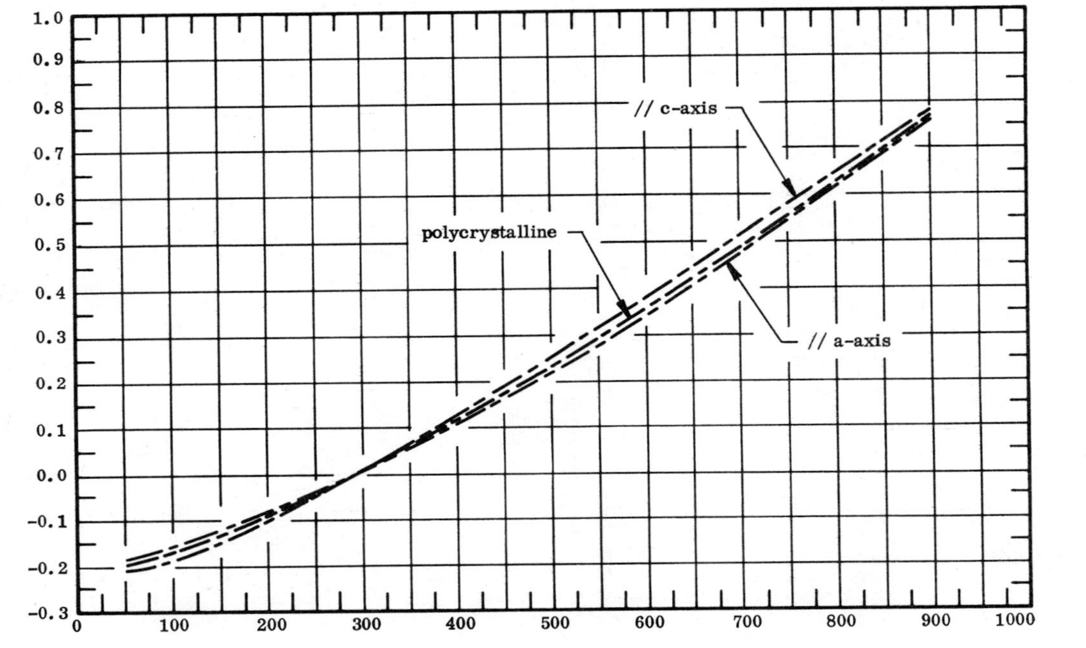

TYPICAL VALUES

[Temperature, T, K; Linear Expansion, $\Delta L/L_0$, %; α, K^{-1}]

T	// a-axis $\Delta L/L_0$	// c-axis $\Delta L/L_0$	polycrystalline $\Delta L/L_0$	polycrystalline $\alpha \times 10^6$
50	-0.189	-0.219	-0.198	6.2
100	-0.157	-0.183	-0.165	7.1
200	-0.081	-0.096	-0.086	8.6
293	0.000	0.000	0.000	9.9
400	0.107	0.124	0.114	11.2
500	0.219	0.253	0.230	12.1
600	0.342	0.381	0.355	12.9
700	0.473	0.516	0.486	13.4
800	0.611	0.650	0.623	13.8
900	0.753	0.779	0.763	13.9

REMARKS

The tabulated values are considered accurate to within ±20%. The values for polycrystalline material in the absence of experimental data are calculated from the single crystal data. These and other values can be represented approximately by the following equations:

// a-axis: $\Delta L/L_0 = -0.217 + 5.237 \times 10^{-4}\,T + 8.076 \times 10^{-7}\,T^2 - 2.114 \times 10^{-10}\,T^3$

// c-axis: $\Delta L/L_0 = -0.249 + 5.541 \times 10^{-4}\,T + 1.184 \times 10^{-6}\,T^2 - 5.890 \times 10^{-10}\,T^3$

polycrystalline: $\Delta L/L_0 = -0.228 + 5.323 \times 10^{-4}\,T + 9.355 \times 10^{-7}\,T^2 - 3.388 \times 10^{-10}\,T^3$

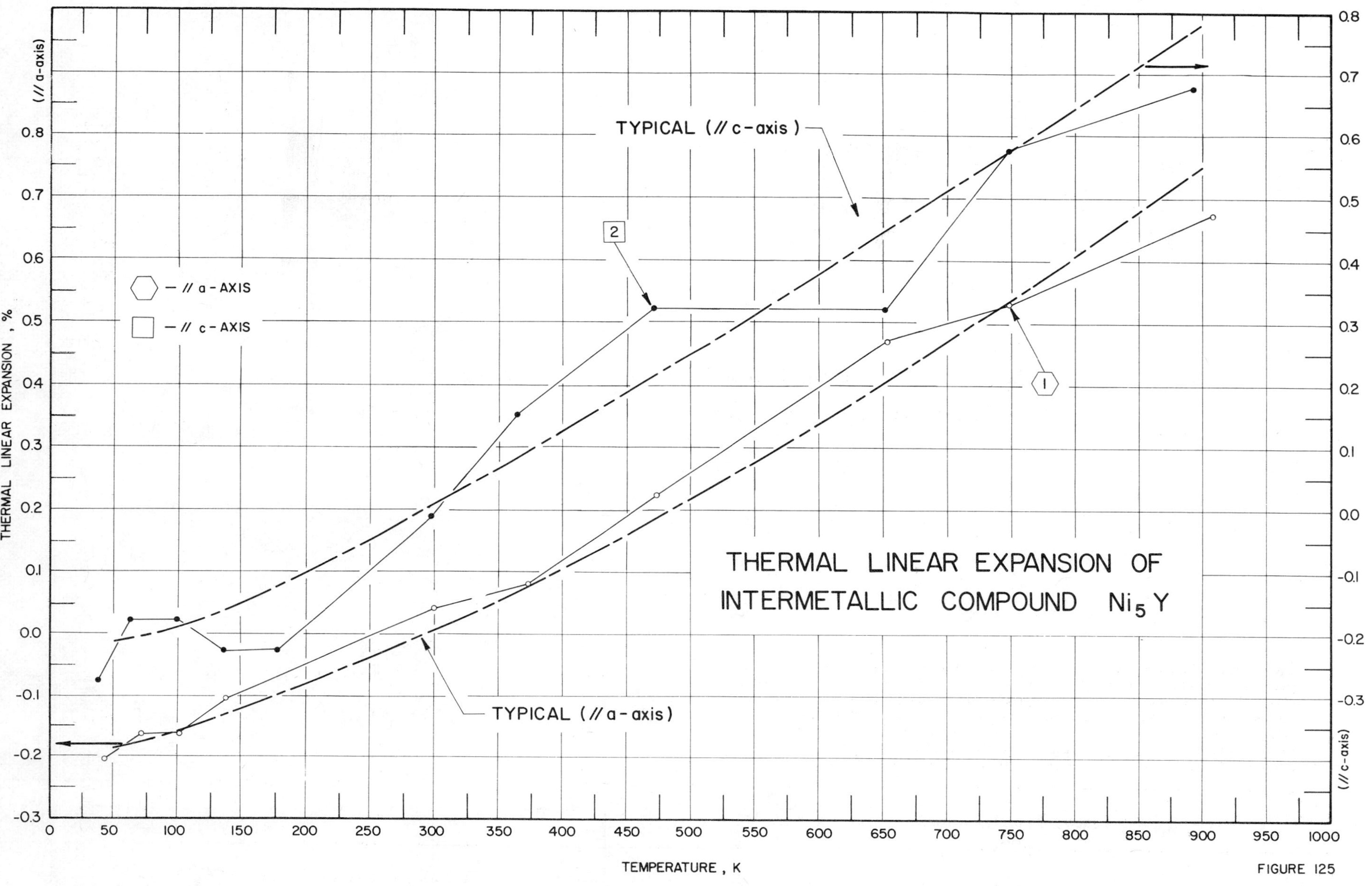
TYPICAL (// c-axis)
2
— // a-AXIS
— // c-AXIS
1
THERMAL LINEAR EXPANSION OF
INTERMETALLIC COMPOUND Ni_5Y
TYPICAL (// a-axis)
THERMAL LINEAR EXPANSION, %
(// a-axis)
(// c-axis)
TEMPERATURE, K

FIGURE 125

SPECIFICATION TABLE 125. THERMAL LINEAR EXPANSION OF Ni_5Y INTERMETALLIC COMPOUND

Cur. No.	Ref. No.	Author(s)	Year	Method Used	Temp. Range, K	Name and Specimen Designation	Composition (weight percent), Specifications, and Remarks
1	709	Shah, J. S.	1971	X	40-900		Samples from Dr. R. Lemaire, C.N.R.S., Grenable France, compounds prepared by melting elements in high-frequency levitation furnace then homogenizing at 1373 K for 3 days, powders were ground in a boron nitride motar then sieved through a special 20 μ sieve, intermetallic compound powder samples were mounted on special quartz fiber, sample structure hexagonal Ca Cu_5 - type, expansion measured along the a-axis, lattice parameter reported as 4.869 Å at 43 K, 4.879 Å at 293 K by graphical interpolation, lattice parameter not corrected for refraction.

DATA TABLE 125. THERMAL LINEAR EXPANSION OF Ni_5Y INTERMETALLIC COMPOUND

[Temperature, T, K; Linear Expansion, $\Delta L/L_0$, %]

T	$\Delta L/L_0$	T	$\Delta L/L_0$
CURVE 1		CURVE 2	
43	-0.205	39	-0.277
72	-0.164	64	-0.176
101	-0.164	97	-0.176
139	-0.102	135	-0.226
183	0.041	176	-0.226
300	0.041	298	-0.050
374	0.082	365	0.151
474	0.225	471	0.327
654	0.471	652	0.327
748	0.532	748	0.579
907	0.676	894	0.680

FIGURE AND TABLE NO. 126R. PROVISIONAL VALUES FOR THERMAL LINEAR EXPANSION OF Nb_3Sn INTERMETALLIC COMPOUND

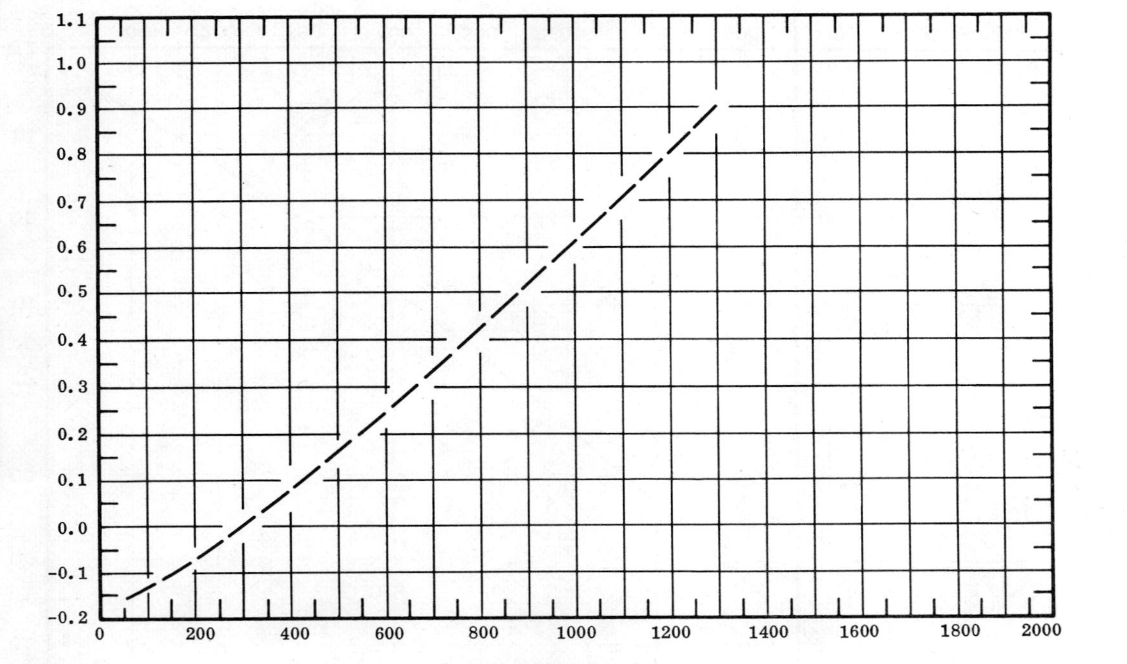

PROVISIONAL VALUES

[Temperature, T, K; Linear Expansion, $\Delta L/L_0$, %; α, K^{-1}]

T	$\Delta L/L_0$	$\alpha \times 10^6$
50	-0.159	5.9
100	-0.130	6.2
200	-0.066	6.7
293	0.000	7.2
400	0.080	7.7
500	0.160	8.2
600	0.243	8.6
700	0.330	8.9
800	0.420	9.2
900	0.513	9.4
1000	0.609	9.6
1100	0.707	9.7
1200	0.804	9.8
1300	0.901	9.9

REMARKS

The tabulated values are considered accurate to within ± 7% over the entire temperature range. These values can be represented approximately by the following equation:

$$\Delta L/L_0 = -0.187 + 5.490 \times 10^{-4} T + 3.296 \times 10^{-7} T^2 - 8.261 \times 10^{-11} T^3$$

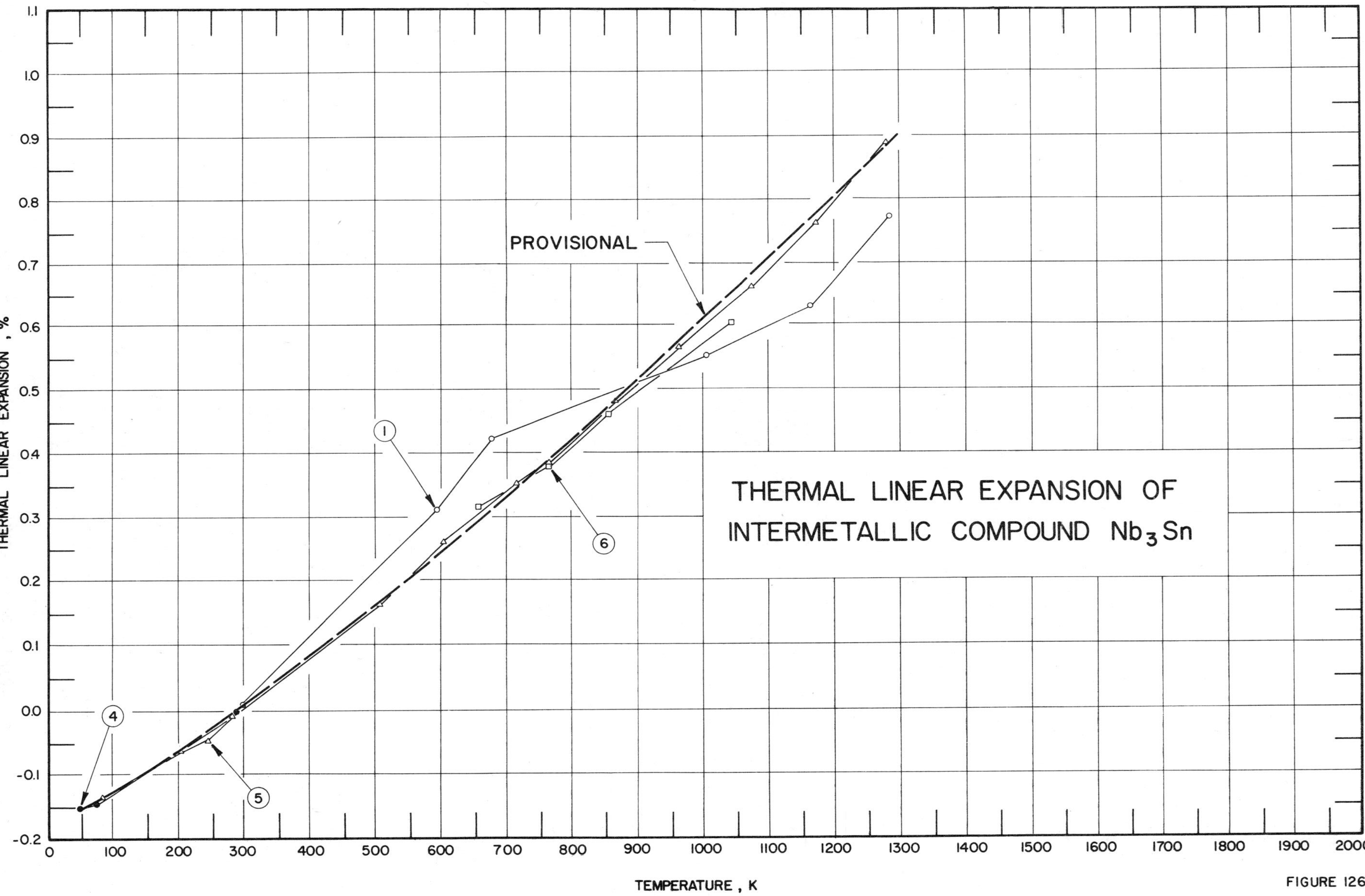
PROVISIONAL
1
6
4
5
THERMAL LINEAR EXPANSION OF
INTERMETALLIC COMPOUND Nb_3Sn
THERMAL LINEAR EXPANSION, %
1.1
1.0
0.9
0.8
0.7
0.6
0.5
0.4
0.3
0.2
0.1
0.0
-0.1
-0.2
0
100
200
300
400
500
600
700
800
900
1000
1100
1200
1300
1400
1500
1600
1700
1800
1900
2000
TEMPERATURE, K

FIGURE 126

SPECIFICATION TABLE 126. THERMAL LINEAR EXPANSION OF Nb_3Sn INTERMETALLIC COMPOUND

Cur. No.	Ref. No.	Author(s)	Year	Method Used	Temp. Range, K	Name and Specimen Designation	Composition (weight percent), Specifications, and Remarks
1	580	Cherry, W.H., Cody, G.D., Cooper, J.L., Gittleman, Y.I., Hanak, J.J., McConville, G.T., Rayl, M., and Rosi, F.D.	1962	X	293-1283		Vapor-deposited crystal.
2*	581	Mailfert, R., Batterman, B.W., and Hanak, J.J.	1969	X	5.5-42.6		Single crystal; expansion measured along a-axis of tetragonal phase; lattice parameter reported at 5.5 K is 5.2942 Å; this curve is here reported using the first given temperature as reference temperature at which $\Delta L/L_0 = 0$.
3*	581	Mailfert, R., et al.	1969	X	5.0-43		The above specimen; expansion measured along c-axis; lattice parameter reported at 5.0 K is 5.2602 Å; this curve is here reported using the first given temperature as reference temperature at which $\Delta L/L_0 = 0$.
4	581	Mailfert, R., et al.	1969	X	50-293		The above specimen; expansion of cubic phase; lattice parameter reported at 293 K is 5.2906 Å.
5	672	Schadler, H.W., Osika, L.M., Salvo, G.P., and DeCarlo, V.J.	1964	X	81-1289		Specimen in the form of Nb_3Sn on Nb sheet prepared by heating 0.005 in. thick sheet of electron beam melted commercial Nb in molten 99.999 Sn from Vulcan Ditinning Co., N.J., at 1220 K for 50 hr; excess Sn removed by etching in 35 HCl; annealed at 1270 K; lattice parameter reported at 81 K is 5.2838 Å; 5.2911 Å at 293 K determined by graphical interpolation; above 1172 K lattice parameter corrected for NbO and NbO_2 using Nelson-Riley extrapolation function.
6	672	Schadler, H.W., et al.	1964	X	288-1047		Single crystal prepared by heating Nb rod in molten 99.999 Sn from Vulcan Detinning Co., N.J., for 2 weeks at 1198 K; pulverized; lattice parameter reported at 288 K is 5.2904 Å; 5.2906 Å at 293 K determined by the graphical interpolation.
7*	672	Schadler, H.W., et al.	1964	X	288-1027		The above specimen; annealed at 870 K; average of two runs; lattice parameter reported at 288 K is 5.2909 Å; 5.2911 Å at 293 K determined by graphical interpolation.
8*	672	Schadler, H.W., et al.	1964	X	294-1047		The above specimen; lattice parameter reported at 294 K is 5.2909 Å; 5.2907 Å at 293 K determined by graphical extrapolation.

*Not shown in figure.

DATA TABLE 126. THERMAL LINEAR EXPANSION OF Nb_3Sn INTERMETALLIC COMPOUND

[Temperature, T, K; Linear Expansion, $\Delta L/L_0$, %]

T	$\Delta L/L_0$
CURVE 1†	
293	0.000
300	0.009
595	0.312
677	0.425
1006	0.558
1163	0.634
1283	0.728
CURVE 2*, †	
6	0.000
6	-0.019
6	-0.032
16	-0.017
27	-0.049
37	-0.087
41	-0.104
42	-0.100
42	-0.127
43	-0.149
CURVE 3*, †	
5	0.000
5	0.017
5	0.034
16	0.021
27	0.044
37	0.095
41	0.177
42	0.177
42	0.194
43	0.405
CURVE 4	
50	-0.151
77	-0.147
293	0.000
CURVE 5	
81	-0.138
201	-0.064
223	-0.047
288	-0.006

T	$\Delta L/L_0$
CURVE 5 (cont.)	
297	0.006*
507	0.164
604	0.261
712	0.355
766	0.387
868	0.486
963	0.580
1076	0.669
1172	0.769
1289	0.896
CURVE 6	
288	-0.004*
656	0.319
763	0.382
856	0.463
1047	0.609
CURVE 7*	
288	-0.002
288	-0.008
661	0.308
719	0.353
756	0.393
856	0.461
951	0.533
1027	0.618
CURVE 8*	
294	0.004
668	0.308
761	0.397
856	0.471
1047	0.624

* Not shown in figure.

† This curve is here reported using the first given temperature as reference temperature at which $\Delta L/L_0 = 0$.

FIGURE AND TABLE NO. 127AR. PROVISIONAL VALUES FOR THERMAL LINEAR EXPANSION OF Pd_3Sn INTERMETALLIC COMPOUND

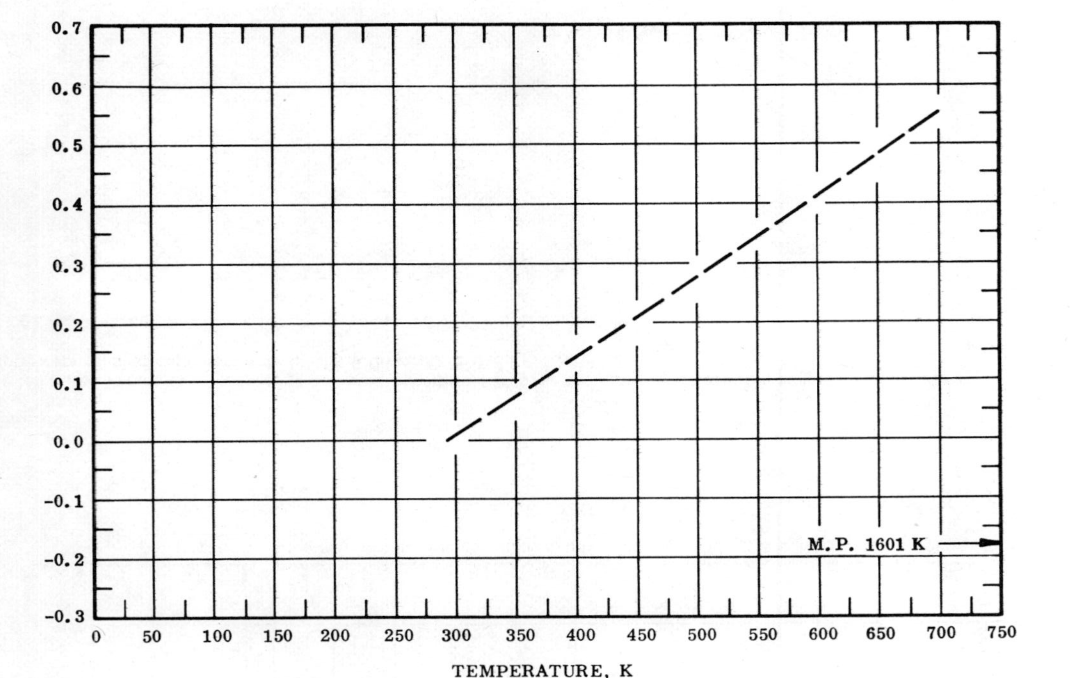

PROVISIONAL VALUES

[Temperature, T, K; Linear Expansion, $\Delta L/L_0$, %; α, K^{-1}]

T	$\Delta L/L_0$	$\alpha \times 10^6$
293	0.000	12.8
400	0.140	13.4
500	0.276	13.8
600	0.414	13.9
700	0.554	14.0

REMARKS

The tabulated values are considered accurate to within ± 7% over the entire temperature range. These values can be represented approximately by the following equation:

$$\Delta L/L_0 = -0.351 + 1.088 \times 10^{-3} T + 4.304 \times 10^{-7} T^2 - 1.961 \times 10^{-10} T^3$$

FIGURE AND TABLE NO. 127BR. PROVISIONAL VALUES FOR THERMAL LINEAR EXPANSION OF Pd_3Yb INTERMETALLIC COMPOUND

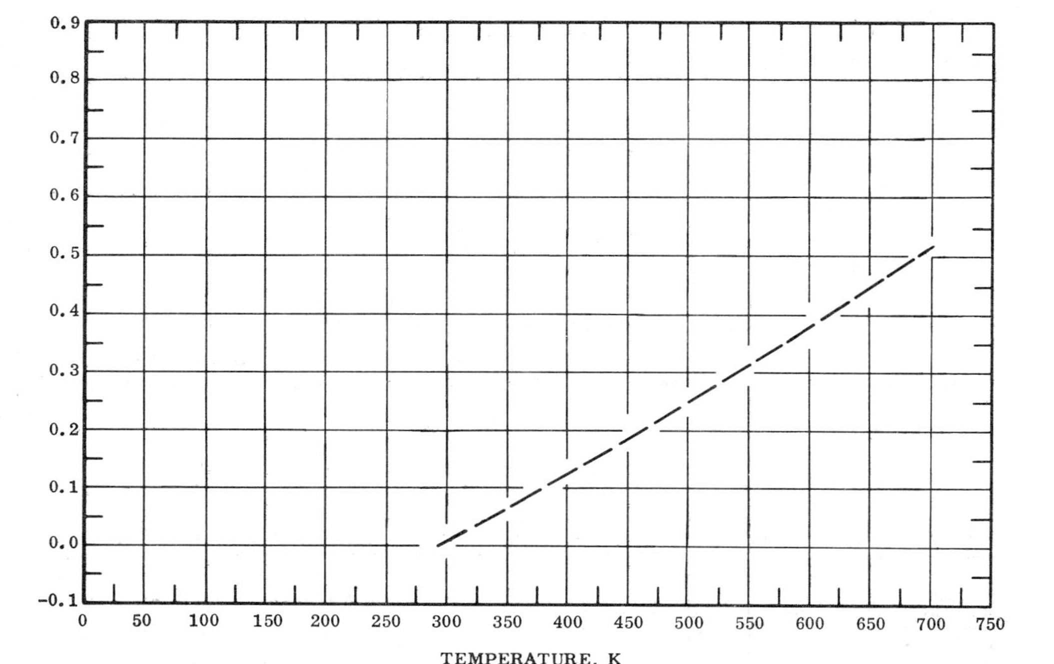

PROVISIONAL VALUES

[Temperature, T, K; Linear Expansion, $\Delta L/L_0$, %; α, K^{-1}]

T	$\Delta L/L_0$	$\alpha \times 10^6$
293	0.000	10.8
400	0.122	12.0
500	0.246	12.8
600	0.376	13.3
700	0.511	13.5

REMARKS

The tabulated values are considered accurate to within ±7% over the entire temperature range. These values can be represented approximately by the following equation:

$$\Delta L/L_0 = -0.248 + 5.681 \times 10^{-4}\, T + 1.093 \times 10^{-6}\, T^2 - 5.100 \times 10^{-10}\, T^3$$

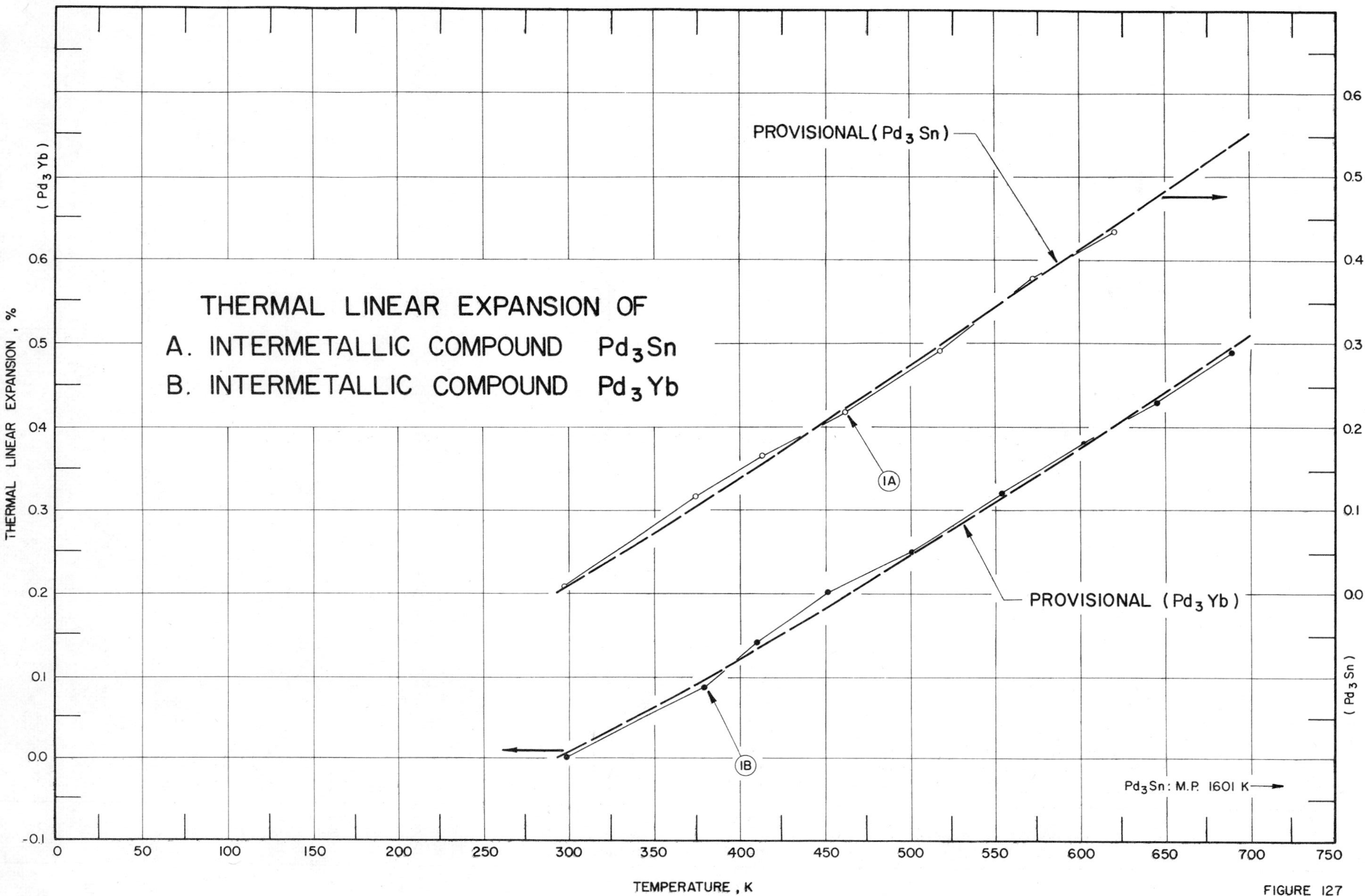
THERMAL LINEAR EXPANSION OF
A. INTERMETALLIC COMPOUND Pd3Sn
B. INTERMETALLIC COMPOUND Pd3Yb
PROVISIONAL (Pd3 Sn)
PROVISIONAL (Pd3 Yb)
1A
1B
Pd3Sn: M.P. 1601 K
THERMAL LINEAR EXPANSION, %
(Pd3 Yb)
(Pd3 Sn)
TEMPERATURE, K
0
50
100
150
200
250
300
350
400
450
500
550
600
650
700
750
-0.1
0.0
0.1
0.2
0.3
0.4
0.5
0.6

FIGURE 127

SPECIFICATION TABLE 127A. THERMAL LINEAR EXPANSION OF Pd_3Sn INTERMETALLIC COMPOUND

Cur. No.	Ref. No.	Author(s)	Year	Method Used	Temp. Range, K	Name and Specimen Designation	Composition (weight percent), Specifications, and Remarks
1	557	Harris, L.R., Raynor, G.V., and Winstanley, C.S.	1967	X	297-620		Specimen prepared from 99.95$^+$ Sn and Pd; obtained from Johnson Matthey and Co.; copper Kα radiation used for analysis; lattice parameter reported at 297 K is 3.9678 Å; 3.9675 Å at 293 K determined by graphical extrapolation.

DATA TABLE 127A. THERMAL LINEAR EXPANSION OF Pd_3Sn INTERMETALLIC COMPOUND

[Temperature, T, K; Linear Expansion, $\Delta L/L_0$, %]

T	$\Delta L/L_0$
CURVE 1	
297	0.007
374	0.118
413	0.166
461	0.221
516	0.294
572	0.380
620	0.436

SPECIFICATION TABLE 127B. THERMAL LINEAR EXPANSION OF Pd_3Yb INTERMETALLIC COMPOUND

Cur. No.	Ref. No.	Author(s)	Year	Method Used	Temp. Range, K	Name and Specimen Designation	Composition (weight percent), Specifications, and Remarks
1	557	Harris, I. R., Raynor, G. V. and Winstanley, C. J.	1967	X	296-734		Specimen prepared from 99.95^+ pure Yb and Pd; obtained from Johnson Matthey and Co.; copper $K\alpha$ radiation used for analysis; lattice parameter reported at 296 K is 4.0248 Å; 4.0247 Å at 293 K determined by graphical extrapolation.

DATA TABLE 127B. THERMAL LINEAR EXPANSION OF Pd_3Yb INTERMETALLIC COMPOUND

[Temperature, T, K; Linear Expansion, $\Delta L/L_0$, %]

T	$\Delta L/L_0$
CURVE 1	
296	0.002
377	0.089
410	0.141
451	0.201
500	0.250
554	0.325
601	0.382
645	0.432
689	0.491
734	0.546

FIGURE AND TABLE NO. 128AR. PROVISIONAL VALUES FOR THERMAL LINEAR EXPANSION OF $PrRu_2$ INTERMETALLIC COMPOUND

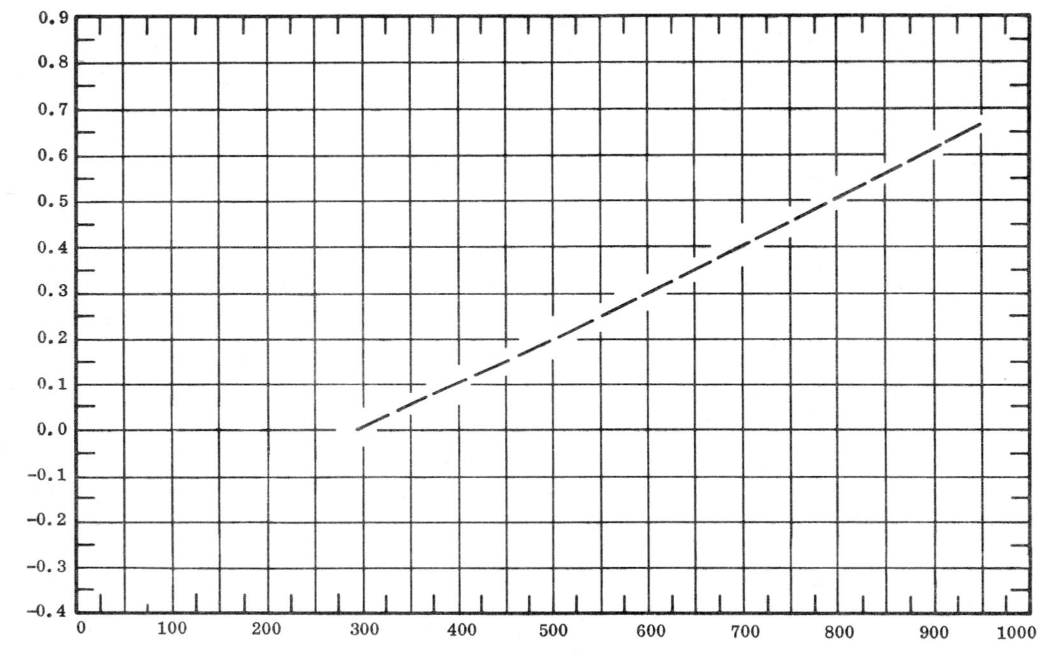

PROVISIONAL VALUES

[Temperature, T, K; Linear Expansion, $\Delta L/L_0$, %; α, K^{-1}]

T	$\Delta L/L_0$	$\alpha \times 10^6$
293	0.000	8.9
400	0.099	9.4
500	0.195	9.9
600	0.296	10.3
700	0.399	10.5
800	0.505	10.6
900	0.611	10.6
950	0.664	10.6

REMARKS

The tabulated values are considered accurate to within ±10% over the entire temperature range. These values can be represented approximately by the following equation:

$$\Delta L/L_0 = -0.230 + 6.651 \times 10^{-4}\, T + 4.580 \times 10^{-7}\, T^2 - 1.772 \times 10^{-10}\, T^3$$

FIGURE AND TABLE NO. 128BR. PROVISIONAL VALUES FOR THERMAL LINEAR EXPANSION OF $PrSn_3$ INTERMETALLIC COMPOUND

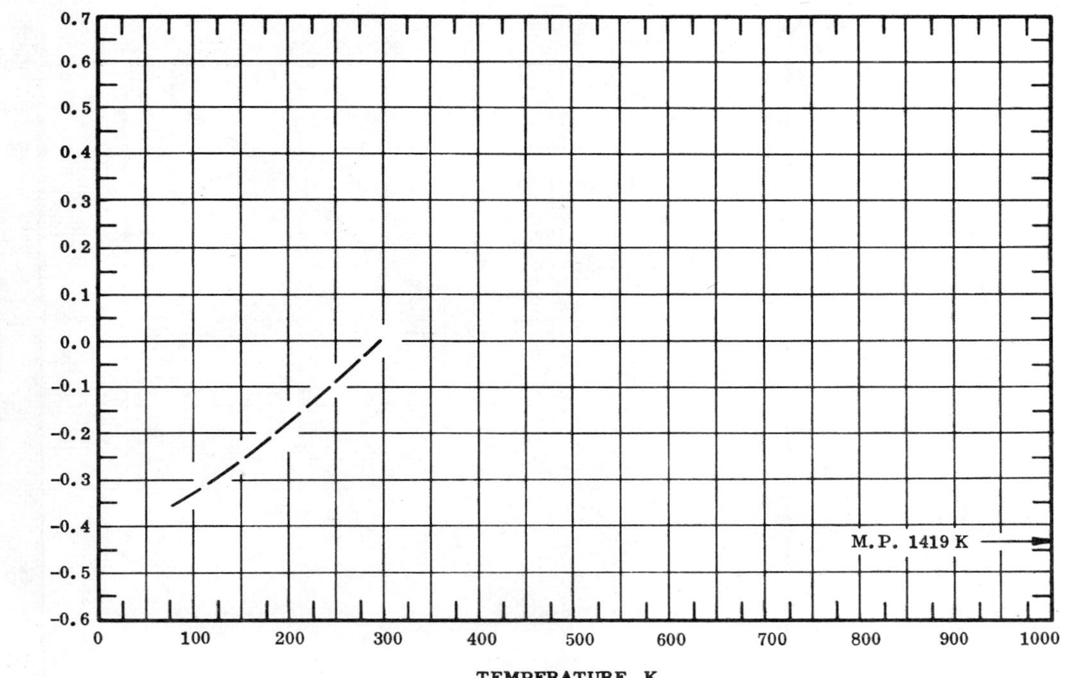

PROVISIONAL VALUES

[Temperature, T, K; Linear Expansion, $\Delta L/L_0$, %; α, K^{-1}]

T	$\Delta L/L_0$	$\alpha \times 10^6$
75	-0.358	12.0
100	-0.327	13.5
200	-0.168	17.5
293	0.000	18.2

REMARKS

The tabulated values are considered accurate to within ± 7% over the entire temperature range. These values can be represented approximately by the following equation:

$$\Delta L/L_0 = -0.428 + 6.162 \times 10^{-4}\, T + 4.539 \times 10^{-6}\, T^2 - 5.672 \times 10^{-9}\, T^3$$

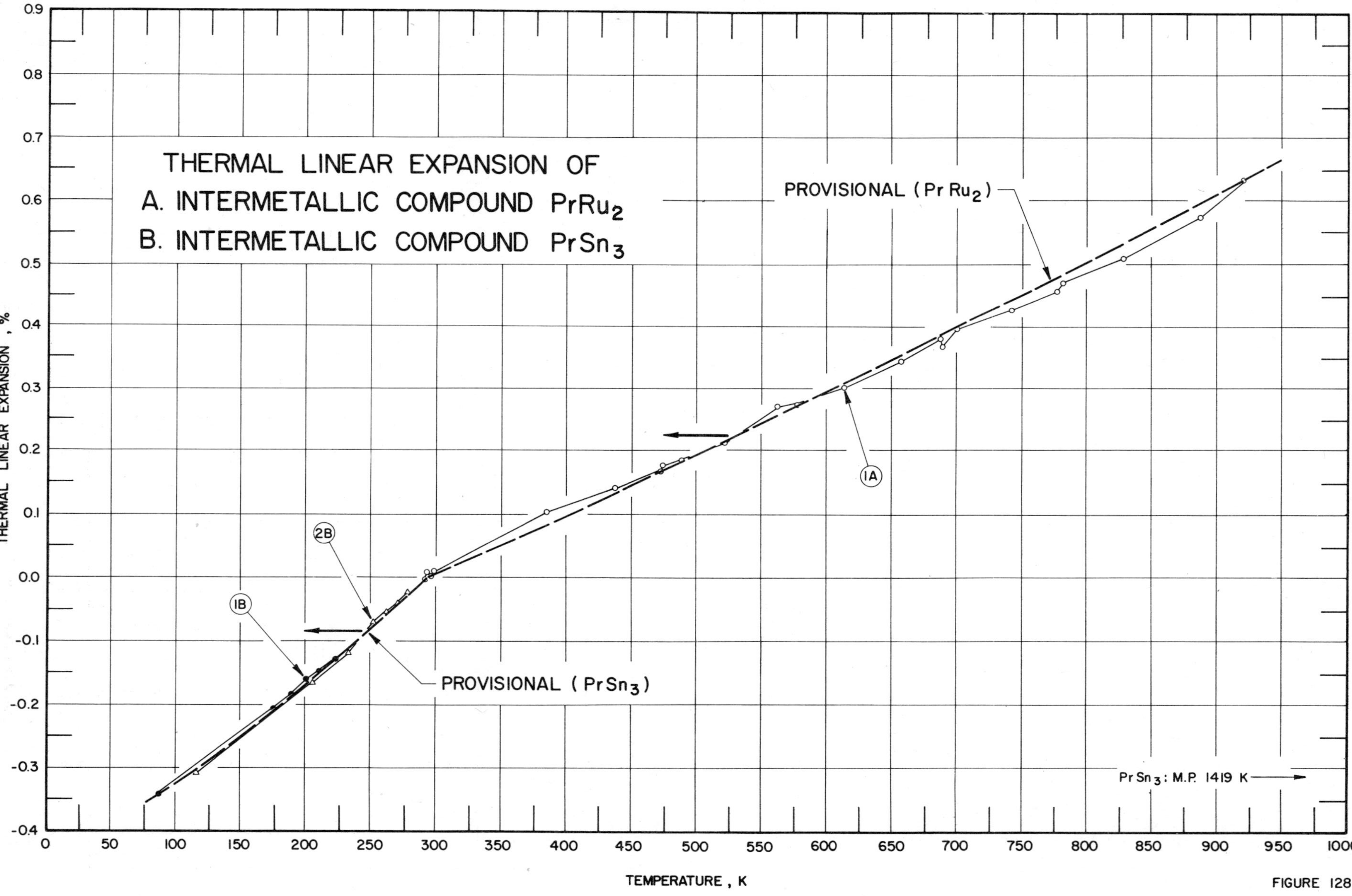
THERMAL LINEAR EXPANSION OF
A. INTERMETALLIC COMPOUND PrRu2
B. INTERMETALLIC COMPOUND PrSn3
PROVISIONAL (Pr Ru2)
PROVISIONAL (PrSn3)
1A
1B
2B
Pr Sn3 : M.P. 1419 K
THERMAL LINEAR EXPANSION, %
TEMPERATURE, K
0.9
0.8
0.7
0.6
0.5
0.4
0.3
0.2
0.1
0.0
-0.1
-0.2
-0.3
-0.4
0
50
100
150
200
250
300
350
400
450
500
550
600
650
700
750
800
850
900
950
1000

FIGURE 128

SPECIFICATION TABLE 128A. THERMAL LINEAR EXPANSION OF $PrRu_2$ INTERMETALLIC COMPOUND

Cur. No.	Ref. No.	Author(s)	Year	Method Used	Temp. Range, K	Name and Specimen Designation	Composition (weight percent), Specifications, and Remarks
1	558	Gschneidner, K.A., Jr., Ellott, R.O., and Cromer, D.T.	1969	X	292-923		Prepared by arc-melting together weighed amounts of components; filings annealed at 720 K for 15 min; CuK_α radiation used; lattice parameter reported at 292 K is 7.6217 Å; 7.6219 Å at 293 K determined by graphical interpolation.

DATA TABLE 128A. THERMAL LINEAR EXPANSION OF $PrRu_2$ INTERMETALLIC COMPOUND

[Temperature, T, K; Linear Expansion, $\Delta L/L_0$, %]

T	$\Delta L/L_0$	T	$\Delta L/L_0$
CURVE 1		CURVE 1 (cont.)	
292	-0.002	784	0.472
292	0.011	828	0.512
295	0.001	888	0.578
297	0.011*	923	0.633
385	0.106		
437	0.143		
473	0.170		
474	0.179*		
489	0.187		
521	0.215		
563	0.271		
578	0.274		
614	0.303		
656	0.345		
675	0.381		
679	0.368		
701	0.400		
741	0.429		
777	0.464		

*Not shown in figure.

SPECIFICATION TABLE 128B. THERMAL LINEAR EXPANSION OF $PrSn_3$ INTERMETALLIC COMPOUND

Cur. No.	Ref. No.	Author(s)	Year	Method Used	Temp. Range, K	Name and Specimen Designation	Composition (weight percent), Specifications, and Remarks
1	682	Harris, I.R. and Raynor, G.V.	1965	X	296-86		Specimen prepared from 99.999 Sn from Messrs. Capper, Pass, Ltd., and Pr containing <0.10 rare earth metals, 0.10 Ta and 0.02 other base metals by the method similar to the one described for $CeIn_3$; measurement with decreasing temperature; lattice parameter reported at 293 K is 4.7064 kX.
2	682	Harris, I.R. and Raynor, G.V.	1965	X	115-278		The above specimen; measurements with increasing temperature; lattice parameter reported at 278 K is 4.7050 kX; 4.7060 kX at 293 K determined by graphical extrapolation.

DATA TABLE 128B. THERMAL LINEAR EXPANSION OF $PrSn_3$ INTERMETALLIC COMPOUND

[Temperature, T, K; Linear Expansion, $\Delta L/L_0$, %]

T	$\Delta L/L_0$	T	$\Delta L/L_0$
CURVE 1		CURVE 2 (cont.)	
296	0.008	272	-0.040
293	0.000	278	-0.021
222	-0.127		
210	-0.149		
200	-0.157		
189	-0.183		
175	-0.206		
86	-0.342		
CURVE 2			
115	-0.308		
205	-0.161		
234	-0.115		
252	-0.070		
262	-0.055		

FIGURE AND TABLE NO. 129R. TYPICAL VALUES FOR THERMAL LINEAR EXPANSION OF $SmAg_3$ INTERMETALLIC COMPOUND

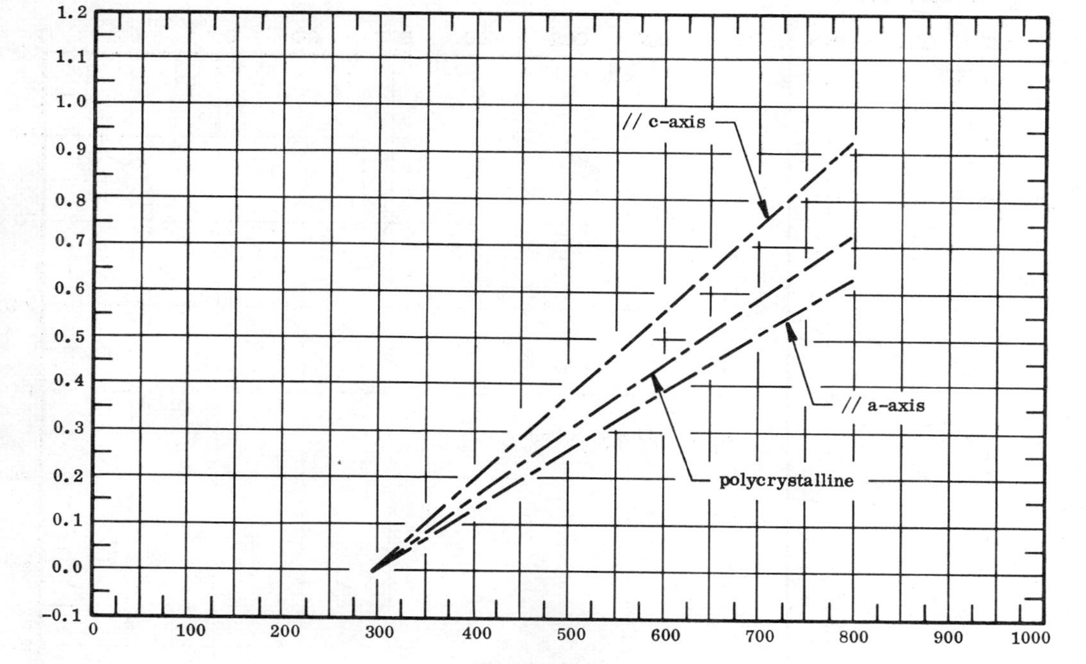

TYPICAL VALUES

[Temperature, T, K; Linear Expansion, $\Delta L/L_0$, %; α, K^{-1}]

T	// a-axis $\Delta L/L_0$	// c-axis $\Delta L/L_0$	polycrystalline $\Delta L/L_0$
293	0.000	0.000	0.000
400	0.133	0.195	0.154
500	0.259	0.377	0.298
600	0.385	0.559	0.442
700	0.509	0.741	0.586
800	0.627	0.923	0.726

// a-axis: $\alpha = 12.5 \times 10^{-6}$

// c-axis: $\alpha = 18.2 \times 10^{-6}$

polycrystalline: $\alpha = 14.4 \times 10^{-6}$

REMARKS

The tabulated values are considered accurate to within ±20%. The values for polycrystalline material in the absence of experimental data are calculated from the single crystal data. These and other values can be represented approximately by the following equations:

// a-axis: $\Delta L/L_0 = -0.321 + 9.452 \times 10^{-4}\, T + 6.453 \times 10^{-7}\, T^2 - 4.345 \times 10^{-10}\, T^3$

// c-axis: $\Delta L/L_0 = -0.535 + 1.832 \times 10^{-3}\, T - 2.102 \times 10^{-8}\, T^2 + 1.170 \times 10^{-11}\, T^3$

polycrystalline: $\Delta L/L_0 = -0.421 + 1.436 \times 10^{-3}\, T + 6.468 \times 10^{-9}\, T^2 - 3.601 \times 10^{-12}\, T^3$

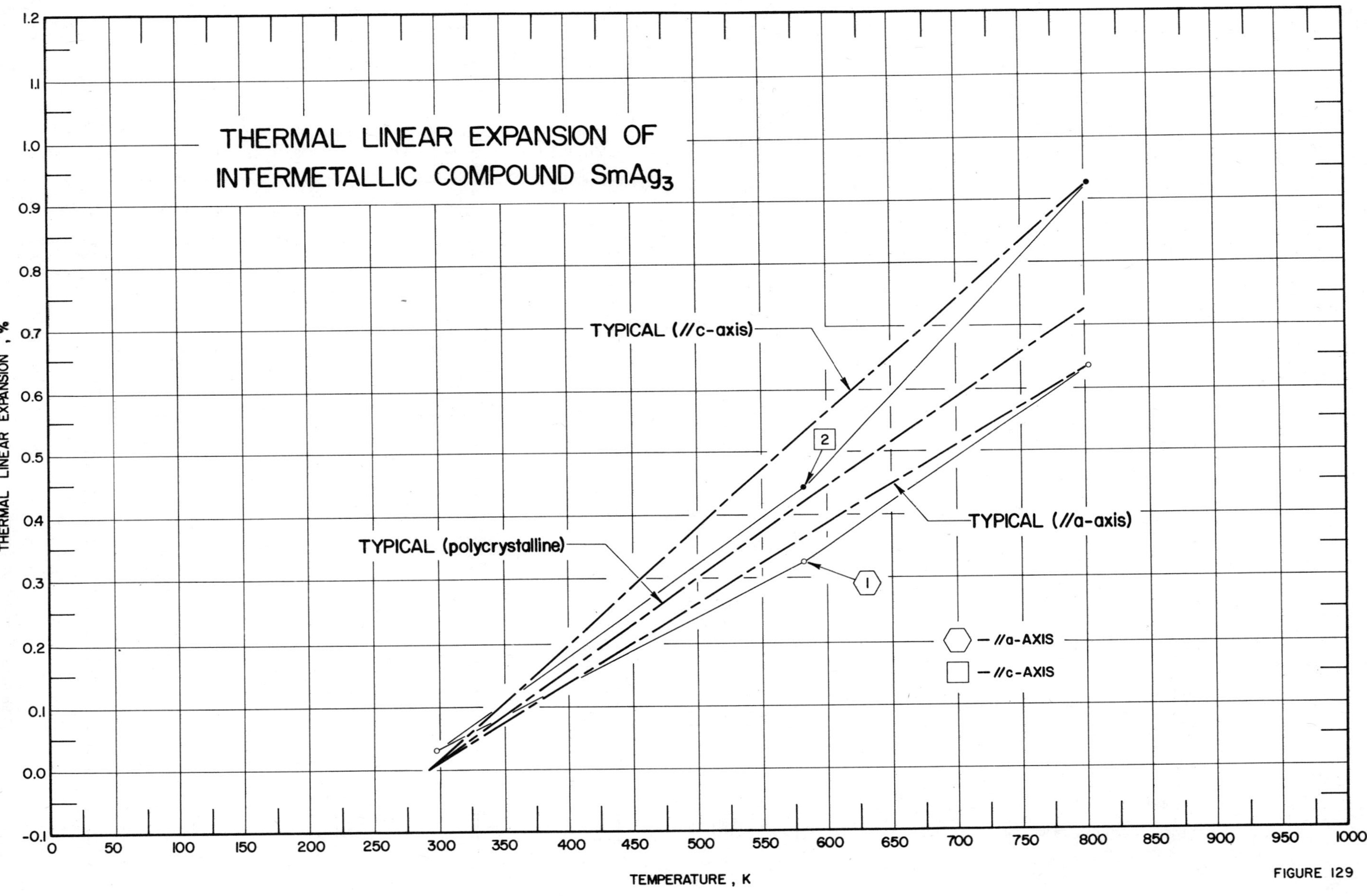
THERMAL LINEAR EXPANSION OF
INTERMETALLIC COMPOUND $SmAg_3$
TYPICAL (//c-axis)
TYPICAL (polycrystalline)
TYPICAL (//a-axis)
2
1
— //a-AXIS
— //c-AXIS
THERMAL LINEAR EXPANSION, %
1.2
1.1
1.0
0.9
0.8
0.7
0.6
0.5
0.4
0.3
0.2
0.1
0.0
-0.1
0
50
100
150
200
250
300
350
400
450
500
550
600
650
700
750
800
850
900
950
1000
TEMPERATURE, K

FIGURE 129

SPECIFICATION TABLE 129. THERMAL LINEAR EXPANSION OF $SmAg_3$ INTERMETALLIC COMPOUND

Cur. No.	Ref. No.	Author(s)	Year	Method Used	Temp. Range, K	Name and Specimen Designation	Composition (weight percent), Specifications, and Remarks
1	587	Steeb, S., Godd, D., and Lohr, C.	1968	X	298-803		99.999$^+$ Ag and 99.0 pure Sm arc-melted in argon atmosphere; placed in quartz capsules, heat-treated several days at 1020 K then quenched in water; expansion along a-axis measured with CuKα x-radiation; lattice parameter reported at 298 K is 12.634 Å; 12.630 Å at 293 K determined by graphical extrapolation.
2	587	Steeb, S., et al.	1968	X	298-803		The above specimen; expansion along c-axis; lattice parameter reported at 298 K is 9.285 Å; 9.282 Å at 293 K determined by graphical extrapolation.

DATA TABLE 129. THERMAL LINEAR EXPANSION OF $SmAg_3$ INTERMETALLIC COMPOUND

[Temperature, T, K; Linear Expansion, $\Delta L/L_0$, %]

T	$\Delta L/L_0$
CURVE 1	
298	0.031
583	0.324
803	0.633
CURVE 2	
298	0.032
583	0.441
803	0.926

SPECIFICATION TABLE 130. THERMAL LINEAR EXPANSION OF Ag_2Tb INTERMETALLIC COMPOUND

Cur. No.	Ref. No.	Author(s)	Year	Method Used	Temp. Range, K	Name and Specimen Designation	Composition (weight percent), Specifications, and Remarks
1*	676	Atoji, M.	1968		5, 300		Specimen prepared from stoichiometric mixture of Tb and Ag, both of 99.9+ pure; heated at 1120 K for one day; unit cell dimensions along a-axis at 5 K and 300 K are 3.698 Å amd 3.710 Å respectively; expansion measured along a-axis.
2*	676	Atoji, M.	1968		5, 300		The above specimen; unit cell dimensions along c-axis at 5 K and 300 K are 9.226 Å and 9.247 Å respectively; expansion measured along c-axis.

DATA TABLE 130. THERMAL LINEAR EXPANSION OF Ag_2Tb INTERMETALLIC COMPOUND

[Temperature, T, K; Linear Expansion, $\Delta L/L_0$, %]

T	$\Delta L/L_0$
CURVE 1*	
5	-0.317
300	0.008
CURVE 2*	
5	-0.230
300	0.006

* No figure given.

FIGURE AND TABLE NO. 131R. PROVISIONAL VALUES FOR THERMAL LINEAR EXPANSION OF **AgZn** INTERMETALLIC COMPOUND

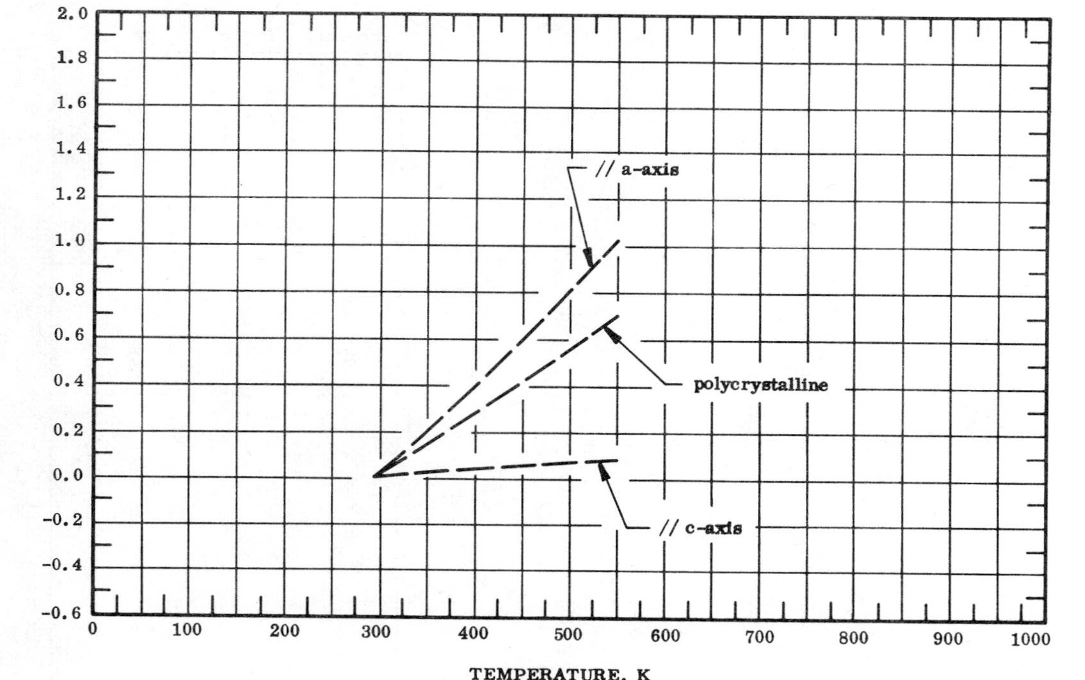

PROVISIONAL VALUES

[Temperature, T, K; Linear Expansion, $\Delta L/L_0$, %; α, K^{-1}]

T	// a-axis $\Delta L/L_0$	// c-axis $\Delta L/L_0$	polycrystalline $\Delta L/L_0$	polycrystalline $\alpha \times 10^6$
293	0.000	0.000	0.000	25.9
400	0.407	0.033	0.282	27.1
500	0.815	0.064	0.566	29.3
550	1.033	0.080	0.715	30.9

REMARKS

The tabulated values are considered accurate to within ±10%. The values for polycrystalline material in the absence of experimental data are calculated from the single crystal data. These and other values can be represented approximately by the following equations:

// a-axis: $\Delta L/L_0 = -1.125 + 4.173 \times 10^{-3}\ T - 1.906 \times 10^{-6}\ T^2 + 2.646 \times 10^{-9}\ T^3$

// c-axis: $\Delta L/L_0 = -0.105 + 4.211 \times 10^{-4}\ T - 2.749 \times 10^{-7}\ T^2 + 2.215 \times 10^{-10}\ T^3$

polycrystalline: $\Delta L/L_0 = -0.784 + 2.911 \times 10^{-3}\ T - 1.333 \times 10^{-6}\ T^2 + 1.810 \times 10^{-9}\ T^3$

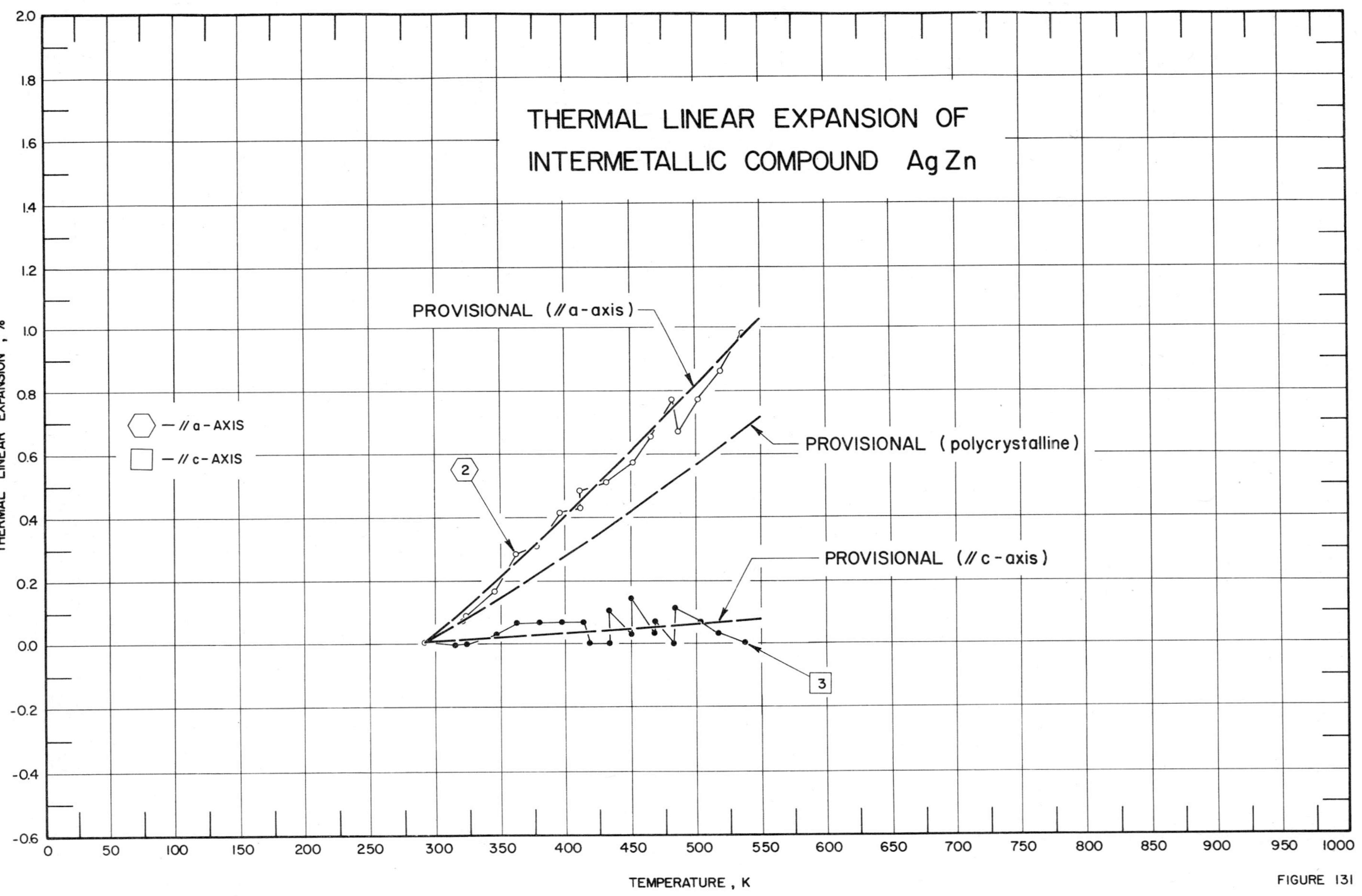
THERMAL LINEAR EXPANSION OF
INTERMETALLIC COMPOUND Ag Zn
PROVISIONAL (//a-axis)
PROVISIONAL (polycrystalline)
PROVISIONAL (//c-axis)
2
3
— //a-AXIS
— //c-AXIS
THERMAL LINEAR EXPANSION, %
2.0
1.8
1.6
1.4
1.2
1.0
0.8
0.6
0.4
0.2
0.0
-0.2
-0.4
-0.6
TEMPERATURE, K
0
50
100
150
200
250
300
350
400
450
500
550
600
650
700
750
800
850
900
950
1000

FIGURE 131

SPECIFICATION TABLE 131. THERMAL LINEAR EXPANSION OF AgZn INTERMETALLIC COMPOUND

Cur. No.	Ref. No.	Author(s)	Year	Method Used	Temp. Range, K	Name and Specimen Designation	Composition (weight percent), Specifications, and Remarks
1*	588	Chaudhuri, A., Clark, H. McL., and Wayman, C. M.	1969	L	293-603		Single crystal rod 0.4 cm diameter, 2.4 cm long, grown by Bridgman technique; quenched from 733 K (β-phase) to produce metastable β' phase; heating rate is 1 C/min; sample changes to stable ζ-phase at 403-423 K and to stable β-phase at 538 K.
2	678	Warlimont, H.	1959	X	293-536		Specimen prepared from 99.9 Ag and 99.99 Zn by melting in carbon resistant oven; measurements along a-axis; lattice parameter reported at 293 K is 7.627 kX.
3	678	Warlimont, H.	1959	X	293-537		The above specimen; measurements along c-axis; lattice parameter reported at 293 K is 2.824 kX.

DATA TABLE 131. THERMAL LINEAR EXPANSION OF AgZn INTERMETALLIC COMPOUND

[Temperature, T, K; Linear Expansion, $\Delta L/L_0$, %]

T	$\Delta L/L_0$	T	$\Delta L/L_0$	T	$\Delta L/L_0$	T	$\Delta L/L_0$
CURVE 1*		CURVE 1 (cont.)*		CURVE 2 (cont.)		CURVE 3 (cont.)	
293	0.000	533	1.266	432	0.511	414	0.071
304	0.009	538	1.276	451	0.577	416	0.000
313	0.026	554	1.345	467	0.655	433	0.000
333	0.101	564	1.393	483	0.773	433	0.106
343	0.152	583	1.487	486	0.669	450	0.035
353	0.207	603	1.583	502	0.773	450	0.142
362	0.253			520	0.865	466	0.035
372	0.313	CURVE 2		536	0.983	468	0.071
393	0.419					483	0.000
403	0.477	293	0.000	CURVE 3		486	0.106
413	0.578	316	0.079			502	0.071
421	0.783	324	0.092	293	0.000	519	0.035
423	0.842	345	0.170	315	0.000	537	0.000
443	0.915	363	0.288	324	0.000		
463	0.995	379	0.315	346	0.035		
483	1.071	396	0.419	364	0.071		
503	1.153	411	0.433	380	0.071		
528	1.249	411	0.485	397	0.071		

* Not shown in figure.

FIGURE AND TABLE NO. 132R. PROVISIONAL VALUES FOR THERMAL LINEAR EXPANSION OF NaTl INTERMETALLIC COMPOUND

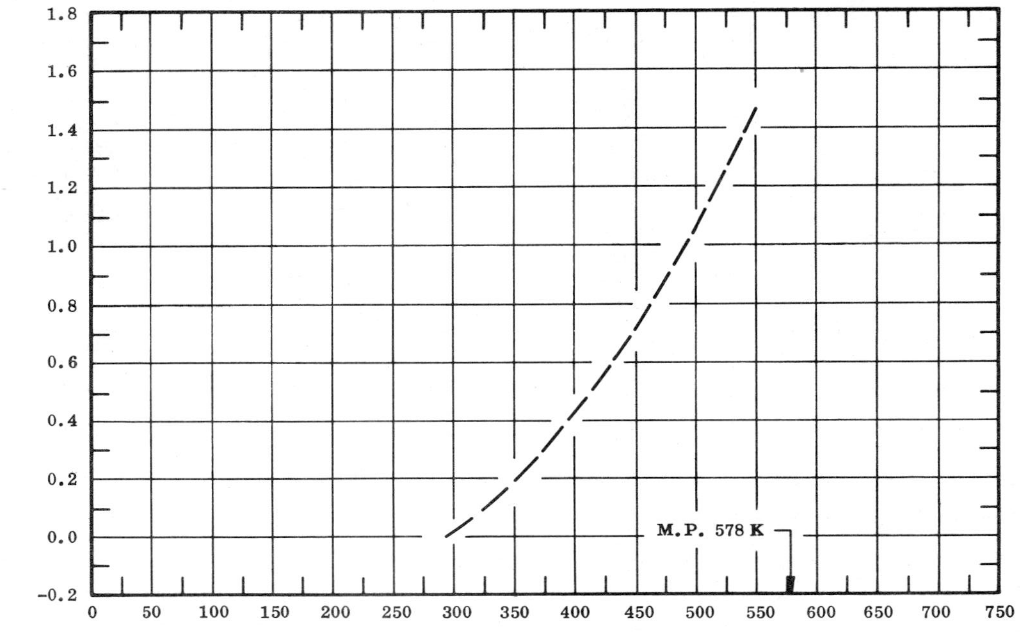

PROVISIONAL VALUES

[Temperature, T, K; Linear Expansion, $\Delta L/L_0$, %; α, K^{-1}]

T	$\Delta L/L_0$	$\alpha \times 10^6$
293	0.000	28
400	0.428	52
500	1.068	76
550	1.477	88

REMARKS

The tabulated values are considered accurate to within ±7% over the entire temperature range. These values can be represented approximately by the following equation:

$$\Delta L/L_0 = -0.123 - 3.563 \times 10^{-3}\, T + 1.039 \times 10^{-5}\, T^2 + 1.007 \times 10^{-9}\, T^3$$

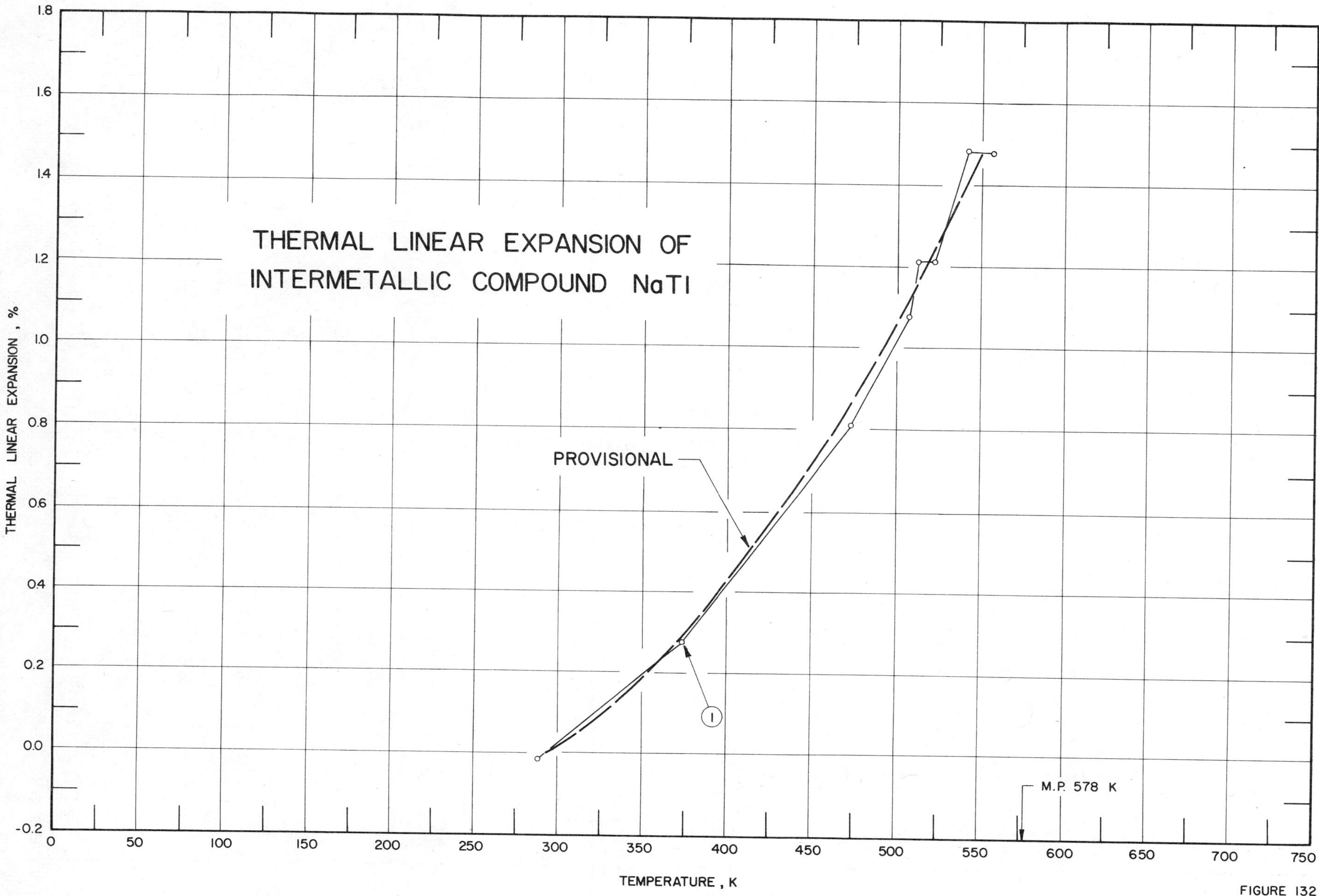
THERMAL LINEAR EXPANSION OF
INTERMETALLIC COMPOUND NaTl
PROVISIONAL
1
M.P. 578 K
THERMAL LINEAR EXPANSION, %
TEMPERATURE, K
1.8
1.6
1.4
1.2
1.0
0.8
0.6
0.4
0.2
0.0
-0.2
0
50
100
150
200
250
300
350
400
450
500
550
600
650
700
750

FIGURE 132

SPECIFICATION TABLE 132. THERMAL LINEAR EXPANSION OF NaTl INTERMETALLIC COMPOUND

Cur. No.	Ref. No.	Author(s)	Year	Method Used	Temp. Range, K	Name and Specimen Designation	Composition (weight percent), Specifications, and Remarks
1	690	Schneider, V.A. and Heymer, G.	1956	X	289-556		Prepared from pure metals; lattice parameter reported at 289 K is 7.488 Å; 7.489 Å at 293 K determined by graphical interpolation.

DATA TABLE 132. THERMAL LINEAR EXPANSION OF NaTl INTERMETALLIC COMPOUND

[Temperature, T, K; Linear Expansion, $\Delta L/L_0$, %]

T	$\Delta L/L_0$
CURVE 1	
289	-0.013
374	0.280
473	0.815
507	1.082
512	1.215
522	1.215
542	1.482
556	1.482

FIGURE AND TABLE NO. 133R. PROVISIONAL VALUES FOR THERMAL LINEAR EXPANSION OF $YbZn_2$ INTERMETALLIC COMPOUND

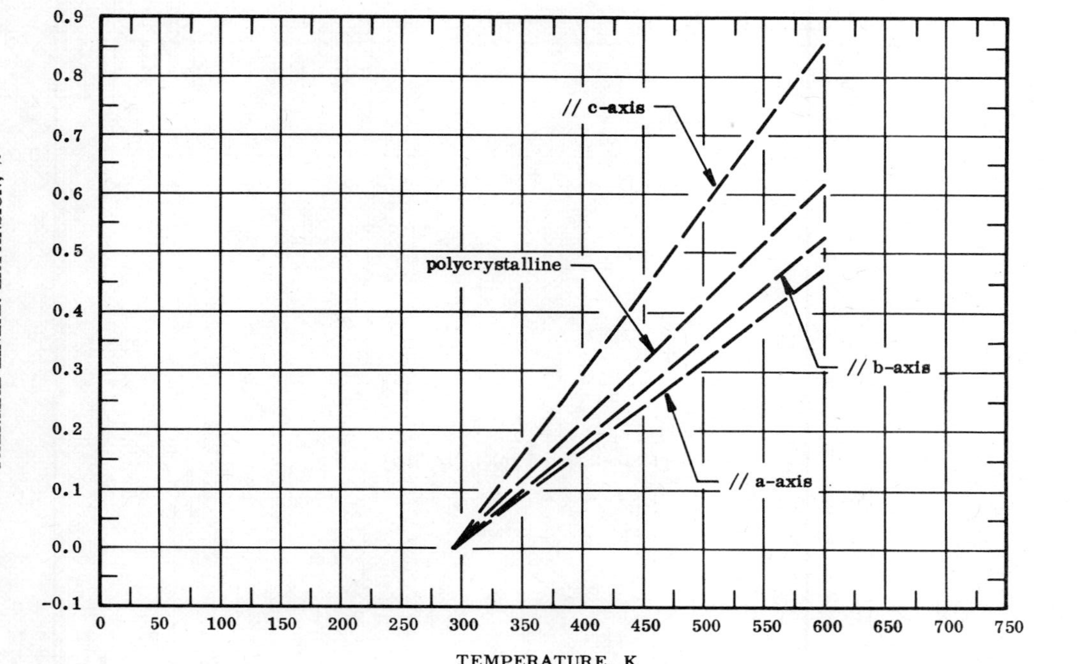

PROVISIONAL VALUES

[Temperature, T, K; Linear Expansion, $\Delta L/L_0$, %; α, K^{-1}]

T	// a-axis $\Delta L/L_0$	// b-axis $\Delta L/L_0$	// c-axis $\Delta L/L_0$	polycrystalline $\Delta L/L_0$
293	0.000	0.000	0.000	0.000
400	0.165	0.184	0.299	0.216
500	0.319	0.355	0.579	0.418
600	0.472	0.527	0.857	0.619

α values

// a-axis: 15.4×10^{-6} (293-600)

// b-axis: 17.1×10^{-6} (293-600)

// c-axis: 28.0×10^{-6} (293-600)

polycrystalline: 20.2×10^{-6} (293-600)

REMARKS

The tabulated values are considered accurate to within ±10%. The values for polycrystalline material in the absence of experimental data are calculated from the single crystal data. These and other values can be represented approximately by the following equations:

// a-axis: $\Delta L/L_0 = -0.476 + 1.716 \times 10^{-3} T - 3.818 \times 10^{-7} T^2 + 2.630 \times 10^{-10} T^3$

// b-axis: $\Delta L/L_0 = -0.504 + 1.720 \times 10^{-3} T - 3.198 \times 10^{-7} T^2 - 8.612 \times 10^{-13} T^3$

// c-axis: $\Delta L/L_0 = -0.764 + 2.392 \times 10^{-3} T + 9.533 \times 10^{-7} T^2 - 7.250 \times 10^{-10} T^3$

polycrystalline: $\Delta L/L_0 = -0.574 + 1.893 \times 10^{-3} T + 2.957 \times 10^{-7} T^2 - 2.253 \times 10^{-10} T^3$

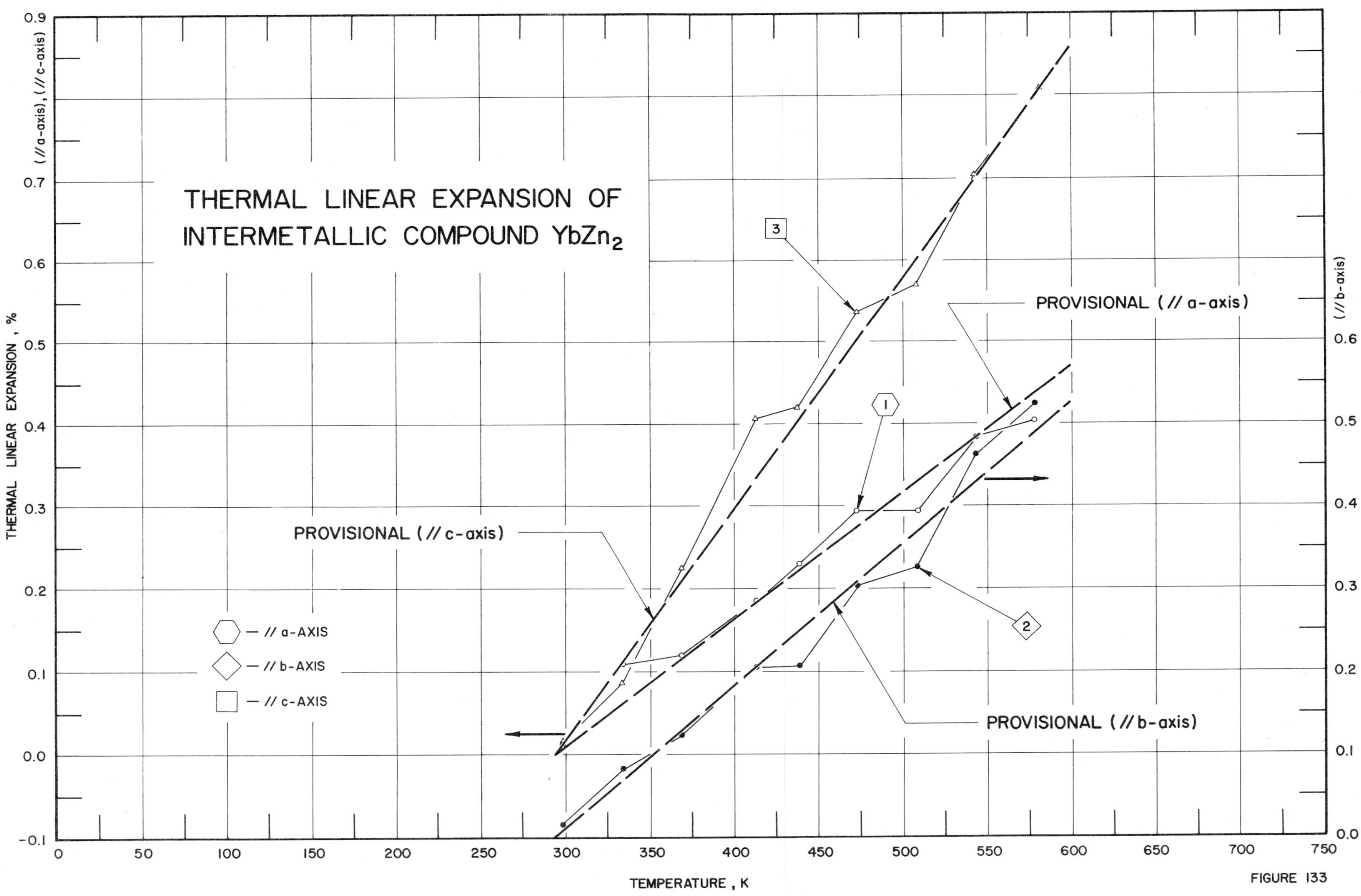
THERMAL LINEAR EXPANSION OF
INTERMETALLIC COMPOUND $YbZn_2$
PROVISIONAL (// a-axis)
PROVISIONAL (// b-axis)
PROVISIONAL (// c-axis)
1
2
3
— // a-AXIS
— // b-AXIS
— // c-AXIS
THERMAL LINEAR EXPANSION, %
(// a-axis), (// c-axis)
(// b-axis)
TEMPERATURE, K
0
50
100
150
200
250
300
350
400
450
500
550
600
650
700
750
-0.1
0.0
0.1
0.2
0.3
0.4
0.5
0.6
0.7
0.9
FIGURE 133

SPECIFICATION TABLE 133. THERMAL LINEAR EXPANSION OF $YbZn_2$ INTERMETALLIC COMPOUND

Cur. No.	Ref. No.	Author(s)	Year	Method Used	Temp. Range, K	Name and Specimen Designation	Composition (weight percent), Specifications, and Remarks
1	591	Michel, D. J. and Ryba, E.	1969	X	298-578		Single phase $YbZn_2$ alloy, prepared in a sealed tantalum can, reduced to -325 mesh, pressed and sealed in 10 mil beryllium foil, CuK_α (λ = 1.5178 Å) radiation employed; expansion measured in a-direction (orthorhombic structure); lattice parameter reported at 298 K is 4.5750 Å; 4.5745 Å at 293 K determined by graphical extrapolation.
2	591	Michel, D. J. and Ryba, E.	1969	X	298-578		Similar to the above specimen except expansion measured in b-direction; lattice parameter reported at 298 K is 7.327 Å; 7.326 Å at 293 K determined by graphical extrapolation.
3	591	Michel, D. J. and Ryba, E.	1969	X	298-578		Similar to the above specimen except expansion measured in c-direction; lattice parameter reported at 298 K is 7.5695 Å; 7.5687 Å at 293 K determined by graphical extrapolation.

DATA TABLE 133. THERMAL LINEAR EXPANSION OF $YbZn_2$ INTERMETALLIC COMPOUND

[Temperature, T, K; Linear Expansion, $\Delta L/L_0$, %]

T	$\Delta L/L_0$	T	$\Delta L/L_0$
CURVE 1		CURVE 2 (cont.)	
298	0.011*	473	0.300
333	0.109	508	0.328
368	0.120	543	0.464
413	0.186	578	0.526
438	0.230		
473	0.295	CURVE 3	
508	0.295		
543	0.383	298	0.011*
578	0.404	333	0.083*
		368	0.176
CURVE 2		413	0.407
		438	0.420
298	0.014*	473	0.539
333	0.082	508	0.572
368	0.123*	543	0.704
413	0.205	578	0.810
438	0.205		

* Not shown in figure.

SPECIFICATION TABLE 134. THERMAL LINEAR EXPANSION OF Zn_2Zr INTERMETALLIC COMPOUND

Cur. No.	Ref. No.	Author(s)	Year	Method Used	Temp. Range, K	Name and Specimen Designation	Composition (weight percent), Specifications, and Remarks
1*	592	Ogawa, S. and Kasai, N.	1969	E	4-45		Polycrystal bar; cylindrical shape of 14.95 mm in length, 13.3 mm in diameter; prepared by sintering method; curie temperature Tc = 18 K, Debye temperature θ_D = 371 ± 5 K; this curve is here reported using the first given temperature as reference temperature at which $\Delta L/L_0 = 0$.
2*	593	Meincke, P.P.M., Fawcett, E., and Knapp, G.S.	1969	L	2-7		Single crystal, dimension about 5 x 4 x 2 mm; this curve is here reported using the first given temperature as reference temperature at which $\Delta L/L_0 = 0$.

DATA TABLE 134. THERMAL LINEAR EXPANSION OF Zn_2Zr INTERMETALLIC COMPOUND

[Temperature, T, K; Linear Expansion, $\Delta L/L_0$, %]

T	$\Delta L/L_0$	T	$\Delta L/L_0$	T	$\Delta L/L_0$	T	$\Delta L/L_0$
CURVE 1*, †, ‡		CURVE 1 (cont.)*, †, ‡		CURVE 2*, †		CURVE 2 (cont.)*, †	
4.0	0.000 x 10^{-3}	19.8	-0.386 x 10^{-3}	1.64	0.000 x 10^{-4}	5.46	-0.966 x 10^{-4}
6.0	-0.043	21.9	-0.271	1.87	-0.040	5.97	-1.179
7.9	-0.099	23.9	-0.114	2.37	-0.112	6.49	-1.418
9.9	-0.175	25.9	0.092	2.88	-0.211	7.00	-1.173
12.0	-0.268	29.7	0.629	3.42	-0.320		
14.0	-0.361	34.9	1.685	3.91	-0.423		
16.0	-0.431	39.8	3.091	4.44	-0.574		
17.9	-0.440	44.9	5.039	4.96	-0.727		

DATA TABLE 134. COEFFICIENT OF THERMAL LINEAR EXPANSION OF Zn_2Zr INTERMETALLIC COMPOUND

[Temperature, T, K; Coefficient of Expansion, α, 10^{-6} K^{-1}]

T	α	T	α
CURVE 1*, ‡		CURVE 1 (cont.)*, †	
4.0	-0.18	21.9	0.67
6.0	-0.25	23.9	0.90
7.9	-0.34	25.9	1.16
9.9	-0.42	29.7	1.67
12.0	-0.47	34.9	2.39
14.0	-0.45	39.8	3.35
16.0	-0.25	44.9	4.29
17.9	0.15		
19.8	0.42		

*No figure given.
†This curve is here reported using the first given temperature as reference temperature at which $\Delta L/L_0 = 0$.
‡Author's data for coefficient of thermal expansion have been integrated by TPRC to obtain $\Delta L/L_0$.

3. BINARY ALLOY SYSTEMS

FIGURE AND TABLE NO. 135R. PROVISIONAL VALUES FOR THERMAL LINEAR EXPANSION OF ALUMINUM-BERYLLIUM SYSTEM Al-Be

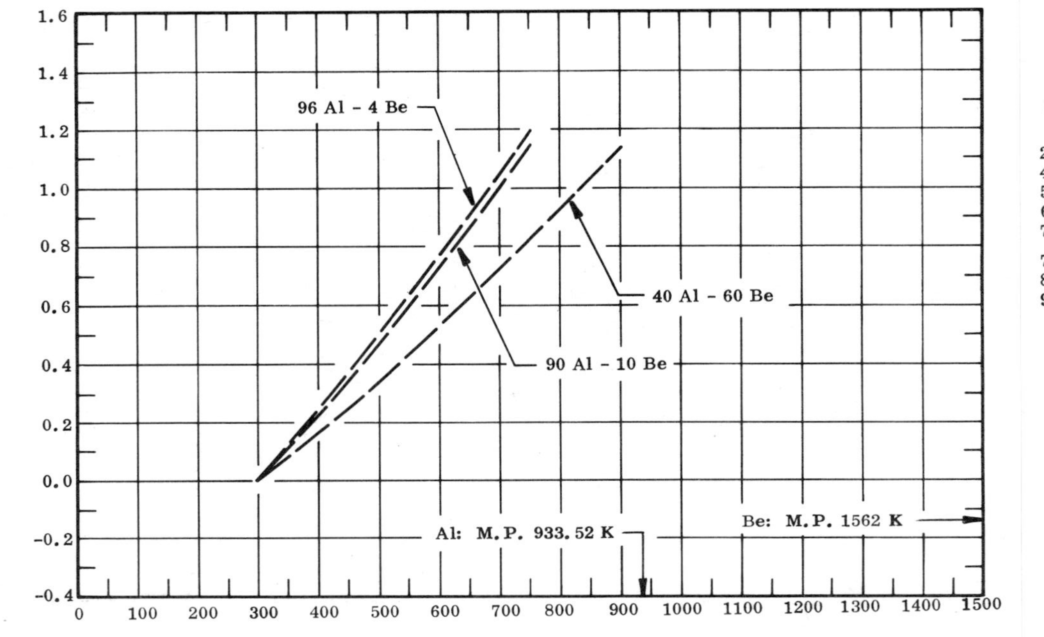

PROVISIONAL VALUES

[Temperature, T, K; Linear Expansion, $\Delta L/L_0$, %; α, K^{-1}]

	(40 Al-60 Be)		(90 Al-10 Be)		(96 Al-4 Be)	
T	$\Delta L/L_0$	$\alpha \times 10^6$	$\Delta L/L_0$	$\alpha \times 10^6$	$\Delta L/L_0$	$\alpha \times 10^6$
293	0.000	14.9	0.000	20.6	0.000	23.3
400	0.170	16.8	0.235	23.4	0.256	24.6
500	0.347	18.4	0.480	25.3	0.509	26.0
600	0.538	19.4	0.739	26.4	0.774	27.2
700	0.734	19.9	1.008	27.3	1.054	28.6
750	0.834	20.2	1.146	27.7	1.199	29.4
800	0.935	20.2				
900	1.139	20.5				

REMARKS

(40 Al-60 Be): The tabulated values for well-annealed alloy are considered accurate to within ± 7% over the entire temperature range. These values can be represented approximately by the following equation:

$$\Delta L/L_0 = -0.332 + 7.237 \times 10^{-4}\ T + 1.600 \times 10^{-6}\ T^2 - 6.537 \times 10^{-10}\ T^3$$

(90 Al-10 Be): The tabulated values for well-annealed alloy are considered accurate to within ± 7% over the entire temperature range. These values can be represented approximately by the following equation:

$$\Delta L/L_0 = -0.455 + 9.442 \times 10^{-4}\ T + 2.376 \times 10^{-6}\ T^2 - 1.055 \times 10^{-9}\ T^3$$

(96 Al-4 Be): The tabulated values for well-annealed alloy are considered accurate to within ± 7% over the entire temperature range. These values can be represented approximately by the following equation:

$$\Delta L/L_0 = -0.638 + 2.042 \times 10^{-3}\ T + 4.183 \times 10^{-7}\ T^2 + 1.658 \times 10^{-10}\ T^3$$

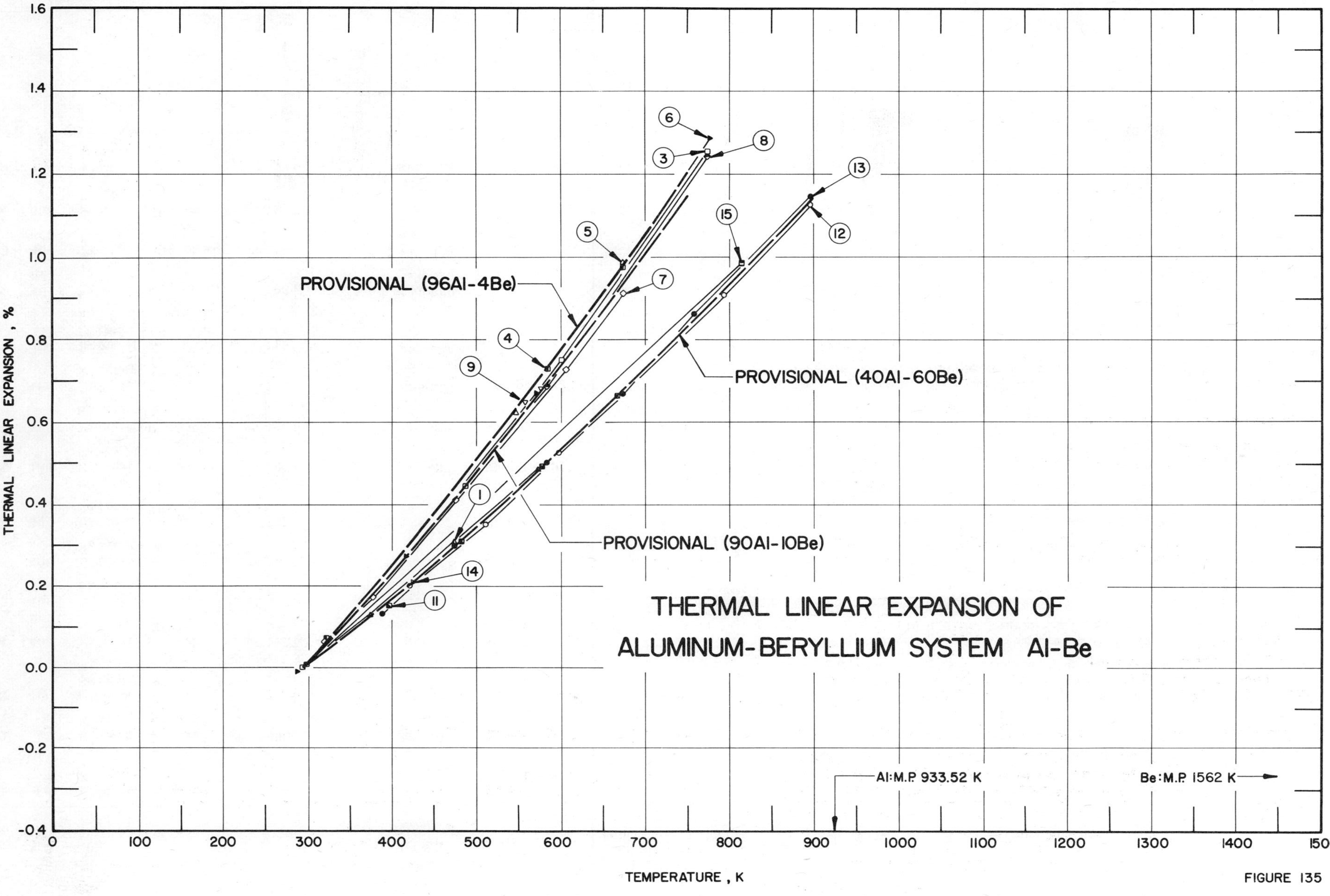
THERMAL LINEAR EXPANSION OF
ALUMINUM-BERYLLIUM SYSTEM Al-Be
PROVISIONAL (96Al-4Be)
PROVISIONAL (40Al-60Be)
PROVISIONAL (90Al-10Be)
Al: M.P. 933.52 K
Be: M.P. 1562 K
THERMAL LINEAR EXPANSION, %
TEMPERATURE, K
1.6
1.4
1.2
1.0
0.8
0.6
0.4
0.2
0.0
-0.2
-0.4
0
100
200
300
400
500
600
700
800
900
1000
1100
1200
1300
1400
1500
1
3
4
5
6
7
8
9
11
12
13
14
15

FIGURE 135

SPECIFICATION TABLE 135. THERMAL LINEAR EXPANSION OF ALUMINUM-BERYLLIUM SYSTEM Al-Be

Cur. No.	Ref. No.	Author(s)	Year	Method Used	Temp. Range, K	Name and Specimen Designation	Composition (weight percent) Al	Be	Composition (continued), Specifications, and Remarks
1	19	Hidnert, P. and Krider, H. S.	1952	L	293-573	Sample 1657	60	40	Specimen cast at 1533 K, heat treated 24 hrs at 825 K; hot forged to 1.91 cm dia, annealed 4 hrs at 825 K; quenched in cold water, cold worked to 1.59 cm dia; expansion measured with increasing temp.
2*	19	Hidnert, P. and Krider, H. S.	1952	L	293-573	Sample 1657	60	40	The above specimen; expansion measured on second test with increasing temp.
3	400	Hidnert, P. and Sweeney, W. T.	1927	L	293-772	Sample 1228	95.7	4.2	0.12 Fe, 0.10 Si, 0.10 Cu, 0.01 Mn; density 2.416 g cm^{-3} at 293 K; about 300 mm long, 10 mm dia, furnished by Beryllium Corp. of America, Cleveland, Ohio; expansion measured on first test with increasing temp.
4	400	Hidnert, P. and Sweeney, W. T.	1927	L	672-297	Sample 1228	95.7	4.2	The above specimen; expansion measured with decreasing temp; zero-point correction of 0.012% determined by graphical extrapolation.
5	400	Hidnert, P. and Sweeney, W. T.	1927	L	296-675	Sample 1228	95.7	4.2	The above specimen; expansion measured on second test with increasing temp; zero-point correction of 0.193% determined by graphical extrapolation.
6	400	Hidnert, P. and Sweeney, W. T.	1927	L	776-288	Sample 1228	95.7	4.2	The above specimen; expansion measured with decreasing temp; zero-point correction of -0.192% determined by graphical interpolation.
7	400	Hidnert, P. and Sweeney, W. T.	1927	L	293-673	Sample 1227	89.7	10.1	0.19 Fe, 0.12 Si, 0.09 Cu, 0.01 Mn, density 2.401 g cm^{-3} at 293 K; obtained from same source as the above specimen; expansion on first test with increasing temp.
8	400	Hidnert, P. and Sweeney, W. T.	1927	L	774-292	Sample 1227	89.7	10.1	The above specimen; expansion measured with decreasing temp; zero-point correction of -0.023% determined by graphical interpolation.
9	400	Hidnert, P. and Sweeney, W. T.	1927	L	293-578	Sample 1227	89.7	10.1	The above specimen; expansion measured on second test with increasing temp; zero-point correction is -0.200%.
10*	400	Hidnert, P. and Sweeney, W. T.	1927	L	778-294	Sample 1227	89.7	10.1	The above specimen; expansion measured with decreasing temp; zero-point correction of -0.205% determined by graphical extrapolation.
11	404 645	Fenn, R. W., Crooks, D. D., Coons, W. C., and Underwood, E. E.	1964	I	396-666		42	58	No details given.
12	404 645	Fenn, R. W., et al.	1964	I	420-896		36	64	No details given.
13	404 645	Fenn, R. W., et al.	1964	I	388-895		33	67	No details given.
14	608	Pinto, N. P. and Burke, E. C.	1971		298,423		38	62	No details given.
15	608	Pinto, N. P. and Burke, E. C.	1971		298,813		38	62	No details given.

* Not shown on figure.

DATA TABLE 135. THERMAL LINEAR EXPANSION OF ALUMINUM-BERYLLIUM SYSTEM Al-Be

[Temperature, T, K; Linear Expansion, $\Delta L/L_0$, %]

T	$\Delta L/L_0$
CURVE 1	
293	0.000*
373	0.133
473	0.311
573	0.487
CURVE 2*	
293	0.000
373	0.133
473	0.308
573	0.498
CURVE 3	
293	0.000
379	0.191
487	0.446
600	0.754
772	1.257
CURVE 4	
672	0.979
585	0.734
404	0.259
321	0.078
297	0.012
CURVE 5	
296	0.007*
366	0.161
546	0.622
588	0.727*
675	0.984
CURVE 6	
776	1.287
571	0.673
413	0.271
288	-0.010
CURVE 7	
293	0.000*
CURVE 7 (cont.)	
320	0.066
376	0.179
476	0.415
606	0.736
673	0.919
CURVE 8	
774	1.243
583	0.691
400	0.244
292	-0.002*
CURVE 9	
293	0.000*
361	0.151
559	0.645
578	0.683
CURVE 10*	
778	1.250
675	0.927
540	0.592
379	0.187
294	0.002
CURVE 11	
396	0.156
482	0.311
577	0.494
666	0.665
CURVE 12	
420	0.201
510	0.352
598	0.525
794	0.914
896	1.128
CURVE 13	
388	0.136
CURVE 13 (cont.)	
473	0.304
582	0.505
671	0.670
758	0.867
895	1.147
CURVE 14	
298	0.0083*
423	0.216
CURVE 15	
298	0.00945*
813	0.983

* Not shown in figure.

FIGURE AND TABLE NO. 136R. PROVISIONAL VALUES FOR THERMAL LINEAR EXPANSION OF ALUMINUM-COPPER SYSTEM Al-Cu

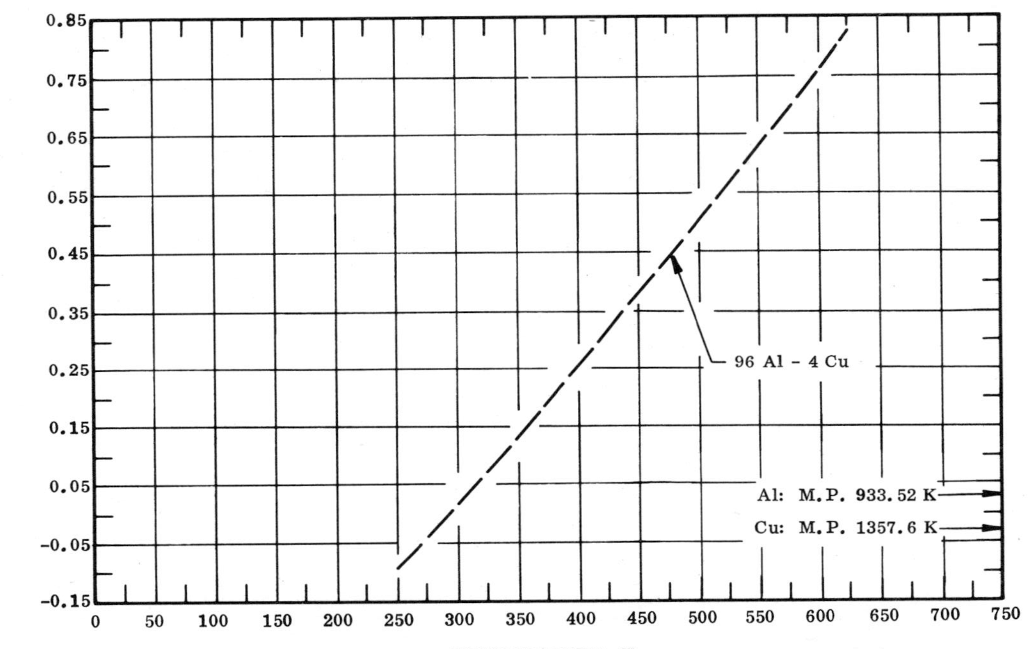

PROVISIONAL VALUES

[Temperature, T, K; Linear Expansion, $\Delta L/L_0$, %; α, K^{-1}]

T	(96 Al-4 Cu) $\Delta L/L_0$	$\alpha \times 10^6$
250	-0.095	22.0
293	0.000	22.8
400	0.254	24.6
500	0.503	25.4
600	0.760	25.9
625	0.827	26.2

REMARKS

(96 Al-4 Cu): The tabulated values for well-annealed alloy are considered accurate to within ± 7% over the entire temperature range. These values can be represented approximately by the following equation:

$$\Delta L/L_0 = -0.575 + 1.594 \times 10^{-3} T + 1.471 \times 10^{-6} T^2 - 6.928 \times 10^{-10} T^3$$

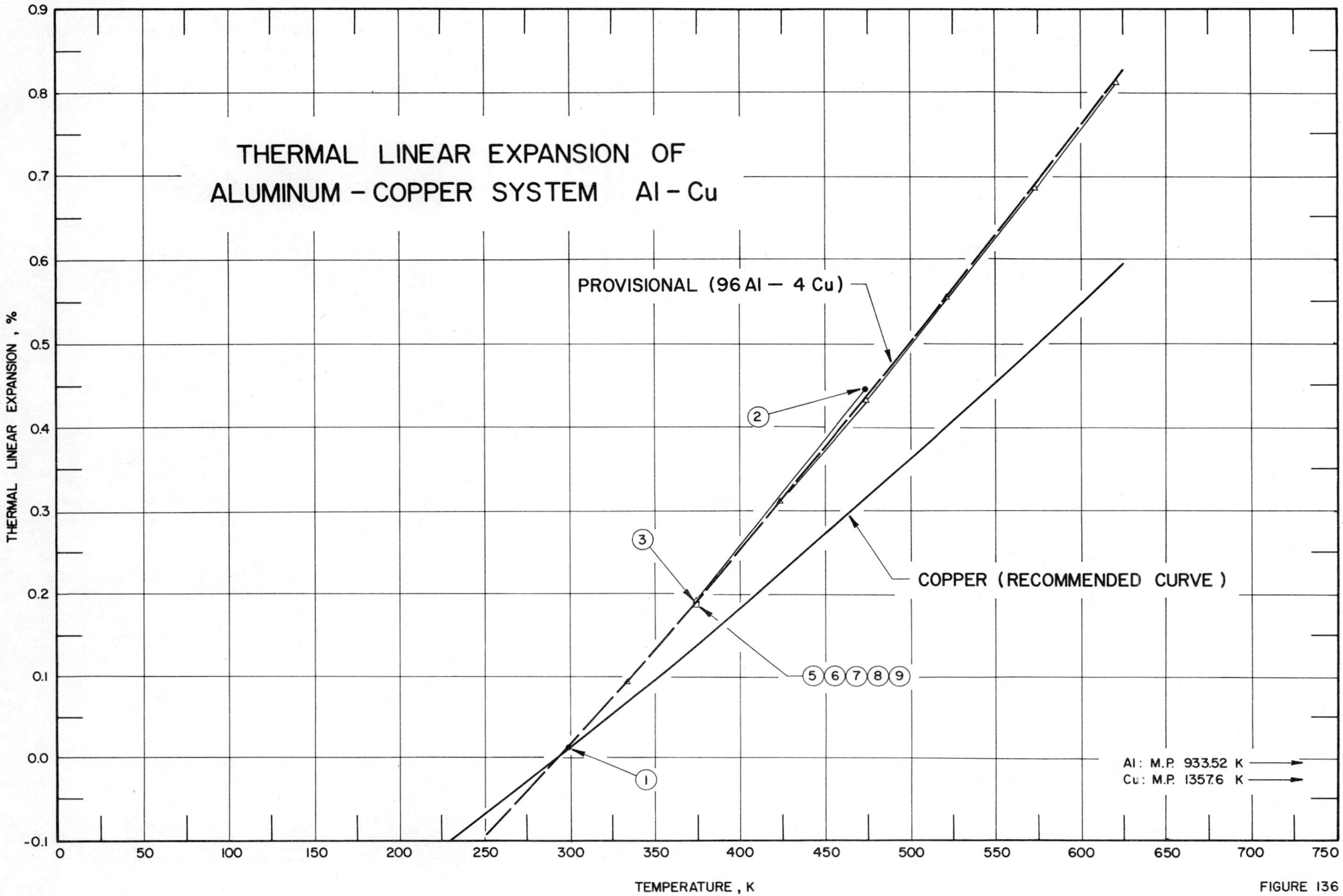
THERMAL LINEAR EXPANSION OF
ALUMINUM - COPPER SYSTEM Al - Cu
PROVISIONAL (96 Al — 4 Cu)
COPPER (RECOMMENDED CURVE)
1
2
3
5 6 7 8 9
Al: M.P. 933.52 K
Cu: M.P. 1357.6 K
THERMAL LINEAR EXPANSION, %
0.9
0.8
0.7
0.6
0.5
0.4
0.3
0.2
0.1
0.0
-0.1
0
50
100
150
200
250
300
350
400
450
500
550
600
650
700
750
TEMPERATURE, K

FIGURE 136

SPECIFICATION TABLE 136. THERMAL LINEAR EXPANSION OF ALUMINUM-COPPER SYSTEM Al-Cu

Cur. No.	Ref. No.	Author(s)	Year	Method Used	Temp. Range, K	Name and Specimen Designation	Composition (weight percent) Al	Cu	Composition (continued), Specifications, and Remarks
1	401	Hume-Rothery, W. and Boultbee, T. H.	1949	X	223-298		96.27	3.73	Prepared from metals of highest purity; Al supplied by British Aluminum Co., Ltd.; expansion determined in several experiments; lattice parameter reported at 223 K is 4.02670 kX (4.03475 Å); 4.03305 kX (4.04111 Å) at 293 K determined by graphical interpolation.
2	401	Hume-Rothery, W. and Boultbee, T. H.	1949	X	298-473		96.27	3.73	Similar to the above specimen; lattice parameter reported at 298 K is 4.03347 kX (4.04153 Å); 4.03305 kX (4.04111 Å) at 293 K determined by graphical extrapolation.
3	35	Honda, K. and Okubo, Y.	1924	L	333-621	L8	88.0	12.0	Cylindrical rod specimen 15.005 cm long, 5 mm diameter.
4*	35	Honda, K. and Okubo, Y.	1924	L	334-622	No. 12	92.0	8.0	Cylindrical rod specimen 14.995 cm long, 5 mm diameter.
5	603	McCollough, E. E.	1931	I	298, 373		97.70	2.30	Specimen prepared by placing required amount of aluminum in graphite crucible and melting in electric furnace previously heated to about 1073 K; after aluminum melted required amount of alloying material added and heating continued for about 10 min.
6	603	McCollough, E. E.	1931	I	298, 373		97.07	2.93	Similar to the above specimen.
7	603	McCollough, E. E.	1931	I	298, 373		95.73	4.27	Similar to the above specimen.
8	603	McCollough, E. E.	1931	I	298, 373		93.90	6.10	Similar to the above specimen.
9	603	McCollough, E. E.	1931	I	298, 373		93.29	6.71	Similar to the above specimen.
10*	603	McCollough, E. E.	1931	I	298, 373		93.0	7.0	Similar to the above specimen.
11*	603	McCollough, E. E.	1931	I	298, 373		92.63	7.37	Similar to the above specimen.
12*	603	McCollough, E. E.	1931	I	298, 373		92.95	7.05	Similar to the above specimen.
13*	603	McCollough, E. E.	1931	I	298, 373		91.13	8.87	Similar to the above specimen.
14*	603	McCollough, E. E.	1931	I	298, 373		91.60	9.40	Similar to the above specimen.
15*	647	Ellwood, E.C. and Silcock, J.M.	1948	X	652-862		99.05	0.95	Alloy prepared by fusing 99.9985 pure Al from British Aluminum Co. Ltd. and 99.9 electrolytically pure cathode copper; homogenized at 819 K for 2 weeks and quenched in water and split across diameter; lattice parameter reported at 652 K is 4.0850 kX (4.0933 Å); this curve is here reported using the first given temperature as reference temperature at which $\Delta L/L_0 = 0$.
16*	647	Ellwood, E.C. and Silcock, J.M.	1948	X	711-858		98.05	1.95	Similar to the above specimen; homogenized at 819 K for 3 weeks; lattice parameter reported at 711 K is 4.0906 kX (4.0989 Å); this curve is here reported using the first given temperature as reference temperature at which $\Delta L/L_0 = 0$.
17*	647	Ellwood, E.C. and Silcock, J.M.	1948	X	756-853		97.02	2.98	Similar to the above specimen; homogenized at 819 K for 4 weeks; lattice parameter reported at 756 K is 4.0941 kX (4.1024 Å); this curve is here reported using the first given temperature as reference temperature at which $\Delta L/L_0 = 0$.
18*	647	Ellwood, E.C. and Silcock, J.M.	1948	X	784-833		96.04	3.96	Similar to the above specimen; homogenized at 819 K for 4 weeks; lattice parameter reported at 784 K is 4.0954 kX (4.1037 Å); this curve is here reported using the first given temperature as reference temperature at which $\Delta L/L_0 = 0$.

* Not shown in figure.

SPECIFICATION TABLE 136. THERMAL LINEAR EXPANSION OF ALUMINUM-COPPER SYSTEM Al-Cu (continued)

Cur. No.	Ref. No.	Author(s)	Year	Method Used	Temp. Range, K	Name and Specimen Designation	Composition (weight percent) Al	Cu	Composition (continued), Specifications, and Remarks
19*	647	Ellwood, E.C. and Silcock, J.M.	1948	X	811-828		95.03	4.97	Similar to the above specimen; homogenized at 819 K for 4 weeks; lattice parameter reported at 811 K is 4.0970 kX (4.1053 Å); this curve is here reported using the first given temperature as reference temperature at which $\Delta L/L_0 = 0$.

* Not shown in figure.

DATA TABLE 136. THERMAL LINEAR EXPANSION OF ALUMINUM-COPPER SYSTEM Al-Cu

[Temperature, T, K; Linear Expansion, $\Delta L/L_0$, %]

T	$\Delta L/L_0$
CURVE 1	
223	-0.157
223	-0.156*
298	0.013
CURVE 2	
298	0.012*
473	0.448
473	0.450*
CURVE 3‡	
333	0.095
373	0.191
423	0.312
473	0.434
521	0.553
573	0.687
621	0.813
CURVE 4*, ‡	
334	0.099
373	0.192
422	0.313
472	0.437
521	0.559
573	0.690
622	0.815
CURVE 5	
298	0.012
373	0.189
CURVE 6	
298	0.012
373	0.189
CURVE 7	
298	0.012
373	0.190
CURVE 8	
298	0.012
373	0.190
CURVE 9	
298	0.012
373	0.192
CURVE 10*	
298	0.012
373	0.194
CURVE 11*	
298	0.012
373	0.195
CURVE 12*	
298	0.023
373	0.197
CURVE 13*	
298	0.012
373	0.186
CURVE 14*	
298	0.011
373	0.181
CURVE 15*, †	
652	0.000
664	0.416
724	0.223
756	0.328
797	0.460
827	0.567
862	0.680
CURVE 16*, †	
711	0.000
742	0.097
783	0.222
815	0.327
842	0.425
858	0.464
CURVE 17*, †	
756	0.000
778	0.061
799	0.129
824	0.217
853	0.303
CURVE 18*, †	
784	0.000
806	0.073
833	0.153
CURVE 19*, †	
811	0.000
822	0.029
828	0.048

DATA TABLE 136. COEFFICIENT OF THERMAL LINEAR EXPANSION OF ALUMINUM-COPPER SYSTEM Al-Cu

[Temperature, T, K; Coefficient of Expansion, α, $10^{-6}\ K^{-1}$]

T	α
CURVE 3*, ‡	
333	23.5
373	24.2
423	24.3
473	24.4
521	25.4
573	26.2
621	26.3
CURVE 4*, ‡	
334	23.5
373	24.6
422	24.7
472	24.8
521	25.0
573	25.4
622	25.5

* Not shown in figure.

† This curve is here reported using the first given temperature as reference temperature at which $\Delta L/L_0 = 0$.

‡ Author's data on coefficient of thermal expansion have been integrated by TPRC to obtain $\Delta L/L_0$.

FIGURE AND TABLE NO. 137R. PROVISIONAL VALUES FOR THERMAL LINEAR EXPANSION OF ALUMINUM-IRON SYSTEM Al-Fe

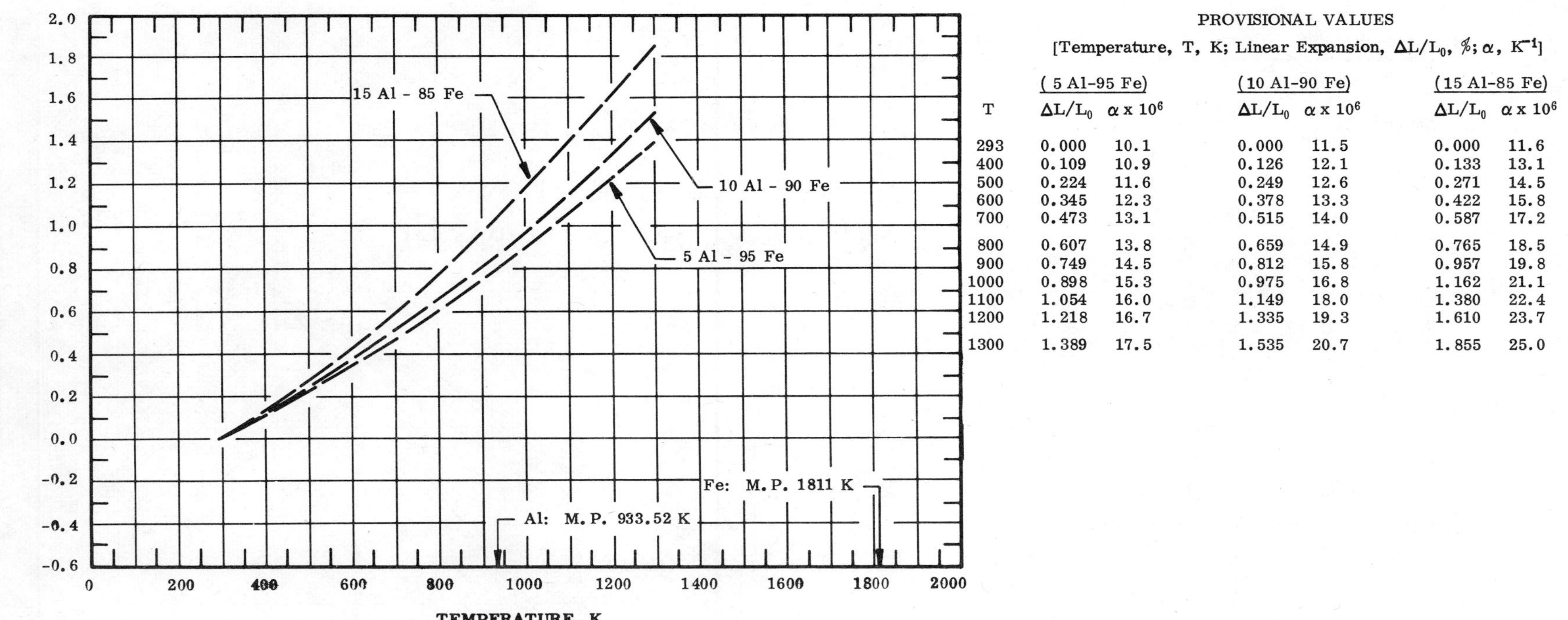

PROVISIONAL VALUES

[Temperature, T, K; Linear Expansion, $\Delta L/L_0$, %; α, K^{-1}]

T	(5 Al-95 Fe) $\Delta L/L_0$	(5 Al-95 Fe) $\alpha \times 10^6$	(10 Al-90 Fe) $\Delta L/L_0$	(10 Al-90 Fe) $\alpha \times 10^6$	(15 Al-85 Fe) $\Delta L/L_0$	(15 Al-85 Fe) $\alpha \times 10^6$
293	0.000	10.1	0.000	11.5	0.000	11.6
400	0.109	10.9	0.126	12.1	0.133	13.1
500	0.224	11.6	0.249	12.6	0.271	14.5
600	0.345	12.3	0.378	13.3	0.422	15.8
700	0.473	13.1	0.515	14.0	0.587	17.2
800	0.607	13.8	0.659	14.9	0.765	18.5
900	0.749	14.5	0.812	15.8	0.957	19.8
1000	0.898	15.3	0.975	16.8	1.162	21.1
1100	1.054	16.0	1.149	18.0	1.380	22.4
1200	1.218	16.7	1.335	19.3	1.610	23.7
1300	1.389	17.5	1.535	20.7	1.855	25.0

REMARKS

(5 Al-95 Fe): The tabulated values for well-annealed alloy are considered accurate to within ± 7% over the entire temperature range. These values can be represented approximately by the following equation:

$$\Delta L/L_0 = -0.264 + 7.910 \times 10^{-4} T + 3.734 \times 10^{-7} T^2 - 3.016 \times 10^{-12} T^3$$

(10 Al-90 Fe): The tabulated values for well-annealed alloy are considered accurate to within ± 7% over the entire temperature range. These values can be represented approximately by the following equation:

$$\Delta L/L_0 = -0.324 + 1.079 \times 10^{-3} T + 5.550 \times 10^{-8} T^2 - 1.650 \times 10^{-10} T^3$$

(15 Al-85 Fe): The tabulated values for well-annealed alloy are considered accurate to within ± 7% over the entire temperature range. These values can be represented approximately by the following equation:

$$\Delta L/L_0 = -0.282 + 7.590 \times 10^{-4} T + 7.002 \times 10^{-7} T^2 - 1.513 \times 10^{-11} T^3$$

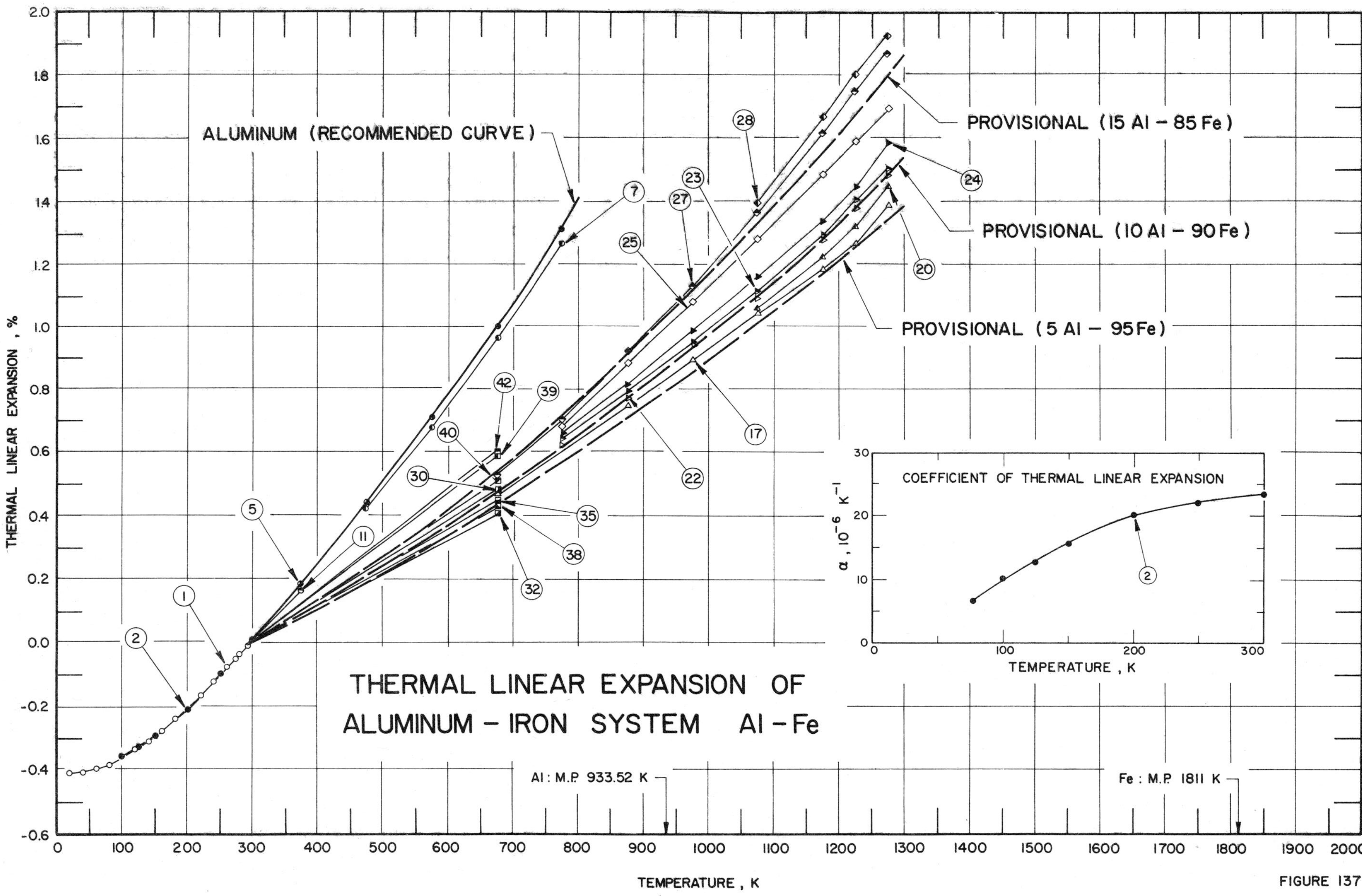
THERMAL LINEAR EXPANSION OF
ALUMINUM - IRON SYSTEM Al - Fe
ALUMINUM (RECOMMENDED CURVE)
PROVISIONAL (15 Al - 85 Fe)
PROVISIONAL (10 Al - 90 Fe)
PROVISIONAL (5 Al - 95 Fe)
COEFFICIENT OF THERMAL LINEAR EXPANSION
α, 10⁻⁶ K⁻¹
TEMPERATURE, K
THERMAL LINEAR EXPANSION, %
Al : M.P. 933.52 K
Fe : M.P. 1811 K
TEMPERATURE, K

FIGURE 137

SPECIFICATION TABLE 137. THERMAL LINEAR EXPANSION OF ALUMINUM-IRON SYSTEM Al-Fe

Cur. No.	Ref. No.	Author(s)	Year	Method Used	Temp. Range, K	Name and Specimen Designation	Composition (weight percent) Al	Fe	Composition (continued), Specifications, and Remarks
1	462	Arp, V., Wilson, J.H., Winrich, L., and Sikora, P.	1962	L	20-293	Al 1100	99.1	0.6	0.2 Cu and 0.1 Si; hardness R_H53, A.S.M. condition O.
2	549	Hordon, M.J., Lement, B.S., and Averbach, B.	1958	L	77-300	2S-O (Al 1100)	Bal.	0.45	0.12 Cu, 0.10 Si, and 0.01 Mn; unstressed specimen.
3*	549	Hordon, M.J., et al.	1958	L	77-300	2S-O (Al 1100)	Bal.	0.45	0.12 Cu, 0.10 Si, and 0.01 Mn; 4000 psi stress applied at 77 K; expansion measured along tensile axis.
4*	549	Hordon, M.J., et al.	1958	L	77-300	2S-O (Al 1100)	Bal.	0.45	0.12 Cu, 0.10 Si, and 0.01 Mn; 6800 psi stress applied at 77 K; expansion measured along tensile axis.
5	463	Willey, L.A. and Fink, W.L.	1945	I	293-773	No. 59410 Alcoa 2S (Al 1100)	Bal.	0.45	1.0 Fe + Si, 0.2 Cu, 0.1 Zn, 0.05 Mn; handbook composition; sheet wrought form, 0.064 in. thick; annealed to 623 K.
6*	463	Willey, L.A. and Fink, W.L.	1945	I	213-373	No. 59410 Alcoa 2S (Al 1100)			Similar to the above specimen.
7	463	Willey, L.A. and Fink, W.L.	1945	I	293-773	No. 59410 Alcoa 2S (Al 1100)			Similar to the above specimen except tested in as rolled condition.
8*	603	McCollough, E.E.	1931	I	298,373		97.50	2.50	Specimen furnished by Aluminum Company of America.
9*	603	McCollough, E.E.	1931	I	298,373		93.88	6.12	Similar to the above specimen.
10*	603	McCollough, E.E.	1931	I	298,373		91.18	8.82	Similar to the above specimen.
11	603	McCollough, E.E.	1931	I	298,373		89.22	10.78	Similar to the above specimen.
12*	523	Sykes, C. and Bampfylde, J.W.	1934		273-1273		0.80	99.20	99.92 pure Fe and 99.0 Al containing 0.53 Fe and 0.47 Si; used raw materials melted together; data taken from smoothed set of curves of α' vs composition.
13*	523	Sykes, C. and Bampfylde, J.W.	1934		273-1273		1.0	Bal.	Similar to the above specimen.
14*	523	Sykes, C. and Bampfylde, J.W.	1934		273-1273		2.0	Bal.	Similar to the above specimen.
15*	523	Sykes, C. and Bampfylde, J.W.	1934		273-1273		3.0	Bal.	Similar to the above specimen.
16*	523	Sykes, C. and Bampfylde, J.W.	1934		273-1273		4.0	Bal.	Similar to the above specimen.
17	523	Sykes, C. and Bampfylde, J.W.	1934		273-1273		5.0	Bal.	Similar to the above specimen.
18*	523	Sykes, C. and Bampfylde, J.W.	1934		273-1273		6.0	Bal.	Similar to the above specimen.
19*	523	Sykes, C. and Bampfylde, J.W.	1934		273-1273		7.0	Bal.	Similar to the above specimen.

* Not shown in figure.

SPECIFICATION TABLE 137. THERMAL LINEAR EXPANSION OF ALUMINUM-IRON SYSTEM Al-Fe (continued)

Cur. No.	Ref. No.	Author(s)	Year	Method Used	Temp. Range, K	Name and Specimen Designation	Composition (weight percent) Al	Fe	Composition (continued), Specifications, and Remarks
20	523	Sykes, C. and Bampfylde, J.W.	1934		273-1273		8.0	Bal.	Similar to the above specimen.
21*	523	Sykes, C. and Bampfylde, J.W.	1934		273-1273		9.0	Bal.	Similar to the above specimen.
22	523	Sykes, C. and Bampfylde, J.W.	1934		273-1273		10.0	Bal.	Similar to the above specimen.
23	523	Sykes, C. and Bampfylde, J.W.	1934		273-1273		11.0	Bal.	Similar to the above specimen.
24	523	Sykes, C. and Bampfylde, J.W.	1934		273-1273		12.0	Bal.	Similar to the above specimen.
25	523	Sykes, C. and Bampfylde, J.W.	1934		273-1273		13.0	Bal.	Similar to the above specimen.
26*	523	Sykes, C. and Bampfylde, J.W.	1934		273-1273		14.0	Bal.	Similar to the above specimen.
27	523	Sykes, C. and Bampfylde, J.W.	1934		273-1273		15.0	Bal.	Similar to the above specimen.
28	523	Sykes, C. and Bampfylde, J.W.	1934		273-1273		15.52	Bal.	Similar to the above specimen.
29*	764	Sinha, A.N. and Balasundaram, L.J.	1967	L	273-673		1.18	Bal.	Samples machined in a lathe or ground to required shape; complete details in earlier publication.
30	764	Sinha, A.N. and Balasundaram, L.J.	1967	L	273-673		3.01	Bal.	Similar to the above specimen.
31*	764	Sinha, A.N. and Balasundaram, L.J.	1967	L	273-673		3.5	Bal.	Similar to the above specimen.
32	764	Sinha, A.N. and Balasundaram, L.J.	1967	L	273-673		4.55	Bal.	Similar to the above specimen.
33*	764	Sinha, A.N. and Balasundaram, L.J.	1967	L	273-673		4.99	Bal.	Similar to the above specimen.
34*	764	Sinha, A.N. and Balasundaram, L.J.	1967	L	273-673		5.37	Bal.	Similar to the above specimen.
35	764	Sinha, A.N. and Balasundaram, L.J.	1967	L	273-673		8.86	Bal.	Similar to the above specimen.
36*	764	Sinha, A.N. and Balasundaram, L.J.	1967	L	273-673		9.43	Bal.	Similar to the above specimen.
37*	764	Sinha, A.N. and Balasundaram, L.J.	1967	L	273-673		10.0	Bal.	Similar to the above specimen.
38	764	Sinha, A.N. and Balasundaram, L.J.	1967	L	273-673		11.11	Bal.	Similar to the above specimen.
39	764	Sinha, A.N. and Balasundaram, L.J.	1967	L	273-673		14.09	Bal.	Similar to the above specimen.

* Not shown in figure.

SPECIFICATION TABLE 137. THERMAL LINEAR EXPANSION OF ALUMINUM-IRON SYSTEM Al-Fe (continued)

Cur. No.	Ref. No.	Author(s)	Year	Method Used	Temp. Range, K	Name and Specimen Designation	Composition (weight percent) Al	Fe	Composition (continued), Specifications, and Remarks
40	764	Sinha, A.N. and Balasundaram, L.J.	1967	L	273-673		14.55	Bal.	Similar to the above specimen.
41*	764	Sinha, A.N. and Balasundaram, L.J.	1967	L	273-673		15.17	Bal.	Similar to the above specimen.
42	764	Sinha, A.N. and Balasundaram, L.J.	1967	L	273-673		16.39	Bal.	Similar to the above specimen.

* Not shown in figure.

DATA TABLE 137. THERMAL LINEAR EXPANSION OF ALUMINUM-IRON SYSTEM Al-Fe

[Temperature, T, K; Linear Expansion, $\Delta L/L_0$, %]

T	$\Delta L/L_0$
CURVE 1	
20	-0.409
40	-0.406*
60	-0.397
80	-0.381*
100	-0.361*
120	-0.336*
140	-0.308
160	-0.275*
180	-0.239
200	-0.202*
220	-0.161
240	-0.119*
260	-0.075
273	-0.045*
280	-0.030
293	0.000
CURVE 2‡	
77	-0.378*
100	-0.358*
125	-0.329
150	-0.294
200	-0.204
250	-0.098
300	0.015*
CURVE 3*, ‡	
77	-0.383
100	-0.363
125	-0.333
150	-0.297
200	-0.206
250	-0.100
300	0.016
CURVE 4*, ‡	
77	-0.376
100	-0.357
125	-0.328
150	-0.292
200	-0.202
250	-0.098
300	0.015
CURVE 5	
293	0.000*
373	0.189
473	0.443
573	0.717
673	1.007
773	1.320
CURVE 6*	
213	-0.174
293	0.000
373	0.187
CURVE 7	
293	0.000*
373	0.184*
473	0.430
573	0.680
673	0.969
773	1.277
CURVE 8*	
298	0.0117
373	0.187
CURVE 9*	
298	0.0114
373	0.182
CURVE 10*	
298	0.0110
373	0.176
CURVE 11	
298	0.0106
373	0.170
CURVE 12*	
273	-0.028
673	0.518
CURVE 12 (cont.)*	
773	0.658
873	0.803
973	0.955
1073	1.104
1173	1.256
1223	1.336
1273	1.453
CURVE 13*	
273	-0.027
673	0.513
773	0.634
873	0.797
973	0.949
1073	1.098
1173	1.251
1223	1.330
1273	1.441
CURVE 14*	
273	-0.026
673	0.494
773	0.633
873	0.772
973	0.928
1073	1.077
1173	1.227
1223	1.310
1273	1.413
CURVE 15*	
273	-0.026
673	0.487
773	0.622
873	0.759
973	0.914
1073	1.062
1173	1.209
1223	1.294
1273	1.393
CURVE 16*	
273	-0.025
673	0.481
773	0.618
873	0.754
973	0.906
1073	1.054
1173	1.199
1223	1.285
1273	1.391
CURVE 17	
273	-0.025*
673	0.474
773	0.616*
873	0.751
973	0.899
1073	1.049
1173	1.194
1223	1.278
1273	1.399
CURVE 18*	
273	-0.026
673	0.467
773	0.610
873	0.748
973	0.898
1073	1.049
1173	1.199
1223	1.285
1273	1.407
CURVE 19*	
273	-0.024
673	0.461
773	0.603
873	0.746
973	0.899
1073	1.054
1173	1.211
1223	1.305
1273	1.420
CURVE 20	
273	-0.024*
673	0.458*
773	0.600*
873	0.746*
973	0.906*
1073	1.066*
1173	1.230*
1223	1.329*
1273	1.439
CURVE 21*	
273	-0.025
673	0.466
773	0.609*
873	0.754*
973	0.924*
1073	1.083
1173	1.262
1223	1.354
1273	1.465
CURVE 22	
273	-0.025*
673	0.467*
773	0.626
873	0.773
973	0.941
1073	1.097
1173	1.283
1223	1.383
1273	1.490
CURVE 23	
273	-0.025*
673	0.469*
773	0.647
873	0.799
973	0.957
1073	1.124
1173	1.307
1223	1.411
1273	1.516
CURVE 24	
273	-0.025*
673	0.473*
773	0.663
873	0.820
973	0.990
1073	1.168
1173	1.349
1223	1.456
1273	1.594
CURVE 25	
273	-0.026*
673	0.500*
773	0.683
873	0.887
973	1.088
1073	1.290
1173	1.492
1223	1.597
1273	1.702
CURVE 26*	
273	-0.027*
673	0.515*
773	0.697*
873	0.906
973	1.115
1073	1.333
1173	1.557
1223	1.669
1273	1.781
CURVE 27	
273	-0.028*
673	0.527
773	0.708
873	0.927
973	1.140
1073	1.378
1173	1.626
1223	1.760
1273	1.878

* Not shown in figure.
‡ Author's data for coefficient of thermal expansion have been integrated by TPRC to obtain $\Delta L/L_0$.

DATA TABLE 137. THERMAL LINEAR EXPANSION OF ALUMINUM-IRON SYSTEM Al-Fe (continued)

T	$\Delta L/L_0$	T	$\Delta L/L_0$	T	$\Delta L/L_0$	T	$\Delta L/L_0$	T	$\Delta L/L_0$	T	$\Delta L/L_0$
CURVE 28		CURVE 30*		CURVE 33*		CURVE 36		CURVE 39		CURVE 42	
273	-0.028*	273	-0.025	273	-0.022	273	-0.023	273	-0.031*	273	-0.032
673	0.530*	293	0.000	293	0.000	293	0.000	293	0.000*	293	0.000
773	0.712*	673	0.481	673	0.423	673	0.433	673	0.589	673	0.601
873	0.936*										
973	1.152*	CURVE 31*		CURVE 34*		CURVE 37*		CURVE 40*			
1073	1.400										
1173	1.665	273	-0.024	273	-0.023	273	-0.024	273	-0.027		
1223	1.809	293	0.000	293	0.000	293	0.000	293	0.000		
1273	1.939	673	0.457	673	0.430	673	0.456	673	0.510		
CURVE 29*		CURVE 32		CURVE 35*		CURVE 38		CURVE 41*			
273	-0.025	273	-0.022	273	-0.024	273	-0.023	273	-0.028		
293	0.000	293	0.000*	293	0.000	293	0.000	293	0.000		
673	0.478	673	0.412	673	0.450	673	0.433	673	0.523		

DATA TABLE 137. COEFFICIENT OF THERMAL LINEAR EXPANSION OF ALUMINUM-IRON SYSTEM Al-Fe

[Temperature, T, K; Coefficient of Expansion, α, 10^{-6} K^{-1}]

T	α	T	α
CURVE 2‡		CURVE 4*,‡	
77	6.8	77	6.8
100	10.2	100	10.1
125	12.9	125	12.9
150	15.8	150	15.8
200	20.1	250	21.9
250	22.1	300	23.3
300	23.4		
CURVE 3*,‡			
77	7.0		
100	10.4		
125	13.1		
150	16.1		
200	20.1		
250	22.4		
300	23.9		

* Not shown in figure.
‡ Author's data for coefficient of thermal expansion have been integrated by TPRC to obtain $\Delta L/L_0$.

FIGURE AND TABLE NO. 138R. PROVISIONAL VALUES FOR THERMAL LINEAR EXPANSION OF ALUMINUM-MAGNESIUM SYSTEM Al-Mg

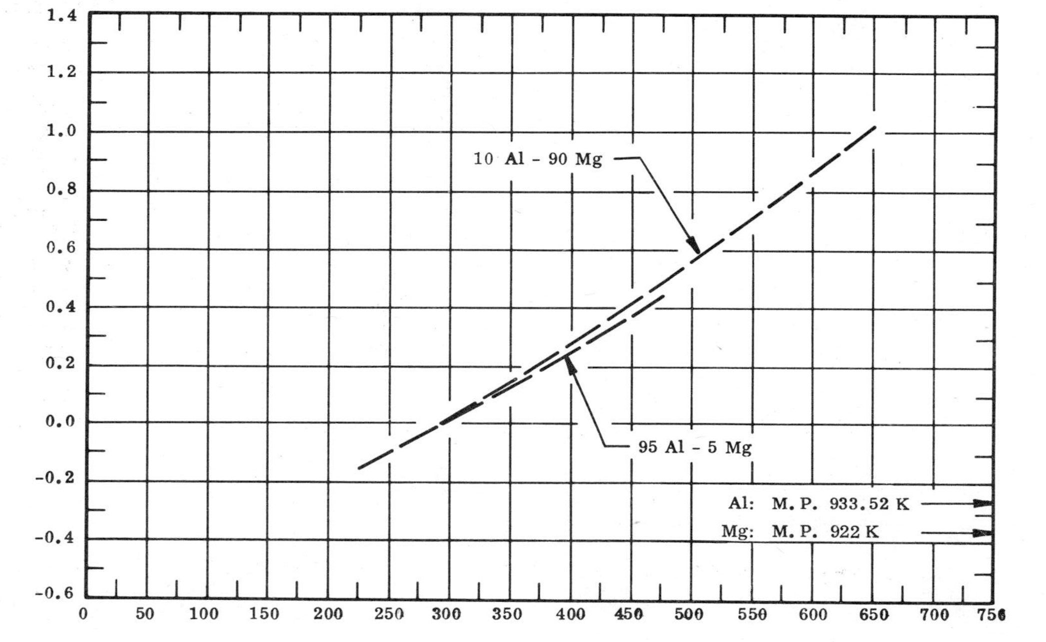

PROVISIONAL VALUES

[Temperature, T, K; Linear Expansion, $\Delta L/L_0$, %; α, K^{-1}]

T	(10 Al-90 Mg) $\Delta L/L_0$	(10 Al-90 Mg) $\alpha \times 10^6$	(95 Al-5 Mg) $\Delta L/L_0$	(95 Al-5 Mg) $\alpha \times 10^6$
225			-0.155	22.2
250			-0.098	22.4
293	0.000	23.6	0.000	22.8
400	0.276	27.7	0.252	24.4
475	0.490	29.5	0.446	26.9
500	0.566	30.1		
600	0.873	31.1		
650	1.028	31.3		

REMARKS

(10 Al-90 Mg): The tabulated values for cast alloy are considered accurate to within ± 7% over entire temperature range. The thermal linear expansion of this alloy is very close to that of pure magnesium. These values can be represented approximately by the following equation:

$$\Delta L/L_0 = -0.715 + 2.861 \times 10^{-3}\ T - 2.352 \times 10^{-6}\ T^2 + 3.107 \times 10^{-9}\ T^3$$

(95 Al-5 Mg): The tabulated values for cast alloy are considered accurate to within ± 7% over entire temperature range. The thermal linear expansion of this alloy is very close to that of pure aluminum. These values can be represented approximately by the following equation:

$$\Delta L/L_0 = -0.440 + 4.468 \times 10^{-4}\ T + 4.278 \times 10^{-6}\ T^2 - 2.293 \times 10^{-9}\ T^3$$

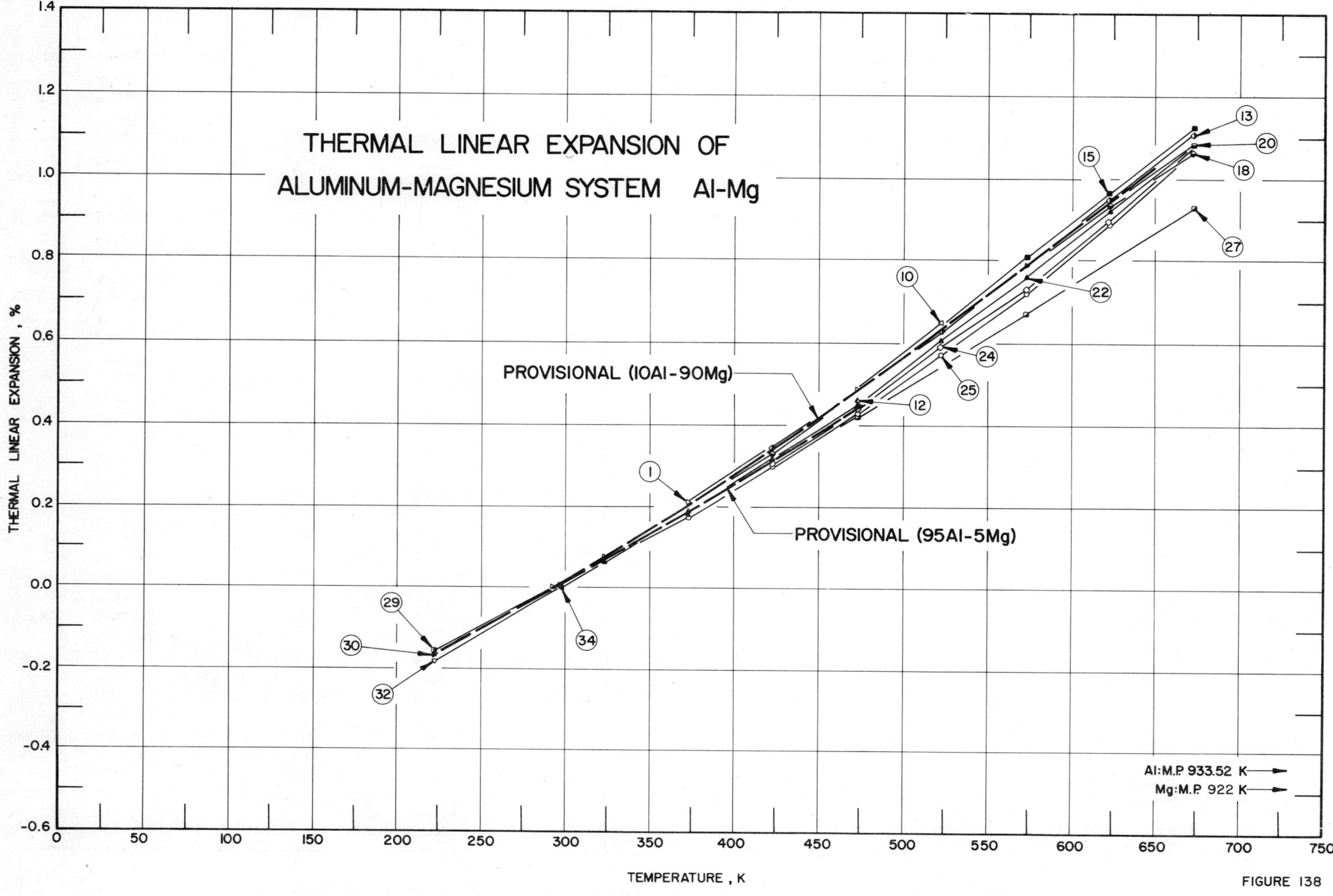

THERMAL LINEAR EXPANSION OF
ALUMINUM-MAGNESIUM SYSTEM Al-Mg
PROVISIONAL (10Al-90Mg)
PROVISIONAL (95Al-5Mg)
Al: M.P. 933.52 K
Mg: M.P. 922 K
13
20
18
15
27
10
22
24
25
12
1
29
30
34
32
THERMAL LINEAR EXPANSION, %
1.4
1.2
1.0
0.8
0.6
0.4
0.2
0.0
-0.2
-0.4
-0.6
0
50
100
150
200
250
300
350
400
450
500
550
600
650
700
750
TEMPERATURE, K

FIGURE 138

SPECIFICATION TABLE 138. THERMAL LINEAR EXPANSION OF ALUMINUM-MAGNESIUM SYSTEM Al-Mg

Cur. No.	Ref. No.	Author(s)	Year	Method Used	Temp. Range, K	Name and Specimen Designation	Composition (weight percent) Al	Mg	Composition (continued), Specifications, and Remarks
1	200	Hidnert, P. and Sweeney, W.T.	1928	T	293-573	Sample 1272	4.44		Specimen cast in vacuum furnace at 953 K.
2*	200	Hidnert, P. and Sweeney, W.T.	1928	T	293-573	Sample 1271	4.36		Specimen extruded at 623 K.
3*	200	Hidnert, P. and Sweeney, W.T.	1928	T	293-573	Sample 1274	6.22		Specimen cast in vacuum furnace at 953 K.
4*	200	Hidnert, P. and Sweeney, W.T.	1928	T	293-573	Sample 1273	6.26		Specimen extruded at 603 K.
5*	200	Hidnert, P. and Sweeney, W.T.	1928	T	293-573	Sample 1277	9.75		Specimen extruded at 603 K, held at 698 K for 8 hrs, quenched, and aged 72 hrs at 448 K.
6*	200	Hidnert, P. and Sweeney, W.T.	1928	T	293-573	Sample 1277	9.75		The above specimen, second test.
7*	200	Hidnert, P. and Sweeney, W.T.	1928	T	293-673	Sample 1277A	10.04		Similar to the above specimen.
8*	200	Hidnert, P. and Sweeney, W.T.	1928	T	293-673	Sample 1277A	10.04		The above specimen, second test.
9*	200	Hidnert, P. and Sweeney, W.T.	1928	T	293-573	Sample 1276	10.19		Specimen cast in vacuum furnace at 953 K.
10	200	Hidnert, P. and Sweeney, W.T.	1928	T	293-573	Sample 1275	10.35		Specimen extruded at 603 K.
11*	200	Hidnert, P. and Sweeney, W.T.	1928	T	293-573	Sample 1275	10.35		The above specimen, second test.
12	427	Souder, W., Hidnert, P. and Fox, J.F.	1934	N	293-573	1273A	6.26	93.70	0.026 Fe, 0.008 Si, 0.008 Cu, 150 mm long with pointed ends, max dia 12 mm, extruded at 603 K.
13	201	Takahasi, K. and Kikuti, R.	1936		273-673		2	Bal.	Rod specimen, about 16 cm long, 5 mm dia, heating rate 2 C/min.
14*	201	Takahasi, K. and Kikuti, R.	1936		273-673		4	Bal.	Similar to the above specimen.
15	201	Takahasi, K. and Kikuti, R.	1936		273-673		6	Bal.	Similar to the above specimen.
16*	201	Takahasi, K. and Kikuti, R.	1936		273-673		8	Bal.	Similar to the above specimen.
17*	201	Takahasi, K. and Kikuti, R.	1936		273-673		10	Bal.	Similar to the above specimen.
18	201	Takahasi, K. and Kikuti, R.	1936		273-673		12	Bal.	Similar to the above specimen.
19*	201	Takahasi, K. and Kikuti, R.	1936		273-673		14	Bal.	Similar to the above specimen.

*Not shown in figure.

SPECIFICATION TABLE 138. THERMAL LINEAR EXPANSION OF ALUMINUM–MAGNESIUM SYSTEM Al–Mg (continued)

Cur. No.	Ref. No.	Author(s)	Year	Method Used	Temp. Range, K	Name and Specimen Designation	Composition (weight percent) Al	Mg	Composition (continued), Specifications, and Remarks
20	201	Takahasi, K. and Kikuti, R.	1936		273-673		16	Bal.	Similar to the above specimen.
21*	201	Takahasi, K. and Kikuti, R.	1936		273-673		20	Bal.	Similar to the above specimen.
22	201	Takahasi, K. and Kikuti, R.	1936		273-673		25	Bal.	Similar to the above specimen.
23*	201	Takahasi, K. and Kikuti, R.	1936		273-673		30	Bal.	Similar to the above specimen.
24	201	Takahasi, K. and Kikuti, R.	1936		273-673		35	Bal.	Similar to the above specimen.
25	201	Takahasi, K. and Kikuti, R.	1936		273-673		40	Bal.	Similar to the above specimen.
26*	201	Takahasi, K. and Kikuti, R.	1936		273-673		43	Bal.	Similar to the above specimen.
27	199	Bell, I.P.	1954	L	293-673	Magnox C	0.9	99.1	
28*	252	Heal, J.T.	1958		293-673	Magnox A-12			Commercial specimen.
29	401	Hume-Rothery, W. and Boultbee, T.H.	1949	X	223-298		97.87	2.13	Prepared from metals of highest purity; Al supplied by British Aluminum Co., Ltd.; data combined from several experiments; lattice spacing reported at 223 K is 4.04275 kX (4.05083 Å); 4.04920 kX (4.05729 Å) at 293 K determined by graphical interpolation.
30	401	Hume-Rothery, W. and Boultbee, T.H.	1949	X	223-298		94.23	5.77	Similar to the above specimen; lattice spacing reported at 298 K is 4.0685 kX (4.07663 Å); 4.06802 kX (4.07615 Å) at 293 K determined by graphical interpolation.
31*	401	Hume-Rothery, W. and Boultbee, T.H.	1949	X	298-473		94.23	5.77	Similar to the above specimen; lattice spacing reported at 298 K is 4.06854 kX (4.07667 Å); 4.06805 kX (4.07618 Å) at 293 K determined by graphical extrapolation.
32	401	Hume-Rothery, W. and Boultbee, T.H.	1949	X	223-298		90.38	9.62	Similar to the above specimen; lattice spacing reported at 298 K is 4.08613 kX (4.09430 Å); 4.08560 kX (4.09377 Å) at 293 K determined by graphical interpolation.
33*	401	Hume-Rothery, W. and Boultbee, T.H.	1949	X	298-473		90.38	9.62	Similar to the above specimen; lattice spacing reported at 298 K is 4.08617 kX (4.09434 Å); 4.08567 kX (4.09384 Å) at 293 K determined by graphical extrapolation.
34	603	McCollough, E.E.	1931	I	297, 373		98.48	1.52	Specimen prepared by placing required amount of Al in graphite crucible and melting in electric furnace previously heated to about 1073 K; after having melted required amount of alloying material was added and heating continued for about 10 min.
35*	603	McCollough, E.E.	1931	I	297, 373		96.64	3.36	Similar to above.
36*	603	McCollough, E.E.	1931	I	297, 373		94.02	5.98	Similar to above.
37*	603	McCollough, E.E.	1931	I	297, 373		92.14	7.86	Similar to above.

*Not shown in figure.

SPECIFICATION TABLE 138. THERMAL LINEAR EXPANSION OF ALUMINUM-MAGNESIUM SYSTEM Al-Mg (continued)

Cur. No.	Ref. No.	Author(s)	Year	Method Used	Temp. Range, K	Name and Specimen Designation	Composition (weight percent) Al	Mg	Composition (continued), Specifications, and Remarks
38*	603	McCollough, E. E.	1931	I	297, 373		91.70	8.30	Similar to above.
39*	603	McCollough, E. E.	1931	I	297, 373		90.00	10.00	Similar to above.
40*	603	McCollough, E. E.	1931	I	297, 373		87.67	12.33	Similar to above.
41*	603	McCollough, E. E.	1931	I	297, 373		85.06	14.94	Similar to above.

* Not shown in figure.

DATA TABLE 138. THERMAL LINEAR EXPANSION OF ALUMINUM-MAGNESIUM SYSTEM Al-Mg

[Temperature, T, K; Linear Expansion, $\Delta L/L_0$, %]

T	$\Delta L/L_0$
CURVE 1	
293	0.000
323	0.0777
373	0.211
473	0.493
573	0.784
CURVE 2*	
293	0.000
323	0.0765
373	0.206
473	0.484
573	0.778
CURVE 3*	
293	0.000
323	0.0759
373	0.211
473	0.490
573	0.784
CURVE 4*	
293	0.000
323	0.0780
373	0.210
473	0.488
573	0.781
CURVE 5	
293	0.000*
323	0.0759*
373	0.203
473	0.477
573	0.621
CURVE 6	
293	0.000*
373	0.206*
473	0.500
573	0.798
CURVE 7	
293	0.000*
373	0.208*
473	0.470*
573	0.641
673	0.858
CURVE 8	
293	0.000*
373	0.206*
473	0.484*
573	0.710
673	0.979
CURVE 9*	
293	0.000
323	0.0759
373	0.210
473	0.491
573	0.787
CURVE 10	
293	0.000*
323	0.0747*
373	0.207*
473	0.490*
523	0.651
CURVE 11*	
293	0.000
323	0.0741
373	0.205
473	0.484
573	0.776
CURVE 12	
373	0.194
473	0.461
573	0.708*
CURVE 13	
273.2	0.000*
323.2	0.076*
373.2	0.208*
423.2	0.344
473.2	0.488*
523.2	0.635
573.2	0.791*
623.2	0.952
673.2	1.115
CURVE 14*	
273.2	0.000
323.2	0.074
373.2	0.205
423.2	0.341
473.2	0.486
523.2	0.637
573.2	0.798
623.2	0.959
673.2	1.120
CURVE 15	
273.2	0.000*
323.2	0.074*
373.2	0.205*
423.2	0.345*
473.2	0.491*
523.2	0.644*
573.2	0.802
623.2	0.965
673.2	1.127
CURVE 16*	
273.2	0.000
323.2	0.071
373.2	0.201
423.2	0.340
473.2	0.490
523.2	0.639
573.2	0.797
623.2	0.955
673.2	1.108
CURVE 17*	
273.2	0.000
323.2	0.071
373.2	0.200
423.2	0.339
473.2	0.485
523.2	0.640
573.2	0.798
623.2	0.946
673.2	1.092
CURVE 18	
273.2	0.000
323.2	0.071*
373.2	0.208*
423.2	0.335
473.2	0.479*
523.2	0.631*
573.2	0.787*
623.2	0.940
673.2	1.065
CURVE 19*	
273.2	0.000
323.2	0.073
373.2	0.200
423.2	0.336
473.2	0.480
523.2	0.631
573.2	0.787
623.2	0.939
673.2	1.079
CURVE 20	
273.2	0.000*
323.2	0.0745*
373.2	0.204*
423.2	0.339*
473.2	0.480*
523.2	0.632*
573.2	0.790*
623.2	0.941*
673.2	1.082
CURVE 21*	
273.2	0.000
323.2	0.071
373.2	0.199
423.2	0.333
473.2	0.475
523.2	0.625
573.2	0.782
623.2	0.940
673.2	1.081
CURVE 22	
273.2	0.000
323.2	0.069
373.2	0.192*
423.2	0.321
473.2	0.459
523.2	0.606
573.2	0.760
623.2	0.922
673.2	1.080*
CURVE 23*	
273.2	0.000
323.2	0.068
373.2	0.191
423.2	0.321
473.2	0.455
523.2	0.599
573.2	0.753
623.2	0.914
673.2	1.077
CURVE 24	
273.2	0.000*
323.2	0.066*
373.2	0.187*
423.2	0.313
473.2	0.448
523.2	0.591
573.2	0.734
623.2	0.900
673.2	1.073*
CURVE 25	
273	0.000*
323	0.063*
373	0.178
423	0.301
473	0.433
523	0.574
573	0.720
623	0.891
673	1.064*
CURVE 26*	
273.2	0.000
323.2	0.063
373.2	0.176
423.2	0.296
473.2	0.428
523.2	0.569
573.2	0.720
623.2	0.889
673.2	1.064
CURVE 27	
293	0.000*
373	0.184*
473	0.423
573	0.672
673	0.935
CURVE 28*	
293	0.000
373	0.203
473	0.472
573	0.756
673	1.060
CURVE 29	
223	-0.159
223	-0.159*
223	-0.159*
298	0.011

*Not shown in figure.

DATA TABLE 138. THERMAL LINEAR EXPANSION OF ALUMINUM-MAGNESIUM SYSTEM Al-Mg (continued)

T	$\Delta L/L_0$
CURVE 30	
223	-0.162
223	-0.162*
298	0.012*
CURVE 31*	
298	0.012
473	0.436
CURVE 32	
223	-0.180
298	0.013*
CURVE 33*	
298	0.012
473	0.445
473	0.441
CURVE 34	
297	0.009
373	0.182
CURVE 35*	
297	0.0092
373	0.184
CURVE 36*	
297	0.0094
373	0.187
CURVE 37*	
297	0.0095
373	0.190
CURVE 38*	
297	0.0096
373	0.191
CURVE 39*	
297	0.0097
373	0.194
CURVE 40*	
297	0.0098
373	0.195
CURVE 41*	
297	0.0098
373	0.196

*Not shown in figure.

SPECIFICATION TABLE 139. THERMAL LINEAR EXPANSION OF ALUMINUM-MANGANESE SYSTEM Al-Mn

Cur. No.	Ref. No.	Author(s)	Year	Method Used	Temp. Range, K	Name and Specimen Designation	Composition (weight percent) Al	Mn	Composition (continued), Specifications, and Remarks
1*	603	McCollough, E. E.	1931	I	298, 373		Bal.	0.52	Specimen furnished by Aluminum Co. of America.
2*	603	McCollough, E. E.	1931	I	298, 373		Bal.	7.12	Similar to the above specimen.
3*	603	McCollough, E. E.	1931	I	298, 373		Bal.	1.48	Similar to the above specimen.

DATA TABLE 139. THERMAL LINEAR EXPANSION OF ALUMINUM-MANGANESE SYSTEM Al-Mn

[Temperature, T, K; Linear Expansion, $\Delta L/L_0$, %]

T	$\Delta L/L_0$
CURVE 1*	
298	0.012
373	0.190
CURVE 2*	
298	0.012
373	0.190
CURVE 3*	
298	0.012
373	0.187

* No figure given.

FIGURE AND TABLE NO. 140R. PROVISIONAL VALUES FOR THERMAL LINEAR EXPANSION OF ALUMINUM-MOLYBDENUM SYSTEM Al-Mo

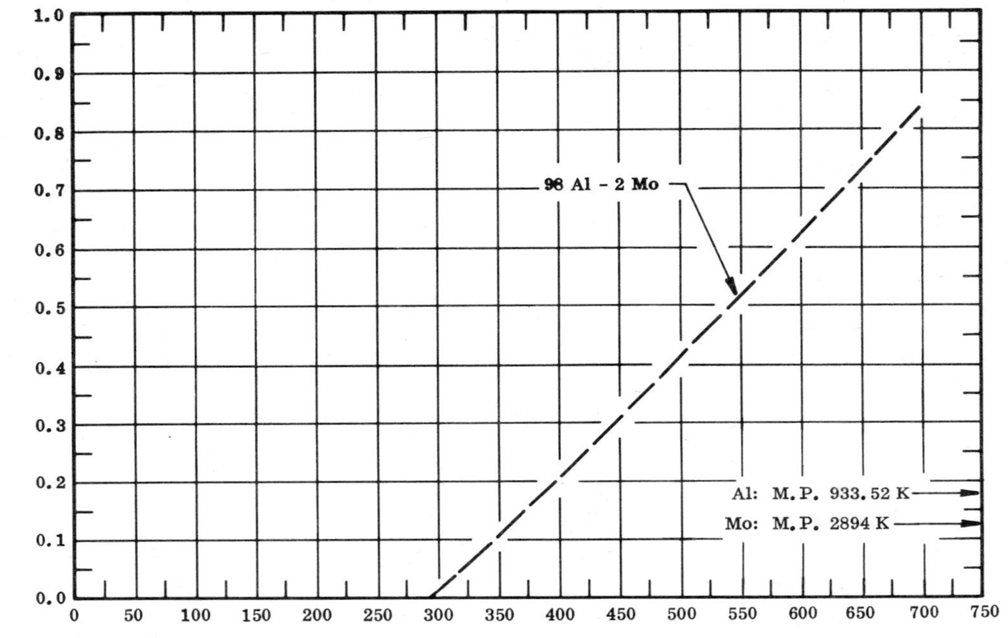

PROVISIONAL VALUES

[Temperature, T, K; Linear Expansion, $\Delta L/L_0$, %; α, K^{-1}]

T	(98 Al-2 Mo) $\Delta L/L_0$	$\alpha \times 10^6$
293	0.000	18.8
400	0.206	20.0
500	0.413	20.7
600	0.620	21.3
700	0.838	22.5

REMARKS

(98 Al-2 Mo): The tabulated values for well-annealed alloy are considered accurate to within ± 7% over the entire temperature range. These values can be represented approximately by the following equation:

$$\Delta L/L_0 = -0.516 + 1.597 \times 10^{-3}\, T + 6.003 \times 10^{-7}\, T^2 - 1.697 \times 10^{-10}\, T^3$$

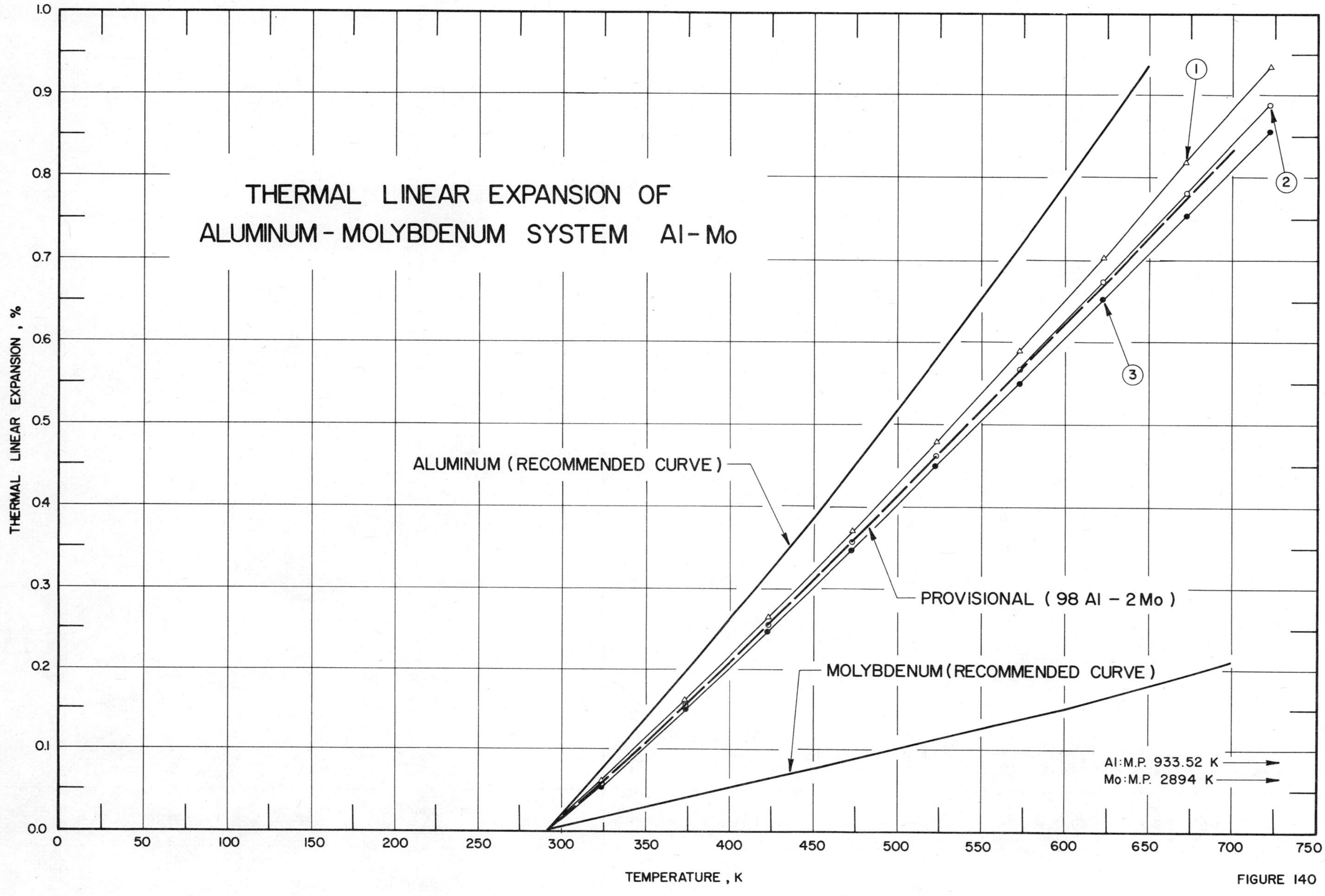

THERMAL LINEAR EXPANSION OF
ALUMINUM-MOLYBDENUM SYSTEM Al-Mo
ALUMINUM (RECOMMENDED CURVE)
PROVISIONAL (98 Al - 2 Mo)
MOLYBDENUM (RECOMMENDED CURVE)
1
2
3
Al: M.P. 933.52 K
Mo: M.P. 2894 K
THERMAL LINEAR EXPANSION, %
TEMPERATURE, K
0.0
0.1
0.2
0.3
0.4
0.5
0.6
0.7
0.8
0.9
1.0
0
50
100
150
200
250
300
350
400
450
500
550
600
650
700
750

FIGURE 140

SPECIFICATION TABLE 140. THERMAL LINEAR EXPANSION OF ALUMINUM-MOLYBDENUM SYSTEM Al-Mo

Cur. No.	Ref. No.	Author(s)	Year	Method Used	Temp. Range, K	Name and Specimen Designation	Composition (weight percent) Al	Mo	Composition (continued), Specifications, and Remarks
1	556	Varich, N.I. and Sheyko, T.I.	1970	X	293-723		Bal.	0.5	Specimen foils 0.09-0.1 mm thick, prepared by method of Dawtz by ejecting drops of melt out rotating copper drum.
2	556	Varich, N.I. and Sheyko, T.I.	1970	X	293-723		Bal.	1.5	Similar to the above specimen.
3	556	Varich, N.I. and Sheyko, T.I.	1970	X	293-723		Bal.	3.2	Similar to the above specimen.

DATA TABLE 140. THERMAL LINEAR EXPANSION OF ALUMINUM-MOLYBDENUM SYSTEM Al-Mo

[Temperature, T, K; Linear Expansion, $\Delta L/L_0$, %]

T	$\Delta L/L_0$	T	$\Delta L/L_0$	T	$\Delta L/L_0$	T	$\Delta L/L_0$	T	$\Delta L/L_0$	T	$\Delta L/L_0$
CURVE 1‡		CURVE 1 (cont.)‡		CURVE 2‡		CURVE 2 (cont.)		CURVE 3‡		CURVE 3 (cont.)	
293	0.000	523	0.479	293	0.000*	523	0.463*	293	0.000*	523	0.451
296	0.006*	573	0.591	296	0.006*	573	0.569	296	0.006*	573	0.552
323	0.060	623	0.704	323	0.058	623	0.676	323	0.056*	623	0.655
373	0.162	673	0.820	373	0.156	673	0.782	373	0.151*	673	0.757
423	0.265	723	0.938	423	0.256*	723	0.889	423	0.249	723	0.859
473	0.371			473	0.359			473	0.349		

DATA TABLE 140. COEFFICIENT OF THERMAL LINEAR EXPANSION OF ALUMINUM-MOLYBDENUM SYSTEM Al-Mo

[Temperature, T, K; Coefficient of Expansion, α, 10^{-6} K^{-1}]

T	α	T	α	T	α	T	α	T	α
CURVE 1*,‡		CURVE 1 (cont.)*,‡		CURVE 2 (cont.)*,‡		CURVE 3*,‡		CURVE 3 (cont.)*,‡	
293	19.84	673	23.34	423	20.37	293	18.64	673	20.49
296	20.00	723	23.82	473	20.83	296	18.66	723	20.44
323	20.17			523	21.00	323	18.83		
373	20.36	CURVE 2*,‡		573	21.29	373	19.23		
423	20.97			623	21.29	423	19.85		
473	21.40	293	19.20	673	21.29	473	20.25		
523	21.98	296	19.29	723	21.41	523	20.26		
573	22.63	323	19.39			573	20.43		
623	22.82	373	19.70			623	20.49		

* Not shown in figure.
‡ Author's data for coefficient of thermal expansion have been integrated by TPRC to obtain $\Delta L/L_0$.

SPECIFICATION TABLE 141. THERMAL LINEAR EXPANSION OF ALUMINUM-NICKEL SYSTEM Al-Ni

Cur. No.	Ref. No.	Author(s)	Year	Method Used	Temp. Range, K	Name and Specimen Designation	Composition (weight percent) Al	Ni	Composition (continued), Specifications, and Remarks
1*	603	McCollough, E. E.	1931	I	298, 373		97.25	2.75	Specimen furnished by Aluminum Co. of America.
2*	603	McCollough, E. E.	1931	I	298, 373		95.50	4.50	Similar to the above.
3*	603	McCollough, E. E.	1931	I	298, 373		94.22	5.78	Similar to the above.
4*	603	McCollough, E. E.	1931	I	298, 373		92.76	7.24	Similar to the above.
5*	603	McCollough, E. E.	1931	I	298, 373		91.78	8.22	Similar to the above.

DATA TABLE 141. THERMAL LINEAR EXPANSION OF ALUMINUM-NICKEL SYSTEM Al-Ni

[Temperature, T, K; Linear Expansion, $\Delta L/L_0$, %]

T	$\Delta L/L_0$	T	$\Delta L/L_0$
CURVE 1*		CURVE 5*	
298	0.012	298	0.011
373	0.187	373	0.179
CURVE 2*			
298	0.012		
373	0.189		
CURVE 3*			
298	0.012		
373	0.186		
CURVE 4*			
298	0.011		
373	0.182		

* No figure given.

FIGURE AND TABLE NO. 142BR. PROVISIONAL VALUES FOR THERMAL LINEAR EXPANSION OF ALUMINUM-SILICON SYSTEM Al-Si

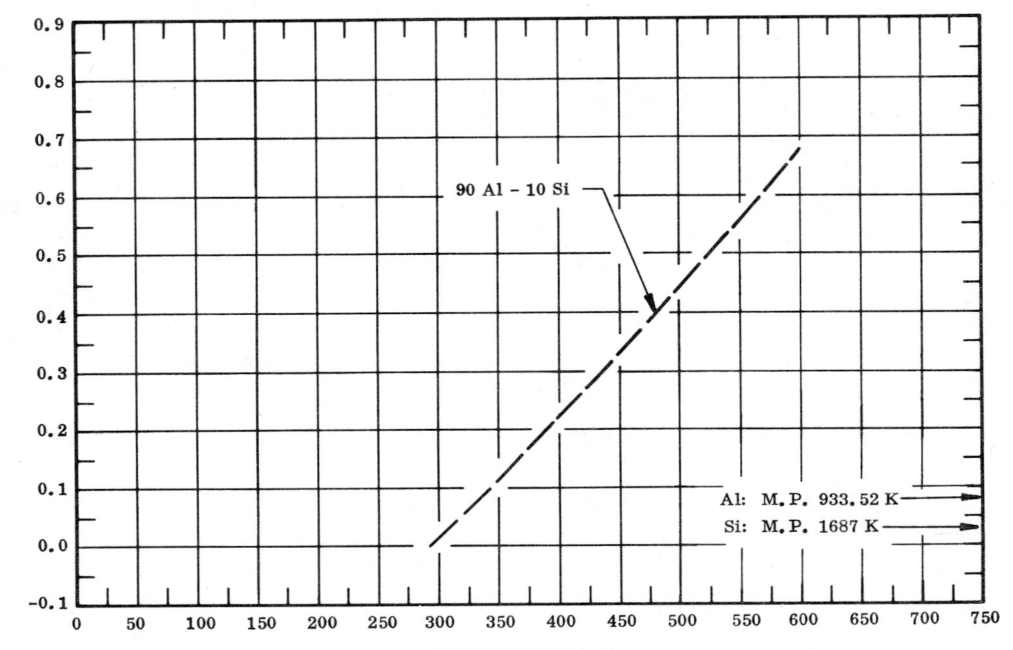

PROVISIONAL VALUES

[Temperature, T, K; Linear Expansion, $\Delta L/L_0$, %; α, K^{-1}]

T	(90 Al-10 Si) $\Delta L/L_0$	$\alpha \times 10^6$
293	0.000	20.1
400	0.222	21.4
500	0.443	22.7
600	0.675	23.7

REMARKS

(90 Al-10 Si): The tabulated values for well-annealed alloy are considered accurate to within ± 7% over the entire temperature range. These values can be represented approximately by the following equation:

$$\Delta L/L_0 = -0.519 + 1.527 \times 10^{-3}\,T + 9.031 \times 10^{-7}\,T^2 - 2.172 \times 10^{-10}\,T^3$$

FIGURE AND TABLE NO. 142CR. PROVISIONAL VALUES FOR THERMAL LINEAR EXPANSION OF ALUMINUM-TITANIUM SYSTEM Al-Ti

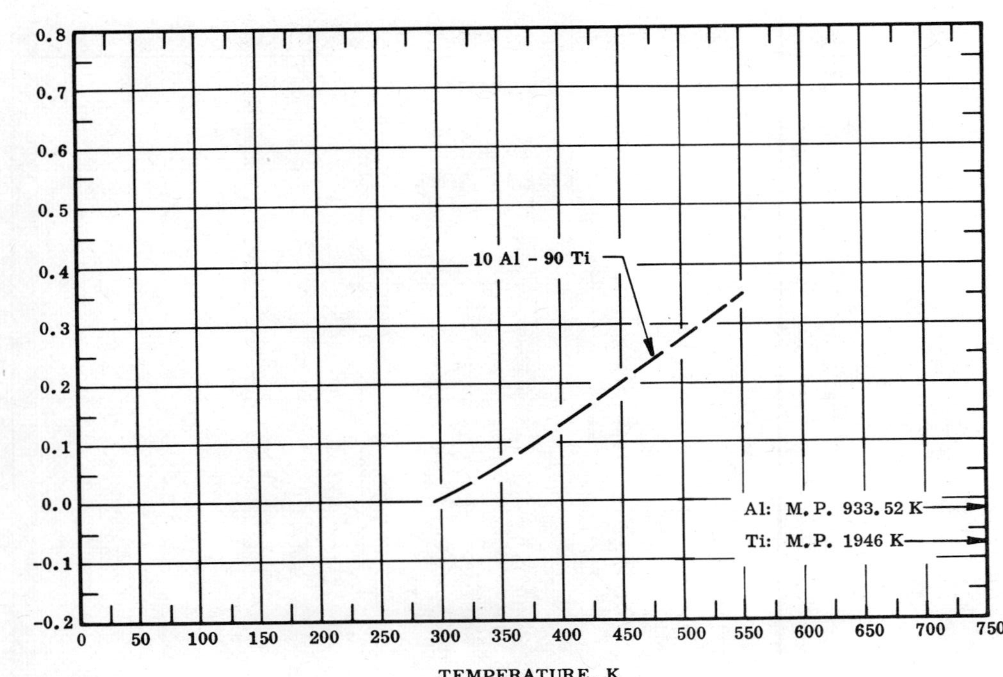

PROVISIONAL VALUES

[Temperature, T, K; Linear Expansion, $\Delta L/L_0$, %; α, K^{-1}]

	(10 Al-90 Ti)	
T	$\Delta L/L_0$	$\alpha \times 10^6$
293	0.000	11.6
400	0.137	13.7
500	0.278	14.4
550	0.350	14.7

REMARKS

(10 Al-90 Ti): The tabulated values for well-annealed alloy are considered accurate to within ± 7% over the entire temperature range. These values can be represented approximately by the following equation:

$$\Delta L/L_0 = -0.172 - 1.663 \times 10^{-4}\, T + 3.176 \times 10^{-6}\, T^2 - 2.092 \times 10^{-9}\, T^3$$

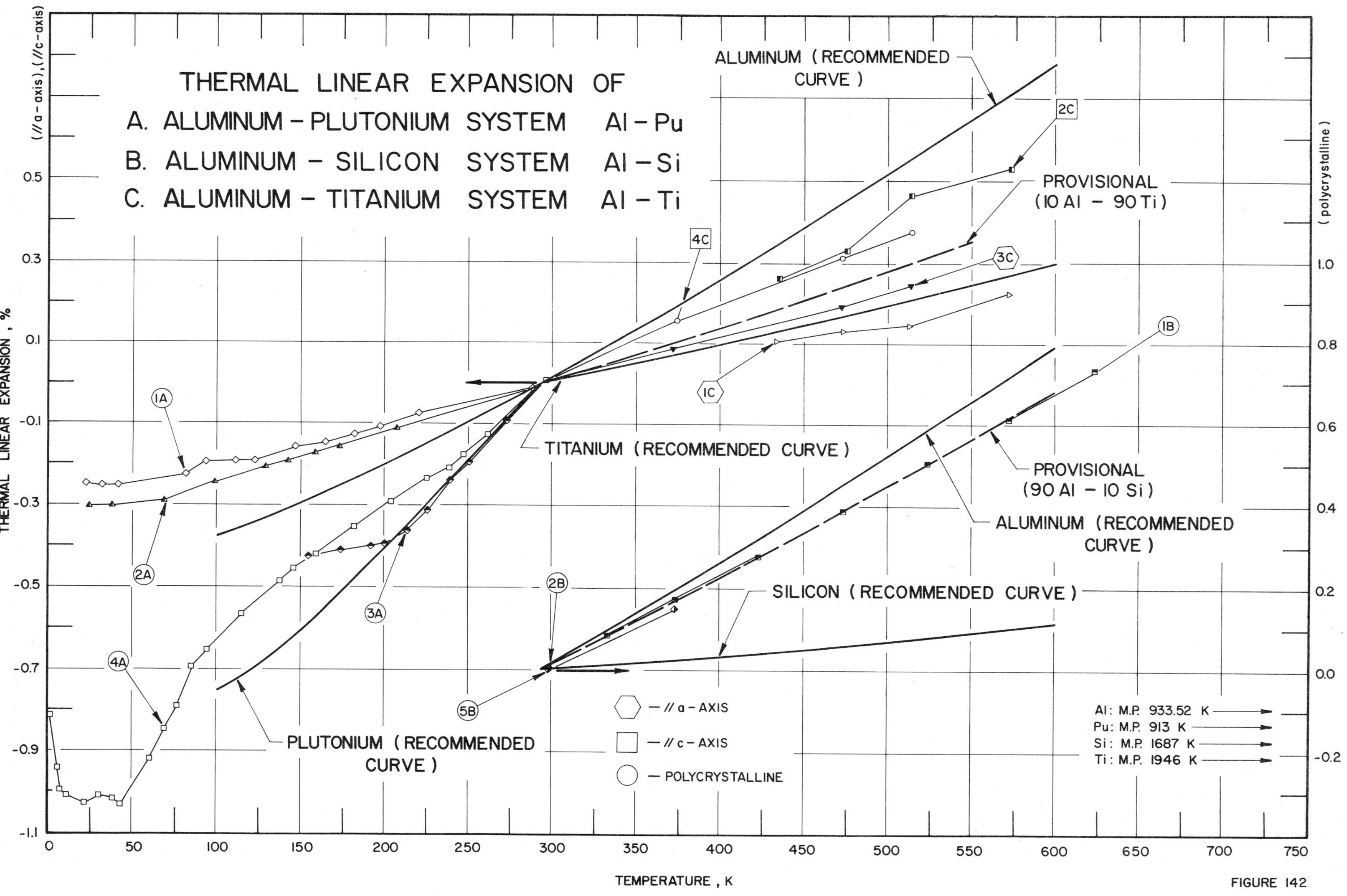

THERMAL LINEAR EXPANSION OF
A. ALUMINUM - PLUTONIUM SYSTEM Al - Pu
B. ALUMINUM - SILICON SYSTEM Al - Si
C. ALUMINUM - TITANIUM SYSTEM Al - Ti
ALUMINUM (RECOMMENDED CURVE)
PROVISIONAL (10 Al - 90 Ti)
TITANIUM (RECOMMENDED CURVE)
PROVISIONAL (90 Al - 10 Si)
ALUMINUM (RECOMMENDED CURVE)
SILICON (RECOMMENDED CURVE)
PLUTONIUM (RECOMMENDED CURVE)
1A
2A
3A
4A
1B
2B
5B
1C
2C
3C
4C
— // a - AXIS
— // c - AXIS
— POLYCRYSTALLINE
Al: M.P. 933.52 K
Pu: M.P. 913 K
Si: M.P. 1687 K
Ti: M.P. 1946 K
THERMAL LINEAR EXPANSION, %
(// a - axis), (// c - axis)
(polycrystalline)
TEMPERATURE, K
0.5
0.3
0.1
-0.1
-0.3
-0.5
-0.7
-0.9
-1.1
1.0
0.8
0.6
0.4
0.2
0.0
-0.2
0
50
100
150
200
250
300
350
400
450
500
550
600
650
700
750

FIGURE 142

SPECIFICATION TABLE 142A. THERMAL LINEAR EXPANSION OF ALUMINUM-PLUTONIUM SYSTEM Al-Pu

Cur. No.	Ref. No.	Author(s)	Year	Method Used	Temp. Range, K	Name and Specimen Designation	Composition (weight percent) Al	Pu	Composition (continued), Specifications, and Remarks
1	278	Solente, P.	1964	X	23-289		4.32	95.68	δ-phase alloy; homogeneous; expansion measured along a-axis; lattice parameter reported at 288.9 K is 4.59815 Å; 4.59835 Å at 293 K determined by graphical extrapolation.
2	278	Solente, P.	1964	X	24-290		8.7	91.3	Similar to the above specimen; lattice parameter reported at 288.9 K is 4.59708 Å; 4.59730 Å at 293 K determined by graphical extrapolation.
3	438	Sandenau, T.A.	1961		155-273		8	92	δ-stabilized alloy; expansion measured on first temperature decrease; zero-point correction is -0.086%.
4	438	Sandenau, T.A.	1961		2-296		8	92	The above specimen; results of several subsequent measurements; zero-point correction is -0.086%.

DATA TABLE 142A. THERMAL LINEAR EXPANSION OF ALUMINUM-PLUTONIUM SYSTEM Al-Pu

[Temperature, T, K; Linear Expansion, $\Delta L/L_0$, %]

T	$\Delta L/L_0$
CURVE 1	
23	-0.250
27	-0.259*
33	-0.253
37	-0.248*
43	-0.252
48	-0.252*
81	-0.221
95	-0.197
104	-0.194*
111	-0.193
123	-0.193
140	-0.177*
146	-0.155
165	-0.145
181	-0.125
197	-0.101
221	-0.078
289	-0.004
293	0.000
CURVE 2	
24	-0.301
30	-0.301*
37	-0.301
49	-0.301*
68	-0.293
79	-0.271*
98	-0.245
109	-0.229*
117	-0.215*
129	-0.203
134	-0.196*
142	-0.190
149	-0.182*
159	-0.171
171	-0.155
208	-0.109
289	-0.005*
CURVE 3	
155	-0.428
175	-0.412
192	-0.399
200	-0.391
207	-0.375*
213	-0.360
218	-0.340*
225	-0.311
240	-0.240
250	-0.195
273	-0.092
CURVE 4	
2	-0.817
6	-0.945
7	-0.972*
8	-0.989*
9	-0.999
CURVE 4 (cont.)	
11	-1.012
14	-1.023*
22	-1.029
28	-1.016*
30	-1.014
35	-1.014*
38	-1.019
43	-1.031
47	-1.010*
61	-0.921
77	-0.788
80	-0.728*
85	-0.699
95	-0.651
115	-0.564
124	-0.529*
137	-0.482
145	-0.458
159	-0.420
CURVE 4 (cont.)	
181	-0.354
204	-0.296
226	-0.239
238	-0.207
247	-0.178
263	-0.128
275	-0.086
296	0.014

*Not shown in figure.

SPECIFICATION TABLE 142B. THERMAL LINEAR EXPANSION OF ALUMINUM-SILICON SYSTEM Al-Si

Cur. No.	Ref. No.	Author(s)	Year	Method Used	Temp. Range, K	Name and Specimen Designation	Composition (weight percent) Al	Si	Composition (continued), Specifications, and Remarks
1	35	Honda, K. and Okubo, Y.	1924	L	332-622	S	Bal.	12	Cylindrical rod specimen 15.025 cm long, 5 mm diameter; data on the coefficient of thermal linear expansion are also reported.
2	603	McCollough, E. E.	1931	I	298, 373		Bal.	22.4	Specimen furnished by Aluminum Co. of America.
3*	603	McCollough, E. E.	1931	I	298, 373		Bal.	21.8	Similar to the above specimen.
4*	603	McCollough, E. E.	1931	I	298, 373		Bal.	20.2	Similar to the above specimen.
5	603	McCollough, E. E.	1931	I	298, 373		Bal.	19.2	Similar to the above specimen.
6*	606	Edwards, J. D.	1922		293, 373		90	10	Specimen of splendid casting qualities.

DATA TABLE 142B. THERMAL LINEAR EXPANSION OF ALUMINUM-SILICON SYSTEM Al-Si

[Temperature, T, K; Linear Expansion, $\Delta L/L_0$, %]

T	$\Delta L/L_0$	T	$\Delta L/L_0$	T	$\Delta L/L_0$	T	$\Delta L/L_0$
CURVE 1‡		CURVE 2		CURVE 4*		CURVE 6*	
332	0.084	298	0.011	298	0.010	293	0.000
374	0.174	373	0.179*	373	0.162	373	0.168
422	0.278	CURVE 3*		CURVE 5			
473	0.390	298	0.011	298	0.009		
523	0.502	373	0.174	373	0.154		
572	0.615						
622	0.737						

DATA TABLE 142B. COEFFICIENT OF THERMAL LINEAR EXPANSION OF ALUMINUM-SILICON SYSTEM Al-Si

[Temperature, T, K; Coefficient of Expansion, α, 10^{-6} K^{-1}]

T	α	T	α
CURVE 1‡		CURVE 1 (cont.)‡	
332	21.0*	523	22.6
374	21.9	572	23.7
422	21.7	622	24.9
473	22.1		

* Not shown in figure.
‡ Author's data for coefficient of thermal expansion have been integrated by TPRC to obtain $\Delta L/L_0$.

SPECIFICATION TABLE 142C. THERMAL LINEAR EXPANSION OF ALUMINUM-TITANIUM SYSTEM Al-Ti

Cur. No.	Ref. No.	Author(s)	Year	Method Used	Temp. Range, K	Name and Specimen Designation	Composition (weight percent) Al	Ti	Composition (continued), Specifications, and Remarks
1	330	Margolin, H. and Portisch, H.	1968	X	433-572		12.5	87.5	0.146 H; polycrystalline needle sample produced from heat-treated rods turned to 2 mm diameter on lathe, then rotation-etched; subjected to hydrogenation at 1173 K to 873 K; 0.2 to 0.5 mm diameter; expansion measured along a-axis.
2	330	Margolin, H. and Portisch, H.	1968	X	435-573		12.5	87.5	The above specimen; expansion measured along c-axis.
3	330	Margolin, H. and Portisch, H.	1968	X	298-513		12.5	87.5	Similar to the above specimen without hydrogenation added; expansion measured along a-axis.
4	330	Margolin, H. and Portisch, H.	1968	X	300-514		12.5	87.5	The above specimen; expansion measured along c-axis.

DATA TABLE 142C. THERMAL LINEAR EXPANSION OF ALUMINUM-TITANIUM SYSTEM Al-Ti

[Temperature, T, K; Linear Expansion, $\Delta L/L_0$, %]

T	$\Delta L/L_0$	T	$\Delta L/L_0$
CURVE 1		CURVE 4	
433	0.105	300	0.013*
474	0.136	374	0.160
513	0.147	473	0.319
572	0.229	514	0.377
CURVE 2			
435	0.244		
475	0.315		
514	0.439		
573	0.504		
CURVE 3			
298	0.007*		
372	0.089		
473	0.200		
513	0.248		

* Not shown in figure.

FIGURE AND TABLE NO. 143AR. PROVISIONAL VALUES FOR THERMAL LINEAR EXPANSION OF ALUMINUM-URANIUM SYSTEM Al-U

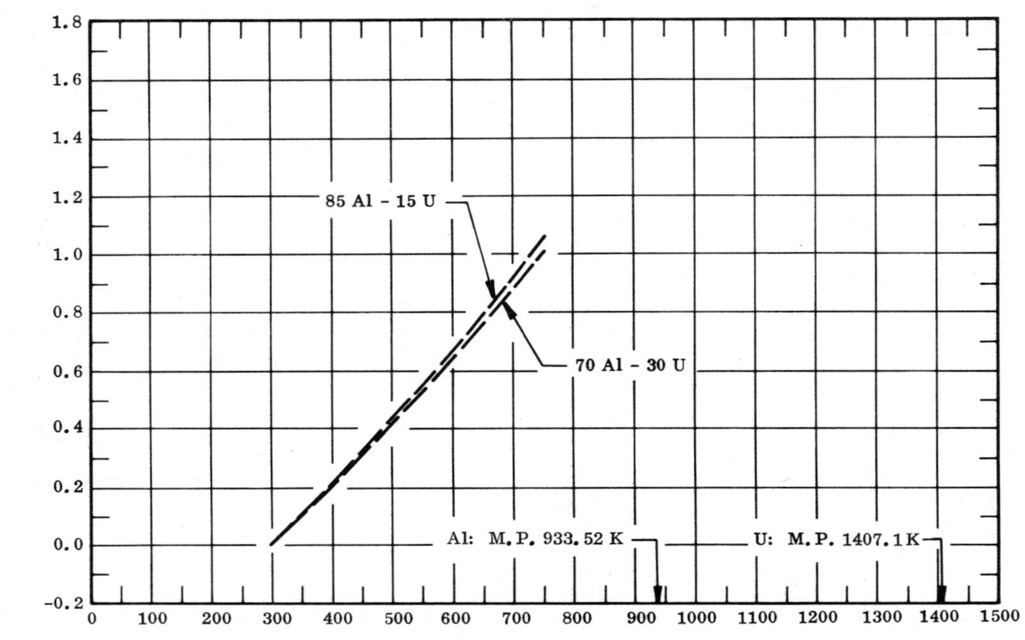

PROVISIONAL VALUES

[Temperature, T, K; Linear Expansion, $\Delta L/L_0$, %; α, K^{-1}]

T	(70 Al-30 U) $\Delta L/L_0$	(70 Al-30 U) $\alpha \times 10^6$	(85 Al-15 U) $\Delta L/L_0$	(85 Al-15 U) $\alpha \times 10^6$
293	0.000	18.0	0.000	19.5
400	0.208	20.5	0.220	21.5
500	0.420	22.2	0.444	23.2
600	0.649	23.3	0.683	24.7
700	0.887	24.0	0.936	25.7
750	1.007	24.3	1.065	25.9

REMARKS

(70 Al-30 U): The tabulated values for well-annealed alloy are considered accurate to within ± 7% over the entire temperature range. These values can be represented approximately by the following equation:

$$\Delta L/L_0 = -0.388 + 7.736 \times 10^{-4}\, T + 2.160 \times 10^{-6}\, T^2 - 9.510 \times 10^{-10}\, T^3$$

(85 Al-15 U): The tabulated values for well-annealed alloy are considered accurate to within ± 7% over the entire temperature range. These values can be represented approximately by the following equation:

$$\Delta L/L_0 = -0.468 + 1.213 \times 10^{-3}\, T + 1.451 \times 10^{-6}\, T^2 - 4.567 \times 10^{-10}\, T^3$$

FIGURE AND TABLE NO. 143BR. PROVISIONAL VALUES FOR THERMAL LINEAR EXPANSION OF ALUMINUM-ZINC SYSTEM Al-Zn

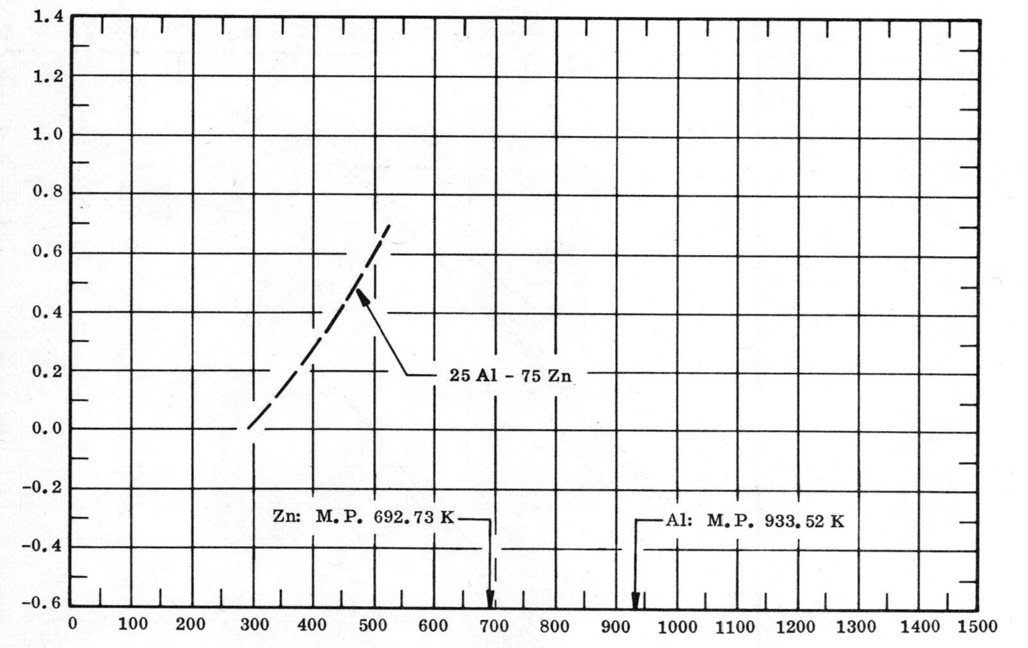

PROVISIONAL VALUES

[Temperature, T, K; Linear Expansion, $\Delta L/L_0$, %; α, K^{-1}]

	(25 Al-75 Zn)	
T	$\Delta L/L_0$	$\alpha \times 10^6$
293	0.000	24.0
400	0.279	28.7
500	0.602	36.9
525	0.699	39.5

REMARKS

(25 Al-75 Zn): The tabulated values for well-annealed alloy are considered accurate to within ± 7% over the entire temperature range. These values can be represented approximately by the following equation:

$$\Delta L/L_0 = -0.752 + 3.260 \times 10^{-3}\, T - 4.170 \times 10^{-6}\, T^2 + 6.136 \times 10^{-9}\, T^3$$

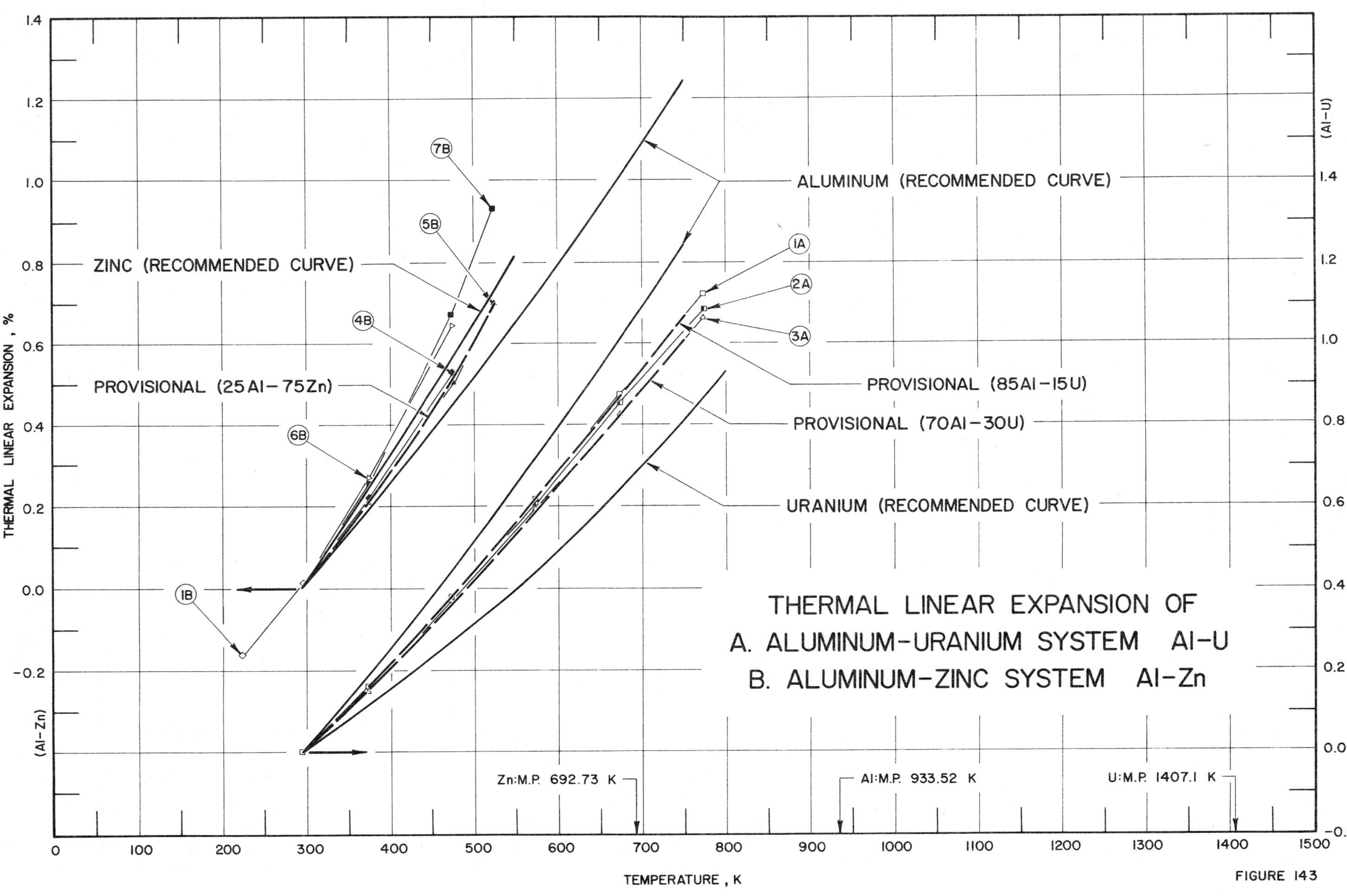

THERMAL LINEAR EXPANSION OF
A. ALUMINUM-URANIUM SYSTEM Al-U
B. ALUMINUM-ZINC SYSTEM Al-Zn
ALUMINUM (RECOMMENDED CURVE)
ZINC (RECOMMENDED CURVE)
PROVISIONAL (25Al-75Zn)
PROVISIONAL (85Al-15U)
PROVISIONAL (70Al-30U)
URANIUM (RECOMMENDED CURVE)
1A
2A
3A
1B
4B
5B
6B
7B
Zn:M.P. 692.73 K
Al:M.P. 933.52 K
U:M.P. 1407.1 K
THERMAL LINEAR EXPANSION, %
(Al-Zn)
(Al-U)
TEMPERATURE, K
1.4
1.2
1.0
0.8
0.6
0.4
0.2
0.0
-0.2
0
100
200
300
400
500
600
700
800
900
1000
1100
1200
1300
1400
1500

FIGURE 143

SPECIFICATION TABLE 143A. THERMAL LINEAR EXPANSION OF ALUMINUM-URANIUM SYSTEM Al-U

Cur. No.	Ref. No.	Author(s)	Year	Method Used	Temp. Range, K	Name and Specimen Designation	Composition (weight percent) Al	U	Composition (continued), Specifications, and Remarks
1	402	Saller, H.A.	1957		293-773		Bal.	12.5	Prepared from forged alloys; 3.25 in. long, 0.5 in. dia.
2	402	Saller, H.A.	1957		293-773		Bal.	22.7	Similar to the above specimen.
3	402	Saller, H.A.	1957		293-773		Bal.	30.5	Similar to the above specimen.

DATA TABLE 143A. THERMAL LINEAR EXPANSION OF ALUMINUM-URANIUM SYSTEM Al-U

[Temperature, T, K; Linear Expansion, $\Delta L/L_0$, %]

T	$\Delta L/L_0$
CURVE 1	
293	0.000
373	0.160
473	0.380
573	0.619
673	0.878
773	1.123
CURVE 2	
293	0.000*
373	0.160*
473	0.382*
573	0.613
673	0.855
773	1.089
CURVE 3	
293	0.000*
373	0.155
473	0.374
573	0.596
673	0.821
773	1.061

* Not shown in figure.

SPECIFICATION TABLE 143B. THERMAL LINEAR EXPANSION OF ALUMINUM-ZINC SYSTEM Al-Zn

Cur. No.	Ref. No.	Author(s)	Year	Method Used	Temp. Range, K	Name and Specimen Designation	Composition (weight percent) Al	Zn	Composition (continued), Specifications, and Remarks
1	401	Hume-Rothery, W. and Boultbee, T.H.	1949	X	223, 298		95.98	4.02	Prepared from metals of highest purity; Al supplied by British Aluminum Co. Ltd.; data obtained in several experiments; lattice parameter reported at 298 K is 4.04036 kX (4.04844 Å); 4.03991 kX (4.04799 Å) at 293 K determined by graphical interpolation.
2*	401	Hume-Rothery, W. and Boultbee, T.H.	1949	X	223, 298		91.64	8.36	Similar to the above specimen; lattice parameter reported at 298 K is 4.03881 kX (4.09627 Å); 4.03822 kX (4.04629 Å) at 293 K determined by graphical interpolation.
3*	401	Hume-Rothery, W. and Boultbee, T.H.	1949	X	223, 298		86.81	13.19	Similar to the above specimen; lattice parameter reported at 298 K is 4.03699 kX (4.04506 Å); 4.03650 kX (4.04457 Å) at 293 K determined by graphical interpolation.
4	20	Hidnert, P.	1925	T	293-373	S389	22.57	77.22	0.11 Fe, 0.05 Si, 0.05 Cu; this alloy indicates marked changes in expansion at 543 K.
5	20	Hidnert, P.	1925	T	293-523		22.57	77.22	The above specimen; expansion measured on second test; this alloy indicates marked changes in expansion at 543 K.
6	20	Hidnert, P.	1925	T	293-473		5.29	94.60	0.02 Cu, 0.02 Fe, 0.01 Si; this alloy indicates marked changes in expansion at 553 K.
7	20	Hidnert, P.	1925	T	293-523		5.29	94.60	The above specimen; expansion measured on second test.

DATA TABLE 143B. THERMAL LINEAR EXPANSION OF ALUMINUM-ZINC SYSTEM Al-Zn

[Temperature, T, K; Linear Expansion, $\Delta L/L_0$, %]

T	$\Delta L/L_0$
CURVE 1	
223	-0.160
298	0.011
CURVE 2*	
223	-0.166
298	0.012
CURVE 3*	
223	-0.175
298	0.012*
CURVE 4	
293	0.000*
373	0.220
473	0.533

T	$\Delta L/L_0$
CURVE 5	
293	0.000*
373	0.208
473	0.509
523	0.702
CURVE 6	
293	0.000*
373	0.266
473	0.643
CURVE 7	
293	0.000*
373	0.256
473	0.670
523	0.936

* Not shown in figure.

FIGURE AND TABLE NO. 144R. PROVISIONAL VALUES FOR THERMAL LINEAR EXPANSION OF ALUMINUM-ZIRCONIUM SYSTEM Al-Zr

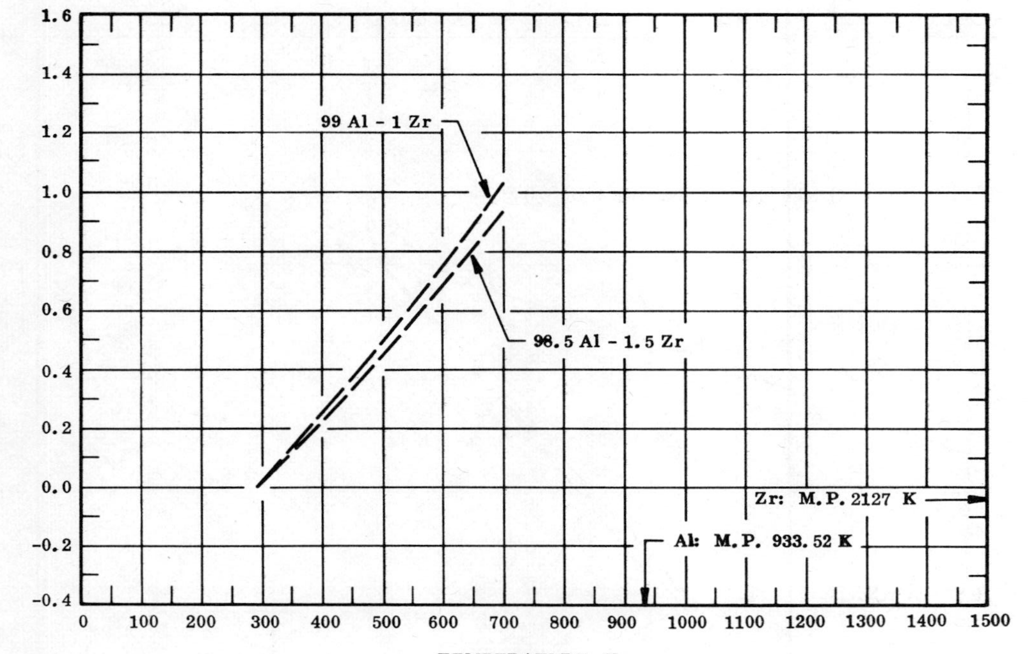

PROVISIONAL VALUES

[Temperature, T, K; Linear Expansion, $\Delta L/L_0$, %; α, K^{-1}]

T	(98.5 Al-1.5 Zr) $\Delta L/L_0$	(98.5 Al-1.5 Zr) $\alpha \times 10^6$	(99 Al-1 Zr) $\Delta L/L_0$	(99 Al-1 Zr) $\alpha \times 10^6$
293	0.000	21.5	0.000	23.3
400	0.229	21.8	0.250	23.6
500	0.452	22.6	0.490	24.6
600	0.685	23.9	0.745	26.7
700	0.941	27.4	1.025	29.4

REMARKS

(98.5 Al-1.5 Zr): The tabulated values for well-annealed alloy are considered accurate to within ±7% over the entire temperature range. These values can be represented approximately by the following equation:

$$\Delta L/L_0 = -0.669 + 2.509 \times 10^{-3}\, T - 1.129 \times 10^{-6}\, T^2 + 1.183 \times 10^{-9}\, T^3$$

(99 Al-1 Zr): The tabulated values for well-annealed alloy are considered accurate to within ±7% over the entire temperature range. These values can be represented approximately by the following equation:

$$\Delta L/L_0 = -0.740 + 2.846 \times 10^{-3}\, T - 1.541 \times 10^{-6}\, T^2 + 1.540 \times 10^{-9}\, T^3$$

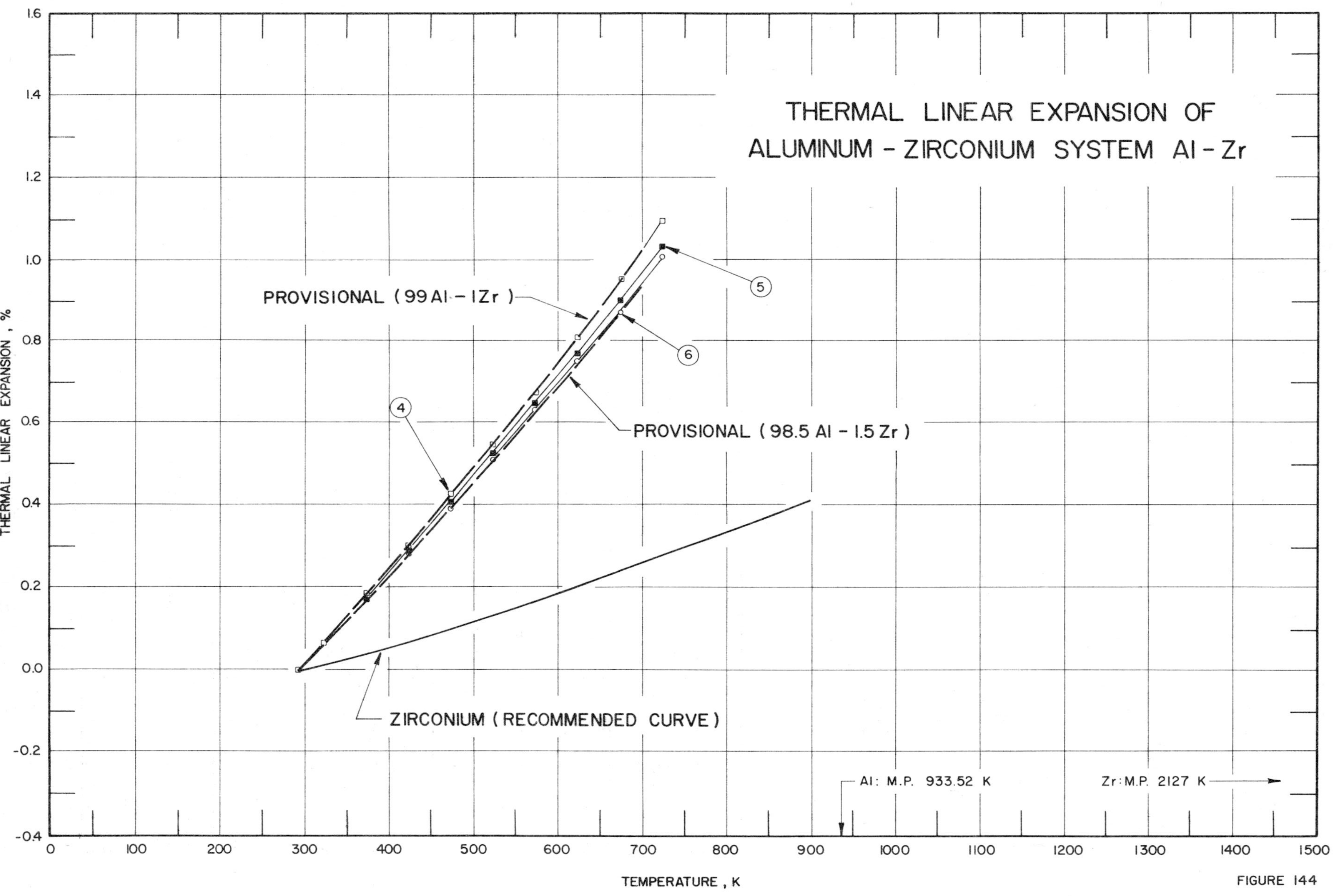
THERMAL LINEAR EXPANSION OF
ALUMINUM - ZIRCONIUM SYSTEM Al - Zr
PROVISIONAL (99 Al - 1 Zr)
PROVISIONAL (98.5 Al - 1.5 Zr)
ZIRCONIUM (RECOMMENDED CURVE)
4
5
6
Al: M.P. 933.52 K
Zr: M.P. 2127 K
THERMAL LINEAR EXPANSION, %
1.6
1.4
1.2
1.0
0.8
0.6
0.4
0.2
0.0
-0.2
-0.4
TEMPERATURE, K
0
100
200
300
400
500
600
700
800
900
1000
1100
1200
1300
1400
1500

FIGURE 144

SPECIFICATION TABLE 144. THERMAL LINEAR EXPANSION OF ALUMINUM-ZIRCONIUM SYSTEM Al-Zr

Cur. No.	Ref. No.	Author(s)	Year	Method Used	Temp. Range, K	Name and Specimen Designation	Composition (weight percent) Al	Zr	Composition (continued), Specifications, and Remarks
1*	403	Varich, N.I., Lyukevich, R.B., Kolomytseva, L.F., Varich, A.N. and Maslov, V.V.	1969	X	1073-1573		99.2	0.8	Specimen prepared as thin film 0.05-0.10 mm thick; lattice parameter expansion measured; lattice parameter reported at 1073 K is 4.04186 kX (4.04994 Å); this curve is here reported using the first given temperature as reference temperature at which $\Delta L/L_0 = 0$.
2*	403	Varich, N.I., et al.	1969	X	1073-1573		98.4	1.6	Similar to the above specimen; lattice parameter reported at 1073 K is 4.04232 kX (4.05040 Å); this curve is here reported using the first given temperature as reference temperature at which $\Delta L/L_0 = 0$.
3*	403	Varich, N.I., et al.	1969	X	1073-1573		97.8	2.2	Similar to the above specimen; lattice parameter expansion measured; lattice parameter reported at 1073 K is 4.04251 kX (4.05059 Å); this curve is here reported using the first given temperature as reference temperature at which $\Delta L/L_0 = 0$.
4	556	Varich, N.I. and Sheyko, T.I.	1970	X	293-723		99.0	1.0	Foil specimen 0.09-0.1 mm thick prepared by ejecting drops of melt onto a rotating copper drum; experiment performed in a KROS apparatus with temperature fluctuations of not more than ±2 deg; the lattice constant was calculated from a (511) line in copper radiation with accuracy of 0.004 Å.
5	556	Varich, N.I. and Sheyko, T.I.	1970	X	293-723		98.8	1.2	Similar to the above specimen.
6	556	Varich, N.I. and Sheyko, T.I.	1970	X	293-723		98.5	1.5	Similar to the above specimen.

* Not shown in figure.

DATA TABLE 144. THERMAL LINEAR EXPANSION OF ALUMINUM-ZIRCONIUM SYSTEM Al-Zr

[Temperature, T, K; Linear Expansion, $\Delta L/L_0$, %]

T	$\Delta L/L_0$
CURVE 1*, †	
1073	0.000
1093	0.003
1113	0.007
1130	0.010
1150	0.013
1173	0.013
1192	0.019
1212	0.021
1235	0.026
1254	0.023
1273	0.027
1373	0.026
1473	0.027
1574	0.028
CURVE 2*, †	
1073	0.000
1091	0.005
1111	0.010
CURVE 2 (cont.)*, †	
1134	0.011
1153	0.017
1173	0.015
1193	0.024
1212	0.023
1233	0.032
1250	0.035
1273	0.041
1292	0.040
1313	0.043
1332	0.052
1374	0.051
1293	0.053
1473	0.056
1573	0.055
CURVE 3*, †	
1073	0.000
1091	0.005
CURVE 3 (cont.)*, †	
1114	0.008
1134	0.012
1152	0.016
1173	0.017
1193	0.024
1209	0.031
1233	0.032
1251	0.039
1273	0.042
1291	0.051
1312	0.060
1333	0.060
1354	0.069
1373	0.068
1393	0.074
1414	0.080
1473	0.082
1573	0.081
CURVE 4‡	
293	0.000*
296	0.007*
323	0.067
373	0.181
423	0.300
473	0.423
523	0.548
573	0.677
623	0.811
673	0.953
723	1.102
CURVE 5‡	
293	0.000*
296	0.006*
323	0.065*
373	0.176
423	0.291*
473	0.406
CURVE 5 (cont.)‡	
523	0.524*
573	0.644
623	0.769
673	0.901*
723	1.038
CURVE 6‡	
293	0.000*
296	0.006*
323	0.063*
373	0.172*
423	0.283
473	0.396*
523	0.512
573	0.629*
623	0.750
673	0.877*
723	1.008

DATA TABLE 144. COEFFICIENT OF THERMAL LINEAR EXPANSION OF ALUMINUM-ZIRCONIUM SYSTEM Al-Zr

[Temperature, T, K; Coefficient of Expansion, α, 10^{-6} K^{-1}]

T	α
CURVE 4*, ‡	
293	21.91
296	22.00
323	22.43
373	23.37
423	24.32
473	24.79
523	25.28
573	26.09
623	27.73
673	29.10
723	30.53
CURVE 5*, ‡	
293	21.43
296	21.47
323	21.93
373	22.63
423	23.04
473	23.31
523	23.79
573	24.32
623	25.56
673	27.06
723	27.95
CURVE 6*, ‡	
293	21.06
296	20.93
323	21.41
373	22.09
423	22.41
473	22.78
523	23.29
573	23.71
623	24.79
673	25.88
723	26.57

*Not shown in figure.
†This curve is here reported using the first given temperature as reference temperature at which $\Delta L/L_0 = 0$.
‡Author's data for coeffficient of thermal expansion have been integrated by TPRC to obtain $\Delta L/L_0$.

FIGURE AND TABLE NO. 145AR. PROVISIONAL VALUES FOR THERMAL LINEAR EXPANSION OF ANTIMONY-BISMUTH SYSTEM Sb-Bi

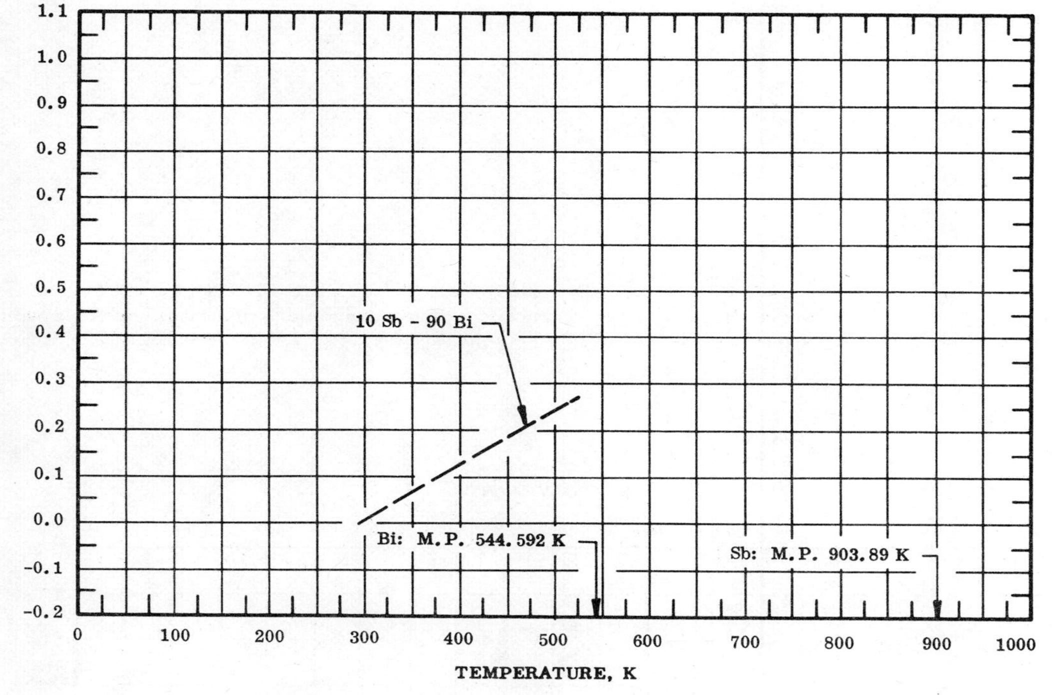

PROVISIONAL VALUES

[Temperature, T, K; Linear Expansion, $\Delta L/L_0$, %; α, K^{-1}]

	(10 Sb-90 Bi)	
T	$\Delta L/L_0$	$\alpha \times 10^6$
293	0.000	11.7
400	0.127	11.7
500	0.240	11.9
525	0.270	12.1

REMARKS

(10 Sb-90 Bi): The tabulated values for well-annealed alloy are considered accurate to within ±7% over the entire temperature range. These values can be represented approximately by the following equation:

$$\Delta L/L_0 = -0.621 + 3.290 \times 10^{-3}\, T - 5.207 \times 10^{-6}\, T^2 + 4.139 \times 10^{-9}\, T^3$$

FIGURE AND TABLE NO. 145BR. PROVISIONAL VALUES FOR THERMAL LINEAR EXPANSION OF ANTIMONY-SILVER SYSTEM Sb-Ag

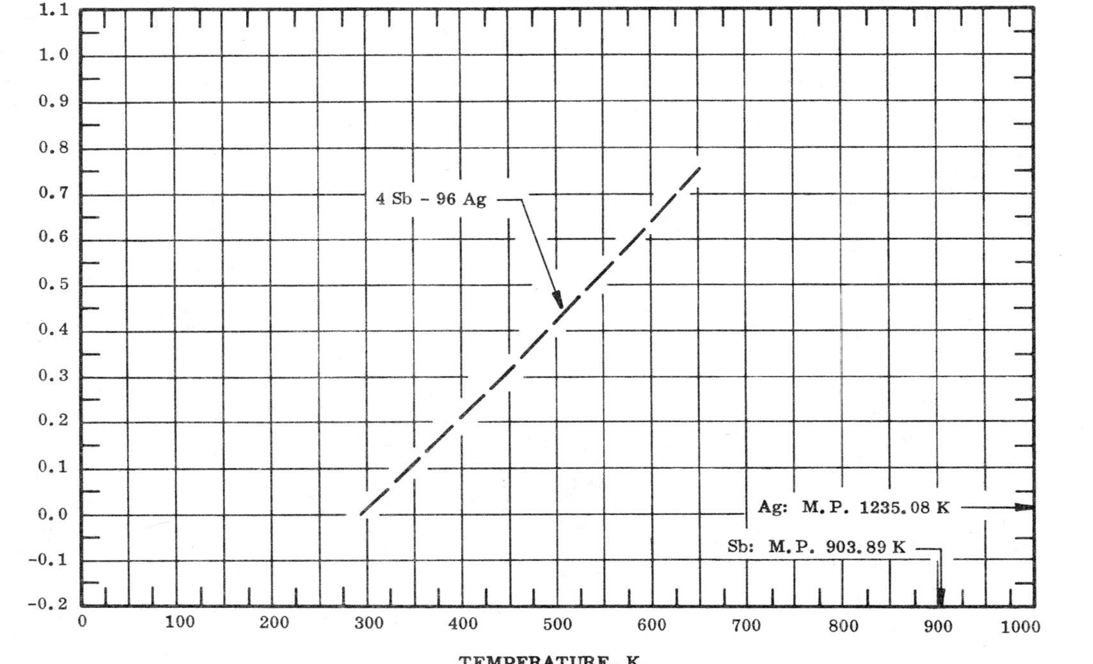

PROVISIONAL VALUES

[Temperature, T, K; Linear Expansion, $\Delta L/L_0$, %; α, K^{-1}]

T	(4 Sb-96 Ag) $\Delta L/L_0$	$\alpha \times 10^6$
293	0.000	20.2
400	0.217	20.5
500	0.425	21.1
600	0.641	21.9
650	0.752	22.4

REMARKS

(4 Sb-96 Ag): The tabulated values for well-annealed alloy are considered accurate to within ± 7% over the entire temperature range. These values can be represented approximately by the following equation:

$$\Delta L/L_0 = -0.596 + 2.093 \times 10^{-3}\, T - 3.388 \times 10^{-7}\, T^2 + 4.762 \times 10^{-10}\, T^3$$

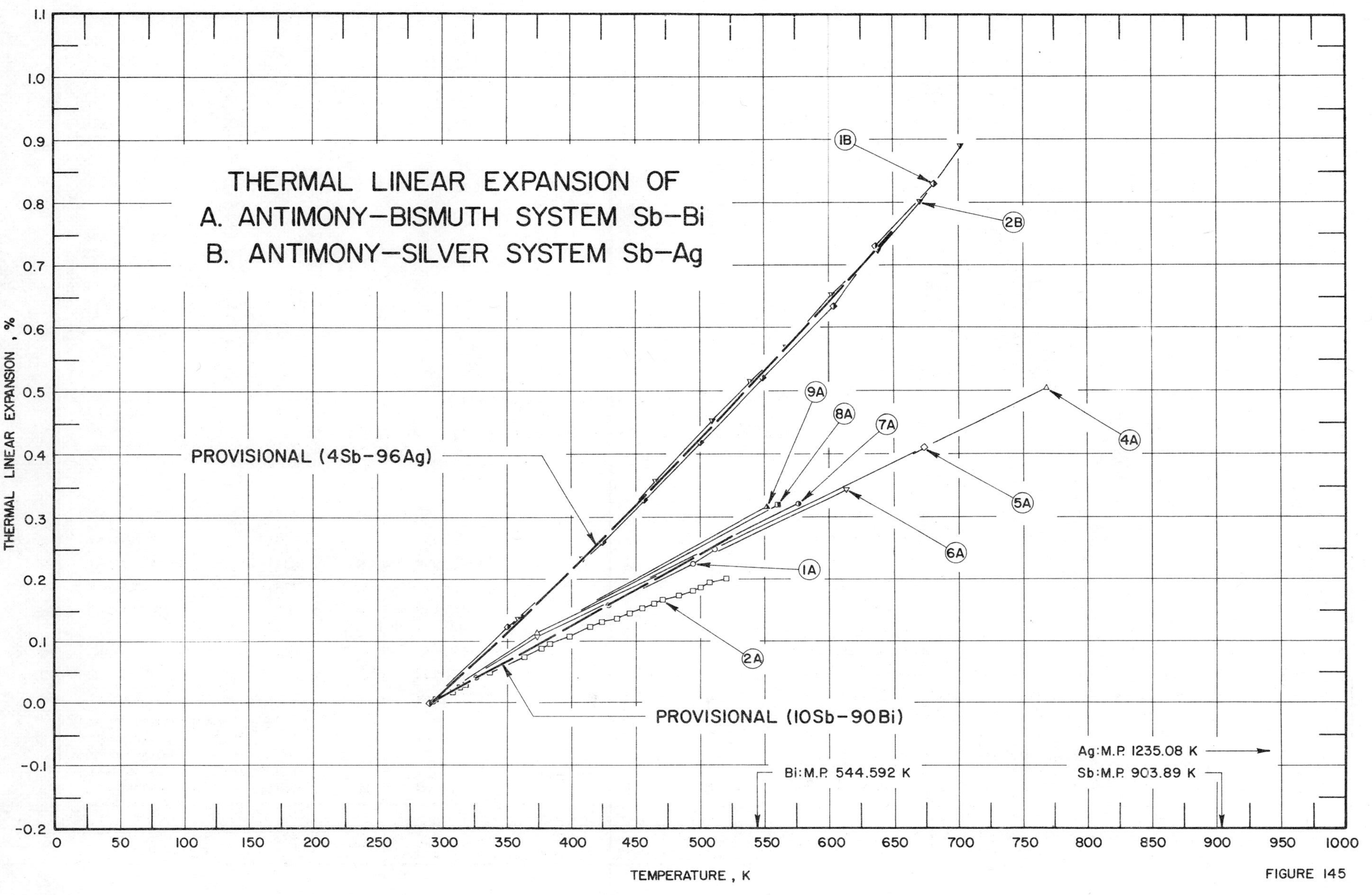
THERMAL LINEAR EXPANSION OF
A. ANTIMONY-BISMUTH SYSTEM Sb-Bi
B. ANTIMONY-SILVER SYSTEM Sb-Ag
PROVISIONAL (4Sb-96Ag)
PROVISIONAL (10Sb-90Bi)
1B
2B
9A
8A
7A
4A
5A
6A
1A
2A
Ag:M.P. 1235.08 K
Sb:M.P. 903.89 K
Bi:M.P. 544.592 K
THERMAL LINEAR EXPANSION , %
1.1
1.0
0.9
0.8
0.7
0.6
0.5
0.4
0.3
0.2
0.1
0.0
-0.1
-0.2
0
50
100
150
200
250
300
350
400
450
500
550
600
650
700
750
800
850
900
950
1000
TEMPERATURE , K

FIGURE 145

SPECIFICATION TABLE 145A. THERMAL LINEAR EXPANSION OF ANTIMONY-BISMUTH SYSTEM Sb-Bi

Cur. No.	Ref. No.	Author(s)	Year	Method Used	Temp. Range, K	Name and Specimen Designation	Composition (weight percent) Sb	Bi	Composition (continued), Specifications, and Remarks
1	649	Zhadanova, V. V., Ivanov, G.A., Inyutkin, A.I., Naletov, V.L., Nikolaev, V.I., Regel, A.R., and Sergeev, V.P.	1971	X	293-512		11.0	89.0	Single crystal by zone method (zone speed 0.05 cm/h) using Bi-000 grade bismuth and extragrade antimony; inhomogeneity was estimated to be 3.7; zero-point correction is -0.023%.
2	649	Zhadanova, V.V., et al.	1971	L	293-520		11.0	89.0	The above specimen; first cycle; zero-point correction is -0.023%.
3*	649	Zhadanova, V.V., et al.	1971	L	293-523		11.0	89.0	The above specimen; second cycle; zero-point correction is -0.023%.
4	669	Pelzel, E.	1959	X	293-768		80	20	Specimen prepared from 99.9 Sb and 99.99 Bi; homogenized for 8 days at various temperatures.
5	669	Pelzel, E.	1959	X	293-673		60	40	Similar to the above specimen.
6	669	Pelzel, E.	1959	X	293-613		40	60	Similar to the above specimen.
7	669	Pelzel, E.	1959	X	293-576		20	80	Similar to the above specimen.
8	669	Pelzel, E.	1959	X	293-560		10	90	Similar to the above specimen.
9	669	Pelzel, E.	1959	X	293-551		5	95	Similar to the above specimen.

DATA TABLE 145A. THERMAL LINEAR EXPANSION OF ANTIMONY-BISMUTH SYSTEM Sb-Bi

[Temperature, T, K; Linear Expansion, $\Delta L/L_0$, %]

T	$\Delta L/L_0$
CURVE 1	
293	0.000
327	0.041
429	0.158
494	0.226
512	0.249
CURVE 2	
293	0.000*
295	0.004
308	0.016
313	0.023
318	0.028
336	0.048
364	0.074
377	0.089
384	0.096
CURVE 2 (cont.)	
399	0.107
415	0.123
424	0.130
435	0.136
445	0.144
455	0.151
464	0.159
471	0.166
483	0.172
493	0.180
500	0.186
507	0.194
520	0.200
CURVE 3*	
293	0.000
CURVE 3 (cont.)*	
344	0.052
358	0.065
379	0.091
390	0.099
399	0.113
406	0.120
417	0.133
430	0.147
441	0.160
448	0.167
454	0.172
459	0.180
464	0.186
472	0.196
476	0.202
481	0.208
489	0.214
CURVE 3 (cont.)*	
513	0.241
523	0.254
CURVE 4	
293	0.000*
373	0.112
768	0.504
CURVE 5	
293	0.000*
373	0.113*
673	0.410
CURVE 6	
293	0.000*
373	0.109
613	0.346
CURVE 7	
293	0.000*
373	0.109*
576	0.321
CURVE 8	
293	0.0000*
373	0.0114*
560	0.0320
CURVE 9	
293	0.0000*
373	0.0118*
551	0.0317

*Not shown in figure.

SPECIFICATION TABLE 145B. THERMAL LINEAR EXPANSION OF ANTIMONY-SILVER SYSTEM Sb-Ag

Cur. No.	Ref. No.	Author(s)	Year	Method Used	Temp. Range, K	Name and Specimen Designation	Composition (weight percent) Sb	Ag	Composition (continued), Specifications, and Remarks
1	303	Owen, E.A. and Roberts, E.W.	1939	X	291-681		3.04	96.96	99.95 pure silver and 99.99 pure antimony starting materials; melted together to form alloy; lattice parameter reported at 291 K is 4.0937 Å; 4.0939 Å at 293 K determined by graphical interpolation.
2	303	Owen, E.A. and Roberts, E.W.	1939	X	291-704		5.38	94.62	Similar to the above specimen; lattice parameter reported at 291 K is 4.1062 Å; 4.1064 Å at 293 K determined by graphical interpolation.

DATA TABLE 145B. THERMAL LINEAR EXPANSION OF ANTIMONY-SILVER SYSTEM Sb-Ag

[Temperature, T, K; Linear Expansion, $\Delta L/L_0$, %]

T	$\Delta L/L_0$	T	$\Delta L/L_0$
CURVE 1		CURVE 2 (cont.)	
291	-0.0042	539	0.514
351	0.122	566	0.569
424	0.261	604	0.652
456	0.328	637	0.725
500	0.420	670	0.801
548	0.521	704	0.890
604	0.636		
636	0.730		
681	0.830		
CURVE 2			
291	-0.0040*		
359	0.133		
409	0.234		
466	0.357		
509	0.452		

* Not shown in figure.

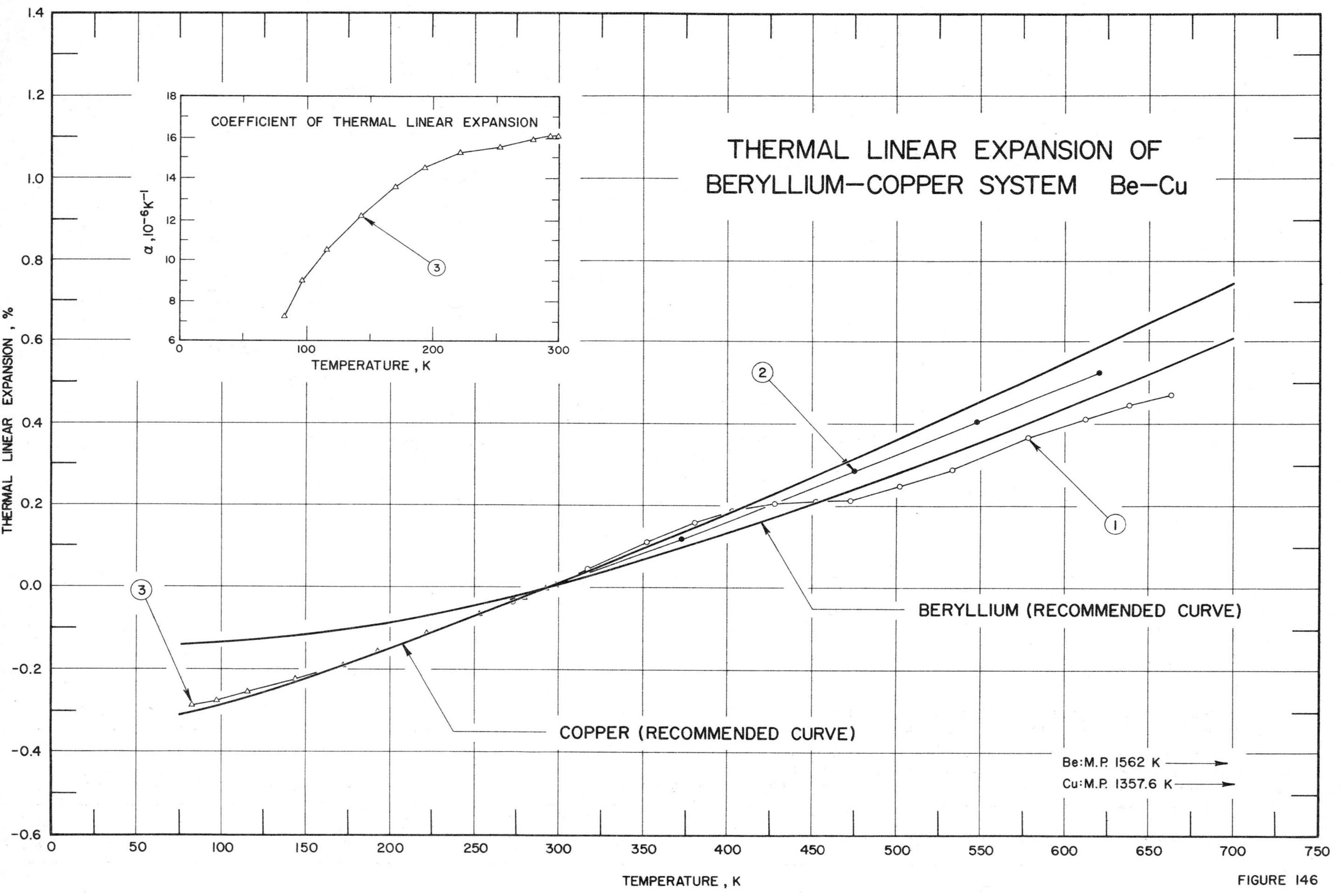

THERMAL LINEAR EXPANSION OF
BERYLLIUM–COPPER SYSTEM Be–Cu
COEFFICIENT OF THERMAL LINEAR EXPANSION
α, 10⁻⁶ K⁻¹
TEMPERATURE, K
BERYLLIUM (RECOMMENDED CURVE)
COPPER (RECOMMENDED CURVE)
Be: M.P. 1562 K
Cu: M.P. 1357.6 K
THERMAL LINEAR EXPANSION, %
TEMPERATURE, K
1
2
3

FIGURE 146

SPECIFICATION TABLE 146. THERMAL LINEAR EXPANSION OF BERYLLIUM-COPPER SYSTEM Be-Cu

Cur. No.	Ref. No.	Author(s)	Year	Method Used	Temp. Range, K	Name and Specimen Designation	Composition (weight percent) Be	Cu	Composition (continued), Specifications, and Remarks
1	408	Wilkes, P. and Barrand, P.	1968	L	273-664		2.0	98	0.01 each of Fe, Mg, Si, and Sn, 0.003 Al, 0.0005 each of Ag, Mn, and Pb, 0.0003 Ni, and 0.0001 Cr; obtained from Johnson Matthey Limited; specimen 0.1 cm diameter and 4 cm long; produced from the billet material by cold swaging and machining, treated in hydrogen atmosphere and water-quenched from 1073 K, holding in nitrogen before measured; isochronal aging after quenching; expansion measured with temperature increasing at 33 C min^{-1}; expansion reported at 273 K is 0.000; 0.0015 cm at 293 K determined by graphical interpolation.
2	408	Wilkes, P. and Barrand, P.	1968	L	620-273		2.0	98	The above specimen; expansion measured with decreasing temperature; expansion reported at 273 K is -0.0060 cm; -0.0048 cm at 293 K determined by graphical interpolation.
3	409	Itskevich, E. S., Voronovskii, A. N., Gavrilov, A. F., and Sukhoparonov, V. A.	1967	L	82-299	Beryllium Bronze	1.0-2.5	Bal.	Heat-treated to Rockwell hardness 38 to 40.

DATA TABLE 146. THERMAL LINEAR EXPANSION OF BERYLLIUM-COPPER SYSTEM Be-Cu

[Temperature, T, K; Linear Expansion, $\Delta L/L_0$, %]

T	$\Delta L/L_0$	T	$\Delta L/L_0$	T	$\Delta L/L_0$	T	$\Delta L/L_0$	T	$\Delta L/L_0$	T	$\Delta L/L_0$
CURVE 1		CURVE 1 (cont.)		CURVE 1 (cont.)		CURVE 2		CURVE 3‡		CURVE 3 (cont.)‡	
273	-0.037	428	0.205	578	0.363	620	0.528	82	-0.283	193	-0.155
316	0.043	451	0.213	611	0.413	548	0.407	96	-0.272	222	-0.111
351	0.115	473	0.218	639	0.443	475	0.287	115	-0.254	253	-0.064
380	0.160	501	0.250	664	0.473	373	0.120	144	-0.220	280	-0.021
403	0.185	533	0.293			273	-0.030	171	-0.186	293	-0.000
										299	0.009

DATA TABLE 146. COEFFICIENT OF THERMAL LINEAR EXPANSION OF BERYLLIUM-COPPER SYSTEM Be-Cu

[Temperature, T, K; Coefficient of Expansion, α, 10^{-6} K^{-1}]

T	α	T	α	T	α
CURVE 3‡		CURVE 3 (cont.)‡		CURVE 3 (cont.)‡	
82	7.22	171	13.60	280	15.98
96	9.01	193	14.56	293	16.15
115	10.53	222	15.24	299	16.15
144	12.28	253	15.50		

‡Author's data for coefficient of thermal expansion have been integrated by TPRC to obtain $\Delta L/L_0$.

SPECIFICATION TABLE 147. THERMAL LINEAR EXPANSION OF BERYLLIUM-IRON SYSTEM Be-Fe

Cur. No.	Ref. No.	Author(s)	Year	Method Used	Temp. Range, K	Name and Specimen Designation	Composition (weight percent) Be	Fe	Composition (continued), Specifications, and Remarks
1*	651	Goggin, W.R. and Paquin, R.A.	1970	L	275-311	FP-175B	99	1	Specimen from Stanford Research Institute; composition of Be unspecified.

DATA TABLE 147. THERMAL LINEAR EXPANSION OF BERYLLIUM-IRON SYSTEM Be-Fe

[Temperature, T, K; Linear Expansion, $\Delta L/L_0$, %]

T	$\Delta L/L_0$
CURVE 1*	
275.2	-0.020
293.2	0.000
311.2	0.020

* No figure given.

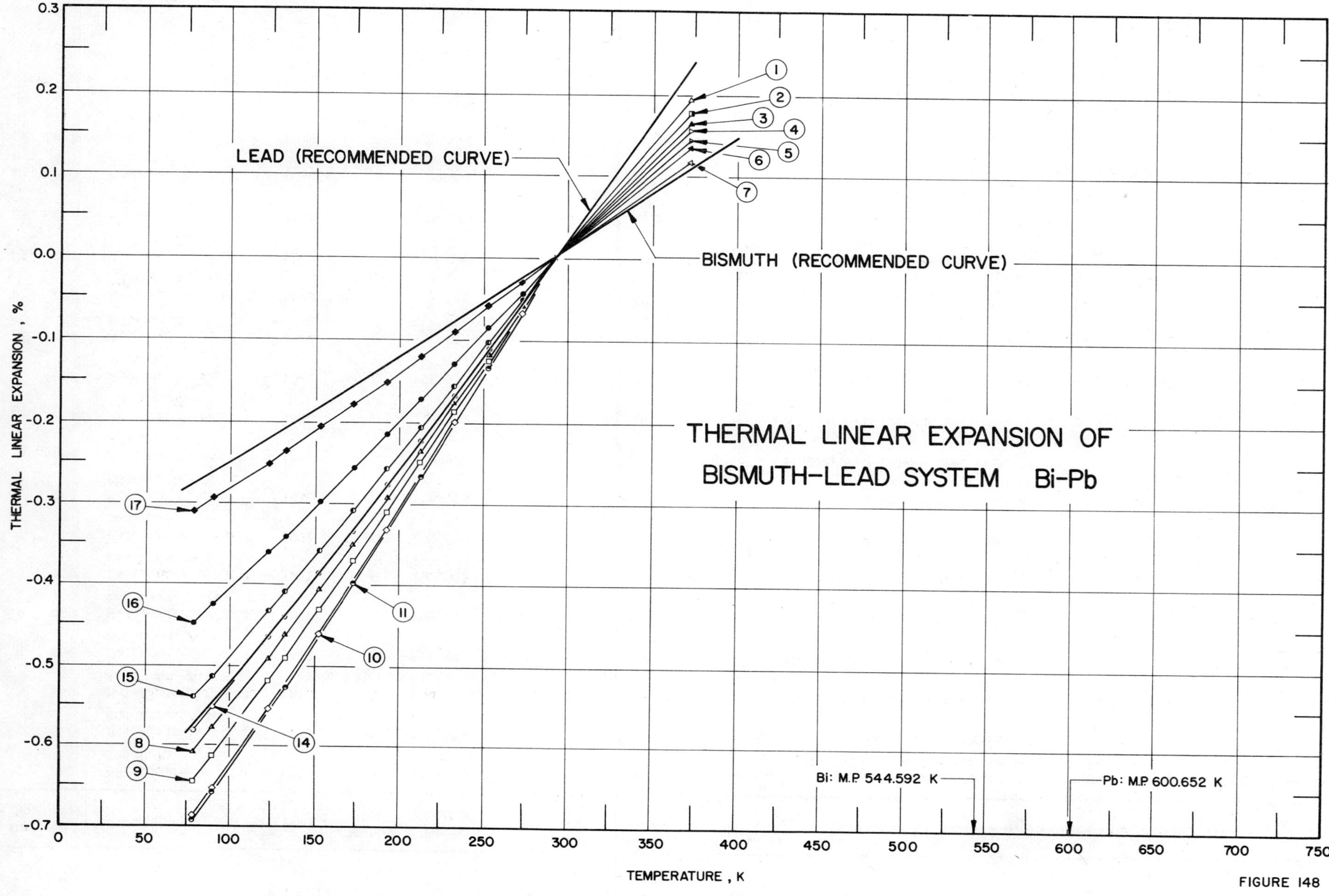
THERMAL LINEAR EXPANSION OF
BISMUTH-LEAD SYSTEM Bi-Pb
LEAD (RECOMMENDED CURVE)
BISMUTH (RECOMMENDED CURVE)
Bi: M.P. 544.592 K
Pb: M.P. 600.652 K
TEMPERATURE , K
THERMAL LINEAR EXPANSION , %
0.3
0.2
0.1
0.0
-0.1
-0.2
-0.3
-0.4
-0.5
-0.6
-0.7
0
50
100
150
200
250
300
350
400
450
500
550
600
650
700
750
1
2
3
4
5
6
7
8
9
10
11
14
15
16
17

FIGURE 148

SPECIFICATION TABLE 148. THERMAL LINEAR EXPANSION OF BISMUTH-LEAD SYSTEM Bi-Pb

Cur. No.	Ref. No.	Author(s)	Year	Method Used	Temp. Range, K	Name and Specimen Designation	Composition (weight percent) Bi	Pb	Composition (continued), Specifications, and Remarks
1	650	Balasundaram, L.J. and Sinha, A.N.	1971	L	293, 373		35	65	Alloys prepared from chemically pure metals; solidified alloys homogenized for 36 hr 50° below eutectic temperature.
2	650	Balasundaram, L.J. and Sinha, A.N.	1971	L	293, 373		45	55	Similar to the above specimen.
3	650	Balasundaram, L.J. and Sinha, A.N.	1971	L	293, 373		55	45	Similar to the above specimen.
4	650	Balasundaram, L.J. and Sinha, A.N.	1971	L	293, 373		60	40	Similar to the above specimen.
5	650	Balasundaram, L.J. and Sinha, A.N.	1971	L	293, 373		70	30	Similar to the above specimen.
6	650	Balasundaram, L.J. and Sinha, A.N.	1971	L	293, 373		80	20	Similar to the above specimen.
7	650	Balasundaram, L.J. and Sinha, A.N.	1971	L	293, 373		90	10	Similar to the above specimen.
8	659	Aoyama, S. and Mikura, Z.	1940	L	78-273		10	90	Commercial specimen; 10 cm rod, both ends smoothed; heated at 503 K for 50 hr.
9	659	Aoyama, S. and Mikura, Z.	1940	L	78-273		20	80	Similar to the above specimen.
10	659	Aoyama, S. and Mikura, Z.	1940	L	78-273		30	70	Similar to the above specimen.
11	659	Aoyama, S. and Mikura, Z.	1940	L	78-273		33	67	Similar to the above specimen; heated at 100° for 80 hr.
12*	659	Aoyama, S. and Mikura, Z.	1940	L	78-273		35	65	Similar to the above specimen.
13*	659	Aoyama, S. and Mikura, Z.	1940	L	78-273		40	60	Similar to the above specimen.
14	659	Aoyama, S. and Mikura, Z.	1940	L	78-273		50	50	Similar to the above specimen.
15	659	Aoyama, S. and Mikura, Z.	1940	L	78-273		56.5	43.5	Similar to the above specimen.
16	659	Aoyama, S. and Mikura, Z.	1940	L	78-273		70	30	Similar to the above specimen.
17	659	Aoyama, S. and Mikura, Z.	1940	L	78-273		90	10	Similar to the above specimen.

*Not shown in figure.

DATA TABLE 148. THERMAL LINEAR EXPANSION OF BISMUTH-LEAD SYSTEM Bi-Pb

[Temperature, T, K; Linear Expansion, $\Delta L/L_0$, %]

T	$\Delta L/L_0$
CURVE 1	
293	0.000
373	0.192
CURVE 2	
293	0.000*
373	0.177
CURVE 3	
293	0.000*
373	0.164
CURVE 4	
293	0.000*
373	0.157
CURVE 5	
293	0.000*
373	0.145
CURVE 6	
293	0.000*
373	0.132
CURVE 7	
293	0.000*
373	0.118
CURVE 8	
93	-0.550
113	-0.519
133	-0.431
153	-0.403
173	-0.347
193	-0.291
213	-0.234
233	-0.176
253	-0.117
273	-0.059
CURVE 9	
93	-0.583
113	-0.550
133	-0.456
153	-0.427
173	-0.368
193	-0.308
213	-0.247
233	-0.186
253	-0.125
273	-0.063
CURVE 10	
93	-0.621
113	-0.587
133	-0.487
153	-0.456
173	-0.393
193	-0.330
213	-0.265
233	-0.200
253	-0.133
273	-0.067*
CURVE 11	
93	-0.627
113	-0.592
133	-0.491
153	-0.460
173	-0.396
193	-0.332
213	-0.267
233	-0.201*
253	-0.134
273	-0.067
CURVE 12*	
93	-0.620
113	-0.585
133	-0.486
153	-0.455
173	-0.392
193	-0.328
CURVE 12 (cont.)*	
213	-0.264
233	-0.197
253	-0.133
273	-0.067
CURVE 13*	
93	-0.581
113	-0.548
133	-0.455
153	-0.426
173	-0.367
193	-0.307
213	-0.247
233	-0.186
253	-0.124
273	-0.062
CURVE 14	
93	-0.525
113	-0.495
133	-0.411
153	-0.385
173	-0.332
193	-0.277
213	-0.223
237	-0.168
253	-0.112
273	-0.056
CURVE 15	
93	-0.488
113	-0.461
133	-0.382
153	-0.357
173	-0.308
193	-0.258
213	-0.207
233	-0.156
253	-0.104
273	-0.052*
CURVE 16	
93	-0.405
113	-0.382
133	-0.317
153	-0.297
173	-0.256
193	-0.214
213	-0.172
233	-0.129
253	-0.086
273	-0.043
CURVE 17	
93	-0.280
113	-0.264
133	-0.220
153	-0.206
173	-0.178
193	-0.149
213	-0.120
233	-0.090
253	-0.060
273	-0.030

*Not shown in figure.

FIGURE AND TABLE NO. 149R. PROVISIONAL VALUES FOR THERMAL LINEAR EXPANSION OF BISMUTH-TIN SYSTEM Bi-Sn

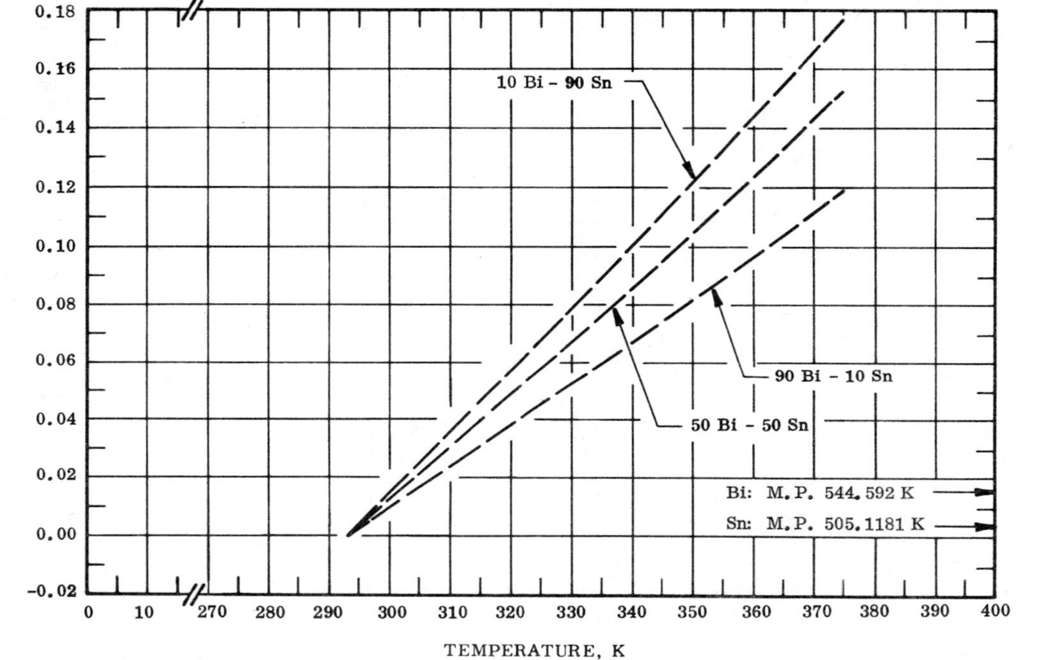

PROVISIONAL VALUES

[Temperature, T, K; Linear Expansion, $\Delta L/L_0$, %; α, K^{-1}]

T	(10 Bi-90 Sn)		(50 Bi-50 Sn)		(90 Bi-10 Sn)	
	$\Delta L/L_0$	$\alpha \times 10^6$	$\Delta L/L_0$	$\alpha \times 10^6$	$\Delta L/L_0$	$\alpha \times 10^6$
293	0.000	21.6	0.000	18.6	0.000	14.0
350	0.124	21.8	0.106	18.7	0.082	14.9
375	0.177	22.0	0.152	18.8	0.119	15.5

REMARKS

(10 Bi-90 Sn): The tabulated values for well-annealed alloy are considered accurate to within ± 7% over the entire temperature range. These values can be represented approximately by the following equation:

$$\Delta L/L_0 = -0.277 - 6.158 \times 10^{-3}\, T + 2.523 \times 10^{-5}\, T^2 - 2.538 \times 10^{-8}\, T^3$$

(50 Bi-50 Sn): The tabulated values for well-annealed alloy are considered accurate to within ± 7% over the entire temperature range. These values can be represented approximately by the following equation:

$$\Delta L/L_0 = -0.051 - 2.594 \times 10^{-3}\, T + 1.333 \times 10^{-5}\, T^2 - 1.325 \times 10^{-8}\, T^3$$

(90 Bi-10 Sn): The tabulated values for well-annealed alloy are considered accurate to within ± 7% over the entire temperature range. These values can be represented approximately by the following equation:

$$\Delta L/L_0 = -1.344 + 9.758 \times 10^{-3}\, T - 2.492 \times 10^{-5}\, T^2 + 2.480 \times 10^{-8}\, T^3$$

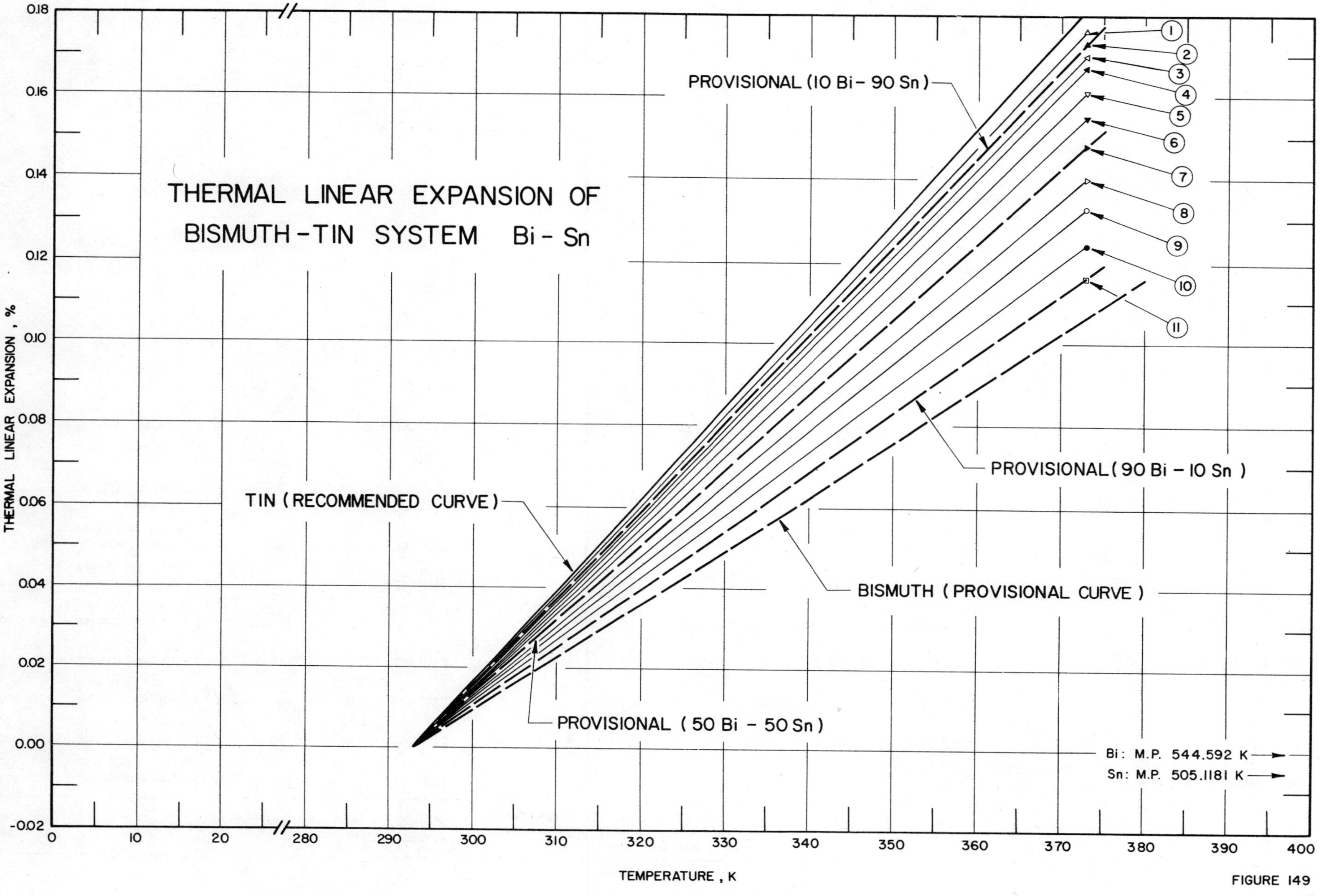

THERMAL LINEAR EXPANSION OF
BISMUTH-TIN SYSTEM Bi - Sn
PROVISIONAL (10 Bi - 90 Sn)
TIN (RECOMMENDED CURVE)
PROVISIONAL (90 Bi - 10 Sn)
BISMUTH (PROVISIONAL CURVE)
PROVISIONAL (50 Bi - 50 Sn)
1
2
3
4
5
6
7
8
9
10
11
Bi: M.P. 544.592 K
Sn: M.P. 505.1181 K
THERMAL LINEAR EXPANSION, %
0.18
0.16
0.14
0.12
0.10
0.08
0.06
0.04
0.02
0.00
-0.02
0
10
20
280
290
300
310
320
330
340
350
360
370
380
390
400
TEMPERATURE, K

FIGURE 149

SPECIFICATION TABLE 149. THERMAL LINEAR EXPANSION OF BISMUTH-TIN SYSTEM Bi-Sn

Cur. No.	Ref. No.	Author(s)	Year	Method Used	Temp. Range, K	Name and Specimen Designation	Composition (weight percent) Bi	Sn	Composition (continued), Specifications, and Remarks
1	650	Balasundarum, L.J. and Sinhi, A.N.	1971	L	293,373		5	95	Alloys prepared from chemically pure metals; solidified alloys homogenized for 36 hr 50° below eutectic temperature.
2	650	Balasundarum, L.J. and Sinhi, A.N.	1971	L	293,373		10	90	Similar to the above specimen.
3	650	Balasundarum, L.J. and Sinhi, A.N.	1971	L	293,373		15	85	Similar to the above specimen.
4	650	Balasundarum, L.J. and Sinhi, A.N.	1971	L	293,373		20	80	Similar to the above specimen.
5	650	Balasundarum, L.J. and Sinhi, A.N.	1971	L	293,373		30	70	Similar to the above specimen.
6	650	Balasundarum, L.J. and Sinhi, A.N.	1971	L	293,373		40	60	Similar to the above specimen.
7	650	Balasundarum, L.J. and Sinhi, A.N.	1971	L	293,373		50	50	Similar to the above specimen.
8	650	Balasundarum, L.J. and Sinhi, A.N.	1971	L	293,373		60	40	Similar to the above specimen.
9	650	Balasundarum, L.J. and Sinhi, A.N.	1971	L	293,373		70	30	Similar to the above specimen.
10	650	Balasundarum, L.J. and Sinhi, A.N.	1971	L	293,373		80	20	Similar to the above specimen.
11	650	Balasundarum, L.J. and Sinhi, A.N.	1971	L	293,373		90	10	Similar to the above specimen.

DATA TABLE 149. THERMAL LINEAR EXPANSION OF BISMUTH-TIN SYSTEM Bi-Sn

[Temperature, T, K; Linear Expansion, $\Delta L/L_0$, %]

T	$\Delta L/L_0$	T	$\Delta L/L_0$
CURVE 1		CURVE 11	
293	0.000	293	0.000
373	0.176	373	0.116
CURVE 2			
293	0.000		
373	0.173		
CURVE 3			
293	0.000		
373	0.170		
CURVE 4			
293	0.000		
373	0.167		
CURVE 5			
293	0.000		
373	0.161		
CURVE 6			
293	0.000		
373	0.155		
CURVE 7			
293	0.000		
373	0.148		
CURVE 8			
293	0.000		
373	0.140		
CURVE 9			
293	0.000		
373	0.133		
CURVE 10			
293	0.000		
373	0.124		

SPECIFICATION TABLE 150. THERMAL LINEAR EXPANSION OF CADMIUM-GOLD SYSTEM Cd-Au

Cur. No.	Ref. No.	Author(s)	Year	Method Used	Temp. Range, K	Name and Specimen Designation	Composition (weight percent) Cd	Au	Composition (continued), Specifications, and Remarks
1*	684	Zirinsky, S.	1956	X	353-473		33.6	66.4	Specimen prepared by melting accurately weighed 99.99 Au from Baker and Co., and 99.99 Cd from N.J. Zinc Co. in evacuated fused quartz tubes; lattice parameter at 353 K is 3.3185 kX; this curve is here reported using the first given temperature as reference temperature at which $\Delta L/L_0 = 0$.
2*	684	Zirinsky, S.	1956	L	353-473		33.6	66.4	The above specimen, but different method of measurement; this curve is here reported using the first given temperature as reference temperature at which $\Delta L/L_0 = 0$.
3*	684	Zirinsky, S.	1956	X	353-473		34.3	65.7	Similar to the above specimen; this curve is here reported using the first given temperature as reference temperature at which $\Delta L/L_0 = 0$.
4*	684	Zirinsky, S.	1956	L	353-473		34.3	65.7	The above specimen, but different method of measurement; this curve is here reported using the first given temperature as reference temperature at which $\Delta L/L_0 = 0$.

DATA TABLE 150. THERMAL LINEAR EXPANSION OF CADMIUM-GOLD SYSTEM Cd-Au

[Temperature, T, K; Linear Expansion, $\Delta L/L_0$, %]

T	$\Delta L/L_0$	T	$\Delta L/L_0$
CURVE 1*, †		CURVE 3 (cont.)*, †	
353	0.000	413	0.134
373	0.042	433	0.177
393	0.081	453	0.221
413	0.120	473	0.265
433	0.163		
453	0.202	CURVE 4*, †	
473	0.244		
		353	0.000
CURVE 2*, †		473	0.257
353	0.000		
473	0.248		
CURVE 3*, †			
353	0.000		
373	0.046		
393	0.088		

* No figure given.
† This curve is here reported using the first given temperature as reference temperature at which $\Delta L/L_0 = 0$.

FIGURE AND TABLE NO. 151AR. PROVISIONAL VALUES FOR THERMAL LINEAR EXPANSION OF CADMIUM-LEAD SYSTEM Cd-Pb

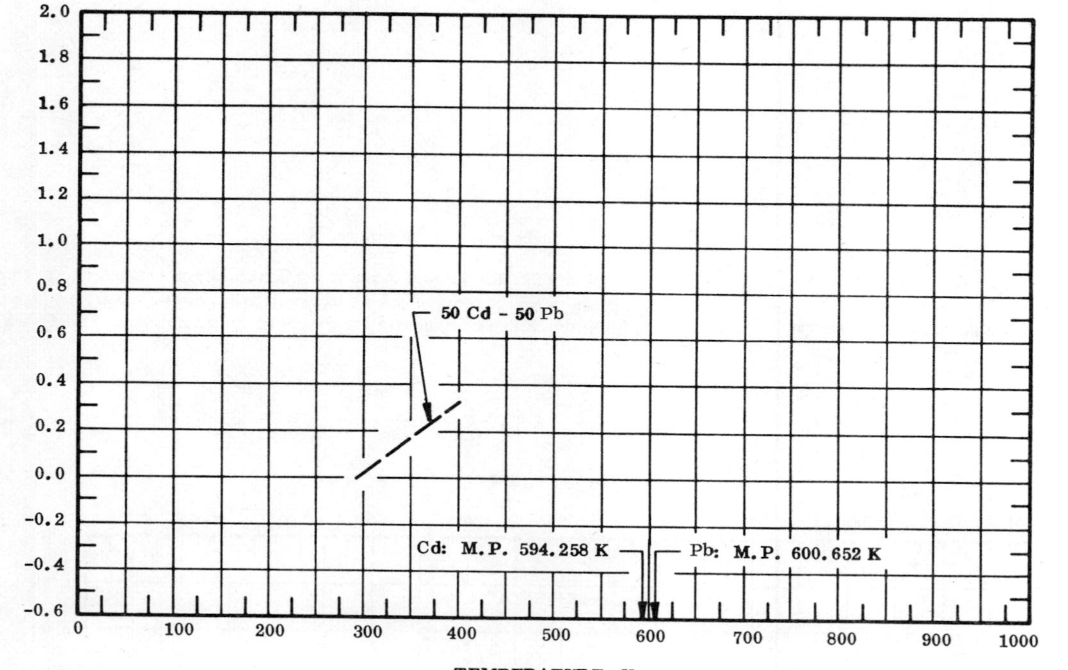

PROVISIONAL VALUES

[Temperature, T, K; Linear Expansion, $\Delta L/L_0$, %; α, K^{-1}]

	(50 Cd-50 Pb)	
T	$\Delta L/L_0$	$\alpha \times 10^6$
293	0.000	29.2
350	0.172	31.0
400	0.330	32.6

REMARKS

(50 Cd-50 Pb): The tabulated values for well-annealed alloy are considered accurate to within ±7% over the entire temperature range. These values can be represented approximately by the following equation:

$$\Delta L/L_0 = -0.617 + 1.034 \times 10^{-3}\, T + 4.532 \times 10^{-6}\, T^2 - 3.006 \times 10^{-9}\, T^3$$

FIGURE AND TABLE NO. 151CR. PROVISIONAL VALUES FOR THERMAL LINEAR EXPANSION OF CADMIUM-SILVER SYSTEM Cd-Ag

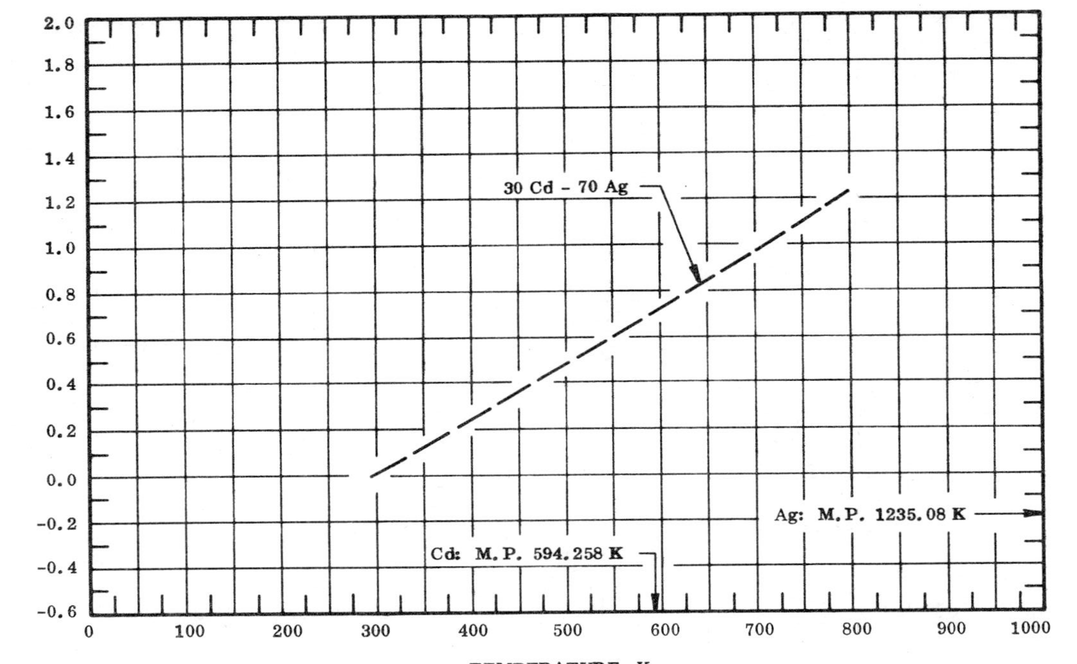

PROVISIONAL VALUES

[Temperature, T, K; Linear Expansion, $\Delta L/L_0$, %; α, K^{-1}]

T	(30 Cd-70 Ag) $\Delta L/L_0$	$\alpha \times 10^6$
293	0.000	23.2
350	0.134	23.3
400	0.250	23.4
500	0.485	23.8
600	0.724	24.4
700	0.971	24.9
800	1.225	25.6

REMARKS

(30 Cd-70 Ag): The tabulated values for well-annealed alloy are considered accurate to within ±7% over the entire temperature range. These values can be represented approximately by the following equation:

$$\Delta L/L_0 = -0.693 + 2.434 \times 10^{-3}\, T - 3.374 \times 10^{-7}\, T^2 + 3.652 \times 10^{-10}\, T^3$$

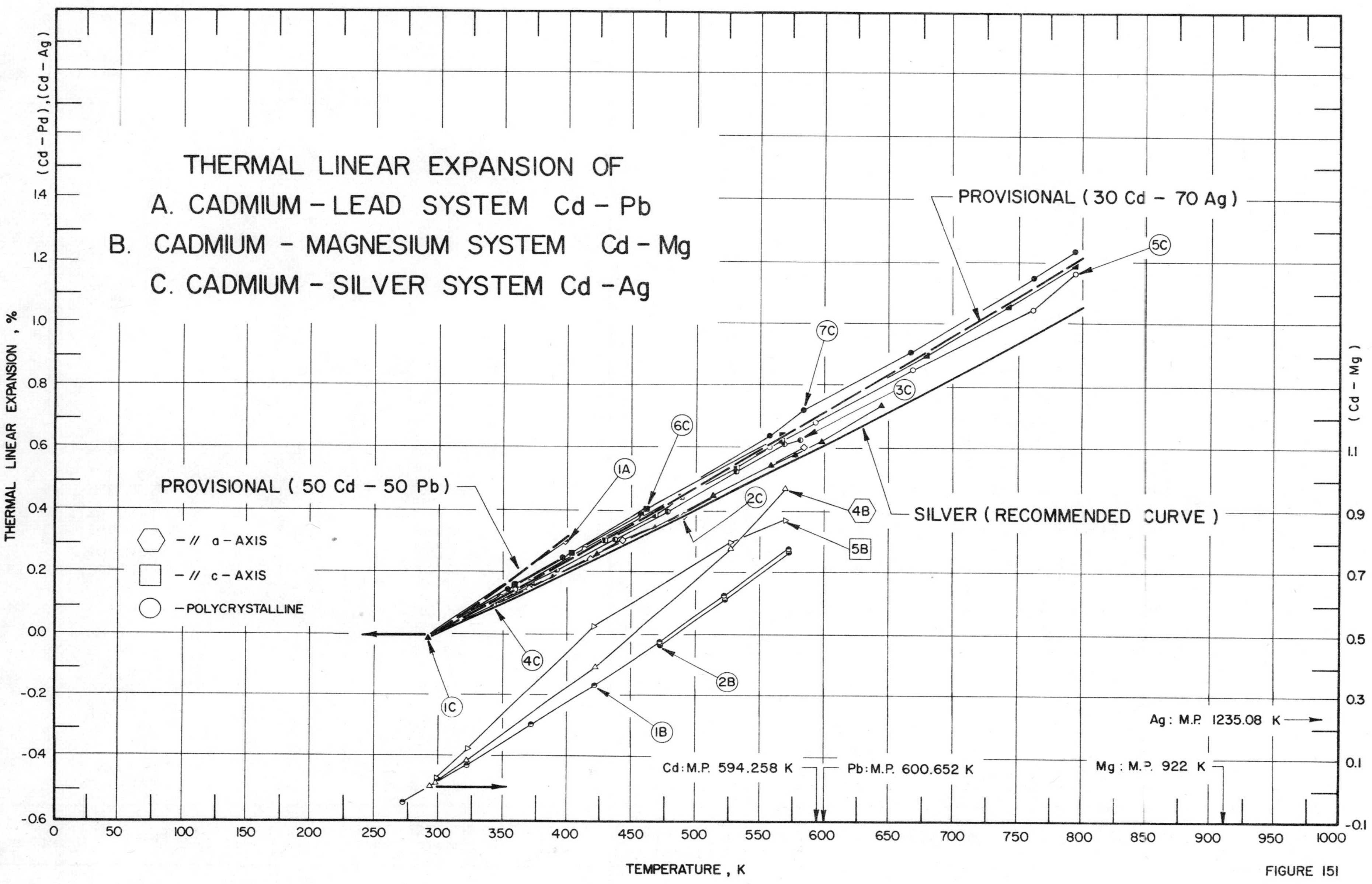
THERMAL LINEAR EXPANSION OF
A. CADMIUM - LEAD SYSTEM Cd - Pb
B. CADMIUM - MAGNESIUM SYSTEM Cd - Mg
C. CADMIUM - SILVER SYSTEM Cd - Ag
— // a - AXIS
— // c - AXIS
— POLYCRYSTALLINE
PROVISIONAL (50 Cd – 50 Pb)
PROVISIONAL (30 Cd – 70 Ag)
SILVER (RECOMMENDED CURVE)
1A
1B
1C
2B
2C
3C
4B
4C
5B
5C
6C
7C
Cd: M.P. 594.258 K
Pb: M.P. 600.652 K
Mg: M.P. 922 K
Ag: M.P. 1235.08 K
THERMAL LINEAR EXPANSION, %
(Cd – Pd), (Cd – Ag)
(Cd – Mg)
TEMPERATURE, K
0
50
100
150
200
250
300
350
400
450
500
550
600
650
700
750
800
850
900
950
1000
-0.6
-0.4
-0.2
0.0
0.2
0.4
0.6
0.8
1.0
1.2
1.4
-0.1
0.1
0.3
0.5
0.7
0.9
1.1

FIGURE 151

SPECIFICATION TABLE 151A. THERMAL LINEAR EXPANSION OF CADMIUM-LEAD SYSTEM Cd-Pb

Cur. No.	Ref. No.	Author(s)	Year	Method Used	Temp. Range, K	Name and Specimen Designation	Composition (weight percent) Cd	Pb	Composition (continued), Specifications, and Remarks
1	665	Balasundaran, L.J. and Sinha, A.N.	1971	L	293-398		80	20	Specimen (17 vol.% Pb) prepared by melting chemically pure metals in pyrex tubes; homogenized and annealed at 50° below the eutectic temperature.
2*	665	Balasundaran, L.J. and Sinha, A.N.	1971	L	293-398		70	30	Similar to the above specimen except 25 vol.% Pb.
3*	665	Balasundaran, L.J. and Sinha, A.N.	1971	L	293-398		55	45	Similar to the above specimen except 39 vol.% Pb.
4*	665	Balasundaran, L.J. and Sinha, A.N.	1971	L	293-398		40	60	Similar to the above specimen except 54 vol.% Pb.
5*	665	Balasundaran, L.J. and Sinha, A.N.	1971	L	293-398		30	70	Similar to the above specimen except 65 vol.% Pb.
6*	665	Balasundaran, L.J. and Sinha, A.N.	1971	L	293-398		23	77	Similar to the above specimen except 73 vol.% Pb.
7*	665	Balasundaran, L.J. and Sinha, A.N.	1971	L	293-398		17.4	82.6	Similar to the above specimen except 79 vol.% Pb.
8*	665	Balasundaran, L.J. and Sinha, A.N.	1971	L	293-398		13	87	Similar to the above specimen except 85 vol.% Pb.
9*	665	Balasundaran, L.J. and Sinha, A.N.	1971	L	293-398		5	95	Similar to the above specimen except 95 vol.% Pb.

DATA TABLE 151A. THERMAL LINEAR EXPANSION OF CADMIUM-LEAD SYSTEM Cd-Pb

[Temperature, T, K; Linear Expansion, $\Delta L/L_0$, %]

T	$\Delta L/L_0$	T	$\Delta L/L_0$	T	$\Delta L/L_0$
CURVE 1		CURVE 4*		CURVE 7*	
293	0.000	293	0.000	293	0.000
398	0.313	398	0.311	398	0.310
CURVE 2*		CURVE 5*		CURVE 8*	
293	0.000	293	0.000	293	0.000
398	0.313	398	0.311	398	0.309
CURVE 3*		CURVE 6*		CURVE 9*	
293	0.000	293	0.000	293	0.000
398	0.313	398	0.310	398	0.308

* Not shown in figure.

SPECIFICATION TABLE 151B. THERMAL LINEAR EXPANSION OF CADMIUM-MAGNESIUM SYSTEM Cd-Mg

Cur. No.	Ref. No.	Author(s)	Year	Method Used	Temp. Range, K	Name and Specimen Designation	Composition (weight percent) Cd	Mg	Composition (continued), Specifications, and Remarks
1	201	Takahasi, K, and K Kikuti, R.	1936		273-573		2	Bal.	Rod specimen, about 16 cm long, 5 mm, diameter, heating rate 2 C/min.
2	201	Takahasi, K. and Kikuti, R.	1936		273-573		4	Bal.	Similar to the above specimen.
3*	201	Takahasi, K. and Kikuti, R.	1936		273-573		6	Bal.	Similar to the above specimen.
4	704	Edwards, D.A., Wallace, W.E., and Craig, R.S.	1952	X	299-570		97.8	2.2	Sample prepared by filing bulk alloy under argon or helium atmosphere, size of filings smaller than 200 mesh, spectroscopic analysis showed only traces of metallic contaminants; magnesium from National Lead Co., Cadmium from Anaconda Copper Mining Co., strain annealed 10 hr at 573 K, aged at room temperature for 9 months, average measurement error of 0.20%, expansion measured along the a-axis, lattice parameter reported at 299 K is 2.9999 Å, 2.9994 Å at 293 K by graphical extrapolation.
5	704	Edwards, D.A., et al.	1952	X	299-570		97.8	2.2	The above specimen, expansion measured along the c-axis, lattice parameter reported at 299 K is 5.5157 Å, 5.5143 Å at 293 K by graphical extrapolation.
6*	704	Edwards, D.A., et al.	1952	X	306-572		94.3	5.7	Similar to the above specimen but not aged; expansion measured along the a-axis; lattice parameter reported at 298 K is 3.0257 Å, two hexagonal phases coexisted between 298 K and 335 K, reported using first temperature as reference temperature at which $\Delta L/L_0 = 0$.
7*	704	Edwards, D.A., et al.	1952	X	306-572		94.3	5.7	The above specimen, expansion measured along the c-axis, lattice parameter reported at 298 K is 5.3971 Å, two hexagonal phases coexisted between 298 K and 335 K, reported using first temperature as reference temperature at which $\Delta L/L_0 = 0$.

* Not shown in figure.

DATA TABLE 151B. THERMAL LINEAR EXPANSION OF CADMIUM-MAGNESIUM SYSTEM   Cd-Mg

[Temperature, T, K; Linear Expansion, $\Delta L/L_0$, %]

T	$\Delta L/L_0$	T	$\Delta L/L_0$
CURVE 1		CURVE 6*, †	
273	-0.048	306	0.000
323	0.072	316	0.198
373	0.201	335	0.346
423	0.337	426	0.679
743	0.478	537	1.273
523	0.625	572	1.303
573	0.776		
		CURVE 7*, †	
CURVE 2			
		306	0.000
273	-0.047*	316	-0.245
323	0.071*	335	-0.371
373	0.198*	426	-0.101
423	0.330*	537	0.007
473	0.469	572	0.120
523	0.614		
573	0.764		
CURVE 3*			
273	-0.048		
323	0.072		
373	0.201		
423	0.338		
473	0.479		
523	0.627		
573	0.777		
CURVE 4			
293	0.000		
299	0.017		
323	0.087		
422	0.394		
528	0.778		
570	0.969		
CURVE 5			
299.2	0.025		
323.2	0.125		
422.2	0.522		
528.2	0.798		
570.2	0.869		

* Not shown in figure.

† This curve is here reported using the first given temperature as reference temperature at which $\Delta L/L_0 = 0$.

SPECIFICATION TABLE 151C. THERMAL LINEAR EXPANSION OF CADMIUM-SILVER SYSTEM Cd-Ag

Cur. No.	Ref. No.	Author(s)	Year	Method Used	Temp. Range, K	Name and Specimen Designation	Composition (weight percent) Cd	Ag	Composition (continued), Specifications, and Remarks
1	303	Owen, E.A. and Roberts, E.W.	1939	X	291-644		8.08	91.92	99.95 pure materials used, fine filings of pure metals heated together in evacuated pyrex tubes; diffusion complete at 723 K; lattice parameter expansion measured; lattice parameter reported at 291 K is 4.0932 Å; 4.0934 Å at 293 K determined by graphical interpolation.
2	303	Owen, E.A. and Roberts, E.W.	1939	X	291-584		14.79	85.21	Similar to the above specimen; lattice parameter expansion measured; lattice parameter reported at 291 K is 4.1074 Å; 4.1076 Å at 293 K determined by graphical interpolation.
3	303	Owen, E.A. and Roberts, E.W.	1939	X	291-581		29.89	70.11	Similar to the above specimen; lattice parameter reported at 291 K is 4.1409 Å; 4.1411 Å at 293 K determined by graphical interpolation.
4	303	Owen, E.A. and Roberts, E.W.	1939	X	291-566		40.97	59.03	Similar to the above specimen; lattice parameter reported at 291 K is 4.1663 Å; 4.1665 Å at 293 K determined by graphical interpolation.
5	642	Quader, Md.A. and Dey, B.N.	1962	X	298-795		24.3	75.7	Alloys prepared from spectroscopically pure Matthey metals homogenized and annealed at 923 K for 6 hr; lattice parameter reported at 298 K is 4.1395 Å; 4.1390 Å at 293 K determined by graphical extrapolation; data on the coefficient of thermal linear expansion also reported.
6	642	Quader, Md.A. and Dey, B.N.	1962	X	299-795		29.3	70.7	Similar to the above specimen; lattice parameter reported at 299 K is 4.1513 Å; 4.1507 Å at 293 K determined by graphical extrapolation; data on the coefficient of thermal linear expansion also reported.
7	642	Quader, Md.A. and Dey, B.N.	1962	X	300-795		33.8	66.2	Similar to the above specimen; lattice parameter reported at 293 K is 4.1631 Å; 4.1624 Å at 293 K determined by graphical extrapolation; data on the coefficient of thermal linear expansion also reported.

DATA TABLE 151C. THERMAL LINEAR EXPANSION OF CADMIUM-SILVER SYSTEM Cd-Ag

[Temperature, T, K; Linear Expansion, $\Delta L/L_0$, %]

T	$\Delta L/L_0$
CURVE 1	
291	-0.0042
350	0.118
388	0.190
423	0.263
468	0.357
514	0.454
559	0.550
577	0.583
598	0.629
644	0.743
CURVE 2	
291	-0.0089*
370	0.155
418	0.244
442	0.304
490	0.398
559	0.557*
584	0.606
CURVE 3	
291	-0.0041*
348	0.124
391	0.219
436	0.313
477	0.400
531	0.532
569	0.620
581	0.637
CURVE 4	
291	-0.0045*
341	0.101
374	0.175
429	0.305
469	0.399
489	0.447
532	0.540
566	0.634
CURVE 5	
298.2	0.0121
310.2	0.0386

T	$\Delta L/L_0$
CURVE 5 (cont.)	
359.2	0.155
413.2	0.280
457.2	0.382
557.2	0.611
592.2	0.691
667.2	0.860
762.2	1.056
795.2	1.167
CURVE 6	
299.2	0.0144*
310.2	0.0409
359.2	0.161
404.2	0.263
462.2	0.403
568.2	0.653
592.2	0.689*
677.2	0.911
741.2	1.067
795.2	1.202
CURVE 7	
300.2	0.0168*
311.2	0.0432*
354.2	0.149
396.2	0.250
457.2	0.399
557.2	0.649
584.2	0.726
667.2	0.918
763.2	1.158
795.2	1.240

* Not shown in figure.

SPECIFICATION TABLE 152. THERMAL LINEAR EXPANSION OF CADMIUM-TIN SYSTEM Cd-Sn

Cur. No.	Ref. No.	Author(s)	Year	Method Used	Temp. Range, K	Name and Specimen Designation	Composition (weight percent) Cd	Sn	Composition (continued), Specifications, and Remarks
1*	327	Matuyama, Y.	1931	L	336-434		1	99	Pure Sn with impurities 0.01 Zn, 0.008 Pb, trace Cd, rolled specimen about 200 mm long, 2.5 mm diameter; expansion measured with increasing temperature; this curve is here reported using the first given temperature as reference temperature at which $\Delta L/L_0 = 0$.
2*	327	Matuyama, Y.	1931	L	434-335		1	99	The above specimen; expansion measured with decreasing temperature; this curve is here reported using the first given temperature as reference temperature at which $\Delta L/L_0 = 0$.
3*	327	Matuyama, Y.	1931	L	336-438		2	98	Similar to the above specimen; expansion measured with increasing temperature; this curve is here reported using the first given temperature as reference temperature at which $\Delta L/L_0 = 0$.
4*	327	Matuyama, Y.	1931	L	436-335		2	98	The above specimen; expansion measured with decreasing temperature; this curve is here reported using the first given temperature as reference temperature at which $\Delta L/L_0 = 0$.
5*	327	Matuyama, Y.	1931	L	334-442		2.5	97.5	Similar to the above specimen; expansion measured with increasing temperature; this curve is here reported using the first given temperature as reference temperature at which $\Delta L/L_0 = 0$.
6*	327	Matuyama, Y.	1931	L	442-335		2.5	97.5	The above specimen; expansion measured with decreasing temperature; this curve is here reported using the first given temperature as reference temperature at which $\Delta L/L_0 = 0$.
7*	327	Matuyama, Y.	1931	L	337-436		3	97	Similar to the above specimen; expansion measured with increasing temperature; this curve is here reported using the first given temperature as reference temperature at which $\Delta L/L_0 = 0$.
8*	327	Matuyama, Y.	1931	L	434-336		3	97	The above specimen; expansion measured with decreasing temperature; this curve is here reported using the first given temperature as reference temperature at which $\Delta L/L_0 = 0$.
9*	327	Matuyama, Y.	1931	L	334-443		3.5	96.5	Similar to the above specimen; expansion measured with increasing temperature; this curve is here reported using the first given temperature as reference temperature at which $\Delta L/L_0 = 0$.
10*	327	Matuyama, Y.	1931	L	438-336		3.5	96.5	The above specimen; expansion measured with decreasing temperature; this curve is here reported using the first given temperature as reference temperature at which $\Delta L/L_0 = 0$.
11*	327	Matuyama, Y.	1931	L	335-433		4	96	Similar to the above specimen; expansion measured with increasing temperature; this curve is here reported using the first given temperature as reference temperature at which $\Delta L/L_0 = 0$.
12*	327	Matuyama, Y.	1931	L	433-335		4	96	The above specimen; expansion measured with decreasing temperature; this curve is here reported using the first given temperature as reference temperature at which $\Delta L/L_0 = 0$.
13*	327	Matuyama, Y.	1931	L	336-434		4.5	95.5	Similar to the above specimen; expansion measured with increasing temperature; this curve is here reported using the first given temperature as reference temperature at which $\Delta L/L_0 = 0$.
14*	327	Matuyama, Y.	1931	L	434-334		4.5	95.5	The above specimen; expansion measured with decreasing temperature; this curve is here reported using the first given temperature as reference temperature at which $\Delta L/L_0 = 0$.

* No figure given.

SPECIFICATION TABLE 152. THERMAL LINEAR EXPANSION OF CADMIUM-TIN SYSTEM Cd-Sn (continued)

Cur. No.	Ref. No.	Author(s)	Year	Method Used	Temp. Range, K	Name and Specimen Designation	Composition (weight percent) Cd	Sn	Composition (continued), Specifications, and Remarks
15*	327	Matuyama, Y.	1931	L	334-440		5	95	Similar to the above specimen; expansion measured with increasing temperature; this curve is here reported using the first given temperature as reference temperature at which $\Delta L/L_0 = 0$.
16*	327	Matuyama, Y.	1931	L	436-333		5	95	The above specimen; expansion measured with decreasing temperature; this curve is here reported using the first given temperature as reference temperature at which $\Delta L/L_0 = 0$.
17*	327	Matuyama, Y.	1931	L	333-440		5.5	94.5	Similar to the above specimen; expansion measured with increasing temperature; this curve is here reported using the first given temperature as reference temperature at which $\Delta L/L_0 = 0$.
18*	327	Matuyama, Y.	1931	L	435-334		5.5	94.5	The above specimen; expansion measured with decreasing temperature; this curve is here reported using the first given temperature as reference temperature at which $\Delta L/L_0 = 0$.
19*	327	Matuyama, Y.	1931	L	333-430		6	94	Similar to the above specimen; expansion measured with increasing temperature; this curve is here reported using the first given temperature as reference temperature at which $\Delta L/L_0 = 0$.
20*	327	Matuyama, Y.	1931	L	425-333		6	94	The above specimen; expansion measured with decreasing temperature; this curve is here reported using the first given temperature as reference temperature at which $\Delta L/L_0 = 0$.
21*	327	Matuyama, Y.	1931	L	336-428		7	93	Similar to the above specimen; expansion measured with increasing temperature; this curve is here reported using the first given temperature as reference temperature at which $\Delta L/L_0 = 0$.
22*	327	Matuyama, Y.	1931	L	423-335		7	93	The above specimen; expansion measured with decreasing temperature; this curve is here reported using the first given temperature as reference temperature at which $\Delta L/L_0 = 0$.

*** No figure given.**

DATA TABLE 152. THERMAL LINEAR EXPANSION OF CADMIUM-TIN SYSTEM Cd-Sn

[Temperature, T, K; Linear Expansion, $\Delta L/L_0$, %]

T	$\Delta L/L_0$
CURVE 1*, †	
336	0.0000
338	0.0006
342	0.0012
346	0.0012
350	0.0012
354	0.0012
360	0.0016
367	0.0016
373	0.0025
379	0.0025
384	0.0037
388	0.0037
392	0.0044
397	0.0053
402	0.0062
406	0.0072
411	0.0081
413	0.0087
415	0.0097
417	0.0103
419	0.0103
422	0.0112
424	0.0119
426	0.0134
429	0.0137
431	0.0137
434	0.0144
CURVE 2*, †	
434	0.0097
432	0.0087
426	0.0082
424	0.0082
421	0.0081
417	0.0082
414	0.0072
412	0.0072
411	0.0072
408	0.0069
406	0.0069
404	0.0066
402	0.0066
399	0.0056
397	0.0050
395	0.0050
CURVE 2 (cont.)*, †	
394	0.0047
393	0.0047
386	0.0044
384	0.0044
381	0.0044
379	0.0044
377	0.0044
375	0.0044
371	0.0044
369	0.0044
367	0.0041
352	0.0025
342	0.0012
337	0.0000
335	0.0000
CURVE 3*, †	
336	0.0000
338	0.0006
342	0.0016
350	0.0028
353	0.0041
359	0.0050
363	0.0062
371	0.0078
375	0.0087
379	0.0097
385	0.0122
388	0.0128
393	0.0150
398	0.0162
402	0.0162
404	0.0168
408	0.0172
412	0.0172
415	0.0172
417	0.0165
419	0.0165
422	0.0165
424	0.0162
426	0.0147
428	0.0131
430	0.0106
431	0.0090
432	0.0059
CURVE 3 (cont.)*, †	
434	0.0031
435	0.0006
436	-0.0016
437	-0.0028
438	-0.0037
CURVE 4*, †	
436	-0.0178
433	-0.0178
431	-0.0178
428	-0.0178
426	-0.0178
424	-0.0178
422	-0.0178
418	-0.0178
415	-0.0178
413	-0.0175
412	-0.0175
409	-0.0175
407	-0.0175
405	-0.0175
403	-0.0175
401	-0.0165
399	-0.0144
398	-0.0125
397	-0.0103
395	-0.0081
394	-0.0050
393	-0.0028
392	-0.0006
391	0.0047
389	0.0069
387	0.0069
385	0.0078
382	0.0078
378	0.0078
372	0.0078
370	0.0078
368	0.0078
366	0.0078
362	0.0069
353	0.0056
348	0.0031
340	0.0019
335	0.0000
CURVE 5*, †	
334	0.0000
340	0.0016
349	0.0041
356	0.0059
362	0.0072
362	0.0078
368	0.0082
369	0.0082
371	0.0094
374	0.0103
379	0.0103
383	0.0125
388	0.0140
393	0.0156
397	0.0172
401	0.0175
406	0.0181
410	0.0181
412	0.0181
415	0.0168
417	0.0162
419	0.0137
421	0.0112
423	0.0078
424	0.0053
426	0.0031
427	-0.0022
429	-0.0069
430	-0.0106
431	-0.0140
433	-0.0184
435	-0.0240
437	-0.0281
438	-0.0300
441	-0.0312
442	-0.0331
CURVE 6*, †	
442	-0.0284
440	-0.0271
438	-0.0262
435	-0.0246
433	-0.0246
432	-0.0246
430	-0.0246
CURVE 6 (cont.)*, †	
428	-0.0246
425	-0.0234
423	-0.0231
421	-0.0231
419	-0.0231
417	-0.0231
412	-0.0231
406	-0.0225
404	-0.0225
403	-0.0215
401	-0.0200
400	-0.0175
399	-0.0153
398	-0.0112
396	-0.0041
395	0.0025
394	0.0082
393	0.0125
391	0.0134
389	0.0134
385	0.0134
380	0.0119
374	0.0115
371	0.0106
369	0.0097
368	0.0097
362	0.0090
362	0.0082
354	0.0062
351	0.0050
341	0.0022
335	0.0000
CURVE 7*, †	
337	0.0000
342	0.0022
345	0.0034
348	0.0047
364	0.0081
371	0.0100
374	0.0100
376	0.0109
381	0.0119
386	0.0140
388	0.0144
CURVE 7 (cont.)*, †	
391	0.0162
395	0.0168
398	0.0187
401	0.0190
404	0.0200
408	0.0209
411	0.0209
414	0.0203
416	0.0175
417	0.0150
419	0.0128
420	0.0103
422	0.0081
423	0.0047
424	-0.0006
425	-0.0059
426	-0.0150
427	-0.0228
428	-0.0290
430	-0.0328
433	-0.0353
436	-0.0362
CURVE 8*, †	
434	-0.0359
432	-0.0359
430	-0.0359
427	-0.0359
425	-0.0359
423	-0.0359
422	-0.0359
421	-0.0359
419	-0.0359
417	-0.0359
415	-0.0359
412	-0.0359
410	-0.0359
410	-0.0359
409	-0.0359
407	-0.0359
405	-0.0359
403	-0.0349
402	-0.0334
401	-0.0306
400	-0.0271

* No figure given.

† This curve is here reported using the first given temperature as reference temperature at which $\Delta L/L_0 = 0$.

DATA TABLE 152. THERMAL LINEAR EXPANSION OF CADMIUM-TIN SYSTEM Cd-Sn (continued)

CURVE 8 (cont.)*, †

T	$\Delta L/L_0$
397	-0.0137
397	-0.0050
396	0.0066
394	0.0131
393	0.0165
389	0.0165
388	0.0168
386	0.0165
383	0.0150
381	0.0144
379	0.0134
369	0.0112
359	0.0078
354	0.0066
349	0.0047
346	0.0037
342	0.0016
337	0.0006
336	0.0000

CURVE 9*, †

T	$\Delta L/L_0$
334	0.0000
335	0.0000
339	0.0009
344	0.0019
347	0.0028
351	0.0034
355	0.0050
358	0.0059
363	0.0069
366	0.0075
373	0.0094
379	0.0112
383	0.0122
387	0.0134
390	0.0144
393	0.0150
397	0.0172
401	0.0197
404	0.0215
407	0.0231
410	0.0231
412	0.0209
415	0.0181
417	0.0153
419	0.0103
421	0.0059

CURVE 9 (cont.)*, †

T	$\Delta L/L_0$
423	0.0022
425	-0.0025
426	-0.0072
427	-0.0106
429	-0.0187
431	-0.0243
432	-0.0303
433	-0.0365
434	-0.0418
436	-0.0496
437	-0.0546
439	-0.0568
441	-0.0568
443	-0.0552

CURVE 10*, †

T	$\Delta L/L_0$
438	-0.0552
435	-0.0558
433	-0.0558
431	-0.0558
430	-0.0558
428	-0.0558
426	-0.0558
423	-0.0558
420	-0.0558
418	-0.0558
417	-0.0558
414	-0.0558
412	-0.0558
409	-0.0558
407	-0.0558
405	-0.0558
404	-0.0521
402	-0.0496
401	-0.0459
400	-0.0421
399	-0.0371
397	-0.0290
396	-0.0190
395	-0.0053
395	-0.0016
395	0.0028
393	0.0081
392	0.0100
388	0.0100
384	0.0100
375	0.0087

CURVE 10 (cont.)*, †

T	$\Delta L/L_0$
371	0.0078
366	0.0069
363	0.0066
359	0.0059
353	0.0047
350	0.0034
339	0.0003
336	0.0000

CURVE 11*, †

T	$\Delta L/L_0$
335	0.0000
342	0.0028
344	0.0034
350	0.0050
352	0.0059
357	0.0069
359	0.0078
365	0.0087
374	0.0112
382	0.0134
384	0.0140
386	0.0147
388	0.0162
391	0.0172
393	0.0184
396	0.0197
398	0.0206
400	0.0218
401	0.0218
402	0.0222
404	0.0222
405	0.0228
407	0.0228
409	0.0222
410	0.0206
411	0.0178
412	0.0131
413	0.0066
414	0.0025
415	-0.0022
416	-0.0053
417	-0.0097
418	-0.0147
419	-0.0187
420	-0.0228
421	-0.0253
423	-0.0275

CURVE 11 (cont.)*, †

T	$\Delta L/L_0$
424	-0.0306
426	-0.0334
429	-0.0353
431	-0.0353
433	-0.0353

CURVE 12*, †

T	$\Delta L/L_0$
433	-0.0540
429	-0.0552
426	-0.0558
422	-0.0574
420	-0.0577
417	-0.0580
415	-0.0587
411	-0.0596
408	-0.0599
407	-0.0599
404	-0.0593
403	-0.0562
402	-0.0534
401	-0.0496
400	-0.0456
399	-0.0406
397	-0.0343
397	-0.0296
396	-0.0250
396	-0.0175
395	0.0041
394	0.0082
392	0.0097
390	0.0103
387	0.0106
385	0.0106
383	0.0103
381	0.0103
376	0.0087
374	0.0087
371	0.0082
368	0.0078
366	0.0072
356	0.0059
351	0.0047
341	0.0012
337	0.0006
335	0.0000

CURVE 13*, †

T	$\Delta L/L_0$
336	0.0000
337	0.0006
343	0.0019
349	0.0031
356	0.0041
363	0.0059
368	0.0069
374	0.0082
378	0.0100
383	0.0115
392	0.0144
399	0.0165
403	0.0178
406	0.0184
408	0.0184
410	0.0159
412	0.0087
413	0.0022
414	-0.0062
416	-0.0190
419	-0.0356
422	-0.0505
425	-0.0552
429	-0.0580
430	-0.0583
432	-0.0583
434	-0.0583

CURVE 14*, †

T	$\Delta L/L_0$
434	-0.0518
432	-0.0518
430	-0.0518
427	-0.0518
425	-0.0518
423	-0.0518
421	-0.0518
418	-0.0527
416	-0.0527
414	-0.0530
412	-0.0530
410	-0.0537
407	-0.0546
406	-0.0546
404	-0.0546
402	-0.0546
401	-0.0534
399	-0.0512

CURVE 14 (cont.)*, †

T	$\Delta L/L_0$
398	-0.0487
397	-0.0449
396	-0.0409
395	-0.0346
394	-0.0259
392	-0.0140
392	-0.0059
391	0.0041
390	0.0081
388	0.0094
387	0.0100
386	0.0115
383	0.0103
378	0.0094
376	0.0094
370	0.0082
368	0.0078
366	0.0075
360	0.0059
350	0.0034
345	0.0022
342	0.0019
337	0.0009
334	0.0000

CURVE 15*, †

T	$\Delta L/L_0$
334	0.0000
341	0.0022
350	0.0034
353	0.0037
363	0.0056
373	0.0072
378	0.0087
383	0.0097
387	0.0115
392	0.0128
394	0.0128
396	0.0137
400	0.0147
403	0.0153
407	0.0153
410	0.0128
412	0.0044
414	-0.0100
416	-0.0190
419	-0.0324
421	-0.0462

* No figure given.
† This curve is here reported using the first given temperature as reference temperature at which $\Delta L/L_0 = 0$.

DATA TABLE 152. THERMAL LINEAR EXPANSION OF CADMIUM-TIN SYSTEM Cd-Sn (continued)

T	$\Delta L/L_0$
CURVE 15 (cont.)*, †	
423	-0.0562
425	-0.0627
427	-0.0686
429	-0.0711
433	-0.0739
435	-0.0771
438	-0.0802
440	-0.0814
CURVE 16*, †	
436	-0.0736
434	-0.0736
432	-0.0736
429	-0.0736
425	-0.0736
423	-0.0736
416	-0.0736
412	-0.0736
407	-0.0736
405	-0.0733
403	-0.0718
401	-0.0699
400	-0.0665
399	-0.0627
397	-0.0593
396	-0.0534
395	-0.0471
394	-0.0399
393	-0.0293
391	-0.0222
390	-0.0062
389	0.0034
388	0.0081
387	0.0087
386	0.0097
384	0.0097
383	0.0094
381	0.0094
379	0.0090
377	0.0082
373	0.0078
365	0.0072
358	0.0062
345	0.0031
342	0.0025
333	0.0000

T	$\Delta L/L_0$
CURVE 17*, †	
333	0.0000
337	0.0006
342	0.0022
351	0.0047
354	0.0050
356	0.0053
364	0.0075
368	0.0094
373	0.0097
378	0.0122
383	0.0144
388	0.0156
401	0.0200
406	0.0209
409	0.0200
410	0.0168
411	0.0115
412	0.0059
414	-0.0031
415	-0.0097
416	-0.0165
418	-0.0218
421	-0.0371
423	-0.0446
425	-0.0521
427	-0.0565
430	-0.0593
432	-0.0615
434	-0.0633
437	-0.0652
440	-0.0665
CURVE 18*, †	
435	-0.0736
432	-0.0724
428	-0.0699
426	-0.0680
424	-0.0674
421	-0.0674
419	-0.0674
418	-0.0674
412	-0.0680
410	-0.0683
409	-0.0683
408	-0.0686
407	-0.0686
405	-0.0680

T	$\Delta L/L_0$
CURVE 18 (cont.)*, †	
403	-0.0680
401	-0.0680
399	-0.0677
397	-0.0668
395	-0.0646
394	-0.0527
393	-0.0362
392	-0.0178
391	-0.0041
389	0.0050
388	0.0082
387	0.0090
386	0.0090
385	0.0090
384	0.0090
383	0.0090
382	0.0090
379	0.0087
371	0.0069
361	0.0056
343	0.0019
340	0.0009
335	0.0000
334	0.0000
CURVE 19*, †	
333	0.0000
337	0.0006
343	0.0012
349	0.0022
353	0.0031
358	0.0041
366	0.0056
371	0.0066
376	0.0075
379	0.0082
386	0.0103
390	0.0115
396	0.0134
401	0.0137
405	0.0134
408	0.0109
410	0.0019
412	-0.0137
414	-0.0300
416	-0.0452
419	-0.0593

T	$\Delta L/L_0$
CURVE 19 (cont.)*, †	
422	-0.0640
424	-0.0665
427	-0.0686
430	-0.0699
CURVE 20*, †	
425	-0.0602
423	-0.0602
421	-0.0602
418	-0.0602
415	-0.0602
412	-0.0602
410	-0.0602
408	-0.0602
407	-0.0602
406	-0.0599
405	-0.0590
404	-0.0574
402	-0.0555
401	-0.0540
400	-0.0524
399	-0.0490
398	-0.0456
397	-0.0396
396	-0.0331
395	-0.0246
394	-0.0137
393	-0.0022
392	0.0047
390	0.0069
387	0.0072
384	0.0072
382	0.0072
378	0.0072
376	0.0062
372	0.0059
368	0.0056
366	0.0053
361	0.0044
358	0.0037
353	0.0031
349	0.0028
347	0.0022
345	0.0019
333	0.0000

T	$\Delta L/L_0$
CURVE 21*, †	
336	0.0000
338	0.0009
341	0.0019
349	0.0059
353	0.0081
359	0.0100
363	0.0112
378	0.0197
383	0.0218
387	0.0253
392	0.0278
396	0.0309
401	0.0337
404	0.0340
407	0.0328
409	0.0250
411	0.0144
414	0.0050
416	-0.0041
420	-0.0128
423	-0.0137
425	-0.0128
428	-0.0106
CURVE 22*, †	
423	-0.0505
419	-0.0509
416	-0.0515
412	-0.0521
409	-0.0521
407	-0.0524
405	-0.0518
404	-0.0505
402	-0.0484
401	-0.0452
399	-0.0406
399	-0.0362
397	-0.0231
394	-0.0003
393	0.0087
392	0.0115
391	0.0122
390	0.0128
389	0.0131
387	0.0131
386	0.0131
384	0.0128

T	$\Delta L/L_0$
CURVE 22 (cont.)*, †	
377	0.0112
374	0.0103
369	0.0090
358	0.0069
353	0.0053
348	0.0041
345	0.0031
343	0.0025
339	0.0009
337	0.0006
335	0.0000

* No figure given.
† This curve is here reported using the first given temperature as reference temperature at which $\Delta L/L_0 = 0$.

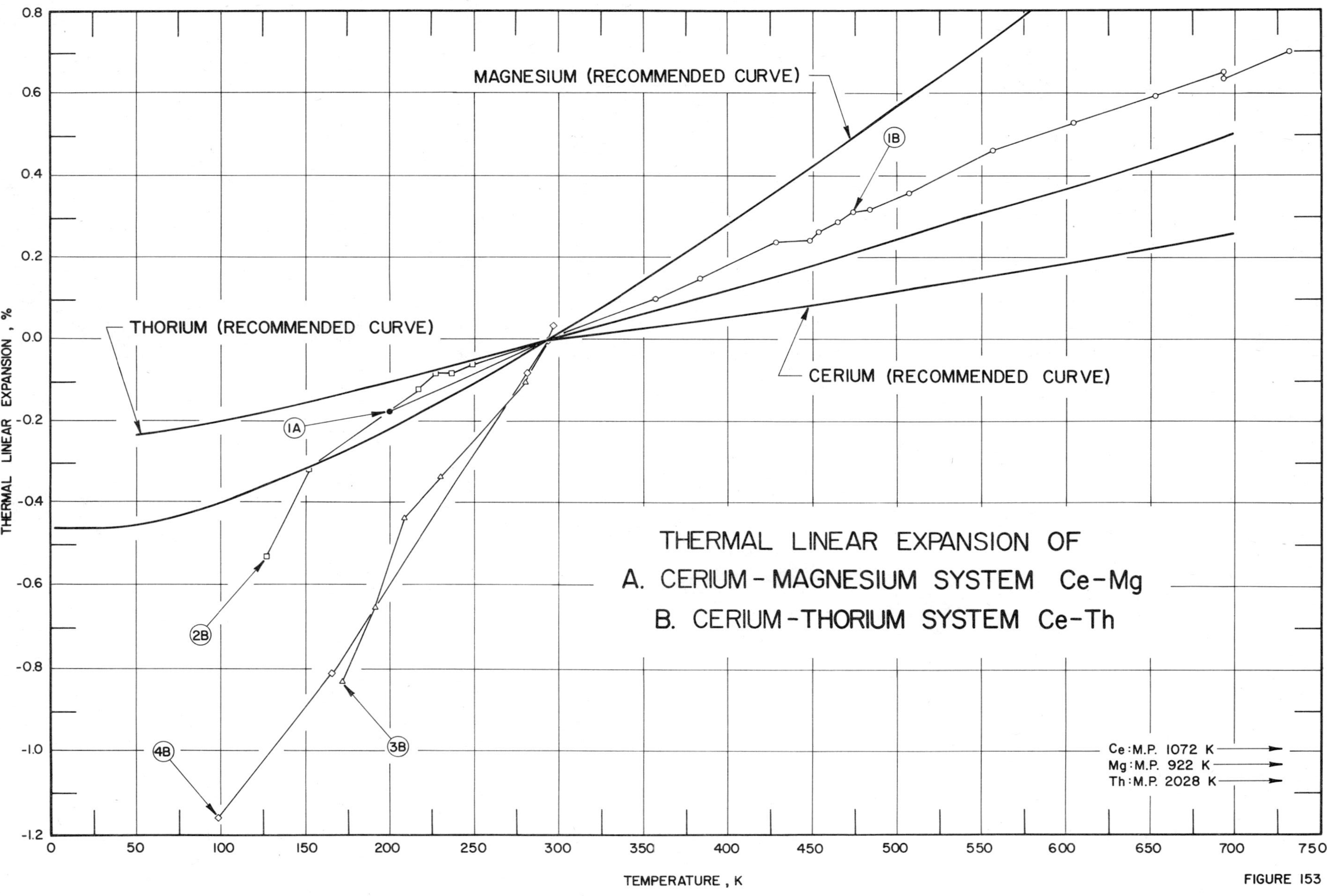
MAGNESIUM (RECOMMENDED CURVE)
1B
THORIUM (RECOMMENDED CURVE)
CERIUM (RECOMMENDED CURVE)
1A
THERMAL LINEAR EXPANSION OF
A. CERIUM - MAGNESIUM SYSTEM Ce-Mg
B. CERIUM - THORIUM SYSTEM Ce-Th
2B
4B
3B
Ce : M.P. 1072 K
Mg : M.P. 922 K
Th : M.P. 2028 K
THERMAL LINEAR EXPANSION, %
0.8
0.6
0.4
0.2
0.0
-0.2
-0.4
-0.6
-0.8
-1.0
-1.2
0
50
100
150
200
250
300
350
400
450
500
550
600
650
700
750
TEMPERATURE, K

FIGURE 153

SPECIFICATION TABLE 153A. THERMAL LINEAR EXPANSION OF CERIUM-MAGNESIUM SYSTEM Ce-Mg

Cur. No.	Ref. No.	Author(s)	Year	Method Used	Temp. Range, K	Name and Specimen Designation	Composition (weight percent) Ce	Mg	Composition (continued), Specifications, and Remarks
1	706	Gscheider, K.A.	1962	X	200, 295		99.13	0.87	Specimen made from filings taken from casting that was heated at 723 K for 200 hr and then air cooled, filings sealed in capillary tubes, heated to 723 K for 10 to 15 min and water quenched.

DATA TABLE 153A. THERMAL LINEAR EXPANSION OF CERIUM-MAGNESIUM SYSTEM Ce-Mg

[Temperature, T, K; Linear Expansion, $\Delta L/L_0$, %]

T	$\Delta L/L_0$
CURVE 1	
200	-0.171
295	0.004*

* Not shown in figure.

SPECIFICATION TABLE 153B. THERMAL LINEAR EXPANSION OF CERIUM-THORIUM SYSTEM Ce-Th

Cur. No.	Ref. No.	Author(s)	Year	Method Used	Temp. Range, K	Name and Specimen Designation	Composition (weight percent) Ce	Th	Composition (continued), Specifications, and Remarks
1	634	Harris, I.R. and Raynor, G.V.	1964	X	299-733		13	87	Prepared from thorium from Atomic Energy Research Establishment, Harwell, England, < 0.1 metallic impurities, 0.02 C, 0.01 N and O, and cerium from Johnson Matthey Co., Ltd, containing < 0.01 rare earth impurities; lattice parameter reported at 299 K is 5.0702 kX (5.0804 Å); 5.0697 kX (5.0799 Å) at 293 K determined by graphical extrapolation.
2	634	Harris, I.R. and Raynor, G.V.	1964	X	127-294		84	16	Similar to the above specimen, lattice parameter reported at 293 K is 5.134 kX (5.144 Å).
3	634	Harris, I.R. and Raynor, G.V.	1964	X	171-293		48	52	Similar to the above specimen, lattice parameter reported at 293 K is 5.101 dX (5.111 Å).
4	634	Harris, I.R. and Raynor, G.V.	1964	X	98-297		38	62	Similar to the above specimen, lattice parameter reported at 297 K is 5.087 dX (5.097 Å), 5.085 dX (5.095 Å) at 293 K determined by graphical interpolation.

DATA TABLE 153B. THERMAL LINEAR EXPANSION OF CERIUM-THORIUM SYSTEM Ce-Th

[Temperature, T, K; Linear Expansion, $\Delta L/L_0$, %]

T	$\Delta L/L_0$
CURVE 1	
299	0.010
301	0.018
357	0.106
383	0.152
429	0.241
449	0.244
453	0.268
465	0.296
474	0.319
484	0.323
507	0.361
558	0.467
605	0.534
653	0.598
693	0.637
693	0.629
733	0.706
CURVE 2	
127	-0.526
151	-0.312
217	-0.117
227	-0.078
229	-0.078*
236	-0.078
248	-0.058
250	-0.058*
293	0.000
294	0.000*
CURVE 3	
171	-0.823
191	-0.549
208	-0.431
230	-0.333
CURVE 3 (cont.)	
280	-0.098
293	0.000
CURVE 4	
98	-1.160
165	-0.806
281	-0.078
293	0.000
297	0.039

* Not shown in figure.

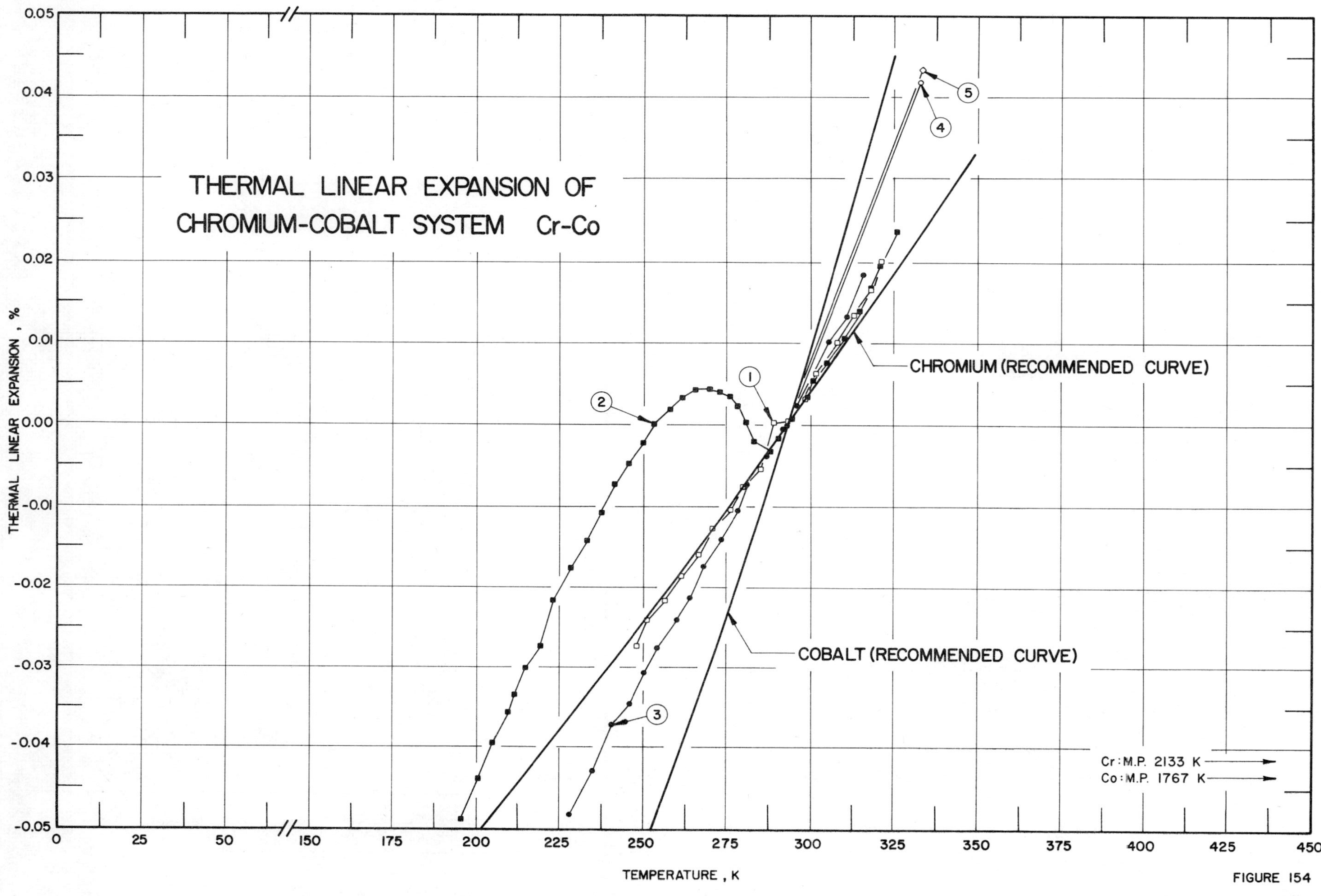

THERMAL LINEAR EXPANSION OF
CHROMIUM-COBALT SYSTEM Cr-Co
CHROMIUM (RECOMMENDED CURVE)
COBALT (RECOMMENDED CURVE)
Cr: M.P. 2133 K
Co: M.P. 1767 K
1
2
3
4
5
THERMAL LINEAR EXPANSION, %
TEMPERATURE, K
0.05
0.04
0.03
0.02
0.01
0.00
-0.01
-0.02
-0.03
-0.04
-0.05
0
25
50
150
175
200
225
250
275
300
325
350
375
400
425
450

FIGURE 154

SPECIFICATION TABLE 154. THERMAL LINEAR EXPANSION OF CHROMIUM-COBALT SYSTEM Cr-Co

Cur. No.	Ref. No.	Author(s)	Year	Method Used	Temp. Range, K	Name and Specimen Designation	Composition (weight percent) Cr	Co	Composition (continued), Specifications, and Remarks
1	397	Endoh, Y., Ishikawa, Y., and Ohno, H.	1968	S	248-322	Dilute Cr alloy with Co	99.094	0.906	Single crystal grown by strain anneal technique from polycrystalline alloys melted several times in argon atmosphere; Néel temperature 290 K; expansion measured in (110) plane along [100]; zero-point correction of 0.018% determined by graphical interpolation.
2	397	Endoh, Y., et al.	1968	S	196-326	Dilute Cr alloy with Co	97.58	2.42	Similar to the above specimen; Néel temperature 300 K; zero-point correction of 0.0054% determined by graphical interpolation.
3	397	Endoh, Y., et al.	1968	S	217-316	Dilute Cr alloy with Co	94.01	5.98	Similar to the above specimen; Néel temperature 297 K; zero-point correction of 0.0287% determined by graphical interpolation.
4	129	Masumoto, H.	1934	L	293, 333		9.0	91.0	Electrolytic cobalt (0.13 Fe, 0.074 Al, 0.05 C, 0.007 P, 0.002 Si, trace Ni, S, Mn) and chromium (1.88 Al, 0.44 Fe, 0.33 Si, 0.03 C) supplied by Scigibayashi & Co. used as raw materials; mixed and melted in alumina crucible, placed in Tammann furnace under hydrogen atm; melt cast into iron mold; test-specimen 10 cm long cut from cast rod; heated for 1 hr at 1273 K in vacuum furnace and slowly cooled.
5	129	Masumoto, H.	1934	L	293, 333		20.0	80.0	Similar to the above specimen.

DATA TABLE 154. THERMAL LINEAR EXPANSION OF CHROMIUM-COBALT SYSTEM Cr-Co

[Temperature, T, K; Linear Expansion, $\Delta L/L_0$, %]

T	$\Delta L/L_0$	T	$\Delta L/L_0$	T	$\Delta L/L_0$	T	$\Delta L/L_0$	T	$\Delta L/L_0$
CURVE 1		CURVE 2		CURVE 2 (cont.)		CURVE 3		CURVE 3 (cont.)	
248	-0.0271	196	-0.0490	266	0.0041*	217	-0.0574	301	0.0054*
252	-0.024	201	-0.0441*	270	0.0042	221	-0.0540*	305	0.0101
257	-0.0216	205	-0.0398*	273	0.0040*	228	-0.0484*	311	0.0132*
262	-0.0186*	210	-0.0356	276	0.0035*	235	-0.0430	316	0.0182
267	-0.0159*	212	-0.0337*	278	0.0021*	241	-0.0373*		
271	-0.0126	215	-0.0303*	281	0.0001	246	-0.0346*	CURVE 4	
276	-0.0102*	219	-0.0274	283	-0.0020*	250	-0.0309		
280	-0.0074	223	-0.0218*	288	-0.0031	254	-0.0275*	293	0.0000
285	-0.0052*	228	-0.0178	291	-0.0017*	260	-0.0240*	333	0.0419
289	-0.0001*	233	-0.0141*	294	0.0009*	264	-0.0212		
294	0.0002	237	-0.0108*	299	0.0032	268	-0.0174	CURVE 5	
298	0.0032*	242	-0.0073	301	0.0054*	273	-0.0140*		
303	0.0063	246	-0.0049*	305	0.0078*	278	-0.0104*	293	0.0000
308	0.0104	250	-0.0022	310	0.0107	281	-0.0072*	333	0.0433
313	0.0135	253	0.0001*	315	0.0140*	287	-0.0039*		
317	0.0168	258	0.0019	318	0.0169	292	-0.0007		
322	0.0202	262	0.0032*	321	0.0196*	296	0.0023*		
				326	0.0237				

*Not shown in figure.

FIGURE AND TABLE NO. 155R. PROVISIONAL VALUES FOR THERMAL LINEAR EXPANSION OF CHROMIUM-IRON SYSTEM Cr-Fe

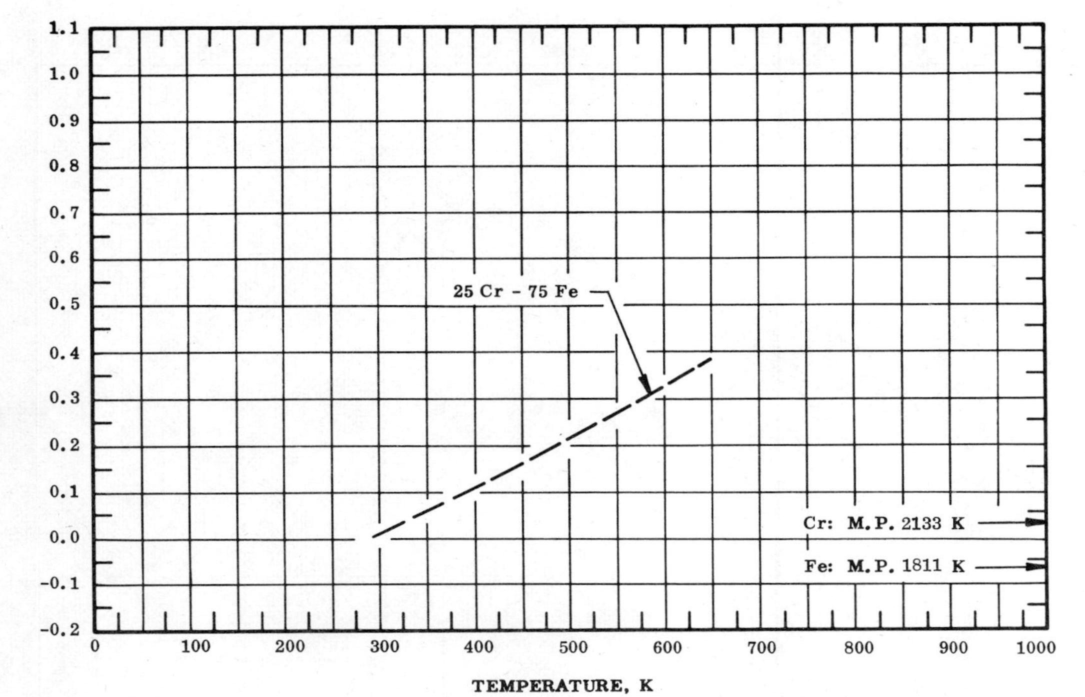

PROVISIONAL VALUES

[Temperature, T, K; Linear Expansion, $\Delta L/L_0$, %; α, K^{-1}]

	(25 Cr-75 Fe)	
T	$\Delta L/L_0$	$\alpha \times 10^6$
293	0.000	9.4
400	0.106	10.3
500	0.212	11.0
600	0.323	11.3
650	0.380	11.5

REMARKS

(25 Cr-75 Fe): The tabulated values for well-annealed alloy are considered accurate to within ±7% over the entire temperature range. The Curie temperature for this alloy is approximately at 870 K. The tabulated values can be represented approximately by the following equation:

$$\Delta L/L_0 = -0.222 + 5.357 \times 10^{-4}\, T + 8.839 \times 10^{-7}\, T^2 - 4.375 \times 10^{-10}\, T^3$$

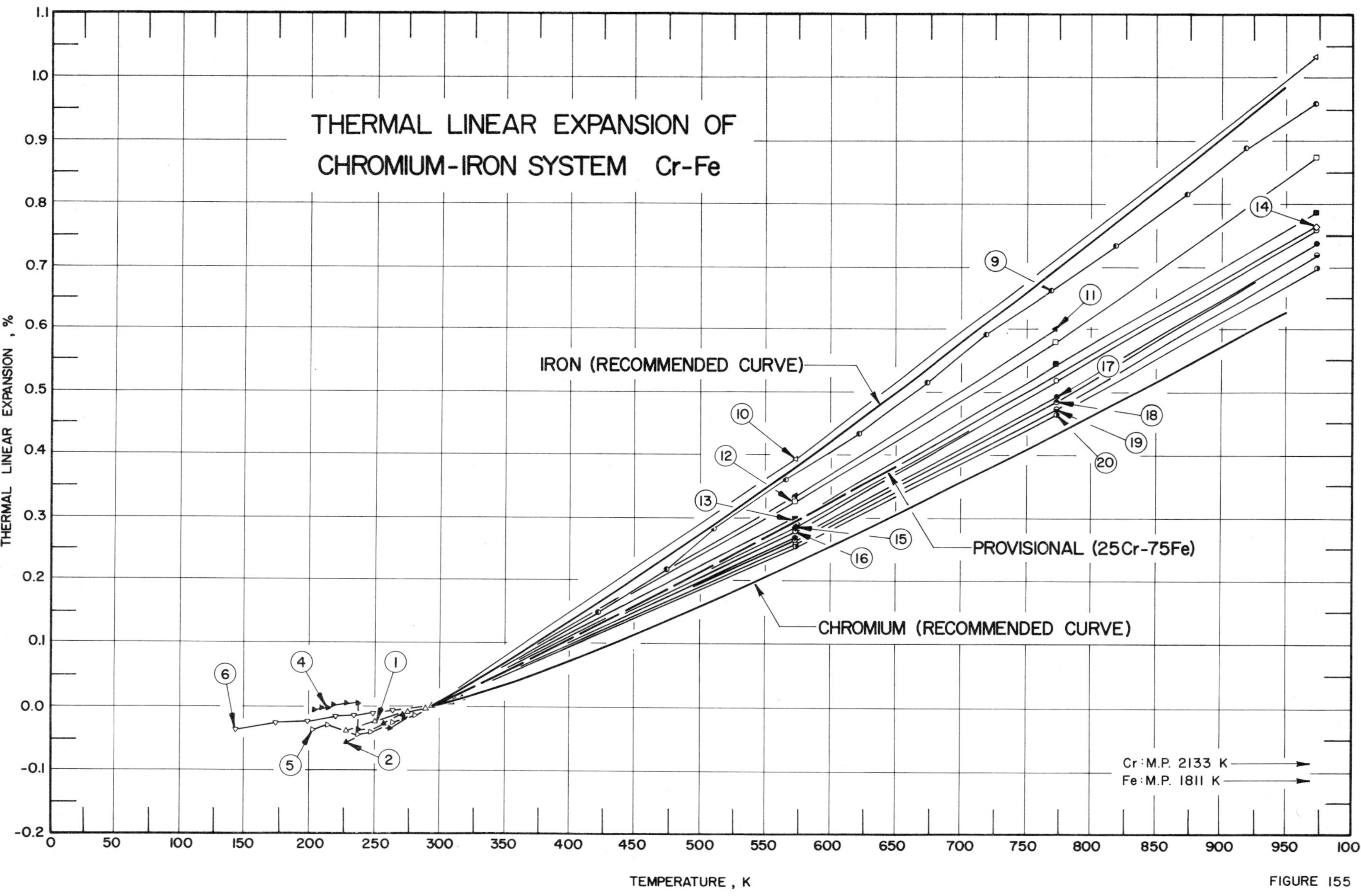
THERMAL LINEAR EXPANSION OF
CHROMIUM-IRON SYSTEM Cr-Fe
IRON (RECOMMENDED CURVE)
PROVISIONAL (25Cr-75Fe)
CHROMIUM (RECOMMENDED CURVE)
Cr: M.P. 2133 K
Fe: M.P. 1811 K
1
2
4
5
6
9
10
11
12
13
14
15
16
17
18
19
20
THERMAL LINEAR EXPANSION, %
TEMPERATURE, K
1.1
1.0
0.9
0.8
0.7
0.6
0.5
0.4
0.3
0.2
0.1
0.0
-0.1
-0.2
0
50
100
150
200
250
300
350
400
450
500
550
600
650
700
750
800
850
900
950
1000

FIGURE 155

SPECIFICATION TABLE 155. THERMAL LINEAR EXPANSION OF CHROMIUM-IRON SYSTEM Cr-Fe

Cur. No.	Ref. No.	Author(s)	Year	Method Used	Temp. Range, K	Name and Specimen Designation	Composition (weight percent) Cr	Fe	Composition (continued), Specifications, and Remarks
1	116	Ishikawa, Y., Hashino, S., and Enden, Y.	1967	S	228-317	S-2	Bal.	0.43	Single crystal prepared by strain anneal technique; Johnson Matthey pure Cr and Fe, arc melted several times in argon atmosphere; cylindrical ingots 5 cm long, 1 cm in diameter, annealed at 1373 K in evacuated silica tubes for 72 hr; pressed at five tons per cm^2, annealed again at 1873 K for 72 hr, in argon atmosphere; disk specimen 7-15 mm in diameter, 1.5 mm thick, oriented with the [011] direction normal to the plane of disk; Néel temperature: 298 K; zero-point correction is 0.0022%.
2	116	Ishikawa, Y., et al.	1967	S	228-318	S-6	Bal.	1.82	Similar to the above specimen; Néel temperature: 288 K; T_{CO}, the temperature where commensurable antiferromagnetic state disappears on cooling, is 266 K; zero-point correction of 0.0070% determined by graphical interpolation.
3*	116	Ishikawa, Y., et al.	1967	S	216-308	S-7	Bal.	2.47	Similar to the above specimen except polycrystalline; Néel temperature: 278 K T_{CO}: 258 K; zero-point correction of 0.0095% determined by graphical interpolation.
4	116	Ishikawa, Y., et al.	1967	S	203-294	S-8	Bal.	4.03	Similar to the above specimen except single crystal; Néel temperature: 261 K; T_{CO}: 240 K; zero-point correction of 0.0026% determined by graphical interpolation.
5	116	Ishikawa, Y., et al.	1967	S	201-298	S-11	Bal.	5.24	Similar to the above specimen; Néel temperature: 245 K, T_{CO}: 223 K; zero-point correction of 0.0074% determined by graphical interpolation.
6	339	Newmann, M.M. and Stevens, K.W.N.	1959		144-290	Alloy 1	Bal.	1.05	Spectroscopically standardized iron and chromium from Johnson Matthey and Co. used; specimen prepared in argon arc furnace at Atomic Energy Research Establishment, Harnwell; sealed in evacuated quartz tube, annealed for about 1 month at 1200 K; water quenched; specimen showed b.c.c. α-phase; machined into rod 2 cm long.
7*	399	Newmann, M.M. and Stevens, K.W.H.	1959		109-294	Alloy 2	Bal.	2.06	Similar to the above specimen.
8*	339	Newmann, M.M. and Stevens, K.W.N.	1959		126-261	Alloy 3	Bal.	5.76	Similar to the above specimen.
9	57	Miller, I.G.	1958		310-1073	X-6	3.97	96.03	Cylindrical specimen, 6.5 mm in diameter, 65 mm long; zero-point correction of 0.017% determined by graphical extrapolation.
10	707	Pridantseva, K.S. and Solov'yeva, N.A.	1965	L	293-973		2	98	Single-phase solid solution powder sample, impurities mainly 1% oxygen, samples melted in an induction furnace in an argon atmosphere, homogenizing anneal at 983 K for 240 hr, 1% measurement error.
11	707	Pridantseva, K.S. and Solov'yeva, N.A.	1965	L	293-973		6	94	Similar to the above specimen.
12	707	Pridantseva, K.S. and Solov'yeva, N.A.	1965	L	293-973		11	89	Similar to the above specimen.
13	707	Pridantseva, K.S. and Solov'yeva, N.A.	1965	L	293-973		21	79	Similar to the above specimen.
14	707	Pridantseva, K.S. and Solov'yeva, N.A.	1965	L	293-973		31	69	Similar to the above specimen.

* Not shown in figure.

SPECIFICATION TABLE 155. THERMAL LINEAR EXPANSION OF CHROMIUM-IRON SYSTEM Cr-Fe (continued)

Cur. No.	Ref. No.	Author(s)	Year	Method Used	Temp. Range, K	Name and Specimen Designation	Composition (weight percent) Cr	Fe	Composition (continued), Specifications, and Remarks
15	707	Pridantseva, K.S. and Solov'yeva, N.A.	1965	L	293-973		40	60	Similar to the above specimen.
16	707	Pridantseva, K.S. and Solov'yeva, N.A.	1965	L	293-973		51	49	Similar to the above specimen.
17	707	Pridantseva, K.S. and Solov'yeva, N.A.	1965	L	293-973		60	40	Similar to the above specimen.
18	707	Pridantseva, K.S. and Solov'yeva, N.A.	1965	L	293-973		70	30	Similar to the above specimen.
19	707	Pridantseva, K.S. and Solov'yeva, N.A.	1965	L	293-973		80	20	Similar to the above specimen.
20	707	Pridantseva, K.S. and Solov'yeva, N.A.	1965	L	293-973		90	10	Similar to the above specimen.
21*	29	Branchereau, M., Navez, M., and Perroux, M.	1962	L	273, 573	Dilver O	25	75	Specimen 100 mm long and 8 mm diameter.

* Not shown in figure.

DATA TABLE 155. THERMAL LINEAR EXPANSION OF CHROMIUM-IRON SYSTEM Cr-Fe

[Temperature, T, K; Linear Expansion, $\Delta L/L_0$, %]

CURVE 1

T	$\Delta L/L_0$
228	-0.0359
231	-0.0342*
233	-0.0332*
237	-0.0302*
241	-0.0276*
246	-0.0249*
250	-0.0220
253	-0.0204*
255	-0.0195*
258	-0.0182*
259	-0.0169*
262	-0.0157*
264	-0.0144*
266	-0.0133*
268	-0.0117*
271	-0.0105*
273	-0.0094*
275	-0.0082
277	-0.0070*
280	-0.0058*
282	-0.0046*
284	-0.0037*
286	-0.0028*
289	-0.0015
291	-0.0009*
293	0.0000
295	0.0001*
297	0.0016*
300	0.0027*
302	0.0042*
304	0.0057*
306	0.0073*
308	0.0080*
311	0.0102*
313	0.0118*
315	0.0135*
317	0.0147*

CURVE 2

T	$\Delta L/L_0$
228	-0.0546
230	-0.0527*
232	-0.0505*
235	-0.0479*
237	-0.0460*
239	-0.0445*

CURVE 2 (cont.)

T	$\Delta L/L_0$
241	-0.0425*
244	-0.0399*
248	-0.0370*
250	-0.0340*
252	-0.0313*
254	-0.0284*
257	-0.0259
259	-0.0210*
262	-0.0175*
264	-0.0155*
266	-0.0141*
268	-0.0126*
271	-0.0108
273	-0.0099*
275	-0.0086*
277	-0.0076*
279	-0.0067*
282	-0.0058*
284	-0.0052*
286	-0.0042*
288	-0.0037*
290	-0.0023*
292	-0.0007*
295	0.0010*
297	0.0025*
300	0.0042*
302	0.0059*
304	0.0087*
306	0.0089*
308	0.0109*
310	0.0122*
312	0.0137*
315	0.0155*
318	0.0177*

CURVE 3*

T	$\Delta L/L_0$
216	-0.0045
220	-0.0024
225	0.0003
230	0.0028
234	0.0064
239	0.0075
241	0.0074
246	0.0086
250	0.0091

CURVE 3 (cont.)

T	$\Delta L/L_0$
252	0.0085
255	0.0085
257	0.0065
258	-0.0338
260	-0.0323
262	-0.0200
264	-0.0280
268	-0.0246
271	-0.0220
273	-0.0201
275	-0.0179
278	-0.0155
380	-0.0145
287	-0.0071
289	-0.0046
291	-0.0024
295	0.0017
300	0.0066
302	0.0093
305	0.0124
308	0.0148

CURVE 4

T	$\Delta L/L_0$
203	-0.0041
209	-0.0013
213	0.0010
218	0.0032
222	0.0048*
227	0.0063
229	0.0068*
232	0.0072*
233	0.0078*
235	0.0077*
237	0.0075
237	0.0238
240	-0.0236*
242	-0.0232*
244	-0.0229*
245	-0.0229*
247	-0.0225*
249	-0.0222*
251	-0.0219*
254	-0.0216*
256	-0.0214*
258	-0.0217*

CURVE 4 (cont.)

T	$\Delta L/L_0$
261	-0.0216
263	-0.0210*
265	-0.0193*
268	-0.0180*
271	-0.0151
277	-0.0120*
281	-0.0087*
286	-0.0055*
290	-0.0021*
294	0.0003*

CURVE 5

T	$\Delta L/L_0$
201	-0.0347
205	-0.0329*
209	-0.0306*
214	-0.0284
218	-0.0273*
222	-0.0273*
227	-0.0296*
230	-0.0330*
232	-0.0365*
234	-0.0393*
237	-0.0406
239	-0.0409*
242	-0.0407*
244	-0.0399*
246	-0.0391
248	-0.0377*
250	-0.0361*
252	-0.0348*
255	-0.0334*
257	-0.0315*
259	-0.0302*
262	-0.0278*
264	-0.0263
266	-0.0244*
268	-0.0228*
270	-0.0203*
273	-0.0182*
274	-0.0169*
276	-0.0161*
277	-0.0143*
280	-0.0121
282	-0.0103*
284	-0.0080*

CURVE 5 (cont.)

T	$\Delta L/L_0$
286	-0.0063*
289	-0.0041*
290	-0.0023*
292	-0.0010*
294	0.0003*
295	0.0021*
298	0.0043*

CURVE 6

T	$\Delta L/L_0$
144	-0.034
174	-0.027
199	-0.022
220	-0.017
236	-0.012
249	-0.009
257	-0.007*
266	-0.005
278	-0.002*
283	-0.001*
290	-0.000*

CURVE 7*

T	$\Delta L/L_0$
109	-0.013
126	-0.011
145	-0.009
160	-0.007
185	-0.003
201	-0.001
225	0.003
240	0.003
250	0.001
260	-0.002
272	-0.004
280	-0.004
290	-0.003
294	0.000

CURVE 8*

T	$\Delta L/L_0$
126	-0.010
145	-0.011
179	-0.007
196	-0.007
224	-0.009

CURVE 8 (cont.)*

T	$\Delta L/L_0$
233	-0.016
240	-0.020
250	-0.016
261	-0.013

CURVE 9

T	$\Delta L/L_0$
310	0.017
322	0.034*
372	0.086*
421	0.148
473	0.217
510	0.282
565	0.360
621	0.434
673	0.515
719	0.589
768	0.663
818	0.734
873	0.815
920	0.888
973	0.960
1021	1.031
1073	1.092

CURVE 10

T	$\Delta L/L_0$
293	0.000*
573	0.391
973	1.030

CURVE 11

T	$\Delta L/L_0$
293	0.000*
573	0.331
773	0.599

CURVE 12

T	$\Delta L/L_0$
293	0.000*
573	0.325
773	0.577
973	0.874

* Not shown in figure.

DATA TABLE 155. THERMAL LINEAR EXPANSION OF CHROMIUM-IRON SYSTEM Cr-Fe (continued)

T	$\Delta L/L_0$
CURVE 13	
293	0.000*
573	0.296
773	0.542
973	0.787
CURVE 14	
293	0.000*
573	0.286
773	0.518*
973	0.762
CURVE 15	
293	0.000*
573	0.288
773	0.518*
973	0.762*
CURVE 16	
293	0.000*
573	0.277
773	0.518
973	0.759
CURVE 17	
293	0.000*
573	0.266
773	0.491
973	0.738
CURVE 18	
293	0.000*
573	0.263
773	0.483
973	0.738*
CURVE 19	
293	0.000*
573	0.256
773	0.471
973	0.719
CURVE 20	
293	0.000*
573	0.251
CURVE 20 (cont.)	
773	0.464
973	0.698
CURVE 21*	
273	-0.0208
573	0.2923

* Not shown in figure.

FIGURE AND TABLE NO. 156BR. PROVISIONAL VALUES FOR THERMAL LINEAR EXPANSION OF CHROMIUM-MOLYBDENUM SYSTEM Cr-Mo

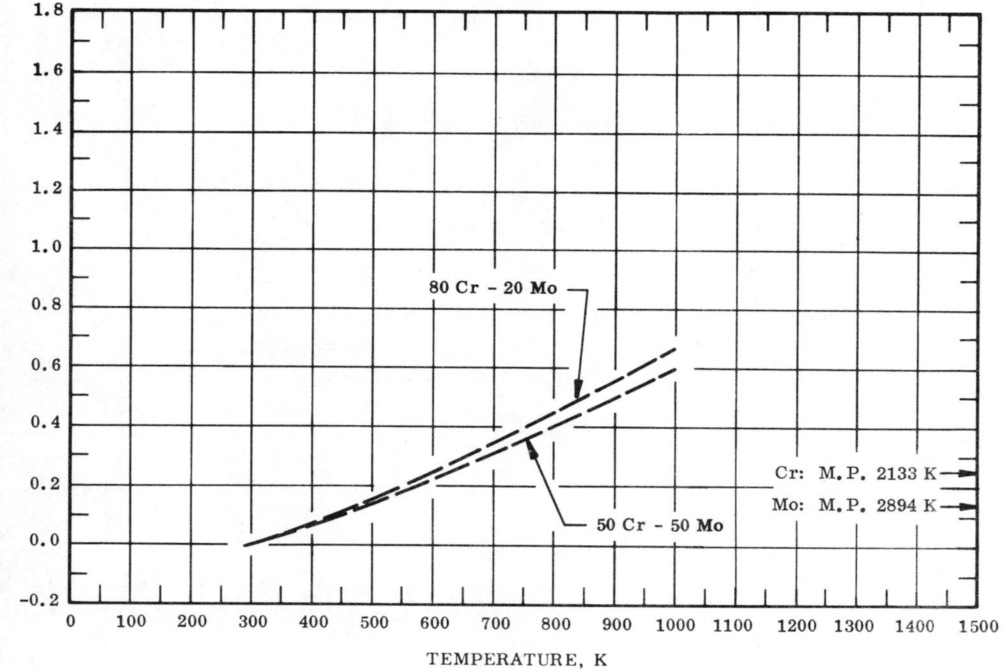

PROVISIONAL VALUES

[Temperature, T, K; Linear Expansion, $\Delta L/L_0$, %; α, K^{-1}]

T	(50 Cr-50 Mo) $\Delta L/L_0$	(50 Cr-50 Mo) $\alpha \times 10^6$	(80 Cr-20 Mo) $\Delta L/L_0$	(80 Cr-20 Mo) $\alpha \times 10^6$
293	0.000	5.7	0.000	6.2
400	0.067	6.9	0.073	7.6
500	0.140	7.8	0.155	8.7
600	0.222	8.6	0.246	9.5
700	0.310	9.1	0.344	10.1
800	0.403	9.5	0.447	10.6
900	0.500	9.9	0.554	10.9
1000	0.601	10.2	0.664	11.2

REMARKS

(50 Cr-50 Mo): The tabulated values for well-annealed alloy are considered accurate to within ±7% over the entire temperature range. These values can be represented approximately by the following equation:

$$\Delta L/L_0 = -0.116 + 1.874 \times 10^{-4} T + 7.716 \times 10^{-7} T^2 - 2.426 \times 10^{-10} T^3$$

(80 Cr-20 Mo): The tabulated values for well-annealed alloy are considered accurate to within ±7% over the entire temperature range. These values can be represented approximately by the following equation:

$$\Delta L/L_0 = -0.117 + 1.455 \times 10^{-4} T + 9.560 \times 10^{-7} T^2 - 3.215 \times 10^{-10} T^3$$

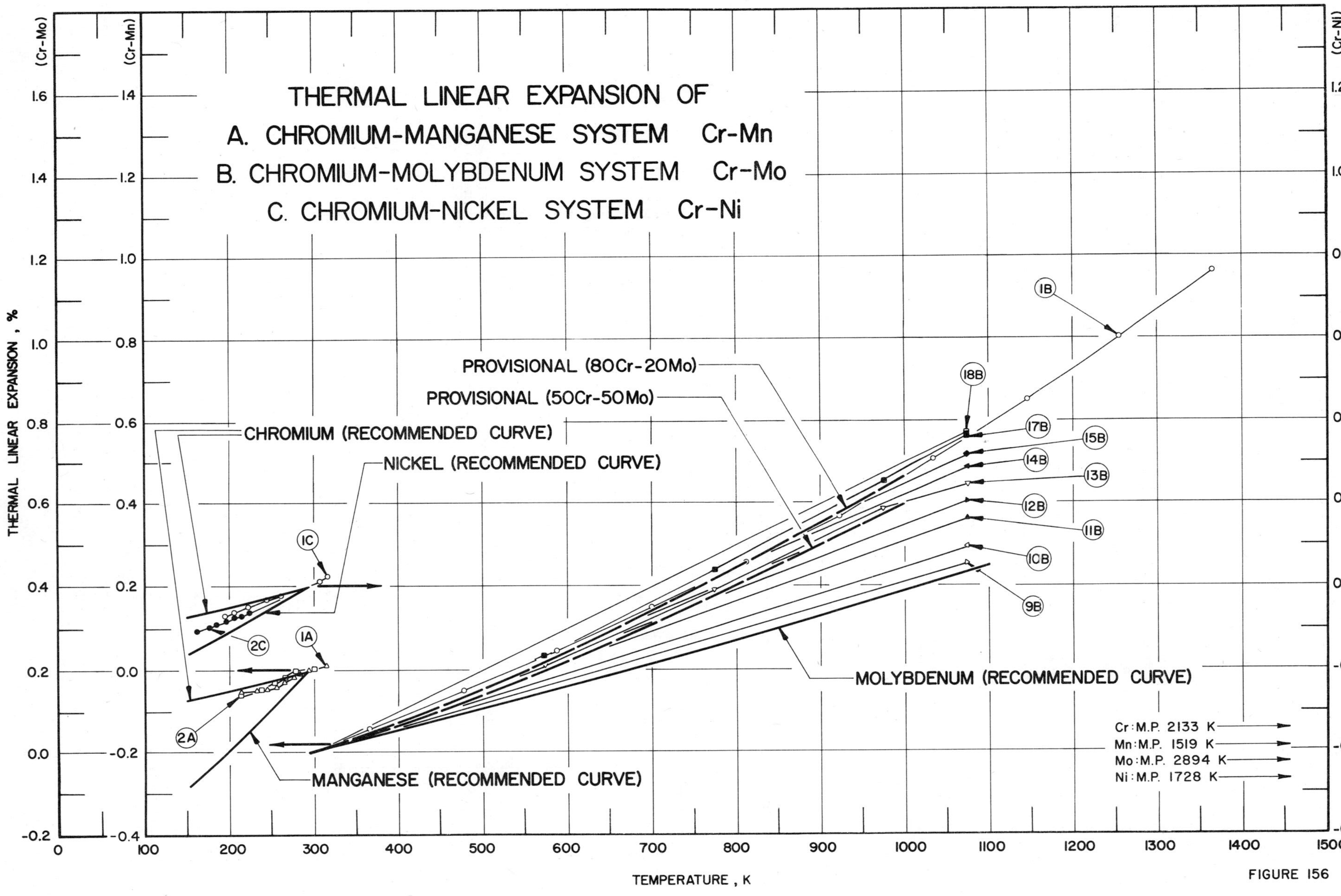

THERMAL LINEAR EXPANSION OF
A. CHROMIUM-MANGANESE SYSTEM Cr-Mn
B. CHROMIUM-MOLYBDENUM SYSTEM Cr-Mo
C. CHROMIUM-NICKEL SYSTEM Cr-Ni
PROVISIONAL (80Cr-20Mo)
PROVISIONAL (50Cr-50Mo)
CHROMIUM (RECOMMENDED CURVE)
NICKEL (RECOMMENDED CURVE)
MOLYBDENUM (RECOMMENDED CURVE)
MANGANESE (RECOMMENDED CURVE)
Cr: M.P. 2133 K
Mn: M.P. 1519 K
Mo: M.P. 2894 K
Ni: M.P. 1728 K
1A
2A
1B
9B
10B
11B
12B
13B
14B
15B
17B
18B
1C
2C
THERMAL LINEAR EXPANSION, %
TEMPERATURE, K
(Cr-Mo)
(Cr-Mn)
(Cr-Ni)
-0.2
0.0
0.2
0.4
0.6
0.8
1.0
1.2
1.4
1.6
-0.4
-0.2
0.0
0.2
0.4
0.6
0.8
1.0
1.2
1.4
-0.6
-0.4
-0.2
0.0
0.2
0.4
0.6
0.8
1.0
1.2
0
100
200
300
400
500
600
700
800
900
1000
1100
1200
1300
1400
1500

FIGURE 156

SPECIFICATION TABLE 156A. THERMAL LINEAR EXPANSION OF CHROMIUM-MANGANESE SYSTEM Cr-Mn

Cur. No.	Ref. No.	Author(s)	Year	Method Used	Temp. Range, K	Name and Specimen Designation	Composition (weight percent) Cr	Mn	Composition (continued), Specifications, and Remarks
1	398	Syono, Y. and Ishikawa, Y.	1967	S	215-317		99.55	0.45	Expansion measured with increasing temperature at 0.5 kb pressure; zero-point correction of -2.46% determined by graphical interpolation.
2	398	Syono, Y. and Ishikawa, Y.	1967	S	318-215		99.55	0.45	The above specimen; expansion measured with decreasing temperature; data measured at pressure of 1.0 and 1.5 kbar also reported by the author; zero-point correction of -2.68% determined by graphical interpolation.

DATA TABLE 156A. THERMAL LINEAR EXPANSION OF CHROMIUM-MANGANESE SYSTEM Cr-Mn

[Temperature, T, K; Linear Expansion, $\Delta L/L_0$, %]

T	$\Delta L/L_0$	T	$\Delta L/L_0$
CURVE 1		CURVE 2	
215	-0.0593	318	0.00990*
231	-0.0525	301	0.00360*
238	-0.0496*	289	-0.00190*
245	-0.0464	279	-0.00830
250	-0.0442*	274	-0.0120*
257	-0.0411	267	-0.0201
263	-0.0365*	258	-0.0312*
266	-0.0330	254	-0.0363
270	-0.0288*	248	-0.0431*
276	-0.0203	243	-0.0475*
279	-0.0146*	238	-0.0512
281	-0.0106	231	-0.0552*
287	-0.0047*	215	-0.0615
292	-0.0007		
302	0.0059*		
317	0.0121		

*Not shown in figure.

SPECIFICATION TABLE 156B. THERMAL LINEAR EXPANSION OF CHROMIUM-MOLYBDENUM SYSTEM Cr-Mo

Cur. No.	Ref. No.	Author(s)	Year	Method Used	Temp. Range, K	Name and Specimen Designation	Composition (weight percent) Cr	Mo	Composition (continued), Specifications, and Remarks
1	120	Shevlin, T.S. and H Houk, C.A.	1954	L	293-1589		80	20	Prepared from electrolytic 99^+ Cr powder, -325 mesh, and hydrogen reduced 99.75^+ Mo powder, -200 mesh, supplied by Charles Hardy, Inc.
2*	708	Ageev, N.V. and Model', M.S.	1963	X	282-893		Bal.	0.4	Sample prepared from electrolytic 99.96 Cr and 99.6 Mo, polycrystalline powder with 40 μ particle size, annealing at 1673 K in argon atmosphere added nitrogen and oxygen impurities to sample, solid solution sample was then pressed into rings, ±0.0001 Å accuracy, lattice parameter at 293 K is 2.8864 Å determined by graphical interpolation.
3*	708	Ageev, N.V. and Model', M.S.	1963	X	282-893		Bal.	0.7	Similar to the above specimen, lattice parameter at 293 K is 2.8869 Å determined by graphical interpolation.
4*	708	Ageev, N.V. and Model', M.S.	1963	X	282-972		Bal.	1.1	Similar to the above specimen, lattice parameter at 293 K is 2.8875 Å determined by graphical interpolation.
5*	708	Ageev, N.V. and Model', M.S.	1963	X	283-972		Bal.	1.5	Similar to the above specimen, lattice parameter at 293 K is 2.8883 Å determined by graphical interpolation.
6*	708	Ageev, N.V. and Model', M.S.	1963	X	285-973		Bal.	4.5	Similar to the above specimen, lattice parameter at 293 K is 2.8930 Å determined by graphical interpolation.
7*	708	Ageev, N.V. and Model', M.S.	1963	X	283-973		Bal.	8.9	Similar to the above specimen, lattice parameter at 293 K is 2.9013 Å determined by graphical interpolation.
8*	708	Ageev, N.V. and Model', M.S.	1963	X	284-967		Bal.	16.2	Similar to the above specimen, lattice parameter at 293 K is 2.9182 Å determined by graphical interpolation.
9	707	Pridantseva, K.S., and Solov'yeva, N.A.	1965	L	293, 1073		2	Bal.	Specimen prepared by powder metallurgy method, impurities mainly 1% oxygen sample was homogenized at temperature which would provide an equilibrium state of the alloy, homogenizing anneal was done in a vacuum of not less than 10^{-4} mm Hg or in argon, single phase solid solution sample, 1% measurement error.
10	707	Pridantseva, K.S., and Solov'yeva, N.A.	1965	L	293, 1073		10	Bal.	Similar to the above specimen.
11	707	Pridantseva, K.S., and Solov'yeva, N.A.	1965	L	293, 1073		24	Bal.	Similar to the above specimen.
12	707	Pridantseva, K.S., and Solov'yeva, N.A.	1965	L	293, 1073		36	Bal.	Similar to the above specimen.
13	707	Pridantseva, K.S., and Solov'yeva, N.A.	1965	L	293, 1073		50	Bal.	Similar to the above specimen.
14	707	Pridantseva, K.S., and Solov'yeva, N.A.	1965	L	293, 1073		60	Bal.	Similar to the above specimen.
15	707	Pridantseva, K.S. and Solov'yeva, N.A.	1965	L	293, 1073		70	Bal.	Similar to the above specimen.
16*	707	Pridantseva, K.S., and Solov'yeva, N.A.	1965	L	293, 1073		80	Bal.	Similar to the above specimen.
17	707	Pridantseva, K.S., and Solov'yeva, N.A.	1965	L	293, 1073		90	Bal.	Similar to the above specimen.

* Not shown in figure.

SPECIFICATION TABLE 156B. THERMAL INEAR EXPANSION OF CHROMIUM-MOLYBDENUM SYSTEM Cr–Mo (continued)

Cur. No.	Ref. No.	Author(s)	Year	Method Used	Temp. Range, K	Name and Specimen Designation	Composition (weight percent) Cr	Mo	Composition (continued), Specifications, and Remarks
18	707	Pridantseva, K. S., and Solov'yeva, N. A.	1965	L	293, 1073		94	Bal.	Similar to the above specimen.

DATA TABLE 156B. THERMAL LINEAR EXPANSION OF CHROMIUM-MOLYBDENUM SYSTEM Cr-Mo

[Temperature, T, K; Linear Expansion, $\Delta L/L_0$, %]

T	$\Delta L/L_0$
CURVE 1	
293	0.000*
366	0.056
478	0.147
589	0.245
700	0.348
811	0.456
922	0.569
1033	0.708
1144	0.852
1255	1.005
1366	1.165
CURVE 2*	
282	-0.007
293	0.000
404	0.076
587	0.243
752	0.399
893	0.541
CURVE 3*	
282	-0.007
293	0.000
404	0.007
587	0.243
752	0.403
893	0.538
CURVE 4*	
282	-0.014
293	0.000
404	0.076
587	0.239
752	0.396
891	0.541
972	0.618
CURVE 5*	
283	-0.007
293	0.000
404	0.083
587	0.250
CURVE 5 (cont.)*	
748	0.416
889	0.562
972	0.652
CURVE 6*	
285	-0.003
293	0.000
401	0.094
587	0.267
739	0.412
893	0.568
973	0.658
CURVE 7*	
283	-0.007
293	0.000
404	0.083
588	0.260
741	0.418
891	0.577
973	0.660
CURVE 8*	
284	-0.003
293	0.000
396	0.089
584	0.261
740	0.409
888	0.560
967	0.639
CURVE 9	
293	0.000
1073	0.458
CURVE 10	
293	0.000
1073	0.498
CURVE 11	
293	0.000
1073	0.565
CURVE 12	
293	0.000
1073	0.609
CURVE 13	
293	0.000
573	0.211
773	0.391
973	0.589
1073	0.647
CURVE 14	
293	0.000*
573	0.218*
773	0.403*
973	0.607*
1073	0.691
CURVE 15	
293	0.000*
573	0.227*
773	0.420*
973	0.633*
1073	0.720
CURVE 16*	
293	0.000
573	0.232
773	0.435
973	0.648
1073	0.742
CURVE 17	
293	0.000
573	0.234
773	0.439
973	0.652
1073	0.767
CURVE 18	
293	0.000
1073	0.775

* Not shown in figure.

SPECIFICATION TABLE 156C. THERMAL LINEAR EXPANSION OF CHROMIUM-NICKEL SYSTEM Cr-Ni

Cur. No.	Ref. No.	Author(s)	Year	Method Used	Temp. Range, K	Name and Specimen Designation	Composition (weight percent) Cr	Ni	Composition (continued), Specifications, and Remarks
1	397	Endoh, Y., Ishikawa, Y., and Ohno, H.	1968	S	195-318	Dilute Cr alloy with Ni	99.46	0.54	Single crystal grown by strain anneal technique from polycrystalline alloys melted several times in argon atmosphere; starting material, Johnson Matthey pure Cr, Ni; expansion measured in (110) plane along [100]; observed at the Néel temperature almost the same as that for pure chromium; Néel temperature 238 K; zero-point correction of 0.009% determined by graphical interpolation.
2	397	Endoh, Y., et al.	1968	S	162-244	Dilute Cr alloy with Ni	98.88	1.12	Similar to the above specimen; Néel temperature 207 K; zero-point correction is 0.000%.
3*	428	Masumoto, H., Sawaya, S., and Nakamura, N.	1969	L	273, 313		4.0	96.0	Electrolytic Ni with 0.180 Co, 0.002 S, 0.002 C, 0.001 Cu, 0.001 Fe, and 99.6 Cr powder; furnace-cooled at rate 300 C/hr after heating at 1273 K for 1 hr; density 8.92 g cm^3 at 293 K.
4*	428	Masumoto, H., et al.	1969	L	273, 313		5.0	95.0	Similar to the above specimen; density 8.88 g cm^{-3} at 293 K.
5*	428	Masumoto, H., et al.	1969	L	273, 313		5.3	94.7	Similar to the above specimen.
6*	428	Masumoto, H., et al.	1969	L	273, 313		5.5	94.5	Similar to the above specimen; density 8.92 g cm^{-3} at 293 K.
7*	428	Masumoto, H., et al.	1969	L	273, 313		6.0	94.0	Similar to the above specimen; density 8.86 g cm^{-3} at 293 K.
8*	428	Masumoto, H., et al.	1969	L	273, 313		6.5	93.5	Similar to the above specimen; density 8.87 g cm^{-3} at 293 K.

*Not shown in figure.

DATA TABLE 156C. THERMAL LINEAR EXPANSION OF CHROMIUM-NICKEL SYSTEM Cr-Ni

[Temperature, T, K; Linear Expansion, $\Delta L/L_0$, %]

T	$\Delta L/L_0$
CURVE 1	
195	-0.073
200	-0.069*
204	-0.066*
208	-0.062
213	-0.058*
217	-0.056*
222	-0.052
227	-0.049*
231	-0.045*
236	-0.043*
241	-0.041*
245	-0.039
249	-0.035*
252	-0.033*
256	-0.029*
261	-0.026
263	-0.023*
268	-0.020*
272	-0.017*
277	-0.013*
281	-0.010*
286	-0.006*
290	-0.002*
295	0.002*
300	0.005*
304	0.009*
309	0.012
313	0.016*
318	0.020
CURVE 2	
162	-0.110
166	-0.107*
171	-0.103*
176	-0.100
179	-0.098*
185	-0.093
189	-0.091*
194	-0.088*
198	-0.084
203	-0.082*
207	-0.079
211	-0.076*
216	-0.072
221	-0.067*

T	$\Delta L/L_0$
CURVE 2 (cont.)	
225	-0.064
230	-0.059*
234	-0.055*
238	-0.051*
243	-0.047*
244	-0.042*
CURVE 3*	
273	-0.027
313	0.027
CURVE 4*	
273	-0.027
313	0.027
CURVE 5*	
273	-0.028
313	0.028
CURVE 6*	
273	-0.029
313	0.029
CURVE 7*	
273	-0.029
313	0.029
CURVE 8*	
273	-0.028
313	0.028

*Not shown in figure.

FIGURE AND TABLE NO. 157R. PROVISIONAL VALUES FOR THERMAL LINEAR EXPANSION OF CHROMIUM-SILICON SYSTEM Cr-Si

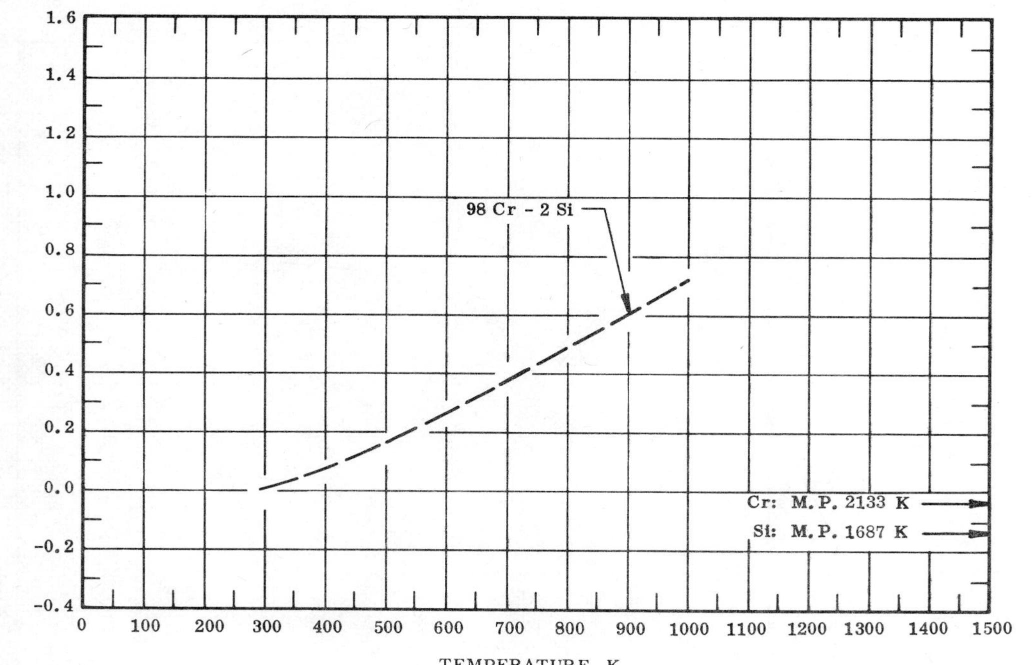

PROVISIONAL VALUES

[Temperature, T, K; Linear Expansion, $\Delta L/L_0$, %; α, K^{-1}]

	(98 Cr-2 Si)	
T	$\Delta L/L_0$	$\alpha \times 10^6$
293	0.000	6.5
400	0.077	8.0
500	0.164	9.4
600	0.264	10.4
700	0.371	11.0
800	0.483	11.5
900	0.600	11.8
1000	0.719	12.1

REMARKS

(98 Cr-2 Si): The tabulated values for well-annealed alloy are considered accurate to within ± 7% over the entire temperature range. These values can be represented approximately by the following equation:

$$\Delta L/L_0 = -0.109 + 5.897 \times 10^{-5}\,T + 1.188 \times 10^{-6}\,T^2 - 4.206 \times 10^{-10}\,T^3$$

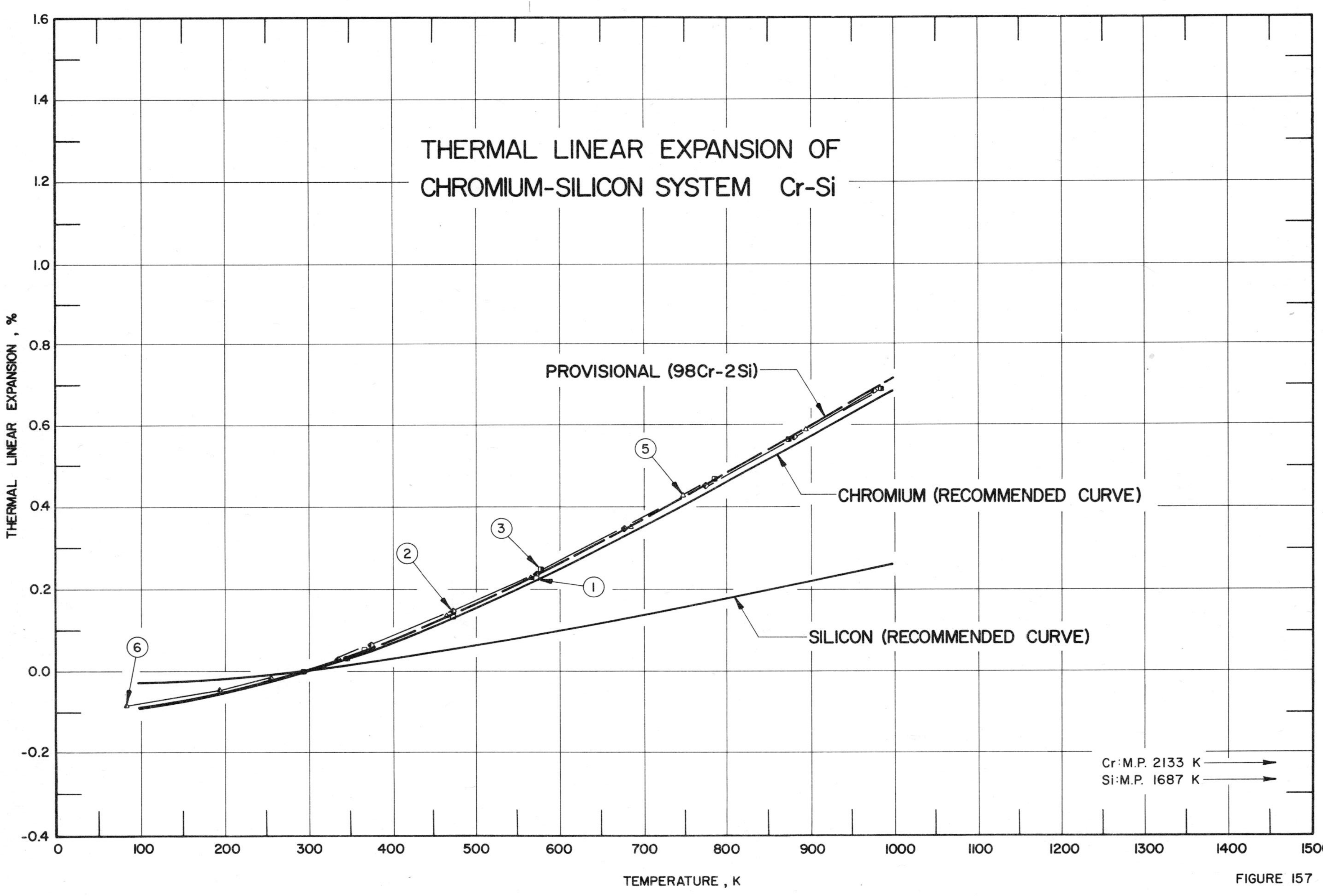

THERMAL LINEAR EXPANSION OF
CHROMIUM-SILICON SYSTEM Cr-Si
PROVISIONAL (98Cr-2Si)
CHROMIUM (RECOMMENDED CURVE)
SILICON (RECOMMENDED CURVE)
1
2
3
5
6
Cr: M.P. 2133 K
Si: M.P. 1687 K
THERMAL LINEAR EXPANSION, %
1.6
1.4
1.2
1.0
0.8
0.6
0.4
0.2
0.0
-0.2
-0.4
0
100
200
300
400
500
600
700
800
900
1000
1100
1200
1300
1400
1500
TEMPERATURE, K

FIGURE 157

SPECIFICATION TABLE 157. THERMAL LINEAR EXPANSION OF CHROMIUM-SILICON SYSTEM Cr-Si

Cur. No.	Ref. No.	Author(s)	Year	Method Used	Temp. Range, K	Name and Specimen Designation	Composition (weight percent) Cr	Si	Composition (continued), Specifications, and Remarks
1	396	Hidnert, P.	1941	I	292-571	1343 I	99.2	0.29	0.05 Mn, 0.01 Fe; hot-swaged rod; 2 mm dia, 6.648 mm long; each sample consisted of three pieces of nearly equal length; prepared by Westinghouse Electric Mfg. Co., Pittsburgh, Pa.; density 7.07 g/cm^3; expansion measured with increasing temp on first test.
2	396	Hidnert, P.	1941	I	292-976	1343 I	99.2	0.29	The above specimen; expansion measured with increasing temp on fourth test.
3	396	Hidnert, P.	1941	I	292-981	1343 I	99.2	0.29	The above specimen; expansion measured with increasing temp on fifth test.
4*	396	Hidnert, P.	1941	I	295-575	1356 I	96.7	0.53	0.09 C, 0.065 N, 0.02 Fe, 0.02 Mn; hot-swaged rod; 4 mm dia, 7.47 mm long; sample consisted of three pieces of nearly equal length; prepared by Research Department, Westinghouse Lamp Division, Bloomfield, N.J.; expansion measured with increasing temp on first test.
5	396	Hidnert, P.	1941	I	293-973	1356 I	96.7	0.53	The above specimen; expansion measured with increasing temp on third test.
6	396	Hidnert, P.	1941	L	81-294	1356	96.7	0.53	0.09 C, 0.065 N, 0.02 Fe, 0.02 Mn; hot-swage rod, 4 mm dia, 170 mm long; prepared by Research Department, Westinghouse Lamp Division, Bloomfield, N.J.; expansion measured with increasing temp on first test.
7*	396	Hidnert, P.	1941	L	92-372	1356	96.7	0.53	The above specimen; expansion measured with increasing temp on third test.

*Not shown in figure.

DATA TABLE 157. THERMAL LINEAR EXPANSION OF CHROMIUM-SILICON SYSTEM Cr-Si

[Temperature, T, K; Linear Expansion, $\Delta L/L_0$, %]

T	$\Delta L/L_0$
CURVE 1	
293	0.000
345	0.036
369	0.054
473	0.137
571	0.228
CURVE 2	
293	0.000*
335	0.030
373	0.061
473	0.145
572	0.239
677	0.349
773	0.456
880	0.575
976	0.691
CURVE 3	
293	0.000*
331	0.027*
373	0.057*
474	0.145*
579	0.246
678	0.352*
785	0.465
876	0.569
981	0.694
CURVE 4*	
293	0.000
334	0.028
369	0.056
475	0.147
575	0.242
CURVE 5	
293	0.000*
331	0.027*
371	0.057*
465	0.139
566	0.227
684	0.353
CURVE 5 (cont.)	
749	0.425
776	0.458*
892	0.593
973	0.693*
CURVE 6	
81	-0.082
192	-0.044
253	-0.016
274	-0.010*
294	0.001*
CURVE 7*	
92	-0.080
195	-0.040
254	-0.013
272	-0.007
299	0.002
302	0.009
334	0.027
372	0.056

*Not shown in figure.

SPECIFICATION TABLE 158. THERMAL LINEAR EXPANSION OF CHROMIUM-TIN Cr-Sn

Cur. No.	Ref. No.	Author(s)	Year	Method Used	Temp. Range, K	Name and Specimen Designation	Composition (weight percent) Cr	Sn	Composition (continued), Specifications, and Remarks
1*	755	Fukamichi, K. and Saito, H.	1972		211-415		Bal.	2.64	Specimen prepared using 99.99 Cr and 99.9 Sn; arc melted together and remelted three times in form of square rod in argon atmosphere and homogenized in vacuum at 1073 K for 80 hr.

DATA TABLE 158. THERMAL LINEAR EXPANSION OF CHROMIUM-TIN Cr-Sn

[Temperature, T, K; Linear Expansion, $\Delta L/L_0$, %]

T	$\Delta L/L_0$	T	$\Delta L/L_0$
CURVE 1*		CURVE 1 (cont.)*	
211	-0.087	368	0.027
231	-0.070	373	0.027
252	-0.056	377	0.028
259	-0.049	382	0.028
263	-0.045	386	0.029
265	-0.041	389	0.031
267	-0.036	392	0.034
268	-0.031	398	0.038
270	-0.027	415	0.052
273	-0.023		
279	-0.014		
293	0.000		
294	0.001		
314	0.012		
337	0.019		
352	0.023		
358	0.024		
363	0.025		

* No figure given.

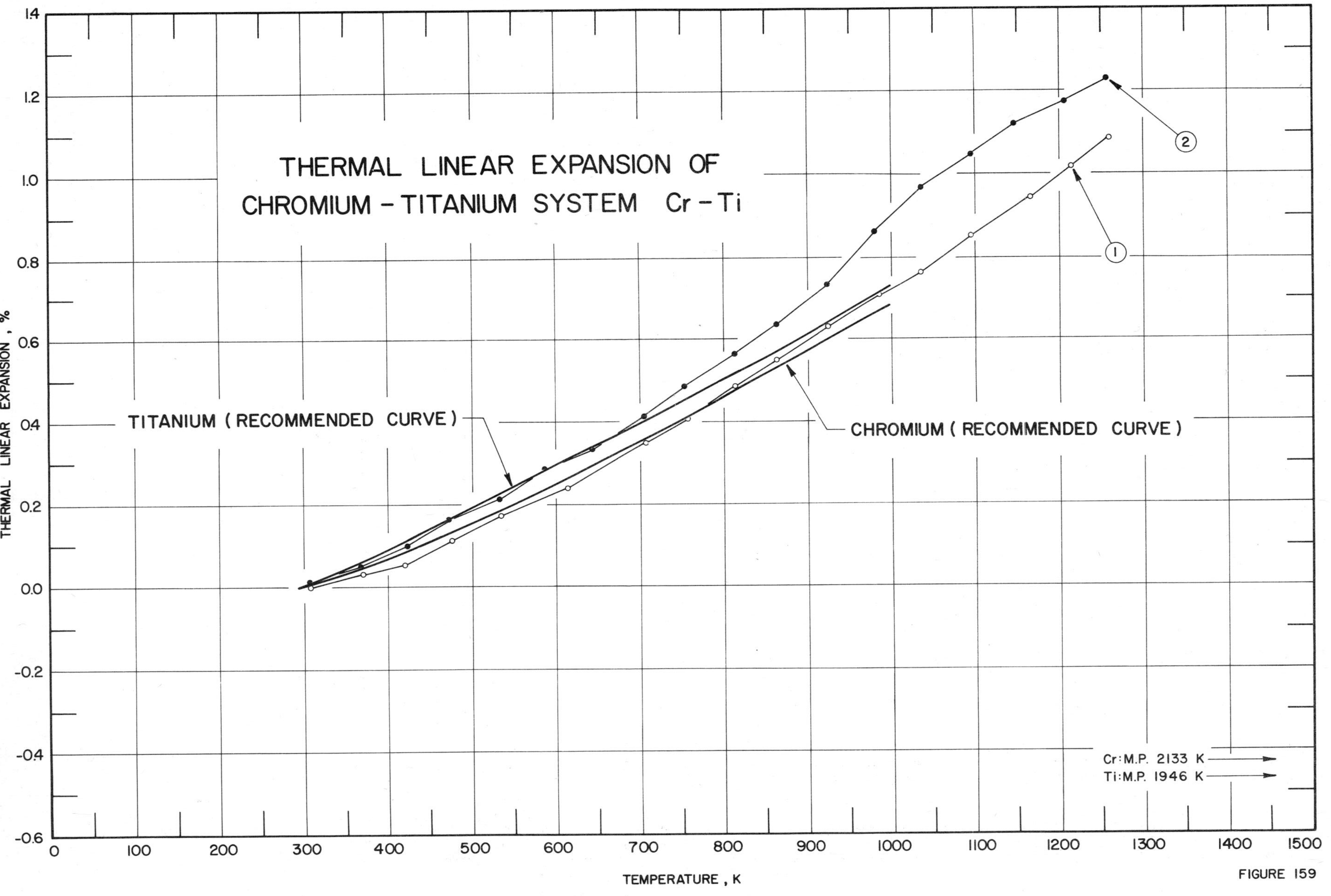
THERMAL LINEAR EXPANSION OF
CHROMIUM - TITANIUM SYSTEM Cr - Ti
TITANIUM (RECOMMENDED CURVE)
CHROMIUM (RECOMMENDED CURVE)
1
2
Cr: M.P. 2133 K
Ti: M.P. 1946 K
THERMAL LINEAR EXPANSION, %
TEMPERATURE, K
1.4
1.2
1.0
0.8
0.6
0.4
0.2
0.0
-0.2
-0.4
-0.6
0
100
200
300
400
500
600
700
800
900
1000
1100
1200
1300
1400
1500

FIGURE 159

SPECIFICATION TABLE 159. THERMAL LINEAR EXPANSION OF CHROMIUM-TITANIUM SYSTEM Cr-Ti

Cur. No.	Ref. No.	Author(s)	Year	Method Used	Temp. Range, K	Name and Specimen Designation	Composition (weight percent) Cr	Ti	Composition (continued), Specifications, and Remarks
1	395	Stetson, A.R. and Metcalfe, A.G.	1967	L	309-1260		77.0	23.0	Specimen prepared by arc-melting components; annealed 16 hrs at 1477 K and cut to 0.6 cm sq, 3 cm long; zero-point correction of 0.007% determined by graphical extrapolation.
2	395	Stetson, A.R. and Metcalfe, A.G.	1967	L	309-1255		28.4	71.6	Similar to the above specimen.

DATA TABLE 159. THERMAL LINEAR EXPANSION OF CHROMIUM-TITANIUM SYSTEM Cr-Ti

[Temperature, T, K; Linear Expansion, $\Delta L/L_0$, %]

T	$\Delta L/L_0$	T	$\Delta L/L_0$
CURVE 1		CURVE 2	
309	0.007	309	0.011
370	0.035	369	0.053
420	0.059	421	0.100
476	0.118	474	0.163
533	0.173	531	0.211
611	0.240	589	0.285
706	0.350	642	0.334
758	0.407	704	0.417
813	0.483	752	0.486
862	0.552	812	0.565
923	0.632	864	0.638
984	0.711	923	0.738
1035	0.765	980	0.861
1097	0.856	1035	0.974
1165	0.949	1097	1.052
1215	1.020	1145	1.122
1260	1.095	1208	1.175
		1255	1.224

FIGURE AND TABLE NO. 160AR. PROVISIONAL VALUES FOR THERMAL LINEAR EXPANSION OF CHROMIUM-URANIUM SYSTEM Cr-U

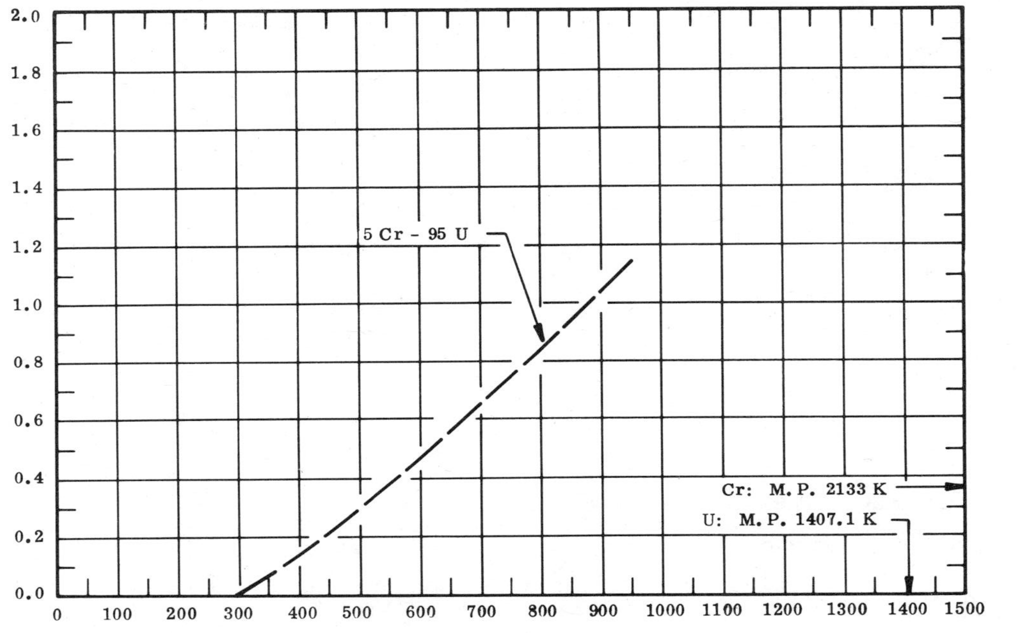

PROVISIONAL VALUES

[Temperature, T, K; Linear Expansion, $\Delta L/L_0$, %; α, K^{-1}]

T	(5 Cr-95 U) $\Delta L/L_0$	$\alpha \times 10^6$
293	0.000	12.8
400	0.147	14.5
500	0.299	16.0
600	0.268	17.4
700	0.648	18.7
800	0.841	20.0
900	1.047	21.2
950	1.155	21.7

REMARKS

(5 Cr-95 U): The tabulated values for well-annealed alloy are considered accurate to within ±7% over the entire temperature range. These values can be represented approximately by the following equation:

$$\Delta L/L_0 = -0.302 + 7.660 \times 10^{-4}\ T + 9.407 \times 10^{-7}\ T^2 - 1.413 \times 10^{-10}\ T^3$$

FIGURE AND TABLE NO. 160BR. PROVISIONAL VALUES FOR THERMAL LINEAR EXPANSION OF CHROMIUM-VANADIUM SYSTEM Cr-V

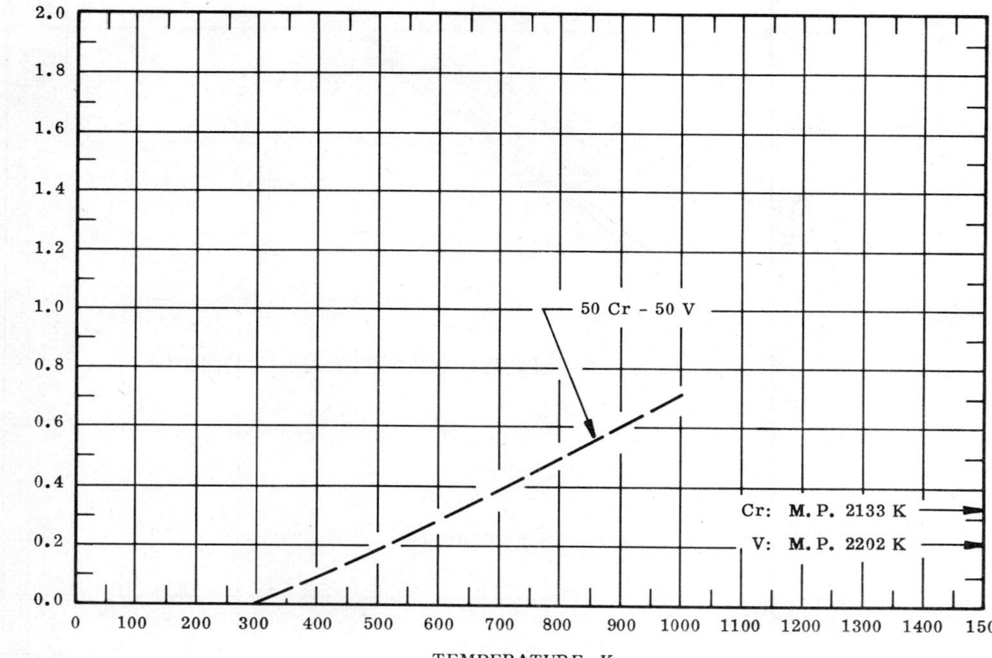

PROVISIONAL VALUES

[Temperature, T, K; Linear Expansion, $\Delta L/L_0$, %; α, K^{-1}]

T	(50 Cr-50 V) $\Delta L/L_0$	$\alpha \times 10^6$
293	0.000	7.9
400	0.089	8.8
500	0.181	9.6
600	0.280	10.2
700	0.384	10.6
800	0.492	10.9
900	0.602	11.1
1000	0.714	11.2

REMARKS

(50 Cr-50 V): The tabulated values for well-annealed alloy are considered accurate to within ±7% over the entire temperature range. These values can be represented approximately by the following equation:

$$\Delta L/L_0 = -0.187 + 4.596 \times 10^{-4}\, T + 6.658 \times 10^{-7}\, T^2 - 2.247 \times 10^{-10}\, T^3$$

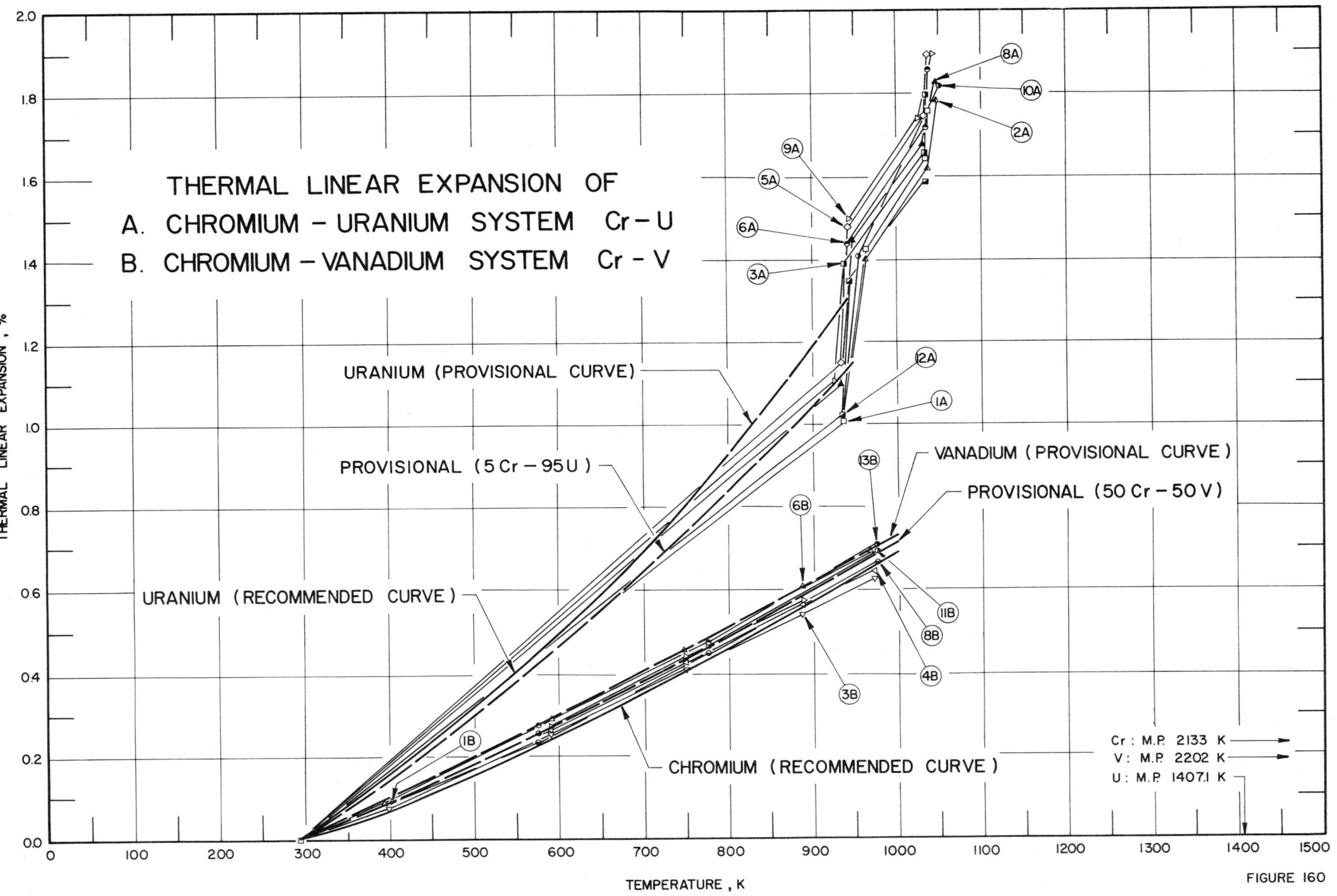
THERMAL LINEAR EXPANSION OF
A. CHROMIUM - URANIUM SYSTEM Cr - U
B. CHROMIUM - VANADIUM SYSTEM Cr - V
URANIUM (PROVISIONAL CURVE)
PROVISIONAL (5 Cr - 95 U)
URANIUM (RECOMMENDED CURVE)
VANADIUM (PROVISIONAL CURVE)
PROVISIONAL (50 Cr - 50 V)
CHROMIUM (RECOMMENDED CURVE)
Cr : M.P. 2133 K
V : M.P. 2202 K
U : M.P. 1407.1 K
8A
10A
2A
9A
5A
6A
3A
12A
1A
13B
6B
11B
8B
4B
3B
1B
THERMAL LINEAR EXPANSION, %
TEMPERATURE, K
0.0
0.2
0.4
0.6
0.8
1.0
1.2
1.4
1.6
1.8
2.0
0
100
200
300
400
500
600
700
800
900
1000
1100
1200
1300
1400
1500
FIGURE 160

SPECIFICATION TABLE 160A. THERMAL LINEAR EXPANSION OF CHROMIUM-URANIUM SYSTEM Cr-U

Cur. No.	Ref. No.	Author(s)	Year	Method Used	Temp. Range, K	Name and Specimen Designation	Composition (weight percent) Cr	U	Composition (continued), Specifications, and Remarks
1	446	Deem, H.W., Winn, R.A. and Lucks, C.F.	1954	L	293-1038	84D	5.2	94.8	Specimen prepared under vacuum in zirconium crucible by induction melting of biscuit of uranium cast in cold graphite, machined to cylinder 7.6 cm long x 0.95 cm dia; expansion measured in vacuum of 2 x 10^{-5} mm Hg under vertical load ~4 psi with temp increasing at 3 C/min.
2	446	Deem, H.W. et al.	1954	L	293-1047	87-D	5.2	94.8	Similar to the above specimen; except using biscuit of uranium cast in cold copper.
3	446	Deem, H.W. et al.	1954	L	293-1035	119-A	5.2	94.8	Similar to the above specimen; heat treated 1 hr at 823 K and cooled slowly.
4*	446	Deem, H.W. et al.	1954	L	293-1043	119-B	5.2	94.8	Similar to the above specimen.
5	446	Deem, H.W. et al.	1954	L	293-1036	120-A	5.2	94.8	Similar to the above specimen.
6	446	Deem, H.W. et al.	1954	L	293-1038	120-B	5.2	94.8	Similar to the above specimen.
7*	446	Deem, H.W. et al.	1954	L	293-1035	661	5.2	94.8	Similar to the above specimen; except using biscuit of uranium cast in warm graphite.
8	446	Deem, H.W. et al.	1954	L	293-1048	787	5.2	94.8	Similar to the above specimen; except using Fernald base uranium cast in graphite; specimen 5.1 cm long, 1.3 cm sq; expansion measured parallel to length of ingot.
9	446	Deem, H.W. et al.	1954	L	293-1041	787-1	5.2	94.8	Similar to the above specimen; expansion measured perpendicular to length of ingot.
10	446	Deem, H.W. et al.	1954	L	293-1051	787-2	5.2	94.8	Similar to the above specimen; expansion measured perpendicular to the two previous directions.
11*	446	Deem, H.W. et al.	1954	L	293-1046	771-1 and 2	5.2	94.8	Cubical specimen; expansion measured parallel to length of ingot.
12	446	Deem, H.W. et al.	1954	L	293-1037	771-1 and 2	5.2	94.8	Cubical specimen; expansion measured perpendicular to length of ingot.
13*	446	Deem, H.W. et al.	1954	L	293-1035	771-1 and 2	5.2	94.8	Cubical specimen; expansion measured perpendicular to the above two directions.

*Not shown in figure.

DATA TABLE 160A. THERMAL LINEAR EXPANSION OF CHROMIUM-URANIUM SYSTEM Cr-U

[Temperature, T, K; Linear Expansion, $\Delta L/L_0$, %]

T	$\Delta L/L_0$	T	$\Delta L/L_0$	T	$\Delta L/L_0$
CURVE 1		CURVE 7*		CURVE 13	
293	0.00	293	0.00	293	0.00
936	1.08	936	1.07	932	1.05
962	1.43	960	1.41	936	1.38
1034	1.65	1031	1.65	1031	1.64
1038	1.76	1035	1.78	1035	1.78
CURVE 2		CURVE 8			
293	0.00*	293	0.00*		
939	1.08*	932	1.10		
966	1.40	948	1.45		
1037	1.62	1030	1.68		
1047	1.79	1048	1.83		
CURVE 3		CURVE 9			
293	0.00*	293	0.00*		
934	1.07	929	1.13		
938	1.39	943	1.50		
1031	1.66	1026	1.74		
1035	1.80	1041	1.90		
CURVE 4*		CURVE 10			
293	0.00	293	0.00*		
935	1.08	933	1.08*		
943	1.40	953	1.41		
1035	1.65	1030	1.65*		
1043	1.80	1051	1.82		
CURVE 5		CURVE 11*			
293	0.00*	293	0.00		
933	1.15	935	1.02		
941	1.48	947	1.37		
1031	1.75	1049	1.61		
1036	1.90	1046	1.75		
CURVE 6		CURVE 12			
293	0.00*	293	0.00*		
937	1.14*	933	1.02		
941	1.44	941	1.35		
1035	1.72	1033	1.59		
1038	1.86	1037	1.72*		

*Not shown in figure.

SPECIFICATION TABLE 160B. THERMAL LINEAR EXPANSION OF CHROMIUM-VANADIUM SYSTEM Cr-V

Cur. No.	Ref. No.	Author(s)	Year	Method Used	Temp. Range, K	Name and Specimen Designation	Composition (weight percent) Cr	V	Composition (continued), Specifications, and Remarks
1	708	Ageev, N.V. and Model', M.S.	1963	X	301-971		Bal.	0.6	Solid solution sample prepared from electrolytic 99.96 Cr and 99.8 V, polycrystalline powder with 40 μ particle size; annealed at 1673 K in argon atmosphere, adding nitrogen and oxygen impurities to sample; powder pressed into rings, ±0.0001 Å accuracy; lattice parameter reported at 293 K is 2.8847 Å determined by graphical extrapolation.
2*	708	Ageev, N.V. and Model', M.S.	1963	X	298-971		Bal.	1.2	Similar to the above specimen; lattice parameter reported at 293 K is 2.8858 Å determined by graphical extrapolation.
3	708	Ageev, N.V. and Model', M.S.	1963	X	296-971		Bal.	2.5	Similar to the above specimen; lattice parameter reported at 293 K is 2.8879 Å determined by graphical extrapolation.
4	708	Ageev, N.V. and Model', M.S.	1963	X	297-971		Bal.	4.9	Similar to the above specimen; lattice parameter reported at 293 K is 2.8900 Å determined by graphical extrapolation.
5*	708	Ageev, N.V. and Model', M.S.	1963	X	296-971		Bal.	6.4	Similar to the above specimen; lattice parameter reported at 293 K is 2.8910 Å determined by graphical extrapolation.
6	708	Ageev, N.V. and Model', M.S.	1963	X	391-967		Bal.	9.8	Similar to the above specimen; lattice parameter reported at 293 K is 2.8960 Å determined by graphical extrapolation.
7*	707	Pridantseva, K.S. and Solov'yeva, N.A.	1965	L	293-973		Bal.	10	Single-phase solid solution powder sample; samples melted in an induction furnace in an argon atmosphere; annealed at 1273 K for 24 hr in a vacuum of not less than 10^{-4} mm Hg or in argon; impurities mainly 1% oxygen; 1% measurement error.
8	707	Pridantseva, K.S. and Solov'yeva, N.A.	1965	L	293-973		Bal.	20	Similar to the above specimen.
9*	707	Pridantseva, K.S. and Solov'yeva, N.A.	1965	L	293-973		Bal.	30	Similar to the above specimen.
10*	707	Pridantseva, K.S. and Solov'yeva, N.A.	1965	L	293-973		Bal.	40	Similar to the above specimen.
11	707	Pridantseva, K.S. and and Solov'yeva, N.A.	1965	L	293-973		Bal.	50	Similar to the above specimen.
12*	707	Pridantseva, K.S. and Solov'yeva, N.A.	1965	L	293-973		Bal.	60	Similar to the above specimen.
13	707	Pridantseva, K.S. and Solov'yeva, N.A.	1965	L	293-973		Bal.	70	Similar to the above specimen.
14*	707	Pridantseva, K.S. and Solov'yeva, N.A.	1965	L	293-973		Bal.	80	Similar to the above specimen.
15*	707	Pridantseva, K.S. and Solov'yeva, N.A.	1965	L	293-973		Bal.	90	Similar to the above specimen.

* Not shown in figure.

DATA TABLE 160B. THERMAL LINEAR EXPANSION OF CHROMIUM-VANADIUM SYSTEM Cr-V

[Temperature, T, K; Linear Expansion, $\Delta L/L_0$, %]

T	$\Delta L/L_0$
CURVE 1	
293	0.000*
301	0.003*
401	0.090
590	0.274
749	0.441
887	0.587
971	0.670*
CURVE 2*	
293	0.000
298	0.003
400	0.076
590	0.253
749	0.420
887	0.560
971	0.642
CURVE 3	
293	0.000*
296	0.000*
399	0.076
590	0.257
749	0.409
887	0.541
971	0.625
CURVE 4	
293	0.000*
297	0.003*
392	0.083*
590	0.267
749	0.426
887	0.562
971	0.645
CURVE 5*	
293	0.000
296	0.007
392	0.087
590	0.284
750	0.461
883	0.600
967	0.693
CURVE 6	
293	0.000*
391	0.086
590	0.291
745	0.457
887	0.609
967	0.695
CURVE 7*	
293	0.000
573	0.240
773	0.468
973	0.690
CURVE 8	
293	0.000*
573	0.235
773	0.452
973	0.673
CURVE 9*	
293	0.000
573	0.255
773	0.454
973	0.675
CURVE 10*	
293	0.000
573	0.259
773	0.459
973	0.676
CURVE 11	
293	0.000*
573	0.257
773	0.470
973	0.694
CURVE 12*	
293	0.000
573	0.258
773	0.471
973	0.702
CURVE 13	
293	0.000*
573	0.263
773	0.475
973	0.709
CURVE 14*	
293	0.000
573	0.243
773	0.479
973	0.709
CURVE 15*	
293	0.000
573	0.251
773	0.457
973	0.704

* Not shown in figure.

SPECIFICATION TABLE 161. THERMAL LINEAR EXPANSION OF COBALT-GOLD SYSTEM Co-Au

Cur. No.	Ref. No.	Author(s)	Year	Method Used	Temp. Range, K	Name and Specimen Designation	Composition (weight percent) Co	Au	Composition (continued), Specifications, and Remarks
1*	628	Krajewski, W., Kruger, J., and Winterhager, H.	1970	L	373-1073		99.54	0.46	Alloy prepared from >99.99 pure cobalt and high purity gold; polished and annealed in vacuum; transformation starts at 722 K on heating and 656 K on cooling; this curve is here reported using the first given temperature as reference temperature at which $\Delta L/L_0 = 0$.
2*	628	Krajewski, W., et al.	1970	L	373-1073		95.95	4.05	Similar to the above specimen; transformation starts at 715 K on heating and 621 K on cooling; this curve is here reported using the first given temperature as reference temperature at which $\Delta L/L_0 = 0$.

DATA TABLE 161. THERMAL LINEAR EXPANSION OF COBALT-GOLD SYSTEM Co-Au

[Temperature, T, K; Linear Expansion, $\Delta L/L_0$, %]

T	$\Delta L/L_0$
CURVE 1*, †	
373	0.000
673	0.411
1073	1.127
CURVE 2*, †	
373	0.000
673	0.396
1073	1.116

*No figure given.
†This curve is here reported using the first given temperature as reference temperature at which $\Delta L/L_0 = 0$.

FIGURE AND TABLE NO. 162R. PROVISIONAL VALUES FOR THERMAL LINEAR EXPANSION OF COBALT-IRON SYSTEM Co-Fe

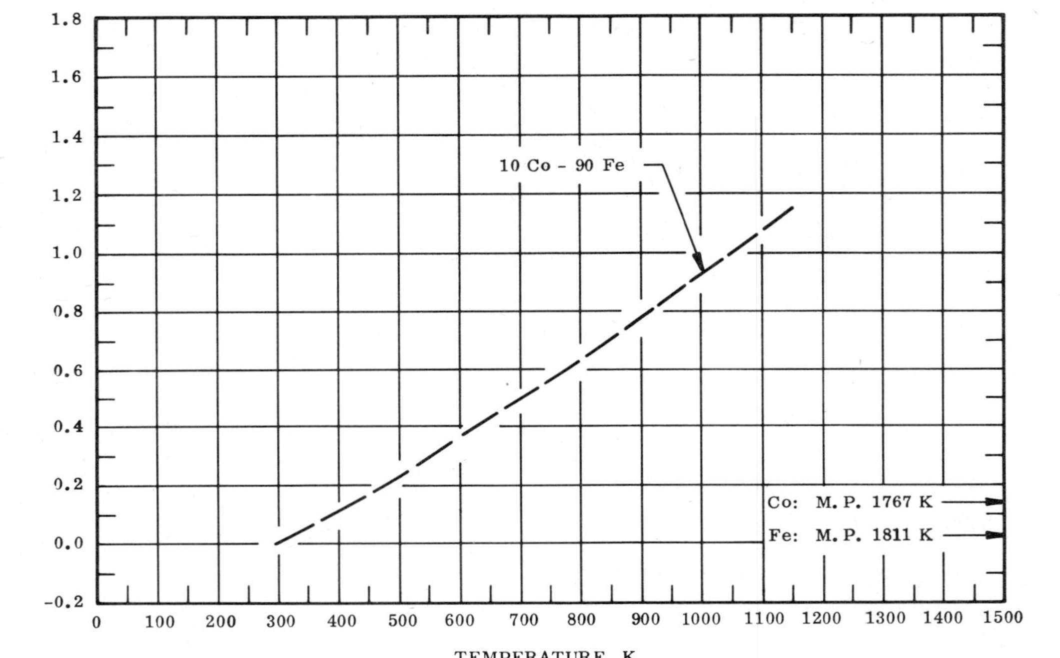

PROVISIONAL VALUES

[Temperature, T, K; Linear Expansion, $\Delta L/L_0$, %; α, K^{-1}]

T	(10 Co-90 Fe) $\Delta L/L_0$	$\alpha \times 10^6$
293	0.000	10.2
400	0.115	11.4
500	0.234	12.4
600	0.362	13.2
700	0.497	13.8
800	0.637	14.2
900	0.781	14.6
1000	0.928	14.8
1100	1.077	15.0
1150	1.152	15.1

REMARKS

(10 Co-90 Fe): The tabulated values for well-annealed alloy are considered accurate to within ±7% over the entire temperature range. These values can be represented approximately by the following equation:

$$\Delta L/L_0 = -0.242 + 6.069 \times 10^{-4}\, T + 8.163 \times 10^{-7}\, T^2 - 2.532 \times 10^{-10}\, T^3$$

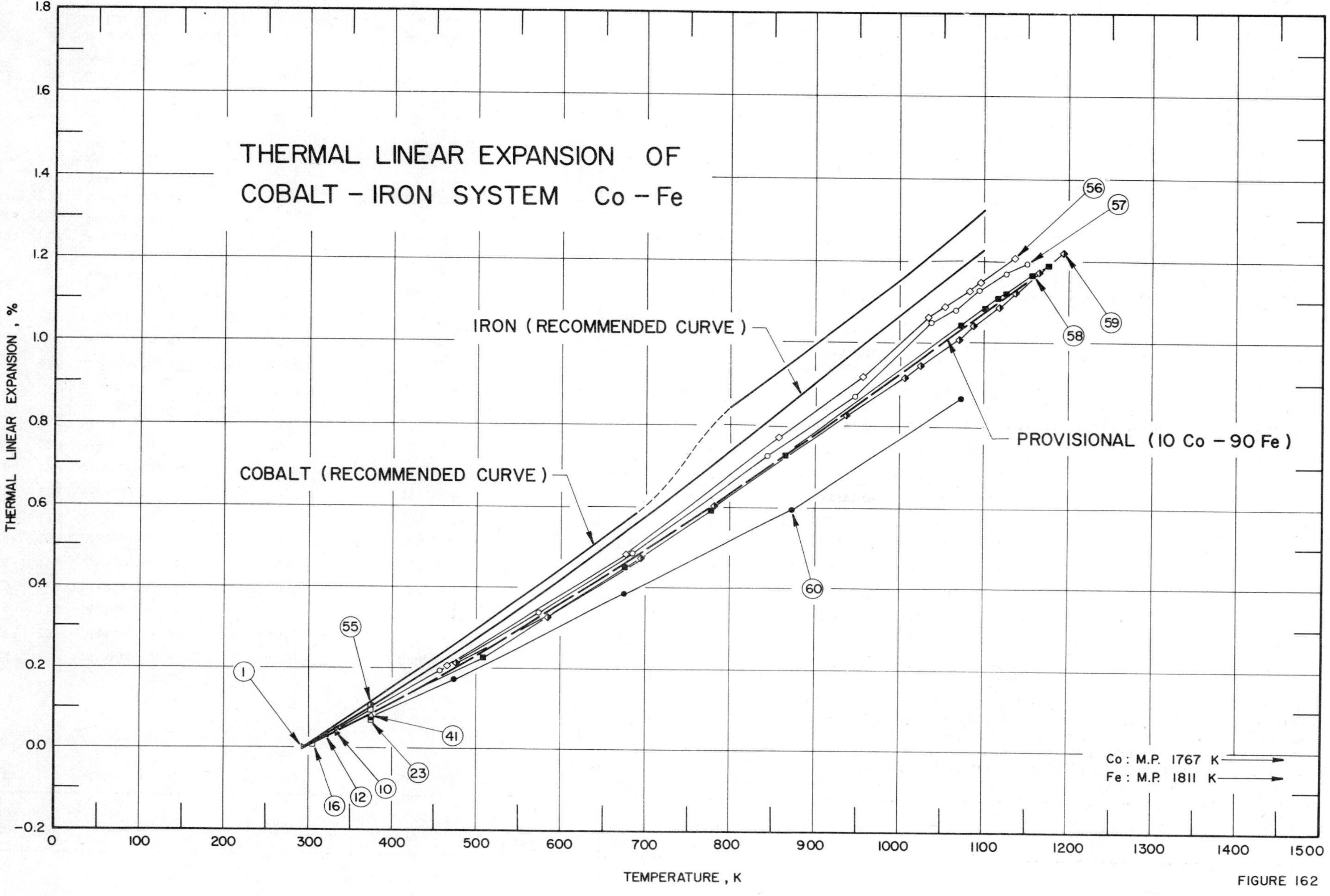
THERMAL LINEAR EXPANSION OF
COBALT - IRON SYSTEM Co - Fe
IRON (RECOMMENDED CURVE)
COBALT (RECOMMENDED CURVE)
PROVISIONAL (10 Co - 90 Fe)
Co : M.P. 1767 K
Fe : M.P. 1811 K
56
57
58
59
60
55
1
41
23
10
12
16
THERMAL LINEAR EXPANSION, %
1.8
1.6
1.4
1.2
1.0
0.8
0.6
0.4
0.2
0.0
-0.2
TEMPERATURE, K
0
100
200
300
400
500
600
700
800
900
1000
1100
1200
1300
1400
1500

FIGURE 162

SPECIFICATION TABLE 162. THERMAL LINEAR EXPANSION OF COBALT-IRON SYSTEM Co-Fe

Cur. No.	Ref. No.	Author(s)	Year	Method Used	Temp. Range, K	Name and Specimen Designation	Composition (weight percent) Co	Fe	Composition (continued), Specifications, and Remarks
1	129	Masumoto, H.	1934	L	293, 333		95.9	4.1	Electrolytic cobalt (0.13 Fe, 0.074 Al, 0.05 C, 0.007 P, 0.002 Si, trace Ni, S, Mn) and electrolytic iron (0.01 C, 0.01 Si, 0.001 P, trace Mn) supplied by Sugibayashi and Co.; used as raw materials; metals mixed in proportion, melted in alumina crucible, placed in Tammann furnace, under hydrogen atmosphere; melt cast into iron mold; test-specimen 10 cm long cut from cast rod; heated for 1 hr at 1273 K in vacuum furnace and slowly cooled.
2*	129	Masumoto, H.	1934	L	293, 333		93.9	6.1	Similar to the above specimen.
3*	129	Masumoto, H.	1934	L	293, 333		90.0	10.0	Similar to the above specimen.
4*	129	Masumoto, H.	1934	L	293, 333		84.1	15.9	Similar to the above specimen.
5*	129	Masumoto, H.	1934	L	293, 333		81.1	18.9	Similar to the above specimen.
6*	129	Masumoto, H.	1934	L	293, 333		79.1	20.9	Similar to the above specimen.
7*	129	Masumoto, H.	1934	L	293, 333		78.1	21.9	Similar to the above specimen.
8*	129	Masumoto, H.	1934	L	293, 333		74.2	25.8	Similar to the above specimen.
9*	129	Masumoto, H.	1934	L	293, 333		69.2	30.8	Similar to the above specimen.
10	129	Masumoto, H.	1934	L	293, 333		59.3	40.7	Similar to the above specimen.
11*	129	Masumoto, H.	1934	L	293, 333		49.4	50.6	Similar to the above specimen.
12	405	Masumoto, H., Saito, H., Sugai, Y., and Kono, T.	1960	L	293, 323		60	40	Electrolytic Co (0.26 Ni, 0.15 Fe, 0.03 C, 0.01 Si, trace Mn, Al, P, S), electrolytic Fe (0.02 Al, trace C, Si, Mn, P, S, Cu) used as raw materials; specimen 3 mm long.
13*	405	Masumoto, H., et al.	1960	L	293, 323		65	35	Similar to the above specimen.
14*	405	Masumoto, H., et al.	1960	L	293, 323		75	25	Similar to the above specimen.
15*	405	Masumoto, H., et al.	1960	L	293, 323		80	20	Similar to the above specimen.
16	56	Masumoto, H. and Nara, S.	1927	L	303, 373		9.98	Bal.	Specimen made by mixing two metals in proportion, melting in alumina crucible in Tammann furnace, cast in iron mold to form round rod 5 mm thick, 20 cm long, annealed at 1173 K for 30 min.
17*	56	Masumoto, H. and Nara, S.	1927	L	303, 373		14.97	Bal.	Similar to the above specimen.
18	56	Masumoto, H. and Nara, S.	1927	L	303, 373		19.96	Bal.	Similar to the above specimen.
19*	56	Masumoto, H. and Nara, S.	1927	L	303, 373		29.95	Bal.	Similar to the above specimen.
20*	56	Masumoto, H. and Nara, S.	1927	L	303, 373		34.94	Bal.	Similar to the above specimen.
21*	56	Masumoto, H. and Nara, S.	1927	L	303, 373		39.97	Bal.	Similar to the above specimen.
22*	56	Masumoto, H. and Nara, S.	1927	L	303, 373		44.92	Bal.	Similar to the above specimen.

* Not shown in figure.

SPECIFICATION TABLE 162. THERMAL LINEAR EXPANSION OF COBALT-IRON SYSTEM Co-Fe (continued)

Cur. No.	Ref. No.	Author(s)	Year	Method Used	Temp. Range, K	Name and Specimen Designation	Composition (weight percent) Co	Fe	Composition (continued), Specifications, and Remarks
23	56	Masumoto, H. and Nara, S.	1927	L	303, 373		49 91	Bal.	Similar to the above specimen.
24*	56	Masumoto, H. and Nara, S.	1927	L	303, 373		59.89	Bal.	Similar to the above specimen.
25*	56	Masumoto, H. and Nara, S.	1927	L	303, 373		64.88	Bal.	Similar to the above specimen.
26	56	Masumoto, H. and Nara, S.	1927	L	303, 373		69.87	Bal.	Similar to the above specimen.
27*	56	Masumoto, H. and Nara, S.	1927	L	303, 373		75.86	Bal.	Similar to the above specimen.
28*	56	Masumoto, H. and Nara, S.	1927	L	303, 373		77.86	Bal.	Similar to the above specimen.
29*	56	Masumoto, H. and Nara, S.	1927	L	303, 373		79.36	Bal.	Similar to the above specimen.
30*	56	Masumoto, H. and Nara, S.	1927	L	303, 373		79.86	Bal.	Similar to the above specimen.
31*	56	Masumoto, H. and Nara, S.	1927	L	303, 373		82.85	Bal.	Similar to the above specimen.
32*	56	Masumoto, H. and Nara, S.	1927	L	303, 373		86.84	Bal.	Similar to the above specimen.
33	56	Masumoto, H. and Nara, S.	1927	L	303, 373		89.84	Bal.	Similar to the above specimen.
34*	56	Masumoto, H. and Nara, S.	1927	L	303, 373		94.33	Bal.	Similar to the above specimen.
35*	56	Masumoto, H. and Nara, S.	1927	L	303, 373		95.83	Bal.	Similar to the above specimen.
36*	56	Masumoto, H. and Nara, S.	1927	L	303, 373		96.83	Bal.	Similar to the above specimen.
37*	56	Masumoto, H. and Nara, S.	1927	L	303, 373		97.82	Bal.	Similar to the above specimen.
38*	56	Masumoto, H. and Nara, S.	1927	L	303, 373		98.62	Bal.	Similar to the above specimen.
39*	56	Masumoto, H. and Nara, S.	1927	L	303, 373		99.32	Bal.	Similar to the above specimen.
40	59	Masumoto, H.	1931	L	303, 373	No. 1	9.9	90.1	Armco Fe, granular Co, supplied by Sugibayashi and Co., refined by electrolysis, mixed, melted in hydrogen atmosphere in alumina crucible in Tammann furnace, cast in iron mold with cylindrical aperture, machined to circular rod 4 mm thick, 10 cm long; heated at 1273 K for 1 hr in electric furnace in hydrogen atmosphere at reduced pressure, slowly cooled; expansion measured in vacuum.

* Not shown in figure.

SPECIFICATION TABLE 162. THERMAL LINEAR EXPANSION OF COBALT-IRON SYSTEM Co-Fe (continued)

Cur. No.	Ref. No.	Author(s)	Year	Method Used	Temp. Range, K	Name and Specimen Designation	Composition (weight percent) Co	Fe	Composition (continued), Specifications, and Remarks
41	59	Masumoto, H.	1931	L	303, 373	No. 2	19.8	80.2	Similar to the above specimen.
42	59	Masumoto, H.	1931	L	303, 373	No. 3	29.7	70.3	Similar to the above specimen.
43	59	Masumoto, H.	1931	L	303, 373	No. 4	34.6	65.4	Similar to the above specimen.
44	59	Masumoto, H.	1931	L	303, 373	No. 5	39.6	60.4	Similar to the above specimen.
45	59	Masumoto, H.	1931	L	303, 373	No. 7	49.4	50.6	Similar to the above specimen.
46*	59	Masumoto, H.	1931	L	303, 373	No. 8	59.3	40.7	Similar to the above specimen.
47*	59	Masumoto, H.	1931	L	303, 373	No. 10	69.2	30.8	Similar to the above specimen.
48	59	Masumoto, H.	1931	L	303, 373	No. 11	74.2	25.8	Similar to the above specimen.
49*	59	Masumoto, H.	1931	L	303, 373	No. 12	78.1	21.9	Similar to the above specimen.
50*	59	Masumoto, H.	1931	L	303, 373	No. 13	79.1	20.9	Similar to the above specimen.
51*	59	Masumoto, H.	1931	L	303, 373	No. 14	81.1	18.9	Similar to the above specimen.
52*	59	Masumoto, H.	1931	L	303, 373	No. 15	84.1	15.9	Similar to the above specimen.
53*	59	Masumoto, H.	1931	L	303, 373	No. 16	90.0	10.0	Similar to the above specimen.
54*	59	Masumoto, H.	1931	L	303, 373	No. 18	93.9	6.1	Similar to the above specimen.
55	59	Masumoto, H.	1931	L	303, 373	No. 20	95.9	4.1	Similar to the above specimen.
56	524	Stuart, H. and Ridley, N.	1969	X	293-1135		3.00	97.00	99.8+ pure Japanese electrolytic iron and 99.998 pure cobalt used; melted together in alumina crucibles under argon atmosphere; swaged to 50% reduction in cross-sectional area; sealed in silica capsule and homogenized 14 days at 1273 K; specimen prepared from filings taken from center of ingot; lattice parameter reported at 293 K is 2.8665 Å.
57	524	Stuart, H. and Ridley, N.	1969	X	293-1150		5.95	94.05	Similar to the above specimen; lattice parameter reported at 293 K is 2.8664 Å.
58	524	Stuart, H. and Ridley, N.	1969	X	293-1176		9.35	90.65	Similar to the above specimen; lattice parameter reported at 293 K is 2.8669 Å.
59	524	Stuart, H. and Ridley, N.	1969	X	293-1193		12.30	87.70	Similar to the above specimen; lattice parameter reported at 293 K is 2.8669 Å.
60	555	Fine, M.E. and Ellis W.C.	1948	L	303-1073	R 390	52.1	Bal.	0.01 Ni; specimen prepared from electrolytic Co and Fe; vacuum melted.

* Not shown in figure.

DATA TABLE 162. THERMAL LINEAR EXPANSION OF COBALT-IRON SYSTEM Co-Fe

[Temperature, T, K; Linear Expansion, $\Delta L/L_0$, %]

T	$\Delta L/L_0$
CURVE 1	
293	0.000
333	0.050
CURVE 2*	
293	0.000
333	0.047
CURVE 3*	
293	0.000
333	0.047
CURVE 4*	
293	0.000
333	0.047
CURVE 5*	
293	0.000
333	0.047
CURVE 6*	
293	0.000
333	0.047
CURVE 7*	
293	0.000
333	0.046
CURVE 8*	
293	0.000
333	0.042
CURVE 9*	
293	0.000
333	0.040
CURVE 10	
293	0.000*
333	0.038
CURVE 11*	
293	0.000
333	0.037
CURVE 12	
293	0.000*
323	0.029
CURVE 13*	
293	0.000
323	0.029
CURVE 14*	
293	0.000
323	0.032
CURVE 15*	
293	0.000
323	0.025
CURVE 16	
303	0.012
373	0.093
CURVE 17*	
303	0.011
373	0.086
CURVE 18	
303	0.010
373	0.079
CURVE 19*	
303	0.009
373	0.075
CURVE 20*	
303	0.011
373	0.084
CURVE 21*	
303	0.010
373	0.079
CURVE 22*	
303	0.009
373	0.072
CURVE 23	
303	0.008*
373	0.068
CURVE 24*	
303	0.009
373	0.071
CURVE 25*	
303	0.010
373	0.079
CURVE 26	
303	0.011
373	0.087
CURVE 27*	
303	0.011
373	0.087
CURVE 28*	
303	0.011
373	0.087
CURVE 29*	
303	0.012
373	0.093
CURVE 30*	
303	0.012
373	0.093
CURVE 31*	
303	0.012
373	0.092
CURVE 32*	
303	0.012
373	0.097
CURVE 33	
303	0.012
373	0.099
CURVE 34*	
303	0.012
373	0.098
CURVE 35*	
303	0.012
373	0.098
CURVE 36*	
303	0.012
373	0.099
CURVE 37*	
303	0.012
373	0.098
CURVE 38*	
303	0.012
373	0.098
CURVE 39*	
303	0.012
373	0.098
CURVE 40*	
303	0.011
373	0.089
CURVE 41	
303	0.010
373	0.081
CURVE 42*	
303	0.009
373	0.078
CURVE 43*	
303	0.010
373	0.078
CURVE 44*	
303	0.010
373	0.077
CURVE 45*	
303	0.009
373	0.075
CURVE 46*	
303	0.010
373	0.076
CURVE 47*	
303	0.010
373	0.080
CURVE 48	
303	0.010*
373	0.084
CURVE 49*	
303	0.011
373	0.092
CURVE 50*	
303	0.012
373	0.093
CURVE 51*	
303	0.012
373	0.094
CURVE 52*	
303	0.012
373	0.094
CURVE 53*	
303	0.012
373	0.093
CURVE 54*	
303	0.012
373	0.094
CURVE 55	
303	0.012*
373	0.100
CURVE 56	
293	0.000
297	0.003*
465	0.206
675	0.481
859	0.771
955	0.921
1032	1.064
1043	1.071
1052	1.099
1059	1.110*
1063	1.117
1071	1.126*
1081	1.129
1087	1.124*
1095	1.150
1112	1.167
1135	1.212
CURVE 57	
293	0.000*
295	0.000

* Not shown in figure.

DATA TABLE 162. THERMAL LINEAR EXPANSION OF COBALT-IRON SYSTEM Co-Fe (continued)

T	$\Delta L/L_0$
CURVE 57 (cont.)	
457	0.192
572	0.338
682	0.485
844	0.726
949	0.876
1037	1.043
1051	1.059
1067	1.086
1074	1.096*
1085	1.120*
1094	1.138
1107	1.138
1118	1.144*
1129	1.158
1137	1.176*
1145	1.178
1150	1.200
CURVE 58	
293	0.000*
295	0.000*
506	0.227
672	0.453
776	0.596
867	0.726*
1071	1.044
1080	1.054*
1091	1.071*
1101	1.090*
1109	1.106
1119	1.115
1128	1.121
1139	1.136*
1147	1.155*
1156	1.165
1166	1.183*
1176	1.198
CURVE 59	
293	0.000*
295	0.000*
374	0.080*
475	0.216
583	0.324
692	0.471
780	0.603
864	0.722*
CURVE 59 (cont.)	
939	0.823
1005	0.921
1022	0.945
1070	1.015
1081	1.029
1088	1.051
1097	1.069
1108	1.082
1119	1.098
1128	1.113
1138	1.131
1149	1.153
1155	1.166
1165	1.178
1174	1.193
1185	1.205*
1193	1.223*
CURVE 60	
303	0.010
473	0.176
673	0.382
873	0.596
1073	0.866

* Not shown in figure.

FIGURE AND TABLE NO. 163BR. PROVISIONAL VALUES FOR THERMAL LINEAR EXPANSION OF COBALT-MOLYBDENUM SYSTEM Co-Mo

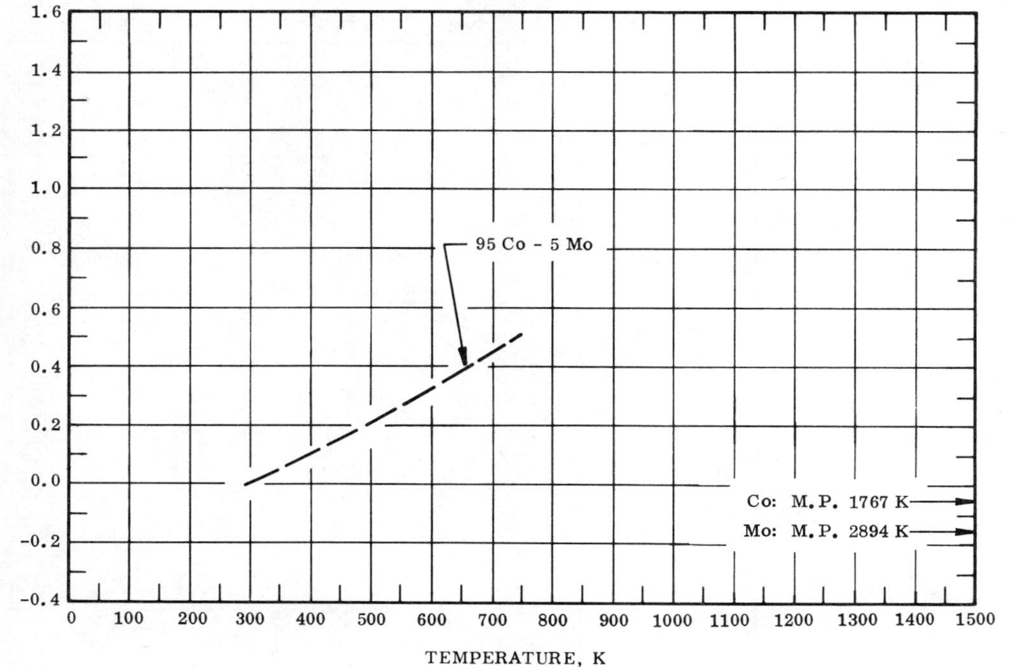

PROVISIONAL VALUES

[Temperature, T, K; Linear Expansion, $\Delta L/L_0$, %; α, K^{-1}]

T	(95 Co-5 Mo) $\Delta L/L_0$	$\alpha \times 10^6$
293	0.000	8.8
400	0.102	10.2
500	0.209	11.3
600	0.325	12.0
700	0.448	12.4
750	0.509	12.5

REMARKS

(95 Co-5 Mo): The tabulated values for well annealed alloy are considered accurate to within ± 7% over the entire temperature range. The phase transformation at about 510 K seems to produce little change in the thermal expansion. The tabulated values can be represented approximately by the following equation:

$$\Delta L/L_0 = -0.163 + 2.172 \times 10^{-4}\, T + 1.337 \times 10^{-6}\, T^2 - 5.743 \times 10^{-10}\, T^3$$

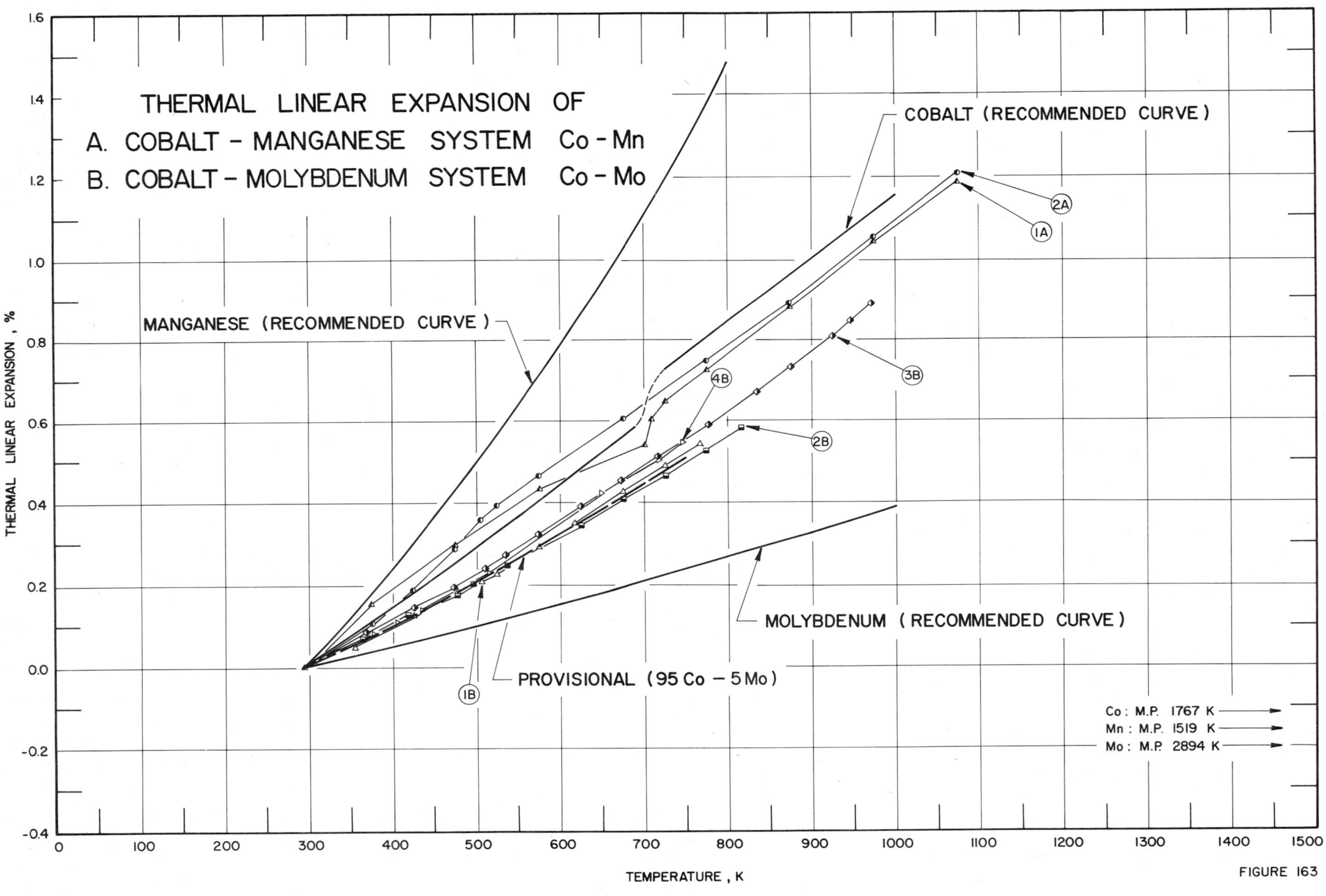
THERMAL LINEAR EXPANSION OF
A. COBALT - MANGANESE SYSTEM Co - Mn
B. COBALT - MOLYBDENUM SYSTEM Co - Mo
COBALT (RECOMMENDED CURVE)
MANGANESE (RECOMMENDED CURVE)
MOLYBDENUM (RECOMMENDED CURVE)
PROVISIONAL (95 Co — 5 Mo)
2A
1A
3B
4B
2B
1B
Co : M.P. 1767 K
Mn : M.P. 1519 K
Mo : M.P. 2894 K
THERMAL LINEAR EXPANSION , %
TEMPERATURE , K
1.6
1.4
1.2
1.0
0.8
0.6
0.4
0.2
0.0
-0.2
-0.4
0
100
200
300
400
500
600
700
800
900
1000
1100
1200
1300
1400
1500

FIGURE 163

SPECIFICATION TABLE 163A. THERMAL LINEAR EXPANSION OF COBALT-MANGANESE SYSTEM Co-Mn

Cur. No.	Ref. No.	Author(s)	Year	Method Used	Temp. Range, K	Name and Specimen Designation	Composition (weight percent) Co	Mn	Composition (continued), Specifications, and Remarks
1	555	Fine, M.E. and Ellis, W.C.	1948	L	293-1073	6013	98.9	0.63	0.1 Fe, 0.09 Si, 0.2 C; specimen prepared from cast ingots by first hot-swaging followed by cold-swaging to 0.18 in. diameter, 2.5 in. long with flat-ground and polished end; before measurement they are annealed in hydrogen for 1 hr at 1173 K; zero-point correction is 0.012%.
2	555	Fine, M.E. and Ellis, W.C.	1948	L	1073-293	6013	98.9	0.63	The above specimen; expansion measured with decreasing temperature; zero-point correction is 0.00%.

DATA TABLE 163A. THERMAL LINEAR EXPANSION OF COBALT-MANGANESE SYSTEM Co-Mn

[Temperature, T, K; Linear Expansion, $\Delta L/L_0$, %]

T	$\Delta L/L_0$	T	$\Delta L/L_0$
CURVE 1		CURVE 2	
293	0.000	293	0.000*
303	0.012	373	0.109
373	0.159	423	0.189
473	0.300	473	0.291
573	0.436	503	0.360
700	0.541	523	0.399
709	0.605	573	0.462
723	0.644	673	0.602
773	0.726	773	0.755
873	0.881	873	0.893
973	1.042	973	1.054
1073	1.191	1073	1.203

* Not shown in figure.

SPECIFICATION TABLE 163B. THERMAL LINEAR EXPANSION OF COBALT-MOLYBDENUM SYSTEM Co-Mo

Cur. No.	Ref. No.	Author(s)	Year	Method Used	Temp. Range, K	Name and Specimen Designation	Composition (weight percent) Co	Mo	Composition (continued), Specifications, and Remarks
1	406	Sykes, W.P. and Graff, H.F.	1935	L	351-766	Sample B	98	2	Pure Co with 0.6 Ni, 0.2 Fe, 99.8 Mo used as raw materials; specimen 3 in. long, 0.625 in. diameter; machined from cast alloy which heated in hydrogen 50 hrs at 1573 K after solification, cooled from 1073 to 573 K in 100 hrs; expansion measured with increasing temperature at 1 to 2 C min^{-1}; expansion increment reported at 351 K is -0.00038 in.; -0.00185 at 293 K determined by graphical extrapolation.
2	406	Sykes, W.P. and Graff, H.F.	1935	L	363-818	Sample C	95	5	Similar to the above specimen; expansion measured with increasing temperature; expansion increment reported at 363 K is -0.00184 in.; -0.00394 at 293 K determined by graphical extrapolation.
3	406	Sykes, W.P. and Graff, H.F.	1935	L	366-970	Sample D	95	5	Similar to the above specimen except previously heated at 1273 K for 25 hrs and quenched; expansion measured with increasing temperature.
4	406	Sykes, W.P. and Graff, H.F.	1935	L	745-365	Sample D	95	5	The above specimen; expansion measured with decreasing temperature; expansion increment reported at 365 K is -0.00596 in.; -0.00809 at 293 K determined by graphical extrapolation.

DATA TABLE 163B. THERMAL LINEAR EXPANSION OF COBALT-MOLYBDENUM SYSTEM Co-Mo

[Temperature, T, K; Linear Expansion, $\Delta L/L_0$, %]

T	$\Delta L/L_0$	T	$\Delta L/L_0$	T	$\Delta L/L_0$
CURVE 1		CURVE 2 (cont.)		CURVE 3 (cont.)	
351	0.049	570	0.290	775	0.598
424	0.127	623	0.349	833	0.675
476	0.182	673	0.414	873	0.732
503	0.213	725	0.468	922	0.807
523	0.232	773	0.533	944	0.843
572	0.299	818	0.586	970	0.892
616	0.351				
672	0.426	CURVE 3		CURVE 4	
723	0.488				
766	0.543	366	0.081	745	0.546
		422	0.144	716	0.506
CURVE 2		472	0.197	649	0.421
		510	0.242	576	0.322*
363	0.070	535	0.275	513	0.239
419	0.127	572	0.328	476	0.193*
475	0.178	624	0.399	433	0.142
496	0.202	670	0.458	406	0.111
536	0.248	713	0.519	376	0.082
				365	0.071*

*Not shown in figure.

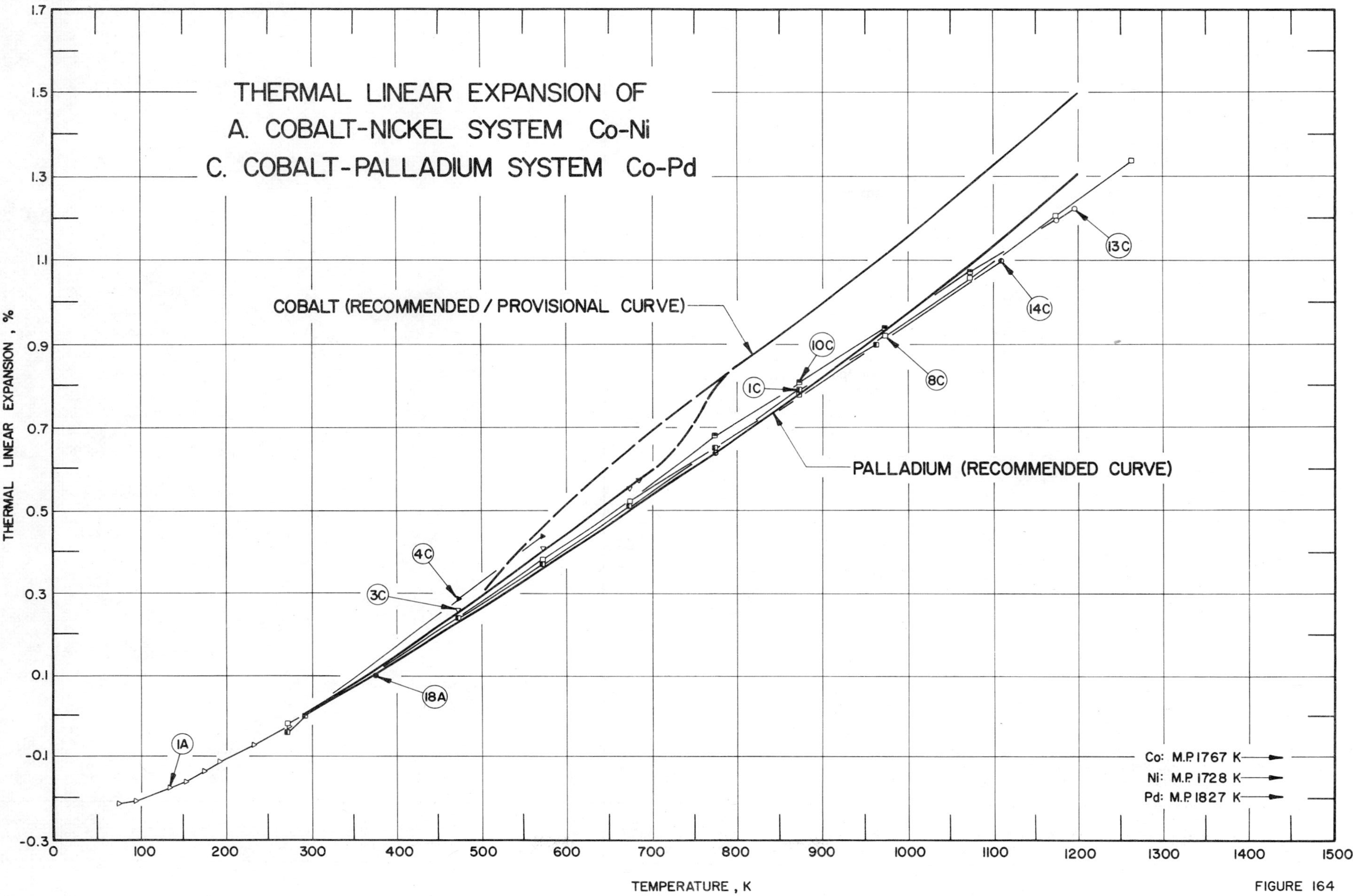
THERMAL LINEAR EXPANSION OF
A. COBALT-NICKEL SYSTEM Co-Ni
C. COBALT-PALLADIUM SYSTEM Co-Pd
COBALT (RECOMMENDED / PROVISIONAL CURVE)
PALLADIUM (RECOMMENDED CURVE)
1A
18A
3C
4C
1C
10C
8C
14C
13C
Co: M.P. 1767 K
Ni: M.P. 1728 K
Pd: M.P. 1827 K
THERMAL LINEAR EXPANSION, %
1.7
1.5
1.3
1.1
0.9
0.7
0.5
0.3
0.1
-0.1
-0.3
0
100
200
300
400
500
600
700
800
900
1000
1100
1200
1300
1400
1500
TEMPERATURE, K

FIGURE 164

SPECIFICATION TABLE 164A. THERMAL LINEAR EXPANSION OF COBALT-NICKEL SYSTEM Co-Ni

Cur. No.	Ref. No.	Author(s)	Year	Method Used	Temp. Range, K	Name and Specimen Designation	Composition (weight percent) Co	Ni	Composition (continued), Specifications, and Remarks
1	142	Aoyama, S. and Ito, T.	1939	L	77-273	Electrolytic Nickel	0.53	99.4	0.05 Fe, 0.02 Al; alloy melted, cast into bars 30 x 30 x 250 mm; hot-forged to bar 8 mm dia, then cut into round bar 5 mm dia, 100 mm long; annealed in vacuum for 3 hr at 1123 K.
2*	56	Masumoto, H. and Nara, S.	1927	L	303, 373		10	90	Specimen made by mixing metals in proportion, melting in alumina crucible in Tammann furnace, cast in iron mold to form round rod 5 mm dia, 20 cm long; annealed at 1173 K for 30 min; solid solution having face-centered cubic lattice.
3*	56	Masumoto, H. and Nara, S.	1927	L	303, 373		20	80	Similar to the above specimen.
4*	56	Masumoto, H. and Nara, S.	1927	L	303, 373		25	75	Similar to the above specimen.
5*	56	Masumoto, H. and Nara, S.	1927	L	303, 373		30	70	Similar to the above specimen.
6*	56	Masumoto, H. and Nara, S.	1927	L	303, 373		40	60	Similar to the above specimen.
7*	56	Masumoto, H. and Nara, S.	1927	L	303, 373		50	50	Similar to the above specimen.
8*	56	Masumoto, H. and Nara, S.	1927	L	303, 373		55	45	Similar to the above specimen.
9*	56	Masumoto, H. and Nara, S.	1927	L	303, 373		60	40	Similar to the above specimen.
10*	56	Masumoto, H. and Nara, S.	1927	L	303, 373		65	35	Similar to the above specimen.
11*	56	Masumoto, H. and Nara, S.	1927	L	303, 373		70	30	Similar to the above specimen; solid solution having hexagonal close-packed lattice.
12*	56	Masumoto, H. and Nara, S.	1927	L	303, 373		80	20	Similar to the above specimen.
13*	56	Masumoto, H. and Nara, S.	1927	L	303, 373		90	10	Similar to the above specimen.
14*	59	Masumoto, H.	1931	L	303, 373	No. 1	10	90	Mond Ni, granular Co supplied by Sugiboyashi & Co.; refined by electrolysis, mixed, melted in hydrogen atm in alumina crucible in Tammann furnace; cast in iron mold with cylindrical aperature machined to circular rod 4 mm thick, 10 cm long; heated to 1273 K for 1 hr in electric furnace in hydrogen atm at reduced pressure, slowly cooled; expansion measured in vacuum.
15*	59	Masumoto, H.	1931	L	303, 373	No. 2	20	80	Similar to the above specimen.
16*	59	Masumoto, H.	1931	L	303, 373	No. 3	30	70	Similar to the above specimen.
17*	59	Masumoto, H.	1931	L	303, 373	No. 4	40	60	Similar to the above specimen.
18	59	Masumoto, H.	1931	L	303, 373	No. 5	50	50	Similar to the above specimen.

* Not shown in figure.

SPECIFICATION TABLE 164A. THERMAL LINEAR EXPANSION OF COBALT-NICKEL SYSTEM Co-Ni (continued)

Cur. No.	Ref. No.	Author(s)	Year	Method Used	Temp. Range, K	Name and Specimen Designation	Composition (weight percent) Co	Ni	Composition (continued), Specifications, and Remarks
19*	59	Masumoto, H.	1931	L	303, 373	No. 6	60	40	Similar to the above specimen.
20*	59	Masumoto, H.	1931	L	303, 373	No. 7	65	35	Similar to the above specimen.
21*	59	Masumoto, H.	1931	L	303, 373	No. 8	70	30	Similar to the above specimen.
22*	59	Masumoto, H.	1931	L	303, 373	No. 9	75	25	Similar to the above specimen.
23*	59	Masumoto, H.	1931	L	303, 373	No. 10	80	20	Similar to the above specimen.
24*	59	Masumoto, H.	1931	L	303, 373	No. 11	85	15	Similar to the above specimen.
25*	59	Masumoto, H.	1931	L	303, 373	No. 13	90	10	Similar to the above specimen.
26*	59	Masumoto, H.	1931	L	303, 373	No. 15	95	5	Similar to the above specimen.
27*	628	Krajewski, W., Kruger, J., and Winterhager, H.	1970	L	373-1073		99.33	0.67	Alloy prepared from >99.99 pure cobalt and high purity nickel; polished and annealed in vacuum; transformation starts at 722 K on heating and 681 K on cooling; this curve is here reported using the first given temperature as reference temperature at which $\Delta L/L_0 = 0$.
28*	628	Krajewski, W., et al.	1970	L	373-1073		91.16	8.84	Similar to the above specimen; transformation starts at 723 K on heating and 663 K on cooling; this curve is here reported using the first given temperature as reference temperature at which $\Delta L/L_0 = 0$.

* Not shown in figure.

DATA TABLE 164A. THERMAL LINEAR EXPANSION OF COBALT-NICKEL SYSTEM Co–Ni

[Temperature, T, K; Linear Expansion, $\Delta L/L_0$, %]

T	$\Delta L/L_0$
CURVE 1	
77	-0.2169
91	-0.2097
133	-0.1784
153	-0.1600
173	-0.1396
193	-0.1183
233	-0.0725
273	-0.0242
CURVE 2*	
303	0.0135
373	0.1076
CURVE 3*	
303	0.0134
373	0.1072
CURVE 4*	
303	0.0132
373	0.1052
CURVE 5*	
303	0.0126
373	0.1010
CURVE 6*	
303	0.0125
373	0.1000
CURVE 7*	
303	0.0132
373	0.1057
CURVE 8*	
303	0.0135
373	0.1081
CURVE 9*	
303	0.0135
373	0.1079
CURVE 10*	
303	0.0129
373	0.1034
CURVE 11*	
303	0.0126
373	0.1005
CURVE 12*	
303	0.0127
373	0.1014
CURVE 13*	
303	0.0128
373	0.1022
CURVE 14*	
303	0.0131
373	0.1045
CURVE 15*	
303	0.0128
373	0.1025
CURVE 16*	
303	0.0128
373	0.1020
CURVE 17*	
303	0.0126
373	0.1005
CURVE 18	
303	0.0127
373	0.1012
CURVE 19*	
303	0.0124
373	0.0991
CURVE 20*	
303	0.0124
373	0.0993
CURVE 21*	
303	0.0121
373	0.0966
CURVE 22*	
303	0.0126
373	0.1007
CURVE 23*	
303	0.0128
373	0.1022
CURVE 24*	
303	0.0125
373	0.0998
CURVE 25*	
303	0.0126
373	0.1009
CURVE 26*	
303	0.0123
373	0.0986
CURVE 27*, †	
373	0.000
673	0.405
1073	1.121
CURVE 28*, †	
373	0.000
673	0.408
1073	1.128

*Not shown in figure.
†This curve is here reported using the first given temperature as reference temperature at which $\Delta L/L_0 = 0$.

SPECIFICATION TABLE 164B. THERMAL LINEAR EXPANSION OF COBALT-NIOBIUM SYSTEM Co-Nb

Cur. No.	Ref. No.	Author(s)	Year	Method Used	Temp. Range, K	Name and Specimen Designation	Composition (weight percent) Co	Nb	Composition (continued), Specifications, and Remarks
1*	628	Krajewski, W., Kruger, J., and Winterhager, H.	1970	L	373-1073		99.57	0.43	Alloy prepared from > 99.99 pure cobalt and high purity niobium; polished and annealed in vacuum; transformation starts at 742 K on heating and 645 K on cooling; this curve is here reported using the first given temperature as reference temperature at which $\Delta L/L_0 = 0$.
2*	628	Krajewski, W., et al.	1970	L	373-1073		96.54	3.46	Similar to above specimen; transformation starts at 796 K on heating and 610 K on cooling; this curve is here reported using the first given temperature as reference temperature at which $\Delta L/L_0 = 0$.

DATA TABLE 164B. THERMAL LINEAR EXPANSION OF COBALT-NIOBIUM SYSTEM Co-Nb

[Temperature, T, K; Linear Expansion, $\Delta L/L_0$, %]

T	$\Delta L/L_0$
	CURVE 1*, †
373	0.000
673	0.384
1073	1.092
	CURVE 2*, †
373	0.000
673	0.366
1073	1.022

*No figure given.
†This curve is here reported using the first given temperature as reference temperature at which $\Delta L/L_0 = 0$.

SPECIFICATION TABLE 164C. THERMAL LINEAR EXPANSION OF COBALT-PALLADIUM SYSTEM Co-Pd

Cur. No.	Ref. No.	Author(s)	Year	Method Used	Temp. Range, K	Name and Specimen Designation	Composition (weight percent) Co	Pd	Composition (continued), Specifications, and Remarks
1	407	Masumoto, H. and Sawaya, S.	1969	L	273-963		36	64	Electrolytic Co with impurities 0.120 Fe, 0.033 Ni, 0.028 C, 0.012 Si, 0.006 S, 0.002 P, 0.001 Al, 0.001 Mn, and 99.9 Pd; mixed and melted in crucible; rod formed, 2 mm diameter, 10 cm long; expansion measured in vacuum with heating rate 1273 K per hr, cooling rate 573 K per hr; same values observed for heating and cooling; density 10.50 g cm^{-3} at 293 K; zero-point correction is -0.04%.
2*	407	Masumoto, H. and Sawaya, S.	1969	L	273-873		30	70	Similar to the above specimen; density 10.70 g cm^{-3} at 293 K; zero-point correction is -0.03%.
3	407	Masumoto, H. and Sawaya, S.	1969	L	273-684		20	80	Similar to the above specimen; density 11.10 g cm^{-3} at 293 K; zero-point correction is -0.03%.
4	407	Masumoto, H. and Sawaya, S.	1969	L	273-573		15	85	Similar to the above specimen; density 11.30 g cm^{-3} at 293 K; zero-point correction is -0.03%.
5*	407	Masumoto, H. and Sawaya, S.	1969	L	273-427		10	90	Similar to the above specimen; density 11.55 g cm^{-3} at 293 K; zero-point correction is -0.03%.
6*	407	Masumoto, H. and Sawaya, S.	1969	L	273-323		6	94	Similar to the above specimen; zero-point correction is -0.02%.
7*	407	Masumoto, H. and Sawaya, S.	1969	L	273-412		4	96	Similar to the above specimen; density 11.80 g cm^{-3} at 293 K; zero-point correction is -0.02%.
8	407	Masumoto, H. and Sawaya, S.	1969	L	273-1263		90	10	The above specimen; zero-point correction is -0.002%.
9*	407	Masumoto, H. and Sawaya, S.	1969	L	1263-273		90	10	The above specimen; expansion measured with cooling rate of 573 K per hr; zero-point correction is -0.002%.
10	407	Masumoto, H. and Sawaya, S.	1969	L	273-1262		80	20	Similar to the above specimen; expansion measured with increasing temperature; zero-point correction is -0.002%.
11*	407	Masumoto, H. and Sawaya, S.	1969	L	1262-273		80	20	The above specimen; expansion measured with decreasing temperature; zero-point correction is -0.002%.
12*	407	Masumoto, H. and Sawaya, S.	1969	L	273-1265		70	30	Similar to the above specimen; no difference in expansion observed between measurements with increasing and decreasing temperature; zero-point correction is -0.003%.
13	407	Masumoto, H. and Sawaya, S.	1969	L	273-1196		60	40	Similar to the above specimen; zero-point correction is -0.003%.
14	407	Masumoto, H. and Sawaya, S.	1969	L	273-1110		50	50	Similar to the above specimen; density 10.00 g cm^{-3} at 293 K; zero-point correction is -0.0034%.
15*	628	Krajewski, W., Kruger, J., and Winterhager, H.	1970	L	373-1073		99.48	0.52	Alloy prepared from > 99.99 pure cobalt and high purity palladium; polished and annealed in vacuum; transformation starts at 724 K on heating and 663 K on cooling; this curve is here reported using the first given temperature as reference temperature at which $\Delta L/L_0 = 0$.
16*	628	Krajewski, W., et al.	1970	L	373-1073		97.10	2.90	Similar to above specimen; transformation starts at 737 K on heating and 611 K on cooling; this curve is here reported using the first given temperature as reference temperature at which $\Delta L/L_0 = 0$.

* Not shown in figure.

DATA TABLE 164C. THERMAL LINEAR EXPANSION OF COBALT-PALLADIUM SYSTEM

[Temperature, T, K; Linear Expansion, $\Delta L/L_0$, %]

T	$\Delta L/L_0$
CURVE 1	
273	-0.04
293	0.00*
373	0.11*
473	0.24
573	0.37*
673	0.51
773	0.65*
873	0.79
963	0.90
CURVE 2*	
273	-0.03
293	0.00
373	0.12
473	0.25
573	0.39
673	0.52
773	0.67
873	0.80
CURVE 3	
273	-0.03*
293	0.00*
373	0.12*
473	0.26*
573	0.41
673	0.55
684	0.57
CURVE 4	
273	-0.03
293	0.00
373	0.14
473	0.29
573	0.44
CURVE 5*	
273	-0.03
293	0.00
373	0.13
427	0.22
CURVE 6*	
273	-0.02
293	0.00
323	0.06
CURVE 7*	
273	-0.02
293	0.00
373	0.12
412	0.16
CURVE 8	
273	-0.02
293	0.00*
373	0.11*
473	0.24
573	0.38
673	0.52
773	0.65
873	0.78
973	0.92
1073	1.06
1173	1.21
1263	1.34
CURVE 9*	
1263	1.34
1173	1.21
1073	1.06
973	0.92
873	0.78
773	0.65
673	0.52
573	0.39
473	0.26
373	0.12
293	0.00
273	-0.02
CURVE 10	
273	-0.02*
293	0.00*
373	0.11*
473	0.25*
CURVE 10 (cont.)	
573	0.38*
673	0.52*
773	0.68
873	0.81
973	0.94
1073	1.07
1173	1.21*
1262	1.33*
CURVE 11*	
1262	1.33
1173	1.21
1073	1.07
973	0.94
873	0.81
773	0.68
673	0.53
573	0.41
473	0.27
373	0.11
293	0.00
273	-0.02
CURVE 12*	
273	-0.03
293	0.00
373	0.11
473	0.24
573	0.38
673	0.50
773	0.64
873	0.77
973	0.93
1073	1.07
1173	1.22
1265	1.33
CURVE 13	
273	-0.03*
293	0.00*
373	0.11*
473	0.25*
573	0.38*
673	0.52*
CURVE 13 (cont.)	
773	0.66*
873	0.79*
973	0.94*
1073	1.07*
1173	1.20
1196	1.22
CURVE 14	
273	-0.03*
293	0.00*
373	0.11*
473	0.24*
573	0.37*
673	0.51*
773	0.64*
873	0.79*
973	0.92*
1073	1.06*
1110	1.10
CURVE 15*, †	
373	0.000
673	0.432
1073	1.176
CURVE 16*, †	
373	0.000
673	0.396
1073	1.092

* Not shown in figure.
† This curve is here reported using the first given temperature as reference temperature at which $\Delta L/L_0 = 0$.

SPECIFICATION TABLE 165. THERMAL LINEAR EXPANSION OF COBALT-PLATINUM SYSTEM Co-Pt

Cur. No.	Ref. No.	Author(s)	Year	Method Used	Temp. Range, K	Name and Specimen Designation	Composition (weight percent) Co	Pt	Composition (continued), Specifications, and Remarks
1*	628	Krajewski, W., Kruger, J., and Winterhager, H.	1970	L	373-1073		99.62	0.38	Alloy prepared from >99.99 pure cobalt and high purity platinum; polished and annealed in vacuum; transformation starts at 766 K on heating and 682 K on cooling; this curve is here reported using the first given temperature as reference temperature at which $\Delta L/L_0 = 0$.
2*	628	Krajewski, W., et al.	1970	L	373-1073		96.44	3.56	Similar to the above specimen; transformation starts at 725 K on heating and 672 K on cooling; this curve is here reported using the first given temperature as reference temperature at which $\Delta L/L_0 = 0$.

DATA TABLE 165. THERMAL LINEAR EXPANSION OF COBALT-PLATINUM SYSTEM Co-Pt

[Temperature, T, K; Linear Expansion, $\Delta L/L_0$, %]

T	$\Delta L/L_0$
CURVE 1*, †	
373	0.000
673	0.447
1073	1.191
CURVE 2*, †	
373	0.000
673	0.408
1073	1.116

*No figure given.

†This curve is here reported using the first given temperature as reference temperature at which $\Delta L/L_0 = 0$.

SPECIFICATION TABLE 166. THERMAL LINEAR EXPANSION OF COBALT-TANTALUM SYSTEM Co-Ta

Cur. No.	Ref. No.	Author(s)	Year	Method Used	Temp. Range, K	Name and Specimen Designation	Composition (weight percent) Co	Ta	Composition (continued), Specifications, and Remarks
1*	628	Krajewski, W., Kruger, J., and Winterhager, H.	1970	L	373-1073		99.49	0.51	Alloy prepared from > 99.99 pure cobalt and high purity tantalum; polished and annealed in vacuum; transformation starts at 725 K on heating and 661 K on cooling; this curve is here reported using the first given temperature as reference temperature at which $\Delta L/L_0 = 0$.
2*	628	Krajewski, W., et al.	1970	L	373-1073		90.51	9.49	Similar to above specimen; transformation starts at 771 K on heating and 593 K on cooling; this curve is here reported using the first given temperature as reference temperature at which $\Delta L/L_0 = 0$.

DATA TABLE 166. THERMAL LINEAR EXPANSION OF COBALT-TANTALUM SYSTEM Co-Ta

[Temperature, T, K; Linear Expansion, $\Delta L/L_0$, %]

T	$\Delta L/L_0$
CURVE 1*,†	
373	0.000
673	0.408
1073	1.124
CURVE 2*,†	
373	0.000
673	0.369
1073	1.009

*No figure given.
†This curve is here reported using the first given temperature as reference temperature at which $\Delta L/L_0 = 0$.

SPECIFICATION TABLE 167. THERMAL LINEAR EXPANSION OF COBALT-TITANIUM SYSTEM Co-Ti

Cur. No.	Ref. No.	Author(s)	Year	Method Used	Temp. Range, K	Name and Specimen Designation	Composition (weight percent) Co	Ti	Composition (continued), Specifications, and Remarks
1*	628	Krajewski, W., Kruger, J., and Winterhager, H.	1970	L	373-1073		99.35	0.65	Alloy prepared from > 99.99 pure cobalt and high purity titanium; polished and annealed in vacuum; transformation starts at 707 K on heating and 634 K on cooling; this curve is here reported using the first given temperature as reference temperature at which $\Delta L/L_0 = 0$.
2*	628	Krajewski, W., et al.	1970	L	373-1073		96.15	3.85	Similar to the above specimen; this curve is here reported using the first given temperature as reference temperature at which $\Delta L/L_0 = 0$.
3*	628	Krajewski, W., et al.	1970	L	373-1073		92.77	7.23	Similar to the above specimen; this curve is here reported using the first given temperature as reference temperature at which $\Delta L/L_0 = 0$.

DATA TABLE 167. THERMAL LINEAR EXPANSION OF COBALT-TITANIUM SYSTEM Co-Ti

[Temperature, T, K; Linear Expansion, $\Delta L/L_0$, %]

T	$\Delta L/L_0$
CURVE 1*,†	
373	0.000
673	0.411
1073	1.115
CURVE 2*,†	
373	0.000
673	0.390
1073	1.038
CURVE 3*,†	
373	0.000
673	0.360
1073	1.000

*No figure given.
†This curve is here reported using the first given temperature as reference temperature at which $\Delta L/L_0 = 0$.

FIGURE AND TABLE NO. 168R. PROVISIONAL VALUES FOR THERMAL LINEAR EXPANSION OF COBALT-TUNGSTEN SYSTEM Co-W

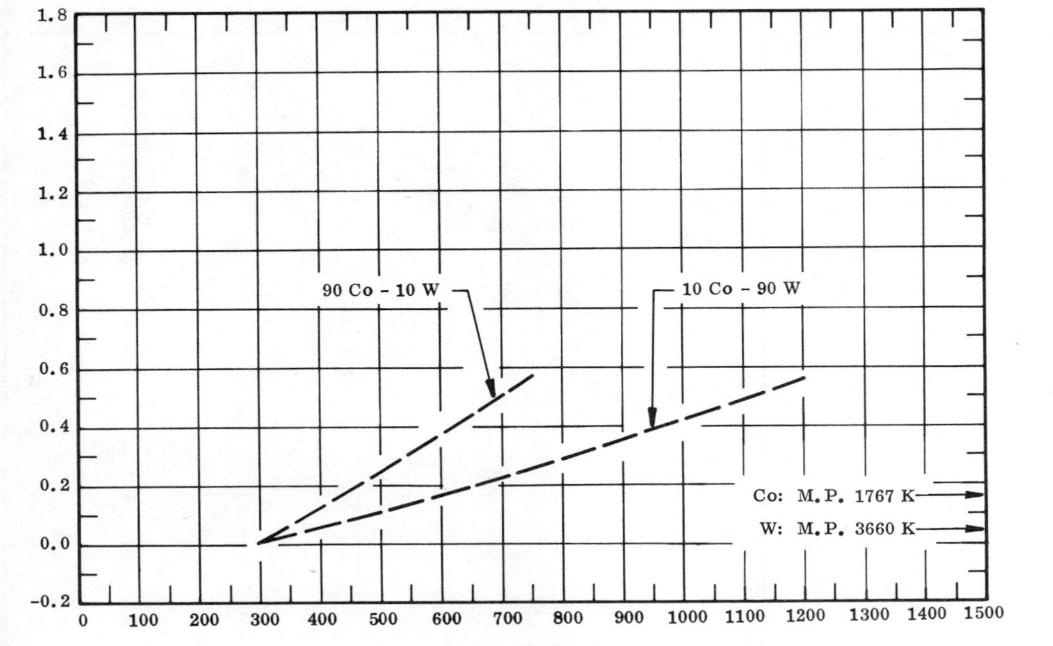

PROVISIONAL VALUES

[Temperature, T, K; Linear Expansion, $\Delta L/L_0$, %; α, K^{-1}]

T	(10 Co-90 W) $\Delta L/L_0$	(10 Co-90 W) $\alpha \times 10^6$	(90 Co-10 W) $\Delta L/L_0$	(90 Co-10 W) $\alpha \times 10^6$
293	0.000	4.8	0.000	11.7
400	0.055	5.3	0.130	12.1
500	0.110	5.6	0.249	12.4
600	0.167	5.9	0.374	12.6
700	0.228	6.1	0.500	12.6
750	0.258	6.2	0.562	12.6
800	0.289	6.3		
900	0.353	6.5		
1000	0.419	6.8		
1100	0.488	7.0		
1200	0.561	7.3		

REMARKS

(10 Co-90 W): The tabulated values for well-annealed alloy are considered accurate to within ±7% over the entire temperature range. These values can be represented approximately by the following equation:

$$\Delta L/L_0 = -0.139 + 4.377 \times 10^{-4}\, T + 1.242 \times 10^{-7}\, T^2 - 2.893 \times 10^{-12}\, T^3$$

(90 Co-10 W): The tabulated values for well-annealed alloy are considered accurate to within ±7% over the entire temperature range. The phase transformation at about 460 K seems to produce little change in the thermal expansion. The tabulated values can be represented approximately by the following equation:

$$\Delta L/L_0 = -0.338 + 1.130 \times 10^{-3}\, T + 9.313 \times 10^{-8}\, T^2 + 2.713 \times 10^{-12}\, T^3$$

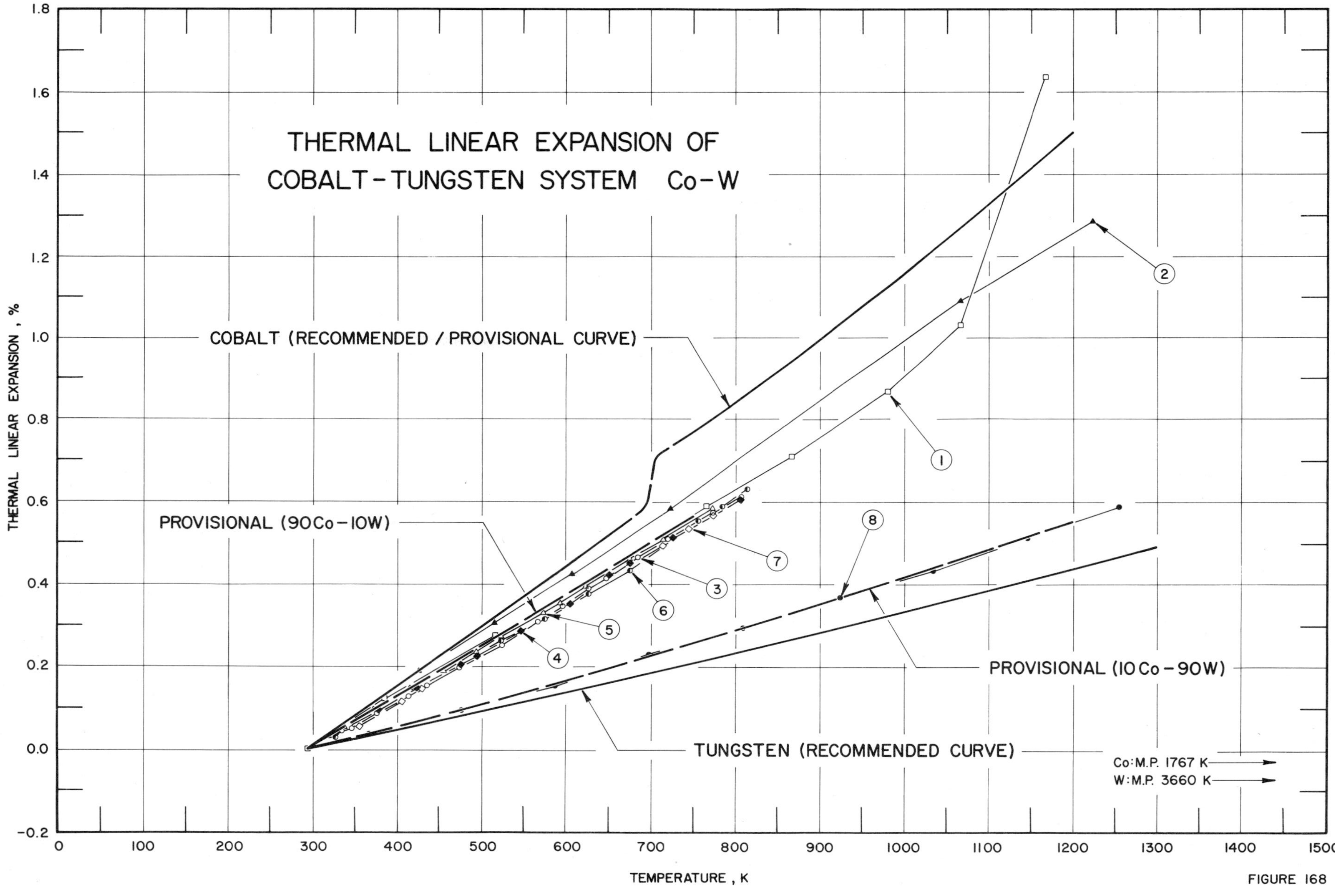
THERMAL LINEAR EXPANSION OF
COBALT - TUNGSTEN SYSTEM Co-W
COBALT (RECOMMENDED / PROVISIONAL CURVE)
PROVISIONAL (90Co-10W)
PROVISIONAL (10Co-90W)
TUNGSTEN (RECOMMENDED CURVE)
Co: M.P. 1767 K
W: M.P. 3660 K
1
2
3
4
5
6
7
8
THERMAL LINEAR EXPANSION, %
-0.2
0.0
0.2
0.4
0.6
0.8
1.0
1.2
1.4
1.6
1.8
TEMPERATURE, K
0
100
200
300
400
500
600
700
800
900
1000
1100
1200
1300
1400
1500

FIGURE 168

SPECIFICATION TABLE 168. THERMAL LINEAR EXPANSION OF COBALT-TUNGSTEN SYSTEM Co-W

Cur. No.	Ref. No.	Author(s)	Year	Method Used	Temp. Range, K	Name and Specimen Designation	Composition (weight percent) Co	W	Composition (continued), Specifications, and Remarks
1	128	Brenna, A., Burkhead, P., and Seegmiller, E.	1947		294-1222		90	10	Alloy electro-deposited on copper bar, bar then removed leaving tube 30 cm long, 5 mm diameter; expansion measured with increasing temperature; zero-point correction of 0.001% determined by graphical extrapolation.
2	128	Brenna, A., et al.	1947		1222-304		90	10	The above specimen; expansion measured with decreasing temperature; zero-point correction of -0.842% determined by graphical extrapolation.
3	406	Sykes, W. P. and Graff, H. F.	1935	L	333-809	Sample B	97	3	Specimen 3 in. long, 0.625 in. diameter; machined from cast alloy which was heated in hydrogen for 50 hr at 1573 K after solification, previously cooled from 1273 to 573 K in 150 hrs; expansion measured with temperature increasing at 1 to 2 C min^{-1}; expansion increment reported at 333 K is -0.00076 in.; -0.00202 in. at 293 K determined by graphical extrapolation.
4	406	Sykes, W. P. and Graff, H. F.	1935	L	328-806	Sample C	95	5	Similar to the above specimen; expansion measured with increasing temperature; expansion increment at 328 K is -0.0020 in.; -0.00322 in. at 293 K determined by graphical extrapolation.
5	406	Sykes, W. P. and Graff, H. F.	1935	L	319-772	Sample D	90	10	Similar to the above specimen; expansion measured with increasing temperature; expansion increment reported at 319 K is -0.00374 in.; -0.00473 in. at 293 K determined by graphical extrapolation.
6	406	Sykes, W. P. and Graff, H. F.	1935	L	326-812	Sample E	90	10	Similar to the above specimen; except previously heated at 1273 K for 25 hrs and quenched; expansion measured with increasing temperature; expansion increment reported at 326 K is -0.0054 in.; -0.00654 in. at 293 K determined by graphical extrapolation.
7	406	Sykes, W. P. and Graff, H. F.	1935	L	803-325	Sample E	90	10	The above specimen; expansion measured with decreasing temperature; expansion increment reported at 325 K is -0.00591 in.; -0.0069 in. at 293 K determined by graphical extrapolation.
8	456	Harrington, L. C. and Rowe, G. H.	1963	O	293-1255	K10	10	90	Machined specimen, 4 mm diameter, 50 mm long, from Kensametal, Inc.; heated in vacuum at the rate of about 1.8 C per min and increased to 2.7 C per min at end of heating; no significant permanent change in length observed after test.

DATA TABLE 168. THERMAL LINEAR EXPANSION OF COBALT-TUNGSTEN SYSTEM Co-W

[Temperature, T, K; Linear Expansion, $\Delta L/L_0$, %]

T	$\Delta L/L_0$
CURVE 1	
294	0.001*
384	0.121
516	0.279
765	0.590
868	0.709
980	0.868
1069	1.031
1167	1.639
1222	2.128*
CURVE 2	
1222	1.285
1067	1.093
721	0.584
608	0.424
515	0.301
425	0.188
304	0.018
CURVE 3	
333	0.042
345	0.054
375	0.087
411	0.128
435	0.156
474	0.200
523	0.252
567	0.301
597	0.341
626	0.385
648	0.419
682	0.464
720	0.510
773	0.574
809	0.617
CURVE 4	
328	0.040
335	0.049*
378	0.097
421	0.146
475	0.202
CURVE 4 (cont.)	
492	0.222
521	0.253*
549	0.283
566	0.302*
601	0.350
623	0.385*
650	0.422
673	0.452
723	0.513
772	0.572*
806	0.612
CURVE 5	
319	0.033
334	0.054
372	0.099
415	0.148
453	0.190
492	0.236
523	0.275
571	0.327
594	0.358
624	0.399
648	0.422*
674	0.460
712	0.510
772	0.584
CURVE 6	
326	0.038
373	0.092*
423	0.146*
524	0.260
574	0.318
625	0.378
672	0.438
722	0.512*
756	0.555
783	0.593
812	0.636
CURVE 7	
803	0.605*
774	0.571
742	0.533
711	0.494
472	0.206*
426	0.145
404	0.117
374	0.083*
352	0.059
325	0.033*
CURVE 8	
293	0.000*
366	0.037
477	0.096
588	0.158
699	0.224
810	0.292
922	0.365
1033	0.438
1144	0.515
1255	0.592

*Not shown in figure.

SPECIFICATION TABLE 169. THERMAL LINEAR EXPANSION OF COBALT-VANADIUM SYSTEM Co-V

Cur. No.	Ref. No.	Author(s)	Year	Method Used	Temp. Range, K	Name and Specimen Designation	Composition (weight percent) Co	V	Composition (continued), Specifications, and Remarks
1*	628	Krajewski, W., Kruger, J., and Winterhager, H.	1970	L	373-1073		99.48	0.52	Alloy prepared from >99.99 pure cobalt and high purity vanadium; transformation starts at 738 K on heating and at 689 K on cooling; this curve is here reported using the first given temperature as reference temperature at which $\Delta L/L_0 = 0$.
2*	628	Krajewski, W., et al.	1970	L	373-1073		97.30	2.70	Similar to the above specimen; transformation starts around 796 K on heating and 673 K on cooling; this curve is here reported using the first given temperature as reference temperature at which $\Delta L/L_0 = 0$.
3*	628	Krajewski, W., et al.	1970	L	373-1073		92.93	7.07	Similar to the above specimen; this curve is here reported using the first given temperature as reference temperature at which $\Delta L/L_0 = 0$.

DATA TABLE 169. THERMAL LINEAR EXPANSION OF COBALT-VANADIUM SYSTEM Co-V

[Temperature, T, K; Linear Expansion, $\Delta L/L_0$, %]

T	$\Delta L/L_0$
CURVE 1*, †	
373	0.000
673	0.405
1073	1.117
CURVE 2*, †	
373	0.000
673	0.378
1073	1.058
CURVE 3*, †	
373	0.000
673	0.372
1073	0.980

*No figure given.
†This curve is here reported using the first given temperature as reference temperature at which $\Delta L/L_0 = 0$.

SPECIFICATION TABLE 170. THERMAL LINEAR EXPANSION OF COBALT-ZIRCONIUM SYSTEM Co-Zr

Cur. No.	Ref. No.	Author(s)	Year	Method Used	Temp. Range, K	Name and Specimen Designation	Composition (weight percent) Co	Zr	Composition (continued), Specifications, and Remarks
1*	628	Krajewski, W., Kruger, J., and Winterhager, H.	1970	L	373-1073		99.15	0.85	Alloy prepared from > 99.99 pure cobalt and high purity zirconium; polished and annealed in vacuum; transformation starts at 719 K on heating and 661 K on cooling; this curve is here reported using the first given temperature as reference temperature at which $\Delta L/L_0 = 0$.
2*	628	Krajewski, W., et al.	1970	L	373-1073		97.65	2.35	Similar to the above specimen; transformation starts at 721 K on heating and 646 K on cooling; this curve is here reported using the first given temperature as reference temperature at which $\Delta L/L_0 = 0$.

DATA TABLE 170. THERMAL LINEAR EXPANSION OF COBALT-ZIRCONIUM SYSTEM Co-Zr

[Temperature, T, K; Linear Expansion, $\Delta L/L_0$, %]

T	$\Delta L/L_0$
CURVE 1*, †	
373	0.000
673	0.405
1073	1.105
CURVE 2*, †	
373	0.000
673	0.396
1073	1.092

*No figure given.
†This curve is here reported using the first given temperature as reference temperature at which $\Delta L/L_0 = 0$.

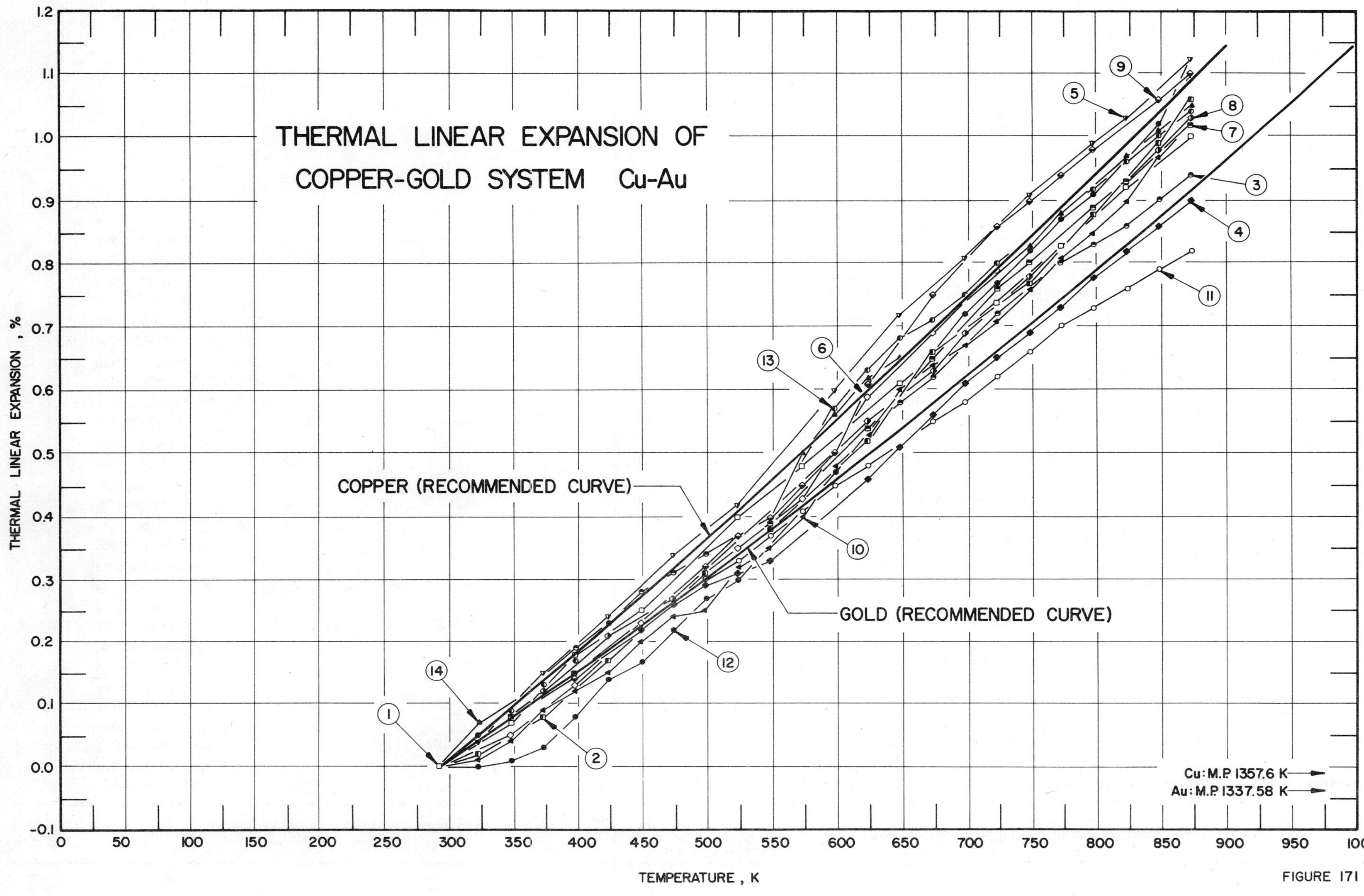
THERMAL LINEAR EXPANSION OF
COPPER-GOLD SYSTEM Cu-Au
COPPER (RECOMMENDED CURVE)
GOLD (RECOMMENDED CURVE)
Cu: M.P. 1357.6 K
Au: M.P. 1337.58 K
THERMAL LINEAR EXPANSION, %
TEMPERATURE, K
1
2
3
4
5
6
7
8
9
10
11
12
13
14

FIGURE 171

SPECIFICATION TABLE 171. THERMAL LINEAR EXPANSION OF COPPER-GOLD SYSTEM Cu-Au

Cur. No.	Ref. No.	Author(s)	Year	Method Used	Temp. Range, K	Name and Specimen Designation	Composition (weight percent) Cu	Au	Composition (continued), Specifications, and Remarks
1	611	Kurnakow, N.S. and Agneew, J.	1931	L	293-873		65.49	34.51	Alloy prepared from chemically pure gold and electrolytic copper; annealed at 1073 K; expansion measured with increasing temperature.
2	611	Kurnakow, N.S. and Agneew, J.	1931	L	873-293		65.49	34.51	The above specimen; expansion measured with decreasing temperature; zero-point correction is 0.02%.
3	611	Kurnakow, N.S. and Agneew, J.	1931	L	293-873		5.38	94.62	Similar to the above specimen; expansion measured with increasing temperature.
4	611	Kurnakow, N.S. and Agneew, J.	1931	L	873-293		5.38	94.62	The above specimen; expansion measured with decreasing temperature; zero-point correction is 0.01%.
5	611	Kurnakow, N.S. and Agneew, J.	1931	L	293-873		43.53	56.47	Similar to the above specimen; expansion measured with increasing temperature.
6	611	Kurnakow, N.S. and Agneew, J.	1931	L	873-293		43.53	56.47	The above specimen; expansion measured with decreasing temperature; zero-point correction is 0.03%.
7	611	Kurnakow, N.S. and Agneew, J.	1931	L	293-873		32.63	67.37	Similar to the above specimen; expansion measured with increasing temperature.
8	611	Kurnakow, N.S. and Agneew, J.	1931	L	873-293		32.63	67.37	The above specimen; expansion measured with decreasing temperature; zero-point correction is 0.01%.
9	611	Kurnakow, N.S. and Agneew, J.	1931	L	293-873		30.49	69.51	Similar to the above specimen; expansion measured with increasing temperature.
10	611	Kurnakow, N.S. and Agneew, J.	1931	L	873-293		30.49	69.51	The above specimen; expansion measured with decreasing temperature; zero-point correction is -0.07%.
11	611	Kurnakow, N.S. and Agneew, J.	1931	L	293-873		30.49	69.51	Alloy prepared from chemically pure gold and electrolytic copper; quenched at 1073 K; expansion measured with increasing temperature.
12	611	Kurnakow, N.S. and Agneew, J.	1931	L	873-293		30.49	69.51	The above specimen; expansion measured with decreasing temperature; zero-point correction is 0.08%.
13	611	Kurnakow, N.S. and Agneew, J.	1931	L	293-873		18.17	81.83	Similar to the above specimen; expansion measured with increasing temperature.
14	611	Kurnakow, N.S. and Agneew, J.	1931	L	873-293		18.17	81.83	The above specimen; expansion measured with decreasing temperature; zero-point correction is 0.03%.

DATA TABLE 171. THERMAL LINEAR EXPANSION OF COPPER-GOLD SYSTEM Cu-Au

[Temperature, T, K; Linear Expansion, $\Delta L/L_0$, %]

T	$\Delta L/L_0$
CURVE 1	
293	0.00
323	0.02*
348	0.07
373	0.12*
398	0.18
423	0.21*
448	0.25
473	0.32*
498	0.36*
523	0.40
548	0.44*
573	0.48
598	0.52*
623	0.56*
648	0.61
673	0.66*
698	0.70*
723	0.74
748	0.79*
773	0.83
798	0.88*
823	0.92
848	0.97*
873	1.00
CURVE 2	
873	1.06
848	0.99
823	0.93*
798	0.88
773	0.82*
748	0.77
723	0.73*
698	0.69*
673	0.66
648	0.60*
623	0.52
598	0.47*
573	0.42*
548	0.38
523	0.35*
498	0.31
473	0.26*
448	0.22*
423	0.17
CURVE 2 (cont.)	
398	0.12*
373	0.08
348	0.04*
323	0.02
293	0.00*
CURVE 3	
293	0.00*
323	0.05*
348	0.10*
373	0.15*
398	0.19
423	0.23
448	0.28
473	0.31
498	0.34
523	0.37*
548	0.41*
573	0.45*
598	0.49*
623	0.53*
648	0.58
673	0.62
698	0.67*
723	0.72
748	0.76*
773	0.80
798	0.83
823	0.86
848	0.90
873	0.94
CURVE 4	
873	0.90
848	0.86
823	0.82
798	0.78
773	0.73
748	0.69
723	0.65
698	0.61
673	0.56
648	0.51
623	0.46
CURVE 4 (cont.)	
598	0.42*
573	0.37*
548	0.33
523	0.31
498	0.29
473	0.26
448	0.22
423	0.19*
398	0.16*
373	0.12*
348	0.09*
323	0.05*
293	0.00*
CURVE 5	
293	0.00*
323	0.04
348	0.10*
373	0.15
398	0.20*
423	0.24
448	0.29*
473	0.34
498	0.38*
523	0.42
548	0.48*
573	0.52*
598	0.60
623	0.67*
648	0.72
673	0.76*
698	0.81
723	0.86*
748	0.91
773	0.95*
798	0.99
823	1.03
848	1.07*
873	1.12
CURVE 6	
873	1.09
848	1.04*
823	0.98*
798	0.93*
CURVE 6 (cont.)	
773	0.89*
748	0.84*
723	0.79
698	0.74*
673	0.69
648	0.64*
623	0.59
598	0.52*
573	0.43
548	0.39*
523	0.35
498	0.30*
473	0.26*
448	0.23
423	0.18*
398	0.13
373	0.09*
348	0.05
323	0.02*
293	0.00*
CURVE 7	
293	0.00*
323	0.04*
348	0.08
373	0.11*
398	0.15
423	0.19*
448	0.23*
473	0.28*
498	0.31*
523	0.35*
548	0.39
573	0.43*
598	0.49*
623	0.54
648	0.61*
673	0.65
698	0.71*
723	0.76
748	0.80
773	0.84*
798	0.89
823	0.93
848	0.98*
873	1.02
CURVE 8	
873	1.03
848	0.98
823	0.93
798	0.88*
773	0.84*
748	0.78
723	0.74*
698	0.69
673	0.63
648	0.59*
623	0.55
598	0.49*
573	0.43*
548	0.39*
523	0.35*
498	0.31*
473	0.27
448	0.24*
423	0.21
398	0.17*
373	0.12
348	0.08*
323	0.03*
293	0.00*
CURVE 9	
293	0.00*
323	0.03*
348	0.07*
373	0.10*
398	0.14
423	0.21*
448	0.25*
473	0.28*
498	0.32
523	0.37
548	0.40
573	0.45
598	0.50
623	0.61
648	0.72*
673	0.75
698	0.81*
723	0.86
748	0.90
773	0.94
798	0.98
CURVE 9 (cont.)	
823	1.03
848	1.06
873	1.10
CURVE 10	
873	1.02*
848	0.97
823	0.90
798	0.85
773	0.81
748	0.76
723	0.71
698	0.67
673	0.64
648	0.60
623	0.53
598	0.48
573	0.40
548	0.35
523	0.32
498	0.25
473	0.24
448	0.20
423	0.15
398	0.12
373	0.09
348	0.04
323	0.01
293	0.00*
CURVE 11	
293	0.00*
323	0.04*
348	0.08*
373	0.12*
398	0.15*
423	0.19*
448	0.23*
473	0.27*
498	0.30
523	0.33
548	0.37
573	0.41
598	0.45

* Not shown in figure.

DATA TABLE 171. THERMAL LINEAR EXPANSION OF COPPER-GOLD SYSTEM Cu-Au (continued)

T	$\Delta L/L_0$
CURVE 11 (cont.)	
623	0.48
648	0.51*
673	0.55
698	0.58
723	0.62
748	0.66
773	0.70
798	0.73
823	0.76
848	0.79
873	0.82
CURVE 12	
873	1.07*
848	1.02
823	0.96*
798	0.91
773	0.87
748	0.82
723	0.77
698	0.72
673	0.65*
648	0.60*
623	0.54*
598	0.47
573	0.40*
548	0.34*
523	0.30
498	0.27
473	0.22
448	0.17
423	0.14
398	0.08
373	0.03
348	0.01
323	0.00
298	0.00*
CURVE 13	
293	0.00*
323	0.05
348	0.09
373	0.13
398	0.17
423	0.21*
448	0.25*
CURVE 13 (cont.)	
473	0.29*
498	0.33*
523	0.37*
548	0.41*
573	0.48*
598	0.57
623	0.63
648	0.68
673	0.71
698	0.75
723	0.80
748	0.84*
773	0.89*
798	0.92
823	0.96
848	1.00
873	1.04
CURVE 14	
873	1.05
848	1.01
823	0.97
798	0.92*
773	0.88
748	0.83
723	0.79*
698	0.74*
673	0.69*
648	0.65
623	0.62
598	0.56
573	0.50
548	0.40*
523	0.36*
498	0.32*
473	0.28*
448	0.25*
423	0.22*
398	0.17*
373	0.13*
348	0.10*
323	0.07
293	0.00*

* Not shown in figure.

FIGURE AND TABLE NO. 172CR. PROVISIONAL VALUES FOR THERMAL LINEAR EXPANSION OF COPPER-MAGNESIUM SYSTEM Cu-Mg

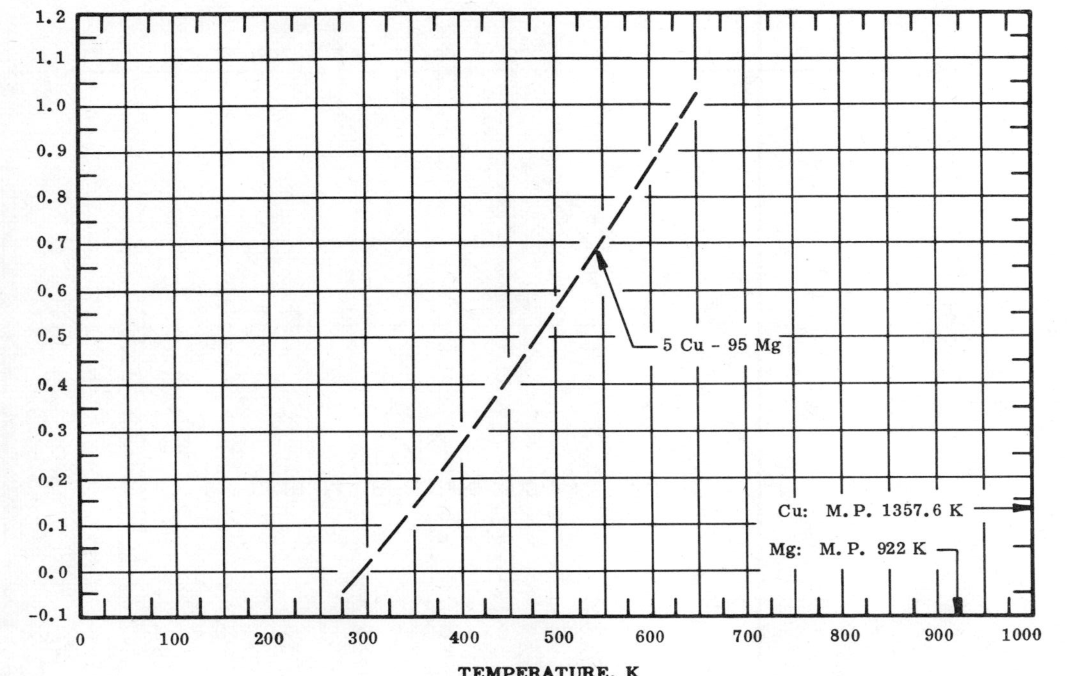

PROVISIONAL VALUES

[Temperature, T, K; Linear Expansion, $\Delta L/L_0$, %; α, K^{-1}]

	(5 Cu-95 Mg)	
T	$\Delta L/L_0$	$\alpha \times 10^6$
275	-0.042	22.8
293	0.000	23.6
400	0.274	27.4
500	0.562	29.7
600	0.868	30.9
650	1.022	31.5

REMARKS

(5 Cu-95 Mg): The tabulated values for well-annealed alloy are considered accurate to within ±7% over the entire temperature range. The thermal linear expansion of this alloy is very close to thermal linear expansion of pure magnesium. These values can be represented approximately by the following equation:

$$\Delta L/L_0 = -0.466 + 6.591 \times 10^{-4}\, T + 3.737 \times 10^{-6}\, T^2 - 1.893 \times 10^{-9}\, T^3$$

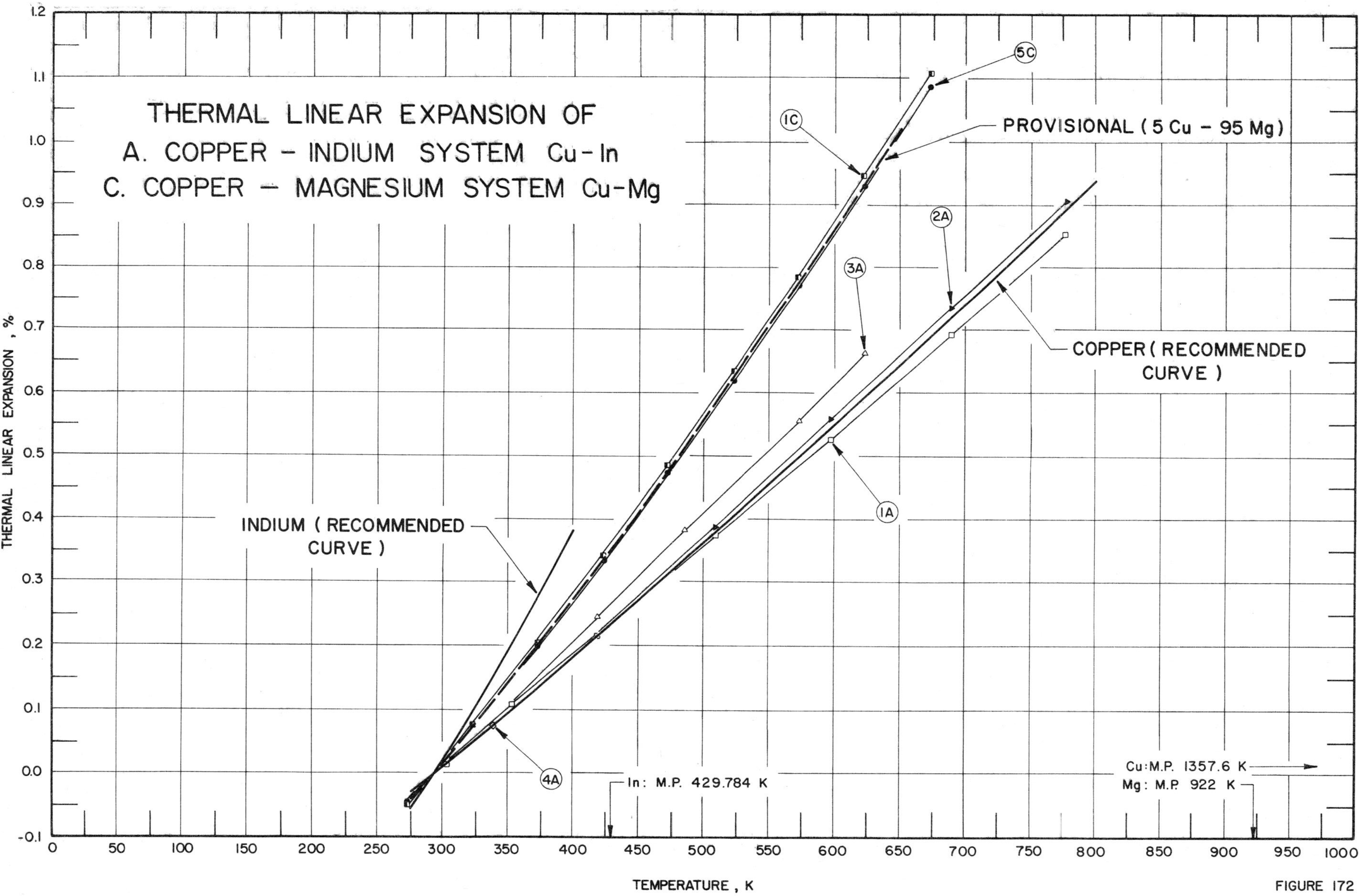
THERMAL LINEAR EXPANSION OF
A. COPPER - INDIUM SYSTEM Cu-In
C. COPPER - MAGNESIUM SYSTEM Cu-Mg
5C
1C
PROVISIONAL (5 Cu - 95 Mg)
2A
3A
COPPER (RECOMMENDED CURVE)
1A
INDIUM (RECOMMENDED CURVE)
4A
In: M.P. 429.784 K
Cu: M.P. 1357.6 K
Mg: M.P. 922 K
THERMAL LINEAR EXPANSION, %
TEMPERATURE, K
1.2
1.1
1.0
0.9
0.8
0.7
0.6
0.5
0.4
0.3
0.2
0.1
0.0
-0.1
0
50
100
150
200
250
300
350
400
450
500
550
600
650
700
750
800
850
900
950
1000

FIGURE 172

SPECIFICATION TABLE 172A. THERMAL LINEAR EXPANSION OF COPPER-INDIUM SYSTEM Cu-In

Cur. No.	Ref. No.	Author(s)	Year	Method Used	Temp. Range, K	Name and Specimen Designation	Composition (weight percent) Cu	In	Composition (continued), Specifications, and Remarks
1	646	De, M.	1969	X	303-778		97.02	2.98	Alloys prepared from spectrographically standardized metals from Messrs. Johnson, Matthey & Co., Ltd., London; homogenized at 1073-1123 K; annealed and treatment terminated by quenching in air; lattice parameter reported at 303 K is 3.6306 Å; 3.6300 Å at 293 K determined by extrapolation; data on the coefficient of thermal linear expansion also reported.
2	646	De, M.	1969	X	303-778		91.05	8.95	Similar to the above specimen; lattice parameter reported at 303 K is 3.6626 Å; 3.6620 Å at 293 K determined by extrapolation; data on the coefficient of thermal linear expansion also reported.
3	646	De, M.	1969	X	303-623		89.35	10.65	Similar to the above specimen; lattice parameter reported at 303 K is 3.6727 Å; 3.6720 Å at 293 K determined by extrapolation; data on the coefficient of thermal linear expansion also reported.
4	743	Straumanis, M.E. and Yu, L.S.	1969	X	288, 338		Bal.	1.0	Alloy prepared from 99.999 Cu and 99.999 In both from ASARCO, samples were placed in quartz tubes which were evacuated, sealed and heated to 1420 K, taken from the melting point down to 670 K in about 8 hr, homogenized at 1073 K for 14 days and then quenched in water, α-phase solid solution, zero-point correction of -0.0083% determined by graphical interpolation.
5*	743	Straumanis, M.E. and Yu, L.S.	1969	X	288, 338		Bal.	2.0	Similar to the above specimen, zero-point correction of -0.0083% determined by graphical interpolation.
6*	743	Straumanis, M.E. and Yu, L.S.	1969	X	288, 338		Bal.	3.0	Similar to the above specimen, zero-point correction of -0.0084% determined by graphical interpolation.
7*	743	Straumanis, M.E. and Yu, L.S.	1969	X	288, 338		Bal.	4.0	Similar to the above specimen, zero-point correction of -0.0085% determined by graphical interpolation.
8*	743	Straumanis, M.E. and Yu, L.S.	1969	X	288, 338		Bal.	8.0	Similar to the above specimen, zero-point correction of -0.0086% determined by graphical interpolation.

* Not shown in figure.

DATA TABLE 172A. THERMAL LINEAR EXPANSION OF COPPER-INDIUM SYSTEM Cu-In

[Temperature, T, K; Linear Expansion, $\Delta L/L_0$, %]

T	$\Delta L/L_0$
CURVE 1	
303	0.016
353	0.105
417	0.215
509	0.375
597	0.529
689	0.694
778	0.857
CURVE 2	
303	0.016*
353	0.104*
417	0.218*
509	0.388
597	0.560
689	0.732
778	0.901
CURVE 3	
303	0.019*
353	0.117*
417	0.242
485	0.381
573	0.556
623	0.662
CURVE 4	
288	-0.0083
338	0.0742
CURVE 5*	
288	-0.0083
338	0.0750
CURVE 6*	
288	-0.0084
338	0.0754
CURVE 7*	
288	-0.0085
338	0.0762
CURVE 8*	
288	-0.0086
338	0.0773

DATA TABLE 172A. COEFFICIENT OF THERMAL LINEAR EXPANSION OF COPPER-INDIUM SYSTEM Cu-In

[Temperature, T, K; Coefficient of Expansion, α, 10^{-6} K^{-1}]

T	α
CURVE 1*	
293	16.69
370	16.79
472	17.14
571	17.35
673	17.81
773	18.29
CURVE 2*	
293	17.60
368	17.76
469	18.00
571	18.56
666	19.01
774	19.50
CURVE 3*	
293	18.92
372	19.29
470	19.89
571	20.34
621	20.62

* Not shown in figure.

SPECIFICATION TABLE 172B. THERMAL LINEAR EXPANSION OF COPPER-IRON SYSTEM Cu-Fe

Cur. No.	Ref. No.	Author(s)	Year	Method Used	Temp. Range, K	Name and Specimen Designation	Composition (weight percent) Cu	Fe	Composition (continued), Specifications, and Remarks
1*	525	Simpson, K.M. and Banister, R.T.	1936		293-1173		50	50	Forged specimen.
2*	780	White, G.K.	1972	E	5-85, 283		Bal.	0.2	Specimen cast from 99.99 pure materials; in form of cylinder 2 cm in diameter and 5 cm long.

DATA TABLE 172B. THERMAL LINEAR EXPANSION OF COPPER-IRON SYSTEM Cu-Fe

[Temperature, T, K; Linear Expansion, $\Delta L/L_0$, %]

T	$\Delta L/L_0$
CURVE 1*	
293	0.000
373	0.148
673	0.627
873	0.969
1073	1.271
1173	1.320

T	$\Delta L/L_0$
CURVE 2*, †, ‡	
5	0.00000
10	0.00001
15	0.00005
20	0.00016
25	0.0004
65	0.014
85	0.029
283	0.281

DATA TABLE 172B. COEFFICIENT OF THERMAL LINEAR EXPANSION OF COPPER-IRON SYSTEM Cu-Fe

[Temperature, T, K; Coefficient of Expansion, α, $10^{-6}\ K^{-1}$]

T	α
CURVE 2*, ‡	
5	0.0132
10	0.044
15	0.122
20	0.28
25	0.57
65	6.30
85	8.96
283	16.50

* No figure given.
† This curve is here reported using the first given temperature as reference temperature at which $\Delta L/L_0 = 0$.
‡ Author's data for coefficient of thermal expansion have been integrated by TPRC to obtain $\Delta L/L_0$.

SPECIFICATION TABLE 172C. THERMAL LINEAR EXPANSION OF COPPER-MAGNESIUM SYSTEM Cu-Mg

Cur. No.	Ref. No.	Author(s)	Year	Method Used	Temp. Range, K	Name and Specimen Designation	Composition (weight percent) Cu	Mg	Composition (continued), Specifications, and Remarks
1	201	Takahasi, K. and Kikuti, R.	1936	L	273-673		2	98	Rod specimen, about 16 cm long, 5 mm dia, heated for 5 hr at 1-2 mm Hg at 620 K, heating rate 2 C/min.
2*	201	Takahasi, K. and Kikuti, R.	1936	L	273-673		4	96	Similar to the above specimen.
3*	201	Takahasi, K. and Kikuti, R.	1936	L	273-673		6	94	Similar to the above specimen.
4*	201	Takahasi, K. and Kikuti, R.	1936	L	273-673		8	92	Similar to the above specimen.
5	201	Takahasi, K. and Kikuti, R.	1936	L	273-673		10	90	Similar to the above specimen.

DATA TABLE 172C. THERMAL LINEAR EXPANSION OF COPPER-MAGNESIUM SYSTEM Cu-Mg

[Temperature, T, K; Linear Expansion, $\Delta L/L_0$, %]

T	$\Delta L/L_0$
CURVE 1	
273	-0.050
323	0.075
373	0.204
423	0.340
473	0.484
523	0.632
573	0.786
623	0.946
673	1.107
CURVE 2*	
273	-0.050
323	0.075
373	0.207
423	0.343
473	0.484
CURVE 2 (cont.)*	
523	0.630
573	0.780
623	0.937
673	1.094
CURVE 3*	
273	-0.05
323	0.074
373	0.202
423	0.337
473	0.478
523	0.625
573	0.777
623	0.733
673	1.094
CURVE 4*	
273	-0.049
323	0.074*
373	0.201
423	0.334
473	0.477
523	0.621
573	0.769
623	0.923
673	1.081
CURVE 5	
273	-0.048*
323	0.071*
373	0.199*
423	0.332
473	0.475*
CURVE 5 (cont.)	
523	0.620
573	0.771
623	0.930
673	1.089

*Not shown in figure.

FIGURE AND TABLE NO. 173R. PROVISIONAL VALUES FOR THERMAL LINEAR EXPANSION OF COPPER-MANGANESE SYSTEM Cu-Mn

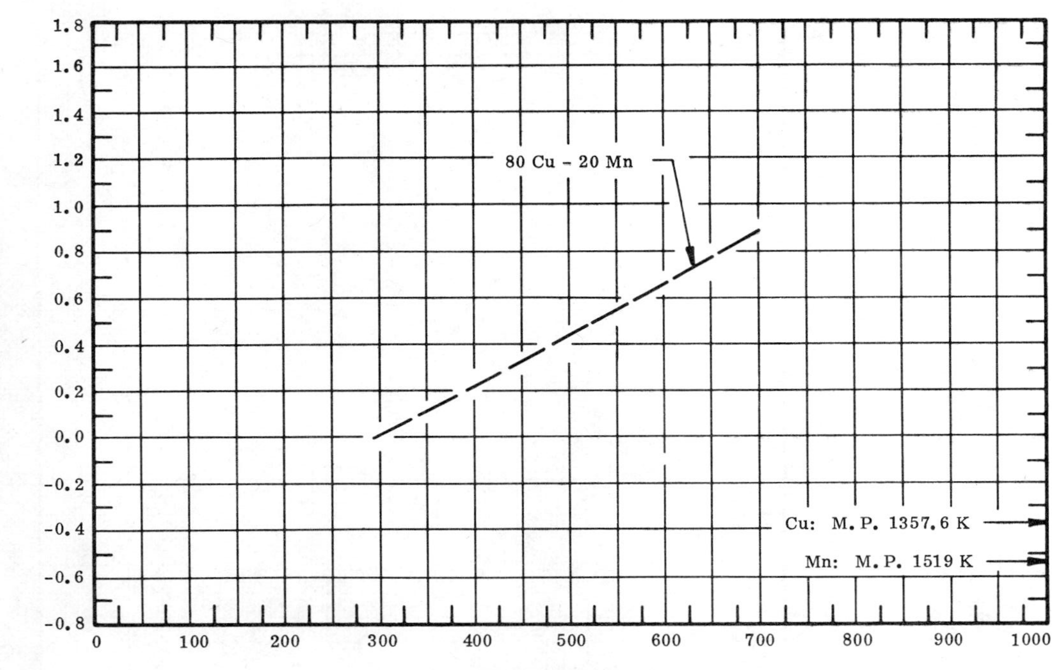

PROVISIONAL VALUES

[Temperature, T, K; Linear Expansion, $\Delta L/L_0$, %; α, K^{-1}]

T	(80 Cu-20 Mn) $\Delta L/L_0$	$\alpha \times 10^6$
293	0.000	20.3
400	0.220	20.8
500	0.435	22.0
600	0.660	23.1
700	0.892	23.4

REMARKS

(80 Cu-20 Mn): The tabulated values for well-annealed alloy are considered accurate to within ± 7% over the entire temperature range. These values can be represented approximately by the following equation:

$$\Delta L/L_0 = -0.513 + 1.505 \times 10^{-3}\, T + 9.455 \times 10^{-7}\, T^2 - 3.285 \times 10^{-10}\, T^3$$

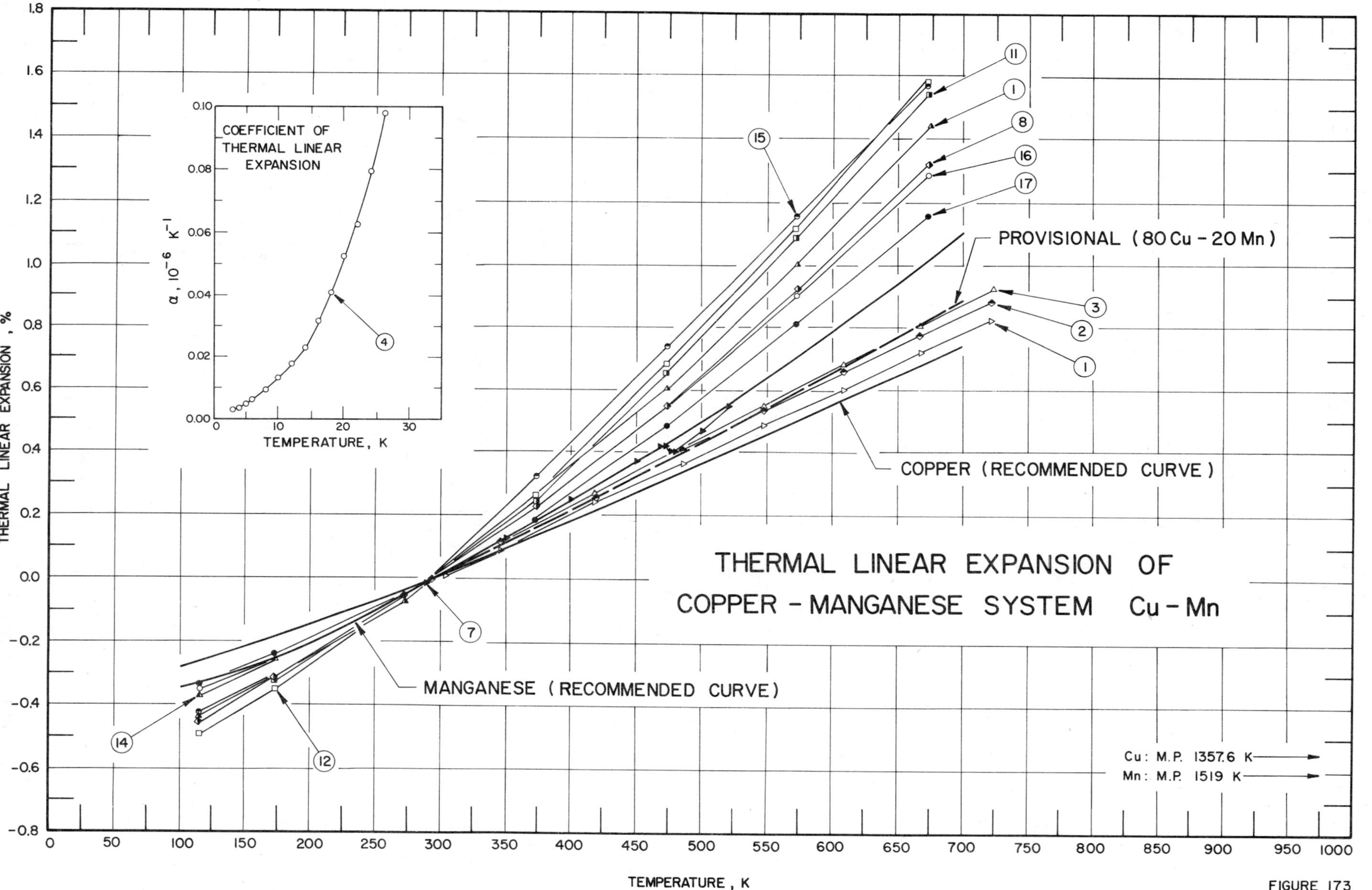
THERMAL LINEAR EXPANSION OF
COPPER - MANGANESE SYSTEM Cu - Mn
THERMAL LINEAR EXPANSION , %
TEMPERATURE , K
PROVISIONAL (80 Cu - 20 Mn)
COPPER (RECOMMENDED CURVE)
MANGANESE (RECOMMENDED CURVE)
Cu : M.P. 1357.6 K
Mn : M.P. 1519 K
COEFFICIENT OF
THERMAL LINEAR
EXPANSION
α, 10⁻⁶ K⁻¹
TEMPERATURE, K
11
1
8
16
17
15
3
2
1
7
4
14
12

FIGURE 173

SPECIFICATION TABLE 173. THERMAL LINEAR EXPANSION OF COPPER-MANGANESE SYSTEM Cu-Mn

Cur. No.	Ref. No.	Author(s)	Year	Method Used	Temp. Range, K	Name and Specimen Designation	Composition (weight percent) Cu	Mn	Composition (continued), Specifications, and Remarks
1	410	De, M.	1967	X	303-723	α-phase	96.06	3.94	Prepared from spectrographically standardized copper and manganese supplied by Messrs. Johnson, Matthey and Co., London; melted together in evacuated and sealed quartz capsules, homogenized for a week at 1073-1123 K; powder sample obtained by hand-filing at room temperature, annealed at 873 K for 6 hr in vacuum; filings sieved through a 250 mesh screen; specimen about 4 mm in length and 0.5 mm in diameter prepared by taking the annealed powder in thin-walled pyrex capillaries with both ends sealed; filtered Cu Kα x-radiation used; data corrected by author by standard extrapolation method; lattice parameter reported at 303 K is 3.6328 Å; 3.6321 Å at 293 K determined by graphical extrapolation.
2	410	De, M.	1967	X	299-723	α-phase	88.06	11.94	Similar to the above specimen; lattice parameter reported at 299 K is 3.6651 Å; 3.6646 Å at 293 K determined by graphical extrapolation.
3	410	De, M.	1967	X	302-723	α-phase	80.08	19.92	Similar to the above specimen; lattice parameter reported at 302 K is 3.6952 Å; 3.6945 Å at 293 K determined by graphical extrapolation.
4*	41	White, G.K.	1965	E	3-26	γ-Mn	17.0	83.0	Prepared from Johnson Matthey specimen 99.99$^+$ Mn and 99.999 Cu; specimen 6 cm long, 2 cm diameter; homogenized for 4 days at 1273 K in argon and quenched; measured in the as-received condition.
5*	421	Makhurane, P. and Gaunt, P.	1969	X	4-437		11.44	88.56	Fe Kα x-radiation used to observe 202 and 220 Bragg reflections; expansion measured along a-axis; lattice parameter reported at 306 K is 3.751 Å; 3.752 Å at 293 K determined by graphical interpolation; Néel temperature is 390 K.
6*	421	Makhurane, P. and Gaunt, P.	1969	X	4-383		11.44	88.56	The above specimen; expansion measured along c-axis; lattice parameter reported at 307 K is 3.646 Å; 3.641 Å at 293 K determined by graphical interpolation; Néel temperature is 390 K.
7	661	Sawaoka, A., Soma, T., Saito, S., and Kndoh, Y.	1971	L	293-521	γ-Mn	5.7	94.3	Prepared from 99.9 Mn and electrolytic Cu by melting in arc furnace in argon atmosphere; annealed at 1273 K for 18 hrs in vacuum and water quenched; zero-point correction -0.078%.
8	664	Masumoto, H., Sawaya, S., and Kikuchi, M.	1972	L	115-673		3.5	Bal.	Prepared from electrolytic Cu with 0.008 Fe, 0.001 Mn, 0.009 Si, and 0.005 S; and electrolytic Mn with 0.007 Fe, 0.008 Si, 0.007 C, and 0.030 S; rod specimen of 2 mm diameter and 11 cm length; heated for 1 hr in argon at 353 K below solidus line; then quenched into water or slow cooled at 100°/hr; average of heating and cooling curves; zero-point correction is -0.452%.
9*	664	Masumoto, H., et al.	1972	L	115-673		5.0	Bal.	Similar to the above specimen; zero-point correction is -0.452%.
10	664	Masumoto, H., et al.	1972	L	115-673		12.65	Bal.	Similar to the above specimen; zero-point correction is -0.440%.
11	664	Masumoto, H., et al.	1972	L	115-673		22.6	Bal.	Similar to the above specimen; zero-point correction is -0.451%.
12	664	Masumoto, H., et al.	1972	L	115-673		30.03	Bal.	Similar to the above specimen; zero-point correction is -0.491%.
13*	664	Masumoto, H., et al.	1972	L	115-673		46.68	Bal.	Similar to the above specimen; zero-point correction is -0.427%.
14	664	Masumoto, H., et al.	1972	L	115-673		55.0	Bal.	Similar to the above specimen; zero-point correction is -0.374%.
15	664	Masumoto, H., et al.	1972	L	115-673		56.3	Bal.	Similar to the above specimen; zero-point correction is -0.420%.
16	664	Masumoto, H., et al.	1972	L	115-673		58.0	Bal.	Similar to the above specimen; zero-point correction is -0.356%.
17	664	Masumoto, H., et al.	1972	L	115-673		81.2	Bal.	Similar to the above specimen; zero-point correction is -0.334%.

* Not shown in figure.

DATA TABLE 173. THERMAL LINEAR EXPANSION OF COPPER-MANGANESE SYSTEM Cu-Mn

[Temperature, T, K; Linear Expansion, $\Delta L/L_0$, %]

T	$\Delta L/L_0$
CURVE 1	
303	0.0193
345	0.0991
417	0.245
485	0.366
547	0.493
609	0.608
666	0.724
723	0.831
CURVE 2	
299	0.0136
345	0.120
417	0.259
485	0.412
547	0.540
609	0.663
666	0.783
723	0.898
CURVE 3	
302	0.0189*
345	0.111*
417	0.273
485	0.419*
547	0.552
609	0.690
666	0.815
723	0.937
CURVE 4*,†,‡	
3	0.0000 x 10^{-4}
4	0.0335
5	0.0770
6	0.132
8	0.287
10	0.516
12	0.831
14	1.25
16	1.80
18	2.53
20	3.47
22	4.64
24	6.09
26	7.87
CURVE 5*	
4	-0.188 †
23	-0.188
64	-0.161
90	-0.108
118	-0.108
158	-0.054
198	-0.001
240	0.026
281	0.026
306	-0.028
335	-0.054
350	-0.374
371	-0.601
379	-0.827
383	-0.961
386	-1.174
T	**a(Å)**
398	3.710 f.c.c.*
418	3.714*
437	3.717*
T	**$\Delta L/L_0$**
CURVE 6*	
4	1.207
22	-1.207
66	-1.207
91	-1.207
119	-1.180
159	-1.070
199	-0.878
240	-0.878
282	-0.109
307	0.138
335	0.358
352	0.605
370	0.935
375	1.182
383	1.402
CURVE 7	
293	0.000
350	0.137
CURVE 7 (cont.)	
399	0.256
450	0.377
468	0.425
471	0.425
477	0.405
480	0.405
500	0.474
521	0.556
CURVE 8	
115	-0.452
173	-0.306
273	-0.051
293	0.000*
373	0.234
473	0.559
573	0.935
673	1.328
CURVE 9*	
115	-0.452
173	-0.306
273	-0.054
293	0.000
373	0.229
473	0.548
573	0.994
673	1.350
CURVE 10	
115	-0.440
173	-0.308*
273	-0.064
293	0.000*
373	0.231*
473	0.601
573	1.010
673	1.445*
CURVE 11	
115	-0.451*
173	-0.320
273	-0.061*
CURVE 11 (cont.)	
293	0.000*
373	0.249
473	0.660
573	1.107
673	1.549*
CURVE 12	
115	-0.491
173	-0.343
273	-0.059*
293	0.000*
373	0.264
473	0.697
573	1.137
673	1.591*
CURVE 13*	
115	-0.427
173	-0.310
273	-0.060
293	0.000
373	0.262
473	0.656
573	1.095
673	1.520
CURVE 14	
115	-0.374
173	-0.256
273	-0.064*
293	0.000*
373	0.266*
473	0.685*
573	1.103*
673	1.544*
CURVE 15	
115	-0.420
173	-0.324*
273	-0.062*
293	0.000*
373	0.325
473	0.744
CURVE 15 (cont.)	
573	1.168
673	1.579*
CURVE 16	
115	-0.356
173	-0.256*
273	-0.046*
293	0.000*
373	0.221*
473	0.546*
573	0.902
673	1.297
CURVE 17	
115	-0.334
173	-0.237
273	-0.046*
293	0.000*
373	0.188
473	0.482
573	0.820
673	1.171

* Not shown in figure.
† This curve is here reported using the first given temperature as reference temperature at which $\Delta L/L_0 = 0$.
‡ Author's data for coefficient of thermal expansion have been integrated by TPRC to obtain $\Delta L/L_0$.

DATA TABLE 173. COEFFICIENT OF THERMAL LINEAR EXPANSION OF COPPER-MANGANESE SYSTEM Cu-Mn

[Temperature, T, K; Coefficient of Expansion, α, 10^{-6} K^{-1}]

T	α
CURVE 4‡	
3	0.029
4	0.038
5	0.049
6	0.061
8	0.094
10	0.135
12	0.180
14	0.235
16	0.320
18	0.410
20	0.525
22	0.625
24	0.800
26	0.980

‡ Author's data for coefficient of thermal expansion have been integrated by TPRC to obtain $\Delta L/L_0$.

FIGURE AND TABLE NO. 174BR. PROVISIONAL VALUES FOR THERMAL LINEAR EXPANSION OF COPPER-NICKEL SYSTEM Cu-Ni

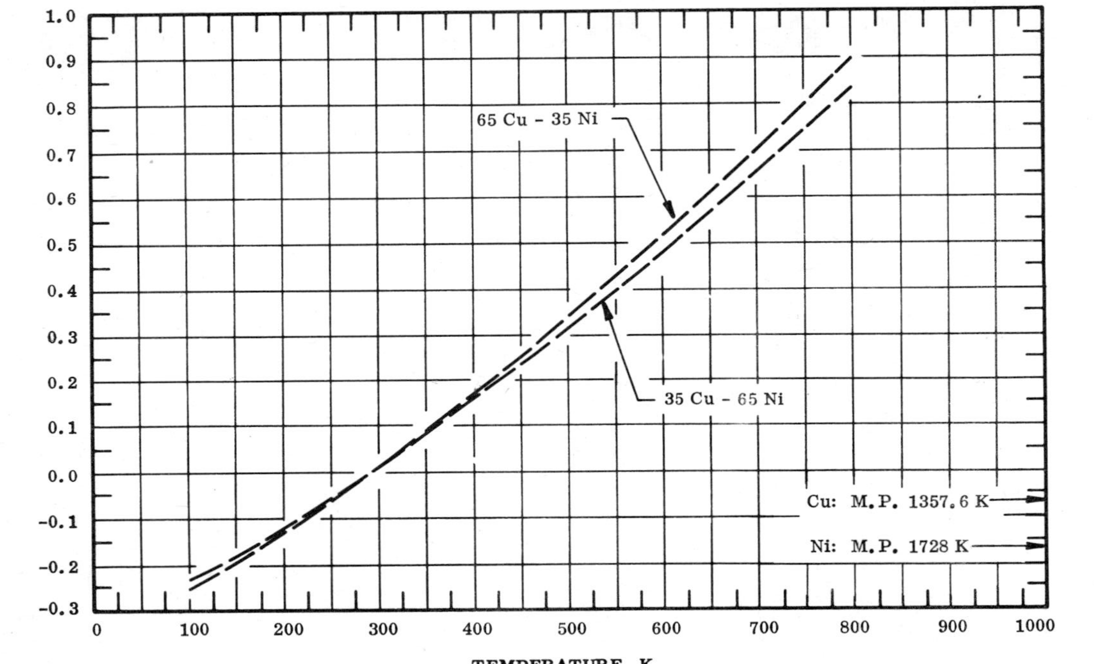

PROVISIONAL VALUES

[Temperature, T, K; Linear Expansion, $\Delta L/L_0$, %; α, K^{-1}]

T	(35 Cu-65 Ni) $\Delta L/L_0$	(35 Cu-65 Ni) $\alpha \times 10^6$	(65 Cu-35 Ni) $\Delta L/L_0$	(65 Cu-35 Ni) $\alpha \times 10^6$
100	-0.232	10.1	-0.256	11.2
200	-0.121	12.2	-0.132	13.4
293	0.000	13.9	0.000	15.0
400	0.156	15.3	0.169	16.4
500	0.314	16.3	0.338	17.4
600	0.482	17.1	0.517	18.2
700	0.655	17.5	0.702	18.8
800	0.832	17.8	0.892	19.2

REMARKS

(35 Cu-65 Ni): The tabulated values for well-annealed alloy are considered accurate to within ±7% over the entire temperature range. These values can be represented approximately by the following equation:

$$\Delta L/L_0 = -0.321 + 7.688 \times 10^{-4}\, T + 1.278 \times 10^{-6}\, T^2 - 5.477 \times 10^{-10}\, T^3$$

(65 Cu-35 Ni): The tabulated values for well-annealed alloy are considered accurate to within ±7% over the entire temperature range. These values can be represented approximately by the following equation:

$$\Delta L/L_0 = -0.360 + 9.188 \times 10^{-4}\, T + 1.196 \times 10^{-6}\, T^2 - 4.883 \times 10^{-10}\, T^3$$

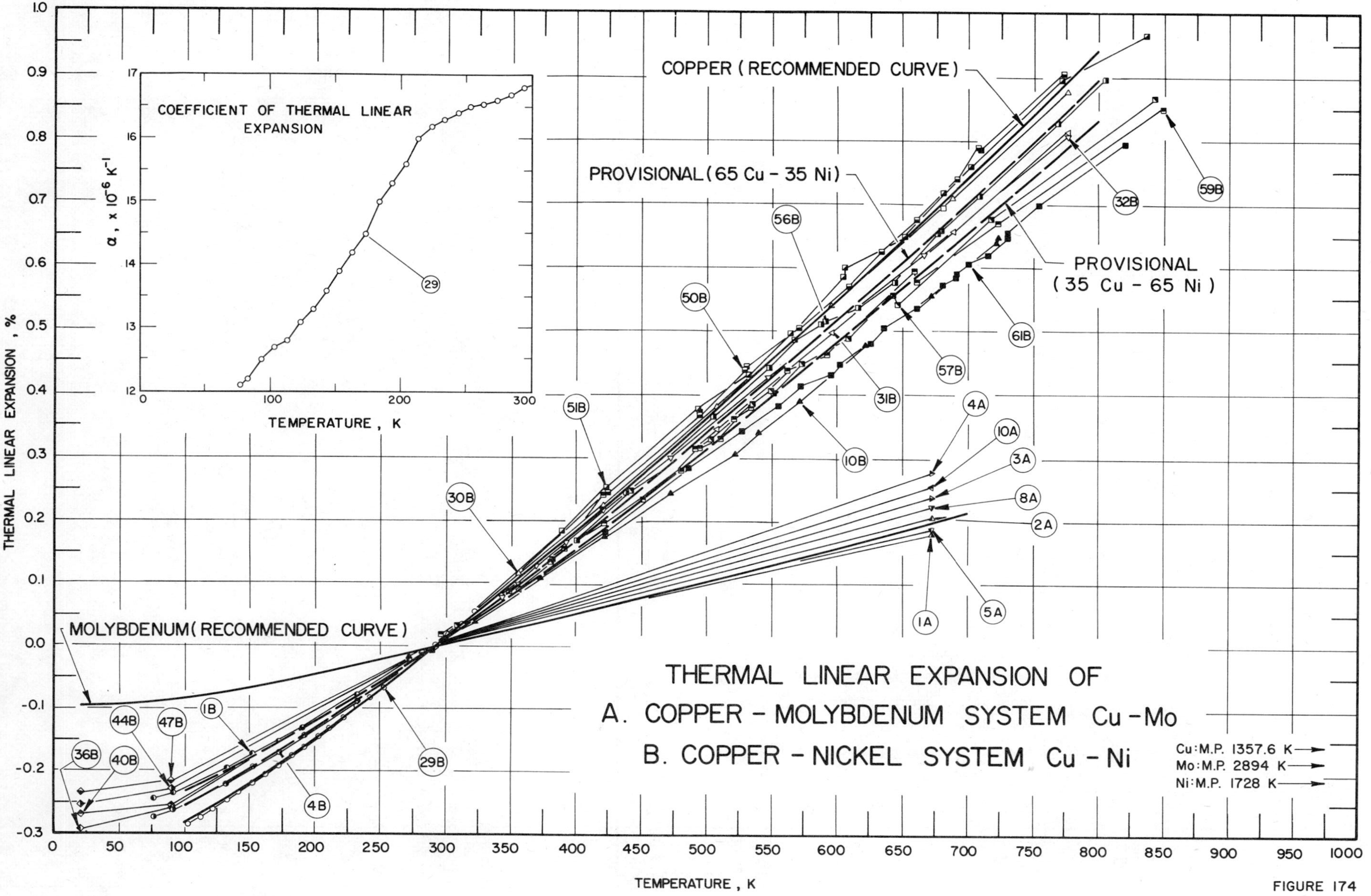
THERMAL LINEAR EXPANSION OF
A. COPPER - MOLYBDENUM SYSTEM Cu - Mo
B. COPPER - NICKEL SYSTEM Cu - Ni
Cu: M.P. 1357.6 K
Mo: M.P. 2894 K
Ni: M.P. 1728 K
COPPER (RECOMMENDED CURVE)
PROVISIONAL (35 Cu - 65 Ni)
PROVISIONAL (65 Cu - 35 Ni)
MOLYBDENUM (RECOMMENDED CURVE)
COEFFICIENT OF THERMAL LINEAR EXPANSION
TEMPERATURE, K
α, x 10⁻⁶ K⁻¹
THERMAL LINEAR EXPANSION, %
TEMPERATURE, K
1A
2A
3A
4A
5A
8A
10A
1B
4B
10B
29B
30B
31B
32B
36B
40B
44B
47B
50B
51B
56B
57B
59B
61B
29

FIGURE 174

SPECIFICATION TABLE 174A. THERMAL LINEAR EXPANSION OF COPPER-MOLYBDENUM SYSTEM Cu-Mo

Cur. No.	Ref. No.	Author(s)	Year	Method Used	Temp. Range, K	Name and Specimen Designation	Composition (weight percent) Cu	Mo	Composition (continued), Specifications, and Remarks
1	228	Baskin, L.M., Savin, A.V., Tumanov, V.I., and Eyduk, Yu. A.	1965	L	291, 673		1.5	98.5	Alloy prepared by powder-metallurgical techniques; sintered at 1973 K, annealed 5 hrs at 1223 K and oil quenched; porosity ~0.6% by volume; only Mo-phase apparent in x-ray analysis.
2	228	Baskin, L.M. et al.	1965	L	291, 673		3.0	97.0	Similar to the above specimen, except two phase apparent.
3	228	Baskin, L.M. et al.	1965	L	291, 673		5.0	95.0	Similar to the above specimen.
4	228	Baskin, L.M. et al.	1965	L	291, 673		6.5	93.5	Similar to the above specimen.
5	228	Baskin, L.M. et al.	1965	L	291, 673		7.0	93.0	Similar to the above specimen.
6*	228	Baskin, L.M. et al.	1965	L	291, 673		7.6	92.4	Similar to the above specimen.
7*	228	Baskin, L.M. et al.	1965	L	291, 673		8.0	92.0	Similar to the above specimen.
8	228	Baskin, L.M. et al.	1965	L	291, 673		9.0	91.0	Similar to the above specimen.
9*	228	Baskin, L.M. et al.	1965	L	291, 673		10.0	90.0	Similar to the above specimen.
10	228	Baskin, L.M. et al.	1965	L	291, 673		14.0	86.0	Similar to the above specimen except sintered at 1873 K.

DATA TABLE 174A. THERMAL LINEAR EXPANSION OF COPPER-MOLYBDENUM SYSTEM Cu-Mo

[Temperature, T, K; Linear Expansion, $\Delta L/L_0$, %]

T	$\Delta L/L_0$	T	$\Delta L/L_0$	T	$\Delta L/L_0$
CURVE 1		CURVE 5		CURVE 9*	
291	-0.001*	291	-0.001	291	-0.001
673	0.179	673	0.186	673	0.205
CURVE 2		CURVE 6*		CURVE 10	
291	-0.001	291	-0.001	291	-0.001
673	0.205	673	0.209	673	0.251
CURVE 3		CURVE 7*			
291	-0.001	291	-0.001		
673	0.236	673	0.201		
CURVE 4		CURVE 8			
291	-0.001	291	-0.001		
673	0.274	673	0.220		

*Not shown in figure.

SPECIFICATION TABLE 174B. THERMAL LINEAR EXPANSION OF COPPER-NICKEL SYSTEM Cu-Ni

Cur. No.	Ref. No.	Author(s)	Year	Method Used	Temp. Range, K	Name and Specimen Designation	Composition (weight percent) Cu	Ni	Composition (continued), Specifications, and Remarks
1	142	Aoyama, S. and Ito, T.	1939	L	77-273	Specimen No. 5	Bal.	59.35	0.07 Fe, 0.04 Mn; alloy melted, cast into bar 30 x 30 x 250 mm; hot-forged to 8 mm diameter bar, then cut into round bar 5 mm diameter, 100 mm length; annealed in vacuum for 3 hr at 1123 K.
2*	142	Aoyama, S. and Ito, T.	1939	L	77-273	Specimen No. 6	Bal.	49.79	0.15 Fe, 0.04 Mn; similar to the above specimen except annealed at 1023 K.
3*	142	Aoyama, S. and Ito, T.	1939	L	77-273	Specimen No. 8	Bal.	29.02	0.19 Fe, 0.02 Mn; similar to the above specimen.
4	142	Aoyama, S. and Ito, T.	1939	L	77-273	Specimen No. 9	Bal.	19.52	0.15 Fe, trace Mn; similar to the above specimen.
5*	142	Aoyama, S. and Ito, T.	1939	L	77-273	Specimen No. 10	Bal.	13.84	0.11 Fe, trace Mn; similar to the above specimen.
6*	142	Aoyama, S. and Ito, T.	1939	L	77-273	Specimen No. 11	Bal.	9.47	0.14 Fe, trace Mn; similar to the above specimen.
7*	142	Aoyama, S. and Ito, T.	1939	L	77-273	Specimen No. 12	Bal.	3.67	0.09 Fe; similar to the above specimen.
8*	142	Aoyama, S. and Ito, T.	1939	L	77-273	Specimen No. 13	Bal.	1.03	0.03 Fe; similar to the above specimen.
9*	60	Holborn, L. and Day, A. L.	1901	T	273-773	Constantan	60	40	No details given.
10	153	Masumoto, H., Saito, H., and Sawaya, S.	1969	L	273-725	No. 2	9.82	90.18	Electrolytic 99.980 Ni with 0.180 Co, 0.002 S, 0.001 Cu, 0.001 Fe, and electrolytic 99.900 Cu with 0.009 Si, 0.008 Fe, 0.005 Sb, 0.005 S, 0.001 Mn; rod specimen 2 mm diameter, 10 cm long; cooled at rate 300 C/hr after heating in vacuum at 1173 K for 30 min; density 8.90 g cm^{-3} at 293 K; zero-point correction is -0.029%.
11*	153	Masumoto, H., et al.	1969	L	273-633	No. 3	19.72	80.27	Similar to the above specimen; zero-point correction is -0.026%.
12*	153	Masumoto, H., et al.	1969	L	273-573	No. 4	24.80	75.28	Similar to the above specimen; zero-point correction is -0.030%.
13*	153	Masumoto, H., et al.	1969	L	273, 313	No. 5	27.65	72.35	Similar to the above specimen.
14*	153	Masumoto, H., et al.	1969	L	273-464	No. 8	29.85	70.15	Similar to the above specimen except density 8.91 g cm^{-3} at 293 K; zero-point correction is -0.032%.
15*	153	Masumoto, H., et al.	1969	L	273-418	No. 10	30.86	69.14	Similar to the above specimen; zero-point correction is -0.040%.
16*	153	Masumoto, H., et al.	1969	L	273-418	No. 11	31.60	68.40	Similar to the above specimen; density 8.92 g cm^{-3} at 293 K.
17*	153	Masumoto, H., et al.	1969	L	273-418	No. 12	32.50	67.50	Similar to the above specimen.
18*	153	Masumoto, H., et al.	1969	L	273-418	No. 13	34.43	65.57	Similar to the above specimen.
19*	153	Masumoto, H., et al.	1969	L	273-418	No. 14	39.68	60.32	Similar to the above specimen.
20*	153	Masumoto, H., et al.	1969	L	273-418	No. 15	44.58	55.42	Similar to the above specimen; density 8.93 g cm^{-3} at 293 K.
21*	153	Masumoto, H., et al.	1969	L	273, 313	No. 16	49.65	50.35	Similar to the above specimen.
22*	153	Masumoto, H., et al.	1969	L	273, 313	No. 17	59.70	40.30	Similar to the above specimen.
23*	153	Masumoto, H., et al.	1969	L	273, 313	No. 18	69.70	30.30	Similar to the above specimen.

* Not shown in figure.

SPECIFICATION TABLE 174B. THERMAL LINEAR EXPANSION OF COPPER-NICKEL SYSTEM Cu-Ni (continued)

Cur. No.	Ref. No.	Author(s)	Year	Method Used	Temp. Range, K	Name and Specimen Designation	Composition (weight percent) Cu	Ni	Composition (continued), Specifications, and Remarks
24*	153	Masumoto, H., et al.	1969	L	273, 313	No. 19	79.46	20.54	Similar to the above specimen.
25*	153	Masumoto, H., et al.	1969	L	273, 313	No. 20	89.70	10.30	Similar to the above specimen.
26*	388	Nakamura, K.	1936	L	293-678		14.2	Bal.	Rod specimen 2.27 mm diameter, 10.16 cm long; density 8.917 g cm^{-3}.
27*	388	Nakamura, K.	1936	L	285-703		25.2	Bal.	Rod specimen 2.28 mm diameter, 10.39 cm long; density 8.926 g cm^{-3}.
28*	388	Nakamura, K.	1936	L	303-703		35.7	Bal.	Rod specimen 2.27 mm diameter, 10.10 cm long; density 8.935 g cm^{-3}.
29	551	Belov, A.K.	1968	L	77-323	M2 Russian Alloy	Bal.	0.2	Cylindrical specimen 3.5 mm in diameter, 50 mm long; sample heated and cooled at a rate of 3-7 deg per min.
30	646	De, M.	1969	X	302-778		74.98	25.02	Alloy from Messrs. Goldsmith Bros. (USA); powder samples of about 4 mm length and 0.5 mm diameter; lattice parameter reported at 302 K is 3.5850 Å; 3.4656 Å at 293 K determined by graphical extrapolation; data on the coefficient of thermal linear expansion also reported.
31	646	De, M.	1969	X	301-778		50.01	49.99	Similar to the above specimen; lattice parameter reported at 301 K is 3.5650 Å; 3.5646 Å at 293 K determined by graphical extrapolation; data on the coefficient of thermal linear expansion also reported.
32	646	De, M.	1969	X	300-778		25.01	74.99	Similar to the above specimen; lattice parameter reported at 300 K is 3.5434 Å; 3.5430 Å at 293 K determined by graphical extrapolation; data on the coefficient of thermal linear expansion also reported.
33*	640	Krupkowski, A. and DeHaas, W.J.	1928	L	20-283		98	2	Alloy prepared using electrolytic copper and nickel; homogenized and annealed for 3 days in vacujm at 1023-1123 K.
34*	640	Krupkowski, A. and DeHaas, W.J.	1928	L	20-283		90	10	Similar to the above specimen.
35*	640	Krupkowski, A. and DeHaas, W.J.	1928	L	20-283		85	15	Similar to the above specimen.
36	640	Krupkowski, A. and DeHaas, W.J.	1928	L	20-283		75	25	Similar to the above specimen.
37*	640	Krupkowski, A. and DeHaas, W.J.	1928	L	20-283		70	30	Similar to the above specimen.
38*	640	Krupkowski, A. and DeHaas, W.J.	1928	L	20-283		65	35	Similar to the above specimen.
39*	640	Krupkowski, A. and DeHaas, W.J.	1928	L	20-283		60	40	Similar to the above specimen.
40	640	Krupkowski, A. and DeHaas, W.J.	1928	L	20-283		55	45	Similar to the above specimen.
41*	640	Krupkowski, A. and DeHaas, W.J.	1928	L	20-283		50	50	Similar to the above specimen.
42*	640	Krupkowski, A. and DeHaas, W.J.	1928	L	20-283		47.5	52.5	Similar to the above specimen.
43*	640	Krupkowski, A. and DeHaas, W.J.	1928	L	20-283		40	60	Similar to the above specimen.

* Not shown in figure.

SPECIFICATION TABLE 174B. THERMAL LINEAR EXPANSION OF COPPER-NICKEL SYSTEM Cu-Ni (continued)

Cur. No.	Ref. No.	Author(s)	Year	Method Used	Temp. Range, K	Name and Specimen Designation	Composition (weight percent) Cu	Ni	Composition (continued), Specifications, and Remarks
44	640	Krupkowski, A. and DeHaas, W.J.	1928	L	20-283		40	60	Similar to the above specimen.
45*	640	Krupkowski, A. and DeHaas, W.J.	1928	L	20-283		35	65	Similar to the above specimen.
46*	640	Krupkowski, A. and DeHaas, W.J.	1928	L	20-283		25	75	Similar to the above specimen.
47	640	Krupkowski, A. and DeHaas, W.J.	1928	L	20-283		20	80	Similar to the above specimen.
48*	640	Krupkowski, A. and DeHaas, W.J.	1928	L	20-283		15	85	Similar to the above specimen.
49*	640	Krupkowski, A. and DeHaas, W.J.	1928	L	20-283		5	95	Similar to the above specimen.
50	738	Kantola, M. and Takola, E.	1967	X	291-843		91	9	Alloy prepared using MD 154 copper shot and MD 101 nickel powder from Metals Disintegrating Corp., N.J.; known weights mixed and heated in furnace in current of hydrogen at 1373 K; alloy then powdered and sifted through sieve of diameter equal to 0.060 mm and slowly cooled from 873 K; lattice parameter at 291 K reported as 3.60275 Å; 3.6029 Å at 293 K determined by graphical interpolation.
51	738	Kantola, M. and Tokola, E.	1967	X	292-837		77	23	Similar to the above specimen; lattice parameter at 292 K reported as 3.58695 Å; 3.5870 Å at 293 K determined by graphical interpolation.
52*	738	Kantola, M. and Tokola, E.	1967	X	294-847		73	27	Similar to the above specimen; lattice parameter at 294 K reported as 3.58318 Å; 3.5829 Å at 293 K determined by graphical extrapolation.
53*	738	Kantola, M. and Tokola, E.	1967	X	291-841		68	32	Similar to the above specimen; lattice parameter at 291 K reported as 3.57860 Å; 3.5787 Å at 293 K determined by graphical interpolation.
54*	738	Kantola, M. and Tokola, E.	1967	X	296-843		58	42	Similar to the above specimen; lattice parameter at 296 K reported as 3.56980 Å; 3.5691 Å at 293 K determined by graphical extrapolation.
55*	738	Kantola, M. and Tokola, E.	1967	X	296-843		53	47	Similar to the above specimen; lattice parameter at 296 K reported as 3.56655 Å; 3.5657 Å at 293 K determined by graphical extrapolation.
56	738	Kantola, M. and Tokola, E.	1967	X	293-832		48	52	Similar to the above specimen; lattice parameter at 293 K reported as 3.56168 Å.
57	738	Kantola, M. and Tokola, E.	1967	X	295-844		39	61	Similar to the above specimen; lattice parameter at 295 K reported as 3.55415 Å; 3.5539 Å at 293 K determined by graphical extrapolation.
58*	738	Kantola, M. and Tokola, E.	1967	X	296-838		33	67	Similar to the above specimen; lattice parameter at 296 K reported as 3.54790 Å; 3.5479 Å at 293 K determined by graphical extrapolation.
59	738	Kantola, M. and Tokola, E.	1967	X	297-851		24	76	Similar to the above specimen; lattice parameter at 297 K reported as 3.54275 Å; 3.5421 Å at 293 K determined by graphical extrapolation.
60*	738	Kantola, M. and Tokola, E.	1967	X	293-909		12	88	Similar to the above specimen; lattice parameter at 293 K reported as 3.5334 Å.
61	738	Kantola, M. and Tokola, E.	1967	X	290-821		5	95	Similar to the above specimen; lattice parameter at 290 K reported as 3.52887 Å; 3.5291 Å at 293 K determined by graphical interpolation.

* Not shown in figure.

DATA TABLE 174B. THERMAL LINEAR EXPANSION OF COPPER-NICKEL SYSTEM Cu-Ni

[Temperature, T, K; Linear Expansion, $\Delta L/L_0$, %]

T	$\Delta L/L_0$
CURVE 1	
77	-0.245
91	-0.236
133	-0.198
153	-0.177
173	-0.155
193	-0.131
233	-0.080
273	-0.027
CURVE 2*	
77	-0.251
91	-0.242
133	-0.202
153	-0.180
173	-0.158
193	-0.134
233	-0.082
273	-0.027
CURVE 3*	
77	-0.267
91	-0.256
133	-0.214
153	-0.191
173	-0.166
193	-0.140
233	-0.086
273	-0.028
CURVE 4	
77	-0.276
91	-0.265
133	-0.221
153	-0.196
173	-0.171
193	-0.144
233	-0.088
273	-0.029
CURVE 5*	
77	-0.283
91	-0.272
133	-0.226
CURVE 5 (cont.)*	
153	-0.201
173	-0.175
193	-0.147
233	-0.090
273	-0.030
CURVE 6*	
77	-0.289
91	-0.277
133	-0.231
153	-0.205
173	-0.178
193	-0.150
233	-0.092
273	-0.030
CURVE 7*	
77	-0.297
91	-0.284
133	-0.236
153	-0.210
173	-0.183
193	-0.154
233	-0.094
273	-0.031
CURVE 8*	
77	-0.301
91	-0.288
133	-0.240
153	-0.213
173	-0.185
193	-0.155
233	-0.095
273	-0.031
CURVE 9*	
273	-0.032
523	0.364
648	0.579
773	0.810
CURVE 10	
273	-0.029
293	0.000*
323	0.039
373	0.108
423	0.174
473	0.241
523	0.306
541	0.339
573	0.389
623	0.476
673	0.556
723	0.639
725	0.648
CURVE 11*	
273	-0.026
293	0.000
323	0.047
373	0.120
423	0.188
452	0.234
473	0.268
523	0.351
573	0.445
623	0.525
633	0.548
CURVE 12*	
273	-0.030
293	0.000
323	0.043
350	0.081
373	0.118
423	0.198
473	0.283
523	0.370
573	0.455
CURVE 13*	
273	-0.029
313	0.029
CURVE 14*	
273	-0.032
293	0.000
323	0.050
373	0.147
423	0.246
464	0.325
CURVE 15*	
273	-0.040
293	0.000
323	0.060
373	0.157
418	0.241
CURVE 16*	
273	-0.029
313	0.029
CURVE 17*	
273	-0.029
313	0.029
CURVE 18*	
273	-0.029
313	0.029
CURVE 19*	
273	-0.030
313	0.030
CURVE 20*	
273	-0.031
313	0.031
CURVE 21*	
273	-0.031
313	0.031
CURVE 22*	
273	-0.031*
313	0.031
CURVE 23*	
273	-0.032
313	0.032
CURVE 24*	
273	-0.032
313	0.032
CURVE 25*	
273	-0.033
313	0.033
CURVE 26*	
293	0.000
308	0.033
328	0.047
353	0.079
378	0.136
403	0.144
428	0.181
453	0.215
478	0.254
503	0.301
528	0.326
553	0.359
578	0.394
603	0.430
628	0.466
653	0.504
678	0.543
CURVE 27*	
285	-0.009
303	0.011
318	0.029
333	0.044
353	0.071
378	0.111
403	0.145
CURVE 27 (cont.)*	
433	0.186
453	0.214
478	0.249
503	0.287
528	0.321
553	0.357
583	0.406
603	0.437
628	0.475
653	0.515
683	0.566
703	0.594
CURVE 28*	
303	0.016
313	0.033
328	0.055
353	0.080
378	0.121
403	0.154
428	0.190
453	0.224
478	0.261
503	0.295
533	0.340
553	0.373
578	0.409
603	0.444
628	0.479
653	0.515
678	0.555
703	0.580
CURVE 29‡	
77	-0.320*
83	-0.312*
93	-0.300*
103	-0.287
113	-0.275
123	-0.262
133	-0.248
143	-0.235
153	-0.221
163	-0.207
173	-0.193

* Not shown in figure.

‡ Author's data for coefficient of thermal expansion have been integrated by TPRC to obtain $\Delta L/L_0$.

DATA TABLE 174B. THERMAL LINEAR EXPANSION OF COPPER-NICKEL SYSTEM Cu-Ni (continued)

T	$\Delta L/L_0$
CURVE 29 (cont.)‡	
183	-0.178
193	-0.163
203	-0.147
213	-0.132
223	-0.116
233	-0.099
243	-0.083
253	-0.066
263	-0.050
273	-0.033
283	-0.017
293	0.000
303	0.017*
313	0.034
323	0.051
CURVE 30	
302.2	0.017
357.2	0.114
421.2	0.223
509.2	0.379
597.2	0.538
689.2	0.709
778.2	0.876
CURVE 31	
301.2	0.011*
357.2	0.090
421.2	0.191
509.2	0.342
597.2	0.494
689.2	0.654
778.2	0.811
CURVE 32	
300.2	0.011*
323.2	0.045
344.2	0.076
357.2	0.096
370.2	0.127
393.2	0.166
421.2	0.217
473.2	0.299
548.2	0.423
666.2	0.618
778.2	0.804
CURVE 33*	
20.6	-0.316
90.2	-0.283
283.2	-0.013
CURVE 34*	
20.6	-0.282
90.2	-0.246
283.2	-0.013
CURVE 35*	
20.6	-0.298
90.2	-0.266
283.2	-0.013
CURVE 36	
20.6	-0.294
90.2	-0.261
283.2	-0.012*
CURVE 37*	
20.6	-0.284
90.2	-0.255
283.2	-0.012
CURVE 38*	
20.6	-0.280
90.2	-0.253
283.2	-0.012
CURVE 39*	
20.6	-0.275
90.2	-0.251
283.2	-0.012
CURVE 40	
20.6	-0.270
90.2	-0.246
283.2	-0.012*
CURVE 41*	
20.6	-0.266
CURVE 41 (cont.)*	
90.2	-0.245
283.2	-0.012
CURVE 42*	
20.6	-0.267
90.2	-0.238
283.2	-0.012
CURVE 43*	
20.6	-0.261
90.2	-0.236
283.2	-0.012
CURVE 44	
20.6	-0.256
90.2	-0.230
283.2	-0.011*
CURVE 45*	
20.6	-0.251
90.2	-0.233
283.2	-0.011
CURVE 46*	
20.6	-0.241
90.2	-0.223
283.2	-0.011
CURVE 47	
20.6	-0.236
90.2	-0.219
283.2	-0.011
CURVE 48*	
20.6	-0.232
90.2	-0.218
283.2	-0.011
CURVE 49*	
20.6	-0.226
90.2	-0.209
283.2	-0.010
CURVE 50	
291	-0.004*
425	0.244
495	0.366
535	0.436
537	0.441
572	0.500
610	0.569
635	0.622
662	0.674
692	0.738
719	0.788
720	0.783
775	0.902
843	1.024
CURVE 51	
292	-0.001*
292	-0.003*
390	0.184
424	0.251
494	0.376
495	0.368
534	0.429
565	0.493
605	0.583
606	0.599
652	0.647
681	0.691
681	0.716
703	0.758
774	0.895
837	0.965
CURVE 52*	
294	0.008
397	0.195
433	0.248
502	0.377
499	0.371
527	0.421
561	0.463
586	0.541
677	0.728
713	0.762
777	0.851
847	0.999
CURVE 53*	
291	-0.003
291	-0.001
395	0.196
430	0.246
498	0.358
497	0.360
538	0.430
567	0.483
598	0.523
601	0.525
622	0.562
648	0.620
674	0.676
676	0.671
709	0.724
774	0.861
841	0.986
CURVE 54*	
296	0.020
409	0.191
448	0.272
510	0.367
512	0.384
559	0.445
584	0.493
616	0.549
613	0.546
633	0.577
675	0.684
713	0.717
714	0.717
769	0.832
843	0.955
CURVE 55*	
296	0.024
400	0.174
435	0.227
501	0.345
505	0.359
573	0.466
638	0.555
713	0.690
714	0.696
774	0.782
843	0.954
CURVE 56	
293	0.000*
294	0.002*
401	0.169
439	0.245
505	0.363
506	0.363
549	0.439
568	0.483
588	0.509
591	0.512
616	0.534
645	0.576
677	0.652
680	0.658
710	0.711
770	0.829
832	0.896
CURVE 57	
295	0.007*
297	0.010*
298	0.011*
309	0.031
381	0.132
383	0.138
442	0.245
504	0.329
550	0.402
551	0.400
574	0.445
609	0.487
643	0.557
646	0.540
718	0.675
844	0.867
CURVE 58*	
296	0.000
298	0.014
301	0.023
306	0.031
308	0.034
311	0.034
317	0.037
323	0.037
323	0.062
383	0.183

* Not shown in figure.

‡ Author's data for coefficient of thermal expansion have been integrated by TPRC to obtain $\Delta L/L_0$.

DATA TABLE 174B. THERMAL LINEAR EXPANSION OF COPPER-NICKEL SYSTEM Cu-Ni (continued)

T	$\Delta L/L_0$
CURVE 58 (cont.)*	
435	0.248
503	0.358
506	0.341
582	0.468
648	0.555
717	0.679
719	0.707
785	0.854
838	0.899
CURVE 59	
297	0.018
349	0.088
391	0.155
421	0.240
421	0.195
451	0.232
481	0.279
491	0.313
495	0.313
511	0.330
522	0.359
535	0.378
536	0.381
563	0.435
593	0.460
661	0.576
660	0.593
724	0.669
851	0.850
CURVE 60*	
293	0.000
299	0.009
419	0.195
455	0.246
471	0.294
505	0.328
531	0.360
564	0.396
573	0.433
586	0.453
598	0.467
610	0.490
624	0.507
636	0.549
651	0.558

T	$\Delta L/L_0$
CURVE 60 (cont.)*	
665	0.580
675	0.589
686	0.603
693	0.597
696	0.603
713	0.654
736	0.696
765	0.747
800	0.824
858	0.892
909	0.960
CURVE 61	
289	-0.007
290	-0.007
423	0.181
487	0.283
528	0.340
556	0.380
574	0.411
596	0.428
603	0.445
626	0.479
637	0.502
662	0.533
681	0.570
691	0.581
691	0.587
702	0.604
716	0.618
731	0.643
731	0.655
755	0.697
821	0.791

* Not shown in figure.

DATA TABLE 174B. COEFFICIENT OF THERMAL LINEAR EXPANSION OF COPPER-NICKEL SYSTEM Cu-Ni

[Temperature, T, K; Coefficient of Expansion, α, 10^{-6} K^{-1}]

T	α	T	α
CURVE 29‡		CURVE 32*	
77	12.20	293	12.85
83	12.30	320	13.61
93	12.50	347	14.68
103	12.70	358	16.90
113	12.79	363	17.75
123	13.10	374	17.54
133	13.30	422	16.91
143	13.59	472	16.40
153	13.89	523	16.36
163	14.20	571	16.36
173	14.50	623	16.52
183	14.99	649	16.52
193	15.30	673	16.60
203	15.60	722	16.60
213	16.00	773	16.41
223	16.20		
233	16.30		
243	16.40		
253	16.49		
263	16.55		
273	16.60		
283	16.70		
293	16.79		
303	16.84*		
313	16.89*		
323	17.00*		
CURVE 30*			
293	16.12		
371	16.72		
474	17.41		
571	17.89		
671	18.33		
772	18.77		
CURVE 31*			
293	13.13		
326	13.45		
376	14.21		
427	15.26		
473	16.10		
571	17.04		
671	17.53		
772	17.61		

*Not shown in figure.

‡Author's data for coefficient of thermal expansion have been integrated by TPRC to obtain $\Delta L/L_0$.

FIGURE AND TABLE NO. 175AR. PROVISIONAL VALUES FOR THERMAL LINEAR EXPANSION OF COPPER-TIN SYSTEM Cu-Sn

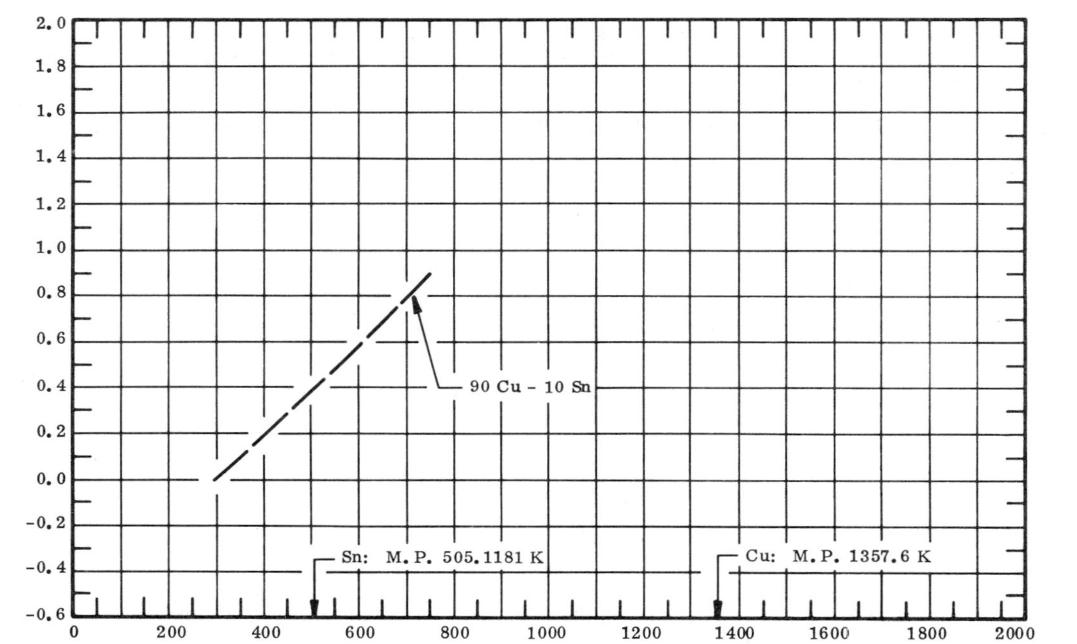

PROVISIONAL VALUES

[Temperature, T, K; Linear Expansion, $\Delta L/L_0$, %; α, K^{-1}]

T	(90 Cu-10 Sn) $\Delta L/L_0$	$\alpha \times 10^6$
293	0.000	17.0
400	0.192	18.8
500	0.337	20.0
600	0.539	20.5
700	0.796	20.8
750	0.901	21.3

REMARKS

(90 Cu-10 Sn): The tabulated values for well-annealed alloy are considered accurate to within ±7% over the entire temperature range. The phase transformation occurs at about 613 K. The tabulated values can be represented approximately by the following equation:

$$\Delta L/L_0 = -0.415 + 1.055 \times 10^{-3}\ T + 1.421 \times 10^{-6}\ T^2 - 6.517 \times 10^{-10}\ T^3$$

FIGURE AND TABLE NO. 175BR. PROVISIONAL VALUES FOR THERMAL LINEAR EXPANSION OF COPPER-TUNGSTEN SYSTEM Cu-W

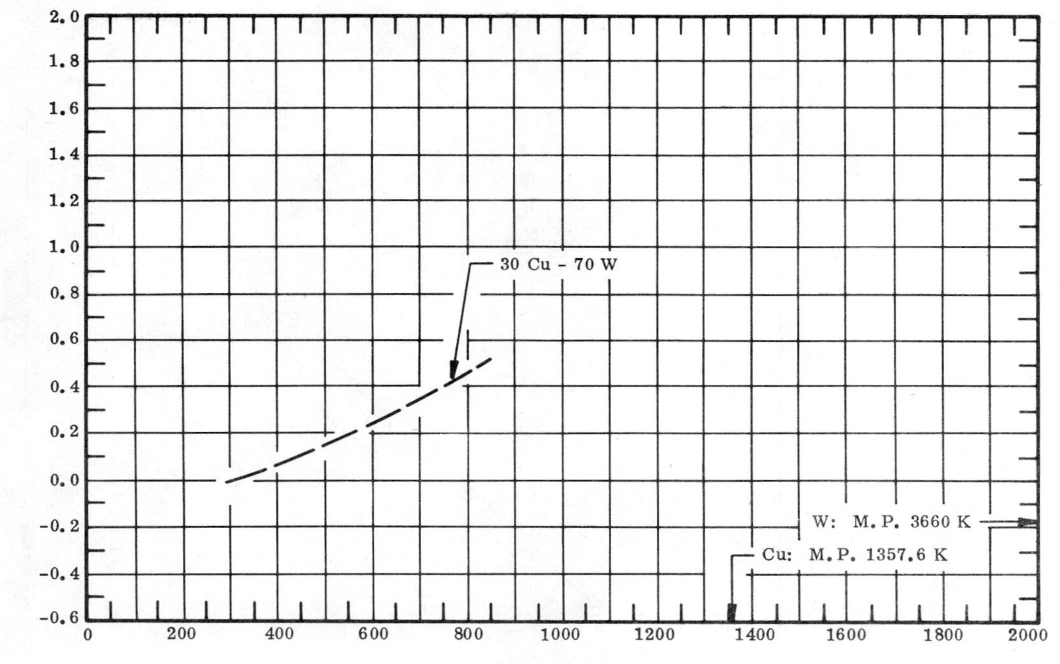

PROVISIONAL VALUES

[Temperature, T, K; Linear Expansion, $\Delta L/L_0$, %; α, K^{-1}]

T	(30 Cu-70 W) $\Delta L/L_0$	$\alpha \times 10^6$
293	0.000	5.9
400	0.073	7.9
500	0.160	9.1
600	0.257	10.1
700	0.360	10.6
800	0.469	10.9
850	0.525	11.2

REMARKS

(30 Cu-70 W): The tabulated values for well-annealed alloy are considered accurate to within ± 7% over the entire temperature range. These values can be represented approximately by the following equation:

$$\Delta L/L_0 = -0.083 - 1.101 \times 10^{-4}\, T + 1.473 \times 10^{-6}\, T^2 - 6.048 \times 10^{-10}\, T^3$$

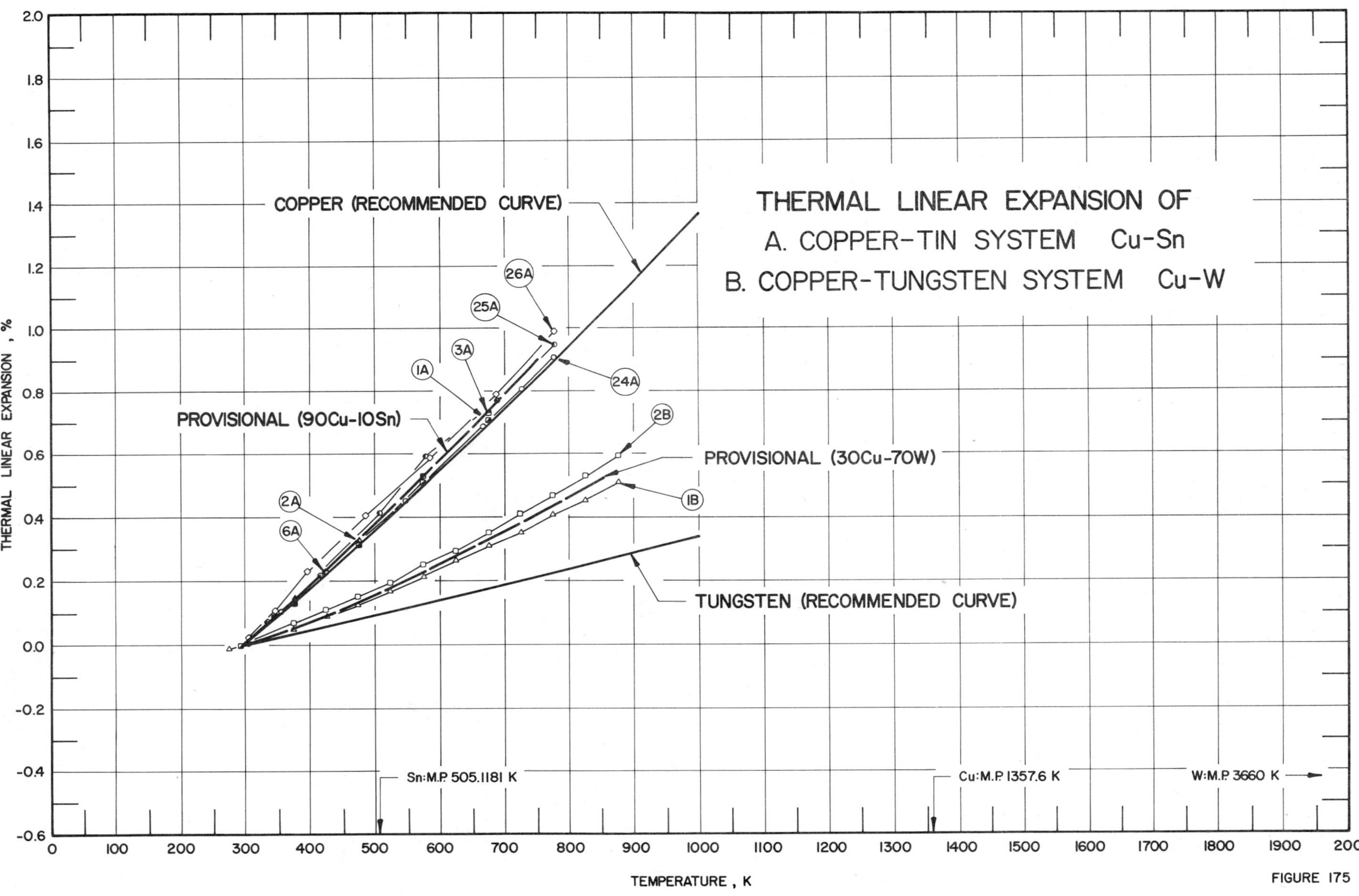
THERMAL LINEAR EXPANSION OF
A. COPPER-TIN SYSTEM Cu-Sn
B. COPPER-TUNGSTEN SYSTEM Cu-W
COPPER (RECOMMENDED CURVE)
PROVISIONAL (90Cu-10Sn)
PROVISIONAL (30Cu-70W)
TUNGSTEN (RECOMMENDED CURVE)
26A
25A
3A
1A
24A
2A
6A
2B
1B
Sn:M.P. 505.1181 K
Cu:M.P. 1357.6 K
W:M.P. 3660 K
THERMAL LINEAR EXPANSION, %
TEMPERATURE, K
2.0
1.8
1.6
1.4
1.2
1.0
0.8
0.6
0.4
0.2
0.0
-0.2
-0.4
-0.6
0
100
200
300
400
500
600
700
800
900
1000
1100
1200
1300
1400
1500
1600
1700
1800
1900
2000
FIGURE 175

SPECIFICATION TABLE 175A. THERMAL LINEAR EXPANSION OF COPPER-TIN SYSTEM Cu-Sn

Cur. No.	Ref. No.	Author(s)	Year	Method Used	Temp. Range, K	Name and Specimen Designation	Composition (weight percent) Cu	Sn	Composition (continued), Specifications, and Remarks
1	412	Cook, M. and Tallis, W.G.	1941	L	293-673	Phosphor Bronze Alloy No. 1	96.84	3.11	0.02 P, < 0.01 each Pb, Fe, Ni; annealed cylindrical specimen 5.1 cm x 1.3 cm dia.
2	412	Cook, M. and Tallis, W.G.	1941	L	293-673	Phosphor Bronze Alloy No. 3	92.60	7.31	0.02 P, < 0.01 each Fe, Pb; similar to the above specimen.
3	412	Cook, M. and Tallis, W.G.	1941	L	293-673	Phosphor Bronze Alloy No. 5	96.16	3.17	0.12 P, < 0.01 each Fe, Pb, Zn; similar to the above specimen.
4*	412	Cook, M. and Tallis, W.G.	1941	L	293-673	Phosphor Bronze Alloy No. 6	94.60	5.27	0.09 P, < 0.01 each Fe, Pb; similar to the above specimen.
5*	412	Cook, M. and Tallis, W.G.	1941	L	293-673	Phosphor Bronze Alloy No. 7	93.19	6.65	0.12 P, < 0.01 each Fe, Pb, Zn; similar to the above specimen.
6	413	Hidnert, P.	1934	L	293-473	No. 1473	87.75	12.25	0.02 P; virgin materials cast at 1422 K, machined to 300 mm long, 9 mm dia.
7*	413	Hidnert, P.	1934	L	293-473		95.40	4.25	Similar to the above specimen.
8*	413	Hidnert, P.	1934	L	293-473		95.40	4.25	Similar to the above specimen; cut from same casting.
9*	413	Hidnert, P.	1934	L	293-473		95.40	4.25	The above specimen; expansion measured on second test.
10*	413	Hidnert, P.	1934	L	293-473		95.40	4.25	Similar to the above specimen; cut from same casting.
11*	413	Hidnert, P.	1934	L	293-473		94.86	4.88	Similar to the above specimen.
12*	413	Hidnert, P.	1934	L	293-473		94.86	4.88	Similar to the above specimen; cut from same casting.
13*	413	Hidnert, P.	1934	L	293-473		94.86	4.88	Similar to the above specimen; cut from same casting.
14*	413	Hidnert, P.	1934	L	293-473		94.86	4.88	Similar to the above specimen; cut from same casting.
15*	413	Hidnert, P.	1934	L	293-473		92.04	7.67	Similar to the above specimen.
16*	413	Hidnert, P.	1934	L	293-473		92.04	7.67	Similar to the above specimen; cut from same casting.
17*	413	Hidnert, P.	1934	L	293-473		92.04	7.67	Similar to the above specimen; cut from same casting.
18*	413	Hidnert, P.	1934	L	293-473		92.04	7.67	Similar to the above specimen; cut from same casting.
19*	413	Hidnert, P.	1934	L	293-473		89.69	10.14	Similar to the above specimen.
20*	413	Hidnert, P.	1934	L	293-473		89.69	10.14	Similar to the above specimen; cut from same casting.
21*	413	Hidnert, P.	1934	L	293-473		89.69	10.14	Similar to the above specimen; cut from same casting.
22*	413	Hidnert, P.	1934	L	293-473		89.69	10.14	Similar to the above specimen; cut from same casting.
23*	413	Hidnert, P.	1934	L	293-473		84.84	14.95	Similar to the above specimen.

* Not shown in figure.

SPECIFICATION TABLE 175A. THERMAL LINEAR EXPANSION OF COPPER-TIN SYSTEM Cu-Sn (continued)

Cur. No.	Ref. No.	Author(s)	Year	Method Used	Temp. Range, K	Name and Specimen Designation	Composition (weight percent) Cu	Sn	Composition (continued), Specifications, and Remarks
24	646	De, M.	1969	X	304-778		94.95	5.05	Alloys prepared from spectrographically standardized metals from Messrs. Johnson, Matthey and Co. Ltd.; homogenized at 1023-1123 K; annealing treatments terminated by quenching in air; lattice parameter reported at 304 K is 3.6432 Å; 3.6425 Å at 293 K determined by graphical extrapolation; coefficient of thermal linear expansion also reported.
25	646	De, M.	1969	X	302-778		89.87	9.13	Similar to the above specimen; lattice parameter reported at 302 K is 3.6644 Å; 3.6638 Å at 293 K determined by graphical extrapolation; data on the coefficient of thermal linear expansion also reported.
26	646	De, M.	1969	X	303-778		86.07	13.93	Similar to the above specimen; lattice parameter reported at 303 K is 3.6940 Å; 3.6932 Å at 293 K determined by graphical extrapolation; data on the coefficient of thermal linear expansion also reported.

DATA TABLE 175A. THERMAL LINEAR EXPANSION OF COPPER-TIN SYSTEM

[Temperature, T, K; Linear Expansion, $\Delta L/L_0$, %]

T	$\Delta L/L_0$
CURVE 1	
293	0.000
373	0.135
473	0.315
573	0.507
673	0.703
CURVE 2	
293	0.000*
373	0.141
473	0.328
573	0.518*
673	0.718*
CURVE 3	
293	0.000*
373	0.141*
473	0.331*
573	0.526
673	0.737
CURVE 4*	
293	0.000
373	0.136
473	0.315
573	0.504
673	0.699
CURVE 5*	
293	0.000
373	0.138
473	0.322
573	0.512
673	0.718
CURVE 6	
293	0.000*
333	0.070
373	0.142*
423	0.234
473	0.326*

T	$\Delta L/L_0$
CURVE 7*	
293	0.000
373	0.139
473	0.320
CURVE 8*	
293	0.000
373	0.140
473	0.322
CURVE 9*	
293	0.000
373	0.138
473	0.319
CURVE 10*	
293	0.000
373	0.139
473	0.320
CURVE 11*	
293	0.000
373	0.139
473	0.319
CURVE 12*	
293	0.000
373	0.138
473	0.319
CURVE 13*	
293	0.000
373	0.139
473	0.320
CURVE 14*	
293	0.000
373	0.138
473	0.317

T	$\Delta L/L_0$
CURVE 15*	
293	0.000
373	0.140
473	0.322
CURVE 16*	
293	0.000
373	0.140
473	0.322
CURVE 17*	
293	0.000
373	0.140
473	0.320
CURVE 18*	
293	0.000
373	0.141
473	0.322
CURVE 19*	
293	0.000
373	0.141
473	0.326
CURVE 20*	
293	0.000
373	0.140
473	0.324
CURVE 21*	
293	0.000
373	0.142
473	0.326
CURVE 22*	
293	0.000
373	0.142
437	0.328

T	$\Delta L/L_0$
CURVE 23*	
293	0.000
373	0.144
473	0.328
CURVE 24	
304	0.019
353	0.104
417	0.220
547	0.458
666	0.692
723	0.804
778	0.914
CURVE 25	
302	0.016*
345	0.093
417	0.235*
509	0.418
579	0.595
689	0.775
778	0.950
CURVE 26	
303	0.022
345	0.108
395	0.233
485	0.406
585	0.588
689	0.799
778	0.994

*Not shown in figure.

DATA TABLE 175A. COEFFICIENT OF THERMAL LINEAR EXPANSION OF COPPER-TIN SYSTEM Cu-Sn

[Temperature, T, K; Coefficient of Expansion, α, 10^{-6} K^{-1}]

T	α
CURVE 24*	
293	17.07
371	17.52
468	18.19
570	18.90
674	19.44
772	19.71
CURVE 25*	
293	18.34
369	18.80
471	19.57
570	19.90
672	20.07
771	20.02
CURVE 26*	
293	19.31
371	19.78
470	20.45
673	20.63
771	20.63

*No figure given.

SPECIFICATION TABLE 175B. THERMAL LINEAR EXPANSION OF COPPER-TUNGSTEN SYSTEM Cu-W

Cur. No.	Ref. No.	Author(s)	Year	Method Used	Temp. Range, K	Name and Specimen Designation	Composition (weight percent) Cu	W	Composition (continued), Specifications, and Remarks
1		Hensel, F.R., Larsen, E.I. and Swazy, E.F.	1942		273-873		20	80	Specimen prepared by powder metallurgy techniques.
2		Hensel, F.R. et al.	1942		273-873		40	60	Similar to the above specimen.

DATA TABLE 175B. THERMAL LINEAR EXPANSION OF COPPER-TUNGSTEN SYSTEM Cu-W

[Temperature, T, K; Linear Expansion, $\Delta L/L_0$, %]

T	$\Delta L/L_0$	T	$\Delta L/L_0$
CURVE 1		CURVE 2	
273	-0.014	273	-0.016*
373	0.058	373	0.064
423	0.097	423	0.106
473	0.134	473	0.150
523	0.175	523	0.196
573	0.218	573	0.244
623	0.262	623	0.294
673	0.309	673	0.347
723	0.357	723	0.403
773	0.407	773	0.463
823	0.459	823	0.525
873	0.513	873	0.592

*Not shown in figure.

FIGURE AND TABLE NO. 176R. PROVISIONAL VALUES FOR THERMAL LINEAR EXPANSION OF COPPER-ZINC SYSTEM Cu-Zn

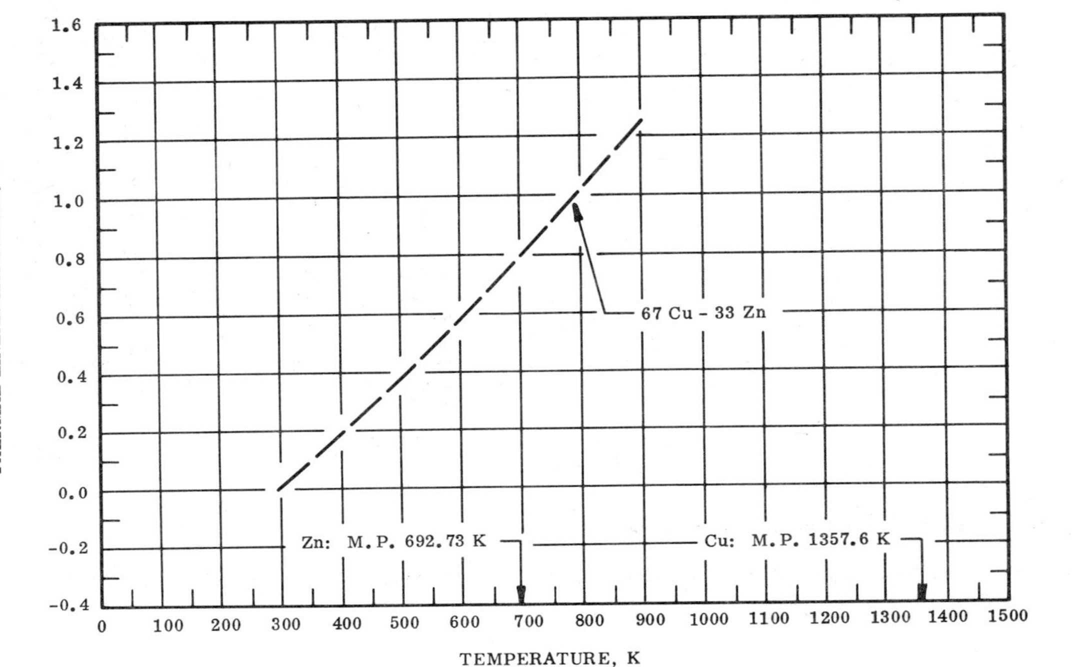

PROVISIONAL VALUES

[Temperature, T, K; Linear Expansion, $\Delta L/L_0$, %; α, K^{-1}]

T	(67 Cu-33 Zn) $\Delta L/L_0$	$\alpha \times 10^6$
293	0.000	17.5
400	0.194	18.9
500	0.389	20.0
600	0.595	20.9
700	0.808	21.7
800	1.028	22.5
900	1.258	23.5

REMARKS

(67 Cu-33 Zn): The tabulated values for well-annealed alloy are considered accurate to within ±7% over the entire temperature range. These values can be represented approximately by the following equation:

$$\Delta L/L_0 = -0.464 + 1.377 \times 10^{-3}\, T + 7.410 \times 10^{-7}\, T^2 - 1.608 \times 10^{-10}\, T^3$$

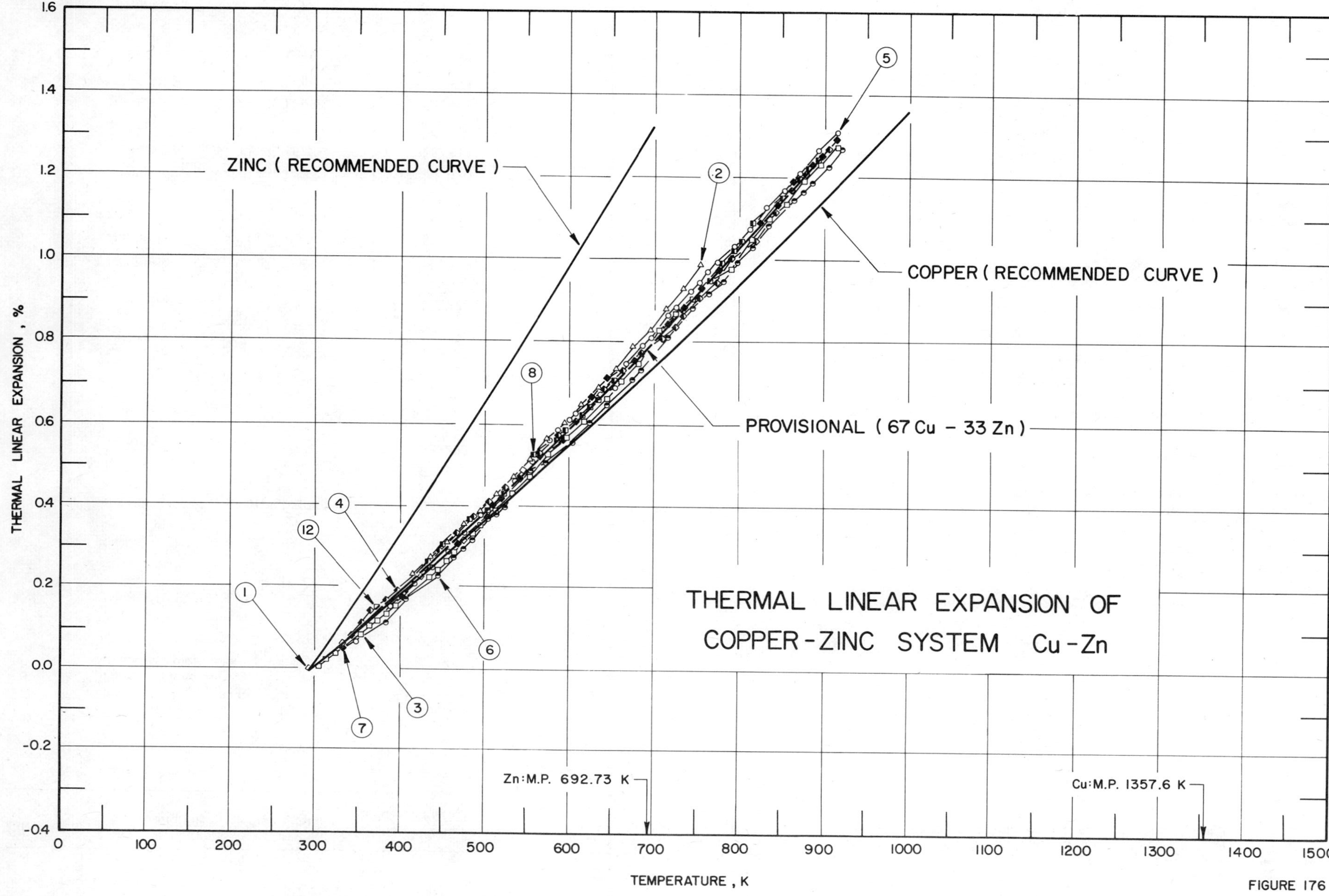
THERMAL LINEAR EXPANSION OF
COPPER-ZINC SYSTEM Cu-Zn
ZINC (RECOMMENDED CURVE)
COPPER (RECOMMENDED CURVE)
PROVISIONAL (67 Cu - 33 Zn)
Zn:M.P. 692.73 K
Cu:M.P. 1357.6 K
THERMAL LINEAR EXPANSION , %
TEMPERATURE , K
1.6
1.4
1.2
1.0
0.8
0.6
0.4
0.2
0.0
-0.2
-0.4
0
100
200
300
400
500
600
700
800
900
1000
1100
1200
1300
1400
1500
1
2
3
4
5
6
7
8
12

FIGURE 176

SPECIFICATION TABLE 176. THERMAL LINEAR EXPANSION OF COPPER-ZINC SYSTEM Cu-Zn

Cur. No.	Ref. No.	Author(s)	Year	Method Used	Temp. Range, K	Name and Specimen Designation	Composition (weight percent) Cu	Zn	Composition (continued), Specifications, and Remarks
1	413	Hidnert, P.	1934	L	293-473	Sample 1472	99.20	0.76	0.04 P; virgin material cast at 1422 K, machined to 300 mm long; 9 mm dia.
2	38	Uffelman, F.L.	1930	I	413-753	Common brass	62	Bal.	Cubic specimen 4 mm on a side.
3	600	Dowdell, R.H. Nagler, C.A., Fine, M.E., Klug, H.P., and Bitsianes, G.	1947	L	293-919		66.87	33.12	0.01 Fe; specimen 5 x 0.5 x 0.25 in.; annealed for 0.5 hr at 573 K, heated to 923 K and cooled immediately; expansion measured with increasing temperature.
4	600	Dowdell, R.H., et al.	1947	L	905-357		66.87	33.12	The above specimen except expansion measured with decreasing temperature.
5	600	Dowdell, R.H., et al.	1947	L	293-916				0.01 Fe; specimen 5 x 0.5 x 0.25 in.; specimen heated 4 hr at 903-923 K; measurements taken every 15 min; gradual decrease as β-phase dissolves; expansion measured with increasing temperature.
6	600	Dowdell, R.H., et al.	1947	L	920-293				The above specimen except expansion measured with decreasing temperature.
7	600	Dowdell, R.H., et al.	1947	L	293-916		66.87	33.12	0.01 Fe; specimen 5 x 0.5 x 0.25 in; second specimen both heating and cooling; after the second step above no β-phase present; expansion measured with increasing temperature.
8	600	Dowdell, R.H., et al.	1947	L	892-293		66.87	33.12	The above specimen except expansion measured with decreasing temperature.
9*	461	Neumeier, L.A., et al.	1969	L	293, 373		0.75	Bal.	0.12 Ti; specimen melted in purified helium atmosphere; cast at 822 K into graphite mold, extruded at 478 K; 0.95 cm rod; expansion measured transverse to extrusion.
10*	461	Neumeier, L.A., et al.	1969	L	293, 373		0.75	Bal.	Similar to the above specimen; extruded at 566 K.
11*	461	Neumeier, L.A., et al.	1969	L	293, 373		0.75	Bal.	Similar to the above specimen; extruded at 655 K.
12	461	Neumeier, L.A., et al.	1969	L	293, 373		0.75	Bal.	Similar to the above specimen; 0.32 cm rod; expansion longitudinal; extruded at 566 K.
13*	461	Neumeier, L.A., et al.	1969	L	293, 373		0.75	Bal.	Similar to the above specimen; extruded at 611 K.
14*	461	Neumeier, L.A., et al.	1969	L	293, 373		0.75	Bal.	Similar to the above specimen.
15*	461	Neumeier, L.A., et al.	1969	L	293, 373		0.75	Bal.	Similar to the above specimen; 0.95 cm rod; extruded at 478 K; expansion measured parallel to extrusion.
16*	461	Neumeier, L.A., et al.	1969	L	293, 373		0.75	Bal.	Similar to the above specimen.
17*	461	Neumeier, L.A., et al.	1969	L	293, 373		0.75	Bal.	Similar to the above specimen; extruded at 500 K.
18*	461	Neumeier, L.A., et al.	1969	L	293, 373		0.75	Bal.	Similar to the above specimen; extruded at 522 K.
19*	461	Neumeier, L.A., et al.	1969	L	293, 373		0.75	Bal.	Similar to the above specimen.
20*	461	Neumeier, L.A., et al.	1969	L	293, 373		0.75	Bal.	Similar to the above specimen; extruded at 544 K.
21*	461	Neumeier, L.A., et al.	1969	L	293, 373		0.75	Bal.	Similar to the above specimen; extruded at 566 K.

* Not shown in figure.

SPECIFICATION TABLE 176. THERMAL LINEAR EXPANSION OF COPPER-ZINC SYSTEM Cu-Zn (continued)

Cur. No.	Ref. No.	Author(s)	Year	Method Used	Temp. Range, K	Name and Specimen Designation	Composition (weight percent) Cu	Zn	Composition (continued), Specifications, and Remarks
22*	461	Neumeier, L.A., et al.	1969	L	293, 373		0.75	99.13	Similar to the above specimen.
23*	461	Neumeier, L.A., et al.	1969	L	293, 373		0.75	99.13	Similar to the above specimen; extruded at 589 K.
24*	461	Neumeier, L.A., et al.	1969	L	293, 373		0.75	99.13	Similar to the above specimen; extruded at 611 K.
25*	461	Neumeier, L.A., et al.	1969	L	293, 373		0.75	99.13	Similar to the above specimen.
26*	461	Neumeier, L.A., et al.	1969	L	293, 373		0.75	99.13	Similar to the above specimen; extruded at 633 K.
27*	461	Neumeier, L.A., et al.	1969	L	293, 373		0.75	99.13	Similar to the above specimen; extruded at 655 K.
28*	461	Neumeier, L.A., et al.	1969	L	293, 373		0.75	99.13	Similar to the above specimen.
29*	643	Owen, E.A. and Pickup, L.	1933	X	623-1073	673 Z			No details given; composition of alloy is in α-region; alloys annealed > 873 K and quenched in water; lattice parameter reported at 623 K is 3.6804 Å; this curve is here reported using the first given temperature as reference temperature at which $\Delta L/L_0 = 0$.
30*	643	Owen, E.A. and Pickup, L.	1933	X	623-973	589 Z α-phase			No details given; similar to above specimen; composition of alloy is in $(\alpha+\beta)$ region; lattice parameter reported at 623 K is 3.694 Å; this curve is here reported using the first given temperature as reference temperature at which $\Delta L/L_0 = 0$.
31*	643	Owen, E.A. and Pickup, L.	1933	X	623-973	β-phase	42.7	57.3	Lump annealed 623 to 1073 K and then rapidly quenched; composition of alloy is in $(\beta+\gamma)$ region; lattice parameter reported at 623 K is 2.950 Å; this curve is here reported using the first given temperature as reference temperature at which $\Delta L/L_0 = 0$.
32*	643	Owen, E.A. and Pickup, L.	1933	X	623-973	589 Z β-phase			No details given; composition of alloy is in $(\alpha+\beta)$ region; lattice parameter reported at 623 K is 2.944 Å; this curve is here reported using the first given temperature as reference temperature at which $\Delta L/L_0 = 0$.
33*	643	Owen, E.A. and Pickup, L.	1933	X	623-1073	561 Z β-phase			Similar to above specimen; composition of alloy is in $(\alpha+\beta)$ region; lattice parameter reported at 623 K is 2.944 Å; this curve is here reported using the first given temperature as reference temperature at which $\Delta L/L_0 = 0$.
34*	643	Owen, E.A. and Pickup, L.	1933	X	623-1073	523 Z			Similar to the above specimen; composition of alloy in β-region; lattice parameter reported at 623 K is 2.946 Å; this curve is here reported using the first given temperature as reference temperature at which $\Delta L/L_0 = 0$.
35*	643	Owen, E.A. and Pickup, L.	1933	X	623-973	453 Z β-phase			Similar to the above specimen; composition of alloy in $(\beta+\gamma)$ region; lattice parameter reported at 623 K is 2.951 Å; this curve is here reported using the first given temperature as reference temperature at which $\Delta L/L_0 = 0$.
36*	643	Owen, E.A. and Pickup, L.	1933	X	623-973	453 Z γ-phase			Similar to the above specimen; composition of alloy is in γ-region; lattice parameter reported at 623 K is 8.828 Å; this curve is here reported using the first given temperature as reference temperature at which $\Delta L/L_0 = 0$.
37*	643	Owen, E.A. and Pickup, L.	1933	X	623-1073	414 Z			Similar to the above specimen; composition of alloy is in γ-region; lattice parameter reported at 623 K is 8.832 Å; this curve is here reported using the first given temperature as reference temperature at which $\Delta L/L_0 = 0$.

*Not shown in figure.

SPECIFICATION TABLE 176. THERMAL LINEAR EXPANSION OF COPPER-ZINC SYSTEM Cu-Zn (continued)

Cur. No.	Ref. No.	Author(s)	Year	Method Used	Temp. Range, K	Name and Specimen Designation	Composition (weight percent) Cu	Zn	Composition (continued), Specifications, and Remarks
38*	643	Owen, E.A. and Pickup, L.	1933	X	623-1073	387 Z			Similar to the above specimen; composition of alloy is in γ-region; lattice parameter reported at 623 K is 8.840 Å; this curve is here reported using the first given temperature as reference temperature at which $\Delta L/L_0 = 0$.
39*	643	Owen, E.A. and Pickup, L.	1933	X	623-1073	γ-phase	42.7	57.3	Lump-annealed 623-1073 K and then rapidly quenched; composition of alloy is in (β+γ) phase; lattice parameter reported at 623 K is 8.834 Å; this curve is here reported using the first given temperature as reference temperature at which $\Delta L/L_0 = 0$.

*Not shown in figure.

DATA TABLE 176. THERMAL LINEAR EXPANSION OF COPPER-ZINC SYSTEM Cu-Zn

[Temperature, T, K; Linear Expansion, $\Delta L/L_0$, %]

T	$\Delta L/L_0$
CURVE 1	
293	0.000
333	0.067
373	0.135
423	0.224
473	0.313
CURVE 2‡	
413	0.234
433	0.273
453	0.313
473	0.353
493	0.394
513	0.435
533	0.477
553	0.519
573	0.562
593	0.605
613	0.649
633	0.694
653	0.739
673	0.786
693	0.833
713	0.882
733	0.933
753	0.984
CURVE 3	
293	0.000*
306	0.009
315	0.023
324	0.037
354	0.083
364	0.104
374	0.119
384	0.133
394	0.154
433	0.223
443	0.240
455	0.263
463	0.283
486	0.333
493	0.350
505	0.372*
514	0.395
CURVE 3 (cont.)	
526	0.417*
535	0.434
546	0.457*
556	0.475
566	0.503*
576	0.526
586	0.548*
597	0.569
608	0.586*
617	0.612
625	0.636*
641	0.664
650	0.684*
660	0.705
673	0.729*
680	0.749
690	0.773*
696	0.810*
706	0.828
716	0.849*
726	0.872
734	0.891*
745	0.905
758	0.930*
765	0.954
779	0.962*
790	0.977
795	1.012*
805	1.035*
815	1.053
828	1.078*
837	1.100
851	1.116*
859	1.139
872	1.164*
879	1.187
888	1.213*
897	1.233
909	1.255*
919	1.278
CURVE 4	
905	1.274
895	1.249*
885	1.227*
CURVE 4 (cont.)	
879	1.208
868	1.190*
861	1.166
850	1.145*
843	1.119
831	1.087*
820	1.045
801	1.022*
790	1.003
782	0.985*
773	0.943
765	0.928*
753	0.908
741	0.888*
732	0.868
728	0.856*
725	0.840
718	0.831*
710	0.811
699	0.792*
681	0.770
673	0.749*
663	0.732
651	0.706*
640	0.689
631	0.666*
607	0.613
596	0.595*
588	0.578
578	0.552*
566	0.533
555	0.513*
548	0.491
538	0.476*
528	0.449
517	0.433*
507	0.411
497	0.394*
487	0.373
478	0.351*
467	0.332
459	0.309*
448	0.287
437	0.267
397	0.187
382	0.162
CURVE 4 (cont.)	
369	0.140
357	0.115
CURVE 5	
293	0.000*
303	0.013
317	0.030*
327	0.040
338	0.052*
350	0.068
358	0.095*
369	0.115
377	0.136*
390	0.155
399	0.176*
417	0.208
430	0.225*
439	0.247
449	0.257*
458	0.285
466	0.309*
478	0.333
488	0.355*
497	0.380
508	0.401*
519	0.427*
528	0.447*
538	0.468
548	0.493*
558	0.516
570	0.535*
578	0.561
588	0.585
600	0.612
609	0.625
618	0.644*
637	0.677
651	0.692
658	0.721*
668	0.747
678	0.767*
689	0.790
697	0.810
706	0.835*
CURVE 5 (cont.)	
716	0.866
728	0.883
742	0.922
754	0.943
763	0.971
773	0.994
782	1.014*
792	1.036
803	1.058*
813	1.079
822	1.099*
832	1.122
842	1.149*
855	1.168
863	1.196*
873	1.218
881	1.241*
894	1.264
903	1.285*
916	1.311
CURVE 6	
920	1.274
915	1.251*
907	1.224
896	1.203*
887	1.183
877	1.164
867	1.144
857	1.128*
849	1.106*
837	1.084
828	1.060*
818	1.036
807	1.017*
796	0.994
789	0.969*
780	0.947
766	0.920
757	0.902*
747	0.884
737	0.862*
727	0.840*
716	0.815
CURVE 6 (cont.)	
694	0.763*
682	0.737
674	0.713
663	0.689*
653	0.664*
641	0.648
632	0.620*
622	0.603
611	0.579*
601	0.555
582	0.525*
573	0.504
554	0.489
543	0.442*
531	0.422
525	0.400
515	0.380
503	0.364
494	0.341*
484	0.318
473	0.297
464	0.276
444	0.227
409	0.165
382	0.113
293	0.000*
CURVE 7	
293	0.000*
308	0.019*
315	0.026*
326	0.043*
334	0.059
345	0.082
364	0.108*
381	0.135*
391	0.154*
401	0.174*
418	0.219*
430	0.243
439	0.263*
451	0.284
468	0.309
478	0.335*

*Not shown in figure.
‡Author's data for coefficient of thermal expansion have been integrated by TPRC to obtain $\Delta L/L_0$.

DATA TABLE 176. THERMAL LINEAR EXPANSION OF COPPER-ZINC SYSTEM Cu-Zn (continued)

T	$\Delta L/L_0$
CURVE 7 (cont.)	
491	0.354*
503	0.372
510	0.401
520	0.420
531	0.442*
541	0.469
551	0.493*
560	0.509*
568	0.529
581	0.549*
591	0.563
596	0.601*
605	0.626*
617	0.653*
627	0.671
636	0.695*
644	0.719
655	0.734*
676	0.758
687	0.785*
699	0.803*
709	0.831*
717	0.848
728	0.868*
737	0.889
750	0.910*
758	0.936
768	0.959*
779	0.976
784	1.006*
799	1.025
807	1.051*
819	1.079*
827	1.093
837	1.112*
849	1.135
867	1.188
879	1.210*
888	1.237
899	1.256
907	1.272*
916	1.290

T	$\Delta L/L_0$
CURVE 8	
892	1.244
881	1.220
874	1.197
860	1.179*
851	1.145
841	1.130*
832	1.109*
819	1.092
814	1.071*
801	1.042
792	1.021
780	0.993
773	0.973*
761	0.955*
754	0.938*
743	0.912*
732	0.887*
722	0.867
711	0.843*
704	0.822
692	0.805*
678	0.786*
669	0.757*
666	0.729*
652	0.707
644	0.688*
633	0.663
624	0.641
613	0.622
602	0.604*
595	0.589
588	0.562
560	0.529
552	0.512*
541	0.488*
532	0.468*
523	0.437
514	0.415*
503	0.395
494	0.383*
483	0.362
474	0.338*
450	0.304
443	0.281*
431	0.263
422	0.232*

T	$\Delta L/L_0$
CURVE 8 (cont.)	
412	0.216*
402	0.196
392	0.177
382	0.154
371	0.130*
293	0.000*
CURVE 9*	
373	0.268
CURVE 10*	
373	0.268
CURVE 11*	
373	0.276
CURVE 12	
373	0.145
CURVE 13*	
373	0.140
CURVE 14*	
373	0.139
CURVE 15*	
373	0.144
CURVE 16*	
373	0.142
CURVE 17*	
373	0.146
CURVE 18*	
373	0.145

T	$\Delta L/L_0$
CURVE 19*	
373	0.143
CURVE 20*	
373	0.141
CURVE 21*	
373	0.142
CURVE 22*	
373	0.140
CURVE 23*	
373	0.141
CURVE 24*	
373	0.140
CURVE 25*	
373	0.136
CURVE 26*	
373	0.134
CURVE 27*	
373	0.134
CURVE 28*	
373	0.132
CURVE 29*, †	
623.2	0.000
673.2	0.027
723.2	0.027
773.2	-0.027
873.2	0.000
1073.2	-0.027

T	$\Delta L/L_0$
CURVE 30*, †	
623.2	0.000
673.2	0.000
723.2	-0.034
773.2	-0.034
823.2	-0.068
873.2	-0.102
923.2	-0.204
973.2	-0.272
1073.2	-0.339
CURVE 31*, †	
623.2	0.000
653.2	-0.034
683.2	-0.034
723.2	-0.034
773.2	-0.034
873.2	-0.068
923.2	-0.034
983.2	-0.034
1073.2	0.000
CURVE 32*, †	
623.2	0.000
673.2	0.000
723.2	-0.034
773.2	-0.034
823.2	-0.068
873.2	-0.102
923.2	-0.204
973.2	-0.272
1073.2	-0.339
CURVE 33*, †	
623.2	0.000
673.2	-0.034
773.2	-0.068
873.2	-0.170
923.2	-0.170
973.2	-0.170
1073.2	-0.170

T	$\Delta L/L_0$
CURVE 34*, †	
623.2	0.000
673.2	0.000
723.2	0.000
773.2	0.000
873.2	0.000
973.2	0.000
1073.2	0.000
CURVE 35*, †	
623.2	0.000
673.2	-0.034
723.2	-0.068
773.2	-0.068
873.2	-0.068
973.2	-0.068
CURVE 36*, †	
623.2	0.000
673.2	-0.056
723.2	-0.102
773.2	-0.102
873.2	-0.102
973.2	-0.045
CURVE 37*, †	
623.2	0.000
653.2	-0.023
673.2	-0.034
723.2	-0.034
773.2	-0.034
873.2	-0.011
973.2	-0.011
1073.2	-0.023
CURVE 38*, †	
623.2	0.000
673.2	0.000
723.2	0.000
773.2	-0.023
873.2	-0.011
1073.2	-0.023

*Not shown in figure.
†This curve is here reported using the first given temperature as reference temperature at which $\Delta L/L_0 = 0$.

DATA TABLE 176. THERMAL LINEAR EXPANSION OF COPPER-ZINC SYSTEM Cu-Zn (continued)

T	$\Delta L/L_0$
CURVE 39*, †	
623.2	0.000
653.2	-0.056
683.2	-0.125
723.2	-0.158
773.2	-0.170
873.2	-0.158
923.2	-0.147
983.2	-0.113
1073.2	-0.045

DATA TABLE 176. COEFFICIENT OF THERMAL LINEAR EXPANSION OF COPPER-ZINC SYSTEM Cu-Zn

[Temperature, T, K; Coefficient of Expansion, α, 10^{-6} K^{-1}]

T	α	T	α
CURVE 2*, ‡		CURVE 2 (cont.)*, ‡	
413	19.3	733	26.0
433	19.7	753	25.0
453	20.0		
473	20.4		
493	20.6		
513	20.8		
533	21.0		
553	21.2		
573	21.5		
593	21.8		
613	22.1		
633	22.5		
653	22.9		
673	23.5		
693	24.0		
713	25.0		

*Not shown in figure.
†This curve is here reported using the first given temperature as reference temperature at which $\Delta L/L_0 = 0$.
‡Author's data for coefficient of thermal expansion have been integrated by TPRC to obtain $\Delta L/L_0$.

FIGURE AND TABLE NO. 177R. PROVISIONAL VALUES FOR THERMAL LINEAR EXPANSION OF DYSPROSIUM-TANTALUM SYSTEM Dy-Ta

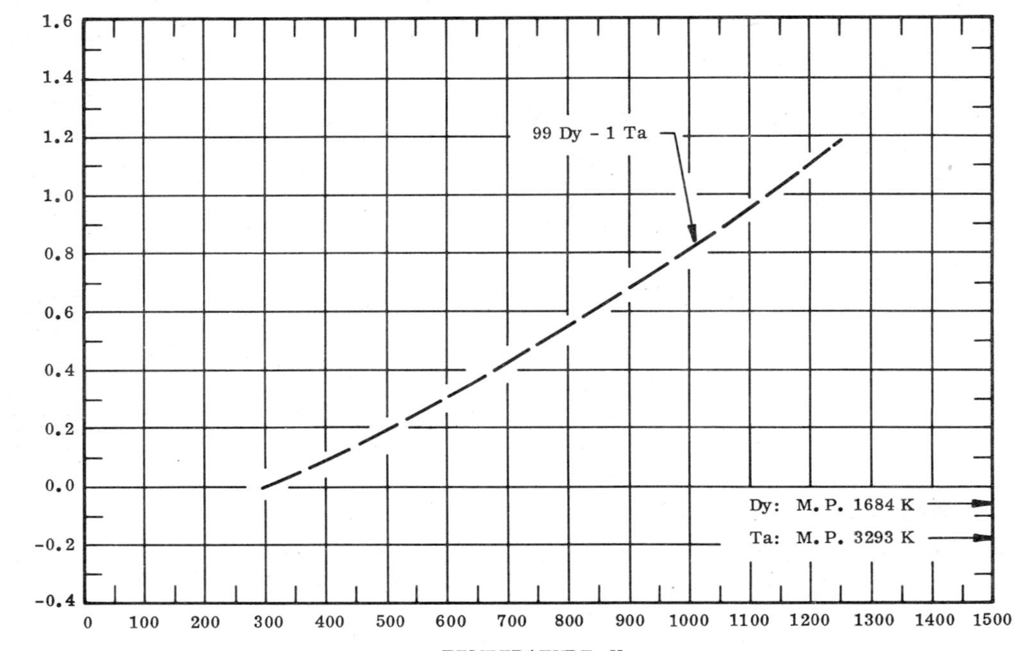

PROVISIONAL VALUES

[Temperature, T, K; Linear Expansion, $\Delta L/L_0$, %; α, K^{-1}]

T	(99 Dy-1 Ta) $\Delta L/L_0$	$\alpha \times 10^6$
293	0.000	8.7
400	0.098	9.6
500	0.199	10.4
600	0.307	11.3
700	0.424	12.0
800	0.548	12.7
900	0.679	13.4
1000	0.816	14.1
1100	0.959	14.7
1200	1.108	15.2
1250	1.183	15.4

REMARKS

(99 Dy-1 Ta): The tabulated values for well-annealed alloy are considered accurate to within ± 7% over the entire temperature range. These values can be represented approximately by the following equation:

$$\Delta L/L_0 = -0.211 + 5.667 \times 10^{-4}\, T + 5.455 \times 10^{-7}\, T^2 - 8.586 \times 10^{-11}\, T^3$$

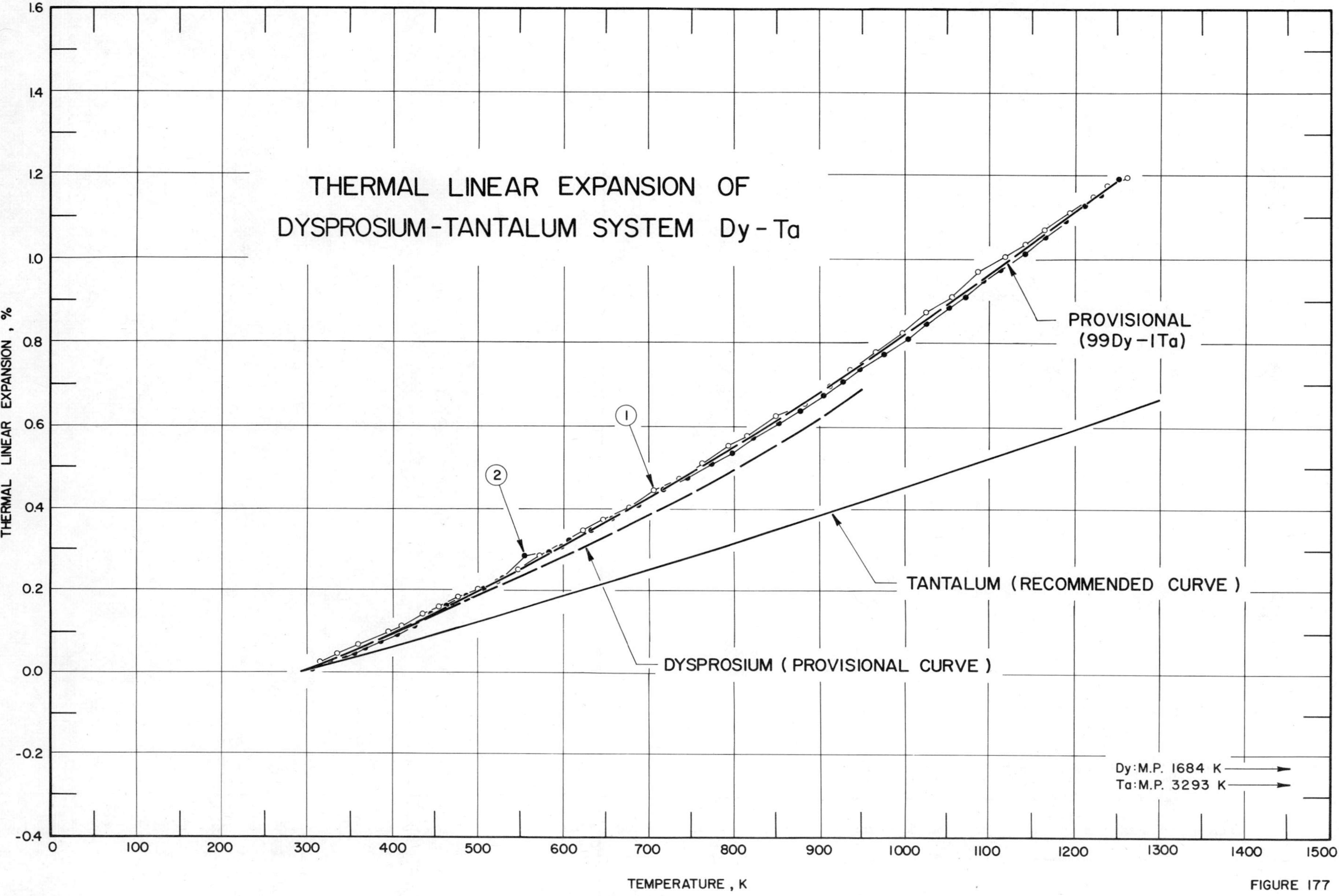
THERMAL LINEAR EXPANSION OF
DYSPROSIUM-TANTALUM SYSTEM Dy-Ta
PROVISIONAL
(99Dy-1Ta)
TANTALUM (RECOMMENDED CURVE)
DYSPROSIUM (PROVISIONAL CURVE)
1
2
Dy:M.P. 1684 K
Ta:M.P. 3293 K
THERMAL LINEAR EXPANSION, %
TEMPERATURE, K
1.6
1.4
1.2
1.0
0.8
0.6
0.4
0.2
0.0
-0.2
-0.4
0
100
200
300
400
500
600
700
800
900
1000
1100
1200
1300
1400
1500

FIGURE 177

SPECIFICATION TABLE 177. THERMAL LINEAR EXPANSION OF DYSPROSIUM - TANTALUM SYSTEM Dy - Ta

Cur. No.	Ref. No.	Author(s)	Year	Method Used	Temp. Range, K	Name and Specimen Designation	Composition (weight percent) Dy	Ta	Composition (continued), Specifications, and Remarks
1	78	Barson, F., Legvold, S. and Spedding, F. H.	1956	L	317-1262		99.0	0.5	0.2 Ca, 0.1 Tb, 0.05 Ho, 0.02 Er, 0.02 Si, 0.095 C, 0.005 Fe, 0.003 N_2; specimen 0.6 cm diameter, 6 cm long; fluorides of Dy bomb reduced with Ca metal to produce compact metallic sample, then vacuum cast, turned to shape; expansion measured with increasing temperature.
2	78	Barson, F., et al.	1956	L	1253-307		99.0	0.5	0.2 Ca; the above specimen; expansion measured with decreasing temperature.

DATA TABLE 177. THERMAL LINEAR EXPANSION OF DYSPROSIUM - TANTALUM SYSTEM Dy - Ta

[Temperature, T, K; Linear Expansion, $\Delta L/L_0$, %]

T	$\Delta L/L_0$
CURVE 1	
317	0.027
335	0.047
346	0.056*
360	0.074
376	0.084*
393	0.104
410	0.120
433	0.144
454	0.160
475	0.185
500	0.204
523	0.228
549	0.257
574	0.289
624	0.342
649	0.375
676	0.403
706	0.442
CURVE 1 (cont.)	
736	0.476
763	0.509
792	0.546
816	0.579
850	0.623
882	0.654
913	0.699
938	0.737
967	0.779
999	0.828
1028	0.876
1056	0.916
1089	0.970
1119	1.002
1142	1.039
1166	1.075
1193	1.111
CURVE 1 (cont.)	
1221	1.145
1240	1.170
1262	1.197
CURVE 2	
1253	1.190
1231	1.158
1214	1.125
1190	1.095
1167	1.057
1142	1.016
1114	0.978
1094	0.946
1073	0.916
1051	0.884
1028	0.848
CURVE 2 (cont.)	
1004	0.812
975	0.768
948	0.735
929	0.702
903	0.674
879	0.638
852	0.606
822	0.568
798	0.537
773	0.507
746	0.479
717	0.444
689	0.411
659	0.379
632	0.346
606	0.321
599	0.312
CURVE 2 (cont.)	
581	0.292
556	0.289
530	0.235
508	0.208
485	0.187
461	0.161
442	0.141
423	0.120
403	0.100
385	0.080
369	0.066
355	0.054
343	0.039
328	0.026
307	0.006

* Not shown in figure.

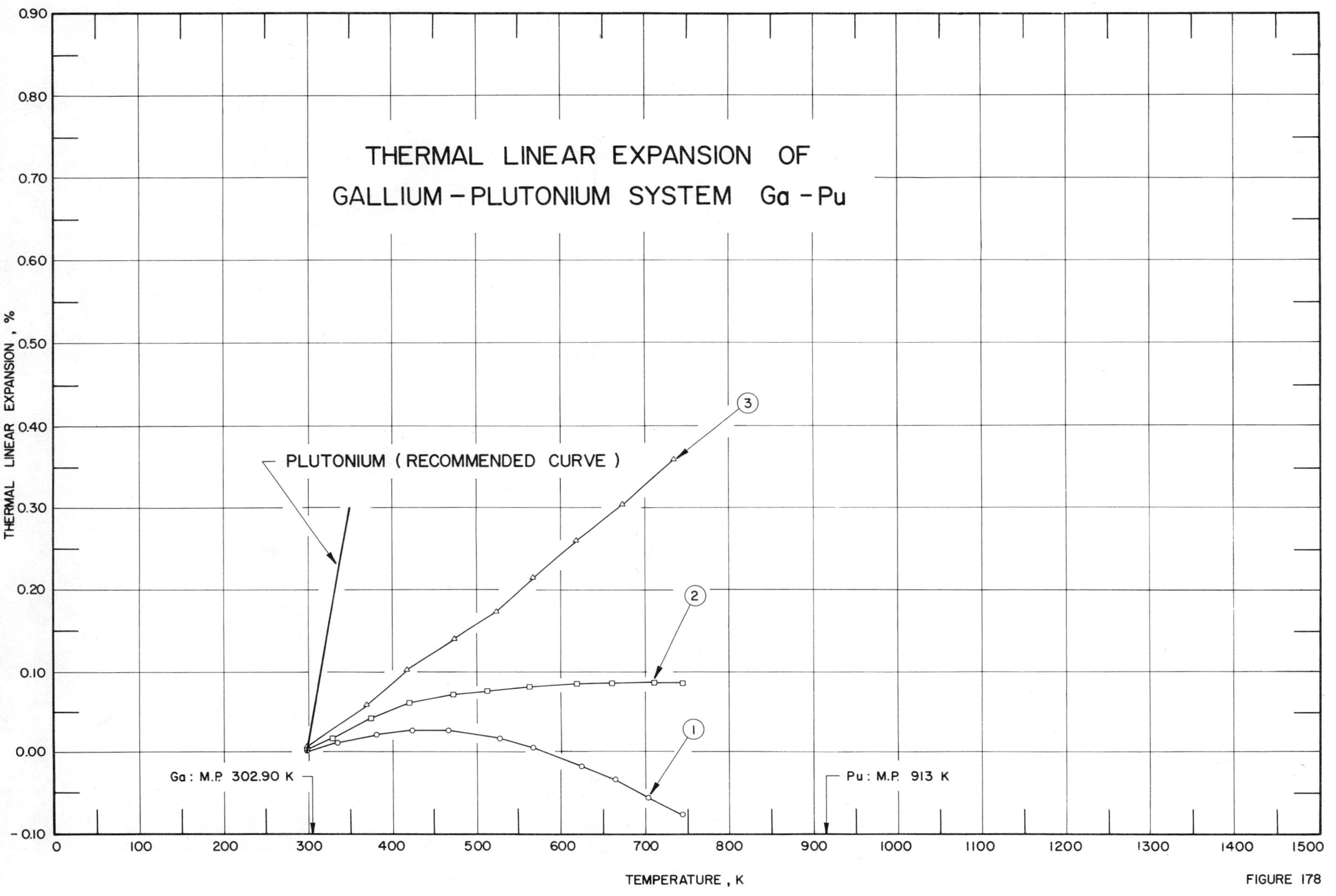
THERMAL LINEAR EXPANSION OF
GALLIUM – PLUTONIUM SYSTEM Ga – Pu
PLUTONIUM (RECOMMENDED CURVE)
3
2
1
Ga : M.P. 302.90 K
Pu : M.P. 913 K
THERMAL LINEAR EXPANSION , %
0.90
0.80
0.70
0.60
0.50
0.40
0.30
0.20
0.10
0.00
– 0.10
0
100
200
300
400
500
600
700
800
900
1000
1100
1200
1300
1400
1500
TEMPERATURE , K

FIGURE 178

SPECIFICATION TABLE 178. THERMAL LINEAR EXPANSION OF GALLIUM-PLUTONIUM SYSTEM Ga-Pu

Cur. No.	Ref. No.	Author(s)	Year	Method Used	Temp. Range, K	Name and Specimen Designation	Composition (weight percent) Ga	Pu	Composition (continued), Specifications, and Remarks
1	439	Taylor, J. M.	1969	O	298-743		0.4	99.6	Specimen prepared by arc melting six times to a side, high purity Pu and Ga (impurity < 0.0001) cast into rod, machined into a right cylinder 9.53 mm in diameter and 50.8 mm long; solid solution sample; zero-point correction is 0.001%.
2	439	Taylor, J. M.	1969	O	298-743		0.85	99.15	Prepared in a manner similar to the above procedure; zero-point correction is 0.002%.
3	439	Taylor, J. M.	1969	O	298-743		1.87	98.13	Prepared in a manner similar to the above procedure; zero-point correction is 0.006%.

DATA TABLE 178. THERMAL LINEAR EXPANSION OF GALLIUM-PLUTONIUM SYSTEM Ga-Pu

[Temperature, T, K; Linear Expansion, $\Delta L/L_0$, %]

T	$\Delta L/L_0$	T	$\Delta L/L_0$	T	$\Delta L/L_0$
CURVE 1		CURVE 2		CURVE 3	
298	0.001	298	0.002*	298	0.006*
333	0.011	329	0.017	369	0.059
380	0.021	373	0.042	418	0.104
423	0.027	420	0.062	473	0.141
468	0.027	473	0.072	523	0.174
528	0.016	513	0.077	568	0.216
568	0.005	563	0.082	619	0.261
623	-0.017	618	0.087	673	0.306
662	-0.033	660	0.087	733	0.361
703	-0.055	710	0.087		
743	-0.077	743	0.087		

*Not shown in figure.

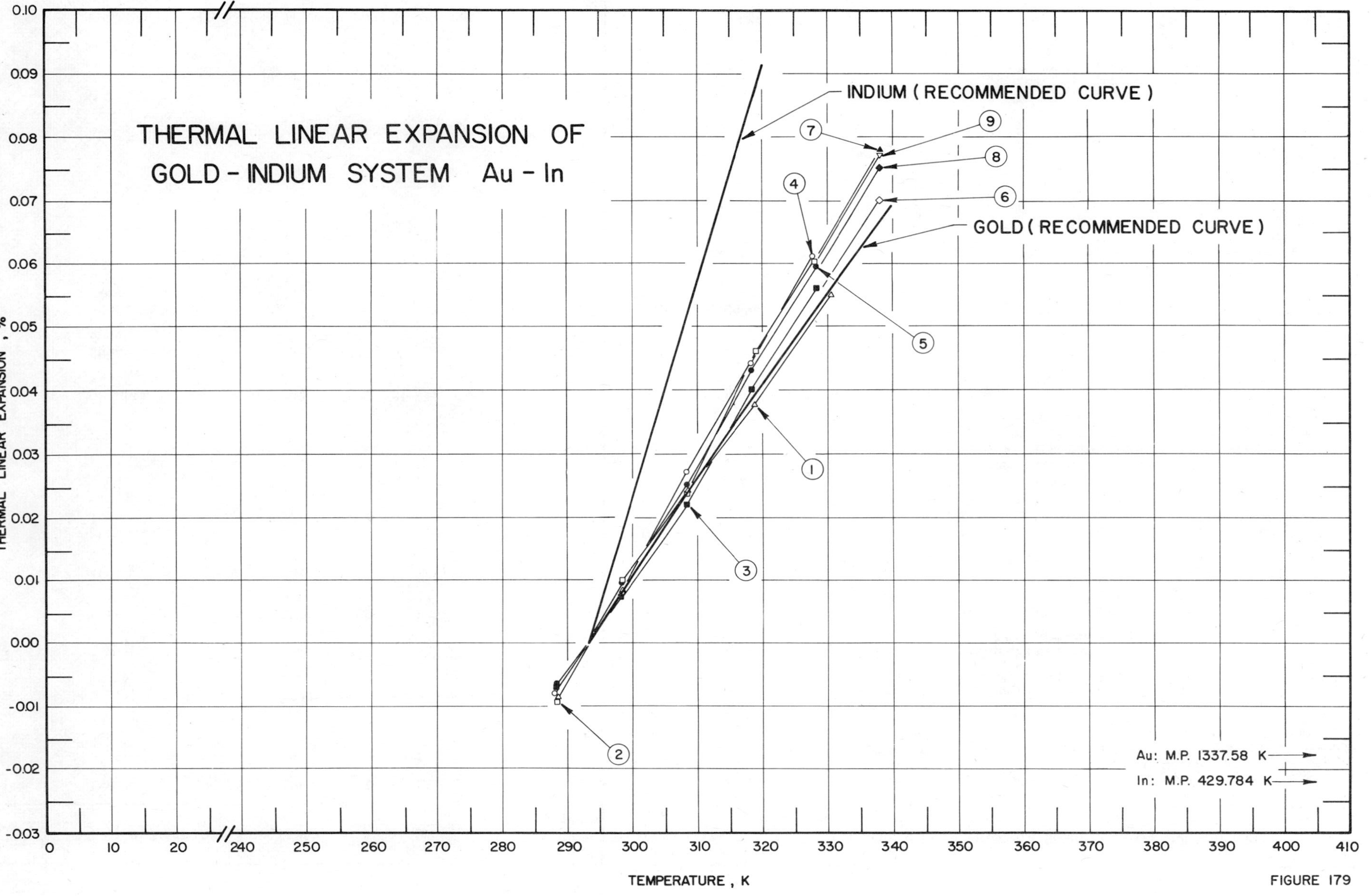
THERMAL LINEAR EXPANSION OF
GOLD - INDIUM SYSTEM Au - In
INDIUM (RECOMMENDED CURVE)
GOLD (RECOMMENDED CURVE)
1
2
3
4
5
6
7
8
9
Au: M.P. 1337.58 K
In: M.P. 429.784 K
THERMAL LINEAR EXPANSION, %
TEMPERATURE, K
0.10
0.09
0.08
0.07
0.06
0.05
0.04
0.03
0.02
0.01
0.00
-0.01
-0.02
-0.03
0
10
20
240
250
260
270
280
290
300
310
320
330
340
350
360
370
380
390
400
410

FIGURE 179

SPECIFICATION TABLE 179. THERMAL LINEAR EXPANSION OF GOLD-INDIUM SYSTEM Au-In

Cur. No.	Ref. No.	Author(s)	Year	Method Used	Temp. Range, K	Name and Specimen Designation	Composition (weight percent) Au	In	Composition (continued), Specifications, and Remarks
1	185	Patel, V.K.	1967	X	288-330		98.8	1.2	Alloy obtained as ingot, homogenized 110 hr at 690 ± 5 C; quenched in ice water; powder specimen filed from ingot and annealed 5 hr at 963 K; expansion measured using $CoK_{\alpha 1}$ x-radiation; lattice parameter reported at 287.6 K is 4.08208 Å; 4.08242 Å at 293 K determined by graphical interpolation.
2	185	Patel, V.K.	1967	X	288-328		97.6	2.4	Similar to the above specimen; lattice parameter reported at 288.3 K is 4.08835 Å; 4.08872 Å at 293 K determined hy graphical interpolation.
3	185	Patel, V.K.	1967	X	288-328		96.0	4.0	Similar to the above specimen; lattice parameter reported at 288.2 K is 4.09630 Å; 4.09658 Å at 293 K determined by graphical interpolation.
4	185	Patel, V.K.	1967	X	288-328		93.9	6.1	Similar to the above specimen; lattice parameter reported at 288.0 K is 4.10590 Å; 4.10622 Å at 293 K determined by graphical interpolation.
5	185	Patel, V.K.	1967	X	288-328		92.6	7.4	Similar to the above specimen; lattice parameter reported at 288.3 K is 4.11248 Å; 4.11284 Å at 293 K determined by graphical interpolation.
6	767	Straumanis, M.E. and Patel, V.K.	1972	X	288-338		Bal.	2.00	Alloy made from 99.999 Au and 99.999 In; impurities in Au were Cu, Mg, Fe, Si, Pb, and Ag (from 0.0001 to 0.0005); impurities in In were Cd, Fe, Pd, and Sn (about 0.0001); samples supplied by American Smelting and Refining Co.; calculated amounts of Au and In were weighed and placed in fuzed quartz tubes, tubes then evacuated and filled with 1/2 atm. He pressure; tubes with samples were heated to 1423 K to liquid, then mixed, quenched in ice water, and homogenized at 963 ± 5 K for 110 hr and quenched again.
7	767	Straumanis, M.E. and Patel, V.K.	1972	X	288-338		Bal.	4.00	Similar to the above specimen.
8	767	Straumanis, M.E. and Patel, V.K.	1972	X	288-338		Bal.	6.65	Similar to the above specimen.
9	767	Straumanis, M.E. and Patel, V.K.	1972	X	288-338		Bal.	10.00	Similar to the above specimen.

DATA TABLE 179. THERMAL LINEAR EXPANSION OF GOLD-INDIUM SYSTEM Au-In

[Temperature, T, K; Linear Expansion, $\Delta L/L_0$, %]

T	$\Delta L/L_0$
CURVE 1	
287.6	-0.0084
298.3	0.0083
308.3	0.0242
318.8	0.0379
329.6	0.0550
CURVE 2	
288.3	-0.0091
298.2	0.0100
308.2	0.0240
319.0	0.0462
328.0	0.0604
CURVE 3	
288.2	-0.0069
298.2	0.0075
308.2	0.0221
318.2	0.0402
328.2	0.0563
CURVE 4	
288.0	-0.0078
298.2	0.0081
308.1	0.0273
318.2	0.0444
327.9	0.0614
CURVE 5	
288.3	-0.0089
298.2	0.0098
308.2	0.0252
318.2	0.0431
328.2	0.0597
CURVE 6	
288	-0.008*
338	0.070
CURVE 7	
288	-0.009*
338	0.078

T	$\Delta L/L_0$
CURVE 8	
288	-0.008*
338	0.075
CURVE 9	
288	-0.009*
338	0.077

* Not shown in figure.

FIGURE AND TABLE NO. 180R. PROVISIONAL VALUES FOR THERMAL LINEAR EXPANSION OF GOLD-PALLADIUM SYSTEM Au-Pd

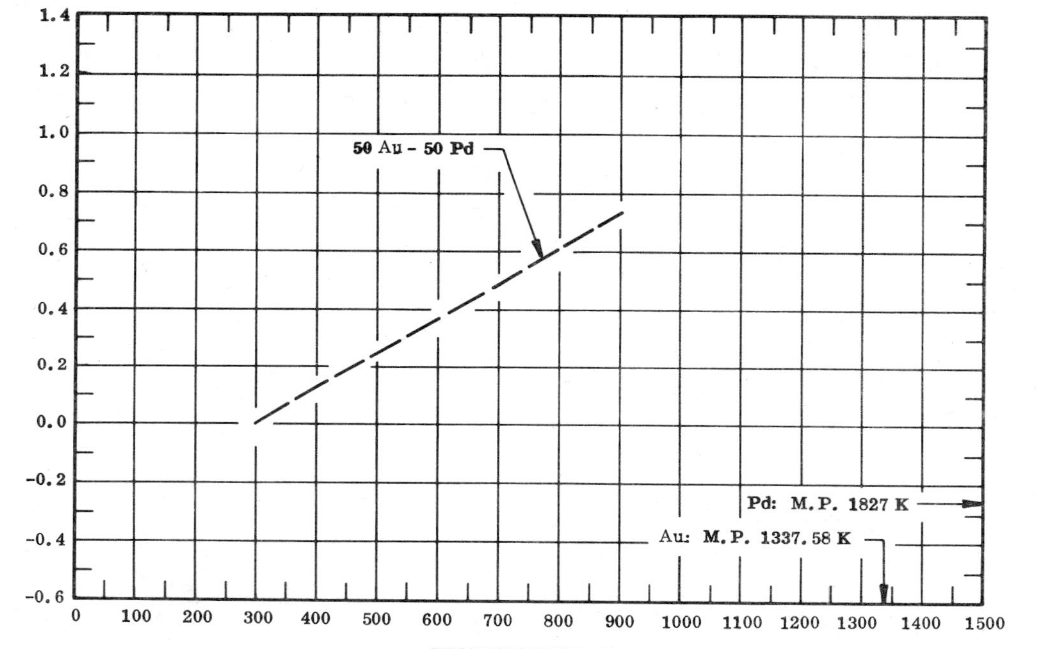

PROVISIONAL VALUES

[Temperature, T, K; Linear Expansion, $\Delta L/L_0$, %; α, K^{-1}]

T	(50 Au-50 Pd) $\Delta L/L_0$	$\alpha \times 10^6$
293	0.000	12
400	0.129	12
500	0.249	12
600	0.369	12
700	0.490	12
800	0.610	12
900	0.730	12

REMARKS

(50 Au-50 Pd): The tabulated values for well-annealed alloy are considered accurate to within ±7% over the entire temperature range. These values can be represented approximately by the following equation:

$$\Delta L/L_0 = -0.346 + 1.168 \times 10^{-3} T + 6.200 \times 10^{-8} T^2 - 3.403 \times 10^{-11} T^3$$

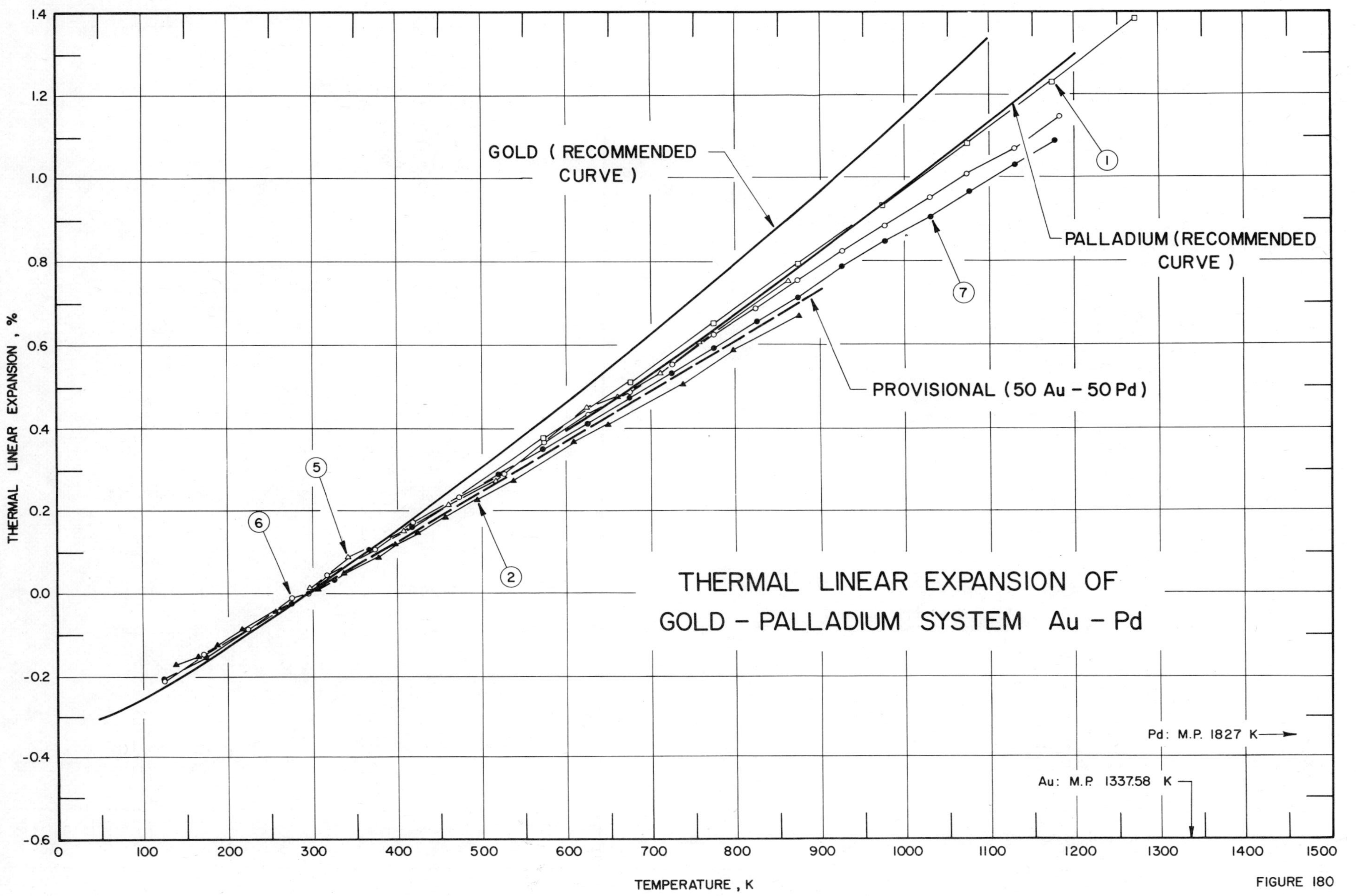

THERMAL LINEAR EXPANSION OF
GOLD - PALLADIUM SYSTEM Au - Pd
GOLD (RECOMMENDED CURVE)
PALLADIUM (RECOMMENDED CURVE)
PROVISIONAL (50 Au - 50 Pd)
1
2
5
6
7
Pd: M.P. 1827 K
Au: M.P. 1337.58 K
THERMAL LINEAR EXPANSION , %
1.4
1.2
1.0
0.8
0.6
0.4
0.2
0.0
-0.2
-0.4
-0.6
0
100
200
300
400
500
600
700
800
900
1000
1100
1200
1300
1400
1500
TEMPERATURE , K

FIGURE 180

SPECIFICATION TABLE 180. THERMAL LINEAR EXPANSION OF GOLD-PALLADIUM SYSTEM Au-Pd

Cur. No.	Ref. No.	Author(s)	Year	Method Used	Temp. Range, K	Name and Specimen Designation	Composition (weight percent) Au	Pd	Composition (continued), Specifications, and Remarks
1	416	Spedding, A. L.	1964	C	273-1273	Palau	80.0	20.0	No details given.
2	417	Masumoto, H., Saito, H., and Kadowaki, S.	1969	L	137-873	No. 1 Pallagold	50	50	Commercial 99.91 Pd and 99.9$^+$ Au; button-melted in argon atmosphere with arc furnace; finished to wire of 2 mm diameter by hot forging, cold swaging and cold drawing; wire cut into pieces, put in alumina crucible, 3 mm inner diameter; melted in vacuum; rod-form single crystal of alloy about 2 mm diameter and 2-10 cm long prepared by Tammann-Bridgman method; heated to 1270 K and cooled at 240 K per hr; expansion measured in <100> direction; zero-point correction is -0.249%.
3*	417	Masumoto, H., et al.	1969	L	130-861	No. 4 Pallagold	50	50	The above specimen; expansion measured in <110> direction; zero-point correction is -0.113%.
4*	417	Masumoto, H., et al.	1969	L	142-871	No. 12 Pallagold	50	50	The above specimen; expansion measured in <111> direction; zero-point correction is 0.027%.
5	601	Devi, U., Rao, C.N., and Rao, K.K.	1965	X	293-862		71.2	28.8	Specimen prepared from pure gold and 99.99 palladium from Johnson and Matthey Co. and W. C. Heraeus and Co., West Germany, respectively were melted together; homogenized for 8 days at 1173 K.
6	271	Masumoto, H. and Sawaya, S.	1969	L	125-1184	Pallagold	62	38	99.9 Pd and 99.9 Au; mixed and melted; rod specimen formed 2 mm diameter, 100 mm long, annealed polycrystalline sample.
7	271	Masumoto, H. and Sawaya, S.	1969	L	124-1177	Pallagold	50	50	Similar to the above specimen.
8*	654	Nagender Naidu, S.V. and Houska, C.R.	1971	X	80-298		61.8	38.2	Prepared from 99.97 Pd and 99.97 Au from Engelhard Industries Inc. by melting in induction furnace under hydrogen atmosphere; homogenized for one week at 1173 K; annealed at 873 K in evacuated quartz tubes after flushing 3 times with argon; lattice parameter reported at 298 K is 3.9377 Å; 3.9376 Å at 293 K determined by graphical interpolation.
9*	654	Nagender Naidu, S.V. and Houska, C.R.	1971	X	80-298		35.1	64.9	Similar to the above specimen; lattice parameter reported at 298 K is 3.9820 Å; 3.9818 Å at 293 K determined by graphical interpolation.
10*	654	Nagender Naidu, S.V. and Houska, C.R.	1971	X	80-298		15.3	84.7	Similar to the above specimen; lattice parameter reported at 298 K is 4.0288 Å; 4.0283 Å at 293 K determined by graphical interpolation.

* Not shown in figure.

DATA TABLE 180. THERMAL LINEAR EXPANSION OF GOLD-PALLADIUM SYSTEM Au-Pd

[Temperature, T, K; Linear Expansion, $\Delta L/L_0$, %]

T	$\Delta L/L_0$
CURVE 1‡	
273	-0.026*
373	0.104*
473	0.237*
573	0.373
673	0.510
773	0.650
873	0.792
973	0.937
1073	1.084
1173	1.234
1273	1.386
CURVE 2	
137	-0.174
163	-0.150
187	-0.121
219	-0.084
255	-0.049
293	0.000*
305	0.014
337	0.047
377	0.082
395	0.120
423	0.151
458	0.189
492	0.225
511	0.246*
539	0.277
577	0.328*
609	0.365
649	0.409
686	0.453*
736	0.504
764	0.547*
798	0.582
834	0.624*
873	0.663
CURVE 3*	
130	-0.183
179	-0.130
214	-0.093
267	-0.034
293	0.000
CURVE 3 (cont.)	
327	0.042
356	0.070
376	0.094
408	0.130
443	0.170
455	0.182
484	0.212
505	0.246
528	0.275
561	0.305
583	0.331
634	0.389
673	0.430
702	0.462
731	0.494
770	0.536
793	0.561
813	0.587
861	0.645
CURVE 4*	
142	-0.174
163	-0.147
197	-0.110
241	-0.060
273	-0.022
293	0.000
316	0.027
375	0.095
412	0.135
454	0.185
485	0.221
524	0.266
550	0.304
590	0.340
620	0.383
656	0.425
701	0.476
728	0.504
771	0.543
793	0.573
826	0.613
871	0.666
CURVE 5	
293	0.000*
296	0.013
340	0.085
408	0.155
415	0.175*
460	0.215
517	0.275
533	0.328*
570	0.350*
624	0.445
660	0.473
676	0.488*
693	0.513*
710	0.533
759	0.605
776	0.625*
822	0.698*
862	0.750
CURVE 6	
125	-0.213
170	-0.147
223	-0.086
274	-0.017
293	0.000
316	0.041
371	0.105
419	0.172
473	0.236
526	0.296
571	0.368
625	0.431
673	0.490
725	0.557
774	0.622
823	0.687
873	0.754
927	0.821
979	0.884
1030	0.951
1073	1.011
1130	1.077
1184	1.146
CURVE 7	
124	-0.205
174	-0.146
223	-0.082*
274	-0.024
293	0.000*
321	0.038
366	0.103
419	0.162
474	0.225
520	0.289
573	0.349
624	0.412
673	0.471
723	0.533
773	0.595
827	0.658
874	0.715
926	0.781
979	0.842
1030	0.903
1077	0.964
1130	1.035
1177	1.094
CURVE 8*	
80	-0.236
195	-0.038
298	0.003
CURVE 9*	
80	-0.244
195	-0.120
298	0.005
CURVE 10*	
80	-0.256
195	-0.174
298	0.012

* Not shown in figure.
‡ Author's data for coefficient of thermal expansion have been integrated by TPRC to obtain $\Delta L/L_0$.

DATA TABLE 180. COEFFICIENT OF THERMAL LINEAR EXPANSION OF GOLD–PALLADIUM SYSTEM Au–Pd

[Temperature, T, K; Coefficient of Expansion, α, 10^{-6} K^{-1}]

T	α
CURVE 1*,‡	
273	13.00
373	13.20
473	13.42
573	13.64
673	13.87
773	14.09
873	14.34
973	14.59
1073	14.84
1173	15.12
1273	15.40

* No figure given.
‡ Author's data for coefficient of thermal expansion have been integrated by TPRC to obtain $\Delta L/L_0$, %.

FIGURE AND TABLE NO. 181R. PROVISIONAL VALUES FOR THERMAL LINEAR EXPANSION OF GOLD-PLATINUM SYSTEM Au-Pt

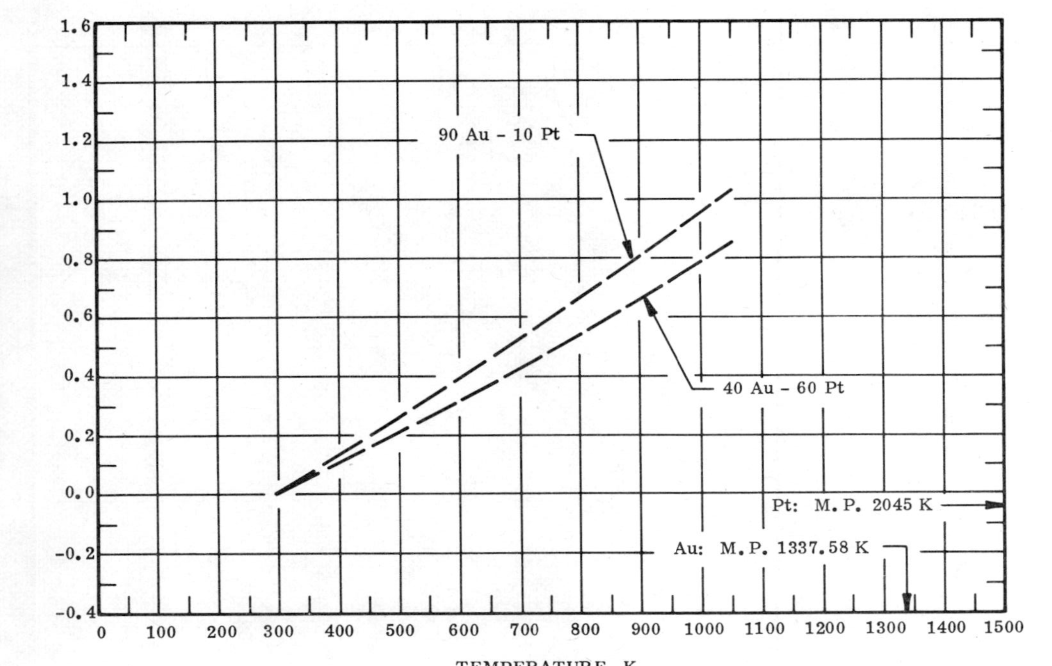

PROVISIONAL VALUES

[Temperature, T, K; Linear Expansion, $\Delta L/L_0$, %; α, K^{-1}]

T	(40 Au-60 Pt) $\Delta L/L_0$	(40 Au-60 Pt) $\alpha \times 10^6$	(90 Au-10 Pt) $\Delta L/L_0$	(90 Au-10 Pt) $\alpha \times 10^6$
293	0.000	9.3	0.000	11.8
400	0.102	10.0	0.130	12.7
500	0.206	10.8	0.261	13.4
600	0.317	11.3	0.398	14.0
700	0.432	11.7	0.538	14.1
800	0.550	11.9	0.678	14.1
900	0.671	12.2	0.818	14.1
1000	0.794	12.5	0.959	14.1
1050	0.857	12.6	1.029	14.1

REMARKS

(40 Au-60 Pt): The tabulated values for well-annealed alloy are considered accurate to within ±7% over the entire temperature range. These values can be represented approximately by the following equation:

$$\Delta L/L_0 = -0.235 + 6.522 \times 10^{-4} T + 5.448 \times 10^{-7} T^2 - 1.674 \times 10^{-10} T^3$$

(90 Au-10 Pt): The tabulated values for well-annealed alloy are considered accurate to within ±7% over the entire temperature range. These values can be represented approximately by the following equation:

$$\Delta L/L_0 = 0.301 + 8.408 \times 10^{-4} T + 7.165 \times 10^{-7} T^2 - 2.971 \times 10^{-10} T^3$$

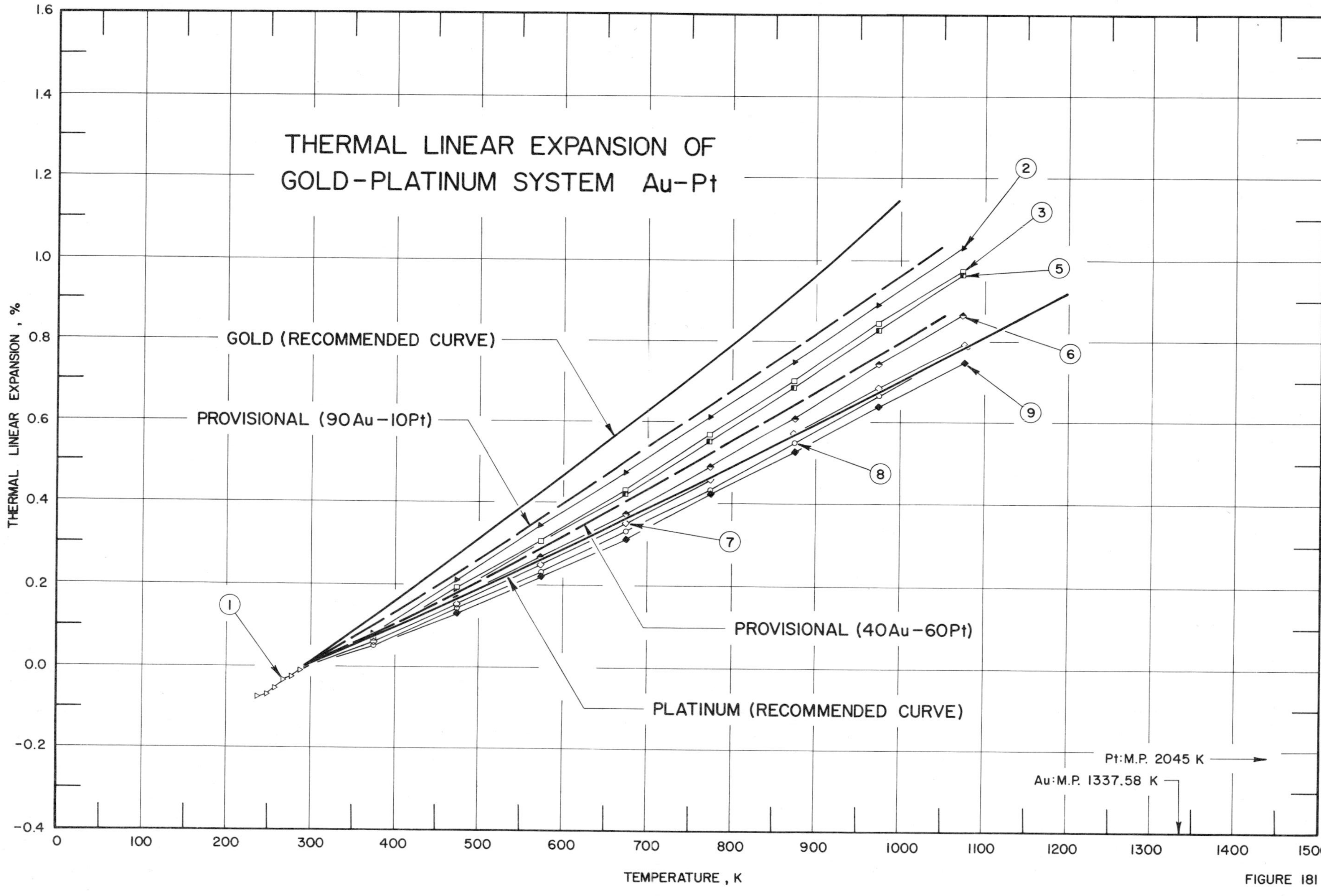
THERMAL LINEAR EXPANSION OF
GOLD-PLATINUM SYSTEM Au-Pt
GOLD (RECOMMENDED CURVE)
PROVISIONAL (90Au-10Pt)
PROVISIONAL (40Au-60Pt)
PLATINUM (RECOMMENDED CURVE)
Pt:M.P. 2045 K
Au:M.P. 1337.58 K
1
2
3
5
6
7
8
9
THERMAL LINEAR EXPANSION, %
1.6
1.4
1.2
1.0
0.8
0.6
0.4
0.2
0.0
-0.2
-0.4
0
100
200
300
400
500
600
700
800
900
1000
1100
1200
1300
1400
1500
TEMPERATURE, K

FIGURE 181

SPECIFICATION TABLE 181. THERMAL LINEAR EXPANSION OF GOLD-PLATINUM SYSTEM Au-Pt

Cur. No.	Ref. No.	Author(s)	Year	Method Used	Temp. Range, K	Name and Specimen Designation	Composition (weight percent) Au	Pt	Composition (continued), Specifications, and Remarks
1	70	Dorsey, H. G.	1908	I	238-288		99	1	99.998 Au used; melted on charcoal, then in graphite mold; specimen 0.6304 cm long.
2	657	Vest, R. W.	1971	L	293-1073		90	10	Specimen prepared from 99.9 pure metals of -352 mesh; data on the coefficient of thermal linear expansion are also reported.
3	657	Vest, R. W.	1971	L	293-1073		80	20	Similar to the above specimen.
4*	657	Vest, R. W.	1971	L	293-1072		75	25	Similar to the above specimen.
5	657	Vest, R. W.	1971	L	293-1072		71	29	Similar to the above specimen.
6	657	Vest, R. W.	1971	L	293-1071		40	60	Similar to the above specimen.
7	657	Vest, R. W.	1971	L	293-1078		30	70	Similar to the above specimen.
8	657	Vest, R. W.	1971	L	293-1080		20	80	Similar to the above specimen.
9	657	Vest, R. W.	1971	L	293-1079		6	94	Similar to the above specimen.

* Not shown in figure.

DATA TABLE 181. THERMAL LINEAR EXPANSION OF GOLD–PLATINUM SYSTEM Au–Pt

[Temperature, T, K; Linear Expansion, $\Delta L/L_0$, %]

T	$\Delta L/L_0$
CURVE 1‡	
238	-0.079
248	-0.065
258	-0.051
268	-0.036
278	-0.022
288	-0.007
CURVE 2	
293	0.000
373	0.080
473	0.214
573	0.344
673	0.473
773	0.611
873	0.749
973	0.892
1073	1.033
CURVE 3	
293	0.000*
373	0.079
473	0.191
573	0.308
673	0.427
773	0.565
873	0.702
973	0.841
1073	0.979
CURVE 4*	
293	0.000
373	0.070
473	0.195
573	0.312
673	0.435
773	0.571
CURVE 4 (cont.)	
873	0.710
973	0.856
1072	0.997
CURVE 5	
293	0.000*
373	0.078*
473	0.189
573	0.302*
673	0.422
773	0.551
873	0.685
973	0.825
1072	0.963
CURVE 6	
293	0.000*
373	0.069*
473	0.165
573	0.269
673	0.373
773	0.492
873	0.613
973	0.742
1071	0.869
CURVE 7	
293	0.000*
373	0.061
473	0.155
573	0.253
673	0.353
773	0.460
CURVE 7 (cont.)	
873	0.570
973	0.685
1078	0.799
CURVE 8	
293	0.000*
373	0.059
473	0.144
573	0.234
673	0.330
773	0.436
873	0.548
973	0.668
1080	0.790
CURVE 9	
293	0.000*
373	0.057*
473	0.135
573	0.223
673	0.317
773	0.423
873	0.527
973	0.640
1079	0.752

DATA TABLE 181. COEFFICIENT OF THERMAL LINEAR EXPANSION OF GOLD–PLATINUM SYSTEM Au–Pt

[Temperature, T, K; Coefficient of Expansion, α, 10^{-6} K^{-1}]

T	α
CURVE 1*, ‡	
238	14.2
248	14.2
258	14.3
268	14.8
278	14.3
288	15.1
CURVE 2*	
373	12.01
473	12.39
573	12.79
673	13.17
773	13.56
873	13.94
973	14.34
1073	14.72
CURVE 3*	
373	10.84
473	11.39
573	11.93
673	12.46
773	13.02
873	13.55
973	14.07
1073	14.61
CURVE 4*	
373	10.83
473	11.43
573	12.03
673	12.65
773	13.25
873	13.86
CURVE 4 (cont.)*	
973	14.45
1073	15.04
CURVE 5*	
373	10.68
473	11.21
573	11.74
673	12.28
773	12.83
873	13.36
973	13.91
1073	14.44
CURVE 6*	
373	9.30
CURVE 6 (cont.)*	
473	9.92
573	10.50
673	11.06
773	11.64
873	12.23
973	13.81
1073	13.37
CURVE 7*	
373	8.73
473	9.22
573	9.71
673	10.16
773	10.68
873	11.15
973	11.63
CURVE 7 (cont.)*	
1073	12.10
CURVE 8*	
373	8.10
473	8.75
573	9.42
673	10.06
773	10.76
873	11.40
973	12.06
1073	12.71
CURVE 9*	
373	7.96
473	8.51
CURVE 9 (cont.)*	
573	9.05
673	9.59
773	10.11
873	10.65
973	11.20
1073	11.73

* Not shown in figure.
‡ Author's data for coefficient of thermal expansion have been integrated by TPRC to obtain $\Delta L/L_0$.

SPECIFICATION TABLE 182. THERMAL LINEAR EXPANSION OF GOLD-SILVER SYSTEM Au-Ag

Cur. No.	Ref. No.	Author(s)	Year	Method Used	Temp. Range, K	Name and Specimen Designation	Composition (weight percent) Au	Ag	Composition (continued), Specifications, and Remarks
1*	654	Nagender Naidu, S.V. and Housha, C.R.	1971	X	80-298		64.6	35.4	Prepared from 99.97 Au and 99.99 Ag from Englehard Industries, Inc., by melting in induction furnace under hydrogen atmosphere; homogenized for 1 wk at 1173 K; annealed at 873 K in evaucated quartz tubes after flushing 3 times with argon; lattice parameter reported at 298 K is 4.0752 Å; 4.0749 Å at 293 K determined by graphical interpolation.

DATA TABLE 182. THERMAL LINEAR EXPANSION OF GOLD-SILVER SYSTEM Au-Ag

[Temperature, T, K; Linear Expansion, $\Delta L/L_0$, %]

T	$\Delta L/L_0$
CURVE 1*	
80	-0.282
195	-0.120
298	0.007*

* No figure given.

FIGURE AND TABLE NO. 183R. PROVISIONAL VALUES FOR THERMAL LINEAR EXPANSION OF HAFNIUM-ZIRCONIUM SYSTEM Hf-Zr

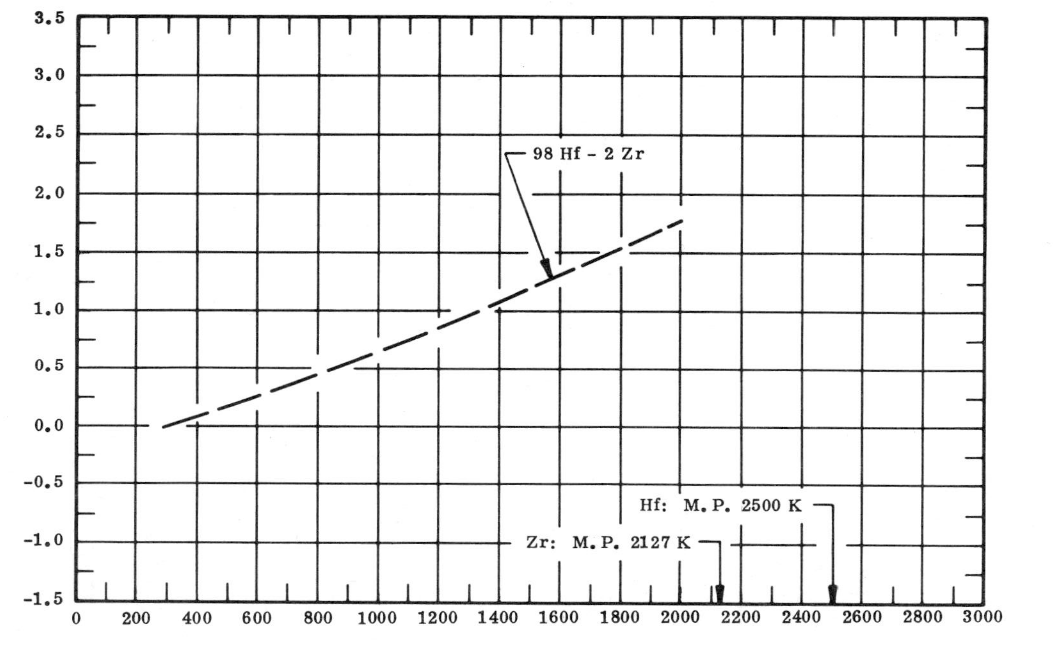

PROVISIONAL VALUES

[Temperature, T, K; Linear Expansion, $\Delta L/L_0$, %; α, K^{-1}]

	(98 Hf-2 Zr)	
T	$\Delta L/L_0$	$\alpha \times 10^6$
293	0.000	7.9
400	0.086	8.3
500	0.171	8.6
600	0.260	9.1
700	0.352	9.4
800	0.448	9.7
900	0.548	10.0
1000	0.650	10.3
1200	0.861	10.8
1400	1.081	11.3
1600	1.308	11.4
1800	1.538	11.5
2000	1.767	11.5

REMARKS

(98 Hf-2 Zr): The tabulated values for well-annealed alloy are considered accurate to within ± 7% over the entire temperature range. These values can be represented approximately by the following equation:

$$\Delta L/L_0 = -0.209 + 6.373 \times 10^{-4}\, T + 2.670 \times 10^{-7}\, T^2 - 4.565 \times 10^{-11}\, T^3$$

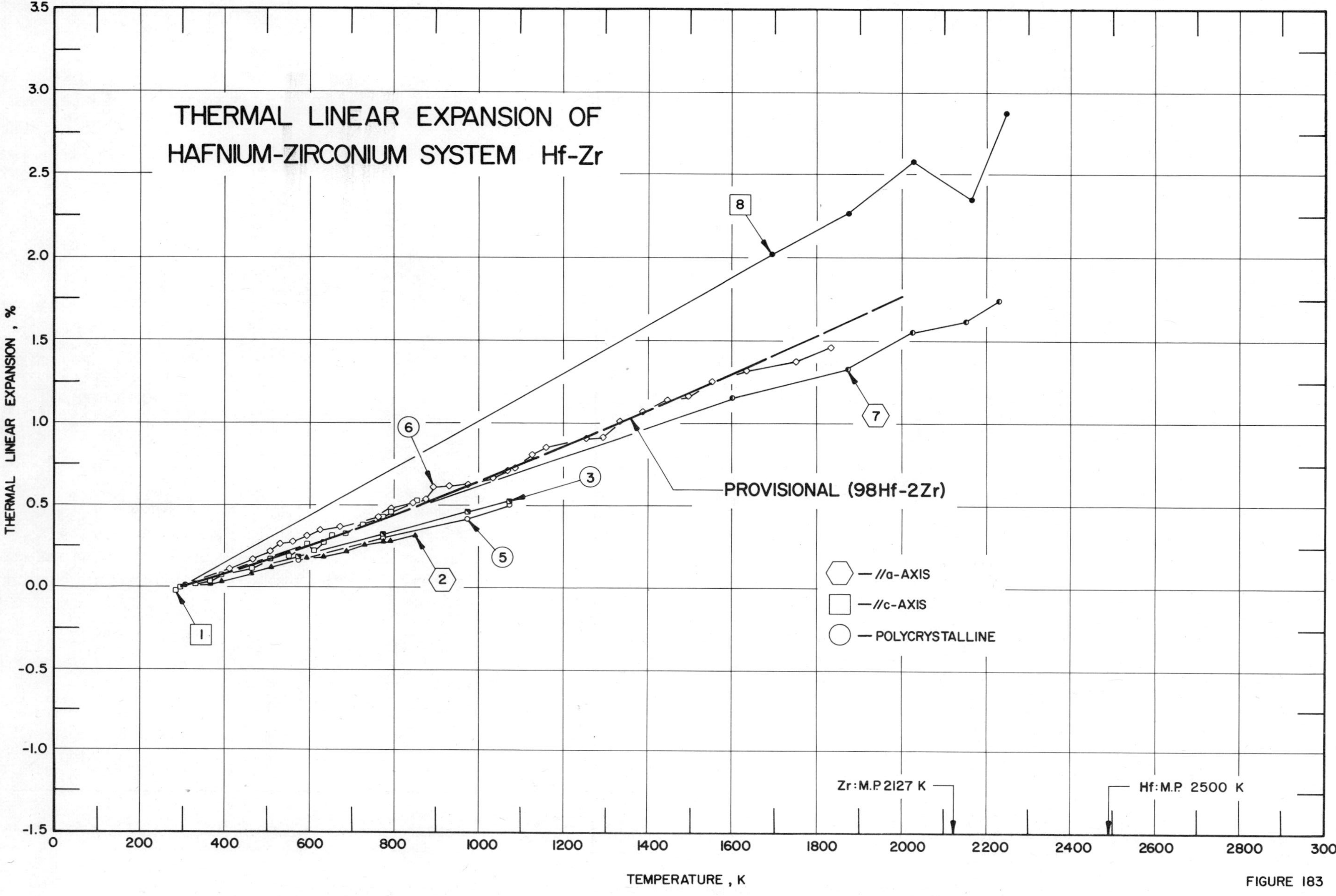
THERMAL LINEAR EXPANSION OF
HAFNIUM-ZIRCONIUM SYSTEM Hf-Zr
PROVISIONAL (98Hf-2Zr)
—//a-AXIS
—//c-AXIS
—POLYCRYSTALLINE
Zr: M.P. 2127 K
Hf: M.P. 2500 K
1
2
3
5
6
7
8
THERMAL LINEAR EXPANSION, %
3.5
3.0
2.5
2.0
1.5
1.0
0.5
0.0
-0.5
-1.0
-1.5
TEMPERATURE, K
0
200
400
600
800
1000
1200
1400
1600
1800
2000
2200
2400
2600
2800
3000
FIGURE 183

SPECIFICATION TABLE 183. THERMAL LINEAR EXPANSION OF HAFNIUM-ZIRCONIUM SYSTEM Hf-Zr

Cur. No.	Ref. No.	Author(s)	Year	Method Used	Temp. Range, K	Name and Specimen Designation	Composition (weight percent) Hf	Zr	Composition (continued), Specifications, and Remarks
1	387	Russell, R. B.	1954	X	281-856		2.32	97.68	Crystal bar machined to 0.471 in. diameter; dehydrogenated at 1673 K for 1 hr in 10^{-6} mm Hg vacuum; swaged to 0.189 in. diameter; chemical analysis taken, and then rolled to 0.012 in. thick sheet; cut into strip about 1 x 0.4 in.; prick-punching with Burgess "Vibro-Tool" before annealing at 773 K in packet in quartz tube sealed off at pressure of 5 x 10^{-6} mm Hg; expansion measured along c-axis; lattice parameter reported at 281.2 K is 5.14494 Å; 5.14545 Å at 293 K determined by graphical interpolation.
2	387	Russell, R. B.	1954	X	281-856		2.32	97.68	The above specimen; expansion measured along a-axis; lattice parameter reported at 281.2 K is 3.23054 Å; 3.23089 Å at 293 K determined by graphical interpolation.
3	338	Pridantseva, K. S.	1964		293-1073	Alloy No. 6	0.65	98.35	Iodide zirconium remelted in arc furnace.
4*	338	Pridantseva, K. S.	1964		293-1073	Alloy No. 8	1.1	98.9	Similar to the above specimen.
5	338	Pridantseva, K. S.	1964		293-1073	Alloy No. 10	1.1	98.8	0.1 Si; similar to the above specimen.
6	358	Fieldhouse, I. B. and Lang, J. I.	1961		293-1833		99	1 max.	0.1 max. each Ti, Si, 0.01 max. each Fe, V, Zn, 0.001 max. each Cu, Mn, Ni, 0.0001 max. Mg; density 13.09 g cm^{-3}.
7	187	Ross, R. G. and Hume-Rothery, W.	1963	X	303-2393		98.4	1.6	Specimen from the Magnesium Eliktron Co., Ltd; expansion measured along a-axis; lattice parameter reported at 303 K is 3.200 Å; 3.1997 Å at 293 K estimated.
8	187	Ross, R. G. and Hume-Rothery, W.	1963	X	303-2243		98.4	1.6	The above specimen; expansion measured along c-axis; lattice parameter reported at 303 K is 5.059 Å; 5.058 Å at 293 K estimated.

*Not shown in figure.

DATA TABLE 183. THERMAL LINEAR EXPANSION OF HAFNIUM-ZIRCONIUM SYSTEM Hf-Zr

[Temperature, T, K; Linear Expansion, $\Delta L/L_0$, %]

T	$\Delta L/L_0$
CURVE 1	
281.2	-0.010
296.7	0.003
298.2	0.026*
299.1	0.016*
299.3	0.018*
299.3	-0.010*
299.5	-0.006*
332.2	0.015
363.2	0.043
399.2	0.074
462.7	0.112
505.0	0.170
553.6	0.190
594.2	0.260
610.4	0.234
637.5	0.289
655.0	0.311
681.7	0.330
729.8	0.396
777.0	0.443
790.7	0.465
856.0	0.533
CURVE 2	
281.2	-0.011*
296.7	0.003*
298.2	0.002*
299.1	-0.010*
299.3	-0.002*
299.3	0.024*
299.5	0.006*
332.2	0.004*
363.2	0.023
399.2	0.041
462.7	0.088
505.0	0.130
553.6	0.147*
594.2	0.186
610.4	0.182*
637.5	0.188
655.0	0.206*
681.7	0.235
729.8	0.256
777.0	0.281
790.7	0.292
856.0	0.329

T	$\Delta L/L_0$
CURVE 3	
293	0.000*
573	0.185
773	0.322
973	0.462
1073	0.537
CURVE 4*	
293	0.000
573	0.183
773	0.314
973	0.454
1073	0.536
CURVE 5	
293	0.000*
573	0.175
773	0.303
973	0.433
1073	0.503
CURVE 6	
293	0.000*
415	0.105
465	0.162
469	0.165*
505	0.227
533	0.260
560	0.277
597	0.315
628	0.350
672	0.370
699	0.382*
762	0.440
794	0.487
842	0.505
875	0.545
894	0.612
932	0.622
951	0.625*
979	0.635
1037	0.680
1065	0.705
1088	0.742
1127	0.807

T	$\Delta L/L_0$
CURVE 6 (cont.)	
1160	0.858
1255	0.902
1295	0.927
1333	1.010
1388	1.070
1444	1.150
1499	1.180
1555	1.260
1638	1.340
1749	1.380
1833	1.460
CURVE 7	
303	0.009*
1603	1.156
1873	1.344
2023	1.563
2153	1.625
2233	1.750
	a(Å)
2273	3.609*
2323	3.616*
2393	3.620*
CURVE 8	
	$\Delta L/L_0$
303	0.015*
1693	2.016
1873	2.273
2023	2.589
2163	2.352
2243	2.886

*Not shown in figure.

FIGURE AND TABLE NO. 184BR. PROVISIONAL VALUES FOR THERMAL LINEAR EXPANSION OF INDIUM-SILVER SYSTEM In-Ag

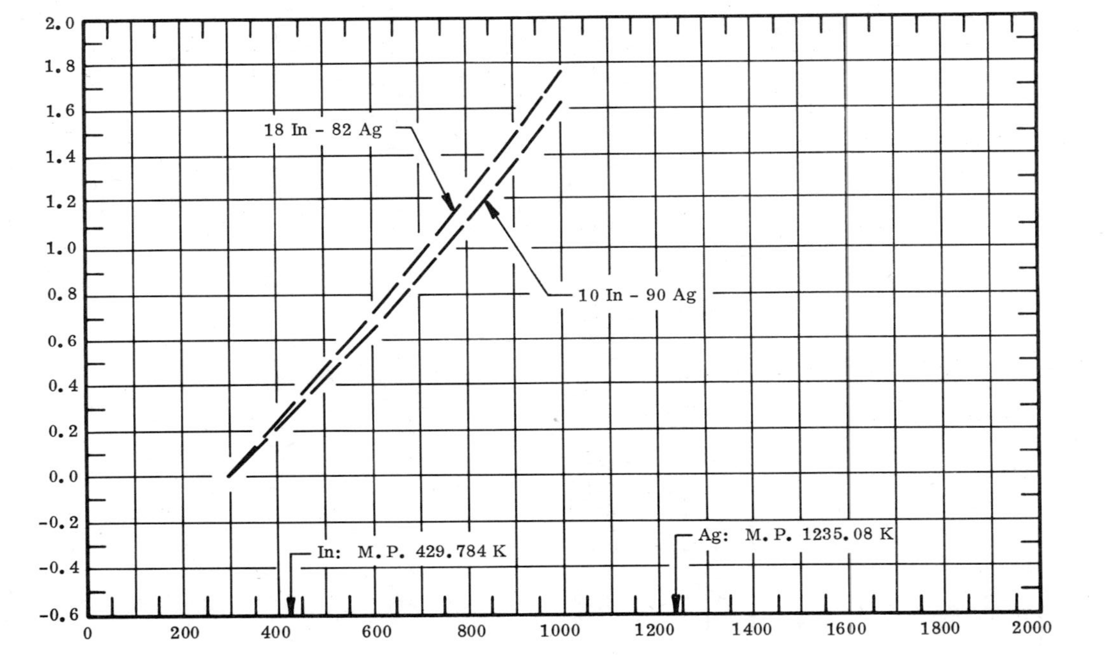

PROVISIONAL VALUES

[Temperature, T, K; Linear Expansion, $\Delta L/L_0$, %; α, K^{-1}]

T	(10 In-90 Ag) $\Delta L/L_0$	(10 In-90 Ag) $\alpha \times 10^6$	(18 In-82 Ag) $\Delta L/L_0$	(18 In-82 Ag) $\alpha \times 10^6$
293	0.000	19.5	0.000	21.5
400	0.216	20.8	0.240	22.8
500	0.429	21.9	0.474	24.0
600	0.652	22.7	0.719	24.9
700	0.882	23.4	0.971	25.6
800	1.120	24.1	1.231	26.2
900	1.366	25.1	1.495	26.7
1000	1.623	26.3	1.765	27.0

REMARKS

(10 In-90 Ag): The tabulated values for well-annealed alloy are considered accurate to within ±7% over the entire temperature range. These values can be represented approximately by the following equation:

$$\Delta L/L_0 = -0.540 + 1.700 \times 10^{-3}\, T + 4.963 \times 10^{-7}\, T^2 - 3.303 \times 10^{-11}\, T^3$$

(18 In-82 Ag): The tabulated values for well-annealed alloy are considered accurate to within ±7% over the entire temperature range. These values can be represented approximately by the following equation:

$$\Delta L/L_0 = -0.572 + 1.721 \times 10^{-3}\, T + 8.610 \times 10^{-7}\, T^2 - 2.478 \times 10^{-10}\, T^3$$

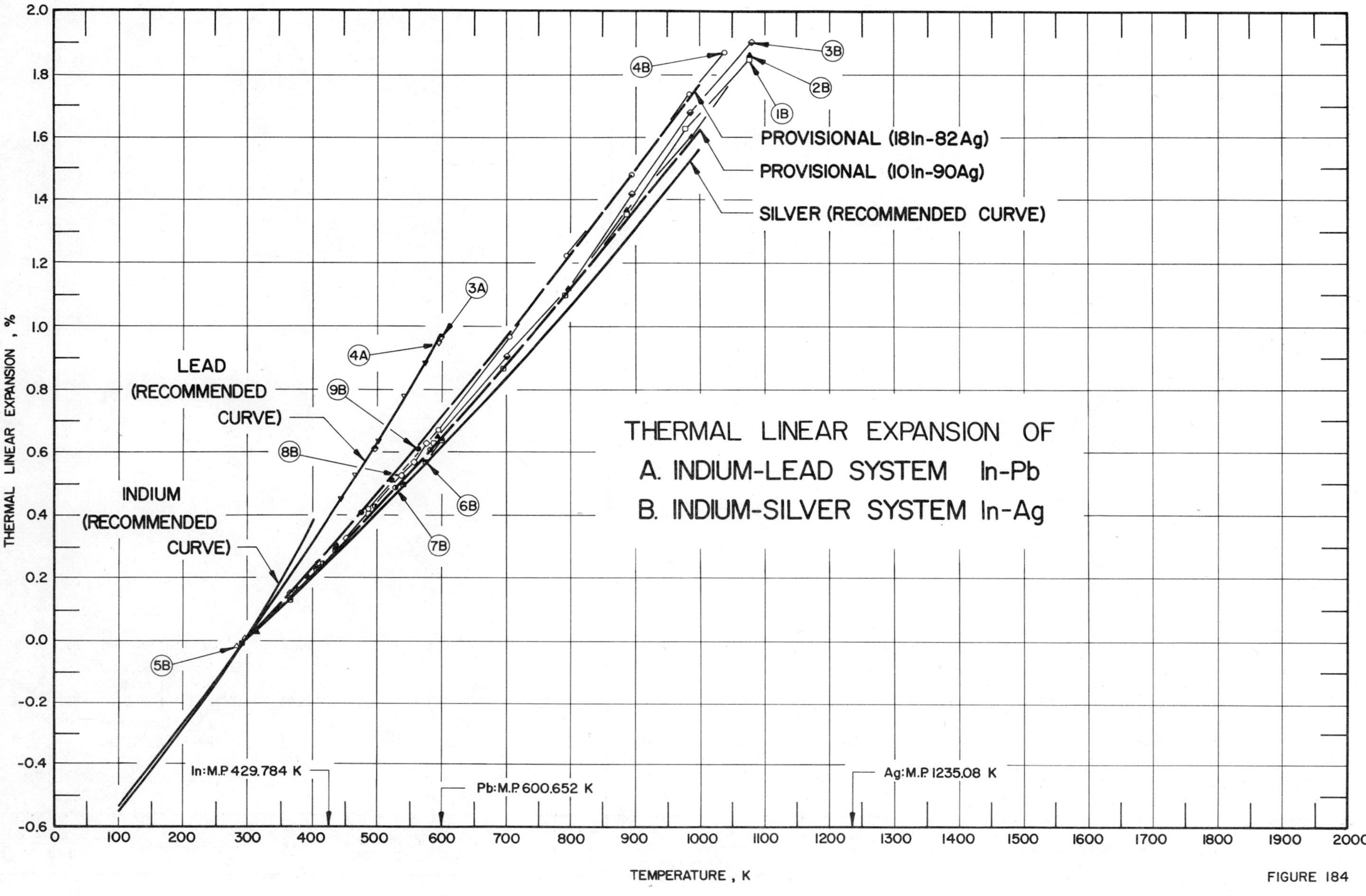
THERMAL LINEAR EXPANSION OF
A. INDIUM-LEAD SYSTEM In-Pb
B. INDIUM-SILVER SYSTEM In-Ag
PROVISIONAL (18In-82Ag)
PROVISIONAL (10In-90Ag)
SILVER (RECOMMENDED CURVE)
LEAD (RECOMMENDED CURVE)
INDIUM (RECOMMENDED CURVE)
In:M.P. 429.784 K
Pb:M.P. 600.652 K
Ag:M.P. 1235.08 K
3B
4B
2B
1B
3A
4A
9B
8B
6B
7B
5B
THERMAL LINEAR EXPANSION, %
2.0
1.8
1.6
1.4
1.2
1.0
0.8
0.6
0.4
0.2
0.0
-0.2
-0.4
-0.6
0
100
200
300
400
500
600
700
800
900
1000
1100
1200
1300
1400
1500
1600
1700
1800
1900
2000
TEMPERATURE, K

FIGURE 184

SPECIFICATION TABLE 184A. THERMAL LINEAR EXPANSION OF INDIUM-LEAD SYSTEM In-Pb

Cur. No.	Ref. No.	Author(s)	Year	Method Used	Temp. Range, K	Name and Specimen Designation	Composition (weight percent) In	Pb	Composition (continued), Specifications, and Remarks
1	193	Smith, J.F. and Schneider, V.L.	1964	X	100-300		99.63	0.36	Dilute solid solution of Pb prepared using 99.9999 In by casting in sealed pyrex container; CuKα x-radiation used; expansion measured perpendicular to unique axis; reported error $\pm 5.0\%$.
2	193	Smith, J.F. and Schneider, V.L.	1964	X	100-300		99.63	0.36	The above specimen; expansion measured parallel to unique axis.
3	742	D' Heurle, F.M., Feder, R., and Nowick, A.S.	1963	X	495-595		0.6	Bal.	Specimen prepared using 99.999 Pb, single crystal sample, expansion measured in argon atmosphere at 2 mm Hg pressure.
4	742	D' Heurle, F.M., et al.	1963	O	444-597		0.6	Bal.	Similar to the above specimen.

DATA TABLE 184A. THERMAL LINEAR EXPANSION OF INDIUM-LEAD SYSTEM In-Pb

[Temperature, T, K; Linear Expansion, $\Delta L/L_0$, %]

T	$\Delta L/L_0$	T	$\Delta L/L_0$
CURVE 1‡		CURVE 4	
100	-0.806*	444	0.456
150	-0.645*	468	0.532
200	-0.456	501	0.641
250	-0.230	542	0.778
300	0.041	573	0.886
		592	0.956
CURVE 2‡		597*	0.974
100	-0.184		
150	-0.086		
200	-0.017		
250	-0.013		
300	-0.006		
CURVE 3			
495	0.618		
596	0.971		

DATA TABLE 184A. COEFFICIENT OF THERMAL LINEAR EXPANSION OF INDIUM-LEAD SYSTEM In-Pb

[Temperature, T, K; Coefficient of Expansion, α, 10^{-6} K^{-1}]

T	α
CURVE 1‡	
100	30.0
150	34.6
200	41.0
250	49.2
300	59.2
CURVE 2‡	
100	22.0
150	17.2
200	10.4
250	1.7
300	-9.0

* Not shown in figure.
‡ Author's data for coefficient of thermal expansion have been integrated by TPRC to obtain $\Delta L/L_0$.

SPECIFICATION TABLE 184B. THERMAL LINEAR EXPANSION OF INDIUM-SILVER SYSTEM In-Ag

Cur. No.	Ref. No.	Author(s)	Year	Method Used	Temp. Range, K	Name and Specimen Designation	Composition (weight percent) In	Ag	Composition (continued), Specifications, and Remarks
1	298	Straumanis, M.E. and Riad, S.M.	1965	X	313-1076		9.0	91.0	99.999+ pure Ag from ASARCO, and In weighed, placed in quartz tubes, sealed, heated to 1273 K and mixed by shaking vigorously 1 hr while melted; tubes quenched, ingots removed, sealed in Pyrex tubes, homogenized 5 hr at 773 K; powder filed from each end; copper Kα x-rays used to observe expansion; lattice parameter expansion measured; lattice parameter reported at 313 K is 4.1151 Å; 4.1134 Å at 293 K determined by graphical extrapolation.
2	298	Straumanis, M.E. and Riad, S.M.	1965	X	311-1079		12.0	88.0	Similar to the above specimen; lattice parameter reported at 311 K is 4.1248 Å; 4.1233 Å at 293 K determined by graphical extrapolation.
3	298	Straumanis, M.E. and Riad, S.M.	1965	X	315-1080		15.0	85.0	Similar to the above specimen; lattice parameter reported at 315 K is 4.1338 Å; 4.1319 Å at 293 K determined by graphical extrapolation.
4	298	Straumanis, M.E. and Riad, S.M.	1965	X	312-1038		18.0	82.0	Similar to the above specimen; lattice parameter reported at 312 K is 4.1443 Å; 4.1426 Å at 293 K determined by graphical extrapolation.
5	298	Straumanis, M.E. and Riad, S.M.	1965	X	286-338		22.0	78.0	Similar to the above specimen; lattice parameter reported at 286.4 K is 4.15035 Å; 4.1509 Å at 293 K determined by graphical interpolation.
6	303	Owen, E.A. and Roberts, E.W.	1939	X	291-599		7.52	92.48	99.95 silver and 99.98 indium used as starting materials; melted together; lattice parameter reported at 291 K is 4.0996 Å; 4.0997 Å at 293 K determined by graphical interpolation.
7	303	Owen, E.A. and Roberts, E.W.	1939	X	291-595		13.02	86.98	Similar to the above specimen; lattice parameter reported at 291 K is 4.1171 Å; 4.1173 Å at 293 K determined by graphical interpolation.
8	303	Owen, E.A. and Roberts, E.W.	1939	X	291-575		15.96	84.04	Similar to the above specimen; lattice parameter reported at 291 K is 4.1265 Å; 4.1267 Å at 293 K determined by graphical interpolation.
9	303	Owen, E.A. and Roberts, E.W.	1939	X	291-562		19.12	80.88	Similar to the above specimen; lattice parameter reported at 291 K is 4.1377 Å; 4.1379 Å at 293 K determined by graphical interpolation.

DATA TABLE 184B. THERMAL LINEAR EXPANSION OF INDIUM-SILVER SYSTEM In-Ag

[Temperature, T, K; Linear Expansion, $\Delta L/L_0$, %]

T	$\Delta L/L_0$
CURVE 1	
313	0.041*
489	0.401
589	0.630
694	0.869
790	1.10
885	1.36
978	1.63
1076	1.85
CURVE 2	
311	0.037
492	0.406*
593	0.648
695	0.864*
795	1.12
887	1.37
987	1.60
1079	1.86
CURVE 3	
315	0.046*
489	0.409*
596	0.649*
700	0.905
794	1.17*
892	1.42
983	1.68
1080	1.91
CURVE 4	
312	0.041*
488	0.420
594	0.674
702	0.963
793	1.22
891	1.48
982	1.74
1038	1.87
CURVE 5	
286.4	-0.014
298.3	0.012
323.2	0.063*
338.3	0.097*
CURVE 6	
291	-0.0019
361	0.133
408	0.227
437	0.293
487	0.392*
540.7	0.500
572.7	0.571
599	0.643
CURVE 7	
291	-0.0055*
374	0.164
417	0.247
473	0.360*
499	0.425
529	0.484
580	0.609
595	0.652*
CURVE 8	
291	-0.0043*
365.7	0.155
412.7	0.245
452.7	0.327
539.7	0.527
557.7	0.577
575	0.632
CURVE 9	
291	-0.0041*
359	0.136*
390.7	0.203
437	0.301
477	0.402
521	0.510
562	0.615

* Not shown in figure.

FIGURE AND TABLE NO. 185R. PROVISIONAL VALUES FOR THERMAL LINEAR EXPANSION OF INDIUM-THALLIUM SYSTEM In-Tl

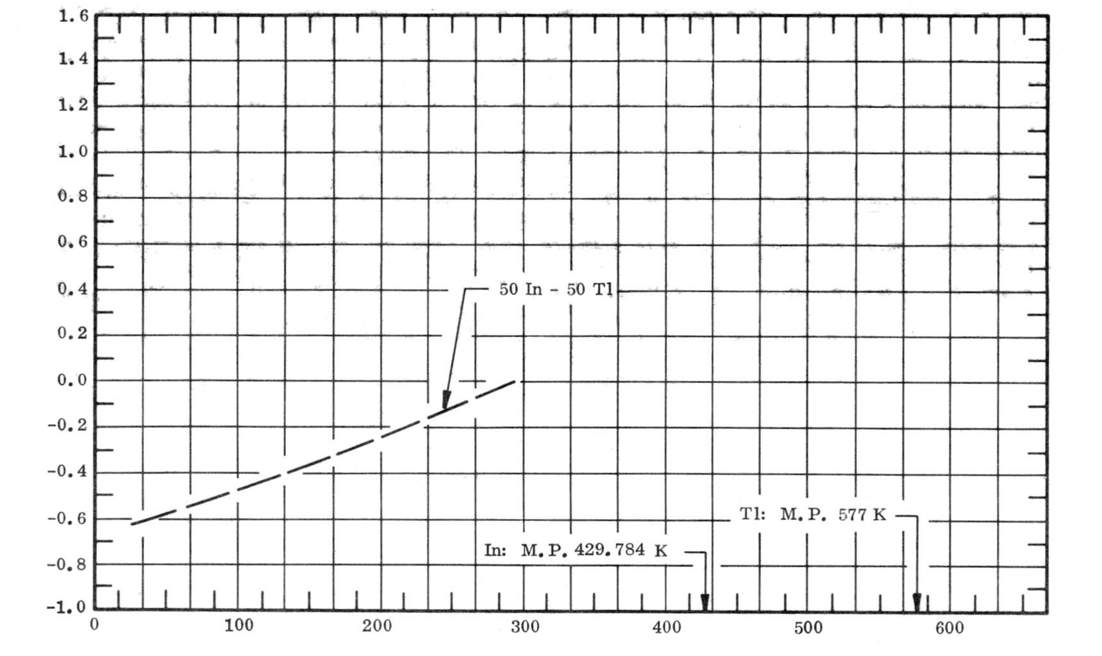

PROVISIONAL VALUES

[Temperature, T, K; Linear Expansion, $\Delta L/L_0$, %; α, K^{-1}]

T	(50 In-50 Tl) $\Delta L/L_0$	$\alpha \times 10^6$
25	-0.621	18.4
50	-0.573	19.7
100	-0.469	21.7
150	-0.356	23.3
200	-0.237	24.5
293	0.000	26.6

REMARKS

(50 In-50 Tl): The tabulated values for well-annealed alloy are considered accurate to within ±7% over the entire temperature range. These values can be represented approximately by the following equation:

$$\Delta L/L_0 = -0.665 + 1.727 \times 10^{-3}\, T + 2.543 \times 10^{-6}\, T^2 - 2.342 \times 10^{-9}\, T^3$$

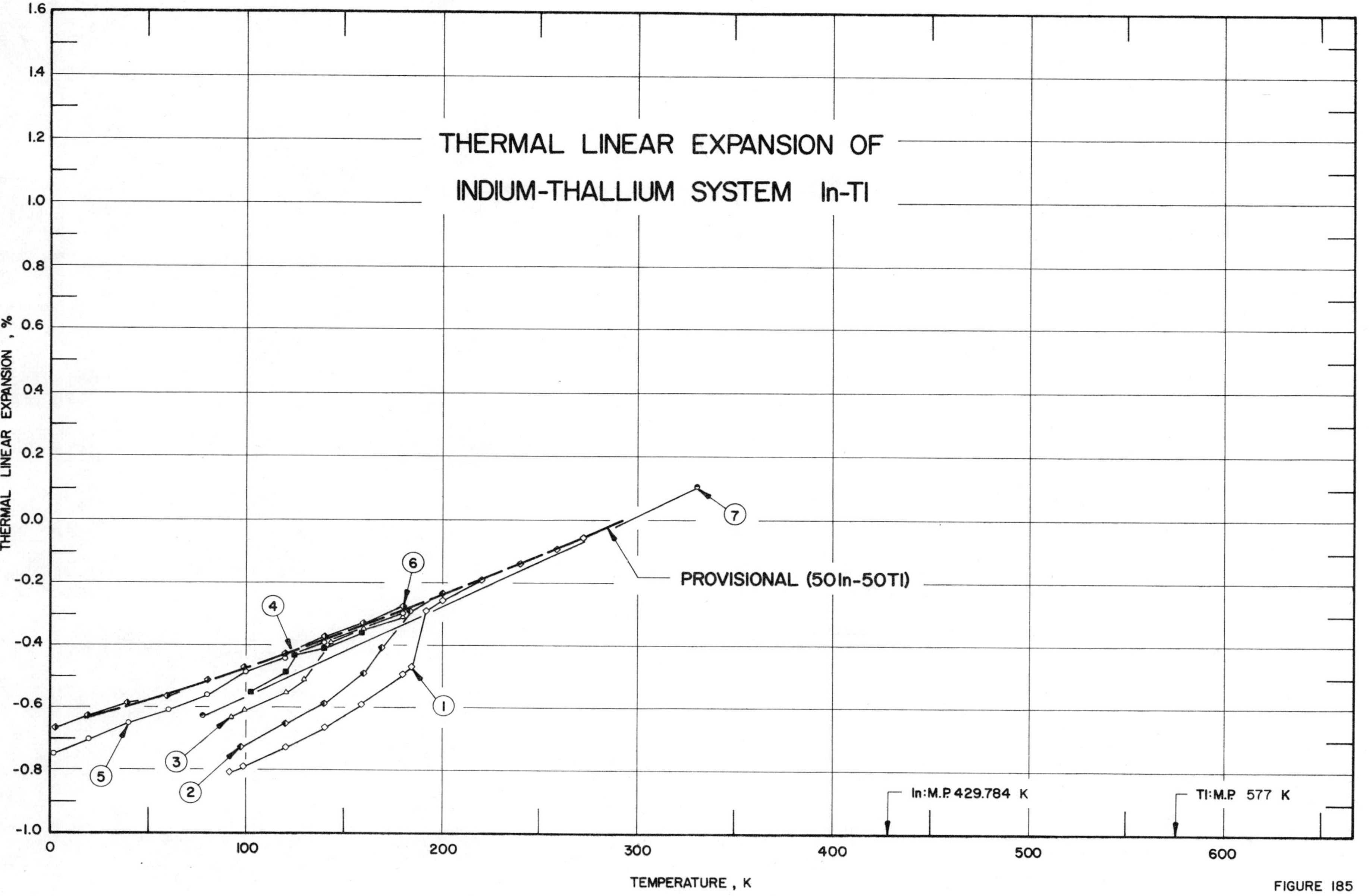
THERMAL LINEAR EXPANSION OF
INDIUM-THALLIUM SYSTEM In-Tl
PROVISIONAL (50In-50Tl)
In: M.P. 429.784 K
Tl: M.P. 577 K
1
2
3
4
5
6
7
THERMAL LINEAR EXPANSION, %
1.6
1.4
1.2
1.0
0.8
0.6
0.4
0.2
0.0
-0.2
-0.4
-0.6
-0.8
-1.0
TEMPERATURE, K
0
100
200
300
400
500
600

FIGURE 185

SPECIFICATION TABLE 185. THERMAL LINEAR EXPANSION OF INDIUM-THALLIUM SYSTEM In-Tl

Cur. No.	Ref. No.	Author(s)	Year	Method Used	Temp. Range, K	Name and Specimen Designation	Composition (weight percent) In	Tl	Composition (continued), Specifications, and Remarks
1	419	Pahlman, J.E. and Smith, J.F.	1968	I	91-273	FCT-FCC	60.83	39.17	99.97 In and 99.95 Tl used; crystal cut to facilitate measurements in [110] direction; expansion measured with increasing temperature; zero-point correction of -0.051% determined by graphical extrapolation.
2	419	Pahlman, J.E. and Smith, J.F.	1968	I	273-97	FCT-FCC	60.83	39.17	The above specimen; expansion measured with decreasing temperature; zero-point correction of -0.050% determined by graphical extrapolation.
3	419	Pahlman, J.E. and Smith, J.F.	1968	I	92-273	FCT-FCC	59.24	40.76	Similar to the above specimen; expansion measured with increasing temperature; zero-point correction of -0.051% determined by graphical extrapolation.
4	419	Pahlman, J.E. and Smith, J.F.	1968	I	273-102	FCT-FCC	59.24	40.76	The above specimen; expansion measured with decreasing temperature; zero-point correction of -0.051% determined by graphical extrapolation.
5	419	Pahlman, J.E. and Smith, J.F.	1968	I	2-272	FCC	55.38	44.62	Similar to the above specimen; zero-point correction of -0.060% determined by graphical extrapolation.
6	419	Pahlman, J.E. and Smith, J.F.	1968	I	3-271	FCC	51.86	48.14	Similar to the above specimen; zero-point correction of -0.050% determined by graphical extrapolation.
7	455	Luo, H.L.	1967		77,330		10.32/ 15.77	89.68/ 84.23	Body-centered cubic structure.

DATA TABLE 185. THERMAL LINEAR EXPANSION OF INDIUM-THALLIUM SYSTEM In-Tl

[Temperature, T, K; Linear Expansion, $\Delta L/L_0$, %]

T	$\Delta L/L_0$
CURVE 1	
91	-0.812
99	-0.794
120	-0.730
140	-0.662
159	-0.589
180	-0.498
184	-0.462
192	-0.283
200	-0.251
220	-0.192
240	-0.139
259	-0.087
273	-0.051
CURVE 2	
273	-0.050*
260	-0.086*
240	-0.138*
220	-0.188*
200	-0.234
180	-0.291
169	-0.412
160	-0.498
140	-0.585
120	-0.654
97	-0.726
CURVE 3	
92	-0.633
99	-0.612
120	-0.547
129	-0.504
140	-0.392
160	-0.347
180	-0.301
200	-0.255*
220	-0.203*
240	-0.150
259	-0.095*
273	-0.055*
CURVE 4	
273	-0.055*
260	-0.095*
240	-0.150*
220	-0.204*
200	-0.259*
180	-0.309*
159	-0.360
140	-0.408
126	-0.431
120	-0.484
102	-0.553
CURVE 5	
2	-0.751
20	-0.705
40	-0.656
60	-0.609
80	-0.560
100	-0.485
120	-0.440
140	-0.395
159	-0.347*
180	-0.300
200	-0.250*
220	-0.198*
240	-0.142*
260	-0.088*
272	-0.054*
CURVE 6	
3	-0.664
19	-0.626
39	-0.584
59	-0.561
80	-0.514
99	-0.473
120	-0.426
140	-0.378
160	-0.328
180	-0.278
200	-0.231*
220	-0.186*
240	-0.140*
260	-0.088*
271	-0.053*
CURVE 7	
77	-0.627
330	0.107

* Not shown in figure.

SPECIFICATION TABLE 186. THERMAL LINEAR EXPANSION OF INDIUM-TIN SYSTEM In-Sn

Cur. No.	Ref. No.	Author(s)	Year	Method Used	Temp. Range, K	Name and Specimen Designation	Composition (weight percent) In	Composition (weight percent) Sn	Composition (continued), Specifications, and Remarks
1*	191	Pollock, D.B.	1969	L	77,293		50	50	Specimen 3.2 cm long.

DATA TABLE 186. THERMAL LINEAR EXPANSION OF INDIUM-TIN SYSTEM In-Sn

[Temperature, T, K; Linear Expansion, $\Delta L/L_0$, %]

T	$\Delta L/L_0$
CURVE 1*	
77	-0.558
293	0.000

* No figure given.

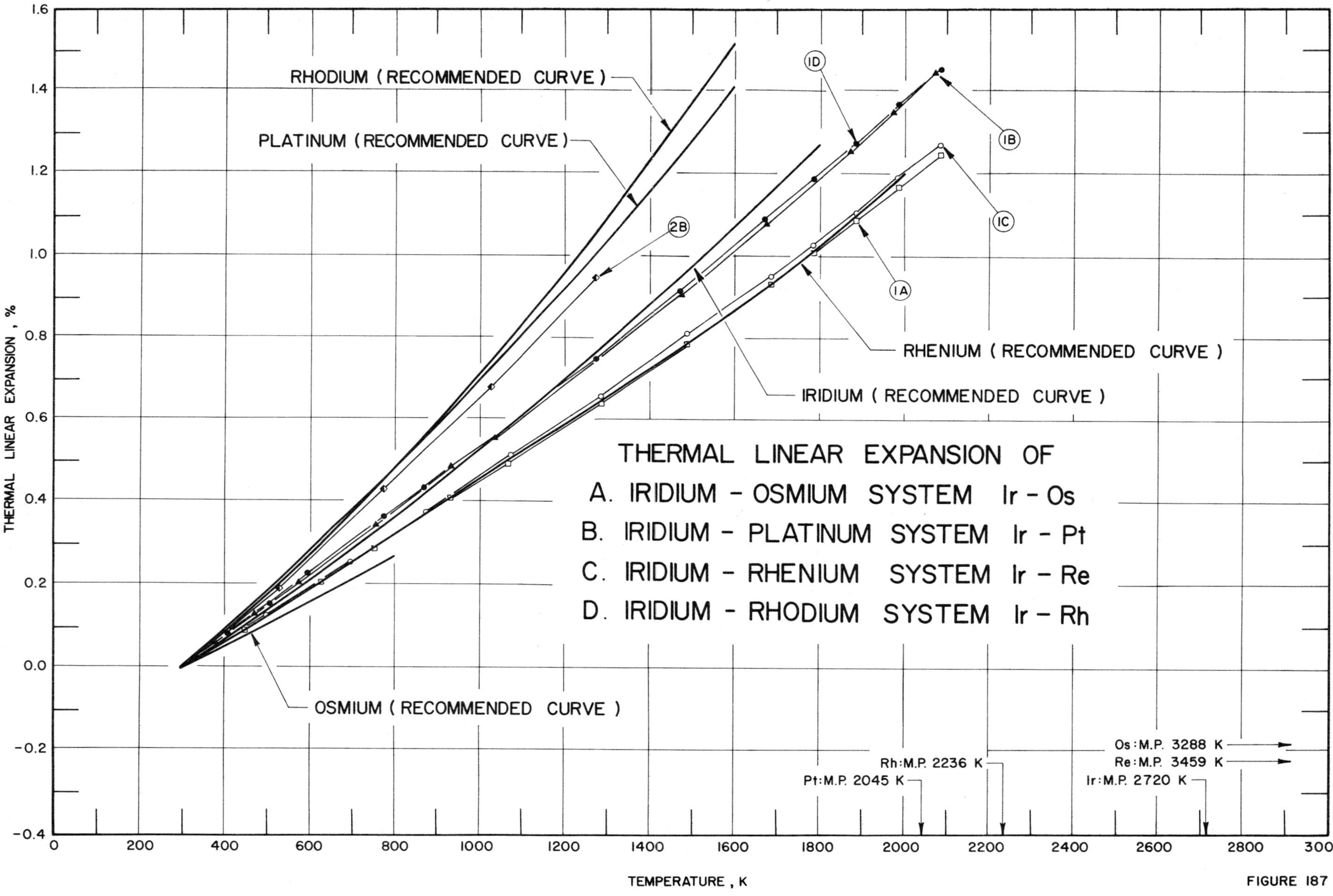

THERMAL LINEAR EXPANSION OF
A. IRIDIUM - OSMIUM SYSTEM Ir - Os
B. IRIDIUM - PLATINUM SYSTEM Ir - Pt
C. IRIDIUM - RHENIUM SYSTEM Ir - Re
D. IRIDIUM - RHODIUM SYSTEM Ir - Rh
RHODIUM (RECOMMENDED CURVE)
PLATINUM (RECOMMENDED CURVE)
RHENIUM (RECOMMENDED CURVE)
IRIDIUM (RECOMMENDED CURVE)
OSMIUM (RECOMMENDED CURVE)
1A
1B
1C
1D
2B
Os : M.P. 3288 K
Re : M.P. 3459 K
Ir : M.P. 2720 K
Rh : M.P. 2236 K
Pt : M.P. 2045 K
THERMAL LINEAR EXPANSION , %
TEMPERATURE , K

FIGURE 187

SPECIFICATION TABLE 187A. THERMAL LINEAR EXPANSION OF IRIDIUM-OSMIUM SYSTEM Ir-Os

Cur. No.	Ref. No.	Author(s)	Year	Method Used	Temp. Range, K	Name and Specimen Designation	Composition (weight percent) Ir	Os	Composition (continued), Specifications, and Remarks
1	420	Harmon, D.P.	1966	X, L	443-2082		75	25	99.96 Ir, obtained as powder from J. Bishop and Co., 99.84 Os, obtained as powder from Englehard Industries, Inc; alloy prepared by standard powder metallurgical techniques.

DATA TABLE 187A. THERMAL LINEAR EXPANSION OF IRIDIUM-OSMIUM SYSTEM Ir-Os

[Temperature, T, K; Linear Expansion, $\Delta L/L_0$, %]

T	$\Delta L/L_0$	T	$\Delta L/L_0$
CURVE 1‡		CURVE 1 (cont.)‡	
443	0.093	1482	0.785
623	0.206	1682	0.934
753	0.287	1782	1.010
933	0.406	1882	1.086
1063	0.492	1982	1.164
1282	0.640	2082	1.244

DATA TABLE 187A. COEFFICIENT OF THERMAL LINEAR EXPANSION OF IRIDIUM-OSMIUM SYSTEM Ir-Os

[Temperature, T, K; Coefficient of Expansion, α, 10^{-6} K^{-1}]

T	α
CURVE 1*, ‡	
443	6.34
623	6.12
753	6.43
933	6.74
1063	6.47
1282	7.12
1482	7.36
1682	7.54
1782	7.58
1882	7.74
1982	7.88
2082	8.09

*Not shown in figure.
‡Author's data for coefficient of thermal expansion have been integrated by TPRC to obtain $\Delta L/L_0$.

SPECIFICATION TABLE 187B. THERMAL LINEAR EXPANSION OF IRIDIUM-PLATINUM SYSTEM Ir-Pt

Cur. No.	Ref. No.	Author(s)	Year	Method Used	Temp. Range, K	Name and Specimen Designation	Composition (weight percent) Ir	Pt	Composition (continued), Specifications, and Remarks
1	420	Harmon, D.P.	1966	X, L	378-2073		75	25	99.96 Ir, obtained as powder from J. Bishop and Co., 99.9$^+$ Pt, obtained as powder from J. Bishop and Co.; alloy prepared by standard powder metallurgical techniques.
2	60	Holborn, L. and Day, A. L.	1901	T	273-1273		20	80	No details given; zero-point correction of -0.015% determined by graphical interpolation.

DATA TABLE 187B. THERMAL LINEAR EXPANSION OF IRIDIUM-PLATINUM SYSTEM Ir-Pt

[Temperature, T, K; Linear Expansion, $\Delta L/L_0$, %]

T	$\Delta L/L_0$	T	$\Delta L/L_0$	T	$\Delta L/L_0$
CURVE 1‡		CURVE 1 (cont.)‡		CURVE 2	
378	0.064	1473	0.910	273	-0.015
468	0.131	1673	1.080	523	0.199
573	0.210	1873	1.257	773	0.430
758	0.347	1973	1.348	1023	0.680
938	0.482	2073	1.443	1273	0.950
1033	0.556				

DATA TABLE 187B. COEFFICIENT OF THERMAL LINEAR EXPANSION OF IRIDIUM-PLATINUM SYSTEM Ir-Pt

[Temperature, T, K; Coefficient of Expansion, α, 10^{-6} K^{-1}]

T	α
CURVE 1*, ‡	
378	7.38
468	7.57
573	7.45
758	7.37
938	7.67
1033	7.80
1473	8.29
1673	8.72
1873	9.00
1973	9.30
2073	9.57

*Not shown in figure.
‡Author's data for coefficient of thermal expansion have been integrated by TPRC to obtain $\Delta L/L_0$.

SPECIFICATION TABLE 187C. THERMAL LINEAR EXPANSION OF IRIDIUM-RHENIUM SYSTEM Ir-Re

Cur. No.	Ref. No.	Author(s)	Year	Method Used	Temp. Range, K	Name and Specimen Designation	Composition (weight percent) Ir	Re	Composition (continued), Specifications, and Remarks
1	420	Harmon, D.P.	1966	X, L	495-2083		90.0	10.0	99.96 Ir, from J. Bishop and Co. in powder form; 99.99$^+$ Re, from Chase Brass and Copper Co., in powder form; alloy prepared by standard powder metallurgical techniques.

DATA TABLE 187C. THERMAL LINEAR EXPANSION OF IRIDIUM-RHENIUM SYSTEM Ir-Re

[Temperature, T, K; Linear Expansion, $\Delta L/L_0$, %]

T	$\Delta L/L_0$	T	$\Delta L/L_0$
CURVE 1‡		CURVE 1 (cont.)‡	
495	0.129	1682	0.954
696	0.258	1782	1.030
873	0.376	1883	1.109
1073	0.513	1983	1.189
1282	0.658	2083	1.271
1482	0.804		

DATA TABLE 187C. COEFFICIENT OF THERMAL LINEAR EXPANSION OF IRIDIUM-RHENIUM SYSTEM Ir-Re

[Temperature, T, K; Coefficient of Expansion, α, 10^{-6} K^{-1}]

T	α
CURVE 1*, ‡	
495	6.34
696	6.47
873	6.83
1073	6.85
1282	7.03
1482	7.56
1682	7.44
1782	7.80
1883	7.91
1983	8.15
2083	8.12

* Not shown in figure.

‡ Author's data for coefficient of thermal expansion have been integrated by TPRC to obtain $\Delta L/L_0$.

SPECIFICATION TABLE 187D. THERMAL LINEAR EXPANSION OF IRIDIUM-RHODIUM SYSTEM Ir-Rh

Cur. No.	Ref. No.	Author(s)	Year	Method Used	Temp. Range, K	Name and Specimen Designation	Composition (weight percent) Ir	Rh	Composition (continued), Specifications, and Remarks
1	420	Harmon, D.P.	1966	L, X	403-2083		75	25	99.96 Ir, obtained as powder from J. Bishop and Co., 99.98 Rh, also powder from J. Bishop and Co.; alloy prepared by standard powder metallurgical techniques.

DATA TABLE 187D. THERMAL LINEAR EXPANSION OF IRIDIUM-RHODIUM SYSTEM Ir-Rh

[Temperature, T, K; Linear Expansion, $\Delta L/L_0$, %]

T	$\Delta L/L_0$	T	$\Delta L/L_0$
CURVE 1‡		CURVE 1 (cont.)‡	
403	0.083	1673	1.089
503	0.158	1783	1.186
598	0.229	1883	1.274
773	0.363	1983	1.364
868	0.436	2083	1.456
1273	0.751		
1473	0.916		

DATA TABLE 187D. COEFFICIENT OF THERMAL LINEAR EXPANSION OF IRIDIUM-RHODIUM SYSTEM Ir-Rh

[Temperature, T, K; Coefficient of Expansion, α, 10^{-6} K^{-1}]

T	α
CURVE 1*, ‡	
403	7.63
503	7.44
598	7.43
773	7.85
868	7.54
1273	8.01
1473	8.55
1673	8.73
1783	8.86
1883	8.87
1983	8.99
2083	9.44

*Not shown in figure.
‡Author's data for coefficient of thermal expansion have been integrated by TPRC to obtain $\Delta L/L_0$.

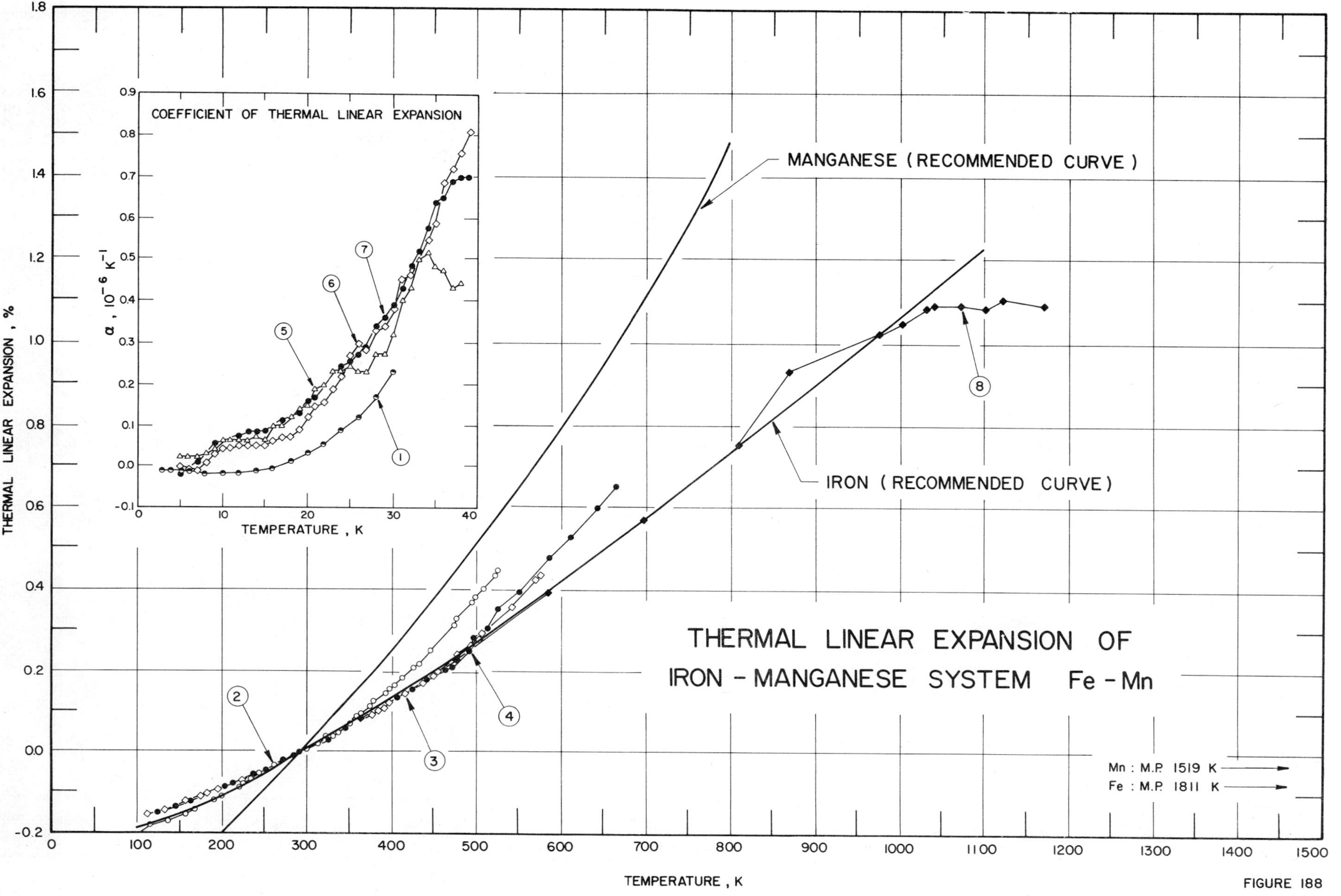
THERMAL LINEAR EXPANSION OF
IRON - MANGANESE SYSTEM Fe - Mn
MANGANESE (RECOMMENDED CURVE)
IRON (RECOMMENDED CURVE)
Mn : M.P. 1519 K
Fe : M.P. 1811 K
THERMAL LINEAR EXPANSION , %
TEMPERATURE , K
COEFFICIENT OF THERMAL LINEAR EXPANSION
α , 10⁻⁶ K⁻¹
TEMPERATURE , K

FIGURE 188

SPECIFICATION TABLE 188. THERMAL LINEAR EXPANSION OF IRON-MANGANESE SYSTEM Fe-Mn

Cur. No.	Ref. No.	Author(s)	Year	Method Used	Temp. Range, K	Name and Specimen Designation	Composition (weight percent) Fe	Mn	Composition (continued), Specifications, and Remarks
1*	41	White, G.K.	1965	E	3-283		50	50	Face-centered cubic structure from Harwell; specimen 6 cm long and 2 cm diameter; measured in the as received condition; this curve is here reported using the first given temperature as reference temperature at which $\Delta L/L_0 = 0$.
2	526	Fujimori, H.	1966	L	100-527		70.34	29.66	70 at.% Fe and 30 at.% Mn; prepared with 99.9 pure electrolytic iron and 99.9 pure electrolytic manganese, mixed and melted in an electric induction furnace and cast in an iron mold in argon; specimen 2 mm diameter and 80 mm long; cut from swaged ingot, lathed to size; zero-point correction of -0.023% determined by graphical interpolation.
3	526	Fujimori, H.	1966	L	114-579		65.37	34.63	Similar to the above specimen except 65 at.% Fe and 35 at.% Mn; zero-point correction of -0.008% determined by graphical interpolation.
4	526	Fujimori, H.	1966	L	126-667		60.39	39.61	Similar to the above specimen except 60 at.% Fe and 40 at.% Mn; zero-point correction of 0.004 % determined by graphical interpolation.
5*	44	Sivasubramanium, K.	1966		5-38		99.69	0.31	Zone melted specimen; Fe from American Iron and Steel Inst.; this curve is here reported using the first given temperature as reference temperature at which $\Delta L/L_0 = 0$.
6*	44	Sivasubramanium, K.	1966		5-39		99.42	0.58	Similar to the above specimen; this curve is here reported using the first given temperature as reference temperature at which $\Delta L/L_0 = 0$.
7*	44	Sivasubramanium, K.	1966		5-39		99.65	0.35	Vacuum melted specimen; Fe from American Iron and Steel Inst.; this curve is here reported using the first given temperature as reference temperature at which $\Delta L/L_0 = 0$.
8	123	Luck, C.F. and Deem, H.W.	1958	L	116-1172	SAE 1010	99.08	0.6	0.13 C, 0.1 Si, 0.05 S, and 0.04 P; from United States Steel Corp.; hot-rolled specimen 0.375 in. diameter by 3 in. long, heating rate of 4 F per min.

* Not shown in figure.

DATA TABLE 188. THERMAL LINEAR EXPANSION OF IRON - MANGANESE SYSTEM Fe - Mn

[Temperature, T, K; Linear Expansion, $\Delta L/L_0$, %]

T	$\Delta L/L_0$
CURVE 1*, †, ‡	
3	0.000 x 10^{-4}
4	-0.011
5	-0.025
6	-0.041
8	-0.078
10	-0.119
12	-0.160
14	-0.195
16	-0.217
18	-0.216
20	-0.183
22	-0.103
24	0.041
26	0.251
28	0.541
30	0.941
75	2.741
85	5.241
283	5.391
CURVE 2	
100	-0.206
119	-0.189
139	-0.169
159	-0.152
162	-0.146*
170	-0.136
191	-0.114
201	-0.103
221	-0.081
226	-0.075
237	-0.063
246	-0.053
250	-0.046*
257	-0.037*
264	-0.030
287	-0.006*
301	0.008
314	0.024
324	0.041
351	0.076
360	0.091
367	0.099
374	0.116

T	$\Delta L/L_0$
CURVE 2 (cont.)	
380	0.126
394	0.148
399	0.159
403	0.167
412	0.184
427	0.212
431	0.221
446	0.253
473	0.313
478	0.330
496	0.370
500	0.381
511	0.401
524	0.435
527	0.446
CURVE 3	
114	-0.151
133	-0.140
150	-0.126*
159	-0.117
176	-0.104
184	-0.096
197	-0.085
225	-0.064
242	-0.045*
288	-0.005*
297	0.004*
303	0.011*
317	0.025*
331	0.041
339	0.050
352	0.064*
378	0.096
383	0.103
391	0.111
399	0.124
416	0.144
429	0.157*
437	0.171
444	0.181*
450	0.193
455	0.202
465	0.216*

T	$\Delta L/L_0$
CURVE 3 (cont.)	
478	0.241
494	0.268
504	0.283
509	0.298
542	0.358
572	0.425
579	0.436
CURVE 4	
126	-0.148
148	-0.132
163	-0.119
186	-0.098*
205	-0.080
214	-0.073
240	-0.052
252	-0.042
273	-0.018
287	-0.006
296	0.003
307	0.015*
318	0.025*
327	0.035
347	0.061
364	0.082*
385	0.113*
391	0.119*
401	0.128*
407	0.138
424	0.156
441	0.180
463	0.205
471	0.217
478	0.230*
491	0.253
498	0.269
508	0.291*
514	0.304
528	0.352
552	0.396
588	0.478
611	0.525
642	0.602
667	0.658

T	$\Delta L/L_0$
CURVE 5*, †, ‡	
5	0.000 x 10^{-4}
6	0.020
7	0.040
8	0.065
9	0.100
10	0.150
11	0.210
12	0.270
13	0.330
14	0.395
15	0.460
16	0.540
17	0.640
18	0.750
19	0.880
20	1.025
21	1.195
22	1.390
23	1.605
24	1.835
25	2.070
26	2.305
27	2.535
28	2.785
29	3.055
30	3.350
31	3.710
32	4.125
33	4.590
34	5.100
35	5.600
36	6.075
37	6.525
38	6.960
CURVE 6*, †, ‡	
5	0.000 x 10^{-4}
6	-0.005
7	-0.015
8	-0.015
9	0.005
10	0.040
11	0.080
12	0.125

T	$\Delta L/L_0$
CURVE 6 (cont.)*, †, ‡	
13	0.175 x 10^{-4}
14	0.225
15	0.275
16	0.330
17	0.395
18	0.465
19	0.545
20	0.650
21	0.785
22	0.940
23	1.115
24	1.320
25	1.565
26	1.850
27	2.140
28	2.445
29	2.780
30	3.140
31	3.555
32	4.010
33	4.490
34	5.015
35	5.585
36	6.225
37	6.930
38	7.670
39	8.455
CURVE 7*, †, ‡	
5	0.000 x 10^{-4}
6	-0.015
7	-0.015
8	0.005
9	0.045
10	0.100
11	0.160
12	0.225
13	0.300
14	0.380
15	0.460
16	0.550
17	0.655
18	0.770
19	0.895

T	$\Delta L/L_0$
CURVE 7 (cont.)*, †, ‡	
20	1.040 x 10^{-4}
21	1.205
22	1.390
23	1.605
24	1.840
25	2.085
26	2.345
27	2.625
28	2.940
29	3.290
30	3.665
31	4.075
32	4.530
33	5.030
34	5.580
35	6.190
36	6.835
37	7.505
38	8.200
39	8.900
CURVE 8	
116	-0.172*
144	-0.148
199	-0.098*
293	0.000
366	0.086
477	0.231
588	0.391
699	0.566
811	0.751
922	0.932
977	1.021
1005	1.056
1033	1.086
1044	1.091
1074	1.093
1102	1.081
1122	1.104
1172	1.092

* Not shown in figure.
† This curve is here reported using the first given temperature as reference temperature at which $\Delta L/L_0 = 0$.
‡ Author's data for coefficient of thermal expansion have been integrated by TPRC to obtain $\Delta L/L_0$.

DATA TABLE 188. COEFFICIENT OF THERMAL LINEAR EXPANSION OF IRON - MANGANESE SYSTEM Fe - Mn

[Temperature, T, K; Coefficient of Expansion, α, 10^{-6} K^{-1}]

T	α
CURVE 1‡	
3	-0.010
4	-0.013
5	-0.015
6	-0.017
8	-0.020
10	-0.021
12	-0.020
14	-0.015
16	-0.007
18	0.008
20	0.025
22	0.055
24	0.090
26	0.120
28	0.170
30	0.230
75	3.85*
85	4.65*
283	10.2*
CURVE 5‡	
5	0.02
6	0.02
7	0.02
8	0.03
9	0.04
10	0.06
11	0.06
12	0.06
13	0.06
14	0.07
15	0.06
16	0.10
17	0.10
18	0.12
19	0.14
20	0.15
21	0.19
22	0.20
23	0.23
24	0.23
25	0.24
26	0.23
27	0.23
CURVE 5 (cont.)‡	
28	0.27
29	0.27
30	0.32
31	0.40
32	0.43
33	0.50
34	0.52
35	0.48
36	0.47
37	0.43
38	0.44
CURVE 6‡	
5	-0.000
6	-0.01
7	-0.01
8	0.01
9	0.03
10	0.04
11	0.04
12	0.05
13	0.05
14	0.05
15	0.05
16	0.06
17	0.07
18	0.07
19	0.09
20	0.12
21	0.15
22	0.16
23	0.19
24	0.22
25	0.27
26	0.30
27	0.28
28	0.33
29	0.34
30	0.38
31	0.45
32	0.46
33	0.50*
34	0.55
35	0.59
CURVE 6 (cont.)‡	
36	0.69
37	0.72
38	0.76
39	0.81
CURVE 7‡	
5	-0.02
6	-0.01*
7	0.01
8	0.03*
9	0.05
10	0.06*
11	0.06*
12	0.07
13	0.08
14	0.08
15	0.08
16	0.10*
17	0.11
18	0.12*
19	0.13
20	0.16
21	0.17
22	0.20*
23	0.23*
24	0.24
25	0.25
26	0.27
27	0.29
28	0.34
29	0.36
30	0.39
31	0.43
32	0.48
33	0.52
34	0.58
35	0.64
36	0.65
37	0.69
38	0.70
39	0.70

*Not shown in figure.
‡Author's data for coefficient of thermal expansion have been integrated by TPRC to obtain $\Delta L/L_0$.

SPECIFICATION TABLE 189. THERMAL LINEAR EXPANSION OF IRON-MOLYBDENUM SYSTEM Fe-Mo

Cur. No.	Ref. No.	Author(s)	Year	Method Used	Temp. Range, K	Name and Specimen Designation	Composition (weight percent) Fe	Mo	Composition (continued), Specifications, and Remarks
1*	44	Krishnan, K. S.	1966	C	5-39		99.66	0.34	Vacuum melted specimen; Fe from American Iron and Steel Inst.; Mo from General Electric Co.; this curve is here reported using the first given temperature as reference temperature at which $\Delta L/L_0 = 0$.
2*	44	Krishnan, K. S.	1966	C	5-39		99.46	0.54	Similar to the above specimen; this curve is here reported using the first given temperature as reference temperature at which $\Delta L/L_0 = 0$.
3*	44	Krishnan, K. S.	1966	C	5-39		99.69	0.31	Zone melted specimen similar to the above specimen; this curve is here reported using the first given temperature as reference temperature at which $\Delta L/L_0 = 0$.
4*	44	Krishnan, K. S.	1966	C	5-39		99.64	0.36	Similar to the above specimen; this curve is here reported using the first given temperature as reference temperature at which $\Delta L/L_0 = 0$.
5*	44	Krishnan, K. S.	1966	C	5-39		99.47	0.53	Similar to the above specimen; this curve is here reported using the first given temperature as reference temperature at which $\Delta L/L_0 = 0$.
6*	44	Krishnan, K. S.	1966	C	5-39		99.25	0.75	Similar to the above specimen; this curve is here reported using the first given temperature as reference temperature at which $\Delta L/L_0 = 0$.
7*	44	Krishnan, K. S.	1966	C	5-39		99.08	0.92	Similar to the above specimen; this curve is here reported using the first given temperature as reference temperature at which $\Delta L/L_0 = 0$.

* No figure given.

DATA TABLE 189. THERMAL LINEAR EXPANSION OF IRON-MOLYBDENUM SYSTEM Fe-Mo

[Temperature, T, K; Linear Expansion, $\Delta L/L_0$, %]

T	$\Delta L/L_0$
CURVE 1*, †, ‡	
5	0.00 x 10^{-4}
6	-0.03
7	-0.04
8	-0.05
9	-0.04
10	-0.02
11	0.01
12	0.04
13	0.07
14	0.12
15	0.17
16	0.23
17	0.31
18	0.39
19	0.49
20	0.60
21	0.74
22	0.89
23	1.06
24	1.25
25	1.46
26	1.69
27	1.95
28	2.22
29	2.53
30	2.86
31	3.23
32	3.62
33	4.02
34	4.46
35	4.95
36	5.50
37	6.09
38	6.72
39	7.34
CURVE 2*, †, ‡	
5	0.00 x 10^{-4}
6	-0.05
7	-0.06
8	-0.02
9	0.03
10	0.05
11	0.04
12	0.05
CURVE 2 (cont.)*, †, ‡	
13	0.09 x 10^{-4}
14	0.15
15	0.23
16	0.34
17	0.46
18	0.58
19	0.69
20	0.81
21	0.93
22	1.08
23	1.25
24	1.45
25	1.66
26	1.90
27	2.16
28	2.45
29	2.77
30	3.09
31	3.44
32	3.82
33	4.25
34	4.75
35	5.33
36	5.95
37	6.57
38	7.20
39	7.84
CURVE 3*, †, ‡	
5	0.00 x 10^{-4}
6	0.01
7	0.06
8	0.10
9	0.13
10	0.14
11	0.14
12	0.14
13	0.16
14	0.18
15	0.22
16	0.28
17	0.38
18	0.50
19	0.62
20	0.74
CURVE 3 (cont.)*, †, ‡	
21	0.88
22	1.02
23	1.19
24	1.37
25	1.58
26	1.82
27	2.07
28	2.31
29	2.56
30	2.85
31	3.18
32	3.54
33	3.95
34	4.42
35	4.95
36	5.54
37	6.18
38	6.85
39	7.58
CURVE 4*, †, ‡	
5	0.00 x 10^{-4}
6	-0.06
7	0.01
8	0.10
9	0.19
10	0.26
11	0.29
12	0.31
13	0.32
14	0.34
15	0.36
16	0.39
17	0.44
18	0.49
19	0.55
20	0.62
21	0.69
22	0.78
23	0.88
24	1.03
25	1.20
26	1.39
27	1.61
28	1.86
CURVE 4 (cont.)*, †, ‡	
29	2.14
30	2.46
31	2.83
32	3.23
33	3.65
34	4.11
35	4.64
36	5.22
37	5.83
38	6.46
39	7.08
CURVE 5*, †, ‡	
5	0.00 x 10^{-4}
6	0.03
7	0.08
8	0.11
9	0.13
10	0.15
11	0.18
12	0.21
13	0.26
14	0.31
15	0.36
16	0.42
17	0.49
18	0.57
19	0.66
20	0.76
21	0.88
22	1.02
23	1.19
24	1.38
25	1.60
26	1.83
27	2.07
28	2.33
29	2.60
30	2.90
31	3.23
32	3.60
33	4.03
34	4.51
35	5.06
36	5.66
CURVE 5 (cont.)*, †, ‡	
37	6.27
38	6.87
39	7.46
CURVE 6*, †, ‡	
5	0.00 x 10^{-4}
6	-0.04
7	-0.04
8	-0.03
9	-0.02
10	-0.01
11	0.02
12	0.05
13	0.09
14	0.15
15	0.21
16	0.28
17	0.36
18	0.46
19	0.56
20	0.68
21	0.83
22	1.00
23	1.19
24	1.42
25	1.67
26	1.93
27	2.19
28	2.45
29	2.71
30	2.98
31	3.28
32	3.60
33	3.97
34	4.41
35	4.93
36	5.54
37	6.17
38	6.79
39	7.41
CURVE 7*, †, ‡	
5	0.00 x 10^{-4}
6	-0.03
CURVE 7 (cont.)*, †, ‡	
7	-0.05
8	-0.05
9	-0.05
10	-0.04
11	-0.01
12	0.06
13	0.14
14	0.23
15	0.33
16	0.43
17	0.53
18	0.64
19	0.75
20	0.87
21	1.01
22	1.17
23	1.35
24	1.56
25	1.79
26	2.05
27	2.32
28	2.62
29	2.93
30	3.25
31	3.60
32	4.00
33	4.45
34	4.94
35	5.51
36	6.16
37	6.83
38	7.49
39	8.11

* No figure given.

† This curve is here reported using the first given temperature as reference temperature at which $\Delta L/L_0 = 0$.

‡ Author's data for coefficient of thermal expansion have been integrated by TPRC to obtain $\Delta L/L_0$.

DATA TABLE 189. COEFFICIENT OF THERMAL LINEAR EXPANSION OF IRON-MOLYBDENUM SYSTEM Fe-Mo

[Temperature, T, K; Coefficient of Expansion, α, 10^{-6} K^{-1}]

T	α	T	α	T	α	T	α	T	α	T	α
CURVE 1*, ‡		CURVE 2 (cont.)*, ‡		CURVE 3 (cont.)*, ‡		CURVE 4 (cont.)*, ‡		CURVE 5 (cont.)*, ‡		CURVE 7*, ‡	
5	-0.05	12	0.02	19	0.12	26	0.21	33	0.46	5	-0.02
6	-0.01	13	0.06	20	0.13	27	0.23	34	0.50	6	-0.04
7	-0.01	14	0.07	21	0.14	28	0.27	35	0.60	7	0.01
8	-0.00	15	0.09	22	0.15	29	0.30	36	0.60	8	-0.01
9	0.02	16	0.12	23	0.18	30	0.34	37	0.61	9	0.00
10	0.02	17	0.12	24	0.19	31	0.39	38	0.60	10	0.02
11	0.03	18	0.12	25	0.23	32	0.41	39	0.58	11	0.05
12	0.03	19	0.11	26	0.25	33	0.43			12	0.07
13	0.03	20	0.12	27	0.24	34	0.49	CURVE 6*, ‡		13	0.09
14	0.06	21	0.13	28	0.25	35	0.58			14	0.10
15	0.05	22	0.16	29	0.25	36	0.58	5	-0.05	15	0.10
16	0.07	23	0.19	30	0.32	37	0.63	6	-0.02	16	0.10
17	0.08	24	0.20	31	0.35	38	0.63	7	0.01	17	0.10
18	0.09	25	0.23	32	0.37	39	0.62	8	0.01	18	0.11
19	0.10	26	0.25	33	0.44			9	0.01	19	0.11
20	0.13	27	0.27	34	0.50	CURVE 5*, ‡		10	0.02	20	0.13
21	0.14	28	0.31	35	0.57			11	0.03	21	0.15
22	0.16	29	0.32	36	0.61	5	0.00	12	0.03	22	0.17
23	0.18	30	0.333	37	0.66	6	0.06	13	0.05	23	0.20
24	0.20	31	0.36	38	0.69	7	0.04	14	0.06	24	0.22
25	0.22	32	0.40	39	0.77	8	0.02	15	0.06	25	0.24
26	0.25	33	0.46			9	0.02	16	0.08	26	0.27
27	0.26	34	0.54	CURVE 4*, ‡		10	0.02	17	0.09	27	0.28
28	0.29	35	0.62			11	0.03	18	0.10	28	0.31
29	0.32	36	0.62	5	-0.14	12	0.04	19	0.11	29	0.32
30	0.35	37	0.63	6	0.03	13	0.05	20	0.13	30	0.32
31	0.38	38	0.62	7	0.09	14	0.06	21	0.16	31	0.37
32	0.40	39	0.66	8	0.10	15	0.04	22	0.18	32	0.43
33	0.41			9	0.08	16	0.07	23	0.21	33	0.47
34	0.46	CURVE 3*, ‡		10	0.05	17	0.08	24	0.24	34	0.52
35	0.53			11	0.02	18	0.08	25	0.26	35	0.62
36	0.56	5	-0.01	12	0.01	19	0.09	26	0.27	36	0.67
37	0.62	6	0.03	13	0.01	20	0.11	27	0.25	37	0.67
38	0.64	7	0.06	14	0.03	21	0.13	28	0.26	38	0.65
39	0.60	8	0.03	15	0.02	22	0.16	29	0.26	39	0.59
		9	0.02	16	0.04	23	0.18	30	0.29		
CURVE 2*, ‡		10	0.00	17	0.05	24	0.20	31	0.30		
		11	0.00	18	0.06	25	0.23	32	0.35		
5	-0.07	12	0.01	19	0.06	26	0.24	33	0.39		
6	-0.03	13	0.02	20	0.07	27	0.24	34	0.48		
7	0.02	14	0.03	21	0.08	28	0.27	35	0.57		
8	0.06	15	0.04	22	0.09	29	0.28	36	0.64		
9	0.03	16	0.09	23	0.12	30	0.31	37	0.62		
10	-0.00	17	0.11	24	0.18	31	0.35	38	0.63		
11	-0.01	18	0.12	25	0.16	32	0.40	39	0.60		

* No figure given.
‡ Author's data for coefficient of thermal expansion have been integrated by TPRC to obtain $\Delta L/L_0$.

FIGURE AND TABLE NO. 190R. PROVISIONAL VALUES FOR THERMAL LINEAR EXPANSION OF IRON-NICKEL SYSTEM Fe-Ni

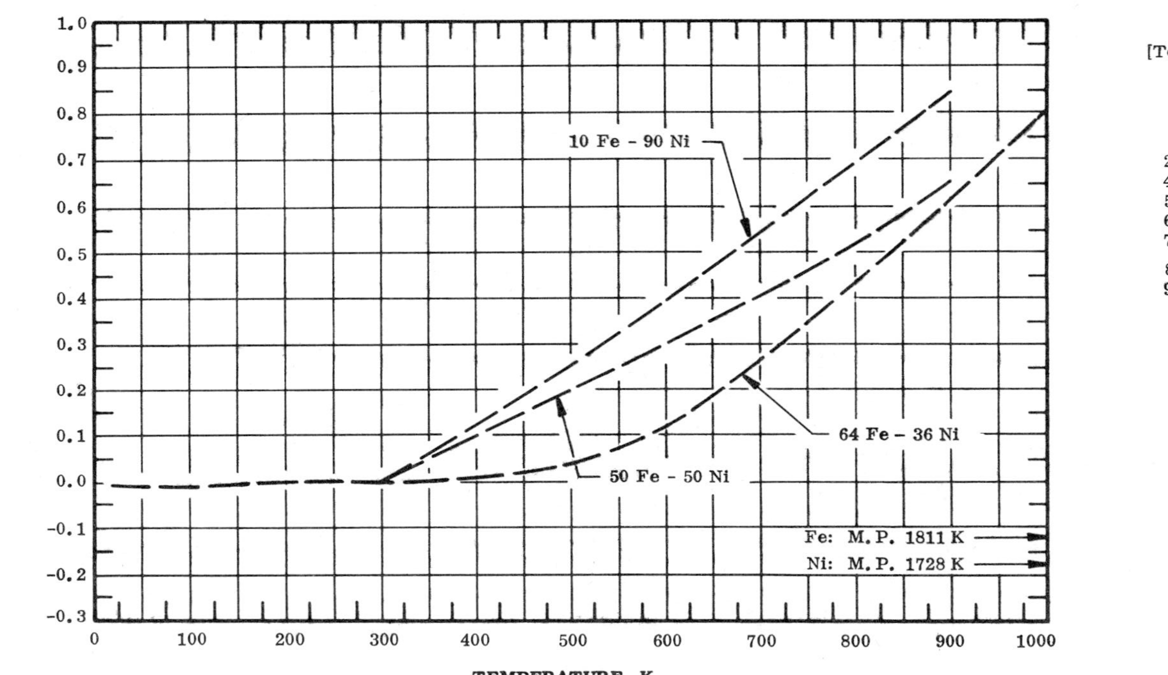

PROVISIONAL VALUES

[Temperature, T, K; Linear Expansion, $\Delta L/L_0$, %; α, K^{-1}]

T	(10 Fe-90 Ni) $\Delta L/L_0$	(10 Fe-90 Ni) $\alpha \times 10^6$	(50 Fe-50 Ni) $\Delta L/L_0$	(50 Fe-50 Ni) $\alpha \times 10^6$
293	0.000	10.4	0.000	9.4
400	0.121	12.2	0.108	9.4
500	0.251	13.6	0.203	9.6
600	0.393	14.6	0.299	9.8
700	0.541	15.2	0.402	10.9
800	0.694	15.4	0.518	12.5
900	0.848	15.4	0.654	14.8

(64 Fe-36 Ni)

T	$\Delta L/L_0$	$\alpha \times 10^6$
10	-0.0116	-0.94
20	-0.0131	-1.69
30	-0.0147	-1.63
40	-0.0161	-1.05
50	-0.0169	-0.48
60	-0.0171	0.05
70	-0.0168	0.57
80	-0.0160	0.94
90	-0.0150	1.18
100	-0.0137	1.41
125	-0.0098	1.60
150	-0.0060	1.40
175	-0.0030	0.99
200	-0.0011	0.53
225	-0.0002	0.18
250	0.0000	0.00
275	-0.0001	-0.01
293	0.0000	0.13
350	0.002	0.6
400	0.006	1.7
450	0.018	2.6
500	0.033	5.1
550	0.070	9.2
600	0.125	13.5
700	0.272	15.6
800	0.437	17.1
900	0.615	18.1
1000	0.800	18.5

REMARKS

(10 Fe-90 Ni): The tabulated values for well-annealed alloy are considered accurate to within ± 7% over the entire temperature range. The Curie temperature of this alloy is approximately at 760 K. These values can be represented approximately by the following equation:

$$\Delta L/L_0 = -0.212 + 3.450 \times 10^{-4}\, T + 1.461 \times 10^{-6}\, T^2 - 5.947 \times 10^{-10}\, T^3$$

(50 Fe-50 Ni): The tabulated values for well-annealed alloy are considered accurate to within ± 7% over the entire temperature range. The Curie temperature of this alloy is approximately at 790 K. These values can be represented approximately by the following equation:

$$\Delta L/L_0 = -0.390 + 1.694 \times 10^{-3}\, T - 1.548 \times 10^{-6}\, T^2 + 1.061 \times 10^{-9}\, T^3$$

(64 Fe-36 Ni): The tabulated values for well-annealed alloy of type Invar are considered accurate to within ± 7% over the entire temperature range.

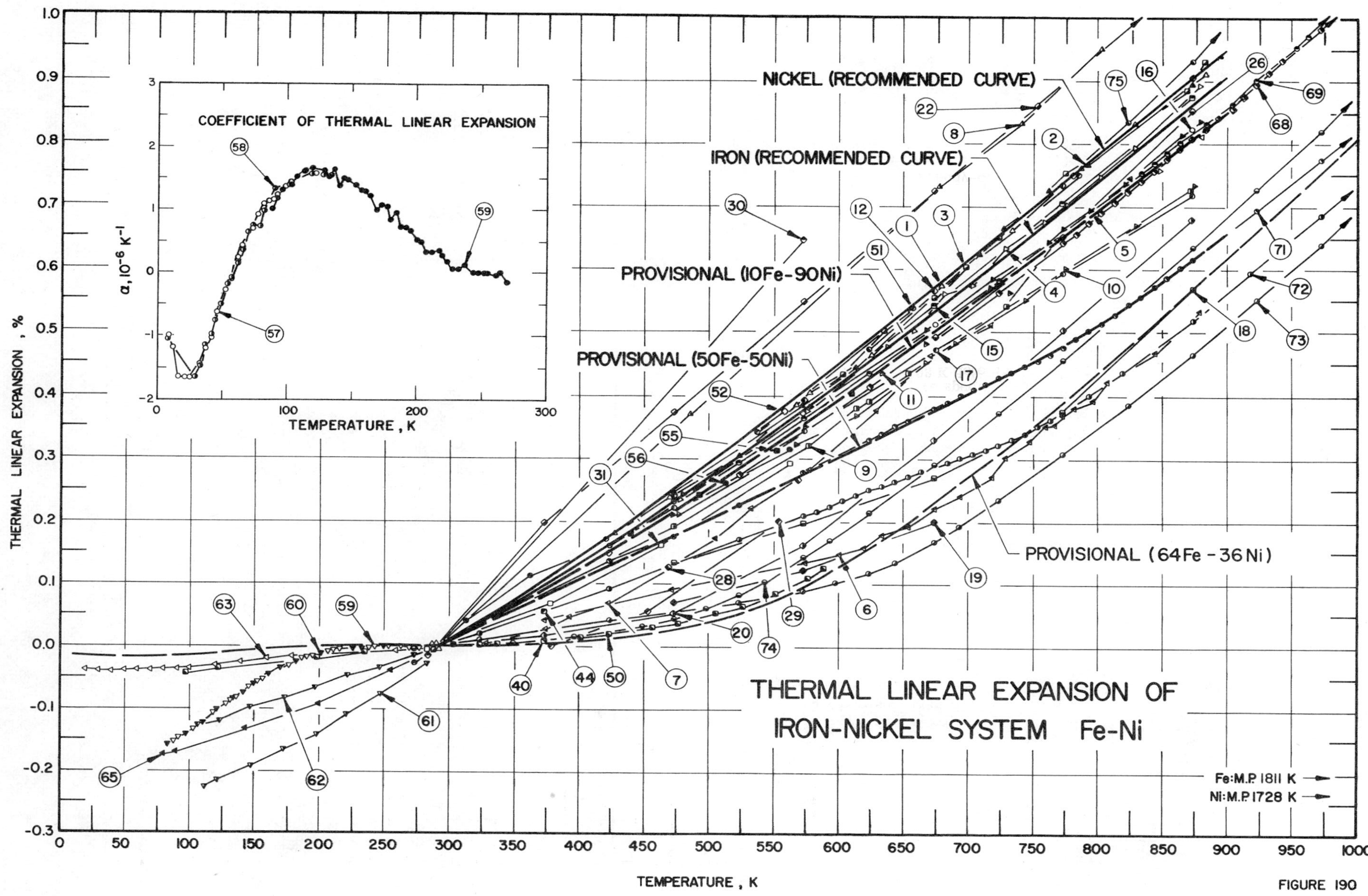
THERMAL LINEAR EXPANSION OF
IRON-NICKEL SYSTEM Fe-Ni
NICKEL (RECOMMENDED CURVE)
IRON (RECOMMENDED CURVE)
PROVISIONAL (10Fe-90Ni)
PROVISIONAL (50Fe-50Ni)
PROVISIONAL (64Fe - 36Ni)
COEFFICIENT OF THERMAL LINEAR EXPANSION
α, 10⁻⁶ K⁻¹
TEMPERATURE, K
THERMAL LINEAR EXPANSION, %
Fe: M.P. 1811 K
Ni: M.P. 1728 K

FIGURE 190

SPECIFICATION TABLE 190. THERMAL LINEAR EXPANSION OF IRON-NICKEL SYSTEM Fe-Ni

Cur. No.	Ref. No.	Author(s)	Year	Method Used	Temp. Range, K	Name and Specimen Designation	Composition (weight percent) Fe	Ni	Composition (continued), Specifications, and Remarks
1	235	Owen, E.A., and Yates, E.L.	1937	X	286-882		Bal.	96.8	Provided by Mond Nickel Company Ltd.; lump annealed at 1423 K for 20 hr in nitrogen atmosphere; lump enclosed in pyrex glass tube, evacuated and sealed; wrapped in asbestos cloth, placed in furnace maintained at 1073 K, allowed to remain at that temperature for several days before cooled slowly to room temperature; lattice parameter expansion measured; lattice parameter reported at 292 K is 3.5207 Å; 3.52075 Å at 293 K determined by graphical interpolation.
2	235	Owen, E.A., and Yates, E.L.	1937	X	293-875		Bal.	94.0	Similar to the above specimen; lattice parameter reported at 293 K is 3.5243 Å.
3	235	Owen, E.A., and Yates, E.L.	1937	X	287-881		Bal.	91.1	Similar to the above specimen; lattice parameter reported at 293 K is 2.5277 Å.
4	235	Owen, E.A., and Yates, E.L.	1937	X	288-879		Bal.	82.9	Similar to the above specimen; lattice parameter reported at 292 is 3.5381 Å, 3.53814 Å at 293 K determined by graphical interpolation.
5	235	Owen, E.A., and Yates, E.L.	1937	X	283-879		Bal.	74.0	Similar to the above specimen; lattice parameter reported at 289 K is 3.5490 Å; 3.5492 Å at 293 K determined by graphical interpolation.
6	530	Owen, E.A., and Yates, E.L.	1937	X	282-878		Bal.	42.7	The alloy possessed a f.c.c. structure and the lines measured in the photographs were doublets reflected from (400) planes; lattice parameter reported at 285 K is 3.5877 Å; 3.5878 at 293 K determined by graphical interpolation.
7	530	Owen, E.A., and Yates, E.L.	1937	X	285-880		Bal.	32.2	This alloy is inside the pure γ' phase and structure should be f.c.c.; lattice parameter reported at 285 K is 3.5813 Å; 3.58135 Å at 293 K determined by graphical interpolation.
8	530	Owen, E.A., and Yates, E.L.	1937	X	288-904		Bal.	24.2	Annealed at 300 C for 24 hr and cooled from this temperature to room temperature in 7 hr; body-centered α phase; lattice parameter reported at 288 K is 3.5735 Å; 3.5738 Å at 293 K determined by graphical interpolation.
9	49	Owen, E.A., and Yates, E.L.	1937	X	286-871		Bal.	16.7	Supplied by the Mond Nickel Company Ltd. and prepared from pure materials. Annealed in powder form at 773 K for 3 hr, b.c.c structure. The purity of iron is 99.966; lattice parameter reported at 291 K is 2.8631 Å; 2.86315 Å at 293 K determined by graphical interpolation.
10	49	Owen, E.A., and Yates, E.L.	1937	X	288-873		Bal.	9.0	Similar to the above specimen; lattice parameter reported at 293 K is 2.8635 Å.
11	49	Owen, E.A., and Yates, E.L.	1937	X	289-848		Bal.	3.1	Similar to the above specimen; lattice parameter reported at 291 K is 2.8622 Å; 2.86226 Å at 293 K determined by graphical interpolation.
12	389	Owen, E.A., Yates, E.L., and Sully, A.H.	1937	X	288-873		Bal.	96.75	Lump material heat treated, then filings taken, heated 14 hr at 873 K; lattice parameter reported at 288 K is 3.52065 Å; 3.52087 Å at 293 determined by graphical interpolation.
13*	389	Owen, E.A., et al.	1937	X	288-873		Bal.	93.95	Similar to the above specimen; lattice parameter reported at 288 K is 3.52405 Å; 3.5243 Å at 293 K determined by graphical interpolation.
14*	389	Owen, E.A., et al.	1937	X	288-873		Bal.	91.04	Similar to the above specimen; lattice parameter reported at 288 K is 3.52775 Å, 3.5280 Å at 293 K determined by graphical interpolation.
15	389	Owen, E.A., et al.	1937	X	288-873		Bal.	82.86	Similar to the above specimen; lattice parameter reported at 288 K is 3.53780 Å; 3.5380 Å at 293 determined by graphical interpolation.

* Not shown in figure.

SPECIFICATION TABLE 190. THERMAL LINEAR EXPANSION OF IRON-NICKEL SYSTEM Fe-Ni (continued)

Cur. No.	Ref. No.	Author(s)	Year	Method Used	Temp. Range, K	Name and Specimen Designation	Composition (weight percent) Fe	Ni	Composition (continued), Specifications, and Remarks
16	389	Owen, E.A., et al.	1937	X	288-873		Bal.	74.00	Similar to the above specimen; lattice parameter reported at 288 is 3.54916 Å; 3.5494 Å at 293 K determined by graphical interpolation.
17	389	Owen, E.A., et al.	1937	X	288-873		Bal.	58.41	Similar to the above specimen; lattice parameter reported at 288 K is 3.56908 Å; 3.5693 Å at 293 K determined by graphical interpolation.
18	389	Owen, E.A., et al.	1937	X	288-873		Bal.	45.50	Lump material heat treated, then filings taken; heated 14 hr at 873 K; specimen in γ-phase; lattice parameter reported at 288 K is 3.58557 Å; 3.5857 Å at 293 K determined by graphical interpolation.
19	389	Owen, E.A., et al.	1937	X	288-873		Bal.	42.71	Similar to the above specimen; lattice parameter reported at 288 K is 3.58827 Å; 3.5883 Å at 293 K determined by graphical interpolation.
20	389	Owen, E.A., et al.	1937	X	288-873		Bal.	35.10	Similar to the above specimen; lattice parameter reported at 288 K is 3.58556 Å; 3.58559 Å at 293 K determined by graphical interpolation.
21*	389	Owen, E.A., et al.	1937	X	288-873		Bal.	32.3	Similar to the above specimen; lattice parameter reported at 288 K is 3.58125 Å; 3.58132 Å at 293 K determined by graphical interpolation.
22	389	Owen, E.A., et al.	1937	X	288-873		Bal.	24.22	Similar to the above specimen; lattice parameter reported at 288 K is 3.5733 Å; 3.5737 Å at 293 determined by graphical interpolation.
23*	389	Owen, E.A., and Yates, E.L.	1937	X	288-573		Bal.	24.2	Similar to the above specimen except α-phase; lattice parameter reported at 288 K is 2.8630 Å; 2.8631 Å at 293 determined by graphical interpolation.
24*	389	Owen, E.A., and Yates, E.L.	1937	X	288-873		Bal.	16.7	Similar to the above specimen; lattice parameter reported at 288 K is 2.86296 Å; 2.8631 Å at 293 determined by graphical interpolation.
25*	389	Owen, E.A., and Yates, E.L.	1937	X	288-873		Bal.	9.0	Similar to the above specimen; lattice parameter reported at 288 K is 2.86338 Å; 2.8635 Å at 293 K determined by graphical interpolation.
26	389	Owen, E.A., and Yates, E.L.	1937	X	288-873		Bal.	3.1	Similar to the above specimen; lattice parameter reported at 288 K is 2.86212 Å; 2.86229 Å at 293 K determined by graphical interpolation.
27*	391	Phragmen, G.	1931	X	283-453	C.F.O. 1586	Bal.	35.52	0.18 Mn, 0.09 Si, 0.02 C; specimen obtained from Société anonyme de Commentry, Fourchambault et Decazeville, Imphy, France; lattice parameter reported at 283 K is 3.5853 Å; 3.5859 Å at 293 K determined by graphical interpolation.
28	391	Phragmen, G.	1931	X	283-568	C.F.O. 1931	Bal.	36.00	0.18 Mn, 0.04 C, trace Si; similar to the above specimen; lattice parameter reported at 283 K is 3.5863 Å; 3.5864 Å at 293 K determined by graphical interpolation.
29	391	Phragmen, G.	1931	X	283-553		Bal.	35.8	0.02 C; alloy prepared by melting pure Ni and electrolytic Fe in vacuum; lattice parameter reported at 283 is 3.5848 Å; 3.5848 Å at 293 K determined by graphical interpolation
30	391	Phragmen, G.	1931	X	283, 573		Bal.	20	Nominal composition, similar to the above specimen; lattice parameter reported at 283 K is 3.572 Å; 3.573 Å at 293 K determined by graphical interpolation.
31	391	Phragmen, G.	1931	X	283-563		Bal.	50.4	Prepared by melting pure Ni and electrolytic Fe in vacuum; lattice parameter reported at 283 K is 3.5794 Å; 3.5797 Å at 293 K determined by graphical interpolation.

* Not shown in figure.

SPECIFICATION TABLE 190. THERMAL LINEAR EXPANSION OF IRON-NICKEL SYSTEM Fe-Ni (continued)

Cur. No.	Ref. No.	Author(s)	Year	Method Used	Temp. Range, K	Name and Specimen Designation	Composition (weight percent) Fe	Ni	Composition (continued), Specifications, and Remarks
32*	531	Asano, H.	1968	X	102-273		Bal.	21.1	Specimen prepared by reduction of iron-nickel oxalate, annealed 30 min at 923 K; lattice parameter reported at 278 K is 3.5832 Å; 3.5834 Å at 293 K determined by graphical extrapolation.
33*	59	Masumoto, H.	1931	L	303, 373	No. 2	95	5	Armco Fe, Mond Ni, supplied by Sugibayashi & Co. refined by electrolysis, mixed and melted in hydrogen atmosphere in alumina crucible in Tammann furnace, cast in iron mould with cylindrical aperture, machined to circular rod 4 mm thick, 10 cm long, heated at 1273 K for 1 hr in electric furnace in hydrogen atmosphere at reduced pressure, slowly cooled, measured in vacuum.
34*	59	Masumoto, H.	1931	L	303, 373	No. 3	90	10	Similar to the above specimen.
35*	59	Masumoto, H.	1931	L	303, 373	No. 4	85	15	Similar to the above specimen.
36*	59	Masumoto, H.	1931	L	303, 373	No. 5	80	20	Similar to the above specimen.
37*	59	Masumoto, H.	1931	L	303, 373	No. 6	75	25	Similar to the above specimen.
38*	59	Masumoto, H.	1931	L	303, 373	No. 8	70	30	Similar to the above specimen.
39*	59	Masumoto, H.	1931	L	303, 373	No. 9	67	33	Similar to the above specimen.
40	59	Masumoto, H.	1931	L	303, 373	No. 11	64	36	Similar to the above specimen.
41*	59	Masumoto, H.	1931	L	303, 373	No. 12	63.5	36.5	Similar to the above specimen.
42*	59	Masumoto, H.	1931	L	303, 373	No. 13	63	37	Similar to the above specimen.
43*	59	Masumoto, H.	1931	L	303, 373	No. 14	60	40	Similar to the above specimen.
44	59	Masumoto, H.	1931	L	303, 373	No. 15	55	45	Similar to the above specimen.
45*	59	Masumoto, H.	1931	L	303, 373	No. 16	50	50	Similar to the above specimen.
46*	59	Masumoto, H.	1931	L	303, 373	No. 17	40	60	Similar to the above specimen.
47*	59	Masumoto, H.	1931	L	303, 373	No. 18	30	70	Similar to the above specimen.
48*	59	Masumoto, H.	1931	L	303, 373	No. 19	20	80	Similar to the above specimen.
49*	59	Masumoto, H.	1931	L	303, 373	No. 20	10	90	Similar to the above specimen.
50	59	Masumoto, H.	1931	L	96-589	No. 24 Invar type	63.5	36.5	Similar to the above specimen; zero-point correction is -0.004%.
51	429	Mochel, N. L.	1927	T	273-732	K5 A (Malleable Nickel)	0.37	99.4	0.12 C, 0.051 Cu, 0.047 Si; cold-drawn material, annealed, cold rolled to 2.5 cm square bar; tested as received; expansion measured with increasing temperature, zero-point correction of -0.0253% determined by graphical interpolation.
52	429	Mochel, N. L.	1927	T	782-273	K5 A (Malleable Nickel)	0.37	99.4	The above specimen; expansion measured with decreasing temperature; zero-point correction of -0.0253% determined by graphical interpolation.
53*	429	Mochel, N. L.	1927	T	273-788	K5 B (Malleable Nickel)	0.32	99.45	0.14 C, 0.056 Si, 0.010 Cu; similar to the above specimen; expansion measured with increasing temperature; zero-point correction of -0.0261% determined by graphical interpolation.

* Not shown in figure.

SPECIFICATION TABLE 190. THERMAL LINEAR EXPANSION OF IRON-NICKEL SYSTEM Fe-Ni (continued)

Cur. No.	Ref. No.	Author(s)	Year	Method Used	Temp. Range, K	Name and Specimen Designation	Composition (weight percent) Fe	Ni	Composition (continued), Specifications, and Remarks
54*	429	Mochel, N. L.	1927	T	788-273	K5 B (Malleable Nickel)	0.32	99.45	The above specimen; expansion measured with decreasing temperature; zero-point correction of -0.0261% determined by graphical interpolation.
55	612	Boiko, B. T., Paltnik, L. S., and Rigachev, A. T.	1967	F	440-610	Permalloy	18	82	Non continuous film of permalloy prepared in electron diffraction camera by evaporation of 99.9 pure metals, measurements with increasing temperatures.
56	612	Boiko, B. T., et al.	1967	F	562-313		18	82	The above specimen; measurements with decreasing temperature.
57*	527	Zakharov, A. L., and Fedotov, L. N.	1967	L	9-121	Invar	Bal.	36	.003 C, .003 S, .002 P, trace Al, Si; produced in vacuum furnace; charge composed of carbonyl iron refined in hydrogen and electrolytic nickel; specimen 5 mm diameter and 43.5 mm long; quenched in water from 1373 K; annealed 8 hr at 588 and 48 hr at 368 K; this curve is here reported using the first temperature as reference temperature at which $\Delta L/L_0 = 0$.
58*	527	Zakharov, A. L., and Fedotov, L. N.	1967	L	9-134				Similar to the above specimen; this curve is here reported using the first temperature as reference temperature at which $\Delta L/L_0 = 0$.
59	527	Zakharov, A. L., and Fedotov, L. N.	1967	L	88-298				Similar to the above specimen.
60	527	Zakharov, A. L., and Fedotov, L. N.	1967	L	83-305				Similar to the above specimen.
61	528	Kachi, S. and Asano, H.	1969	X	112-283	Invar alloy	Bal.	25	Specimen prepared by hydrogen reduction of Fe-Ni oxalates; lattice parameter reported at 283 K is 3.5828 Å; 3.5837 Å at 293 K determined by graphical extrapolation.
62	528	Kachi, S. and Asano, H.	1969	X	106-288	Invar alloy	Bal.	31	Similar to the above specimen; lattice parameter reported at 288 K is 3.5880 Å; 3.5881 Å at 293 determined by graphical extrapolation.
63	467	Clark, A. F.	1968	L	0-300	LR35	Bal.	36.04	<0.1 each C, P, S, N, O; specimen machined to 0.64 cm square x 20.32 cm long; hot rolled specimen; Rockwell hardness B-75.
64*	191	Pollock, D. B.	1969	L	77, 293	Niromet-42	Bal.	42	Specimen 3.2 cm long; alloy from W. B. Driver Co. (expansion of fused silica taken as zero).
65	529	Janser, G. R.	1966		80-293		Bal.	18	0.090 in. thick sheet of maraging steel; air and vacuum melted, solution annealed.
66*	529	Janser, G. R.	1966		80-293		Bal.	18	Similar to the above specimen; solution annealed and aged.
67*	741	Tanji, Y.	1971	L	273-1072		10.38	89.60	0.007 Cu, 0.006 Si, 0.003 Mn, 0.00049 C, 0.00048 S, 0.00019 P; specimen prepared with electrolytic Fe and electrolytic Ni, mixed and melted in high frequency induction furnace; forged, rolled, faced to finished flat rectangular bar of length 11.995 cm, width 1.1990 cm, and thickness 0.1480 cm, annealed in vacuum at 1273 K for 10 hr and cooled in furnace, density: 8.757 g/cm^3 at room temperature.
68	741	Tanji, Y.	1971	L	273-1072		21.48	78.50	Similar to the above specimen; 0.006 Cu, 0.006 Si, 0.004 Mn, 0.0010 C, 0.0010 S, 0.00039 P; length 11.995 cm, width 1.2007 cm, thickness 0.1495 cm; density 8.600 g/cm^3 at room temperature.

* Not shown in figure.

SPECIFICATION TABLE 190. THERMAL LINEAR EXPANSION OF IRON-NICKEL SYSTEM Fe-Ni (continued)

Cur. No.	Ref. No.	Author(s)	Year	Method Used	Temp. Range, K	Name and Specimen Designation	Composition (weight percent) Fe	Ni	Composition (continued), Specifications, and Remarks
69	741	Tanji, Y.	1971	L	273-1072		29.96	70.02	Similar to the above specimen; 0.006 Si, 0.006 Cu, 0.004 Mn, 0.0014 C, 0.0014 S, 0.00054 P; length 11.995 cm, width 1.1989 cm, thickness 0.1484 cm; density 8.476 g/cm^3 at room temperature.
70*	741	Tanji, Y.	1971	L	273-1072		39.28	60.70	Similar to the above specimen; 0.006 Si, 0.005 Cu, 0.005 Mn, 0.0018 C, 0.0018 S, 0.00071 P; length 11.988 cm, width 1.1860 cm, thickness 0.1483 cm; density 8.334 g/cm^3 at room temperature.
71	741	Tanji, Y.	1971	L	273-1072		50.02	49.96	Similar to the above specimen; 0.005 Si, 0.005 Mn, 0.004 Cu, 0.0024 C, 0.0023 S, 0.00090 P; length 11.990 cm, width 1.1995 cm, thickness 0.1477 cm; density 8.205 g/cm^3 at room temperature.
72	741	Tanji, Y.	1971	L	273-1072		55.56	44.43	Similar to the above specimen; 0.005 Mn, 0.005 Si, 0.004 Cu, 0.0026 C, 0.0026 S, 0.0010 P; length 12.000 cm, width 1.1997 cm, thickness 0.1277 cm; density 8.165 g/cm^3 at room temperature.
73	741	Tanji, Y.	1971	L	273-1072		60.36	39.62	Similar to the above specimen; 0.005 Mn, 0.005 Si, 0.003 Cu, 0.0028 C, 0.0028 S, 0.0011 P; length 11.993 cm, width 1.1990 cm, thickness 0.1493 cm; density 8.142 g/cm^3 at room temperature.
74	741	Tanji, Y.	1971	L	273-1072		64.28	35.70	Similar to the above specimen; 0.006 Mn, 0.005 Si, 0.0030 C, 0.003 Cu, 0.0030 S, 0.0012 P; length 11.992 cm, width 1.1987 cm, thickness 0.1258 cm; density 8.138 g/cm^3 at room temperature.
75	741	Tanji, Y.	1971	L	273-1072		70.16	29.82	Similar to the above specimen; 0.006 Mn, 0.005 Si, 0.0033 C, 0.0032 S, 0.003 Cu, 0.0013 P; length 11.990 cm, width 1.1983, thickness 0.1492 cm; density 8.165 g/cm^3 at room temperature.
76*	14	Palatnik, L.S., Pugachev, A.T., Boiko, B.T., and Bratsykhin, V.M.	1967	X	438-608	Permalloy	18	82	Underlayer of NaCl condensed on base previous to formation of film; specimen film 400 to 600 Å thick; expansion measured with increasing temperature.
77*	14	Palatnik, L.S., et al.	1967	X	561-312	Permalloy	18	82	The above specimen; expansion measured with decreasing temperature.

* Not shown in figure.

DATA TABLE 190. THERMAL LINEAR EXPANSION OF IRON-NICKEL SYSTEM Fe-Ni

[Temperature, T, K; Linear Expansion, $\Delta L/L_0$, %]

T	$\Delta L/L_0$
CURVE 1	
286	0.001
290	0.004
292	-0.001
375	0.112*
427	0.183
470	0.240
524	0.308
568	0.388
578	0.408
625	0.467
678	0.575
732	0.663
786	0.760
827	0.831
882	0.910
CURVE 2	
293	0.000*
373	0.105*
473	0.247
571	0.386
612	0.448
680	0.573*
707	0.621
712	0.636*
726	0.664
762	0.724
784	0.749*
790	0.758*
793	0.769
829	0.828*
873	0.897
CURVE 3	
287	-0.006
293	0.000*
375	0.113*
474	0.249*
525	0.315*
576	0.400*
604	0.436
635	0.502
672	0.556
681	0.575*
697	0.607
CURVE 3 (cont.)	
730	0.672*
775	0.754
822	0.836*
867	0.887
881	0.930
CURVE 4	
288	-0.010*
292	-0.001*
375	0.098*
381	0.098*
478	0.236
575	0.378
636	0.479
657	0.502
680	0.561
702	0.578
726	0.632
757	0.677
775	0.702
802	0.745
829	0.796
879	0.892
CURVE 5	
283	-0.008*
289	-0.005*
377	0.108*
422	0.170
480	0.243
579	0.384
666	0.505
718	0.584
774	0.668
820	0.736
865	0.815
879	0.837
CURVE 6	
282	-0.010*
285	-0.004*
374	0.041
472	0.091
CURVE 6 (cont.)	
575	0.135
601	0.149
620	0.160
633	0.180
654	0.194
674	0.222
694	0.241
719	0.272
728	0.302
747	0.330
755	0.328*
760	0.353
766	0.356
773	0.372
782	0.381
796	0.394
807	0.425
844	0.470
878	0.537
CURVE 7	
285	-0.002*
328	0.007
375	0.029
393	0.049
423	0.071
468	0.121
505	0.177
533	0.222
578	0.283
628	0.378
670	0.465
717	0.551
769	0.649
825	0.747
880	0.816
CURVE 8	
288	-0.010*
485	0.371
677	0.732
740	0.832
803	0.953
869	1.059*
CURVE 8 (cont.)	
904	1.124*
904	1.140*
CURVE 9	
286	-0.013*
287	-0.009*
288	-0.006*
289	-0.013*
290	-0.006*
373	0.082*
473	0.194
577	0.323
579	0.337*
613	0.382
618	0.382*
623	0.393
653	0.438
691	0.487*
729	0.543
774	0.599*
871	0.728
CURVE 10	
288	-0.003*
289	-0.003*
290	-0.003*
291	-0.007*
293	0.000*
373	0.112*
476	0.213
567	0.325
568	0.321*
569	0.328*
613	0.370
667	0.454
689	0.485
743	0.552
775	0.601
829	0.671
873	0.733
CURVE 11	
289	-0.009*
290	-0.006*
291	-0.002*
374	0.095*
472	0.211
574	0.351
633	0.438
666	0.483
717	0.571
730	0.567
762	0.647
795	0.696
848	0.759
CURVE 12	
288	-0.007*
373	0.106*
473	0.242
573	0.392
673	0.569
773	0.736
873	0.903
CURVE 13*	
288	-0.007
373	0.105
473	0.244
573	0.394
673	0.569
773	0.741
873	0.905
CURVE 14*	
288	-0.007
373	0.105
473	0.242
573	0.392
673	0.562
773	0.748
873	0.912
CURVE 15	
288	-0.006*
373	0.104*
473	0.243*
573	0.389*
673	0.542
773	0.706
873	0.878
CURVE 16	
288	-0.006*
373	0.101*
473	0.232
573	0.372
673	0.513
773	0.666*
873	0.826
CURVE 17	
288	-0.006*
373	0.098*
473	0.221
573	0.345
673	0.471
773	0.595
873	0.720
CURVE 18	
288	-0.004*
373	0.061*
473	0.136
573	0.200
673	0.272
773	0.379
873	0.570
CURVE 19	
288	-0.002*
373	0.028*
473	0.074
573	0.122
673	0.203
773	0.351*
873	0.513

* Not shown in figure.

DATA TABLE 190. THERMAL LINEAR EXPANSION OF IRON-NICKEL SYSTEM Fe-Ni (continued)

[Temperature, T, K; Linear Expansion, $\Delta L/L_0$, %]

T	$\Delta L/L_0$
CURVE 20	
288	-0.001*
373	0.019
473	0.054
573	0.164
673	0.331
773	0.504
873	0.680
CURVE 21*	
288	-0.002
373	0.035
473	0.132
573	0.289
673	0.463
773	0.639
873	0.816
CURVE 22	
288	-0.018*
373	0.200
473	0.376
573	0.550
673	0.726
773	0.901
873	1.076
CURVE 23*	
288	-0.005
373	0.082
473	0.190
573	0.313
CURVE 24*	
288	-0.005
373	0.084
473	0.203
573	0.328
673	0.463
773	0.603
873	0.744
CURVE 25*	
288	-0.005
CURVE 25 (cont.)*	
373	0.092
473	0.209
573	0.332
673	0.463
773	0.597
873	0.734
CURVE 26	
288	-0.006*
373	0.093*
473	0.215*
573	0.349*
673	0.493
773	0.651
873	0.853
CURVE 27*	
283	-0.002
378	0.017
453	0.051
CURVE 28	
283	-0.003*
383	0.025*
468	0.128
568	0.268
CURVE 29	
283	0.000*
378	0.000
453	0.056
553	0.204
CURVE 30	
283	-0.023
573	0.649
CURVE 31	
283	-0.008*
378	0.070
463	0.162
563	0.291
CURVE 32*	
102	-0.232
110	-0.229
120	-0.221
143	-0.209
155	-0.198
161	-0.190
175	-0.179
185	-0.179
188	-0.167
197	-0.151
208	-0.137
212	-0.112
228	-0.120
230	-0.114
237	-0.087
246	-0.064
252	-0.047
257	-0.064
276	-0.042
278	-0.006
CURVE 33*	
303	0.011
373	0.089
CURVE 34*	
303	0.011
373	00.086
CURVE 35*	
303	0.009
373	0.079
CURVE 36*	
303	0.011
373	0.088
CURVE 37*	
303	0.013
373	0.108
CURVE 38*	
303	0.012
373	0.096
CURVE 39*	
303	0.003
373	0.028
CURVE 40	
303	0.001
373	0.010
CURVE 41*	
303	0.001
373	0.010
CURVE 42*	
303	0.001
373	0.012
CURVE 43*	
303	0.004
373	0.033
CURVE 44	
303	0.007
373	0.057
CURVE 45*	
303	0.009
373	0.077
CURVE 46*	
303	0.0113
373	0.0902
CURVE 47*	
303	0.0122
373	0.0979
CURVE 48*	
303	0.0128
373	0.1022
CURVE 49*	
303	0.0129
373	0.1031
CURVE 50	
96	-0.041
122	-0.034
164	-0.025*
198	-0.016
235	-0.009
273	-0.004*
293	0.000*
307	0.002*
323	0.003
337	0.005
373	0.010*
402	0.016
424	0.021
457	0.031
476	0.038
505	0.053
527	0.067
551	0.085
576	0.110
589	0.126
CURVE 51	
273	-0.0253
298	0.0063*
420	0.1703
537	0.3443
656	0.5408
785	0.7503
CURVE 52	
782	0.7503
558	0.3783
298	0.0103
CURVE 53*	
273	-0.0261
298	0.0065
419	0.1645
537	0.3395
660	0.5435
788	0.7539
CURVE 54*	
788	0.7539
515	0.3035
293	0.0065
CURVE 55	
440	0.181
491	0.243
552	0.312
610	0.401
CURVE 56	
313	0.040
363	0.112
431	0.168
513	0.261
562	0.318
CURVE 57*, †, ‡	
9	0.000
12	-0.003
16	-0.008
21	-0.017
25	-0.023
32	-0.034
37	-0.041
41	-0.045
45	-0.048
48	-0.050
53	-0.052
57	-0.052
61	-0.051
65	-0.050
73	-0.045
77	-0.042
81	-0.038

* Not shown in figure.

† This curve is here reported using the first given temperature as reference temperature at which $\Delta L/L_0 = 0$.

‡ Author's data for coefficient of thermal expansion have been integrated by TPRC to obtain $\Delta L/L_0$.

DATA TABLE 190. THERMAL LINEAR EXPANSION OF IRON-NICKEL SYSTEM Fe-Ni (continued)

[Temperature, T, K; Linear Expansion, $\Delta L/L_0$, %]

T	$\Delta L/L_0$
CURVE 57 (cont.)*, †, ‡	
85	-0.034
89	-0.029
92	-0.025
98	-0.017
102	-0.012
114	0.006
121	0.017
CURVE 58*, †, ‡	
9	0.000
29	-0.027
33	-0.033
37	-0.038
41	-0.042
44	-0.045
49	-0.048
52	-0.049
57	-0.050
61	-0.050
65	-0.049
69	-0.047
73	-0.044
78	-0.040
81	-0.038
90	-0.027
101	-0.012
110	0.000
118	0.012
126	0.025
134	0.037
CURVE 59‡	
88	-0.0153
92	-0.0148
96	-0.0143*
104	-0.0133
107	-0.0128*
112	-0.0120*
119	-0.0109
128	-0.0094
132	-0.0088
136	-0.0082
140	-0.0076*

T	$\Delta L/L_0$
CURVE 59 (cont.)‡	
143	-0.0071*
147	-0.0065
153	-0.0057
157	-0.0051
160	-0.0047*
164	-0.0042*
168	-0.0038*
172	-0.0034*
176	-0.0029
179	-0.0026*
184	-0.0022*
187	-0.0019
191	-0.0017
195	-0.0014
200	-0.0011*
203	-0.0009*
207	-0.0007
211	-0.0006
216	-0.0004
219	-0.0003*
223	-0.0002*
227	-0.0002*
231	-0.0001*
236	-0.0001*
239	-0.0001
243	-0.0000
247	-0.0000*
251	-0.0000*
254	-0.0000
260	-0.0000*
263	-0.0000
269	-0.0001
277	-0.0001
287	-0.0001*
298	0.0001*
CURVE 60‡	
83	-0.00159
86	-0.00156*
94	-0.0147*
98	-0.0141
100	-0.0138*
102	-0.0135*
106	-0.0129*

T	$\Delta L/L_0$
CURVE 60 (cont.)‡	
110	-0.0122
114	-0.0116*
118	-0.0110*
122	-0.0103
126	-0.0097*
129	-0.0092*
134	-0.0084
138	-0.0078*
142	-0.0072
146	-0.0066*
150	-0.0060
155	-0.0054*
159	-0.0048*
162	-0.0045
166	-0.0040*
170	-0.0036
174	-0.0032*
178	-0.0028*
182	-0.0024
186	-0.0020*
190	-0.0017*
194	-0.0014*
198	-0.0012*
201	-0.0010
205	-0.0008*
210	-0.0006*
214	-0.0005*
217	-0.0004
222	-0.0003
226	-0.0002*
230	-0.0001
233	-0.0001*
235	-0.0000*
237	0.0000*
241	0.0000*
245	0.0000
250	0.0000
253	0.0000*
258	0.0000*
262	0.0000*
270	0.0000*
275	-0.0001*
286	0.0000
291	0.0000*
295	0.0000*
305	0.0002

T	$\Delta L/L_0$
CURVE 61	
112	-0.226
122	-0.215
148	-0.193
173	-0.167
199	-0.140
221	-0.109
247	-0.075
274	-0.045*
283	-0.025
CURVE 62	
106	-0.128
123	-0.120
147	-0.100
173	-0.084
197	-0.067
222	-0.047
246	-0.031
273	-0.011
288	-0.003*
CURVE 63	
0	-
10	-
20	-0.038
30	-0.039
40	-0.040
50	-0.039
60	-0.038
70	-0.037
80	-0.036*
90	-0.035
100	-0.033
120	-0.029
140	-0.024
160	-0.019
180	-0.015
200	-0.011*
220	-0.008*
240	-0.006*
260	-0.005
273	-0.004*
280	-0.004*
293	0.000*

T	$\Delta L/L_0$
CURVE 64*	
77	-0.118
293	0.000*
CURVE 65	
293	0.000*
255	-0.0378
200	-0.0910
144	-0.1324
89	-0.1704
80	-0.1739
CURVE 66*	
293	0.000
255	-0.032
200	-0.078
144	-0.122
89	-0.158
80	-0.165
CURVE 67*, ‡	
273	-0.021
293	0.000
323	0.033
373	0.090
423	0.149
473	0.211
523	0.275
573	0.341
583	0.354
593	0.368
603	0.382
613	0.396
623	0.410
633	0.424
643	0.438
653	0.453
663	0.467
673	0.482
683	0.497
693	0.512
703	0.528
713	0.544
723	0.560

T	$\Delta L/L_0$
CURVE 67 (cont.)*, ‡	
733	0.576
743	0.593
753	0.608
763	0.624
773	0.639
783	0.655
793	0.670
803	0.686
813	0.702
823	0.718
873	0.798
923	0.882
973	0.967
1023	1.055
1073	1.145
CURVE 68‡	
273	-0.021*
293	0.000*
323	0.033*
373	0.090*
423	0.150
473	0.212*
523	0.277
573	0.345*
623	0.416
673	0.489*
723	0.565
773	0.643
783	0.659
793	0.675
803	0.691
813	0.708
823	0.724
833	0.741
843	0.758
853	0.776
863	0.793
873	0.810
883	0.827
893	0.844
903	0.861
913	0.878
923	0.896
933	0.913

* Not shown in figure.
† This curve is here reported using the first given temperature as reference temperature at which $\Delta L/L_0 = 0$.
‡ Author's data for coefficient of thermal expansion have been integrated by TPRC to obtain $\Delta L/L_0$.

DATA TABLE 190. THERMAL LINEAR EXPANSION OF IRON-NICKEL SYSTEM Fe-Ni (continued)

[Temperature, T, K; Linear Expansion, $\Delta L/L_0$, %]

T	$\Delta L/L_0$
CURVE 68 (cont.)‡	
943	0.931
973	0.984
1023	1.076*
1073	1.170*
CURVE 69‡	
273	-0.024*
293	0.000*
323	0.036*
373	0.098*
423	0.162
473	0.227*
523	0.295
573	0.365
623	0.436
673	0.509*
723	0.583
773	0.660*
823	0.737*
833	0.753
843	0.769
853	0.785
863	0.801
873	0.817*
883	0.833
893	0.850*
903	0.867
913	0.884*
923	0.902
933	0.919*
943	0.936*
953	0.954
963	0.972
973	0.989*
1023	1.080*
1073	1.172*
CURVE 70*	
273	-0.023
293	0.000
323	0.035
373	0.094
423	0.155
473	0.219

T	$\Delta L/L_0$
CURVE 70 (cont.)*	
523	0.286
573	0.355
623	0.426
673	0.499
683	0.514
693	0.528
703	0.543
713	0.558
723	0.573
733	0.588
743	0.602
753	0.617
763	0.632
773	0.646
783	0.661
793	0.675
803	0.689
813	0.703
823	0.717
833	0.730
843	0.745
853	0.760
863	0.776
873	0.792
883	0.809
893	0.826
903	0.843
913	0.859
923	0.876
933	0.894
973	0.963
1023	1.052
1073	1.143
CURVE 71‡	
273	-0.020*
293	0.000*
323	0.030*
373	0.079*
423	0.128
473	0.178
523	0.229
573	0.280
623	0.331
633	0.341

T	$\Delta L/L_0$
CURVE 71 (cont.)‡	
643	0.352
653	0.362
673	0.381
683	0.391
693	0.401
703	0.411
713	0.420
723	0.430
733	0.439
743	0.448
753	0.457
763	0.467
773	0.478
783	0.490
793	0.503
803	0.516
913	0.530
823	0.544*
833	0.559
843	0.573
853	0.588
863	0.603
873	0.618
923	0.695
973	0.775
1023	0.858*
1073	0.942*
CURVE 72‡	
273	-0.013*
293	0.000*
323	0.020
373	0.055*
423	0.093
473	0.133*
523	0.172
573	0.212
583	0.219
593	0.227
603	0.235
613	0.243
623	0.251
633	0.258
643	0.266
653	0.273

T	$\Delta L/L_0$
CURVE 72 (cont.)‡	
663	0.281
673	0.288*
683	0.296
693	0.303
703	0.310
713	0.317
723	0.324
733	0.332
743	0.341
753	0.352
763	0.363
773	0.375*
793	0.401
819	0.439
873	0.522
893	0.554
918	0.596
973	0.689
1018	0.769*
1073	0.870*
CURVE 73‡	
273	-0.008*
293	0.000*
323	0.011
373	0.026*
423	0.041
473	0.055*
523	0.071
573	0.090
597	0.101
623	0.118
643	0.136
673	0.170
693	0.196
723	0.237
773	0.310
823	0.388
873	0.469
923	0.553
973	0.641
1023	0.731*
1073	0.825*

T	$\Delta L/L_0$
CURVE 74‡	
273	-0.002
293	0.000*
303	0.001*
323	0.003*
349	0.007
373	0.010*
397	0.016
423	0.023*
449	0.033
473	0.045
498	0.062
523	0.084
543	0.106
573	0.144
593	0.171
623	0.216
673	0.293
723	0.375
773	0.459
823	0.546
873	0.636
923	0.728
973	0.822
1023	0.919
1073	1.018
CURVE 75‡	
273	-0.008*
293	0.000*
299	0.004*
323	0.024*
373	0.083*
423	0.152*
473	0.228*
523	0.308*
573	0.390*
623	0.475
673	0.562*
723	0.651
773	0.743*
823	0.836
973	0.931
923	1.028*
973	1.127*
1023	1.227*
1073	1.330*

T	$\Delta L/L_0$
CURVE 76*	
438	0.165
490	0.224
551	0.299
608	0.387
CURVE 77*	
561	0.303
512	0.243
431	0.146
362	0.094
312	0.018

* Not shown in figure.
‡ Author's data for coefficient of thermal expansion have been integrated by TPRC to obtain $\Delta L/L_0$.

DATA TABLE 190. COEFFICIENT OF THERMAL LINEAR EXPANSION OF IRON-NICKEL SYSTEM Fe-Ni

[Temperature, T, K; Coefficient of Expansion, α, 10^{-6} K^{-1}]

T	α
CURVE 57‡	
9	-1.00
12	-1.18
16	-1.63
21	-1.67
25	-1.67
32	-1.42
37	-1.20
41	-1.03
45	-0.62
48	-0.50
53	-0.19
57	0.09
61	0.24
65	0.45
73	0.71
77	0.93
81	1.10
85	1.14*
89	1.17*
92	1.24
98	1.37
102	1.45*
114	1.62
121	1.58
CURVE 58‡	
9	-1.05
29	-1.65
33	-1.48
37	-1.14
41	-0.99
44	-0.75*
49	-0.55*
52	-0.27
57	-0.10
61	0.17
65	0.38
69	0.65
73	0.76*
78	0.74
81	0.98
90	1.33
101	1.40
110	1.50
118	1.58
CURVE 58 (cont.)‡	
126	1.56
134	1.55
CURVE 59‡	
88	1.00
92	1.19
96	1.32
104	1.40*
107	1.51
112	1.61
119	1.67
128	1.61
132	1.51
136	1.64
140	1.39
143	1.50
147	1.46
153	1.40
157	1.31
160	1.30*
164	1.22
168	1.00
172	1.09
176	1.07*
179	0.85
184	0.97
187	0.72
191	0.72
195	0.69*
200	0.54
203	0.50*
207	0.34
211	0.34*
216	0.37
219	0.29*
223	0.18
227	0.09*
231	0.08
236	0.13
239	0.10*
243	0.00
247	0.00*
251	0.00*
254	0.00
260	-0.03
CURVE 59 (cont.)‡	
263	0.00
269	-0.13
277	0.02
287	0.09
298	0.17
CURVE 60‡	
83	1.02
86	1.09
94	1.28
98	1.45
100	1.52
102	1.54
106	1.60
110	1.54
114	1.57
118	1.63
122	1.65
126	1.61
129	1.50
134	1.63
138	1.46
142	1.50
146	1.49
150	1.38
155	1.37
159	1.26
162	1.21
166	1.08
170	1.08
174	1.00
178	1.01
182	0.89
186	0.84
190	0.69
194	0.71
198	0.57
201	0.57
205	0.38
210	0.37
214	0.37
217	0.26
222	0.29
226	0.17
230	0.15
CURVE 60 (cont.)‡	
233	0.09
235	0.09
237	0.09
241	0.07
245	0.00
250	0.00
253	-0.04
258	0.00
262	-0.04
270	-0.05
275	0.00
286	0.03
291	0.05
295	0.15
305	0.31
CURVE 67*,‡	
273	10.65
293	10.74
323	11.14
373	11.67
423	12.08
473	12.57
523	12.97
573	13.45
583	13.62
593	13.75
603	13.83
613	13.88
623	14.18
633	14.26
643	14.36
653	14.52
663	14.64
673	14.98
683	15.12
693	15.39
703	15.63
713	15.82
723	16.27
733	17.10
743	15.89
753	15.39
763	15.39
773	15.50
CURVE 67 (cont.)*,‡	
783	15.54
793	15.58
803	15.68
813	15.77
823	15.96
873	16.40
923	16.83
973	17.34
1023	17.83
1073	18.23
CURVE 68‡	
273	10.63
293	10.78
323	11.06
373	11.73
423	12.24
473	12.78
523	13.31
573	13.83
623	14.36
673	14.93
723	15.38
773	15.88
783	15.96
793	16.08
803	16.33
813	16.57
823	16.78
833	16.99
843	17.23
853	17.51
863	17.16
873	16.97
883	16.97
893	17.11
903	17.24
913	17.36
923	17.49
933	17.49
943	17.68
973	17.98
1023	18.51
1073	19.04
CURVE 69‡	
273	11.71
293	11.98
323	12.17
373	12.52
423	12.93
473	13.36
523	13.80
573	14.00
623	14.47
673	14.69
723	15.06
773	15.49
823	15.70
833	15.70
843	15.82
853	15.92
863	16.00
873	16.34
883	16.67
893	16.95
903	17.03
913	17.17
923	17.31
933	17.34
943	17.48
953	17.59
963	17.70
973	17.86
1023	18.30
1073	18.79
CURVE 70*,‡	
273	11.15
293	11.40
323	11.64
373	12.09
423	12.49
473	13.06
523	13.55
573	13.97
623	14.44
673	14.88
683	14.79
693	14.79
703	14.79

* Not shown in figure.

‡ Author's data for coefficient of thermal expansion have been integrated by TPRC to obtain $\Delta L/L_0$.

DATA TABLE 190. COEFFICIENT OF THERMAL LINEAR EXPANSION OF IRON-NICKEL SYSTEM Fe-Ni (continued)

T	α
CURVE 70 (cont.)*, ‡	
713	14.79
723	14.79
733	14.72
743	14.71
753	14.71
763	14.61
773	14.50
783	14.36
793	14.18
803	14.10
813	13.85
823	13.64
833	13.44
843	15.04
853	15.77
863	16.35
873	16.47
883	16.64
893	16.78
903	16.84
913	16.96
923	17.06
933	17.16
973	17.53
1023	17.99
1073	18.49
CURVE 71‡	
273	9.91
293	9.91
323	9.91
373	9.80
423	9.91
473	10.13
523	10.21
573	10.21
623	10.21
633	10.13
643	10.02
653	9.97
673	9.89
683	9.80
693	9.64
703	9.64
713	9.43
723	9.29
733	9.21

T	α
CURVE 71 (cont.)‡	
743	9.04
753	9.13
763	10.05
773	12.10
783	12.64
793	13.20
803	13.73
813	13.99
823	14.23
833	14.36
843	14.68
853	14.93
863	15.05
873	15.17
923	15.70
973	16.27
1023	16.68
1073	17.24
CURVE 72‡	
273	6.17
293	6.56
323	6.81
373	7.33
423	7.79
473	7.98
523	7.92
573	7.79
583	7.79
593	7.79
603	7.79
613	7.79
623	7.60
633	7.61
643	7.60
653	7.55
663	7.46
673	7.37
683	7.31
693	7.10
703	7.04
713	7.04
723	7.04
733	8.52
743	10.05
753	11.02
763	11.70

T	α
CURVE 72 (cont.)‡	
773	12.48
793	13.96
819	14.84
873	16.01
893	16.31
918	16.69
973	17.40
1018	18.04
1073	18.68
CURVE 73‡	
273	3.90
293	3.74
323	3.36
373	2.95
423	2.76
473	2.95
523	3.37
573	4.23
597	5.41
623	7.62
643	10.36
673	12.21
693	13.42
723	14.07
773	15.22
823	15.88
873	16.52
923	17.15
973	17.84
1023	18.47
1073	19.12
CURVE 74‡	
273	0.80
293	0.86
303	1.00
323	1.22
349	1.44
373	1.80
397	2.45
423	3.17
449	4.34
473	5.80
498	7.75
523	9.95

T	α
CURVE 74 (cont.)‡	
543	11.93
573	13.39
593	14.33
623	15.03
673	16.04
723	16.61
773	17.17
823	17.65
873	18.13
923	18.65
973	19.14
1023	19.62
1073	20.01
CURVE 75‡	
273	2.48
293	5.28
299	7.07
323	10.23
373	13.15
423	14.63
473	15.65
523	16.28
573	16.73
623	17.12
673	17.67
723	18.12
773	18.48
823	18.81
873	19.20
923	19.53
973	19.91
1023	20.35
1073	20.74

* Not shown in figure.
‡ Author's data for coefficient of thermal expansion have been integrated by TPRC to obtain $\Delta L/L_0$.

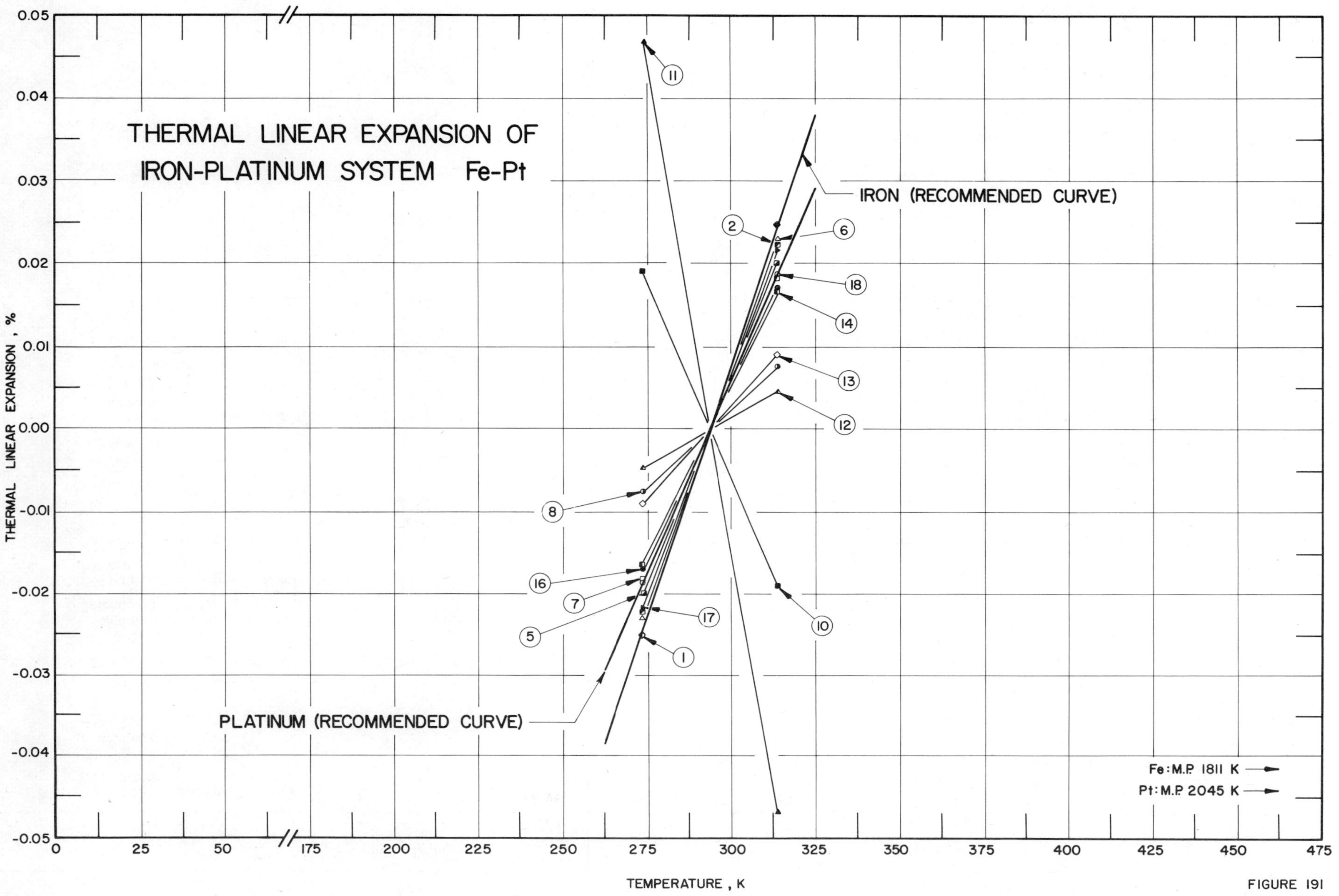
THERMAL LINEAR EXPANSION OF
IRON-PLATINUM SYSTEM Fe-Pt
IRON (RECOMMENDED CURVE)
PLATINUM (RECOMMENDED CURVE)
Fe: M.P. 1811 K
Pt: M.P. 2045 K
THERMAL LINEAR EXPANSION, %
TEMPERATURE, K
0.05
0.04
0.03
0.02
0.01
0.00
-0.01
-0.02
-0.03
-0.04
-0.05
0
25
50
175
200
225
250
275
300
325
350
375
400
425
450
475
11
2
6
18
14
13
12
8
16
7
5
17
1
10

FIGURE 191

SPECIFICATION TABLE 191. THERMAL LINEAR EXPANSION OF IRON-PLATINUM SYSTEM Fe-Pt

Cur. No.	Ref. No.	Author(s)	Year	Method Used	Temp. Range, K	Name and Specimen Designation	Composition (weight percent) Fe	Pt	Composition (continued), Specifications, and Remarks
1	631	Masumoto, H. and Kobayashi, T.	1965	L	273-313		90.0	10.0	Alloy prepared using 99.99 pure iron and platinum; melted in alumina crucible in Tammann furnace in hydrogen atmosphere; heated at 1273 K for 1 hr in vacuum furnace and cooled in it.
2	631	Masumoto, H. and Kobayashi, T.	1965	L	273-313		80.0	20.0	Similar to the above specimen.
3*	631	Masumoto, H. and Kobayashi, T.	1965	L	273-313		70.0	30.0	Similar to the above specimen.
4*	631	Masumoto, H. and Kobayashi, T.	1965	L	273-313		60.0	40.0	Similar to the above specimen.
5	631	Masumoto, H. and Kobayashi, T.	1965	L	273-313		55.0	45.0	Similar to the above specimen.
6	631	Masumoto, H. and Kobayashi, T.	1965	L	273-313		50.0	50.0	Similar to the above specimen.
7	631	Masumoto, H. and Kobayashi, T.	1965	L	273-313		48.0	52.0	Similar to the above specimen.
8	631	Masumoto, H. and Kobayashi, T.	1965	L	273-313		47.0	53.0	Similar to the above specimen.
9*	631	Masumoto, H. and Kobayashi, T.	1965	L	273-313		46.0	54.0	Similar to the above specimen; negative value of the mean temperature coefficient is explained by the theory of invar.
10	631	Masumoto, H. and Kobayashi, T.	1965	L	273-313		45.0	55.0	Similar to the above specimen; negative value of the mean temperature coefficient is explained by the theory of invar.
11	631	Masumoto, H. and Kobayashi, T.	1965	L	273-313		44.0	56.0	Similar to the above specimen; negative value of the mean temperature coefficient is explained by the theory of invar.
12	631	Masumoto, H. and Kobayashi, T.	1965	L	273-313		43.0	57.0	Similar to the above specimen.
13	631	Masumoto, H. and Kobayashi, T.	1965	L	273-313		42.0	58.0	Similar to the above specimen.
14	631	Masumoto, H. and Kobayashi, T.	1965	L	273-313		35.0	65.0	Similar to the above specimen.
15*	631	Masumoto, H. and Kobayashi, T.	1965	L	273-313		32.0	68.0	Similar to the above specimen.
16	631	Masumoto, H. and Kobayashi, T.	1965	L	273-313		30.0	70.0	Similar to the above specimen.
17	631	Masumoto, H. and Kobayashi, T.	1965	L	273-313		20.0	80.0	Similar to the above specimen.
18	631	Masumoto, H. and Kobayashi, T.	1965	L	273-313		10.0	90.0	Similar to the above specimen.

*Not shown in figure.

DATA TABLE 191. THERMAL LINEAR EXPANSION OF IRON-PLATINUM SYSTEM Fe-Pt

[Temperature, T, K; Linear Expansion, $\Delta L/L_0$, %]

T	$\Delta L/L_0$	T	$\Delta L/L_0$
CURVE 1		CURVE 11	
273	-0.025	273	0.047
313	0.025	313	-0.047
CURVE 2		CURVE 12	
273	-0.022	273	-0.004
313	0.022	313	0.004
CURVE 3		CURVE 13	
273	-0.023	273	-0.009
313	0.023	313	0.009
CURVE 4*		CURVE 14	
273	-0.018	273	-0.016
313	0.018	313	0.016
CURVE 5		CURVE 15*	
273	-0.020	273	-0.016
313	0.020	313	0.016
CURVE 6		CURVE 16*	
273	-0.023	273	-0.017
313	0.023	313	0.017
CURVE 7		CURVE 17	
273	-0.018	273	-0.021
313	0.018	313	0.021
CURVE 8		CURVE 18	
273	-0.007	273	-0.018
313	0.007	313	0.018
CURVE 9			
273	0.011		
313	-0.011		
CURVE 10*			
273	0.019		
313	-0.019		

*Not shown in figure.

FIGURE AND TABLE NO. 192R. PROVISIONAL VALUES FOR THERMAL LINEAR EXPANSION OF IRON-RHODIUM SYSTEM Fe-Rh

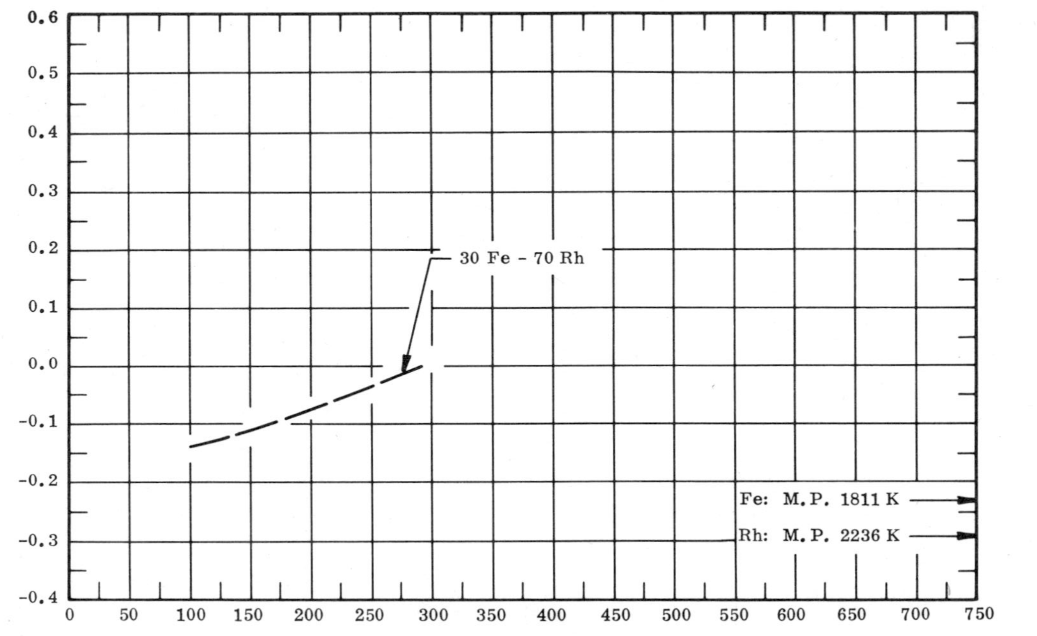

PROVISIONAL VALUES

[Temperature, T, K; Linear Expansion, $\Delta L/L_0$, %; α, K^{-1}]

T	(30 Fe-70 Rh) $\Delta L/L_0$	$\alpha \times 10^6$
100	-0.140	4.6
150	-0.113	6.3
200	-0.077	7.8
293	0.000	9.0

REMARKS

(30 Fe-70 Rh): The tabulated values for well-annealed alloy are considered accurate to within ±7% over the entire temperature range. These values can be represented approximately by the following equation:

$$\Delta L/L_0 = -0.154 - 2.020 \times 10^{-4}\, T + 3.896 \times 10^{-6}\, T^2 - 4.817 \times 10^{-9}\, T^3$$

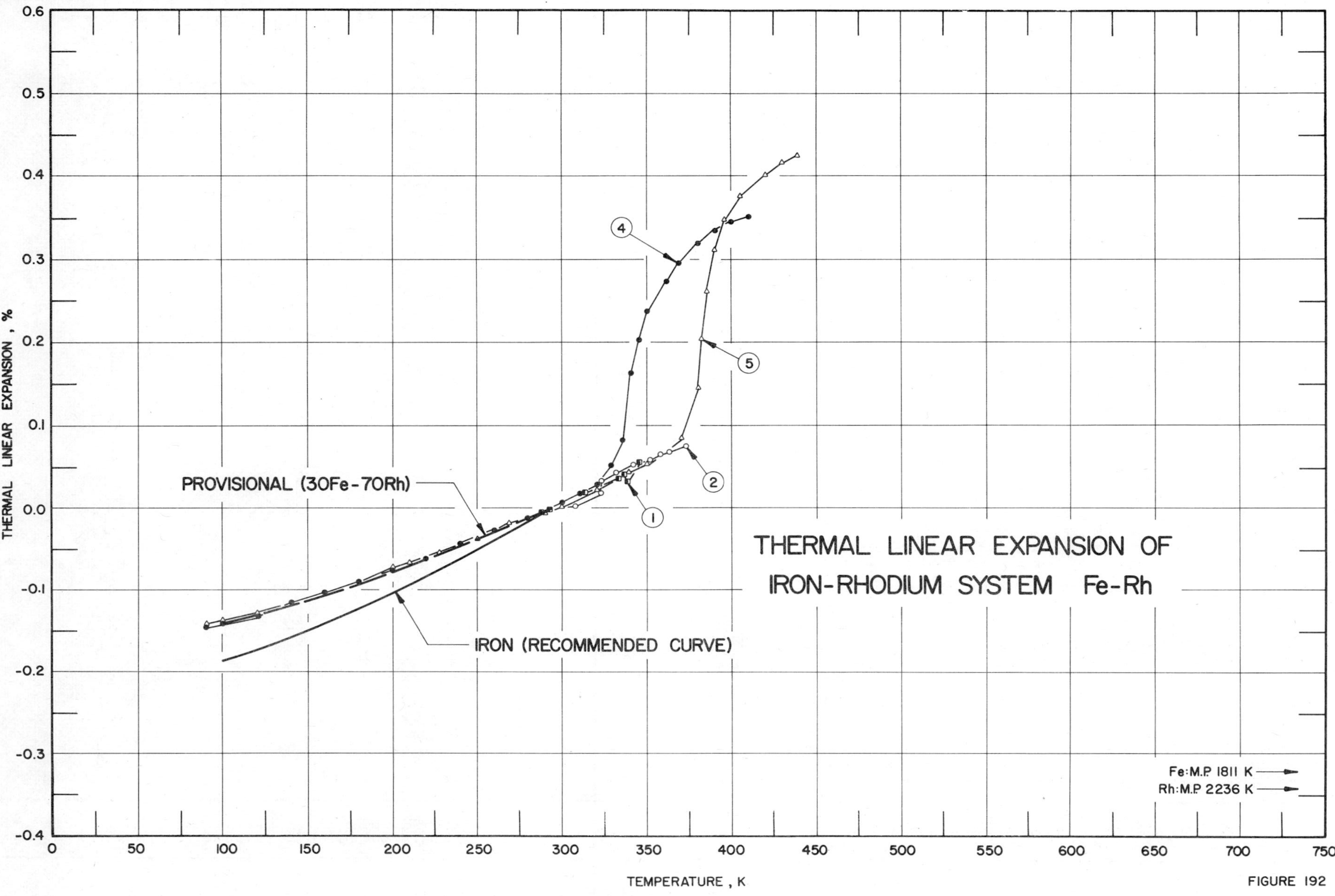

THERMAL LINEAR EXPANSION OF
IRON-RHODIUM SYSTEM Fe-Rh
PROVISIONAL (30Fe-70Rh)
IRON (RECOMMENDED CURVE)
1
2
4
5
Fe:M.P. 1811 K
Rh:M.P. 2236 K
THERMAL LINEAR EXPANSION, %
TEMPERATURE, K
0.6
0.5
0.4
0.3
0.2
0.1
0.0
-0.1
-0.2
-0.3
-0.4
0
50
100
150
200
250
300
350
400
450
500
550
600
650
700
750

FIGURE 192

SPECIFICATION TABLE 192. THERMAL LINEAR EXPANSION OF IRON-RHODIUM SYSTEM Fe-Rh

Cur. No.	Ref. No.	Author(s)	Year	Method Used	Temp. Range, K	Name and Specimen Designation	Composition (weight percent) Fe	Rh	Composition (continued), Specifications, and Remarks
1	440	Zsoldes, L.	1967	X	289-363		31.70	68.30	Specimen prepared on fused and drawn wires; contained a considerable amount of the γ-phase; received from Dr. L. Pál of Central Research Institute of Physics; average of two or more independent cooling and heating measurements; phase transition near 346 K; lattice parameter reported at 289 K is 2.9862 Å; 2.9863 Å at 293 K determined by graphical interpolation.
2	440	Zsoldes, L.	1967	X	295-423		35.18	64.82	Chemically prepared sample; all nearly pure α' (B_2-types) structures; phase transition between 358-372 K; lattice parameter reported at 295 K is 2.9857 Å; 2.9856 Å at 293 K determined by graphical extrapolation.
3*	440	Zsoldes, L.	1967	X	295-362		35.54	64.46	Specimen prepared on fused and drawn wires containing an amount of γ-phase; phase transition near 323 K; lattice parameter reported at 294 K is 2.9860 Å; 2.9861 Å at 293 K determined by graphical extrapolation.
4	441	McKinnon, J.B., Melville, D., and Lee, E.W.	1970	L	90-410	Fe-Rh II	32.28	67.71	Sample prepared in argon arc furnace; vacuum annealed for 24 hr at 1273 K; zero-point correction is -0.247%.
5	441	McKinnon, J.B., et al.	1970	L	90-439	Fe-Rh III	25.05	74.93	Similar to the above specimen; zero-point correction is -0.143%.

*Not shown in figure.

DATA TABLE 192. THERMAL LINEAR EXPANSION OF IRON-RHODIUM SYSTEM Fe-Rh

[Temperature, T, K; Linear Expansion, $\Delta L/L_0$, %]

T	$\Delta L/L_0$
CURVE 1	
289	-0.003
292	-0.003*
293	0.000
293	0.003*
314	0.020
333	0.037
336	0.040
337	0.033
340	0.040*
345	0.057
	a(Å)
346	2.9873*
346	2.9874*
348	2.9873*
358	2.9871*
358	2.9877*
363	2.9872*
CURVE 2	
	$\Delta L/L_0$
293	0.000*
295	0.003*
308	0.003
323	0.020
323	0.033
332	0.044
342	0.053
352	0.060
352	0.064*
358	0.067
358	0.070*
363	0.070
372	0.077
	a(Å)
358	2.9867*
358	2.9868*
362	2.9868*
372	2.9869*
382	2.9872*
387	2.9873*

T	a(Å)
CURVE 2 (cont.)	
392	2.9874*
402	2.9876*
423	2.9880*
CURVE 3*	
	$\Delta L/L_0$
293	0.000
294	-0.003
294	0.007
311	0.010
322	0.037
323	0.033
323	0.047
	a(Å)
323	2.9860
323	2.9857
326	2.9860
329	2.9862
332	2.9862
336	2.9860
336	2.9861
362	2.9868
CURVE 4	
	$\Delta L/L_0$
90	-0.147
100	-0.142
121	-0.132
140	-0.118
160	-0.104
180	-0.090
200	-0.076
220	-0.062
240	-0.043
260	-0.028
280	-0.011
293	0.000*
300	0.009
310	0.019
320	0.030

T	$\Delta L/L_0$
CURVE 4 (cont.)	
329	0.053
335	0.084
340	0.164
345	0.204
350	0.237
361	0.272
369	0.299
380	0.320
390	0.335
400	0.347
410	0.354
CURVE 5	
90	-0.143
100	-0.138
120	-0.129
140	-0.118*
160	-0.104*
180	-0.090*
200	-0.071
210	-0.066
220	-0.059*
229	-0.052
239	-0.045*
250	-0.036
260	-0.028*
269	-0.019
280	-0.011*
290	-0.004
293	0.000*
300	0.003
310	0.014*
320	0.022
339	0.044
350	0.054
359	0.066*
370	0.087
380	0.149
382	0.205
385	0.262
386	0.283*
390	0.312
392	0.335*
396	0.350
405	0.376

T	$\Delta L/L_0$
CURVE 5 (cont.)	
420	0.402
430	0.419
439	0.428

*Not shown in figure.

FIGURE AND TABLE NO. 193R. PROVISIONAL VALUES FOR THERMAL LINEAR EXPANSION OF IRON-SILICON SYSTEM Fe-Si

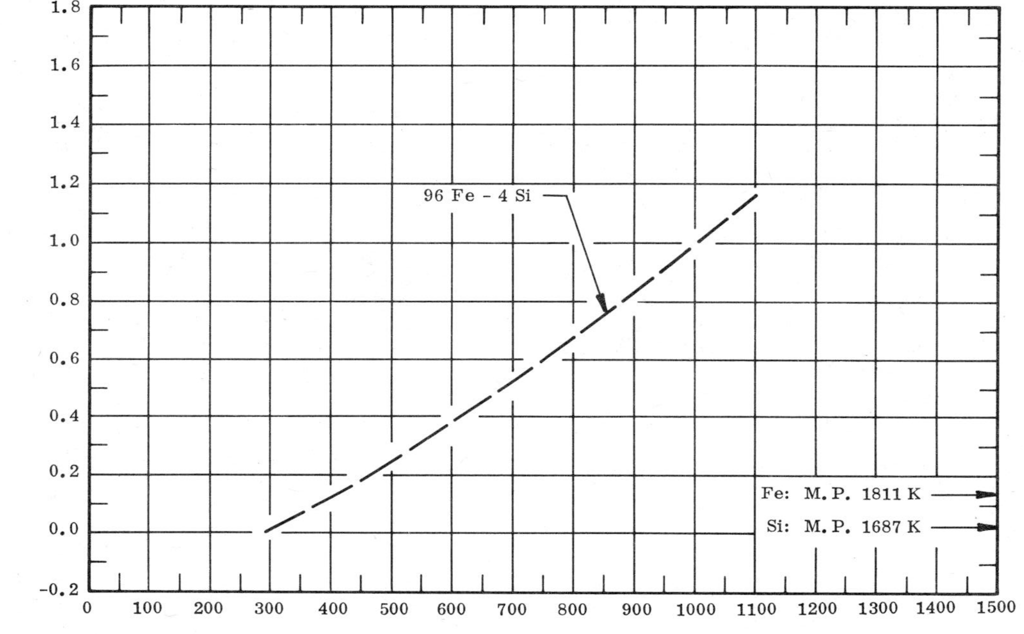

PROVISIONAL VALUES

[Temperature, T, K; Linear Expansion, $\Delta L/L_0$, %; α, K^{-1}]

	(96 Fe-4 Si)	
T	$\Delta L/L_0$	$\alpha \times 10^6$
293	0.000	10.8
400	0.121	11.9
500	0.246	13.0
600	0.381	13.9
700	0.523	14.7
800	0.674	15.5
900	0.833	16.1
1000	0.997	16.5
1100	1.162	16.6

REMARKS

(96 Fe-4 Si): The tabulated values for well-annealed alloy are considered accurate to within ± 7% over the entire temperature range. These values can be represented approximately by the following equation:

$$\Delta L/L_0 = -0.254 + 6.492 \times 10^{-4}\, T + 7.989 \times 10^{-7}\, T^2 - 1.984 \times 10^{-10}\, T^3$$

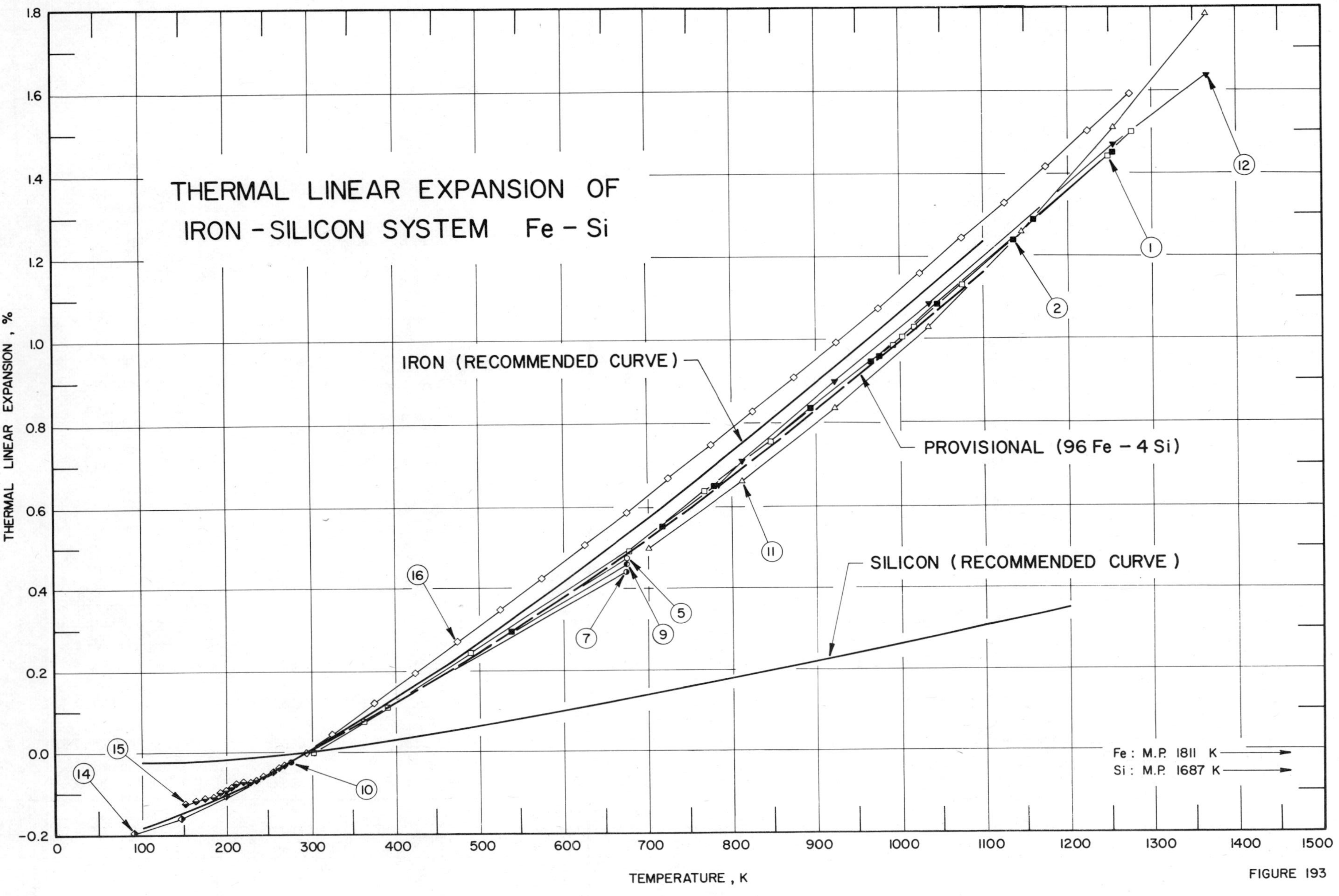
THERMAL LINEAR EXPANSION OF
IRON - SILICON SYSTEM Fe - Si
IRON (RECOMMENDED CURVE)
PROVISIONAL (96 Fe - 4 Si)
SILICON (RECOMMENDED CURVE)
Fe : M.P. 1811 K
Si : M.P. 1687 K
THERMAL LINEAR EXPANSION, %
TEMPERATURE, K
1
2
5
7
9
10
11
12
14
15
16
1.8
1.6
1.4
1.2
1.0
0.8
0.6
0.4
0.2
0.0
-0.2
0
100
200
300
400
500
600
700
800
900
1000
1100
1200
1300
1400
1500

FIGURE 193

SPECIFICATION TABLE 193. THERMAL LINEAR EXPANSION OF IRON-SILICON SYSTEM Fe-Si

Cur. No.	Ref. No.	Author(s)	Year	Method Used	Temp. Range, K	Name and Specimen Designation	Composition (weight percent) Fe	Si	Composition (continued), Specifications, and Remarks
1	51	Souder, W. and Hidnert, P.	1922	T	302-1276	S 457	Bal.	3.70	0.19 Mn, 0.176 Cr, 0.09 C, 0.05⁻ Ni, 0.005⁻ V; annealed specimen; expansion measured with increasing temperature.
2	51	Souder, W. and Hidnert, P.	1922	T	1276-538	S 457	Bal.	3.70	The above specimen; expansion measured with decreasing temperature.
3*	764	Sinhi, A.N. and Balasundaram, L.J.	1967	L	273-673		Bal.	0.28	Samples were machined on a lathe or ground to required shape.
4*	764	Sinhi, A.N. and Balasundaram, L.J.	1967	L	273-673		Bal.	1.586	Similar to the above specimen.
5	764	Sinhi, A.N. and Balasundaram, L.J.	1967	L	273-673		Bal.	3.32	Similar to the above specimen.
6*	764	Sinhi, A.N. and Balasundaram, L.J.	1967	L	273-673		Bal.	3.98	Similar to the above specimen.
7	764	Sinhi, A.N. and Balasundaram, L.J.	1967	L	273-673		Bal.	7.10	Similar to the above specimen.
8*	764	Sinhi, A.N. and Balasundaram, L.J.	1967	L	273-673		Bal.	7.17	Similar to the above specimen.
9	764	Sinhi, A.N. and Balasundaram, L.J.	1967	L	273-673		Bal.	7.41	Similar to the above specimen.
10	764	Sinhi, A.N. and Balasundaram, L.J.	1967	L	273-673		Bal.	11.21	Similar to the above specimen.
11	225	Valentich, J.	1965	L	700-1366	Cubex	Bal.	3.5	Doubly oriented; expansion measured across grain.
12	225	Valentich, J.	1965	L	783-1366	Cubex	Bal.	3.5	Similar to the above specimen; expansion measured with grain; expansion measured with increasing temperature.
13*	225	Valentich, J.	1965	L	1366-737	Cubex	Bal.	3.5	Similar to the above specimen; expansion measured with grain; expansion measured with decreasing temperature.
14	717	Technology Utilization Division, NASA	1969	L	20-293	No. 5 Relay	Bal.	2.5	No details given.
15	775	Kuz'min, R.N.; Nikitina, S.V., Tyuteva, N.d., and Golovnin, V.A.	1972	X	150-293	Steel St.10	Bal.	0.24	0.12 C, 0.025 S, 0.021 P; specimen prepared from bar of hot-rolled steel 18-20 mm diameter; forged to plates 5-6 mm thick; annealed 3 hr in oxidation-free atmosphere; ground, chemically etched, and electro-polished to discs 20 mm in diameter and 20-30 μ thick.
16	781	Kohlhaas, R. and Pandey, R.K.	1965	X	293-1273		Bal.	3.34	0.012 C, 0.07 Mn, 0.016 P, 0.014 S, lattice parameter calculated from the equation: $a_T = a_0 + bT + cT^2$ where $a_0 = 2.8565$ Å, $b = 4.26 \times 10^{-5}$ Å/C, $C = 3.75 \times 10^{-9}$ Å/C², Curie temperature = 1009 K.

* Not shown in figure.

DATA TABLE 193. THERMAL LINEAR EXPANSION OF IRON-SILICON SYSTEM Fe-Si

[Temperature, T, K; Linear Expansion, $\Delta L/L_0$, %]

T	$\Delta L/L_0$
CURVE 1	
302	0.000
362	0.075
390	0.112
489	0.244
677	0.496
766	0.639
846	0.758
991	0.992
1002	1.009
1016	1.033
1033	1.056
1074	1.137
1162	1.294*
1249	1.442
1276	1.501
CURVE 2	
1276	1.501*
1203	1.354
1161	1.292
1135	1.241
1047	1.091
974	0.962
965	0.950
892	0.836
779	0.651
717	0.554
538	0.298
CURVE 3*	
273	-0.025
293	0.000
673	0.477
CURVE 4*	
273	-0.024
293	0.000
673	0.464
CURVE 5	
273	-0.025*
293	0.000*
673	0.476
CURVE 6*	
273	-0.024
293	0.000
673	0.463
CURVE 7	
273	-0.023*
293	0.000*
673	0.442
CURVE 8*	
273	-0.023
293	0.000
673	0.445
CURVE 9	
273	-0.024*
293	0.000*
673	0.460
CURVE 10	
273	-0.025
293	0.000*
673	0.471*
CURVE 11	
700	0.500
811	0.661
922	0.838
1033	1.033
1144	1.263
1255	1.515
1366	1.786
CURVE 12	
783	0.653
811	0.712
922	0.902
1033	1.090
1144	1.275*
1255	1.470
1366	1.637
CURVE 13*	
1366	1.788
1255	1.470
1144	1.234
1033	1.040
922	0.840
811	0.678
737	0.580
CURVE 14	
20	-0.214*
33	-0.215*
88	-0.197
144	-0.160
199	-0.107
254	-0.046
293	-0.000*
CURVE 15	
150	-0.125
155	-0.122*
162	-0.119
166	-0.116*
172	-0.113
175	-0.111*
179	-0.108*
180	-0.108*
183	-0.105
186	-0.103*
188	-0.101*
191	-0.097
195	-0.093*
198	-0.090
200	-0.087*
205	-0.082
208	-0.079*
211	-0.076
215	-0.073*
217	-0.072*
219	-0.072
221	-0.072*
225	-0.072*
225	-0.072*
228	-0.071
230	-0.070*
233	-0.068*
CURVE 15 (cont.)	
234	-0.067
238	-0.062*
242	-0.058
245	-0.055*
251	-0.048*
260	-0.038
267	-0.030
293	0.000*
CURVE 16	
293	0.000
323	0.045
373	0.121
423	0.197
473	0.274
523	0.351
573	0.429
623	0.508
673	0.587
723	0.668
773	0.748
823	0.830
873	0.912
923	0.995
973	1.078
1023	1.162
1073	1.247
1123	1.332
1173	1.418
1223	1.505
1273	1.592

* Not shown in figure.

FIGURE AND TABLE NO. 194R. PROVISIONAL VALUES FOR THERMAL LINEAR EXPANSION OF LEAD-TIN SYSTEM Pb-Sn

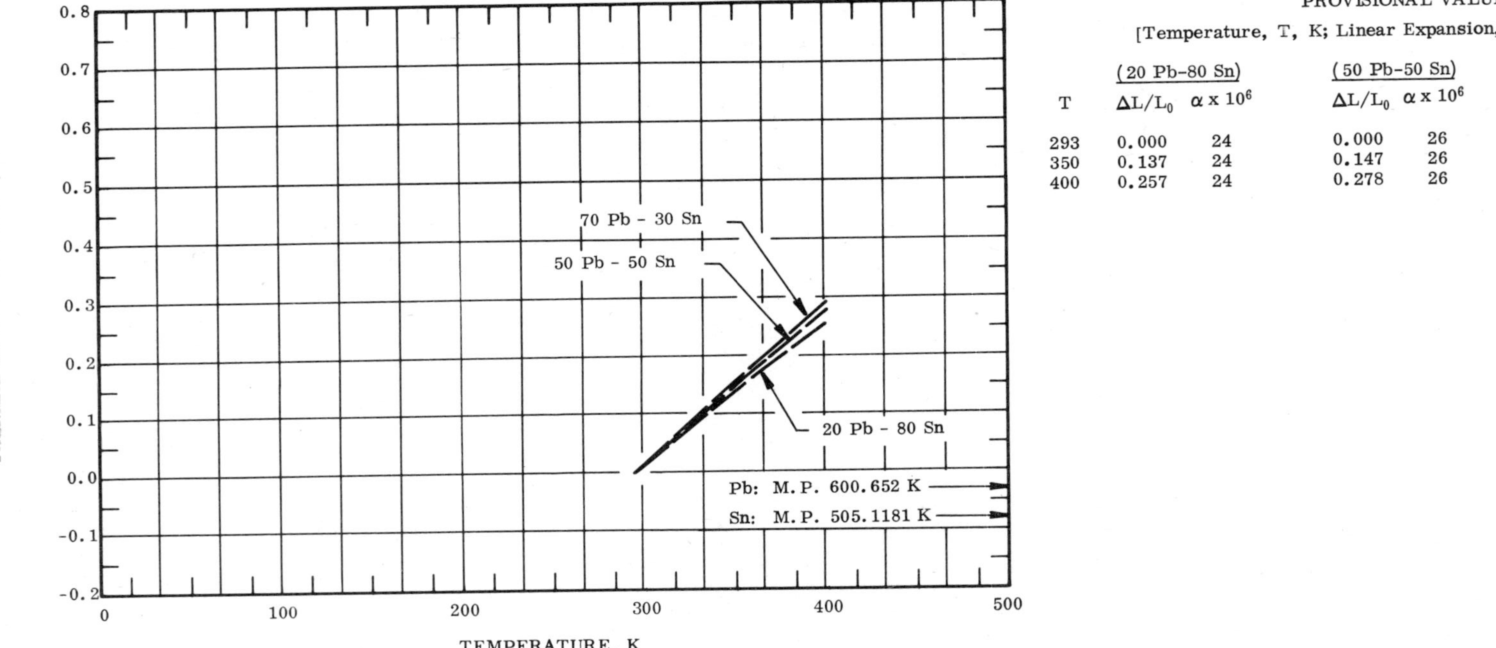

PROVISIONAL VALUES

[Temperature, T, K; Linear Expansion, $\Delta L/L_0$, %; α, K^{-1}]

T	(20 Pb-80 Sn) $\Delta L/L_0$	(20 Pb-80 Sn) $\alpha \times 10^6$	(50 Pb-50 Sn) $\Delta L/L_0$	(50 Pb-50 Sn) $\alpha \times 10^6$	(70 Pb-30 Sn) $\Delta L/L_0$	(70 Pb-30 Sn) $\alpha \times 10^6$
293	0.000	24	0.000	26	0.000	27
350	0.137	24	0.147	26	0.155	27
400	0.257	24	0.278	26	0.290	27

REMARKS

The tabulated values for well-annealed alloy are considered accurate to within ± 7% over the entire temperature range.

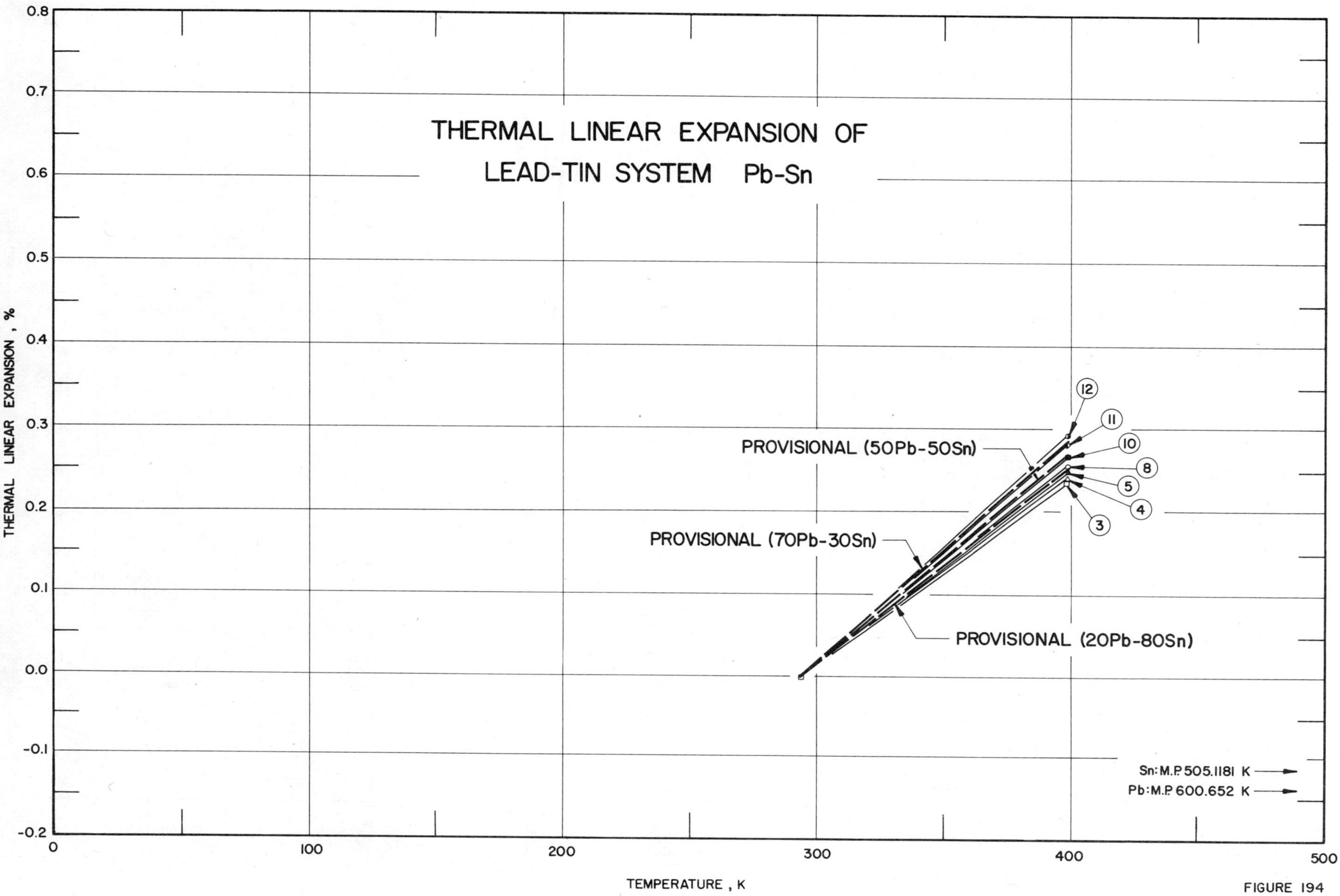
THERMAL LINEAR EXPANSION OF
LEAD-TIN SYSTEM Pb-Sn
PROVISIONAL (50Pb-50Sn)
PROVISIONAL (70Pb-30Sn)
PROVISIONAL (20Pb-80Sn)
12
11
10
8
5
4
3
Sn: M.P. 505.1181 K
Pb: M.P. 600.652 K
THERMAL LINEAR EXPANSION, %
0.8
0.7
0.6
0.5
0.4
0.3
0.2
0.1
0.0
-0.1
-0.2
0
100
200
300
400
500
TEMPERATURE, K

FIGURE 194

SPECIFICATION TABLE 194. THERMAL LINEAR EXPANSION OF LEAD-TIN SYSTEM Pb-Sn

Cur. No.	Ref. No.	Author(s)	Year	Method Used	Temp. Range, K	Name and Specimen Designation	Composition (weight percent) Pb	Sn	Composition (continued), Specifications, and Remarks
1*	327	Matuyama, Y.	1931	L	357-465		0.6	99.4	Pure Sn with impurities 0.01 Zn, 0.008 Pb, trace Cd; rolled specimen about 200 mm long, 1.5 mm diameter; this curve is here reported using the first given temperature as reference temperature at which $\Delta L/L_0 = 0$.
2*	327	Matuyama, Y.	1931	L	361-466		1	99	Similar to the above specimen; this curve is here reported using the first given temperature as reference temperature at which $\Delta L/L_0 = 0$.
3	665	Balasundaram, L.J. and Sinha, A.N.	1971	L	293-398		5	95	Specimen (3 volume percent of Pb) prepared by melting chemically pure metals in Pyrex tube; homogenized and annealed 50° below eutectic temperature.
4	665	Balasundaram, L.J. and Sinha, A.N.	1971	L	293-398		15	85	Similar to the above specimen except 11 volume percent of Pb.
5	665	Balasundaram, L.J. and Sinha, A.N.	1971	L	293-398		25	75	Similar to the above specimen except 18 volume percent of Pb.
6*	665	Balasundaram, L.J. and Sinha, A.N.	1971	L	293-398		35	65	Similar to the above specimen except 26 volume percent of Pb.
7*	665	Balasundaram, L.J. and Sinha, A.N.	1971	L	293-398		40	60	Similar to the above specimen except 31 volume percent of Pb.
8	665	Balasundaram, L.J. and Sinha, A.N.	1971	L	293-398		45	55	Similar to the above specimen except 36 volume percent of Pb.
9*	665	Balasundaram, L.J. and Sinha, A.N.	1971	L	293-398		50	50	Similar to the above specimen except 40 volume percent of Pb.
10	665	Balasundaram, L.J. and Sinha, A.N.	1971	L	293-398		60	40	Similar to the above specimen except 51 volume percent of Pb.
11	665	Balasundaram, L.J. and Sinha, A.N.	1971	L	293-398		70	30	Similar to the above specimen except 62 volume percent of Pb.
12	665	Balasundaram, L.J. and Sinha, A.N.	1971	L	293-398		80	20	Similar to the above specimen except 75 volume percent of Pb.

* Not shown in figure.

DATA TABLE 194. THERMAL LINEAR EXPANSION OF LEAD-TIN SYSTEM Pb-Sn

[Temperature, T, K; Linear Expansion, $\Delta L/L_0$, %]

T	$\Delta L/L_0$
CURVE 1*,†	
357	0.0000
359	0.0000
364	0.0002
368	0.0002
370	0.0002
372	0.0004
373	0.0004
375	0.0004
378	0.0004
380	0.0006
382	0.0006
383	0.0006
385	0.0006
387	0.0008
390	0.0010
394	0.0011
396	0.0011
399	0.0013
401	0.0013
404	0.0015
406	0.0017
408	0.0019
410	0.0019
412	0.0021
416	0.0025
418	0.0025
420	0.0027
423	0.0029
425	0.0029
427	0.0030
429	0.0032
431	0.0034
433	0.0036
435	0.0038
436	0.0040
437	0.0042
439	0.0044
441	0.0046
443	0.0048
448	0.0051
452	0.0055
455	0.0057
458	0.0061
460	0.0063
463	0.0076
465	0.0080
CURVE 2*,†	
361	0.0000
363	0.0002
366	0.0002
369	0.0002
376	0.0002
379	0.0002
381	0.0002
383	0.0002
385	0.0004
388	0.0004
390	0.0006
393	0.0006
395	0.0008
398	0.0010
401	0.0010
404	0.0011
407	0.0013
409	0.0015
412	0.0017
415	0.0017
417	0.0019
420	0.0021
423	0.0023
426	0.0023
429	0.0027
432	0.0029
434	0.0029
437	0.0032
439	0.0034
442	0.0036
445	0.0038
447	0.0042
449	0.0044
450	0.0046
451	0.0048
453	0.0049
455	0.0051
458	0.0053
460	0.0057
463	0.0061
466	0.0065
CURVE 3	
293	0.000
398	0.236
CURVE 4	
293	0.000*
398	0.241
CURVE 5	
293	0.000*
398	0.250
CURVE 6*	
293	0.000
398	0.252
CURVE 7*	
293	0.000
398	0.253
CURVE 8	
293	0.000*
398	0.257
CURVE 9*	
293	0.000
398	0.264
CURVE 10	
293	0.000*
398	0.269
CURVE 11	
293	0.000*
398	0.281
CURVE 12	
293	0.000
398	0.293

* Not shown in figure.
† This curve is here reported using the first given temperature as reference temperature at which $\Delta L/L_0 = 0$.

SPECIFICATION TABLE 195. THERMAL LINEAR EXPANSION OF LEAD-THALLIUM SYSTEM Pb-Tl

Cur. No.	Ref. No.	Author(s)	Year	Method Used	Temp. Range, K	Name and Specimen Designation	Composition (weight percent) Pb	Tl	Composition (continued), Specifications, and Remarks
1*	742	D'Heurle, F.M., Feder, R., and Nowick, A.S.	1963	X	446-598		Bal.	0.5	Specimen prepared using 99.999 Pb, single crystal sample, expansion measured in argon atmosphere at 2 mm Hg pressure.
2*	742	D'Heurle, F.M., et al.	1963	O	467-599		Bal.	0.5	Similar to the above specimen.

DATA TABLE 195. THERMAL LINEAR EXPANSION OF LEAD-THALLIUM SYSTEM Pb-Tl

[Temperature, T, K; Linear Expansion, $\Delta L/L_0$, %]

T	$\Delta L/L_0$
CURVE 1*	
446	0.464
492	0.609
500	0.637
551	0.811
580	0.913
593	0.957
595	0.965
598	0.977
CURVE 2*	
467	0.528
558	0.834
585	0.931
591	0.953
599	0.984

* No figure given.

FIGURE AND TABLE NO. 196BR. PROVISIONAL VALUES FOR THERMAL LINEAR EXPANSION OF MAGNESIUM-NICKEL SYSTEM Mg-Ni

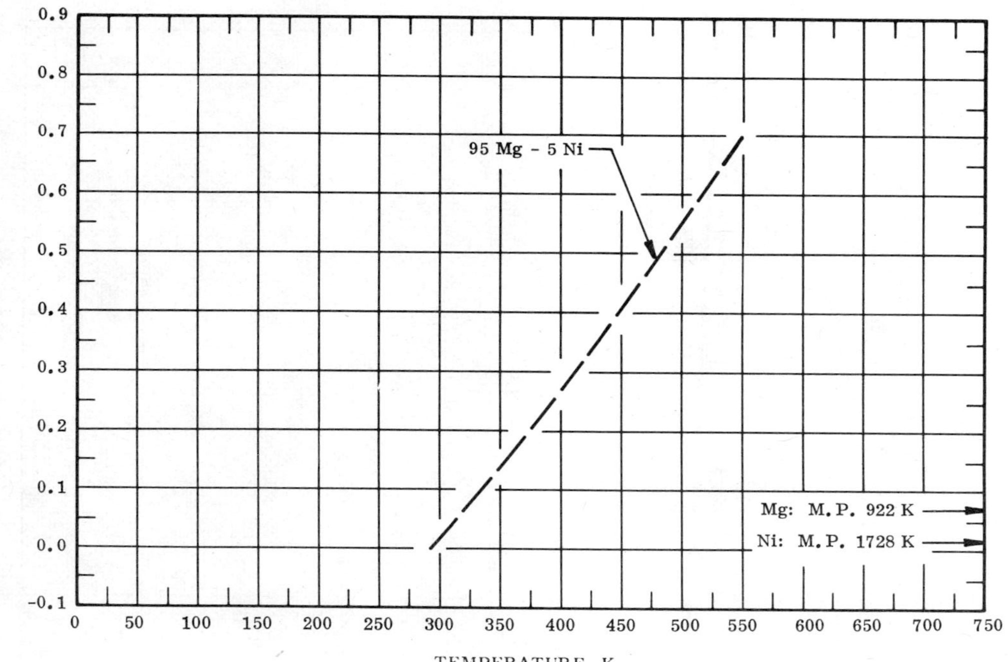

PROVISIONAL VALUES

[Temperature, T, K; Linear Expansion, $\Delta L/L_0$, %; α, K^{-1}]

T	(95 Mg-5 Ni) $\Delta L/L_0$	$\alpha \times 10^6$
293	0.000	23.3
400	0.272	27.1
500	0.555	29.3
550	0.700	29.5

REMARKS

(95 Mg-5 Ni): The tabulated values for well-annealed alloy are considered accurate to within ± 7% over the entire temperature range. These values can be represented approximately by the following equation:

$$\Delta L/L_0 = -0.415 + 2.590 \times 10^{-4}\, T + 4.795 \times 10^{-6}\, T^2 - 2.870 \times 10^{-9}\, T^3$$

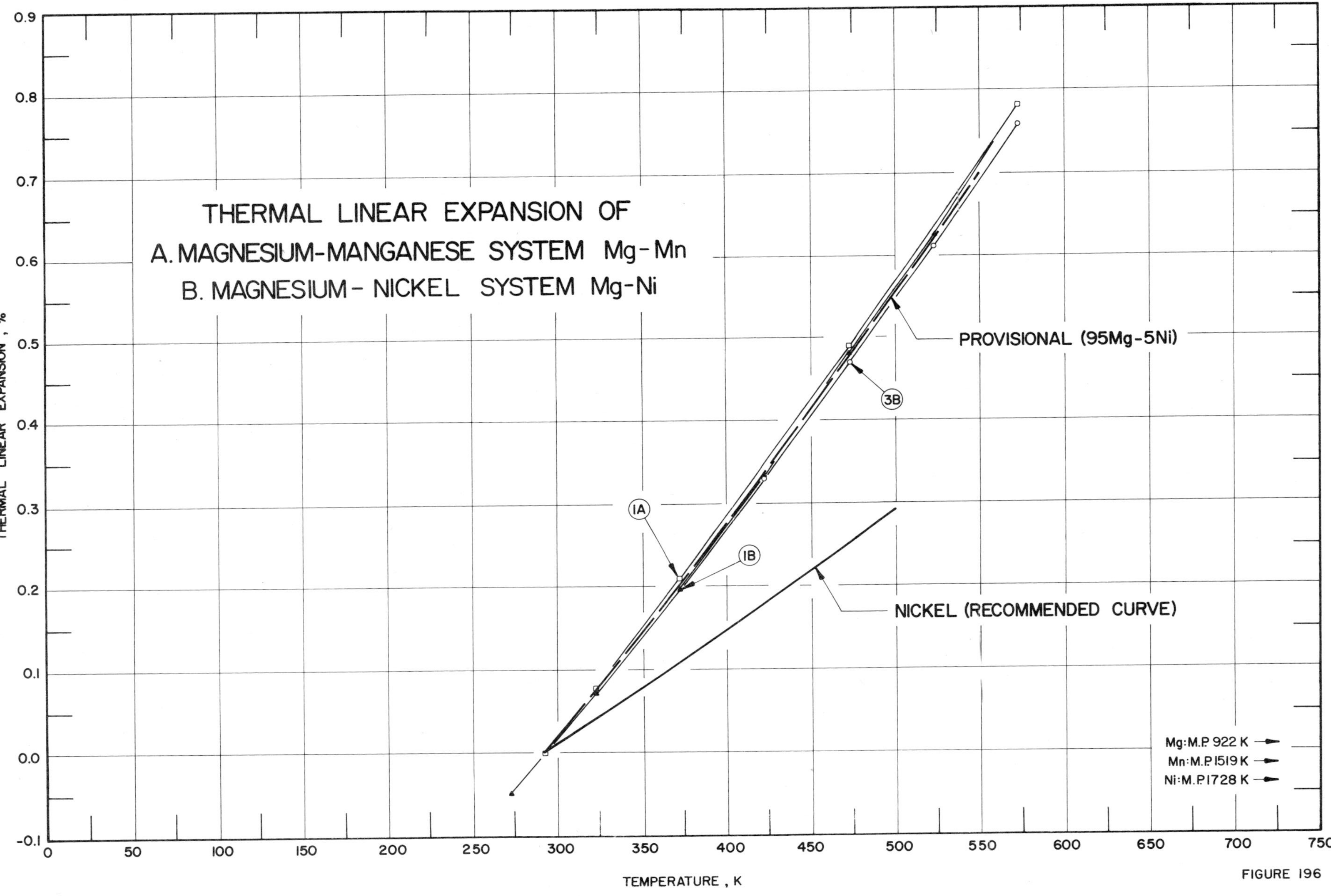
THERMAL LINEAR EXPANSION OF
A. MAGNESIUM-MANGANESE SYSTEM Mg-Mn
B. MAGNESIUM-NICKEL SYSTEM Mg-Ni
PROVISIONAL (95Mg-5Ni)
3B
1A
1B
NICKEL (RECOMMENDED CURVE)
Mg: M.P. 922 K
Mn: M.P. 1519 K
Ni: M.P. 1728 K
THERMAL LINEAR EXPANSION, %
0.9
0.8
0.7
0.6
0.5
0.4
0.3
0.2
0.1
0.0
-0.1
0
50
100
150
200
250
300
350
400
450
500
550
600
650
700
750
TEMPERATURE, K
FIGURE 196

SPECIFICATION TABLE 196A. THERMAL LINEAR EXPANSION OF MAGNESIUM-MANGANESE SYSTEM Mg-Mn

Cur. No.	Ref. No.	Author(s)	Year	Method Used	Temp. Range, K	Name and Specimen Designation	Composition (weight percent) Mg	Mn	Composition (continued), Specifications, and Remarks
1	200	Hidnert, P. and Sweeney, W.T.	1928	T	293-573	Sample 1280	Bal.	0.91	0.01 Al; specimen cast at 943 K.

DATA TABLE 196A. THERMAL LINEAR EXPANSION OF MAGNESIUM-MANGANESE SYSTEM Mg-Mn

[Temperature, T, K; Linear Expansion, $\Delta L/L_0$, %]

T	$\Delta L/L_0$
CURVE 1	
293	0.000
323	0.078
373	0.210
473	0.490
573	0.781

SPECIFICATION TABLE 196B. THERMAL LINEAR EXPANSION OF MAGNESIUM-NICKEL SYSTEM Mg-Ni

Cur. No.	Ref. No.	Author(s)	Year	Method Used	Temp. Range, K	Name and Specimen Designation	Composition (weight percent) Mg	Ni	Composition (continued), Specifications, and Remarks
1	201	Takahasi, K. and Kikuti, R.	1936		273-573		Bal.	2	Rod specimen, about 16 cm long, 5 mm dia, heating rate 2 C/min.
2*	201	Takahasi, K. and Kikuti, R.	1936		273-573		Bal.	4	Similar to the above specimen.
3	201	Takahasi, K. and Kikuti, R.	1936		273-573		Bal.	6	Similar to the above specimen.

DATA TABLE 196B. THERMAL LINEAR EXPANSION OF MAGNESIUM-NICKEL SYSTEM Mg-Ni

[Temperature, T, K; Linear Expansion, $\Delta L/L_0$, %]

T	$\Delta L/L_0$	T	$\Delta L/L_0$
CURVE 1		CURVE 3	
273	-0.048	273	-0.048*
323	0.072	323	0.072*
373	0.199	373	0.198*
423	0.336	423	0.331
473	0.480	473	0.470
523	0.628	523	0.612
573	0.780*	573	0.760
CURVE 2*			
273	-0.048		
323	0.072		
373	0.198		
423	0.331		
473	0.472		
523	0.620		
573	0.771		

*Not shown in figure.

FIGURE AND TABLE NO. 197R. PROVISIONAL VALUES FOR THERMAL LINEAR EXPANSION OF MAGNESIUM-SILVER SYSTEM Mg-Ag

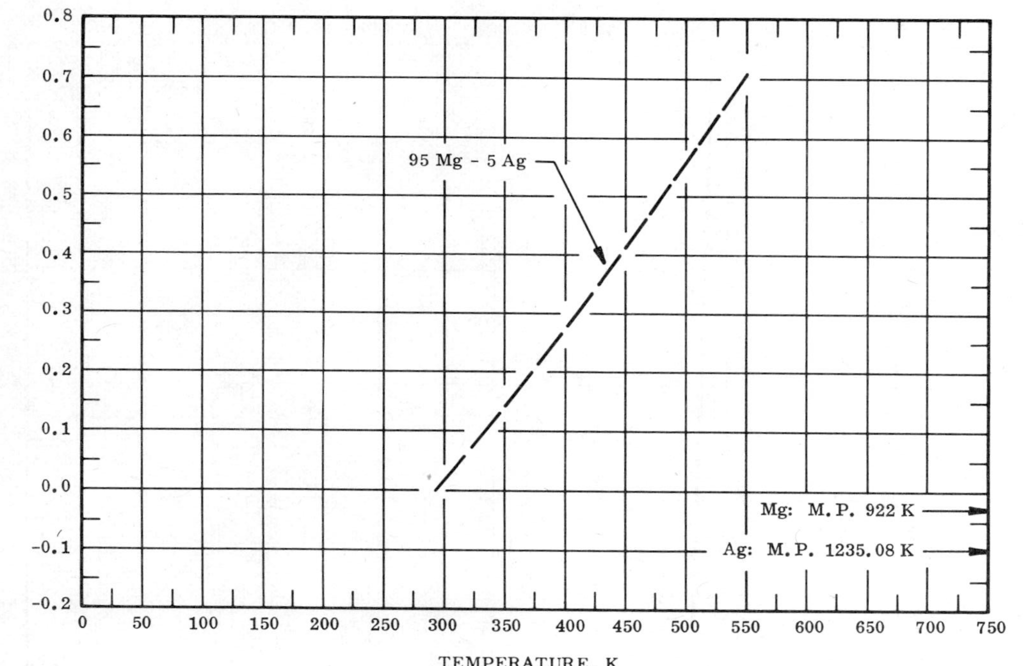

PROVISIONAL VALUES

[Temperature, T, K; Linear Expansion, $\Delta L/L_0$, %; α, K^{-1}]

T	(95 Mg-5 Ag) $\Delta L/L_0$	$\alpha \times 10^6$
293	0.000	23.4
400	0.273	27.7
500	0.560	29.6
550	0.710	30.3

REMARKS

(95 Mg-5 Ag): The tabulated values for well-annealed alloy are considered accurate to within ±7% over the entire temperature range. These values can be represented approximately by the following equation:

$$\Delta L/L_0 = -0.375 - 2.592 \times 10^{-5}\, T + 5.396 \times 10^{-6}\, T^2 - 3.204 \times 10^{-9}\, T^3$$

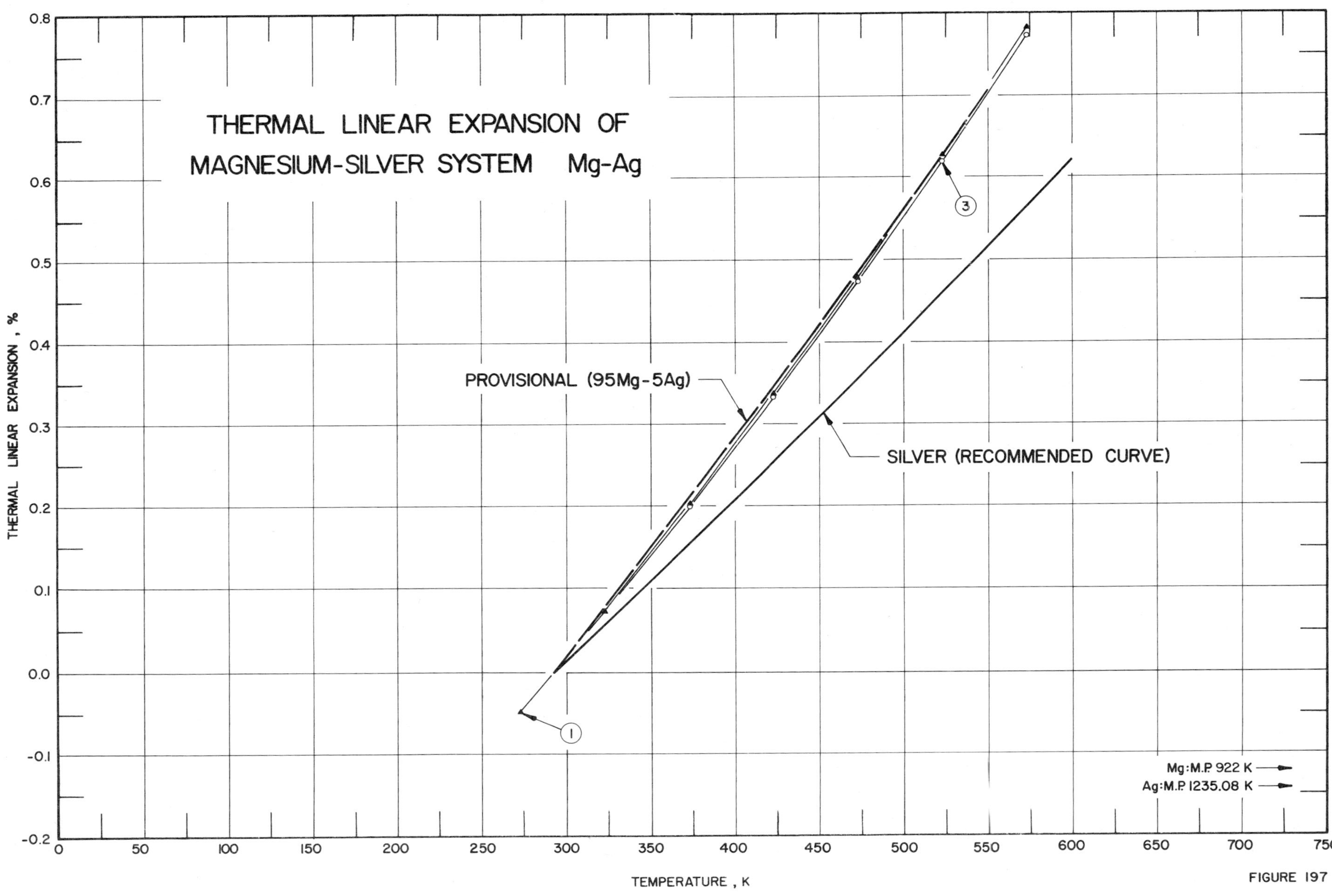
THERMAL LINEAR EXPANSION OF
MAGNESIUM-SILVER SYSTEM Mg-Ag
PROVISIONAL (95Mg-5Ag)
SILVER (RECOMMENDED CURVE)
3
1
Mg: M.P. 922 K
Ag: M.P. 1235.08 K
THERMAL LINEAR EXPANSION, %
TEMPERATURE, K
0.8
0.7
0.6
0.5
0.4
0.3
0.2
0.1
0.0
-0.1
-0.2
0
50
100
150
200
250
300
350
400
450
500
550
600
650
700
750

FIGURE 197

SPECIFICATION TABLE 197. THERMAL LINEAR EXPANSION OF MAGNESIUM-SILVER SYSTEM Mg-Ag

Cur. No.	Ref. No.	Author(s)	Year	Method Used	Temp. Range, K	Name and Specimen Designation	Composition (weight percent) Mg	Ag	Composition (continued), Specifications, and Remarks
1	201	Takahasi, K. and Kikuti, R.	1936		273-573		Bal.	2	Rod specimen, about 16 cm long, 5 mm dia, heating rate 2 C/min.
2*	201	Takahasi, K. and Kikuti, R.	1936		273-573		Bal.	4	Similar to the above specimen.
3	201	Takahasi, K. and Kikuti, R.	1936		273-573		Bal.	6	Similar to the above specimen.

DATA TABLE 197. THERMAL LINEAR EXPANSION OF MAGNESIUM-SILVER SYSTEM Mg-Ag

[Temperature, T, K; Linear Expansion, $\Delta L/L_0$, %]

T	$\Delta L/L_0$	T	$\Delta L/L_0$
CURVE 1		CURVE 3	
273	-0.049	273	-0.048*
323	0.073	323	0.072*
373	0.202	373	0.200
423	0.356	423	0.334
473	0.479	473	0.474
523	0.629	523	0.620
573	0.784	573	0.773
CURVE 2*			
273	-0.049		
323	0.073		
373	0.202		
423	0.339		
473	0.482		
523	0.630		
573	0.784		

*Not shown in figure.

FIGURE AND TABLE NO. 198R. PROVISIONAL VALUES FOR THERMAL LINEAR EXPANSION OF MAGNESIUM-TIN SYSTEM Mg-Sn

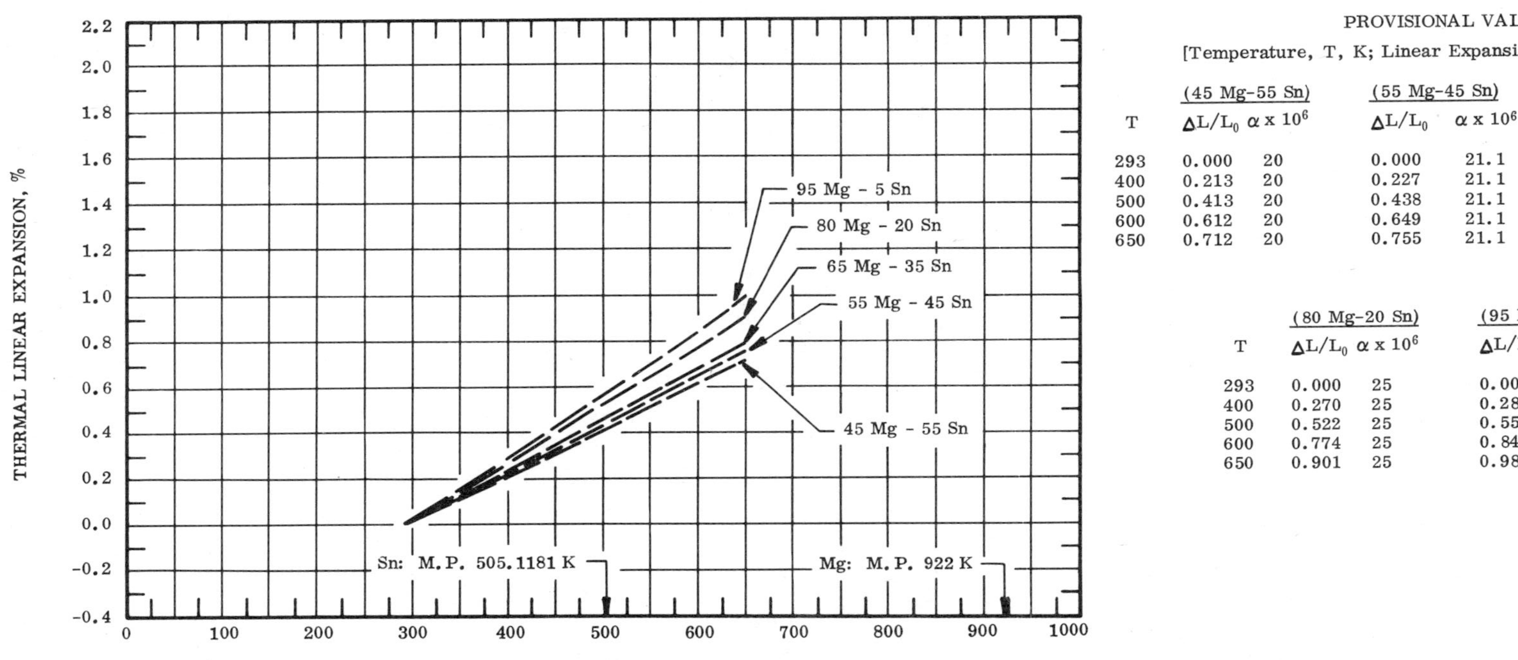

PROVISIONAL VALUES

[Temperature, T, K; Linear Expansion, $\Delta L/L_0$, %; α, K^{-1}]

T	(45 Mg-55 Sn)		(55 Mg-45 Sn)		(65 Mg-35 Sn)	
	$\Delta L/L_0$	$\alpha \times 10^6$	$\Delta L/L_0$	$\alpha \times 10^6$	$\Delta L/L_0$	$\alpha \times 10^6$
293	0.000	20	0.000	21.1	0.000	22
400	0.213	20	0.227	21.1	0.237	22
500	0.413	20	0.438	21.1	0.459	22
600	0.612	20	0.649	21.1	0.680	22
650	0.712	20	0.755	21.1	0.791	22

T	(80 Mg-20 Sn)		(95 Mg-5 Sn)	
	$\Delta L/L_0$	$\alpha \times 10^6$	$\Delta L/L_0$	$\alpha \times 10^6$
293	0.000	25	0.000	26
400	0.270	25	0.285	27
500	0.522	25	0.559	28
600	0.774	25	0.841	29
650	0.901	25	0.988	30

REMARKS

(45 Mg-55 Sn): The tabulated values for well-annealed alloy are considered accurate to within ±7% over the entire temperature range. These values can be represented approximately by the following equation:

$$\Delta L/L_0 = -0.578 + 1.950 \times 10^{-3}\, T + 9.503 \times 10^{-8}\, T^2 - 6.548 \times 10^{-11}\, T^3$$

(55 Mg-45 Sn): The tabulated values for well-annealed alloy are considered accurate to within ±7% over the entire temperature range. These values can be represented approximately by the following equation:

$$\Delta L/L_0 = -0.641 + 2.259 \times 10^{-3}\, T - 3.067 \times 10^{-7}\, T^2 + 2.079 \times 10^{-10}\, T^3$$

(65 Mg-35 Sn): The tabulated values for well-annealed alloy are considered accurate to within ±7% over the entire temperature range. These values can be represented approximately by the following equation:

$$\Delta L/L_0 = -0.648 + 2.209 \times 10^{-3}\, T + 2.142 \times 10^{-8}\, T^2 - 1.912 \times 10^{-11}\, T^3$$

(80 Mg-20 Sn): The tabulated values for well-annealed alloy are considered accurate to within ±7% over the entire temperature range. These values can be represented approximately by the following equation:

$$\Delta L/L_0 = -0.756 + 2.643 \times 10^{-3}\, T - 2.773 \times 10^{-7}\, T^2 + 2.041 \times 10^{-10}\, T^3$$

(95 Mg-5 Sn): The tabulated values for well-annealed alloy are considered accurate to within ±7% over the entire temperature range. These values can be represented approximately by the following equation:

$$\Delta L/L_0 = -0.778 + 2.714 \times 10^{-3}\, T - 3.678 \times 10^{-7}\, T^2 + 5.721 \times 10^{-10}\, T^3$$

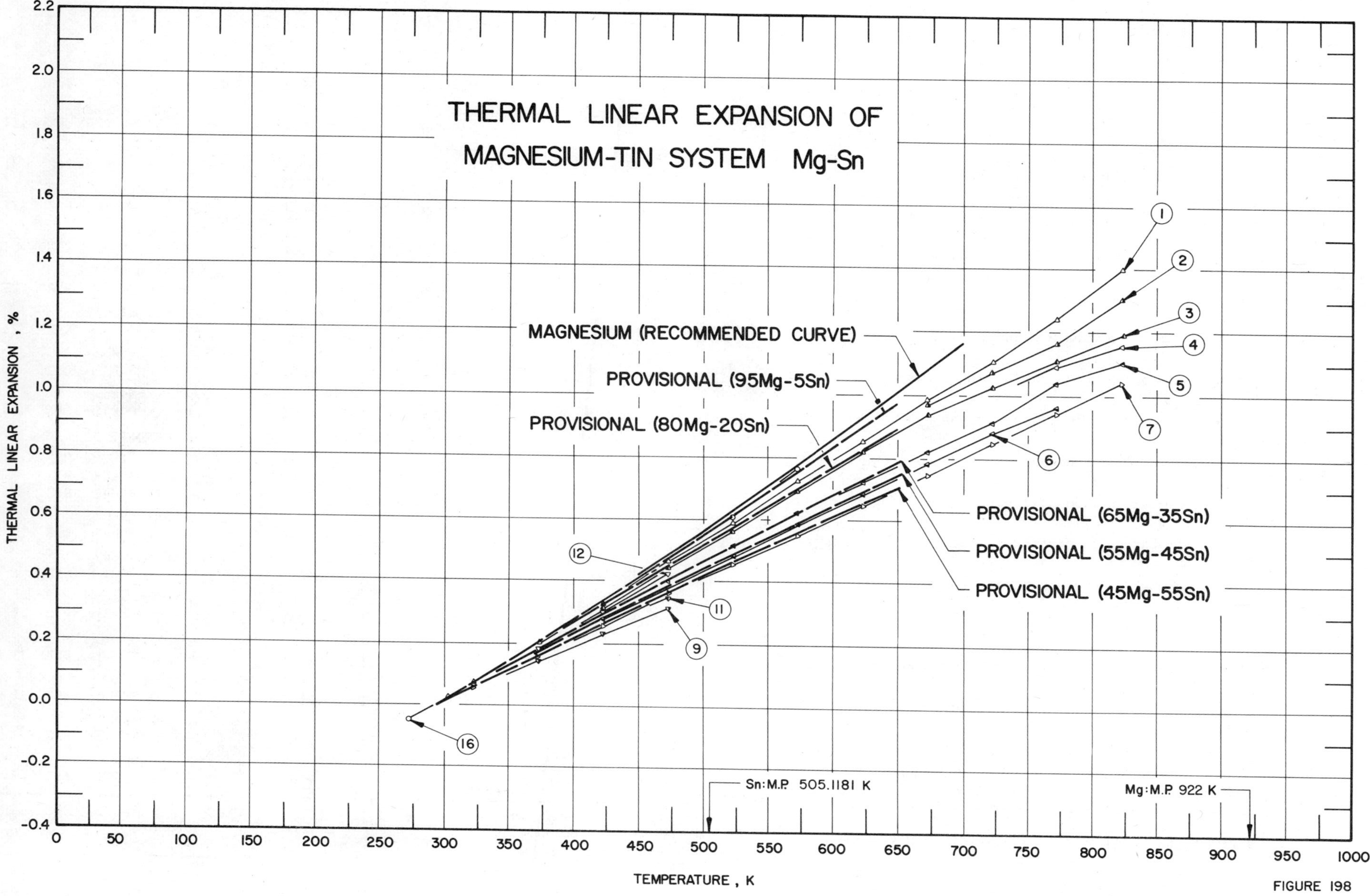
THERMAL LINEAR EXPANSION OF
MAGNESIUM-TIN SYSTEM Mg-Sn
MAGNESIUM (RECOMMENDED CURVE)
PROVISIONAL (95Mg-5Sn)
PROVISIONAL (80Mg-20Sn)
PROVISIONAL (65Mg-35Sn)
PROVISIONAL (55Mg-45Sn)
PROVISIONAL (45Mg-55Sn)
Sn:M.P. 505.1181 K
Mg:M.P. 922 K
1
2
3
4
5
6
7
9
11
12
16
THERMAL LINEAR EXPANSION, %
2.2
2.0
1.8
1.6
1.4
1.2
1.0
0.8
0.6
0.4
0.2
0.0
-0.2
-0.4
0
50
100
150
200
250
300
350
400
450
500
550
600
650
700
750
800
850
900
950
1000
TEMPERATURE, K

FIGURE 198

SPECIFICATION TABLE 198. THERMAL LINEAR EXPANSION OF MAGNESIUM-TIN SYSTEM Mg-Sn

Cur. No.	Ref. No.	Author(s)	Year	Method Used	Temp. Range, K	Name and Specimen Designation	Composition (weight percent) Mg	Sn	Composition (continued), Specifications, and Remarks
1	263	Grube, G. and Vosskühler, H.	1934	L	303-823		Bal.	4.7	Cylindrical specimen of 8 cm long and 5 mm diameter at 303 K; zero-point correction of 0.025%.
2	263	Grube, G. and Vosskühler, H.	1934	L	303-823		Bal.	9.1	Similar to the above specimen; zero-point correction of 0.026%.
3	263	Grube, G. and Vosskühler, H.	1934	L	303-823		Bal.	13.1	Similar to the above specimen; zero-point correction of 0.024%.
4	263	Grube, G. and Vosskühler, H.	1934	L	303-823		Bal.	20.5	Similar to the above specimen; zero-point correction of 0.025%.
5	263	Grube, G. and Vosskühler, H.	1934	L	303-823		Bal.	35.2	Similar to the above specimen; zero-point correction of 0.022%.
6	263	Grube, G. and Vosskühler, H.	1934	L	303-823		Bal.	46.3	Similar to the above specimen; zero-point correction of 0.021%.
7	263	Grube, G. and Vosskühler, H.	1934	L	303-473		Bal.	55.0	Similar to the above specimen; zero-point correction of 0.020%.
8*	263	Grube, G. and Vosskühler, H.	1934	L	303-473		Bal.	83.0	Similar to the above specimen; zero-point correction of 0.018%.
9	263	Grube, G. and Vosskühler, H.	1934	L	303-473		Bal.	85.1	Similar to the above specimen; zero-point correction of 0.017%.
10*	263	Grube, G. and Vosskühler, H.	1934	L	303-473		Bal.	90.1	Similar to the above specimen; zero-point correction of 0.017%.
11	263	Grube, G. and Vosskühler, H.	1934	L	303-473		Bal.	91.9	Similar to the above specimen; zero-point correction of 0.018%.
12	263	Grube, G. and Vosskühler, H.	1934	L	303-473		Bal.	97.8	Similar to the above specimen; zero-point correction of 0.022%.
13*	263	Grube, G. and Vosskühler, H.	1934	L	303-473		Bal.	99.2	Similar to the above specimen; zero-point correction of 0.025%.
14*	263	Grube, G. and Vosskühler, H.	1934	L	303-473		Bal.	99.6	Similar to the above specimen; zero-point correction of 0.025%.
15*	263	Grube, G. and Vosskühler, H.	1934	L	303-473		Bal.	99.8	Similar to the above specimen; zero-point correction of 0.027%.
16	201	Takahasi, K. and Kikuti, R.	1936		273-573		Bal.	2	Rod specimen about 16 cm long, 5 mm diameter; heating rate 2 C/min.
17*	201	Takahasi, K. and Kikuti, R.	1936		273-573		Bal.	4	Similar to the above specimen.
18*	201	Takahasi, K. and Kikuti, R.	1936		273-573		Bal.	6	Similar to the above specimen.

* Not shown in figure.

DATA TABLE 198. THERMAL LINEAR EXPANSION OF MAGNESIUM–TIN SYSTEM Mg–Sn

[Temperature, T, K; Linear Expansion, $\Delta L/L_0$, %]

T	$\Delta L/L_0$
CURVE 1	
303	0.025
323	0.075
373	0.201
423	0.331
473	0.464
523	0.593
573	0.726
623	0.858
673	0.988
723	1.107
773	1.247
823	1.407
CURVE 2	
303	0.026*
323	0.079*
373	0.205*
423	0.335*
473	0.470*
523	0.597*
573	0.731*
623	0.853*
673	0.976
723	1.075
773	1.172
823	1.301
CURVE 3	
303	0.024*
323	0.073*
373	0.193*
423	0.318
473	0.445
523	0.566
573	0.696
623	0.821
673	0.941
723	1.029
773	1.117
823	1.198
CURVE 4	
303	0.025*
323	0.075*
CURVE 4 (cont.)	
373	0.195*
423	0.318*
473	0.446*
523	0.567*
573	0.693*
623	0.812*
673	0.927*
723	1.018*
773	1.096
823	1.160
CURVE 5	
303	0.022*
323	0.066*
373	0.180*
423	0.294*
473	0.401
523	0.520
573	0.630
623	0.727
673	0.823
723	0.919
773	1.042
823	1.104
CURVE 6	
303	0.021*
323	0.003*
373	0.169*
423	0.280
473	0.384*
523	0.488
573	0.592
623	0.690
673	0.786
723	0.883
773	0.965
823	1.053*
CURVE 7	
303	0.020*
323	0.059
373	0.160
423	0.258
CURVE 7 (cont.)	
473	0.362
523	0.458
573	0.555
623	0.652
673	0.748
723	0.850
773	0.943
823	1.042
CURVE 8*	
303	0.018
323	0.053
373	0.139
423	0.225
473	0.312
CURVE 9	
303	0.017*
323	0.050*
373	0.142
423	0.231
473	0.317
CURVE 10*	
303	0.017
323	0.057
373	0.144
423	0.243
473	0.342
CURVE 11	
303	0.018*
323	0.054*
373	0.149*
423	0.278*
473	0.347
CURVE 12	
303	0.022*
323	0.065*
373	0.184
423	0.301
473	0.423
CURVE 13*	
303	0.025
323	0.075
373	0.206
423	0.332
473	0.459
CURVE 14*	
303	0.025
323	0.075
373	0.201
423	0.331
473	0.468
CURVE 15*	
303	0.027
323	0.080
373	0.212
423	0.344
473	0.487
CURVE 16	
273	-0.047
323	0.071
373	0.197
423	0.329
473	0.467
523	0.612
573	0.763
CURVE 17*	
273	-0.049
323	0.073
373	0.204
423	0.341
473	0.484
523	0.631
573	0.785
CURVE 18*	
273	-0.049
323	0.073
373	0.202
CURVE 18 (cont.)*	
423	0.339
473	0.479
523	0.625
573	0.776

* Not shown in figure.

FIGURE AND TABLE NO. 199BR. PROVISIONAL VALUES FOR THERMAL LINEAR EXPANSION OF MAGNESIUM-ZINC SYSTEM Mg-Zn

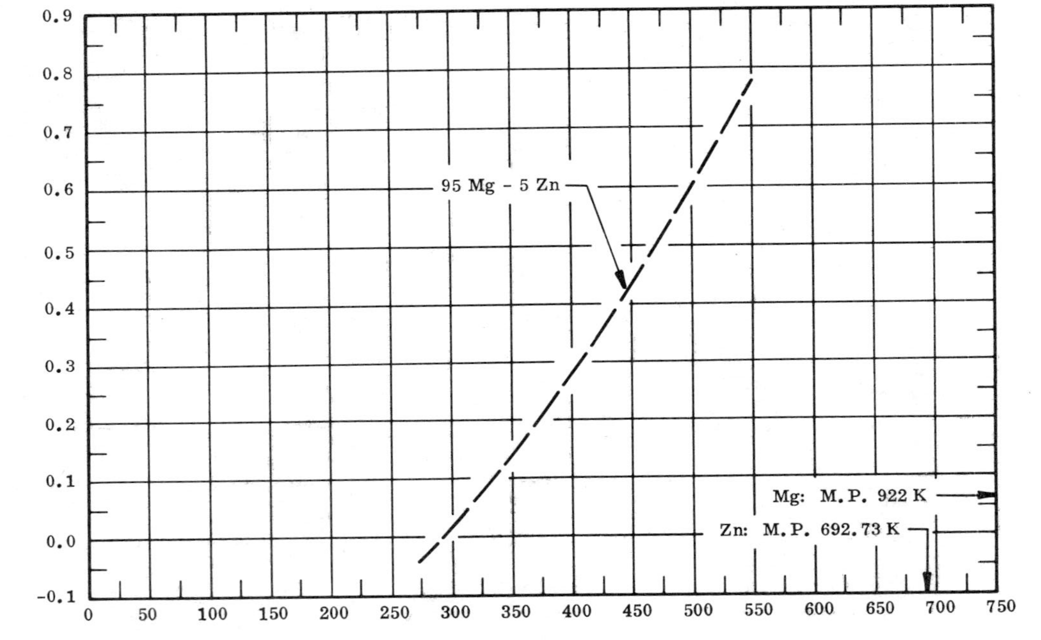

PROVISIONAL VALUES

[Temperature, T, K; Linear Expansion, $\Delta L/L_0$, %; α, K^{-1}]

T	(95 Mg-5 Zn) $\Delta L/L_0$	$\alpha \times 10^6$
275	-0.042	23.0
293	0.000	24.1
400	0.286	29.5
500	0.608	34.2
550	0.782	36.5

REMARKS

(95 Mg-5 Zn): The tabulated values for well-annealed alloy are considered accurate to within ± 7% over the entire temperature range. These values can be represented approximately by the following equation:

$$\Delta L/L_0 = -0.475 + 8.011 \times 10^{-4}\, T + 2.922 \times 10^{-6}\, T^2 - 4.025 \times 10^{-10}\, T^3$$

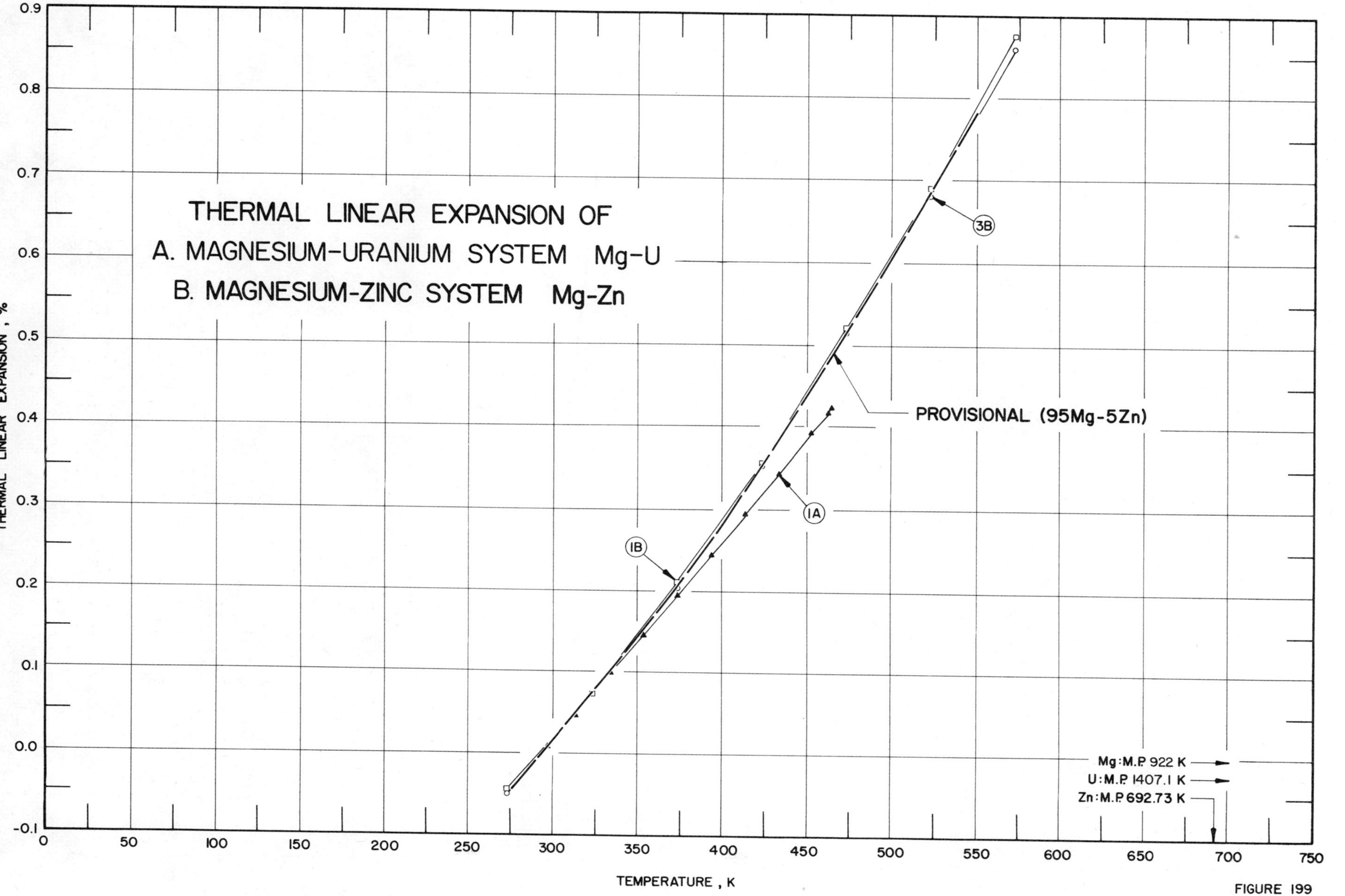
THERMAL LINEAR EXPANSION OF
A. MAGNESIUM-URANIUM SYSTEM Mg-U
B. MAGNESIUM-ZINC SYSTEM Mg-Zn
PROVISIONAL (95Mg-5Zn)
3B
1A
1B
Mg: M.P. 922 K
U: M.P. 1407.1 K
Zn: M.P. 692.73 K
THERMAL LINEAR EXPANSION, %
TEMPERATURE, K
-0.1
0.0
0.1
0.2
0.3
0.4
0.5
0.6
0.7
0.8
0.9
0
50
100
150
200
250
300
350
400
450
500
550
600
650
700
750

FIGURE 199

SPECIFICATION TABLE 199A. THERMAL LINEAR EXPANSION OF MAGNESIUM-URANIUM SYSTEM Mg-U

Cur. No.	Ref. No.	Author(s)	Year	Method Used	Temp. Range, K	Name and Specimen Designation	Composition (weight percent) Mg	U	Composition (continued), Specifications, and Remarks
1	447	McCreight, L.R.	1952	I	296-464		29.5	70.5	Prepared by cold compacting powder mixture, then hot extruding; zero-point correction is ± 0.000%.

DATA TABLE 199A. THERMAL LINEAR EXPANSION OF MAGNESIUM-URANIUM SYSTEM Mg-U

[Temperature, T, K; Linear Expansion, $\Delta L/L_0$, %]

T	$\Delta L/L_0$
CURVE 1	
296	0.007
313	0.049
333	0.100
352	0.147
373	0.197
393	0.247
413	0.296
432	0.345
452	0.395
462	0.418
464	0.423

SPECIFICATION TABLE 199B. THERMAL LINEAR EXPANSION OF MAGNESIUM-ZINC SYSTEM Mg-Zn

Cur. No.	Ref. No.	Author(s)	Year	Method Used	Temp. Range, K	Name and Specimen Designation	Composition (weight percent) Mg	Zn	Composition (continued), Specifications, and Remarks
1	201	Takahasi, K. and Kikuti, R.	1936		273-573		Bal.	2	Rod specimen, about 16 cm long, 5 mm dia, heating rate 2 C/min.
2*	201	Takahasi, K. and Kikuti, R.	1936		273-573		Bal.	4	Similar to the above specimen.
3	201	Takahasi, K. and Kikuti, R.	1936		273-573		Bal.	6	Similar to the above specimen.

DATA TABLE 199B. THERMAL LINEAR EXPANSION OF MAGNESIUM-ZINC SYSTEM Mg-Zn

[Temperature, T, K; Linear Expansion, $\Delta L/L_0$, %]

T	$\Delta L/L_0$	T	$\Delta L/L_0$
CURVE 1		CURVE 3	
273	-0.049	273	-0.049
323	0.074	323	0.074*
373	0.211	373	0.206
423	0.359	423	0.357
473	0.521	473	0.517
523	0.691	523	0.681
573	0.877	573	0.860
CURVE 2*			
273	-0.049		
323	0.074		
373	0.212		
423	0.357		
473	0.521		
523	0.686		
573	0.866		

*Not shown in figure.

FIGURE AND TABLE NO. 200AR. PROVISIONAL VALUES FOR THERMAL LINEAR EXPANSION OF MANGANESE-NICKEL SYSTEM Mn-Ni

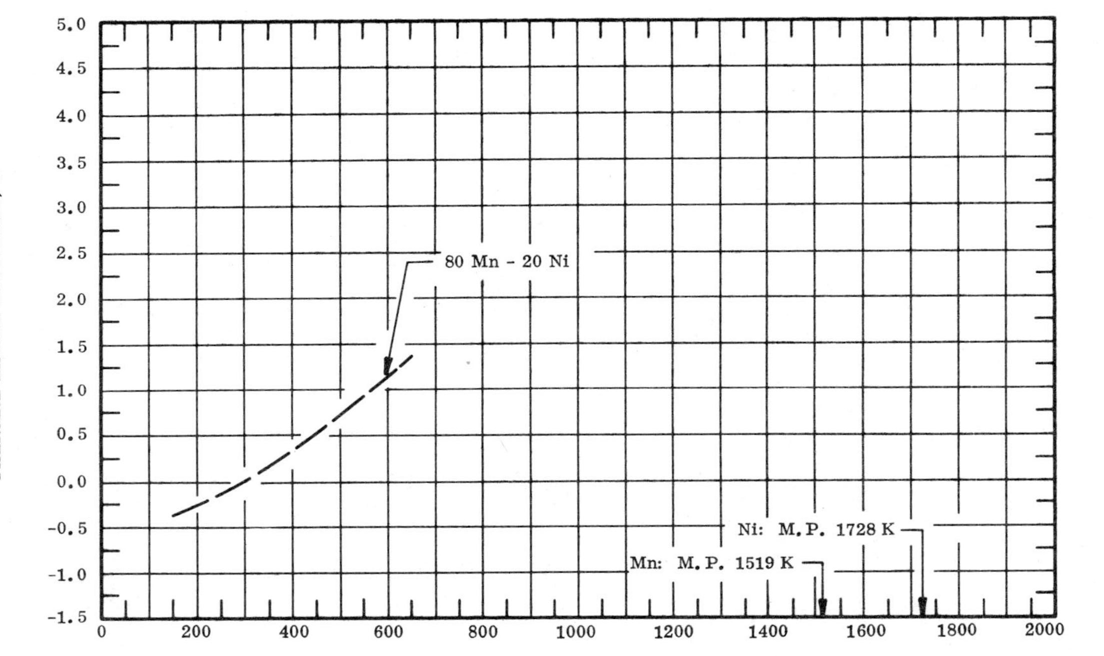

PROVISIONAL VALUES

[Temperature, T, K; Linear Expansion, $\Delta L/L_0$, %; α, K^{-1}]

(80 Mn-20 Ni)

T	$\Delta L/L_0$	$\alpha \times 10^6$
150	-0.354	20.7
200	-0.244	23.6
293	0.000	28.9
400	0.341	34.8
500	0.716	40.2
600	1.144	45.5
650	1.379	48.1

REMARKS

(80 Mn-20 Ni): The tabulated values for well-annealed alloy are considered accurate to within ±7% over the entire temperature range. These values can be represented approximately by the following equation:

$$\Delta L/L_0 = -0.596 + 1.172 \times 10^{-3}\, T + 3.033 \times 10^{-6}\, T^2 - 2.418 \times 10^{-10}\, T^3$$

FIGURE AND TABLE NO. 200BR. PROVISIONAL VALUES FOR THERMAL LINEAR EXPANSION OF MANGANESE-SILVER SYSTEM Mn-Ag

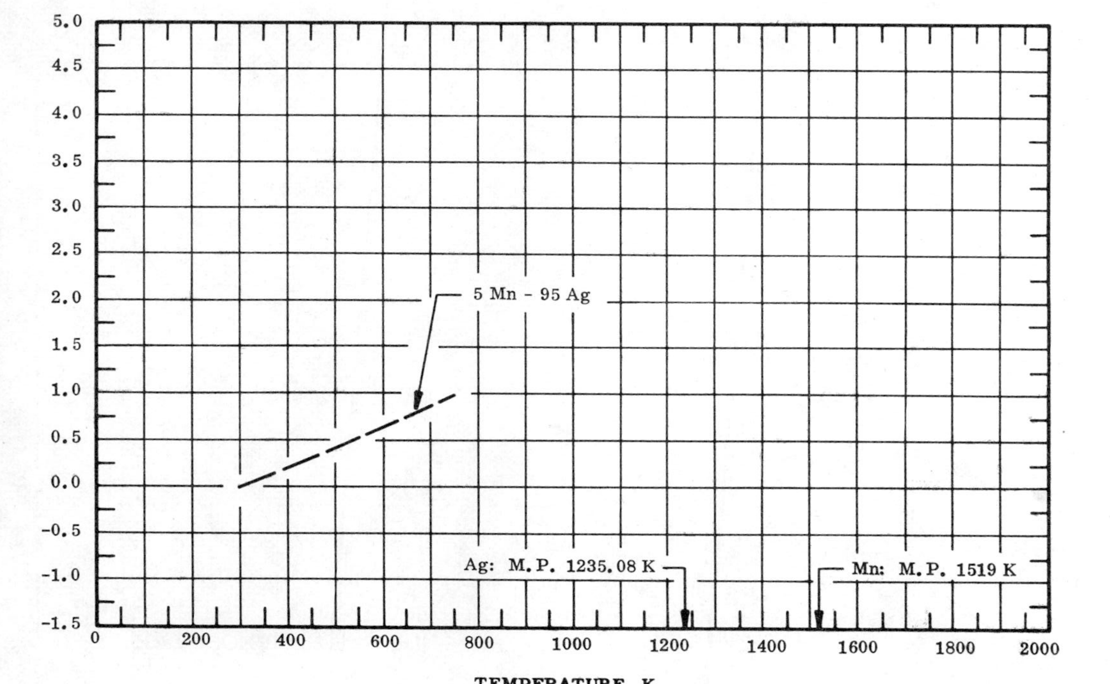

PROVISIONAL VALUES

[Temperature, T, K; Linear Expansion, $\Delta L/L_0$, %; α, K^{-1}]

T	(5 Mn-95 Ag) $\Delta L/L_0$	$\alpha \times 10^6$
293	0.000	18.6
400	0.209	20.5
500	0.420	22.0
600	0.646	23.0
700	0.880	23.7
750	0.999	24.0

REMARKS

(5 Mn-95 Ag): The tabulated values for well-annealed alloy are considered accurate to within ± 7% over the entire temperature range. These values can be represented approximately by the following equation:

$$\Delta L/L_0 = -0.428 + 1.036 \times 10^{-3} T + 1.648 \times 10^{-6} T^2 - 6.552 \times 10^{-10} T^3$$

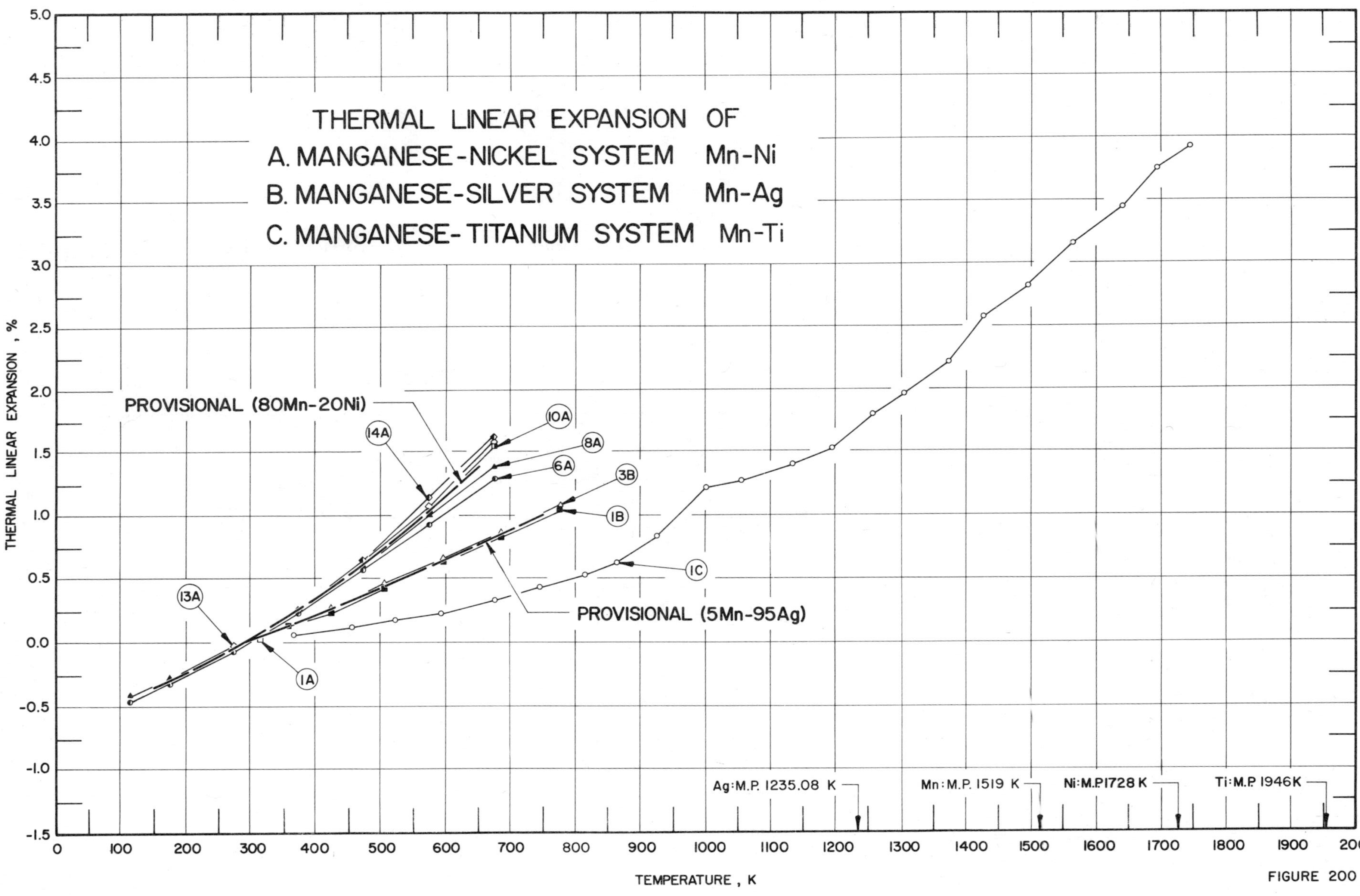
THERMAL LINEAR EXPANSION OF
A. MANGANESE-NICKEL SYSTEM Mn-Ni
B. MANGANESE-SILVER SYSTEM Mn-Ag
C. MANGANESE-TITANIUM SYSTEM Mn-Ti
PROVISIONAL (80Mn-20Ni)
PROVISIONAL (5Mn-95Ag)
10A
14A
8A
6A
3B
1B
1C
13A
1A
Ag: M.P. 1235.08 K
Mn: M.P. 1519 K
Ni: M.P. 1728 K
Ti: M.P. 1946 K
THERMAL LINEAR EXPANSION, %
5.0
4.5
4.0
3.5
3.0
2.5
2.0
1.5
1.0
0.5
0.0
-0.5
-1.0
-1.5
TEMPERATURE, K
0
100
200
300
400
500
600
700
800
900
1000
1100
1200
1300
1400
1500
1600
1700
1800
1900
2000

FIGURE 200

SPECIFICATION TABLE 200A. THERMAL LINEAR EXPANSION OF MANGANESE-NICKEL SYSTEM Mn-Ni

Cur. No.	Ref. No.	Author(s)	Year	Method Used	Temp. Range, K	Name and Specimen Designation	Composition (weight percent) Mn	Ni	Composition (continued), Specifications, and Remarks
1	241	Masumoto, H., Sawaya, S., and Kodowaki, S.	1969	L	273, 313		4.65	95.35	Electrolytic Ni with 0.180 Co, 0.002 S, 0.002 C, 0.001 Cu, 0.001 Fe, and 99.9 electrolytic Mn; rod specimen 2 mm diameter, 10 cm long; cooled at rate of 300 C/hr after heating in vacuum at 1273 K for 1 hr; density 8.81 g cm^{-3} at 293 K.
2*	241	Masumoto, H., et al.	1969	L	273, 313		9.40	90.60	Similar to the above specimen; density 8.74 g cm^{-3} at 293 K.
3*	241	Masumoto, H., et al.	1969	L	273, 313		12.20	87.80	Similar to the above specimen; density 8.61 g cm^{-3} at 293 K.
4*	241	Masumoto, H., et al.	1969	L	273, 313		14.20	85.80	Similar to the above specimen; density 8.56 g cm^{-3} at 293 K.
5*	241	Masumoto, H., et al.	1969	L	273, 313		16.10	83.90	Similar to the above specimen; density 8.55 g cm^{-3} at 293 K.
6	663	Masumoto, H., Sawaya, S., and Kikuchi, M.	1971	L	115-673		Bal.	5.1	Alloy prepared from electrolytic Ni with 0.18 Co, 0.001 Fe, 0.001 C, and 0.001 S, and electrolytic Mn with 0.007 Fe, 0.008 Si, 0.007 C, and 0.030 S; rod specimen 2 mm diameter and 11 cm length; cooled at rate of 100 degrees/hr after heating to 1223 K for 1 hr; average of heating and cooling curves; zero-point correction is -0.454%.
7*	663	Masumoto, H., et al.	1971	L	115-673		Bal.	10.15	Similar to the above specimen; zero-point correction is -0.439%.
8	663	Masumoto, H., et al.	1971	L	115-673		Bal.	15.22	Similar to the above specimen; zero-point correction is -0.422%.
9*	663	Masumoto, H., et al.	1971	L	115-673		Bal.	18.30	Similar to the above specimen; zero-point correction is -0.430%.
10	663	Masumoto, H., et al.	1971	L	115-673		Bal.	20.30	Similar to the above specimen; zero-point correction is -0.413%.
11*	663	Masumoto, H., et al.	1971	L	115-673		Bal.	22.35	Similar to the above specimen; zero-point correction is -0.405%.
12*	663	Masumoto, H., et al.	1971	L	115-673		Bal.	25.8	Similar to the above specimen; zero-point correction is -0.446%.
13	663	Masumoto, H., et al.	1971	L	115-673		Bal.	30.36	Similar to the above specimen; zero-point correction is -0.426%.
14	663	Masumoto, H., et al.	1971	L	115-673		Bal.	35.30	Similar to the above specimen; zero-point correction is -0.442%.
15*	663	Masumoto, H., et al.	1971	L	115-673		Bal.	40.42	Similar to the above specimen; zero-point correction is -0.425%.

*Not shown in figure.

DATA TABLE 200A. THERMAL LINEAR EXPANSION OF MANGANESE-NICKEL SYSTEM Mn-Ni

[Temperature, T, K; Linear Expansion, $\Delta L/L_0$, %]

T	$\Delta L/L_0$
CURVE 1	
273	-0.028
313	0.028
CURVE 2*	
273	-0.028
313	0.028
CURVE 3*	
273	-0.027
313	0.027
CURVE 4*	
273	-0.028
313	0.028
CURVE 5*	
273	-0.028
313	0.028
CURVE 6	
115	-0.454
173	-0.309
273	-0.054
293	0.000*
373	0.223
473	0.565
573	0.912
673	1.278
CURVE 7*	
115	-0.439
173	-0.305
273	-0.057
293	0.000
373	0.228
473	0.575
573	0.935
673	1.318
CURVE 8	
115	-0.422
173	-0.289
273	-0.051*
293	0.000*
373	0.226*
473	0.605
573	1.005
673	1.395
CURVE 9*	
115	-0.430
173	-0.293
273	-0.056
293	0.000
373	0.206
473	0.553
573	0.990
673	1.469
CURVE 10	
115	-0.413*
173	-0.297*
273	-0.049*
293	0.000*
373	0.210*
473	0.559*
573	1.030
673	1.552
CURVE 11*	
115	-0.405
173	-0.292
273	-0.049
293	0.000
373	0.200
473	0.525
573	1.007
673	1.523
CURVE 12*	
115	-0.446
173	-0.322
273	-0.054
293	0.000
373	0.239
473	0.595
573	1.061
673	1.552
CURVE 13	
115	-0.426*
173	-0.304*
273	-0.046
293	0.000*
373	0.221*
473	0.595*
573	1.054
673	1.574
CURVE 14	
115	-0.442*
173	-0.322*
273	-0.058*
293	0.000*
373	0.245
473	0.629
573	1.140
673	1.616
CURVE 15*	
115	-0.425
173	-0.319
273	-0.052
293	0.000
373	0.271
473	0.670
573	1.173
673	1.637

*Not shown in figure.

SPECIFICATION TABLE 200B. THERMAL LINEAR EXPANSION OF MANGANESE-SILVER SYSTEM Mn-Ag

Cur. No.	Ref. No.	Author(s)	Year	Method Used	Temp. Range, K	Name and Specimen Designation	Composition (weight percent) Mn	Composition (weight percent) Ag	Composition (continued), Specifications, and Remarks
1	646	De, M.	1969	X	293-774		2.90	97.10	Alloys prepared from spectrographically standardized metal from Messrs. Johnson, Matthey and Co., Ltd, London; homogenized at 1023-1123 K; annealed and quenched in air; lattice parameter reported at 297 K is 4.0860 Å; 4.0857 Å at 293 K determined by graphical extrapolation; data on coefficient of thermal linear expansion also reported.
2*	646	De, M.	1969	X	293-772		5.94	94.06	Similar to the above specimen; lattice parameter reported at 297 K is 4.0858 Å; 4.0855 Å at 293 K determined by graphical extrapolation; data on coefficient of thermal linear expansion also reported.
3	646	De, M.	1969	X	293-773		8.98	91.02	Similar to the above specimen; lattice parameter reported at 297 K is 4.0854 Å; 4.0851 Å at 293 K determined by graphical extrapolation; data on coefficient of thermal linear expansion also reported.

DATA TABLE 200B. THERMAL LINEAR EXPANSION OF MANGANESE-SILVER SYSTEM Mn-Ag

[Temperature, T, K; Linear Expansion, $\Delta L/L_0$, %]

T	$\Delta L/L_0$	T	$\Delta L/L_0$	T	$\Delta L/L_0$
CURVE 1		CURVE 2*		CURVE 3	
297	0.007	297	0.007	297	0.007*
357	0.115	357	0.117	357	0.120*
421	0.232	421	0.257	421	0.272
509	0.428	509	0.458	509	0.463
597	0.644	597	0.844	597	0.668
689	0.827	689	0.854	689	0.864
778	1.050	778	1.050	778	1.070

DATA TABLE 200B. COEFFICIENT OF THERMAL LINEAR EXPANSION OF MANGANESE-SILVER SYSTEM Mn-Ag

[Temperature, T, K; Coefficient of Expansion, α, 10^{-6} K^{-1}]

T	α	T	α	T	α
CURVE 1*		CURVE 2*		CURVE 3*	
293	17.26	293	18.03	293	18.25
376	18.12	375	19.35	372	19.70
474	19.69	472	21.20	472	21.88
573	21.91	571	22.24	572	22.70
673	23.14	772	22.67	673	22.73
774	24.51			773	22.33

*Not shown in figure.

SPECIFICATION TABLE 200C. THERMAL LINEAR EXPANSION OF MANGANESE-TITANIUM SYSTEM Mn-Ti

Cur. No.	Ref. No.	Author(s)	Year	Method Used	Temp. Range, K	Name and Specimen Designation	Composition (weight percent) Mn	Ti	Composition (continued), Specifications, and Remarks
1	358	Fieldhouse, I.B. and Lang, J.I.	1961		366-1743	Titanium C110M	7.9	91.81	0.15 O, 0.03 C, 0.01 W; density 4.59 g cm^{-3}; zero-point correction of -0.02% determined by graphical extrapolation.

DATA TABLE 200C. THERMAL LINEAR EXPANSION OF MANGANESE-TITANIUM SYSTEM Mn-Ti

[Temperature, T, K; Linear Expansion, $\Delta L/L_0$, %]

T	$\Delta L/L_0$	T	$\Delta L/L_0$
CURVE 1		CURVE 1 (cont.)	
366	0.058	1567	3.153
457	0.125	1640	3.453
521	0.173	1694	3.755
591	0.223	1743	3.948
675	0.348		
744	0.446		
813	0.529		
863	0.638		
927	0.849		
1002	1.226		
1056	1.258		
1135	1.403		
1194	1.521		
1257	1.803		
1306	1.970		
1372	2.215		
1429	2.586		
1495	2.830		

SPECIFICATION TABLE 201. THERMAL LINEAR EXPANSION OF MOLYBDENUM-NICKEL SYSTEM Mo-Ni

Cur. No.	Ref. No.	Author(s)	Year	Method Used	Temp. Range, K	Name and Specimen Designation	Composition (weight percent) Mo	Ni	Composition (continued), Specifications, and Remarks
1*	428	Masumoto, H., Sawaya, S. and Nakamura, N.	1969	L	273,313		7.0	93.0	Electrolytic Ni with 0.180 Co, 0.002 S, 0.002 C, 0.001 Cu, 0.001 Fe, and 99.80 Mo; powder mixed and pressed; melted to 10 cm long, 2 mm dia, furnace-cooled at rate 300 C/hr after heating at 1273 K for 1 hr; density 9.01 g/cm^3 at 293 K.
2*	428	Masumoto, H. et al.	1969	L	273,313		7.8	92.2	Similar to the above specimen; density 9.06 g/cm^3 at 293 K.
3*	428	Masumoto, H. et al.	1969	L	273,313		8.0	92.0	Similar to the above specimen; density 9.08 g/cm^3 at 293 K.
4*	428	Masumoto, H. et al.	1969	L	273,313		8.2	91.8	Similar to the above specimen.
5*	428	Masumoto, H. et al.	1969	L	273,313		8.5	91.5	Similar to the above specimen; density 9.10 g/cm^3 at 293 K.
6*	428	Masumoto, H. et al.	1969	L	273,313		10.0	90.0	Similar to the above specimen.

DATA TABLE 201. THERMAL LINEAR EXPANSION OF MOLYBDENUM-NICKEL SYSTEM Mo-Ni

[Temperature, T, K; Linear Expansion, $\Delta L/L_0$, %]

T	$\Delta L/L_0$	T	$\Delta L/L_0$
CURVE 1*		CURVE 4*	
273	-0.025	273	-0.026
313	0.025	313	0.026
CURVE 2*		CURVE 5*	
273	-0.024	273	-0.026
313	0.024	313	0.026
CURVE 3*		CURVE 6*	
273	-0.025	273	-0.026
313	0.025	313	0.026

*No figure given.

FIGURE AND TABLE NO. 202R. PROVISIONAL VALUES FOR THERMAL LINEAR EXPANSION OF MOLYBDENUM-NIOBIUM SYSTEM Mo-Nb

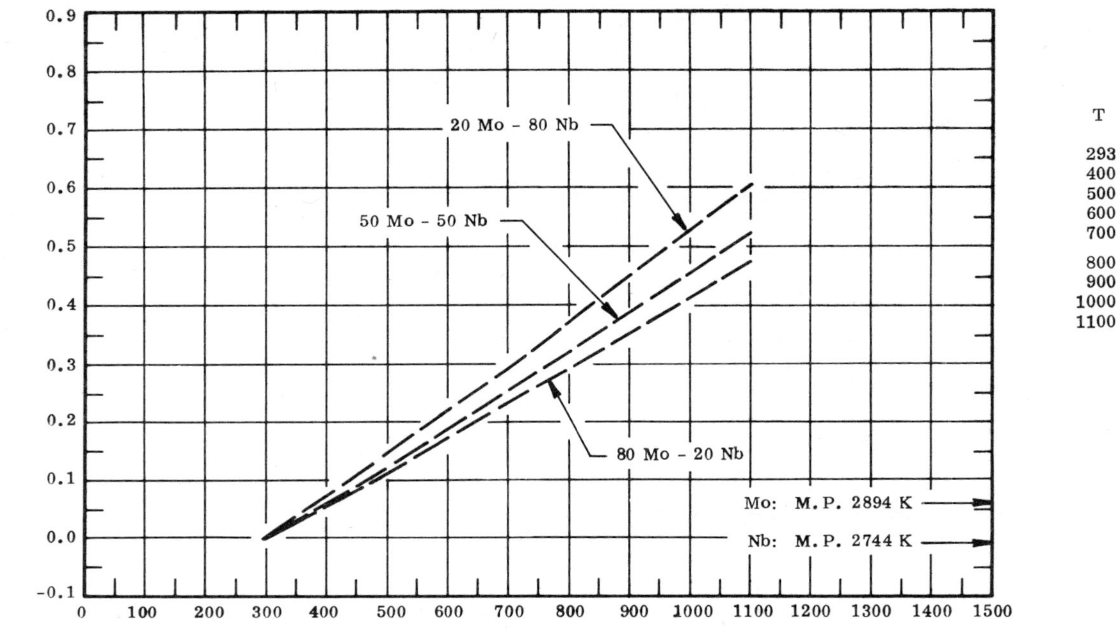

PROVISIONAL VALUES

[Temperature, T, K; Linear Expansion, $\Delta L/L_0$, %; α, K^{-1}]

T	(20 Mo-80 Nb) $\Delta L/L_0$	(20 Mo-80 Nb) $\alpha \times 10^6$	(50 Mo-50 Nb) $\Delta L/L_0$	(50 Mo-50 Nb) $\alpha \times 10^6$	(80 Mo-20 Nb) $\Delta L/L_0$	(80 Mo-20 Nb) $\alpha \times 10^6$
293	0.000	7.1	0.000	7.2	0.000	5.7
400	0.076	7.2	0.060	7.2	0.062	5.7
500	0.149	7.3	0.124	7.3	0.120	5.7
600	0.223	7.5	0.189	7.5	0.176	5.7
700	0.298	7.6	0.255	7.7	0.234	5.7
800	0.375	7.7	0.323	7.7	0.292	5.9
900	0.453	7.8	0.391	7.7	0.352	6.2
1000	0.531	7.8	0.459	7.7	0.413	6.2
1100	0.608	7.8	0.525	7.8	0.475	6.2

REMARKS

(20 Mo-80 Nb): The tabulated values for well-annealed alloy are considered accurate to within ± 7% over the entire temperature range. These values can be represented approximately by the following equation:

$$\Delta L/L_0 = -0.193 + 6.201 \times 10^{-4}\,T + 1.532 \times 10^{-7}\,T^2 - 4.961 \times 10^{-11}\,T^3$$

(50 Mo-50 Nb): The tabulated values for well-annealed alloy are considered accurate to within ± 7% over the entire temperature range. These values can be represented approximately by the following equation:

$$\Delta L/L_0 = -0.137 + 3.690 \times 10^{-4}\,T + 3.838 \times 10^{-7}\,T^2 - 1.570 \times 10^{-10}\,T^3$$

(80 Mo-20 Nb): The tabulated values for well-annealed alloy are considered accurate to within ± 7% over the entire temperature range. These values can be represented approximately by the following equation:

$$\Delta L/L_0 = -0.181 + 6.480 \times 10^{-4}\,T - 1.399 \times 10^{-7}\,T^2 + 8.444 \times 10^{-11}\,T^3$$

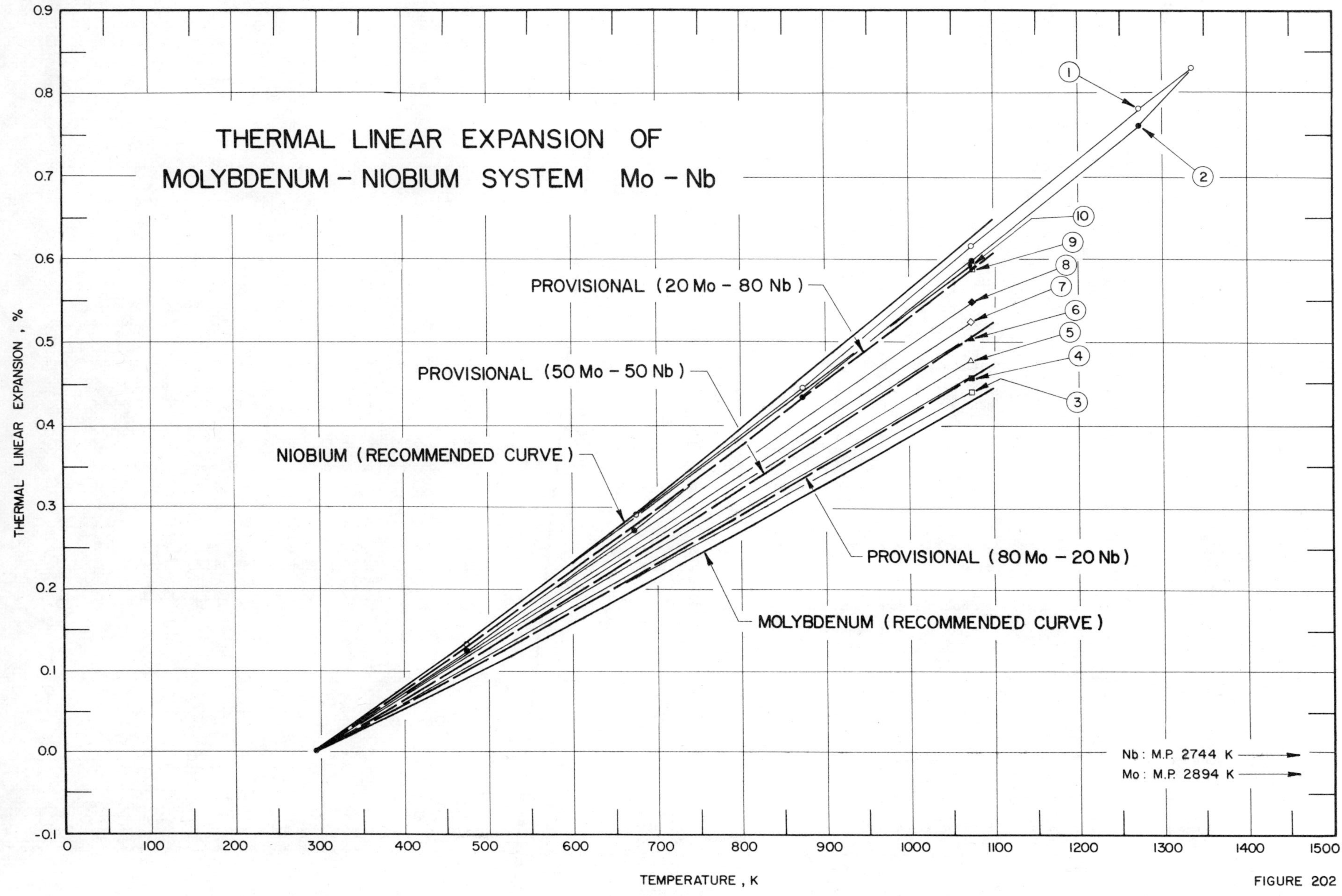
THERMAL LINEAR EXPANSION OF
MOLYBDENUM - NIOBIUM SYSTEM Mo - Nb
1
2
10
9
8
7
6
5
4
3
PROVISIONAL (20 Mo – 80 Nb)
PROVISIONAL (50 Mo – 50 Nb)
NIOBIUM (RECOMMENDED CURVE)
PROVISIONAL (80 Mo – 20 Nb)
MOLYBDENUM (RECOMMENDED CURVE)
Nb : M.P. 2744 K
Mo : M.P. 2894 K
THERMAL LINEAR EXPANSION , %
0.9
0.8
0.7
0.6
0.5
0.4
0.3
0.2
0.1
0.0
-0.1
0
100
200
300
400
500
600
700
800
900
1000
1100
1200
1300
1400
1500
TEMPERATURE , K
FIGURE 202

SPECIFICATION TABLE 202. THERMAL LINEAR EXPANSION OF MOLYBDENUM-NIOBIUM SYSTEM Mo-Nb

Cur. No.	Ref. No.	Author(s)	Year	Method Used	Temp. Range, K	Name and Specimen Designation	Composition (weight percent) Mo	Nb	Composition (continued), Specifications, and Remarks
1	253	Ul'yanov, R.A., Tarasov, N.D., and Mikhaylov, Ya.D.	1964	L	293-1339		2.71	97.29	Starting material 98.7 pure Nb, then vacuum refined; alloy prepared by arc smelting in argon atmosphere; expansion measured under 10^{-5} mm Hg vacuum.
2	253	Ul'yanov, R.A., et al.	1964	L	293-1339		4.95	95.05	Similar to the above specimen.
3	707	Pridantseva, K.S. and Solov'yeva, N.A.	1965	L	293, 1073		Bal.	10	Specimen prepared by powder metallurgy method, impurities mainly 2% oxygen; sample was homogenized at temperature which would provide an equilibrium state of the alloy; homogenizing anneal was done in a vacuum of not less than 10^{-4} mm Hg or in argon; single phase solid solution sample; 2% measurement error.
4	707	Pridantseva, K.S. and Solov'yeva, N.A.	1965	L	293, 1073		Bal.	20	Similar to the above specimen.
5	707	Pridantseva, K.S. and Solov'yeva, N.A.	1965	L	293, 1073		Bal.	30	Similar to the above specimen.
6	707	Pridantseva, K.S. and Solov'yeva, N.A.	1965	L	293, 1073		Bal.	40	Similar to the above specimen.
7	707	Pridantseva, K.S. and Solov'yeva, N.A.	1965	L	293, 1073		Bal.	50	Similar to the above specimen.
8	707	Pridantseva, K.S. and Solov'yeva, N.A.	1965	L	293, 1073		Bal.	60	Similar to the above specimen.
9	707	Pridantseva, K.S. and Solov'yeva, N.A.	1965	L	293, 1073		Bal.	70	Similar to the above specimen.
10	707	Pridantseva, K.S. and Solov'yeva, N.A.	1965	L	293, 1073		Bal.	90	Similar to the above specimen.
11	748	Hubbell, W.C. and Brotzen, F.R.	1972	L	83, 373		17.3	Bal.	Prepared from reactor grade Nb and Mo powders mixed and compacted into 6.2 mm rounds approx. 200 mm long, sample melted by electron-beam floating-zone in vacuum of 10^{-6} Torr to transform powder compacts into solid solutions. Cylinders about 10 mm long were spark sliced from sample and end faces polished, crystal axes orientation determined by Laue back-reflection method, composition determined from density measurements, $\pm 1.5 \times 10^{-6}$ K^{-1} uncertainty in α, 8.860 gm cm^{-3} at 298 K.
12*	748	Hubbell, W.C. and Brotzen, F.R.	1972	L	83, 373		23.9	Bal.	Similar to the above specimen, 8.968 gm cm^{-3} at 298 K.
13*	748	Hubbell, W.C. and Brotzen, F.R.	1972	L	83, 373		34.6	Bal.	Similar to the above specimen, 9.127 gm cm^{-3} at 298 K.
14	748	Hubbell, W.C. and Brotzen, F.R.	1972	L	83, 373		52.4	Bal.	Similar to the above specimen, 9.440 gm cm^{-3} at 298 K.
15*	748	Hubbell, W.C. and Brotzen, F.R.	1972	L	83, 373		75.8	Bal.	Similar to the above specimen, 9.828 gm cm^{-3} at 298 K.
16	748	Hubbell, W.C. and Brotzen, F.R.	1972	L	83, 373		92.3	Bal.	Similar to the above specimen, 10.095 gm cm^{-3} at 298 K.

* Not shown in figure.

DATA TABLE 202. THERMAL LINEAR EXPANSION OF MOLYBDENUM-NIOBIUM SYSTEM Mo-Nb

[Temperature, T, K; Linear Expansion, $\Delta L/L_0$, %]

T	$\Delta L/L_0$	T	$\Delta L/L_0$
CURVE 1		CURVE 8	
293	0.000	293	0.000*
473	0.132	1073	0.550
673	0.293	CURVE 9	
873	0.447	293	0.000*
1073	0.619	1073	0.589
1273	0.781	CURVE 10	
1339	0.832	293	0.000*
CURVE 2		1073	0.595
293	0.000*	CURVE 11	
473	0.124	83	-0.160
673	0.271	373	0.061
873	0.433	CURVE 12*	
1073	0.598	83	-0.153
1273	0.761	373	0.058
1339	0.832*	CURVE 13*	
CURVE 3		83	-0.147
293	0.000*	373	0.056
1073	0.440	CURVE 14	
CURVE 4		83	-0.137
293	0.000*	373	0.052
1073	0.456	CURVE 15*	
CURVE 5		83	-0.120
293	0.000*	373	0.046
1073	0.479	CURVE 16	
CURVE 6		83	-0.109
293	0.000*	373	0.042
1073	0.505		
CURVE 7			
293	0.000*		
1073	0.525		

* Not shown in figure.

FIGURE AND TABLE NO. 203R. PROVISIONAL VALUES FOR THERMAL LINEAR EXPANSION OF MOLYBDENUM-RHENIUM SYSTEM Mo-Re

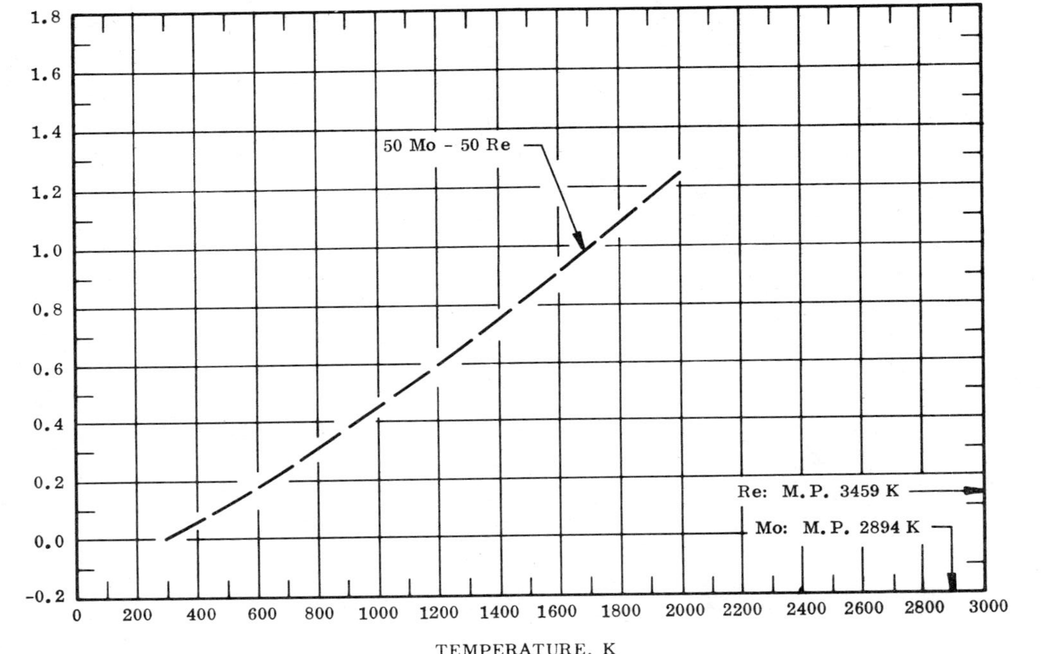

PROVISIONAL VALUES

[Temperature, T, K; Linear Expansion, $\Delta L/L_0$, %; α, K^{-1}]

T	(50 Mo-50 Re) $\Delta L/L_0$	$\alpha \times 10^6$
293	0.000	5.5
400	0.059	5.8
500	0.118	6.0
600	0.179	6.2
700	0.242	6.4
800	0.307	6.6
900	0.375	6.8
1000	0.444	7.1
1200	0.591	7.6
1400	0.747	8.0
1600	0.910	8.3
1800	1.078	8.5
2000	1.250	8.6

REMARKS

(50 Mo-50 Re): The tabulated values for well-annealed alloy are considered accurate to within ±7% over the entire temperature range. These values can be represented approximately by the following equation:

$$\Delta L/L_0 = -0.144 + 4.460 \times 10^{-4}\, T + 1.603 \times 10^{-7}\, T^2 - 1.726 \times 10^{-11}\, T^3$$

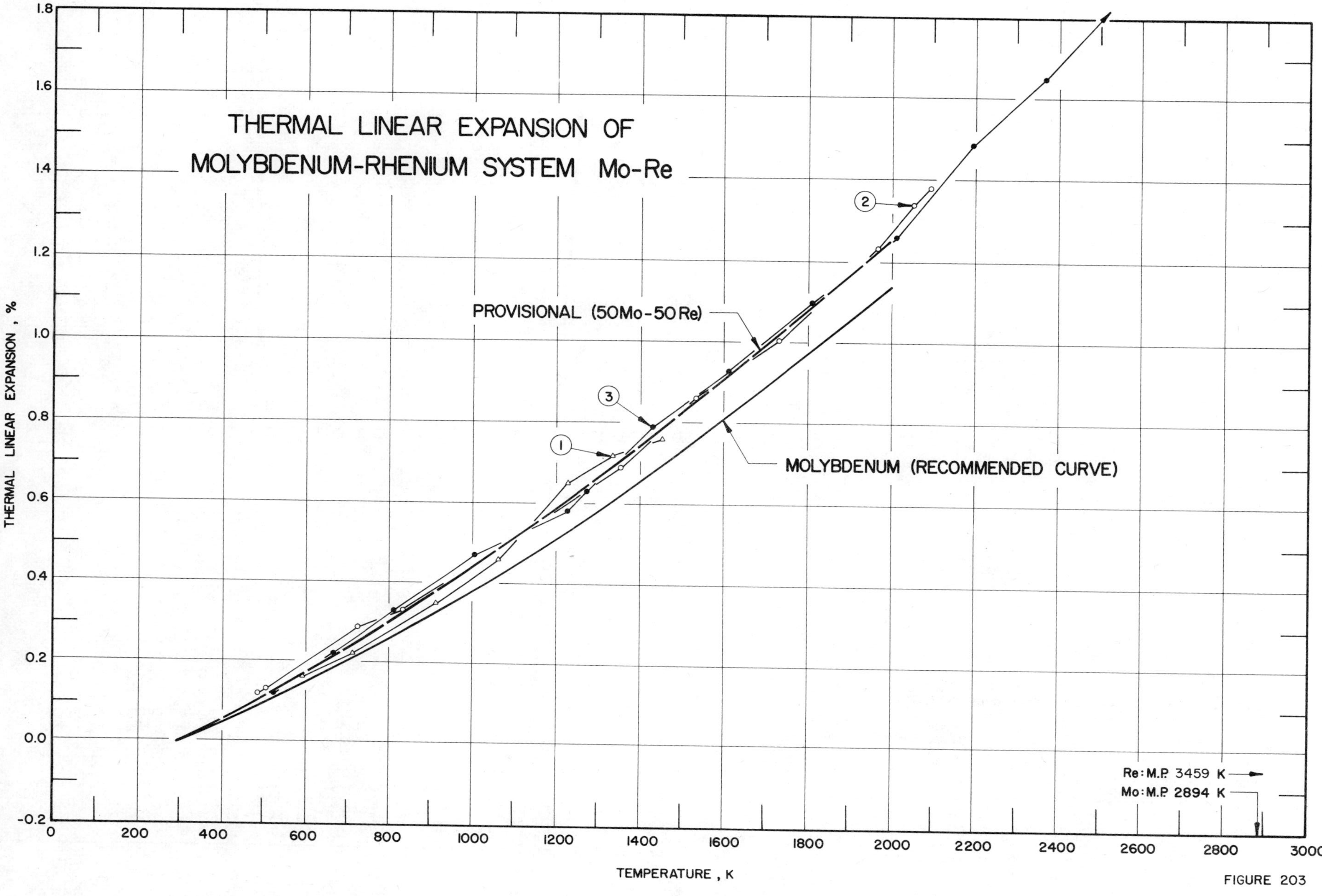
THERMAL LINEAR EXPANSION OF
MOLYBDENUM-RHENIUM SYSTEM Mo-Re
PROVISIONAL (50Mo-50Re)
MOLYBDENUM (RECOMMENDED CURVE)
1
2
3
Re: M.P. 3459 K
Mo: M.P. 2894 K
THERMAL LINEAR EXPANSION, %
TEMPERATURE, K
-0.2
0.0
0.2
0.4
0.6
0.8
1.0
1.2
1.4
1.6
1.8
0
200
400
600
800
1000
1200
1400
1600
1800
2000
2200
2400
2600
2800
3000

FIGURE 203

SPECIFICATION TABLE 203. THERMAL LINEAR EXPANSION OF MOLYBDENUM-RHENIUM SYSTEM Mo-Re

Cur. No.	Ref. No.	Author(s)	Year	Method Used	Temp. Range, K	Name and Specimen Designation	Composition (weight percent) Mo	Re	Composition (continued), Specifications, and Remarks
1	210	Conway, J.B. and Losekamp, A.C.	1966	T	599-1455		50	50	Sheet 20 mil thick formed by powder metallurgical techniques; expansion measured in helium atm on first test; zero-point correction is ±0.00%.
2	210	Conway, J.B. and Losekamp, A.C.	1966	T	486-2099		50	50	The above specimen; expansion measured on second test; zero-point correction is ±0.00%.
3	210	Conway, J.B. and Losekamp, A.C.	1966	T	522-2511		50	50	The above specimen; expansion measured on third test; zero-point correction is ±0.00%.

DATA TABLE 203. THERMAL LINEAR EXPANSION OF MOLYBDENUM-RHENIUM SYSTEM Mo-Re

[Temperature, T, K; Linear Expansion, $\Delta L/L_0$, %]

T	$\Delta L/L_0$
CURVE 1	
599	0.16
712	0.22
916	0.35
1061	0.46
1226	0.65
1336	0.72
1455	0.76
CURVE 2	
486	0.12
502	0.13
722	0.29
839	0.33
1351	0.69
1530	0.86
1732	1.01
1961	1.23
2052	1.34
2099	1.38
CURVE 3	
522	0.12
662	0.22
813	0.33
1001	0.47
1222	0.58
1271	0.63
1429	0.79
1610	0.93
1817	1.10
2017	1.26
2194	1.49
2365	1.65
2511	1.83*

*Not shown in figure.

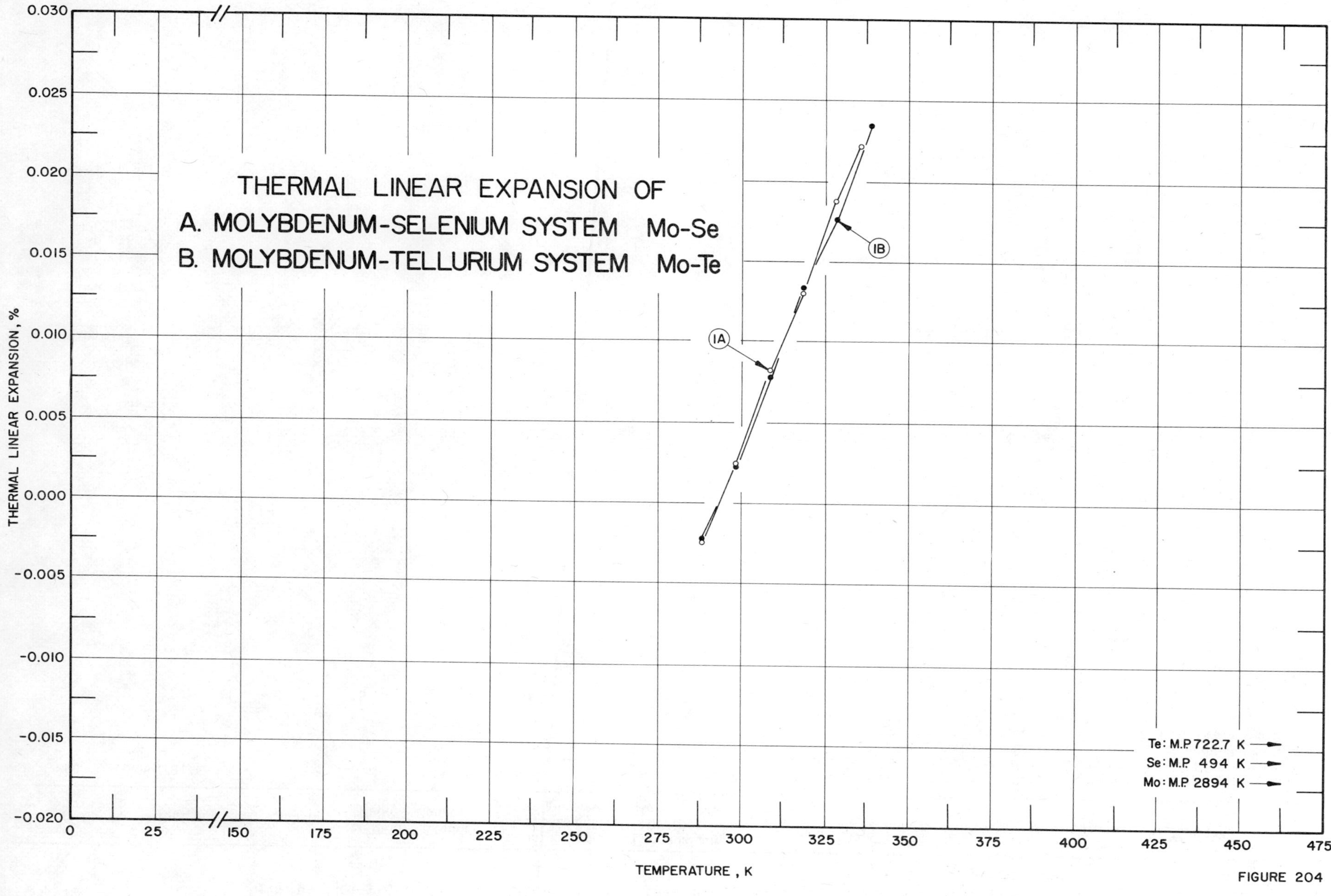
THERMAL LINEAR EXPANSION OF
A. MOLYBDENUM-SELENIUM SYSTEM Mo-Se
B. MOLYBDENUM-TELLURIUM SYSTEM Mo-Te
IA
IB
Te: M.P. 722.7 K
Se: M.P. 494 K
Mo: M.P. 2894 K
THERMAL LINEAR EXPANSION, %
0.030
0.025
0.020
0.015
0.010
0.005
0.000
-0.005
-0.010
-0.015
-0.020
TEMPERATURE, K
0
25
150
175
200
225
250
275
300
325
350
375
400
425
450
475

FIGURE 204

SPECIFICATION TABLE 204A. THERMAL LINEAR EXPANSION OF MOLYBDENUM-SELENIUM SYSTEM Mo-Se

Cur. No.	Ref. No.	Author(s)	Year	Method Used	Temp. Range, K	Name and Specimen Designation	Composition (weight percent) Mo	Se	Composition (continued), Specifications, and Remarks
1	209	Shodhn, R. P.	1965	X	288-336		98.35	1.65	99.9935 Mo from Hilger Co., 99.999 Se from ASARCO; powders ground intimately in mortar, placed into clean, dry quartz tube, evacuated and sealed; placed in furnace for 144 hrs at 1373 K, then quenched in water; lattice parameter reported at 288 K is 3.14697 Å; 3.14705 Å at 293 K determined by graphical interpolation.

DATA TABLE 204A. THERMAL LINEAR EXPANSION OF MOLYBDENUM-SELENIUM SYSTEM Mo-Se

[Temperature, T, K; Linear Expansion, $\Delta L/L_0$, %]

T	$\Delta L/L_0$
CURVE 1	
288	-0.0025
298	0.0025
308	0.0083
318	0.0130
328	0.0187
336	0.0222

SPECIFICATION TABLE 204B. THERMAL LINEAR EXPANSION OF MOLYBDENUM-TELLURIUM SYSTEM Mo-Te

Cur. No.	Ref. No.	Author(s)	Year	Method Used	Temp. Range, K	Name and Specimen Designation	Composition (weight percent) Mo	Te	Composition (continued), Specifications, and Remarks
1	209	Shodhn, R.P.	1965	X	288-338		97.37	2.63	99.9935 Mo from Hilger Co.; 99.999 Te from ASARCO; ground intimately in mortar, placed in clean, dry quartz tube, evacuated and sealed; heated 144 hrs at 1373 K, then quenched in water; lattice parameter reported at 288 K is 3.14700 Å; 3.14708 Å at 293 K determined by graphical interpolation.

DATA TABLE 204B. THERMAL LINEAR EXPANSION OF MOLYBDENUM-TELLURIUM SYSTEM Mo-Te

[Temperature, T, K; Linear Expansion, $\Delta L/L_0$, %]

T	$\Delta L/L_0$
CURVE 1	
288	-0.0024
298	0.0024
308	0.0078
318	0.0132
328	0.0176
338	0.0234

FIGURE AND TABLE NO. 205R. PROVISIONAL VALUES FOR THERMAL LINEAR EXPANSION OF MOLYBDENUM-TITANIUM SYSTEM Mo-Ti

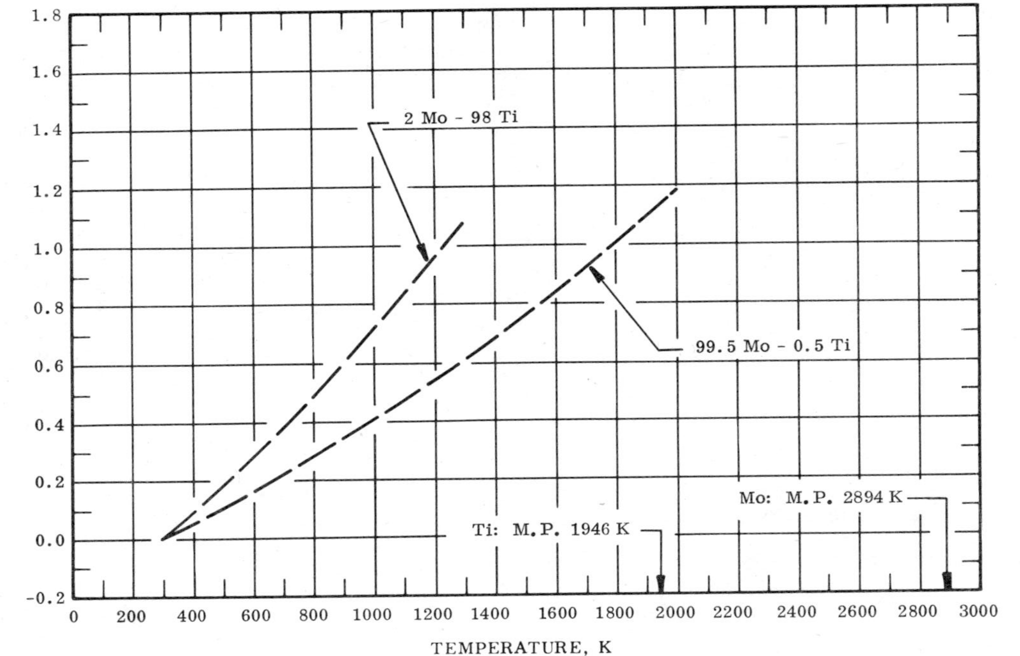

PROVISIONAL VALUES

[Temperature, T, K; Linear Expansion, $\Delta L/L_0$, %; α, K^{-1}]

T	(2 Mo-98 Ti) $\Delta L/L_0$	(2 Mo-98 Ti) $\alpha \times 10^6$	(99.5 Mo-0.5 Ti) $\Delta L/L_0$	(99.5 Mo-0.5 Ti) $\alpha \times 10^6$
293	0.000	7.8	0.000	5.1
400	0.089	8.7	0.056	5.2
500	0.180	9.5	0.109	5.4
600	0.278	10.1	0.164	5.7
700	0.382	10.7	0.222	5.9
800	0.491	11.1	0.281	6.1
900	0.604	11.5	0.343	6.3
1000	0.721	11.8	0.406	6.5
1200	0.960	12.1	0.538	6.9
1300	1.080	12.1		
1400			0.681	7.2
1600			0.835	7.9
1800			1.000	8.7
2000			1.181	9.3

REMARKS

(2 Mo-98 Ti): The tabulated values for well-annealed alloy are considered accurate to within ± 7% over the entire temperature range. These values can be represented approximately by the following equation:

$$\Delta L/L_0 = -0.187 + 4.802 \times 10^{-4}\ T + 5.849 \times 10^{-7}\ T^2 - 1.571 \times 10^{-10}\ T^3$$

(99.5 Mo-0.5 Ti): The tabulated values for well-annealed alloy are considered accurate to within ± 7% over the entire temperature range. These values can be represented approximately by the following equation:

$$\Delta L/L_0 = -0.145 + 4.783 \times 10^{-4}\ T + 5.118 \times 10^{-8}\ T^2 + 2.035 \times 10^{-11}\ T^3$$

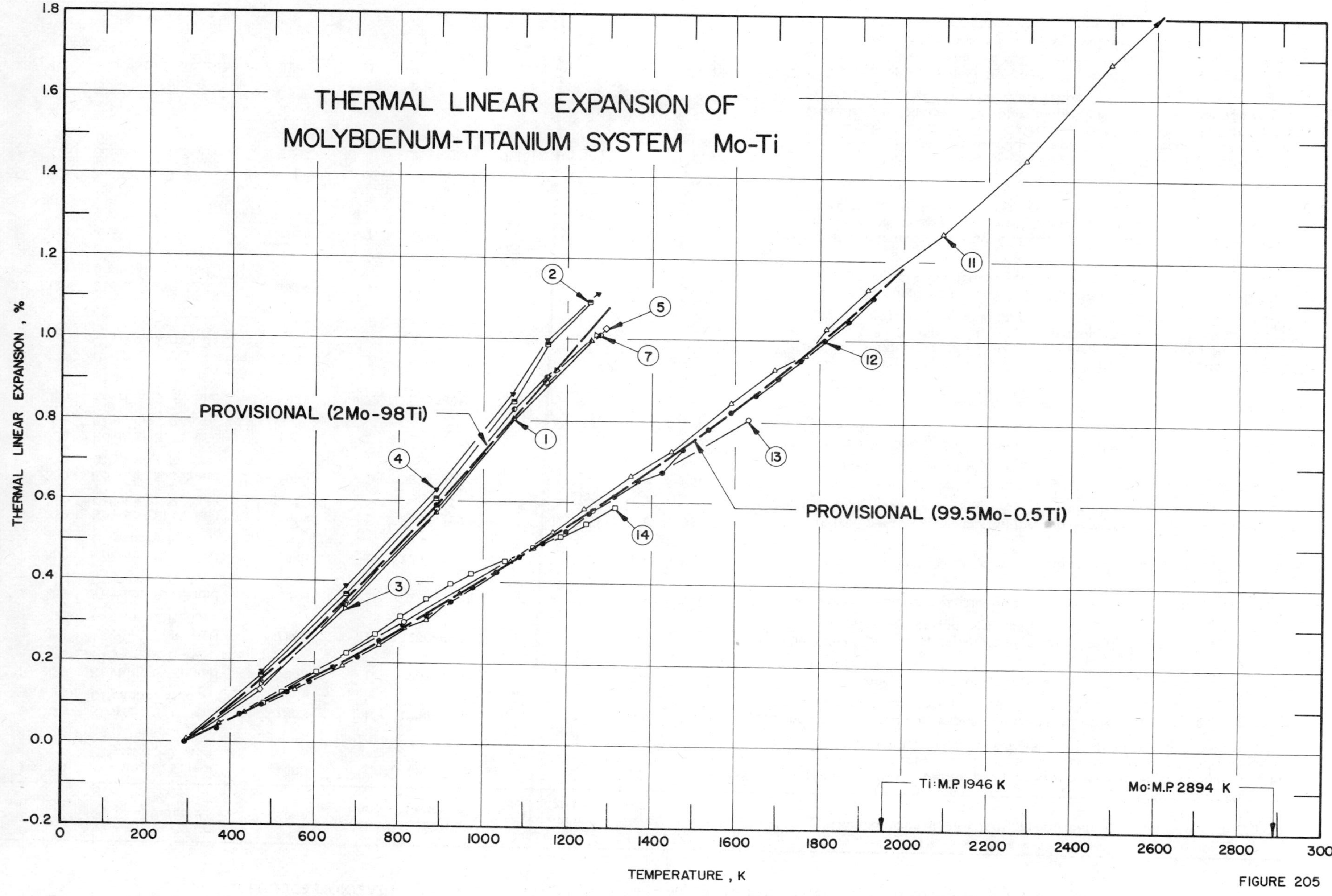
THERMAL LINEAR EXPANSION OF
MOLYBDENUM-TITANIUM SYSTEM Mo-Ti
PROVISIONAL (2Mo-98Ti)
PROVISIONAL (99.5Mo-0.5Ti)
Ti:M.P 1946 K
Mo:M.P 2894 K
1
2
3
4
5
7
11
12
13
14
THERMAL LINEAR EXPANSION, %
1.8
1.6
1.4
1.2
1.0
0.8
0.6
0.4
0.2
0.0
-0.2
TEMPERATURE, K
0
200
400
600
800
1000
1200
1400
1600
1800
2000
2200
2400
2600
2800
3000
FIGURE 205

SPECIFICATION TABLE 205. THERMAL LINEAR EXPANSION OF MOLYBDENUM-TITANIUM SYSTEM Mo-Ti

Cur. No.	Ref. No.	Author(s)	Year	Method Used	Temp. Range, K	Name and Specimen Designation	Composition (weight percent) Mo	Ti	Composition (continued), Specifications, and Remarks
1	330	Kovtun, S.F. and Ul'yanov, R.A.	1963	L	293-1255		2.0	98.0	Pure materials melted together in argon atmosphere, cooled, remelted 4 times; rolled into rods in vacuum; annealed 40 hr at 1473 K; expansion measured with increasing temperature on first test.
2	330	Kovtun, S.F. and Ul'yanov, R.A.	1963	L	1255-295		2.0	98.0	The above specimen; expansion measured with decreasing temperature; zero-point correction is 0.09%.
3	330	Kovtun, S.F. and Ul'yanov, R.A.	1963	L	296-1265		2.0	98.0	The above specimen; expansion measured with increasing temperature on second test.
4	330	Kovtun, S.F. and Ul'yanov, R.A.	1963	L	1265-296		2.0	98.0	The above specimen; expansion measured with decreasing temperature; zero-point correction is 0.186%.
5	330	Kovtun, S.F. and Ul'yanov, R.A.	1963	L	296-1288		2.0	98.0	The above specimen; expansion measured with increasing temperature on third test.
6*	330	Kovtun, S.F. and Ul'yanov, R.A.	1963	L	1288-296		2.0	98.0	The above specimen; expansion measured with decreasing temperature; zero-point correction is 0.26%.
7	330	Kovtun, S.F. and Ul'yanov, R.A.	1963	L	295-1272		2.0	98.0	The above specimen; expansion measured with increasing temperature on fourth test.
8*	330	Kovtun, S.F. and Ul'yanov, R.A.	1963	L	1272-299		2.0	98.0	The above specimen; expansion measured with decreasing temperature; zero-point correction is 0.345%.
9*	330	Kovtun, S.F. and Ul'yanov, R.A.	1963	L	299-1270		2.0	98.0	The above specimen; expansion measured with increasing temperature on fifth test.
10*	330	Kovtun, S.F. and Ul'yanov, R.A.	1963	L	1270-297		2.0	98.0	The above specimen; expansion measured with decreasing temperature; zero-point correction is 0.426%.
11	425	Hedge, J.C., Kostenko, C., and Lang, J.I.	1963	L	300-2749		Bal.	0.50	0.07 Zr, < 0.002 Fe, < 0.001 Ni, < 0.005 Si, 0.0290 C, and trace of O_2, N_2, H_2; obtained from Climax Molybdenum Co.; specimen dimensions 0.5 in. diameter by 6 in. long; expansion measured in argon atmosphere with temperature increasing at ~ 2.7 C min^{-1}; reported error 2%.
12	423	Fieldhouse, I.B., Long, J.I., and Blau, H.H. Jr.	1960	L	293-1922		Bal.	0.50	0.01-0.03 C, supplied by General Electric Co., Cleveland, Ohio; recrystallized for 35 min at 1866 K; reported error < 3%.
13	348	Anthony, F.M. and Pearl, H.A.	1960	L	299-1630		Bal.	0.50	Arc cast billet from American Metal Climax Corp.; recyrstallized at 1866 K; reduced approx. 60% by direct rolling, then machined into 3 in. long, 0.375 in. diameter specimen; reported error < 5%.
14	424	Burman, R.W.	1964	L	480-1313		99.4	0.50	0.08 Zr, 0.01 C; no details given.
15*	459	Southern Research Inst.	1966	L	298-588	TZM Alloy	Bal.	0.42	0.088 Zr, 0.032 C, 0.0042 O_2, <0.0035 Si, 0.0018 Fe, <0.001 Ni, 0.0003 N_2, and 0.0001 H_2; sheet material obtained from Refractomet Division of Universal-Cyclops Steel Corp; initial and final length 2.999 in; expansion measured in Quartz-tube dilatometer; zero-point correction of 0.003% determined by graphical extrapolation.
16*	459	Southern Research Inst.	1966	L	301-671	TZM Alloy	Bal.	10.49	0.104 Zr, 0.030 C, 0.002 Ni, <0.0015 Fe, <0.001 Si, 0.0012 O_2, 0.0003 N_2, and 0.00017 H_2; similar to the above specimen; zero-point correction of 0.003% determined by graphical extrapolation.

* Not shown in figure.

SPECIFICATION TABLE 205. THERMAL LINEAR EXPANSION OF MOLYBDENUM-TITANIUM SYSTEM Mo-Ti (continued)

Cur. No.	Ref. No.	Author(s)	Year	Method Used	Temp. Range, K	Name and Specimen Designation	Composition (weight percent) Mo	Ti	Composition (continued), Specifications, and Remarks
17*	459	Southern Research Inst.	1966	L	293-1655	TZM Alloy	Bal.	0.50	0.119 Zr, 0.031 C, <0.0035 Si, 0.0015 O_2, <0.0015 Fe, <0.001 Ni, 0.00024 H_2 and 0.0002 N_2; similar to the above specimen except using graphite dilatometer.
18*	459	Southern Research Inst.	1966	L	293-1655	TZM Alloy	Bal.	0.50	0.083 Zr, 0.024 C, <0.0035 Si, 0.0022 O_2, <0.0015 Fe, <0.001 Ni, 0.0010 N_2 and 0.0002 H_2; similar to the above specimen.

* Not shown in figure.

DATA TABLE 205. THERMAL LINEAR EXPANSION OF MOLYBDENUM-TITANIUM SYSTEM Mo-Ti

[Temperature, T, K; Linear Expansion, $\Delta L/L_0$, %]

T	$\Delta L/L_0$
CURVE 1	
293	0.000*
471	0.158
674	0.350
886	0.577
1073	0.802
1255	1.006
CURVE 2	
1255	1.096
1152	0.986
1071	0.847
885	0.604
672	0.368
471	0.164
295	0.006*
CURVE 3	
296	0.002*
474	0.142
673	0.335
885	0.572*
1071	0.801*
1171	0.922
1265	1.016
CURVE 4	
1265	1.113
1157	0.999
1071	0.863
883	0.622
672	0.388
471	0.170
296	0.002*
CURVE 5	
296	0.003*
472	0.216
673	0.338*
884	0.580
1069	0.804*
1152	0.900
1288	1.028
CURVE 6*	
1288	1.114
1157	0.990
1070	0.854
885	0.621
673	0.378
471	0.166
296	0.004
CURVE 7	
295	0.002*
472	0.141*
672	0.341
886	0.589
1071	0.823
1151	0.904
1272	1.011
CURVE 8*	
1272	1.093
1147	0.981
1070	0.863
883	0.623
674	0.374
471	0.170
299	0.008
CURVE 9*	
299	0.004
472	0.134
672	0.331
887	0.575
1070	0.817
1175	0.923
1270	1.017
CURVE 10*	
1270	1.097
1159	0.995
1072	0.856
881	0.608
671	0.362
472	0.158
297	0.005
CURVE 11	
300	0.004
378	0.047
436	0.078
554	0.133
664	0.193
757	0.244
815	0.286
862	0.308
945	0.375
1066	0.455
1162	0.525
1240	0.584
1348	0.666
1442	0.731
1588	0.845
1693	0.932
1762	0.967
1811	1.036
1907	1.135
2095	1.276
2284	1.459
2491	1.694
2647	1.892
2749	2.045
CURVE 12	
293	0.000
366	0.036
422	0.068
477	0.095
533	0.127
588	0.158
644	0.189
700	0.219
755	0.253
811	0.186
866	0.318
922	0.353
977	0.388
1033	0.427
1088	0.463
1144	0.500
1200	0.529
1255	0.577
1311	0.618
CURVE 12 (cont.)	
1366	0.658
1422	0.678
1477	0.737
1533	0.783
1588	0.825
1644	0.869
1700	0.913
1755	0.958
1811	1.005
1866	1.056
1922	1.110
CURVE 13	
299	0.003
816	0.298
1261	0.578
1630	0.805
CURVE 14	
480	0.098
526	0.124
605	0.178
675	0.226
742	0.271
805	0.314
869	0.359
922	0.399
972	0.420
1054	0.454
1120	0.488
1187	0.518
1246	0.549
1313	0.592
CURVE 15*	
298	0.003
339	0.022
421	0.067
505	0.113
588	0.154
CURVE 16*	
301	0.003
CURVE 16 (cont.)*	
339	0.022
423	0.062
505	0.111
591	0.156
671	0.200
CURVE 17*	
293	0.000
729	0.240
848	0.304
975	0.381
1114	0.458
1119	0.458
1222	0.522
1297	0.561
1419	0.646
1561	0.728
1655	0.789
CURVE 18*	
293	0.000
966	0.391
1130	0.478
1136	0.479
1241	0.539
1269	0.544
1422	0.652
1536	0.722
1655	0.794

* Not shown in figure.

FIGURE AND TABLE NO. 206R. PROVISIONAL VALUES FOR THERMAL LINEAR EXPANSION OF MOLYBDENUM-TUNGSTEN SYSTEM Mo-W

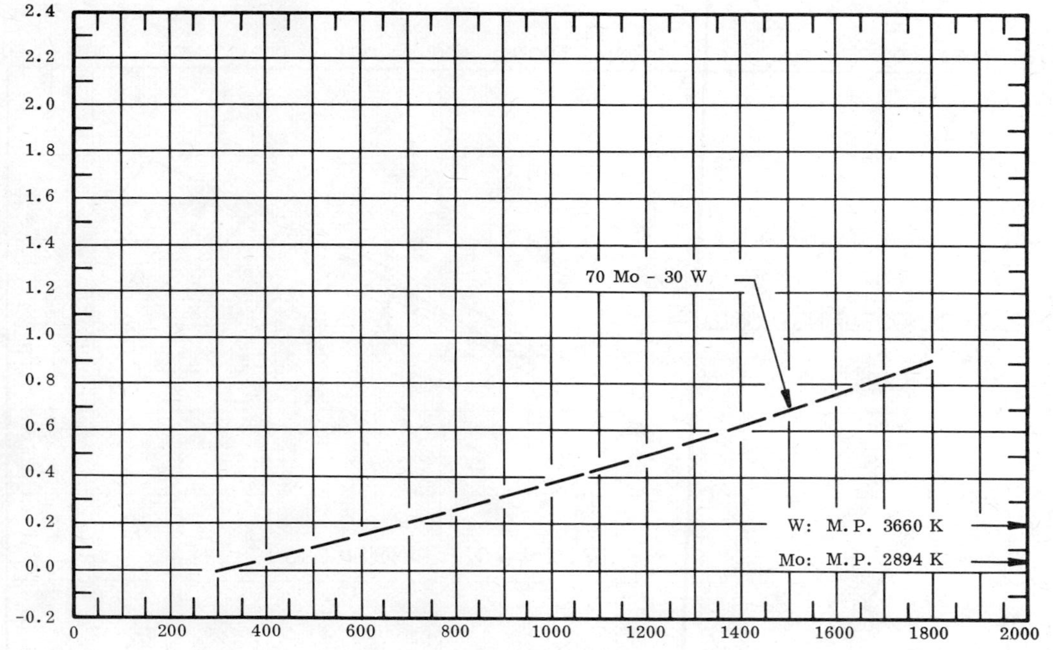

PROVISIONAL VALUES

[Temperature, T, K; Linear Expansion, $\Delta L/L_0$, %; α, K^{-1}]

T	(70 Mo-30 W) $\Delta L/L_0$	$\alpha \times 10^6$
293	0.000	4.9
400	0.052	5.0
500	0.101	5.1
600	0.153	5.2
700	0.205	5.4
800	0.259	5.5
900	0.315	5.7
1000	0.373	5.8
1200	0.492	6.2
1400	0.622	6.7
1600	0.762	7.4
1800	0.917	8.2

REMARKS

(70 Mo-30 W): The tabulated values for well-annealed alloy are considered accurate to within ±7% over the entire temperature range. These values can be represented approximately by the following equation:

$$\Delta L/L_0 = -0.143 + 4.817 \times 10^{-4}\, T + 4.874 \times 10^{-10}\, T^2 + 3.253 \times 10^{-11}\, T^3$$

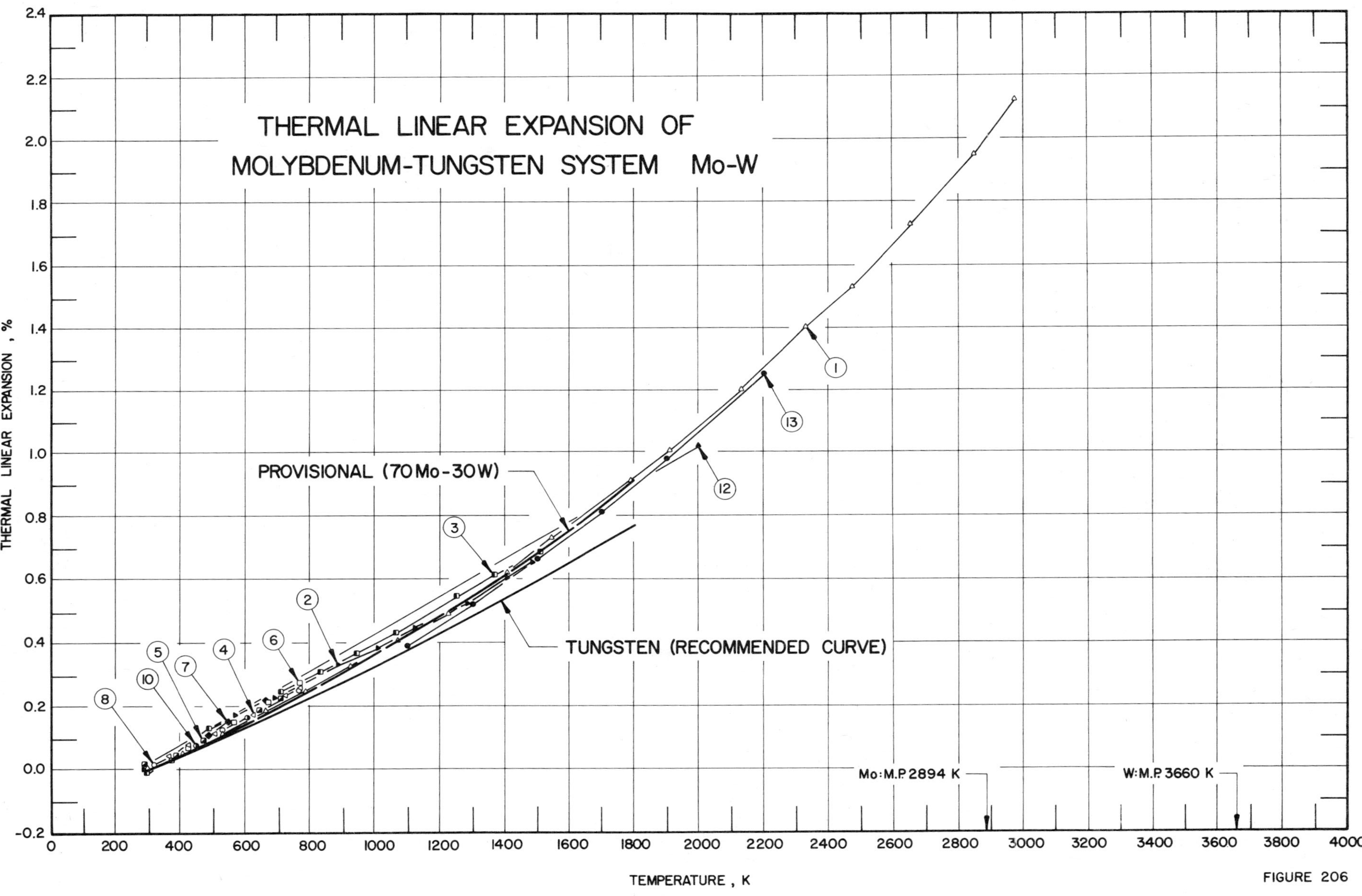
THERMAL LINEAR EXPANSION OF
MOLYBDENUM-TUNGSTEN SYSTEM Mo-W
THERMAL LINEAR EXPANSION, %
TEMPERATURE, K
PROVISIONAL (70 Mo-30 W)
TUNGSTEN (RECOMMENDED CURVE)
Mo: M.P. 2894 K
W: M.P. 3660 K
1
2
3
4
5
6
7
8
10
12
13
2.4
2.2
2.0
1.8
1.6
1.4
1.2
1.0
0.8
0.6
0.4
0.2
0.0
-0.2
0
200
400
600
800
1000
1200
1400
1600
1800
2000
2200
2400
2600
2800
3000
3200
3400
3600
3800
4000

FIGURE 206

SPECIFICATION TABLE 206. THERMAL LINEAR EXPANSION OF MOLYBDENUM-TUNGSTEN SYSTEM Mo-W

Cur. No.	Ref. No.	Author(s)	Year	Method Used	Temp. Range, K	Name and Specimen Designation	Composition (weight percent) Mo	W	Composition (continued), Specifications, and Remarks
1	425	Hedge, J.C., Kostenko, C., and Lang, J.I.	1963	L	300-2973		70.09	29.83	0.07 Zr and 0.012 C; specimen dimensions 0.5 in. diameter by 6 in. long; obtained from Climax Molybdenum Co.; expansion measured in argon atmosphere with temperature increasing at ~ 2.8 C min^{-1}; reported error 2%; zero-point correction is 0.002%.
2	224	Clark, D. and Knight, D.	1965	X	293-1489		50.1	49.9	Alloy prepared by powder metallurgy techniques; lattice parameter reported at 308 K is 3.1524 Å; 3.1527 Å at 293 K determined by graphical extrapolation.
3	224	Clark, D. and Knight, D.	1965	X	293-1515		75	25	Similar to the above specimen; lattice parameter reported at 293 K is 3.1500 Å.
4	226	Hidnert, P. and Gero, W.B.	1924	T	300-773	S516	Bal.	1.85	Hexagonal ingot 1.59 cm diameter; prepared from fine grained powder; expansion measured with increasing temperature; zero-point correction of 0.003% determined by graphical extrapolation.
5	226	Hidnert, P. and Gero, W.B.	1924	T	710-298	S516	Bal.	1.85	The above specimen; expansion measured with decreasing temperature; zero-point correction of 0.003% determined by graphical extrapolation.
6	226	Hidnert, P. and Gero, W.B.	1924	T	293-770	S516A	Bal.	1.85	0.03 Fe, 0.03 Si; the above specimen, swaged to 0.635 cm diameter; expansion measured with increasing temperature.
7	226	Hidnert, P. and Gero, W.B.	1924	T	669-385	S516A	Bal.	1.85	The above specimen; expansion measured with decreasing temperature; zero-point correction of -0.001% determined by graphical extrapolation.
8	226	Hidnert, P. and Gero, W.B.	1924	T	293-773	S516B	Bal.	1.80	0.05 Fe, 0.014 Si; the above specimen, swaged to 0.445 cm diameter; expansion measured with increasing temperature.
9*	226	Hidnert, P. and Gero, W.B.	1924	T	656-293	S516B	Bal.	1.80	The above specimen; expansion measured with decreasing temperature; zero-point correction of 0.006% determined by graphical extrapolation.
10	226	Hidnert, P. and Gero, W.B.	1924	T	293-777	S516C	Bal.	1.80	The above specimen; swaged to 0.254 cm diameter; expansion measured with increasing temperature.
11*	226	Hidnert, P. and Gero, W.B.	1924	T	667-293	S516C	Bal.	1.80	The above specimen; expansion measured with decreasing temperature; zero-point correction of -0.022% determined by graphical extrapolation.
12	458	Bossart, P.N.	1936	L	293, 2000		62.5	37.5	Alloy prepared by powder metallurgy techniques, sintered at 1873 K for 1 hr.
13	458	Bossart, P.N.	1936	L	1100-2200		62.5	37.5	Data computed by TPRC from equation given by author as $\Delta L/L = 3.5 \times 10^{-6} (T_C-20) + 1.6 \times 10^{-9} (T_C-20)^2$; equation valid from 1100 K to 2200 K.

*Not shown in figure.

DATA TABLE 206. THERMAL LINEAR EXPANSION OF MOLYBDENUM-TUNGSTEN SYSTEM Mo-W

[Temperature, T, K; Linear Expansion, $\Delta L/L_0$, %]

T	$\Delta L/L_0$
CURVE 1	
300	0.003
406	0.043
538	0.114
663	0.182
793	0.247
927	0.322
1074	0.410
1223	0.494
1405	0.620
1545	0.725
1790	0.907
1908	1.003
2138	1.206
2333	1.402
2473	1.536
2655	1.730
2846	1.954
2973	2.122
CURVE 2	
293	0.000
308	0.009*
573	0.171
692	0.225
709	0.246*
886	0.330
1012	0.385
1124	0.445
1288	0.527
1403	0.601
1489	0.653
CURVE 3	
293	0.000*
300	-0.012
496	0.126
710	0.241
836	0.303
946	0.362
1069	0.430
1256	0.542
1379	0.611
1423	0.623*
1515	0.685
CURVE 4	
300	0.003*
368	0.040
429	0.072
470	0.094*
515	0.117
589	0.155*
623	0.176
655	0.193*
666	0.199*
723	0.231
773	0.260
CURVE 5	
710	0.223
641	0.183
474	0.087
386	0.040
298	0.010
CURVE 6	
293	0.000*
315	0.012*
372	0.039
428	0.068
510	0.111*
569	0.143
624	0.172*
677	0.204
728	0.240*
770	0.272
CURVE 7	
669	0.211
552	0.147
489	0.106
385	0.047*
CURVE 8	
293	0.000*
320	0.015
381	0.044*
431	0.068*
CURVE 8 (cont.)	
475	0.091*
531	0.121
566	0.137*
630	0.172*
722	0.226*
773	0.254
CURVE 9*	
656	0.195
459	0.086
369	0.037
293	0.000
CURVE 10	
293	0.000*
329	0.017*
389	0.046*
450	0.078
476	0.090*
522	0.113*
574	0.143*
615	0.160
668	0.195*
731	0.236*
777	0.262*
CURVE 11*	
667	0.182
573	0.130
498	0.095
385	0.041
293	0.000
CURVE 12	
293	0.00*
2000	1.02
CURVE 13	
1100	0.39
1300	0.52
1500	0.66
CURVE 13 (cont.)	
1700	0.81
1900	0.98
2000	1.06*
2200	1.25

*Not shown in figure.

FIGURE AND TABLE NO. 207R. PROVISIONAL VALUES FOR THERMAL LINEAR EXPANSION OF MOLYBDENUM-URANIUM SYSTEM Mo-U

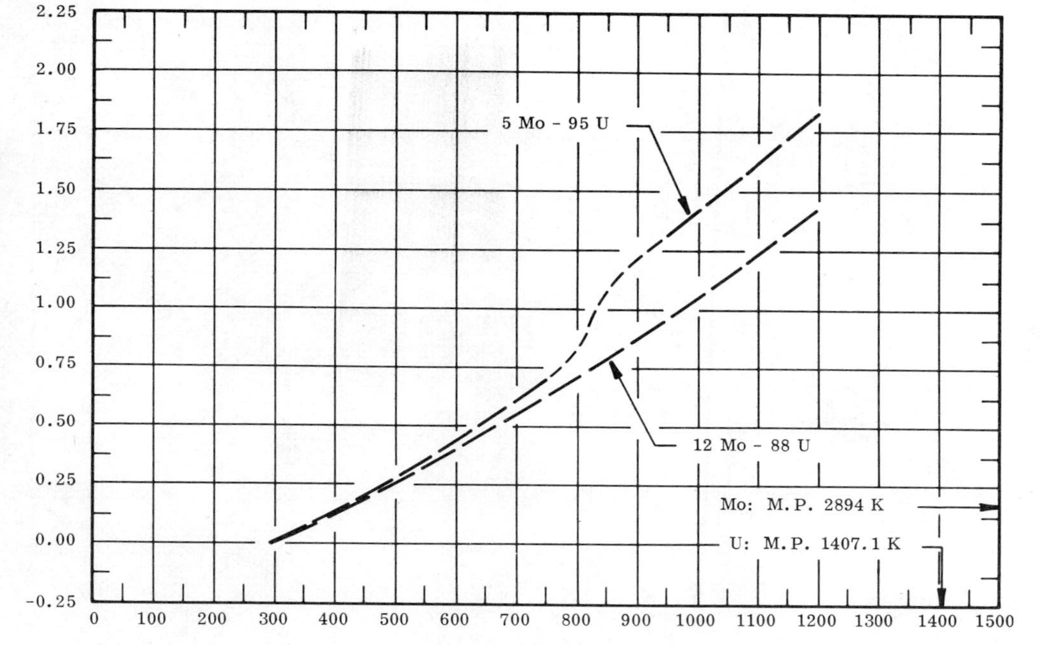

PROVISIONAL VALUES

[Temperature, T, K; Linear Expansion, $\Delta L/L_0$, %; α, K^{-1}]

T	(5 Mo-95 U) $\Delta L/L_0$	(5 Mo-95 U) $\alpha \times 10^6$	(12 Mo-88 U) $\Delta L/L_0$	(12 Mo-88 U) $\alpha \times 10^6$
293	0.000	13.2	0.000	12.4
400	0.145	14.2	0.137	13.1
500	0.292	15.1	0.272	13.8
600	0.446	16.0	0.414	14.6
700	0.611	17.0	0.562	15.3
800			0.718	16.0
900			0.882	16.7
950	1.312	20.2	0.968	17.1
1000	1.414	20.5	1.055	17.5
1100	1.623	20.8	1.233	18.3
1200	1.832	20.9	1.420	19.1

REMARKS

(5 Mo-95 U): The tabulated values considered accurate to within ±7% for an alloy whose heat treatment resulted in the transformation of part of the phase into other intermediate phases. A cooling and heating hysteresis is observed in the region of the phase transformations (720-950 K). The thermal linear expansion of this alloy is slightly lower than for an alloy which has been water quenched from the γ-uranium phase. The tabulated values can be represented approximately by the following equations:

$$\Delta L/L_0 = -0.352 + 1.088 \times 10^{-3}\, T + 3.589 \times 10^{-7}\, T^2 + 7.467 \times 10^{-11}\, T^3 \quad (293 < T < 700)$$

$$\Delta L/L_0 = -0.779 + 2.467 \times 10^{-3}\, T - 4.206 \times 10^{-7}\, T^2 + 1.481 \times 10^{-10}\, T^3 \quad (950 < T < 1200)$$

(12 Mo-88 U): The tabulated values considered accurate to within ±7% are for an alloy whose heat treatment has resulted into the transformation of part of the phase into other intermediate phases. The thermal linear expansion of this alloy is slightly higher than for an alloy which has been water quenched from the γ-uranium phase. The tabulated values can be represented approximately by the following equation:

$$\Delta L/L_0 = -0.335 + 1.050 \times 10^{-3}\, T + 3.175 \times 10^{-7}\, T^2 + 2.242 \times 10^{-11}\, T^3$$

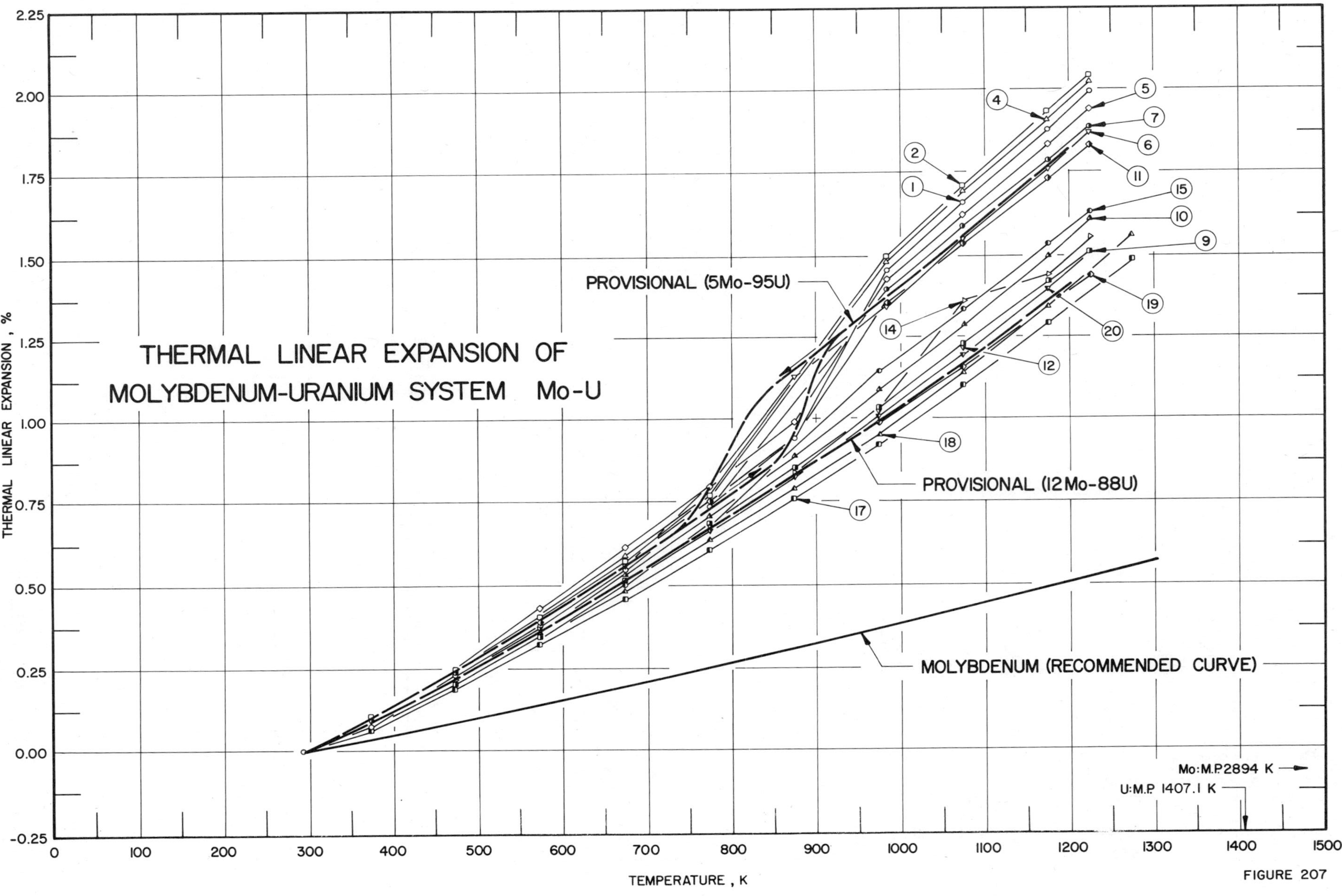
THERMAL LINEAR EXPANSION OF
MOLYBDENUM-URANIUM SYSTEM Mo-U
PROVISIONAL (5Mo-95U)
PROVISIONAL (12Mo-88U)
MOLYBDENUM (RECOMMENDED CURVE)
Mo:M.P. 2894 K
U:M.P. 1407.1 K
THERMAL LINEAR EXPANSION, %
TEMPERATURE, K
2.25
2.00
1.75
1.50
1.25
1.00
0.75
0.50
0.25
0.00
-0.25
0
100
200
300
400
500
600
700
800
900
1000
1100
1200
1300
1400
1500

FIGURE 207

SPECIFICATION TABLE 207. THERMAL LINEAR EXPANSION OF MOLYBDENUM-URANIUM SYSTEM Mo-U

Cur. No.	Ref. No.	Author(s)	Year	Method Used	Temp. Range, K	Name and Specimen Designation	Composition (weight percent) Mo	U	Composition (continued), Specifications, and Remarks
1	448	Saller, H.A., Dickerson, R.F. and Murr, W.E.	1956	L	293-1223		3.34	96.66	Prepared by combined induction and consumable-electrode arc-melting processes; heat treated 1 hr at 1073 K, water quenched; expansion measured with increasing temp under 2 x 10^{-5} mm Hg vacuum.
2	448	Saller, H.A. et al.	1956	L	1223-293		3.34	96.66	The above specimen; expansion measured with decreasing temp.
3*	448	Saller, H.A. et al.	1956	L	293-1223		3.34	96.66	Similar to the above specimen; heat treated 1 hr at 1073 K, air cooled; specimen protected in evacuated glass envelope; expansion measured with increasing temp.
4	448	Saller, H.A. et al.	1956	L	1223-293		3.34	96.66	The above specimen; expansion measured with decreasing temp at 298 K per min under vacuum of 2 x 10^{-5} mm Hg.
5	448	Saller, H.A. et al.	1956	L	293-1223		4.99	95.01	Prepared by combined induction and consumable-electrode arc-melting processes; heat treated 1 hr at 1073 K, water quenched; expansion measured with increasing temp under vacuum of 2 x 10^{-5} mm Hg.
6	448	Saller, H.A. et al.	1956	L	1223-293		4.99	95.01	The above specimen; expansion measured with decreasing temp.
7	448	Saller, H.A. et al.	1956	L	293-1223		4.99	95.01	Prepared by combined induction and consumable electrode arc-melting processes; heat treated 1 hr at 1073 K, furnace cooled; expansion measured with increasing temp.
8*	448	Saller, H.A. et al.	1956	L	1223-293		4.99	95.01	The above specimen; expansion measured with decreasing temp.
9	448	Saller, H.A. et al.	1956	L	293-1223		7.18	92.82	Similar to the above specimen; heat treated 1 hr at 1073 K, water quenched; expansion measured with increasing temp under 2 x 10^{-5} mm Hg vacuum.
10	448	Saller, H.A. et al.	1956	L	1223-293		7.18	92.82	The above specimen; expansion measured with decreasing temp.
11	448	Saller, H.A. et al.	1956	L	293-1223		7.18	92.82	Similar to the above specimen; heat treated 1 hr at 1073 K, furnace cooled to 773 K, held 100 hrs, furnace cooled; expansion measured with increasing temp.
12	448	Saller, H.A. et al.	1956	L	1223-293		7.18	92.82	The above specimen; expansion measured with decreasing temp.
13*	448	Saller, H.A. et al.	1956	L	293-1223		9.36	90.64	Similar to the above specimen; heat treated 1 hr at 1073 K, water quenched; expansion measured with increasing temp under 2 x 10^{-5} mm Hg vacuum.
14	448	Saller, H.A. et al.	1956	L	1223-293		9.36	90.64	The above specimen; expansion measured with decreasing temp.
15	448	Saller, H.A. et al.	1956	L	293-1223		9.36	90.64	Similar to the above specimen; heat treated 1 hr at 1073 K, furnace cooled to 773 K, held 200 hrs, furnace cooled; expansion measured with increasing temp under 2 x 10^{-5} mm Hg vacuum.
16*	448	Saller, H.A. et al.	1956	L	1223-293		9.36	90.64	The above specimen; expansion measured with decreasing temp.
17	448	Saller, H.A. et al.	1956	L	293-1273		12.1	87.9	Prepared by combined induction and consumable electrode arc-melting processes; heat treated 1 hr at 1073 K, water quenched; expansion measured with increasing temp.
18	448	Saller, H.A. et al.	1956	L	1273-293		12.1	87.9	The above specimen; expansion measured with decreasing temp.
19	448	Saller, H.A. et al.	1956	L	293-1223		12.1	87.9	Prepared by combined induction and consumable electrode arc-melting processes; heat treated 1 hr at 1073 K, furnace cooled to 773 K, held 2 weeks, furnace cooled; expansion measured with increasing temp.
20	448	Saller, H.A. et al.	1956	L	1223-293		12.1	87.9	The above specimen; expansion measured with decreasing temp.

* Not shown in figure.

DATA TABLE 207. THERMAL LINEAR EXPANSION OF MOLYBDENUM-URANIUM SYSTEM Mo-U

[Temperature, T, K; Linear Expansion, $\Delta L/L_0$, %]

T	$\Delta L/L_0$
CURVE 1	
293	0.000
373	0.087
473	0.228
573	0.384
673	0.550
773	0.729
873	0.944
973	1.462
1073	1.666
1173	1.880
1223	1.993
CURVE 2	
1223	2.044
1173	1.934
1073	1.717
973	1.506
773	0.766
673	0.576
573	0.404
473	0.249
373	0.106
293	0.000*
CURVE 3*	
293	0.000
373	0.095
473	0.241
573	0.400
673	0.570
773	0.757
873	0.971
1073	1.700
1173	1.906
1223	2.012
CURVE 4	
1223	2.028
1173	1.916
1073	1.693
973	1.482
773	0.766*
673	0.582
573	0.414*
CURVE 4 (cont.)	
473	0.251*
373	0.115*
293	0.000*
CURVE 5	
293	0.000*
373	0.106*
473	0.253*
573	0.433
673	0.618
773	0.795
873	0.991
973	1.432
1073	1.626
1173	1.835
1223	1.941
CURVE 6	
1223	1.874
1173	1.767
1073	1.550
973	1.347
873	1.158
773	0.796*
673	0.566
573	0.406*
473	0.256*
373	0.113*
293	0.000*
CURVE 7	
293	0.000*
373	0.093*
473	0.237
573	0.396*
673	0.557*
773	0.734*
973	1.397
1073	1.583
1173	1.784
1223	1.886
CURVE 8*	
1223	1.866
1173	1.762
1073	1.545
973	1.339
873	1.146
673	0.566
573	0.403
473	0.248
373	0.103
293	0.000
CURVE 9	
293	0.000*
373	0.083*
473	0.203
573	0.348
673	0.517
773	0.683
873	0.854
973	1.038
1073	1.231
1173	1.423
1223	1.510
CURVE 10	
1223	1.613
1173	1.500
1073	1.283
973	1.085
873	0.837
773	0.714
673	0.539
573	0.354
473	0.236*
373	0.102*
293	0.000*
CURVE 11	
293	0.000*
373	0.098*
473	0.236*
573	0.396
673	0.565*
773	0.752
CURVE 11 (cont.)	
873	0.949*
973	1.358
1073	1.540
1173	1.734
1223	1.838
CURVE 12	
1223	1.522*
1173	1.420*
1073	1.222
973	1.021
873	0.834*
773	0.666
673	0.517*
573	0.366
473	0.224*
373	0.094
293	0.000*
CURVE 13*	
293	0.000
373	0.103
473	0.242
573	0.395
673	0.558
773	0.730
873	0.913
973	1.093
1073	1.269
1173	1.442
1223	1.521
CURVE 14	
1223	1.557
1173	1.443
1073	1.382
973	1.023*
873	0.842
773	0.676
673	0.517*
573	0.360*
473	0.218
373	0.094*
293	0.000*
CURVE 15	
293	0.000*
373	0.080*
473	0.207*
573	0.354*
673	0.508
773	0.674*
973	1.150
1073	1.336
1173	1.535
1223	1.628
CURVE 16*	
1223	1.535
1173	1.431
1073	1.218
973	1.026
873	0.845
773	0.676
673	0.515
573	0.364
473	0.226
373	0.095
293	0.000
CURVE 17	
293	0.000*
373	0.072
473	0.188
573	0.324
673	0.460
773	0.603
873	0.759
973	0.922
1073	1.103
1173	1.294
1273	1.494
CURVE 18	
1273	1.558
1173	1.344
1073	1.146
973	0,961
873	0.794
773	0.633
CURVE 18 (cont.)	
673	0.484
573	0.342*
473	0.211*
373	0.091*
293	0.000*
CURVE 19	
293	0.000*
373	0.110*
473	0.248*
573	0.392*
673	0.536*
773	0.684*
873	0.836*
973	0.992
1073	1.156
1173	1.343*
1223	1.440
CURVE 20	
1223	1.511*
1173	1.398
1073	1.192
973	1.003
873	0.825
773	0.663*
673	0.507*
573	0.357*
473	0.224*
373	0.094*
293	0.000*

* Not shown in figure.

FIGURE AND TABLE NO. 208R. PROVISIONAL VALUES FOR THERMAL LINEAR EXPANSION OF MOLYBDENUM-VANADIUM SYSTEM Mo-V

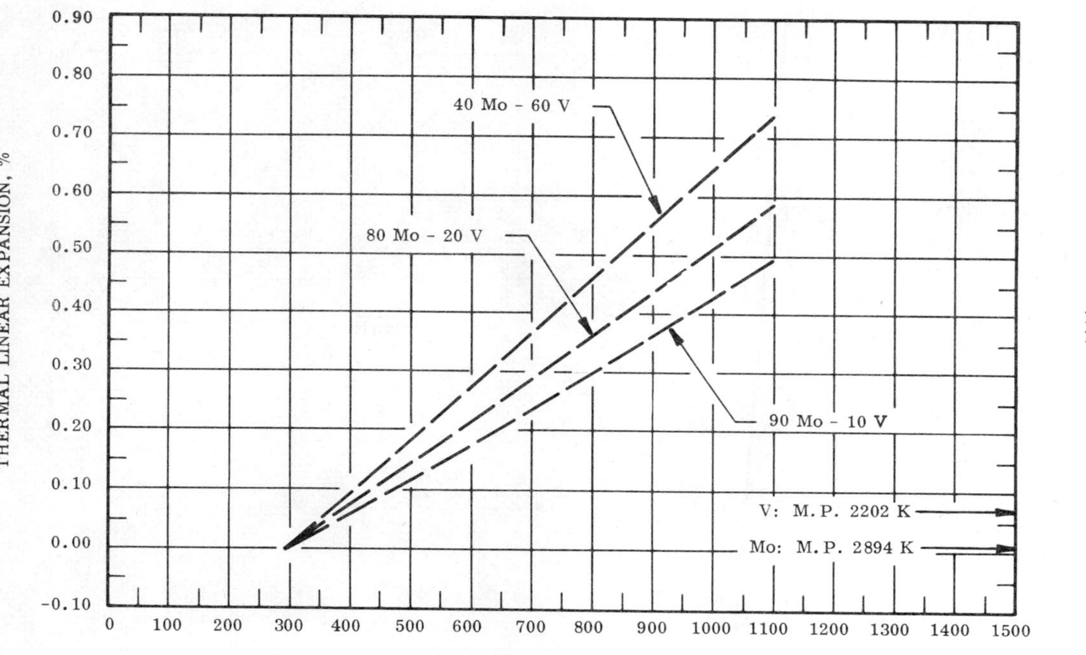

PROVISIONAL VALUES

[Temperature, T, K; Linear Expansion, $\Delta L/L_0$, %; α, K^{-1}]

T	(40 Mo-60 V) $\Delta L/L_0$	(40 Mo-60 V) $\alpha \times 10^6$	(80 Mo-20 V) $\Delta L/L_0$	(80 Mo-20 V) $\alpha \times 10^6$	(90 Mo-10V) $\Delta L/L_0$	(90 Mo-10V) $\alpha \times 10^6$
293	0.000	8.2	0.000	6.5	0.000	5.6
400	0.091	8.9	0.072	6.9	0.060	5.7
500	0.182	9.2	0.143	7.2	0.118	5.8
600	0.275	9.3	0.216	7.4	0.176	6.0
700	0.369	9.3	0.290	7.4	0.237	6.2
800	0.462	9.3	0.363	7.4	0.300	6.4
900	0.554	9.3	0.437	7.4	0.364	6.5
1000	0.646	9.3	0.511	7.4	0.430	6.6
1100	0.739	9.3	0.585	7.4	0.497	6.7

REMARKS

(40 Mo-60 V): The tabulated values for well-annealed alloy are considered accurate to within ±7% over the entire temperature range. These values can be represented approximately by the following equation:

$$\Delta L/L_0 = -0.231 + 7.105 \times 10^{-4} T + 2.959 \times 10^{-7} T^2 - 1.282 \times 10^{-10} T^3$$

(80 Mo-20 V): The tabulated values for well-annealed alloy are considered accurate to within ±7% over the entire temperature range. These values can be represented approximately by the following equation:

$$\Delta L/L_0 = -0.188 + 5.978 \times 10^{-4} T + 1.622 \times 10^{-7} T^2 - 6.052 \times 10^{-11} T^3$$

(90 Mo-10 V): The tabulated values for well-annealed alloy are considered accurate to within ±7% over the entire temperature range. These values can be represented approximately by the following equation:

$$\Delta L/L_0 = -0.153 + 5.005 \times 10^{-4} T + 8.227 \times 10^{-8} T^2 + 4.180 \times 10^{-13} T^3$$

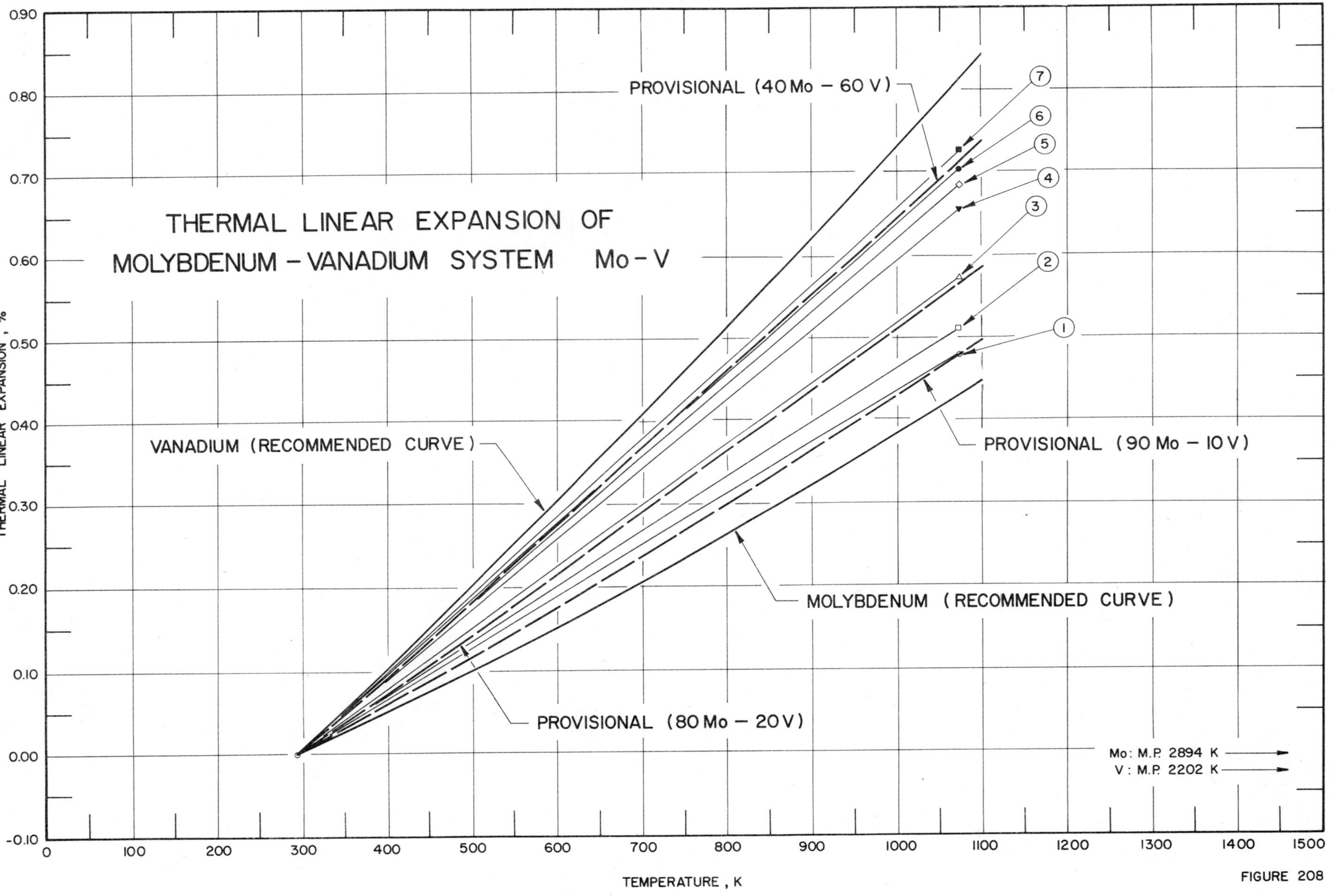
THERMAL LINEAR EXPANSION OF
MOLYBDENUM - VANADIUM SYSTEM Mo - V
PROVISIONAL (40 Mo – 60 V)
VANADIUM (RECOMMENDED CURVE)
PROVISIONAL (90 Mo – 10 V)
MOLYBDENUM (RECOMMENDED CURVE)
PROVISIONAL (80 Mo – 20 V)
7
6
5
4
3
2
1
Mo : M.P. 2894 K
V : M.P. 2202 K
THERMAL LINEAR EXPANSION , %
0.90
0.80
0.70
0.60
0.50
0.40
0.30
0.20
0.10
0.00
-0.10
0
100
200
300
400
500
600
700
800
900
1000
1100
1200
1300
1400
1500
TEMPERATURE , K

FIGURE 208

SPECIFICATION TABLE 208. THERMAL LINEAR EXPANSION OF MOLYBDENUM-VANADIUM SYSTEM Mo-V

Cur. No.	Ref. No.	Author(s)	Year	Method Used	Temp. Range, K	Name and Specimen Designation	Composition (weight percent) Mo	V	Composition (continued), Specifications, and Remarks
1	707	Pridantseva, K.S. and Solov'yeva, N.A.	1965	L	293-1073		Bal.	10	Single-phase solid solution powder sample, samples melted in an induction furnace in an argon atmosphere, impurities mainly 1% oxygen, sample homogenized at temperature which would provide an equilibrium state of the alloy, homogenizing anneal was done in a vacuum of not less than 10^{-4} mm Hg or in argon, 1% measurement error.
2	707	Pridantseva, K.S. and Solov'yeva, N.A.	1965	L	293-1073		Bal.	15	Similar to the above specimen.
3	707	Pridantseva, K.S. and Solov'yeva, N.A.	1965	L	293-1073		Bal.	22	Similar to the above specimen.
4	707	Pridantseva, K.S. and Solov'yeva, N.A.	1965	L	293-1073		Bal.	38	Similar to the above specimen.
5	707	Pridantseva, K.S. and Solov'yeva, N.A.	1965	L	293-1073		Bal.	44	Similar to the above specimen.
6	707	Pridantseva, K.S. and Solov'yeva, N.A.	1965	L	293-1073		Bal.	57	Similar to the above specimen.
7	707	Pridantseva, K.S. and Solov'yeva, N.A.	1965	L	293-1073		Bal.	67	Similar to the above specimen.

DATA TABLE 208. THERMAL LINEAR EXPANSION OF MOLYBDENUM-VANADIUM SYSTEM Mo-V

[Temperature, T, K; Linear Expansion, $\Delta L/L_0$, %]

T	$\Delta L/L_0$	T	$\Delta L/L_0$
CURVE 1		CURVE 5	
293	0.000	293	0.000*
1073	0.480	1073	0.686
CURVE 2		CURVE 6	
293	0.000*	293	0.000*
1073	0.512	1073	0.703
CURVE 3		CURVE 7	
293	0.000*	293	0.000*
1073	0.574	1073	0.729
CURVE 4			
293	0.000*		
1073	0.657		

* Not shown in figure.

FIGURE AND TABLE NO. 209R. PROVISIONAL VALUES FOR THERMAL LINEAR EXPANSION OF NICKEL-PALLADIUM SYSTEM Ni-Pd

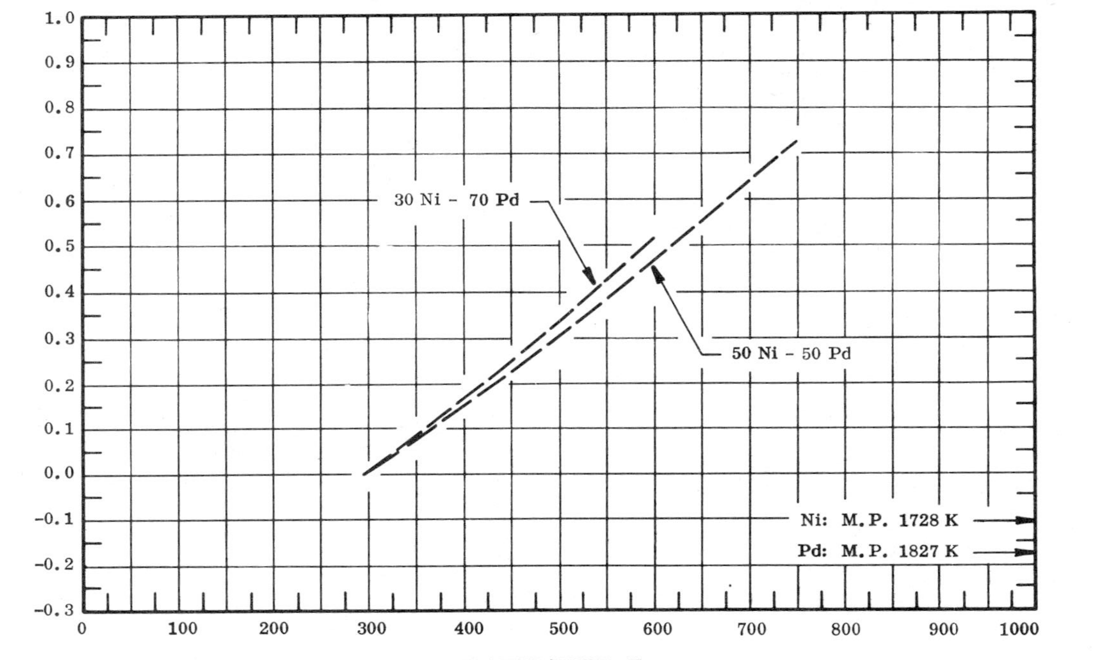

PROVISIONAL VALUES

[Temperature, T, K; Linear Expansion, $\Delta L/L_0$, %; α, K^{-1}]

	(30 Ni-70 Pd)		(50 Ni-50 Pd)	
T	$\Delta L/L_0$	$\alpha \times 10^6$	$\Delta L/L_0$	$\alpha \times 10^6$
293	0.000	14.7	0.000	13.2
400	0.167	16.3	0.150	14.6
500	0.336	17.4	0.302	15.8
600	0.510	17.9	0.466	16.8
700			0.636	17.3
750			0.723	17.4

REMARKS

(30 Ni-70 Pd): The tabulated values for well-annealed alloy are considered accurate to within ±7% over the entire temperature range. These values can be represented approximately by the following equation:

$$\Delta L/L_0 = -0.331 + 7.082 \times 10^{-4}\, T + 1.714 \times 10^{-6}\, T^2 - 9.294 \times 10^{-10}\, T^3$$

(50 Ni-50 Pd): The tabulated values for well-annealed alloy are considered accurate to within ±7% over the entire temperature range. These values can be represented approximately by the following equation:

$$\Delta L/L_0 = -0.312 + 7.719 \times 10^{-4}\, T + 1.128 \times 10^{-6}\, T^2 - 4.229 \times 10^{-10}\, T^3$$

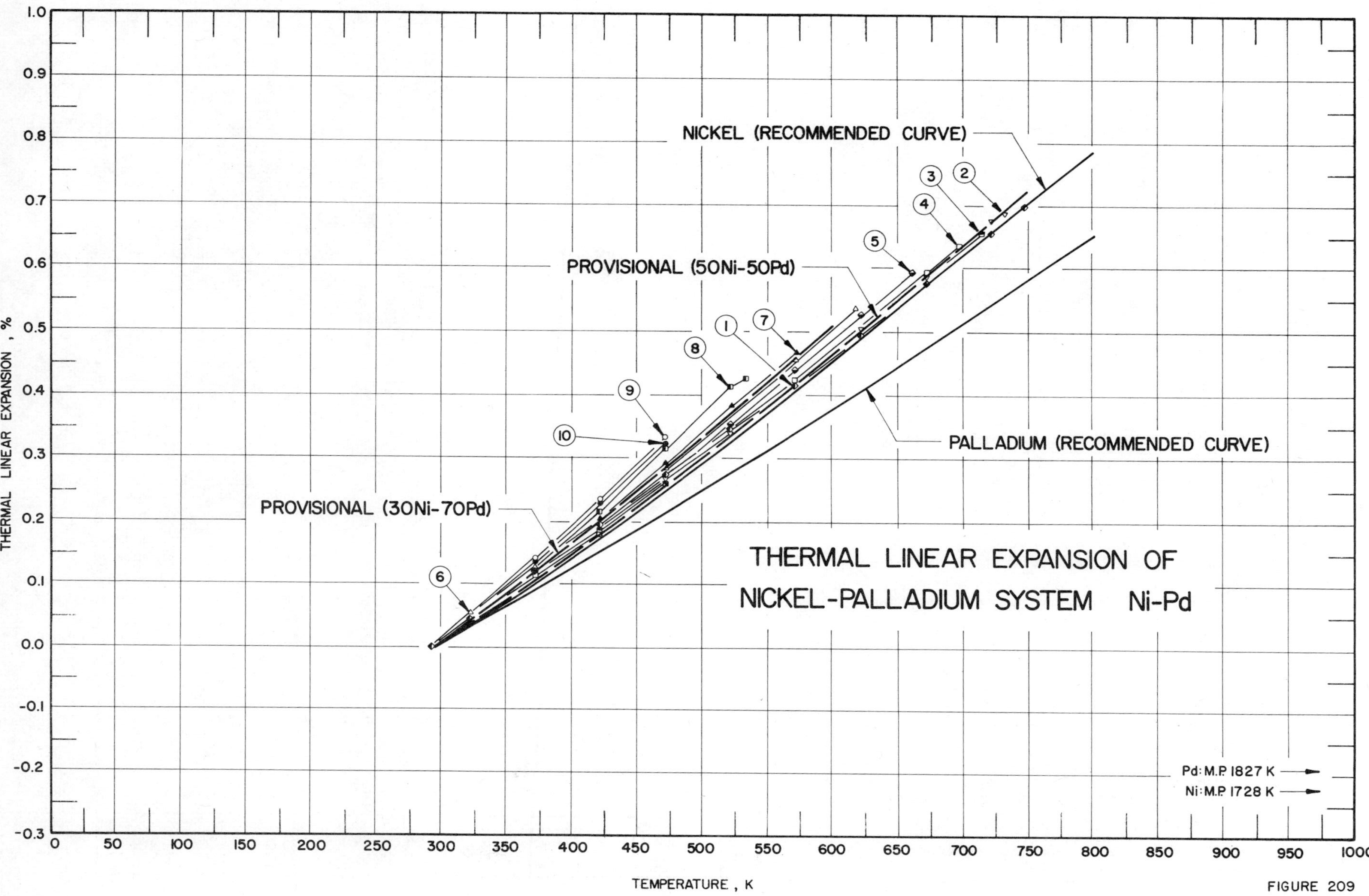
THERMAL LINEAR EXPANSION OF
NICKEL-PALLADIUM SYSTEM Ni-Pd
NICKEL (RECOMMENDED CURVE)
PALLADIUM (RECOMMENDED CURVE)
PROVISIONAL (50Ni-50Pd)
PROVISIONAL (30Ni-70Pd)
1
2
3
4
5
6
7
8
9
10
Pd: M.P. 1827 K
Ni: M.P. 1728 K
THERMAL LINEAR EXPANSION, %
TEMPERATURE, K
1.0
0.9
0.8
0.7
0.6
0.5
0.4
0.3
0.2
0.1
0.0
-0.1
-0.2
-0.3
0
50
100
150
200
250
300
350
400
450
500
550
600
650
700
750
800
850
900
950
1000

FIGURE 209

SPECIFICATION TABLE 209. THERMAL LINEAR EXPANSION OF NICKEL-PALLADIUM SYSTEM Ni-Pd

Cur. No.	Ref. No.	Author(s)	Year	Method Used	Temp. Range, K	Name and Specimen Designation	Composition (weight percent) Ni	Pd	Composition (continued), Specifications, and Remarks
1	658	Masumoto, H. and Sawaya, S.	1970	L	293-747		80	20	Alloy prepared from electrolytic Ni and 99.9 Pd by melting in alumina crucible in H_2 atmosphere; annealed in vacuum at 1273 K for 1 hr and cooled at rate of 300 degrees per hr; zero-point correction is -0.032%; magnetic transition point near 630 K.
2	658	Masumoto, H. and Sawaya, S.	1970	L	293-732		60	40	Similar to the above specimen; zero-point correction is -0.033%; magnetic transition point near 601 K.
3	658	Masumoto, H. and Sawaya, S.	1970	L	293-715		52.5	47.5	Similar to the above specimen; zero-point correction is -0.030%; magnetic transition point near 550 K.
4	658	Masumoto, H. and Sawaya, S.	1970	L	293-698		50	50	Similar to the above specimen; zero-point correction is -0.033%; magnetic transition point near 537 K.
5	658	Masumoto, H. and Sawaya, S.	1970	L	293-661		45	55	Similar to the above specimen; zero-point correction is -0.033%; magnetic transition point near 522 K.
6	658	Masumoto, H. and Sawaya, S.	1970	L	293-617		36	64	Similar to the above specimen; zero-point correction is -0.032%; magnetic transition point near 506 K.
7	658	Masumoto, H. and Sawaya, S.	1970	L	293-573		27	73	Similar to the above specimen; zero-point correction is -0.031%; magnetic transition point near 467 K.
8	658	Masumoto, H. and Sawaya, S.	1970	L	293-535		17	83	Similar to the above specimen; zero-point correction is -0.024%; magnetic transition point near 418 K.
9	658	Masumoto, H. and Sawaya, S.	1970	L	293-475		15	85	Similar to the above specimen; zero-point correction is -0.024%; magnetic transition point near 336 K.
10	658	Masumoto, H. and Sawaya, S.	1970	L	293-473		10	90	Similar to the above specimen; zero-point correction is -0.035%; magnetic transition point near 310 K.

DATA TABLE 209. THERMAL LINEAR EXPANSION OF NICKEL-PALLADIUM SYSTEM Ni-Pd

[Temperature, T, K; Linear Expansion, $\Delta L/L_0$, %]

T	$\Delta L/L_0$
CURVE 1	
293	0.000
323	0.044
373	0.120
423	0.197
473	0.273
523	0.347
573	0.417
623	0.491
673	0.577
723	0.658
747	0.699
CURVE 2	
293	0.000*
323	0.046*
373	0.121*
423	0.196*
473	0.269
523	0.343
573	0.417*
623	0.501
673	0.589
723	0.675
732	0.688
CURVE 3	
293	0.000*
323	0.042*
373	0.115*
423	0.185
473	0.261
523	0.340*
573	0.416*
623	0.501*
673	0.593*
715	0.658
CURVE 4	
293	0.000*
323	0.045*
373	0.113
423	0.180
473	0.265*
CURVE 4 (cont.)	
523	0.343*
573	0.422
623	0.503*
673	0.594
698	0.632
CURVE 5	
293	0.000*
323	0.045*
373	0.122*
423	0.193*
473	0.269*
523	0.355
573	0.440
623	0.528
661	0.594
CURVE 6	
293	0.000*
323	0.051
373	0.126
423	0.199*
473	0.285
523	0.364*
573	0.459
617	0.538
CURVE 7	
293	0.000*
323	0.048*
373	0.120*
423	0.204
473	0.292
523	0.383
573	0.469
CURVE 8	
293	0.000*
323	0.035*
373	0.120*
423	0.216
473	0.317
523	0.413
535	0.427
CURVE 9	
293	0.000*
323	0.050*
373	0.140
423	0.237
473	0.335
CURVE 10	
293	0.000*
323	0.050*
373	0.136
423	0.230
473	0.322

*Not shown in figure.

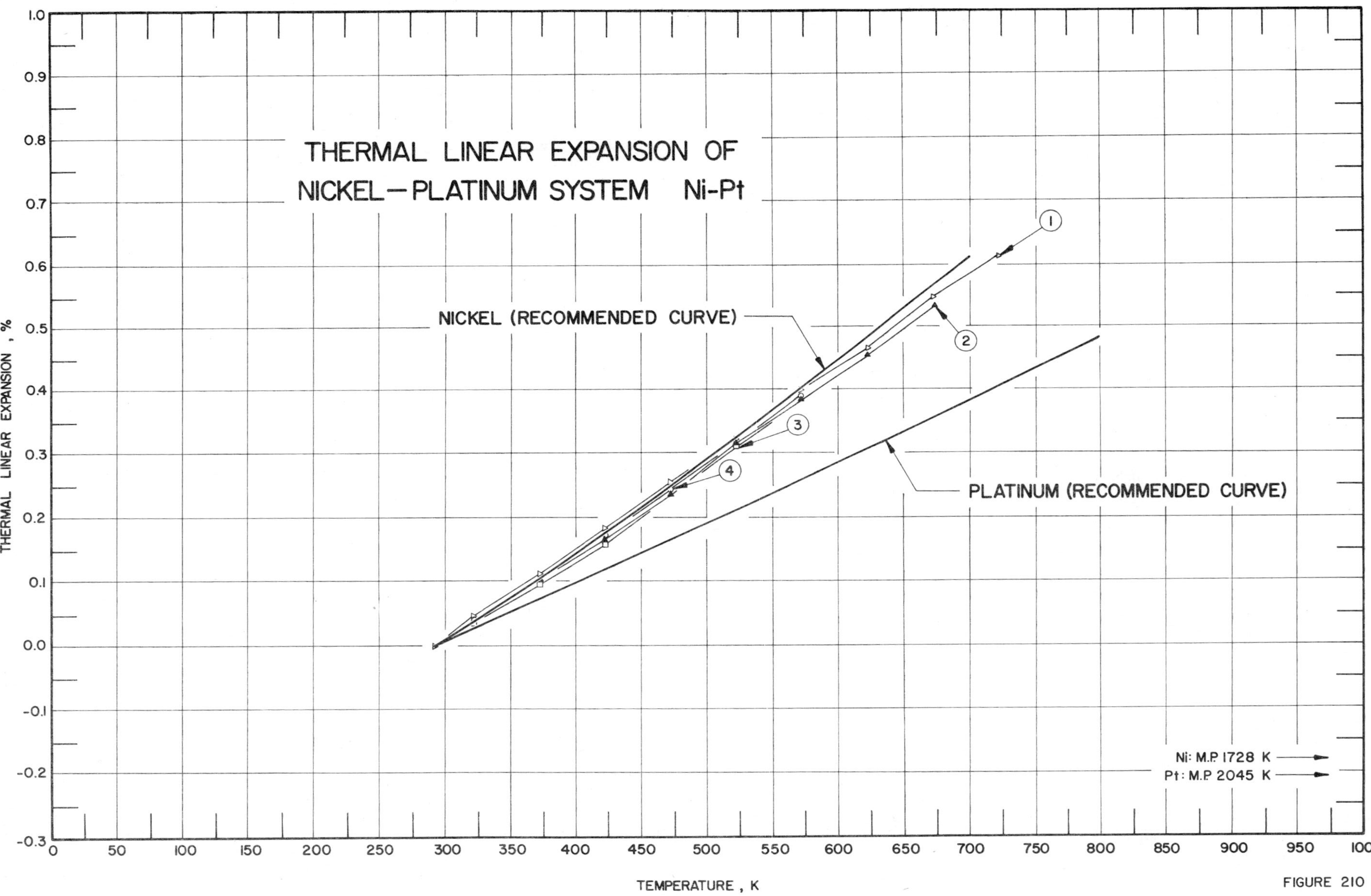
THERMAL LINEAR EXPANSION OF
NICKEL—PLATINUM SYSTEM Ni-Pt
1
NICKEL (RECOMMENDED CURVE)
2
3
4
PLATINUM (RECOMMENDED CURVE)
Ni: M.P. 1728 K
Pt: M.P. 2045 K
THERMAL LINEAR EXPANSION, %
1.0
0.9
0.8
0.7
0.6
0.5
0.4
0.3
0.2
0.1
0.0
-0.1
-0.2
-0.3
0
50
100
150
200
250
300
350
400
450
500
550
600
650
700
750
800
850
900
950
1000
TEMPERATURE, K

FIGURE 210

SPECIFICATION TABLE 210. THERMAL LINEAR EXPANSION OF NICKEL-PLATINUM SYSTEM Ni-Pt

Cur. No.	Ref. No.	Author(s)	Year	Method Used	Temp. Range, K	Name and Specimen Designation	Composition (weight percent) Ni	Pt	Composition (continued), Specifications, and Remarks
1	658	Masumoto, H. and Sawaya, S.	1970	L	293-723		80	20	Alloy prepared from electrolytic Ni and 99.99 Pt by melting in alumina crucible in H_2 atmosphere; annealed in vacuum at 1273 K for 1 hr and cooled at rate of 300 degrees per hr; zero-point correction is -0.032%; magnetic transition point near 572 K.
2	658	Masumoto, H. and Sawaya, S.	1970	L	293-673		55	45	Similar to the above specimen; zero-point correction is -0.024%; magnetic transition point near 458 K.
3	658	Masumoto, H. and Sawaya, S.	1970	L	293-623		52	48	Similar to the above specimen; zero-point correction is -0.031%; magnetic transition point near 432 K.
4	658	Masumoto, H. and Sawaya, S.	1970	L	293-573		42	58	Similar to the above specimen; zero-point correction is -0.026%; magnetic transition point near 353 K.
5*	658	Masumoto, H. and Sawaya, S.	1970	L	293-523		38	62	Similar to the above specimen; zero-point correction is -0.028%; magnetic transition point near 318 K.
6*	658	Masumoto, H. and Sawaya, S.	1970	L	293-520		32	68	Similar to the above specimen; zero-point correction is -0.025%.

DATA TABLE 210. THERMAL LINEAR EXPANSION OF NICKEL-PLATINUM SYSTEM Ni-Pt

[Temperature, T, K; Linear Expansion, $\Delta L/L_0$, %]

T	$\Delta L/L_0$	T	$\Delta L/L_0$	T	$\Delta L/L_0$	T	$\Delta L/L_0$
CURVE 1		CURVE 2 (cont.)		CURVE 4		CURVE 6*	
293	0.000	473	0.237	293	0.000*	293	0.000
323	0.047	523	0.314	323	0.042	323	0.047
373	0.111	573	0.385	373	0.106*	373	0.111
423	0.183	623	0.458	423	0.171	423	0.176
473	0.256	673	0.531	473	0.244	473	0.249
523	0.322			523	0.318*	520	0.314
573	0.400	CURVE 3		573	0.380		
623	0.469						
673	0.548	293	0.000*	CURVE 5*			
723	0.616	323	0.038				
		373	0.099	293	0.000		
CURVE 2		423	0.160	323	0.041		
		473	0.235*	373	0.108		
293	0.000*	523	0.312	423	0.180		
323	0.041*	573	0.386*	473	0.253		
373	0.104	623	0.457*	523	0.323		
423	0.167						

*Not shown in figure.

FIGURE AND TABLE NO. 211R. PROVISIONAL VALUES FOR THERMAL LINEAR EXPANSION OF NICKEL-SILICON SYSTEM Ni-Si

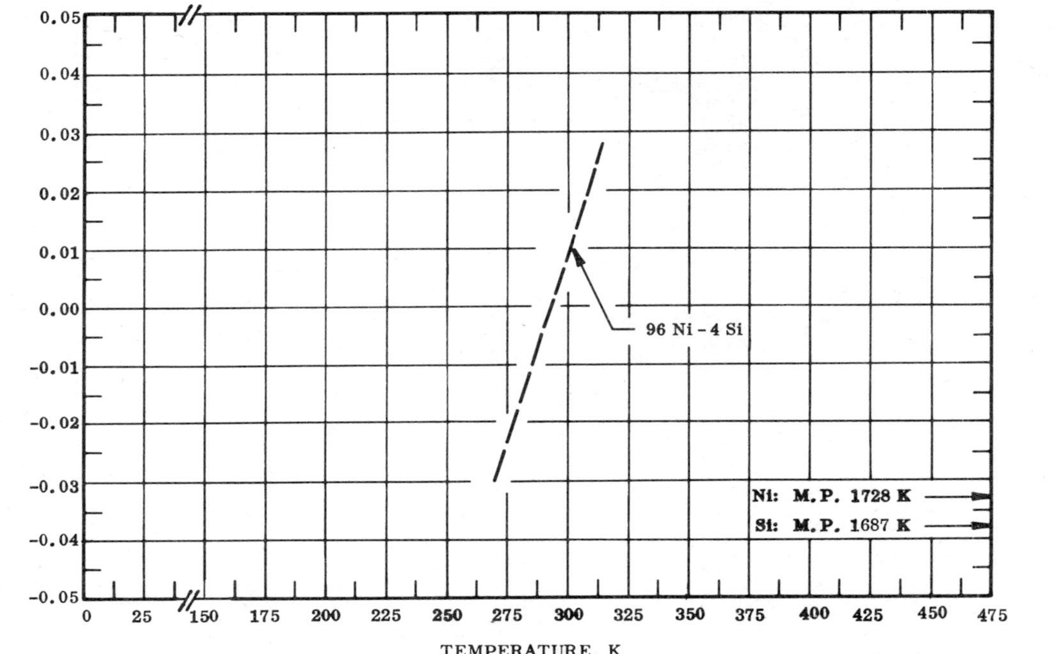

PROVISIONAL VALUES

[Temperature, T, K; Linear Expansion, $\Delta L/L_0$, %; α, K^{-1}]

T	(96 Ni-4 Si) $\Delta L/L_0$	$\alpha \times 10^6$
270	-0.030	13.0
293	0.000	12.9
315	0.028	12.7

REMARKS

(96 Ni-4 Si): The tabulated values for well-annealed alloy are considered accurate to within ± 7% over the entire temperature range. These values can be represented approximately by the following equation:

$$\Delta L/L_0 = -1.312 + 1.086 \times 10^{-2}\ T - 3.263 \times 10^{-5}\ T^2 + 3.701 \times 10^{-8}\ T^3$$

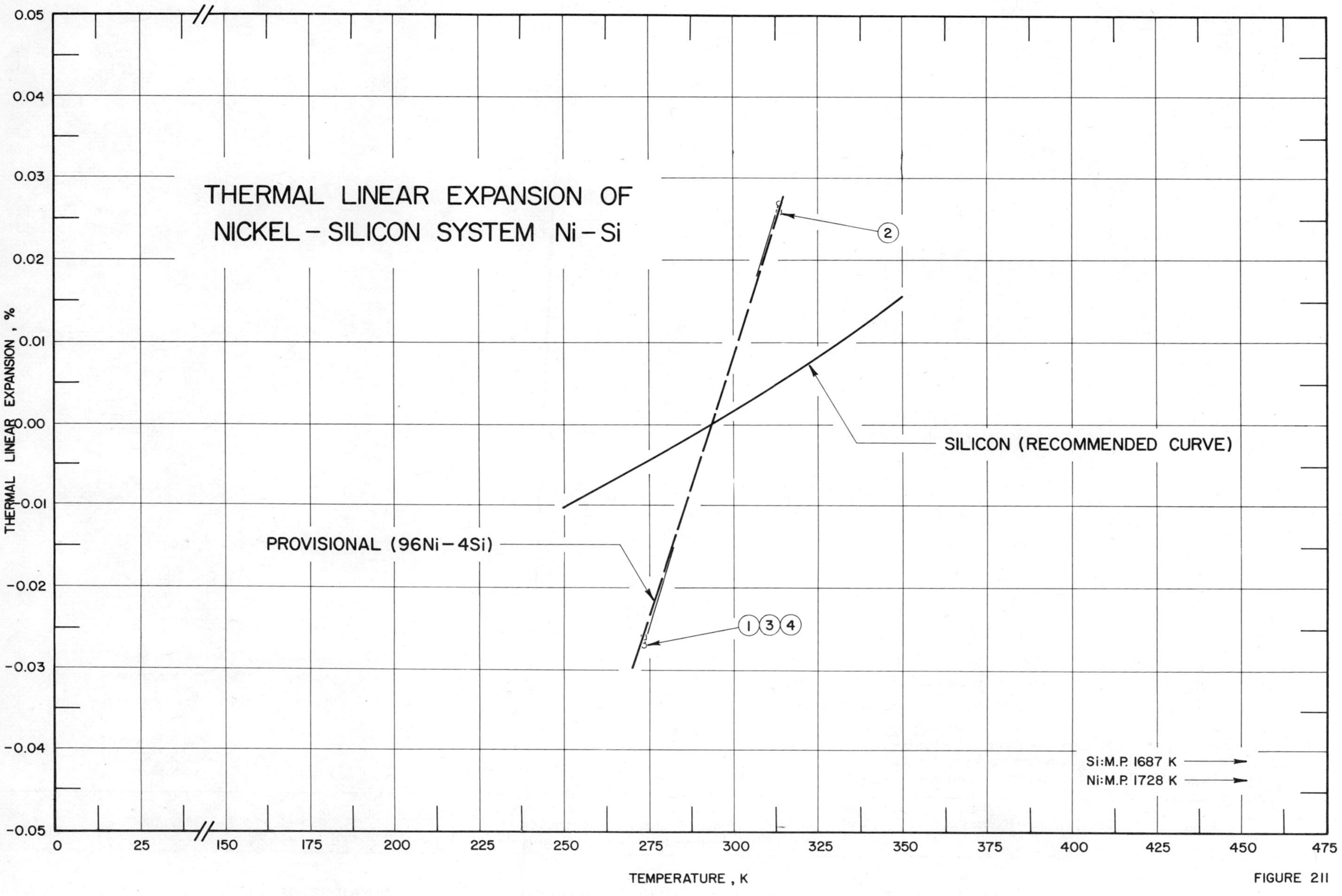
THERMAL LINEAR EXPANSION OF
NICKEL - SILICON SYSTEM Ni - Si
2
SILICON (RECOMMENDED CURVE)
PROVISIONAL (96Ni - 4Si)
1 3 4
Si: M.P. 1687 K
Ni: M.P. 1728 K
THERMAL LINEAR EXPANSION, %
0.05
0.04
0.03
0.02
0.01
0.00
-0.01
-0.02
-0.03
-0.04
-0.05
0
25
150
175
200
225
250
275
300
325
350
375
400
425
450
475
TEMPERATURE, K
FIGURE 211

SPECIFICATION TABLE 211. THERMAL LINEAR EXPANSION OF NICKEL-SILICON SYSTEM Ni-Si

Cur. No.	Ref. No.	Author(s)	Year	Method Used	Temp. Range, K	Name and Specimen Designation	Composition (weight percent) Ni	Si	Composition (continued), Specifications, and Remarks
1	430	Masumoto, H., Sawaya, S. and Nakamura, N.	1969	L	273, 313		97.1	2.9	Electrolytic Ni, metallic silicon with 1.990 Fe, 0.113 Mn, 0.044 Cu, 0.026 P, 0.025 C, 0.003 S; rod specimen 3.6 mm dia, 12 cm long, furnace-cooled at 300 C/hr after heating at 1273 K for 1 hr; density 8.72 g cm^{-3} at 293 K.
2	430	Masumoto, H. et al.	1969	L	273, 313		96.1	3.9	Similar to the above specimen; density 8.66 g cm^{-3} at 293 K.
3	430	Masumoto, H. et al.	1969	L	273, 313		95.7	4.3	Similar to the above specimen; density 8.60 g cm^{-3} at 293 K.
4	430	Masumoto, H. et al.	1969	L	273, 313		95.1	4.9	Similar to the above specimen; density 8.56 g cm^{-3} at 293 K.

DATA TABLE 211. THERMAL LINEAR EXPANSION OF NICKEL-SILICON SYSTEM Ni-Si

[Temperature, T, K; Linear Expansion, $\Delta L/L_0$, %]

T	$\Delta L/L_0$
CURVE 1	
273	-0.027
313	0.027
CURVE 2	
273	-0.026
313	0.026
CURVE 3	
273	-0.027
313	0.027
CURVE 4	
273	-0.027
313	0.027

FIGURE AND TABLE NO. 212R. PROVISIONAL VALUES FOR THERMAL LINEAR EXPANSION OF NICKEL-TIN SYSTEM Ni-Sn

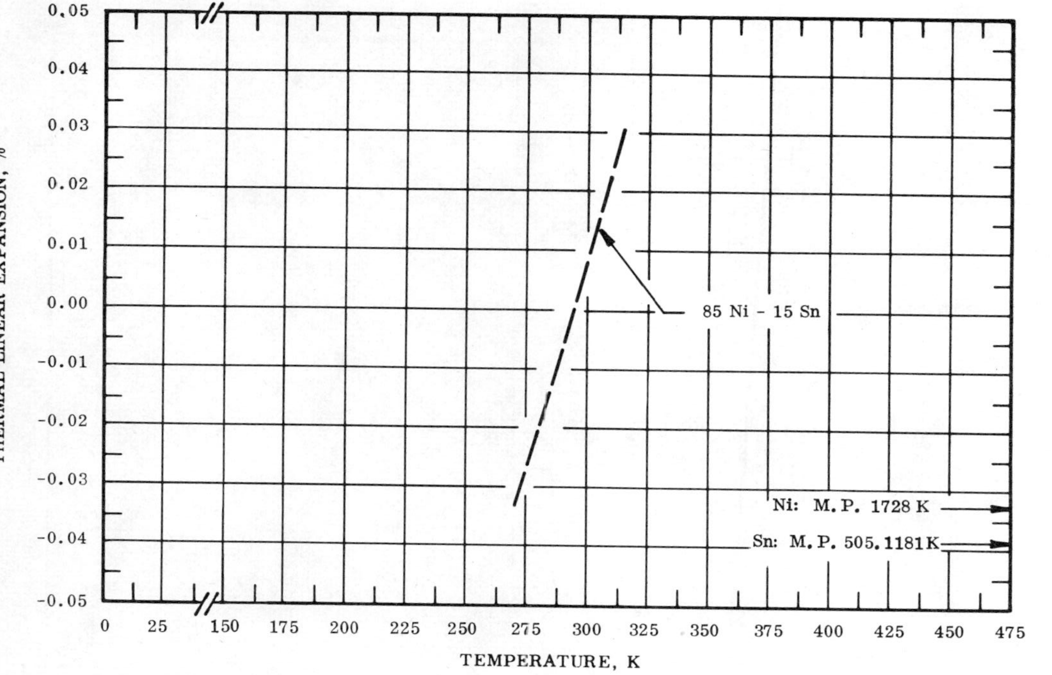

PROVISIONAL VALUES

[Temperature, T, K; Linear Expansion, $\Delta L/L_0$, %; α, K^{-1}]

	(85 Ni-15 Sn)	
T	$\Delta L/L_0$	$\alpha \times 10^6$
270	-0.033	14.3
293	0.000	14.2
315	0.031	14.0

REMARKS

(85 Ni-15 Sn): The tabulated values for well-annealed alloy are considered accurate to within ±7% over the entire temperature range. The Curie temperature occurs at about 630 K. These values can be represented approximately by the following equation:

$$\Delta L/L_0 = -1.369 + 1.144 \times 10^{-2} T - 3.510 \times 10^{-5} T^2 + 4.090 \times 10^{-8} T^3$$

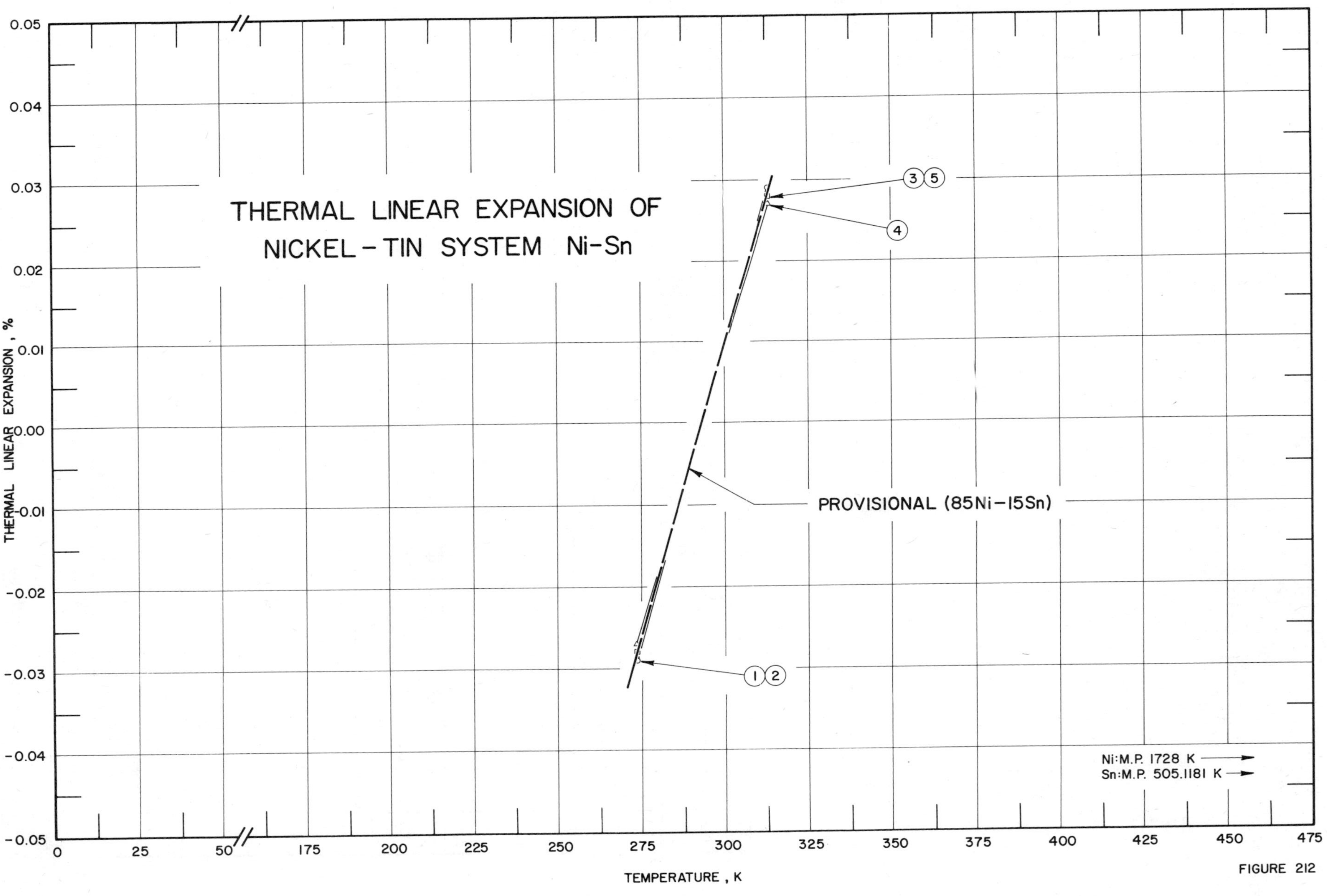

THERMAL LINEAR EXPANSION OF
NICKEL - TIN SYSTEM Ni-Sn
PROVISIONAL (85Ni-15Sn)
1
2
3
5
4
Ni:M.P. 1728 K
Sn:M.P. 505.1181 K
THERMAL LINEAR EXPANSION, %
0.05
0.04
0.03
0.02
0.01
0.00
-0.01
-0.02
-0.03
-0.04
-0.05
0
25
50
175
200
225
250
275
300
325
350
375
400
425
450
475
TEMPERATURE, K

FIGURE 212

SPECIFICATION TABLE 212. THERMAL LINEAR EXPANSION OF NICKEL-TIN SYSTEM Ni-Sn

Cur. No.	Ref. No.	Author(s)	Year	Method Used	Temp. Range, K	Name and Specimen Designation	Composition (weight percent) Ni	Sn	Composition (continued), Specifications, and Remarks
1	430	Masumoto, H., Sawaya, S. and Nakamura, N.	1969	L	273, 313		87.0	13.0	Electrolytic Ni, 99.9 Sn; rod specimen 3.6 mm dia, 12 cm long; furnace-cooled at 300 C/hr after heating at 1273 K for 1 hr; density 8.90 g cm^{-3} at 293 K.
2	430	Masumoto, H. et al.	1969	L	273, 313		85.2	14.8	Similar to the above specimen; density 8.88 g cm^{-3} at 293 K.
3	430	Masumoto, H. et al.	1969	L	273, 313		84.6	15.4	Similar to the above specimen; density 8.90 g cm^{-3} at 293 K.
4	430	Masumoto, H. et al.	1969	L	273, 313		84.0	16.0	Similar to the above specimen; density 8.87 g cm^{-3} at 293 K.
5	430	Masumoto, H. et al.	1969	L	273, 313		82.2	17.8	Similar to the above specimen; density 8.89 g cm^{-3} at 293 K.

DATA TABLE 212. THERMAL LINEAR EXPANSION OF NICKEL-TIN SYSTEM Ni-Sn

[Temperature, T, K; Linear Expansion, $\Delta L/L_0$, %]

T	$\Delta L/L_0$	T	$\Delta L/L_0$
CURVE 1		CURVE 4	
273	-0.029	273	-0.027
313	0.029	313	0.027
CURVE 2		CURVE 5	
273	-0.029	273	-0.028
313	0.029	313	0.028
CURVE 3			
273	-0.028		
313	0.028		

FIGURE AND TABLE NO. 213R. PROVISIONAL VALUES FOR THERMAL LINEAR EXPANSION OF NICKEL-TITANIUM SYSTEM Ni-Ti

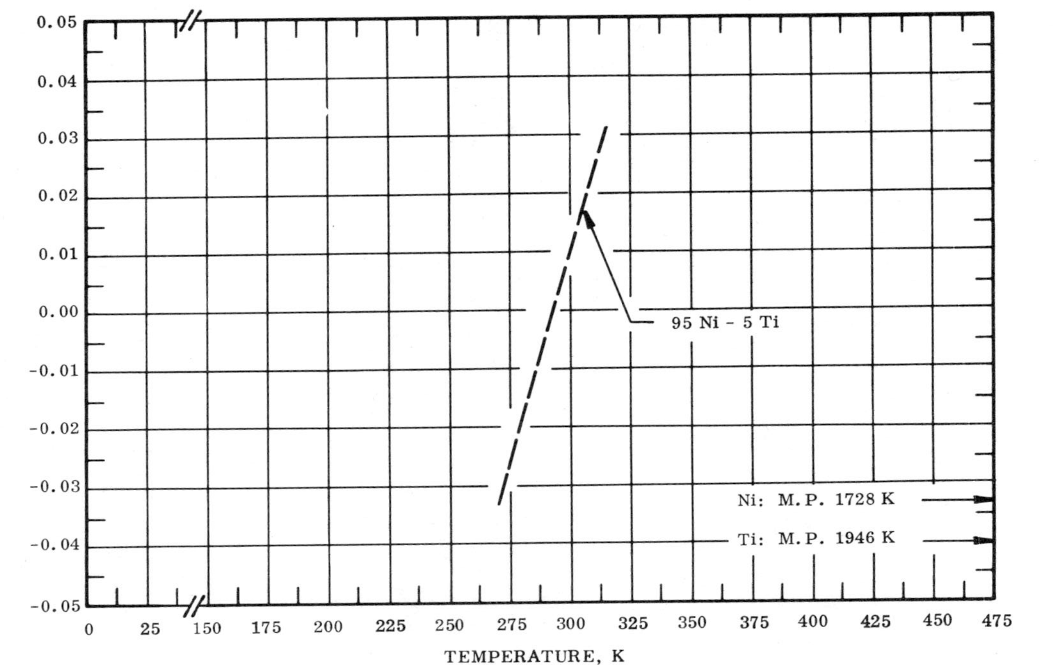

PROVISIONAL VALUES

[Temperature, T, K; Linear Expansion, $\Delta L/L_0$, %; α, K^{-1}]

	(95 Ni-5 Ti)	
T	$\Delta L/L_0$	$\alpha \times 10^6$
270	-0.034	14.7
293	0.000	14.4
315	0.031	14.1

REMARKS

(95 Ni-5 Ti): The tabulated values for well-annealed alloy are considered accurate to within ± 7% over the entire temperature range. These values can be represented approximately by the following equation:

$$\Delta L/L_0 = 1.296 - 1.648 \times 10^{-2}\, T + 6.215 \times 10^{-5}\, T^2 - 7.168 \times 10^{-8}\, T^3$$

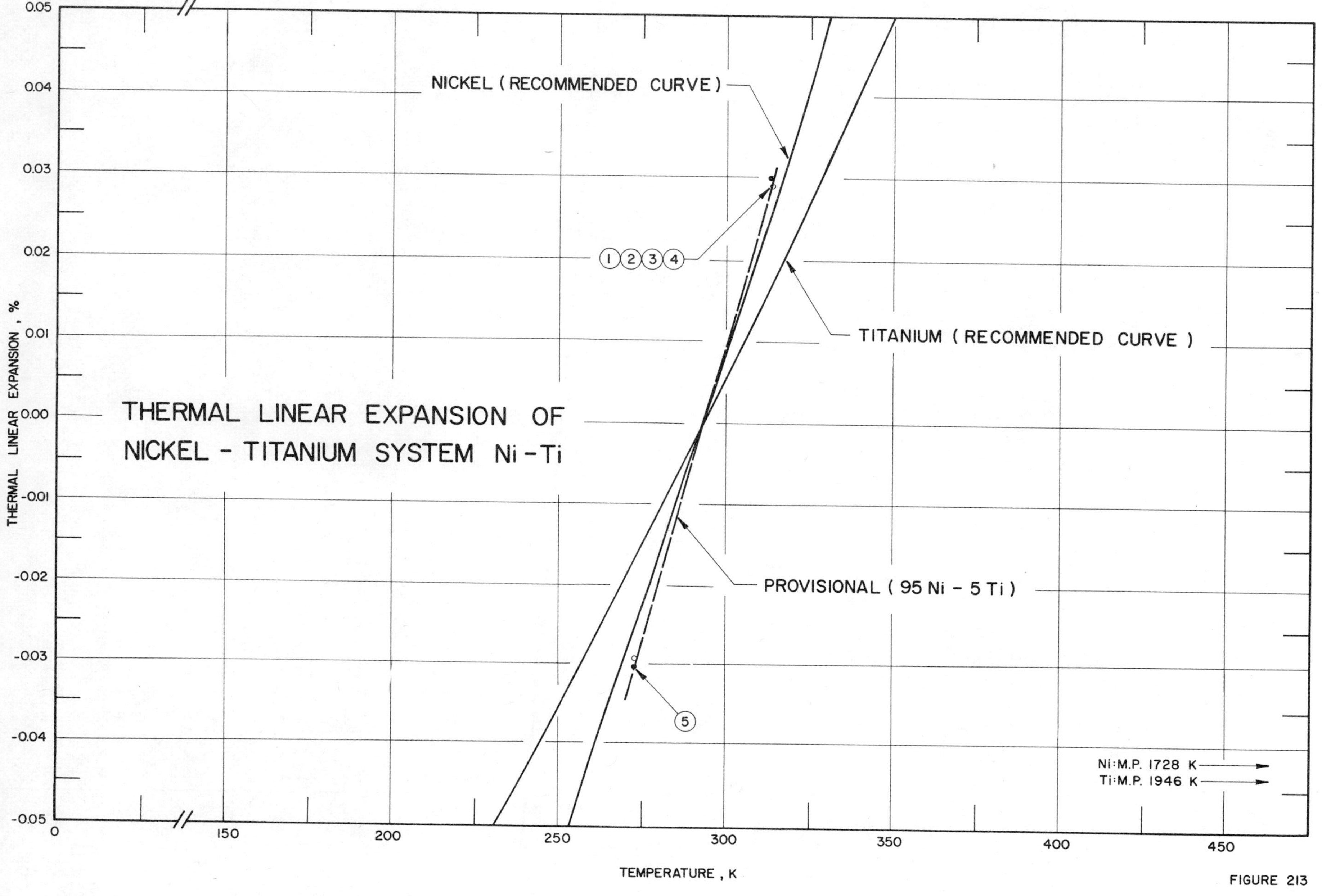
THERMAL LINEAR EXPANSION OF
NICKEL - TITANIUM SYSTEM Ni-Ti
NICKEL (RECOMMENDED CURVE)
TITANIUM (RECOMMENDED CURVE)
PROVISIONAL (95 Ni - 5 Ti)
Ni: M.P. 1728 K
Ti: M.P. 1946 K
TEMPERATURE , K
THERMAL LINEAR EXPANSION , %

FIGURE 213

SPECIFICATION TABLE 213. THERMAL LINEAR EXPANSION OF NICKEL-TITANIUM SYSTEM Ni-Ti

Cur. No.	Ref. No.	Author(s)	Year	Method Used	Temp. Range, K	Name and Specimen Designation	Composition (weight percent) Ni	Ti	Composition (continued), Specifications, and Remarks
1	241	Masumoto, H., Sawaya, S. and Kadowaki, S.	1969	L	273, 313		95.23	4.77	Electrolytic Ni with 0.180 Co, 0.002 S, 0.002 C, 0.001 Cu, 0.001 Fe, and 99.5+ Ti; rod specimen 2 mm dia, 10 cm long; cooled at rate 300 C/hr after heating in vacuum at 1273 K for 1 hr; density 8.77 g cm^{-3} at 293 K.
2	241	Masumoto, H. et al.	1969	L	273, 313		94.75	5.25	Similar to the above specimen; density 8.70 g cm^{-3} at 293 K.
3	241	Masumoto, H. et al.	1969	L	273, 313		94.26	5.74	Similar to the above specimen; density 8.67 g cm^{-3} at 293 K.
4	241	Masumoto, H. et al.	1969	L	273, 313		93.78	6.22	Similar to the above specimen; density 8.64 g cm^{-3} at 293 K.
5	241	Masumoto, H. et al.	1969	L	273, 313		93.31	6.69	Similar to the above specimen; density 8.61 g cm^{-3} at 293 K.

DATA TABLE 213. THERMAL LINEAR EXPANSION OF NICKEL-TITANIUM SYSTEM Ni-Ti

[Temperature, T, K; Linear Expansion, $\Delta L/L_0$, %]

T	$\Delta L/L_0$	T	$\Delta L/L_0$
CURVE 1		CURVE 4	
273	-0.029	273	-0.029
313	0.029	313	0.029
CURVE 2		CURVE 5	
273	-0.029	273	-0.030
313	0.029	313	0.030
CURVE 3			
273	-0.029		
313	0.029		

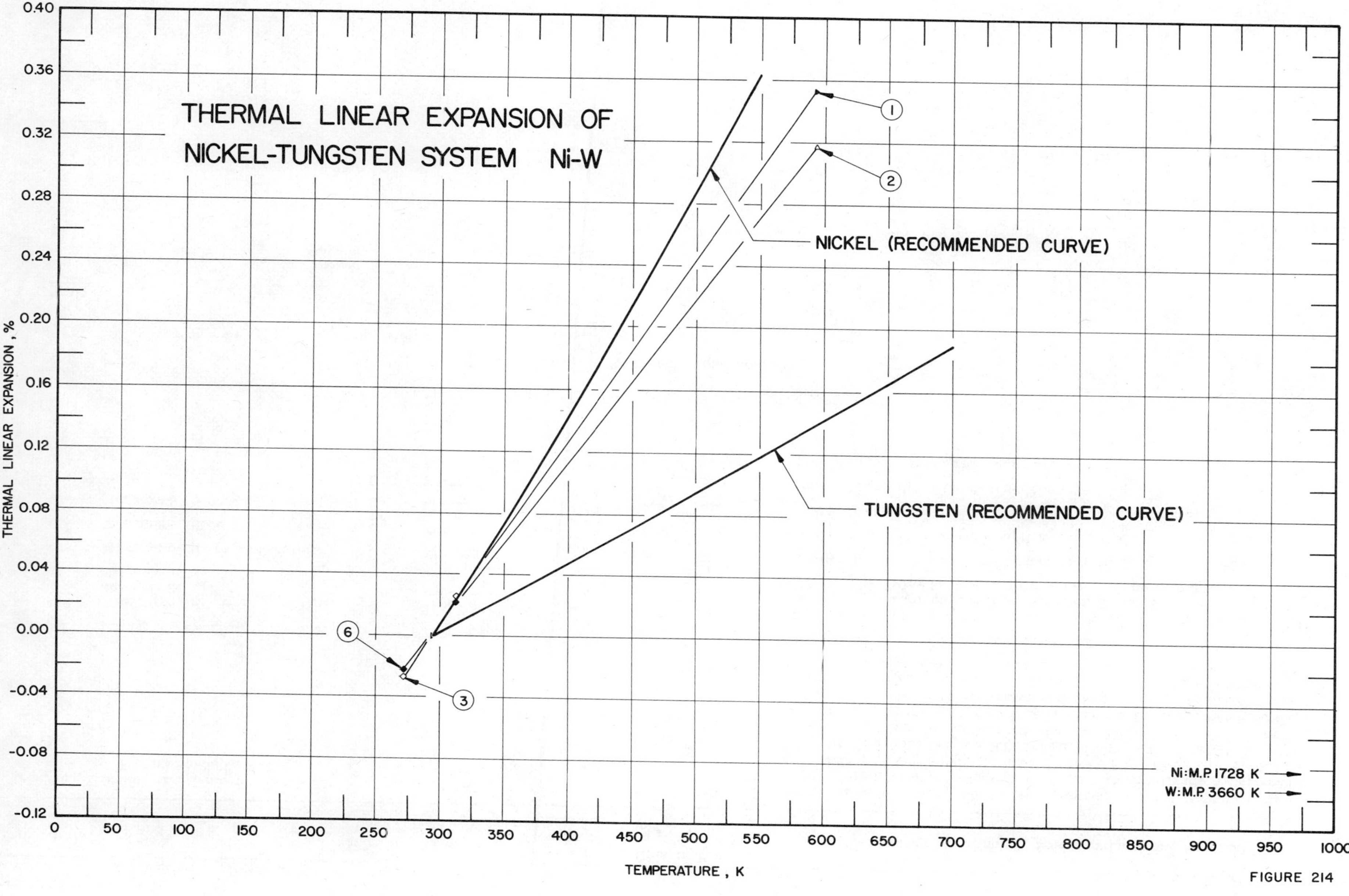
THERMAL LINEAR EXPANSION OF
NICKEL-TUNGSTEN SYSTEM Ni-W
1
2
NICKEL (RECOMMENDED CURVE)
TUNGSTEN (RECOMMENDED CURVE)
6
3
Ni: M.P. 1728 K
W: M.P. 3660 K
THERMAL LINEAR EXPANSION, %
0.40
0.36
0.32
0.28
0.24
0.20
0.16
0.12
0.08
0.04
0.00
-0.04
-0.08
-0.12
0
50
100
150
200
250
300
350
400
450
500
550
600
650
700
750
800
850
900
950
1000
TEMPERATURE, K

FIGURE 214

SPECIFICATION TABLE 214. THERMAL LINEAR EXPANSION OF NICKEL-TUNGSTEN SYSTEM Ni-W

Cur. No.	Ref. No.	Author(s)	Year	Method Used	Temp. Range, K	Name and Specimen Designation	Composition (weight percent) Ni	W	Composition (continued), Specifications, and Remarks
1	236	Davis, M., Densen, C.E. and Rendall, J.H.	1955-56		293, 593		70	30	Prepared by powder metallurgy techniques; Grade R.P. x 99.99 W supplied by British Thomson-Houston Co. Ltd., Rugby; Grade A Carbonyl Ni powder, having impurity, 0.07 C; 0.016 Si; 0.013 Fe; 0.01-0.2 O; 0.003 S; 0.0005 Mn; 0.0003 Mg; supplied by Mond Nickel Co., Ltd.; mixed powder compacted in thin-walled latex rubber sheaths by applying hydrostatic pressure; sintering carried out in platinum-wound electric muffle-furnace with protective atm of commercial hydrogen; raised to temp of 1673 K in 3-4 hrs, held for 2-3 hrs, hot forged down to 0.1 in. dia in rotary swaging machines.
2	236	Davis, M. et al.	1955-56		293, 593		60	40	Similar to the above specimen; solid solution.
3	428	Masumoto, H., Sawaya, S. and Nakamura, N.	1969	L	273, 313		92.0	8.0	Electrolytic Ni with 0.180 Co, 0.002 S, 0.002 C, 0.001 Cu, 0.001 Fe, and 99.9 W; powder melted to 10 cm long, 2 mm dia; furnace-cooled at rate 300 C/hr, after heating at 1273 K for 1 hr; density 9.28 g cm^{-3} at 293 K.
4*	428	Masumoto, H. et al.	1969	L	273, 313		89.0	11.0	Similar to the above specimen; density 9.40 g cm^{-3} at 293 K.
5*	428	Masumoto, H. et al.	1969	L	273, 313		86.0	14.0	Similar to the above specimen; density 9.54 g cm^{-3} at 293 K.
6	428	Masumoto, H. et al.	1969	L	273, 313		85.3	14.7	Similar to the above specimen; density 9.55 g cm^{-3} at 293 K.
7*	428	Masumoto, H. et al.	1969	L	273, 313		84.7	15.3	Similar to the above specimen; density 9.57 g cm^{-3} at 293 K.
8*	428	Masumoto, H. et al.	1969	L	273, 313		84.0	16.0	Similar to the above specimen; density 9.55 g cm^{-3} at 293 K.

DATA TABLE 214. THERMAL LINEAR EXPANSION OF NICKEL-TUNGSTEN SYSTEM Ni-W

[Temperature, T, K; Linear Expansion, $\Delta L/L_0$, %]

T	$\Delta L/L_0$	T	$\Delta L/L_0$	T	$\Delta L/L_0$
CURVE 1		CURVE 4*		CURVE 7*	
293	0.000	273	-0.026	273	-0.025
593	0.354	313	0.026	313	0.025
CURVE 2		CURVE 5*		CURVE 8*	
293	0.000*	273	-0.027	273	-0.026
593	0.319	313	0.027	313	0.026
CURVE 3		CURVE 6			
273	-0.026	273	-0.022		
313	0.026	313	0.022		

*Not shown in figure.

FIGURE AND TABLE NO. 215R. PROVISIONAL VALUES FOR THERMAL LINEAR EXPANSION OF NICKEL-VANADIUM SYSTEM Ni-V

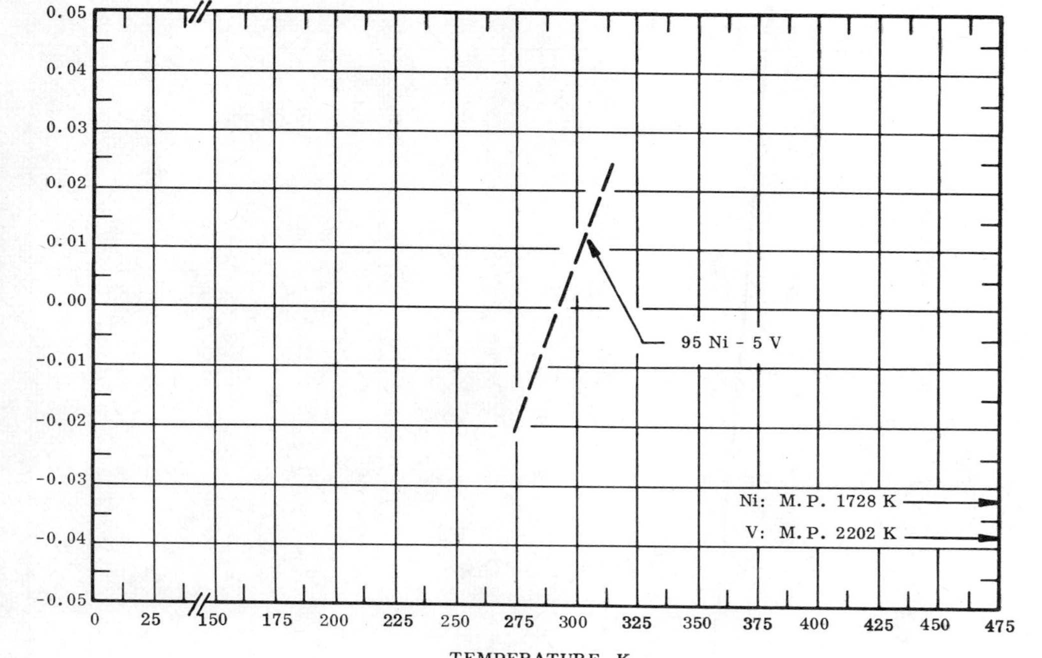

PROVISIONAL VALUES

[Temperature, T, K; Linear Expansion, $\Delta L/L_0$, %; α, K^{-1}]

T	(95 Ni-5 V) $\Delta L/L_0$	$\alpha \times 10^6$
275	-0.021	11.5
293	0.000	11.5
315	0.024	11.5

REMARKS

(95 Ni-5 V): The tabulated values for well-annealed are considered accurate to within ± 7% over the entire temperature range. These values can be represented approximately by the following equation:

$$\Delta L/L_0 = -1.140 + 9.093 \times 10^{-3}\,T - 2.608 \times 10^{-5}\,T^2 + 2.838 \times 10^{-8}\,T^3$$

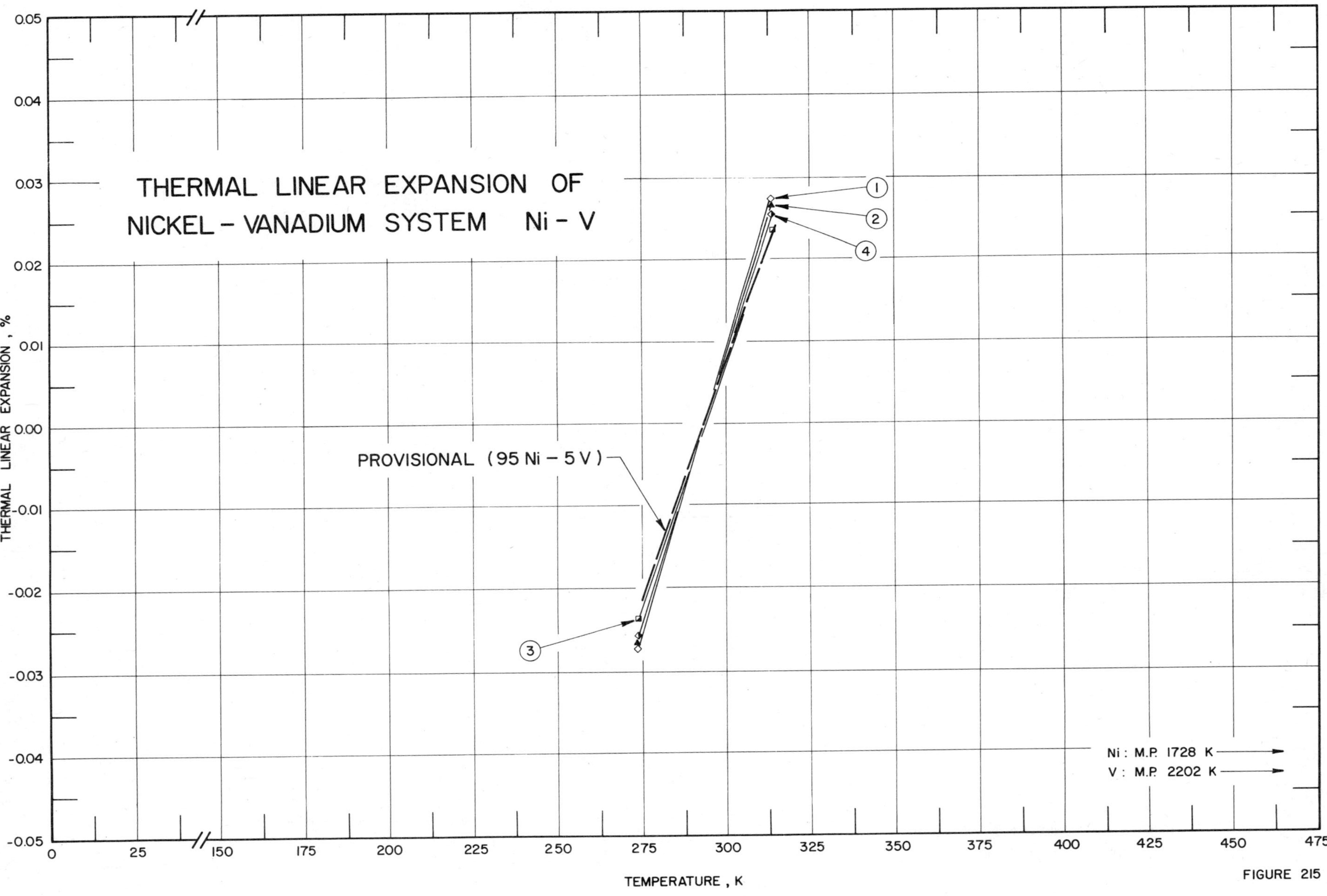
THERMAL LINEAR EXPANSION OF
NICKEL - VANADIUM SYSTEM Ni - V
1
2
4
PROVISIONAL (95 Ni – 5 V)
3
Ni : M.P. 1728 K
V : M.P. 2202 K
THERMAL LINEAR EXPANSION, %
0.05
0.04
0.03
0.02
0.01
0.00
-0.01
-0.02
-0.03
-0.04
-0.05
0
25
150
175
200
225
250
275
300
325
350
375
400
425
450
475
TEMPERATURE, K
FIGURE 215

SPECIFICATION TABLE 215. THERMAL LINEAR EXPANSION OF NICKEL-VANADIUM SYSTEM Ni-V

Cur. No.	Ref. No.	Author(s)	Year	Method Used	Temp. Range, K	Name and Specimen Designation	Composition (weight percent) Ni	V	Composition (continued), Specifications, and Remarks
1	430	Masumoto, H., Sawaya, S. and Nakamura, N.	1969	L	273, 313		97.6	2.4	Electrolytic Ni, 99.9 V; rod specimen 3.6 mm dia, 12 cm long; furnace-cooled at 300 C/hr after heating at 1273 K for 1 hr; density 8.89 g cm^{-3} at 293 K.
2	430	Masumoto, H. et al.	1969	L	273, 313		96.8	3.2	Similar to the above specimen; density 8.87 g cm^{-3} at 293 K.
3	430	Masumoto, H. et al.	1969	L	273, 313		95.7	4.3	Similar to the above specimen; density 8.85 g cm^{-3} at 293 K.
4	430	Masumoto, H. et al.	1969	L	273, 313		95.5	4.5	Similar to the above specimen; density 8.86 g cm^{-3} at 293 K.
5*	430	Masumoto, H. et al.	1969	L	273, 313		95.0	5.0	Similar to the above specimen; density 8.82 g cm^{-3} at 293 K.
6*	430	Masumoto, H. et al.	1969	L	273, 313		94.6	5.4	Similar to the above specimen; density 8.80 g cm^{-3} at 293 K.

DATA TABLE 215. THERMAL LINEAR EXPANSION OF NICKEL-VANADIUM SYSTEM Ni-V

[Temperature, T, K; Linear Expansion, $\Delta L/L_0$, %]

T	$\Delta L/L_0$	T	$\Delta L/L_0$
CURVE 1		CURVE 4	
273	-0.027	273	-0.025
313	0.027	313	0.025
CURVE 2		CURVE 5*	
273	-0.026	273	-0.026
313	0.026	313	0.026
CURVE 3		CURVE 6*	
273	-0.022	273	-0.026
313	0.022	313	0.026

* Not shown in figure.

FIGURE AND TABLE NO. 216R. PROVISIONAL VALUES FOR THERMAL LINEAR EXPANSION OF NICKEL-ZINC SYSTEM Ni-Zn

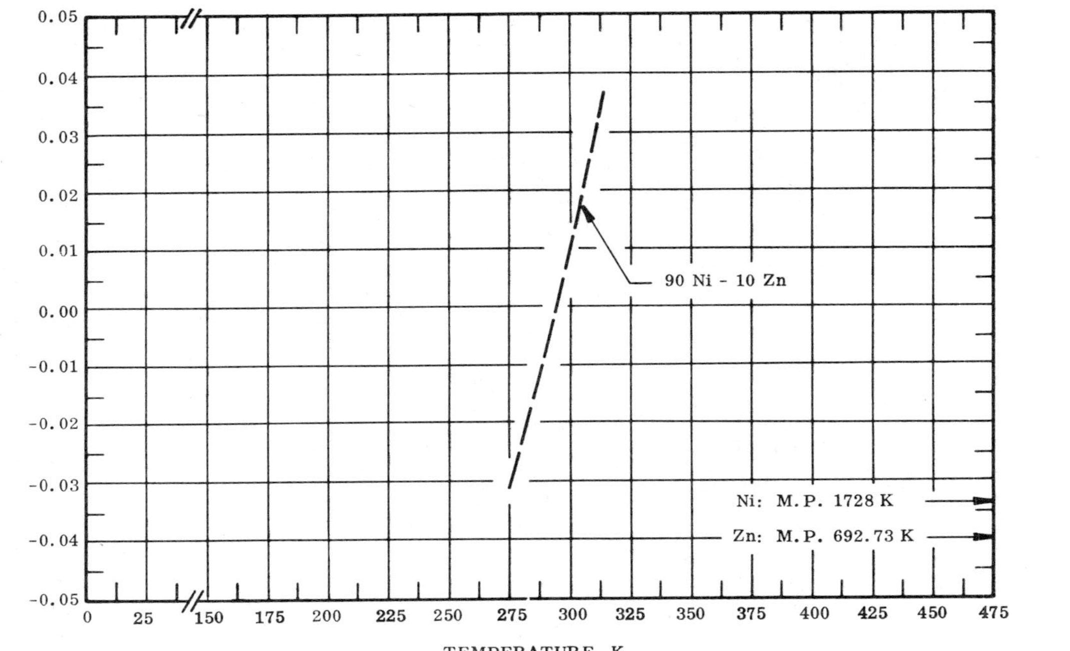

PROVISIONAL VALUES

[Temperature, T, K; Linear Expansion, $\Delta L/L_0$, %; α, K^{-1}]

T	(90 Ni-10 Zn) $\Delta L/L_0$	$\alpha \times 10^6$
275	-0.031	17.0
293	0.000	17.2
315	0.037	17.6

REMARKS

(90 Ni-10 Zn): The tabulated values for well-annealed alloy are considered accurate to within ±7% over the entire temperature range. These values can be represented approximately by the following equation:

$$\Delta L/L_0 = -1.542 + 1.239 \times 10^{-2}\, T - 3.640 \times 10^{-5}\, T^2 + 4.119 \times 10^{-8}\, T^3$$

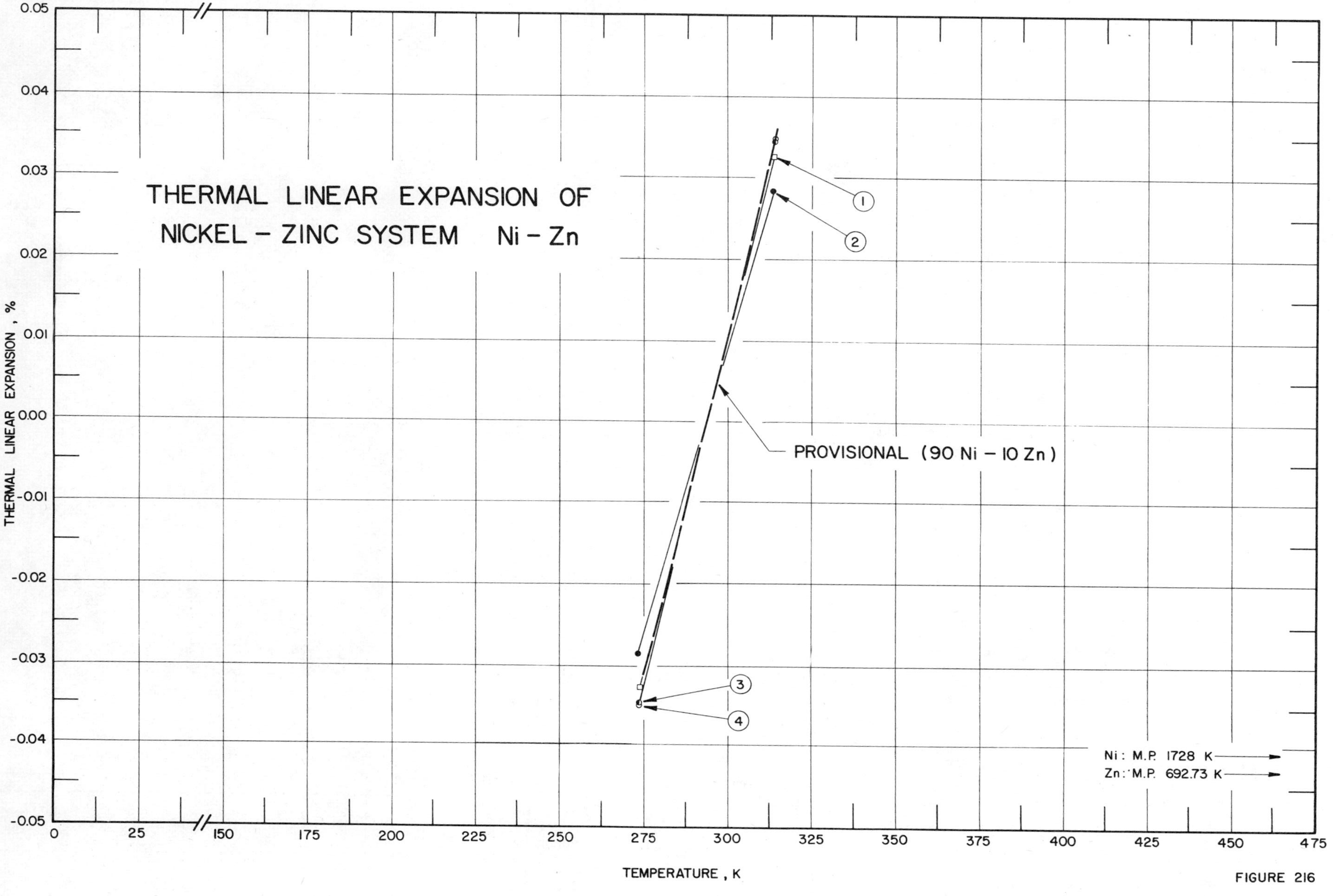
THERMAL LINEAR EXPANSION OF
NICKEL - ZINC SYSTEM Ni - Zn
PROVISIONAL (90 Ni - 10 Zn)
1
2
3
4
Ni : M.P. 1728 K
Zn : M.P. 692.73 K
THERMAL LINEAR EXPANSION , %
TEMPERATURE , K

FIGURE 216

SPECIFICATION TABLE 216. THERMAL LINEAR EXPANSION OF NICKEL-ZINC SYSTEM Ni-Zn

Cur. No.	Ref. No.	Author(s)	Year	Method Used	Temp. Range, K	Name and Specimen Designation	Composition (weight percent) Ni	Zn	Composition (continued), Specifications, and Remarks
1	241	Masumoto, H., Sawaya, S. and Kadowaki, S.	1969	L	273, 313		91.20	8.80	Electrolytic Ni with 0.180 Co, 0.002 S, 0.002 C, 0.001 Cu, 0.001 Fe, and 99.5+ Zn; rod specimen 2 mm dia, 10 cm long; cooled at rate 300 C/hr after heating in vacuum at 1273 K for 1 hr, density 8.94 g cm^{-3} at 293 K.
2	241	Masumoto, H. et al.	1969	L	273, 313		90.60	9.40	Similar to the above specimen.
3	241	Masumoto, H. et al.	1969	L	273, 313		90.06	9.94	Similar to the above specimen; density 8.93 g cm^{-3} at 293 K.
4	241	Masumoto, H. et al.	1969	L	273, 313		89.40	10.60	Similar to the above specimen; density 8.94 g cm^{-3} at 293 K.
5*	241	Masumoto, H. et al.	1969	L	273, 313		88.70	11.30	Similar to the above specimen; density 8.92 g cm^{-3} at 293 K.

DATA TABLE 216. THERMAL LINEAR EXPANSION OF NICKEL-ZINC SYSTEM Ni-Zn

[Temperature, T, K; Linear Expansion, $\Delta L/L_0$, %]

T	$\Delta L/L_0$	T	$\Delta L/L_0$
CURVE 1		CURVE 4	
273	-0.032	273	-0.035
313	0.032	313	0.035
CURVE 2		CURVE 5*	
273	-0.028	273	-0.032
313	0.028	313	0.032
CURVE 3			
273	-0.034		
313	0.034		

*Not shown in figure.

FIGURE AND TABLE NO. 217AR. PROVISIONAL VALUES FOR THERMAL LINEAR EXPANSION OF NIOBIUM-RHENIUM SYSTEM Nb-Re

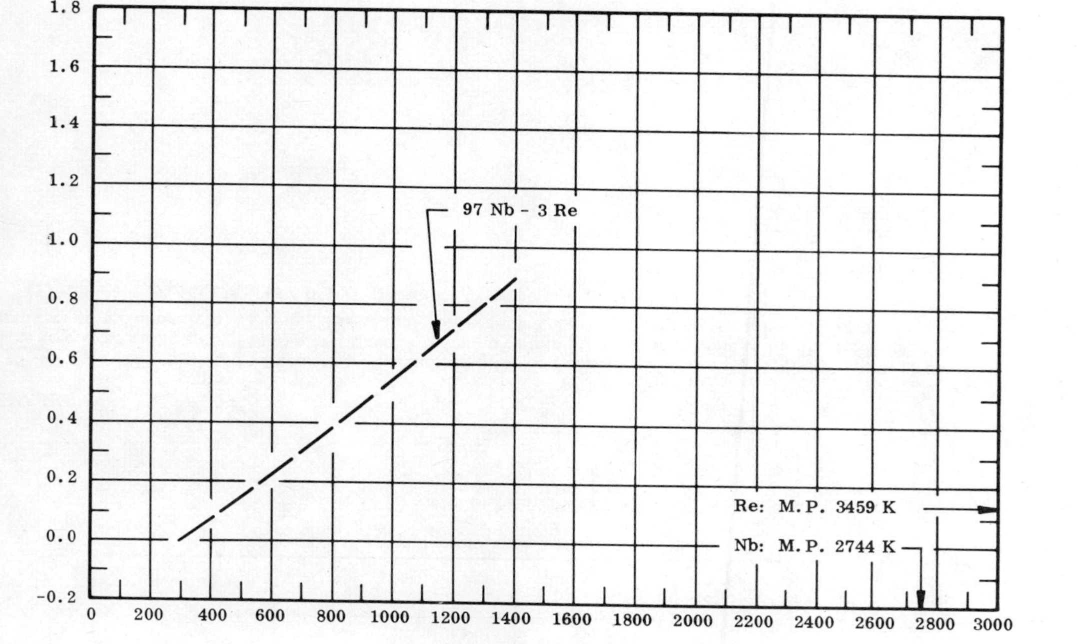

PROVISIONAL VALUES

[Temperature, T, K; Linear Expansion, $\Delta L/L_0$, %; α, K^{-1}]

T	(97 Nb-3 Re) $\Delta L/L_0$	$\alpha \times 10^6$
293	0.000	7.2
400	0.079	7.5
500	0.155	7.7
600	0.232	7.8
700	0.311	8.0
800	0.392	8.1
900	0.474	8.2
1000	0.557	8.3
1200	0.722	8.4
1400	0.892	8.4

REMARKS

(97 Nb-3 Re): The tabulated values for well-annealed alloy are considered accurate to within ± 7% over the entire temperature range. These values can be represented approximately by the following equation:

$$\Delta L/L_0 = -0.201 + 6.467 \times 10^{-4}\, T + 1.480 \times 10^{-7}\, T^2 - 3.767 \times 10^{-11}\, T^3$$

FIGURE AND TABLE NO. 217BR. PROVISIONAL VALUES FOR THERMAL LINEAR EXPANSION OF NIOBIUM-TANTALUM SYSTEM Nb-Ta

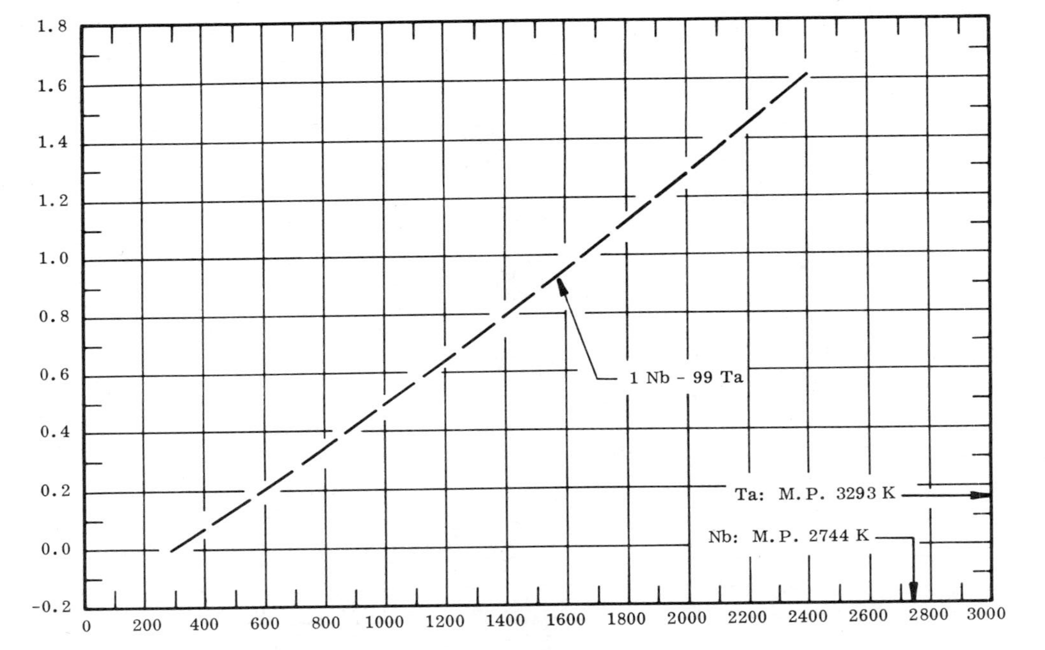

PROVISIONAL VALUES

[Temperature, T, K; Linear Expansion, $\Delta L/L_0$, %; α, K^{-1}]

T	(1 Nb-99 Ta) $\Delta L/L_0$	$\alpha \times 10^6$
293	0.000	6.5
400	0.070	6.6
500	0.137	6.7
600	0.205	6.8
700	0.274	6.9
800	0.344	7.1
900	0.415	7.2
1000	0.488	7.3
1200	0.636	7.5
1400	0.788	7.7
1600	0.945	7.9
1800	1.106	8.2
2000	1.271	8.4
2200	1.441	8.6
2400	1.615	8.8

REMARKS

(1 Nb-99 Ta): The tabulated values for well-annealed alloy are considered accurate to within ± 7% over the entire temperature range. The thermal linear expansion of this alloy is very close to the thermal linear expansion of pure tantalum. These values can be represented approximately by the following equation:

$$\Delta L/L_0 = -0.186 + 6.175 \times 10^{-4}\, T + 5.625 \times 10^{-8}\, T^2 - 3.739 \times 10^{-13}\, T^3$$

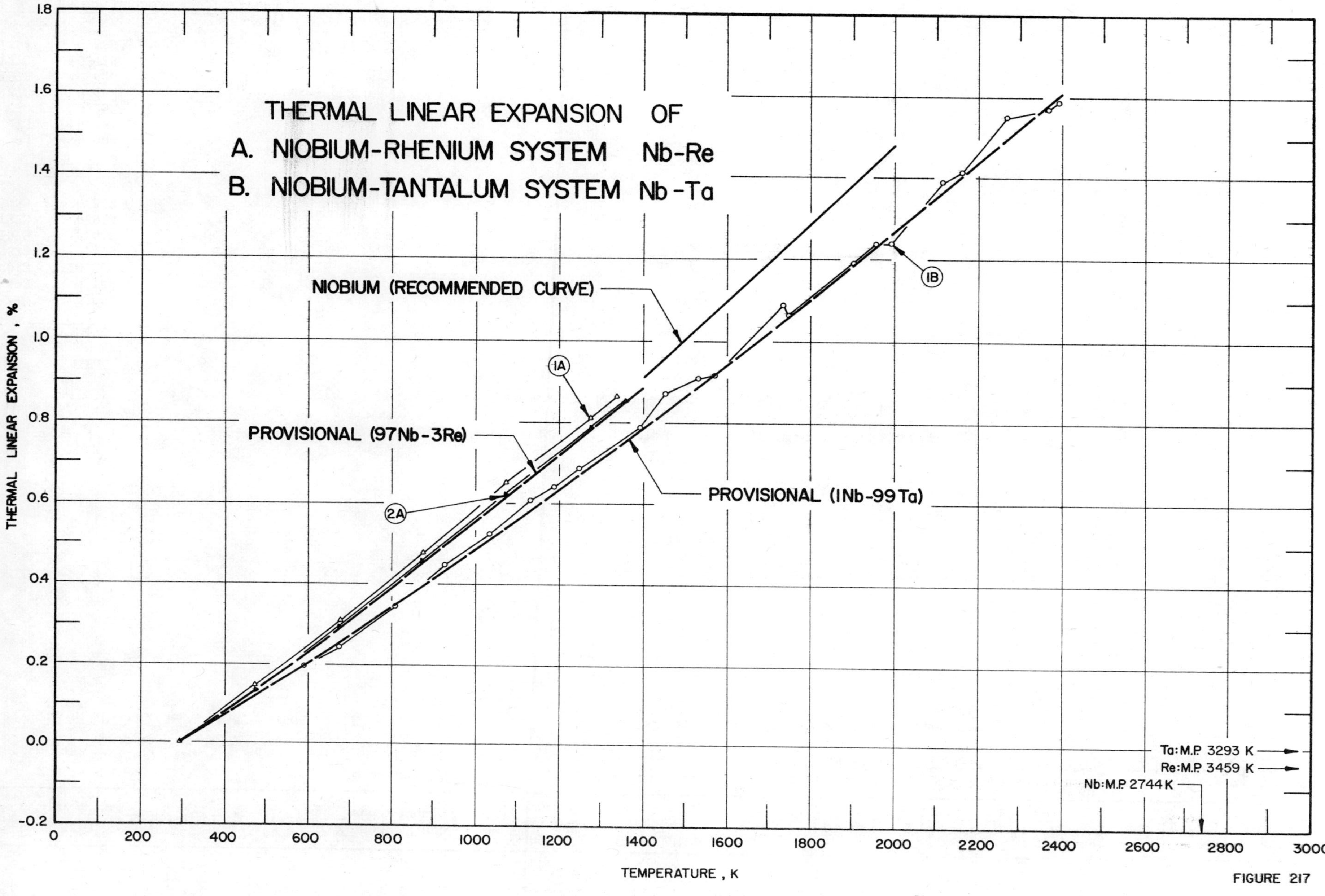
THERMAL LINEAR EXPANSION OF
A. NIOBIUM-RHENIUM SYSTEM Nb-Re
B. NIOBIUM-TANTALUM SYSTEM Nb-Ta
NIOBIUM (RECOMMENDED CURVE)
1A
PROVISIONAL (97Nb-3Re)
2A
PROVISIONAL (1Nb-99Ta)
1B
Ta: M.P. 3293 K
Re: M.P. 3459 K
Nb: M.P. 2744 K
THERMAL LINEAR EXPANSION, %
1.8
1.6
1.4
1.2
1.0
0.8
0.6
0.4
0.2
0.0
-0.2
0
200
400
600
800
1000
1200
1400
1600
1800
2000
2200
2400
2600
2800
3000
TEMPERATURE, K
FIGURE 217

SPECIFICATION TABLE 217A. THERMAL LINEAR EXPANSION OF NIOBIUM-RHENIUM SYSTEM Nb-Re

Cur. No.	Ref. No.	Author(s)	Year	Method Used	Temp. Range, K	Name and Specimen Designation	Composition (weight percent) Nb	Re	Composition (continued), Specifications, and Remarks
1	253	Ul'yanov, R.A., Tarasov, N.D. and Mikhaylov, Ya.D.	1964	L	293-1338		99.0	1.0	Starting material 98.7 pure Nb; then vacuum refined; alloy prepared by arc smelting in argon atm; expansion measured under 10^{-5} mm Hg vacuum.
2	253	Ul'yanov, R.A. et al.	1964	L	293-1354		97.44	2.56	Similar to the above specimen.

DATA TABLE 217A. THERMAL LINEAR EXPANSION OF NIOBIUM-RHENIUM SYSTEM Nb-Re

[Temperature, T, K; Linear Expansion, $\Delta L/L_0$, %]

T	$\Delta L/L_0$
CURVE 1	
293	0.000
473	0.142
673	0.309
873	0.479
1073	0.647
1273	0.813
1338	0.862
CURVE 2	
293	0.000*
473	0.134
673	0.297
873	0.458
1073	0.620
1273	0.788
1354	0.856

* Not shown in figure.

SPECIFICATION TABLE 217B. THERMAL LINEAR EXPANSION OF NIOBIUM-TANTALUM SYSTEM Nb-Ta

Cur. No.	Ref. No.	Author(s)	Year	Method Used	Temp. Range, K	Name and Specimen Designation	Composition (weight percent) Nb	Ta	Composition (continued), Specifications, and Remarks
1	623	Mochalov, G.A. and Ivanov, O.S.	1969	L	592-2395		1.2	98.6	Specimen cut from a bar prepared by powder metallurgy; machined, heated in vacuum for stabilization and held at maximum dilatometer temperature for 5 min; dilatometer measurements below 1273 K, cathetometer measurements above 1273 K.

DATA TABLE 217B. THERMAL LINEAR EXPANSION OF NIOBIUM-TANTALUM SYSTEM Nb-Ta

[Temperature, T, K; Linear Expansion, $\Delta L/L_0$, %]

T	$\Delta L/L_0$	T	$\Delta L/L_0$
CURVE 1		CURVE 1 (cont.)	
592	0.199	2160	1.419
672	0.241	2266	1.552
805	0.342	2361	1.565
923	0.442	2395	1.587
1036	0.521		
1130	0.605		
1186	0.640		
1243	0.685		
1390	0.791		
1454	0.871		
1531	0.913		
1577	0.920		
1736	1.094		
1743	1.065		
1903	1.197		
1958	1.240		
1998	1.240		
2118	1.396		

FIGURE AND TABLE NO. 218AR. PROVISIONAL VALUES FOR THERMAL LINEAR EXPANSION OF NIOBIUM-URANIUM SYSTEM Nb-U

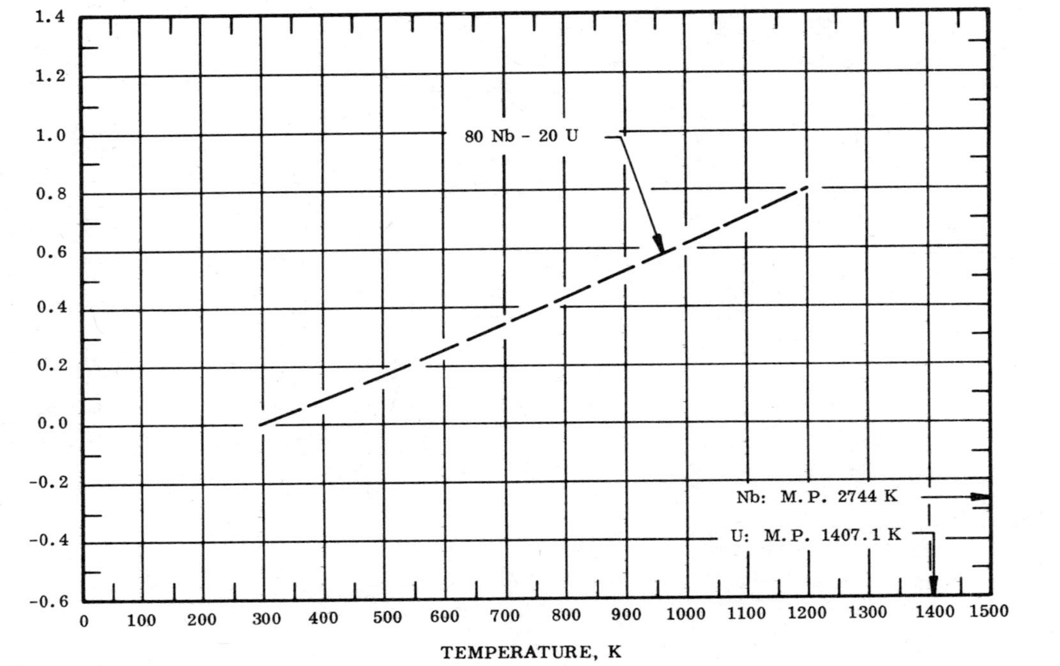

PROVISIONAL VALUES

[Temperature, T, K; Linear Expansion, $\Delta L/L_0$, %; α, K^{-1}]

T	(80 Nb-20 U) $\Delta L/L_0$	$\alpha \times 10^6$
293	0.000	7.8
400	0.086	8.3
500	0.169	8.5
600	0.254	8.8
700	0.344	9.0
800	0.436	9.1
900	0.527	9.2
1000	0.621	9.4
1200	0.810	9.7

REMARKS

(80 Nb-20 U): The tabulated values for well-annealed alloy are considered accurate to within ±7% over the entire temperature range. These values can be represented approximately by the following equation:

$$\Delta L/L_0 = -0.213 + 6.670 \times 10^{-4}\, T + 2.276 \times 10^{-7}\, T^2 - 6.064 \times 10^{-11}\, T^3$$

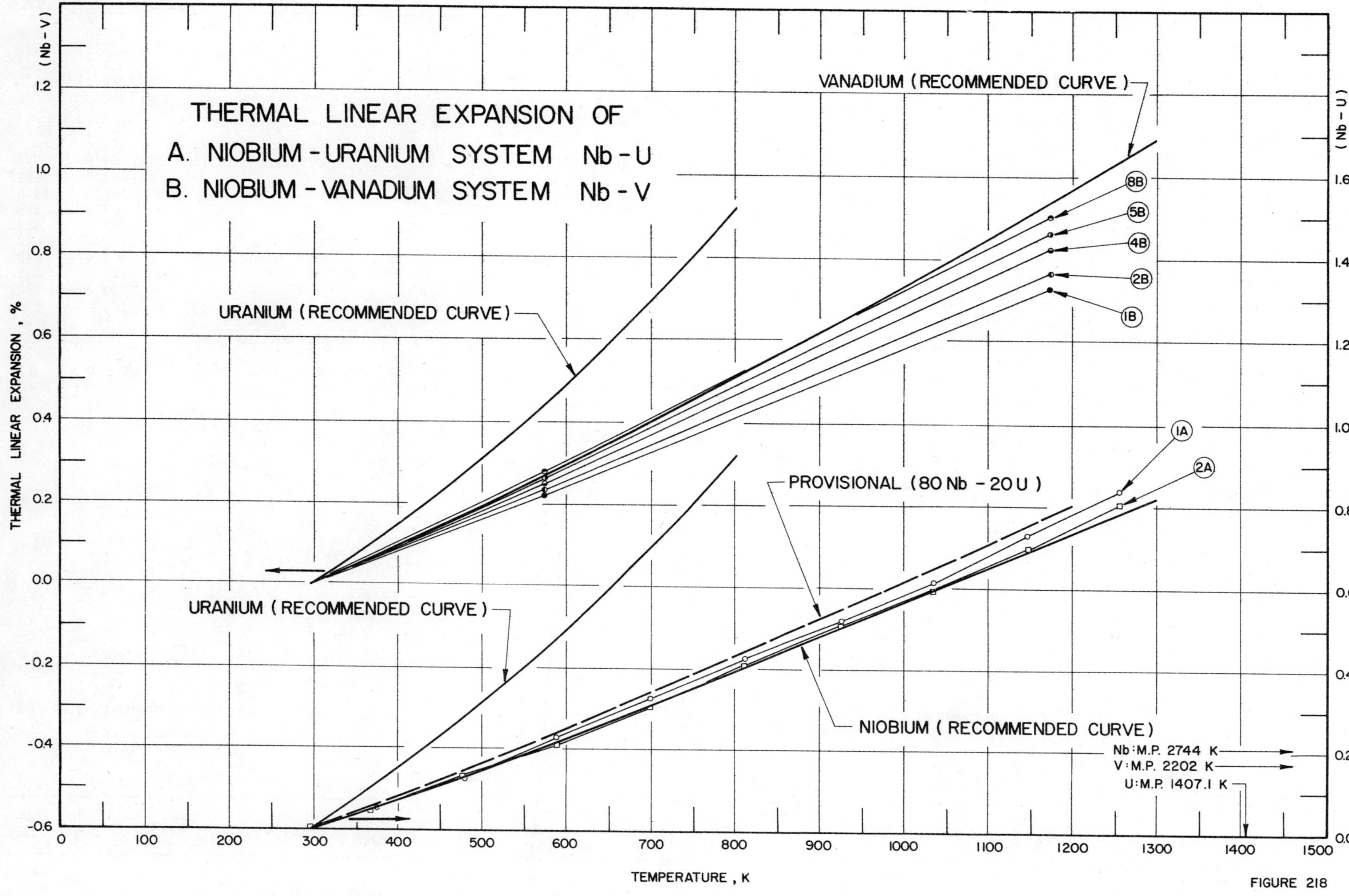
THERMAL LINEAR EXPANSION OF
A. NIOBIUM - URANIUM SYSTEM Nb - U
B. NIOBIUM - VANADIUM SYSTEM Nb - V
VANADIUM (RECOMMENDED CURVE)
8B
5B
4B
2B
1B
URANIUM (RECOMMENDED CURVE)
1A
2A
PROVISIONAL (80 Nb - 20 U)
URANIUM (RECOMMENDED CURVE)
NIOBIUM (RECOMMENDED CURVE)
Nb: M.P. 2744 K
V: M.P. 2202 K
U: M.P. 1407.1 K
THERMAL LINEAR EXPANSION, %
(Nb - V)
(Nb - U)
TEMPERATURE, K

FIGURE 218

SPECIFICATION TABLE 218A. THERMAL LINEAR EXPANSION OF NIOBIUM-URANIUM SYSTEM Nb-U

Cur. No.	Ref. No.	Author(s)	Year	Method Used	Temp. Range, K	Name and Specimen Designation	Composition (weight percent) Nb	U	Composition (continued), Specifications, and Remarks
1	431	DeMastry, J.A., Mook, D.P., Epstein, S.G., Bauer, A.A. and Dickerson, R.F.	1961	L	293-1255		90	10	Thermal cycled twice from room temp to about 1033 K and back to room temp; measured in vacuum of at least 5 x 10^{-5} mm Hg.
2	431	DeMastry, J.A. et al.	1961	L	293-1255		80	20	Similar to the above specimen.

DATA TABLE 218A. THERMAL LINEAR EXPANSION OF NIOBIUM-URANIUM SYSTEM Nb-U

[Temperature, T, K; Linear Expansion, $\Delta L/L_0$, %]

T	$\Delta L/L_0$	T	$\Delta L/L_0$
CURVE 1		CURVE 2 (cont.)	
293	0.000	810	0.422
366	0.042	922	0.520
477	0.136	1033	0.617
588	0.214	1144	0.735
699	0.306	1255	0.840
810	0.403		
922	0.505		
1033	0.598		
1144	0.700		
1255	0.804		
CURVE 2			
293	0.000*		
372	0.060		
477	0.128		
588	0.229		
699	0.323		

* Not shown in figure.

SPECIFICATION TABLE 218B. THERMAL LINEAR EXPANSION OF NIOBIUM-VANADIUM SYSTEM Nb-V

Cur. No.	Ref. No.	Author(s)	Year	Method Used	Temp. Range, K	Name and Specimen Designation	Composition (weight percent) Nb	V	Composition (continued), Specifications, and Remarks
1	707	Pridantseva, K.S. and Solov'yeva, N.A.	1965	L	293, 1173		Bal.	10	Specimen prepared by powder metallurgy method, impurities mainly 1% oxygen, sample was homogenized at temperature which would provede an equilibrium state of the alloy, homogenizing anneal was done in a vacuum of up to 10^{-4} mm Hg or in argon, single phase solid solution sample, 1% measurement error.
2	707	Pridantseva, K.S. and Solov'yeva, N.A.	1965	L	293, 1173		Bal.	20	Similar to the above specimen.
3*	707	Pridantseva, K.S. and Solov'yeva, N.A.	1965	L	293, 1173		Bal.	25	Similar to the above specimen.
4	707	Pridantseva, K.S. and Solov'yeva, N.A.	1965	L	293, 1173		Bal.	33	Similar to the above specimen.
5	707	Pridantseva, K.S. and Solov'yeva, N.A.	1965	L	293, 1173		Bal.	42	Similar to the above specimen.
6*	707	Pridantseva, K.S. and Solov'yeva, N.A.	1965	L	293, 1173		Bal.	53	Similar to the above specimen.
7*	707	Pridantseva, K.S. and Solov'yeva, N.A.	1965	L	293, 1173		Bal.	70	Similar to the above specimen.
8	707	Pridantseva, K.S. and Solov'yeva, N.A.	1965	L	293, 1173		Bal.	80	Similar to the above specimen.

DATA TABLE 218B. THERMAL LINEAR EXPANSION OF NIOBIUM-VANADIUM SYSTEM Nb-V

[Temperature, T, K; Linear Expansion, $\Delta L/L_0$, %]

T	$\Delta L/L_0$	T	$\Delta L/L_0$	T	$\Delta L/L_0$
CURVE 1		CURVE 4		CURVE 7*	
293	0.000	293	0.000*	293	0.000
573	0.223	573	0.251	573	0.278
1173	0.730	1173	0.825	1173	0.895
CURVE 2		CURVE 5		CURVE 8	
293	0.000*	293	0.000*	293	0.000*
573	0.237	573	0.263	573	0.280
1173	0.766	1173	0.864	1173	0.906
CURVE 3*		CURVE 6*			
293	0.000	293	0.000		
573	0.242	573	0.270		
1173	0.792	1173	0.882		

* Not shown in figure.

FIGURE AND TABLE NO. 219R. PROVISIONAL VALUES FOR THERMAL LINEAR EXPANSION OF NIOBIUM-ZIRCONIUM SYSTEM Nb-Zr

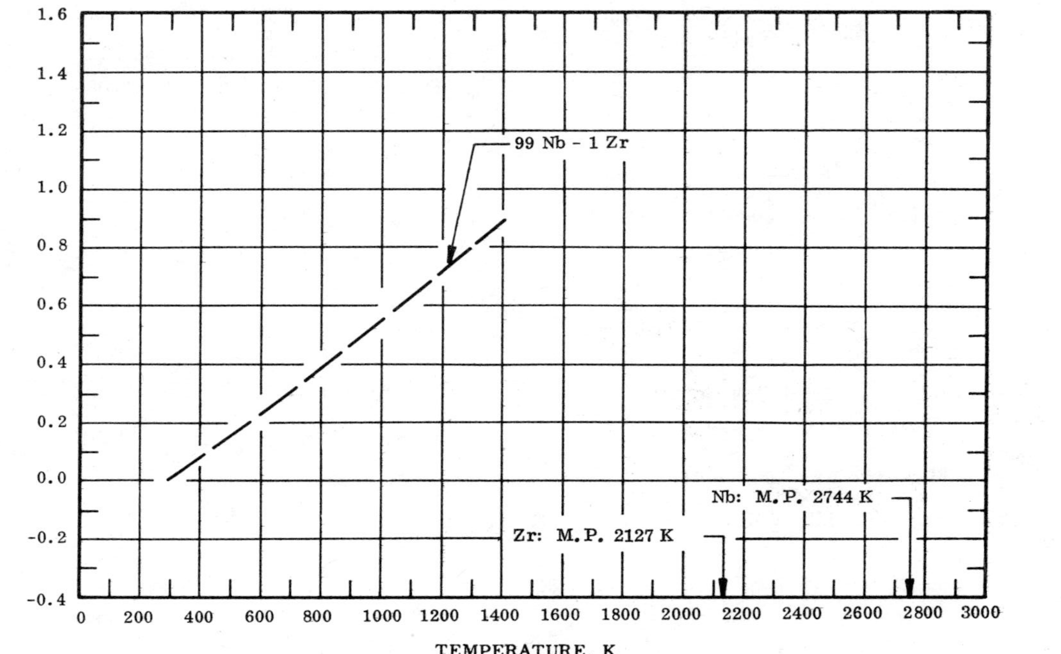

PROVISIONAL VALUES

[Temperature, T, K; Linear Expansion, $\Delta L/L_0$, %; α, K^{-1}]

T	(99 Nb-1 Zr) $\Delta L/L_0$	$\alpha \times 10^6$
293	0.000	7.0
400	0.077	7.3
500	0.152	7.6
600	0.229	7.8
700	0.308	8.0
800	0.390	8.2
900	0.472	8.3
1000	0.555	8.4
1200	0.725	8.6
1400	0.899	8.7

REMARKS

(99 Nb-1 Zr): The tabulated values for well-annealed alloy are considered accurate to within ± 7% over the entire temperature range. These values can be represented approximately by the following equation:

$$\Delta L/L_0 = -0.194 + 6.146 \times 10^{-4}\ T + 1.765 \times 10^{-7}\ T^2 - 4.153 \times 10^{-11}\ T^3$$

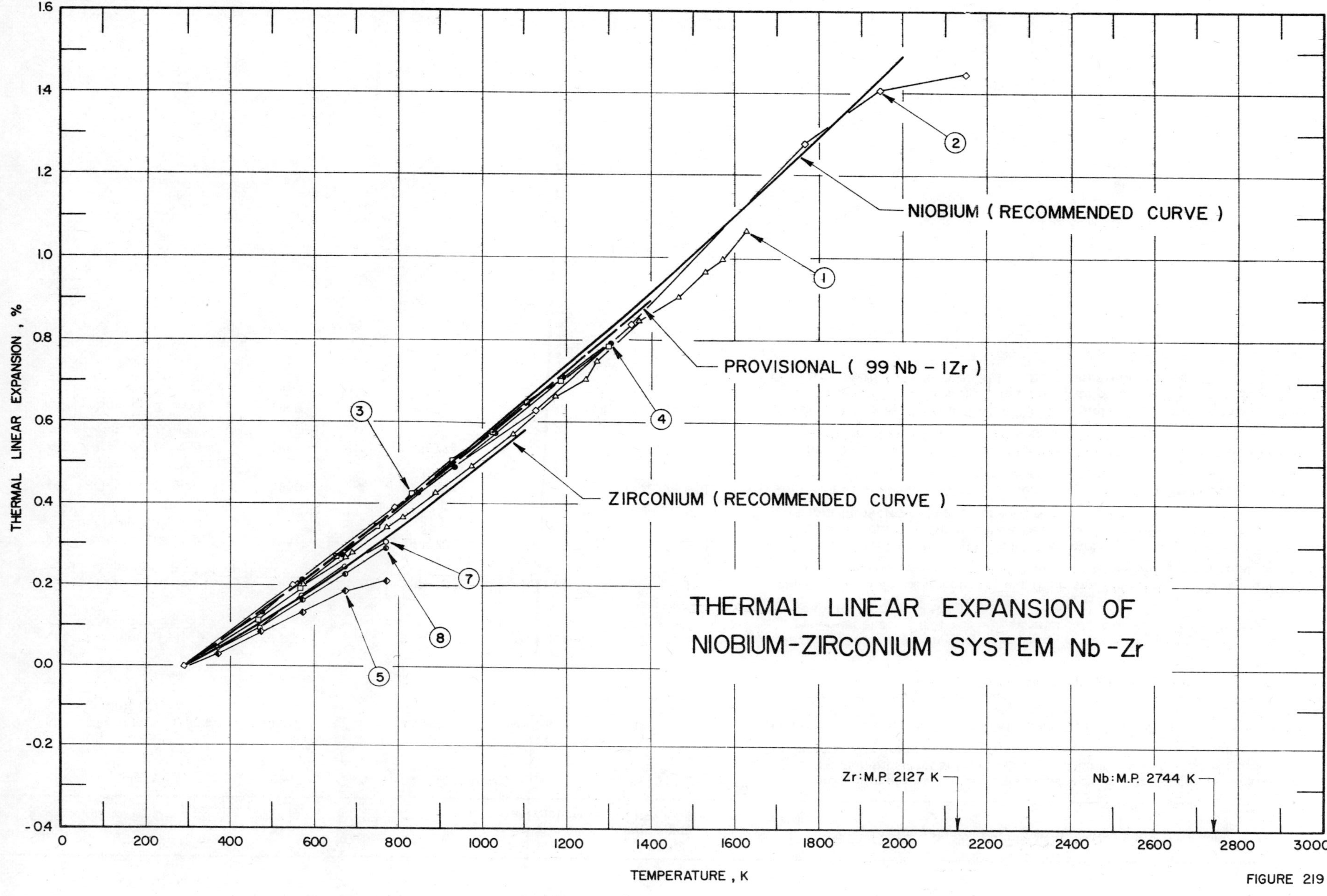
THERMAL LINEAR EXPANSION OF
NIOBIUM-ZIRCONIUM SYSTEM Nb-Zr
NIOBIUM (RECOMMENDED CURVE)
PROVISIONAL (99 Nb - 1Zr)
ZIRCONIUM (RECOMMENDED CURVE)
Zr: M.P. 2127 K
Nb: M.P. 2744 K
1
2
3
4
5
7
8
THERMAL LINEAR EXPANSION , %
1.6
1.4
1.2
1.0
0.8
0.6
0.4
0.2
0.0
-0.2
-0.4
0
200
400
600
800
1000
1200
1400
1600
1800
2000
2200
2400
2600
2800
3000
TEMPERATURE , K

FIGURE 219

SPECIFICATION TABLE 219. THERMAL LINEAR EXPANSION OF NIOBIUM-ZIRCONIUM SYSTEM Nb-Zr

Cur. No.	Ref. No.	Author(s)	Year	Method Used	Temp. Range, K	Name and Specimen Designation	Composition (weight percent) Nb	Zr	Composition (continued), Specifications, and Remarks
1	432	Ewing, C. T.	1965	L	575-1627		99.0	1.0	99.8 starting materials; tested as received; data for three separate experiments combined by author; zero-point correction of 0.006% determined by graphical extrapolation.
2	433	Pears, C.D. and Oglesby, S.	1962		293-2628	Run No. E9	99.2	0.5	Obtained from General Astrometals Corp.; after exposure 99.5 Nb and 0.41 C; hot pressed; specimen melted; density at 298 K before exposure 7.88 g cm^{-3}, density after exposure 8.32 g cm^{-3}; initial length 2.982 in.; expansion measured in helium atmosphere; reported error 5%.
3	434	Fisher, C.R. and Achener, P.Y.	1965	L	297-1300	Specimen A	99	1	Specimen cut to 50 mm length; expansion measured in argon atmosphere; reported error <3%; zero-point correction of 0.003% determined by graphical extrapolation.
4	434	Fisher, C.R. and Achener, P.Y.	1965	L	297-1305	Specimen B	99	1	Similar to the above specimen; cut from same material, but at right angles to it; zero-point correction of 0.003% determined by graphical extrapolation.
5	449	Beyer, B. and Juknat, R.	1966	L	293-774		1.17	Bal.	0.035 C, 0.0167 O, 0.005 N, 0.0008 H; melted in vacuum electric arc-furnace from Van Arkel Zr bar and reactorgrade Nb; wire specimen 4 mm diameter; zero-point correction is -0.007%.
6*	449	Beyer, B. and Juknat, R.	1966	L	293-774		1.17	Bal.	Similar to the above specimen; specimen from hot-rolled sheet parallel to the direction of rolling; zero-point correction is -0.007%.
7	449	Beyer, B. and Juknat, R.	1966	L	293-774		1.17	Bal.	Similar to the above specimen; specimen from hot-rolled sheet transverse to the direction of rolling; zero-point correction is -0.001%.
8	449	Beyer, B. and Juknat, R.	1966	L	293-773		2.54	Bal.	0.0375 O, 0.006 C, 0.0035 N, 0.0015 H; melted in vacuum electric arc-furnace from Van Arkel Zr bar and reactor grade Nb; heat-treated at 773 K, water-cooled for 8 hr at 323 K; zero-point correction is -0.007%.

* Not shown in figure.

DATA TABLE 219. THERMAL LINEAR EXPANSION OF NIOBIUM-ZIRCONIUM SYSTEM Nb-Zr

[Temperature, T, K; Linear Expansion, $\Delta L/L_0$, %]

T	$\Delta L/L_0$
CURVE 1	
575	0.200
591	0.209*
670	0.269
689	0.280
774	0.341
808	0.366
809	0.369*
885	0.425
896	0.437*
978	0.495
988	0.507*
1075	0.570
1077	0.579*
1081	0.576*
1173	0.663
1175	0.654*
1242	0.707
CURVE 1 (cont.)	
1275	0.753
1277	0.755*
1370	0.847
1465	0.906
1466	0.900*
1533	0.967
1573	1.008
1627	1.069
CURVE 2	
293	0.00
549	0.20
796	0.39
1122	0.63
1356	0.84
1769	1.28
CURVE 2 (cont.)	
1944	1.41
2158	1.45
2158	1.37*
2349	1.15*
2388	1.09*
2544	0.77*
2627	-0.40*
CURVE 3	
297	0.003*
465	0.115
561	0.197
655	0.270
744	0.344
833	0.421
925	0.504
CURVE 3 (cont.)	
1021	0.579
1108	0.648
1189	0.704
1300	0.788
CURVE 4	
297	0.003*
478	0.138
571	0.207
663	0.275
753	0.352*
843	0.424
939	0.495
1030	0.579
1116	0.649*
1193	0.714
1305	0.798
CURVE 5	
293	0.000*
376	0.037
474	0.084
575	0.133
674	0.186
774	0.214
CURVE 6*	
293	0.000
376	0.037
474	0.084
575	0.133
674	0.186
774	0.214
CURVE 7	
293	0.000*
375	0.044
474	0.097
575	0.173
674	0.244
774	0.305
CURVE 8	
293	0.000*
374	0.035*
473	0.096*
575	0.162
673	0.229
773	0.291

DATA TABLE 219. COEFFICIENT OF THERMAL LINEAR EXPANSION OF NIOBIUM-ZIRCONIUM SYSTEM Nb-Zr

[Temperature, T, K; Coefficient of Expansion, α, $10^{-6}\ K^{-1}$]

T	α
CURVE 3	
465	3.70
561	4.09
655	4.14
744	4.24
833	4.33
925	4.43
1021	4.42
1108	4.42
1189	4.37
1300	4.35
CURVE 4	
478	4.14
571	4.13
663	4.13
753	4.26
CURVE 4 (cont.)	
843	4.28
939	4.26
1030	4.36
1116	4.38
1193	4.41
1305	4.38

* Not shown in figure.

FIGURE AND TABLE NO. 220R. PROVISIONAL VALUES FOR THERMAL LINEAR EXPANSION OF LEAD-SILVER SYSTEM Pd-Ag

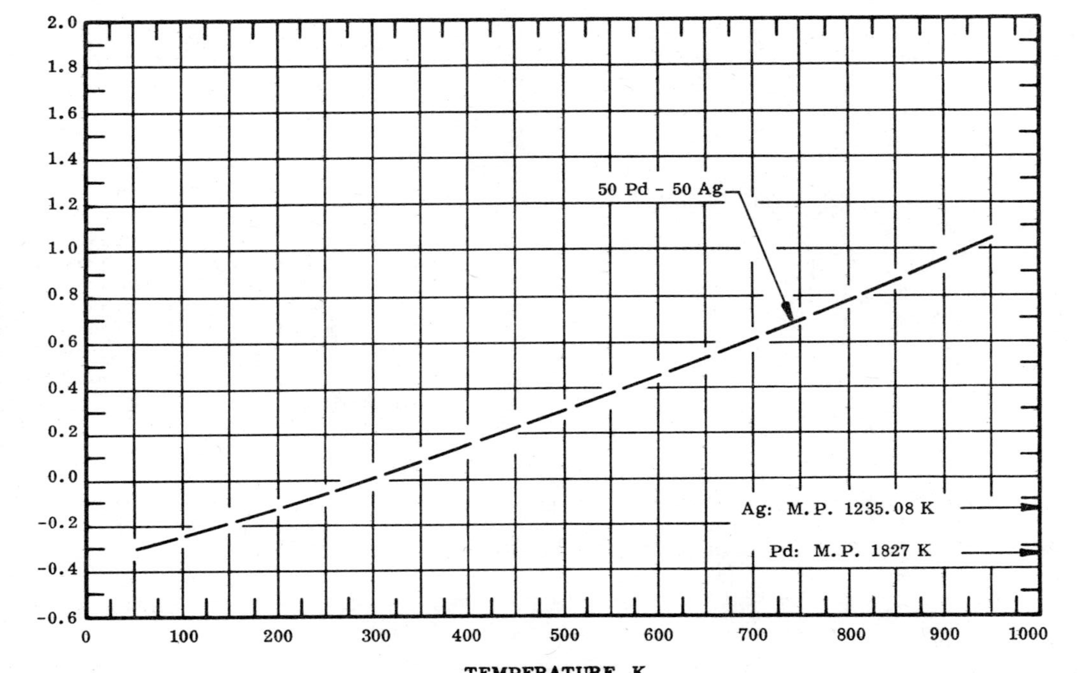

PROVISIONAL VALUES

[Temperature, T, K; Linear Expansion, $\Delta L/L_0$, %; α, K^{-1}]

(50 Pd-50 Ag)

T	$\Delta L/L_0$	$\alpha \times 10^6$
50	-0.305	11.6
100	-0.247	11.9
200	-0.124	13.0
293	0.000	13.4
400	0.148	14.2
500	0.294	15.0
600	0.447	15.7
700	0.609	16.5
800	0.781	17.5
900	0.961	18.8
950	1.053	19.6

REMARKS

(50 Pd-50 Ag): The tabulated values for well-annealed alloy are considered accurate to within ± 7% over the entire temperature range. These values can be represented approximately by the following equation:

$$\Delta L/L_0 = -0.360 + 1.120 \times 10^{-3}\, T + 3.660 \times 10^{-7}\, T^2 + 2.061 \times 10^{-11}\, T^3$$

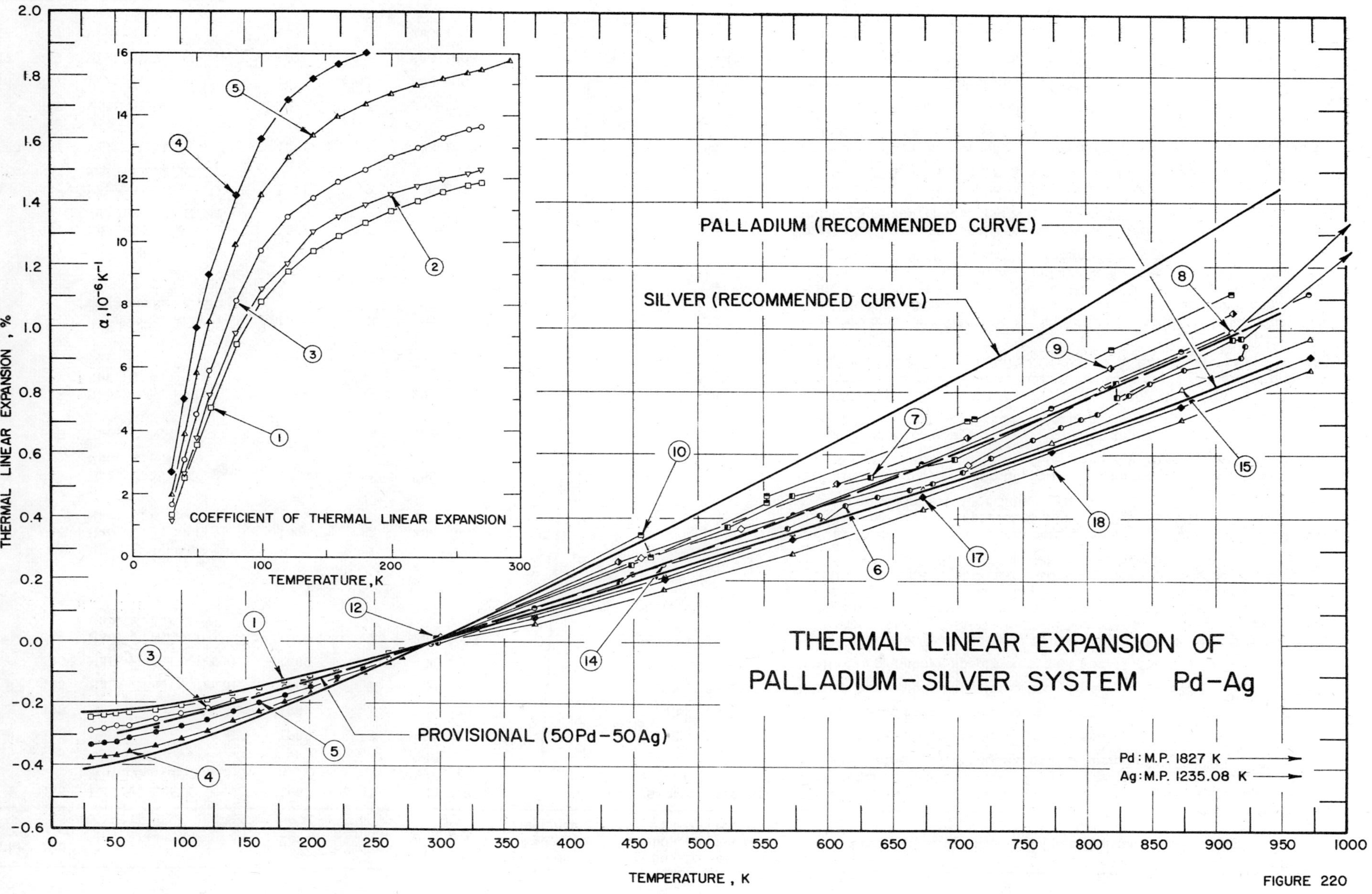
THERMAL LINEAR EXPANSION OF
PALLADIUM-SILVER SYSTEM Pd-Ag
PALLADIUM (RECOMMENDED CURVE)
SILVER (RECOMMENDED CURVE)
PROVISIONAL (50Pd-50Ag)
Pd: M.P. 1827 K
Ag: M.P. 1235.08 K
THERMAL LINEAR EXPANSION, %
TEMPERATURE, K
COEFFICIENT OF THERMAL LINEAR EXPANSION
$\alpha, 10^{-6} K^{-1}$
TEMPERATURE, K

FIGURE 220

SPECIFICATION TABLE 220. THERMAL LINEAR EXPANSION OF PALLADIUM-SILVER SYSTEM Pd-Ag

Cur. No.	Ref. No.	Author(s)	Year	Method Used	Temp. Range, K	Name and Specimen Designation	Composition (weight percent) Pd	Ag	Composition (continued), Specifications, and Remarks
1	435	Bailey, A.C., Waterhouse, N., and Yates, B.	1969	I	30-270		95.73	4.27	Hollow cylinder with 2.6 cm O.D., 2.3 cm I.D. and 1 cm high; supplied by Johnson, Matthey and Co., Ltd.; annealed in vacuum for between 30 and 60 min at 1173 K; cooled over a period of 24 hr.
2*	435	Bailey, A.C., et al.	1969	I	30-270		82.33	17.67	Similar to the above specimen except annealed at 1123 K.
3	435	Bailey, A.C., et al.	1969	I	30-270		54.43	45.57	Similar to the above specimen except annealed at 1073 K.
4	435	Bailey, A.C., et al.	1969	I	30-293		11.78	88.22	Similar to the above specimen except annealed at 873 K.
5	435	Bailey, A.C., et al.	1969	I	30-293		27.77	72.23	Similar to the above specimen except annealed at 973 K.
6	648	Rao, C.N. and Rao, K.K.	1964	X	299-1053		25	75	0.008 Si, 0.008 Fe, 0.004 Cu, 0.004 Al, and 0.002 B; specimen from Bishop and Co., Malverin, Pa.; homogenized in vacuum for a week at 1173 K; measurements also on specimen quenched from 613 K, 793 K, and 923 K for 1, 3, and 8 hr, respectively; constant room temperature lattice parameter obtained; lattice parameter reported at 299 K is 3.9328 Å; 3.9327 Å at 293 K determined by extrapolation.
7	648	Rao, C.N. and Rao, K.K.	1964	X	298-1058		39	61	Similar to the above specimen; homogenized in vacuum for a week at 1173 K; lattice parameter reported at 298 K is 3.9590 Å; 3.9587 Å at 293 K determined by extrapolation.
8	648	Rao, C.N. and Rao, K.K.	1964	X	300-1063		50	50	Similar to the above specimen; lattice parameter reported at 300 K is 3.9772 Å; 3.9768 Å at 293 K determined by extrapolation.
9	648	Rao, C.N. and Rao, K.K.	1964	X	298-913		63.1	36.9	Similar to the above specimen; lattice parameter reported at 298 K is 4.0035 Å; 4.0031 Å at 293 K determined by extrapolation.
10	648	Rao, C.N. and Rao, K.K.	1964	X	298-913		80	20	0.010 Si, 0.008 Fe, 0.003 Cu, 0.002 Al, 0.007 B; specimen from the above source; homogenized in vacuum for a week at 1173 K; lattice parameter reported at 296 K is 4.0376 Å; 4.0371 Å at 293 K determined by extrapolation.
11*	654	Nagender Naidu, S.V. and Houska, C.R.	1971	X	80-298		75	25	Prepared from 99.97 Pd and 99.99 Ag from Engelhard Industries, Inc. by melting in induction furnace under hydrogen atmosphere; homogenized for 1 week at 1173 K; annealed at 873 K in evacuated quartz tubes after flushing 3 times with argon; lattice parameter reported at 298 K is 3.9348 Å; 3.9346 Å at 293 K determined by graphical interpolation.
12	654	Nagender Naidu, S.V. and Houska, C.R.	1971	X	80-298		50	50	Similar to the above specimen; lattice parameter reported at 298 K is 3.9785 Å; 3.9782 Å at 293 K determined by graphical interpolation.
13*	654	Nagender Naidu, S.V. and Houska, C.R.	1971	X	80-298		25	75	Similar to the above specimen; lattice parameter reported at 298 K is 4.0280 Å; 4.0275 Å at 293 K determined by graphical interpolation.
14	657	Vest, R.W.	1971	L	293-1067		20	80	Specimen prepared from 99.9 pure metals; data on the coefficient of thermal linear expansion also reported.
15	657	Vest, R.W.	1971	L	293-1063		40	60	Similar to the above specimen.
16*	657	Vest, R.W.	1971	L	293-1067		60	40	Similar to the above specimen.
17	657	Vest, R.W.	1971	L	293-1068		70	30	Similar to the above specimen.
18	657	Vest, R.W.	1971	L	293-1067		80	20	Similar to the above specimen.

* Not shown in figure.

DATA TABLE 220. THERMAL LINEAR EXPANSION OF PALLADIUM-SILVER SYSTEM Pd-Ag

[Temperature, T, K; Linear Expansion, $\Delta L/L_0$, %]

T	$\Delta L/L_0$
CURVE 1‡	
30	-0.241
40	-0.239
50	-0.236
60	-0.232
80	-0.221
100	-0.206
120	-0.189
140	-0.170
160	-0.150
180	-0.129
200	-0.108
220	-0.085
240	-0.063
260	-0.039
270	-0.027
CURVE 2*,‡	
30	-0.252
40	-0.250
50	-0.247
60	-0.243
80	-0.231
100	-0.215
120	-0.197
140	-0.178
160	-0.156
180	-0.134
200	-0.112
220	-0.088
240	-0.065
260	-0.040
270	-0.028
CURVE 3‡	
30	-0.283
40	-0.280
50	-0.276
60	-0.271
80	-0.257
100	-0.239
120	-0.219
140	-0.196
160	-0.173
180	-0.149
CURVE 3 (cont.)‡	
200	-0.124
220	-0.098
240	-0.072
260	-0.045
270	-0.031*
CURVE 4‡	
30	-0.377
40	-0.373
50	-0.367
60	-0.359
80	-0.338
100	-0.313
120	-0.286
140	-0.256
160	-0.225
180	-0.193
200	-0.161
220	-0.127
240	-0.093
260	-0.059
270	-0.041
293	0.000*
CURVE 5‡	
30	-0.331
40	-0.328
50	-0.323
60	-0.316
80	-0.299
100	-0.277
120	-0.253
140	-0.227
160	-0.200
180	-0.171
200	-0.142
220	-0.113
240	-0.082
260	-0.052*
270	-0.036*
293	0.000*
CURVE 6	
299	0.003
438	0.198
448	0.221
568	0.374
568	0.376*
593	0.402
595	0.394
613	0.440
613	0.442*
636	0.463
663	0.491*
663	0.496
680	0.511
680	0.516*
703	0.542
703	0.542*
706	0.537*
708	0.552*
723	0.592*
725	0.595
758	0.648
758	0.646
778	0.681*
780	0.684
795	0.715*
797	0.720*
808	0.732*
809	0.732
818	0.769*
823	0.783
833	0.796
833	0.791*
848	0.821
848	0.826*
875	0.872
920	0.910
923	0.948
1053	1.139*
1053	1.203*
CURVE 7	
298	0.008*
448	0.250
448	0.242*
523	0.374
CURVE 7 (cont.)	
633	0.525
698	0.584
823	0.829
823	0.781
913	0.967
920	0.965
1053	1.233*
1058	1.223*
CURVE 8	
300	0.010
455	0.272
536	0.362
706	0.568
811	0.805
913	0.998
1063	1.323*
CURVE 9	
298	0.010*
438	0.260
606	0.505
706	0.659
818	0.872
913	1.052
CURVE 10	
298	0.012*
455	0.344
463	0.275
551	0.443
553	0.468
706	0.701
713	0.716
816	0.924*
818	0.939
913	1.117
CURVE 11*	
80	-0.218
195	-0.107
298	0.005
CURVE 12	
80	-0.264
195	-0.133
298	0.007
CURVE 13*	
80	-0.285
195	-0.159
298	0.012
CURVE 14	
293	0.000
373	0.110
473	0.251
573	0.409
673	0.574
773	0.744
873	0.925
973	1.113
1067	1.297*
CURVE 15	
293	0.000*
373	0.097
473	0.218
573	0.355
673	0.494
773	0.640
873	0.802
973	0.973
1063	1.145*
CURVE 16*	
293	0.000
373	0.090
473	0.205
573	0.339
673	0.488
773	0.628
873	0.781
973	0.922
1067	1.103
CURVE 17	
293	0.000*
373	0.089
473	0.203
573	0.336
673	0.468
773	0.608
873	0.758
973	0.917
1068	1.106*
CURVE 18	
293	0.000*
373	0.060
473	0.170
573	0.288
673	0.425
773	0.560
873	0.709
973	0.870
1067	1.028*

* Not shown in figure.
‡ Author's data for coefficient of thermal expansion have been integrated by TPRC to obtain $\Delta L/L_0$.

DATA TABLE 220. COEFFICIENT OF THERMAL LINEAR EXPANSION OF PALLADIUM-SILVER SYSTEM Pd-Ag

[Temperature, T, K; Coefficient of Expansion, α, 10^{-6} K^{-1}]

T	α
CURVE 1‡	
30	1.36
40	2.45
50	3.59
60	4.76
80	6.68
100	8.06
120	9.03
140	9.71
160	10.20
180	10.60
200	11.00
220	11.30
240	11.60
260	11.80
270	11.90
CURVE 2‡	
30	1.16
40	2.46
50	3.77
60	5.03
80	7.09
100	8.56
120	9.35
140	10.30
160	10.80
180	11.20
200	11.50
220	11.80
240	12.00
260	12.20
270	12.30
CURVE 3‡	
30	1.67
40	3.02
50	4.52
60	5.90
80	8.19
100	9.76
120	10.80
140	11.40
160	11.90
180	12.30
CURVE 3 (cont.)‡	
200	12.70
220	13.00
240	13.30
260	13.60
270	13.70
CURVE 4‡	
30	2.7
40	5.0
50	7.3
60	9.0
80	11.5
100	13.3
120	14.5
140	15.2
160	15.7
180	16.1
200	16.5*
220	16.8*
240	17.1*
260	17.4*
270	17.6*
293	18.1*
CURVE 5‡	
30	2.1
40	4.0
50	5.9
60	7.5
80	10.0
100	11.5
120	12.7
140	13.4
160	14.0
180	14.4
200	14.7
220	15.0
240	15.2
260	15.4
270	15.5
293	15.8
CURVE 14*	
373	13.77
473	14.58
573	15.51
673	16.34
773	17.22
873	18.06
973	18.91
1073	19.73
CURVE 15*	
373	11.31
473	12.30
573	13.32
673	14.30
773	15.34
873	16.39
973	17.35
1073	18.36
CURVE 16*	
373	11.81
473	12.51
573	13.21
673	13.87
773	14.50
873	15.20
973	15.84
1073	16.52
CURVE 17*	
373	10.26
473	11.14
573	11.95
673	12.81
773	13.64
873	14.47
973	15.33
1073	16.13
CURVE 18*	
373	11.14
473	11.87
CURVE 18 (cont.)*	
573	12.62
673	13.33
773	14.03
873	14.76
973	15.45
1073	16.18

* Not shown in figure.
‡ Author's data for coefficient of thermal expansion have been integrated by TPRC to obtain $\Delta L/L_0$.

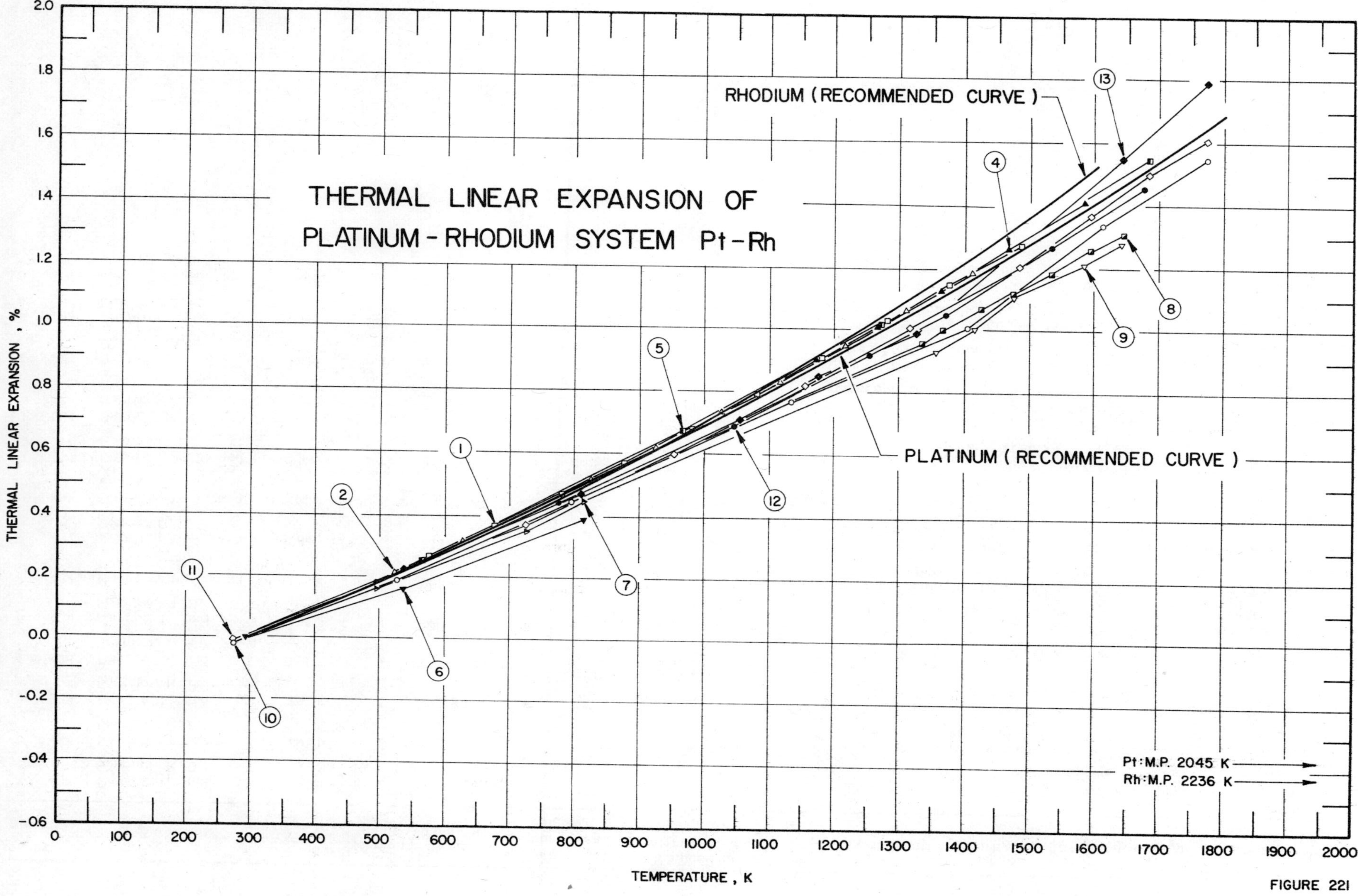
THERMAL LINEAR EXPANSION OF
PLATINUM - RHODIUM SYSTEM Pt-Rh
RHODIUM (RECOMMENDED CURVE)
PLATINUM (RECOMMENDED CURVE)
1
2
4
5
6
7
8
9
10
11
12
13
Pt: M.P. 2045 K
Rh: M.P. 2236 K
THERMAL LINEAR EXPANSION, %
TEMPERATURE, K
2.0
1.8
1.6
1.4
1.2
1.0
0.8
0.6
0.4
0.2
0.0
-0.2
-0.4
-0.6
0
100
200
300
400
500
600
700
800
900
1000
1100
1200
1300
1400
1500
1600
1700
1800
1900
2000

FIGURE 221

SPECIFICATION TABLE 221. THERMAL LINEAR EXPANSION OF PLATINUM-RHODIUM SYSTEM Pt-Rh

Cur. No.	Ref. No.	Author(s)	Year	Method Used	Temp. Range, K	Name and Specimen Designation	Composition (weight percent) Pt	Rh	Composition (continued), Specifications, and Remarks
1	436	Day, A.L. and Sosman, R.B.	1910	T	575-1486		80	20	Bar specimen 500 mm x 6 mm diameter; zero-point correction of -0.018% determined by graphical extrapolation.
2	436	Day, A.L. and Sosman, R.B.	1910	T	522-1410		80	20	The above specimen; expansion measured on second test; zero-point correction of -0.018% determined by graphical extrapolation.
3*	436	Day, A.L. and Sosman, R.B.	1910	T	573-1480		80	20	The above specimen; expansion measured on third test; zero-point correction of -0.018% determined by graphical extrapolation.
4	436	Day, A.L. and Sosman, R.B.	1910	T	1171-1583		80	20	The above specimen; expansion measured on fourth test; zero-point correction of -0.022% determined by graphical extrapolation.
5	436	Day, A.L. and Sosman, R.B.	1910	T	566-1686		80	20	Similar to the above specimen; expansion measured a year later; zero-point correction of -0.018% determined by graphical extrapolation.
6	437	Engelke, W.T. and Pears, C.D.	1962	L	293-814		60	40	Specimen annealed 30 min. at 1800 F; hollow cylinder 3.140 in. long, 0.040 in. thick walls.
7	437	Engelke, W.T. and Pears, C.D.	1962	L	814-293		60	40	The above specimen; expansion measured with decreasing temperature; zero-point correction is 0.04%.
8	437	Engelke, W.T. and Pears, C.D.	1962	L	293-1647	Run No. 1410-E1	96	4	Similar to the above specimen.
9	437	Engelke, W.T. and Pears, C.D.	1962	L	1647-293	Run No. 1410-E1	96	4	Similar to the above specimen; expansion measured with decreasing temperature; zero-point correction is -0.03%.
10	554	Barter, B. and Darling, A.S.	1960	T	273-1773		90	10	Specimen was in the form of 0.125 in. diameter drawn rod; measurements carried out in air; zero-point correction of -0.45% determined by graphical interpolation.
11	554	Barter, B. and Darling, A.S.	1960	T	273-1772		80	20	Similar to the above specimen; zero-point correction of -1.04% determined by graphical interpolation.
12	554	Barter, B. and Darling, A.S.	1960	T	273-1773		70	30	Similar to the above specimen; zero-point correction of -1.22% determined by graphical interpolation.
13	554	Barter, B. and Darling, A.S.	1960	T	273-1773		5	95	Similar to the above specimen; zero-point correction of -1.43% determined by graphical interpolation.

* Not shown in figure.

DATA TABLE 221. THERMAL LINEAR EXPANSION OF PLATINUM-RHODIUM SYSTEM Pt-Rh

[Temperature, T, K; Linear Expansion, $\Delta L/L_0$, %]

T	$\Delta L/L_0$
CURVE 1	
575	0.263
678	0.364
780	0.469
878	0.572
980	0.682
1080	0.795
1180	0.910
1280	1.030
1377	1.148
1486	1.276
CURVE 2	
522	0.213
626	0.315
725	0.414
823	0.516
922	0.620
1025	0.733
1116	0.834
1214	0.947
1310	1.063
1410	1.184
CURVE 3*	
573	0.259
675	0.362
778	0.468
878	0.575
979	0.684
1078	0.796
1179	0.911
1278	1.028
1376	1.148
1480	1.275
CURVE 4	
1171	0.904
1266	1.014
1361	1.130
1469	1.260
1583	1.411

T	$\Delta L/L_0$
CURVE 5	
566	0.252
969	0.672
1271	1.020
1276	1.022*
1375	1.144*
1469	1.264*
1582	1.413*
1686	1.548
CURVE 6	
293	0.000
537	0.160
814	0.390
CURVE 7	
814	0.43
725	0.35
493	0.16
293	0.00*
CURVE 8	
293	0.00*
1333	0.96
1366	1.00
1425	1.07
1478	1.12
1533	1.18
1597	1.26
1647	1.31
CURVE 9	
1647	1.28
1589	1.21
1475	1.11
1416	1.00
1355	0.93
293	0.00*
CURVE 10	
273	-0.02
529	0.19
787	0.44
1133	0.77

T	$\Delta L/L_0$
CURVE 10 (cont.)	
1402	1.05
1614	1.34
1773	1.55
CURVE 11	
273	-0.017
293	0.00*
524	0.19*
722	0.37
952	0.60
1156	0.82
1318	1.01
1482	1.21
1595	1.37
1684	1.50
1772	1.61
CURVE 12	
273	-0.01*
293	0.00*
499	0.18
779	0.44
1044	0.69
1251	0.92
1370	1.05
1538	1.27
1674	1.46
1773	1.61*
CURVE 13	
273	-0.01*
293	0.00
539	0.22
810	0.47
1053	0.71
1177	0.85
1327	0.99
1488	1.27
1647	1.55
1773	1.79

* Not shown in figure.

FIGURE AND TABLE NO. 222R. PROVISIONAL VALUES FOR THERMAL LINEAR EXPANSION OF PLATINUM-RUTHENIUM SYSTEM Pt-Ru

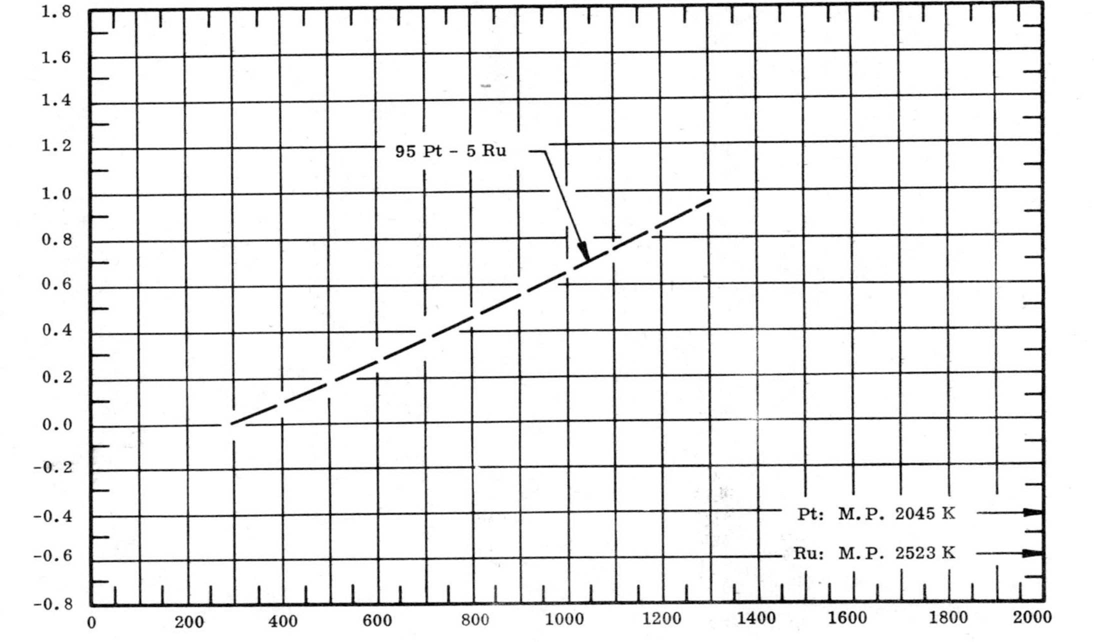

PROVISIONAL VALUES

[Temperature, T, K; Linear Expansion, $\Delta L/L_0$, %; α, K^{-1}]

T	(95 Pt-5 Ru) $\Delta L/L_0$	$\alpha \times 10^6$
293	0.000	8.6
400	0.093	8.7
500	0.182	8.9
600	0.271	9.0
700	0.361	9.2
800	0.453	9.3
900	0.546	9.4
1000	0.642	9.8
1200	0.842	10.3
1300	0.949	10.9

REMARKS

(95 Pt-5 Ru): The tabulated values for well-annealed alloy are considered accurate to within ± 7% over the entire temperature range. These values can be represented approximately by the following equation:

$$\Delta L/L_0 = -0.261 + 9.021 \times 10^{-4}\, T - 6.843 \times 10^{-8}\, T^2 + 6.950 \times 10^{-11}\, T^3$$

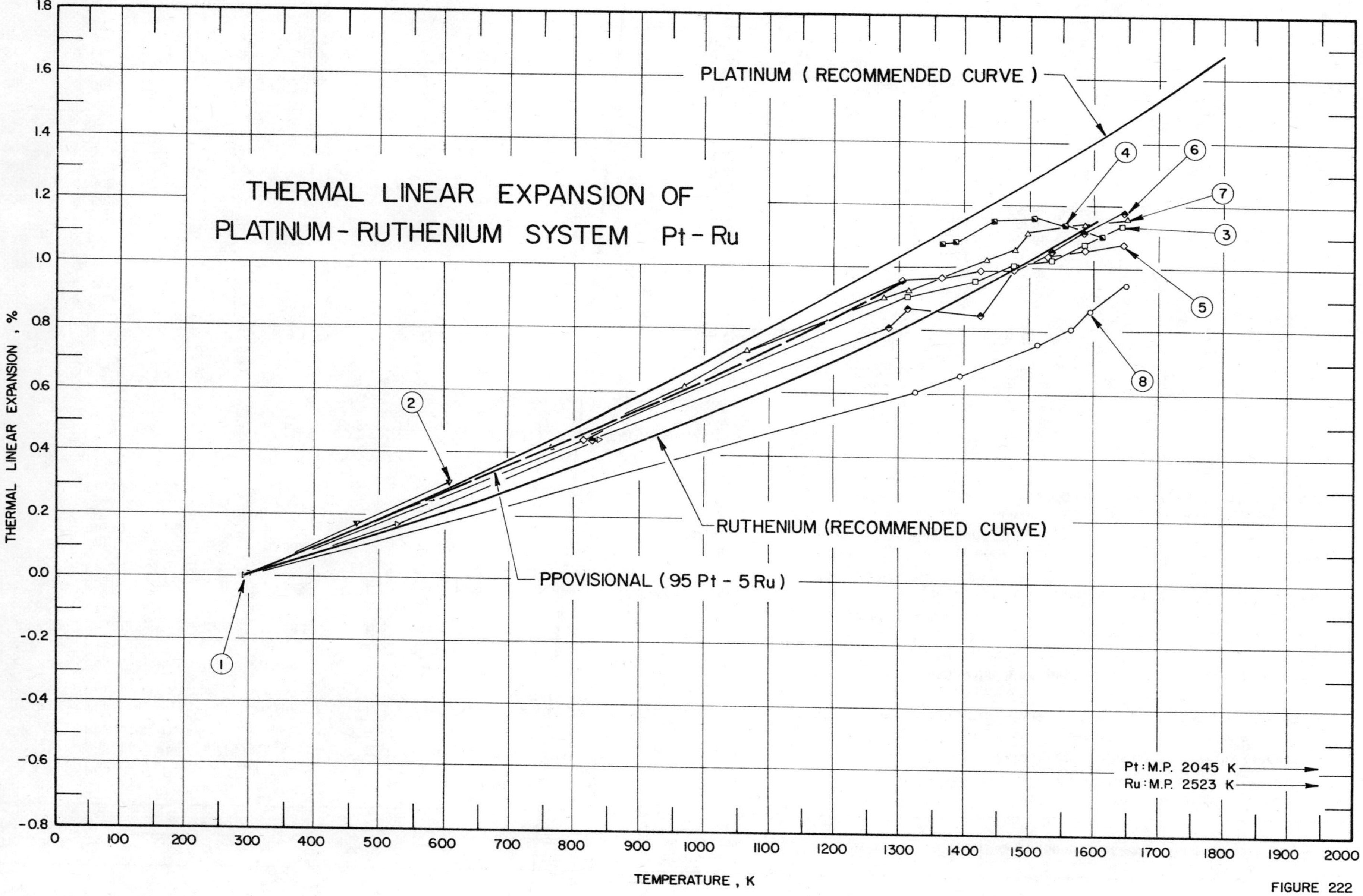

THERMAL LINEAR EXPANSION OF
PLATINUM - RUTHENIUM SYSTEM Pt - Ru
PLATINUM (RECOMMENDED CURVE)
RUTHENIUM (RECOMMENDED CURVE)
PPOVISIONAL (95 Pt - 5 Ru)
Pt : M.P. 2045 K
Ru : M.P. 2523 K
THERMAL LINEAR EXPANSION , %
TEMPERATURE , K
1.8
1.6
1.4
1.2
1.0
0.8
0.6
0.4
0.2
0.0
-0.2
-0.4
-0.6
-0.8
0
100
200
300
400
500
600
700
800
900
1000
1100
1200
1300
1400
1500
1600
1700
1800
1900
2000
1
2
3
4
5
6
7
8

FIGURE 222

SPECIFICATION TABLE 222. THERMAL LINEAR EXPANSION OF PLATINUM-RUBIDIUM SYSTEM Pt-Ru

Cur. No.	Ref. No.	Author(s)	Year	Method Used	Temp. Range, K	Name and Specimen Designation	Composition (weight percent) Pt	Ru	Composition (continued), Specifications, and Remarks
1	437	Engelke, W.T. and Pears, C.D.	1962	L	293-839		95	5	Specimen annealed 30 min at 1033 K; hollow cylinders 3.011 in. long, 0.040 in. thick walls; expansion measured with increasing temperature.
2	437	Engelke, W.T. and Pears, C.D.	1962	L	839-304		95	5	The above specimen; expansion measured with decreasing temperature; zero-point correction of 0.052% determined by graphical interpolation.
3	437	Engelke, W.T. and Pears, C.D.	1962	L	293-1644	Run No. 1409-E1	95	5	The above specimen; expansion measured with increasing temperature.
4	437	Engelke, W.T. and Pears, C.D.	1962	L	1614-1366	Run No. 1409-E1	95	5	The above specimen; expansion measured with decreasing temperature.
5	437	Engelke, W.T. and Pears, C.D.	1962	L	293-1647	Run No. 1409-E2	95	5	The above specimen except 3.015 in. long; expansion measured with increasing temperature.
6	437	Engelke, W.T. and Pears, C.D.	1962	L	1647-828	Run No. 1409-E2	95	5	The above specimen; expansion measured with decreasing temperature; zero-point correction is 0.10%.
7	437	Engelke, W.T. and Pears, C.D.	1962	L	293-1650	Run No. 1409-E3	95	5	Similar to the above specimen except 3.000 in. long; expansion measured with increasing temperature.
8	437	Engelke, W.T. and Pears, C.D.	1962	L	1650-293	Run No. 1409-E3	95	5	The above specimen; expansion measured with decreasing temperature; zero-point correction is -0.21%; authors feel that the data from this run is the most reliable.

DATA TABLE 222. THERMAL LINEAR EXPANSION OF PLATINUM-RUBIDIUM SYSTEM Pt-Ru

[Temperature, T, K; Linear Expansion, $\Delta L/L_0$, %]

T	$\Delta L/L_0$
CURVE 1	
293	0.000
529	0.170
839	0.450
CURVE 2	
839	0.502
609	0.302
465	0.172
304	0.012
CURVE 3	
293	0.000*
1311	0.910
1419	0.96
1478	1.01
CURVE 3 (cont.)	
1533	1.03
1589	1.08
1644	1.14
CURVE 4	
1614	1.11
1558	1.14
1503	1.16
1441	1.15
1389	1.09
1366	1.08
CURVE 5	
293	0.00*
815	0.45
CURVE 5 (cont.)	
1302	0.96
1366	0.97
1425	0.90
1478	1.00
1530	1.04
1589	1.06
1647	1.08
CURVE 6	
1647	1.18
1586	1.12
1536	1.06
1472	0.99
1425	0.85
1311	0.87
1283	0.81
CURVE 6 (cont.)	
828	0.44
CURVE 7	
293	0.00*
574	0.25
764	0.42
970	0.62
1065	0.73
1278	0.90
1311	0.93
1366	0.97*
1436	1.03
1478	1.06
1494	1.11
1589	1.14
1650	1.16
CURVE 8	
1650	0.95
1594	0.87
1566	0.81
1511	0.76
1394	0.66
1325	0.61
293	0.00*

* Not shown in figure.

FIGURE AND TABLE NO. 223R. PROVISIONAL VALUES FOR THERMAL LINEAR EXPANSION OF RHENIUM-TUNGSTEN SYSTEM Re-W

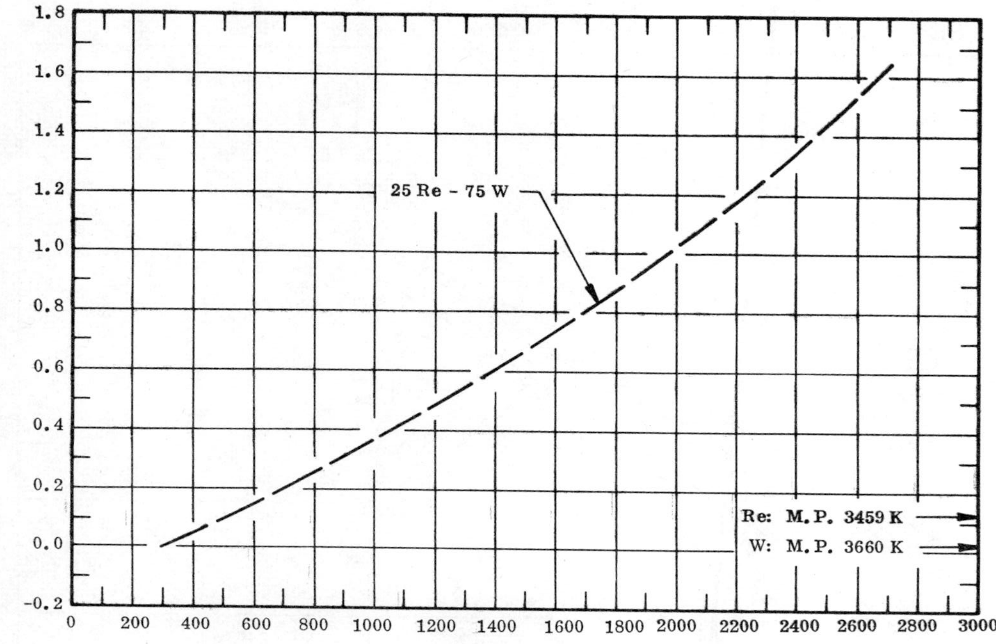

PROVISIONAL VALUES

[Temperature, T, K; Linear Expansion, $\Delta L/L_0$, %; α, K^{-1}]

T	(25 Re-75 W) $\Delta L/L_0$	$\alpha \times 10^6$
293	0.000	4.9
400	0.053	5.0
500	0.102	5.0
600	0.153	5.1
700	0.205	5.2
800	0.258	5.3
900	0.312	5.5
1000	0.368	5.7
1200	0.485	6.1
1400	0.609	6.4
1600	0.742	6.9
1800	0.883	7.2
2000	1.034	7.8
2200	1.196	8.4
2400	1.369	9.0
2600	1.555	9.5
2700	1.654	10.5

REMARKS

(25 Re-75 W): The tabulated values for well-annealed alloy are considered accurate to within ±7% over the entire temperature range. These values can be represented approximately by the following equation:

$$\Delta L/L_0 = -0.135 + 4.568 \times 10^{-4}\, T + 2.960 \times 10^{-8}\, T^2 + 1.722 \times 10^{-11}\, T^3$$

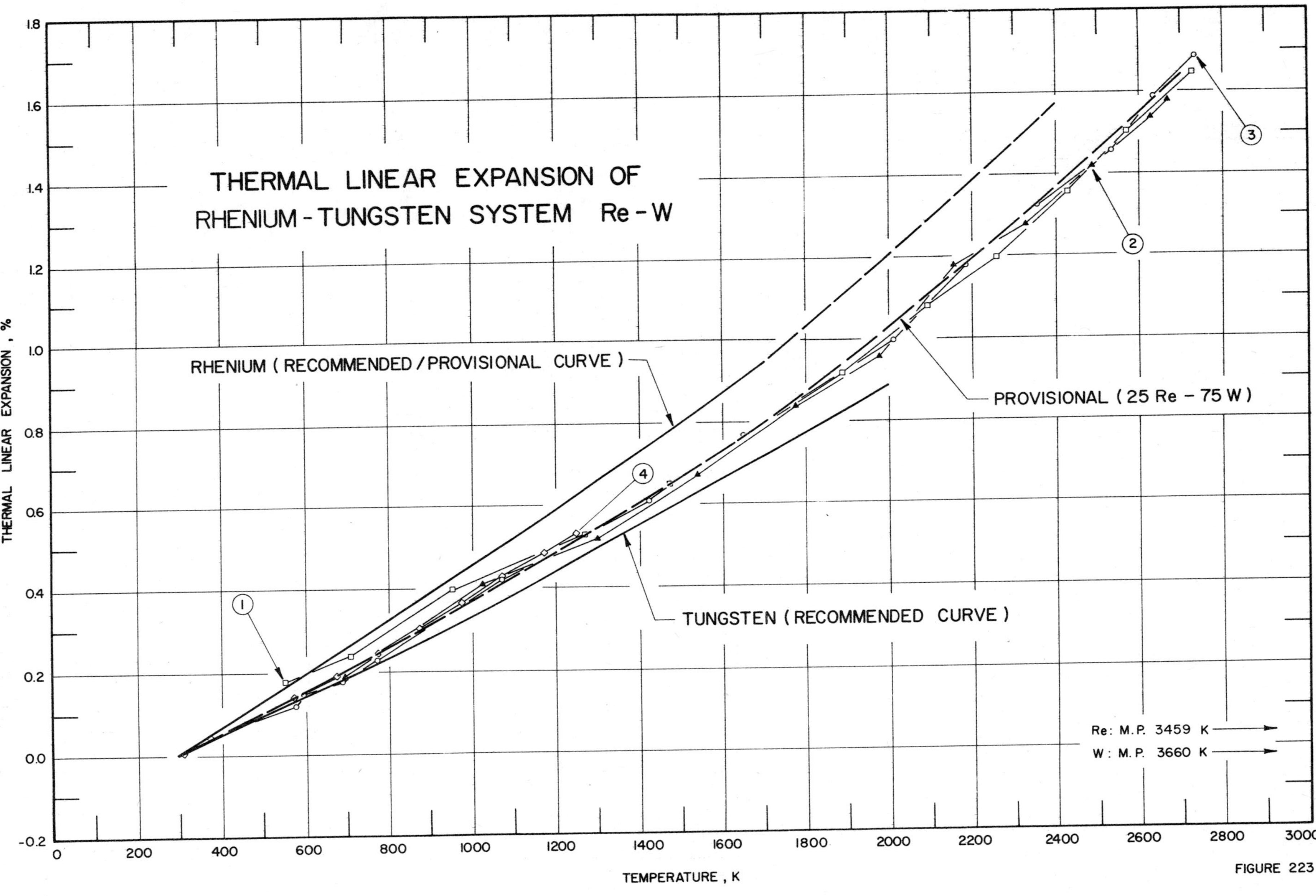

THERMAL LINEAR EXPANSION OF
RHENIUM-TUNGSTEN SYSTEM Re-W
RHENIUM (RECOMMENDED / PROVISIONAL CURVE)
PROVISIONAL (25 Re – 75 W)
TUNGSTEN (RECOMMENDED CURVE)
Re : M.P. 3459 K
W : M.P. 3660 K
1
2
3
4
THERMAL LINEAR EXPANSION , %
1.8
1.6
1.4
1.2
1.0
0.8
0.6
0.4
0.2
0.0
-0.2
0
200
400
600
800
1000
1200
1400
1600
1800
2000
2200
2400
2600
2800
3000
TEMPERATURE , K

FIGURE 223

SPECIFICATION TABLE 223. THERMAL LINEAR EXPANSION OF RHENIUM-TUNGSTEN SYSTEM Re-W

Cur. No.	Ref. No.	Author(s)	Year	Method Used	Temp. Range, K	Name and Specimen Designation	Composition (weight percent) Re	W	Composition (continued), Specifications, and Remarks
1	210	Conway, J.B. and Losekamp, A.C.	1966	T	554-2721		25	75	Sheet 20 mil thick, formed by powder metallurgical techniques; expansion measured in helium atm on first run.
2	210	Conway, J.B. and Losekamp, A.C.	1966	T	691-2665		25	75	The above specimen; expansion measured on second run.
3	210	Conway, J.B. and Losekamp, A.C.	1966	T	455-2739		25	75	The above specimen; expansion measured on third run.
4	211	Yaggee, F.L. and Styles, J.W.	1965	L	304-1248		26	74	No details given; zero-point correction of 0.006% determined by graphical extrapolation.

DATA TABLE 223. THERMAL LINEAR EXPANSION OF RHENIUM-TUNGSTEN SYSTEM Re-W

[Temperature, T, K; Linear Expansion, $\Delta L/L_0$, %]

T	$\Delta L/L_0$
CURVE 1	
554	0.18
707	0.24
958	0.40
1265	0.53
1474	0.65
1884	0.92
2091	1.08
2258	1.20
2425	1.36
2569	1.51
2721	1.65
CURVE 2	
691	0.19
1027	0.41
1300	0.52
1540	0.67
1779	0.84
CURVE 2 (cont.)	
1979	0.96
2156	1.18
2321	1.28
2488	1.42
2625	1.54
2665	1.58
CURVE 3	
455	0.08
575	0.12
597	0.15
685	0.18
773	0.23
1073	0.42
1420	0.61
1653	0.77
2010	1.00
2181	1.18
CURVE 3 (cont.)	
2359	1.33
2534	1.46
2638	1.59
2739	1.69
CURVE 4	
304	0.006
373	0.041
473	0.088
574	0.140
673	0.192
772	0.248
871	0.304
973	0.365
1072	0.424
1172	0.487
1248	0.534

FIGURE AND TABLE NO. 224R. PROVISIONAL VALUES FOR THERMAL LINEAR EXPANSION OF SILVER-TIN SYSTEM Ag-Sn

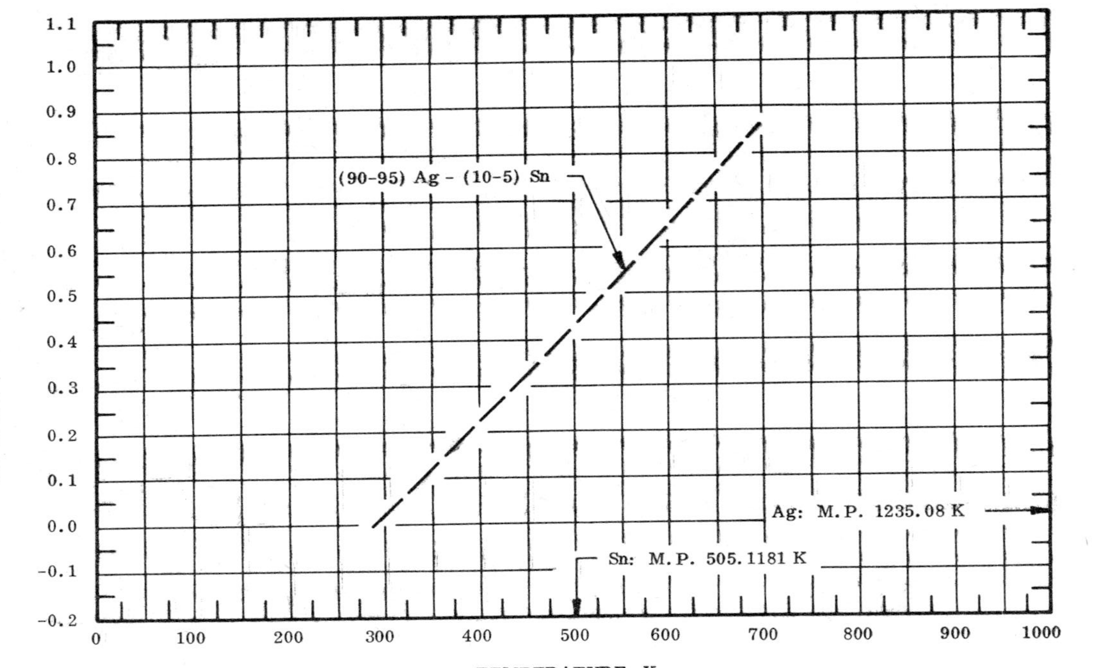

PROVISIONAL VALUES

[Temperature, T, K; Linear Expansion, $\Delta L/L_0$, %; α, K^{-1}]

	([90-95] Ag - [10-5] Sn)	
T	$\Delta L/L_0$	$\alpha \times 10^6$
293	0.000	20.0
400	0.217	20.3
500	0.424	21.2
600	0.641	22.3
700	0.868	22.9

REMARKS

([90-95] Ag - [10-5] Sn): The tabulated values for well-annealed alloy are considered accurate to within ±7% over the entire temperature range. These values can be represented approximately by the following equation:

$$\Delta L/L_0 = -0.585 + 2.042 \times 10^{-3} T - 2.862 \times 10^{-7} T^2 + 4.812 \times 10^{-10} T^3$$

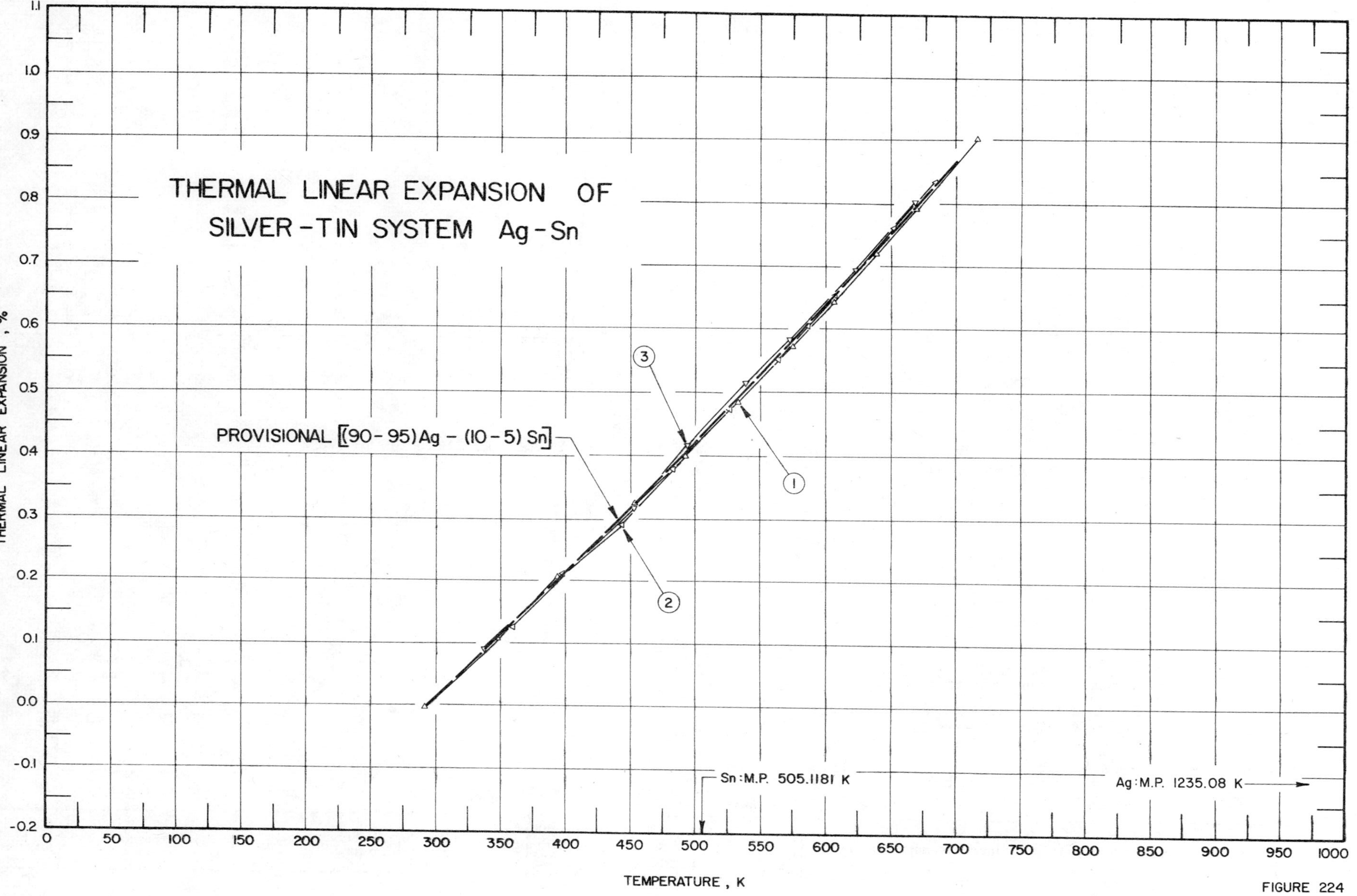

THERMAL LINEAR EXPANSION OF
SILVER - TIN SYSTEM Ag - Sn
PROVISIONAL [(90 - 95) Ag - (10 - 5) Sn]
3
1
2
Sn : M.P. 505.1181 K
Ag : M.P. 1235.08 K
THERMAL LINEAR EXPANSION , %
1.1
1.0
0.9
0.8
0.7
0.6
0.5
0.4
0.3
0.2
0.1
0.0
-0.1
-0.2
0
50
100
150
200
250
300
350
400
450
500
550
600
650
700
750
800
850
900
950
1000
TEMPERATURE , K

FIGURE 224

SPECIFICATION TABLE 224. THERMAL LINEAR EXPANSION OF SILVER-TIN SYSTEM Ag-Sn

Cur. No.	Ref. No.	Author(s)	Year	Method Used	Temp. Range, K	Name and Specimen Designation	Composition (weight percent) Ag	Sn	Composition (continued), Specifications, and Remarks
1	303	Owen, E.A. and Roberts, E.W.	1939	X	291-716		95.32	4.68	99.95 pure silver and 99.98 tin starting materials; melted together to form alloy; lattice parameter expansion measured; lattice parameter reported at 291 K is 4.0954 Å; 4.0956 Å at 293 K determined by graphical interpolation.
2	303	Owen, E.A. and Roberts, E.W.	1939	X	291-685		92.75	7.25	Similar to the above specimen; lattice parameter reported at 291 K is 4.1044 Å; 4.1046 Å at 293 K determined by graphical interpolation.
3	303	Owen, E.A. and Roberts, E.W.	1939	X	291-669		90.91	9.09	Similar to the above specimen; lattice parameter reported at 291 K is 4.1133 Å; 4.1135 Å at 293 K determined by graphical interpolation.

DATA TABLE 224. THERMAL LINEAR EXPANSION OF SILVER-TIN SYSTEM Ag-Sn

[Temperature, T, K; Linear Expansion, $\Delta L/L_0$, %]

T	$\Delta L/L_0$	T	$\Delta L/L_0$	T	$\Delta L/L_0$
CURVE 1		CURVE 2 (cont.)		CURVE 3 (cont.)	
291	-0.004	482	0.380	669	0.801
347	0.108	526	0.473		
394	0.206	563	0.551		
453	0.325	586	0.608		
491	0.400	607	0.659*		
533	0.486	653	0.763		
575	0.575	685	0.836		
606	0.645				
639	0.722	CURVE 3			
669	0.794				
716	0.905	291	-0.004*		
		336	0.091		
CURVE 2		395	0.203*		
		452	0.318		
291	-0.004*	493	0.416		
359	0.126	539	0.514		
398	0.212	573	0.584		
443	0.291	623	0.694		

* Not shown in figure.

FIGURE AND TABLE NO. 225R. PROVISIONAL VALUES FOR THERMAL LINEAR EXPANSION OF TANTALUM-TUNGSTEN SYSTEM Ta-W

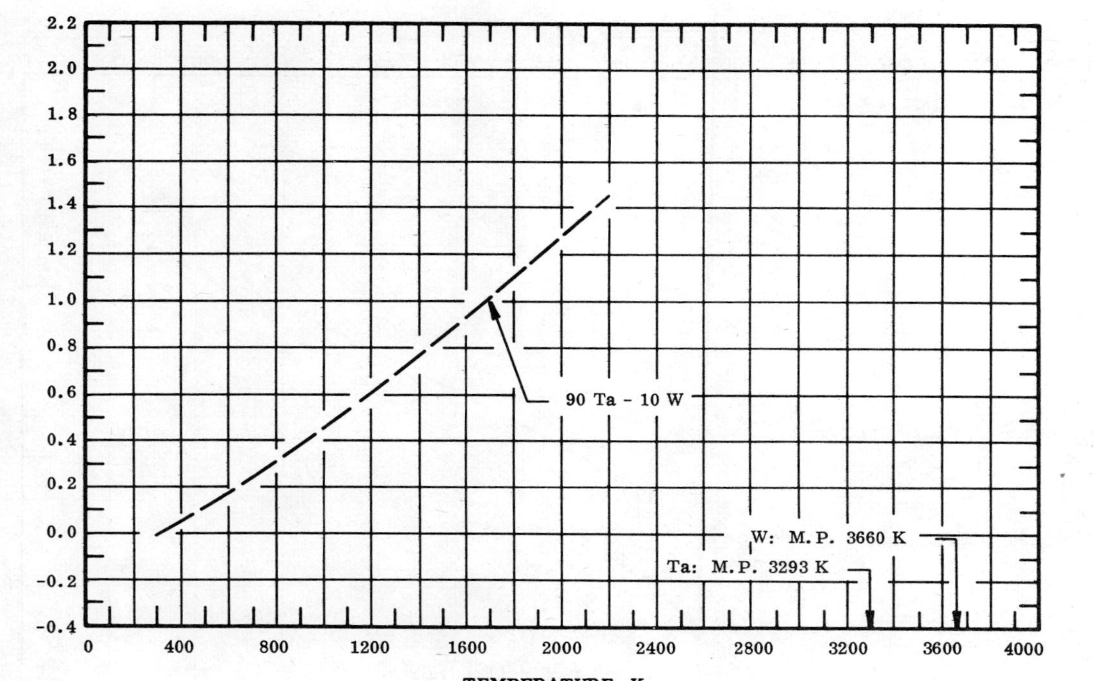

PROVISIONAL VALUES

[Temperature, T, K; Linear Expansion, $\Delta L/L_0$, %; α, K^{-1}]

T	(90 Ta-10 W) $\Delta L/L_0$	$\alpha \times 10^6$
293	0.000	5.9
400	0.064	6.0
500	0.125	6.2
600	0.188	6.4
700	0.254	6.6
800	0.321	6.8
900	0.390	7.0
1000	0.461	7.2
1200	0.610	7.6
1400	0.760	8.0
1600	0.930	8.4
1800	1.101	8.7
2000	1.279	9.1
2200	1.464	9.4

REMARKS

(90 Ta-10 W): The tabulated values for well-annealed alloy are considered accurate to within ±7%. The thermal linear expansion of this alloy is very close to the thermal linear expansion of pure tantalum. The tabulated values can be represented approximately by the following equation:

$$\Delta L/L_0 = -0.160 + 5.174 \times 10^{-4}\ T + 1.077 \times 10^{-7}\ T^2 \times 10^2 - 3.343 \times 10^{-12}\ T^3$$

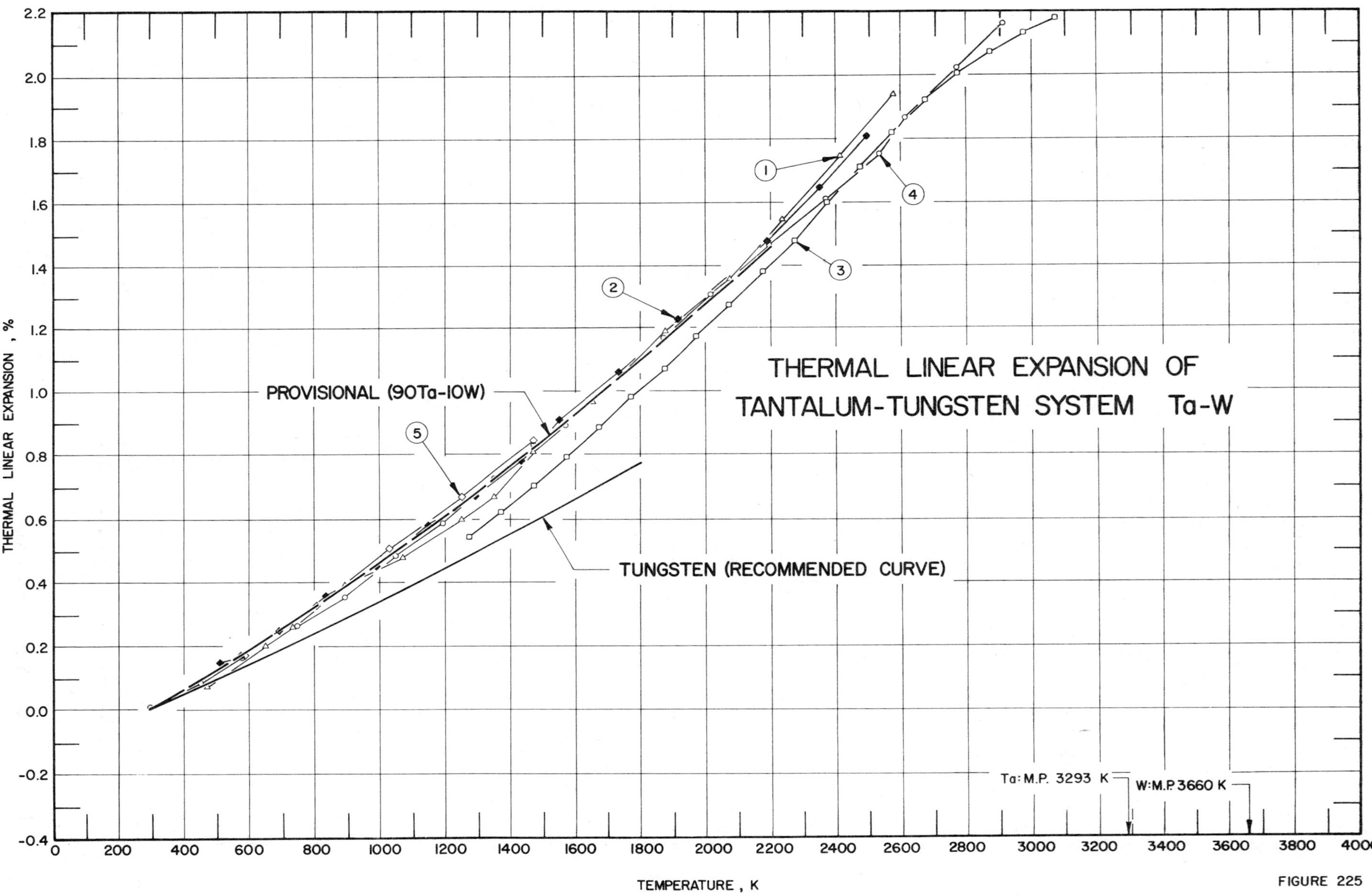
THERMAL LINEAR EXPANSION OF
TANTALUM-TUNGSTEN SYSTEM Ta-W
PROVISIONAL (90Ta-10W)
TUNGSTEN (RECOMMENDED CURVE)
1
2
3
4
5
Ta: M.P. 3293 K
W: M.P. 3660 K
THERMAL LINEAR EXPANSION, %
-0.4
-0.2
0.0
0.2
0.4
0.6
0.8
1.0
1.2
1.4
1.6
1.8
2.0
2.2
0
200
400
600
800
1000
1200
1400
1600
1800
2000
2200
2400
2600
2800
3000
3200
3400
3600
3800
4000
TEMPERATURE, K

FIGURE 225

SPECIFICATION TABLE 225. THERMAL LINEAR EXPANSION OF TANTALUM-TUNGSTEN SYSTEM Ta-W

Cur. No.	Ref. No.	Author(s)	Year	Method Used	Temp. Range, K	Name and Specimen Designation	Composition (weight percent) Ta	W	Composition (continued), Specifications, and Remarks
1	210	Conway, J. B. and Losekamp, A. C.	1966	T	473-2579		90	10	Arc cast sheet 20 mil thick; expansion measured in helium atm on first run.
2	210	Conway, J. B. and Losekamp, A. C.	1966	T	514-2499		90	10	The above specimen; expansion measured on second run.
3	453	Torti, M. L.	1960		1273-3173		90	10	No details given; zero-point correction is 0.2%.
4	425	Hedge, J. C., Kosteuko, C. and Lang, J. I.	1963	T	299-2980		90.34	9.50	0.087 Nb, 0.015 Mo, 0.02 Ti, 0.005 Fe, 0.02 Si, 0.001 C, 0.005 O_2, 0.003 N_2; obtained from Fansteel Metallurgical Corp.; specimen dimensions 1/2 in. diameter, 6 in. long; expansion measured in argon atm with heating rate of approx. 2.7 K per min; zero-point correction is 0.004%.
5	454	Powers, D. J.	1963	L	299-1475		90	10	Specimen placed in dilatometer which fitted into Pt wound tube furnace; the test equipment enclosed in vacuum chamber of pressure 0.5 in Hg; argon gas purge initiated when specimen temperature reached 360-420 K; reported error ±5%.
6*	621	Fitzer, E.	1972	L	293-1573	Participant # 5	9.45	Bal.	0.0044 O, 0.0014 N, 0.0014 C, <0.0025 Nb, sample arc cast from Norton Co. distributed by AFML, Wright Patterson AFB, annealed specimen from Heat Number 8437.
7*	621	Fitzer, E.	1972	L	293-1573	Participant # 8	9.45	Bal.	Similar to the above specimen.
8*	621	Fitzer, E.	1972	O	293-1673	Participant #46	9.45	Bal.	Similar to the above specimen.

* Not shown in figure.

DATA TABLE 225. THERMAL LINEAR EXPANSION OF TANTALUM-TUNGSTEN SYSTEM Ta-W

[Temperature, T, K; Linear Expansion, $\Delta L/L_0$, %]

T	$\Delta L/L_0$
CURVE 1	
473	0.07
643	0.20
738	0.26
893	0.39
1066	0.48
1249	0.60
1349	0.67
1465	0.81
1648	0.97
1879	1.19
2074	1.36
2239	1.55
2412	1.75
2579	1.94
CURVE 2	
514	0.15
575	0.17
684	0.25
837	0.36
988	0.45
1142	0.58
1295	0.67
1434	0.78
1554	0.91
1736	1.06
1916	1.23
2185	1.48
2342	1.65
2499	1.81
CURVE 3	
1273	0.54
1373	0.62
1473	0.71
1573	0.79
1673	0.88
1773	0.98
1873	1.07
1973	1.17
2073	1.27
2173	1.38
2273	1.48
2373	1.60
CURVE 3 (cont.)	
2473	1.71
2573	1.82
2673	1.92
2773	2.00*
2873	2.07*
2973	2.13*
3073	2.18*
3173	2.22*
CURVE 4	
299	0.004
449	0.088
580	0.165
745	0.261
897	0.357
1044	0.481
1191	0.588
1347	0.726
1472	0.835
1563	0.898
1735	1.063*
1868	1.171
2018	1.306
2198	1.475
2363	1.603
2528	1.757
2618	1.865
2761	2.020
2915	2.161
2980	2.243*
CURVE 5	
299	0.004*
588	0.171
807	0.323
1030	0.501
1250	0.674
1475	0.841
CURVE 6*	
293	0.000
373	0.060
473	0.124
CURVE 6 (cont.)*	
573	0.185
673	0.250
773	0.310
873	0.385
973	0.450
1073	0.521
1173	0.584
1273	0.655
1373	0.725
1473	0.792
1573	0.890
CURVE 7*	
293	0.000
473	0.129
573	0.188
673	0.257
773	0.325
873	0.388
973	0.454
1073	0.530
1173	0.604
1273	0.649
1373	0.720
1473	0.785
1573	0.870
CURVE 8*	
293	0.000
373	0.050
473	0.115
573	0.181
673	0.249
773	0.316
873	0.384
973	0.453
1073	0.522
1173	0.593
1273	0.668
1373	0.747
1473	0.832
1573	0.926
1673	1.032

* Not shown in figure.

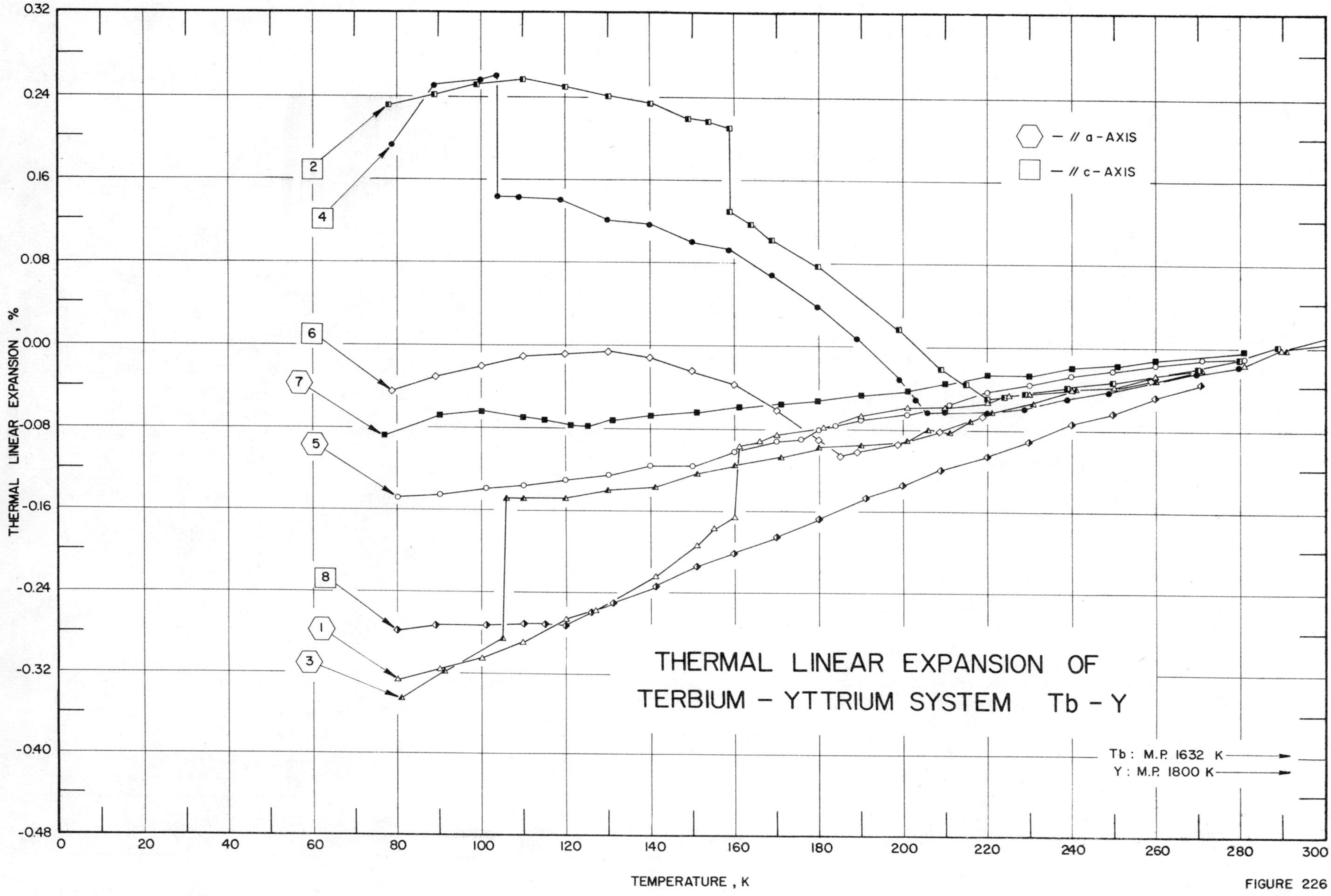
THERMAL LINEAR EXPANSION OF
TERBIUM - YTTRIUM SYSTEM Tb - Y
— // a - AXIS
— // c - AXIS
Tb : M.P. 1632 K
Y : M.P. 1800 K
THERMAL LINEAR EXPANSION, %
TEMPERATURE, K
0.32
0.24
0.16
0.08
0.00
-0.08
-0.16
-0.24
-0.32
-0.40
-0.48
0
20
40
60
80
100
120
140
160
180
200
220
240
260
280
300
1
2
3
4
5
6
7
8

FIGURE 226

SPECIFICATION TABLE 226. THERMAL LINEAR EXPANSION OF TERBIUM-YTTRIUM SYSTEM Tb-Y

Cur. No.	Ref. No.	Author(s)	Year	Method Used	Temp. Range, K	Name and Specimen Designation	Composition (weight percent) Tb	Y	Composition (continued), Specifications, and Remarks
1	317	Finkel', V.A. and Vorob'ev, V.V.	1968	X	80-300		95.00	5.00	99.5 Tb and 99.8 Y melted together in arc furnace at 0.5 to 0.6 atmosphere; specimen 10 x 20 x 2 mm cut from casting; annealed 10-50 hr in 2 x 10^{-6} mm Hg vacuum at 1473 K; expansion measured along a-axis; magnetic transition temperatures are T_c = 168 K and T_N = 213 K; lattice parameter reported at 300 K is 3.6071 Å; 3.6067 Å at 293 K determined by graphical interpolation.
2	317	Finkel', V.A. and Vorob'ev, V.V.	1968	X	78-300		95.00	5.00	The above specimen; expansion measured along c-axis; lattice parameter reported at 300 K is 5.6908 Å; 5.6907 Å at 293 K determined by graphical interpolation.
3	317	Finkel', V.A. and Vorob'ev, V.V.	1968	X	81-300		90.00	10.00	Similar to the above specimen; expansion measured along a-axis; magnetic transition temperatures are T_c = 100 K and T_N = 202 K; lattice parameter reported at 300 K is 3.6139 Å; 3.6136 Å at 293 K determined by graphical interpolation.
4	317	Finkel', V.A. and Vorob'ev, V.V.	1968	X	79-300		90.00	10.00	The above specimen; expansion measured along c-axis; lattice parameter reported at 300 K is 5.7024 Å; 5.7021 Å at 293 K determined by graphical interpolation.
5	317	Finkel', V.A. and Vorob'ev, V.V.	1968	X	80-300		80.00	20.00	Similar to the above specimen; expansion measured along a-axis; magnetic transition temperature is T_N = 183 K; lattice parameter reported at 300 K is 3.6209 Å; 3.6208 Å at 293 K determined by graphical interpolation.
6	317	Finkel', V.A. and Vorob'ev, V.V.	1968	X	79-300		80.00	20.00	The above specimen; expansion measured along c-axis; lattice parameter reported at 300 K is 5.7104 Å; 5.7101 Å at 293 K determined by graphical interpolation.
7	317	Finkel', V.A. and Vorob'ev, V.V.	1968	X	77-300		50.01	49.99	Similar to the above specimen; expansion measured along a-axis; magnetic transition temperature is T_N = 120 K; lattice parameter reported at 300 K is 3.6370 · 3.63693 Å at 293 K determined by graphical interpolation.
8	317	Finkel', V.A. and Vorob'ev, V.V.	1968	X	80-300		50.01	49.99	The above specimen; expansion measured along c-axis; lattice parameter reported at 300 K is 5.7093 Å; 5.9709 Å at 293 K determined by graphical interpolation.

DATA TABLE 226. THERMAL LINEAR EXPANSION OF TERBIUM-YTTRIUM SYSTEM Tb-Y

[Temperature, T, K; Linear Expansion, $\Delta L/L_0$, %]

T	$\Delta L/L_0$
CURVE 1	
80	-0.327
90	-0.316
100	-0.305
110	-0.289
120	-0.267
127	-0.258
141	-0.225
151	-0.194
155	-0.178
160	-0.167
161	-0.097
166	-0.092*
170	-0.086
181	-0.096
190	-0.067
201	-0.059
210	-0.059
220	-0.053
225	-0.047
230	-0.045
240	-0.042
250	-0.039
260	-0.028
270	-0.023
281	-0.017
290	-0.003
300	0.009
CURVE 2	
78	0.232
89	0.243
99	0.252
110	0.257
120	0.250
130	0.243
140	0.236
149	0.220
154	0.217
159	0.210*
159	0.129
164	0.118
169	0.102
180	0.078
199	0.016
209	-0.021
215	-0.036
CURVE 2 (cont.)	
220	-0.050
224	-0.049*
229	-0.045
239	-0.040
250	-0.035
260	-0.029
270	-0.021
280	-0.012
289	-0.001
300	0.002
CURVE 3	
81	-0.346
91	-0.318
105	-0.285
106	-0.149*
110	-0.149
120	-0.149
130	-0.141
141	-0.138
151	-0.124
160	-0.116
171	-0.108
180	-0.099
190	-0.096
201	-0.091
206	-0.080
211	-0.083
216	-0.072
221	-0.063
231	-0.055
241	-0.041
250	-0.041
260	-0.033
271	-0.022
280	-0.011*
291	-0.002
300	0.009*
CURVE 4	
79	0.194
89	0.252
100	0.257*
104	0.261
104	0.143
CURVE 4 (cont.)	
109	0.143
119	0.141
130	0.121
140	0.117
150	0.100
159	0.092
169	0.068
180	0.038
189	0.008
199	-0.032
203	-0.052
206	-0.063
210	-0.063
220	-0.063
229	-0.060
239	-0.051
249	-0.044
260	-0.035
270	-0.025
280	-0.018
290	-0.002*
300	0.005
CURVE 5	
80	-0.148
90	-0.145
101	-0.139
110	-0.136
120	-0.131
130	-0.126
140	-0.117
150	-0.117
160	-0.103*
170	-0.092
176	-0.090
180	-0.081
184	-0.076
190	-0.070
201	-0.065
211	-0.056
220	-0.043
230	-0.037
240	-0.028
250	-0.023
260	-0.018
271	-0.012
CURVE 5 (cont.)	
281	-0.010
291	-0.001*
300	0.004
CURVE 6	
79	-0.045
89	-0.031
100	-0.020
110	-0.011
120	-0.008
130	-0.006
140	-0.011
150	-0.024
160	-0.038
170	-0.062
180	-0.090
183	-0.106
189	-0.102
199	-0.094
209	-0.081
219	-0.067
229	-0.062
239	-0.052
249	-0.045*
259	-0.031
270	-0.057
280	-0.023
290	-0.003*
300	0.006*
CURVE 7	
77	-0.088
90	-0.067
100	-0.064
110	-0.069
115	-0.072
121	-0.075
125	-0.075
131	-0.072
140	-0.067
151	-0.064
161	-0.058
171	-0.056
180	-0.053
190	-0.047
CURVE 7 (cont.)	
201	-0.042
210	-0.036
220	-0.028
230	-0.028
240	-0.020
251	-0.017
260	-0.012
271	-0.012*
281	-0.003
291	-0.001*
300	0.002*
CURVE 8	
80	-0.278
89	-0.273
101	-0.273
110	-0.271
115	-0.271
120	-0.273*
126	-0.259
131	-0.250
141	-0.234
151	-0.215
160	-0.201
170	-0.185
180	-0.168
191	-0.147
200	-0.134
209	-0.119
220	-0.106
230	-0.091
240	-0.074
250	-0.064
260	-0.049
271	-0.036
281	-0.019
290	-0.003*
300	0.007*

* Not shown in figure.

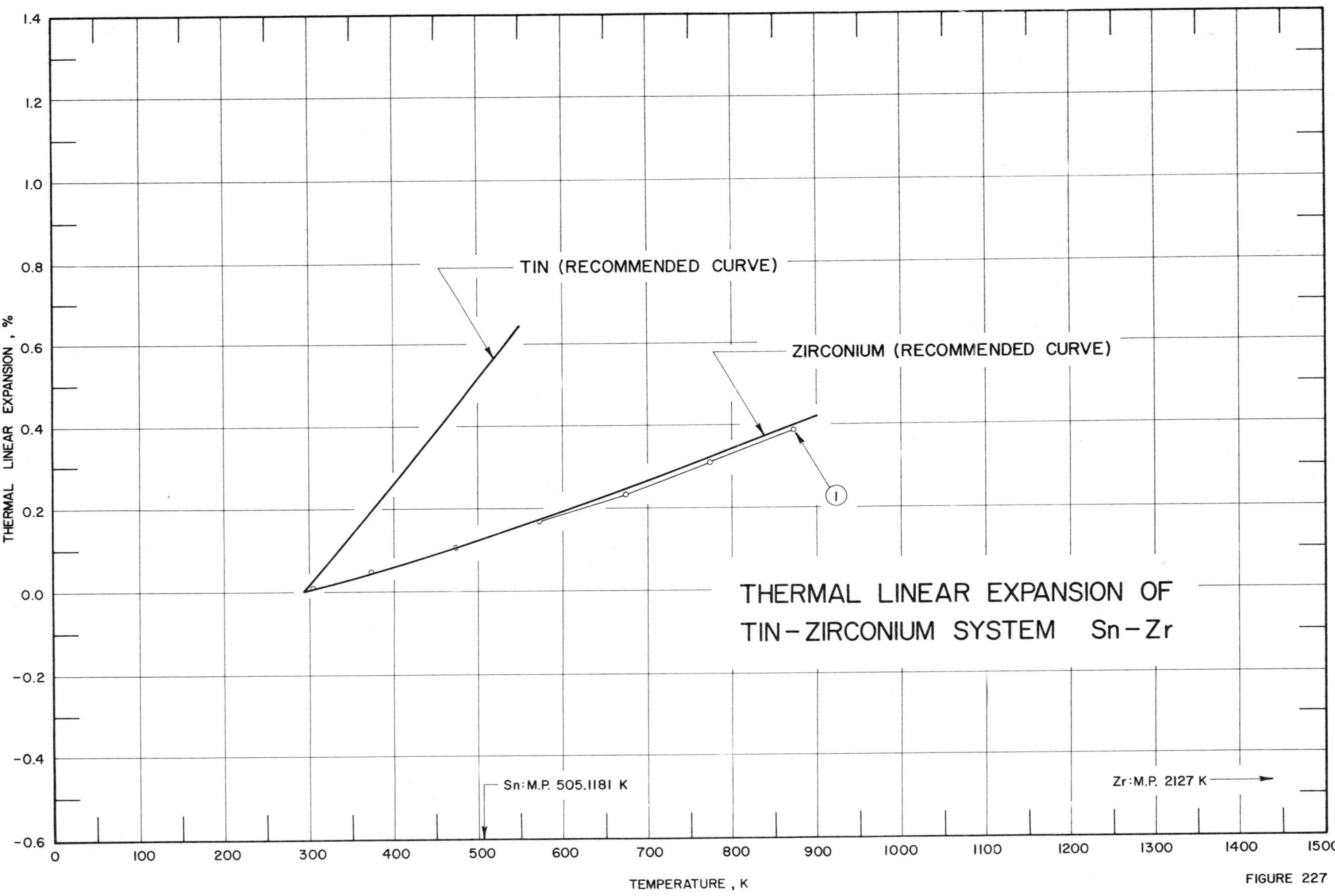

TIN (RECOMMENDED CURVE)
ZIRCONIUM (RECOMMENDED CURVE)
1
THERMAL LINEAR EXPANSION OF
TIN-ZIRCONIUM SYSTEM Sn-Zr
Sn: M.P. 505.1181 K
Zr: M.P. 2127 K
THERMAL LINEAR EXPANSION, %
1.4
1.2
1.0
0.8
0.6
0.4
0.2
0.0
-0.2
-0.4
-0.6
0
100
200
300
400
500
600
700
800
900
1000
1100
1200
1300
1400
1500
TEMPERATURE, K

FIGURE 227

SPECIFICATION TABLE 227. THERMAL LINEAR EXPANSION OF TIN-ZIRCONIUM SYSTEM Sn-Zr

Cur. No.	Ref. No.	Author(s)	Year	Method Used	Temp. Range, K	Name and Specimen Designation	Composition (weight percent) Sn	Zr	Composition (continued), Specifications, and Remarks
1	450	Mehan, R.L. and Cutler, G.L.	1958	L	303-873	Zircaloy-2	1.53	98.018	0.155 Fe, 0.0883 O, 0.0863 Cr, 0.0493 Ni, 0.04 C, 0.006 Al, 0.004 N, <0.004 Hf, 0.0038 W, 0.0029 Cu, 0.0028 Pb, 0.0028 Si, <0.002 Ti, 0.0009 Mn, 0.0005 Mo; 25% cold-rolled and annealed plate; material from KAPL; expansion in the three principle directions of rolling measured; possible difference for transverse specimen observed from 773 K up; 10^{-5} mm Hg vacuum maintained during each measurement; reported values avgd by author from three directions.

DATA TABLE 227. THERMAL LINEAR EXPANSION OF TIN-ZIRCONIUM SYSTEM Sn-Zr

[Temperature, T, K; Linear Expansion, $\Delta L/L_0$, %]

T	$\Delta L/L_0$
CURVE 1	
303	0.005
373	0.046
473	0.102
573	0.171
673	0.239
773	0.310
873	0.382

SPECIFICATION TABLE 288. THERMAL LINEAR EXPANSION OF TITANIUM-TUNGSTEN SYSTEM Ti-W

Cur. No.	Ref. No.	Author(s)	Year	Method Used	Temp. Range, K	Name and Specimen Designation	Composition (weight percent) Ti	W	Composition (continued), Specifications, and Remarks
1*	395	Stetson, A.R. and Metcalfe, A.G.	1967	L	293-922		18.01	81.99	Specimen approximately 1 to 1.5 in. long by 0.25 in. square; prepared by arc melting the specific composition in nonconsumable arc melter, then annealed at 1178 K for 16 hr; specimen eloxed from arc melt; expansion measured in argon.
2*	395	Stetson, A.R. and Metcalfe, A.G.	1967	L	293-922		53.11	46.89	Similar to the above specimen.

DATA TABLE 228. THERMAL LINEAR EXPANSION OF TITANIUM-TUNGSTEN SYSTEM Ti-W

[Temperature, T, K; Linear Expansion, $\Delta L/L_0$, %]

T	$\Delta L/L_0$
CURVE 1*	
293	0.000
478	0.119
922	0.589
CURVE 2*	
293	0.000
478	0.128
922	0.595

* No figure given.

FIGURE AND TABLE NO. 229R. PROVISIONAL VALUES FOR THERMAL LINEAR EXPANSION OF TITANIUM-VANADIUM SYSTEM Ti-V

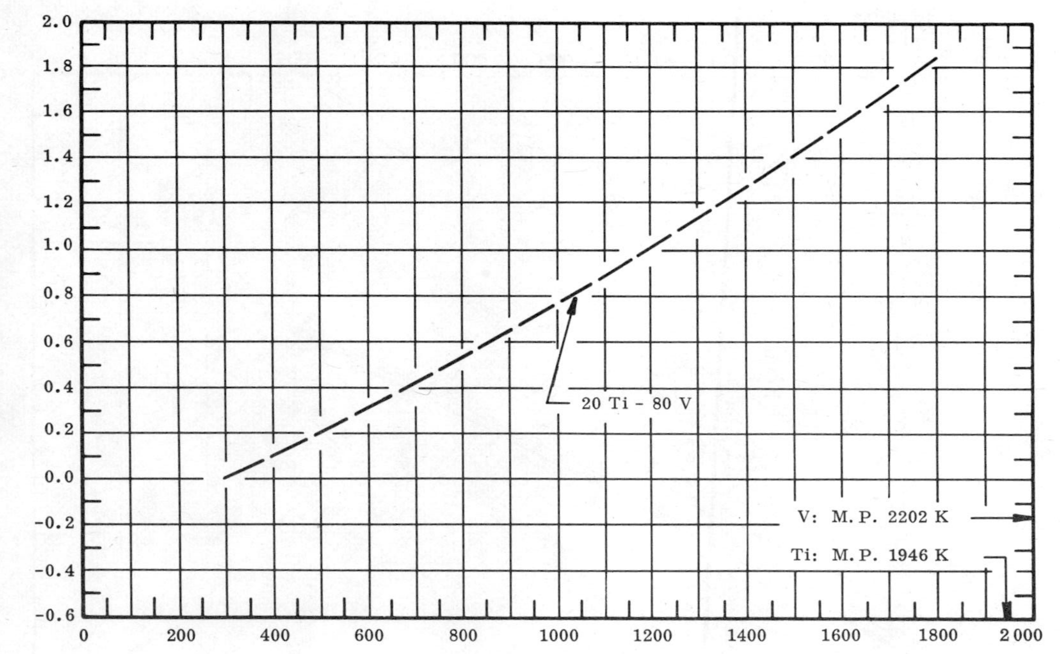

PROVISIONAL VALUES

[Temperature, T, K; Linear Expansion, $\Delta L/L_0$, %; α, K^{-1}]

T	(20 Ti-80 V) $\Delta L/L_0$	$\alpha \times 10^6$
293	0.000	10.2
400	0.111	10.4
500	0.216	10.6
600	0.323	10.8
700	0.432	11.0
800	0.543	11.3
900	0.657	11.5
1000	0.774	11.8
1200	1.017	12.5
1400	1.276	13.4
1600	1.553	14.4
1800	1.854	15.7

REMARKS

(20 Ti-80 V): The tabulated values for well-annealed alloy are considered accurate to within ±7% over the entire temperature range. These values can be represented approximately by the following equation:

$$\Delta L/L_0 = -0.300 + 1.019 \times 10^{-3} T - 6.916 \times 10^{-10} T^2 + 5.493 \times 10^{-11} T^3$$

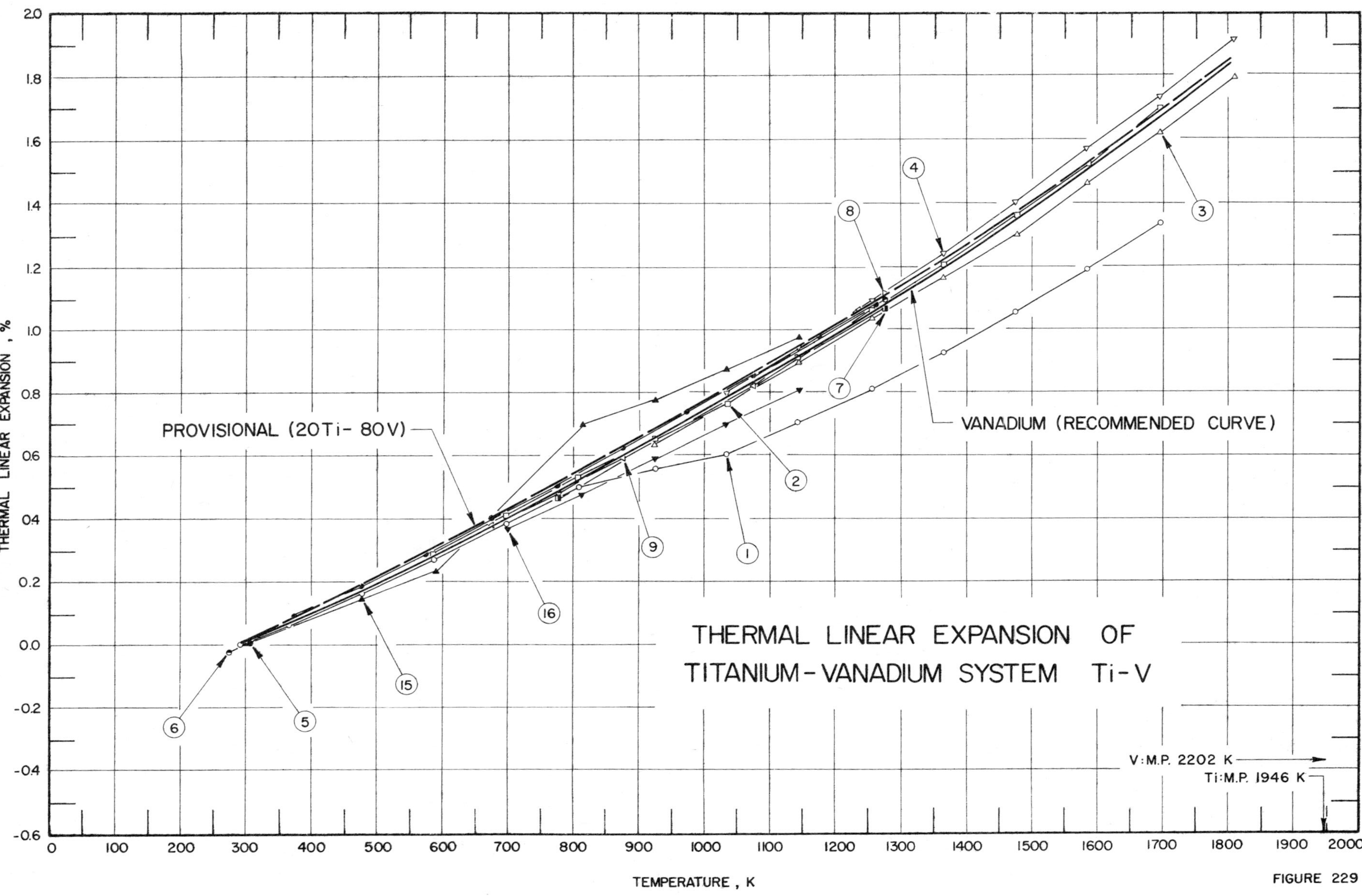
THERMAL LINEAR EXPANSION OF
TITANIUM-VANADIUM SYSTEM Ti-V
VANADIUM (RECOMMENDED CURVE)
PROVISIONAL (20Ti- 80V)
V: M.P. 2202 K
Ti: M.P. 1946 K
1
2
3
4
5
6
7
8
9
15
16
THERMAL LINEAR EXPANSION, %
TEMPERATURE, K
2.0
1.8
1.6
1.4
1.2
1.0
0.8
0.6
0.4
0.2
0.0
-0.2
-0.4
-0.6
0
100
200
300
400
500
600
700
800
900
1000
1100
1200
1300
1400
1500
1600
1700
1800
1900
2000
FIGURE 229

SPECIFICATION TABLE 229. THERMAL LINEAR EXPANSION OF TITANIUM-VANADIUM SYSTEM Ti-V

Cur. No.	Ref. No.	Author(s)	Year	Method Used	Temp. Range, K	Name and Specimen Designation	Composition (weight percent) Ti	V	Composition (continued), Specifications, and Remarks
1	333	Williams, D.N.	1961	L	293-1699	Ti-8 V			Specimen prepared from 140 Bhn sponge; annealed; expansion measured in vacuum of 3×10^{-4} mm Hg with increasing temperature; discontinuity at 1044 K due to α-β transition in Ti.
2	333	Williams, D.N.	1961	L	1699-293	Ti-8 V			The above specimen; expansion measured with decreasing temperature.
3	333	Williams, D.N.	1961	L	293-1810	Ti-20 V			Similar to the above specimen; expansion measured with increasing temperature; discontinuity at 922 K due to α-β transition in Ti.
4	333	Williams, D.N.	1961	L	1810-293	Ti-20 V			The above specimen; expansion measured with decreasing temperature.
5	460	Rhude, H.V., Yaggee, F.L., Savage, H., and Dunworth, R.J.	1964		304-1260		80.0	20.0	No details given; zero-point correction of 0.012% determined by graphical extrapolation.
6	117	Yaggee, F.L. and Styles, J.W.	1966	L	273-1273		70.0	30.0	Density 5.53 ± 0.01 g cm^{-3}.
7	117	Yaggee, F.L. and Styles, J.W.	1966	L	273-1273		60.0	40.0	Density 5.36 ± 0.01 g cm^{-3}.
8	117	Yaggee, F.L. and Styles, J.W.	1966	L	273-1273		50.0	50.0	Density 5.20 g cm^{-3}.
9	62	Yaggee, F.L. and Styles, J.W.	1969	L	273-1233		80	20	Reported error $< \pm 2\%$.
10*	602	Adenstedt, H.K., Pequignot, J.R., and Raymer, J.M.	1951	L	293-922		Bal.	10	Specimen prepared by Armour Research Foundation using DuPont Process A sponge 99.9^+ Ti, and Union Carbide and Carbon Corp. 99.8^+ V; cast samples approximately 0.125 in. x 0.125 in. x 1.25 in. were cold rolled with intermittent vacuum anneals.
11*	602	Adenstedt, H.K., et al.	1951	L	293-700		Bal.	30	Similar to the above specimen.
12*	602	Adenstedt, H.K., et al.	1951	L	293-700		Bal.	60	Similar to the above specimen.
13*	602	Adenstedt, H.K., et al.	1951	L	293-922		Bal.	70	Similar to the above specimen.
14*	602	Adenstedt, H.K., et al.	1951	L	293-922		Bal.	90	Similar to the above specimen.
15	602	Adenstedt, H.K., et al.	1951	L	300-1144		Bal.	15	Similar to the above specimen except kept at 1061 K for 21 hr and water quenched (beta-treated); length: 33.2 mm; expansion measured with increasing temperature; heating rate: 1 F/min.
16	602	Adenstedt, H.K., et al.	1951	L	1144-478		Bal.	15	Similar to the above specimen; expansion measured with decreasing temperature; cooling rate: 1 F/min.

* Not shown in figure.

DATA TABLE 229. THERMAL LINEAR EXPANSION OF TITANIUM-VANADIUM SYSTEM Ti-V

[Temperature, T, K; Linear Expansion, $\Delta L/L_0$, %]

T	$\Delta L/L_0$
CURVE 1	
293	0.000
366	0.063
477	0.163
588	0.271
699	0.383
810	0.499
922	0.559
1033	0.601
1144	0.701
1255	0.809
1366	0.925
1477	1.057
1588	1.191
1699	1.337
CURVE 2	
1699	1.696
1588	1.522
1477	1.366
1366	1.210
1255	1.060
1144	0.906
1033	0.762
922	0.650
810	0.530
699	0.410
588	0.292
477	0.174*
366	0.066*
293	0.000*
CURVE 3	
293	0.000*
366	0.066*
477	0.170*
588	0.276*
699	0.388*
810	0.508*
922	0.632
1033	0.762*
1144	0.896
1255	1.028
1366	1.162
1477	1.308
1588	1.462
CURVE 3 (cont.)	
1699	1.626
1810	1.794
CURVE 4	
1810	1.917
1699	1.739
1588	1.571
1477	1.407
1366	1.245
1255	1.091
1144	0.941
1033	0.793
922	0.653*
810	0.529*
699	0.401*
588	0.283*
477	0.173*
366	0.067*
293	0.000*
CURVE 5	
304	0.012
373	0.088
473	0.189
573	0.292
673	0.401
773	0.511
873	0.623
972	0.740
1073	0.858
1172	0.980
1260	1.079
CURVE 6	
273	-0.020
773	0.482
1273	1.096
CURVE 7	
273	-0.020*
773	0.469
1273	1.074
CURVE 8	
273	-0.021*
773	0.504*
1273	1.104
CURVE 9	
273	-0.02*
473	0.18*
673	0.38
873	0.59
1073	0.82
1233	1.06
CURVE 10*	
293	0.000
478	0.173
700	0.390
922	0.616
CURVE 11*	
293	0.000
478	0.175
700	0.392
CURVE 12*	
293	0.000
700	0.405
CURVE 13*	
293	0.000
922	0.623
CURVE 14*	
293	0.000
478	0.176
700	0.395
922	0.628
CURVE 15	
300	0.006
366	0.069*
CURVE 15 (cont.)	
478	0.145
589	0.229
700	0.380*
811	0.696
922	0.777
1033	0.870
1144	0.973
CURVE 16	
1144	0.801
1033	0.693
922	0.581
811	0.473
700	0.370
589	0.271*
478	0.169*

* Not shown in figure.

FIGURE AND TABLE NO. 230R. PROVISIONAL VALUES FOR THERMAL LINEAR EXPANSION OF TITANIUM-ZIRCONIUM SYSTEM Ti-Zr

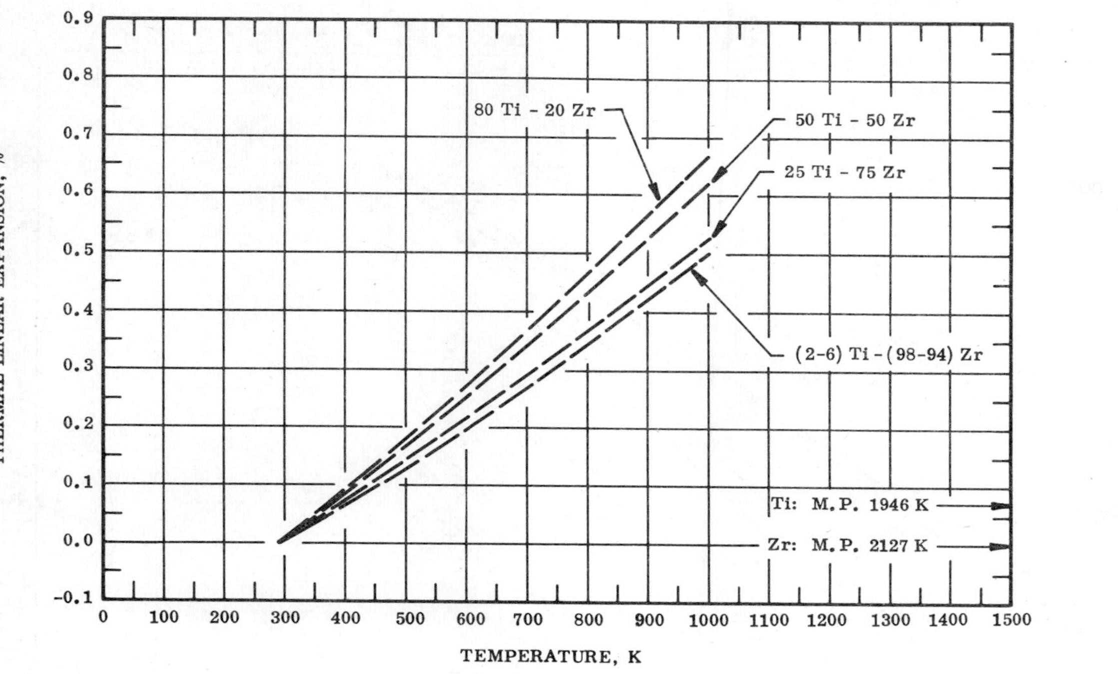

PROVISIONAL VALUES

[Temperature, T, K; Linear Expansion, $\Delta L/L_0$, %; α, K^{-1}]

T	([2-6] Ti-[98-94] Zr) $\Delta L/L_0$	$\alpha \times 10^6$	(25 Ti-75 Zr) $\Delta L/L_0$	$\alpha \times 10^6$
293	0.000	5.5	0.000	6.8
400	0.062	6.1	0.073	6.9
500	0.127	6.7	0.143	7.1
600	0.197	7.2	0.216†	7.4
700	0.271	7.6	0.290†	7.6
800	0.347	7.7	0.367†	7.7
900	0.424	7.8	0.446†	7.9
1000	0.500	7.8	0.524†	7.9

T	(50 Ti-50 Zr) $\Delta L/L_0$	$\alpha \times 10^6$	(80 Ti-20 Zr) $\Delta L/L_0$	$\alpha \times 10^6$
293	0.000	7.1	0.000	8.3
400	0.081	8.1	0.091	8.8
500	0.166	8.6	0.180	9.0
600	0.254†	8.9	0.272†	9.3
700	0.343†	9.0	0.368†	9.7
800	0.434†	9.2	0.466†	10.1
900	0.525†	9.3	0.568†	10.3
1000	0.620†	9.4	0.670†	10.4

† Uncertainties in these values are slightly higher.

REMARKS

([2-6] Ti-[98-94] Zr): The tabulated values for well-annealed alloy are considered accurate to within ± 7% over the entire temperature range. These values can be represented approximately by the following equation:

$$\Delta L/L_0 = -0.119 + 2.573 \times 10^{-4} T + 5.824 \times 10^{-7} T^2 - 2.195 \times 10^{-10} T^3$$

(25 Ti-75 Zr): The tabulated values for well-annealed alloy are considered accurate to within ± 7% over the entire temperature range. These values can be represented approximately by the following equation:

$$\Delta L/L_0 = -0.181 + 5.682 \times 10^{-4} T + 1.818 \times 10^{-7} T^2 - 4.509 \times 10^{-11} T^3$$

(50 Ti-50 Zr): The tabulated values for well-annealed alloy are considered accurate to within ± 7% over the entire temperature range. These values can be represented approximately by the following equation:

$$\Delta L/L_0 = -0.199 + 5.845 \times 10^{-4} T + 3.622 \times 10^{-7} T^2 - 1.276 \times 10^{-10} T^3$$

(80 Ti-20 Zr): The tabulated values for well-annealed alloy are considered accurate to within ± 7% over the entire temperature range. These values can be represented approximately by the following equation:

$$\Delta L/L_0 = -0.222 + 6.889 \times 10^{-4} T + 2.589 \times 10^{-7} T^2 + 5.544 \times 10^{-11} T^3$$

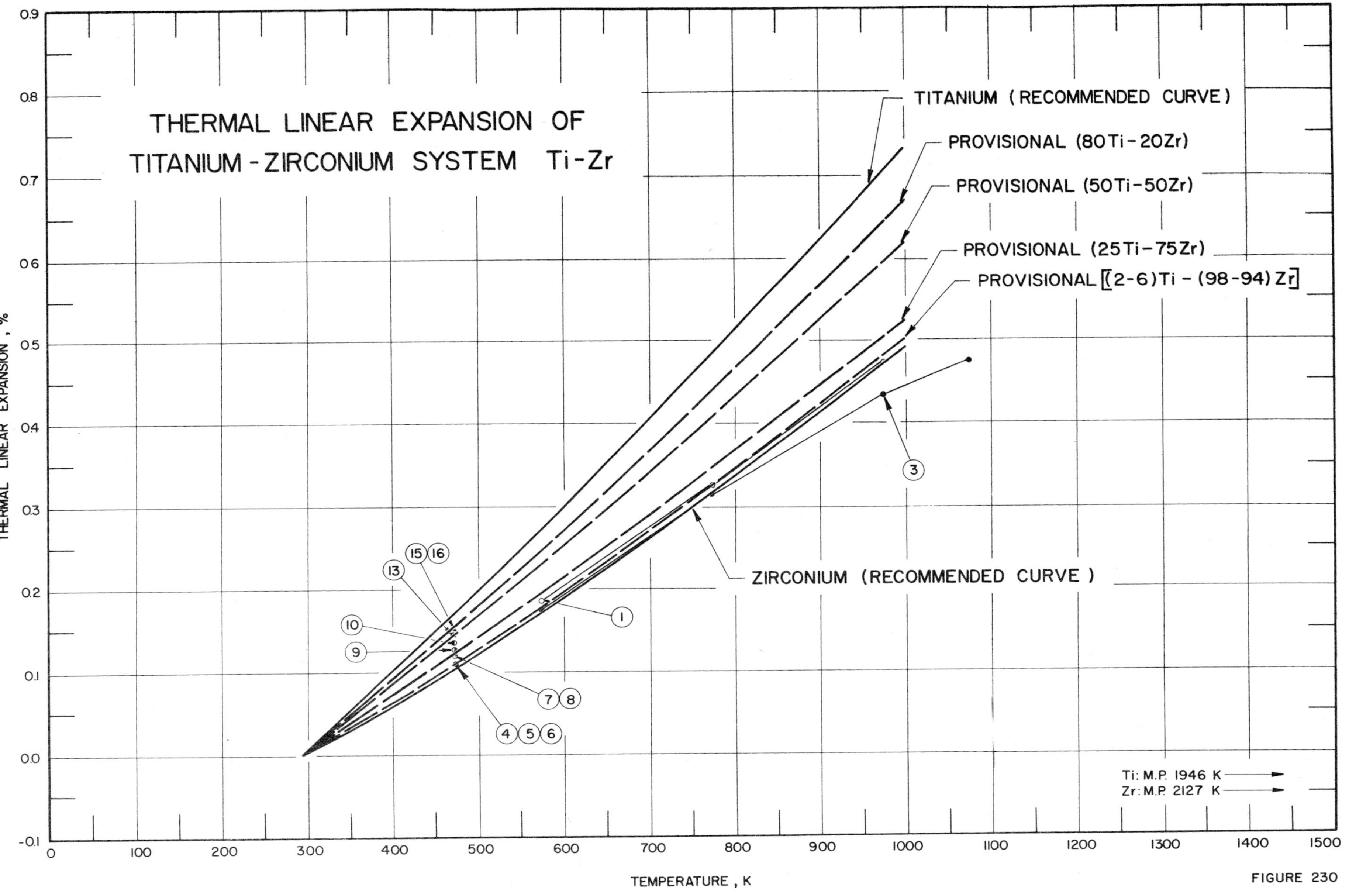
THERMAL LINEAR EXPANSION OF
TITANIUM-ZIRCONIUM SYSTEM Ti-Zr
TITANIUM (RECOMMENDED CURVE)
PROVISIONAL (80Ti-20Zr)
PROVISIONAL (50Ti-50Zr)
PROVISIONAL (25Ti-75Zr)
PROVISIONAL [(2-6)Ti-(98-94)Zr]
ZIRCONIUM (RECOMMENDED CURVE)
1
3
4 5 6
7 8
9
10
13
15 16
Ti: M.P. 1946 K
Zr: M.P. 2127 K
THERMAL LINEAR EXPANSION, %
-0.1
0.0
0.1
0.2
0.3
0.4
0.5
0.6
0.7
0.8
0.9
0
100
200
300
400
500
600
700
800
900
1000
1100
1200
1300
1400
1500
TEMPERATURE, K

FIGURE 230

SPECIFICATION TABLE 230. THERMAL LINEAR EXPANSION OF TITANIUM-ZIRCONIUM SYSTEM Ti-Zr

Cur. No.	Ref. No.	Author(s)	Year	Method Used	Temp. Range, K	Name and Specimen Designation	Composition (weight percent) Ti	Zr	Composition (continued), Specifications, and Remarks
1	338	Pridantseva, K.S.	1969	L	293-1073	Alloy No. 2	2	97.97	0.03 Hf; iodide zirconium, remelted in arc furnace.
2*	338	Pridantseva, K.S.	1969		293-1073	Alloy No. 2	2	97.97	0.03 Hf; similar to the above specimen; quintuple thermal shock of alloy: 573 K to 213 K; then 293 K to 90 K.
3	338	Pridantseva, K.S.	1969		293-1073	Alloy No. 3	2	97.82	0.05 each Al, Mo, V, 0.03 Hf; iodide zirconium remelted in arc furnace.
4	707	Pridantseva, K.S. and Solov'yeva, N.A.	1965	L	293,473		2	Bal.	Specimen prepared by powder metallurgy method, impurities mainly 1% oxygen, sample homogenized at temperature which would provide an equilibrium state of the alloy, homogenizing anneal was done in a vacuum of up to 10^{-4} mm Hg or in argon, single phase solid solution sample, 1% measurement error.
5	707	Pridantseva, K.S. and Solov'yeva, N.A.	1965	L	293,473		5	Bal.	Similar to the above specimen.
6	707	Pridantseva, K.S. and Solov'yeva, N.A.	1965	L	293,473		6	Bal.	Similar to the above specimen.
7	707	Pridantseva, K.S. and Solov'yeva, N.A.	1965	L	293,473		8	Bal.	Similar to the above specimen.
8	707	Pridantseva, K.S. and Solov'yeva, N.A.	1965	L	293,473		11	Bal.	Similar to the above specimen.
9	707	Pridantseva, K.S. and Solov'yeva, N.A.	1965	L	293,473		21	Bal.	Similar to the above specimen.
10	707	Pridantseva, K.S. and Solov'yeva, N.A.	1965	L	293,473		31	Bal.	Similar to the above specimen.
11*	707	Pridantseva, K.S. and Solov'yeva, N.A.	1965	L	293,473		36	Bal.	Similar to the above specimen.
12*	707	Pridantseva, K.S. and Solov'yeva, N.A.	1965	L	293,473		41	Bal.	Similar to the above specimen.
13	707	Pridantseva, K.S. and Solov'yeva, N.A.	1965	L	293,473		50	Bal.	Similar to the above specimen.
14*	707	Pridantseva, K.S. and Solov'yeva, N.A.	1965	L	293,473		60	Bal.	Similar to the above specimen.
15	707	Pridantseva, K.S. and Solov'yeva, N.A.	1965	L	293,473		70	Bal.	Similar to the above specimen.
16	707	Pridantseva, K.S. and Solov'yeva, N.A.	1965	L	293,473		80	Bal.	Similar to the above specimen.
17*	707	Pridantseva, K.S. and Solov'yeva, N.A.	1965	L	293,473		90	Bal.	Similar to the above specimen.
18*	707	Pridantseva, K.S. and Solov'yeva, N.A.	1965	L	293,473		11	Bal.	Powder sample; 1% measurement error.
19*	707	Pridantseva, K.S. and Solov'yeva, N.A.	1965	L	293,473		21	Bal.	Similar to the above specimen.

* Not shown in figure.

SPECIFICATION TABLE 230. THERMAL LINEAR EXPANSION OF TITANIUM-ZIRCONIUM SYSTEM Ti-Zr (continued)

Cur. No.	Ref. No.	Author(s)	Year	Method Used	Temp. Range, K	Name and Specimen Designation	Composition (weight percent) Ti	Zr	Composition (continued), Specifications, and Remarks
20*	707	Pridantseva, K.S. and Solov'yeva, N.A.	1965	L	293,473		27	Bal.	Similar to the above specimen.
21*	707	Pridantseva, K.S. and Solov'yeva, N.A.	1965	L	293,473		35	Bal.	Similar to the above specimen.
22*	707	Pridantseva, K.S. and Solov'yeva, N.A.	1965	L	293,473		45	Bal.	Similar to the above specimen.
23*	707	Pridantseva, K.S. and Solov'yeva, N.A.	1965	L	293,473		55	Bal.	Similar to the above specimen.
24*	707	Pridantseva, K.S. and Solov'yeva, N.A.	1965	L	293,473		65	Bal.	Similar to the above specimen.
25*	707	Pridantseva, K.S. and Solov'yeva, N.A.	1965	L	293,473		75	Bal.	Similar to the above specimen.
26*	707	Pridantseva, K.S. and Solov'yeva, N.A.	1965	L	293,473		80	Bal.	Similar to the above specimen.

DATA TABLE 230. THERMAL LINEAR EXPANSION OF TITANIUM-ZIRCONIUM SYSTEM Ti-Zr

[Temperature, T, K; Linear Expansion, $\Delta L/L_0$, %]

T	$\Delta L/L_0$
CURVE 1	
293	0.000*
573	0.187
773	0.327
973	0.475
1073	0.520
CURVE 2*	
293	0.000
573	0.185
773	0.322
973	0.468
1073	0.520
CURVE 3	
293	0.000*
573	0.174
773	0.313
973	0.434
1073	0.475
CURVE 4	
293	0.000
473	0.111
CURVE 5	
293	0.000
473	0.113
CURVE 6	
293	0.000
473	0.115
CURVE 7	
293	0.000
473	0.121
CURVE 8	
293	0.000
473	0.121
CURVE 9	
293	0.000
473	0.128
CURVE 10	
293	0.000
473	0.133
CURVE 11*	
293	0.000
473	0.138
CURVE 12*	
293	0.000
473	0.142
CURVE 13	
293	0.000
473	0.147
CURVE 14*	
293	0.000
473	0.151
CURVE 15	
293	0.000
473	0.154
CURVE 16	
293	0.000
473	0.156
CURVE 17*	
293	0.000
473	0.160
CURVE 18*	
293	0.000
473	0.115
CURVE 19*	
293	0.000
473	0.121
CURVE 20*	
293	0.000
473	0.124
CURVE 21*	
293	0.000
473	0.130
CURVE 22*	
293	0.000
473	0.135
CURVE 23*	
293	0.000
473	0.140
CURVE 24*	
293	0.000
473	0.144
CURVE 25*	
293	0.000
473	0.146
CURVE 26*	
293	0.000
473	0.148

* Not shown in figure.

FIGURE AND TABLE NO. 231R. PROVISIONAL VALUES FOR THERMAL LINEAR EXPANSION OF URANIUM-ZIRCONIUM SYSTEM U-Zr

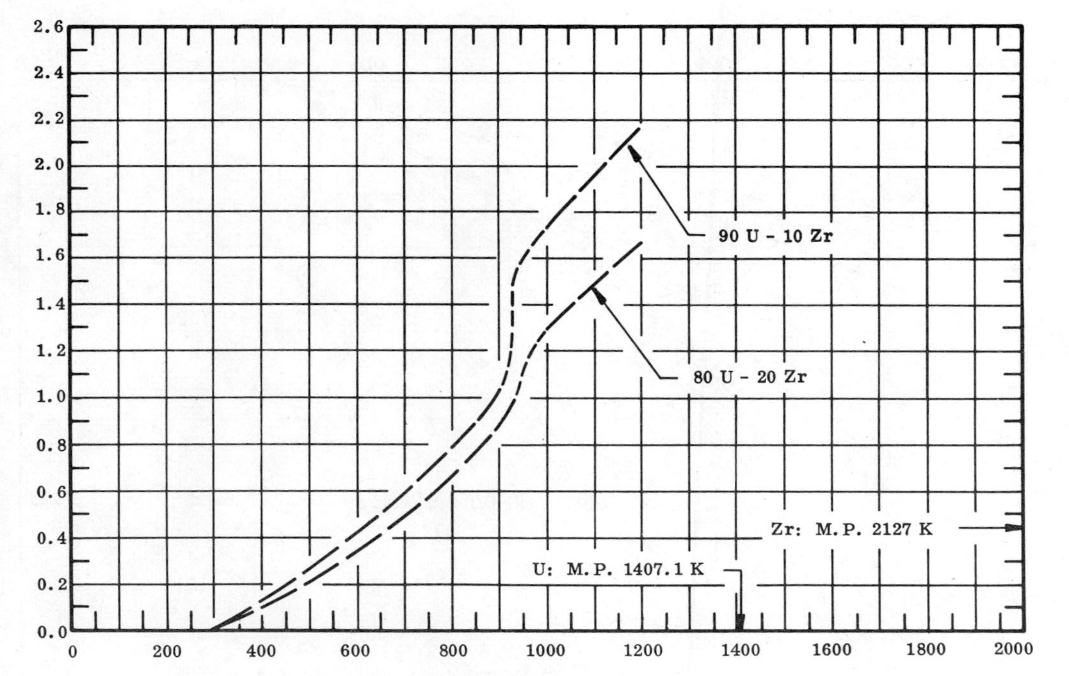

PROVISIONAL VALUES

[Temperature, T, K; Linear Expansion, $\Delta L/L_0$, %; α, K^{-1}]

	(80 U-20 Zr)		(90 U-10 Zr)	
T	$\Delta L/L_0$	$\alpha \times 10^6$	$\Delta L/L_0$	$\alpha \times 10^6$
293	0.000	9.8	0.000	13.2
400	0.107	10.5	0.142	13.5
500	0.219	11.9	0.281	14.2
600	0.348	13.9	0.433	16.0
700	0.500	16.6	0.603	18.2
800	0.681	19.9	0.799	21.2
900†	0.899	24.1	1.027	24.7
1000†	1.300	18.7	1.725	22.5
1100	1.487	18.7	1.950	22.5
1200	1.674	18.7	2.175	22.5

† Phase transformation between 900-1000 K.

REMARKS

(80 U-20 Zr): The tabulated values for well-annealed alloy are considered accurate to within ±7% over the entire temperature range. These values can be represented approximately by the following equation:

$$\Delta L/L_0 = -0.301 + 1.160 \times 10^{-3}\, T - 7.790 \times 10^{-7}\, T^2 + 1.080 \times 10^{-9}\, T^3 \qquad (293 < T < 900)$$

(90 U-10 Zr): The tabulated values for well-annealed alloy are considered accurate to within ±7% over the entire temperature range. These values can be represented approximately by the following equation:

$$\Delta L/L_0 = -0.424 + 1.658 \times 10^{-3}\, T - 1.052 \times 10^{-6}\, T^2 + 1.115 \times 10^{-9}\, T^3 \qquad (293 < T < 900)$$

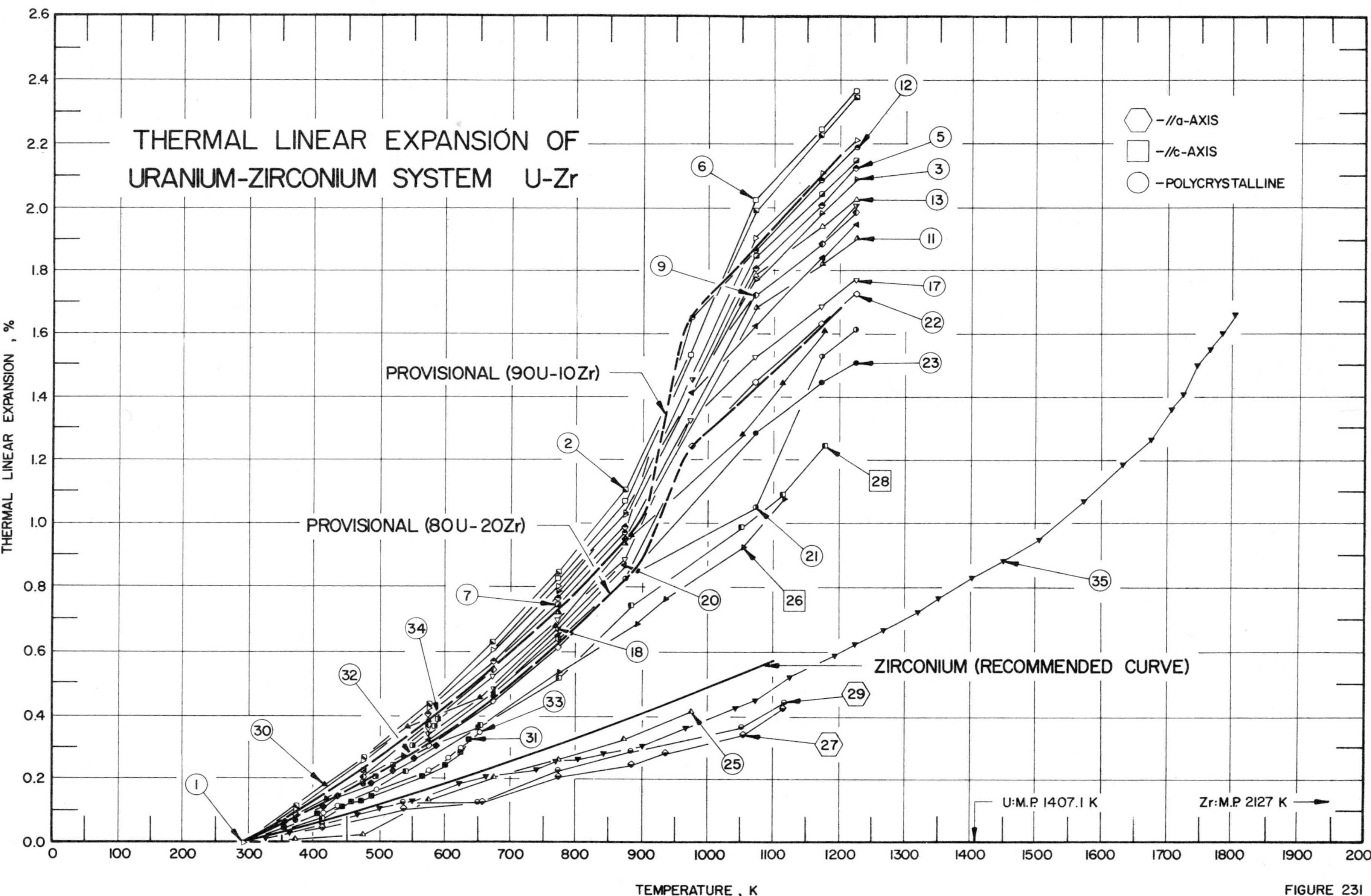
THERMAL LINEAR EXPANSIÓN OF
URANIUM-ZIRCONIUM SYSTEM U-Zr
-//a-AXIS
-//c-AXIS
-POLYCRYSTALLINE
PROVISIONAL (90U-10Zr)
PROVISIONAL (80U-20Zr)
ZIRCONIUM (RECOMMENDED CURVE)
U:M.P. 1407.1 K
Zr:M.P. 2127 K
THERMAL LINEAR EXPANSION, %
TEMPERATURE, K
1
2
3
5
6
7
9
11
12
13
17
18
20
21
22
23
25
26
27
28
29
30
31
32
33
34
35
0.0
0.2
0.4
0.6
0.8
1.0
1.2
1.4
1.6
1.8
2.0
2.2
2.4
2.6
0
100
200
300
400
500
600
700
800
900
1000
1100
1200
1300
1400
1500
1600
1700
1800
1900
2000

FIGURE 231

SPECIFICATION TABLE 231. THERMAL LINEAR EXPANSION OF URANIUM-ZIRCONIUM SYSTEM U-Zr

Cur. No.	Ref. No.	Author(s)	Year	Method Used	Temp. Range, K	Name and Specimen Designation	Composition (weight percent) U	Zr	Composition (continued), Specifications, and Remarks
1	448	Saller, H.A., Dickerson, R.F. and Murr, W.E.	1956	L	293-1223		Bal.	2.95	Prepared by combined induction and consumable-electrode arc-melting; heat treated 1 hr at 1073 K, held 24 hrs at 943 K, furnace cooled; expansion measured with increasing temp under 2 x 10^{-5} mm Hg vacuum.
2	448	Saller, H.A. et al.	1956	L	1223-293		Bal.	2.95	The above specimen; expansion measured with decreasing temp.
3	448	Saller, H.A. et al.	1956	L	293-1223		Bal.	2.95	Similar to the above specimen and conditions except heat treated 1 hr at 1073 K, isothermally transformed 2 hrs at 823 K; then 5 min at 1053 K, isothermally transformed 2 hrs at 823 K, water quenched, then repeat 5 min at 1053 K, 2 hr at 823 K; expansion measured with increasing temp.
4*	448	Saller, H.A. et al.	1956	L	1223-293		Bal.	2.95	The above specimen; expansion measured with decreasing temp.
5	448	Saller, H.A. et al.	1956	L	293-1223		Bal.	4.96	Prepared by combined induction and consumable-electrode arc-melting; heat treated 1 hr at 1073 K, held 24 hrs at 943 K, furnace cooled; expansion measured with increasing temp under 2 x 10^{-5} mm Hg vacuum.
6	448	Saller, H.A. et al.	1956	L	1223-293		Bal.	4.96	The above specimen; expansion measured with decreasing temp.
7	448	Saller, H.A. et al.	1956	L	293-1223		Bal.	4.96	Similar to the above specimen and conditions except heat treated 1 hr at 1073 K, isothermally transformed 2 hrs at 823 K; then 5 min at 1053 K; isothermally transformed 2 hrs at 823 K, water quenched, then repeat 5 min at 1053 K, 2 hrs at 823 K; expansion measured with increasing temp.
8*	448	Saller, H.A. et al.	1956	L	1223-293		Bal.	4.96	The above specimen; expansion measured with decreasing temp.
9	448	Saller, H.A. et al.	1956	L	293-1223		Bal.	7.07	Prepared by combined induction and consumable-electrode arc-melting; heat treated 1 hr at 1073 K, held 24 hrs at 943 K, furnace cooled; expansion measured with increasing temp under 2 x 10^{-5} mm Hg vacuum.
10*	448	Saller, H.A. et al.	1956	L	1223-293		Bal.	7.07	The above specimen; expansion measured with decreasing temp.
11	448	Saller, H.A. et al.	1956	L	293-1223		Bal.	7.07	Similar to the above specimen and conditions except heat treated 1 hr at 1073 K, isothermally transformed 2 hrs at 823 K; then 5 min at 1053 K; isothermally transformed 2 hrs at 823 K, water quenched, then repeat 5 min at 1053 K, 2 hrs at 823 K; expansion measured with increasing temp.
12	448	Saller, H.A. et al.	1956	L	1223-293		Bal.	7.07	The above specimen; expansion measured with decreasing temp.
13	448	Saller, H.A. et al.	1956	L	293-1223		Bal.	9.86	Prepared by combined induction and consumable-electrode arc-melting; heat treated 1 hr at 1073 K, held 24 hrs at 943 K, furnace cooled; expansion measured with increasing temp under 2 x 10^{-5} mm Hg vacuum.
14*	448	Saller, H.A. et al.	1956	L	1223-293		Bal.	9.86	The above specimen; expansion measured with decreasing temp.
15*	448	Saller, H.A. et al.	1956	L	293-1223		Bal.	9.86	Similar to the above specimen and conditions except heat treated 1 hr at 1073 K, furnace cooled to 1023 K, furnace cooled from 1023 K to 843 K at 0.5 C per min; expansion measured with increasing temp.
16*	448	Saller, H.A. et al.	1956	L	1223-293		Bal.	9.86	The above specimen; expansion measured with decreasing temp.
17	448	Saller, H.A. et al.	1956	L	293-1223		Bal.	15.5	Similar to the above specimen; expansion measured with increasing temp.
18	448	Saller, H.A. et al.	1956	L	1223-293		Bal.	15.5	The above specimen; expansion measured with decreasing temp.

* Not shown in figure.

SPECIFICATION TABLE 231. THERMAL LINEAR EXPANSION OF URANIUM-ZIRCONIUM SYSTEM U-Zr (continued)

Cur. No.	Ref. No.	Author(s)	Year	Method Used	Temp. Range, K	Name and Specimen Designation	Composition (weight percent) U	Zr	Composition (continued), Specifications, and Remarks
19*	448	Saller, H.A., Dickerson, R.F. and Murr, W.E.	1956	L	293-1223		Bal.	15.5	Prepared by combined induction and consumable-electrode arc-melting; heat treated 1 hr at 1073 K, held 24 hrs at 943 K, furnace cooled; expansion measured with increasing temp under 2 x 10^{-5} mm Hg vacuum.
20	448	Saller, H.A. et al.	1956	L	1223-293		Bal.	15.5	The above specimen; expansion measured with decreasing temp.
21	448	Saller, H.A. et al.	1956	L	293-1223		Bal.	20.0	Prepared by combined induction and consumable-electrode arc-melting; heat treated 1 hr at 1073 K, isothermally transformed 823 K for 2 hrs, water quenched; expansion measured with increasing temp.
22	448	Saller, H.A. et al.	1956	L	1223-293		Bal.	20.0	The above specimen; expansion measured with decreasing temp.
23	448	Saller, H.A. et al.	1956	L	293-1223		Bal.	20.0	Similar to the above specimen; heat treated 1 hr at 1073 K, furnace cooled to 1023 K; furnace cooled from 1023 K to 843 K at 0.5 C per min; expansion measured with increasing temp.
24*	448	Saller, H.A. et al.	1956	L	1223-293		Bal.	20.0	The above specimen; expansion measured with decreasing temp.
25	382	Loy, S.M. and Vetrano, J.B.	1960		293-973		7	93	Bulk density 6.80 g cm^{-3}.
26	387	Russell, R.B.	1954	X	416-1115		2.11	97.89	Hexagonal (α) crystal bar machined to 0.471 in. diameter; dehydrogenated at 1400 C for 1 hr in vacuum of 10^{-6} mm Hg, swaged to 0.189 in. diameter; chemical analysis taken; rolled to 0.012 in. thick sheet, cut into strip about 1 x 0.4 in.; expansion measured along c-axis; lattice parameter reported at 416 K is 5.1542 x 10^{-8} cm; 5.1470 x 10^{-8} cm at 293 K determined by graphical extrapolation.
27	387	Russell, R.B.	1954	X	411-1114		2.11	97.89	The above specimen; expansion measured along a-axis; lattice parameter reported at 414 K is 5.1375 x 10^{-8} cm; 5.1348 x 10^{-8} cm at 293 K determined by graphical extrapolation.
28	387	Russell, R.B.	1954	X	414-1180		4.11	95.89	Similar to the above specimen; expansion measured along c-axis; lattice parameter reported at 414 K is 5.1490 x 10^{-8} cm; 5.1450 x 10^{-8} cm at 293 K determined by graphical extrapolation.
29	387	Russell, R.B.	1954	X	412-1119		4.11	95.89	The above specimen; expansion measured along a-axis; lattice parameter reported at 412 K is 5.1362 x 10^{-8} cm; 5.1331 x 10^{-8} cm at 293 K determined by graphical extrapolation.
30	387	Russell, R.B.	1954	X	415-1178		4.11	95.89	Similar to the above specimen except cubic (β) form of crystal; lattice parameter reported at 415 K is 3.5632 x 10^{-8} cm; 3.5567 x 10^{-8} cm at 293 K determined by graphical extrapolation.
31	447	McCreight, L.R.	1952	I	302-639		31	69	Prepared by mixing respective hydrides, then decomposing and sintering in one operation; extruded at 1273 K.
32	447	McCreight, L.R.	1952	I	303-586		31	69	The above specimen; expansion measured on second test; zero-point correction of 0.006% determined by graphical extrapolation.
33	447	McCreight, L.R.	1952	I	302-657		31	69	The above specimen; expansion measured on third test; zero-point correction of 0.001% determined by graphical extrapolation.

* Not shown in figure.

SPECIFICATION TABLE 231. THERMAL LINEAR EXPANSION OF URANIUM-ZIRCONIUM SYSTEM U-Zr (continued)

Cur. No.	Ref. No.	Author(s)	Year	Method Used	Temp. Range, K	Name and Specimen Designation	Composition (weight percent) U	Zr	Composition (continued), Specifications, and Remarks
34	447	McCreight, L.R.	1952	I	299-589		31	69	The above specimen; expansion measured on fourth test; zero-point correction is ± 0.000%.
35	451	Hedge, J.C., Lang, J.I., Howe, E., and Elliot, R.	1963	T	300-1805		10.48	89.52	Specimen machined into rod 0.5 in. diameter, 6 in. long; expansion measured in argon atmosphere; density 430 lb ft^{-3}; zero-point correction of 0.003% determined by graphical extrapolation.
36*	451	Hedge, J.C., et al.	1963	T	300-1803		10.58	Bal.	Similar to the above specimen except hydrided to 1.5% H_2, density 383 lb ft^{-3}; shrinkage explained by evolution of hydrogen from sample; expansion measured in hydrogen atmosphere; zero-point correction of 0.002% determined by graphical extrapolation.

*Not shown in figure.

DATA TABLE 231. THERMAL LINEAR EXPANSION OF URANIUM-ZIRCONIUM SYSTEM U-Zr

[Temperature, T, K; Linear Expansion, $\Delta L/L_0$, %]

T	$\Delta L/L_0$
CURVE 1	
293	0.000
373	0.108
473	0.261
573	0.425
673	0.604
773	0.804
873	1.034
1073	1.902
1173	2.108
1223	2.216
CURVE 2	
1223	2.345
1173	2.229
1073	1.996
873	1.110
773	0.849
673	0.633
573	0.440
473	0.274
373	0.118
293	0.000*
CURVE 3	
293	0.000*
373	0.109*
473	0.267*
573	0.432*
673	0.605*
773	0.798
873	1.038
1073	1.778
1173	1.986
1223	2.098
CURVE 4*	
1223	2.336
1173	2.221
1073	1.996
873	1.154
773	0.850
673	0.630
573	0.442
473	0.275
CURVE 4 (cont.)*	
373	0.114
293	0.000
CURVE 5	
293	0.000*
373	0.107*
473	0.244*
573	0.407
673	0.578
773	0.771
873	0.995
1073	1.807
1173	2.015
1223	2.124
CURVE 6	
1223	2.362
1173	2.247
1073	2.024
973	1.537
873	1.078
773	0.826
673	0.615*
573	0.435*
473	0.272*
373	0.116*
293	0.000*
CURVE 7	
293	0.000*
373	0.092
473	0.237
573	0.402*
673	0.570*
773	0.754
873	0.993*
1073	1.851
1173	2.044
1223	2.148
CURVE 8*	
1223	2.329
1173	2.215
CURVE 8 (cont.)*	
1073	1.986
973	1.768
873	1.061
773	0.807
673	0.602
573	0.420
473	0.261
373	0.108
293	0.000
CURVE 9	
293	0.000*
373	0.082*
473	0.214
573	0.379
673	0.554
773	0.745*
873	0.955
1073	1.722
1173	1.893
1223	1.988
CURVE 10*	
1223	2.239
1173	2.123
1073	1.894
973	1.680
873	0.995
773	0.757
673	0.565
573	0.397
473	0.246
373	0.104
293	0.000
CURVE 11	
293	0.000*
373	0.090*
473	0.226*
573	0.387
673	0.544*
773	0.723
873	0.943
1073	1.682
CURVE 11 (cont.)	
1173	1.828
1223	1.903
CURVE 12	
1223	2.199
1173	2.088
1073	1.866
973	1.652
873	0.977
773	0.740*
673	0.549*
573	0.388*
473	0.238*
373	0.099*
293	0.000*
CURVE 13	
293	0.000*
373	0.084*
473	0.216*
573	0.380*
673	0.564*
773	0.766*
873	0.953*
1073	1.786
1173	1.943
1223	2.026
CURVE 14*	
1223	2.228
1173	2.112
1073	1.885
973	1.663
873	0.966
773	0.732
673	0.544
573	0.390
473	0.250
373	0.111
293	0.000
CURVE 15*	
293	0.000
CURVE 15 (cont.)*	
373	0.111
473	0.257
573	0.423
673	0.601
773	0.788
873	0.996
1073	1.831
1173	2.012
1223	2.106
CURVE 16*	
1223	2.182
1173	2.071
1073	1.848
973	1.632
873	0.946
773	0.717
673	0.530
573	0.374
473	0.235
373	0.099
293	0.000
CURVE 17	
293	0.000*
373	0.098*
473	0.211*
573	0.361
673	0.524
773	0.700
873	0.894
973	1.331
1073	1.531
1173	1.688
1223	1.772
CURVE 18	
1223	2.012
1173	1.896
1073	1.730*
973	1.459
873	0.894*
773	0.672
673	0.495
CURVE 18 (cont.)	
573	0.347
473	0.211*
373	0.086*
293	0.000*
CURVE 19*	
293	0.000
373	0.086
473	0.207
573	0.351
673	0.506
773	0.687
873	0.882
1073	1.539
1173	1.693
1273	1.762
CURVE 20	
1223	1.956
1173	1.841
1073	1.627
973	1.418
873	0.872
773	0.659
673	0.481
573	0.330
473	0.202*
373	0.084*
293	0.000*
CURVE 21	
293	0.000*
373	0.082*
473	0.200
573	0.337*
673	0.482*
773	0.645
873	0.836
1073	1.058
1173	1.537
1223	1.619

* Not shown in figure.

DATA TABLE 231. THERMAL LINEAR EXPANSION OF URANIUM-ZIRCONIUM SYSTEM U-Zr (continued)

T	$\Delta L/L_0$
CURVE 22	
1223	1.734
1173	1.639
1073	1.444
973	1.248
873	0.825*
773	0.620
673	0.455
573	0.316
473	0.193*
373	0.080*
293	0.000*
CURVE 23	
293	0.000*
373	0.074
473	0.191
573	0.321*
673	0.462
773	0.619*
1073	1.299
1173	1.441
1223	1.506
CURVE 24*	
1223	1.730
1173	1.628
1073	1.434
973	1.246
873	0.848
773	0.645
673	0.473
573	0.349
473	0.199
373	0.081
293	0.000
CURVE 25	
293	0.000*
373	0.017
473	0.025
573	0.139
673	0.201
773	0.263
873	0.334
973	0.411
CURVE 26	
416	0.140
539	0.280
650	0.367
772	0.542
890	0.690
934	0.767
1057	0.936
1115	1.080
CURVE 27	
411	0.052
538	0.109
652	0.136
773	0.208
885	0.255
932	0.290
1057	0.344
1114	0.428
CURVE 28	
414	0.077
418	0.193
540	0.223
657	0.371
773	0.524
884	0.746
1052	0.989
1115	1.096
1180	1.245
CURVE 29	
412	0.060
539	0.124
651	0.126
773	0.231
881	0.296
1053	0.366
1114	0.413*
1119	0.446
CURVE 30	
415	0.182
541	0.370
655	0.460
CURVE 30 (cont.)	
770	0.683
883	0.966
1052	1.296
1112	1.444
1116	1.444*
1178	1.619
CURVE 31	
302	0.007*
333	0.033*
352	0.048
408	0.090
420	0.099*
446	0.117
459	0.126
471	0.137
486	0.147
567	0.207
576	0.219*
582	0.228*
600	0.250
609	0.261*
624	0.292
625	0.299*
639	0.337
CURVE 32	
303	0.010*
333	0.039*
357	0.062
371	0.077*
415	0.118
439	0.143
488	0.185
520	0.231
527	0.239*
553	0.268
567	0.286*
579	0.305*
586	0.313
CURVE 33	
302	0.007*
324	0.026*
333	0.033*
CURVE 33 (cont.)	
352	0.050*
415	0.098
427	0.107*
439	0.118
448	0.129*
458	0.139*
494	0.168
571	0.231
600	0.258*
608	0.272
613	0.280*
625	0.300
657	0.358
CURVE 34	
299	0.007*
309	0.016*
438	0.143*
460	0.164*
474	0.187*
492	0.212
509	0.236*
520	0.258
535	0.282*
550	0.305
573	0.350*
581	0.374
589	0.397
CURVE 35	
300	0.003*
329	0.017
361	0.037
419	0.054*
467	0.089
500	0.110
550	0.139
622	0.188
663	0.206
740	0.234
770	0.261
805	0.267
822	0.267*
843	0.284
872	0.289*
900	0.312
CURVE 35 (cont.)	
927	0.327*
968	0.362
998	0.382*
1042	0.424
1078	0.457
1129	0.524
1193	0.594
1223	0.626
1270	0.675
1320	0.724
1353	0.769
1400	0.836
1450	0.888
1501	0.959
1572	1.073
1633	1.183
1677	1.263
1708	1.363
1726	1.413
1747	1.503
1765	1.553
1782	1.603
1805	1.663
CURVE 36*	
300	0.002
340	0.025
392	0.072
445	0.110
468	0.154
498	0.182
557	0.239
625	0.324
703	0.422
759	0.534
819	0.635
857	0.722
892	0.774
953	0.880
1023	0.949
1071	1.002
1112	0.987
1120	0.947
1129	0.889
1141	0.795
1152	0.604
1186	0.285
CURVE 36 (cont.)*	
1222	0.009
1259	-0.175
1313	-0.488
1352	-0.673
1382	-0.776
1423	-0.860
1466	-0.960
1509	-1.008
1530	-1.048
1567	-1.068
1593	-1.068
1613	-1.038
1630	-1.008
1647	-0.988
1685	-0.938
1722	-0.883
1760	-0.832
1803	-0.784

* Not shown in figure.

4. MULTIPLE ALLOYS

FIGURE AND TABLE NO. 232R. PROVISIONAL VALUES FOR THERMAL LINEAR EXPANSION OF ALUMINUM + BERYLLIUM + ΣX_i ALLOYS Al + Be + ΣX_i

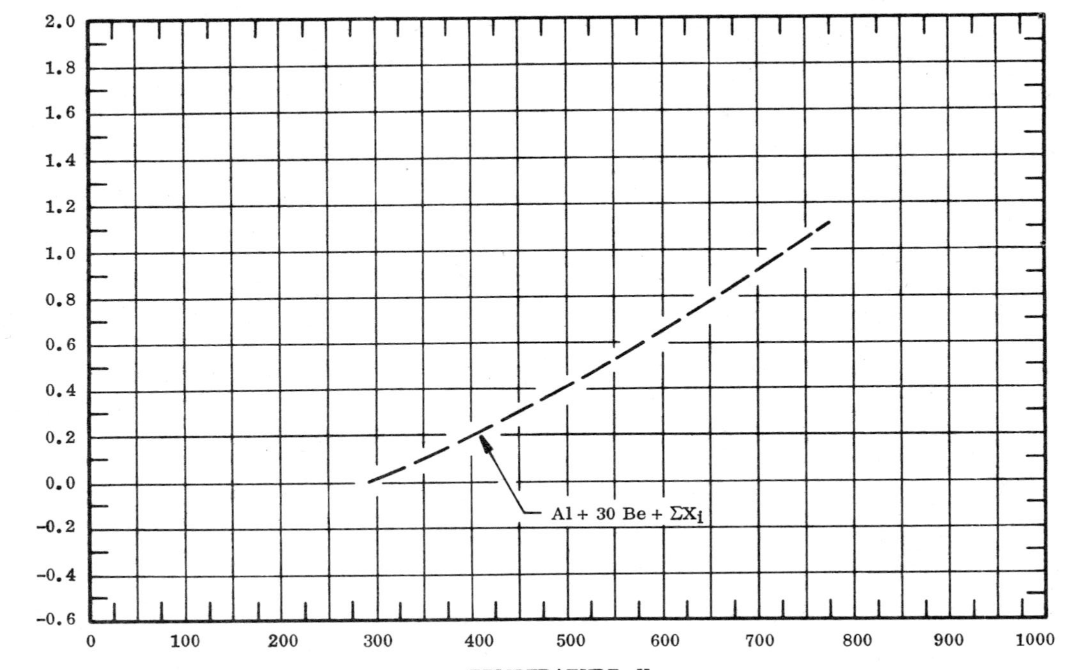

PROVISIONAL VALUES

[Temperature, T, K; Linear Expansion, $\Delta L/L_0$, %; α, K^{-1}]

T	(Al + 30 Be + ΣX_i) $\Delta L/L_0$	$\alpha \times 10^6$
293	0.000	15.9
400	0.193	19.9
500	0.407	22.9
600	0.649	25.0
700	0.905	26.3
775	1.106	26.8

REMARKS

(Al + 30 Be + ΣX_i) : The tabulated values are for well-annealed alloy and are considered accurate to within ± 7% over the entire temperature range. These values can be represented approximately by the following equation:

$$\Delta L/L_0 = -0.254 + 2.437 \times 10^{-5}\ T + 3.268 \times 10^{-6}\ T^2 - 1.339 \times 10^{-9}\ T^3$$

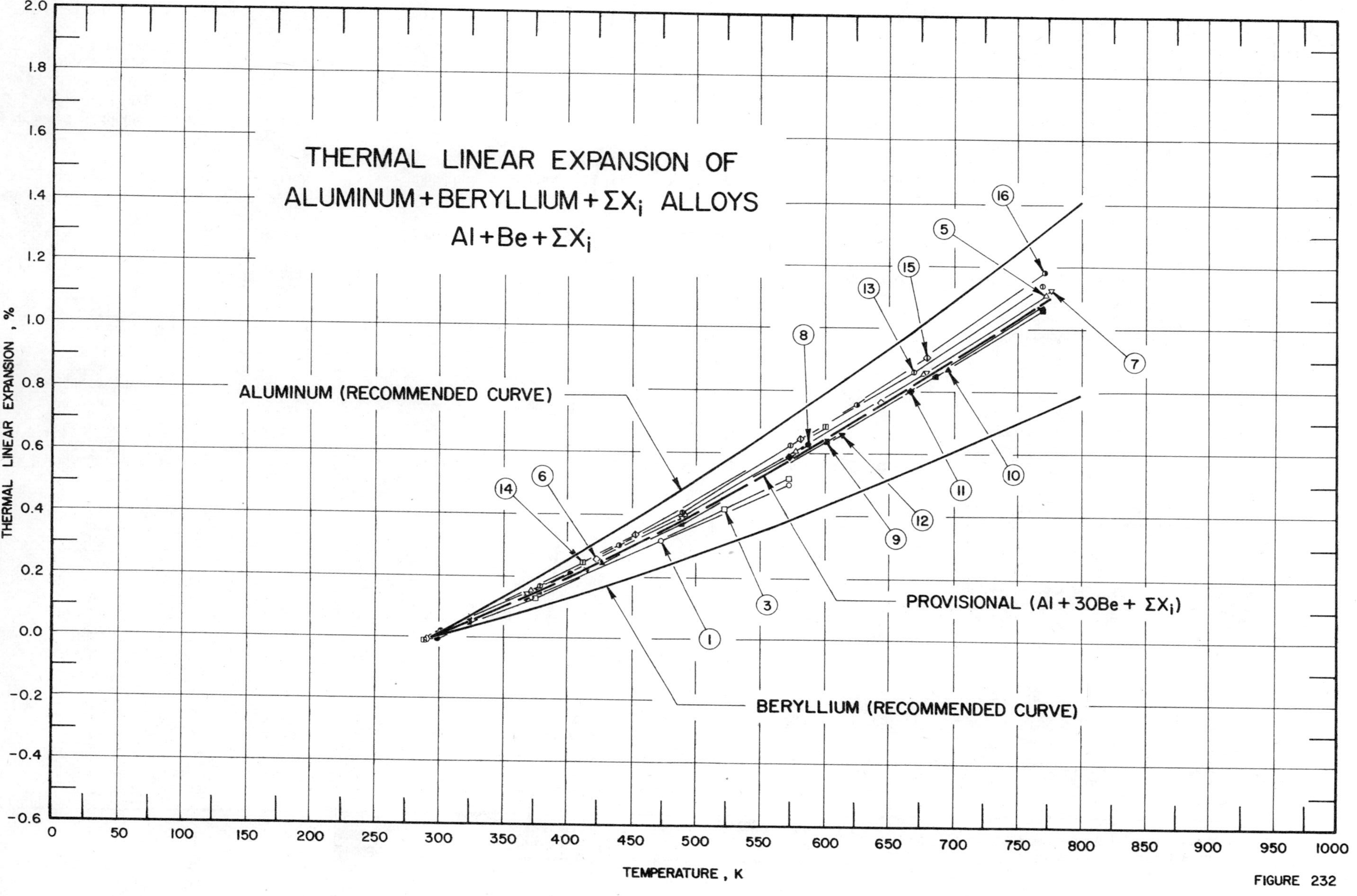
THERMAL LINEAR EXPANSION OF
ALUMINUM + BERYLLIUM + ΣXi ALLOYS
Al + Be + ΣXi
ALUMINUM (RECOMMENDED CURVE)
PROVISIONAL (Al + 30Be + ΣXi)
BERYLLIUM (RECOMMENDED CURVE)
THERMAL LINEAR EXPANSION, %
TEMPERATURE, K
2.0
1.8
1.6
1.4
1.2
1.0
0.8
0.6
0.4
0.2
0.0
-0.2
-0.4
-0.6
0
50
100
150
200
250
300
350
400
450
500
550
600
650
700
750
800
850
900
950
1000
1
3
5
6
7
8
9
10
11
12
13
14
15
16

FIGURE 232

SPECIFICATION TABLE 232. THERMAL LINEAR EXPANSION OF ALUMINUM + BERYLLIUM + ΣX_i ALLOYS Al + Be + ΣX_i

Cur. No.	Ref. No.	Author(s)	Year	Method Used	Temp. Range, K	Name and Specimen Designation	Composition (weight percent) Al	Be	Si	Mg	Cu	Fe	Mn	Composition (continued), Specifications, and Remarks
1	19	Hidnert, P. and Krider, H.S.	1952	L	293-573	Sample 1671	Bal.	35	0.5-1.0	0.5-1.0				Forged specimen, solution heat-treated at 589 K, quenched and aged at 422 K; expansion measured with increasing temperature.
2*	19	Hidnert, P. and Krider, H.S.	1952	L	573-293	Sample 1671	Bal.	35	0.5-1.0	0.5-1.0				The above specimen; expansion measured with decreasing temperature.
3	19	Hidnert, P. and Krider, H.S.	1952	L	293-573	Sample 1671	Bal.	35	0.5-1.0	0.5-1.0				The above specimen; expansion measured with increasing temperature on second test.
4*	19	Hidnert, P. and Krider, H.S.	1952	L	573-293	Sample 1671	Bal.	35	0.5-1.0	0.5-1.0				The above specimen; expansion measured with decreasing temperature.
5	400	Hidnert, P. and Sweeney, W.T.	1927	L	298-774	Sample 1225	71.9	27.5	0.10		0.08	0.58	0.05	Rod specimen about 300 mm long, 10 mm diameter; furnished by Beryllium Corp. of America; prepared using virgin Al and Be containing about 99% beryllium; cast alloys prepared by heating metals to about 1373 to 1473 K with small amount of fluoride-chloride; density 2.202 g cm^{-3} at 293 K; expansion measured with increasing temp; zero-point correction of 0.012% determined by graphical extrapolation.
6	400	Hidnert, P. and Sweeney, W.T.	1927	L	643-291	Sample 1225	71.9	27.5	0.10		0.08	0.58	0.05	The above specimen; expansion measured with decreasing temp; zero-point correction of 0.022% determined by graphical interpolation.
7	400	Hidnert, P. and Sweeney, W.T.	1927	L	294-777	Sample 1225	71.9	27.5	0.10		0.08	0.58	0.05	The above specimen; expansion measured with increasing temp on second test; zero-point correction of -0.199% determined by graphical extrapolation.
8	400	Hidnert, P. and Sweeney, W.T.	1927	L	589-298	Sample 1225	71.9	27.5	0.10		0.08	0.58	0.05	The above specimen; expansion measured with decreasing temp on second test; zero-point correction of -0.199% determined by graphical extrapolation.
9	400	Hidnert, P. and Sweeney, W.T.	1927	L	293-771	Sample 1224	66.3	32.7	0.11		0.09	0.84	0.06	Similar to the above specimen; density 2.225 g cm^{-3} at 293 K; expansion measured with increasing temp.
10	400	Hidnert, P. and Sweeney, W.T.	1927	L	696-296	Sample 1224	66.3	32.7	0.11		0.09	0.84	0.06	The above specimen; expansion measured with decreasing temp; zero-point correction of 0.0101% determined by graphical extrapolation.
11	400	Hidnert, P. and Sweeney, W.T.	1927	L	296-769	Sample 1224	66.3	32.7	0.11		0.09	0.84	0.06	The above specimen; expansion measured with increasing temp on second test; zero-point correction of -0.192% determined by graphical extrapolation.
12	400	Hidnert, P. and Sweeney, W.T.	1927	L	613-290	Sample 1224	66.3	32.7	0.11		0.09	0.84	0.06	The above specimen; expansion measured with decreasing temp on second test; zero-point correction of -0.201% determined by graphical interpolation.
13	400	Hidnert, P. and Sweeney, W.T.	1927	L	293-770	Sample 1226	81.0	18.6	0.12		0.11	0.36	0.05	Obtained from same source as above specimen; density 2.242 g cm^{-3} at 293 K; expansion measured on first test with increasing temp.

* Not shown in figure.

SPECIFICATION TABLE 232. THERMAL LINEAR EXPANSION OF ALUMINUM + BERYLLIUM + ΣX_i ALLOYS Al + Be + ΣX_i (continued)

Cur. No.	Ref. No.	Author(s)	Year	Method Used	Temp. Range, K	Name and Specimen Designation	Composition (weight percent) Al	Be	Si	Mg	Cu	Fe	Mn	Composition (continued), Specifications, and Remarks
14	400	Hidnert, P. and Sweeney, W.T.	1927	L	600-289	Sample 1226	81.0	18.6	0.12		0.11	0.36	0.05	The above specimen; expansion measured with decreasing temp; zero-point correction of 0.0364% determined by graphical interpolation.
15	400	Hidnert, P. and Sweeney, W.T.	1927	L	290-679	Sample 1226	81.0	18.6	0.12		0.11	0.36	0.05	The above specimen; expansion measured with increasing temp on second test; zero-point correction of -0.206% determined by graphical interpolation.
16	400	Hidnert, P. and Sweeney, W.T.	1927	L	772-301	Sample 1226	81.0	18.6	0.12		0.11	0.36	0.05	The above specimen; expansion measured with decreasing temp; zero-point correction of -0.199% determined by graphical extrapolation.

DATA TABLE 232. THERMAL LINEAR EXPANSION OF ALUMINUM + BERYLLIUM + ΣX_i ALLOYS   Al + Be + ΣX_i

[Temperature, T, K; Linear Expansion, $\Delta L/L_0$, %]

T	$\Delta L/L_0$
CURVE 1	
293	0.000
373	0.134
473	0.322
573	0.501
CURVE 2*	
573	0.518
523	0.421
373	0.133
293	0.000
CURVE 3	
293	0.000
373	0.130
523	0.421
573	0.521
CURVE 4*	
573	0.518
523	0.419
373	0.133
293	0.000
CURVE 5	
298	0.012
324	0.067
373	0.157
491	0.405
579	0.610
678	0.856
774	1.117
CURVE 6	
643	0.768
424	0.260
291	-0.004
CURVE 7	
294	0.002*
369	0.144
490	0.392
580	0.608*
680	0.859
777	1.128
CURVE 8	
589	0.626
403	0.216
298	0.007
CURVE 9	
293	0.000*
324	0.054
378	0.156
487	0.375
601	0.640
687	0.845
771	1.064
CURVE 10	
696	0.867
427	0.251
296	0.005
CURVE 11	
296	0.005
368	0.138
574	0.591
666	0.802
769	1.070
CURVE 12	
613	0.662
417	0.230
290	-0.005
CURVE 13	
293	0.000*
323	0.054
379	0.170
489	0.408
575	0.626
669	0.861
770	1.143
CURVE 14	
600	0.693
412	0.250
289	-0.008
CURVE 15	
290	-0.005
370	0.151
453	0.340
581	0.644
679	0.909
CURVE 16	
772	1.183
626	0.761
440	0.306
301	0.016

* Not shown in figure.

FIGURE AND TABLE NO. 233R. PROVISIONAL VALUES FOR THERMAL LINEAR EXPANSION OF ALUMINUM + COPPER + ΣX_i ALLOYS Al + Cu + ΣX_i

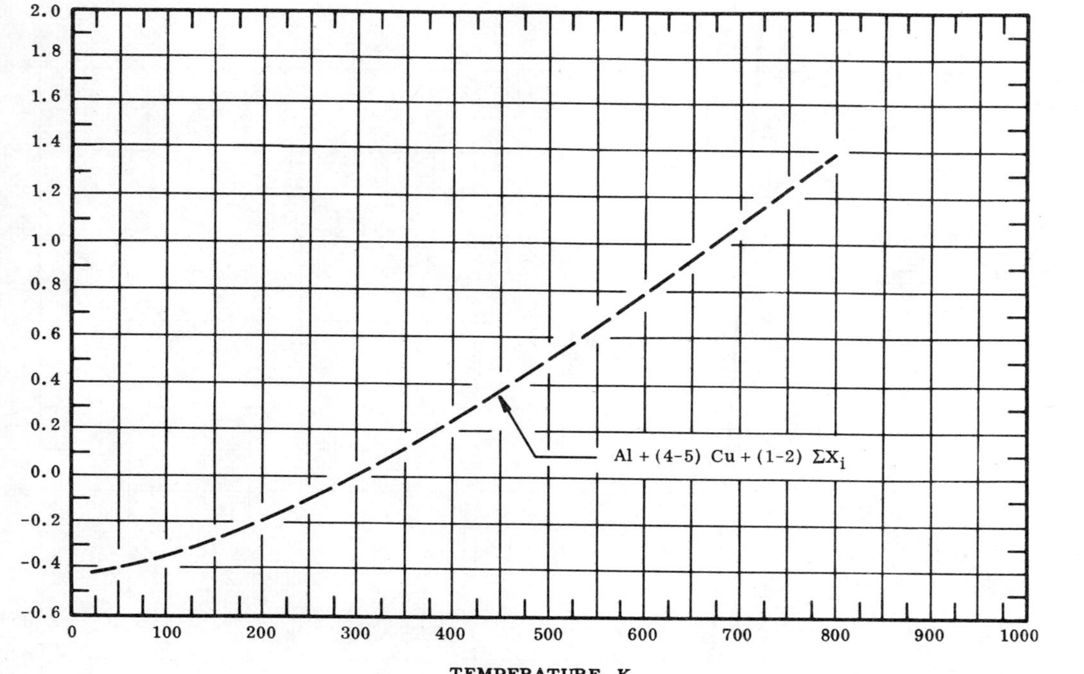

PROVISIONAL VALUES

[Temperature, T, K; Linear Expansion, $\Delta L/L_0$, %; α, K^{-1}]

[Al + (4-5) Cu + (1-2) ΣX_i]

T	$\Delta L/L_0$	$\alpha \times 10^6$
20	-0.428	8.7
25	-0.420	9.0
50	-0.400	10.4
75	-0.371	11.8
100	-0.340	13.1
150	-0.268	15.6
200	-0.184	17.9
250	-0.090	20.0
293	0.000	21.6
400	0.249	25.1
450	0.378	26.4
500	0.515	27.5
550	0.655	28.4
600	0.798	29.1
650	0.946	29.7
700	1.095	30.0
750	1.245	30.1
800	1.396	30.1

REMARKS

[Al + (4-5) Cu + (1-2) ΣX_i]: The tabulated values for well-annealed alloy of type Al 2018, Al 2017, Al 2014, Al 2024, Al 2025, Al 2020, Y-alloy, Russian V-65, and Russian D16, are considered accurate to within ±7% over the entire temperature range. These values can be represented approximately by the following equation:

$$\Delta L/L_0 = -0.451 + 7.585 \times 10^{-4} T + 2.971 \times 10^{-6} T^2 - 1.304 \times 10^{-9} T^3$$

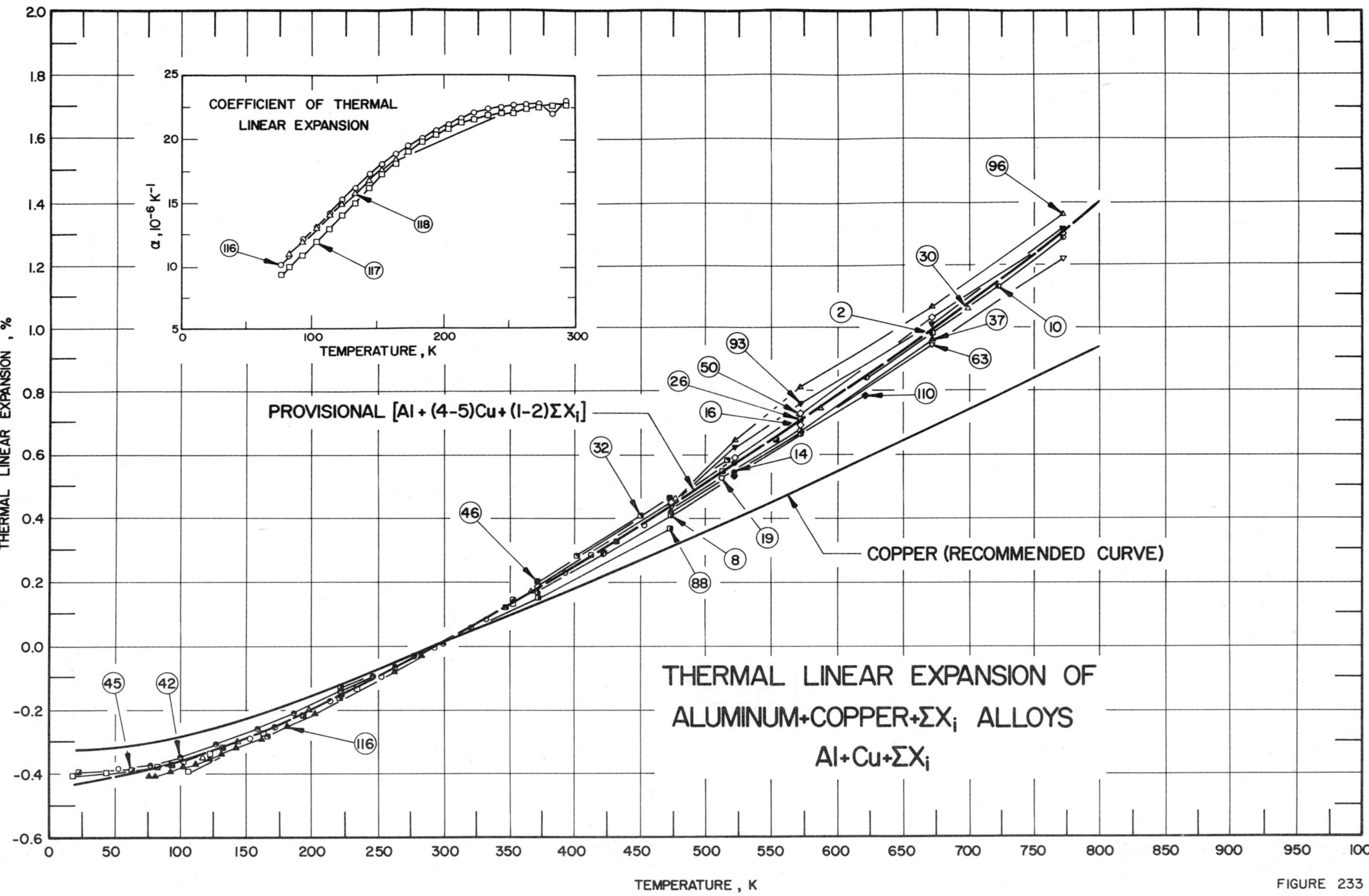
THERMAL LINEAR EXPANSION OF
ALUMINUM+COPPER+ΣXi ALLOYS
Al+Cu+ΣXi
THERMAL LINEAR EXPANSION, %
TEMPERATURE, K
PROVISIONAL [Al + (4-5)Cu + (1-2)ΣXi]
COPPER (RECOMMENDED CURVE)
COEFFICIENT OF THERMAL
LINEAR EXPANSION
α, 10⁻⁶ K⁻¹
TEMPERATURE, K
2.0
1.8
1.6
1.4
1.2
1.0
0.8
0.6
0.4
0.2
0.0
-0.2
-0.4
-0.6
0
50
100
150
200
250
300
350
400
450
500
550
600
650
700
750
800
850
900
950
1000
96
30
2
37
10
93
63
50
26
110
16
14
32
46
19
8
88
45
42
116
118
117
25
20
15
10
5

FIGURE 233

SPECIFICATION TABLE 233. THERMAL LINEAR EXPANSION OF ALUMINUM + COPPER + ΣX_i ALLOYS $Al + Cu + \Sigma X_i$

Cur. No.	Ref. No.	Author(s)	Year	Method Used	Temp. Range, K	Name and Specimen Designation	Composition (weight percent) Al	Cu	Fe	Si	Mn	Mg	Ni	Composition (continued), Specifications, and Remarks
1*	19	Hidnert, P. and Krider, H.S.	1952	L	293-673	Alcoa 18S	Bal.	3.98	0.45	0.58	0.05	0.64	2.01	0.02 Zr, 0.01 each Bi, Cr, Pb, Ti; specimen solution heat treated 1 hr at 789 K; quenched in water and aged 10 hrs at 444 K; expansion measured with increasing temp.
2*	19	Hidnert, P. and Krider, H.S.	1952	L	673-293	Alcoa 18S	Bal.	3.98	0.45	0.58				0.2 Zr, 0.1 each Bi, Cr, Pb, Ti; the above specimen; expansion measured with decreasing temp.
3*	19	Hidnert, P. and Krider, H.S.	1952	L	293-673	Alcoa 18S	Bal.	3.98	0.45	0.58				0.02 Zr, 0.1 each Bi, Cr, Pb, Ti; similar to the above specimen except aged 100 hrs at 644 K; expansion measured with increasing temp.
4*	19	Hidnert, P. and Krider, H.S.	1952	L	673-293	Alcoa 18S	Bal.	3.98	0.45	0.58				0.2 Zr, 0.1 each Bi, Cr, Pb, Ti; the above specimen; expansion measured with decreasing temp.
5*	19	Hidnert, P. and Krider, H.S.	1952	L	293-673	Alcoa 18S	Bal.	3.98	0.45					0.2 Zr, 0.1 each Bi, Cr, Pb, Ti; similar to the above specimen, except aged 500 hrs at 700 K; expansion measured with increasing temp.
6*	19	Hidnert, P. and Krider, H.S.	1952	L	673-293	Alcoa 18S	Bal.	3.98	0.45					0.2 Zr, 0.1 each Bi, Cr, Pb, Ti; the above specimen; expansion measured with decreasing temp.
7*	463	Willey, L.A. and Fink, W.L.	1945	I	293-573	Alcoa 18S	Bal.	4.0				0.5	2.0	Bar wrought form, forged to 1 in. sq; tested in the as-forged condition.
8	463	Willey, L.A. and Fink, W.L.	1945	I	293-573	Alcoa 18S	Bal.	4.0	2.0	0.5		0.5	2.0	Similar to the above specimen, except annealed to 623 K.
9*	463	Willey, L.A. and Fink, W.L.	1945	I	213-373	Alcoa 18S		4.0	2.0	0.5				Similar to the above specimen, except heat-treated and aged according to standard practice.
10	466	Lucks, C.F., Thompson, H.B., Smith, A.R., Curry, F.P., Deem, H.W. and Bing, G.F.	1951	L	83-773	24S-T4 aluminum	93.4	4.5			0.6	1.5		Heat treatment T4; material supplied by Aluminum Co. of America; density 2.78 g cm^{-3} at 293 K.
11*	463	Willey, L.A. and Fink, W.L.	1945	I	293-573	Alcoa 25S	Bal.	4.5		0.8	0.8			Bar wrought form, forged to 1 in. sq; tested in as-forged condition.
12*	463	Willey, L.A. and Fink, W.L.	1945	I	293-573	Alcoa 25S	Bal.	4.5	0.8	0.8				Similar to the above specimen, except annealed to 623 K.
13*	463	Willey, L.A. and Fink, W.L.	1945	I	213-373	Alcoa 25S	Bal.	4.5	0.8	0.8				Similar to the above specimen, except heat-treated and aged according to standard practice.
14	19	Hidnert, P. and Krider, H.S.	1952	L	223-573	Al 122	Bal.	9.0	1.0	0.4	0.3	0.03		Specimen heated 20 hrs at 498 K and air cooled; expansion measured with increasing temp.
15*	19	Hidnert, P. and Krider, H.S.	1952	L	293-523	Al 31	Bal.	1.8	0.6	0.2	0.02	0.6		1.3 Sn, 1.1 Zn, 0.23 Ti, 0.2 Cr; sand cast specimen; expansion measured with increasing temp.
16	19	Hidnert, P. and Krider, H.S.	1952	L	293-523	Al 31	Bal.	1.8	0.6	0.2				1.3 Sn, 1.1 Zn, 0.23 Ti, 0.2 Cr; similar to the above specimen, except heated 2 hrs at 616 K and cooled slowly; expansion measured with increasing temp.
17*	19	Hidnert, P. and Krider, H.S.	1952	L	573-223	Al 31	Bal.	1.8	0.6	0.2				1.3 Sn, 1.1 Zn, 0.23 Ti, 0.2 Cr; the above specimen; expansion measured with decreasing temp.

* Not shown in figure.

SPECIFICATION TABLE 233. THERMAL LINEAR EXPANSION OF ALUMINUM + COPPER + ΣX_i ALLOYS Al + Cu + ΣX_i (continued)

Cur. No.	Ref. No.	Author(s)	Year	Method Used	Temp. Range, K	Name and Specimen Designation	Al	Cu	Fe	Si	Mn	Mg	Ni	Composition (continued), Specifications, and Remarks
							Composition (weight percent)							
18*	191	Pollock, D.B.	1969	L	77, 293	Al 2014 T-6	93.6	4.4		0.8	0.8	0.4		Specimen 3.2 cm long.
19	15	Rhodes, B.L., Moeller, C.E., Hopkins, V. and Marx, T.I.	1963	L	18-573	Al 2014 T-6	Bal.	4.5	1.0	0.8	0.8	0.4		0.25 Zn, 0.15 Ti, 0.10 Cr, 0.15 ΣX_i; specimen 10.16 cm long, 6.2 mm dia.
20*	463	Willey, L.A. and Fink, W.L.	1945	I	293-573	(Al 2014) Alcoa 14S	Bal.	4.4		0.8	0.8	0.4		Bar wrought form, forged to 1 in. sq; tested in as-forged condition.
21*	463	Willey, L.A. and Fink, W.L.	1945	I	293-573	(Al 2014) Alcoa 14S	Bal.	4.4		0.8	0.8	0.4		Similar to the above specimen, except annealed to 623 K.
22*	463	Willey, L.A. and Fink, W.L.	1945	I	213-373	(Al 2014) Alcoa 14S	Bal.	4.4		0.8	0.8			Similar to the above specimen, except heat-treated and aged according to standard practice.
23*	463	Willey, L.A. and Fink, W.L.	1945	I	293-573	(Al 2017) Alcoa 17S	Bal.	4.0			0.5	0.5		Rod specimen, 0.75 in. dia rolled and drawn; tested in as-drawn condition.
24*	463	Willey, L.A. and Fink, W.L.	1945	I	293-573	(Al 2017) Alcoa 17S	Bal.	4.0			0.5	0.5		Similar to the above specimen, except annealed to 623 K.
25*	463	Willey, L.A. and Fink, W.L.	1945	I	213-373	(Al 2017) Alcoa 17S	Bal.	4.0			0.5	0.5		Similar to the above specimen, except heat-treated and aged according to standard practice.
26	15	Rhodes, B.L., Moeller, C.E., Hopkins, V. and Marx, T.I.	1963	I/L	18-573	X2020-T6	Bal.	4.5	0.40	0.40	0.55	0.03		0.25 Zn, 0.10 Ti, 0.15 ΣX_i; specimen 10.16 cm long, 6.2 mm dia.
27*	462	Arp, V., Wilson, J.H., Winrich, L. and Sikora, P.	1962	L	20-293	Al 2020	Bal.	4.3	0.1	0.1	0.5			1.1 Li, 0.2 Cd; hardness R_B91.
28*	467	Clark, A.F.	1968	L	0-300	Al 2024	93.6	4.1	0.2	0.1	0.5	1.4		0.1 Zn; specimen machined to 20.32 cm long, 0.64 cm sq; Rockwell hardness B-25.
29*	467	Clark, A.F.	1968	L	0-300	Al 2024	Bal.	4.1	0.2					0.1 Zn; similar to the above specimen; hardness B-83.
30	123	Lucks, C.F. and Deem, H.W.	1958	L	116-699	Al 2024-T4								0.375 in. dia, 3 in. long; material from Aluminum Co. of America.
31*	462	Arp, V., Wilson, J.H., Winrich, L. and Sikora, P.	1962	L	20-293	Al 2024	93.6	4.1	1.4	0.5	0.2	0.1	0.1	0.1 Zn; hardness R_B83, A.S.M. condition T-86.
32	225	Valentich, J.	1965	L	106-516	Al 2024								No details given; zero-point correction is -0.1%.

* Not shown in figure.

SPECIFICATION TABLE 233. THERMAL LINEAR EXPANSION OF ALUMINUM + COPPER + ΣX_i ALLOYS Al + Cu + ΣX_i (continued)

Cur. No.	Ref. No.	Author(s)	Year	Method Used	Temp. Range, K	Name and Specimen Designation	Composition (weight percent) Al	Cu	Fe	Si	Mn	Mg	Ni	Composition (continued), Specifications, and Remarks
33*	19	Hidnert, P. and Krider, H.S.	1952	L	293-673	Al 24S (Al 2024)	Bal.	4.41	0.25	0.10	0.67	1.41	0.01	0.02 Zn, 0.01 each Bi, Cr, Pb, Ti; specimen solution heat treated 1 hr at 766 K, quenched in water and aged to room temp; expansion measured with increasing temp.
34*	19	Hidnert, P. and Krider, H.S.	1952	L	673-293	Al 24S (Al 2024)	Bal.	4.41	0.25	0.10				0.02 Zn, 0.01 each Bi, Cr, Pb, Ti; the above specimen; expansion measured with decreasing temp.
35*	19	Hidnert, P. and Krider, H.S.	1952	L	293-673	Al 24S (Al 2024)	Bal.	4.41	0.25	0.10				0.02 Zn, 0.01 each Bi, Cr, Pb, Ti; similar to the above specimen, then aged 100 hrs at 644 K; expansion measured with increasing temp.
36*	19	Hidnert, P. and Krider, H.S.	1952	L	673-293	Al 24S (Al 2024)	Bal.	4.41	0.25	0.10				0.02 Zn, 0.01 each Bi, Cr, Pb, Ti; the above specimen; expansion measured with decreasing temp.
37	19	Hidnert, P. and Krider, H.S.	1952	L	293-673	Al 24S (Al 2024)	Bal.	4.41	0.25	0.10				0.02 Zn, 0.01 each Bi, Cr, Pb, Ti; similar to the above specimen, except aged 500 hrs at 700 K; expansion measured with increasing temp.
38*	19	Hidnert, P. and Krider, H.S.	1952	L	673-293	Al 24S (Al 2024)	Bal.	4.41	0.25	0.10				0.02 Zn, 0.01 each Bi, Cr, Pb, Ti; the above specimen; expansion measured with decreasing temp.
39*	463	Willey, L.A. and Fink, W.L.	1945	I	293-573	(Al 2024) Alcoa 24S		4.5	1.5	0.6				Sheet wrought form cold-rolled to 0.064 in. thick; tested in as-rolled condition.
40*	463	Willey, L.A. and Fink, W.L.	1945	I	293-573	(Al 2024) Alcoa 24S	Bal.	4.5			0.6	1.5		Similar to the above specimen, except annealed to 623 K.
41*	463	Willey, L.A. and Fink, W.L.	1945	I	213-373	(Al 2024) Alcoa 24S	Bal.	4.5						Similar to the above specimen, except heat-treated and aged according to standard practice.
42	511	Hertz, J.	1962		77-300	Al 2024 T-3								No details given; zero-point correction is 0.014%.
43*	463	Willey, L.A. and Fink, W.L.	1945	I	213-373	(Al 2117) Alcoa A17S	Bal.	2.5				0.3		Rod wrought form, rolled and drawn to 0.75 in. dia, heat-treated and aged according to standard practice.
44*	199	Bell, I.P.	1954	L	293-673	Al 2219 RR 57	92.3	6.8	0.3	0.2	0.4			No details given.
45	15	Rhodes, B.L., Moeller, C.E., Hopkins, V. and Marx, T.I.	1963	L	18-573	Al 2219 T81	Bal.	6.3	0.30	0.20	0.30	0.02		0.10 Zn, 0.06 Ti, 0.15 ΣX_i; specimen 10.16 cm long, 6.2 mm dia.
46	199	Bell, I.P.	1954	L	293-673	Al 2618 RR 58	92.75	2.7	1.3	0.25		1.8	1.2	No details given.
47*	20	Hidnert, P.	1925	T	293-773	Duralumin S835	Bal.	3.68	0.35	0.25	0.57	0.36		Sand cast specimen.

* Not shown in figure.

SPECIFICATION TABLE 233. THERMAL LINEAR EXPANSION OF ALUMINUM + COPPER + ΣX_i ALLOYS Al + Cu + ΣX_i (continued)

Cur. No.	Ref. No.	Author(s)	Year	Method Used	Temp. Range, K	Name and Specimen Designation	Composition (weight percent) Al	Cu	Fe	Si	Mn	Mg	Ni	Composition (continued), Specifications, and Remarks
48*	20	Hidnert, P.	1925	T	293-773	Duralumin S836	Bal.	3.68	0.35	0.25				Cut from same bar as above specimen.
49*	20	Hidnert, P.	1925	T	293-773	Duralumin S456	Bal.	3.74	0.52	0.30		1.08		Hot rolled at 683 K from 8.89 cm to 0.66 cm thick.
50	20	Hidnert, P.	1925	T	293-773	Duralumin S430	Bal.	3.74	0.52		0.57			Duplicate of the above specimen, cold rolled to 0.23 cm and quenched in water from 793 K, then aged two days at 393 K.
51*	20	Hidnert, P.	1925	T	293-573	Duralumin S935	Bal.	3.66	0.37	0.16	0.51	0.52		0.20 Ca; hard rolled specimen.
52*	20	Hidnert, P.	1925	T	293-773	Duralumin S935	Bal.	3.66	0.37	0.16				0.20 Ca; the above specimen; expansion measured on second test.
53*	20	Hidnert, P.	1925	T	293-573	Duralumin S936	Bal.	3.66	0.37	0.16				0.20 Ca; duplicate of the above specimen.
54*	20	Hidnert, P.	1925	T	293-773	Duralumin S936	Bal.	3.66	0.37	0.16				0.20 Ca; the above specimen; expansion measured on second test.
55*	20	Hidnert, P.	1925	T	293-573	Duralumin S937	Bal.	3.66	0.37	0.16				0.20 Ca; duplicate of the above specimen, heated to 773 K and quenched.
56*	20	Hidnert, P.	1925	T	293-773	Duralumin S937	Bal.	3.66	0.37	0.16				0.20 Ca; the above specimen; expansion measured on second test.
57*	20	Hidnert, P.	1925	T	293-573	Duralumin S938	Bal.	3.66	0.37	0.16	0.51	0.52	0.20	0.20 Ca; duplicate of the above specimen.
58*	20	Hidnert, P.	1925	T	293-773	Duralumin S938	Bal.	3.66	0.37	0.16				0.20 Ca; the above specimen; expansion measured on second test.
59*	19	Hidnert, P. and Krider, H.S.	1952	L	293-673	Aluminum alloy XB-18S	Bal.	3.89	0.31	0.58	0.01	1.43	2.14	0.02 Zn, 0.01 each Bi, Cr, Pb, Ti; specimen solution heat treated 1 hr at 789 K, quenched in water and aged 10 hrs at 444 K; expansion measured with increasing temp.
60*	19	Hidnert, P. and Krider, H.S.	1952	L	673-293	Aluminum alloy XB-18S	Bal.	3.89	0.31	0.58				0.02 Zn, 0.01 each Bi, Cr, Pb, Ti; the above specimen; expansion measured with decreasing temp.
61*	19	Hidnert, P. and Krider, H.S.	1952	L	293-673	Aluminum alloy XB-18S	Bal.	3.89	0.31	0.58				0.02 Zn, 0.01 each Bi, Cr, Pb, Ti; similar to the above specimen except aged 100 hrs at 644 K; expansion measured with increasing temp.
62*	19	Hidnert, P. and Krider, H.S.	1952	L	673-293	Aluminum alloy XB-18S	Bal.	3.89	0.31	0.58				0.02 Zn, 0.01 each Bi, Cr, Pb, Ti; the above specimen; expansion measured with decreasing temp.
63	13	Irmann, R.	1955		293-773	Y-alloy	92.5	4				1.5	2	High heat-strength alloy.
64*	464	Barber, C.R.	1949	L	293-573	Y-alloy	Bal.	3.76	0.40	0.15		1.33	1.85	Rod specimen 1.4 cm dia, 30 cm long; tested as received.
65*	464	Barber, C.R.	1949	L	293-573	Y-alloy	Bal.	3.76	0.40	0.15		1.33	1.85	Similar to the above specimen, except heat treated at 784 K, quenched in hot water and aged at room temp.

* Not shown in figure.

SPECIFICATION TABLE 233. THERMAL LINEAR EXPANSION OF ALUMINUM + COPPER + ΣX_i ALLOYS Al + Cu + ΣX_i (continued)

Cur. No.	Ref. No.	Author(s)	Year	Method Used	Temp. Range, K	Name and Specimen Designation	Composition (weight percent) Al	Cu	Fe	Si	Mn	Mg	Ni	Composition (continued), Specifications, and Remarks
66*	19	Hidnert, P. and Krider, H.S.	1952	L	293-573	Sample 967AN	Bal.	5.81	0.42	0.36				Cast specimen, heated to 773 K and cooled slowly, reheated to 573 K and cooled slowly; expansion measured with increasing temp.
67*	19	Hidnert, P. and Krider, H.S.	1952	L	573-223	Sample 967AN	Bal.	5.81	0.42	0.36				The above specimen; expansion measured with decreasing temp.
68*	19	Hidnert, P. and Krider, H.S.	1952	L	293-573	Sample 968AN	Bal.	5.81	0.42	0.36				Similar to the above specimen; expansion measured with increasing temp.
69*	19	Hidnert, P. and Krider, H.S.	1952	L	573-293	Sample 968AN	Bal.	5.81	0.42	0.36				The above specimen; expansion measured with decreasing temp.
70*	19	Hidnert, P. and Krider, H.S.	1952	L	293-573	Sample 830AN	Bal.	7.68	0.46	0.39	0.33			Similar to the above specimen; expansion measured with increasing temp.
71*	19	Hidnert, P. and Krider, H.S.	1952	L	573-223	Sample 830AN	Bal.	7.68	0.46	0.39				The above specimen; expansion measured with decreasing temp.
72*	19	Hidnert, P. and Krider, H.S.	1952	L	293-573	Sample 831AN	Bal.	7.87	0.45	0.33	0.22			Similar to the above specimen; expansion measured with increasing temp.
73*	19	Hidnert, P. and Krider, H.S.	1952	L	573-223	Sample 831AN	Bal.	7.87	0.45	0.33				The above specimen; expansion measured with decreasing temp.
74*	19	Hidnert, P. and Krider, H.S.	1952	L	293-573	Sample 832AN	Bal.	7.87	0.45	0.33				Similar to the above specimen, except heated to 573 K and slowly cooled twice; expansion measured with increasing temp.
75*	19	Hidnert, P. and Krider, H.S.	1952	L	573-293	Sample 832AN	Bal.	7.87	0.45	0.33	0.22			The above specimen; expansion measured with decreasing temp.
76*	19	Hidnert, P. and Krider, H.S.	1952	L	293-573	Sample 969AN	Bal.	9.95	0.44	0.39				Cast specimen, heated to 773 K and slowly cooled, reheated to 573 K and slowly cooled; expansion measured with increasing temp.
77*	19	Hidnert, P. and Krider, H.S.	1952	L	573-223	Sample 969AN	Bal.	9.95	0.44	0.39				The above specimen; expansion measured with decreasing temp.
78*	19	Hidnert, P. and Krider, H.S.	1952	L	293-573	Sample 970AN	Bal.	9.95	0.44	0.39				Similar to the above specimen; expansion measured with increasing temp.
79*	19	Hidnert, P. and Krider, H.S.	1952	L	573-293	Sample 970AN	Bal.	9.95	0.44	0.39				The above specimen; expansion measured with decreasing temp.
80*	19	Hidnert, P. and Krider, H.S.	1952	L	293-573	Sample 971AN	Bal.	11.88	0.43	0.39				Similar to the above specimen; expansion measured with increasing temp.
81*	19	Hidnert, P. and Krider, H.S.	1952	L	573-223	Sample 971AN	Bal.	11.88	0.43	0.39				The above specimen; expansion measured with decreasing temp.
82*	19	Hidnert, P. and Krider, H.S.	1952	L	293-573	Sample 972AN	Bal.	11.88	0.43	0.39				Similar to the above specimen; expansion measured with increasing temp.
83*	19	Hidnert, P. and Krider, H.S.	1952	L	573-293	Sample 972AN	Bal.	11.88	0.43	0.39				The above specimen; expansion measured with decreasing temp.
84*	19	Hidnert, P. and Krider, H.S.	1952	L	293-473	Sample 1096	Bal.	10.09	1.14	0.21		0.26		Specimen cast in iron mold; expansion measured with increasing temp.

* Not shown in figure.

SPECIFICATION TABLE 233. THERMAL LINEAR EXPANSION OF ALUMINUM + COPPER + ΣX_i ALLOYS Al + Cu + ΣX_i (continued)

Cur. No.	Ref. No.	Author(s)	Year	Method Used	Temp. Range, K	Name and Specimen Designation	Composition (weight percent) Al	Cu	Fe	Si	Mn	Mg	Ni	Composition (continued), Specifications, and Remarks
85*	19	Hidnert, P. and Krider, H.S.	1952	L	473-293	Sample 1096	Bal.	10.09	1.14					The above specimen; expansion measured with decreasing temp.
86*	19	Hidnert, P. and Krider, H.S.	1952	L	293-473	Sample 1097	Bal.	10.09	1.14					Specimen heated to 472 K and cooled very slowly; expansion measured with increasing temp.
87*	19	Hidnert, P. and Krider, H.S.	1952	L	473-293	Sample 1097	Bal.	10.09	1.14					The above specimen; expansion measured with decreasing temp.
88	19	Hidnert, P. and Krider, H.S.	1952	L	223-473	Sample 1267	Bal.	7.21	0.84	6.78			7.18	Specimen annealed; expansion measured with increasing temp.
89*	19	Hidnert, P. and Krider, H.S.	1952	L	293-473	Sample 1100	Bal.	9.86	0.30	9.79			4.03	Specimen cast in iron mold; expansion measured with increasing temp.
90*	19	Hidnert, P. and Krider, H.S.	1952	L	473-293	Sample 1100	Bal.	9.86	0.30	9.79				The above specimen; expansion measured with decreasing temp.
91*	19	Hidnert, P. and Krider, H.S.	1952	L	293-473	Sample 1101	Bal.	9.86	0.30	9.79				Specimen heated to 672 K and cooled very slowly; expansion measured with increasing temp.
92*	19	Hidnert, P. and Krider, H.S.	1952	L	473-293	Sample 1101	Bal.	9.86	0.30	9.79				The above specimen; expansion measured with decreasing temp.
93	20	Hidnert, P.	1925	T	293-773	S833	Bal.	3.75	0.36	0.30	0.18			Sand cast specimen.
94*	20	Hidnert, P.	1925	T	293-573	S834	Bal.	3.75	0.36	0.30				Specimen cut from same bar as the above.
95*	20	Hidnert, P.	1925	T	293-773	S967	Bal.	5.81	0.42	0.36				Specimen cast in green sand; similar to specimen used for curves 66 and 67 except not heat treated.
96	20	Hidnert, P.	1925	T	293-773	S968	Bal.	5.81	0.42	0.36				Specimen cut from same bar as the above.
97*	20	Hidnert, P.	1925	T	293-573	S829	Bal.	7.68	0.46	0.39	0.33			Specimen cast in green sand.
98*	20	Hidnert, P.	1925	T	293-773	S830	Bal.	7.68	0.46	0.39				Cut from same bar as the above specimen; similar to specimen used for curves 70, 71 except not heat treated.
99*	20	Hidnert, P.	1925	T	293-773	S831	Bal.	7.87	0.45	0.33	0.22			Specimen cast in green sand; similar to specimen used for curves 72, 73 except not heat treated.
100*	20	Hidnert, P.	1925	T	293-573	S832	Bal.	7.87	0.45	0.33				Cut from same rod as above specimen; similar to specimen used for curves 74, 75 except not heat treated.
101*	20	Hidnert, P.	1925	T	293-773	S969	Bal.	9.95	0.44	0.39				Specimen cast in green sand; similar to specimen used for curves 76, 77 except not heat treated.
102*	20	Hidnert, P.	1925	T	293-773	S970	Bal.	9.95	0.44	0.39				Cut from same bar as the above specimen; similar to specimen used for curves 78, 79 except not heat treated.
103*	20	Hidnert, P.	1925	T	293-773	S971	Bal.	11.88	0.43	0.39				Specimen cast in green sand; similar to specimen used for curves 80, 81 except not heat treated.
104*	20	Hidnert, P.	1925	T	293-773	S972	Bal.	11.88	0.43	0.39				Cut from same bar as the above specimen; similar to specimen used for curves 82, 83 except not heat treated.

* Not shown in figure.

SPECIFICATION TABLE 233. THERMAL LINEAR EXPANSION OF ALUMINUM + COPPER + ΣX_i ALLOYS Al + Cu + ΣX_i (continued)

Cur. No.	Ref. No.	Author(s)	Year	Method Used	Temp. Range, K	Name and Specimen Designation	Composition (weight percent) Al	Cu	Fe	Si	Mn	Mg	Ni	Composition (continued), Specifications, and Remarks
105*	20	Hidnert, P.	1925	T	293-773	S839	Bal.	1.91	0.51	0.30	1.08			Sand cast specimen.
106*	20	Hidnert, P.	1925	T	293-573	S840	Bal.	1.91	0.51	0.30				Cut from same bar as above specimen.
107*	20	Hidnert, P.	1925	T	293-573	S872	Bal.	4.41	0.57	3.75				Sand cast specimen, heated to 673 K and cooled in furnace before testing.
108*	20	Hidnert, P.	1925	T	293-573	S875	Bal.	6.62	0.64	4.08				Similar to the above specimen.
109*	464	Barber, C.R.	1949	L	293-573	RR59	Bal.	3.31	1.23	0.88		1.46	1.20	0.07 Ti; rod specimen 1.4 cm dia, 30 cm long; heat treated 2 hrs at 798 K, quenched; then 16 hrs at 443 K and quenched.
110	35	Honda, K. and Okubo, Y.	1924	L	333-622	2E8	Bal.	12			1			Cylindrical rod specimen, 15.025 cm long, 5 mm thick; zero-point correction of 0.092% obtained by extrapolation.
111*	35	Honda, K. and Okubo, Y.	1924	L	331-622	2L11		7						1 Sn; similar to the above specimen, 15.015 cm long; zero-point correction of 0.089% obtained by extrapolation.
112*	35	Honda, K. and Okubo, Y.	1924	L	332-624	R		2.0						1 Cr; similar to the above specimen, 15.035 cm long; zero-point correction of 0.092% obtained by extrapolation.
113*	35	Honda, K. and Okubo, Y.	1924	L	332-621	Y		4				1.5	2	Similar to the above specimen, 15.000 cm long; zero-point correction of 0.091% obtained by extrapolation.
114*	35	Honda, K. and Okubo, Y.	1924	L	333-623	N		2				1.5		Similar to the above specimen, 15.045 cm long; zero-point correction of 0.096% obtained by extrapolation.
115*	35	Honda, K. and Okubo, Y.	1924	L	329-620	D1		4			0.5	0.5		Similar to the above specimen, 15.025 cm long; zero-point correction of 0.086% obtained by extrapolation.
116	551	Belov, A.K.	1968	L	77-293	V65 Russian alloy	Bal.	4.5	0.2	0.25	0.5	0.3		Cylindrical specimen 3.5 mm in diameter x 50 mm long; sample heated and cooled at a rate of 3-7 deg. per min.
117	551	Belov, A.K.	1968	L	77-293	D16 Russian alloy	Bal.	4.9	0.8	0.8	0.9	1.8		0.3 Zn; similar to the above specimen.
118	551	Belov, A.K.	1968	L	77-293	D18 Russian alloy	Bal.	3	0.5	0.5	0.2	0.5		Similar to the above specimen.

* Not shown in figure.

SPECIFICATION TABLE 233. THERMAL LINEAR EXPANSION OF ALUMINUM + COPPER + ΣX_i ALLOYS Al + Cu + ΣX_i (continued)

Cur. No.	Ref. No.	Author(s)	Year	Method Used	Temp. Range, K	Name and Specimen Designation	Composition (weight percent) Al	Cu	Fe	Si	Mn	Mg	Ni	Composition (continued), Specifications, and Remarks
119	603	McCollough, E. E.	1931	I	298-373		Bal.	4.00		0.30	0.55			The material for the preparation of the alloy was furnished by the Aluminum Co. of America.
120	603	McCollough, E. E.	1931	I	298-373		Bal.	4.14		0.22	0.59			Same as above.
121	603	McCollough, E. E.	1931	I	298-373		Bal.	4.37		0.74	0.81			Same as above.

DATA TABLE 233. THERMAL LINEAR EXPANSION OF ALUMINUM + COPPER + ΣX_i ALLOYS Al + Cu + ΣX_i

[Temperature, T, K; Linear Expansion, $\Delta L/L_0$, %]

T	$\Delta L/L_0$
CURVE 1*	
293	0.000
373	0.184
473	0.425
573	0.706
673	0.984
CURVE 2*	
673	0.954
473	0.421*
373	0.179*
293	0.000
CURVE 3*	
293	0.000
373	0.179
473	0.418
573	0.675
673	0.942
CURVE 4*	
673	0.958
473	0.421
373	0.178
293	0.000
CURVE 5*	
293	0.000
373	0.184
473	0.427
573	0.686
673	0.965
CURVE 6*	
673	0.958
473	0.421
373	0.178
293	0.000
CURVE 7*	
293	0.000
CURVE 7 (cont.)*	
373	0.178
473	0.421
573	0.694
CURVE 8	
293	0.000*
373	0.179
473	0.419
573	0.678
CURVE 9*	
213	-0.167
293	0.000
373	0.181
CURVE 10	
83	-0.3797*
123	-0.3281*
173	-0.2464
223	-0.1494*
323	0.0663
373	0.1763*
423	0.2973
473	0.4304*
523	0.5773
573	0.7139*
623	0.8423
673	0.9758*
723	1.1375
773	1.2948
CURVE 11*	
293	0.000
373	0.182
473	0.428
573	0.722
CURVE 12*	
293	0.000
373	0.182
473	0.427
CURVE 12 (cont.)*	
573	0.686
CURVE 13*	
213	-0.173
293	0.000
373	0.186
CURVE 14	
223	-0.145
293	0.000*
373	0.178*
473	0.419*
523	0.547
573	0.675*
CURVE 15*	
293	0.000
373	0.186
473	0.443
523	0.584
CURVE 16	
293	0.000*
373	0.182*
473	0.432*
573	0.703
CURVE 17*	
573	0.703
473	0.436
373	0.186
293	0.000
223	-0.151
CURVE 18*	
77	-0.378
293	0.000
CURVE 19	
53	-0.380
CURVE 19 (cont.)	
63	-0.380*
73	-0.380*
83	-0.372*
93	-0.363*
103	-0.353
113	-0.341*
123	-0.327*
133	-0.312*
143	-0.297*
153	-0.281
163	-0.264*
173	-0.246*
193	-0.209*
213	-0.171
233	-0.130*
253	-0.088
273	-0.044*
293	0.000
313	0.045*
333	0.091
353	0.045*
333	0.091
353	0.138*
373	0.186*
393	0.235
413	0.284*
433	0.333*
453	0.381
473	0.431*
493	0.482*
513	0.532
533	0.583*
553	0.634*
573	0.684
CURVE 20*	
293	0.000
373	0.182
473	0.428
573	0.722
CURVE 21*	
293	0.000
373	0.180
CURVE 21 (cont.)*	
473	0.425
573	0.686
CURVE 22*	
213	-0.171
293	0.000
373	0.184
CURVE 23*	
293	0.000
373	0.180
473	0.427
573	0.675
CURVE 24*	
293	0.000
373	0.184
473	0.432
573	0.700
CURVE 25*	
213	-0.173
293	0.000
373	0.188
CURVE 26	
18	-0.402
23	-0.400*
33	-0.398*
43	-0.396
53	-0.394*
63	-0.391*
73	-0.385*
83	-0.377
93	-0.368*
103	-0.356*
113	-0.343*
123	-0.330
133	-0.315*
143	-0.300*
153	-0.284*
CURVE 26 (cont.)	
163	-0.267
173	-0.249*
193	-0.211*
213	-0.172*
233	-0.132
253	-0.089*
273	-0.043*
293	0.000*
313	0.046*
333	0.093*
353	0.140
373	0.190*
393	0.240*
413	0.290
433	0.341*
453	0.393*
473	0.446*
493	0.500
513	0.554*
533	0.609*
553	0.665*
573	0.721
CURVE 27*	
20	-0.406
40	-0.404
60	-0.397
80	-0.383
100	-0.361
120	-0.335
140	-0.305
160	-0.271
180	-0.235
200	-0.197
220	-0.158
240	-0.117
260	-0.073
273	-0.043
280	-0.029
293	0.000
CURVE 28*	
0	-0.392
10	-0.392

* Not shown in figure.

DATA TABLE 233. THERMAL LINEAR EXPANSION OF ALUMINUM + COPPER + ΣX_i ALLOYS Al + Cu + ΣX_i (continued)

T	$\Delta L/L_0$
CURVE 28 (cont.)*	
20	-0.392
30	-0.391
40	-0.389
50	-0.386
60	-0.382
70	-0.375
80	-0.367
90	-0.358
100	-0.347
120	-0.323
140	-0.294
160	-0.262
180	-0.228
200	-0.191
220	-0.153
240	-0.113
260	-0.071
273	-0.044
280	-0.029
293	0.000
300	0.015
CURVE 29*	
0	-0.396
10	-0.396
20	-0.396
30	-0.396
40	-0.394
50	-0.391
60	-0.387
70	-0.380
80	-0.372
90	-0.363
100	-0.351
120	-0.325
140	-0.295
160	-0.262
180	-0.227
200	-0.190
220	-0.151
240	-0.111
260	-0.070
273	-0.043
280	-0.028
293	0.000
300	0.015

T	$\Delta L/L_0$
CURVE 30	
116	-0.345
144	-0.294
199	-0.194
293	0.000*
366	0.174
477	0.464
588	0.755
699	1.062
CURVE 31*	
20	-0.411
40	-0.408
60	-0.400
80	-0.385
100	-0.364
120	-0.338
140	-0.308
160	-0.275
180	-0.237
200	-0.198
220	-0.158
240	-0.116
260	-0.073
273	-0.044
280	-0.029
293	0.000
CURVE 32	
106	-0.394
166	-0.278
223	-0.153
293	0.000
347	0.132
401	0.286
450	0.413
516	0.583
CURVE 33*	
293	0.000
373	0.185
473	0.421
573	0.666
673	0.946

T	$\Delta L/L_0$
CURVE 34*	
673	0.996
473	0.423
373	0.179
293	0.000
CURVE 35*	
293	0.000
373	0.184
473	0.425
573	0.686
673	0.984
CURVE 36*	
673	0.988
473	0.427
373	0.182
293	0.000
CURVE 37	
293	0.000*
373	0.182
473	0.427
573	0.680
673	0.969
CURVE 38*	
673	0.980
473	0.430
373	0.184
293	0.000
CURVE 39*	
293	0.000
373	0.182
473	0.419
573	0.661
CURVE 40*	
293	0.000
373	0.182
473	0.430
573	0.692

T	$\Delta L/L_0$
CURVE 41*	
213	-0.172
293	0.000
373	0.186
CURVE 42	
77	-0.371
97	-0.351*
100	-0.341
109	-0.332*
118	-0.317*
127	-0.304
137	-0.292*
145	-0.273*
158	-0.258
178	-0.224*
186	-0.208
196	-0.188*
207	-0.168*
224	-0.133
235	-0.116*
245	-0.096
252	-0.078*
263	-0.060
272	-0.039*
277	-0.026
291	-0.007*
293	0.000*
300	0.014
CURVE 43*	
213	-0.174
293	0.000
373	0.190
CURVE 44*	
293	0.000
373	0.196
473	0.443
573	0.692
673	0.942
CURVE 45	
18	-0.394*
23	-0.394

T	$\Delta L/L_0$
CURVE 45 (cont.)	
33	-0.393*
43	-0.392*
53	-0.389*
63	-0.384
73	-0.379*
83	-0.372*
93	-0.364
103	-0.354*
113	-0.343*
123	-0.329*
133	-0.314
143	-0.298*
153	-0.282*
163	-0.265*
173	-0.247*
193	-0.210
213	-0.170*
233	-0.128*
253	-0.087*
273	-0.043*
293	0.000*
313	0.047*
333	0.094*
353	0.141
373	0.189*
393	0.238*
413	0.288*
433	0.338
453	0.389*
473	0.441*
493	0.494*
513	0.548
533	0.601*
553	0.656
573	0.712*
CURVE 46	
293	0.000*
373	0.208
473	0.466*
573	0.722*
673	0.997*
CURVE 47*	
293	0.000
373	0.189

T	$\Delta L/L_0$
CURVE 47 (cont.)*	
473	0.443
523	0.591
573	0.728
673	1.015
773	1.310
CURVE 48*	
293	0.000
373	0.186
473	0.446
523	0.593
573	0.728
673	1.018
773	1.325
CURVE 49*	
293	0.000
373	0.190
473	0.445
523	0.582
573	0.720
673	0.999
773	1.306
CURVE 50	
293	0.000*
373	0.190
473	0.454
523	0.593
573	0.739
673	1.037
773	1.310
CURVE 51*	
293	0.000
373	0.185
473	0.468
523	0.610
573	0.748
CURVE 52*	
293	0.000
373	0.180

* Not shown in figure.

DATA TABLE 233. THERMAL LINEAR EXPANSION OF ALUMINUM + COPPER + ΣX_i ALLOYS Al + Cu + ΣX_i (continued)

T	$\Delta L/L_0$
CURVE 52 (cont.)*	
473	0.443
523	0.570
573	0.708
673	0.996
773	1.253
CURVE 53*	
293	0.000
373	0.185
473	0.468
523	0.614
573	0.753
CURVE 54*	
293	0.000
373	0.178
473	0.423
523	0.554
573	0.692
673	0.977
773	1.219
CURVE 55*	
293	0.000
373	0.186
473	0.432
523	0.582
573	0.725
CURVE 56*	
293	0.000
373	0.175
473	0.412
523	0.545
573	0.692
673	0.977
773	1.267
CURVE 57*	
293	0.000
373	0.186
473	0.436
523	0.584
573	0.734
CURVE 58*	
293	0.000
373	0.181
473	0.434
523	0.573
573	0.717
673	1.011
773	1.291
CURVE 59*	
293	0.000
373	0.184
473	0.421
573	0.683
673	0.984
CURVE 60*	
673	0.954
473	0.425
373	0.181
293	0.000
CURVE 61*	
293	0.000
373	0.181
473	0.434
573	0.680
673	0.954
CURVE 62*	
673	0.950
573	0.680
373	0.181
293	0.000
CURVE 63	
293	0.000*
373	0.194*
473	0.432
573	0.686
673	0.950
773	1.224
CURVE 64*	
293	0.000
373	0.171
473	0.405
573	0.666
CURVE 65*	
293	0.000
373	0.172
473	0.409
573	0.661
CURVE 66*	
293	0.000
373	0.183
473	0.430
573	0.689
CURVE 67*	
573	0.689
473	0.425
373	0.177
293	0.000
223	-0.152
CURVE 68*	
293	0.000
373	0.184
473	0.432
573	0.694
CURVE 69*	
573	0.694
473	0.427
373	0.181
293	0.000
223	-0.150
CURVE 70*	
293	0.000
373	0.182
473	0.427
573	0.686
CURVE 71*	
573	0.689
473	0.427
373	0.182
293	0.000
223	-0.145
CURVE 72*	
293	0.000
373	0.182
473	0.427
573	0.683
CURVE 73*	
573	0.689
473	0.427
373	0.182
293	0.000
223	-0.144
CURVE 74*	
293	0.000
373	0.179
473	0.423
573	0.683
CURVE 75*	
573	0.686
473	0.423
373	0.179
293	0.000
CURVE 76*	
293	0.000
373	0.180
473	0.425
573	0.680
CURVE 77*	
573	0.683
473	0.425
373	0.179
293	0.000
223	-0.146
CURVE 78*	
293	0.000
373	0.180
473	0.425
573	0.683
CURVE 79*	
573	0.686
473	0.427
373	0.180
293	0.000
CURVE 80*	
293	0.000
373	0.178
473	0.418
573	0.675
CURVE 81*	
573	0.678
473	0.419
373	0.179
293	0.000
223	-0.145
CURVE 82*	
293	0.000
373	0.180
473	0.423
573	0.678
CURVE 83*	
573	0.680
473	0.423
373	0.179
293	0.000
CURVE 84*	
293	0.000
373	0.180
473	0.457
CURVE 85*	
473	0.409
373	0.182
293	0.000
CURVE 86*	
293	0.000
373	0.178
473	0.419
CURVE 87*	
473	0.419
373	0.180
293	0.000
CURVE 88	
223	-0.125
293	0.000*
373	0.160
473	0.378
CURVE 89*	
293	0.000
373	0.155
473	0.392
CURVE 90*	
473	0.353
373	0.157
293	0.000
CURVE 91*	
293	0.000
373	0.154
473	0.360
CURVE 92*	
473	0.360
373	0.158
293	0.000

* Not shown in figure.

DATA TABLE 233. THERMAL LINEAR EXPANSION OF ALUMINUM + COPPER + ΣX_i ALLOYS Al + Cu + ΣX_i (continued)

T	$\Delta L/L_0$
CURVE 93	
293	0.000*
373	0.190*
473	0.443*
523	0.621
573	0.762
673	1.015
773	1.320
CURVE 94*	
293	0.000
373	0.182
473	0.439
523	0.584
573	0.739
CURVE 95*	
293	0.000
373	0.190
473	0.448
523	0.610
573	0.778
673	1.056
773	1.344
CURVE 96	
293	0.000*
373	0.186*
473	0.454*
523	0.646
573	0.818
673	1.064
773	1.368
CURVE 97*	
293	0.000
373	0.197
473	0.455
523	0.600
573	0.759
CURVE 98*	
293	0.000
373	0.190
473	0.473
523	0.639
573	0.784
673	1.011
773	1.315
CURVE 99*	
293	0.000
373	0.187
473	0.482
523	0.637
573	0.784
673	1.011
773	1.306
CURVE 100*	
293	0.000
373	0.187
473	0.443
523	0.607
573	0.767
CURVE 101*	
293	0.000
373	0.179
473	0.436
523	0.616
573	0.801
673	1.045
773	1.325
CURVE 102*	
293	0.000
373	0.179
523	0.639
573	0.778
673	1.015
773	1.296
CURVE 103*	
293	0.000
373	0.179
473	0.434
523	0.616
573	0.801
673	1.045
773	1.325
CURVE 104*	
293	0.000
373	0.178
473	0.425
523	0.656
573	0.801
673	1.018
773	1.296
CURVE 105*	
293	0.000
373	0.189
473	0.454
523	0.614
573	0.753
673	1.018
773	1.320
CURVE 106*	
293	0.000
373	0.190
473	0.436
523	0.591
573	0.750
CURVE 107*	
293	0.000
373	0.179
473	0.421
523	0.547
573	0.675
CURVE 108*	
293	0.000
373	0.174
473	0.412
523	0.538
573	0.661
CURVE 109*	
293	0.000
373	0.171
473	0.401
573	0.661
CURVE 110‡	
333	0.092
373	0.184
422	0.300
472	0.419
522	0.540
573	0.667
622	0.791
CURVE 111*, ‡	
331	0.089
373	0.188
422	0.304
473	0.427
522	0.547
573	0.673
622	0.797
CURVE 112*, ‡	
332	0.092
373	0.188
422	0.309
472	0.435
523	0.564
573	0.692
624	0.827
CURVE 113*, ‡	
332	0.091
374	0.188
423	0.304
472	0.422
521	0.542
571	0.667
621	0.795
CURVE 114*, ‡	
333	0.096
374	0.195
421	0.311
472	0.438
520	0.559
571	0.691
623	0.829
CURVE 115*, ‡	
329	0.086
371	0.186
421	0.311
471	0.439
520	0.566
570	0.696
620	0.827
CURVE 116‡	
77	-0.409
83	-0.402
93	-0.391
103	-0.378
113	-0.364
123	-0.350
133	-0.334
143	-0.317
153	-0.299*
163	-0.281
173	-0.262*
183	-0.242
193	-0.221*
203	-0.201
213	-0.179*
223	-0.157
233	-0.135
CURVE 116 (cont.)‡	
243	-0.113
253	-0.090*
263	-0.068
273	-0.045
283	-0.022
293	0.000*
CURVE 117‡	
73	-0.396
83	-0.390
93	-0.379
103	-0.368
113	-0.355
123	-0.342
133	-0.327
143	-0.311
153	-0.295
163	-0.277
173	-0.258
183	-0.239
193	-0.219
203	-0.198
213	-0.177
223	-0.155
233	-0.134
243	-0.112
253	-0.090
263	-0.068
273	-0.045
283	-0.023
293	0.000
CURVE 118‡	
77	-0.403
83	-0.396
93	-0.385
103	-0.372
113	-0.359
123	-0.344
133	-0.329
143	-0.312
153	-0.295
163	-0.277
173	-0.258
183	-0.239

*Not shown in figure.
‡Author's data for coefficient of thermal expansion have been integrated by TPRC to obtain $\Delta L/L_0$.

DATA TABLE 233. THERMAL LINEAR EXPANSION OF ALUMINUM + COPPER + ΣX_i ALLOYS Al + Cu + ΣX_i (continued)

T	$\Delta L/L_0$
CURVE 118 (cont.)‡	
193	-0.219
203	-0.198
213	-0.177
223	-0.156
233	-0.134
243	-0.112
253	-0.090
263	-0.068
273	-0.045
283	-0.023
293	0.000

T	$\Delta L/L_0$
CURVE 119	
298	0.011
373	0.187
CURVE 120	
298	0.011
373	0.181
CURVE 121	
298	0.011
373	0.182

DATA TABLE 233. COEFFICIENT OF THERMAL LINEAR EXPANSION OF ALUMINUM + COPPER + ΣX_i ALLOYS Al + Cu + ΣX_i

[Temperature, T, K; Coefficient of Expansion, α, 10^{-6} K^{-1}]

T	α
CURVE 110*, ‡	
333	22.4
422	23.6
472	24.0
573	25.3
622	25.5
CURVE 111*, ‡	
331	23.2
373	23.7
473	24.2
522	24.6
622	25.3
CURVE 112*, ‡	
332	22.9
373	24.2
472	25.2
523	25.4
624	26.6

T	α
CURVE 113*, ‡	
332	23.1
374	23.4
472	24.3
521	24.8
621	25.7
CURVE 114*, ‡	
333	23.7
374	24.5
472	24.9
520	25.5
623	26.6
CURVE 115*, ‡	
329	23.2
371	24.5
471	25.9
520	26.0
620	26.1

T	α
CURVE 116‡	
77	10.20
83	11.00
93	12.10
103	13.20
113	14.30
123	15.30
133	16.30
143	17.30
153	18.10
163	18.90
173	19.50
183	20.10
193	20.60
203	21.20
213	21.60
223	22.00
233	22.30
243	22.50
253	22.60
263	22.70
273	22.80
283	22.00
293	23.00

T	α
CURVE 117‡	
77	9.40
83	10.00
93	10.90
103	12.00
113	13.00
123	14.10
133	15.20
143	16.30
153	17.30
163	18.20
173	19.10
183	19.80
193	20.40
203	20.90
213	21.30
223	21.60
233	21.80
243	22.00
253	22.00
263	22.40
273	22.50
283	22.60
293	22.70

T	α
CURVE 118‡	
77	10.20
83	11.10
93	12.00
103	13.10
113	14.10
123	15.00
133	15.90
143	16.80
153	17.60
163	18.40
173	19.20*
183	19.80*
193	20.40*
203	20.90*
213	21.30*
223	21.60*
233	21.80*
243	22.00*
253	22.20
263	22.40*
273	22.50*
283	22.60*
293	22.70*

*Not shown in figure.
‡Author's data for coefficient of thermal expansion have been integrated by TPRC to obtain $\Delta L/L_0$.

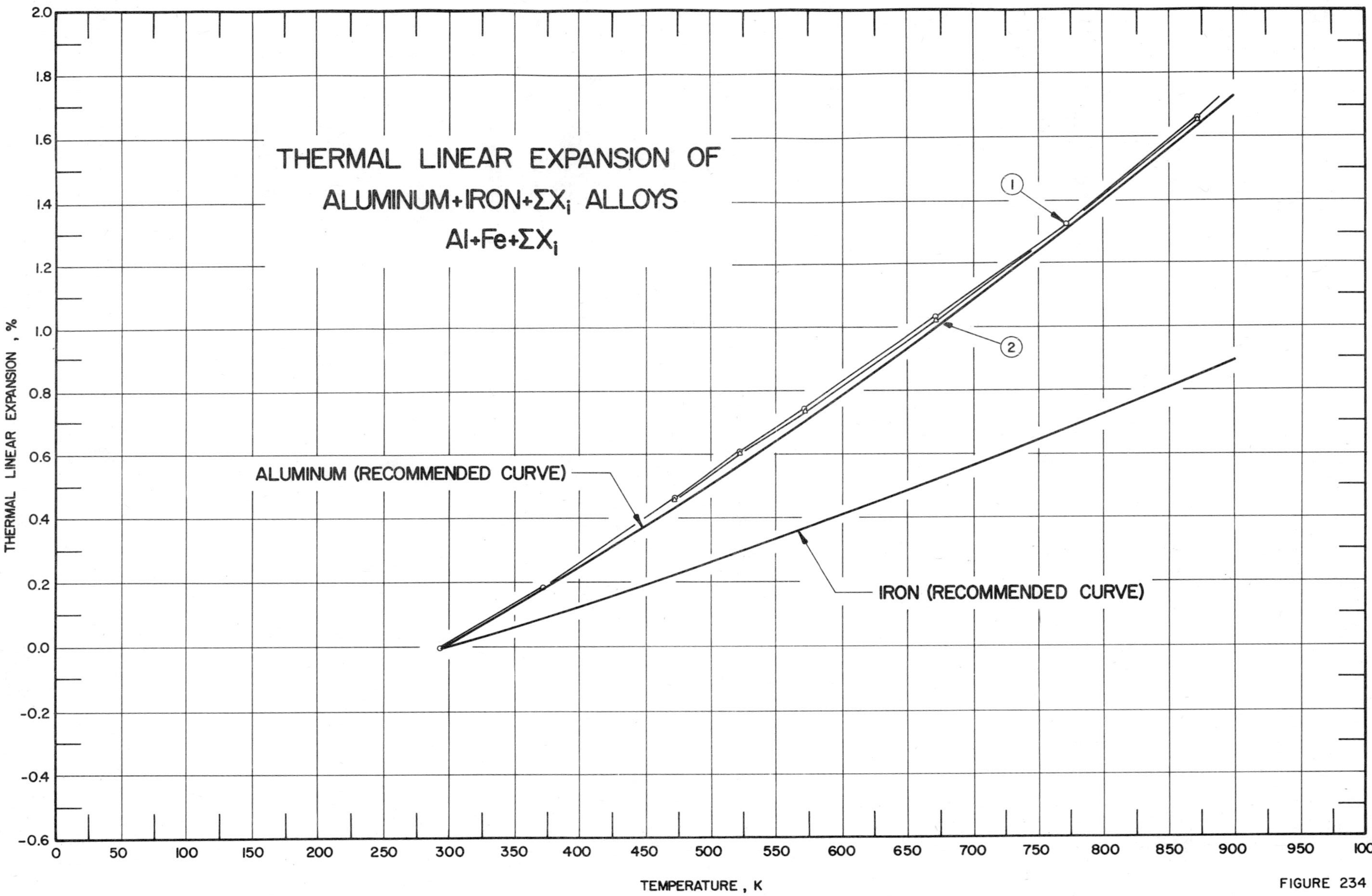

THERMAL LINEAR EXPANSION OF
ALUMINUM+IRON+ΣX_i ALLOYS
Al+Fe+ΣX_i
1
2
ALUMINUM (RECOMMENDED CURVE)
IRON (RECOMMENDED CURVE)
THERMAL LINEAR EXPANSION , %
2.0
1.8
1.6
1.4
1.2
1.0
0.8
0.6
0.4
0.2
0.0
-0.2
-0.4
-0.6
0
50
100
150
200
250
300
350
400
450
500
550
600
650
700
750
800
850
900
950
1000
TEMPERATURE , K

FIGURE 234

SPECIFICATION TABLE 234. THERMAL LINEAR EXPANSION OF ALUMINUM + IRON + ΣX_i ALLOYS $Al + Fe + \Sigma X_i$

Cur. No.	Ref. No.	Author(s)	Year	Method Used	Temp. Range, K	Name and Specimen Designation	Composition (weight percent) Al	Fe	Cu	Si	Mn	Composition (continued), Specifications, and Remarks
1	20	Hidnert, P.	1925	T	293-873	S843	99.15	0.36	0.10	0.32	0.07	Commercial aluminum specimen cut from sheet rolled to 0.66 cm thick.
2	20	Hidnert, P.	1925	T	293-873	S844	99.15	0.36	0.10	0.32		Similar (duplicate) to the above specimen.

DATA TABLE 234. THERMAL LINEAR EXPANSION OF ALUMINUM + IRON + ΣX_i ALLOYS $Al + Fe + \Sigma X_i$

[Temperature, T, K; Linear Expansion, $\Delta L/L_0$, %]

T	$\Delta L/L_0$
CURVE 1	
293	0.000
373	0.191
473	0.470
523	0.612
573	0.750
673	1.037
773	1.344
873	1.665
CURVE 2	
293	0.000*
373	0.192*
473	0.463
523	0.607
573	0.745
673	1.026

T	$\Delta L/L_0$
CURVE 2 (cont.)	
773	1.334*
873	1.659

* Not shown in figure.

FIGURE AND TABLE NO. 235R. PROVISIONAL VALUES FOR THERMAL LINEAR EXPANSION OF ALUMINUM + MAGNESIUM + ΣX_i ALLOYS Al + Mg + ΣX_i

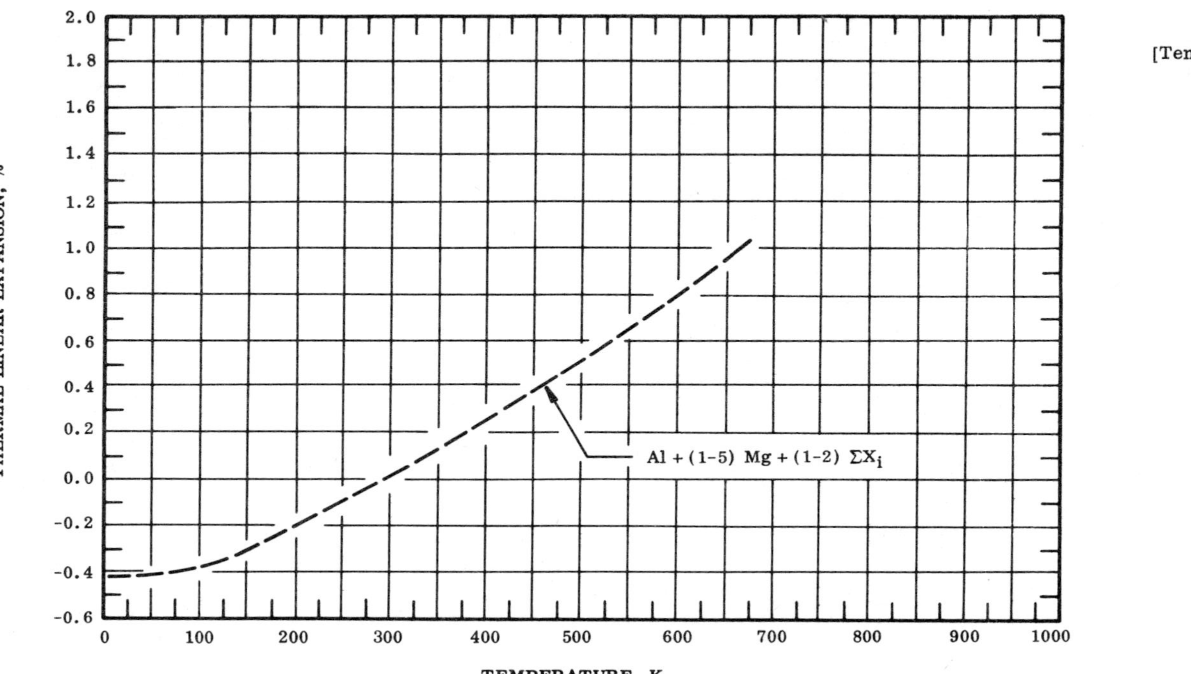

PROVISIONAL VALUES

[Temperature, T, K; Linear Expansion, $\Delta L/L_0$, %; α, K^{-1}]

[Al + (1 - 5) Mg + (1 - 2) ΣX_i]

T	$\Delta L/L_0$	$\alpha \times 10^6$
10	-0.418	
20	-0.416	
30	-0.415	
40	-0.413	
50	-0.410	
75	-0.400	
100	-0.392	12.2
125	-0.345	18.7
150	-0.300	19.3
175	-0.250	19.8
200	-0.200	20.3
225	-0.150	20.9
250	-0.096	21.5
293	0.000	22.5
350	0.131	23.8
400	0.254	25.0
450	0.382	26.3
500	0.517	27.5
550	0.657	28.8
600	0.800	30.1
650	0.958	31.5
675	1.039	32.1

REMARKS

[Al + (1 - 5) Mg + (1 - 2) ΣX_i] : The tabulated values for well-annealed alloy of type Al6061, 6053, 5052, 5083, and 5456 are considered accurate to within ± 7% over the entire temperature range. These values can be represented approximately by the following equation:

$$\Delta L/L_0 = -0.563 + 1.613 \times 10^{-3} T + 1.003 \times 10^{-6} T^2 + 1.812 \times 10^{-10} T^3 \qquad (100 < T < 675)$$

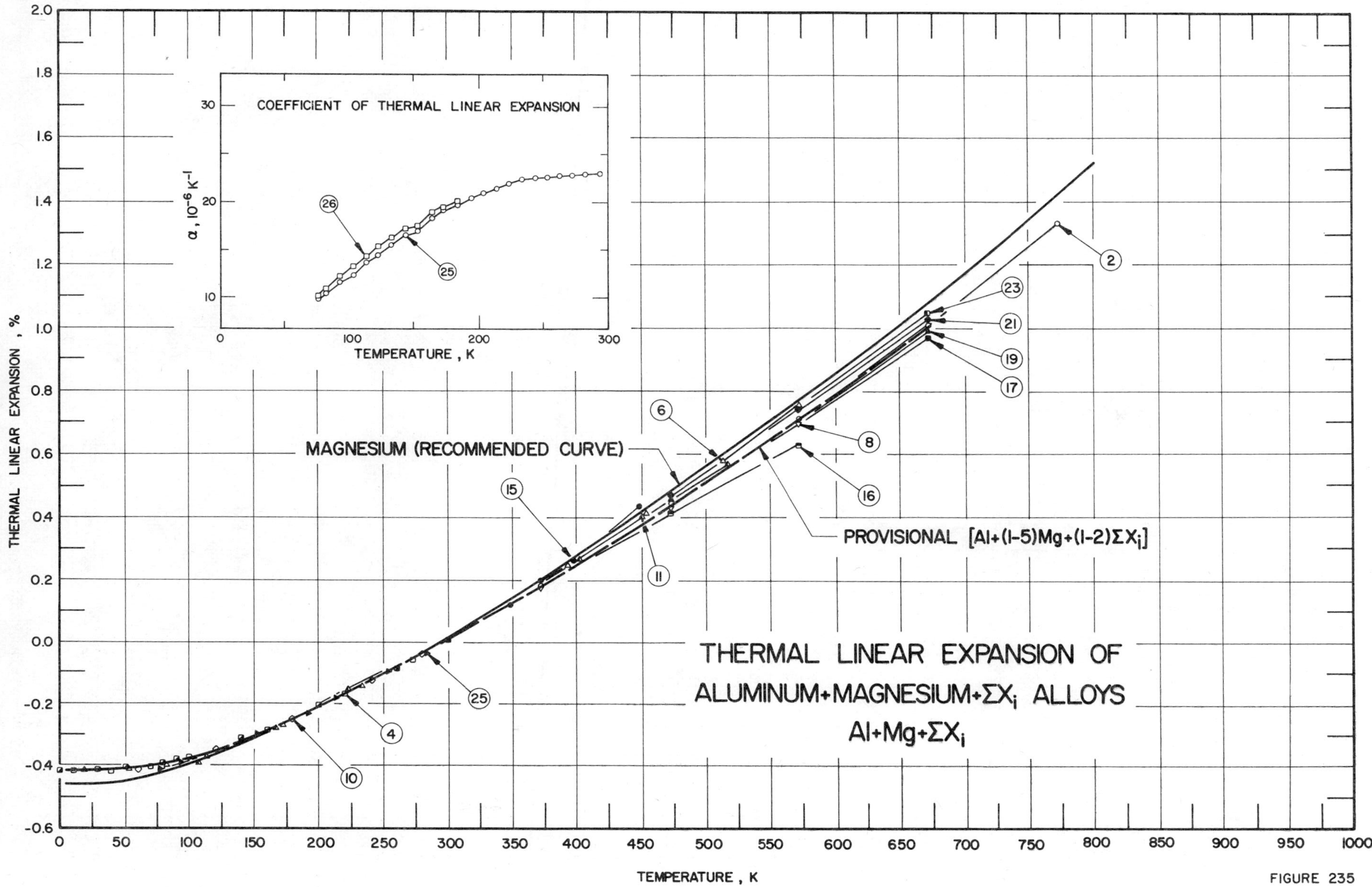

COEFFICIENT OF THERMAL LINEAR EXPANSION
α, 10^{-6} K^{-1}
30
20
10
0
100
200
300
TEMPERATURE , K
26
25
THERMAL LINEAR EXPANSION , %
2.0
1.8
1.6
1.4
1.2
1.0
0.8
0.6
0.4
0.2
0.0
-0.2
-0.4
-0.6
2
23
21
19
17
6
8
16
MAGNESIUM (RECOMMENDED CURVE)
15
11
PROVISIONAL [Al+(1-5)Mg+(1-2)ΣX$_i$]
THERMAL LINEAR EXPANSION OF
ALUMINUM+MAGNESIUM+ΣX$_i$ ALLOYS
Al+Mg+ΣX$_i$
25
4
10
0
50
100
150
200
250
300
350
400
450
500
550
600
650
700
750
800
850
900
950
1000
TEMPERATURE , K

FIGURE 235

SPECIFICATION TABLE 235. THERMAL LINEAR EXPANSION OF ALUMINUM + MAGNESIUM + ΣX_i ALLOYS Al + Mg + ΣX_i

Cur. No.	Ref. No.	Author(s)	Year	Method Used	Temp. Range, K	Name and Specimen Designation	Composition (weight percent) Al	Mg	Ni	Si	Zn	Cu	Fe	Composition (continued), Specifications, and Remarks
1*	463	Willey, L.A. and Fink, W.L.	1945	I	293-773	Al 5052	Bal.	2.5						0.25 Cr; sheet wrought form, 0.064 in. thick; tested in as-rolled condition.
2	463	Willey, L.A. and Fink, W.L.	1945	I	293-773	Al 5052	Bal.	2.5						0.25 Cr; similar to the above specimen, except annealed to 623 K.
3*	463	Willey, L.A. and Fink, W.L.	1945	I	213-373	Al 5052	Bal.	2.5						0.25 Cr; similar to the above specimen.
4	467	Clark, A.F.	1962	L	0-300	Al 5083	94.3	4.75					0.19	0.63 Mn, 0.13 Cr, Trace (< 0.1) Si, Cu, Zn, Ti, V, Ni; specimen 0.64 cm sq x 20.32 cm long; Rockwell hardness B-43, expansion measured across the specimen.
5	467	Clark, A.F.	1962	L	0-300	Al 5083	94.3	4.75					0.19	The above specimen; expansion measured along specimen.
6	15	Rhodes, B.L., Moeller, C.E., Hopkins, V. and Marx, T.I.	1963	L	18-573	Al-5456	Bal.	5.1		0.20	0.25	0.10	0.20	0.08 Mn, 0.13 Cr, 0.15 ΣX_i; specimen 10.16 cm long, 6.2 mm dia.
7*	463	Willey, L.A. and Fink, W.L.	1945	I	293-573	Al 6053	Bal.	1.3		0.7				0.25 Cr; rod wrought form, 3/4 in. dia rolled and drawn, tested in as-drawn condition.
8	463	Willey, L.A. and Fink, W.L.	1945	I	293-573	Al 6053	Bal.	1.3		0.7				Similar to the above specimen, except annealed to 623 K.
9*	463	Willey, L.A. and Fink, W.L.	1945	I	213-373	Al 6053	Bal.	1.3		0.7				Similar to the above specimen, except heat-treated and aged according to standard practice.
10	462	Arp, V., Wilson, J.H., Winrich, L. and Sikora, P.	1962	L	20-293	Al 6061	Bal.	1.0		0.6	0.3	0.3	0.7	0.3 Cr, 0.2 Mn, 0.2 Ti; hardness R_B51, A.S.M. condition T6.
11	225	Valentich, J.	1965	L	106-516	Al 6061								No details given; zero-point correction is -0.1%.
12*	463	Willey, L.A. and Fink, W.L.	1945	I	293-573	Alcoa 615 Al 6061	Bal.	1.0		0.6		0.25		0.25 Cr, sheet wrought form; cold rolled to 0.064 in.; tested in as-rolled condition.
13*	463	Willey, L.A. and Fink, W.L.	1945	I	293-573	Alcoa 615 Al 6061	Bal.	1.0		0.6		0.25		Similar to the above specimen, except annealed to 623 K.
14*	463	Willey, L.A. and Fink, W.L.	1945	I	213-373	Alcoa 615 Al 6061	Bal.	1.0		0.6		0.25		Similar to the above specimen, except heat-treated and aged according to standard practice.
15	34	Asay, J.R., Urzendowski, S.R. and Guenther, A.H.	1967	L	173-473	Al 6061 T6	98.0							Principal impurities Mg, Cr, Si, Cu; zero-point correction of 0.01% determined by graphical interpolation.
16	464	Barber, C.R.	1949	L	293-573	RR131D	Bal.	1.39	1.20	0.50	0.45	0.30	0.30	0.25 Co, 0.18 Cr, 0.12 Ti, 0.11 Mn, rod specimen 1.4 cm dia, 30 cm long; heat treated 10 hrs at 433-443 K and air cooled.
17	465	Losano, L.	1930	L	293-673	No. 16	93.40	6.03		0.26		0.08	0.32	No details given.
18*	465	Losano, L.	1930	L	293-673	No. 17	89.82	9.36		0.28		0.08	0.38	No details given.

*Not shown in figure.

SPECIFICATION TABLE 235. THERMAL LINEAR EXPANSION OF ALUMINUM + MAGNESIUM + ΣX_i ALLOYS Al + Mg + ΣX_i (continued)

Cur. No.	Ref. No.	Author(s)	Year	Method Used	Temp. Range, K	Name and Specimen Designation	Composition (weight percent) Al	Mg	Ni	Si	Zn	Cu	Fe	Composition (continued), Specifications, and Remarks
19	465	Losano, L.	1930	L	293-673	No. 18	78.72	20.75		0.30		0.02	0.20	No details given.
20*	465	Losano, L.	1930	L	293-673	No. 19	67.96	31.08		0.30		0.06	0.46	No details given.
21	465	Losano, L.	1930	L	293-673	No. 20	58.12	40.92		0.38		0.03	0.52	No details given.
22*	465	Losano, L.	1930	L	293-673	No. 21	55.84	43.22		0.35		0.05	0.54	No details given.
23	465	Losano, L.	1930	L	293-673	No. 22	50.12	49.10		0.40		0.07	0.28	No details given.
24*	550	Edel'man, N.M.	1965	L	293-673	AD 31 Russian alloy	Bal.	0.4/ 0.9		0.3/ 0.7		0.1		0.1 Cr, 0.1 Mn; no details given.
25	551	Belov, A.K.	1968	L	77-293	AMg5 Russian alloy	Bal.	5.7	0.2	0.4		0.2	0.4	0.6 Mn; cylindrical specimen 3.5 mm in diameter x 50mm long; sample heated and cooled at a rate of 3-7 deg. per min.
26*	551	Belov, A.K.	1968	L	77-293	AMg6 Russian alloy	Bal.	6.8		0.4	0.2		0.4	0.8 Mn; similar to the above specimen.

*Not shown in figure.

DATA TABLE 235. THERMAL LINEAR EXPANSION OF ALUMINUM + MAGNESIUM + ΣX_i ALLOYS Al + Mg + ΣX_i

[Temperature, T, K; Linear Expansion, $\Delta L/L_0$, %]

T	$\Delta L/L_0$
CURVE 1*	
293	0.000
373	0.190
473	0.439
573	0.703
673	0.992
773	1.296
CURVE 2	
293	0.000
373	0.191
473	0.448
573	0.722
673	1.015
773	1.339
CURVE 3*	
213	-0.176
293	0.000
373	0.190
CURVE 4	
0	-0.418*
10	-0.418
20	-0.418*
30	-0.417*
40	-0.416
50	-0.413*
60	-0.408*
70	-0.401
80	-0.393*
90	-0.383*
100	-0.371
120	-0.344*
140	-0.313*
160	-0.279
180	-0.242*
200	-0.203*
220	-0.162
240	-0.119*
260	-0.076*
273	-0.047
280	-0.031*
293	0.000*
300	0.015*
CURVE 5	
0	-0.411
10	-0.411*
20	-0.411*
30	-0.410
40	-0.409*
50	-0.406
60	-0.402*
70	-0.395*
80	-0.387
90	-0.378
100	-0.366*
120	-0.340*
140	-0.309
160	-0.275*
180	-0.239*
200	-0.200
220	-0.160*
240	-0.118*
260	-0.075
273	-0.046*
280	-0.030*
293	0.000*
300	0.015
CURVE 6	
18	-0.414
23	-0.414*
33	-0.413*
43	-0.412*
53	-0.410
63	-0.408*
73	-0.404*
83	-0.396
93	-0.386*
103	-0.375*
113	-0.362
123	-0.348*
133	-0.333*
143	-0.316
153	-0.300*
163	-0.282*
173	-0.263
193	-0.224*
213	-0.182*
233	-0.138
253	-0.093*
CURVE 6 (cont.)	
273	-0.047*
293	0.000*
313	0.048*
333	0.098
353	0.149*
373	0.201*
393	0.255
413	0.309*
433	0.363*
453	0.416
473	0.471*
495	0.525*
513	0.581
533	0.640*
553	0.700*
573	0.761
CURVE 7*	
293	0.000
373	0.173
473	0.427
573	0.689
CURVE 8	
293	0.000*
373	0.183
473	0.434
573	0.703
CURVE 9*	
213	-0.174
293	0.000
373	0.188
CURVE 10	
20	-0.419*
40	-0.416*
60	-0.408
80	-0.393*
100	-0.370*
120	-0.343
140	-0.312*
160	-0.278*
CURVE 10 (cont.)	
180	-0.241
200	-0.202*
220	-0.161*
240	-0.118
260	-0.074*
273	-0.045*
280	-0.030
293	0.000*
CURVE 11	
106	-0.394
166	-0.278
223	-0.153
293	0.000*
347	0.132*
401	0.276
450	0.403
516	0.573
CURVE 12*	
293	0.000
373	0.182
473	0.427
573	0.686
CURVE 13*	
293	0.000
373	0.188
473	0.437
573	0.711
CURVE 14*	
213	-0.173
293	0.000
373	0.188
CURVE 15	
173	-0.191*
198	-0.160*
223	-0.126*
248	-0.090*
273	-0.041*
CURVE 15 (cont.)	
298	0.010*
323	0.066*
348	0.129
373	0.203
398	0.269
423	0.351*
448	0.438
473	0.462*
CURVE 16	
293	0.000*
373	0.190*
473	0.418
573	0.633
CURVE 17	
293	0.000*
373	0.193*
473	0.441*
573	0.705*
673	0.973
CURVE 18*	
293	0.000
373	0.194
473	0.447
573	0.710
673	0.981
CURVE 19	
293	0.000*
373	0.197*
473	0.454
573	0.722*
673	0.999
CURVE 20*	
293	0.000
373	0.198
473	0.459
573	0.734
673	1.017
CURVE 21	
293	0.000*
373	0.202*
473	0.468
573	0.748
673	1.035
CURVE 22*	
293	0.000
373	0.204
473	0.472
573	0.752
673	1.040
CURVE 23	
293	0.000*
373	0.207*
473	0.478*
573	0.762*
673	1.053
CURVE 24*	
293	0.000
373	0.187
473	0.437
573	0.722
673	1.015
CURVE 25‡	
77	-0.403
83	-0.397*
93	-0.386
103	-0.374*
113	-0.361*
123	-0.347
133	-0.332*
143	-0.316*
153	-0.299
163	-0.281*
173	-0.262*
183	-0.242*
193	-0.222
203	-0.201*
213	-0.180

* Not shown in figure.
‡ Author's data for coefficient of thermal expansion have been integrated by TPRC to obtain $\Delta L/L_0$.

DATA TABLE 235. THERMAL LINEAR EXPANSION OF ALUMINUM + MAGNESIUM + ΣX_i ALLOYS Al + Mg + ΣX_i (continued)

T	$\Delta L/L_0$
CURVE 25 (cont.)‡	
223	-0.158*
233	-0.136*
243	-0.114*
253	-0.091
263	-0.069*
273	-0.046*
283	-0.023
293	0.000*
CURVE 26*, ‡	
77	-0.408
83	-0.402
93	-0.391
103	-0.378
113	-0.364
123	-0.349
133	-0.334
143	-0.317
CURVE 26 (cont.)‡	
153	-0.299
163	-0.281
173	-0.262
183	-0.242
193	-0.222
203	-0.201
213	-0.180
223	-0.158
233	-0.136
243	-0.114
253	-0.091
263	-0.069
273	-0.046
283	-0.023
293	0.000

DATA TABLE 235. COEFFICIENT OF THERMAL LINEAR EXPANSION OF ALUMINUM + MAGNESIUM + ΣX_i ALLOYS Al + Mg + ΣX_i

[Temperature, T, K; Coefficient of Expansion, **α**, 10^{-6} K^{-1}]

T	α
CURVE 25‡	
77	9.90
83	10.50
93	11.50
103	12.50
113	13.59
123	14.60
133	15.60
143	16.60
153	17.50
163	18.40
173	19.20
183	19.90
193	20.50
203	21.10
213	21.59
CURVE 25 (cont.)‡	
223	21.99
233	22.29
243	22.50
253	22.60
263	22.70
273	22.80
283	22.90
293	23.00
CURVE 26‡	
77	10.20
83	10.99
93	12.09
103	13.19
113	14.29
CURVE 26 (cont.)‡	
123	15.30
133	16.20
143	17.10
153	17.90
163	18.69
173	19.30
183	20.00*
193	20.50*
203	21.10*
213	21.59*
223	21.99*
233	22.29*
243	22.50*
253	22.60*
263	22.70*
273	22.80*
CURVE 26 (cont.)‡	
283	22.90
293	23.00*

* Not shown in figure.
‡ Author's data for coefficient of thermal expansion have been integrated by TPRC to obtain $\Delta L/L_0 = 0$.

FIGURE AND TABLE NO. 236R. PROVISIONAL VALUES FOR THERMAL LINEAR EXPANSION OF ALUMINUM + MANGANESE + ΣX_i ALLOYS Al + Mn + ΣX_i

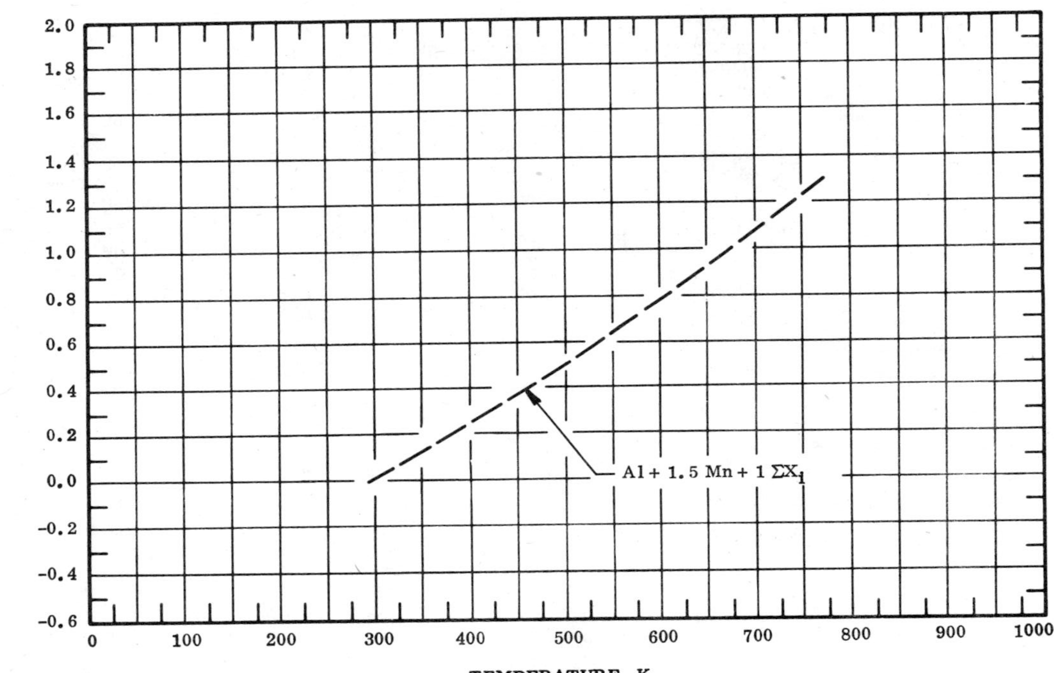

PROVISIONAL VALUES

[Temperature, T, K; Linear Expansion, $\Delta L/L_0$, %; α, K^{-1}]

T	($Al + 1.5\ Mn + 1\ \Sigma X_i$) $\Delta L/L_0$	$\alpha \times 10^6$
293	0.000	23.0
400	0.256	24.8
500	0.513	26.4
600	0.784	27.9
700	1.070	29.4
775	1.296	30.5

REMARKS

($Al + 1.5\ Mn + 1\ \Sigma X_i$): The tabulated values are for well-annealed alloy of type Al 3003 (Alcoa 3S) and are considered accurate to within ± 7% over the entire temperature range. The thermal linear expansion of this alloy is very close to that of pure Aluminum. The tabulated values can be represented approximately by the following equation:

$$\Delta L/L_0 = -0.605 + 1.819 \times 10^{-3}\ T + 8.597 \times 10^{-7}\ T^2 - 5.550 \times 10^{-11}\ T^3$$

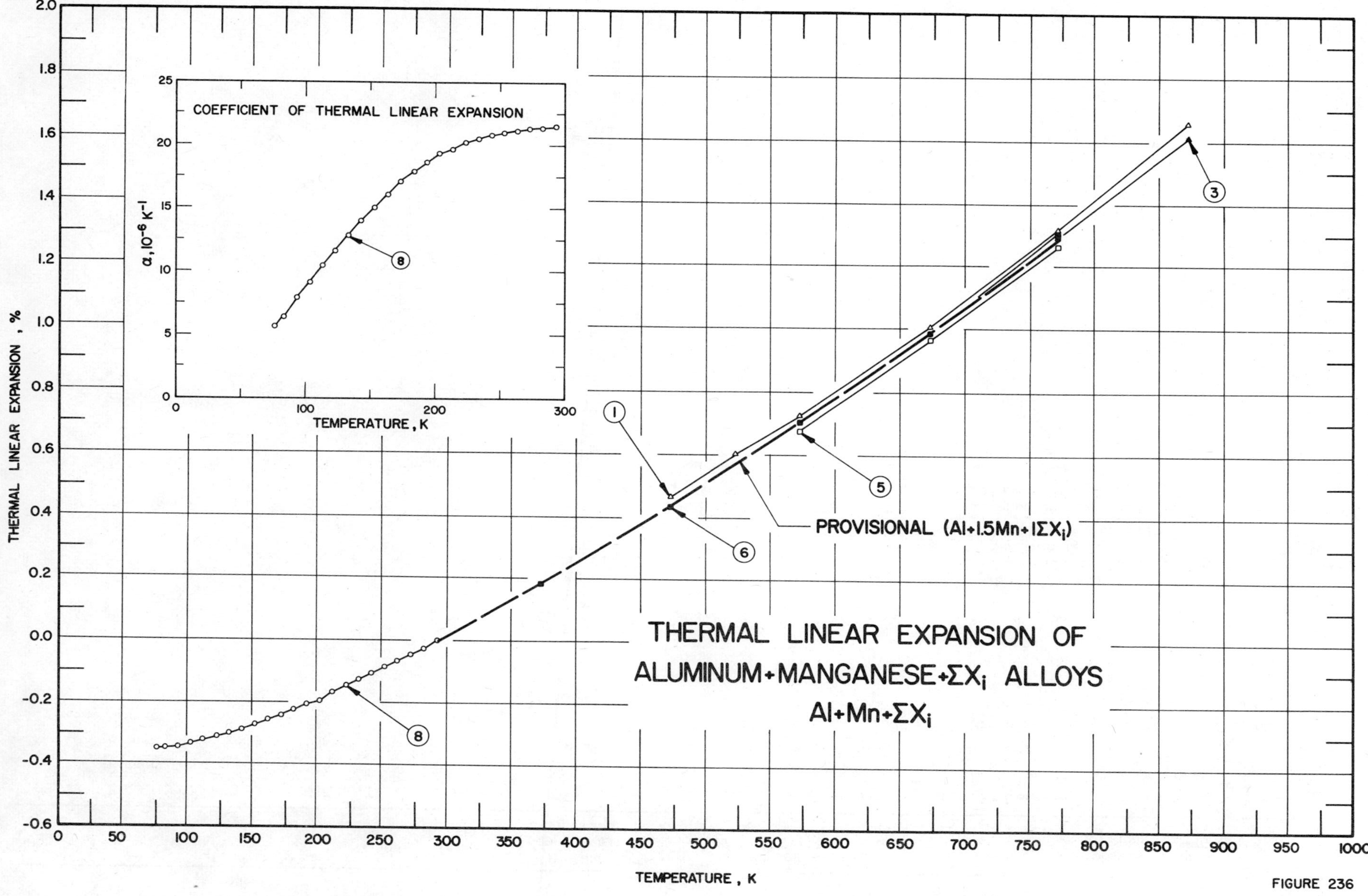
COEFFICIENT OF THERMAL LINEAR EXPANSION
α, 10⁻⁶ K⁻¹
TEMPERATURE, K
THERMAL LINEAR EXPANSION, %
PROVISIONAL (Al+1.5Mn+1ΣXi)
THERMAL LINEAR EXPANSION OF
ALUMINUM+MANGANESE+ΣXi ALLOYS
Al+Mn+ΣXi
TEMPERATURE, K

FIGURE 236

SPECIFICATION TABLE 236. THERMAL LINEAR EXPANSION OF ALUMINUM + MANGANESE + ΣX_i ALLOYS $Al+Mn+\Sigma X_i$

Cur. No.	Ref. No.	Author(s)	Year	Method Used	Temp. Range, K	Name and Specimen Designation	Composition (weight percent) Al	Mn	Fe	Si	Cu	Composition (continued), Specifications, and Remarks
1	20	Hidnert, P.	1925	T	293-873	3S	Bal.	1.05	0.57	0.41	0.19	3S sheet rolled to 0.66 cm.
2*	20	Hidnert, P.	1925	T	293-873	3S	Bal.	1.05	0.57	0.41		Duplicate of the above specimen.
3	20	Hidnert, P.	1925	T	293-873	3S	Bal.	1.80	0.84	0.40	0.23	Sand cast alloy.
4*	20	Hidnert, P.	1925	T	293-773	3S	Bal.	1.80	0.84	0.40		Cut from same bar as the above specimen.
5	463	Willey, L.A. and Fink, W.L.	1945	I	293-773	Alcoa 3S	Bal.	1.2				Sheet specimen, cold-rolled to 0.064 in.; tested in as-rolled condition.
6	463	Willey, L.A. and Fink, W.L.	1945	I	293-773	Alcoa 3S	Bal.	1.2				Similar to the above specimen, except annealed to 623 K.
7*	463	Willey, L.A. and Fink, W.L.	1945	I	213-373	Alcoa 3S	Bal.	1.2				Similar to the above specimen.
8	551	Belov, A.K.	1968	L	77-293	AMtS Russian alloy	Bal.	1.6	0.7	0.6	0.2	Cylindrical specimen 3.5 mm in diameter x 50 mm long; sample heated and cooled at a rate of 3-7 deg. per min; data on the coefficient of thermal linear expansion are also reported.
9*	603	McCollough, E.E.	1931	I	298, 373		Bal.	1.07		0.36	0.14	Material for the preparation of the alloy was furnished by Aluminum Co. of America.
10*	603	McCollough, E.E.	1931	I	298, 373		Bal.	1.29		0.38	0.11	Same as above.

*Not shown in figure.

DATA TABLE 236. THERMAL LINEAR EXPANSION OF ALUMINUM + MANGANESE + ΣX_i ALLOYS Al+Mn+ΣX_i

[Temperature, T, K; Linear Expansion, $\Delta L/L_0$, %]

T	$\Delta L/L_0$
CURVE 1	
293	0.000
373	0.190
473	0.463
523	0.600
573	0.725
673	1.007
773	1.320
873	1.659
CURVE 2*	
293	0.000
373	0.190
473	0.461
523	0.596
573	0.714
673	0.992
773	1.315
873	1.647
CURVE 3	
293	0.000*
373	0.185
473	0.436
523	0.568
573	0.714
673	0.984
773	1.296
873	1.618
CURVE 4*	
293	0.000
373	0.185
473	0.439
523	0.580
573	0.722
673	1.003
773	1.320
CURVE 5	
293	0.000
373	0.185
473	0.428
573	0.678
673	0.961
773	1.262
CURVE 6	
293	0.000*
373	0.186*
473	0.436*
573	0.703
673	0.988
773	1.301
CURVE 7*	
213	-0.171
293	0.000
373	0.185
CURVE 8‡	
77	-0.353
83	-0.350
93	-0.343
103	-0.334
113	-0.324
123	-0.313
133	-0.301
143	-0.288
153	-0.273
163	-0.258
173	-0.241
183	-0.224
193	-0.205
CURVE 8 (cont.)‡	
203	-0.186
213	-0.167
223	-0.147
233	-0.127
243	-0.106
253	-0.085
263	-0.064
273	-0.043
283	-0.021
293	0.000
CURVE 9*	
298	0.012
373	0.186
CURVE 10*	
298	0.012
373	0.186

DATA TABLE 236. COEFFICIENT OF THERMAL LINEAR EXPANSION OF ALUMINUM + MANGANESE + ΣX_i ALLOYS Al+Mn+ΣX_i

[Temperature, T, K; Coefficient of Expansion, α, 10^{-6} K^{-1}]

T	α
CURVE 8‡	
77	5.60
83	6.40
93	7.80
103	9.10
113	10.40
123	11.60
133	12.79
143	14.00
153	14.99
163	16.09
173	17.10
183	17.90
193	18.59
203	19.30
CURVE 8 (cont.)‡	
213	19.70
223	20.19
233	20.50
243	20.79
253	21.00
263	21.20
273	21.29
283	21.40
293	21.50

*Not shown in figure.
‡Author's data for coefficient of thermal expansion have been integrated by TPRC to obtain $\Delta L/L_0$.

FIGURE AND TABLE NO. 237R. PROVISIONAL VALUES FOR THERMAL LINEAR EXPANSION OF ALUMINUM + NICKEL + ΣX_i ALLOYS Al + Ni + ΣX_i

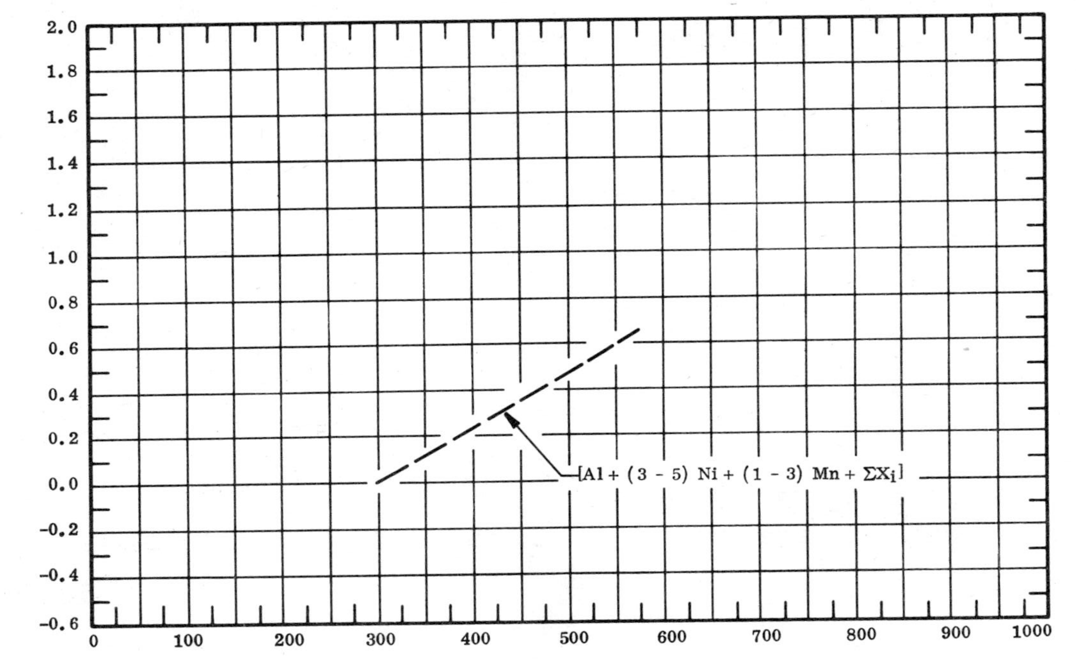

PROVISIONAL VALUES

[Temperature, T, K; Linear Expansion, $\Delta L/L_0$, %; α, K^{-1}]

T	[Al + (3 - 5) Ni + (1 - 3) Mn + ΣX_i] $\Delta L/L_0$	$\alpha \times 10^6$
293	0.000	20.8
400	0.235	23.0
500	0.471	24.6
575	0.662	25.6

REMARKS

[Al + (3 - 5) Ni + (1 - 3) Mn + ΣX_i]: The tabulated values for well-annealed alloy are considered accurate to within ± 7% over the entire temperature range. These values can be represented approximately by the following equation:

$$\Delta L/L_0 = -0.493 + 1.231 \times 10^{-3}\, T + 1.740 \times 10^{-6}\, T^2 - 6.804 \times 10^{-10}\, T^3$$

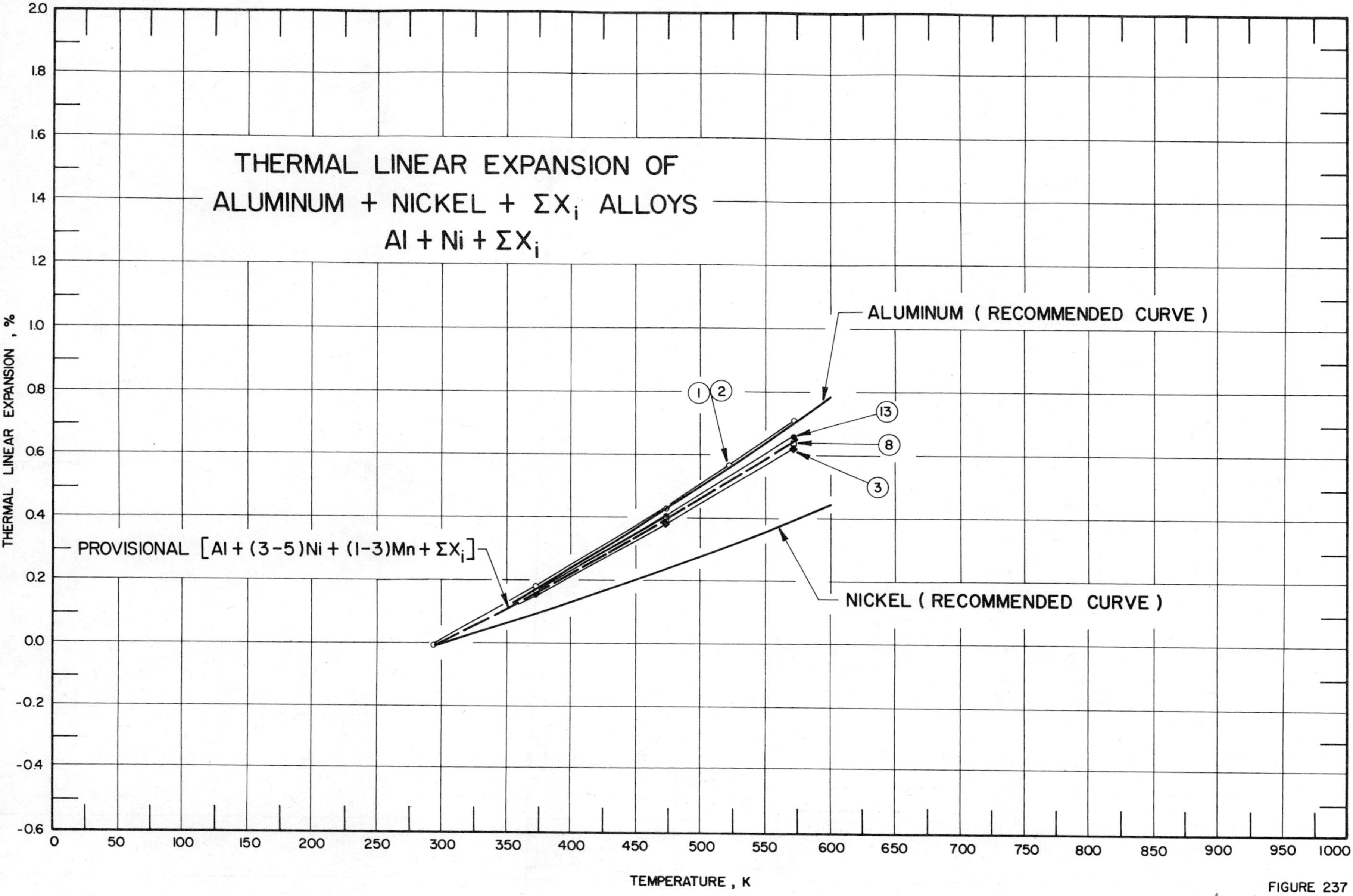

THERMAL LINEAR EXPANSION OF
ALUMINUM + NICKEL + ΣXi ALLOYS
Al + Ni + ΣXi
ALUMINUM (RECOMMENDED CURVE)
1
2
13
8
3
PROVISIONAL [Al + (3-5) Ni + (1-3) Mn + ΣXi]
NICKEL (RECOMMENDED CURVE)
THERMAL LINEAR EXPANSION, %
2.0
1.8
1.6
1.4
1.2
1.0
0.8
0.6
0.4
0.2
0.0
-0.2
-0.4
-0.6
0
50
100
150
200
250
300
350
400
450
500
550
600
650
700
750
800
850
900
950
1000
TEMPERATURE, K

FIGURE 237

SPECIFICATION TABLE 237. THERMAL LINEAR EXPANSION OF ALUMINUM + NICKEL + ΣX_i ALLOYS $Al+Ni+\Sigma X_i$

Cur. No.	Ref. No.	Author(s)	Year	Method Used	Temp. Range, K	Name and Specimen Designation	Composition (weight percent) Al	Ni	Mn	Cu	Cr	Fe	Mg	Composition (continued), Specifications, and Remarks
1	20	Hidnert, P.	1925	T	293-573	Verilite S425	95.5	1.5	0.5	1.0	1.5			Cast specimen.
2	20	Hidnert, P.	1925	T	293-573	Verilite S426	95.5	1.5	0.5	1.0				Similar to the above specimen.
3	464	Barber, C.R.	1949	L	293-573	RAE-40C	Bal.	5.0	3.0	2.0	0.5	0.5	0.5	0.4 Be, 0.3 Si; wrought rod specimen 1.4 cm dia, 30 cm long; tested as received.
4*	464	Barber, C.R.	1949	L	293-573	RAE-40C	Bal.	5.0	3.0	2.0				Similar to the above specimen, heat treated 6 hrs at 843 K and cold water quenched; 20 hrs at 423 K and air cooled.
5*	464	Barber, C.R.	1949	L	293-573	RAE-47D	Bal.	4.0	3.0	1.0		0.5	0.5	0.3 Si; wrought rod specimen, 1.4 cm dia x 30 cm long; tested as received.
6*	464	Barber, C.R.	1949	L	293-573	RAE-47D	Bal.	4.0	3.0	1.0				Similar to the above specimen; heat treated 6 hrs at 843 K and cold water quenched; then 20 hrs at 433 K and air cooled.
7*	464	Barber, C.R.	1949	L	293-573	RAE-47D	Bal.	4.0	3.0	1.0				Similar to the above specimen, except chill cast; tested with no heat treatment.
8	464	Barber, C.R.	1949	L	293-573	RAE-55 No. 29A	Bal.	3.05	1.98	1.68	0.45	0.39	0.50	0.19 Si, 0.08 Ti; wrought rod specimen 1.4 cm dia x 30 cm long; tested as received.
9*	464	Barber, C.R.	1949	L	293-573	RAE-55 No. 39A	Bal.	3.01	1.41	1.68	0.17	0.10	0.49	0.15 Si, 0.03 Ti; similar to the above specimen.
10*	464	Barber, C.R.	1949	L	293-573	RAE-40C	Bal.	5.0	3.0	2.0	0.5	0.5	0.5	0.4 Be, 0.3 Si; cast specimen similar to specimen used for curve 1.
11*	464	Barber, C.R.	1949	L	293-573	RAE-40C	Bal.	5.0	3.0	2.0				Cast specimen similar to specimen used for curve 2.
12*	464	Barber, C.R.	1949	L	293-573	RAE-47B	Bal.	4.0	3.0	1.0		0.5	0.5	0.2 Si, 0.2 Ti; specimen tested in sand cast conditions.
13	464	Barber, C.R.	1949	L	293-573	RAE-55	Bal.	2.90	1.55	1.89	0.15	0.43	0.56	0.21 Si, 0.07 Ti; chill cast rod specimen 1.4 cm dia, 30 cm long; tested as received.

* Not shown in figure.

DATA TABLE 237. THERMAL LINEAR EXPANSION OF ALUMINUM + NICKEL + ΣX_i ALLOYS Al+Ni+ ΣX_i

[Temperature, T, K; Linear Expansion, $\Delta L/L_0$, %]

T	$\Delta L/L_0$
CURVE 1	
293	0.000
373	0.186
473	0.436
523	0.575
573	0.722
CURVE 2*	
293	0.000
373	0.186
473	0.436
523	0.575
573	0.722
CURVE 3	
293	0.000*
373	0.162
473	0.387
573	0.630
CURVE 4*	
293	0.000
373	0.161
473	0.382
573	0.616
CURVE 5*	
293	0.000
373	0.170
473	0.396
573	0.633
CURVE 6*	
293	0.000
373	0.167
473	0.391
573	0.633
CURVE 7*	
293	0.000
373	0.168

T	$\Delta L/L_0$
CURVE 7 (cont.)	
473	0.391
573	0.633
CURVE 8	
293	0.000*
373	0.175
473	0.401
573	0.644
CURVE 9*	
293	0.000
373	0.177
473	0.405
573	0.647
CURVE 10*	
293	0.000
373	0.168
473	0.396
573	0.636
CURVE 11*	
293	0.000
373	0.166
473	0.394
573	0.636
CURVE 12*	
293	0.000
373	0.175
473	0.403
573	0.655
CURVE 13	
293	0.000*
373	0.178*
473	0.410
573	0.666

* Not shown in figure.

FIGURE AND TABLE NO. 238R. PROVISIONAL VALUES FOR THERMAL LINEAR EXPANSION OF ALUMINUM + SILICON + ΣX_i ALLOYS Al + Si + ΣX_i

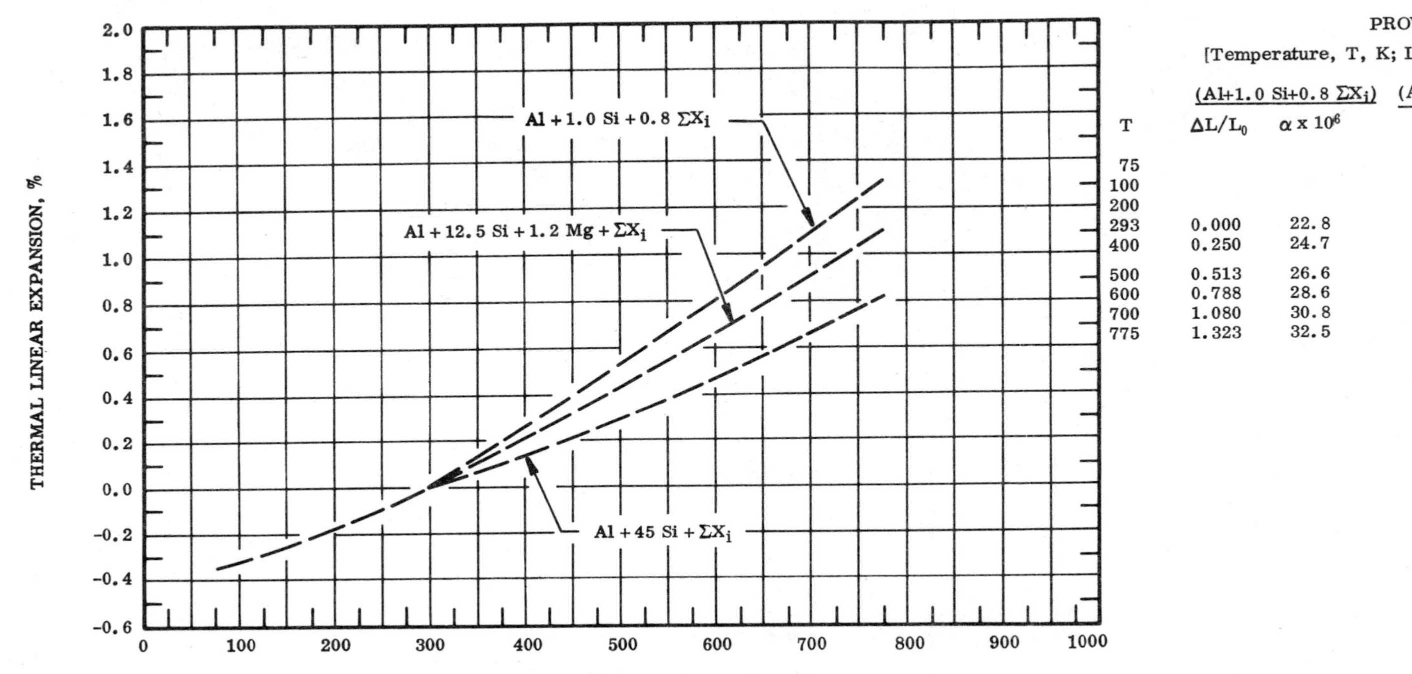

PROVISIONAL VALUES

[Temperature, T, K; Linear Expansion, $\Delta L/L_0$, %; α, K^{-1}]

	(Al+1.0 Si+0.8 ΣX_i)		(Al+12.5 Si+1.2 Mg+ΣX_i)		(Al+45 Si+ΣX_i)	
T	$\Delta L/L_0$	$\alpha \times 10^6$	$\Delta L/L_0$	$\alpha \times 10^6$	$\Delta L/L_0$	$\alpha \times 10^6$
75			-0.339	11.3		
100			-0.310	12.4		
200			-0.165	16.2		
293	0.000	22.8	0.000	19.1	0.000	12.4
400	0.250	24.7	0.218	21.6	0.144	14.5
500	0.513	26.6	0.443	23.2	0.299	16.4
600	0.788	28.6	0.679	24.1	0.473	18.2
700	1.080	30.8	0.921	24.2	0.664	20.0
775	1.323	32.5	1.103	24.3	0.819	21.3

REMARKS

(Al + 1.0 Si + 0.8 ΣX_i): The tabulated values for well-annealed alloy of type Al 6151 are considered accurate to within ±7% over the entire temperature range. The thermal linear expansion of this alloy is very close to that of pure aluminum. These values can be represented approximately by the following equation:

$$\Delta L/L_0 = -0.604 + 1.859 \times 10^{-3} T + 6.239 \times 10^{-7} T^2 + 2.383 \times 10^{-10} T^3$$

(Al + 12.5 Si + 1.2 Mg + ΣX_i): The tabulated values for well-annealed alloy of type Al 4032 are considered accurate to within ±7% over the entire temperature range. These values can be represented approximately by the following equation:

$$\Delta L/L_0 = -0.410 + 7.804 \times 10^{-4} T + 2.481 \times 10^{-6} T^2 - 1.249 \times 10^{-9} T^3$$

(Al + 45 Si + ΣX_i): The tabulated values for well annealed alloy are considered accurate to within ±7% over the entire temperature range. These values can be represented approximately by the following equation:

$$\Delta L/L_0 = -0.275 + 6.281 \times 10^{-4} T + 1.086 \times 10^{-6} T^2 - 9.896 \times 10^{-11} T^3$$

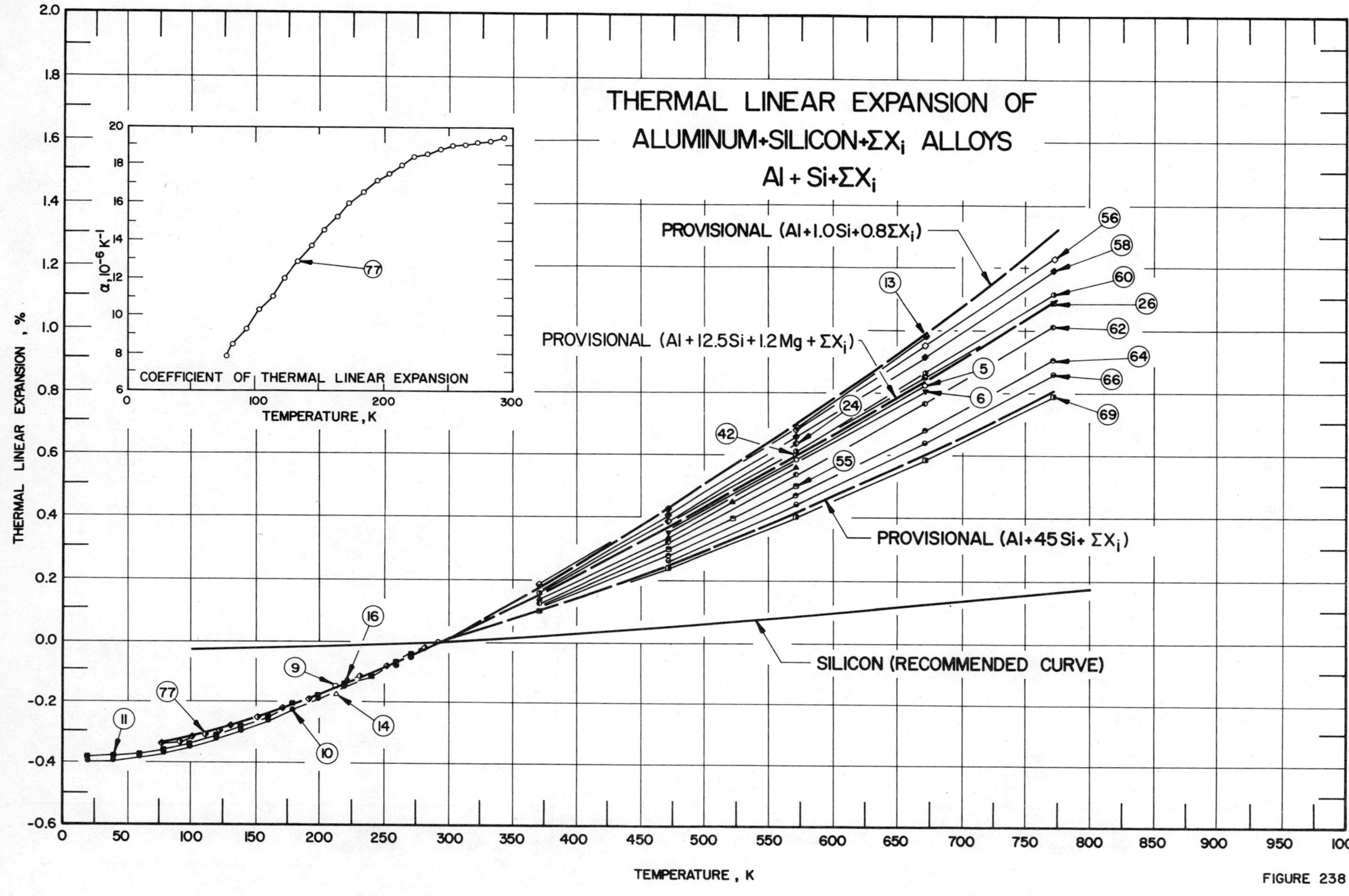

THERMAL LINEAR EXPANSION OF
ALUMINUM+SILICON+ΣXi ALLOYS
Al + Si+ΣXi
PROVISIONAL (Al+1.0Si+0.8ΣXi)
PROVISIONAL (Al+12.5Si+1.2Mg+ΣXi)
PROVISIONAL (Al+45Si+ΣXi)
SILICON (RECOMMENDED CURVE)
THERMAL LINEAR EXPANSION, %
TEMPERATURE, K
COEFFICIENT OF THERMAL LINEAR EXPANSION
α, 10⁻⁶ K⁻¹

FIGURE 238

SPECIFICATION TABLE 238. THERMAL LINEAR EXPANSION OF ALUMINUM + SILICON + ΣX_i ALLOYS $Al + Si + \Sigma X_i$

Cur. No.	Ref. No.	Author(s)	Year	Method Used	Temp. Range, K	Name and Specimen Designation	Composition (weight percent) Al	Si	Fe	Cu	Na	Mn	Ni	Composition (continued), Specifications, and Remarks
1*	19	Hidnert, P. and Krider, H.S.	1952	L	293-673	Al-4032	Bal.	12.18	0.41	0.89		0.01	0.87	1.20 Mg, 0.02 Zn, 0.01 each Cr, Tr; specimen solution heat-treated 1 hr at 789 K, quenched in water; aged 12 hrs at 444 K; expansion measured with increasing temp.
2*	19	Hidnert, P. and Krider, H.S.	1952	L	673-293	Al-4032	Bal.	12.18	0.41	0.89		0.01	0.87	1.20 Mg, 0.02 Zn; the above specimen; expansion measured with decreasing temperature.
3*	19	Hidnert, P. and Krider, H.S.	1952	L	293-673	Al-4032	Bal.	12.18	0.41	0.89		0.01	0.87	1.20 Mg, 0.02 Zn; similar to the above specimen, except aged 100 hrs at 644 K; expansion measured with increasing temperature.
4*	19	Hidnert, P. and Krider, H.S.	1952	L	673-293	Al-4032	Bal.	12.18	0.41	0.89		0.01	0.87	1.20 Mg, 0.02 Zn; the above specimen; expansion measured with decreasing temperature.
5	19	Hidnert, P. and Krider, H.S.	1952	L	293-673	Al-4032	Bal.	12.18	0.41	0.89		0.01	0.87	1.20 Mg, 0.02 Zn; similar to the above specimen; expansion measured with increasing temperature.
6	19	Hidnert, P. and Krider, H.S.	1952	L	673-293	Al-4032	Bal.	12.18	0.41	0.89		0.01	0.87	1.20 Mg, 0.02 Zn; the above specimen; expansion measured with decreasing temperature.
7*	463	Willey, L.A. and Fink, W.L.	1945	I	293-573	Al-4032	Bal.	12.5		0.9			0.9	1.0 Mg; specimen forged to 1 in. square; temperature measured without further treatment.
8*	463	Willey, L.A. and Fink, W.L.	1945	I	293-573	Al-4032	Bal.	12.5		0.9			0.9	1.0 Mg; similar to the above specimen except annealed to 623 K.
9	463	Willey, L.A. and Fink, W.L.	1945	I	213-373	Al-4032	Bal.	12.5		0.9			0.9	1.0 Mg; similar to the above specimen except heat-treated and aged according to standard practice.
10	462	Arp, V., Wilson, J.H., Winrich, L., and Sikora, P.	1962	L	20-293	Al-356	Bal.	6.9	0.2	0.1				0.2 Mg, 0.1 Ti; hardness R_B41, A.S.M. condition T6.
11	462	Arp, V., et al.	1962	L	20-293	Tens-50	Bal.	8.2	0.2					0.4 Mg, 0.2 Ti, 0.1 Be; hardness R_B60, A.S.M. condition T6.
12*	463	Willey, L.A. and Fink, W.L.	1945	I	293-673	Al-6151		1.0						0.6 Mg, 0.25 Cr; forged to 1 in. square; expansion measured without further treatment.
13	463	Willey, L.A. and Fink, W.L.	1945	I	293-673	Al-6151		1.0						0.6 Mg, 0.25 Cr; similar to the above specimen except annealed to 623 K.
14	463	Willey, L.A. and Fink, W.L.	1945	I	293-373	Al-6151		1.0						0.6 Mg, 0.25 Cr; similar to the above specimen except heat-treated and aged according to standard practice.
15*	19	Hidnert, P. and Krider, H.S.	1952	L	223, 293	Sample 1203A	Bal.	13.08	0.76	0.15				Specimen normalized 1 hr at 673 K and cooled slowly; expansion measured with increasing temperature.

*Not shown in figure.

SPECIFICATION TABLE 238. THERMAL LINEAR EXPANSION OF ALUMINUM + SILICON + ΣX_i ALLOYS Al + Si + ΣX_i (continued)

Cur. No.	Ref. No.	Author(s)	Year	Method Used	Temp. Range, K	Name and Specimen Designation	Al	Composition (weight percent) Si	Fe	Cu	Na	Mn	Ni	Composition (continued), Specifications, and Remarks
16	19	Hidnert, P. and Krider, H.S.	1952	L	223-573	Sample 1204	Bal.	17.27	0.81	0.12				Similar to the above specimen; expansion measured with increasing temperature.
17*	19	Hidnert, P. and Krider, H.S.	1952	L	293, 573	Sample 1198	Bal.	13.35	0.82	4.03				Similar to the above specimen; expansion measured with increasing temperature.
18*	19	Hidnert, P. and Krider, H.S.	1952	L	573-223	Sample 1198	Bal.	13.35	0.82	4.03				Similar to the above specimen; expansion measured with decreasing temperature.
19*	19	Hidnert, P. and Krider, H.S.	1952	L	293-473	Sample 1099	Bal.	10.16	0.31	9.08				Specimen cast in iron mold heated to 1023 K and cooled very slowly; expansion measured with increasing temperature.
20*	19	Hidnert, P. and Krider, H.S.	1952	L	473-293	Sample 1099	Bal.	10.16	0.31	9.08				The above specimen; expansion measured with decreasing temperature.
21*	19	Hidnert, P. and Krider, H.S.	1952	L	473-293	Sample 1098	Bal.	10.16	0.31	9.08				Specimen cast in iron mold.
22*	20	Hidnert, P.	1925	T	293-573	S866	Bal.	4.15	0.52	0.33				Sand cast specimen, heated to 673 K and cooled in furnace before testing.
23*	20	Hidnert, P.	1925	T	293-573	S871	Bal.	7.28	0.47	0.27				Similar to the above specimen.
24	20	Hidnert, P.	1925	T	293-573	S868	Bal.	9.81	0.50	0.22				Similar to the above specimen.
25*	20	Hidnert, P.	1925	T	293-773	S939	Bal.	12.55	0.56	0.08				Specimen cast in green sand.
26	20	Hidnert, P.	1925	T	293-773	S939	Bal.	12.55	0.56	0.08				The above specimen; expansion measured on second test.
27*	20	Hidnert, P.	1925	T	293-773	S940	Bal.	12.55	0.56	0.08				Specimen cut from same bar as the above specimen.
28*	20	Hidnert, P.	1925	T	293-773	S940	Bal.	12.55	0.56	0.08				The above specimen; expansion measured on second test.
29*	20	Hidnert, P.	1925	T	293-773	S941	Bal.	12.54	0.56	0.08	0.10			Similar to the above specimen except sodium added in molten state.
30*	20	Hidnert, P.	1925	T	293-773	S942	Bal.	12.54	0.56	0.08	0.10			Cut from same bar as the above specimen.
31*	20	Hidnert, P.	1925	T	293-573	S869	Bal.	3.33	0.55	2.2		0.01		Sand cast specimen, heated to 673 K and cooled in furnace before testing.
32*	20	Hidnert, P.	1925	T	293-573	S867	Bal.	7.42	0.53	2.43				Similar to the above specimen.
33*	20	Hidnert, P.	1925	T	293-573	S870	Bal.	9.96	0.60	2.33				Similar to the above specimen.
34*	20	Hidnert, P.	1925	T	293-523	S873	Bal.	6.61	0.57	4.53				Similar to the above specimen.
35*	20	Hidnert, P.	1925	T	293-573	S873	Bal.	6.61	0.57	4.53				The above specimen; expansion measured on second test.
36*	20	Hidnert, P.	1925	T	293-573	S876	Bal.	10.28	0.54	4.58				Similar to the above specimen.
37*	20	Hidnert, P.	1925	T	293-573	S877	Bal.	9.45	0.53	6.29				Similar to the above specimen.

*Not shown in figure.

SPECIFICATION TABLE 238. THERMAL LINEAR EXPANSION OF ALUMINUM + SILICON + ΣX_i ALLOYS Al + Si + ΣX_i (continued)

Cur. No.	Ref. No.	Author(s)	Year	Method Used	Temp. Range, K	Name and Specimen Designation	Composition (weight percent) Al	Si	Fe	Cu	Na	Mn	Ni	Composition (continued), Specifications, and Remarks
38*	20	Hidnert, P.	1925	T	293-573	S878	Bal.	3.12	0.55	2.4		0.93		Similar to the above specimen.
39*	20	Hidnert, P.	1925	T	293-573	S880	Bal.	9.97	0.50	2.32		0.82		Similar to the above specimen.
40*	20	Hidnert, P.	1925	T	293-573	S879	Bal.	10.22	0.56	2.49		0.89		Similar to the above specimen.
41*	20	Hidnert, P.	1925	T	293-573	S874	Bal.	10.18	0.70	2.47		1.17		Similar to the above specimen.
42	464	Barber, C.R.	1949	L	293-573	Lo-Ex	Bal.	11.80	0.50	1.03		0.03	1.02	0.91 Mg, 0.02 Ti; rod specimen 1.4 cm dia, 30 cm long; heat-treated 12 hr at 795 K, quenched; aged 4 hr at 428 K, air cooled; aged 4 hr at 473 K, air cooled.
43*	464	Barber, C.R.	1949	L	293, 373	SA-1	Bal.	11.0	0.5	5.0				0.6 Mg, 0.2 Ce, 0.05 Ti; rod specimen 1.4 cm dia, 30 cm long; tested as received.
44*	464	Barber, C.R.	1949	L	293-573	SA-1	Bal.	11.0	0.5	5.0				Similar to the above specimen except heat-treated 3 hr at 773 K, quenched in cold water; aged 16 hr at 438 K, air cooled.
45*	464	Barber, C.R.	1949	L	293, 373	SA-14	Bal.	11.0	0.5	5.0		0.4		0.5 Mg, 0.3 Ce, 0.1 Ti; rod specimen 1.4 cm dia, 30 cm long; tested as received.
46*	464	Barber, C.R.	1949	L	293-573	SA-14	Bal.	11.0	0.5	5.0		0.4		Similar to the above specimen except heat-treated 3 hr at 773 K, quenched in cold water; aged 16 hr at 438 K, air cooled.
47	464	Barber, C.R.	1949	L	293-573	RR-50	Bal.	2.25	1.18	1.40			0.90	0.12 Mg, 0.19 Ti; rod specimen 1.4 cm dia, 30 cm long; heat-treated 10 hr at 433-443 K; air cooled.
48*	464	Barber, C.R.	1949	L	293-573	RR-53C	Bal.	2.42	1.12	1.33			0.87	0.50 Mg, 0.15 Ti; rod specimen 1.4 cm dia, 30 cm long; heat-treated 2 hr at 803 K, water quenched; aged 15 hr at 433-443 K.
49*	464	Barber, C.R.	1949	L	293, 373	Alpax-Gamma	Bal.	12.0	0.28			0.29		0.35 Mg; rod specimen 1.4 cm dia, 30 cm long; tested as received.
50*	464	Barber, C.R.	1949	L	293-573	Alpax-Gamma	Bal.	12.0	0.28			0.29		Similar to the above specimen; heat-treated 1 hr at 783-791 K, cold water quenched; aged 16 hr at 423-438 K.
51*	464	Barber, C.R.	1949	L	293, 373	SA-1	Bal.	11.0	0.5	5.0				0.6 Mg, 0.2 Ce, 0.05 Ti; similar to specimen for curve 43 except chill cast.
52*	464	Barber, C.R.	1949	L	293-573	SA-1	Bal.	11.0	0.5	5.0				Similar to specimen for curve 44 except chill cast.
53*	464	Barber, C.R.	1949	L	293, 373	SA-14	Bal.	11.0	0.5	5.0		0.1		0.5 Mg, 0.3 Ce, 0.1 Ti; similar to specimen for curve 45 except chill cast.
54*	464	Barber, C.R.	1949	L	293-573	SA-14	Bal.	11.0	0.5	5.0		0.1		Similar to specimen for curve 46 except chill cast.

*Not shown in figure.

SPECIFICATION TABLE 238. THERMAL LINEAR EXPANSION OF ALUMINUM + SILICON + ΣX_i ALLOYS Al + Si + ΣX_i (continued)

Cur. No.	Ref. No.	Author(s)	Year	Method Used	Temp. Range, K	Name and Specimen Designation	Composition (weight percent) Al	Si	Fe	Cu	Na	Mn	Ni	Composition (continued), Specifications, and Remarks
55	427	Souder, W., Hidnert, P., and Fox, J. F.	1934	O	373-573	1206A	Bal.	19.30	0.84	3.14		1.08	4.18	150 mm long with pointed ends, maximum dia 12 mm; annealed at 673 K.
56	465	Losano, L.	1930		293-773	No. 1	99.70	0.18	0.11	trace				No details given.
57*	465	Losano, L.	1930		293-773	No. 2	98.76	0.84	0.21	0.03				No details given.
58	465	Losano, L.	1930		293-773	No. 3	94.82	4.88	0.23	0.01				No details given.
59*	465	Losano, L.	1930		293-773	No. 4	92.08	7.56	0.28	0.03				No details given.
60	465	Losano, L.	1930		293-773	No. 5	88.10	11.62	0.21	0.01				No details given.
61*	465	Losano, L.	1930		293-773	No. 6	84.26	15.23	0.29	0.07				No details given.
62	465	Losano, L.	1930		293-773	No. 7	78.32	21.34	0.30	0.02				No details given.
63*	465	Losano, L.	1930		293-773	No. 8	72.46	27.30	0.18	0.04				No details given.
64	465	Losano, L.	1930		293-773	No. 9	67.31	32.34	0.30	0.03				No details given.
65*	465	Losano, L.	1930		293-773	No. 10	65.44	33.98	0.48	0.05				No details given.
66	465	Losano, L.	1930		293-773	No. 11	61.26	38.48	0.19	0.07				No details given.
67*	465	Losano, L.	1930		293-773	No. 12	58.62	41.02	0.38	trace				No details given.
68*	465	Losano, L.	1930		293-773	No. 13	53.20	46.30	0.41	0.03				No details given.
69	465	Losano, L.	1930		293-773	No. 14	49.72	49.70	0.45	0.06				No details given.
70*	552	Kempf, L. W.	1933	L	293-573	Sample 5542	Bal.	13	0.76	0.15				Specimen approximately 4 in. long, 0.475 in. dia; annealed 20 hr at 225, air cooled to room temperature; specimen enclosed in a nickel chamber and the expansion of this cylinder relative to quartz was calibrated in terms of temperature.
71*	552	Kempf, L. W.	1933	L	293-573	Sample 5542	Bal.	13	0.76	0.15				The above specimen.
72*	552	Kempf, L. W.	1933	L	293-573	Sample 5542	Bal.	13	0.76	0.15				The above specimen.
73*	552	Kempf, L. W.	1933	L	293-573	Sample 5542	Bal.	13	0.76	0.15				The above specimen.
74*	552	Kempf, L. W.	1933	L	293-573	Sample 5542	Bal.	13	0.76	0.15				The above specimen.
75*	552	Kempf, L. W.	1933	L	293-573	Sample 5542	Bal.	13	0.76	0.15				The above specimen.
76*	552	Kempf, L. W.	1933	L	293-573	Sample 5542	Bal.	13	0.76	0.15				The above specimen.
77	551	Belov, A. K.	1968	L	77-293	Al2 Russian alloy	Bal.	13		0.6		0.5		0.3 Zn; cylindrical specimen 3.5 mm in dia, 50 mm long; sample heated and cooled at a rate of 3-7 deg per min; data on the coefficient of thermal linear expansion are also reported.

* Not shown in figure.

DATA TABLE 238. THERMAL LINEAR EXPANSION OF ALUMINUM + SILICON + ΣX_i ALLOYS Al + Si + ΣX_i

[Temperature, T, K; Linear Expansion, $\Delta L/L_0$, %]

T	$\Delta L/L_0$
CURVE 1*	
293	0.000
373	0.160
473	0.380
573	0.627
673	0.882
CURVE 2*	
673	0.821
473	0.356
373	0.149
293	0.000
CURVE 3*	
293	0.000
373	0.158
473	0.367
573	0.582
673	0.817
CURVE 4*	
673	0.809
573	0.568
293	0.000
CURVE 5	
293	0.000
373	0.155
473	0.369
573	0.588
673	0.828
CURVE 6	
673	0.809
473	0.353
373	0.149*
293	0.000*
CURVE 7*	
293	0.000
373	0.159
CURVE 7 (cont.)*	
473	0.374
573	0.610
CURVE 8*	
293	0.000
373	0.155
473	0.365
573	0.591
CURVE 9	
213	-0.147
293	0.000*
373	0.159
CURVE 10	
20	-0.388
40	-0.386
60	-0.378
80	-0.364
100	-0.343
120	-0.316
140	-0.286
160	-0.255
180	-0.220
200	-0.184
220	-0.146
240	-0.108
260	-0.068
273	-0.041
280	-0.027
293	0.000*
CURVE 11	
20	-0.377
40	-0.374
60	-0.366
80	-0.352
100	-0.331
120	-0.305
140	-0.275
160	-0.242
180	-0.206
200	-0.170
CURVE 11 (cont.)	
220	-0.134
240	-0.097
260	-0.060
273	-0.037
280	-0.023*
293	0.000*
CURVE 12*	
293	0.000
373	0.183
473	0.428
573	0.694
673	0.977
CURVE 13	
293	0.000
373	0.185
473	0.436
573	0.700
673	0.984
CURVE 14	
213	-0.173
293	0.000*
373	0.186
CURVE 15*	
223	-0.128
CURVE 16	
223	-0.120
293	0.000*
373	0.150
473	0.344
523	0.453
573	0.563
CURVE 17*	
523	0.483
CURVE 18*	
573	0.605
293	0.000
223	-0.124
CURVE 19*	
293	0.000
373	0.160
473	0.378
CURVE 20*	
473	0.378
373	0.163
293	0.000
CURVE 21*	
293	0.000
373	0.163
473	0.401
CURVE 22*	
293	0.000
373	0.178
473	0.418
573	0.675
CURVE 23*	
293	0.000
373	0.174
473	0.410
573	0.658
CURVE 24	
293	0.000
373	0.169
473	0.394
573	0.641
CURVE 25*	
293	0.000
373	0.155
CURVE 25 (cont.)*	
473	0.387
573	0.694
673	0.942
773	1.186
CURVE 26	
293	0.000*
373	0.161*
473	0.385*
573	0.619*
673	0.859
773	1.099
CURVE 27*	
293	0.000
373	0.155
473	0.378
573	0.683
673	0.923
773	1.162
CURVE 28*	
293	0.000
373	0.178
473	0.398
573	0.638
673	0.882
773	1.114
CURVE 29*	
293	0.000
373	0.158
473	0.374
573	0.622
673	0.874
773	1.104
CURVE 30*	
293	0.000
373	0.154
473	0.364
573	0.622
CURVE 30 (cont.)	
673	0.870
773	1.104
CURVE 31*	
293	0.000
373	0.187
473	0.430
523	0.557
573	0.683
CURVE 32*	
293	0.000
373	0.174
473	0.405
523	0.529
573	0.655
CURVE 33*	
293	0.000
373	0.166
473	0.391
523	0.511
573	0.636
CURVE 34*	
293	0.000
373	0.171
473	0.405
523	0.531
CURVE 35*	
293	0.000
373	0.172
473	0.401
523	0.524
573	0.647
CURVE 36*	
293	0.000
373	0.163
473	0.383

* Not shown in figure.

DATA TABLE 238. THERMAL LINEAR EXPANSION OF ALUMINUM + SILICON + ΣX_i ALLOYS   Al + Si + ΣX_i (continued)

T	$\Delta L/L_0$
CURVE 36 (cont.)*	
523	0.501
573	0.619
CURVE 37*	
293	0.000
373	0.165
473	0.380
523	0.506
573	0.622
CURVE 38*	
293	0.000
373	0.178
473	0.421
523	0.545
573	0.666
CURVE 39*	
293	0.000
373	0.163
473	0.387
523	0.506
573	0.627
CURVE 40*	
293	0.000
373	0.166
473	0.387
523	0.506
573	0.624
CURVE 41*	
293	0.000
373	0.163
473	0.387
523	0.506
573	0.624
CURVE 42	
293	0.000*
373	0.155*
CURVE 42 (cont.)	
473	0.371
573	0.602
CURVE 43*	
373	0.157
CURVE 44*	
293	0.000
373	0.155
473	0.382
573	0.605
CURVE 45*	
373	0.153
CURVE 46*	
293	0.000
373	0.152
473	0.364
573	0.577
CURVE 47	
293	0.000*
373	0.177
473	0.412
573	0.661
CURVE 48*	
293	0.000
373	0.196
473	0.443
573	0.683
CURVE 49*	
373	0.160
CURVE 50*	
293	0.000
373	0.160
CURVE 50 (cont.)*	
473	0.378
573	0.610
CURVE 51*	
373	0.161
CURVE 52*	
293	0.000
373	0.159
473	0.376
573	0.605
CURVE 53*	
373	0.158
CURVE 54*	
293	0.000
373	0.158
473	0.371
573	0.596
CURVE 55	
373	0.132
473	0.308
523	0.400
573	0.501
CURVE 56	
293	0.000*
373	0.190
473	0.438
573	0.695
673	0.959
773	1.233
CURVE 57*	
293	0.000
373	0.190
473	0.437
573	0.694
CURVE 57 (cont.)*	
673	0.960
773	1.232
CURVE 58	
293	0.000*
373	0.179*
473	0.419
573	0.669
673	0.928
773	1.196
CURVE 59*	
293	0.000
373	0.173
473	0.405
573	0.648
673	0.903
773	1.166
CURVE 60	
293	0.000*
373	0.162
473	0.381
573	0.617
673	0.866
773	1.127
CURVE 61*	
293	0.000
373	0.150
473	0.357
573	0.584
673	0.827
773	1.082
CURVE 62	
293	0.000*
373	0.134
473	0.327
573	0.540
673	0.772
773	1.017
CURVE 63*	
293	0.000
373	0.123
473	0.300
573	0.500
673	0.722
773	0.956
CURVE 64	
293	0.000*
373	0.116
473	0.283
573	0.474
673	0.686
773	0.913
CURVE 65*	
293	0.000
373	0.113
473	0.277
573	0.465
673	0.675
773	0.900
CURVE 66	
293	0.000*
373	0.109*
473	0.266
573	0.445
673	0.648
773	0.865
CURVE 67*	
293	0.000
373	0.106
473	0.259
573	0.434
673	0.633
773	0.848
CURVE 68*	
293	0.000
373	0.103
CURVE 68 (cont.)*	
473	0.250
573	0.417
673	0.607
773	0.814
CURVE 69	
293	0.000*
373	0.102
473	0.245
573	0.407
673	0.589
773	0.792
CURVE 70*	
293	0.000
373	0.157
473	0.365
573	0.599
CURVE 71*	
293	0.000
373	0.157
473	0.367
573	0.602
CURVE 72*	
293	0.000
373	0.154
473	0.362
573	0.591
CURVE 73*	
293	0.000
373	0.157
473	0.364
573	0.588
CURVE 74*	
293	0.000
373	0.156
473	0.364
573	0.594

* Not shown in figure.

DATA TABLE 238. THERMAL LINEAR EXPANSION OF ALUMINUM + SILICON + ΣX_i ALLOYS $Al + Si + \Sigma X_i$ (continued)

T	$\Delta L/L_0$	T	$\Delta L/L_0$
CURVE 75*		CURVE 77 (cont.)‡	
293	0.000	113	-0.302
373	0.153	123	-0.290
473	0.358	133	-0.278
573	0.588	143	-0.264*
		153	-0.250
CURVE 76*		163	-0.235*
		173	-0.219
293	0.000	183	-0.203*
373	0.153	193	-0.186
473	0.360	203	-0.169*
573	0.591	213	-0.151*
		223	-0.133*
CURVE 77‡		233	-0.114
		243	-0.096*
77	-0.336	253	-0.077
83	-0.331*	263	-0.058*
93	-0.322	273	-0.039*
103	-0.312	283	-0.019
		293	0.000

DATA TABLE 238. COEFFICIENT OF THERMAL LINEAR EXPANSION OF ALUMINUM + SILICON + ΣX_i ALLOYS $Al + Si + \Sigma X_i$

[Temperature, T, K; Coefficient of Expansion, α, 10^{-6} K^{-1}]

T	α	T	α
CURVE 77‡		CURVE 77 (cont.)‡	
77	7.80	233	18.59
83	8.50	243	18.80
93	9.30	253	19.00
103	10.30	263	19.09
113	11.10	273	19.20
123	12.00	283	19.30
133	12.90	293	19.40
143	13.70		
153	14.60		
163	15.30		
173	16.00		
183	16.60		
193	17.19		
203	17.59		
213	17.99		
223	18.40		

* Not shown in figure.

‡Author's data for coefficient of thermal expansion have been integrated by TPRC to obtain $\Delta L/L_0$.

FIGURE AND TABLE NO. 239R. PROVISIONAL VALUES FOR THERMAL LINEAR EXPANSION OF ALUMINUM + ZINC + ΣX_i ALLOYS Al + Zn + ΣX_i

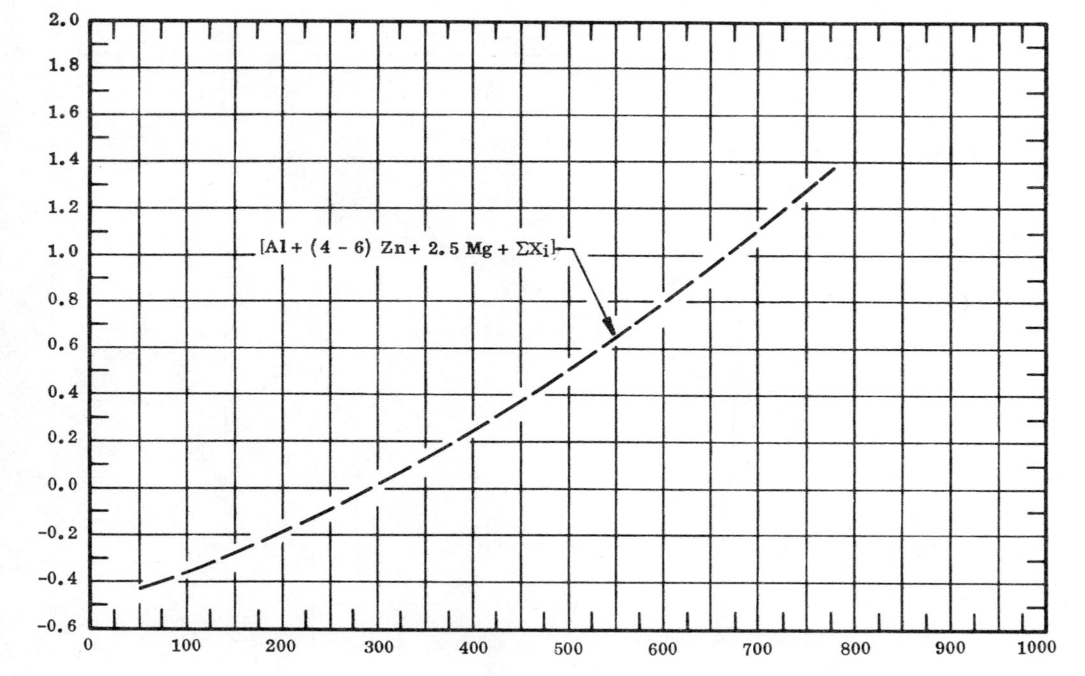

PROVISIONAL VALUES

[Temperature, T, K; Linear Expansion, $\Delta L/L_0$, %; α, K^{-1}]

T	[Al + (4 - 6) Zn + 2.5 Mg + ΣX_i] $\Delta L/L_0$	$\alpha \times 10^6$
50	-0.425	12.0
100	-0.359	14.5
200	-0.192	18.8
293	0.000	22.3
400	0.257	25.6
500	0.527	28.1
600	0.818	29.9
700	1.124	31.1
775	1.359	31.6

REMARKS

[Al + (4 - 6) Zn + 2.5 Mg + ΣX_i]: The tabulated values for well-annealed alloy of types Al 7039, Al 7075, and Al 7079 alloys are considered accurate to within ± 7% over the entire temperature range. These values can be represented approximately by the following equation:

$$\Delta L/L_0 = -0.478 + 9.368 \times 10^{-4} T + 2.688 \times 10^{-6} T^2 - 1.082 \times 10^{-9} T^3$$

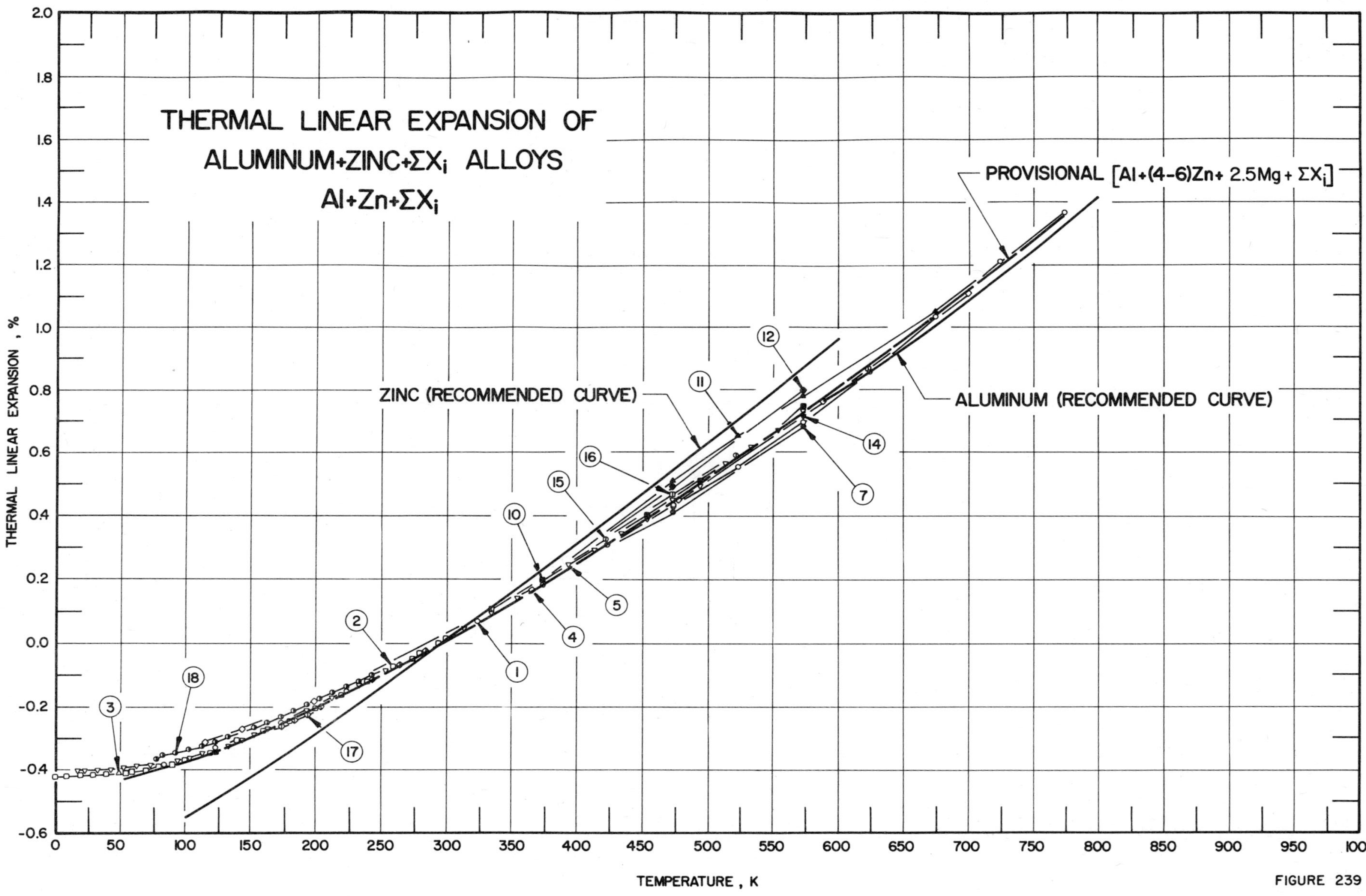
THERMAL LINEAR EXPANSION OF
ALUMINUM+ZINC+ΣX_i ALLOYS
Al+Zn+ΣX_i
PROVISIONAL [Al+(4-6)Zn+ 2.5Mg+ ΣX_i]
ZINC (RECOMMENDED CURVE)
ALUMINUM (RECOMMENDED CURVE)
12
11
14
16
15
7
10
5
2
4
1
18
3
17
THERMAL LINEAR EXPANSION, %
2.0
1.8
1.6
1.4
1.2
1.0
0.8
0.6
0.4
0.2
0.0
-0.2
-0.4
-0.6
0
50
100
150
200
250
300
350
400
450
500
550
600
650
700
750
800
850
900
950
1000
TEMPERATURE, K
FIGURE 239

SPECIFICATION TABLE 239. THERMAL LINEAR EXPANSION OF ALUMINUM + ZINC + ΣX_i ALLOYS Al + Zn + ΣX_i

Cur. No.	Ref. No.	Author(s)	Year	Method Used	Temp. Range, K	Name and Specimen Designation	Al	Zn	Cu	Fe	Si	Mn	Mg	Composition (continued), Specifications, and Remarks
							Composition (weight percent)							
1	466	Lucks, C.F., Thompson, H.B., Smith, A.R., Curry, F.P., Deem, H.W. and Bing, G.F.	1951	L	83-773	75S-T6 aluminum	90	5.6	1.6				2.5	0.3 Cr; heat treatment T6; supplied by Aluminum Co. of America; density 2.80 g/cm³ at 293 K; zero-point correction of -0.046% determined by graphical interpolation.
2	467	Clark, A.F.	1968	L	0-300	Al 7039	93.4	3.60				0.23	2.55	0.20 Cr; trace (< 0.1) Fe, Cu, Si, Ti, Be; machined to 0.64 cm sq x 20.3 cm long; Rockwell hardness B-75.
3	467	Clark, A.F.	1968	L	0-300	Al 7075-T73	90.6	5.3	1.4	0.14	0.10		2.2	0.22 Cr; trace (< 0.1) Ti, Mn, Ni; machined to 0.64 cm sq x 20.3 cm long; Rockwell hardness B-73.
4	123	Lucks, C.F. and Deem, H.W.	1958	L	116-669	Al 7075-T6								0.375 in. dia by 3 in. long; heating and cooling rate 3 F per min; specimen produced by Aluminum Co. of America.
5	15	Rhodes, B.L., Moeller, C.E., Hopkins, V. and Marx, T.I.	1963	L	18-573	Al 7075-T6	Bal.	5.6	1.6	0.7	0.5	0.3	2.5	0.3 Cr, 0.2 Ti, 0.15 ΣX_i, specimen 10.16 cm long by 6.2 mm; reported error < 1.5%.
6	462	Arp, V., Wilson, J.H., Winrich, L. and Sikora, P.	1962	L	20-293	Al 7075	Bal.	5.6	1.6	0.7	0.5	0.3	2.5	0.3 Cr, 0.2 Ti; hardness R_B90, A.S.M. condition T6.
7	463	Willey, L.A. and Fink, W.L.	1945	I	293-573	75S	Bal.	5.75	1.6			0.15	2.5	0.25 Cr; standard alloy for extrusion, sheet wrought form, 0.064 in cold-rolled, temp as rolled.
8*	463	Willey, L.A. and Fink, W.L.	1945	I	293-573	75S	Bal.	5.75	1.6			0.15	2.5	0.25 Cr; similar to the above specimen, except annealed through heating to temp 623 K.
9	463	Willey, L.A. and Fink, W.L.	1945	I	213-373	75S	Bal.	5.75	1.6			0.15	2.5	0.25 Cr; similar to the above specimen, except heat-treated and aged according to standard practice.
10	15	Rhodes, B.L., Moeller, C.E., Hopkins, V. and Marx, T.I.	1963	L	18-573	Al 7079-T6	Bal.	4.3	0.6	0.40	0.30	0.20	0.33	0.18 Cr, 0.10 Ti, 0.15 ΣX_i; specimen 10.16 cm long; 6.2 mm dia.
11	20	Hidnert, P.	1925	T	293-773	S841	Bal.	12.17	1.47	0.31	0.21	0.01		Specimen rolled and drawn to hexagonal shape.
12*	20	Hidnert, P.	1925	T	293-573	S842	Bal.	12.17	1.47					Cut from same bar as above specimen.
13*	464	Barber, C.R.	1949	L	293,373	RR-77	Bal.	4.96	2.20	0.31	0.26	0.54	2.54	Rod specimen 1.4 cm dia, 30 cm long; tested as received.
14	464	Barber, C.R.	1949	L	293-573	RR-77	Bal.	4.96	2.20	0.31	0.26	0.54	2.54	Similar to the above specimen; 2 hr solution heat treated at 723 K, quenched in water at 343 K; 4 hr aged at 408 K, air cooled.
15	35	Honda, K. and Okubo, Y.	1924	L	333-622	L5		13.5	2.5					Cylindrical rod specimens 15.000 cm long, 5 mm thick; zero-point correction of 0.100 determined by extrapolation.

* Not shown in figure.

SPECIFICATION TABLE 239. THERMAL LINEAR EXPANSION OF ALUMINUM + ZINC + ΣX_i ALLOYS Al + Zn + ΣX_i (continued)

Cur. No.	Ref. No.	Author(s)	Year	Method Used	Temp. Range, K	Name and Specimen Designation	Composition (weight percent) Al	Zn	Cu	Fe	Se	Mn	Mg	Composition (continued), Specifications, and Remarks
16	35	Honda, K. and Okubo, Y.	1924	L	334-622	B	Bal.	20	3					Similar to the above specimen; zero-point correction of 0.104% determined by extrapolation.
17*	551	Belov, A.K.	1968	L	77-293	V-94 Russian alloy	Bal	6.7	2				1.6	Cylindrical specimen 3.5 mm in diameter, 50 mm long; sample heated and cooled at a rate of 3-7 deg per min.
18*	551	Belov, A.K.	1968	L	77-293	AL9 Russian alloy	Bal.	10	0.8	1.5	6	0.4		Similar to the above specimen.

* Not shown in figure.

DATA TABLE 239. THERMAL LINEAR EXPANSION OF ALUMINUM + ZINC + ΣX_i ALLOYS Al + Zn + ΣX_i

[Temperature, T, K; Linear Expansion, $\Delta L/L_0$, %]

T	$\Delta L/L_0$
CURVE 1	
83	-0.386
123	-0.334
173	-0.251
223	-0.153
323	0.068
373	0.184
423	0.308
473	0.434
523	0.556
573	0.697
623	0.857
673	1.033
723	1.212
773	1.366
CURVE 2	
0	-0.421
10	-0.421
20	-0.420
30	-0.420
40	-0.418
56	-0.414
60	-0.409
70	-0.402
80	-0.393
90	-0.383
100	-0.371
120	-0.344
140	-0.313
160	-0.279
180	-0.242
200	-0.203
220	-0.162
240	-0.120
260	-0.076
273	-0.047
280	-0.031
293	0.000
300	0.015
CURVE 3	
0	-0.419*
10	-0.419*
20	-0.419*
CURVE 3 (cont.)	
30	-0.418*
40	-0.417*
50	-0.413*
60	-0.408*
70	-0.401*
80	-0.392*
90	-0.382*
100	-0.370*
120	-0.343
140	-0.312*
160	-0.278*
180	-0.241*
200	-0.202*
220	-0.161*
240	-0.119*
260	-0.076*
273	-0.047*
280	-0.031*
293	0.000*
300	0.015*
CURVE 4	
116	-0.314
144	-0.276
199	-0.185
366	0.163
477	0.449
588	0.765
699	1.107
CURVE 5	
18	-0.404
23	-0.404
33	-0.403
43	-0.401
53	-0.398
63	-0.395
73	-0.390
83	-0.383
93	-0.374
103	-0.364
113	-0.352
123	-0.338*
133	-0.324
CURVE 5 (cont.)	
143	-0.308
153	-0.291
163	-0.274
173	-0.255
193	-0.216
213	-0.176
233	-0.134
253	-0.091
273	-0.046
293	0.000*
313	0.046
333	0.093
353	0.141
373	0.191
393	0.242
413	0.292
433	0.342
453	0.393
473	0.444
493	0.496
513	0.562
533	0.612
553	0.672
573	0.734
CURVE 6	
20	-0.416
40	-0.413
60	-0.405
80	-0.391
100	-0.368
120	-0.340*
140	-0.309*
160	-0.275
180	-0.241*
200	-0.202*
220	-0.161
240	-0.119
260	-0.076
273	-0.046*
280	-0.030*
293	0.000*
CURVE 7	
293	0.000*
373	0.189*
473	0.410
573	0.686
CURVE 8*	
293	0.000
373	0.186
473	0.436
573	0.728
CURVE 9	
213	-0.173
293	0.000*
373	0.189*
CURVE 10	
18	-0.408
23	-0.408
33	-0.408
43	-0.407
53	-0.405
63	-0.402
73	-0.398*
83	-0.391*
93	-0.382*
103	-0.370*
113	-0.357*
123	-0.343*
133	-0.328*
143	-0.312*
153	-0.295*
163	-0.277*
173	-0.259*
193	-0.220*
213	-0.178*
233	-0.136*
253	-0.092*
273	-0.046*
293	0.000*
313	0.047*
333	0.096
353	0.145
CURVE 10 (cont.)	
373	0.195*
393	0.246*
413	0.297*
433	0.348
453	0.401
473	0.454*
493	0.507
513	0.562*
533	0.620
553	0.681
573	0.743
CURVE 11	
293	0.000*
373	0.194*
473	0.506
523	0.651
573	0.781
673	1.049
773	1.373
CURVE 12*	
293	0.000
373	0.204
473	0.491
523	0.656
573	0.792
CURVE 13*	
373	0.189
CURVE 14	
293	0.000*
373	0.183*
473	0.425
573	0.711
CURVE 15‡	
333	0.100
373	0.200
422	0.326
CURVE 15 (cont.)‡	
472	0.456
522	0.587
573	0.722
622	0.851
CURVE 16‡	
334	0.104
373	0.203
423	0.333
473	0.466
522	0.596
573	0.733*
622	0.866
CURVE 17*, ‡	
77	-0.399
83	-0.393
93	-0.383
103	-0.371
113	-0.359
123	-0.345
133	-0.330
143	-0.315
153	-0.298
163	-0.280
173	-0.262
183	-0.242
193	-0.222
203	-0.201
213	-0.180
223	-0.158
233	-0.136
243	-0.114
253	-0.091
263	-0.069
273	-0.046
283	-0.023
293	0.000
CURVE 18*, ‡	
77	-0.364
83	-0.359
93	-0.350

* Not shown in figure.

‡ Author's values for coefficient of expansion integrated by TPRC to obtain $\Delta L/L_0$.

DATA TABLE 239. THERMAL LINEAR EXPANSION OF ALUMINUM + ZINC + ΣX_i ALLOYS $Al + Zn + \Sigma X_i$ (continued)

T	$\Delta L/L_0$
CURVE 18 (cont.)‡	
103	-0.339
113	-0.327
123	-0.314
133	-0.299
143	-0.284
153	-0.268
163	-0.251
173	-0.234
183	-0.215
193	-0.197
203	-0.178
213	-0.159
223	-0.139
233	-0.120
243	-0.100
253	-0.080
263	-0.060
273	-0.040
283	-0.020
293	0.000

DATA TABLE 239. COEFFICIENT OF THERMAL LINEAR EXPANSION OF ALUMINUM + ZINC + ΣX_i ALLOYS $Al + Zn + \Sigma X_i$

[Temperature, T, K; Coefficient of Expansion, α, 10^{-6} K^{-1}]

T	α
CURVE 15*, ‡	
333	24.6
373	25.4
422	26.0
472	26.2
522	26.2
573	26.4
622	26.4
CURVE 16*, ‡	
334	24.9
373	25.8
423	26.4
473	26.6
522	26.7
573	27.0
622	27.0
CURVE 17‡	
77	9.30
83	10.00
93	10.99
103	12.09
113	13.10
123	14.10
133	15.20
143	16.30
153	17.19
163	18.10
173	19.00
183	19.79
193	20.50
203	21.10
213	21.50
223	21.99
233	22.29
243	22.50
253	22.60
263	22.70
273	22.80
283	22.90
293	23.00
CURVE 18‡	
77	8.00
83	8.80
93	10.20
103	11.40
113	12.60
123	13.70
133	14.69
143	15.60
153	16.60
163	17.30
173	17.99
183	18.40
193	18.69
203	19.00
213	19.30
223	19.40
233	19.60
243	19.79
253	19.90
263	20.00
273	20.09
283	20.19
293	20.30

*Not shown in figure.

‡Author's data for coefficient of thermal expansion have been integrated by TPRC to obtain $\Delta L/L_0$.

FIGURE AND TABLE NO. 240R. PROVISIONAL VALUES FOR THERMAL LINEAR EXPANSION OF BERYLLIUM + ALUMINUM + ΣX_i ALLOYS   Be + Al + ΣX_i

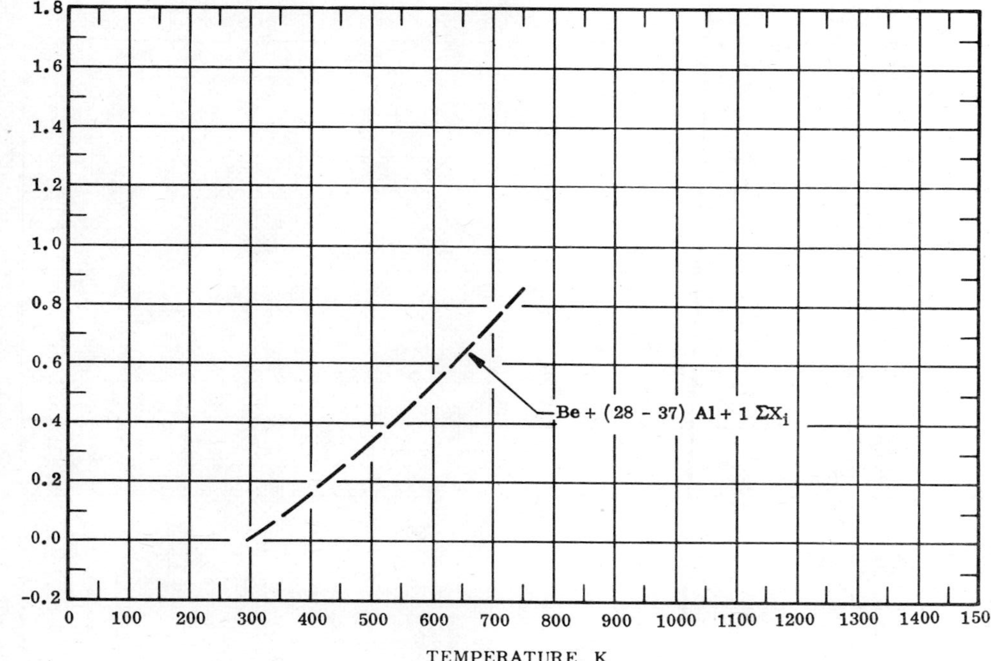

PROVISIONAL VALUES

[Temperature, T, K; Linear Expansion, $\Delta L/L_0$, %; α, K^{-1}]

	[Be + (28-37) Al + 1 ΣX_i]	
T	$\Delta L/L_0$	$\alpha \times 10^6$
293	0.000	14.4
400	0.168	16.7
500	0.343	18.6
600	0.539	20.2
700	0.748	21.5
750	0.856	22.0

REMARKS

[Be + (28-37) Al + 1 ΣX_i]: The tabulated values are for well-annealed alloys of type Lockalloy and are considered accurate to within ±7% over the entire temperature range. These values can be represented approximately by the following equation:

$$\Delta L/L_0 + -0.306 + 6.090 \times 10^{-4}\ T + 1.641 \times 10^{-6}\ T^2 - 5.150 \times 10^{-10}\ T^3$$

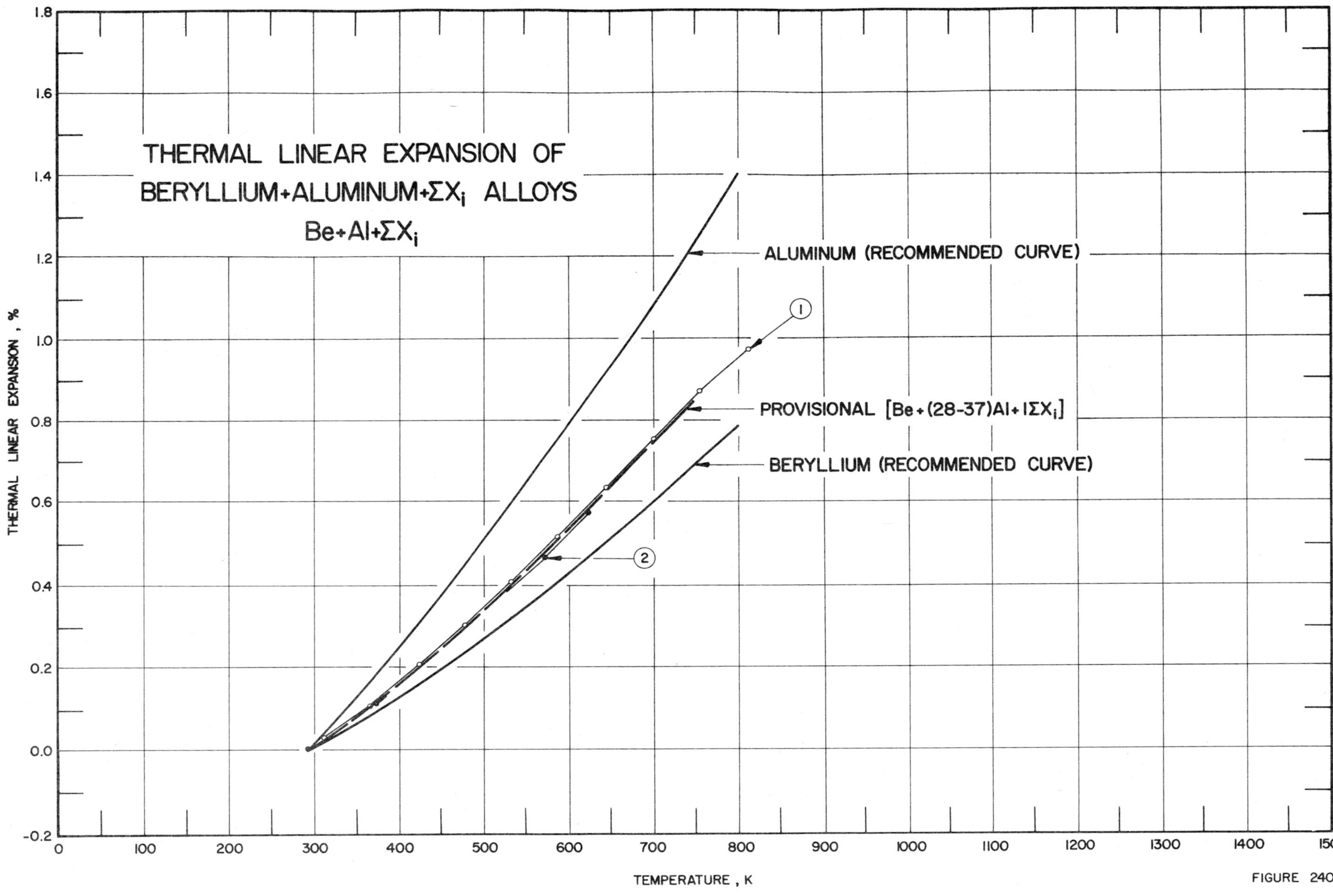
THERMAL LINEAR EXPANSION OF
BERYLLIUM+ALUMINUM+ΣX$_i$ ALLOYS
Be+Al+ΣX$_i$
ALUMINUM (RECOMMENDED CURVE)
PROVISIONAL [Be+(28-37)Al+1ΣX$_i$]
BERYLLIUM (RECOMMENDED CURVE)
1
2
THERMAL LINEAR EXPANSION, %
1.8
1.6
1.4
1.2
1.0
0.8
0.6
0.4
0.2
0.0
-0.2
0
100
200
300
400
500
600
700
800
900
1000
1100
1200
1300
1400
1500
TEMPERATURE, K

FIGURE 240

SPECIFICATION TABLE 240. THERMAL LINEAR EXPANSION OF BERYLLIUM + ALUMINUM + ΣX_i ALLOYS Be + Al + ΣX_i

Cur. No.	Ref. No.	Author(s)	Year	Method Used	Temp. Range, K	Name and Specimen Designation	Composition (weight percent) Be	Al	Si	Fe	C	Mg	Composition (continued), Specifications, and Remarks
1	468	Santschi, W.H. and Marz, W.G.	1966		310-811	Lockalloy	62	36.86	0.10	0.08	0.04	0.02	Specimen tested as-rolled; 0.80 BeO + AlO; density 2.09 g cm^{-3}; specific heat 0.39 cal/gm c.
2	19	Hidnert, P. and Krider, H.S.	1952	L	293-623	Sample 1630	71.3	27.9	<0.04	0.25		0.5	Specimen cast in iron mold.

DATA TABLE 240. THERMAL LINEAR EXPANSION OF BERYLLIUM + ALUMINUM + ΣX_i ALLOYS Be + Al + ΣX_i

[Temperature, T, K; Linear Expansion, $\Delta L/L_0$, %]

T	$\Delta L/L_0$
CURVE 1	
311	0.027
367	0.114
422	0.207
478	0.305
533	0.411
589	0.518
644	0.633
700	0.747
755	0.866
811	0.977

T	$\Delta L/L_0$
CURVE 2	
293	0.000
373	0.118
573	0.465
623	0.571

FIGURE AND TABLE NO. 241R. PROVISIONAL VALUES FOR THERMAL LINEAR EXPANSION OF CHROMIUM + IRON + ΣX_i ALLOYS Cr + Fe + ΣX_i

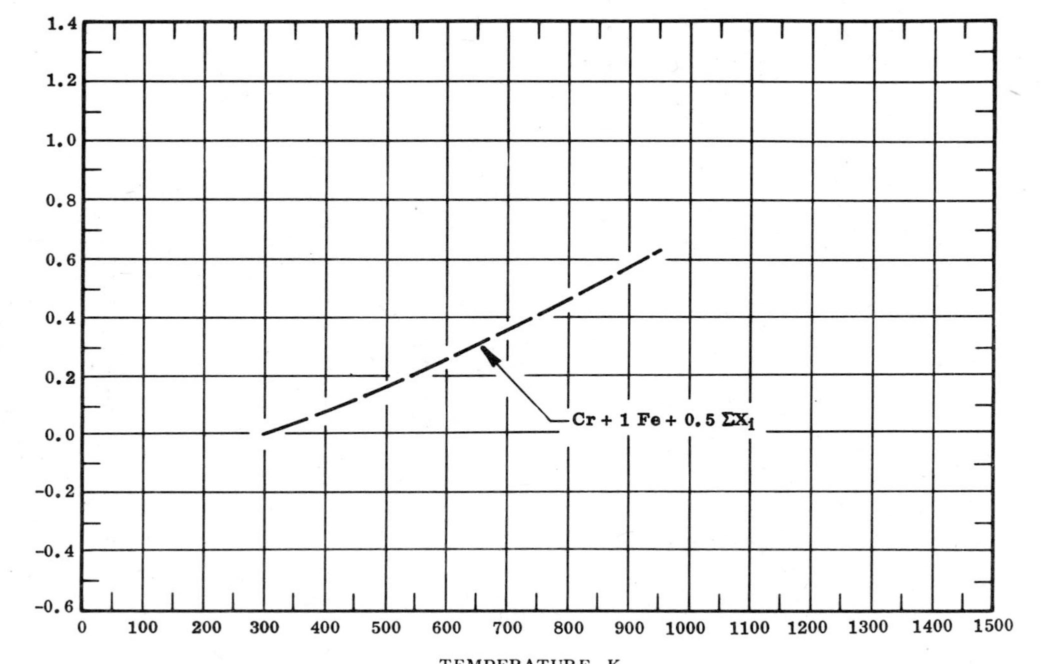

PROVISIONAL VALUES

[Temperature, T, K; Linear Expansion, $\Delta L/L_0$, %; α, K^{-1}]

T	(Cr + 1 Fe + 0.5 ΣX_i)	
	$\Delta L/L_0$	$\alpha \times 10^6$
293	0.000	6.3
400	0.075	7.9
500	0.162	9.1
600	0.259	10.0
700	0.361	10.6
800	0.468	10.8
900	0.577	10.8
950	0.630	10.8

REMARKS

(Cr + 1 Fe + 0.5 ΣX_i): The tabulated values for well-annealed alloy are considered accurate to within ±7% over the entire temperature range. These values can be represented approximately by the following equation:

$$\Delta L/L_0 = -0.100 + 2.439 \times 10^{-6}\, T + 1.305 \times 10^{-6}\, T^2 - 5.249 \times 10^{-10}\, T^3$$

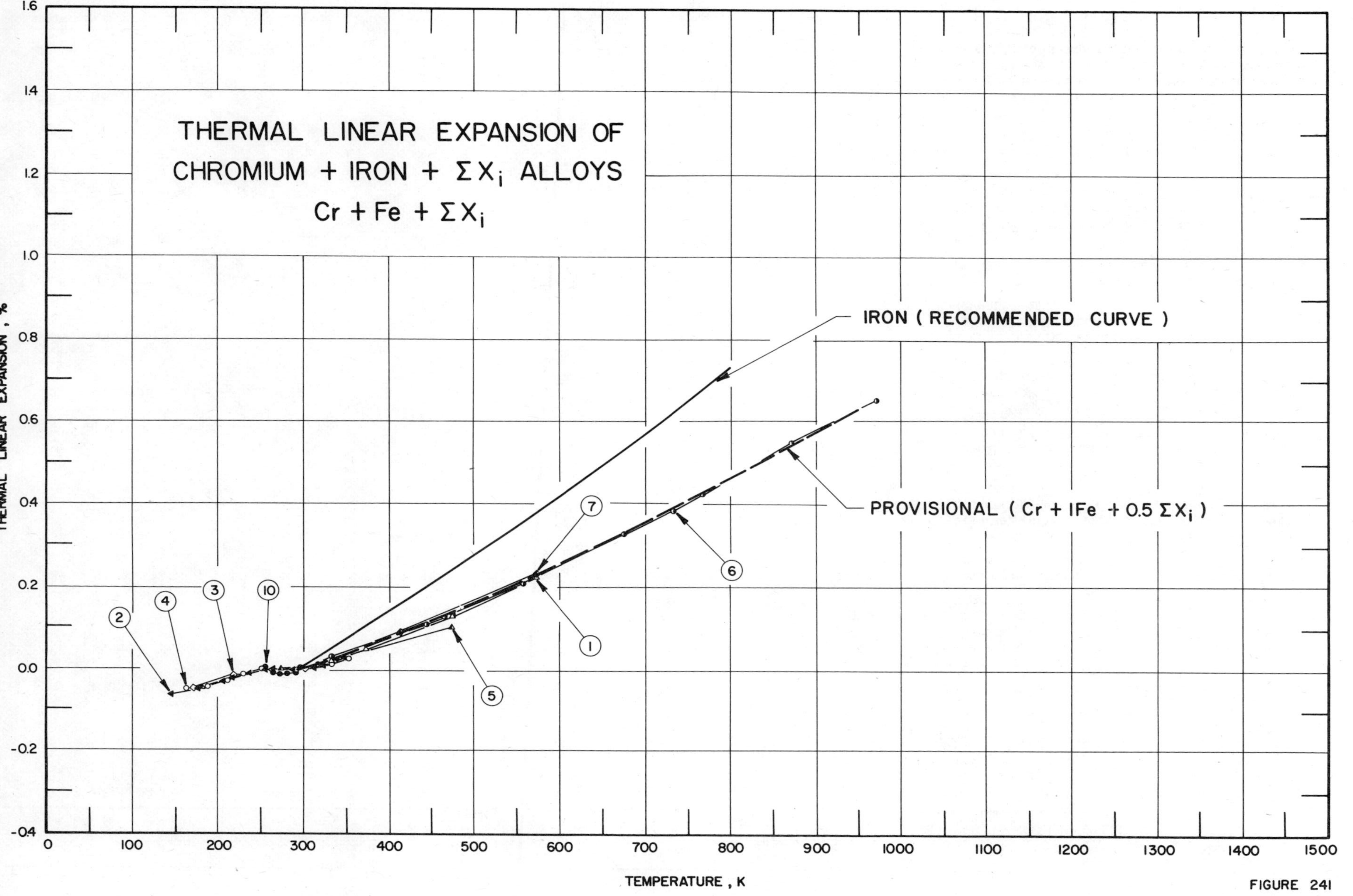
THERMAL LINEAR EXPANSION OF
CHROMIUM + IRON + Σ X_i ALLOYS
Cr + Fe + Σ X_i
IRON (RECOMMENDED CURVE)
PROVISIONAL (Cr + 1Fe + 0.5 Σ X_i)
1
2
3
4
5
6
7
10
THERMAL LINEAR EXPANSION, %
1.6
1.4
1.2
1.0
0.8
0.6
0.4
0.2
0.0
-0.2
-0.4
TEMPERATURE, K
0
100
200
300
400
500
600
700
800
900
1000
1100
1200
1300
1400
1500

FIGURE 241

SPECIFICATION TABLE 241. THERMAL LINEAR EXPANSION OF CHROMIUM + IRON + ΣX_i ALLOYS $Cr + Fe + \Sigma X_i$

Cur. No.	Ref. No.	Author(s)	Year	Method Used	Temp. Range, K	Name and Specimen Designation	Composition (weight percent) Cr	Fe	Si	C	Mn	N	Composition (continued), Specifications, and Remarks
1	396	Hidnert, P.	1941	T	300-574	1284	98.3	0.7	0.34	0.08	0.06	0.014	Hot-swaged rod, 3 mm in diameter, 298 mm long; prepared by Research Dept., Westinghouse Lamp Div., Bloomfield, N.J.; density 7.136 g cm^{-3} (at 298 K); expansion measured with increasing temperature; first test; zero-point correction of 0.004% determined by graphical extrapolation.
2	396	Hidnert, P.	1941	T	146-572	1284							The above specimen; expansion measured with increasing temperature; third test.
3	396	Hidnert, P.	1941	T	302-171	1284							The above specimen; expansion measured with decreasing temperature; third test; zero-point correction of -0.003% determined by graphical interpolation.
4	396	Hidnert, P.	1941	T	146-573	1284							The above specimen; expansion measured with increasing temperature; fourth test.
5	396	Hidnert, P.	1941	T	437-181	1284							The above specimen; expansion measured with decreasing temperature; fourth test; zero-point correction of -0.003% determined by graphical interpolation.
6	396	Hidnert, P.	1941	T	296-970	1284							The above specimen; expansion measured with increasing temperature; sixth test; zero-point correction of 0.003% determined by graphical extrapolation.
7	396	Hidnert, P.	1941	T	574-293	1284	98.3	0.7	0.34	0.08	0.06	0.016	The above specimen; expansion measured with decreasing temperature; sixth test; zero-point correction is 0.005%.
8*	396	Hidnert, P.	1941	L	256-304	1284							Similar to the above specimen; 201 mm long specimen; expansion measured with increasing temperature; seventh test.
9*	396	Hidnert, P.	1941		257-331	1284							The above specimen; expansion measured with increasing temperature; eighth test.
10	396	Hidnert, P.	1941	I	297-257	1443I	97.4	2.01	0.54	0.04			Cast piece; 8.380 mm long; sample consisted of three pieces of nearly equal length; expansion measured with decreasing temperature; first test; zero-point correction of -0.0267% determined by graphical interpolation.
11*	396	Hidnert, P.	1941	I	303-286	1443I							The above specimen; expansion measured with decreasing temperature; second test; zero-point correction is -0.0526%.
12*	396	Hidnert, P.	1941	I	272-156	1443I							The above specimen; expansion measured with decreasing temperature; third test; this curve is here reported using the first given temperature as reference temperature at which $\Delta L/L_0 = 0$.

* Not shown in figure.

DATA TABLE 241. THERMAL LINEAR EXPANSION OF CHROMIUM + IRON + ΣX_i ALLOYS $Cr + Fe + \Sigma X_i$

[Temperature, T, K; Linear Expansion, $\Delta L/L_0$, %]

T	$\Delta L/L_0$
CURVE 1	
300	0.004
332	0.022
372	0.050
474	0.135
574	0.225
CURVE 2	
146	-0.064
178	-0.050
208	-0.030
219	-0.022
238	-0.009
256	0.000
266	0.002
277	0.000*
289	0.000
311	0.008
323	0.015
348	0.033
372	0.049*
474	0.132
572	0.224*
CURVE 3	
302	0.001
220	-0.018
171	-0.054
CURVE 4	
146	-0.063*
164	-0.055
188	-0.042
203	-0.033*
211	-0.026
221	-0.019*
230	-0.013
252	0.001
268	0.000*
292	0.000
313	0.010*
322	0.015*
333	0.020
351	0.034
CURVE 4 (cont.)	
373	0.049*
474	0.134*
573	0.226*
CURVE 5	
437	0.101
302	0.001*
273	-0.003
233	-0.011*
181	-0.044
CURVE 6	
296	0.003*
319	0.016
335	0.029*
377	0.056
441	0.108
464	0.129
557	0.215
675	0.327
732	0.387
768	0.426
872	0.543
970	0.654
CURVE 7	
574	0.230
474	0.139*
414	0.087
293	0.000*
CURVE 8*	
256	-0.002
285	-0.001
296	0.000
304	0.002
CURVE 9*	
257	-0.003
283	-0.002
309	0.004
CURVE 9 (cont.)*	
331	0.015
CURVE 10	
297	0.004
291	-0.002
286	-0.006*
281	-0.010
276	-0.013
266	-0.009
263	-0.004*
261	-0.001*
257	0.005
CURVE 11*	
303	0.008
297	0.004
293	0.000
290	-0.004
286	-0.008
CURVE 12*, †	
156	0.0000
166	0.0018
177	0.0050
186	0.0057
203	0.0090
218	0.0131
233	0.0154
241	0.0162
247	0.0174
254	0.0182
260	0.0176
266	0.0173
272	0.0109

* Not shown in figure.

† This curve is here reported using the first given temperature as reference temperature at which $\Delta L/L_0 = 0$.

FIGURE AND TABLE NO. 242R. PROVISIONAL VALUES FOR THERMAL LINEAR EXPANSION OF CHROMIUM + SILICON + ΣX_i ALLOYS Cr + Si + ΣX_i

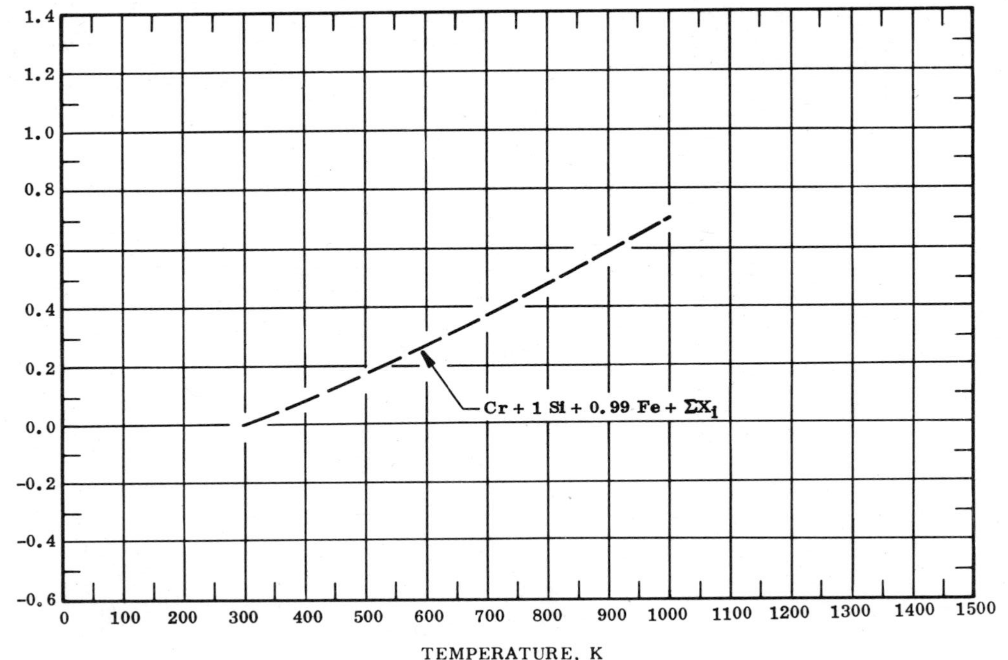

PROVISIONAL VALUES

[Temperature, T, K; Linear Expansion, $\Delta L/L_0$, %; α, K^{-1}]

	(Cr + 1 Si + 0.99 Fe + ΣX_i)	
T	$\Delta L/L_0$	$\alpha \times 10^6$
293	0.000	7.6
400	0.087	8.6
500	0.176	9.3
600	0.273	10.0
700	0.376	10.6
800	0.484	10.9
900	0.594	11.0
1000	0.703	11.1

REMARKS

(Cr + 1 Si + 0.99 Fe + ΣX_i): The tabulated values are for well-annealed alloy with primary impurity being carbon. These are considered accurate to within ±7% over the entire temperature range and can be represented approximately by the following equation:

$$\Delta L/L_0 = -0.164 + 3.455 \times 10^{-4} T + 8.158 \times 10^{-7} T^2 - 2.948 \times 10^{-10} T^3$$

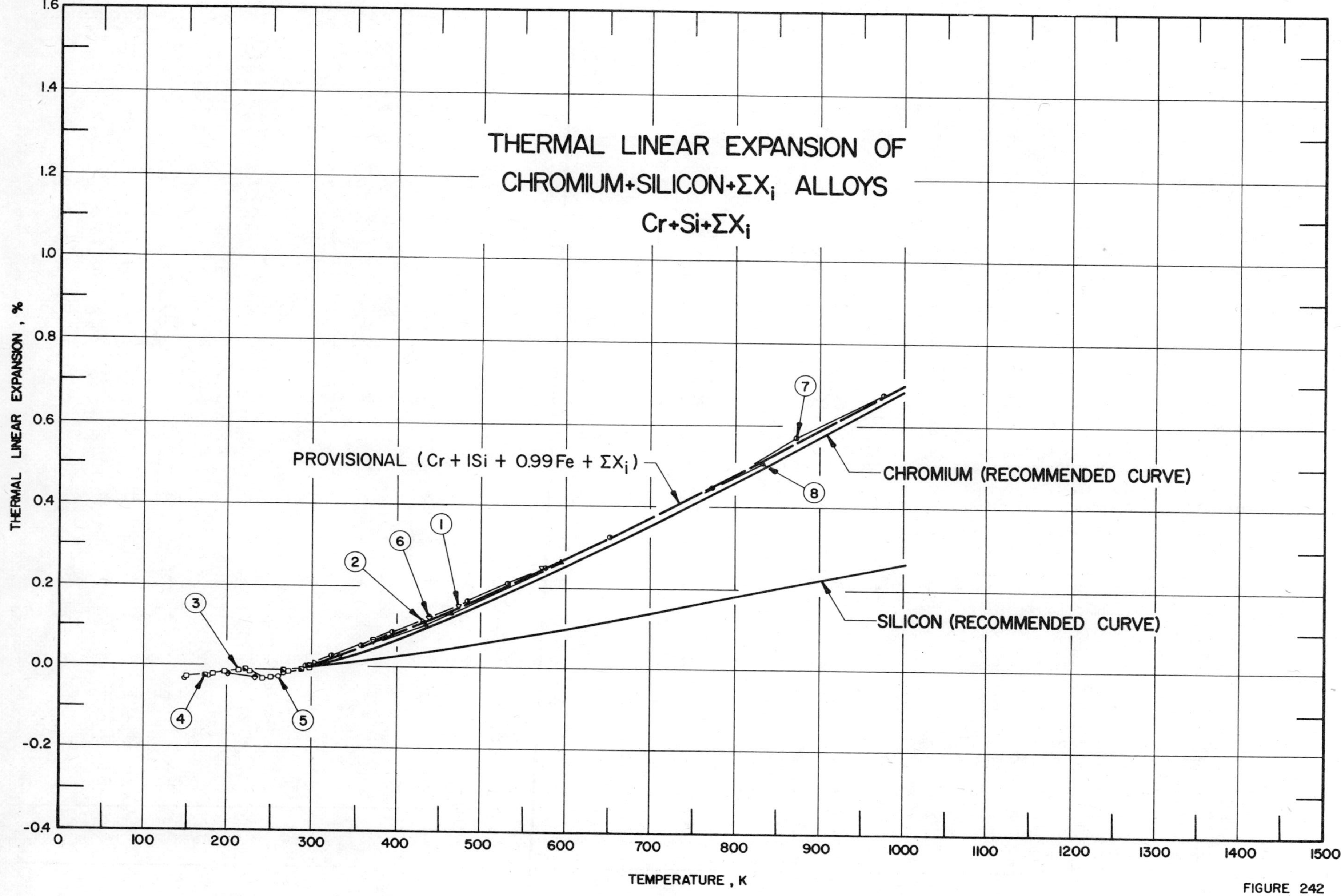

THERMAL LINEAR EXPANSION OF
CHROMIUM+SILICON+ΣX_i ALLOYS
Cr+Si+ΣX_i
PROVISIONAL (Cr + 1Si + 0.99 Fe + ΣX_i)
CHROMIUM (RECOMMENDED CURVE)
SILICON (RECOMMENDED CURVE)
1
2
3
4
5
6
7
8
THERMAL LINEAR EXPANSION, %
1.6
1.4
1.2
1.0
0.8
0.6
0.4
0.2
0.0
-0.2
-0.4
0
100
200
300
400
500
600
700
800
900
1000
1100
1200
1300
1400
1500
TEMPERATURE, K

FIGURE 242

SPECIFICATION TABLE 242. THERMAL LINEAR EXPANSION OF CHROMIUM + SILICON + ΣX_i ALLOYS $Cr + Si + \Sigma X_i$

Cur. No.	Ref. No.	Author(s)	Year	Method Used	Temp. Range, K	Name and Specimen Designation	Cr	Si	Fe	C	Mn	N	Composition (continued), Specifications, and Remarks
							Composition (weight percent)						
1	396	Hidnert, P.	1941	T	298-573	1285	96.3	1.0	0.9	0.53	0.07	0.03	Cast bar, 9 x 9 mm cross section, 300 mm long; prepared by the Union Carbide & Carbon Research Laboratories, Inc., New York; density 6.974 g cm^{-3}; expansion measured with increasing temperature; first test.
2	396	Hidnert, P.	1941	T	430-303	1285	96.3	1.0	0.9	0.53	0.07	0.03	The above specimen; first test, expansion measured with decreasing temperature.
3	396	Hidnert, P.	1941	T	151-575	1285	96.3	1.0	0.9	0.53	0.07	0.03	The above specimen; expansion measured with increasing temperature; third test.
4	396	Hidnert, P.	1941	T	296-174	1285	96.3	1.0	0.9	0.53	0.07	0.03	The above specimen; expansion measured with decreasing temperature; third test.
5	396	Hidnert, P.	1941	T	150-576	1285	96.3	1.0	0.9	0.53	0.07	0.03	The above specimen; expansion measured with increasing temperature; fourth test.
6	396	Hidnert, P.	1941	T	438-200	1285	96.3	1.0	0.9	0.53	0.07	0.03	The above specimen; expansion measured with decreasing temperature; fourth test.
7	396	Hidnert, P.	1941	T	293-975	1285	96.3	1.0	0.9	0.53	0.07	0.03	The above specimen; expansion measured with increasing temperature; sixth test.
8	396	Hidnert, P.	1941	T	831-294	1285	96.3	1.0	0.9	0.53	0.07	0.03	The above specimen; expansion measured with decreasing temperature; sixth test.

DATA TABLE 242. THERMAL LINEAR EXPANSION OF CHROMIUM + SILICON + ΣX_i ALLOYS $Cr + Si + \Sigma X_i$

[Temperature, T, K; Linear Expansion, $\Delta L/L_0$, %]

T	$\Delta L/L_0$
CURVE 1	
298	0.002
331	0.030
372	0.064
473	0.155
573	0.249
CURVE 2	
430	0.117
303	0.008
CURVE 3	
151	-0.027
183	-0.020
196	-0.018
214	-0.016
229	-0.018
241	-0.030
253	-0.027
272	-0.014
288	-0.002
334	0.034*
373	0.067
474	0.155*
575	0.251*
CURVE 4	
296	0.004*
289	-0.004
267	-0.015
224	-0.015
174	-0.022
CURVE 5	
150	-0.032
177	-0.025
186	-0.023*
214	-0.017*
224	-0.017*
244	-0.034*
262	-0.024
268	-0.019
332	0.033*
373	0.066*
474	0.155*
576	0.252*
CURVE 6	
438	0.124
296	0.003*
288	-0.005*
253	-0.029*
234	-0.024
200	-0.019
CURVE 7	
293	0.000
324	0.027
359	0.056
393	0.086
481	0.164
531	0.210
575	0.251
651	0.323
771	0.446
875	0.562
975	0.678
CURVE 8	
831	0.507
593	0.261
462	0.138
432	0.112
294	0.000*

* Not shown in figure.

FIGURE AND TABLE NO. 243R. PROVISIONAL VALUES FOR THERMAL LINEAR EXPANSION OF COBALT + CHROMIUM + ΣX_i ALLOYS Co + Cr + ΣX_i

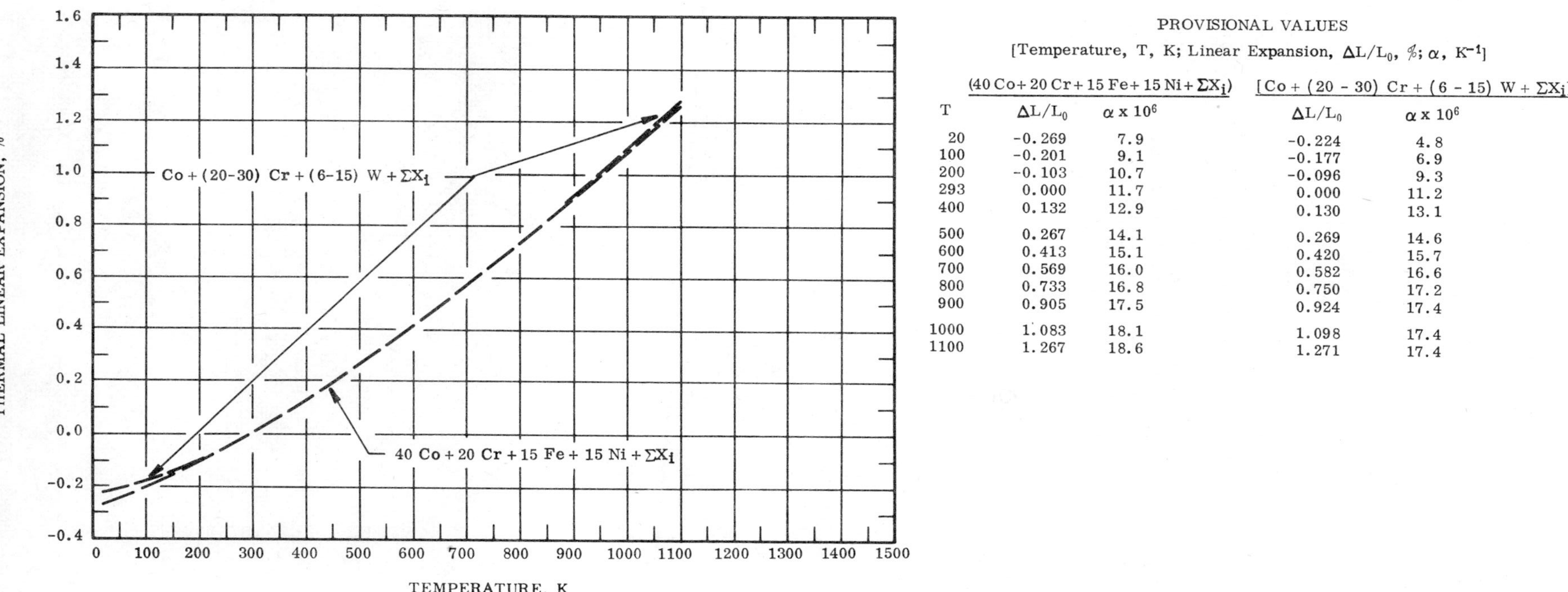

PROVISIONAL VALUES

[Temperature, T, K; Linear Expansion, $\Delta L/L_0$, %; α, K^{-1}]

T	(40 Co+ 20 Cr+ 15 Fe+ 15 Ni+ ΣX_i) $\Delta L/L_0$	$\alpha \times 10^6$	[Co + (20 - 30) Cr + (6 - 15) W + ΣX_i] $\Delta L/L_0$	$\alpha \times 10^6$
20	-0.269	7.9	-0.224	4.8
100	-0.201	9.1	-0.177	6.9
200	-0.103	10.7	-0.096	9.3
293	0.000	11.7	0.000	11.2
400	0.132	12.9	0.130	13.1
500	0.267	14.1	0.269	14.6
600	0.413	15.1	0.420	15.7
700	0.569	16.0	0.582	16.6
800	0.733	16.8	0.750	17.2
900	0.905	17.5	0.924	17.4
1000	1.083	18.1	1.098	17.4
1100	1.267	18.6	1.271	17.4

REMARKS

(40 Co+ 20 Cr+ 15 Fe+ 15 Ni+ ΣX_i): The tabulated values for well-annealed alloy of type Elgilog are considered accurate to within ± 7% over the entire temperature range. These values can be represented approximately by the following equation:

$$\Delta L/L_0 = -0.286 + 7.575 \times 10^{-4} T + 7.743 \times 10^{-7} T^2 - 1.642 \times 10^{-10} T^3$$

[Co + (20 - 30) Cr + (6 - 15) W + ΣX_i]: The tabulated values for well-annealed alloy of type Stellites are considered accurate to within ± 7% over the entire temperature range. These values can be represented approximately by the following equation:

$$\Delta L/L_0 = -0.233 + 4.234 \times 10^{-4} T + 1.414 \times 10^{-6} T^2 - 5.050 \times 10^{-10} T^3$$

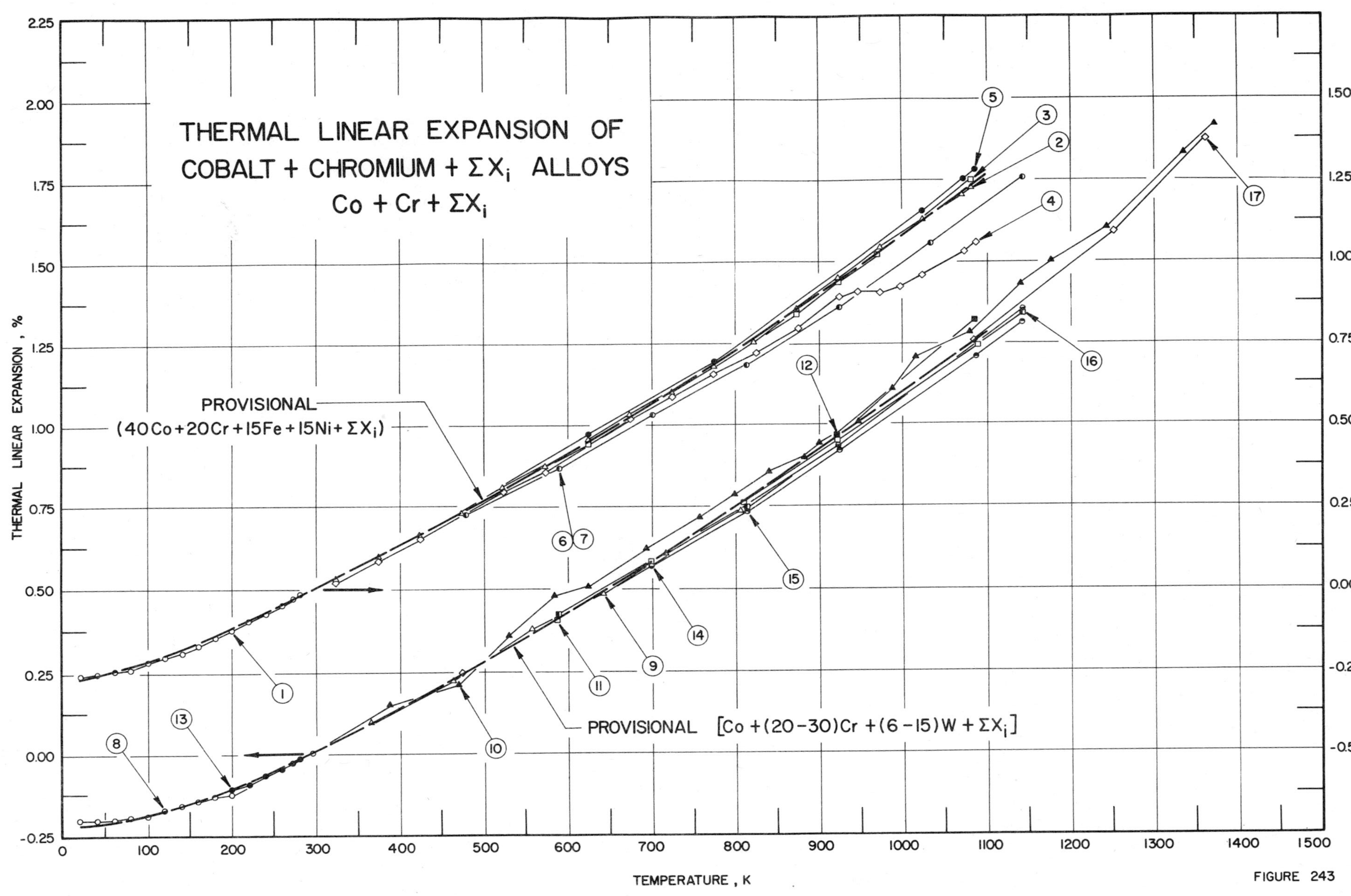
THERMAL LINEAR EXPANSION OF
COBALT + CHROMIUM + ΣX_i ALLOYS
Co + Cr + ΣX_i
PROVISIONAL
(40Co+20Cr+15Fe+15Ni+ΣX_i)
PROVISIONAL [Co+(20-30)Cr+(6-15)W+ΣX_i]
THERMAL LINEAR EXPANSION, %
TEMPERATURE, K
2.25
2.00
1.75
1.50
1.25
1.00
0.75
0.50
0.25
0.00
-0.25
1.50
1.25
1.00
0.75
0.50
0.25
0.00
-0.25
-0.50
0
100
200
300
400
500
600
700
800
900
1000
1100
1200
1300
1400
1500
1
2
3
4
5
6
7
8
9
10
11
12
13
14
15
16
17

FIGURE 243

SPECIFICATION TABLE 243. THERMAL LINEAR EXPANSION OF COBALT + CHROMIUM + ΣX_i ALLOYS $Co + Cr + \Sigma X_i$

Cur. No.	Ref. No.	Author(s)	Year	Method Used	Temp. Range, K	Name and Specimen Designation	Composition (weight percent) Co	Cr	Fe	Ni	Mo	Mn	C	Composition (continued), Specifications, and Remarks
1	462	Arp, V., Wilson, J.H., Winrich, L., and Sikora, P.	1962	L	20-293	Elgiloy	40.0	20.0	16.0	15.0	7.0	2.0	0.15	Hardness R_c46, cold reduced 45%.
2	470	Henmi, Z., Okada, M., and Nagai, T.	1967	X	293-1082	Elgiloy	41.00	20.37	14.43	15.00	7.00	1.96	0.17	0.03 Cu, 0.02 Be, 0.018 P, 0.004 S; 0.05 mm permanent change, quenched sample; expansion measured with increasing temp; zero-point correction is -0.173%.
3	470	Henmi, Z., et al.	1967	X	1082-293	Elgiloy	41.00	20.37	14.43	15.00	7.00	1.96	0.17	The above specimen; expansion measured with decreasing temp; zero-point correction is -0.15%.
4	470	Henmi, Z., et al.	1967	X	293-1088	Elgiloy	41.00	20.37	14.43	15.00	7.00	1.96	0.17	Similar to the above specimen after drawn 48.6%; expansion measured with increasing temp; zero-point correction is -0.321%.
5	470	Henmi, Z., et al.	1967	X	1088-293	Elgiloy	41.00	20.37	14.43	15.00	7.00	1.96	0.17	The above specimen; expansion measured with decreasing temp; zero-point correction is -0.09%.
6	471	Nejedlik, J.F.	1966		293-1144	L-605; L4-1701	Bal.	20.08	2.01	10.21		1.41	0.010	14.82 W, 0.14 Si, 0.015 P, 0.010 S; rod specimen 0.318 cm diameter x 5.00 cm long; machined from annealed stock.
7*	471	Nejedlik, J.F.	1966		293-1144	L-605; 186-4-1702	Bal.	19.63	2.11	10.42		1.43	0.012	14.78 W, 0.15 Si, 0.014 P, 0.012 S; similar to the above specimen.
8	462	Arp, V., Wilson, J.H., Winrich, L., and Sikora, P.	1962	L	20-293	Stellite 3	Bal.	30.5	3.0	3.0			2.45	12.5 W; sand cast specimen, hardness R_c55.
9	225	Valentich, J.	1965	L	293-807	Stellite 6	Bal.	33	3.5	3			1.1	6 W; no information given.
10	472	Fieldhouse, I.B., Hedge, J.C., Lang, J.I., Waterman, T.E.	1958	T	299-1524	Stellite 21	60.49	26.69	1.54	2.38	5.42		0.258	Zero-point correction of 0.0103% determined by graphical extrapolation.
11	473	Sweeney, W.O.	1947		294-1089	Stellite 21	Bal.	25/30	2 Max	1.5/3.5	4.5/6.5		0.3/0.35	Density 8.30 g cm^{-3}.
12	473	Sweeney, W.O.	1947		294-1089	Stellite 23	Bal.	23/29	2 Max	1.5 Max			0.35/0.50	4/7 W; density 8.54 g cm^{-3}.
13	462	Arp, V., Wilson, J.H., Winrich, L., and Sikora, P.	1962	L	20-293	Stellite 25	Bal.	20.2	2.4	10.0		1.6	0.07	15.2 W, 0.6 Si; hardness R_c41; cold reduced 26%.
14	473	Sweeney, W.O.	1947		294-1144	Stellite 27	30 Min	23/29	2 Max	Bal.	5/7		0.35/0.50	Density 8.21 g cm^{-3}.

* Not shown in figure.

SPECIFICATION TABLE 243. THERMAL LINEAR EXPANSION OF COBALT + CHROMIUM + ΣX_i ALLOYS Co + Cr + ΣX_i (continued)

Cur. No.	Ref. No.	Author(s)	Year	Method Used	Temp. Range, K	Name and Specimen Designation	Composition (weight percent) Co	Cr	Fe	Ni	Mo	Mn	C	Composition (continued), Specifications, and Remarks
15	473	Sweeney, W.O.	1947		294-1144	Stellite 30	Bal.	23/ 29	2 Max	13/ 17	5/ 7		0.35/ 0.50	Density 8.31 g cm^{-3}.
16*	473	Sweeney, W.O.	1947		294-1144	Stellite 31	Bal.	23/ 28	1.50 Max	9/ 12			0.45/ 0.60	6/9 W; density 8.61 g cm^{-3}.
17	474	Morral, F.R. and Wagner, H.J.	1960		293-1367	PWA653-A	Bal.	20/ 22	1.0/ 2.5	1.0 Max.		0.5 Max.	0.4/ 0.5	10-12 W, 1.5-2.5 Nb + Ta, 0.50 Max. Si, 0.040 Max. P, 0.040 Max. S; as cast; M.P. 2400-2450; density 0.321 lb in.$^{-3}$.
18*	358	Fieldhouse, I.B. and Lang, J.I.	1961		300-1395	Haynes Stellite HE 1049	43.6	26.0	3.0			0.8	0.40	15.0 W, 10.0 N, 0.8 Si, 0.4 B; density 8.85 g cm^{-3}.
19*	129	Masumoto, H.	1934	L	293-333		75.0	15.0	10.0					The materials used were electrolytic cobalt (0.13 Fe, 0.074 Al, 0.05 C, 0.007 P, 0.002 Si, trace Ni, S, Mn), electrolytic iron (0.01 C, 0.01 Si, 0.001 P, trace Mn), chromium (1.88 Al, 0.44 Fe, 0.33 Si, 0.03 C); supplied by Sugibayashi & Co.; these metals first mixed and melted in an alumina crucible, placed in a Tammann furnace under hydrogen atmosphere; the melt then cast into an iron mold, then from the cast rod, a test-specimen, 10 cm long was cut; heated for 1 hr at 1000 in a vacuum furnace and then slowly cooled.
20*	129	Masumoto, H.	1934	L	293-333		70.0	20.0	10.0					The same treatment as above specimen.
21*	129	Masumoto, H.	1934	L	293-333		65.0	20.0	15.0					The same treatment as above specimen.

* Not shown in figure.

DATA TABLE 243. THERMAL LINEAR EXPANSION OF COBALT + CHROMIUM + ΣX_i ALLOYS $Co + Cr + \Sigma X_i$

[Temperature, T, K; Linear Expansion, $\Delta L/L_0$, %]

T	$\Delta L/L_0$
CURVE 1	
20	-0.259
40	-0.255
60	-0.248
80	-0.237
100	-0.221
120	-0.205
140	-0.186
160	-0.166
180	-0.143
200	-0.119
220	-0.094
240	-0.069
260	-0.042
273	-0.025
280	-0.016
293	0.000
CURVE 2	
293	0.000*
323	0.030
373	0.096
423	0.165
473	0.231
523	0.303
573	0.372
623	0.452
673	0.528
723	0.600
773	0.680
823	0.757
873	0.865
923	0.949
973	1.047
1023	1.131
1073	1.201
1082	1.227
CURVE 3	
1082	1.250
1073	1.224
1023	1.130*
973	1.025
923	0.940
CURVE 3 (cont.)	
873	0.843
823	0.753*
773	0.681*
723	0.602*
673	0.526*
623	0.448
573	0.370*
523	0.307*
473	0.233*
423	0.166*
373	0.094*
323	0.036*
293	0.000*
CURVE 4	
293	0.000*
323	0.025
373	0.084
423	0.154
473	0.228*
523	0.299
573	0.363
623	0.446*
673	0.515
723	0.588
773	0.655
823	0.717
873	0.796
923	0.885
948	0.906
973	0.906
998	0.921
1023	0.961
1073	1.029
1088	1.054
CURVE 5	
1088	1.285
1073	1.254
1023	1.157
973	1.051*
923	0.954*
873	0.856*
CURVE 5 (cont.)	
823	0.769*
773	0.685
723	0.612*
673	0.535*
623	0.467
573	0.382*
523	0.308*
473	0.234*
423	0.166*
373	0.100*
323	0.035*
293	0.000*
CURVE 6	
293	0.000*
477	0.227
589	0.365
700	0.529
811	0.682
922	0.866
1033	1.067
1144	1.264
CURVE 7	
293	0.000
477	0.227
589	0.365
700	0.529
811	0.682
922	0.866
1033	1.067
1144	1.264
CURVE 8	
20	-0.208
40	-0.206
60	-0.203
80	-0.196
100	-0.186
120	-0.174
140	-0.160
160	-0.145
CURVE 8 (cont.)	
180	-0.128
200	-0.111
220	-0.089*
240	-0.066*
260	-0.041*
273	-0.025*
280	-0.016*
293	0.000*
CURVE 9	
293	0.000*
367	0.092
465	0.221
558	0.356
641	0.486
718	0.606
807	0.733
CURVE 10	
299	0.010*
389	0.146
470	0.205
530	0.355
585	0.475
625	0.505
692	0.619
758	0.712
800	0.789
840	0.851
883	0.898
900	0.934
944	1.008
986	1.106
1015	1.209
1080	1.288
1140	1.429
1178	1.505
1247	1.610
1339	1.828
1372	1.920
1425	2.051
1474	2.208*
1524	2.342*
CURVE 11	
294	0.002*
589	0.417*
700	0.583
811	0.762
922	0.949
1089	1.243
CURVE 12	
294	0.002*
589	0.406
700	0.583*
811	0.762*
922	0.960
1089	1.323
CURVE 13	
20	-0.206*
40	-0.203*
60	-0.198*
80	-0.191*
100	-0.181*
120	-0.169*
140	-0.156*
160	-0.142*
180	-0.124*
200	-0.105
220	-0.086
240	-0.064
260	-0.042
273	-0.025
280	-0.017
293	0.000
CURVE 14	
294	0.002*
589	0.401*
700	0.570
811	0.749*
922	0.938
1089	1.242*
1144	1.354
CURVE 15	
294	0.002*
589	0.410*
700	0.575*
811	0.737
922	0.914
1089	1.206
1144	1.308
CURVE 16	
294	0.002
589	0.424
700	0.577*
811	0.744
922	0.926*
1089	1.207*
1144	1.347
CURVE 17	
293	0.000
477	0.249
699	0.571*
922	0.940*
1088	1.260
1255	1.593
1366	1.874
CURVE 18*	
299	0.002
421	0.042
465	0.069
532	0.137
644	0.285
699	0.433
753	0.567
809	0.662
868	0.850
973	0.985
1029	1.146
1143	1.496
1199	1.604
1318	1.941
1394	2.012

* Not shown in figure.

DATA TABLE 243. THERMAL LINEAR EXPANSION OF COBALT + CHROMIUM + ΣX_i ALLOYS Co + Cr + ΣX_i (continued)

T	$\Delta L/L_0$
CURVE 19*	
293	0.000
333	0.041
CURVE 20*	
293	0.000
333	0.042
CURVE 21*	
293	0.000
333	0.053

* Not shown in figure.

FIGURE AND TABLE NO. 244R. PROVISIONAL VALUES FOR THERMAL LINEAR EXPANSION OF COBALT + IRON + ΣX_i ALLOYS Co + Fe + ΣX_i

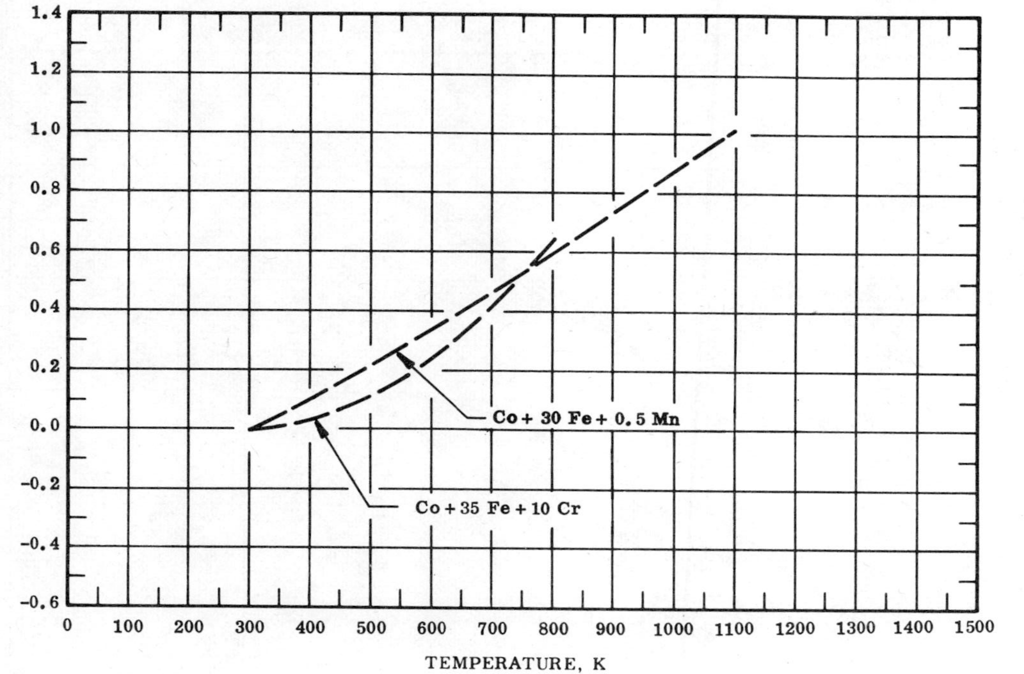

PROVISIONAL VALUES

[Temperature, T, K; Linear Expansion, $\Delta L/L_0$, %; α, K^{-1}]

	(Co + 30 Fe + 0.5 Mn)		(Co + 35 Fe + 10 Cr)	
T	$\Delta L/L_0$	$\alpha \times 10^6$	$\Delta L/L_0$	$\alpha \times 10^6$
293	0.000	10.7	0.000	0.1
400	0.115	10.8	0.028	6.6
500	0.225	11.2	0.124	12.3
600	0.339	11.6	0.270	16.4
700	0.458	12.2	0.448	18.9
800	0.584	12.9	0.641	19.6
900	0.717	13.8		
1000	0.860	14.8		
1100	1.014	15.9		

REMARKS

(Co + 30 Fe + 0.5 Mn): The tabulated values are for well-annealed alloy are are considered accurate to within ± 7% over the entire temperature range. These values can be represented approximately by the following equation:

$$\Delta L/L_0 = -0.314 + 1.101 \times 10^{-3} T - 1.587 \times 10^{-7} T^2 + 2.328 \times 10^{-10} T^3$$

(Co + 35 Fe + 10 Cr): The tabulated values for well-annealed alloy are considered accurate to within ± 7% over the entire temperature range. These values can be represented approximately by the following equation:

$$\Delta L/L_0 = 0.476 - 3.343 \times 10^{-3} T + 6.681 \times 10^{-6} T^2 - 2.805 \times 10^{-9} T^3$$

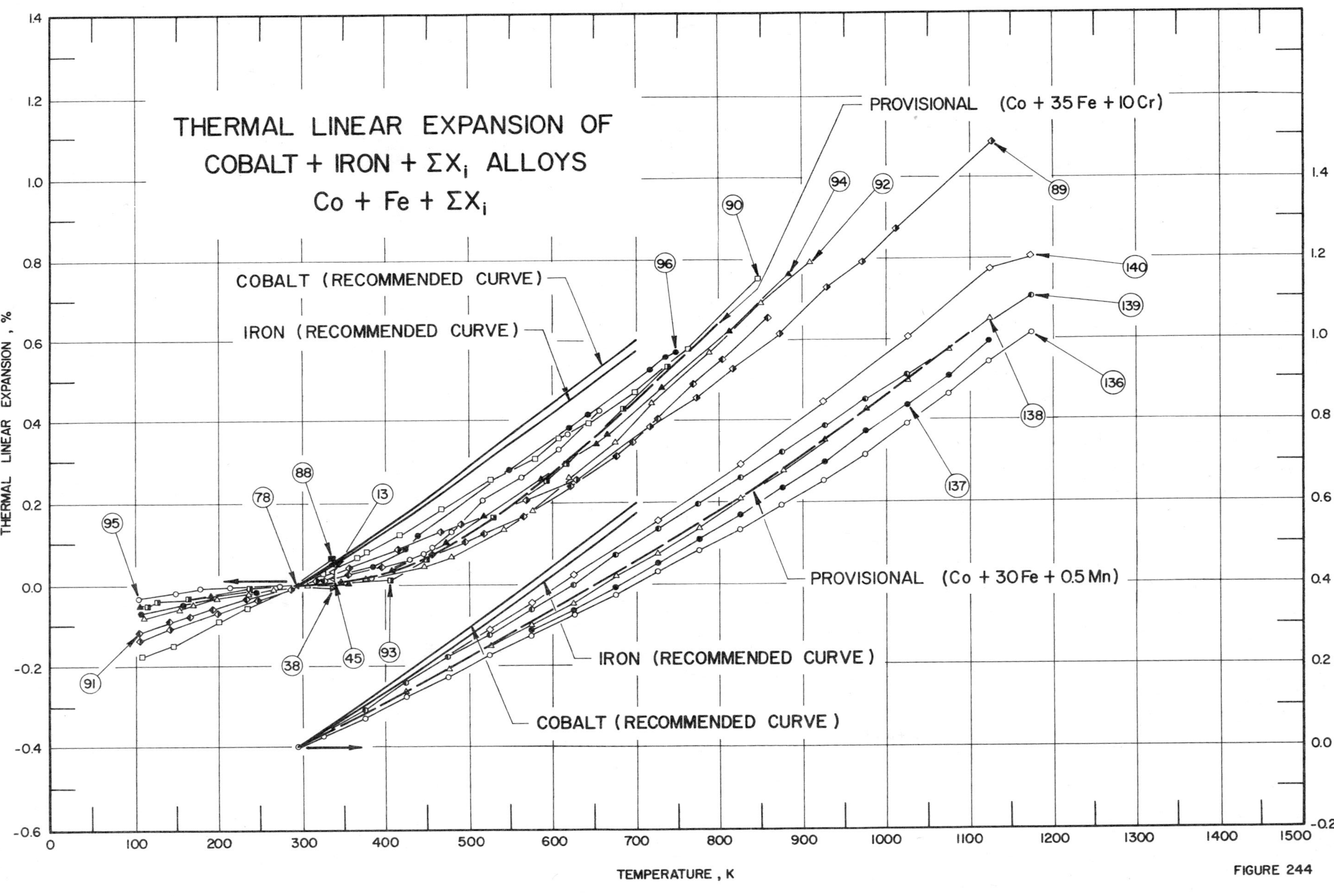

THERMAL LINEAR EXPANSION OF
COBALT + IRON + ΣXi ALLOYS
Co + Fe + ΣXi
PROVISIONAL (Co + 35 Fe + 10 Cr)
COBALT (RECOMMENDED CURVE)
IRON (RECOMMENDED CURVE)
PROVISIONAL (Co + 30 Fe + 0.5 Mn)
IRON (RECOMMENDED CURVE)
COBALT (RECOMMENDED CURVE)
95
78
88
13
91
38
45
93
96
90
94
92
89
140
139
136
138
137
THERMAL LINEAR EXPANSION, %
TEMPERATURE, K
FIGURE 244

SPECIFICATION TABLE 244. THERMAL LINEAR EXPANSION OF COBALT + IRON + ΣX_i ALLOYS $Co + Fe + \Sigma X_i$

Cur. No.	Ref. No.	Author(s)	Year	Method Used	Temp. Range, K	Name and Specimen Designation	Co	Fe	Cr	Ni	W	Mo	Mn	Composition (continued), Specifications, and Remarks
							Composition (weight percent)							
1	129	Masumoto, H.	1934	L	293-333		75.0	20.0	5.0					Electrolytic cobalt (0.13 Fe, 0.074 Al, 0.05 C, 0.007 P, 0.002 Si, trace Ni, S, Mn), electrolytic iron (0.01 C, 0.01 Si, 0.001 P, trace Mn), chromium (1.88 Al, 0.44 Fe, 0.33 Si, 0.03 C); supplied by Sugibayashi & Co.; these metals mixed in a suitable proportion and melted in an alumina crucible, placed in a Tammann furnace under hydrogen atmosphere; cast into an iron mold with a cylindrical aperture 6 mm thick and 26 cm long; from the rod a specimen, 10 cm long was cut; heat for 1 hr at 1273 K in a vacuum furnace and slowly cooled within it.
2*	129	Masumoto, H.	1934	L	293-333		70.0	25.0	5.0					The same treatment as above specimen.
3*	129	Masumoto, H.	1934	L	293-333		65.0	30.0	5.0					The same treatment as above specimen.
4*	129	Masumoto, H.	1934	L	293-333		60.0	35.0	5.0					The same treatment as above specimen.
5*	129	Masumoto, H.	1934	L	293-333		55.0	40.0	5.0					The same treatment as above specimen.
6*	129	Masumoto, H.	1934	L	293-333		64.5	28.0	7.5					The same treatment as above specimen.
7*	129	Masumoto, H.	1934	L	293-333		62.5	30.0	7.5					The same treatment as above specimen.
8*	129	Masumoto, H.	1934	L	293-333		60.5	32.0	7.5					The same treatment as above specimen.
9*	129	Masumoto, H.	1934	L	293-333		58.5	34.0	7.5					The same treatment as above specimen.
10*	129	Masumoto, H.	1934	L	293-333		56.5	36.0	7.5					The same treatment as above specimen.
11*	129	Masumoto, H.	1934	L	293-333		54.5	38.0	7.5					The same treatment as above specimen.
12*	129	Masumoto, H.	1934	L	293-333		52.0	40.5	7.5					The same treatment as above specimen.
13	129	Masumoto, H.	1934	L	293-333		49.5	43.0	7.5					The same treatment as above specimen.
14*	129	Masumoto, H.	1934	L	293-333		57.5	34.5	8.0					The same treatment as above specimen.
15*	129	Masumoto, H.	1934	L	293-333		57.0	35.0	8.0					The same treatment as above specimen.
16*	129	Masumoto, H.	1934	L	293-333		56.5	35.5	8.0					The same treatment as above specimen.
17*	129	Masumoto, H.	1934	L	293-333		60.0	31.5	8.5					The same treatment as above specimen.
18*	129	Masumoto, H.	1934	L	293-333		59.0	32.5	8.5					The same treatment as above specimen.
19*	129	Masumoto, H.	1934	L	293-333		58.0	33.5	8.5					The same treatment as above specimen.
20*	129	Masumoto, H.	1934	L	293-333		57.0	34.5	8.5					The same treatment as above specimen.
21*	129	Masumoto, H.	1934	L	293-333		56.5	35.0	8.5					The same treatment as above specimen.
22*	129	Masumoto, H.	1934	L	293-333		56.0	35.5	8.5					The same treatment as above specimen.
23*	129	Masumoto, H.	1934	L	293-333		55.0	36.5	8.5					The same treatment as above specimen.
24*	129	Masumoto, H.	1934	L	293-333		54.5	37.0	8.5					The same treatment as above specimen.

*Not shown in figure.

SPECIFICATION TABLE 244. THERMAL LINEAR EXPANSION OF COBALT + IRON + ΣX_i ALLOYS Co + Fe + ΣX_i (continued)

Cur. No.	Ref. No.	Author(s)	Year	Method Used	Temp. Range, K	Name and Specimen Designation	Composition (weight percent) Co	Fe	Cr	Ni	W	Mo	Mn	Composition (continued), Specifications, and Remarks
25	129	Masumoto, H.	1934	L	293-333		54.0	37.5	8.5					The same treatment as above specimen.
26*	129	Masumoto, H.	1934	L	293-333		80.0	11.0	9.0					The same treatment as above specimen.
27*	129	Masumoto, H.	1934	L	293-333		70.0	21.0	9.0					The same treatment as above specimen.
28*	129	Masumoto, H.	1934	L	293-333		65.0	26.0	9.0					The same treatment as above specimen.
29*	129	Masumoto, H.	1934	L	293-333		61.0	30.0	9.0					The same treatment as above specimen.
30*	129	Masumoto, H.	1934	L	293-333		60.0	31.0	9.0					The same treatment as above specimen.
31*	129	Masumoto, H.	1934	L	293-333		59.0	32.0	9.0					The same treatment as above specimen.
32*	129	Masumoto, H.	1934	L	293-333		58.0	33.0	9.0					The same treatment as above specimen.
33*	129	Masumoto, H.	1934	L	293-333		57.0	34.0	9.0					The same treatment as above specimen.
34*	129	Masumoto, H.	1934	L	293-333		56.0	35.0	9.0					The same treatment as above specimen.
35*	129	Masumoto, H.	1934	L	293-333		55.5	35.5	9.0					The same treatment as above specimen.
36*	129	Masumoto, H.	1934	L	293-333		55.0	36.0	9.0					The same treatment as above specimen.
37*	129	Masumoto, H.	1934	L	293-333		54.5	36.5	9.0					The same treatment as above specimen.
38	129	Masumoto, H.	1934	L	293-333		54.0	37.0	9.0					The same treatment as above specimen.
39	129	Masumoto, H.	1934	L	293-333		53.0	38.0	9.0					The same treatment as above specimen.
40*	129	Masumoto, H.	1934	L	293-333		49.5	41.5	9.0					The same treatment as above specimen.
41*	129	Masumoto, H.	1934	L	293-333		74.5	16.0	9.5					The same treatment as above specimen.
42*	129	Masumoto, H.	1934	L	293-333		69.5	21.0	9.5					The same treatment as above specimen.
43*	129	Masumoto, H.	1934	L	293-333		64.5	26.0	9.5					The same treatment as above specimen.
44*	129	Masumoto, H.	1934	L	293-333		61.5	29.0	9.5					The same treatment as above specimen.
45	129	Masumoto, H.	1934	L	293-333		59.5	31.0	9.5					The same treatment as above specimen.
46*	129	Masumoto, H.	1934	L	293-333		57.5	33.0	9.5					The same treatment as above specimen.
47*	129	Masumoto, H.	1934	L	293-333		56.5	34.0	9.5					The same treatment as above specimen.
48*	129	Masumoto, H.	1934	L	293-333		56.0	34.5	9.5					The same treatment as above specimen.
49*	129	Masumoto, H.	1934	L	293-333		55.0	35.5	9.5					The same treatment as above specimen.
50*	129	Masumoto, H.	1934	L	293-333		54.5	36.0	9.5					The same treatment as above specimen.
51*	129	Masumoto, H.	1934	L	293-333		54.0	36.5	9.5					The same treatment as above specimen.
52*	129	Masumoto, H.	1934	L	293-333		53.5	37.0	9.5					The same treatment as above specimen.
53*	129	Masumoto, H.	1934	L	293-333		53.0	37.5	9.5					The same treatment as above specimen.
54*	129	Masumoto, H.	1934	L	293-333		52.5	38.0	9.5					The same treatment as above specimen.
55*	129	Masumoto, H.	1934	L	293-333		60.0	30.0	10.0					The same treatment as above specimen.
56*	129	Masumoto, H.	1934	L	293-333		59.0	31.0	10.0					The same treatment as above specimen.

*Not shown in figure.

SPECIFICATION TABLE 244. THERMAL LINEAR EXPANSION OF COBALT + IRON + ΣX_i ALLOYS $Co + Fe + \Sigma X_i$ (continued)

Cur. No.	Ref. No.	Author(s)	Year	Method Used	Temp. Range, K	Name and Specimen Designation	Composition (weight percent) Co	Fe	Cr	Ni	W	Mo	Mn	Composition (continued), Specifications, and Remarks
57*	129	Masumoto, H.	1934	L	293-333		57.0	33.0	10.0					The same treatment as above specimen.
58	129	Masumoto, H.	1934	L	293-333		55.0	35.0	10.0					The same treatment as above specimen.
59*	129	Masumoto, H.	1934	L	293-333		54.5	35.5	10.0					The same treatment as above specimen.
60*	129	Masumoto, H.	1934	L	293-333		54.0	36.0	10.0					The same treatment as above specimen.
61*	129	Masumoto, H.	1934	L	293-333		53.5	36.5	10.0					The same treatment as above specimen.
62*	129	Masumoto, H.	1934	L	293-333		53.0	37.0	10.0					The same treatment as above specimen.
63*	129	Masumoto, H.	1934	L	293-333		52.5	37.5	10.0					The same treatment as above specimen.
64*	129	Masumoto, H.	1934	L	293-333		59.5	30.0	10.5					The same treatment as above specimen.
65*	129	Masumoto, H.	1934	L	293-333		58.5	31.0	10.5					The same treatment as above specimen.
66*	129	Masumoto, H.	1934	L	293-333		57.5	32.0	10.5					The same treatment as above specimen.
67*	129	Masumoto, H.	1934	L	293-333		55.0	34.5	10.5					The same treatment as above specimen.
68*	129	Masumoto, H.	1934	L	293-333		54.5	35.0	10.5					The same treatment as above specimen.
69*	129	Masumoto, H.	1934	L	293-333		52.5	37.0	10.5					The same treatment as above specimen.
70*	129	Masumoto, H.	1934	L	293-333		51.5	38.0	10.5					The same treatment as above specimen.
71*	129	Masumoto, H.	1934	L	293-333		53.5	35.5	11.0					The same treatment as above specimen.
72*	129	Masumoto, H.	1934	L	293-333		51.5	37.5	11.0					The same treatment as above specimen.
73*	129	Masumoto, H.	1934	L	293-333		67.5	21.0	11.5					The same treatment as above specimen.
74*	129	Masumoto, H.	1934	L	293-333		62.5	26.0	11.5					The same treatment as above specimen.
75*	129	Masumoto, H.	1934	L	293-333		58.5	30.0	11.5					The same treatment as above specimen.
76*	129	Masumoto, H.	1934	L	293-333		56.5	32.0	11.5					The same treatment as above specimen.
77*	129	Masumoto, H.	1934	L	293-333		54.5	34.0	11.5					The same treatment as above specimen.
78	129	Masumoto, H.	1934	L	293-333		49.5	39.0	11.5					The same treatment as above specimen.
79*	129	Masumoto, H.	1934	L	293-333		52.0	36.0	12.0					The same treatment as above specimen.
80*	129	Masumoto, H.	1934	L	293-333		70.0	15.0	15.0					The same treatment as above specimen.
81*	129	Masumoto, H.	1934	L	293-333		65.0	20.0	15.0					The same treatment as above specimen.
82*	129	Masumoto, H.	1934	L	293-333		62.5	22.5	15.0					The same treatment as above specimen.
83*	129	Masumoto, H.	1934	L	293-333		60.0	25.0	15.0					The same treatment as above specimen.
84*	129	Masumoto, H.	1934	L	293-333		55.0	30.0	15.0					The same treatment as above specimen.
85*	129	Masumoto, H.	1934	L	293-333		50.0	35.0	15.0					The same treatment as above specimen.
86*	129	Masumoto, H.	1934	L	293-333		60.0	20.0	20.0					The same treatment as above specimen.
87*	129	Masumoto, H.	1934	L	293-333		55.0	25.0	20.0					The same treatment as above specimen.
88	129	Masumoto, H.	1934	L	293-333		50.0	30.0	20.0					The same treatment as above specimen.

* Not shown in figure.

SPECIFICATION TABLE 244. THERMAL LINEAR EXPANSION OF COBALT + IRON + ΣX_i ALLOYS $Co + Fe + \Sigma X_i$ (continued)

Cur. No.	Ref. No.	Author(s)	Year	Method Used	Temp. Range, K	Name and Specimen Designation	Composition (weight percent) Co	Fe	Cr	Ni	W	Mo	Mn	Composition (continued), Specifications, and Remarks
89	475	Saito, H., Fujimori, H., and Saito, T.	1969	X	104-1128	$(F_{1-x}Co_x)_{0.89}Cr_{0.11}$ x = 0.831	75.5	14.5	10.0					99.9 purity Fe, 99.5 purity Co, 99.1 purity Cr; mixed and melted in vacuum; disc form 2 mm x 8 cm; annealed in vacuum at 1273 K for 5 hr; zero-point correction is -0.106%.
90	475	Saito, H., et al.	1969	X	108-847	$(F_{1-x}Co_x)_{0.89}Cr_{0.11}$ x = 0.691	63.23	26.79	9.98					Similar to the above specimen; zero-point correction is 0.094%.
91	475	Saito, H., et al.	1969	X	104-859	$(F_{1-x}Co_x)_{0.89}Cr_{0.11}$ x = 0.663	60.75	29.26	9.99					Similar to the above specimen.
92	475	Saito, H., et al.	1969	X	110-908	$(F_{1-x}Co_x)_{0.89}Cr_{0.11}$ x = 0.635	58.25	31.75	10.0					Similar to the above specimen.
93	475	Saito, H., et al.	1969	X	113-739	$(F_{1-x}Co_x)_{0.89}Cr_{0.11}$ x = 0.601	55.24	34.74	10.02					Similar to the above specimen; zero-point correction is -0.003%.
94	475	Saito, H., et al.	1969	X	105-882	$(F_{1-x}Co_x)_{0.89}Cr_{0.11}$ x = 0.596	54.79	35.17	10.04					Similar to the above specimen; zero-point correction is -0.007%.
95	475	Saito, H., et al.	1969	X	103-762	$(F_{1-x}Co_x)_{0.89}Cr_{0.11}$ x = 0.590	54.24	35.73	10.03					Similar to the above specimen.
96	475	Saito, H., et al.	1969	X	107-746	$(F_{1-x}Co_x)_{0.89}Cr_{0.11}$ x = 0.584	53.73	36.24	10.03					Similar to the above specimen; zero-point correction is -0.007%.
97*	476	Masumoto, H., Saito, H., and Kikuchi, M.	1967	L	273-1073	No. 1	38.0	22.4	12.0	16.5	4.0	4.0	1.2	1.0 Ti, 0.8 Si; single crystal with 15-20 cm long, 3 mm diameter; transition temp 831 K; zero-point correction is -0.028%.
98*	476	Masumoto, H., et al.	1967	L	273-1073	No. 5	38.0	22.4	12.0	16.5	4.0	4.0	1.2	Similar to the above specimen; zero-point correction is -0.023%.
99*	476	Masumoto, H., et al.	1967	L	274-1073	No. 10	38.0	22.4	12.0	16.5	4.0	4.0	1.2	Similar to the above specimen; zero-point correction is -0.026%.
100*	405	Masumoto, H., Saito, H., Sugai, Y., and Kono, T.	1960	L	293-323		55	40				5		Electrolytic Co (0.26 Ni, 0.15 Fe, 0.03 C, 0.01 Si, trace Mn, Al, P, S), electrolytic Fe (0.02 Al, trace C, Si, Mn, P, S, Cu), and powder Mo with purity 99.94; specimen 3 mm long.
101*	405	Masumoto, H., et al.	1960	L	293-323		65	30				5		Similar to the above specimen.
102*	405	Masumoto, H., et al.	1960	L	293-323		75	20				5		Similar to the above specimen.
103*	405	Masumoto, H., et al.	1960	L	293-323		50	40				10		Similar to the above specimen.
104*	405	Masumoto, H., et al.	1960	L	293-323		55	35				10		Similar to the above specimen.

* Not shown in figure.

SPECIFICATION TABLE 244. THERMAL LINEAR EXPANSION OF COBALT + IRON + ΣX_i ALLOYS Co + Fe + ΣX_i (continued)

Cur. No.	Ref. No.	Author(s)	Year	Method Used	Temp. Range, K	Name and Specimen Designation	Co	Fe	Cr	Ni	W	Mo	Mn	Composition (continued), Specifications, and Remarks
105*	405	Masumoto, H., et al.	1960	L	293-323		60	30				10		Similar to the above specimen.
106*	405	Masumoto, H., et al.	1960	L	293-323		65	25				10		Similar to the above specimen.
107*	405	Masumoto, H., et al.	1960	L	293-323		70	20				10		Similar to the above specimen.
108*	405	Masumoto, H., et al.	1960	L	293-323		55	32.5				12.5		Similar to the above specimen.
109*	405	Masumoto, H., et al.	1960	L	293-323		57.5	30				12.5		Similar to the above specimen.
110*	405	Masumoto, H., et al.	1960	L	293-323		60	27.5				12.5		Similar to the above specimen.
111*	405	Masumoto, H., et al.	1960	L	293-323		45	40				15		Similar to the above specimen.
112*	405	Masumoto, H., et al.	1960	L	293-323		50	35				15		Similar to the above specimen.
113*	405	Masumoto, H., et al.	1960	L	293-323		55	30				15		Similar to the above specimen.
114*	405	Masumoto, H., et al.	1960	L	293-323		57.5	27.5				15		Similar to the above specimen.
115*	405	Masumoto, H., et al.	1960	L	293-323		60	25				15		Similar to the above specimen.
116*	405	Masumoto, H., et al.	1960	L	293-323		50	32.5				17.5		Similar to the above specimen.
117*	405	Masumoto, H., et al.	1960	L	293-323		55	27.5				17.5		Similar to the above specimen.
118*	405	Masumoto, H., et al.	1960	L	293-323		50	30				20		Similar to the above specimen.
119*	405	Masumoto, H., et al.	1960	L	293-323		55	25				20		Similar to the above specimen.
120*	405	Masumoto, H., et al.	1960	L	293-323		60	20				20		Similar to the above specimen.
121*	405	Masumoto, H., et al.	1960	L	293-323		45	30				25		Similar to the above specimen.
122*	405	Masumoto, H., et al.	1960	L	293-323		50	25				25		Similar to the above specimen.
123*	59	Masumoto, H.	1931	L	303-373	No. 41	39.6	35.4		25				Granular Co, Armco Fe, Mond Ni; mixed and melted in hydrogen atmosphere in alumina crucible in Tammann furnace, cast in iron mold with cylindrical aperture, machined to circular rod 4 mm thick, 10 cm long; heated at 1273 K for 1 hr in electric furnace in hydrogen atmosphere at reduced pressure; expansion measured in vacuum.
124*	59	Masumoto, H.	1931	L	303-373	No. 42	39.9	30.1		30				Similar to the above specimen.
125*	59	Masumoto, H.	1931	L	303-373	No. 45	49.8	40.2		10				Similar to the above specimen.
126*	59	Masumoto, H.	1931	L	303-373	No. 46	49.5	35.5		15				Similar to the above specimen.
127*	59	Masumoto, H.	1931	L	303-373	No. 47	49.8	30.2		20				Similar to the above specimen.
128*	59	Masumoto, H.	1931	L	303-373	No. 51	59.8	30.2		10				Similar to the above specimen.
129*	59	Masumoto, H.	1931	L	303-373	No. 52	59.4	25.6		15				Similar to the above specimen.
130*	59	Masumoto, H.	1931	L	303-373	No. 53	59.8	20.2		20				Similar to the above specimen.
131*	59	Masumoto, H.	1931	L	303-373	No. 55	69.7	25.3		5				Similar to the above specimen.

* Not shown in figure.

SPECIFICATION TABLE 244. THERMAL LINEAR EXPANSION OF COBALT + IRON + ΣX_i ALLOYS Co + Fe + ΣX_i (continued)

Cur. No.	Ref. No.	Author(s)	Year	Method Used	Temp. Range, K	Name and Specimen Designation	Composition (weight percent) Co	Fe	Cr	Ni	W	Mo	Mn	Composition (continued), Specifications, and Remarks
132*	59	Masumoto, H.	1931	L	303-373	No. 56	69.7	20.3		10				Similar to the above specimen.
133*	59	Masumoto, H.	1931	L	303-373	No. 60	79.7	10.3		10				Similar to the above specimen.
134*	59	Masumoto, H.	1931	L	303-373	No. 65	89.9	5.1		5				Similar to the above specimen.
135	59	Masumoto, H.	1931	L	303-373	No. 67	94.9	3.1		2				Similar to the above specimen.
136	555	Fine, M.E. and Ellis, W.C.	1948	L	293-1173	6072	50.1	49.2					0.47	< 0.01 Si; specimens were prepared from cast ingots by first hot-swaging followed by cold-swaging to 0.180 in. dia. rod 2.5 in. long with flat-ground and polished end; before measurement they are annealed in hydrogen 1 hr at 1173 K; zero-point correction is 0.006%.
137	555	Fine, M.E. and Ellis, W.C.	1948	L	293-1123	6034	59.8	39.4					0.58	< 0.01 Si; similar to the above specimen; zero-point correction is 0.005%.
138	555	Fine, M.E. and Ellis, W.C.	1948	L	293-1123	6035	69.6	29.8					0.39	< 0.01 Si; similar to the above specimen; zero-point correction is 0.002%.
139	555	Fine, M.E. and Ellis, W.C.	1948	L	293-1173	6036	79.4	20.2					0.48	0.01 Si; similar to the above specimen; zero-point correction is 0.006%.
140	555	Fine, M.E. and Ellis, W.C.	1948	L	293-1148	6028	89.6	9.6					0.63	0.03 Si; similar to the above specimen; zero-point correction is 0.009%.

* Not shown in figure.

DATA TABLE 244. THERMAL LINEAR EXPANSION OF COBALT + IRON + ΣX_i ALLOYS $Co + Fe + \Sigma X_i$

[Temperature, T, K; Linear Expansion, $\Delta L/L_0$, %]

T	$\Delta L/L_0$	T	$\Delta L/L_0$	T	$\Delta L/L_0$	T	$\Delta L/L_0$	T	$\Delta L/L_0$	T	$\Delta L/L_0$
CURVE 1*		CURVE 10*		CURVE 19*		CURVE 28*		CURVE 37*		CURVE 46*	
293	0.000	293	0.000	293	0.000	293	0.000	293	0.000	293	0.000
333	0.043	333	0.035	333	0.016	333	0.028	333	0.001	333	0.016
CURVE 2*		CURVE 11*		CURVE 20*		CURVE 29*		CURVE 38		CURVE 47*	
293	0.000	293	0.000	293	0.000	293	0.000	293	0.000	293	0.000
333	0.039	333	0.037	333	0.013	333	0.023	333	-0.004	333	0.013
CURVE 3*		CURVE 12*		CURVE 21*		CURVE 30*		CURVE 39*		CURVE 48*	
293	0.000	293	0.000	293	0.000	293	0.000	293	0.000	293	0.000
333	0.038	333	0.036	333	0.012	333	0.021	333	0.016	333	0.009
CURVE 4*		CURVE 13		CURVE 22*		CURVE 31*		CURVE 40*		CURVE 49*	
293	0.000	293	0.000*	293	0.000	293	0.000	293	0.000	293	0.000
333	0.038	333	0.038	333	0.013	333	0.020	333	0.033	333	0.003
CURVE 5*		CURVE 14*		CURVE 23*		CURVE 32*		CURVE 41*		CURVE 50*	
293	0.000	293	0.000	293	0.000	293	0.000	293	0.000	293	0.000
333	0.038	333	0.015	333	0.017	333	0.015	333	0.039	333	0.002
CURVE 6*		CURVE 15*		CURVE 24*		CURVE 33*		CURVE 42*		CURVE 51*	
293	0.000	293	0.000	293	0.000	293	0.000	293	0.000	293	0.000
333	0.033	333	0.012	333	0.019	333	0.013	333	0.033	333	0.000
CURVE 7*		CURVE 16*		CURVE 25*		CURVE 34*		CURVE 43*		CURVE 52*	
293	0.000	293	0.000	293	0.000	293	0.000	293	0.000	293	0.000
333	0.030	333	0.014	333	0.028	333	0.011	333	0.028	333	0.002
CURVE 8*		CURVE 17*		CURVE 26*		CURVE 35*		CURVE 44*		CURVE 53*	
293	0.000	293	0.000	293	0.000	293	0.000	293	0.000	293	0.000
333	0.029	333	0.026	333	0.040	333	0.006	333	0.024	333	0.000
CURVE 9*		CURVE 18*		CURVE 27*		CURVE 36*		CURVE 45		CURVE 54*	
293	0.000	293	0.000	293	0.000	293	0.000	293	0.000	293	0.000
333	0.030	333	0.024	333	0.035	333	0.004	333	0.020	333	0.029

* Not shown in figure.

DATA TABLE 244. THERMAL LINEAR EXPANSION OF COBALT + IRON + ΣX_i ALLOYS $Co + Fe + \Sigma X_i$ (continued)

T	$\Delta L/L_0$
CURVE 55*	
293	0.000
333	0.020
CURVE 56*	
293	0.000
333	0.018
CURVE 57*	
293	0.000
333	0.015
CURVE 58*	
293	0.000
333	0.006
CURVE 59*	
293	0.000
333	0.003
CURVE 60*	
293	0.000
333	0.001
CURVE 61*	
293	0.000
333	-0.001
CURVE 62*	
293	0.000
333	0.001
CURVE 63*	
293	0.000
333	0.011
CURVE 64*	
293	0.000
333	0.023
CURVE 65*	
293	0.000
333	0.021
CURVE 66*	
293	0.000
333	0.018
CURVE 67*	
293	0.000
333	0.008
CURVE 68*	
293	0.000
333	0.007
CURVE 69*	
293	0.000
333	0.035
CURVE 70*	
293	0.000
333	0.043
CURVE 71*	
293	0.000
333	0.048
CURVE 72*	
293	0.000
333	0.044
CURVE 73*	
293	0.000
333	0.035
CURVE 74*	
293	0.000
333	0.029
CURVE 75*	
293	0.000
333	0.024
CURVE 76*	
293	0.000
333	0.041
CURVE 77*	
293	0.000
333	0.049
CURVE 78	
293	0.000
333	0.054
CURVE 79*	
293	0.000
333	0.052
CURVE 80*	
293	0.000
333	0.041
CURVE 81*	
293	0.000
333	0.038
CURVE 82*	
293	0.000
333	0.049
CURVE 83*	
293	0.000
333	0.056
CURVE 84*	
293	0.000
333	0.062
CURVE 85*	
293	0.000
333	0.064
CURVE 86*	
293	0.000
333	0.059
CURVE 87*	
293	0.000
333	0.063
CURVE 88	
293	0.000
333	0.065
CURVE 89	
104	-0.135
140	-0.103
199	-0.064
244	-0.035
286	-0.006
293	0.000*
315	0.017
354	0.044
413	0.089
CURVE 89 (cont.)	
467	0.129
490	0.147
568	0.206
628	0.257
714	0.384
771	0.457
816	0.527
871	0.618
928	0.726
970	0.796
1011	0.876
1128	1.083
CURVE 90	
108	-0.177
144	-0.144
200	-0.087
232	-0.056
285	-0.005*
293	0.000*
323	0.031
363	0.067
377	0.081
415	0.121
467	0.185
525	0.256
578	0.306
606	0.355
641	0.396
698	0.467
762	0.579
847	0.749
CURVE 91	
104	-0.116
140	-0.086
164	-0.075
192	-0.056
232	-0.030
293	0.000*
324	0.014
354	0.028
CURVE 91 (cont.)	
393	0.044
455	0.078
494	0.105
518	0.123
564	0.164
620	0.240
673	0.313
693	0.343
727	0.407
769	0.489
802	0.550
859	0.656
CURVE 92	
110	-0.080
151	-0.058
169	-0.045
197	-0.032
235	-0.018
265	-0.007
293	0.000*
304	0.000
331	0.010*
344	0.012
373	0.019
444	0.047
478	0.072
541	0.136
577	0.181
619	0.260
672	0.345
718	0.444
789	0.569
850	0.692
908	0.798
CURVE 93	
113	-0.049
125	-0.038
161	-0.028
235	-0.003
285	0.000*

* Not shown in figure.

DATA TABLE 244. THERMAL LINEAR EXPANSION OF COBALT + IRON + ΣX_i ALLOYS Co + Fe + ΣX_i (continued)

T	$\Delta L/L_0$
CURVE 93 (cont.)	
293	0.000*
309	0.000*
335	0.003*
403	0.017
448	0.061
493	0.113*
529	0.165
591	0.254
617	0.297
683	0.426
739	0.534
CURVE 94	
105	-0.049
167	-0.028*
190	-0.022
238	-0.014*
288	-0.001*
293	0.000*
316	0.001*
357	0.006
407	0.037
472	0.104
517	0.166
584	0.258
651	0.341
663	0.364
730	0.481
812	0.620
882	0.761
CURVE 95	
103	-0.030
148	-0.016
177	-0.006
211	-0.002
271	0.000
293	0.000*
324	0.004*
381	0.018
387	0.028*
403	0.041*
428	0.061
443	0.076

T	$\Delta L/L_0$
CURVE 95 (cont.)	
455	0.092
478	0.131
490	0.157*
516	0.204
562	0.271
607	0.330
618	0.367
653	0.421
762	0.574*
CURVE 96	
107	-0.066
156	-0.044
195	-0.027*
243	-0.016
286	-0.004*
293	0.000*
307	0.006*
332	0.018*
384	0.047
423	0.090
438	0.121
462	0.139*
511	0.206*
548	0.280
620	0.382
641	0.417
716	0.523
735	0.559
746	0.569
CURVE 97*	
273	-0.024
293	0.000*
322	0.041
374	0.105
409	0.160
449	0.206
502	0.267
573	0.365
626	0.433
671	0.491
710	0.538
749	0.591

T	$\Delta L/L_0$
CURVE 97 (cont.)*	
777	0.629
809	0.673
835	0.701
879	0.774
952	0.884
979	0.932*
1008	0.976
1047	1.037
1073	1.088
CURVE 98*	
273	-0.025
293	0.000
324	0.039
374	0.108
415	0.157
445	0.210
499	0.272
573	0.369
624	0.438
674	0.502
708	0.548
752	0.598
773	0.627
809	0.671
833	0.706
877	0.775
949	0.885
978	0.935
1011	0.979
1047	1.044
1073	1.093
CURVE 99*	
274	-0.03
293	0.00
322	0.04
374	0.10
413	0.16
447	0.20
496	0.27
576	0.37
624	0.43
672	0.50

T	$\Delta L/L_0$
CURVE 99 (cont.)*	
710	0.54
748	0.60
779	0.63
807	0.67
834	0.70
879	0.77
949	0.88
980	0.93
1013	0.98
1050	1.04
1073	1.09
CURVE 100*	
293	0.000
323	0.032
CURVE 101*	
293	0.000
323	0.032
CURVE 102*	
293	0.000
323	0.037
CURVE 103*	
293	0.000
323	0.030
CURVE 104*	
293	0.000
323	0.030
CURVE 105*	
293	0.000
323	0.026
CURVE 106*	
293	0.000
323	0.031

T	$\Delta L/L_0$
CURVE 107*	
293	0.000
323	0.034
CURVE 108*	
293	0.000
323	0.017
CURVE 109*	
293	0.000
323	0.021
CURVE 110*	
293	0.000
323	0.025
CURVE 111*	
293	0.000
323	0.029
CURVE 112*	
293	0.000
323	0.028
CURVE 113*	
293	0.000
323	0.022
CURVE 114*	
293	0.000
323	0.023
CURVE 115*	
293	0.000
323	0.027
CURVE 116*	
293	0.000
323	0.029

T	$\Delta L/L_0$
CURVE 117*	
293	0.000
323	0.023
CURVE 118*	
293	0.000
323	0.026
CURVE 119*	
293	0.000
323	0.026
CURVE 120*	
293	0.000
323	0.029
CURVE 121*	
293	0.000
323	0.025
CURVE 122*	
293	0.000
323	0.024
CURVE 123*	
303	0.015
373	0.092
CURVE 124*	
303	0.012
373	0.095
CURVE 125*	
303	0.010
373	0.077
CURVE 126*	
303	0.011
373	0.089

* Not shown in figure.

DATA TABLE 244. THERMAL LINEAR EXPANSION OF COBALT + IRON + ΣX_i ALLOYS $Co + Fe + \Sigma X_i$ (continued)

T	$\Delta L/L_0$
CURVE 127*	
303	0.012
373	0.094
CURVE 128*	
303	0.011
373	0.091
CURVE 129*	
303	0.012
373	0.094
CURVE 130*	
303	0.012
373	0.096
CURVE 131*	
303	0.011
373	0.091
CURVE 132*	
303	0.012
373	0.095
CURVE 133*	
303	0.012
373	0.095
CURVE 134*	
303	0.012
373	0.096
CURVE 135*	
303	0.012
373	0.098
CURVE 136	
293	0.000
323	0.028
373	0.074
423	0.123
473	0.173
523	0.226
573	0.277
623	0.327
673	0.377
723	0.432
773	0.483
823	0.537
873	0.592
923	0.652
973	0.718
1023	0.789
1073	0.861
1123	0.940
1173	1.016
CURVE 137	
293	0.000*
323	0.032*
373	0.080*
423	0.132*
473	0.184*
523	0.231*
573	0.289
623	0.338
673	0.394
723	0.453
773	0.514
823	0.570
873	0.636
923	0.698
973	0.770
1023	0.837
1073	0.907
1123	0.996
CURVE 138	
293	0.000*
323	0.031*
CURVE 138 (cont.)	
373	0.082*
423	0.135
473	0.194
523	0.247
573	0.303
623	0.357
673	0.421
723	0.478
773	0.539
823	0.609
873	0.676
923	0.751
973	0.826
1023	0.898
1073	0.972
1123	1.048
CURVE 139	
293	0.000*
323	0.037*
373	0.096
423	0.159
473	0.221
523	0.278
573	0.340
623	0.400
673	0.473
723	0.533
773	0.596
823	0.660
873	0.721
923	0.784
973	0.850
1023	0.910
1073	0.978*
1123	1.044*
1173	1.106
CURVE 140	
293	0.000*
323	0.040*
373	0.098*
423	0.162*
CURVE 140 (cont.)	
473	0.229*
523	0.289
573	0.356
623	0.423
673	0.492*
723	0.555
773	0.629*
823	0.695
873	0.766*
923	0.843
973	0.924*
1023	1.006
1073	1.081*
1128	1.174
1148	1.209

* Not shown in figure.

FIGURE AND TABLE NO. 245R. PROVISIONAL VALUES FOR THERMAL LINEAR EXPANSION OF COBALT + NICKEL + ΣX_i ALLOYS Co + Ni + ΣX_i

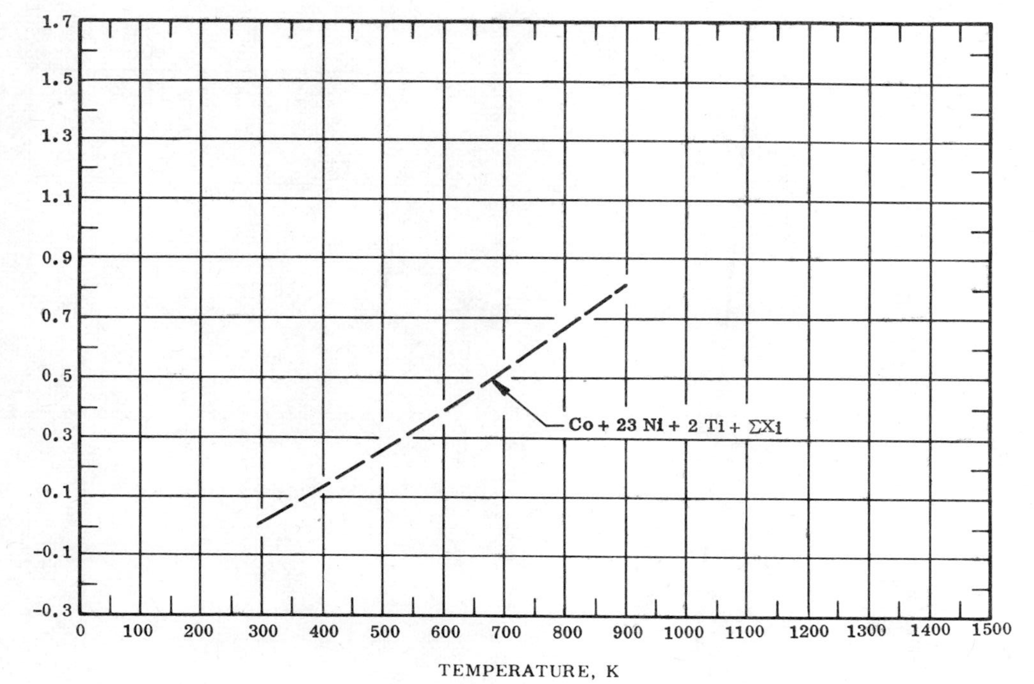

PROVISIONAL VALUES

[Temperature, T, K; Linear Expansion, $\Delta L/L_0$, %; α, K^{-1}]

T	(Co + 23 Ni + 2 Ti + ΣX_i) $\Delta L/L_0$	$\alpha \times 10^6$
293	0.000	12.4
400	0.135	12.5
500	0.260	12.6
600	0.389	13.0
700	0.520	13.4
800	0.658	14.1
900	0.804	15.0

REMARKS

(Co + 23 Ni + 2 Ti + ΣX_i): The tabulated values for well-annealed alloy of type Nivco are considered accurate to within ±7% over the entire temperature range. These values can be represented approximately by the following equation:

$$\Delta L/L_0 = -0.381 + 1.380 \times 10^{-3}\,T - 3.487 \times 10^{-7}\,T^2 + 3.090 \times 10^{-10}\,T^3$$

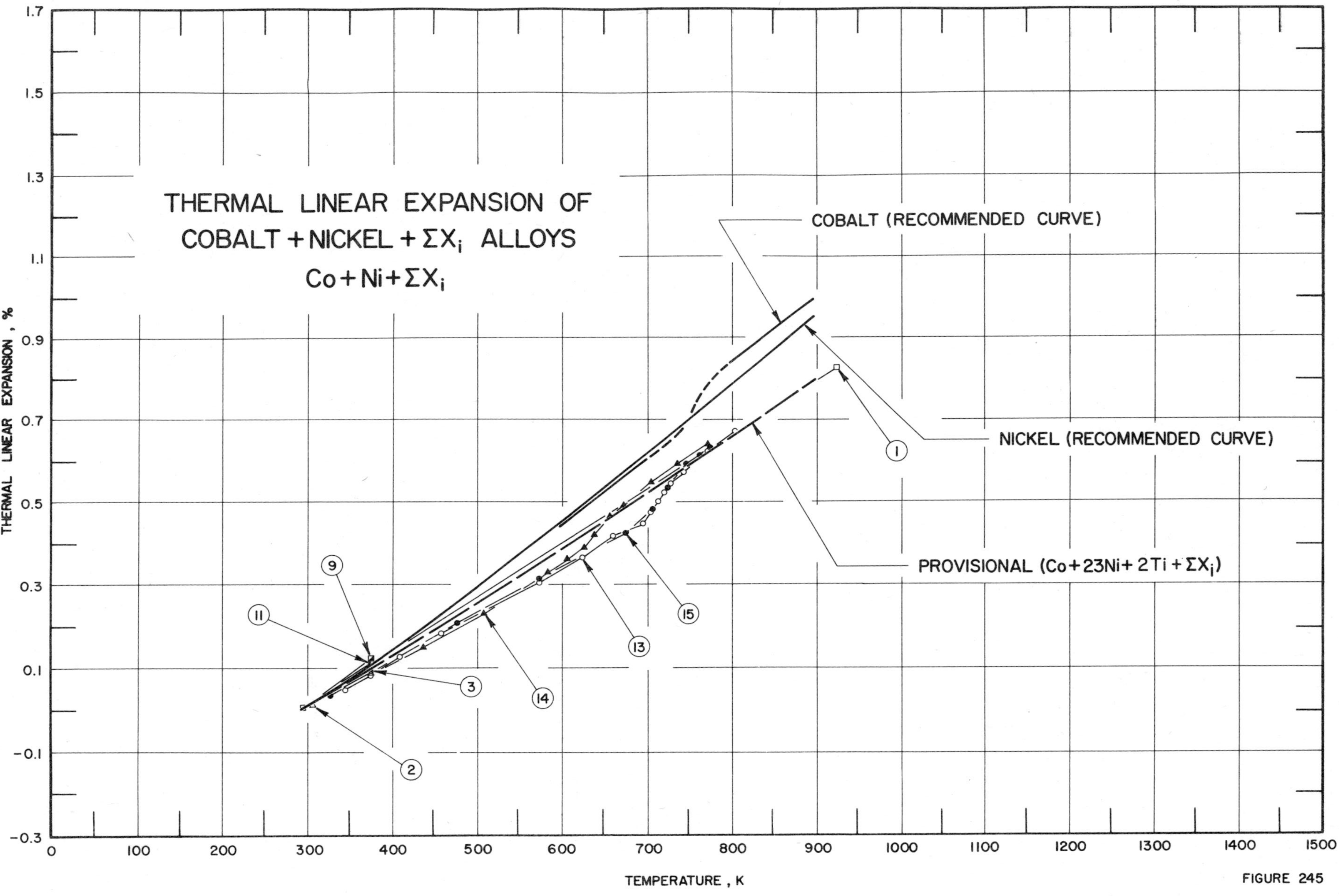
THERMAL LINEAR EXPANSION OF
COBALT + NICKEL + ΣXi ALLOYS
Co + Ni + ΣXi
COBALT (RECOMMENDED CURVE)
NICKEL (RECOMMENDED CURVE)
PROVISIONAL (Co+23Ni+2Ti+ΣXi)
1
2
3
9
11
13
14
15
THERMAL LINEAR EXPANSION, %
1.7
1.5
1.3
1.1
0.9
0.7
0.5
0.3
0.1
-0.1
-0.3
TEMPERATURE, K
0
100
200
300
400
500
600
700
800
900
1000
1100
1200
1300
1400
1500

FIGURE 245

SPECIFICATION TABLE 245. THERMAL LINEAR EXPANSION OF COBALT + NICKEL + ΣX_i ALLOYS Co + Ni + ΣX_i

Cur. No.	Ref. No.	Author(s)	Year	Method Used	Temp. Range, K	Name and Specimen Designation	Composition (weight percent) Co	Ni	Ti	Zr	Al	Fe	Composition (continued), Specifications, and Remarks
1	477	Kueser, P.E., Pavlovic, D.M., Lane, D.H., Clark, J.J., and Spewock, M.	1967	L	295-922	Nivco Alloy	73.18	23.5	2.04	1.03	0.25		Specimen heated to 1725±15 F in an air atmosphere, held at temperature 1 hr, water-quenched, then age-hardened at 1225±5 F for 25 hrs in air to a minimum hardness of Rockwell C36; a steam-turbine blading alloy manufactured by the Westinghouse Electric Corp, Blairsville, Pennsylvania; 2 in. long bar.
2	59	Masumoto, H.	1931	L	303-373	No. 48	49.8	30				20.2	Granular Co, Mond Ni, Armco Fe; mixed and melted in hydrogen atmosphere in alumina crucible in Tammann furnace, cast in iron mold with cylindrical aperture, machined to circular rod 4 mm thick, 10 cm long; heated at 1273 K for 1 hr in electric furnace in hydrogen atmosphere at reduced pressure; expansion measured in vacuum.
3	59	Masumoto, H.	1931	L	303-373	No. 49	49.8	40				10.2	Similar to the above specimen.
4*	59	Masumoto, H.	1931	L	303-373	No. 54	59.8	30				10.2	Similar to the above specimen.
5*	59	Masumoto, H.	1931	L	303-373	No. 57	69.7	20				10.3	Similar to the above specimen.
6*	59	Masumoto, H.	1931	L	303-373	No. 58	74.9	22				3.1	Similar to the above specimen.
7*	59	Masumoto, H.	1931	L	303-373	No. 59	74.9	24				1.1	Similar to the above specimen.
8*	59	Masumoto, H.	1931	L	303-373	No. 61	79.9	16				4.1	Similar to the above specimen.
9	59	Masumoto, H.	1931	L	303-373	No. 62	79.9	18				2.1	Similar to the above specimen.
10*	59	Masumoto, H.	1931	L	303-373	No. 63	84.9	10				5.1	Similar to the above specimen.
11	59	Masumoto, H.	1931	L	303-373	No. 64	84.9	13				2.1	Similar to the above specimen.
12*	59	Masumoto, H.	1931	L	303-373	No. 66	89.9	7				3.1	Similar to the above specimen.
13	406	Sykes, W.P. and Graff, H.F.	1935	L	344-804	Sample A	Bal.	0.6				0.2	Specimen 3 in. long x 0.625 in. diameter; machined from hydrogen-reduced powder; cooled from 1073 to 573 K in 100 hrs; heating rate 1 to 2 C/min.
14	406	Sykes, W.P. and Graff, H.F.	1935	L	804-344	Sample A	Bal.	0.6				0.2	The above specimen; expansion measured with decreasing temperature.
15	406	Sykes, W.P. and Graff, H.F.	1935	L	344-804	Sample A	Bal.	0.6				0.2	Similar to the above specimen except previously cooled from 1273 to 573 K in 150 hrs; expansion measured with increasing temperature; zero-point correction of 0.0145% determined by graphical extrapolation.

* Not shown in figure.

DATA TABLE 245. THERMAL LINEAR EXPANSION OF COBALT + NICKEL + ΣX_i ALLOYS Co + Ni + ΣX_i

[Temperature, T, K; Linear Expansion, $\Delta L/L_0$, %]

T	$\Delta L/L_0$
CURVE 1	
295.42	0.003
922.09	0.831
CURVE 2	
303	0.0122
373	0.0975
CURVE 3	
303	0.0125
373	0.1001
CURVE 4*	
303	0.0125
373	0.0993
CURVE 5*	
303	0.0125
373	0.0998
CURVE 6*	
303	0.0121
373	0.0965
CURVE 7*	
303	0.0124
373	0.0995
CURVE 8*	
303	0.0123
373	0.0981
CURVE 9	
303	0.0128*
373	0.1025
CURVE 10*	
303	0.0121
373	0.0966
CURVE 11	
303	0.0127*
373	0.1018
CURVE 12*	
303	0.0126
373	0.1004
CURVE 13	
344	0.052
374	0.085
409	0.129
458	0.182
574	0.306
623	0.363
660	0.420
693	0.452
703	0.480
712	0.502
720	0.524
727	0.543
742	0.570
773	0.621
804	0.670
CURVE 14	
771	0.639
737	0.593
705	0.544
672	0.497
656	0.467
638	0.422
625	0.397
605	0.361
581	0.330
509	0.237
435	0.155
344	0.052*
CURVE 15	
327	0.037
376	0.093*
475	0.206
574	0.317
674	0.425
707	0.482
723	0.535
748	0.590
761	0.615
774	0.634

* Not shown in figure.

FIGURE AND TABLE NO. 246R. PROVISIONAL VALUES FOR THERMAL LINEAR EXPANSION OF COPPER + ALUMINUM + ΣX_i ALLOYS Cu + Al + ΣX_i

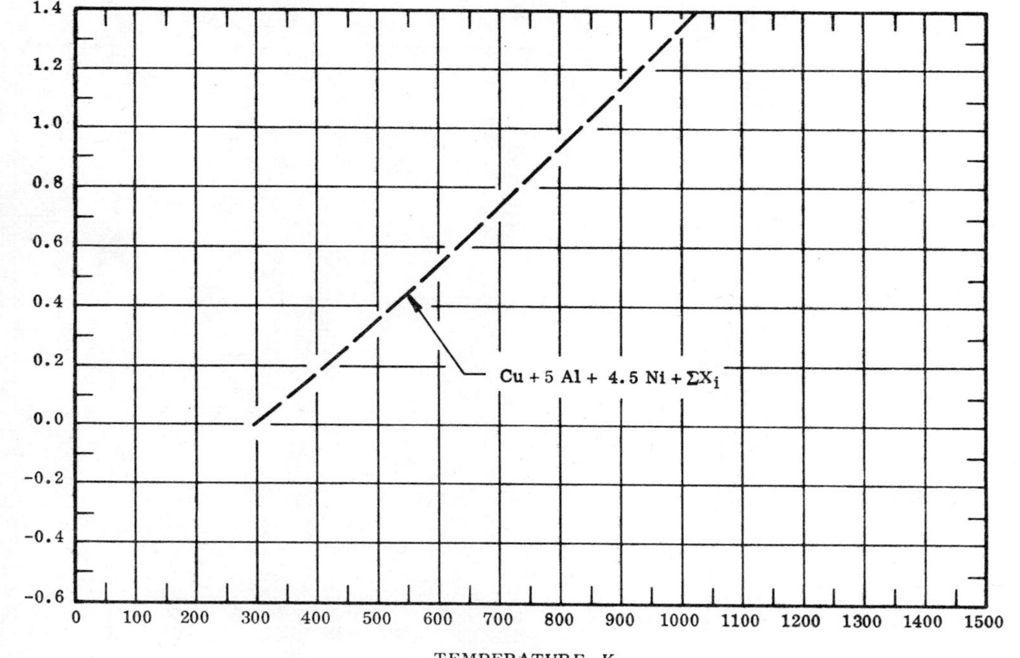

PROVISIONAL VALUES

[Temperature, T, K; Linear Expansion, $\Delta L/L_0$, %; α, K^{-1}]

(Cu + 5 Al + 4.5 Ni + ΣX_i)		
T	$\Delta L/L_0$	$\alpha \times 10^6$
293	0.000	15.9
400	0.177	17.1
500	0.353	18.1
600	0.538	18.9
700	0.731	19.6
800	0.930	20.3
900	1.136	20.8
1000	1.346	21.2
1050	1.451	21.3

REMARKS

(Cu + 5 Al + 4.5 Ni + ΣX_i): Tabulated values are for well-annealed alloy of type Tempaloy 841 and are considered accurate to within ±7% over the entire temperature range. The thermal linear expansion of this alloy is very close to that of pure copper. These values can be represented approximately by the following equation:

$$\Delta L/L_0 = -0.408 + 1.185 \times 10^{-3}\, T + 7.789 \times 10^{-7}\, T^2 - 2.099 \times 10^{-10}\, T^3$$

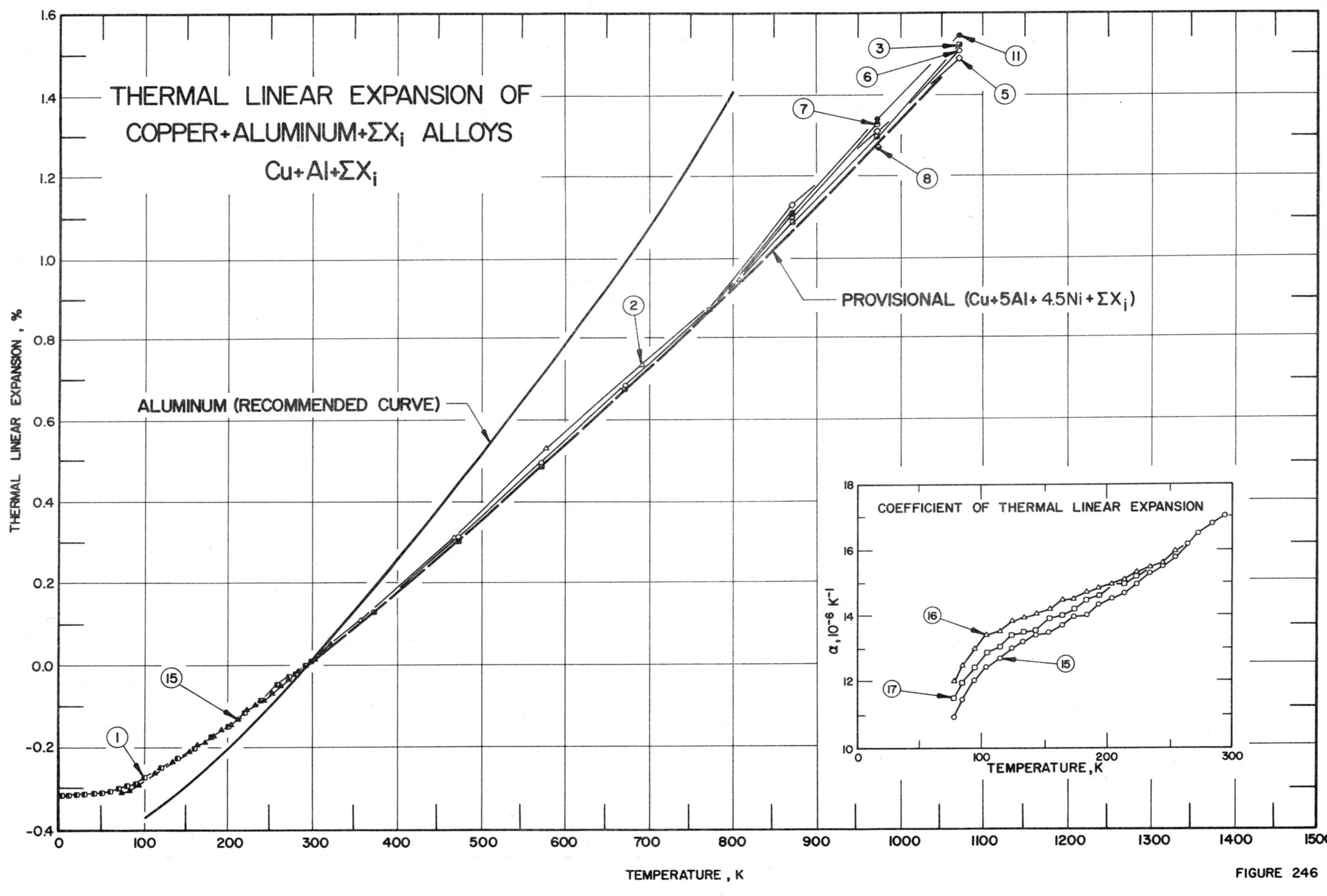

THERMAL LINEAR EXPANSION OF
COPPER+ALUMINUM+ΣXi ALLOYS
Cu+Al+ΣXi
THERMAL LINEAR EXPANSION, %
TEMPERATURE, K
ALUMINUM (RECOMMENDED CURVE)
PROVISIONAL (Cu+5Al+4.5Ni+ΣXi)
COEFFICIENT OF THERMAL LINEAR EXPANSION
α, 10⁻⁶ K⁻¹
TEMPERATURE, K
1
2
3
5
6
7
8
11
15
16
17

FIGURE 246

SPECIFICATION TABLE 246. THERMAL LINEAR EXPANSION OF COPPER + ALUMINUM + ΣX_i ALLOYS Cu + Al + ΣX_i

Cur. No.	Ref. No.	Author(s)	Year	Method Used	Temp. Range, K	Name and Specimen Designation	Composition (weight percent) Cu	Al	Fe	Ni	Si	Composition (continued), Specifications, and Remarks
1	467	Clark, A. F.	1968	L	0-300	Aluminum Bronze D	90.95	6.57	2.13			Specimen machined to 0.64 cm square x 20.32 cm long; Rockwell hardness B-90.
2	225	Valentich, J.	1965	L	293-806	Aluminum Bronze						No information given.
3	478	Hidnert, P. and Dickson, G.	1943	L	293-1073	Tempaloy 841	89.67	5.04		4.47	0.82	Specimen cast at 1473 K and machined to 0.95 cm diameter; expansion measured with increasing temperature.
4*	478	Hidnert, P. and Dickson, G.	1943	L	1073-293	Tempaloy 841	89.67	5.04		4.47	0.82	The above specimen; expansion measured with decreasing temperature.
5	478	Hidnert, P. and Dickson, G.	1943	L	293-1073	Tempaloy 841	89.67	5.04		4.47	0.82	Specimen cast at 1473 K; annealed 3 hrs at 1123 K, quenched in water, heat treated at 723 K for 3 hrs and cooled slowly in air; machined to 0.95 cm diameter; expansion measured with increasing temperature.
6	478	Hidnert, P. and Dickson, G.	1943	L	1073-293	Tempaloy 841	89.67	5.04		4.47	0.82	The above specimen; expansion measured with decreasing temperature.
7	478	Hidnert, P. and Dickson, G.	1943	L	293-973	Tempaloy 841	89.67	5.04		4.47	0.82	Specimen hot rolled at 1023 K from 3.81 cm to 1.91 cm diameter; annealed at 1073 K, quenched in water, drawn to 1.59 cm diameter, annealed, drawn to 1.27 cm diameter, annealed, drawn to 0.95 cm diameter, annealed 3 hrs at 1123 K and quenched; expansion measured with increasing temperature.
8	478	Hidnert, P. and Dickson, G.	1943	L	973-293	Tempaloy 841	89.67	5.04		4.47	0.82	The above specimen; expansion measured with decreasing temperature.
9*	478	Hidnert, P. and Dickson, G.	1943	L	293-1073	Tempaloy 841	89.67	5.04		4.47	0.82	The above specimen; heat treated at 973 K and slowly cooled; expansion measured with increasing temperature.
10*	478	Hidnert, P. and Dickson, G.	1943	L	1073-293	Tempaloy 841	89.67	5.04		4.47	0.82	The above specimen; expansion measured with decreasing temperature.
11	478	Hidnert, P. and Dickson, G.	1943	L	293-1073	Tempaloy 841	89.67	5.04		4.47	0.82	Similar to specimen used for curves 5 and 6, heat treated 3 hrs at 723 K and cooled slowly in air; expansion measured with increasing temperature.
12*	478	Hidnert, P. and Dickson, G.	1943	L	1073-293	Tempaloy 841	89.67	5.04		4.47	0.82	The above specimen; expansion measured with decreasing temperature.
13*	478	Hidnert, P. and Dickson, G.	1943	L	293-1073	Tempaloy 841	89.67	5.04		4.47	0.82	Similar to the above specimen; heated to 1073 K and cooled slowly; expansion measured with increasing temperature.
14*	478	Hidnert, P. and Dickson, G.	1943	L	1073-293	Tempaloy 841	89.67	5.04		4.47	0.82	The above specimen; expansion measured with decreasing temperature.

* Not shown in figure.

SPECIFICATION TABLE 246. THERMAL LINEAR EXPANSION OF COPPER + ALUMINUM + ΣX_i ALLOYS Cu + Al + ΣX_i (continued)

Cur. No.	Ref. No.	Author(s)	Year	Method Used	Temp. Range, K	Name and Specimen Designation	Composition (weight percent) Cu	Al	Fe	Ni	Si	Composition (continued), Specifications, and Remarks
15	551	Belov, A.K.	1968	L	77-323	Br AZh 9-4 Russian Cu alloy	Bal.	10	4.0	0.5		1.0 Zn, 0.5 Mn; cylindrical specimen 3.5 mm in diameter, 50 mm long; sample heated and cooled at a rate of 3-7 deg per min.
16*	551	Belov, A.K.	1968	L	77-323	AZhMts 10-3-1.5 Russian Cu alloy	Bal.	11	4.0	0.5		2.0 Mn. 0.5 Zn; similar to the above specimen.
17*	551	Belov, A.K.	1968	L	77-323	BrAZhN 10-4-4 Russian Cu alloy	Bal.	11	5.5	5.5		0.5 Mn, 0.3 Zn; similar to the above specimen.

* Not shown in figure.

DATA TABLE 246. THERMAL LINEAR EXPANSION OF COPPER + ALUMINUM + ΣX_i ALLOYS $Cu + Al + \Sigma X_i$

[Temperature, T, K; Linear Expansion, $\Delta L/L_0$, %]

T	$\Delta L/L_0$
CURVE 1	
0	-0.316
10	-0.316
20	-0.315
30	-0.315
40	-0.313
50	-0.310
60	-0.305
70	-0.299
80	-0.291
90	-0.283
100	-0.273
120	-0.252
140	-0.227
160	-0.201
180	-0.174
200	-0.145
220	-0.116
240	-0.085
260	-0.054
273	-0.033
280	-0.022
293	0.000
300	0.010
CURVE 2	
293	0.000*
358	0.109
468	0.314
578	0.527
690	0.737
806	0.954
CURVE 3	
293	0.000*
373	0.131
473	0.302
573	0.487
673	0.676
773	0.878
873	1.090
973	1.299
1073	1.521
CURVE 4*	
1073	1.529
973	1.306
673	0.684
473	0.306
373	0.134
293	0.000
CURVE 5	
293	0.000*
373	0.134*
473	0.308*
573	0.490*
673	0.688*
773	0.893*
873	1.131
973	1.306
1073	1.490
CURVE 6	
1073	1.513
973	1.292*
673	0.680
573	0.493
473	0.308
373	0.134*
293	0.000*
CURVE 7	
293	0.000*
373	0.136*
473	0.302*
573	0.482*
673	0.676*
773	0.878*
873	1.102
973	1.326
CURVE 8	
973	1.272
773	0.874*
673	0.680*
573	0.490*
CURVE 8 (cont.)	
473	0.308*
373	0.133*
293	0.000*
CURVE 9*	
293	0.000
373	0.134
473	0.306
573	0.487
673	0.680
773	0.878
873	1.073
973	1.272
1073	1.490
CURVE 10*	
1073	1.513
673	0.680
573	0.487
473	0.306
373	0.132
293	0.000
CURVE 11	
293	0.000*
373	0.134*
473	0.306*
573	0.487*
673	0.680*
773	0.878*
873	1.108
973	1.340
1073	1.544
CURVE 12*	
1073	1.513
973	1.292
673	0.680
573	0.490
473	0.306
373	0.133
293	0.000
CURVE 13*	
293	0.000
373	0.133
473	0.304
573	0.484
673	0.680
773	0.874
873	1.085
973	1.292
1073	1.505
CURVE 14*	
1073	1.521
973	1.300
673	0.680
573	0.490
473	0.308
373	0.131
293	0.000
CURVE 15‡	
77	-0.307
83	-0.301
93	-0.289
103	-0.277*
113	-0.264
123	-0.251*
133	-0.238
143	-0.225*
153	-0.211
163	-0.198
173	-0.184
183	-0.170
193	-0.156
203	-0.142
213	-0.127
223	-0.112
233	-0.097
243	-0.082
253	-0.066
263	-0.050
273	-0.034
283	-0.017
293	0.000*
303	0.017
CURVE 15 (cont.)‡	
313	0.035
323	0.052
CURVE 16*, ‡	
77	-0.318
83	-0.311
93	-0.298
103	-0.285
113	-0.271
123	-0.258
133	-0.244
143	-0.230
153	-0.216
163	-0.201
173	-0.187
183	-0.172
193	-0.158
203	-0.143
213	-0.128
223	-0.113
233	-0.097
243	-0.082
253	-0.066
263	-0.050
273	-0.034
283	-0.017
293	0.000
303	0.017
313	0.034
323	0.052
CURVE 17*, ‡	
77	-0.312
83	-0.305
93	-0.293
103	-0.281
113	-0.268
123	-0.255
133	-0.241
143	-0.228
153	-0.214
163	-0.200
173	-0.186
183	-0.172
CURVE 17 (cont.)*, ‡	
193	-0.157
203	-0.142
213	-0.128
223	-0.112
233	-0.097
243	-0.082
253	-0.066
263	-0.050
273	-0.034
283	-0.017
293	0.000
303	0.017
313	0.035
323	0.052

* Not shown in figure.
‡ Author's data for coefficient of thermal expansion have been integrated by TPRC to obtain $\Delta L/L_0$.

DATA TABLE 246. COEFFICIENT OF THERMAL LINEAR EXPANSION OF COPPER + ALUMINUM + ΣX_i ALLOYS $Cu + Al + \Sigma X_i$

[Temperature, T, K; Coefficient of Expansion, α, 10^{-6} K^{-1}]

T	α
CURVE 15‡	
77	10.99
83	11.40
93	12.00
103	12.39
113	12.70
123	13.00
133	13.19
143	13.40
153	13.49
163	13.70
173	13.89
183	14.00
193	14.29
203	14.50
213	14.69
223	14.99
233	15.30
243	15.50
253	15.79
263	16.20
273	16.49
283	16.79
293	17.00
303	17.30*
313	17.50*
323	17.70*
CURVE 16‡	
77	12.00
83	12.50
93	13.00
103	13.40
113	13.59
123	13.80
133	13.89
143	14.10
153	14.20
163	14.40
173	14.50
183	14.69
193	14.80
203	14.99
213	15.10
223	15.30
CURVE 16 (cont.)‡	
233	15.39
243	15.60
253	15.90
263	16.20*
273	16.49*
283	16.79*
293	17.00*
303	17.24*
313	17.50*
323	17.70*
CURVE 17‡	
77	11.5
83	11.9
93	12.4
103	12.8
113	13.1
123	13.3
133	13.5
143	13.6
153	13.8
163	14.0
173	14.2
183	14.4
193	14.6
203	14.8
213	14.9
223	15.2
233	15.4
243	15.6*
253	15.9*
263	16.2*
273	16.5*
283	16.8*
293	17.0*
303	17.3*
313	17.5*
323	17.7*

* Not shown in figure.

‡ Author's data for coefficient of thermal expansion have been integrated by TPRC to obtain $\Delta L/L_0$.

FIGURE AND TABLE NO. 247R. PROVISIONAL VALUES FOR THERMAL LINEAR EXPANSION OF COPPER + BERYLLIUM + ΣX_i ALLOYS Cu + Be + ΣX_i

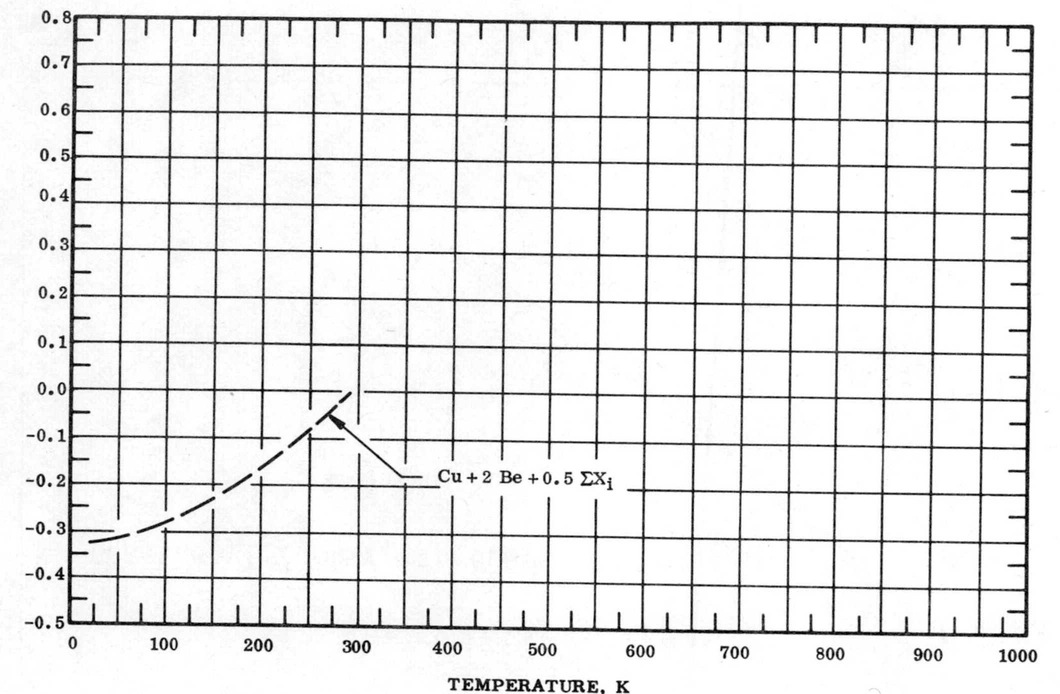

PROVISIONAL VALUES

[Temperature, T, K; Linear Expansion, $\Delta L/L_0$, %; α, K^{-1}]

T	(Cu + 2 Be + 0.5 ΣX_i)	
	$\Delta L/L_0$	$\alpha \times 10^6$
20	-0.323	0.4
50	-0.317	4.2
100	-0.281	9.5
200	-0.151	15.5
293	0.000	15.9

REMARKS

(Cu + 2 Be + 0.5 ΣX_i): The tabulated values are for well-annealed alloy of type Copper Beryllium and are accurate to within ± 7% over the entire temperature range. The thermal linear expansion of this alloy is very close to that of pure copper. The tabulated values can be represented approximately by the following equation:

$$\Delta L/L_0 = -0.321 - 2.482 \times 10^{-4}\, T + 7.442 \times 10^{-6}\, T^2 - 9.788 \times 10^{-9}\, T^3$$

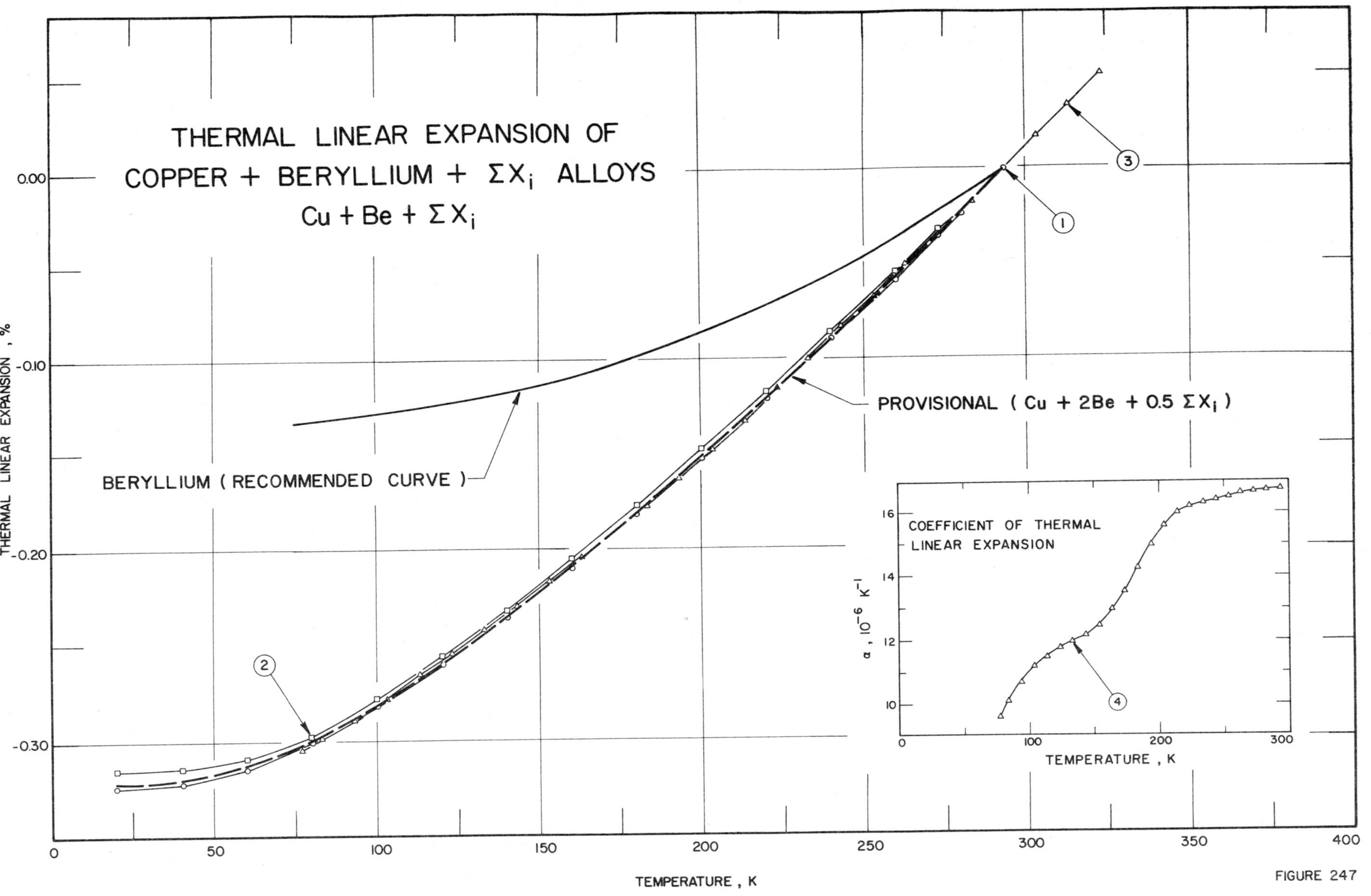
THERMAL LINEAR EXPANSION OF
COPPER + BERYLLIUM + ΣX_i ALLOYS
Cu + Be + ΣX_i
PROVISIONAL (Cu + 2Be + 0.5 ΣX_i)
BERYLLIUM (RECOMMENDED CURVE)
COEFFICIENT OF THERMAL
LINEAR EXPANSION
α, 10^-6 K^-1
TEMPERATURE , K
THERMAL LINEAR EXPANSION , %
TEMPERATURE , K
1
2
3
4
0.00
-0.10
-0.20
-0.30
0
50
100
150
200
250
300
350
400
0
100
200
300
10
12
14
16

FIGURE 247

SPECIFICATION TABLE 247. THERMAL LINEAR EXPANSION OF COPPER + BERYLLIUM + ΣX_i ALLOYS Cu + Be + ΣX_i

Cur. No.	Ref. No.	Author(s)	Year	Method Used	Temp. Range, K	Name and Specimen Designation	Composition (weight percent) Cu	Be	Co	Al	Fe	Si	Ni	Composition (continued), Specifications, and Remarks
1	462	Arp, V., Wilson, J.H., Winrich, I. and Sikora, P.	1962	L	20-293	Berylco 25	Bal.	1.8	0.2	0.1	0.1	0.1		Hardness R_B55; specimen annealed.
2	462	Arp, V., et al.	1962	L	20-293	Berylco 25	Bal.	1.8	0.2	0.1	0.1	0.1		Hardness R_B95; hard.
3	551	Belov, A.K.	1968	L	77-323	BrB_2 Russian alloy	Bal.	2.2					0.5	Cylindrical specimen 3.5 mm in diameter, 50 mm long; sample heated and cooled at a rate of 3-7 deg. per min.

DATA TABLE 247. THERMAL LINEAR EXPANSION OF COPPER + BERYLLIUM + ΣX_i ALLOYS Cu + Be + ΣX_i

[Temperature, T, K; Linear Expansion, $\Delta L/L_0$, %]

CURVE 1

T	$\Delta L/L_0$
20	-0.324
40	-0.322
60	-0.315
80	-0.300
100	-0.281
120	-0.259
140	-0.235
160	-0.210
180	-0.182
200	-0.152
220	-0.121
240	-0.089
260	-0.057
273	-0.035
280	-0.023
293	0.000

CURVE 2

T	$\Delta L/L_0$
20	-0.316
40	-0.314
60	-0.309
80	-0.297
100	-0.277
120	-0.256
140	-0.232
160	-0.205
180	-0.177
200	-0.147
220	-0.117
240	-0.086
260	-0.054
273	-0.032
280	-0.022*
293	0.000*

CURVE 3‡

T	$\Delta L/L_0$
77	-0.304
83	-0.298
93	-0.288
103	-0.277
113	-0.265
123	-0.254
133	-0.242
143	-0.230
153	-0.217
163	-0.205
173	-0.191
183	-0.177
193	-0.163
203	-0.148
213	-0.132
223	-0.116
233	-0.099
243	-0.083
253	-0.067
263	-0.050
273	-0.033*
283	-0.017
293	0.000*
303	0.017
313	0.034
323	0.051

DATA TABLE 247. COEFFICIENT OF THERMAL LINEAR EXPANSION OF COPPER + BERYLLIUM + ΣX_i ALLOYS Cu + Be + ΣX_i

[Temperature, T, K; Coefficient of Expansion, α, 10^{-6} K^{-1}]

CURVE 3‡

T	α
77	9.60
83	10.10
93	10.70
103	11.20
113	11.50
123	11.80
133	12.00
143	12.20
153	12.50
163	13.00
173	13.59
183	14.29
193	14.99
203	15.60
213	16.00
223	16.20
233	16.30
243	16.40
253	16.49
263	16.60
273	16.65
283	16.70
293	16.75
303	16.79*
313	16.89*
323	16.95*

*Not shown in figure.

‡Author's data for coefficient of thermal expansion have been integrated by TPRC to obtain $\Delta L/L_0$.

FIGURE AND TABLE NO. 248R. PROVISIONAL VALUES FOR THERMAL LINEAR EXPANSION OF COPPER + LEAD + ΣX_i ALLOYS $Cu + Pb + \Sigma X_i$

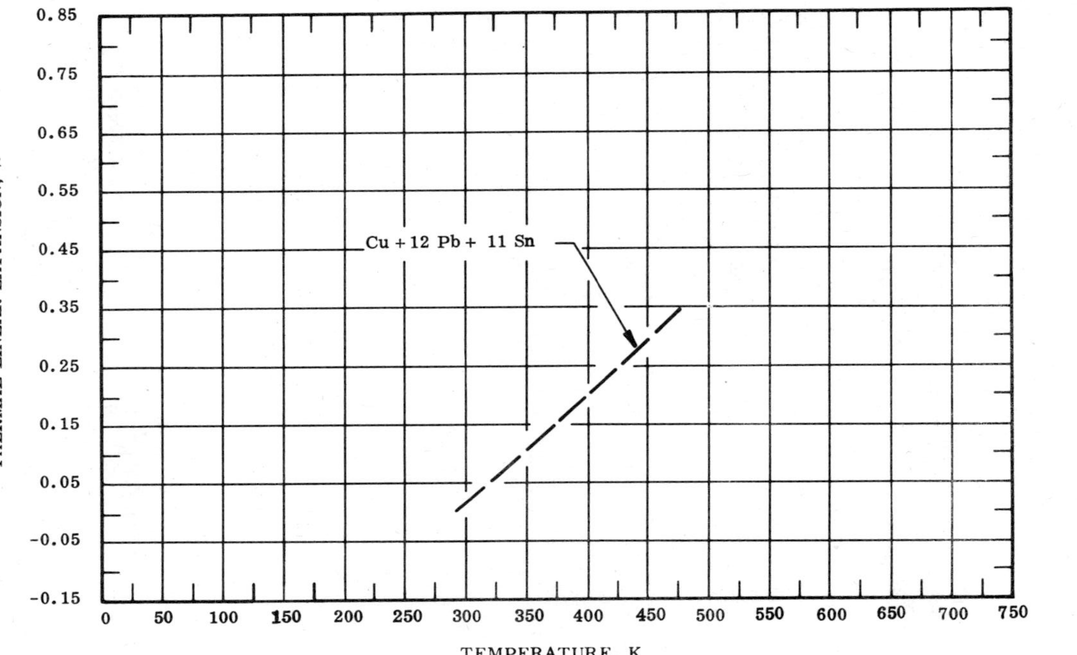

PROVISIONAL VALUES

[Temperature, T, K; Linear Expansion, $\Delta L/L_0$, %; α, K^{-1}]

T	(Cu + 12 Pb + 11 Sn) $\Delta L/L_0$	$\alpha \times 10^6$
293	0.000	18.5
400	0.200	18.8
475	0.340	18.9

REMARKS

(Cu + 12 Pb + 11 Sn): The tabulated values are for well-annealed alloy and are considered accurate to within ± 7% over the entire temperature range. These values can be represented approximately by the following equation:

$$\Delta L/L_0 = -0.736 + 3.372 \times 10^{-3} T - 3.906 \times 10^{-6} T^2 + 3.321 \times 10^{-9} T^3$$

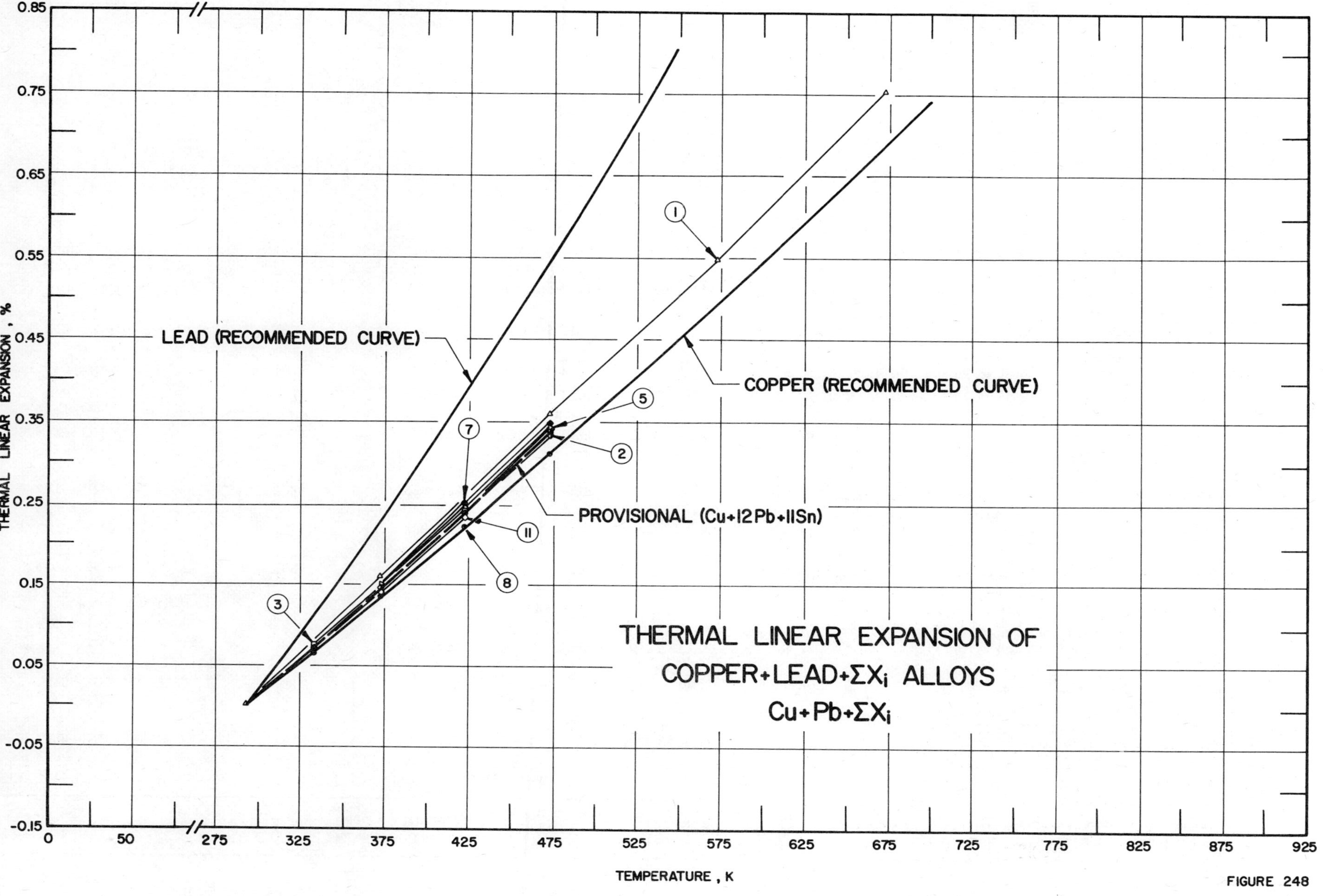
THERMAL LINEAR EXPANSION OF
COPPER+LEAD+ΣXi ALLOYS
Cu+Pb+ΣXi
LEAD (RECOMMENDED CURVE)
COPPER (RECOMMENDED CURVE)
PROVISIONAL (Cu+12Pb+11Sn)
1
2
3
5
7
8
11
THERMAL LINEAR EXPANSION, %
TEMPERATURE, K
0.85
0.75
0.65
0.55
0.45
0.35
0.25
0.15
0.05
-0.05
-0.15
0
50
275
325
375
425
475
525
575
625
675
725
775
825
875
925

FIGURE 248

SPECIFICATION TABLE 248. THERMAL LINEAR EXPANSION OF COPPER + LEAD + ΣX_i ALLOYS $Cu + Pb + \Sigma X_i$

Cur. No.	Ref. No.	Author(s)	Year	Method Used	Temp. Range, K	Name and Specimen Designation	Composition (weight percent) Cu	Pb	Sb	Sn	Ni	Zn	Composition (continued), Specifications, and Remarks
1	478	Hidnert, P. and Dickson, G.	1943	L	293-773	Sample 934	64	34	2				Rod 0.95 cm diameter machined from sand casting.
2	479	Hidnert, P.	1934	L	293-473	Sample 1475	77.00	11.90	2	11.08			0.02 P; prepared from virgin materials; cast at 1422 K, machined to 300 mm long x 9 mm diameter; expansion measured on first test.
3	479	Hidnert, P.	1934	L	293-473	Sample 1475	77.00	11.90	2	11.08			The above specimen; expansion measured on second test.
4*	479	Hidnert, P.	1934	L	293-473	Sample 1476	69.82	20.50	2	9.72			0.02 P; similar to the above specimen; expansion measured on first test.
5	479	Hidnert, P.	1934	L	293-473	Sample 1476	69.82	20.50	2	9.72			The above specimen; expansion measured on second test.
6*	479	Hidnert, P.	1934	L	293-473	Sample 1477	65.40	26.62	2	7.75	0.25		Similar to the above specimen; expansion measured on first test.
7	479	Hidnert, P.	1934	L	293-473	Sample 1477	65.40	26.62	2	7.75	0.25		The above specimen; expansion measured on second test.
8	479	Hidnert, P.	1934	L	293-473	Sample 1478	95.70	2.32	2	1.95			0.03 P; similar to the above specimen except prepared from commercial materials; expansion measured on first test.
9*	479	Hidnert, P.	1934	L	293-473	Sample 1478	95.70	2.32	2	1.95			The above specimen; expansion measured on second test.
10*	479	Hidnert, P.	1934	L	293-473	Sample 1479	75.86	12.00	0.15	11.57	0.20	0.25	0.04 P; similar to the above specimen.
11	479	Hidnert, P.	1934	L	293-423	Sample 1481	82.28	7.48	0.15	6.70	0.15	3.20	0.05 P; similar to the above specimen; expansion measured on first test.
12*	479	Hidnert, P.	1934	L	293-473	Sample 1481	82.28	7.48	0.15	6.70	0.15	3.20	The above specimen; expansion measured on second test.
13*	479	Hidnert, P.	1934	L	293-473	Sample 1484	73.00	21.79	0.15	4.45	0.35	0.25	0.05 P; similar to the above specimen; expansion measured on first test.
14*	479	Hidnert, P.	1934	L	293-473	Sample 1484	73.00	21.79	0.15	4.45	0.35	0.25	The above specimen; expansion measured on second test.

* Not shown in figure.

DATA TABLE 248. THERMAL LINEAR EXPANSION OF COPPER + LEAD + ΣX_i ALLOYS $Cu + Pb + \Sigma X_i$

[Temperature, T, K; Linear Expansion, $\Delta L/L_0$, %]

T	$\Delta L/L_0$
CURVE 1	
293	0.000
373	0.161
473	0.360
573	0.560
673	0.756
773	0.950*
CURVE 2	
293	0.000*
333	0.071
373	0.149
423	0.240
473	0.335
CURVE 3	
293	0.000
333	0.074
373	0.148*
423	0.243
473	0.338
CURVE 4*	
293	0.000
333	0.074
373	0.149
423	0.242
473	0.338
CURVE 5	
293	0.000*
333	0.076*
373	0.151
423	0.247
473	0.344
CURVE 6*	
293	0.000
333	0.076
373	0.152
423	0.246
473	0.342
CURVE 7	
293	0.000*
333	0.075*
373	0.153
423	0.251
473	0.349
CURVE 8	
293	0.000*
333	0.068
373	0.138
423	0.225
473	0.311
CURVE 9*	
293	0.000
333	0.068
373	0.137
473	0.317
CURVE 10*	
293	0.000
333	0.072
373	0.146
423	0.242
473	0.337
CURVE 11	
293	0.000*
333	0.071*
373	0.145
423	0.237
CURVE 12*	
293	0.000
333	0.071
373	0.144
423	0.237
473	0.331
CURVE 13*	
293	0.000
333	0.074
373	0.151
423	0.244
473	0.338
CURVE 14*	
293	0.000
333	0.075
373	0.154
423	0.250
473	0.346

*Not shown in figure.

SPECIFICATION TABLE 249. THERMAL LINEAR EXPANSION OF COPPER + MANGANESE + ΣX_i ALLOYS $Cu + Mn + \Sigma X_i$

Cur. No.	Ref. No.	Author(s)	Year	Method Used	Temp. Range, K	Name and Specimen Designation	Composition (weight percent) Cu	Mn	Al	Fe	Composition (continued), Specifications, and Remarks
1*	478	Hidnert, P. and Dickson, G.	1943	L	373-293	Sample 1575	84.93	9.4	5.5	0.22	Cold drawn wire 1 mm diameter; baked in air 18 hrs at 413 K; expansion measured with decreasing temperature.

DATA TABLE THERMAL LINEAR EXPANSION OF COPPER + MANGANESE + ΣX_i ALLOYS $Cu + Mn + \Sigma X_i$

[Temperature, T, K; Linear Expansion, $\Delta L/L_0$, %]

T	$\Delta L/L_0$
CURVE 1*	
373	0.139
293	0.000

* No figure given.

FIGURE AND TABLE NO. 250R. PROVISIONAL VALUES FOR THERMAL LINEAR EXPANSION OF COPPER + NICKEL + ΣX_i ALLOYS   Cu + Ni + ΣX_i

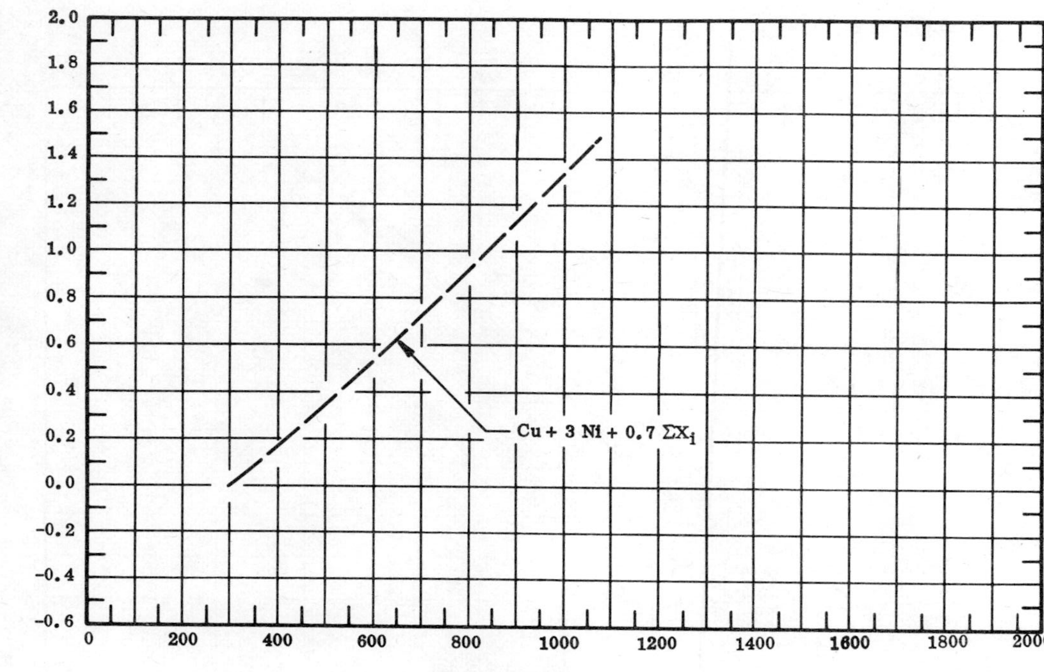

PROVISIONAL VALUES

[Temperature, T, K; Linear Expansion, $\Delta L/L_0$, %; α, K^{-1}]

T	(Cu + 3 Ni + 0.7 ΣX_i) $\Delta L/L_0$	$\alpha \times 10^6$
293	0.000	16.3
400	0.174	16.7
500	0.348	17.8
600	0.528	18.7
700	0.720	19.5
800	0.919	20.2
900	1.124	20.9
1000	1.334	21.3
1073	1.491	21.6

REMARKS

(Cu + 3 Ni + 0.7 ΣX_i): The tabulated values for well-annealed alloy of type Tempaloy 836 are considered accurate to within ±7% over the entire temperature range. These values can be represented approximately by the following equation:

$$\Delta L/L_0 = -0.403 + 1.180 \times 10^{-3}\, T + 7.210 \times 10^{-7}\, T^2 - 1.639 \times 10^{-10}\, T^3$$

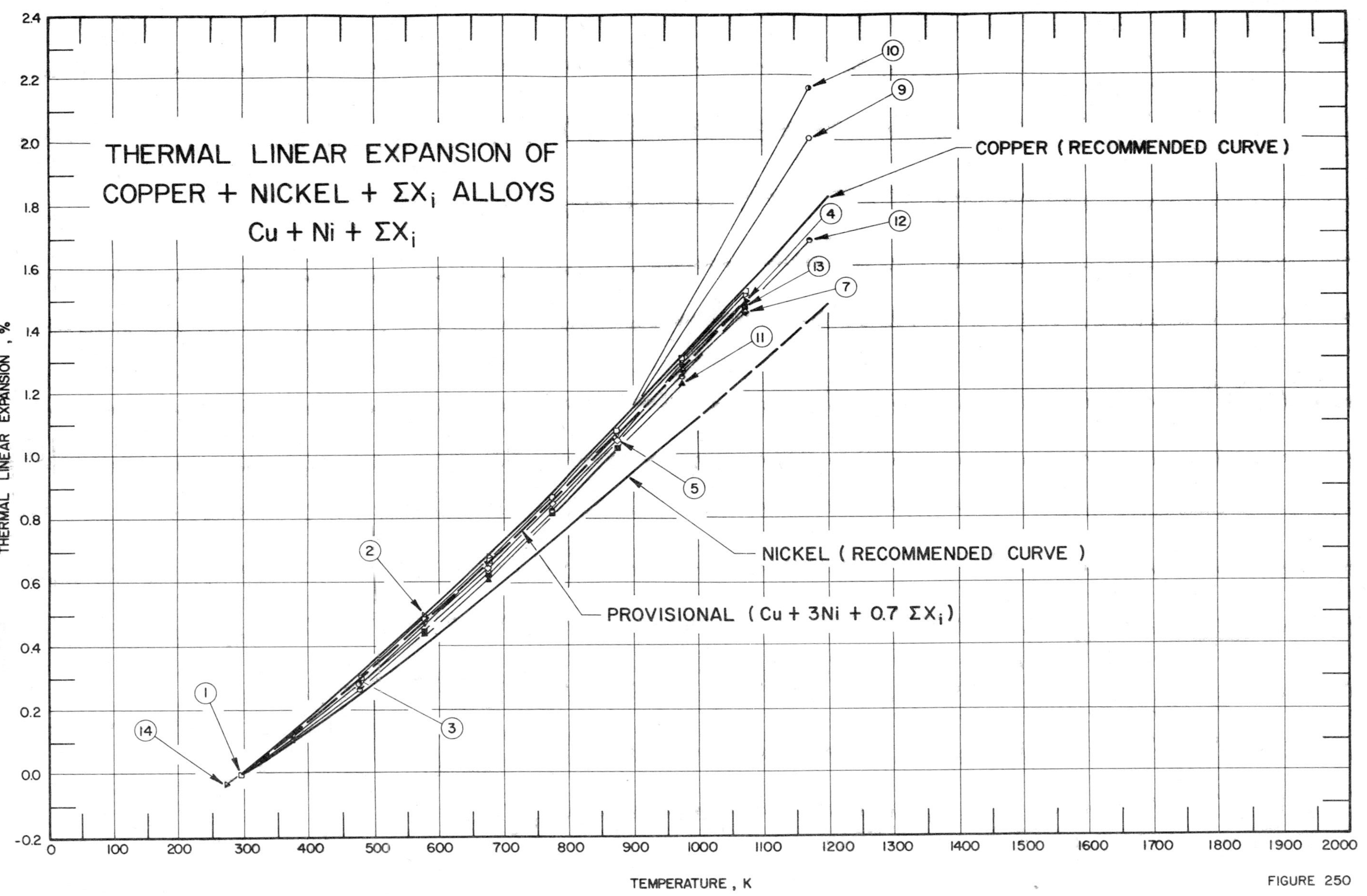

THERMAL LINEAR EXPANSION OF
COPPER + NICKEL + ΣX_i ALLOYS
Cu + Ni + ΣX_i
COPPER (RECOMMENDED CURVE)
NICKEL (RECOMMENDED CURVE)
PROVISIONAL (Cu + 3Ni + 0.7 ΣX_i)
1
2
3
4
5
7
9
10
11
12
13
14
THERMAL LINEAR EXPANSION, %
-0.2
0.0
0.2
0.4
0.6
0.8
1.0
1.2
1.4
1.6
1.8
2.0
2.2
2.4
TEMPERATURE, K
0
100
200
300
400
500
600
700
800
900
1000
1100
1200
1300
1400
1500
1600
1700
1800
1900
2000

FIGURE 250

SPECIFICATION TABLE 250. THERMAL LINEAR EXPANSION OF COPPER + NICKEL + ΣX_i ALLOYS Cu + Ni + ΣX_i

Cur. No.	Ref. No.	Author(s)	Year	Method Used	Temp. Range, K	Name and Specimen Designation	Composition (weight percent) Cu	Ni	Sn	Si	Fe	Pb	Zn	Composition (continued), Specifications, and Remarks
1	478	Hidnert, P. and Dickson, G.	1943	L	293-1073	Tempaloy 836	96.06	3.02		0.61	0.12			Specimen cast at 1473 K and machined to 0.95 cm diameter; expansion measured with increasing temperature.
2	478	Hidnert, P. and Dickson, G.	1943	L	1073-293	Tempaloy 836	96.06	3.02		0.61	0.12			The above specimen; expansion measured with decreasing temperature.
3	478	Hidnert, P. and Dickson, G.	1943	L	293-1073	Tempaloy 836	96.06	3.02		0.61	0.12			Specimen cast at 1473 K, annealed at 1123 K for 3 hrs, quenched in water, heat-treated 3 hrs at 723 K, and cooled slowly in air; machined to 0.95 cm diameter; expansion measured with increasing temperature.
4	478	Hidnert, P. and Dickson, G.	1943	L	1073-293	Tempaloy 836	96.06	3.02		0.61	0.12			The above specimen; expansion measured with decreasing temperature.
5	478	Hidnert, P. and Dickson, G.	1943	L	293-1073	Tempaloy 836	96.00	3.14		0.86				0.794 cm rod annealed at 1123 K and quenched; expansion measured with increasing temperature.
6*	478	Hidnert, P. and Dickson, G.	1943	L	293-1073	Tempaloy 836	95.71	3.22		0.69				Specimen drawn from 0.95 cm to 0.79 cm; expansion measured with increasing temperature.
7	478	Hidnert, P. and Dickson, G.	1943	L	1073-293	Tempaloy 836	95.71	3.22		0.69				The above specimen; expansion measured with decreasing temperature.
8*	427	Souder, W., Hidnert, P., and Fox, J. F.	1934	O	293-773	Tempaloy (soft)	96.00	3.14		0.86				150 mm long with pointed ends, maximum diameter 12 mm; soft, annealed at 850 C and quenched.
9	478	Hidnert, P. and Dickson, G.	1943	L	293-1173	Sample 1085	68	20	12					Bar 0.635 cm square, cast.
10	478	Hidnert, P. and Dickson, G.	1943	L	293-1173	Sample 1084	71	18	6			4	1	Cast bar 0.635 cm square.
11	478	Hidnert, P. and Dickson, G.	1943	L	293-973	Sample 1026	64.9	28.5			4.9			0.9 Mn; cast bar 1.27 cm x 1.11 cm cross section.
12	478	Hidnert, P. and Dickson, G.	1943	L	293-1173	Aterite No. 9	55	32			6		7	Cast bar 1.27 cm square cross section.
13	478	Hidnert, P. and Dickson, G.	1943	L	293-1073	Aterite No. 4	48.80	31.26		0.66	8.83	2.23	8.93	0.41 C, 0.16 Mn; cast bar 0.953 cm square.
14	480	Cook, M.	1936	L	273-673	Nickel Silver Alloy No. 5	62.16	20.22			0.05	0.005	17.44	0.13 Mn, 0.012 C, <0.005 any others; annealed rod 12.7 cm x 1.6 cm diameter.
15*	480	Cook, M.	1936	L	273-673	Nickel Silver Alloy No. 6	61.96	25.56			0.07	0.004	12.31	0.10 Mn, 0.020 C, <0.005 any others; similar to the above specimen.
16*	480	Cook, M.	1936	L	273-673	Nickel Silver Alloy No. 7	62.02	29.77			0.09	0.003	7.93	0.14 Mn, 0.019 C, <0.01 any others; similar to the above specimen.
17*	142	Aoyama, S. and Ito, T.	1939	L	77-273		Bal.	39.31			0.27			0.03 Mn, alloy melted, cast into bar 30 x 30 x 250 mm; hot-forged to 8 mm diameter bar, then cut into round bar 5 mm diameter, 100 mm length; annealed in vacuum for three hr at 1023 C.

* Not shown in figure.

DATA TABLE 250. THERMAL LINEAR EXPANSION OF COPPER + NICKEL + ΣX_i ALLOYS Cu + Ni + ΣX_i

[Temperature, T, K; Linear Expansion, $\Delta L/L_0$, %]

T	$\Delta L/L_0$
CURVE 1	
293	0.000
373	0.133
473	0.308
573	0.490
673	0.676
773	0.874
873	1.085
973	1.306
1073	1.521
CURVE 2	
1073	1.513
973	1.292*
673	0.680
573	0.496
473	0.311*
373	0.136*
293	0.000*
CURVE 3	
293	0.000*
373	0.125
473	0.297
573	0.479
673	0.669
773	0.864*
873	1.079
973	1.312*
1073	1.513*
CURVE 4	
1073	1.490
973	1.285
673	0.680*
573	0.493*
473	0.310*
373	0.138
293	0.000*
CURVE 5	
293	0.000*
373	0.133*
473	0.301*
573	0.482*
673	0.657
773	0.854
873	1.056
973	1.285*
1073	1.505*
CURVE 6*	
293	0.000
373	0.135
473	0.306
573	0.496
673	0.684
773	0.878
873	1.085
973	1.299
1073	1.552
CURVE 7	
1073	1.459
973	1.265
873	1.061*
673	0.676*
573	0.490*
473	0.308*
293	0.000*
CURVE 8*	
373	0.134
473	0.306
573	0.479
673	0.661
773	0.850
CURVE 9	
293	0.000*
473	0.292
673	0.642
873	1.050*
1173	2.015
CURVE 10	
293	0.000*
473	0.295*
673	0.646*
873	1.073*
1173	2.165
CURVE 11	
293	0.000
573	0.442
673	0.616
973	1.224
CURVE 12	
293	0.000
373	0.119
473	0.277
573	0.448*
773	0.830
973	1.244
1073	1.451
1173	1.690
CURVE 13	
293	0.000*
473	0.277*
573	0.451
673	0.631
773	0.821
873	1.021
973	1.238*
1073	1.474
CURVE 14	
273	-0.031
373	0.123*
473	0.292*
573	0.470*
673	0.661*
CURVE 15*	
273	-0.032
373	0.126
473	0.295
573	0.476
673	0.669
CURVE 16*	
273	-0.032
373	0.130
473	0.299
573	0.476
673	0.676
CURVE 17*	
77	-0.258
91	-0.248
133	-0.207
153	-0.185
173	-0.161
193	-0.136
233	-0.083
273	-0.027

* Not shown in figure.

FIGURE AND TABLE NO. 251R. PROVISIONAL VALUES FOR THERMAL LINEAR EXPANSION OF COPPER + SILICON + ΣX_i ALLOYS Cu + Si + ΣX_i

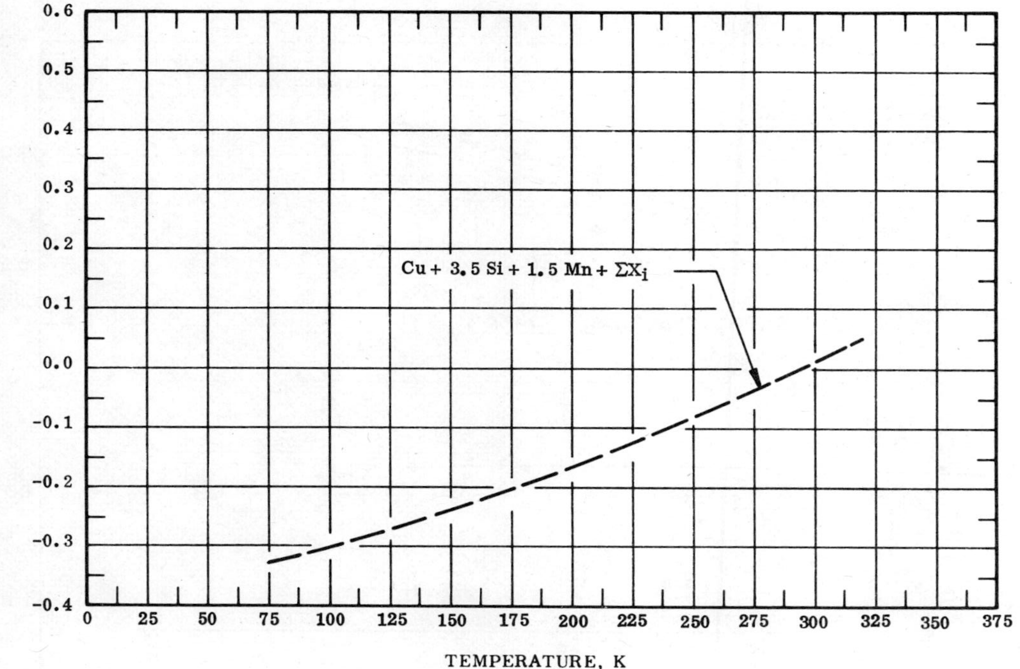

PROVISIONAL VALUES

[Temperature, T, K; Linear Expansion, $\Delta L/L_0$, %; α, K^{-1}]

(Cu + 3.5 Si + 1.5 Mn + ΣX_i)		
T	$\Delta L/L_0$	$\alpha \times 10^6$
75	-0.328	10.9
100	-0.300	11.8
200	-0.162	15.6
293	0.000	19.4
320	0.054	20.6

REMARKS

(Cu + 3.5 Si + 1.5 Mn + ΣX_i): The tabulated values are for well-annealed alloy of type Russian Br. KMts 3-1 and are considered accurate to within ± 7% over the entire temperature range. These values can be represented approximately by the following equation:

$$\Delta L/L_0 = -0.400 + 8.441 \times 10^{-4}\, T + 1.576 \times 10^{-6}\, T^2 + 6.732 \times 10^{-10}\, T^3$$

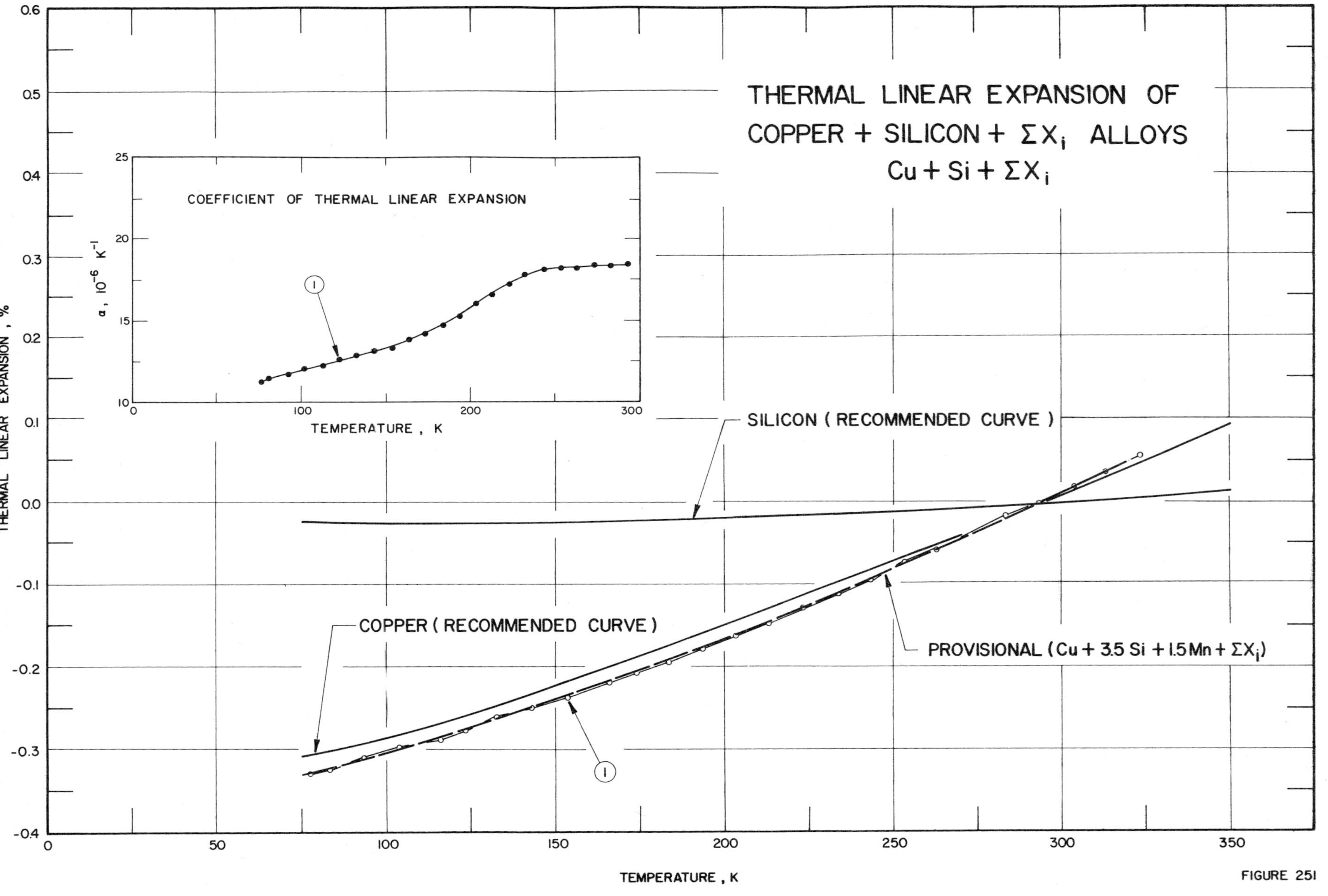

THERMAL LINEAR EXPANSION OF
COPPER + SILICON + ΣX_i ALLOYS
Cu + Si + ΣX_i
COEFFICIENT OF THERMAL LINEAR EXPANSION
α, 10^{-6} K^{-1}
25
20
15
10
0
100
200
300
TEMPERATURE , K
1
SILICON (RECOMMENDED CURVE)
COPPER (RECOMMENDED CURVE)
PROVISIONAL (Cu + 3.5 Si + 1.5 Mn + ΣX_i)
1
THERMAL LINEAR EXPANSION , %
0.6
0.5
0.4
0.3
0.2
0.1
0.0
-0.1
-0.2
-0.3
-0.4
0
50
100
150
200
250
300
350
TEMPERATURE , K

FIGURE 251

SPECIFICATION TABLE 251. THERMAL LINEAR EXPANSION OF COPPER + SILICON + ΣX_i ALLOYS $Cu + Si + \Sigma X_i$

Cur. No.	Ref. No.	Author(s)	Year	Method Used	Temp. Range, K	Name and Specimen Designation	Composition (weight percent) Cu	Si	Mn	Sn	Zn	Composition (continued), Specifications, and Remarks
1	551	Belov, A.K.	1968	L	77-323	Br. KMts 3-1 Russian alloy	Bal.	3.5	1.5	0.25	0.5	Cylindrical specimen 3.5 mm in diameter by 50 mm long; sample heated and cooled at a rate of 3-7 deg per min.

DATA TABLE 251. THERMAL LINEAR EXPANSION OF COPPER + SILICON + ΣX_i ALLOYS $Cu + Si + \Sigma X_i$

[Temperature, T, K; Linear Expansion, $\Delta L/L_0$, %]

T	$\Delta L/L_0$	T	$\Delta L/L_0$
CURVE 1‡		CURVE 1 (cont.)‡	
77	-0.326	203	-0.160
83	-0.320	213	-0.144
93	-0.308	223	-0.127
103	-0.296	233	-0.110
113	-0.284	243	-0.092
123	-0.272	253	-0.073
133	-0.259	263	-0.055
143	-0.246	273	-0.037
153	-0.233	283	-0.018
163	-0.219	293	0.000
173	-0.205	303	0.019
183	-0.191	313	0.037
193	-0.176	323	0.056

DATA TABLE 251. COEFFICIENT OF THERMAL LINEAR EXPANSION OF COPPER + SILICON + ΣX_i ALLOYS $Cu + Si + \Sigma X_i$

[Temperature, T, K; Coefficient of Expansion, α, 10^{-6} K^{-1}]

T	α	T	α	T	α	T	α
CURVE 1‡		CURVE 1 (cont.)‡		CURVE 1 (cont.)‡		CURVE 1 (cont.)‡	
77	11.2	143	13.1	213	16.6	283	18.4
83	11.4	153	13.4	223	17.3	293	18.5
93	11.7	163	13.6	233	17.8	303	18.6*
103	12.0	173	14.1	243	18.1	313	18.7*
113	12.3	183	14.6	253	18.2	323	18.8*
123	12.6	193	15.3	263	18.3		
133	12.8	203	16.0	273	18.4		

*Not shown in figure.
‡Author's data for coefficient of thermal expansion have been integrated by TPRC to obtain $\Delta L/L_0$.

FIGURE AND TABLE NO. 252R. PROVISIONAL VALUES FOR THERMAL LINEAR EXPANSION OF COPPER + TIN + ΣX_i ALLOYS   Cu + Sn + ΣX_i

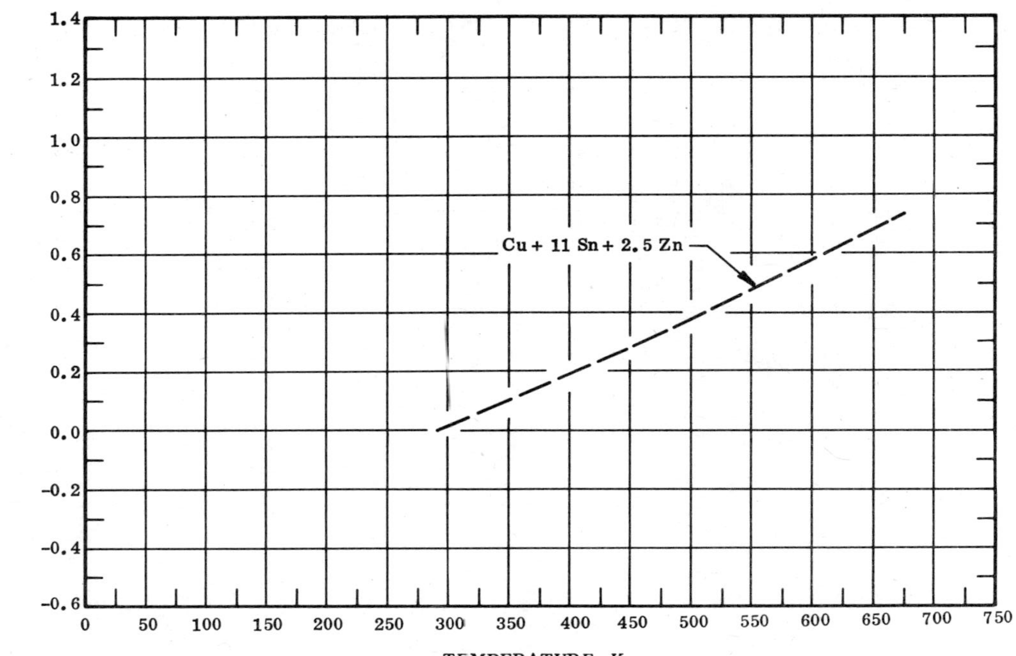

PROVISIONAL VALUES

[Temperature, T, K; Linear Expansion, $\Delta L/L_0$, %; α, K^{-1}]

T	(Cu + 11 Sn + 2.5 Zn) $\Delta L/L_0$	$\alpha \times 10^6$
293	0.000	17.0
400	0.190	18.3
500	0.377	19.3
600	0.574	19.9
675	0.724	20.2

REMARKS

(Cu + 11 Sn + 2.5 Zn): The tabulated values are for well-annealed alloy of type bronze and are considered accurate to within ±7% over the entire temperature range. These values can be represented approximately by the following equation:

$$\Delta L/L_0 = -0.429 + 1.187 \times 10^{-3} T + 1.081 \times 10^{-6} T^2 - 4.566 \times 10^{-10} T^3$$

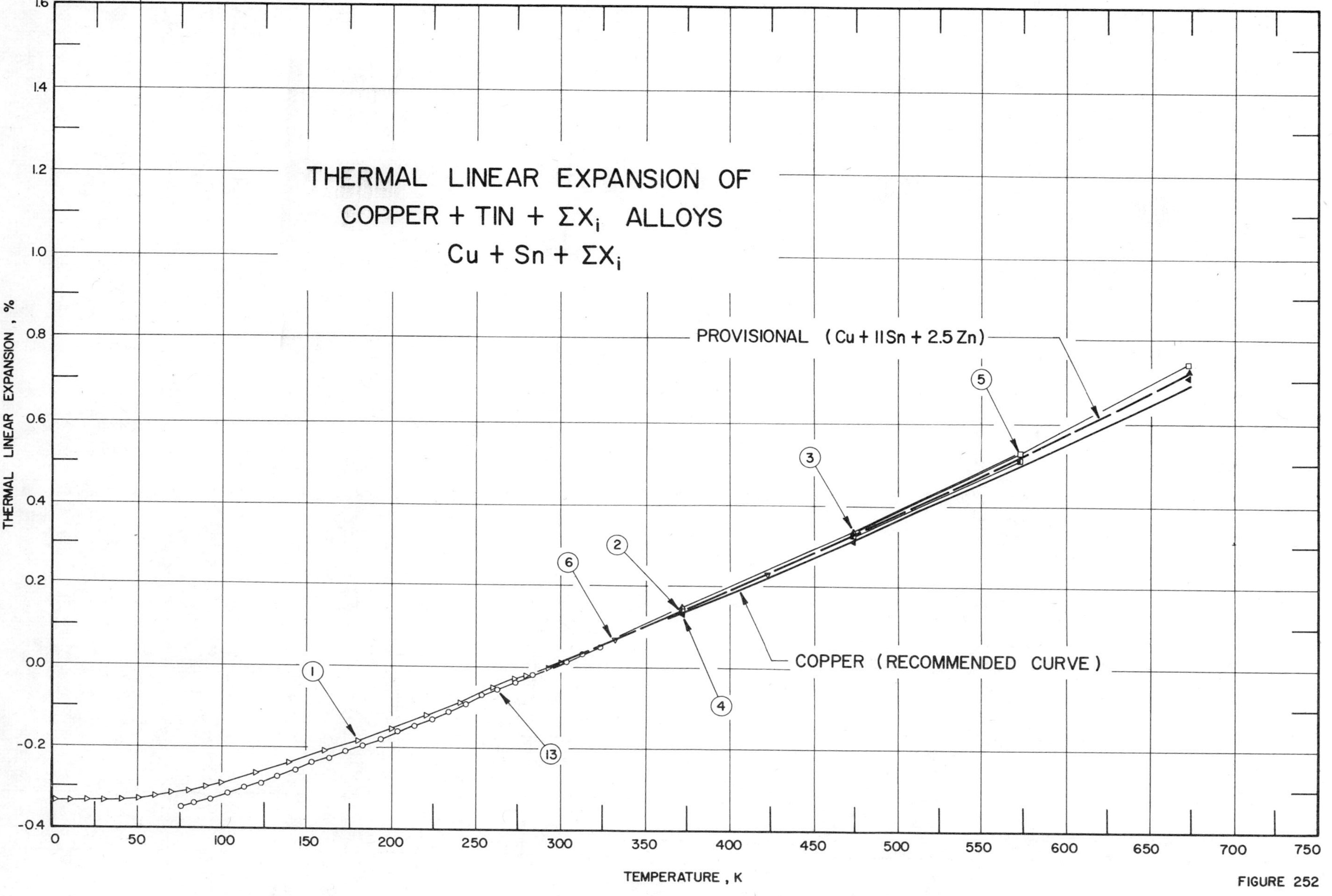
THERMAL LINEAR EXPANSION OF
COPPER + TIN + ΣXi ALLOYS
Cu + Sn + ΣXi
PROVISIONAL (Cu + 11Sn + 2.5 Zn)
COPPER (RECOMMENDED CURVE)
1
2
3
4
5
6
13
THERMAL LINEAR EXPANSION, %
1.6
1.4
1.2
1.0
0.8
0.6
0.4
0.2
0.0
-0.2
-0.4
0
50
100
150
200
250
300
350
400
450
500
550
600
650
700
750
TEMPERATURE, K

FIGURE 252

SPECIFICATION TABLE 252. THERMAL LINEAR EXPANSION OF COPPER + TIN + ΣX_i ALLOYS $Cu + Sn + \Sigma X_i$

Cur. No.	Ref. No.	Author(s)	Year	Method Used	Temp. Range, K	Name and Specimen Designation	Composition (weight percent) Cu	Sn	P	Zn	Pb	Ni	Composition (continued), Specifications, and Remarks
1	467	Clark, A. F.	1968	L	0-300	Phosphor Bronze A	95.93	4.85	0.18	<0.1	<0.1		<0.1 Fe; specimen machined to 0.64 cm square x 20.32 cm long; Rockwell hardness B-91.
2	427	Souder, W., Hidnert, P., and Fox, J. F.	1934	N	293-573	Bronze Sample 894A	85.5/87.5	10.5/11.5		2.0/3.0			0.2 maximum impurities; 150 mm long with pointed ends, maximum diameter 12 mm.
3	478	Hidnert, P. and Dickson, G.	1943	L	873-293	Bronze Sample 1444A	84.84	14.95			0.21		Cast rod with 2.22 cm diameter; expansion measured with decreasing temperature.
4	412	Cook, M. and Tallis, W. G.	1941	L	293-673	Phosphor Bronze Alloy No. 2	96.50	3.09	0.39		<0.005	0.01	0.01 Fe; annealed cylinder 5.1 cm x 1.3 cm diameter.
5	412	Cook, M. and Tallis, W. G.	1941	L	293-673	Phosphor Bronze Alloy No. 4	92.20	7.41	0.38		0.01		0.02 Fe; similar to the above specimen.
6	413	Hidnert, P.	1934	L	293-473	No. 1474	83.80	11.96	0.02		4.22		Specimen prepared from virgin materials, cast at 1422 K, machined to 300 mm long x 9 mm diameter.
7*	413	Hidnert, P.	1934	L	293-473	Sample 1480	80.02	9.58	0.04	0.60	9.45	0.15	0.15 Sb; similar to the above specimen except from commercial materials.
8*	413	Hidnert, P.	1934	L	293-473	Sample 1482	88.63	8.58		0.34	2.08	0.25	0.15 Sb; similar to the above specimen.
9*	413	Hidnert, P.	1934	L	293-423	Sample 1483	86.75	9.97	0.03	2.75	0.53		Similar to the above specimen; expansion measured on first test.
10*	413	Hidnert, P.	1934	L	293-333	Sample 1483	86.75	9.97	0.03	2.75	0.53		The above specimen; expansion measured on second test.
11*	413	Hidnert, P.	1934	L	293-423	Sample 1483	86.75	9.97	0.03	2.75	0.53		The above specimen; expansion measured on third test.
12*	413	Hidnert, P.	1934	L	293-473	Sample 1483	86.75	9.97	0.03	2.75	0.53		The above specimen; expansion measured on fourth test.
13	551	Belov, A. K.	1968	L	77-323	Br OF 10-1 Russian Cu alloy	Bal.	11	1.2	0.3	0.2		0.5 Sb, 0.2 Fe; cylindrical specimen 3.5 mm in diameter; sample heated and cooled at a rate of 3-7 deg per min; data on the coefficient of thermal linear expansion are also reported.

* Not shown in figure.

DATA TABLE 252. THERMAL LINEAR EXPANSION OF COPPER + TIN + ΣX_i ALLOYS Cu + Sn + ΣX_i

[Temperature, T, K; Linear Expansion, $\Delta L/L_0$, %]

T	$\Delta L/L_0$
CURVE 1	
0	-0.330
10	-0.330
20	-0.330
30	-0.329
40	-0.327
50	-0.324
60	-0.319
70	-0.312
80	-0.304
90	-0.295
100	-0.285
120	-0.262
140	-0.237
160	-0.209
180	-0.180
200	-0.150
220	-0.119
240	-0.087
260	-0.054
273	-0.033
280	-0.021
293	-0.000
300	0.013
CURVE 2	
373	0.136
473	0.326
573	0.515
CURVE 3	
873	1.160*
773	0.941
673	0.726
573	0.524*
473	0.328
373	0.142
293	0.000*
CURVE 4	
293	0.000*
373	0.133
473	0.315
573	0.510*
673	0.714
CURVE 5	
293	0.000*
373	0.139*
473	0.329*
573	0.532
673	0.741
CURVE 6	
293	0.000*
333	0.070
373	0.143*
423	0.235
473	0.329*
CURVE 7*	
293	0.000
333	0.072
373	0.147
423	0.239
473	0.335
CURVE 8*	
293	0.000
333	0.069
373	0.140
423	0.229
473	0.322
CURVE 9*	
293	0.000
333	0.070
373	0.142
423	0.235
CURVE 10*	
293	0.000
333	0.071
CURVE 11*	
293	0.000
333	0.070
373	0.142
423	0.234
CURVE 12*	
293	0.000
333	0.070
373	0.143
423	0.235
473	0.329
CURVE 13‡	
77	-0.345
83	-0.338
93	-0.325
103	-0.311
113	-0.297
123	-0.283
133	-0.268
143	-0.254
153	-0.239
163	-0.224
173	-0.208
183	-0.193
193	-0.176
203	-0.159
213	-0.142
223	-0.124
233	-0.107
243	-0.089
253	-0.071
263	-0.054
273	-0.036
283	-0.018
293	0.000*
303	0.018
313	0.036
323	0.054

DATA TABLE 252. COEFFICIENT OF THERMAL LINEAR EXPANSION OF COPPER + TIN + ΣX_i ALLOYS Cu + Sn + ΣX_i

[Temperature, T, K; Coefficient of Expansion, α, 10^{-6} K^{-1}]

T	α
CURVE 13‡	
77	12.39
83	12.70
93	13.19
103	13.70
113	14.10
123	14.50
133	14.10
143	14.80
153	14.99
163	15.20
173	15.50
183	16.00
193	16.70
203	17.19
213	17.50
223	17.59
233	17.65
243	17.70
253	17.75
263	17.80
273	17.85
283	17.90
293	17.94
303	17.99
313	18.05
323	18.10

*Not shown in figure.
‡Author's data for coefficient of thermal expansion have been integrated by TPRC to obtain $\Delta L/L_0$.

FIGURE AND TABLE NO. 253R. PROVISIONAL VALUES FOR THERMAL LINEAR EXPANSION OF COPPER + ZINC + ΣX_i ALLOYS   Cu + Zn + ΣX_i

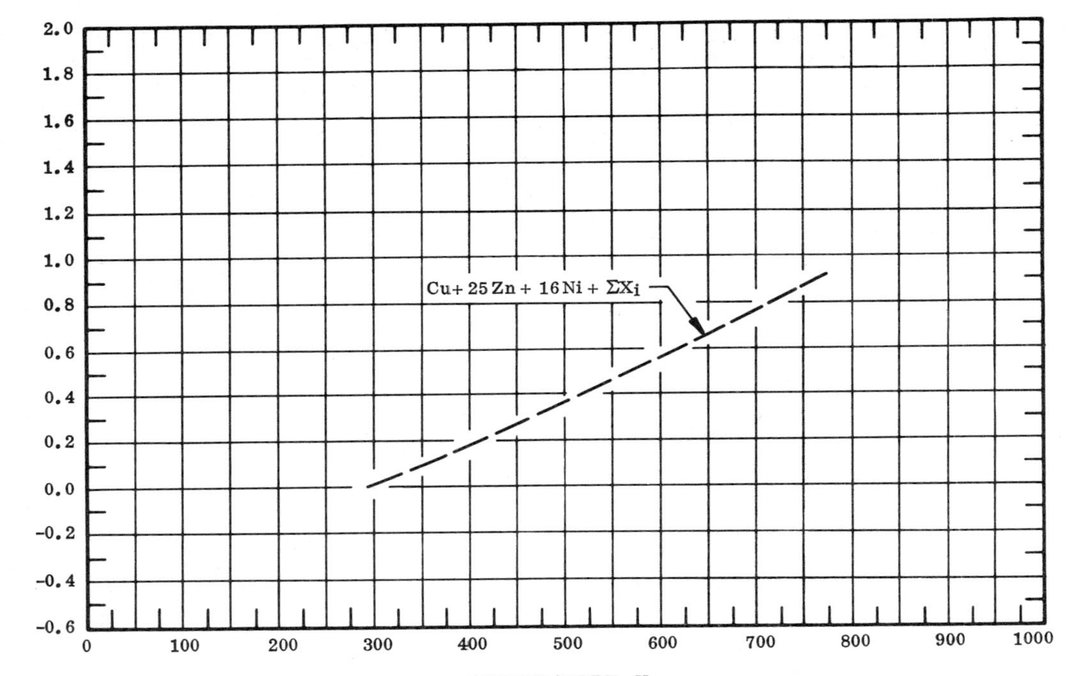

PROVISIONAL VALUES

[Temperature, T, K; Linear Expansion, $\Delta L/L_0$, %; α, K^{-1}]

(Cu + 25 Zn + 16 Ni + ΣX_i)		
T	$\Delta L/L_0$	$\alpha \times 10^6$
293	0.000	17.3
400	0.186	17.5
500	0.365	18.2
600	0.551	19.4
700	0.756	21.4
775	0.922	23.2

REMARKS

(Cu + 25 Zn + 16 Ni + ΣX_i): The tabulated values for well-annealed alloy of type Nickel Silver are considered accurate to within ± 7% over the entire temperature range. These values can be represented approximately by the following equation:

$$\Delta L/L_0 = -0.555 + 2.127 \times 10^{-3}\, T - 1.106 \times 10^{-6}\, T^2 + 1.060 \times 10^{-9}\, T^3$$

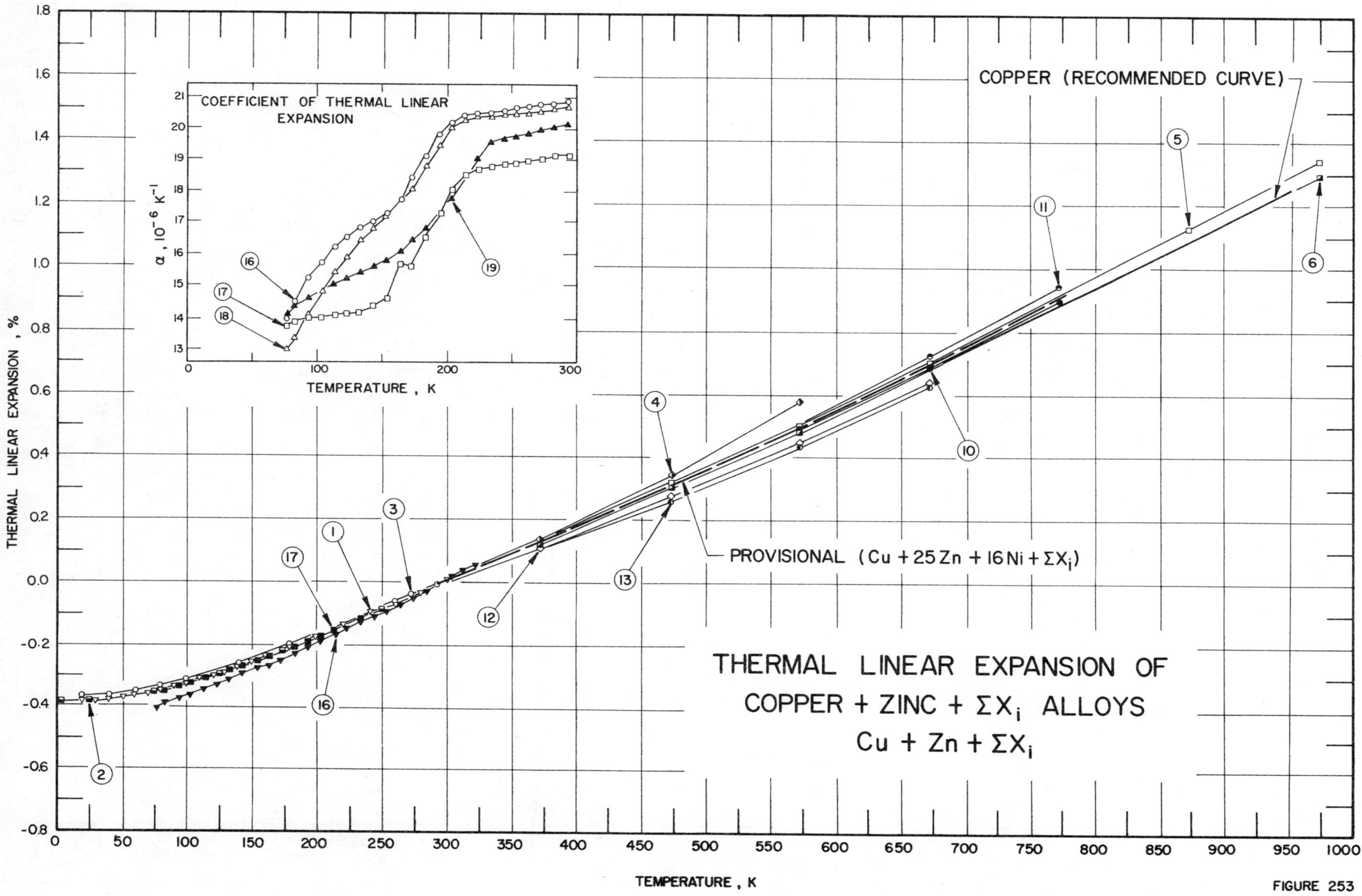
THERMAL LINEAR EXPANSION OF
COPPER + ZINC + ΣX_i ALLOYS
Cu + Zn + ΣX_i
COPPER (RECOMMENDED CURVE)
PROVISIONAL (Cu + 25 Zn + 16 Ni + ΣX_i)
THERMAL LINEAR EXPANSION, %
TEMPERATURE, K
COEFFICIENT OF THERMAL LINEAR EXPANSION
α, 10⁻⁶ K⁻¹
TEMPERATURE, K
1
2
3
4
5
6
10
11
12
13
16
17
18
19
1.8
1.6
1.4
1.2
1.0
0.8
0.6
0.4
0.2
0.0
-0.2
-0.4
-0.6
-0.8
0
50
100
150
200
250
300
350
400
450
500
550
600
650
700
750
800
850
900
950
1000
21
20
19
18
17
16
15
14
13
0
100
200
300

FIGURE 253

SPECIFICATION TABLE 253. THERMAL LINEAR EXPANSION OF COPPER + ZINC + ΣX_i ALLOYS $Cu + Zn + \Sigma X_i$

Cur. No.	Ref. No.	Author(s)	Year	Method Used	Temp. Range, K	Name and Specimen Designation	Composition (weight percent) Cu	Zn	Al	Mn	Fe	Sn	Pb	Composition (continued), Specifications, and Remarks
1	481	Johnston, H. L., Altman, H. W., and Rubin, T.	1965	I	18-300	ASTM B 16 Yellow Brass								Specimen from Williams and Co. consisted of three sectors cut from bar, filed to length to within 0.20 of wavelength of sodium D radiation; data smoothed by author.
2	302	Beenakker, Y. M. and Swenson, C. A.	1955	L	4.2-300	Brass								Technical grade.
3	462	Arp, V., Wilson, J. H., Winrich, L. and Sikora, P.	1962	L	20-293	Brass 70/30	70.3	29.6						Hardness R_B88; 0.750 hard.
4	478	Hidnert, P. and Dickson, G.	1943	L	293-573	Manganese Bronze	66.2	Bal.	4.8	3.5	2.6			Rod 0.95 cm diameter machined from 3.49 cm diameter casting.
5	478	Hidnert, P. and Dickson, G.	1943	L	973-293	Red Brass	84.96	5.15				5.02	4.87	Cast rod; expansion measured with decreasing temperature.
6	478	Hidnert, P. and Dickson, G.	1943	L	293-1073	Sample 683	65	22		0.5	1.5			11 Ni; cold rolled, cold drawn rod 0.635 cm diameter.
7*	478	Hidnert, P. and Dickson, G.	1943	L	293-573	Sample 280	63.60	Bal.			1.72		0.05	13.18 Ni; rod 1.27 cm diameter.
8*	478	Hidnert, P. and Dickson, G.	1943	L	293-473	Sample 281	63.60	Bal.			1.72		0.05	13.18 Ni; rod 1.27 cm diameter; expansion measured with increasing temperature.
9*	478	Hidnert, P. and Dickson, G.	1943	L	473-293	Sample 281	63.60	Bal.			1.72		0.05	13.18 Ni; the above specimen; expansion measured with decreasing temperature.
10	478	Hidnert, P. and Dickson, G.	1943	L	293-773	(Nickel Silver)	58.4	Bal.		0.15	0.22			15.7 Ni; tube specimen 0.476 cm outside diameter.
11	478	Hidnert, P. and Dickson, G.	1943	L	773-293	(Nickel Silver)	58.4	Bal.		0.15	0.22			15.7 Ni; the above specimen; expansion measured with decreasing temperature.
12	480	Cook, M.	1936	L	273-673	(Nickel Silver)	62.62	27.14		0.13	0.04		0.005	10.05 Ni, <0.1 each any others; annealed rod 12.7 cm x 1.6 cm diameter.
13	480	Cook, M.	1936	L	273-673	Alloy No. 2	63.47	24.31		0.13	0.04		0.005	12.33 Ni; similar to the above specimen.
14*	480	Cook, M.	1936	L	273-673	Alloy No. 3	62.43	22.08		0.10	0.04		0.003	15.35 Ni, 0.014 C, <0.005 any others; similar to the above specimen.
15*	480	Cook, M.	1936	L	273-673	Alloy No. 4	62.05	19.36		0.12	0.07		0.004	18.40 Ni, <0.01 each any others; similar to the above specimen.
16	551	Belov, A. K.	1968	L	77-323	LS 59-1 Russian alloy	Bal.	37			0.5		1.9	Cylindrical specimen 3.5 mm in diameter, 50 mm long; sample heated and cooled at a rate of 3-7 deg per min.

* Not shown in figure.

SPECIFICATION TABLE 253. THERMAL LINEAR EXPANSION OF COPPER + ZINC + ΣX_i ALLOYS $Cu + Zn + \Sigma X_i$ (continued)

Cur. No.	Ref. No.	Author(s)	Year	Method Used	Temp. Range, K	Name and Specimen Designation	Composition (weight percent) Cu	Zn	Al	Mn	Fe	Sn	Pb	Composition (continued), Specifications, and Remarks
17	551	Belov, A.K.	1968	L	77-323	LK 80-3L Russian alloy	Bal.	12		1.0	0.6	0.3	0.5	4.5 Si; similar to the above specimen.
18	551	Belov, A.K.	1968	L	77-323	LZhMts 59-1-1 Russian alloy	Bal.	40	0.2	0.8	1.2	0.7	0.2	Similar to the above specimen.
19	551	Belov, A.K.	1968	L	77-323	L62 Russian alloy	Bal.	39			0.2			Similar to the above specimen.

DATA TABLE 253. THERMAL LINEAR EXPANSION OF COPPER + ZINC + ΣX_i ALLOYS Cu + Zn + ΣX_i

[Temperature, T, K; Linear Expansion, $\Delta L/L_0$, %]

T	$\Delta L/L_0$
CURVE 1	
20	-0.383
30	-0.382
40	-0.380
50	-0.375
60	-0.368
70	-0.360
80	-0.350
90	-0.339
100	-0.326
110	-0.313
120	-0.299
130	-0.284
140	-0.269
150	-0.253
160	-0.237
170	-0.221*
180	-0.204
190	-0.187*
200	-0.170
210	-0.152*
220	-0.134
230	-0.117*
240	-0.099
250	-0.080*
260	-0.062
270	-0.043*
280	-0.025
290	-0.006*
300	0.013
CURVE 2	
4.2	-0.383
25	-0.383
50	-0.375*
75	-0.354
100	-0.326*
125	-0.292
150	-0.254*
175	-0.215
200	-0.172*
250	-0.080
300	0.012*
CURVE 3	
20	-0.369
40	-0.366
60	-0.355
80	-0.337
100	-0.313
120	-0.288*
140	-0.260
160	-0.229*
180	-0.196
200	-0.163*
220	-0.128*
240	-0.093*
260	-0.058
273	-0.035
280	-0.023*
293	0.000
CURVE 4	
293	0.000*
373	0.150
473	0.356
573	0.582
CURVE 5	
973	1.346
873	1.125
673	0.707
573	0.510
473	0.324
373	0.142*
293	0.000*
CURVE 6	
293	0.000*
373	0.132
473	0.306
573	0.487
973	1.300
1073	1.537
CURVE 7*	
293	0.000
373	0.138
473	0.315
573	0.504
CURVE 8*	
293	0.000
373	0.138
473	0.317
CURVE 9*	
473	0.320
373	0.139
293	0.000
CURVE 10	
293	0.000*
373	0.139*
573	0.498
673	0.695
773	0.907
CURVE 11	
773	0.946
673	0.726
473	0.313*
293	0.000*
CURVE 12	
273	-0.030*
373	0.120
473	0.284
573	0.459
673	0.646
CURVE 13	
273	-0.030*
373	0.118*
473	0.277
573	0.454
673	0.638
CURVE 14*	
273	-0.030
373	0.120
473	0.281
573	0.454
673	0.638
CURVE 15*	
273	-0.030
373	0.118
473	0.277
573	0.448
673	0.638
CURVE 16‡	
77	-0.401
83	-0.393
93	-0.378
103	-0.362
113	-0.346
123	-0.330
133	-0.313
143	-0.297
153	-0.279
163	-0.262
173	-0.244
183	-0.225
193	-0.206
203	-0.186
213	-0.165
223	-0.145
233	-0.124
243	-0.104
253	-0.083
CURVE 16 (cont.)‡	
263	-0.062
273	-0.042
283	-0.021
293	0.000*
303	0.021
313	0.042
323	0.063
CURVE 17‡	
77	-0.359*
83	-0.350
93	-0.336
103	-0.322
113	-0.308
123	-0.294*
133	-0.280
143	-0.266
153	-0.251
163	-0.236
173	-0.220*
183	-0.204
193	-0.187
203	-0.170
213	-0.151
223	-0.133*
233	-0.114
243	-0.095*
253	-0.076*
263	-0.057*
273	-0.038*
283	-0.019*
293	0.000*
303	0.019*
313	0.038*
323	0.058*
CURVE 18*, ‡	
77	-0.394
83	-0.386
93	-0.372
103	-0.357
CURVE 18 (cont.)*, ‡	
113	-0.342
123	-0.327
133	-0.311
143	-0.294
153	-0.277
163	-0.259
173	-0.242
183	-0.223
193	-0.204
203	-0.184
213	-0.164
223	-0.144
233	-0.123
243	-0.103
253	-0.082
263	-0.062
273	-0.041
283	-0.021
293	0.000
303	0.021
313	0.042
323	0.063
CURVE 19*, ‡	
77	-0.373
83	-0.365
93	-0.350
103	-0.336
113	-0.321
123	-0.306
133	-0.290
143	-0.275
153	-0.259
163	-0.243
173	-0.227
183	-0.210
193	-0.193
203	-0.176
213	-0.158
223	-0.139
233	-0.119
243	-0.100

* Not shown in figure.

‡ Author's data for coefficient of thermal expansion have been integrated by TPRC to obtain $\Delta L/L_0$.

DATA TABLE 253. THERMAL LINEAR EXPANSION OF COPPER + ZINC + ΣX_i ALLOYS Cu + Zn + ΣX_i (continued)

T	$\Delta L/L_0$
CURVE 19 (cont.)*, ‡	
253	-0.080
263	-0.060
273	-0.040
283	-0.020
293	0.000
303	0.020
313	0.041
323	0.061

DATA TABLE COEFFICIENT OF THERMAL LINEAR EXPANSION OF COPPER + ZINC + ΣX_i ALLOYS Cu + Zn + ΣX_i

[Temperature, T, K; Coefficient of Expansion, α, 10^{-6} K^{-1}]

T	α
CURVE 16‡	
77	14.00
83	14.50
93	15.20
103	15.70
113	16.20
123	16.49
133	16.79
143	17.00
153	17.30
163	17.70
173	18.40
183	19.09
193	19.79
203	20.19
213	20.40
223	20.50
233	20.54
CURVE 16 (cont.)‡	
243	20.60
253	20.70
263	20.75
273	20.79
283	20.84
293	20.89
303	20.94*
313	21.00*
323	21.10*
CURVE 17‡	
77	13.80
83	13.89
93	14.00
103	14.05
113	14.10
123	14.15
CURVE 17 (cont.)‡	
133	14.20
143	14.40
153	14.60
163	15.70
173	15.60
183	16.49
193	17.30
203	18.10
213	18.50
223	18.69
233	18.80
243	18.90
253	18.95
263	19.00
273	19.05
283	19.19
293	19.15
303	19.20*
313	19.30*
323	19.40*
CURVE 18‡	
77	13.00
83	13.40
93	14.10
103	14.80
113	15.39
123	15.90
133	16.40
143	16.79
153	17.19
163	17.70*
173	18.10
183	18.80
193	19.50
203	20.09
213	20.30
223	20.35
233	20.40
243	20.45
253	20.50
263	20.54
CURVE 18 (cont.)‡	
273	20.60
283	20.65
293	20.70
303	20.79*
313	20.89*
323	21.00*
CURVE 19‡	
77	14.20
83	14.40
93	14.60
103	14.80*
113	14.99
123	15.20
133	15.39
143	15.60
153	15.79
163	16.09
173	16.40
CURVE 19 (cont.)‡	
183	16.79
193	17.30*
203	17.80
213	18.50*
223	19.09
233	19.60
243	19.70
253	19.79
263	19.90
273	20.00
283	20.09
293	20.19
303	20.30*
313	20.35*
323	20.40*

* Not shown in figure.
‡ Author's data for coefficient of thermal expansion have been integrated by TPRC to obtain $\Delta L/L_0$.

SPECIFICATION TABLE 254. THERMAL LINEAR EXPANSION OF DYSPROSIUM + TANTALUM + ΣX_i ALLOYS $Dy + Ta + \Sigma X_i$

Cur. No.	Ref. No.	Author(s)	Year	Method Used	Temp. Range, K	Name and Specimen Designation	Composition (weight percent) Dy	Ta	Ca	Composition (continued), Specifications, and Remarks
1*	78	Barson, F., Legvold, S., and Spedding, F.H.	1956	L	317-1262		99.0	0.5	0.2	0.1 Tb, 0.05 Ho, 0.02 Er, 0.02 Si, 0.0095 C, 0.005 Fe, 0.003 N_2; specimen 0.6 cm diameter, 6 cm long; fluorides of Dy bomb reduced with Ca metal to produce compact metallic sample, then vacuum cast, turned to shape; expansion measured with increasing temperature.
2*	78	Barson, F., et al.	1956	L	1253-307		99.0	0.5	0.2	The above specimen; expansion measured with decreasing temperature.

DATA TABLE 254. THERMAL LINEAR EXPANSION OF DYSPROSIUM + TANTALUM + ΣX_i ALLOYS $Dy + Ta + \Sigma X_i$

[Temperature, T, K; Linear Expansion, $\Delta L/L_0$, %]

T	$\Delta L/L_0$	T	$\Delta L/L_0$	T	$\Delta L/L_0$	T	$\Delta L/L_0$	T	$\Delta L/L_0$
CURVE 1*		CURVE 1 (cont.)*		CURVE 1 (cont.)*		CURVE 2 (cont.)*		CURVE 2 (cont.)*	
317	0.027	736	0.476	1240	1.170	948	0.735	508	0.208
335	0.047	763	0.509	1262	1.197	929	0.702	485	0.187
346	0.056	792	0.546			903	0.674	461	0.161
360	0.074	816	0.579	CURVE 2*		879	0.638	442	0.141
376	0.084	850	0.623			852	0.606	423	0.120
393	0.104	882	0.654	1253	1.190	822	0.568	403	0.100
410	0.120	913	0.699	1231	1.158	798	0.537	385	0.080
433	0.144	938	0.737	1214	1.125	773	0.507	369	0.066
454	0.160	967	0.779	1190	1.095	746	0.479	355	0.054
475	0.185	999	0.828	1167	1.057	717	0.444	343	0.039
500	0.204	1028	0.876	1142	1.016	689	0.411	328	0.026
523	0.228	1056	0.916	1114	0.978	659	0.379	307	0.006
549	0.257	1089	0.970	1094	0.946	632	0.346		
574	0.289	1119	1.002	1073	0.916	606	0.321		
624	0.342	1142	1.039	1051	0.884	599	0.312		
649	0.375	1166	1.075	1028	0.848	581	0.292		
676	0.403	1193	1.111	1004	0.812	556	0.289		
706	0.442	1221	1.145	975	0.768	530	0.235		

* No figure given.

SPECIFICATION TABLE 255. THERMAL LINEAR EXPANSION OF GOLD + SILVER + ΣX_i ALLOYS $Au + Ag + \Sigma X_i$

Cur. No.	Ref. No.	Author(s)	Year	Method Used	Temp. Range, K	Name and Specimen Designation	Composition (weight percent) Au	Ag	Pd	Composition (continued), Specifications, and Remarks
1*	654	Nagender Naidu, S.V. and Houska, C.R.	1971	X	80-298		52	29	19	Prepared from 99.97 Au, 99.99 Ag, and 99.97 Pd from Engelhard Industries Inc. by melting in induction furnace under hydrogen atmosphere; homogenized for 1 week at 1173 K; annealed at 873 K in evacuated quartz tubes after flushing 3 times with argon; lattice parameter reported at 298 K is 4.0253 Å; 4.0249 Å at 293 K determined by graphical interpolation.

DATA TABLE 255. THERMAL LINEAR EXPANSION OF GOLD + SILVER + ΣX_i ALLOYS $Au + Ag + \Sigma X_i$

[Temperature, T, K; Linear Expansion, $\Delta L/L_0$, %]

T	$\Delta L/L_0$
CURVE 1*	
80	-0.283
195	-0.159
298	0.010

* No figure given.

FIGURE AND TABLE NO. 256AR. PROVISIONAL VALUES FOR THERMAL LINEAR EXPANSION OF IRIDIUM + OSMIUM + ΣX_i ALLOYS Ir + Os + ΣX_i

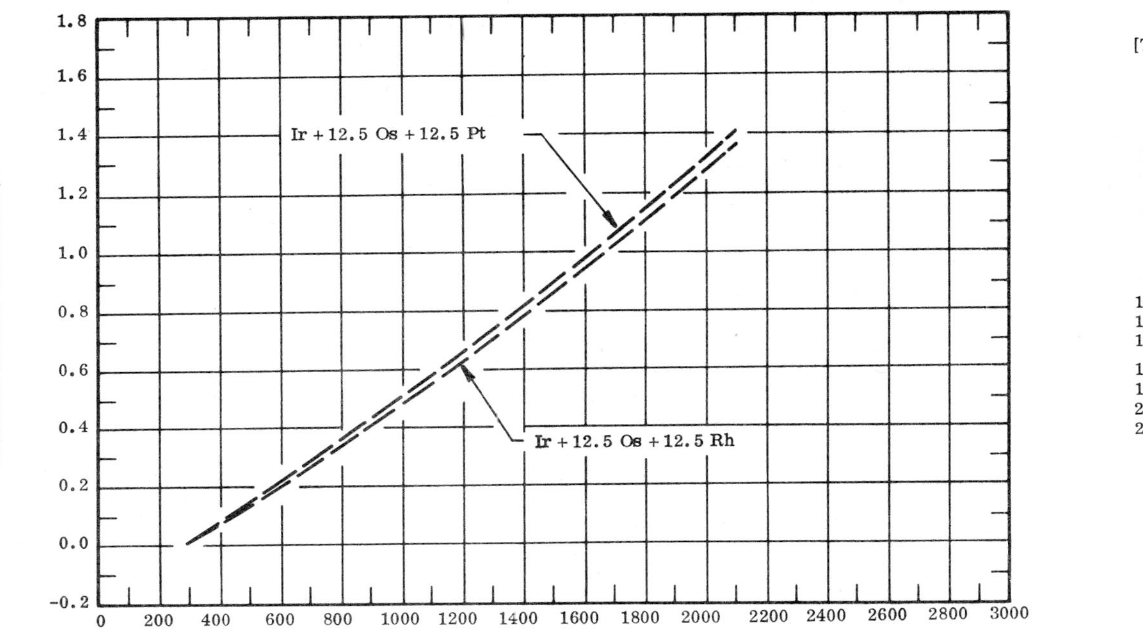

PROVISIONAL VALUES

[Temperature, T, K; Linear Expansion, $\Delta L/L_0$, %; α, K^{-1}]

T	(Ir + 12.5 Os + 12.5 Pt) $\Delta L/L_0$	$\alpha \times 10^6$	(Ir + 12.5 Os + 12.5 Rh) $\Delta L/L_0$	$\alpha \times 10^6$
293	0.000	6.7	0.000	6.5
400	0.072	6.8	0.069	6.6
500	0.141	7.0	0.136	6.7
600	0.213	7.1	0.204	6.8
700	0.284	7.2	0.272	6.9
800	0.355	7.3	0.341	7.0
900	0.429	7.4	0.411	7.1
1000	0.503	7.5	0.482	7.2
1200	0.655	7.7	0.629	7.5
1400	0.811	7.9	0.782	7.8
1600	0.973	8.2	0.941	8.1
1800	1.141	8.6	1.107	8.4
2000	1.317	9.1	1.278	8.7
2100	1.407	9.2	1.365	8.8

REMARKS

(Ir + 12.5 Os + 12.5 Pt): The tabulated values for well-annealed alloy are considered accurate to within ± 7% over the entire temperature range. These values can be represented approximately by the following equation:

$$\Delta L/L_0 = -0.198 + 6.634 \times 10^{-4} T + 2.728 \times 10^{-8} T^2 + 9.832 \times 10^{-12} T^3$$

(Ir + 12.5 Os + 12.5 Rh): The tabulated values for well-annealed alloy are considered accurate to within ± 7% over the entire temperature range. These values can be represented approximately by the following equation:

$$\Delta L/L_0 = -0.187 + 6.229 \times 10^{-4} T + 3.975 \times 10^{-8} T^2 + 7.254 \times 10^{-12} T^3$$

FIGURE AND TABLE NO. 256BR. PROVISIONAL VALUES FOR THERMAL LINEAR EXPANSION OF IRIDIUM + PLATINUM + ΣX_i Ir + Pt + ΣX_i

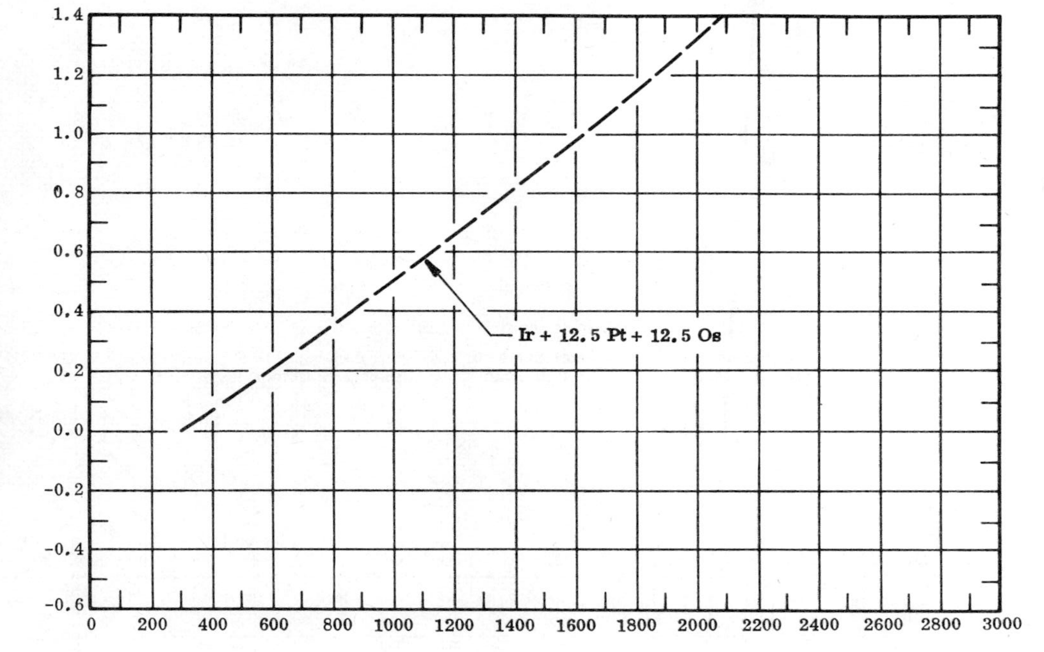

PROVISIONAL VALUES

[Temperature, T, K; Linear Expansion, $\Delta L/L_0$, %; α, K^{-1}]

(Ir + 12.5 Pt + 12.5 Os)

T	$\Delta L/L_0$	$\alpha \times 10^6$
293	0.000	6.7
400	0.072	6.8
500	0.141	7.0
600	0.213	7.1
700	0.284	7.2
800	0.355	7.3
900	0.429	7.4
1000	0.503	7.5
1200	0.655	7.7
1400	0.811	7.9
1600	0.973	8.2
1800	1.141	8.6
2000	1.317	9.1
2100	1.407	9.2

REMARKS

(Ir + 12.5 Pt + 12.5 Os): The tabulated values for well-annealed alloy are considered accurate to within ±7% over the entire temperature range. These values can be represented approximately by the following equation:

$$\Delta L/L_0 = -0.198 + 6.634 \times 10^{-4}\,T + 2.728 \times 10^{-6}\,T^2 + 9.832 \times 10^{12}\,T^3$$

FIGURE AND TABLE NO. 256CR. PROVISIONAL VALUES FOR THERMAL LINEAR EXPANSION OF IRIDIUM + RHENIUM + ΣX_i ALLOYS Ir + Re + ΣX_i

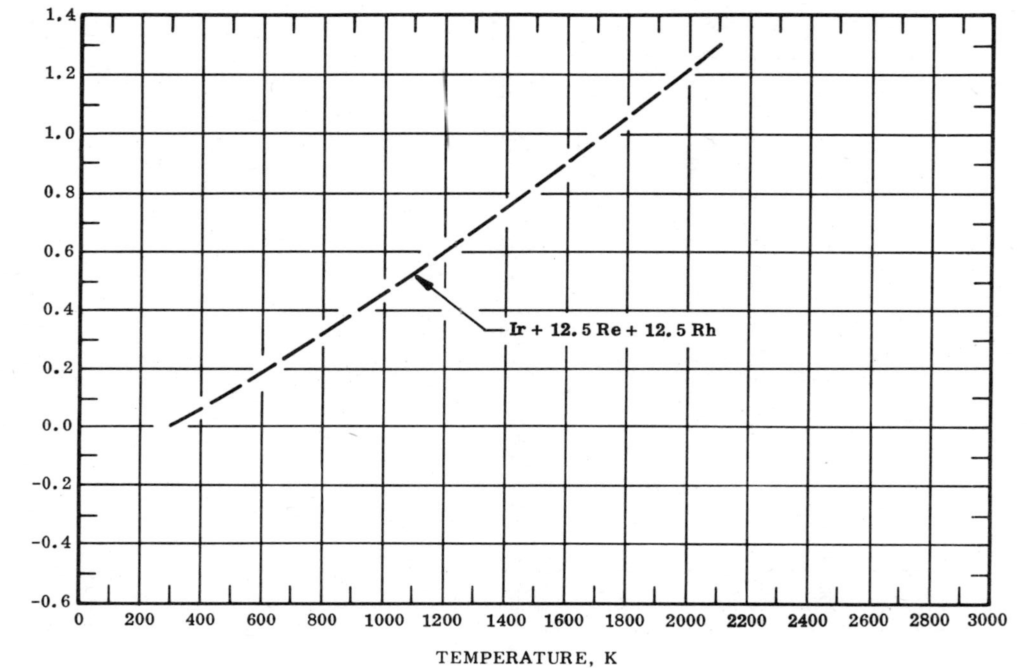

PROVISIONAL VALUES

[Temperature, T, K; Linear Expansion, $\Delta L/L_0$, %; α, K^{-1}]

(Ir + 12.5 Re + 12.5 Rh)

T	$\Delta L/L_0$	$\alpha \times 10^6$
293	0.000	6.0
400	0.065	6.2
500	0.128	6.4
600	0.193	6.6
700	0.260	6.7
800	0.327	6.8
900	0.396	6.9
1000	0.466	7.1
1200	0.608	7.2
1400	0.754	7.4
1600	0.904	7.6
1800	1.059	7.9
2000	1.221	8.2
2100	1.304	8.4

REMARKS

(Ir + 12.5 Re + 12.5 Rh): The tabulated values for well-annealed alloy are considered accurate to within ± 7% over the entire temperature range. These values can be represented approximately by the following equation:

$$\Delta L/L_0 = -0.180 + 5.930 \times 10^{-4}\, T + 5.080 \times 10^{-8}\, T^2 + 1.567 \times 10^{-12}\, T^3$$

FIGURE AND TABLE NO. 256DR. PROVISIONAL VALUES FOR THERMAL LINEAR EXPANSION OF IRIDIUM + RHODIUM + ΣX_i ALLOYS   Ir + Rh + ΣX_i

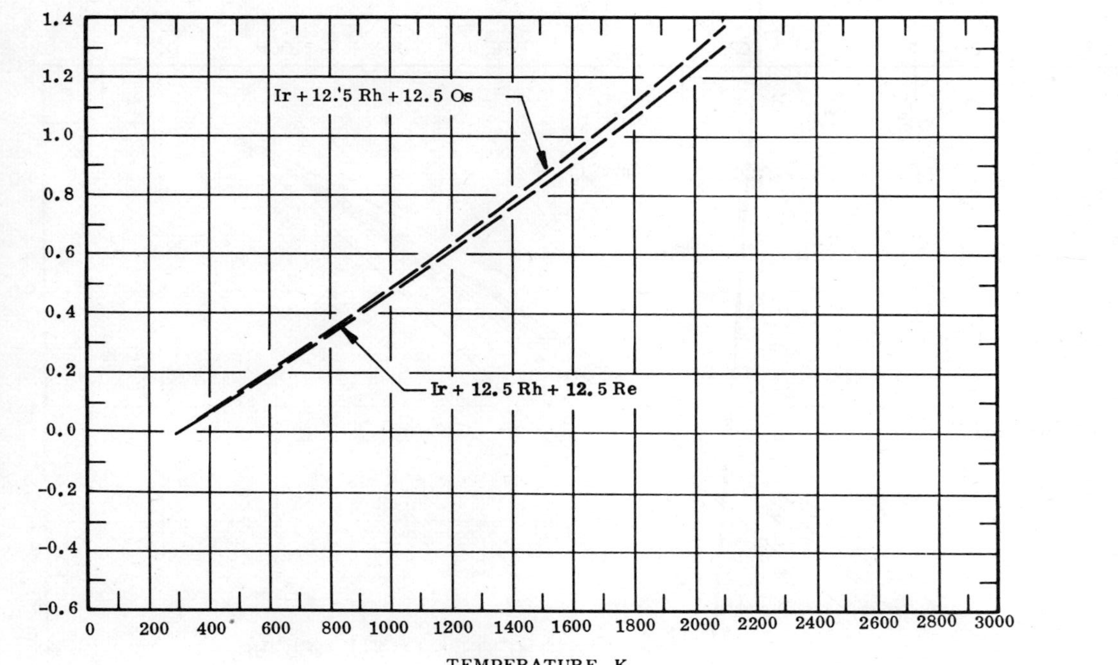

PROVISIONAL VALUES

[Temperature, T, K; Linear Expansion, $\Delta L/L_0$, %; α, K^{-1}]

T	(Ir + 12.5 Rh + 12.5 Os) $\Delta L/L_0$	$\alpha \times 10^6$	(Ir + 12.5 Rh + 12.5 Re) $\Delta L/L_0$	$\alpha \times 10^6$
293	0.000	6.5	0.000	6.0
400	0.069	6.6	0.065	6.2
500	0.136	6.7	0.128	6.4
600	0.204	6.8	0.193	6.6
700	0.272	6.9	0.260	6.7
800	0.341	7.0	0.327	6.8
900	0.411	7.1	0.396	6.9
1000	0.482	7.2	0.466	7.1
1200	0.629	7.5	0.608	7.2
1400	0.782	7.8	0.754	7.4
1600	0.941	8.1	0.904	7.6
1800	1.107	8.4	1.059	7.9
2000	1.278	8.7	1.221	8.2
2100	1.365	8.8	1.304	8.4

REMARKS

(Ir + 12.5 Rh + 12.5 Os): The tabulated values for well-annealed alloy are considered accurate to within ±7% over the entire temperature range. These values can be represented approximately by the following equation:

$$\Delta L/L_0 = -0.187 + 6.229 \times 10^{-4}\, T + 3.975 \times 10^{-8}\, T^2 + 7.254 \times 10^{-12}\, T^3$$

(Ir + 12.5 Rh + 12.5 Re): The tabulated values for well-annealed alloy are considered accurate to within ±7% over the entire temperature range. These values can be represented approximately by the following equation:

$$\Delta L/L_0 = -0.180 + 5.930 \times 10^{-4}\, T + 5.080 \times 10^{-8}\, T^2 + 1.567 \times 10^{-12}\, T^3$$

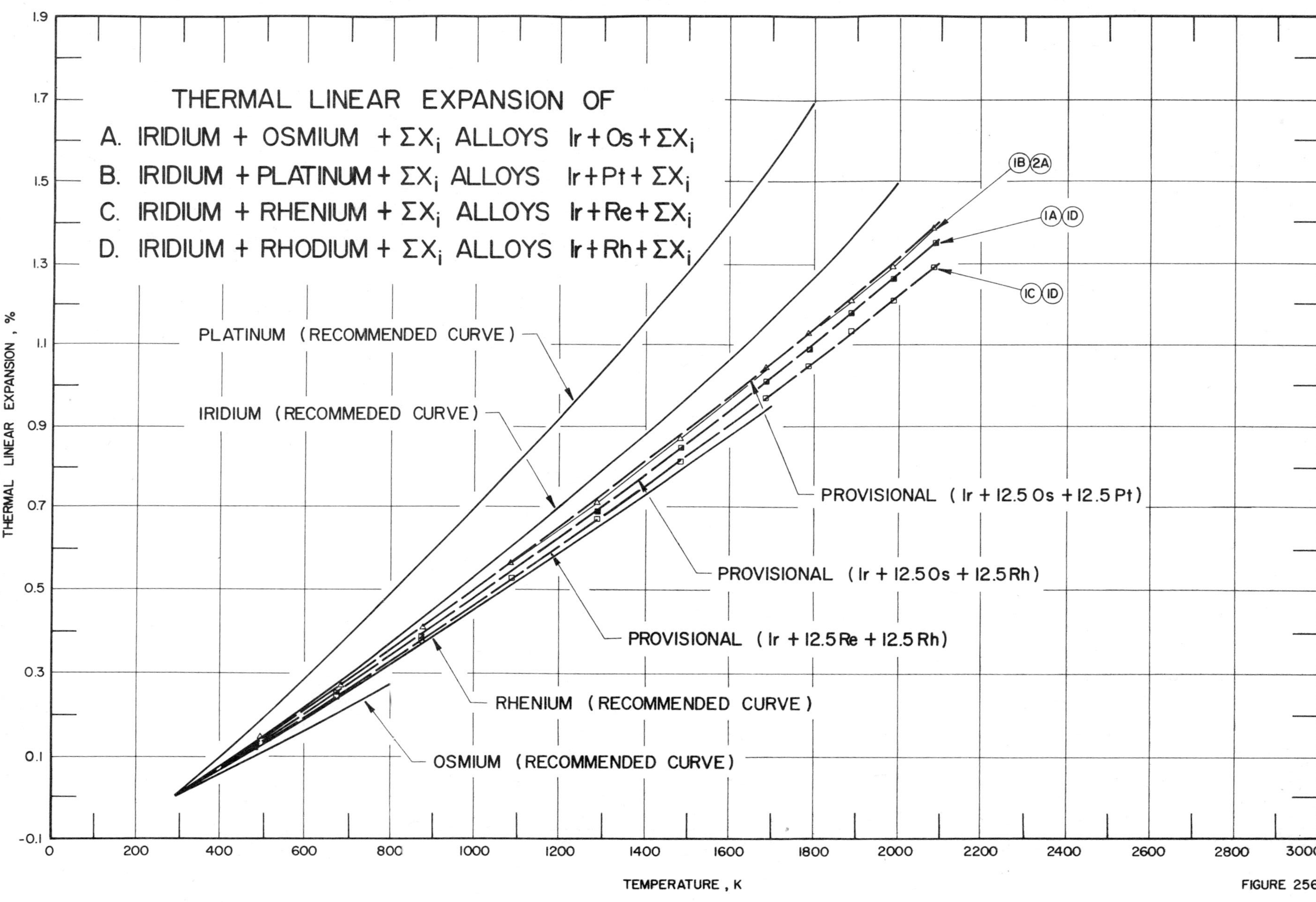

THERMAL LINEAR EXPANSION OF
A. IRIDIUM + OSMIUM + ΣXi ALLOYS Ir + Os + ΣXi
B. IRIDIUM + PLATINUM + ΣXi ALLOYS Ir + Pt + ΣXi
C. IRIDIUM + RHENIUM + ΣXi ALLOYS Ir + Re + ΣXi
D. IRIDIUM + RHODIUM + ΣXi ALLOYS Ir + Rh + ΣXi
PLATINUM (RECOMMENDED CURVE)
IRIDIUM (RECOMMEDED CURVE)
1B 2A
1A 1D
1C 1D
PROVISIONAL (Ir + 12.5 Os + 12.5 Pt)
PROVISIONAL (Ir + 12.5 Os + 12.5 Rh)
PROVISIONAL (Ir + 12.5 Re + 12.5 Rh)
RHENIUM (RECOMMENDED CURVE)
OSMIUM (RECOMMENDED CURVE)
THERMAL LINEAR EXPANSION, %
1.9
1.7
1.5
1.3
1.1
0.9
0.7
0.5
0.3
0.1
-0.1
0
200
400
600
800
1000
1200
1400
1600
1800
2000
2200
2400
2600
2800
3000
TEMPERATURE, K

FIGURE 256

SPECIFICATION TABLE 256A. THERMAL LINEAR EXPANSION OF IRIDIUM + OSMIUM + ΣX_i ALLOYS Ir + Os + ΣX_i

Cur. No.	Ref. No.	Author(s)	Year	Method Used	Temp. Range, K	Name and Specimen Designation	Composition (weight percent) Ir	Os	Rh	Pt	Composition (continued), Specifications, and Remarks
1	420	Harmon, D.P.	1966	X, L	480-2082		75	12.5	12.5		99.96 Ir powder from J. Bishop and Co., 99.84 Os powder from Englehard Industries, 99.98 Rh powder from J. Bishop and Co.; alloy prepared by standard powder metallurgical techniques.
2	420	Harmon, D.P.	1966	X, L	498-2083		75	12.5		12.5	99.96 Ir powder from J. Bishop and Co., 99.84 Os powder from Englehard Industries, 99.99+ Pt powder from J. Bishop and Co.; alloy prepared by standard powder metallurgical techniques.

DATA TABLE 256A. THERMAL LINEAR EXPANSION OF IRIDIUM + OSMIUM + ΣX_i ALLOYS Ir + Os + ΣX_i

[Temperature, T, K; Linear Expansion, $\Delta L/L_0$, %]

T	$\Delta L/L_0$	T	$\Delta L/L_0$	T	$\Delta L/L_0$	T	$\Delta L/L_0$
CURVE 1‡		CURVE 1 (cont.)‡		CURVE 2‡		CURVE 2 (cont.)‡	
480	0.124	1482	0.851	498	0.144	1483	0.875
673	0.251	1682	1.013	684	0.271	1683	1.045
873	0.388	1782	1.095	878	0.408	1783	1.126
1073	0.532	1882	1.178	1083	0.562	1883	1.211
1282	0.691	1982	1.263	1283	0.717	1983	1.298
		2082	1.350			2083	1.386

DATA TABLE 256A. COEFFICIENT OF THERMAL LINEAR EXPANSION OF IRIDIUM + OSMIUM + ΣX_i ALLOYS Ir + Os + ΣX_i

[Temperature, T, K; Coefficient of Expansion, α, 10^{-6} K^{-1}]

T	α	T	α	T	α	T	α
CURVE 1*,‡		CURVE 1 (cont.)*,‡		CURVE 2*,‡		CURVE 2 (cont.)*,‡	
480	6.65	1782	8.22	498	7.00	1783	8.10
673	6.55	1882	8.26	684	6.84	1883	8.58
873	6.88	1982	8.56	878	7.09	1983	8.69
1073	7.19	2082	8.64	1083	7.50	2083	8.80
1282	7.58			1283	7.73		
1482	8.03			1483	7.94		
1682	8.08			1683	8.46		

* Not shown in figure.

‡ Author's data for coefficient of thermal expansion have been integrated by TPRC to obtain $\Delta L/L_0$.

SPECIFICATION TABLE 256B. THERMAL LINEAR EXPANSION OF IRIDIUM + PLATINUM + ΣX_i ALLOYS Ir + Pt + ΣX_i

Cur. No.	Ref. No.	Author(s)	Year	Method Used	Temp. Range, K	Name and Specimen Designation	Composition (weight percent) Ir	Pt	Os	Composition (continued), Specifications, and Remarks
1	420	Harmon, D.P.	1966	X, L	498-2083		75	12.5	12.5	99.96 Ir powder from J. Bishop and Co., 99.99⁺ Pt powder from J. Bishop and Co., 99.84 Os powder from Englehard Industries; alloy prepared by standard metallurgical techniques.

DATA TABLE 256B. THERMAL LINEAR EXPANSION OF IRIDIUM + PLATINUM + ΣX_i ALLOYS Ir + Pt + ΣX_i

[Temperature, T, K; Linear Expansion, $\Delta L/L_0$, %]

T	$\Delta L/L_0$	T	$\Delta L/L_0$	T	$\Delta L/L_0$
CURVE 1‡		CURVE 1 (cont.)‡		CURVE 1 (cont.)‡	
498	0.144	1483	0.875	2083	1.386
684	0.271	1683	1.045		
878	0.408	1783	1.126		
1083	0.562	1883	1.211		
1283	0.717	1983	1.298		

DATA TABLE 256B. COEFFICIENT OF THERMAL LINEAR EXPANSION OF IRIDIUM + PLATINUM + ΣX_i ALLOYS Ir + Pt + ΣX_i

[Temperature, T, K; Coefficient of Expansion, α, 10^{-6} K^{-1}]

T	α
CURVE 1*, ‡	
498	7.00
684	6.84
878	7.09
1083	7.50
1283	7.73
1483	7.94
1683	8.46
1783	8.10
1883	8.58
1983	8.69
2083	8.80

* No figure given.
‡ Author's data for coefficient of thermal expansion have been integrated by TPRC to obtain $\Delta L/L_0$.

SPECIFICATION TABLE 256C. THERMAL LINEAR EXPANSION OF IRIDIUM + RHENIUM + ΣX_i ALLOYS Ir + Re + ΣX_i

Cur. No.	Ref. No.	Author(s)	Year	Method Used	Temp. Range, K	Name and Specimen Designation	Composition (weight percent) Ir	Re	Rh	Composition (continued), Specifications, and Remarks
1	420	Harmon, D.P.	1966	X, L	518-2083		75	12.5	12.5	99.96 Ir powder from J. Bishop and Co., 99.98 Rh powder from J. Bishop and Co., 99.99$^+$ Re powder from Chase Brass and Copper Co.; alloy prepared by standard powder metallurgical techniques.

DATA TABLE 256C. THERMAL LINEAR EXPANSION OF IRIDIUM + RHENIUM + ΣX_i ALLOYS Ir + Re + ΣX_i

[Temperature, T, K; Linear Expansion, $\Delta L/L_0$, %]

T	$\Delta L/L_0$
CURVE 1‡	
518	0.145
673	0.245
873	0.380
1083	0.531
1283	0.673
1483	0.819
1683	0.974
1783	1.051
1883	1.131
1983	1.213
2083	1.296

DATA TABLE 256C. COEFFICIENT OF THERMAL LINEAR EXPANSION OF IRIDIUM + RHENIUM + ΣX_i ALLOYS Ir + Re + ΣX_i

[Temperature, T, K; Coefficient of Expansion, α, 10^{-6} K^{-1}]

T	α	T	α
CURVE 1*, ‡		CURVE 1 (cont.)*, ‡	
518	6.44	1783	7.66
673	6.45	1883	8.03
873	6.78	1983	8.21
1083	7.18	2083	8.25
1283	7.08		
1483	7.33		
1683	7.74		

* No figure given.
‡ Author's data for coefficient of thermal expansion have been integrated by TPRC to obtain $\Delta L/L_0$.

SPECIFICATION TABLE 256D. THERMAL LINEAR EXPANSION OF IRIDIUM + RHODIUM + ΣX_i ALLOYS   Ir + Rh + ΣX_i

Cur. No.	Ref. No.	Author(s)	Year	Method Used	Temp. Range, K	Name and Specimen Designation	Composition (weight percent) Ir	Rh	Re	Os	Composition (continued), Specifications, and Remarks
1	420	Harmon, D.P.	1966	X, L	518-2083		75	12.5	12.5		99.96 Ir powder from J. Bishop and Co., 99.98 Rh powder from J. Bishop and Co., 99.99+ Re powder from Chase Brass and Copper Co.; alloy prepared by standard powder metallurgical techniques.
2	420	Harmon, D.P.	1966	X, L	480-2082		75	12.5		12.5	99.96 Ir powder from J. Bishop and Co., 99.98 Rh powder from J. Bishop and Co., 99.84 Os powder from Englehard Industries; alloy prepared by standard powder metallurgical techniques.

DATA TABLE 256D. THERMAL LINEAR EXPANSION OF IRIDIUM + RHODIUM + ΣX_i ALLOYS   Ir + Rh + ΣX_i

[Temperature, T, K; Linear Expansion, $\Delta L/L_0$, %]

T	$\Delta L/L_0$	T	$\Delta L/L_0$	T	$\Delta L/L_0$
CURVE 1‡		CURVE 1 (cont.)‡		CURVE 2 (cont.)‡	
518	0.145	1883	1.131	873	0.388
673	0.245	1983	1.213	1073	0.532*
873	0.380	2083	1.296	1282	0.691
1083	0.531			1482	0.851
1283	0.673	CURVE 2‡		1682	1.013
1483	0.819			1782	1.095
1683	0.974	480	0.124	1882	1.178
1783	1.051	673	0.251	1982	1.263
				2082	1.350

DATA TABLE 256D. COEFFICIENT OF THERMAL LINEAR EXPANSION OF IRIDIUM + RHODIUM + ΣX_i ALLOYS   Ir + Rh + ΣX_i

[Temperature, T, K; Coefficient of Expansion, α, 10^{-6} K^{-1}]

T	α	T	α	T	α	T	α	T	α	T	α
CURVE 1*, ‡		CURVE 1 (cont.)*, ‡		CURVE 1 (cont.)*, ‡		CURVE 2*, ‡		CURVE 2 (cont.)*, ‡		CURVE 2 (cont.)*, ‡	
518	6.44	1283	7.08	1883	8.03	480	6.65	1282	7.58	1882	8.26
673	6.45	1483	7.33	1983	8.21	673	6.55	1482	8.03	1982	8.56
873	6.78	1683	7.74	2083	8.25	873	6.88	1682	8.08	2082	8.64
1083	7.18	1783	7.66			1073	7.19	1782	8.22		

* Not shown in figure.
‡ Author's data for coefficient of thermal expansion have been integrated by TPRC to obtain $\Delta L/L_0$.

FIGURE AND TABLE NO. 257R. PROVISIONAL VALUES FOR THERMAL LINEAR EXPANSION OF IRON + CARBON + ΣX_i ALLOYS Fe + C + ΣX_i

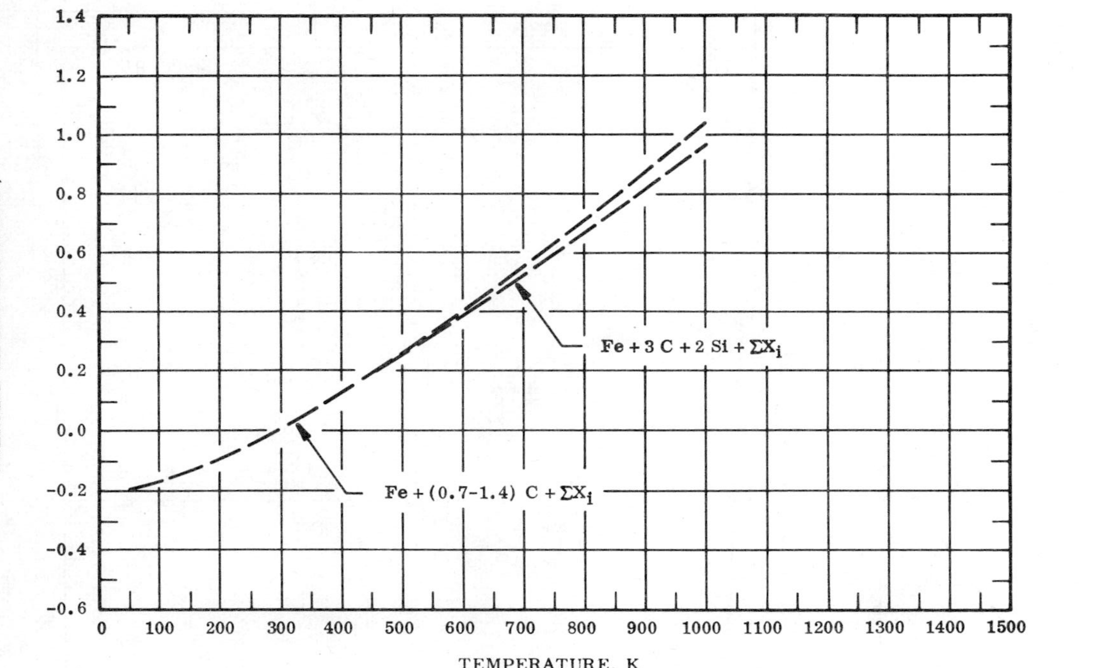

PROVISIONAL VALUES

[Temperature, T, K; Linear Expansion, $\Delta L/L_0$, %; α, K^{-1}]

T	[Fe + (0.7 - 1.4) C + ΣX_i] $\Delta L/L_0$	$\alpha \times 10^6$	(Fe + 3 C + 2 Si + ΣX_i) $\Delta L/L_0$	$\alpha \times 10^6$
50	-0.199	5.6		
100	-0.169	6.7		
200	-0.091	8.9		
293	0.000	10.7	0.000	11.9
400	0.127	12.4	0.128	12.4
500	0.258	13.7	0.255	13.1
600	0.401	14.8	0.389	13.7
700	0.554	15.6	0.528	14.1
800	0.713	16.2	0.671	14.5
900	0.876	16.4	0.817	14.7
1000	1.041	16.5	0.965	14.9

REMARKS

[Fe + (0.7 - 1.4) C + ΣX_i]: The tabulated values for well-annealed alloy of type Carbon Steel are considered accurate to within ± 7% over the entire temperature range. These values can be represented approximately by the following equation:

$$\Delta L/L_0 = -0.223 + 4.337 \times 10^{-4}\, T + 1.273 \times 10^{-6}\, T^2 - 4.446 \times 10^{-10}\, T^3$$

(Fe + 3 C + 2 Si + ΣX_i): The tabulated values for well-annealed alloy of type Cast Iron are considered accurate to within ± 7% over the entire temperature range. These values can be represented approximately by the following equation:

$$\Delta L/L_0 = -0.296 + 8.422 \times 10^{-4}\, T + 6.081 \times 10^{-7}\, T^2 - 1.910 \times 10^{-10}\, T^3$$

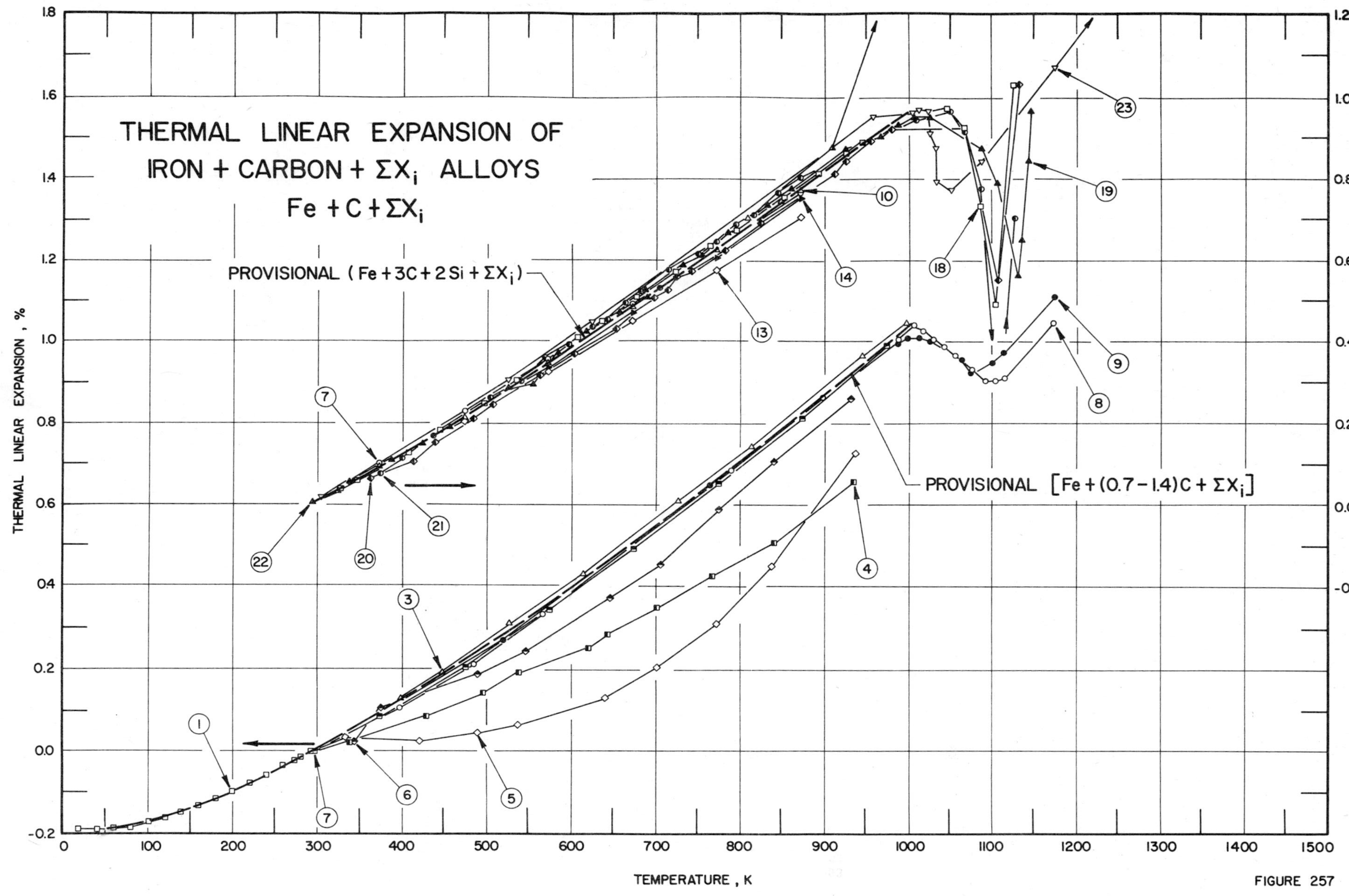
THERMAL LINEAR EXPANSION OF
IRON + CARBON + ΣX_i ALLOYS
Fe + C + ΣX_i
PROVISIONAL (Fe + 3C + 2 Si + ΣX_i)
PROVISIONAL [Fe + (0.7 – 1.4) C + ΣX_i]
THERMAL LINEAR EXPANSION, %
TEMPERATURE, K
1
3
4
5
6
7
8
9
10
13
14
18
19
20
21
22
23
0
100
200
300
400
500
600
700
800
900
1000
1100
1200
1300
1400
1500
-0.2
0.0
0.2
0.4
0.6
0.8
1.0
1.2
1.4
1.6
1.8

FIGURE 257

SPECIFICATION TABLE 257. THERMAL LINEAR EXPANSION OF IRON + CARBON + ΣX_i ALLOYS Fe + C + ΣX_i

Cur. No.	Ref. No.	Author(s)	Year	Method Used	Temp. Range, K	Name and Specimen Designation	Fe	Composition (weight percent) C	Mn	Si	P	S	Composition (continued), Specifications, and Remarks
1	462	Arp, V., Wilson, J.H., Winrich, L., and Sikora, P.	1962	L	20-293	Carbon Steel 1075	Bal.	0.80	0.30	0.15			Hardness R_c 43, quench hardened and tempered.
2	42	Stuart, H. and Ridley, N.	1966	X	1018-1209	Carbon Steel	99.16	0.81	0.01	0.01	0.002	0.008	Alloy produced by vacuum melting; austenitized at 1233 K for 30 min; formed into wire 0.5 mm diameter, polished and sealed in silica tubes, then coated with graphite; expansion studied in austenite range; this curve is here reported using the first given temperature as reference temperature at which $\Delta L/L_0 = 0$; lattice parameter reported at 1018 K is 3.6607 Å.
3	42	Stuart, H. and Ridley, N.	1966	X	293-998	Carbon Steel	98.53	1.44	0.01	0.01	0.001	0.006	Alloy produced by vacuum melting; austenitized in inert atm at 960 C for 30 min; placed in evacuated silica tube, quenched in water and tempered at 973 K for 30 hr; formed into wire, 5 mm diameter, polished and sealed in silica tube; expansion measured where ferrite and cementite phases coexist in specimen; lattice parameter reported at 293 K is 2.8660 Å.
4	42	Stuart, H. and Ridley, N.	1966	X	293-933	Carbon Steel	98.53	1.44	0.01	0.006	0.001		Cementite phase electrolytically extracted from the above specimen; expansion measured along a-axis; lattice parameter reported at 293 K is 4.5251 Å.
5	42	Stuart, H. and Ridley, N.	1966	X	293-935	Carbon Steel	98.53	1.44	0.01	0.006	0.001		Same specimen as above; expansion measured along b-axis; lattice parameter reported at 293 K is 5.0890 Å.
6	42	Stuart, H. and Ridley, N.	1966	X	293-930	Carbon Steel	98.53	1.44	0.01	0.006	0.001		Same specimen as above; expansion measured along c-axis; lattice parameter reported at 293 K is 6.7422 Å.
7	51	Souder, W. and Hidnert, P.	1922	T	293-973	Carbon Steel S482	Bal.	1.28	0.37				0.19 Cr; annealed specimen.
8	51	Souder, W. and Hidnert, P.	1922	T	293-1172	Carbon Steel S556	Bal.	0.252	0.06	0.007	0.012	0.035	Annealed specimen; expansion measured with increasing temperature.
9	51	Souder, W. and Hidnert, P.	1922	T	1172-300	Carbon Steel S556	Bal.	0.252	0.06	0.007	0.012	0.035	The above specimen; expansion measured with decreasing temperature; zero-point correction is 0.061%.
10	521	Fitzgeorge, D. and Pope, J.A.	1959		373-873	Iron N	Bal.	3.09	1.09	0.55	0.18	0.103	Material supplied by North Eastern Marine Engineering Co. (1938) Ltd., Wallsend-on-Tyne; specimen 0.25 in. in diameter and 4 in. long.
11*	521	Fitzgeorge, D. and Pope, J.A.	1959		373-873	Iron L	Bal.	3.30	1.56	1.18	0.345	0.069	0.05 Ni, 0.02 Mo, 0.02 V, and trace T; material supplied by The Wallsend Slipway and Engineering Co., Ltd., Wallsend-on-Tyne; specimen 0.25 in. in diameter and 4 in. long.

* Not shown in figure.

SPECIFICATION TABLE 257. THERMAL LINEAR EXPANSION OF IRON + CARBON + ΣX_i ALLOYS Fe + C + ΣX_i (continued)

Cur. No.	Ref. No.	Author(s)	Year	Method Used	Temp. Range, K	Name and Specimen Designation	Composition (weight percent) Fe	C	Mn	Si	P	S	Composition (continued), Specifications, and Remarks
12*	521	Fitzgeorge, D. and Pope, J.A.	1959		373-873	Iron D	Bal.	3.16	0.78	0.99	0.28	0.097	0.13 V, 0.09 Ni, 0.04 Ti, and 0.02 Mo; material supplied by Wm. Denny and Brothers, Ltd., Dumbarton; specimen 0.25 in. in diameter and 4 in. long.
13	521	Fitzgeorge, D. and Pope, J.A.	1959		373-873	Iron E	Bal.	3.23	0.81	1.3	0.163	0.114	1.43 Ni, 0.44 Mo, 0.02 Ti, and trace V; material supplied by Vickers-Armstrongs, Ltd., Barrow in Furnes; specimen 0.25 in. in diameter, and 4 in. long.
14	521	Fitzgeorge, D. and Pope, J.A.	1959		373-873	Iron V	Bal.	2.93	0.70	1.11	0.073	0.112	1.25 Cu, 0.42 Mo, 0.05 Ni, and 0.01 Cr; material supplied by Richards (Leicester), Ltd., Leicester; specimen 0.25 in. in diameter and 4 in. long.
15*	521	Fitzgeorge, D. and Pope, J.A.	1959		373-873	Iron R	Bal.	2.87	0.86	2.08	0.138	0.139	0.15 Cr, 0.132 Cu, 0.92 Ni, and 0.024 Mo; material supplied by Sheepbridge Engineering, Ltd., Chesterfield; specimen 0.25 in. in diameter and 4 in. long.
16*	521	Fitzgeorge, D. and Pope, J.A.	1959		373-873	Iron F	Bal.	3.51	0.37	2.46	0.019	0.01	2.06 Ni, 0.068 Cu, 0.016 Mo, and trace Cr; material supplied by Sheepbridge Engineering, Ltd., Chesterfield; specimen 0.25 in. in diameter and 4 in. long.
17	521	Fitzgeorge, D. and Pope, J.A.	1959		373-873	Iron G	Bal.	3.50	0.37	2.46	0.017	0.01	2.06 Ni, 0.068 Cu, 0.016 Mo, and trace Cr; material supplied by Sheepbridge Engineering, Ltd., Chesterfield; specimen 0.25 in. in diameter and 4 in. long.
18	522	Neimark, B.E., Monina, E.F., Kartuzova, L.M., and Kainova, R.A.	1967	L	327-1129	Cast Iron 43	Bal.	3.72	0.24	2.60	0.054		0.13 Mg, 0.1 Cr; structure of iron, globular graphite plus plate graphite π + 50% ϕ; sample 200 mm long, measurements made every 5 to 10 C on a vacuum quartz dilatometer; zero-point correction of 0.012% determined by graphical extrapolation.
19	522	Neimark, B.E., et al.	1967	L	338-1147	Cast Iron 35	Bal.	3.05	0.59	3.47			0.05 Mg, 0.02 Ce, 0.04 Ca, 0.04 Ti; containing globular graphite finally ground plus plate graphite 3-5% ϕ + 10-15 % π; sample 200 mm long, vacuum quartz dilatometer used; zero-point correction of 0.012% determined by graphical extrapolation.
20	522	Neimark, B.E., et al.	1967	L	329-1132	Cast Iron No. 45	94.24	2.65	0.28	2.59			0.11 Mg, 0.11 Ti, 0.02 Ca, 0.019 Ce; containing globular graphite ϕ + 20 - 30% π; sample 200 mm long, vacuum quartz dilatometer used; zero-point correction of -0.016% determined by graphical extrapolation.
21	522	Neimark, B.E., et al.	1967	L	326-1127	Cast Iron No. 31	Bal.	3.45	0.29	2.65		0.008	0.08 Mg, 0.5 Ti; bonded carbon 0.61%, having plate graphite ϕ + 20% π; sample 200 mm long; vacuum quartz dilatometer used; zero-point correction of -0.006% determined by graphical extrapolation.

* Not shown in figure.

SPECIFICATION TABLE 257. THERMAL LINEAR EXPANSION OF IRON + CARBON + ΣX_i ALLOYS $Fe + C + \Sigma X_i$ (continued)

Cur. No.	Ref. No.	Author(s)	Year	Method Used	Temp. Range, K	Name and Specimen Designation	Composition (weight percent) Fe	C	Mn	Si	P	S	Composition (continued), Specifications, and Remarks
22	51	Souder, W. and Hidnert, P.	1922	T	293-1175	S483	Bal.	3.08		1.68			Annealed specimen; expansion measured with increasing temperature.
23	51	Souder, W. and Hidnert, P.	1922	T	1175-304	S483	Bal.	3.08		1.68			The above specimen; expansion measured with decreasing temperature; zero-point correction of -0.903% determined by graphical extrapolation.

DATA TABLE 257. THERMAL LINEAR EXPANSION OF IRON + CARBON + ΣX_i ALLOYS $Fe + C + \Sigma X_i$

[Temperature, T, K; Linear Expansion, $\Delta L/L_0$, %]

T	$\Delta L/L_0$
CURVE 1	
20	-0.198
40	-0.196
60	-0.192
80	-0.187
100	-0.177
120	-0.164
140	-0.150
160	-0.133
180	-0.116
200	-0.098
220	-0.079
240	-0.059
260	-0.037
273	-0.023
280	-0.015
293	0.000
CURVE 2*, †	
1018	0.000
1067	0.131
1106	0.235
1158	0.369
1209	0.492
CURVE 3	
293	0.000
399	0.133
445	0.199
528	0.311
612	0.429
726	0.611
812	0.740
943	0.963
998	1.050
CURVE 4	
293	0.000*
339	0.022
426	0.086
494	0.141
539	0.197
620	0.254
641	0.285
700	0.349
CURVE 4 (cont.)	
769	0.422
841	0.502
933	0.659
CURVE 5	
293	0.000*
334	0.004
420	0.026
488	0.049
538	0.065
640	0.130
700	0.206
772	0.312
839	0.442
935	0.727
CURVE 6	
293	0.000*
343	0.027
425	0.105
487	0.187
545	0.242
644	0.375
704	0.457
773	0.592
840	0.705
930	0.862
CURVE 7	
298	0.006
373	0.088
473	0.204
573	0.340
673	0.491
773	0.654
873	0.815
973	0.985
CURVE 8	
293	0.000*
396	0.107
482	0.214
568	0.338
CURVE 8 (cont.)	
789	0.682
897	0.862
974	0.991*
985	1.008
1009	1.042
1016	1.036
1030	1.004
1043	0.984
1055	0.963
1073	0.934
1090	0.904
1102	0.903
1113	0.912
1172	1.057
CURVE 9	
1172	1.118
1113	0.973
1100	0.945
1074	0.920
1061	0.958
1044	0.984*
1027	1.000
1013	1.006
999	1.003
986	0.999
767	0.646
520	0.271
300	0.008*
CURVE 10‡	
373	0.097*
473	0.218
573	0.347
673	0.483
773	0.624
873	0.770
CURVE 11*, ‡	
373	0.097
473	0.219
573	0.349
673	0.485
773	0.625
CURVE 11 (cont.)*, ‡	
873	0.768
CURVE 12*, ‡	
373	0.094
473	0.212
573	0.338
673	0.470
773	0.606
873	0.746
CURVE 13‡	
373	0.091*
473	0.204
573	0.323
673	0.448
773	0.578
873	0.712
CURVE 14‡	
373	0.094*
473	0.211*
573	0.336
673	0.467
773	0.603
873	0.744
CURVE 15*, ‡	
373	0.091
473	0.205
573	0.326
673	0.453
773	0.585
873	0.720
CURVE 16*, ‡	
373	0.097
473	0.218
573	0.343
673	0.473
773	0.608
873	0.747
CURVE 17*, ‡	
373	0.100
473	0.226
573	0.356
673	0.489
773	0.627*
873	0.768
CURVE 18	
327	0.038
347	0.059
408	0.123
441	0.180
471	0.211*
495	0.244
536	0.303
608	0.407
636	0.449
679	0.506
724	0.573
765	0.636
795	0.677
851	0.757
892	0.815
927	0.865
944	0.891
1047	0.971
1069	0.923
1084	0.733
1105	0.487
1129	1.033
CURVE 19	
338	0.053
386	0.111
421	0.145
456	0.190
527	0.288
556	0.295
582	0.374
619	0.421
659	0.475
688	0.526
731	0.588
757	0.617
785	0.667
CURVE 19 (cont.)	
810	0.702
834	0.737
850	0.742
863	0.779
925	0.875
966	0.913
988	0.931
1007	0.947
1028	0.955
1087	0.878
1107	0.783
1130	0.559
1135	0.644
1144	0.842
1147	0.968
CURVE 20	
329	0.032
362	0.060
412	0.105
438	0.149
483	0.216
509	0.242
565	0.319
602	0.367
652	0.432
698	0.501
715	0.525
724	0.558
743	0.573
782	0.622
826	0.691
873	0.770
911	0.814
927	0.844
951	0.897
980	0.920
1067	0.921
1106	0.545
1132	1.037
CURVE 21	
326	0.029
373	0.076
400	0.115

* Not shown in figure.

† This curve is here reported using the first given temperature as reference temperature at which $\Delta L/L_0 = 0$.

DATA TABLE 257. THERMAL LINEAR EXPANSION OF IRON + CARBON + ΣX_i ALLOYS Fe + C + ΣX_i (continued)

T	$\Delta L/L_0$
CURVE 21 (cont.)	
435	0.165
472	0.212
505	0.262
541	0.308
570	0.360
599	0.393
625	0.437
641	0.456
665	0.492
683	0.521
704	0.533
706	0.558
717	0.574
733	0.594
751	0.619
772	0.642
798	0.686
817	0.714
847	0.761
866	0.783
873	0.803
898	0.815
1010	0.941
1030	0.957
1050	0.962
1067	0.919
1087	0.770
1104	0.176
1127	0.700
CURVE 22	
293	0.000
689	0.509
910	0.880
980	1.306*
1005	1.511*
1013	1.539*
1079	1.662*
1081	1.655*
1087	1.655*
1128	1.786*
1175	1.978*
CURVE 23	
1175	1.075
1089	0.842
1050	0.764
CURVE 23 (cont.)	
1035	0.769*
1032	0.793
1032	0.825*
1032	0.851*
1032	0.873
1028	0.898*
1026	0.917
1023	0.941*
1023	0.963
1018	0.978*
1012	0.969
1005	0.960
959	0.950
622	0.442
527	0.307
304	0.013

* Not shown in figure.

FIGURE AND TABLE NO. 258R. PROVISIONAL VALUES FOR THERMAL LINEAR EXPANSION OF IRON + CHROMIUM + ΣX_i ALLOYS Fe + Cr + ΣX_i

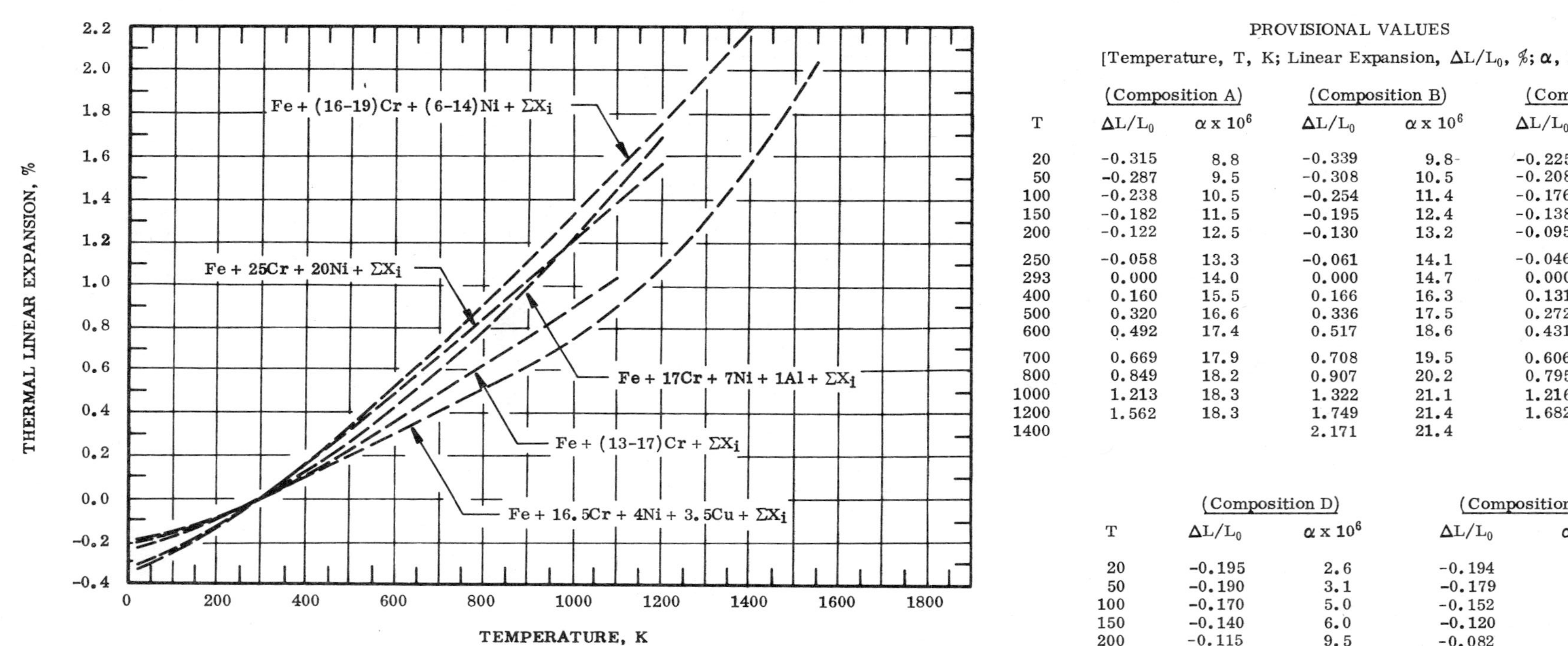

PROVISIONAL VALUES

[Temperature, T, K; Linear Expansion, $\Delta L/L_0$, %; α, K^{-1}]

T	(Composition A) $\Delta L/L_0$	$\alpha \times 10^6$	(Composition B) $\Delta L/L_0$	$\alpha \times 10^6$	(Composition C) $\Delta L/L_0$	$\alpha \times 10^6$
20	-0.315	8.8	-0.339	9.8	-0.225	5.2
50	-0.287	9.5	-0.308	10.5	-0.208	5.9
100	-0.238	10.5	-0.254	11.4	-0.176	7.0
150	-0.182	11.5	-0.195	12.4	-0.138	8.1
200	-0.122	12.5	-0.130	13.2	-0.095	9.2
250	-0.058	13.3	-0.061	14.1	-0.046	10.3
293	0.000	14.0	0.000	14.7	0.000	11.1
400	0.160	15.5	0.166	16.3	0.131	13.2
500	0.320	16.6	0.336	17.5	0.272	15.0
600	0.492	17.4	0.517	18.6	0.431	16.7
700	0.669	17.9	0.708	19.5	0.606	18.3
800	0.849	18.2	0.907	20.2	0.795	19.7
1000	1.213	18.3	1.322	21.1	1.216	22.2
1200	1.562	18.3	1.749	21.4	1.682	24.3
1400			2.171	21.4		

T	(Composition D) $\Delta L/L_0$	$\alpha \times 10^6$	(Composition E) $\Delta L/L_0$	$\alpha \times 10^6$
20	-0.195	2.6	-0.194	4.3
50	-0.190	3.1	-0.179	4.9
100	-0.170	5.0	-0.152	6.0
150	-0.140	6.0	-0.120	7.0
200	-0.115	9.5	-0.082	7.9
250	-0.051	10.0	-0.041	8.8
293	0.000	10.5	0.000	9.5
400	0.115	10.6	0.108	10.9
500	0.214	10.7	0.225	12.1
600	0.308	10.7	0.350	12.9
700	0.403	10.7	0.483	13.5
800	0.505	10.7	0.619	13.8
900	0.619	12.2	0.760	13.9
1000	0.750	14.2	0.899	13.9
1100	0.905	16.8	1.034	13.9
1200	1.088	19.9		
1300	1.306	23.6		
1400	1.563	27.8		
1500	1.865	32.6		
1550	2.034	35.2		

REMARKS

The tabulated values for various Fe + Cr + ΣX_i alloys are for well-annealed alloys and are considered accurate to within ± 7% over the entire temperature range. These values can be represented approximately by the equations given below:

Composition A (Fe + 25Cr + 20 Ni + ΣX_i): This is an alloy of type stainless steel SS310.

$$\Delta L/L_0 = -0.334 + 8.362 \times 10^{-4} T + 1.169 \times 10^{-6} T^2 - 4.586 \times 10^{-10} T^3$$

Composition B [Fe + (16-19)Cr + (6-14)Ni + ΣX_i]: This is an alloy of type stainless steels SS301, 302, 303, 304L, 316, 321, 347, Kh18N9T Russian alloy, Kh17N7Yu Russian alloy, EI-572 Russian steel, EI-606 Russian steel, and 18 Cr - 10 Ni Austenite.

$$\Delta L/L_0 = -0.358 + 9.472 \times 10^{-4} T + 1.031 \times 10^{-6} T^2 - 2.978 \times 10^{-10} T^3$$

Composition C (Fe + 17Cr + 7Ni + 1Al + ΣX_i): This is an alloy of type 17-7PH.

$$\Delta L/L_0 = -0.235 + 4.749 \times 10^{-4} T + 1.181 \times 10^{-6} T^2 - 2.041 \times 10^{-10} T^3$$

Composition D (Fe + 16.5Cr + 4Ni + 3.5Cu + ΣX_i): This is an alloy of type 17-4PH.

$$\Delta L/L_0 = -0.427 + 1.842 \times 10^{-3} T - 1.575 \times 10^{-6} T^2 + 9.106 \times 10^{-10} T^3 \quad (200 < T < 1550)$$

Composition E [Fe + (13-17)Cr + ΣX_i]: This is an alloy of type stainless steels, SS406, 410, 416, 420, 422, 430, 440, and 446.

$$\Delta L/L_0 = -0.201 + 3.865 \times 10^{-4} T + 1.152 \times 10^{-6} T^2 - 4.379 \times 10^{-10} T^3$$

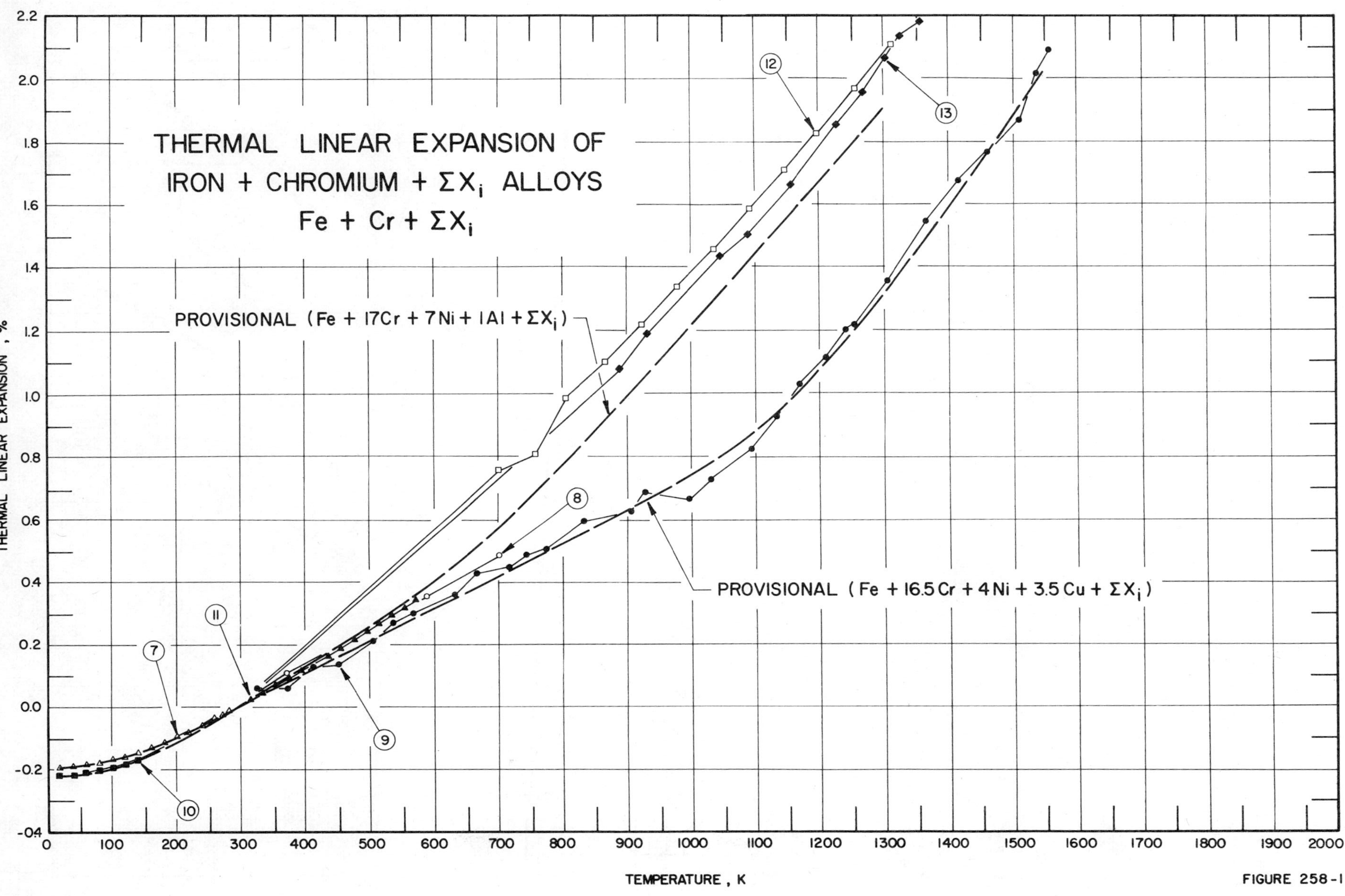
THERMAL LINEAR EXPANSION OF
IRON + CHROMIUM + ΣX$_i$ ALLOYS
Fe + Cr + ΣX$_i$
PROVISIONAL (Fe + 17Cr + 7Ni + 1Al + ΣX$_i$)
PROVISIONAL (Fe + 16.5Cr + 4Ni + 3.5Cu + ΣX$_i$)
12
13
8
11
7
9
10
THERMAL LINEAR EXPANSION, %
2.2
2.0
1.8
1.6
1.4
1.2
1.0
0.8
0.6
0.4
0.2
0.0
-0.2
-0.4
0
100
200
300
400
500
600
700
800
900
1000
1100
1200
1300
1400
1500
1600
1700
1800
1900
2000
TEMPERATURE, K

FIGURE 258-1

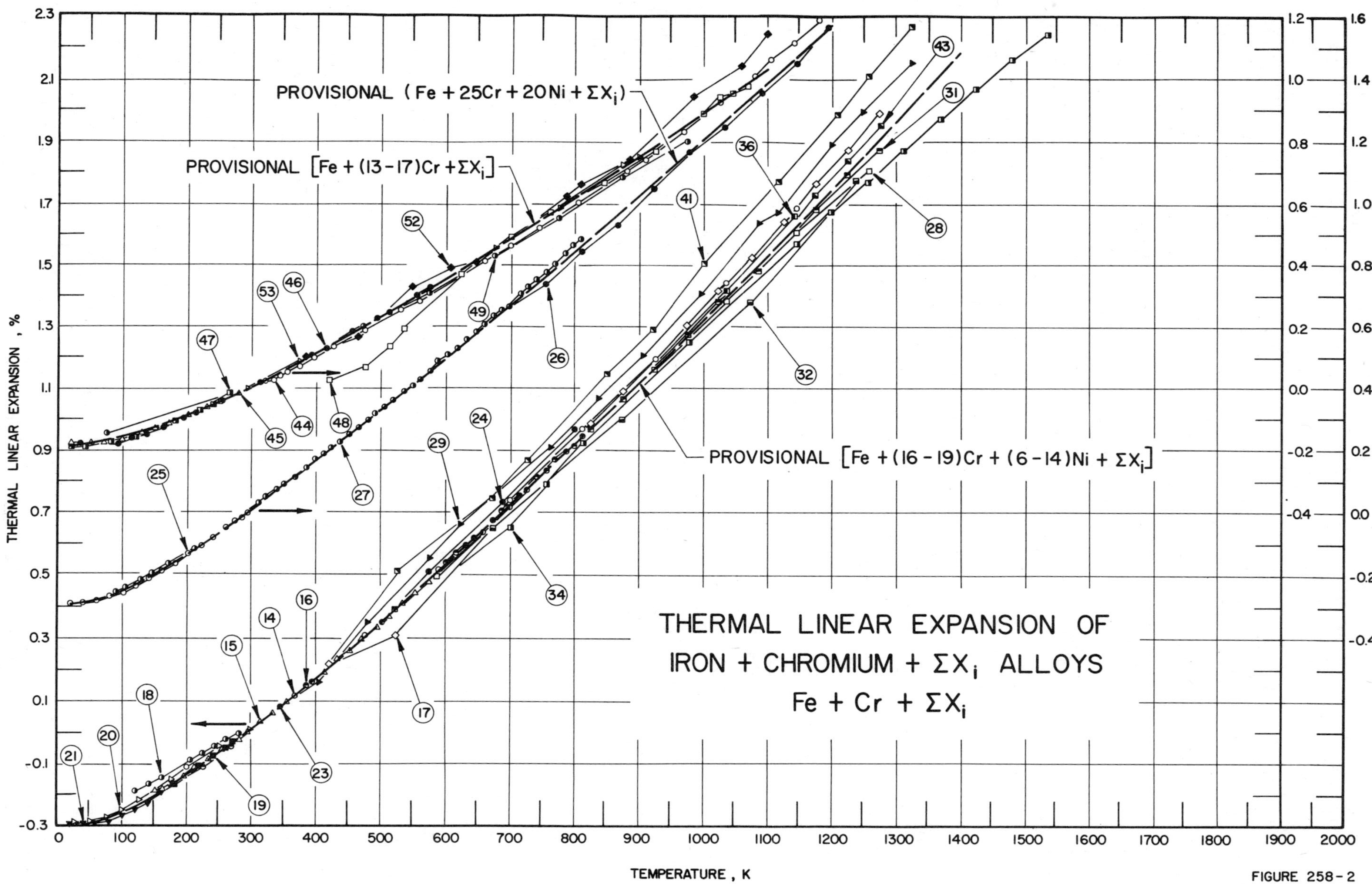
THERMAL LINEAR EXPANSION OF
IRON + CHROMIUM + ΣX_i ALLOYS
Fe + Cr + ΣX_i
PROVISIONAL (Fe + 25Cr + 20Ni + ΣX_i)
PROVISIONAL [Fe + (13 - 17)Cr + ΣX_i]
PROVISIONAL [Fe + (16 - 19)Cr + (6 - 14)Ni + ΣX_i]
THERMAL LINEAR EXPANSION, %
TEMPERATURE, K

FIGURE 258-2

SPECIFICATION TABLE 258. THERMAL LINEAR EXPANSION OF IRON + CHROMIUM + ΣX_i ALLOYS $Fe + Cr + \Sigma X_i$

Cur. No.	Ref. No.	Author(s)	Year	Method Used	Temp. Range, K	Name and Specimen Designation	Composition (weight percent) Fe	Cr	Mn	Si	C	Ni	P	Composition (continued), Specifications, and Remarks
1*	217	Totskii, E.E.	1964	L	79-1373	1Kh18N9T Russian Alloy	Bal.	18-20			0.12	8-10		1.0 Ti; specimen annealed rod 200 mm long.
2*	551	Belov, A.K.	1968	L	77-323	Kh18N9T Russian Alloy	Bal.	20	2	0.8		11		0.8 Ti; cylindrical specimen 3.5 mm in diameter, 50 mm long; sample heated and cooled at a rate of 3-7 deg/min; data on the coefficient of thermal linear expansion also reported.
3*	533	Hoenie, A.F. and Roach, D.B.	1966		293-699	15-5 PH Steel	Bal.	14.0-15.5	1.0	1.0	0.07	3.5-5.5	0.04	2.5 to 4.5 Cu, 0.15 to 0.45 Nb + Ta, 0.03 S; martensitic precipitation hardenable steel from Armco Steel Corp; solution annealed by heating 30 min at 1311 K, then air cooled and oil quenched; specimen condition A.
4*	533	Hoenie, A.F. and Roach, D.B.	1966		200-699	15-5 PH Steel	Bal.	14.0-15.5	1.0	1.0	0.07	3.5-5.5	0.04	Similar to the above specimen; precipitation hardened 1 hr at 755 K; specimen condition H-900.
5*	533	Hoenie, A.F. and Roach, D.B.	1966		293-699	15-5 PH Steel	Bal.	14.0-15.5	1.0	1.0	0.07	3.5-5.5	0.04	Similar to the above specimen; averaged 4 hr at 852 K; specimen condition H-1075.
6*	217	Totskii, E.E.	1964		273-1373	15KhM Russian Steel	Bal.	0.80-1.10	0.40-0.70					0.40 to 0.60 Mo, 0.20 Cu; specimen annealed rod 200 mm long.
7	462	Arp, V., Wilson, J.H., Winrich, L., and Sikora, P.	1962	L	20-293	17-4PH	Bal.	16.0	0.2	0.5	0.03	4.3	0.02	3.6 Cu, 0.2 Nb, 0.01 S; hardness R_c38; condition H 1100.
8	533	Hoenie, A.F. and Roach, D.B.	1966		293-699	17-4PH								Martensitic precipitation hardenable steel developed by Armco Steel Corp.
9	358	Fieldhouse, I.B. and Lang, J.I.	1961	T	293-1560	17-4PH	72.9	16.4	1.0	1.0	0.07	4.2	0.04	4.1 Cu, 0.3 Nb + Ta; heat treatment H900; density 7.74 g cm^{-3}.
10	462	Arp, V., Wilson, J.H., Winrich, L., and Sikora, P.	1962	L	20-293	17-7PH	Bal.	17.2	0.7	0.4	0.07	7.4	0.04	1.2 Al, 0.01 S; hardness R_c42; condition TH 1050.
11	15	Rhodes, B.L., Moeller, C.T., Hopkins, V., and Marx, T.I.	1963	I, L	18-573	17-7PH	Bal.	17.0	0.7	0.4	0.07	7.0		1.2 Al; 10.16 cm long x 6.2 mm.
12	495	Seibel, R.D. and Mason, G.L.	1958		293-1644	17-7PH	72.62	17.08	0.71	0.45	0.70	7.21	0.024	1.19 Al, 0.017 S; heating rate at 1.7 to 2.8 K per min in vacuum of 10^{-5} to 10^{-6} mm Hg; density 7.43 g cm^{-3}.
13	472	Fieldhouse, I.B., Hedge, J.C., Lang, J.I., and Waterman, T.E.	1958	T	300-1400	17-7PH	72.21	17.30	0.60	0.49	0.074	7.06		1.11 Al; zero-point correction of 0.01% determined by graphical extrapolation.
14	123	Lucks, C.F. and Deem, H.W.	1958	L	116-1144	301 Stainless Steel	71	18	2	1		8		Handbook composition; 0.375 in. diameter by 3 in. long; hot rolled material from Republic Steel Corp; annealed 1 hr at 1310.9 K and water quenched; heating rate 2.2 K per min.

* Not shown in figure.

SPECIFICATION TABLE 258. THERMAL LINEAR EXPANSION OF IRON + CHROMIUM + ΣX_i ALLOYS Fe + Cr + ΣX_i (continued)

Cur. No.	Ref. No.	Author(s)	Year	Method Used	Temp. Range, K	Name and Specimen Designation	Composition (weight percent) Fe	Cr	Mn	Si	C	Ni	P	Composition (continued), Specifications, and Remarks
15	15	Rhodes, B.L., Moeller, C.I., Hopkins, V., and Marx, T.I.	1963	I, L	18-573	301 Stainless Steel	Bal.	17	2.0	1.0	0.15	7.0		10.16 cm long x 6.2 mm.
16	534	Furman, D.E.	1950		89-811	301 Stainless Steel	Bal.	16.91	0.80	0.54	0.13	7.25		Specimen made in a 10# high frequency furnace given an anneal at 1339 K for 30 min and water quenched prior to machining the expansion specimen; 0.250 in. diameter by 4.000 in. long; zero-point correction of 0.02% determined by graphical interpolation.
17	466	Lucks, C.F., Thompson, H.B., Smith, A.R., Curry, F.P., and Deem, H.W.	1951	L	83-1273	301 Stainless Steel	Bal.	16-18	2.0		0.08-0.2	6-8		Supplied by Republic Steel Corp. (nominal composition indicated); hot rolled, annealed 1 hr at 1310.9 K and water quenched; density 7.85 g cm^{-3}.
18	511	Hertz, J.	1962		76-288	301-FH								No details given; zero-point correction of -0.004% determined by graphical extrapolation.
19	462	Arp, V., Wilson, J.H., Winrich, L., and Sikora, P.	1962	L	20-293	302 Stainless Steel	Bal.	18.6	0.6	0.6	0.08	8.7	0.02	0.01 S; hardness R_c 31, cold drawn specimen.
20	302	Beenakker, J.J.M. and Swenson, C.A.	1955	L	4-300	302 Stainless Steel								Technical grade.
21	462	Arp, V., Wilson, J.H., Winrich, L., and Sikora, P.	1962	L	20-293	303 Stainless Steel	Bal.	17.6	1.2	0.6	0.10	8.7	0.03	0.4 Cu, 0.4 Mo, 0.29 S; hardness R_B95; specimen annealed.
22*	462	Arp, V., et al.	1962	L	20-293	304L Stainless Steel	Bal.	18.4	1.4	0.6	0.02	9.7	0.02	0.01 S; hardness R_B94; annealed specimen.
23	534	Furman, D.E.	1950		89-810	304L Stainless Steel	Bal.	19.19	0.65	0.53	0.068	8.49	0.024	0.007 S; 0.250 in.diameter by 4.000 in. long; prior to machining the expansion specimen, the commercial grade steel was given an anneal at 1339 K for 30 min and water quenched.
24	225	Valentich, J.	1965	L	293-806	304L Stainless Steel								No details given.
25	462	Arp, V., Wilson, J.H., Winrich, L., and Sikora, P.	1962	L	20-293	310 Stainless Steel	Bal.	24.8	1.7	0.7	0.08	20.8	0.02	0.1 Cu, 0.02 S; hardness R_B79; specimen annealed.
26	495	Seibel, R.D. and Mason, G.L.	1958		293-1644	310 Stainless Steel	46.58	24.94	1.57	0.37	0.062	19.60	0.018	0.018 S; heating rate at 1.7 to 2.8 K per min in vacuum of 10^{-5} to 10^{-6} mm Hg; density 7.90 g cm^{-3}.
27	534	Furman, D.E.	1950		89-811	310 Stainless Steel	Bal.	27.22	1.51	0.42	0.111	21.64	0.022	0.010 S; 0.250 in. diameter by 4.000 in. long; commercial grade steel was given an anneal at 1340 K for 30 min and water quenched to machine the specimen; zero-point correction of -0.254% determined by graphical interpolation.

* Not shown in figure.

SPECIFICATION TABLE 258. THERMAL LINEAR EXPANSION OF IRON + CHROMIUM + ΣX_i ALLOYS $Fe + Cr + \Sigma X_i$ (continued)

Cur. No.	Ref. No.	Author(s)	Year	Method Used	Temp. Range, K	Name and Specimen Designation	Composition (weight percent) Fe	Cr	Mn	Si	C	Ni	P	Composition (continued), Specifications, and Remarks
28	123	Lucks, C.F. and Deem, H.W.	1958	L	116-1255	316 Stainless Steel	65	19	2	1		13		Handbook composition; 0.375 in. diameter by 3 in. long; hot rolled material from Timken Roller Bearing Co.; annealed 1 hr at 1367 K and water quenched; heating rate 2.2 K per min.
29	250	Fieldhouse, I.B., Hedge, J.C., and Lang, J.I.	1958	T	300-1616	316 Stainless Steel								Zero-point correction of 0.01% determined by graphical extrapolation.
30*	534	Furman, D.E.	1950		89-811	316 Stainless Steel	Bal.	17.78	1.77	0.58	0.057	12.70	0.012	2.38 Mo, 0.012 S; 0.250 in. diameter by 4.000 in. long; commercial grade steel was given an anneal at 1339 K for 30 min and water quenched to machine the specimen; zero-point correction of -0.264% determined by graphical interpolation.
31	466	Lucks, C.F., Thompson, H.B., Smith, A.R., Curry, F.P., Deem, H.W., and Bing, G.F.	1951	L	83-1273	316 Stainless Steel	Bal.	16.82	1.59	0.26	0.108	11.66	0.018	2.18 Mo, 0.023 S; hot-rolled, annealed 1 hr at 1366.49 K and water quenched; supplied by Timken Roller Bearing Co.; density 7.96 g cm^{-3}.
32	62	Yaggee, F.L. and Styles, R.W.	1969	L	273-1233	AISI-316 Stainless Steel								Reported error $< \pm 2\%$.
33*	462	Arp, V., Wilson, J.H., Winrich, L., and Sikora, P.	1962	L	20-293	321 Stainless Steel	Bal.	17.9	1.4	0.6	0.06	9.8	0.02	0.3 Cu, 0.2 Mo, 0.02 S; hardness R_B97; specimen annealed.
34	495	Seibel, R.D. and Mason, G.L.	1958	L	293-1644	321 Stainless Steel	69.05	17.59	1.53	0.71	0.091	9.85	trace	1.17 Ti, 0.009 S; heating rate 1.7 to 2.8 K per min in vacuum of 10^{-5} to 10^{-6} mm of Hg; density 7.89 g cm^{-3}.
35*	462	Arp, V., Wilson, J.H., Winrich, L., and Sikora, P.	1962	L	20-293	347 Stainless Steel	Bal.	18.0	1.5	0.6	0.06	10.3	0.02	0.9 Nb, 0.2 Cu, 0.2 Mo, 0.02 S; hardness R_B95; specimen annealed.
36	123	Lucks, C.F. and Deem, H.W.	1958	L	116-1144	347 Stainless Steel	62	18	2	1		14		3.0 Mo; handbook composition; 0.375 in. diameter by 3 in. long; hot-rolled material from Timken Roller Bearing Co.; annealed 1 hr at 1367 K and water quenched; heating rate 2.2 K per min.
37*	499	Aerojet-General Corp.	1963	L	367-1311	347 Stainless Steel								Specimen from stock material.
38*	15	Rhodes, B.L., Moeller, C.E., Hopkins, V., and Marx, T.I.	1963	I, L	18-573	347 Stainless Steel	Bal.	18.0	2.0	1.0	0.08	11.0		10.16 cm long by 6.2 mm.
39*	96	Watrous, J.D.	1961	L	298-1080	347 Stainless Steel								Heating rate 3 C per min; run 1; zero-point correction of 0.007% determined by graphical extrapolation.

* Not shown in figure.

SPECIFICATION TABLE 258. THERMAL LINEAR EXPANSION OF IRON + CHROMIUM + ΣX_i ALLOYS Fe + Cr + ΣX_i (continued)

Cur. No.	Ref. No.	Author(s)	Year	Method Used	Temp. Range, K	Name and Specimen Designation	Composition (weight percent) Fe	Cr	Mn	Si	C	Ni	P	Composition (continued), Specifications, and Remarks
40*	96	Watrous, J.D.	1961	L	298-1078	347 Stainless Steel								The above specimen; run 2; zero-point correction of 0.007% determined by graphical extrapolation.
41	250	Fieldhouse, I.B., Hedge, J.C., and Lang, J.I.	1958	T	300-1494	347 Stainless Steel								Zero-point correction of 0.010% determined by graphical extrapolation.
42*	534	Furman, D.E.	1950		89-811	347 Stainless Steel	Bal.	18.65	1.74	0.56	0.068	11.30	0.019	0.77 Nb, 0.006 S; 0.250 in. diameter by 4.000 in. long; commercial grade steel was given an anneal at 1340 K for 30 min and water quenched to machine the specimen; zero-point correction of -0.272% determined by graphical interpolation.
43	466	Lucks, C.F., Thompson, H.B., Smith, A.R., Curry, F.P., Deem, H.W., and Bing, G.F.	1951	L	83-1273	347 Stainless Steel	Bal.	17.65	1.64	0.58	0.06	10.94	0.013	0.73 Nb, 0.09 Cu, 0.02 Mo, 0.017 S; supplied by Timken Roller Bearing Co.; hot-rolled, annealed 1 hr at 1366 K and water quenched; density 7.90 g cm^{-3} (at 293 K).
44	535	McCoy, H.E., Jr.	1968	L	302-1267	406 Stainless Steel	Bal.	14.1	0.41	0.35	0.079	0.44		3.91 Al, 0.32 Ti, 0.11 Zr, 0.0038 O, 0.019 N; as-received material annealed at undesignated temperature, cold swaged to 0.75 in. before specimen made; 0.250 in. diameter, 2 in. long; thermocouple Pt-10% Rh Vs Pt attached outside of quartz tube approx. 0.010 in. from specimen; heated and cooled at linear rate of 1 C/min; zero-point correction of 0.0074% determined by graphical extrapolation.
45	481	Johnston, H.L., Altman, H.W., and Rubin, T.	1965	I	20-300	410 Stainless Steel	84	14	1	1				Composition handbook values; supplied by Carnegie-Illinois Steel Co.; specimen consisted of three sectors cut from bar, filed to length, smooth to within 1/5 of wavelength of sodium D radiation; data smoothed by author; data on the coefficient of thermal linear expansion are also reported.
46	15	Rhodes, B.L., Moeller, C.E., Hopkins, V., and Marx, T.I.	1963	I/L	18-573	410 Stainless Steel	Bal.	12.5	1.0	1.0	0.15			Specimen 10.16 cm long by 6.2 mm.
47	462	Arp, V., Wilson, J.H., Winrich, L., and Sikora, P.	1962	L	20-293	416 Stainless Steel	Bal.	12.6	0.5	0.6	0.13		0.02	0.4 Mo, 0.22 S; hardness R_c 41; hot rolled specimen.
48	358	Fieldhouse, I.B. and Lang, J.I.	1961	T	299-1448	420 Stainless Steel	84.99	13.10	0.48	0.41	0.30	0.50	0.02	0.12 Cu, 0.06 Mo, 0.011 S; density 7.71 g cm^{-3}.

* Not shown in figure.

SPECIFICATION TABLE 258. THERMAL LINEAR EXPANSION OF IRON + CHROMIUM + ΣX_i ALLOYS Fe + Cr + ΣX_i (continued)

Cur. No.	Ref. No.	Author(s)	Year	Method Used	Temp. Range, K	Name and Specimen Designation	Fe	Cr	Mn	Si	C	Ni	P	Composition (continued), Specifications, and Remarks
							Composition (weight percent)							
49	217	Totskii, E. E.	1964	L	79-1273	E1-802 Russian Steel, 422 Stainless Steel								Cylindrical specimen 200 mm long, annealed.
50*	467	Clark, A. F.	1968	L	0-300	AISI-430F Steel	81.2	17.07	0.52	0.14	<0.1	0.32	<0.1	0.48 Mo, 0.28 S; machined to 0.64 cm square by 20.32 cm long, cold drawn; Rockwell hardness B-92.
51*	462	Arp, V., Wilson, J.H., Winrich, L., and Sikora, P.	1962	L	20-293	440 C Steel	Bal.	17.3	0.5	0.4	1.08		0.02	0.6 Mo, 0.01 S; hardness R_c 40; hot rolled specimen.
52	472	Fieldhouse, I. B., Hedge, J.C., Lang, J.I., and Waterman, T.E.	1958	T	300-1493	446 Stainless Steel	70.55	27.61			0.086			0.01 Mo; zero-point correction of 0.0074% determined by graphical extrapolation.
53	533	Hoenie, A.F. and Roach, D.B.	1966		293-894	455 Stainless Steel	Bal.	11.0-13.0	0.5	0.5	0.03	7.0-10.0		1.0-3.0 Cu, 0.9-1.4 Ti, 0.25-0.5 Nb + Ta, 0.005 B; low carbon martensitic precipitation-hardenable steel from Carpenter Steel Co.
54*	533	Hoenie, A.F. and Roach, D.B.	1966		298-922	AFC-77 Steel	Bal.	14.5			0.15			13.5 Co, 5.0 Mo, 0.5 V, 0.05 N; bar stock precipitation hardenable stainless from Crucible Steel Co.; austenitized 1 hr at 1366 K, oil quenched 1/2 hr at 200 K, tempered 4 hr at 866 K.
55*	536	Battelle Memorial Lab.	1966		298, 811	AFC-77 Steel	Bal.	15						13.5 Co, 5.0 Mo, 0.5 V; handbook values; tempered at 867 K.
56*	358	Fieldhouse, I.B. and Lang, J.I.	1961		299-1583	AM-335 Steel	75.5	15.66	0.94	0.05	0.12	4.27	0.02	2.82 Mo; density 7.78 g cm^{-3}.
57*	467	Clark, A. F.	1968	L	0-300	AM-35 Steel (AISI-633)	75.1	16.2	0.82	0.15	0.12	4.5	<0.1	2.8 Mo, 0.23 Cu, <0.1 each N, S, V, Ti, Nb; machined to 0.64 cm square by 20.32 cm long; Rockwell hardness B-96.
58*	467	Clark, A. F.	1968	L	0-300	AM-35 Steel (AISI-633)	75.1	16.2	0.82	0.15	0.12	4.5	<0.1	2.8 Mo, 0.12 Cu, <0.1 each N, S, V, Ti, Nb; similar to the above specimen except precipitation hardened to hardness C-44.
59*	533	Hoenie, A.F. and Roach, D.B.	1966		293-873	AM-362 Steel	Bal.	14.5	0.3	0.2	0.03	6.5	0.015	0.8 Ti, 0.015 S; precipitation hardenable martensitic stainless steel developed by Allegheny Ludlum Steel Corp.
60*	533	Hoenie, A.F. and Roach, D.B.	1966		293-773	AM-363 Steel	Bal.	11.0-12.0	0.3	0.15	0.05	4.0-5.0		>0.5 Ti; low strength martensitic steel from Allegheny Ludlum Steel Corp; solution annealed 5 min at 1144 K and aged 5 min at 811 K.
61*	537	Smirnov, V.V., Pokhumurskii, V.I., and Boltarovich, A.V.	1966		293-973	Russian EP-479 Stainless Steel	Bal.	15.0-16.6	<0.6	<0.6	0.12-0.18	2.0-2.5	<0.03	1.2-1.5 Mo, 0.05-0.1 N_2, <0.03 S; specific gravity 7.74 g cm^{-3}.

* Not shown in figure.

SPECIFICATION TABLE 258. THERMAL LINEAR EXPANSION OF IRON + CHROMIUM + ΣX_i ALLOYS Fe + Cr + ΣX_i (continued)

Cur. No.	Ref. No.	Author(s)	Year	Method Used	Temp. Range, K	Name and Specimen Designation	Composition (weight percent) Fe	Cr	Mn	Si	C	Ni	P	Composition (continued), Specifications, and Remarks
62*	477	Kueser, P. E., Pavlovic, D. M., Lane, D. H., Clark, J. J., and Spewock, M.	1967	L	295, 755	H-11 Steel (AMS-6487)	91.70	4.89	0.20	0.87	0.40		0.01	1.30 Mo, 0.53 V, 0.006 S; premium quality steel from Universal Cyclops Steel Corp., AISI condition H-11; 2 in. long specimen.
63*	538	Yershov, V. M. and Oslon, N. L.	1968	X	301-1351	Kh12M Russian Steel	Bal.	12.21	0.40	0.30	1.57		0.020	0.42 Mo, 0.033 S; produced in induction furnace forged to square bar 12 x 12 mm, after annealing; cross-section 12 x 12 and thickness 0.8-0.9 mm; measured of austenite line (200) from 1050 C on cooling; lattice parameter reported at 301 K is 3.5777 kX.
64*	537	Smirnov, V. V., Pokhumurskii, V. I., and Boltarovich, A. V.	1966		293-973	Kh17N2 Russian Alloy	Bal.	17				2		Nominal composition.
65*	539	Neimark, B. E., Lyusternik, V. E., and Korytina, S. F.	1964		300-1092	Kh17N7Yu Russian Alloy Steel	Bal.	17.26	0.64	0.58	0.08	7.27		1.17 Al, 0.012 S; specimen quenched in air from 1323 K; data on the coefficient of thermal linear expansion are also reported.
66*	473	Sweeney, W. O.	1947		294-1144	Low C Multimet	Bal.	19/23			0.12/ 0.20	19/23		19/23 Co, 2.5/4 Mo, 0.75/1.50 Nb; density 8.20 g cm^{-3}, wrought alloy.
67*	473	Sweeney, W. O.	1947		294-1089	Med. C Multimet	Bal.	19/23			0.30/ 0.50	19/23		19/23 Co, 2.5/4 Mo, 0.75/1.50 Nb; density 8.20 g cm^{-3}, wrought alloy.
68*	533	Hoenie, A. F. and Roach, D. M.	1966		294-810	PH14-8Mo Steel	Bal.	13.5-15.5	1.0	1.0	0.02-0.05	7.5-9.5	0.015	2.0-3.0 Mo, 0.75-1.5 Al, 0.01 S; developed by Armco Steel Corp.; high strength reached by heat treatment; condition SRH-950; density 7.71 g cm^{-3}.
69*	540	Trehan, Y. N.	1968	L	373-1273		Bal.	31.4	5.67	0.65	0.04			0.35 N; 5 cm long; solution heat treated for 1/2 hr at 1323 K and water quenched before machining.
70*	540	Trehan, Y. N.	1968	L	373-1273		Bal.	30.0	10.91	0.75	0.03			0.89 N; similar to the above specimen.
71*	540	Trehan, Y. N.	1968	L	373-1273		Bal.	29.59	5.26	1.21	0.07			0.61 N; similar to the above specimen.
72*	540	Trehan, Y. N.	1968	L	373-1273		Bal.	26.09	20.45	0.43	0.06			0.99 N; similar to the above specimen.
73*	540	Trehan, Y. N.	1968	L	773-1273		Bal.	23.54	12.85	1.02	0.03			0.86 N; similar to the above specimen.
74*	540	Trehan, Y. N.	1968	L	373-1273		Bal.	22.67	14.88	0.58	0.05			0.67 N; similar to the above specimen.
75*	540	Trehan, Y. N.	1968	L	373-1273		Bal.	21.51	17.86	0.80	0.07			0.75 N; similar to the above specimen.
76*	540	Trehan, Y. N.	1968	L	573-1273		Bal.	21.2	13.06	0.36	0.05			0.63 N; 5 cm long; solution heat treated for 1/2 hr at 1050 C and water quenched before machining.
77*	540	Trehan, Y. N.	1968	L	373-1273		Bal.	20.98	14.9	0.86	0.06			0.72 N; similar to the above specimen.
78*	540	Trehan, Y. N.	1968	L	473-1273		Bal.	20.5	19.66	0.96	0.05			0.86 N; similar to the above specimen.
79*	540	Trehan, Y. N.	1968	L	473-1273		Bal.	16.83	11.64	0.69	0.03			0.42 N; similar to the above specimen.

* Not shown in figure.

SPECIFICATION TABLE 258. THERMAL LINEAR EXPANSION OF IRON + CHROMIUM + ΣX_i ALLOYS Fe + Cr + ΣX_i (continued)

Cur. No.	Ref. No.	Author(s)	Year	Method Used	Temp. Range, K	Name and Specimen Designation	Composition (weight percent) Fe	Cr	Mn	Si	C	Ni	P	Composition (continued), Specifications, and Remarks
80*	540	Trehan, Y.N.	1968	L	473-1273		Bal.	13.1	5.51	0.78	0.04			0.24 N; similar to the above specimen.
81*	51	Souder, W. and Hidnert, P.	1922	T	298-973	S 552	Bal.	0.57	0.46	0.09	0.36		0.011	0.12 V, 0.029 S; annealed specimen.
82*	51	Souder, W. and Hidnert, P.	1922	T	298-973	S 565	Bal.	13.0			0.30-0.40			Annealed specimen.
83*	51	Souder, W. and Hidnert, P.	1922	T	302-1184	S 551	Bal.	1.17	0.08	0.110	0.35		0.010	0.14 V, 0.027 S; annealed specimen; expansion measured with increasing temperature; zero-point correction of 0.010% determined by graphical extrapolation.
84*	51	Souder, W. and Hidnert, P.	1922	T	1184-303	S 551	Bal.	1.17	0.08	0.110				The above specimen; expansion measured with decreasing temperature; zero-point correction of 0.045% determined by graphical extrapolation.
85*	51	Souder, W. and Hidnert, P.	1922	T	302-1173	S 558	Bal.	0.85	0.05	0.846	0.122		0.020	0.23 V, 0.040 S; annealed specimen; expansion measured with increasing temperature; zero-point correction of 0.010% determined by graphical extrapolation.
86*	51	Souder, W. and Hidnert, P.	1922	T	1173-300	S 558	Bal.	0.85	0.05	0.846				The above specimen; expansion measured with decreasing temperature; zero-point correction of -0.028% determined by graphical extrapolation.
87*	51	Souder, W. and Hidnert, P.	1922	T	293-1190	S 557	Bal.	0.92	0.08	0.038	0.168		0.010	0.24 V, 0.64 Mo, 0.029 S; annealed specimen; expansion measured with increasing temperature; zero-point correction is -0.002%.
88*	51	Souder, W. and Hidnert, P.	1922	T	1190-1001	S 557	Bal.	0.92	0.08	0.038				The above specimen; expansion measured with decreasing temperature.
89*	217	Totskii, E.E.	1964	L	79-1373	EI-572 Russian Steel	Bal.	18-20	0.75-1.5		0.28-0.35	8-10		1-1.5 Mo, W, 0.2-0.5 Nb, Ti; cylindrical annealed rod 200 cm long.
90*	217	Totskii, E.E.	1964	L	273-1373	EI-606 Russian Steel	Bal.	18-20	1-2	1.3-1.8	0.07	8-10		2.2-2.7 V; similar to the above specimen.
91*	541	Stuart, H. and Ridley, N.	1969	X	293-1310	18Cr-10Ni Austenite	72.39	17.7				9.91		Melting together iron (99.986 pure); chromium (>99.95) and nickel (>99.998) in recrystallized alumina boats under a purified argon atmosphere; fillings size 200-350 mesh (British Standard sieve), given a strain relieving anneal at 1073 K for 1 hr before measurement; lattice reproducible to ± 0.0002 Å, temperature control ± 2 C; lattice parameter reported at 293 K is 3.5901 Å.
92*	541	Stuart, H. and Ridley, N.	1969	X	375-1274	18Cr-10Ni Austenite	72.39	17.7				9.91		The above specimen.

* Not shown in figure.

SPECIFICATION TABLE 258. THERMAL LINEAR EXPANSION OF IRON + CHROMIUM + ΣX_i ALLOYS Fe + Cr + ΣX_i

Cur. No.	Ref. No.	Author(s)	Year	Method Used	Temp. Range, K	Name and Specimen Designation	Fe	Cr	Composition (weight percent) Mn	Si	C	Ni	P	Composition (continued), Specifications, and Remarks
93*	429	Mochel, N. L.	1927	T	273-795	K3	Bal.	0.87	0.51	0.19	0.39		0.015	0.21 Mo, 0.029 S; specimen rolled to 2.85 cm diameter, tested as received; expansion measured with increasing temperature; zero-point correction of -0.021% determined by graphical interpolation.
94*	429	Mochel, N. L.	1927	T	795-297	K3	Bal.	0.87	0.51					The above specimen; expansion measured with decreasing temperature; zero-point correction of -0.021% determined by graphical extrapolation.
95*	429	Mochel, N. L.	1927	T	273-787	K3A	Bal.	0.87	0.51	0.19	0.39			Similar to the above specimen; expansion measured with increasing temperature; zero-point correction of -0.022% determined by graphical interpolation.
96*	429	Mochel, N. L.	1927	T	787-297	K3A	Bal.	0.87	0.51					The above specimen; expansion measured with decreasing temperature; zero-point correction of -0.022% determined by graphical extrapolation.
97*	551	Belov, A. K.	1968	L	77-323	Kh21G7ANS Russian Alloy	Bal.	21	7.8	0.29		5.9		Cylindrical specimen 3.5 mm in diameter; sample heated and cooled at a rate of 3-7 deg per min.
98*	149	Burger, E. E.	1934	L	298-678	Ascoloy No. 55	Bal.	26-30	≤1.0		≤0.25	≤0.60		Chrome-iron alloy; handbook composition; specimen tested in form of rod 12 in. long and 0.25 in. in diameter; zero-point correction of 0.005% determined by graphical extrapolation.
99*	717	Technology Utilization Division, NASA	1969	L	19-293	Armco 21-6-9	Bal.	22	10			7.5		Bar specimen; annealed at 1070 C/hr; water quenched.

* Not shown in figure.

DATA TABLE 258. THERMAL LINEAR EXPANSION OF IRON + CHROMIUM + ΣX_i ALLOYS $Fe + Cr + \Sigma X_i$

[Temperature, T, K; Linear Expansion, $\Delta L/L_0$, %]

T	$\Delta L/L_0$
CURVE 1*	
79	-0.272
273	-0.032
373	0.134
473	0.311
573	0.496
673	0.685
773	0.879
873	1.080
973	1.288
1073	1.503
1173	1.725
1273	1.957
1373	2.198
CURVE 2*, ‡	
79	-0.279
83	-0.276
93	-0.268
103	-0.261
113	-0.252
123	-0.242
133	-0.231
143	-0.220
153	-0.207
163	-0.194
173	-0.181
183	-0.168
193	-0.154
203	-0.139
213	-0.125
223	-0.110
233	-0.095
243	-0.080
253	-0.064
263	-0.049
273	-0.033
283	-0.016
293	0.000
303	0.017
313	0.033
323	0.050
CURVE 3*	
293	0.000
366	0.079
471	0.199
CURVE 3 (cont.)*	
588	0.329
699	0.461
CURVE 4*	
200	-0.097
293	0.000
366	0.079
471	0.199
588	0.335
699	0.476
CURVE 5*	
293	0.000
366	0.083
471	0.216
588	0.351
699	0.497
CURVE 6*	
273	-0.021
323	0.034
373	0.098
423	0.165
473	0.233
573	0.266
673	0.519
773	0.670
873	0.827
973	0.988
998	1.023
1023	1.049
1048	1.061
1073	1.043
1123	0.963
1148	0.934
1153	0.937
1173	0.982
1273	1.221
1373	1.442
CURVE 7	
20	-0.197
40	-0.195
60	-0.190
CURVE 7 (cont.)	
80	-0.183
100	-0.173
120	-0.161
140	-0.148
160	-0.133
180	-0.117
200	-0.099
220	-0.080
240	-0.061
260	-0.039
273	-0.024
280	-0.016
293	0.000*
CURVE 8	
293	0.000
373	0.108
588	0.358
699	0.490
CURVE 9	
293	0.00*
326	0.06
371	0.06
412	0.13
453	0.14
503	0.21
536	0.27
567	0.30
587	0.33*
632	0.36
668	0.43
718	0.45
742	0.49
774	0.51
830	0.60
905	0.63
929	0.69
997	0.67
1030	0.73
1094	0.83
1134	0.93
1169	1.03
1209	1.11
1240	1.20
1253	1.22
CURVE 9 (cont.)	
1305	1.36
1368	1.45
1415	1.58
1460	1.77
1511	1.87
1538	2.02
1560	2.10
CURVE 10	
20	-0.222
40	-0.221
60	-0.218
80	-0.211
100	-0.199
120	-0.185
140	-0.167
160	-0.149*
180	-0.129*
200	-0.109*
220	-0.087*
240	-0.064*
260	-0.041*
273	-0.025*
280	-0.017*
293	0.000*
CURVE 11	
18	-0.190*
23	-0.190*
33	-0.190*
43	-0.190*
53	-0.190*
63	-0.188*
73	-0.186*
83	-0.184*
93	-0.180*
103	-0.175*
113	-0.169*
123	-0.163*
133	-0.156*
143	-0.148*
153	-0.140*
163	-0.132*
173	-0.124*
193	-0.105*
213	-0.086*
CURVE 11 (cont.)	
233	-0.065*
253	-0.044*
273	-0.022*
293	0.000*
313	0.023*
333	0.046*
353	0.070*
373	0.094*
393	0.118*
413	0.142*
433	0.167
453	0.192
473	0.217
493	0.242
513	0.268
533	0.294
553	0.319
573	0.345
CURVE 12	
293	0.000
699	0.760
755	0.810
810	0.990
866	1.100
922	1.220
977	1.340
1033	1.460
1088	1.590
1144	1.710
1199	1.830
1255	1.970
1310	2.110
1366	2.260*
1422	2.400*
1477	2.540*
1533	2.680*
1588	2.850*
1644	3.040*
CURVE 13	
300	0.01*
889	1.08
931	1.19
981	1.27*
1045	1.44
CURVE 13 (cont.)	
1089	1.51
1156	1.67
1159	1.72*
1194	1.81*
1226	1.86*
1267	1.96*
1267	1.98*
1302	2.07
1326	2.14
1353	2.19
1375	2.29*
1400	2.35*
CURVE 14	
116	-0.239*
144	-0.207*
199	-0.133
366	0.119
477	0.311
588	0.521
699	0.739
810	0.969
922	1.199
1033	1.439
1144	1.691
CURVE 15	
18	-0.262*
23	-0.262*
33	-0.262*
43	-0.262*
53	-0.261*
63	-0.259*
73	-0.256*
83	-0.252*
93	-0.246*
103	-0.238*
113	-0.231*
123	-0.222*
133	-0.213*
143	-0.202*
153	-0.192
163	-0.180
173	-0.168*
193	-0.142
213	-0.116

* Not shown in figure.
‡ Author's data for coefficient of thermal expansion have been integrated by TPRC to obtain $\Delta L/L_0$.

DATA TABLE 258. THERMAL LINEAR EXPANSION OF IRON + CHROMIUM + ΣX_i ALLOYS Fe + Cr + ΣX_i (continued)

T	$\Delta L/L_0$
CURVE 15 (cont.)	
233	-0.088
253	-0.059
273	-0.030
293	0.000
313	0.032
333	0.063
353	0.094
373	0.127
393	0.160
413	0.194
433	0.228
453	0.262
473	0.298
493	0.334
513	0.370
533	0.407
553	0.444
573	0.480
CURVE 16	
89	-0.282*
94	-0.273*
100	-0.270*
105	-0.262*
111	-0.253*
117	-0.247*
122	-0.239*
128	-0.233*
133	-0.228*
139	-0.223*
144	-0.214*
150	-0.209*
155	-0.200*
161	-0.193*
167	-0.187*
172	-0.180*
178	-0.170
183	-0.164*
189	-0.159*
194	-0.150*
200	-0.142*
205	-0.136*
211	-0.124*
217	-0.121*
222	-0.112*
228	-0.108
233	-0.096*
239	-0.094*
244	-0.082*
CURVE 16 (cont.)	
250	-0.072*
255	-0.070*
261	-0.062*
267	-0.052*
272	-0.035*
278	-0.028*
283	-0.020*
289	-0.007*
294	0.002*
300	0.008*
305	0.021*
311	0.028*
317	0.040*
322	0.048*
328	0.059*
333	0.067*
338	0.077*
344	0.081*
349	0.091*
355	0.096*
360	0.109*
366	0.120*
372	0.133*
377	0.142*
383	0.151*
388	0.158*
394	0.169*
399	0.178*
405	0.187*
410	0.194*
416	0.204*
422	0.217*
435	0.238*
499	0.267*
463	0.288*
477	0.313*
491	0.342*
505	0.372*
519	0.393*
533	0.416*
547	0.447*
560	0.469*
574	0.495*
588	0.512*
602	0.545
616	0.572
630	0.594
644	0.620
658	0.647*
672	0.676
CURVE 16 (cont.)	
685	0.706
699	0.730*
713	0.756
727	0.786
741	0.818
755	0.838
769	0.869
783	0.892
797	0.917
811	0.949
CURVE 17	
83	-0.270*
123	-0.233*
173	-0.175*
223	-0.107*
323	0.048*
373	0.135*
423	0.222
473	0.307*
523	0.310
573	0.489*
623	0.586*
673	0.686*
723	0.784*
773	0.884*
823	0.988
873	1.092
923	1.195*
973	1.308
1023	1.422
1073	1.531
1123	1.647
1173	1.764
1223	1.880
1273	1.994
CURVE 18	
76	-0.225*
99	-0.208*
111	-0.196*
120	-0.187
131	-0.175*
140	-0.167
160	-0.148
171	-0.134*
190	-0.116*
196	-0.105*
CURVE 18 (cont.)	
207	-0.093
224	-0.072
234	-0.058*
242	-0.045
253	-0.037*
262	-0.028*
266	-0.024
278	-0.012
288	-0.004*
CURVE 19	
20	-0.305
40	-0.302
60	-0.297
80	-0.287
100	-0.270
120	-0.247
140	-0.222
160	-0.195
180	-0.166
200	-0.137
220	-0.108
240	-0.078
260	-0.048
273	-0.030
280	-0.018
293	0.000
CURVE 20	
4	-0.284
25	-0.284
50	-0.281
75	-0.269
100	-0.246
125	-0.217
150	-0.185
175	-0.152
200	-0.120
250	-0.042
300	0.047
CURVE 21	
20	-0.299
40	-0.298
60	-0.293
80	-0.282
100	-0.265
CURVE 21 (cont.)	
120	-0.245
140	-0.222
160	-0.196
180	-0.169
200	-0.141
220	-0.113
240	-0.085
260	-0.054
273	-0.032
280	-0.022
293	0.000
CURVE 22*	
20	-0.306
40	-0.303
60	-0.294
80	-0.281
100	-0.265
120	-0.245
140	-0.222
160	-0.195
180	-0.168
200	-0.140
220	-0.113
240	-0.083
260	-0.052
273	-0.031
280	-0.021
293	0.000
CURVE 23	
89	-0.271*
94	-0.263*
100	-0.257*
105	-0.249*
111	-0.243*
116	-0.238*
122	-0.231*
128	-0.225*
133	-0.217*
139	-0.213*
144	-0.206*
150	-0.201*
155	-0.192*
161	-0.184*
166	-0.177*
172	-0.172*
178	-0.168*
CURVE 23 (cont.)	
183	-0.160*
189	-0.152*
194	-0.143*
200	-0.140*
205	-0.130*
211	-0.126*
216	-0.114*
222	-0.104*
228	-0.095*
233	-0.089*
239	-0.083*
244	-0.076*
250	-0.071*
255	-0.058*
261	-0.046*
266	-0.040*
272	-0.029*
278	-0.024*
283	-0.013*
289	-0.005*
294	0.002*
300	0.012*
305	0.018*
311	0.028*
316	0.037*
322	0.044*
328	0.055*
333	0.063*
339	0.073*
344	0.082
350	0.091*
355	0.100*
361	0.107*
366	0.118*
372	0.128*
378	0.134*
383	0.145*
389	0.151*
394	0.161
400	0.172*
405	0.182*
411	0.191*
416	0.199*
422	0.207*
436	0.236*
450	0.259*
464	0.281*
478	0.309*
491	0.334*
505	0.358

* Not shown in figure.

DATA TABLE 258. THERMAL LINEAR EXPANSION OF IRON + CHROMIUM + ΣX_i ALLOYS $Fe + Cr + \Sigma X_i$ (continued)

CURVE 23 (cont.)

T	$\Delta L/L_0$
519	0.383*
533	0.402*
547	0.429*
561	0.455*
575	0.484*
589	0.507*
603	0.536*
616	0.563*
630	0.588*
644	0.614*
658	0.636
672	0.667*
686	0.695*
700	0.724
714	0.741*
728	0.768
741	0.809*
755	0.831*
769	0.858*
783	0.887*
797	0.917*
810	0.945*

CURVE 24

T	$\Delta L/L_0$
293	0.000*
358	0.109*
468	0.314*
578	0.527
690	0.737
806	0.954

CURVE 25

T	$\Delta L/L_0$
20	-0.288
40	-0.286
60	-0.281
80	-0.271
100	-0.256
120	-0.235
140	-0.212
160	-0.188
180	-0.163
200	-0.135
220	-0.106
240	-0.077
260	-0.048
273	-0.029
280	-0.019
293	0.000

CURVE 26

T	$\Delta L/L_0$
293	0.000*
699	0.669
755	0.741
810	0.842
866	0.935
922	1.045
977	1.150
1033	1.250
1088	1.360
1144	1.460
1199	1.570
1255	1.680*
1310	1.790*
1366	1.910*
1422	2.020*
1477	2.140*
1533	2.240*
1588	2.360*
1644	2.510*

CURVE 27

T	$\Delta L/L_0$
89	-0.254
94	-0.246*
99	-0.241*
105	-0.235
110	-0.232*
116	-0.227*
122	-0.221*
127	-0.216
133	-0.211*
138	-0.203*
144	-0.199
149	-0.190*
155	-0.183*
160	-0.177*
166	-0.175*
172	-0.165
177	-0.158*
183	-0.154*
188	-0.144*
194	-0.137*
199	-0.127*
205	-0.122*
210	-0.113*
216	-0.104*
222	-0.097*
227	-0.090*
233	-0.083*
238	-0.076*

* Not shown in figure.

CURVE 27 (cont.)

T	$\Delta L/L_0$
244	-0.067*
249	-0.061*
255	-0.054*
260	-0.042*
266	-0.037*
272	-0.026*
277	-0.019*
283	-0.010*
289	-0.004*
294	0.001*
300	0.010*
305	0.016*
310	0.031
316	0.039*
322	0.046
327	0.054*
333	0.058*
338	0.070
344	0.079*
349	0.089
355	0.095*
360	0.104*
366	0.112
372	0.124*
377	0.129*
383	0.142
388	0.151*
394	0.160*
399	0.171
405	0.177*
410	0.187
416	0.196*
422	0.209
435	0.227
449	0.253
463	0.277
477	0.300
491	0.321
505	0.345
519	0.366
533	0.394
547	0.412
560	0.431
574	0.461
588	0.488
602	0.510
616	0.539
630	0.560
644	0.584
658	0.613

CURVE 27 (cont.)

T	$\Delta L/L_0$
672	0.640
685	0.659
699	0.688*
713	0.711
727	0.737
741	0.761
755	0.785
769	0.810
783	0.843
797	0.867
811	0.892

CURVE 28

T	$\Delta L/L_0$
116	-0.261*
144	-0.221*
199	-0.141*
366	0.114*
477	0.298*
588	0.498
699	0.718*
810	0.939*
922	1.164
1033	1.388
1144	1.610
1255	1.850

CURVE 29

T	$\Delta L/L_0$
300	0.010*
407	0.162
481	0.352
579	0.554
623	0.662
765	0.917
837	1.048
908	1.213
996	1.402
1085	1.638
1119	1.674
1199	1.893
1249	1.997
1325	2.159
1392	2.347*
1442	2.447*
1472	2.524*
1499	2.580*
1542	2.679*
1592	2.825*
1616	2.932*

CURVE 30*

T	$\Delta L/L_0$
89	-0.264
94	-0.256
99	-0.249
105	-0.245
110	-0.237
116	-0.231
122	-0.225
127	-0.219
133	-0.212
138	-0.206
144	-0.197
149	-0.193
155	-0.186
160	-0.179
166	-0.171
172	-0.167
177	-0.161
183	-0.153
188	-0.145
194	-0.141
199	-0.133
205	-0.124
210	-0.119
216	-0.113
222	-0.105
227	-0.095
233	-0.088
238	-0.081
244	-0.075
249	-0.067
255	-0.061
260	-0.053
266	-0.043
272	-0.036
277	-0.026
283	-0.016
288	-0.007
294	0.002
299	0.011
305	0.016
310	0.026
316	0.030
322	0.043
327	0.054
333	0.062
338	0.071
344	0.080
349	0.086
355	0.095
360	0.103

CURVE 30 (cont.)*

T	$\Delta L/L_0$
366	0.118
372	0.121
377	0.136
383	0.142
388	0.149
394	0.161
399	0.170
405	0.182
410	0.189
416	0.199
422	0.208
435	0.230
449	0.257
463	0.282
477	0.308
491	0.334
505	0.356
519	0.387
533	0.408
547	0.433
560	0.456
574	0.484
588	0.510
602	0.538
616	0.561
630	0.588
644	0.614
658	0.639
672	0.665
685	0.694
699	0.717
713	0.745
727	0.769
741	0.796
755	0.825
769	0.851
783	0.879
797	0.908
811	0.934

CURVE 31

T	$\Delta L/L_0$
83	-0.277*
123	-0.237*
173	-0.178*
223	-0.107*
323	0.043*
373	0.128*
423	0.212*
473	0.299*

DATA TABLE 258. THERMAL LINEAR EXPANSION OF IRON + CHROMIUM + ΣX_i ALLOYS Fe + Cr + ΣX_i ALLOYS (continued)

T	$\Delta L/L_0$
CURVE 31 (cont.)	
523	0.392
573	0.486*
623	0.593*
673	0.683*
723	0.779*
773	0.873*
823	0.969
873	1.071
923	1.171*
973	1.279
1023	1.382
1073	1.483
1123	1.667*
1173	1.683
1223	1.786
1273	1.896
CURVE 32	
273	-0.03*
473	0.30*
673	0.65*
873	1.01
1073	1.38
1233	1.77
CURVE 33*	
20	-0.294
40	-0.293
60	-0.290
80	-0.281
100	-0.267
120	-0.249
140	-0.227
160	-0.201
180	-0.173
200	-0.144
220	-0.115
240	-0.085
260	-0.053
273	-0.032
280	-0.021
293	0.000
CURVE 34	
293	0.000*
699	0.650
755	0.790
CURVE 34 (cont.)	
810	0.925
866	1.040*
922	1.160*
977	1.260
1033	1.370*
1088	1.490*
1144	1.570
1199	1.670
1255	1.770
1310	1.870
1366	1.970
1422	2.070
1477	2.160
1533	2.240
1588	2.340*
1644	2.420*
CURVE 35*	
20	-0.298
40	-0.297
60	-0.294
80	-0.282
100	-0.267
120	-0.248
140	-0.225
160	-0.200
180	-0.172
200	-0.143
220	-0.113
240	-0.083
260	-0.052
273	-0.031
280	-0.020
293	0.000
CURVE 36	
116	-0.258*
144	-0.224*
199	-0.147*
366	0.122*
477	0.313*
588	0.523*
699	0.734*
810	0.953*
922	1.182*
1033	1.417
1144	1.664
CURVE 37*	
367	0.128
477	0.321
588	0.520
699	0.727
810	0.937
922	1.152
1033	1.371
1144	1.594
1255	1.820
1311	1.935
CURVE 38*	
18	-0.294
23	-0.294
33	-0.293
43	-0.292
53	-0.290
63	-0.287
73	-0.283
83	-0.277
93	-0.269
103	-0.261
113	-0.252
123	-0.242
133	-0.232
143	-0.220
153	-0.208
163	-0.196
173	-0.183
193	-0.155
213	-0.126
233	-0.095
253	-0.064
273	-0.032
293	0.000
313	0.034
333	0.068
353	0.102
373	0.138
393	0.174
413	0.211
433	0.247
453	0.284
473	0.321
493	0.359
513	0.397
533	0.434
553	0.472
573	0.509
CURVE 39*	
298	0.007
391	0.145
510	0.352
673	0.663
872	1.079
1080	1.525
CURVE 40*	
298	0.007
376	0.113
508	0.329
673	0.647
872	1.076
1078	1.533
CURVE 41	
300	0.010*
377	0.125*
435	0.206
524	0.509
669	0.745
728	0.869
850	1.147
920	1.298
1001	1.506
1114	1.769
1208	1.982
1260	2.114
1322	2.275
1400	2.449*
1443	2.575*
1494	2.685*
CURVE 42*	
89	-0.272
94	-0.264
99	-0.260
105	-0.254
110	-0.250
116	-0.244
122	-0.238
127	-0.231
133	-0.226
138	-0.222
144	-0.214
149	-0.203
155	-0.197
CURVE 42 (cont.)*	
160	-0.192
166	-0.184
172	-0.174
177	-0.170
183	-0.164
188	-0.154
194	-0.147
199	-0.139
205	-0.131
210	-0.126
216	-0.116
222	-0.108
227	-0.101
233	-0.094
238	-0.081
244	-0.073
249	-0.069
255	-0.058
260	-0.050
266	-0.039
272	-0.031
277	-0.022
283	-0.018
289	-0.008
294	0.002
300	0.011
305	0.018
310	0.028
316	0.037
322	0.047
327	0.058
333	0.066
338	0.077
344	0.086
349	0.095
355	0.104
360	0.112
366	0.120
372	0.128
377	0.136
383	0.147
388	0.154
394	0.165
399	0.176
405	0.185
410	0.193
416	0.205
422	0.217
435	0.238
449	0.269
CURVE 42 (cont.)*	
463	0.290
477	0.316
491	0.343
505	0.367
519	0.392
533	0.414
547	0.444
560	0.465
574	0.497
588	0.522
602	0.547
616	0.578
630	0.603
644	0.622
658	0.645
672	0.677
685	0.710
699	0.736
713	0.764
727	0.789
741	0.822
755	0.847
769	0.872
783	0.900
797	0.927
811	0.957
CURVE 43	
83	-0.280*
123	-0.238*
173	-0.178*
223	-0.108*
323	0.048*
373	0.132*
423	0.218*
473	0.307*
523	0.402*
573	0.498*
623	0.593*
673	0.690*
723	0.786*
773	0.882*
823	0.982*
873	1.088*
923	1.197*
973	1.304*
1023	1.412*
1073	1.517*
1123	1.625*

* Not shown in figure.

DATA TABLE 258. THERMAL LINEAR EXPANSION OF IRON + CHROMIUM + ΣX_i ALLOYS Fe + Cr + ΣX_i (continued)

T	$\Delta L/L_0$
CURVE 43 (cont.)	
1173	1.731
1223	1.840
1273	1.953
CURVE 44	
302	0.007*
319	0.021
333	0.030
341	0.018*
341	0.044
353	0.055
373	0.077
384	0.055*
396	0.101
422	0.097*
428	0.139
473	0.168*
473	0.188
494	0.197*
530	0.257
561	0.280
581	0.317*
645	0.388*
660	0.416
681	0.433*
700	0.463
707	0.509*
738	0.553*
746	0.522
798	0.582*
804	0.605
836	0.635*
882	0.708
911	0.740*
941	0.796
969	0.835*
1014	0.904
1024	0.929
1080	1.013*
1107	1.073*
1145	1.129*
1180	1.197*
1216	1.265*
1234	1.314*
1267	1.374*
1267	1.383*

T	$\Delta L/L_0$
CURVE 45‡	
20	-0.176
30	-0.176*
40	-0.176*
50	-0.175
60	-0.174*
70	-0.172
80	-0.169*
90	-0.165**
100	-0.161
110	-0.156*
120	-0.150*
130	-0.144
140	-0.137*
150	-0.130
160	-0.123*
170	-0.115
180	-0.107*
190	-0.098*
200	-0.089
210	-0.081*
220	-0.071*
230	-0.062
240	-0.053*
250	-0.043
260	-0.033*
270	-0.023*
280	-0.013
290	-0.003*
300	0.007*
CURVE 46	
18	-0.178*
23	-0.178*
33	-0.178
43	-0.178*
53	-0.178*
63	-0.177
73	-0.176*
83	-0.173
93	-0.170
103	-0.165*
113	-0.160
123	-0.154*
133	-0.147*
143	-0.140*
153	-0.133
163	-0.124*
173	-0.116*
193	-0.099

T	$\Delta L/L_0$
CURVE 46 (cont.)	
213	-0.080
233	-0.060
253	-0.040
273	-0.020*
293	0.000
313	0.022
333	0.045*
353	0.068*
373	0.090*
393	0.114
413	0.138
433	0.162*
453	0.186
473	0.210*
493	0.234
513	0.258
533	0.282*
553	0.306
573	0.330
CURVE 47	
20	-0.186
40	-0.185
60	-0.182*
80	-0.176
100	-0.167*
120	-0.155
140	-0.141*
160	-0.125*
180	-0.108
200	-0.090*
220	-0.072
240	-0.053
260	-0.033
273	-0.019
280	-0.013
293	0.000
CURVE 48	
299	0.001*
420	0.023
478	0.069
513	0.140
534	0.196
620	0.370
698	0.493
759	0.575
843	0.667

T	$\Delta L/L_0$
CURVE 48 (cont.)	
922	0.773
1005	0.898
1025	0.941
1042	0.954
1070	0.980
1098	0.874*
1118	0.804*
1141	0.870*
1202	1.025*
1258	1.167*
1285	1.236*
1308	1.320*
1362	1.501*
1400	1.596*
1423	1.648*
1448	1.706*
CURVE 49	
79	-0.148
273	-0.020
373	0.084
473	0.195
573	0.312
673	0.433
773	0.559
873	0.687
973	0.815
1073	0.939*
1083	0.951*
1093	0.961*
1103	0.968*
1113	0.956*
1123	0.933*
1133	0.914*
1143	0.915*
1153	0.922*
1163	0.931*
1173	0.942*
1198	0.974*
1223	1.016*
1273	1.140*
CURVE 50*	
0	-0.178
10	-0.178
20	-0.177
30	-0.177
40	-0.177

T	$\Delta L/L_0$
CURVE 50 (cont.)*	
50	-0.176
60	-0.174
70	-0.171
80	-0.168
90	-0.164
100	-0.160
120	-0.149
140	-0.136
160	-0.122
180	-0.106
200	-0.089
220	-0.071
240	-0.053
260	-0.034
273	-0.021
280	-0.014
293	0.000
300	0.007
CURVE 51*	
20	-0.185
40	-0.184
60	-0.182
80	-0.177
100	-0.168
120	-0.157
140	-0.143
160	-0.131
180	-0.115
200	-0.095
220	-0.076
240	-0.056
260	-0.035
273	-0.021
280	-0.013
293	0.000
CURVE 52	
300	0.007
387	0.101
466	0.170
504	0.249
548	0.339
604	0.397
645	0.417
783	0.627
810	0.665
884	0.748

T	$\Delta L/L_0$
CURVE 52 (cont.)	
930	0.834
986	0.944
1060	1.049
1101	1.149
1128	1.223*
1173	1.311*
1208	1.291*
1260	1.452*
1291	1.587*
1301	1.575*
1317	1.567*
1372	1.742*
1373	1.777*
1449	1.994
1493	2.061*
CURVE 53	
293	0.000
373	0.087
473	0.196
573	0.324
673	0.451
773	0.586
873	0.727
894	0.754
CURVE 54*	
298	0.005
366	0.069
422	0.130
477	0.186
533	0.249
588	0.312
644	0.377
699	0.445
755	0.512
810	0.592
866	0.660
922	0.725
CURVE 55*	
298	0.006
811	0.594
CURVE 56*	
299	0.004

* Not shown in figure.
‡ Author's data for coefficient of thermal expansion have been integrated by TPRC to obtain $\Delta L/L_0$.

DATA TABLE 258. THERMAL LINEAR EXPANSION OF IRON + CHROMIUM + ΣX_i ALLOYS Fe + Cr + ΣX_i (continued)

T	$\Delta L/L_0$
CURVE 56 (cont.)*	
384	0.027
419	0.057
463	0.088
560	0.175
619	0.258
662	0.325
703	0.394
755	0.509
792	0.562
835	0.615
867	0.662
892	0.681
920	0.706
949	0.719
978	0.686
997	0.647
1008	0.642
1032	0.637
1050	0.658
1060	0.669
1115	0.749
1172	0.852
1227	0.950
1255	1.010
1315	1.150
1357	1.287
1424	1.433
1478	1.562
1535	1.695
1583	1.822
CURVE 57*	
0	-0.282
10	-0.282
20	-0.282
30	-0.281
40	-0.280
50	-0.279
60	-0.276
70	-0.272
80	-0.267
90	-0.260
100	-0.253
120	-0.235
140	-0.214
160	-0.190
180	-0.164
200	-0.137
220	-0.109
CURVE 57 (cont.)*	
240	-0.081
260	-0.051
273	-0.032
280	-0.021
293	0.000
300	0.010
CURVE 58*	
0	-0.193
10	-0.193
20	-0.193
30	-0.193
40	-0.192
50	-0.191
60	-0.189
70	-0.187
80	-0.183
90	-0.179
100	-0.174
120	-0.162
140	-0.148
160	-0.132
180	-0.115
200	-0.096
220	-0.077
240	-0.057
260	-0.036
273	-0.023
280	-0.015
293	0.000
300	0.006
CURVE 59*	
293	0.000
373	0.082
473	0.194
573	0.307
673	0.424
775	0.547
873	0.668
CURVE 60*	
293	0.000
373	0.082
573	0.307
773	0.544
CURVE 61*	
293	0.000
373	0.084
473	0.196
573	0.316
673	0.441
773	0.571
873	0.702
973	0.830
CURVE 62*	
295	0.003
755	0.564
CURVE 63*	
301	0.045
321	0.156
339	0.232
361	0.280
376	0.330
394	0.358
410	0.347
420	0.372
437	0.352
462	0.372
465	0.394
481	0.422
503	0.436
507	0.500
526	0.551
548	0.551
557	0.635
589	0.652
648	0.772
673	0.878
707	0.942
747	1.051
793	1.155
1351	2.394
CURVE 64*	
293	0.000
373	0.084
473	0.203
573	0.336
673	0.486
773	0.643
873	0.748
CURVE 64 (cont.)*	
973	0.864
CURVE 65*, ‡	
300	0.011
320	0.042
347	0.085
380	0.139
421	0.209
462	0.282
514	0.377
558	0.461
581	0.508
594	0.534
613	0.573
631	0.608
644	0.633
662	0.668
685	0.711
704	0.746
723	0.781
749	0.830
764	0.858
783	0.894
809	0.945
847	1.020
866	1.059
889	1.105
906	1.138
928	1.179
950	1.220
967	1.252
986	1.288
1005	1.325
1008	1.331
1036	1.386
1057	1.429
1084	1.489
1092	1.507
CURVE 66*	
294	0.002
589	0.463
700	0.651
811	0.848
922	1.064
1089	1.399
1144	1.517
CURVE 67*	
294	0.002
589	0.423
700	0.608
811	0.788
922	0.971
1089	1.290
CURVE 68*	
294	0.001
366	0.069
477	0.196
588	0.329
699	0.460
810	0.596
CURVE 69*	
373	0.071
473	0.182
573	0.294
673	0.421
773	0.553
873	0.663
973	0.856
1073	0.964
1173	1.142
1273	1.346
CURVE 70*	
373	0.092
473	0.252
573	0.394
673	0.540
773	0.710
873	0.880
973	1.042
1073	1.152
1173	1.340
1273	1.583
CURVE 71*	
373	0.059
473	0.171
573	0.282
673	0.404
773	0.521
873	0.641
CURVE 71 (cont.)*	
973	0.783
1073	0.942
1173	1.096
1273	1.243
CURVE 72*	
373	0.114
473	0.290
573	0.456
673	0.633
773	0.807
873	1.007
973	1.204
1073	1.404
1173	1.598
1273	1.819
CURVE 73*	
773	0.820
873	1.041
973	1.252
1073	1.441
1173	1.663
1273	1.903
CURVE 74*	
373	0.126
473	0.296
573	0.477
673	0.668
773	0.868
873	1.079
973	1.306
1073	1.540
1173	1.770
1273	2.324
CURVE 75*	
373	0.132
473	0.312
573	0.492
673	0.696
773	0.885
873	1.096
973	1.313
1073	1.533

* Not shown in figure.

‡ Author's data for coefficient of thermal expansion have been integrated by TPRC to obtain $\Delta L/L_0$.

DATA TABLE 258. THERMAL LINEAR EXPANSION OF IRON + CHROMIUM + ΣX_i ALLOYS Fe + Cr + ΣX_i (continued)

T	$\Delta L/L_0$
CURVE 75 (cont.)*	
1173	1.744
1273	1.974
CURVE 76*	
573	0.472
673	0.654
773	0.850
873	1.054
973	1.265
1073	1.480
1173	1.712
1273	1.945
CURVE 77*	
373	0.122
473	0.295
573	0.466
673	0.641
773	0.848
873	1.039
973	1.251
1073	1.473
1173	1.704
1273	1.918
CURVE 78*	
473	0.297
573	0.473
673	0.674
773	0.886
873	1.095
973	1.258
1073	1.524
1173	1.749
1273	1.886
CURVE 79*	
473	0.266
573	0.441
673	0.611
773	0.810
873	0.982
1073	1.362
1173	1.552
1273	1.712
CURVE 80*	
473	0.218
573	0.361
673	0.493
773	0.631
873	0.775
973	0.818
1073	0.913
1173	1.153
1273	1.309
CURVE 81*	
298	0.006
373	0.094
473	0.220
573	0.364
673	0.515
773	0.675
873	0.841
973	1.009
CURVE 82*	
298	0.005
373	0.080
473	0.186
573	0.306
673	0.432
773	0.567
873	0.706
973	0.843
CURVE 83*	
302	0.010
365	0.081
445	0.177
599	0.405
678	0.528
692	0.548
768	0.665
880	0.855
996	1.040
1035	1.095
1044	1.059
1050	1.016
1057	0.964
1063	0.931
1073	0.908
CURVE 83 (cont.)*	
1088	0.908
1099	0.931
1142	1.033
1184	1.131
CURVE 84*	
1184	1.166
1135	1.053
1084	0.941
1079	0.929
1062	0.890
1043	0.857
1035	0.871
1028	0.903
1019	0.929
1006	0.969
1000	0.993
993	1.031
990	1.032
981	1.021
971	1.005
869	0.834
648	0.483
514	0.287
303	0.024
CURVE 85*	
302	0.010
361	0.080
603	0.403
706	0.554
795	0.692
886	0.841
964	0.962
1007	1.030
1037	1.072
1056	1.104
1082	1.125
1099	1.133
1111	1.143
1126	1.149
1154	1.164
1173	1.169
CURVE 86*	
1173	1.131
CURVE 86 (cont.)*	
1152	1.107
1140	1.104
1126	1.101
1116	1.094
1096	1.088
1084	1.086
1057	1.063
1042	1.049
1024	1.037
1005	1.023
996	1.007
979	0.970
855	0.775
668	0.489
526	0.286
347	0.066
300	0.008
CURVE 87*	
293	0.000
305	0.019
460	0.189
573	0.355
637	0.444
693	0.525
779	0.665
847	0.775
877	0.832
1054	1.096
1067	1.087
1078	1.080
1089	1.071
1101	1.064
1113	1.048
1121	1.037
1131	1.028
1143	1.028
1154	1.035
1190	1.088
CURVE 88*	
1190	1.090
1138	0.978
1113	0.947
1106	0.964
1092	0.973
1079	1.000
CURVE 88 (cont.)*	
1067	1.013
1053	1.019
1041	1.025
1033	1.024
1019	1.023
1006	1.019
1001	1.023
CURVE 89*	
79	-0.255
273	-0.033
373	0.133
473	0.305
573	0.482
673	0.665
773	0.855
873	1.053
973	1.259
1073	1.474
1173	1.697
1273	1.929
1373	2.172
CURVE 90*	
273	-0.033
373	0.133
473	0.306
573	0.486
673	0.672
773	0.866
873	1.067
973	1.275
1073	1.488
1173	1.708
1273	1.935
1373	2.169
CURVE 91*	
293	0.000
298	0.011
414	0.203
512	0.373
595	0.507
606	0.529
721	0.749
892	1.058
CURVE 91 (cont.)*	
946	1.173
992	1.264
1097	1.493
1163	1.624
1209	1.749
1310	1.997
CURVE 92*	
375	0.150
472	0.303
573	0.471
674	0.652
772	0.834
872	1.031
972	1.235
1072	1.452
1173	1.685
1274	1.920
CURVE 93*	
273	-0.021
298	0.005
421	0.143
541	0.305
662	0.481
795	0.696
CURVE 94*	
795	0.696
622	0.427
488	0.233
297	0.006
CURVE 95*	
273	-0.022
298	0.005
447	0.152
539	0.304
661	0.483
787	0.683
CURVE 96*	
787	0.683
618	0.418

* Not shown in figure.

DATA TABLE 258. THERMAL LINEAR EXPANSION OF IRON + CHROMIUM + ΣX_i ALLOYS $Fe + Cr + \Sigma X_i$ (continued)

T	$\Delta L/L_0$
CURVE 96 (cont.)*	
475	0.215
297	0.004
CURVE 97*, ‡	
77	-0.243
83	-0.241
93	-0.237
103	-0.231
113	-0.225
123	-0.218
133	-0.210
143	-0.201
153	-0.191
163	-0.180
173	-0.169
183	-0.157
193	-0.144
203	-0.131
213	-0.118
223	-0.104
233	-0.090
243	-0.075
253	-0.061
263	-0.046
273	-0.031
283	-0.015
293	0.000
303	0.016
313	0.031
323	0.048
CURVE 98*	
298	0.005
323	0.033
348	0.059
373	0.085
398	0.110
423	0.136
445	0.161
473	0.189
498	0.214
523	0.243
548	0.270
573	0.296
598	0.320
623	0.346
648	0.374

T	$\Delta L/L_0$
CURVE 98 (cont.)*	
673	0.401
678	0.405
CURVE 99*	
19	-0.285
33	-0.280
88	-0.265
144	-0.220
200	-0.155
255	-0.065
293	0.000

* Not shown in figure.

‡ Author's data for coefficient of thermal expansion have been integrated by TPRC to obtain $\Delta L/L_0$.

DATA TABLE 258. COEFFICIENT OF THERMAL LINEAR EXPANSION OF IRON + CHROMIUM + ΣX_i ALLOYS $Fe + Cr + \Sigma X_i$

[Temperature, T, K; Coefficient of Expansion, α, 10^{-6} K^{-1}]

T	α
CURVE 2‡	
77	6.00
83	6.60
93	7.50
103	8.40
113	9.30
123	10.30
133	11.20
143	12.00
153	12.60
163	13.10
173	13.40
183	13.80
193	14.00
203	14.30
213	14.70
223	14.90
233	15.10
243	15.40
253	15.70
263	15.90
273	16.10
283	16.30
293	16.50
303	16.70*
313	16.80*
323	17.00*
CURVE 45‡	
20	10.36
30	10.24
40	10.11
50	9.98
60	9.84
70	9.70
80	9.55
90	9.39
100	9.22
110	9.03
120	8.82
130	8.59
140	8.34
150	8.06
160	7.74
170	7.38
180	6.97
190	6.50
CURVE 45 (cont.)‡	
200	5.96
210	5.34
220	4.66
230	3.95
240	3.21
250	2.44
260	1.64
270	0.92
280	0.43
290	0.17
300	0.06
CURVE 65*, ‡	
300	15.42
320	15.71
347	16.02
380	16.73
421	17.64
462	17.86
514	18.56
558	19.78
581	20.57
594	20.46
613	20.00
631	19.54
644	19.23
662	19.04
685	18.24
704	18.54
723	18.60
749	18.90
764	19.00
783	19.07
809	19.57
847	20.31
866	20.31
889	19.68
906	18.81
928	18.67
950	18.60
967	18.98
986	19.19
1005	19.89
1008	19.37
1036	19.93
1057	21.16
CURVE 65 (cont.)*, ‡	
1084	22.93
1092	23.76
CURVE 97‡	
77	3.20
83	3.80
93	4.70
103	5.80
113	6.70
123	7.70
133	8.60
143	9.40
153	10.30
163	10.99
173	11.69
183	12.30
193	12.79
203	13.30
213	13.70
223	14.00
233	14.29
243	14.60
253	14.80
263	14.99
273	15.20
283	15.39
293	15.60
303	15.70*
313	16.00*
323	16.09*

* Not shown in figure.

‡ Author's data for coefficient of thermal expansion have been integrated by TPRC to obtain $\Delta L/L_0$.

FIGURE AND TABLE NO. 259R. PROVISIONAL VALUES FOR THERMAL LINEAR EXPANSION OF IRON + COBALT + ΣX_i ALLOYS Fe + Co + ΣX_i

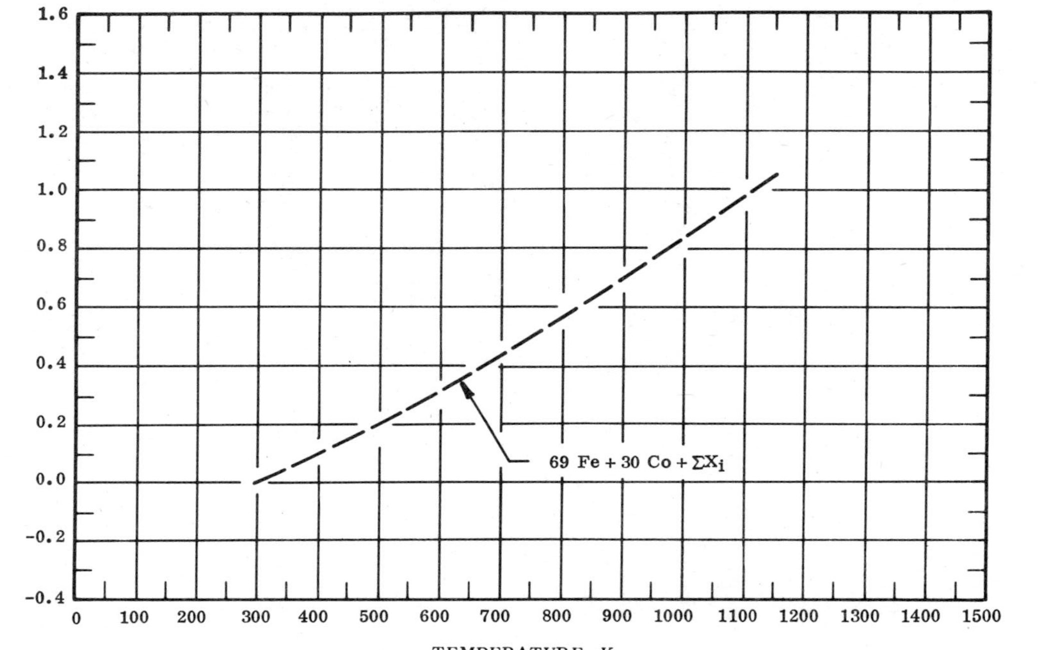

PROVISIONAL VALUES

[Temperature, T, K; Linear Expansion, $\Delta L/L_0$, %; α, K^{-1}]

(69 Fe + 30 Co + ΣX_i)		
T	$\Delta L/L_0$	$\alpha \times 10^6$
293	0.000	9.4
400	0.105	10.3
500	0.211	11.0
600	0.325	11.7
700	0.446	12.4
800	0.572	12.9
900	0.705	13.4
1000	0.842	13.9
1100	0.982	14.2
1150	1.054	14.4

REMARKS

(69 Fe + 30 Co + ΣX_i): The tabulated values for well-annealed alloy are considered accurate to within ± 7% over the entire temperature range. These values can be represented approximately by the following equation:

$$\Delta L/L_0 = -0.235 + 6.536 \times 10^{-4}\, T + 5.353 \times 10^{-7}\, T^2 - 1.119 \times 10^{-10}\, T^3$$

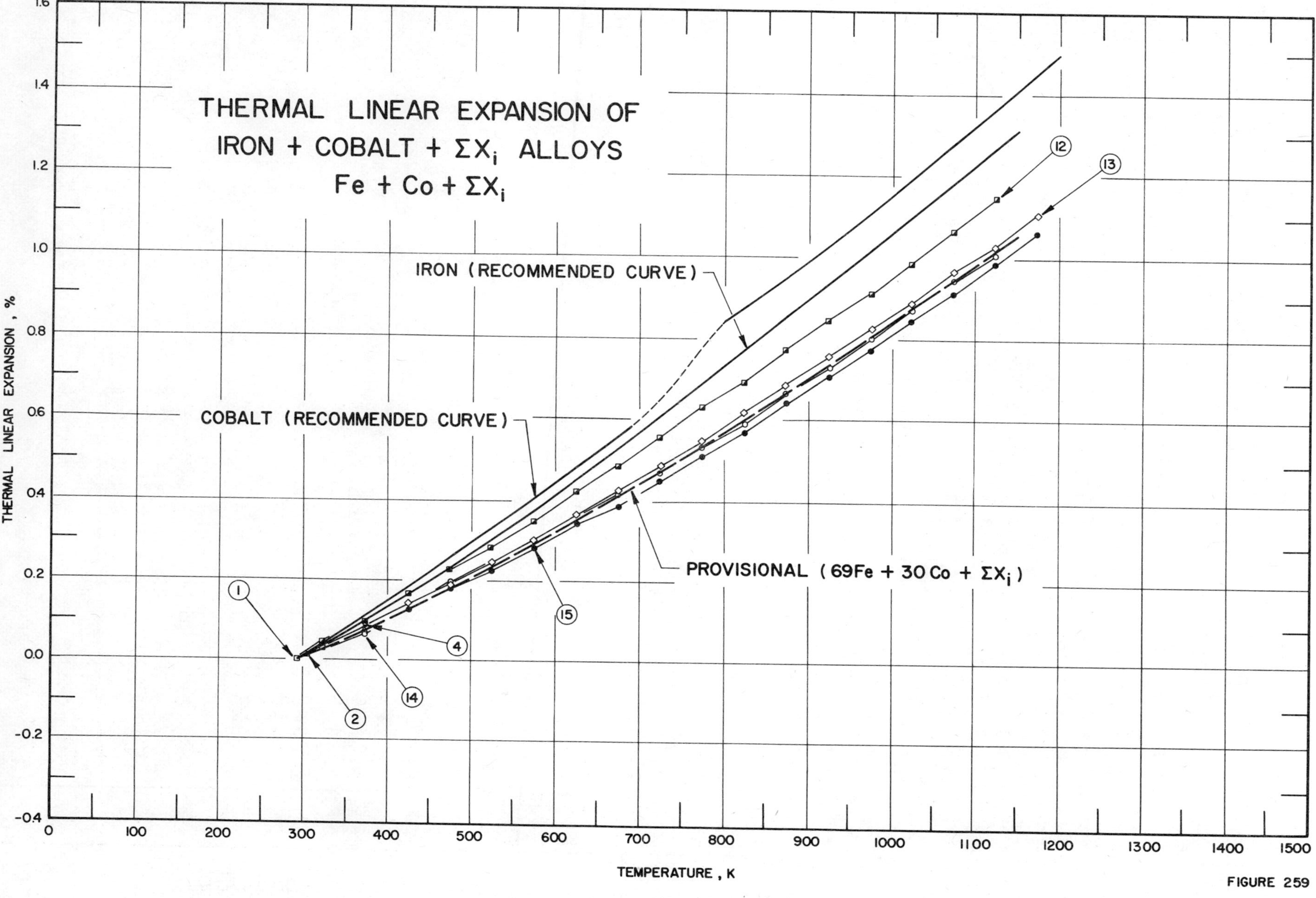

THERMAL LINEAR EXPANSION OF
IRON + COBALT + ΣX$_i$ ALLOYS
Fe + Co + ΣX$_i$
IRON (RECOMMENDED CURVE)
COBALT (RECOMMENDED CURVE)
PROVISIONAL (69Fe + 30Co + ΣX$_i$)
1
2
4
12
13
14
15
THERMAL LINEAR EXPANSION, %
1.6
1.4
1.2
1.0
0.8
0.6
0.4
0.2
0.0
-0.2
-0.4
TEMPERATURE, K
0
100
200
300
400
500
600
700
800
900
1000
1100
1200
1300
1400
1500

FIGURE 259

SPECIFICATION TABLE 259. THERMAL LINEAR EXPANSION OF IRON + COBALT + ΣX_i ALLOYS Fe + Co + ΣX_i

Cur. No.	Ref. No.	Author(s)	Year	Method Used	Temp. Range, K	Name and Specimen Designation	Composition (weight percent) Fe	Co	Mo	Ni	Mn	Si	Composition (continued), Specifications, and Remarks
1	405	Masumoto, H., Saito, H., Sugai, Y. and Kono, T.	1960	L	293-323		50	45	5				Electrolyte Fe (0.02 Al, trace C, Si, Mn, P, S), electrolytic Co (0.26 Ni, 0.15 Fe, 0.03 C, 0.01 Si, trace Mn, Al, P, S) and powder Mo with purity 99.94; specimen 3 mm long.
2	59	Masumoto, H.	1931	L	303-373	No. 15	70.1	19.9		10			Armco Fe, granular Co, Mond Ni, mixed and melted in hydrogen atmosphere in alumina crucible in Tammann furnace, cast in iron mould with cylindrical aperture, machined to circular rod 4 mm thick, 10 cm long, heated at 1000 K for 1 hr in electric furnace in hydrogen atmosphere at reduced pressure; expansion measured in vacuum.
3*	59	Masumoto, H.	1931	L	303-373	No. 27	60.1	29.9		10			Similar to the above specimen.
4	59	Masumoto, H.	1931	L	303-373	No. 28	55.1	29.9		15			Similar to the above specimen.
5*	59	Masumoto, H.	1931	L	303-373	No. 29	50.1	29.9		20			Similar to the above specimen.
6*	59	Masumoto, H.	1931	L	303-373	No. 30	47.3	29.7		23			Similar to the above specimen.
7*	59	Masumoto, H.	1931	L	303-373	No. 31	45.1	29.9		25			Similar to the above specimen.
8*	59	Masumoto, H.	1931	L	303-373	No. 37	50.1	39.9		10			Similar to the above specimen.
9*	59	Masumoto, H.	1931	L	303-373	No. 38	45.4	39.6		15			Similar to the above specimen.
10*	59	Masumoto, H.	1931	L	303-373	No. 39	42.4	39.6		18			Similar to the above specimen.
11*	59	Masumoto, H.	1931	L	303-373	No. 40	40.1	39.9		20			Similar to the above specimen.
12	555	Fine, M. E. and Ellis, W. C.	1948	L	293-1123	5931	89.4	10.2			0.31	<0.01	Specimen prepared from cast ingots by first hot-swaging followed by cold-swaging to 0.18 in. diameter x 2.50 in. long with flat-ground and polished ends; before measurement they are annealed in hydrogen for 1 hr at 900 C; zero-point correction is 0.014%.
13	555	Fine, M. E. and Ellis, W. C.	1948	L	293-1173	5932	79.2	20.0			0.47		Similar to the above specimen; zero-point correction is 0.009%.
14	555	Fine, M. E. and Ellis, W. C.	1948	L	293-1123	5933	69.4	30.2			0.45		Similar to the above specimen; zero-point correction is 0.012%.
15	555	Fine, M. E. and Ellis, W. C.	1948	L	293-1173	5934	59.4	40.1			0.44		Similar to the above specimen; zero-point correction is 0.005%.

* Not shown in figure.

DATA TABLE 259. THERMAL LINEAR EXPANSION OF IRON + COBALT + ΣX_i ALLOYS Fe + Co + ΣX_i

[Temperature, T, K; Linear Expansion, $\Delta L/L_0$, %]

T	$\Delta L/L_0$
CURVE 1	
293	0.000
323	0.026
CURVE 2	
303	0.0099
373	0.0788
CURVE 3*	
303	0.0099
373	0.0793
CURVE 4	
303	0.0104*
373	0.0835
CURVE 5*	
303	0.0103
373	0.0821
CURVE 6*	
303	0.0099
373	0.0793
CURVE 7*	
303	0.0094
373	0.0754
CURVE 8*	
303	0.0099
373	0.0791
CURVE 9*	
303	0.0099
373	0.0794
CURVE 10*	
303	0.0099
373	0.0795
CURVE 11*	
303	0.0102
373	0.0814
CURVE 12	
293	0.000*
323	0.042
373	0.097
423	0.161
473	0.229
523	0.285
573	0.352
623	0.422
673	0.486
723	0.559
773	0.630
823	0.692
873	0.772
923	0.848
973	0.919
1023	0.991
1073	1.069
1123	1.147
CURVE 13	
293	0.000*
323	0.037*
373	0.085*
423	0.141
473	0.194
523	0.252
573	0.309
623	0.367
673	0.431
723	0.494
773	0.555
823	0.620
873	0.689
923	0.759
973	0.831
1023	0.894
1073	0.969
1123	1.035
1173	1.109
CURVE 14	
293	0.000*
323	0.033*
373	0.089*
423	0.142*
473	0.197
523	0.252*
573	0.302*
623	0.362*
673	0.420
723	0.477
773	0.538
823	0.596
873	0.663
923	0.733
973	0.801
1023	0.876
1073	0.942
1123	1.012
CURVE 15	
293	0.000*
323	0.028*
373	0.075
423	0.131
473	0.181
523	0.231
573	0.286
623	0.342
673	0.396
723	0.458
773	0.517
823	0.576
873	0.642
923	0.709
973	0.775
1023	0.847
1073	0.919
1123	0.993
1173	1.063

* Not shown in figure.

FIGURE AND TABLE NO. 260R. PROVISIONAL VALUES FOR THERMAL LINEAR EXPANSION OF IRON + COPPER + ΣX_i ALLOYS   Fe + Cu + ΣX_i

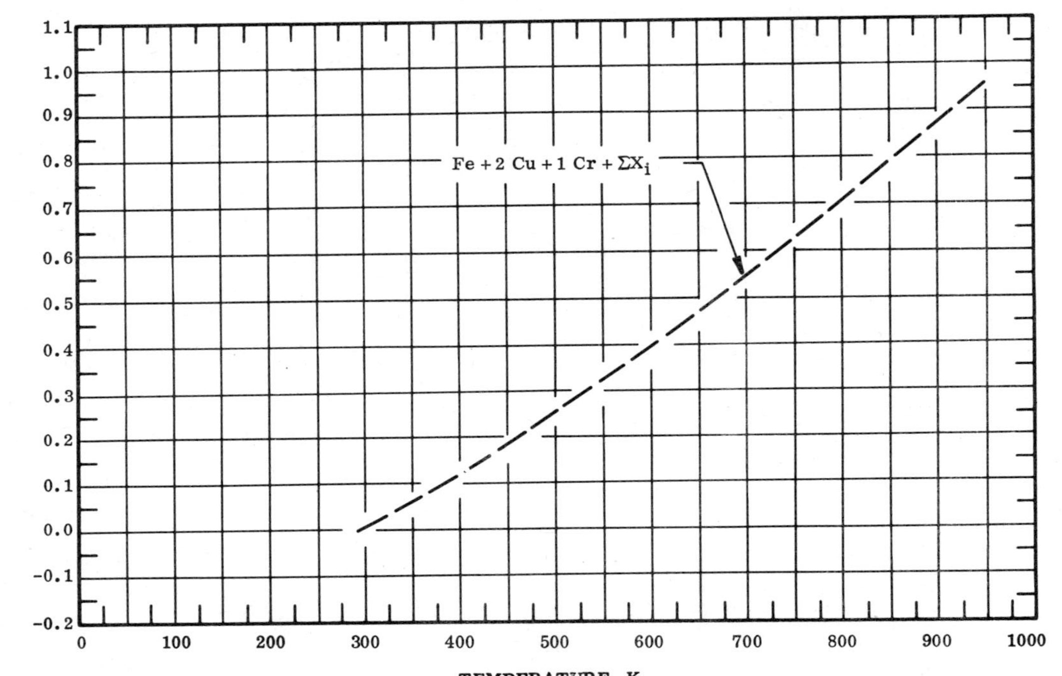

PROVISIONAL VALUES

[Temperature, T, K; Linear Expansion, $\Delta L/L_0$, %; α, K^{-1}]

(Fe + 2 Cu + 1 Cr + ΣX_i)

T	$\Delta L/L_0$	$\alpha \times 10^6$
293	0.000	10.7
350	0.064	11.6
400	0.124	12.4
450	0.188	13.1
500	0.255	13.7
550	0.324	14.3
600	0.397	14.8
650	0.472	15.2
700	0.550	15.6
750	0.629	15.9
800	0.709	16.2
850	0.790	16.3
900	0.872	16.4
950	0.955	16.5

REMARKS

(Fe + 2 Cu + 1 Cr + ΣX_i): The tabulated values for well-annealed alloy are considered accurate to within ± 7% over the entire temperature range. These values can be represented approximately by the following equation:

$$\Delta L/L_0 = -0.227 + 4.437 \times 10^{-4}\ T + 1.258 \times 10^{-6}\ T^2 - 4.377 \times 10^{-10}\ T^3$$

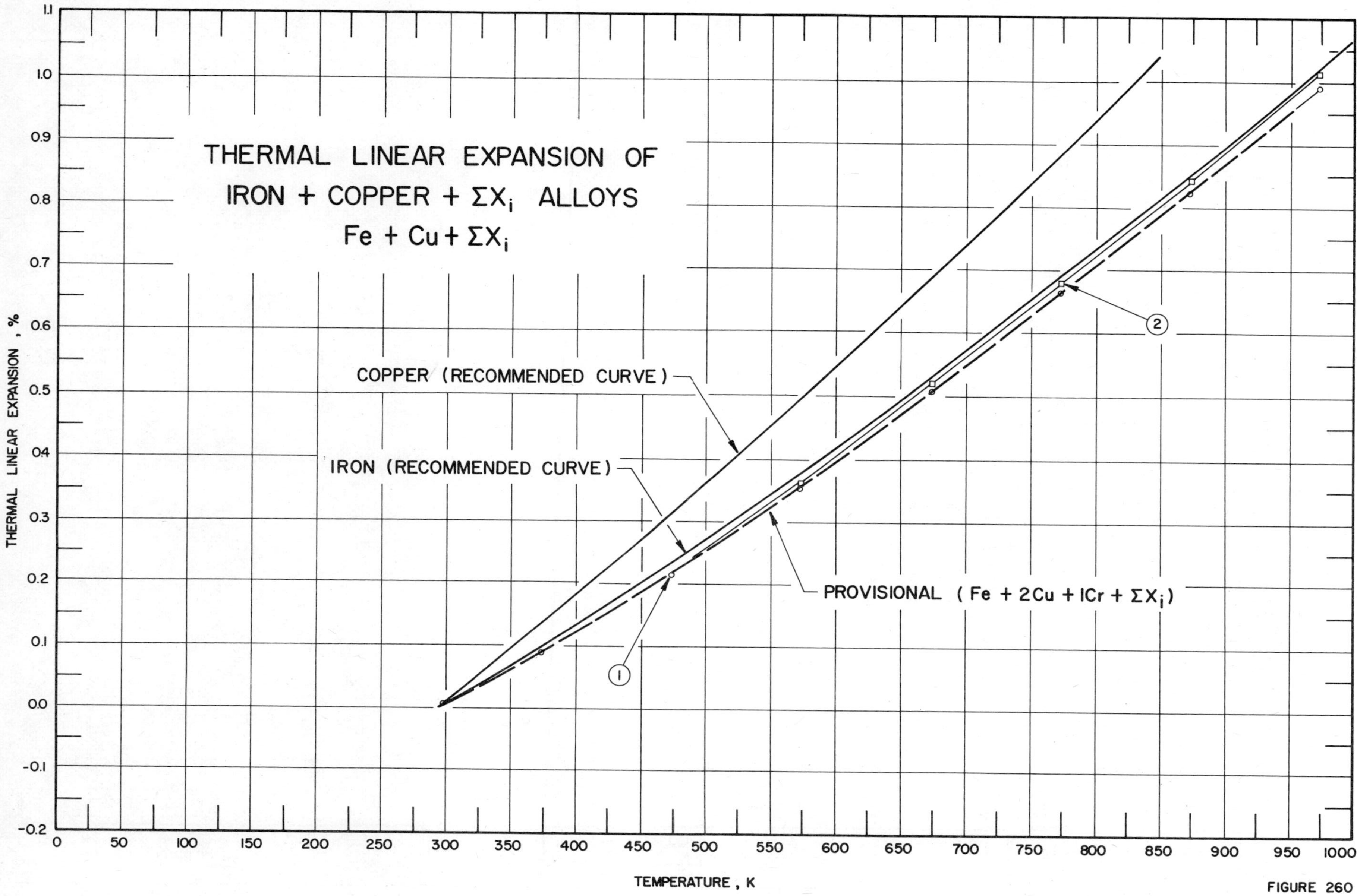
THERMAL LINEAR EXPANSION OF
IRON + COPPER + ΣXi ALLOYS
Fe + Cu + ΣXi
COPPER (RECOMMENDED CURVE)
IRON (RECOMMENDED CURVE)
PROVISIONAL (Fe + 2Cu + 1Cr + ΣXi)
1
2
THERMAL LINEAR EXPANSION, %
1.1
1.0
0.9
0.8
0.7
0.6
0.5
0.4
0.3
0.2
0.1
0.0
-0.1
-0.2
0
50
100
150
200
250
300
350
400
450
500
550
600
650
700
750
800
850
900
950
1000
TEMPERATURE, K

FIGURE 260

SPECIFICATION TABLE 260. THERMAL LINEAR EXPANSION OF IRON + COPPER + ΣX_i ALLOYS Fe + Cu + ΣX_i

Cur. No.	Ref. No.	Author(s)	Year	Method Used	Temp. Range, K	Name and Specimen Designation	Composition (weight percent) Fe	Cu	Cr	V	C	Mn	Composition (continued), Specifications, and Remarks
1	51	Souder, W. and Hidnert, P.	1922	T	298-973	S 555	Bal.	1.85	1.15	0.21	0.144	0.10	0.035 S, 0.034 Si, 0.03 P; annealed specimen.
2	51	Souder, W. and Hidnert, P.	1922	T	298-973	S 560	Bal.	2.70	0.82	0.26	0.342	0.28	0.094 Si, 0.043 S, 0.01 P; annealed specimen.

DATA TABLE 260. THERMAL LINEAR EXPANSION OF IRON + COPPER + ΣX_i ALLOYS Fe + Cu + ΣX_i

[Temperature, T, K; Linear Expansion, $\Delta L/L_0$, %]

T	$\Delta L/L_0$	T	$\Delta L/L_0$
CURVE 1		CURVE 2	
298	0.006	298	0.006*
373	0.090	373	0.093*
473	0.216	473	0.219*
573	0.354	573	0.361
673	0.510	673	0.521
773	0.666	773	0.680
873	0.826	873	0.844
973	0.993	973	1.013

*Not shown in figure.

FIGURE AND TABLE NO. 261R. PROVISIONAL VALUES FOR THERMAL LINEAR EXPANSION OF IRON + MANGANESE + ΣX_i ALLOYS Fe + Mn + ΣX_i

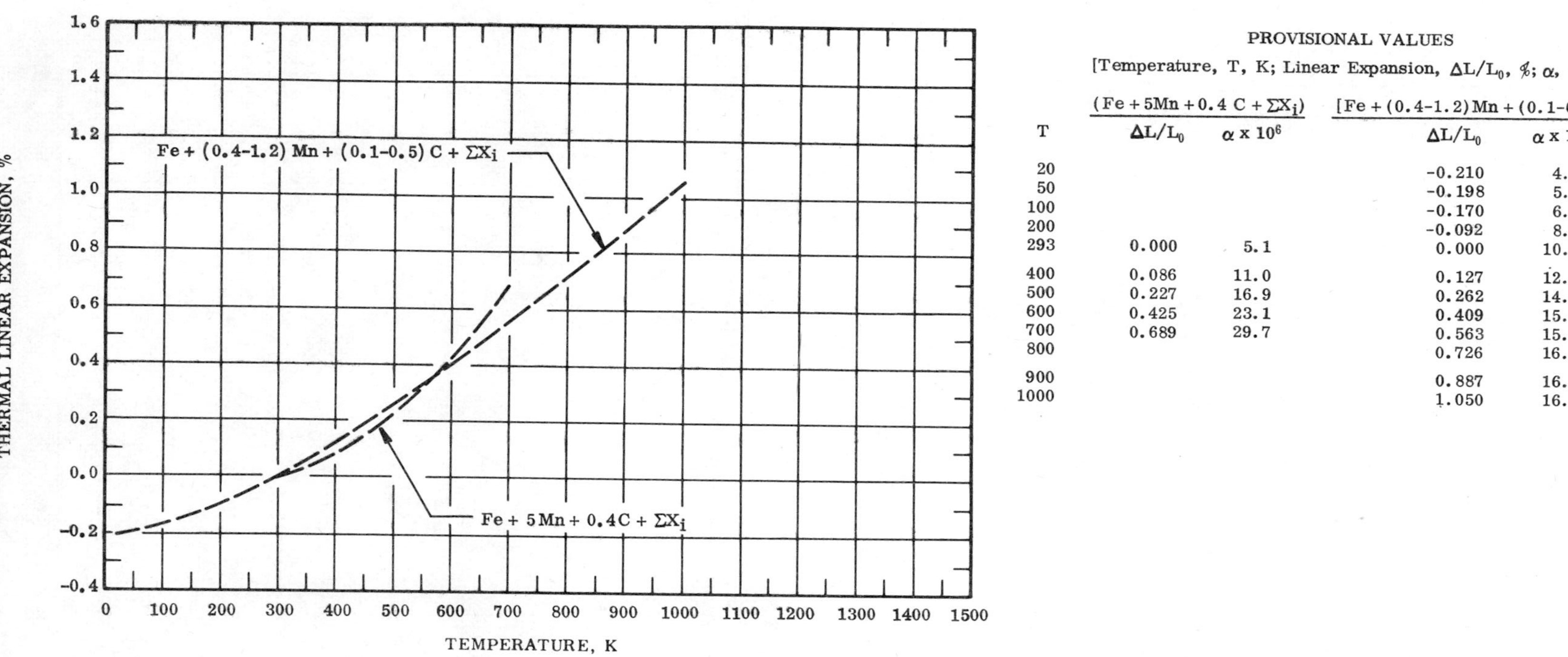

PROVISIONAL VALUES

[Temperature, T, K; Linear Expansion, $\Delta L/L_0$, %; α, K^{-1}]

T	(Fe + 5Mn + 0.4 C + ΣX_i)		[Fe + (0.4-1.2) Mn + (0.1-0.5) C + ΣX_i]	
	$\Delta L/L_0$	$\alpha \times 10^6$	$\Delta L/L_0$	$\alpha \times 10^6$
20			-0.210	4.1
50			-0.198	5.0
100			-0.170	6.4
200			-0.092	8.9
293	0.000	5.1	0.000	10.9
400	0.086	11.0	0.127	12.8
500	0.227	16.9	0.262	14.2
600	0.425	23.1	0.409	15.2
700	0.689	29.7	0.563	15.9
800			0.726	16.2
900			0.887	16.2
1000			1.050	16.3

REMARKS

(Fe + 5 Mn + 0.4 C + ΣX_i): The tabulated values for well-annealed alloy of type steel 40G5 are considered accurate to within ±7% over the entire temperature range. These values can be represented approximately by the following equation:

$$\Delta L/L_0 = 0.069 - 9.291 \times 10^{-4} T + 2.213 \times 10^{-6} T^2 + 5.434 \times 10^{-10} T^3$$

[Fe + (0.4-1.2) Mn + (0.1-0.5) C + ΣX_i]: The tabulated values for well-annealed alloy of type SAE 1010, SAE 1020 are considered accurate to within ±7% over the entire temperature range. These values can be represented approximately by the following equation:

$$\Delta L/L_0 = -0.219 + 3.528 \times 10^{-4} T + 1.523 \times 10^{-6} T^2 - 6.093 \times 10^{-10} T^3$$

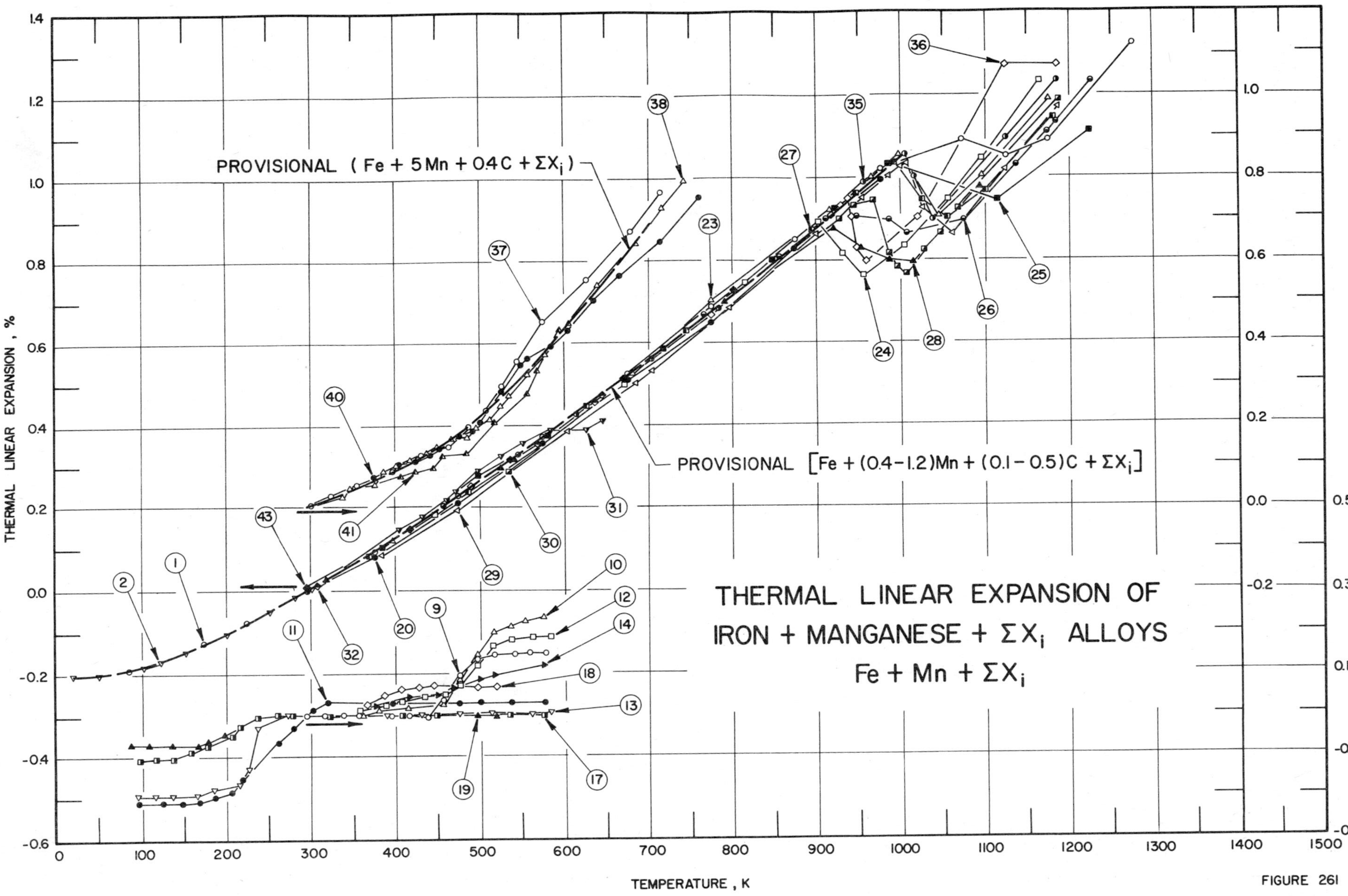
THERMAL LINEAR EXPANSION OF
IRON + MANGANESE + ΣXi ALLOYS
Fe + Mn + ΣXi
PROVISIONAL (Fe + 5 Mn + 0.4 C + ΣXi)
PROVISIONAL [Fe + (0.4 – 1.2) Mn + (0.1 – 0.5) C + ΣXi]
THERMAL LINEAR EXPANSION, %
TEMPERATURE, K
0
100
200
300
400
500
600
700
800
900
1000
1100
1200
1300
1400
1500
-0.6
-0.4
-0.2
0.0
0.2
0.4
0.6
0.8
1.0
1.2
1.4
-0.3
-0.1
0.1
0.3
0.5
1
2
9
10
11
12
13
14
17
18
19
20
23
24
25
26
27
28
29
30
31
32
35
36
37
38
40
41
43

FIGURE 261

SPECIFICATION TABLE 261. THERMAL LINEAR EXPANSION OF IRON + MANGANESE + ΣX_i ALLOYS $Fe + Mn + \Sigma X_i$

Cur. No.	Ref. No.	Author(s)	Year	Method Used	Temp. Range, K	Name and Specimen Designation	Composition (weight percent) Fe	Mn	C	Si	P	Cr	Composition (continued), Specifications, and Remarks
1	466	Lucks, C. F., Thompson, H. B., Smith, A.R., Curry, F.P., Deem, H.W., and Bing, G.F.	1951	L	83-1273	SAE 1010 Steel (AISI C1010)	Bal.	0.42	0.10		0.008		0.028 S; hot-rolled; material supplied by United States Steel Corporation; density 7.85 g/cm^3 at 293 K; zero-point correction of -0.023% determined by graphical interpolation.
2	481	Johnston, H. L., H.W., and Rubin, T.	1965	I	20-300	SAE 1020 Low Carbon Steel	99.17	0.60	0.23				Handbook composition; supplied by Carnegie-Illinois Steel Co., used without further heat treatment; specimen consisted of three sectors cut from bar; filed to length to within 1/5 wave-length of sodium D radiation; data smoothed by author; zero-point correction of 0.200% determined by graphical interpolation.
3*	302	Beenakker, Y.Y.M., and Swenson, C.A.	1955	L	4.2-300	SAE 1020 Low Carbon Steel							Nominal composition; cold rolled.
4*	540	Trehan, Y.N.	1968	L	473-1273		Bal.	19.65	0.04	0.59		15.2	0.57 N; 5 cm long; solution heat treated for 1/2 hr at 1323 K and water quenched before machining.
5*	540	Trehan, Y.N.	1968	L	373-1173		Bal.	18.48	0.08	0.23		10.06	0.35 N; similar to the above specimen.
6*	542	Lysok, L.I., Makogon, Yu.N., and Nikolior, B.I.	1968	L	386-92	45G10	Bal.	10	0.45				Fine-grain cylindrical specimen; 4-5 mm diameter, 10-15 mm long; made from 45G10 steel, austentic structure at room temperature after quenching in water from 1373 K; expansion measured with decreasing temperature; $\gamma \rightarrow \epsilon'$ transformation; zero-point correction is -0.0527%.
7*	542	Lysok, L.I., et al.	1968	L	92-386	45G10	Bal.	10	0.45				Above specimen; expansion measured with increasing temperature; $\epsilon \rightarrow \gamma$ martensitic transformation; zero-point correction is -0.0068%.
8	542	Lysok, L.I., et al.	1968	L	574-94	35G18	Bal.	18	0.35				Fine-grain cylindrical specimen; 4-5 mm diameter, 10-15 mm long; made from 35G18 steel; austentic structure at room temperature after quenching in water from 1373 K; expansion measured with decreasing temperature after one cycle; zero-point correction is -0.126%.
9	542	Lysok, L.I., et al.	1968	L	293-574	35G18							Above specimen; expansion measured with increasing temperature after one cycle; $\epsilon \rightarrow \gamma$ transformation.
10	542	Lysok, L.I., et al.	1968	L	293-574	35G18							Above specimen; expansion measured with increasing temperature after three (3) cycles; zero-point correction is -0.006%.
11	542	Lysok, L.I., et al.	1968	L	573-97	35G18							Above specimen; expansion measured with decreasing temperature after 3 cycles; zero-point correction is -0.212%.
12	542	Lysok, L.I., et al.	1968	L	293-581	35G18							Above specimen; expansion measured with increasing temperature after 6 cycles.

* Not shown in figure.

SPECIFICATION TABLE 261. THERMAL LINEAR EXPANSION OF IRON + MANGANESE + ΣX_i ALLOYS Fe + Mn + ΣX_i (continued)

Cur. No.	Ref. No.	Author(s)	Year	Method Used	Temp. Range, K	Name and Specimen Designation	Composition (weight percent) Fe	Mn	C	Si	P	Cr	Composition (continued), Specifications, and Remarks
13	542	Lysok, L.I., Makogon, Yu.N., and Nikolior, B.I.	1968	L	581-97	35G18							Above specimen; expansion measured with decreasing temperature after 6 cycles; zero-point correction is -0.189%
14	542	Lysok, L.I., et al.	1968	L	293-575	35G18							Above specimen; expansion measured with increasing temperature after 15 cycles; zero-point correction is -0.002%.
15*	542	Lysok, L.I., et al.	1968	L	575-98	35G18							Above specimen; expansion measured with decreasing temperature after 15 cycles; zero-point correction is -0.124%.
16*	542	Lysok, L.I., et al.	1968	L	300-571	35G18							Above specimen; expansion measured with increasing temperature after 25 cycles; zero-point correction is ±0.000%.
17	542	Lysok, L.I., et al.	1968	L	574-94	35G18							Above specimen; expansion measured with decreasing temperature after 25 cycles; zero-point correction is -0.104%.
18	542	Lysok, L.I., et al.	1968	L	299-519	35G18							Above specimen; expansion measured with increasing temperature after 70-100 cycles; zero-point correction is ±0.000%.
19	542	Lysok, L.I., et al.	1968	L	519-89	35G18							Above specimen; expansion measured with decreasing temperature after 70-100 cycles; zero-point correction is -0.073%.
20	51	Souder, W. and Hidnert, P.	1922	T	298-973	S 548	Bal.	1.21	0.41	0.12	0.05		0.05 S; annealed specimen.
21*	51	Souder, W. and Hidnert, P.	1922	T	293-1173	S 546	Bal.	1.42	0.35	0.20	0.013	1.00	0.11 V, 0.057 S; annealed specimen; expansion measured with increasing temperature.
22*	51	Souder, W. and Hidnert, P.	1922	T	1173-293	S 546	Bal.	1.42		0.20		1.00	The above specimen; expansion with decreasing temperature; zero-point correction is 0.114%.
23	51	Souder, W. and Hidnert, P.	1922	T	296-1174	S 547	Bal.	1.21	0.49	0.12	0.05		0.050 S; annealed specimen; expansion measured with increasing temperature; zero-point correction of 0.004% determined by graphical extrapolation.
24	51	Souder, W. and Hidnert, P.	1922	T	1174-301	S 547	Bal.	1.21		0.12			The above specimen; expansion measured with decreasing temperature; zero-point correction of 0.044% determined by graphical extrapolation.
25	51	Souder, W. and Hidnert, P.	1922	T	290-1225	S 504	Bal.	1.10	0.20		0.05	0.5	0.5 Ni, 0.05 S; annealed specimen; expansion measured with increasing temperature; zero-point correction is -0.008%.
26	51	Souder, W. and Hidnert, P.	1922	T	1225-292	S 504	Bal.	1.10	0.20			0.5	The above specimen; expansion measured with decreasing temperature; zero-point correction is 0.032%.

* Not shown in figure.

SPECIFICATION TABLE 261. THERMAL LINEAR EXPANSION OF IRON + MANGANESE + ΣX_i ALLOYS $Fe + Mn + \Sigma X_i$ (continued)

Cur. No.	Ref. No.	Author(s)	Year	Method Used	Temp. Range, K	Name and Specimen Designation	Composition (weight percent) Fe	Mn	C	Si	P	Cr	Composition (continued), Specifications, and Remarks
27	51	Souder, W. and Hidnert, P.	1922	T	293-1179	S 562	Bal.	1.17	0.380	0.10	0.055		0.81 Ni, 0.067 S; annealed specimen; expansion measured with increasing temperature.
28	51	Souder, W. and Hidnert, P.	1922	T	1179-318	S 562	Bal.	1.17		0.10			The above specimen; expansion measured with decreasing temperature; zero-point correction of 0.021% determined by graphical extrapolation.
29	51	Souder, W. and Hidnert, P.	1922	T	302-1187	S 566	Bal.	0.68	0.418	0.23	0.012		0.025 S; annealed specimen; expansion measured with increasing temperature; zero-point correction of 0.008% determined by graphical extrapolation.
30	51	Souder, W. and Hidnert, P.	1922	T	1187-312	S 566	Bal.	0.68	0.418	0.23			The above specimen; expansion measured with decreasing temperature; zero-point correction of 0.031% determined by graphical extrapolation.
31	51	Souder, W. and Hidnert, P.	1922	T	301-644	S 614	Bal.	1.00	0.85	0.25	<0.02	0.50	0.45 W; hardened from 1123 K; expansion measured with increasing temperature; zero-point correction of 0.050% determined by graphical extrapolation.
32	51	Souder, W. and Hidnert, P.	1922	T	644-292	S 614	Bal.	1.00		0.25		0.50	The above specimen; expansion measured with decreasing temperature; zero-point correction is 0.12%.
33*	51	Souder, W. and Hidnert, P.	1922	T	295-1179	S 549	Bal.	0.57	0.44	0.161	0.013		0.14 V, 0.033 S; annealed specimen; expansion measured with increasing temperature; zero-point correction is ±0.000%.
34*	51	Souder, W. and Hidnert, P.	1922	T	1179-791	S 549	Bal.	0.57		0.161			The above specimen; expansion measured with decreasing temperature.
35	51	Souder, W. and Hidnert, P.	1922	T	298-1181	S 550	Bal.	0.92	0.59	0.25	0.024		0.033 S; annealed specimen; expansion measured with increasing temperature; zero-point correction of 0.005% determined by graphical extrapolation.
36	51	Souder, W. and Hidnert, P.	1922	T	1181-306	S 550	Bal.	0.92		0.25			The above specimen; expansion measured with decreasing temperature; zero-point correction of 0.045% determined by graphical extrapolation.
37	538	Yershov, V.M. and Oslon, N.L.	1968	X	298-958	Steel 40G5	Bal.	4.55	0.41	0.25	0.021	0.10	0.030 S; produced in induction furnace; forged to square bar 12 x 12 mm, after annealing; made with cross-section of 12 x 12 and thickness of 0.8-0.9 mm; measured of austentic line (200) on cooling from 953 K; lattice parameter reported at 298 K is 3.5893 kX (35965 Å); 359815 kX (3.5963 Å) at 293 K determined by graphical extrapolation.
38	538	Yershov, V.M. and Oslon, N.L.	1968	X	740-298	Steel 40G5	Bal.	4.55	0.41	0.25	0.021	0.10	Similar to the above specimen, except on cooling from 993 K; lattice parameter reported at 298 K is 3.5883 kX (3.5955 Å); 3.5881 kX (3.5953 Å) at 293 K determined by graphical extrapolation.

* Not shown in figure.

SPECIFICATION TABLE 261. THERMAL LINEAR EXPANSION OF IRON + MANGANESE + ΣX_i ALLOYS Fe + Mn + ΣX_i (continued)

Cur. No.	Ref. No.	Author(s)	Year	Method Used	Temp. Range, K	Name and Specimen Designation	Composition (weight percent) Fe	Mn	C	Si	P	Cr	Composition (continued), Specifications, and Remarks
39*	538	Yershov, V.M. and Oslon, N. L.	1968	X	378-1273	Steel 40G5	Bal.	4.55	0.41	0.25	0.021	0.10	Similar to the above specimen; except on cooling from 1273 K; lattice parameter reported at 378 K is 3.5898 kX (3.5970 Å); this curve is here reported using the first given temperature as reference temperature at which $\Delta L/L_0 = 0$.
40	538	Yershov, V.M. and Oslon, N. L.	1968	X	760-373	Steel 40G5	Bal.	4.55	0.41	0.25	0.021	0.10	Similar to the above specimen; measured on cooling from 973 to 373 K; lattice parameter reported at 373 K is 3.5929 kX (3.6001 Å); 3.5902 kX (3.5974 Å) at 293 K determined by graphical extrapolation.
41	538	Yershov, V.M. and Oslon, N. L.	1968	X	407-594	Steel 40G5	Bal.	4.55	0.41	0.25	0.021	0.10	The above specimen; followed by heating to 623 K; lattice parameter reported at 407 K is 3.5936 kX (3.6001 Å); 3.5907 kX (3.5979 Å) at 293 K determined by graphical extrapolation.
42*	538	Yershov, V.M. and Oslon, N. L.	1968	X	562-301	Steel 40G5	Bal.	4.55	0.41	0.25	0.021	0.10	The above specimen; cooling to room temperature; lattice parameter reported at 301 K is 3.5810 kX (3.5882 Å); 3.5802 kX (3.5874 Å) at 293 K determined by graphical extrapolation.
43	429	Mochel, N. L.	1927	T	273-800	K1	Bal.	0.42	0.17		0.012		0.035 S; specimen rolled to 2.5 cm round; tested as received; expansion measured with increasing temperature; zero-point correction of -0.023% determined by graphical interpolation.
44*	429	Mochel, N. L.	1927	T	800-300	K1	Bal.	0.42	0.17		0.012		The above specimen; expansion measured with decreasing temperature; zero-point correction of -0.028% determined by graphical extrapolation.
45*	429	Mochel, N. L.	1927	T	273-789	K1A	Bal.	0.42	0.17		0.012		Similar to the above specimen; expansion measured with increasing temperature; zero-point correction of -0.0235% determined by graphical interpolation.
46*	429	Mochel, N. L.	1927	T	789-297	K1A	Bal.	0.42	0.17		0.012		The above specimen; expansion measured with decreasing temperature; zero-point correction of -0.015% determined by graphical extrapolation.

* Not shown in figure.

DATA TABLE 261. THERMAL LINEAR EXPANSION OF IRON + MANGANESE + ΣX_i ALLOYS $Fe + Mn + \Sigma X_i$

[Temperature, T, K; Linear Expansion, $\Delta L/L_0$, %]

T	$\Delta L/L_0$
CURVE 1	
83	-0.190
123	-0.164
173	-0.123*
223	-0.076
323	0.036*
373	0.097
423	0.160*
473	0.226
523	0.297*
573	0.369
623	0.446*
673	0.525
723	0.607*
773	0.690
823	0.771*
873	0.854
923	0.936*
973	1.020
1023	1.080*
1073	1.094
1123	1.051*
1173	1.093
1223	1.208*
1273	1.323
CURVE 2‡	
20	-0.201
30	-0.201*
40	-0.201*
50	-0.200
60	-0.198*
70	-0.195*
80	-0.191
90	-0.187*
100	-0.182
110	-0.176*
120	-0.170
130	-0.163*
140	-0.155*
150	-0.147
160	-0.138*
170	-0.130*
180	-0.120*
190	-0.111*
200	-0.101
CURVE 2 (cont.)‡	
210	-0.091*
220	-0.081*
230	-0.070*
240	-0.060*
250	-0.049
260	-0.038*
270	-0.027*
280	-0.015
290	-0.004*
300	0.008
CURVE 3*	
4.2	-0.171
25	-0.171
50	-0.171
75	-0.166
100	-0.154
125	-0.139
150	-0.121
175	-0.099
200	-0.075
250	-0.025
300	0.030
CURVE 4*	
473	0.297
573	0.498
673	0.687
773	0.898
873	1.115
973	1.338
1073	1.570
1273	1.979
CURVE 5*	
373	0.114
473	0.301
573	0.479
673	0.670
773	0.876
873	1.103
973	1.329
1073	1.562
1173	1.804
CURVE 6*	
386	0.001
373	0.001
354	0.000
331	0.000
311	0.000
293	0.000
273	0.000
250	0.000
230	-0.001
209	-0.013
191	-0.038
173	-0.051
152	-0.053
132	-0.053
110	-0.053
92	-0.053
CURVE 7*	
92	-0.007
110	-0.007
132	-0.007
152	-0.007
193	-0.007
212	-0.007
232	-0.007
251	-0.007
273	-0.006
293	0.000
310	0.021
331	0.042
354	0.046
373	0.047
386	0.047
CURVE 8	
574	0.018
558	0.019
540	0.021
516	0.018
493	0.017
474	0.018
454	0.018
434	0.015
418	0.018
CURVE 8 (cont.)	
395	0.017
374	0.018
356	0.020
333	0.016
309	0.008
293	0.000
289	-0.007
268	-0.053
252	-0.108
241	-0.126
213	-0.142
173	-0.148
153	-0.149
135	-0.147
115	-0.147
94	-0.147
CURVE 9	
293	0.000
314	0.000
336	0.000
356	-0.001
374	0.000
395	0.000
415	0.000
436	0.006
454	0.035
472	0.094
496	0.136
516	0.144
540	0.147
558	0.145
574	0.144
CURVE 10	
293	0.000
314	0.000
334	0.000
360	0.000
378	0.007
395	0.012
414	0.017
456	0.028
477	0.088
CURVE 10 (cont.)	
495	0.144
515	0.202
535	0.219
551	0.229
574	0.236
CURVE 11	
573	0.030
550	0.030*
534	0.030
511	0.029*
492	0.030*
474	0.030
456	0.030*
434	0.030*
416	0.029*
394	0.030
373	0.030*
354	0.030*
340	0.030*
319	0.030
298	0.012
293	0.000
278	-0.027
260	-0.067
239	-0.113*
219	-0.153
207	-0.182
185	-0.196
167	-0.203
143	-0.207
121	-0.207
97	-0.209
CURVE 12	
293	0.000*
312	0.000*
336	0.003*
358	0.011
378	0.021
394	0.026*
408	0.035
431	0.045
457	0.054
CURVE 12 (cont.)	
478	0.077
495	0.120
516	0.168
536	0.180
561	0.190
581	0.190
CURVE 13	
581	0.001
561	0.001
538	0.001*
516	0.001
493	0.001*
474	0.001
452	0.001*
431	0.001*
411	0.000
382	0.000*
359	0.000*
338	0.000
316	0.000*
300	0.000*
293	0.000*
273	0.000
255	-0.012
236	-0.035
217	-0.130
198	-0.165
181	-0.176
163	-0.189
140	-0.189
116	-0.189
97	-0.189
CURVE 14	
293	0.000*
318	-0.003*
338	-0.002*
356	0.006*
376	0.020
392	0.029
413	0.041
450	0.052
468	0.065

* Not shown in figure.

‡ Author's data for coefficient of thermal expansion have been integrated by TPRC to obtain $\Delta L/L_0$.

DATA TABLE 261. THERMAL LINEAR EXPANSION OF IRON + MANGANESE + ΣX_i ALLOYS $Fe + Mn + \Sigma X_i$ (continued)

T	$\Delta L/L_0$
CURVE 14 (cont.)	
498	0.086
516	0.099
575	0.118
CURVE 15*	
575	-0.004
552	-0.003
535	-0.003
518	-0.003
496	-0.004
472	-0.003
451	-0.003
434	-0.001
415	-0.003
399	-0.001
383	-0.001
363	-0.002
341	-0.002
318	-0.002
297	0.000
293	0.000
279	0.000
257	0.000
240	-0.007
219	-0.022
191	-0.069
176	-0.098
157	-0.115
137	-0.121
114	-0.125
98	-0.124
CURVE 16*	
300	0.000
318	0.000
333	0.000
350	0.012
370	0.027
387	0.040
402	0.048
413	0.052
433	0.063
454	0.071
472	0.076
496	0.079
511	0.085

T	$\Delta L/L_0$
CURVE 16 (cont.)*	
534	0.090
555	0.097
571	0.104
CURVE 17	
574	0.000
555	0.000*
535	0.000
512	0.000*
491	0.000*
472	0.000*
449	0.000
430	0.000*
408	0.000
392	0.000*
373	0.000*
353	0.000*
337	0.000*
320	0.000
295	0.000*
276	0.000
259	0.000
239	-0.003
216	-0.027
203	-0.049
179	-0.078
158	-0.093
134	-0.104
114	-0.104
94	-0.104
CURVE 18	
299	0.000*
318	0.000*
339	0.003*
354	0.010*
368	0.023
383	0.046
404	0.060
424	0.067
446	0.073
474	0.073*
495	0.073
519	0.072*

T	$\Delta L/L_0$
CURVE 19	
519	-0.001
495	0.000
474	0.000
446	0.000*
430	0.002*
409	0.000*
387	0.001*
366	0.001*
350	0.000*
331	0.003*
310	0.002*
296	0.000*
293	0.000*
277	0.002*
258	0.003*
234	-0.020*
212	-0.023*
197	-0.048
173	-0.066
156	-0.074
136	-0.071
109	-0.073
89	-0.073
CURVE 20	
298	0.005*
373	0.089
473	0.211
573	0.354
673	0.512
773	0.669
873	0.829
973	0.995
CURVE 21	
293	0.000
369	0.096
461	0.209
578	0.377
670	0.521
778	0.691
916	0.914
964	0.988
988	1.024
997	1.040
1018	1.067

T	$\Delta L/L_0$
CURVE 21 (cont.)	
1027	1.055
1032	1.041
1033	1.022
1037	1.010
1042	0.988
1047	0.960
1053	0.945
1066	0.943
1086	0.952
1132	1.037
1173	1.144
CURVE 22	
1173	1.258
1132	1.151
1091	1.065
1061	0.999
1030	0.930
1016	0.899
1002	0.863
995	0.851
985	0.829
967	0.799
958	0.798
950	0.805
946	0.821
944	0.881
941	0.940
930	0.935
850	0.797
721	0.590
602	0.413
536	0.312
356	0.072
293	0.000
CURVE 23	
296	0.004*
395	0.119
479	0.225*
576	0.365*
678	0.529
774	0.704
916	0.921
962	1.005
997	1.056

T	$\Delta L/L_0$
CURVE 23 (cont.)	
1000	1.052*
1014	0.972*
1024	0.934*
1043	0.911
1097	1.006
1174	1.192
CURVE 24	
1174	1.232
1097	1.046
1054	0.948
1003	0.835
967	0.764*
954	0.764
941	0.770*
930	0.815
916	0.913*
909	0.910*
901	0.895
877	0.853*
814	0.748
670	0.506
446	0.181
301	0.009*
CURVE 25	
290	-0.008*
293	0.000*
382	0.108
496	0.258
580	0.378
715	0.590
847	0.807
920	0.923
983	1.037
1112	0.948
1120	0.949*
1225	1.199
CURVE 26	
1225	1.239
1182	1.139
1173	1.117
1137	1.036
1133	1.032*

T	$\Delta L/L_0$
CURVE 26 (cont.)	
1074	0.899
1019	0.823*
1018	0.845*
1018	0.861*
1009	0.866
998	0.884*
988	0.899
956	0.911*
946	0.909
937	0.902*
928	0.916*
918	0.916*
910	0.903
855	0.808
767	0.669
545	0.324
454	0.202
293	0.000*
292	-0.003*
CURVE 27	
293	0.000*
370	0.088
488	0.239
627	0.443
742	0.624
892	0.877
949	0.967
990	1.034
996	1.032*
1007	0.982*
1014	0.954*
1024	0.942
1037	0.926*
1047	0.915*
1055	0.907
1097	0.954*
1101	0.966
1179	1.147
CURVE 28	
1179	1.168*
1101	0.987*
1097	0.975
1013	0.798
1008	0.792*

* Not shown in figure.

DATA TABLE 261. THERMAL LINEAR EXPANSION OF IRON + MANGANESE + ΣX_i ALLOYS $Fe + Mn + \Sigma X_i$ (continued)

T	$\Delta L/L_0$
CURVE 28 (cont.)	
998	0.793*
988	0.799
970	0.820*
962	0.825*
951	0.830
941	0.833*
931	0.835*
920	0.879
913	0.908*
903	0.892*
789	0.700
540	0.317
525	0.295
318	0.029
CURVE 29	
302	0.008*
381	0.086
472	0.191
606	0.385
684	0.507
701	0.535
794	0.683
900	0.868
955	0.954
983	1.003
999	1.030
1006	1.034
1009	1.032*
1015	0.998*
1026	0.931
1038	0.882*
1047	0.871*
1060	0.886
1122	1.020
1187	1.173
CURVE 30	
1187	1.196
1068	0.923
1046	0.870
1027	0.827
1005	0.776
996	0.773*
991	0.788
982	0.814
CURVE 30 (cont.)	
966	0.842
941	0.939
925	0.908
798	0.696*
532	0.285
312	0.007*
CURVE 31	
301	0.009
402	0.146
430	0.179
461	0.216
471	0.238
475	0.238*
479	0.250*
498	0.282
526	0.329
550	0.354
583	0.385
599	0.384*
625	0.389
641	0.403
644	0.404*
CURVE 32	
644	0.474
635	0.455
624	0.437*
577	0.372*
569	0.358
465	0.212*
311	0.020*
303	0.010
293	0.000*
292	-0.003*
CURVE 33*	
295	0.000
303	0.000
374	0.095
482	0.223
590	0.384
678	0.522
761	0.656
869	0.831
CURVE 33 (cont.)*	
969	1.005
998	1.050
1008	1.060
1024	0.992
1036	0.956
1045	0.930
1057	0.915
1069	0.919
1105	0.998
1131	1.053
1179	1.171
CURVE 34*	
1179	1.171
1131	1.053
1105	0.998
1069	0.919
1055	0.887
1025	0.828
1019	0.827
1009	0.854
996	0.882
983	0.897
970	0.908
957	0.927
952	0.952
940	0.943
931	0.924
791	0.686
CURVE 35	
298	0.005
371	0.086*
471	0.214*
615	0.426
701	0.561
781	0.686
899	0.884*
964	0.994
982	1.023*
1002	1.054
1007	1.052*
1013	1.008
1027	0.924*
1036	0.901
1046	0.918*
CURVE 35 (cont.)	
1122	1.098
1181	1.238
CURVE 36	
1181	1.278
1122	1.278
1034	0.944*
1019	0.902
982	0.817*
958	0.794
947	0.835
940	0.910
933	0.951
927	0.933
915	0.913
771	0.674
489	0.241
306	0.015*
CURVE 37	
298	0.005
322	0.027
352	0.052
362	0.060*
379	0.072*
394	0.091
407	0.105*
429	0.124
443	0.138*
461	0.147
473	0.169*
484	0.197
497	0.228*
509	0.239
518	0.266*
527	0.300
542	0.356
577	0.411
596	0.456
626	0.556
677	0.670
711	0.765
958	1.386
CURVE 38	
740	0.796
715	0.730
681	0.640
638	0.540
602	0.448
577	0.376
554	0.328
535	0.270
522	0.245
512	0.217
492	0.189
481	0.172
467	0.167
448	0.142
437	0.133
431	0.128*
418	0.119
400	0.102
385	0.088
374	0.055
357	0.055*
347	0.047
336	0.027
326	0.033*
298	0.005*
CURVE 39*	
378	0.000
393	0.653
394	0.731
406	0.768
428	0.770
447	0.770
463	0.770
473	0.796
492	0.801
517	0.857
542	0.924
575	1.011
606	1.036
625	1.084
657	1.171
746	1.263
779	1.347
801	1.404
833	1.484
1188	2.308
1273	2.521
CURVE 40	
760	0.754
713	0.642
662	0.567
631	0.506
601	0.431
581	0.394
559	0.367
549	0.353
526	0.280
499	0.208
489	0.186
473	0.172
461	0.152*
450	0.141
439	0.130
422	0.116
402	0.102
388	0.088*
373	0.094*
373	0.074
CURVE 41	
407	0.079
423	0.090
444	0.096
455	0.121
481	0.135
516	0.207
554	0.280*
557	0.277
568	0.336
594	0.433
CURVE 42	
562	0.612
536	0.603
518	0.528
509	0.508
474	0.416
437	0.338
415	0.327
398	0.273
373	0.201
362	0.142
348	0.140
335	0.112

* Not shown in figure.

DATA TABLE 261. THERMAL LINEAR EXPANSION OF IRON + MANGANESE + ΣX_i ALLOYS Fe + Mn + ΣX_i (continued)

T	$\Delta L/L_0$	T	$\Delta L/L_0$
CURVE 42 (cont.)		CURVE 45*	
317	0.067	273	-0.023
301	0.022	295	0.002
		421	0.159
CURVE 43		539	0.318
		661	0.505
273	-0.023	789	0.716
300	0.008		
414	0.147	CURVE 46*	
538	0.315		
669	0.515	789	0.725
800	0.730	617	0.446
		483	0.249
CURVE 44*		297	0.008
800	0.725		
623	0.436		
300	0.009		

DATA TABLE COEFFICIENT OF THERMAL LINEAR EXPANSION OF IRON + MANGANESE + ΣX_i ALLOYS Fe + Mn + ΣX_i

[Temperature, T, K; Coefficient of Expansion, α, 10^{-6} K^{-1}]

T	α	T	α
CURVE 2‡		CURVE 2 (cont.)‡	
20	11.85	180	7.84
30	11.67	190	7.33
40	11.49	200	6.77
50	11.31	210	6.16
60	11.13	220	5.50
70	10.95	230	4.78
80	10.76	240	3.99
90	10.56	250	3.13
100	10.35	260	2.27
110	10.14	270	1.44
120	9.92	280	0.76
130	9.67	290	0.34
140	9.39	300	0.12
150	9.07		
160	8.71		
170	8.30		

* Not shown in figure.
‡ Author's data for coefficient of thermal expansion have been integrated by TPRC to obtain $\Delta L/L_0$.

FIGURE AND TABLE NO. 262R. PROVISIONAL VALUES FOR THERMAL LINEAR EXPANSION OF IRON + NICKEL + ΣX_i ALLOYS   Fe + Ni + ΣX_i

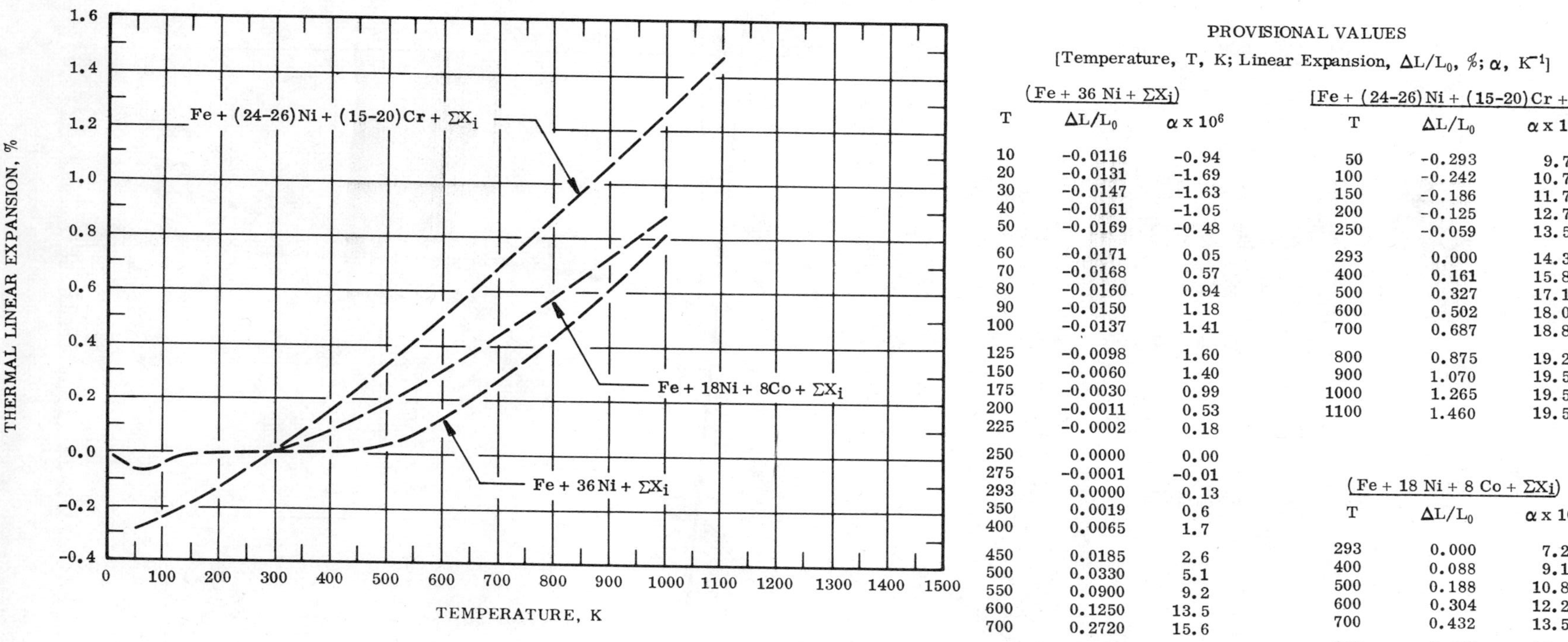

PROVISIONAL VALUES

[Temperature, T, K; Linear Expansion, $\Delta L/L_0$, %; α, K^{-1}]

(Fe + 36 Ni + ΣX_i)

T	$\Delta L/L_0$	$\alpha \times 10^6$
10	-0.0116	-0.94
20	-0.0131	-1.69
30	-0.0147	-1.63
40	-0.0161	-1.05
50	-0.0169	-0.48
60	-0.0171	0.05
70	-0.0168	0.57
80	-0.0160	0.94
90	-0.0150	1.18
100	-0.0137	1.41
125	-0.0098	1.60
150	-0.0060	1.40
175	-0.0030	0.99
200	-0.0011	0.53
225	-0.0002	0.18
250	0.0000	0.00
275	-0.0001	-0.01
293	0.0000	0.13
350	0.0019	0.6
400	0.0065	1.7
450	0.0185	2.6
500	0.0330	5.1
550	0.0900	9.2
600	0.1250	13.5
700	0.2720	15.6
800	0.4370	17.1
900	0.6150	18.1
1000	0.8000	18.5

[Fe + (24-26) Ni + (15-20) Cr + ΣX_i]

T	$\Delta L/L_0$	$\alpha \times 10^6$
50	-0.293	9.7
100	-0.242	10.7
150	-0.186	11.7
200	-0.125	12.7
250	-0.059	13.5
293	0.000	14.3
400	0.161	15.8
500	0.327	17.1
600	0.502	18.0
700	0.687	18.8
800	0.875	19.2
900	1.070	19.5
1000	1.265	19.5
1100	1.460	19.5

(Fe + 18 Ni + 8 Co + ΣX_i)

T	$\Delta L/L_0$	$\alpha \times 10^6$
293	0.000	7.2
400	0.088	9.1
500	0.188	10.8
600	0.304	12.2
700	0.432	13.5
800	0.573	14.5
900	0.723	15.4
1000	0.882	16.1

REMARKS

(Fe + 36Ni + ΣX_i): The tabulated values for well-annealed alloy of type Invar are considered accurate to within ±7% over the entire temperature range.

[Fe + (24-26) Ni + (15-20) Cr + ΣX_i]: The tabulated values for well-annealed alloy of type A-286 steel, 20Cr-25Ni alloy, Kh15N24T2 Russian alloy are considered accurate to within ±7% over the entire temperature range. These values can be represented approximately by the following equation:

$$\Delta L/L_0 = -0.339 + 8.574 \times 10^{-4}\, T + 1.148 \times 10^{-6}\, T^2 - 4.002 \times 10^{-10}\, T^3$$

(Fe + 18Ni + 8Co + ΣX_i): The tabulated values for well-annealed alloy of type 18NiCoMo steel (Grade 250), 18NiCoMo steel (Grade 300) are considered accurate to within ±7% over the entire temperature range. These values can be represented approximately by the following equation:

$$\Delta L/L_0 = -0.118 + 6.201 \times 10^{-5}\, T + 1.261 \times 10^{-6}\, T^2 - 3.250 \times 10^{-10}\, T^3$$

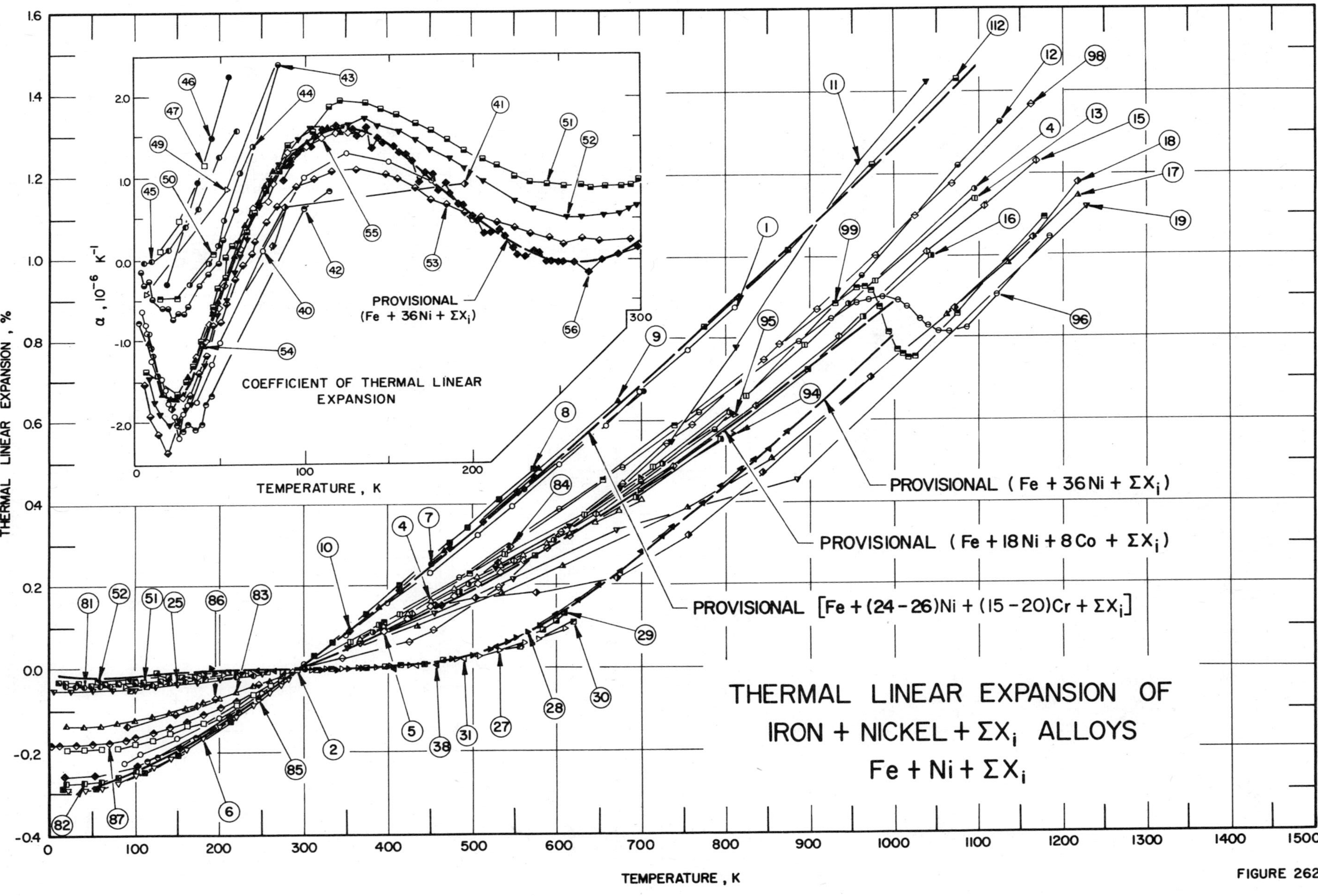
THERMAL LINEAR EXPANSION OF
IRON + NICKEL + ΣX_i ALLOYS
Fe + Ni + ΣX_i
PROVISIONAL (Fe + 36 Ni + ΣX_i)
PROVISIONAL (Fe + 18 Ni + 8 Co + ΣX_i)
PROVISIONAL [Fe + (24 - 26)Ni + (15 - 20)Cr + ΣX_i]
THERMAL LINEAR EXPANSION, %
TEMPERATURE, K
COEFFICIENT OF THERMAL LINEAR EXPANSION
PROVISIONAL (Fe + 36Ni + ΣX_i)
α, 10^{-6} K^{-1}
TEMPERATURE, K

FIGURE 262

SPECIFICATION TABLE 262. THERMAL LINEAR EXPANSION OF IRON + NICKEL + ΣX_i ALLOYS Fe + Ni + ΣX_i

Cur. No.	Ref. No.	Author(s)	Year	Method Used	Temp. Range, K	Name and Specimen Designation	Composition (weight percent) Fe	Ni	Cr	Mn	Si	Co	C	Composition (continued), Specifications, and Remarks
1	534	Furman, D.F.	1950		88.7-810.9	330 S.S.	Bal.	35.19	15.30	1.81	0.62		0.052	0.006 S, P, 0.250 in. diameter by 4.000 in. long; commercial grade steel was given an anneal at 134 K for 30 min and water quenched to machine the specimen; zero-point correction of -0.226% determined by graphical interpolation.
2	462	Arp. V., Wilson, S.H., Winrich, L., and Sikara, P.	1962	L	20-293	2800 Steel	Bal.	8.80	0.2	0.71	0.1		0.09	0.02 S, P; hardness R_c 29; tempered specimen.
3*	462	Arp. V., et al.	1962	L	20-293	4340 Steel	Bal.	1.8	0.8	0.7	0.3		0.39	0.02 S, P, 0.3 Mo; hardness R_c 32; annealed specimen.
4	543	Hoenie, A.F.	1965	L	366-700	Grade 250 18NiCoMo steel	Bal.	18.5		0.1	0.1	8.0	0.03	4.9 Mo, 0.4 Ti, 0.1 P, Al, 0.05 Al, 0.02 Zr, 0.01 S, 0.003 B; specimen finish machined; solution annealed at 1089 K, aged 3 hr at 756 K; expansion measured on 3 in gage length with heating rate 1.1 K per minute.
5	543	Hoenie, A.F.	1965	L	366-700	Grade 300 18NiCoMo steel	Bal.	18.5		0.01	0.1	9.0		4.9 Mo, 0.62 Ti, 0.1 P, Al, 0.05 Ca, 0.02 Zr, 0.01 S, 0.003 B; similar 'o the above specimen.
6	462	Arp, V., Wilson, S.H., Winrich, L., and Sikara, P.	1962	L	20-293	A-286 steel	Bal.	25.4	14.8	1.4	0.6		0.04	1.2 Mo, 2.1 Ti, 0.3 V, 0.2 Al, 0.01 P, S; hardness R_c 30, age-hardened specimen.
7	454	Powers, D.J.	1963		299-702	A-286 steel	Bal.	26	15	1.35	0.95		0.05	1.25 Mo, 1.95 Ti, 0.3 V, 0.2 Al.
8	15	Rhodes, B.L., Moeller, C.E., Hopkins, V., and Marx, T.I.	1963	I+ L	18-573	A-286 steel	Bal.	26.0	15.0	1.35	0.5		0.08	1.25 Mo, 2.0 Ti, 0.3 V, 0.25 Al; 10.16 cm long x 6.2 mm.
9	225	Valentich, S.	1965	L	293-673	A-286 steel								No details given.
10	15	Rhodes, B.L., Moeller, C.E., Hopkins, V., and Marx, T.I.	1963	I+ L	18-573	Carpenter 20-CB steel	Bal.	29.0	20.0	0.75	1.0		0.07	3.0 Cu, 2.0 Mo; 10.16 cm long x 6.2 mm.
11	232	Shevlin, T.S. and Newkirk, H.W.	1956	T	301-1128	Fernichrome	37	30	8			25		Rough rod formed by hydrostatia pressing; metal powders contained in rubber envelope at 3500 psi; pressed with carbowax binder, then dewaxed using 48 hr schedule from room temperature to 408 K; specimen 0.5 in diameter x 12.0 in long; expansion measured with increasing temperature; zero-point correction of -0.095% determined by graphical extrapolation.
12	232	Shevlin, T.S. and Newkirk, H.W.	1956	T	1128-295	Fernichrome	37	30	8			25		The above specimen; expansion measured with decreasing temperature; zero-point correction of -0.404% determined by graphical extrapolation.

* Not shown in figure.

SPECIFICATION TABLE 262. THERMAL LINEAR EXPANSION OF IRON + NICKEL + ΣX_i ALLOYS Fe + Ni + ΣX_i (continued)

Cur. No.	Ref. No.	Author(s)	Year	Method Used	Temp. Range, K	Name and Specimen Designation	Fe	Ni	Composition (weight percent) Cr	Mn	Si	Co	C	Composition (continued), Specifications, and Remarks
13	232	Shevlin, T.S. and Newkirk, H.W.	1956	T	301-1096	Fernichrome	37	30	8			25		The above specimen; run two, expansion measured with increasing temperature; zero-point correction of 0.600% determined by graphical extrapolation.
14	232	Shevlin, T.S. and Newkirk, H.W.	1956	T	1096-301	Fernichrome	37	30	8			25		The above specimen; expansion measured with decreasing temperature; zero-point correction of -0.575% determined by graphical extrapolation.
15	232	Shevlin, T.S. and Newkirk, H.W.	1956	T	298-1170	Fernichrome	37	30	8			25		The above specimen; run three; expansion measured with increasing temperature; zero-point correction of -0.008% determined by graphical extrapolation.
16	232	Shevlin, T.S. and Newkirk, H.W.	1956	T	1170-298	Fernichrome	37	30	8			25		The above specimen; expansion measured with decreasing temperature; zero-point correction of -0.006% determined by graphical extrapolation.
17	232	Shevlin, T.S. and Newkirk, H.W.	1956	T	292-1220	Fernico steel	54.0	28.0				18.0		Rough rod formed by hydrastatic pressing metal powders contained in rubber envelope at 35,000 psi; pressed with carbowax binder, then dewaxed using 48 hr schedule from room temperature to 408 K; specimen 0.5 in. diameter x 12.0 in. long; expansion measured with increasing temperature; zero-point correction of -0.203% determined by graphical interpolation.
18	232	Shevlin, T.S. and Newkirk, H.W.	1956	T	1220-299	Fernico steel	54.0	28.0				18.0		The above specimen; expansion measured with decreasing temperature; zero-point correction of -0.170% determined by graphical extrapolation.
19	232	Shevlin, T.S. and Newkirk, H.W.	1956	T	290-1230	Fernico steel	54.0	28.0				18.0		The above specimen; run two; expansion measured with increasing temperature; zero-point correction of -0.009%.
20*	232	Shevlin, T.S. and Newkirk, H.W.	1956	T	1230-470	Fernico steel	54.0	28.0				18.0		The above specimen; expansion measured with decreasing temperature; this curve is here reported using the first given temperature as reference temperature at which $\Delta L/L_0 = 0$.
21*	544	Foster, J.D. and Finnie, I.	1968	L	293-311	Invar								High purity; heat treated; 10 cm long.
22*	544	Foster, J.D. and Finnie, I.	1968	L	293-313	Invar								Similar to the above specimen.
23*	544	Foster, J.D. and Finnie, I.	1968	L	293-312	Invar								Similar to the above specimen.

* Not shown in figure.

SPECIFICATION TABLE 262. THERMAL LINEAR EXPANSION OF IRON + NICKEL + ΣX_i ALLOYS Fe + Ni + ΣX_i (continued)

Cur. No.	Ref. No.	Author(s)	Year	Method Used	Temp. Range, K	Name and Specimen Designation	Fe	Ni	Cr	Mn	Si	Co	C	Composition (continued), Specifications, and Remarks
24*	544	Foster, J.D. and Finnie, I.	1968	L	293-300	Invar								Commercial specimen, annealed; 15 cm long.
25	302	Beenakker, J.J. and Swenson, C.A.	1955	L	4-300	Invar								Of the same stock as used by General Electric Laboratory at Schenactady for transfer tubes.
26*	225	Valentich, S.	1965	L	293-562									No information given.
27	59	Masumoto, H.	1931	L	89-608	No. 124 Invar	62.5	34.0				3.5		Armco Fe, Mond Ni, Granular Co., mixed and melted in hydrogen atm in alumina crucible in Tammann furnace, cast in iron mould with cylindrical aperature machined to circular rod 4 mm thick, 10 cm long, heated at 1273 K for 1 hr in electric furnace in hydrogen atm at reduced pressure, expansion measured in vacuum; zero-point correction is -0.002%.
28	59	Masumoto, H.	1931	L	191-579	No. 95 Invar	63.5	32.5				4.0		Similar to the above specimen.
29	59	Masumoto, H.	1931	L	124-608	No. 96 Invar	63.0	33.0				4.0		Similar to the above specimen.
30	59	Masumoto, H.	1931	L	95-619	No. 97 Invar	62.5	33.5				4.0		Similar to the above specimen; zero-point correction is 0.003%.
31	59	Masumoto, H.	1931	L	250-595	No. 72 Invar	63.5	31.5				5.0		Similar to the above specimen.
32*	59	Masumoto, H.	1931	L	176-623	No. 71 Invar	62.5	32.5				5.0		Similar to the above specimen.
33*	59	Masumoto, H.	1931	L	266-598	No. 100 Invar	63.5	30.5				6.0		Similar to the above specimen.
34*	59	Masumoto, H.	1931	L	210-626	No. 103 Invar	62.5	31.5				6.0		Similar to the above specimen.
35*	59	Masumoto, H.	1931	L	203-604	No. 6 Invar	64.0	31.0		0.35		5.0		Similar to the above specimen.
36*	59	Masumoto, H.	1931	L	227-642	No. 7 Invar	63.0	31.0		0.38		6.0		Similar to the above specimen.
37*	59	Masumoto, H., Kikucki, M., and Sawaya, S.	1969	L	244-873	No. 1 Invar	63	32				5		Single crystal bar, about 12 cm long; 3 mm diameter; heated in vacuum at 1273 K for 1 hr cooled at rate 300 C/hr; expansion measured in <100> direction.
38	59	Masumoto, H., et al.	1969	L	243-873	No. 7 Invar	63	32				5		Similar to the above specimen; expansion measured in <110> direction.
39*	59	Masumoto, H., et al.	1969	L	243-873	No. 10 Invar	63	32				5		Similar to the above specimen; expansion measured in <111> direction.

* Not shown in figure.

SPECIFICATION TABLE 262. THERMAL LINEAR EXPANSION OF IRON + NICKEL + ΣX_i ALLOYS Fe + Ni + ΣX_i (continued)

Cur. No.	Ref. No.	Author(s)	Year	Method Used	Temp. Range, K	Name and Specimen Designation	Composition (weight percent) Fe	Ni	Cr	Mn	Si	Co	C	Composition (continued), Specifications, and Remarks
40*	752	Schlosser, W.F., Graham, G.M., and Meincke, P.P.M.	1971	E	5-200	Invar	64.45	35.06	0.08	0.90	0.36	0.05	0.09	0.01 Se; 2.7 cm long, 0.5 cm diameter; cylindrical specimen of polycrystalline material from the rod of Commercial Free Cut Invar; made by Carpenter Steel Co. Reading, Pa; annealed sample heat treated at 1373 K for one hr in helium atm, and water quenched; then held at 580 K for 8 hr in air cooled to 368 K, and held at this temperature for 2 days; thermal expansion measured in magnetic field of 21.6 kOe; this curve is here reported using the first given temperature as reference temperature at which $\Delta L/L_0 = 0$.
41*	752	Schlosser, W.F., et al.	1971	E	23-195	Invar	64.45	35.06	0.08	0.90	0.36	0.05	0.09	Similar to the above specimen; thermal expansion measured in zero magnetic field; this curve is here reported using the first given temperature as reference temperature at which $\Delta L/L_0 = 0$.
42*	752	Schlosser, W.F., et al.	1971	E	7-115	Invar	64.45	35.06	0.08	0.90	0.36	0.05	0.09	Similar to the above specimen; except cold-worked Invar; thermal expansion measured in magnetic field of 21.6 kOe; this curve is here reported using the first given temperature as reference temperature at which $\Delta L/L_0 = 0$.
43*	753	Schlosser, W.F., Latal, E., Meincke, P.P.M., Graham, G.M. and Colling, D.A.	1972	E	4-231	Invar LM46ANN	57.93	33.60				8.47		Cylindrical specimen; 2.29 cm long, 0.3 cm diameter; levitation melted, homogenized, machined and cold reduced 75% in area, then annealed for one hr at 1373 K; thermal expansion measured in zero magnetic field; this curve is here reported using the first given temperature as reference temperature at which $\Delta L/L_0 = 0$.
44*	753	Schlosser, W.F., et al.	1972	E	32-68	Invar LM46ANN	57.93	33.60				8.47		Similar to the above specimen; except thermal expansion measured in 2.16 T magnetic field; this curve is here reported using the first given temperature as reference temperature at which $\Delta L/L_0 = 0$.
45*	753	Schlosser, W.F., et al.	1972	E	6-96	Invar LM46ANN	52.00	30.30				17.70		Similar to the above specimen; except thermal expansion measured in zero magnetic field; this curve is here reported using the first given temperature as reference temperature at which $\Delta L/L_0 = 0$.
46*	753	Schlosser, W.F., et al.	1972	E	6-55	Invar LM54ANN	52.00	30.30				17.70		Similar to the above specimen; except thermal expansion measured in 2.16 T magnetic field; this curve is here reported using the first given temperature as reference temperature at which $\Delta L/L_0 = 0$.

* Not shown in figure.

SPECIFICATION TABLE 262. THERMAL LINEAR EXPANSION OF IRON + NICKEL + ΣX_i ALLOYS Fe + Ni + ΣX_i (continued)

Cur. No.	Ref. No.	Author(s)	Year	Method Used	Temp. Range, K	Name and Specimen Designation	Fe	Ni	Composition (weight percent) Cr	Mn	Si	Co	C	Composition (continued), Specifications, and Remarks
47*	753	Schlosser, W.F., et al.	1972	E	5-40	Invar LM50ANN	51.97	38.50				9.53		Similar to the above specimen; except thermal expansion measured in zero magnetic field; this curve is here reported using the first given temperature as reference temperature at which $\Delta L/L_0 = 0$.
48*	753	Schlosser, W.F., et al.	1972	E	6-46	Invar LM50ANN	51.97	38.50				9.53		Similar to the above specimen; except thermal expansion measured in 2.16 T magnetic field; this curve is here reported using the first given temperature as reference temperature at which $\Delta L/L_0 = 0$.
49*	753	Schlosser, W.F., et al.	1972	E	6-100	Invar LM50CW	51.97	38.50				9.53		Similar to the above specimen; except without annealing process; cold-worked; thermal expansion measured in zero magnetic field; this curve is here reported using the first given temperature as reference temperature at which $\Delta L/L_0 = 0$.
50*	753	Schlosser, W.F., et al.	1972	E	10-46	Invar LM50CW	51.97	38.50				9.53		Similar to the above specimen; except thermal expansion measured in 2.16 T magnetic field; this curve is here reported using the first given temperature as reference temperature at which $\Delta L/L_0 = 0$.
51	754	Bol'shakov, Yu, V., Zakharov, A.I., Pozvonkov, F.M., Solov'eva, N.A., and Fridman, V.G.	1971	L	11-301	Invar 36NKh	Bal.	36.2	0.56	0.27	0.09		0.03	0.006 S; alloy melted in open induction furnace and cast in ingots weighing 60 kg; specimen cut from strip 5 mm thick obtained by cold rolling with 36% reduction; heated at 1173 K for 15 min, cooled in air; thermal linear coefficient also reported; zero-point correction of -0.024%.
52	754	Bol'shakov, Yu, V., et al.	1971	L	10-322	Invar 36NKh	Bal.	36.2	0.56	0.27	0.09		0.03	Similar to the above specimen; except heating at 1103 K for 30 min, cooling in air; tempering at 588 K for 1 hr; zero-point correction of -0.016%.
53*	754	Bol'shakov, Yu, V., et al.	1971	L	5-322	Invar 36NKh	Bal.	36.2	0.56	0.27	0.09		0.03	Similar to the above specimen; except plastic deformation with 36% reduction; zero-point correction of -0.007%.
54*	527	Zakharov, A.L. and Fedotov, L.N.	1967	L	9-121	Invar	Bal.	36			trace		0.003	0.003 S, 0.002 P, trace Al; produced in vacuum furnace; charge composed of carbonyl iron refined in hydrogen and electrolytic nickel; specimen 5 mm diameter and 43.5 mm long; quenched in water from 1373 K; annealed 8 hr at 588 K and 48 hr at 368 K; this curve is here reported using the first given temperature as reference temperature at which $\Delta L/L_0 = 0$.

* Not shown in figure.

SPECIFICATION TABLE 262. THERMAL LINEAR EXPANSION OF IRON + NICKEL + ΣX_i ALLOYS Fe + Ni + ΣX_i (continued)

Cur. No.	Ref. No.	Author(s)	Year	Method Used	Temp. Range, K	Name and Specimen Designation	Fe	Ni	Composition (weight percent) Cr	Mn	Si	Co	C	Composition (continued), Specifications, and Remarks
55*	527	Zakharov, A.L. and Fedotov, L.N.	1967	L	9-134	Invar	Bal.	36			trace		0.003	Similar to the above specimen.
56*	527	Zakharov, A.L. and Fedotov, L.N.	1967	L	88-298	Invar	Bal.	36			trace		0.003	Similar to the above specimen.
57*	527	Zakharov, A.L. and Fedotov, L.N.	1967	L	83-305	Invar	Bal.	36			trace		0.003	Similar to the above specimen.
58*	749	Zakharov, A.I., Perepelkina, A.M., and Shiryaeva, A.N.	1972	L	293, 353	Superinvar	Bal.	25.16		0.4	0.1	4.3	≤0.01	≤0.009 S, ≤0.003 P, ≤0.005 O, and ≤0.002 N; melted PVD-40 Fe, electrolytic N-1 Ni and K-1 Co in open induction furnace with basic crucible; water quenching from 1123 K (30 min); tempering at 588 K for 1 hr.
59*	749	Zakharov, A.I., et al.	1972	L	293, 353	Superinvar	Bal.	32.0		0.1	0.1	4.0		Similar to the above specimen; $\gamma - \alpha$ phase transformation temperature at 218 K.
60*	749	Zakharov, A.I., et al.	1972	L	293, 353	Superinvar	Bal.	32.5		0.18	0.4	4.1		Similar to the above specimen.
61*	749	Zakharov, A.I., et al.	1972	L	293, 353	Superinvar	Bal.	32.2		0.3	0.2	3.2		Similar to the above specimen; $\gamma - \alpha$ phase transformation temperature at 208 K.
62*	749	Zakharov, A.I., et al.	1972	L	293, 353	Superinvar	Bal.	32.0		0.2	0.4	6.0		Similar to the above specimen; $\gamma - \alpha$ phase transformation temperature at 193 K.
63*	749	Zakharov, A.I., et al.	1972	L	293, 353	Superinvar	Bal.	31.5		0.4	0.2	5.0		Similar to the above specimen.
64*	749	Zakharov, A.I., et al.	1972	L	293, 353	Superinvar	Bal.	31.0			0.03	5.4		0.36 Tl; similar to the above specimen; $\gamma - \alpha$ phase transformation temperature at 238 K.
65*	749	Zakharov, A.I., et al.	1972	L	293, 353	Superinvar	Bal.	31.1			0.05	5.0		0.73 Tl; similar to the above specimen; $\gamma - \alpha$ phase transformation temperature at 263 K.
66*	749	Zakharov, A.I., et al.	1972	L	293, 353	Superinvar	Bal.	32.0	0.03			5.0		Similar to the above specimen; $\gamma - \alpha$ phase transformation temperature at 223 K.
67*	749	Zakharov, A.I., et al.	1972	L	293, 353	Superinvar	Bal.	31.4	0.09			4.3		Similar to the above specimen; $\gamma - \alpha$ phase transformation temperature at 208 K.
68*	749	Zakharov, A.I., et al.	1972	L	293, 353	Superinvar	Bal.	31.4	0.2			4.3		Similar to the above specimen; $\gamma - \alpha$ phase transformation temperature at 198 K.
69*	749	Zakharov, A.I., et al.	1972	L	293, 353	Superinvar	Bal.	32.2			0.06	4.2		Similar to the above specimen; $\gamma - \alpha$ phase transformation temperature at 233 K.
70*	749	Zakharov, A.I., et al.	1972	L	293, 353	Superinvar	Bal.	32.2			0.08	4.2		Similar to the above specimen; $\gamma - \alpha$ phase transformation temperature at 243 K.
71*	749	Zakharov, A.I., et al.	1972	L	293, 353	Superinvar	Bal.	32.2			0.1	4.2		Similar to the above specimen; $\gamma - \alpha$ phase transformation temperature at 273 K.
72*	749	Zakharov, A.I., et al.	1972	L	293, 353	Superinvar	Bal.	32.2		0.4		4.2		Similar to the above specimen; $\gamma - \alpha$ phase transformation temperature at 258 K.
73*	749	Zakharov, A.I., et al.	1972	L	293, 353	Superinvar	Bal.	32.2		0.9		4.2		Similar to the above specimen; $\gamma - \alpha$ phase transformation temperature at 218 K.

* Not shown in figure.

SPECIFICATION TABLE 262. THERMAL LINEAR EXPANSION OF IRON + NICKEL + ΣX_i ALLOYS Fe + Ni + ΣX_i (continued)

Cur. No.	Ref. No.	Author(s)	Year	Method Used	Temp. Range, K	Name and Specimen Designation	Fe	Ni	Composition (weight percent) Cr	Mn	Si	Co	C	Composition (continued), Specifications, and Remarks
74*	749	Zakharov, A.I., et al.	1972	L	293, 353	Superinvar	Bal.	32.2		0.9		4.2		Similar to the above specimen; $\gamma - \alpha$ phase transformation temperature at 178 K.
75*	749	Zakharov, A.I., et al.	1972	L	293, 353	Superinvar	Bal.	32.2		0.22	0.1	4.2		Similar to the above specimen; $\gamma - \alpha$ phase transformation temperature at 203 K.
76*	749	Zakharov, A.I., et al.	1972	L	293, 353	Superinvar	Bal.	32.2		0.22	0.3	4.2		Similar to the above specimen; $\gamma - \alpha$ phase transformation temperature at 208 K.
77*	749	Zakharov, A.I., et al.	1972	L	293, 353	Superinvar	Bal.	32.2		0.22	0.4	4.2		Similar to the above specimen; $\gamma - \alpha$ phase transformation temperature at 213 K.
78*	749	Zakharov, A.I., et al.	1972	L	293, 353	Superinvar	Bal.	32.4		0.25	0.16	4.2		Similar to the above specimen except Armco iron used in the charge; $\gamma - \alpha$ phase transformation temperature at 163 K.
79*	749	Zakharov, A.I., et al.	1972	L	293, 353	Superinvar	Bal.	32.4		0.25	0.15	4.2		Similar to the above specimen; $\gamma - \alpha$ phase transformation temperature at 143 K.
80*	749	Zakharov, A.I., et al.	1972	L	293, 353	Superinvar	Bal.	32.4		0.25	0.009	4.2		Similar to the above specimen; $\gamma - \alpha$ phase transformation temperature at 133 K.
81	467	Clark, A.F.	1968	L	20-300	Invar 36	Bal.	35.99		0.81	0.35		<0.1	0.17 Cu, <0.1 S, P; machined to 0.64 cm square x 20.32 cm long; 12-15% cold drown; fully martensitio; Rockwell hardness B-98.
82	462	Arp, V., Wilson, J.H., Winrich, L., and Sikora, P.	1962	L	20-293	Kromare 55	Bal.	20.76	16.01	8.76	0.22		0.05	2.12 Mo, 0.011 S, 0.003 P; hardness R_B61, cast specimen.
83	462	Arp, V., et al.	1962	L	20-293	Ni span C	Bal.	42.7	5.1	0.5			0.03	2.5 Ti, 0.4 Al, 0.1 Cu, S; hardness R_c35; age hardened specimen.
84	225	Valentich, J.	1965	L	91-546	Ni span								No information given.
85	545	Kachi, S., Asano, N., and Nakanashi, N.	1968	X	113-282		Bal.	25.66	18.11		0.23			1.30 Al; lattice parameter expansion measured; lattice parameter reported at 282 K is 3.58286 Å; 3.58346 Å at 293 K determined by graphical extrapolation.
86	545	Kachi, S., et al.	1968	X	108-286		Bal.	31.98	18.05		0.22			1.30 Al; lattice parameter expansion measured; lattice parameter reported at 286 K is 3.58802; 3.58818 Å at 293 K determined by graphical extrapolation.
87	467	Clark, A.F.	1968	L	0-300	Fe-29% Ni	Bal.	28.45		0.51	0.91		<0.01	<0.01 S; machined to 0.64 cm square x 20.32 cm long; fully martensitia; Rockwell hardness B-92.
88*	477	Kueser, P.E., Pavlovic, D.M., Lane, D.H., Clark, J.J., and Spavock, M.	1967	L	295-755	Maraging steel	Bal.	15.0				9.0		5.0 Mo, 0.7 Al, Ti; ultra-high strength, martensitia age-hardenable steef from Allegheny-Cudlum Steel Corp.
89*	477	Kueser, P.E., et al.	1967	L	295-755	Maraging steel	Bal.	18.0				8.0		4.0 Mo, 0.8 Ti; similar to the above specimen.

* Not shown in figure.

SPECIFICATION TABLE 262. THERMAL LINEAR EXPANSION OF IRON + NICKEL + ΣX_i ALLOYS Fe + Ni + ΣX_i (continued)

Cur. No.	Ref. No.	Author(s)	Year	Method Used	Temp. Range, K	Name and Specimen Designation	Fe	Ni	Cr	Mn	Si	Co	C	Composition (continued), Specifications, and Remarks
							Composition (weight percent)							
90*	546	Eiselstein, H. L.	1967	L	77-1028		Bal.	39.56		0.22	0.12		0.01	0.22 Al, 0.04 Cu, 2.90 (Nb + Ta), 0.007 S; specimen hot rolled; annealed 0.5 hr at 1255 K; aged 8 hr at 936 K; expansion measured with increasing temperature; zero-point correction of 0.002% determined by graphical interpolation.
91*	546	Eiselstein, H. L.	1967	L	1044-300		Bal.	39.56		0.22	0.12		0.01	0.22 Al, 0.04 Cu; the above specimen; expansion measured with decreasing temperature; zero-point correction of -0.041% determined by graphical extrapolation.
92*	546	Eiselstein, H. L.	1967	L	76-1224		Bal.	39.56		0.22	0.12			0.22 Al, 0.04 Cu; similar to the above specimen except cold rolled; zero-point correction of 0.002% determined by graphical interpolation.
93*	546	Eiselstein, H. L.	1967	L	1242-297		Bal.	39.56		0.22	0.12			0.22 Al, 0.04 Cu; the above specimen; expansion measured with decreasing temperature; zero-point correction of -0.076% determined by graphical extrapolation.
94	547	Dobkowski, D.S., Porter, L.F., and Loveday, G.E.	1964		299-810	Heat no. X 53014	Bal.	18.30		0.032	0.022	8.09	0.014	0.18 Al, 2.60 Mo, 0.15 Ti, 0.012 S, <0.01 each N, P, B, and O; specimen 0.125 in. diameter x 2 in. long; fully solution - annealed at 1200 K for at least 1 hr, aged, air cooled, and machined.
95	547	Dobkowski, D.S., et al.	1964		299-810	Heat no. X 14689	Bal.	12	5					3.0 Mo; specimen 0.125 in diameter x 2 in. long; similar to the above specimen.
96	51	Souder, W. and Hidnert, P.	1922	T	301-1185	S 553	Bal.	3.94	2.50	0.01	0.135		0.168	0.39 V, 0.026 S, 0.010 P; annealed specimen; expansion with increasing temperature; zero-point correction of 0.018% determined by graphical extrapolation.
97*	51	Souder, W. and Hidnert, P.	1922	T	1185-299	S 553	Bal.	3.94	2.50		0.135			The above specimen; expansion measured with decreasing temperature; zero-point correction of 0.080% determined by graphical extrapolation.
98	51	Souder, W. and Hidnert, P.	1922	T	289-1163	S 484	Bal.	34.52					0.14	Annealed specimen; expansion measured with increasing temperature; zero-point correction is -0.004%.
99	51	Souder, W. and Hidnert, P.	1922	T	296-1180	S 559	Bal.	3.59		0.78	0.094		0.326	0.035 S, 0.014 P; annealed specimen; expansion measured with increasing temperature; zero-point correction of 0.002% determined by graphical extrapolation.
100*	51	Souder, W. and Hidnert, P.	1922	T	1180-299	S 559	Bal.	3.59		0.78	0.094			The above specimen; expansion measured with decreasing temperature; zero-point correction of 0.077% determined by graphical extrapolation.

* Not shown in figure.

SPECIFICATION TABLE 262. THERMAL LINEAR EXPANSION OF IRON + NICKEL + ΣX_i ALLOYS Fe + Ni + ΣX_i (continued)

Cur. No.	Ref. No.	Author(s)	Year	Method Used	Temp. Range, K	Name and Specimen Designation	Fe	Ni	Composition (weight percent) Cr	Mn	Si	Co	C	Composition (continued), Specifications, and Remarks
101*	51	Souder, W. and Hidnert, P.	1922	T	300-1177	S 563	Bal.	3.67		1.21	1.04		0.388	0.043 S, 0.010 P; annealed specimen; expansion measured with increasing temperature; zero-point correction of 0.007% determined by graphical extrapolation.
102*	51	Souder, W. and Hidnert, P.	1922	T	1177-298	S 563	Bal.	3.67		1.21	1.04			The above specimen; expansion measured with decreasing temperature; zero-point correction of 0.098% determined by graphical extrapolation.
103*	51	Souder, W. and Hidnert, P.	1922	T	297-1177	S 554	Bal.	2.00		1.11	0.115		0.410	0.053 P, 0.049 S; annealed specimen; expansion measured with increasing temperature; zero-point correction of 0.007% determined by graphical extrapolation.
104*	51	Souder, W. and Hidnert, P.	1922	T	1177-308	S 554	Bal.	2.00		1.11	0.115			The above specimen; expansion measured with decreasing temperature; zero-point correction of 0.022% determined by graphical extrapolation.
105*	548	Zemtsova, N.D., Vasilevskaya, M.M., and Malyshev, K.A.	1964	X	272-1024	N25Kh2T2	Bal.	24.50	1.73	0.37	0.20		0.04	2.32 Ti; α-phase specimen soaked 15 min at 573 to 1173 K; lattice parameter reported at 293 is 2.87200 Å.
106*	548	Zemtsova, N.D., et al.	1964	X	273-1262	N25Kh2T2	Bal.	24.50	1.73					The above specimen in γ-phase; lattice parameter reported at 293 K is 3.59105 Å.
107*	548	Zemtsova, N.D., et al.	1964	X	276-1027	N24Kh2T3	Bal.	23.50	1.77	0.53	0.34		0.05	3.28 Ti; similar to the above specimen in α-phase; α-phase lattice parameter reported at 293 K is 2.87490 Å.
108*	548	Zemtsova, N.D., et al.	1964	X	288-1268	N24Kh2T3	Bal.	23.50	1.77					The above specimen in γ-phase; γ-phase lattice parameter reported at 293 K is 3.58613 Å.
109*	538	Yershov, V.M. and Oslon, N.L.	1968	X	628-209	Steel 40N7	Bal.	7.25	0.09	0.57	0.20		0.38	0.028 S, 0.023 P; produced in induction furnace, forged to square bar 12 x 12 mm, after annealing, made with cross-section of 12 x 12 and thickness of 0.8-0.9 mm measured of austenite line (200) on cooling from 963 K; lattice parameter reported at 293 K is 3.5957 Å.
110*	538	Yershov, V.M. and Oslon, N.L.	1968	X	653-280	Steel 40N7	Bal.	7.25	0.09	0.57	0.20			Similar to the above specimen, except cooling from 1373 K; lattice parameter reported at 280 K is 3.5891 Å; 3.5901 Å at 293 K determined by graphical interpolation.

* Not shown in figure.

SPECIFICATION TABLE 262. THERMAL LINEAR EXPANSION OF IRON + NICKEL + ΣX_i ALLOYS Fe + Ni + ΣX_i (continued)

Cur. No.	Ref. No.	Author(s)	Year	Method Used	Temp. Range, K	Name and Specimen Designation	Composition (weight percent) Fe	Ni	Cr	Mn	Si	Co	C	Composition (continued), Specifications, and Remarks
111*	541	Stuart, H. and Ridley, N.	1969	X	293-1292	20 Cr-25 Ni	Bal.	24.74	19.72					Fillings for x-ray powder specimen; size 200-350 mesh (British Standard Sieve), given a strain relieve anneal at 1073 K for 1 hr before measurement; alloy prepared by melting together iron (99.986 pure); chromium (>99.95) and nickel (>99.998) in recrystallized alumina boats under a purified argon atm, lattice reproducible to ±0.0002 Å, temperature control ±2 C; lattice parameter reported at 293 K is 3.5847 Å.
112	541	Stuart, H. and Ridley, N.	1969	X	375-1271	20 Cr-25 Ni	Bal.	24.74	19.72					The above specimen; mean linear expansion coefficient recorded.
113*	59	Masumoto, H.	1931	L	303, 373	No. 2	80.1	10				9.9		Armco Fe, Mond Ni, Granular Co., mixed and melted in hydrogen atm in alumina crucible in Tammann furnace, cast in iron mould with cylindrical aperature, machined to circular rod 4 mm thick, 10 cm long, heated at 1273 K for 1 hr in electric furnace in hydrogen atm at reduced pressure; expansion measured in vacuum.
114*	59	Masumoto, H.	1931	L	303, 373	No. 3	70.1	20				9.9		Similar to the above specimen.
115*	59	Masumoto, H.	1931	L	303, 373	No. 4	65.0	25				10.0		Similar to the above specimen.
116*	59	Masumoto, H.	1931	L	303, 373	No. 5	62.0	28				10.0		Similar to the above specimen.
117*	59	Masumoto, H.	1931	L	303, 373	No. 6	60.1	30				9.9		Similar to the above specimen.
118*	59	Masumoto, H.	1931	L	303, 373	No. 7	58.0	32				10.0		Similar to the above specimen.
119*	59	Masumoto, H.	1931	L	303, 373	No. 8	55.0	35				10.0		Similar to the above specimen.
120*	59	Masumoto, H.	1931	L	303, 373	No. 9	50.1	40				9.9		Similar to the above specimen.
121*	59	Masumoto, H.	1931	L	303, 373	No. 16	60.2	20				19.8		Similar to the above specimen.
122*	59	Masumoto, H.	1931	L	303, 373	No. 17	55.1	25				19.9		Similar to the above specimen.
123*	59	Masumoto, H.	1931	L	303, 373	No. 18	52.2	28				19.8		Similar to the above specimen.
124*	59	Masumoto, H.	1931	L	303, 373	No. 19	50.1	30				19.9		Similar to the above specimen.
125*	59	Masumoto, H.	1931	L	303, 373	No. 20	45.1	35				19.9		Similar to the above specimen.
126*	59	Masumoto, H.	1931	L	303, 373	No. 21	40.1	40				19.9		Similar to the above specimen.
127*	59	Masumoto, H.	1931	L	303, 373	No. 32	40.1	30				29.9		Similar to the above specimen.
128*	551	Belov, A. K.	1968	L	77-323	Kh15N24T2 Russian Alloy	Bal.	24	15					2 Ti; cylindrical specimen 3.5 mm in diameter, 50 mm long; sample heated and cooled at a rate of 3-7 deg per min; data on the coefficient of thermal linear expansion are also reported.

* Not shown in figure.

SPECIFICATION TABLE 262. THERMAL LINEAR EXPANSION OF IRON + NICKEL + ΣX_i ALLOYS Fe + Ni + ΣX_i (continued)

Cur. No.	Ref. No.	Author(s)	Year	Method Used	Temp. Range, K	Name and Specimen Designation	Composition (weight percent) Fe	Ni	Cr	Mn	Si	Co	C	Composition (continued), Specifications, and Remarks
129*	551	Belov, A.K.	1968	L	77-323	OKhN3M	Bal.	3.3	0.9	0.8	0.37			0.3 Mo; similar to the above specimen; data on the coefficient of thermal linear expansion are also reported.
130*	551	Belov, A.K.	1968	L	77-323	18Kh2N4VA Russian Alloy	Bal.	4.5	1.7	0.55	0.37		0.21	1.2 W; similar to the above specimen; data on the coefficient of thermal linear expansion are also reported.
131*	530	White, G.K.	1965	E	3-30	Nilo 36	63.1	36.2		0.3		0.1		0.1 Cu; from Henry Wiggin Ltd.; specimen 2 cm diameter and 6 cm long; specimen tested as received; this curve is here reported using the first given temperature as reference temperature at which $\Delta L/L_0 = 0$.
132*	530	White, G.K.	1965	E	4-30	Nilo 40	56.3	43.1		0.3		0.1		0.1 Cu; similar to the above specimen; this curve is here reported using the first given temperature as reference temperature at which $\Delta L/L_0 = 0$.
133*	532	Rohde, R.W. and Graham, R.A.	1969	L	297-685		69.8	29.6		0.4	0.1		0.1	Commercial alloy; annealed 2 hr at 923 K and 10^{-6} torr; austenitic structure; zero-point correction of 0.006% determined by graphical extrapolation.
134*	532	Rohde, R.W. and Graham, R.A.	1969	L	297-685		69.8	29.6		0.4	0.1			Similar to the above specimen; martensitic structure; zero-point correction of 0.025% determined by graphical extrapolation.
135*	391	Phragmén, G.	1931	X	283-568	C.F.O. 242N29	Bal.	29.44		0.35	0.10		0.05	Specimen obtained from Société Anonyme de Commentry, Fourchombault et Decazeville, Imphy France; lattice parameter reported at 283 K is 3.5753 Å; 3.5759 Å at 293 K determined by graphical interpolation.
136*	391	Phragmén, G.	1931	X	283-568	C.F.O. 1303	Bal.	34.28		0.43	0.18		0.05	Similar to the above specimen; lattice parameter reported at 283 K is 3.5753 Å; 3.5835 Å at 293 determined by graphical interpolation.
137*	391	Phragmén, G.	1931	X	283-453	C.F.O. 1932	Bal.	36.40		0.25	0.04		0.05	Similar to the above specimen; lattice parameter reported at 283 K is 3.5853 Å; 3.5854 Å at 293 K determined by graphical interpolation.
138*	391	Phragmén, G.	1931	X	283-568	C.F.O. 261N38	Bal.	37.92		0.32	trace		0.05	Similar to the above specimen; lattice parameter reported at 283 K is 3.5883 Å; 3.5884 Å at 293 K determined by graphical interpolation.

* Not shown in figure.

DATA TABLE 262. THERMAL LINEAR EXPANSION OF IRON + NICKEL + ΣX_i ALLOYS Fe + Ni + ΣX_i

[Temperature, T, K; Linear Expansion, $\Delta L/L_0$, %]

T	$\Delta L/L_0$
CURVE 1	
88.7	-0.226
94.3	-0.209*
99.8	-0.205*
105.4	-0.202
110.9	-0.197*
116.5	-0.194*
122.0	-0.189*
127.6	-0.187*
133.2	-0.182*
138.7	-0.176*
144.3	-0.174*
149.8	-0.165
155.4	-0.160*
160.9	-0.155*
166.5	-0.151*
172.0	-0.149*
177.6	-0.141*
183.2	-0.135*
188.7	-0.131*
194.3	-0.125*
199.8	-0.117
205.4	-0.112*
210.9	-0.108*
216.5	-0.099*
222.0	-0.093*
227.6	-0.088*
233.2	-0.078*
238.7	-0.074*
244.3	-0.068*
249.8	-0.059
255.4	-0.051*
260.9	-0.044*
266.5	-0.038*
272.0	-0.031*
277.6	-0.023*
283.2	-0.012*
288.7	-0.007*
294.3	0.002*
299.8	0.012
305.4	0.016*
310.9	0.026*
316.5	0.032*
322.1	0.044*
327.6	0.050*
333.2	0.056*
338.7	0.062*
344.3	0.074*

T	$\Delta L/L_0$
CURVE 1 (cont.)	
349.8	0.083
355.4	0.090*
360.9	0.098*
366.5	0.105*
372.0	0.114*
377.6	0.121*
383.2	0.130*
388.7	0.141*
394.3	0.146*
399.8	0.159
405.4	0.166*
410.9	0.174*
416.5	0.186*
422.1	0.190*
435.9	0.214*
449.8	0.233
463.7	0.257*
477.6	0.284*
491.5	0.304*
505.4	0.324
519.3	0.352*
533.2	0.374*
547.1	0.395
560.9	0.425*
574.8	0.450*
588.7	0.473*
602.6	0.497
616.5	0.521*
630.4	0.543*
644.3	0.569*
658.2	0.592
672.1	0.620*
685.9	0.644*
699.8	0.671
713.7	0.695*
727.6	0.718*
741.5	0.744*
755.4	0.773
769.3	0.797*
783.2	0.824*
797.1	0.849*
810.9	0.874
CURVE 2	
20	-0.195
40	-0.193

T	$\Delta L/L_0$
CURVE 2 (cont.)	
60	-0.191
80	-0.187
100	-0.180
120	-0.170
140	-0.155
160	-0.137
180	-0.119
200	-0.100*
220	-0.081
240	-0.060
260	-0.038
273	-0.023
280	-0.015
293	0.000
CURVE 3*	
20	-0.197
40	-0.196
60	-0.193
80	-0.187
100	-0.177
120	-0.165
140	-0.151
160	-0.136
180	-0.119
200	-0.099
220	-0.079
240	-0.059
260	-0.037
273	-0.023
280	-0.015
293	0.000
CURVE 4	
366	0.0742
394	0.1038
422	0.1340
450	0.164
478	0.193
505	0.221
533	0.250
561	0.277
589	0.303
616	0.328
644	0.354

T	$\Delta L/L_0$
CURVE 4 (cont.)	
672	0.381
700	0.410
CURVE 5	
366	0.0629
394	0.0929
422	0.123
450	0.151
478	0.178
505	0.205
533	0.232
561	0.260
589	0.291
616	0.322
644	0.352*
672	0.381*
700	0.409*
CURVE 6	
20	-0.291
40	-0.290
60	-0.285
80	-0.274
100	-0.257
120	-0.237
140	-0.214
160	-0.191
180	-0.166
200	-0.138
220	-0.110
240	-0.082
260	-0.032
273	-0.031
280	-0.021
293	0.000*
CURVE 7	
299	0.011*
450	0.254
561	0.432
702	0.674

T	$\Delta L/L_0$
CURVE 8	
18	-0.286
23	-0.286*
33	-0.286*
43	-0.285*
53	-0.284
63	-0.281*
73	-0.276*
83	-0.270*
93	-0.264*
103	-0.256*
113	-0.246
123	-0.237*
133	-0.227*
143	-0.215*
153	-0.203
163	-0.190*
173	-0.177*
193	-0.150*
213	-0.122
233	-0.094*
253	-0.064*
273	-0.033*
293	0.000*
313	0.033
333	0.066
353	0.100*
373	0.134
393	0.167*
413	0.202
433	0.236*
453	0.271*
473	0.306
493	0.342
513	0.378*
533	0.414
553	0.451*
573	0.487
CURVE 9	
293	0.000*
388	0.149
467	0.287
579	0.486
673	0.648

T	$\Delta L/L_0$
CURVE 10	
18	-0.258
23	-0.258*
33	-0.258*
43	-0.258*
53	-0.257
63	-0.255*
73	-0.251*
83	-0.246*
93	-0.240*
103	-0.233
113	-0.225*
123	-0.217*
133	-0.208*
143	-0.199*
153	-0.189
163	-0.178*
173	-0.166*
193	-0.140*
213	-0.114
233	-0.087*
253	-0.060*
273	-0.031*
293	0.000*
313	0.030*
333	0.062*
353	0.094
373	0.126*
393	0.158*
413	0.191
433	0.224*
453	0.257*
473	0.291
493	0.325*
513	0.359
533	0.393
553	0.427*
573	0.463*
CURVE 11	
301	0.007*
350	0.053
394	0.100*
410	0.185*
529	0.248
615	0.346
676	0.435

* Not shown in figure.

DATA TABLE 262. THERMAL LINEAR EXPANSION OF IRON + NICKEL + ΣX_i ALLOYS Fe + Ni + ΣX_i (continued)

T	$\Delta L/L_0$
CURVE 11 (cont.)	
733	0.549
811	0.780
959	1.239
1039	1.429
1128	1.636*
CURVE 12	
1128	1.327
1078	1.222
961	0.952
788	0.580
700	0.455
295	0.001*
CURVE 13	
301	0.008*
355	0.069
414	0.132
538	0.275
631	0.371
713	0.490
824	0.664
893	0.783
979	0.940
1096	1.141
CURVE 14	
1096	1.166
946	0.888
804	0.621
724	0.499
597	0.314
495	0.205
380	0.094
301	0.006
CURVE 15	
298	0.004*
391	0.090*
428	0.136
549	0.260
646	0.374
738	0.492
838	0.638
CURVE 15 (cont.)	
934	0.804
1040	1.010
1109	1.121
1170	1.232
CURVE 16	
1170	1.234*
1044	1.003
962	0.852
899	0.721
793	0.554
693	0.412
574	0.275
484	0.187
389	0.108
298	0.006*
CURVE 17	
292	-0.001*
347	0.046
433	0.104
534	0.184
607	0.249
753	0.389
855	0.510
1065	0.859
1138	0.981
1220	1.149
CURVE 18	
1220	1.182
1166	1.046
1071	0.872
972	0.706
844	0.477
756	0.322
671	0.219
576	0.181
502	0.170
454	0.159
299	0.009*
CURVE 19	
290	-0.002*
CURVE 19 (cont.)	
365	0.057*
453	0.133
549	0.217
670	0.334
884	0.453
1230	1.120
CURVE 20*, †	
470	0.000
680	0.067
782	0.176
938	0.511
1149	0.880
1230	0.989
CURVE 21*	
293	0.0000
299	0.0002
304	0.0004
311	0.0007
CURVE 22*	
293	0.0000
301	0.0002
305	0.0003
310	0.0003
313	0.0004
CURVE 23*	
293	0.0000
299	0.0001
304	0.0002
308	0.0003
312	0.0005
CURVE 24*	
293	0.0000
299	0.0004
300	0.0005
301	0.0007
303	0.0011
306	0.0017
309	0.0022
CURVE 25	
4.2	-0.046
25	-0.046
50	-0.046
75	-0.046
100	-0.042
125	-0.037
150	-0.031
175	-0.025
200	-0.018
250	-0.006
300	0.007*
CURVE 26*	
293	0.000
339	0.000
403	0.019
465	0.056
500	0.091
532	0.135
562	0.200
CURVE 27	
89	-0.037
136	-0.026
175	-0.018
205	-0.009
236	-0.005
273	-0.002*
289	0.000*
293	0.000*
309	0.000
324	0.001
344	0.003
363	0.005
386	0.008
402	0.010
425	0.012
452	0.018
481	0.025
509	0.035
532	0.045
561	0.063
579	0.076
608	0.101
CURVE 28	
191	0.008
193	-0.003
204	-0.005*
219	-0.003
249	-0.001*
274	0.000*
289	0.000*
293	0.000*
305	0.000*
320	0.000
338	0.000
353	0.001
369	0.002
386	0.004*
404	0.006*
422	0.009*
453	0.017*
477	0.025
500	0.035
521	0.049
541	0.066
553	0.078
565	0.090
579	0.106
CURVE 29	
124	-0.002*
124	-0.008
128	-0.017*
136	-0.017
144	-0.016*
154	-0.015
163	-0.013*
178	-0.011*
188	-0.009*
219	-0.005*
247	-0.003*
273	-0.002
293	0.000*
311	0.000*
325	0.001*
341	0.002*
361	0.003*
374	0.004
396	0.006
421	0.011*
441	0.015
CURVE 29 (cont.)	
465	0.022
477	0.024*
502	0.033*
528	0.046*
557	0.069*
583	0.099
598	0.117
608	0.137
CURVE 30	
95	-0.045
114	-0.038
132	-0.034*
154	-0.026*
182	-0.020
205	-0.014*
228	-0.008
255	-0.004
273	-0.002*
290	-0.001*
293	0.000*
314	0.001
331	0.002
347	0.004*
367	0.006*
389	0.007*
415	0.010
443	0.015*
473	0.022*
502	0.030*
529	0.040*
558	0.056
580	0.073*
606	0.098*
619	0.113
CURVE 31	
250	-0.001*
273	0.000*
291	0.000*
293	0.000*
309	0.000*
332	0.001*
346	0.001*
362	0.002*
387	0.004*

* Not shown in figure.

† This curve is here reported using the first given temperature as reference temperature at which $\Delta L/L_0 = 0$.

DATA TABLE 262. THERMAL LINEAR EXPANSION OF IRON + NICKEL + ΣX_i ALLOYS Fe + Ni + ΣX_i (continued)

T	$\Delta L/L_0$
CURVE 31 (cont.)	
410	0.007*
432	0.011
463	0.018*
488	0.027
517	0.041*
543	0.063*
570	0.089*
595	0.120*
CURVE 32*	
176	-0.015
191	-0.013
208	-0.010
232	-0.007
251	-0.003
277	-0.002
291	-0.001
293	-0.000
307	0.000
326	0.002
345	0.003
363	0.005
380	0.006
401	0.007
413	0.010
426	0.011
445	0.015
470	0.020
493	0.028
544	0.050
576	0.075
598	0.098
623	0.130
CURVE 33*	
266	0.000
277	0.000
290	0.000
293	0.000
310	0.000
325	0.000
340	0.000
355	0.001
372	0.002
393	0.003
417	0.006
CURVE 33 (cont.)*	
432	0.011
453	0.015
482	0.027
502	0.038
531	0.060
557	0.084
580	0.110
598	0.135
CURVE 34*	
210	0.004
226	0.003
250	0.002
291	0.000
293	0.000
307	0.000
331	0.002
348	0.002
369	0.003
389	0.005
408	0.007
427	0.008
460	0.013
481	0.019
509	0.027
536	0.038
565	0.055
589	0.073
612	0.096
626	0.118
CURVE 35*	
203	0.002
222	0.001
252	0.001
276	0.000
290	0.000
293	0.000
316	0.000
344	0.000
370	0.002
389	0.004
415	0.008
437	0.015
464	0.020
495	0.036
CURVE 35 (cont.)*	
532	0.066
568	0.094
604	0.141
CURVE 36*	
227	0.000
263	-0.001
289	0.000
293	0.000
315	0.000
340	0.001
371	0.002
398	0.004
427	0.008
478	0.016
534	0.036
566	0.058
609	0.103
642	0.142
CURVE 37*	
244	0.000
263	0.000
273	0.000
293	0.000
311	0.000
323	0.000
334	0.000
352	0.000
372	0.000
391	0.000
406	0.000
429	0.007
443	0.010
457	0.014
473	0.019
498	0.043
529	0.059
553	0.083
574	0.108
598	0.135
623	0.163
650	0.202
673	0.240
698	0.279
724	0.322
CURVE 37 (cont.)*	
738	0.345
756	0.373
773	0.400
796	0.441
814	0.470
831	0.503
850	0.532
873	0.565
CURVE 38	
243	0.00
264	0.00
273	0.00*
293	0.00*
311	0.00*
321	0.00*
332	0.00*
352	0.00*
373	0.00*
393	0.00*
407	0.00
427	0.005*
441	0.008*
458	0.016
473	0.022*
498	0.036*
527	0.056
551	0.078*
575	0.100*
599	0.130
623	0.165
649	0.203
673	0.237
700	0.282
723	0.320
736	0.345
773	0.401
794	0.442
818	0.482
833	0.508
851	0.536
873	0.571
CURVE 39*	
243	0.00
263	0.00
CURVE 39 (cont.)*	
273	0.00
293	0.00
311	0.00
322	0.00
333	0.00
351	0.00
373	0.00
392	0.00
407	0.00
428	0.005
442	0.012
457	0.017
473	0.026
498	0.038
528	0.057
553	0.080
573	0.103
597	0.135
621	0.167
647	0.202
673	0.244
698	0.288
721	0.324
735	0.343
773	0.400
793	0.440
817	0.472
834	0.509
852	0.537
873	0.567
CURVE 40*, †, ‡	
5	0.0000
6	0.0001
7	0.0002
8	0.0002
10	0.0003
10	-0.0005
11	-0.0006
14	-0.0008
16	-0.0011
18	-0.0014
20	-0.0017
24	-0.0025
26	-0.0028
28	-0.0033
32	-0.0040
CURVE 40 (cont.)*, †, ‡	
36	-0.0049
45	-0.0062
50	-0.0068
75	-0.0078
100	-0.0063
125	-0.0034
150	-0.0002
175	0.0026
200	0.0049
CURVE 41*, †, ‡	
23	0.0000
24	-0.0002
82	-0.0046
89	-0.0043
195	0.0044
CURVE 42*, †, ‡	
7	0.0000
9	-0.0002
11	-0.0004
13	-0.0006
19	-0.0016
25	-0.0028
28	-0.0034
31	-0.0040
32	-0.0042
36	-0.0048
37	-0.0053
39	-0.0055
42	-0.0062
45	-0.0067
50	-0.0075
75	-0.0090
100	-0.0080
115	-0.0069
CURVE 43*, †, ‡	
4	0.0000
6	-0.0000
8	-0.0001
10	-0.0002
13	-0.0003
13	-0.0003
15	-0.0004

* Not shown in figure.

† This curve is here reported using the first given temperature as reference temperature at which $\Delta L/L_0 = 0$.

‡ Author's data for coefficient of thermal expansion have been integrated by TPRC to obtain $\Delta L/L_0$.

DATA TABLE 262. THERMAL LINEAR EXPANSION OF IRON + NICKEL + ΣX_i ALLOYS Fe + Ni + ΣX_i (continued)

T	$\Delta L/L_0$
CURVE 43 (cont.)*, †, ‡	
16	-0.0004
18	-0.0005
20	-0.0006
21	-0.0007
23	-0.0008
25	-0.0009
29	-0.0012
32	-0.0014
37	-0.0016
43	-0.0017
48	-0.0017
52	-0.0017
58	-0.0014
85	0.0027
99	0.0065
108	0.0094
124	0.0150
158	0.0281
176	0.0357
201	0.0461
231	0.0574
CURVE 44*, †, ‡	
32	0.0000
37	-0.0001
43	-0.0002
48	-0.0001
53	0.0000
63	0.0008
68	0.0015
CURVE 45*, †, ‡	
6	0.0000
8	0.0000
10	0.0000
19	0.0001
30	0.0004
37	0.0008
50	0.0021
60	0.0036
80	0.0092
86	0.0118
96	0.0167
CURVE 46*, †, ‡	
6	0.0000
19	0.0002
36	0.0012
45	0.0024
55	0.0043
CURVE 47*, †, ‡	
5	0.0000
8	0.0000
10	0.0000
15	0.0000
26	0.0004
40	0.0016
CURVE 48*, †, ‡	
6	0.0000
9	0.0000
35	0.0013
46	0.0026
CURVE 49*, †, ‡	
6	0.0000
8	-0.0001
10	-0.0002
17	-0.0004
23	-0.0009
48	-0.0018
54	-0.0015
79	0.0029
89	0.0056
100	0.0087
CURVE 50*, †, ‡	
10	0.0000
15	-0.0002
25	-0.0007
35	-0.0010
46	-0.0011
CURVE 51	
11	-0.025
15	-0.025*
18	-0.026*
20	-0.026
25	-0.027*
29	-0.028*
33	-0.028
39	-0.029*
49	-0.030
56	-0.030*
64	-0.030
73	-0.030*
81	-0.029
89	-0.028*
100	-0.027
110	-0.025
119	-0.023*
130	-0.021
140	-0.019*
149	-0.018*
159	-0.016*
170	-0.014
179	-0.012*
189	-0.011
198	-0.010*
208	-0.009*
218	-0.007*
229	-0.006*
239	-0.005*
249	-0.004*
258	-0.003*
269	-0.002*
282	-0.001*
292	0.000*
293	0.000*
301	0.001*
CURVE 52	
10	-0.017*
14	-0.018*
18	-0.018*
23	-0.020*
28	-0.020*
35	-0.022*
CURVE 52 (cont.)	
43	-0.022*
48	-0.023*
56	-0.023
61	-0.023*
68	-0.023*
76	-0.023*
81	-0.022*
91	-0.021
101	-0.020*
111	-0.018*
120	-0.016*
130	-0.015*
140	-0.013*
150	-0.012*
161	-0.010*
171	-0.008*
180	-0.007
190	-0.006*
200	-0.005*
210	-0.004
221	-0.003*
231	-0.002*
240	-0.002*
250	-0.001*
260	-0.001*
273	-0.001*
283	0.000*
293	0.000*
303	0.001*
313	0.002*
322	0.003*
CURVE 53*	
5	-0.007
8	-0.007
12	-0.009
19	-0.010
28	-0.011
33	-0.012
36	-0.013
41	-0.014
45	-0.014
57	-0.015
64	-0.015
CURVE 53 (cont.)	
72	-0.015
80	-0.014
89	-0.014
100	-0.013
109	-0.012
119	-0.011
130	-0.009
139	-0.008
149	-0.007
159	-0.006
170	-0.005
179	-0.004
188	-0.004
200	-0.003
209	-0.002
219	-0.002
230	-0.002
240	-0.001
250	-0.001
259	-0.001
270	0.000
281	0.000
293	0.000
302	0.001
312	0.001
322	0.001
CURVE 54*, †, ‡	
9	0.000
12	-0.003
16	-0.008
21	-0.017
25	-0.023
32	-0.034
37	-0.041
41	-0.045
45	-0.048
48	-0.050
53	-0.052
57	-0.052
61	-0.051
65	-0.050
73	-0.045
77	-0.042
CURVE 54 (cont.)*, †, ‡	
81	-0.038
85	-0.034
89	-0.029
92	-0.025
98	-0.017
102	-0.012
114	0.006
121	0.017
CURVE 55*, †, ‡	
9	0.000
29	-0.027
33	-0.033
37	-0.038
41	-0.042
44	-0.045
49	-0.048
52	-0.049
57	-0.050
61	-0.050
65	-0.049
69	-0.047
73	-0.044
78	-0.040
81	-0.038
90	-0.027
101	-0.012
110	0.000
118	0.012
126	0.025
134	0.037
CURVE 56*, †, ‡	
88	-0.0153
92	-0.0148
96	-0.0143
104	-0.0133
107	-0.0128
112	-0.0120
119	-0.0109
128	-0.0094
132	-0.0088
136	-0.0082

* Not shown in figure.
† This curve is here reported using the first given temperature as reference temperature at which $\Delta L/L_0 = 0$.
‡ Author's data for coefficient of thermal expansion have been integrated by TPRC to obtain $\Delta L/L_0$.

DATA TABLE 262. THERMAL LINEAR EXPANSION OF IRON + NICKEL + ΣX_i ALLOYS Fe + Ni + ΣX_i (continued)

T	$\Delta L/L_0$
CURVE 56 (cont.)*, ‡	
140	-0.0076
143	-0.0071
147	-0.0065
153	-0.0057
157	-0.0051
160	-0.0047
164	-0.0043
168	-0.0039
172	-0.0034
176	-0.0029
179	-0.0026
184	-0.0022
187	-0.0019
191	-0.0017
195	-0.0014
200	-0.0011
203	-0.0009
207	-0.0007
211	-0.0006
216	-0.0004
219	-0.0003
223	-0.0002
227	-0.0002
231	-0.0001
236	-0.0001
239	0.0001
243	0.0000
247	0.0000
251	0.0000
254	0.0000
260	0.0000
263	0.0000
269	-0.0001
277	-0.0001
287	-0.0001
298	0.0001
CURVE 57*, ‡	
83	-0.0159
86	-0.0156
94	-0.0147
98	-0.0141
100	-0.0138
102	-0.0135
106	-0.0129
110	-0.0122
CURVE 57 (cont.)*, ‡	
114	-0.0116
118	-0.0110
122	-0.0103
126	-0.0097
129	-0.0092
134	-0.0084
138	-0.0078
142	-0.0072
146	-0.0066
150	-0.0060
155	-0.0054
159	-0.0048
162	-0.0045
166	-0.0040
170	-0.0036
174	-0.0032
178	-0.0028
182	-0.0024
186	-0.0020
190	-0.0017
194	-0.0014
198	-0.0012
201	-0.0010
205	-0.0008
210	-0.0006
214	-0.0005
217	-0.0004
222	-0.0003
226	-0.0002
230	-0.0001
233	-0.0001
235	0.0000
237	0.0000
241	0.0000
245	0.0000
250	0.0000
253	0.0000
258	0.0000
262	0.0000
270	0.0000
275	-0.0001
286	0.0000
291	0.0000
295	0.0000
305	0.0002
CURVE 58*	
293	0.000
353	0.062
CURVE 59*	
293	0.000
353	-0.001
CURVE 60*	
293	0.000
353	0.043
CURVE 61*	
293	0.000
353	0.008
CURVE 62*	
293	0.000
353	0.007
CURVE 63*	
293	0.000
353	0.003
CURVE 64*	
293	0.000
353	0.001
CURVE 65*	
293	0.000
353	0.004
CURVE 66*	
293	0.000
353	0.001
CURVE 67*	
293	0.000
353	-0.001
CURVE 68*	
293	0.0000
353	0.0003
CURVE 69*	
293	0.000
353	-0.001
CURVE 70*	
293	0.000
353	0.001
CURVE 71*	
293	0.000
353	0.013
CURVE 72*	
293	0.000
353	0.002
CURVE 73*	
293	0.000
353	0.004
CURVE 74*	
293	0.000
353	0.006
CURVE 75*	
293	0.000
353	0.002
CURVE 76*	
293	0.000
353	0.003
CURVE 77*	
293	0.000
353	0.003
CURVE 78*	
293	0.000
353	0.001
CURVE 79*	
293	0.000
353	0.001
CURVE 80*	
293	0.000
353	-0.001
CURVE 81	
20	-0.037
30	-0.039
40	-0.040
50	-0.040*
60	-0.039
70	-0.039
80	-0.038*
90	-0.037*
100	-0.036*
120	-0.033*
140	-0.029
160	-0.025
180	-0.020*
200	-0.016*
220	-0.013
240	-0.010
260	-0.008*
273	-0.006*
280	-0.004*
293	0.000*
CURVE 82	
20	-0.276
40	-0.272
60	-0.266
80	-0.259
100	-0.247
120	-0.232*
140	-0.211*
160	-0.188*
180	-0.164*
CURVE 82 (cont.)	
200	-0.137*
220	-0.109*
240	-0.081*
260	-0.053*
273	-0.032*
280	-0.022*
293	0.000*
CURVE 83	
20	-0.135
40	-0.135
60	-0.133
80	-0.128
100	-0.121
120	-0.112
140	-0.102
160	-0.091
180	-0.079
200	-0.067
220	-0.054
240	-0.040
260	-0.026
273	-0.015
280	-0.010*
293	0.000*
CURVE 84	
91	-0.135
174	-0.083
293	0.000*
370	0.063*
464	0.154
502	0.207*
532	0.260
546	0.299
CURVE 85	
113	-0.220
125	-0.209
148	-0.186*
171	-0.161
198	-0.131*
222	-0.101
246	-0.068

* Not shown in figure.
‡ Author's data for coefficient of thermal expansion have been integrated by TPRC to obtain $\Delta L/L_0$.

DATA TABLE 262. THERMAL LINEAR EXPANSION OF IRON + NICKEL + ΣX_i ALLOYS $Fe + Ni + \Sigma X_i$ (continued)

T	$\Delta L/L_0$
CURVE 85 (cont.)	
271	-0.034*
282	-0.017*
CURVE 86	
108	-0.131*
123	-0.119*
147	-0.102
172	-0.081*
196	-0.068
221	-0.050*
245	-0.033
270	-0.014*
286	-0.004
CURVE 87	
0	0.180
10	0.180
20	0.180*
30	0.179
40	0.179*
50	0.178
60	0.176*
70	0.173
80	0.170*
90	0.166
100	0.161
120	0.150
140	0.137
160	0.122
180	0.106
200	0.089
220	0.071
240	0.053*
260	0.033*
273	0.021*
280	0.014*
293	0.000*
CURVE 88*	
295	0.002
755	0.461
CURVE 89*	
295	0.002
755	0.461
CURVE 90*	
77	-0.061
297	0.002
369	0.035
480	0.134
589	0.295
700	0.472
812	0.661
859	0.745
923	0.875
1028	1.150
CURVE 91*	
1044	1.106
922	0.874
814	0.650
700	0.470
589	0.287
478	0.138
367	0.038
300	0.003
CURVE 92*	
76	-0.060
295	0.002
367	0.034
478	0.124
589	0.285
700	0.460
811	0.647
922	0.868
1033	1.150
1140	1.402
1224	1.574
CURVE 93*	
1242	1.500
1159	1.324
1033	1.111
922	0.883
811	0.661
700	0.474
589	0.301
478	0.108
367	0.037
297	0.002
CURVE 94	
299	0.007*
699	0.432
810	0.569
CURVE 95	
299	0.008*
699	0.461
810	0.615
CURVE 96	
301	0.008*
365	0.081*
484	0.221
602	0.388
678	0.488
769	0.622
886	0.792
923	0.845
956	0.884
970	0.895
987	0.901
1003	0.891
1017	0.875
1031	0.848
1041	0.829
1054	0.818
1067	0.816
1088	0.824
1123	0.902
1185	1.045
CURVE 97*	
1185	1.117
1123	0.974
1088	0.896
1075	0.869
1030	0.765
996	0.686
954	0.596
909	0.498
861	0.398
813	0.288
798	0.247
776	0.230
764	0.222
CURVE 97 (cont.)*	
746	0.206
737	0.196
722	0.183
707	0.167
693	0.210
691	0.248
684	0.267
668	0.276
659	0.307
646	0.311
625	0.311
611	0.319
596	0.319
580	0.313
564	0.305
299	0.011
CURVE 98	
289	-0.004
293	0.000
345	0.028
424	0.067
453	0.098
531	0.196
589	0.286*
603	0.330
678	0.445
728	0.544
760	0.593
846	0.749
863	0.787
909	0.869
979	1.004
1027	1.103
1069	1.180
1163	1.367
CURVE 99	
296	0.002
396	0.111
496	0.228
527	0.259*
654	0.458
739	0.589
867	0.788*
931	0.885
CURVE 99 (cont.)	
956	0.923
967	0.925
972	0.919
982	0.878
993	0.818
1004	0.773
1012	0.761
1018	0.754
1025	0.757
1031	0.764*
1076	0.861
1106	0.929
1121	0.955
1180	1.100
CURVE 100*	
1180	1.175
1121	1.030
1106	1.004
1076	0.936
1008	0.778
954	0.660
905	0.563
897	0.561
893	0.562
880	0.577
860	0.596
849	0.625
838	0.661
826	0.680
816	0.683
807	0.675
631	0.412
529	0.283
299	0.011
CURVE 101*	
300	0.007
405	0.128
475	0.222
570	0.342
719	0.551
783	0.654
831	0.724
918	0.863
943	0.902
CURVE 101 (cont.)*	
965	0.931
983	0.928
989	0.908
990	0.890
993	0.868
1003	0.840
1014	0.847
1026	0.858
1035	0.870
1044	0.888
1104	1.021
1127	1.076
1177	1.184
CURVE 102*	
1177	1.275
1127	1.167
1104	1.112
1076	1.055
1049	0.998
1038	0.980
1018	0.940
977	0.849
957	0.797
945	0.771
923	0.729
907	0.694
873	0.636
848	0.599
834	0.580
820	0.573
813	0.573
798	0.564
784	0.541
774	0.524
480	0.159
298	0.004
CURVE 103*	
297	0.007
472	0.207
580	0.371
665	0.511
1001	1.052
1007	1.047
1015	1.030

* Not shown in figure.

DATA TABLE 262. THERMAL LINEAR EXPANSION OF IRON + NICKEL + ΣX_i ALLOYS Fe + Ni + ΣX_i (continued)

T	$\Delta L/L_0$
CURVE 103 (cont.)*	
1027	1.026
1039	1.018
1047	1.008
1059	0.994
1085	0.981
1105	0.984
1122	1.006
1177	1.136
CURVE 104*	
1177	1.151
1122	1.021
1063	0.896
1051	0.896
1040	0.910
1027	0.931
1015	0.944
1001	0.958
988	0.958
963	0.953
935	0.943
823	0.756
068	0.375
472	0.223
308	0.022
CURVE 105*	
272	0.000
293	0.000
574	0.000
676	-0.034
770	-0.172
824	-0.139
870	-0.071
897	-0.060
927	-0.069
1024	-0.135
CURVE 106*	
273	-0.001
293	0.000
573	-0.004
670	-0.028
816	-0.134
CURVE 106 (cont.)*	
874	-0.173
898	-0.082
928	-0.185
972	-0.189
1023	-0.167
1076	-0.059
1125	-0.030
1262	0.000
CURVE 107*	
276	0.009
293	0.000
575	-0.063
677	-0.095
782	-0.165
832	-0.164
876	-0.272
901	-0.131
935	-0.094
975	-0.095
1027	-0.095
CURVE 108*	
288	-0.004
293	0.000
577	-0.004
679	-0.004
774	-0.164
826	-0.166
877	-0.088
903	-0.096
928	-0.087
979	-0.144
1027	-0.145
1080	-0.091
1133	-0.001
1268	-0.004
CURVE 109*	
628	0.738
566	0.588
543	0.499
514	0.429
491	0.384
CURVE 109 (cont.)*	
467	0.334
458	0.256
430	0.220
416	0.170
400	0.150
392	0.117
378	0.100
367	0.058
349	0.044
333	0.022
304	0.014
293	0.000
281	-0.008
271	-0.036
255	-0.017
244	-0.056
224	-0.053
209	-0.070
CURVE 110*	
280	-0.054
291	-0.006
300	0.022
307	0.050
323	0.066
329	0.122
339	0.136
347	0.184
368	0.189
387	0.203
402	0.211
417	0.211
427	0.223
451	0.231
467	0.273
482	0.323
499	0.370
518	0.426
542	0.479
568	0.549
617	0.680
653	0.772
CURVE 111*	
293	0.000
CURVE 111 (cont.)*	
296	0.014
454	0.276
531	0.404
582	0.502
630	0.583
707	0.722
835	0.954
925	1.116
981	1.233
1120	1.506
1174	1.645
1205	1.713
1292	1.928
CURVE 112	
375	0.148
474	0.301*
572	0.463*
675	0.642*
775	0.826
872	1.017
972	1.222
1073	1.434
1173	1.661*
1271	1.891*
CURVE 113*	
303	0.0095
373	0.0762
CURVE 114*	
303	0.0104
373	0.0833
CURVE 115*	
303	0.0096
373	0.0766
CURVE 116*	
303	0.0081
373	0.0645
CURVE 117*	
303	0.0012
373	0.0097
CURVE 118*	
303	0.0032
373	0.0259
CURVE 119*	
303	0.0060
373	0.0482
CURVE 120*	
303	0.0087
373	0.0698
CURVE 121*	
303	0.0104
373	0.0829
CURVE 122*	
303	0.0098
373	0.0785
CURVE 123*	
303	0.0070
373	0.0563
CURVE 124*	
303	0.0082
373	0.0658
CURVE 125*	
303	0.0099
373	0.0795
CURVE 126*	
303	0.0114
373	0.0911
CURVE 127*	
303	0.0110
373	0.0883
CURVE 128*,‡	
77	-0.253
83	-0.251
93	-0.246
103	-0.241
113	-0.234
123	-0.227
133	-0.219
143	-0.210
153	-0.200
163	-0.190
173	-0.179
183	-0.167
193	-0.154
203	-0.140
213	-0.125
223	-0.110
233	-0.095
243	-0.079
253	-0.064
263	-0.048
273	-0.032
283	-0.016
293	0.000
303	0.016
313	0.033
323	0.049
CURVE 129*,‡	
77	-0.198
83	-0.195
93	-0.188
103	-0.181
113	-0.173
123	-0.165
133	-0.156
143	-0.147
153	-0.139
163	-0.130
173	-0.121
183	-0.112
193	-0.103

* Not shown in figure.

‡ Author's data for coefficient of thermal expansion have been integrated by TPRC to obtain $\Delta L/L_0$.

DATA TABLE 262. THERMAL LINEAR EXPANSION OF IRON + NICKEL + ΣX_i ALLOYS Fe + Ni + ΣX_i (continued)

T	$\Delta L/L_0$
CURVE 129 (cont.)*, ‡	
203	-0.093
213	-0.084
223	-0.074
233	-0.064
243	-0.054
253	-0.044
263	-0.033
273	-0.022
283	-0.011
293	0.000
303	0.012
313	0.024
323	0.036
CURVE 130*, ‡	
77	-0.207
83	-0.203
93	-0.196
103	-0.188
113	-0.180
123	-0.171
133	-0.162
143	-0.153
153	-0.144
163	-0.135
173	-0.126
183	-0.116
193	-0.106
203	-0.096
213	-0.086
223	-0.076
233	-0.066
243	-0.055
253	-0.044
263	-0.033
273	-0.022
283	-0.011
293	0.000
303	0.011
313	0.023
323	0.035
CURVE 131*, †, ‡	
3	0.0000
4	-0.0003

T	$\Delta L/L_0$
CURVE 131 (cont.)*, †, ‡	
5	-0.0003
6	0.0002
8	0.0015
10	0.0032
12	0.0053
14	0.0076
16	0.0102
18	0.0130
20	0.0159
22	0.0189
24	0.0220
26	0.0250
28	0.0251
30	0.0221
CURVE 132*, †, ‡	
4	0.0000
5	-0.0001
6	-0.0003
8	-0.0007
10	-0.0012
12	-0.0019
14	-0.0026
16	-0.0033
18	-0.0041
20	-0.0050
22	-0.0058
24	-0.0066
26	-0.0074
28	-0.0080
30	-0.0086
CURVE 133*	
297	0.004
301	0.006
323	0.029
348	0.057
373	0.087
423	0.161
473	0.244
523	0.329
573	0.416
623	0.503
673	0.589
685	0.607

T	$\Delta L/L_0$
CURVE 134*	
308	0.025
323	0.052
338	0.071
373	0.105
423	0.163
473	0.228
523	0.297
531	0.301
573	0.362
623	0.430
648	0.466
654	0.455
661	0.455
673	0.449
693	0.430
705	0.410
723	0.403
732	0.388
741	0.369
750	0.363
760	0.356
773	0.356
785	0.362
794	0.373
809	0.395
CURVE 135*	
283	-0.017
398	0.174
468	0.353
568	0.599
CURVE 136*	
283	-0.004
383	0.035
468	0.175
568	0.364
CURVE 137*	
283	-0.002
378	0.017
453	0.062

T	$\Delta L/L_0$
CURVE 138*	
283	-0.003
383	0.030
463	0.066
568	0.256

* Not shown in figure.
† This curve is here reported using the first given temperature as reference temperature at which $\Delta L/L_0 = 0$.
‡ Author's data for coefficient of thermal expansion have been integrated by TPRC to obtain $\Delta L/L_0$.

DATA TABLE 262. COEFFICIENT OF THERMAL LINEAR EXPANSION OF IRON + NICKEL + ΣX_i ALLOYS Fe + Ni + ΣX_i

[Temperature, T, K; Coefficient of Expansion, α, 10^{-6} K^{-1}]

T	α
CURVE 5*	
366	4.79
394	5.11
422	5.29
450	5.35
478	5.35
505	5.37
533	5.38
561	5.40
589	5.47
616	5.53
644	5.57
672	5.59
700	5.58
CURVE 40‡	
5	-0.61
6	-0.78
7	-0.88
8	-0.89
10	-1.03
10	-1.23
11	-1.16
14	-1.40
16	-1.45
18	-1.58
20	-1.72
24	-1.89
26	-2.17
28	-1.86
32	-1.75
36	-1.72
45	-1.28
50	-1.00
75	0.17
100	1.04
125	1.32
150	1.23
175	1.02
200	0.75
CURVE 41‡	
23	-1.78
24	-1.77*
82	0.20
CURVE 41 (cont.)‡	
89	0.69
195	0.95
CURVE 42‡	
7	-0.74
9	-1.10*
11	-1.29*
13	-1.46*
19	-1.76*
25	-2.00
28	-2.08
31	-1.99
32	-2.15*
36	-2.07
37	-2.11*
39	-2.00
42	-1.77
45	-1.64
50	-1.37
75	0.18*
100	0.64
115	0.85
CURVE 43‡	
4	-0.10
6	-0.27
8	-0.22
10	-0.47
13	-0.26*
13	-0.45*
15	-0.55
16	-0.47*
18	-0.47*
20	-0.55
21	-0.62*
23	-0.69
25	-0.64
29	-0.64
32	-0.55
37	-0.27
43	-0.12
48	0.00
52	0.28
58	0.63
CURVE 43 (cont.)‡	
85	2.42
99	3.01*
108	3.32*
124	3.73*
158	4.01*
176	4.35*
201	3.99*
231	3.53*
CURVE 44‡	
32	-0.26
37	-0.24*
43	-0.01
48	0.21
53	0.52
63	1.13
68	1.41
CURVE 45‡	
6	0.00
8	0.02*
10	0.03
19	0.16
30	0.44
37	0.64
50	1.29
60	1.60
80	4.05*
86	4.53*
96	5.35*
CURVE 46‡	
6	0.02*
19	0.27
36	0.99
45	1.51
55	2.24
CURVE 47‡	
5	0.00*
8	0.00*
10	0.06*
CURVE 47 (cont.)‡	
15	0.14
26	0.50
40	1.17
CURVE 48*, ‡	
6	0.00
9	0.06
35	0.93
46	1.64
CURVE 49‡	
6	-0.28*
8	-0.37
10	-0.45*
17	-0.58*
23	-0.71*
43	-0.27*
54	0.90
79	2.46
89	3.06*
100	3.04*
CURVE 50‡	
10	-0.41*
15	-0.45
25	-0.45
35	-0.30*
46	0.10
CURVE 51	
7	-1.35
18	-1.94*
25	-1.63
30	-1.48
34	-1.26
39	-0.96
46	-0.55
48	-0.38
53	-0.19
58	0.07
62	0.20
66	0.35
CURVE 51 (cont.)	
70	0.59
77	0.87
85	1.11
96	1.41
106	1.59
115	1.87
126	1.94
135	1.94
146	1.85
155	1.77
166	1.69
175	1.58
185	1.48
195	1.35
206	1.25
215	1.17
225	1.06
234	0.99
244	0.94
255	0.91
265	0.91
272	0.87
278	0.90
288	0.90
293	0.92
297	1.02
308	1.04*
316	1.14*
CURVE 52	
7	-1.40
8	-1.45
11	-1.62
12	-1.73
15	-1.89
21	-2.00
27	-1.95
30	-1.81
33	-1.72
36	-1.38*
41	-1.27
49	-0.66
53	-0.47
58	-0.12
61	0.08
CURVE 52 (cont.)	
65	0.27
69	0.53
74	0.68
78	0.79
79	1.01
85	1.17
96	1.49
104	1.63
126	1.67
135	1.73
144	1.63
156	1.57
166	1.47
175	1.36
186	1.25
195	1.13
206	0.99
225	0.71
236	0.61
256	0.52
265	0.52
278	0.57
288	0.57
293	0.63
298	0.68
308	0.74*
318	0.87*
CURVE 53	
6	-1.50
9	-1.92
14	-2.14
19	-2.36
26	-1.97*
30	-1.82*
33	-1.63
38	-1.37
42	-1.15
46	-0.79
50	-0.75
54	-0.50
59	-0.21
63	-0.02
67	0.10
70	0.25

* Not shown in figure.
‡ Author's data for coefficient of thermal expansion have been integrated by TPRC to obtain $\Delta L/L_0$.

DATA TABLE 262. COEFFICIENT OF THERMAL LINEAR EXPANSION OF IRON + NICKEL + ΣX_i ALLOYS Fe + Ni + ΣX_i (continued)

T	α
CURVE 53 (cont.)	
77	0.49
84	0.68
95	0.91
106	1.02
116	1.11
136	1.12
147	1.07
155	1.01
165	0.90
175	0.74
184	0.70
195	0.59
205	0.55
214	0.46
224	0.40
234	0.33
245	0.28
254	0.20
265	0.26
276	0.24
293	0.26
298	0.24*
307	0.37*
CURVE 54‡	
9	-1.00
12	-1.18*
16	-1.63
21	-1.67
25	-1.67
32	-1.42
37	-1.20
41	-1.03
45	-0.62
48	-0.50
53	-0.19*
57	0.09
61	0.24
65	0.45
73	0.71
77	0.93
81	1.10
85	1.14*
89	1.17
92	1.24
98	1.37
102	1.45

T	α
CURVE 54 (cont.)‡	
114	1.62
121	1.58
CURVE 55‡	
9	-1.05*
29	-1.65
33	-1.48*
37	-1.14
41	-0.99
44	-0.75
49	-0.55
52	-0.27
57	-0.10*
61	0.17
65	0.38*
69	0.65
73	0.76*
78	0.74
81	0.98
90	1.33
101	1.40
110	1.50
118	1.58
126	1.56
134	1.55
CURVE 56‡	
99	1.00
92	1.19
96	1.32
104	1.40
107	1.51
112	1.61
119	1.67
129	1.61
132	1.51
136	1.64
140	1.39
143	1.50
147	1.46
153	1.40
157	1.31
160	1.30
164	1.22
168	1.00
172	1.09

T	α
CURVE 56 (cont.)‡	
176	1.07
179	0.85
184	0.97
187	0.72
191	0.72
195	0.69
200	0.54
203	0.50
207	0.34
211	0.34
216	0.37
219	0.29
223	0.18
227	0.09
231	0.08
236	0.13
239	0.10
243	0.00
247	0.00
251	0.00
254	0.00
260	-0.03
263	0.00
269	-0.13
277	0.02
287	0.09
298	0.17
CURVE 57*, ‡	
83	1.02
86	1.09
94	1.28
98	1.45
100	1.52
102	1.54
106	1.60
110	1.54
114	1.57
118	1.63
122	1.65
126	1.61
129	1.50
134	1.63
138	1.46
142	1.50
146	1.49
150	1.38

T	α
CURVE 57 (cont.)*, ‡	
155	1.37
159	1.26
162	1.21
166	1.08
170	1.08
174	1.00
178	1.01
182	0.89
186	0.84
190	0.69
194	0.71
198	0.57
201	0.57
205	0.38
210	0.37
214	0.37
217	0.26
222	0.29
226	0.17
230	0.15
233	0.09
235	0.09
237	0.09
241	0.07
245	0.00
250	0.00
253	-0.04
258	0.00
262	-0.04
270	-0.05
275	0.00
286	0.03
291	0.05
295	0.15
305	0.31
CURVE 128*, ‡	
77	3.7
83	4.2
93	5.0
103	6.0
113	6.9
123	7.7
133	8.5
143	9.2
153	9.9
163	10.7

T	α
CURVE 128 (cont.)*, ‡	
173	11.6
183	12.5
193	13.4
203	14.3
213	14.9
223	15.3
233	15.5
243	15.6
253	15.7
263	15.8
273	15.9
283	16.1
293	16.2
303	16.4
313	16.6
323	16.8
CURVE 129*, ‡	
77	5.4
83	6.0
93	7.0
103	7.7
113	8.3
123	8.4
133	8.5
143	8.7
153	8.8
163	8.9
173	9.0
183	9.1
193	9.3
203	9.5
213	9.7
223	9.8
233	10.0
243	10.2
253	10.4
263	10.6
273	10.9
283	11.2
293	11.5
303	11.8
313	12.2
323	12.5

T	α
CURVE 130*, ‡	
77	6.6
83	7.0
93	7.7
103	8.2
113	8.4
123	8.6
133	8.8
143	9.0
153	9.2
163	9.3
173	9.5
183	9.7
193	9.8
203	10.0
213	10.2
223	10.4
233	10.6
243	10.7
253	10.8
263	10.9
273	11.0
283	11.2
293	11.4
303	11.5
313	11.6
323	11.7
CURVE 131*, ‡	
3	-0.28
4	-0.37
5	-0.47
6	-0.56
8	-0.75
10	-0.93
12	-1.10
14	-1.24
16	-1.35
18	-1.45
20	-1.51
22	-1.52
24	-1.52
26	-1.50
28	-1.48
30	-1.45

* Not shown in figure.
‡ Author's data for coefficient of thermal expansion have been integrated by TPRC to obtain $\Delta L/L_0$.

DATA TABLE 262. COEFFICIENT OF THERMAL LINEAR EXPANSION OF IRON + NICKEL + ΣX_i ALLOYS Fe + Ni + ΣX_i (continued)

T	α
CURVE 132*, ‡	
4	-0.12
5	-0.15
6	-0.17
8	-0.23
10	-0.29
12	-0.35
14	-0.37
16	-0.39
18	-0.41
20	-0.42
22	-0.41
24	-0.40
26	-0.36
28	-0.31
30	-0.25

* Not shown in figure.
‡ Author's data for coefficient of thermal expansion have been integrated by TPRC to obtain $\Delta L/L_0$.

FIGURE AND TABLE NO. 263R. PROVISIONAL VALUES FOR THERMAL LINEAR EXPANSION OF IRON + TUNGSTEN + ΣX_i ALLOYS Fe + W + ΣX_i

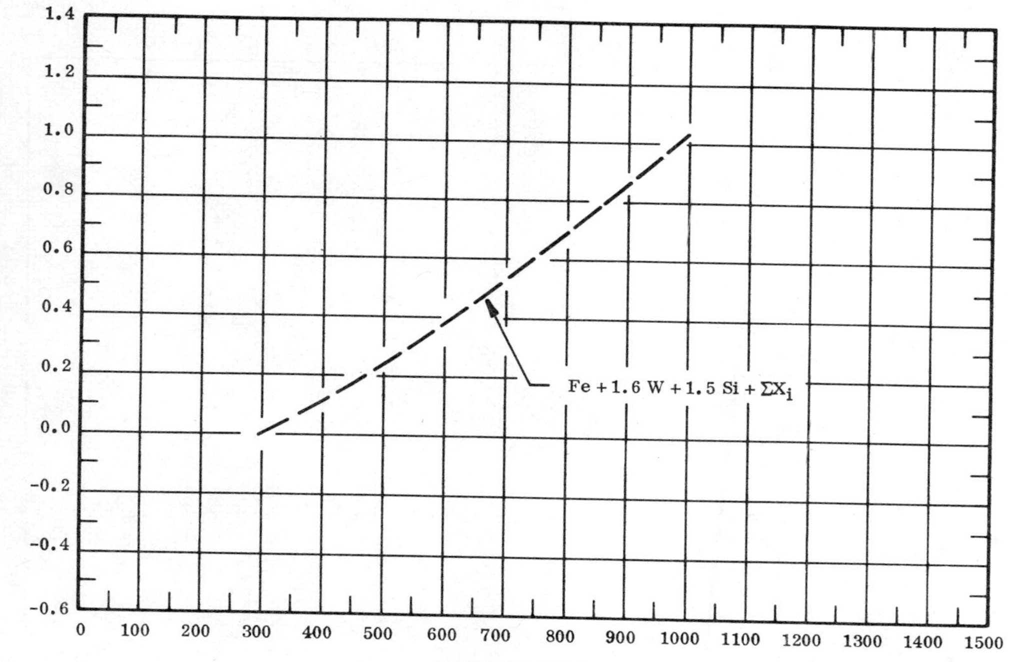

PROVISIONAL VALUES

[Temperature, T, K; Linear Expansion, $\Delta L/L_0$, %; α, K^{-1}]

(Fe + 1.6 W + 1.5 Si + ΣX_i)		
T	$\Delta L/L_0$	$\alpha \times 10^6$
293	0.000	9.8
400	0.117	11.8
500	0.243	13.4
600	0.382	14.6
700	0.536	15.7
800	0.695	16.4
900	0.862	16.9
1000	1.032	17.0

REMARKS

(Fe + 1.6 W + 1.5 Si + ΣX_i): The tabulated values for well-annealed alloy are considered accurate to within ±7% over the entire temperature range. These values can be represented approximately by the following equation:

$$\Delta L/L_0 = -0.191 + 2.839 \times 10^{-4}\, T + 1.396 \times 10^{-6}\, T^2 - 4.567 \times 10^{-10}\, T^3$$

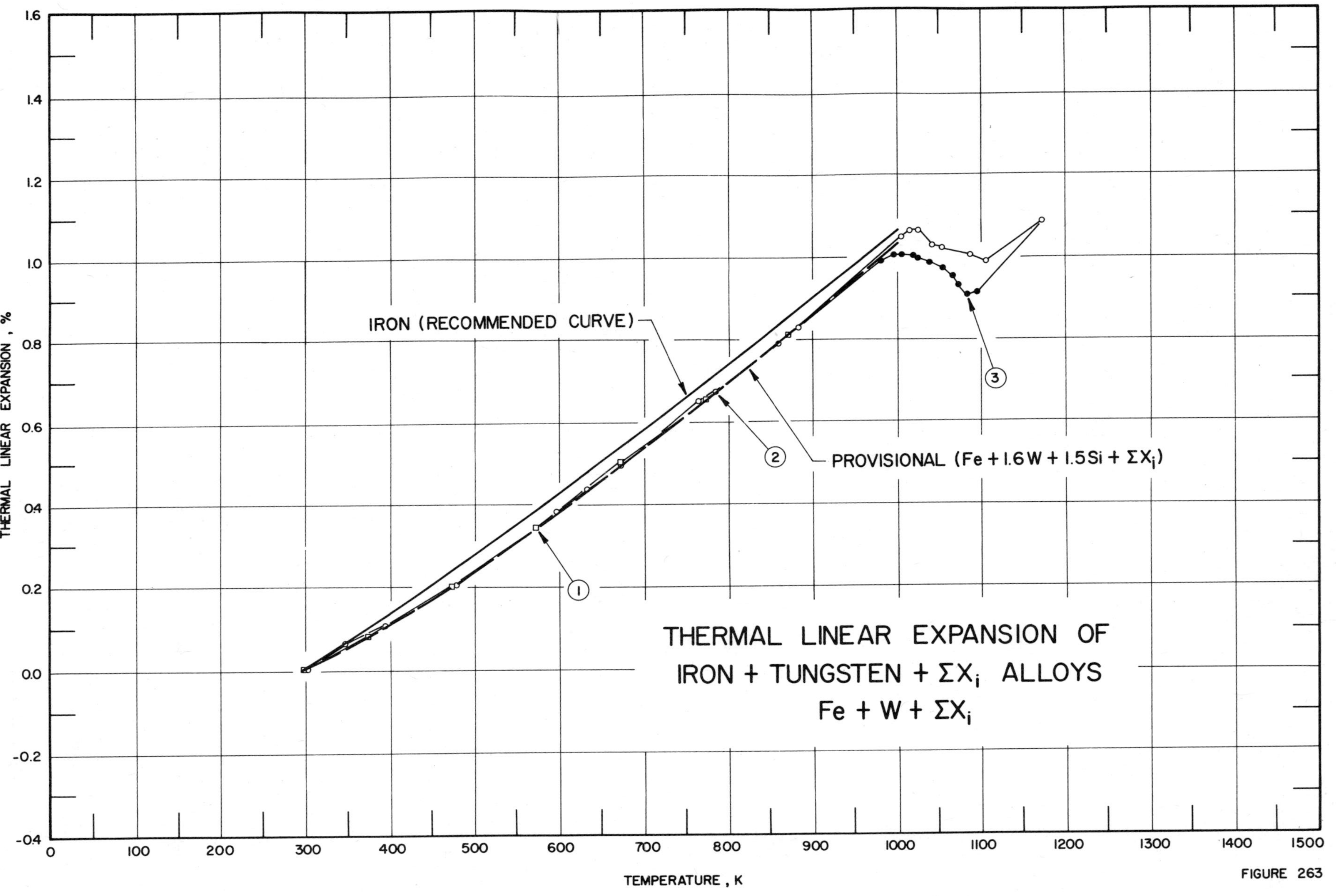
THERMAL LINEAR EXPANSION OF
IRON + TUNGSTEN + ΣXi ALLOYS
Fe + W + ΣXi
IRON (RECOMMENDED CURVE)
PROVISIONAL (Fe + 1.6 W + 1.5 Si + ΣXi)
1
2
3
THERMAL LINEAR EXPANSION, %
TEMPERATURE, K
1.6
1.4
1.2
1.0
0.8
0.6
0.4
0.2
0.0
-0.2
-0.4
0
100
200
300
400
500
600
700
800
900
1000
1100
1200
1300
1400
1500

FIGURE 263

SPECIFICATION TABLE 263. THERMAL LINEAR EXPANSION OF IRON + TUNGSTEN + ΣX_i ALLOYS $Fe + W + \Sigma X_i$

Cur. No.	Ref. No.	Author(s)	Year	Method Used	Temp. Range, K	Name and Specimen Designation	Composition (weight percent) Fe	W	Si	C	Mn	S	P	Composition (continued), Specifications, and Remarks
1	51	Souder, W. and Hidnert, P.	1922	T	298-873	S 564	Bal.	1.58	1.45	0.512	0.42	0.021	0.016	Annealed specimen.
2	51	Souder, W. and Hidnert, P.	1922	T	293-1173	S 561	Bal.	3.96	0.095	0.396	0.25	0.023	0.012	Annealed specimen; expansion measured with increasing temperature.
3	51	Souder, W. and Hidnert, P.	1922	T	1173-980	S 561	Bal.	3.96	0.095	0.396	0.25	0.023	0.012	The above specimen; expansion measured with decreasing temperature.

DATA TABLE 263. THERMAL LINEAR EXPANSION OF IRON + TUNGSTEN + ΣX_i ALLOYS $Fe + W + \Sigma X_i$

[Temperature, T, K; Linear Expansion, $\Delta L/L_0$, %]

T	$\Delta L/L_0$	T	$\Delta L/L_0$	T	$\Delta L/L_0$
CURVE 1		CURVE 2 (cont.)		CURVE 3 (cont.)	
298	0.005	765	0.647	1082	0.906
373	0.083	783	0.672	1074	0.927
473	0.204	860	0.798	1066	0.949
573	0.341	881	0.836	1055	0.973
673	0.500	1007	1.053	1040	0.987
773	0.657	1019	1.065	1029	0.997
873	0.818	1027	1.067	1020	1.003
		1042	1.037	1007	1.003
CURVE 2		1052	1.023	995	1.003
		1088	1.006	980	0.995
293	0.000*	1096	1.003*		
304	0.008	1108	0.990		
348	0.062	1173	1.091		
393	0.110				
476	0.212	CURVE 3			
598	0.381				
631	0.438	1173	1.091*		
671	0.491	1096	0.917		

* Not shown in figure.

FIGURE AND TABLE NO. 264R. PROVISIONAL VALUES FOR THERMAL LINEAR EXPANSION OF MAGNESIUM + ALUMINUM + ΣX_i ALLOYS Mg + Al + ΣX_i

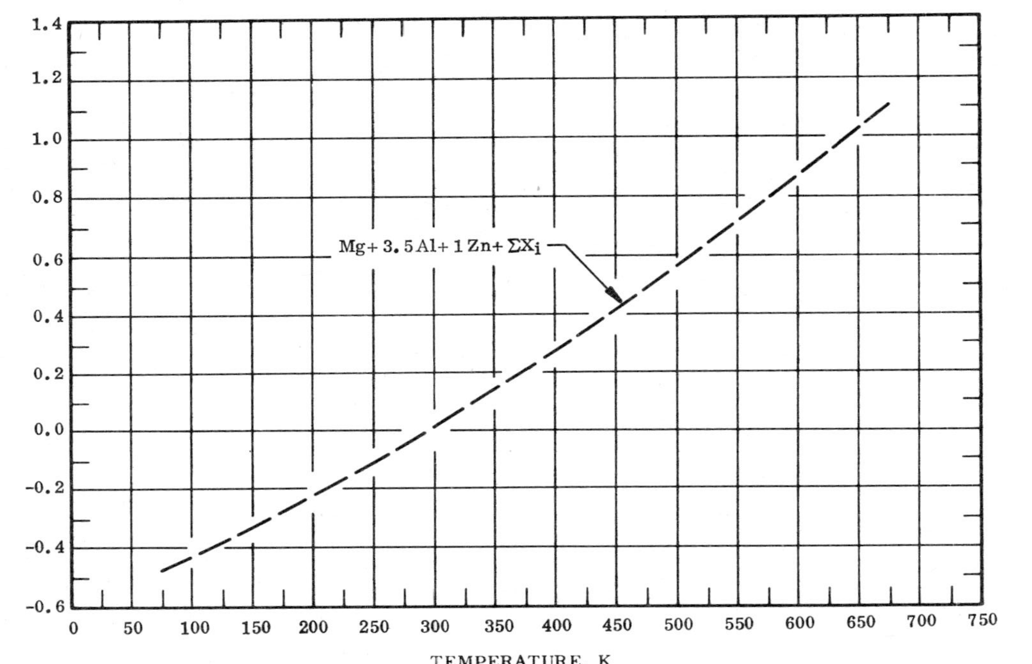

PROVISIONAL VALUES

[Temperature, T, K; Linear Expansion, $\Delta L/L_0$, %; α, K^{-1}]

T	$\Delta L/L_0$ (Mg + 3.5 Al + 1 Zn + ΣX_i)	$\alpha \times 10^6$ (Mg + 3.5 Al + 1 Zn + ΣX_i)
75	-0.476	18.4
100	-0.430	19.2
200	-0.221	22.4
293	0.000	25.0
400	0.281	27.5
500	0.566	29.6
600	0.868	30.8
675	1.102	31.6

REMARKS

(Mg+ 3.5 Al+ 1 Zn+ ΣX_i): The tabulated values for well-annealed alloy of type AZ 31A alloy are considered accurate to within ± 7% over the entire temperature range. The thermal linear expansion of this alloy is very close to that of pure magnesium. These values can be represented approximately by the following equation:

$$\Delta L/L_0 = -0.607 + 1.562 \times 10^{-3}\, T + 1.924 \times 10^{-6}\, T^2 - 7.297 \times 10^{-10}\, T^3$$

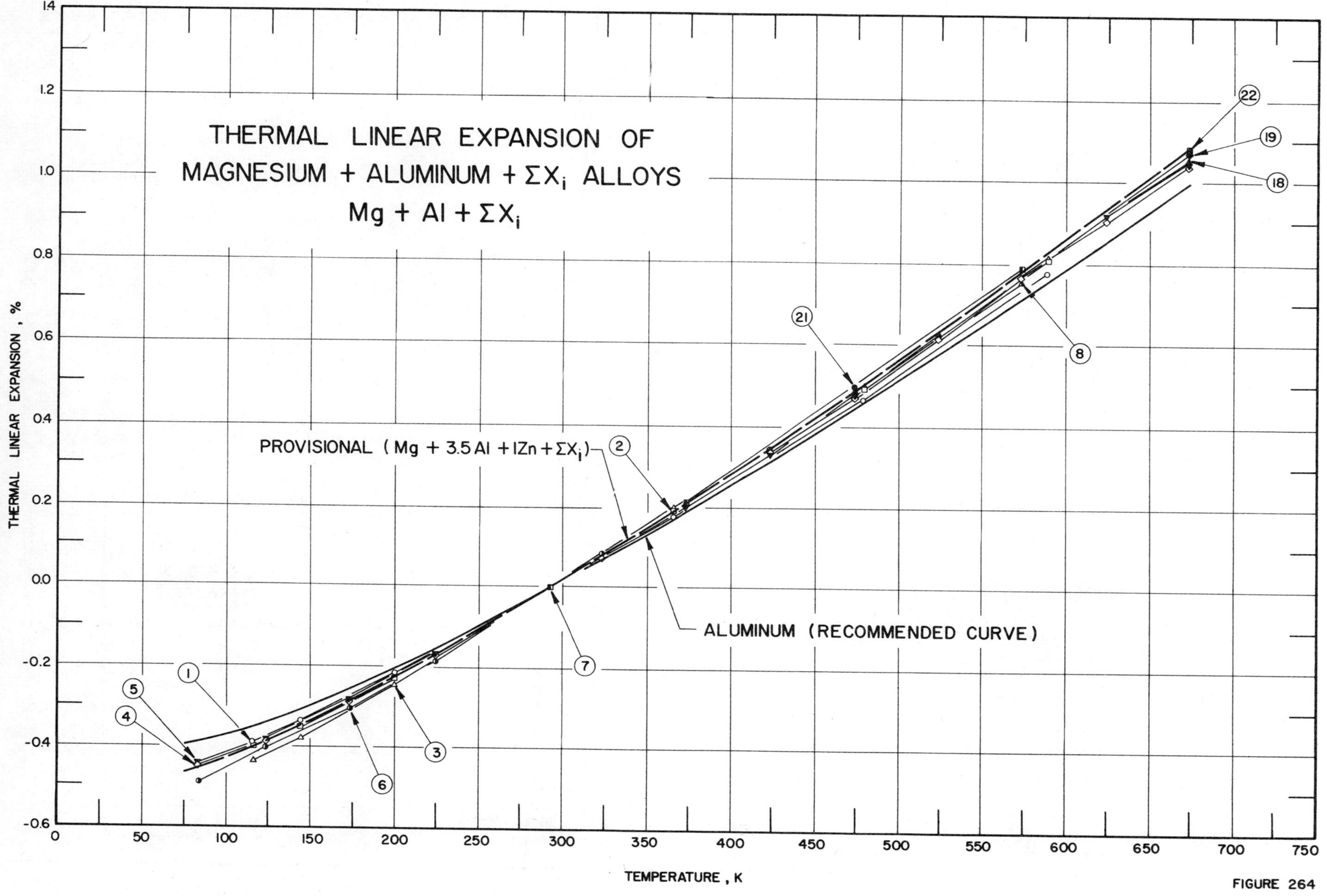
THERMAL LINEAR EXPANSION OF
MAGNESIUM + ALUMINUM + ΣX_i ALLOYS
Mg + Al + ΣX_i
PROVISIONAL (Mg + 3.5 Al + 1Zn + ΣX_i)
ALUMINUM (RECOMMENDED CURVE)
THERMAL LINEAR EXPANSION, %
TEMPERATURE, K
1.4
1.2
1.0
0.8
0.6
0.4
0.2
0.0
-0.2
-0.4
-0.6
0
50
100
150
200
250
300
350
400
450
500
550
600
650
700
750
1
2
3
4
5
6
7
8
18
19
21
22

FIGURE 264

SPECIFICATION TABLE 264. THERMAL LINEAR EXPANSION OF MAGNESIUM + ALUMINUM + ΣX_i ALLOYS $Mg + Al + \Sigma X_i$

Cur. No.	Ref. No.	Author(s)	Year	Method Used	Temp. Range, K	Name and Specimen Designation	Composition (weight percent) Mg	Al	Mn	Zn	Cu	Cd	Sn	Composition (continued), Specifications, and Remarks
1	123	Lucks, C.F. and Deem, H.W.	1958	L	116-588	AZ 31 A (AN-M-29)	Bal.	3.5	0.2	1.3				0.3 Si; 0.375 in. in diameter by 3 in. long; material hot-rolled; from Dow Chemical Co.; specimen annealed 1 hr at 589 K and furnace cooled; heating rate 3 F per min; mean linear expansion transverse to the direction of rolling.
2	123	Lucks, C.F. and Deem, H.W.	1958	L	116-588	AZ 31 A (AN-M-29)	Bal.	3.5	0.2	1.3				The above specimen; expansion longitudinal to the direction of rolling.
3	123	Lucks, C.F. and Deem, H.W.	1958	L	116-588	AZ 31 A (AN-M-29)	Bal.	3.5	0.2	1.3				The above specimen; expansion vertical (normal) to the direction of rolling.
4	466	Lucks, C.F., Thompson, H.B., Smith, A.R., Curry, F.P., Deem, H.W., and Bing, G.F.	1951	L	83-673	AZ 31 A (AN-M-29)	Bal.	2.5/ 3.5	0.2	0.7/ 1.3	0.05			0.3 Si, 0.005 Fe, 0.005 Ni, 0.3 other elements; hot-rolled; annealed 1 hr at 588.72 K and furnace cooled; density 1.78 g cm^{-3}; mean linear expansion transverse to the direction of rolling.
5	466	Lucks, C.F., et al.	1951	L	83-673	AZ 31 A (AN-M-29)	Bal.	2.5/ 3.5	0.2	0.7/ 1.3	0.05			Similar to the above specimen; measured in the longitudinal direction; material supplied by Dow Chemical Co.
6	466	Lucks, C.F., et al.	1951	L	83-673	AZ 31 A (AN-M-29)	Bal.	2.5/ 3.5	0.2	0.7/ 1.3	0.05			Similar to the above specimen; measured in the vertical direction.
7	200	Hidnert, P. and Sweeney, W.T.	1928	T	293-573	Sample 1279	Bal.	4.14	0.27					Specimen cast in vacuum furnace at 963 K.
8	200	Hidnert, P. and Sweeney, W.T.	1928	T	293-573	Sample 1278	Bal.	4.10	0.27					Specimen extruded at 663 K.
9*	201	Takahasi, K. and Kikuti, R.	1936	T	273-573	A 77	Bal.	4	0.4	3				Rod specimen about 16 cm long, 5 mm diameter; heating rate 2 C/min.
10*	201	Takahasi, K. and Kikuti, R.	1936	T	273-573	A 78	Bal.	6	0.4	3				Similar to the above specimen.
11*	201	Takahasi, K. and Kikuti, R.	1936	T	273-573	A 79	Bal.	8.5	0.15	0.5	2	1		Similar to the above specimen.
12*	201	Takahasi, K. and Kikuti, R.	1936	T	273-573	A 82	Bal.	5	0.4	3.0		3		Similar to the above specimen.
13*	201	Takahasi, K. and Kikuti, R.	1936	T	273-573	A 100	Bal.	4		1		1	1	Similar to the above specimen.
14*	201	Takahasi, K. and Kikuti, R.	1936	T	273-573	A 103	Bal.	5	0.4	3		3		Similar to the above specimen.
15*	201	Takahasi, K. and Kikuti, R.	1936	T	273-573	A 111	Bal.	4		0.5		2	1	Similar to the above specimen.
16*	201	Takahasi, K. and Kikuti, R.	1936	T	273-573	A 112	Bal.	4				3	1	Similar to the above specimen.
17*	201	Takahasi, K. and Kikuti, R.	1936	T	273-573	A 114	Bal.	4				2	2	Similar to the above specimen.

* Not shown in figure.

SPECIFICATION TABLE 264. THERMAL LINEAR EXPANSION OF MAGNESIUM + ALUMINUM + ΣX_i ALLOYS Mg + Al + ΣX_i (continued)

Cur. No.	Ref. No.	Author(s)	Year	Method Used	Temp. Range, K	Name and Specimen Designation	Composition (weight percent) Mg	Al	Mn	Zn	Cu	Cd	Sn	Composition (continued), Specifications, and Remarks
18	465	Losano, L.	1930		293-673	No. 23	55.05	43.86			0.08			0.58 Fe, 0.41 Si.
19	465	Losano, L.	1930		293-673	No. 24	59.20	40.00			0.03			0.41 Fe, 0.30 Si.
20*	465	Losano, L.	1930		293-673	No. 25	64.30	34.70			0.05			0.50 Fe, 0.38 Si.
21	465	Losano, L.	1930		293-673	No. 26	69.82	28.98			0.09			0.60 Fe, 0.45 Si.
22	465	Losano, L.	1930		293-673	No. 27	80.24	18.76			0.04			0.52 Fe, 0.46 Si.
23*	465	Losano, L.	1930		293-673	No. 28	85.22	13.82			0.08			0.49 Fe, 0.42 Si.
24*	465	Losano, L.	1930		293-673	No. 29	91.18	7.96			0.03			0.40 Fe, 0.40 Si.
25*	465	Losano, L.	1930		293-673	No. 30	97.02	2.32			Trace			0.29 Fe, 0.36 Si.

* Not shown in figure.

DATA TABLE 264. THERMAL LINEAR EXPANSION OF MAGNESIUM + ALUMINUM + ΣX_i ALLOYS Mg + Al + ΣX_i

[Temperature, T, K; Linear Expansion, $\Delta L/L_0$, %]

T	$\Delta L/L_0$
CURVE 1	
116	-0.388
144	-0.336
199	-0.218
366	0.177
477	0.467
588	0.781
CURVE 2	
116	-0.400
144	-0.350
199	-0.228
366	0.193
477	0.493
588	0.813
CURVE 3	
116	-0.435
144	-0.376
199	-0.246
366	0.193
477	0.493
588	0.813
CURVE 4	
83	-0.447
123	-0.384
173	-0.287
223	-0.173
323	0.075
373	0.205*
423	0.336
473	0.474
523	0.619
573	0.767
623	0.908
673	1.045
CURVE 5	
83	-0.445
123	-0.384
173	-0.286
CURVE 5 (cont.)	
223	-0.173
323	0.076*
373	0.201
423	0.336
473	0.478
523	0.621
573	0.780
623	0.920
673	1.060
CURVE 6	
83	-0.473
123	-0.405
173	-0.301
223	-0.183
323	0.079
373	0.213*
423	0.343
473	0.481
523	0.625
573	0.767
623	0.914
673	1.063
CURVE 7	
293	0.000
323	0.078
373	0.213
473	0.491
573	0.787
CURVE 8	
293	0.000*
323	0.075*
373	0.205*
473	0.475*
573	0.762
CURVE 9*	
273	-0.050
323	0.075
CURVE 9 (cont.)	
373	0.206
423	0.346
473	0.490
523	0.638
573	0.791
CURVE 10*	
273	-0.052
323	0.078
373	0.211
423	0.350
473	0.492
523	0.643
573	0.799
CURVE 11*	
273	-0.049
323	0.074
373	0.204
423	0.343
473	0.483
523	0.629
573	0.782
CURVE 12*	
273	-0.049
323	0.074
373	0.205
423	0.345
473	0.495
523	0.647
573	0.804
CURVE 13*	
273	-0.048
323	0.073
373	0.202
423	0.337
473	0.477
523	0.625
573	0.779
CURVE 14*	
273	-0.049
323	0.074
373	0.211
423	0.350
473	0.496
523	0.648
573	0.804
CURVE 15*	
273	-0.049
323	0.073
373	0.202
423	0.339
473	0.482
523	0.630
573	0.788
CURVE 16*	
273	-0.050
323	0.075
373	0.205
423	0.342
473	0.486
523	0.636
573	0.793
CURVE 17*	
273	-0.049
323	0.073
373	0.202
423	0.388
473	0.480
523	0.629
573	0.784
CURVE 18	
293	0.000*
373	0.210*
473	0.483*
573	0.770*
673	1.063
CURVE 19	
293	0.000*
373	0.212*
473	0.487*
573	0.776*
673	1.070
CURVE 20*	
293	0.000
373	0.213
473	0.489
573	0.779
673	1.074
CURVE 21	
293	0.000*
373	0.214*
473	0.491
573	0.783*
673	1.079*
CURVE 22	
293	0.000*
373	0.214*
473	0.492*
573	0.783*
673	1.080
CURVE 23*	
293	0.000
373	0.215
473	0.493
573	0.784
673	1.082
CURVE 24*	
293	0.000
373	0.215
473	0.493
573	0.784
673	1.083
CURVE 25*	
293	0.000
373	0.216
473	0.494
573	0.785
673	1.084

* Not shown in figure.

FIGURE AND TABLE NO. 265AR. PROVISIONAL VALUES FOR THERMAL LINEAR EXPANSION OF MAGNESIUM + COPPER + ΣX_i ALLOYS   Mg + Cu + ΣX_i

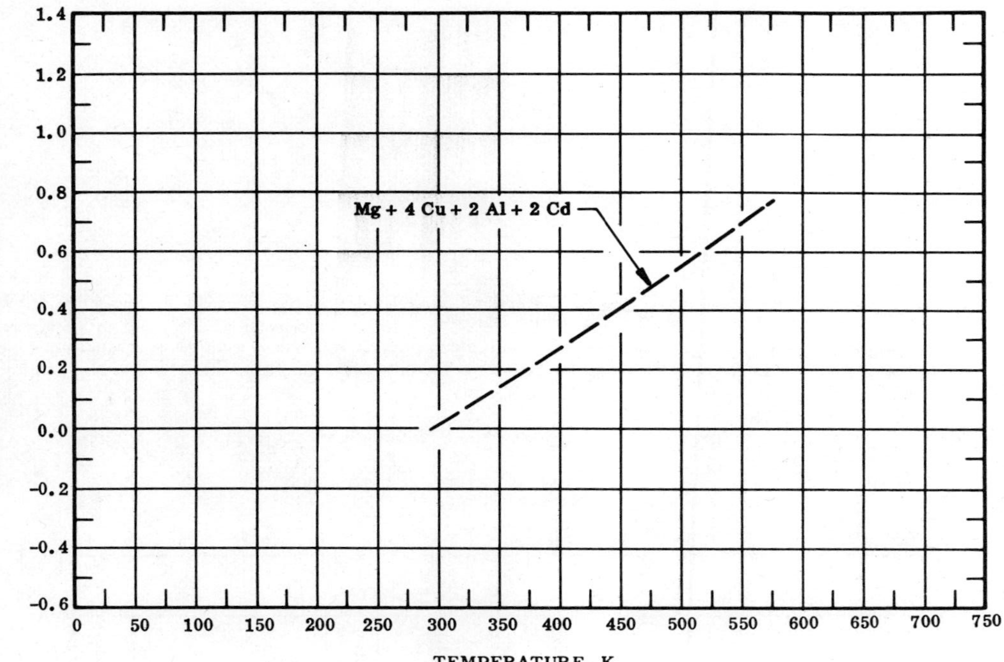

PROVISIONAL VALUES

[Temperature, T, K; Linear Expansion, $\Delta L/L_0$, %; α, K^{-1}]

(Mg + 4 Cu + 2 Al + 2 Cd)

T	$\Delta L/L_0$	$\alpha \times 10^6$
293	0.000	26.9
400	0.273	27.6
500	0.558	28.8
575	0.772	28.8

REMARKS

(Mg + 4 Cu + 2 Al + 2 Cd): The tabulated values are for well-annealed alloy and are considered accurate to within ± 7% over the entire temperature range. The thermal linear expansion of this alloy is very close to that of pure magnesium. These values can be represented approximately by the following equation:

$$\Delta L/L_0 = -0.294 - 7.106 \times 10^{-4}\, T + 7.295 \times 10^{-6}\, T^2 - 4.934 \times 10^{-9}\, T^3$$

FIGURE AND TABLE NO. 265BR. PROVISIONAL VALUES FOR THERMAL LINEAR EXPANSION OF MAGNESIUM + THORIUM + ΣX_i ALLOYS Mg + Th + ΣX_i

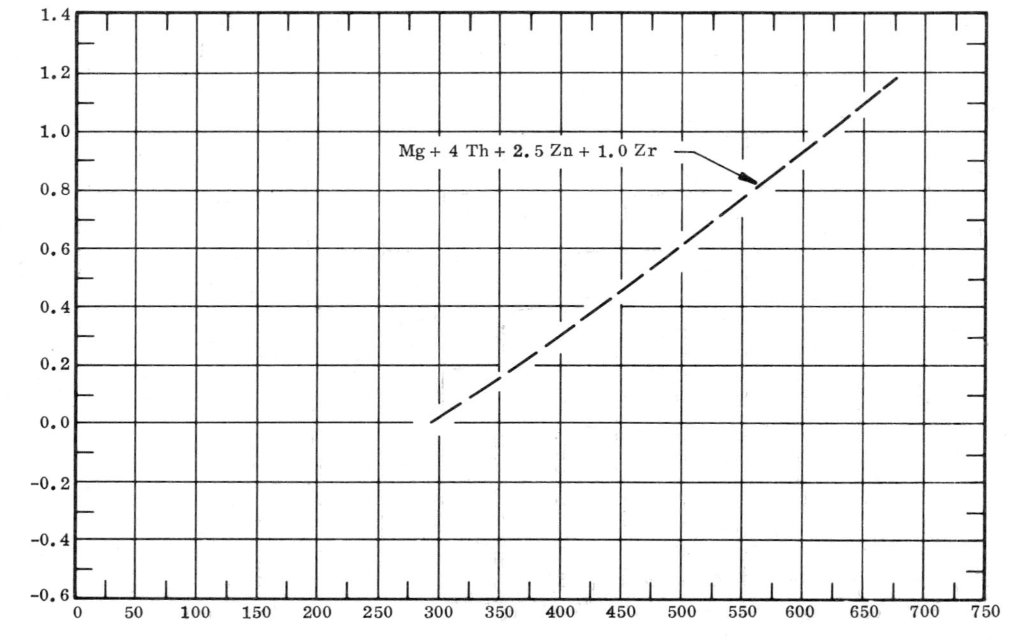

PROVISIONAL VALUES

[Temperature, T, K; Linear Expansion, $\Delta L/L_0$, %; α, K^{-1}]

(Mg + 4 Th + 2.5 Zn + 1.0 Zr)		
T	$\Delta L/L_0$	$\alpha \times 10^6$
293	0.000	28.5
400	0.310	29.5
500	0.612	30.9
600	0.930	32.6
675	1.179	34.0

REMARKS

(Mg + 4 Th + 2.5 Zn + 1.0 Zr): The tabulated values are for well-annealed alloy of type HZ 32A and are considered accurate to within ±7% over the entire temperature range. These values can be represented approximately by the following equation:

$$\Delta L/L_0 = -0.805 + 2.700 \times 10^{-3}\ T + 2.041 \times 10^{-6}\ T^2 + 4.980 \times 10^{-10}\ T^3$$

FIGURE AND TABLE NO. 265CR. PROVISIONAL VALUES FOR THERMAL LINEAR EXPANSION OF MAGNESIUM + ZINC + ΣX_i ALLOYS   Mg + Zn + ΣX_i

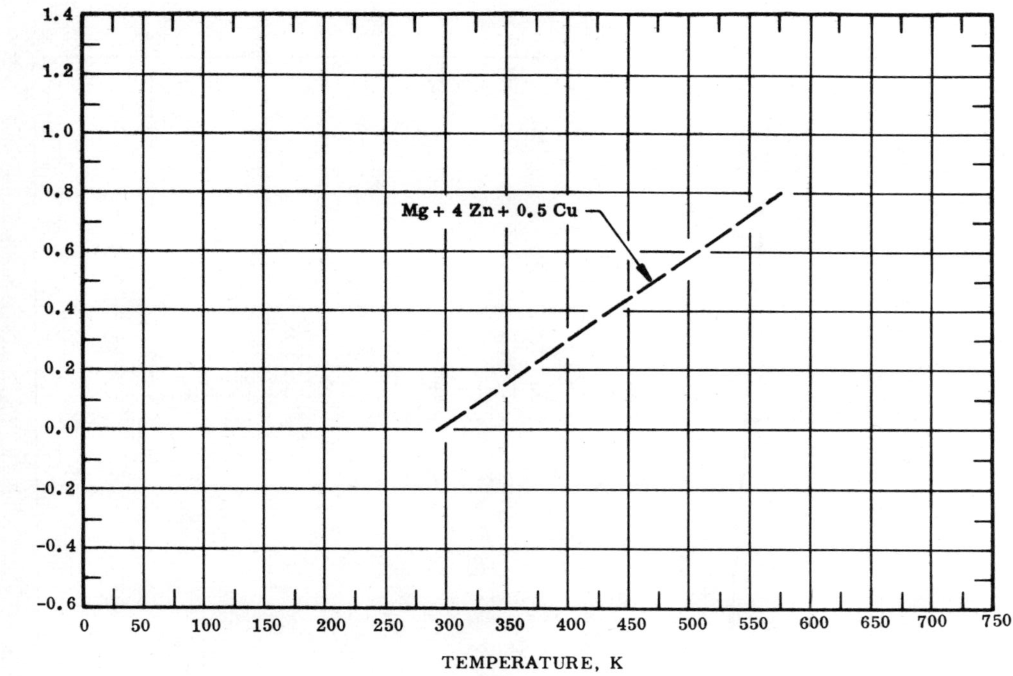

PROVISIONAL VALUES

[Temperature, T, K; Linear Expansion, $\Delta L/L_0$, %; α, K^{-1}]

T	(Mg + 4 Zn + 0.5 Cu) $\Delta L/L_0$	$\alpha \times 10^6$
293	0.000	25.1
400	0.285	28.2
500	0.577	29.8
575	0.802	30.2

REMARKS

(Mg + 4 Zn + 0.5 Cu): The tabulated values are for well-annealed alloy of Type Elektron Amt (Mat) and are considered accurate to within ± 7% over the entire temperature range. The thermal line at expansion of this alloy is very close to that of pure magnesium. The tabulated values can be represented approximately by the following equation:

$$\Delta L/L_0 = -0.547 + 1.070 \times 10^{-3} T + 3.243 \times 10^{-6} T^2 - 1.784 \times 10^{-9} T^3$$

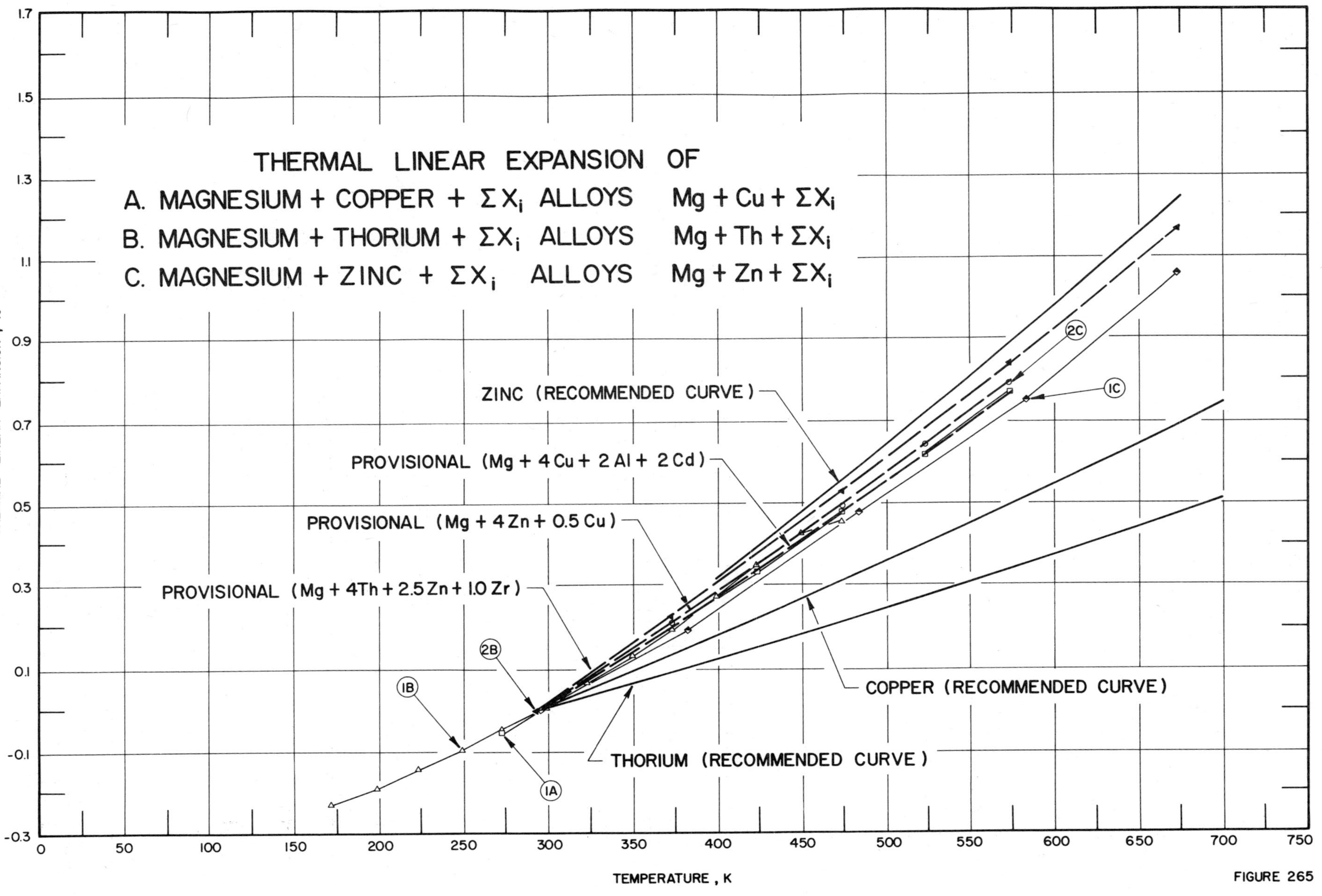

THERMAL LINEAR EXPANSION OF
A. MAGNESIUM + COPPER + ΣX$_i$ ALLOYS Mg + Cu + ΣX$_i$
B. MAGNESIUM + THORIUM + ΣX$_i$ ALLOYS Mg + Th + ΣX$_i$
C. MAGNESIUM + ZINC + ΣX$_i$ ALLOYS Mg + Zn + ΣX$_i$
ZINC (RECOMMENDED CURVE)
PROVISIONAL (Mg + 4 Cu + 2 Al + 2 Cd)
PROVISIONAL (Mg + 4 Zn + 0.5 Cu)
PROVISIONAL (Mg + 4Th + 2.5 Zn + 1.0 Zr)
2C
1C
2B
1B
1A
COPPER (RECOMMENDED CURVE)
THORIUM (RECOMMENDED CURVE)
THERMAL LINEAR EXPANSION, %
1.7
1.5
1.3
1.1
0.9
0.7
0.5
0.3
0.1
-0.1
-0.3
0
50
100
150
200
250
300
350
400
450
500
550
600
650
700
750
TEMPERATURE, K

FIGURE 265

SPECIFICATION TABLE 265A. THERMAL LINEAR EXPANSION OF MAGNESIUM + COPPER + ΣX_i ALLOYS $Mg + Cu + \Sigma X_i$

Cur. No.	Ref. No.	Author(s)	Year	Method Used	Temp. Range, K	Name and Specimen Designation	Composition (weight percent) Mg	Cu	Al	Cd	Mn	Composition (continued), Specifications, and Remarks
1	201	Takahasi, K. and Kikuti, R.	1936	T	273-573	A80	Bal.	4	2	2	0.2	Rod specimen about 16 cm long, 5 mm diameter; heating rate 2 C/min.

DATA TABLE 265A. THERMAL LINEAR EXPANSION OF MAGNESIUM + COPPER + ΣX_i ALLOYS $Mg + Cu + \Sigma X_i$

[Temperature, T, K; Linear Expansion, $\Delta L/L_0$, %]

T	$\Delta L/L_0$
CURVE 1	
273	-0.048
323	0.072*
373	0.199
423	0.333
473	0.474
523	0.619
573	0.774

* Not shown in figure.

SPECIFICATION TABLE 265B. THERMAL LINEAR EXPANSION OF MAGNESIUM + THORIUM + ΣX_i ALLOYS   Mg + Th + ΣX_i

Cur. No.	Ref. No.	Author(s)	Year	Method Used	Temp. Range, K	Name and Specimen Designation	Composition (weight percent) Mg	Th	Mn	Zn	Zr	Composition (continued), Specifications, and Remarks
1	34	Asay, J.R.	1968	L	173-473	ASTM B 90 HM-21A	Bal.	1.5/ 2.5	0.45/ 1.1			0.30 impurities; density 1.782 g cm^{-3}; zero-point correction is 0.011%.
2	199	Bell, I. P.	1954	L	293-673	ASTM B 80 HZ-32A	92.5	4.0		2.5	1.0	No details given.

DATA TABLE 265B. THERMAL LINEAR EXPANSION OF MAGNESIUM + THORIUM + ΣX_i ALLOYS   Mg + Th + ΣX_i

[Temperature, T, K; Linear Expansion, $\Delta L/L_0$, %]

T	$\Delta L/L_0$
CURVE 1	
173	-0.224
298	-0.184
223	-0.140
248	-0.093
273	-0.044
298	0.011
323	0.074
348	0.133
373	0.199
398	0.271
423	0.346
448	0.422
473	0.453

T	$\Delta L/L_0$
CURVE 2	
293	0.000
373	0.226
473	0.524
573	0.840
673	1.170

SPECIFICATION TABLE 265C. THERMAL LINEAR EXPANSION OF MAGNESIUM + ZINC + ΣX_i ALLOYS $Mg + Zn + \Sigma X_i$

Cur. No.	Ref. No.	Author(s)	Year	Method Used	Temp. Range, K	Name and Specimen Designation	Composition (weight percent) Mg	Zn	Th	Zr	Cu	Al	Composition (continued), Specifications, and Remarks
1	469	Shih, C.	1966		296-783	ASTM B 80 HZ-62A	90.6	6.2	2.2	1			No details given; zero-point correction is ±0.007%.
2	201	Takahasi, K. and Kikuti, R.	1936	T	323-573	A83	Bal.	4			0.5		Rod specimen, about 16 cm long, 5 mm diameter, heating rate 2 C/min.
3*	201	Takahasi, K. and Kikuti, R.	1936	T	323-573	A84	Bal.	4			0.5	1.5	Similar to the above specimen.

DATA TABLE 265C. THERMAL LINEAR EXPANSION OF MAGNESIUM + ZINC + ΣX_i ALLOYS $Mg + Zn + \Sigma X_i$

[Temperature, T, K; Linear Expansion, $\Delta L/L_0$, %]

T	$\Delta L/L_0$	T	$\Delta L/L_0$
CURVE 1		CURVE 3*	
296	0.007	273	-0.050
383	0.191	323	0.075
483	0.478	373	0.206
583	0.756	423	0.343
683	1.060	473	0.485
783	1.336*	523	0.635
		573	0.791
CURVE 2			
273	-0.050*		
323	0.075*		
373	0.208		
423	0.347*		
473	0.491		
523	0.640		
573	0.793		

* Not shown in figure.

SPECIFICATION TABLE 266. THERMAL LINEAR EXPANSION OF MOLYBDENUM + TITANIUM + ΣX_i ALLOYS Mo + Ti + ΣX_i

Cur. No.	Ref. No.	Author(s)	Year	Method Used	Temp. Range, K	Name and Specimen Designation	Composition (weight percent) Mo	Ti	Fe	Al	Ni	Si	Composition (continued), Specifications, and Remarks
1*	433	Pears, C.D. and Oglesby, S.	1962		294-2539		98.6	0.7	0.3	0.2	0.2	0.1	Elements found by semi-quantitative emission spectrography; after exposure 98.3 Mo and 0.20 C; initial length 2.980 in.; General Astrometals Corp; hot pressed; specimen melted on post inspection; run No. E10; density in g cm^{-3} at 298 K (25 C) before exposure 9.49 by wax coating specimen and 9.39 by immersion in xylene, after exposure 8.64 by ASTM method B 311-58, 8.68 by wax coating specimen and 9.05 by immersion in xylene; expansion measured in helium atmosphere.

DATA TABLE 266. THERMAL LINEAR EXPANSION OF MOLYBDENUM + TITANIUM + ΣX_i ALLOYS Mo + Ti + ΣX_i

[Temperature, T, K; Linear Expansion, $\Delta L/L_0$, %]

T	$\Delta L/L_0$
CURVE 1*	
293	0.00
781	0.24
1078	0.39
1085	0.39
1410	0.54
1685	0.64
1899	0.89
1938	0.76
2105	0.02
2133	-0.44
2294	-1.44
2360	-2.14
2383	-2.61
2477	-2.85
2538	-1.04

* No figure given.

SPECIFICATION TABLE 267. THERMAL LINEAR EXPANSION OF NICKEL + ALUMINUM + ΣX_i ALLOYS Ni + Al + ΣX_i

Cur. No.	Ref. No.	Author(s)	Year	Method Used	Temp. Range, K	Name and Specimen Designation	Composition (weight percent) Ni	Al	Cr	Composition (continued), Specifications, and Remarks
1*	301	Arbuzov, M.P. and Zelenkov, I.A.	1964	L	293-793		85.6	11.8	2.6	Alloy prepared from pure metals.

DATA TABLE 267. THERMAL LINEAR EXPANSION OF NICKEL + ALUMINUM + ΣX_i ALLOYS Ni + Al + ΣX_i

[Temperature, T, K; Linear Expansion, $\Delta L/L_0$, %]

T	$\Delta L/L_0$	T	$\Delta L/L_0$
CURVE 1*, ‡		CURVE 1 (cont.)*, ‡	
293	0.000	673	0.488
372	0.093	725	0.563
429	0.162	775	0.638
476	0.221	793	0.665
524	0.283	828	0.716
575	0.351	874	0.785
624	0.419	925	0.864
		973	0.941

DATA TABLE 267. COEFFICIENT OF THERMAL LINEAR EXPANSION OF NICKEL + ALUMINUM + ΣX_i ALLOYS Ni + Al + ΣX_i

[Temperature, T, K; Coefficient of Expansion, α, 10^{-6} K^{-1}]

T	α	T	α
CURVE 1*, ‡		CURVE 1 (cont.)*, ‡	
293	11.45	725	14.75
372	11.99	775	15.11
429	12.39	793	14.46
476	12.73	828	14.83
524	13.17	874	15.27
575	13.53	925	15.79
624	13.93	973	16.27
673	14.29		

* No figure given.
‡ Author's data for coefficient of thermal expansion have been integrated by TPRC to obtain $\Delta L/L_0$.

FIGURE AND TABLE NO. 268R. PROVISIONAL VALUES FOR THERMAL LINEAR EXPANSION OF NICKEL + CHROMIUM + ΣX_i ALLOYS Ni + Cr + ΣX_i

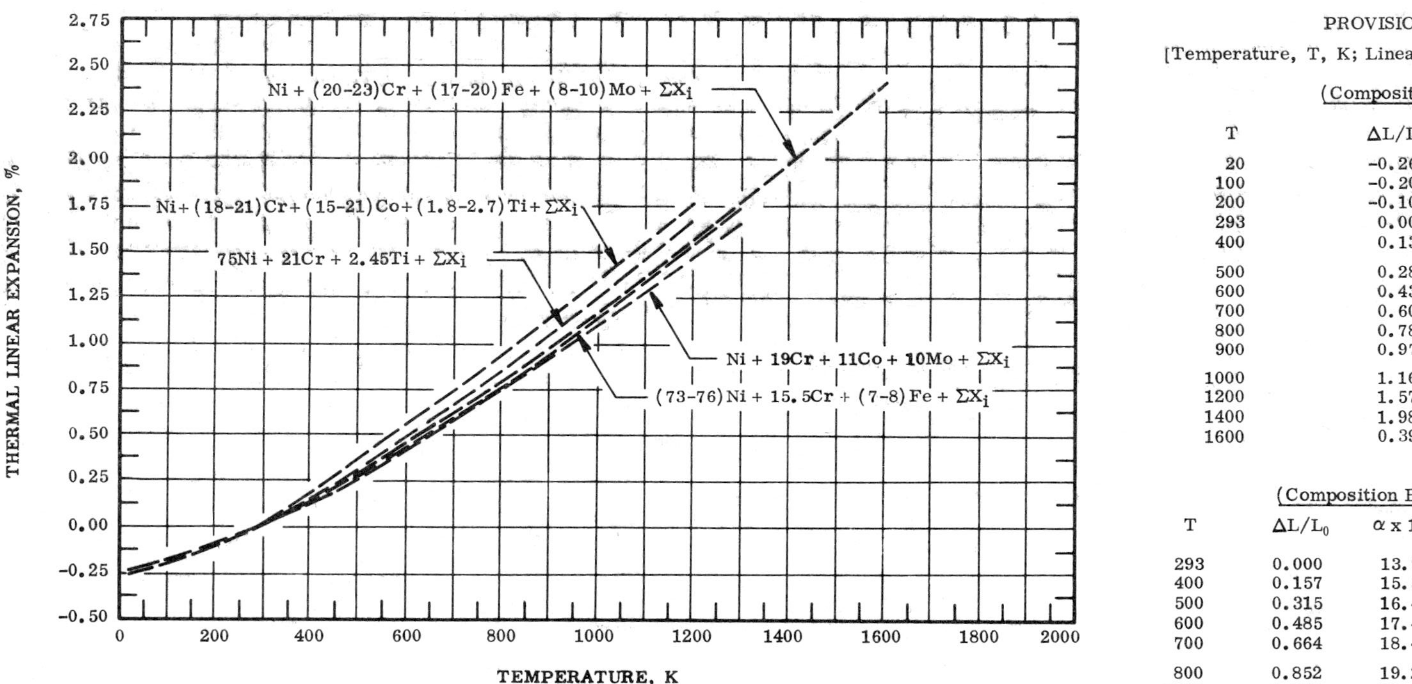

PROVISIONAL VALUES

[Temperature, T, K; Linear Expansion, $\Delta L/L_0$, %; α, K^{-1}]

(Composition A)

T	$\Delta L/L_0$	$\alpha \times 10^6$
20	-0.262	6.9
100	-0.200	8.5
200	-0.105	10.4
293	0.000	12.0
400	0.137	13.7
500	0.281	15.1
600	0.439	16.3
700	0.608	17.4
800	0.787	18.4
900	0.975	19.1
1000	1.169	19.7
1200	1.573	20.5
1400	1.984	20.6
1600	0.391	20.6

T	(Composition B) $\Delta L/L_0$	$\alpha \times 10^6$	(Composition C) $\Delta L/L_0$	$\alpha \times 10^6$
293	0.000	13.9	0.000	16.9
400	0.157	15.3	0.186	17.6
500	0.315	16.4	0.365	18.3
600	0.485	17.4	0.552	18.8
700	0.664	18.4	0.742	19.3
800	0.852	19.2	0.938	19.7
1000	1.249	20.4	1.340	20.4
1200	1.667	21.3	1.753	20.8

T	(Composition D) $\Delta L/L_0$	$\alpha \times 10^6$	(Composition E) $\Delta L/L_0$	$\alpha \times 10^6$
20	-0.238	5.7	-0.262	7.4
100	-0.185	7.6	-0.197	8.7
200	-0.098	9.7	-0.102	10.3
293	0.000	11.4	0.000	11.6
400	0.132	13.2	0.133	13.1
500	0.271	14.6	0.270	14.4
600	0.423	15.7	0.420	15.6
700	0.585	16.7	0.581	16.7
800	0.756	17.4	0.753	17.6
1000	1.112	18.1	1.124	19.4
1200	1.474	18.2	1.525	20.7
1300	1.650	18.2	1.735	21.2

REMARKS

The tabulated values for various Ni + Cr + ΣX_i alloys are for well-annealed alloys and are considered accurate to within ± 7% over the entire temperature range. These values can be represented approximately by the equations given below:

Composition A [Ni + (20-23) Cr + (17-20) Fe + (8-10) Mo + ΣX_i]: This is an alloy of type Hastelloy X.

$$\Delta L/L_0 = -0.275 + 6.475 \times 10^{-4}\ T + 1.064 \times 10^{-6}\ T^2 - 2.672 \times 10^{-10}\ T^3$$

Composition B (75 Ni + 21 Cr + 2.45 Ti + ΣX_i): This is an alloy of type Nimonic 80.

$$\Delta L/L_0 = -0.345 + 9.531 \times 10^{-4}\ T + 8.314 \times 10^{-7}\ T^2 - 1.902 \times 10^{-10}\ T^3$$

Composition C [Ni + (18-21) Cr + (15-21) Co + (1.8-2.7) Ti + ΣX]: This is an alloy of type Nimonic 90.

$$\Delta L/L_0 = -0.459 + 1.447 \times 10^{-3}\ T + 4.634 \times 10^{-7}\ T^2 - 1.108 \times 10^{-10}\ T^3$$

Composition D (Ni + 19 Cr + 11 Co + 10 Mo + ΣX_i): This is an alloy of type Rene 41.

$$\Delta L/L_0 = -0.251 + 5.257 \times 10^{-4}\ T + 1.222 \times 10^{-6}\ T^2 - 3.868 \times 10^{-10}\ T^3$$

Composition E [(73-76) Ni + 15.5 Cr + (7-8) Fe + ΣX_i]: This is an alloy of type Inconel 600 and Inconel X 750.

$$\Delta L/L_0 = -0.277 + 7.105 \times 10^{-4}\ T + 8.451 \times 10^{-7}\ T^2 - 1.548 \times 10^{-10}\ T^3$$

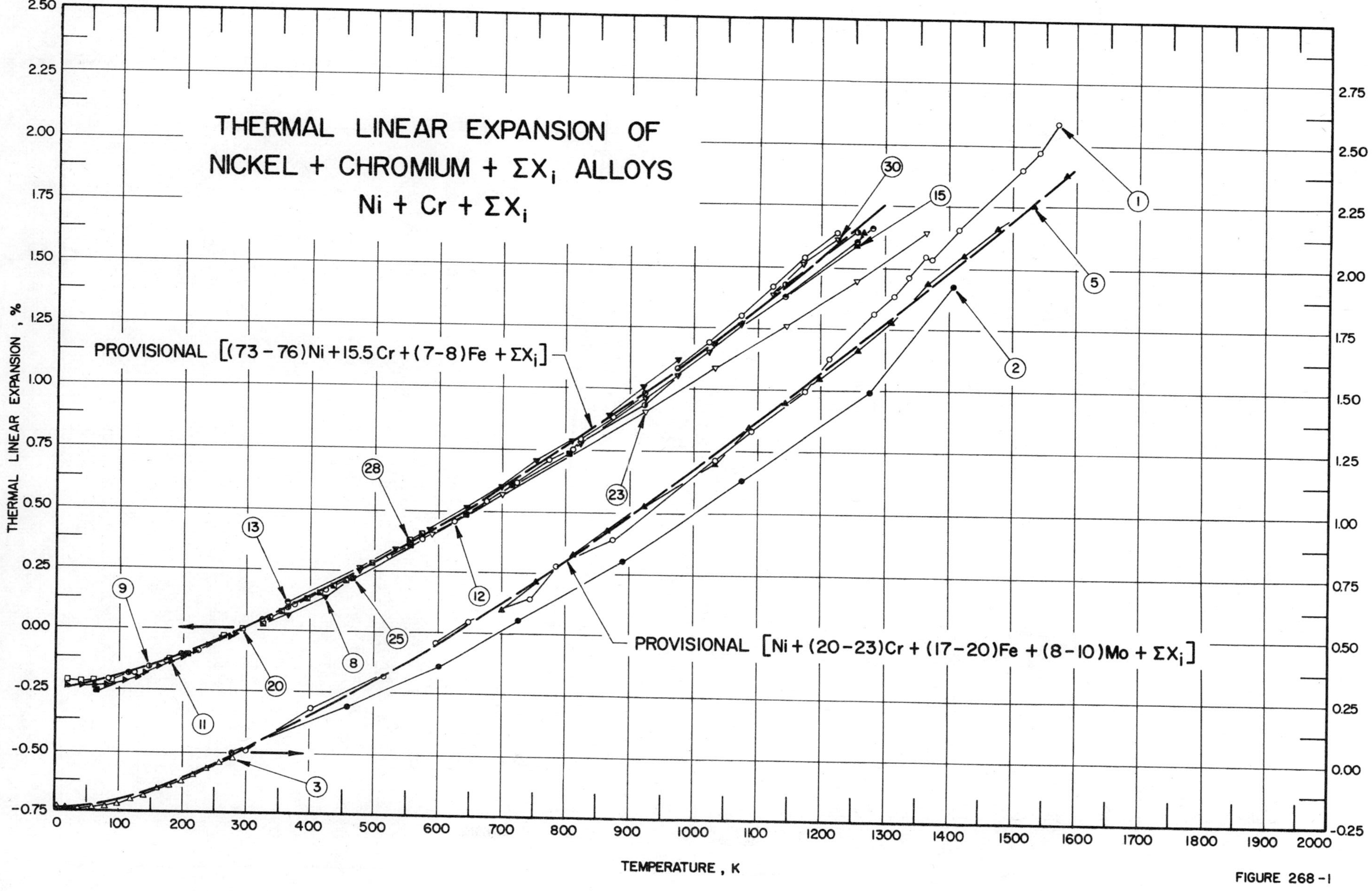
THERMAL LINEAR EXPANSION OF
NICKEL + CHROMIUM + ΣXi ALLOYS
Ni + Cr + ΣXi
PROVISIONAL [(73-76) Ni + 15.5 Cr + (7-8) Fe + ΣXi]
PROVISIONAL [Ni + (20-23) Cr + (17-20) Fe + (8-10) Mo + ΣXi]
THERMAL LINEAR EXPANSION, %
TEMPERATURE, K
0
100
200
300
400
500
600
700
800
900
1000
1100
1200
1300
1400
1500
1600
1700
1800
1900
2000
-0.75
-0.50
-0.25
0.00
0.25
0.50
0.75
1.00
1.25
1.50
1.75
2.00
2.25
2.50
-0.25
0.00
0.25
0.50
0.75
1.00
1.25
1.50
1.75
2.00
2.25
2.50
2.75
1
2
3
5
8
9
11
12
13
15
20
23
25
28
30

FIGURE 268-1

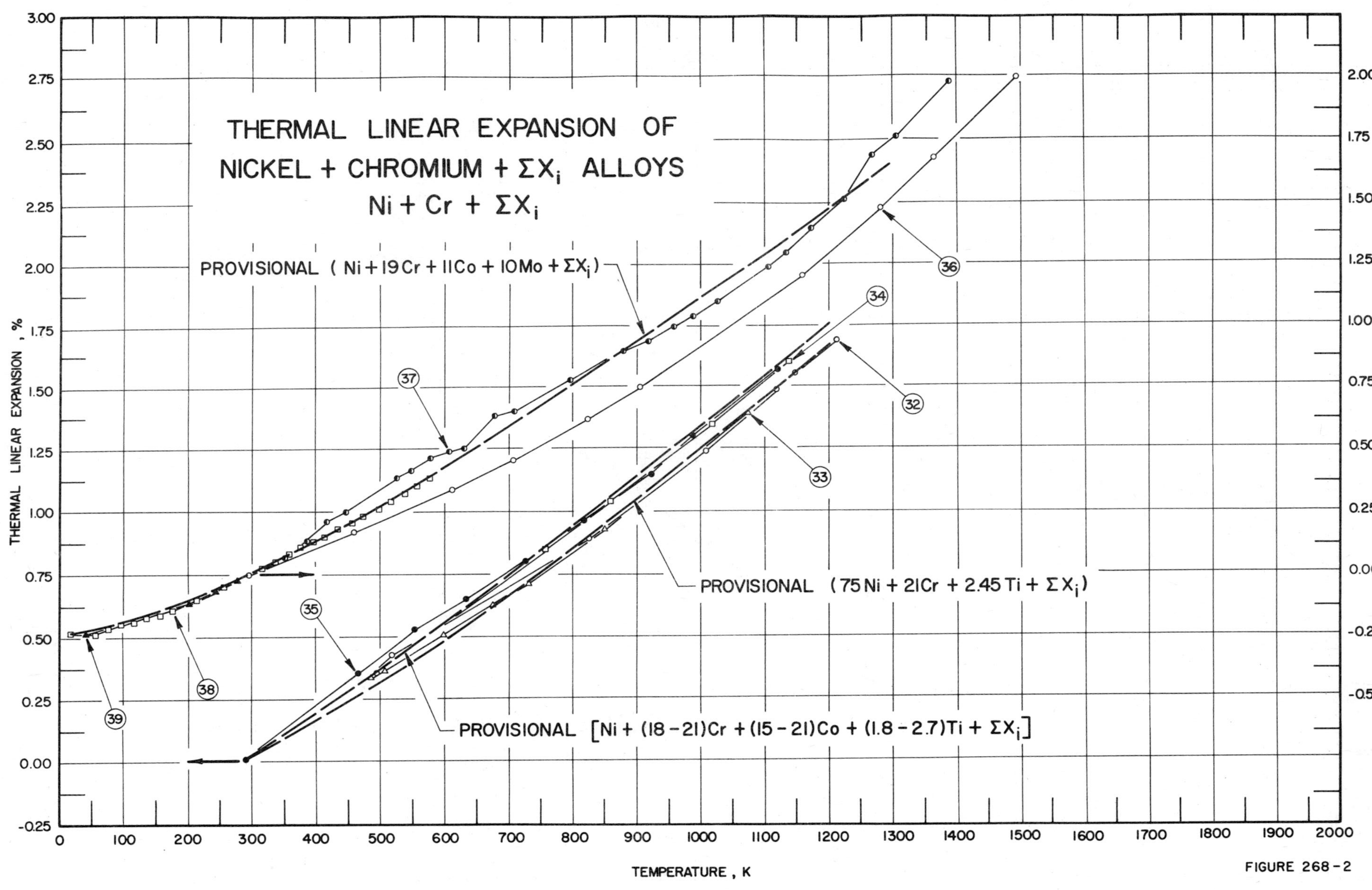

THERMAL LINEAR EXPANSION OF
NICKEL + CHROMIUM + ΣX_i ALLOYS
Ni + Cr + ΣX_i
PROVISIONAL (Ni + 19 Cr + 11 Co + 10 Mo + ΣX_i)
PROVISIONAL (75 Ni + 21 Cr + 2.45 Ti + ΣX_i)
PROVISIONAL [Ni + (18 - 21) Cr + (15 - 21) Co + (1.8 - 2.7) Ti + ΣX_i]
32
33
34
35
36
37
38
39
THERMAL LINEAR EXPANSION , %
3.00
2.75
2.50
2.25
2.00
1.75
1.50
1.25
1.00
0.75
0.50
0.25
0.00
-0.25
2.00
1.75
1.50
1.25
1.00
0.75
0.50
0.25
0.00
-0.25
-0.50
TEMPERATURE , K
0
100
200
300
400
500
600
700
800
900
1000
1100
1200
1300
1400
1500
1600
1700
1800
1900
2000

FIGURE 268-2

SPECIFICATION TABLE 268. THERMAL LINEAR EXPANSION OF NICKEL + CHROMIUM + ΣX_i ALLOYS $Ni + Cr + \Sigma X_i$

Cur. No.	Ref. No.	Author(s)	Year	Method Used	Temp. Range, K	Name and Specimen Designation	Composition (weight percent) Ni	Cr	Fe	Mo	Co	Ti	Al	Composition (continued), Specifications, and Remarks
1	250	Fieldhouse, I.B., Hedge, J.C., and Lang, J.I.	1958	T	300-1571	Hastelloy R-235	Bal.	15	10	5.5	2.5	2.5	2	0.15 C; zero-point correction of 0.012% determined by graphical extrapolation.
2	482	Leggett, H., Cook, J. L., Schwab, D.E., and Powers, C.T.	1965		278-1505	Hastelloy X	41.5	23	20	10	2.5			1 Mn, 1 Si, 1.0 W; material supplied by Union Carbide, Stellite Div.; machined by Douglas for identification purposes; data average of results on several runs on different specimens; zero-point correction of -0.015% determined by graphical interpolation.
3	467	Clark, A.F.	1968	L	0-300	Hastelloy X	46.1	22.17	18.6	9.2	1.76			0.84 Mn, 0.71 W, 0.38 Si, 0.20 Al, trace (< 0.1) C, Cu, V, S, P, Ti, B; machined to 0.64 cm square x 20.32 cm long; Rockwell hardness B-90.
4*	467	Clark, A.F.	1968	L	0-300	Hastelloy X								Similar to the above specimen, composition not specified; Rockwell hardness C-19.
5	495	Seibel, R.D. and Mason, G. L.	1958		293-1644	Hastelloy X	51.15	19.79	17.95	7.43	1.58	0.19		0.81 Mn, 0.86 Si, 0.13 W, 0.11 C; heated at 3 to 5 F per min in vacuum of 10^{-5} to 10^{-6} mm Hg; density 8.15 g cm^{-3}.
6*	358	Fieldhouse, I.B. and Lang, J.I.	1956		300-1562	Inco 713 C	71.53	11.0	5.0	3.5	0.25	0.25	6.5	1.0 Si, 1.0 Mn, 1.0 Cb + Ta, 0.2 C; density 9.23 g cm^{-3}.
7*	496	Heyer, B.A. and Marlatt, J.W.	1966		293-1366	Inco 713 C	Bal.	13.24	0.012	4.32		0.088	5.85	2.10 (Nb + Ta), 0.07 Si, 0.05 Zr, 0.07 C, 0.011 S, 0.011 B; low carbon alloy, cast, heat-treated 2 hr at 1450 K, air cooled.
8	497	Smoke, E.J., et al.	1960	L	366-917	Inconel	76	16	6					No details given.
9	123	Lucks, C.F. and Deem, H.W.	1958	L	116-1255	Inconel								0.375 in diameter x 3 in. long; hot-rolled material from International Nickel Co., Inc; annealed 3 hr at 1600 F, 15 min at 1800 F, and air cooled; heating rate 2.2 C per min.
10*	495	Siebel, R.D. and Mason, G. L.	1958		293-1644	Inconel	75.54	15.15	8.24		0.094	0.35		0.30 Mn, 0.23 Si, 0.077 C; heated at rate of 2 C per min in vacuum of 10^{-5} to 10^{-6} mm Hg; density 8.40 g cm^{-3}.
11*	462	Arp, V., Wilson, J.H., Winrich, L., and Sikora, P.	1962	L	20-293	Inconel	Bal.	15.5	7.3					0.2 Mn, 0.2 Si, 0.04 C, 0.01 Cu, 0.01 S; hardness R_C27; cold drawn and tempered specimen.
12	466	Lucks, C.F., Thompson, H.B., Smith, A.R., Curry, F.P., Deem, H.W., and Bing, G.F.	1951	L	83-1223	Inconel	78.92	14.62	5.80					0.23 Mn, 0.19 Si, 0.12 Cu, 0.09 C, and 0.007 S; supplied by International Nickel Co.; hot-rolled; annealed 3 hr at 1144.3 K, 15 min at 1255.4 K, and air cooled.

*Not shown in figure.

SPECIFICATION TABLE 268. THERMAL LINEAR EXPANSION OF NICKEL + CHROMIUM + ΣX_i ALLOYS Ni + Cr + ΣX_i (continued)

Cur. No.	Ref. No.	Author(s)	Year	Method Used	Temp. Range, K	Name and Specimen Designation	Composition (weight percent) Ni	Cr	Fe	Mo	Co	Ti	Al	Composition (continued), Specifications, and Remarks
13	498	O'Sullivan, W.J.	1955		298-1280	Low C Inconel	75.99	14.42	8.87					0.28 Mn, 0.22 Cu, 0.17 Si, 0.007 S, 0.02 C; expansion measured with increasing temp; zero-point correction of -0.008% determined by graphical extrapolation.
14*	498	O'Sullivan, W.J.	1955		870-298	Low C Inconel	75.99	14.42	8.87					The above specimen; expansion measured with decreasing temperature; zero-point correction of -0.008% determined by graphical extrapolation.
15	498	O'Sullivan, W.J.	1955		298-1266	Med. C. Inconel	76.45	14.96	7.89					0.26 Mn, 0.19 Si, 0.15 Cu, 0.007 S, 0.07 C; expansion measured with increasing temp; zero-point correction of -0.014% determined by graphical extrapolation.
16*	498	O'Sullivan, W.J.	1955		764-298	Med. C Inconel	76.45	14.96	7.89					The above specimen; expansion measured with decreasing temp; zero-point correction of -0.014% determined by graphical extrapolation.
17*	498	O'Sullivan, W.J.	1955		298-1255	Hi C Inconel	75.64	15.32	8.17					0.33 Mn, 0.21 Si, 0.19 Cu, 0.007 S, 0.11 C; expansion measured with increasing temp; zero-point correction of -0.013% determined by graphical extrapolation.
18*	498	O'Sullivan, W.J.	1955		1069-298	Hi C Inconel	75.64	15.32	8.17					0.33 Mn; the above specimen; expansion measured with decreasing temp; zero-point correction of -0.013% determined by graphical extrapolation.
19*	225	Valentich, J.	1965	L	66-805	Inconel 600	Bal.	14/17	6/8					1.75-2.25 Nb, 0.1 C; No details given.
20	467	Clark, A.F.	1968	L	0-300	Inconel 625								Specimen machined to 0.64 cm square x 20.32 cm long; annealed to Rockwell hardness C-25.
21*	467	Clark, A.F.	1968	L	0-300	Inconel 718	54.57	18.06	17.08	3.18		0.85	0.44	5.12 Nb + Ta, 0.29 Mn, 0.24 Si, trace (< 0.1) Cu, C, S; machined to 0.64 cm square x 20.32 cm long; specimen age-hardened; Rockwell hardness C-39.
22	225	Valentich, J.	1965	L	66-807	Inconel 718								No details given.
23*	499	Aerojet-General Corp., Sacramento, California	1963	L	366-1366	Inconel 718								Specimen from stock material.

*Not shown in figure.

SPECIFICATION TABLE 268. THERMAL LINEAR EXPANSION OF NICKEL + CHROMIUM + ΣX_i ALLOYS Ni + Cr + ΣX_i (continued)

Cur. No.	Ref. No.	Author(s)	Year	Method Used	Temp. Range, K	Name and Specimen Designation	Ni	Cr	Fe	Mo	Co	Ti	Al	Composition (continued), Specifications, and Remarks
24*	500	Heathman, J.H.	1968		18-811	Inconel 718								Typical results; composition of several specimens varied; tests made after various heat treatments; no information specified; zero-point correction of 0.006% determined by graphical interpolation.
25*	225	Valentich, J.	1965	L	66-807	Inconel 750 (Inconel X750)		15	6.75			2.5	0.80	0.85 Nb, 0.70 Mn, 0.30 Si, 0.05 Cu, 0.04 C, 0.007 S; no details given.
26*	123	Lucks, C.F. and Deem, H.W.	1958	L	116-1255	Inconel X (Inconel X750)		15	6.8			2.5	0.75	0.85 Nb, 0.04 C; 0.375 in. diameter x 3 in. long; hot-rolled material from International Nickel Co., Inc; solution treated 3 hr at 2100 F, air cooled; double aged 24 hr at 1550 F, air cooled; heated at 2.2 C per min.
27*	495	Seibel, R.D. and Mason, G.L.	1958		293-1644	Inconel X (Inconel X750)	73.03	14.04	7.93			2.73	0.56	0.67 Mn, 0.57 Nb, 0.41 Si; heated at 2 C per min in vacuum of 10^{-5} to 10^{-6} mm Hg; density 8.20 g cm^{-3}.
28	15	Rhodes, B.L., Moeller, C.E., Hopkins, V., and Marx, T.I.	1963	L	18-573	Inconel X (Inconel X750)		15.0	7.0			2.5	0.9	1.02 Nb, 0.7 Mn, 0.3 Si, 0.04 C; specimen 10.16 cm long x 6.2 mm.
29*	462	Arp, V., Wilson, J.H., Winrich, L., and Sikora, P.	1962		20-293	Inconel X (Inconel X750)		15.4	7.0			2.5	0.9	0.7 Nb, 0.6 Mn, 0.3 Si, 0.05 C, 0.01 S; hardness R_C39; specimen hot rolled, direct aged and tempered.
30	466	Lucks, C.F., Thompson, H.B., Smith, A.R., Curry, F.P., Deem, H.W., and Bing, G.F.	1951	L	83-1273	Inconel X (Inconel X750)	72.94	14.65	6.97			2.44	0.93	1.01 Cb, 0.54 Mn, 0.46 Si, 0.03 C, 0.02 Cu, 0.007 S; hot-rolled; solution heat-treated 3 hr at 1422.05 K; air cooled, double aged; 24 hr at 1116.49 K, air cooled; 20 hr at 977.6 K and air cooled.
31	358	Fieldhouse, I.B. and Lang, J.I.	1956		293-1552	M252 (GEJ1500)	57.15	18.65	<0.30	9.98		2.74	1.17	9.75 Cu, 0.12 C, 0.07 Mn, 0.06 Si; dissolved at 2223 K and air cooled; density 8.22 g cm^{-3}.
32	313	Makin, S.M., Standring, J., and Hunter, P.M.	1953	I	492-1210	Nimonic 80								Specimen precipitation hardened; expansion measured with increasing temperature.
33	313	Makin, S.M., et al.	1953	I	1070-478	Nimonic 80								The above specimen; expansion measured with decreasing temperature.
34	313	Makin, S.M., et al.	1953	I	754-1133	Nimonic 90								Specimen precipitation hardened; expansion measured with increasing temperature.
35	313	Makin, S.M., et al.	1953	I	1120-452	Nimonic 90								The above specimen; expansion with decreasing temperature.

*Not shown in figure.

SPECIFICATION TABLE 268. THERMAL LINEAR EXPANSION OF NICKEL + CHROMIUM + ΣX_i ALLOYS Ni + Cr + ΣX_i (continued)

Cur. No.	Ref. No.	Author(s)	Year	Method Used	Temp. Range, K	Name and Specimen Designation	Composition (weight percent) Ni	Cr	Fe	Mo	Co	Ti	Al	Composition (continued), Specifications, and Remarks
36	482	Leggett, H., Cook, J. L., Schwab, D. E., and Powers, C. T.	1965		293-1495	Rene 41	52.5	19	3	10	11	3	1.5	Material supplied by Union Carbide, Stellite Div.; machined by Douglas for identification purposes; data average of results on several runs on different specimens.
37	358	Fieldhouse, I. B. and Lang, J. I.	1961		293-1552	Rene 41 (GE-J1610)	54.60	18.60	1.54	9.63	10.73	3.14	1.49	0.11 C, 0.08 Mn, 0.07 Si; solutioned at 1352 K and water quenched; density 8.08 g cm^{-3}.
38	15	Rhodes, B. L., Moeller, C. E., Hopkins, V., and Marx, T. I.	1963	L	18-573	Rene 41 (GE-J1610)		19.0	3.0	10.0	11.0	3.0	1.5	0.10 C; 10.16 cm long x 6.2 mm.
39	462	Arp, V., Wilson, J. H., Winrich, L., and Sikora, P.	1962	L	20-293	Rene 41		18.8	1.3	9.7	10.5	3.2	1.4	0.2 Si, 0.09 C, 0.01 S; hardness R_C 39; solution treated specimen.
40*	467	Clark, A. F.	1968	L	0-300	Udimet 630 C-29	52.1	17.6	17.1	2.75		0.89	0.63	6.05 (Nb + Ta), 2.8 W, trace (< 0.1) S, B; machined to 0.64 cm square x 20.32 cm long; Rockwell hardness C-29.
41*	467	Clark, A. F.	1968	L	0-300	Udimet 630 C-32								Similar to the above specimen except Rockwell hardness C-40.
42*	467	Clark, A. F.	1968	L	0-300	Udimet 630	59.4	18.8	0.69	4.01	12.67	2.92	1.45	Trace (< 0.1) each C, Zr, Cu, Mn, Si, B, S, P; specimen annealed and machined to 0.64 cm square x 20.32 cm long; Rockwell hardness C-29.
43*	467	Clark, A. F.	1968	L	0-300	Udimet 630								Similar to the above specimen except age-hardened to Rockwell hardness C-32.
44*	217	Totskii, E. E.	1964	L	79-1373	EI-607 Russian alloy	Bal.	15.25				1.9		0.04 C; specimen annealed rod 200 mm long.
45*	217	Totskii, E. E.	1964	L	273-1273	EI-617 Russian alloy	Bal.	13/ 18		2/ 4		1.8/ 2.3	1.7/ 2.3	5-7 W, 0.5-1.0 V, 0.2 Ce, 0.02 B; similar to the above specimen.
46*	429	Mochel, N. L.	1927	T	273-787	K4 (Nichrome)	76.0	18.8	0.80					2.79 Mn, 0.66 Cu, 0.31 C, 0.27 Si; rolled to 1.9 cm diameter rod, tested as received; expansion measured with increasing temp; zero-point correction of -0.024% determined by graphical interpolation.
47*	429	Mochel, N. L.	1927	T	787-273	K4 (Nichrome)	76.0	18.8						The above specimen; expansion measured with decreasing temp; zero-point correction of -0.015% determined by graphical extrapolation.
48*	217	Mochel, N. L.	1927	L	273-787	K4 A (Nichrome)	76.0	18.8						Similar to the above specimen; expansion measured with increasing temperature; zero-point correction of -0.0244% determined by graphical interpolation.
49*	217	Mochel, N. L.	1927	L	787-273	K4 A (Nichrome)	76.0	18.8						The above specimen; expansion measured with decreasing temperature; zero-point correction of -0.015% determined by graphical extrapolation.

* Not shown in figure.

DATA TABLE 268. THERMAL LINEAR EXPANSION OF NICKEL + CHROMIUM + ΣX_i ALLOYS $Ni + Cr + \Sigma X_i$

[Temperature, T, K; Linear Expansion, $\Delta L/L_0$, %]

T	$\Delta L/L_0$
CURVE 1	
300	0.012
401	0.188
514	0.312
594	0.462
648	0.549
742	0.642
785	0.779
872	0.897
1033	1.217
1092	1.328
1175	1.489
1212	1.629
1283	1.817
1315	1.887
1336	1.969
1366	2.051
1382	2.044
1419	2.162
1480	2.313
1519	2.407
1542	2.473
1571	2.597
CURVE 2	
278	0.000
458	0.195
599	0.371
727	0.560
890	0.811
1075	1.135
1274	1.497
1505	1.939
CURVE 3	
0	-0.243
10	-0.243*
20	-0.243
30	-0.243*
40	-0.242
50	-0.239*
60	-0.236
70	-0.232*
80	-0.227
90	-0.221*

T	$\Delta L/L_0$
CURVE 3 (cont.)	
100	-0.214
120	-0.198
140	-0.180
160	-0.160
180	-0.138
200	-0.116
220	-0.093
240	-0.068
260	-0.043
273	-0.027*
280	-0.018
293	0.000
300	0.008*
CURVE 4*	
0	-0.237
10	-0.237
20	-0.237
30	-0.236
40	-0.235
50	-0.233
60	-0.231
70	-0.227
80	-0.222
90	-0.217
100	-0.210
120	-0.195
140	-0.177
160	-0.158
180	-0.136
200	-0.114
220	-0.091
240	-0.067
260	-0.042
273	-0.025
280	-0.016
293	0.000
300	0.010
CURVE 5	
293	0.000*
699	0.600
755	0.715
810	0.835

T	$\Delta L/L_0$
CURVE 5 (cont.)	
866	0.930
922	1.030
977	1.130
1033	1.230
1088	1.340
1144	1.450
1199	1.550
1255	1.670
1310	1.790
1366	1.940
1422	2.060
1477	2.170
1533	2.260
1588	2.380*
CURVE 6*	
299	0.000
368	0.041
450	0.126
501	0.207
566	0.307
614	0.389
672	0.482
727	0.570
758	0.609
812	0.680
866	0.769
920	0.863
978	0.975
1012	1.046
1089	1.180
1119	1.236
1179	1.385
1230	1.517
1295	1.673
1363	1.917
1372	2.114
1455	2.216
1503	2.358
1562	2.578
CURVE 7*	
293	0.000
366	0.074

T	$\Delta L/L_0$
CURVE 7 (cont.)*	
477	0.235
589	0.405
600	0.616
711	0.806
822	1.006
933	1.868
1144	1.437
1255	1.697
1366	2.038
CURVE 8	
366	0.068
422	0.155
477	0.248
533	0.331
588	0.421
644	0.508
699	0.598
755	0.678
810	0.786
866	0.886
922	1.009
977	1.093
CURVE 9	
116	-0.182
144	-0.159
199	-0.105
366	0.090*
477	0.240*
588	0.402*
699	0.571
810	0.752
922	0.953
1033	1.181
1144	1.430*
1255	1.690*
CURVE 10*	
293	0.000
699	0.800
755	0.920
810	1.030

T	$\Delta L/L_0$
CURVE 10 (cont.)	
866	1.140
922	1.250
977	1.350
1033	1.460
1088	1.570
1144	1.680
1199	1.770
1255	1.870
1310	1.970
1366	2.080
1422	2.180
1477	2.227
1533	2.340
1588	2.450
1644	2.540
CURVE 11*	
20	-0.242
40	-0.240
60	-0.235
80	-0.228
100	-0.216
120	-0.203
140	-0.184
160	-0.164
180	-0.143
200	-0.121
220	-0.098
240	-0.073
260	-0.046
273	-0.028
280	-0.019
293	0.000
CURVE 12	
83	-0.207
123	-0.181
173	-0.137
223	-0.084
323	0.037
373	0.103
423	0.172
473	0.242*
523	0.313

T	$\Delta L/L_0$
CURVE 12 (cont.)	
573	0.385
623	0.460
673	0.539
723	0.620
773	0.711
823	0.797
873	0.885
923	0.977
973	1.077
1023	1.185
1073	1.299
1123	1.423
1173	1.535
1223	1.639
CURVE 13	
298	0.008*
366	0.116
478	0.264
589	0.429*
700	0.602
811	0.794*
922	0.963*
1033	1.180*
1144	1.384
1255	1.612
1280	1.662
CURVE 14*	
870	0.884
811	0.787
700	0.603
589	0.419
478	0.253
366	0.095
298	0.010
CURVE 15	
298	0.006*
366	0.100
478	0.252*
589	0.414*
700	0.592*

*Not shown in figure.

DATA TABLE 268. THERMAL LINEAR EXPANSION OF NICKEL + CHROMIUM + ΣX_i ALLOYS $Ni + Cr + \Sigma X_i$ (continued)

T	$\Delta L/L_0$
CURVE 15 (cont.)	
811	0.780*
922	0.962*
1033	1.167
1144	1.366
1255	1.587
1266	1.642
CURVE 16*	
764	0.701
700	0.589
589	0.405
478	0.237
366	0.091
298	0.012
CURVE 17*	
298	0.007
366	0.096
478	0.259
589	0.413
700	0.583
811	0.770
922	0.976
1033	1.175
1144	1.365
1255	1.634
CURVE 18*	
1069	1.263
1033	1.186
922	0.980
811	0.772
700	0.586
589	0.417
478	0.250
366	0.092
298	0.011
CURVE 19*	
66	-0.259
138	-0.185
207	-0.108
CURVE 19 (cont.)*	
293	0.000
365	0.101
466	0.239
579	0.425
688	0.613
805	0.804
CURVE 20	
0	-0.221
10	-0.221*
20	-0.221
30	-0.221*
40	-0.220
50	-0.218*
60	-0.216
70	-0.212*
80	-0.208*
90	-0.202
100	-0.196*
120	-0.182*
140	-0.166*
160	-0.148*
180	-0.128
200	-0.108*
220	-0.086*
240	-0.064*
260	-0.040*
273	-0.025
280	-0.017*
293	0.000*
300	0.008*
CURVE 21*	
0	-0.238
10	-0.238
20	-0.238
30	-0.237
40	-0.236
50	-0.235
60	-0.232
70	-0.228
80	-0.223
90	-0.217
100	-0.211
CURVE 21 (cont.)*	
120	-0.195
140	-0.177
160	-0.157
180	-0.136
200	-0.114
220	-0.091
240	-0.067
260	-0.042
273	-0.026
280	-0.017
293	0.000
300	0.009
CURVE 22	
66	-0.259
138	-0.185*
207	-0.108
293	0.000*
367	0.092*
465	0.221
558	0.356
641	0.486
718	0.606
807	0.733
CURVE 23‡	
367	0.100*
478	0.252
589	0.410
700	0.574
811	0.743*
922	0.913
1033	1.087
1144	1.267
1255	1.454
1367	1.645
CURVE 24*	
18	-0.230
33	-0.228
61	-0.217
89	-0.202
116	-0.185
CURVE 24 (cont.)*	
144	-0.162
200	-0.107
255	-0.045
311	0.021
366	0.091
422	0.170
478	0.247
533	0.333
589	0.415
644	0.502
700	0.588
755	0.681
811	0.765
CURVE 25*	
66	-0.259
138	-0.185
207	-0.108
293	0.000
367	0.092
465	0.221
558	0.356
641	0.486
718	0.606
807	0.733
CURVE 26*	
116	-0.206
144	-0.176
199	-0.113
366	0.092
477	0.237
588	0.397
699	0.568
810	0.755
922	0.952
1033	1.177
1144	1.430
1255	1.688
CURVE 27	
293	0.000*
699	0.440
CURVE 27 (cont.)	
755	0.550
810	0.665
866	0.765
922	0.865
977	0.970
1033	1.070
1088	1.180
1144	1.290
1199	1.390
1255	1.510
1310	1.610
1366	1.720
1422	1.810
1477	1.910
1533	1.980
1588	2.100
1644	2.200
CURVE 28	
18	-0.221*
23	-0.221*
33	-0.221*
43	-0.221*
53	-0.220*
63	-0.218*
73	-0.216*
83	-0.212*
93	-0.207*
103	-0.201*
113	-0.195*
123	-0.188*
133	-0.180*
143	-0.171*
153	-0.162*
163	-0.152*
173	-0.142*
193	-0.121*
213	-0.099*
233	-0.075*
253	-0.051*
273	-0.026*
293	0.000*
313	0.026
333	0.052
353	0.080
CURVE 28 (cont.)	
373	0.107*
393	0.135
413	0.164
433	0.193
453	0.223
473	0.254*
493	0.285
513	0.316*
533	0.349*
553	0.385
573	0.422
CURVE 29*	
20	-0.218
40	-0.217
60	-0.213
80	-0.205
100	-0.195
120	-0.180
140	-0.163
160	-0.146
180	-0.127
200	-0.106
220	-0.084
240	-0.061
260	-0.038
273	-0.023
280	-0.015
293	0.000
CURVE 30	
83	-0.211*
123	-0.181*
173	-0.137*
223	-0.084*
323	0.037*
373	0.103*
423	0.172*
473	0.242*
523	0.313*
573	0.385*
623	0.460*
673	0.539*
723	0.620*

*Not shown in figure.
‡Author's data for coefficient of thermal expansion have been integrated by TPRC to obtain $\Delta L/L_0$.

DATA TABLE 268. THERMAL LINEAR EXPANSION OF NICKEL + CHROMIUM + ΣX_i ALLOYS $Ni + Cr + \Sigma X_i$ (continued)

T	$\Delta L/L_0$
CURVE 30 (cont.)	
773	0.701*
823	0.784
873	0.871*
923	0.960
973	1.054
1023	1.156
1073	1.265
1123	1.382
1173	1.509
1223	1.621
1273	1.727
CURVE 31	
293	0.00*
340	0.09
384	0.14
414	0.18
444	0.25
525	0.41
549	0.43
576	0.48
603	0.51
627	0.54
679	0.66
708	0.67
794	0.79
874	0.88*
917	0.93
958	0.97
985	1.04
1025	1.08
1103	1.24
1131	1.33
1171	1.42
1214	1.53
1265	1.67
1303	1.78
1390	1.94
1460	2.13
1508	2.26
1539	2.46*
1551	2.54*
CURVE 32	
492	0.358
517	0.403
821	0.888
1007	1.243
1119	1.493
1145	1.553
1210	1.694
CURVE 33	
1070	1.398
888	1.006
850	0.929
730	0.724
672	0.627
596	0.509
506	0.372
478	0.340
CURVE 34	
754	0.858
855	1.046
1019	1.361
1133	1.611
CURVE 35	
1120	1.573
985	1.304
920	1.161
818	0.973
727	0.813
629	0.651
550	0.528
452	0.364
CURVE 36	
293	0.000
465	0.173
610	0.337
705	0.455
CURVE 36 (cont.)	
826	0.622
906	0.746
1011	0.930
1160	1.218
1279	1.477
1365	1.678
1495	2.009
CURVE 37	
293	0.00
340	0.07
384	0.14
414	0.22
444	0.25
525	0.38
549	0.42
576	0.47
603	0.49
627	0.50
679	0.64
708	0.65
794	0.78
874	0.89
917	0.94
958	1.00
985	1.04
1025	1.10
1103	1.24
1131	1.31
1171	1.40
1214	1.52
1265	1.68
1303	1.76
1390	1.92
1460	2.12*
1508	2.25*
1539	2.42*
1551	2.54*
CURVE 38	
18	-0.215
23	-0.215*
33	-0.215*
43	-0.215*
53	-0.214
CURVE 38 (cont.)	
63	-0.212*
73	-0.208
83	-0.204
93	-0.199
103	-0.193*
113	-0.186
123	-0.179*
133	-0.172
143	-0.164*
153	-0.155
163	-0.146*
173	-0.136
193	-0.115*
213	-0.094
233	-0.072*
253	-0.048
273	-0.025*
293	0.000
313	0.026
333	0.051
353	0.077
373	0.103
393	0.129
413	0.156
433	0.182
453	0.209
473	0.236
493	0.264
513	0.292
533	0.321
553	0.349
573	0.379
CURVE 39	
20	-0.217*
40	-0.215
60	-0.212*
80	-0.205*
100	-0.193*
120	-0.178*
140	-0.160*
160	-0.141*
180	-0.122*
200	-0.103*
220	-0.081*
CURVE 39 (cont.)	
240	-0.059
260	-0.037*
273	-0.022
280	-0.014*
293	0.000*
CURVE 40*	
0	-0.237
10	-0.237
20	-0.237
30	-0.236
40	-0.235
50	-0.233
60	-0.231
70	-0.227
80	-0.222
90	-0.217
100	-0.210
120	-0.195
140	-0.178
160	-0.159
180	-0.138
200	-0.116
220	-0.093
240	-0.069
260	-0.044
273	-0.027
280	-0.018
293	0.000
300	0.008
CURVE 41*	
0	-0.234
10	-0.234
20	-0.234
30	-0.233
40	-0.232
50	-0.231
60	-0.228
70	-0.225
80	-0.220
90	-0.214
100	-0.208
120	-0.193
CURVE 41 (cont.)*	
140	-0.176
160	-0.156
180	-0.136
200	-0.114
220	-0.091
240	-0.067
260	-0.043
273	-0.027
280	-0.018
293	0.000
300	0.008
CURVE 42*	
0	-0.222
10	-0.222
20	-0.222
30	-0.222
40	-0.221
50	-0.219
60	-0.217
70	-0.213
80	-0.209
90	-0.203
100	-0.197
120	-0.183
140	-0.166
160	-0.148
180	-0.128
200	-0.108
220	-0.086
240	-0.063
260	-0.040
273	-0.025
280	-0.016
293	0.000
300	0.008
CURVE 43*	
0	-0.221
10	-0.221
20	-0.221
30	-0.220
40	-0.219
50	-0.217

*Not shown in figure.

DATA TABLE 268. THERMAL LINEAR EXPANSION OF NICKEL + CHROMIUM + ΣX_i ALLOYS Ni + Cr + ΣX_i (continued)

T	$\Delta L/L_0$
CURVE 43 (cont.)*	
60	-0.215
70	-0.211
80	-0.207
90	-0.202
100	-0.197
120	-0.183
140	-0.167
160	-0.149
180	-0.129
200	-0.109
220	-0.087
240	-0.065
260	-0.041
273	-0.025
280	-0.017
293	0.000
300	0.009
CURVE 44	
79	-0.207
273	-0.026*
373	0.106*
473	0.244*
573	0.388*
673	0.540*
773	0.697
873	0.865
973	1.057
1073	1.259
1173	1.472
1273	1.692*
1373	1.925
CURVE 45	
273	-0.024*
373	0.099*
473	0.230
573	0.365
673	0.505
773	0.652
873	0.813
973	0.991
1073	1.190
1173	1.418*
1273	1.701

T	$\Delta L/L_0$
CURVE 46	
273	-0.024*
298	0.006*
416	0.157
538	0.330
661	0.521
787	0.734
CURVE 47	
787	0.743
623	0.470
465	0.234
297	0.006*
CURVE 48*	
273	-0.0244*
297	0.0049*
420	0.1569
537	0.3249
661	0.5149
787	0.7319
CURVE 49*	
787	0.741
622	0.465
470	0.237
297	0.008

* Not shown in figure.

FIGURE AND TABLE NO. 269R. PROVISIONAL VALUES FOR THERMAL LINEAR EXPANSION OF NICKEL + COBALT + ΣX_i ALLOYS   Ni + Co + ΣX_i

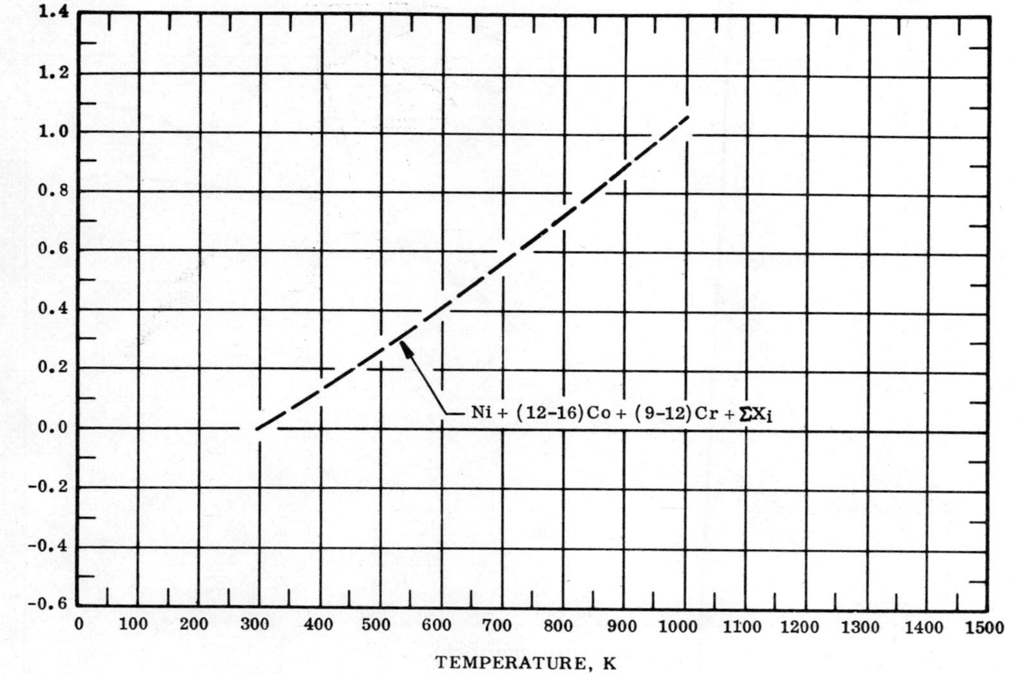

PROVISIONAL VALUES

[Temperature, T, K; Linear Expansion, $\Delta L/L_0$, %; α, K^{-1}]

[Ni + (12 - 16) Co + (9 - 12) Cr + ΣX_i]		
T	$\Delta L/L_0$	$\alpha \times 10^6$
293	0.000	12.6
400	0.137	13.2
500	0.272	13.8
600	0.414	14.5
700	0.563	15.2
800	0.719	16.1
900	0.884	17.0
1000	1.060	18.1

REMARKS

[Ni + (12 - 16) Co + (9 - 12) Cr + ΣX_i]: The tabulated values for well-annealed alloy of type Russian EI 929 are considered accurate to within ± 7% over the entire temperature range. These values can be represented approximately by the following equation:

$$\Delta L/L_0 = -0.351 + 1.145 \times 10^{-3}\, T + 1.366 \times 10^{-7}\, T^2 + 1.292 \times 10^{-10}\, T^3$$

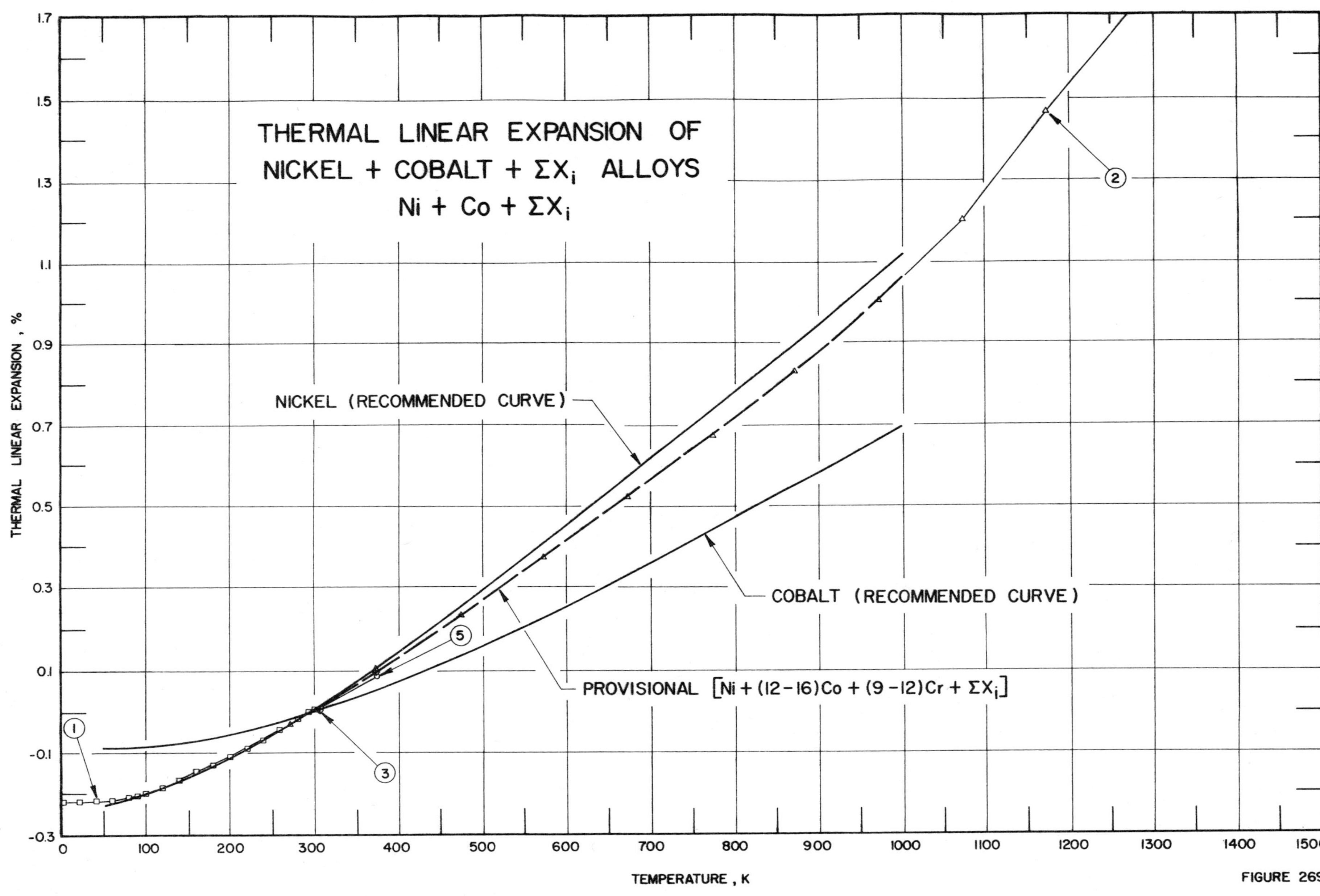

THERMAL LINEAR EXPANSION OF
NICKEL + COBALT + ΣXi ALLOYS
Ni + Co + ΣXi
NICKEL (RECOMMENDED CURVE)
COBALT (RECOMMENDED CURVE)
PROVISIONAL [Ni + (12 - 16) Co + (9 - 12) Cr + ΣXi]
1
2
3
5
THERMAL LINEAR EXPANSION, %
1.7
1.5
1.3
1.1
0.9
0.7
0.5
0.3
0.1
-0.1
-0.3
TEMPERATURE, K
0
100
200
300
400
500
600
700
800
900
1000
1100
1200
1300
1400
1500

FIGURE 269

SPECIFICATION TABLE 269. THERMAL LINEAR EXPANSION OF NICKEL + COBALT + ΣX_i ALLOYS Ni + Co + ΣX_i

Cur. No.	Ref. No.	Author(s)	Year	Method Used	Temp. Range, K	Name and Specimen Designation	Ni	Co	Cr	Mo	Al	Ti	Fe	Composition (continued), Specifications, and Remarks
							Composition (weight percent)							
1	467	Clark, A. F.	1968	L	0-300	Udimet 700	52.8	19.0	15.2	4.95	4.40	3.43	0.13	< 0.1 each Cu, Mn, Si, C, Zr, B, S; specimen machined to 0.64 cm square x 20.32 cm long; Rockwell hardness C-38.
2	217	Totskii, E. E.	1964	L	273-1273	EI-929 Russian alloy	Bal.	12/ 16	9/ 12	4.5 6.5	4/ 6	5.0	3.6/ 4.5	1.4 to 2 Ti, 0.2 to 0.8 V, 0.50 Mn, 0.50 Si, 0.12 C, 0.02 B; specimen annealed rod 200 mm long.
3	59	Masumoto, H.	1931	L	303-373	No. 24	70	19.9					10.1	Mond Ni, granular Co, Armco Fe, mixed and melted in hydrogen atmosphere in alumina crucible in Tammann furnace, cast in iron mold with cylindrical aperture; machined to circular rod 4 mm thick, 10 cm long; heated at 1000 for 1 hr in electric furnace in hydrogen atmosphere at reduced pressure; expansion measured in vacuum.
4*	59	Masumoto, H.	1931	L	303-373	No. 34	50	29.9					20.1	Similar to the above specimen.
5	59	Masumoto, H.	1931	L	303-373	No. 35	60	29.9					10.1	Similar to the above specimen.
6*	59	Masumoto, H.	1931	L	303-373	No. 43	40	39.9					20.1	Similar to the above specimen.
7*	59	Masumoto, H.	1931	L	303-373	No. 44	50	39.9					10.1	Similar to the above specimen.

DATA TABLE 269. THERMAL LINEAR EXPANSION OF NICKEL + COBALT + ΣX_i ALLOYS Ni + Co + ΣX_i

[Temperature, T, K; Linear Expansion, $\Delta L/L_0$, %]

T	$\Delta L/L_0$	T	$\Delta L/L_0$	T	$\Delta L/L_0$	T	$\Delta L/L_0$
CURVE 1		CURVE 1 (cont.)		CURVE 2 (cont.)		CURVE 5	
0	-0.219	240	-0.063	1073	1.209	303	0.012 *
10	-0.219*	260	-0.040	1173	1.470	373	0.099
20	-0.219	273	-0.024*	1273	1.779		
30	-0.218*	280	-0.016			CURVE 6*	
40	-0.217	293	0.000	CURVE 3			
50	-0.216*	300	0.008			303	0.012
60	-0.213			303	0.013	373	0.097
70	-0.210*	CURVE 2		373	0.101		
80	-0.205					CURVE 7*	
90	-0.200	273	-0.026	CURVE 4*			
100	-0.194	373	0.104			303	0.012
120	-0.180	473	0.238	303	0.012	373	0.097
140	-0.164	573	0.377	373	0.098		
160	-0.146	673	0.521				
180	-0.127	773	0.672				
200	-0.106	873	0.835				
220	-0.085	973	1.013				

* Not shown in figure.

FIGURE AND TABLE NO. 270R. PROVISIONAL VALUES FOR THERMAL LINEAR EXPANSION OF NICKEL + COPPER + ΣX_i ALLOYS   Ni + Cu + ΣX_i

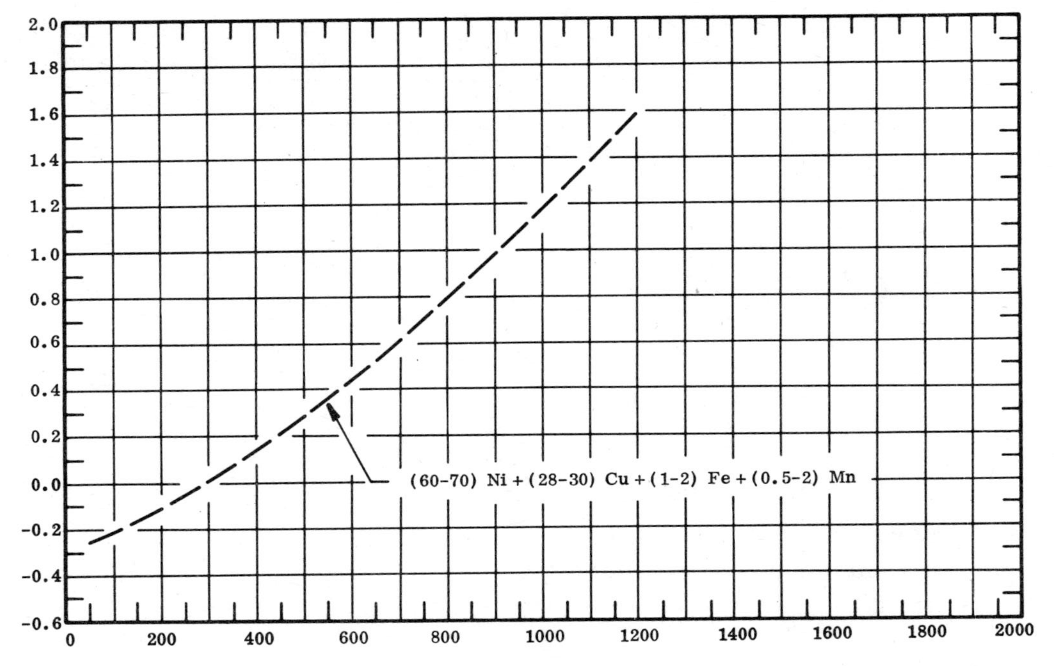

PROVISIONAL VALUES

[Temperature, T, K; Linear Expansion, $\Delta L/L_0$, %; α, K^{-1}]

[(60-70) Ni + (28-30) Cu + (1-2) Fe + (0.5-2) Mn]		
T	$\Delta L/L_0$	$\alpha \times 10^6$
50	-0.264	8.9
100	-0.217	9.8
200	-0.111	11.3
293	0.000	12.7
400	0.144	14.1
500	0.291	15.4
600	0.450	16.4
700	0.620	17.4
800	0.798	18.2
900	0.985	19.0
1000	1.179	19.7
1100	1.378	20.1
1200	1.582	20.6

REMARKS

[(60-70) Ni + (28-30) Cu + (1-2) Fe + (0.5-2) Mn]: The tabulated values are for well-annealed alloy of the type Monel and are considered accurate to within ±7% over the entire temperature range. These values can be represented approximately by the following equation:

$$\Delta L/L_0 = -0.302 + 8.074 \times 10^{-4}\, T + 8.740 \times 10^{-7}\, T^2 - 1.962 \times 10^{-10}\, T^3$$

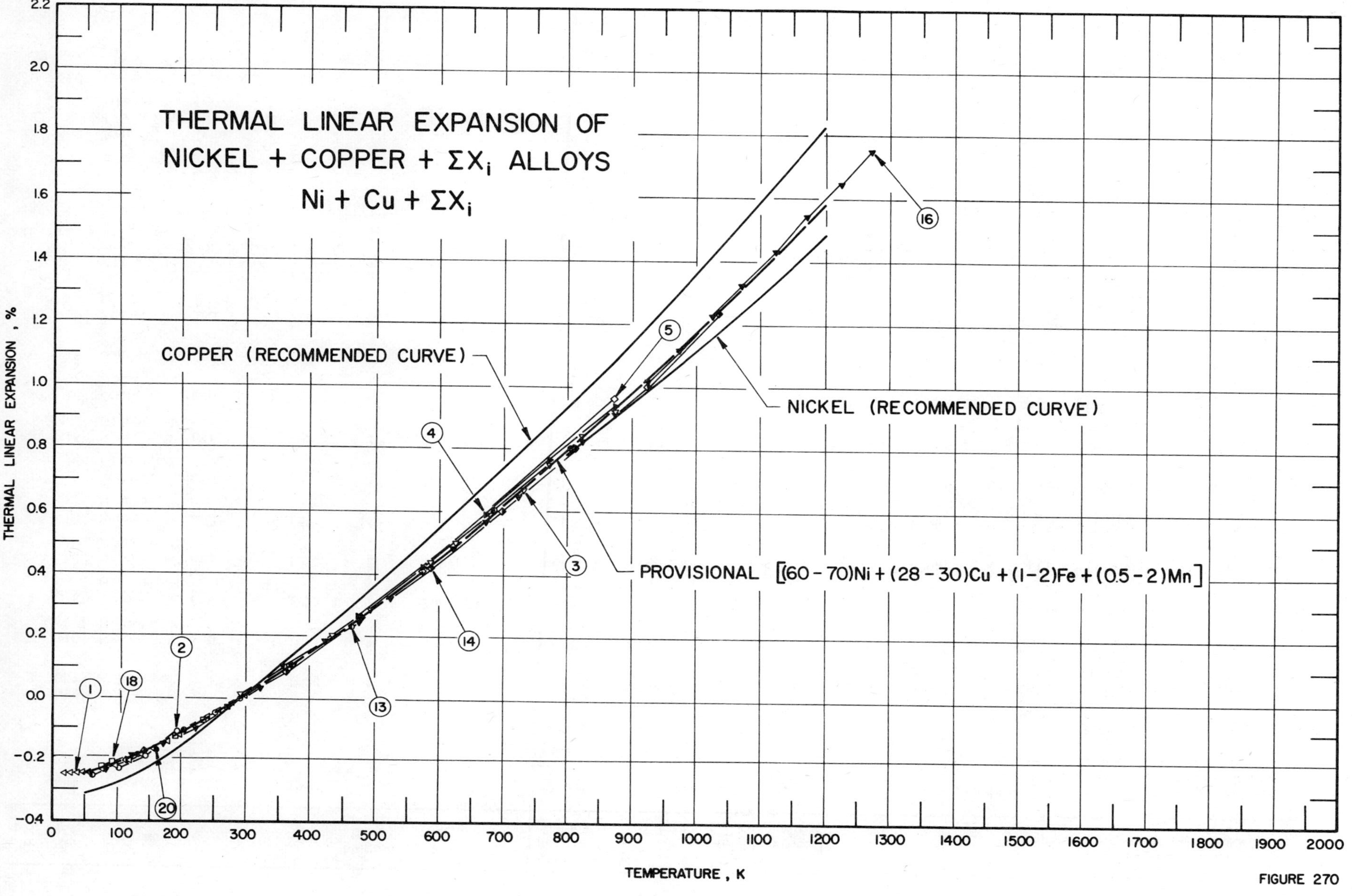
THERMAL LINEAR EXPANSION OF
NICKEL + COPPER + ΣX_i ALLOYS
Ni + Cu + ΣX_i
COPPER (RECOMMENDED CURVE)
NICKEL (RECOMMENDED CURVE)
PROVISIONAL [(60 - 70)Ni + (28 - 30)Cu + (1 - 2)Fe + (0.5 - 2)Mn]
1
18
2
20
13
14
3
4
5
16
THERMAL LINEAR EXPANSION, %
2.2
2.0
1.8
1.6
1.4
1.2
1.0
0.8
0.6
0.4
0.2
0.0
-0.2
-0.4
0
100
200
300
400
500
600
700
800
900
1000
1100
1200
1300
1400
1500
1600
1700
1800
1900
2000
TEMPERATURE, K

FIGURE 270

SPECIFICATION TABLE 270. THERMAL LINEAR EXPANSION OF NICKEL + COPPER + ΣX_i ALLOYS Ni + Cu + ΣX_i

Cur. No.	Ref. No.	Author(s)	Year	Method Used	Temp. Range, K	Name and Specimen Designation	Composition (weight percent) Ni	Cu	Fe	Mn	C	Si	S	Composition (continued), Specifications, and Remarks
1	481	Johnston, H. L., Altman, H. W., and Rubin, T.	1965	I	18-300	Monel	67.5	30	1.5	1.0				Alloy obtained from Williams and Co., Columbus, Ohio; specimen consisted of three sectors cut from bar, filed to length within 0.20 of wave-length of sodium D radiation; cold rolled without further heat treatment; data smoothed by author.
2	501	Ackerman, D. E.	1936	L	140-299	Monel	67.74	29.66	1.3	1.07	0.14	0.06	0.005	Specimen 4 in. long from an annealed 1.5 in. square forging; fused silica dilatometer used.
3	16	Souder, W. and Hidnert, P.	1921	T	298-873	Monel S 641	66.58	29.57	1.79	1.78	0.15	0.09	0.030	Hot rolled wire rod, 0.64 cm round.
4	16	Souder, W. and Hidnert, P.	1921	T	298-873	Monel S 688	66.58	29.57	1.79	1.78	0.15	0.09	0.030	The above specimen annealed 1 hr at 1144 K.
5	16	Souder, W. and Hidnert, P.	1921	T	298-873	Monel S 648	60.05	32.46	2.21	2.00	0.15	0.87	0.035	2.22 Pb; leaded Monel; cast 0.95 cm square.
6*	16	Souder, W. and Hidnert, P.	1921	T	298-873	Monel S 649	60.05	32.46	2.21	2.00	0.15	0.87	0.035	The above specimen; annealed 1 hr at 1171 K.
7*	16	Souder, W. and Hidnert, P.	1921	T	298-873	Monel S 663	66.18	28.42	2.37	2.10	0.18	0.70	0.038	Cast, nonleaded specimen; 0.95 cm square.
8*	16	Souder, W. and Hidnert, P.	1921	T	298-873	Monel S 664	66.18	28.42	2.37	2.10	0.18	0.70	0.038	The above specimen; annealed 1 hr at 1171 K and slowly cooled.
9*	16	Souder, W. and Hidnert, P.	1921	T	298-873	Monel S 650	67.32	28.73	1.74	1.66	0.31	0.19	0.35	Hot rolled, high carbon 0.95 cm rod.
10*	16	Souder, W. and Hidnert, P.	1921	T	298-873	Monel S 651	67.32	28.73	1.74	1.66	0.31	0.19	0.35	The above specimen, annealed 1 hr at 1171 K.
11*	16	Souder, W. and Hidnert, P.	1921	T	298-873	Monel S 661	68.87	29.63	1.60	0.18	0.13	0.15	0.027	Hot rolled sheet 0.31 cm thick x 0.95 cm wide.
12*	16	Souder, W. and Hidnert, P.	1921	T	298-873	Monel S 662	68.87	29.63	1.60	0.18	0.13	0.15	0.027	The above specimen; annealed 1 hr at 1144 K.
13	225	Valentich, J.	1965	L		Monel 400	Bal.	31.5	1.35	0.90	0.12	0.15	0.005	No information given.
14	123	Lucks, C. F. and Deem, H. W.	1958	L	116-1033	Monel K								0.375 in. diameter x 3 in. long; hot rolled specimen from International Nickel Co., Inc; annealed 1 hr at 1172 K; heating rate 2.2 K per minute.
15*	462	Arp, V., Wilson, J. H., Winrich, L., and Sikora, P.	1962	L	20-293	Monel K	Bal.	30.9	1.2	0.4	0.15	0.3	0.01	2.9 Al, 0.5 Ti; hardness R_C35; age hardened.
16	466	Lucks, C. F. and Thompson, H. B.	1951	L	83-1273	Monel K	65.51	29.23	0.86	0.60		0.09		3.0 Al; hot-rolled; annealed 1 hr at 1172.05 K and water quenched; material supplied by International Nickel Co.; density 8.46 g cm^{-3} at 293 K.

* Not shown in figure.

SPECIFICATION TABLE 270. THERMAL LINEAR EXPANSION OF NICKEL + COPPER + ΣX_i ALLOYS $Ni + Cu + \Sigma X_i$ (continued)

Cur. No.	Ref. No.	Author(s)	Year	Method Used	Temp. Range, K	Name and Specimen Designation	Composition (weight percent) Ni	Cu	Fe	Mn	C	Si	S	Composition (continued), Specifications, and Remarks
17*	462	Arp, V., Wilson, J.H., Winrich, L., and Sikora, P.	1962	L	20-293	S Monel	Bal.	27.7	1.8	0.7	0.03	4.1		Hardness R_C25; annealed specimen.
18	142	Aoyama, S. and Ito, T.	1939	L	77-273	Sample No. 1	88.92	10.75	0.3	0.03				Alloy melted, cast into bars of 30 x 30 x 250 mm; hot-forged to 8 mm diameter bar, then cut into round bar 5 mm diameter, 100 mm length; annealed in vacuum for 3 hrs at 1123 K.
19*	142	Aoyama, S. and Ito, T.	1939	L	77-273	Sample No. 2	84.54	14.96	0.45	0.05				Similar to the above specimen.
20	142	Aoyama, S. and Ito, T.	1939	L	77-273	Sample No. 3	78.99	20.64	0.33	0.04				Similar to the above specimen.
21*	142	Aoyama, S. and Ito, T.	1939	L	77-273	Sample No. 4	68.86	30.88	0.21	0.05				Similar to the above specimen.

* Not shown in figure.

DATA TABLE 270. THERMAL LINEAR EXPANSION OF NICKEL + COPPER + ΣX_i ALLOYS Ni + Cu + ΣX_i

[Temperature, T, K; Linear Expansion, $\Delta L/L_0$, %]

T	$\Delta L/L_0$
CURVE 1	
20	-0.251
30	-0.251
40	-0.250
50	-0.248
60	-0.245
70	-0.241*
80	-0.236
90	-0.230*
100	-0.223
110	-0.215*
120	-0.206
130	-0.197*
140	-0.187
150	-0.177*
160	-0.167
170	-0.156*
180	-0.144
190	-0.133*
200	-0.121
210	-0.109*
220	-0.096
230	-0.084*
240	-0.061
250	-0.058*
260	-0.045
270	-0.031*
280	-0.018
290	-0.004*
300	0.010
CURVE 2	
105	-0.229
144	-0.189
199	-0.121
255	-0.049
293	0.000
299	0.008 *
CURVE 3	
296.6	0.0051
333.4	0.0568
375.8	0.1176
438.4	0.2105
CURVE 3 (cont.)	
492.0	0.2924
587.8	0.4426
627.1	0.5074
680.8	0.5970
731.1	0.6724
774.4	0.7586
821.7	0.8421
874.5	0.9385
CURVE 4	
298	0.007*
373	0.114*
473	0.267
573	0.425
673	0.590
773	0.764
873	0.941
CURVE 5	
298	0.007*
373	0.111*
473	0.261*
573	0.419
673	0.587*
773	0.764*
873	0.970
CURVE 6*	
298	0.007
373	0.114
473	0.264
573	0.421
673	0.588
773	0.769
873	0.952
CURVE 7*	
298	0.006
373	0.110
473	0.261
573	0.418
CURVE 7 (cont.)*	
673	0.584
773	0.761
873	0.946
CURVE 8*	
298	0.007
373	0.111
473	0.261
573	0.418
673	0.584
773	0.762
873	0.949
CURVE 9*	
298	0.007
373	0.116
473	0.267
573	0.426
673	0.590
773	0.761
873	0.923
CURVE 10*	
298	0.007
373	0.114
473	0.264
573	0.422
673	0.587
773	0.761
873	0.938
CURVE 11*	
298	0.007
373	0.114
473	0.261
573	0.417
673	0.581
773	0.757
873	0.936
CURVE 12*	
298	0.007
373	0.112
473	0.262
573	0.421
673	0.585
773	0.758
873	0.939
CURVE 13	
66	-0.259
138	-0.185
207	-0.108
293	0.000*
365	0.101
466	0.239
579	0.425*
688	0.613
805	0.804
CURVE 14	
116	-0.206
144	-0.176
199	-0.115*
366	0.097
477	0.261
588	0.429
699	0.609
810	0.805
922	1.015
1033	1.240
CURVE 15*	
20	-0.249
40	-0.248
60	-0.244
80	-0.235
100	-0.221
120	-0.205
140	-0.187
160	-0.167
180	-0.144
CURVE 15 (cont.)*	
200	-0.121
220	-0.096
240	-0.070
260	-0.043
273	-0.026
280	-0.017
293	0.000
CURVE 16	
83	-0.236
123	-0.199
173	-0.149
223	-0.091
323	0.039
373	0.110
423	0.184
473	0.256
523	0.332
573	0.410*
623	0.492
673	0.574
723	0.656
773	0.743*
823	0.831
873	0.922
923	1.019
973	1.120
1023	1.226
1073	1.336
1123	1.439
1173	1.545
1223	1.651
1273	1.759
CURVE 17*	
20	-0.252
40	-0.250
60	-0.246
80	-0.237
100	-0.223
120	-0.207
140	-0.188
160	-0.168
CURVE 17 (cont.)*	
180	-0.146
200	-0.123
220	-0.098
240	-0.072
260	-0.046
273	-0.029
280	-0.018
293	0.000
CURVE 18	
77	-0.225
91	-0.218
133	-0.184*
153	-0.165
173	-0.146*
193	-0.123
233	-0.076
273	-0.026
CURVE 19*	
77	-0.227
91	-0.219
133	-0.185
153	-0.166
173	-0.145
193	-0.123
233	-0.076
273	-0.026
CURVE 20	
77	-0.232*
91	-0.222
133	-0.188*
153	-0.168
173	-0.147*
193	-0.125*
233	-0.077*
273	-0.026*

* Not shown in figure.

DATA TABLE 270. THERMAL LINEAR EXPANSION OF NICKEL + COPPER + ΣX_i ALLOYS Ni + Cu + ΣX_i (continued)

T	$\Delta L/L_0$
CURVE 21*	
77	-0.236
91	-0.228
133	-0.191
153	-0.171
173	-0.149
193	-0.126
233	-0.076
273	-0.026

* Not shown in figure.

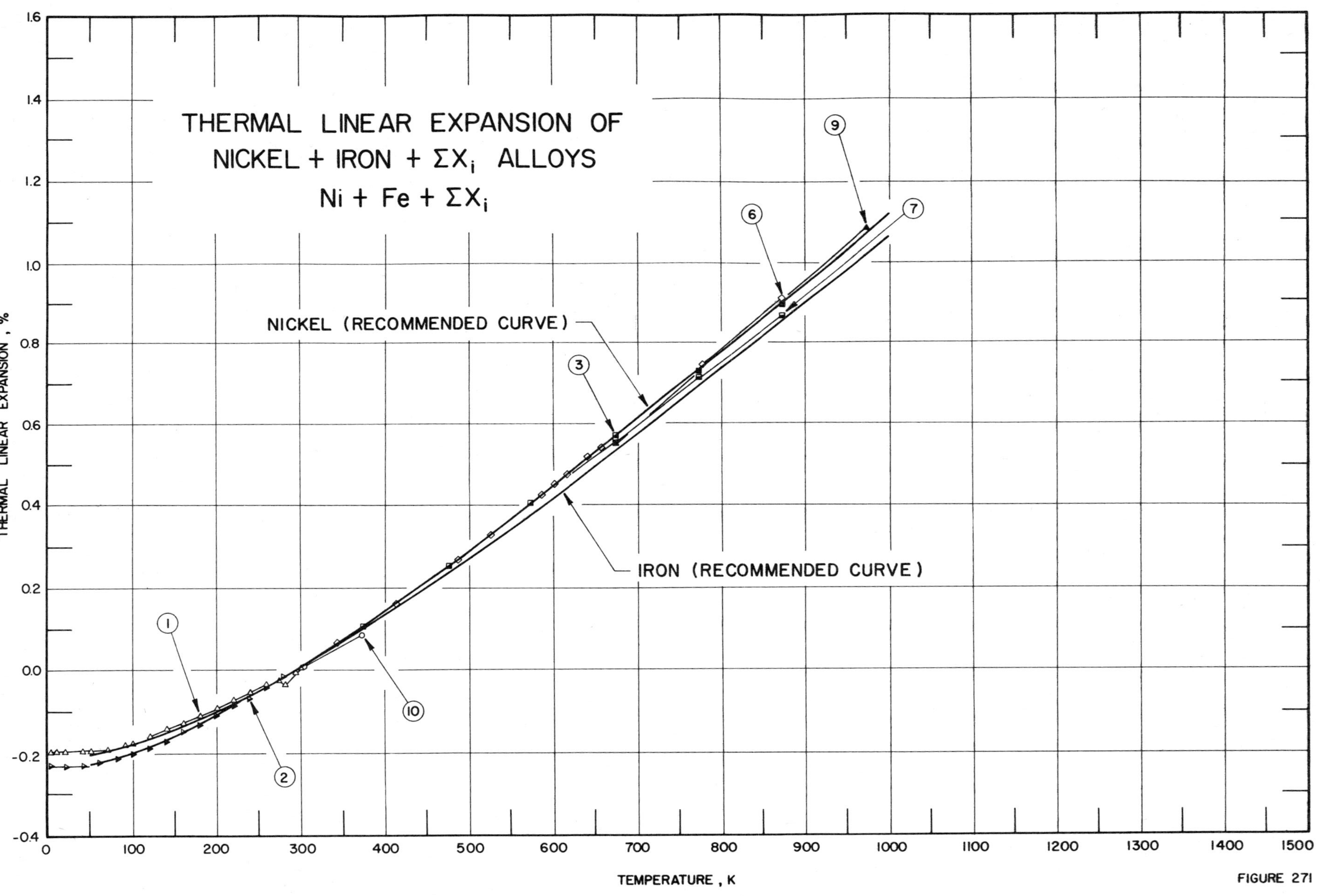
THERMAL LINEAR EXPANSION OF
NICKEL + IRON + ΣX_i ALLOYS
Ni + Fe + ΣX_i
NICKEL (RECOMMENDED CURVE)
IRON (RECOMMENDED CURVE)
1
2
3
6
7
9
10
THERMAL LINEAR EXPANSION, %
1.6
1.4
1.2
1.0
0.8
0.6
0.4
0.2
0.0
-0.2
-0.4
0
100
200
300
400
500
600
700
800
900
1000
1100
1200
1300
1400
1500
TEMPERATURE, K
FIGURE 271

SPECIFICATION TABLE 271. THERMAL LINEAR EXPANSION OF NICKEL + IRON + ΣX_i ALLOYS Ni + Fe + ΣX_i

Cur. No.	Ref. No.	Author(s)	Year	Method Used	Temp. Range, K	Name and Specimen Designation	Composition (weight percent) Ni	Fe	Si	Mn	Cu	C	S	Composition (continued), Specifications, and Remarks
1	467	Clark, A. F.	1968	L	0-300	Fe-50% Ni	50.39	48.93	0.28	0.40		<0.1		<0.1 each Cr, Co, Mo; specimen machined to 0.64 cm sq. x 20.32 cm long; R_H B-47.
2	467	Clark, A. F.	1968	L	0-300	Fe-79% Ni	78.63	20.67	0.19	0.51		<0.1	<0.1	<0.1 each Mo, P; specimen machined to 0.64 cm sq. x 20.32 cm long; R_H B-54.
3	16	Souder, W. and Hidnert, P.	1921	T	298-873	S 635	99.05	0.37	0.12	0.19	0.12	0.12	0.027	Hot rolled, 0.95 cm round specimen of commercial nickel; data on the mean coefficient of thermal linear expansion are also reported.
4*	16	Souder, W. and Hidnert, P.	1921	T	298-873	S 636	99.05	0.37	0.12					The above specimen; annealed 1 hr at 1144 K.
5*	16	Souder, W. and Hidnert, P.	1921	T	298-873	S 652	99.02	0.37	0.16	0.22	0.12	0.08	0.020	Hot rolled, low carbon 0.95 cm round specimen of commercial nickel.
6	16	Souder, W. and Hidnert, P.	1921	T	295-875	S 653	99.02	0.37	0.16					The above specimen; annealed 1 hr at 1171 K; zero-point correction of 0.0026% determined by graphical extrapolation.
7	16	Souder, W. and Hidnert, P.	1921	T	298-873	S 654	98.76	0.38	0.26	0.18	0.17	0.22	0.020	Hot rolled, 0.95 cm round specimen of high carbon commercial nickel.
8*	16	Souder, W. and Hidnert, P.	1921	T	298-873	S 655	98.76	0.38	0.26					The above specimen; annealed 1 hr at 1171 K.
9	427	Souder, W., Hidnert, P., and Fox, J. F.	1934	N	373-973	1403 A	63.0	18.1	0.78	2.14		0.08		15.9 Cr, 0.013 P; 150 mm long with pointed ends; max. diameter 12 mm; hot-rolled.
10	59	Masumoto, H.	1931	L	303-373	No. 10	50	40.0						10.0 Co, Mond Ni, Armco Fe, granular Co, mixed and melted in hydrogen atmosphere in alumina crucible in Tammann furnace; cast in iron mold with cylindrical aperture, machined to circular rod 4 mm thick, 10 cm long, heated at 1000 K for 1 hr in electric furnace in hydrogen atmosphere at reduced pressure; expansion measured in vacuum.
11*	59	Masumoto, H.	1931	L	303-373	No. 11	60	30.0						10.0 Co; similar to the above specimen.
12*	59	Masumoto, H.	1931	L	303-373	No. 12	70	20.0						10.0 Co; similar to the above specimen.
13*	59	Masumoto, H.	1931	L	303-373	No. 13	80	10.0						10.0 Co; similar to the above specimen.
14*	59	Masumoto, H.	1931	L	303-373	No. 22	50	30.1						19.9 Co; similar to the above specimen.
15*	59	Masumoto, H.	1931	L	303-373	No. 23	60	20.1						19.9 Co; similar to the above specimen.
16*	59	Masumoto, H.	1931	L	303-373	No. 33	40	30.1						29.9 Co; similar to the above specimen.
17*	662	Ibragimov, E. A., Selisskiy, Ya. P., and Sorokin, M. N.	1966	X	526-867		Bal.	17.05		6.35		<0.005	<0.002	0.15-0.3 Si, <0.002 P; specimen prepared in induction furnace from electrolytic Ni and Mn, and commercially pure Fe; 1 mm diameter wire specimen; prior to measurements specimens were subjected to various types of heat treatment and finally cold drawn and annealed at 723 K between 15 min to 256 hr; phase transition at 792 K; this curve is here reported using the first given temperature as reference temperature at which $\Delta L/L_0 = 0$.

* Not shown in figure.

SPECIFICATION TABLE 271. THERMAL LINEAR EXPANSION OF NICKEL + IRON + ΣX_i ALLOYS Ni + Fe + ΣX_i (continued)

Cur. No.	Ref. No.	Author(s)	Year	Method Used	Temp. Range, K	Name and Specimen Designation	Ni	Fe	Composition (weight percent) Si	Mn	Cu	C	S	Composition (continued), Specifications, and Remarks
18*	662	Ibragimov, E.A., Selisskiy, Ya. P., and Sorokin, M. N.	1966	X	525-898		Bal.	15.85		8.1		<0.005	<0.002	Similar to the above specimen; phase transition at 789 K; this curve is here reported using the first given temperature as reference temperature at which $\Delta L/L_0 = 0$.
19*	662	Ibragimov, E.A., et al.	1966	X	524-865		Bal.	12.9		10.25		<0.005	<0.002	Similar to the above specimen; phase transition at 801 K; this curve is here reported using the first given temperature as reference temperature at which $\Delta L/L_0 = 0$.
20*	41	White, G.K.	1965	E	4-85	Nilo 50	51.1	47.8		0.4	0.2			0.4 Co; from Henry Wiggin Ltd.; specimen 2 cm in diameter and 5 cm long; measured in the as-received condition; this curve is here reported using the first given temperature as reference temperature at which $\Delta L/L_0 = 0$.

* Not shown in figure.

DATA TABLE 271. THERMAL LINEAR EXPANSION OF NICKEL + IRON + ΣX_i ALLOYS Ni + Fe + ΣX_i

[Temperature, T, K; Linear Expansion, $\Delta L/L_0$, %]

T	$\Delta L/L_0$
CURVE 1	
0	-0.196
10	-0.196
20	-0.196
30	-0.195*
40	-0.195
50	-0.193
60	-0.191*
70	-0.187
80	-0.183*
90	-0.179
100	-0.172
120	-0.159
140	-0.144
160	-0.127
180	-0.110
200	-0.092
220	-0.073
240	-0.053
260	-0.034
273	-0.021
280	-0.031
293	0.000
300	0.007
CURVE 2	
0	-0.226
10	-0.226*
20	-0.225
30	-0.225*
40	-0.224
50	-0.222*
60	-0.220
70	-0.216*
80	-0.211
90	-0.206*
100	-0.199
120	-0.184
140	-0.167
160	-0.148
180	-0.128
200	-0.107
220	-0.085
240	-0.063
260	-0.039
273	-0.024*
280	-0.016
CURVE 2 (cont.)	
293	0.000*
300	0.009*
CURVE 3	
298	0.007*
373	0.106
473	0.250
573	0.403
673	0.571
773	0.726
873	0.895
CURVE 4*	
298	0.007
373	0.106
473	0.251
573	0.405
673	0.569
773	0.735
873	0.898
CURVE 5*	
298	0.007
373	0.103
473	0.248
573	0.402
673	0.571
773	0.736
873	0.904
CURVE 6	
295.0	0.0026*
342.8	0.0654
412.1	0.1617
485.4	0.2693
528.0	0.3353
574.0	0.4073*
585.2	0.4265
600.0	0.4517
614.7	0.4782
621.3	0.4884*
640.9	0.5206
656.6	0.5456
CURVE 6 (cont.)	
661.2	0.5535*
777.6	0.7436
875.4	0.9089
CURVE 7	
298	0.006*
373	0.104
473	0.246
573	0.403
673	0.562
773	0.719
873	0.863
CURVE 8*	
298	0.006
373	0.106
473	0.248
573	0.404
673	0.567
773	0.729
873	0.892
CURVE 9	
373	0.107*
473	0.257*
573	0.406*
673	0.559
773	0.720*
873	0.905*
973	1.081
CURVE 10	
303	0.0111
373	0.0887
CURVE 11*	
303	0.012
373	0.096
CURVE 12*	
303	0.012
373	0.102
CURVE 13*	
303	0.012
373	0.102
CURVE 14*	
303	0.012
373	0.097
CURVE 15*	
303	0.012
373	0.100
CURVE 16*	
303	0.012
373	0.095
CURVE 17*, †, ‡	
526	0.000
757	0.359
792	0.427
853	0.542
911	0.635
CURVE 18*, †, ‡	
525	0.000
732	0.330
789	0.462
843	0.583
898	0.670
CURVE 19*, †, ‡	
524	0.000
748	0.395
801	0.568
823	0.637
865	0.712
CURVE 20*, †, ‡	
4	0.0000 x 10^{-2}
5	0.0001
6	0.0001
CURVE 20 (cont.)*, †, ‡	
8	0.0003 x 10^{-2}
10	0.0004
12	0.0004
14	0.0005
16	0.0009
18	0.0017
20	0.0030
22	0.0050
24	0.0078
26	0.0116
28	0.0171
30	0.0252
65	0.7042
75	1.0892
85	1.5667
283	17.1592

* Not shown in figure.
† This curve is here reported using the first given temperature as reference temperature at which $\Delta L/L_0 = 0$.
‡ Author's data for coefficient of thermal expansion have been integrated by TPRC to obtain $\Delta L/L_0$.

DATA TABLE 271. COEFFICIENT OF THERMAL LINEAR EXPANSION OF NICKEL + IRON + ΣX_i ALLOYS   Ni + Fe + ΣX_i

[Temperature, T, K; Coefficient of Expansion, α, 10^{-6} K^{-1}]

T	α
CURVE 17*, ‡	
526	13.70
757	17.40
792	21.70
853	16.00
911	16.00
CURVE 18*, ‡	
525	14.00
732	17.90
789	28.70
843	15.90
898	15.90
CURVE 19*, ‡	
524	14.80
748	20.50
801	44.70
823	18.10
865	17.80
CURVE 20*, ‡	
4	0.006
5	0.007
6	0.007
8	0.007
10	0.004
12	0.000
14	0.010
16	0.030
18	0.050
20	0.080
22	0.120
24	0.160
26	0.220
28	0.330
30	0.480
65	3.400
75	4.300
85	5.250
283	10.500

* Not shown in figure.
‡ Author's data on coefficient of thermal expansion have been integrated by TPRC to obtain $\Delta L/L_0$.

FIGURE AND TABLE NO. 272R. PROVISIONAL VALUES FOR THERMAL LINEAR EXPANSION OF NICKEL + MANGANESE + ΣX_i ALLOYS   Ni + Mn + ΣX_i

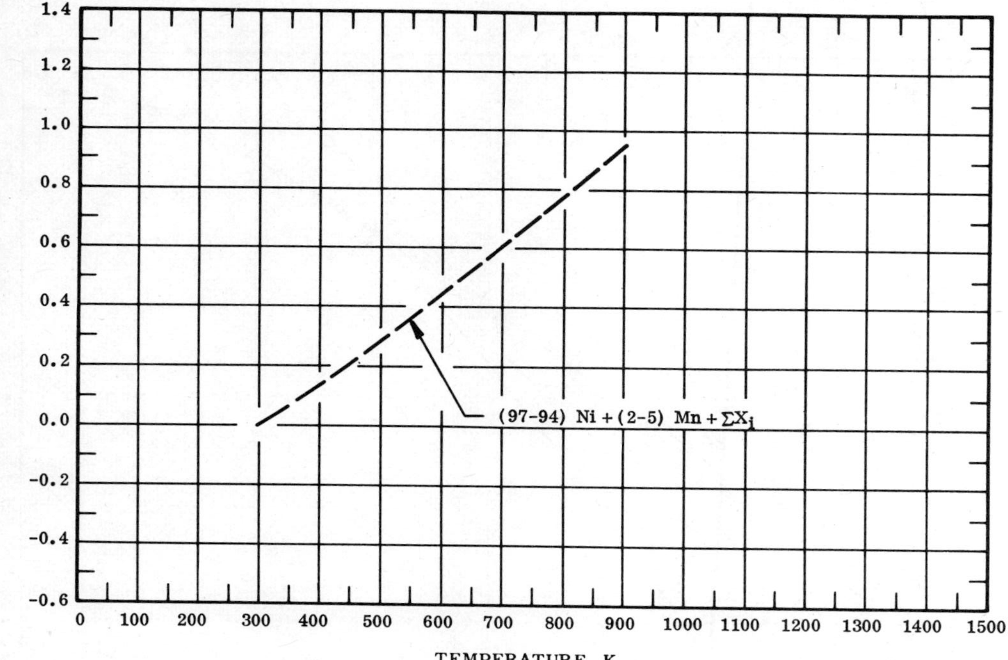

PROVISIONAL VALUES

[Temperature, T, K; Linear Expansion, $\Delta L/L_0$, %; α, K^{-1}]

[(97-94) Ni + (2-5) Mn + ΣX_i]		
T	$\Delta L/L_0$	$\alpha \times 10^6$
293	0.000	12.3
400	0.141	14.1
500	0.288	15.3
600	0.446	16.2
700	0.612	17.0
800	0.784	17.4
900	0.959	17.6

REMARKS

[(97-94) Ni + (2-5) Mn + ΣX_i] : The tabulated values are for well-annealed alloy and are considered accurate to within ±7% over the entire temperature range. The thermal linear expansion for this alloy is very close to that of pure nickel. These values can be represented approximately by the following equation:

$$\Delta L/L_0 = -0.282 + 6.454 \times 10^{-4} T + 1.208 \times 10^{-6} T^2 - 4.367 \times 10^{-10} T^3$$

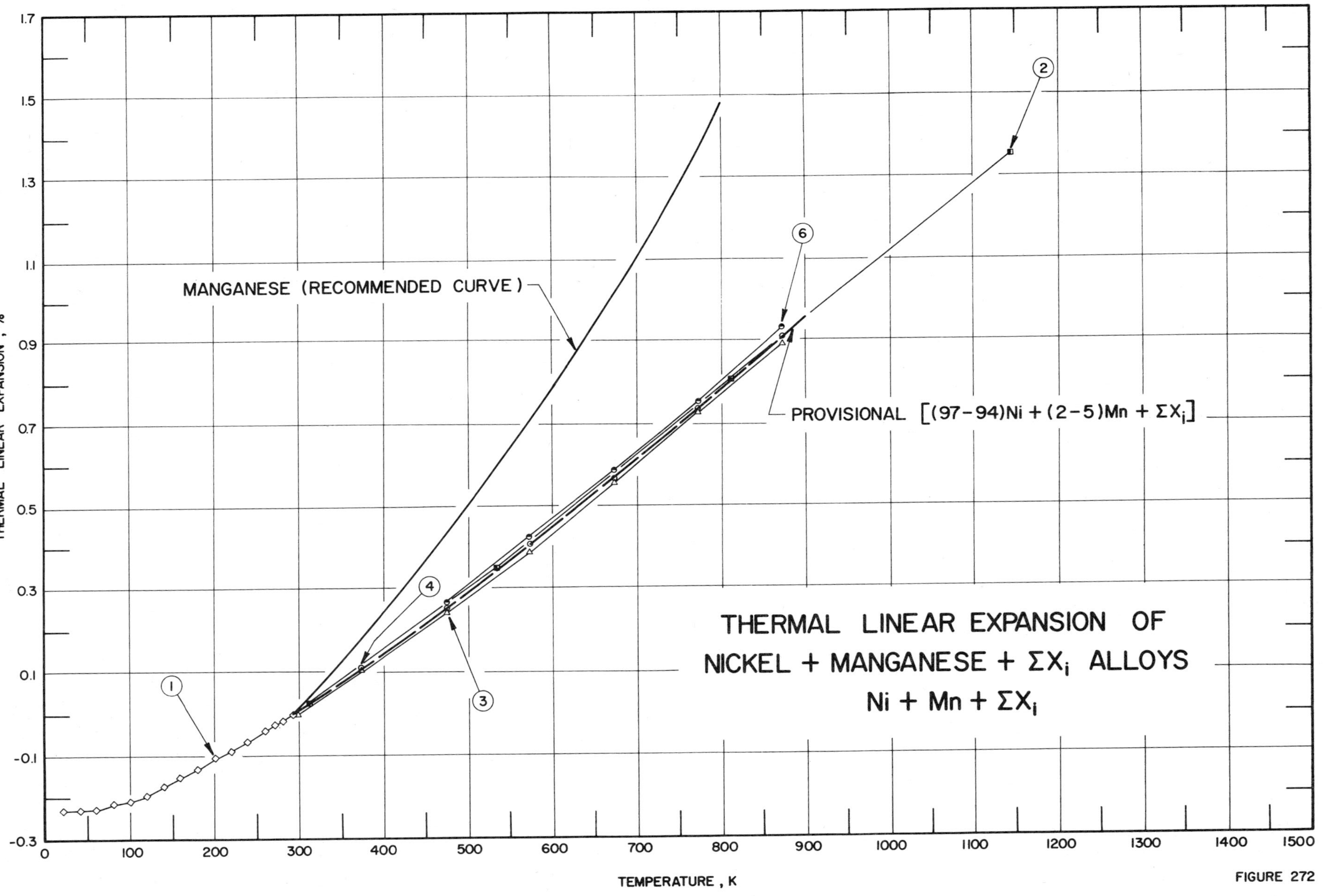

THERMAL LINEAR EXPANSION OF
NICKEL + MANGANESE + ΣX_i ALLOYS
Ni + Mn + ΣX_i
MANGANESE (RECOMMENDED CURVE)
PROVISIONAL [(97-94)Ni + (2-5)Mn + ΣX_i]
1
2
3
4
6
THERMAL LINEAR EXPANSION, %
1.7
1.5
1.3
1.1
0.9
0.7
0.5
0.3
0.1
-0.1
-0.3
0
100
200
300
400
500
600
700
800
900
1000
1100
1200
1300
1400
1500
TEMPERATURE, K

FIGURE 272

SPECIFICATION TABLE 272. THERMAL LINEAR EXPANSION OF NICKEL + MANGANESE + ΣX_i ALLOYS Ni + Mn + ΣX_i

Cur. No.	Ref. No.	Author(s)	Year	Method Used	Temp. Range, K	Name and Specimen Designation	Composition (weight percent) Ni	Mn	Zr	Fe	Cu	Si	C	Composition (continued), Specifications, and Remarks
1	462	Arp, V., Wilson, J.H., Winrich, L., and Sikora, P.	1962	L	60-293	Nickel A	Bal.	0.3		0.1			0.06	0.01 S; hardness R_B, annealed.
2	502	Skinner, E.N.	1951		311-1143	Y-4839	99.02	0.30	0.27	0.08	0.02	0.11	0.01	0.066 Al, 0.056 Ti, 0.038 Mg, 0.005 S; specimen cut from 5000 lb induction heat of material.
3	16	Souder, W. and Hidnert, P.	1921	T	298-873	S 637	97.05	2.08		0.44	0.15	0.15	0.09	0.029 S; hot rolled; 0.64 cm round rod specimen of commercial nickel.
4	16	Souder, W. and Hidnert, P.	1921	T	298-873	S 638	97.05	2.08		0.44				The above specimen; annealed 1 hr at 1144 K.
5*	16	Souder, W. and Hidnert, P.	1921	T	298-873	S 639	94.21	4.92		0.46	0.14	0.10	0.12	0.030 S; similar to the above specimen; unannealed.
6	16	Souder, W. and Hidnert, P.	1921	T	298-873	S 640	94.21	4.92		0.46				The above specimen; annealed 1 hr at 1144 K.
7*	662	Ibragimov, E.A., Selisskiy, Ya.P., and Sorokin, M.N.	1966	X	518-897		Bal.	13.4		9.4		0.15-0.3	<0.005	<0.002 S, P, wire specimen of 1 mm diameter prepared in induction furnace from electrolytic Ni, Mn, and commercially pure Fe; prior to measurements specimen underwent series of heat treatment and finally cold drawn and annealed at 450 C between 15 min and 256 hr; phase transition at 788 K; this curve is here reported using the first given temperature as reference temperature at which $\Delta L/L_0 = 0$.
8*	662	Ibragimov, E.A., et al.	1966	X	531-867		Bal.	15.3		7.4		0.15-0.3	<0.005	Similar to the above specimen; phase transition at 777 K; this curve is here reported using the first given temperature as reference temperature at which $\Delta L/L_0 = 0$.
9*	662	Ibragimov, E.A., et al.	1966	X	530-905		Bal.	17.0		5.15		0.15-0.3	<0.005	Similar to the above specimen; phase transition at 771 K; this curve is here reported using the first given temperature as reference temperature at which $\Delta L/L_0 = 0$.
10*	662	Ibragimov, E.A., et al.	1966	X	531-841		Bal.	23.5		0.30		0.15-0.3	<0.005	Similar to the above specimen; phase transition at 751 K; this curve is here reported using the first given temperature as reference temperature at which $\Delta L/L_0 = 0$.

* Not shown in figure.

DATA TABLE 272. THERMAL LINEAR EXPANSION OF NICKEL + MANGANESE + ΣX_i ALLOYS Ni + Mn + ΣX_i

[Temperature, T, K; Linear Expansion, $\Delta L/L_0$, %]

T	$\Delta L/L_0$
CURVE 1	
20	-0.230
40	-0.228
60	-0.223
80	-0.215
100	-0.203
120	-0.188
140	-0.170
160	-0.149
180	-0.129
200	-0.107
220	-0.085
240	-0.063
260	-0.039
273	-0.025
280	-0.016
293	0.000
CURVE 2	
311	0.026
CURVE 2 (cont.)	
533	0.348
811	0.803
1143	1.360
CURVE 3	
298	0.006
373	0.105
473	0.240
573	0.388
673	0.560
773	0.726
873	0.897
CURVE 4	
298	0.069*
373	0.108
473	0.251
573	0.409
CURVE 4 (cont.)	
673	0.569
773	0.735
873	0.908
CURVE 5*	
298	0.007
373	0.105
473	0.246
573	0.400
673	0.559
773	0.726
873	0.899
CURVE 6	
298	0.007*
373	0.112*
473	0.262
573	0.421
CURVE 6 (cont.)	
673	0.585
773	0.758
873	0.939
CURVE 7*, †, ‡	
518	0.000
713	0.329
788	0.530
821	0.615
897	0.746
CURVE 8*, †, ‡	
531	0.000
746	0.390
777	0.475
811	0.565
867	0.671
CURVE 9*, †, ‡	
530	0.000
733	0.358
771	0.458
813	0.567
905	0.738
CURVE 10*, †, ‡	
531	0.000
689	0.311
751	0.467
791	0.563
841	0.662

DATA TABLE 272. COEFFICIENT OF THERMAL LINEAR EXPANSION OF NICKEL + MANGANESE + ΣX_i ALLOYS Ni + Mn + ΣX_i

[Temperature, T, K; Coefficient of Expansion, α, 10^{-6} K^{-1}]

T	α
CURVE 7*, ‡	
518	14.80
713	19.00
788	34.70
821	16.50
897	18.10
CURVE 8*, ‡	
531	15.10
746	21.20
777	33.70
811	19.30
867	18.50
CURVE 9*, ‡	
530	15.20
733	20.10
771	32.90
813	18.80
905	18.30
CURVE 10*, ‡	
531	17.70
689	21.70
751	28.60
791	19.50
841	20.10

* Not shown in figure.
† This curve is here reported using the first given temperature as reference temperature at which $\Delta L/L_0 = 0$.
‡ Author's data on coefficient of thermal expansion have been integrated by TPRC to obtain $\Delta L/L_0$.

FIGURE AND TABLE NO. 273R. PROVISIONAL VALUES FOR THERMAL LINEAR EXPANSION OF NICKEL + MOLYBDENUM + ΣX_i ALLOYS Ni + Mo + ΣX_i

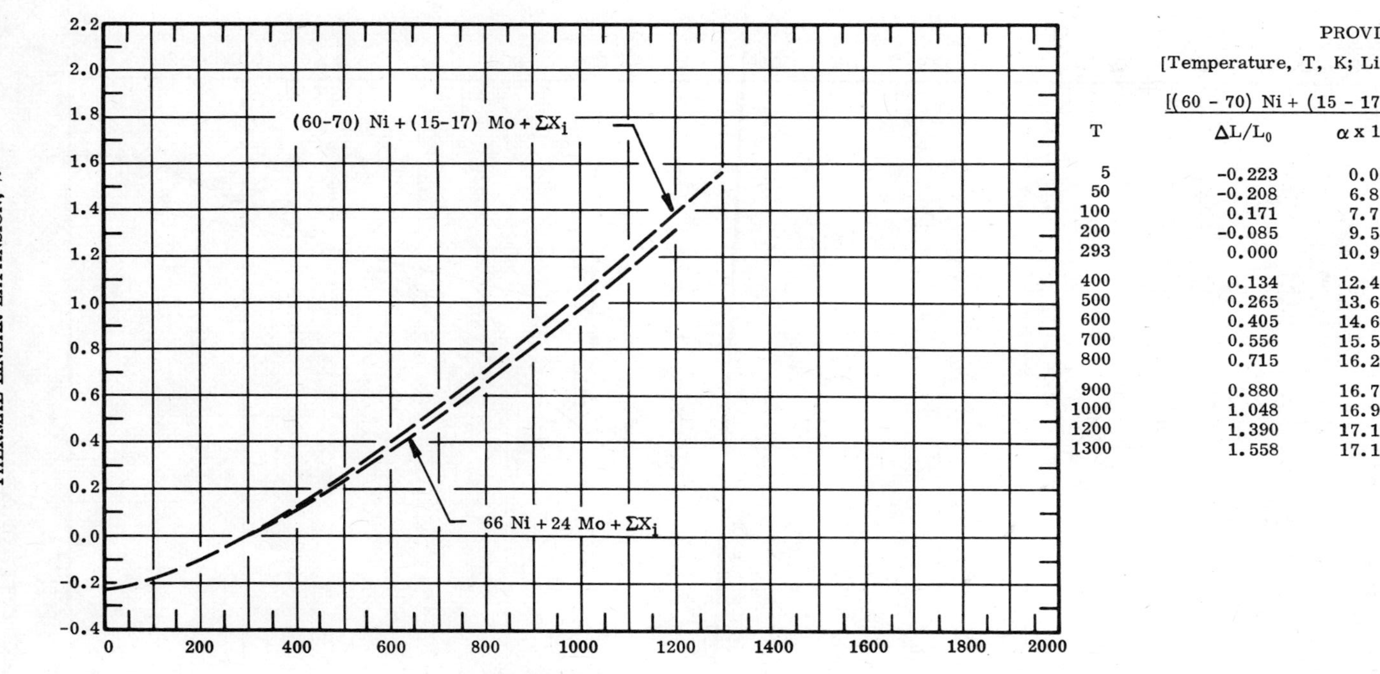

PROVISIONAL VALUES

[Temperature, T, K; Linear Expansion, $\Delta L/L_0$, %; α, K^{-1}]

T	[(60 - 70) Ni + (15 - 17) Mo + ΣX_i] $\Delta L/L_0$	$\alpha \times 10^6$	(66 Ni + 24 Mo + ΣX_i) $\Delta L/L_0$	$\alpha \times 10^6$
5	-0.223	0.05		
50	-0.208	6.8		
100	0.171	7.7		
200	-0.085	9.5		
293	0.000	10.9	0.000	9.3
400	0.134	12.4	0.109	11.1
500	0.265	13.6	0.228	12.6
600	0.405	14.6	0.359	13.9
700	0.556	15.5	0.505	14.9
800	0.715	16.2	0.658	15.7
900	0.880	16.7	0.817	16.2
1000	1.048	16.9	0.979	16.5
1200	1.390	17.1	1.310	16.5
1300	1.558	17.1		

REMARKS

[(60 - 70) Ni + (15 - 17) Mo + ΣX_i]: The tabulated values for well-annealed alloy of type Hastellay C and Hastelloy N are considered accurate to within ± 7% over the entire temperature range. These values can be represented approximately by the following equation:

$$\Delta L/L_0 = -0.244 + 5.800 \times 10^{-4}\, T + 1.004 \times 10^{-6}\, T^2 - 2.969 \times 10^{-10}\, T^3$$

(66 Ni + 24 Mo + ΣX_i): The tabulated values for well-annealed alloy of type Hastellay B are considered accurate to within ± 7% over the entire temperature range. These values can be represented approximately by the following equation:

$$\Delta L/L_0 = -0.184 + 2.743 \times 10^{-4}\, T + 1.292 \times 10^{-6}\, T^2 - 4.044 \times 10^{-10}\, T^3$$

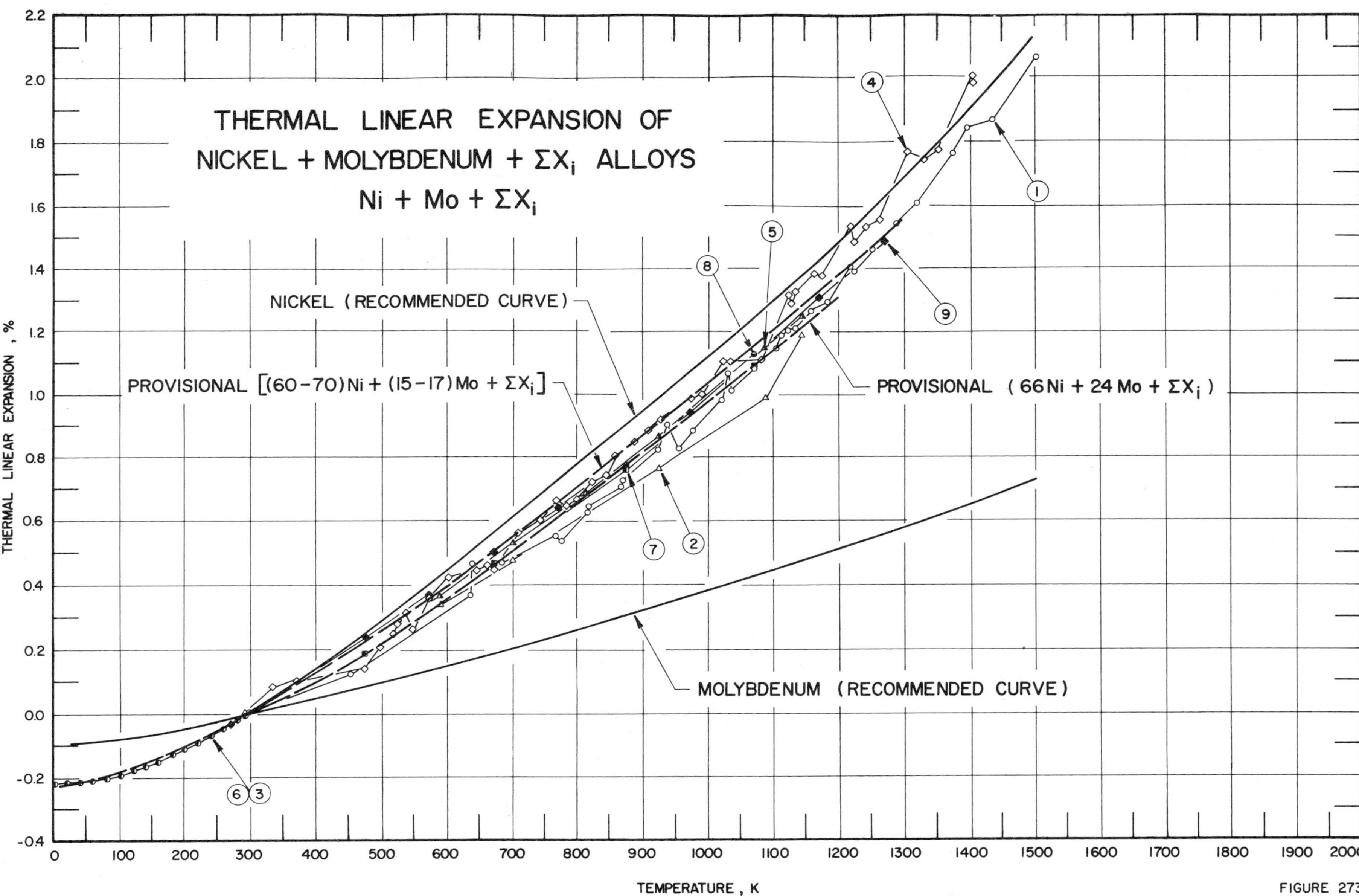
THERMAL LINEAR EXPANSION OF
NICKEL + MOLYBDENUM + ΣX$_i$ ALLOYS
Ni + Mo + ΣX$_i$
NICKEL (RECOMMENDED CURVE)
PROVISIONAL [(60-70) Ni + (15-17) Mo + ΣX$_i$]
PROVISIONAL (66 Ni + 24 Mo + ΣX$_i$)
MOLYBDENUM (RECOMMENDED CURVE)
1
2
3
4
5
6
7
8
9
THERMAL LINEAR EXPANSION, %
2.2
2.0
1.8
1.6
1.4
1.2
1.0
0.8
0.6
0.4
0.2
0.0
-0.2
-0.4
TEMPERATURE, K
0
100
200
300
400
500
600
700
800
900
1000
1100
1200
1300
1400
1500
1600
1700
1800
1900
2000

FIGURE 273

SPECIFICATION TABLE 273. THERMAL LINEAR EXPANSION OF NICKEL + MOLYBDENUM + ΣX_i ALLOYS Ni + Mo + ΣX_i

Cur. No.	Ref. No.	Author(s)	Year	Method Used	Temp. Range, K	Name and Specimen Designation	Composition (weight percent) Ni	Mo	Fe	Cr	Mn	Si	Composition (continued), Specifications, and Remarks
1	472	Fieldhouse, I. B., Hedge, J. C., Lang, J. I. and Waterman, T. E.	1958	T	300-1502	Hastelloy B	65. 56	23. 89	5. 00				0. 022 C; no details given.
2	473	Sweeney, W. O.	1947		293-1144	Hastelloy B	62. 5/ 66. 5	26/ 30	4/ 6				0. 02-0. 12 C; density 9. 24 g cm^{-3}, wrought specimen.
3	467	Clark, A. F.	1968	L	0-300	Hastelloy C	58. 42	15. 91	5. 07	15. 35	0. 45	0. 76	3. 44 W, 0. 35 Co, 0. 25 V, trace (<0. 1) each C, P, S; machined to 0. 64 cm square x 20. 32 cm long; Rockwell hardness C-22.
4	472	Fieldhouse, I. B., Hedge, J. C., Lang, J. I. and Waterman, T. E.	1958	T	300-1405	Hastelloy C	56. 04	14. 55	4. 99	15. 82			4. 45 W, 0. 069 C; zero-point correction of 0. 016% determined by graphical extrapolation.
5	473	Sweeney, W. O.	1947		293-1144	Hastelloy C	Bal.	15/ 18	4. 5/ 7	13/ 15. 5			3. 75-5. 25 W, 0. 20-0. 40 V, 0. 04-0. 15 C; density 8. 94 g cm^{-3}.
6*	467	Clark, A. F.	1968	L	0-300	Hastelloy N	70. 33	16. 02	4. 54	7. 75	0. 47	0. 50	0. 28 V, 0. 11 Co, trace (<0. 1) each C, W, Cu, S, B, P; machined to 0. 64 cm square x 20. 32 cm long; Rockwell hardness B-90.
7	373	Watrous, J. D.	1961	L	293-1073	Hastelloy N							Rod purchased from Haynes-Stellite Co., Kokomo, Indiana; reactor cladding material; heating rate 3 C per minute; run 1.
8	373	Watrous, J. D.	1961	L	293-1073	Hastelloy N							The above specimen; run 2.
9	349	Fulkerson, S. D.	1959	L	273-1273		70. 82	16. 90	4. 21	6. 86	0. 84	0. 23	0. 14 C; zero-point correction of -0. 028% determined by graphical interpolation.

* Not shown in figure.

DATA TABLE 273. THERMAL LINEAR EXPANSION OF NICKEL + MOLYBDENUM + ΣX_i ALLOYS Ni + Mo + ΣX_i

[Temperature, T, K; Linear Expansion, $\Delta L/L_0$, %]

T	$\Delta L/L_0$
CURVE 1	
300	0.006
451	0.132
638	0.375
640	0.468
681	0.479
768	0.559
775	0.540
816	0.627
819	0.642
869	0.718
870	0.728
924	0.825
939	0.916
955	0.830
975	0.893
1020	0.986
1022	1.068*
1030	1.068
1033	1.016
1072	1.081
1105	1.151
1111	1.194
1124	1.202
1135	1.215
1160	1.267
1163	1.312*
1182	1.300
1217	1.404
1222	1.393
1252	1.464
1260	1.469*
1288	1.547
1320	1.613
1377	1.767
1395	1.841
1438	1.876
1502	2.069
CURVE 2	
293	0.000
589	0.341
700	0.481
811	0.621
922	0.762
1089	0.997
1144	1.192
CURVE 3	
0	-0.218
10	-0.218*
20	-0.218
30	-0.217*
40	-0.216
50	-0.214*
60	-0.211
70	-0.208*
80	-0.204
90	-0.199*
100	-0.193
120	-0.179
140	-0.163
160	-0.145
180	-0.125
200	-0.105
220	-0.084
240	-0.063
260	-0.040
273	-0.025*
280	-0.017
293	0.000
300	0.009*
CURVE 4	
300	0.016*
334	0.096
473	0.143
497	0.213
518	0.257
523	0.286
539	0.320
549	0.265
572	0.365
604	0.425
648	0.449
663	0.466
673	0.454
709	0.566
743	0.604
769	0.662
782	0.650
800	0.672
822	0.726
844	0.743
857	0.801
CURVE 4 (cont.)	
887	0.859
908	0.896
928	0.931
972	0.994
990	1.05
1023	1.12
1035	1.11
1040	1.12*
1084	1.19
1125	1.32
1128	1.29
1137	1.33
1167	1.39
1177	1.38
1218	1.54
1223	1.49
1241	1.54
1266	1.56
1308	1.77
1330	1.75
1357	1.78
1405	2.01
1405	1.96
CURVE 5	
293	0.000*
589	0.373
700	0.538
811	0.693
922	0.875
1089	1.156
1144	1.256
CURVE 6*	
0	-0.219
10	-0.219
20	-0.219
30	-0.218
40	-0.217
50	-0.215
60	-0.212
70	-0.208
80	-0.204
90	-0.199
100	-0.193
CURVE 6 (cont.)*	
120	-0.179
140	-0.162
160	-0.145
180	-0.126
200	-0.106
220	-0.085
240	-0.064
260	-0.041
273	-0.026
280	-0.017
293	0.000
300	0.010
CURVE 7	
293	0.000*
472	0.196
673	0.470
873	0.761
1073	1.097
CURVE 8	
293	0.000*
472	0.196*
673	0.470*
873	0.777
1073	1.133
CURVE 9	
273	-0.028
373	0.110
473	0.242
573	0.376
673	0.509
773	0.644
873	0.779*
973	0.950
1073	1.129*
1173	1.311
1273	1.491

* Not shown in figure.

SPECIFICATION TABLE 274. THERMAL LINEAR EXPANSION OF NICKEL + SILICON + ΣX_i ALLOYS Ni + Si + ΣX_i

Cur. No.	Ref. No.	Author(s)	Year	Method Used	Temp. Range, K	Name and Specimen Designation	Composition (weight percent) Ni	Si	Fe	Mn	Cu	C	Composition (continued), Specifications, and Remarks
1*	347	Neel, D.S., Pears, C.D. and Oglesby, S. Jr.	1962		293-1339	Grade "A"	98.18	0.5	0.5	0.35	0.25	0.2	0.02 S; J. M. Tully Metal and Supply Co.; cold rolled from melt; expansion given in increasing temperatures; density 546 lb ft^{-3} by volume displacement; measured in argon.
2*	347	Neel, D. S., et al.	1962		1339-760	Grade "A"							Similar to the above specimen except expansion measured with decreasing temperature.
3*	347	Neel, D. S., et al.	1962		760-1377	Grade "A"							Similar to the above specimen except expansion measured with increasing temperature in second thermal cycle.
4*	347	Neel, D. S., et al.	1962		1377-486	Grade "A"							Similar to the above specimen except expansion measured with decreasing temperature in second thermal cycle.
5*	347	Neel, D. S., et al.	1962		486-1422	Grade "A"							Similar to the above specimen except expansion measured with increasing temperature in third thermal cycle.
6*	347	Neel, D. S., et al.	1962		1422-293	Grade "A"							Similar to the above specimen except expansion measured with decreasing temperature in third thermal cycle.

DATA TABLE 274. THERMAL LINEAR EXPANSION OF NICKEL + SILICON + ΣX_i ALLOYS Ni + Si + ΣX_i

[Temperature, T, K; Linear Expansion, $\Delta L/L_0$, %]

T	$\Delta L/L_0$
CURVE 1*	
293	0.00
370	0.05
555	0.23
732	0.50
761	0.52
1007	0.91
1254	1.35
1339	1.50
CURVE 2*	
1339	1.50
760	0.46
CURVE 3*	
760	0.46
968	0.78
1377	1.50
CURVE 4*	
1377	1.50
486	0.12
CURVE 5*	
486	0.12
629	0.24
849	0.55
1093	0.96
1349	1.40
1422	1.51
CURVE 6*	
1422	1.58
293	0.00

* No figure given.

FIGURE AND TABLE NO. 275R. PROVISIONAL VALUES FOR THERMAL LINEAR EXPANSION OF NIOBIUM + MOLYBDENUM + ΣX_i ALLOYS Nb + Mo + ΣX_i

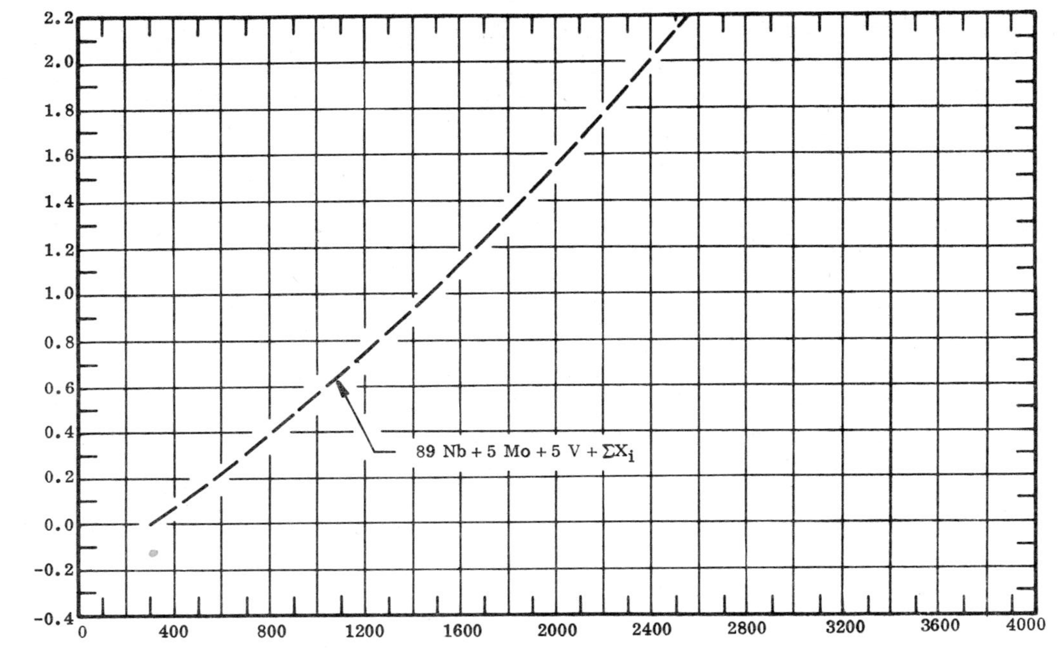

PROVISIONAL VALUES

[Temperature, T, K; Linear Expansion, $\Delta L/L_0$, %; α, K^{-1}]

T	(89 Nb + 5 Mo + 5 V + ΣX_i) $\Delta L/L_0$	$\alpha \times 10^6$
293	0.000	7.4
400	0.080	7.6
500	0.157	7.8
600	0.237	8.0
700	0.317	8.2
800	0.400	8.4
900	0.485	8.6
1000	0.572	8.8
1200	0.752	9.2
1400	0.939	9.6
1600	1.134	9.9
1800	1.337	10.3
2000	1.549	10.9
2200	1.769	11.2
2400	1.999	11.7
2600	2.238	12.2

REMARKS

(89 Nb + 5 Mo + 5 V + ΣX_i): The tabulated values for well-annealed alloy of type Nb - 5 Mo - 5 V - 1 Zr are considered accurate to within ± 7% over the entire temperature range. These values can be represented approximately by the following equation:

$$\Delta L/L_0 = -0.212 + 6.963 \times 10^{-4} T + 8.348 \times 10^{-8} T^2 + 4.250 \times 10^{-12} T^3$$

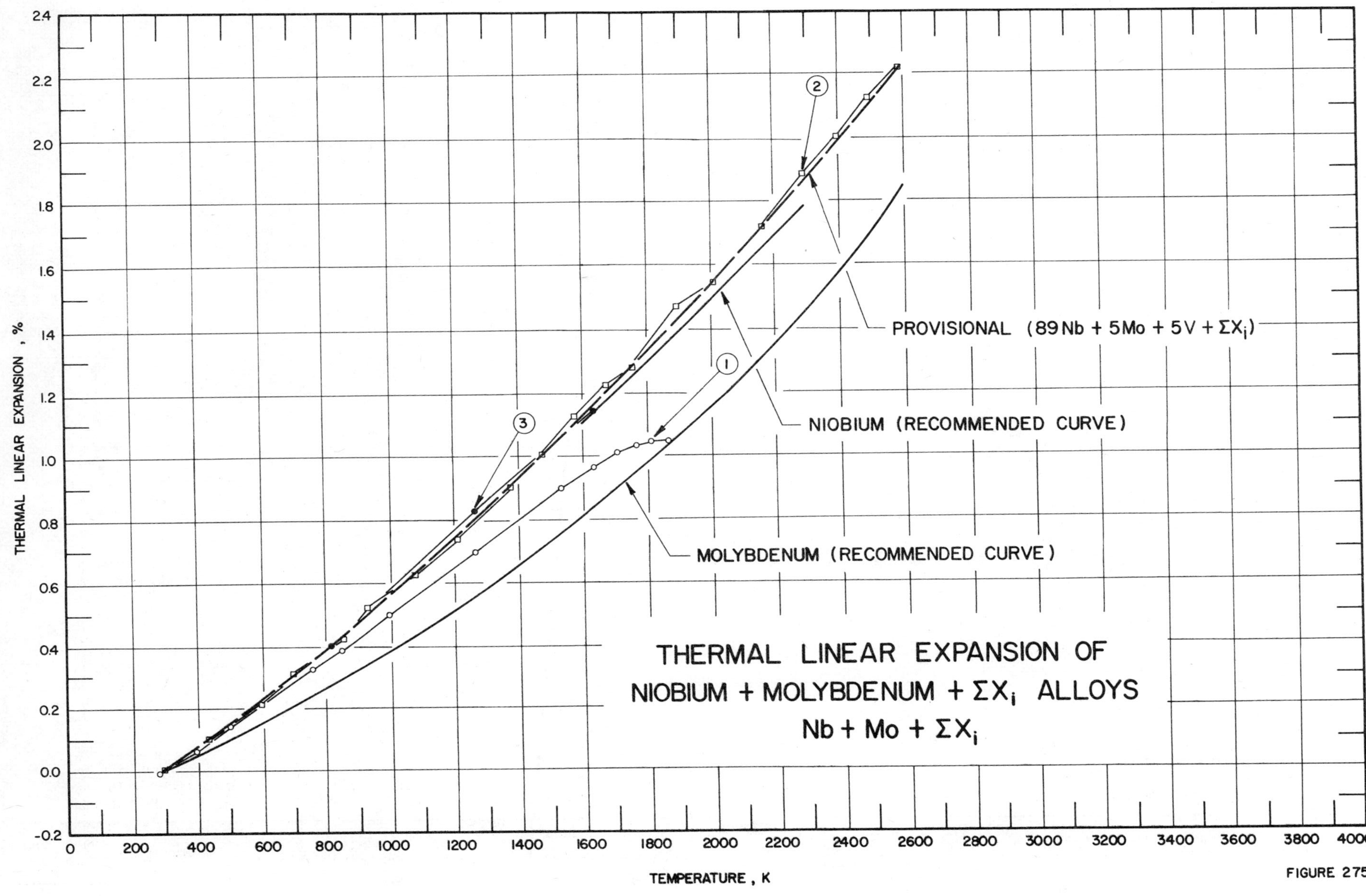

THERMAL LINEAR EXPANSION OF
NIOBIUM + MOLYBDENUM + ΣX$_i$ ALLOYS
Nb + Mo + ΣX$_i$
PROVISIONAL (89 Nb + 5 Mo + 5 V + ΣX$_i$)
NIOBIUM (RECOMMENDED CURVE)
MOLYBDENUM (RECOMMENDED CURVE)
1
2
3
THERMAL LINEAR EXPANSION, %
TEMPERATURE, K
-0.2
0.0
0.2
0.4
0.6
0.8
1.0
1.2
1.4
1.6
1.8
2.0
2.2
2.4
0
200
400
600
800
1000
1200
1400
1600
1800
2000
2200
2400
2600
2800
3000
3200
3400
3600
3800
4000

FIGURE 275

SPECIFICATION TABLE 275. THERMAL LINEAR EXPANSION OF NIOBIUM + MOLYBDENUM + ΣX_i ALLOYS $Nb + Mo + \Sigma X_i$

Cur. No.	Ref. No.	Author(s)	Year	Method Used	Temp. Range, K	Name and Specimen Designation	Composition (weight percent) Nb	Mo	V	Zr	C	N_2	O_2	Composition (continued), Specifications, and Remarks
1	482	Leggett, H., Cook, J.L., Schwab, D.E., and Powers, C.T.	1965		280-1867	Nb-5Mo-5V-1Zr	89	5	5	1				Material supplied by Metals and Controls Div. of Texas Instruments and by Westinghouse, machined by Douglas for identification; data average of results of several runs on materials from both suppliers; zero-point correction of -0.008% determined by graphical interpolation.
2	425	Hedge, J.C., Kostenko, C., and Lang, J.I.	1963	T	299-2586		88.77	5.03	5.02	1.13	0.028	0.0136	0.0093	Specimen dimensions 0.5 in. diameter x 6 in. long; Westinghouse Electric Corp; density 538 lb ft^{-3}; measured in argon atmosphere with heating rate of approximately 5 F per min.
3	348	Anthony, F.M. and Pearl, H.A.	1960		299-1633			10						10 Ti; double arc melted in vacuum by E.I. duPont de Nemours and Co., extruded in air and in argon to 1589-1644 K (2400-2500 F), recrystallized at 1533 K (2300 F) for 1 hr in an inert atmosphere, and then machined into 3 in. long, 0.375 in. diameter specimen by Thompson Products Inc.

DATA TABLE 275. THERMAL LINEAR EXPANSION OF NIOBIUM + MOLYBDENUM + ΣX_i ALLOYS $Nb + Mo + \Sigma X_i$

[Temperature, T, K; Linear Expansion, $\Delta L/L_0$, %]

T	$\Delta L/L_0$	T	$\Delta L/L_0$	T	$\Delta L/L_0$
CURVE 1		CURVE 2 (cont.)		CURVE 3	
280	-0.008	598	0.219	299	0.008*
396	0.067	699	0.319	819	0.410
499	0.144	845	0.425	1260	0.835
758	0.336	942	0.529	1633	1.142
843	0.398	1079	0.637		
992	0.505	1205	0.743		
1264	0.701	1368	0.901		
1538	0.905	1474	1.014		
1628	0.970	1579	1.129		
1703	1.013	1677	1.222		
1765	1.037	1752	1.284		
1810	1.048	1898	1.474		
1867	1.053	2008	1.554		
		2163	1.725		
CURVE 2		2293	1.895		
		2393	2.004		
299	0.005	2498	2.136		
430	0.101	2586	2.245		

* Not shown in figure.

FIGURE AND TABLE NO. 276R. PROVISIONAL VALUES FOR THERMAL LINEAR EXPANSION OF NIOBIUM + TANTALUM + ΣX_i ALLOYS Nb + Ta + ΣX_i

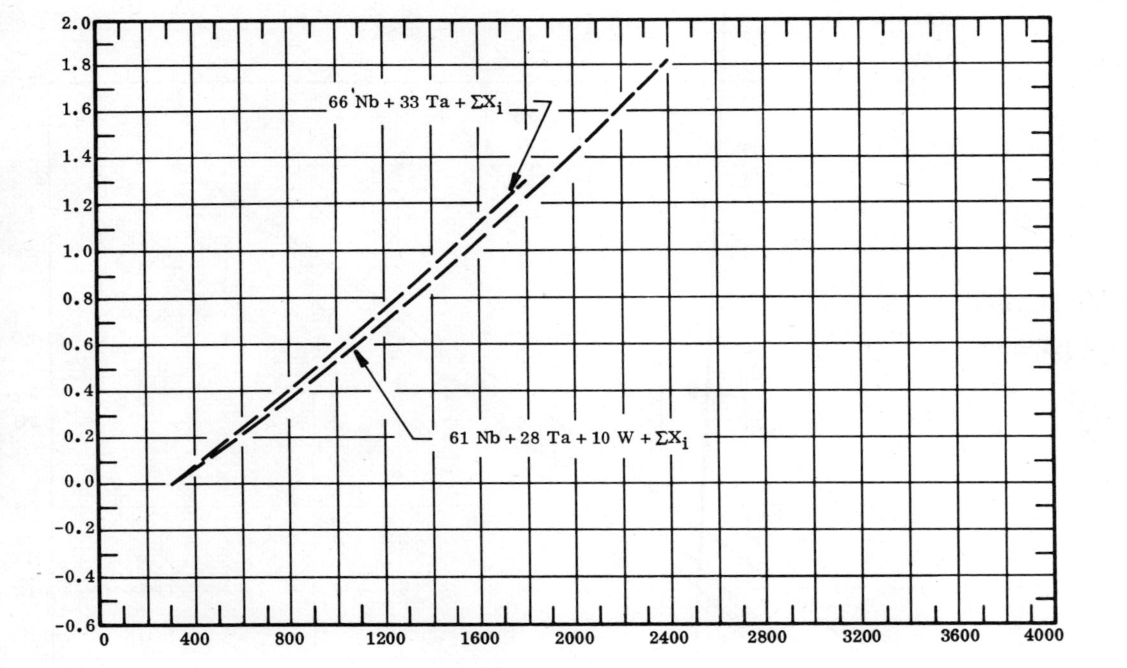

PROVISIONAL VALUES

[Temperature, T, K; Linear Expansion, $\Delta L/L_0$, %; α, K^{-1}]

	(61 Nb + 28 Ta + 10 W + ΣX_i)		(66 Nb + 33 Ta + ΣX_i)	
T	$\Delta L/L_0$	$\alpha \times 10^6$	$\Delta L/L_0$	$\alpha \times 10^6$
293	0.000	7.1	0.000	7.3
400	0.075	7.2	0.080	7.6
500	0.149	7.3	0.157	7.8
600	0.222	7.4	0.236	8.0
700	0.298	7.6	0.317	8.2
800	0.375	7.8	0.401	8.4
900	0.454	8.0	0.485	8.6
1000	0.534	8.2	0.573	8.8
1200	0.699	8.4	0.751	9.0
1400	0.870	8.7	0.932	9.1
1600	1.048	9.0	1.114	9.1
1800	1.232	9.4	1.295	9.1
2000	1.422	9.7		
2200	1.619	10.0		
2400	1.822	10.3		

REMARKS

61 Nb + 28 Ta + 10 W + ΣX_i): The tabulated values for well-annealed alloy are considered accurate to within ±7% over the entire temperature range. These values can be represented approximately by the following equation:

$$\Delta L/L_0 = -0.202 + 6.632 \times 10^{-4}\, T + 7.073 \times 10^{-8}\, T^2 + 1.807 \times 10^{-12}\, T^3$$

66 Nb + 33 Ta + ΣX_i): The tabulated values for well-annealed alloy of Type FS82 alloy are considered accurate to within ±7% over the entire temperature range. These values can be represented approximately by the following equation:

$$\Delta L/L_0 = -0.197 + 6.229 \times 10^{-4}\, T + 1.886 \times 10^{-7}\, T^2 - 4.112 \times 10^{-11}\, T^3$$

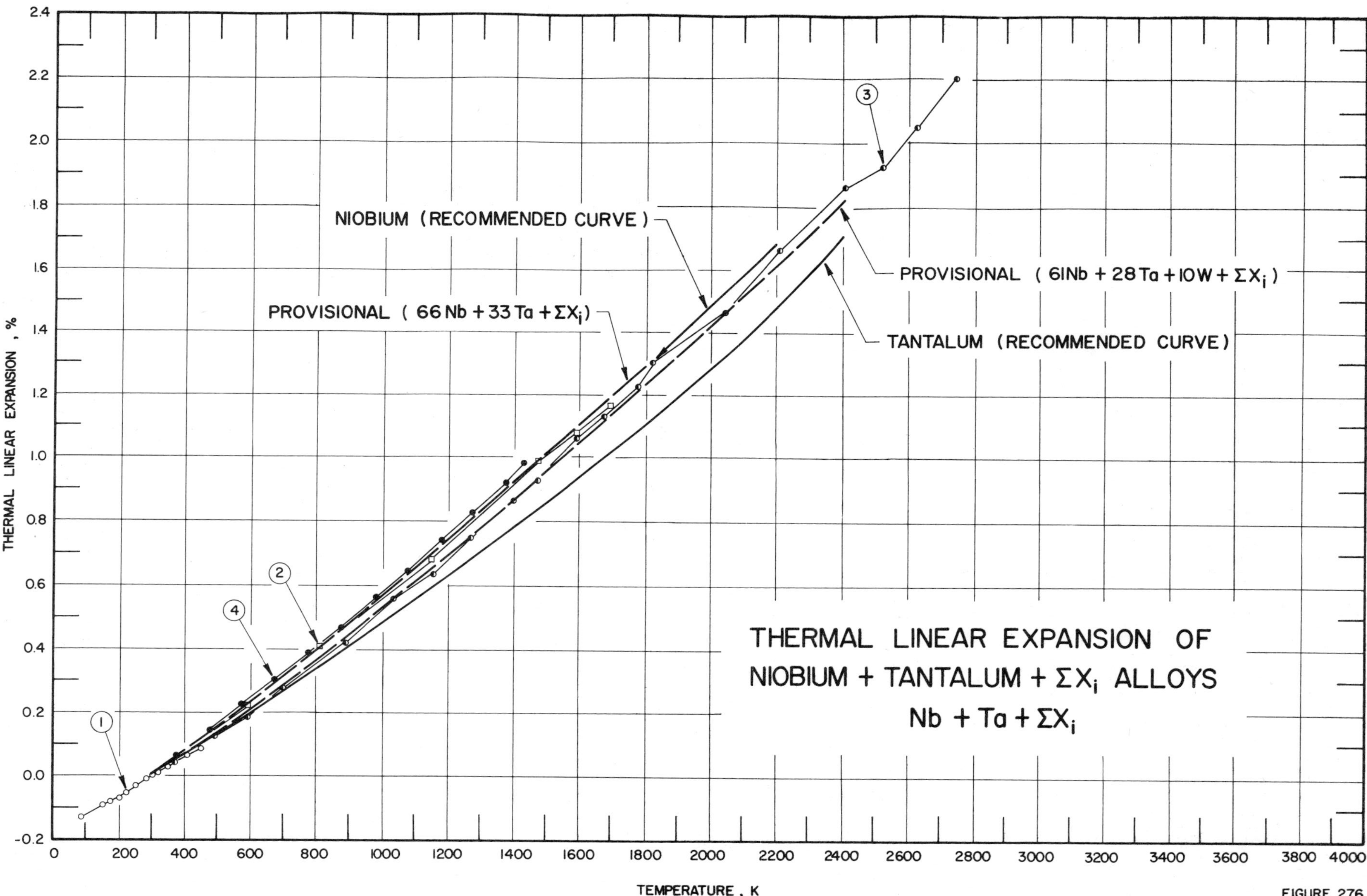
THERMAL LINEAR EXPANSION OF
NIOBIUM + TANTALUM + ΣX$_i$ ALLOYS
Nb + Ta + ΣX$_i$
NIOBIUM (RECOMMENDED CURVE)
PROVISIONAL (66 Nb + 33 Ta + ΣX$_i$)
PROVISIONAL (61Nb + 28 Ta + 10 W + ΣX$_i$)
TANTALUM (RECOMMENDED CURVE)
1
2
3
4
THERMAL LINEAR EXPANSION , %
2.4
2.2
2.0
1.8
1.6
1.4
1.2
1.0
0.8
0.6
0.4
0.2
0.0
-0.2
TEMPERATURE , K
0
200
400
600
800
1000
1200
1400
1600
1800
2000
2200
2400
2600
2800
3000
3200
3400
3600
3800
4000
FIGURE 276

SPECIFICATION TABLE 276. THERMAL LINEAR EXPANSION OF NIOBIUM + TANTALUM + ΣX_i ALLOYS Nb + Ta + ΣX_i

Cur. No.	Ref. No.	Author(s)	Year	Method Used	Temp. Range, K	Name and Specimen Designation	Composition (weight percent) Nb	Ta	W	Si	Zr	Mo	Composition (continued), Specifications, and Remarks
1	244	Harris, B. and Peacock, D.E.	1966	L	82-444	Nb-132	60.113	20.13	14.89			4.867	Three different rod specimens, two annealed for 1 hr at 1773 K and the third annealed for 1 hr at 2273 K; measured in liquid nitrogen, liquid bromoethane, and liquid polyalkylene; curve composed of results from several runs on three specimens; anomaly in the thermal linear expansion observed below 300 K; zero-point correction of -0.123% determined by graphical interpolation.
2	483	Jones, R.L.	1961		293-1699	FS-82 Niobium Alloy	66.3	33			0.7		Sheet (0.040 gage) from the Fansteel Metallurgical Co., supplied in the stress-relieved condition (1 hr at 1311 K); thermal expansion gotten as a by-product of the data obtained from the tensile tests performed on the high heating rate, high strain rate testing machine.
3	425	Hedge, J.C., Kostenko, C., and Lang, J.I.	1963	T	300-2747		60.8	27.84	10.40	0.01	0.92		0.007 Fe, 0.009 Ni, 0.005 Ti, 0.004 C, 0.005 O_2, and 0.002 N_2; specimen dimensions 0.50 in. diameter by 6 in. long; density 669 lb ft^{-3}; Fansteel Metallurgical Corp.; measured in argon atmosphere with heating rate of 5 F per min.
4	484	Neff, C.W., Frank, R.G., and Luft, L.	1961		298-1423	FS-82B							Tested in argon atmosphere.

DATA TABLE 276. THERMAL LINEAR EXPANSION OF NIOBIUM + TANTALUM + ΣX_i ALLOYS Nb + Ta + ΣX_i

[Temperature, T, K; Linear Expansion, $\Delta L/L_0$, %]

T	$\Delta L/L_0$
CURVE 1	
82	-0.123
155	-0.088
155	-0.086*
157	-0.085
159	-0.083*
163	-0.085*
167	-0.083*
170	-0.080
174	-0.078*
174	-0.076*
185	-0.073*
188	-0.068*
190	-0.070*
190	-0.067*
194	-0.067*
198	-0.064*
200	-0.062
205	-0.065*
205	-0.059*
208	-0.061*
219	-0.055*
219	-0.051*
221	-0.051
227	-0.046*
229	-0.046*
231	-0.045*
234	-0.042*
240	-0.043*
240	-0.038*
245	-0.037*
248	-0.030*
252	-0.026
254	-0.024*
259	-0.026*
259	-0.023*
260	-0.020*
265	-0.017*
273	-0.013*
281	-0.011
283	-0.007*
287	-0.003*
291	0.001*
295	-0.001*
297	0.002*
300	0.002
CURVE 1 (cont.)	
300	0.005*
304	0.005*
312	0.010*
320	0.013
323	0.017*
327	0.019*
337	0.023*
350	0.030
362	0.039*
369	0.041*
382	0.051*
392	0.053*
397	0.059*
408	0.066
411	0.066*
422	0.073*
429	0.078*
434	0.076*
441	0.084*
444	0.086
CURVE 2	
293	0.000*
588	0.221
810	0.408
1144	0.683
1366	0.900
1477	0.993
1588	1.083
1699	1.173
CURVE 3	
299	0.004*
369	0.045
484	0.125
592	0.190
700	0.287
885	0.421
1033	0.561
1150	0.640
1266	0.756
1393	0.873
1474	0.938
CURVE 3 (cont.)	
1595	1.078
1673	1.140
1778	1.227
1817	1.308
2040	1.467
2213	1.672
2403	1.865
2521	1.934
2622	2.048
2747	2.203
CURVE 4	
298	0.004*
373	0.070
473	0.146
573	0.224
673	0.310
773	0.391
873	0.477
973	0.569
1073	0.654
1173	0.746
1273	0.834
1373	0.927
1423	0.982

* Not shown in figure.

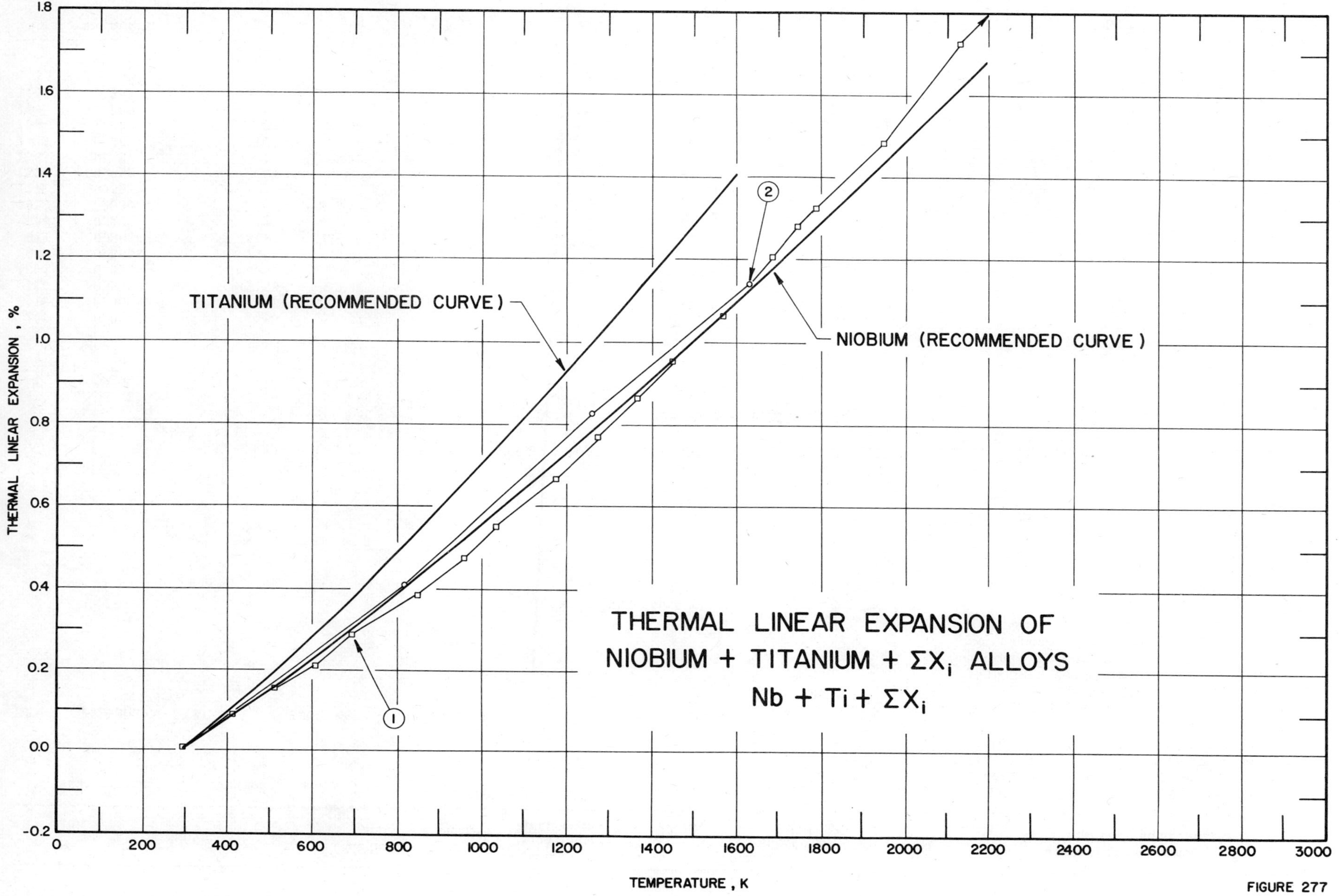
THERMAL LINEAR EXPANSION OF
NIOBIUM + TITANIUM + ΣX_i ALLOYS
Nb + Ti + ΣX_i
TITANIUM (RECOMMENDED CURVE)
NIOBIUM (RECOMMENDED CURVE)
1
2
THERMAL LINEAR EXPANSION, %
1.8
1.6
1.4
1.2
1.0
0.8
0.6
0.4
0.2
0.0
-0.2
0
200
400
600
800
1000
1200
1400
1600
1800
2000
2200
2400
2600
2800
3000
TEMPERATURE, K

FIGURE 277

SPECIFICATION TABLE 277. THERMAL LINEAR EXPANSION OF NIOBIUM + TITANIUM + ΣX_i ALLOYS Nb + Ti + ΣX_i

Cur. No.	Ref. No.	Author(s)	Year	Method Used	Temp. Range, K	Name and Specimen Designation	Composition (weight percent) Nb	Ti	Zr	Mo	Composition (continued), Specifications, and Remarks
1	425	Hedge, J.C., Kostenko, C., and Lang, J.I.	1963	T	300-2578		85.08	10.0	4.9		0.0014 C, 0.0244 O_2, 0.0014 H_2, and 0.0024 N_2; specimen dimensions 0.5 in. diameter x 6 in. long; DuPont; measured in argon atmosphere with heating rate of approximately 5 F per min.
2	348	Anthony, F.M. and Pearl, H.A.	1960		298-1633			10		10	Double arc melted in a vacuum by E.I. DuPont de Nemours and Co.; extruded in air and in argon up to 1589-1644 K (2400-2500 F), recrystallized at 1533 K (2300 F) for 1 hr in an inert atmosphere, and then machined into 3 in. long, 0.375 in. diameter specimen by Thompson Products Inc.

DATA TABLE 277. THERMAL LINEAR EXPANSION OF NIOBIUM + TITANIUM + ΣX_i ALLOYS Nb + Ti + ΣX_i

[Temperature, T, K; Linear Expansion, $\Delta L/L_0$, %]

T	$\Delta L/L_0$	T	$\Delta L/L_0$
CURVE 1		CURVE 1 (cont.)	
299	0.005	1783	1.327
415	0.093	1948	1.485
510	0.156	2133	1.737
604	0.214	2258	1.861*
690	0.284	2405	1.993*
842	0.384	2577	2.180*
958	0.479		
1033	0.550	CURVE 2	
1172	0.665		
1273	0.764	299	0.008*
1365	0.867	819	0.410
1455	0.958	1260	0.835
1564	1.065	1633	1.142
1682	1.205		
1744	1.283		

*Not shown in figure.

FIGURE AND TABLE NO. 278R. PROVISIONAL VALUES FOR THERMAL LINEAR EXPANSION OF NIOBIUM + TUNGSTEN + ΣX_i ALLOYS   Nb + W + ΣX_i

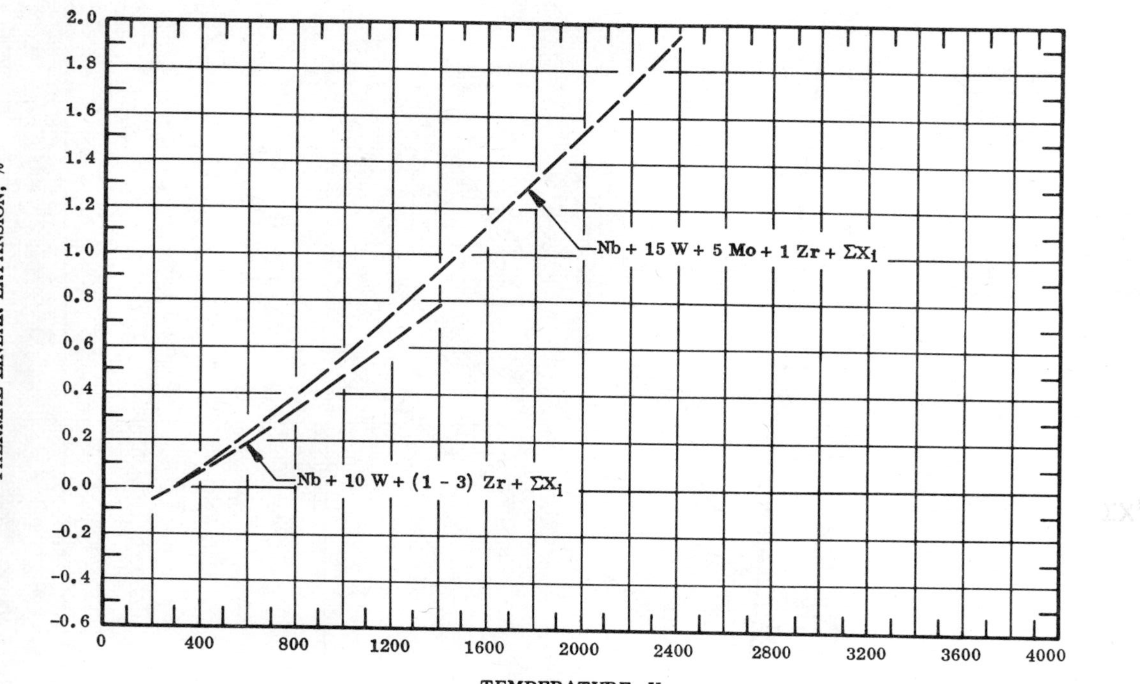

PROVISIONAL VALUES

[Temperature, T, K; Linear Expansion, $\Delta L/L_0$, %; α, K^{-1}]

T	[Nb + 10 W + (1-3) Zr + ΣX_i] $\Delta L/L_0$	$\alpha \times 10^6$	(Nb + 15 W + 5 Mo + 1 Zr + ΣX_i) $\Delta L/L_0$	$\alpha \times 10^6$
200	-0.055	6.0		
293	0.000	6.2	0.000	7.6
400	0.067	6.3	0.081	7.7
500	0.131	6.4	0.159	7.8
600	0.195	6.6	0.238	7.9
700	0.262	6.7	0.317	8.0
800	0.330	6.9	0.399	8.2
900	0.400	7.1	0.480	8.4
1000	0.472	7.3	0.567	8.6
1200	0.620	7.6	0.740	8.8
1400	0.778	8.0	0.921	9.2
1600			1.109	9.6
1800			1.306	10.0
2000			1.511	10.5
2200			1.727	11.0
2400			1.952	11.5

REMARKS

[Nb + 10 W + (1-3) Zr + ΣX_i] : The tabulated values for well-annealed alloy of types D-43 and Nb-752 alloys are considered accurate to within ±7% over the entire temperature range. These values can be represented approximately by the following equation:

$$\Delta L/L_0 = -0.173 + 5.759 \times 10^{-4}\, T + 5.540 \times 10^{-8}\, T^2 + 1.309 \times 10^{-11}\, T^3$$

(Nb + 15 W + 5 Mo + 1 Zr + ΣX_i): The tabulated values for well-annealed alloy are considered accurate to within ±7% over the entire temperature range. These values can be represented approximately by the following equation:

$$\Delta L/L_0 = -0.219 + 7.319 \times 10^{-4}\, T + 3.984 \times 10^{-8}\, T^2 + 1.336 \times 10^{-11}\, T^3$$

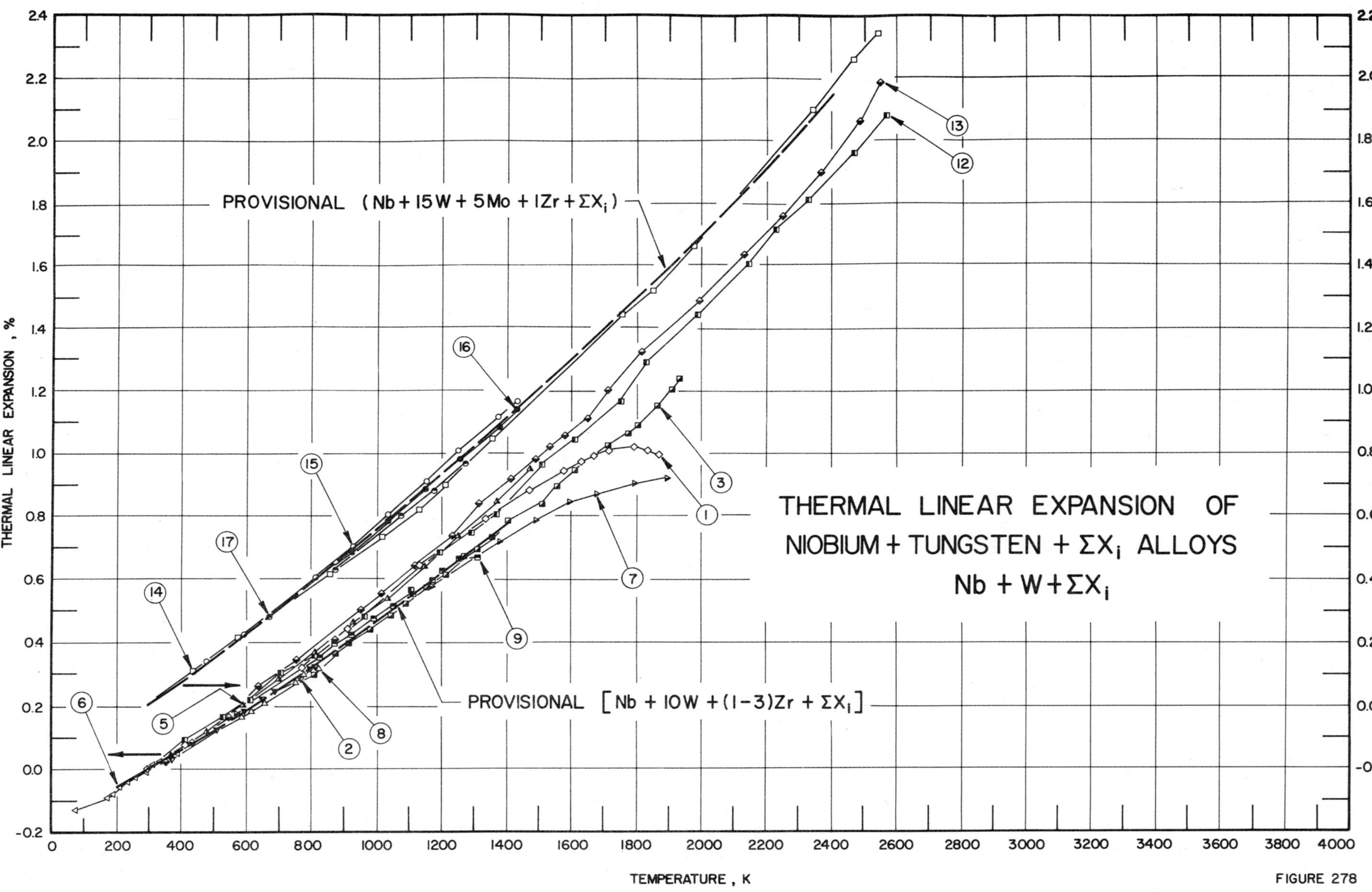
THERMAL LINEAR EXPANSION OF
NIOBIUM + TUNGSTEN + ΣXi ALLOYS
Nb + W + ΣXi
PROVISIONAL (Nb + 15 W + 5 Mo + 1 Zr + ΣXi)
PROVISIONAL [Nb + 10 W + (1 - 3) Zr + ΣXi]
THERMAL LINEAR EXPANSION, %
TEMPERATURE, K
0
200
400
600
800
1000
1200
1400
1600
1800
2000
2200
2400
2600
2800
3000
3200
3400
3600
3800
4000
-0.2
0.0
0.2
0.4
0.6
0.8
1.0
1.2
1.4
1.6
1.8
2.0
2.2
2.4
1
2
3
5
6
7
8
9
12
13
14
15
16
17

FIGURE 278

SPECIFICATION TABLE 278. THERMAL LINEAR EXPANSION OF NIOBIUM + TUNGSTEN + ΣX_i ALLOYS Nb + W + ΣX_i

Cur. No.	Ref. No.	Author(s)	Year	Method Used	Temp. Range, K	Name and Specimen Designation	Nb	W	Zr	C	Mo	Composition (continued), Specifications, and Remarks
1	482	Leggett, H., Cook, J. L., Schwab, D. E., and Powers, C. T.	1965		293-1863	Nb-752	87.5	10	2.5			Material supplied by Union Carbide Stellite Div. and Metals and Controls Div. of Texas Instruments; machined by Douglas for identification; data average of results of several runs on both materials.
2	485	Beck, E.J. and Schwartzberg, F.R.	1965	L	293-813	Nb-752	87.58	9.5	2.9	0.002		0.0095 N, 0.0065 O, 0.0007 H; supplied by Union Carbide Stellite Div.; specimen about 8.9 cm x 1.3 cm diameter with 7.8 cm radius sphere on one end, 0.60 cm radius sphere on other; melted, forged, cold-rolled, and milled to size; annealed 1 hr at 2300 F; initial length 8.910 cm.
3	485	Beck, E.J. and Schwartzberg, F.R.	1965	L	293-1922	Nb-752						The above specimen except initial length 7.617 cm and 7.62 cm radius sphere at each end.
4*	485	Beck, E.J. and Schwartzberg, F.R.	1965	L	293-1922	Nb-752						Similar to the above specimen except initial length 7.605 cm.
5	486	Burney, J.D.	1952		293-1477	Nb-752	87.48	10	2.5			40 ppm C, 60 ppm O_2, <100 ppm N_2, and 6 ppm H_2; density 9.02 g cm^{-3} at 295 K (22 C).
6	244	Harris, B. and Peacock, D.E.	1966	L	80-400	D-43 alloy	88.98	9.98	1.03	0.01		Rod specimen; curve composed of three sets of results, two from specimens annealed 1 hr at 1770 K, third from specimen in cold-worked condition; expansion measured in liquid nitrogen, liquid bromoethane, and liquid polyalkylene; zero-point correction of -0.122% determined by graphical interpolation.
7	482	Leggett, H., Cook, J. L., Schwab, D. E., and Powers, C. T.	1965		293-1887	D-43 alloy	89	10	1			Material supplied by DuPont Metal Products Div. and by Metals and Controls Div. of Texas Instruments, machined by Douglas for identification; data average of results on several runs on both materials.
8	485	Beck, E.J. and Schwartzberg, F.R.	1965	L	293-820	Specimen 2	89.4	9.5	0.99	0.0886		0.0034 N, 0.0163 O, 0.0012 H; specimen cylinder with 1.27 cm diameter with 7.5 cm radius sphere on one end, 0.67 cm diameter on other; prepared as 20 cm ingot by double consummable vacuum arc melting, extruded to a 5 x 15 cm sheet bar, conditioned and cross-rolled to 0.64 cm plate, flattened, surface-conditioned and vacuum annealed at 1470 K for 1 hr, cold-rolled to 0.0254 cm, re-annealed 1 hr at 1470 K, cold-rolled to approximate gage, annealed at 1920 K for 10 min, cold-rolled to size and vacuum annealed 1 hr at 1700 K, then flattened; specimen 8.905 cm long.
9	485	Beck, E.J. and Schwartzberg, F.R.	1965	L	293-1312	Specimen 2 D-43 alloy						The above specimen, initial length 7.623 cm; expansion on run 2.
10	485	Beck, E.J. and Schwartzberg, F.R.	1965	L	293-1916	Specimen 2 D-43 alloy						The above specimen; initial length 7.625 cm; expansion on run 3.

* Not shown in figure.

SPECIFICATION TABLE 278. THERMAL LINEAR EXPANSION OF NIOBIUM + TUNGSTEN + ΣX_i ALLOYS Nb + W + ΣX_i (continued)

Cur. No.	Ref. No.	Author(s)	Year	Method Used	Temp. Range, K	Name and Specimen Designation	Composition (weight percent) Nb	W	Zr	C	Mo	Composition (continued), Specifications, and Remarks
11*	485	Beck, E.J. and Schwartzberg, F.R.	1965	L	293-1933	Specimen 1 D-43 alloy	89.4	9.5	0.99	0.0886		0.0034 N, 0.0163 O, 0.0012 H; similar to the above specimen except with 7.62 cm radius sphere on each end; initial length 7.623 cm.
12	425	Hedge, J.C., Kostenko, C., and Lang, J.I.	1963		300-2566		87.47	9.93	2.58	0.002		0.0120 O_2, 0.0009 H_2, and 0.0060 N_2; specimen dimensions 0.5 in. diameter x 6 in. long; Haynes Stellite Co.; measured in argon atmosphere with heating rate of approximately 5F per min.
13	425	Hedge, J.C., et al.	1963		300-2558		89.26	9.7	0.88	0.0810		0.0052 O_2, 0.0004 H_2, and 0.0033 N_2; specimen dimensions 0.5 in. diameter x 6 in. long; DuPont; measured in argon atmosphere with heating rate of approximately 5 F per min.
14	425	Hedge, J.C., et al.	1963		300-2543		78.29	15.3	1.08	0.0340	5.26	0.0167 O_2, 0.0061 H_2, and 0.0211 N_2; specimen dimensions 0.5 in. diameter x 6 in. long; measured in argon atmosphere with heating rate of approximately 5 F per min.
15	484	Neff, C.W., Frank, R.G., and Luft, L.	1961		298-1422	F-48 (heat No. W-5-T)		13.5/ 16.5	0.85/ 1.15	0.02/ 0.04	4.5/ 5.5	0.1 max. Ta, 0.02-0.05 O, 0.4 max. N, 0.0015 max. H; heating; 0.5 in. gage material; measured in argon.
16	484	Neff, C.W., et al.	1961		1422-298	F-48 (heat No. W-5-T)						Expansion given in decreasing temperatures in argon.
17	487	Savitskii, E.M., Baron, V.V., and Ivanova, K.N.	1968		373-1273	RN-6 Russian alloy	Bal.	4.5/ 6.0	1.0/ 1.5		4.5/ 6.0	Samples under investigation were prepared by the usual vacuum-arc and electron-beam melting processes with subsequent hot and cold working.

* Not shown in figure.

DATA TABLE 278. THERMAL LINEAR EXPANSION OF NIOBIUM + TUNGSTEN + ΣX_i ALLOYS $Nb + W + \Sigma X_i$

[Temperature, T, K; Linear Expansion, $\Delta L/L_0$, %]

T	$\Delta L/L_0$
CURVE 1	
293	0.000
412	0.084
769	0.336
907	0.455
1136	0.646
1338	0.799
1465	0.883
1578	0.949
1630	0.979
1671	0.999
1715	1.014
1784	1.022
1832	1.017
1863	1.000
CURVE 2	
293	0.0005*
374	0.0376
376	0.0718*
509	0.1288
533	0.1459*
598	0.1915
648	0.2229
671	0.2343*
691	0.2485
762	0.2970
800	0.3255
813	0.3341
CURVE 3	
293	0.00*
806	0.31
873	0.37
918	0.40
975	0.44
1040	0.49
1089	0.53
1151	0.58
1205	0.62
1270	0.68
1314	0.70
1353	0.74
1400	0.79
1505	0.84
CURVE 3 (cont.)	
1555	0.90
1566	0.90*
1613	0.95
1666	0.99*
1711	1.03
1772	1.07
1800	1.10*
1861	1.16
1872	1.17*
1905	1.21
1922	1.24
CURVE 4*	
293	0.00
789	0.33
874	0.39
935	0.44
995	0.48
1026	0.51
1083	0.55
1153	0.59
1166	0.60
1194	0.63
1241	0.67
1308	0.71
1355	0.74
1403	0.78
1472	0.83
1553	0.88
1589	0.91
1644	0.95
1700	1.00
1755	1.04
1803	1.08
1855	1.13
1922	1.20
CURVE 5	
293	0.000*
366	0.049
477	0.129
589	0.208
700	0.293
811	0.373
CURVE 5 (cont.)	
922	0.464
1033	0.546
1144	0.643
1255	0.744
1366	0.850
1477	0.959
CURVE 6	
77	-0.122
166	-0.083
168	-0.081*
181	-0.072
182	-0.073*
189	-0.065*
202	-0.062*
208	-0.053*
211	-0.054
225	-0.044*
236	-0.037
246	-0.031*
257	-0.026*
260	-0.022
270	-0.015*
291	-0.006*
291	-0.003
317	0.020
338	0.033
363	0.048*
383	0.059
417	0.082*
CURVE 7	
293	0.000*
569	0.183
1161	0.580
1374	0.723
1484	0.791
1595	0.844
1675	0.878
1791	0.910
1887	0.926
CURVE 8	
293	0.000*
586	0.17401
593	0.18257*
613	0.19112
657	0.21965
687	0.24532*
753	0.28811
778	0.30523
790	0.31379*
820	0.32805
CURVE 9	
293	0.00*
821	0.36
869	0.40
920	0.43
985	0.48
1043	0.52
1109	0.56
1169	0.60
1200	0.63
1257	0.66
1312	0.69
CURVE 10*	
293	0.00
818	0.36
871	0.39
920	0.42
985	0.47
1030	0.50
1094	0.54
1105	0.54
1139	0.57
1197	0.61
1258	0.65
1325	0.71
1363	0.72
1414	0.77
1475	0.82
1536	0.87
1583	0.88
1644	0.92
1705	0.98
CURVE 10 (cont.)*	
1761	1.02
1811	1.07
1869	1.12
1916	1.17
CURVE 11*	
293	0.00
814	0.32
867	0.37
921	0.41
982	0.45
1040	0.50
1100	0.55
1168	0.59
1209	0.62
1268	0.66
1306	0.69
1385	0.75
1419	0.78
1472	0.82
1533	0.85
1585	0.90
1644	0.94
1694	0.99
1747	1.05
1811	1.10
1861	1.17
1933	1.24
CURVE 12	
299	0.005*
418	0.097
532	0.172
616	0.221
704	0.306
814	0.368
958	0.488
1104	0.575
1197	0.687
1293	0.757
1365	0.806
1506	0.965
1613	1.044
1753	1.176
CURVE 12 (cont.)	
1828	1.299
1993	1.458
2148	1.616
2263	1.738
2363	1.820
2473	1.978
2566	2.085
CURVE 13	
299	0.003*
348	0.027
432	0.093
542	0.171
634	0.272
755	0.348
857	0.410
948	0.508
1012	0.560
1118	0.651
1234	0.741
1318	0.843
1410	0.928
1497	0.990
1534	1.023
1578	1.065
1647	1.123
1712	1.204
1813	1.332
1988	1.495
2138	1.643
2253	1.770
2373	1.906
2493	2.078
2558	2.194
CURVE 14	
299	0.005*
437	0.117
578	0.220
700	0.302
859	0.420
907	0.453*
1019	0.543
1124	0.630

*Not shown in figure.

DATA TABLE 278. THERMAL LINEAR EXPANSION OF NIOBIUM + TUNGSTEN + ΣX_i ALLOYS Nb + W + ΣX_i (continued)

T	$\Delta L/L_0$
CURVE 14 (cont.)	
1208	0.709
1355	0.855
1483	0.990*
1580	1.068
1649	1.130
1757	1.246
1843	1.325
1973	1.475
2128	1.646
2248	1.769*
2343	1.912
2468	2.062
2543	2.154
CURVE 15	
298	0.004*
366	0.057*
477	0.144
588	0.229
699	0.317*
810	0.415
922	0.515
1033	0.615
1144	0.720
1255	0.818
1366	0.923
1422	0.973
CURVE 16	
1422	0.953
1366	0.895
1255	0.788
1144	0.686
1033	0.586
922	0.494
810	0.402*
699	0.315*
588	0.222*
477	0.137*
366	0.051*
298	0.003*
CURVE 17	
373	0.058*
473	0.132*
673	0.282
873	0.436
1073	0.601
1173	0.687
1273	0.776

*Not shown in figure.

SPECIFICATION TABLE 279. THERMAL LINEAR EXPANSION OF PALLADIUM + GOLD + ΣX_i ALLOYS Pd + Au + ΣX_i

Cur. No.	Ref. No.	Author(s)	Year	Method Used	Temp. Range, K	Name and Specimen Designation	Composition (weight percent) Pd	Au	Ag	Composition (continued), Specifications, and Remarks
1*	654	Nagender Naidu, S.V. and Houska, C.R.	1971	X	80-298		68	21	11	Prepared from 99.97 Pd, 99.97 Au, and 99.99 Ag from Engelhard Industries Inc. by melting in induction furnace under hydrogen atmosphere; homogenized for 1 week at 1173 K; annealed at 873 K in evacuated quartz tubes after flushing 3 times with argon; lattice parameter reported at 298 K is 3.9356 Å; 3.9353 Å at 293 K determined by graphical interpolation.
2*	654	Nagender, Naidu, S.V. and Houska, C.R.	1971	X	80-298		41	38	21	Similar to the above specimen; lattice parameter reported at 298 K is 3.9792 Å; 3.9789 Å at 293 K determined by graphical interpolation.

DATA TABLE 279. THERMAL LINEAR EXPANSION OF PALLADIUM + GOLD + ΣX_i ALLOYS Pd + Au + ΣX_i

[Temperature, T, K; Linear Expansion, $\Delta L/L_0$, %]

T	$\Delta L/L_0$
CURVE 1*	
80	-0.236
195	-0.125
298	0.008
CURVE 2*	
80	-0.261
195	-0.133
298	0.008

* No figure given.

FIGURE AND TABLE NO. 280AR. PROVISIONAL VALUES FOR THERMAL LINEAR EXPANSION OF TANTALUM + NIOBIUM + ΣX_i ALLOYS Ta + Nb + ΣX_i

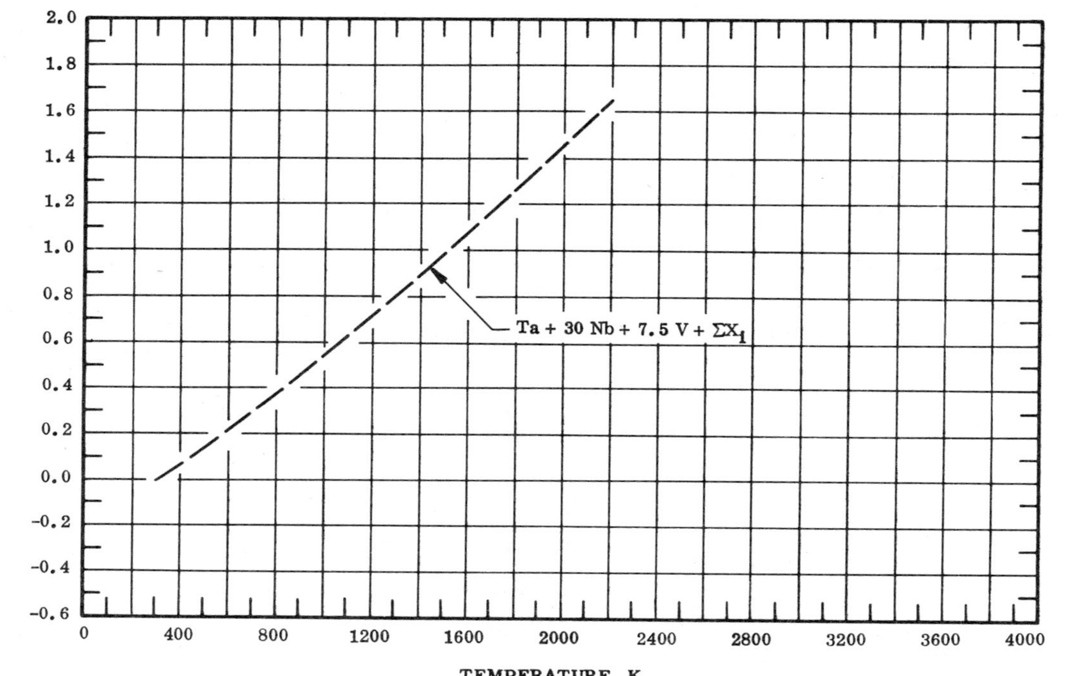

PROVISIONAL VALUES

[Temperature, T, K; Linear Expansion, $\Delta L/L_0$, %; α, K^{-1}]

(Ta + 30 Nb + 7.5 V + ΣX_i)		
T	$\Delta L/L_0$	$\alpha \times 10^6$
293	0.000	6.5
400	0.073	6.9
600	0.215	7.5
800	0.371	8.0
1000	0.536	8.5
1200	0.710	8.9
1400	0.892	9.3
1600	1.080	9.5
1800	1.271	9.6
2000	1.466	9.7
2200	1.660	9.7

REMARKS

(Ta + 30 Nb + 7.5 V + ΣX_i): The tabulated values for well-annealed alloy are considered accurate to within ±7% over the entire temperature range. These values can be represented approximately by the following equation:

$$\Delta L/L_0 = -0.171 + 5.310 \times 10^{-4}\, T + 2.101 \times 10^{-7}\, T^2 - 3.318 \times 10^{-11}\, T^3$$

FIGURE AND TABLE NO. 280BR. PROVISIONAL VALUES FOR THERMAL LINEAR EXPANSION OF TANTALUM + TUNGSTEN + ΣX_i ALLOYS   Ta + W + ΣX_i

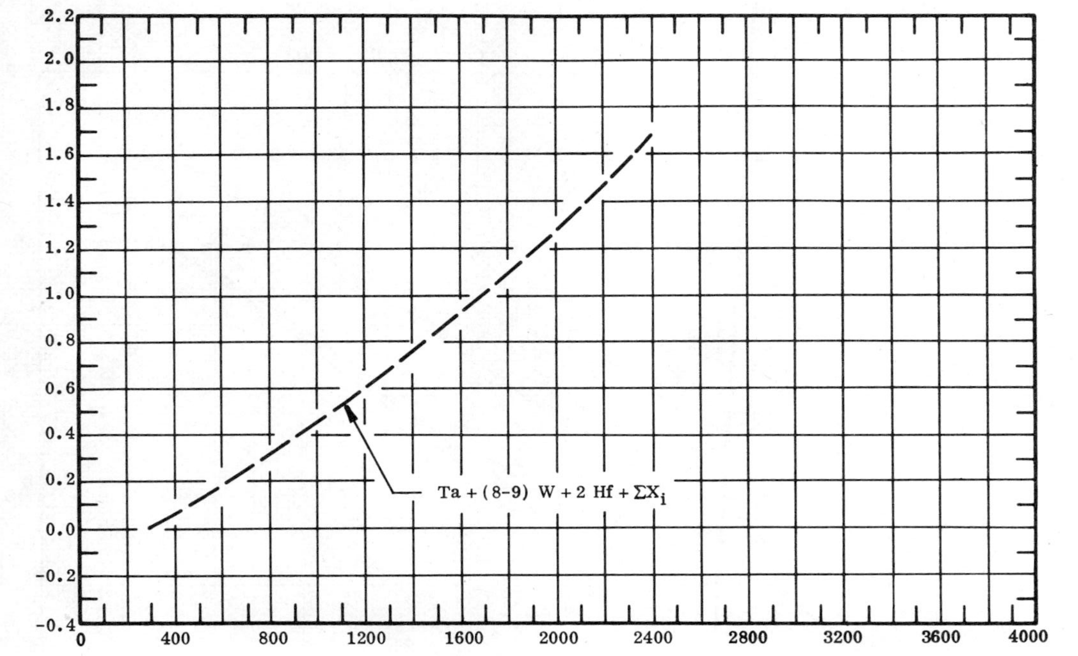

PROVISIONAL VALUES

[Temperature, T, K; Linear Expansion, $\Delta L/L_0$, %; α, K^{-1}]

[Ta + (8-9) W + 2 Hf + ΣX_i]		
T	$\Delta L/L_0$	$\alpha \times 10^6$
293	0.000	5.9
400	0.065	6.0
500	0.123	6.2
600	0.191	6.5
700	0.252	6.7
800	0.323	6.8
900	0.389	7.0
1000	0.463	7.2
1200	0.610	7.5
1400	0.765	8.0
1600	0.929	8.4
1800	1.102	8.8
2000	1.284	9.4
2200	1.477	10.0
2400	1.682	10.4

REMARKS

[Ta + (8-9) W + 2 Hf + ΣX_i]: The tabulated values for well-annealed alloy of type T-111 alloy are considered accurate to within ±7% over the entire temperature range. These values can be represented approximately by the following equation:

$$\Delta L/L_0 = -0.173 + 5.588 \times 10^{-4}\ T + 6.166 \times 10^{-8}\ T^2 + 1.120 \times 10^{-11}\ T^3$$

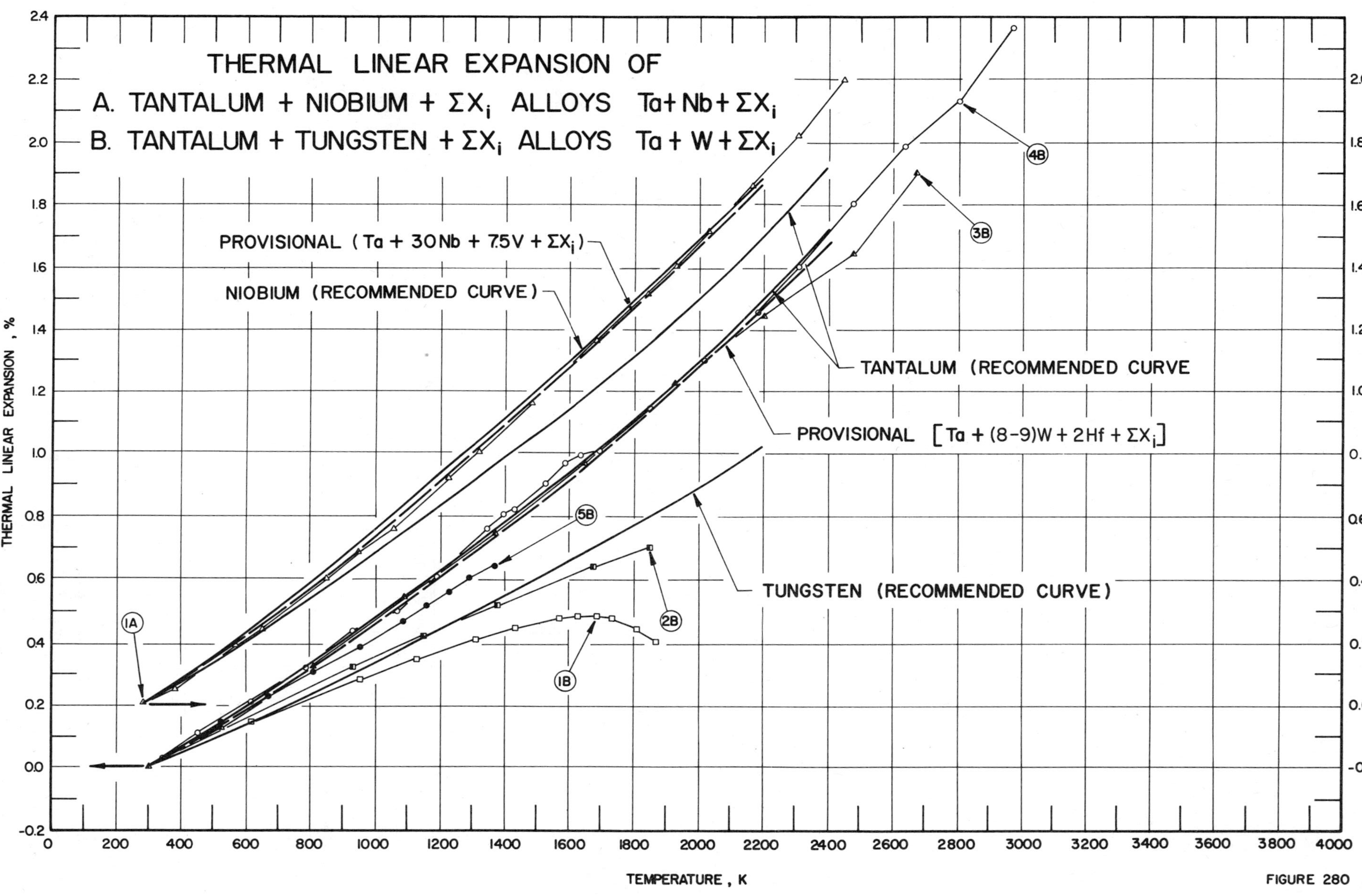
THERMAL LINEAR EXPANSION OF
A. TANTALUM + NIOBIUM + ΣX_i ALLOYS Ta+ Nb + ΣX_i
B. TANTALUM + TUNGSTEN + ΣX_i ALLOYS Ta + W + ΣX_i
PROVISIONAL (Ta + 30Nb + 7.5V + ΣX_i)
NIOBIUM (RECOMMENDED CURVE)
TANTALUM (RECOMMENDED CURVE
PROVISIONAL [Ta + (8-9)W + 2Hf + ΣX_i]
TUNGSTEN (RECOMMENDED CURVE)
1A
1B
2B
3B
4B
5B
THERMAL LINEAR EXPANSION, %
TEMPERATURE, K
2.4
2.2
2.0
1.8
1.6
1.4
1.2
1.0
0.8
0.6
0.4
0.2
0.0
-0.2
0
200
400
600
800
1000
1200
1400
1600
1800
2000
2200
2400
2600
2800
3000
3200
3400
3600
3800
4000
FIGURE 280

SPECIFICATION TABLE 280A. THERMAL LINEAR EXPANSION OF TANTALUM + NIOBIUM + ΣX_i ALLOYS   Ta + Nb + ΣX_i

Cur. No.	Ref. No.	Author(s)	Year	Method Used	Temp. Range, K	Name and Specimen Designation	Composition (weight percent) Ta	Nb	V	Composition (continued), Specifications, and Remarks
1	425	Hedge, J.C., Kostenko, C., and Lang, J.I.	1963	T	300-2443		62.12	30.3	7.47	0.09 C, 0.015 O_2, and 0.0065 N_2; specimen dimensions 0.50 in. diameter by 6 in. long; Wah Chang Corp; density 721 lb ft^{-3}; measured in argon atmosphere with heating rate of approximately 5 F per min.

DATA TABLE 280A. THERMAL LINEAR EXPANSION OF TANTALUM + NIOBIUM + ΣX_i ALLOYS   Ta + Nb + ΣX_i

[Temperature, T, K; Linear Expansion, $\Delta L/L_0$, %]

T	$\Delta L/L_0$
CURVE 1	
300	0.004
388	0.050
579	0.198
658	0.247
848	0.405
946	0.495
1060	0.567
1233	0.723
1323	0.801
1484	0.965
1688	1.166
1853	1.323
1937	1.411
2035	1.524
2176	1.668
CURVE 1 (cont.)	
2316	1.823
2443	2.014

SPECIFICATION TABLE 280B. THERMAL LINEAR EXPANSION OF TANTALUM + TUNGSTEN + ΣX_i ALLOYS Ta + W + ΣX_i

Cur. No.	Ref. No.	Author(s)	Year	Method Used	Temp. Range, K	Name and Specimen Designation	Composition (weight percent) Ta	W	Hf	Nb	Composition (continued), Specifications, and Remarks
1	482	Leggett, H., Cook, J.L., Schwab, D.E., and Powers, C.T.	1965		298-1866	T-111	90	8	2		Specimen supplied by Westinghouse, machined for identification by Douglas Aircraft; data average of results on several specimens; run 1; annealed.
2	482	Leggett, H., et al.	1965		298-1848	T-111	90	8	2		Similar to the above specimen; run 2.
3	488	Ammon, R.L. and Begley, R.T.	1963	L	300-2672	T-111	90	8	2		99.94 Ta from Kennametal, Inc; W from Fansteel Kulite Tungsten Co.; Hf from Bettis Atomic Power Lab. and Wah Chang; specimen prepared by melting components together; 2 in. x 0.25 in. in diameter.
4	425	Hedge, J.C., Kostenko, C., and Lang, J.I.	1963	T	300-2972	T-111	88.79	9	2.2		0.0041 C, 0.0040 O_2, 0.0023 N_2; specimen dimensions 0.5 in. diameter x 6 in. long; Westinghouse Electric Corp; density 1058 lb ft^{-3}; measured in argon atmosphere with heating rate approximately 5 F per min.
5	489	Baker, W.H.	1966	L	522-1366	T-222	Bal.	9.2	2.5	0.05	No details given.

DATA TABLE 280B. THERMAL LINEAR EXPANSION OF TANTALUM + TUNGSTEN + ΣX_i ALLOYS Ta + W + ΣX_i

[Temperature, T, K; Linear Expansion, $\Delta L/L_0$, %]

T	$\Delta L/L_0$
CURVE 1	
298	0.000*
618	0.148
952	0.289
1128	0.357
1319	0.417
1436	0.455
1567	0.482
1635	0.493
1684	0.493
1739	0.483
1802	0.454
1866	0.411
CURVE 2	
298	0.000*
929	0.325
1144	0.421
CURVE 2 (cont.)	
1379	0.530
1677	0.650
1848	0.708
CURVE 3	
300	0.004
533	0.134
811	0.326
1089	0.556
1367	0.753
1644	0.973
1922	1.231
2200	1.441
2478	1.651
2672	1.911
CURVE 4	
300	0.004*
344	0.032
458	0.107
619	0.212
788	0.324
930	0.441
1068	0.507
1183	0.616
1341	0.764
1396	0.812
1425	0.822
1523	0.912
1585	0.974
1635	0.992
1690	1.006
1841	1.151
2019	1.309
2184	1.459
CURVE 4 (cont.)	
2316	1.604
2479	1.804
2639	1.987
2800	2.138
2973	2.369
CURVE 5‡	
522	0.136*
533	0.142
672	0.225
811	0.309
950	0.394
1089	0.479
1158	0.521
1228	0.564
1297	0.606
1366	0.647

DATA TABLE 280B. COEFFICIENT OF THERMAL LINEAR EXPANSION OF TANTALUM + TUNGSTEN + ΣX_i ALLOYS Ta + W + ΣX_i

[Temperature, T, K; Coefficient of Expansion, α, 10^{-6} K^{-1}]

T	α
CURVE 5‡	
522	5.92
533	5.94
672	6.03
811	6.07
950	6.12
1089	6.12
1158	6.10
1228	6.07
1297	6.01
1366	5.90

* Not shown in figure.
‡ Author's data for coefficient of thermal expansion have been integrated by TPRC to obtain $\Delta L/L_0$.

FIGURE AND TABLE NO. 281R. PROVISIONAL VALUES FOR THERMAL LINEAR EXPANSION OF TITANIUM + ALUMINUM + ΣX_i ALLOYS Ti + Al + ΣX_i

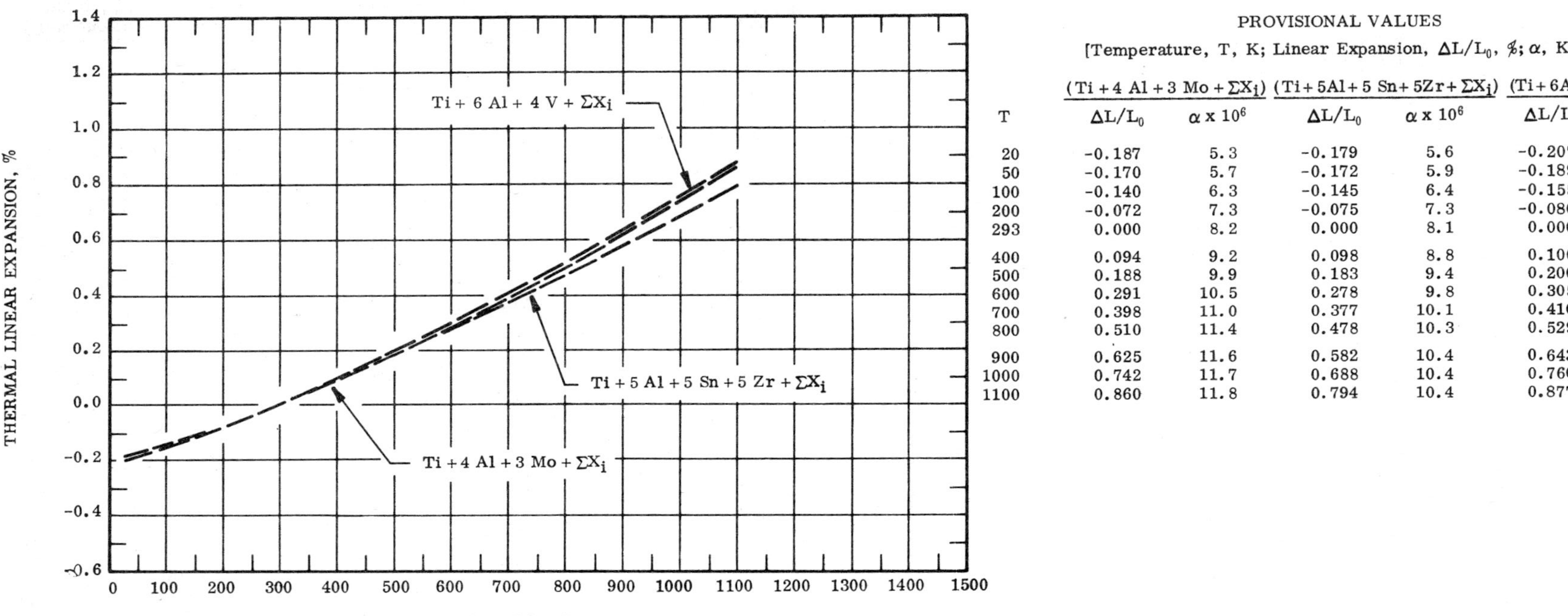

PROVISIONAL VALUES

[Temperature, T, K; Linear Expansion, $\Delta L/L_0$, %; α, K^{-1}]

T	(Ti + 4 Al + 3 Mo + ΣX_i) $\Delta L/L_0$	$\alpha \times 10^6$	(Ti + 5Al + 5 Sn + 5Zr + ΣX_i) $\Delta L/L_0$	$\alpha \times 10^6$	(Ti + 6Al + 4V + ΣX_i) $\Delta L/L_0$	$\alpha \times 10^6$
20	-0.187	5.3	-0.179	5.6	-0.207	6.2
50	-0.170	5.7	-0.172	5.9	-0.189	6.5
100	-0.140	6.3	-0.145	6.4	-0.155	7.1
200	-0.072	7.3	-0.075	7.3	-0.080	8.1
293	0.000	8.2	0.000	8.1	0.000	8.9
400	0.094	9.2	0.098	8.8	0.100	9.7
500	0.188	9.9	0.183	9.4	0.200	10.3
600	0.291	10.5	0.278	9.8	0.305	10.8
700	0.398	11.0	0.377	10.1	0.410	11.2
800	0.510	11.4	0.478	10.3	0.529	11.4
900	0.625	11.6	0.582	10.4	0.643	11.6
1000	0.742	11.7	0.688	10.4	0.760	11.6
1100	0.860	11.8	0.794	10.4	0.877	11.6

REMARKS

(Ti + 4Al + 3Mo + ΣX_i): The tabulated values for well-annealed alloy are considered accurate to within ± 7% over the entire temperature range. These values can be represented approximately by the following equation:

$$\Delta L/L_0 = -0.197 + 5.082 \times 10^{-4}\, T + 6.251 \times 10^{-7}\, T^2 - 1.939 \times 10^{-10}\, T^3$$

(Ti + 5Al + 5Sn + 5Zr + ΣX_i): The tabulated values for well-annealed alloy are considered accurate to within ± 7% over the entire temperature range. These values can be represented approximately by the following equation:

$$\Delta L/L_0 = -0.200 + 5.408 \times 10^{-4}\, T + 5.442 \times 10^{-7}\, T^2 - 1.965 \times 10^{-10}\, T^3$$

(Ti + 6Al + 4 V + ΣX_i): The tabulated values for well-annealed alloy are considered accurate to within ± 7% over the entire temperature range. These values can be represented approximately by the following equation:

$$\Delta L/L_0 = -0.220 + 5.992 \times 10^{-4}\, T + 5.807 \times 10^{-7}\, T^2 - 1.994 \times 10^{-10}\, T^3$$

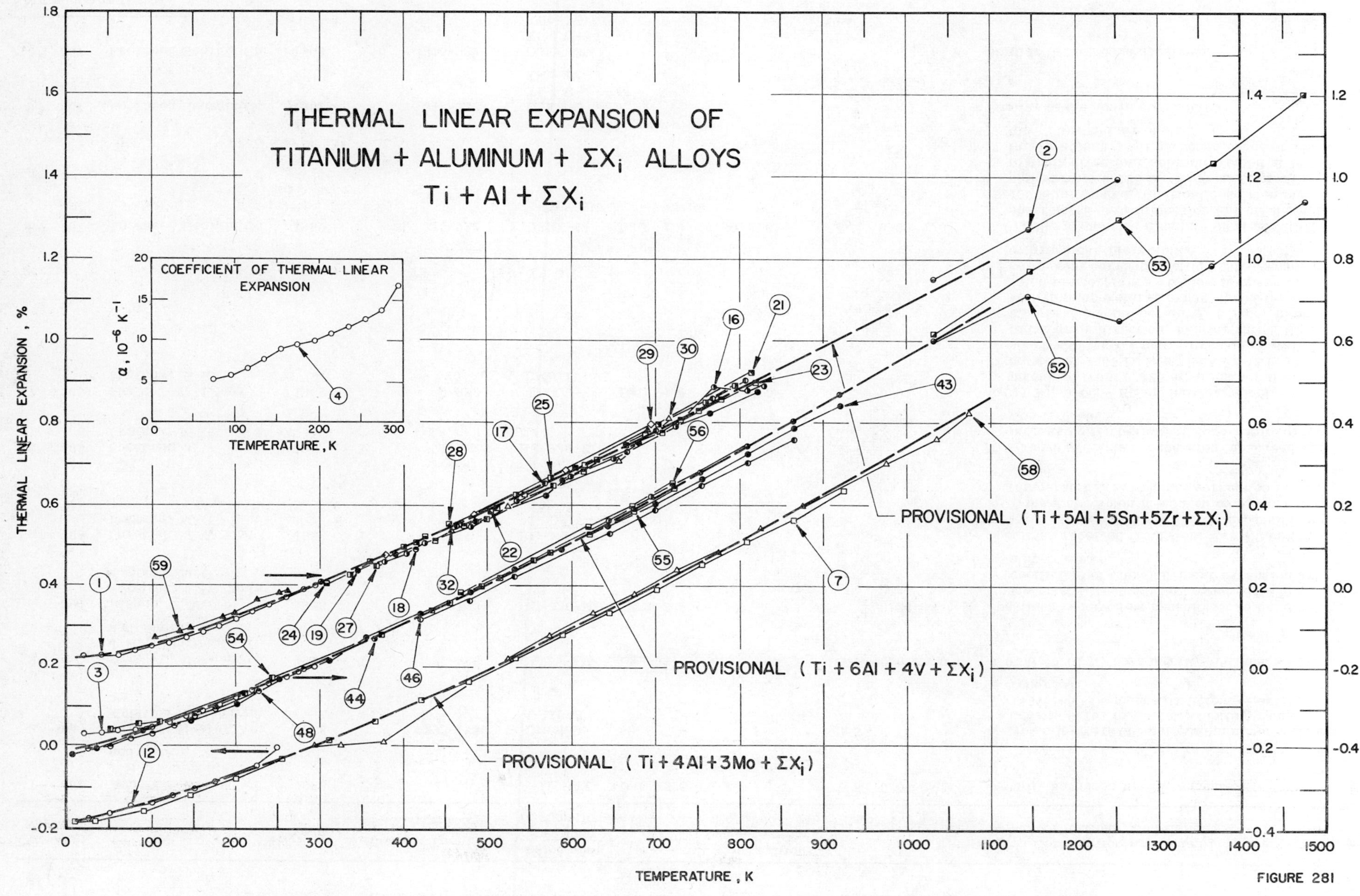
THERMAL LINEAR EXPANSION OF
TITANIUM + ALUMINUM + ΣXi ALLOYS
Ti + Al + ΣXi
COEFFICIENT OF THERMAL LINEAR EXPANSION
α, 10^-6 K^-1
TEMPERATURE, K
PROVISIONAL (Ti + 5Al + 5Sn + 5Zr + ΣXi)
PROVISIONAL (Ti + 6Al + 4V + ΣXi)
PROVISIONAL (Ti + 4Al + 3Mo + ΣXi)
THERMAL LINEAR EXPANSION, %
TEMPERATURE, K

FIGURE 281

SPECIFICATION TABLE 281. THERMAL LINEAR EXPANSION OF TITANIUM + ALUMINUM + ΣX_i Ti + Al + ΣX_i

Cur. No.	Ref. No.	Author(s)	Year	Method Used	Temp. Range, K	Name and Specimen Designation	Composition (weight percent) Ti	Al	Fe	Mo	Mn	Sn	C	V	Composition (continued), Specifications, and Remarks
1	462	Arp, V., Wilson, J. H., Winrich, L. and Sikora, P.	1962	L	20-293	A-110-AT	Bal.	5.5	0.2			2.5	0.07		0.02 H; hardness R_C 35; annealed.
2	503	Crucible Steel Co. of America (Pittsburgh, Pa.)	1958		293-1255	Crucible A-110AT	92.5	5.0				2.5			α-type alloy; 110 000 psi minimum yield strength; density 0.161 lb/cu. in.; melting range 1550-1650 C; mean coefficient of expansion recorded.
3*	462	Arp, V., Wilson, J. H., Winrich, L. and Sikora, P.	1962	L	20-293	C-120-AV	Bal.	6.2	0.1				0.01	4.0	0.01 H, 0.01 N; hardness R_C 36; annealed.
4	504	Bishop, S. M., Spretnak, J. W. and Fontana, M. G.	1953	L	73-293	RC-130-B	92.2	3.8			3.8		0.24		Specimen received as annealed 1.59 cm diameter rod; tested as received; data for coefficient of thermal linear expansion are also reported.
5*	505	Spretnak, J. W. and Fontana, M. G.	1952	L	297-77	RC-130-B									Specimen received from Rem-Cru Titanium, Inc.; annealed 1 hr at 978 K; expansion of 7.6 cm specimen measured with decreasing temperature; zero-point correction is 0.006%.
6*	505	Spretnak, J. W. and Fontana, M. G.	1952	L	77-297	RC-130-B									The above specimen; expansion measured with increasing temperature; zero-point correction is 0.006%.
7	506	McGee, W. M. and Matthew, B. R.	1962	L	9-922	Ti-4Al-3Mo-1V	Bal.	4.4	0.10	3.0			0.03	1.0	0.011 N_2, 0.0057 H_2; cylindrical specimen machined from 0.125 in. thick solution treated at 1655 F, aged (925 F, 12 hrs) sheet with their longer dimension parallel to the grain direction; heating rate 2 F per minute at low temperature, 5 F per minute at high temperature; error <3.0% for elevated temperature, <5% low temperature; zero-point correction of 0.011% determined by graphical interpolation.
8*	507	Lockheed Georgia Co.	1962		310-922	D3EE-1 Ti-4Al-3Mo-1V	Bal.	4.4	0.10	3.0			0.03	1.0	0.011 N_2, 57 ppm H_2; Crucible (heat no. R6736, sheet no. B-32); dimension 0.1 in. diameter, 1.968 in. long; cylindrical specimen machined from 0.125 in. thick sheet parallel to grain direction; solution treated at 1174 K for 15-30 min., oil quenched, aged at 769 K for 12 hrs and air cooled.
9*	507	Lockheed Georgia Co.	1962		310-922	D3EE-2 Ti-4Al-3Mo-1V									Similar to the above specimen.
10*	507	Lockheed Georgia Co.	1962		310-922	D3EE-3 Ti-4Al-3Mo-1V									Similar to the above specimen.

* Not shown in figure.

SPECIFICATION TABLE 281. THERMAL LINEAR EXPANSION OF TITANIUM + ALUMINUM + ΣX_i Ti + Al + ΣX_i (continued)

Cur. No.	Ref. No.	Author(s)	Year	Method Used	Temp. Range, K	Name and Specimen Designation	Composition (weight percent) Ti	Al	Fe	Mo	Mn	Sn	C	V	Composition (continued), Specifications, and Remarks
11*	507	Lockheed Georgia Co.	1962		310-922	Ti-4Al-3Mo-1V									Average value of D3EE-1, -2, -3.
12	507	Lockheed Georgia Co.	1962		3.7-250	D3EL-4 Ti-4Al-3Mo-1V									Similar to specimen D3EE-1 except the value at 3.7K is extrapolated; zero-point correction of -0.01% determined by graphical extrapolation.
13*	507	Lockheed Georgia Co.	1962		3.7-250	D3EL-5 Ti-4Al-3Mo-1V									Similar to the above specimen; zero-point correction of -0.01% determined by graphical extrapolation.
14*	507	Lockheed Georgia Co.	1962		3.7-250	D3EL-6 Ti-4Al-3Mo-1V									Similar to the above specimen; zero-point correction of -0.01% determined by graphical extrapolation.
15*	507	Lockheed Georgia Co.	1962		3.7-250	Ti-4Al-3Mo-1V									Average value of D3EL-4, -5, -6; zero-point correction of -0.01% determined by graphical extrapolation.
16	508	Dotson, C. L.	1967	L	293-818	Ti-5Al-5Sn-5Zr	84.1	5.3	0.05			5.1	0.025		5.3 Zr, 0.10 O, 0.011 N, 0.005 H; ingots press forged 16 in. square from 2050 F and to 3 x 12 in. sheet bars; slabs contained and ultrasonically inspected; sheet bars cut and rolled to intermediate size, from 1880-1900 F, descaled, acid pickled, and surfaces conditioned to remove defects; material finish-rolled from 1750 F, descaled and acid pickled, anneal-flattened at 1350 F for 8 hrs and rough ground; sheets final annealed at 1650 F for 0.50 hr, descaled finish ground, acid pickled, and tested as 0.040 in. gauge; expansion measured parallel to rolling direction; data for Run 2 on specimen; zero-point correction is ±0.00%.
17	508	Dotson, C. L.	1967	L	299-818	Ti-5Al-5Sn-5Zr	84.1	5.3				5.1			5.3 Zr; the above specimen; data for Run 3; zero-point correction of 0.005% determined by graphical extrapolation.
18	508	Dotson, C. L.	1967	L	492-296	Ti-5Al-5Sn-5Zr									The above specimen; expansion measured with decreasing temperature; zero-point correction of 0.002% determined by graphical extrapolation.
19	508	Dotson, C. L.	1967	L	297-825	Ti-5Al-5Sn-5Zr									The above specimen; expansion measured transverse to rolling direction; data for Run 1; zero-point correction of 0.003% determined by graphical extrapolation.
20	508	Dotson, C. L.	1967	L	772-300	Ti-5Al-5Sn-5Zr									The above specimen; expansion measured with decreasing temperature; zero-point correction of 0.03% determined by graphical extrapolation.

* Not shown in figure.

SPECIFICATION TABLE 281. THERMAL LINEAR EXPANSION OF TITANIUM + ALUMINUM + ΣX_i Ti + Al + ΣX_i (continued)

Cur. No.	Ref. No.	Author(s)	Year	Method Used	Temp. Range, K	Name and Specimen Designation	Composition (weight percent) Ti	Al	Fe	Mo	Mn	Sn	C	V	Composition (continued), Specifications, and Remarks
21*	508	Dotson, C. L.	1967	L	300-814	Ti-5Al-5Sn-5Zr	84.1	5.3				5.1			5.3 Zr; the above specimen; data for Run 2; zero-point correction of 0.005% determined by graphical extrapolation.
22*	508	Dotson, C. L.	1967	L	504-296	Ti-5Al-5Sn-5Zr									The above specimen; expansion measured with decreasing temperature; zero-point correction of 0.001% determined by graphical extrapolation.
23*	508	Dotson, C. L.	1967	L	302-816	Ti-5Al-5Sn-5Zr	84.1	5.3	0.05			5.1	0.025		5.3 Zr, 0.10 O, 0.011 N, 0.010 H; ingots press forged 16 in. square from 2050 F and to 4 in. square billets from 1950 F; billets thoroughly conditioned and ultrasonically inspected; billets re-cogged to 2.50 in. square from 1950 F, conditioned, rolled to 0.50 x 1.125 in. bars from 1925 F and descaled; rolled bars annealed-straightened at 1650 F for 2 hrs, A. C., descaled and tested; expansion measured parallel to rolling direction; data for Run 1 on specimen; zero-point correction of 0.006% determined by graphical extrapolation.
24*	508	Dotson, C. L.	1967	L	425-295	Ti-5Al-5Sn-5Zr									The above specimen; expansion measured with decreasing temperature; zero-point correction of 0.001% determined by graphical extrapolation.
25	508	Dotson, C. L.	1967	L	294-813	Ti-5Al-5Sn-5Zr	84.1	5.3				5.1			5.3 Zr, 0.10 O; the above specimen; data for Run 2; zero-point correction of 0.003% determined by graphical extrapolation.
26*	508	Dotson, C. L.	1967	L	684-294	Ti-5Al-5Sn-5Zr									The above specimen; expansion measured with decreasing temperature; zero-point correction of 0.006% determined by graphical extrapolation.
27	508	Dotson, C. L.	1967	L	299-813	Ti-5Al-5Sn-5Zr-1Mo-1V	83.5	4.9	0.12	1.0		4.7	0.031	1.0	4.7 Zr, 0.08 O, 0.012 H, 0.012 N; ingots press forged to 7 x 11 in. sections from 2050 F and to 3 x 12 in. slabs from 1950 F, conditioned, and ultrasonically inspected; sheet bars cut and rolled to intermediate size from 1860 F, descaled, acid-pickled, and the surfaces conditioned to remove defects; material finish rolled from 1700 F, descaled and acid pickled, anneal flattened at 1350 F (8 hrs) and rough ground; sheets were annealed at 1550 F (0.50 hrs) A. C., descaled and rough ground, annealed at 1400 F for 0.25 hrs A. C., descaled, finish ground, acid pickled, and tested at 0.040 in. gauge; expansion measured parallel to rolling direction; data for Run 1 on specimen; zero-point correction of 0.004% determined by graphical extrapolation.

* Not shown in figure.

SPECIFICATION TABLE 281. THERMAL LINEAR EXPANSION OF TITANIUM + ALUMINUM + ΣX_i Ti + Al + ΣX_i (continued)

Cur. No.	Ref. No.	Author(s)	Year	Method Used	Temp. Range, K	Name and Specimen Designation	Composition (weight percent) Ti	Al	Fe	Mo	Mn	Sn	C	V	Composition (continued), Specifications, and Remarks
28*	508	Dotson, C. L.	1967	L	630-297	Ti-5Al-5Sn-5Zr-1Mo-1V	83.5	4.9	0.12	1.0		4.7	0.031	1.0	4.7 Zr, 0.08 O, 0.012 H, 0.012 N; the above specimen; expansion measured with decreasing temperature; zero-point correction of 0.023% determined by graphical extrapolation.
29*	508	Dotson, C. L.	1967	L	299-816	Ti-5Al-5Sn-5Zr-1Mo-1V	83.5	4.9		1.0		4.7	0.031	1.0	4.7 Zr, 0.08 O, 0.012 H, 0.012 N; the above specimen; data for Run 2; zero-point correction of 0.006% determined by graphical extrapolation.
30	508	Dotson, C. L.	1967	L	719-298	Ti-5Al-5Sn-5Zr-1Mo-1V									The above specimen; expansion measured with decreasing temperature; zero-point correction of -0.002% determined by graphical extrapolation.
31*	508	Dotson, C. L.	1967	L	297-815	Ti-5Al-5Sn-5Zr-1Mo-1V									The above specimen; expansion measured transverse to rolling direction; data for Run 1; zero-point correction of 0.004% determined by graphical extrapolation.
32*	508	Dotson, C. L.	1967	L	694-298	Ti-5Al-5Sn-5Zr-1Mo-1V									The above specimen; expansion measured with decreasing temperature; zero-point correction of 0.019% determined by graphical extrapolation.
33*	508	Dotson, C. L.	1967	L	299-815	Ti-5Al-5Sn-5Zr-1Mo-1V									The above specimen; data for Run 2; zero-point correction of 0.004% determined by graphical extrapolation.
34*	508	Dotson, C. L.	1967	L	652-299	Ti-5Al-5Sn-5Zr-1Mo-1V									The above specimen; expansion measured with decreasing temperature; zero-point correction of 0.001% determined by graphical extrapolation.
35*	508	Dotson, C. L.	1967	L	300-816	Ti-6Al-2Sn-4Zr-2Mo	85.4	6.2	0.05	2.0					4.2 Zr, 0.08 O, 0.006 N, 0.005 H; ingots press forged to 7 x 11 in. section from 2050 F and to 3 x 12 in. section from 1950 F, conditioned and ultrasonically inspected; sheet bars cut and rolled to intermediate size from 1790-1800 F, descaled, acid pickled, and surfaces conditioned to remove defects; material finish rolled at 1750 F, descaled and acid pickled, anneal flattened at 1350 F for 8 hrs and rough ground; sheets annealed at 1650 F for 0.50 hrs A. C., descaled and rough ground, annealed at 1450 F for 0.25 hrs A. C., descaled, finish ground, acid pickled, and tested at 0.040 in. gauge; expansion measured parallel to rolling direction; data for Run 2 on specimen; zero-point correction of 0.005% determined by graphical extrapolation.

* Not shown in figure.

SPECIFICATION TABLE 281. THERMAL LINEAR EXPANSION OF TITANIUM + ALUMINUM + ΣX_i Ti + Al + ΣX_i (continued)

Cur. No.	Ref. No.	Author(s)	Year	Method Used	Temp. Range, K	Name and Specimen Designation	Composition (weight percent) Ti	Al	Fe	Mo	Mn	Sn	C	V	Composition (continued), Specifications, and Remarks
36	508	Dotson, C. L.	1967	L	629-299	Ti-6Al-2Sn-4Zr-2Mo	85.4	6.2	0.05	2.0		2.0	0.022		4.2 Zr, 0.08 O, 0.006 N, 0.005 H; the above specimen; expansion measured with decreasing temperature; zero-point correction of 0.005% determined by graphical extrapolation.
37*	508	Dotson, C. L.	1967	L	299-816	Ti-6Al-2Sn-4Zr-2Mo	85.4	6.2		2.0		2.0			The above specimen; data for Run 3; zero-point correction of 0.005% determined by graphical extrapolation.
38*	508	Dotson, C. L.	1967	L	554-299	Ti-6Al-2Sn-4Zr-2Mo									The above specimen; expansion measured with decreasing temperature; zero-point correction of -0.005% determined by graphical extrapolation.
39*	508	Dotson, C. L.	1967	L	300-813	Ti-6Al-2Sn-4Zr-2Mo									The above specimen; expansion measured transverse to rolling direction; data for Run 2; zero-point correction of 0.006% determined by graphical extrapolation.
40*	508	Dotson, C. L.	1967	L	636-299	Ti-6Al-2Sn-4Zr-2Mo									The above specimen; expansion measured with decreasing temperature; zero-point correction of 0.001% determined by graphical extrapolation.
41*	508	Dotson, C. L.	1967	L	299-538	Ti-6Al-2Sn-4Zr-2Mo	85.5	6.0	0.05	2.1		2.2	0.025		4.1 Zr, 0.010 N; similar to the above specimen; expansion measured transverse to rolling direction; data for Run 1 on specimen; zero-point correction of 0.005% determined by graphical extrapolation.
42*	508	Dotson, C. L.	1967	L	689-305	Ti-6Al-2Sn-4Zr-2Mo									The above specimen; expansion measured with decreasing temperature; zero-point correction of 0.01% determined by graphical extrapolation.
43	506	McGee, W. M. and Matthew, B. R.	1962	L	4-922	Ti-6Al-4V	Bal.	6.03	0.15				0.043	4.0	0.009N_2; cylindrical specimen machined from 0.125 in. thick solution treated at 1700 F, aged (900 F, 4 hrs) sheet with their longer dimension parallel to the grain direction; heating rate 2 F per minute at low temperature, 5 F per minute at high temperature; error <3.0% for elevated temperature, <5.0% for low temperature; zero-point correction of 0.016% determined by graphical interpolation.
44	507	Lockheed Georgia Co.	1962		310-922	B9EE-1 Ti-6Al-4V	Bal.	6.03	0.15				0.043	4.0	0.009 N_2; 800 ppm O, 63 ppm H; dimensions 0.1 in. diameter, 1.968 in. long; Reactive Metals (heat no. 32167, sheet no. 1777A-1); cylindrical specimen machined from 0.125 in. thick sheet parallel to grain direction; solution treated at 1199 K for 20 minutes, oil quenched, aged at 755 K for 4 hrs, and air cooled; specimen increased 0.025% in length after cooled from 922 K to 310 K.

* Not shown in figure.

SPECIFICATION TABLE 281. THERMAL LINEAR EXPANSION OF TITANIUM + ALUMINUM + ΣX_i Ti + Al + ΣX_i (continued)

Cur. No.	Ref. No.	Author(s)	Year	Method Used	Temp. Range, K	Name and Specimen Designation	Composition (weight percent) Ti	Al	Fe	Mo	Mn	Sn	C	V	Composition (continued), Specifications, and Remarks
45*	507	Lockheed Georgia Co.	1962		310-922	B9EE-2 Ti-6Al-4V	Bal.	6.03	0.15				0.043	4.0	0.009 N_2; similar to the above specimen except specimen increased 0.011%.
46*	507	Lockheed Georgia Co.	1962		310-922	B9EE-3 Ti-6Al-4V									Similar to the above specimen except no increment in length mentioned.
47*	507	Lockheed Georgia Co.	1962		310-922	Ti-6Al-4V									Average value of B9EE-1, -2, -3.
48	507	Lockheed Georgia Co.	1962		3.7-250	B9EL-4 Ti-6Al-4V	Bal.	6.03	0.15				0.043	4.0	0.009 N_2; similar to the specimen B9EE-3; the value at 3.7 K is extrapolated; zero-point correction of -0.018% determined by graphical extrapolation.
49*	507	Lockheed Georgia Co.	1962		3.7-250	B9EL-5 Ti-6Al-4V									Similar to the above specimen; zero-point correction of -0.018% determined by graphical extrapolation.
50*	507	Lockheed Georgia Co.	1962		3.7-250	B9EL-6 Ti-6Al-4V									Similar to the above specimen; zero-point correction of -0.018% determined by graphical extrapolation.
51*	507	Lockheed Georgia Co.	1962		3.7-250	Ti-6Al-4V									Average value of B9EL-4, -5, -6; zero-point correction of -0.018% determined by graphical extrapolation.
52	333	Williams, D. N.	1961		293-1699		Bal.	6.03						4.0	Specimen 0.625 in. diameter rod; prepared from 140 Bhn sponge; annealed; expansion given in increasing temperatures; beta transus temperature 1258 K; melting point 1866 K; measured in vacuum of about 3 x 10^{-4} mm Hg.
53	333	Williams, D. N.	1961		1699-293		Bal.	6						4	Similar to the above specimen except expansion given in decreasing temperatures.
54	15	Rhodes, B.L., Moeller, C. E., Hopkins, V. and Marx, T. I.	1963	I/L	18-573		Bal.	6.1	0.40				0.10	4.0	0.02 H, 0.07 N; specimen 10.16 cm long x 6.2 mm diameter.
55*	509	Deel, O. L. and Hyler, W. S.	1968	L	294-755	Ti-6Al-4V-3Co	Bal.	6.11	0.17				0.015	3.88	3.28 Co, 0.12 O_2; specimen from Cobalt Information Center; solution treated 45 min. at 1061 K, water quenched, aged 4 hrs at 755 K; density 0.164 lb in.$^{-3}$; expansion measured under 2 x 10^{-5} mm Hg vacuum with increasing temperature; zero-point correction is ±0.000%.
56	509	Deel, O. L. and Hyler, W. S.	1963	L	755-294	Ti-6Al-4V-3Co									The above specimen; expansion measured with decreasing temperature; zero-point correction of 0.019% determined by graphical extrapolation.

* Not shown in figure.

SPECIFICATION TABLE 281. THERMAL LINEAR EXPANSION OF TITANIUM + ALUMINUM + ΣX_i Ti + Al + ΣX_i (continued)

Cur. No.	Ref. No.	Author(s)	Year	Method Used	Temp. Range, K	Name and Specimen Designation	Composition (weight percent) Ti	Al	Fe	Mo	Mn	Sn	C	V	Composition (continued), Specifications, and Remarks
57*	510	Neimark, B. E., Korytina, S. F. and Monina, E. F.	1969		293-973	Russian Alloy VT-5	Bal.	5.0	0.3				0.05		0.15 Si, 0.15 O_2, 0.04 N_2, 0.015 H_2; annealed at 1073 K; expansion measured under 10^{-3} to 10^{-4} mm Hg vacuum.
58	510	Neimark, B. E., et al.	1969		293-1073	Russian Alloy VT-8	Bal.	5.8/ 6.0	0.4	2.8/ 3.8			0.1		0.2 Si, 0.2 O_2, 0.05 N_2, 0.01 H_2; hardening in water at temperature of 1223 K and subsequently tempered at 873 K.
59	511	Hertz, J.	1962		107-292	5.0 Al-2.5 Sn Ti Alloy	Bal.	5.0				2.5			No details given; zero-point correction of -0.001% determined by graphical extrapolation.

* Not shown in figure.

DATA TABLE 281. THERMAL LINEAR EXPANSION OF TITANIUM + ALUMINUM + ΣX_i Ti + Al + ΣX_i

[Temperature, T, K; Linear Expansion, $\Delta L/L_0$, %]

T	$\Delta L/L_0$
CURVE 1	
20	-0.175
40	-0.172
60	-0.168
80	-0.162
100	-0.153
120	-0.143
140	-0.130
160	-0.116
180	-0.100
200	-0.084
220	-0.067
240	-0.050
260	-0.031
273	-0.018*
280	-0.012
293	0.000
CURVE 2	
293	0.000*
366	0.069*
478	0.173*
589	0.282*
700	0.388*
811	0.503*
922	0.623*
1033	0.746
1144	0.873
1255	0.987
CURVE 3*	
20	-0.173
40	-0.171
60	-0.167
80	-0.162
100	-0.154
120	-0.141
140	-0.126
160	-0.110
180	-0.094
200	-0.078
220	-0.062
240	-0.045
260	-0.027
273	-0.016
CURVE 3 (cont.)*	
280	-0.009
293	0.000
CURVE 4‡	
73	-2.16
93	-2.05
113	-1.93
133	-1.79
153	-1.62
173	-1.44
193	-1.24
213	-1.04
233	-0.81
253	-0.57
273	-0.20
293	0.00*
CURVE 5*	
77	-0.210
93	-0.202
113	-0.190
133	-0.176
153	-0.159
173	-0.141
193	-0.122
213	-0.100
233	-0.078
253	-0.053
273	-0.028
293	0.000
297	0.006
CURVE 6*	
77	-0.210
93	-0.203
113	-0.191
133	-0.178
153	-0.163
173	-0.146
193	-0.126
213	-0.105
233	-0.082
253	-0.056
CURVE 6 (cont.)*	
273	-0.029
293	0.000
297	0.006
CURVE 7	
9	-0.186
33	-0.185
89	-0.160
144	-0.120
200	-0.079
255	-0.033
311	0.016
366	0.061
422	0.117
478	0.168
533	0.223
589	0.278
644	0.333
700	0.390
755	0.448
811	0.505
866	0.562
922	0.621
CURVE 8*	
310	0.017
366	0.069
422	0.118
477	0.172
533	0.227
588	0.281
644	0.334
699	0.390
755	0.445
810	0.502
866	0.559
922	0.617
CURVE 9*	
310	0.015
366	0.063
422	0.119
477	0.174
CURVE 9 (cont.)*	
533	0.230
588	0.285
644	0.340
699	0.397
755	0.457
810	0.511
866	0.569
922	0.631
CURVE 10*	
310	0.016
366	0.067
422	0.119
477	0.174
533	0.231
588	0.283
644	0.339
699	0.397
755	0.451
810	0.510
866	0.569
922	0.631
CURVE 11*	
310	0.016
366	0.067
422	0.119
477	0.174
533	0.230
588	0.284
644	0.338
699	0.395
755	0.452
810	0.508
866	0.566
922	0.627
CURVE 12	
3.7	-0.18*
25	-0.18
50	-0.17
75	-0.15
100	-0.14
CURVE 12 (cont.)	
125	-0.12
150	-0.11
175	-0.09
250	-0.07*
225	-0.05
250	-0.03
CURVE 13*	
3.7	-0.18
25	-0.17
50	-0.16
75	-0.15
100	-0.13
125	-0.12
150	-0.10
175	-0.08
200	-0.07
225	-0.05
250	-0.03
CURVE 14*	
3.7	-0.18
25	-0.18
50	-0.17
75	-0.16
100	-0.15
125	-0.13
150	-0.11
175	-0.10
200	-0.08
225	-0.06
250	-0.03
CURVE 15*	
3.7	-0.18
25	-0.18
50	-0.17
75	-0.15
100	-0.14
125	-0.12
150	-0.10
175	-0.09
200	-0.07
CURVE 15 (cont.)*	
225	-0.05
250	-0.03
CURVE 16	
294	0.000*
297	0.003*
376	0.058
403	0.079
426	0.102
480	0.153
543	0.211
601	0.271
650	0.315
664	0.328
723	0.389
771	0.487
810	0.479
818	0.489
CURVE 17	
299	0.005*
384	0.067*
463	0.148
536	0.222*
569	0.255
583	0.263*
699	0.380
769	0.460
810	0.507
CURVE 18	
492	0.159*
413	0.091
296	0.002*
CURVE 19	
297	0.003*
344	0.038
376	0.068*
389	0.078
486	0.164
572	0.222

* Not shown in figure.
‡ Author's data for coefficient of thermal expansion have been integrated by TPRC to obtain $\Delta L/L_0$.

DATA TABLE 281. THERMAL LINEAR EXPANSION OF TITANIUM + ALUMINUM + ΣX_i Ti + Al + ΣX_i (continued)

T	$\Delta L/L_0$
CURVE 19 (cont.)	
589	0.262
681	0.350
766	0.425
817	0.479
822	0.483*
825	0.483*
CURVE 20	
772	0.459*
687	0.372
616	0.300
545	0.223
300	0.004*
CURVE 21	
300	0.005*
371	0.063*
482	0.164*
580	0.271*
705	0.399
814	0.514
CURVE 22	
504	0.182
461	0.144*
296	0.001*
CURVE 23	
302	0.006*
402	0.079*
439	0.114
471	0.149
497	0.174*
550	0.223*
585	0.259*
616	0.291
664	0.343
705	0.381
729	0.403
752	0.427
776	0.455
816	0.501
CURVE 24	
425	0.125
410	0.104*
306	0.081
295	0.001*
CURVE 25	
294	0.003*
297	0.003*
335	0.028
337	0.035*
359	0.057
360	0.057*
399	0.092
414	0.106
476	0.160*
480	0.163*
535	0.219
539	0.222*
576	0.265
643	0.332*
694	0.378*
761	0.446
796	0.488
813	0.501*
CURVE 26	
684	0.367*
610	0.290
534	0.206*
510	0.188
302	0.006*
294	0.003*
CURVE 27	
299	0.004*
366	0.052
492	0.167
577	0.255
701	0.375*
813	0.486*
CURVE 28	
630	0.309
469	0.147
297	0.004*
CURVE 29	
299	0.006*
375	0.070
485	0.172
594	0.284
697	0.394
816	0.518*
CURVE 30	
719	0.406
654	0.333
524	0.199
298	0.004*
CURVE 31*	
297	0.004
385	0.079
489	0.171
535	0.216
575	0.253
646	0.324
726	0.401
815	0.497
CURVE 32	
694	0.379
458	0.140
298	0.003*
CURVE 33*	
299	0.004
370	0.062
481	0.163
576	0.262
653	0.345
706	0.399
815	0.515
CURVE 34*	
652	0.329
529	0.205
436	0.123
299	0.007
CURVE 35*	
300	0.005
306	0.005
379	0.066
388	0.070
489	0.168
587	0.256
706	0.405
816	0.487
CURVE 36*	
629	0.290
524	0.186
299	0.005
CURVE 37*	
299	0.005
380	0.067
483	0.164
583	0.262
704	0.390
816	0.513
CURVE 38*	
554	0.228
299	0.005
CURVE 39*	
300	0.006
376	0.062
483	0.169
578	0.264
697	0.381
813	0.501
CURVE 40*	
636	0.322
463	0.169
299	0.007
CURVE 41*	
299	0.005
391	0.080
480	0.165
CURVE 41 (cont.)*	
594	0.277
696	0.358
811	0.470
CURVE 42*	
689	0.354
442	0.114
305	0.01
CURVE 43	
4	-0.217
33	-0.207
89	-0.179
144	-0.137
200	-0.087
255	-0.033
311	0.016
366	0.072
422	0.127
478	0.182
533	0.237
589	0.292
644	0.348
700	0.405
755	0.464
811	0.524
866	0.584
922	0.641
CURVE 44	
310	0.017*
366	0.071
422	0.127
477	0.185
533	0.241*
588	0.302*
644	0.360
699	0.420
755	0.480
810	0.543
866	0.605
922	0.670
CURVE 45*	
310	0.016
366	0.068
422	0.122
477	0.178
533	0.232
588	0.288
644	0.345
699	0.404
755	0.462
810	0.523
866	0.582
922	0.645
CURVE 46	
310	0.016*
366	0.068*
422	0.120
477	0.174
533	0.227
588	0.281*
644	0.336
699	0.391
755	0.446
810	0.504
866	0.558
922	0.617*
CURVE 47*	
310	0.017
366	0.070
422	0.124
477	0.180
533	0.234
588	0.291
644	0.348
699	0.406
755	0.463
810	0.524
866	0.582
922	0.645
CURVE 48	
3.7	-0.213*
25	-0.208
50	-0.200
75	-0.184

* Not shown in figure.

DATA TABLE 281. THERMAL LINEAR EXPANSION OF TITANIUM + ALUMINUM + ΣX_i Ti + Al + ΣX_i (continued)

T	$\Delta L/L_0$
CURVE 48 (cont.)	
100	-0.170
125	-0.152
150	-0.132
175	-0.111
200	-0.089*
225	-0.065
250	-0.041
CURVE 49*	
3.7	-0.212
25	-0.206
50	-0.195
75	-0.181
100	-0.164
125	-0.146
150	-0.127
175	-0.107
200	-0.085
225	-0.063
250	-0.040
CURVE 50*	
3.7	-0.214
25	-0.209
50	-0.203
75	-0.185
100	-0.168
125	-0.152
150	-0.130
175	-0.109
200	-0.087
225	-0.064
250	-0.041
CURVE 51*	
3.7	-0.213
25	-0.208
50	-0.199
75	-0.183
100	-0.167
125	-0.150
150	-0.130
175	-0.109
200	-0.087
225	-0.064
250	-0.041
CURVE 52	
293	0.000*
366	0.070*
477	0.178*
588	0.292*
699	0.410*
810	0.534*
922	0.666*
1033	0.804
1144	0.912
1255	0.854
1366	0.988
1477	1.142
1588	1.296*
1699	1.450*
CURVE 53	
1699	1.739*
1588	1.563*
1477	1.399
1366	1.239
1255	1.099
1144	0.965
1033	0.813
922	0.667*
810	0.531*
699	0.407*
588	0.291*
477	0.179*
366	0.071*
293	0.000*
CURVE 54	
18	-0.158*
23	-0.158*
33	-0.157*
43	-0.156*
53	-0.155
63	-0.154*
73	-0.152*
83	-0.149
93	-0.146*
103	-0.142*
113	-0.138
123	-0.133*
133	-0.128*
143	-0.122*
153	-0.166*
163	-0.109*
CURVE 54 (cont.)	
173	-0.102*
193	-0.087*
213	-0.072
233	-0.054
253	-0.036*
273	-0.019*
293	0.000*
313	0.018*
333	0.037*
353	0.056*
373	0.076
393	0.097*
413	0.117*
433	0.138
453	0.158
473	0.180*
493	0.200
513	0.221
533	0.242*
553	0.264
573	0.285
CURVE 55	
294	0.000
373	0.071
422	0.119
472	0.171
519	0.223
575	0.277
622	0.331
672	0.386
722	0.440
755	0.473
CURVE 56	
755	0.492
722	0.457
672	0.402
622	0.348
575	0.291*
519	0.238*
472	0.185
422	0.132*
373	0.082
294	0.002*
CURVE 57*	
293	0.000
323	0.0246
373	0.0719
423	0.120
473	0.170
523	0.221
573	0.274
623	0.329
673	0.382
723	0.433
773	0.491
823	0.547
873	0.603
923	0.660
973	0.716
CURVE 58	
293	0.000
323	0.0235
373	0.0661
423	0.115*
473	0.167*
523	0.220
573	0.274
623	0.327
673	0.381
723	0.438
773	0.489*
823	0.544
873	0.594
923	0.642*
973	0.696
1023	0.760
1073	0.824
CURVE 59	
107	-0.129
132	-0.116
145	-0.108
186	-0.078
198	-0.068
224	-0.045
252	-0.021
261	-0.015
282	-0.001*
292	-0.001*

* Not shown in figure.

DATA TABLE 281. COEFFICIENT OF THERMAL LINEAR EXPANSION OF TITANIUM + ALUMINUM + ΣX_i Ti + Al + ΣX_i

[Temperature, T, K; Coefficient of Expansion, α, 10^{-6} K^{-1}]

T	α
CURVE 4‡	
73	5.3
93	5.7
113	6.7
133	7.7
153	8.9
173	9.4
193	9.9
213	10.9
233	11.6
253	12.7
273	13.7
293	16.7

‡ Author's data for coefficient of thermal expansion have been integrated by TPRC to obtain $\Delta L/L_0$.

FIGURE AND TABLE NO. 282R. PROVISIONAL VALUES FOR THERMAL LINEAR EXPANSION OF TITANIUM + CHROMIUM + ΣX_i ALLOYS $Ti + Cr + \Sigma X_i$

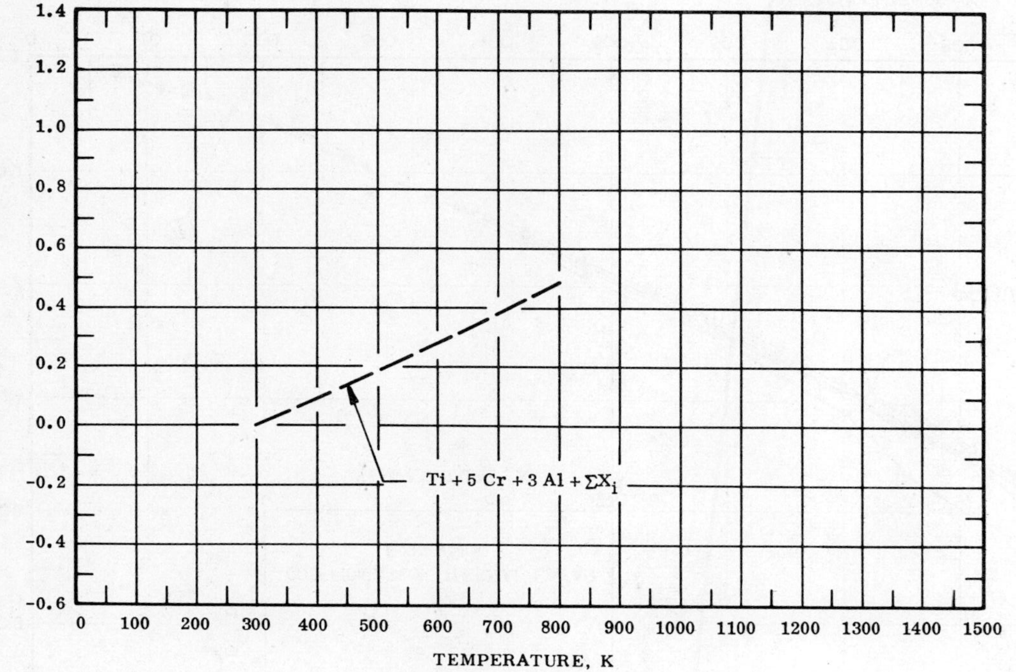

PROVISIONAL VALUES

[Temperature, T, K; Linear Expansion, $\Delta L/L_0$, %; α, K^{-1}]

T	$\Delta L/L_0$	$\alpha \times 10^6$
($Ti + 5Cr + 3Al + \Sigma X_i$)		
293	0.000	8.5
400	0.095	9.1
500	0.187	9.6
600	0.286	10.0
700	0.388	10.3
800	0.492	10.6

REMARKS

($Ti + 5Cr + 3Al + \Sigma X_i$): The tabulated values for well-annealed alloy are considered accurate to within ± 7% over the entire temperature range. These values can be represented approximately by the following equation:

$$\Delta L/L_0 = -0.219 + 6.355 \times 10^{-4}\, T + 4.173 \times 10^{-7}\, T^2 - 1.269 \times 10^{-10}\, T^3$$

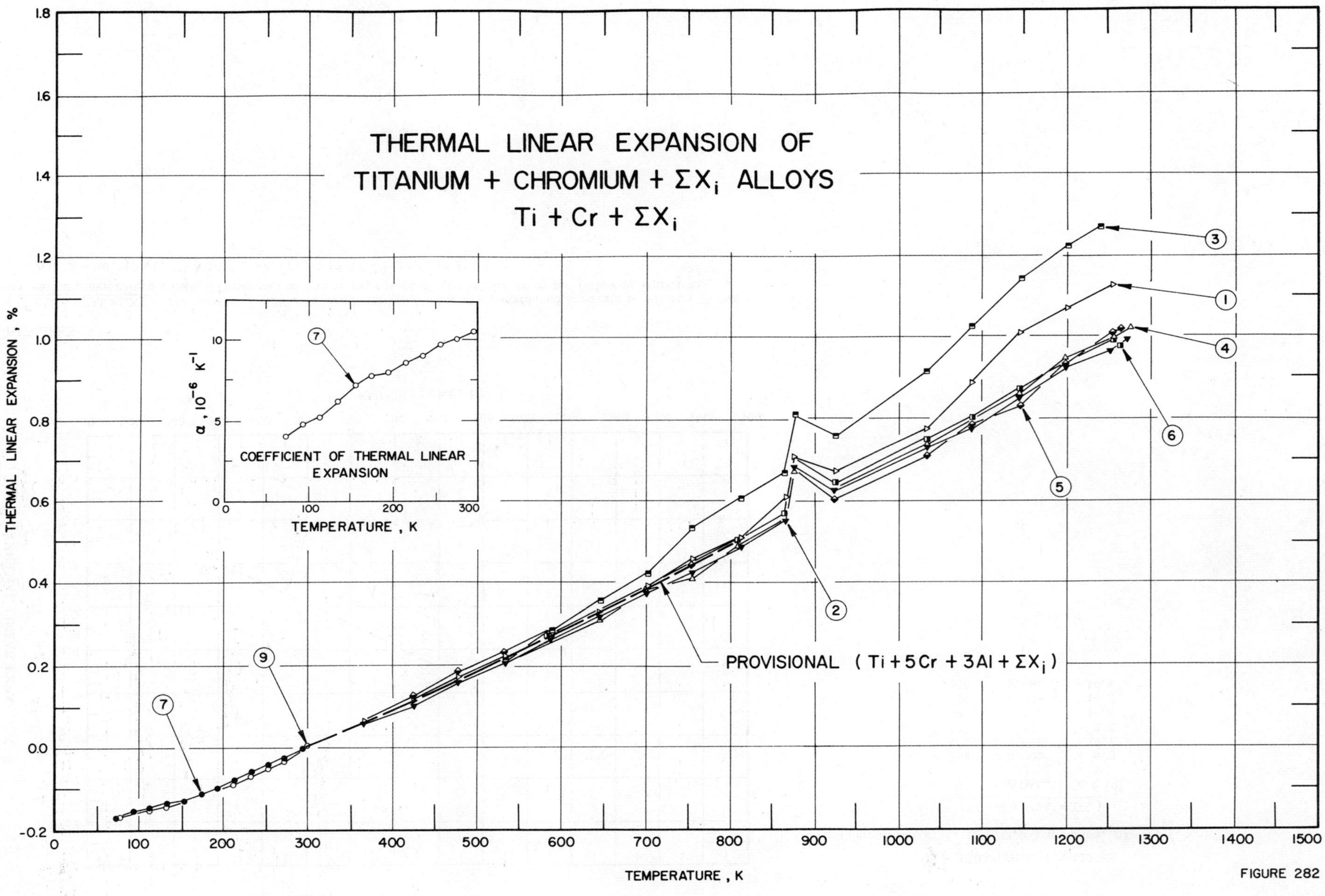
THERMAL LINEAR EXPANSION OF
TITANIUM + CHROMIUM + ΣXi ALLOYS
Ti + Cr + ΣXi
PROVISIONAL (Ti + 5 Cr + 3 Al + ΣXi)
COEFFICIENT OF THERMAL LINEAR EXPANSION
TEMPERATURE , K
α , 10⁻⁶ K⁻¹
THERMAL LINEAR EXPANSION , %
TEMPERATURE , K
1
2
3
4
5
6
7
9

FIGURE 282

SPECIFICATION TABLE 282. THERMAL LINEAR EXPANSION OF TITANIUM + CHROMIUM + ΣX_i ALLOYS $Ti + Cr + \Sigma X_i$

Cur No.	Ref. No.	Author(s)	Year	Method Used	Temp. Range, K	Name and Specimen Designation	Composition (weight percent) Ti	Cr	Al	Fe	C	N	H_2	Composition (continued), Specifications, and Remarks
1	512	Materials Lab., Wright-Patterson AFB, Ohio	1958	L	367-1255	1st run - A Ti-3Al-5Cr	Bal.	4.94	3.47	0.25	0.05	0.036	0.0244	0.625 in. bar stock from Mallory-Sharon Titanium Corp; machined into 0.125 in. diameter, 1.75 in. long; mill annealed at 1199 K 10 min, water quenched; tested under pressure of 0.8 micron; anomaly for this and the following specimens between 860-920 K possibly due to evolution of H_2; between 930-1000 K possibly due to formation of $TiCr_2$ observed; expansion measured with increasing temperature.
2	512	Materials Lab., Wright-Patterson AFB, Ohio	1958	L	1272-367	2nd run - A Ti-3Al-5Cr								The above specimen; expansion measured with decreasing temperature.
3	512	Materials Lab., Wright-Patterson AFB, Ohio	1958	L	367-1241	1st run - B Ti-3Al-5Cr	Bal.	4.94	3.47	0.25	0.05	0.036	0.0188	Similar to the above specimen; tested under pressure; expansion measured with increasing temperature.
4	512	Materials Lab., Wright-Patterson AFB, Ohio	1958	L	1278-367	2nd run - B Ti-3Al-5Cr								The above specimen; vacuum cooled from 1241 K; expansion measured with decreasing temperature.
5	512	Materials Lab., Wright-Patterson AFB, Ohio	1958	L	367-1264	1st run - C Ti-3Al-5Cr	Bal.	4.94	3.47	0.25	0.05	0.036	0.0168	0.625 in. bar stock from Mallory-Sharon Titanium Corp; machined into 0.125 in. diameter, 1.75 in. long, following heat treatment; mill annealed, 811 K, 26 hr, air cooled; in heating cycle, 1 F per min; expansion measured with increasing temperature.
6	512	Materials Lab., Wright-Patterson AFB, Ohio	1958	L	1264-367	2nd run - C Ti-3Al-5Cr								The above specimen; vacuum cooled from 1260 K; expansion measured with decreasing temperature.
7	504	Bishop, S.M., Spretnak, J.W., and Fontana, M.G.	1953	L	73-293	Ti-150-A	95.9	2.7		1.3	<0.05	<0.08		Specimen received as annealed 1.59 cm diameter rod; tested as received; data on the coefficient of thermal linear expansion are also reported.
8*	505	Spretnak, J.W. and Fontana, M.G.	1952	L	297-77	Ti-150-A								Specimen received from Titanium Metals Corp. of America; annealed 6 hr at 922 K; expansion measured with decreasing temperature; zero-point correction is 0.006%.
9	505	Spretnak, J.W. and Fontana, M.G.	1952	L	77-297	Ti-150-A								The above specimen; expansion measured with increasing temperature; zero-point correction is 0.006%.

* Not shown in figure.

DATA TABLE 282. THERMAL LINEAR EXPANSION OF TITANIUM + CHROMIUM + ΣX_i ALLOYS Ti + Cr + ΣX_i

[Temperature, T, K; Linear Expansion, $\Delta L/L_0$, %]

T	$\Delta L/L_0$
CURVE 1	
367	0.068
422	0.119
478	0.172
533	0.223
589	0.274
644	0.331
700	0.398
755	0.460
811	0.515
866	0.611
922	0.670
878	0.705
1033	0.775
1089	0.890
1144	1.013
1200	1.079*
1255	1.127
CURVE 2	
1272	0.995
1255	0.961
1200	0.921
1144	0.849
1089	0.779
1033	0.723
878	0.681
922	0.625
866	0.548
811	0.485
755	0.424
700	0.372
644	0.320
589	0.263
533	0.208
478	0.161
422	0.108
367	0.061
CURVE 3	
367	0.065*
422	0.114*
478	0.163*
CURVE 3 (cont.)	
533	0.216
589	0.287
644	0.360
700	0.424
755	0.531
811	0.604
866	0.669
922	0.757
878	0.812
1033	0.917
1089	1.029
1144	1.147
1200	1.222
1241	1.278
CURVE 4	
1278	1.030
1255	1.006
1200	0.948
1144	0.860
1089	0.804
1033	0.735
878	0.680
922	0.625*
866	0.549*
811	0.487
755	0.419
700	0.383*
644	0.319
589	0.269*
533	0.210
478	0.165
422	0.116
367	0.066
CURVE 5	
367	0.073*
422	0.129
478	0.184
533	0.233
589	0.285*
644	0.337*
CURVE 5 (cont.)	
700	0.389*
755	0.441
811	0.493*
866	0.545*
922	0.608
878	0.674*
1033	0.714
1089	0.782
1144	0.836
1200	0.938
1255	1.013
1264	1.021
CURVE 6	
1264	0.978
1255	0.987
1200	0.930*
1144	0.873
1089	0.801
1033	0.745
922	0.645
878	0.702*
866	0.567
811	0.503
755	0.440*
700	0.380*
644	0.328*
589	0.271
533	0.216*
478	0.166*
422	0.116*
367	0.066*
CURVE 7‡	
73	-0.167
93	-0.158
113	-0.148
133	-0.137
153	-0.123
173	-0.108
193	-0.093
213	-0.076
CURVE 7 (cont.)‡	
233	-0.059
253	-0.040
273	-0.021
293	0.000
CURVE 8*	
297	0.006
293	0.000
273	-0.022
253	-0.042
233	-0.062
213	-0.078
193	-0.095
173	-0.111
153	-0.126
133	-0.139
113	-0.151
93	-0.161
77	-0.167
CURVE 9	
77	-0.167
93	-0.162*
113	-0.153
133	-0.141
153	-0.129*
173	-0.114*
193	-0.099*
213	-0.083
233	-0.065
253	-0.046
273	-0.024
293	0.000*
297	0.006

*Not shown in figure.
‡Author's data for coefficient of thermal expansion have been integrated by TPRC to obtain $\Delta L/L_0$.

DATA TABLE 282. COEFFICIENT OF THERMAL LINEAR EXPANSION OF TITANIUM + CHROMIUM + ΣX_i ALLOYS $Ti + Cr + \Sigma X_i$

[Temperature, T, K; Coefficient of Expansion, α, 10^{-6} K^{-1}]

T	α
CURVE 7‡	
73	4.0
93	4.8
113	5.2
133	6.2
153	7.2
173	7.7
193	8.0
213	8.5
233	8.9
253	9.6
273	10.0
293	10.6

‡Author's data for coefficient of thermal expansion have been integrated by TPRC to obtain $\Delta L/L_0$.

FIGURE AND TABLE NO. 283AR. PROVISIONAL VALUES FOR THERMAL LINEAR EXPANSION OF TITANIUM + TIN + ΣX_i ALLOYS Ti + Sn + ΣX_i

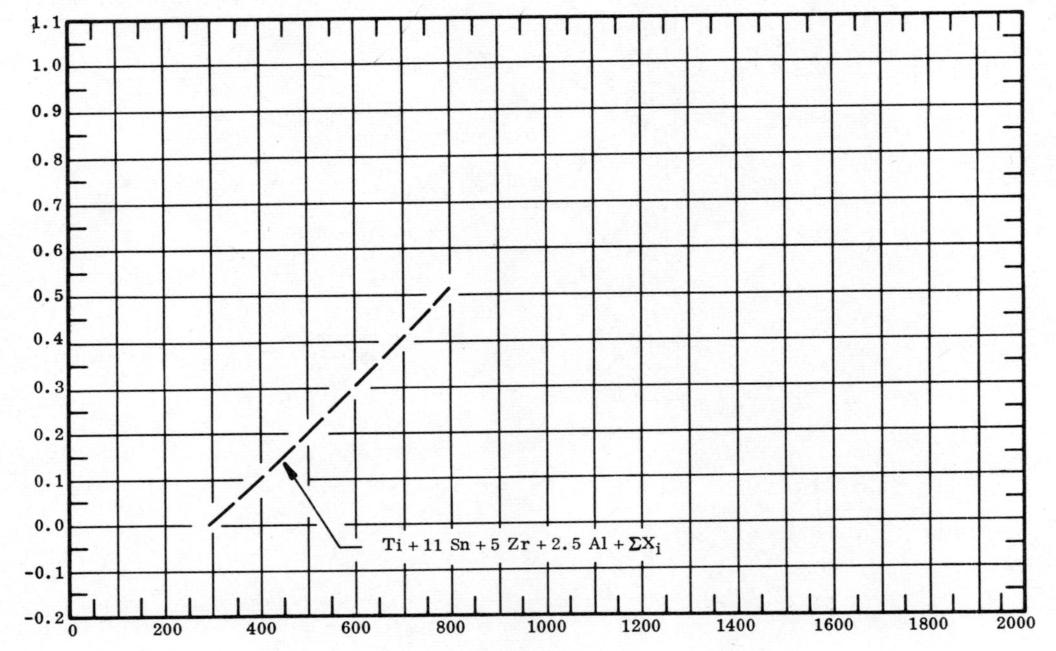

PROVISIONAL VALUES

[Temperature, T, K; Linear Expansion, $\Delta L/L_0$, %; α, K^{-1}]

(Ti + 11 Sn + 5 Zr + 2.5 Al + ΣX_i)

T	$\Delta L/L_0$	$\alpha \times 10^6$
293	0.000	8.4
400	0.094	9.4
500	0.193	10.1
600	0.296	10.6
700	0.403	10.8
800	0.512	10.9

REMARKS

(Ti + 11 Sn + 5 Zr + 2.5 Al + ΣX_i): The tabulated values for well-annealed alloy are considered accurate to within ±7% over the entire temperature range. These values can be represented approximately by the following equation:

$$\Delta L/L_0 = -0.191 + 4.314 \times 10^{-4} T + 8.597 \times 10^{-7} T^2 - 3.741 \times 10^{-10} T^3$$

FIGURE AND TABLE NO. 283BR. PROVISIONAL VALUES FOR THERMAL LINEAR EXPANSION OF TITANIUM + VANADIUM + ΣX_i ALLOYS Ti + V + ΣX_i

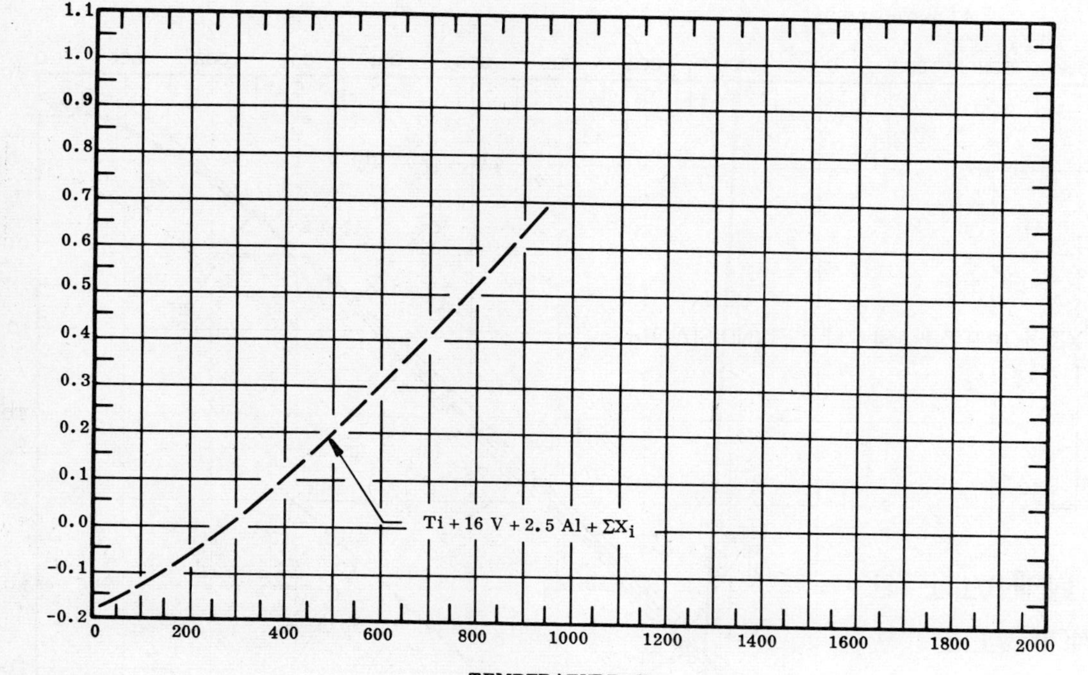

PROVISIONAL VALUES

[Temperature, T, K; Linear Expansion, $\Delta L/L_0$, %; α, K^{-1}]

(Ti + 16 V + 2.5 Al + ΣX_i)		
T	$\Delta L/L_0$	$\alpha \times 10^6$
10	-0.177	4.2
50	-0.159	4.8
100	-0.133	5.6
200	-0.069	7.1
293	0.000	8.2
400	0.095	9.3
500	0.193	10.1
600	0.297	10.7
700	0.407	11.1
800	0.520	11.4
900	0.634	11.4
950	0.691	11.4

REMARKS

(Ti + 16 V + 2.5 Al + ΣX_i): The tabulated values for well-annealed alloy are considered accurate to within ± 7% over the entire temperature range. These values can be represented approximately by the following equation:

$$\Delta L/L_0 = -0.183 + 4.005 \times 10^{-4}\, T + 8.636 \times 10^{-7}\, T^2 - 3.358 \times 10^{-10}\, T^3$$

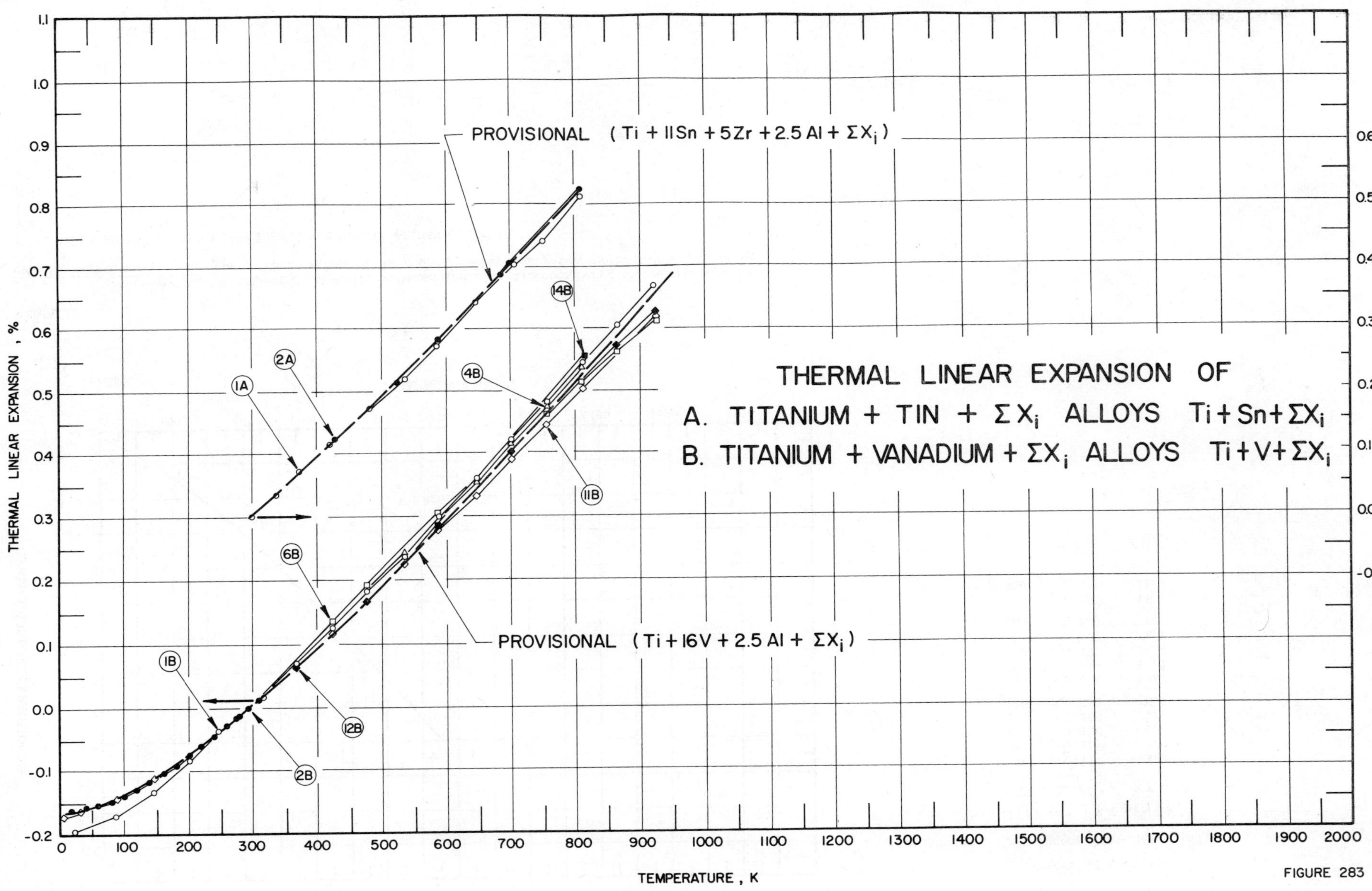
PROVISIONAL (Ti + 11Sn + 5Zr + 2.5 Al + ΣXi)
THERMAL LINEAR EXPANSION OF
A. TITANIUM + TIN + Σ Xi ALLOYS Ti + Sn + ΣXi
B. TITANIUM + VANADIUM + ΣXi ALLOYS Ti + V + ΣXi
PROVISIONAL (Ti + 16V + 2.5 Al + ΣXi)
1A
2A
14B
4B
11B
6B
1B
12B
2B
THERMAL LINEAR EXPANSION, %
1.1
1.0
0.9
0.8
0.7
0.6
0.5
0.4
0.3
0.2
0.1
0.0
-0.1
-0.2
0.6
0.5
0.4
0.3
0.2
0.1
0.0
-0.1
0
100
200
300
400
500
600
700
800
900
1000
1100
1200
1300
1400
1500
1600
1700
1800
1900
2000
TEMPERATURE, K

FIGURE 283

SPECIFICATION TABLE 283A. THERMAL LINEAR EXPANSION OF TITANIUM + TIN + ΣX_i ALLOYS   Ti + Sn + ΣX_i

Cur. No.	Ref. No.	Author(s)	Year	Method Used	Temp. Range, K	Name and Specimen Designation	Composition (weight percent) Ti	Sn	Zr	Al	Mo	Si	O	Composition (continued), Specifications, and Remarks
1	508	Dotson, C. L.	1967	L	295-810	Ti-679	80.7	10.8	4.7	2.4	0.97	0.23	0.127	0.08 Fe, 0.023 C, 0.012 N, and 0.007 H; billet stock 8 in. round press-forged to 4 in. square billets from 1185 K, conditioned, and ultrasonically inspected; billets were re-cogged to 2.5 in. square from 1185-1200 K, conditioned, rolled to 0.5 x 1.125 in. bars from 1185 K and descaled; rolled bars annealed-straightened at 1170 K for 2 hr, A.C. and 770 K 24 hr, A.C., descaled and tested; expansion measured parallel to rolling direction; data for specimen on run 2; zero-point correction of 0.005% determined by graphical extrapolation.
2	508	Dotson, C. L.	1967	L	299-810	Ti-679								The above specimen; data for run 3; zero-point correction of 0.007% determined by graphical extrapolation.

DATA TABLE 283A. THERMAL LINEAR EXPANSION OF TITANIUM + TIN + ΣX_i ALLOYS   Ti + Sn + ΣX_i

[Temperature, T, K; Linear Expansion, $\Delta L/L_0$, %]

T	$\Delta L/L_0$	T	$\Delta L/L_0$
CURVE 1		CURVE 2	
297	0.005	300	0.007
337	0.037	373	0.071*
372	0.072	429	0.121
420	0.115	481	0.177*
482	0.172	525	0.217
535	0.221	588	0.286
585	0.274	685	0.389
646	0.342	698	0.409
708	0.404	810	0.528
750	0.446		
810	0.516		

*Not shown in figure.

SPECIFICATION TABLE 283B. THERMAL LINEAR EXPANSION OF TITANIUM + VANADIUM + ΣX_i Ti + V + ΣX_i

Cur. No.	Ref. No.	Author(s)	Year	Method Used	Temp. Range, K	Name and Specimen Designation	Composition (weight percent) Ti	V	Cr	Al	Fe	N	C	Composition (continued), Specifications, and Remarks
1	506	McGee, W. M. and Matthew, B. R.	1962	L	2-922	B-120-VCA Ti Alloy	Bal.	13.9	10.4	3.5	0.25		0.04	0.025 N_2, 0.0114 H_2; cylindrical specimen machined from 0.125 in. thick solution treated at 1450 F, aged (900 F, 60 hrs) sheet with their longer dimension parallel to grain direction; heating rate 2 F per minute at low temperature, 5 F per minute at high temperature; error <3.0% for elevated temperature, <5.0% low temperature; zero-point correction of 0.016% determined by graphical interpolation.
2	462	Arp, V., Wilson, J. H., Winrich, L. and Sikora, P.	1962	L	20-293	B-120-VCA	Bal.	13.5	10.8	3.0	0.2	0.02	0.03	0.01 H; hardness R_C 34, annealed.
3*	507	Lockheed Georgia Co.	1962		311-922	B-120-VCA A3EE-2	Bal.	13.9	10.4	3.5	0.25		0.04	114 ppm H; dimension 0.1 in. diameter, 1.968 in. long; cylindrical specimen machined from 0.125 in. thick sheet parallel to grain direction; solution treated at 1060 K for 20 minutes, air cooled, aged at 755 K for 60 hrs, and air cooled.
4	507	Lockheed Georgia Co.	1962		311-922	A3EE-3								Similar to the above specimen.
5*	507	Lockheed Georgia Co.	1962		311-922									Average value of A3EE-2, -3.
6	507	Lockheed Georgia Co.	1962		311-922	A3EE-1								Similar to the specimen A3EE-2 except decreased 0.045% in length after cooled from 922 K to 311 K.
7*	507	Lockheed Georgia Co.	1962		307-250	A3EL-4								Similar to the above specimen except the value at -453 F is extrapolated; zero-point correction of -0.02% determined by graphical extrapolation.
8*	507	Lockheed Georgia Co.	1962		307-250	A3EL-5								Similar to the above specimen; zero-point correction of -0.02% determined by graphical extrapolation.
9*	507	Lockheed Georgia Co.	1962		307-250	A3EL-6								Similar to the above specimen; zero-point correction of -0.02% determined by graphical extrapolation.
10*	507	Lockheed Georgia Co.	1962		307-250									Average value of A3EL-4, -5, -6; zero-point correction of -0.02% determined by graphical extrapolation.
11	506	McGee, W. M. and Matthew, B. R.	1962	L	5-922	Ti-2.5Al-16V	Bal.	15.84		2.31	0.21		0.041	0.018 N_2, 0.065 O_2, 0.0067 H_2; cylindrical specimen machined from 0.125 in. thick solution treated at 1038 K, aged 4 hrs at 805 K; heating rate 2 F per minute at low temperatures, 5 F per minute at high temperatures; zero-point correction of 0.017% determined by graphical interpolation.

* Not shown in figure.

SPECIFICATION TABLE 283B. THERMAL LINEAR EXPANSION OF TITANIUM + VANADIUM + ΣX_i Ti + V + ΣX_i (continued)

Cur. No.	Ref. No.	Author(s)	Year	Method Used	Temp. Range, K	Name and Specimen Designation	Ti	V	Cr	Al	Fe	N	C	Composition (continued), Specifications, and Remarks
12	507	Lockheed Georgia Co.	1962		311-922	C9EE-1 Ti-2. 5Al-16V	Bal.	15. 84		2. 31	0. 21	0. 018	0. 41	650 ppm O, 67 ppm H; dimension 0. 1 in. diameter, 1. 968 in. long; cylindrical specimen machined from 0. 125 in. thick sheet, parallel to grain direction; solution treated at 1039 K for 30 min., water quenched, aged at 805 K for 4 hrs, and air cooled; specimen decreased 0. 016% in length after cooled from 922 K to 311 K.
13*	507	Lockheed Georgia Co.	1962		311-922	C9EE-2 Ti-2. 5Al-16V								Similar to the above specimen except increased 0. 045%.
14	507	Lockheed Georgia Co.	1962		311-922	C9EE-3 Ti-2. 5Al-16V								Similar to the above specimen except increased 0. 084%.
15*	507	Lockheed Georgia Co.	1962		311-922									Average value of C9EE-1, -2, -3; average increment 0. 0508%.
16*	507	Lockheed Georgia Co.	1962		3. 7-250	C9EL-4 Ti-2. 5Al-16V								Similar to the C9EE-1 except no increment mentioned and the value at -453 F is extrapolated; zero-point correction of -0. 01% determined by graphical extrapolation.
17*	507	Lockheed Georgia Co.	1962		3. 7-250	C9EL-5 Ti-2. 5Al-16V								Similar to the above specimen; zero-point correction of -0. 01% determined by graphical extrapolation.
18*	507	Lockheed Georgia Co.	1962		3. 7-250	C9EL-6 Ti-2. 5Al-16V								Similar to the above specimen; zero-point correction of -0. 01% determined by graphical extrapolation.
19*	507	Lockheed Georgia Co.	1962		3. 7-250	Ti-2. 5Al-16V								Average value of C9EL-4, -5, -6; zero-point correction of -0. 01% determined by graphical extrapolation.

* Not shown in figure.

DATA TABLE 283B. THERMAL LINEAR EXPANSION OF TITANIUM + VANADIUM + ΣXi   Ti + V + ΣXi

[Temperature, T, K; Linear Expansion, $\Delta L/L_0$, %]

T	$\Delta L/L_0$
CURVE 1	
2	-0.206
33	-0.200
89	-0.176
144	-0.134
200	-0.084
255	-0.034
311	0.016
366	0.070
422	0.126
478	0.182
533	0.239
589	0.300
644	0.360
700	0.420
755	0.483
811	0.545
866	0.608
922	0.670
CURVE 2	
20	-0.161
40	-0.159
60	-0.155
80	-0.149
100	-0.140
120	-0.130
140	-0.118
160	-0.105
180	-0.092
200	-0.077
220	-0.061
240	-0.045
260	-0.028
273	-0.016
280	-0.011
293	0.000
CURVE 3*	
310	0.017
366	0.068
422	0.122
477	0.180
533	0.235
588	0.294
644	0.355
699	0.418
CURVE 3 (cont.)*	
755	0.482
810	0.548
866	0.615
922	0.686
CURVE 4	
310	0.017
366	0.071*
422	0.127*
477	0.184*
533	0.242
588	0.299*
644	0.358
699	0.418
755	0.478
810	0.540
866	0.602*
922	0.665*
CURVE 5*	
310	0.017
366	0.070
422	0.125
477	0.182
533	0.239
588	0.297
644	0.357
699	0.418
755	0.480
810	0.544
866	0.609
922	0.676
CURVE 6	
310	0.019*
366	0.076*
422	0.135
477	0.193
533	0.247*
588	0.304
644	0.357*
699	0.411*
755	0.462
810	0.516
866	0.565
CURVE 6 (cont.)	
922	0.614
CURVE 7*	
3.7	-0.20
25	-0.20
50	-0.19
75	-0.18
100	-0.17
125	-0.15
150	-0.13
175	-0.11
200	-0.09
225	-0.07
250	-0.04
CURVE 8*	
3.7	-0.21
25	-0.20
50	-0.20
75	-0.18
100	-0.17
125	-0.15
150	-0.13
175	-0.11
200	-0.09
225	-0.07
250	-0.04
CURVE 9*	
3.7	-0.20
25	-0.21
50	-0.20
75	-0.18
100	-0.17
125	-0.15
150	-0.13
175	-0.11
200	-0.09
225	-0.07
250	-0.04
CURVE 10*	
3.7	-0.20
25	-0.20
CURVE 10 (cont.)*	
50	-0.19
75	-0.18
100	-0.17
125	-0.15
150	-0.13
175	-0.11
200	-0.09
225	-0.07
250	-0.04
CURVE 11	
5	-0.173
33	-0.167
89	-0.146
144	-0.114
200	-0.073*
255	-0.028
311	0.020*
366	0.067*
422	0.120
478	0.171*
533	0.223
589	0.277
644	0.332
700	0.388
755	0.446
811	0.502
866	0.567*
922	0.619
CURVE 12	
310	0.016*
366	0.064
422	0.114
477	0.168
533	0.220*
588	0.276*
644	0.331*
699	0.400
755	0.449*
810	0.508*
866	0.572
922	0.635
CURVE 13*	
310	0.017
366	0.068
422	0.122
477	0.174
533	0.229
588	0.283
644	0.337
699	0.393
755	0.448
810	0.504
866	0.560
922	0.618
CURVE 14	
310	0.016*
366	0.066*
422	0.117*
477	0.171*
533	0.223*
588	0.283
644	0.330*
699	0.388*
755	0.444*
810	0.501*
866	0.558
922	0.617*
CURVE 15*	
310	0.016
366	0.066
422	0.118
477	0.171
533	0.224
588	0.279
644	0.333
699	0.394
755	0.447
810	0.505
866	0.564
922	0.624
CURVE 16*	
3.7	-0.17
25	-0.17
50	-0.16
CURVE 16 (cont.)*	
75	-0.16
100	-0.14
125	-0.13
150	-0.11
175	-0.09
200	-0.07
225	-0.05
250	-0.03
CURVE 17*	
3.7	-0.18
25	-0.17
50	-0.17
75	-0.15
100	-0.14
125	-0.12
150	-0.11
175	-0.09
200	-0.07
225	-0.05
250	-0.03
CURVE 18*	
3.7	-0.17
25	-0.18
50	-0.16
75	-0.15
100	-0.14
125	-0.12
150	-0.11
175	-0.09
200	-0.07
225	-0.05
250	-0.03
CURVE 19*	
3.7	-0.17
25	-0.17
50	-0.16
75	-0.15
100	-0.14
125	-0.12
150	-0.11
175	-0.09
200	-0.07
225	-0.05
250	-0.02

* Not shown in figure.

FIGURE AND TABLE NO. 284R. PROVISIONAL VALUES FOR THERMAL LINEAR EXPANSION OF TUNGSTEN + NICKEL + ΣX_i ALLOYS W + Ni + ΣX_i

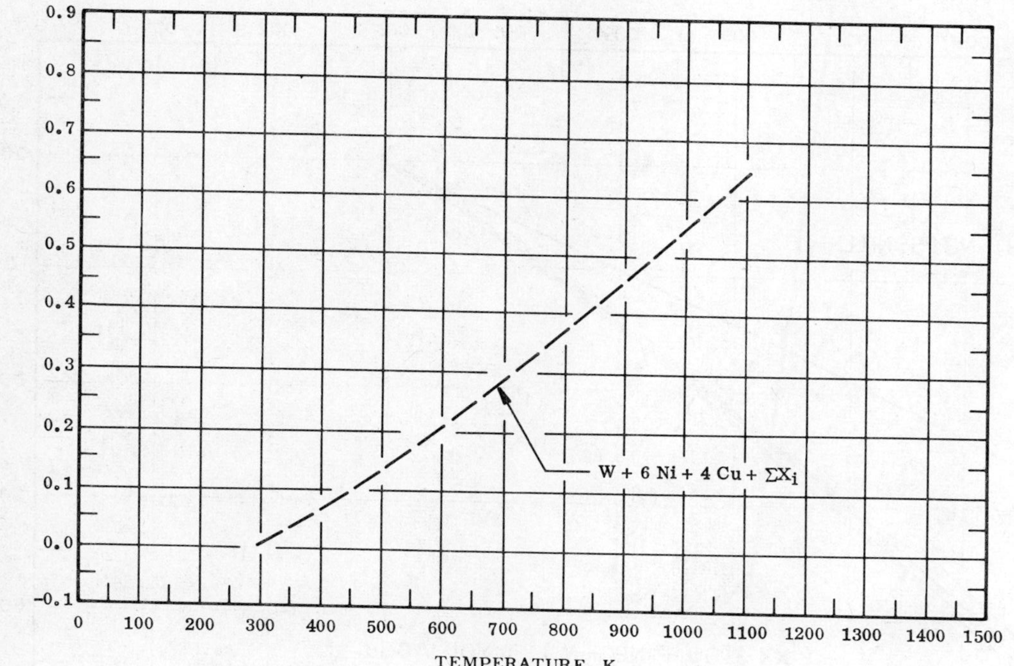

PROVISIONAL VALUES

[Temperature, T, K; Linear Expansion, $\Delta L/L_0$, %; α, K^{-1}]

(W + 6 Ni + 4 Cu + ΣX_i)

T	$\Delta L/L_0$	$\alpha \times 10^6$
293	0.000	6.2
400	0.068	6.8
500	0.140	7.4
600	0.218	7.9
700	0.300	8.3
800	0.383	8.6
900	0.470	8.8
1000	0.559	8.9
1100	0.650	9.0

REMARKS

(W + 6 Ni + 4 Cu + ΣX_i): The tabulated values for well-annealed alloy of type Mallory Metal 1000 are considered accurate to within ±10% over the entire temperature range. Addition of Boron up to 0.2% to this alloy does not change the thermal linear expansion. These values can be represented approximately by the following equation:

$$\Delta L/L_0 = -0.146 + 3.708 \times 10^{-4}\,T + 4.824 \times 10^{-7}\,T^2 - 1.465 \times 10^{-10}\,T^3$$

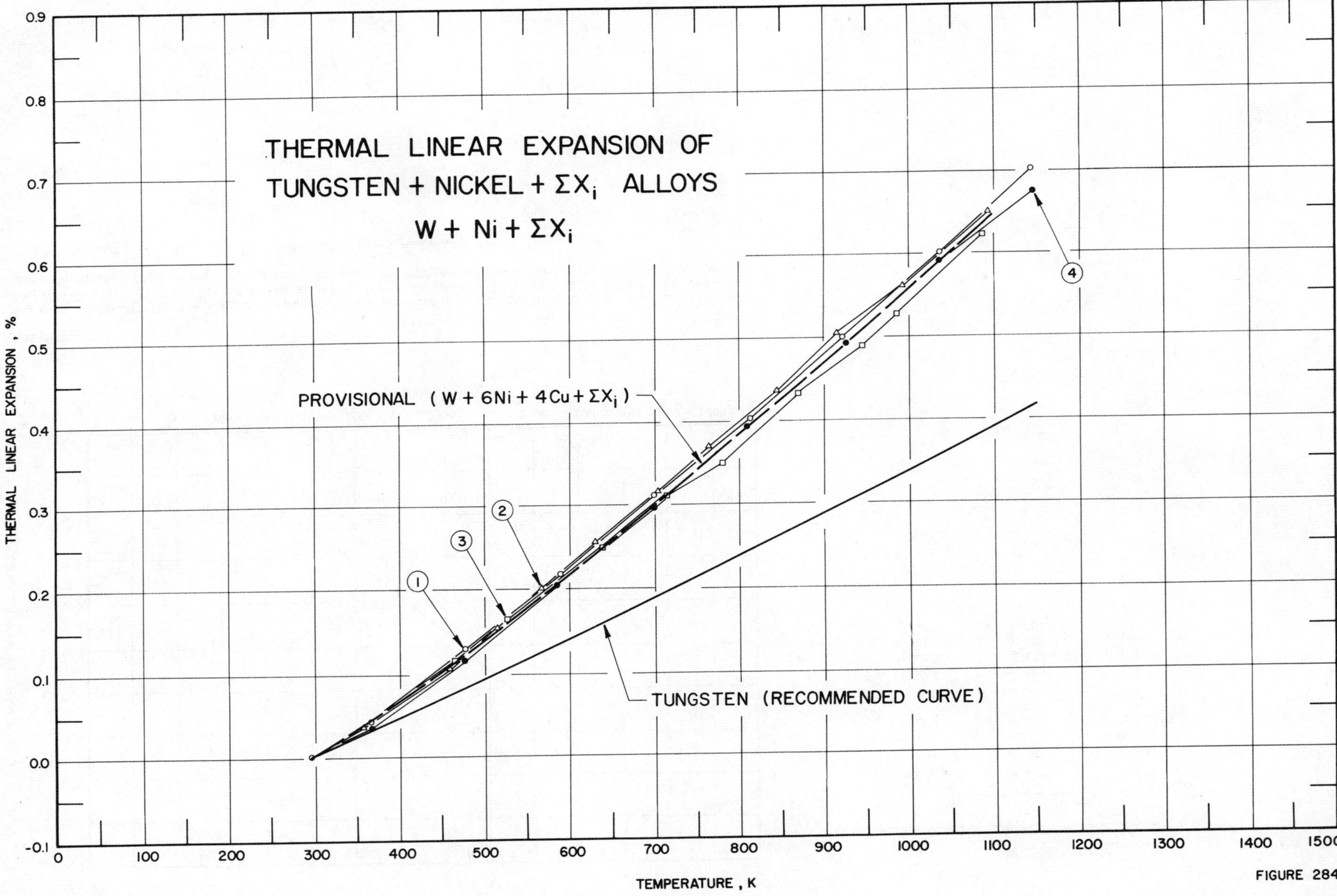
THERMAL LINEAR EXPANSION OF
TUNGSTEN + NICKEL + ΣX_i ALLOYS
W + Ni + ΣX_i
PROVISIONAL (W + 6Ni + 4Cu + ΣX_i)
TUNGSTEN (RECOMMENDED CURVE)
1
2
3
4
THERMAL LINEAR EXPANSION, %
0.9
0.8
0.7
0.6
0.5
0.4
0.3
0.2
0.1
0.0
-0.1
TEMPERATURE, K
0
100
200
300
400
500
600
700
800
900
1000
1100
1200
1300
1400
1500

FIGURE 284

SPECIFICATION TABLE 284. THERMAL LINEAR EXPANSION OF TUNGSTEN + NICKEL + ΣX_i W + Ni + ΣX_i

Cur. No.	Ref. No.	Author(s)	Year	Method Used	Temp. Range, K	Name and Specimen Designation	Composition (weight percent) W	Ni	Cu	B	Composition (continued), Specifications, and Remarks
1	490	Baxter, W. G. and Welch, F. M.	1961		294-1144	Mallory 1000 metal	90	6	4		Measured in argon.
2	490	Baxter, W. G. and Welch, F. M.	1961		294-1094	Vendor A	90.1	5.81	3.76	0.23	92 B^{10} enrichment; density 16.58 g cm^{-3}; measured in argon.
3	490	Baxter, W. G. and Welch, F. M.	1961		294-1085	Vendor B	90.83	6.32	1.98	0.15	Density 16.38 g cm^{-3}; measured in argon.
4	490	Baxter, W. G. and Welch, F. M.	1961		294-1142	B50YA12B	90	5/6	3/4	0.20/0.25	92 min. B^{10}, 0.1 max. Co, 0.01 max. Cd, 0.05 max. total rare earth; density 16.7 ± 0.2 g cm^{-3}; measured in argon.

DATA TABLE 284. THERMAL LINEAR EXPANSION OF TUNGSTEN + NICKEL + ΣX_i W + Ni + ΣX_i

[Temperature, T, K; Linear Expansion, $\Delta L/L_0$, %]

T	$\Delta L/L_0$	T	$\Delta L/L_0$	T	$\Delta L/L_0$
CURVE 1		CURVE 2 (cont.)		CURVE 3 (cont.)	
294	0.000*	630	0.256	942	0.490
366	0.042	702	0.315	981	0.528
479	0.129	763	0.370	1085	0.622
590	0.218	842	0.437		
700	0.311	914	0.508	CURVE 4	
813	0.404	992	0.562	294	0.000*
924	0.501	1093	0.652	364	0.039
1037	0.604			474	0.116
1143	0.704	CURVE 3		587	0.204
		294	0.000*	699	0.299
CURVE 2		462	0.118	809	0.396
294	0.000*	525	0.166	922	0.493
359	0.038	635	0.253	1033	0.591
460	0.107	713	0.311	1141	0.677
515	0.155	778	0.350		
567	0.201	868	0.433		

* Not shown in figure.

SPECIFICATION TABLE 285. THERMAL LINEAR EXPANSION OF TUNGSTEN + RHENIUM + ΣX_i W + Re + ΣX_i

Cur. No.	Ref. No.	Author(s)	Year	Method Used	Temp. Range, K	Name and Specimen Designation	Composition (weight percent) W	Re	Mo	Composition (continued), Specifications, and Remarks
1*	211	Yaggee, F. L. and Styles, J. W.	1965	L	296-1242		46.49	35.31	18.20	No details given.

DATA TABLE 285. THERMAL LINEAR EXPANSION OF TUNGSTEN + RHENIUM + ΣX_i W + Re + ΣX_i

[Temperature, T, K; Linear Expansion, $\Delta L/L_0$, %]

T	$\Delta L/L_0$
CURVE 1*	
296	0.002
374	0.043
472	0.097
573	0.153
673	0.213
773	0.279
873	0.347
973	0.416
1073	0.491
1173	0.560
1242	0.614

* No figure given.

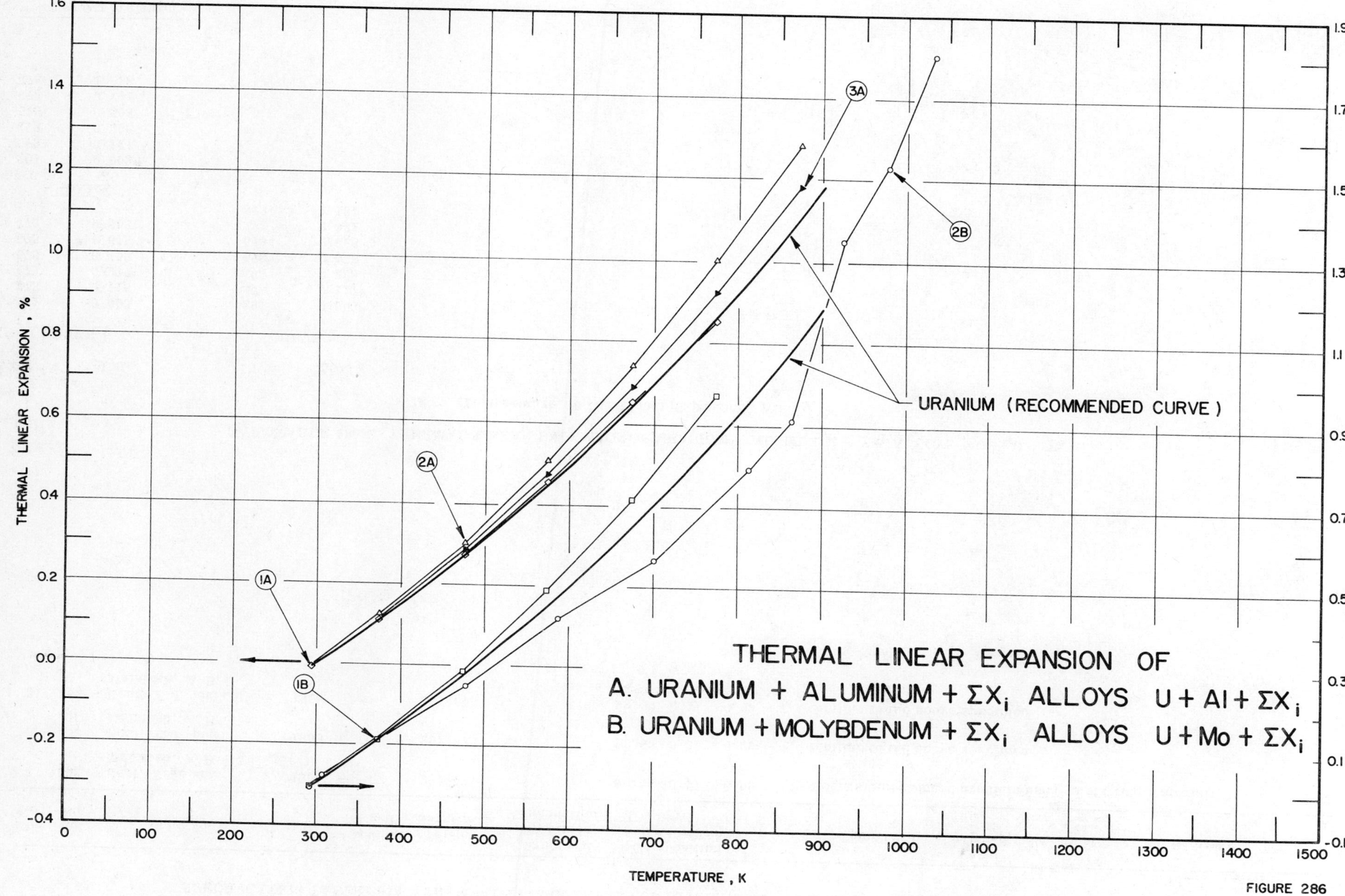
THERMAL LINEAR EXPANSION OF
A. URANIUM + ALUMINUM + ΣX_i ALLOYS U + Al + ΣX_i
B. URANIUM + MOLYBDENUM + ΣX_i ALLOYS U + Mo + ΣX_i
URANIUM (RECOMMENDED CURVE)
1A
2A
3A
1B
2B
THERMAL LINEAR EXPANSION, %
TEMPERATURE, K
0
100
200
300
400
500
600
700
800
900
1000
1100
1200
1300
1400
1500
1.6
1.4
1.2
1.0
0.8
0.6
0.4
0.2
0.0
-0.2
-0.4
1.9
1.7
1.5
1.3
1.1
0.9
0.7
0.5
0.3
0.1
-0.1

FIGURE 286

SPECIFICATION TABLE 286A. THERMAL LINEAR EXPANSION OF URANIUM + ALUMINUM + ΣX_i ALLOYS U + Al + ΣX_i

Cur. No.	Ref. No.	Author(s)	Year	Method Used	Temp. Range, K	Name and Specimen Designation	Composition (weight percent) U	Al	C	Composition (continued), Specifications, and Remarks
1	362	Heal, T.J. and McIntosh, A.B.	1958		291-773		99.44	0.47	~0.09	Magnesium-reduced uranium used as starting material.
2	362	Heal, T.J. and McIntosh, A.B.	1958		291-873		99.68	0.23	~0.09	Similar to the above specimen.
3	362	Heal, T.J. and McIntosh, A.B.	1958		291-873		99.80	0.11	~0.09	Similar to the above specimen.

DATA TABLE 286A. THERMAL LINEAR EXPANSION OF URANIUM + ALUMINUM + ΣX_i ALLOYS U + Al + ΣX_i

[Temperature, T, K; Linear Expansion, $\Delta L/L_0$, %]

T	$\Delta L/L_0$	T	$\Delta L/L_0$
CURVE 1		CURVE 3	
291	-0.003	291	-0.003*
373	0.116	373	0.117*
473	0.277	473	0.283
573	0.453	573	0.473
673	0.645	673	0.686
773	0.853	773	0.922
		873	1.181
CURVE 2			
291	-0.003*		
373	0.124		
473	0.303		
573	0.508		
673	0.741		
873	1.288		

* Not shown in figure.

SPECIFICATION TABLE 286B. THERMAL LINEAR EXPANSION OF URANIUM + MOLYBDENUM + ΣX_i ALLOYS U + Mo + ΣX_i

Cur. No.	Ref. No.	Author(s)	Year	Method Used	Temp. Range, K	Name and Specimen Designation	Composition (weight percent) U	Mo	C	Composition (continued), Specifications, and Remarks
1	362	Heal, T. J. and McIntosh, A. B.	1958		291-773		98.13	1.65	0.22	Magnesium reduced U used as starting material.
2	513	Corni, S.	1967	L	310-1033		92	8		No details given; zero-point correction of 0.013% determined by graphical extrapolation.

DATA TABLE 286B. THERMAL LINEAR EXPANSION OF URANIUM + MOLYBDENUM + ΣX_i ALLOYS U + Mo + ΣX_i

[Temperature, T, K; Linear Expansion, $\Delta L/L_0$, %]

T	$\Delta L/L_0$
CURVE 1	
291	-0.003
373	0.116
473	0.287
573	0.488
673	0.717
773	0.976
CURVE 2	
310	0.022
478	0.247
589	0.420
700	0.567
811	0.794
868	0.915
922	1.356
979	1.539
1033	1.813

* Not shown in figure.

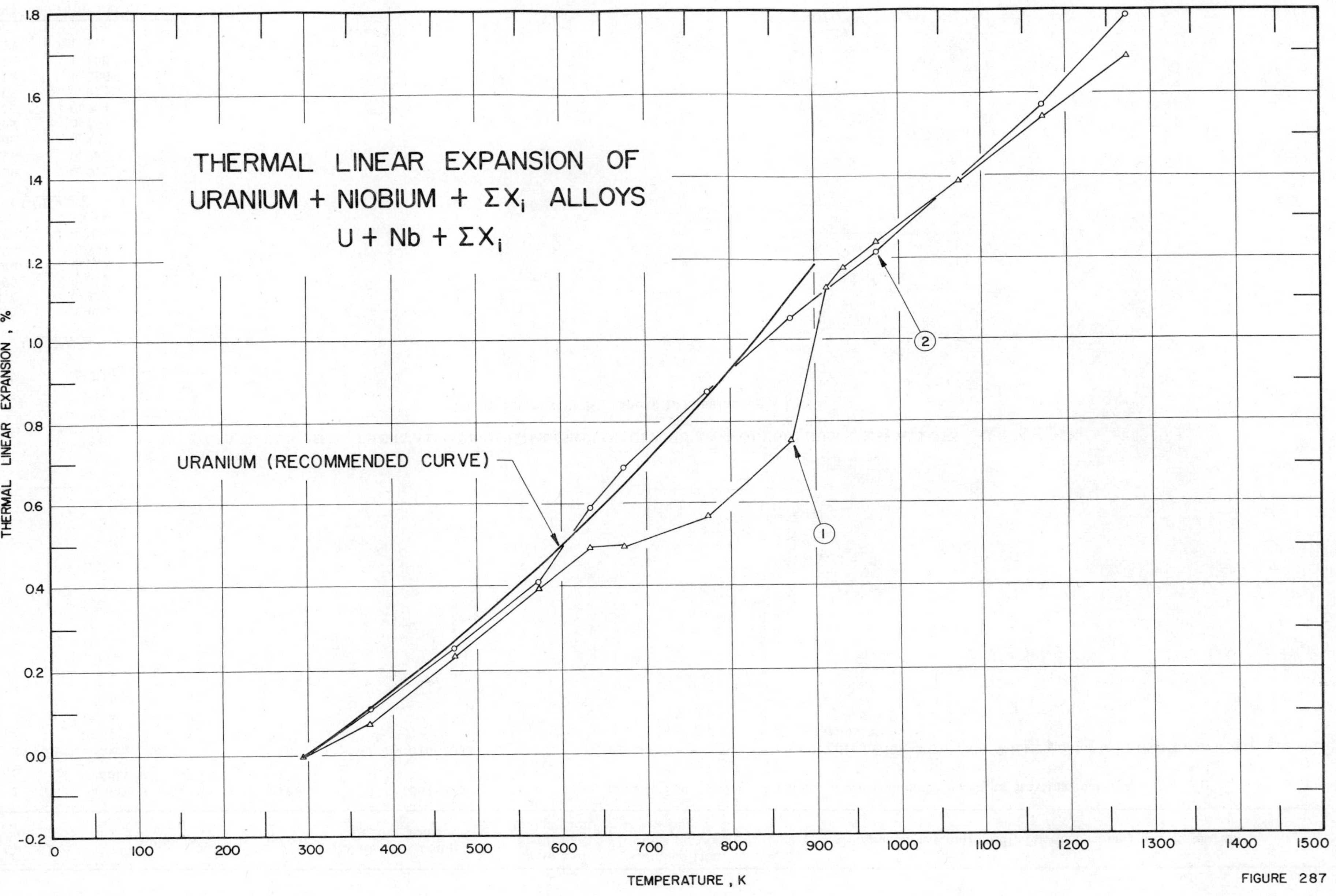
THERMAL LINEAR EXPANSION OF
URANIUM + NIOBIUM + ΣX_i ALLOYS
U + Nb + ΣX_i
URANIUM (RECOMMENDED CURVE)
1
2
THERMAL LINEAR EXPANSION, %
1.8
1.6
1.4
1.2
1.0
0.8
0.6
0.4
0.2
0.0
-0.2
0
100
200
300
400
500
600
700
800
900
1000
1100
1200
1300
1400
1500
TEMPERATURE, K

FIGURE 287

SPECIFICATION TABLE 287. THERMAL LINEAR EXPANSION OF URANIUM + NIOBIUM + ΣX_i ALLOYS U + Nb + ΣX_i

Cur. No.	Ref. No.	Author(s)	Year	Method Used	Temp. Range, K	Name and Specimen Designation	Composition (weight percent) U	Nb	Zr	Composition (continued), Specifications, and Remarks
1	514	Patterson, C. A. W. and Vandervoort, R. R.	1964		293-1273		90. 0	7. 5	2. 5	99.96^+ U, 99.66^+ Nb, 99.75^+ Zr; alloy remelted five times and homogenized 72 hrs at 850 C; specimen calculated to be 2. 881 in. long; expansion measured with increasing temperature.
2	514	Patterson, C. A. W. and Vandervoort, R. R.	1964		1273-293					The above specimen; expansion measured with decreasing temperature; zero-point correction is 0. 125%.

DATA TABLE 287. THERMAL LINEAR EXPANSION OF URANIUM + NIOBIUM + ΣX_i ALLOYS U + Nb + ΣX_i

[Temperature, T, K; Linear Expansion, $\Delta L/L_0$, %]

T	$\Delta L/L_0$	T	$\Delta L/L_0$
CURVE 1		CURVE 2	
293	0. 00	1273	1. 784
373	0. 073	1173	1. 572
473	0. 236	1074	1. 388*
573	0. 399	973	1. 218
631	0. 493	873	1. 045
673	0. 493	774	0. 868
773	0. 569	673	0. 683
873	0. 757	635	0. 590
913	1. 125	573	0. 410
935	1. 180	473	0. 253
973	1. 124	373	0. 104
1073	1. 392	293	0. 000*
1173	1. 541		
1273	1. 690		

* Not shown in figure.

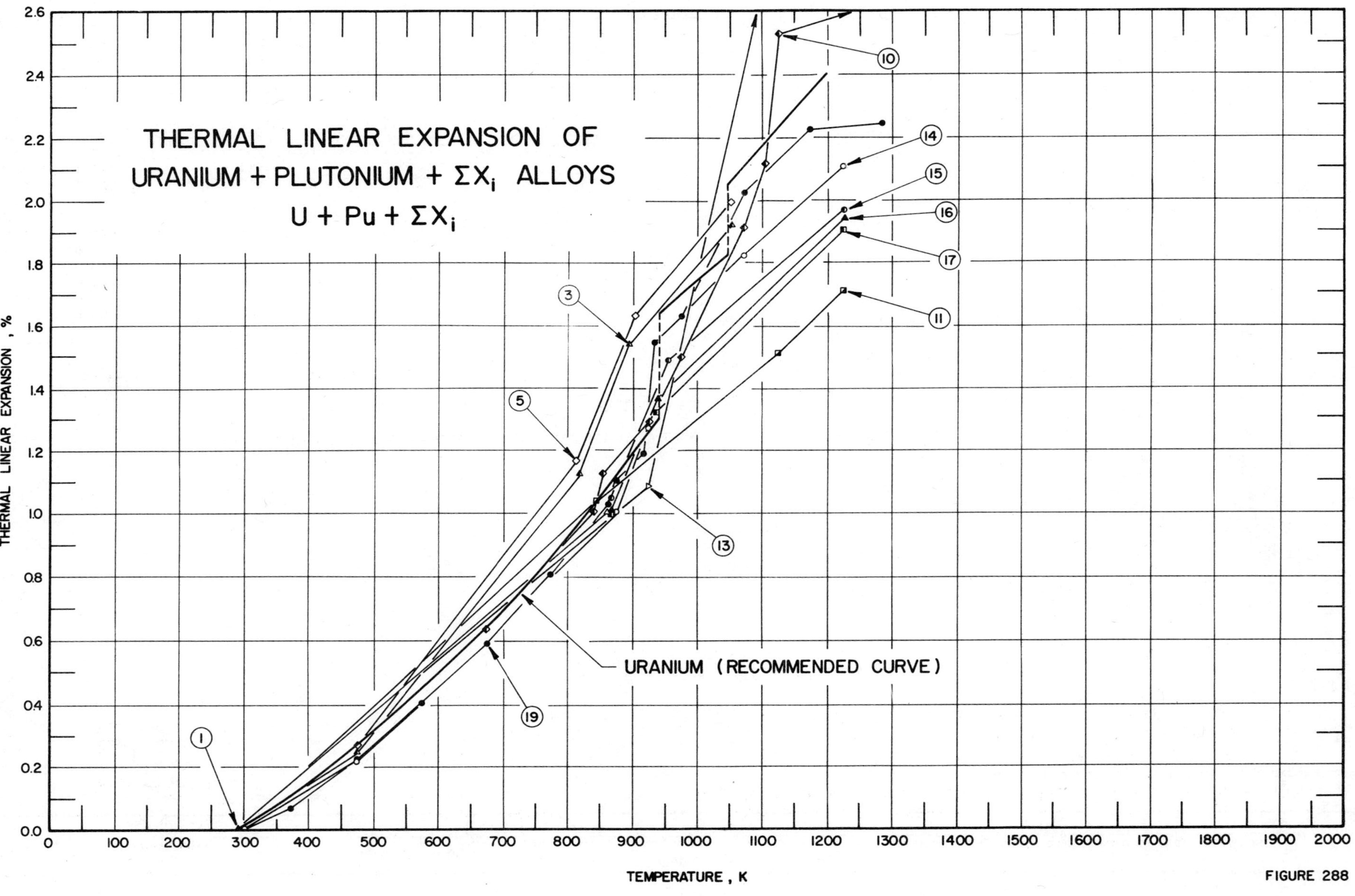

THERMAL LINEAR EXPANSION OF
URANIUM + PLUTONIUM + ΣX_i ALLOYS
$U + Pu + \Sigma X_i$
URANIUM (RECOMMENDED CURVE)
THERMAL LINEAR EXPANSION, %
TEMPERATURE, K
0.0
0.2
0.4
0.6
0.8
1.0
1.2
1.4
1.6
1.8
2.0
2.2
2.4
2.6
0
100
200
300
400
500
600
700
800
900
1000
1100
1200
1300
1400
1500
1600
1700
1800
1900
2000
1
3
5
10
11
13
14
15
16
17
19

FIGURE 288

SPECIFICATION TABLE 288. THERMAL LINEAR EXPANSION OF URANIUM + PLUTONIUM + ΣX_i ALLOYS $U + Pu + \Sigma X_i$

Cur. No.	Ref. No.	Author(s)	Year	Method Used	Temp. Range, K	Name and Specimen Designation	Composition (weight percent) U	Pu	Fe	Mo	Ti	Zr	Composition (continued), Specifications, and Remarks
1	515	Boucher, R., Barthelemy, P., and Milet, C.	1965	L	293-863		74	25	1				No other details given.
2*	515	Boucher, R., et al.	1965	L	293-863		73.5	25	1.5				No other details given.
3	516	Rhude, H. V.	1966	L	298-1053		72.5	25		2.5			Expansion measured in rolling direction.
4*	516	Rhude, H. V.	1966	L	298-1053		72.5	25		2.5			Expansion measured transverse to rolling direction.
5	516	Rhude, H. V.	1966	L	298-1053		72.5	25		2.5			Expansion measured on specimen as cast.
6*	517	Smotritskii, G.S., Chebotarev, N.T., Kutaitsev, V.I., and Petrov, P.N.	1967	L	373, 803		82	10		8			Melting temperature 1353-1441 K; transition point (γ-phase) 823-880 K.
7*	517	Smotritskii, G.S., et al.	1967	L	415, 855		80	12		8			Melting temperature 1323-1383 K; transition point (γ-phase) 823-863 K; density 17.3 g cm^{-3}.
8*	517	Smotritskii, G.S., et al.	1967	L	373, 873		78	14		8			Melting temperature 1298-1398 K; transition point (γ-phase) 948-873 K.
9*	517	Smotritskii, G.S., et al.	1967	L	378, 768		71	20		9			Melting temperature 1263-1405 K; transition point (γ-phase) 723-803 K; density 17.73 g cm^{-3}.
10	518	Rhude, H. V.	1965	L	293-1240		78.5	15			6.5		Specimen cylinder 6 mm diameter, 25 mm long; heating rate 2 C/min.
11	519	Kelman, L.R.	1965	L	298-1223		78.5	15			6.5		No other details given.
12*	519	Kelman, L.R.	1965	L	298-1223		75	15			10		No other details given.
13	520	Kelman, L.R., Savage, H., and Walter, C.M.	1967	L	298-1223		66.2	22			11.8		Specimen prepared from reactor-grade U and Pu, crystal bar Ti.
14	518	Rhude, H. V.	1965	L	301-1223		75	15				10	Specimen cylinder 6 mm diameter, 25 mm long; heating rate 2 C/min; zero-point correction of 0.001% determined by graphical extrapolation.
15	519	Kelman, L.R.	1965	L	298-1223		82.6	11.1				6.3	No other details given.
16	519	Kelman, L.R.	1965	L	298-1223		75	15				10	No other details given.
17	519	Kelman, L.R.	1965	L	298-1223		67.5	18.4				14.1	No other details given.
18*	519	Kelman, L.R.	1965	L	298-1133		79.5	17.1				3.4	No other details given.
19	62	Yaggee, F.L. and Styles, J.W.	1969	L	273-1286		63.3	22.2				14.5	Transformations observed at 868 K and 933 K; reported error <± 2%; zero-point correction is -0.025%.

* Not shown in figure.

DATA TABLE 288. THERMAL LINEAR EXPANSION OF URANIUM + PLUTONIUM + ΣX_i ALLOYS $U + Pu + \Sigma X_i$

[Temperature, T, K; Linear Expansion, $\Delta L/L_0$, %]

T	$\Delta L/L_0$
CURVE 1	
293	0.000
863	1.003
CURVE 2*	
293	0.00
863	1.14
CURVE 3	
298	0.007
473	0.256
818	1.135
893	1.545
1053	1.925
CURVE 4*	
298	0.008
473	0.275
818	1.131
893	1.541
1053	1.928
CURVE 5	
298	0.008*
473	0.277
813	1.168
903	1.628
1053	2.008
CURVE 6*	
373	0.122
803	0.780
CURVE 7*	
415	0.239
855	1.102
CURVE 8*	
373	0.136
848	0.944
CURVE 9*	
378	0.173
768	0.969
CURVE 10	
293	0.00*
473	0.25*
673	0.64
840	1.04
853	1.13
924	1.30
928	1.37*
973	1.50
1074	1.92
1112	2.22
1128	2.53
1240	2.76*
CURVE 11	
298	0.010*
843	1.040
1123	1.516
1223	1.716
CURVE 12*	
298	0.010
953	1.268
1123	3.596
1223	3.788
CURVE 13	
298	0.008*
923	1.094
1088	3.057*
1223	3.347*
CURVE 14	
301.2	0.01
473.2	0.22
673.2	0.59*
873.2	1.03
922.2	1.28
CURVE 14 (cont.)	
935.2	1.55*
1073.2	1.83
1223.2	2.14
CURVE 15	
298	0.009*
868	1.053
953	1.486
1223	1.974
CURVE 16	
298	0.009*
868	1.012
938	1.376
1223	1.949
CURVE 17	
298	0.009*
868	1.007
933	1.331
1223	1.911
CURVE 18*	
298	0.011
843	1.166
1053	3.119
1133	3.250
CURVE 19	
273	-0.025*
293	0.000*
373	0.075
473	0.231
573	0.402
673	0.595
773	0.816
864	1.033
873	1.104
916	1.188
931	1.549
973	1.633
CURVE 19 (cont.)	
1073	2.028
1173	2.234
1286	2.258

* Not shown in figure.

SPECIFICATION TABLE 289. THERMAL LINEAR EXPANSION OF URANIUM + VANADIUM + ΣX_i ALLOYS $U + V + \Sigma X_i$

Cur. No.	Ref. No.	Author(s)	Year	Method Used	Temp. Range, K	Name and Specimen Designation	Composition (weight percent) U	V	C	Composition (continued), Specifications, and Remarks
1*	362	Heal, T. J. and McIntosh, A. B.	1958		291-873		99.38	0.43	0.19	Magnesium reduced U used as starting materials.

DATA TABLE 289. THERMAL LINEAR EXPANSION OF URANIUM + VANADIUM + ΣX_i ALLOYS $U + V + \Sigma X_i$

[Temperature, T, K; Linear Expansion, $\Delta L/L_0$, %]

T	$\Delta L/L_0$
CURVE 1*	
291	-0.003
373	0.115
473	0.289
573	0.500
673	0.745
773	1.025
873	1.341

* No figure given.

SPECIFICATION TABLE 290. THERMAL LINEAR EXPANSION OF VANADIUM + CHROMIUM + ΣX_i V + Cr + ΣX_i

Cur. No.	Ref. No.	Author(s)	Year	Method Used	Temp. Range, K	Name and Specimen Designation	Composition (weight percent) V	Cr	Ti	Composition (continued), Specifications, and Remarks
1*	117	Yaggee, F. L. and Styles, J. W.	1966	L	273, 773		80.0	15.0	5.0	Density 6.16 g cm^{-3}.
2*	117	Yaggee, F. L. and Styles, J. W.	1966	L	273, 1273		80.0	15.0	5.0	The above specimen.

DATA TABLE 290. THERMAL LINEAR EXPANSION OF VANADIUM + CHROMIUM + ΣX_i V + Cr + ΣX_i

[Temperature, T, K; Linear Expansion, $\Delta L/L_0$, %]

T	$\Delta L/L_0$
CURVE 1*	
273	-0.0207
773	0.498
CURVE 2*	
273	-0.0230
1273	1.127

* No figure given.

SPECIFICATION TABLE 291. THERMAL LINEAR EXPANSION OF VANADIUM + TITANIUM + ΣX_i V + Ti + ΣX_i

Cur. No.	Ref. No.	Author(s)	Year	Method Used	Temp. Range, K	Name and Specimen Designation	Composition (weight percent) V	Ti	Cr	Composition (continued), Specifications, and Remarks
1*	117	Yaggee, F. L. and Styles, J. W.	1966	L	273, 773		77.5	15.0	7.5	Density 5.88 ± 0.02 g cm^{-3}.
2*	117	Yaggee, F. L. and Styles, J. W.	1966	L	273, 1273		77.5	15.0	7.5	The above specimen.

DATA TABLE 291. THERMAL LINEAR EXPANSION OF VANADIUM + TITANIUM + ΣX_i V + Ti + ΣX_i

[Temperature, T, K; Linear Expansion, $\Delta L/L_0$, %]

T	$\Delta L/L_0$
CURVE 1*	
273	-0.0196
773	0.469
CURVE 2*	
273	-0.0215
1273	1.054

* No figure given.

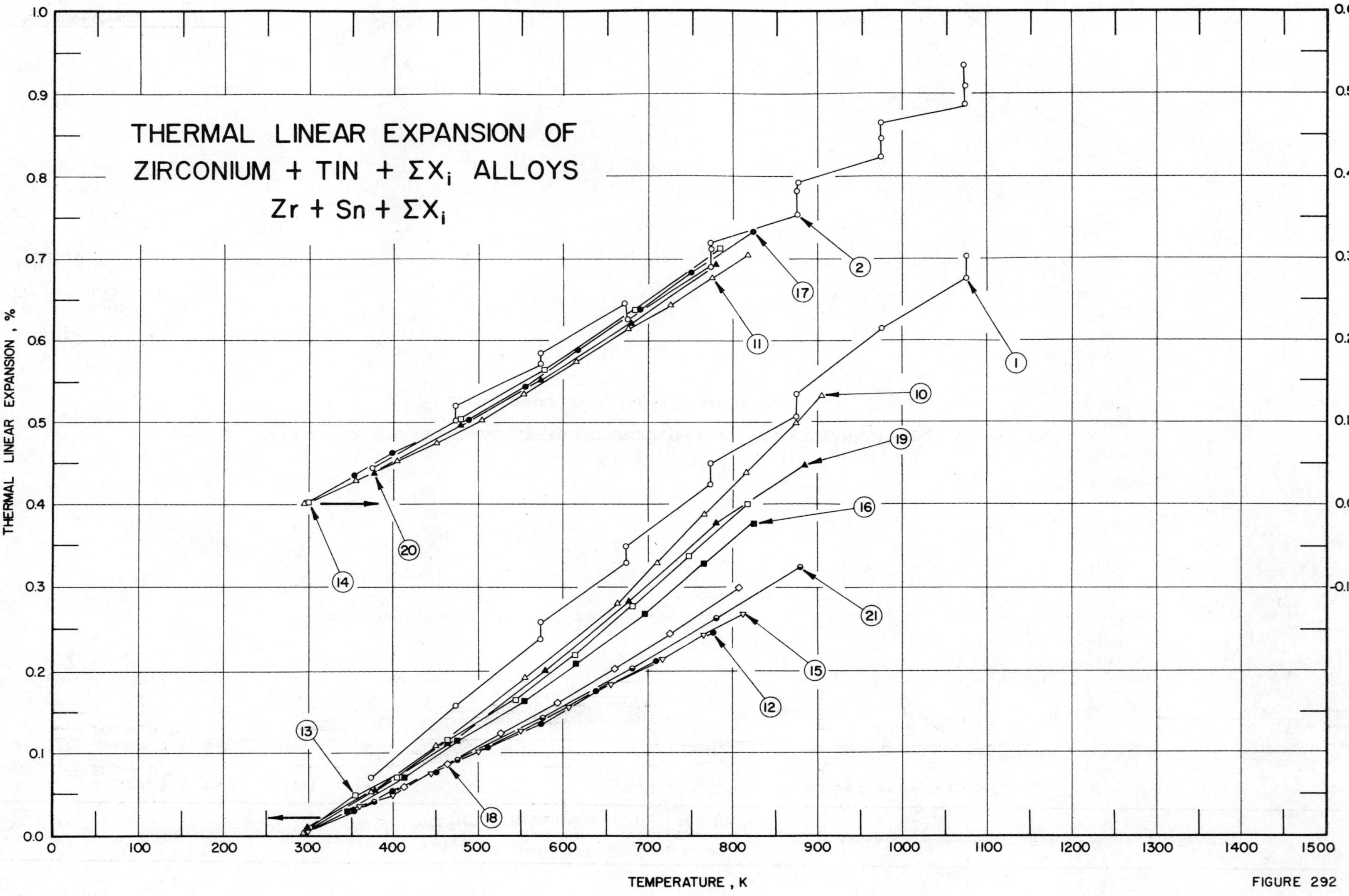
THERMAL LINEAR EXPANSION OF
ZIRCONIUM + TIN + ΣX$_i$ ALLOYS
Zr + Sn + ΣX$_i$
THERMAL LINEAR EXPANSION, %
TEMPERATURE, K
0
100
200
300
400
500
600
700
800
900
1000
1100
1200
1300
1400
1500
0.0
0.1
0.2
0.3
0.4
0.5
0.6
0.7
0.8
0.9
1.0
-0.1
0.0
0.1
0.2
0.3
0.4
0.5
0.6
1
2
10
11
12
13
14
15
16
17
18
19
20
21

FIGURE 292

SPECIFICATION TABLE 292. THERMAL LINEAR EXPANSION OF ZIRCONIUM + TIN + ΣX_i ALLOYS $Zr + Sn + \Sigma X_i$

Cur. No.	Ref. No.	Author(s)	Year	Method Used	Temp. Range, K	Name and Specimen Designation	Composition (weight percent) Zr	Sn	Fe	Cr	O	H_2	N	Composition (continued), Specifications, and Remarks
1	491	Mehan, R. L. and Wiesinger, R. W.	1961	V	373-1074	Zircaloy 2								Expansion measured in direction normal to rolling.
2	491	Mehan, R. L. and Wiesinger, R. W.	1961	V	373-1074	Zircaloy 2								Expansion measured transverse to rolling.
3*	491	Mehan, R. L. and Wiesinger, R. W.	1961	V	373-1075	Zircaloy 2								Expansion measured longitudinal to rolling.
4*	492	Slattery, G. F.	1966	L	390-753	Zircaloy 4	Bal.	1.44	0.16	0.09	0.1	0.026	0.004	Specimen hydrided 16 hr at 363 ± 3 K in 0.5 percent solution H_2SO_4, annealed at 573 K to 773 K for between 100 and 700 hr, machined to 0.406 cm diameter by 5.00 cm long; heating rate 90 C per hr.
5*	492	Slattery, G. F.	1966	L	312-873	Zircaloy 4	Bal.	1.44	0.16	0.09	0.1	0.0585	0.004	Similar to the above specimen; expansion measured with increasing temperature.
6*	492	Slattery, G. F.	1966	L	855-359	Zircaloy 4	Bal.	1.44	0.16	0.09				The above specimen; expansion measured with decreasing temperature.
7*	493	Scott, D. B.	1965	L	295-1273	Zircaloy 4	Bal.	1.2/ 1.7	0.18/ 0.24	0.07/ 0.13				Specimen 3.69 cm long before heating; expansion measured in axial direction and with increasing temperature.
8*	493	Scott, D. B.	1965	L	1234-295	Zircaloy 4	Bal.	1.2/ 1.7	0.18/ 0.24	0.07/ 0.13				The above specimen; expansion measured with decreasing temperature; zero-point correction of 0.551% determined by graphical extrapolation.
9*	493	Scott, D. B.	1965	L	300-871	Zircaloy 4	Bal.	1.2/ 1.7	0.18/ 0.24	0.07/ 0.13				The above specimen; reinforced with Zircaloy 4 rings; deformed above 872 K; expansion measured along radial direction; zero-point correction of 0.0025% determined by graphical extrapolation.
10	494	Kearns, J. J.	1965	L	293-903	Zircaloy 4								From raw mix, extruded 95% at 1144 K, cold swaged 84% and annealed 1 hr at 922 K; hot rolled to 1.91 cm thick at 1102 K; cold rolled 3% per pass with 90 deg. rotation between each pass until total reduction of 45% reached, recrystallized 1 hr at 922 K; expansion measured in vacuum normal to direction of rolling.
11	494	Kearns, J. J.	1965	L	293-818	Zircaloy 4								The above specimen; expansion measured transverse to rolling direction; zero-point correction is -0.006%.
12	494	Kearns, J. J.	1965	L	293-776	Zircaloy 4								The above specimen; expansion measured in rolling direction; zero-point correction is 0.009%.

* Not shown in figure.

SPECIFICATION TABLE 292. THERMAL LINEAR EXPANSION OF ZIRCONIUM + TIN + ΣX_i ALLOYS $Zr + Sn + \Sigma X_i$ (continued)

Cur. No.	Ref. No.	Author(s)	Year	Method Used	Temp. Range, K	Name and Specimen Designation	Composition (weight percent) Zr	Sn	Fe	Cr	O	H_2	N	Composition (continued), Specifications, and Remarks
13	494	Kearns, J.J.	1965	L	293-816	Zircaloy 4								Similar to the above specimen except after annealing unidirectionally rolled to 57% reduction at 1061 K, rolled 10% at 825 K, annealed 3 hr at 1033 K; expansion measured in vacuum normal to rolling direction; zero-point correction is -0.01%.
14	494	Kearns, J.J.	1965	L	293-878	Zircaloy 4								The above specimen; expansion measured transverse to rolling direction; zero-point correction is -0.009%.
15	494	Kearns, J.J.	1965	L	293-812	Zircaloy 4								The above specimen; expansion measured in rolling direction; zero-point correction is -0.009%.
16	494	Kearns, J.J.	1965	L	293-823	Zircaloy 4								Similar to the above specimen; further treated with beta-phase; annealed 15 min. at 1322 K and air cooled; expansion measured in vacuum normal to rolling direction; zero-point correction is -0.011%.
17	494	Kearns, J.J.	1965	L	293-822	Zircaloy 4								The above specimen; expansion measured transverse to rolling direction; zero-point correction is -0.010%.
18	494	Kearns, J.J.	1965	L	293-809	Zircaloy 4								The above specimen; expansion measured in rolling direction; zero-point correction is -0.008%.
19	494	Kearns, J.J.	1965	L	293-883	Zircaloy 4								From raw mix, extruded 95% at 1144 K, cold swaged 84% and annealed 1 hr at 922 K; hydrogenated to 4300 ppm at 994 K rolled in beta-phase condition to 57% reduction at 1061 K, vacuum degassed 51 hr at 922 K, and 24 hr at 944 K; final hydrogen content < 20 ppm; expansion measured in vacuum normal to rolling direction; zero-point correction is -0.011%.
20	494	Kearns, J.J.	1965	L	293-872	Zircaloy 4								The above specimen; expansion measured transverse to rolling direction; zero-point correction is -0.008%.
21	494	Kearns, J.J.	1965	L	293-879	Zircaloy 4								The above specimen; expansion measured in rolling direction; zero-point correction is -0.009%.

DATA TABLE 292. THERMAL LINEAR EXPANSION OF ZIRCONIUM + TIN + ΣX_i ALLOYS $Zr + Sn + \Sigma X_i$

[Temperature, T, K; Linear Expansion, $\Delta L/L_0$, %]

T	$\Delta L/L_0$
CURVE 1	
373	0.069
473	0.155
473	0.159*
573	0.238
573	0.258
673	0.330
673	0.349
773	0.424
773	0.450
875	0.507
875	0.533
975	0.593
974	0.614
1074	0.677
1074	0.703
CURVE 2	
373	0.043
472	0.101
472	0.109*
472	0.120
572	0.172
572	0.185
669	0.247
673	0.242*
673	0.225
772	0.290
772	0.311
772	0.320
874	0.355
876	0.383
876	0.397
974	0.423
974	0.448
974	0.465
1073	0.488
1073	0.510
1073	0.537
CURVE 3	
373	0.029
373	0.035*
473	0.074*
473	0.092
CURVE 3 (cont.)*	
572	0.118
571	0.143
673	0.166
673	0.188
774	0.211
774	0.228
875	0.263
875	0.274
973	0.312
973	0.321
1074	0.347
1075	0.367
CURVE 4*, ‡	
390	0.054
433	0.078
485	0.107
516	0.125
559	0.150
599	0.177
637	0.206
675	0.240
698	0.263
701	0.266
705	0.269
710	0.273
715	0.277
720	0.280
725	0.283
733	0.287
753	0.297
CURVE 5*, ‡	
312	0.010
354	0.031
397	0.054
450	0.086
456	0.090
490	0.110
501	0.116
530	0.134
559	0.153
573	0.161
601	0.179
CURVE 5 (cont.)*, ‡	
612	0.187
650	0.213
688	0.243
698	0.252
709	0.262
718	0.271
725	0.278
731	0.285
736	0.290
743	0.298
754	0.312
766	0.327
774	0.337
784	0.349
793	0.361
801	0.370
806	0.376
811	0.382
815	0.386
820	0.391
822	0.392
829	0.397
838	0.402
846	0.407
857	0.413
873	0.422
CURVE 6*, ‡	
855	0.392
841	0.384
816	0.369
791	0.354
777	0.345
769	0.337
762	0.328
751	0.314
746	0.308
740	0.301
732	0.293
724	0.285
711	0.272
697	0.259
683	0.246
660	0.226
644	0.213
CURVE 6 (cont.)*, ‡	
628	0.200
610	0.187
593	0.175
542	0.142
417	0.068
383	0.049
359	0.036
CURVE 7*	
295	0.001
374	0.035
473	0.080
570	0.128
670	0.172
771	0.211
869	0.247
968	0.281
1068	0.318
1084	0.323
1105	0.324
1169	0.249
1230	0.113
1245	0.120
1273	0.144
CURVE 8*	
1234	0.607
1137	0.543
1106	0.555
1087	0.550
1070	0.543
974	0.483
872	0.395
771	0.323
675	0.250
574	0.180
371	0.052
295	0.002
CURVE 9*	
300	0.002
372	0.029
476	0.078
CURVE 9 (cont.)*	
576	0.152
675	0.241
773	0.342
871	0.449
CURVE 10	
293	0.000
374	0.051
450	0.108
555	0.191
656	0.281
710	0.329
767	0.387
817	0.439
873	0.498
903	0.531
CURVE 11	
293	0.000
351	0.028
402	0.052
449	0.074
502	0.103
553	0.136
615	0.177
676	0.213
724	0.245
774	0.278
818	0.307
CURVE 12	
293	0.000
295	0.000
352	0.028
399	0.051
449	0.075
510	0.104
572	0.136
639	0.174
709	0.211
776	0.249
CURVE 13	
293	0.000
354	0.047
402	0.069
467	0.115
541	0.165
611	0.219
680	0.277
749	0.336
816	0.400
CURVE 14	
293	0.000
299	0.003
374	0.043*
477	0.103
579	0.167
682	0.239
781	0.313
878	0.387*
CURVE 15	
293	0.000
297	0.001
360	0.032
400	0.049
447	0.074
499	0.100
550	0.126
609	0.156
654	0.183
713	0.214
766	0.244
812	0.269
CURVE 16	
293	0.000
349	0.029
413	0.069
477	0.112
552	0.161
616	0.209
694	0.268
765	0.328
823	0.376

* Not shown in figure.
‡ Author's data for coefficient of thermal expansion have been integrated by TPRC to obtain $\Delta L/L_0$.

DATA TABLE 292. THERMAL LINEAR EXPANSION OF ZIRCONIUM + TIN + ΣX_i ALLOYS $Zr + Sn + \Sigma X_i$ (continued)

T	$\Delta L/L_0$
CURVE 17	
293	0.000
295	0.001
355	0.031
398	0.063
484	0.103
555	0.146
618	0.189
690	0.238
750	0.283
822	0.334
CURVE 18	
293	0.000
358	0.030
413	0.056
465	0.086
528	0.123
593	0.160

T	$\Delta L/L_0$
CURVE 18 (cont.)	
660	0.203
725	0.247
809	0.299
CURVE 19	
293	0.000
299	0.004
376	0.052
462	0.111
580	0.201
677	0.281
779	0.377
883	0.477
CURVE 20	
293	0.000*
300	0.006*

T	$\Delta L/L_0$
CURVE 20 (cont.)	
377	0.038
478	0.096
574	0.153
677	0.221
777	0.292
872	0.356*
CURVE 21	
293	0.000*
300	0.003*
379	0.041
477	0.090
578	0.147
679	0.204
779	0.265
879	0.324

DATA TABLE 292. COEFFICIENT OF THERMAL LINEAR EXPANSION OF ZIRCONIUM + TIN + ΣX_i ALLOYS $Zr + Sn + \Sigma X_i$

[Temperature, T, K; Coefficient of Expansion, α, $10^{-6}\ K^{-1}$]

T	α
CURVE 4‡	
390	5.64
433	5.55
485	5.74
516	5.69
559	5.97
599	7.22
637	8.28
675	9.61
698	10.11
701	9.60
705	8.42
710	7.38
715	6.57
720	5.79
725	5.44
733	5.03
753	4.80

T	α
CURVE 5‡	
312	4.96
354	5.13
397	5.84
450	6.12
456	5.83
490	5.97
501	6.26
530	6.12
559	6.43
573	6.30
601	6.47
612	6.65
650	7.32
688	8.41
698	9.03
709	9.61
718	10.15

T	α
CURVE 5 (cont.)‡	
725	10.68
731	11.18
736	11.42
743	11.66
754	12.28
766	12.68
774	12.72
784	12.61
793	12.21
801	12.01
806	11.89
811	11.34
815	9.86
820	8.19
822	7.20
829	6.21
838	5.76

T	α
CURVE 5 (cont.)‡	
846	5.63
857	5.59
873	5.58
CURVE 6‡	
855	6.00
841	5.94
816	5.87
791	6.03
777	7.74
769	12.02
762	12.90
751	12.45
746	11.61
740	11.00
732	10.38

T	α
CURVE 6 (cont.)‡	
724	10.03
711	9.33
697	9.04
683	9.00
660	8.71
644	8.18
628	7.64
610	7.16
593	6.63
542	6.12
417	5.73
383	5.50
359	5.49

* Not shown in figure.

‡ Author's data for coefficient of thermal expansion have been integrated by TPRC to obtain $\Delta L/L_0$.

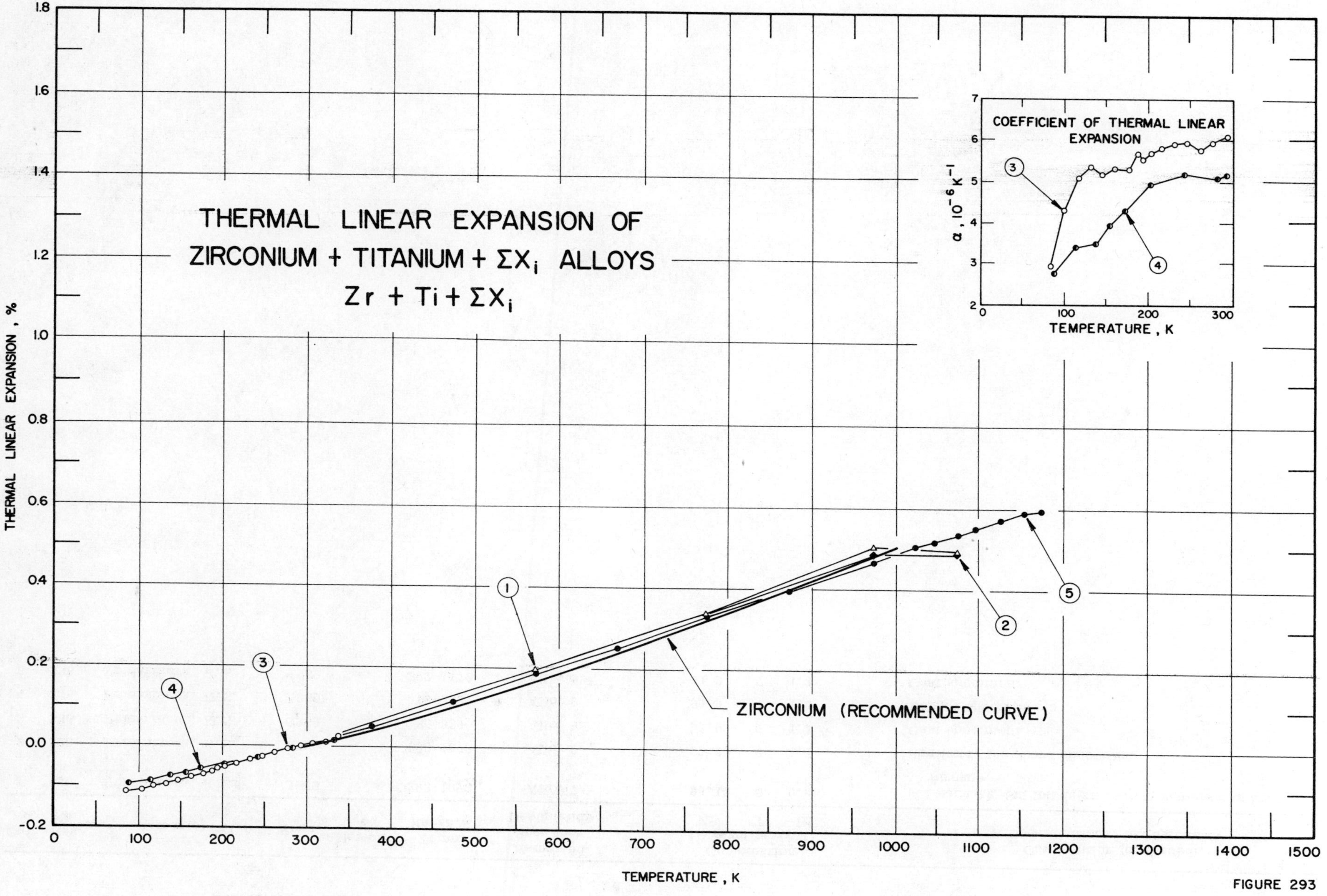
THERMAL LINEAR EXPANSION OF
ZIRCONIUM + TITANIUM + ΣX$_i$ ALLOYS
Zr + Ti + ΣX$_i$
COEFFICIENT OF THERMAL LINEAR EXPANSION
α, 10^{-6} K^{-1}
TEMPERATURE, K
ZIRCONIUM (RECOMMENDED CURVE)
THERMAL LINEAR EXPANSION, %
TEMPERATURE, K
1
2
3
4
5

FIGURE 293

SPECIFICATION TABLE 293. THERMAL LINEAR EXPANSION OF ZIRCONIUM + TITANIUM + ΣX_i ALLOYS $Zr + Ti + \Sigma X_i$

Cur. No.	Ref. No.	Author(s)	Year	Method Used	Temp. Range, K	Name and Specimen Designation	Composition (weight percent) Zr	Ti	Hf	Composition (continued), Specifications, and Remarks
1	338	Pridantseva, K. S.	1969		293-1073	Alloy 7	94.15	5	0.65	0.1 each Al, Sn; iodide zirconium remelted in arc furnace.
2	338	Pridantseva, K. S.	1969		293-1073	Alloy 9	91.9	7	1.1	Similar to the above specimen.
3	338	Pridantseva, K. S.	1969		85-339	Alloy 9	91.9	7	1.1	Cast specimen.
4	338	Pridantseva, K. S.	1969		89-332	Alloy 9	91.9	7	1.1	Forged specimen.
5	338	Pridantseva, K. S.	1969		377-1172	Alloy 9	91.9	7	1.1	Cast specimen.

DATA TABLE 293. THERMAL LINEAR EXPANSION OF ZIRCONIUM + TITANIUM + ΣX_i ALLOYS Zr + Ti + ΣX_i

[Temperature, T, K; Linear Expansion, $\Delta L/L_0$, %]

T	$\Delta L/L_0$
CURVE 1	
293	0.000*
573	0.195
773	0.339
973	0.503
1073	0.496
CURVE 2	
293	0.000*
573	0.189*
773	0.338*
973	0.484
1073	0.488
CURVE 3‡	
85	-0.112
100	-0.106
116	-0.099
131	-0.091
147	-0.083
162	-0.075
179	-0.066
188	-0.061
193	-0.058
203	-0.053
217	-0.044
233	-0.035
248	-0.026
262	-0.018
279	-0.008
293	-0.000
294	0.001*
308	0.009
323	0.018
339	0.028
CURVE 4‡	
89	-0.089
113	-0.082
139	-0.073
156	-0.066
173	-0.059
203	-0.046
242	-0.026
282	-0.005
293	-0.000*
308	0.008*
332	0.020
CURVE 5‡	
377	0.054
473	0.116
572	0.183
670	0.250
775	0.324
873	0.394
974	0.466
1024	0.502
1047	0.517
1075	0.536
1096	0.549
1125	0.568
1152	0.585
1172	0.599

DATA TABLE 293. COEFFICIENT OF THERMAL LINEAR EXPANSION OF ZIRCONIUM + TITANIUM + ΣX_i ALLOYS Zr + Ti + ΣX_i

[Temperature, T, K; Coefficient of Expansion, α, 10^{-6} K^{-1}]

T	α
CURVE 3‡	
85	2.94
100	4.24
116	5.09
131	5.35
147	5.19
162	5.27
179	5.27
188	5.67
193	5.54
203	5.68
217	5.80
233	5.88
248	5.91
262	5.78
279	5.90
293	6.05*
294	6.02
308	6.07
323	6.14
339	6.29
CURVE 4	
89	2.72
113	3.40
139	3.45
156	3.97
173	4.21
203	4.96
242	5.18
282	5.04
293	5.19
308	5.38
332	5.07
CURVE 5‡	
377	6.27
473	6.66
572	6.79
670	6.96
775	7.06
873	7.22
974	7.13
1024	7.00
1047	6.70
1075	6.50
1096	6.40
1125	6.43
1152	6.51
1172	6.61

*Not shown in figure.
‡Author's data for coefficient of thermal expansion have been integrated by TPRC to obtain $\Delta L/L_0$.

References to Data Sources

Ref. No.	TPRC No.	
1	43613	Nix, F.C. and MacNair, D., "The Thermal Expansion of Pure Metals: Copper, Gold, Aluminum, Nickel, and Iron," Phys. Rev., 60, 597-605, 1941.
2	40154	King, A.D., Cornish, A.J., and Burke, J., "Technique for Measuring Vacancy Concentrations in Metals at the Melting Point," J. Appl. Phys., 37(13), 4717-22, 1966.
3	46546	Fraser, D.B. and Hollis-Hallett, A.C., "The Coefficient of Thermal Expansion of Various Cubic Metals Below 100 K," Can. J. Phys., 43(2), 193-219, 1965.
4	54124	Figgins, B.F., Jones, G.O., and Riley, D.P., "The Thermal Expansion of Aluminium at Low Temperatures as Measured by an X-Ray Diffraction Method," Phil. Mag., 1(8), 747-58, 1956.
5	52439	Gerchikova, N.S., Kolobnev, N.I., Stepanova, M.G., and Fridlyander, I.N., "The Influence of the Aluminum Oxide Content on the Structure and Properties of Pressed Objects Made of Sap," Teploprochn. Mater. Spechennoy Alyum. Pudry, 5-16, 1961; English Translation: Joint Publications Research Service Rept. JPRS-17818, 15 pp., 1963.
6	52473	Matveyev, B.I., Kishnev, P.V., and Khanova, I.R., "Properties of Semifinished Products Made of Sintered Aluminum Powder," Teploprochn. Mater. Spechennoy Alyum. Pudry, 108-12, 1961; English Translation: Joint Publications Research Service Rept. JPRS-17818, 123-8, 1963.
7	43608	Bijl, D. and Pullan, H., "A New Method for Measuring the Thermal Expansion of Solids at Low Temperatures, the Thermal Expansion of Copper and Aluminum and the Grüneisen Rule," Physica, 21, 285-98, 1955.
8	52573	Altman, H.W., Rubin, T., and Johnston, H.L., "Coefficient of Thermal Expansion of Solids at Low Temperature. III. The Thermal Expansion of Pure Metals, with the Data for Aluminum, Nickel, Titanium, and Zirconium," Ohio State University, Cryogenic Laboratory Rept. OSU-TR-264-27, 10 pp., 1954. [AD 26 970]
9	51609	Gibbons, D.F., "Thermal Expansion of Some Crystals with the Diamond Structure," Phys. Rev., 112(1), 136-40, 1958.
10	43626	Wilson, A.J.C., "The Thermal Expansion of Aluminum. Further Experiments," Proc. Phys. Soc. (London), 54, 487-91, 1942.
11	6348	Taylor, C.S., Willey, L.A., Smith, D.W., and Edwards, J.D., "The Properties of High-Purity Aluminium," Metals and Alloys, 7(8), 189-92, 1938.
12	43612	Ayres, H.D., "Coefficients of Linear Expansion at Low Temperatures," Phys. Rev., 20, 38-51, 1905.
13	52254	Irmann, R., "Properties and Uses of Sintered Aluminium," Metal Treat., 22, 245-50, 1955.
14	45423	Palatnik, L.S., Pugachev, A.T., Boiko, B.T., and Bratsykhin, V.M., "Electron Diffraction Determination of Thermal Expansion Coefficients of Thin Films," Izv. Akad. Nauk SSR Ser. Fiz., 31(3), 478-81, 1967.
15	46067	Rhodes, B.L., Moeller, C.E., Hopkins, V., and Marx, T.I., "Thermal Expansion of Several Technical Metals from -255 to 300 C," Advances Cryog. Engr., 8, 278-86, 1963; Proc. Cryog. Eng. Conf. (University of California, August 14-16, 1962), Paper E-7, 20 pp., 1962.
16	35401	Souder, W.H. and Hidnert, P., "Thermal Expansion of Ni, Monel Metal, Stellite, Stainless Steel, and Al," Nat. Bur. Stand. Tech. News Bull., 17, 497-519, 1921.
17	35426	Simmons, R.O. and Balluffi, R.W., "Measurements of Equilibrium Vacancy Concentrations in Aluminum," Phys. Rev., 117(1), 52-61, 1960.
18	35449	Zubenko, V.V. and Umansky, M.M., "X-Ray Determination of the Thermal Expansion Coefficient of Polycrystalline Substances in the Interval -50 to 100 C," Kristallografiya, 1(4), 436-41, 1956.
19	35448	Hidnert, P. and Krider, H.S., "Thermal Expansion of Aluminum and Some Aluminum Alloys," J. Res. Nat. Bur. Stand., 48(3), 209-19, 1952.
20	35453	Hidnert, P., "Thermal Expansion of Aluminum and Various Important Aluminum Alloys," J. Res. Nat. Bur. Stand., 19, 697-731, 1925.
21	36215	Schmidt, F.J., and Hess, I.J., "Electroforming Aluminum for Solar Energy Concentrators," NASA-CR-197, 1, 1965.
22	55961	Fraser, D.B. and Hollis-Hallett, A.C., "The Coefficient of Linear Expansion and Grüneisen Gamma of Cu, Ag, Au, Fe, Ni, and Al from 4 to 300 K," Proc. Int. Conf. Low Temperature Phys., Seventh, 689-92, 1961.
23	55481	Carr, R.H., "Use of Mutual Inductance Techniques to Measure Thermal Expansion at Low Temperatures," Iowa State University, Ph.D. Thesis, 105 pp., 1963. [Univ. Microfilms No. 63-5171]
24	56406	Nicklow, R.M. and Young, R.A., "Thermal Expansion of Silver Chloride," Phys. Rev., 129(5), 1936-43, 1963.
25	56413	Wilson, A.J.C., "The Thermal Expansion of Aluminium from 0 to 650 C," Proc. Phys. Soc., 53, 235-44, 1941.

Ref. No.	TPRC No.	
26	50877	Buffington, R.M. and Latimer, W.M., "The Measurement of Coefficients of Expansion at Low Temperatures. Some Thermodynamic Applications of Expansion Data," J. Amer. Chem. Soc., 48, 2305-19, 1926.
27	57244	Pathak, P.D. and Vasavada, N.G., "Thermal Expansion and the Law of Corresponding States," J. Phys. C (Solid State Phys.), 3(2), L44-8, 1970.
28	55978	Abbiss, C.P., Huzan, E., and Jones, G.O., "Thermal Expansion of Aluminum at Low Temperatures," Proc. Int. Conf. Low Temperature Phys., Seventh, 688-9, 1961.
29	53245	Branchereau, M., Navez, M., and Perroux, M., "Precision Industrial Dilatometer," Verres Refract., 16(3), 159-67, 1962.
30	56780	Strelkov, P.G. and Novikova, S.I., "Silica Dilatometer for Low Temperatures. 1. Thermal Expansion of Copper and Aluminum," Prib. Tekh. Eksp. (USSR), 5, 105-10, 1957; English Translation: Sandia Laboratories, Library Operations Division, Technical Translation Service, SC-T-69-1054, 13 pp., 1969. [PB-187836T]
31	55989	Kagan, A.S., "Measurement of the Thermal Expansion Coefficient by X-Ray Methods," Zavodskaya Lab., 30(4), 459-61, 1964.
32	57757	Kochanovska, A., "Investigation of Thermal Dilatation of Cubic Metals," Physica, 15(1/2), 191-6, 1949.
33	58065	Richards, J.W., "The Over-All Linear Expansion of Three Face-Centered Cubic Metals (Al, Cu, Pb) from -190 C to Near Their Melting Points," Trans. Am. Soc. Metals, 30, 326-36, 1942.
34	52896	Asay, J.R., Urzendowski, S.R., and Guenther, A.H., "Ultrasonic and Thermal Studies of Selected Plastics, Laminated Materials, and Metals," U.S. Air Force Rept. AFWL-TR-67-91, 266 pp., 1968. [AD 827 596]
35	58039	Honda, K. and Okubo, Y., "On the Measurement of the Coefficients of Thermal Expansion for Aluminium Alloys and Alloys of Nickel-Iron and Cobalt-Iron," Sci. Reports Tohoku Imp. Univ., 13, 101-7, 1924.
36	55703	Otte, H.M., Montague, W.G., and Welch, D.O., "X-Ray Diffractometer Determination of the Thermal Expansion Coefficient of Aluminum Near Room Temperature," J. Appl. Phys., 34(10), 3149-50, 1963.
37	58923	Cornish, A.J. and Burke, J., "A High Temperature Attachment for an X-Ray Diffractometer for Precision Lattice Parameter Measurements," J. Sci. Instrum., 42, 212-18, 1965.
38	58940	Uffelmann, F.L., "The Expansion of Metals at High Temperatures," Phil. Mag., 7, 10(65), 633-59, 1930.
39	58409	Petrov, Yu.I., "Thermal Expansion and Melting Anomalies of Small Aluminum Crystals," Fiz. Tverd. Tela, 5(9), 2462-76, 1963; English Translation: Soviet Physics-Solid State, 5(9), 1793-805, 1964.
40	44135	Gorton, A.T., Bitsianes, G., and Joseph, T.L., "Thermal Expansion Coefficients for Iron and Its Oxides from X-Ray Diffraction Measurements at Elevated Temperatures," Trans. Met. Soc. AIME, 233(8), 1519-25, 1965.
41	39762	White, G.K., "Thermal Expansion of Magnetic Metals at Low Temperatures," Proc. Phys. Soc., 86, 159-69, 1965.
42	44975	Stuart, H. and Ridley, N., "Thermal Expansion of Cementite and Other Phases," J. Iron Steel Inst. (London), 204(7), 711-17, 1966.
43	36986	Fasiska, E.J. and Zwell, L., "Thermal Expansion of Iron Phosphide," Trans. Met. Soc. AIME, 239(6), 924-5, 1967.
44	50931	Krishnan, K.S., "Thermal Expansion in Dilute Solid Solutions," University of Florida, Ph.D. Thesis, 210 pp., 1966. [Univ. Microfilms No. 67-12932]
45	50901	Ridley, N. and Stuart, H., "Lattice Parameter Anomalies at the Curie Point of Pure Iron," Brit. J. Appl. Phys., 1(10), 1291-5, 1968.
46	54140	Basinski, Z.S., Hume-Rothery, W., and Sutton, A.L., "The Lattice Expansion of Iron," 229A, 459-67, 1955.
47	46145	Allen, R.D., "The Thermal Expansion of Synthetic Graphites at Temperature Intervals between 80 and 2000 F," Calif. Inst. of Tech., Jet Propulsion Lab. Progress Rept. No. 30-20, 17 pp., 1959. [AD 235 454]
48	49396	Dorsey, H.G., "Coefficient of Linear Expansion at Low Temperatures," Phys. Rev., 25, 88-102, 1907.
49	55969	Owen, E.A. and Yates, E.L., "An X-Ray Investigation of Pure Iron-Nickel Alloys. Part 3. The Thermal Expansion of Alloys Rich in Iron," Proc. Phys. Soc., 49, 307-14, 1937.
50	7668	Jaeger, F.M., Rosenbohm, E., and Zuithoff, A.J., "The Exact Measurement of the Specific Heat and Other Physical Properties of Solid Substances at High Temperatures," Rec. Trav. Chim., 57, 1313-40, 1938.
51	41306	Souder, W. and Hidnert, P., "Thermal Expansion of a Few Steels," Natl. Bur. Stand. Tech. News Bull., 17, 611-26, 1922.
52	28538	National Physical Laboratory Med., "Physical Constants of Pure Metals," H. M. Stationary Office (London), Pamphlet, 27 pp., 1936.
53	57700	Von Batchelder, F.W. and Raeuchle, R.F., "Re-Examination of the Symmetries of Iron and Nickel by the Powder Method," Acta Cryst., 7, 464, 1954.
54	54956	Straumanis, M.E. and Kim, D.C., "Lattice Constants, Thermal Expansion Coefficients, Densities, and Perfection of Structure of Pure Iron and of Iron Loaded with Hydrogen," Z. Metallk., 60(4), 272-7, 1969.

Ref. No.	TPRC No.	
55	57074	Richter, F. and Lotter, U., "On the Volume Magnetostriction of Nickel, Iron, and Cobalt," Phys. Stat. Solidi, 34(2), K149-52, 1969.
56	58043	Masumoto, H. and Nara, S., "On the Coefficient of Thermal Expansion in Nickel-Cobalt and Iron-Cobalt Alloys, and the Magnetostriction of Iron-Nickel Alloys," Sci. Reports Tohoku Univ., 16, 333-41, 1927.
57	58051	Miller, Yu.G., "Coefficients of Linear Expansion of Dilute Alpha-Solid Solutions of Fe with Cr, Mo + W," Sov. Physics Dokl., 3(2), 409-10, 1958.
58	58062	Austin, J.B. and Pierce, R.H.H., Jr., "The Linear Thermal Expansion and Alpha-Gamma Transformation Temperature (A_3 Point) of Pure Iron," Trans. Am. Soc. Metals, 22, 447-70, 1934.
59	58045	Masumoto, H., "On the Thermal Expansion of the Alloys of Iron, Nickel, and Cobalt, and the Cause of the Small Expansibility of Alloys of the Invar Type," Sci. Reports Tohoku Univ., 20, 101-23, 1931.
60	37540	Holborn, L. and Day, A.L., "On the Expansion of Certain Metals at High Temperatures," Ann. Physik, 4, 104-22, 1901; English Translation: Amer. J. Sci., 32(4), 374-90, 1901.
61	38389	Austin, J.B. and Pierce, R.H.H., Jr., "The Linear Thermal Expansion and Alpha-Gamma Transformation Temperature of Pure Iron," Physics, 4(12), 409-10, 1933.
62	58105	Yaggee, F.L. and Styles, J.W., "Autographic Dilatometer for Use with Pyrophoric and Alpha-Active Materials," Argonne Natl. Lab. Rept. ANL-7643, 19 pp., 1969.
63	52388	Bunton, G.V. and Weintroub, S., "The Thermal Expansion of Antimony and Bismuth at Low Temperatures," J. Phys. C (Solid State Phys.), 2(1), 116-23, 1969.
64	53414	Deshpande, V.T. and Pawar, R.R., "Anisotropic Lattice Thermal Expansion of Antimony," Curr. Sci., 38(1), 9, 1969.
65	35472	Hidnert, P., "Thermal Expansion of Monocrystalline and Polycrystalline Antimony," J. Res. Nat. Bur. Stand., 14, 523-44, 1935.
66	38669	Feder, R. and Nowick, A.S., "Use of Thermal Expansion Measurements to Detect Lattice Vacancies Near the Melting Point of Pure Lead and Aluminum," Phys. Rev., 109(6), 1959-63, 1958.
67	58707	Richter, V.F., "The Thermal Expansion of Technical Pure Iron (Armco-Iron) and Pure Iron Between -190 and 1450 C," 41(7), 709-14, 1970.
68	41321	Canning, D.A. and Weintroub, S., "Thermal Expansion of Lead at Low Temperatures," Can. J. Phys., 43(5), 955-9, 1965.
69	42330	Feder, R. and Nowick, A.S., "Equilibrium Vacancy Concentration in Pure Pb and Dilute Pb-Ti and Pb-In Alloys," Phil. Mag., 15, 805-12, 1967.
70	40328	Dorsey, H.G., "Further Measurements of the Coefficient of Linear Expansion at Low Temperatures," Phys. Rev., 27(1), 1-10, 1908.
71	54125	Dheer, P.N. and Surange, S.L., "Thermal Expansion of Lead at Low Temperatures," Phil. Mag., 8, 3(31), 665-74, 1958.
72	40337	Olsen, J.L. and Rohrer, H., "The Volume Change at the Superconducting Transition," Helv. Phys. Acta, 30, 49-65, 1957.
73	41390	Hidnert, P. and Sweeney, W.T., "Thermal Expansion of Lead," J. Res. Nat. Bur. Stand., 9, 703-9, 1932.
74	55931	White, G.K., "Thermal Expansion at Low Temperatures. IV. Normal and Superconducting Lead," Phil. Mag., 7, 271-8, 1962.
75	44496	Nic, F.C. and MacNair, D., "The Thermal Expansion of Pure Metals. II. Molybdenum, Palladium, Silver, Tantalum, Tungsten, Platinum, and Lead," Phys. Rev., 61, 74-8, 1942.
76	57759	Van Duijn, J. and Van Galen, J., "Influence of Vacancies on the Thermal Expansion of Lead Near the Melting Point," Physica, 23(7), 622-4, 1957.
77	48148	Andres, K., "Thermal Expansion and Magnetostriction of Lanthanum Metal at Low Temperatures," Phys. Rev., 168(3), 708-14, 1968.
78	52506	Barson, F., Legvold, S., and Spedding, F.H., "Thermal Expansion of Rare Earth Metals," USAEC Rept. ISC-831 (U.S. Air Force Rept. WADC-56-88), 71 pp., 1956.
79	52761	Barson, F., Legvold, S., and Spedding, F.H., "A Low Temperature Dilatometric Study of Some Rare Earth Metals," USAEC Rept. ISC-424, 64 pp., 1953.
80	39550	Andres, K., "Giant Thermal Expansion Coefficients in Rare Earth Metals at Low Temperatures," Phys. Rev. Lett., 10(6), 223-5, 1963.
81	52486	Trombe, F. and Foex, M., "Anomalies of Dilation of Metals of the Ferromagnetic Rare Earths Gadolinium and Dysprosium," J. Res. Centre Natl. Recherche Sci. Lab., Bellevue (Paris), 23, 71-5, 1953.
82	59158	Hanak, J.J., "High Temperature Allotropy of the Rare-Earth Elements," Iowa State University of Science and Technology, Ph.D. Thesis, 167 pp., 1959. [Univ. Microfilms No. 59-5668]
83	50208	Laquer, H.L., "Low Temperature Thermal Expansion of Various Materials," USAEC Rept. AECD-3076, 58 pp., 1952.

Ref. No.	TPRC No.	
84	40731	Snyder, D.D., Zimmerman, W.B., and Kuhlman, H., "Thermal Expansion of Isotopically-Enriched Lithium, 4.2 to 300 K," USAEC Rept. TID-20252, 24 pp., 1964.
85	56791	Kogan, V.S. and Khotkevich, V.I., "Temperature Dependence of the Effect of Isotopic Composition on the Size of the Lattice Constant in Lithium," Zhur. Eksp. Theoret. Fiz., 42, 916-17, 1962; English Translation: Sov. Phys. JETP, 15, 632-3, 1962.
86	60290	Lynch, R.W. and Edwards, L.R., "Thermal Expansion Coefficients and Grüneisen Parameters of bcc Li-Mg Alloys," J. Appl. Phys., 41(13), 5135-7, 1970.
87	58017	Childs, B.G. and Weintroub, S., "The Measurement of the Thermal Expansion of Single Crystals of Tin by an Interferometric Method," Proc. Phys. Soc. (London), 63B, 267-77, 1950.
88	58467	Swenson, C.A., "Lithium Metal. An Experimental Equation of State," J. Phys. Chem. Solids, 27(1), 33-8, 1966.
89	51009	Andres, K., "The Measurement of Thermal Expansion of Metals at Low Temperatures," Cryogenics, 2, 93-7, 1961.
90	52511	Gordon, P., "A High Temperature Precision X-Ray Camera. Some Measurements of the Thermal Coefficients of Expansion of Beryllium," J. Appl. Phys., 20, 908-17, 1949.
91	17600	Zelikman, A.N., Kislyakov, I.P., and Balshin, M.Yu., "Handbook on Machine-Building Materials. Vol. 2. Ferrous Metals and Their Alloys (Selected Articles)," Spravochnik po Mashinostr. Mater. Tsvet. Metal. Ikh. Splavy, Moscow, 1959; English Translation: Wright-Patterson Air Force Base, Technical Document Liaison Office Rept. MCL-681/1, 210 pp., 1961. [AD256155]
92	47057	Martin, A.J. and Moore, A., "The Structure of Beryllium with Particular Reference to Temperatures above 1200 C," J. Less-Common Metals, 1, 85-93, 1959.
93	52483	Meyerhoff, R.W. and Smith, J.F., "Anisotropic Thermal Expansion of Single Crystals of Thallium, Yttrium, Beryllium, and Zinc at Low Temperatures," J. Appl. Phys., 33(1), 219-24, 1962.
94	42396	Chirkin, V.S., "Thermal Diffusivity and Thermal Conductivity of Metallic Beryllium," At. Energ. (USSR), 20(1), 80-2, 1966.
95	46880	Brett, N.H. and Russell, L.E., "Physical Properties of the Oxide. The Thermal Expansion of PuO_2 and Some Other Actinide Oxides Between Room Temperature and 1000 C (5)," Plutonium, 397-410, 1960.
96	52423	Watrous, J.D., "Thermal Expansion of Snap Materials," USAEC Rept. NAA-SR-6047, 19 pp., 1961.
97	57897	Finkel, V.A. and Papirov, I.I., "Crystal Structure of Beryllium at 77-300 K," Fiz. Metal. Metalloved., 6, 1108-10, 1968; English Translation: Phys. Metals Metallogr., 26(6), 150-2, 1968.
98	57752	Owen, E.A. and Richards, T.L., "On the Thermal Expansion of Beryllium," Phil. Mag., 22, 304-11, 1936.
99	58061	Treco, R.M., "Thermal Expansion Characteristics of Beryllium," Trans. Am. Inst. Mining Met. Eng., 188, 1274-6, 1950.
100	52396	White, G.K., "Thermal Expansion of Bismuth at Low Temperature," J. Phys. C (Solid State Phys.), 2(3), 575-6, 1969.
101	56540	Cave, E.F., "The Anisotropic Thermal Expansion of Bismuth," University of Missouri, Ph.D. Thesis, 96 pp., 1958. [Univ. Microfilms No. 58-5247]
102	58024	Roberts, J.K., "The Thermal Expansion of Crystals of Metallic Bismuth," Proc. Roy. Soc. London, 106, 385-99, 1924.
103	58025	Jay, A.H., "The Thermal Expansion of Bismuth by X-Ray Measurements," Proc. Roy. Soc., 143, 465-72, 1934.
104	46534	McCammon, R.D. and White, G.K., "Thermal Expansion at Low Temperatures of Hexagonal Metals: Mg, Zn, and Cd," Phil. Mag., 11(114), 1125-34, 1965.
105	41295	Madaiah, N. and Graham, G.M., "Thermal Expansion of Cadmium and Indium," Can. J. Phys., 42(1), 221-5, 1964.
106	55962	White, G.K., "Thermal Expansion of Solids at Low Temperature," Proc. Int. Conf. Low Temperature Phys., Eighth, 394-6, 1963.
107	52346	Kotelnikov, V.A. and Petrov, Yu.I., "Anomalous Thermal Expansion and Melting of Small Cadmium Crystals," Fiz. Tverd. Tela, 11(5), 1391-3, 1969; English Translation: Sov. Phys. Solid State, 11(5), 1128-9, 1969.
108	57751	Owen, E.A. and Roberts, E.W., "The Thermal Expansion of the Crystal Lattices of Cadmium, Osmium, and Ruthenium," Phil. Mag., 22, 290-304, 1936.
109	57740	Hachkovsky, W.F. and Strelkov, P.G., "Anomalous Expansion of Zinc and Cadmium Near the Melting Point," Nature, 139, 715-16, 1937.
110	52677	Bernstein, B.T. and Smith, J.F., "Coefficients of Thermal Expansion for Face-Centered Cubic and Body-Centered Cubic Calcium," Acta Cryst., 12, 419-20, 1959.
111	52422	Smith, R.D. and Morrice, E., "Electrical Resistivity of Cerium Metal from 4 to 300 K," U.S. Bureau of Mines Rept. BM-RI-6480, 13 pp., 1964.
112	52675	Pavlov, V.S. and Finkel, V.A., "Low-Temperature Crystal Structure of Cerium," Fiz. Metal. Metalloved., 24(6), 1123-4, 1967; English Translation: Phys. Metals Metallogr. USSR, 24(6), 137-8, 1967.

Ref. No.	TPRC No.	
113	53234	Trombe, F. and Foex, M., "Dilatometric Study and Determination of the Allotropic States of Lanthanum and Metallic Cerium," Compt. Rend., 217, 501-3, 1943.
114	52880	Rashid, M.S. and Altstetter, C.J., "Allotropic Transformations in Cerium," Trans. Met. Soc. AIME, 236, 1649-56, 1966. [AD 656 798]
115	53998	Anderson, M.S., Gutman, E.J., Packard, J.R., and Swenson, C.A., "Equation of State for Cesium Metal to 23 Kbar," J. Phys. Chem. Solids, 30(6), 1587-601, 1969.
116	43672	Ishikawa, Y., Hoshino, S., and Endoh, Y., "Antiferromagnetism in Dilute Iron Chromium Alloys," J. Phys. Soc. Japan, 22(5), 1221-32, 1967.
117	51109	Yaggee, F.L. and Styles, J.W., "Linear Thermal Expansion of Vanadium, Titanium, Chromium, and Some Vanadium-Base Binary and Ternary Alloys," in Argonne National Laboratory Annual Progress Report for 1966, USAEC Rept. ANL-7299, 77-8, 1966.
118	47073	Straumanis, M.E. and Weng, C.C., "The Precise Lattice Constant and the Expansion Coefficient of Chromium between 10 and 60 C," Acta Cryst., 8, 367-71, 1955.
119	55154	Yaggee, F.L., Gilbert, E.R., and Styles, J.W., "Thermal Expansivities, Thermal Conductivities, and Densities of Vanadium, Titanium, Chromium, and Some Vanadium-Base Alloys (A Comparison with Austenitic Stainless Steel)," J. Less-Common Metals, 19(1), 39-51, 1969.
120	52567	Shevlin, T.S. and Hauck, C.A., "Alumina-Base Cermets," U.S. Air Force Rept. WADC-TR-54-173 (Pt. 1), 45 pp., 1954. [AD 49 092]
121	6560	Lucks, C.F. and Deem, H.W., "Thermal Conductivities, Heat Capacities, and Linear Thermal Expansion of Five Materials," U.S. Air Force Rept. WADC-TR-55-496, 65 pp., 1956. [AD 97 185]
122	46661	Muller, S. and Dunner, P., "Determination of the Lattice Parameter of Chromium in the Temperature Region from 20 to 1500 C," Z. Naturforsch., 20A(9), 1225-6, 1965.
123	9736	Lucks, C.F. and Deem, H.W., "Thermal Properties of Thirteen Metals," Am. Soc. Test. Mater. Spec. Tech. Publ. 227, 29 pp., 1958.
124	39751	White, G.K., "The Anomalous Thermal Expansion of Chromium," Aust. J. Phys., 14, 359-67, 1961.
125	39564	Hidnert, P., "Thermal Expansion of Electrolytic Chromium," J. Res. Nat. Bur. Stand., 26, 81-91, 1941.
126	39749	James, W.J., Straumanis, M.E., and Rao, P.B., "The Anomaly in the Expansivity Curve of Chromium," J. Inst. Metals, 90, 176-7, 1961-2.
127	33951	Bolgov, I.S., Smirnov, Yu.N., and Finkel, V.A., "Phase Transformations in Cobalt," Fiz. Metal. Metalloved., 17(6), 877-80, 1964; English Translation: Phys. Metals Metallogr., 17(6), 76-9, 1964.
128	43584	Brenner, A., Burkhead, P., and Seegmiller, E., "Electrodeposition of Tungsten Alloys Containing Iron, Nickel, and Cobalt," J. Res. Natl. Bur. Stand., 39, 351-83, 1947.
129	55975	Masumoto, H., "On the Thermal Expansion of Alloys of Cobalt, Iron, and Chromium, and a New Alloy Stainless-Invar," Sci. Rept. Tohoku Imp. Univ., 23, 265-80, 1934.
130	44455	Petrov, Yu.I., "Structure and Thermal Expansion of Small Cobalt Particles," Kristallografiya, 11(6), 931-2, 1966.
131	53633	Kulesko, G.I. and Seryugin, A.L., "Geometrical Shape of Martensite Plates in Cobalt," Fiz. Metal. Metalloved., 26(2), 327-30, 1968; English Translation: Phys. Metals Metallogr., 26(2), 140-3, 1968.
132	45786	Masumoto, H., Saito, H., and Kikuchi, M., "The Thermal Expansion and the Temperature Change of Young's Modulus of Single Crystals of Hexagonal Cobalt," Nippon Kinzoku Gakkaishi, 31(1), 25-30, 1967.
133	44229	Ullrich, H.-J., "Precision Lattice Parameter Measurements by Interferences from Lattice Sources (Kossel Lines) and Divergent Beam X-Ray Diffraction (Pseudo-Kossel-Lines) in Back Reflection," Phys. Stat. Sol., 20(2), K113-17, 1967.
134	54133	Simmons, R.O. and Balluffi, R.W., "Low-Temperature Thermal Expansion of Copper," Phys. Rev., 2, 108(2), 278-80, 1957.
135	10880	Paine, R.M., Stonehouse, A.J., and Beaver, W.W., "An Investigation of Intermetallic Compounds for Very High Temperature Applications," U.S. Air Force Rept. WADC TR 59-29 (Pt. 2), 119 pp., 1959. [AD 244 758]
136	53331	Gehlen, P.C., "A 77-1300 K Single Crystal X-Ray Specimen Chamber," Rev. Sci. Instr., 40(5), 715-18, 1969.
137	52580	Carr, R.H., McCammon, R.D., and White, G.K., "The Thermal Expansion of Copper at Low Temperatures," Proc. Roy. Soc., 280A, 72-84, 1964.
138	6502	Fieldhouse, I.B., Hedge, J.C., Lang, J.T., Takata, A.N., and Waterman, T.E., "Measurements of Thermal Properties," U.S. Air Force Rept. WADC TR 55-495 (Pt. I), 64 pp., 1956. [AD 110 404]
139	52966	Bunton, G.V. and Weintroub, S., "Sensitive Optical Lever Dilatometer for Use at Low Temperatures and the Thermal Expansion of Copper," Cryogenics, 8(6), 354-60, 1968.
140	52490	Pandey, H.D. and Dayal, B., "Bradburn-Furth Equation of State and Thermal Expansion of Face-Centered Metals," Phys. Stat. Sol., 5, 273-7, 1964.
141	55731	Simmons, R.O. and Balluffi, R.W., "Measurement of Equilibrium Concentrations of Vacancies in Copper," Phys. Rev., 129(4), 1533-44, 1963.

Ref. No.	TPRC No.	
142	52119	Aoyama, S. and Ito, T., "Thermal Expansion of Nickel-Copper Alloys at Low Temperatures (Part I)," Sci. Repts. Tohoku Univ., 27(1), 348-64, 1939.
143	45520	Nasekovskii, A.P., "On the Temperature Dependence of the Thermal Expansion Coefficient for Cubiform Crystals," Ukr. Fiz. Zh., 12(8), 1352-5, 1967.
144	41507	Keesom, W.H., Van Agt, F.P.G.A.J., and Jansen, A.F.J., "The Thermal Expansion of Copper between 101 and 253 C," Verslag. Gewone Vergadering Afdel. Natuurk. Konink. Akad. Wetenschappen, 35, 262-7, 1926; English Translation: Proc. Roy. Acad. Sci. Amsterdam, 29, 786-91, 1926; Communications Kamerlingh Onnes Lab. Univ. Leiden, 182A, 2-9, 1926.
145	19989	Sandenaw, T.A., "The Thermal Expansion of Plutonium Metal Below 300 K," USAEC Rept. LA-2394, 23 pp., 1960.
146	55207	Rubin, T., Altman, H.W., and Johnston, H.L., "Coefficients of Thermal Expansion of Solids at Low Temperatures. 1. The Thermal Expansion of Copper from 15 to 300 K," Ohio State University, Cryogenic Lab. Rept. OSU-TR-264-11, 20 pp., 1954.
147	48613	Novikova, S.I. and Strelkov, P.G., "Thermal Expansion of Silicon at Low Temperatures," Fiz. Tverd. Tela, 1(12), 1841-3, 1960; English Translation: Sov. Phys.-Solid State, 1(12), 1687-9, 1960.
148	55594	Kos, J.F. and Lamarche, J.L.G., "Thermal Expansion of the Noble Metals Below 15 K," Can. J. Phys., 47(22), 2509-18, 1969.
149	57719	Burger, E.E., "The Expansion Characteristics of Some Common Glasses and Metals," Gen. Elec. Rev., 37(2), 93-6, 1934.
150	57211	Eisenstein, A., "A Study of Oxide Cathodes by X-Ray Diffraction Methods. Part I. Methods, Conversion Studies, and Thermal Expansion Coefficients," J. Appl. Phys., 17, 434-43, 1946.
151	54991	Kasai, N., "Measurement of Thermal Expansion at Low Temperatures," Denki Shikenjo Iho, 32(12), 1157-61, 1968.
152	57758	Eppelsheimer, D.S. and Penman, R.R., "Thermal Dilation of Copper," Physica, 16(10), 792-4, 1950.
153	55005	Masumoto, H., Saito, H., and Sawaya, S., "Thermal Expansion Coefficient and the Temperature Coefficient of Young's Modulus of Nickel-Copper Alloys," Nippon Kinzoku Gakkaishi, 33(5), 593-5, 1969.
154	58873	Shapiro, J.M., Taylor, D.R., and Graham, G.M., "A Sensitive Dilatometer for Use at Low Temperatures," Can. J. Phys., 42(5), 835-46, 1964.
155	58917	Hume-Rothery, W. and Andrews, K.W., "The Lattice Spacing and Thermal Expansion of Copper," J. Inst. Metals, 68, 19-26, 1942.
156	58410	Mitra, G.B. and Mitra, S.K., "X-Ray Diffraction Study of Copper at High Temperatures," Indian J. Phys., 37, 462-72, 1963.
157	46555	Clark, A.E., DeSavage, B.F., and Bozorth, R., "Anomalous Thermal Expansion and Magnetostriction of Single-Crystal Dysprosium," Phys. Rev., 138A(1), A216-24, 1965.
158	42838	Finkel, V.A. and Vorobev, V.V., "Crystal Structure of Dysprosium at 77 to 300 K," Zh. Eksptl. Teoret. Fiz., 51(3), 786-90, 1966; English Translation: Sov. Phys.-JETP, 24(3), 524-6, 1966.
159	46005	Banister, J.R., Legvold, S., and Spedding, F.H., "Structure of Gd, Dy, and Er at Low Temperatures," Phys. Rev., 94(5), 1140-2, 1954.
160	39031	Darnell, F.J., "Temperature Dependence of Lattice Parameters for Gd, Dy, and Ho," Phys. Rev., 130(5), 1825-8, 1963.
161	60280	Hahn, T.A., "Thermal Expansion of Copper from 20 to 800 K - Standard Reference Material 736," J. Appl. Phys., 41(13), 5096-101, 1970.
162	47380	Spedding, F.H., Hanak, J.J., and Daane, A.H., "The Preparation and Properties of Europium," Trans. Met. Soc. AIME, 212, 379-83, 1958.
163	46698	Ergin, Yu.V., "Anomalies in the Temperature Dependence of the Thermal Expansion Coefficient of a Gadolinium Single Crystal," J. Exptl. Theoret. Phys. (USSR), 48, 1062-4, 1965; English Translation: Sov. Phys.-JETP, 21(4), 709-10, 1965.
164	40004	Vorobev, V.V., Smirnov, Yu.N., and Finkel, V.A., "The Crystal Structure of Gadolinium at 120-370 K," Zh. Eksptl. Teoret. Fiz., 49(6), 1774-8, 1965; English Translation: Sov. Phys.-JETP, 22(6), 1212-15, 1966.
165	56055	Cadieu, F.J. and Douglass, D.H., Jr., "Effects of Impurities on Higher Order Phase Transitions," Phys. Rev. Lett., 21(10), 680-2, 1968.
166	58030	Birss, R.R., "The Thermal Expansion Anomaly of Gadolinium," Proc. Roy. Soc., 255A, 398-406, 1960.
167	46777	Zhdanova, V.V. and Kontorova, T.A., "Thermal Expansion of Doped Germanium," Fiz. Tverd. Tela, 7(11), 3331-8, 1965; English Translation: Sov. Phys.-Solid State, 7(11), 2685-9, 1966.
168	47108	Zhdanova, V.V., Kekua, M.G., and Samadashvili, T.Z., "Thermal Expansion of Alloys in the Si-Ge System," Izv. Akad. Nauk SSSR, Neorg. Mater., 3(7), 1263-4, 1967; English Translation: Inorg. Mater. (USSR), 3(7), 1112-14, 1967.
169	46535	Carr, R.H., McCammon, R.D., and White, G.K., "Thermal Expansion of Germanium and Silicon at Low Temperatures," Phil. Mag., 12(115), 157-63, 1965.
170	39437	Sparks, P.W. and Swenson, C.A., "Thermal Expansions from 2 to 40 K of Germanium, Silicon, and Four III-V Compounds," Phys. Rev., 163(3), 779-90, 1967.

Ref. No.	TPRC No.	
171	47406	Sparks, P.W., "Thermal Expansion of Tetrahedrally Bonded Solids at Low Temperatures," Iowa State University of Science and Technology, Ph.D. Thesis, 114 pp., 1967. [Univ. Microfilms No. 67-5628]
172	50513	Janot, C., Bianchi, G., and George, B., "Dilatometric Study of the Exact Defects at Thermodynamic Equilibrium. The Case of Germanium," Compt. Rend., 267B(5), 336-9, 1968.
173	54096	Novikova, S.I., "Thermal Expansion of Germanium at Low Temperatures," Fiz. Tverd. Tela, 2(1), 43-4, 1960; English Translation: Sov. Phys.-Solid State, 2(1), 37-8, 1960.
174	50034	Singh, H.P., "Determination of Thermal Expansion of Germanium, Rhodium, and Iridium by X-Rays," Acta Crystal., Sect. A, 24(4), 469-71, 1968.
175	55976	Nan, S. and Yi-Huan, L., "X-Ray Measurement of the Thermal Expansion of Germanium, Silicon, Indium Antimonide, and Gallium Arsenide," Sci. Sinica, 14(11), 1582-9, 1965.
176	50224	Baldwin, T.O., "The Effect of Fast Neutron Irradiation at Ambient Temperature on the Lattice Parameters of Silicon and Germanium Crystals," Phys. Rev. Lett., 21(13), 901-3, 1968.
177	38804	Zhdanova, V.V., "Effect of Impurities on the Coefficient of Thermal Expansion of P-Ge," Fiz. Tverd. Tela, 5(11), 3341-3, 1963; English Translation: Sov. Phys. Solid State, 5(11), 2450-1, 1964.
178	55139	Fine, M.E., "Elasticity and Thermal Expansion of Germanium Between -195 and 275 C," J. Appl. Phys., 24(3), 338-40, 1953.
179	58464	Straumanis, M.E. and Aka, E.Z., "Lattice Parameters, Coefficients of Thermal Expansion, and Atomic Weights of Purest Silicon and Germanium," J. Appl. Phys., 23(3), 330-4, 1952.
180	70158	Kirby, R.K., "Thermal Expansion of Tungsten from 293 to 1800 K," High Temp.-High Pressures, 4(4), 459-62, 1972.
181	46220	Merryman, R.G. and Kempter, C.P., "Precise Temperature Measurement in Debye-Scherrer Specimens at Elevated Temperatures," J. Am. Ceram. Soc., 48(4), 202-5, 1965.
182	49030	Merryman, R.G., "A Study of Temperature Measurement Precision in Debye-Scherrer Specimens During High Temperature X-Ray Diffraction Measurement of Thermal Expansion," USAEC Rept. LA-2687, 98 pp., 1962.
183	43617	Dutta, B.N. and Dayal, B., "Lattice Constants and Thermal Expansion of Gold Up to 878 C by X-Ray Method," Phys. Stat. Sol., 3, 473-7, 1963.
184	51089	Vermaak, J.S., Kuhlmann-Wilsdorf, D., "Measurement of the Average Surface Stress of Gold as a Function of Temperature in the Temperature Range 50-985°," J. Phys. Chem., 72(12), 4150-4, 1968.
185	51137	Patel, V.K., "Lattice Constants, Thermal Expansion Coefficients, Densities, and Imperfections in Gold and the Alpha-Phase of the Gold-Indium System," University of Missouri at Rolla, M.S. Thesis, 109 pp., 1967.
186	43615	Simmons, R.O. and Balluffi, R.W., "Measurement of Equilibrium Concentration of Lattice Vacancies in Gold," Phys. Rev., 125(3), 862-72, 1962.
187	47075	Ross, R.G. and Hume-Rothery, W., "High Temperature X-Ray Metallography," J. Less-Common Metals, 5, 258-70, 1963.
188	52760	Baldwin, E.E., "The Thermal Expansion and Elevated Temperature Mechanical Strength of Hafnium. Report No. 1," USAEC Rept. KAPL-M-EEB-7, 14 pp., 1954.
189	7009	Adenstedt, H.K., "Physical, Thermal, and Electrical Properties of Hafnium and High Purity Zirconium," Trans. Am. Soc. Metals, 44, 949-73, 1949.
190	41164	Rhyne, J.J., Legvold, S., and Rodine, E.T., "Anomalous Thermal Expansion and Magnetostriction of Holmium Single Crystals," Phys. Rev., 154(2), 266-9, 1967.
191	54716	Pollock, D.B., "Thermal Expansion Values for Installation of an Irtran-6 Window," Appl. Opt., 8(4), 837-8, 1969.
192	52487	Shinoda, G., "X-Ray Investigations on the Thermal Expansion of Solids. Part 1," Mem. Coll. Sci. Kyoto Imp. Univ., 16A, 193-201, 1933.
193	52513	Smith, J.F. and Schneider, V.L., "Anisotropic Thermal Expansion of Indium," J. Less-Common Metals, 7, 17-22, 1964.
194	40337	Olsen, J.L. and Rohrer, H., "The Volume Change at the Superconducting Transition," Helv. Phys. Acta, 30, 49-65, 1957.
195	40345	Hidnert, P. and Blair, M.G., "Thermal Expansivity and Density of Indium," J. Res. Nat. Bur. Stand., 30, 427-33, 1943.
196	40993	Swenson, C.A., "Properties of Indium and Thallium at Low Temperatures," Phys. Rev., 100(6), 1607-13, 1955.
197	58018	Vernon, E.V. and Weintroub, S., "The Measurement of the Thermal Expansion of Single Crystals of Indium and Tin with a Photoelectric Recording Dilatometer," Proc. Phys. Soc. London, 66B, 887-94, 1953.
198	49102	Schaake, H.F., "Thermal Expansion of Iridium from 4.2 to 300 K," J. Less-Common Metals, 15(1), 103-5, 1968.
199	47254	Bell, I.P., "Thermal Reactor-Physical Properties of Materials of Construction," UKAEA R. & D.B.(C) Tech. Memo. 225, 5 pp., 1954.

Ref. No.	TPRC No.	
200	41428	Hidnert, P. and Sweeney, W.T., "Thermal Expansion of Magnesium and Some of Its Alloys," J. Res. Nat. Bur. Stand., 1, 771-92, 1928.
201	57748	Takahasi, K. and Kikuti, R., "Thermal Expansion Coefficients of Magnesium and Its Alloys," Nippon Kinzoku No Kenkyu, 13, 401-14, 1936.
202	58084	Hanawalt, J.D. and Frevel, L.K., "X-Ray Measurement of the Thermal Expansion of Magnesium," Zeit. Kristallogr., 98A, 84-8, 1938.
203	40942	Marples, J.A.C., "The Lattice Parameters of Alpha-Manganese at Low Temperatures," Phys. Lett., 24A(4), 207-8, 1967.
204	44766	Gazzara, C.P., Middleton, R.M., Weiss, R.J., and Hall, E.O., "A Refinement of the Parameters of Alpha Manganese," U.S. Army Rept. AMRA-TR-67-08, 15 pp., 1967. [AD 651 205]
205	58029	Basinski, Z.S. and Christian, J.W., "A Pressurized High-Temperature Debye-Scherrer Camera, and Its Use to Determine the Structures and Coefficients of Expansion of Gamma and Delta Manganese," Proc. Roy. Soc., 223A, 554-60, 1954.
206	49398	Hill, D.M., "The Principal Expansion Coefficients of Single Crystals of Mercury," Phys. Rev., 48, 620-4, 1935.
207	58411	Carpenter, L.G. and Oakley, F.H., "The Thermal Expansion and Atomic Heat of Solid Mercury," Phil. Mag., 7, 12(77), 511-22, 1931.
208	48587	Frantsevich, I.N., Zhurakovskii, E.A., and Lyashchenko, A.B., "Elastic Constants and Characteristics of the Electron Structure of Certain Classes of Refractory Compounds Obtained by the Metal-Powder Method," Izv. Akad. Nauk SSSR, Neorg. Mater., 3(1), 8-16, 1967; English Translation: Inorg. Mater. (USSR), 3(1), 6-12, 1967.
209	35122	Shodhan, R.P., "Solid Solubility of Sulfur, Selenium, and Tellurium in Molybdenum at 1100 C, Lattice Constants, Thermal Expansion Coefficients, and Densities of Molybdenum and the Molybdenum-Base Alloys," University of Missouri at Rolla, M.S. Thesis, 78 pp., 1965.
210	41642	Conway, J.B. and Losekamp, A.C., "Thermal Expansion Characteristics of Several Refractory Metals to 2500 C," Trans. Met. Soc. AIME, 236(5), 702-9, 1966.
211	51435	Yaggee, F.L. and Styles, J.W., "Linear Thermal Expansion of Refractory Metals and Alloys," in Argonne National Laboratory Annual Progress Report for 1965, USAEC Rept. ANL-7155, 148-9, 1965.
212	47077	Pawar, R.R., "Lattice Expansion of Molybdenum," Curr. Sci., 36(16), 428, 1967.
213	35079	Amonenko, V.M., Vyugov, P.N., and Gumenyuk, V.S., "Thermal Expansion of Tungsten, Molybdenum, Tantalum, Niobium, and Zirconium at High Temperatures," Teplofiz. Vys. Temp., Akad. Nauk SSSR, 2(1), 29-31, 1964; English Translation: High Temp. (USSR), 2(2), 22-4, 1964.
214	49357	Mulyakaev, L.M., Dubinin, G.N., Ryumin, V.P., and Golubeva, A.S., "Diffusion Chrome Plating of Molybdenum," Izv. Akad. Nauk SSSR, Neorg. Mater., 3(11), 2114-7, 1967; English Translation: Inorg. Mater. (USSR), 3(11), 1842-5, 1967.
215	45358	Edwards, J.W., Speiser, R., and Johnston, H.L., "High Temperature Structure and Thermal Expansion of Some Metals as Determined by X-Ray Diffraction Data. I. Platinum, Tantalum, Niobium, and Molybdenum," J. Appl. Phys., 22(4), 424-8, 1951.
216	52127	Eisenlohr, A., "Recent Advances in Arc-Plasma Metallizing," in Symposium on Processing Materials for Re-Entry Structures, 1960, U.S. Air Force Rept. WADD-TR-60-58, 315-32, 1960. [AD 241 597]
217	52494	Totskii, E.E., "Experimental Determination of the Coefficient of Linear Expansion of Metals and Alloys," Teplofiz. Vys. Temp., 2(2), 205-14, 1964; English Translation: High Temp. (USSR), 2(2), 181-9, 1964.
218	21359	Levinstein, M.A., "Properties of Plasma Sprayed Materials," U.S. Air Force Rept. WADD-TR-60-654, 91 pp., 1961. [AD 264 223]
219	52747	Denman, G.L., "An Automatic Recording Dilatometer for Thermal Expansion Measurements to 2000 F," U.S. Air Force Rept. ASD-TDR-62-315, 19 pp., 1962. [AD 282 838]
220	16437	Worthing, A.G., "Physical Properties of Well-Seasoned Molybdenum and Tantalum as a Function of Temperature," Phys. Rev., 28, 190-201, 1926.
221	6980	Rasor, N.S. and McClelland, J.D., "Thermal Properties of Materials. Part I. Properties of Graphite, Molybdenum, and Tantalum to Their Destruction Temperatures," U.S. Air Force Rept. WADC-TR-56-400, Pt. I, 53 pp., 1957. [AD 118 144]
222	41672	Schad, L.W. and Hidnert, P., "Preliminary Determination of the Thermal Expansion of Molybdenum," Nat. Bur. Stand. Tech. News Bull., 15, 31-40, 1919.
223	37030	Valentich, J., "Thermal Expansion Measurement," Instrum. Control Systems, 42, 91-4, 1969.
224	36366	Clark, D. and Knight, D., "Thermal Expansion of Beta-Silicon Carbide and Some Tungsten-Molybdenum Alloys," U.S. Air Force Rept. RAE-TR-65049, 13 pp., 1965. [AD 464 397]
225	37995	Valentich, J., "New Values for Thermal Coefficients," Prod. Eng., 63-71, 1965.
226	41675	Hidnert, P. and Gero, W.B., "Thermal Expansion of Molybdenum," Nat. Bur. Stand. Tech. News Bull., 19, 429-44, 1924.
227	45129	Davidson, D.L., "The Properties of Solid Solution, Molybdenum-Rich, Molybdenum-Rhenium Alloys from -190 to 100 C. Part I. Elastic Properties. Part II. Plastic Properties, Rice University, Ph.D. Thesis, 169 pp., 1968. [Univ. Microfilms No. 68-15612]

Ref. No.	TPRC No.	
228	53527	Baskin, M.L., Savin, A.V., Tumanov, V.T., and Eyduk, Yu.A., "Mutual Solubility of Copper and Molybdenum and Certain Properties of Alloys of Molybdenum with Copper," Izv. Akad. Nauk SSSR, Otdel. Tekh. Nauk Met. i Toplivo., 4, 111-14, 1961; English Translation: U.S. Air Force Rept. FTD-TT-65-1009, 11 pp., 1965. [AD 624 908]
229	53235	Trombe, F. and Foex, M., "Dilatometric Study of Metallic Neodymium," Compt. Rend., 232, 63-5, 1951.
230	6386	Jaeger, F.M., Bottema, J.A., and Rosenbohm, E., "The Exact Measurements of the Specific Heat of Solid Substances at High Temperatures. X. Specific Heat, Electrical Resistance, Thermoelectrical Behaviour, and Thermal Expansion of Some Rare Earth Metals," Rec. Trav. Chim., 57(11), 1137-82, 1938.
231	46765	Petrov, Yu.I., "Anomalous Thermal Expansion and X-Ray Scattering in Small Nickel Particles Near the Curie Temperature," Fiz. Tverdogo Tela., 6(7), 2155-9, 1964; English Translation: Sov. Phys.-Solid State, 6(7), 1697-700, 1965.
232	47542	Shevlin, T.S., Newkirk, H.W., Jr., Stevens, E.G., and Greenhouse, H.M., "Preliminary Microscopic Studies of Cermets at High Temperatures. Part 2," U.S. Air Force Rept. WADC-TR-54-33 (Pt. 2), 127 pp., 1956. [AD 90 041]
233	43600	Owen, E.A. and Yates, E.L., "X-Ray Measurement of the Thermal Expansion of Pure Nickel," Phil. Mag., 21, 809-19, 1936.
234	45423	Palatnik, L.S., Pugachev, A.T., Boiko, B.T., and Bratsykhin, V.M., "Electron Diffraction Determination of Thermal Expansion Coefficients of Thin Films," Izv. Akad. Nauk SSR Ser. Fiz., 31(3), 478-81, 1967.
235	55967	Yates, E.L. and Owen, E.A., "X-Ray Investigation of Pure Iron-Nickel Alloys. Part 1. Thermal Expansion of Alloys Rich in Nickel," Proc. Phys. Soc., 49, 17-28, 1937.
236	54109	Davis, M., Densem, C.E., and Rendall, J.H., "The Manufacture and Properties of High-Strength Nickel-Tungsten Alloys," J. Inst. Metals, 84, 160-4, 1955.
237	43134	Jordan, L. and Swanger, W.H., "The Properties of Pure Nickel," J. Res. Nat. Bur. Stand., 5, 1291-307, 1930.
238	51181	Masumoto, H., Saito, H., Murakami, Y., and Kikuchi, M., "Crystal Anisotropy and Temperature Dependence of Young's Modulus of Nickel Single Crystals," Nippon Kinzoku Gakkaishi, 32(6), 525-8, 1968.
239	56507	Newkirk, H.W., Jr. and Sisler, H.H., "Determination of Residual Stresses in Titanium Carbide-Base Cermets by High-Temperature X-Ray Diffraction," J. Amer. Cer. Soc., 41(3), 93-103, 1958.
240	55831	Pathak, P.D., Gupta, M.C., and Trivedi, J.M., "Recording Instrument for Measurement of Thermal Expansion," Ind. J. Phys., 43(2), 104-7, 1969.
241	56258	Masumoto, H., Sawaya, S., and Kadowaki, S., "Effect of Addition of Manganese, Titanium, or Zinc on the Elinvar Property of Nickel at High Temperatures," Nippon Kinzoku Gakkaishi, 33(10), 1138-40, 1969.
242	48346	Hidnert, P. and Krider, H.S., "Thermal Expansion of Columbium," J. Res. Natl. Bur. Stand., 11, 279-84, 1933.
243	42845	Vasyutinskiy, B.M., Kartmazov, G.N., Smirnov, Yu.M., and Finkel, V.A., "High-Temperature Crystalline Structure of Niobium and Vanadium," Fiz. Metal. Metalloved., 21(4), 620-1, 1966; English Translation: Phys. Metals Metallogr. (USSR), 21(4), 134-5, 1966.
244	39030	Harris, B. and Peacock, D.E., "Physical Properties of Some Niobium (Columbium) Alloys at Low Temperature," Trans. Met. Soc. AIME, 236(4), 471-3, 1966.
245	46142	Nowotny, H., Laube, E., Auer-Welsbach, H., Schob, E., Boller, H., Gorelzki, H., and Geist, F., "Advanced Experiments on Metal-Silicon-Boron and Metal-Carbon-Boron Systems," University of Vienna, Final Tech. Rept., June 1960-May 1961, 49 pp., 1961. [AD 261 390]
246	46697	Smirnov, Yu.M. and Finkel, V.A., "Crystal Structure of Tantalum, Niobium, and Vanadium at 110-400 K," J. Exptl. Theoret. Phys. (USSR), 49, 1077-82, 1965; English Translation: Sov. Phys.-JETP, 22(4), 750-3, 1966.
247	48189	Johnson, P.M., Lincoln, R.L., and McClure, E.R., "Development of a High-Temperature Interferometric Dilatometer Using a Laser Light Source," U.S. Bureau of Mines Rept. BMRI-7142, 13 pp., 1968.
248	52514	Conway, J.B., Fincel, R.M., Jr., and Losekamp, A.C., "The Linear Thermal Expansion of Niobium (Columbium)," Trans. Met. Soc. AIME, 233, 844-5, 1965.
249	6656	Tottle, C.R., "The Physical and Mechanical Properties of Niobium," J. Inst. Metals, 85, 375-8, 1957.
250	7689	Fieldhouse, I.B., Hedge, J.C., and Lang, J.T., "Measurements of Thermal Properties," U.S. Air Force Rept. WADC-TR-58-274, 79 pp., 1958. [AD 206 892]
251	43249	White, G.K., "Thermal Expansion of Vanadium, Niobium, and Tantalum at Low Temperatures," Cryogenics, 2, 292-6, 1962.
252	19417	Heal, T.J., "Mechanical and Physical Properties of Magnesium and Niobium Canning Materials," Proc. U.N. Intern. Conf. Peaceful Uses At. Energy, 2nd, Geneva, 5, 208-19, 1958.
253	52408	Ulyanov, R.A., Tarasov, N.D., and Mikhaylov, Ya.D., "The Effect of Alloying on the Thermal Expansion of Niobium," U.S. Air Force Rept. FTD-MT-64-16, 148-54, 1964. [AD 604 600]

Ref. No.	TPRC No.	
254	57756	Rosenbohm, E., "An Apparatus for the Photographic Record of the Linear Expansion of Metals. Expansion Coefficients of Copper and Nickel," Physica, 5, 385-98, 1938.
255	56787	Bollenrath, F., "A New Optical Dilatometer," Zeit. Metallk., 26(3), 62-5, 1934.
256	38545	Jacobs, R.B. and Goetz, A., "The Thermal Expansion of the Bismuth Lattice Between 25 and 530 K," Phys. Rev., 51, 159-64, 1937.
257	38712	Bolef, D.I. and DeKlerk, J., "Anomalies in the Elastic Constants and Thermal Expansion of Chromium Single Crystals," Phys. Rev., 129(3), 1063-7, 1963.
258	52584	Matsumoto, T. and Mitsui, T., "Thermal Expansion of Chromium Single Crystals at the Néel Temperature," J. Phys. Soc. Japan, 27(3), 786, 1969.
259	58072	Klemm, W., "Note on the Expansion Coefficient of Gallium and the Product αT_s for Elements," Z. Anorg. Allgem. Chem., 198, 178-83, 1931.
260	56568	Richards, T.W. and Boyer, S., "Further Studies Concerning Gallium: Its Electrolytic Behavior, Purification, Melting Point, Density, Coefficient of Expansion, Compressibility, Surface Tension, and Latent Heat of Fusion," J. Amer. Chem. Soc., 43, 274-94, 1921.
261	44678	Shinoda, G., "Further Investigations on the Thermal Expansion of Solids by X-Ray Method," Proc. Phys.-Math. Soc. Japan, 16(3), 436-8, 1934.
262	57865	Goens, E. and Schmid, E., "Elastic Constants, Electrical Resistivity and Thermal Expansion of Magnesium Crystals," Physik. Z., 37(11), 385-91, 1936.
263	58076	Grube, G. and Vosskuhler, H., "Electric Conductivity and Phase Diagram of Binary Alloys," Z. Elektrochem., 40(8), 566-70, 1934.
264	56734	Janot, C., Mallejac, D., and George, B., "Determination of the Parameters of Formation of Defects in Magnesium by Dilatometry," Compt. Rend., 270B(6), 404-6, 1970.
265	58083	Owen, E.A. and Roberts, E.W., "The Crystal Parameters of Osmium and Ruthenium at Different Temperatures," Zeit. Kristall., 96, 497-8, 1937.
266	38668	Smith, C.S., "Properties of Plutonium Metal," Phys. Rev., 94, 1068-9, 1954.
267	58021	Jaeger, F.M. and Zanstra, J.E., "On the Allotropism of Rhodium and on Some Phenomena Observed in the X-Ray-Analysis of Heated Metal-Wires," Proc. Royal Acad., Amsterdam, 34, 15-32, 1931.
268	43618	Dutta, B.N. and Dayal, B., "Lattice Constants and Thermal Expansion of Palladium and Tungsten Up to 878 C by X-Ray Method," Phys. Stat. Sol., 3, 2253-9, 1963.
269	52963	Waterhouse, N. and Yates, B., "Interferometric Measurement of the Thermal Expansion of Silver and Palladium at Low Temperatures," Cryogenics, 8(5), 267-71, 1968.
270	51182	Masumoto, H., Saito, H., and Kadowaki, S., "Young's Modulus of Palladium Single Crystals at High Temperatures," Nippon Kinzoku Gakkaishi, 32(6), 529-32, 1968.
271	53650	Masumoto, H. and Sawaya, S., "Nonmagnetic Elinvar-Type Alloy Pallagold in the Palladium-Gold System," Nippon Kinzoku Gakkaishi, 33(1), 121-5, 1969.
272	56412	Owen, E.A. and Jones, J.T., "The Effect of Pressure and Temperature on the Occlusion of Hydrogen by Palladium," Proc. Phys. Soc., 49, 587-602, 1937.
273	45165	Mauer, F.A. and Bolz, L.H., "Measurement of Thermal Expansion of Cermet Components by High Temperature X-Ray Diffraction," U.S. Air Force Rept. WADC-TR-55-473, 57 pp., 1955. [AD 95 329]; Supplement No. 1, NBS-5837, 47 pp., 1957. [AD 155 555]
274	43582	Brand. J.A. and Goldschmidt, H.J., "Temperature Calibration of a High-Temperature X-Ray Diffraction Camera," J. Sci. Instr., 33(2), 41-5, 1956.
275	55926	Owen, E.A. and Yates, E.L., "The Thermal Expansion of the Crystal Lattices of Silver, Platinum, and Zinc," Phil. Mag., 17(110), 113-31, 1934.
276	55963	Andres, K., "The Thermal Expansion of Some Metals at Low Temperatures," Proc. Int. Conf. Low Temperatures Physics, Eighth, 397-8, 1963.
277	58833	Liu, L.-G., Takahashi, T., and Bassett, W.A., "Effect of Pressure and Temperature on the Lattice Parameters of Rhenium," J. Phys. Chem. Solids, 31(6), 1345-51, 1970.
278	34222	Solente, P., "Crystallographic Study and Self-Irradiation Effects of Plutonium at Low Temperature," Paris University, France, Ph.D. Thesis, 57 pp., 1965; USAEC Rept. CEA-R-2735.
279	48424	Lee, J.A., Marples, J.A.C., Mendelssohn, K., and Sutcliffe, P.W., "Some Physical Properties of Plutonium Metal at Low Temperatures," Plutonium 1965 Proc. Int. Conf. Plutonium, 3rd, London, 176-88, 1967.
280	53237	Lallement, R., "Expansion and Thermoelectric Power of Alpha-Plutonium at Low Temperature," J. Phys. Chem. Solids, 24, 1617-24, 1963.
281	46908	Grove, G.R., "Reactor Fuels and Materials Development Plutonium Research," USAEC Rept. MLM-1402, 50 pp., 1967.
282	52764	Elliott, R.E. and Tate, R.E., "A Determination of the Coefficient of Thermal Expansion of Alpha Plutonium," USAEC Rept. LA-1390, 20 pp., 1952.
283	37536	Abramson, R., Boucher, R., Fabre, R., Monti, H., Pascard, R., Anselin, F., and Grison, E., "Some Properties of Plutonium and Its Alloys," French AEC Rept. A/Conf. 15(P)327, 30 pp., 1958.

Ref. No.	TPRC No.	
284	57701	Zachariasen, W.H., "Unit Cell and Thermal Expansion of Beta-Plutonium Metal," Acta Cryst., 12, 175-6, 1959.
285	56476	Zachariasen, W.H. and Ellinger, F.H., "Crystal Chemical Studies of the 5F-Series of Elements. XXIV. The Crystal Structure and Thermal Expansion of Gamma-Plutonium," Acta Cryst., 8(7), 431-3, 1955.
286	45523	Ellinger, F.H., "Crystal Structure of δ' Plutonium and the Thermal Expansion Characteristics of δ, δ', and ε Plutonium," J. Metals, 206, 1256-9, 1956.
287	55716	Brocklehurst, R.E., Goode, J.M., and Vassamillet, L.F., "Coefficient of Expansion of Polonium," J. Chem. Phys., 27, 985, 1957.
288	46613	Stokes, R.H., "The Molar Volumes and Thermal Expansion Coefficients of Solid and Liquid Potassium from 0-85 C," J. Phys. Chem. Solids, 27(1), 51-6, 1966.
289	54112	Monfort, C.E., III, and Swenson, C.A., "An Experimental Equation of State for Potassium Metal," J. Phys. Chem. Solids, 26, 291-301, 1965.
290	41147	Marples, J.A.C., "Thermal Expansion of Protactinium Metal," Acta Cryst., 18(4), 815-16, 1965.
291	54129	Andres, K., "The Thermal Expansion of Molybdenum, Tungsten, Rhenium, Platinum, and Cadmium at Low Temperatures," Phys. Letters, 7(5), 315-6, 1963.
292	44160	Pawar, R.R. and Deshpande, V.T., "X-Ray Determination of the Thermal Expansion of Rhenium," Curr. Sci., 36(5), 120-1, 1967.
293	58069	Wasilewski, R.J., "Axial Thermal Expansion of Rhenium," Trans. Met. Soc. AIME, 221, 1081-2, 1961.
294	49319	Swanger, W.H., "Melting, Mechanical Working, and Some Physical Properties of Rhodium," J. Res. Natl. Bur. Stand., 3, 1029-40, 1929.
295	55956	Bale, E.S., "The Structure of Rhodium," Platinum Metals Rev., 2, 61-3, 1958.
296	34084	Mardon, P.G., Nichols, J.L., Pearce, J.H., and Poole, D.M., "Some Properties of Scandium Metal," Nature (London), 189, 566-8, 1961.
297	48797	Geiselman, D., "The Metallurgy of Scandium," U.S. Air Force Rept. WADD-TR-60-894, 54 pp., 1961. [AD268188]
298	46674	Straumanis, M.E. and Riad, S.M., "Solubility Limit of Indium in Silver and Thermal Expansion Coefficients of the Solid Solutions," Trans. Met. Soc. AIME, 233(5), 964-7, 1965.
299	47523	Petrov, Yu.I. and Fedorov, Yu.I., "Thermal Expansion and Melting of Small Crystals of Silver," Dokl. Akad. Nauk SSSR, 175(6), 1325-7, 1967; English Translation: Dokl. Phys. Chem., 175(6), 644-6, 1967.
300	53402	Michel, D.J., "Some Thermal and Mechanical Properties of the Intermetallic Compound $HoZn_2$," Pennsylvania State University, Ph.D. Thesis, 77 pp., 1968. [Univ. Microfilms No. 69-14543]
301	59597	Arbuzov, M.P. and Zelenkov, I.A., "Thermal Expansion of Certain Transition Metals and Alloys on Their Base," Fiz. Metal. Metalloved. Akad. Nauk SSSR, Ural Filial, 18(2), 311-12, 1964; English Translation: Phys. Metals Metallogr., 18(2), 149-50, 1964.
302	52492	Beenakker, J.J.M. and Swenson, C.A., "Total Thermal Contractions of Some Technical Metals to 4.2 K," Rev. Sci. Instr., 26, 1204-5, 1955.
303	43601	Owen, E.A. and Roberts, E.W., "Factors Affecting the Limit of Solubility of Elements in Copper and Silver," Phil. Mag., 27, 294-327, 1939.
304	41611	Evans, D.J. and Winstanley, C.J., "Measurement of the Coefficient of Thermal Expansion of Solids and Its Precise Variation with Temperature," J. Sci. Instr., 43(10), 772-3, 1966.
305	44529	Simmons, R.O. and Balluffi, R.W., "Measurement of the Equilibrium Concentration of Lattice Vacancies in Silver Near the Melting Point," Phys. Rev., 119(2), 600-5, 1960.
306	56415	Hume-Rothery, W. and Reynolds, P.W., "A High-Temperature Debye-Scherrer Camera, and Its Application to the Study of the Lattice Spacing of Silver," Proc. Roy. Soc. (London), 167A, 25-34, 1938.
307	58922	Spreadborough, J. and Christian, J.W., "High-Temperature X-Ray Diffractometer," J. Sci. Instrum., 36, 116-18, 1959.
308	40776	Feder, R. and Charbnau, H.P., "Equilibrium Defect Concentration in Crystalline Sodium," Phys. Rev., 149(2), 464-71, 1966.
309	54130	Sifgel, S. and Quimby, S.L., "The Thermal Expansion of Crystalline Sodium Between 80 K and 290 K," Phys. Rev., 54(1), 76-8, 1938.
310	46673	Conway, J.B., Fincel, R.M., Jr., and Losekamp, A.C., "Effects of Contaminants on the Thermal Expansion of Tantalum," Trans. Met. Soc. AIME, 233(4), 841-2, 1965.
311	40096	Vyugov, P.N. and Gumenyuk, V.S., "Thermal Expansion of Tungsten and Tantalum in the Range 1500-3000 C," Teplofiz. Vysok. Temp., 3(6), 936-7, 1965; English Translation: High Temp. (USSR), 3(6), 879-80, 1965.
312	52096	Hidnert, P., "Thermal Expansion of Tantalum," J. Res. Natl. Bur. Stand., 2, 887-96, 1929.
313	42965	Makin, S.M., Standring, J., and Hunter, P.M., "Determination of the Coefficient of Linear Thermal Expansion of Metals at T°C. A Review of Progress from 1 December 1952-1 June 1953," UKAEC Rept. RDB-(C)-TN-45, 8 pp., 1953.
314	52421	Saldinger, I.L. and Glasier, L.F., Jr., "Mechanical and Physical Properties of Tantalum and Carburized Tantalum Above Approximately 3500 F," Aerojet-General Corp., Materials and Processes Dept., Rept. M-1795, 66 pp., 1959.

Ref. No.	TPRC No.	
315	42833	Finkel, V.A., Smirnov, Yu.N., and Vorobev, V.V., "Crystal Structure of Terbium at 120-300 K," Zh. Eksptl. i Teoret. Fiz., 51(1), 32-7, 1966; English Translation: Sov. Phys.-JETP, 24(1), 21-4, 1967.
316	47781	Alberts, L., DuPlessis, P.De.V., "Thermal Expansion and Forced Magnetostriction in a Terbium Single Crystal," J. Appl. Phys., 39(2), 581-2, 1968.
317	50398	Finkel, V.A. and Vorobev, V.V., "Crystal Structure of Terbium-Yttrium Alloys at 77-330 K," Zh. Eksptl. i Teoret. Fiz., 53(6), 1913-9, 1967; English Translation: Sov. Phys.-JETP, 26(6), 1086-9, 1968.
318	39001	Larikov, L.N., Falchenko, V.M., and Koblova, E.A., "Thermodynamic Properties of Thallium," Ukr. Fiz. Zh., 11(2), 212-16, 1966.
319	43569	Armstrong, P.E., Carlson, O.N., and Smith, J.F., "Elastic Constants of Thorium Single Crystals in the Range 77-400 K," J. Appl. Phys., 30(1), 36-41, 1959.
320	36105	Thompson, J.G., "Some Physical Properties of Commercial Thorium," Metals Alloys, 4, 114-18, 1933.
321	36877	Danielson, G.C., Murphy, G., Peterson, D., and Rogers, B.A., "Progress Report of an Investigation of the Properties of Thorium and Some of Its Alloys. Progress Report Feb. 1, 1952," USAEC Rept. ISC-208, 38 pp., 1952.
322	9233	Smith, J.F., "Physical Constants, Crystal Structure, and Thermodynamic Properties," in The Metal Thorium, Amer. Soc. Metals, 133-47, 1958.
323	54101	Novikova, S.I., "Study of the Thermal Expansion of Alpha-Tin, Indium Antimonide, and Cadmium Telluride," Fiz. Tverd. Tela, 2(9), 2341-4, 1960; English Translation: Sov. Phys.-Solid State, 2(9), 2087-9, 1961.
324	57702	Deshpande, V.T. and Sirdeshmukh, D.B., "Thermal Expansion of Tetragonal Tin," Acta Cryst., 14(4), 355-6, 1961.
325	57704	Deshpande, V.T. and Sirdeshmukh, D.B., "Thermal Expansion of Tin in the Beta-Gamma Transition Region," Acta Cryst., 15, 294-5, 1962.
326	57743	Thewlis, J. and Davey, A.R., "Thermal Expansion of Grey Tin," Nature, 174, 1011, 1954.
327	58046	Matuyama, Y., "On the Question of the Allotropy of White Tin and the Equilibrium Diagram of the System Tin-Cadmium," Sci. Reports, Tohoku Univ., 20, 649-80, 1931.
328	35521	Packwood, R.H. and Black, P.J., "The Absence of Pre-Melting Anomalies in Tin," Proc. Phys. Soc. (London), 86, 653-66, 1965.
329	43792	Kornilov, I.I. and Boriskina, N.G. (Editors), "Mechanical and Engineering Properties of Titanium Alloys," Novyye Issled. Splavov, Pt. III (Moscow), 167-333, 1965; English Translation: Joint Publications Research Service Rept. JPRS-36680, 235 pp., 1966.
330	46128	Koftun, S.F. and Ulyanov, R.A., "Effect of Alloying on the Physicochemical Properties of Titanium," in Works of the Fifth Conf. on Metallurgy, Physical Metallurgy, and Application of Titanium and Its Alloys, 1963, NASA Rept. NASA-TT-F-338, 120-36, 1965.
331	42170	Margolin, H. and Portisch, H., "Hydrogen-Induced Expansions in Titanium-Aluminum Alloys," Trans. Met. Soc. AIME, 242(9), 1901-13, 1968.
332	48242	Greiner, E.S. and Ellis, W.C., "Thermal and Electrical Properties of Ductile Titanium," Am. Inst. Mining Met. Eng. Tech. Pub. TP 2466, 9 pp., 1948.
333	43639	Williams, D.N., "Thermal Expansion of Beta-Titanium," Trans. Met. Soc. AIME, 221, 411-12, 1961.
334	52791	Willens, R.H., "A Vacuum X-Ray Diffractometer for High Temperature Studies and an Investigation of the Allotropic Transformation of Titanium," U.S. Air Force Rept. AFOSR-1839, 131 pp., 1961. [AD 270 632]
335	50047	Cowan, J.A., Pawlowicz, A.T., and White, G.K., "Thermal Expansion of Polycrystalline Titanium and Zircondium," Cryogenics, 8(3), 155-7, 1968.
336	44985	Hidnert, P., "Thermal Expansion of Titanium," J. Res. Nat. Bur. Stand., 30, 101-5, 1943.
337	52542	Wasilewski, R.J., "Thermal Expansion of Titanium and Some TiO Alloys," Trans. Met. Soc. AIME, 221, 1231-5, 1961.
338	56358	Pridantseva, K.S., "Manufacture, Properties and Application of Zirconium," Nauch.-Issled. Inst. Chern. Met., 51, 139-52, 1967; U.S. Air Force Rept. FTD-MT-24-457-68, 21 pp., 1969. [AD 694 926]
339	58032	Berry, R.L.P. and Raynor, G.V., "A Note on the Lattice Spacings of Titanium at Elevated Temperatures," Research, London, 6(4), (Suppl.-1), 21S-23S, 1953.
340	58059	Greiner, E.S. and Ellis, W.C., "Thermal and Electrical Properties of Ductile Titanium," Trans. Am. Inst. Mining Met. Engr., 180, 657-65, 1949.
341	43082	Houska, C.R., "Thermal Expansion and Atomic Vibration Amplitudes for Titanium Carbide, Titanium Nitride, Zirconium Carbide, Zirconium Nitride, and Pure Tungsten," U.S. Air Force Rept. TR-C-17, 16 pp., 1963.
342	45040	Steele, S.R., Pappis, J., Schilling, H., and Hagen, L., "Chemical Vapor Deposited Materials for Electron Tubes," Raytheon Co., Waltham, Mass. Rept. ECOM-01343-7, 57 pp., 1967. [AD 651 021]

Ref. No.	TPRC No.	
343	36875	Takamori, T. and Tomozawa, M., "Behavior of Interlayers of Glass-to-Tungsten Seals," J. Am. Ceram. Soc., 48(8), 405-9, 1965.
344	43748	Rausch, J.J., "Protective Coatings for Refractory Metals in Rocket Engines," Illinois Institute of Technology Rept. IITRI-B237-45, 117 pp., 1966; NASA Rept. NASA-CR-71317.
345	50536	Brizes, W.F., "Mechanical Properties of the Group IV-B and V-B Transition Metal Monocarbides," NASA Rept. NASA-CR-95887, 45 pp., 1968.
346	6722	Forsythe, W.E. and Worthing, A.G., "The Properties of Tungsten and the Characteristics of Tungsten Lamps," Astrophys. J., 61, 146-85, 1925.
347	26074	Neel, D.S., Pears, C.D., and Oglesby, S., Jr., "The Thermal Properties of Thirteen Solid Materials to 5000 F or Their Destruction Temperatures," U.S. Air Force Rept. WADD-TR-60-924, 216 pp., 1962. [AD 275 536]
348	16727	Anthony, F.M. and Pearl, H.A., "Investigation of Feasibility of Utilizing Available Heat Resistant Materials for Hypersonic Leading Edge Applications. Vol. 3. Screening Test Results and Selection of Materials," U.S. Air Force Rept. WADC-TR-59-744(Vol. 3), 347 pp., 1960.
349	52756	Fulkerson, S.D., "Apparatus for Determining Linear Thermal Expansions of Materials in Vacuum of Controlled Atmosphere," USAEC Rept. ORNL-2856, 39 pp., 1960.
350	40762	Mauer, F.A. and Bolz, L.H., "Thermal Expansion of Cermet Components by High Temperature X-Ray Diffraction," Natl. Bur. Stand. Rept. NBS-3148, 38 pp., 1959.
351	55939	Hidnert, P. and Sweeney, W.T., "Thermal Expansion of Tungsten," J. Res. Natl. Bur. Stand., 20, 483-7, 1925.
352	56403	Worthing, A.G., "The Thermal Expansion of Tungsten at Incandescent Temperatures," Phys. Rev., 10(2), 638-41, 1917.
353	45127	Hidnert, P. and Sweeney, W.T., "Thermal Expansion of Tungsten," Nat. Bur. Stand. Tech. News Bull., 20, 483-7, 1925.
354	52745	Baun, W.L., "A High Temperature X-Ray Diffractometer Specimen Mount," U.S. Air Force Rept. WADC-TN-59-139, 11 pp., 1959.
355	54480	Knibbs, R.H., "Measurement of Thermal Expansion Coefficient of Tungsten at Elevated Temperatures," J. Sci. Instrum., 2, 2(6), 515-17, 1969.
356	57749	Goucher, F.S., "On the Strength of Tungsten Single Crystals and Its Variation With Temperature," Phil. Mag., 48(284), 229-49, 1924.
357	57724	Benedicks, C., Berlin, D.W., and Phragmen, G., "A Method for the Determination of the Specific Gravity of Liquid Iron and Other Metals of High Melting Point," Iron Steel Inst. (London), Carnegie Scholarship Mem., 13, 129-74, 1924.
358	24812	Fieldhouse, I.B. and Lang, J.I., "Measurement of Thermal Properties," U.S. Air Force Rept. WADD-TR-60-904, 119 pp., 1961. [AD 268 304]
359	58097	James, W.J. and Straumanis, M.E., "Lattice Parameter and Coefficient of Thermal Expansion of Vanadium," Z. Physik. Chem., 29, 134-42, 1961.
360	45633	Bridge, J.R., Schwartz, C.M., and Vaughan, D.A., "X-Ray Diffraction Determination of the Coefficients of Expansion of Alpha Uranium," J. Metals, 206, 1282-5, 1956.
361	52504	Lloyd, L.T., "Thermal Expansion of Alpha-Uranium Single Crystals," USAEC Rept. ANL-5972, 48 pp., 1959.
362	47103	Heal, T.J. and McIntosh, A.B., "High Temperature Properties of Uranium and Its Alloys," UKAEC Rept. A/Conf. 15(P)49, 48 pp., 1958.
363	49282	Andres, K., "Thermal Expansion of Alpha-Uranium Below 10 K," Phys. Rev., 170(3), 614-17, 1968.
364	40743	Laquer, H.L. and Schuch, A.F., "Low Temperature Dilatometry of Uranium," USAEC Rept. LAMS-1358, 13 pp., 1952.
365	52531	Thewlis, J., "An X-Ray Powder Study of Beta-Uranium," Acta Cryst., 5, 790-4, 1952.
366	55950	Barrett, C.S., Mueller, M.H., and Hitterman, R.L., "Crystal Structure Variations in Alpha Uranium at Low Temperatures," Phys. Rev., 129(2), 625-9, 1963.
367	7406	Schuch, A.F. and Laquer, H.L., "Low-Temperature Thermal Expansion of Uranium," Phys. Rev., 2, 86(5), 803, 1952.
368	52484	Ibrahim, E.F., "Thermal Expansion Measurements on Irradiated Uranium," J. Inst. Metals, 93(4), 117-23, 1964-5.
369	52409	Saller, H.A., Rough, F.A., and Chubb, W., "The Properties of Uranium Containing Minor Additions of Chromium, Silicon, or Titanium," USAEC Rept. BMI-1068(Del.), 32 pp., 1956.
370	57196	Battelle Memorial Institute, "Metallurgy of Tuballoy," USAEC Rept. CT-2144, 214-30, 1944.
371	57213	Kelpfer, H.H. and Chiotti, P., "Characteristics of the Solid State Transformations in Uranium," USAEC Rept. ISC-893, 35 pp., 1957.
372	42771	Finkel, V.A. and Vorobev, V.V., "Temperature Dependence of the Periods of the Yttrium Crystal Lattice at 77 to 300 K," Kristallografiya, 13(3), 550-1, 1968; English Translation: Sov. Phys.-Crystallogr., 13(3), 457-8, 1968.

Ref. No.	TPRC No.	
373	52432	Nolting, H.J., Simmons, C.R., and Klingenberg, J.J., "Preparation and Properties of High Purity Yttrium Metal," J. Inorg. Nucl. Chem., 14, 208-16, 1960.
374	46547	Channing, D.A. and Weintroub, S., "Thermal Expansion of Single Crsytals of Zinc at Low Temperatures," Can. J. Phys., 43(7), 1328-33, 1965.
375	54875	Gilder, H.M. and Wallmark, G.N., "Thermal Expansion Measurements of Vacancy Formation Parameters in Zinc Single Crystals," Phys. Rev., 182(3), 771-7, 1969.
376	55927	Owen, E.A. and Iball, J., "Thermal Expansion of Zinc by the X-Ray Method," Phil. Mag., 16, 479-88, 1933.
377	49991	Freeman, J.R. and Brandt, P.F., "Pure Zinc at Normal and Elevated Temperatures," Natl. Bur. Stand. Tech. News Bull., 20, 661-95, 1926.
378	57215	Claus, K. and Lohberg, K., "Irreversible Changes in Length of Zinc Under Thermal Stress," Zeit. Metallk., 46, 582-8, 1955; English Translation: USAEC Rept. IGRL-T/C-43, 11 pp., 1957.
379	46627	Goldak, J., Lloyd, L.T., and Barrett, C.S., "Lattice Parameters, Thermal Expansions, and Grüneisen Coefficients of Zirconium, 4.2 to 1130 K," Phys. Rev., 144(2), 478-84, 1966.
380	46825	Couterne, J.C. and Cizeron, G., "Determination of the Principal Coefficient of Thermal Expansion of Alpha-Zirconium," J. Nucl. Mater. (Amsterdam), 20, 75-82, 1966; English Translation: USAEC Rept. WAPD-TRANS-45, 14 pp., 1966.
381	52497	Balluffi, R.W., Resnick, R., and Timper, A.J., "Dilatometric Studies of Zirconium and Zirconium Tin Alloys Between 25 and 1100 C," USAEC Rept. SEP-90, 13 pp., 1952.
382	52268	Toy, S.M. and Vetrano, J.B., "Properties of Zirconium Hydride and Zirconium-Uranium Alloy Hydrides," USAEC Rept. NAA-SR-4244-DEL, 18 pp., 1960.
383	52583	Lloyd, L.T., "Thermal Expansion of Alpha-Zirconium Single Crystals," USAEC Rept. ANL-6591, 41 pp., 1963.
384	47211	Zwikker, C., "Modification Changes of Zirconium and Hafnium," Physica, 6, 361-5, 1926.
385	1546	Squire, C.F. and Kaufmann, A.R., "The Magnetic Susceptibility of Titanium and Zirconium," J. Chem. Phys., 9, 673-7, 1941.
386	56494	Skinner, G.B. and Johnston, H.L., "Thermal Expansion of Zirconium Between 298 K and 1600 K," J. Chem. Phys., 21(8), 1383-4, 1953.
387	56784	Russell, R.B., "Coefficients of Thermal Expansion for Zirconium," Trans. Am. Inst. Mining Met. Engrs., 200, 1045-52, 1954.
388	58049	Nakamura, K., "Effect of Temperature on Young's Modulus of Elasticity in Nickel-Copper Alloys," Sci. Reports Tohoku Univ., 25, 415-25, 1936.
389	57985	Owen, E.A., Yates, E.L., and Sully, A.H., "An X-Ray Investigation of Pure Iron-Nickel Alloys. Part 4. The Variation of Lattice-Parameter with Composition," Proc. Phys. Soc., 49, 315-22, 1937.
390	57696	Suzuki, H. and Miyahara, S., "The Crystal Distortion of Vanadium," J. Phys. Soc. Japan, 21(12), 2735, 1966.
391	57694	Phragmen, G., "X-Ray Investigation of Certain Nickel Steels of Low Thermal Expansion," J. Iron Steel Inst., 123, 465-77, 1931.
392	57687	Raynor, G.V. and Hume-Rothery, W., "A Technique for the X-Ray Powder Photography of Reactive Metals and Alloys, with Special Reference to the Lattice Spacing of Magnesium at High Temperatures," J. Inst. Metals, 65, 379-87, 1939.
393	60588	Pavese, F., Righini, F., and Ruffino, G., "Push-Rod Dilatometer with Interferometric Transducer and the Thermal Expansion of Copper," Rev. Int. Hautes Temp. Refract., 7(3), 252-6, 1970.
394	61398	James, W.J. and Straumanis, M.E., "Lattice Parameter and Coefficient of Thermal Expansion of Thorium," Acta Cryst., 9, 376-9, 1956.
395	46702	Stetson, A.R. and Metcalfe, A.G., "Development of Coatings for Columbium Base Alloys. Part 1 - Basic Property Measurements and Coating System Development," U.S. Air Force Rept. AFML-TR-67-139 (Pt. 1), 301 pp., 1967. [AD 821 628]
396	39565	Hidnert, P., "Thermal Expansion of Cast and of Swaged Chromium," J. Res. Natl. Bur. Stand., 27, 113-24, 1941.
397	48166	Endoh, Y., Ishikawa, Y., and Ohno, H., "Anitferromagnetism of Dilute Chromium Alloys with Cobalt and Nickel," J. Phys. Soc. Japan, 24(2), 263-70, 1968.
398	45835	Syono, Y. and Ishikawa, Y., "Pressure Effect on the First-Order Transition Between Commensurable and Incommensurable Spin States of Dilute Chromium Alloys," Phys. Rev. Letters, 19(13), 747-9, 1967.
399	41769	Newmann, M.M. and Stevens, K.W.H., "Magnetic Susceptibilities of Iron-Chromium Alloys," Proc. Phys. Soc., 74, 290-6, 1959.
400	55941	Hidnert, P. and Sweeney, W.T., "Thermal Expansion of Beryllium and Aluminum-Beryllium Alloys," J. Res. Natl. Bur. Stand., 22, 533-45, 1927; Metal Industry, 32, 397-400, 423, 1928.
401	57753	Hume-Rothery, W. and Boultbee, T.H., "The Coefficients of Expansion of Some Solid Solutions in Aluminium," Phil. Mag., 40, 71-80, 1949.

Ref. No.	TPRC No.	
402	23756	Saller, H.A., "Preparation and Properties of the Aluminum-Uranium Alloys," in Nuclear Science and Technology, (Extracts from J. Metallurgy and Ceramics, Issue Nos. 1-6, TID-65-69, July 1948-January 1951), USAEC Rept. TID-2501(DEL), 12 pp., 1957.
403	58652	Varich, N.I., Lyukevich, R.B., Kolomytseva, L.F., Varich, A.N., and Maslov, V.V., "Effect of Superheating the Melt on the Structure and Properties of Rapidly Cooled Al-Zr Alloys," Fiz. Metal. Metalloved., 27(2), 361-4, 1969; English Translation: Phys. Metals Metallogr., 27(2), 176-9, 1969.
404	36115	Fenn, R.W., Jr., Crooks, D.D., Coons, W.C., and Underwood, E.E., "Properties and Behavior of Beryllium-Aluminum Alloys," AD 453 918, 39 pp., 1964.
405	45695	Masumoto, H., Saito, H., Sugai, Y., and Kono, T., "New Elinvar-Type Alloy Moelinvar in the Ternary System of Cobalt, Iron, and Molybdenum," J. Japan Inst. Metals, 24, 44-6, 1960.
406	58063	Sykes, W.P. and Graff, H.F., "The Cobalt-Molybdenum System," Trans. Am. Soc. Metals, 23, 249-85, 1935.
407	55132	Masumoto, H. and Sawaya, S., "Thermal Expansion Coefficient and the Temperature Coefficient of Young's Modulus of Cobalt and Palladium Alloys," Nippon Kinzoku Gakkaishi, 33(6), 685-7, 1969.
408	47825	Wilkes, P. and Barrand, P., "On the Role of Vacancies in Solute Clustering in Quenched and Aged Copper-Beryllium Alloys," Acta Met., 16(2), 159-66, 1968.
409	52249	Itskevich, E.S., Voronovskii, A.N., Gavrilov, A.F., and Sukhoparov, V.A., "High Pressure (Up to 18 Kilobars) Chamber for Operating at Helium Temperatures," Prib. Tekh. Eksp., 11(6), 161-4, 1966; English Translation: Instrum. Exp. Tech. (USSR), 6, 1452-4, 1966.
410	45149	De, M., "The Thermal Expansion of Gamma-Phase Cu-Mn Alloys at High Temperatures," Indian J. Phys., 41(2), 79-86, 1967.
411	43718	Mukherjee, K.P., Bandyopadhyaya, J., and Gupta, K.P., "Phase Relationship and Crystal Structure of Intermediate Phases in the Cu-Si System in the Composition Range of 17 to 5 At. Pct. Silicon," Trans. Met. Soc. AIME, 245(10), 2335-8, 1969.
412	16288	Cook, M. and Tallis, W.G., "The Physical Properties and Annealing Characteristics of Standard Phosphor-Bronze Alloys," J. Inst. Metals, 67, 49-65, 1941.
413	40315	Hidnert, P., "Thermal Expansion of Bearing Bronzes," J. Res. Natl. Bur. Stand., 12, 391-400, 1934.
414	53092	Lehr, P. and Langeron, J.P., "On the Expansion Behavior of Single Crystals of Alpha-Uranium," Compt. Rend., 241, 1130-3, 1955.
415	50415	Saur, E. and Wenkowitsch, V., "Thermal Lattice Expansion and Microscopic Expansion of Nickel in the Region of the Curie Temperature," Ber. Oberhess Ges. Natur.-Heilk. Giessen Naturwiss Abt., 29, 14-23, 1958.
416	52817	Spedding, P.L., "Linear Expansion Coefficient of Palau Alloy," J. Less-Common Metals, 7, 395-6, 1964.
417	54511	Masumoto, H., Saito, H., Kadowaki, S., "Temperature Dependence of Young's Modulus of Single Crystals of Pd-50% Au Alloy Pallagold," Nippon Kinzoku Gakkaishi, 33(1), 126-9, 1969.
418	51933	Iwasaki, H. and Uesugi, T., "X-Ray Measurement of Order in the β' Phases of Noble Metal Alloys. I. β' AuZn Intermetallic," J. Phys. Soc. Japan, 25(6), 1640-6, 1968.
419	51751	Pahlman, J.E. and Smith, J.F., "Thermal Expansion of Indium-Thallium Alloys in the Vicinity of the f.c.c.-f.c.t. Transition," J. Less-Common Metals, 16(4), 397-405, 1968.
420	40276	Harmon, D.P., "Iridium-Base Alloys and Their Behavior in the Presence of Carbon," U.S. Air Force Rept. AFML-TR-66-290, 115 pp., 1966. [AD 802 109]
421	54708	Makhurane, P. and Gaunt, P., "Lattice Distortion, Elasticity and Antiferromagnetic Order in Copper-Manganese Alloys," J. Phys. C (Solid State Phys.), 2(6), 959-65, 1969.
422	47777	Pal, L., Kren, E., Kadar, G., Szabo, P., and Tarnoczi, T., "Magnetic Structures and Phase Transformations in Mn-Based Copper Gold-I Type Alloys," J. Appl. Phys., 39(2), 538-44, 1968.
423	16590	Fieldhouse, I.B., Lang, J.I., and Blau, H.H., "Investigation of Feasibility of Utilizing Available Heat Resistant Materials for Hypersonic Leading Edge Applications. Vol. IV. Thermal Properties of Molybdenum Alloy and Graphite," U.S. Air Force Rept. WADC-TR-59-744(Vol. 4), 78 pp., 1960. [AD 249 166]
424	37565	Burman, R.W., "Performance of New Molybdenum-Base Alloy Tools in Casting Processes," Modern Castings, 46(2), 471-80, 1964.
425	25961	Hedge, J.C., Kopec, J.W., Kostenko, C., and Lang, J.I., "Thermal Properties of Refractory Alloys," U.S. Air Force Rept. ASD-TDR-63-597, 128 pp., 1963. [AD 424 375]
426	37205	Lifanov, I.I. and Sherstyukov, N.G., "Thermal Expansion of Copper in the Temperature Range -185 to 300 C," Izmer. Tekh., 12, 39-44, 1968; English Translation: Meas. Tech. (USSR), 12, 1653-9, 1968.
427	56075	Souder, W., Hidnert, P., and Fox, J.F., "Autographic Thermal Expansion Apparatus," J. Res. Natl. Bur. Stand., 13(4), 497-513, 1934.
428	55855	Masumoto, H., Sawaya, S., and Nakamura, N., "Effect of Addition of Chromium, Molybdenum, or Tungsten on the Elinvar Property of Nickel at High Temperature," Nippon Kinziku Gakkaishi, 33(9), 1003-6, 1969.
429	37599	Mochel, N.L., "Report on Thermal Expansion of Materials," Amer. Soc. Test. Mater. Proc., 27(1), 153-62, 1927.
430	56257	Masumoto, H., Sawaya, S., and Nakamura, N., "Effect of Addition of Silicon, Tin, or Vanadium on the Elinvar Property of Nickel at High Temperature," Nippon Kinzoku Gakkaishi, 33(10), 1134-7, 1969.

Ref. No.	TPRC No.	
431	24656	Demastry, J.A., Moak, D.P., Epstein, S.G., Bauer, A.A., and Dickerson, R.F., "Development of Niobium-Uranium Alloys for Elevated-Temperature Fuel Applications," USAEC Rept. BMI-1536, 64 pp., 1961.
432	34236	Ewing, C.T., Stone, J.P., Spann, J.R., Steinkuller, E.W., Williams, D.D., and Miller, R.R., "High-Temperature Properties of Potassium," NASA Rept. NRL-6233, 87 pp., 1965. [AD 622 190]
433	26008	Pears, C.D., "The Thermal Properties of Twenty-Six Solid Materials to 5000 F or Their Destruction Temperatures," U.S. Air Force Rept. ASD-TDR-62-765, 420 pp., 1962. [AD 298 061]
434	35842	Fisher, C.R. and Achener, P.Y., "Alkali Metals Evaluation Program," USAEC Rept. AGN-8131, 28 pp., 1965.
435	53845	Batley, A.C., Waterhouse, N., and Yates, B., "The Thermal Expansion of Palladium-Silver Alloys at Low Temperatures," J. Phys. C (Solid State Phys.), 2(5), 769-76, 1969.
436	43333	Day, A.L. and Sosman, R.B., "The Nitrogen Thermometer from Zinc to Palladium," Am. J. Sci., 29, 93-161, 1910.
437	27344	Engelke, W.T. and Pears, C.D., "Thermal Conductivity and Expansion of Two Platinum Alloys," USAEC Rept. UCRL-13061, 39 pp., 1962.
438	21358	Sandenaw, T.A., "Heat Capacity, Thermal Expansion, and Electrical Resistivity of an 8 Atomic Percent Aluminum-Plutonium (Delta-Phase Stabilized) Alloy Below 300 K," Phys. and Chem. Solids, 16, 329-36, 1960.
439	54269	Taylor, J.M., "Thermal Expansion of Some Plutonium-Gallium Solid Solution Alloys," J. Nucl. Mater., 31(3), 339-41, 1969.
440	41807	Zsoldos, L., "Lattice Parameter Change of Iron-Rhodium Alloys Due to Antiferromagnetic-Ferromagnetic Transformation," Phys. Status Solidi, 20(1), K25-8, 1967.
441	58267	McKinnon, J.B., Melville, D., and Lee, E.W., "The Antiferromagnetic-Ferromagnetic Transition in Iron-Rhodium Alloys," J. Phys. C (Suppl. to Solid State Physics) Metal Physics, 3(1), S46-S58, 1970.
442	44456	Ponyatovskii, E.G., Kutsar, A.R., and Dubovka, G.T., "Possible Presence of a Special Triple Point in the P-T Diagram of a Fe-Rh Alloy," Kristallografiya, 12(1), 79-83, 1967.
443	41176	Neumann, J.P. and Chang, Y.A., "Thermal Expansion and Elastic Constants of β'-Silver Magnesium Intermetallic. I. The Coefficient of Thermal Expansion from 77 to 800 K," J. Appl. Phys., 38(2), 647-9, 1967.
444	35662	Riad, S.M., "Lattice Parameters, Thermal Expansion Coefficients, Densities, and Imperfections in Silver and in the Alpha-Phase of the Silver-Indium System," University of Missouri, Ph.D. Thesis, 158 pp., 1964. [Univ. Microfilms No. 65-2542]
445	48062	Eremenko, V.N. and Listovnichii, V.E., "Structure and Properties of Sulphur-Titanium Alloys," in Chalcogenides, Akademia Nauk of the Ukrainian SSR Materials Science Institute, Kiev, 69-78, 1967.
446	10088	Deem, H.W., Winn, R.A., and Lucks, C.F., "Thermal Conductivity and Linear Expansion of the Eutectic Uranium-Chromium Alloy," USAEC Rept. BMI-900, 16 pp., 1954. [AD 85 812]
447	52759	McCreight, L.R., "Thermal Expansion Measurements of Six Fuel Materials," USAEC Rept. KAPL-M-LRM-7, 9 pp., 1952.
448	52578	Saller, H.A., Dickerson, R.F., and Murr, W.F., "Uranium Alloys for High-Temperature Application," USAEC Rept. BMI-1098, 45 pp., 1956.
449	46188	Beyer, B. and Juknat, R., "Physical and Mechanical Properties of Zirconium-Niobium Alloys Containing 1 and 2.5 Wt. % Niobium," AEC German Democratic Republic AEC Rept. CONF-661059(Vol. 1), 55-75, 1966.
450	52496	Mehan, R.L. and Cutler, G.L., "Thermal Expansion of Zircaloy-2 Between Room Temperature and 1000 C," USAEC Rept. KAPL-M-RLM-15, 22 pp., 1958.
451	27792	Hedge, J.C., Lang, J.I., Howe, E., and Elliott, R., "High Temperature Property Study for Re-entering NAP," U.S. Air Force Rept. AFSWC-TDR-63-17, 91 pp., 1963. [AD 412 586]
452	61545	Leksina, I.E. and Novikova, S.I., "Thermal Expansion of Copper, Silver, and Gold Within a Wide Range of Temperatures," Fiz. Tverd. Tela, 5(4), 1094-9, 1963; English Translation: Sov. Phys.-Solid State, 5(4), 798-801, 1963.
453	46146	Torti, M.L., "Development of Tantalum-Tungsten Alloys for High Performance Propulsion System Components," National Research Corp., Cambridge, Mass. Rept. NRC-11-1-032, 34 pp., 1960. [AD 252 217L]
454	52799	Powers, D.J., "Thermal Expansion Determinations on Tantalum 90-Tungsten 10 Alloy, A-286 Steel, and Bico-Loy Steel," Bell Aerosystems Company Rept. BLR-63-3-(M), 14 pp., 1963. [AD 401 292]
455	41171	Luo, H.L. and Willens, R.H., "Superconducting Transitions in Body-Centered Cubic Thallium-Indium Alloys," Phys. Rev., 154(2), 436-8, 1967.
456	46859	Harrington, L. and Rowe, G.H., "Thermal Expansion of Thirteen Tungsten Carbide Cermets from 68 to 1800 F," Amer. Soc. Test. Mater. Proc., 63, 633-45, 1963; USAEC Rept. CNLM-4479, 33 pp., 1963.
457	46826	Hensel, F.R., Larsen, E.I., and Swazy, E.F., "Physical Properties of Metal Compositions With a Refractory Metal Base," Powder Met. Bull., 483-92, 1942.

Ref. No.	TPRC No.	
458	6349	Bossart, P.N., "Spectral Emissivities, Resistivity, and Thermal Expansion of Tungsten-Molybdenum Alloys," Physics, 7(2), 50-4, 1936.
459	40463	Southern Research Inst., Birmingham, Ala., "The Mechanical and Thermal Properties of Tungsten and TZM Sheet Produced in the Refractory Metal Sheet Rolling Program, Part I," AD638631, 200 pp., 1966.
460	51419	Rhude, H.V., Yaggee, F.L., Savage, H., and Dunworth, R.J., "High-Temperature Physical and Thermal Properties of V-20 Wt. % Ti Alloy," in Argonne National Laboratory Annual Progress Report for 1964, USAEC Rept. ANL-7000, 11-12, 1964.
461	54262	Neumeier, L.A. and Risbeck, J.S., "Effect of Varied Extrusion Temperature on the Properties of a Zinc-Copper-Titanium Alloy," Bureau of Mines Rept. BM-RI-7229, 26 pp., 1969. [PB-183378]
462	47107	Arp, V., Wilson, J.H., Winrich, L., and Sikora, P., "Thermal Expansion of Some Engineering Materials from 20 to 293 K," Cryogenics, 2, 230-5, 1962.
463	56783	Willey, L.A. and Fink, W.L., "An Interferometer Type of Dilatometer, and Some Typical Results," Trans. Am. Inst. Mining Met. Engrs., 162, 642-55, 1945.
464	6548	Powell, R.W., Hickman, M.J., and Barber, C.R., "Some Physical Properties of Aluminium Alloys at Elevated Temperatures. I. Thermal Conductivity and Electrical Resistivity. II. Linear Thermal Expansion," Metallurgia, 41(241), 15-21, 1949.
465	16277	Losana, L., "Contribution to the Knowledge of Some Light Alloys," Ind. Chim. (Milan), 5, 145-50, 1930.
466	6940	Lucks, C.F., Thompson, H.B., Smith, A.R., Curry, F.P., Deem, H.W., and Bing, G.F., "The Experimental Measurement of Thermal Conductivities, Specific Heats, and Densities of Metallic, Transparent, and Protective Materials. Part I," U.S. Air Force Rept. AF-TR-6145(Pt. I), 127 pp., 1951.
467	50161	Clark, A.F., "Low Temperature Thermal Expansion of Some Metallic Alloys," Cryogenics, 8(5), 282-9, 1968.
468	44471	Santschi, W.H. and Marz, W.G., "Manufacturing and Properties of Beryllium-Aluminum Composite Materials," Met. Soc. Conf. 1964, 33, 523-37, 1966.
469	35023	Shih, C., "Snap 19 Test Report Thermal Performance of Materials and Mechanical Joints," Martin Nuclear Co. Rept. MND-3607-112, 52 pp., 1966.
470	48419	Henmi, Z., Okada, M., and Nagai, T., "The Mechanism of Age Hardening in Elgiloy," Nippon Kinzoku Gakkaishi, 31(12), 1351-5, 1967.
471	46931	Nejedlik, J.F., "The Embrittlement Characteristics of a Low-Silicon Modified Cobalt-Base Alloy (L-605) at 1200 and 1600 F," USAEC Rept. TID-23826, 77 pp., 1966.
472	6970	Fieldhouse, I.B., Hedge, J.C., Lang, J.I., and Waterman, T.E., "Thermal Properties of High Temperature Materials," U.S. Air Force Rept. WADC-TR-57-487, 78 pp., 1957.
473	20404	Sweeny, W.O., "Haynes Alloys for High-Temperature Service," Trans. Am. Soc. Mech. Engrs., 69, 569-81, 1947.
474	16620	Morral, F.R. and Wagner, H.J., "Physical and Mechanical Properties of the Cobalt-Chromium-Tungsten Alloy WI-52," Defense Metals Information Center, Memo 66, 19 pp., 1960. [AD243903]
475	55408	Satto, H., Fujimori, H., and Saito, T., "Spontaneous Volume Magnetostriction of (Fe(1-x) Co(x)) 0.89 Cr(0.11) Alloys. I. Invar Characteristics of (Fe(1-x) Co(x)) 0.89 Cr(0.11) Alloys," Nippon Kinzoku Gakkaishi, 33(2), 231-4, 1969.
476	45787	Masumoto, H., Saito, H., and Kikuchi, M., "Elastic Anisotropies and Their Temperature Dependence of Single Crystals of High Elasticity Alloy Dia-Flex in the Co-Fe-Ni-Cr-W-Mo System," Nippon Kinzoku Gakkaishi, 31(3), 263-8, 1967.
477	47409	Kueser, P.E., Pavlovic, D.M., Lane, D.H., Clark, J.J., and Spewock, M., "Properties of Magnetic Materials for Use in High-Temperature Space Power Systems," NASA Rept. NASA-SP-3043, 318 pp., 1967.
478	40273	Hidnert, P. and Dickson, G., "Thermal Expansion of Some Industrial Copper Alloys," J. Res. Natl. Bur. Stand., 31, 77-82, 1943.
479	37389	Bocker, S., "The Stabilization of Mixed Carbides of Uranium-Plutonium by Zirconium. Part 1. Uranium Carbide with Small Additions of Zirconium," NASA Rept. CEA-R-3765, 65 pp., 1969.
480	6280	Cook, M., "The Physical Properties and Annealing Characteristics of Standard Nickel Silver Alloys," J. Inst. Metals, 58, 151-71, 1936.
481	46623	Johnston, H.L., Altman, H.W., and Rubin, T., "Coefficients of Thermal Expansions of Alloys at Low Temperatures," J. Chem. Eng. Data, 10(3), 241-2, 1965.
482	34059	Leggett, H., Cook, J.L., Schwab, D.E., and Powers, C.T., "Mechanical and Physical Properties of Super Alloy and Coated Refractory Alloy Foils," U.S. Air Force Rept. AFML-TR-65-147, 954 pp., 1965. [AD468607]
483	52259	Jones, R.L., "The Elevated Temperature Tensile and Creep-Rupture Properties of the FS-82 Columbium Alloy," General Dynamics Corporation, Convair Astronautics Div. Rept. GDA-ERR-AN-049, 68 pp., 1961.
484	33326	Neff, C.W., Frank, F.G., and Luft, L., "Refractory Metals Structural Development Program. Vol. 2. Refractory Alloy and Coating Development," U.S. Air Force Rept. ASD-TR-61-392(V2), 342 pp., 1961. [AD267844]

Ref. No.	TPRC No.	
485	34121	Beck, E.J., "Determination of Mechanical and Thermophysical Properties of Refractory Metals," U.S. Air Force Rept. RTD-CR-65-1, 192 pp., 1965. [AD 471 505]
486	52411	Burney, J.D., "Cermets Containing Molybdenum Disilicide and Aluminum Oxide," in Proceedings of the WADC Ceramic Conference on Cermets, U.S. Air Force Rept. WADC-TR-52-327, 103-6, 1952. [AD 1 183]
487	47058	Savitskii, E.M., Baron, V.V., and Ivanova, K.N., "Niobium-Base Alloys and Their Properties," Sampe Journal, 4(2), 21-5, 1968.
488	52105	Ammon, R.L. and Begley, R.T., "Pilot Production and Evaluation of Tantalum Alloy Sheet," Westinghouse Electric Corp., Astronuclear Lab. Rept. WANL-PR-M-004, 83 pp., 1963.
489	52894	Baker, W.H., Jr., "Demonstration of a Submerged Cooled Nozzle," U.S. Air Force Rept. AFRPL-TR-66-91, 120 pp., 1966. [AD 481 829]
490	20484	Baxter, W.G. and Welch, F.H., "Physical and Mechanical Properties and Oxidation-Resistant Coatings for a Tungsten-Base Alloy," USAEC Rept. APEX-623, 89 pp., 1961.
491	36756	Mehan, R.L. and Wiesinger, R.W., "Mechanical Properties of Zircaloy-2," USAEC Rept. KAPL-2110, 47 pp., 1961.
492	44978	Slattery, G.F., "The Effect of Hydride on the Thermal Expansivity of Zirconium Alloys," J. Less-Common Metals, 11(2), 89-98, 1966.
493	42318	Scott, D.B., "Physical and Mechanical Properties of Zircaloy 2 and 4," USAEC Rept. WCAP-3269-41, 68 pp., 1965.
494	43535	Kearns, J.J., "Thermal Expansion and Preferred Orientation in Zircaloy," USAEC Rept. WAPD-TM-472, 39 pp., 1965.
495	10616	Seibel, R.D. and Mason, G.L., "Thermal Properties of High Temperature Materials," U.S. Air Force Rept. WADC-TR-57-468, 58 pp., 1958. [AD 155 605]
496	52884	Heyer, B.A., Marlatt, J.W., and Avery, H.S., "Manufacturing Process Development for Super-Alloy Cast Parts," ABEX Corp. Research Center Rept. XL-1360-4, IR-8-297(IV), 72 pp., 1966. [AD 804 266]
497	52665	Smoke, E.J., Rooda, J., Sirkis, M.D., Holsberg, P.J., Molony, D.A., and Choa, C., "Study of High Temperature Materials," U.S. Air Force Rept. RADC-TR-60-233, 35 pp., 1960. [AD 248 105]
498	20970	O'Sullivan, W.J., Jr., "Some Thermal and Mechanical Properties of Inconel at Temperatures for Use in Aerodynamic Heating Research," Am. Soc. Test. Mater. Proc., 55, 757-64, 1955.
499	52518	Aerojet-General Corporation, Liquid Rocket Plant, Sacramento, Calif., "Coefficient of Thermal Expansion of Inconel 718 and 347," Aerojet-General Corp. Rept. AGC-4-433, DVR-63-566, 2 pp., 1963. [AD 431 387]
500	56351	Heathman, J.H., "Hydrogen Tankage Application to Manned Aerospace Systems, Phases II and III. Volume I. Design and Analytical Investigations," U.S. Air Force Rept. AFFDL-TR-68-75-Vol.I, 171 pp., 1968. [AD 833 232]
501	52488	Ackerman, D.E., "Properties of Monel at Low Temperatures," Metal Progress, 30(5), 56-60, 1936.
502	46733	Skinner, E.N., "Properties of a Nickel-Zirconium Alloy at Elevated Temperature," in Proc. of Metall. and Materials Inform. Meeting, Volume 1, International Nickel Co. Rept. B8I-82, TID-5061, 497-508, 1951.
503	10938	Crucible Steel Co. of America, Pittsburgh, Pa., "Crucible A-110AT Titanium Base Alloy," Data Sheet A-110AT, 8 pp., 1958.
504	43638	Bishop, S.M., Spretnak, J.W., and Fontana, M.G., "Mechanical Properties, Including Fatigue of Titanium-Base Alloys RC-130-B and TI-150-A at Very Low Temperatures," Trans. ASM, 45, 993-1007, 1953.
505	45320	Spretnak, J.W. and Fontana, M.G., "Low-Temperature Mechanical Properties, Including Fatigue, of Titanium-Base Alloys RC-130-B and TI-150-A," in Investigation of Mechanical Properties and Physical Metallurgy of Aircraft Alloys at Very Low Temperatures, U.S. Air Force Rept. AF-TR-5662(Pt.4), 35 pp., 1952.
506	31540	McGee, W.M. and Matthews, B.R., "Determination of Design Data for Heat Treated Titanium Alloy Sheet. Vol. 2A. Details of Data Collection Program. Test Techniques and Results for Tension, Compression, Bearing, Shear, Crippling, Joints, and Physical Properties," U.S. Air Force Rept. ASD-TDR-62-335(2A), 413 pp., 1962. [AD 298 765]
507	31460	Lockheed-Georgia Co., Marietta, Ga., "Determination of Design Data for Heat Treated Titanium Alloy Sheet. Vol. 3. Tables of Data Collected," U.S. Air Force Rept. ASD-TDR-62-335, 326 pp., 1962. [AD 297 803]
508	43119	Dotson, C.L., "Mechanical and Thermal Properties of High-Temperature Titanium Alloys," U.S. Air Force Rept. AFML-TR-67-41, 338 pp., 1967. [AD 814 022]
509	53091	Deel, O.L. and Hyler, W.S., "Engineering Data on Newly Developed Structural Materials," U.S. Air Force Rept. AFML-TR-68-211, 139 pp., 1968. [AD 840 065]
510	56467	Neimark, B.E., Korytina, S.F., and Monina, E.F., "A Detailed Study of the Physical Properties of Titanium Alloys VT5 and VT8," Teploenergetika, 16(6), 52-5, 1969; English Translation: Ther. Eng. (USSR), 16(6), 80-3, 1969.
511	56354	Hertz, J., "An Evaluation of the Mechanical Properties of Adhesives at Cryogenic Temperatures and Their Correlation with Molecular Structure," General Dynamics/Astronautics, San Diego, Calif. Rept. GDA-ERR-AN-196, 54 pp., 1962. [AD 831 537]

Ref. No.	TPRC No.	
512	52790	Wright-Patterson Air Force Base, Materials Laboratory, Ohio, "Dilatation Curves of Titanium Alloys," U.S. Air Force Rept. WCRT-TM-56-85 (Pt. 3), 12 pp., 1958.
513	49250	Cerni, S., "Phase 1," NASA Rept. WANL-3800-9, WANL-PR-(SS)-010, 119 pp., 1967.
514	51223	Peterson, C.A.W. and Vandervoort, R.R., "The Properties of a Metastable Gamma-Phase Uranium-Base Alloy: U-7.5 Nb-2.5 Zr," USAEC Rept. UCRL-7869, 36 pp., 1964.
515	48427	Boucher, R., Barthelemy, P., and Milet, C., "A Study of Plutonium-Based Alloys Carried Out at Fontenay-Aux-Roses," Plutonium 1965 Proc. Int. Conf. Plutonium, 3rd, London, 485-509, 1967.
516	51110	Rhude, H.V., "Properties of Zero-Power Uranium-Plutonium Metal Fuels," in Argonne National Laboratory Annual Progress Report for 1966, USAEC Rept. ANL-7299, 92, 1966.
517	49036	Smotritskii, G.S., Chebotarev, N.T., Kutaitsev, V.I., and Petrov, P.N., "Multicomponent Alloys Based on Uranium-Plutonium as a Fuel for Fast Reactors," Plutonium Reactor Fuel Proc. Symp. Brussels, 165-75, 1967.
518	51426	Rhude, H.V., "Thermal Expansion of U-Pu-Zr and U-Pu-Ti Alloys," in Argonne National Laboratory Annual Progress Report for 1965, USAEC Rept. ANL-7155, 18-20, 1965.
519	51425	Kelman, L.R., "Properties of Uranium-Plutonium-Base Metallic Fuel Alloys. Review of Developments," in Argonne National Laboratory Annual Progress Report for 1965, USAEC Rept. ANL-7155, 14-15, 1965.
520	48426	Kelman, L.R., Savage, H., Walter, C.M., Blumenthal, B., Dunworth, R.J., and Rhude, H.V., "Status of Metallic Plutonium Fast Power-Breeder Fuels," Plutonium 1965 Proc. Int. Conf. Plutonium, 3rd, London, 458-84, 1967.
521	34458	Fitzgeorge, D. and Pope, J.A., "The Thermal and Elastic Properties of Eight Cast-Irons," Trans. Northeast Coast Inst. of Engrgs. and Shipbuilders, 75(6), 285-330, 1959.
522	46177	Neimark, B.E., Monina, E.F., Kartuzova, L.M., and Kainova, R.A., "Physical Properties of Cast Irons as a Function of Temperature," Liteinoe Prozvod, 9, 33-5, 1967.
523	9349	Sykes, C. and Bampfylde, J.W., "The Physical Properties of Iron-Aluminium Alloys," J. Iron Steel Inst. (London), 129(2), 389-418, 1934.
524	52358	Stuart, H. and Ridley, N., "Lattice Parameters and Curie-Point Anomalies of Iron-Cobalt Alloys," Brit. J. Appl. Phys. (J. Phys. D), 2(4), 485-91, 1969.
525	57736	Simpson, K.M. and Banister, R.T., "Alloys of Copper and Iron," Metals and Alloys, 7, 88-94, 1936.
526	39962	Fujimori, H., "Thermal Expansion, Electrical Resistance and the Effect of Hydrostatic Pressure on the Néel Temperature in Fe-Mn Alloys," J. Phys. Soc. Japan, 21(10), 1860-5, 1966.
527	50390	Zakharov, A.I. and Fedotov, L.N., "Low-Temperature Thermal Expansion of Invar," Fiz. Metal. Met Metalloved., 23(4), 759-60, 1967; English Translation: Phys. Metals Metallogr. (USSR), 23(4), 201-3, 1967.
528	46149	Kachi, S. and Asano, H., "Concentration Fluctuations and Anomalous Properties of the Invar Alloy," J. Phys. Soc. Japan, 27(3), 536-41, 1969.
529	48096	Janser, G.R., "Summary of Material Technology of M-1 Engine," NASA Rept. NASA-CR-54961, 76 pp., 1966.
530	55968	Owen, E.A. and Yates, E.L., "X-Ray Investigation of Pure Iron-Nickel Alloys. Part 2. Thermal Expansion of Some Further Alloys," Proc. Phys. Soc., 49, 178-88, 1937.
531	51174	Asano, H., "Magnetism of Gamma Iron-Nickel Invar Alloys with Low Nickel Concentration," J. Phys. Soc. Japan, 27(3), 542-53, 1969.
532	40435	Rohde, R.W. and Graham, R.A., "The Effect of Hydrostatic Pressure on the Martensitic Reversal of an Iron-Nickel-Carbon Alloy," Trans. Met. Soc. AIME, 245(11), 2441-5, 1969.
533	40367	Hoenie, A.F. and Roach, D.B., "New Developments in High-Strength Stainless Steels," U.S. Air Force Rept. DMIC-223, 40 pp., 1966.
534	55979	Furman, D.E., "Thermal Expansion Characteristics of Stainless Steels, Between -300 and 1000 F," Trans. AIME, 188, 688-91, 1950.
535	54202	McCoy, H.E., Jr., "Physical and Mechanical Properties of Type 406 Stainless Steel," USAEC Rept. ORNL-TM-2263, 38 pp., 1968.
536	52858	Battelle Memorial Inst., Columbus, Ohio, "Mechanical-Property Data. AFC-77 Steel. Tempered (1100 F) Sheet," AD 803545, 7 pp., 1966.
537	44276	Smirnov, V.V., Pokhmurskii, V.I., and Boltarovich, A.V., "Physico-Mechanical and Corrosion Properties of Heat-Resistant Stainless Steel EP-479," Fiz.-Khim. Mekh. Materialov, 2(3), 304-7, 1966; English Translation: Sov. Mater. Science, 2(3), 218-20, 1966.
538	53994	Yershov, V.M. and Olson, N.L., "Change in the Linear Expansion Coefficient of Austenite on Transformation to Martensite," Fiz. Metal. Metalloved., 25(5), 874-81, 1968; English Translation: Phys. Metals Metallogr. (USSR), 25(5), 118-23, 1968.
539	36780	Neimark, B.E., Lyusternik, V.E., and Korytina, S.F., "Physical Properties of DH17 N7 Yu Steel," Teplofiz. Vys. Temp., 2(5), 725-9, 1964; English Translation: High Temp. (USSR), 2(5), 652-5, 1964.
540	53160	Trehan, Y.N., "Thermal Expansion Characteristics of Some Iron-Chromium-Manganese-Nitrogen Alloys," Nat. Met. Lab.-Tech. J., 10(2), 31-4, 1968.

Ref. No.	TPRC No.	
541	54475	Stuart, H. and Ridley, N., "Lattice Parameters and Thermal Expansion Coefficients of 18Cr-10Ni and 20Cr-25Ni Austenites," J. Iron Steel Inst. (London), 207(3), 368, 1969.
542	53993	Lysak, L.I., Makogon, Yu.N., and Nikolin, B.I., "Dilatometric and X-Ray Study of Gamma in Equilibrium with ϵ' and γ in Equilibrium with ϵ Transformations," Fiz. Metal. i Metalloved., 25(3), 562-5, 1968; English Translation: Phys. Metals Metallogr. (USSR), 25(3), 197-200, 1968.
543	34046	Hoenie, A.F., Lumm, J.A., Shelton, R.J., and Wallace, R.A., "Determination of Mechanical Property Design Values for 18 Nicomo 250 and 300 Grade Maraging Steels," U.S. Air Force Rept. NA64H-423-4, ML-TR-65-197, 214 pp., 1965. [AD 470 742]
544	49267	Foster, J.D. and Finnie, I., "Method for Measuring Small Thermal Expansion with a Single Frequency Helium-Neon Laser," Rev. Sci. Instr., 39(5), 654-7, 1968.
545	50560	Kachi, S., Asano, H., and Nakanishi, N., "Lattice Constants of Low Nickel Invar Alloy and Its Anomalous Low Expansion Coefficient," J. Phys. Soc. Japan, 25(3), 909, 1968.
546	45045	Eiselstein, H.L., "An Age-Hardenable, Low Expansion Alloy for Cryogenic Service," Advan. Cryog. Eng., 12, 508-19, 1967.
547	52779	Darkowski, D.S., Porter, L.J., and Loveday, G.E., "The Production and Properties of 180 KSI Minimum-Yield-Strength Maraging Steels," AD 604 874, 49 pp., 1964.
548	52167	Zemtsova, N.D., Vasilevskaya, M.M., and Malyshev, K.A., "Aging of Fe-Ni-Ti Alloys in the Process of the Reverse Alpha-Gamma Martensitic Transformation," Fiz. Metal. i Metalloved., 24(2), 293-8, 1967; English Translation: Phys. Metals Metallogr. (USSR), 24(2), 89-94, 1967.
549	35422	Hordon, M.J., Lement, B.S., and Averbach, B.L., "Influence of Plastic Deformation on Expansivity and Elastic Modulus of Aluminum," Acta Met., 6, 446-53, 1958.
550	39748	Edelman, N.M., "Aluminum Alloys in Civil Construction," in Building Structures from Aluminum Alloys, (English Translation of a Russian Reference Book for Construction Workers), U.S. Air Force Rept. FTD-MT-64-156, 53-68, 1965. [AD 620 816]
551	59797	Belov, A.K., "Expansion Coefficients of Structural Materials at Low Temperatures," Metalloved. Term. Obrab. Metal., 4, 20-2, 1968; English Translation: Metal Sci. Heat Treat. Metals (USSR), 4, 267-9, 1968.
552	60359	Kempf, L.W., "Thermal Expansivity of Aluminum Alloys," Trans. Am. Inst. Mining Met. Engrs., 104, 308-24, 1933.
553	47149	Coursey, B.M. and Heric, E.L., "Congruence Principle Applied to Viscosity of n-Alkane Mixtures," Mol. Phys., 13(3), 287-91, 1967.
554	60520	Barter, B. and Darling, A.S., "Thermal Expansion of Rhodium-Platinum Alloys," Platinum Metals Rev., 4(4), 138-40, 1960.
555	49562	Fine, M.E. and Ellis, W.C., "Thermal Expansion Properties of Iron-Cobalt Alloys," Metals Tech. Publication 2320, 13 pp., 1948.
556	67149	Varich, N.I. and Sheyko, T.I., "Thermal Expansion of Aluminum-Molybdenum, Aluminum-Zirconium Alloys Prepared at High Cooling Rates," Fiz. Metal. Metalloved., Akad. Nauk SSSR, Ural. Filial, 30(2), 443-5, 1970; English Translation: Phys. Metals Metallogr. (USSR), 30(2), 231-2, 1970.
557	40921	Harris, I.R., Raynor, G.V., and Winstanley, C.J., "Rare Earth Intermediate Phases. IV. High-Temperature Lattice Spacings of Some R.E.PD3 Phases," J. Less-Common Metals, 12(1), 69-74, 1967.
558	44405	Gschneider, K.A., Jr., Elliott, R.O., and Cromer, D.T., "Thermal Expansion of $LaRu_2$, $CeRu_2$, and $PrRu_2$ from 20 to 900°," J. Less-Common Metals, 8(4), 217-21, 1965.
559	56778	Novikova, S.I. and Abrikosov, N.Kh., "Thermal Expansion of Aluminum Antimonide, Gallium Antimonide, Zinc Telluride, and Mercury Telluride at Low Temperatures," Fiz. Tverd. Tela, 5(8), 2138-40, 1963; English Translation: Sov. Phys.-Solid State, 5(8), 1558-9, 1964.
560	49217	Mansey, R.C., Raynor, G.V., and Harris, I.R., "Rare-Earth Intermediate Phases. V. The Cubic Laves Phases Formed by Rare-Earth Metals with Iron and Nickel," J. Less-Common Metals, 14(3), 329-36, 1968.
561	42850	Straumanis, M.E. and Kim, C.D., "Lattice Parameters, Thermal Expansion Coefficients, Phase Width, and Perfection of the Structure of Gallium Antimonide and Indium Antimonide," J. Appl. Phys., 36(12), 3822-5, 1965.
562	39368	Kaneko, T. and Matsumoto, M., "Anomalous Thermal Expansion and Specific Heat of the Ferromagnetic Compound Au_4Mn," J. Phys. Soc. Japan, 27(5), 1141-3, 1969.
563	50107	Matsumoto, M., Kaneko, T., and Kamigaki, K., "On the Pressure Effect of the Néel Temperature in the Compound AuMn," J. Phys. Soc. Japan, 25(2), 631, 1968.
564	52743	Booker, J., Paine, R.M., and Stonehouse, A.J., "Investigation of Intermetallic Compounds for Very High Temperature Applications," U.S. Air Force Rept. WADC-TR-176, 23 pp., 1960.
565	47669	Ryba, E., "Transformations in AB2 Intermetallic Compounds," Pennsylvania State Univ., Dept. of Metallurgy, Rept. NYO-3560-7, 24 pp., 1967.
566	51773	Goncharova, E.V., Zhuze, V.P., Zhdanova, V.V., Zhukova, T.B., Smirnov, I.A., and Shadrichev, E.V., "Thermal and Electrical Properties of Lanthanum Antimonide," Fiz. Tverd. Tela, 10(5), 1322-9, 1968; English Translation: Sov. Phys.-Solid State, 10(5), 1052-7, 1968.

Ref. No.	TPRC No.	
567	53188	Samsonov, G.V., Paderno, Yu.B., and Rud, B.M., "Physical Properties of Germanides of Ceric Group and Yttrium Rare Earth Metals," Rev. Intern. Hautes Temp. Refract., 5(2), 105-10, 1968.
568	34467	Chung, P.L., Whitten, W.B., and Danielson, G.C., "Lattice Dynamics of Mg_2Ge," J. Phys. Chem. Solids, 26(12), 1753-60, 1965.
569	52728	Novikova, S.I., "Thermal Expansion of Substances Having a Defect Structure of the A2 (Group III) B3 (Group VI) Type and the Antifluorite Structure," Fiz. Tverd. Tela, 10(10), 3141-2, 1968; English Translation: Sov. Phys.-Solid State, 10(10), 2481-2, 1969.
570	48198	Gordienko, V.A. and Nikolaev, V.I., "Spontaneous Magnetostriction of the Compound $MnAu_2$," Fiz. Tverd. Tela, 9(10), 2820-2, 1967; English Translation: Sov. Phys.-Solid State, 9(10), 2216-18, 1968.
571	48564	Kren, E., Kadar, G., and Szabo, P., "Effect of Fe Substitution on the Magnetic Structure of Mn_3Pt," Phys. Lett., 26A(11), 556-7, 1968.
572	40707	Kishii, T., "Thermal Expansion of Dumet Wires," Toshiba Rev. Intern. Ed. 12, 3, 265-72, 1957; English Translation: Special Libraries Association Rept. TT-66-12506, 39 pp., 1966.
573	52873	Owen, L., Jr., "Advanced, Fiber-Reinforced Tungsten Nozzle," 3rd Quarterly Rept. Oct. 1966-Dec. 1966, AD 379 508, 40 pp., 1967.
574	55664	Lee, H.L., Jr., Swartz, M.L., and Smith, F.F., "Physical Properties of Four Thermosetting Dental Restorative Resins," J. Dent. Res., 48(4), 526-35, 1969.
575	52774	Glaser, F.W., "Elevated Temperature Properties of Zirconium Boride Alloys," in Cemented Borides, U.S. Air Force Rept. WADC-52-40 (Appendix II), 18 pp., 1953. [AD 9 866]
576	31473	Dickerson, R.F., "Studies of Fuels (AEC-DRD)," in Progress Relating to Civilian Applications During July, 1963, USAEC Rept. BMI-1642, B1-2, 1963.
577	37506	Griesenauer, N.M., Farkas, M.S., and Rough, F.A., "Thorium and Thorium-Uranium Compounds as Potential Thermal Breeder Fuels," USAEC Rept. BMI-1680, 35 pp., 1964.
578	54667	Farkas, M.S., "Preparation and Evaluation of Nonoxide Thorium-Base Fuel, in Proc. of Second International Thorium Fuel Cycle Symposium (Gatlinburg, Tennessee, May 3-6, 1966), USAEC Rept. AEC Symp. Ser. 12, CONF-660524, 445-62, 1968.
579	53524	Arbuzov, M.P. and Chuprina, V.G., "Thermal Expansion of Ni_3Al-Ni_3Nb Alloys," Izv. Vuz, Fizika, 9(2), 106-10, 1966; English Translation: Sov. Phys. J., 9(2), 70-3, 1966.
580	41664	Cherry, W.H., Cody, G.D., Cooper, J.L., Gittleman, J.I., Hanak, J.J., McConville, G.T., Rayl, M., and Rosi, F.D., "Superconductivity in Metals and Alloys," U.S. Air Force Rept. ASD-TDR-62-269, 86 pp., 1962. [AD 286 456]
581	54010	Mailfert, R., Batterman, B.W., and Hanak, J.J., "Observations Related to the Order of the Low Temperature Structural Transformation in V_3Si and Nb_3Sn," Phys. Stat. Sol., 32(1), K67-9, 1969.
582	34457	Wasilewski, R.J., "Thermal Vacancies in NiAl," Acta Met., 15(11), 1757-9, 1967.
583	47099	Singleton, R.H., Wallace, A.V., and Miller, D.G., "Nickel Aluminide Leading Edge for a Turbine Vane," in Summary of the Eleventh Refractory Composites Working Group Meeting, U.S. Air Force Rept. AFML-TR-66-179, 717-38, 1966. [AD 804 083]
584	50376	Wasolewski, R.J., Butler, S.R., and Hanlon, J.E., "Constitutional and Thermal Structure Defects in Nickel-Gallium," J. Appl. Phys., 39(9), 4234-41, 1968.
585	57158	Rud, B.M., Lynchak, K.A., and Paderno, Yu.B., "Some Properties of Praseodymium Germanides," Izv. Akad. Nauk SSSR, Neorg. Mater., 5(8), 1350-3, 1969; English Translation: Inorg. Mater. (USSR), 5(8), 1152-5, 1969.
586	48558	Neumann, J.P. and Chang, Y.A., "The Influence of Temperature on the Lattice Parameters of the Intermetallic Compound Ag_2Al," Trans. Met. Soc. AIME, 242(4), 700-2, 1968.
587	48867	Steeb, S., Godel, D., and Lohr, C., "On the Structure of the Compounds Ag_3, R.E. (R.E. = Y, La, Ce, Sm, Gd, Dy, Ho, Er)," J. Less-Common Metals, 15(2), 137-41, 1968.
588	49159	Chaudhuri, A., Clark, H.M., and Wayman, C.M., "The Effect of Stress on the β'-ζ° Transformation in Equiatomic AgZn," Acta Met., 17(6), 735-44, 1969.
589	24810	Booker, J., Paine, R.M., and Stonehouse, A.J., "Investigation of Intermetallic Compounds for Very High Temperature Applications," U.S. Air Force Rept. WADD-TR-60-889, 133 pp., 1961. [AD 265 625]
590	9980	Tripler, A.B., Jr., Snyder, M.J., and Duckworth, W.H., "Further Studies of Sintered Refractory Uranium Compounds," USAEC Rept. BMI-1313, 54 pp., 1959.
591	53856	Michel, D.J. and Ryba, E., "The Thermal Expansion Behavior of $YbZn_2$ and Other KHg_2-Type Intermetallic Compounds," J. Less-Common Metals, 18(2), 159-65, 1969.
592	54699	Ogawa, S. and Kasai, N., "Thermal Expansion Anomaly in $ZrZn_2$," J. Phys. Soc. Japan, 27(3), 789, 1969.
593	56886	Meincke, P.P.M., Fawcett, E., and Knapp, G.S., "Thermal Expansion and Magnetostriction of Zirconium Zinc," Solid State Commun., 7(22), 1643-5, 1969.
594	60319	Gertsriken, S.D. and Slyusar, B.F., "On the Determination of the Energy of Vacancy Formations and Their Number in Pure Metals," Fiz. Metal Metalloved., 6(6), 1061-9, 1958; English Translation: Phys. Metal Metallogr. (USSR), 6(6), 103-11, 1958.

Ref. No.	TPRC No.	
595	57690	Sully, A.H., Brandes, E.A., and Mitchell, K.W., "The Effect of Temperature and Purity on the Ductility and Other Properties of Chromium," J. Inst. Metals, 81, 585-97, 1952-3.
596	49975	Conway, J.B. and Flagella, P.N., "High-Temperature Reactor Materials Research," in AEC Fuels and Materials Development Program Progress Report No. 67, USAEC Rept. GEMP-67, 11-41, 1967.
597	49784	Pawar, R.R., "Thermal Expansion of Rhodium," Curr. Sci., 37(8), 224-5, 1968.
598	59303	White, G.K. and Pawlowicz, A.T., "Thermal Expansion of Rhodium, Iridium, and Palladium at Low Temperatures," J. Low Temp. Phys., 2(5/6), 631-9, 1970.
599	59529	Hall, E.O. and Crangle, J., "An X-Ray Investigation of the Reported High-Temperature Allotropy of Ruthenium," Acta Cryst., 10, 240-2, 1957.
600	60545	Dowdell, R.L., Nagler, C.A., Fine, M.E., Klug, H.P., and Bitsianes, G., "Beta Laminations in Cartridge Brass," Trans. Am. Soc. Metals, 41, 985-1000, 1949.
601	59819	Devi, U., Rao, C.N., and Rao, K.K., "Effect of Temperature on the Lattice Parameter of a Gold-41.02 At.% Palladium Alloy," Acta Met., 13(1), 44-5, 1965.
602	60551	Adenstedt, H., Pequignot, J.R., and Raymer, J.M., "The Titanium-Vanadium System," Trans. Am. Soc. Metals, 44, 990-1003, 1952.
603	38358	McCollough, E.E., "The Thermal Expansion of Some Aluminum Alloys," Physics, 1, 334-9, 1931.
604	60283	Fahmy, A.A. and Ragai, A.N., "Thermal-Expansion Behavior of Two-Phase Solids," J. Appl. Phys., 41(13), 5108-11, 1970.
605	60474	Haas, M., "The Dilatometric Study of Light Metals," J. Inst. Metals, 39, 233-54, 1928.
606	59808	Edwards, J.D., "Properties and Manufacture of Aluminum-Silicon Alloys," Chem. Met. Eng., 27(13), 654-5, 1922.
607	61070	Varich, N.I. and Sheiko, T.I., "Thermal Expansion of Aluminum-Molybdenum and Aluminum-Zirconium Alloys Obtained at a High Rate of Cooling," Fiz. Metal Metalloved., 30(2), 443-5, 1970.
608	45824	Pinto, N.P. and Burke, E.C., "Characteristics of Be-38% Al Sheet," Conf. Int. Met. Beryllium, Grenoble, France, p. 533, 1965.
609	58362	Chistov, S.F., Chernov, A.P., and Dombovskii, S.A., "Study of the Linear Expansion of Vitreous and Polycrystalline Selenium and As_2Se_3," Izv. Akad. Nauk SSSR, Neorg. Mater., 4(12), 2085-8, 1968; English Translation: Inorg. Mater. (USSR), 4(12), 1814-16, 1968.
610	57411	Stokes, A.R. and Wilson, A.J.C., "The Thermal Expansion of Lead from 0 to 320 C," Proc. Phys. Soc. London, 53, 658-61, 1941.
611	57686	Kurnakow, N.S. and Ageew, N.W., "Physico-Chemical Study of the Gold-Copper Solid Solutions," J. Inst. Metals, 46, 481-501, 1931.
612	56850	Boiko, B.T., Palatnik, L.S., and Pugachev, A.T., "Determination of the Thermal Expansion Coefficient of Thin Evaporated Films by Electron Diffraction," Proc. Colloq. Thin Films, 2nd, 1967, 474-82, 1968.
613	67281	Fitzer, E. and Weisenburger, S., "Cooperative Measurement of the Thermal Expansion Behavior of Different Materials Up to 1000 Degrees by Pushrod Dilatometers," AIP Conf. Proc., 3, 25-35, 1972.
614	65394	McLean, K.O., Swenson, C.A., and Case, C.R., "Thermal Expansion of Copper, Silver, and Gold Below 30 K," J. Low Temp. Phys., 7(1-2), 77-98, 1972.
615	65393	White, G.K. and Collins, J.G., "Thermal Expansion of Copper, Silver, and Gold at Low Temperatures," J. Low Temp. Phys., 7(1-2), 43-75, 1972.
616	65027	Straumanis, M.E., "Redetermination of the Lattice Parameters, Densities, and Thermal Expansion Coefficients of Silver and Gold, and Perfection of the Structure," Monatsh. Chem., 102(5), 1377-86, 1971.
617	64985	Straumanis, M.E. and Woodward, C.L., "Lattice Parameters and Thermal Expansion Coefficients of Aluminum, Silver, and Molybdenum at Low Temperatures. Comparison with Dilatometric Data," Acta Crystallogr., Sect. A, 27 (Pt. 6), 549-51, 1971.
618	64243	Straumanis, M.E., Rao, P.B., and James, W.J., "Lattice Parameters, Expansion Coefficients, and Densities of Indium and Indium-Cadmium Alloys," Z. Metallk., 62(6), 493-8, 1971.
619	38863	Deshpande, V.T. and Pawar, R., "Anisotropic Thermal Expansion of Indium," Acta Crystallogr., 25A(3), 415-6, 1969.
620	58430	Finkel, V.A., Glamazda, V.I., and Kovtun, G.P., "A Phase Transition in Vanadium," Zh. Eksptl. i Teor. Fiz., 57(4), 1065-8, 1969; English Translation: Sov. Phys.-JETP, 30(4), 581-3, 1970.
621	66321	Fitzer, E., "Thermophysical Properties of Solid Materials. Project Section 1B-Thermal Expansion Measurements from 1000-2600 C," Advisory Group for Aeronautical Research and Development Rept. AGARD-AR-38, 52 pp., 1972.
622	59267	Westlake, D.G. and Ockers, S.T., "Thermal Expansion of Vanadium and Vanadium Hydride at Low Temperatures," J. Less-Common Metals, 22(2), 225-30, 1970.
623	59604	Mochalov, G.A. and Ivanov, O.S., "High-Temperature Vacuum Dilatometer with Optical Measurement of Specimen Length," Zavod. Lab., 35(1), 116-18, 1969; English Translation: Ind. Lab. USSR, 35(1), 139-41, 1969.

Ref. No.	TPRC No.	
624	66320	Halvorsen, J.J., "The Electrical Resistivity and Thermal Expansion of Iridium at High Temperatures," Montana State University, M.S. Thesis, 40 pp., 1971.
625	65075	Feder, R. and Nowick, A.S., "Dilatometric and X-Ray Thermal Expansion in Non-Cubic Crystals. II. Experiments on Cadmium," Phys. Rev., 5B(4), 1244-53, 1972.
626	65203	Stebler, B., Andersson, C.G., and Kristensson, O., "Anomalous Thermal Expansion of Chromium Near T(N) (Néel Temperature)," Phys. Scr., 1(5-6), 281-5, 1970.
627	69015	Gordienko, V.A. and Nikolaev, V.I., "Magnetic Anomalies of the Thermal Expansion of Chromium," Zh. Eksp. Teor. Fiz. Pisma Red., 14(1), 6-9, 1971; English Translation: JETP Letters, 14(1), 3-5, 1971.
628	59564	Krajewski, W., Krueger, J., and Winterhager, H., "Allotropic Transformation and Thermal Expansion of Cobalt Binary Alloys Between 100 and 800 Degrees. I. II," Cobalt (Eng. Ed.), 47, 81-8, 1970, 48, 120-8, 1970.
629	62970	Vertogradskii, V.A., "Arrangement for Measuring Linear Expansion Coefficient of Wire Specimens Over a Broad Temperature Range," Zavod. Lab., 35(4), 515-17, 1969; English Translation: Ind. Lab. USSR, 35(4), 619-20, 1969.
630	60285	Cowder, L.R., Zocher, R.W., Kerrisk, J.F., and Lyon, L.L., "Thermal Expansion of Extruded Graphite-Zirconium Carbide Composites," J. Appl. Phys., 41(13), 5118-21, 1970.
631	60563	Masumoto, H. and Kobayashi, T., "On the Thermal Expansion Coefficient and the Temperature Coefficient of Young's Modulus of Iron-Platinum Alloys," Trans. Japan Inst. Metals, 6(2), 113-15, 1965.
632	61887	Weisenburger, S., "Analysis of Push-Rod Dilatometer Measurements," Rev. Int. Hautes Temp. Refract., 7(4), 410-5, 1970.
633	47076	Collins, J.G., Cowan, J.A., and White, G.K., "Thermal Expansion at Low Temperatures of Anisotropic Metals: Indium," Cryogenics, 7(4), 219-24, 1967.
634	60411	Harris, I.R. and Raynor, G.V., "The Lattice Spacings of Thorium Cerium-Alloys at Elevated Temperatures," J. Less-Common Metals, 7, 11-16, 1964.
635	65540	Tonnies, J.J., "Preparation of Lanthanide Single Crystals. Elastic Moduli and Thermal Expansion of Lutetium Single Crystals from 4.2 to 300 K," Iowa State University, Ph.D. Thesis, 79 pp., 1971. [Univ. Microfilms No. 71-14267]
636	67518	Evans, D.L. and Fischer, G.R., "Determination of the Thermal Expansion of Platinum by X-Ray Diffraction," AIP Conf. Proc., 3, 97-104, 1972.
637	66961	White, G.K., "Thermal Expansion of Platinum at Low Temperatures," J. Phys., 2F(2), L30-1, 1972.
638	67287	Hahn, T.A. and Kirby, R.K., "Thermal Expansion of Platinum from 293 to 1900°K," AIP Conf. Proc., 3, 87-95, 1972.
639	63396	Fitzer, E., "Thermophysical Properties of Solid Materials. Project Section 1A. Cooperative Thermal Expansion Measurements Up to 1000 C," Advisory Group for Aerospace Research and Development Corp. Rept. AGARD-AR-31-71, 63 pp., 1971. [AD723 581]
640	38065	Krupkowski, A. and DeHaas, W.J., "Thermo-Electric and Dilation Determinations with Ni-Cu Alloys at Low Temperatures," Ver. Gewone Vergad. Afdel Nat. Konink Akad Wetensch., 37, 810-18, 1928; English Translation: Communications Kamerlingh Onnes Lab. Univ. Leiden, 194B, 14-24, 1928.
641	61010	Marples, J.A.C., "The Effect of Alloying Additions of Plutonium and Neptunium on the Lattice Parameters of Alpha-Uranium Between 4 and 300 K," J. Phys. Chem. Solids, 31(11), 2421-30, 1970.
642	71367	Quader, Md.A. and Dey, B.N., "Lattice Expansion and Debye Temperature of A-Phase Silver-Cadmium Alloys," Ind. J. Phys., 36, 43-54, 1962.
643	60347	Owen, E.A. and Pickup, L., "Variation of Mean Atomic Volume with Temperature in Copper-Zinc Alloys, with Observations on the Beta-Transformation," Proc. Roy. Soc. London, 140, 191-204, 1933.
644	71369	Graham, J., Moore, A., and Raynor, G.V., "The Effect of Temperature on the Lattice Spacings of Indium," J. Inst. Metals, 84, 86-7, 1955.
645	35722	Fenn, R.W., Jr., Glass, R.A., Needham, R.A., and Steinberg, M.A., "Beryllium-Aluminum Alloys," Spacecraft Rockets J., 2(1), 87-93, 1965.
646	57989	De, M., "Thermal Expansion of Some Copper- and Silver-Base Alloys at High Temperatures," Indian J. Phys., 43(7), 367-76, 1969.
647	57689	Ellwood, E.C. and Silcock, J.M., "The Lattice Spacings of the Solid Solution of Copper in Aluminium," J. Inst. Metals, 74, 457-67, 1948.
648	71368	Rao, C.N. and Rao, K.K., "Effect of Temperature on the Lattice Parameters of Some Silver-Palladium Alloys," Can. J. Phys., 42(7), 1336-42, 1964.
649	62733	Zhdanova, V.V., Ivanov, G.A., Inyutkin, A.I., Naletov, V.L., Nikolaev, V.I., Regel, A.R., and Sergeev, V.P., "Effects of Twinned Layers on the Thermal Expansion of Bismuth-Antimony Alloys," Fiz. Tverd. Tela, 13(1), 199-204, 1971; English Translation: Sov. Phys.-Solid State, 13(1), 157-60, 1971.
650	66871	Balasundaram, L.J. and Sinha, A.N., "Thermal Expansion of Bismuth-Tin and Bismuth-Lead Alloys," Curr. Sci., 41(6), 211-2, 1972.

Ref. No.	TPRC No.	
651	63400	Goggin, W.R. and Paquin, R.A., "Optical Materials Study Program," AD 865 842L, 79 pp., 1970.
652	57190	Schaake, H.F., "Study of Thermal and Mechanical Properties of Selected Solids from 4° to the Melting Point," U.S. Air Force Rept. AFCRL-69-0538, 54 pp., 1969. [AD 699 579]
653	66733	Halvorson, J.J. and Wimber, R.T., "Thermal Expansion of Iridium at High Temperatures," J. Appl. Phys., 43(6), 2519-22, 1972.
654	64616	Naidu, S.V.N. and Houska, C.R., "X-Ray Determinations of the Debye Temperatures and Thermal Expansions for the Palladium-Silver-Gold System," J. Appl. Phys., 42(12), 4971-5, 1971.
655	68965	Epanchintsev, O.G., Konyukhova, M.A., and Vertman, A.A., "Use of Hydrostatic Microbalance for Determining the Coefficient of Volume Expansion of Metals," Zavod. Lab., 38(1), 53-4, 1972; English Translation: Ind. Lab. (USSR), 38(1), 66-7, 1972.
656	69667	Smirnov, Yu.N. and Timoshenko, V.M., "Magnetic Contribution to Thermal Expansion of Palladium," Pis'ma Zh. Eksp. Teor. Fiz., 15(8), 473-7, 1972; English Translation: JETP Letters, 15(8), 334-7, 1972.
657	65829	Vest, R.W., "Conduction Mechanisms in Thick Film Microcircuits," AD 727 988, 47 pp., 1971.
658	62592	Masumoto, H. and Sawaya, S., "Thermal Expansion Coefficient and the Temperature Coefficient of Young's Modulus of Nickel-Palladium and Nickel-Platinum Alloys," Trans. Jap. Inst. Metals, 11(6), 391-4, 1970.
659	60383	Aoyama, S. and Mikura, Z., "Thermal Expansion of Pb-Bi Alloys at Low Temperatures," Nippon Kinzoku Gakkaishi, 4(12), 397-8, 1940.
660	69892	Finkel, V.A., Palatnik, M.I., and Kovtun, G.P., "X-Ray Diffraction Study of the Thermal Expansion of Ruthenium, Osmium and Rhenium at 77-300 K," Fiz. Metal. Metalloved., 32(1), 212-6, 1971; English Translation: Phys. Metals Metallogr. (USSR), 32(1), 231-5, 1971.
661	64883	Sawaoka, A., Soma, T., Saito, S., and Endoh, Y., "Effect of High Pressure on the Néel Temperature of Gamma-Manganese," Phys. Stat. Sol., 47B(2), K99-101, 1971.
662	41561	Ibragimov, E.A., Selisskii, Ya.P., and Sorokin, M.N., "Investigation of the Effects of Ordering in Ternary Ni-Fe-Mn Solid Solutions," Izv. Akad. Nauk, SSSR, Metally, 5, 152-8, 1966; English Translation: Russ. Met. (Metal.), 5, 82-7, 1966.
663	65738	Masumoto, H., Sawaya, S., and Kikuchi, M., "Nonmagnetic Elinvar-Type Alloy in Manganese-Nickel System," Nippon Kinzoku Gakkaishi, 35(12), 1143-9, 1971.
664	69499	Masumoto, H., Sawaya, S., and Kikuchi, M., "New Nonmagnetic Elinvar-Type Alloys in the Manganese-Copper System," Trans. Japan Inst. Metals, 13(1), 21-7, 1972.
665	64621	Balasundaram, L.J. and Sinha, A.N., "Thermal Expansion of Lead-Tin and Lead-Cadmium Alloys," J. Appl. Phys., 42(12), 5207, 1971.
666	71679	Zalkin, A. and Ramsey, W.J., "Intermetallic Compounds Between Lithium and Lead. III. The β'-β Transition in Lithium-Lead Intermetallic," J. Phys. Chem., 61, 1413-5, 1957.
667	71680	Koster, W. and Schneider, A., "Elastic Modulus and Vaporization of Intermediate Phases in Gold-Cadmium System," Z. Metallk., 32(6), 156-9, 1940.
668	71681	Schubert, K. and Wall, E., "Crystal Structure of High Temperature and Phase of the Copper-Zinc System," Z. Metallk., 40, 383-5, 1949.
669	71682	Pelzel, E., "Specific Volumes of Solid and Liquid Antimony Bismuth Alloys," Z. Metallk., 50, 392-5, 1959.
670	71683	Cave, E.F. and Holroyd, L.V., "Thermal Expansion Coefficients of Bismuth," J. Appl. Phys., 31(8), 1357-8, 1960.
671	71684	Muldawer, L., "X-Ray Study of Ternary Ordering of the Noble Metals in Silver-Gold-Zinc and Copper-Gold-Zinc Intermetallic Compounds," J. Appl. Phys., 37(5), 2062-6, 1966.
672	71688	Schadler, H.W., Osika, L.M., Salvo, G.P., and DeCarlo, V.J., "The Thermal Expansion of Niobium-Tin Intermetallic," Trans. Met. Soc. AIME, 20, 1074-7, 1964.
673	71689	Taylor, A. and Floyd, R.W., "The Constitution of Nickel-Rich Alloys of the Nickel-Titanium-Aluminum System," J. Inst. Metals, 81, 25-32, 1952-3.
674	71690	Flinn, P.A., McManus, G.M., and Rayne, J.A., "Elastic Constants of Ordered and Disordered Copper-Gold Intermetallic from 4.2 to 300 K," J. Phys. Chem. Solids, 15, 189-95, 1960.
675	71691	Cheng, C.H., "The Elastic Constants of Magnesium-Silver and Magnesium-Copper Intermetallic Single Crystals," J. Phys. Chem. Solids, 28, 413-6, 1967.
676	71692	Atoji, M., "Magnetic Structure of Terbium-Silver Intermetallic," J. Chem. Phys., 48(8), 3380-3, 1968.
677	71693	Buehler, W.J. and Wiley, R.C., "Titanium-Nickel, Ductile Intermetallic Compound," Trans. Amer. Soc. Metals, 55, 269-76, 1962.
678	71719	Warlimont, H., "Method Standardization and Examples of Continuous Structure-Analysis with X-Ray," Z. Metallk., 50(12), 708-16, 1959.
679	71696	Zhuravlev, N.N. and Stepanova, A.A., "An X-Ray Investigation of Superconducting Alloys of Bismuth with Platinum in the Range 20-640 C," Kristallografiya, 7(2), 310-1, 1962; English Translation: Sov. Phys.-Crystallogr., 7(2), 241-2, 1962.

Ref. No.	TPRC No.	
680	71697	Nakagawa, Y. and Hori, T., "Magnetic Transition and Crystal Distortion in Manganese-Mercury and Manganese-Zinc Intermetallics," J. Phys. Soc. Japan, 17, 1313-14, 1962.
681	71698	Bergevin, F. and Muldawer, L., "Crystallographic Study of Certain Alloys of Iron-Rhodium," Compt. Rend., 252, 1347-9, 1961.
682	71714	Harris, I.R. and Raynor, G.V., "Rare Earth Intermediate Phases. I. Phases Formed with Tin and Indium," J. Less-Common Metals, 9, 7-19, 1965.
683	71715	Harris, I.R., Norman, M., and Bryant, A.W., "A Study of Some Palladium-Indium, Platinum-Indium and Platinum-Tin Alloys," J. Less-Common Metals, 16, 427-40, 1968.
684	71716	Zirinsky, S., "The Temperature Dependence of the Elastic Constants of Gold-Cadmium Alloys," Acta Met., 4, 164-71, 1956.
685	71717	Smith, J.H. and Gaunt, P., "The Martensitic Transformations in Gold-Manganese Alloys Near the Equiatomic Composition," Acta Met., 9, 819-24, 1961.
686	71718	Hume-Rothery, W. and Andrews, K.W., "The Equilibrium Diagram of the System Silver-Gallium. II. The Region from 20 to 33 At.% Gallium," Z. Metallk., 50(11), 661-2, 1959.
687	71773	Straumanis, M.E. and Chopra, K.S., "Lattice Parameters, Expansion Coefficients and Extent of the Dialuminum Gold Phase," Z. Phys. Chem (Frankfort Am. Main), 42, 344-50, 1964.
688	56527	Smith, J.F. and Ogren, J.R., "Electrical Properties and Thermal Expansion of the Laves Phases, $CaMg_2$ and $MgCu_2$," J. Appl. Phys., 29(11), 1523-5, 1958.
689	31847	Snyder, M.J. and Tripler, A.B., Jr., "Some Refractory Uranium Compounds Preparation and Properties," Am. Soc. Test. Mater. Spec. Tech. Publ. 276, 293-300, 1960.
690	60015	Schneider, A. and Heymer, G., "The Temperature Dependence of the Molvolumina of the Phases Sodium Thallium and Lithium Cadmium Intermetallic," Zeit. Anorg. Allgem. Chem., 286, 118-35, 1956.
691	44360	Chang, L.-C., "Coefficients of Thermal Expansion of Au-Cd Alloys Containing 47.5 At.% Cd," J. Appl. Phys., 22, 525-6, 1951.
692	60402	Owen, E.A. and Liu, Y.H., "The Thermal Expansion of the Gold-Copper Alloy $AuCu_3$," Phil. Mag., 38, 354-60, 1947.
693	59454	Ryabov, V.R., Lozovskaya, A.V., and Vasilyev, V.G., "Properties of the Intermetallic Compounds of the System Iron-Aluminium," Fiz. Metal. Metalloved., 27(4), 668-73, 1969; English Translation: Phys. Metals Metallogr., 27(4), 98-103, 1969.
694	59047	Pearce, R.J., "The Thermal Expansion of Uranium Trialuminum Intermetallic in the Temperature Range 20 to 750 C," J. Nucl. Mater., 17(2), 201-2, 1965.
695	49218	Michel, D.J. and Ryba, E., "The High-Temperature Lattice Parameters of $HoZn_2$," J. Less-Common Metals, 14(3), 367-9, 1968.
696	64012	Awad, F.G. and Gugan, D., "The Thermal Expansion of Copper, Aluminum, Potassium Chloride, and Potassium Iodide Between 10 and 80 K," Cryogenics, 11(5), 414-5, 1971.
697	60672	Gupta, M.L. and Singh, S., "Thermal Expansion of Cerium, Holmium, and Lutetium Oxide from 100 to 300 K by an X-Ray Method," J. Am. Ceram. Soc., 53(12), 663-5, 1970.
698	63413	Rabkin, D.M., Ryabov, V.R., Lozovskaya, A.V., and Dovzhenko, V.A., "Preparation and Properties of Copper-Aluminum Intermetallic Compounds," Porosh Met., USSR, 10(8), 101-7, 1970; English Translation: Sov. Powder Met. Metal. Ceram., 8, 695-700, 1970.
699	58220	McLean, K.O., "Low Temperature Thermal Expansion of Copper, Silver, Gold, and Aluminum," Iowa State University, Ph.D. Thesis, 97 pp., 1969.
700	61047	Gachkovskii, V.F. and Strelkov, P.G., "Thermal Constants at High Temperatures," Zhur. Eksp. Teor. Fiz., 7(4), 532-48, 1937; English Translation: U.S. Air Force Rept. FTD-MT-71-1599, 57 pp., 1971.
701	63566	Woodard, C.L., "X-Ray Determination of Lattice Parameters and Thermal Expansion Coefficients of Aluminum, Silver, and Molybdenum at Cryogenic Temperatures," University of Missouri at Rolla, Ph.D. Thesis, 142 pp., 1969. [Univ. Microfilms No. 70-11768]
702	38132	Goetz, A. and Hergenrother, R.C., "Macroscopic and Lattice Expansion of Bismuth Single Crystals," Phys. Rev., 38, 2075-7, 1931.
703	59588	Goetz, A. and Hergenrother, R.C., "X-Ray Studies of the Thermal Expansion of Bismuth Single Crystals," Phys. Rev., 40(5), 643-61, 1932.
704	41926	Edwards, D.A., Wallace, W.E., and Craig, R.S., "Magnesium-Cadmium Alloys. IV. The Cadmium-Rich Alloys, Some Lattice Parameters and Phase Relationships Between 25 and 300°. Structure of the $MgCd_3$ Superlattice. Schottky Defects and the Anomalous Entropy," J. Amer. Chem. Soc., 74, 5256-61, 1952.
705	39328	Bridgman, P.W., "Electrical Resistance Under Pressure, Including Certain Liquid Metals," Proc. Amer. Acad. Arts Sci., 56(3), 61-154, 1921.
706	57219	Gschneidner, K.A., Jr., "Influence of Magnesium on Some of the Physical Properties of Cerium," USAEC Rept. LADC-5615, 13 pp., 1962.
707	65500	Pridantseva, K.S. and Solveva, N.A., "Thermal Expansion of Solid Solutions in Heat-Resistant Metals of the 4th, 5th, and 6th Groups of the Periodic System," Vysok. Neorg. Soedin. Akad. Nauk Ukr. SSR, Inst. Probl. Mater., 41-7, 1965; English Translation: U.S. Air Force Rept. FTD-HC-23-57-72, FTD-MT-24, 1396-71, 8 pp., 1972. [AD 745 847]

Ref. No.	TPRC No.	
708	62682	Ageev, N.V. and Medel, M.S., "Thermal Expansion of Chromium and Solid Solutions Based on Chromium," Issled. Zharoproch. Splavam. Akad. Nauk SSSR, Inst. Met., 10, 15-22, 1963; English Translation: U.S. Air Force Rept. FTD-MT-24-1398-71, 26 pp., 1971.
709	66170	Shah, J.S., "Thermal Lattice Expansion of Various Types of Solids," University of Missouri, Ph.D. Thesis, 126 pp., 1971. [Univ. Microfilms No. 71-25776]
710	48022	Erfling, H.D., "Studies to the Thermal Expansion of Solids at Low Temperatures. II. (Cr, B-Mn, Mo, Rh, Be, Graphite, Tl, Zr, Bi, Sb, Sn, and Beryll)," Ann. Physik., 34(5), 136-60, 1939.
711	68779	White, G.K., "Thermal Expansion of Trigonal Elements at Low Temperatures. Arsenic, Antimony, and Bismuth," J. Phys., 5C(19), 2731-45, 1972.
712	37704	Klemm, W., Spitzer, H., and Niermann, H., "Investigation of Some Semi-Metals," Angew. Chem., 72(24), 985-94, 1960.
713	52942	Esser, H. and Eusterbrock, H., "Investigation of the Thermal Expansion of Some Metals and Alloys with Improved Dilatometer," Archiv. Eisenhutten., 14(7), 341-55, 1941.
714	34153	Andres, K., "Thermal Expansion of Metals at Low Temperature," Phys. Kondens Mater., 2, 294-333, 1964.
715	57733	Shinoda, G., "X-Ray Investigations on the Thermal Expansion of Solids. Part 2," Kyoto Imp. Univ. Memoirs College Sci., 17A, 27-31, 1934.
716	35406	Andres, K. and Rohrer, H., "Thermal Expansion at Low Temperatures," Helv. Phys. Acta, 34, 398-401, 1961.
717	63999	Technology Utilization Division, NASA, "Technical Support Package for Tech. Brief 69-10055 - Thermal Expansion Properties of Aerospace Materials," NTIS Rept. 174 pp., 1969. [PB-184749]
718	64530	Vorob'ev, V.V., Palatnik, M.I., and Finkel, V.A., "Structural Effects at the Curie Point of Gadolinium," Phys. Stat. Sol., 47B(1), K53-8, 1971.
719	73087	Austin, J.B., "A Vacuum Apparatus for Measuring Thermal Expansion at Elevated Temperatures with Measurements on Platinum, Gold, Magnesium, and Zinc," Physics, 3, 240-67, 1932.
720	68046	Schuerch, H.U., "Thermally Stable Macrocomposite Structures," NASA Rept. NASA-CR-1973, 22 pp., 1972.
721	73179	Bibring, H. and Sebilleau, F., "Structure of Allotropic Transformation of Cobalt," Rev. Met., 52(7), 569-78, 1955.
722	73178	Owen, E.A. and Madoc Jones, D., "Effect of Grain Size on the Crystal Structure of Cobalt," Proc. Phys. Soc. (London), 67B, 656-66, 1954.
723	63116	Griessen, R. and Ott, H.R., "The Volume Change Between Normal and Superconducting Aluminum Above 0.3 K," Phys. Lett., 36A(2), 113-4, 1971.
724	39635	Fine, M.E., Greiner, E.S., and Ellis, W.C., "Transitions in Chromium," Trans. AIME, 189, 56-8, 1951.
725	73287	Marick, L., "Variation of Resistance and Structure of Cobalt with Temperature and a Discussion of Its Photoelectric Emission," Phys. Rev., 49, 831-7, 1936.
726	68642	White, G.K., "Thermal Expansion of Calcium, Strontium, and Barium at Low Temperatures," J. Phys., 2F(5), 865-72, 1972.
727	58101	Cath, P.G. and Von Steenis, O.L., "The Expansion Coefficient of Barium and Calcium and Allotropy," Z. Tech. Physik., 17(7), 239-41, 1936.
728	48885	Erfling, H.D., "Studies on the Thermal Expansion of Solid Materials at Low Temperature. III. (Calcium, Niobium, Thorium, Vanadium, Silicon, Titanium, Zirconium)," Ann. Phys., 41(5), 467-75, 1942.
729	39260	Schulze, A., "Allotropy Investigations on Very Pure Calcium," Physik. Z., 36(18), 595-600, 1935.
730	63615	Kondorskii, E.I., Kostina, T.I., and Ekonomova, L.N., "Magnetostriction and Thermal Expansion of Chromium," J. Phys. (Paris), Colloq., 1 (Part 1), C1-417 - C1-8, 1971.
731	45398	Cohen, R.L., Hüfner, S., and West, K.W., "First-Order Phase Transition in Europium Metal," Phys. Rev., 184(2), 263-70, 1969.
732	65959	Bhalla, A.S. and White, E.W., "Coefficient of Linear Expansion of Silicon and Germanium by Double Crystal X-Ray Spectrometer," Phys. Stat. Sol., 5A(1), K51-3, 1971.
733	62928	Finkel, V.A. and Palatnik, M.I., "Crystal Structure of Holmium and Erbium between 77 and 300 K," Zh. Eksp. Teor. Fiz., 59(5), 1518-23, 1970; English Translation: Sov. Phys.-JETP, 32(5), 828-31, 1971.
734	37679	Feder, R., "Equilibrium Defect Concentration in Crystalline Lithium," Phys. Rev., 2B(4), 828-34, 1970.
735	67164	Golutvin, Yu.M. and Maslennikova, E.G., "The Heat Capacity of Metallic Hafnium," Izv. Akad. Nauk SSSR, Metal., 5, 174-83, 1970; English Translation: Russ. Met. (USSR), 5, 129-35, 1970.
736	62585	Singh, S., Khanduri, N.C., and Tsang, T., "Thermal Expansions of Thulium and Lutetium, 90-300 K," Scr. Met., 5(3), 167-73, 1971.
737	71924	Rubin, T., Johnston, H.L., and Altman, H.W., "The Thermal Expansion of Lead," J. Phys. Chem., 66, 266-8, 1962.

Ref. No.	TPRC No.	
738	45391	Kantola, M. and Tokola, E., "X-Ray Studies on the Thermal Expansion of Copper-Nickel Alloys," Ann. Acad. Sci. Fenn, A VI (223), 11 pp., 1967.
739	71453	Petukhov, V.A. and Chekhovskoi, V.Ya., "Thermal Expansion of Molybdenum and Tungsten at High Temperatures," High Temp.-High Pressures (USSR), 4(6), 671-7, 1972.
740	67875	Gurevich, M.E. and Larikov, L.N., "An Automatic Dilatometer," Faz. Prevraschen. Metal. Splavakh, Akad. Nauk Ukr. SSR, 27, 230-4, 1970; English Translation: U.S. Air Force Rept. FTD-MT-24-893-71, 12 pp., 1971. [AD 737 886]
741	64130	Tanji, Y., "Thermal Expansion Coefficient and Spontaneous Volume Magnetostriction of Iron-Nickel (FCC) Alloys," J. Phys. Soc. Japan, 31(5), 1366-73, 1971.
742	71918	d'Heurle, F.M., Feder, R., and Nowick, A.S., "Equilibrium Concentration of Lattice Vacancies in Lead and Lead Alloys," J. Phys. Soc. Japan, 18, Suppl. II, 184-90, 1963.
743	56155	Straumanis, M.E. and Yu, L.S., "Lattice Parameters, Densities, Expansion Coefficients, and Perfection of Structure of Copper and of Copper-Indium α-Phase," Acta Crystallogr., Sect. A, 25(6), 676-82, 1969.
744	57880	Linkoaho, M. and Rantavuori, E., "Debye Temperature and the Coefficient of Thermal Expansion for Vanadium and Niobium by X-Ray Diffraction," Phys. Stat. Sol., 37(2), 1970.
745	57039	Lebedev, V.P., Mamalui, A.A., Pervakov, V.A., Petrenko, N.S., Popov, V.P., and Khotkevich, V.I., "Thermal Expansion of Niobium, Molybdenum, and Their Alloy at Low Temperatures," Ukr. Fiz. Zh., 14(5), 746-50, 1969.
746	57592	Straumanis, M.E. and Zyszezynski, S., "Lattice Parameters, Thermal Expansion Coefficients, and Densities of Niobium, and of Solid Solutions Niobium Oxygen and Niobium-Nitrogen-Oxygen and Their Defect Structure," J. Appl. Crystallogr., 3(Pt. 1), 1-6, 1970.
747	69430	Lisovskii, Yu.A., "Analysis of Thermal Expansion of Transition Metals in the Quasiharmonic Approximation," Fiz. Tverd. Tela, 14(8), 2329-33, 1972; English Translation: Sov. Phys.-Solid State, 14(8), 2015-8, 1973.
748	66740	Hubbell, W.C. and Brotzen, F.R., "Elastic Constants of Niobium-Molybdenum Alloys in the Temperature Range -190 to 100 C," J. Appl. Phys., 43(8), 3306-12, 1972.
749	69442	Zakharov, A.I., Perepelkina, A.M., and Shiryaeva, A.N., "Effect of Alloying on Thermal Expansion of Supervinvar Alloy," Metalloved. Term. Obrab. Metal., 6, 62-4, 1972; English Translation: Metal Sci. Heat Treat. Metals (USSR), 6, 539-41, 1972.
750	65022	Chatterjee, D. and Taylor, K.N.R., "Lattice Expansion in Some Rare-Earth Laves Phase Compounds," J. Less-Common Metals, 25(4), 423-6, 1971.
751	61521	Kasai, N. and Ogawa, S., "Thermal Expansion of Au_4V," J. Phys. Soc. Japan, 30(3), 736-41, 1971.
752	63492	Schlosser, W.F., Graham, G.M., and Meincke, P.P.M., "The Temperature and Magnetic Field Dependence of the Forced Magnetostriction and Thermal Expansion of Invar," J. Phys. Chem. Solids, 32(5), 927-38, 1971.
753	68015	Schlosser, W.F., Latal, E., Meincke, P.P.M., Graham, G.M., and Colling, D.A., "Low Temperature Invar Anomalies in Fe-Ni-Co Alloys," Amer. Inst. Phys. Conf. Proc., 3, 195-202, 1972.
754	66980	Bol'shakov, Yu.V., Zakharov, A.I., Pozvonkov, F.M., Solov'eva, N.S., and Fridman, V.G., "Thermal Expansion of Alloy 36 NKh at 4.2 to 300 K," Metalloved Term. Obrab. Metal., 3, 57, 1971; English Translation: Metal Sci. Heat Treat., USSR, 3, 234, 1971.
755	68426	Fukamichi, K. and Saito, H., "Anomalous Thermal Expansion of Chromium-Rich Chromium-Tin Alloys," J. Phys. Soc. Jap, 33(5), 1485, 1972.
756	45652	Woolley, J.C., "Thermal Expansion of GaSb at High Temperatures," J. Electrochem. Soc., 142(4), 461, 1965.
757	63410	Mal'ko, P.I., Arensburger, D.S., Pugin, V.S., Nemchenko, V.F., and L'vov, S.N., "Thermal and Electrical Properties of Porous Titanium," Porosh. Met., USSR, 10(8), 35-8, 1970; English Translation: Sov. Powder Met. Metal Ceram., 8, 642-44, 1970.
758	66796	Kraftmakher, Ya.A., "Vacancy Equilibrium and Thermal Expansion of Tungsten at High Temperatures," Fiz. Tverd. Tela, 14(2), 392-4, 1972; English Translation: Sov. Phys.-Solid State, 14(2), 325-7, 1972.
759	62137	Bolef, D.I., Smith, R.E., and Miller, J.G., "Elastic Properties of Vanadium. I. Temperature Dependence of the Elastic Constants and the Thermal Expansion," Phys. Rev. B, 3B(12), 4100-8, 1971. [AD 727 587]
760	62138	Bolef, D.I., Smith, R.E., and Miller, J.G., "Elastic Properties of Vanadium. II. The Role of Interstitial Hydrogen," Phys. Rev., 3B(12), 4108-15, 1971. [AD 727 388]
761	74095	Brodskiy, B.R. and Neymark, B.E., "Thermal Expansion of Zirconium and Vanadium," Izv. Akad. Nauk SSSR, Metal., 221-4, 1971; English Translation: Russ. Met. (Metally), 6, 156-60, 1971.
762	61106	Apostolou, S.F., "X-Ray Thermal Expansion Measurements of Dislocation Climb in Zinc," Rensselaer Polytechnic Inst., M.S. Thesis, 20 pp., 1970.
763	58355	Wallmark, G.N., "Thermal Expansion Measurements of Vacancy Formation Parameters and Anisotropic Atomic Rearrangements in Zinc Single Crystals," Rensselaer Polytechnic Inst., M.S. Thesis, 35 pp., 1969.
764	60212	Sinha, A.N. and Balasundaram, L.J., "Coefficients of Thermal Expansion of Iron-Aluminum and Iron-Silicon Alloys," Indian Inst. Metals, 23(1), 5-7, 1970.

Ref. No.	TPRC No.	
765	67285	Willemsen, H.W., Vittoratos, E., and Meincke, P.P.M., "Low-Temperature Thermal Expansion of Pure Zinc and a Dilute Zinc-Manganese Alloy," Amer. Inst. Phys., (Conf. Proc., Univ. Toronto), 3, 72-6, 1972.
766	60649	Lysak, L.I. and Andrushchik, L.O., "Dilatometric Analysis of Low-Temperature Phase Transformations in Rhenium Steels," Fiz. Metal. Metalloved., 28(3), 478-81, 1969; English Translation: Phys. Metals Metallogr. (USSR), 28(3), 97-101, 1969.
767	66384	Straumanis, M. E. and Patel, V.K., "Perfection of the Structure of the Gold-Indium Alpha Phase. Lattice Parameters, Solubility Limit, Expansion Coefficients, Densities, and Voids," Z. Metallk., 63(1), 33-7, 1972.
768	63814	Andreeva, L.P. and Gel'd, P.V., "Coefficients of Thermal and Module of Elasticity of Iron Silicides," Izv. Vyssh. Ucheb. Zaved., Chern. Met., 8(2), 111-17, 1965; English Translation: Foreign Technology Division Rept., 27 pp., 1971. [AD 745 825]
769	41588	Gen, M.Ya. and Petrov, Yu.I., "Thermal Expansion and Structural Anomalies of Small Iron Particles," Dokl. Akad. Nauk SSSR, 179(6), 1311-13, 1968; English Translation: Soviet Phys.-Doklady, 13(4), 358-60, 1968.
770	74337	Bernstein, L. and Beals, R.J., "Thermal Expansion and Related Bonding Problems of Some III-V Compound Semiconductors," J. Appl. Phys., 32, 122-3, 1961.
771	74608	Darnell, F.J., "Lattice Parameters of Terbium and Erbium at Low Temperatures," Phys. Rev., 132(3), 1098-1100, 1963.
772	61787	Goldberg, L.R. and Matlock, K., "Delta and Epsilon Thermal Expansion Coefficients and the Delta-to-Epsilon Contraction for Some Plutonium-Rich Alloys," USAEC Rept. UCRL-7223, 18 pp., 1970.
773	68291	Fournier, J.-M., "Magnetic Properties and Electronic Structure of Plutonium," J. Phys., Radium, 33(7), 699-706, 1972.
774	37992	White, G.K., "Thermal Expansion of Anisotropic Metals at Low Temperatures," Phys. Letters, 8(8), 294, 1964.
775	70209	Kun'min, R.N., Nikitina, S.V., Ovchinnikov, A.N., Ryuteva, N.D., and Golovnin, V.A., "Investigation of the Ductile-to-Brittle Transition in Carbon Steels," Metalloved. Term. Obrab. Metal., 8, 38,43-4, 1972; English Translation: Metal Sci. Heat Treat. Metals, (USSR), 8, 696-8, 1972.
776	51212	Schmitz-Pranghe, N. and Duenner, P., "Crystal Structure and Thermal Expansion of Scandium, Titanium, Vanadium, and Manganese," Z. Metallk., 59(5), 377-82, 1968.
777	59236	Chekhovskoi, V.Ya. and Petukhov, V.A., "Measurement of Thermal Expansion of Solids at High Temperatures," Proc. Symp. Thermophys. Prop., 5th, 366-72, 1970.
778	58914	Jette, E.R., "Some Physical Properties of Plutonium Metal," J. Chem. Phys., 23(2), 365-68, 1955.
779	57741	Powell, R.W., "Some Anisotropic Properties of Gallium," Nature, 164(4160), 153-4, 1949.
780	67327	White, G.K., "Expansion Coefficient of Coppers at 283 K and Low Temperatures," Amer. Inst. Phys. Conf. Proc., 3, 59-64, 1972.
781	42188	Kohlhaas, R. and Pandey, R.K., "The Physical Properties of an Iron-Silicon Alloy with 3.34% Silicon in the High-Temperature Region," Z. Angew. Phys., 21(4), 365-7, 1966.

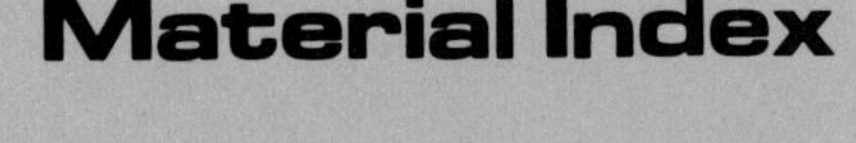

Material Index

Material Index